The Comprehensive Sourcebook of Bacterial Protein Toxins

The Comprehensive Sourcebook of Bacterial Protein Toxins

Second Edition

Editors

Joseph E. Alouf
Institut Pasteur, Lille, France

and

John H. Freer
University of Glasgow, Glasgow, UK

ACADEMIC PRESS

London San Diego Boston New York Sydney Tokyo Toronto

Produced and typeset by Gray Publishing, Tunbridge Wells, Kent
Printed in China
99 00 01 02 03 04 AW 9 8 7 6 5 4 3 2 1

Contents

SECTION II: MEMBRANE-DAMAGING TOXINS

SECTION III: OTHER TOXINS OF CLINICAL, PHARMACOLOGICAL, IMMUNOLOGICAL AND THERAPEUTIC INTEREST

A colour plate section appears between pp. 336 and 337

Contributors

David W.K. Acheson
Tupper Research Institute, New England Medical
Center, 750, Washington Street, Boston, MA 02111,
USA

Joseph E. Alouf
Institut Pasteur, 1 Rue du Professeur A. Calmette–BP
245, 59019 Lille Cedex, France

Andreas Ambrosch
Otto-von-Guericke-Universität Magdeburg, Institut
für Medizinische Mikrobiologie, Leipziger Strasse 44,
D-39120 Magdeburg, Germany

Rudy Antoine
Laboratoire de Microbiologie Genetique et
Moleculaire, INSERM U447, Institut Pasteur de Lille,
1 Rue du Professeur Calmette, F-59019 Lille Cedex,
France

Marina de Bernard
Dipartimento di Scienzie Biomediche, Università di
Padova, Via G. Colombo 3, I-35121 Padova,
Italy

Gregory A. Bohach
Department of Microbiology, Molecular Biology and
Biochemistry, University of Idaho, Moscow, ID 83844,
USA

Patrice Boquet
INSERM U452–Faculté de Médecine, 28 Avenue de
Valombrose, 06107 Nice, Cedex 15, France

Volkmar Braun
Institut für Mikrobiologie II Auf der Morgenstelle 28,
D-72076 Tübingen, Germany

Amy E. Bryant
Veterans Affairs Medical Centre, Infectious Diseases
Section (Bldg 45), 500 West Fort Street, Boise,
ID 83702-4598, USA

Sigrid Brynestad
Norwegian College of Veterinary Medicine,
P.O. Box 8146 Dep, N-0033 Oslo, Norway

J. Thomas Buckley
Department of Biochemistry and Microbiology,
University of Victoria, Box 3055, Victoria, British
Columbia V8W 3P6, Canada

Michelle C. Callegan
University of Oklahoma Health Sciences Center,
Department of Ophthalmology, 608 S.L. Young Blvd,
Oklahoma City, OK 73104, USA

Christophe Carnoy
Equipe Mixte Inserm E99-19 Université JE 2225,
Departement de Pathogènése des Maladies
Infectieuses et Parasitaires, Institut de Biologie de
Lille, 1 rue du Professeur Calmette, 13P–245-59021,
Lille Cedex, France

Sabine Castano
Centre de Recherche Paul Pascal–CNRS, Avenue A.
Schweitzer, 33600 Pessac, France

Trinad Chakraborty
Institute for Medical Microbiology, Justus Liebig-
University Giessen, D-35392 Giessen, Germany

Esteban Chaves-Olarte
Microbiology and Tumorbiology Center, Karolinska
Institute, Box 280, S-17177 Stockholm, Sweden

Ayub Darji
Department of Cell Biology and Immunology, National Center of Biotechnology, D-38124 Braunschweig, Germany

Ulrich Dobrindt
Institut für Molekulare Infektionsbiologie, Röntgenring 11, D-97070 Würzburg, Germany

Andreas Drynda
Otto-von-Guericke-Universität Magdeburg, Institut für Medizinische Mikrobiologie, Leipziger Strasse 44, D-39120 Magdeburg, Germany

J. Daniel Dubreuil
Faculty of Veterinary Medicine, University of Montreal, 3200 Sicotte, P.O. Box 5000, Saint-Hyacinthe, Quebec J2S 7C6, Canada

Jean Dufourcq
Centre de Recherche Paul Pascal–CNRS, Avenue A. Schweitzer, 33600 Pessac, France

Pål Ø. Falnes
Department of Biochemistry, Institute of Cancer Research, The Norwegian Radium Hospital, Montebello, N-0310 Oslo, Norway

John H. Freer
Division of Infection and Immunity, University of Glasgow, Joseph Black Building, Glasgow G12 8QQ, UK

Michael S. Gilmore
University Oklahoma Health Sciences Center, Departments of Microbiology, and Immunology and Opthalmology, P.O. Box 26901, Oklahoma City, OK 73190, USA

Werner Goebel
Theodor-Boveri-Institut (Biozentrum), LS Mikrobiologie, Universität Würzburg, Am Hubland, D-97074 Würzburg, Germany

Joanne Goranson-Siekierke
University of Colorado Health Science Center, Department of Microbiology – Campus Box B-175, 4200 East Ninth Avenue, Denver, CO 80262, USA

Per Einar Granum
Norwegian College of Veterinary Medicine, P.O. Box 8146 Dep., N-0033 Oslo, Norway

Jörg Hacker
Institut für Molekulare Infektionsbiologie, Röntgenring 11, D-97070 Würzburg, Germany

Judit Herreros
Imperial Cancer Research Fund, 44 Lincoln's Inn Fields, Room 614-5, London WC2A 3PX, UK

Ralf Hertle
Institut für Mikro-/Membranbiologie, Universität Tübingen Auf der Morgenstelle 28, D-72076 Tübingen, Germany

Timothy R. Hirst
Department of Pathology and Microbiology, School of Medical Sciences, University of Bristol, University Walk, Bristol BS8 1TD, UK

Randall K. Holmes
University of Colorado Health Science Center, Department of Microbiology – Campus Box B-175, 4200 East Ninth Avenue, Denver, CO 80262, USA

Jan Holmgren
University of Göteborg, Guldhedsgatan 10, S-413 46 Göteborg, Sweden

Thomas Jacobs
Department of Cell Biology and Immunology, National Center of Biotechnology, D-38124 Braunschweig, Germany

Bradley D. Jett
University of Oklahoma Health Sciences Center, Department of Opthalmology, 608 S.L. Young Blvd. Oklahoma City, OK 73104, USA

Gerald T. Keusch
Tupper Research Institute, New England Medical Center, 750 Washington Street, Boston, MA 02111, USA

Werner Köhler
Institute für experimentelle Mikrobiologie, Friedrich Schiller University, Winzerlaer Straße 10, D-07745 Jena, Germany

Brigitte König
Otto-von-Guericke-Universität Magdeburg, Institut für Medizinische Mikrobiologie, Leipziger Strasse 44, D-39120 Magdeburg, Germany

Wolfgang König
Otto-von-Guericke-Universität Magdeburg, Institut
für Medizinische Mikrobiologie, Leipziger Strasse 44,
D-39120 Magdeburg, Germany

Giovanna Lalli
Imperial Cancer Research Fund, 44 Lincoln's Inn
Fields, London WC2A 3PX, UK

Stephen A. Leppla
NIDCR NIH, Bldg 30, Room 316, Bethesda,
MD 20892-4350, USA

Camille Locht
Laboratoire de Microbiologie Genetique et
Moleculaire, INSERM U447, Institut Pasteur de Lille,
1 Rue du Professeur Calmette, F-59019 Lille Cedex,
France

Albrecht Ludwig
Theodor-Boveri-Institut (Biozentrum),
LS Mikrobiologie, Universität Würzburg,
D-97074 Würzburg, Germany

Jean-Christophe Marvaud
Unité des Interactions Bactéries Cellules,
Institut Pasteur, 28 Rue du Docteur Roux,
75724 Paris Cedex 15, France

Vega Masignani
IRIS, Chiron SpA, Via Fiorentina 1, I-53100 Sienna,
Italy

Gianfranco Menestrina
CNR-ITC Centro di Fisica degli Stati Aggregati,
Via Sommarive 14, I-38050 Povo, Trento, Italy

Timothy J. Mitchell
Division of Infection and Immunity, Joseph Black
Building, University of Glasgow, Glasgow G12 8QQ,
UK

Steven R. Monday
Department of Microbiology, Molecular Biology and
Biochemistry, University of Idaho, Moscow, ID 83844,
USA

Cesare Montecucco
Dipartimento di Scienzie Biomediche, Università di
Padova, Via G. Colombo 3, I-35121 Padova, Italy

Michael Moos
Institut für Med. Mikrobiologie und Hygiene,
Verfügungsgebäude für Forschung und Entwicklung,
Obere Zahlbacher Strasse 63, D-55101 Mainz,
Germany

Heide Müller-Alouf
Institut Pasteur, 1 Rue du Professeur A. Calmette –
BP 245, 59019 Lille Cedex, France

John R. Murphy
Section of Biomolecular Medicine, Boston University
School of Medicine, Boston, MA 02118-2340, USA

G. Balakrish Nair
National Institute of Cholera and Enteric
Diseases P-33, CIT Road, Scheme XM, Beliaghata,
P.O. Box 177, Calcutta 700 010, India

Sjur Olsnes
Department of Biochemistry, Institute of Cancer
Research, The Norwegian Radium Hospital,
Montebello, N-0310 Oslo, Norway

Michael Palmer
Institute for Medical Microbiology and Hygiene,
Hocchaus Augustusplatz, D-55101 Mainz, Germany

Emanuele Papini
Dipartimento di Scienze Biomediche ed Oncologia
Umana, Università di Bari, 70124 Bari, Italy

Michael W. Parker
St Vincent's Institute of Medical Research, 41 Victoria
Parade, Fitzroy, Victoria 3065, Australia

Yves Piémont
Institut de Bactériologie, 3 Rue Koeberlé,
67000 Strasbourg, France

Mariagrazia Pizza
IRIS, Chiron, SpA, Via Fiorentina 1, I-53100 Sienna,
Italy

Michel R. Popoff
Unité des Interactions Bactéries Cellules,
Institut Pasteur, 28 Rue du Docteur Roux,
75724 Paris Cedex 15, France

Gilles Prévost
LTAB–UPRES EA 1318, Institut de Bactériologie de la
Faculté de Médicine, Université Louis Pasteur,
Hopitaux Universitaires de Strasbourg,
3 rue Koeberlé, F-67000 Strasbourg, France

Thandavarayan Ramamurthy
National Institute of Cholera and Enteric
Diseases P-33, CIT Road, Scheme XM, Beliaghata,
P.O. Box 177, Calcutta 700 010, India

Rino Rappuoli
IRIS, Chiron SpA, Via Fiorentina 1, I-53100 Siena,
Italy

Julian I. Rood
Department of Microbiology, Monash University,
Clayton, Victoria 3168, Australia

Jamie Rossjohn
St Vincent's Institute of Medical Research, 41 Victoria
Parade, Fitzroy, Victoria 3065, Australia

Giampietro Schiavo
Molecular NeuroPathobiology Laboratory,
Imperial Cancer Research Fund, 44 Lincoln's
Inn Fields, Room 614-5, London WC2A 3PX,
UK

Cynthia L. Sears
Johns Hopkins University School of Medicine, Ross
Building Room 933, 720 Rutland Avenue, Baltimore,
MD 21205, USA

Bret R. Sellman
Microbiology and Molecular Genetics, Harvard
Medical School, Boston, MA 02115, USA

Sumio Shinoda
Faculty of Pharmaceutical Sciences, Okayama
University, Tsushima-Naka, Okayama 700–8530,
Japan

Michel Simonet
Equipe Mixte Inserm E99-19 Université JE 2225,
Departement de Pathogènése des Maladies
Infectieuses et Parasitaires, Institut de Biologie de
Lille, 1 rue du Professeur Calmette, 13P–245-59021,
Lille Cedex, France

Dennis L. Stevens
Veterans Affairs Medical Center, Infectious Diseases
Section (Bldg 45), 500 West Fort Street, Boise,
ID 83702-4598, USA

Ann-Mari Svennerholm
Department of Medical Microbiology and
Immunology, Göteborg University, S-41346 Göteborg,
Sweden

Tae Takeda
Department of Infectious Diseases Research
National, Children's Medical Research Center,
Taishido 3-35-31, Setagaya Ku, Tokyo-154, Japan

Jean-Claude Talbot
Centre de Recherche Paul Pascal–CNRS, Avenue A.
Schweitzer, 33600 Pessac, France

John L. Telford
IRIS, Chiron SpA, I-53100 Sienna, Italy

Monica Thelestam
Microbiology and Tumorbiology Center, Karolinska
Institute, Box 280, S-17177 Stockholm, Sweden

Richard W. Titball
Defence Evaluation and Research Agency, CBD,
Porton Down, Salisbury SP4 0JQ, UK

Rodney K. Tweten
The University of Oklahoma Health Sciences Center,
Microbiology and Immunology, BMSB-1053, 940 S.L.
Young Blvd, Oklahoma City, OK 73104, USA

Johanna vanderSpek
Section of Biomolecular Medicine, Boston University
School of Medicine, Boston, MA 02118-2340, USA

Beatrix Vécsey Semjén
CNR-ITC Centro di Fisica degli Stati Aggregati,
Via Sommarive 14, I-38050 Povo, Trento, Italy

Christoph von Eichel-Streiber
Institut für Med. Mikrobiologie und Hygiene,
Verfügungsgebäude für Forschung und Entwicklung,
Obere Zahlbacher Straße 63, D-55101 Mainz,
Germany

Siegfried Weiss
Department of Cell Biology and Immunology,
National Center of Biotechnology, D-38124
Braunschweig, Germany

Jørgen Wesche
Department of Biochemistry, Institute of Cancer
Research, The Norwegian Radium Hospital,
Montebello, N-0310 Oslo, Norway

Ken-ichi Yoshino
Department of Infectious Diseases Research,
National Children's Medical Research Center,
Taishido 3-35-31, Setagaya Ku, Tokyo-154, Japan

Preface to Second Edition

It is now eight years since the publication of the first edition of the *Sourcebook of Bacterial Protein Toxins*, and during that period many previously well-recognized toxins now constitute the prototypes of toxin families which share common mechanisms and, most probably, common evolutionary origins (e.g. the RTX toxins, STs and LTs of enterobacteria). Of great significance in this context was the recent recognition of pathogenicity islands which are responsible for the mobility and horizontal spread of toxins and other virulence genes between closely related species. Further remarkable progress has been made in the definition of molecular mechanisms of a wide range of toxins, with increasing numbers having enzymic mechanisms revealed, including ADP-ribosylation and glycosylation of novel targets (e.g. small Ras G-proteins). Among the most exciting discoveries of the 1990s, simply because it explained the potency of botulinum and tetanus toxins which have challenged toxinologists for over 100 years, was the demonstration that they consist of Zn-proteinases of exquisite specificity, cleaving proteins associated with vesicle fusion at presynaptic membranes. We now have the three-dimensional structure of botulinum toxin, which should yield further detailed information on the functional domains of these fascinating molecules.

The host immune system is not only the primary defence against colonization, and sometimes invasion, by toxigenic bacteria but it also constitutes the major target for a growing repertoire of bacterial toxins which can act either directly by cytoxicity towards immune effector cells (e.g. the leucocidins of *Staphylococci* and the leucotoxins of *Pasteurellae*) or by more subtle routes involving deregulation of cytokine production (e.g. the superantigenic toxins of *Staphylococci* and *Streptococci*).

Genome analysis promises to identify many more toxins over the next decade and should provide rich pastures for functional analysis in the next century. These analyses undoubtedly offer new opportunities for use as tools to dissect cellular processes as well as novel potential therapeutic agents.

A volume such as the *Sourcebook*, although attempting to cover the major groups of toxins at a relatively detailed level, inevitably has some gaps, and if these occur in your favourite area of research, then we can only apologize. The content of the book always reflects a compromise between what the editors would like and what they are able to include, yet still meet the deadlines set for the production of the book. We are greatly indebted to all the contributors, especially to those who met the first submission deadline. We wish to record our thanks to Tessa Picknett at Academic Press, who initiated the second edition, and to Lilian Leung who saw the project through to a successful conclusion with persistence and good humour. We also thank Mrs Patricia Paul for invaluable secretarial work in compiling chapters and correspondence with authors.

Joseph E. Alouf and John H. Freer

Preface to First Edition

Great strides have been made in the depth of our understanding of the structure and mechanisms of action of bacterial toxins over the last decade. The current pace of this advance in knowledge is particularly impressive, and results largely from the power that gene manipulation techniques have offered in experimental biology.

Recent research achievements in the field of bacterial toxins, which consist of about 240 protein toxins as well as a relatively small number of non-protein toxins, reflect the extensive and productive blending of disciplines such as molecular genetics, protein chemistry and crystallography, immunology, neurobiology, pharmacology and biophysics. Furthermore, the exciting developments in many areas of cell biology, and particularly in membrane-associated mechanisms relating to signalling and communication, export and import of proteins and to cytoskeletal functions, have been facilitated because critical steps in these processes constitute the targets for bacterial toxins. Thus, we have toxins available which can be used to probe many fundamental aspects of eukaryotic cell biology.

Disruption of these same central cellular processes *in vivo* can also be the critical event in the pathogenesis of infectious diseases for man or domestic animals. Many such infectious diseases have major social or economic impacts on man, and such considerations have quickened the pace of the search for therapeutic agents. Currently, a number of physically inactivated bacterial or hybrid engineered toxoids are used as immunogens in vaccination programmes, and there is a major international effort to develop new and more effective vaccines based on our deeper understanding of the molecular events in pathogenesis and the host response to infection.

Since the publication of the excellent multi-volume treatise on *Bacterial Toxins* edited by S. Ajl, S. Kadis and T. Montie (Academic Press) in the early 1970s, most of the books published in the past twenty years have covered the subject by presenting individual toxins or groups of toxins in separate chapters. This is not the main approach followed in this book. Our aim is not to give an exhaustive review of the wide spectrum of the protein toxin repertoire but rather to give an 'in depth' critical review of the original and the newly expanding body of information accumulated during the past decade or so. The multifaceted aspects of toxin research and the multidisciplinary approaches adopted suggested to us that 'state of the art' toxin research might best be presented by putting together in several chapters the common structural and/or functional aspects of toxin 'families'. Other chapters highlight the various physiological or genetic mechanisms regulating toxin expression and the therapeutic or vaccine applications of genetically engineered toxins.

The 22 chapters of this book have been written by 44 internationally known specialists who have significantly contributed to the progress in the domains covered. It is hoped that this book will appeal to a wide readership, including microbiologists, biochemists, cell biologists and physicians. Also, we hope it will arouse the interest of students and scientists in other disciplines who see the power of these fascinating biological agents, either as exquisitely specific probes of cellular processes or as extremely potent agents of infectious disease.

Finally, we would like to thank all the authors for their contributions, and particularly to those who delivered their manuscripts by the first deadline. We also express our appreciation to the editorial staff at Academic Press for their help and patience throughout the preparation of this book.

J.E. Alouf and J.H. Freer

INTRODUCTORY SECTION

1

Plasmids, phages and pathogenicity islands: lessons on the evolution of bacterial toxins

Ulrich Dobrindt and Jörg Hacker

INTRODUCTION: THE GENOMIC STRUCTURE OF PROKARYOTES

The advent of new molecular biological techniques and the speed at which sequence information of entire genomes has become available have shed light on many new aspects of microbiology. Knowledge of the sequences and their organization on the genome has proven particularly useful for comparative studies of different bacterial species and the evolutionary processes that drive their differentiation.

A compilation of information available from the literature shows that a striking correlation exists between the distribution of bacterial toxins and moveable genetic elements in bacteria. Indeed, a great number of bacterial toxins is structurally or functionally related to plasmids, phages or pathogenicity islands. In this review, we will attempt to demonstrate that there is a structural and functional interdependence between bacterial protein toxins and the mechanisms of genetic exchange of microbial pathogenesis.

Most prokaryotes contain a single circular 'core' chromosome, which harbours essential housekeeping genes and which is not transferable *per se*. The arrangement of genes on the chromosome is characterized by a frequently clustered organization or a close link between functionally related genes. Genes located on the 'core' chromosome exhibit a relatively homogeneous G + C content and a specific codon usage. The

genomic organization of closely related bacteria is very similar (Holm, 1986; Andersson and Kurland, 1990; Sharp, 1991). The structure of the 'core' genome, however, is often interrupted by the presence of DNA stretches harbouring genes with a G + C content and a codon usage that differs from those of the 'core' genome. Such regions, which we term 'genomic islands', may code for functions important for the survival of bacteria under specific conditions. Genomic islands, which comprise toxin-specific genes or other virulence determinants, have been termed 'pathogenicity islands' (Pais). Genomic islands of non-pathogenic bacteria may also code for gene products involved in symbiosis ('symbiosis islands') (Sullivan and Ronson, 1998), metabolism, drug resistance, secretion, degradation of xenobiotics or other important processes (Hacker *et al.*, 1997). It can be concluded, therefore, that the modular composition of the bacterial genome is a common feature of prokaryotes not restricted to pathogenic organisms. Pathogenicity islands, however, belong to the best-known genomic islands of prokaryotes and are therefore considered as paradigms for genomic structures in bacteria.

Bacterial genomes also include repetitive extragenic palindromic sequences (Higgins *et al.*, 1988) and short interspersed repetitive sequence elements (Lupski and Weinstock, 1992). Such elements presumably influence gene regulation, chromosome structure and chromosomal rearrangements. Other extragenic components have been described, such as the transposable insertion sequence (IS) elements and transposons (Deonier, 1996). They are restricted to moving themselves, and sometimes additional sequences, by recombination events from one site of their genome to other sites of the same genome (Berg, 1989). In addition, many prokaryotes can contain genetic information on plasmids that are extrachromosomal DNA elements with the ability of autonomous replication. The genetic information encoded by plasmids can contain valuable genes, which may be advantageous under certain conditions, such as toxin genes, but also resistance determinants and genes coding for specific metabolic properties. Many plasmids can be transferred from cell to cell. Some have the ability to become integrated into the chromosome, thereby losing the control of their own replication (Hardy, 1986). Other genetic elements frequently found in association with the genome are integrons, retrophages and lysogenic bacteriophages (Berg, 1989; Inouye and Inouye, 1991; Hall and Collis, 1995; Travisano and Inouye, 1995; Jones *et al.*, 1997). They are able to integrate into various sites of the host cell chromosome as well. The excision of the phage genome out of the bacterial chromosome during the lytic cycle can be accompanied by the transfer of chromosomal sequences. Interestingly, toxin-specific genes are often located on bacteriophage genomes.

With respect to the above-mentioned structure of bacterial chromosomes and the different classes of mobile elements frequently found in prokaryotic genomes, it can be concluded that the genome of prokaryotes constantly undergoes structural variations due to the potential of mobile elements to integrate into different sites on the chromosome and to promote chromosomal rearrangements via recombinational mechanisms. Genetic information can be lost by deletion or can be acquired by horizontal gene transfer from other species. These genetic variations result in phenotypic changes of the organisms and in the selection of new variants. This is an ongoing evolutionary process, which also affects the existence, expression and distribution of virulence-associated genes, including toxin determinants. We will attempt to demonstrate in the following chapter that genes encoding bacterial protein toxins are constantly exposed to the possibility of genetic rearrangements and that these mechanisms belong to the driving forces of the evolution of microbial pathogens.

PROTEIN TOXINS ENCODED BY MOBILE GENETIC ELEMENTS

Protein toxins encoded by plasmids

Many bacterial pathogens harbour plasmids, which carry protein toxin determinants. They frequently contribute to specific combinations of virulence factors, which argues for a coevolution of specific factors in different pathotypes. Some of these plasmids integrate into the chromosome (Zagaglia *et al.*, 1991; Colonna *et al.*, 1995). Important plasmid-encoded toxins are given in Table 1.1.

Gram-negative bacteria

Intestinal *Escherichia coli* bacteria may cause different types of diarrhoeal diseases. The differences in the clinical pictures reflect the different pathotypes of intestinal *E. coli*, such as enterotoxigenic *E. coli* (ETEC), enteropathogenic *E. coli* (EPEC) and others (see Table 1.1). A main feature of the different intestinal *E. coli* pathotypes is the presence of pathotype-specific plasmids, which often encode protein toxins. ETEC strains cause diarrhoea through the action of two different plasmid-encoded types of enterotoxins, the heat-labile enterotoxin (LT) and the heat-stable enterotoxin (ST) (see Chapters 6 and 29). These strains usually contain the determinant for a LT only, a ST only, or both toxin types (Gyles *et al.*, 1974; Levine, 1987; Gyles, 1992). Heat-labile toxins (LTs) are closely related to the cholera toxin of *Vibrio cholerae* (Spangler, 1992; Sixma *et al.*, 1993). Two unrelated classes of plasmid-encoded heat-stable toxins without sequence homology exist (STa and STb). The STb encoding gene (*estB*) is found on heterogeneous plasmids, which may also contain other properties (Harnett and Gyles, 1985; Echeverria *et al.*, 1985; Dubreuil, 1997).

The low-molecular-mass ST called EAST1 of enteroaggregative *E. coli* (EAEC) strains (Savarino *et al.*, 1991, 1993) exhibits 50% protein identity to STa and is encoded on so-called EAF-plasmids, which range from 50 to 70 MDa in size (Nataro *et al.*, 1987). This enterotoxin can also be expressed in addition to STa from some ETEC, EHEC and many EPEC strains (Nataro *et al.*, 1987; Savarino *et al.*, 1996; Yamamoto and Echeverria, 1996).

Enteroinvasive *E. coli* produce an enterotoxin that is also expressed by *Shigella flexneri*. Therefore, it has been designated ShET2 (*Shigella* enterotoxin 2) as well. The toxin-encoding gene (*sen*) is located on the 140-MDa pInv-plasmid (Nataro *et al.*, 1995; Sears and Kaper, 1996).

ETEC strains and extraintestinal *E. coli* from animals often carry the structural gene for CNF2 (*cnf2*) on the F-like Vir-plasmid (Oswald and De Rycke, 1990; Oswald *et al.*, 1994). The Vir-plasmid can also contain sequences that are homologous to adhesins as well as cytolethal distending toxin (CDT) determinants (Mainil *et al.*, 1997; Peres *et al.*, 1997). Interestingly, *cnf1*, which is a homologous gene of *cnf2*, is located on Pais of uropathogenic *E. coli* strains (Blum *et al.*, 1995; Donnenberg and Welch, 1995; Swenson *et al.*, 1996).

TABLE 1.1 Toxins encoded by bacterial plasmids

Organism	Pathotype[a]	Plasmid-encoded property	Gene symbol	Other plasmid-encoded factors
Escherichia coli	ETEC	Heat-labile enterotoxin (LT)	*elt, etx*	ST, CFAs, drug resistance colicins
	ETEC	Heat-stable enterotoxin (ST)[b]	*est*	ST, LT, CFAs, drug resistance, colicins
	EAEC, ETEC, EHEC, EPEC	EAST1	*ast*	Type-IV adhesin 'bundle forming pili'
	Extraintestinal *E. coli*	Alpha-haemolysin (Hly)	*hly*	–
	EHEC	Enterohaemolysin (Ehx)	*ehx*	–
	Extraintestinal *E. coli*, ETEC	Cytotoxic necrotizing factor 2 (CNF2)	*cnf2*	F17-, AFA/Dr adhesins, CDT
	EIEC	EIEC enterotoxin (=*Shigella* enterotoxin 2)	*sen*	Genes required for invasion, Ipa proteins, type-III secretion system
Yersinia spp.		Yop proteins	*yop*	Genes required for invasion, Yop proteins, type-III secretion system
Shigella spp.		Ipa proteins	*ipa*	
		Shigella enterotoxin 2	*sen*	Genes required for invasion, Ipa proteins, type-III secretion system
Salmonella spp.		Invasins	*spv*	Genes required for intracellular growth
Enterococcus faecalis		Cytolysin	*cyl*	–
Staphylococcus aureus		Enterotoxin type D	*entD*	Penicillin and cadmium resistance
		Exfoliative toxin B	*etb*	Cadmium resistance, Bacteriocin
Clostridium tetani		Tetanus neurotoxin (TeTx)	*tet*	–
Clostridium botulinum		Botulinum neurotoxin type G (BoNT/G)	*botG*	Bacteriocin (Boticin G)
Bacillus anthracis		Anthrax toxin (LF, EF, PA)	*lef, cya, pag*	–

[a]EAEC, enteroaggregative *E. coli*; EHEC, enterohaemorrhagic *E. coli*; EPEC, enteropathogenic *E. coli*; ETEC, enterotoxigenic *E. coli*.
[b]Genes encoding for STa (*estA*) and STb (*estB*) have been found on Tn*1681* (So and McCarthy, 1980) and Tn*4521* (Lee *et al.*, 1985; Hu *et al.*, 1987; Hu and Lee, 1988), respectively.

Extraintestinal *E. coli*, a major cause of urinary tract infections (UTI), sepsis and newborn meningitis (NBM), often carry Pais which contain the α-haemolysin gene cluster (*hly*) (Hacker and Hughes, 1985). Some *E. coli* strains contain *hly* determinants on plasmids that are heterogeneous in size (50–60 kb), conjugational behaviour and incompatibility group (De la Cruz *et al.*, 1980). The plasmid-encoded and chromosomally encoded *hly* determinants show a high overall homology (De la Cruz *et al.*, 1980; Müller *et al.*, 1983). Nevertheless, the sequences of the upstream regions differ significantly (Knapp *et al.*, 1985; Hess *et al.*, 1986). The enterohaemolysin determinant (*ehxABD*) of enterohaemorrhagic *E. coli* strains is encoded on plasmids (pO157) which range in size from 93 to 104 kb (Schmidt *et al.*, 1995, 1996). Although this toxin has a high overall similarity to α-haemolysin, the amino- and carboxy-termini of both proteins are different (Schmidt and Karch, 1996; Bauer and Welch, 1996).

Virulence plasmids of pathogenic *Yersinia*, *Shigella* and *Salmonella* species share the absence of genes encoding conventional protein toxins. The *Yersinia* and *Shigella* virulence plasmids code for a conserved type-III protein secretion system (Sasakawa *et al.*, 1989; Haddix and Straley, 1992; Venkatesan *et al.*, 1992; Sasakawa *et al.*, 1993; Allaoui *et al.*, 1995). All pathogenic species of *Yersinia* (*Y. pestis*, *Y. pseudotuberculosis* and *Y. enterocolitica*) harbour a virulence plasmid (pYV) of about 70 kb in size. This plasmid encodes the Yop virulon that organizes the secretion of several proteins termed Yops (*Yersinia* outer proteins) which are essential for virulence (Portnoy and Falkow, 1981; Galyov *et al.*, 1993; Straley *et al.*, 1993). Some of the Yops are considered as 'unconventional toxins', i.e. the cytotoxin YopE, YopH and YopO (Bliska *et al.*, 1992) as well as YopB, which shows sequence similarity to contact haemolysins, such as IpaB of *Shigella* species (Beuscher *et al.*, 1995).

All virulent *Shigellae* and enteroinvasive *E. coli* strains harbour a 220-kb pInv-plasmid which encodes all genes that are essential for epithelial cell invasion (Parsot and Sansonetti, 1996). The effectors of the invasion process, termed invasion plasmid antigens (Ipa), are encoded on this plasmid (Venkatesan *et al.*, 1988; Sasakawa *et al.*, 1989; Venkatesan *et al.*, 1991) as well as the invasin IpaB, which has homology to pore-forming toxins (Ménard *et al.*, 1994; Beuscher *et al.*, 1995). In addition, the invasion plasmid contains the *sen* gene encoding the *Shigella* enterotoxin 2, which is also called EIEC enterotoxin. The *sen* gene from *Shigella flexneri* 2a and EIEC share 99% identity (Nataro *et al.*, 1995).

Many clinical isolates of invasive as well as of non-typhoidal *S. typhimurium* and *S. enteritidis* contain virulence plasmids (Montenegro *et al.*, 1991; Fierer *et al.*, 1992; Gulig *et al.*, 1993). A highly conserved 8-kb region of these virulence plasmids, which carries the *spv* genes, is responsible for mediating systemic infections and increases the growth rate of *Salmonella* in mice (Gulig *et al.*, 1992; Gulig and Doyle, 1993).

Gram-positive bacteria

The *Clostridium tetani* structural gene encoding the tetanus neurotoxin (TeTx) is located on a 75-kb plasmid (pCL1) (Laird *et al.*, 1980; Finn *et al.*, 1984). All *Clostridium botulinum* type-G strains investigated contain an 81-MDa plasmid (Strom *et al.*, 1984), which presumably encodes either the structural gene (*botG*) of the botulinum neurotoxin type-G (BoNT/G) or regulatory genes important for BoNT/G expression (Eklund *et al.*, 1988; Minton, 1995).

Virulent *Bacillus anthracis* strains contain two large plasmids, pX01 (170–185 kb) and pX02 (90–95 kb), which are both required for virulence (Kaspar and Robertson, 1987). The genes for the three-component anthrax exotoxin are located on pX01 (Mikesell *et al.*, 1983; Thorne, 1985). The plasmid pX02 contains the genes for the expression of another virulence factor of

B. anthracis, the D-glutamic acid-composed capsule (Green *et al.*, 1985; Uchida *et al.*, 1985).

Cytolytic strains of *Enterococcus faecalis* produce a cytolysin, which shows a homology to lantibiotics. The cytolysin determinant (*cyl*) consists of six tandemly arranged genes. These are encoded in one operon on large (*ca.* 60 kb), transmissible and pheromone-responsive plasmids (Jett *et al.*, 1994; Gilmore *et al.*, 1994). Cytolysin determinants have also been located on the chromosome of *Enterococcus* (Ike and Clewell, 1992).

The structural gene (*entD*) coding for the staphylococcal enterotoxin D (SED) has been located on the 27.6-kb penicillinase plasmid pIB485 (Bayles and Iandolo, 1989). The two exfoliative toxin determinants responsible for the staphylococcal scalded skin syndrome have been assigned to different genetic loci. The gene for exfoliative toxin A (*eta*) has been mapped on the chromosome and that for the exfoliative toxin B (*etb*) on the plasmid pRW0019 (Warren *et al.*, 1975; Jackson and Iandolo, 1985).

Protein toxins encoded by bacteriophages

Bacteriophages encode a variety of toxin genes of pathogenic bacteria. The toxin genes are frequently located next to the bacteriophage attachment site, which argues for acquisition by a transduction mechanism. The bacteriophage-encoded toxins are given in Table 1.2.

Gram-negative bacteria

Full virulence of *Vibrio cholerae* depends on two co-ordinately expressed factors: the cholera toxin (CT) and toxin-co-regulated pili (TCP). The CT encoding genes (*ctxAB*) are located on the CTX element, which ranges from 7 to 9.7 kb in size and occurs frequently in multiple tandemly arranged copies. This element has been identified as a filamentous bacteriophage designated CTXφ (Mekalanos, 1983; Waldor and Mekalanos, 1996)

TABLE 1.2 Toxins encoded by bacteriophages

Host organism	Bacteriophage-encoded toxin	Gene symbol	Phage designation
Vibrio cholerae	Cholera toxin (CT)	*ctx*	CTXφ
E. coli (EHEC)	Shiga toxin (Stx)	*stx*	H19, 933
Pseudomonas aeruginosa	Cytotoxin (CTX)	*ctx*	φCTX[a]
Corynebacterium diphtheriae	Diphtheria toxin (DT)	*tox*	β, ω[b]
Clostridium botulinum	Botulinum neurotoxin type C1 and D (BoNT/C1, BoNT/D)	*botC, botD*	cI
Streptococcus pyogenes	Erythrogenic toxin type A and C (SPEA, SPEC)	*speA, speC*	SPE-phage, T12, CS112
Staphylococcus aureus	Enterotoxin A (SEA)	*entA*	PS42D
	Staphylokinase	*sak*	

[a]The chromosomal attachment site (*attB*) of φCTX has been mapped to the 3′-end of the tRNA^{Ser} (Hayashi *et al.*, 1993).
[b]The chromosomal attachment sites (*attB1* and *attB2*) of β and ω overlap with a duplicate tRNA gene encoding the tRNA$_2$^{Arg} (Ratti *et al.*, 1997).

and is restricted to toxigenic strains (Miller and Mekalanos, 1984; Kovach et al., 1996). Structurally, the CTX element resembles a compound transposon (Pearson et al., 1993). The core region of the CTX element encodes several toxins such as CT (ctxAB), zonula occludens toxin (zot) and the accessory cholera toxin (ace). The complete CTX element is self-transmissible and can replicate as a plasmid or lead to the production of extracellular virions (Waldor and Mekalanos, 1996). Transmission of CTXφ requires the expression of TCP pili. The tcp genes are located on a separate pathogenicity island (Kovach et al., 1996).

Shiga toxin (Stx, formerly called Shiga-like toxin or Vero toxin) is the major virulence factor of enterohaemorrhagic E. coli strains. Two immunologically non-cross-reactive groups of Stx can be distinguished (Stx1 and Stx2). One EHEC strain expresses Stx1 only, Stx2 only, both toxin types or multiple forms of Stx2 (O'Brien et al., 1992; Nataro and Kaper, 1998). The Stx1 of EHEC is essentially identical to Shiga toxin from Shigella dysenteriae (Takeda, 1995) and shows no significant sequence variation within the Stx1 group. In contrast, sequence variation does exist among the members of the Stx2 group (Jackson et al., 1987; O'Brien et al., 1992). The identically organized structural genes for Stx1 and Stx2 are located on lysogenic lambdoid phages, whereas those for Stx2v are encoded on the chromosome (Jackson et al., 1987; Marques et al., 1987; Rietra et al., 1989). Several variant Stx genes from human Shiga toxin-producing E. coli strains do not appear to be phage encoded, but they may be carried on defective bacteriophages (Paton et al., 1992, 1993). Some Citrobacter freundii and Enterobacter spp. strains produce a Stx2 toxin and contain a stx2 gene, which is homologous to that of E. coli (Schmidt et al., 1993).

Certain Pseudomonas aeruginosa strains produce a cytotoxin (Ctx). The corresponding structural gene (ctx) is carried by a temperate bacteriophage (φCTX) which is able to convert non-toxigenic into toxigenic strains. The chromosomal attachment site (attB) of φCTX has been mapped to the 3'-end of the tRNA^{Ser} (Hayashi et al., 1993).

Gram-positive bacteria

The determinant encoding diphtheria toxin (tox) is part of the genome of corynephage β, which converts non-toxigenic Corynebacterium diphtheriae strains into toxinogenic strains (Barksdale and Pappenheimer, 1954; Collier, 1982). Various different corynephages are known, including some non-converting phages. The corynephages ω and β form polylysogens by inserting into two different bacterial attachment sites, attB1 and attB2. As a result, the level of toxigenicity is increased as the tox gene dose also influences the toxin produc-

tion (Rappuoli and Ratti, 1984). Both attachment sites overlap with a duplicate tRNA gene encoding the tRNA$_2^{Arg}$ (Ratti et al., 1997).

Although many different C. botulinum strains carry bacteriophages, only in case of the structural genes of BoNT/C1 and BoNT/D (botC and botD) has it been proven that they are part of the genomes of bacteriophages (Inoue and Iida, 1970; Eklund et al., 1971).

Four different serological types of erythrogenic toxins (SPE types A, B, C and D) have been identified for Streptococcus pyogenes (Watson, 1960; McMillan et al., 1987). Depending on the strain and culture conditions, group A streptococci produce a single type of these toxins only, a combination thereof, or all of them simultaneously (Alouf et al., 1991). The structural genes of SPE type A and C (speA, speC) are phage encoded (Zabriskie, 1964; Colon-Whitt et al., 1979). SPE-converting phages produce the toxin in the lysogenic state, the virulent state (obligate lytic growth) or the pseudolysogenic state. In this case, their replication remains extrachromosomal (Johnson et al., 1980; McKane and Ferretti, 1981; Nida and Ferretti, 1982).

The staphylococcal bacteriophage PS42D carries the Staphylococcus aureus enterotoxin A (SEA) encoding gene entA and has a tendency to form defective lysogens. The S. aureus chromosome contains several bacterial attachment sites (attB) for the bacteriophage (Betley and Mekalanos, 1985). The structural gene (sak) for staphylokinase of S. aureus is encoded by a bacteriophage (Sako and Tsuchida, 1983).

Protein toxin genes and other mobile genetic elements

Plasmids and bacteriophages are elements that increase the genetic flexibility of bacteria. They contribute to the evolution of pathogens upon horizontal gene transfer followed by integration into the chromosome. The fact that toxin genes have the capacity to spread among bacterial strains, and even between species, is underlined by the occurrence of identical toxin determinants, or those with similar functions, on different genetic elements, such as chromosomes, phages and plasmids (see also Table 1.3). Thus, the E. coli heat-labile enterotoxin (LT) genes (elt, etx) are located on plasmids (Nataro and Kaper, 1998), while the related cholera toxin (CT) structural gene (ctxAB) of toxigenic V. cholerae strains is phage associated (Waldor and Mekalanos, 1996). A similar situation is reported for the Shiga toxin structural genes of S. dysenteriae (chromosomally encoded) (Hale, 1991) and E. coli (phage-encoded). The genes encoding α-haemolysin and CNF-toxins of pathogenic E. coli may be located

TABLE 1.3 The location of protein toxin encoding genes and their homologues[a]

Toxin	Location	Organism	Homologous toxin	Location	Organism
Cholera toxin (CT)	Phage	*V. cholerae*	Heat-labile enterotoxin (LT)	Plasmid, chromosome	*E. coli*
α-Haemolysin	Chromosome, Pai	*E. coli*	α-Haemolysin	Plasmid	*E. coli*
CNF 1	Chromosome, Pai	*E. coli*	CNF 2	Plasmid	*E. coli*
TeTx	Plasmid	*C. tetani*	BoNT/C, BoNT/D	Phage	*C. botulinum*
BoNT/G	Plasmid	*C. botulinum*	BoNT/A, BoNT/F	Chromosome?	*C. botulinum*
			BoNT/E	Plasmid, phage?	*C. butyricum*
SEA	Phage	*S. aureus*	SED	Plasmid	*S. aureus*

[a]Not all homologues are listed for each toxin.

either on plasmids (Waalwijk *et al.*, 1984; Hess *et al.*, 1986; Oswald *et al.*, 1994) or as parts of Pais on the chromosome (Knapp *et al.*, 1984, 1986; Falbo *et al.*, 1993; Blum *et al.*, 1995; Swenson *et al.*, 1996). Similar findings have been reported for clostridial neurotoxin genes (Hauser *et al.*, 1992; Whelan *et al.*, 1994; Minton, 1995) and for the enterotoxins of staphylococci (Altboum *et al.*, 1985; Betley and Mekalanos, 1985; Couch *et al.*, 1988; Johns and Khan, 1988; Bayles and Iandolo, 1989; Iandolo, 1989).

The location of identical or closely related toxin-encoding genes on different genetic elements raises the question of whether toxin determinants may be part of transposons or other genetic structures that have the capacity to jump between different genetic elements. A transposon location was only reported for the ST enterotoxin structural genes of *E. coli* that have been found on Tn*1681* (So and McCarthy, 1980) and on Tn*4521* (Lee *et al.*, 1985; Hu *et al.*, 1987; Hu and Lee, 1988). However, many toxin genes such as those encoding α-haemolysin, CNF1, LT enterotoxins, cholera toxin and others are located next to intact IS elements, which may form complex transposons similar to Tn*5* or Tn*10*. In a recent article, Mazel *et al.* (1998) described the presence of a distinct class of integrons in the genome of *V. cholerae*. The integron is composed of repeated sequences (VCRs) and genes encoding virulence factors, such as a lipoprotein and a haemagglutinin. Interestingly, the *sto* gene coding for the ST enterotoxin is also flanked by two VCR sequences, which raises the possibility that toxin genes may also be part of an integron structure with the consequence of transfer and expression in new host organisms.

TOXINS ENCODED BY PATHOGENICITY ISLANDS

Pathogenicity islands

As mentioned above, determinants encoding pathogenicity factors, including toxins and other properties (i.e. bacteriocins, antibiotic resistance factors, secretion systems and enzymes required for the degradation of xenobiotics) of pathogenic as well as of non-pathogenic bacteria, can be located on mobile genetic elements such as plasmids, transposons and bacteriophages (see Tables 1.1 and 1.2). In addition, such determinants have been mapped on chromosomes where they may be part of genomic islands. Virulence-associated genes are integral parts of the Pais (Hacker *et al.*, 1990; Blum *et al.*, 1994; Lee, 1996), which may carry genes coding for toxins, adhesins, invasins, secretion or iron-uptake systems (see Table 1.4). Pathogenicity islands represent rather complex genetic units, which usually share several characteristics (Hacker *et al.*, 1997):

(a) carriage of (often many) virulence genes
(b) presence in pathogenic strains, and absence or sporadic occurrence in less pathogenic strains of one species or a related species
(c) different G+C content and unusual codon usage in comparison with the host genome, indicating an acquisition by horizontal gene transfer
(d) occupation of large genomic regions (often >30 kb) preferentially, but not exclusively, on the chromosome
(e) compact, distinct genetic units, often flanked by direct repeats
(f) association with tRNA genes and/or IS elements at their boundaries
(g) presence of (often cryptic) 'mobility' genes (IS elements, integrases, transposases, origins of plasmid replication)
(h) instability.

Figure 1.1 shows two examples of prototypes of Pais located in the chromosome of the uropathogenic *E. coli* strain 536, which match all the features of Pais mentioned above. Obviously, not all DNA fragments known as Pais exhibit all eight characteristics. However, they should show most of them to be considered as a Pai.

TABLE 1.4 Virulence properties encoded by pathogenicity islands

Virulence property		Gene symbol	Organism
Toxins		*hly, cnfI, cnfII*	*E. coli* (UPEC)
		set1A, set1B	*Sh. flexneri 2a*
		ptx	*B. pertussis*
Adhesins		*sfa, pap, prf, prs*	*E. coli* (UPEC)
		eaeA	*E. coli* (EPEC, EHEC)
		tcpA	*V. cholerae*
Invasins		*tir, esp*	*E. coli* (EPEC, EHEC)
		sip	*S. typhimurium*
Secretion systems	Type-III system	(SPI-1: *inv, spa*)	*S. typhimurium*
		(SPI-2: *spi, ssa*)	
	Type-III system	(*sep*)	*E. coli* (EPEC, EHEC)
	Type-IV system	*virB, virD*	*H. pylori*
Iron uptake		*irp2, fyuA*	*E. coli* (UPEC)
		irp1, irp2, fyuA	*Yersinia* spp.

Pathogenicity island-encoded toxins

Enterobacteria

Pathogenicity islands have been described for many Gram-positive and Gram-negative bacteria. The main properties of Pais of Gram-negative bacteria are given in Table 1.5. The first Pais that encode α-haemolysin as a protein toxin were discovered in the genome of uropathogenic *E. coli* by Goebel, Hacker and co-workers. The uropathogenic *E. coli* strains 536, J96 and CFT073 carry Pais ranging from 50 to 190 kb in size (see Table 1.5). The Pais I and II of *E. coli* 536 both encode α-haemolysin. Additionally, Pai II carries the *hly* operon, which is associated with the *prf* determinant encoding P-related fimbriae. On Pai I, the up- and downstream regions of the *cnfI* determinant have been mapped, but not the *cnfI* structural gene itself (Hacker *et al.*, 1990; Blum

et al., 1994; Blum-Oehler *et al.*, 1998). In the uropathogenic *E. coli* strain J96, the Pais carry an *hly* determinant linked to a *pap* determinant, or a *prs* determinant linked to a *cnfI* determinant, respectively (Donnenberg and Welch, 1995; Swenson *et al.*, 1996). In both strains (536 and J96), the Pais are associated with tRNA genes. Pai I and II of *E. coli* 536 are located at the 3′-end of *selC* and *leuX*, respectively (Blum *et al.*, 1994; see also Figure 1.1), whereas the Pais of *E. coli* J96 are associated with the 3′-ends of *pheV* or *pheR*. The Pais I, II of *E. coli* 536 and one Pai of *E. coli* J96 delete at high frequency from the chromosome (Hacker *et al.*, 1990; Blum *et al.*, 1995; Swenson *et al.*, 1996). Recently, two additional Pais have been detected in the *E. coli* strain 536. Pai III contains an *sfa* determinant, whereas Pai IV contains the *fyuA-irp* gene cluster. Both Pais are also associated with minor tRNA encoding genes (see Table 1.5) (Blum-Oehler, unpublished).

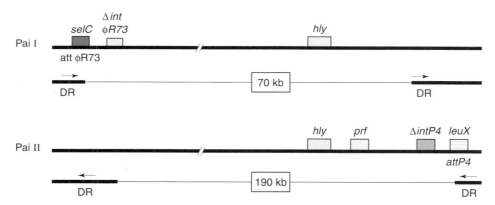

FIGURE 1.1 Diagram of the Pais I and II of the uropathogenic *E. coli* strain 536. Both Pais are flanked by 16 bp (Pai I) and 18 bp (Pai II) direct repeats (DRs). Upon deletion of the Pais, one copy of the DRs remains in the chromosome of the deletion strains. The thinner lines represent deleted Pai sequences. Abbreviations: *hly*, α-haemolysin determinant; *prf*, P-related fimbriae determinant; *intP4*, pseudo gene of the P4 integrase; *intA*, pseudo gene of the φR73 integrase gene; *selC*, gene encoding the tRNA^Sec; *att*φR73, attachment site of the φR73 phage; *leuX*, gene encoding the tRNA_5^Leu; *att*P4, attachment site of the P4 phage.

TABLE 1.5 Pathogenicity islands (Pais) and Pai-like structures of Gram-negative bacteria

Organism	Pai designation	Encoded toxins	Other Pai-encoded genes	Boundary sequences	Associated gene
E. coli 536 (UPEC)	Pai I	α-Haemolysin	–	DR 16 bp	selC
	Pai II	α-Haemolysin	prf	DR 18 bp	leuX
	Pai III	–	sfa	–	thrW
	Pai IV	–	fyuA, irp2	–	asnT
E. coli J96 (UPEC)	Pai IV	α-Haemolysin	pap		pheV
	Pai V	α-Haemolysin, CNF 1	prs	DR 135 bp	pheR
E. coli CFT073 (UPEC)	–	α-Haemolysin	pap	DR 9 bp	metV
E. coli K1	kps Pai	–	kps		pheV
E. coli E2348/69 (EPEC)	LEE	EspA, B, D ?	eaeA, sepA-I	–	selC
Y. pestis	HPI (pgm locus)	–	hms, HFRS, fyuA, irpB-D	IS100	–
Y. enterocolitica	HPI	–	fyuA, irp2 (irp1)	–	asnT
S. typhimurium	SPI-1	Sip/SspB, C, D ?	inf, spa, hil	–	–
	SPI-2	–	spi, ssa	–	valV
	SPI-3	–	mgt	–	selC
	SPI-4	–	Toxin secretion genes?	–	putative tRNA gene
S. flexneri 2a	she locus	Shigella enterotoxin 1	she	IS elements	–
V. cholerae	TCP–ACF element	–	acf, tcp, tox, int	att sites	ssrA
H. pylori	cag Pai	–	cagA-T	DR 31 bp	glr
D. nodosus	vap region	–	vapA-E	DR 19 bp	serV
	vrl region		vrl	–	ssrA
B. pertussis	ptx–ptl locus	Pertussis toxin (PTX)	ptl	–	tRNA(Asp)
B. fragilis	Pathogenicity islet	Fragilysin	orf1 (metalloprotease MP II encoding gene)	DR 12 bp	–

The pyelonephritogenic *E. coli* strain CFT073 contains a 58-kb Pai carrying an α-haemolysin determinant together with a *pap* operon. This Pai is flanked by direct repeats, which differ from those reported for the *E. coli* strains 536 and J96. The insertion site of this Pai is 75 bp downstream of the tRNA encoding gene *metV* (Kao *et al.*, 1997). In addition to the genes encoding α-haemolysin and P-fimbriae, the Pai of the strain CFT073 contains 44 open reading frames (ORFs). Interestingly, some ORFs show homology to iron-uptake systems as well as to ORFs already identified in the K-12 genome (Guyer *et al.*, 1998). The latter finding argues for a stepwise acquisition of Pai-DNA fragments.

As only extraintestinal *E. coli* strains carry capsule genes, it is expected that the corresponding genes are also a part of Pais. A neuroinvasive *E. coli* K1 strain indeed contains a 20-kb *kps* 'pathogenicity island' coding for genes that are required for the regulation, synthesis and export of the polysialic acid capsule components. This region is inserted next to a phenyl-alanine-tRNA gene, *pheV* (Cieslwicz and Vimr, 1997).

In enteropathogenic and enterohaemorrhagic *E. coli*, the 35.5-kb 'locus of enterocyte effacement' (LEE) forms a Pai and is sufficient to induce the 'attaching and effacing' (A/E) phenotype when transferred on a plasmid into *E. coli* K-12 (McDaniel and Kaper, 1997). This region contains no conventional toxin genes but genes encoding the adhesin intimin (*eaeA*), several secreted proteins essential for the A/E phenotype (*espADB*) and, thirdly, the *sep* genes encoding for a type-III secretion system (Jarvis *et al.*, 1995; Elliott *et al.*, 1998). In addition, the gene (*tir*) encoding the receptor required for intimate attachment to eukaryotic cells is encoded by LEE as well. The Tir protein is secreted by the type-III secretion system and concomitantly inserted into the eukaryotic membrane (Kenny *et al.*, 1997). Various *sep* genes exhibit a high homology with chromosomally or plasmid-encoded type-III secretion system components of other enteric pathogens such as *Salmonella*, *Yersinia* and *Shigella*. Sequence homologues of this region are also present in other Enterobacteriaceae that cause A/E lesions, such as several EHEC strains or diarrhoeagenic *Hafnia alvei*, and *Citrobacter rodentium* (McDaniel *et al.*, 1995; Nataro and Kaper, 1998). Interestingly, the LEE locus is inserted near the 3'-end of *selC* into the chromosome of the EPEC strain E2348/69 and in EHEC O157:H7, which is also the integration site for the retronphage φR73, the Pai I of the UPEC strain 536 as well as the SPI-3 of *S. typhimurium* (Blum *et al.*, 1994; McDaniel *et al.*, 1995; Blanc-Potard and Groisman, 1997). As indicated in Figure 1.2, *selC* serves as the integration site of several genetic elements in different organisms.

Yersinia pestis, *Y. pseudotuberculosis* serotype O1 and *Y. enterocolitica* biotype 1b contain an unstable chromosomal region termed a 'high pathogenicity island' (HPI), which is involved in iron acquisition by the siderophore yersiniabactin and required for the expression of virulence in a mouse model (Fetherston *et al.*, 1992). In *Y. pestis*, the entire 102-kb HPI region carries a haemin storage locus (*hms*), genes encoding proteins that are involved in the production of yersiniabactin (*irp1* and *irp2*), the *fyuA* gene coding for the yersiniabactin receptor, which cross-reacts with pesticin, but no protein toxin gene (Perry *et al.*, 1990; Heesemann *et al.*, 1993; Lucier *et al.*, 1996). The HPI locus of *Y. pestis* is flanked by single IS*100* insertion ele-

ments, which confers a great flexibility on this region. Some strains have lost the entire HPI upon homologous recombination between both IS*100* elements. Various strains of *Y. enterocolitica* and *Y. pseudotuberculosis* lack the *hms* locus but contain a 45-kb region including the *fyuA-irp* gene cluster (Heesemann *et al.*, 1993; Carniel *et al.*, 1996). The 45-kb HPI region of *Y. enterocolitica* strain Ye8081 is associated with the asparagine-specific tRNA gene *asnT* on one side and includes IS elements and other 'repeat sequences' (Carniel *et al.*, 1996).

Gene sequences nearly identical to the *irp2-fyuA* gene cluster of *Yersinia* spp. have been detected in several pathogenic species of Enterobacteriaceae, such as enteroaggregative (EAEC) and uropathogenic *E. coli* strains (Schubert *et al.*, 1998; see above). This underlines the spread of these Pais by horizontal gene transfer.

In *Salmonella typhimurium*, two chromosomal regions designated 'Salmonella pathogenicity islands' (SPI-1 and SPI-2) have been described (Mills *et al.*, 1995; Shea *et al.*, 1996). Both SPIs encode structurally similar but functionally distinct type-III secretion systems. The 40-kb SPI-1 is involved in epithelial cell entry and comprises at least 25 genes, which are predominantly coding for a type-III secretion system (*inv*, *spa*) and its effectors (Mills *et al.*, 1995; Galán, 1996). The organization, size and sequence of the *inv* and *spa* genes in *Salmonella* resemble a set of genes of the *Shigella* virulence

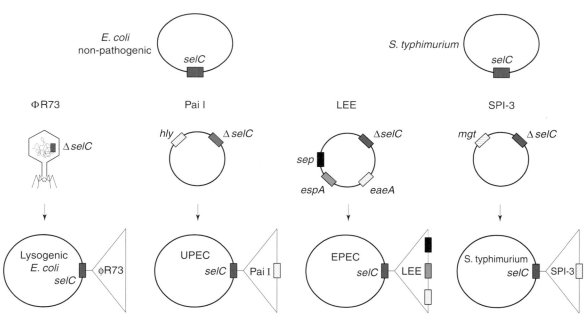

FIGURE 1.2 Integration site of Pais of EHEC, UPEC and *Salmonella* into *selC* illustrating the generation of different pathotypes upon the acquisition of Pais. The tRNA locus *selC* acts as a target sequence for the integration of the bacteriophage φR73; Pai I (UPEC) and LEE (EPEC and EHEC) as well as SPI-3 (*S. typhimurium*). Abbreviations: *selC*, gene encoding the tRNA^{Sec}; *hly*, α-haemolysin determinant; LEE, locus of enterocyte effacement; *sep*, secretion of *E. coli* protein genes; *espA*, enteropathogenic *E. coli* secreted protein A gene; *eaeA*, *E. coli* attaching and effacing gene; *mgt*, *S. typhimurium* magnesium transport gene.

plasmid. This implies that these sequences are transmissible (Groisman and Ochman, 1993). Moreover, homologous sequences to the Inv/Spa system have been detected in other enteric bacteria, such as *Salmonella, Yersinia, Erwinia, Xanthomonas* and *Pseudomonas* (Groisman and Ochman, 1997).

SPI-2 is about 40 kb in size and is important for intramacrophage survival (Ochman *et al.*, 1996; Shea *et al.*, 1996). This pathogenicity island is associated with a tRNA gene, *valV* (Hensel *et al.*, 1997). SPI-2 contains at least 17 genes encoding a two-component regulatory system and a type-III secretion system (Spi/Ssa system) which differs from the SPI-1 encoded Inv/Spa system (Ochman *et al.*, 1996; Hensel *et al.*, 1997). Several genes of the *ssa* operon have extensive similarity to *ysc* genes of *Yersinia* spp. In contrast, the SPI-2 is only present in the genus *Salmonella* (Groisman and Ochman, 1993). The *stpA* gene, located between the *sip* and *iag* loci on the *Salmonella typhi* chromosome, codes for a protein with high similarity to the *Yersinia enterocolitica* YopE and YopH proteins (Arricau *et al.*, 1997).

A 17-kb pathogenicity island (SPI-3), harbouring the *mgtBC* operon required for intramacrophage survival, is inserted near the *selC* gene of *S. typhimurium* chromosome (Blanc-Potard and Groisman, 1997) (see Figure 1.2).

Recently, a fourth *Salmonella*-Pai (SPI-4) has been identified in *Salmonella typhimurium*. This Pai is about 25 kb in size and comprises 18 putative open reading frames, some of which encode proteins that have homology to proteins involved in toxin secretion. SPI-4 is required for intramacrophage survival and is associated with a putative tRNA gene sequence downstream of the *ssb* gene (Wong *et al.*, 1998).

Another toxin-specific gene is located on a pathogenicity island of *Shigella flexneri*, termed *she* locus. The *she* locus of *S. flexneri* 2a codes for the protein ShMu, which has haemagglutinin and mucinase activities. ShMu shares homology with the virulence-related immunoglobulin A protease-like family of secreted proteins. Within the *she* locus, the two *Shigella* enterotoxin 1 encoding genes, *set1B* and *set1A*, are tandemly located in the opposite orientation. This element is part of a 51-kb deletable chromosomal element, which has been termed *she* Pai. The flanking regions of this Pai contain several IS elements, such as IS2, IS600 and a copy of IS629 that was disrupted by the insertion of a bacterial group-II intron (Rajamukar *et al.*, 1997).

Other Gram-negative bacteria

Vibrio cholerae strains O1 and O139 contain a 39.5-kb Pai (VPI) which is present in pathogenic but absent from non-pathogenic strains. The VPI comprises genes like *tcpA*, encoding an important adhesin and the receptor

for CTXϕ. The VPI may have originated from a bacteriophage as it is flanked by *att* sites and integrates near a 10Sa RNA, *ssrA* (Kovach *et al.*, 1996; Karaolis *et al.*, 1998). VPI includes a sequence, orf1 and its adjacent region, which exhibits homology to transposase-encoding genes of other bacteria, such as *Arthrobacter nicotinovorans, Vibrio anguillarum* or Tn903 of *E. coli* (Karaolis *et al.*, 1998). The fact that a cholera toxin-converting bacteriophage infection requires the VPI-encoded Tcp adhesin as a receptor suggests a co-evolution of the VPI and the CTXϕ-encoded *ctx* genes (Waldor and Mekalanos, 1996).

Helicobacter pylori strains can be divided into two families, depending on the production of a functional vacuolating cytotoxin, VacA, and of the cytotoxin-associated antigen CagA (type-I strains) or the lack thereof (type II) (Xiang *et al.*, 1995). Only type-I strains contain a 40-kb Pai with more than 40 potential ORFs, including the gene *cagA*, which codes for the cytotoxin-associated antigen A. This Pai is associated with the gene encoding the glutamate racemase (*glr*). The gene *vacA*, encoding the vacuolating cytotoxin, is not located on the Pai. It is present in both type-I and type-II strains, although active VacA is only produced in type-I strains (Atherton *et al.*, 1995; Covacci *et al.*, 1997). The presence of a functional Pai is responsible for the increased virulence of these strains. The *cag* Pai is flanked by 31-bp direct repeats and is disrupted in some strains by IS605. These structures seem to be involved in partial deletions of the Pai-generating strains with intermediate virulence (Censini *et al.*, 1996; Akopyants *et al.*, 1998). Some of the Pai-encoded proteins are related to similar molecules, which are part of type-IV secretion systems found in *Agrobacterium tumefaciens* (Vir), *Bordetella pertussis* (Ptl) involved in the transfer of pertussis toxin or tumour-inducing DNA into host eukaryotic cells. Ptl/Vir homologues encoded on conjugative plasmids of *E. coli* have a comparable function for the DNA transfer to recipient bacterial cells (Weiss *et al.*, 1993; Winans *et al.*, 1996; Christie, 1997). Therefore, it is speculated that the expression of genes located on the *cag* Pai is required for the delivery of proteins involved in virulence and host–cell interaction of *H. pylori* (Covacci *et al.*, 1997).

Strains of *Dichelobacter nodosus*, the principal causative agent of ovine foot-rot in sheep, can be grouped according to the severity of their virulence. This correlates with the presence of multiple copies of a DNA segment of 11.9 kb in size (*vap*) and that of another functionally unknown locus (*vrl*) (Cheetham *et al.*, 1995; Billington *et al.*, 1996a), both of which exhibit characteristics of Pais. *Dichelobacter nodosus* strain A198 contains three copies of the *vap* region. Regions 1 and 3 are adjacent on the chromosome, whereas region 2 is located elsewhere on the chromosome. The *vap* regions 1 and 2 contain several *vap* genes and an *intA* gene (Katz *et al.*, 1992; Cheetham *et al.*, 1995). Both regions are associated with tRNA genes, the latter being identified as *serV*. The tRNA gene next to region 1 differs from *serV*, indicating that both regions integrated independently into two different tRNA loci. Three genes located downstream of region 3, with similarities to genes found on mobile genetic elements, could indicate the integration of a mobile genetic element such as an integrated bacteriophage or a conjugative transposon (Cheetham *et al.*, 1995; Bloomfield *et al.*, 1997). A plasmid harbouring the *vap* genes has been isolated from a *D. nodosus* strain that is capable of integrating into the chromosome, resulting in Pai formation (Billington *et al.*, 1996b). The 27-kb *vrl* region has inserted into the chromosome at the 3'-end of a gene which is homologous to the *ssrA* gene of *E. coli* (Billington *et al.*, 1996a).

Bordetella pertussis, the causative agent of whooping cough, secretes several toxins implicated in this disease. The major virulence factor produced by *B. pertussis* is the pertussis toxin PTX (Tamura *et al.*, 1982). The five genes encoding the different subunits of the toxin (*ptx*) and the nine accessory genes required for the PTX secretion system (*ptl*) are organized as a polycistronic operon (Locht and Keith, 1986; Weiss *et al.*, 1993; Farizo *et al.*, 1996), which has some characteristics of a pathogenicity island. In *B. pertussis*, the upstream region of the *ptx–ptl* locus contains a truncated IS element similar to IS*481* of *B. pertussis*. The last gene of *ptx–ptl* locus (*ptlH*) is followed by a tRNA gene coding for an asparagine-specific tRNA and by an inverted repeat. When the chromosomal insertion point of the *ptx–ptl* locus of *B. pertussis* is compared with that of *B. parapertussis* and *B. bronchiseptica*, this site appears to be different in *B. pertussis* from those in *B. parapertussis* and *B. bronchiseptica* (Antoine *et al.*, 1998).

Enterotoxigenic strains of *Bacteroides fragilis* produce a metalloprotease toxin designated fragilysin. The encoding gene is located on a DNA region termed a 'fragilysin pathogenicity islet' (see below), which contains an ORF coding for a second metalloprotease (MPII), and another ORF, which encodes a protein with homology to a snake cytotoxin. The fragilysin pathogenicity islet is 6033 bp in size, contains nearly perfect 12-bp direct repeats near both ends of the islet and is not present in non-toxigenic strains of *B. fragilis* (Moncrief *et al.*, 1998).

A Pai-like region has been suggested for *Neisseria meningitidis*, which comprises capsule-related genes. Representational difference analysis revealed that the 'region 1' is specific for *N. meningitidis* and not for *N. gonorrhoeae* (Tinsley and Nassif, 1996).

Gram-positive bacteria

The majority of the toxins encoded by Pais has been described for Gram-negative bacteria. In the last few years, however, an increasing number of Gram-positive bacteria with Pais has been detected (see Table 1.6).

The toxic shock syndrome toxin 1 (TSST-1) of *Staphylococcus aureus* is encoded by a gene (*tst*) which is carried by related pathogenicity islands designated SaPI1 and SaPI2 of the *S. aureus* strains RN4282 and RN3984, respectively. These genetic elements are absent in TSST-1-negative strains of *S. aureus* and can be mobilized by staphylococcal phages. SaPI1 is 15.2 kb in size and is flanked by 17 bp direct repeats. SaPI1 integrates site- and orientation-specific into the chromosome. This Pai also contains a gene with homology to another superantigen toxin (*ent*), a *D. nodosus vapE* homologue and a putative integrase gene (Lindsay *et al.*, 1998).

The pathogenic species of *Listeria* (*L. monocytogenes*, *L. ivanovii*) as well as the non-pathogenic *L. seeligeri* harbour a 10-kb virulence cluster containing the listeriolysin gene (*hly*), flanked on one side by the *plcA–prfA* region and on the other side by the lecithinase operon, which is required for intercellular spread during infection (Portnoy *et al.*, 1992; Gouin *et al.*, 1994). In *L. seeligeri*, this gene cluster is not properly expressed, owing to a reduced transcription of *prfA*, which encodes the transcriptional activator of *Listeria* virulence genes. The same genes that are located at both ends of the virulence cluster of *Listeria* (*ldh*, *prs*) are present in avirulent strains but, instead of the virulence gene cluster, housekeeping genes are located between

TABLE 1.6 Pathogenicity islands of gram-positive bacteria

Organism	Pai designation	Encoded toxins	Other Pai-encoded genes	Boundary sequences	Associated gene
S. aureus	SaPI 1 SaPI 2	Toxic shock syndrome toxin 1	*ent*, *int*, ORF11	DR 17 bp	–
L. monocytogenes	Virulence cluster	Listeriolysin	*prfA*, *plcA*, *mpl*, *actA*, *plcB*	–	–
L. ivanovii	*i-inlC* gene cluster	–	–	–	tRNA(Thr)
C. difficile	Pathogenicity locus	Enterotoxin Cytotoxin	*tcdC-E*	–	–

ldh and *prs* (J. Kreft, personal communication). Another putative pathogenicity region of *Listeria* has recently been described. The *inlC* gene of *L. monocytogenes* and the *i-inlC* gene of *L. ivanovii* show high sequence similarities, but they differ with respect to their chromosomal localization. Whereas *inlC* of *L. monocytogenes* is a monocistronic gene, located between two housekeeping genes (*rplS* and *infC*), the *i-inlC* of *L. ivanovii* is associated with another small internalin gene (*i-inlD*) and has been inserted into a minor threonine-specific tRNA as part of a larger DNA fragment. This tRNA belongs to a rRNA/tRNA locus with a similar organization to the *rrnB* locus of *B. subtilis* (Engelbrecht *et al.*, 1998). The *i-inlC* gene cluster shares some features with Pais.

The 19-kb *Clostridium difficile* pathogenicity locus (PaLoc), which exhibits several features of a Pai, is specific for virulent strains of *C. difficile* and comprises the genes encoding the enterotoxin (*tcdA*), and the cytotoxin (*tcdB*) as well as the accessory genes *tcdC-E* (Braun *et al.*, 1996). The integration site of the PaLoc was defined as a 115-bp fragment, which forms a 20-bp hairpin loop. This stretch is found only in non-toxinogenic strains and is replaced by the PaLoc in toxinogenic strains (Braun *et al.*, 1996).

Pathogenicity islets

Pathogenicity islands represent complex pieces of DNA, which exhibit different features. The genomes of Gram-positive as well as Gram-negative bacteria often contain small DNA fragments (up to 5–7 kb) with a different G + C content compared with the host genome and which are quite often flanked by direct repeats. This is an indication for an acquisition by horizontal gene transfer. According to B. Finlay, such regions, if present in pathogenic but absent in apathogenic strains and if encoding virulence factors, were termed 'pathogenicity islets' (Stein *et al.*, 1996). Putative 'pathogenicity islets' have already been described in the sections on pathogenicity islands and pathogenicity island-encoded toxins. In this section, we will describe some of these 'islets', often toxin-encoding, of a selected number of pathogens.

Two regions with a G + C content differing from that of the average *Salmonella* genome have been localized on the *S. enterica* sv. Typhimurium chromosome at 25 and 27 min, respectively. They are restricted to *Salmonella* spp. and required for virulence. The 1.6-kb *sifA* locus is inserted into the chromosome at 25 min between the *potB* and *potC* genes (Stein *et al.*, 1996). In contrast to the *pot* genes, *sifA* has no sequence homologues among non-pathogenic bacteria. The *sifA* segment is flanked by 14-bp direct repeats. This may indicate that acquisition of this *sifA* segment occurred by site-specific

recombination (Stein *et al.*, 1996). The region located at 27 min comprises two virulence genes, *msgA* and *pagC*, which are required for intramacrophage survival (Gunn *et al.*, 1995). The PagC outer membrane protein exhibits sequence similarities to several enterobacterial proteins including the *Yersinia* Ail protein (Pulkkinen and Miller, 1991).

Several other examples of chromosomal regions with characteristics of pathogenicity islets have been described. They will be mentioned briefly, although no link to encoded protein toxins exists. *Erwinia amylovora*, a plant pathogen, carries a 6.6-kb disease-specific region (*dspEF*), which is required for pathogenicity next to the *hrp* gene cluster. Both proteins encoded by *dspEF* show homology to proteins expressed in *Pseudomonas syringae* pv. tomato. DspF resembles chaperones for virulence factors secreted by type-III secretion systems of animal pathogens (Bogdanove *et al.*, 1998). *Haemophilus influenza* type b can catabolize tryptophan. This ability is more frequently found among pathogenic *H. influenza* strains and is linked to the possession of the tryptophanase operon (*tna*). *Haemophilus influenza* type b contains genes homologous to the *tna* operon of *E. coli*, which are part of a 3.1-kb fragment inserted into the *H. influenza* type b chromosome next to *mutS*. This gene cluster is flanked by 43-bp direct repeats and is absent in most of the *Haemophilus* species (Martin *et al.*, 1998). *Streptococcus pyogenes* is the only species of *Streptococcus* that contains the 6-kb *vir*-regulon. Therefore, this region has some similarity with a Pai and encodes M and M-related proteins (*emm*, *mrp*, *enn*), a major regulator (*mga*) and a C5b peptidase (*scpA*) (Podbielski *et al.*, 1996).

Instability of pathogenicity islands

The flanking regions of several Pais are often characterized by the occurrence of direct repeats, which can be involved in recombinational events leading to the deletion of the Pai (see Tables 1.5 and 1.6). In the case of Pais I and II of the UPEC strain 536 and one Pai of J96, respectively, they can spontaneously delete from the chromosome at high frequencies upon site-specific recombination between short direct repeats (Blum *et al.*, 1994, 1995; Swenson *et al.*, 1996). IS elements or RS-sequences can also be responsible for the excision and possibly integration of genetic elements. In *Y. pestis*, homologous recombination between IS*100* elements flanking the HPI leads to the deletion of this Pai (Fetherston *et al.*, 1992). The excision of these Pais is accompanied by the loss of Pai-encoded determinants (*cnfI*, *hly*, *pap*, *pgm* locus) and reduced virulence of the resulting strains (Fetherston *et al.*, 1992; Blum *et al.*, 1994, 1995; Swenson *et al.*, 1996).

Deletion events leading to the modulation of virulence have also been described for *H. pylori* strains. It is suggested that such deletion events may represent a specific form of adaptation of pathogens to certain niches, hosts or tissues.

ROLE OF tRNA GENES

In various bacterial species as well as in eukaryotes, tRNA genes are commonly used as integration sites for foreign DNA including bacteriophages, plasmids or retroviruses (Cheetham and Katz, 1995). As already mentioned, many Pais are also associated with genes coding for tRNAs (see Tables 1.5 and 1.6). As indicated in Figure 1.2, the 3'-end of the gene coding for the selenocysteine-specific tRNA (*selC*) has been described as the attachment site of the retronphage φR73 (Inouye *et al.*, 1991). The insertion of Pais of uropathogenic and EPEC strains as well as of *Salmonella typhimurium*, which all encode different virulence factors, next to the tRNA gene *selC* underlines the specificity of the underlying recombination process (Blum *et al.*, 1995; McDaniel *et al.*, 1995; Blanc-Potard and Groisman, 1997). Several other tRNA genes serve as integration sites for Pais (see Tables 1.5 and 1.6). Usually, their expression is not influenced by the integration of a mobile genetic element. Another class of RNA genes has been reported to function as a chromosomal insertion target site. The 10Sa RNA encoding gene has been located adjacent to the *V. cholerae* VPI and to the *D. nodosus vrl* region (Billington *et al.*, 1996a; Karaolis *et al.*, 1998).

The presence of multiple copies of tRNA genes and their conserved nucleotide sequences may be the reason for the frequent insertion of extragenic elements next to tRNA genes. They consequently provide multiple stable integration sites, and their sequence similarities among different species could be advantageous for horizontal gene transfer.

Apart from that, tRNAs also have an impact on the expression of certain genes. The availability of minor tRNAs can modulate gene expression via facilitated translation of mRNAs containing high amounts of rare codons. In the uropathogenic *E. coli* strain 536, the *leuX*-encoded tRNA$_5^{Leu}$ influences the expression of various virulence traits, including toxins and metabolic activities (Ritter *et al.* 1995, 1997; Susa *et al.*, 1996; Dobrindt *et al.*, 1998). In the case of type-I fimbriation, this minor leucyl-tRNA enhances the *fim* expression by facilitated translation of the *fimB* gene. FimB acts as a positive regulator necessary for the production of the major fimbrial subunit (Ritter *et al.*, 1997).

Recent investigations revealed that in the *E. coli* strain 536 the tRNA$_5^{Leu}$ also influences the expression of α-haemolysin (see Figure 1.3). The availability of this tRNA has an impact on the transcription of the *hly*

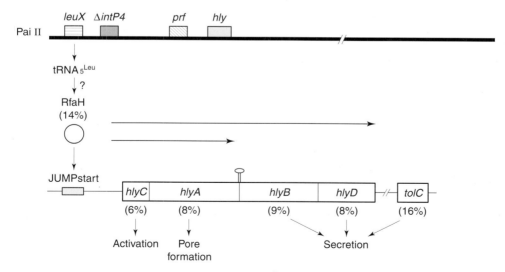

FIGURE 1.3 Model illustrating the influence of the *leuX* encoded tRNA$_5^{Leu}$ on the *expression* of α-haemolysin. A regulatory protein encoded by *rfaH* regulates the transcription of the *hly* determinant. The RfaH protein and a DNA element (JUMPstart) upstream of *hlyC* function together and enhance the transcription of the *hly* operon and in particular transcriptional readthrough beyond *hlyD*. The availability of functional tRNA$_5^{Leu}$ could influence the expression of RfaH and therefore expression of HlyA. The horizontal arrows indicate the length of the *hly* transcripts. Numbers in brackets indicate the percentage of *leuX*-specific codons among all leucine-specific codons of the corresponding gene. Abbreviations: *prf*, P-related fimbriae determinant; *intP4*, pseudo gene of the P4 integrase; *leuX*, gene encoding the tRNA$_5^{Leu}$; *hly*, genes of the α-haemolysin determinant; *tolC*, gene encoding the *E. coli* outer membrane protein TolC; Pai II, pathogenicity island II.

operon. In a *leuX* negative background, the secretion of α-haemolysin is delayed and the total concentration of the secreted toxin is reduced compared with the wild-type. Western- and Northern-blot analysis revealed that the transcription of the *hly* operon is reduced. The expression of *rfaH*, encoding a transcriptional activator of the *hly* operon, may be impaired in the absence of a functional $tRNA_5^{Leu}$, as this gene contains a high amount of *leuX*-specific codons (Blum-Oehler *et al.*, 1998; Dobrindt, unpublished). Interestingly, one of the *hly* determinants of the UPEC strain 536 is encoded by Pai II that is associated with *leuX* (Blum *et al.*, 1994).

In enterohaemorrhagic *E. coli* O157 and non-O157 strains, a tRNA gene, *ileX*, is located about 260 bp upstream of the translational start codon of *stx2A*, encoding the Shiga toxin 2, on the lysogenic bacteriophage genome. This tRNA gene has not been found in the corresponding region upstream of *stx1A*, which has fewer *ileX*-specific codons than *stx2a* (Schmidt *et al.*, 1997). There are several other examples of gene expression in prokaryotes, which is dependent on the presence of minor tRNAs (Fernández-Moreno *et al.*, 1991; Sauer and Dürre, 1992). Non-virulent strains of *Shigella flexneri*, which contain a defective virR gene, can regain virulence upon the transfer of $tRNA_1^{Tyr}$- or $tRNA_1^{Leu}$-encoding DNA (Hromockyj *et al.*, 1992). The *fimU*-encoded tRNA is necessary for a proper expression of type-1 fimbriae of *Salmonella typhimurium* and *S. enteritidis* (Swenson *et al.*, 1994; Clouthier *et al.*, 1998).

From an evolutionary point of view, the presence of tRNA-encoding genes on extrachromosomal genetic elements, also on those that encode toxins, may reflect their function as targets for the integration of foreign DNA. In addition, the presence of an additional copy of a putative minor tRNA gene may facilitate a proper expression of transferred genes, especially in a new host organism with a different codon usage.

CONCLUSIONS

Evolution of new pathogenic variants caused by pathogenicity islands and mobile genetic elements

Genes encoding toxins and other virulence factors can be located on Pais and/or on mobilizable genetic elements such as plasmids, transposons and bacteriophages. This is especially true for micro-organisms which lack the capacity to take up foreign DNA because of natural competence. The presence of so-called 'mobility sequences', such as direct repeats, sequences with homology to integrases or to plasmid origins of

replication near the termini of Pais (see Table 1.4), imply that Pais are derived from integrated mobile elements such as bacteriophages and plasmids (Groisman and Ochman, 1996; Lee, 1996; Hacker *et al.*, 1997). As indicated in Figure 1.4, plasmids and phages are able to form co-integrates. This has been found frequently in *Streptomyces* (Leblond and Decaris, 1994) and may be an explanation for the existence of features of plasmids and phages on Pais. Therefore, integrated plasmids and bacteriophages have been considered as 'Pai precursors' or 'Pre-Pais' (Lee, 1996; Hacker *et al.*, 1997). This underlines the strong mutual dependence between plasmids, bacteriophages and Pais. This interdependence contributes not only to the genetic flexibility of bacteria carrying these genetic elements, but also to the fast evolution of new pathogenic variants (Groisman and Ochman, 1996).

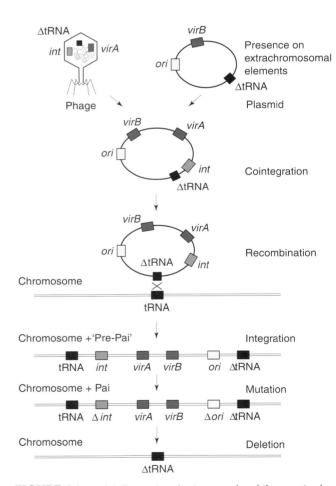

FIGURE 1.4 Model illustrating the impact of mobile genetic elements on the evolution of Pais. A chromosomal integration of mobile genetic elements into the chromosome can occur upon recombination of homologous sites. A stabilization of the integrated DNA may occur by mutations in integrase genes (*int*) and/or origins of replication (*ori*). The deletion of Pais may lead to avirulent variants.

The spread of specific genomic regions through lateral gene transfer followed by integration into the bacterial genome is a mechanism by which new variants of microbes could be generated. This can occur by lysogenic conversion in which the prophage confers specific changes in the bacterial phenotype or by the uptake of plasmids or DNA fragments by conjugation or transformation. A genomic island or its more specified variant (Pai) may be the result of such transfer processes and subsequently generated point mutations. Such spontaneous point mutations in 'mobility genes' may lead to a fixation of laterally acquired genes in specific strains. Under specific selective pressure, a 'homing' of new variants by inactivation of mobility genes could be advantageous, resulting in some structures of Pais described in this article.

Horizontal gene transfer and the evolution of toxin families

The existence of toxin families with common properties and sequence homologies in the relevant genes argues that such a distribution of genetic information is a result of lateral gene transfer. The most prominent examples of genetically related toxin families are ADP ribosylating toxins (diphtheria toxin, pertussis toxin, cholera toxin, *E. coli* LT, *Pseudomonas* exotoxin A and S and others) and other AB toxins (Shiga toxins), pore-forming RTX toxins (i.e. various haemolysins, high pathogenicity island *a* leucotoxin, *B. pertussis* adenylate cyclase-haemolysin), clostridial neurotoxins (*C. botulinum* and *C. tetani* neurotoxins), and proteins secreted by a type-III secretion system. Many of them are encoded on plasmids, phages or Pais. Members of these toxin families can be transferred by mobile genetic elements or may have formerly been transferred by them. Other examples, such as the occurrence of many different proteins, which share functional features of superantigens, suggest convergent evolution of proteins with similar toxic characteristics in different bacteria. This is less probable in the case of toxin families than the acquisition by horizontal gene transfer, because of the complexity of mutational and evolutionary events on the DNA level that would have to occur in order to result in functionally and structurally similar proteins.

The great similarity in the protein sequences of the *E. coli* LT and *V. cholerae* CT led to the assumption that CT, LT I and LT II originally derived from the same ancestral gene. Other toxins immunologically related to CT can be found in *Salmonella*, *Pseudomonas*, *Campylobacter* and *Aeromonas*. This may be the result of different evolutionary pathways that this gene followed in different species (Rappuoli and Pizza, 1991).

The pattern of distribution of genes involved in toxin expression on plasmids relative to the chromosome indicates that they may derive from a common ancestor such as the type-III protein secretion systems of various bacterial animal or plant pathogens (*Yersinia* spp., *Shigella flexneri*, *Salmonella typhimurium*, EPEC, *Pseudomonas aeruginosa*, *Chlamydia* spp., *Erwinia* spp., *Pseudomonas syringae*, *Ralstonia solanacearum*, *Xanthomonas campestris*) (Bonas, 1994; Collmer and Bauer, 1994; Mecsas and Strauss, 1996; Yahr *et al.*, 1996; Hsia *et al.*, 1997). Although the pathogenicity mechanisms and, therefore, the secreted effector molecules of these bacteria differ, the type-III secretion apparatus is conserved and effector molecules from one pathogen can be translocated by another system when the appropriate chaperones are present (Mecsas and Strauss, 1996).

Taken together, virulence determinants in general and toxin genes in particular are frequently encoded by mobile genetic elements. They are encoded on large genomic regions as well-designated Pais, which often contain clusters of virulence-associated genes. These genetic elements with the capacity to be spread by horizontal gene transfer contribute to the rapid evolution of bacterial pathogens as the rearrangement, excision and acquisition of large genomic regions creates new pathogenic variants. The occurrence of toxin-encoding genes on various interdepending mobile genetic elements, their ability to delete from and integrate into chromosomal DNA and the existence of toxin families among a wide variety of bacterial species show that the continuing process of the evolution of new pathogens is significantly connected with the transfer of foreign DNA harbouring toxin determinants.

ACKNOWLEDGEMENTS

Our work related to the subject of this article is supported by the Boehringer Ingelheim Fonds, the Sonderforschungsbereich 479 and the Fonds der Chemischen Industrie. We thank G. Blum-Oehler and U. Hentschel for carefully reading the manuscript.

REFERENCES

Akopyants, N.S., Clifton, S.W., Kersulyte, D., Crabtree, J.E., Youree, B.E., Reece, C.A., Bukanov, N.O., Drazek, E.S., Roe, B.A. and Berg, D.E. (1998). Analyses of the *cag* pathogenicity island of *Helicobacter pylori*. Mol. Microbiol. **28**, 37–53.

Allaoui, A., Schulte, R. and Cornelis, G.R. (1995). Mutational analysis of the *Yersinia enterocolitica virC* operon: characterization of *yscE, F, G, I, J, K* required for Yop secretion and *yscH* encoding YopR. Mol. Microbiol. **18**, 343–55.

Alouf, J.E., Knöll, H. and Köhler, W. (1991). The family of mitogenic, shock-inducing and superantigenic toxins from staphylococci and streptococci. In: *Sourcebook of Bacterial Protein Toxins* (eds J.E. Alouf and J.H. Freer), pp. 367–414. Academic Press, London.

Altboum, Z., Hertman, I. and Sadrid, S. (1985). Penicillinase plasmid-linked genetic determinants for enterotoxin B and C_1 production in *Staphylococcus aureus*. *Infect. Immun.* **47**, 514–21.

Andersson, S.G.E. and Kurland, C.G. (1990). Codon preference in free-living microorganisms. *Microbiol. Rev.* **54**, 198–210.

Antoine, R., Raze, D. and Locht, C. (1998). The pertussis toxin locus has some features of a pathogenicity island. *Zentralbl. Bakteriol. [Suppl.]* **29**, 393–4.

Arricau, N., Hermant, D., Waxin, H. and Popoff, M.Y. (1997). Molecular characterization of the *Salmonella typhi* StpA protein that is related to both *Yersinia* YopE cytotoxin and YopH tyrosine phosphatase. *Res. Microbiol.* **148**, 21–6.

Atherton, J.C., Cao, P., Peek, R.M., Jr, Tummuru, M.K., Blaser, M.J. and Cover, T.L. (1995). Mosaicism in vacuolating cytotoxin alleles of *Helicobacter pylori*. Association of specific *vacA* types with cytotoxin production and peptic ulceration. *J. Biol. Chem.* **279**, 17771–7.

Barksdale, W.L. and Pappenheimer, A.M., Jr (1954). Phage–host relationships in nontoxinogenic and toxinogenic diphtheria bacilli. *J. Bacteriol.* **67**, 220–32.

Bauer, M.E. and Welch, R.A. (1996). Characterization of an RTX-toxin from enterohemorrhagic *Escherichia coli* O157:H7. *Infect. Immun.* **64**, 167–75.

Bayles, K.W. and Iandolo, J.J. (1989). Genetic and molecular analyses of the gene encoding staphylococcal enterotoxin D. *J. Bacteriol.* **171**, 4799–806.

Berg, D.E. (1989). *Mobile DNA*. ASM Press, Washington, DC.

Betley, M.J. and Mekalanos, J.J. (1985). Staphylococcal enterotoxin A is encoded by phage. *Science* **229**, 185–7.

Beuscher, H.U., Rödel, F., Forsberg, Å. and Röllinghoff, M. (1995). Bacterial evasion of host immune response: *Yersinia enterocolitica* encodes a suppressor for tumor necrosis factor alpha expression. *Infect. Immun.* **63**, 1270–7.

Billington, S.J., Johnston, J.L. and Rood, J.I. (1996a). Virulence regions and virulence factors of the ovine footrot pathogen, *Dichelobacter nodosus*. *FEMS Microbiol. Lett.* **145**, 147–56.

Billington, S.J., Sinistaj, M., Cheetham, B.F., Ayres, A., Moses, E.K., Katz, M.E. and Rood, J.I. (1996b). Identification of a native *Dichelobacter nodosus* plasmid and implications for the evolution of the *vap* regions. *Gene* **172**, 111–16.

Blanc-Potard, A.-B. and Groisman, E.A. (1997). The *Salmonella selC* locus contains a pathogenicity island mediating intra-macrophage survival. *EMBO J.* **16**, 5376–85.

Bliska, J.B., Clemens, J.C., Dixon, J.E. and Falkow, S. (1992). The *Yersinia* tyrosine phosphatase: specificity of a bacterial virulence determinant for phosphoproteins in the J774A.1 macrophage. *J. Exp. Med.* **176**, 1625–30.

Bloomfield, G.A., Whittle, G., McDonagh, M.B., Katz, M.E. and Cheetham, B.F. (1997). Analysis of sequences flanking the *vap* regions of *Dichelobacter nodosus*: evidence for multiple integration events, a killer system, and a new genetic element. *Microbiology* **143**, 553–62.

Blum, G., Ott, M., Lischewski, A., Ritter, A., Imrich, H., Tschäpe, H. and Hacker, J. (1994). Excision of large DNA regions termed pathogenicity islands from tRNA-specific loci in the chromosome of an *Escherichia coli* wild-type pathogen. *Infect. Immun.* **62**, 606–14.

Blum, G., Falbo, V., Caprioli, A. and Hacker, J. (1995). Gene clusters encoding the cytotoxic necrotizing factor 1, Prs-fimbriae and α-hemolysin form the pathogenicity island II of the uropathogenic *Escherichia coli* strain J96. *FEMS Microbiol. Lett.* **126**, 189–96.

Blum-Oehler, G., Dobrindt, U., Weiß, N., Janke, B., Schubert, S., Rakin, A., Heesemann, J., Marre, R. and Hacker, J. (1998). Pathogenicity islands of uropathogenic *Escherichia coli*: implications for the evolution of virulence. *Zentralbl. Bakteriol. [Suppl.]* **29**, 380–6.

Bogdanove, A.J., Kim, J.W., Wie, Z., Kolchinsky, P., Charkowski, A.O., Conlin, A.K., Collmer, A. and Beer, S.V. (1998). Homology and functional similarity of an *hrp*-linked pathogenicity locus, *dspEF*, of *Erwinia amylovora* and the avirulence locus, *avrE* of *Pseudomonas syringae* pathovar tomato. *Proc. Natl Acad. Sci. USA* **95**, 1325–30.

Bonas, U. (1994). *hrp* genes of phytopathogenic bacteria. *Curr. Top. Microbiol. Immunol.* **192**, 79–98.

Braun, V., Hundsberger, T., Leukel, P., Sauerborn, M. and von Eichel Streiber, C. (1996). Definition of the single integration site of the pathogenicity locus in *Clostridium difficile*. *Gene* **181**, 29–38.

Carniel, E., Guilvout, I. and Prentice, M. (1996). Characterization of a large chromosomal 'high-pathogenicity-island' in biotype 1B *Yersinia enterocolitica*. *J. Bacteriol.* **178**, 6743–7751.

Censini, S., Lange, C., Xiang, Z.Y., Crabtree, T.E., Ghiara, P., Borodovsky, M., Rappuoli, R. and Covacci, A. (1996). *cag*, a pathogenicity island of *Helicobacter pylori*, encodes type I-specific and disease-associated virulence factors. *Proc. Natl Acad. Sci. USA* **93**, 14648–53.

Cheetham, B.F. and Katz, M.E. (1995). A role for bacteriophages in the evolution and transfer of bacterial virulence determinants. *Mol. Microbiol.* **18**, 201–8.

Cheetham, B.F., Tattersall, D.B., Bloomfield, G.A., Rood, J.I. and Katz, M.E. (1995). Identification of a bacteriophage-related integrase gene in a *vap* region of the *Dichelobacter nodosus* genome. *Gene* **162**, 53–8.

Christie, P.J. (1997). *Agrobacterium tumefaciens* T-complex transport apparatus: a paradigm for a new family of multifunctional transporters in eubacteria. *J. Bacteriol.* **179**, 3085–4.

Cieslewicz, M. and Vimr, E. (1997). Reduced polysialic acid capsule expression in *Escherichia coli* K1 mutants with chromosomal defects in *kpsF*. *Mol. Microbiol.* **26**, 237–49.

Clouthier, S.C., Collinson, S.K., White, A.P., Banser, P.A. and Kay, W.W. (1998). tRNA^{Arg} (*fimU*) and expression of SEF14 and SEF21 in *Salmonella enteritidis*. *J. Bacteriol.* **180**, 840–5.

Collier, R.J. (1982). Structure and activity of diphtheria toxin. In: *ADP-ribosylation Reactions* (eds D. Hayashi and K. Ueda), pp. 575–92. Academic Press, New York.

Collmer, A. and Bauer, D.W. (1994). *Erwinia chrysanthemi* and *Pseudomonas syringae*: plant pathogens trafficking in extracellular virulence proteins. *Curr. Top. Microbiol. Immunol.* **192**, 43–78.

Colonna, B., Casalino, M., Fradiani, P.A., Zagaglia, C., Naitza, S., Leoni, L., Prosseda, G., Coppo, A., Gherlardini, P. and Nicoletti, M. (1995). H-NS regulation of virulence gene expression in enteroinvasive *Escherichia coli* harboring the virulence plasmid integrated into the host chromosome. *J. Bacteriol.* **177**, 4703–12.

Colon-Whitt, A., Whitt, R.S. and Cole, R.M. (1979). Production of an erythrogenic toxin (streptococcal pyrogenic exotoxin) by a nonlysogenized group-A streptococcus. In: *Pathogenic Streptococci* (ed. M.T. Parker), pp. 64–5. Reedbooks, Chertsey.

Couch, J.L., Soltis, M.T. and Betley, M.J. (1988). Cloning and nucleotide sequence of the type E staphylococcal enterotoxin gene. *J. Bacteriol.* **170**, 2954–60.

Covacci, A., Falkow, S., Berg, D.E. and Rappuoli, R. (1997). Did the inheritance of a pathogenicity island modify the virulence of *Helicobacter pylori*? *Trends Microbiol.* **5**, 205–8.

De la Cruz, F., Müller, D., Ortiz, J.M. and Goebel, W. (1980). Hemolysis determinant common to *Escherichia coli* hemolytic plasmids of different incompatibility groups. *J. Bacteriol.* **143**, 825–33.

Deonier, R.C. (1996). Native insertion sequence elements: locations, distributions, and sequence relationships. In: Escherichia coli *and* Salmonella typhimurium. *Cellular and Molecular Biology* (eds F.C. Neidhardt, R. Curtis III, J.L. Ingraham, E.C.C. Lin, K. Brooks Low Jr, B. Magasanik, W.S. Reznikoff, M. Riley, M. Schaechter and H.E. Umbarger), pp. 2000–11. ASM Press, Washington, DC.

Dobrindt, U., Cohen, P.S., Utley, M., Mühldorfer, I. and Hacker, J. (1998). The *leuX*-encoded tRNA$_5$^Leu but not the pathogenicity islands I and II influence the survival of the uropathogenic *Escherichia coli* strain 536 in CD-1 mouse bladder mucus in the stationary phase. *FEMS Microbiol. Lett.* **162**, 135–41.

Donnenberg, M.S. and Welch, R.A. (1995). Virulence determinants of uropathogenic *Escherichia coli*. In: *Urinary Tract Infections: Molecular Pathogenesis and Clinical Management* (eds H.L.T. Mobley and J.W. Warren), pp. 135–74. ASM Press, Washington, DC.

Dubreuil, J.D. (1997). *Escherichia coli* STb enterotoxin. *Microbiology* **143**, 1783–95.

Echeverria, P., Seriwatana, J., Taylor, D.N., Tirapat, C. and Rowe, B. (1985). *Escherichia coli* contains plasmids coding for heat-stable b, other enterotoxins, and antibiotic resistance. *Infect. Immun.* **48**, 843–6.

Eklund, M.W., Poysky, F.T., Reed, S.M. and Smith, C.A. (1971). Bacteriophages and toxigenicity of *Clostridium botulinum* type D. *Nature (London) New Biol.* **235**, 16–18.

Eklund, M.W., Poysky, F.T., Mseitif, L.M. and Strom, M.S. (1988). Evidence for plasmid-mediated toxin production and bacteriocin production in *Clostridium botulinum* type G. *Appl. Environ. Microbiol.* **54**, 1405–8.

Elliott, S.J., Wainwright, L.A., McDaniel, T.K., Jarvis, K.G., Deng, Y., Lai, L.-C., McNamara, B.P., Donnenberg, M.S. and Kaper, J.B. (1998). The complete sequence of the locus of enterocyte effacement (LEE) from enteropathogenic *Escherichia coli* E2348/69. *Mol. Microbiol.* **28**, 1–4.

Engelbrecht, F., Dickneite, C., Lampidis, R., Götz, M., DasGupta, U. and Goebel, W. (1998). Sequence comparison of the chromosomal regions encompassing the internalin C genes (*inlC*) of *Listeria monocytogenes* and *L. ivanovii*. *Mol. Gen. Genet.* **257**, 186–97.

Falbo, V., Pace, T., Picci, L., Pizzi, E. and Caprioli, A. (1993). Isolation and nucleotide sequence of the gene encoding cytotoxic necrotizing factor 1 of *Escherichia coli*. *Infect. Immun.* **61**, 4904–14.

Farizo, K.M., Cafarella, T.G. and Burns, D.L. (1996). Evidence for a ninth gene, *ptlI*, in the locus encoding the pertussis toxin secretion system of *Bordetella pertussis* and formation of a PtlI–PtlF complex. *J. Biol. Chem.* **271**, 31643–9.

Fernández-Moreno, M.A., Caballero, J.L., Hopwood, D.A. and Malpartida, F. (1991). The *act* cluster contains regulatory and antibiotic export genes, direct targets for translational control by the *bldA* transfer RNA gene of *Streptomyces*. *Cell* **66**, 769–80.

Fetherston, J.D., Schuetze, P. and Perry, R.D. (1992). Loss of pigmentation phenotype in *Yersinia pestis* is due to the spontaneous deletion of 102 kb of chromosomal DNA which is flanked by a repetitive element. *Mol. Microbiol.* **6**, 2693–704.

Fierer, J., Krause, M., Tauxe, R. and Guiney, D. (1992). *Salmonella typhimurium* bacteremia: association with the virulence plasmid. *J. Infect. Dis.* **166**, 639–42.

Finn, C.W., Jr, Silver, R.P., Habig, W.H. and Hardegree, M.C. (1984). The structural gene for tetanus neurotoxin is on a plasmid. *Science* **224**, 881–4.

Galán, J.E. (1996). Molecular genetic bases of *Salmonella* entry into host cells. *Mol. Microbiol.* **20**, 263–72.

Galyov, E.E., Håkansson, S., Forsberg, Å. and Wolf-Watz, H. (1993). A secreted protein kinase of *Yersinia pseudotuberculosis* is an indispensable virulence determinant. *Nature* **361**, 730–1.

Gilmore, M.S., Segarra, R.A., Booth, M.C., Bogie, C.P., Hall, L.R. and Clewell, D.B. (1994). Genetic structure of the *Enterococcus faecalis* plasmid pAD1-encoded cytolytic toxin system and its relationship to lantibiotic determinants. *J. Bacteriol.* **176**, 7335–44.

Gouin, E., Mengaud, J. and Cossart, P. (1994). The virulence gene cluster of *Listeria monocytogenes* is also present in *Listeria ivanovii*, an animal pathogen, and *Listeria seeligeri*, a non-pathogenic species. *Infect. Immun.* **62**, 3550–3.

Green, B.D., Battisti, L., Koehler, T.M., Thorne, C.B. and Ivins, B.E. (1985). Demonstration of a capsule plasmid in *Bacillus anthracis*. *Infect. Immun.* **49**, 291–7.

Groisman, E.A. and Ochman, H. (1993). Cognate gene clusters govern invasion of host epithelial cells by *Salmonella typhimurium* and *Shigella flexneri*. *EMBO J.* **12**, 3779–87.

Groisman, E.A. and Ochman, H. (1996). Pathogenicity islands: bacterial evolution in quantum leaps. *Cell* **87**, 791–4.

Groisman, E.A. and Ochman, H. (1997). How *Salmonella* became a pathogen. *Trends Microbiol.* **5**, 343–9.

Gulig, P.A. and Doyle, T.J. (1993). The *Salmonella typhimurium* virulence plasmid increases the growth rate of salmonellae in mice. *Infect. Immun.* **61**, 504–11.

Gulig, P.A., Caldwell, A.L. and Chiodo, V.A. (1992). Identification, genetic analysis, and DNA sequence of a 7.8-kilobase virulence region of the *Salmonella typhimurium* virulence plasmid. *Mol. Microbiol.* **6**, 1395–411.

Gulig, P.A., Danbara, H., Guiney, D.G., Lax, A.J., Norel, F. and Rhen, M. (1993). Molecular analysis of virulence genes of the *Salmonella* virulence plasmids. *Mol. Microbiol.* **7**, 825–30.

Gunn, J.S., Alpuche-Aranda, C.M., Loomis, W.P., Belden, W.J. and Miller, S.I. (1995). Characterization of the *Salmonella typhimurium pagC/pagD* chromosomal region. *J. Bacteriol.* **177**, 5040–7.

Guyer, D.M., Kao, J.-S. and Mobley, H.L.T. (1998). Genomic analysis of a pathogenesis island in uropathogenic *Escherichia coli* CFT073: distribution of homologous sequences among pyelonephritis, cystitis, catheter-associated bacteriuria, and fecal isolates. *Infect. Immun.* **66**, 4411–17.

Gyles, C.M. (1992). *Escherichia coli* cytotoxins and enterotoxins. *Can. J. Microbiol.* **38**, 734–46.

Gyles, C.M., So, M. and Falkow, S. (1974). The enterotoxin plasmids of *Escherichia coli*. *J. Infect. Dis.* **130**, 40–8.

Hacker, J. and Hughes, C. (1985). Genetics of *Escherichia coli* hemolysin. *Curr. Top. Microbiol. Immunol.* **118**, 139–62.

Hacker, J., Bender, L., Ott, M., Wingender, J., Lund, B., Marre, R. and Goebel, W. (1990). Deletions of chromosomal regions coding for fimbriae and hemolysins occur *in vivo* and *in vitro* in various extraintestinal *Escherichia coli* isolates. *Microb. Pathogen.* **8**, 213–25.

Hacker, J., Blum-Oehler, G., Mühldorfer, I. and Tschäpe, H. (1997). Pathogenicity islands of virulent bacteria: structure, function and impact on microbial evolution. *Mol. Microbiol.* **23**, 1089–97.

Haddix, P.L. and Straley, S.C. (1992). Structure and regulation of the *Yersinia pestis yscBCDEF* operon. *J. Bacteriol.* **174**, 4820–8.

Hale, T.L. (1991). Genetic basis of virulence in *Shigella* species. *Microbiol. Rev.* **55**, 206–24.

Hall, R.M. and Collis, C.M. (1995). Mobile gene cassettes and integrons: capture and spread of genes by site-specific recombination. *Mol. Microbiol.* **15**, 593–600.

Hardy, K. (1986). *Bacterial Plasmids*, 2nd edn. ASM Press, Washington, DC.

Harnett, N.M. and Gyles, C.L. (1985). Linkage of genes for heat-stable enterotoxin, drug resistance, K99 antigen, and colicin in bovine and porcine strains of enterotoxigenic *Escherichia coli*. *Am. J. Vet. Res.* **46**, 428–33.

Hauser, D., Gibert, M., Boquet, P. and Popoff, M.R. (1992). Plasmid localization of a type E botulinal neurotoxin gene homologue in toxigenic *Clostridium butyricum* strains, and absence of this gene in non-toxigenic *C. butyricum* strains. *FEMS Microbiol. Lett.* **99**, 251–6.

Hayashi, T., Matsumoto, H., Ohnishi, M. and Terawaki, Y. (1993). Molecular analysis of a cytotoxin-converting phage, φCTX, of *Pseudomonas aeruginosa*: structure of the *attP-cos-ctx*-region and integration into the serine tRNA gene. *Mol. Microbiol.* **7**, 657–67.

Heesemann, J., Hantke, K., Vocke, T., Saken, E., Rakin, A., Stojiljkovic, I. and Berner, R. (1993). Virulence of *Yersinia enterocolitica* is closely associated with siderophore production, expression of an iron-repressible outer membrane protein of 65 000 Da and pesticin sensitivity. *Mol. Microbiol.* **8**, 397–408.

Hensel, M., Shea, J.E., Bäumler, A.J., Gleeson, C., Blattner, F. and Holden, D.W. (1997). Analysis of the boundaries of *Salmonella* pathogenicity island 2 and the corresponding chromosomal region of *Escherichia coli* K-12. *J. Bacteriol.* **179**, 1105–11.

Hess, J., Wels, W., Vogel, M. and Goebel, W. (1986). Nucleotide sequence of a plasmid-encoded hemolysin determinant and its comparison with a corresponding chromosomal hemolysin sequence. *FEMS Microbiol. Lett.* **34**, 1–11.

Higgins, C.F., McLaren, R.S. and Newbury, S.F. (1988). Repetitive extragenic palindromic sequences, mRNA stability and gene expression: evolution by gene conversion – a review. *Gene* **72**, 3–14.

Holm, L. (1986). Codon usage and gene expression. *Nucl. Acids Research* **14**, 3075–87.

Hromockyj, A.E., Tucker, S.C. and Maurelli, A.T. (1992). Temperature regulation of *Shigella* virulence: identification of the repressor gene *virR*, an analogue of *hns*, and partial complementation by tyrosyl transfer RNA (tRNA$_1^{Tyr}$). *Mol. Microbiol.* **6**, 2113–24.

Hsia, R.-C., Pannekoek, Y., Ingerowski, E. and Bavoil, P.M. (1997). Type III secretion genes identify a putative virulence locus of *Chlamydia*. *Mol. Microbiol.* **25**, 351–9.

Hu, S.T. and Lee, C.H. (1988). Characterization of the transposon carrying the STII gene of enterotoxigenic *Escherichia coli*. *Mol. Gen. Genet.* **214**, 490–5.

Hu, S.T., Yang, M.K., Spandau, D.F. and Lee, C.H. (1987). Characterization of the terminal sequences flanking the transposon that carries the *Escherichia coli* enterotoxin STII gene. *Gene* **55**, 157–67.

Iandolo, J.J. (1989). Genetic analysis of extracellular toxins of *Staphylococcus aureus*. *Annu. Rev. Microbiol.* **43**, 375–402.

Ike, Y. and Clewell, D.B. (1992). Evidence that the hemolysin bacteriocin phenotype of *Enterococcus faecalis* subsp. *zymogenes* can be determined by plasmids in different incompatibility groups as well as by the chromosome. *J. Bacteriol.* **174**, 8172–7.

Inoue, K. and Iida, H. (1970). Conversion to toxigenicity in *Clostridium botulinum* type C. *Jpn J. Microbiol.* **14**, 87–9.

Inouye, M. and Inouye, S. (1991). Retroelements in bacteria. *Trends Biochem. Sci.* **16**, 18–21.

Inouye, S., Sunshine, M.G., Six, E.W. and Inouye, M. (1991). Retronphage φR73: an *E. coli* phage that contains a retroelement and integrates into a tRNA gene. *Science* **252**, 969–71.

Jackson, M.P. and Iandolo, J.J. (1985). Cloning and expression of the exfoliative toxin B gene from *Staphylococcus aureus*. *J. Bacteriol.* **166**, 574–80.

Jackson, M.P., Neill, R.J., O'Brien, A.D., Holmes, R.K. and Newland, J.W. (1987). Nucleotide sequence analysis and comparison of the structural genes for Shiga-like toxin I and Shiga-like toxin II encoded by bacteriophages from *Escherichia coli* 933. *FEMS Microbiol. Lett.* **44**, 109–14.

Jarvis, K.G., Girón, J.A., Jerse, A.E., McDaniel, T.K., Donnenberg, M.S. and Kaper, J.B. (1995). Enteropathogenic *Escherichia coli* contains a putative type III system necessary for the export of proteins involved in attaching and effacing lesion formation. *Proc. Natl Acad. Sci. USA* **92**, 7996–8000.

Jett, B.D., Huycke, M.M. and Gilmore, M.S. (1994). Virulence of enterococci. *Clin. Microbiol. Rev.* **7**, 462–78.

Johns, B.M. and Khan, S.A. (1988). Staphylococcal enterotoxin B gene is associated with a discrete genetic element. *J. Bacteriol.* **170**, 4033–9.

Johnson, L.P., Schlievert, P.M. and Watson, D.W. (1980). Transfer of group A streptococcal pyrogenic exotoxin production to non-toxinogenic strains by lysogenic conversion. *Infect. Immun.* **28**, 254–7.

Jones, M.E., Peters, E., Weersink, A.M., Fluit, A. and Verhoef, J. (1997). Widespread occurrence of integrons causing multiple antibiotic resistance in bacteria. *Lancet* **349**, 1742–3.

Kao, J.-S., Stucker, D.M., Warren, J.W. and Mobley, H.L.T. (1997). Pathogenicity island sequences of pyelonephritogenic *Escherichia coli* CFT073 are associated with virulent uropathogenic strains. *Infect. Immun.* **65**, 2812–20.

Karaolis, D.K.R., Johnson, J.A., Bailey, J.C., Boedeker, E.C., Kaper, J.B. and Reeves, P.R. (1998). A *Vibrio cholerae* pathogenicity island associated with epidemic and pandemic strains. *Proc. Natl Acad. Sci. USA* **95**, 3134–9.

Kaspar, R.L. and Robertson, D.L. (1987). Purification and physical analysis of *Bacillus anthracis* plasmids pX01 and pX02. *Biochem. Biophys. Res. Commun.* **149**, 362–8.

Katz, M.E., Strugnell, R.A. and Rood, J.I. (1992). Molecular characterization of a genomic region associated with virulence in *Dichelobacter nodosus*. *Infect. Immun.* **60**, 4586–92.

Kenny, B., DeVinney, R., Stein, M., Reinscheid, D.J., Frey, E.A. and Finlay, B.B. (1997). Enteropathogenic *E. coli* (EPEC) transfers its receptor for intimate adherence into mammalian cells. *Cell* **91**, 511–20.

Knapp, S., Hacker, J., Then, I., Müller, D. and Goebel, W. (1984). Multiple copies of hemolysin genes and associated sequences in the chromosomes of uropathogenic *Escherichia coli* strains. *J. Bacteriol.* **159**, 1027–33.

Knapp, S., Then, I., Wels, W., Michel, G., Tschäpe, H., Hacker, J. and Goebel, W. (1985). Analysis of the flanking regions from different haemolysin determinants of *Escherichia coli*. *Mol. Gen. Genet.* **200**, 385–92.

Knapp, S., Hacker, J., Jarchau, T. and Goebel, W. (1986). Large, unstable inserts in the chromosome affect virulence properties of uropathogenic *Escherichia coli* O6 strain 536. *J. Bacteriol.* **168**, 22–30.

Kovach, M.E., Shaffer, M.D. and Peterson, K.M. (1996). A putative integrase gene defines the distal end of a large cluster of ToxR-regulated colonization genes in *Vibrio cholerae*. *Microbiology* **142**, 2165–74.

Laird, W.J., Aaronson, W., Silver, R.P., Habig, W.H. and Hardegree, M.C. (1980). Plasmid-associated toxigenicity in *Clostridium tetani*. *J. Infect. Dis.* **142**, 623.

Leblond, P. and Decaris, B. (1994). New insights into the genetic instability of *Streptomyces*. *FEMS Microbiol. Lett.* **123**, 225–32.

Lee, C.A. (1996). Pathogenicity islands and the evolution of bacterial pathogens. *Infect. Agents Dis.* **5**, 1–7.

Lee, C.H., Hu, S.T., Swiatek, P.J., Moseley, S.L., Allen, S.D. and So, M. (1985). Isolation of a novel transposon which carries the *Escherichia coli* enterotoxin STII gene. *J. Bacteriol.* **162**, 615–20.

Levine, M.M. (1987). *Escherichia coli* that cause diarrhea: enterotoxigenic, enteropathogenic, enteroinvasive, enterohemorrhagic, and enteroadherent. *J. Infect. Dis.* **155**, 377–89.

Lindsay, J.A., Ruzin, A., Ross, H.F., Kurepina, N. and Novick, R.P. (1998). The gene for toxic shock toxin is carried by a family of mobile pathogenicity islands in *Staphylococcus aureus*. *Mol. Microbiol.* **29**, 527–43.

Locht, C. and Keith, J.M. (1986). Pertussis toxin gene: nucleotide sequence and genetic organization. *Science* **232**, 1258–64.

Lucier, T.S., Fetherston, J.D., Brubaker, R.R. and Perry, R.D. (1996). Iron uptake and iron-repressible polypeptides in *Yersinia pestis*. *Infect. Immun.* **64**, 3023–31.

Lupski, J.R. and Weinstock, G.M. (1992). Short interspersed repetitive DNA sequences in prokaryotic genomes. *J. Bacteriol.* **174**, 4525–9.

Mainil, J.G., Jacquemin, E., Herault, F. and Oswald, E. (1997). Presence of *pap-*, *sfa-*, and *afa*-related sequences in necrotoxigenic *Escherichia coli* isolates from cattle: evidence for new variants of the AFA family. *Can. J. Vet. Res.* **61**, 193–9.

Marques, L.R.M., Peiris, J.S.M., Cryz, S.J. and O'Brien, A.D. (1987). *Escherichia coli* strains isolated from pigs with edema disease produce a variant of Shiga-like toxin II. *FEMS Microbiol Lett.* **44**, 33–8.

Martin, K., Morlin, G., Smith, A., Nordyke, A., Eisenstark, A. and Golomb, M. (1998). The tryptophanase gene cluster of *Haemophilus influenzae* type b: evidence for horizontal gene transfer. *J. Bacteriol.* **180**, 107–18.

Mazel, D., Dychino, B., Webb, V.A. and Davies, J. (1998). A distinct class of integron in the *Vibrio cholerae* genome. *Science* **280**, 605–8.

McDaniel, T.K. and Kaper, J.B. (1997). A cloned pathogenicity island from enteropathogenic *Escherichia coli* confers the attaching and effacing phenotype on *E. coli* K-12. *Mol. Microbiol.* **23**, 399–407.

McDaniel, T.K., Jarvis, K.G., Donnenberg, M.S. and Kaper, J.B. (1995). A genetic locus of enterocyte effacement conserved among diverse enterobacterial pathogens. *Proc. Nat. Acad. Sci. USA* **92**, 1664–8.

McKane, L. and Ferretti, J.J. (1981). Phage–host interactions and the production of type A streptococcal exotoxin in group A streptococci. *Infect. Immun.* **34**, 915–19.

McMillan, R.A., Bloomster, T.A., Saeed, A.M., Henderson, K.L., Zinn, N.E., Abernathy, R., Watson, D.W. and Greenberg, R.N. (1987). Characterization of a fourth streptococcal pyrogenic exotoxin (SPED). *FEMS Microbiol. Lett.* **44**, 317–22.

Mecsas, J. and Strauss, E.J. (1996). Molecular mechanisms of bacterial virulence: type III secretion and pathogenicity islands. *Emerg. Infect. Dis.* **2**, 271–88.

Mekalanos, J.J. (1983). Duplication and amplification of toxin genes in *Vibrio cholerae*. *Cell* **35**, 253–63.

Ménard, R., Sansonetti, P.J. and Parsot, C. (1994). The secretion of the *Shigella flexneri* Ipa invasins is induced by the epithelial cell and controlled by IpaB and IpaD. *EMBO J.* **13**, 555–68.

Mikesell, P., Ivins, B.E., Ristroph, J.D. and Dreier, T.M. (1983). Evidence for plasmid-mediated toxin production in *Bacillus anthracis*. *Infect. Immun.* **39**, 371–6.

Miller, V.L. and Mekalanos, J.J. (1984). Synthesis of cholera toxin is positively regulated at the transcriptional level by *toxR*. *Proc. Natl Acad. Sci. USA* **81**, 3471–5.

Mills, D.M., Bajaj, V. and Lee, C.A. (1995). A 40 kb chromosomal fragment encoding *Salmonella typhimurium* invasion genes is absent from the corresponding region of the *Escherichia coli* K-12 chromosome. *Mol. Microbiol.* **15**, 749–59.

Minton, N.P. (1995). Molecular genetics of clostridial neurotoxins. *Curr. Top. Microbiol. Immunol.* **195**, 161–94.

Moncrief, J.S., Duncan, A.J., Wright, R.L., Barroso, L.A. and Wilkins, T.D. (1998). Molecular characterization of the fragilysin pathogenicity islet of enterotoxigenic *Bacteroides fragilis*. *Infect. Immun.* **66**, 1735–9.

Montenegro, M.A., Morelli, G. and Helmuth, R. (1991). Heteroduplex analysis of *Salmonella* virulence plasmids and their prevalence in isolates of defined sources. *Microb. Pathog.* **11**, 391–7.

Müller, D., Hughes, C. and Goebel, W. (1983). Relationship between plasmid and chromosomal hemolysin determinants in *Escherichia coli*. *J. Bacteriol.* **153**, 846–51.

Nataro, J.P. and Kaper, J.B. (1998). Diarrheagenic *Escherichia coli*. *Clin. Microbiol. Rev.* **11**, 142–201.

Nataro, J.P., Maher, K.O., Mackie, P. and Kaper, J.B. (1987). Characterization of plasmids encoding the adherence factor of enteropathogenic *Escherichia coli*. *Infect. Immun.* **55**, 2370–7.

Nataro, J.P., Seriwatana, J., Fasano, A., Maneval, D.R., Guers, L.D., Noriega, F., Dubovsky, F., Levine, M.M. and Morris, R.G., Jr (1995). Identification and cloning of a novel plasmid-encoded enterotoxin of enteroinvasive *E. coli* and *Shigella* strains. *Infect. Immun.* **63**, 4721–8.

Nida, S.K. and Ferretti, J.J. (1982). Phage influence on the synthesis of extracellular toxins in group A streptococci. *Infect. Immun.* **36**, 745–50.

O'Brien, A.D., Tesh, V.L., Donohue-Rolfe, A., Jackson, M.P., Olsnes, S., Sandvig, K., Lindberg, A.A. and Keusch, G.T. (1992). Shiga toxin: biochemistry, genetics, mode of action, and role in pathogenesis. *Curr. Top. Microbiol. Immunol.* **180**, 65–94.

Ochman, H., Soncini, F.C., Solomon, F. and Groisman, E.A. (1996). Identification of a pathogenicity island required for *Salmonella* survival in host cells. *Proc. Natl Acad. Sci. USA* **93**, 7800–4.

Oswald, E. and De Rycke, J. (1990). A single protein of 110 kDa is associated with the multinucleating and necrotizing activity encoded by the Vir plasmid of *Escherichia coli*. *FEMS Microbiol. Lett.* **56**, 279–84.

Oswald, E., Sugai, M., Labigne, A., Wu, H.C., Fiorentini, C., Boquet, P. and O'Brien, A.D. (1994). Cytotoxic necrotizing factor type 2 produced by virulent *Escherichia coli* modifies the small GTP-binding proteins Rho involved in assembly of actin stress fibers. *Proc. Natl Acad. Sci. USA* **91**, 3814–18.

Parsot, C. and Sansonetti, P.J. (1996). Invasion and the pathogenesis of *Shigella* infections. *Curr. Top. Microbiol. Immunol.* **209**, 25–42.

Paton, A.W., Paton, J.C., Heuzenroeder, M.W., Goldwater, P.N. and Manning, P.A. (1992). Cloning and nucleotide sequence of a variant Shiga-like toxin II gene from *Escherichia coli* OX3:H21 isolated from a case of sudden infant death syndrome. *Microb. Pathog.* **13**, 225–36.

Paton, A.W., Paton, J.C., Goldwater, P.N., Heuzenroeder, M.W. and Manning, P.A. (1993). Sequence of a variant Shiga-like toxin type-I operon of *Escherichia coli* O111:H-. *Gene* **129**, 87–92.

Pearson, G.D.N., Woods, A., Chiang, S.L. and Mekalanos, J.J. (1993). CTX genetic element encodes a site-specific recombination system and an intestinal colonization factor. *Proc. Natl Acad. Sci. USA* **90**, 3750–4.

Peres, S.Y., Marches, O., Daigle, F., Nougayrede, J.P., Herault, F., Tasca, C., De Rycke, J. and Oswald, E. (1997). A new cytolethal distending toxin (CDT) from *Escherichia coli* producing CNF2 blocks HeLa cell division in G2/M phase. *Mol. Microbiol.* **24**, 1095–107.

Perry, R.D., Pendrak, M.L. and Schuetze, P. (1990). Identification and cloning of a hemin storage locus involved in the pigmentation phenotype of *Yersinia pestis*. *J. Bacteriol.* **172**, 5929–37.

Podbielski, A., Woischnik, M., Pohl, B. and Schmidt, K.H. (1996). What is the size of the group A streptococcal *vir* regulon? The Mga regulator affects expression of secreted and surface virulence factors. *Med. Microbiol. Immunol.* **185**, 171–81.

Portnoy, D.A. and Falkow, S. (1981). Virulence-associated plasmids from *Yersina enterocolitica* and *Yersinia pestis*. *J. Bacteriol.* **148**, 877–3.

Portnoy, D.A., Chakraborty, T., Goebel, W. and Cossart, P. (1992). Molecular determinants of *Listeria monocytogenes* pathogenesis. *Infect. Immun.* **60**, 1263–7.

Pulkkinen, W.S. and Miller, S.I. (1991). A *Salmonella typhimurium* virulence protein is similar to a *Yersinia enterocolitica* invasion protein and a bacteriophage lambda outer membrane protein. *J. Bacteriol.* **173**, 86–93.

Rajamukar, J., Sasakawa, C. and Adler, B. (1997). Use of a novel approach, termed island probing, identifies the *Shigella flexneri she* pathogenicity island which encodes a homolog of the immunoglobulin A protease-like family of proteins. *Infect. Immun.* **65**, 4604–14.

Rappuoli, R. and Pizza, M. (1991). Structure and evolutionary aspects of ADP-ribosylating toxins. In: *Sourcebook of Bacterial Protein Toxins* (eds J.E. Alouf and J.H. Freer), pp. 1–22. Academic Press, London.

Rappuoli, R. and Ratti, G. (1984). Physical map of the chromosomal region of *Corynebacterium diphtheriae* containing corynephage attachment sites *attB1* and *attB2*. *J. Bacteriol.* **158**, 325–30.

Ratti, G., Covacci, A. and Rappuoli, R. (1997). A tRNA$_2^{Arg}$ gene of *Corynebacterium diphtheriae* is the chromosomal integration site for toxinogenic bacteriophages. *Mol. Microbiol.* **25**, 1179–81.

Rietra, P.J.U.G.M., Willshaw, G.A., Smith, H.R., Field, A.M., Scotland, S.M. and Rowe, B. (1989). Comparison of Vero cytotoxin-encoding phages from *Escherichia coli* of human and bovine origin. *J. Gen. Microbiol.* **135**, 2307–18.

Ritter, A., Blum, G., Emödy, L., Kerenyi, M., Böck, A., Neuhierl, B., Rabsch, W., Scheutz, F. and Hacker, J. (1995). tRNA genes and pathogenicity islands: influence on virulence and metabolic properties of uropathogenic *Escherichia coli*. *Mol. Microbiol.* **17**, 109–21.

Ritter, A., Gally, D.L., Olsen, P.B., Dobrindt, U., Friedrich, A., Klemm, P. and Hacker, J. (1997). The Pai-associated *leuX* specific tRNA$_5^{Leu}$ affects type 1 fimbriation in pathogenic *Escherichia coli* by control of FimB recombinase expression. *Mol. Microbiol.* **25**, 871–2.

Sako, T. and Tsuchida, N. (1983). Nucleotide sequence of the staphylokinase gene from *Staphylococcus aureus*. *Nucleic Acids Res.* **11**, 7679–93.

Sasakawa, C., Adler, B., Tobe, T., Okada, N., Nagai, S., Komatsu, K. and Yoshikawa, M. (1989). Functional organization and nucleotide sequence of virulence region-2 on the large virulence plasmid of *Shigella flexneri* 2a. *Mol. Microbiol.* **3**, 1191–201.

Sasakawa, C., Komatsu, K., Tobe, T., Suzuki, T. and Yoshikawa, M. (1993). Eight genes in region 5 that form an operon are essential for invasion of epithelial cells by *Shigella flexneri* 2a. *J. Bacteriol.* **175**, 2334–46.

Sauer, U. and Dürre, P. (1992). Possible function of tRNA$_{ACG}^{Tyr}$ in regulation of solvent formation in *Clostridium acetobutylicum*. *FEMS Microbiol. Lett.* **100**, 147–54.

Savarino, S.J., Fasano, A., Robertson, C. and Levine, M.M. (1991). Enteroaggregative *Escherichia coli* elaborate a heat-stable enterotoxin demonstrable in an *in vitro* rabbit intestinal model. *J. Clin. Invest.* **87**, 1450–5.

Savarino, S.J., Fasano, A., Watson, J., Martin, B.M., Levine, M.M., Guandalini, S. and Guerry, P. (1993). Enteroaggregative *Escherichia coli* heat-stable enterotoxin 1 represents another subfamily of *E. coli* heat-stable toxin. *Proc. Natl Acad. Sci. USA* **90**, 3093–7.

Savarino, S.J., McVeigh, A., Watson, J., Molina, J., Cravioto, A., Echeverria, P., Bhan, M.K., Levine, M.M. and Fasano, A. (1996). Enteroaggregative *Escherichia coli* heat-stable enterotoxin is not restricted to enteroaggregative *E. coli*. *J. Inf. Dis.* **173**, 1019–22.

Schmidt, H. and Karch, H. (1996). Enterohemolytic phenotypes and genotypes of Shiga toxin-producing *Escherichia coli* O111 strains from patients with diarrhea and hemolytic uremic syndrome. *J. Clin. Microbiol.* **34**, 2364–7.

Schmidt, H., Montag, M., Bockemühl, J., Heesemann, J. and Karch, H. (1993). Shiga-like toxin II-related cytotoxins in *Citrobacter freundii* strains from humans and beef samples. *Infect. Immun.* **61**, 534–43.

Schmidt, H., Beutin, L. and Karch, H. (1995). Molecular analysis of the plasmid-encoded hemolysin of *Escherichia coli* O157:H7 strain EDL933. *Infect. Immun.* **63**, 1055–61.

Schmidt, H., Kernbach, C. and Karch, H. (1996). Analysis of the EHEC *hly* operon and its location in the physical map of the large plasmid of enterhaemorrhagic *Escherichia coli* O157:H7. *Microbiology* **142**, 907–14.

Schmidt, H., Scheef, J., Janetzki-Mittmann, C., Datz, M. and Karch, H. (1997). An *ileX* tRNA gene is located close to the Shiga toxin II operon in enterohemorrhagic *Escherichia coli* O157 and non-O157 strains. *FEMS Microbiol. Lett.* **149**, 39–44.

Schubert, S., Rakin, A., Karch, H., Carniel, E. and Heesemann, J. (1998). Prevalence of the 'high-pathogenicity island' of *Yersinia* species among *Escherichia coli* strains that are pathogenic to humans. *Infect. Immun.* **66**, 480–5.

Sears, C.L. and Kaper, J.B. (1996). Enteric bacterial toxins: mechanims of action and linkage to intestinal secretion. *Microbiol. Rev.* **60**, 167–215.

Sharp, P.M. (1991). Determinants of DNA sequence divergence between *Escherichia coli* and *Salmonella typhinurium*: codon usage, map position, and concerted evolution. *J. Mol. Evol.* **33**, 23–33.

Shea, J.E., Hensel, M., Gleeson, C. and Holden, D.W. (1996). Identification of a virulence locus encoding a second type III secretion system in *Salmonella typhimurium*. *Proc. Natl Acad. Sci. USA* **93**, 2593–7.

Sixma, T.K., Kalk, K.H., van Zanten, B.A., Dauter, Z., Kingma, J., Witholt, B. and Hol, W.G. (1993). Refined structure of *Escherichia coli* heat-labile enterotoxin, a close relative of cholera toxin. *J. Mol. Biol.* **230**, 890–918.

So, M. and McCarthy, B.J. (1980). Nucleotide sequence of transposon Tn1681 encoding a heat-stable toxin (ST) and its identification in enterotoxigenic *Escherichia coli* strains. *Proc. Natl Acad. Sci. USA* **77**, 4011–15.

Spangler, B.D. (1992). Structure and function of cholera toxin and the related *Escherichia coli* heat-labile enterotoxin. *Microbiol. Rev.* **56**, 622–47.

Stein, M.A., Leung, K.Y., Zwick, M., Garcia-del Portillo, F. and Finlay, B.B. (1996). Identification of a *Salmonella* virulence gene required for formation of filamentous structures containing lysosomal membrane glycoproteins within epithelial cells. *Mol. Microbiol.* **20**, 151–64.

Straley, S.C., Plano, G.V., Skrzypek, E. and Bliska, J.B. (1993). Yops of *Yersinia* spp. pathogenic for humans. *Infect. Immun.* **61**, 3105–10.

Strom, M.S., Eklund, M.W. and Poysky, F.T. (1984). Plasmids in *Clostridium botulinum* and related *Clostridium* species. *Appl. Environ. Microbiol.* **48**, 956–63.

Sullivan, J.T. and Ronson, C.W. (1998). Evolution of rhizobia by acquisition of a 500-kb symbiosis island that integrates into a phe-tRNA gene. *Proc. Natl Acad. Sci. USA* **95**, 5145–9.

Susa, M., Kreft, B., Wasenauer, G., Ritter, A., Hacker, J. and Marre, R. (1996). Influence of cloned tRNA genes from a uropathogenic *Escherichia coli* strain on adherence to primary human renal tubular epithelial cells and nephropathogenicity in rats. *Infect. Immun.* **64**, 5390–4.

Swenson, D.L., Kim, K.-J., Six, E.W. and Clegg, S. (1994). The gene *fimU* affects expression of *Salmonella typhimurium* type 1 fimbriae and is related to the *Escherichia coli* tRNA gene *argU*. *Mol. Gen. Genet.* **244**, 216–18.

Swenson, D.L., Bukanov, N.O., Berg, D.E. and Welch, R.A. (1996). Two pathogenicity islands in uropathogenic *Escherichia coli* J96: cosmid cloning and sample sequencing. *Infect. Immun.* **64**, 3736–43.

Takeda, Y. (1995). Shiga and Shiga-like (Vero) toxins. In: *Bacterial Toxins and Virulence Factors in Disease* (eds J. Moss, B. Iglewski, M. Vaughan and T.A. Tu), pp. 313–26. Marcel Dekker, New York.

Tamura, M., Nogimori, K., Murai, S., Yajima, M., Ito, K., Katada, T., Ui, M. and Ishii, S. (1982). Subunit structure of islet-activating protein, pertussis toxin, in conformity with the A-B model. *Biochemistry* **21**, 5516–22.

Thorne, C.B. (1985). Genetics of *Bacillus anthracis*. In: *Microbiology-85* (eds L. Lieve, P.F. Bonventre, J.A. Morello, S. Schlessinger, S.D. Silver and H.C. Wu), p. 56. ASM Press, Washington, DC.

Tinsley, C.R. and Nassif, X. (1996). Analysis of the genetic differences between *Neisseria meningitidis* and *Neisseria gonorrhoeae*: two closely related bacteria expressing two different pathogenicities. *Proc. Natl Acad. Sci. USA* **93**, 11109–14.

Travisano, M. and Inouye, M. (1995). Retrons: retroelements of no known function. *Trends Microbiol.* **3**, 209–11.

Uchida, I., Sekizaki, T., Hashimoto, K. and Terakado, N. (1985). Association of the encapsulation of *Bacillus anthracis* with a 60 megadalton plasmid. *J. Gen. Microbiol.* **131**, 363–7.

Venkatesan, M.M. and Buysse, J.M. (1991). Nucleotide sequence of invasion plasmid antigen gene *ipaA* from *Shigella flexneri* 5. *Nucleic Acids Res.* **18**, 1648.

Venkatesan, M.M., Buysse, J.M. and Kopecko, D.J. (1988). Characterization of invasion plasmid antigen (*ipaBCD*) genes from *Shigella flexneri*. *Proc. Natl Acad. Sci. USA* **85**, 9317–21.

Venkatesan, M.M., Buysse, J.M. and Oaks, E.V. (1992). Surface presentation of *Shigella flexneri* invasion plasmid antigens requires the products of the *spa* locus. *J. Bacteriol.* **174**, 1990–2001.

Waalwijk, C., de Graaff, J. and MacLaren, D.M. (1984). Physical mapping of hemolysin plasmid pCW2, which codes for virulence of a nephropathogenic *Escherichia coli* strain. *J. Bacteriol.* **159**, 424–6.

Waldor, M.K. and Mekalanos, J.J. (1996). Lysogenic conversion by a filamentous phage encoding cholera toxin. *Science* **272**, 1910–14.

Warren, R., Rogolsky, M., Wiley, B.B. and Glasgow, L.A. (1975). Isolation of extrachromosomal deoxyribonucleic acid for exfoliative toxin production from phage group II *Staphylococcus aureus*. *J. Bacteriol.* **122**, 99–105.

Watson, D.W. (1960). Host–parasite factors in group A streptococcal infection. Pyrogenic and other effects of immunologic distinct exotoxins related to scarlet fever toxins. *J. Exp. Med.* **111**, 255–84.

Weiss, A.A., Johnson, F.D. and Burns, D.L. (1993). Molecular characterization of an operon required for pertussis toxin secretion. *Proc. Natl Acad. Sci. USA* **90**, 2970–4.

Whelan, S.M., Garcia, J.L., Elmore, M.J. and Minton, N.P. (1994). The botulinum neurotoxin gene of the type A *Clostridium botulinum* strain NCTC 2916 is followed by a gene (*lycA*) encoding a lysozyme. *Zentralbl. Bakteriol. [Suppl.]* **24**, 162–3.

Winans, S.C., Burns, D.L. and Christie, P.J. (1996). Adaptation of a conjugal transfer system for the export of pathogenic molecules. *Trends Microbiol.* **4**, 64–8.

Wong, K.-K., McClelland, M., Stillwell, L.C., Sisk, E.C., Thurston, S.J. and Saffer, J.D. (1998). Identification and sequence analysis of a 27-kilobase chromosomal fragment containing a *Salmonella* pathogenicity island located at 92 minutes on the chromosome map of *Salmonella enterica* serovar *Typhimurium* LT2. *Infect. Immun.* **66**, 3365–71.

Xiang, Z.Y., Censini, S., Baeyli, P.F., Telford, J.L., Figura, N., Rappuoli, R. and Covacci, A. (1995). Analysis of expression of CagA and VacA virulence factors in 43 strains of *Helicobacter pylori* reveals that clinical isolates can be divided into two major types and that CagA is not necessary for expression of the vacuolating cytotoxin. *Infect. Immun.* **63**, 94–8.

Yahr, T.L., Goranson, J. and Frank, D.W. (1996). Exoenzyme S of *Pseudomonas aeruginosa* is secreted by a type III pathway. *Mol. Microbiol.* **22**, 991–1103.

Yamamoto, T. and Echeverria, P. (1996). Detection of the enteroaggregative *Escherichia coli* heat-stable enterotoxin 1 gene sequences in enterotoxigenic *E. coli* strains pathogenic for humans. *Infect. Immun.* **64**, 1441–5.

Zabriskie, J.B. (1964). The role of temperate bacteriophage in the production of erythrogenic toxin by group A streptococci. *J. Exp. Med.* **119**, 761–80.

Zagaglia, C., Casalino, M., Colonna, B., Conti, C., Calconi, A. and Nicoletti, M. (1991). Virulence plasmids of enteroinvasive *Escherichia coli* and *Shigella flexneri* integrate into a specific site on the host chromosome: integration greatly reduces expression of plasmid-carried virulence genes. *Infect. Immun.* **59**, 792–9.

SECTION I

TOXINS ACTING ON CYTOSOLIC TARGETS

2

The Ras superfamily of small GTP-binding proteins as targets for bacterial toxins

Patrice Boquet

INTRODUCTION

From the early 1980s to the late 1990s, important breakthroughs took place in the research fields of both signal transduction and bacterial toxinology. Indeed, a new family of small-molecular-mass GTP-binding proteins, namely the Ras superfamily, was progressively discovered and its cellular functions unravelled. Almost concomitantly, it was shown that, in their armamentarium, certain bacteria had protein toxins that selectively modify these newly discovered GTP-binding proteins.

To date, more than 70 different small GTP-binding proteins (also named GTPases owing to their ability to hydrolyse GTP into GDP) have been described. Also, more than 12 bacterial toxins or bacterial enzymes modifying some of these GTPases have been isolated and their precise activities at the molecular level elucidated. Some of these newly discovered bacterial toxins also had new toxic enzymic activities (such as glucosyl-transferases or deamidases). Amazingly, the small GTP-binding protein, Rho, was found to be the substrate of three different toxins: *Clostridium botulinum* exoenzyme C3 (C3), *C. difficile* toxins B (TCdB) and *Escherichia coli* cytotoxic necrotizing factor 1 (CNF1). Each of them exhibited a different enzymic activity (ADP-ribosyltransferase, glucosyl-transferase and deamidase, respectively) able either to inactivate or to activate Rho.

The aims of this chapter are to describe the molecular activities of the Ras superfamily proteins, and to understand how toxins interfere with these molecules and why some of these GTP-binding proteins have been selected as targets for toxins.

THE RAS PARADIGM

Ras is the prototype of a regulatory protein involved in transmembrane signalling. A major feature of Ras, as well as of the other small GTP-binding proteins, is the absolute requirement for membrane attachment in order to exert their activities (reviewed by Downward, 1990). Owing to the covalent attachment of a farnesyl (a precursor of cholesterol) and a palmitate molecule to its carboxy-terminal domain (named the CAAX box) (Hanckock *et al.*, 1989), Ras is permanently located on the cytoplasmic face of the cell membrane.

The p21 Ras molecule was first reported as a transforming protein associated with animal tumour viruses, but it was rapidly demonstrated that Ras had a normal counterpart in cells (Ellis *et al.*, 1981). The difference between the tumour-inducing molecule and its normal cellular analogue was a point mutation at glycine 12, glycine 13 or glutamine 61 (reviewed by Barbacid, 1987). Mutations lead to the inability of the protein to hydrolyse GTP to GDP, thus leaving the molecule permanently in its active state.

The crystal structure of Ras has been solved to high resolution (Pai *et al.*, 1990). It shows the structural domains involved in nucleotide binding and in GTP

27

hydrolysis. Two domains of Ras undergo conformational change from the GTP-bound to GDP-bound state. These domains are so-called 'switches' (reviewed by Wittinghofer and Pai, 1991; Wittinghofer and Valencia, 1995; Wittinghofer and Nassar, 1996). Switch 1 (residues 32–40) is involved in the interaction with the Ras downstream effector, Raf (Nassar et al., 1995). Switch 2 covers residues 60–76 and is implicated in the hydrolysis of GTP. Membrane localization, binding of GTP, hydrolysis of GTP into GDP, switch 1 and 2 functions and conformational changes are the key features of the p21Ras superfamily GTPases.

One of the most interesting features of Ras is its activation/deactivation cycle, central to understanding its role as an intracellular regulator. By its ability to turn on, just for a certain period, a cascade of events, Ras is an intracellular timer for biochemical reactions. To achieve this task, Ras uses the gamma phosphate of GTP to induce structural deformations in both Ras switch 1 and 2 domains. This is performed by attracting, through ionic interactions, tyrosine 32 (Y32), threonine 35 (T35) of switch 1 and glycine 60 (G60) of switch 2 on to the gamma phosphate of GTP, pulling these two domains toward the gamma phosphate of GTP in the Ras molecule. This subsequent deformation leads to the activated state of the GTPase. Hydrolysis of GTP into GDP releases the internal tension exerted by the gamma phosphate on Y32, T35 and G60, allowing both switches 1 and 2 of Ras to return back, like a spring, to the inactive state of the GTPase (Wittinghofer and Nassar, 1996).

In addition to this basic mechanism, two groups of molecules interact with Ras to regulate the activation/deactivation cycle.

Ras has a high affinity for GDP and GTP in the presence of Mg^{2+} ($K_m = 10^{-10}$ M^{-1}). Therefore, a protein factor (named GEF, for guanine exchange factor) must remove the bound GDP from inactive Ras. Following removal of GDP, a rapid binding of GTP to the GTPase takes place, since there is a higher concentration of GTP than GDP in the cytosol. The mechanism by which RasGEF (named SOS, from a mutation found in *Drosophila* called son of sevenless; Simon et al., 1991) promotes GDP release has recently been established (Boriack-Sjodin et al., 1998). SOS first inserts into Ras an α-helix, which results in the displacement of switch 1, opening up the nucleotide binding site. Then sidechains belonging to that α-helix act at the level of switch 2 of Ras, altering the chemical environment of the β phosphate group of GDP and its associated magnesium ion, resulting in GDP release. SOS does not block the binding sites for the ribose and base moieties of GTP, allowing nucleotides to be released and to rebind to the GTPase (reviewed by Wittinghofer, 1998).

As in all GEFs isolated, Ras GEF contains two important functional polypeptidic regions: one involved in the exchange of GDP to GTP, the DH domain (for Dbl homology) (reviewed by Cerione and Zheng, 1996) and one involved in phospholipid binding, the PH domain (for plesktrin homology) (reviewed by Lemmon et al., 1996). The PH domain probably plays a pivotal role in recruiting GEFs to membranes, thus in the close vicinity of small GTPases. The PH domain binds phosphatidylinositol-3,4,5-phosphate (PIP3), a lipid generated by phosphorylation of the phosphatidylinositol phospholipid (PI), located on the cytosolic face of the lipid membrane. PI3-kinase, which performs the phosphorylation of PI to PIP3, is activated by tyrosine kinase receptors such as growth factor receptors (reviewed by Toker and Cantley, 1997). Activation of the PI3-kinase, creating areas rich in PIP3, might target GEFs to membranes in close proximity to small GTPases. For Ras, however, a more direct link between growth factor receptors of the tyrosine kinase family and SOS takes place. By binding to a protein (Grb2) named 'adaptor' (reviewed by Pawson, 1995), SOS recognizes phosphorylated tyrosines of activated tyrosine kinase receptors (Chardin et al., 1993). In fact, a dual system, consisting of the binding of growth factor receptor, directly or adaptator-mediated, and a membrane attachment by protein–lipid interactions, may target small GTPase GEFs to membranes.

To turn active Ras into the inactive state, the Ras-bound GTP molecule must be hydrolysed into GDP. The rate of hydrolysis of GTP by the sole intrinsic GTPase activity of Ras is very slow. It can be accelerated several thousand times through interaction with a protein named GTPase activating protein (RasGAP) (Trahey and McCormick, 1987). The Ras switch 2 domain is implicated in catalysing the hydrolysis of GTP (Wittinghofer and Nassar, 1996). The binding of a water molecule by Ras glutamine 61 (Q61) in the switch 2 domain is central to performing GTP hydrolysis leading to the release of inorganic phosphate. Switch 2 of Ras is a highly mobile polypeptide and therefore inefficient for achieving the transition state required for a high rate of hydrolytic activity (Scheffzek et al., 1997). The RasGAP reduces the mobility of switch 2, and by introducing within Ras an arginine residue (named the arginine finger; Bourne, 1997) in close proximity to the Ras Q61 residue, this also stabilizes the water molecule bound to Q61 in the transition state, therefore increasing tremendously the catalytic activity of the GTPase (Scheffzek et al., 1997).

There is a great similarity between mechanisms inducing GTP hydrolysis in small and heterotrimeric GTPases. Indeed, both types of GTP-binding protein require a switch 2-stabilizing domain and an arginine finger to achieve GTP hydrolysis (Bourne, 1997). The switch 2-stabilizing domain and the arginine finger are integrated in heterotrimeric GTPases and localized in a separate protein (GAP) in small GTPases (Bourne, 1997). Bacterial toxins have taken advantage of this mechanism of GTP hydrolysis to manipulate GTP-binding proteins. Cholera toxin ADP-ribosylates the arginine 201 residue of αGs, which plays the role of arginine finger in this heterotrimeric G-protein subunit. This hinders αGs from hydrolysing GTP via destabilization of the water molecule required for the hydrolysis step (reviewed by Bourne *et al.*, 1991). Conversely, cytotoxic necrotizing factor 1 (CNF1), a toxin from *E. coli*, deamidates Rho glutamine 63 (homologous to glutamine 61 of Ras) into glutamic acid, impairing the binding of the water molecule required for Rho GTP hydrolysis (Flatau *et al.*, 1997; Schmidt *et al.*, 1997). It is remarkable to observe that cholera toxin and CNF1 achieve the same goal, which is to activate permanently the GTP-binding proteins, by blocking GTP hydrolysis activities through manipulation of two different residues involved in the same reaction.

The Ras paradigm represents the basic mechanism of small GTPases. This applies to other members of the Ras superfamily although, as described above, the Ras cycle is modified in other small GTPases. Indeed, a large majority of these regulatory molecules do not reside permanently on membranes like Ras and, therefore, must be kept in an inactive state within the cytosol.

AP

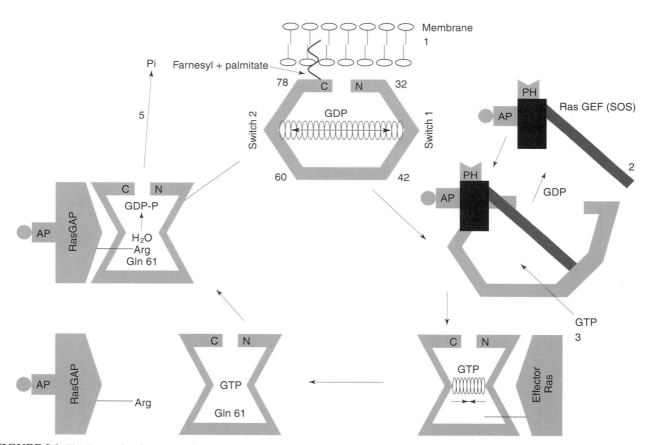

FIGURE 2.1 The Ras cycle of activation/inactivation is described through steps 1 to 5. (1) Ras is associated with the plasma membrane by lipidic modifications at its *C*-terminus end. RasGDP is inactive owing to the fact that switches 1 and 2 are exposed at the surface of the molecule (the relaxed spring indicates that switches 1 and 2 are pushed in opposite directions by structural constraint). (2) Ras GEF (SOS) (with a PH and adaptor AD domains) binds to switch 2 and opens switch 1 by pushing on this domain through a helix structure (shaded grey). This opens the Ras molecule, allowing GDP to dissociate. (3) GTP enters Ras. Owing to the attraction of switches 1 and 2 on the γ-phosphate of GTP, these domains are pulled one to each other (the compressed spring indicates this mechanism). Under this conformation, Ras binds and activates its downstream effector (Raf). (4) RasGAP [with an adaptor (AD) domain] stabilizes the switch 2 domain. (5) By introducing an arginine finger in the vicinity of Ras glutamine 61, RasGAP stabilizes the transition state and allows rapid hydrolysis of GTP into GDP. The reaction then returns to step 1.

As pointed out above, bacterial toxins which interfere with small GTPases have taken advantage of the structure and the mechanism of activation of these regulatory molecules. Indeed, toxins which inhibit small GTPases will modify their switch 1 domain, whereas those activating them will act on their switch 2 domain. Figure 2.1 summarizes the Ras paradigm.

FROM RAS TO THE RAS SUPERFAMILY

The discovery of the Rho protein from the marine snail *Aplysia californica*, then of *Saccharomyces cerevisiae* YPT1, and sec4 (two so-called Rab-like GTPases) immediately after Ras, started a long list of incoming members of the Ras superfamily. Later on, this super-family was divided into several groups on the basis of their sequence homologies and their cellular functions. Ras, Rho and Rab subfamilies were shown to control each of three apparently non-overlapping activities in cells (reviewed by Hall, 1990): the growth control by Ras, the intracellular vesicular trafficking by Rab and the actin cytoskeleton organization by Rho. This distinction is actually less clear and, apparently, the Ras and Rho subfamilies can be regrouped. Indeed, the Ras subfamily, in addition to its role in controlling cell growth, is able to regulate actin cytoskeleton organization (in fact via activation of Rho GTPases), whereas the Rho subfamily, in addition to its control of the actin cytoskeleton, regulates cell growth (reviewed by Hall, 1998a). A major difference between Ras subfamily and the Rho and Rab subfamilies must be emphasized. Rho and Rab proteins, in contrast to Ras, are not permanently associated with membranes, but are transferred to membranes only when they need to perform their tasks. In the cytosol, Rho and Rab are associated with a protein named guanine dissociation inhibitor (GDI) (Fukumoto *et al.*, 1990). There are different GDIs for Rho and Rab GTPases, although they have a broad specificity. For instance, RhoGDI can associate with Rho, Rac and Cdc42. RhoGDI or RabGDI maintain Rho or Rab GTPases in an inactive state (bound to GDP) within the cytosol. In addition, GDIs may direct or withdraw the GTPases to or from membranes.

Two other groups of small GTPases then joined the Ras superfamily, namely ARF and Ran. ARF and Ran regulate two other basic cellular mechanisms. Certain ARF molecules activate phospholipase D (reviewed by Cockcroft, 1997) and subsequently control the pro-cessing of lipids on the cytosolic face of membranes and the formation of vesicles from membrane donor compartments, such as the Golgi apparatus or the endoplasmic reticulum (reviewed by Alb *et al.*, 1996). Ran controls the transport of proteins containing a nuclear localization signal (NLS) through nuclear pores (Moore and Blobel, 1993).

The Ras subfamily encompasses three main mem-bers: Ras, Rap and Ral. All three are substrates for the lethal toxin (LT) of *Clostridium sordellii* (Popoff *et al.*, 1996; Just *et al.*, 1996; Hofmann *et al.*, 1996). Ras is also one of the preferred substrates for *Pseudomonas aerugina* exoenzyme S (McGuffie *et al.*, 1998).

The Rho subfamily encompasses 9 members (Rho A, B, C, G, E, Rac1, Rac2, Cdc42/G25K and Rnd1). In Swiss 3T3 mouse cells there is a clear hierarchy among Rho, Rac and Cdc42 (reviewed by Tapon and Hall, 1997). Cdc42 activates Rac which in turn activates Rho (Nobes and Hall, 1995). Ras can also activate Rac and Rho (Ridley *et al.*, 1992). Rho and Rac are not farnesy-lated, like Ras, but geranyl-geranylated at their *C*-ter-minal end (Katayama *et al.*, 1991) (except for Rho B which can be either farnesylated or geranyl-geranyl-lated) (Adamson *et al.*, 1992). RhoGDP and RacGTP 3D structures are known (Wei *et al.*, 1997; Hirshberg *et al.*, 1997). The 3D structure of the RhoGAP domain inter-acting with Rho (Barrett *et al.*, 1997) and the 3D Rho structure in association with its RhoGAP (Rittinger *et al.*, 1997) have also been established. Structures of Rho and Rac are, as expected, similar to that of Ras. However, one major deviation from the Ras structure is the presence of an insert (residues 124–136 in Rho, 122–134 in Rac) which is a short three-turn helix. The role of this helix is unknown.

Members of the Rho subfamily are clearly the pre-ferred targets of bacterial toxins.

The Rab subfamily has now expanded to more than 50 members. It is noteworthy that no bacterial toxin has been found that specifically modifies a Rab GTPase. Tetanus and botulinum neurotoxins (reviewed by Boquet *et al.*, 1998) and *Helicobacter pylori* VacA (reviewed by Cover, 1996), which interfere with intra-cellular vesicular trafficking, exert their toxic activities through strategies different from the direct modifica-tion of small Rab GTPases. This might indicate that Rabs may be too redundant and, thus, will not be reliable targets for bacterial toxins.

The ARF subfamily contains six main subtypes (ARF1–ARF6) (reviewed by Kahn, 1995), whereas Ran is still a lonesome molecule in higher eukaryotes (reviewed by Bischoff and Coutavas, 1995). ARFs and Ran are the most divergent members of the Ras superfamily. ADP-ribosylating factor (ARF) was first isolated as a factor required for ADP-ribosylation of the αGs subunit by cholera toxin (Kahn and Gilman, 1986). Curiously, ARF also binds the complex formed between αGi and the ADP-ribosylating pertussis toxin

(PTX) but, in contrast to cholera toxin, is not required for the enzymic activity of PTX. Unlike Ras, Rho and Rac GTPases, ARF is attached to the membrane by its amino-terminus via a covalently linked myristic acid (Kahn *et al.*, 1988). ARFs GTPases are regulated by a GEF and a GAP but, to date, no GDI for ARFs has been isolated. Certain ARF exchange factors are the target of the fungal toxin brefeldin A (Donaldson *et al.*, 1992), which inactivates these GEFs.

To date, no bacterial toxins have been found that modify ARFs or Ran.

CELLULAR DOWNSTREAM EFFECTORS AND EFFECTS OF THE RAS GTPASES SUPERFAMILY

Owing to the abundance of data on this topic, it has become quite difficult to give a short overview on this aspect of small GTPases. Since this book is devoted to bacterial toxins, this section shall focus on those small GTPases that are substrates of bacterial toxins. As the Rho GTP-binding proteins are the most usual targets of toxins, downstream effectors and effects of Rho GTPases will be described in detail. However, a short description of the Ras downstream effectors will first be given. Indeed, a large clostridial toxin (the LT from *C. sordellii*) affects members of the Ras family of proteins (Popoff *et al.*, 1996; Just *et al.*, 1996) and the lethal factor (LF) of *Bacillus anthracis* was shown recently to proteolyse and to inactivate two well-known downstream effectors of Ras (Duesbery *et al.*, 1998).

Ras GEF (SOS) is activated by growth factors when they bind to their tyrosine kinase receptors such as EGF or PDGF receptors (reviewed by van der Geer, 1994). The Ras downstream effector pathway (named the Ras/ERK pathway) starts with the serine/threonine kinase Raf (reviewed by Wood 1995). Raf is activated by RasGTP by attachment to the plasma membrane (Leevers *et al.*, 1994). Raf phosphorylates the double-specificity threonine/tyrosine kinases Mek1 and Mek2, which are then active via their combined phosphorylation activities (reviewed by Marshall, 1994). Mek1 and 2 in turn activate the MAP kinases (ERK1 and ERK2). ERK1 and 2 are serine/threonine kinases with broad-range target specificities (reviewed by Leevers and Marshall, 1995). Upon phosphorylation, ERKs can translocate to the nucleus and activate transcription factors such as c-jun (Pulverer *et al.*, 1991). In addition, ERK1 and ERK2 phosphorylate and activate cytoplasmic proteins such as microtubule associated proteins, phospholipase A2 and the S6 kinase.

Activation of the Ras/Raf/Mek/ERK pathway leads to either cell proliferation (Meloche *et al.*, 1992) or cell differentiation (Noda *et al.*, 1988; Traverse *et al.*, 1992; Wood *et al.*, 1993). Signals that differentiate proliferation from differentiation are not totally understood, but the length of time during which ERKs are phosphorylated, and thus can be transferred and act in the nucleus, seems to play a pivotal role (Marshall, 1995). *Clostridium sordellii* LT toxin inactivates Ras, Rap and Rac (Popoff *et al.*, 1996; Just *et al.*, 1996). Treatment of NIH 3T3 cells with LT abrogates the ability of EGF to activate MAP kinases; cells then resume growth and die (Popoff *et al.*, 1996; Just *et al.*, 1996). The LF of the tripartite toxin of *B. anthracis* (reviewed elsewhere in this book; see Chapter 12) contains a consensus sequence for zinc metalloprotease (Klimpel *et al.*, 1994). The targets for this metalloprotease activity are Mek1 and Mek2. LF releases a small peptide from the amino-terminus of Mek1 and Mek2, inactivating these proteins and inhibiting the downstream signal transduction pathway (Duesbery *et al.*, 1998).

An example of how bacterial toxins can be useful tools in cell biology is illustrated by the discovery of the Rho cellular functions. Indeed, the first clue for Rho cellular activity was found when it was observed that *C. botulinum* C3 transferase exoenzyme, shown to inactivate Rho, caused dissociation of actin stress fibres in cultured Vero cells (Chardin *et al.*, 1989). Based on this observation, it was later shown that microinjection of activated recombinant Rho protein into quiescent Swiss 3T3 fibroblasts induced the assembly of focal adhesions and actin stress fibres (Paterson *et al.*, 1990). The identification of three new signal transduction pathways linking growth factors, their receptors and the actin cytoskeleton organization via Rho, Rac and Cdc42 (Ridley and Hall, 1992; Ridley *et al.*, 1992; Nobes and Hall, 1995; Kozma *et al.*, 1996) then initiated a new field of investigations in the scientific community. Rho links membrane receptors activated by extracellular factors such as lysophosphatidic acid (LPA), bombesin or thrombin to the formation of actin stress fibres and focal adhesion contacts (Ridley and Hall, 1992). LPA, bombesin and thrombin are growth factors acting through signalling via heterotrimeric G-protein receptors. For a long time, the link between heterotrimeric G-protein receptors and small GTPases of the Rho family was not fully understood. Very recently, a first clue to this link was found. Heterotrimeric G-proteins of the $G\alpha13$ family were shown directly to activate a RhoGEF (p115RhoGEF) The p115RhoGEF is a GAP for $G\alpha13$ (Kozasa *et al.*, 1998) since it contains a consensus sequence of RGS proteins that play the role of GAPs for heterotrimeric G-proteins (reviewed by Hall, 1998b).

The GTP-bound form of Gα13 binds to the RGS domain of p115 RhoGEF (Hart *et al.*, 1998). Binding of Gα13GTP to p115RhoGEF unfolds the molecule, exposing the DH domain, thereby activating its exchange activity for Rho (Hart *et al.*, 1998).

Rac1 links tyrosine kinase receptors, such as the EGF or PDGF receptors, probably by activation of the PI-3 kinase to actin polymerization at the level of the plasma membrane (Ridley *et al.*, 1992) in a process termed 'membrane ruffling', leading to the formation of a thin cellular membrane extension (lamellipodium) made from actin filaments. Lamellipodia can be observed best in migrating fibroblasts or during neuronal growth cone extension (reviewed by Machesky and Hall, 1996; Hall, 1998b). In addition, a specialized function for Rac has been described in the control of NADPH oxidase activation implicated in the generation of superoxide radicals in professional phagocytic cells (Abo *et al.*, 1991).

Cdc42 links membrane receptors (activated by bradykinin, for instance) to the formation of finger-like tips extensions, termed filopodia, formed by actin filaments (Nobes and Hall, 1995; Kozma *et al.*, 1996). Filopodia, as lamellipodia, are also involved in cell spreading and migration.

Over the past few years, some of the events leading from Rho, Rac or Cdc42 activation to the formation of actin stress fibres, lamellipodia or filopodia have been elucidated. As for Ras GTPases, Rho, Rac and Cdc42 activate downstream kinases, which, in turn, control activities involved in either actin contraction (Rho) or actin polymerization (Rho, Rac and Cdc42). These mechanisms will be briefly described.

Rho-GTP binds and activates a serine/threonine kinase, the Rho kinase (a 160-kDa protein termed ROK) (Matsui *et al.*, 1996; Ishizaki *et al.*, 1996), which subsequently phosphorylates the myosin-binding subunit of myosin-light chain phosphatase, causing inactivation of the phosphatase and, therefore, accumulation of phosphorylated myosin light chain (Kimura *et al.*, 1996). This activity results in an increase in phosphorylated myosin-light chains which interact with actin filaments and induce stress fibre formation by bundling and contractility (Chrzanowska-Wodnicka and Burridge, 1996). In addition, it was shown that Rho (although probably not directly by ROK) could activate ERM proteins (for ezrin, radixin and moesin) (reviewed by Tsukita *et al.*, 1997) and link transmembrane receptor proteins (such as CD44) to actin filaments (Hirao *et al.*, 1996). Binding assays revealed that ezrin interacts with its membrane receptor CD44 only when Rho is in the GTP-bound form (Hirao *et al.*, 1996). In permeabilized cells, moesin was essential for the formation of focal contacts induced by activated Rho (Mackay *et al.*, 1997).

Phosphatidylinositol-4-phosphate 5-kinase (PIP-5 kinase), which converts phosphatidylinositol-phosphate (PIP) to PIP2, is also a target for Rho (Chong *et al.*, 1994). How Rho activates this enzyme is not clear, since Rho seems to interact with PIP-5 kinase in both the GDP and GTP-bound forms (Ren *et al.*, 1996). The toxin CNF1, by activating Rho, activates a PIP-5 kinase associated with the cytoskeleton (Fiorentini *et al.*, 1997a). PIP2 is a potent regulator of several actin binding proteins, such as vinculin (Gilmore and Burridge, 1996). PIP2 also, by lowering the affinity of the capping protein CapZ for the actin barbed end filaments, allows reinitiation of actin polymerization (Schafer *et al.*, 1996). Thus, Rho may also regulate the actin cytoskeleton by PIP2-dependent mechanisms.

Rac affects the actin cytoskeleton probably via activation of a kinase. Recently, the kinase LIMK-1 has been shown to phosphorylate and to inactivate the actin depolymerizing filament factor, cofilin, leading to accumulation of actin filaments (Arber *et al.*, 1998; Yang *et al.*, 1998). Rac also activates the PIP-5 kinase in platelets (Hartwig *et al.*, 1995) and the actin-severing protein gelsolin in cultivated fibroblasts (Azuma *et al.*, 1998).

Downstream targets of Cdc42 are still less well defined than those of Rac. The product of the Wiskott–Aldrich syndrome gene, WASP, is a target for Cdc42 (Symons *et al.*, 1996). WASP has no kinase activity and is probably a simple adaptor molecule, which can bind profilin by a proline-rich motif. Profilin is an actin-binding protein able to both sequester actin monomers and promote actin polymerization by exchanging ADP to ATP on G-actin, allowing rapid growth of filaments (Pantaloni and Carlier, 1993; Theriot and Mitchison, 1993). In addition, a 65-kDa protein isolated from brain, homologous to WASP (named N-WASP for neuronal WASP) constitutes a downstream target for Cdc42 (Miki *et al.*, 1996). Unlike WASP, N-WASP induces long filopodia when overexpressed in cells (Miki *et al.*, 1998). Very interestingly, N-WASP contains at its carboxy-terminal end a region homologous to cofilin which is hidden in the resting state of N-WASP. Upon association with GTP-bound Cdc42, the cofilin-like domain of N-WASP is exposed and can depolymerize (at least partially) actin filaments, exposing uncapped barbed ends. At the level of the barbed end, filaments can induce actin polymerization, using profilin, which loads G-actin monomers with ATP, therefore producing filopodia (Miki *et al.*, 1998). Cdc42 can probably also modulate the activity of the PIP-5 kinase, at least in platelets (Hartwig *et al.*, 1995).

It must be emphasized that Rac and Cdc42, by their ability to polymerize actin at the barbed ends, can elongate pre-existing filaments but apparently cannot initiate the polymerization of the first actin monomers, a step called nucleation. The nucleation of actin filaments was recently attributed to molecules named Arps (actin related proteins). In the close vicinity of the cell membrane, Arps induce nucleation of actin by binding to the pointed end of an actin dimer, allowing polymerization at the barbed end (reviewed by Machesky and Way, 1998). By their activities on actin-binding proteins, Rac and Cdc42 might elongate the first nucleus of polymerized actin, resulting in thin filament formation. Cross-linking of these thin actin filaments via actin cross-linker proteins (such as fimbrin or α-actinin) (reviewed by Matsudeira, 1991) will then result in the formation of either filopodia or lamellipodia.

Members of the Rho subfamily of GTPases monitor gene transcription, as does Ras, in addition to its effect on the control of actin cytoskeleton organization. These two processes are apparently accomplished by two different non-overlapping parts of their effector domains (Lamarche et al., 1996). Rac and Cdc42 control the JNK/P38 MAP kinase pathways (Minden et al., 1995) by phosphorylating (like ERKs) transcription factors (Bagrodia et al., 1995). Rho controls the binding of the transcription factor SRF to the DNA enhancer element SRE by a still unknown phosphorylation-independent mechanism (Hill et al., 1995).

By their activities on the control of both the actin cytoskeleton organization and gene transcription, the Rho family of GTPases is involved in a large variety of cellular processes, such as cell polarity determination, phagocytosis, epithelial and endothelial cell junctions, intestinal cell migration, cell cycle, apoptosis and neuronal plasticity.

Cell polarity is a process by which a cell will be divided into two domains, the apical and the basolateral, each exhibiting separate functions and a different pattern of membrane proteins and lipids. The first event occurring when a non-polarized cell undergoes polarization to become an epithelial, endothelial or nerve cell, is the organization of its actin cytoskeleton together with its genes transcription program in response to specific cues (reviewed by Drubin and Nelson, 1996). Rho GTPases can co-ordinate these two events. The best example of cell polarity control by Rho GTPases is the formation of the bud site in the yeast *Saccharomyces cerevisiae*. This site determines cell polarity with actin cables organized along the axis of the cell, actin patches in the bud loosely associated with the inner face of the membrane, and a ring of actin spots forming a collar around the bud. Cdc42 is essential for bud formation, since mutation in either Cdc42 or the Cdc42 GEF (Cdc43) blocks budding and leads to uni-

form cell surface growth and disorganization of the actin cytoskeleton (reviewed by Hall, 1998a).

Phagocytosis of antibody-opsonized particles by macrophages is a sequential process triggered by the interaction of particles with surface Fc receptors. This results in the development of membrane extensions, forming the phagocytic cup, which engulf the particles (reviewed by Greenberg and Silverstein, 1993). Actin filaments are thought to provide the driving force for phagocytosis. The fact that polymerized actin and phosphorylated tyrosine accumulates around the phagocytic cup (Greenberg et al., 1993) indicates that this process might depend on Rho GTPases. Indeed, Rho was essential for accumulation of phosphotyrosine and F-actin filaments around the phagocytic cup (Hackam et al., 1997). In addition to the role of Rho in F-actin and phosphotyrosine accumulation, Rac and Cdc42 both exert a pivotal activity for closing the phagocytic cup (Massol et al., 1998). Studies on human macrophages indicated that phagocytosis involving complement receptor (CR3) was down-regulated when Rho was permanently activated (Capo et al., 1998). Rho and Rac are also required for macrophage chemotaxis, whereas Cdc42 is required for cells to respond to a gradient of the colony-stimulating factor 1 (CSF-1) but not essential for locomotion (Allen et al., 1998).

Epithelial cells are linked together by a junctional complex comprising adherens junctions, tight junctions and desmosomes (reviewed by Gumbiner, 1996). Adherens and tight junctions are linked to actin filaments, whereas desmosomes are linked to intermediate filaments. Cadherins are responsible for the Ca^{2+}-dependent cell–cell adhesion in adherens junctions. Occludins are transmembrane proteins localized in tight junctions (Furuse et al., 1993). Their cytoplasmic tail interacts with proteins such as ZO1, ZO2 and actin filaments (reviewed by Mitic and Anderson, 1998). Epithelial cell–cell adhesion is controlled by the Rho GTPases, which regulate the formation of the tight junctions in polarized intestinal epithelial T84 and HT29 cells (Nusrat et al., 1995). The *Drosophila* dRac regulates actin assembly at the adherens junctions of the wing disc epithelium (Eaton et al., 1995). Rho and Rac are required for the establishment of cadherin-dependent cell–cell contacts in human keratinocytes (Braga et al., 1997), and Rac and Rho are involved in the regulation of cell–cell adhesion of MDCK cells (Takaishi et al., 1997). Rho, by inducing smooth acto-myosin contraction, might drive the opening/closing mechanism of occludin molecules in tight junctions, probably via contraction/relaxation of the perijunctional actin ring. Rac could allow the homophilic binding of cadherins by inducing the formation of lamellipodia at the membrane.

In the intestine, endogenous growth factors stimulate the migration of epithelial crypt cells to villi, thus maintaining tissue integrity. Rho GTPases might play an essential role in the initial phase of cell migration during mucosal restitution (Santos *et al.*, 1997).

RasGTP, by activating Raf, activates the MAP kinases pathway and triggers progression from the G1 to the S phase of the cell cycle. Excessive signalling from Ras/Raf induces a cyclin-dependent-kinase inhibitor (p21Waf1/Cip1), thereby blocking the entry of cells into the S phase. When activated, Rho blocks the inhibitory effect of Waf1/Cip1 on the cell cycle entry (Olson *et al.*, 1998).

Rho may also be important in the control of cell survival or apoptosis. For instance, Rho plays a selective role in early thymic development and its inactivation leads to apoptosis in murine T-lymphoma cells (Moorman *et al.*, 1996; Henning *et al.*, 1997). A link between the functional state of Rho and the anti-apoptotic factor Bcl2 has been reported (Gomez *et al.*, 1997; Fiorentini *et al.*, 1998a).

Neuronal plasticity requires the regulation of synaptic efficacy involving control of the delayed rectifier Shaker family of potassium channels (Meiri *et al.*, 1997). Rho regulates the Kv1.1 channel activity of the Shaker family by controlling its tyrosine phosphorylation state or by a direct physical association with this channel (Cachero *et al.*, 1998).

Finally, it has been claimed that Rho and Rac regulate receptor-mediated endocytosis (Lamaze *et al.*, 1996) and that a new class of Rho GTPases, namely RhoD (Murphy *et al.*, 1996), might be involved in vesicular transport governing early endosome mobility and distribution (Murphy *et al.*, 1996). However, these data await confirmation.

COVALENT MODIFICATIONS OF SMALL GTP-BINDING PROTEINS INDUCED BY BACTERIAL TOXINS

Two types of bacterial toxins acting on small GTPases can be identified, depending on whether they inactivate [exoenzyme C3 from *Clostridium botulinum* and related proteins, *Clostridium difficile* toxin A and B, *Clostridium sordellii* lethal (LT) and haemorrhagic (HT) toxins and *Clostridium novyi* α-toxin] (reviewed by von Eichel-Streiber *et al.*, 1996; Aktories, 1997; Boquet *et al.*, 1998) or activate [cytotoxic necrotizing factor 1 (CNF1) from *Escherichia coli*, dermonecrotic toxin (DNT) from *Bordetella pertussis*) small GTP binding proteins (reviewed by Aktories, 1997; Fiorentini *et al.*, 1998b].

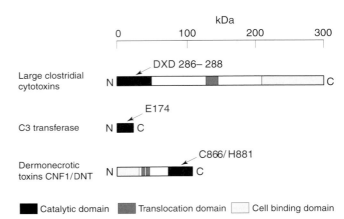

FIGURE 2.2 Comparison of the primary structure of toxins inactivating or activating small GTP-binding proteins. The lengths of the whole toxin and the domains are proportional. The function of each domain is explained at the bottom of the figure. The amino acids DXD 286–288, E174, C886 and H881 refer to toxin catalytic sites. Relative molecular masses are in kilodaltons (kDa). Large clostridial cytotoxins are: *C. difficile* toxins A and B, *C. sordellii* LT and HT toxins and *C. novyi* α-toxin.

Toxins inactivating the Rho subfamily of GTPases, particularly toxins A and B from *C. difficile* and *C. sordellii*, are reviewed elsewhere in this book. Readers are therefore referred to Chapter 8 for details. As a general outline concerning the molecular mechanisms of these toxins, we can say that (a) toxins inactivating small GTP-binding proteins do so by modifying an amino acid residue of the effector domain (switch 1) of the protein, (b) switch 1, being exposed on the surface of the GTPases when they are bound to GDP, will be a better substrate for toxins in this conformation, (c) modification of the switch 1 domain will impair the binding of GTPases to their downstream effectors by steric hindrance (see Figure 2.1).

Two different types of enzymic activities are borne by bacterial toxins that inactivate small GTP-binding proteins. The first enzymic toxin shown to be directed to an identified small GTP-binding protein was the *C. botulinum* ADP-ribosyl transferase exoenzyme C3 (Chardin *et al.*, 1989). For more details about the molecular mode of action of ADP-ribosyltransferases, readers must consult Chapter 3 on this topic in this book. Exoenzyme C3 is not a toxin, since it does not contain or bind to a polypeptide domain able to allow its translocation into the cytosol. Therefore, C3 must be either incubated with cells at high concentration (5–40 µg ml^{-1} depending on the cell species) or fused to a protein allowing membrane translocation such as a diphtheria toxin B fragment (C3DTB) (Boquet *et al.*, 1995). Microinjection of C3 (Paterson *et al.*, 1990) or direct expression of the C3 DNA by transfection (Hill *et al.*, 1995) is also commonly used to introduce this enzyme to cultured cells. C3 ADP-ribosylates specifically Rho and exhibits only 1/100 of its activity on Rac and Cdc42. C3 modifies Rho on asparagine 41 (Sekine *et al.*, 1989), which belongs to the switch 1 domain of Rho. Asparagine 41 of Rho is solvent accessible in the RhoGDP structure (Wei *et al.*, 1997) but this residue is masked when the GTPase is associated to RhoGDI (Bourmeyster *et al.*, 1992). Since 90% of the cellular Rho is associated with RhoGDI in the cytosol, *in vivo* ADP-ribosylation of Rho by C3 is probably rate-limited by the slow, spontaneous equilibrium dissociation of the RhoGDP-GDI complex that makes asparagine 41 accessible for C3 (Fujihara *et al.*, 1997). Upon ADP-ribosylation, Rho can reassociate with its GDI (Fujihara *et al.*, 1997). When GTPγS is added to permeabilized cells, cytosolic Rho is translocated from the cytosol to the membrane (Gong *et al.*, 1997). ADP-ribosylation of Rho by C3 totally blocks the membrane translocation of Rho induced by GTPγS (Fujihara *et al.*, 1997). Therefore, inhibition of Rho-mediated effects by ADP-ribosylation is due to the impediment of the GTPase to associate with a membrane-bound protein.

The second enzymic activity exerted by a bacterial toxin to inhibit small GTP-binding proteins is the glucosyl-transferase activity. This activity was first found associated with *C. difficile* B and A toxins (Just *et al.*, 1995a, b). It was later extended to *C. sordellii* LT, HT toxins and *C. novyi* α-toxin (Popoff *et al.*, 1996; Just *et al.*, 1996; Genth *et al.*, 1996; Selzer *et al.*, 1996). This activity consists of the transfer of a glucose (or *N*-acetyl glucosamine in the case of *C. novyi* α-toxin) moiety from UDP-glucose to the threonine 37 of Rho (or threonine 35 of Rac, Ras, Rap and Cdc42). The structure of Rho bound to GDP shows that threonine 37 is exposed outside the molecule and that the switch 1 domain is well ordered and can thus be recognized easily by glycosyltransferase toxins (Wei *et al.*, 1997). Glycosylation of residues within the switch 1 domain renders the Rho and Ras GTPases inactive owing to impairment of the binding of modified GTPases to their effectors (Herrmann *et al.*, 1998). Effects of glucosylation of Ras by LT show that the modification does not modify the nucleotide binding ability of Ras (Herrmann *et al.*, 1998) but reduces its intrinsic GTPase activity and completely blocks the p120RAsGAP GTPase activity owing to the failure of RasGAP to bind to the glucosylated GTP-binding protein (Herrmann *et al.*, 1998). The enzymic activity of both toxin B from *C. difficile* and *C. sordellii* LT toxin is localized in the *N*-terminal polypeptide of the molecule (residues 1–546) (Hofmann *et al.*, 1997, 1998). Asparagines 286 and 288 of LT toxin, forming a DXD motif conserved in a large number of glycosyltransferases, is essential for the LT enzymic activity and probably for all large clostridial cytotoxins (Busch *et al.*, 1998).

The third enzymic activity found in toxins interfering with small GTP-binding protein surprisingly activates these molecules. As already pointed out, this enzymic modification occurs at the level of the Rho switch 2 domain. Modification of this switch 2 domain will modify the intrinsic or GAP-assisted GTP hydrolysis activity of the GTP-binding molecules, resulting in their permanent activation. Toxins CNF1 from *E. coli* and DNT from *Bordetella bronchiseptica* activate Rho (Flatau *et al.*, 1997; Schmidt *et al.*, 1997; Horiguchi *et al.*, 1997). The first hint of this activity was shown when HEp-2 cells where incubated with CNF1 and then their cytosol ADP-ribosylated with exoenzyme C3. Under such conditions, the molecular mass of the Rho protein shifted to a slightly higher value (Fiorentini *et al.*, 1994; Oswald *et al.*, 1994). Later, the same observation was made in the case of DNT (Horiguchi *et al.*, 1995). This result indicated a possible post-translational modification of the GTP-binding protein after CNF1 or DNT treatment. CNF1 was then shown to modify Rho directly *in vitro* without the need for cellular cofactors (Flatau *et al.*, 1997; Schmidt *et al.*, 1997). Microsequencing of CNF1-modified Rho revealed a single change in amino acid composition consisting of the replacement of glutamine 63 by a glutamic acid (Flatau *et al.*, 1997). The same modification induced by CNF1 on Rho was found concomitantly by mass spectrometry (Schmidt *et al.*, 1997) and later for DNT (Horiguchi *et al.*, 1997). Therefore, by converting glutamine to glutamic acid, CNF1 and DNT exert a deamidase activity. As indicated in this chapter, the equivalent amino acid of Rho glutamine 63 in Ras is glutamine 61. This amino acid has a pivotal role in the GTPase activity of small GTP-binding proteins. Ras glutamine 61 and Rho glutamine 63 mediate intrinsic and GAP-assisted GTP hydrolysis activities of these GTPases (Scheffzek *et al.*, 1997; Rittinger *et al.*, 1997). Deamidation of glutamine 63 of Rho blocks both the intrinsic and GAP-dependent GTP hydrolysis activity of the small GTP-binding protein, leaving Rho permanently activated (Flatau *et al.*, 1997 Schmidt *et al.*, 1997). CNF1 enzymic activity is borne by its 30-kDa carboxy-terminal end, whereas the cell binding domain of the toxin is located on its amino-terminal domain (Lemichez *et al.*, 1997). Cysteine 866 and histidine 881 have been implicated in the deamidase activity of CNF1 (thus like cysteine proteases) (Schmidt *et al.*, 1998). According to this work, CNF1 can also behave *in vitro* as a transglutaminase.

Figure 2.2 compares the primary structures of toxins acting on small GTPases and Figure 2.4 shows the different impact of toxins on the downstream effects of small GTPases.

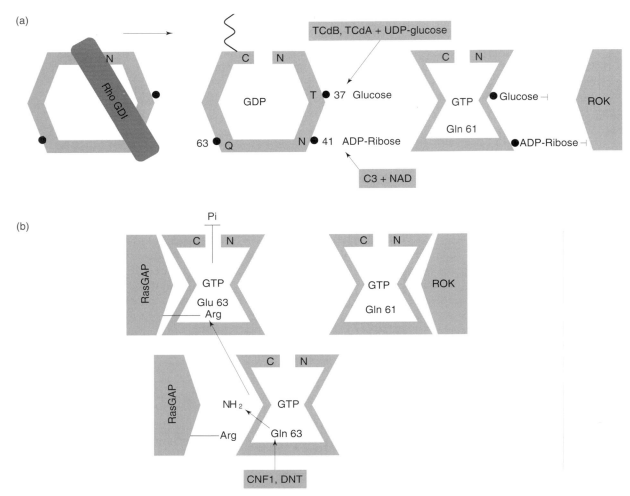

FIGURE 2.3 Modifications induced by toxins on the Rho small GTPases. Toxins inactivating Rho (a) modify the switch 1 domain, usually when it is exposed to the surface of the GTPase (under the GDP bound form). Modification of the switch 1 domain impedes the binding of the small GTPase to its effector. Toxins activating Rho (b) deamidate the Rho glutamine 63 (analogous to Ras glutamine 61) into glutamic acid, which is not suited for the binding of the water molecule required for RhoGAP GTP hydrolysis. Rho stays in the GTP bound form and activates permanently its downstream effector ROK. Association of RhoGDI with Rho hinders asparagine 41, and thus blocks the C3-catalysed ADP-ribosylation of this residue. RhoGDI may not block threonine 37 and glutamine 63, thereby allowing large clostridial cytotoxins or dermonecrotic toxins to interact with their substrate residues. This may explain why large clostridial cytotoxins and dermonecrotic toxins are much more powerful than exoenzyme C3 on cells.

CONCLUSION: WHY ARE SMALL GTPASES TARGETS FOR BACTERIAL TOXINS?

Bacterial toxins modifying small GTP-binding proteins are produced mainly by bacteria growing in internal cavities of the body. For instance, *C. difficile* toxins A and B operate in the intestine, CNF1 in the urinary tract and DNT in the respiratory tract. Therefore, the manipulation of small GTPases by toxins produced by these organisms must perform an effect that alters a specialized function of cells lining these cavities, namely epithelial cells. The main feature of epithelia is that they are made of cells sealed one to another by specialized structures: the tight and adherens junctions. As reported above, small GTPases play an important role in the regulation of these junctions. Therefore, it may be assumed that bacterial toxins affecting small GTP-binding proteins do so mostly to modulate the permeability properties of epithelia. The best case for such a modulation of epithelial permeability is exemplified by *C. difficile* toxins A and B. These toxins, which glucosylate Rho, Rac and Cdc42 (Just *et al.*, 1995a, b), decrease in a dose-dependent manner the trans-epithelial resistance (TER) of intestinal T84 cells cultivated in a polarized fashion on filters, clearly demonstrating that they open tight junctions (Hecht *et al.*, 1988, 1992). The chimeric toxin C3DTB, which also selectively inactivates Rho in intact cells (but by ADP-ribosylation), is able to decrease the TER of T84 cells (thus also opening tight junctions), demonstrating that Rho controls tight junction permeability (Nusrat *et al.*, 1995). Interestingly, inactivation of Rho by C3 in T84 cells involves a selective modification of the apical actin cytoskeleton with additional delocalization of the pro-

tein ZO1 from tight junctions (Nusrat *et al.*, 1995). Conversely, it was shown that CNF1, which activates Rho, does not decrease the TER of T84 cells and does not modify the ZO1 localization from tight junctions (Hofman *et al.*, 1998). Very interestingly, CNF1 reduces the epithelial transmigration of polymorphonuclear leucocytes (PMNs) through T84 monolayers (Hofman *et al.*, 1998). Thus CNF1, by activating Rho, may enforce the tight junction permeability of T84 cells.

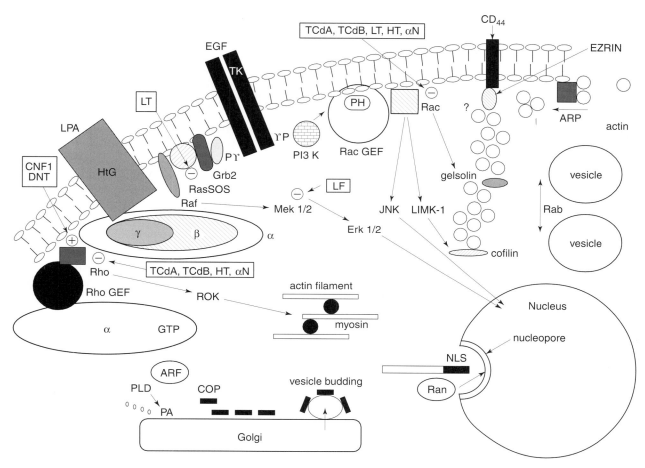

FIGURE 2.4 Impact of bacterial toxins on the downstream effects of small GTPases. The Ras signalling pathway is activated by tyrosine kinase receptors (TK) and controls the cell division via activation of the Mek and Erk kinases. The lethal toxin of *C. sordellii* (LT) glucosylates Ras and inactivates it, blocking the Rho downstream cascade. The lethal factor of *Bacillus anthracis* (LF) proteolyses Mek1/2, blocking the downstream cascade of kinases. The Rho signalling pathway is activated by a heterotrimeric G-protein receptor (HtG). Upon activation by its ligand lysophosphatidic acid (LPA), HtG stimulates the binding of GTP on Gα. The GαGTP binds to the RhoGEF and activates its nucleotide exchange domain (DH), stimulating the formation of RhoGTP. RhoGTP activates ROK which, by increasing myosin light chain phosphorylation, induces actin filament bundling and contractility. *Clostridium difficile* toxins A and B (TCdA, TCdB), *C. sordellii* haemorrhagic toxin (HT) and *C. novyi* α-toxin (αN) inactivate Rho by glycosylation, blocking the downstream effectors of the small GTPase. Rac is activated by tyrosine kinase receptors but probably via stimulation of the PI3-kinase activity, which produces PIP3. PIP3 allows binding of the RacGEF PH domain to the membrane in close proximity to Rac. Activated Rac can elongate actin filaments through direct activation of gelsolin or by stimulating the LIMK-1 kinase, which phosphorylates and inactivates the actin depolymerizing factor cofilin. Nucleation of actin is made by ARP proteins, activated by the cell membrane. TCdA, TCdB, LT HT and αN inactivate Rac by glycosylation and therefore block the downstream effects of this GTPase (actin depolymerization). Rab GTPases are implicated in vesicular fusion. By activating phospholipase D (PLD), ARF, which produces phosphatidic acid (PA) on the Golgi membrane, allows coat proteins (COP) to bind and to induce vesicle budding from this compartment. Ran allows protein with a nuclear localization signal (NLS) to enter the nucleus through the nucleopore.

Opening or closing tight and adherens junctions by manipulation of small GTP binding proteins of the Rho family may have some important implications in terms of bacterial virulence. Some of the most important players in the primary host defence against bacterial infection are PMNs. When alerted to a bacterial infection outbreak in the digestive, urinary or respiratory tracts, PMNs move from the bloodstream to the basolateral side of epithelial cells through adhesion to integrins. Then PMNs crawl on the surface of epithelial cells and cross the epithelium in the space between two adjacent cells (space named the paracellular pathway) (Madara, 1988). The paracellular pathway permeability is controlled by tight junctions (Madara, 1988). PMNs cross the paracellular pathway by brute force, using their own motility. Tight junctions may restrict (if they are tightly closed) or increase (if they are opened) the ability and the number of PMNs crossing the paracellular pathway and reaching the luminal side of the epithelium. Bacterial toxins might modulate the barrier effect of tight junctions to PMNs through activation/deactivation of Rho small GTP binding proteins. For instance, by opening intestinal tight junctions, *C. difficile* toxins A and B will facilitate the trans-epithelial migration of PMNs, allowing accumulation of PMNs in the intestinal lumen. Accumulation of PMNs in the lumen will start an inflammatory process by releasing cytokines, reactive oxygen species and proteolytic enzymes (among other products), which will lyse intestinal cells. This inflammatory process may be beneficial to the growth of *C. difficile* bacteria in the intestine, since it will provide micro-organisms with nutrients and iron, released by injured cells. However, a toxin such as CNF1, which activates the Rho protein and therefore increases the tight junction barrier efficiency to PMNs, may reduce the ability of PMNs to cross the paracellular pathway. Subsequently, the number of PMNs reaching the luminal side of the epithelium will be decreased, facilitating the multiplication of CNF1-producing bacteria.

Beside their effects on epithelial barrier permeability, bacterial toxins acting on small GTPases may also be important in controlling phagocytosis of bacteria by professional or non-professional phagocytes. Rho, Rac and Cdc42 are pivotal for the induction of phagocytosis, as described above. Therefore, it is easy to imagine that toxins inhibiting these GTP-binding proteins in professional phagocytes will block the engulfment of bacteria and, therefore, their destruction. The fact that Rac, a substrate of *C. difficile* toxins A and B, is involved in the control of NADPH oxidase activity (Abo *et al.*, 1991) underlines the role of these toxins as virulence factor for macrophages. However, the most interesting point is that toxins activating small GTPases such as CNF1 are able to induce phagocytosis in non-professional phagocytic cells, such as epithelial cells (Falzano *et al.*, 1993; Fiorentini *et al.*, 1997b). To induce their entry into cells, invasive bacteria such as *Shigella* or *Salmonella* also need to activate small GTPases of the Rho subfamily (Adam *et al.*, 1996; Chen *et al.*, 1996; Watarai *et al.*, 1997; Hardt *et al.*, 1998). They do so because Rho GTPases control most of the molecular steps required for phagocytosis. Two molecular strategies have therefore been set up by bacteria to activate Rho GTPases of host cells. The first one consists of transferring activated Rho GTPases exchange factor directly through the cell membrane into the cytosol by a type-III secretion mechanism (reviewed by Hueck, 1998). This has been described in the case of *Salmonella typhimurium*, which injects into the host cell the SopE protein, an exchange factor for Cdc42 and Rac GTPases (Hardt *et al.*, 1998). The second strategy utilizes CNF1 which, upon its entry into the cytosol by receptor-mediated endocytosis and membrane translocation, induces an activating point mutation on the GTP-binding protein (Flatau *et al.*, 1997; Schmidt *et al.*, 1997). These two strategies will induce membrane ruffling, leading to a process called 'triggered phagocytosis' (reviewed by Finlay and Cossart, 1997).

Finally, the influence of Rho on the control of apoptosis may also explain why some bacterial toxins target this GTPase. It is now clear that inhibition of Rho activity provokes apoptosis (Moorman *et al.*, 1996; Henning *et al.*, 1997). Recently, it was shown that TCdA and TCdB induce apoptosis (Mahida *et al.*, 1996; Fiorentini *et al.*, 1998c). The hypothesis may be proposed that TCdA and TCdB can also induce apoptosis of macrophages, inducing the release of the pro-inflammatory cytokine interleukin-1, as described for *Shigella flexneri* (Zychinsky and Sansonetti, 1997). Release of interleukin-1 will, in turn, recruit other macrophages and PMNs, initiating an inflammatory process (as reported for *C. difficile* infections; reviewed by von Eichel-Streiber *et al.*, 1996). In contrast, activation of Rho protects cells from apoptosis (Fiorentini *et al.*, 1997b). In epithelia, crypt cells differentiated into villus cells have a short lifetime: they undergo apoptosis and disappear via phagocytosis, a process termed shedding. This ensures a constant removal of villus cells on which bacteria have bound, preventing bacterial proliferation or invasion. By activating Rho, CNF1 may block the apoptosis of villus cells and therefore their shedding, and will allow bacteria bound to or invading these cells to multiply actively (Fiorentini *et al.*, 1997b). This clever mechanism will be finally beneficial for bacterial virulence.

Figures 2.5 and 2.6 summarize why bacterial toxins may affect small GTP-binding proteins.

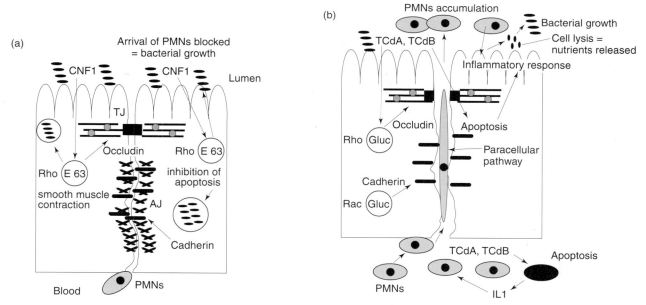

FIGURE 2.5 Effects of bacterial toxins activating or inactivating Rho on the epithelial barrier. Toxins activating Rho (CNF1, DNT) (a) will close tight junctions (TJ), decreasing access of polymorphonuclear leucocytes (PMNs) to the lumen. With fewer PMNs in the lumen, bacteria will proliferate. CNF1 will also induce invasion of bacteria through induction of their phagocytosis by Rho activation. CNF1, by blocking apoptosis of epithelial cells, will allow them to stay alive longer in the epithelia, thereby favouring the intracellular proliferation of invading bacteria or the multiplication of cell surface attached bacteria. Toxins inactivating Rho and Rac (TCDA, TCdB) (b) open tight and adherens junctions (AJ), thus leaving the paracellular pathway free for access of PMNs, which can accumulate in the lumen. Accumulation of PMNs leads to an inflammatory response, which injures cells, releasing nutrients (especially iron). This allows bacteria to grow better. In addition, induction of apoptosis by TCdA and TCdB (when these toxins reach the bloodstream by open tight junctions) will kill macrophages, inducing the release of interleukin-1 (IL1). IL1, in turn, will recruit more macrophages and PMNs, increasing the inflammatory response.

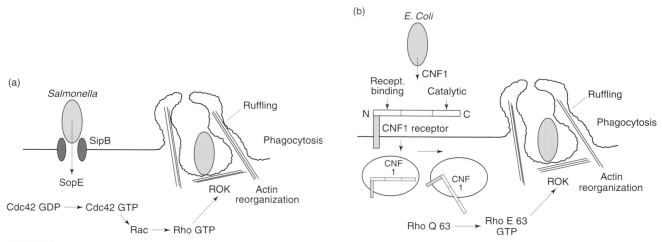

FIGURE 2.6 Two mechanisms activating small GTP-binding proteins, elaborated by bacteria to invade cells: (a) to induce its own phagocytosis, *Salmonella* injects through SipB (a pore-forming protein by which *Salmonella* virulence factors are injected into the cell by the type-III secretion mechanism: for a review see Hueck, 1998), a Cdc42 exchange factor (SopE). SopE accelerates the exchange of GDP to GTP on Cdc42, which is activated and can stimulate Rac and then Rho. Rho activates the Rho kinase, which allows actin filament bundling, contractility, membrane ruffling and phagocytosis. (b) CNF1 produced by *E. coli* will bind to its cell receptor, through its *N*-terminal end. CNF1 then enters the cell by receptor-mediated endocytosis and its catalytic domain (*C*-terminal end) is translocated from the lumen of the endocytic vesicles to the cytosol. In the cytosol, the catalytic domain of CNF1 deamidates Rho on glutamine 63 (Q63) into glutamic acid (RhoE63). Rho E63 does not hydrolyse GTP and, thus, is activated permanently. Rho E63 stimulates ROK, triggering phagocytosis similar to the effect in (a) for *Salmonella*.

ACKNOWLEDGEMENT

I would like to thank Michel Gauthier (INSERM U452, Nice, France) for critical reading of the manuscript.

REFERENCES

Abo, A., Pick, E., Hall, A., Totty, N., Teahan C.G. and Segal, A.W. (1991). The small GTP-binding protein Rac is involved in the activation of the phagocyte NADPH oxidase. *Nature* **353**, 668–70.

Adam, T., Giry, M., Boquet, P. and Sansonetti, P. (1996). Rho dependent membrane folding causes *Shigella* entry into epithelial cells. *EMBO J.* **15**, 3315–21.

Adamson, P., Marshall, C.J., Hall, A. and Tillbrook, P.A. (1992). Post-translational modifications of the p21 Rho proteins. *J. Biol. Chem.* **267**, 20033–8.

Aktories, K. (1997). Rho proteins: targets for bacterial toxins. *Trends Microbiol.* **5**, 282–8.

Alb, J.G., Kearns, M.A. and Bankaitis, V.A. (1996). Phospholipids metabolism and membrane dynamics. *Curr. Opin. Cell Biol.* **8**, 534–41.

Allen, W.E., Zicha, D., Ridley, A.J. and Jones, G. (1998). A role for Cdc42 in macrophage chemotaxis. *J. Cell Biol.* **141**, 1147–57.

Arber, S., Barbayannis, F.A., Hanser, M., Schneider, C., Stanyon, C. A., Bernard, O. and Caroni, P. (1998). Regulation of actin dynamics through phosphorylation of cofilin by Lim-kinase. *Nature* **393**, 805–9.

Azuma, T., Witke, W., Stossel, T.P., Hartwig, J.H. and Kwiatkowski, D.J. (1998). Gelsolin is a downstream effector of Rac for fibroblast motility. *EMBO J.* **17**, 1362–70.

Bagrodia, S., Derijard, B., Davis, J. and Cerione, R.A. (1995). Cdc42 and PAK-mediated signalling leads to Jun kinase and p38 mitogen-activated protein kinase activation. *J. Biol. Chem.* **270**, 27995–8.

Barbacid, M. (1987). Ras gene. *Annu. Rev. Biochem.* **56**, 779–827.

Barrett, T., Xiao, B., Dodson, G., Ludbrook, S.B., Nurmahomed K., Gamblin, S., Musacchio, A., Smerdon, S.J. and Eccleston, J.(1997). The structure of the GTPase-activating from p50RhoGAP. *Nature* **385**, 458–61.

Bischoff, F.R. and Coutavas, E. (1995). Ran/TC4. In: *Guidebook to Small GTPases* (eds M. Zerial and L.A. Huber), pp. 457–60. Oxford University Press, New York.

Boquet, P., Popoff, M.R., Giry, M., Lemichez, E. and Bergez-Aullo, P. (1995). Inhibition of p21Rho in intact cells by C3 diphtheria toxin chimera proteins. *Methods Enzymol.* **256**, 297–306.

Boquet, P., Munro, P., Fiorentini, C. and Just, I. (1998). Toxins from anaerobic bacteria: specificity and molecular mechanisms of action. *Curr. Opin. Microbiol.* **1**, 66–74.

Boriack-Sjodin, P.A., Margarit, S.M., Bar-Sagi, D. and Kuriyan, J. (1998). The structural basis of the activation of Ras by SOS. *Nature* **394**, 337–43.

Bourmeyster, N., Stasia, M.J., Garin, J., Gagnon, J., Boquet, P. and Vignais, P. (1992). Copurification of Rho protein and the Rho-GDP dissociation inhibitor from bovine neutrophil cytosol. Effect of phosphoinositides on Rho ADP-ribosylation by the C3 exoenzyme of *Clostridium botulinum*. *Biochemistry* **31**, 12863–9.

Bourne, H.R. (1997). The arginine finger strikes again. *Nature* **389**, 673–4.

Bourne, H.R., Sanders, D. and McCormick, F. (1991). The GTPase superfamily: conserved structure and molecular mechanism. *Nature* **349**, 117–27.

Braga, V.M.M., Machesky, L.M., Hall, A. and Hotchin, N.A. (1997). The small GTPases Rho and Rac are required for the establishment of cadherin-dependent cell–cell contacts. *J. Cell Biol.* **137**, 1421–31.

Busch, C., Hofmann, F., Selzer, J., Munro, S., Jeckel, D. and Aktories, K. (1998). A common motif of eukaryotic glycosyltransferases is essential for the enzyme activity of large clostridial cytotoxins. *J. Biol. Chem.* **273**, 19566–72.

Cachero, T.G., Morielli, A.D. and Peralta, E.G. (1998). The small GTP-binding protein RhoA regulates a delayed rectifier potassium channel. *Cell* **93**, 1077–85.

Capo, C., Sangueldoce, M.V., Meconi, S., Flatau, G., Boquet, P. and Mége, J.L. (1998). Effect of cytotoxic necrotizing factor 1 on actin cytoskeleton in human monocytes: role in the regulation of integrin-dependent phagocytosis. *J. Immunol.* **161**, 4301–8.

Cerione, R.A. and Zheng, Y. (1996). The Dbl family of oncogenes. *Curr. Opin. Cell Biol.* **8**, 216–22.

Chardin, P., Boquet, P., Madaule, P., Popoff, M.R., Rubin, E.J. and Gill, D.M. (1989). The mammalian G-protein RhoC is ADP-ribosylated by *Clostridium botulinum* exoenzyme C3 and affects actin microfilament in Vero cells. *EMBO J.* **8**, 1087–92.

Chardin, P., Camonis, J.H., Gale, N.W., VanAelst, L., Schlessinger, J. and Wigler, M.H. (1993). Human SOS1: a guanine nucleotide exchange factor for Ras that binds Grb2. *Science* **260**, 1338–43.

Chen, L.M., Hobbie, S. and Galan, J.E. (1996). Requirement for Cdc42 for *Salmonella*-induced cytoskeletal and nuclear responses. *Science* **274**, 2115–18.

Chong, L.D., Traynor-Kaplan, A., Bokoch, G.M. and Schwartz, M.A. (1994). The small GTP-binding protein Rho regulates a phospholipidinositol4-phosphate5-kinase in mammalian cells. *Cell* **79**, 507–13.

Chrzanowska-Wodnicka, M. and Burridge, K. (1996). Rho-stimulated contractility drives the formation of stress fibres and focal adhesions. *J. Cell Biol.* **6**, 1403–15.

Cockcroft, S. (1997). Phospholipase D: regulation by GTPases and protein kinase C and physiological relevance. *Prog. Lipid Res* **35**, 345–70.

Cover, T.A. (1996). The vacuolating cytotoxin of *Helicobacter pylori*. *Mol. Microbiol.* **20**, 241–6.

Donaldson, J.G., Finnazi, D. and Klausner, R.D. (1992). Brefeldin A inhibits Golgi membrane-catalyzed exchange of guanine nucleotide onto ARF protein. *Nature* **360**, 350–2.

Downward, J. (1990). The Ras superfamily of small GTP binding proteins. *Trends Biochem. Sci.* **15**, 449–77.

Drubin, D.G. and Nelson, W.J. (1996). Origins of cell polarity. *Cell* **84**, 335–44.

Duesbery, N.S., Webb C.P., Leppla, S.H., Gordon, V.M., Klimpel, K.R., Copeland, T.D., Ahn, N.G., Oskarsson, M.K., Fukusawa, K., Paull, K.D. and VandeWoude, G.F. (1998). Proteolytic inactivation of Map kinase-kinase by anthrax lethal factor. *Science* **280**, 734–7.

Eaton, S.P., Auvinen, L., Luo, Y., Jan, N. and Simons, K. (1995). Cdc42 and rac1 control different actin-dependent processes in the *Drosophila* wing disc epithelium. *J. Cell Biol.* **131**, 151–64.

Ellis, R.W., Defeo, D., Shih, T.Y., Gonda, M.A., Young, H.A., Tsuchida,N., Lowy, D.R. and Scolnick, E. (1981). The p21 src genes of Harvey and Kirsten sarcoma viruses originate from divergent members of a family of normal vertebrate genes. *Nature* **292**, 506–11.

Falzano, L., Fiorentini, C., Donelli, G., Michel, E., Kocks, C., Cossart, P., Cabanié, L., Oswald, E. and Boquet, P. (1993). Induction of phagocytic behaviour in human epithelial cells by *Escherichia coli* cytotoxic necrotizing factor 1. *Mol. Microbiol.* **9**, 1247–54.

Fiorentini, C., Giry, M., Donelli, G., Falzano, L., Aullo, P. and Boquet, P. (1994). *Escherichia coli* cytotoxic necrotizing factor 1 increases actin assembly via the p21 Rho GTPase. *Zentrabl. Bakteriol Suppl.* **24**, 404–5.

Fiorentini, C., Fabbri, A., Flatau, G., Donelli, G., Matarrese, P., Lemichez, E., Falzano, L. and Boquet, P. (1997a). *Escherichia coli* cytotoxic necrotizing factor 1 (CNF1) a toxin that activates the Rho GTPase. *J. Biol. Chem.* **272**, 19532–7.

Fiorentini, C., Fabbri, A., Matarrese, P., Falzano, L., Boquet, P. and Malorni, W. (1997b). Hinderance of apoptosis and phagocytic behaviour: induced by *Escherica coli* necrotizing factor (CNF1): two related activities in epithelial cells. *Biochem. Biophys. Res. Commun.* **241**, 341–6.

Fiorentini, C., Matarrese, P., Straface, E., Falzano, L., Fabbri, A., Donelli, G., Cossarizza, A., Boquet, P. and Malorni, W. (1998a). Toxin-induced activation of Rho GTP-binding protein increases Bcl-2 expression and influences mitochondrial homeostasis. *Exp. Cell Res.* **242**, 341–50.

Fiorentini, C., Gauthier, M., Donelli, G. and Boquet, P. (1998b). Bacterial toxins and the Rho GTP-binding protein: what microbes teach us about cell regulation. *Cell Death Diff.* **5**, 720–8.

Fiorentini, C., Fabbri, A., Falzano, L., Fattorossi, A., Matarrese, P., Rivabene, R. and Donelli, G. (1998c). *Clostridium difficile* toxin B induces apoptosis in intestinal cultured cells. *Infect. Immun.* **66**, 2660–5.

Finlay, B. and Cossart, P. (1997). Exploitation of the mammalian host cell functions by bacterial pathogens. *Science* **276**, 718–25.

Flatau, G., Lemichez, E., Gauthier, M., Chardin, P., Paris, S., Fiorentini, C. and Boquet, P. (1997). Toxin-induced activation of the G-protein p21 Rho by deamidation of glutamine. *Nature* **387**, 729–33.

Fujihara, H., Walker, L.A., Gong, M.C., Lemichez, E., Boquet, P., Somlyo, A.V. and Somlyo, A.P. (1997). Inhibition of RhoA translocation and calcium sensitization by *in vivo* ADP-ribosylation of Rho with the chimeric DC3B. *Mol. Cell Biol.* **8**, 2437–47.

Fukumoto, Y., Kaibuchi, K., Hori, Y., Fujioka H., Azaku, S., Ueda, T., Kikuchi, A. and Takai, Y. (1990). Molecular cloning and characterization of a novel type of regulatory protein (GDI) for the Rho protein, ras p21-like small GTP binding protein. *Oncogene* **5**, 1321–8.

Furuse, M., Hirase, T., Itoh, M., Nagafuchi, A., Yonemura, S., Tsukita, S. and Tsukita, S. (1993). Occludin: a novel integral membrane protein localized in tight junction. *J. Cell Biol.* **123**, 1777–88.

Genth, M., Hofmann, F., Selzer, J., Rex, G., Aktories, K. and Just, I. (1996). Difference in protein substrate specificity between hemorrhagic toxin and lethal toxin from *Clostrium sordellii*. *Biochem. Biophys. Res. Commun.* **229**, 370–4.

Gilmore, A.P. and Burridge, K. (1996). Regulation of vinculin binding to talin and actin by phosphatidyl-inositol 4-5 biphosphate. *Nature* **306**, 704–7.

Gomez, J., Martinez, C., Giry, M., Garcia, A. and Rebollo, A. (1997). Rho prevent apoptosis through Bcl-2 expression: implications for interleukin-2 receptor signal transduction. *Eur. J. Immunol.* **27**, 2793–9.

Gong, C., Fujihara, H., Somlyo, A.V. and Somlyo, A.P. (1997). Translocation of RhoA associated with Ca^{2+} sensitization of smooth muscle. *J. Biol. Chem.* **272**, 10704–9.

Greenberg, S. and Silverstein, S.C. (1993). Phagocytosis. In: *Fundamental Immunology* (ed. W.E. Paul), pp. 941–64. Raven Press, New York.

Greenberg, S., Chang, P. and Silverstein, S.C. (1993). Tyrosine phosphorylation is required for Fc receptor mediated phagocytosis in mouse macrophages. *J. Exp. Med.* **177**, 529–34.

Gumbiner, B.M. (1996). Cell adhesion: the molecular basis of tissue architecture and morphogenesis. *Cell* **84**, 345–57.

Hackam, D.J., Rotstein, O., Schreiber, A., Zhang, W.J. and Grinstein, S. (1997). Rho is required for the initiation of calcium signalling and phagocytosis by Fcg receptors in macrophages. *J. Exp. Med.* **186**, 955–66.

Hall, A. (1990). The cellular functions of small GTP-binding proteins. *Science* **249**, 635–40.

Hall, A. (1998a). Rho GTPases and the actin cytoskeleton. *Science* **279**, 509–14.

Hall, A. (1998b). G-proteins and small GTPases: distant relatives keep in touch. *Science* **280**, 2074–5.

Hanckock, J.C., Paterson, H.F. and Marshall, C.J. (1989). All Ras proteins are polyisoprenylated but only some are palmitoylated. *Cell* **57**, 1167–77.

Hardt, W.D., Chen, L.H., Schuebel, K.E., Bustelo, X.R. and Galan, J.E. (1998). *S. typhimurium* encodes an activator of Rho GTPases that induces membrane ruffling and nuclear response in host cells. *Cell* **93**, 815–26.

Hart, M.J., Jiang, X., Kozasa, T., Roscoe, W., Singer, W.D. Gilman, A.G., Sternweiss, P.C. and Bollag, G. (1998). Direct stimulation of the nucleotide exchange activity of the p115RhoGEF by Ga13. *Science* **280**, 2112–14.

Hartwig, J.H., Bokoch, G.M., Carpenter, C.L., Jeanmey, P.A., Taylor, L.A., Token, A. and Stossel, T.P. (1995). Thrombin receptor ligation and activated Rac uncap actin filament barbed ends through phosphorylation synthesis in permeabilized platelets. *Cell* **82**, 643–53.

Hecht, G., Pothoulakis, C., Lamont, J.T. and Madara, J.L. (1988). *Clostridium difficile* toxin A perturbs cytoskeletal structures qnd tight junction permeability of cultured human intestinal epithelial cells. *J. Clin. Invest.* **56**, 1053–61.

Hecht, G., Koutsouris, A., Pothoulakis, C., Lamont, T.J. and Madara, J.L. (1992). *Clostridium difficile* toxin B disrupts the barrier function of T84 monolayers. *Gastroenterology* **102**, 416–23.

Henning, S.W., Galandrini, R., Hall, A. and Cantrell, D.A. (1997). The GTPase Rho has a critical role in thymus development. *EMBO J.* **16**, 2397–407.

Herrmann, C., Ahmadian, M.R., Hofmann, F. and Just, I. (1998). Functional consequences of monoglucosylation of Ha-Ras at effector domain amino acid threonine 35. *J. Biol. Chem.* **273**, 16134–9.

Hill, C.S., Wynne, J. and Treisman, R. (1995). The Rho family GTPases RhoA, Rac1 and Cdc42Hs regulate transcriptional activation by SRF. *Cell* **81**, 1159–70.

Hirao, M., Sato, N., Kondo, T., Yonemura, S., Monden M., Sasaki, T., Takai, Y., Tsukita, S. and Tsukita, S. (1996). Regulation mechanism of ERM (ezrin/radixin/moesin) protein/plasma association: possible involvement of phosphatidylinositol turnover and Rho-dependent signaling pathway. *J. Cell Biol.* **135**, 37–51.

Hirshberg, M., Stockley, R.W., Dodson, G. and Webb, N. (1997). The crystal structure of human Rac1, a member of the Rho-family complexed with a GTP analogue. *Nature Struct. Biol.* **4**, 147–51.

Hofman, P., Flatau, G., Selva, E., Gauthier, M., Le Negrate, G., Fiorentini, C., Rossi, B. and Boquet, P. (1998). *Escherichia coli* cytotoxic necrotizing factor 1 effaces microvilli and decreases transmigration of polymorphonuclear leukocytes in intestinal T84 epithelial cell monolayers. *Infect. Immun.* **66**, 2494–500.

Hofmann, F., Rex, G., Aktories, K. and Just, I. (1996). The Ras-related protein Ral is monoglucosylated by *Clostridium sordellii* lethal toxin. *Biochem. Biophys. Res. Commun.* **227**, 77–81.

Hofmann, F., Busch, C., Prepens, U., Just, I. and Aktories, K. (1997). Localization of the glucosyltransferase activity of *Clostridium difficile* toxin B to the N-terminal part of the holotoxin. *J. Biol. Chem.* **272**, 11074–8.

Hofmann, F., Busch, C. and Aktories, K. (1998). Chimeric clostridial cytotoxins: identification of the N-terminal region involved in protein substrate recognition. *Infect. Immun.* **66**, 1076–81.

Horiguchi, Y., Senda, T., Sugimoto, N., Katahira, J. and Matsuda, M. (1995). *Bordetella bronchiseptica* dermonecrotizing toxin stimulates assembly of actin stress fibres and focal adhesions by modifying the small GTP-binding protein Rho. *J. Cell Science* **108**, 3243–51.

Horiguchi, Y., Inoue, N., Masuda, M., Kashimoto, M., Katahira, T., Sujimoto, N. and Matsuda, M. (1997). *Bordetella bronchiseptica* dermonecrotizing toxin induces reorganization of actin stress fibres through deamidation of Gln-63 of the GTP-binding protein Rho. *Proc. Natl Acad. Sci. USA* **94**, 11623–6.

Hueck, C.J. (1998). Type III secretion systems in bacterial pathogens of animals and plants. *Microbiol. Mol. Biol. Rev.* **62**, 379–433.

Ishizaki, T., Mackawa, M., Fujisawa, K., Okawa, K., Iwamatsu, A., Fujita, A., Watanabe, N., Saito, Y., Kakizuka, Y., Morii, N. and Narumiya, S. (1996). The small GTP-binding protein Rho binds to and activates a 160 kDa ser/thr protein kinase homologous to myotonic dystrophy kinase. *EMBO J.* **15**, 1885–93.

Just, I., Selzer, J., Wilm, M., von Eichel-Streiber, C., Mann, M. and Aktories, K. (1995a). Glucosylation of Rho proteins by *Clostridium difficile* toxin B. *Nature* **375**, 500–3.

Just, I., Wilm, M., Selzer, J., Rex, G., von Eichel-Streiber, C., Mann, M. and Aktories, K. (1995b). The enterotoxin from *Clostridium difficile* (ToxA) monoglucosylates the Rho proteins. *J. Biol. Chem.* **270**, 13932–6.

Just, I., Selzer, J., Hofman, F., Green, G.A. and Aktories, K. (1996). Inactivation of Ras by *Clostridium sordelli* lethal toxin-catalyzed glucosylation. *J. Biol. Chem.* **271**, 10149–53.

Kahn, R.A. (1995). The ARF subfamily. In: *Guidebook to the Small GTPases* (eds M. Zerial and L.A. Huber), pp. 429–32. Oxford University Press, New York.

Kahn, R.A. and Gilman, A.G. (1986). The protein cofactor necessary for ADP-ribosylation of Gs by Cholera toxin is itself a GTP-binding protein. *J. Biol. Chem.* **261**, 7906–11.

Kahn, R.A., Goddard, C. and Neukirch, G.M. (1988). Chemical and immunological characterization of the 21 kD ADP-ribosylation factor (ARF) of adenylate cyclase. *J. Biol. Chem.* **263**, 8282–7.

Katayama, M., Kawata, M., Yoshida, Y., Horiuchi, H., Yamamoto, T., Matsuura, Y. and Takai, Y. (1991). The post-translationally modified structure of bovine smooth aortic muscle RhoA. p21. *J. Biol. Chem.* **266**, 12639–45.

Kimura, K., Ito, M., Amano, M., Chihara, K., Fukuta, Y., Nakafuku, M., Yamamori, B., Feng, J., Nakano, T., Okawa, K., Iwamatsu, A. and Kaibuchi, I. (1996). Regulation of myosin phosphatase by Rho and Rho-associated kinase (Rho-kinase). *Science* **273**, 245–7.

Klimpel, K.R., Arora, M. and Leppla, S.H. (1994). Anthrax toxin lethal factor contains a zinc metallo-protease consensus sequence which is required for lethal activity. *Mol. Microbiol.* **13**, 1093–1100.

Kozasa, T., Jiang, X., Hart, M., Sternweiss, P.M, Singer, W.O., Gilman, A.G., Bollag, G. and Sternweiss, P.C. (1998). p115RhoGEF, a GTPase activating protein for Ga12 and Ga13. *Science* **280**, 2109–11.

Kozma, R., Ahmed, S., Best, A. and Lim, L. (1996). The GTPase-activating protein n-chimaerin cooperates with Rac1 and Cdc42Hs to induce the formation of lamellipodia and filopodia. *Mol. Cell Biol.* **16**, 5069–80.

Lamarche, N., Tapon, N., Stower, L., Burbelo, P.D., Aspenström P., Bridges, T., Chan, J. and Hall, A. (1996). Rac and Cdc42 induce actin polymerization and G1 cell cycle progression independently of p65PAK and the JNK/SAPK MAP kinase cascade. *Cell* **87**, 519–29.

Lamaze, C., Chuang, T.H., Terlecky, L.J., Bokoch, G.M. and Schmidt, S.L. (1996). Regulation of receptor-mediated endocytosis by Rho and Rac. *Nature* **382**, 177–9.

Lemichez, E., Flatau, G., Bruzzone, M., Boquet, P. and Gauthier, M. (1997). Molecular localization of the *Escherichia coli* cytotoxic necrotizing factor 1 cell binding and catalytic domains. *Mol. Microbiol.* **24**, 1061–70.

Lemmon, M.A., Ferguson, K.M. and Schlessinger, J. (1996). PH domains: diverse sequence with a common field recruit signaling molecules to the cell surface. *Cell* **85**, 621–4.

Leveers, S.J. and Marshall, C.J. (1995). Extracellular signalling-regulated kinases (ERKs) In: *Guidebook to the Small GTPases* (eds M. Zerial and L.A. Huber), pp. 160–4. Oxford University Press, New York.

Leveers, S.J., Paterson, H.F. and Marshall, C.J. (1994). Requirement for Ras in Raf activation is overcome by targetting Raf to the plasma membrane. *Nature* **369**, 411–14.

Machesky, L.M. and Hall, A. (1996). Rho: a connexion between membrane receptor signalling and the cytoskeleton. *Trends Cell Biol.* **6**, 304–10.

Machesky, L.M. and Way, M. (1998). Actin branches out. *Nature* **394**, 125–6.

Mackay, D.J.G., Esch, F., Furthmayr, M. and Hall, A. (1997). Rho and Rac-dependent assembly of focal adhesion complexes and actin filaments in permeabilized fibroblasts: an essential role of ezrin/radixin/moesin proteins. *J. Cell Biol.* **138**, 927–38.

Madara, J.L. (1988). Tight junction dynamics: is the paracellular transport regulated? *Cell* **53**, 497–8.

Mahida, Y.R., Makh, S., Hyde, S., Gray, T. and Borriello, S.P. (1996). Effect of *Clostridium difficile* toxin A on human intestinal cells: induction of interleukin 8 production and apoptosis. *Gut* **38**, 337–47.

Marshall, C.J. (1994). MAP-kinase-kinase-kinase, MAP-kinase-kinase and MAP-kinase. *Curr. Opin. Gene Dev.* **4**, 82–9.

Marshall, C.J. (1995). Specificity of receptor tyrosine kinase signalling: transient versus sustained extracellular signal-regulated kinase activation. *Cell* **80**, 179–85.

Massol, P., Montcourrier, P., Guillemot, J.C. and Chavrier, P. (1998). Completion of FcR-mediated phagocytosis requires the sequential function of Cdc42 and Rac1 and depends on protein phosphatase activity. *EMBO J.* **17**, 6219–29.

Matsudeira, P. (1991). Molecular organization of actin crosslinking proteins. *Trends Biol. Sci.* **16**, 87–92.

Matsui, T., Amano, M., Yamamoto, T., Chihara, K., Nakafuka, M., Ito, M., Nakano, T., Okawa, K., Iwamatsu, A. and Kaibuchi, K. (1996). Rho-associated kinase, a novel serine/threonine kinase, as putative target for the small GTP-binding protein Rho. *EMBO J.* **15**, 2208–16.

McGuffie, E.M., Franck, D.W., Vincent, T.S. and Olson, J.C. (1989). Modification of Ras in eukaryotic cells by *Pseudomonas aeruginosa* exoenzyme S. *Infect. Immun.* **66**, 2607–13.

Meiri, N., Ghelardini, C., Tesco, G., Galeotti, N., Dahl, D., Tomsic, D., Cavallero, S., Quattrone, A., Cappaciolini, S., Bartolini, A. and Alkon. D.L. (1997). Reversible antisense inhibition of shaker-like Kv1.1 potassium channel expression impairs associative memory in mouse and rat. *Proc. Natl Acad. Sci. USA* **94**, 4430–4.

Meloche, S., Seuwen, K., Pagés, G. and Pouysségur, J. (1992). Biphasic and synergistic activation of p44 (mapK) (ERK1) by growth factors: correlation between late phase activation and mitogenicity. *Mol. Endocrinol.* **658**, 845–54.

Miki, M., Miura, K. and Takenawa, T.N. (1996). N-WASP, a novel actin-depolymerizing protein, regulates the cortical cytoskeleton rearrangment in a PIP2-dependent manner downstream of tyrosine kinases. *EMBO J.* **15**, 5326–35.

Miki, M., Sasaki, T., Takai, Y. and Takenawa, T. (1998). Induction of Filopodia formation by a WASP-related actin-depolymerizing factor N-WASP. *Nature*, **391**, 93–6.

Minden, A., Lin, A., Claret, F.X., Abo, A. and Karin M. (1995). Selective activation of the JAK signalling cascade and C-Jun transcriptional activity by the small GTPase Rac and Cdc42. *Cell* **81**, 1147–57.

Mitic, L.L. and Anderson, J.M. (1998). Molecular architecture of tight junctions. *Annu. Rev. Physiol.* **60**, 121–42.

Moore, M.S. and Blobel, G. (1993). The GTP-binding protein Ran/TC4 is required for protein import into the nucleus. *Nature* **365**, 661–3.

Moorman, J.P., Bobak, D.A. and Hahn, C.S. (1996). Inactivation of the small GTP-binding protein Rho induces multinucleate cell formation and apoptosis in murine T lymphocytes EL4. *J. Immunol.* **22**, 4147–53.

Murphy, C., Saffrich, R., Grummt, M., Gournier, H., Rybin, V., Rubino, M., Auvinen, P., Lütcke, A., Parton, R.G. and Zerial, M. (1996). Endosome dynamics regulated by a Rho protein. *Nature* **384**, 427–32.

Nassar, N., Horn, G., Heurmann, C., Scherrer, A., McCormick, F. and Wittinghofer, A. (1995). The 2,2 A crystal structure of the Ras-binding domain of the serine/threonine kinase C-Raf1 in complex with Rap 1A and a GTP analogue. *Nature* **375**, 554–60.

Nobes, C.D. and Hall, A. (1995). Rho, Rac and Cdc42 GTPases regulate the assembly of multimolecular focal complexes associated with actin stress fibres lamellipodia and filopodia. *Cell* **81**, 1–20.

Noda, M., Ko, M., Ogura, A., Liu, D.G., Amano, T., Takano, T. and Ikawa, Y. (1988). Sarcoma virus carrying ras oncogene induce differentiation-associated properties in a neuronal cell. *Nature* **318**, 73–5.

Nusrat, A., Giry, M., Turner, J.R., Colgan, S.P., Parkos, S.A., Carnes, D., Lemichez, E., Boquet, P. and Madara, J.L. (1995). Rho protein regulates tight junction and perijunctional actin organization in polarized epithelia. *Proc. Natl Acad. Sci. USA* **92**, 10629–33.

Olson, M.F., Paterson, H.F. and Marshall, C.J. (1998). Signals from Ras and Rho GTPases interact to regulate expression of p21 Waf/Cip1. *Nature* **394**, 295–9.

Oswald, E., Sugai, M., Labigne, A., Wu, H.C., Fiorentini, C., Boquet, P. and O'Brien, A.D. (1994). Cytotoxic necrotizing factor type 2 produced by virulent *Escherichia coli* modifies the small GTP-binding protein Rho involved in assembly of actin stress fibres. *Proc. Natl Acad. Sci. USA* **91**, 3814–18.

Pai, E.F., Krengel, U., Petsko, G.A., Goody, R.S., Kabsch, W. and Wittinghofer, A. (1990). Refined crystal structure of the triphosphate conformation of the H-Ras p21 at 1,35A resolution: implication for the mechanism of GTP hydrolysis. *EMBO J.* **9**, 2351–9.

Pantaloni, D. and Carlier, M.F. (1993). How profilin promotes actin filament assembly in the presence of thymosin b4. *Cell* **75**, 1007–14.

Paterson, H.F., Self, A.J., Garret, M.D., Just, I., Aktories, K. and Hall, A. (1990). Microinjection of recombinant p21 Rho induces rapid changes in morphology. *J. Cell Biol.* **111**, 1001–7.

Pawson, T. (1995). Protein modules and the signalling networks. *Nature* **373**, 573–80.

Popoff, M.R., Chaves-Olarte, E., Lemichez, E., von Eichel-Streiber, C., Thelestam, M., Chardin P., Cussac, D., Antonny, B., Chavrier, P., Flatau, G. and Boquet, P. (1996). Ras, Rap and Rac small GTP-binding proteins are target for *Clostridium sordellii* lethal toxin glucosylation. *J. Biol. Chem.* **271**, 10217–24.

Pulverer, B.J., Kyriakis, J.M., Avruch, J., Nikolakaki, E. and Woodgett, J.R. (1991). Phosphorylation of C-Jun mediated by MAP kinases. *Nature* **353**, 670–4.

Ren, Bokoch, G.M., Traynor-Kaplan A., Jenkins, G.H., Anderson R.A. and Schwartz, M.A. (1996). Physical association of the small GTPase Rho with a 68 kDa phosphatidylinositol 4-phosphate 5 kinase in Swiss 3T3 cells. *Mol. Biol. Cell* **7**, 435–42.

Ridley, A.J., Paterson, H.F., Johnston, C.L., Dickmann, O. and Hall, A. (1992). The small GTP-binding protein Rac regulates growth factor-induced membrane ruffling. *Cell* **70**, 401–10.

Ridley, A.J. and Hall, A. (1992). The small GTP-binding protein Rho regulates the assembly of focal adhesion and actin stress fibres in response to growth factor. *Cell* **70**, 389–99.

Rittinger, K., Walker, P.A., Eccleston, J.F., Smerdon, S.J. and Gamblin, S.J. (1997). Structure at 1,65 A of RhoA and its GTPase-activating protein in complex with a transition state analogue. *Nature* **389**, 753–62.

Santos, M.F., McCormack, S.A., Guo, Z., Okolicany, J., Zheng, Y., Johnson, L.R. and Tigyi, G. (1997). Rho proteins play a critical role in cell migration during the early phase of mucosal restitution. *J. Clin. Invest.* **100**, 216–25.

Schafer, D.A., Jennings, P.B. and Cooper, J.A. (1996). Dynamics of capping proteins and actin assembly *in vitro*: uncapping barbed ends by polyphosphoinositides. *J. Cell Biol.* **135**, 169–79.

Scheffzek, K., Ahmadian, M.R., Kabsch, W., Wiesmüller, L., Lautwein, A., Schmitz, F. and Wittinghofer, A. (1997). The Ras–RasGAP complex: structural basis for GTPase activation and its loss in oncogenic Ras mutants. *Science* **277**, 333–8.

Schmidt, G., Sehr, P., Wilm, M., Selzer, J., Mann, M. and Aktories, K. (1997). Rho gln 63 is deamidated by *Escherichia coli* cytotoxic necrotizing factor 1. *Nature* **387**, 725–9.

Schmidt, G., Selzer, J., Lerm, M. and Aktories, K. (1998). The Rho-deamidating cytotoxic necrotizing factor 1 from *Escherichia coli* possesses transglutaminase activity. *J. Biol. Chem.* **273**, 13669–74.

Sekine, A., Fujiwara, M. and Narumyia, S. (1989). Asparagine residue in the Rho gene product is the modification site for Botulinum ADP-ribosyltransferase. *J. Biol. Chem.* **264**, 8602–5.

Selzer, J., Hofmann, F., Rex, G., Wilm, M., Mann, M., Just, I. and Aktories, K. (1996). *Clostridium novyi* a-toxin-catalyzed incorporation of GlcNac into Rho subfamily of proteins. *J. Biol. Chem.* **271**, 25173–7.

Simon, M.A., Bowtell, D.D.L., Dodson, G.S., Laverty, T.R. and Rubin, G.M. (1991). Ras-1 and a putative guanine nucleotide exchange factor perform crucial steps in signalling by the sevenless protein tyrosine kinase. *Cell* **67**, 701–16.

Symons, M., Derry, J.M., Karlak, B., Jiang, S., Lemahieu, V., McCormick, F., Francke, U. and Abo, A. (1996). Wiskott–Aldrich syndrome protein, a novel effector for the GTPase Cdc42Hs is implicated in actin polymerization. *Cell* **84**, 723–34.

Takaishi, K., Sasaki, T., Kotani, M., Nishioka, M. and Takai, Y. (1997). Regulation of cell–cell adhesion by Rac and Rho small GTP-binding protein in MDCK cells. *J. Cell Biol.* **139**, 1047–59.

Tapon, N. and Hall, A. (1997). Rho, Rac and Cdc42 GTPases regulate the organization of the actin cytoskeleton. *Curr. Opin. Cell Biol.* **9**, 86–92.

Theriot, J.A. and Mitchison, T.J. (1993). The three faces of profilin. *Cell* **75**, 835–8.

Toker, A. and Cantley, L.C. (1997). Signalling through the lipid products of phosphoinositide-3OH-kinase. *Nature* **387**, 673–6.

Trahey, M. and McCormick, F. (1987). A cytoplasmic protein stimulates normal H-Ras p21 GTPase but does not affect oncogenic mutants. *Science* **238**, 542–5.

Traverse, S., Gomez, N., Paterson, H.F., Marshall, C.J. and Cohen, P. (1992). Sustained activation of the mitogen-activated protein (MAP) kinase cascade may be required for differentiation of PC12. Comparison of the effects of nerve growth factor and epidermal growth factor. *Biochem. J.* **288**, 351–5.

Tsukita, S., Yonemura, Y. and Tuskita, S. (1997). ERM (ezrin/radixin/moesin) family: from cytoskeleton to signal transduction. *Curr. Opin. Cell Biol.* **9**, 70–5.

van der Geer, P., Hunter, T. and Linberg, R.A. (1994). Receptor protein-kinases and their signal transduction. *Annu. Rev. Cell. Biol.* **10**, 251–338.

von Eichel-Streiber, C., Boquet, P., Sauerborn, P. and Thelestam, M. (1996). Large clostridial cytotoxins: a family of glycosyl-transferases modifying small GTP-binding proteins. *Trends Microbiol.* **4**, 375–82.

Watarai, M., Kamata, Y., Kosaki, S. and Sasakawa, C. (1997). Rho, a small GTP-binding protein, is essential for *Shigella* invasion of epithelial cells. *J. Exp. Med.* **185**, 281–92.

Wei, Y., Zhang, Y., Derewenda, U., Liu, X., Minor, W., Nakamoto, R.K., Somlyo, A.V., Somlyo, A.P. and Derewenda, Z.S. (1997). Crystal structure of RhoAGDP and its functional implications. *Nature Struct. Biol.* **4**, 699–703.

Wittinghofer, A. (1998). Caught in the act of switch on. *Nature* **394**, 317–20.

Wittinghofer, A. and Nassar, N. (1996). How Ras-related proteins talk to their effectors. *Trends Biochem. Sci.* **21**, 488–91.

Wittinghofer, A. and Pai, E.F. (1991). The structure of Ras protein: a model for universal molecular switch. *Trends Biochem. Sci.* **16**, 382–7.

Wittinghofer, A. and Valencia, A. (1995). Three-dimensional structure of Ras and Ras-related proteins. In: *Guidebook to the Small GTPases* (eds M. Zerial and L.A. Huber), pp. 20–9. Oxford University Press, New York.

Wood, K.W. (1995). Raf1. In: *Guidebook to the Small GTPases* (eds M. Zerial and L.A. Huber), pp. 150–5. Oxford University Press, New York.

Wood, K.W., Qi, H., D'Arcangelo, G., Armstrong, R.C., Roberts, T.M. and Halegoua, S., (1993). The cytoplasmic raf oncogene induces a neuronal phenotype in PC12 cells: a potential role for cellular raf kinases in neuronal growth factor signal transduction. *Proc. Natl Acad. Sci. USA* **90**, 5016–20.

Yang, N., Miguchi, O., Ohashi, K., Nagata, K., Wada, A., Kangawa, K., Nishida, E. and Mizuno, K. (1998). Cofilin phosphorylation by LIM-kinase 1 and its role in Rac-mediated actin reorganization. *Nature* **393**, 809–12.

Zychinsky, A. and Sansonetti, P. (1997). Apoptosis in bacterial pathogenesis. *J. Clin. Invest.* **100**, 493–6.

3

Molecular, functional and evolutionary aspects of ADP-ribosylating toxins

Mariagrazia Pizza, Vega Masignani and Rino Rappuoli

INTRODUCTION

The mono-ADP-ribosylation reaction is an enzymic activity that was initially discovered and studied at the molecular level in bacterial toxins. For a while this reaction was believed to be a peculiarity of bacteria and no role was known for it on normal metabolism of eukaryotic cells. Recently, many enzymes with ADP-ribosylating activity have been discovered in eukaryotic cells. Examples are mammalian ART1–5 and the rodent RT6. This growing family of enzymes shows that ADP-ribosylation is also an enzymic reaction with an important role in the post-translational modification of the eukaryotic cells.

ADP-ribosylating toxins are a variety of bacterial proteins with totally unrelated structures that have in common only one feature: they behave as enzymes with ADP-ribosyltransferase activity. The toxins with this activity are shown in Figure 3.1. The enzymic activity, which is represented by the black 'A' domain in Figure 3.1, is responsible for the toxic effect. This is contained in a protein that usually has a size of approximately 20–25 kDa, but in some proteins can be higher. Based on their overall structure the toxins can be divided into A–B toxins, binary toxins or A-only toxins. A–B and binary toxins are molecules released by bacteria in the periplasm or in the extracellular environment. The A domain contains the enzymic toxic activity, while the B domain is a non-toxic part that functions as a carrier or delivery system for the A

domain: it binds the receptor on the surface of eukaryotic cells and helps its translocation within the cell.

As shown in Figure 3.1, the A–B toxins are the best studied and we know the three-dimensional structure of most of them. In some cases, the B domain can be divided into the B (cell binding) and T (membrane translocation) domains.

The binary toxins have a similar organization; however, in this case the A and B domains are separately secreted in the culture medium. The B domain then binds to the receptor on the surface of target eukaryotic cells, and only then is able to bind the A domain and help its translocation into the cytosol.

Toxins with this structure are the C2 toxin of *Clostridium botulinium* and the related toxins shown in Figure 3.1.

The A-only toxins are those toxins constituted uniquely by the A domain with the enzymic activity. A typical example of this type of toxin is the exoenzyme S of *Pseudomonas aeruginosa* that is directly injected by bacteria into eukaryotic cells by a specialized secretion system. Other toxins with this structure have a still unknown mechanism of cell entry (exoenzyme C3 of *C. botulinium*).

THE ENZYMIC REACTION AND THE SUBSTRATES

Two types of ADP-ribosylation reactions occur in nature: poly- and mono-ADP-ribosylation (Ueda and

---

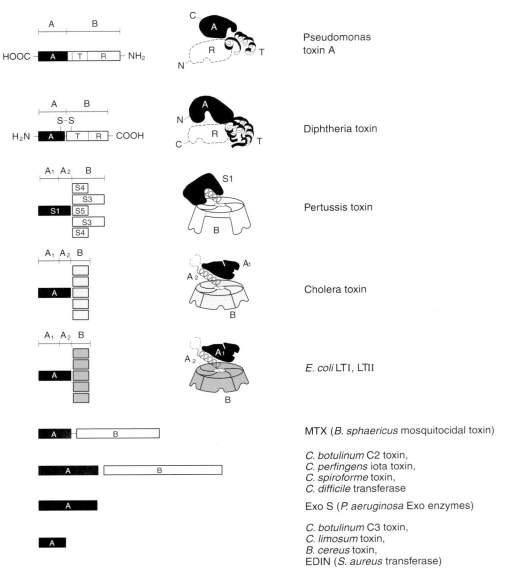

FIGURE 3.1 Structural organization of bacterial ADP-ribosylating toxins. The enzymically active, toxic moiety (A) is black. The carrier domain (B), when present, is white (left). When crystallographic data are available, a schematic representation of the quaternary structures of the domains is reported (right).

Hayaishi, 1985; Althaus and Richter, 1987). Poly-ADP-ribosylation occurs mostly in the nucleus of eukaryotic cells where histones and other nuclear proteins are post-translationally modified first by the covalent attachment of an ADP-ribose group to the carboxyl group of the C-terminal amino acid, and then by the elongation of the ADP-ribose chain through the addition of further ADP-ribose groups. Here we are interested in mono-ADP-ribosylation, an enzymic reaction which is mediated by bacterial toxins, phage and *Escherichia coli* proteins, and

a growing family of cytoplasmic and membrane-associated eukaryotic enzymes.

Mono-ADP-ribosyltransferases bind NAD and transfer the ADP-ribose group to a specific target protein that, following the post-translational modification, usually changes dramatically its function. The ADP-ribose group is transferred to a nitrogen atom in the side-chain of amino acids such as diphthamide, arginine, asparagine or cysteine, according to the reaction shown in Figure 3.2.

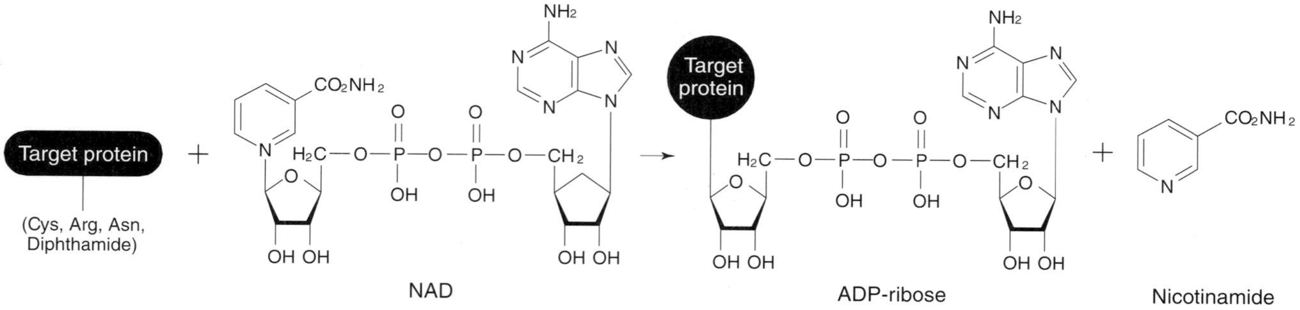

FIGURE 3.2 Mechanism of the mono-ADP-ribosylation reaction catalysed by ADP-ribosyltransferases: an ADP-ribose moiety is transferred to a specific target protein and a nicotinamide group is released. Following the post-translational modification the target protein changes its properties.

Interestingly, with the exception of actin, all of the eukaryotic proteins that are ADP-ribosylated by bacterial toxins belong to a quite restricted family: they are GTP-binding proteins (G-proteins). These proteins are molecular switches involved in a variety of cell functions, including cell proliferation, tissue differentiation, signal transduction, protein synthesis, protein translocation, vesicular trafficking and cytoskeleton structure (see Chapter 2 of this book). These proteins consist of two parts: a common core structure involved in GTP-binding, and one or more different domains that act as effectors, or are involved in the interactions with the molecules regulated by these molecular switches (Figure 3.3, and Kjeldgaard et al., 1996). G-proteins have four regions of sequence homology (named $\Sigma 1$–$\Sigma 4$ in Figure 3.3) and a structurally conserved core that binds GTP. This domain is the core of the molecular switch that is in the 'on' conformation when GTP-bound. Hydrolysis of GTP to GDP causes a conformational change of an α-helix in this core structure that turns the switch to the 'off' conformation (Hamm and Gilchrist, 1996; Sunahara et al., 1997; Gether and Kobilka, 1998; Gilchrist et al., 1998; Ji et al., 1998; Lefkowitz, 1998; Vaughan, 1998). The GTP-binding proteins are divided into four groups: the Ras-like G-proteins, the heterotrimeric G-proteins, the translation factors and tubulins. Tubulins do not contain the conserved core and are not substrates of ADP-ribosylation. All other groups of G-proteins have one or more components that are substrates of ADP-ribosyltransferases. One question that remains unanswered is whether ADP-ribosylating toxins derive from a common ancestor that recognized the common core structure of G-proteins, or whether during evolution they have independently selected G-proteins many times as targets just because they regulate very important cellular circuits and, therefore, are ideal

targets for molecules that intend to intoxicate cells or tissues.

The eukaryotic GTP-binding proteins which are mono-ADP-ribosylated by bacterial toxins are as follows.

Elongation factor 2

Eukaryotic elongation factor 2 (EF-2) is a protein of 95 700 Da which is involved in protein synthesis (Kohno et al., 1986). It contains a post-transcriptionally modified histidine residue (diphthamide 715) (Van Ness et al., 1980a, b), which is ADP-ribosylated by diphtheria (DT) and Pseudomonas exotoxin A (PAETA) (Honjo et al., 1968; Gill et al., 1969). Following ADP-ribosylation, EF-2 is unable to carry out protein synthesis and this causes rapid cell death. The region containing diphthamide 715 is very close to the anticodon recognition domain of EF-2. This suggests that ADP-ribosylation interferes with EF-2 binding to the tRNA. The homologous bacterial EFG contains a lysine in this position and therefore it is not a substrate for these toxins (see Figure 3.3) (Kohno et al., 1986). Therefore, bacteria are resistant to diphtheria and Pseudomonas toxins. In evolution, the first organisms containing an EF-2, which is susceptible to DT and PAETA, are the archaebacteria.

Heterotrimeric G-proteins

Heterotrimeric G-proteins are involved in transduction of signals from surface-exposed receptors to intracellular effectors. They generally consist of an α subunit, containing the GTP-binding domain and the intrinsic GTPase activity, and the $\beta\gamma$ subunit complex, containing two protein chains. In the 'off' position, the α subunit binds ADP and is found in a membrane-bound

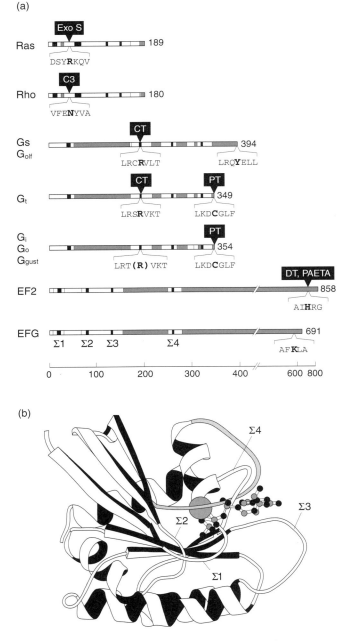

FIGURE 3.3 (a) schematic representation of GTP-binding proteins. The $\Sigma 1$–$\Sigma 4$ are the regions that form the core of the GTP-binding domain. They are present in all G-proteins. In the figure they are indicated by black boxes. The amino acids that are ADP-ribosylated by each toxin are indicated in bold. (b) Three-dimensional structure of the GTP-binding domain common to all G-proteins and formed by the homologous regions $\Sigma 1$–$\Sigma 4$.

complex together with the β and γ subunits. A positive signal from the receptor will cause the exchange of the GDP with GTP. The GTP-bound form of α dissociates from the $\beta\gamma$ complex and interacts with a downstream effector molecule, thus transducing an extracellular event in a change of the intracellular chemical environment. They usually regulate the activity of enzymes such as adenylate cyclase, phospholipase C and cyclic GMP-phosphodiesterase, which release secondary messengers into the cytoplasm as a response to external stimuli (Stryer and Bourne, 1986; Neer and Clapham, 1988). Adenyl cyclase, for instance, is regulated by two GTP-binding proteins: G_s and G_i. G_s receives signals from stimulatory receptors located on the surface of eukaryotic cells and stimulates the activity of adenyl cyclase. G_i, in contrast, receives signals from the inhibitory receptors and inhibits the adenyl cyclase activity (Gilman, 1984). In addition to G_s and G_i, the family of heterotrimeric G-proteins contains G_o, G_t, G_g, G_{olf}, and many other less characterized proteins with similar function. As shown in Figure 3.3, cholera toxin (CT) ADP-ribosylates Arg201 of the α subunit of G_s, G_{olf}, G_t. The corresponding Arg residues in G_i and G_o are ADP-ribosylated only when they are activated by the receptor. Arg201 is located in the core region that is involved in binding GTP. ADP-ribosylation has therefore a direct consequence on the enzymic activity: it slows down GTP hydrolysis, thus keeping G_s in a permanent 'on' position. Pertussis toxin (PT) ADP-ribosylates Cys352 of the α subunit of G_i, G_o, G_{gust}, G_t and other G-proteins, but is unable to ADP-ribosylate G_s and G_{olf} that have a tyrosine instead of a cysteine in this position. This carboxy-terminal region of G_α is involved in interaction with the receptor and therefore ADP-ribosylation causes receptor uncoupling. ADP-ribosylation of G-proteins by cholera and pertussis toxins causes a variety of effects in different tissues. In the case of adenyl cyclase, treatment with cholera toxin causes constitutive activation of the enzyme and accumulation of the second messenger cAMP, while treatment with pertussis toxin uncouples G_i from its receptor so that it becomes unable to inactivate adenyl cyclase.

Ras, Rho and the small GTP-binding proteins

The Ras proteins are a large family of membrane-associated GTP-binding proteins of approximately 21 kDa. They function in transmembrane-signalling systems that control a variety of cellular processes including growth, proliferation and differentiation of cells. The human Ras family consists of three highly homologous members: N-, K- and K-Ras. They contain a 'CAAX' carboxy-terminal box that can be modified by the addition of a farnesyl lipid moiety, thus increasing their hydrophobicity. In the active conformation the GTP-bound Ras interacts with the GTPase activating proteins (GAPs) effector molecules that turn it off by stimulating GTP hydrolysis. Many Ras proteins are proto-oncogenes, requiring a single amino acid change to turn them into oncogenes. The mutations that are mostly found in tumours are in position 12 in Σ1 and in position 61 in Σ3. The oncogenic proteins exhibit a low GTPase activity following interaction with GAP, and therefore are blocked mostly in the 'on' position.

Rho (Ras homology) are a subgroup of the Ras family GTPases containing several members including RhoA, B, C, D, E and G, Rac1, 2 and 3, and Cdc42. Like Ras, they contain the 'CAAX' carboxy-terminal box and can be post-translationally modified. However, while Ras is only plasma membrane-associated, Rho can be found in a variety of locations, including the cytoplasm (RhoA and C, Rac1 and 2), the endosomes (RhoB), or the Golgi and endoplasmic reticulum (Cdc42). Rho proteins are controlled by three groups of regulatory proteins. Guanine nucleotide exchange factors (GEFs) induce activation of Rho by facilitating the GDP/GTP exchange, GTPase-activating proteins (GAPs) stimulate the hydrolysis of GTP, thus turning off Rho proteins, while guanine dissociation inhibitors (GDIs) keep Rho in the inactive form. Rho, Rac and Cdc42 are involved in the regulation of the actin cytoskeleton. Rho induces formation of focal adhesion and actin stress fibres. Rac is involved in lamellopodia formation and induces adhesion complexes. Cdc42 induces the formation of microspikes. In summary, the Rho family of GTPases is involved in focal adhesion, integrin function, cell movement and cell division. Ras and Rho are ADP-ribosylated at Asn41 by the exoenzyme S of *Pseudomonas aeruginosa* and the C3 enzyme of *C. botulinum*, respectively. ADP-ribosylation causes inactivation of Ras and Rho, and inability to stimulate actin polymerization, with consequent loss of cell shape and cell rounding.

Actin

Actin, a polypetide of 375 amino acid residues, is an ATP-binding protein that, in addition to contraction of muscle cells, in non-muscle cells controls the shape and the spatial organization of the cells, the cell movement, endo- and exocytosis, vesicle transport, cell contact and mitosis. Microfilaments, which are the major structure of the cytoskeleton, are filamentous structures 7–9 nm in diameter, composed of polymerized actin. Rapid

changes in cell shape are based on the ability of microfilaments to polymerize and depolymerize. Several bacterial toxins ADP-ribosylate monomeric actin at Arg177. This Arg is located at the contact site between actin monomers, and therefore it can be modified in monomeric G-actin, but it is not available for ADP-ribosylation in polymerized actin. Once ADP-ribosylated, monomeric actin is unable to make contact with the other actin monomers and therefore it cannot polymerize. As a consequence, toxin-treated cells are not able to build microfilaments and their cytoskeletal structure is rapidly destroyed. The toxins that ADP-ribosylate actin are *C. botulinum* C2 toxin, *Clostridium perfringens* iota toxin, *Clostridium spiriforme* toxin and the ADP-ribosylating toxin of *Clostridium difficile*.

CELL ENTRY

Since the target proteins of ADP-ribosylating toxins are all located in the cytosol or in the inner face of the cytoplasmic membrane, the toxins need to cross the cell membrane in order to reach their intracellular targets. This is done mainly in two ways: (a) receptor-mediated endocytosis (see Figure 3.4, paths 1 and 2); or (b) direct injection of the toxin from the bacterium into the cytosol of the eukaryotic cell by a type-III secretion system (see Figure 3.4, path 3).

Receptor-mediated endocytosis

Receptor-mediated endocytosis is used by soluble toxins with an A–B structure and by binary toxins where the A and B domains are physically separated. The toxins with this structure are released by bacteria in the culture medium and bind to the surface of eukaryotic cells by their B domain, which contains a receptor-binding site. Following binding the A–B toxins are internalized and located in membrane-bound vesicles (early endosomes). In the case of binary toxins, the B domain binds first to the cell receptor and then captures the A domain to the cell surface. The A–B complex is then internalized. Following internalization, two quite different pathways are used by different toxins to translocate their A domain into the cytosol. The two pathways have been best studied for diphtheria and cholera toxin, respectively. Diphtheria toxin A domain crosses the membrane early after internalization (see Figure 3.4, path 1a). As soon as the pH of the endosomes decreases to 5.5 units, the B domain changes conformation and exposes hydrophobic α-helices that are no longer soluble in water and therefore penetrate the lipid bilayer of the membrane (Figure 3.4, paths 1b and 1c). This initiates a process that favours the translocation of the A subunit across the membrane (Figure 3.4, path 1d). (See the subsection on diphtheria toxin later in this chapter for a more detailed description of this process.)

The toxins known to cross the membrane by a mechanism similar to diphtheria toxin are botulinum and tetanus toxins.

Conversely, cholera toxin has a much more complicated intracellular route before it reaches the cytoplasm (Figure 3.4, paths 2a and 2b). Following internalization in early endosomes, it undergoes a retrograde transport back to the Golgi apparatus (Figure 3.4, path 2c), and across it until it reaches the endoplasmic reticulum, from where the A subunit is finally translocated into the cytosol (Figure 3.4, path 2c). The routing of the A subunit to this pathway is believed to be mediated by an amino acid motif (KDEL) that is similar to the endoplasmic reticulum retention domain of eukaryotic proteins.

The toxins that follow the intracellular route of cholera toxin are: Shiga toxin, the related Verotoxins, pertussis toxin, *E. coli* heat-labile enterotoxin and *Pseudomonas* exotoxin A (Johannes and Goud, 1998).

Direct transfer of toxins from bacteria to eukaryotic cells

Some bacteria have evolved sophisticated transport systems to transfer directly their weapons into

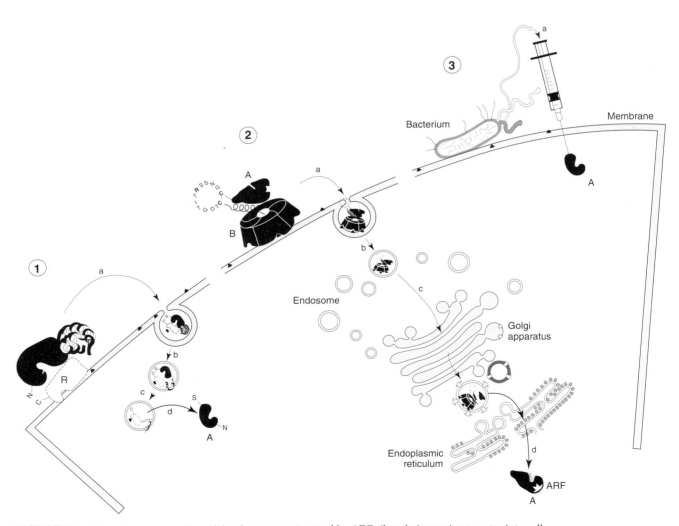

FIGURE 3.4 Schematic representation of the three strategies used by ADP-ribosylating toxins to enter into cells.

eukaryotic cells. The most famous are the type-III and type-IV secretion systems of Gram-negative bacteria. In these cases the transmembrane machinery of flagella and conjugative pili, respectively, has evolved into contact-dependent delivery systems able to translocate macromolecules into eukaryotic cells. The toxins translocated by these mechanisms do not need a B domain as they are injected directly into eukaryotic cells. (In Figure 3.4, path 3, they are, in fact, represented as being injected into the cells using a syringe.) An ADP-ribosylating toxin injected into cells by a type-III secretion system is exoenzyme S of *P. aeruginosa*. This is made up of a single polypeptide chain containing the enzymically active portion in the carboxy-terminal region and a part recognized by the secretion system in the amino-terminal domain.

BACTERIAL TOXINS

Toxins acting on protein synthesis

Diphtheria toxin

Diphtheria toxin (DT) is a protein molecule, of 58 350 Da, that is released into the supernatant by toxinogenic *Corynebacterium diphtheriae* strains (Pappenheimer, 1977; Collier, 1982). The toxin is synthesized as a single polypeptide chain that, after mild trypsin treatment and reduction of a disulfide bond, can be divided into two functionally different moieties: fragment A and fragment B of 21 150 and 37 200 Da, respectively (Pappenheimer, 1977) (Figure 3.5a). The molecule contains four cysteines and two disulfide bridges: the first one joins Cys186 to Cys201 and links fragment A to fragment B, whereas the other is contained within fragment B and joins Cys461 to Cys471. The determination of the three-dimensional structure of the molecule (Choe *et al.*, 1992; Bennett *et al.*, 1994) has shown that the toxin can be divided into three separate domains: the C (catalytic) domain, corresponding to fragment A; the T (transmembrane or translocation) domain, composed of nine α-helices (TH1–TH9) (Figure 3.5c); and the carboxy-terminal R (receptor-binding) domain.

Fragment A is a NAD$^+$-binding enzyme that catalyses the transfer of the ADP-ribosyl group to a post-translationally modified histidine residue (diphthamide) present in the cytoplasmic elongation factor 2 (EF-2) of eukaryotic cells (Brown and Bodley, 1979; Van Ness *et al.*, 1980a). The EF-2–ADP-ribose complex is inactive and, therefore, diphtheria toxin causes inhibition of protein synthesis and cell death. It has been shown that *in vitro* a single molecule of fragment

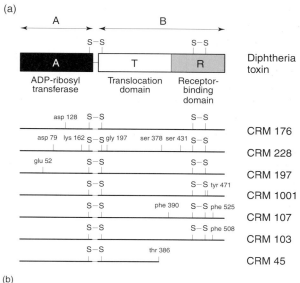

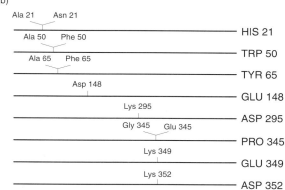

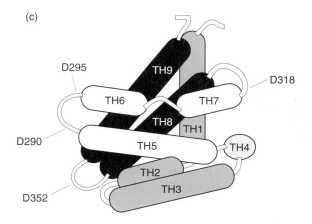

FIGURE 3.5 Diphtheria toxin: (a) structure of the toxin and mutant CRMs obtained by random chemical mutagenesis; (b) mutants obtained by site-directed mutagenesis; (c) three-dimensional structure of the seven α-helices that form the hydrophilic T domain. The position of the mutations described in this region are shown.

A is enough to kill one eukaryotic cell (Yamaizumi *et al.*, 1978). This observation correlates well with the *in vivo* studies that show that, on a weight basis, diphtheria toxin is one of the most potent bacterial toxins; the minimal lethal dose for humans and sensitive animals is below $0.1 \mu g\ kg^{-1}$ of body weight (Pappenheimer, 1984).

Although the entire lethal activity of diphtheria toxin is located in fragment A, fragment B is required for receptor binding and for translocation of fragment A across the cell membrane. The COOH-terminal domain, involved in receptor binding (R), and the NH$_2$-terminal hydrophobic domain (T), involved in the interaction with the cell membrane, mediate this process. The toxin receptor is the heparin-binding EGF-like growth factor precursor (Naglich *et al.*, 1992; Brown *et al.*, 1993; Iwamoto *et al.*, 1994; Hooper and Eidels, 1995) that is present on most animal cells. Murine cells contain amino acid changes in the receptor that make rodents insensitive to diphtheria toxin (Mitamura *et al.*, 1995; Brooke *et al.*, 1998). During the cell intoxication process, the toxin binds the cells through its receptor-binding domain and is internalized by receptor-mediated endocytosis. When the pH of the endosomes decreases below pH 5.5, fragment B undergoes a conformational change that exposes the hydrophobic regions of the T domain, allowing interaction with the endosomal membrane and translocation of fragment A from the endosome into the cytoplasm (Sandvig and Olsnes, 1981; Neville and Hudson, 1986; Cieplak *et al.*, 1987; Papini *et al.*, 1987a, b). Therefore, whereas fragment A is very stable over a wide range of pH, temperature and denaturing agents (it can be boiled, and treated with urea or various detergents without affecting its enzymic activity), fragment B is very sensitive to low pH and is unstable in many buffers. Exposure of fragment B to pH below 5.5 causes non-reversible denaturation of fragment B.

The diphtheria toxin gene is carried by a family of closely related bacteriophages (corynebacteriophages) that integrate into the bacterial chromosome and convert non-toxinogenic, non-virulent *C. diphtheriae* strains into toxinogenic, highly virulent species (Freeman, 1951; Uchida *et al.*, 1971). Lysogenic *C. diphtheriae* strains produce diphtheria toxin only when the medium is depleted of iron. The iron regulation of DT synthesis had already been discovered by A.M. Pappenheimer (1938), and today is well understood at the molecular level. The region upstream of the DT gene contains a sequence that is bound by an iron-binding repressor that dissociates from DNA only in the absence of iron (Qiu *et al.*, 1995; Ding *et al.*, 1996; Pohl *et al.*, 1998; White *et al.*, 1998).

The best-known toxinogenic corynephages are βtox$^+$ (Costa *et al.*, 1981), isolated in 1950 during an outbreak of diphtheria in Canada, and ωtox$^+$, isolated from the hypertoxinogenic PW8 strain (Rappuoli *et al.*, 1983a). The process of integration of corynephage DNA into the bacterial chromosome closely resembles the mechanism of lysogenization described by Campbell (1962) for bacteriophages: the phage DNA contains a locus called the phage attachment site (*att*P), which is homologous to a sequence in the bacterial chromosome (bacterial attachment site or *att*B). Following site-specific recombination between these two sites, the phage DNA becomes stably integrated into the bacterial chromosome. Unlike *E. coli*, the chromosome of *C. diphtheriae* contains two primary attachment sites (attB1 and attB2) (Rappuoli *et al.*, 1983b; Rappuoli and Ratti, 1984). This allows the stable integration of two copies of corynephage DNA into the bacterial chromosome (Rappuoli *et al.*, 1983b). Because each phage carries one copy of the *tox* gene, double lysogens produce twice the toxin produced by monolysogens. This property has been very important for obtaining strains hyper-producing CRM197, a non-toxic form of diphtheria toxin, which is used as a carrier for *Haemophilus influenzae* type B (Anderson *et al.*, 1987; Egan *et al.*, 1995; Rothbrock *et al.*, 1995) and for meningococcal (Costantino *et al.*, 1992) conjugate vaccines, and has been proposed as a new vaccine against diphtheria (Rappuoli, 1983).

The gene coding for diphtheria toxin is contained within a 1850 base pair (bp) *EcoRI–HindIII* DNA fragment, which has been cloned and entirely sequenced from phages βtox$^+$ and ωtox$^+$ (Greenfield *et al.*, 1983; Ratti *et al.*, 1983). The sequences of the two *tox* genes obtained from phages isolated more than 50 years apart are identical, showing the remarkable conservation of the diphtheria toxin molecule. DT was the first toxin for which structure–function relationships were elucidated, and it has been a model for all the other toxins.

Mutants of diphtheria toxin

Most of the functional and structural properties of DT were initially deduced from the analysis of a number of non-toxic DT mutants (cross-reacting materials or CRMs), encoded by corynephages which had been mutagenized by nitrosguanidine (Uchida *et al.*, 1971, 1973a–c; Laird and Groman, 1976). The analysis of the properties of these mutants and the sequence of their genes was the first tool available for mapping the functional domains of diphtheria toxin, and to identify some of the amino acids that play an important role in the enzymic activity of fragment

A, the receptor-binding domain of fragment B and the translocation of fragment A across the eukaryotic cell membrane. The most relevant diphtheria toxin mutants are described below and reported in Figure 3.5a.

CRM 176 contains a Gly128 to aspartic acid mutation that reduces the enzymic activity of fragment A by a factor of 10 (Comanducci et al., 1987). CRM 228 contains two mutations in fragment A and three in fragment B, which result in inactive A and B fragments (Kaczorek et al., 1983). The mutation Gly79 → Asp was later shown to be responsible alone for the lack of enzymic activity of fragment A (Johnson and Nicholls, 1994a).

CRM 197 contains a single glycine to glutamic acid change in position 52 that makes the fragment A unable to bind NAD$^+$ and, therefore, enzymically inactive (Pappenheimer et al., 1972; Giannini et al., 1984). Being enzymically inactive, and therefore non-toxic, but otherwise identical to diphtheria toxin, CRM 197 has been proposed as a natural candidate to develop a new vaccine against diphtheria. However, some subtle structural differences between diphtheria toxin and CRM 197 have been detected (Bigio et al., 1987), making this mutant more susceptible to proteases and less immunogenic than diphtheria toxoid (Pappenheimer et al., 1972). After stabilization with formalin, CRM 197 becomes immunogenic and able to induce protective antibody titres against diphtheria (Porro et al., 1980); however, its potency per μg of protein is still slightly lower than that of diphtheria toxoid.

CRM 1001 has an enzymically active fragment A and a fragment B that binds the toxin receptors, but is unable to translocate fragment A across the eukaryotic cell membrane. The nucleotide sequence of its gene has shown that Cys471 has been replaced by tyrosine, thus inhibiting the formation of the disulfide bridge between Cys461 and Cys471 (Dell'Arciprete et al., 1988).

CRM 103 and CRM 107 are two mutants that are unable to bind the receptors on the surface of eukaryotic cells (Laird and Groman, 1976; Greenfield et al., 1987). Sequence analysis has shown that both molecules contain mutations in the R domain. In CRM 103, Ser508 has been changed to phenylalanine, whereas in CRM 107 both Leu390 and Ser525 have been replaced by a phenylalanine. CRM 45 is a shorter molecule prematurely terminated because of a non-sense mutation that introduces a stop codon at position 387; therefore, CRM 45 lacks entirely the R domain and is unable to bind the toxin receptors (Giannini et al., 1984).

The characterization of the structure–function relationships, initially performed using the above mutants produced by nitrosoguanidine mutagenesis, has progressed more recently by studying mutant proteins constructed through site-directed mutagenesis, and containing amino acid changes in those positions that the crystal structure and biochemical studies had suggested as being important. Among the mutations produced by site-directed mutagenesis, the most relevant are described below and shown in Figure 3.5b.

Histidine 21, initially identified as important for catalysis by biochemical methods (Papini et al., 1989), was later confirmed to be so important that it could not be substituted with other amino acids without abolishing the enzymic activity. Some activity was only maintained when His was replaced by Asn (Blanke et al., 1994a; Johnson and Nicholls, 1994b).

Substitution of Try50 with alanine decreased ADP-ribosyltransferase by 10^5-fold, while substitution with phenalanine had only minimal effects, suggesting an important role for an aromatic residue in this position for NAD affinity (Wilson et al., 1994).

Tyrosine 65 was initially identified by photolabelling studies as important for binding the nicotinamide ring (Papini et al., 1991), and later confirmed to be relevant by mutagenesis studies. Alanine substitution caused a 350-fold decrease in enzymic activity, while phenalanine substitution caused only a small decrease (Blanke et al., 1994b).

Glutamic acid 148 was also initially identified as being near the catalytic site by photoaffinity labelling (Carroll et al., 1985), and was later shown to be so important for catalysis that it could not be substituted even with the closely related aspartic acid (Tweten et al., 1985).

Aspartic acid 295, which is located in the hairpin loop between the α-helices TH5 and TH6 of the T domain, was important for membrane interaction of fragment B, because its substitution with a lysine residue caused a reduction in toxicity (Silverman et al., 1994). Finally, several amino acids, located between α-helices TH8 and TH9, were also very important for DT translocation across the endosomal membrane. Substitution of Pro345 with glycine or glutamic acid caused a 99% reduction in toxicity (Johnson et al., 1993). Substitution of Glu349 or Asp352 with lysine, caused more than 100-fold reduction in toxicity (Silverman et al., 1994).

Mechanisms of the catalytic and translocation processes

The information deriving from biochemical, structural and mutagenesis studies has increased our knowledge on the way in which the catalytic and translocation domains work. The present state of understanding is described below.

The active site (Figure 3.11) is formed by an α-helix bent over a β-strand, which form the ceiling and the floor of the NAD-binding cavity, respectively. Within the cavity Tyr54 and Tyr65 are parallel, and between their aromatic rings there is just enough space to allow the entrance of the nicotinamide ring of NAD, which is then held firmly sandwiched between the two tyrosines. At each side of the NAD-binding cavity, there are two amino acids which are essential for catalysis: His21 and Glu148.

Histidine 21 is likely to be involved in positioning the NAD into the cavity, so that it is available for the nucleophilic attack of the substrate. Glutamic acid 148 is likely to be involved in interaction with the incoming substrate molecule. The structure of the active site, shown in Figure 3.11, is conserved in all ADP-ribosylating enzymes (Domenighini *et al.*, 1991, 1994, 1995a).

The T domain (Figure 3.5c) is formed by nine α-helices arranged in three layers. The internal layer contains helices TH8 and TH9, which are hydrophobic and amphipathic, respectively. The intermediate layer contains helices TH5, TH6 and TH7, which form a second hydrophobic shell. Finally, the external layer contains helices TH1, TH2 and TH3, which are rich in charged residues and help in maintaining the T domain and the toxin in a soluble form at neutral pH. When the pH decreases within the endosomes during toxin internalization, the charged residues located at the hairpin between two hydrophobic α-helices, such as Asp295, Glu349 and Asp352, lose their charge, thus making the entire structure hydrophobic. This triggers the interaction of the T domain with the membrane and the consequent translocation of the A domain in the cytoplasmic site.

The genetic organization of DT, with the region encoding for the receptor-binding domain at the 3'-end of the gene, has allowed the development of a number of hybrid toxins in which the receptor-binding domain has been replaced by interleukin 2, or by the melanocyte hormone MSH (Murphy *et al.*, 1986; Williams *et al.*, 1987; Bacha *et al.*, 1988). These new molecules kill activated T-cells and MSH receptor-bearing cells, respectively. They might find therapeutic applications for the treatment of allograft rejection and melanomas.

Pseudomonas *exotoxin A*

Pseudomonas exotoxin A (PAETA) is secreted in the culture medium as a single polypeptide chain of 613 amino acids of which the sequence and the three-dimensional structure to 3 Å resolution is known (Gray *et al.*, 1984; Allured *et al.*, 1986). According to X-ray crystallography studies, the molecule can be divided into three domains (Hwang *et al.*, 1987; Pastan and FitzGerald, 1989; Siegall *et al.*, 1989). Domain I is composed of two non-contiguous regions: Ia comprising amino acids 1–252 and Ib composed of amino acids 365–404.

Domain II is composed of amino acids 253–364, while domain III comprises amino acids 405–613 (see Figures 3.1 and 3.6). Four disulfide bridges are present, two located in domain Ia, one in domain Ib and one in domain II. Genetic studies, based mainly on the expression of mutated forms of the PAETA gene in *E. coli*, have shown that the deletion of domain Ia results in non-toxic, enzymically active molecules which cannot bind the cells. A similar result can be obtained by mutating Lys57 to Glu (Jinno *et al.*, 1988). Deletions in domain II result in molecules which bind to the

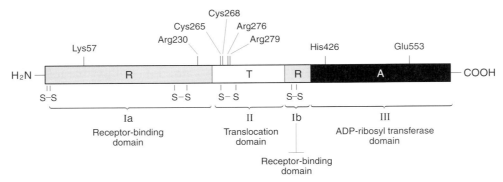

FIGURE 3.6 Domain organization of *Pseudomonas aeruginosa* exotoxin A (PAETA). The enzymic domain (A) is shown in black, while the receptor-binding (R, Ia and Ib) and the translocation (T) domains are coloured grey and white, respectively. The more relevant mutants affecting either cell binding or enzymic activity are reported in the upper portion of the figure. Disulfide bridges are also shown.

cells and which are enzymically active but not toxic. A similar result can be obtained by mutagenizing Arg276, Arg279 and Arg230 (Jinno et al., 1989) or by converting Cys265 and Cys268 to other amino acids, thus changing the structure of domain II (Siegall et al., 1989). Deletions or mutations in domain III result in enzymically inactive molecules (Siegall et al., 1989).

Based on these observations it can be concluded that the three domains observed in the X-ray structure correspond exactly to three functional domains involved in cell recognition (domain I), membrane translocation (domain II) and ADP-ribosylation of elongation factor 2 (domain III). PAETA can be described as a typical bacterial toxin with an A–B structure, having a mechanism of action identical to diphtheria toxin. Although identical in function, DT and PAETA have an opposite structural organization: the enzymically active domain is at the C-terminus of PAETA and at the N-terminus of DT. While the B domains of DT and PAETA do not have any structural similarity, the enzymes share a common structure of the catalytic site (see Figure 3.11 for details) (Brandhuber et al., 1988; Carroll and Collier, 1988). The structure of this site, well described by X-ray crystallography, has been further corroborated by functional studies which have identified amino acids playing a key role in enzymic activity. Glutamic acid 553 was initially shown to be at the catalytic site, because it was the only amino acid photoaffinity-labelled by NAD^+ (Carroll and Collier, 1987). Later, it was shown that substitution of Glu553 with any amino acid, including Asp, decreased the enzymic activity by a factor of 1000 (Douglas and Collier, 1990), and that deletion of Glu553 completely abolished the toxicity of PAETA. Similarly, iodination of Tyr481, which is also at the catalytic site, abolished enzymic activity (Brandhuber et al., 1988). Other amino acids which, although not located at the catalytic site, are shown to be essential for enzymic activity are His426 and residues 405–408 (Galloway et al., 1989). Histidine 426 has been proposed to be necessary for the interaction between PAETA and EF-2.

The well-defined structural divisions in separated domains make PAETA an ideal candidate for the development of chimeric toxins by replacing the gene parts encoding for domain I with others encoding for cell-binding domains with different specificities. So far, nucleotides encoding domain I have been replaced by sequences encoding interleukin-2, interleukin-6, interleukin-4 and T-cell antigen CD4. In all instances, expression of these genes in E. coli has given new toxins which specifically kill the cells bearing the receptor recognized by the new domain I. Such molecules are promising candidates for the treatment of arthritis and allograft rejection (PAETA–IL-2), AIDS (PAETA–CD4) and other diseases (Chaudhary et al., 1987, 1988; Lorberboum-Galski et al., 1988; Siegall et al., 1988; Ogata et al., 1989; Baldwin et al., 1996; Mori et al., 1997; Zimmermann et al., 1997; Essand and Pastan, 1998).

Toxins acting on signal transduction – action on large G-proteins

Pertussis toxin

Pertussis toxin (PT) is a protein of 105 000 Da released into the extracellular medium by Bordetella pertussis, the aetiological agent of whooping cough (see Chapter 7 of this book). PT is a complex bacterial toxin composed of five different subunits which have been named S1 (21 220 Da), S2 (21 920 Da), S3 (21 860 Da), S4 (12 060 Da) and S5 (11 770 Da) (see Figure 3.1), according to their electrophoretic mobility (Tamura et al., 1982; Sekura et al., 1985). Exposure of PT to 2 M urea disassembles the PT into the monomer A (subunit S1) and the oligomer B (which comprises the subunits S2, S3, S4 and S5). Upon exposure to 5 M urea, the B oligomer can be dissociated into two dimers: dimer 1 (comprising S2 and S4) and dimer 2 (comprising S3 and S4), and the monomer S5. With 8 M urea PT dissociates into five monomeric subunits (Tamura et al., 1982).

As in the case for the other ADP-ribosylating toxins, the B oligomer of PT binds the receptors on the surface of eukaryotic cells and allows the toxic subunit S1 to reach its intracellular target proteins. In Chinese hamster ovary (CHO) cells, the PT receptor is a 165-kDa glycoprotein which binds the PT B oligomer through a branched mannose structure containing sialic acid. Dimers S2–S4 and S3–S4 also bind the same receptor (Witvliet et al., 1989; Stein et al., 1994a).

S1 ADP-ribosylates the Cys353 of the α subunit of protein G_i, and the corresponding cysteine in G_o, G_{gust} and G_t (see Figure 3.3) (Katada et al., 1983; West et al., 1985). G_{sand}, which contain a Tyr in place of the Cys residue, are not ADP-ribosylated by PT. ADP-ribosylation of the above G-proteins causes alteration in the response of eukaryotic cells to exogenous stimuli and results in a variety of phenotypes. In vivo, the most relevant consequences of PT intoxication are: leucocytosis, histamine sensitization, increased insulin production with consequent hypoglycaemia and potentiation of anaphylaxis (Sekura et al., 1985). In vitro, PT has a number of different activities, the most relevant of which is the change in cell morphology

TABLE 3.1 Toxic and non-toxic properties of PT and PT–9K/129G mutant

Property	Native PT	PT–9K/129G mutant	Reference
Toxic properties of PT			
CHO cell-clustered growth (ng ml^{-1})	0.005	>5000[a]	Pizza et al. (1989)
Histamine sensitization (μg mouse^{-1})	0.1–0.5	>50[a]	Nencioni et al. (1990)
Leucocytosis stimulation (μg mouse^{-1})	0.02	>50[a]	Nencioni et al. (1990)
Anaphylaxis potentiation (μg mouse^{-1})	0.04	>7.5[a]	Nencioni et al. (1990)
Enhanced insulin secretion (μg mouse^{-1})	<1	>25[a]	Nencioni et al. (1990)
IgE induction (in vitro) (ng ml^{-1})	0.8	>100[a]	van der Pouw-Kraan et al. (1995)
IgE induction (in vivo) (ng rat^{-1})	10	>200[a]	van der Pouw-Kraan et al. (1995)
Long-lasting enhancement of nerve-mediated intestinal permeabilization of antigen uptake (ng rat^{-1})	1	>200[a]	Kesecka et al. (1994)
Inhibition of IL-1-induced IL-2 release in ELA 6.1 cells (μg ml^{-1})	0.1	>100[a]	Zumbihl et al. (1995)
Inhibition of neutrophil migration (μg rat^{-1})	0.2	>1.2[a]	Brito et al. (1997)
Lethal dose (μg kg^{-1})	15	>1500[a]	Unpublished data
ADP-ribosylation (ng)	1	>20 000[a]	Pizza et al. (1989)
Non-toxic properties of PT			
T-cell mitogenicity (μg ml^{-1})	0.1–0.3	0.1–0.3	Nencioni et al. (1991)
Haemagglutination (μg well^{-1})	0.1–0.5	0.1–0.5	Nencioni et al. (1991)
Mitogenicity for PT-specific T-cells (μg ml^{-1})	3	3	Pizza et al. (1989)
Platelet activation (μg ml^{-1})	5	5	Sindt et al. (1994)
Mucosal adjuvanticity (μg mouse^{-1})	3	3	Roberts et al. (1995)
Affinity constant (monoclonal 1B7, anti-S$_1$)	2.4×10^8	6.1×10^8	Nencioni et al. (1990)
Affinity constant (polyclonal anti-PT)	2.0×10^{10}	9.8×10^9	Nencioni et al. (1991)

[a]No effect was observed at the highest dose reported used in the assay.

in CHO cells, a phenotype which is able to detect as little as 10 pg of active PT (Hewlett et al., 1983). Further in vitro activities are haemagglutination, T-cell mitogenicity, inhibition of migration of peritoneal macrophages, enhancement of receptor-mediated accumulation of cAMP and many others (Sekura et al., 1985). In contrast to the other bacterial toxins where all the activities are mediated by the enzymically active subunit, in the case of PT several biological activities are due to the B oligomer (Tamura et al., 1983). In fact, the B subunit of pertussis toxin is a polyclonal mitogen for T-cells. In vitro the effect requires at least 0.3 μg of pertussis toxin or of its B subunit. The effect has not been observed in vivo, possibly because the dose required for activity is never achieved. It also has haemagglutination activity. In vitro, the effect requires 0.3 μg ml^{-1} of wild-type PT, or of its B subunit. In addition, the B oligomer is able to induce signal transduction through the inositol phosphate pathway. The numerous in vitro and in vivo activities of PT are listed in Table 3.1. As shown in the table, the toxic activities are all abolished when the active site is inactivated by site-directed mutagenesis, such as in mutant PT–9K/129G. In contrast, the non-toxic biological activities that are mediated by the receptor-

binding site present in the B oligomer are all maintained, even in the absence of ADP-ribosyltransferase activity.

The genes encoding for the five subunits of pertussis toxin are clustered in a fragment of DNA of 3200 base pairs, organized in an operon structure and in the following order: S1, S2, S5, S4 and S3 (Locht and Keith, 1986; Nicosia et al., 1986). Each of the five subunits is co-translationally exported into the periplasmic space where the holotoxin is assembled. The release of the toxin into the extracellular medium requires the products of the ptl locus, an operon containing 11 genes which is homologous to the tra operon of E. coli and the VirB operon of Agrobacterium tumefaciens. The ptl operon is located downstream from the PT operon, is transcribed by the PT promoter and codes for a type-IV secretion system (Covacci and Rappuoli, 1993; Weiss et al., 1993; Ricci et al., 1996).

The amino acid sequence of subunit S1 shows a significant homology with the A1 promoter of cholera and E. coli LT toxins (see Figure 3.10) (Locht and Keith, 1986; Nicosia et al., 1986). The crystal structure of the toxin has revealed a structural similarity of the catalytic subunits to the other ADP-ribosylating toxins (Figure

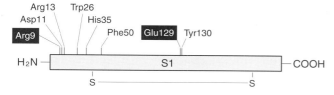

FIGURE 3.7 Toxic subunit S_1 of the pertussis toxin (PT) and illustration of some of the residues which have been mutagenized. In particular, double substitution of Arg9 and Glu129 (black boxed) produces the most well-known mutant which has been used for the construction of an acellular vaccine against pertussis.

3.11) and a structure of the B subunit that resembles that of cholera toxin organized around a central cavity composed by the α-helices of each subunit (Stein *et al.*, 1994b).

Mutants of pertussis toxin

The role of many amino acids of the S1 subunit has been tested by site-directed mutagensis initially performed to produce non-toxic mutants of PT to be used as vaccines. The minimal region still enzymically active contains amino acids 4–179 (Barbieri and Cortina, 1988; Cieplak *et al.*, 1988; Pizza *et al.*, 1988). Within this region, many amino acids have been changed by site-directed mutagenesis (Figure 3.7) (Barbieri and Cortina, 1988; Black *et al.*, 1988; Burnette *et al.*, 1988; Pizza *et al.*, 1988; Kaslow *et al.*, 1989; Lobet *et al.*, 1989; Locht *et al.*, 1989). Arginine 9, Asp11, Arg13, Trp26, His35, Phe50, Glu129 and Tyr130 were essential for enzymic activity. In fact, their replacement with other amino acids reduced the activity of recombinant S1 molecules to levels equal to or below 1%. Most of these amino acids could be replaced without altering the overall structure of the recombinant S1 molecule which was still recognized by toxin-neutralizing monoclonal antibodies specific for a conformational epitope. In addition to the amino acids described above, Cys41 was close to the active site as its alkylation decreased enzymic activity (Kaslow *et al.*, 1989). By photoaffinity labelling with NAD$^+$, Glu129 was shown to be equivalent to Glu148 of DT and Glu553 of PAETA (Barbieri *et al.*, 1989). When the above amino acid changes were introduced either alone or in combination into the PT operon in the *B. pertussis* chromosome, a number of mutant PT molecules was obtained; most of them had a polyacrylamide gel electrophoretic pattern indistinguishable from the wild-type PT and a reduced toxicity. Some mutants were unable to assemble the S1 subunit and released into the extracellular medium only the B oligomer. These mutants contain mutations in any of the two cysteines, in regions homologous to cholera toxin (8D–9G, 50E, 88E–89S), or contain three amino acid changes as in mutant 13L–26I–129G. The above mutations are believed to alter the structure of the S1 subunit and therefore prevent its assembly into the holotoxin (Pizza *et al.*, 1990). In general, the molecules containing single amino acids mutations had a toxicity reduced from 4- to 1000-fold, but none of them was completely non-toxic. The mutant that has become most famous contains two amino acid substitutions: Arg9 $\rightarrow$ Lys and Glu129 $\rightarrow$ Gly (PT–9K/129G). This mutant has a structure indistinguishable from wild-type but is completely free of any toxicity (see Table 3.1). It has been used for the construction of a vaccine against pertussis that has been extensively tested in clinical trials and shown to induce protection from disease. The vaccine containing this mutant is presently licensed in several countries (Pizza *et al.*, 1989; Rappuoli, 1997).

Mutants in the B oligomer have also been constructed. The most relevant is the one containing deletion of Asn105 in S2 and Lys105 in S3, which resulted in drastic reduction of binding (Lobet *et al.*, 1993).

Cholera toxin and Escherichia coli enterotoxin

Cholera toxin (CT) and *E. coli* heat-labile enterotoxin (LT) are A–B toxins sharing high homology (80% identity) in their primary structure (Dallas and Falkow, 1980; Spicer *et al.*, 1981) and superimposable tertiary structures (Sixma *et al.*, 1991). Both toxins are composed of a pentameric B oligomer that binds the receptor(s) on the surface eukaryotic cells and an enzymically active A subunit that is responsible for the toxicity (Holmgren, 1981; Moss and Vaughan, 1988) (see Figure 3.1) (see Chapters 6 and 40 of this book).

The A subunit is composed of a globular structure linked to the B oligomer by a trypsin-sensitive loop and a long α-helix whose carboxy-terminus enters into the central cavity of the B oligomer, thus anchoring the A subunit to the B pentamer (Sixma *et al.*, 1991). Following protease cleavage of the loop, the A subunit is divided into the globular (enzymically active) A_1 and the carboxy-terminal A_2 fragments that remain linked by a disulfide bridge between the A_1–Cys187 and the A_2–Cys199. Proteolytic cleavage of the loop and reduction of the disulfide bridge are both necessary in order to generate the enzymic activity (Gill and Rappaport, 1979). The loop is intact when the molecules are produced in *E. coli*, while it is already cleaved when molecules are produced in *Vibrio cholerae*, which produces a specific protease to cleave this loop. The A_1

subunit contains the ADP-ribosylating activity and transfers the ADP-ribose group to the stimulatory α subunit of G_s. Once this G protein is ADP-ribosylated the adenylate cyclase is permanently activated, causing abnormal intracellular accumulation of cAMP (Field *et al.*, 1989a, b). While cAMP accumulation is believed to be responsible for the toxicity of CT and LT, we cannot exclude a contribution of interactions with less well-characterized G proteins to the toxicity.

The active site of the A subunit has a structure similar to all enzymes with mono-ADP-ribosylating activity (see Figure 3.11; Domenighini *et al.*, 1994; Domenighini and Rappuoli, 1996). The catalytic site is formed by a cavity with a β-strand (the floor of the cavity) followed by an α-helix (the ceiling of the cavity) that contains two amino acids that are essential for catalysis: these are Arg7 (Burnette *et al.*, 1991; Lobet *et al.*, 1991) and Glu110–Glu112 (Tsuji *et al.*, 1990, 1991). A peculiar feature of CT and LT is that the basal ADP-ribosyltransferase activity is enhanced by interaction with 20-kDa guanine-nucleotide binding proteins, known as ADP-ribosylation factors or ARFs (Tsai *et al.*, 1988).

ARF proteins not only function as activators of CT and LT, but also play a crucial role in vesicular membrane trafficking in both endocytic and exocytic pathways, maintenance of organelle integrity and assembly of coat proteins in eukaryotic cells. The most abundant and the best-known ARF family members are class I ARF, which are involved in the endoplasmic reticulum (ER)–Golgi and intra-Golgi transport. Like all GTP-binding proteins, ARF can bind GDP or GTP. The GDP-bound form is present in the cytosol. Initiation of vesicle formation occurs when a Golgi membrane enzyme interacts with the ARF–GDP complex and catalyses the release of GDP and the binding of GTP. The ARF–GTP form mediates the assembly of cytosolic coat proteins (coatomers) to Golgi membranes, inducing budding of the transported vesicles (Donaldson and Klausner, 1994; Boman and Kahn, 1995; Moss and Vaughan, 1995). Therefore, we can see the A_1 fragment as having at least two independent functions: enzymic activity and ARF binding (see Figure 3.4, path 2). The enzymic activity takes place within the eukaryotic cells and can be measured *in vitro*, whereas whether ARF activates CT and LT in the eukaryotic cells still remains unknown. While the region of LT and CT carrying the catalytic site is well characterized, so far we have no idea of where the ARF binding site is localized on the A_1 subunit. However, it is known that both trypsinization and reduction of the A subunit are needed for expression of a functional ARF binding site (Moss *et al.*, 1993), and in a recent study on non-toxic derivatives of LT it has been shown that the two sites

(catalytic and ARF binding) are independent and located on different regions of the A_1 domain (Stevens *et al.*, 1999).

The B oligomer is a pentameric molecule of 55 kDa, containing five identical polypeptide monomers. The structure is compact, resistant to trypsin and requires boiling in the presence of sodium dodecyl sulfate to be dissociated. The five subunits are arranged in a cylinder-like structure with a central cavity that, on one side, exposes five symmetrical cavities that are responsible for binding to the eukaryotic cell receptor (Sixma *et al.*, 1991, 1993). The receptor binding site is specific for a variety of galactose-containing molecules and shows a different fine specificity between LT and CT. CT binds mostly to the ganglioside GM1, which is believed to be the major toxin receptor (Holmgren, 1973), while LT binds not only GM1 but also to other glycosphingolipids (Teneberg *et al.*, 1994), to glycoprotein receptors present in the intestine of rabbits and humans (Holmgren *et al.*, 1982, 1985; Griffiths *et al.*, 1986), to polyglycosilceramides (PCGs) (Karlsson *et al.*, 1996) and to paragloboside (Teneberg *et al.*, 1994). Furthermore, the two variants of LT, human LT (hLT) and porcine LT (pLT), which differ only by four amino acids (Domenighini *et al.*, 1995b), are identical in their binding to glycoproteins and PCGs, but different in binding to paragloboside. Only pLT, and not hLT, is able to bind paragloboside and the reason seems to be due to the differences in the residue in position 13, which is an Arg in pLT and an His in hLT (Karlsson *et al.*, 1996). The different receptor-binding activities of the molecules may be significant for the qualitatively different immunological properties exhibited by LT and CT. The B oligomer can be produced in great quantities from *E. coli* transformed with the B subunit gene under the control of a strong promoter (Lebens *et al.*, 1993). The monomeric subunits are individually secreted into the periplasmic space where they are assembled into the pentameric structure (Hofstra and Witholt, 1985). CTB and LTB are systemic and mucosal immunogens (see Chapter 6), and this property is dependent on the binding to the GM1 receptor. In addition, they are strong inhibitors of T-cell activation and are able to induce apoptosis of CD8$^+$ T cells (Nashar *et al.*, 1996) and, to a lesser extent, of CD4$^+$ T cells (Truitt *et al.*, 1998). On the basis of results of different studies, Figure 3.4 summarizes the sequence of events that take place during cell intoxication (Bastiaens *et al.*, 1996; Majoul *et al.*, 1996). The sequence is as follows.

The holotoxin binds the receptor located on the eukaryotic cell membrane (Figure 3.4, path 2a), it is internalized into vesicles that, instead of taking the usual endocytic pathway leading to lysosomes, are

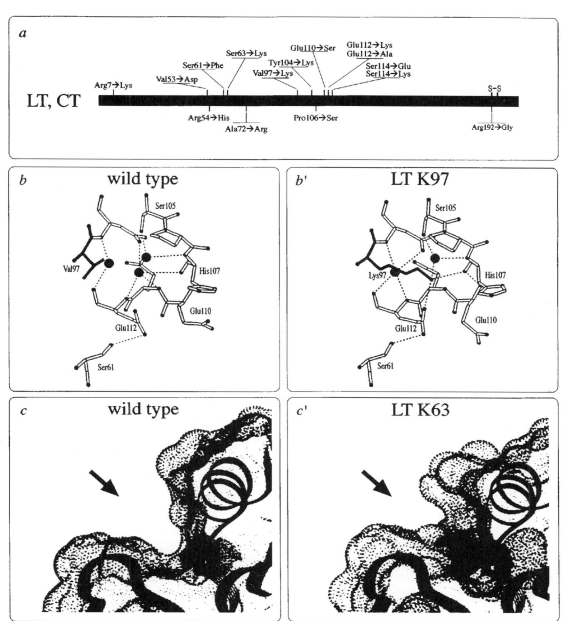

FIGURE 3.8 (a) Schematic representation of LT–CT toxins, where the most relevant site-directed mutations are reported. Two of them are illustrated in more detail in (b) and (c). (b) Structure of the wild-type LT and of the mutant LT K97 (b′). Glu112 is involved in one hydrogen bond in the native structure and in two in the mutant. (c) Three-dimensional structure of the enzymic cavity of the wild-type LT and (c′) of the mutant LT K63. The arrows show how this single mutation can affect the dimensions of the pocket and thus the entrance of the NAD molecule.

transported retrogradely to the Golgi compartment (Figure 3.4, paths 2b and 2c). At this stage the toxin is still in the A–B$_5$ form. Subsequently, the A and B subunits are dissociated, the A subunit is transported from the Golgi to the ER, while the B subunit persists in the Golgi and is later degraded. The A subunit or the A$_1$ is translocated, by an unknown mechanism, from the ER to the cytosol where, according to the *in vitro* data, it could interact with the soluble ARF and be activated (Figure 3.4, path 2d). Then, the active A$_1$ subunit ADP-ribosylates the α subunit of G$_s$ located on the plasma membrane.

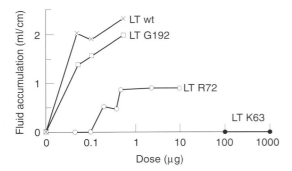

FIGURE 3.9 Toxicity in the rabbit ileal loop of wild-type LT and of the most important mutants.

Mutants of cholera and Escherichia coli *heat-labile enterotoxin*

In order to study the structure–function of CT and LT, and to find molecules that are non-toxic but still active as mucosal adjuvants and immunogens, more than 50 different site-directed mutants have been produced (Okamoto *et al.*, 1988; Tsuji *et al.*, 1990, 1991; Burnette *et al.*, 1991; Lobet *et al.*, 1991; Lycke *et al.*, 1992; Pizza *et al.*, 1994a, b; Fontana *et al.*, 1995; Douce *et al.*, 1995; de Haan *et al.*, 1996; Guidry *et al.*, 1997; Yamamoto *et al.*, 1997). The most important ones are reported in Figure 3.8. Among these, the ones that have been best studied are LTK63, LTK97 and LTK7, for which the three-dimensional structure has been determined (Merritt *et al.*, 1995; van den Akker *et al.*, 1995, 1997). One of the most interesting mutants is LTK97, where the Val → Lys mutation does not change at all the three-dimensional structure of the molecule but introduces a salt bridge between the charged amino group of Lys97, and the carboxylate of the Glu112, thus making it unavailable for further interactions (Figure 3.8, panels b and b'). The observation that a simple hydrogen bond inactivates the enzymic activity suggests an important role for the negative charge of the glutamic acid in the enzymic activity. Other interesting mutants are LTK63 and CTK63 (containing a serine-to-lysine substitution in position 63 of the A subunit). These holotoxoids have no detectable enzymic activity and no toxicity *in vitro* or *in vivo* (Figure 3.9), even when huge amounts of them are used (Giannelli *et al.*, 1997; Giuliani *et al.*, 1998). Therefore, they can be considered complete knockouts of enzymic activity but otherwise indistinguishable from wild-type. In fact, many other biological properties measured to date are maintained intact, including receptor and ARF-binding (Stevens *et al.*, 1998). The X-ray structure of LTK63 has shown complete identity to the wild-type LT across the entire molecule, with the exception of the active site, where the bulky side-chain of Lys63 fills the catalytic cavity,

thus making it unsuitable for the enzymic activity (see Figure 3.8, panels c and c') (van den Akker *et al.*, 1997). LTK63 is an excellent mucosal adjuvant, although the activity is reproducibly reduced in comparison to LT (Di Tommaso *et al.*, 1996; Partidos *et al.*, 1996; Douce *et al.*, 1997; Ghiara *et al.*, 1997; Giuliani *et al.*, 1998; Marchetti *et al.*, 1998), while CTK63 is a less active mucosal adjuvant (Douce *et al.*, 1997).

A second class of mutant molecules contains LTR72 (containing an alanine-to-arginine substitution in position 72 of the A subunit) and CTS106 (containing a proline-to-serine substitution in position 106 of the A subunit). These mutants have approximately 1% of the wild-type ADP-ribosylating activity, a toxicity *in vitro* in Y1 cells reduced by a factor of 10^4–10^5 and a toxicity *in vivo*, in the rabbit ileal loop, reduced by 25–100-fold (Figure 3.9). Both LTR72 and CTS106 are excellent mucosal adjuvants, being as effective as LT and CT, respectively (Douce *et al.*, 1997; Giuliani *et al.*, 1998).

Mutants in the protease-sensitive loop have also been constructed with the aim of making the loop insensitive to proteases, and therefore the toxin not susceptible to the activation process that is necessary for enzymic activity and toxicity. Among the many mutants constructed, LTG192 is the best characterized. In this LT mutant, the arginine in position 192 is replaced by a glycine (Grant *et al.*, 1994; Dickinson and Clements, 1995; Giannelli *et al.*, 1997). *In vitro*, the mutant is completely resistant to trypsin treatment. *In vivo*, proteases other than trypsin may partially cleave the loop and activate the toxin because toxicity is detectable. The toxicity observed in Y1 cells is approximately 10^3 times lower than that of wild-type toxin during the first 8 h of incubation and becomes only 5–10 times lower than wild-type following longer incubation (Giannelli *et al.*, 1997). In practice, this molecule takes longer to be activated but delivers approximately the same total enzymic activity as wild-type. The difference is that the delivery of the active toxin is diluted over a longer time. *In vivo*, in the rabbit ileal loop, almost no difference in toxicity is observed between LTG192 and wild-type LT (Figure 3.9) (Giannelli *et al.*, 1997). Ongoing human trials are expected to establish the safety profile of these molecules (*Vaccine Weekly*, 1997).

Mutant derivatives of LTB have also been constructed. Those which are defective in receptor binding (for example LTB–D33, containing a glycine-to-aspartic acid substitution in position 33) were almost completely non-immunogenic at mucosal surfaces, suggesting that an intact receptor-binding site is necessary for both binding and the immunogenicity associated with the molecule (Nashar *et al.*, 1996).

Other studies with a non-receptor-binding mutant of LT (LTD33) have confirmed that receptor binding is also necessary for adjuvanticity (Guidry *et al.*, 1997). Interestingly, non-binding LTB mutants also lose other immune-modulating activities including their ability to induce apoptosis of CD4[+] and CD8[+] cells (Nashar *et al.*, 1996; Truitt *et al.*, 1998).

Binary toxins and toxins that are directly delivered to the cytoplasm of eukaryotic cells

C2 toxin

Clostridium botulinum C2 toxin is a member of a family of 'binary' cytotoxins that ADP-ribosylate monomeric G-actin at an arginine in position 177 (Vandekerckhove *et al.*, 1988). Being the arginine at the contact site between actin monomers, the binding of the ADP-ribose makes actin unable to polymerize (Aktories *et al.*, 1986). C2 is composed of two separate molecules, the 50-kDa enzymically active toxin and a binding component that is synthesized as a 100-kDa precursor, proteolytically cleaved to generate a 75-kDa fragment that binds both the cell receptor on the cell surface and the 50-kDa active component (Aktories and Wegner, 1992; Popoff, 1998). This organization resembles closely that of the EF, LF and PA of *Bacillus anthracis*. Toxins related to C2 are *C. perfringens* iota toxin, *C. spiriforme* toxin and a similar molecule produced by *C. difficile* (see Chapter 8 of this book).

Exoenzyme S

Some toxins do not have a receptor-binding component and a translocation domain, and are directly injected by bacteria into the cytoplasm of eukaryotic cells. In this case, bacteria intoxicate individual eukaryotic cells by using a contact-dependent secretion system to inject or deliver toxic proteins into the cytoplasm of eukaryotic cells. This is done by using specialized secretion systems that in Gram-negative bacteria have been called type III or type IV, depending on whether they use a transmembrane structure similar to flagella or conjugative pili. The only ADP-ribosylating toxin that is delivered by a type-III secretion system is exoenzyme S (ExoS) of *P. aeruginosa*. For convenience we also include here the C3 toxin of *C. botulinum* for which a receptor-binding and translocation domain is not known.

The exoenzyme S of *P. aeruginosa* is a 49-kDa protein with ADP-ribosylating activity that ADP-ribosylates the small G-protein Ras at position 41 (Ganesan *et al.*, 1998). In order to become enzymically active, ExoS requires the interaction with a cytoplasmic factor named FAS or 14.3.3 (Coburn *et al.*, 1991).

The toxin is injected into eukaryotic cells by a type-III secretion system (Frithz-Lindsten *et al.*, 1997). When cells are transfected with the *exoS* gene under the control of an eukaryotic cell promoter, collapse of the cytoskeleton and a change in the cell morphology that results in the rounding up of the cells are observed.

C3

Exoenzyme C3 is a protein of 211 amino acids that is produced by *C. botulinum* (Aktories *et al.*, 1989; Chardin *et al.*, 1989) and that *in vitro* ADP-ribosylates the small regulatory protein Rho at Asn41 (Sekine *et al.*, 1989), thus inactivating its function. If the protein is microinjected into cells or the cells are transfected with the C3 gene under a eukaryotic promoter, this causes disruption of actin stress fibres, rounding of the cells and formation of arborescent protrusions (Hill *et al.*, 1995). However, for the moment we do not know whether C3 alone is able to enter cells and intoxicate them because no mechanism of cell entry has been found. Toxins with activity similar to C3 have been identified in *Staphylococcus aureus* (EDIN), *Clostridium limosum*, *B. cereus* and *Legionella pneumophila*.

EUKARYOTIC MONO-ADP-RIBOSYLTRANSFERASES

Eukaryotic mono-ADP-ribosyltransferases represent a growing class of enzymes. Their deduced amino acid sequences have similarities to those of viral and bacterial toxin enzymes in the region of the active site cleft, which is consistent with a common mechanism of NAD-binding and ADP-ribose transfer (Domenighini and Rappuoli, 1996).

Recent findings suggest that in vertebrates, this post-translational protein modification may be used to control important endogenous physiological functions such as the induction of long-term potentiation in the brain, terminal muscle cell differentiation and the cytotoxic activity of killer T-cells (McMahon *et al.*, 1993; Zolkiewska and Moss, 1993; Schuman *et al.*, 1994; Wang *et al.*, 1994).

The first vertebrate ribosyltransferases were purified and sequenced from chicken bone marrow and from rabbit and human skeletal muscle, and their specific target proteins have been identified. The majority of eukaryotic enzymes are arginine-specific transferases, none the less ADP-ribosylation of cysteines was reported in bovine and human erythrocytes and platelet membranes (Tanuma *et al.*, 1988; Saxty and van Heyningen, 1995).

The family of mammalian enzymes consists of five proteins (ART1–5) which share extensive similarities in their gene structure and amino acid sequence (Okazaki and Moss, 1998).

The rabbit ART1 is a 36-kDa protein, and its deduced amino acid sequence possesses hydrophobic amino- and carboxy-terminal signal peptides that are characteristic of GPI-linked proteins. There is roughly 75% sequence identity among ART1 muscle enzymes isolated from humans, rats and rabbits, this feature being consistent with considerable conservation across species. Like CT and LT, the muscle transferases specifically use the guanidino group of arginine as an ADP-ribose acceptor. The $\alpha 7$ integrin has been shown to be the target protein for cell-surface mono-ADP-ribosylation in muscle cells (Zolkiewska and Moss, 1993).

The ART1 enzymes have significant amino acid sequence identity to the RT6 (ART2) family of rodent T-cell differentiation and activation antigens (Takada et al., 1995). The expression of RT6 proteins appears in post-thymic lymphocytes and is restricted to peripheral T cells and intestinal intraepithelial lymphocytes. In the mouse, there are two functional copies of the RT6 gene (Rt6-1 and Rt6-2) located on chromosome 7. Rat and mouse ART2 sequences are roughly 80% identical, while in humans and chimpanzees the ART2 genes contain three premature stop codons and thus appear not to be expressed. In humans the role of ART2 may be assumed by other related ADP-ribosyltransferases.

ART3 and ART4 were recently cloned from human testis and spleen, respectively, and they possess several regions of sequence similarity with ART1. Moreover, the hydropathy profiles of the amino- and carboxyl-terminal sequences of ART3 and ART4 demonstrate hydrophobic signal sequences consistent with the possibility that these enzymes, like ART1, may be GPI-linked.

An ART5 cDNA was cloned from Yac-1 murine lymphoma cells, and its deduced amino acid sequence has similarities to other ART proteins in regions believed to be involved in catalytic activity.

Other mammalian ADP-ribosyltransferases have been purified from rat brain and adrenal medulla (Fujita et al., 1995). These enzymes modify, to different degrees, β/γ-actin, smooth muscle γ-actin, G_s, G_i and G_o. The in vitro modification of brain and adrenal G-proteins suggests potential mechanisms for cell signalling, similar to those observed with the bacterial toxins.

Previous studies have shown that the bacterial ADP-ribosyltransferases share three separate regions of similarity in their amino acid sequences, which also seem to be present in the muscle transferases and RT6 enzymes. Alignment of these domains highlights conserved residues within the catalytic sites of the mammalian and bacterial toxin ADP-ribosyltransferases (Figure 3.10) (Okazaki and Moss, 1994).

In computer modelling studies of the mouse ART2, Arg126 on a β-strand (region 1), Ser147 on a β-strand followed by an α-helix (region 2) and the active site Glu184 on a β-strand (region 3) are positioned in the catalytic cleft in a manner similar to that found in the crystal structure of the bacterial CT, LT and PT. In the alignment of deduced amino acid sequences of ART1, ART4 and ART5, a region 1 arginine and region 2 serine similar to those in LT and PT appear to be conserved. Based on site-directed mutagenesis and amino acid sequence alignment, the region containing Glu–X–Glu present in ART1 and ART5 is postulated to be analogous to Glu110 and Glu112 of LT and CT.

Moreover, the deduced amino acid sequences of human poly(ADP-ribose) polymerases (PARPs) and, perhaps, ART3 appear to have regions of similarity that align with DT and PAETA. To support this observation, crystal structure of the chicken PARP (Ruf et al., 1996) and mutagenesis of human PARP (Marsischky et al., 1995; Trucco et al. 1996) demonstrated that Glu988, which is essential for ADP-ribose chain elongation, is positioned in a cleft similar to that found in bacterial toxins.

These data are consistent with the hypothesis that several of the bacterial toxins and vertebrate transferases possess a common mechanism of NAD binding and ADP-ribose transfer, and that differences observed in the three-dimensional structures may reflect differences in substrate proteins.

A COMMON STRUCTURE OF THE CATALYTIC SITE

The analysis of the sequences of ADP-ribosylating enzymes known to date shows that all enzymes can be divided into two groups (Figure 3.10): the DT-like group, mainly composed of DT, PAETA and human poly-ADP-ribosylpolymerases (PARPs); and the CT-like group comprising CT, LT, PT, MTX (Bacillus sphaericus mosquitocidal toxin), ExoS, other bacterial ADP-ribosyltransferases and eukaryotic enzymes such as RT6. The obvious question is whether all these enzymes that perform the same reaction on different substrates have something in common.

While some homology is present within the CT family, overall no significant and extended sequence

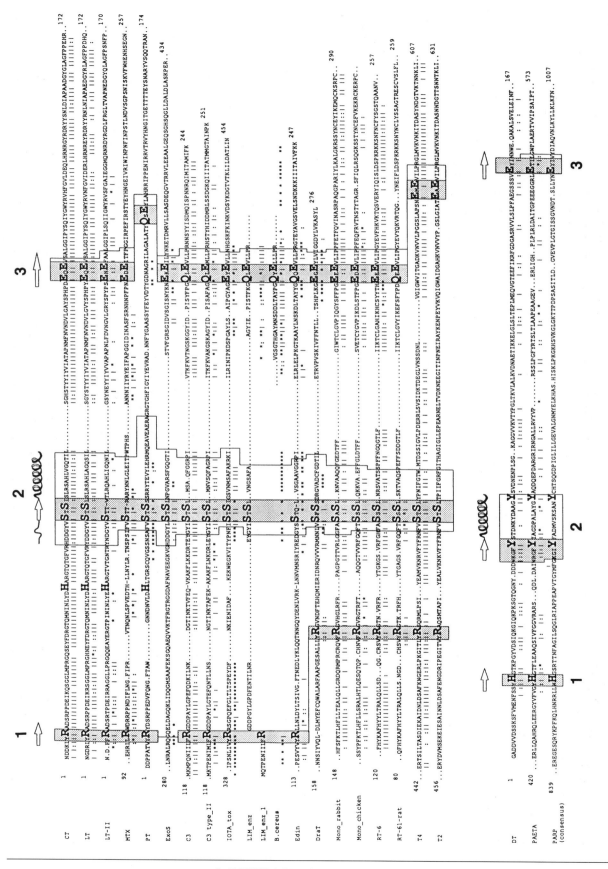

FIGURE 3.10 Sequence alignment of the ADP-ribosylating enzymes. Linear tracts (|) and columns (:) stand for identity and conservative substitution, respectively, while asterisks (*) indicate identity with at least one other amino acid among the sequences of the group. Conserved regions 1, 2 and 3, which are relevant to enzymic activity, are boxed.

similarity can be detected to justify the observed common mechanism of catalysis. Nevertheless, biochemical experiments of photoaffinity labelling and studies of site-directed mutagenesis had previously demonstrated that all the toxins possessed a catalytic glutamic acid so important for enzymic activity that not even a conservative substitution with an aspartic acid could be tolerated without loss or drastic decrease of toxicity. The structural identity of the active site was later confirmed by the crystal structure of several enzymes and computer modelling studies. The catalytic cavity is formed by a β-strand followed by a slanted α-helix that has a different length in the various toxins. The β-strand and the α-helix represent, respectively, the lower and the upper face of the cavity in which the nicotinamide ring of NAD enters and is anchored during the enzymic reaction (Figure 3.11, see colour plate section). Although all the toxins share a similar folding in the region of the catalytic domain, at the amino acid level, the only residues which appear to be conserved among all representatives of the DT and CT subgroups are a glutamic acid and an histidine or an arginine; X-ray data have shown that they are carried by two independent β-strands and retain an equivalent spatial position and orientation in all the enzymes. The knowledge of the structure of the active site, and of the amino acids that form it, allowed the primary sequences to be looked at again and similarities to be found that had previously been disregarded.

Briefly, all the mentioned toxins share three common features, namely, the cavity formed by the β–α structure and the two catalytic residues (identified by numbers 2, 1 and 3, respectively, in Figure 3.10). A closer look at these regions shows that the β–α structure (region 2) has very little or no amino acid sequence homology, however it is identified by an …S–X–S… motif in most of the enzymes of the CT group and by an …Y–(X)$_{10}$–Y… in the DT group. The glutamic acid shown in Figure 3.10, region 3 (namely Glu148 of DT, Glu553 of PAETA, Glu112 of CT and LT, Glu129 of PT, etc.) is characterized by a Glu/Gln–X–Glu in the CT group. The other conserved amino acid (region 1) is an arginine in the CT group and an histidine in the DT group (His21 of DT, His440 of PAETA, Arg7 of CT, Arg9 of PT, etc.). This region is usually identified by the Aro–Arg/His motif (region 1). The essential role of the amino acids in regions 1 and 3, and of the structure of the cavity in region 2, have been confirmed experimentally by site-directed mutagenesis. In the case of DT, it has been established that the carboxylate group of the side chain of Glu148 lies near the nicotinamide ring of NAD, at a distance of 4 Å from the N-glycosidic bond. Extending this observation to all the toxins, the more likely explanation, in terms of enzymic activity, is that the essen-

tial glutamic acid could play a role in stabilizing a chemical intermediate of NAD during a Sn1-type mechanism in which the nicotinamide ring might be dissociated because of the following interaction with the nucleophilic residue of the incoming substrate (diphtamide in the case of DT and PAETA, arginine in the case of LT and CT, and cysteine in the case of PT) (Figures 3.2 and 3.4). Although several models have been proposed to explain the possible function of the conserved histidine/arginine, it seems now widely accepted that this residue does not play a direct role in catalysis (Figure 3.11 in colour plate section); it very probably has the function of maintaining the integrity of the active-site pocket upon formation of structurally stabilizing hydrogen bonds.

This block of homology β–α domain includes a number of amino acids that, while maintaining the same secondary structure in both DT and CT families, still strongly differ in terms of sequence (Figure 3.10, region 2). This region corresponds to the core of the active site cleft which is devoted to the docking of NAD. The difference detected in terms of primary structure between the two subgroups suggests that the toxins may adopt a diverse mechanism of binding. The consensus sequence generated for the DT group is characterized by two conserved tyrosines spaced by 10 amino acids, and located on the middle portion of the β-strand and on the internal face of the α-helix, respectively. Tyrosine 54 and Tyr65 of DT (Figure 3.11), and Tyr470 and Tyr481 of PAETA play a very important role in catalysis, as they anchor the nicotinamide ring creating a π pile of three aromatic rings which strengthen the overall binding of NAD and stabilize the complex. In PT, a similar role is likely to be played by Tyr59 and Tyr63, which have a similar spatial orientation and distance from each other. This observation is supported by the fact that in CT and LT, where the stacking interactions produced by the two tyrosines are lacking, the affinity for NAD is 1000-fold lower. In the case of the CT group, the box β–α is centred on a conserved core region characterized by the consensus Ser–Thr–Ser that is observed and predicted to fold in a β-strand which represents the floor of the cavity. No conclusive information is available to define the precise role of these small, polar residues in catalysis, but experiments of site-directed mutagenesis confirm that they are extremely important in maintaining the cavity available for NAD entrance and docking. Substitutions of Ser61 and Ser63 of LT with Phe and Lys, respectively, produce non-toxic mutants.

Another amino acid that has been proposed as being important in catalysis is His35 of PT, which is located near the beginning of the β-strand which forms the floor of the cavity, in a position equivalent to that of His44 of LT and CT; a functionally homologous His is also present in MTX, but absent in DT and PAETA.

In the three-dimensional structure, this residue appears to be sufficiently close to the oxygen atom of the ribose ring of NAD to interact with it and increase the electrophilicity of the adjacent anomeric carbon atom. The absence of an equivalent residue in DT and PAETA again supports the idea that the two groups of toxins perform the same enzymic activity in a slightly different fashion.

An additional feature that is common to all ADP-ribosylating toxins is the need for a conformational rearrangement in order to achieve enzymic activity. In the native structure, the NAD-binding site of LT and CT is obstructed by a loop comprising amino acids 47–56 which needs to be displaced in order to obtain a functional NAD-binding cavity. A functionally homologous region is also present in PT where the loop comprises residues 199–207. In the case of DT, where the crystallographic data of the complex are available, the observation that the active site loop consisting of amino acids 39–46 changes structure upon NAD binding suggests that these residues may be important for the recognition of the ADP-ribose acceptor substrate, EF-2.

This proposal is supported by at least two lines of evidence: first, DT and PAETA have a high degree of sequence similarity in this loop region with a number of identical or highly conservative substitutions, supporting the idea that these residues have some essential function; second, antibodies raised against a peptide corresponding to this loop sequence were able to block the catalytic domain of DT from catalysing the ADP-ribosylation of EF-2.

The recent publication of the crystallographic data of the DT–NAD complex, and the presence of common features within all ADP-ribosylating toxins, allow speculation on a possible common mechanism of catalysis (Bell and Eisenberg, 1997).

The best hypothesis is that NAD enters the cavity upon displacement of the mobile loop which is then made available for the recognition of the substrate; NAD is subsequently docked at the bottom of the pocket where a small residue (the conserved serine in the β–α box of the CT-group, the Thr56 of DT and the Ala472 of PAETA) is required to allow good positioning. The nicotinamide moiety of NAD is then blocked in a suitable position by means of stacking interactions provided by a couple of aromatic rings (Tyr54 and Tyr65 of DT, Tyr470 and Tyr481 of PAETA and, possibly, Tyr59 and Tyr63 of PT, respectively). In this context the conserved arginine/histidine might display its key role in main-

taining the correct shape of the active site pocket via hydrogen bonds formed with the backbone of the structure and possibly with one of the ribose moiety.

The enzymic reaction is then catalysed by the essential glutamic acid which is likely to stabilize a positively charged oxocarbonium intermediate of NAD, in order to favour its subsequent interaction with the nucleophilic residue of the incoming substrate (Figure 3.11, in colour plate section).

REFERENCES

Aktories, K., Bärmann, M., Ohishi, I., Tsuyama, S., Jakobs, K.H. and Habermann, E. (1986). Botulinum C2 toxin ADP-ribosylates actin. *Nature* **322**, 390–2.

Aktories, K., Braun, U., Rösener, S., Just, I. and Hall, A. (1989). The *rho* gene product expressed in *E. coli* is a substrate of botulinum ADP-ribosyltransferase C3. *Biochem. Biophys. Res. Comm.* **158**, 209–13.

Aktories, K. and Wegner, A. (1992). Mechanisms of the cytopathic action of actin–ADP-ribosylating toxins. *Mol. Microbiol.* **6**, 2905–8.

Allured, V.S., Collier, R.J., Carroll, S.F. and McKay, D.B. (1986). Structure of exotoxin A of *Pseudomonas aeruginosa* at 3.0-Angstrom resolution. *Proc. Natl Acad. Sci. USA* **83**, 1320–4.

Althaus, F.R. and Richter, C. (1987). ADP-ribosylation of proteins. Enzymology and biological significance. *Mol. Biol. Biochem. Biophys.* **37**, 1–237.

Anderson, P., Pichichero, M., Edwards, K., Porch, C.R. and Insel, R. (1987). Priming and induction of *Haemophilus influenzae* type b capsular antibodies in early infancy by Dpo20, an oligosaccharide–protein conjugate vaccine. *J. Pediatr.* **111**, 644–50.

Bacha, P., Williams, D.P., Waters, C., Williams, J.M., Murphy, J.R. and Strom, T.B. (1988). Interleukin 2 receptor-targeted cytotoxicity. Interleukin 2 receptor-mediated action of a diphtheria toxin-related interleukin 2 fusion protein. *J. Exp. Med.* **167**, 612–22.

Baldwin, R.L., Kobrin, M.S., Tran, T., Pastan, I. and Korc, M. (1996). Cytotoxic effects of TGF-alpha-Pseudomonas exotoxin A fusion protein in human pancreatic carcinoma cells. *Pancreas* **13**, 16–21.

Barbieri, J.T. and Cortina, G. (1988). ADP-ribosyltransferase mutations in the catalytic S-1 subunit of pertussis toxin. *Infect. Immun.* **56**, 1934–41.

Barbieri, J.T., Mende-Mueller, L.M., Rappuoli, R. and Collier, R.J. (1989). Photolabeling of Glu-129 of the S-1 subunit of pertussis toxin with NAD. *Infect. Immun.* **57**, 3549–54.

Bastiaens, P.I., Majoul, I.V., Verveer, P.J., Soling, H.D. and Jovin, T.M. (1996). Imaging the intracellular trafficking and state of the AB$_5$ quaternary structure of cholera toxin. *EMBO J.* **15**, 4246–53.

Bell, C.E. and Eisenberg, D. (1997). Crystal structure of diphtheria toxin bound to nicotinamide adenine dinucleotide. *Adv. Exp. Med. Biol.* **419**, 35–43.

Bennett, M.J., Choe, S. and Eisenberg, D. (1994). Refined structure of dimeric diphtheria toxin at 2.0 Angstrom resolution. *Protein Sci.* **3**, 1444–63.

Bigio, M., Rossi, R., Nucci, D., Antoni, G., Rappuoli, R. and Ratti, G. (1987). Conformational changes in diphtheria toxoids. Analysis with monoclonal antibodies. *FEBS Lett.* **218**, 271–6.

Black, W.J., Munoz, J.J., Peacock, M.G., Schad, P.A., Cowell, J.L., Burchall, J.J., Lim, M., Kent, A., Steinman, L. and Falkow, S. (1988). ADP-ribosyltransferase activity of pertussis toxin and immunomodulation by *Bordetella pertussis*. *Science* **240**, 656–9.

Blanke, S.R., Huang, K., Wilson, B.A., Papini, E., Covacci, A. and Collier, R.J. (1994a). Active-site mutations of the diphtheria toxin catalytic domain – role of Histidine-21 in nicotinamide adenine dinucleotide binding and ADP-ribosylation of elongation factor 2. *Biochemistry* **33**, 5155–61.

Blanke, S.R., Huang, K. and Collier, R.J. (1994b). Active-site mutations of diphtheria toxin: role of tyrosine-65 in NAD binding and ADP-ribosylation. *Biochemistry* **33**, 15494–500.

Boman, A.L. and Kahn, R.A. (1995). Arf proteins: the membrane traffic police? *Trends Biochem. Sci.* **20**, 147–50.

Brandhuber, B.J., Allured, V.S., Falbel, T.G. and McKay, D.B. (1988). Mapping the enzymic active site of *Pseudomonas aeruginosa* exotoxin A. *Proteins* **3**, 146–54.

Brito, G.A.C., Souza, M.H.L.P., Melo-Filho, A.A., Hewlett, E.L., Lima, A.A.M., Flores, C.A. and Ribiero, R.A. (1997). Role of pertussis toxin A subunit in neutrophil migration and vascular permeability. *Infect. Immun.* **65**, 1114–18.

Brooke, J.S., Cha, J.H. and Eidels, L. (1998). Diphtheria toxin:receptor interaction: association, dissociation, and effect of pH. *Biochem. Biophys. Res. Commun.* **248**, 297–302.

Brown, B.A. and Bodley, J.W. (1979). Primary structure at the site in beef and wheat elongation factor 2 of ADP-ribosylation by diphtheria toxin. *FEBS Lett.* **103**, 253–5.

Brown, J.G., Almond, B.D., Naglich, J.G. and Eidels, L. (1993). Hypersensitivity to diphtheria toxin by mouse cells expressing both diphtheria toxin receptor and CD9 antigen. *Proc. Natl Acad. Sci. USA* **90**, 8184–8.

Burnette, W.N., Cieplak, W., Mar, V.L., Kaljot, K.T., Sato, H. and Keith, J.M. (1988). Pertussis toxin S1 mutant with reduced enzyme activity and a conserved protective epitope. *Science* **242**, 72–4.

Burnette, W.N., Mar, V.L., Platler, B.W., Schlotterbeck, J.D., McGinley, M.D., Stoney, K.S., Rohde, M.F. and Kaslow, H.R. (1991). Site-specific mutagenesis of the catalytic subunit of cholera toxin: substituting lysine for arginine 7 causes loss of activity. *Infect. Immun.* **59**, 4266–70.

Campbell, A.M. (1962). Episomes. *Adv. Genet.* **11**, 101–45.

Carroll, S.F. and Collier, R.J. (1987). Active site of *Pseudomonas aeruginosa* exotoxin A. Glutamic acid 553 is photolabeled by NAD and shows functional homology with glutamic acid 148 of diphtheria toxin. *J. Biol. Chem.* **262**, 8707–11.

Carroll, S.F. and Collier, R.J. (1988). Amino acid sequence homology between the enzymic domains of diphtheria toxin and *Pseudomonas aeruginosa* exotoxin A. *Mol Microbiol.* **2**, 293–6.

Carroll, S.F., McCloskey, J.A., Crain, P.F., Oppenheimer, N.J., Marschner, T.M. and Collier, R.J. (1985). Photoaffinity labeling of diphtheria toxin fragment A with NAD: structure of the photoproduct at position 148. *Proc. Natl Acad. Sci. USA* **82**, 7237–41.

Chardin, P., Boquet, P., Madaule, P., Popoff, M.R., Rubin, E.J. and Gill, D.M. (1989). The mammalian G-protein Rho C is ADP-ribosylated by *Clostridium botulinum* exoenzyme C3 and affects actin microfilaments in Vero cells. *EMBO J.* **8**, 1087–92.

Chaudhary, V.K., FitzGerald, D.J., Adhya, S. and Pastan, I. (1987). Activity of a recombinant fusion protein between transforming growth factor type alpha and *Pseudomonas* toxin. *Proc. Natl Acad. Sci. USA* **84**, 4538–42.

Chaudhary, V.K., Xu, Y.H., FitzGerald, D., Adhya, S. and Pastan, I. (1988). Role of domain II of *Pseudomonas* exotoxin in the secretion of proteins into the periplasm and medium by *Escherichia coli*. *Proc. Natl Acad. Sci. USA* **85**, 2939–43.

Choe, S., Bennett, M.J., Fujii, G., Curmi, P.M., Kantardjieff, K.A., Collier, R.J. and Eisenberg. D. (1992). The crystal structure of diphtheria toxin. *Nature* **357**, 216–22.

Cieplak, W., Burnette, W.N., Mar, V.L., Kaljot, K.T., Morris, C.F., Chen, K.K., Sato, H. and Keith, J.M. (1988). Identification of a region in the S1 subunit of pertussis toxin that is required for enzymic activity and that contributes to the formation of a neutralizing antigenic determinant. *Proc. Natl Acad. Sci. USA* **85**, 4667–71.

Cieplak, W., Gaudin, H.M. and Eidels, L. (1987). Diphtheria toxin receptor. Identification of specific diphtheria toxin-binding proteins on the surface of Vero and BS-C-1 cells. *J. Biol. Chem.* **262**, 13246–53.

Coburn, J., Kane, A.V., Feig, L. and Gill, D.M. (1991). *Pseudomonas aeruginosa* exoenzyme S requires a eukaryotic protein for ADP-ribosyltransferase activity. *J. Biol. Chem.* **266**, 6438–46.

Collier, R.J. (1982). Structure and activity of diphtheria toxin. In: *ADP-ribosylation Reactions* (eds D. Hayashi and K. Ueda), p. 575. Academic Press, New York.

Comanducci, M., Ricci, S., Rappuoli, R. and Ratti, G. (1987). The nucleotide sequence of the gene coding for diphtheria toxoid CRM 176. *Nucleic Acids Res.* **15**, 5897.

Costa, J., Michel, J.L., Rappuoli, R. and Murphy, J. (1981). Restriction map of corynebacteriophages βc and βvir and physical localization of diphtheria tox operon. *J. Bacteriol.* **148**, 124–30.

Costantino, P., Viti, S., Podda, A., Velmonte, M.A., Nencioni, L. and Rappuoli, R. (1992). Development and phase I clinical testing of a conjugate vaccine against meningococcus A and C. *Vaccine* **10**, 691–8.

Covacci, A. and Rappuoli, R. (1993). Pertussis toxin export requires accessory genes located downstream from the pertussis toxin operon. *Mol. Microbiol.* **8**, 429–34.

Dallas, W.S. and Falkow, S. (1980). Amino acid sequence homology between cholera toxin and *Escherichia coli* heat-labile toxin. *Nature* **288**, 499–501.

de Haan, L., Verweij, W.R., Feil, I.K., Lijnema, T.H., Hol, W.G., Agsteribbe, E. and Wilschut, J. (1996). Mutants of the *Escherichia coli* heat-labile enterotoxin with reduced ADP-ribosylation activity or no activity retain the immunogenic properties of the native holotoxin. *Infect. Immun.* **64**, 5413–16.

Dell'Arciprete, L., Colombatti, M., Rappuoli, R. and Tridente, G. (1988). A C terminus cysteine of diphtheria toxin B chain involved in immunotoxin cell penetration and cytotoxicity. *J. Immunol.* **140**, 2466–71.

Dickinson, B.L. and Clements, J.D. (1995). Dissociation of *Escherichia coli* heat-labile enterotoxin adjuvanticity from ADP-ribosyltransferase activity. *Infect. Immun.* **63**, 1617–23.

Ding, X., Zeng, H., Schiering, N., Ringe, D. and Murphy, J.R. (1996). Identification of the primary metal ion-activation sites of the diphtheria tox repressor by X-ray crystallography and site-directed mutational analysis. *Nat. Struct. Biol.* **3**, 382–7.

Di Tommaso, A., Saletti, G., Pizza, M., Rappuoli, R., Dougan, G., Abrignani, S., Douce, G. and De Magistris, M.T. (1996). Induction of antigen-specific antibodies in vaginal secretions by using a non-toxic mutant of heat-labile enterotoxin as a mucosal adjuvant. *Infect. Immun.* **64**, 974–9.

Domenighini, M. and Rappuoli, R. (1996). Three conserved consensus sequences identify the NAD binding site of ADP-ribosylating enzymes, expressed by eukaryotes, bacteria and T-even bacteriophages. *Mol. Microbiol.* **21**, 667–74.

Domenighini, M., Montecucco, C., Ripka, W.C. and Rappuoli, R. (1991). Computer modelling of the NAD binding site of ADP-ribosylating toxins: active-site structure and mechanism of NAD binding. *Mol. Microbiol.* **5**, 23–31.

Domenighini, M., Magagnoli, C., Pizza, M. and Rappuoli, R. (1994) Common features of the NAD-binding and catalytic site of ADP-ribosylating toxins. *Mol. Microbiol.* **14**, 41–50.

Domenighini, M., Pizza, M. and Rappuoli, R. (1995a). Bacterial ADP-ribosyltransferases. In: *Bacterial Toxins and Virulence Factors in Disease* (eds J. Moss, B. Iglewski, M. Vaughan, A.T.), pp. 59–80. Marcel Dekker, New York.

Domenighini, M., Pizza, M., Jobling, M.G., Holmes, R.K. and Rappuoli, R. (1995b). Identification of errors among database sequence entries and comparison of correct amino acid sequences for the heat-labile enterotoxins of *Escherichia coli* and *Vibrio cholerae*. *Mol. Microbiol.* **15**, 1165–7.

Donaldson, J.G. and Klausner, R.D. (1994). ARF: a key regulatory switch in membrane traffic and organelle structure. *Curr. Opin. Cell Biol.* **6**, 527–32.

Douce, G., Turcotte, C., Cropley, I., Roberts, M., Pizza, M., Domenighini, M., Rappuoli R. and Dougan, G. (1995). Mutants of *Escherichia coli* heat-labile toxin lacking ADP-ribosyltransferase activity act as nontoxic, mucosal adjuvants. *Proc. Natl Acad. Sci. USA* **92**, 1644–8.

Douce, G., Fontana, M., Pizza, M., Rappuoli, R. and Dougan, G. (1997). Intranasal immunogenicity and adjuvanticity of site-directed mutant derivatives of cholera toxin. *Infect. Immun.* **65**, 2821–8.

Douglas, C.M. and Collier, R.J. (1990). *Pseudomonas aeruginosa* exotoxin A: alterations of biological and biochemical properties resulting from mutation of glutamic acid 553 to aspartic acid. *Biochemistry* **29**, 5043–9.

Egan, W., Frasch, C.E. and Anthony, B.F. (1995). Lot-release criteria, postlicensure quality control, and the *Haemophilus influenzae* type b conjugate vaccines. *J. Am. Med. Assoc.* **273**, 888–9.

Essand, M. and Pastan, I. (1998). Anti-prostate immunotoxins: cytotoxicity of E4 antibody-*Pseudomonas* exotoxin constructs. *Int. J. Cancer* **77**, 123–7.

Field, M., Rao, M.C. and Chang, E.B. (1989a). Intestinal electrolyte transport and diarrheal disease (1). *N. Engl. J. Med.* **321**, 800–6.

Field, M., Rao, M.C. and Chang, E.B. (1989b). Intestinal electrolyte transport and diarrheal disease (2). *N. Engl. J. Med.* **321**, 879–83.

Fontana, M.R., Manetti, R., Giannelli, V., Magagnoli, C., Marchini, A., Olivieri, R., Domenighini, M., Rappuoli, R. and Pizza, M. (1995). Construction of nontoxic derivatives of cholera toxin and characterization of the immunological response against the A subunit. *Infect. Immun.* **63**, 2356–60.

Freeman, V.J. (1951). Studies on the virulence of bacteriophage infected strains of *Corynebacterium diphtheriae*. *J. Bacteriol.* **61**, 675–88.

Frithz-Lindsten, E., Du, Y., Rosqvist, R. and Forsberg, A. (1997). Intracellular targeting of exoenzyme S of *Pseudomonas aeruginosa* via type III-dependent translocation induces phagocytosis resistance, cytotoxicity and disruption of actin microfilaments. *Mol. Microbiol.* **25**, 1125–39.

Fujita, H., Okamoto, H. and Tsuyama, S. (1995). ADP-ribosylation in adrenal glands: purification and characterization of mono-ADP-ribosyltransferases and ADP-ribosylhydrolase affecting cytoskeletal actin. *Int. J. Biochem. Cell. Biol.* **27**, 1065–78.

Galloway, D.R., Hedstrom, R.C., McGowan, J.L., Kessler, S.P. and Wozniak, D.J. (1989). Biochemical analysis of CRM 66. A nonfunctional *Pseudomonas aeruginosa* exotoxin A. *J. Biol. Chem.* **264**, 14869–73.

Ganesan, A.K., Frank, D.W., Misra, R.P., Schmidt, G. and Barbieri, J.T. (1998). *Pseudomonas aeruginosa* exoenzyme S ADP-ribosylates Ras at multiple sites. *J. Biol. Chem.* **273**, 7332–7.

Gether, U. and Kobilka, B.K. (1998). G protein-coupled receptors. II. Mechanism of agonist activation. *J. Biol. Chem.* **273**, 17979–82.

Ghiara, P., Rossi, M., Marchetti, M., Di Tommaso, A., Vindigni, C., Ciampolini, F., Covacci, A., Telford, J.L., De Magistris, M.T., Pizza, M., Rappuoli, R. and Del Giudice, G. (1997). Therapeutic intragastric vaccination against *Helicobacter pylori* in mice eradicates an otherwise chronic infection and confers protection against reinfection. *Infect. Immun.* **65**, 4996–5002.

Giannelli, V., Fontana, M.R., Giuliani, M.M., Guangcai, D., Rappuoli, R. and Pizza, M. (1997). Protease susceptibility and toxicity of heat-labile enterotoxins with a mutation in the active site or in the protease-sensitive loop. *Infect. Immun.* **65**, 331–4.

Giannini, G., Rappuoli, R. and Ratti, G. (1984). The amino acid sequence of two nontoxic mutants of diphtheria toxin: CRM 45 and CRM 197. *Nucleic Acids Res.* **12**, 4063–9.

Gilchrist, A., Mazzoni, M.R., Dineen, B., Dice, A., Linden, J., Proctor, W.R., Lupica, C.R., Dunwiddie, T.V. and Hamm, H.E. (1998). Antagonists of the receptor–G protein interface block Gi-coupled signal transduction. *J. Biol. Chem.* **273**, 14912–19.

Gill, D.M. and Rappaport, R.S. (1979). Origin of the enzymically active A1 fragment of cholera toxin. *J. Infect. Dis.* **139**, 674–80.

Gill, D.M., Pappenheimer, A.M., Jr, Brown, R. and Kurnick, J.T. (1969). Studies on the mode of action of diphtheria toxin. VII. Toxin-stimulated hydrolysis of nicotinamide adenine dinucleotide in mammalian cell extracts. *J. Exp. Med.* **129**, 1–21.

Gilman, A.G. (1984). G proteins and dual control of adenylate cyclase. *Cell* **36**, 577–9.

Giuliani, M.M., Del Giudice, G., Giannelli, V., Dougan, G., Douce, G., Rappuoli, R. and Pizza, M. (1998). Mucosal adjuvanticity and immunogenicity of LTR72, a novel mutant of *Escherichia coli* heat-labile enterotoxin with partial knockout of ADP-ribosyltransferase activity. *J. Exp. Med.* **187**, 1123–32.

Grant, C.C., Messer, R.J. and Cieplak, W., Jr (1994). Role of trypsin-like cleavage at arginine 192 in the enzymic and cytotonic activities of *Escherichia coli* heat-labile enterotoxin. *Infect. Immun.* **62**, 4270–8.

Gray, G.L., Smith, D.H., Baldridge, J.S., Harkins, R.N., Vasil, M.L., Chen, E.Y. and Heyneker, H.L. (1984). Cloning nucleotide sequence, and expression in *Escherichia coli* of the exotoxin. A structural gene of *Pseudomonas aeruginosa*. *Proc. Natl Acad. Sci. USA* **81**, 2645–9.

Greenfield, L., Bjorn, M.J., Horn, G., Fong, D., Buck, G.A., Collier, R.J. and Kaplan, D.A. (1983). Nucleotide sequence of the structural gene for diphtheria toxin carried by corynephage 0. *Proc. Natl Acad. Sci. USA* **80**, 6853–7.

Greenfield L., Johnson, V.G. and Youle, R.J. (1987). Mutations in diphtheria toxin separate binding from entry and amplify immunotoxin selectivity. *Science* **238**, 536–9.

Griffiths, S.L., Finkelstein, R.A. and Critchley, D.R. (1986). Characterization of the receptor for cholera toxin and *Escherichia coli* heat-labile toxin in rabbit intestinal brush borders. *Biochem. J.* **238**, 313–22.

Guidry, J.J., Cardenas, L., Cheng, E. and Clements, J.D. (1997). Role of receptor binding in toxicity, immunogenicity, and adjuvanticity of *Escherichia coli* heat-labile enterotoxin. *Infect. Immun.* **65**, 4943–50.

Hamm, H.E. and Gilchrist, A. (1996). Heterotrimeric G proteins. *Curr. Opin. Cell Biol.* **8**, 189–96.

Hewlett, E.L., Sauer, K.T., Myers, G.A., Cowell, J.L. and Guerrant, R.L. (1983). Induction of a novel morphological response in Chinese hamster ovary cells by pertussis toxin. *Infect. Immun.* **40**, 1198–203.

Hill, C.S., Wynne, J. and Treisman, R. (1995). The Rho family GTPases RhoA, Rac1, and Cdc42Hs regulate transcriptional activation by SRF. *Cell* **81**, 1159–70.

Hofstra, H. and Witholt, B. (1985). Heat-labile enterotoxin in *Escherichia coli*. Kinetics of association of subunits into periplasmic holotoxin. *J. Biol. Chem.* **260**, 16037–44.

Holmgren, J. (1973). Comparison of the tissue receptors for *Vibrio cholerae* and *Escherichia coli* enterotoxins by means of gangliosides and natural cholera toxoid. *Infect. Immun.* **8**, 851–9.

Holmgren, J. (1981). Actions of cholera toxin and the prevention and treatment of cholera. *Nature* **292**, 413–17.

Holmgren, J., Fredman, P., Lindblad, M., Svennerholm, A.M. and Svennerholm, L. (1982). Rabbit intestinal glycoprotein receptor for *Escherichia coli* heat-labile enterotoxin lacking affinity for cholera toxin. *Infect. Immun.* **38**, 424–33.

Holmgren, J., Lindblad, M., Fredman, P., Svennerholm, L. and Myrvold, H. (1985). Comparison of receptors for cholera and *Escherichia coli* enterotoxins in human intestine. *Gastroenterology* **89**, 27–35.

Honjo, T., Nishizuka, Y. and Hayaishi, O. (1968). Diphtheria toxin-dependent adenosine diphosphate ribosylation of aminoacyl transferase II and inhibition of protein synthesis. *J. Biol. Chem.* **243**, 3553–5.

Hooper, K.P. and Eidels, L. (1995). Localization of a critical diphtheria toxin-binding domain to the C-terminus of the mature heparin-binding EGF-like growth factor region of the diphtheria toxin receptor. *Biochem. Biophys. Res. Commun.* **206**, 710–17.

Hwang, J., Fitzgerald, D.J., Adhya, S. and Pastan, I. (1987). Functional domains of *Pseudomonas* exotoxin identified by deletion analysis of the gene expressed in *E. coli*. *Cell* **48**, 129–36.

Iwamoto, R., Higashiyama, S., Mitamura, T., Taniguchi, N., Klagsbrun, M. and Mekada, E. (1994). Heparin-binding EGF-like growth factor, which acts as the diphtheria toxin receptor, forms a complex with membrane protein DRAP27/CD9, which up-regulates functional receptors and diphtheria toxin sensitivity. *EMBO J.* **13**, 2322–30.

Ji, T.H., Grossmann, M. and Ji, I. (1998). G protein-coupled receptors. I. Diversity of receptor–ligand interactions. *J. Biol. Chem.* **273**, 17299–302.

Jinno, Y., Chaudhary, V.K., Kondo, T., Adhya, S., FitzGerald, D.J. and Pastan, I. (1988). Mutational analysis of domain I of *Pseudomonas* exotoxin. Mutations in domain I of *Pseudomonas* exotoxin which reduce cell binding and animal toxicity. *J. Biol. Chem.* **263**, 13203–7.

Jinno, Y., Ogata, M., Chaudhary, V.K., Willingham, M.C., Adhya, S., FitzGerald, D. and Pastan, I. (1989). Domain II mutants of *Pseudomonas* exotoxin deficient in translocation. *J. Biol. Chem.* **264**, 15953–9.

Johannes, L. and Goud, B. (1998). Surfing on a retrograde wave: how does Shiga toxin reach the endoplasmic reticulum? *Trends Cell. Biol.* **8**, 158–62.

Johnson, V.G. and Nicholls, P.J. (1994a). Identification of a single amino acid substitution in the diphtheria toxin A chain of CRM 228 responsible for the loss of enzymic activity. *J. Bacteriol.* **176**, 4766–9.

Johnson, V.G. and Nicholls, P.J. (1994b). Histidine 21 does not play a major role in diphtheria toxin catalysis. *J. Biol. Chem.* **269**, 4349–54.

Johnson V.G., Nicholls, P.J., Habig, W.H. and Youle, R.J. (1993). The role of proline 345 in diphtheria toxin translocation. *J. Biol. Chem.* **268**, 3514–19.

Kaczorek, M., Delpeyroux, F., Chenciner, N., Streeck, R.E., Murphy, J.R., Boquet, P. and Tiollais, P. (1983). Nucleotide sequence and expression of the diphtheria tox 228 gene in *Escherichia coli*. *Science* **221**, 855–8.

Karlsson, K.A., Teneberg, S., Angstrom, J., Kjellberg, A., Hirst, T.R., Berstrom, J. and Miller-Podraza, H. (1996). Unexpected carbohydrate cross-binding by *Escherichia coli* heat-labile enterotoxin. Recognition of human and rabbit target cell glycoconjugates in comparison with cholera toxin. *Bioorg. Med. Chem.* **11**, 1919–28.

Kaslow, H.R., Schlotterbeck, J.D., Mar, V.L. and Burnette, W.N. (1989). Alkylation of cysteine 41, but not cysteine 200, decreases the ADP-ribosyltransferase activity of the S1 subunit of pertussis toxin. *J. Biol. Chem.* **264**, 6386–90.

Katada, T., Tamura, M. and Ui, M. (1983). The A protomer of islet-activating protein, pertussis toxin, as an active peptide catalyzing ADP-ribosylation of a membrane protein. *Arch. Biochem. Biophys.* **224**, 290–8.

Kjeldgaard, M., Nyborg, J. and Clark, B.F. (1996). The GTP binding motif: variations on a theme. *FASEB J.* **10**, 1347–68.

Kohno, K., Uchida, T., Ohkubo, H., Nakanishi, S., Nakanishi, T., Fukui, T., Ohtsuka, E., Ikehara, M. and Okada, Y. (1986). Amino acid sequence of mammalian elongation factor 2 deduced from the cDNA sequence: homology with GTP-binding proteins. *Proc. Natl Acad. Sci. USA* **83**, 4978–82.

Kosecka, U., Marshall, J. S., Crowe, S. E., Bienenstock J. and Perdue, M.H. (1994). Pertussis toxin stimulates hypersensitivity and enhances nerve-mediated antigen uptake in rat intestine. *Am. J. Physiol.* **267**, G745–53.

Laird, W. and Groman, N. (1976). Isolation and characterization of tox mutants of corynebacteriophage beta. *J. Virol.* **19**, 220–7.

Lebens, M., Johansson, S., Osek, J., Lindblad, M. and Holmgren, J. (1993). Large-scale production of *Vibrio cholerae* toxin B subunit for use in oral vaccines. *Biotechnology* **11**, 1574–8.

Lefkowitz, R.J. (1998). G protein-coupled receptors. III. New roles for receptor kinases and beta-arrestins in receptor signaling and desensitization. *J. Biol. Chem.* **273**, 18677–80.

Lobet, Y., Cieplak, W., Jr, Smith, S.G. and Keith, J.M. (1989). Effects of mutations on enzyme activity and immunoreactivity of the S1 subunit of pertussis toxin. *Infect. Immun.* **57**, 3660–2.

Lobet, Y., Cluff, C.W. and Cieplak, W., Jr (1991). Effect of site-directed mutagenic alterations on ADP-ribosyltransferase activity of the A subunit of *Escherichia coli* heat-labile enterotoxin. *Infect. Immun.* **59**, 2870–9.

Lobet, Y., Feron, C., Dequesne, G., Simoen, E., Hauser, P. and Locht, C. (1993). Site-specific alterations in the B oligomer that affect receptor-binding activities and mitogenicity of pertussis toxin. *J. Exp. Med.* **177**, 79–87.

Locht, C. and Keith, J.M. (1986). Pertussis toxin gene: nucleotide sequence and genetic organization. *Science* **232**, 1258–64.

Locht, C., Capiau, C. and Feron, C. (1989). Identification of amino acid residues essential for the enzymic activities of pertussis toxin. *Proc. Natl Acad. Sci. USA* **86**, 3075–9.

Lorberboum-Galski, H., FitzGerald, D., Chaudhary, V., Adhya, S. and Pastan, I. (1988). Cytotoxic activity of an interleukin 2–*Pseudomonas* exotoxin chimeric protein produced in *Escherichia coli. Proc. Natl Acad. Sci. USA* **85**, 1922–6.

Lycke, N., Tsuji, T. and Holmgren, J. (1992). The adjuvant effect of *Vibrio cholerae* and *Escherichia coli* heat-labile enterotoxins is linked to their ADP-ribosyltransferase activity. *Eur. J. Immunol.* **22**, 2277–81.

Majoul, I.V., Bastiaens, P.I. and Soling, H.D. (1996). Transport of an external Lys–Asp–Glu–Leu (KDEL) protein from the plasma membrane to the endoplasmic reticulum: studies with cholera toxin in Vero cells. *J. Cell Biol.* **133**, 777–89.

Marchetti, M., Rossi, M., Giannelli, V., Giuliani, M.M., Pizza, M., Censini, S., Covacci, A., Massari, P., Pagliaccia, C., Manetti, R., Telford, J.L., Douce, G., Dougan, G., Rappuoli, R. and Ghiara, P. (1998). Protection against *Helicobacter pylori* infection in mice by intragastric vaccination with *H. pylori* antigens is achieved using a non-toxic mutant of *E. coli* heat-labile enterotoxin (LT) as adjuvant. *Vaccine* **16**, 33–7.

Marsischky, C.T., Wilson, B.A. and Collier, R.J. (1995). Role of glutamic acid 988 of human poly-ADP-ribose polymerase in polymer formation. Evidence for active site similarities to the ADP-ribosylating toxins. *J. Biol. Chem.* **270**, 3247–54.

McMahon, K.K., Piron, K.J., Ha, V.T. and Fullerton, A.T. (1993). Developmental and biochemical characteristics of the cardiac membrane-bound arginine-specific mono-ADP-ribosyltransferase. *Biochem. J.* **293**, 789–93.

Merritt, E.A., Sarfaty, S., Pizza, M., Domenighini, M., Rappuoli, R. and Hol, W.G. (1995). Mutation of a buried residue causes loss of activity but no conformational change in the heat-labile enterotoxin of *Escherichia coli. Nat. Struct. Biol.* **2**, 269–72.

Mitamura, T., Higashiyama, S., Taniguchi, N., Klagsbrun, M. and Mekada, E. (1995). Diphtheria toxin binds to the epidermal growth gactor (EGF)-like domain of human heparin-binding EGF-like growth factor/diphtheria toxin receptor and inhibits specifically its mitogenic activity. *J. Biol. Chem.* **270**, 1015–19.

Mori, T., Shoemaker, R.H., McMahon, J.B., Gulakowski, R.J., Gustafson, K.R. and Boyd, M.R. (1997). Construction and enhanced cytotoxicity of a [cyanovirin-N]-[*Pseudomonas* exotoxin] conjugate against human immunodeficiency virus-infected cells. *Biochem. Biophys. Res. Commun.* **239**, 884–8.

Moss, J. and Vaughan, M. (1988). Cholera toxin and *E. coli* enterotoxins and their mechanisms of action. In: *Handbook of Natural Toxins*, Vol. 4, *Bacterial Toxins* (eds M. Hardegree and A.T. Ru), pp. 39–87. Marcel Dekker, New York.

Moss, J. and Vaughan, M. (1995). Structure and function of ARF proteins: activators of cholera toxin and critical components of intracellular vesicular transport processes. *J. Biol. Chem.* **270**, 12327–30.

Moss, J., Stanley, S.J., Vaughan, M. and Tsuji, T. (1993). Interaction of ADP-ribosylation factor with *Escherichia coli* enterotoxin that contains an inactivating lysine 112 substitution. *J. Biol. Chem.* **268**, 6383–7.

Murphy, J.R., Bishai, W., Borowski, M., Miyanohara, A., Boyd, J. and Nagle, S. (1986). Genetic construction, expression, and melanoma-selective cytotoxicity of a diphtheria toxin-related alpha-melanocyte-stimulating hormone fusion protein. *Proc. Natl Acad. Sci. USA* **83**, 8258–62.

Naglich, J.G., Metherall, J.E., Russell, D.W. and Eidels, L. (1992). Expression cloning of a diphtheria toxin receptor: identity with a heparin-binding EGF-like growth factor precursor. *Cell* **69**, 1051–61.

Nashar, T.O., Webb, H.M., Eaglestone, S., Williams, N.A. and Hirst, T.R. (1996). Potent immunogenicity of the B subunits of *Escherichia coli* heat-labile enterotoxin: receptor binding is essential and induces differential modulation of lymphocyte subsets. *Proc. Natl Acad. Sci. USA* **93**, 226–30.

Neer, E.J. and Clapham, D.E. (1988). Roles of G protein subunits in transmembrane signalling. *Nature* **333**, 129–34.

Nencioni, L., Pizza, M., Bugnoli, M., De Magistris, T., Di Tommaso, A., Giovannoni, F., Manetti, R., Marsili, I., Matteucci, G., Nucci, D., Olivieri, R., Pileri, P., Presentini, R., Villa, L., Kreeftenberg, H., Silvestri, S., Tagliabue, A. and Rappuoli, R. (1990). Characterization of genetically inactivated pertussis toxin mutants: candidates for a new vaccine against whooping cough. *Infect. Immun.* **58**, 1308–15.

Nencioni, L., Volpini, G., Peppoloni, S., de Magistris, M.T., Marsili, I. and Rappuoli, R. (1991). Properties of the pertussis toxin mutant PT–9K/129G after formaldehyde treatment. *Infect. Immun.* **59**, 625–30.

Neville, D.M., Jr and Hudson, T.H. (1986). Transmembrane transport of diphtheria toxin, related toxins, and colicins. *Annu. Rev. Biochem.* **55**, 195–224.

Nicosia, A., Perugini, M., Franzini, C., Casagli, M.C., Borri, M.G., Antoni, G., Almoni, M., Neri, P., Ratti, G. and Rappuoli, R. (1986). Cloning and sequencing of the pertussis toxin genes: operon structure and gene duplication. *Proc. Natl Acad. Sci. USA* **83**, 4631–5.

Ogata, M., Chaudhary, V.K., FitzGerald, D.J. and Pastan, I. (1989). Cytotoxic activity of a recombinant fusion protein between interleukin 4 and *Pseudomonas* exotoxin. *Proc. Natl Acad. Sci. USA* **86**, 4215–19.

Okamoto, K., Okamoto, K., Miyama, A., Tsuji, T., Honda, T. and Miwatani, T. (1988). Effect of substitution of glycine for arginine at position 146 of the A1 subunit on biological activity of *Escherichia coli* heat-labile enterotoxin. *J. Bacteriol.* **170**, 2208–11.

Okazaki, I.J. and Moss, J. (1994). Common structure of the catalytic sites of mammalian and bacterial toxin ADP-ribosyltransferases. *Mol. Cell. Biochem.* **138**, 177–81.

Okazaki, I.J. and Moss, J. (1998). Glycosylphosphatidylinositol-anchored and secretory isoforms of mono-ADP-ribosyltransferases. *J. Biol. Chem.* **273**, 23617–20.

Papini, E., Colonna, R., Cusinato, F., Montecucco, C., Tomasi, M. and Rappuoli, R. (1987a). Lipid interaction of diphtheria toxin and mutants with altered fragment B. 1. Liposome aggregation and fusion. *Eur. J. Biochem.* **169**, 629–35.

Papini, E., Schiavo, G., Tomasi, M., Colombatti, M., Rappuoli, R. and Montecucco, C. (1987b). Lipid interaction of diphtheria toxin and mutants with altered fragment B. 2. Hydrophobic photolabelling and cell intoxications. *Eur. J. Biochem.* **169**, 637–44.

Papini, E., Schiavo, G., Sandona, D., Rappuoli, R. and Montecucco, C. (1989). Histidine 21 is at the NAD$^+$ binding site of diphtheria toxin. *J. Biol. Chem.* **264**, 12385–8.

Papini, E., Santucci, A., Schiavo, G., Domenighini, M., Neri, P., Rappuoli, R. and Montecucco, C. (1991). Tyr-65 is photolabeled by 8-azido adenine and 8-azido-adenosine at the NAD binding site of diphtheria toxin. *J. Biol. Chem.* **266**, 2494–8.

Pappenheimer, A.M. (1938). Diphtheria toxin. II. The action of ketene and formaldehyde. *J. Biol. Chem.* **125**, 201–8.

Pappenheimer, A.M., Jr (1977). Diphtheria toxin. *Annu. Rev. Biochem.* **46**, 69–94.

Pappenheimer, A.M., Jr (1984). Diphtheria. In: *Bacterial Vaccines* (ed. R. Germanier), pp. 1–16. Academic Press, New York.

Pappenheimer, A.M., Jr, Uchida, T. and Harper, A.A. (1972). An immunological study of the diphtheria toxin molecule. *Immunochemistry* **9**, 891–906.

Partidos, C.D., Pizza, M., Rappuoli, R. and Steward, M.W. (1996). The adjuvant effect of a non-toxic mutant of heat-labile enterotoxin of *Escherichia coli* for the induction of measles virus-specific CTL responses after intranasal co-immunization with a synthetic peptide. *Immunology* **89**, 483–7.

Pastan, I. and FitzGerald, D. (1989). *Pseudomonas* exotoxin: chimeric toxins. *J. Biol. Chem.* **264**, 15157–60.

Pizza, M., Bartoloni, A., Prugnola, A., Silvestri, S. and Rappuoli, R. (1988). Subunit S1 of pertussis toxin: mapping of the regions essential for ADP-ribosyltransferase activity. *Proc. Natl Acad. Sci. USA* **85**, 7521–5.

Pizza, M., Covacci, A., Bartoloni, A., Perugini, M., Nencioni, L., De Magistris, M.T., Villa, L., Nucci, D., Manetti, R., Bugnoli, M., Giovannoni, F., Olivieri, R., Barbieri, J.T., Sato, H. and Rappuoli, R. (1989). Mutants of pertussis toxin suitable for vaccine development. *Science* **246**, 497–500.

Pizza, M., Bugnoli, M., Manetti, R., Covacci, A. and Rappuoli, R. (1990). The subunit S1 is important for pertussis toxin secretion. *J. Biol. Chem.* **265**, 17759–63.

Pizza, M., Domenighini, M., Hol, W., Giannelli, V., Fontana, M.R., Giuliani, M.M., Magagnoli, C., Peppoloni, S., Manetti, R. and Rappuoli, R. (1994a). Probing the structure–activity relationship of *Escherichia coli* LT-A by site-directed mutagenesis. *Mol. Microbiol.* **14**, 51–60.

Pizza, M., Fontana, M.R., Giuliani, M.M., Domenighini, M., Magagnoli, C., Giannelli, V., Nucci, D., Hol, W., Manetti, R. and Rappuoli, R. (1994b). A genetically detoxified derivative of heat-labile *Escherichia coli* enterotoxin induces neutralizing antibodies against the A subunit. *J. Exp. Med.* **180**, 2147–53.

Pohl, E., Holmes, R.K. and Hol, W.G. (1998). Motion of the DNA-binding domain with respect to the core of the diphtheria toxin repressor (DtxR) revealed in the crystal structures of apo- and holo-DtxR. *J. Biol. Chem.* **273**, 22420–7.

Popoff, M.R. (1998). Interactions between bacterial toxins and intestinal cells. *Toxicon* **36**, 665–85.

Porro, M., Saletti, M., Nencioni, L., Tagliaferri, L. and Marsili, I. (1980). Immunogenic correlation between cross-reacting material (CRM 197) produced by a mutant of *Corynebacterium diphtheriae* and diphtheria toxoid. *J. Infect. Dis.* **142**, 716–24.

Qiu, X., Verlinde, C.L., Zhang, S., Schmitt, M.P., Holmes, R.K. and Hol, W.G. (1995). Three-dimensional structure of the diphtheria toxin repressor in complex with divalent cation co-repressors. *Structure* **3**, 87–100.

Rappuoli, R. (1983). Isolation and characterization of *Corynebacterium diphtheriae* nontandem double lysogens hyperproducing CRM 197. *Appl. Enzym. Microbiol.* **45**, 560–4.

Rappuoli, R. (1997). Rational design of vaccines. *Nat. Med.* **3**, 374–6.

Rappuoli, R. and Ratti, G. (1984). Physical map of the chromosomal region of *Corynebacterium diphtheriae* containing corynephage attachment sites attB1 and attB2. *J. Bacteriol.* **158**, 325–30.

Rappuoli, R., Michel, J.L. and Murphy, J.R. (1983a). Restriction endonuclease map of corynebacteriophage ωctox⁺ isolated from the Park Williams no. 8 strain of *Corynebacterium diphtheriae*. *J. Virol.* **45**, 524–30.

Rappuoli, R., Michel, J.L. and Murphy, J.R. (1983b). Integration of corynebacterophages βtox⁺, ωtox⁺, and γtox⁻ into two attachment sites on the *Corynebacterium diphtheriae* chromosome. *J. Bacteriol.* **153**, 1202–10.

Ratti, G., Rappuoli, R. and Giannini, G. (1983). Complete nucleotide sequence of the gene coding for diphtheria toxin in the corynephage omegatox⁺ genome. *Nucleic Acids Res.* **11**, 6589–95.

Ricci, S., Rappuoli, R. and Scarlato, V. (1996). The pertussis toxin liberation genes of *Bordetella* pertussis are transcriptionally linked to the pertussis toxin operon. *Infect. Immun.* **64**, 1458–60.

Roberts, M., Bacon, A., Rappuoli, R., Pizza, M., Cropley, I., Douce, G., Dougan, G., Marinaro, M., McGhee, J. and Chatfield, S. (1995). A mutant pertussis toxin molecule that lacks ADP-ribosyltransferase activity, PT–9K/129G, is an effective mucosal adjuvant for intranasally delivered proteins. *Infect. Immun.* **63**, 2100–8.

Rothbrock, G., Smithee, L., Rados, M. and Baughman, W. (1995). Progress toward elimination of *Haemophilus influenzae* type b disease among infants and children – United States, 1993–1994 (Reprinted from MMWR, vol. 44, pg 545–550, 1995). *J. Am. Med. Assoc.* **274**, 1334–5.

Ruf, A., Mennissier de Murcia, J., de Murcia, G. and Schulz, G.E. (1996). Structure of the catalytic fragment of poly(ADP-ribose) polymerase from chicken. *Proc. Natl Acad. Sci. USA* **93**, 7481–5.

Sandvig, K. and Olsnes, S. (1981). Rapid entry of nicked diphtheria toxin into cells at low pH. Characterization of the entry process and effects of low pH on the toxin molecule. *J. Biol. Chem.* **256**, 9068–76.

Saxty, B.A. and van Heyningen, S. (1995). The purification of a cysteine-dependent NAD⁺ glycohydrolase activity from bovine erythrocytes and evidence that it exhibits a novel ADP-ribosyltransferase activity. *Biochem. J.* **310**, 931–7.

Schuman, E.M., Meffert, M.K., Schulman, H. and Madison, D.V. (1994). An ADP-ribosyltransferase as a potential target for nitric oxide action in hippocampal long-term potentiation. *Proc. Natl Acad. Sci. USA* **91**, 11958–62.

Sekine, A., Fujiwara, M. Narumiya, S. (1989). Asparagine residue in the *rho* gene product is the modification site for botulinum ADP-ribosyl transferase. *J. Biol. Chem.* **264**, 8602–5.

Sekura, R.D., Moss, J. and Vaughan, M. (1985). In: *Pertussis Toxins*, pp. 1–250. Academic Press, New York.

Siegall, C.B., Chaudhary, V.K., FitzGerald, D.J. and Pastan, I. (1988). Cytotoxic activity of an interleukin 6-Pseudomonas exotoxin fusion protein on human myeloma cells. *Proc. Natl Acad. Sci. USA* **85**, 9738–42.

Siegall, C.B., Chaudhary, V.K., FitzGerald, D.J. and Pastan, I. (1989). Functional analysis of domains II, Ib, and III of *Pseudomonas* exotoxin. *J. Biol. Chem.* **264**, 14256–61.

Silverman, J.A., Mindell, J.A., Finkelstein, A., Shen, W.H. and Collier, R.J. (1994). Mutational analysis of the helical hairpin region of diphtheria toxin transmembrane domain. *J. Biol. Chem.* **269**, 22524–32.

Sindt, K.A., Hewlett, E.L., Redpath, G.T., Rappuoli, R., Gray, L.S. and Vandenberg, S. R. (1994). Pertussis toxin activates platelets through an interaction with platelet glycoprotein Ib. *Infect. Immun.* **62**, 3108–14.

Sixma, T.K., Pronk, S.E., Kalk, K.H., Wartna, E.S., van Zanten, B.A., Witholt. B. and Hol, W.G. (1991). Crystal structure of a cholera toxin-related heat-labile enterotoxin from *E. coli*. *Nature* **351**, 371–7.

Sixma, T.K., Kalk, K.H., van Zanten, B.A., Dauter, Z., Kingma, J., Witholt, B. and Hol, W.G. (1993). Refined structure of *Escherichia coli* heat-labile enterotoxin, a close relative of cholera toxin. *J. Mol. Biol.* **230**, 890–918.

Spicer, E.K., Kavanaugh, W.M., Dallas, W.S., Falkow, S., Konigsberg, W.H. and Schafer D.E. (1981). Sequence homologies between A subunits of *Escherichia coli* and *Vibrio cholerae* enterotoxins. *Proc. Natl Acad. Sci. USA* **78**, 50–4.

Stein, P.E., Boodhoo, A., Armstrong, G.D., Heerze, L.D., Cockle, S.A., Klein, M.H. and Read, R.J. (1994a). Structure of a pertussis toxin–sugar complex as a model for receptor binding. *Nat. Struct. Biol.* **1**, 591–6.

Stein, P.E., Boodhoo, A., Armstrong, G.D., Cockle, S.A., Klein, M.H. and Read, R.J. (1994b). The crystal structure of pertussis toxin. *Structure* **2**, 45–57.

Stevens, L., Moss J., Vaughan M., Pizza M. and Rappuoli, R. (1999). Effects of site-directed mutagenesis of *E. coli* heat labile enterotoxin on ADP-ribosyltransferase activity and interaction with ADP-ribosylation factors (ARFs). *Infect. Immun.* **67**, 259–65.

Stryer, L. and Bourne, H.R. (1986). G proteins: a family of signal transducers. *Annu. Rev. Cell. Biol.* **2**, 391–419.

Sunahara, R.K., Tesmer, J.J., Gilman, A.G. and Sprang, S.R. (1997). Crystal structure of the adenylyl cyclase activator Gs alpha. *Science* **278**, 1943–7.

Takada, T., Iida, K. and Moss, J. (1995). Conservation of a common motif in enzymes catalyzing ADP-ribose transfer. Identification of domains in mammalian transferases. *J. Biol. Chem.* **270**, 541–4.

Tamura, M., Nogimori, K., Murai, S., Yajima, M., Ito, K., Katada, T., Ui, M. and Ishii, S. (1982). Subunit structure of islet-activating protein, pertussis toxin, in conformity with the A–B model. *Biochemistry* **21**, 5516–22.

Tamura, M., Nogimori, K., Yajima, M., Ase, K. and Ui, M. (1983). A role of the B-oligomer moiety of islet-activating protein, pertussis toxin, in development of the biological effects on intact cells. *J. Biol. Chem.* **258**, 6756–61.

Tanuma, S., Kawashima, K. and Endo, H. (1988). Eukaryotic mono(ADP-ribosyl)transferase that ADP-ribosylates GTP-binding regulatory Gi protein. *J. Biol. Chem.* **263**, 5485–9.

Teneberg, S., Hirst, T.R., Angstrom, J. and Karlsson, K.A. (1994). Comparison of the glycolipid-binding specificities of cholera toxin and porcine *Escherichia coli* heat-labile enterotoxin: identification of a receptor-active non-ganglioside glycolipid for the heat-labile toxin in infant rabbit small intestine. *Glycoconj. J.* **11**, 533–40.

Trucco, C., Flatter, E., Fribourg, S., de Murcia, G. and Menissier-de Murcia, J. (1996). Mutations in the amino-terminal domain of the human poly(ADP-ribose) polymerase that affect its catalytic activity but not its DNA binding capacity. *FEBS Lett.* **399**, 313–16.

Truitt, R.L., Hanke, C., Radke, J., Mueller, R. and Barbieri, J.T. (1998). Glycosphingolipids as novel targets for T-cell suppression by the B subunit of recombinant heat-labile enterotoxin. *Infect. Immun.* **66**, 1299–308.

Tsai, S.C., Adamik, R., Moss, J. and Aktories, K. (1988). Separation of the 24 kDa substrate for botulinum C3 ADP-ribosyltransferase and the cholera toxin ADP-ribosylation factor. *Biochem. Biophys. Res. Commun.* **152**, 957–61.

Tsuji, T., Inoue, T., Miyama, A., Okamoto, K., Honda, T. and Miwatani, T. (1990). A single amino acid substitution in the A subunit of *Escherichia coli* enterotoxin results in a loss of its toxic activity. *J. Biol. Chem.* **265**, 22520–5.

Tsuji, T., Inoue, T., Miyama, A. and Noda, M. (1991). Glutamic acid-112 of the A subunit of heat-labile enterotoxin from enterotoxi-genic *Escherichia coli* is important for ADP-ribosyltransferase activity. *FEBS Lett.* **291**, 319–21.

Tweten, R.K., Barbieri, J.T. and Collier, R.J. (1985). Diphtheria toxin. Effect of substituting aspartic acid for glutamic acid 148 on ADP-ribosyltransferase activity. *J. Biol. Chem.* **260**, 10392–4.

Uchida, T., Gill, D.M. and Pappenheimer, A.M., Jr (1971). Mutation in the structural gene for diphtheria toxin carried by temperate phage β. *Nature New Biol.* **233**, 8–11.

Uchida, T., Pappenheimer, A.M., Jr and Greany, R. (1973a). Diphtheria toxin and related proteins I. Isolation and properties of mutant proteins serologically related to diphtheria toxin. *J. Biol. Chem.* **248**, 3838–44.

Uchida, T., Pappenheimer, A.M., Jr and Harper, A.A. (1973b). Diphtheria toxin and related proteins. II. Kinetic studies on intoxication of HeLa cells by diphtheria toxin and related proteins. *J. Biol. Chem.* **248**, 3845–50.

Uchida, T., Pappenheimer, A.M., Jr and Harper, A.A. (1973c). Diphtheria toxin and related proteins. III. Reconstruction of hybrid 'diphtheria toxin' from nontoxic mutant proteins. *J. Biol. Chem.* **248**, 3851–4.

Ueda, K. and Hayaishi, O. (1985). ADP-ribosylation. *Annu. Rev. Biochem.* **54**, 73–100.

Vandekerckhove, J., Schering, B., Bärmann, M. and Aktories, K. (1988). Botulinum C2 toxin ADP-ribosylates cytoplasmic b/v-actin in arginine 177. *J. Biol. Chem.* **263**, 13739–42.

van den Akker, F., Merritt, E.A., Pizza, M., Domenighini, M., Rappuoli, R. and Hol, W.G. (1995). The Arg7Lys mutant of heat-labile enterotoxin exhibits great flexibility of active site loop 47–56 of the A subunit. *Biochemistry* **34**, 10996–1004.

van den Akker, F., Pizza, M., Rappuoli, R. and Hol, W.G. (1997). Crystal structure of a non-toxic mutant of heat-labile enterotoxin, which is a potent mucosal adjuvant. *Protein Sci.* **12**, 2650–4.

van der Pouw-Kraan, T., Rensink, I., Rappuoli, R. and Aarden, L. (1995). Co-stimulation of T cells via CD28 inhibits human IgE production. Reversal by pertussis toxin. *Clin. Exp. Immunol.* **99**, 473–8.

Van Ness, B.G., Howard, J.B. and Bodley, J.W. (1980a). ADP-ribosylation of elongation factor 2 by diphtheria toxin. NMR spectra and proposed structures of ribosyl-diphthamide and its hydrolysis products. *J. Biol. Chem.* **255**, 10710–16.

Van Ness, B.G., Howard, J.B. and Bodley, J.W. (1980b). ADP-ribosylation of elongation factor 2 by diphtheria toxin. Isolation and properties of the novel ribosyl-amino acid and its hydrolysis products. *J. Biol. Chem.* **255**, 10717–20.

Vaughan, M. (1998). G protein-coupled receptors minireview series. *J. Biol. Chem.* **273**, 17297.

Wang, J., Nemoto, E., Kots, A.Y., Kaslow, H.R. and Dennert, G. (1994). Regulation of cytotoxic T cells by ecto-nicotinamide adenine dinucleotide (NAD) correlates with cell surface GPI-anchored/arginine ADP-ribosyltransferase. *J. Immunol.* **153**, 4048–58.

Weiss, A.A., Johnson, F.D. and Burns, D.L. (1993). Molecular characterization of an operon required for pertussis toxin secretion. *Proc. Natl Acad. Sci. USA* **90**, 2970–4.

West, R.E., Jr, Moss, J., Vaughan, M., Liu, T. and Liu, T.Y. (1985). Pertussis toxin-catalyzed ADP-ribosylation of transducin. Cysteine 347 is the ADP-ribose acceptor site. *J. Biol. Chem.* **260**, 14428–30.

White, A., Ding, X., vanderSpek, J.C., Murphy, J.R. and Ringe, D. (1998). Structure of the metal-ion-activated diphtheria toxin repressor/tox operator complex. *Nature* **394**, 502–6.

Williams, D.P., Parker, K., Bacha, P., Bishai, W., Borowski, M., Genbauffe, F., Strom, T.B. and Murphy, J.R (1987).

Diphtheria toxin receptor binding domain substitution with inter-leukin-2: genetic construction and properties of a diphtheria toxin-related interleukin-2 fusion protein. *Protein Eng.* **1**, 493–8.

Wilson, B.A., Blanke, S.R., Reich, K.A. and Collier, R.J. (1994). Active-site mutations of diphtheria toxin – tryptophan 50 is a major determinant of NAD affinity. *J. Biol. Chem.* **269**, 23296–301.

Witvliet, M.H., Burns, D.L., Brennan, M.J., Poolman, J.T. and Manclark, C.R. (1989). Binding of pertussis toxin to eucaryotic cells and glycoproteins. *Infect. Immun.* **57**, 3324–30.

Yamaizumi, M., Mekada, E., Uchida, T. and Okada, Y. (1978) One molecule of diphtheria toxin fragment A introduced into a cell can kill the cell. *Cell* **15**, 245–50.

Yamamoto, S., Kiyono, H., Yamamoto, M., Imaoka, K., Fujihashi, K., Van Ginkel, F.W., Noda, M., Takeda, Y. and McGhee, J.R. (1997).

A nontoxic mutant of cholera toxin elicits Th2-type responses for enhanced mucosal immunity. *Proc. Natl Acad. Sci.USA* **94**, 5267–72.

Zimmermann, S., Wels, W., Froesch, B.A., Gerstmayer, B., Stahel, R.A. and Zangemeister-Wittke, U. (1997). A novel immunotoxin recognising the epithelial glycoprotein-2 has potent anti-tumoural activity on chemotherapy-resistant lung cancer. *Cancer Immunol. Immunother.* **44**, 1–9.

Zolkiewska, A. and Moss, J. (1993). Integrin alpha 7 as substrate for a glycosylphosphatidylinositol-anchored ADP-ribosyltransferase on the surface of skeletal muscle cells. *J. Biol. Chem.* **268**, 25273–6.

Zumbihl, R., Dornand, J., Fischer, T., Cabane, S., Rappuoli, R., Bouaboula, M., Casellas, P. and Rouot, B. (1995). IL-1 stimulates a diverging signaling pathway in EL4 6.1 thymoma cells. *J. Immunol.* **155**, 181–9.

4

Binding, uptake, routing and translocation of toxins with intracellular sites of action

Sjur Olsnes, Jørgen Wesche and Pål Ø. Falnes

INTRODUCTION

An increasing number of bacterial protein toxins is found to exert their action on targets that are located in the cytosol. The different toxins act enzymically on very different cytosolic components and the consequences for the cell vary from minor effects on cell physiology to cell death. Common to all these effects is that they have consequences for the host organism that promote the multiplication and spreading of the pathogenic bacterium. Although some toxins may be lethal to the host organism, it is questionable whether this is the intended effect in most cases. It is more likely that lethality is an overshoot reaction of the toxic effect that would otherwise give the bacterium a survival advantage. Toxins from *Clostridia* may be exceptions as these bacteria require anaerobic conditions for growth and may therefore profit from killing the host.

In some cases the toxic effect consists of inhibition of protein synthesis, whereas in other cases modulation of signalling mechanisms in the cells, making the cell unable to cope properly with the infection. Some toxins inhibit exocytic processes, which has most serious consequences for the nervous system by blocking synaptic transmission.

For many of the toxins the intracellular mechanisms of action are known at the molecular, and even at the atomic, level but it is still not known in detail how the toxins penetrate cellular membranes to access their targets. However, even in this field there has been much progress in recent years. There appear to be two main categories of toxins concerning entry; those toxins that are equipped with their own translocation device and those that must rely on mechanisms provided by the target cells. Valuable complementary information on the toxins discribed here can be found in the various chapters of this book dealing more specifically with these toxins.

INTRACELLULAR TARGETS OF TOXINS

The toxins discussed here all act enzymically on targets in the cytosol. The toxins act directly on their cytoplasmic substrates without intermediary molecules. We are therefore confronted with the problem of transporting enzymically active proteins across cellular membranes to their cytosolic targets. Before discussing what is known about the translocation mechanisms, the intracellular actions of the different toxins will be described briefly. The toxins are listed here in groups having the same or closely related intracellular targets.

Toxins modifying elongation factor 2

Diphtheria toxin is the pathogenicity factor in diphtheria (Pappenheimer, 1977), and exotoxin A plays a decisive role in the pathogenicity of *Pseudomonas*

73

aeruginosa in both animals and plants (Rahme *et al.*, 1995). Both toxins act by modifying elongation factor 2 (EF-2) involved in protein synthesis (Collier, 1967, 1975; Iglewski and Kabat, 1975; Pappenheimer, 1977). This protein contains an unusual amino acid, diphthamide, which is formed by post-translational modification of a histidine residue (Van Ness *et al.*, 1980a, b). The toxins transfer the ADP-ribose moiety from NAD to this unique residue found only in EF-2. As a result of the modification, the elongation factor cannot interact properly with the ribosomes and protein synthesis is blocked. The cell will die after a time lag of several hours.

Toxins modifying the α subunit of trimeric G-proteins

Cholera toxin, *Escherichia coli* heat-labile toxin (causing less severe diarrhoea than in cholera) and pertussis toxin also act by ADP-ribosylating their intracellular targets, but in these cases the target is the α subunit of various heterotrimeric G-proteins (Gill, 1975; Gill and Meren, 1978; Katada and Ui, 1982; Katada *et al.*, 1986), and the amino acid modified is arginine (cholera toxin and *E. coli* heat-labile toxin) or cysteine (pertussis toxin). Whereas cholera toxin and *E. coli* heat-labile toxin inactivate the GTPase of the α subunit by ADP-ribosylating an arginine close to the active site, pertussis toxin ADP-ribosylates a cysteine residue close to the C-terminal end of the α subunit of the G$_i$ subfamily and thereby prevents interaction of the G-protein with receptors. The ADP-ribosylation by cholera toxin is stimulated by a cytoplasmic protein, ARF, which is an essential regulator of vesicular transport (Kahn and Gilman, 1984; Moss and Vaughan, 1995).

Toxins with actin as targets

The C2 toxin from *Clostridium botulinum*, *C. perfringens iota* toxin and *C. spiriforme* toxin act by ADP-ribosylating the ATP-binding protein actin. There are at least six mammalian actin isoforms, but *C. botulinum* C2 toxin ADP-ribosylates only the cytoplasmic form (Aktories *et al.*, 1986, 1992). The other toxins are less selective. The target for the toxins is the monomeric G-actin, while the polymerized F-actin cannot be ADP-ribosylated by the toxins. The modified site is located at the position where actin–actin interactions occur during polymerization. Therefore, the ADP-ribosylated actin is not able to participate in the polymerization process. The modified actin will, in fact, act as a stop signal for polymerization – a 'capping' agent. Also, the ATPase activity of the actin monomer is blocked. As actin is in a constant process of polymerization and depolymer-

ization, inhibition of polymerization at the growing end of the polymer will eventually lead to actin depolymerization.

Toxins that act on the small G-protein Rho

An export protein from *C. botulinum*, the exoenzyme C3, is able to ADP-ribosylate the small G-protein Rho, which regulates actin polymerization and intracellular vesicular transport (Chardin *et al.*, 1989). Similar enzymes are found in a wide variety of bacteria (Just *et al.*, 1992a, b), including *Staphylococcus aureus* (Inoue *et al.*, 1991), although no bacterial substrate is known. The use of exoenzyme C3 has played an important role in the recent elucidation of the mechanism of action of certain microbial toxins, the *Clostridium difficile* toxins A and B (Hofmann *et al.*, 1997) and the *E. coli* necrotizing toxin (Lemichez *et al.*, 1997). When Rho had been pretreated with *C. difficile* toxins it could no longer be ADP-ribosylated by exoenzyme C3, clearly indicating that the *C. difficile* toxin had modified Rho. This modification was later found to lie in the transfer of glucose from the precursor UDP-Glc to a threonine residue in Rho. In the toxin-treated cells the stress fibres disassemble. The related rac and Cdc-42 were also modified (Just *et al.*, 1995a, b). Lethal toxin from *C. sordellii* has the same activity (Popoff *et al.*, 1996). The toxins did not modify the more distantly related G-proteins ras, rab and ARF. *Clostridium novyi* toxin incorporates *N*-acetylglucosamine rather than glucose into Rho (Selzer *et al.*, 1996).

Cytotoxic necrotizing factor (CNF1) from certain strains of *E. coli* is another toxin acting on Rho. The toxin consists of a single polypeptide chain (137 kDa) with a putative *N*-terminal receptor-binding domain, a central hydrophobic domain and a *C*-terminal catalytic domain (Lemichez *et al.*, 1997). The toxin induces deamidation of a glutamine residue (Flatau *et al.*, 1997; Schmidt *et al.*, 1997), leading to permanent activation of Rho. As a result, there is a strong overactivity in actin polymerization, and formation of multinucleated cells. Furthermore, the toxin induces phagocytosis in non-phagocytic cells, enabling toxigenic bacteria to enter and multiply in cells and to be transcytosed across epithelia. It is apparently part of the same strategy that the toxin inhibits apoptosis in the intoxicated cells (Fiorentini *et al.*, 1997).

Toxins with protease activity

Tetanus and botulinum neurotoxins are often considered as the most toxic compounds known. While tetanus toxin induces spastic paralysis, the botulinum toxins induce flaccid paralysis. This apparent difference in mode of action is, however, only due to different target neurons for the toxins (see Chapter 10). Tetanus toxin acts primarily on inhibitory glycinergic neurons in the central nervous system, whereas botulinum toxins prevent release of acetylcholine at the neuromuscular junction (Burgen *et al.*, 1949). At the cellular level all these neurotoxins inhibit exocytosis of neurotransmitters from synaptic vesicles. The toxins consist of two functional parts: a receptor-binding part and an enzymic part. The enzymic part, the light (L) chain, is in each case a metalloprotease which differs slightly in specificity between the different toxins. The L-chain enters the cytosol and cleaves the cytosolic tail of certain proteins (synaptobrevin, syntaxin and Snap-25) involved in vesicle fusion (Schiavo *et al.*, 1992; Montecucco and Schiavo, 1993), including fusion with the plasma membrane in the case of exocytosis. When artificially introduced into the cytosol, the toxins can also inhibit exocytosis in many non-neuronal cells that do not have receptors for these toxins.

Recently, the lethal factor of anthrax toxin was also found to be a specific protease. It cleaves the amino terminus of mitogen-activated protein kinase kinases 1 and 2 (MAPKK1 and MAPKK2), and thereby inactivates the enzymes and disrupts signal transduction in the cells (Duesbery *et al.*, 1998).

Invasive adenylate cyclases

The oedema factor (EF) of anthrax toxin is an invasive adenylate cyclase which is only active after it has bound the cytosolic protein calmodulin (Leppla, 1982). Another invasive and calmodulin-activated adenylate cyclase (CyaA) is produced by *Bordetella pertussis* (Rogel *et al.*, 1991). The 177-kDa CyaA–protein can be divided into an *N*-terminal catalytic domain and a *C*-terminal hemolytic domain. Upon activation, both EF and CyaA elevate the level of cAMP in the cytosol, thus interfering with the bacteriocidal capacity of the intoxicated cell.

Toxins acting on ribosomes

Shiga toxins and Shiga-like toxins (SLT), as well as the plant toxins ricin, abrin, modeccin, volkensin and viscumin, kill target cells by inactivating the ribosomes (Olsnes *et al.*, 1973, 1975; Refsnes *et al.*, 1977; Reisbig *et al.*, 1981; Stirpe *et al.*, 1982). In each case the enzymic activity is carried by a polypeptide chain of approximately 30 kDa that is disulfide-linked to the rest of the toxin. The A-chain is a specific glycohydrolase that cleaves off an adenine residue from an exposed loop of the 28S RNA in the large ribosomal subunit (Endo and Tsurugi, 1987; Endo *et al.*, 1987, 1988). This residue

is involved in the binding of the elongation factors to the ribosome (Moazed *et al.*, 1988; Holmberg and Nygård, 1996; Orita *et al.*, 1996) and, as a result, protein synthesis is blocked.

From the nature of the enzymic activity of the active parts of the toxins discussed here and the topology of their cellular targets, it is clear that the toxins must reach the cytosol in order to act. In some cases cytoplasmically located co-factors, such as calmodulin and ARF, are required. The striking variety in toxin structure, target and mechanism of action indicates that the toxins have developed independently, and that transfer of an enzymic moiety to the cytosol is an efficient way of interfering with cellular processes.

TOXIN BINDING

The first essential step in the intoxication by all toxins described here is their binding to structures ('receptors') at the cell surface. All toxins discussed here consist of two functionally different parts, commonly designated A and B. A and B can be two different proteins or two domains of a single polypeptide chain. While the A part is the effector part discussed above, the B part binds to cell-surface receptors. Cells lacking the relevant receptor are resistant to the toxin in question. The identification of receptors has been difficult and, up to now, only a few toxin receptors have been identified. These belong to very different groups of cell-surface structures.

Receptors for cholera toxin and *Escherichia coli* heat-labile toxin

The first identified receptor for a protein toxin was that for cholera toxin, the protein that causes the extensive and often lethal diarrhoea seen in cholera. Cholera toxin and the closely related *E. coli* heat-labile toxin consist of a doughnut-shaped pentameric B subunit linked non-covalently to a single A subunit (Gill, 1976; Merritt *et al.*, 1994a). Several toxins have this architecture (A–B$_5$ toxins). Each of the five poly-peptides of the B subunit of cholera toxin and *E. coli* heat-labile toxin binds one glycolipid molecule (Sixma *et al.*, 1992), the ganglioside GM$_1$ (van Heyningen, 1974). The toxin binds to the carbohydrate part of the ganglioside, and the binding is such that the A subunit sits on top of the pentameric ring and points away from the membrane (Merritt *et al.*, 1994a, b). Cells that lack the receptor ganglioside are resistant to cholera toxin, but they can be rendered sensitive when incubated with GM$_1$ which is able to insert itself into the membrane. The heat-labile toxin from *E. coli*, which has a similar mode of action, also binds to gangliosides, but in this case it is less restricted and can also bind to asialo-GM$_1$ (MacKenzie *et al.*, 1997).

Receptors for Shiga and Shiga-like toxins

Shiga toxin is produced by the most virulent form of the dysentery bacterium, *Shigella dysenteriae 1*. An almost identical toxin is produced by certain pathogenic forms of *E. coli*. Shiga toxin also belongs to the A–B$_5$ toxins (Olsnes *et al.*, 1981; Donohue-Rolfe *et al.*, 1984; Stein *et al.*, 1992). The B subunit resembles that of cholera toxin, but its constituent peptide chains are smaller. The toxin binds to a glycolipid, but in this case not to a ganglioside but to a neutral glycolipid, the globoside Gb$_3$. Each of the five polypeptides in the pentameric ring appears to be able to bind three glycolipid molecules (Ling *et al.*, 1998). The essentially identical SLT1 and slightly different SLT2 bind to the same receptor, whereas the related, but somewhat different, SLT2e binds to another glycolipid receptor, the globoside Gb$_4$ (DeGrandis *et al.*, 1989). In the case of the receptors for Shiga and SLT toxins, the chain length of the ceramide group plays a role for the efficiency of the receptor (Sandvig *et al.*, 1994).

Receptors for pertussis toxin

Pertussis toxin has an architecture resembling that of the A–B$_5$ toxins, but in this case the pentameric ring is constructed from three different and two identical polypeptides, which are all related (Saukkonen *et al.*, 1992). The receptor has not been identified with certainty, but several proteins ranging from 70 to 165 kDa bind the toxin (Brennan *et al.*, 1988; Clark and Armstrong, 1990). The toxin also binds to several soluble glycoproteins, which could mean that the pentameric ring binds to carbohydrates. The binding could be restricted to S2 and S3 subunits which have lectin properties and bind to oligosaccharides with α(2–6)-linked sialic acid residues (Saukkonen *et al.*, 1992). Differently from cholera and Shiga toxins, the receptor-binding domain appears to be close to the enzymic S1 subunit, positioning this subunit in close proximity to the membrane (Stein *et al.*, 1994).

Diphtheria toxin receptor

Diphtheria toxin is synthesized as a single polypeptide chain which is easily cleaved by trypsin or furin into two fragments designated A and B. The B fragment consists of two parts, a T ('transmembrane') domain which is involved in translocation and an R ('receptor-binding') domain that binds the toxin to cell-surface receptors. Growth factors of the epidermal growth factor (EGF) family are synthesized as transmembrane proteins and are subsequently cleaved close to the external membrane leaflet to release the growth factor. In some cases the precursor is not cleaved and can act by stimulating receptors on adjacent cells (juxta-cellular stimulation). Diphtheria toxin receptor is the uncleaved precursor of one growth factor in this group, viz. the precursor of heparin-binding EGF-like growth factor (Naglich *et al.*, 1992).

The receptor also binds heparin, and only after heparin is bound is diphtheria toxin able to interact with the receptor (Shishido *et al.*, 1995; Valdizan *et al.*, 1995). The R domain of the toxin binds to the EGF-like domain of the receptor. Certain amino acid differences in this region between the murine and human sequences are the reason why murine cells do not bind the toxin and therefore mice are highly resistant to diphtheria toxin (Hooper and Eidels, 1996). The transmembrane and intracellular part of the receptor do not appear to have any role in the receptor function as they could be removed and replaced by a glycophosphoinositol (GPI) anchor without eliminating the ability of the receptor to support toxin entry (Lanzrein *et al.*, 1996).

Receptor for exotoxin A from *Pseudomonas aeruginosa*

This toxin consists of a single polypeptide chain with three different domains (Allured *et al.*, 1986). The N-terminal domain, Ia, binds to a cell-surface protein that also acts as binding site for α_2-macroglobulin and for low-density lipoprotein (Kounnas *et al.*, 1992). Domain II is almost entirely α-helical and may be involved in translocation (Wick *et al.*, 1990). The C-terminal domain III is the enzymically active part of the toxin.

Binary toxins

A number of toxins is produced as two different proteins that associate at the level of the cell surface. Among these are anthrax toxin, *Clostridium botulinum* C2 toxin, *C. perfringens iota* toxin and *C. spiroforme* toxin (Ensoli *et al.*, 1993). Anthrax toxin is the best-studied member of this group. It consists of three parts: a protein termed the protective antigen (PA), because antibodies against it protect against the toxin, the oedema factor (EF) and the lethal factor (LF). Intact PA (83 kDa) is a monomer that binds to receptors at the cell surface. The receptor so far has not been identified, but it may be a protein of 85–90 kDa (Escuyer and Collier, 1991). PA binds with a K_d = 0.9 nM to approximately 10 000 receptors per CHO cell. It is easily cleaved by proteases, such as trypsin and furin, to release a 20-kDa fragment (Figure 4.1). The emerging PA_{63} is able to bind EF (an adenylate cyclase) or LF (a metalloprotease) in a mutually exclusive manner and with high affinity ($K_d = 10$ pM) (Leppla *et al.*, 1988). Furthermore, upon exposure to low pH, PA_{63} forms a heptameric ring which inserts into the membrane, presumably forming a β-barrel (Petosa *et al.*, 1997), and forms a cation-conducting channel in the membrane (Blaustein *et al.*, 1989; Milne and Collier, 1993). Exposure to low pH is a requirement for translocation of EF and LF to the cytosol (Friedlander, 1986). The other toxins in this group also consist of a 80–100-kDa binding part corresponding to PA which is cleaved to release a 20-kDa fragment. Subsequently, the active subunit is bound in a similar way as found in anthrax toxin.

Several toxins that do not translocate an active subunit to the cytosol also form heptameric rings that insert into membranes. Among these are staphylococcal α-toxin and aerolysin (Song *et al.*, 1996). In all cases that have been studied, the heptameric protein forms an ion-conducting channel and appears to insert into the membrane as a β-barrel, as was indeed demonstrated in the case of the staphylococcal α-toxin.

Bordetella pertussis invasive adenylate cyclase

In addition to pertussis toxin mentioned above, *B. pertussis* produces an invasive adenylate cyclase. In contrast to anthrax toxin, where the separate EF moiety is an adenylate cyclase, the adenylate cyclase from *B. per-*

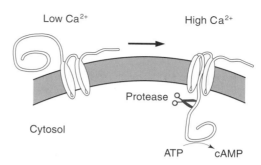

FIGURE 4.2 Translocation across the plasma membrane of the invasive adenylate cyclase, CyaA, from *Bordetella pertussis*. Calcium is not required for binding, but it is necessary for translocation of the enzymically active part. The membrane potential is also required. After translocation the enzymic part may be cleaved off and activated by calmodulin.

tussis is a single-chain 177-kDa molecule which also contains the receptor-binding part (Figure 4.2). It is a major pathogenicity factor in pertussis (Gross *et al.*, 1992; see Chapters 7 and 16, this volume). A receptor has not been identified, but the second domain, which consists of 38 glycine and aspartate-rich nonapeptide repeats, may be involved in binding to the membrane. Such repeats are also found in *E. coli* haemolysin and other toxins (RTX toxins) that bind to membranes in a calcium-dependent manner (Benz *et al.*, 1994).

The invasive adenylate cyclase also has haemolytic activity. Interestingly, like the *E. coli* haemolysin, it must be palmitoylated to obtain both haemolytic and toxic activity, but this modification is not necessary for the enzymic activity (Hackett *et al.*, 1994).

Clostridial neurotoxins

The extreme toxicity to animals and the high specificity of binding and action of tetanus and botulinum toxins on neuronal tissues indicate that the toxins must bind to highly specific receptors. The observation that tetanus toxin binds to polysialogangliosides (van Heyningen, 1974), which are widely distributed, has therefore not satisfied the expectations of the receptor. There has been an extensive search for protein receptors, but up to now only one has been found, namely that for botulinum toxin B. This receptor is a transmembrane protein, synaptotagmin, found on synaptic and other vesicles (Nishiki *et al.*, 1994). The possibility of a glycolipid receptor for clostridial neurotoxins should, however, not be excluded. Thus, Shiga toxin, which, at least in some animals, is equally as toxic as clostridial neurotoxins, binds to a widely distributed glycolipid, Gb_3. In spite of this, the toxic effect in animals is mainly seen in the microvasculature of the kidney and the central nervous system, which are particularly rich in Gb_3.

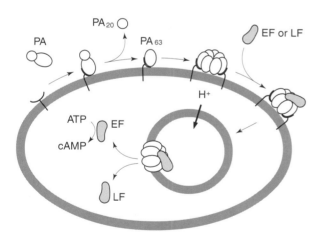

FIGURE 4.1 Schematic representation of the entry of anthrax toxin. The protective antigen (PA) binds to receptors at the cell surface and is proteolytically cleaved to release a 20-kDa fragment. The remaining part aggregates into a heptamer and forms a binding site for the oedema factor (EF) and the lethal factor (LF) that bind in a mutually exclusive manner. The complex is endocytosed and, after acidification of the vesicle, EF or LF is translocated to the cytosol.

Binding of plant toxins

A number of plant toxins, such as ricin, abrin, modeccin, viscumin and volkensin, transfers an enzymically active polypeptide into the cytosol in a manner similar to the bacterial toxins mentioned above (Olsnes *et al.*, 1974; Refsnes *et al.*, 1977; Stirpe *et al.*, 1982, 1985). They all consist of two polypeptide chains linked by a single disulfide bond. The B-chain has lectin properties and binds to carbohydrates containing terminal galactose. Defined receptors have not been identified, and it is likely that several surface glycolipids and glycoproteins can act as functional receptors for these toxins (Hughes and Gardas, 1976; Meager *et al.*, 1976; Olsnes and Refsnes, 1978).

ENDOCYTIC UPTAKE OF TOXINS

With the exception of the invasive adenylate cyclase from *Bordetella pertussis*, all toxins that act on intracellular targets appear to require endocytic uptake. The first indication for this was obtained with the plant toxins abrin and ricin, which are highly efficient in inactivating ribosomes in a cell-free system from rabbit reticulocytes. The toxins bind extensively to reticulocytes, but there was no evidence for the inhibition of protein synthesis in the intact cells. The toxins were highly active on nucleated mammalian cells. As the reticulocytes have little or no endocytosis compared to the nucleated cells, it was suggested that endocytosis is required for the toxins to enter the cytosol (Olsnes *et al.*, 1974).

That toxins are, indeed, taken up by endocytosis was observed early on. Thus, ricin was efficiently endocytosed and it was claimed that the intoxication occurs when endosomes disrupt (Nicolson, 1974). Later experiments have not confirmed this possibility. Also, diphtheria toxin was demonstrated to be endocytosed (Morris *et al.*, 1985). The problem with most of the early experiments was that it could not be decided whether the endocytic uptake was a process involved in the intoxication process, rather than an epiphenomenon.

The first indication that endocytosis is directly involved came with the observation that the toxic effect of diphtheria toxin can be inhibited by NH$_4$Cl (Kim and Groman, 1965) and that the protection can be overcome by a short exposure to medium with a low pH (Draper and Simon, 1980; Sandvig and Olsnes, 1980). NH$_4$Cl prevents acidification of intracellular vesicles at neutral, but not at acidic, pH. In addition, ionophores that are able to prevent acidification of intracellular vesicles protect the cells (Sandvig and Olsnes, 1982). Because low pH is normally found in endosomes and lysosomes, this indicated that transport to one of these organelles is required for intoxication. It was also found that protein synthesis was inhibited more rapidly in toxin-treated cells that were briefly exposed to low pH than when the low pH pulse was omitted (Sandvig and Olsnes, 1980, 1981). Further experiments demonstrated that the low pH pulse induces translocation of the toxin directly from the cell surface to the cytosol (Moskaug *et al.*, 1988), and thus induces at the level of the cell surface the process that normally occurs in the endosomes.

Endocytosis occurs partly by a process involving coated pits and coated vesicles, and partly by other less well-characterized mechanisms. In the case of diphtheria toxin (Morris *et al.*, 1985), Shiga toxin (Sandvig *et al.*, 1989), *Pseudomonas aeruginosa* exotoxin A (FitzGerald *et al.*, 1980; Morris and Saelinger, 1986) and *C. difficile* toxin (Kushnaryov and Sedmak, 1989) the major part of the uptake appears to take place by endocytosis from coated pits. In the case of both exotoxin A and Shiga toxin the receptors are diffusely distributed over the surface and, upon toxin binding and incubation at 37°C, they accumulate in coated pits and are rapidly endocytosed (FitzGerald *et al.*, 1980; Sandvig *et al.*, 1989). Such uptake can be inhibited in different ways. Treatment of the cells with hypertonic medium (Daukas and Zigmond, 1985) or by depleting the cells of potassium (Larkin *et al.*, 1983) inhibits or blocks uptake from coated pits. Another method consists of adjusting the cytoplasmic pH to one pH-unit lower than that of the medium (Sandvig *et al.*, 1987). It was found that all of these treatments are able to inhibit the toxic effect of diphtheria toxin, whereas the toxicity of ricin was affected very little (Moya *et al.*, 1985; Sandvig *et al.*, 1987), indicating that the productive uptake of ricin does not, to a large extent, occur from coated pits. In addition, cholera toxin and tetanus toxin have been reported to enter cells from non-coated areas of the surface membrane (Montesano *et al.*, 1982), possibly from caveolae (Orlandi and Fishman, 1998).

INTRACELLULAR ROUTING

As will be discussed below, diphtheria toxin and anthrax toxin are equipped with their own translocation apparatus and their translocation across the endosomal membrane is triggered by the low pH in this organelle.

Most toxins acting in the cytosol do not appear to be equipped with a translocation device and they require retrograde transport further into the cell to reach a compartment where they can translocate (Figure 4.3). Retrograde transport to the trans-Golgi network was

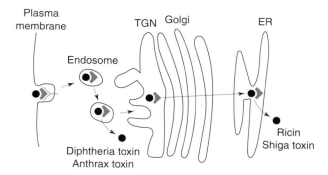

FIGURE 4.3 Intracellular transport of toxins. The toxins are taken up by endocytosis and transported retrogradely along the exocytic pathway. Diphtheria toxin and anthrax toxin are able to translocate from endosomes, whereas other toxins, such as ricin and Shiga toxin, are transported to the Golgi apparatus and further to the ER where translocation appears to take place.

first demonstrated in the case of cholera toxin (Joseph *et al.*, 1978; Gonatas *et al.*, 1980) and ricin (van Deurs *et al.*, 1986, 1987). Earlier, such transport had been observed with certain lipids, but not with proteins. Quantitation demonstrated that approximately 5% of the total endocytosed ricin reached the trans-Golgi network (van Deurs *et al.*, 1988). Later, similar findings were made with cholera toxin and pertussis toxin (el Baya *et al.*, 1997). In each case, treatment with brefeldin A of cells that are sensitive to this drug protected them against intoxication, indicating that the toxins must be transported to or beyond the Golgi apparatus.

It was observed early on that cells differ strongly in their sensitivity to Shiga toxin (Eiklid and Olsnes, 1980). A resistant cell line became sensitive to Shiga toxin after treatment overnight with butyric acid (Sandvig *et al.*, 1991). Such treatment has been extensively used to induce various genes. To check whether the treatment resulted in altered intracellular transport of the toxin, the authors carried out ultrastructural studies. Not only was the toxin visible in the Golgi region, but a considerable amount of toxin was found in the endoplasmic reticulum (ER), including the perinuclear space (Sandvig *et al.*, 1992). This was the first demonstration of protein transport from the cell surface to the ER.

That toxin transport to the ER may occur had been suggested earlier. Cholera toxin was found to contain the amino acid sequence Lys–Asp–Glu–Leu (KDEL) in the *C*-terminus of its A-chain (Chaudhary *et al.*, 1990). This sequence is recognized by a receptor found in the ER and in the intermediate or ERGIC compartment between the ER and the Golgi apparatus (Scheel and Pelham, 1996). The role of the KDEL receptor is to ensure that proteins that are marked with this sequence do not leave the ER by the exocytic pathway.

The presence of a KDEL sequence in the *C*-terminus of the A-chain of cholera toxin therefore suggested that the toxin must reach the ER to act.

Exotoxin A from *P. aeruginosa* contains a sequence that resembles KDEL, viz. RDELK. The final Lys residue is easily cleaved off by exoproteases, leaving an RDEL sequence at the *C*-terminus (Fiorentini *et al.*, 1997). This sequence is able to bind to the KDEL receptor, although with reduced affinity. When the *C*-terminus was modified such that it would not bind to the KDEL-receptor, the toxic effect was lost. Replacement with the sequence KDEL restored the toxicity (Chaudhary *et al.*, 1990). These findings suggested that exotoxin A must reach the ER to act. It had earlier been found that acidification of intracellular compartments is required (FitzGerald *et al.*, 1980; Ray and Wu, 1981). The reason for this could be that binding to the KDEL receptor requires a low pH in the Golgi apparatus. Similar to ricin, low concentrations of monensin (50 nM), that do not increase the pH of intracellular organelles, sensitized the cells to PEA (Ray and Wu, 1981). With ricin it can be shown that transport to the ER was increased under these conditions (Wesche *et al.*, in preparation).

As Shiga toxin, which is known to enter the ER, does not contain a KDEL-like sequence, it is likely that this toxin or its receptor, Gb_3, binds to a molecule that is normally transported retrograde to the ER (Johannes and Goud, 1998). This may also be the case with other toxins. We therefore studied whether ricin, which does not have a KDEL-like sequence, is transported to the ER. Although ricin is easily seen to be transported retrogradely from the cell surface to the trans-Golgi network (van Deurs *et al.*, 1986), it could never be visualized in the ER ultrastructurally. To test in a more sensitive system whether ricin is transported to the Golgi and ER, the ricin A-chain was modified to contain a tyrosine sulfation signal (Rapak *et al.*, 1997). Export proteins containing such a signal are sulfated on tyrosine residues when they reach the Golgi apparatus on their way out of the cell. Ricin coming in from the exterior and transported retrogradely to the Golgi apparatus should therefore also be sulfated if it contained such a sequence. Addition of radioactive sulfate to the medium should, in that case, result in labelling of the toxin (Figure 4.4a, b). This proved to be the case (Rapak *et al.*, 1997).

To test whether the labelled ricin is transported further backwards to the ER, glycosylation signals were also engineered into the toxin A-chain. Glycosylation can be visualized as an increase in the relative molecular mass of the protein and a corresponding reduction in the migration rate in sodium dodecyl sulfate–polyacrylamide gel electrophoresis (SDS–PAGE). Such a

(a)

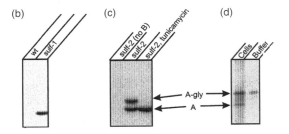

Ricin A (wt):	-CAPPPSSQF
Ricin A-sulf-1:	-CAPPP**SAEDYEYPS**
Ricin A-sulf-2:	-CAPPP**SAEDYEYPS**NGTKNNTSQ

FIGURE 4.4 Labelling of ricin with sulfate in the Golgi apparatus and subsequent toxin transport to the ER. (a) The *C*-terminus of the ricin A-chain was in one case modified to contain a tyrosine sulfation site (in bold type). In another case it was modified to contain a sulfation site as well as glycosylation sites (underlined). (b) Upon incubation with cells in the presence of radioactive sulfate, the A-chain was labelled. (c) When the A-chain contained glycosylation sites, part of it was glycosylated, indicating transport to the ER. The inhibitor of glycosylation, tunicamycin, prevented the glycosylation. (d) When the cells were permeabilized with streptolysin O, part of the glycosylated A-chain was released from the cell, indicating that it had been translocated to the cytosol.

shift was, indeed, obtained in good yield (Rapak *et al.*, 1997). The data therefore demonstrate that ricin is also transported retrograde through the Golgi apparatus to the ER. Addition of a *C*-terminal KDEL sequence to the ricin A-chain (Tagge *et al.*, 1996; Wesche *et al.*, in preparation) or Shiga toxin (Kim *et al.*, 1998) increased the rate and efficiency of transport to the ER.

It has earlier been found that treatments that disrupt the Golgi apparatus, such as brefeldin A and ilimaquinone, protect cells against ricin. This is also the case with Shiga toxin and other toxins that do not have their own translocation apparatus. Therefore, transport through the Golgi apparatus appears to be required for intoxication.

Transport from endosomes to the trans-Golgi network is interrupted at temperatures below 18°C. Under these conditions the toxic effect of ricin, Shiga toxin, *Pseudomonas* toxin and others is prevented (Eiklid and Olsnes, 1983; Sandvig *et al.*, 1984). Transient expression of mutant Sar1 and ARF1 to inhibit retrograde transport reduced the toxicity of ricin (Simpson *et al.*, 1995). In contrast, diphtheria toxin is able to intoxicate cells even under conditions that prevent vesicular transport beyond the endosomes.

The most extensive vesicular toxin transport known is that of tetanus toxin. The toxin is taken up by peripheral nerves by endocytosis and then transported retrogradely to the body of the neuron in the cen-

tral nervous system and further on to synapses that the neuron makes with other neurons. The toxin passes the synaptic cleft and is taken up by the secondary neurons (Coen *et al.*, 1997). This trans-synaptic movement apparently takes place by exocytosis of the toxin together with neurotransmitters from the primary neuron and endocytic uptake by the secondary neuron, along with emptied synaptic vesicles. It is even possible that the toxin can, in this way, pass on into a third neuron. Thus the toxin reaches its targets, which are inhibitory glycinergic neurons in the central nervous system.

Apparently, it is only in the glycinergic target neurons that the toxin is able to penetrate into the cytosol and carry out its proteolytic effect on synaptobrevin or VAMP, a small transmembrane protein involved in exocytosis. It is not clear what prevents it from intoxicating the primary motor neuron (if it did, the toxin would not be exocytosed and delivered to the secondary neuron) and how it finally gets into the cytosol of the target neuron. If, in order to translocate, the toxin must reach the ER or another intracellular organelle different from the endosome, the protein guiding it to this location could be lacking in the non-sensitive cells.

TRANSLOCATION TO THE CYTOSOL

Translocation of toxins to the cytosol appears to occur from at least three different locations: the cell surface, the endosomes and the ER. Many toxins have the ability to form ion-conducting channels through biological membranes. Such a property is observed in the case of diphtheria toxin (Kagan *et al.*, 1981), *P. aeruginosa* exotoxin A (Gambale *et al.*, 1992), tetanus toxin (Boquet and Duflot, 1982; Beise *et al.*, 1994), botulinum toxin (Donovan and Middlebrook, 1986; Blaustein *et al.*, 1987), anthrax PA (Koehler and Collier, 1991), *C. botulinum* B2 toxin (Schmid *et al.*, 1994) and *B. pertussis* invasive adenylate cyclase, CyaA (Benz *et al.*, 1994). Ion channels could not be found with Shiga toxin, cholera toxin or ricin. The latter group of toxins appears to require transport back to the ER for translocation to take place. Here they can take advantage of an established machinery designed to transport misfolded proteins from the ER to the cytosol (Wiertz *et al.*, 1996). The channels formed by toxins described here have similar properties, they are selective for cations and have low conductivity (a few tens of pS), much lower than the channel of the sec61p complex in the ER (220 pS at 45 mM potassium glutamate) involved in the export of proteins across the membrane (Simon and Blobel, 1991, 1992).

There has been much discussion as to whether the enzymically active A-moity of the toxins is translocated through the channel. The estimated diameter of the toxin channels is such that a folded protein is unlikely to pass, but after complete unfolding, passage of the protein could be possible (Kagan *et al.*, 1981). In the case of diphtheria toxin, the A fragment apparently blocks the channel, as free B fragment alone induces higher conductivity than the whole toxin (Lanzrein *et al.*, 1997).

The finding that mutations of diphtheria toxin that largely eliminate the channel properties do not prevent translocation (Falnes *et al.*, 1992) suggests that the A-moiety is not translocated through the channel. Possibly, the translocation device used by these toxins is a remodelled pore-forming toxin that has retained some of its former pore-forming capacity. In fact, the translocation domain of several toxins in this group resembles membrane-inserting domains of pore-forming toxins.

If a toxin enters from the surface, one would expect that there is no real lag time from receptor binding to the appearance of an intracellular response. This is indeed the case with the invasive adenylate cyclase from *B. pertussis*. A certain lag period, as well as protection from intoxication by drugs that inhibit organellar acidification and by drugs or temperatures (below 19°C) that prevent retrograde transport within the cells, are indications for translocation of the toxin from intracellular organelles. It should be noted that the ability of various inhibitors of organellar acidification, such as monensin and NH_4Cl, to protect against a toxin does not necessarily mean that translocation takes place from an acidified organelle. Alternatively, the toxin may have to pass through an acidified organelle to become modified before it is transported further on to the organelle (that may not be acidified) where the translocation takes place. The drugs may also interfere with retrograde transport with the result that the toxin does not reach the organelle in question. Inability to bind to the KDEL receptor at neutral pH is a possible mechanism. Also, the drugs could interfere with the antegrade transport of the receptor. Therefore, studies with inhibitors should always be supported by other kinds of evidence.

Criteria for translocation

To decide whether a toxic protein is translocated to the cytosol, assessment of the toxic effect is sufficient. However, to study the efficiency of translocation, or if more than the enzymically active part of the toxin is translocated, it is necessary to use more direct methods. For such purposes the protein can be supplied with a signal recognized by enzymes or other molecules exclusively present in the cytosol, resulting in a detectable modification of the protein. Some methods will be summarized here.

1. The protein can be supplied with a *C*-terminal farnesylation signal, a CAAX (C, cysteine; A, an aliphatic amino acid; X, methionine or certain other amino acids) box (Falnes *et al.*, 1995; Wiedlocha *et al.*, 1995). If the four *C*-terminal amino acids of a cytosolic protein form a CAAX box, enzymes in the cytosol will link a farnesyl group on to the cysteine residue with subsequent removal of the three terminal amino acids and carboxylmethylation of the exposed terminal cysteine residue (Figure 4.5). Farnesylating enzymes are present in the cytosol of eukaryotic cells and the modification can be assessed in different ways:
   (a) labelling the cells with radioactive mevalonic acid, a precursor of the farnesyl group, will label all cytosolic proteins containing a CAAX box. The

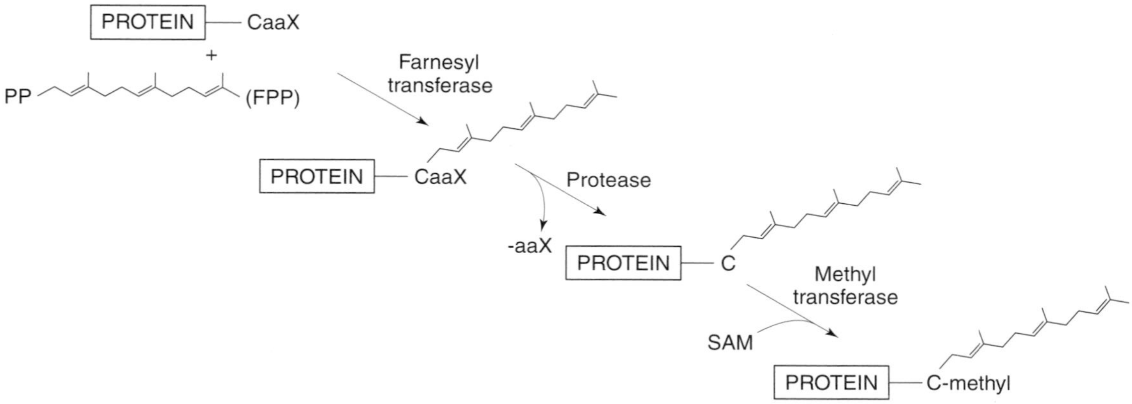

FIGURE 4.5 Farnesylation of proteins containing a *C*-terminal CAAX box. Farnesyl transferase transfers the farnesyl group from the precursor, farnesyl pyrophosphate (FPP), on to the cysteine residue of the CAAX box. The last three amino acids are cleaved off and the cysteine is subsequently carboxymethylated.

translocated protein can then be visualized by subsequent solubilization of the cells and immunoprecipitation with specific antibody, followed by SDS–PAGE

(b) the modified protein usually migrates slightly more rapidly in SDS–PAGE than the unmodified molecule (Figure 4.6) (Falnes *et al.*, 1995)

(c) the modified protein tends to partition into Triton X-114, even if the unmodified molecule does not (Falnes *et al.*, 1995).

2. A nuclear localization signal can be added to the protein. After being translocated to the cytosol, the protein may then be transported to the nucleus (Wiedlocha *et al.*, 1994). Cell fractionation followed by SDS–PAGE can be used to assess how much is present in the nucleus and, by inference, how much has been translocated into the cytosol. A caveat here is that many labelled proteins bind unspecifically to the plastic of the cell culture vessel and can be released when the cells are solubilized with a non-ionic detergent. In the subsequent centrifugation to separate the nuclear and cytosolic fractions they will often precipitate with the nuclei. To avoid this problem, treatment of the cells with trypsin or pronase is recommended, followed by addition of a protease inhibitor, before the cells are solubilized.

3. A signal for degradation by the proteasome system can be added (Falnes and Olsnes, 1998). A control lacking this signal will also be required. If the protein with the signal is rapidly degraded after translocation has been induced, this indicates a cytosolic localization, provided the control without the degradation signal is stable. The simplest modification is to place a destabilizing amino acid at the *N*-terminus according to the *N*-end rule (Figure 4.7). This can be done by expressing the protein as a fusion protein behind a cleavage site specific for a rare-cutting protease, such as factor X. In this case the *N*-terminal amino acid can be any amino acid except for proline.

4. Tyrosine phosphorylation (but not necessarily phosphorylation of serine and threonine) appears to occur selectively in the cytosol and in the nuclei. If an externally added protein carrying a tyrosine phosphorylation site becomes phosphorylated at this site, it may be taken as evidence that the protein has crossed the membrane to gain access to the cytosol or to the nucleoplasm. Even phosphorylation in a site specific for a particular intracellular kinase that phosphorylates in serine and threonine can be used if this is the only phosphorylation site in the protein.

5. Direct demonstration of the protein in the cytosol. If the plasma membrane is selectively permeabilized (e.g. with streptolysin O) proteins in the cytosol may leak out into the medium and can be recovered by immunoprecipitation (Rapak *et al.*, 1997). This is the most difficult procedure to evaluate and it is necessary to ensure that intracellular organelles, such as the ER, are not disrupted or permeabilized by the treatment.

Toxin translocation at the level of the surface membrane

Only in the case of one toxin is there evidence, so far, that the translocation occurs at the level of the plasma membrane, viz. the invasive adenylate cyclase from *B. pertussis*, CyaA. This 177-kDa protein, provided it is palmitoylated (Hackett *et al.*, 1994), binds to and apparently inserts into membranes, even at 4°C, as it cannot be washed off the membranes afterwards by carbonate at alkaline pH. At low concentrations of calcium the toxin is able to bind firmly to the membrane, but there is no evidence for intoxication of the cells until the extracellular calcium concentration is increased to millimolar levels (Rogel and Hanski, 1992). Furthermore, it is necessary to increase the temperature above 20°C for intoxication to occur. This indicates that translocation of the toxin to the cytosol requires millimolar concentrations of calcium, and that it is strongly temperature dependent (Rogel and Hanski, 1992). Interestingly, if the cells are depolarized, the toxin will not be translocated (Otero *et al.*, 1995; Szabo *et al.*, 1994), opening up the possibility that the membrane potential provides energy for the translocation.

The following sequence of events may therefore occur: after binding of the toxin to cell-surface structures, the haemolysin-like domain near the *C*-terminus inserts into the membrane and forms an ion-conducting channel (Benz *et al.*, 1994). The presence of calcium leads to a partial unfolding of the remaining part of the toxin which is then translocated to the cytoplasmic side, possibly driven by the electrical

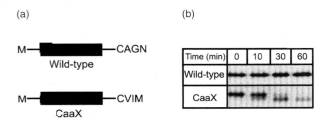

FIGURE 4.6 Farnesylation of a diphtheria toxin A fragment modified to contain a CAAX box. Upon translocation to the cytosol the wild-type A fragment remains unchanged. The A fragment with a CAAX box is modified by farnesylation, which can be detected by the more rapid migration in SDS–PAGE. It also becomes more unstable and is eventually degraded.

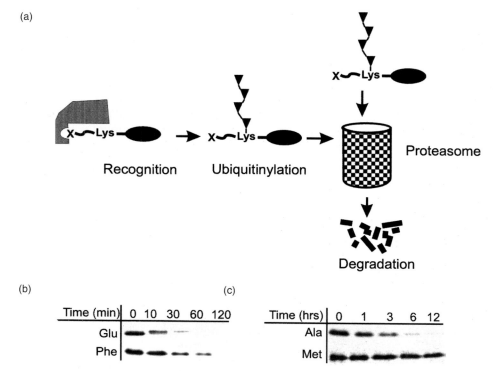

FIGURE 4.7 Degradation of the diphtheria toxin A fragment with a destabilizing *N*-terminal amino acid. (a) Enzymes in the cytosol transfer ubiquitin residues on to proteins having a destabilizing amino acid at their *N*-terminus. The ubiquitinylated proteins are degraded by the proteasomes which are also located in the cytosol. (b, c) The translocated A fragment with the stabilizing amino acid methionine is stable in the cytosol for more than 12 h. A fragments with the destabilizing amino acids phenylalanine or glutamic acid at their *N*-terminus are degraded quickly. *N*-terminal alanine provides intermediate stability.

potential across the membrane (Figure 4.2). It is not clear whether translocation occurs through the ion-conducting pore. Once in the cytosol, the catalytic part binds calmodulin, and it is possible that this binding is involved in pulling the toxin across the membrane (Oldenburg *et al.*, 1992). Finally, a 45-kDa enzymically active fragment, which is dependent on calmodulin for activity, may be released by a cytosolic protease (Rogel and Hanski, 1992).

It is interesting that the invasive adenylate cyclase is also active on sheep erythrocytes that do not have endocytic activity and that in these cells, as in other cells, the cAMP level starts to accumulate immediately upon addition of the toxin at physiological temperature and in the presence of calcium (Rogel *et al.*, 1991). In the case of cholera toxin (Gill and King, 1975) and anthrax EF (Gordon *et al.*, 1989) there is always a lag period before cAMP accumulation can be observed. Addition of calmodulin to the exterior of the erythrocytes increased the haemolytic activity of CyaA, but apparently prevented translocation to the cytosol (Rogel *et al.*, 1991).This probably means that the catalytic part of the toxin must unfold to be able to translocate, and that binding of calmodulin at the exterior inhibits the unfolding.

Translocation from endosomes: diphtheria toxin and anthrax toxin

The mechanism of translocation is best understood in the case of diphtheria toxin and anthrax toxin. As these toxins are equipped with their own translocation apparatus, the only requirement for translocation is binding to the specific receptors followed by exposure to low pH (Figures 4.1 and 4.8). Normally the low pH is obtained in endosomes and translocation therefore occurs as soon as proton pumps present in the endosome membrane have lowered the intravesicular pH sufficiently (i.e. to a value below pH 5.3) (Moskaug *et al.*, 1987, 1988; Koehler and Collier, 1991; Milne and Collier, 1993).

The two toxins, which are structurally unrelated, have adapted different, but related, mechanisms of translocation. The three-dimensional structures of diphtheria toxin and of the PA of anthrax toxin have been resolved (Choe *et al.*, 1992; Bennett *et al.*, 1994; Petosa *et al.*, 1997). The B fragment of diphtheria toxin consists of two domains: the receptor-binding domain (the R domain) and the transmembrane domain (the T domain), which resembles the presumed pore-forming

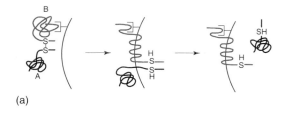

(a)

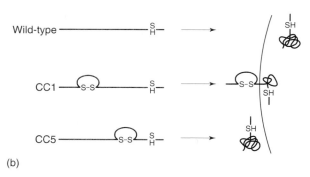

(b)

FIGURE 4.8 Model of diphtheria toxin translocation. (a) The toxin binds to its receptor by the B fragment and, upon exposure to low pH, the B fragment inserts into the membrane pulling the C-terminus of the A fragment across the membrane. Upon exposure to the cytosol, the disulfide bridge linking the two fragments is reduced and the A fragment can be fully translocated to the cytosol. (b) Insertion of an artificial disulfide in the A fragment prevents translocation. If the disulfide loop is located in the N-terminal part of the A fragment, translocation proceeds up to this point and the A fragment becomes stuck in the membrane. If it is located close to the C-terminus, the translocated part is too short to ensure anchoring of the A fragment in the membrane, which is released into the external medium.

domains of *Bacillus thuringiensis* endotoxin (Li *et al.*, 1991) and certain colicins (Parker and Pattus, 1993). Recent crystallization data of the toxin bound to its receptor demonstrated that the R domain swings out after binding, thus exposing the T domain (Louie *et al.*, 1997). The T domain is an 'inside out membrane protein' (Lesieur *et al.*, 1997) that, upon exposure to low pH, undergoes a conformational change to expose normally hidden hydrophobic helices that insert into the membrane. Helices 8 and 9, which form a hairpin, initiate the process by forming two antiparallel transmembrane helices. A glutamic acid and an aspartic acid residue in the loop between the two helices are central in this process. Being protonated at low pH, they allow the insertion of the helices into the membrane. When they reach the cytoplasmic side of the membrane, where the pH is neutral, they lose the protons and in this way lock the hairpin in the membrane (Mindell *et al.*, 1992). Mutation of these residues to lysine strongly reduced the toxicity of the molecule.

As a result of the insertion of the T domain, a cation-selective channel is formed in the membrane. The A fragment appears partly to block this channel, because when B fragment alone is used the channel has a larger conductance than that obtained with whole toxin (Lanzrein *et al.*, 1997).

Apparently initiated by the insertion of the B fragment, the A fragment is translocated across the membrane (Madshus, 1994; Madshus *et al.*, 1994; Falnes and Olsnes, 1995). When cells with diphtheria toxin bound to receptors at the cell surface are exposed to low pH, the translocation takes place across the plasma membrane. Compared to the normal situation where the A fragment is translocated from the endosomes, translocation across the plasma membrane has a number of advantages. Thus, it is possible to induce synchronous translocation, to alter the conditions of the external medium, and to differentiate between toxin that is translocated and toxin that is not. This enables tests to be carried out on how a number of conditions influences the translocation.

When cells with surface-bound labelled diphtheria toxin were exposed to low pH to induce translocation, and the cells were subsequently treated with pronase to remove untranslocated material, two protected bands were obtained. One band (21 kDa) represents the A fragment, whereas the other band (25 kDa) represents the C-terminal part of the B fragment (Moskaug *et al.*, 1988). This indicates that only part of the B fragment is translocated to a compartment where it cannot be reached by pronase. Later, when the structure was resolved (Choe *et al.*, 1992), the helical nature of the T domain suggested that it spans the membranes several times and would therefore be partly accessible to the pronase. It is more difficult to explain why the R domain is protected (Moskaug *et al.*, 1991), which could mean that this domain is translocated to the inside of the membrane. In fact, experiments with insertion of diphtheria toxin into artificial lipid membranes also indicate that the R domain is inserted into the membrane (Tortorella *et al.*, 1995; Quertenmont *et al.*, 1996).

Early experiments indicated that the translocation of diphtheria toxin requires not only a low pH in the medium, but also that the pH of the cytosol is neutral, such that a transmembrane pH gradient is present (Sandvig *et al.*, 1986; Moskaug *et al.*, 1987). An inwardly directed proton gradient of at least one pH-unit was required for efficient translocation. A pH gradient was not required for insertion of the B fragment into the membrane as low extracellular pH was sufficient for this process (Moskaug *et al.*, 1987). However, for the subsequent translocation of the A fragment an inwardly directed pH gradient was required, as was an inwardly directed anion gradient (Moskaug *et al.*, 1989).

When a disulfide bridge was introduced into the A fragment, translocation was prevented (Falnes *et al.*, 1994). Depending on where in the A fragment the disulfide was located, shorter or larger parts of the A fragment were translocated, indicating that the translocation occurred until it was stopped by the disulfide (Figure 4.8). It therefore appears that the A fragment must unfold to a considerable extent to translocate.

Possibly, the translocation occurs in the following way: insertion of the B fragment into the membrane pulls with it the A fragment, which is still disulfide-linked to the B fragment, and thus starts the translocation. Upon exposure of the disulfide to the reducing conditions in the cytosol, the disulfide is cleaved and the A fragment is free to translocate. It has been found that the A fragment partially unfolds at low pH (Blewitt *et al.*, 1985). Refolding in the neutral cytosol could pull the rest of the A fragment across. In this way, the inwardly directed pH gradient could be the driving force for the translocation.

Also, in the case of anthrax toxin, unfolding appears to be required. Although fusion proteins of the PA-binding domain of anthrax toxin lethal factor (LF_N) and diphtheria toxin A fragment were efficiently translocated at low pH, this was not the case when the fusion protein contained a disulfide in the diphtheria toxin A fragment part (Wesche *et al.*, 1998). Furthermore, a fusion protein of LF_N and dihydrofolate reductase was translocated in the absence, but not in the presence, of methotrexate. Methotrexate binds strongly to dihydrofolate reductase and prevents its unfolding. A fusion protein of LF_N and a mutant dihydrofolate reductase, that does not efficiently bind methotrexate, was translocated even in the presence of the drug. Similar data were obtained with a fusion protein of DTA and dihydrofolate reductase (Klingenberg and Olsnes, 1996).

Translocation from compartments beyond endosomes

With the exception of reports from one laboratory (Taupiac *et al.*, 1996; Beaumelle *et al.*, 1997; Alami *et al.*, 1998), most data indicate that cholera toxin, Shiga toxin, *P. aeruginosa* exotoxin A, ricin, abrin and others are not able to translocate from endosomes. Thus, the toxic effect of these proteins is inhibited when the cells are kept at temperatures below 18°C to prevent transport to the trans-Golgi network, or when the Golgi apparatus is disrupted by incubation of the cells with brefeldin A (Yoshida *et al.*, 1990) or ilimaquinone (Nambiar and Wu, 1995). There is evidence that these toxins are translocated from the ER. This, so far, has only been demonstrated in the case of ricin.

When ricin, modified to contain a tyrosine sulfation signal and glycosylation signals, was added to cells and incubated in the presence of radioactive sulfate (Figure 4.4a, c) the labelled toxin was glysosylated, indicating retrograde transport to the ER (Rapak *et al.*, 1997). To study to what extent the sulfate-labelled toxin was translocated to the cytosol, the cells were permeabilized with streptolysin O. To avoid disruption of intracellular organelles, streptolysin O was added at 0°C and the cells were then washed. Upon subsequent incubation at 37°C, pores are formed in the plasma membrane allowing cytosolic proteins to leak out into the medium (Bhakdi *et al.*, 1993). Control experiments demonstrated that intracellular organelles including the ER were not disrupted. When the medium was analysed for the presence of ricin A-chain (Figure 4.4d), it was found that mainly the glycosylated form was obtained (Rapak *et al.*, 1997). This indicates that only A-chain that has reached the ER is able to translocate to the cytosol. Presumably, the translocation takes place across the membrane of the ER, but so far it is not known by what mechanism it occurs. It is possible that unfolding is also required here. Thus, when a disulfide bridge is introduced into ricin A-chain, the toxic effect is reduced (Argent *et al.*, 1994).

PEA contains a *C*-terminal sequence (RDELK) that may bind to the KDEL receptor after removal of the terminal lysine residue. It may therefore be transported to the ER. The toxin is cleaved by furin to generate a 37-kDa fragment that is disulfide-linked to the receptor-binding domain. Upon reduction, this fragment may be released. In homogenized cells it was found in the cytosolic fraction, consistent with the possibility that it had been translocated *in vivo* (Ogata *et al.*, 1990). However, the extensive disruption of cellular organelles, such as the ER, during mechanical disruption of cells (Rapak *et al.*, 1997) makes it difficult to make this conclusion definitive.

In experiments where a signal sequence was placed in front of the 37-kDa fragment of PEA, the toxin was not translocated fully into microsomes, but only far enough to allow the signal sequence to be cleaved off by the signal peptidase that has its active site facing the ER lumen (Theuer *et al.*, 1993). The ability to abort the translocation resided in the *N*-terminal part of the protein, which represents part of the α-helical domain II believed to be important for translocation (Theuer *et al.*, 1994).

Cholera toxin and the related *E. coli* heat-labile toxin contain a KDEL and an RDEL sequence, respectively, and several reports have demonstrated transport of the toxin to the ER (Majoul *et al.*, 1996; Sandvig *et al.*, 1996). However, it appears that sequences binding to the KDEL receptor are not required for toxicity, at least not in all cases. Thus, in *E. coli* heat-labile toxin, mutations of the C-terminal RDEL sequence such that it would not be recognized by the KDEL receptor did not result in loss of toxicity (Cieplak *et al.*, 1995). Because Shiga toxin and ricin, which are both transported to the ER, do not contain a sequence that would be recognized by the KDEL receptor, it is also possible that cholera and *E. coli* heat-labile toxin are transported retrograde to the ER independent of the C-terminal sequence that could either be a back-up system for transport or have an entirely different role.

A major function of the ER is to allow translocated proteins to fold correctly before export. For this purpose the organelle contains large amounts of chaperone proteins (Helenius *et al.*, 1992). Mutated proteins that are unable to fold correctly are degraded. It was long believed that the degradation takes place in the ER, but recent data have indicated that the misfolded proteins are instead translocated back into the cytosol where they are degraded by the proteasomes (Hiller *et al.*, 1996; Knop *et al.*, 1996; Werner *et al.*, 1996; Wiertz *et al.*, 1996; Suzuki *et al.*, 1998). The reverse translocation appears to take place through the same sec61p channel that is used for translocation of proteins from the cytosol to the ER. It thus appears that the channel can operate in both directions. It is a tempting idea that toxins that reach the ER by retrograde transport are treated like misfolded proteins and translocated to the cytosol where they somehow avoid the proteasomes.

Most toxins that enter the cytosol contain a disulfide bridge that links the enzymically active polypeptide to the binding part. In the case of diphtheria toxin, this disulfide is reduced concomitantly with translocation, apparently due to exposure of the disulfide to the reducing conditions of the cytosol (Falnes and Olsnes, 1995). In the case of toxins believed to be translocated from the ER, it is not clear where and how the disulfide is reduced. One possibility is that the whole protein is translocated to the cytosol and reduced there. Alternatively, the reduction could occur in the lumen of the ER before translocation takes place. A resident ER enzyme, protein-disulfide isomerase, has been shown to reduce the disulfide of the A-chain of cholera toxin *in vitro* (Majoul *et al.*, 1997; Orlandi, 1997). It is probable that the reduction does not occur before the toxin has reached the compartment where translocation takes place. Thus, in the case of cholera toxin (Majoul *et al.*, 1996) and Shiga toxin (Garred *et al.*, 1997), it was found that when the disulfide is reduced the enzymically active A_1 fragment dissociates from the remaining part of the toxin.

A number of other toxins may also enter the cytosol from a post-endosomal compartment, but so far there is only circumstantial evidence that this is the case and the possibility should remain open that they are translocated from endosomes. The toxic effect of *C. difficile* toxin B was inhibited by monensin and ammonium chloride, indicating that transport to an acidic compartment is necessary. However, the effect of NH_4Cl could not be overcome by a short exposure to low pH (Florin and Thelestam, 1983; Caspar *et al.*, 1987). This could mean that the translocation is not induced by the low pH in the endosomes.

The binary toxin from *Clostridium spiriforme* requires endocytic uptake from coated pits in order to act. The finding that intoxication cannot be blocked by preventing acidification of the vesicles demonstrates that the translocation mechanism is different to that of diphtheria and anthrax toxins. Incubation at 15°C, which prevents traffic from the endosomes to the trans-Golgi network, did not protect against the *C. difficile* toxin, and the possibility must be considered that the toxin translocates from the endosomes, triggered by something other than low pH (Popoff *et al.*, 1989).

In the case of pertussis toxin, reducing the acidification of intracellular organelles, as well as treatment with brefeldin A, inhibited the toxic effect, suggesting that the toxin must pass through an acidified compartment and, probably, through the Golgi apparatus to be able to translocate to the cytosol (Xu and Barbieri, 1995). The release of the A-chain from the pentameric B subunit is facilitated by the presence of ATP, indicating that it occurs in an ATP-containing compartment, possibly the ER (Hazes *et al.*, 1996). A KDEL-like sequence is not present in pertussis toxin.

TOXINS AS VEHICLES TO CARRY PEPTIDES AND PROTEINS INTO THE CYTOSOL

As toxins have the unusual property of penetrating cellular membranes and entering the cytosol, the possibility of using them as vehicles to take peptides and proteins into cells has been considered by several authors. With the elucidation of the mechanism of major histocompatibility complex (MHC) class I presentation of antigens (Townsend and Bodmer, 1989), the toxins appeared as interesting candidates for directing virus and tumour antigens into the class I pathway in order to obtain MHC class I immunity (Stenmark *et al.*, 1991). A number of peptides was fused to the N-terminus of the diphtheria toxin A fragment, and the constructs were dialysed together with the B fragment to form holotoxin and then given to cells. Toxicity experiments indicated that the constructs entered the cytosol.

However, the possibility existed that the fused peptide had been cleaved off the A fragment before translocation occurred, and it was therefore necessary to demonstrate in a more direct way that the whole fusion protein had been translocated. This was carried out by incubating the cells with the reconstituted fusion protein in the cold to let binding occur, the cells were then exposed to low pH to induce translocation and, finally, non-translocated toxin was removed from the cells with pronase. It was found that the whole fusion protein was, indeed, protected. Furthermore, when the cells were permeabilized to release cytosolic proteins, the fusion protein was released, indicating that it was free in the cytosol (Stenmark et al., 1991).

Later studies also showed that whole proteins can be translocated with diphtheria toxin as a vehicle. Acidic fibroblast growth factor fused to the N-terminus of the diphtheria toxin A fragment was efficiently translocated (Wiedlocha et al., 1994). This translocation was prevented when the A fragment or the growth factor was engineered to contain an internal disulfide, or when the unfolding of the growth factor was prevented by heparin which binds tightly to the protein (Wiedlocha et al., 1994; Wesche et al., 1998). An additional diphtheria toxin A fragment added to the N-terminus of the primary A fragment was also translocated (Madshus et al., 1992).

It appears that it is also necessary for the passenger protein to be able to unfold to a considerable extent for translocation to occur. This was first observed with a fusion protein of the diphtheria toxin A fragment (DTA) and acidic fibroblast growth factor (aFGF) that was fused to the N-terminus of the toxin (Wiedlocha et al., 1994). It was found that when this construct was dialysed together with the diphtheria toxin B fragment (DTB), to allow a disulfide bridge to be formed, it could be translocated to the cytosol by the diphtheria toxin pathway upon exposure to low pH. The translocation was inhibited by various polyanions that induce tight folding of the aFGF. Also, when a disulfide bridge was introduced into the aFGF (the growth factor normally has none) translocation was inhibited (Wesche et al., submitted).

Interestingly, acidic fibroblast growth factor, which was easily translocated to the cytosol as a fusion protein with diphtheria toxin, could not be translocated when fused to LF$_N$ of anthrax toxin (Wesche et al., 1998). In fact, the majority of fusion proteins tested could not be demonstrated to reach the cytosol in intact form. Apparently, for transport by the toxin pathway there are strict requirements as to the folding of the construct and to its ability to unfold at low pH. Furthermore, the results demonstrate how important it is to study, by direct methods, that the whole fusion construct has been translocated and not only the toxic part of it.

Fusion of LF$_N$ to the enzymically active parts of diphtheria or Shiga toxin resulted in a protein that was toxic in the presence, but not in the absence, of PA (Arora and Leppla, 1994). This indicated that the fusion protein had entered the cytosol by the anthrax toxin pathway. The fusion could be at either end of LF$_N$, and in both cases highly toxic products were obtained.

When the enzymically active domain I of PEA was replaced by the bacterial RNAse, barnase, the construct exhibited toxicity at 1–10 µg/ml concentrations in cells that were resistant to PEA (Prior et al., 1992). Exoenzyme C3 from C. botulinum fused in front of diphtheria toxin B fragment was able to induce ADP-ribosylation of Rho in living cells at lower concentrations than those required for the exoenzyme alone (Aullo et al., 1993). In both cases it is not known how the translocation occurred, and it was in both cases rather inefficient.

In the invasive adenylate cyclase from B. pertussis, CyaA, it is possible to insert heterologous peptides without disturbing the translocation function. Peptides from the virus of lymphocytic choriomeningitis were inserted into this locus and, when the resulting protein was given to animals, an MHC class I response was obtained (Saron et al., 1997). It has not been directly established, so far, that peptides translocated to the cytosol by the toxin were indeed the reason for the response, as MHC class I molecules can be loaded with peptides generated in the ER. Production of cytotoxic T-lymphocytes was dependent on the injection of the fusion protein together with alum, which could mean that another mechanism is involved.

A modified PEA, where the catalytic domain had been removed and replaced by MHC-restricted epitopes from influenza virus, was able to sensitize cells to cytotoxic T-lymphocytes with the corresponding specificity (Donnelly et al., 1993). A mutated form of the toxin that had reduced ability to be cleaved by furin and therefore was less able to translocate to the cytosol was also less active in sensitizing target cells to the lymphocytes. However, later studies with various inhibitors have thrown doubt on whether the epitopes had really been translocated to the cytosol, rather than being cleaved off in the endosomes and loaded on to internalized MHC class I molecules that are recycled back to the cell surface (Ulmer et al., 1994). This again focuses attention on the importance of visualizing, by direct means, that the fusion protein is translocated to the cytosol.

A decapeptide fused to the N-terminus of cholera toxin A-chain did not prevent reassociation with the B subunit, but the toxicity was reduced approximately 10-fold, possibly owing to interference of the decapeptide with the active site of the A-chain (Sanchez *et al.*, 1997). It was not formally shown, however, that the modified A-chain did, in fact, reach the cytosol. The possibility exists that the moderate toxicity was due to molecules that had lost this epitope.

So far, the anthrax toxin system is the most promising one for the transport of MHC class I epitopes into the cytosol. A fusion protein of LF_N and gp120 from HIV was effective in sensitizing target cells for lysis by cytotoxic T-lymphocytes (Goletz *et al.*, 1997). The sensitizing effect was abolished by lactacystin, which inhibits the activity of the proteasomes, indicating that processing of the antigen by these cytoplasmic organelles is necessary. It was, however, not demonstrated that the whole fusion protein was translocated to the cytosol, and from other experiments that indicate that the protein has to unfold to translocate, some problems are presented by the fact that gp120 contains several disulfide bridges. In fact, only a few of the fusion proteins that have been tested are efficiently translocated to the cytosol. Therefore, in this case, too, more direct methods for assaying translocation are warranted.

Another approach was taken by Ballard *et al.* (1996), who fused a nonapeptide from *Listeria monocytogenes* to LF_N and demonstrated protection against infection.

It appears from the data discussed here that it is possible to use toxins as carriers to take passenger peptides and proteins into cells. The requirements for unfolding and non-interference of the passenger polypeptide with the translocation mechanisms restrict, however, the usefulness of toxins for this purpose. It is therefore necessary to ensure that the fusion protein in question really is translocated to the cytosol before firm conclusions can be made.

ACKNOWLEDGEMENTS

This work was supported by the Norwegian Cancer Society, the Novo Nordisk Foundation, the Norwegian Research Council, Blix Fund for the Promotion of Medical Research, Rachel and Otto Kr. Bruun's legat, and the Jahre Foundation.

REFERENCES

Aktories, K., Bärmann, M., Ohishi, I., Tsuyama, S., Jakobs, K.H. and Habermann, E. (1986). Botulinum C2 toxin ADP-ribosylates actin. *Nature* **322**, 390–2.

Aktories, K., Mohr, C. and Koch, G. (1992). *Clostridium botulinum* C3 ADP-ribosyltransferase. *Curr. Top. Microbiol. Immunol.* **175**, 115–31.

Alami, M., Taupiac, M.P., Reggio, H., Bienvenue, A. and Beaumelle, B. (1998). Involvement of ATP-dependent *Pseudomonas* exotoxin translocation from a late recycling compartment in lymphocyte intoxication procedure. *Mol. Biol. Cell* **9**, 387–402.

Allured, V.S., Collier, R.J., Carroll, S.F. and McKay, D.B. (1986). Structure of exotoxin A of *Pseudomonas aeruginosa* at 3.0-Angstrom resolution. *Proc. Natl Acad. Sci. USA* **83**, 1320–4.

Argent, R.H., Roberts, L.M., Wales, R., Robertus, J.D. and Lord, J.M. (1994). Introduction of a disulfide bond into ricin A chain decreases the cytotoxicity of the ricin holotoxin. *J. Biol. Chem.* **269**, 26705–10.

Arora, N. and Leppla, S.H. (1994). Fusions of anthrax toxin lethal factor with shiga toxin and diphtheria toxin enzymic domains are toxic to mammalian cells. *Infect. Immun.* **62**, 4955–61.

Aullo, P., Giry, M., Olsnes, S., Popoff, M.R., Kocks, C. and Boquet, P. (1993). A chimeric toxin to study the role of the 21 kDa GTP binding protein rho in the control of actin microfilament assembly. *EMBO J.* **12**, 921–31.

Ballard, J.D., Collier, R.J. and Starnbach, M.N. (1996). Anthrax toxin-mediated delivery of a cytotoxic T-cell epitope *in vivo*. *Proc. Natl Acad. Sci. USA* **93**, 12531–4.

Beaumelle, B., Taupiac, M.P., Lord, J.M. and Roberts, L.M. (1997). Ricin A chain can transport unfolded dihydrofolate reductase into the cytosol. *J. Biol. Chem.* **272**, 22097–102.

Beise, J., Hahnen, J., Andersen-Beckh, B. and Dreyer, F. (1994). Pore formation by tetanus toxin, its chain and fragments in neuronal membranes and evaluation of the underlying motifs in the structure of the toxin molecule. *Naunyn-Schmiedebergs Arch. Pharmacol.* **349**, 66–73.

Bennett, M.J., Choe, S. and Eisenberg, D. (1994). Refined structure of dimeric diphtheria toxin at 2.0 Å resolution. *Protein Sci.* **3**, 1444–63.

Benz, R., Maier, E., Ladant, D., Ullmann, A. and Sebo, P. (1994). Adenylate cyclase toxin (CyaA) of *Bordetella pertussis*. Evidence for the formation of small ion-permeable channels and comparison with HlyA of *Escherichia coli*. *J. Biol. Chem.* **269**, 27231–9.

Bhakdi, S., Weller, U., Walev, I., Martin, E., Jonas, D. and Palmer, M. (1993). A guide to the use of pore-forming toxins for controlled permeabilization of cell membranes. *Med. Microbiol. Immunol. (Berl.)* **182**, 167–75.

Blaustein, R.O., Germann, W.J., Finkelstein, A. and DasGupta, B.R. (1987). The N-terminal half of the heavy chain of botulinum type A neurotoxin forms channels in planar phospholipid bilayers. *FEBS Lett.* **226**, 115–20.

Blaustein, R.O., Koehler, T.M., Collier, R.J. and Finkelstein, A. (1989). Anthrax toxin: channel-forming activity of protective antigen in planar phospholipid bilayers. *Proc. Natl Acad. Sci. USA* **86**, 2209–13.

Blewitt, M.G., Chung, L.A. and London, E. (1985). Effect of pH on the conformation of diphtheria toxin and its implications for membrane penetration. *Biochemistry* **24**, 5458–64.

Boquet, P. and Duflot, E. (1982). Tetanus toxin fragment forms channels in lipid vesicles at low pH. *Proc. Natl Acad. Sci. USA* **79**, 7614–18.

Brennan, M.J., David, J.L., Kenimer, J.G. and Manclark, C.R. (1988). Lectin-like binding of pertussis toxin to a 165-kilodalton Chinese hamster ovary cell glycoprotein. *J. Biol. Chem.* **263**, 4895–9.

Burgen, A.S.V., Dickens, F. and Zatman, L.J. (1949). The action of botulinum toxin on the neuro-muscular junction. *J. Physiol. (Lond.)* **109**, 10–24.

Caspar, M., Florin, I. and Thelestam, M. (1987). Calcium and calmodulin in cellular intoxication with *Clostridium difficile* toxin B. *J. Cell Physiol.* **132**, 168–72.

Chardin, P., Boquet, P., Madaule, P., Popoff, M.R., Rubin, E.J. and Gill, D.M. (1989). The mammalian G protein rhoC is ADP-ribosylated by *Clostridium botulinum* exoenzyme C3 and affects actin microfilaments in Vero cells. *EMBO J.* **8**, 1087–92.

Chaudhary, V.K., Jinno, Y., FitzGerald, D. and Pastan, I. (1990). *Pseudomonas* exotoxin contains a specific sequence at the carboxyl terminus that is required for cytotoxicity. *Proc. Natl Acad. Sci. USA* **87**, 308–12.

Choe, S., Bennett, M.J., Fujii, G., Curmi, P.M., Kantardjieff, K.A., Collier, R.J. and Eisenberg, D. (1992). The crystal structure of diphtheria toxin. *Nature* **357**, 216–22.

Cieplak, W., Messer, R.J., Konkel, M.E. and Grant, C.C.R. (1995). Role of a potential endoplasmic reticulum retention sequence (RDEL) and the Golgi complex in the cytotonic activity of *Escherichia coli* heat-labile enterotoxin. *Mol. Microbiol.* **16**, 789–800.

Clark, C.G. and Armstrong, G.D. (1990). Lymphocyte receptors for pertussis toxin. *Infect. Immun.* **58**, 3840–6.

Coen, L., Osta, R., Maury, M. and Brulet, P. (1997). Construction of hybrid proteins that migrate retrogradely and transynaptically into the central nervous system. *Proc. Natl Acad. Sci. USA* **94**, 9400–5.

Collier, R.J. (1967). Effect of diphtheria toxin on protein synthesis: inactivation of one of the transfer factors. *J. Mol. Biol.* **25**, 83–98.

Collier, R.J. (1975). Diphtheria toxin: mode of action and structure. *Bacteriol. Rev.* **39**, 54–85.

Daukas, A. and Zigmond, S.H. (1985). Inhibition of receptor-mediated but not fluid-phase endocytosis in polymorphonuclear leukocytes. *J. Cell Biol.* **101**, 1673–9.

DeGrandis, S., Law, H., Brunton, J., Gyles, C. and Lingwood, C.A. (1989). Globotetraosylceramide is recognized by the pig edema disease toxin. *J. Biol. Chem.* **264**, 12520–5.

Donnelly, J.J., Ulmer, J.B., Hawe, L.A., Friedman, A., Shi, X.P., Leander, K.R., Shiver, J.W., Oliff, A.I., Martinez, D., Montgomery, D. and Liu, M.A. (1993). Targeted delivery of peptide epitopes to class I major histocompatibility molecules by a modified *Pseudomonas* exotoxin. *Proc. Natl Acad. Sci. USA* **90**, 3530–4.

Donohue-Rolfe, A., Keusch, G.T., Edson, C., Thorley-Lawson, D. and Jacewicz, M. (1984). Pathogenesis of Shigella diarrhea. IX. Simplified high yield purification of Shigella toxin and characterization of subunit composition and function by the use of subunit-specific monoclonal and polyclonal antibodies. *J. Exp. Med.* **160**, 1767–81.

Donovan, J.J. and Middlebrook, J.L. (1986). Ion-conducting channels produced by botulinum toxin in planar lipid membranes. *Biochemistry* **25**, 2872–6.

Draper, R.K. and Simon, M.I. (1980). The entry of diphtheria toxin into the mammalian cell cytoplasm: evidence for lysosomal involvement. *J. Cell Biol.* **87**, 849–54.

Duesbery, N.S., Webb, C.P., Leppla, S.H., Gordon, V.M., Klimpel, K.R., Copeland, T.D., Ahn, N.G., Oskarsson, M.K., Fukasawa, K., Paull, K.D. and Vande Woude, G.F. (1998). Proteolytic inactivation of MAP-kinase-kinase by anthrax lethal factor. *Science* **280**, 734–7.

Eiklid, K. and Olsnes, S. (1980). Interaction of *Shigella shigae* cytotoxin with receptors on sensitive and insensitive cells. *J. Recept. Res.* **1**, 199–213.

Eiklid, K. and Olsnes, S. (1983). Entry of *Shigella dysenteriae* toxin into HeLa cells. *Infect. Immun.* **42**, 771–7.

el Baya, A., Linnemann, R., von Olleschik-Elbheim, L., Robenek, H. and Schmidt, M.A. (1997). Endocytosis and retrograde transport of pertussis toxin to the Golgi complex as a prerequisite for cellular intoxication. *Eur. J. Cell Biol.* **73**, 40–8.

Endo, Y. and Tsurugi, K. (1987). RNA *N*-glycosidase activity of ricin A-chain. Mechanism of action of the toxic lectin ricin on eukaryotic ribosomes. *J. Biol. Chem.* **262**, 8128–30.

Endo, Y., Mitsui, K., Motizuki, M. and Tsurugi, K. (1987). The mechanism of action of ricin and related toxic lectins on eukaryotic ribosomes. The site and the characteristics of the modification in 28 S ribosomal RNA caused by the toxins. *J. Biol. Chem.* **262**, 5908–12.

Endo, Y., Tsurugi, K., Yutsudo, T., Takeda, Y., Ogasawara, T. and Igarashi, K. (1988). Site of action of a Vero toxin (VT2) from *Escherichia coli* O157: H7 and of Shiga toxin on eukaryotic ribosomes. RNA *N*-glycosidase activity of the toxins. *Eur. J. Biochem.* **171**, 45–50.

Ensoli, B., Buonaguro, L., Barillari, G., Fiorelli, V., Gendelman, R., Morgan, R.A., Wingfield, P. and Gallo, R.C. (1993). Release, uptake, and effects of extracellular human immunodeficiency virus type 1 Tat protein on cell growth and viral transactivation. *J. Virol.* **67**, 277–87.

Escuyer, V. and Collier, R.J. (1991). Anthrax protective antigen interacts with a specific receptor on the surface of CHO-K1 cells. *Infect. Immun.* **59**, 3381–6.

Falnes, P.O. and Olsnes, S. (1995). Cell-mediated reduction and incomplete membrane translocation of diphtheria toxin mutants with internal disulfides in the A fragment. *J. Biol. Chem.* **270**, 20787–93.

Falnes, P.O. and Olsnes, S. (1998). Modulation of the intracellular stability and toxicity of diphtheria toxin through degradation by the N-end rule pathway. *EMBO J.* **17**, 615–25.

Falnes, P.O., Madshus, I.H., Sandvig, K. and Olsnes, S. (1992). Replacement of negative by positive charges in the presumed membrane-inserted part of diphtheria toxin B fragment. Effect on membrane translocation and on formation of cation channels. *J. Biol. Chem.* **267**, 12284–90.

Falnes, P.O., Choe, S., Madshus, I.H., Wilson, B.A. and Olsnes, S. (1994). Inhibition of membrane translocation of diphtheria toxin A-fragment by internal disulfide bridges. *J. Biol. Chem.* **269**, 8402–7.

Falnes, P.O., Wiedlocha, A., Rapak, A. and Olsnes, S. (1995). Farnesylation of CaaX-tagged diphtheria toxin A-fragment as a measure of transfer to the cytosol. *Biochemistry* **34**, 11152–9.

Fiorentini, C., Fabbri, A., Matarrese, P., Falzano, L., Boquet, P. and Malorni, W. (1997). Hindrance of apoptosis and phagocytic behavior induced by *Escherichia coli* cytotoxic necrotizing factor 1: two related activities in epithelial cells. *Biochem. Biophys. Res. Commun.* **241**, 341–6.

FitzGerald, D., Morris, R.E. and Saelinger, C.B. (1980). Receptor-mediated internalization of *Pseudomonas* toxin by mouse fibroblasts. *Cell* **21**, 867–73.

Flatau, G., Lemichez, E., Gauthier, M., Chardin, P., Paris, S., Florentini, C. and Boquet, P. (1997). Toxin-induced activation of the G protein p21 Rho by deamidation of glutamine. *Nature* **387**, 729–33.

Florin, I. and Thelestam, M. (1983). Internalization of *Clostridium difficile* cytotoxin into cultured human lung fibroblasts. *Biochim. Biophys. Acta* **763**, 383–92.

Friedlander, A.M. (1986). Macrophages are sensitive to anthrax lethal toxin through an acid-dependent process. *J. Biol. Chem.* **261**, 7123–6.

Gambale, F., Rauch, G., Belmonte, G. and Menestrina, G. (1992). Properties of *Pseudomonas aeruginosa* exotoxin A ionic channel incorporated in planar lipid bilayers. *FEBS Lett.* **306**, 41–5.

Garred, O., Dubinina, E., Polesskaya, A., Olsnes, S., Kozlov, J. and Sandvig, K. (1998). Role of the disulfide bond in Shiga toxin A-chain for toxin entry into cells. *J. Biol. Chem.* **272**, 11414–19.

Gill, D.M. (1975). Involvement of nicotinamide adenine dinucleotide in the action of cholera toxin *in vitro*. *Proc. Natl Acad. Sci. USA* **72**, 2064–8.

Gill, D.M. (1976). The arrangement of subunits in cholera toxin. *Biochemistry* **15**, 1242–8.

Gill, D.M. and King, C.A. (1975). The mechanism of action of cholera toxin in pigeon erythrocyte lysates. *J. Biol. Chem.* **250**, 6424–32.

Gill, D.M. and Meren, R. (1978). ADP-ribosylation of membrane proteins catalyzed by cholera toxin: basis of the activation of adenylate cyclase. *Proc. Natl Acad. Sci. USA* **75**, 3050–4.

Goletz, T.J., Klimpel, K.R., Arora, N., Leppla, S.H., Keith, J.M. and Berzofsky, J.A. (1997). Targeting HIV proteins to the major histocompatibility complex class I processing pathway with a novel gp120-anthrax toxin fusion protein. *Proc. Natl Acad. Sci. USA* **94**, 12059–64.

Gonatas, J., Stieber, A., Olsnes, S. and Gonatas, N.K. (1980). Pathways involved in fluid phase and adsorptive endocytosis in neuroblastoma. *J. Cell Biol.* **87**, 579–88.

Gordon, V.M., Young, W.W., Jr, Lechler, S.M., Gray, M.C., Leppla, S.H. and Hewlett, E.L. (1989). Adenylate cyclase toxins from *Bacillus anthracis* and *Bordetella pertussis*. Different processes for interaction with and entry into target cells. *J. Biol. Chem.* **264**, 14792–6.

Gross, M.K., Au, D.C., Smith, A.L. and Storm, D.R. (1992). Targeted mutations that ablate either the adenylate cyclase or hemolysin function of the bifunctional cyaA toxin of *Bordetella pertussis* abolish virulence. *Proc. Natl Acad. Sci. USA* **89**, 4898–902.

Hackett, M., Guo, L., Shabanowitz, J., Hunt, D.F. and Hewlett, E.L. (1994). Internal lysine palmitoylation in adenylate cyclase toxin from *Bordetella pertussis*. *Science* **266**, 433–5.

Hazes, B., Boodhoo, A., Cockle, S.A. and Read, R.J. (1996). Crystal structure of the pertussis toxin–ATP complex: a molecular sensor. *J. Mol. Biol.* **258**, 661–71.

Helenius, A., Marquardt, T. and Braakman, I. (1992). The endoplasmic reticulum as a protein-folding compartment. *Trends Cell Biol.* **2**, 227–31.

Hiller, M., Finger, A., Schweiger, M. and Wolf, D. (1996). ER degradation of a misfolded luminal protein by the cytosolic ubiquitin–proteasome pathway. *Science* **273**, 1725–8.

Hofmann, F., Busch, C., Prepens, U., Just, I. and Aktories, K. (1997). Localization of the glucosyltransferase activity of *Clostridium difficile* toxin B to the N-terminal part of the holotoxin. *J. Biol. Chem.* **272**, 11074–8.

Holmberg, L. and Nygård, O. (1996). Depurination of A4256 in 28 S rRNA by the ribosome-inactivating proteins from barley and ricin results in different ribosome conformations. *J. Mol. Biol.* **259**, 81–94.

Hooper, K.P. and Eidels, L. (1996). Glutamic acid 141 of the diphtheria toxin receptor (HB-EGF precursor) is critical for toxin binding and toxin sensitivity. *Biochem. Biophys. Res. Commun.* **220**, 675–80.

Hughes, R.C. and Gardas, A. (1976). Phenotypic reversion of ricin-resistant hamster fibroblasts to a sensitive state after coating with glycolipid receptors. *Nature* **264**, 63–6.

Iglewski, B.H. and Kabat, D. (1975). NAD-dependent inhibition of protein synthesis by *Pseudomonas aeruginosa* toxin. *Proc. Natl Acad. Sci. USA* **72**, 2284–8.

Inoue, S., Sugai, M., Murooka, Y., Paik, S.Y., Hong, Y.M., Ohgai, H. and Suginaka, H. (1991). Molecular cloning and sequencing of the epidermal cell differentiation inhibitor gene from *Staphylococcus aureus*. *Biochem. Biophys. Res. Commun.* **174**, 459–64.

Johannes, L. and Goud, B. (1998). Surfing on a retrograde wave: how does Shiga toxin reach the endoplasmic reiculum? *Trends Cell Biol.* **8**, 158–62.

Joseph, K.C., Kim, S.U., Stieber, A. and Gonatas, N.K. (1978). Endocytosis of cholera toxin into neuronal GERL. *Proc. Natl Acad. Sci. USA* **75**, 2815–19.

Just, I., Mohr, C., Schallehn, G., Menard, L., Didsbury, J.R., Vandekerckhove, J., van Damme, J. and Aktories, K. (1992a). Purification and characterization of an ADP-ribosyltransferase produced by *Clostridium limosum*. *J. Biol. Chem.* **267**, 10274–80.

Just, I., Schallehn, G. and Aktories, K. (1992b). ADP-ribosylation of small GTP-binding proteins by *Bacillus cereus*. *Biochem. Biophys. Res. Commun.* **183**, 931–6.

Just, I., Selzer, J., Wilm, M., von Eichel-Streiber, C., Mann, M. and Aktories, K. (1995a). Glucosylation of Rho proteins by *Clostridium difficile* toxin B. *Nature* **375**, 500–3.

Just, I., Wilm, M., Selzer, J., Rex, G., von Eichel-Streiber, C., Mann, M. and Aktories, K. (1995b). The enterotoxin from *Clostridium difficile* (ToxA) monoglucosylates the Rho proteins. *J. Biol. Chem.* **270**, 13932–6.

Kagan, B.L., Finkelstein, A. and Colombini, M. (1981). Diphtheria toxin fragment forms large pores in phospholipid bilayer membranes. *Proc. Natl Acad. Sci. USA* **78**, 4950–4.

Kahn, R.A. and Gilman, A.G. (1984). Purification of a protein cofactor required for ADP-ribosylation of the stimulatory regulatory component of adenylate cyclase by cholera toxin. *J. Biol. Chem.* **259**, 6228–34.

Katada, T. and Ui, M. (1982). Direct modification of the membrane adenylate cyclase system by islet-activating protein due to ADP-ribosylation of a membrane protein. *Proc. Natl Acad. Sci. USA* **79**, 3129–33.

Katada, T., Oinuma, M. and Ui, M. (1986). Two guanine nucleotide-binding proteins in rat brain serving as the specific substrate of islet-activating protein, pertussis toxin. Interaction of the alpha-subunits with beta gamma-subunits in development of their biological activities. *J. Biol. Chem.* **261**, 8182–91.

Kim, J.H., Johannes, L., Goud, B., Antony, C., Lingwood, C.A., Daneman, R. and Grinstein, S. (1998). Noninvasive measurement of the pH of the endoplasmic reticulum at rest and during calcium release. *Proc. Natl Acad. Sci. USA* **95**, 2997–3002.

Kim, K. and Groman, N.B. (1965). *In vitro* inhibition of diphtheria toxin action by ammonium salts and amines. *J. Bacteriol.* **90**, 1552–6.

Klingenberg, O. and Olsnes, S. (1996). Ability of methotrexate to inhibit translocation to the cytosol of dihydrofolate reductase fused to diphtheria toxin. *Biochem. J.* **313**, 647–53.

Knop, M., Finger, A., Braun, T., Hellmuth, K. and Wolf, D.H. (1996). Der1, a novel protein specifically required for endoplasmic reticulum degradation in yeast. *EMBO J.* **15**, 753–63.

Koehler, T.M. and Collier, R.J. (1991). Anthrax toxin protective antigen: low-pH-induced hydrophobicity and channel formation in liposomes. *Mol. Microbiol.* **5**, 1501–6.

Kounnas, M.Z., Morris, R.E., Thompson, M.R., FitzGerald, D.J., Strickland, D.K. and Saelinger, C.B. (1992). The alpha 2-macroglobulin receptor/low density lipoprotein receptor-related protein binds and internalizes *Pseudomonas* exotoxin A. *J. Biol. Chem.* **267**, 12420–3.

Kushnaryov, V.M. and Sedmak, J.J. (1989). Effect of *Clostridium difficile* enterotoxin A on ultrastructure of Chinese hamster ovary cells. *Infect. Immun.* **57**, 3914–21.

Lanzrein, M., Sand, O. and Olsnes, S. (1996). GPI-anchored diphtheria toxin receptor allows membrane translocation of the toxin without detectable ion channel activity. *EMBO J.* **15**, 725–34.

Lanzrein, M., Falnes, P., Sand, O. and Olsnes, S. (1997). Structure–function relationship of the ion channel formed by diphtheria toxin in Vero cell membranes. *J. Membr. Biol.* **156**, 141–8.

Larkin, J.M., Brown, M.S., Goldstein, J.L. and Anderson, R.G.W. (1983). Depletion of intracellular potassium arrests coated pit formation and receptor-mediated endocytosis in fibroblasts. *Cell* **33**, 273–85.

Lemichez, E., Flatau, G., Bruzzone, M., Boquet, P. and Gauthier, M. (1997). Molecular localization of the *Escherichia coli* cytotoxic necrotizing factor CNF1 cell-binding and catalytic domains. *Mol. Microbiol.* **24**, 1061–70.

Leppla, S.H. (1982). Anthrax toxin edema factor: a bacterial adenylate cyclase that increases cyclic AMP concentrations of eukaryotic cells. *Proc. Natl Acad. Sci. USA* **79**, 3162–6.

Leppla, S.H., Friedlander, A.M. and Cora, E.M. (1988). Proteolytic activation of anthrax toxin bound to cellular receptors. *Zentbl. Bakteriol.* Suppl. 17, 111–12.

Lesieur, C., Vecsey-Semjen, B., Abrami, L., Fivaz, M. and Gisou van der Goot, F. (1997). Membrane insertion: the strategies of toxins. *Mol. Membr. Biol.* **14**, 45–64.

Li, J.D., Carroll, J. and Ellar, D.J. (1991). Crystal structure of insecticidal delta-endotoxin from *Bacillus thuringiensis* at 2.5 Å resolution. *Nature* **353**, 815–21.

Ling, H., Boodhoo, A., Hazes, B., Cummings, M.D., Armstrong, G.D., Brunton, J.L. and Read, R.J. (1998). Structure of the Shiga-like toxin-I B-pentamer complexed with an analog of its receptor GB3. *Biochemistry* **37**, 1777–88.

Louie, G.V., Yang, W., Bowman, M.E. and Choe, S. (1997). Crystal structure of the complex of diphtheria toxin with an extracellular fragment of its receptor. *Mol. Cell* **1**, 67–78.

MacKenzie, C.R., Hirama, T., Lee, K.K., Altman, E. and Young, N.M. (1997). Quantitative analysis of bacterial toxin affinity and specificity for glycolipid receptors by surface plasmon resonance. *J. Biol. Chem.* **272**, 5533–8.

Madshus, I.H. (1994). The *N*-terminal alpha-helix of fragment B of diphtheria toxin promotes translocation of fragment A into the cytoplasm of eukaryotic cells. *J. Biol. Chem.* **269**, 17723–9.

Madshus, I.H., Olsnes, S. and Stenmark, H. (1992). Membrane translocation of diphtheria toxin carrying passenger protein domains. *Infect. Immun.* **60**, 3296–302.

Madshus, I.H., Wiedlocha, A. and Sandvig, K. (1994). Intermediates in translocation of diphtheria toxin across the plasma membrane. *J. Biol. Chem.* **269**, 4648–52.

Majoul, I.V., Bastiaens, P.I.H. and Söling, H.D. (1996). Transport of an external Lys–Asp–Glu–Leu (KDEL) protein from the plasma membrane to the endoplasmic reticulum: studies with cholera toxin in Vero cells. *J. Cell Biol.* **133**, 777–89.

Majoul, I., Ferrari, D. and Söling, H.D. (1997). Reduction of protein disulfide bonds in an oxidizing environment – the disulfide bridge of cholera toxin A-subunit is reduced in the endoplasmic reticulum. *FEBS Lett.* **401**, 104–8.

Meager, A., Ungkitchanukit, A. and Hughes, R.C. (1976). Variants of hamster fibroblasts resistant to *Ricinus communis* toxin (ricin). *Biochem. J.* **154**, 113–24.

Merritt, E.A., Sarfaty, S., van den Akker, F., L'Hoir, C., Martial, J.A. and Hol, W.G. (1994a). Crystal structure of cholera toxin B-pentamer bound to receptor GM1 pentasaccharide. *Protein Sci.* **3**, 166–75.

Merritt, E.A., Sixma, T.K., Kalk, K.H., van Zanten, B.A. and Hol, W.G. (1994b). Galactose-binding site in *Escherichia coli* heat-labile enterotoxin (LT) and cholera toxin (CT). *Mol. Microbiol.* **13**, 745–53.

Milne, J.C. and Collier, R.J. (1993). pH-dependent permeabilization of the plasma membrane of mammalian cells by anthrax protective antigen. *Mol. Microbiol.* **10**, 647–53.

Mindell, J.A., Silverman, J.A., Collier, R.J. and Finkelstein, A. (1992). Locating a residue in the diphtheria toxin channel. *Biophys. J.* **62**, 41–4.

Moazed, D., Robertson, J.M. and Noller, H.F. (1988). Interaction of elongation factors EF-G and EF-Tu with a conserved loop in 23S RNA. *Nature* **334**, 362–4.

Montecucco, C. and Schiavo, G. (1993). Tetanus and botulism neurotoxins: a new group of zinc proteases. *Trends Biochem. Sci.* **18**, 324–7.

Montesano, R., Roth, J., Robert, A. and Orci, L. (1982). Non-coated membrane invaginations are involved in binding internalization of cholera and tetanus toxins. *Nature* **296**, 651–3.

Morris, R.E. and Saelinger, C.B. (1986). Reduced temperature alters *Pseudomonas* exotoxin A entry into the mouse LM cell. *Infect. Immun.* **52**, 445–53.

Morris, R.E., Gerstein, A.S., Bonventre, P.F. and Saelinger, C.B. (1985). Receptor-mediated entry of diphtheria toxin into monkey kidney (Vero) cells: electron microscopic evaluation. *Infect. Immun.* **50**, 721–7.

Moskaug, J.O., Sandvig, K. and Olsnes, S. (1987). Cell-mediated reduction of the interfragment disulfide in nicked diphtheria toxin. A new system to study toxin entry at low pH. *J. Biol. Chem.* **262**, 10339–45.

Moskaug, J.O., Sandvig, K. and Olsnes, S. (1988). Low pH-induced release of diphtheria toxin A-fragment in Vero cells. Biochemical evidence for transfer to the cytosol. *J. Biol. Chem.* **263**, 2518–25.

Moskaug, J.O., Sandvig, K. and Olsnes, S. (1989). Role of anions in low pH-induced translocation of diphtheria toxin. *J. Biol. Chem.* **264**, 11367–72.

Moskaug, J.O., Stenmark, H. and Olsnes, S. (1991). Insertion of diphtheria toxin B-fragment into the plasma membrane at low pH. Characterization and topology of inserted regions. *J. Biol. Chem.* **266**, 2652–9.

Moss, J. and Vaughan, M. (1995). Structure and function of ARF proteins: activators of cholera toxin and critical components of intracellular vesicular transport processes. *J. Biol. Chem.* **270**, 12327–30.

Moya, M., Dautry-Varsat, A., Goud, B., Louvard, D. and Boquet, P. (1985). Inhibition of coated pit formation in Hep2 cells blocks the cytotoxicity of diphtheria toxin but not that of ricin toxin. *J. Cell Biol.* **101**, 548–59.

Naglich, J.G., Metherall, J.E., Russell, D.W. and Eidels, L. (1992). Expression cloning of a diphtheria toxin receptor: identity with a heparin-binding EGF-like growth factor precursor. *Cell* **69**, 1051–61.

Nambiar, M.P. and Wu, H.C. (1995). Ilimaquinone inhibits the cytotoxicities of ricin, diphtheria toxin, and other protein toxins in Vero cells. *Exp. Cell Res.* **219**, 671–8.

Nicolson, G.L. (1974). Ultrastructural analysis of toxin binding and entry into mammalian cells. *Nature* **251**, 628–30.

Nishiki, T., Kamata, Y., Nemoto, Y., Omori, A., Ito, T., Takahashi, M. and Kozaki, S. (1994). Identification of protein receptor for *Clostridium botulinum* type B neurotoxin in rat brain synaptosomes. *J. Biol. Chem.* **269**, 10498–503.

Ogata, M., Chaudhary, V.K., Pastan, I. and FitzGerald, D.J. (1990). Processing of *Pseudomonas* exotoxin by a cellular protease results in the generation of a 37,000-Da toxin fragment that is translocated to the cytosol. *J. Biol. Chem.* **265**, 20678–85.

Oldenburg, D.J., Gross, M.K., Wong, C.S. and Storm, D.R. (1992). High-affinity calmodulin binding is required for the rapid entry of *Bordetella pertussis* adenylyl cyclase into neuroblastoma cells. *Biochemistry* **31**, 8884–91.

Olsnes, S. and Refsnes, K. (1978). On the mechanism of toxin resistance in cell variants resistant to abrin and ricin. *Eur. J. Biochem.* **88**, 7–15.

Olsnes, S., Heiberg, R. and Pihl, A. (1973). Inactivation of eucaryotic ribosomes by the toxic plant proteins abrin and ricin. *Mol. Biol. Rep.* **1**, 15–20.

Olsnes, S., Refsnes, K. and Pihl, A. (1974). Mechanism of action of the toxic lectins abrin and ricin. *Nature* **249**, 627–31.

Olsnes, S., Fernandez-Puentes, C., Carrasco, L. and Vazquez, D. (1975). Ribosome inactivation by the toxic lectins abrin and ricin. Kinetics of the enzymic activity of the toxin A-chains. *Eur. J. Biochem.* **60**, 281–8.

Olsnes, S., Reisbig, R. and Eiklid, K. (1981). Subunit structure of Shigella cytotoxin. *J. Biol. Chem.* **256**, 8732–8.

Orita, M., Nishikawa, F., Kohno, Senda, T., Mitsui, Y., Yaeta, E., Kazunari, T. and Nishikawa, S. (1996). High-resolution NMR study of a GdAGA tetranucleotide loop that is an improved substrate for ricin, a cytotoxic plant protein. *Nucleic Acids Res.* **24**, 611–18.

Orlandi, P.A. (1997). Protein-disulfide isomerase-mediated reduction of the A subunit of cholera toxin in a human intestinal cell line. *J. Biol. Chem.* **272**, 4591–9.

Orlandi, P.A. and Fishman, P.H. (1998). Filipin-dependent inhibition of cholera toxin: evidence for toxin internalization and activation through caveolae-like domains. *J. Cell Biol.* **141**, 905–15.

Otero, A.S., Yi, X.B., Gray, M.C., Szabo, G. and Hewlett, E.L. (1995). Membrane depolarization prevents cell invasion by *Bordetella pertussis* adenylate cyclase toxin. *J. Biol. Chem.* **270**, 9695–7.

Pappenheimer, A.M., Jr (1977). Diphtheria toxin. *Annu. Rev. Biochem.* **46**, 69–94.

Parker, M.W. and Pattus, F. (1993). Rendering a membrane protein soluble in water: a common packing motif in bacterial protein toxins. *Trends Biochem. Sci.* **18**, 391–5.

Petosa, C., Collier, R.J., Klimpel, K.R., Leppla, S.H. and Liddington, R.C. (1997). Crystal structure of the anthrax toxin protective antigen. *Nature* **385**, 833–8.

Popoff, M.R., Milward, F.W., Bancillon, B. and Boquet, P. (1989). Purification of the *Clostridium spiroforme* binary toxin and activity of the toxin on HEp-2 cells. *Infect. Immun.* **57**, 2462–9.

Popoff, M.R., Chaves-Olarte, E., Lemichez, E., von Eichel-Streiber, C., Thelestam, M., Chardin, P., Cussac, D., Antonny, B., Chavrier, P., Flatau, G., Giry, M., de Gunzburg, J. and Boquet, P. (1996). Ras, Rap, and Rac small GTP-binding proteins are targets for *Clostridium sordellii* lethal toxin glucosylation. *J. Biol. Chem.* **271**, 10217–24.

Prior, T.I., FitzGerald, D.J. and Pastan, I. (1992). Translocation mediated by domain II of *Pseudomonas* exotoxin A: transport of barnase into the cytosol. *Biochemistry* **31**, 3555–9.

Quertenmont, P., Wattiez, R., Falmagne, P., Ruysschaert, J.M. and Cabiaux, V. (1996). Topology of diphtheria toxin in lipid vesicle membranes: a proteolysis study. *Mol. Microbiol.* **21**, 1283–96.

Rahme, L.G., Stevens, E.J., Wolfort, S.F., Shao, J., Tompkins, R.G. and Ausubel, F.M. (1995). Common virulence factors for bacterial pathogenicity in plants and animals. *Science* **268**, 1899–902.

Rapak, A., Falnes, P.O. and Olsnes, S. (1997). Retrograde transport of mutant ricin to the endoplasmic reticulum with subsequent translocation to cytosol. *Proc. Natl Acad. Sci. USA* **94**, 3783–8.

Ray, B. and Wu, H.C. (1981). Enhancement of cytotoxicities of ricin and *Pseudomonas* toxin in Chinese hamster ovary cells by nigericin. *Mol. Cell Biol.* **1**, 552–9.

Refsnes, K., Haylett, T., Sandvig, K. and Olsnes, S. (1977). Modeccin – a plant toxin inhibiting protein synthesis. *Biochem. Biophys. Res. Commun.* **79**, 1176–83.

Reisbig, R., Olsnes, S. and Eiklid, K. (1981). The cytotoxic activity of Shigella toxin. Evidence for catalytic inactivation of the 60 S ribosomal subunit. *J. Biol. Chem.* **256**, 8739–44.

Rogel, A. and Hanski, E. (1992). Distinct steps in the penetration of adenylate cyclase toxin of *Bordetella pertussis* into sheep erythrocytes. Translocation of the toxin across the membrane. *J. Biol. Chem.* **267**, 22599–605.

Rogel, A., Meller, R. and Hanski, E. (1991). Adenylate cyclase toxin from *Bordetella pertussis*. The relationship between induction of cAMP and hemolysis. *J. Biol. Chem.* **266**, 3154–61.

Sanchez, J., Argotte, R. and Buelna, A. (1997). Engineering of cholera toxin A-subunit for carriage of epitopes at its amino end. *FEBS Lett.* **401**, 95–7.

Sandvig, K. and Olsnes, S. (1980). Diphtheria toxin entry into cells is facilitated by low pH. *J. Cell Biol.* **87**, 828–32.

Sandvig, K. and Olsnes, S. (1981). Rapid entry of nicked diphtheria toxin into cells at low pH. Characterization of the entry process and effects of low pH on the toxin molecule. *J. Biol. Chem.* **256**, 9068–76.

Sandvig, K. and Olsnes, S. (1982). Entry of the toxic proteins abrin, modeccin, ricin, and diphtheria toxin into cells. I. Requirement for calcium. *J. Biol. Chem.* **257**, 7495–503.

Sandvig, K., Sundan, A. and Olsnes, S. (1984). Evidence that diphtheria toxin and modeccin enter the cytosol from different vesicular compartments. *J. Cell Biol.* **98**, 963–70.

Sandvig, K., Tonnessen, T.I., Sand, O. and Olsnes, S. (1986). Requirement of a transmembrane pH gradient for the entry of diphtheria toxin into cells at low pH. *J. Biol. Chem.* **261**, 11639–44.

Sandvig, K., Olsnes, S., Petersen, O.W. and van Deurs, B. (1987). Acidification of the cytosol inhibits endocytosis from coated pits. *J. Cell Biol.* **105**, 679–89.

Sandvig, K., Olsnes, S., Brown, J.E., Petersen, O.W. and van Deurs, B. (1989). Endocytosis from coated pits of Shiga toxin: a glycolipid-binding protein from *Shigella dysenteriae* 1. *J. Cell Biol.* **108**, 1331–43.

Sandvig, K., Prydz, K., Ryd, M. and van Deurs, B. (1991). Endocytosis and intracellular transport of the glycolipid-binding ligand Shiga toxin in polarized MDCK cells. *J. Cell Biol.* **113**, 553–62.

Sandvig, K., Garred, O., Prydz, K., Kozlov, J.V., Hansen, S.H. and van Deurs, B. (1992). Retrograde transport of endocytosed Shiga toxin to the endoplasmic reticulum. *Nature* **358**, 510–12.

Sandvig, K., Ryd, M., Garred, O., Schweda, E., Holm, P.K. and van Deurs, B. (1994). Retrograde transport from the Golgi complex to the ER of both Shiga toxin and the nontoxic Shiga B-fragment is regulated by butyric acid and cAMP. *J. Cell Biol.* **126**, 53–64.

Sandvig, K., Garred, O. and van Deurs, B. (1996). Thapsigargin-induced transport of cholera toxin to the endoplasmic reticulum. *Proc. Natl Acad. Sci. USA* **93**, 12339–43.

Saron, M.F., Fayolle, C., Sebo, P., Ladant, D., Ullmann, A. and Leclerc, C. (1997). Anti-viral protection conferred by recombinant adenylate cyclase toxins from *Bordetella pertussis* carrying a CD8[+] T cell epitope from lymphocytic choriomeningitis virus. *Proc. Natl Acad. Sci. USA* **94**, 3314–19.

Saukkonen, K., Burnette, W.N., Mar, V.L., Masure, H.R. and Tuomanen, E.I. (1992). Pertussis toxin has eukaryotic-like carbohydrate recognition domains. *Proc. Natl Acad. Sci. USA* **89**, 118–22.

Scheel, A. and Pelham, R. (1996). Purification and characterization of the human KDEL receptor. *Biochemistry* **35**, 10203–9.

Schiavo, G., Poulain, B., Rossetto, O., Benfenati, F., Tauc, L. and Montecucco, C. (1992). Tetanus toxin is a zinc protein and its inhibition of neurotransmitter release and protease activity depend on zinc. *EMBO J.* **11**, 3577–83.

Schmid, A., Benz, R., Just, I. and Aktories, K. (1994). Interaction of *Clostridium botulinum* C2 toxin with lipid bilayer membranes. Formation of cation-selective channels and inhibition of channel function by chloroquine. *J. Biol. Chem.* **269**, 16706–11.

Schmidt, G., Sehr, P., Wilm, M., Selzer, J., Mann, M. and Aktories, K. (1997). Gln63 of Rho is deamidated by *Escherichia coli* cytotoxic necrotizing factor-1. *Nature* **387**, 725–9.

Selzer, J., Hofmann, F., Rex, G., Wilm, M., Mann, M., Just, I. and Aktories, K. (1996). *Clostridium novyi* alpha-toxin-catalyzed incorporation of GlcNAc into Rho subfamily proteins. *J. Biol. Chem.* **271**, 25173–7.

Shishido, Y., Sharma, K.D., Higashiyama, S., Klagsbrun, M. and Mekada, E. (1995). Heparin-like molecules on the cell surface potentiate binding of diphtheria toxin to the diphtheria toxin receptor membrane-anchored heparin-binding epidermal growth factor-like growth factor. *J. Biol. Chem.* **270**, 29578–85.

Simon, S.M. and Blobel, G. (1991). A protein-conducting channel in the endoplasmic reticulum. *Cell* **65**, 371–80.

Simon, S.M. and Blobel, G. (1992). Signal peptides open protein-conducting channels in *E. coli. Cell* **69**, 677–84.

Simpson, J.C., Dascher, C., Roberts, L.M., Lord, J.M. and Balch, W.E. (1995). Ricin cytotoxicity is sensitive to recycling between the endoplasmic reticulum and the Golgi complex. *J. Biol. Chem.* **270**, 20078–83.

Sixma, T.K., Pronk, S.E., Kalk, K.H., van Zanten, B.A., Berghuis, A.M. and Hol, W.G. (1992). Lactose binding to heat-labile enterotoxin revealed by X-ray crystallography. *Nature* **355**, 561–4.

Song, L., Hobaugh, M.R., Shustak, C., Cheley, S., Bayley, H. and Gouaux, J.E. (1996). Structure of staphylococcal alpha-hemolysin, a heptameric transmembrane pore. *Science* **274**, 1859–66.

Stein, P.E., Boodhoo, A., Tyrrell, G.J., Brunton, J.L. and Read, R.J. (1992). Crystal structure of the cell-binding B oligomer of vero-toxin-1 from *E. coli*. *Nature* **355**, 748–50.

Stein, P.E., Boodhoo, A., Armstrong, G.D., Cockle, S.A., Klein, M.H. and Read, R.J. (1994). The crystal structure of pertussis toxin. *Structure* **2**, 45–57.

Stenmark, H., Moskaug, J.O., Madshus, I.H., Sandvig, K. and Olsnes, S. (1991). Peptides fused to the amino-terminal end of diphtheria toxin are translocated to the cytosol. *J. Cell Biol.* **113**, 1025–32.

Stirpe, F., Barbieri, L., Abbondanza, A., Falasca, A.I., Brown, A.N., Sandvig, K., Olsnes, S. and Pihl, A. (1985). Properties of volkensin, a toxic lectin from *Adenia volkensii*. *J. Biol. Chem.* **260**, 14589–95.

Stirpe, F., Sandvig, K., Olsnes, S. and Pihl, A. (1982). Action of vis-cumin, a toxic lectin from mistletoe, on cells in culture. *J. Biol. Chem.* **257**, 13271–7.

Suzuki, T., Yan, Q. and Lennarz, W.J. (1998). Complex, two-way traffic of molecules across the membrane of the endoplasmic reticulum. *J. Biol. Chem.* **273**, 10083–6.

Szabo, G., Gray, M.C. and Hewlett, E.L. (1994). Adenylate cyclase toxin from *Bordetella pertussis* produces ion conductance across artificial lipid bilayers in a calcium- and polarity-dependent manner. *J. Biol. Chem.* **269**, 22496–9.

Tagge, E., Chandler, J., Tang, B.L., Hong, W.J., Willingham, M.C. and Frankel, A. (1996). Cytotoxicity of KDEL-terminated ricin toxins correlates with distribution of the KDEL receptor in the Golgi. *J. Histochem. Cytochem.* **44**, 159–65.

Taupiac, M.P., Alami, M. and Beaumelle, B. (1996). Translocation of full-length *Pseudomonas* exotoxin from endosomes is driven by ATP hydrolysis but requires prior exposure to acidic pH. *J. Biol. Chem.* **271**, 26170–3.

Theuer, C.P., Buchner, J., FitzGerald, D. and Pastan, I. (1993). The *N*-terminal region of the 37-kDa translocated fragment of *Pseudomonas* exotoxin A aborts translocation by promoting its own export after microsomal membrane insertion. *Proc. Natl Acad. Sci. USA* **90**, 7774–8.

Theuer, C., Kasturi, S. and Pastan, I. (1994). Domain II of *Pseudomonas* exotoxin A arrests the transfer of translocating nascent chains into mammalian microsomes. *Biochemistry* **33**, 5894–900.

Tortorella, D., Sesardic, D., Dawes, C.S. and London, E. (1995). Immunochemical analysis shows all three domains of diphtheria toxin penetrate across model membranes. *J. Biol. Chem.* **270**, 27446–52.

Townsend, A. and Bodmer, H. (1989). Antigen recognition by class I-restricted T lymphocytes. (Review.) *Annu. Rev. Immunol.* **7**, 601–24.

Ulmer, J.B., Donnelly, J.J. and Liu, M.A. (1994). Presentation of an exogenous antigen by major histocompatibility complex class I molecules. *Eur. J. Immunol.* **24**, 1590–6.

Valdizan, E.M., Loukianov, E.V. and Olsnes, S. (1995). Induction of toxin sensitivity in insect cells by infection with baculovirus encoding diphtheria toxin receptor. *J. Biol. Chem.* **270**, 16879–85.

van Deurs, B., Tonnessen, T.I., Petersen, O.W., Sandvig, K. and Olsnes, S. (1986). Routing of internalized ricin and ricin conjugates to the Golgi complex. *J. Cell Biol.* **102**, 37–47.

van Deurs, B., Petersen, O.W., Olsnes, S. and Sandvig, K. (1987). Delivery of internalized ricin from endosomes to cisternal Golgi elements is a discontinuous, temperature-sensitive process. *Exp. Cell Res.* **171**, 137–52.

van Deurs, B., Sandvig, K., Petersen, O.W., Olsnes, S., Simons, K. and Griffiths, G. (1988). Estimation of the amount of internalized ricin that reaches the trans-Golgi network. *J. Cell Biol.* **106**, 253–67.

van Heyningen, W.E. (1974). Gangliosides as membrane receptors for tetanus toxin, cholera toxin and serotonin. *Nature* **249**, 415–17.

Van Ness, B.G., Howard, J.B. and Bodley, J.W. (1980a). ADP-ribosylation of elongation factor 2 by diphtheria toxin. Isolation and properties of the novel ribosyl-amino acid and its hydrolysis products. *J. Biol. Chem.* **255**, 10717–20.

Van Ness, B.G., Howard, J.B. and Bodley, J.W. (1980b). ADP-ribosylation of elongation factor 2 by diphtheria toxin. NMR spectra and proposed structures of ribosyl-diphthamide and its hydrolysis products. *J. Biol. Chem.* **255**, 10710–16.

Werner, E.D., Brodsky, J.L. and McCracken, A.A. (1996). Proteasome-dependent endoplasmic reticulum-associated protein degradation: an unconventional route to a familiar fate. *Proc. Natl Acad. Sci. USA* **93**, 13797–801.

Wesche, J., Elliot, J.L., Falnes, P.Ø., Olnes, S. and Collier, J. (1998). Characterization of membrane translocation by anthrax protective antigen. *Biochemistry* **37**, 15737–46.

Wick, M.J., Hamood, A.N. and Iglewski, B.H. (1990). Analysis of the structure–function relationship of *Pseudomonas aeruginosa* exotoxin A. *Mol. Microbiol.* **4**, 527–35.

Wiedlocha, A., Falnes, P., Madshus, I.H., Sandvig, K. and Olsnes, S. (1994). Dual mode of signal transduction by externally added acidic fibroblast growth factor. *Cell* **76**, 1039–51.

Wiedlocha, A., Falnes, P.O., Rapak, A., Klingenberg, O., Muñoz, R. and Olsnes, S. (1995). Translocation to cytosol of exogenous, *CAAX*-tagged acidic fibroblast growth factor. *J. Biol. Chem.* **270**, 30680–85.

Wiertz, E.J., Jones, T.R., Sun, L., Bogyo, M., Geuze, H.J. and Ploegh, H.L. (1996). The human cytomegalovirus US11 gene product dislocates MHC class I heavy chains from the endoplasmic reticulum to the cytosol. *Cell* **84**, 769–79.

Xu, Y. and Barbieri, J.T. (1995). Pertussis toxin-mediated ADP-ribosylation of target proteins in Chinese hamster ovary cells involves a vesicle trafficking mechanism. *Infect. Immun.* **63**, 825–32.

Yoshida, T., Chen, C., Zhang, M. and Wu, H.C. (1990). Increased cytotoxicity of ricin in a putative Golgi-defective mutant of Chinese hamster ovary cell. *Exp. Cell Res.* **190**, 11–16.

Regulation of diphtheria toxin production: characterization of the role of iron and the diphtheria toxin repressor

Joanne Goranson-Siekierke and Randall K. Holmes

INTRODUCTION

Several recent reviews of clinical diphtheria are available (Rakhmanova *et al.*, 1996; Bisgard *et al.*, 1998; Holmes, 1998). The major virulence determinant of *Corynebacterium diphtheriae* is diphtheria toxin (DT), the cause of the systemic complications seen with diphtheria. Frequently, the onset of diphtheria is insidious, with only a low-grade fever and mild sore throat. DT causes local necrosis of mucosal epithelial cells, resulting in isolated greyish spots that begin to coalesce within 24 h and develop into a tough pseudomembrane that can extend into the trachea and cause suffocation. The systemic effects of the toxin are also serious and can cause damage to heart muscle, peripheral nerves and kidneys.

Treatment is by administration of diphtheria antitoxin. The outcome in diphtheria depends on multiple factors such as the age of the patient, the location and extent of the membrane, and the promptness of administration of antitoxin. Although diphtheria has become a clinical rarity in the USA owing to the widespread use of an effective toxoid vaccine (Committee on Infectious Diseases, 1997), *C. diphtheriae* remains an important bacterial pathogen. Humans are the only reservoir of the disease and, world-wide, there is a significant mortality rate that can reach 5–10% during epidemics in poorly immunized or unimmunized populations (Chen *et al.*, 1985; Bisgard *et al.*, 1998; Vitek and Wharton, 1998).

Both small and large epidemics of diphtheria have occurred in developed countries during the past 20 years, and it is not necessary to travel to underdeveloped countries to risk exposure to diphtheria. Alcoholism, low socio-economic status, crowded living conditions and Native American ethnic background were risk factors in significant outbreaks of diphtheria that occurred in the USA during the 1970s and 1980s (Chen *et al.*, 1985; Harnisch *et al.*, 1989), and a recent report documents the continuing carriage of toxinogenic strains of *C. diphtheriae* among Native Americans in North Dakota [Centers for Disease Control and Prevention (CDC), 1997]. A small diphtheria outbreak in Sweden between 1984 and 1986, mainly among alcoholics, resulted in 17 cases, three deaths and six episodes of reversible paralysis (Rappuoli *et al.*, 1988). Since 1969, a number of small outbreaks in England and Wales was due to contact between susceptible individuals and asymptomatic carriers with imported infections caused by both toxigenic and non-toxigenic *C. diphtheriae* strains (Begg and Balraj, 1995). Since 1990, epidemic diphtheria on a massive scale has re-emerged in the New Independent States of the former Soviet Union, with a cumulative total of more than 100 000 cases through 1995 (CDC, 1995a, b, 1996; Nakao *et al.*, 1996). In 1994, at least 20 imported cases of diphtheria were reported in eastern and western European countries neighbouring the New Independent States (CDC, 1995a, b). These reports demonstrate the continuing potential for the spread of diphtheria and indicate the need for maintaining adequate immunization of adults. Most populations in developed countries are highly immunized in childhood, but adults in many developed countries are likely to have inadequate levels of circulating antibody, making introductions of virulent *C. diphtheriae* a potentially serious risk (Karzon and Edwards, 1988).

Diphtheria, first identified as a clinical entity by Brettoneau in 1826, was shown in 1888 by Roux and Yersin to be caused by an extracellular toxin produced by *C. diphtheriae* (reviewed in Pappenheimer and Gill, 1973; Holmes, 1998, 1999). DT production by *C. diphtheriae* is mediated by phage conversion (Freeman, 1951; Groman, 1953a, b; Barksdale and Pappenheimer, 1954), and the structural gene, *tox*, is present in the genomes of several corynebacteriophages such as the well-studied phage β (Barksdale and Arden, 1974; Uchida *et al.*, 1971). With certain strains of *C. diphtheriae*, such as PW8, DT comprises as much as 5% of the protein synthesized and 75% of all the protein that is secreted (Pappenheimer and Gill, 1973; Murphy *et al.*, 1974). DT is a member of the A:B family of toxins and is an ADP-ribosyltransferase that must be processed from its propeptide form to be enzymatically active (reviewed in Krueger and Barbieri, 1995; Holmes, 1999). DT transfers the ADP ribose moiety from NAD to elongation factor 2 (EF-2) (Honjo *et al.*, 1969), resulting in inactivation of EF-2 and inhibition of protein synthesis in the cytoplasm of target cells (Collier, 1967; Goor and Pappenheimer, 1967). DT has been purified, analysed biochemically and crystallized (Choe *et al.*, 1992; Wilson and Collier, 1992; Bell and Eisenberg, 1997). The crystal structure reveals three functional domains of DT: a receptor-binding domain that allows DT to bind to cells and gain entry by receptor-mediated endocytosis; a central translocation domain required for delivery of the A domain from endocytic vesicles to the cytoplasm; and the amino-terminal A domain, which has the ADP-ribosyltransferase activity responsible for covalent modification of EF-2 (Choe *et al.*, 1992). Intoxication of eukaryotic cells is due directly to the ADP-ribosylation of EF-2, and introduction of a single molecule of fragment A of DT into the cytoplasm results in the death of the eukaryotic cell (Yamaizumi *et al.*, 1978).

Transmission of the *tox* gene in a population occurs directly via respiratory droplets containing toxigenic *C. diphtheriae* or indirectly via transmission of *tox*[+] phages and subsequent lysogenization of non-toxigenic *C. diphtheriae* already present in the nasopharynx (Pappenheimer and Murphy, 1983). DT is not essential for the metabolism of either the host bacterium or the bacteriophage. Although the *tox* gene does not confer a selective advantage *in vitro*, it is believed to be of benefit to the bacteria in the infected host by mechanisms such as releasing nutrients (haem, etc.) from killed cells or by affecting the probability that *C. diphtheriae* will be spread from an infected individual to other susceptible individuals (Pappenheimer, 1982; Holmes, 1998, 1999).

REGULATION OF PRODUCTION OF DIPHTHERIA TOXIN

Successful pathogenic bacteria must be able to adapt to changes in their environment and co-ordinate the expression of genes necessary for their survival. Iron is an essential element both for bacterial pathogens and for the human host, and the ability to compete for iron in the host environment is important for the survival and growth of most pathogenic bacteria. The concentration of free soluble iron in the mammalian host is severely limited, and the expression of many virulence determinants, including toxins, is induced by low-iron conditions in several bacterial pathogens including *C. diphtheriae* (reviewed in Litwin and Calderwood, 1993).

Iron-mediated regulation of diptheria toxin production

It has long been known that DT production by *C. diphtheriae* is influenced by the amount of iron in the growth medium. A culture of toxinogenic diphtheria bacilli will not produce toxin unless the iron content of the growth medium is low, and DT synthesis in low-iron medium is maximal during the declining phase of bacterial growth, when iron becomes the rate-limiting nutrient (Locke and Main, 1931; Pappenheimer and Johnson, 1936; Mueller and Miller, 1941; Pappenheimer, 1955; Edwards and Seamer, 1960; Barksdale, 1970). Thus, toxin production is maximal under conditions of iron starvation and is severely inhibited under high-iron growth conditions. Under iron-limiting conditions, the *tox* gene can be expressed from the genome of a *tox*[+] prophage (Groman, 1953a, b; Barksdale and Pappenheimer, 1954; Barksdale and Arden, 1974), from a vegetatively replicating *tox*[+] phage (Matsuda and Barksdale, 1966, 1967), and from a repressed but non-integrated superinfecting *tox*[+] phage genome (Gill *et al.*, 1972). Control of toxin production is, therefore, independent of the regulatory systems that govern the vegetative replication and the prophage state of *tox*[+] corynephages.

After it was established that DT is encoded by the *tox* gene of corynephages and that regulation of *tox* expression is controlled by the physiological state of *C.*

diphtheriae, an *in vitro* transcription/translation system in *Escherichia coli* was used to investigate whether the regulatory factors involved in *tox* expression are encoded by the *tox*⁺ corynephage or the genome of *C. diphtheriae* (Murphy *et al.*, 1974). DNA from the *tox*⁺ corynephage β was found to direct the synthesis of DT and other phage-encoded proteins in such extracts regardless of the presence of iron, but DT synthesis was specifically and selectively inhibited by the addition of a crude extract from *C. diphtheriae*. Mutants of *C. diphtheriae* C7 (β) were later identified that were insensitive to the inhibitory effect of iron on DT production, and the phenotype was regulated by the bacterial genome and not by the genome of corynephage β (Kanei *et al.*, 1977). Mutant β phages that allowed for the production of DT by *C. diphtheriae* lysogens under high-iron growth conditions were also isolated (Murphy *et al.*, 1976; Welkos and Holmes, 1981a, b), and the phenotype was due to a *cis*-acting phage regulatory element (Murphy *et al.*, 1976; Welkos and Holmes, 1981b) that mapped immediately upstream from the *tox* gene (Welkos and Holmes, 1981b). It was also shown that the regulation of DT expression by iron occurred at the level of *tox* transcription (Murphy *et al.*, 1978; Welkos and Holmes, 1981a). Taken together, these data supported the hypothesis that an iron-dependent repressor with the ability to inhibit transcription of the *tox* gene of phage β under high-iron growth conditions was encoded by the *C. diphtheriae* genome (Holmes, 1975; Murphy *et al.*, 1976; Murphy and Bacha, 1979).

Regulatory sequences for the *tox* gene

The structural gene for diphtheria toxin (*tox*) and upstream regulatory sequences were cloned and sequenced (Greenfield *et al.*, 1983; Kaczorek *et al.*, 1983; Ratti *et al.*, 1983), and the *tox* promoter was characterized (Leong and Murphy, 1985). S1 nuclease mapping of the *tox* mRNA in both *E. coli* and *C. diphtheriae* showed that transcription initiated at nucleotide positions −41 and −40 upstream from the GTG translation start signal (Leong and Murphy, 1985). A sequence similar to the consensus for the −35 region of σ^{70} promoters of *E. coli* started at position −74 upstream from the *tox* structural gene, and two putative −10 regions were located starting at positions −54 and −48 (Leong and Murphy, 1985). Using site-directed mutagenesis, it was later shown that the sequence beginning at −48 was the primary −10 promoter region and, upon inactivation of the primary region, the sequence starting at position −54 was able to function as a −10 promoter region, although less efficiently (Boyd and Murphy, 1988). In addition to the −35 and −10 regions of the promoter, a 27-bp interrupted palindromic sequence was also present (Kaczorek *et al.*, 1983; Greenfield *et al.*, 1983; Ratti *et al.*, 1983). Since many prokaryotic operators are proximal to their corresponding promoter regions and exhibit dyad symmetry, this palindrome was hypothesized to be the *tox* operator.

Further evidence for the existence of a chromosomally encoded repressor of DT expression was provided by the observation that protein(s) in crude extracts from *C. diphtheriae* grown under high-iron conditions were able to bind specifically to the predicted *tox* operator and protect it from DNase I digestion (Fourel *et al.*, 1989). Since protection of the putative *tox* operator required the presence of iron in the reaction, it was hypothesized that the binding factor was the diphtheria toxin repressor.

THE DIPHTHERIA TOXIN REPRESSOR

Cloning of the *dtxR* gene and characterization of DtxR

The gene for the diphtheria toxin repressor (*dtxR*) was subsequently cloned independently by two groups by screening *C. diphtheriae* C7 genomic libraries in *E. coli* to identify clones that caused iron-dependent inhibition of expression of a β-galactosidase reporter gene that was transcribed under control of the *tox* operator/promoter (Boyd *et al.*, 1990; Schmitt and Holmes, 1991a). The *dtxR* gene encodes an M_r 25 316 polypeptide consisting of 226 amino acids, which is transcribed constitutively in *C. diphtheriae* at a low level under both high- and low-iron conditions (Schmitt and Holmes, 1991a). In Gram-negative bacteria, expression of genes that encode iron-regulated virulence determinants is regulated in a similar manner by the Fur (ferric uptake regulator) protein (Hantke, 1984; Litwin and Calderwood, 1993). Although the diphtheria toxin repressor (DtxR) and Fur are both iron-dependent repressors, they share little amino acid sequence identity and their regulatory actions are specific for different operators (Boyd *et al.*, 1990; Schmitt and Holmes, 1991a). DtxR and Fur are the prototypes, therefore, for two different families of bacterial iron-dependent regulatory proteins.

In wild-type *E. coli*, the cloned *tox* promoter is constitutively expressed regardless of the iron concentration (Boyd *et al.*, 1990; Schmitt and Holmes, 1991a). In the presence of the cloned *dtxR* gene, however, expression from the *tox* promoter in *E. coli* is regulated in an iron-dependent fashion (Boyd *et al.*, 1990; Schmitt and Holmes, 1991a). Recombinant DtxR was expressed in *E. coli* and purified using various methods, including metal ion affinity chromatography (Tao *et al.*, 1992; Schmitt *et al.*, 1992). Using gel mobility shift analysis, it was shown that binding of purified DtxR to a radiolabelled diphtheria *tox* operator probe is specific and requires the presence of Fe^{2+} or another appropriate divalent transition metal ion (Tao *et al.*, 1992; Schmitt *et al.*, 1992). DtxR was shown by DNase I footprinting to bind to an approximately 30-bp region upstream of the *tox* gene (Schmitt *et al.*, 1992; Tao and Murphy, 1992; Schmitt and Holmes, 1993). The DtxR binding site in the *tox* operator/promoter contains the 27-bp region of interrupted dyad symmetry previously postulated to be the *tox* operator, and it overlaps with the −10 promoter region. DNase I footprinting analysis of additional DtxR-regulated operators indicated that their common feature is a 19-bp palindromic core with the consensus sequence 5′-TTAGGTTAGCCTAACCTAA-3′ (Schmitt and Holmes, 1994; Tao and Murphy, 1994; Lee *et al.*, 1997). *In vitro* sequence-specific binding of DtxR to the *tox* operator requires activation by any of the transition metals Cd^{2+}, Co^{2+}, Fe^{2+}, Mn^{2+}, Ni^{2+} or Zn^{2+}, although Fe^{2+} is the physiologically relevant activator *in vivo* (Schmitt *et al.*, 1992; Tao *et al.*, 1992, 1994; Schmitt and Holmes, 1993). Hydroxyl radical footprinting indicated that DtxR binds in a symmetrical manner about the dyad axis of the *tox* operator and suggested that DtxR binds to DNA as a dimer or multimer (Schmitt and Holmes, 1993). Binding of Ni^{2+} by DtxR has an apparent K_d of 9×10^{-7} M to 2×10^{-6} M, is cooperative, and results in the quenching of the intrinsic fluorescence of W104 in DtxR (Wang *et al.*, 1994; Tao *et al.*, 1995). Monomeric apo-DtxR is in weak equilibrium with dimers, but the dimers are stabilized by interaction with the divalent cations that are required for activation of the repressor activity (Tao *et al.*, 1995).

Structure and function of DtxR

The relationships between structure and function of DtxR have been investigated by a combination of biochemical, molecular genetic and X-ray crystallographic methods. A number of early genetic studies focused on defining the metal-binding and DNA-binding sites of DtxR and the mechanism by which binding of divalent cations activates the dimeric repressor protein. DtxR contains a single Cys residue in DtxR at position 102. The sulfhydryl group of C102 in active DtxR is in the reduced form, and formation of a disulfide-linked dimer inactivates repressor activity (Schmitt *et al.*, 1992). To determine whether C102 was essential for DtxR activity, saturation mutagenesis of the C102 codon was performed (Tao and Murphy, 1993). Replacement of C102 with any amino acid except aspartate resulted in the loss of DtxR

repressor activity in an *E. coli* reporter strain carrying a transcriptional fusion of the *tox* operator/promoter region with *lacZ*. DtxR with the C102D substitution required appropriate divalent metal ions for activation and specific binding to the *tox* promoter, but it had a lower metal-binding affinity than wild-type DtxR (Tao and Murphy, 1993). These results indicated that Cys102 plays an important role in DtxR activity, and it was concluded that Cys102 is positioned in the metal ion-activation site of DtxR.

Random bisulfite mutagenesis of cloned *dtxR* in *E. coli* resulted in the identification of 20 mutations in single codons, 18 of which resulted in single amino acid substitutions and two of which were chain-terminating (Wang *et al.*, 1994). The phenotypes of these mutants provided early evidence that a sequence in the middle third of DtxR including residues His98-XXX-Cys102-XXX-His106 is involved in metal binding and that a predicted helix-turn-helix motif in the amino-terminal third of DtxR (residues 27–50) is involved in DNA binding (Wang *et al.*, 1994). In addition, sequencing of a mutant *dtxR* allele from a strain of *C. diphtheriae* deficient in repressor activity revealed an R47H substitution in the deduced amino acid sequence (Schmitt and Holmes, 1991b; Boyd *et al.*, 1992). Subsequent molecular genetic studies were designed to test specific structural models of DtxR, and further discussion of genetic analyses will be integrated with the results of crystallographic studies of DtxR described below.

The structure of holo-DtxR in complex with six different divalent transition metals (Cd^{2+}, Co^{2+}, Fe^{2+}, Mn^{2+}, Ni^{2+} and Zn^{2+}) was determined by X-ray diffraction analysis at 2.8 Å resolution (Qiu *et al.*, 1995). Each DtxR monomer possesses three distinct domains. The amino terminal domain 1 (residues 1–73) contains a helix-turn-helix motif, required for binding DNA. Domain 2 (residues 74–144) is responsible for dimerization and possesses two distinct metal binding sites, which lie 10 Å apart. The formation of DtxR homodimers is mediated by protein–protein interactions between the monomers, and binding of divalent cations is required for activation of repressor activity (Tao *et al.*, 1995). Furthermore, deletion of the amino-terminal 47 amino acids from domain 1 abolishes the DNA-binding activity of DtxR without affecting its capacity to form dimers (Tao *et al.*, 1995). The carboxy-terminal domain 3 (residues 145–226) has greater flexibility than domains 1 and 2 (Qiu *et al.*, 1995). Domain 3 is not resolved in most crystal structures of DtxR, and its function is unknown.

Metal-binding site 1 was fully occupied in all of the DtxR–cation complexes and was shown to involve His79, Glu83, His98 and a solvent molecule as ligands

(Qiu *et al.*, 1995). Metal-binding site 2 was highly occupied only when DtxR was in complex with CdCl$_2$ and included the carbonyl group of Cys102 and the side-chains of Glu105 and His106 and a solvent molecule as ligands. The side-chain of Cys102 did not interact with the metal at site 2, but rather with the imidazole ring of His98, one of the direct ligands of metal-binding site 1 (Qiu *et al.*, 1995). In these crystals, however, the SH- group of Cys102 appeared to be oxidized (Qiu *et al.*, 1995, 1996) or covalently modified (Pohl *et al.*, 1997), which may have altered its interaction with the cation bound at site 2.

Other crystal structures of DtxR solved at 3 Å resolution (Schiering *et al.*, 1995), and at 2.4 Å resolution for the variant of DtxR with Asp replacing Cys at position 102 (Ding *et al.*, 1996), confirmed that amino acid residues Cys102, Glu105 and His106 co-ordinate with metal at site 2, as seen by Qiu *et al.* (1995, 1996), and indicated additional co-ordination of the metal at site 2 by Met10. Replacement of Met10, Cys102, Glu105 or His106 with alanine dramatically decreased repressor activity of DtxR, whereas replacement of the metal-binding site 1 ligands His79, Glu83 or His98 with alanine has less effect on repressor activity (Tao and Murphy, 1993; Ding *et al.*, 1996), leading to the proposal that metal-binding site 2 is the primary metal ion-activation site of the DtxR (Ding *et al.*, 1996). Based on small differences between the structures of apo-DtxR and holo-DtxR dimers at relatively low resolution, activation as a consequence of binding metal ions was proposed to occur by a calliper-like rotation of the monomers relative to each other, resulting in an alteration of the quaternary structure of DtxR (Schiering *et al.*, 1995; Ding *et al.*, 1996).

High-resolution structures of DtxR complexed with cobalt (2.0 Å refinement) and manganese (2.2 Å refinement) indicated that metal binding site 1 was well occupied in both structures and showed that a sulfate ion served as a fourth ligand for the divalent cation at this site (Qiu *et al.*, 1996). The sulfate anion participates in an extensive network of hydrogen bonds with other ligands in DtxR, including Arg80, Ser126 and Asn130 (Qiu *et al.*, 1996). Under physiological conditions, phosphate may occupy the anion site instead of sulfate and may function as a co-corepressor for DtxR. These analyses also revealed that the third domain of DtxR exhibits an SH3-like fold and does not appear to function in either DNA- or metal-binding by the repressor (Qiu *et al.*, 1996).

The structure of DtxR with cobalt was also determined at 100 K with 1.85 Å resolution, providing the highest resolution to date, and at room temperature with zinc at 2.4 Å resolution (Pohl *et al.*, 1997). Comparison of the two structures revealed no signifi-

cant differences, but provided the most accurate view of the anion–cation binding site 1 (Pohl *et al.*, 1997). In the 1.85 Å cobalt–DtxR structure, the metal ion at site 1 was co-ordinated tetrahedrally by His79, Glu83, His98 and a sulfate ion; and Arg80, Ser126 and Asn130 were involved in co-ordination of the sulfate ion. Metal binding site 2 was not occupied in this cobalt–DtxR structure, but as in previous studies, the sulfhydryl group of Cys102 appeared to be covalently modified (Pohl *et al.*, 1997). The zinc–DtxR crystal was prepared in the absence of sulfate, yet a molecule was present in the structure at the anion binding site (Pohl *et al.*, 1997). This was presumed to be a phosphate anion, since it exhibited tetrahedral density, and phosphate buffers were used during purification of the protein. As with the cobalt–DtxR structure, this zinc–DtxR structure showed no cation at metal-binding site 2. Thus, the cation–anion binding site 1 appears to be a structurally important feature of DtxR, and it is noteworthy that the anion-binding ligands Arg80, Ser126 and Asn130 are conserved in homologues of DtxR from several other bacterial species (Doukhan *et al.*, 1995; Gunter-Seeboth and Schupp, 1995; Oguiza *et al.*, 1995; Schmitt *et al.*, 1995).

The hypothesis that the anion-binding ligands at site 1 are important for DtxR activity was tested by site-directed mutagenesis of *dtxR* (Goranson-Siekierke *et al.*, 1999). Alanine was substituted for each of the amino acids that functions as a ligand for binding of metal ions at sites 1 and 2 and for anion binding at site 1. These studies confirmed that substitutions of alanine for the residues involved in metal co-ordination at site 2 (C102, E105 and H106) had a more dramatically deleterious effect on DtxR activity, as measured using a *tox* operator–*lacZ* reporter construct in *E. coli*, than substitutions of the metal-co-ordinating residues at site 1 (H79, E83 and H98). Remarkably, however, alanine substitutions for the anion-co-ordinating ligands of site 1 (R80, S126 and N130) resulted in inactivation of DtxR repressor activity to an extent comparable with the alanine substitutions for ligands at site 2. A direct interaction between residue R80 of the anion–cation binding site and E20 of the DNA binding motif in domain 1 was also analysed in this study and was shown to be essential for DtxR activity (Goranson-Siekierke *et al.*, 1999). Thus, both the anion–cation binding site 1 and the cation binding site 2 are required for DtxR activity.

When high-resolution (2.2–2.4 Å) structures of form 1 and form 2 crystals of apo-DtxR and holo-DtxR in the presence of zinc were compared, it was found that the N-terminal DNA-binding domain and the last 20 amino acids of the dimerization domain of each subunit show significant movement with respect to the immobile dimer core as a consequence of binding divalent transition metals (Pohl *et al.*, 1998a). Thus, activation of DtxR appears to involve a change in the tertiary structure of DtxR, rather than a change in quaternary structure as previously proposed (Schiering *et al.*, 1995; Ding *et al.*, 1996). In addition, the structure of the iron-dependent regulator (IdeR) from *Mycobacterium tuberculosis*, which is a homologue of DtxR, was recently solved in complex with Zn^{2+} (Pohl *et al.*, 1998b). In this structure, both metal-binding sites are fully occupied. C102 is not oxidized or covalently modified, and the sulfur atoms of both C102 and Met10 serve as ligands of the metal ion at site 2 (in addition to the carbonyl group of C102 and the side-chains of E105 and H106). The change in tertiary structure relative to apo-DtxR is significantly greater than that observed previously for holo-DtxR, bringing the DNA-binding helices of the helix-turn-helix motifs in domain 1 closer together than in holo-DtxR. This is the first structure of an activated wild-type iron-dependent regulator of the DtxR family with both metal-binding sites fully occupied.

The C102D variant of DtxR was recently crystallized in complex with an oligonucleotide corresponding in sequence to the *tox* operator, and the structure of the complex was determined at 3 Å resolution (White *et al.*, 1998). Surprisingly, the ternary complex contains two dimers of DtxR-C102D, which bind to opposite faces of the oligonucleotide, without protein–protein interactions between the separate dimers of DtxR-C102D. The two DtxR-C102D dimers bind symmetrically to partially overlapping regions of the interrupted palindromic core region of the operator. Additional studies are needed, however, to provide a full explanation for the structural basis of the sequence-specific binding of DtxR to its cognate operators.

Global regulation by DtxR

DtxR is not only involved in the regulation of expression of DT, but is an iron-dependent global regulator of metabolism in *C. diphtheriae*. Production of DT and siderophore is co-ordinately regulated during the transition of *C. diphtheriae* from high- to low-iron growth conditions (Tai *et al.*, 1990). Production of both DT and siderophore is derepressed in *C. diphtheriae* C7(*β*)hm723, which has reduced DtxR function, and complementation with a wild-type *dtxR* allele restores the repressible phenotype under high-iron conditions (Schmitt and Holmes, 1991a).

Screening of promoters on DNA fragments from the chromosome of *C. diphtheriae* strain C7 in an *E. coli* functional assay led to the identification of five iron- and DtxR-regulated operator/promoters from the C.

diphtheriae chromosome, named IRP1 to IRP5 (Schmitt and Holmes, 1994; Lee *et al.*, 1997; Schmitt *et al.*, 1997), in addition to the previously characterized *tox* operator/promoter from phage *β*. A gene encoding a 38-kDa lipoprotein that may function as a ferric siderophore receptor lies downstream of IRP1 (Schmitt *et al.*, 1997); a 15-kDa homologue of the AraC family of transcriptional activators is encoded downstream of IRP3 (Lee *et al.*, 1997); and downstream of IRP4 and IRP5 are open reading frames encoding polypeptides that have no homology to other known sequences (Lee *et al.*, 1997). In addition, a haem oxygenase encoded by the *hmuO* gene of *C. diphtheriae* was recently characterized, and it was also shown to be regulated by DtxR and iron (Schmitt, 1997a, b; Wilks and Schmitt, 1998). A total of 14 proteins that are expressed preferentially by *C. diphtheriae* in low-iron conditions has also been identified by sodium dodecyl sulfate–polyacrylamide gel electrophoresis (Tai and Zhu, 1995), but the genes that encode them have not been determined and it has not yet been established whether DtxR regulates their production.

Homologues of DtxR in other bacteria

DtxR is the prototype for a family of iron-dependent regulatory proteins, IdeRs, that have been identified in a number of bacteria other than *C. diphtheriae*, including *Brevibacterium lactofermentum* (Oguiza *et al.*, 1995), *Mycobacterium smegmatis* and *M. tuberculosis* (Doukhan *et al.*, 1995; Dussurget *et al.*, 1996; Schmitt *et al.*, 1995), *Streptomyces lividans* and *S. pilosus* (Gunter-Seeboth and Schupp, 1995) and *Staphylococcus epidermidis* and *S. aureus* (Hill *et al.*, 1998). The cloned DtxR homologue from *M. tuberculosis* complements defects in a *dtxR* mutant of *C. diphtheriae* (Schmitt *et al.*, 1995). These homologues are highly conserved, having approximately 30–60% overall identity with DtxR, although domains 1 and 2 exhibit up to 90% homology, while domain 3 shows much greater variability. Strikingly, the amino acids in the anion–cation binding site 1 and in metal-binding site 2 that have been implicated as being important for the activity of DtxR are identical among these DtxR homologues, except for a deduced H79D substitution in IdeR encoded by *B. lactofermentum* (Oguiza *et al.*, 1995) and a deduced C102E substitution in SirR encoded by *S. epidermidis* (Hill *et al.*, 1998).

SUMMARY AND CONCLUSIONS

The fact that production of DT by *C. diphtheriae* is inhibited by excess iron in the growth medium was established during the 1930s. Evidence that this effect is mediated by a diphtheria toxin repressor that requires iron for its activity was obtained during the 1970s. It is only within the last decade, however, that the *dtxR* gene has been cloned and that rapid advances have been made in characterizing purified wild-type and mutant forms of DtxR by biochemical and X-ray crystallographic techniques.

A molecular understanding of the structure and function of DtxR is rapidly emerging. DtxR is a homodimer, and each monomer has three distinct domains. The amino-terminal domain contains a helix-turn-helix motif responsible for sequence-specific binding of DtxR to DNA. The central domain mediates dimerization and contains two binding sites for Fe^{2+} or other divalent transition metal ions. Site 1 is a complex anion–cation binding site, at which the ligands for the cation are the side-chains of residues H79, E83 and H98 plus a sulfate or phosphate anion, and the ligands for the anion are the side-chains of residues R80, S126 and N130 plus the cation. At site 2, the ligands for the cation are the carbonyl oxygen of C102, the side-chains of C102, E105 and H106, and the side-chain of Met10. The carboxyl-terminal domain has an SH3-like fold, suggesting that it may be involved in protein–protein interactions, but its function is not yet proven. Binding of metal at both site 1 and site 2 appears to be required for full activation of DtxR, although the relative importance of metal binding at site 1 and site 2 remains controversial. Activation of DtxR as a consequence of metal binding is accomplished by changes in the tertiary structure of each monomer, consisting of slight motion of domain 1 and the carboxyl-terminal segment of domain 2 with respect to the immobile core of the dimer in domain 2. As a consequence of this conformational change, binding of holo-DtxR to a 19-bp core region of DtxR-regulated operators is optimized. The structure of a ternary complex consisting of two dimers of holo-DtxR with an oligonucleotide corresponding to the *tox* operator sequence was reported during the past year, but a detailed structural explanation for the sequence specificity of DtxR binding to DNA awaits further investigation.

DtxR is involved in regulating not only the *tox* gene of specific corynebacteriophages but also a diverse group of genes in *C. diphtheriae* that encode a corynebacterial siderophore, components of the siderophore-dependent iron-uptake system, a haem oxygenase that is required for assimilation of iron from haem, and several other recently identified proteins of unknown function. DtxR functions as a negative regulator for transcription of these genes, which are expressed either individually or as components of operons. Under high-iron conditions holo-DtxR binds to the operator regions of

DtxR-regulated operator/promoters and prevents transcription. Under low-iron conditions, the Fe^{2+} co-repressor dissociates from holo-DtxR, and the resulting apo-DtxR lacks repressor activity. As a consequence, the genes controlled by the set of DtxR-regulated operator/promoters are expressed.

DtxR is now recognized as the prototype for a family of iron-activated global regulatory proteins in prokaryotes, which are called iron-dependent regulators. These IdeRs are found in several pathogenic and non-pathogenic bacteria. It is quite common for specific toxins and virulence factors in pathogenic bacteria to be produced preferentially under low-iron conditions as components of regulons controlled by either DtxR and its homologues or Fur and its homologues. Future studies of global gene regulation by DtxR and related IdeRs are likely, therefore, to provide important new insights into physiological processes, toxins and other virulence factors that contribute to growth within the host environment of *C. diphtheriae* and other important bacterial pathogens including *M. tuberculosis*, *M. leprae* and *S. aureus*.

ACKNOWLEDGEMENTS

Research on diphtheria and the diphtheria toxin repressor in the authors' laboratory is funded in part by research grant AI-14107 and national research service award AI-10038 from the National Institute of Allergy and Infectious Diseases, National Institutes of Health, Bethesda, MD, USA.

REFERENCES

Barksdale, L. (1970). *Corynebacterium diphtheriae* and its relatives. *Bacteriol. Rev.* **4**, 378–422.

Barksdale, L. and Arden, S.B. (1974). Persisting bacteriophage infections, lysogeny, and phage conversion. *Annu. Rev. Microbiol.* **28**, 265–99.

Barksdale, W.L. and Pappenheimer, A.M., Jr (1954). Phage–host relationships in nontoxigenic and toxigenic diphtheria bacilli. *J. Bacteriol.* **56**, 220–32.

Begg, N. and Balraj, V. (1995). Diphtheria: are we ready for it? *Arch. Dis. Child.* **73**, 568–72.

Bell, C.E. and Eisenberg, D. (1997). Crystal structure of nucleotide-free diphtheria toxin. *Biochem.* **36**, 481–8.

Bisgard, K.M., Hardy, I.R., Popovic, T., Strebel, P.M., Wharton, M., Chen, R.T. and Hadler, S.C. (1998). Respiratory diphtheria in the United States, 1980 through 1995. *Am. J. Public Health* **88**, 787–91.

Boyd, J. and Murphy, J.R. (1988). Analysis of the diphtheria *tox* promoter by site-directed mutagenesis. *J. Bacteriol.* **170**, 5949–52.

Boyd, J.M., Manish, O.N. and Murphy, J.R. (1990). Molecular cloning and DNA sequence of a diphtheria *tox* iron-dependent regulatory element (*dtxR*) from *Corynebacterium diphtheriae*. *Proc.*

Natl Acad. Sci. USA **87**, 5968–72.

Boyd, J.M., Hall, K.C. and Murphy, J.R. (1992). DNA sequencing and characterization of *dtxR* alleles from *Corynebacterium diphtheriae* PW8(–), 1030(–), and C7hm723(–). *J. Bacteriol.* **174**, 1268–72.

Centers for Disease Control and Prevention (1995a). Diphtheria acquired by U.S. citizens in the Russian Federation and Ukraine – 1994. *Morb. Mortal. Wkly Rep.* **44**, 177–81.

Centers for Disease Control and Prevention (1995b). Diphtheria epidemic – new independent states of the former Soviet Union, 1990–1994. *J. Am. Med. Assoc.* **273**, 1250–2.

Centers for Disease Control and Prevention (1996). Update: diphtheria epidemic – New Independent States of the Former Soviet Union, January 1995–March 1996. *Morb. Mortal. Wkly Rep.* **45**, 693–7.

Centers for Disease Control and Prevention (1997). Toxigenic *Corynebacterium diphtheriae* – Northern Plains Indian Community, August–October 1996. *Morb. Mortal. Wkly Rep.* **46**, 506–10.

Chen, R.T., Broome, C.V., Weinstein, R.A., Weaver, R. and Tsai, T.F. (1985). Diphtheria in the United States, 1971–81. *Am. J. Public Health* **75**, 1393–7.

Choe, S., Bennett, M.J., Fujii, G., Curmi, P.M., Kantardjieff, K.A., Collier, R.J. and Eisenberg, D. (1992). The crystal structure of diphtheria toxin. *Nature* **357**, 216–22.

Collier, R.J. (1967). Effect of diphtheria toxin on protein synthesis: inactivation of one of the transfer factors. *J. Mol. Biol.* **25**, 83–98.

Committee on Infectious Diseases (1997). Diphtheria. In: *1997 Red Book: Report of the Committee on Infectious Diseases*, 24th edn, pp. 191–5. American Academy of Pediatrics, Elk Grove Village, IL.

Ding, X., Zeng, H., Schiering, N., Ringe, D. and Murphy, J.R. (1996). Identification of the primary metal ion-activation sites of the diphtheria *tox* repressor by X-ray crystallography and site-directed mutational analysis. *Nat. Struct. Biol.* **3**, 382–7.

Doukhan, L., Predich, M., Nair, G., Dussurget, O., Mandic-Mulec, I., Cole, S.T., Smith, D.R. and Smith, I. (1995). Genomic organization of the mycobacterial sigma gene cluster. *Gene* **165**, 67–70.

Dussurget, O., Rodriguez, M. and Smith, I. (1996). An *ideR* mutant of *Mycobacterium smegmatis* has derepressed siderophore production and an altered oxidative stress response. *Mol. Microbiol.* **22**, 535–44.

Edwards, D.C. and Seamer, P.A. (1960). The uptake of iron by *Corynebacterium diphtheriae* growing in submerged medium. *J. Gen. Microbiol.* **22**, 715–22.

Fourel, G., Phalipon, A. and Kaczorek, A. (1989). Evidence for direct regulation of diphtheria toxin gene transcription by an Fe^{2+}-dependent DNA-binding repressor, DtoxR, in *Corynebacterium diphtheriae*. *Infect. Immun.* **57**, 3221–5.

Freeman, V.J. (1951). Studies on the virulence of bacteriophage-infected strains of *Corynebacterium diphtheriae*. *J. Bacteriol.* **61**, 675–88.

Gill, D.M., Uchida, T. and Singer, R.A. (1972). Expression of diphtheria toxin genes carried by integrated and nonintegrated phage beta. *Virology* **50**, 665–8.

Goor, R.S. and Pappenheimer, A.M., Jr (1967). Studies on the mode of action of diphtheria toxin. II. Site of toxin action in cell-free extracts. *J. Exp. Med.* **126**, 899–912.

Goranson-Siekierke, J., Pohl, E., Hol, W.G.J. and Holmes, R.K. (1999). Anion-coordinating residues at binding site 1 are essential for the biological activity of the diphtheria toxin repressor (DtxR). *Infect. Immun.* **67**, 1806–11.

Greenfield, L., Bjorn, M.J., Horn, G., Fong, D., Buck, G.A., Collier, R.J. and Kaplan, D.A. (1983). Nucleotide sequence of the structural gene for diphtheria toxin carried by corynebacteriophage beta. *Proc. Natl Acad. Sci. USA* **80**, 6853–7.

Groman, N.B. (1953a). The relation of bacteriophage to the change of *Corynebacterium diphtheriae* from avirulence to virulence. *Science* **117**, 297–9.

Groman, N.B. (1953b). Evidence for the induced nature of the change from nontoxigenicity in *Corynebacterium diphtheriae* as a result of exposure to specific bacteriophage. *J. Bacteriol.* **66**, 134–91.

Gunter-Seeboth, K. and Schupp, T. (1995). Cloning and sequence analysis of *the Corynebacterium diphtheriae dtxR* homologue from *Streptomyces lividans* and *S. pilosus* encoding a putative iron repressor protein. *Gene* **166**, 117–19.

Hantke, K. (1984). Cloning of the repressor protein gene of iron-regulated systems in *Escherichia coli* K-12. *Mol. Gen. Genet.* **197**, 337–41.

Harnisch, J.P., Tronca, E., Nolan, C.M., Turck, M. and Holmes, K.K. (1989). Diphtheria among alcoholic urban adults. *Ann. Intern. Med.* **111**, 71–82.

Hill, P.J., Cockayne, A., Landers, P., Jorrissey, J.A., Sims, C.M. and Williams, P. (1998). SirR, a novel iron-dependent repressor in *Staphylococcus epidermidis*. *Infect. Immun.* **66**, 4123–9.

Holmes, R.K. (1975). Genetic aspects of toxinogenesis in bacteria. In: *Microbiology – 1975* (ed. D. Schlessinger), pp. 296–301. American Society for Microbiology, Washington, DC.

Holmes, R.K. (1998). Diphtheria, other corynebacterial infections, and anthrax. In: *Harrison's Principles of Internal Medicine* (eds A.S. Fauci *et al.*), pp. 892–9. McGraw-Hill, New York.

Holmes, R.K. (1999). Biology and molecular epidemiology of diphtheria toxin and the *tox* gene. *J. Infect. Dis. (Supplement on Diphtheria)*, in press.

Honjo, T., Nishizuka, Y. and Hayaishi, O. (1969). Adenosine diphosphoribosylation of aminoacyl transferase II by diphtheria toxin. *Cold Spring Harb. Symp. Quant. Biol.* **34**, 603–68.

Kaczorek, M., Delpeyroux, F., Chenciner, N., Streeck, R.E., Murphy, J.R., Boquet, P. and Tiollais, P. (1983). Nucleotide sequence and expression of the diphtheria *tox228* gene in *Escherichia coli*. *Science* **221**, 855–8.

Kanei, C., Uchida, T. and Yoneda, M. (1977). Isolation from *Corynebacterium diphtheriae* C7 (beta) of bacterial mutants that produce toxin in medium with excess iron. *Infect. Immun.* **18**, 203–9.

Karzon, D.T. and Edwards, K.M. (1988). Diphtheria outbreaks in immunized populations. *N. Engl. J. Med.* **318**, 41–3.

Krueger, K.M. and Barbieri, J.T. (1995). The family of bacterial ADP-ribosylating exotoxins. *Clin. Microbiol. Rev.* **8**, 34–47.

Lee, J.H., Wang, T., Ault, K., Liu, J., Schmitt, M.P. and Holmes, R.K. (1997). Identification and characterization of three new promoter/operators from *Corynebacterium diphtheriae* that are regulated by the diphtheria toxin repressor (DtxR) and iron. *Infect. Immun.* **65**, 4273–80.

Leong, D. and Murphy, J.R. (1985). Characterization of the diphtheria *tox* transcript in *Corynebacterium diphtheriae* and *Escherichia coli*. *J. Bacteriol.* **163**, 1114–19.

Litwin, C.M. and Calderwood, S.B. (1993). Role of iron in regulation of virulence genes. *Clin. Microbiol. Rev.* **6**, 137–49.

Locke, A. and Main, E.R. (1931). The relation of copper and iron to the production of toxin and enzyme action. *J. Infect. Dis.* **48**, 419–35.

Matsuda, M. and Barksdale, L. (1966). Phage-directed synthesis of diphtherial toxin in non-toxinogenic *Corynebacterium diphtheriae*. *Nature* **210**, 911–13.

Matsuda, M. and Barksdale, L. (1967). System for the investigation of the bacteriophage-directed synthesis of diphtherial toxin. *J. Bacteriol.* **93**, 722–30.

Mueller, J.H. and Miller, P.A. (1941). Production of diphtheria toxin of high potency (100 LF) on a reproducible medium. *J. Immunol.* **40**, 21–32.

Murphy, J.R. and Bacha, P. (1979). Regulation of diphtheria toxin production. In: *Microbiology – 1979* (ed. D. Schlessinger), pp. 181–6. American Society for Microbiology, Washington, DC.

Murphy, J.R., Pappenheimer, A.M., Jr and de Borms, S.T. (1974). Synthesis of diphtheria *tox*-gene products in *Escherichia coli* extracts. *Proc. Natl Acad. Sci. USA* **71**, 11–15.

Murphy, J.R., Skiver, J. and McBride, G. (1976). Isolation and partial characterization of a corynebacteriophage beta, *tox* operator constitutive-like mutant lysogen of *Corynebacterium diphtheriae*. *J. Virol.* **18**, 235–44.

Murphy, J.R., Michel, J.L. and Teng, M. (1978). Evidence that the regulation of diphtheria toxin production is directed at the level of transcription. *J. Bacteriol.* **135**, 511–16.

Nakao, H., Pruckler, J.M., Mazurova, I.K., Narvskaia, O.V., Glushkevich, T., Marijevski, V.F., Kravetz, A.N., Fields, B.S., Wachsmuth, I.K. and Popovic, T. (1996). Heterogeneity of diphtheria toxin gene, *tox*, and its regulatory element, *dtxR*, in *Corynebacterium diphtheriae* strains causing epidemic diphtheria in Russia and Ukraine. *J. Clin. Microbiol.* **34**, 1711–16.

Oguiza, J.A., Tao, X., Marcos, A.T., Martin, J.F. and Murphy, J.R. (1995). Molecular cloning, DNA sequence analysis, and characterization of the *Corynebacterium diphtheriae dtxR* homolog from *Brevibacterium lactofermentum*. *J. Bacteriol.* **177**, 465–7.

Pappenheimer, A.M., Jr (1955). The pathogenesis of diphtheria. In: *Mechanisms of Microbial Pathogenicity. Fifth Symposium of the Society for General Microbiology*, pp. 40–56. Cambridge University Press, Cambridge.

Pappenheimer, A.M., Jr (1982). Diphtheria: studies on the biology of an infectious disease. *Harvey Lect.* **76**, 45–73.

Pappenheimer, A.M., Jr and Gill, D.M. (1973). Diphtheria. *Science* **182**, 353–8.

Pappenheimer, A.M., Jr and Johnson, S.J. (1936). Studies in diphtheria toxin production. I. The effect of iron and copper. *Br. J. Exp. Pathol.* **17**, 335–41.

Pappenheimer, A.M., Jr and Murphy, J.R. (1983). Studies on the molecular epidemiology of diphtheria. *Lancet* **ii**, 923–6.

Pohl, E., Qiu, X., Must, L.M., Holmes, R.K. and Hol, W.G.J. (1997). Comparison of high-resolution structures of the diphtheria toxin repressor in complex with cobalt and zinc at the cation–anion binding site. *Protein Sci.* **6**, 1114–18.

Pohl, E., Holmes, R.K. and Hol, W.G.J. (1998a). Motion of the DNA-binding domain with respect to the core of the diphtheria toxin repressor revealed in the crystal structures of apo- and holo-DtxR. *J. Biol. Chem.* **273**, 22420–7.

Pohl, E., Holmes, R.K. and Hol, W.G.J. (1998b). Crystal structure of the iron-dependent regulator (IdeR) from *Mycobacterium tuberculosis* shows both metal binding sites fully occupied. *J. Mol. Biol.*, **285**, 1145–56.

Qiu, X., Verlinde, C.L.M.J., Zhang, S., Schmitt, M.P., Holmes, R.K. and Hol, W.G.J. (1995). Three-dimensional structure of the diphtheria toxin repressor in complex with divalent cation co-repressors. *Structure* **3**, 87–100.

Qiu, X., Pohl, E., Holmes, R.K. and Hol, W.G.J. (1996). High-resolution structure of the diphtheria toxin repressor complexed with cobalt and manganese reveals an SH3-like third domain and suggests a possible role of phosphate as co-corepressor. *Biochemistry* **35**, 12292–302.

Rakhmanova, A.G., Lumio, J., Groundstroem, K., Valova, E., Nosikova, E., Tanasijchuk, T. and Saikku, J. (1996). Diphtheria outbreak in St. Petersburg: clinical characteristics of 1860 adult patients. *Scand. J. Infect. Dis.* **28**, 37–40.

Rappuoli, R., Perugini, M. and Falsen, E. (1988). Molecular epidemiology of the 1984–1986 outbreak of diphtheria in Sweden. *N. Engl. J. Med.* **318**, 12–14.

Ratti, G., Rappuoli, R. and Giannini, G. (1983). The complete nucleotide sequence of the gene coding for diphtheria toxin in the corynephage omega (*tox$^+$*) genome. *Nucleic Acids Res.* **11**, 6589–95.

Schiering, N., Tao, X., Zeng, H., Murphy, J.R., Petsko, G.A. and Ringe, D. (1995). Structures of the apo- and the metal ion-activated forms of the diphtheria *tox* repressor from *Corynebacterium diphtheriae*. *Proc. Natl Acad. Sci. USA* **92**, 9843–50.

Schmitt, M.P. (1997a). Transcription of the *Corynebacterium diphtheriae hmuO* gene is regulated by iron and heme. *Infect. Immun.* **65**, 4634–41.

Schmitt, M.P. (1997b). Utilization of host iron sources by *Corynebacterium diphtheriae*: identification of a gene whose product is homologous to eukaryotic heme oxygenases and is required for acquisition of iron from heme and hemoglobin. *J. Bacteriol.* **179**, 838–45.

Schmitt, M.P. and Holmes, R.K. (1991a). Iron-dependent regulation of diphtheria toxin and siderophore expression by the cloned *Corynebacterium diphtheriae* repressor gene *dtxR* in *C. diphtheriae* C7 strains. *Infect. Immun.* **59**, 1899–904.

Schmitt, M.P. and Holmes, R.K. (1991b). Characterization of a defective diphtheria toxin repressor (*dtxR*) allele and analysis of *dtxR* transcription in wild-type and mutant strains of *Corynebacterium diphtheriae*. *Infect. Immun.* **59**, 3903–8.

Schmitt, M.P. and Holmes, R.K. (1993). Analysis of diphtheria toxin repressor–operator interactions and characterization of a mutant repressor with decreased binding activity for divalent metals. *Mol. Microbiol.* **9**, 173–81.

Schmitt, M.P. and Holmes, R.K. (1994). Cloning, sequence, and footprint analysis of two promoter/operators from *Corynebacterium diphtheriae* that are regulated by the diphtheria toxin repressor and iron. *J. Bacteriol.* **176**, 1141–9.

Schmitt, M.P., Twiddy, E.M. and Holmes, R.K. (1992). Purification and characterization of the diphtheria toxin repressor. *Proc. Natl Acad. Sci. USA* **89**, 7576–80.

Schmitt, M.P., Predich, M., Doukhan, L., Smith, I. and Holmes, R.K. (1995). Characterization of an iron-dependent regulatory protein (IdeR) of *Mycobacterium tuberculosis* as a functional homolog of the diphtheria toxin repressor (DtxR) from *Corynebacterium diphtheriae*. *Infect. Immun.* **63**, 4284–9.

Schmitt, M.P., Talley, B.G. and Holmes, R.K. (1997). Characterization of lipoprotein IRP1 from *Corynebacterium diphtheriae*, which is regulated by the diphtheria toxin repressor (DtxR) and iron. *Infect. Immun.* **65**, 5364–7.

Tai, S.S. and Zhu, Y.Y. (1995). Cloning of a *Corynebacterium diphtheriae* iron-repressible gene that shares sequence homology with the AhpC subunit of alkyl hydroperoxide reductase of *Salmonella typhimurium*. *J. Bacteriol.* **177**, 3512–17.

Tai, S.-P.S., Kraft, A.E., Nootheti, P. and Holmes, R.K. (1990). Coordinate regulation of siderophore and diphtheria toxin production by iron in *Corynebacterium diphtheriae*. *Microb. Pathog.* **9**, 267–73.

Tao, X. and Murphy, J.R. (1992). Binding of the metalloregulatory protein DtxR to the diphtheria *tox* operator requires a divalent heavy metal ion and protects the palindromic sequence from DNase I digestion. *J. Biol. Chem.* **267**, 21761–4.

Tao, X. and Murphy, J.R. (1993). Cysteine-102 is positioned in the metal binding activation site of the *Corynebacterium diphtheriae* regulatory element DtxR. *Proc. Natl Acad. Sci. USA* **90**, 8524–8.

Tao, X. and Murphy, J.R. (1994). Determination of the minimal essential nucleotide sequence for diphtheria tox repressor binding by *in vitro* affinity selection. *Proc. Natl Acad. Sci. USA* **91**, 9646–50.

Tao, X., Boyd, J. and Murphy, J.R. (1992). Specific binding of the diphtheria *tox* regulatory element DtxR to the *tox* operator requires divalent heavy metal ions and a 9-base-pair interrupted palindromic sequence. *Proc. Natl Acad. Sci. USA* **89**, 5897–901.

Tao, X., Nikolaus, S., Zeng, H., Ringe, D. and Murphy, J.R. (1994). Iron, DtxR and the regulation of diphtheria toxin expression. *Mol. Microbiol.* **14**, 191–7.

Tao, X., Zeng, H. and Murphy, J.R. (1995). Transition metal ion activation of DNA binding by the diphtheria *tox* repressor requires the formation of stable homodimers. *Proc. Natl Acad. Sci. USA* **92**, 6803–7.

Uchida, T., Gill, D.M. and Pappenheimer, A.M., Jr (1971). Mutation in the structural gene for diphtheria toxin carried by temperate phage *β*. *Nature (London) New Biol.* **233**, 8–11.

Vitek, C.R. and Wharton, M. (1998). Diphtheria in the former Soviet Union: reemergence of a pandemic disease. *Emerg. Infect. Dis.* **4**, 539–50.

Wang, Z., Schmitt, M.P. and Holmes, R.K. (1994). Characterization of mutations that inactivate the diphtheria toxin repressor gene (*dtxR*). *Infect. Immun.* **62**, 1600–8.

Welkos, S.L. and Holmes, R.K. (1981a). Regulation of toxinogenesis in *Corynebacterium diphtheriae*. I Mutations in bacteriophage beta that alter the effects of iron on toxin production. *J. Virol.* **37**, 936–45.

Welkos, S.L. and Holmes, R.K. (1981b). Regulation of toxinogenesis in *Corynebacterium diphtheriae*. II Genetic mapping of a *tox* regulatory mutation in bacteriophage beta. *J. Virol.* **37**, 946–54.

White, A., Ding, X., vanderSpek, J.C., Murphy, J.R. and Ringe, D.R. (1998). Structure of the metal-iron-activated diphtheria toxin repressor/*tox* operator complex. *Nature* **394**, 502–6.

Wilks, A. and Schmitt, M.P. (1998). Expression and characterization of a heme oxygenase (HmuO) from *Corynebacterium diphtheriae*. Iron acquisition requires oxidative cleavage of the heme macrocycle. *J. Biol. Chem.* **273**, 837–41.

Wilson, B.A. and Collier, R.J. (1992). Diphtheria toxin and *Pseudomonas aeruginosa* exotoxin A: active-site structure and enzymic mechanism. *Curr. Top. Microbiol. Immunol.* **175**, 27–41.

Yamaizumi, M., Mekada, E., Uchida, T. and Okada, Y. (1978). One molecule of diphtheria toxin fragment A introduced into a cell can kill the cell. *Cell* **15**, 245–50.

6

Cholera toxin and *Escherichia coli* heat-labile enterotoxin

Timothy R. Hirst

INTRODUCTION

In past few years, there have been over 1000 publications citing studies on cholera toxin and related enterotoxins. Consequently, cholera toxin (Ctx) from *Vibrio cholerae* and the related heat-labile enterotoxins (Etx) from certain toxinogenic strains of *Escherichia coli* have emerged as the most well-characterized virulence factors produced by pathogenic micro-organisms.

Studies on the biogenesis of Ctx in *V. cholerae* have provided important insights into the mechanisms of gene regulation and protein secretion in bacteria, and resulted in Ctx being a paradigm of virulence factor expression. Studies on the interaction, uptake and action of Ctx and Etx in epithelial cells have illuminated the mechanisms by which the toxins cause severe, and at times, life-threatening diarrhoeal disease. They have also provided a greater understanding of cellular events associated with vesicular movement and targeting, a central feature of eukaryotic cell biology.

However, perhaps one of the most surprising features of Ctx and Etx is their exceptional immunogenicity. During natural infection, or upon artificial administration at mucosal sites, the toxins elicit such a potent anti-toxin antibody response that they have been heralded as the most potent mucosal immunogens yet identified. Moreover, addition of minute, subtoxic quantities of Ctx or Etx augments the immunogenicity of other antigens, a finding that has spawned considerable interest in the potential of the toxins as adjuvants for use in mucosal vaccines.

More recently, the B-subunit component (see below) of Ctx and Etx has been reported to down-regulate pro-inflammatory immune responses, which has led to their successful testing in animal models as therapeutic or prophylactic agents to treat or prevent inflammatory autoimmune disorders such as arthritis, diabetes or experimental autoimmune encephalomyelitis (a model of multiple sclerosis). These developments should continue the growing interest in Ctx and Etx, provide an opportunity for a greater understanding of immune regulation and offer new therapeutic possibilities in human medicine.

This chapter contains what might be considered an overview of the 'life and times' of these remarkable toxins, from the genetics of toxin expression, assembly and secretion in bacteria to their interaction, uptake and action in disease, to their capacity to orchestrate dramatic effects on immune systems, and to their use as novel therapeutic agents.

CHOLERA AND RELATED ENTEROPATHIES

Robert Koch, the great German bacteriologist, established that a 'comma-bacillus', now called *Vibrio cholerae* serogroup *O1*, was the causative agent of asiatic cholera (Koch, 1884). The organism, which is transmitted in contaminated water, colonizes the surface of the small bowel and induces the production of copious quantities of watery diarrhoea. The rapid spread of the organism and its capacity to reduce a previously healthy individual to the point of death in as little as 6–8 h made cholera a disease to be feared. However, with the widespread introduction of fluid and electrolyte replacement therapies (intravenous or oral), the mortality rate can be reduced to less than 1%.

None the less, the devastating impact of cholera is all too evident, as seen in the Rwandan refugee camps in 1994 when an estimated 600 000 cases of cholera occurred over a 3-week period, including 45 000 deaths (Waldman, 1998). The world-wide incidence of cholera is difficult to evaluate, although it has been estimated that there are approximately 8 million cases per annum, including 124 000 deaths (Black, 1986). In 1992, a new serogroup of *V. cholerae* emerged that caused a large number of cholera cases and deaths in India and Bangladesh, with over 100 000 cases of cholera and 1473 deaths reported in Bangladesh alone during the first 3 months of 1993 (Albert *et al.*, 1993; Sack, 1996). The strain did not react with antibodies against the *O1*-antigen or any of the other 137 non-*O1* serogroups, and has been designated *O139*. Although *V. cholerae O139* produces a different O-antigen, it shares many pathogenic features with *V. cholerae O1* (biotype El Tor), including production of Ctx (Nakashima *et al.*, 1995).

In 1956, S.N. De noted that the common gut bacterium *E. coli* was also capable of causing a cholera-like disease (De *et al.*, 1956). By the early 1970s it had become clear that enterotoxigenic *E. coli* (ETEC) were of global significance as agents of diarrhoeal disease in both humans and domestic animals. ETEC have a world-wide distribution, but their significance as human pathogens is most evident in developing countries, where it has been estimated that approximately 20% of all acute (life-threatening) diarrhoeal cases in children under 5 years of age are due to ETEC (Evans *et al.*, 1977; Black *et al.*, 1980; Agbonlahor and Odugbemi, 1982; DeMol *et al.*, 1983; Echeverria *et al.*, 1985; Steffen, 1986). A community-based study of children under 5 years of age in rural villages in Bangladesh revealed an incidence of 1–2 ETEC episodes per child per year (Black *et al.*, 1981). For travellers to developing countries, ETEC are the most commonly identified cause of diarrhoea (Black, 1986; Steffen, 1986). The secretory diarrhoea produced by ETEC is due to their ability to produce either an oligomeric heat-labile enterotoxin (Etx) that is structurally and functionally homologous to Ctx (see below) or a heat-stable enterotoxin (ST) or both (for a review, see Nataro and Kaper, 1998).

Enterotoxigenic strains of *E. coli* also cause diarrhoeal disease (colibacillosis) in neonatal pigs, calves, lambs and chickens with concomitant losses to the farming community (Smith and Halls, 1967; Tsuji *et al.*, 1988; Inoue *et al.*, 1993; Moon and Bunn, 1993). Most of the detailed studies of the toxins produced by veterinary ETEC strains have been from pigs. This has revealed that porcine heat-labile Etx (pEtx) shares 98% sequence identity with hEtx produced by ETEC strains of human origin (Leong *et al.*, 1985; Yamamoto *et al.*, 1987; Inoue *et al.*, 1993).

STRUCTURE OF CHOLERA TOXIN AND RELATED ENTEROTOXINS

The structural analysis of Ctx was initiated by Finkelstein and LoSpalluto with the purification of Ctx from sterile culture filtrates of *V. cholerae* (Finkelstein *et al.*, 1966; Finkelstein and LoSpalluto, 1969). The successful crystallization and X-ray analyses of Etx and Ctx by W.G. Hol and co-workers (Sixma *et al.*, 1991) and E. Westbrook and associates (Zhang *et al.*, 1995a, b) respectively, confirmed the predictions of earlier biochemical studies that the toxins are comprised of six non-covalently associated subunits: one A-subunit of *ca.* 28 000 Da arranged on a toroidal ring of five B-subunits, each with a relative molecular mass of approximately 12 000 (Gill, 1976; Clements and Finkelstein, 1979; Kunkel and Robertson, 1979; Gill *et al.*, 1981; Geary *et al.*, 1982). When Ctx was analysed by sodium dodecyl sulfate–polyacrylamide gel electrophoresis (SDS–PAGE) under non-reducing conditions, the A-subunit migrated as a single polypeptide with an apparent electrophoretic mobility of *ca.* 28 000 Da, but under reducing conditions, separated into two polypeptides, A1 (M_r = *ca.* 22 000) and A2 (M_r = *ca.* 5500), corresponding to the *N*-terminal and *C*-terminal fragments of the A-subunit, respectively (Gill, 1976).

Primary sequence

The primary amino acid sequences of the A- and B-subunits from hEtx, pEtx, chicken Etx (cEtx) and Ctx (from classical and *El Tor* strains of *V. cholerae O1* and from *V. cholerae O139*) have been deduced from cloned DNA sequences (Dallas and Falkow, 1980; Gennaro *et al.*, 1982; Spicer and Noble, 1982; Dallas, 1983; Gennaro and Greenaway, 1983; Lockman and Kaper, 1983; Mekalanos *et al.*, 1983; Yamamoto and Yokota, 1983; Lockman *et al.*, 1984; Yamamoto *et al.*, 1984, 1987; Dykes *et al.*, 1985; Leong *et al.*, 1985; Takao *et al.*, 1985; Sanchez and Holmgren, 1989; Brickman *et al.*, 1990; Dams *et al.*,

1991; Inoue *et al.*, 1993). Figure 6.1 shows a comparison of the complete amino acid sequences of the A- and B-subunits of Ctx from the classical *V. cholerae O1* strain 569B and those of hEtx from an *E. coli* strain H74–114 of human origin (Leong *et al.*, 1985; Dams *et al.*, 1991).

From the DNA sequence it was apparent that both the A- and B-subunits of Ctx and Etx are synthesized as longer precursors with amino-terminal signal sequence extensions of 18 and 21 amino acids, respectively (denoted by negative numbers in Figure 6.1).

A-subunit

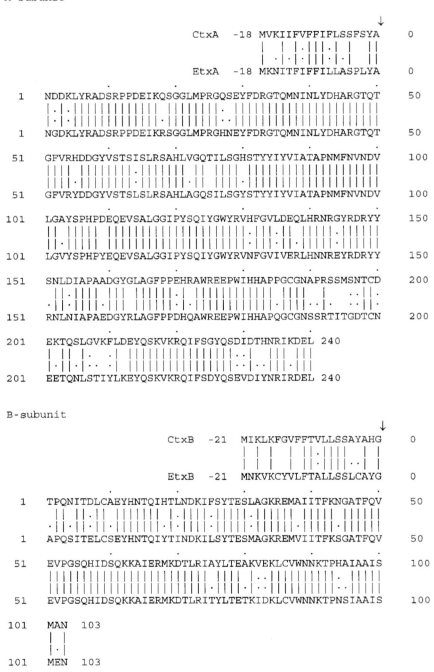

FIGURE 6.1 Primary amino acid sequence alignments of the A- and B-subunits of cholera toxin and *E. coli* heat-labile enterotoxin. Cholera toxin sequences are from the classical biotype *V. cholerae O1* strain 569B (Dams *et al.*, 1991); heat-labile enterotoxin sequences are derived from *E. coli* strain H74–114 of human origin (Leong *et al.*, 1985). Amino acids comprising the signal sequences of the A- and B-subunits are denoted by negative numbers. The arrows indicate the positions of cleavage between the signal sequences and the mature polypeptides.

Such signal sequences are involved in targeting precursors to the cytoplasmic membrane and are cleaved off during translocation to yield a mature polypeptide.

The mature A-subunits of the Ctx (CtxA) and Etx (EtxA), shown in Figure 6.1, are 240 amino acids in length ($M_r = ca.$ 27 200), and share considerable sequence identity, with 197 identical residues (82.1%). Similarly, the mature B-subunits (CtxB and EtxB) are comprised of 103 amino acids ($M_r = ca.$ 11 800) of which 86 residues (83.5%) are identical. The only region where there is little homology between Etx and Ctx is between amino acid residues 190 and 212 of the A-subunits, where the sequence identity is 34.8%. This region contains an exposed loop formed by the generation of a disulfide bond between two cysteine residues at positions 187 and 199. The loop can be easily 'nicked' by exogenous proteases, thereby cleaving the A-subunit to its A1/A2 polypeptides.

Commercially available preparations of Ctx contain A-subunits that have already undergone proteolytic nicking, with cleavage occurring between residues 192 and 193 (Xia *et al.*, 1984). In contrast, Etx purified from *E. coli* is unnicked, although the A-subunit can be easily cleaved by exogenously added trypsin, yielding A1 and A2 polypeptides similar to those of Ctx (Clements and Finkelstein, 1979; Clements *et al.*, 1980).

Recent studies have also revealed that enterocytes possess an apically located endoprotease capable of nicking the A-subunits of *E. coli*-derived Etx or Ctx (Lencer *et al.*, 1997). The nicking of the A-subunit activates the ADP-ribosylating activity of the A1-polypeptide (see below) and is thus important for toxin action. The primary sequence of the A1 polypeptide also reveals that only two out of the 192 residues are lysine – an exceptionally low proportion compared with most other proteins. This may have arisen to limit the possibility of ubiquitination and proteasome-mediated degradation of the A1-fragment during its entry into the eukaryotic cell (see below).

The primary sequences of the B-subunits of Ctx and Etx reveal a number of other distinctive features. Firstly, there is a pair of conserved cysteine residues (at positions 9 and 86), which were subsequently shown to form a single intramolecular disulfide bridge (Hardy *et al.*, 1988; Sixma *et al.*, 1991), the formation of which is essential for toxin assembly (see below). Secondly, they have a high content of charged residues, which for each B-subunit of Ctx include 3 Asp, 7 Glu, 9 Lys and 3 Arg residues (Figure 6.1). Thirdly, each B-subunit possesses a single conserved tryptophan residue (at position 88), which forms the floor of the receptor binding pocket.

Analyses of a variety of *ctx* genes from different strains of *V. cholerae* have revealed a small number of amino acid differences in the B-subunit sequences compared with that shown in Figure 6.1, and also in the subunits of Etx from human and porcine origin (Tables 6.1 and 6.2). These minor differences have no known effect on the 3D structure of the toxin or functional properties, but they do appear to give rise to significant novel antigenic specificities.

Quaternary structure

The successful determination of the X-ray structure of porcine heat-labile enterotoxin from *E. coli* by W.G.J. Hol

TABLE 6.1 Amino acid sequence heterogeneity in B-subunits of cholera toxin

V. cholerae strain	Biotype	Amino acid number[a]			Reference
		B-subunit			
		+18	+47	+54	
569B	Classical	His	Thr	Gly	Dams *et al.* (1991)
41	Classical	His	Thr	Gly	Dams *et al.* (1991)
O395	Classical	His	Thr	Gly	Dams *et al.* (1991)
2125	El Tor	Tyr	Ile	Gly	Dams *et al.* (1991)
62746	El Tor	Tyr	Ile	Gly	Dams *et al.* (1991)
3083	El Tor	Tyr	Ile	(Ser)	Dams *et al.* (1991), Brickman *et al.* (1990)

[a]Corresponds to the amino acid number in the mature subunit sequence.

TABLE 6.2 Amino acid sequence heterogeneity in A- and B-subunits of Etx from *E. coli* of human and porcine origin

E. coli strain	Origin	Amino acid number[a]				Reference[b]
		A-subunit				
		+4	+212	+213	+238	
H74–114	Human	Lys	Lys	Glu	Asp	1
H10407[c]	Human	Lys	Arg	Lys	Asn	2, 3
P307	Porcine	Arg	Arg	Glu	Asp	4
		B-subunit				
		+4	+13	+46	+102	
H74–114	Human	Ser	His	Ala	Glu	5
H10407c	Human	Ser	Arg	Ala	Glu	3, 6
P307	Porcine	Thr	Arg	Glu	Lys	5

[a]Corresponds to the amino acid number in the mature subunit sequence.
[b]1, Webb and Hirst, unpublished results; 2, Yamamoto *et al.* (1984); 3, Inoue *et al.* (1993); 4, Dykes *et al.* (1985); 5, Leong *et al.* (1985); 6, Yamamoto and Yokota (1983).
[c]The amino acid sequences of the A- and B-subunits of cEtx from chicken ETEC isolate were found to be identical to those of strain H10407 (Inoue *et al.*, 1993).

and co-workers in 1991 provided an exquisite insight into the tertiary and quaternary structure of this toxin family (Sixma *et al.*, 1991, 1993).

This revealed that the A1-fragment had a single-domain structure with a somewhat triangular shape, with one side abutting a 23 residue α-helix formed by the A2-polypeptide (Figure 6.2). The *C*-terminal segment of the A2-polypeptide was also revealed to extend downwards into the central pore of the doughnut-shaped ring of five B-subunits (Figure 6.2). Each B-subunit in the pentameric structure was folded into two three-stranded anti-parallel β-sheets (with one sheet on each side of the monomer facing an adjacent B-subunit) and a large central helix positioned at the wall of the central pore, with the five helices (one from each subunit) forming a pentagonal helix barrel. This forms a pore, 30 Å long, with a diameter ranging from *ca.* 11 Å near the surface at which the A-subunit is positioned to *ca.* 15 Å on the lower surface (see Figure 6.2). The three-stranded anti-parallel β-sheets on each side of the B-subunit interact directly with the β-sheets of the adjacent B-subunits (via multiple hydrogen bonds) – a feature that contributes to the remarkable stability of the B-pentamer. The B-subunit receptor binding site is found on the lower convoluted surface, with each B-monomer possessing a single receptor-site (Hol *et al.*, 1995; Sixma *et al.*, 1992).

The structural fold of EtxB has also been found in other proteins; staphylococcal nuclease, the anticodon binding domain of asp-tRNA synthase, which binds oligonucleotides, and verotoxin B-subunit pentamer, which binds a different glycolipid receptor, Gb3 (Arnone *et al.*, 1971; Murzin, 1993; Sixma *et al.*, 1993). The similarity of the B-subunits of Etx (and Ctx) to these other oligosaccharide and oligonucleotide binding proteins has led to the coining of the term 'OB-fold' for proteins that possess this overall structure, even though they do not share any homology in their primary sequences.

The porcine variant of Etx used in the crystallographic determination was purified from a recombinant *E. coli K-12* strain and, as a consequence, the A-subunit was unnicked. The residues 189–195 corresponding to the exposed loop and nicking site between the A1- and A2-polypeptides was not revealed in the electron density map, presumably because of the flexibility of this region. Likewise, the three amino-terminal and four carboxy-terminal amino acids of the A-subunit were not visible within the refined structure.

The subsequent X-ray crystallographic determination of the Ctx structure by Westbrook and co-workers revealed a molecule with a similar overall fold to that of pEtx. The only region of major structural dissimi-larity was found in the A2-segment from residue 227, which in Ctx appears to enter the central pore of the B-pentamer as an α-helix rather than as an extended chain (Zhang *et al.*, 1995a). In addition, the last four (KDEL) residues were clearly visible in the electron density of Ctx and may reflect the more compact structure of the CtxA2- when compared with the EtxA2-fragment.

The A1-polypeptide (amino acids 1–194) contains the enzyme active site capable of ADP-ribosylation (for a fuller description see below). When the crystal structure of Etx was compared with that of another ADP-ribosylating toxin, exotoxin A (ETA) from *Pseudomonas aeruginosa*, it was apparent that 44 residues in the A1-polypeptide could be superimposed with residues in the enzymic domain of ETA (Sixma *et al.*, 1993). Of these, only three amino acids were identical: Tyr 6, Ala 69 and Glu 102 in Etx, with the latter corresponding to the active site residue Glu 553 of ETA.

FIGURE 6.2 Ribbon diagram of the crystal structure of the porcine variant of *E. coli* heat-labile enterotoxin. The A-subunit (comprising the A1- and A2-fragments) forms a triangular-shaped structure; the long α-helix of the A2-fragment is clearly visible to the left-hand side. The B-subunits form a pentameric ring with a central pore in which the *C*-terminal portion of the A2-fragment is situated. The GM1-receptor binding sites in the B-pentamer are located on the lower convoluted side of the molecule.

BIOGENESIS OF CHOLERA TOXIN AND RELATED ENTEROTOXINS

Toxin gene organization

One of the important discoveries in cholera research in the past few years was made by Waldor and Mekalanos, who found that the structural genes, *ctxA* and *ctxB*, encoding the A- and B-subunits of Ctx are located on a lysogenic filamentous bacteriophage, designated CTXΦ (Waldor and Mekalanos, 1996).

The *ctxAB* genes are organized as an operon located near one end of the phage genome (Figure 6.3). The phage genome is comprised of a 7-kb DNA segment containing a 4.5-kb central core region and a 2.7-kb repetitive sequence (RS), which is often duplicated. The RS sequences are capable of undergoing RecA-dependent homologous recombination, to give tandem amplification of the entire phage genome and flanking RS1 repeats. The core region of the CTXΦ contains four genes in addition to the *ctxAB* operon, namely *zot* (encoding the so-called zonula occludens toxin), *ace* (encoding an accessory cholera enterotoxin), *cep* (encoding a core-encoded pilin) and *or fU* (encoding a product of unknown function) (Mekalanos, 1983; Mekalanos *et al.*, 1983; Goldberg and Mekalanos, 1986; Fasano *et al.*, 1991; Baudry *et al.*, 1992; Pearson *et al.*, 1993; Trucksis *et al.*, 1993). The RS sequences also have at least four open reading frames (*rstABCR*), which determine expression of a site-specific recombination system, capable of catalysing the integration of the CTXΦ into the chromosomes of non-toxinogenic *V. cholerae* strains at a particular 17-bp target sequence, termed *attRS1* (Pearson *et al.*, 1993).

Waldor and Mekalanos showed that the CTXΦ can undergo excision from the chromosome and replicate as a plasmid yielding a single-stranded phage DNA, which is incorporated into the filamentous CTXΦ particle (Waldor and Mekalanos, 1996). The similarity in gene organization of the *cep, orfU, ace* and *zot* genes to those found in the filamentous bacteriophage M13 suggests that their biological function and properties need to be reassessed. The sequence identity between Cep and gene VIII, encoding the major capsid protein of M13, suggests that Cep may be a major structural component of the phage particle rather than a pilus, as originally proposed (Ho *et al.*, 1990).

Several filamentous bacteriophages use pili as receptors for infection of host bacterial cells. In *V. cholerae*, a toxin co-regulated pilus (termed TCP) necessary for bacterial colonization of the human intestine has also been found to be the functional receptor for the CTXΦ (Waldor and Mekalanos, 1996). Expression of TCP is optimal under conditions found in the intestine, suggesting that the CTXΦ may have evolved to be a highly efficient transmissible agent within the gastrointestinal environment.

In contrast to the location of the *ctx* genes on a lysogenic phage, the genes specifying the production of Etx are found on large naturally occurring plasmids, called *ENT* plasmids, present in enterotoxinogenic strains of *E. coli* (Gyles *et al.*, 1977). In many instances, these plasmids have been shown to be transmissible by conjugation, and often to possess additional genes for drug resistance and colonization antigens (Smith and Linggood, 1971; Gyles *et al.*, 1977).

Operon structure

Because the *E. coli* Etx genes are plasmid-located, Falkow and co-workers were able to use the emerging techniques of genetic engineering in the mid-1970s to isolate and clone the Etx genes from an *ENT* plasmid (P307) derived from an ETEC strain of porcine origin (So *et al.*, 1978). The genes encoding the A- and

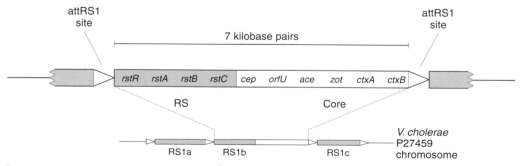

FIGURE 6.3 Structure of the CTX DNA element and CTXΦ. The 7.0-kb CTX element is comprised of three distinct subrepeats: the 4.5-kb core region (not shaded), the 2.5–2.7-kb RS sequences (shaded) and the 17–18-bp attRS1 sites (open triangles). The 10 identified genes in the core region and RS sequences are shown. (Adapted from Waldor and Mekalanos, 1996.)

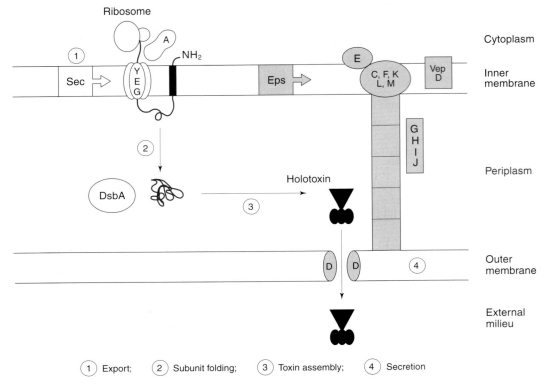

1 Export;   2 Subunit folding;   3 Toxin assembly;   4 Secretion

FIGURE 6.4 Schematic model of the pathway of toxin biogenesis in *E. coli* (steps 1–3 only) and in *V. cholerae* (steps 1–4). Step 1, precursor A- and B-subunits are exported across the cytoplasmic membrane via a Sec-dependent pathway involving SecA, SecE, SecY and SecG (SecD, SecF, YajC not shown), followed by their cleavage by LepB (not shown) to yield mature A- and B-subunits that are released into the periplasm. Step 2, folding of the mature subunits is mediated both by intramolecular interactions within the polypeptide chain and by intermolecular interactions with factors that may facilitate or catalyse folding events, such as DsbA and peptidyl prolyl *cis–trans* isomerase. Step 3, holotoxin assembly occurs via subunit–subunit interactions to give rise to an AB$_5$ structure. In *E. coli* the holotoxin is thought to remain entrapped in the periplasm, only being released owing to non-specific disruption of the bacterial outer membrane by bile salts or intestinal proteases. Step 4, holotoxin is secreted across the outer membrane of *V. cholerae* via the Eps-secretion machinery, comprising an outer membrane pore (EpsD) and at least 10 other proteins (EpsC, EpsE-N). It is likely that the secretion pore is gated to prevent the non-specific loss of resident periplasmic proteins or the influx of low molecular mass solutes. Energy in the form of ATP and pmf is required for holotoxin translocation across the outer membrane.

B-subunits of Etx have been variously designated by early investigators as either *eltA/eltB* or *toxA/toxB*, with the use of subscripts, H or P, in some instances to denote the origin of the gene as deriving from a human or porcine strain of *E. coli* . Because of these disparities in nomenclature, the author has chosen to use a general mnemonic, *etxA* and *etxB*, for the genes encoding the A- and B-subunits of *E. coli* heat-labile enterotoxin, irrespective of the source of the ETEC infection.

The *etx* genes from P307, and two human ETEC isolates, H10407 and H74–114, have been sequenced (Dallas and Falkow, 1980; Spicer and Noble, 1982; Yamamoto and Yokota, 1982, 1983; Yamamoto *et al.*, 1984, 1987; Leong *et al.*, 1985; Webb and Hirst, unpublished). This revealed that *etxA* and *etxB* genes overlap one another by four nucleotides, with the last four nucleotides of *etxA* (including the stop codon for

the A-subunit) being the first four nucleotides of *etxB*. When Mekalanos *et al.* (1983) sequenced the *ctxAB* operon, an identical 4-bp overlap between the *ctxA* and *ctxB* genes was found (Lockman and Kaper, 1983; Mekalanos *et al.*, 1983).

A consensus promoter sequence for RNA polymerase binding is located upstream of *etxA*, whereas no such sequence is found in the nucleotides proximal to *etxB* (Dallas and Falkow, 1979; Dallas *et al.*, 1979). Consequently, the enterotoxin genes are transcribed as a polycistronic mRNA from a single promoter proximal to *etxA*.

Upstream of the single promoter site proximal to the *ctxA* gene is a series of three to eight tandemly repeated copies of a TTTTGAT motif, involved in ToxR-mediated activation of Ctx expression (see below) (Mekalanos *et al.*, 1983; Miller and Mekalanos, 1984; Miller *et al.*, 1987).

Regulation of toxin expression

In *V. cholerae*, several genes required for virulence, including Ctx gene expression, are under the control of ToxR (for an excellent review, see DiRita, 1992). This protein influences the transcription of the *ctx* operon in response to changes in several environmental factors, including aeration, osmolarity, pH, temperature and the availability of several amino acids (DiRita, 1992; Miller and Mekalanos, 1988; Mekalanos, 1992). ToxR is a 32.5-kDa integral cytoplasmic membrane protein, which is essential for pathogenesis of *V. cholerae* in humans (Miller and Mekalanos, 1984; Miller *et al.*, 1987; Herrington *et al.*, 1988). It directly regulates Ctx production as well as another control protein, ToxT, which in turn regulates the expression of toxin-coregulated pili (TCP), various outer membrane proteins and putative pathogenicity factors (Miller and Mekalanos, 1984; Taylor *et al.*, 1987). The ToxR protein directly increases the transcription of the *ctxAB* promoter, by binding to the TTTTGAT repeat motif found upstream of the −35 region (Miller *et al.*, 1987).

Approximately one-third of the ToxR protein is located in the periplasm, where it is presumed to sense environmental signals, such as pH and temperature, whilst the remaining two-thirds of the molecule is located in the cytoplasm, where it acts as a DNA binding protein (Miller *et al.*, 1987). An additional protein, ToxS, is transcribed from the same operon as ToxR, and is also an integral membrane protein with a large periplasmic domain. At present, it is thought that ToxS stabilizes ToxR as a homodimer in the membrane, facilitating attainment of an activated state. The amino-terminal, cytoplasmic domain of ToxR shows some homology with DNA binding proteins of the OmpR-subclass of the two-component regulators, namely OmpR, VirG and PhoP involved in transcriptional activation (Miller *et al.*, 1987; Albright *et al.*, 1989).

However, ToxR cannot be viewed as a typical member of the OmpR-subclass of the two-component regulators, because these are usually cytoplasmic proteins that are phosphorylated upon activation of a sensor kinase protein. Since ToxR does not appear to have the phosphoacceptor domain characteristic of members of this regulator family, it may not be activated by phosphorylation. Indeed, the fact that ToxR is a transmembrane protein may have circumvented the need to transduce the environmental signal by phosphorylation.

In a *toxR* mutant strain of *V. cholerae*, no Ctx expression occurs. Somewhat surprisingly, constitutive expression of ToxT overcomes this defect, suggesting either that ToxT as well as ToxR may modulate the activity of the *ctx* promoter or, alternatively, that ToxT induces expression of hitherto unidentified regulatory factors that influence toxin expression (DiRita, 1992; DiRita *et al.*, 1996). Recent studies by DiRita and co-workers have highlighted further the complexity of Ctx expression. By generating a *toxT* mutant of *V. cholerae*, which possessed a functional ToxRS system, they found that ToxT was required for *ctxAB* expression in *V. cholerae* (Champion *et al.*, 1997). This was a surprising finding given that ToxRS alone had been shown to be sufficient for activation of *ctxAB* expression, when these genes were cloned and investigated in *E. coli*. These results imply that ToxT may be the primary activator of *ctxAB* expression and not ToxR, as previously reported.

The expression of *E. coli* heat-labile enterotoxin is influenced by growth temperatures, with an approximate fourfold elevation in Etx production at 37°C compared with 30°C (T.R. Hirst and B.-E. Uhlin, unpublished observations). While this could be due to an influence on any one of the multiple steps in toxin biogenesis, recent evidence suggests that this is mediated at the level of *etx* mRNA transcription as a result of changes in the local topology of DNA caused by a histone-like protein H1 (also called H-NS).

Translational control of subunit stoichiometry

Translation of the polycistronic *ctx* or *etx* mRNAs that arise from transcriptional initiation upstream of the *ctxA* or *etxA* cistrons is the start of events that lead to the synthesis of the A- and B-subunits.

Initially, the subunits are produced as precursor polypeptides (preA- and preB-) which then undergo export across the cytoplasmic membrane (see below). Since the subunit structure of the assembled toxins is 1A:5B-subunits, this implies that the synthesis of the A- and B-subunits may be stoichiometrically controlled in some way. A careful examination of the relative levels of A- and B-subunit synthesis in *E. coli* expressing either porcine or human Etx indicated that one to two A-subunits are synthesized per five B-subunits (Hofstra and Witholt, 1984; Hirst and Hardy, unpublished observations). The molecular mechanisms responsible for this have, as yet, not been comprehensively investigated.

The fact that both subunits are expressed from a polycistronic mRNA means that the relative levels of A- and B-subunit synthesis must be governed by events that occur after transcriptional initiation. The most likely explanation is that differences in the efficiency of the Shine–Dalgarno (SD) sequences just upstream of the start codons of the A- and B-cistrons influence translational initiation, although other explanations, such as the presence of local stable secondary mRNA structures interfering with or promoting ribosomal recognition of the SD sequence or ribosome movement along the mRNA, cannot be excluded. An analysis of the SD sequences upstream of the initiation codons of the A- and B-subunits of Ctx and Etx revealed clear differences in the number of nucleotides of the A- and B-cistrons which share complementarity with the SD consensus sequence (5'-AGGAGG-3'), which would be consistent with the higher level of B-subunit synthesis being due to the presence of a more efficient ribosome binding site adjacent to the B-subunit cistron.

Toxin export across the cytoplasmic membrane

Protein export across the cytoplasmic membranes of bacteria, especially in *E. coli*, has been extensively investigated over the past 25 years (Driessen *et al.*, 1998). Protein translocation is dependent on both distinct structural properties of exported proteins, such as the presence of amino-terminal signal peptides, and a highly efficient translocation machinery that not only delivers exported proteins to the membrane but also achieves their translocation across it (Figure 6.4).

Our current perception of how the A- and B-subunits of Etx and Ctx cross the cytoplasmic membranes of *E. coli* and *V. cholerae* is based primarily on an understanding of protein export in general, in laboratory strains of *E. coli* K-12. The precise molecular mechanism by which polypeptides translocate across the cytoplasmic membrane remains to be elucidated but, based on current concepts, it seems most likely the polypeptide is threaded like a 'string of beads' through a Sec translocation channel in an energy-dependent process requiring both ATP and proton motive force (pmf).

The signal peptides present at the amino-termini of the precursor A- and B-subunits of Ctx/Etx are 18 and 21 amino acids long, respectively (Figure 6.1). A comparison of the signal peptides encoded by the *ctx* and *etx* genes, and between *etx* genes of human and porcine origin shows that the signal sequences are less well conserved than the mature sequences (Figure 6.1) (Mekalanos *et al.*, 1983; Yamamoto *et al.*, 1987; Dam *et al.*, 1991). This is not too unexpected, since there is considerable sequence diversity among signal peptides in general, even though they perform a conserved function (Randall and Hardy, 1989).

To date, eight Sec proteins are known to be involved in the targeting and translocation of precursor proteins across bacterial cytoplasmic membranes (reviewed in Driessen *et al.*, 1998). These are SecA (a translocation

ATPase), SecB (a molecular chaperone), SecE, G and Y (which form a trimeric translocase channel in the cytoplasmic membrane), SecD, F and the largely uncharacterized YajC (also located in the cytoplasmic membrane, although SecD and SecF have large periplasmic loops) and LepB (leader peptidase II – responsible for cleaving off the signal peptide from the precursor molecule). Similar proteins have been identified in other Gram-negative and Gram-positive bacterial species (Pugsley, 1993; Simonen and Palva, 1993), although, as yet, the respective homologues in the Vibrionaceae, including *V. cholerae*, have not been characterized.

Protein export is an energy-dependent process, requiring both ATP and pmf (for reviews, see Driessen *et al.*, 1998; Geller, 1991). Early studies on the biogenesis of Etx in *E. coli* showed that in the presence of uncouplers of pmf, unprocessed precursor B-subunits accumulated (Palva *et al.*, 1981). The ATP requirement for export across the cytoplasmic membrane is presumed to be due solely to the activity of SecA. It is now accepted that ATP and pmf function at different stages of the translocation process, with ATP playing an essential role in the early events of precursor insertion and translocation, whilst pmf completes the process. However, the exact function of pmf in translocation remains to be precisely determined, especially in light of the findings that different precursors appear to require different levels of pmf for efficient export (Daniels *et al.*, 1981; Driessen and Wicker, 1991; Driessen, 1992).

The last step in the export of the toxin subunits across the cytoplasmic membrane of both *E. coli* and *V. cholerae* involves their release from the translocation channel into the aqueous environment of the periplasm. This step coincides with the important events of subunit folding and the acquisition of tertiary structure. Interactions between secondary structural elements are likely to begin as soon the amino-terminal portion of the polypeptide emerges from the translocation channel, and will lead to the formation of loosely packed tertiary domains. If the polypeptide is exported cotranslationally, it is conceivable that a significant degree of folding may have occurred on the periplasmic face of the cytoplasmic membrane, before the carboxy-terminal portion of the polypeptide has been synthesized. The release of polypeptides from the membrane may, in fact, be dependent on the acquisition of a soluble tertiary-like structure, since parameters, such as low temperature (which slow down folding) or the use of truncated proteins that do not fold correctly, can result in the failure to release the polypeptide from the membrane (Ito and Beckwith, 1981; Koshland and Botstein, 1982; Hengge and Boos,

1985; Minsky *et al.*, 1986; Fitts *et al.*, 1987; Sandkvist *et al.*, 1987, 1990).

Pulse-chase experiments, using [^{35}S]-methionine, have been used to follow the kinetics of toxin subunit export and release into the periplasm of *E. coli* and *V. cholerae* (Hirst *et al.*, 1983; Hofstra and Witholt, 1984, 1985; Hirst and Holmgren, 1987a; Witholt *et al.*, 1988). This revealed that, in common with other proteins, export of the B-subunits of porcine and human Etx was very fast: release was complete within 10 s. In contrast, Hofstra and Witholt (1984) showed that release of mature A-subunits occurred much more slowly, with up to a third remaining membrane-associated 3 min after the initiation of the pulse-chase. Given that these studies used a recombinant plasmid, EWD299, which produced porcine EtxA- and EtxB-subunits in a molar ratio of approximately 2:5, it is conceivable that if A-subunits are expelled from the translocase but fail to associate with B-subunits, they may then irreversibly reassociate with the membrane. This certainly occurs for mutant B-subunits with minor deletions or substitutions at their carboxy-termini or mutations in the Cys-residues, which prevent correct assembly (Sandkvist *et al.*, 1990; Hardy and Hedges, 1996).

Toxin folding and assembly

The folding and assembly of the A- and B-subunits to form a final toxin structure clearly represent key molecular processes in toxin biogenesis. While the events of folding and assembly can be thought of as distinct processes, it is perhaps more appropriate to view them as a pathway of events involving both intra- and intermolecular protein interactions.

It has been observed that the denaturation of both purified Ctx and Etx by various denaturants (under non-reducing conditions) results in the reversible disassembly and unfolding of the toxins, which reassemble into biologically active AB$_5$ complexes upon their return to non-denaturing conditions (Finkelstein *et al.*, 1974a; Hardy *et al.*, 1988; Ruddock *et al.*, 1995, 1996). However, the folding and assembly of proteins within cells can be markedly affected by interactions with other proteins such as chaperones or enzymes that catalyse distinct folding steps, as well as membrane surfaces. Thus, to understand toxin folding and assembly, it is necessary to consider not only the contribution of the amino acids in the A- and B-subunits in folding and assembly, but also the possible role played by cellular folding and assembly factors.

During assembly, the intramolecular interactions that give rise to tertiary structure in the individual toxin subunits will create interfaces which will allow specific intermolecular interactions that ultimately lead to stable quaternary complexes. The nature of the interactions between B-subunits, and between the A- and B-subunits, affords the possibility that these associations may affect late folding events. It is also important to consider that the folded B-subunit monomer contains two surfaces that extensively pack against adjacent B-subunits in the assembled toxin (Figure 6.2) (Sixma *et al.*, 1993). Since these surfaces are particularly hydrophobic, it is possible that the folded monomer may require 'shielding' by chaperone-like proteins to prevent them from aggregating or non-specifically associating with the membrane.

The A- and B-subunit polypeptides of Ctx and Etx undoubtedly begin to fold as soon as they emerge from the translocase channel, although it is not yet clear to what extent they achieve a folded tertiary structure before being fully released into the periplasm. Recently, it was demonstrated that the formation of the disulfide bond in the B-subunit was critically dependent on a periplasmic enzyme, DsbA (Peek and Taylor, 1992; Yu *et al.*, 1992). The periplasmic location of this enzyme suggests that this late step in folding of the B-subunit occurs in the periplasm. Furthermore, it has been shown by pulse-chase experiments that the release of toxin subunits into the periplasm precedes the appearance of assembled oligomers (Hirst *et al.*, 1983; Hofstra and Witholt, 1984; Hirst and Holmgren, 1987a). As a result of these observations, it has been proposed that toxin assembly occurs largely within the periplasmic compartment of the bacterial envelope (Figure 6.4), although the possibility remains that certain subunit–subunit interactions may start at the membrane (Hirst, 1991).

In common with many other exported and secreted polypeptides, the A- and B-subunits of Etx and Ctx possess cysteine residues that must oxidize to form specific intrachain disulfide bonds during folding. Each of the toxin subunits possesses a single disulfide bond, between Cys 187 and Cys 199 in the A-subunit and between Cys 9 and Cys 86 in each B-subunit (Sixma *et al.*, 1993). The importance of the formation of disulfide bonds in toxin biogenesis was first demonstrated by Hardy *et al.* (1988), who showed that the addition of the sulfhydryl reducing reagent, dithiothreitol (DTT), to an EtxB-producing strain of *E. coli* immediately stopped the assembly of new EtxB-pentamers, although it had no effect on the pentamers that had already assembled. The subsequent demonstration that substitution of Cys 9 or Cys 86 by Ser in the B-subunit of Ctx or Etx (Jobling and Holmes, 1991; Hardy and Hedges, 1996) abolished B-subunit assembly into stable oligomers confirmed that disulfide bond formation in the B-subunit is an essential step in toxin biogenesis.

Until recently, it was presumed that the more oxidizing environment of the periplasm would allow the spontaneous (air) oxidation of juxtaposed cysteine residues in secretory proteins. However, in common with other exported proteins, it is now known that the formation of disulfide bonds in the enterotoxin molecule is catalysed by a periplasmic thiol-disulfide oxidoreductase (Yu *et al.*, 1992; Bardwell, 1994; Missiakas and Raina, 1997). Genes encoding analogous enzymes responsible for disulfide bond formation (*dsbA*) have been identified by others in *E. coli*, *V. cholerae* and *Haemophilus influenzae* (where the mnemonic *ppfA*, *tcpG* and *por* have also been used) (Bardwell *et al.*, 1991; Kamitani *et al.*, 1992; Peek and Taylor, 1992; Tomb, 1992; Yu *et al.*, 1992).

The *cis–trans* isomerization of peptidylprolyl bonds is recognized to be a slow, rate-limiting step in the refolding of proteins *in vitro* (Jaenicke, 1987, 1991; Lang *et al.*, 1987). Peptidylprolyl *cis–trans* isomerases are located in the cytoplasm and periplasm of bacteria and are thought to accelerate *cis–trans* prolyl isomerization *in vivo* (Liu and Walsh, 1990; Hayano *et al.*, 1991; Missiakas and Raina, 1997; Dartigalongue and Raina, 1998). Since the folded B-subunits, revealed by X-ray crystallography, contain a single *cis*-Pro at position 93, it is tempting to speculate that the B-subunit interacts with a peptidylprolyl *cis–trans* isomerize as it folds in the periplasm.

The possibility that toxin folding and assembly may be aided by interaction with a general molecular chaperone in the periplasm has not yet been fully explored. The intriguing observation by Schonberger *et al.* (1991) that the heterologous expression of hEtxB in *Saccharomyces cerevisiae* was dependent on a functional *KAR2* gene, a homologue of the Hsp70 chaperone BiP (Schonberger *et al.*, 1991), raised the possibility that efficient folding and assembly of the B-subunit in the periplasm might also be dependent on bacterial chaperones.

The precise pathway of subunit interactions and the identity of assembly intermediates that lead to the formation of an AB$_5$ holotoxin complex are not yet known. The B-subunits of Ctx and Etx readily assemble in the periplasm (in the absence of concomitant A-subunit synthesis) to form stable pentamers (Hirst *et al.*, 1984a). This should not be taken to imply that assembly involves the formation of the B-pentamer, before association with the A-subunit, since there is compelling evidence that the A-subunit interacts with B-subunits before pentamerization is complete.

The kinetics of assembly of radiolabelled toxin subunits *in vivo* has been investigated in several laboratories. It has emerged that the attainment of an SDS-stable B-pentamer structure (the most convenient

measure of toxin assembly) occurs with a half-time of approximately 15–60 s, depending on the bacterial strain, growth temperature, Etx plasmid and level of toxin expression (Hirst *et al.*, 1983, 1984a, b; Hofstra and Witholt, 1985; Hirst and Holmgren, 1987b; Hardy *et al.*, 1988).

While the exact tertiary conformation of B-subunit monomers at the time they begin to interact with one another is not known, the monomer intermediate probably bears at least some resemblance to the B-subunit revealed in the X-ray crystallographic analysis of the AB$_5$ complex. Consequently, much of the tertiary fold and the interfaces of the putative B-monomer are likely to have formed before assembly begins. However, the attainment of the final B-monomer conformation may depend on structural rearrangements that occur when the subunit interfaces combine. From the X-ray structure of the holotoxin, it is clear that in the final state each B-subunit forms extensive hydrophobic interactions with its neighbour via β-sheet and multiple inter-subunit salt bridges (Sixma *et al.*, 1993). It is not known which, if any, of these interactions are necessary to initiate assembly or to stabilize assembly intermediates.

Although the exact pathway of B-subunit pentamerization is not known, it has been speculated that a preferred pathway might involve the initial dimerization of two B-subunit monomers, followed by incorporation of a further monomer, and culminating in the association of a further dimer to form a stable pentameric structure (Hirst, 1995).

The association of the A- and B-subunits to form an AB$_5$ holotoxin adds another level of complexity to the assembly process. The crystal structure of the holotoxin gives a picture of the final outcome of these interactions. The major contacts between the A- and B-subunits are clustered towards the C-terminal portion of the A2-polypeptide which is inserted into the central pore of the B-pentamer, where there are five salt-bridge interactions (Sixma *et al.*, 1993) (see Figure 6.2). Within the central pore there are surprisingly few specific contacts: just two regions of hydrophobic contact at each end of the pore and three salt bridges between three adjacently positioned B-subunits (Sixma *et al.*, 1993).

In vivo studies on Etx assembly in *E. coli* have revealed that the rate at which the B-subunits attain a pentameric structure in the periplasm is increased by *ca.* fourfold when the A-subunit is co-expressed (Hardy *et al.*, 1988). This implied that the A-subunit stabilizes an intermediate in B-subunit assembly. Although several pathways for A-/B-subunit assembly can be envisaged, it seems plausible to hypothesize that assembly of the A- and B-subunits proceeds via the formation of an AB$_3$ intermediate. Indeed, it may be the stabilization of the B-trimer by the A-subunit that leads to the enhancement in the rate of B-subunit pentamerization observed by Hardy *et al.* (1988). If the generation of the AB_3 intermediate is part of the normal pathway of toxin assembly, then its subsequent association with a preformed B-dimer would form the holotoxin complex.

Jobling and Holmes (1992) have provided unequivocal evidence that only the A2-polypeptide of Ctx is necessary and sufficient for association with the B-subunit to form a stable oligomer: with the A2-polypeptide serving to anchor A1- to the B-pentamer. By constructing translational fusions in which the entire A2-fragment (from amino acid residues *ca.* 197–240) was fused to the C-terminus of alkaline phosphatase (PhoA-A2), β-lactamase (Bla-A2) or maltose binding protein (MBP-A2), they were able to demonstrate that the A2 component of the fusion retained its capacity to assemble with CtxB-subunits to form holotoxin-like chimeras (Jobling and Holmes, 1992).

The effect of alterations in the A2-fragment on its capacity to assemble with B-subunits was also tested. This revealed that changing the last four carboxyl terminal amino acids from -Lys-Asp-Glu-Leu-COOH ($-_{237}$KDEL$_{240}$) to -Gln-Asp-Glu-Leu-COOH ($-_{237}$QDEL$_{240}$) did not affect the capacity of the otherwise native A-subunit to assemble with B-subunits. Likewise, the replacement of the KDEL sequence by -Arg-Gly-Gly-Ala-Arg-COOH ($-_{237}$RGGAR$_{241}$) in a PhoA–A2 fusion did not prevent its assembly into a holotoxin-like chimera (Jobling and Holmes, 1992). Thus, the identity of the last four amino acids in the A2-fragment does not appear to be important in ensuring the formation of stable A-/B-subunit interactions. However, the presence of amino acids at those positions is essential for holotoxin formation, since the deletion of these residues in *E. coli* EtxA resulted in an A-subunit that was unable to form a stable AB_5 complex (Streatfield *et al.*, 1992).

Several mutations in the B-subunits have been obtained that prevent A- and B-subunit association during assembly, but which do not appear to inhibit B-pentamer formation (Sandkvist *et al.*, 1987; Jobling and Holmes, 1991). One example is the introduction of extensions at the C-terminus of EtxB (Sandkvist *et al.*, 1987). This is probably due to steric hindrance, since both the C-terminus of the B-subunit and the A-subunit are positioned on the upper face of the B-pentamer.

Another mutant in which Arg35 in the B-subunit of Ctx was substituted for negatively charged residues (either Glu or Asp) resulted in the B-subunit's being unable to form Ctx holotoxin, even though the B-subunit assembled into pentamers (Jobling and Holmes, 1991). This finding is more difficult to explain, since the crystal structure did not show any interaction

between Arg35 and the A-subunit. Several site-specific point mutations in the A1-fragment of porcine EtxA have resulted in the A-subunit failing to assemble into a stable holotoxin complex. The majority of these substitutions introduce either a large hydrophobic amino acid or Pro or Gly, which is likely to affect the folding of the A1-fragment.

Although all of the above studies have provided a significant insight into the assembly of Ctx and related enterotoxins *in vivo*, the precise pathway of subunit–subunit interactions, and the relative importance of particular residues in ensuring the formation of an AB_5 holotoxin complex, remain to be solved.

Toxin secretion across the outer membrane

The final step of toxin biogenesis in *V. cholerae* is the translocation of the periplasmically located holotoxin complex across the bacterial outer membrane, in contrast to enterotoxinogenic strains of *E. coli*, where the assembled toxin appears to remain entrapped within the periplasm (Hirst *et al.*, 1984a, b; Hofstra and Witholt, 1984). ETEC may, therefore, have to rely on damage to their outer membranes by host intestinal factors, such as by bile salts and proteases, to cause the non-specific release of toxin to the gut milieu (Hunt and Hardy, 1991), or possibly, to depend on the expression of a normally cryptic outer membrane secretion system (Francetic and Pugsley, 1996).

In *V. cholerae*, an active toxin secretion system exists that selectively secretes Ctx across the outer membrane to the external milieu (Neill *et al.*, 1983; Hirst *et al.*, 1984a). The difference in secretory behaviour of *V. cholerae* and *E. coli* is not attributable to the differences in Ctx and Etx *per se*, since the expression of the cloned *ctx* genes in *E. coli* resulted in Ctx remaining in the periplasm, whilst expression of Etx in *V. cholerae* resulted in its efficient secretion to the medium (Pearson and Mekalanos, 1982; Mekalanos *et al.*, 1983; Neill *et al.*, 1983; Hirst *et al.*, 1984a; Hirst, 1991). *Vibrio cholerae*, therefore, has a distinct secretory apparatus able to translocate a fully folded and assembled protein, such as Ctx, across its outer membrane. The apparatus must function in a fundamentally different manner from the 'thread of beads' mechanism postulated to be involved in protein export across the bacterial cytoplasmic membrane.

The role of the various subunits in toxin secretion has been investigated by studying engineered bacterial strains carrying either a wild-type operon encoding both the A- and B-subunits, or mutant operons in which the A- or B-subunits are expressed separately (Hirst *et al.*, 1984a). This led to the conclusion that the B-subunit (or the assembled B-pentamer) contains the relevant structural determinants to permit secretion across the *V. cholerae* outer membrane. Thus, *V. cholerae* strains expressing *E. coli* Etx or the individual EtxA or EtxB subunits were able to secrete the assembled EtxB subunits as efficiently as the fully assembled AB$_5$ toxin.

In contrast, when expressed alone, the A-subunit remained cell-associated (Hirst *et al.*, 1984a). This finding led to the hypothesis that the A-subunits must be associated with the B-pentamer prior to the step at which the B-subunits engage the secretory machinery, if it is to be translocated across the outer membrane. It also implicated a need for the assembly of the A- and B-subunits to be well co-ordinated in order to avoid secretion of the B-pentamer (a potential competitive inhibitor of toxin action in the gut). As indicated above, the ability of the A-subunit to stabilize a B-subunit intermediate during assembly should ensure production of AB$_5$ complexes rather than B-pentamers, and thereby avoid this problem.

An analysis of the rate of efflux of labelled holotoxin from the periplasm showed that it resembled a simple first-order kinetic process. Thus, once the toxin molecules have assembled in the periplasm all of them have an equal probability of engaging the secretory machinery and being successfully translocated to the medium (Hirst and Holmgren, 1987a).

The transient entry of secreted proteins into the periplasm prior to their secretion across the bacterial outer membrane is now recognized as a common step in the secretion of most (but not all) extracellularly located proteins that are produced by Gram-negative bacteria (for reviews, see Hirst and Welch, 1988; Pugsley, 1993).

The first indication that *V. cholerae* had specific gene(s) required for the process of toxin translocation through the outer membrane was obtained in the 1970s, when non-toxinogenic isolates of *V. cholerae* strain 569B were obtained by NTG chemical mutagenesis (Finkelstein *et al.*, 1974b; Holmes *et al.*, 1975). This led to identification of a mutant, designated M14, which accumulated Ctx in the periplasm and failed to secrete it into the growth medium (Holmes *et al.*, 1975; Hirst and Holmgren, 1987a).

Extensive studies have been undertaken in recent years on the genetic basis of secretion of a range of extracellular proteins from Gram-negative bacteria. Pugsley (1993) has termed these secretion systems the main terminal branch of the general secretion pathway (GSP) of Gram-negative bacteria, whereas others refer to them as type-II secretion systems (for reviews, see Pugsley *et al.*, 1990, 1997; Pugsley, 1993).

The best-characterized example of this pathway is that required for secretion of pullulanase (PulA) by *Klebsiella oxytoca* (Figure 6.5). Pullulanase is a lipoprotein that is synthesized as a longer precursor with an amino-terminal signal sequence. It is exported across the cytoplasmic membrane with the aid of the Sec-proteins, and undergoes proteolytic processing to remove its signal sequence (Pugsley *et al.*, 1991). However, secretion of pullalanase across the outer membrane of *Klebsiella* (and also when expressed heterologously in *E. coli*) is dependent on 14 out of 15 genes that flank *pulA*, namely *pulC-O* and *pulS* (Figure 6.5). Although 13 of these genes encode proteins that are exported across the cytoplasmic membrane, only PulD appears to be integrated into the outer membrane (Pugsley *et al.*, 1990; Pugsley, 1993). Since several of the Pul proteins (PulG-J; Figure 6.5) are homologous to type-IV pilin proteins, it has been suggested that they may assemble into a pseudopilus that provides a scaffold that couples the cytoplasmic and outer membranes together (Figure 6.4).

Uniquely, PulE, one of the Pul proteins that is essential for pullanase secretion, is not synthesized with a recognizable signal sequence for export, and appears to be located in the cytoplasm (Pugsley, 1993). The PulE protein and its various homologues from other Gram-negative bacteria have a conserved ATP-binding motif similar to that of ATPases and kinases (Walker *et al.*, 1982; Pugsley, 1993). It is possible that this protein may either transduce energy into the process of protein secretion, or provide energy for the assembly of Pul proteins into a complex. When the *exeE* gene of *Aeromonas hydrophila* (which shares extensive sequence homology with *pulE*) was mutated by Jiang and Howard (1992), aerolysin secretion from this organism was prevented. The introduction of a plasmid encoding hEtxB into the *exeE* mutant strain of *A. hydrophila* resulted in the accumulation of correctly assembled B-subunits in the periplasm, whereas the parental strain secreted Etx into the medium (Yu and Hirst, 1995).

Sandkvist *et al.* (1993) have obtained direct evidence that toxin secretion from *V. cholerae* is dependent on the main terminal branch of the GSP of Gram-negative bacteria. They identified a fragment of chromosomal DNA from an *El Tor* strain, which complemented the toxin secretion defect in M14. This study took advantage of the observation that the M14 mutant exhibits a pleiotropic defect in several cellular functions, including the secretion of proteases. Thus, screening for restored protease secretion provided a convenient means of testing for transcomplementing DNA (Sandkvist *et al.*, 1993). Various positive clones were identified and tested in a GM1-ELISA to verify that the cloned DNA restored toxin secretion as well as protease secretion.

One complementing clone was identified (containing a DNA fragment of 15 kb) (Sandkvist *et al.*, 1993).

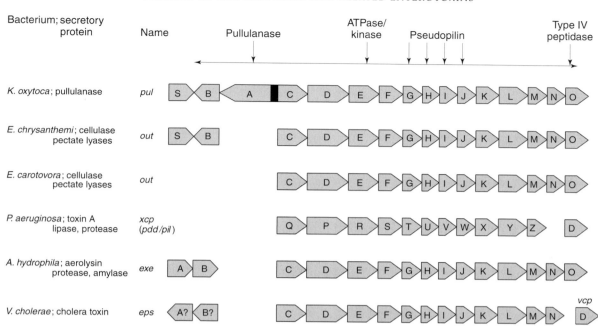

FIGURE 6.5 Schematic representation of the organization of genes involved in the secretion of proteins across the outer membranes of Gram-negative bacteria, including the prototypic *pul*-operon from *Klebsiella oxytoca*, the *out*-operon from *Erwinia chrysanthemi* and *E. carotovora*, the *xcp* (*pil*)-operon from *Pseudomonas aeruginosa*, the *exe*-operon from *Aeromonas hydrophila* and the *eps*-operon from *Vibrio cholerae*.

Sequence analysis revealed that this clone contained an open reading frame, which showed a high degree of homology (45–60% identity) to the PulE/ExeE proteins involved in the secretion of extracellular proteins by other Gram-negative bacteria (Sandkvist *et al.*, 1993) and was designated *epsE*. Overbye *et al.* (1993) located additional genes downstream of *epsE* required for toxin secretion by *V. cholerae* (Figure 6.5). These genes, designated *epsLMN*, shared sequence homology with *pulLMN* (Overbye *et al.*, 1993).

An analysis of the entire *eps* gene cluster revealed that it contained 12 genes, designated *epsC-N*, which have a similar organization and sequence homology to the Pul- and related secretion systems (Sandkvist *et al.*, 1997). In most GSP operons, the distal gene (i.e. *pulO* and homologues) encodes a pre-pilin peptidase that processes type-IV prepilins. This is necessary for construction of a functional secretory apparatus. Although there is no *pulO*-like gene linked to the *esp* operon of *V. cholerae*, recent studies by Marsh and Taylor have revealed that an unlinked gene, designated *vcpD*, encodes a type-IV prepilin peptidase required for Ctx secretion (Marsh and Taylor, 1998).

In common with other extracellularly located proteins, the exact mechanism by which Ctx is secreted remains something of an enigma. However, the discovery that the 'D' proteins of the secretory system are assembled into the outer membrane as multimeric oligomers with a central pore of sufficient size for direct transport of a folded secretory protein (Koster *et al.*, 1997) has provided a conceptual framework around which to consider the mechanism of toxin secretion from *V. cholerae*.

One such model, the 'gated-pore model', assumes that components of the secretory machinery form a channel or pore through which toxin molecules are translocated without interacting with the lipid phase of the outer membrane. Several features common to most of the proteins that use the main terminal branch of the general secretion pathway suggest that these proteins may prefer to translocate via hydrophilic (water-filled) pores. Most of the proteins appear to adopt folded structures prior to translocation, and more importantly, many of them have hydrophilic surfaces and are very water-soluble proteins.

Thus, in the absence of some trigger to expose buried hydrophobic regions, these proteins are unlikely to be able to insert into the lipid phase of the outer membrane. However, such secretion pores (secretins) must be able to exclude small resident periplasmic proteins and small solutes, while at the same time permitting the secretion of bulky proteins such as Ctx, aerolysin or pullulanase. Although the Pul-like secretion machinery in different Gram-negative bacteria appears to be able to translocate several proteins, these homologous secretion systems none the less discriminate between different secreted proteins. This argues for some kind of specific recognition between a particular secreted protein and components of the secretory machinery. Such a mechanism would provide a convenient means of discriminating between resident periplasmic proteins and those passing through the periplasm *en route* for secretion through outer membrane. Consequently, secreted proteins may possess some kind of common structural motif for the targeting of secretory proteins across the bacterial envelope (Lu and Lory, 1996).

Since no common linear sequence motifs have been identified that could fulfil this role, it may be necessary to search for shared three-dimensional motifs once sufficient crystal structures of secreted proteins have been determined. If such pores are not permanently occupied with a secretory protein, their opening would presumably have to be regulated in order to maintain the limited permeability properties of the outer membrane. This regulation could operate by secretory proteins triggering the pores to open by direct interaction. This could also require a source of energy, either from pmf or ATP hydrolysis to ensure that the pore opens. Thus, in this model the pore would act as a gate, opening upon interaction with any secreted protein that possesses the postulated structural motif.

The role of energy in toxin secretion has recently been studied further, using specific inhibitors to assess the requirement for pmf and ATP. Both pulse-chase experiments and quantitative enzyme-linked immunosorbent assay (ELISA) were used to monitor the fate of toxin in the periplasm (poised for secretion) after addition of the proton ionophore CCCP (Hillary, 1998). This revealed that CCCP blocked toxin secretion, and that it did so without causing a cycle of futile ATP hydrolysis to occur, thereby providing strong circumstantial evidence for a direct role for pmf in the toxin secretory event. The use of 50 mM sodium arsenate (which reduces ATP levels in the cell) also blocked toxin secretion and is consistent with the view that ATP may also have an important role, which could involve EpsE. Studies on *A. hydrophila* have revealed that *exeAB*, unlinked to the main GSP operon (*exeC-N*), is also necessary for aerolysin secretion and that ExeA has a putative ATP-binding site. It has been speculated that ATP hydrolysis by ExeA, rather than ExeE, is responsible the ATP dependence of aerolysin secretion by *A. hydrophila* (Pugsley *et al.*, 1997). Two homologues of *exeA* have been found in *V. cholerae* and these may serve a similar function (Hillary and Hirst, unpublished observations). However, the exact role of energy and how this is coupled to the secretory event still need to be explored.

ACTION OF CHOLERA TOXIN AND RELATED ENTEROTOXINS IN MEDIATING DIARRHOEA

Once released from the micro-organism, either by secretion or by non-specific lysis, the toxin is free to interact with the epithelium of the gut. The precise action of Ctx and related enterotoxins on eukaryotic cells depends on a complex sequence of events that eventually leads to alterations in ion fluxes and a concomitant loss of water, characteristic of cholera and related diarrhoeal diseases.

Receptor binding

Toxin action begins with the binding of the B-subunit moiety to cell surface receptors. The principal receptor for Ctx and Etx is GM1-ganglioside (Gal(β1–3)GalNAc(β1–4){NeuAc(α2–3)}Gal(β1–4)Glc(β1–1)ceramide), a glycosphingolipid found ubiquitously on the surface of eukaryotic cells (van Heyningen *et al.*, 1971; Cuatrecasas, 1973; Holmgren, 1973). The toxins bind via the pentasaccharide moiety of GM1, and recently, the X-ray structure of CtxB bound to the pentasaccharide was determined (Merritt *et al.*, 1994, 1998). This revealed that each GM1 binding site is primarily located within a single B-monomer, and is situated on the side of the toxin molecule away from the A-subunit (Figure 6.2). The GM1 binding pocket is formed by several loop regions in the B-subunit, notably a large fully conserved loop comprising residues 51–58, and loops 10–14 and 89–93. In addition, at one end of the binding pocket is a loop from an adjacent monomer (residues 31–36), which is not directly involved in saccharide binding; but substitutions in this region can block receptor interaction (Jobling and Holmes, 1991; Nashar *et al.*, 1996a).

The most important interactions between the B-subunit and the pentasaccharide result from hydrogen bond interactions with the terminal Gal and sialic acid (NeuAc) residues, and to a more limited extent with GalNAc. This results in high-affinity binding with reported dissociation constants for interaction with GM1 of 7.3×10^{-10} M for CtxB and 5.7×10^{-10} M for EtxB (Kuziemko *et al.*, 1996). The B-subunits of Ctx and Etx also show some affinity for GD1b, with a dissociation constant approximately one order of magnitude lower than for GM1. In addition, EtxB interacts with asialo-GM1, lactosylceramide and certain galacto-proteins (Orlandi *et al.*, 1994; Teneberg *et al.*, 1994; Karlsson *et al.*, 1996; Backstrom *et al.*, 1997). The consequences of these differential binding specificities on the toxicity and immunological properties of Ctx and Etx have yet to be fully evaluated.

Uptake and trafficking

The use of polarized human colonic epithelial T84 cells (Lencer *et al.*, 1992) has greatly facilitated studies of Ctx and Etx toxin action, since toxicity can be readily monitored as the induction of electrogenic Cl⁻ secretion (Figure 6.6). Recently, it was demonstrated that following binding of Ctx to T84 cells, Ctx–G_{M1} complexes cluster in caveolae-like detergent-insoluble subdomains of the plasma membrane (Wolf *et al.*, 1998).

Invagination and internalization of these membrane domains results in the formation of smooth apical endocytic vesicles (AE) that enter vesicular trafficking pathways, leading to transport of the toxin to the trans-Golgi network (TGN) (Figure 6.6) (Lencer *et al.*, 1992, 1995, 1993; Nambiar *et al.*, 1993; Bastiaens *et al.*, 1996; Majoul *et al.*, 1996; Sandvig *et al.*, 1996). The observation that Brefeldin A (which causes dissolution of the endoplasmic reticulum, ER) inhibits Ctx action (Lencer *et al.*, 1993; Nambiar *et al.*, 1993; Orlandi and Fishman, 1993) led to the conclusion that the toxin must traffic into the endoplasmic reticulum (ER). This was further substantiated by the finding that an intact - KDEL sequence at the *C*-terminus of CtxA (RDEL in EtxA) was necessary for efficient toxin action (Lencer *et al.*, 1995). As -KDEL sequences are normally found on ER-resident proteins, such as BiP and protein disulfide isomerase (PDI), which are retrieved from the TGN back to the ER by a -KDEL binding protein Erd-2, it has been hypothesized that the A-subunit binds to Erd-2, thereby facilitating toxin entry into the ER (Lencer *et al.*, 1995).

It has been speculated the A-subunit may detach from the B-subunits in the TGN, since only the A-subunit (A1–A2 fragments) has been detected in the ER (Bastiaens *et al.*, 1996; Majoul *et al.*, 1996; Sandvig *et al.*, 1996), although transport of the holotoxin from the Golgi to the ER, followed by rapid dissociation and anterograde transport of the B-subunit back to the Golgi, has not been excluded. Reduction of the disulfide bond between the A1- and A2-fragments is thought to be catalysed by protein disulfide isomerase resident in the ER (Majoul *et al.*, 1997; Orlandi, 1997), followed by translocation of the A1-fragment across the ER-membrane to the cytosolic compartment.

If translocation of the toxin occurs in the ER, one possible mechanism might involve reverse transport via the Sec61p secretion channel, which has recently been shown to occur for aberrantly folded MHC molecules that are returned to the cytosol for targeted destruction by the proteasome. The possibility that the A-subunits of Ctx and Etx (as well as other toxins such as Shiga toxin) might utilize this pathway is particularly attractive, given the absence (or near absence) of lysine residues in the A1-fragments, which would preclude them from being ubiquitinated as they emerge from the Sec61p channel. Because of its hydrophobicity, the A1-fragment may remain associated with the cytosolic face of the ER-membrane, rather than being released free into the cytosol.

The subsequent trafficking, and presumed delivery of the A1-fragment to the basolateral membrane, is less well understood, although the finding that it interacts with so-called ADP-ribosylation factors (ARFs) (Moss and Vaughan, 1992), involved in vesicular transport, may facilitate its anterograde targeting via baslolaterally directed vesicles (BE) to the basolateral membrane (Figure 6.6). ADP-ribosylation factors have also been demonstrated directly to increase toxin activity *in vitro* (Moss *et al.*, 1994), a finding that may be important in determining the magnitude of toxin action *in vivo*. The A1-fragments of Ctx and Etx accomplish their toxic effects by ADP-ribosylating $G_{s\alpha}$, a component of the

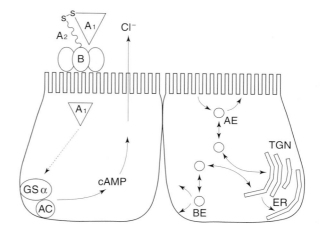

FIGURE 6.6 Schematic model of toxin uptake and action in polarized epithelia. The left-hand side shows the functionally defined requirements for toxin action, including toxin binding at the apical surface, uptake of the A1-fragment, A1-fragment-mediated ADP-ribosylation of Gsα leading to activation of adenylate cyclase (AC) and a concomitant rise in cAMP levels which leads to chloride-ion efflux , the primary event responsible for causing diarrhoea. The right-hand side shows the uptake and trafficking of toxin in polarized cells. The toxin binds to GM1 within caveolae at the apical surface which invaginate to form apically derived endosomes (AE), which enter trafficking pathways leading to the trans-Golgi network (TGN). In the TGN, the *C*-terminal K(R)DEL-motif in the A-subunit enables it to move retrograde via the Golgi cisternae to the endoplasmic reticulum (ER). Translocation of the A1-peptide to cytosolic face of the ER is postulated to occur via the Sec61p secretion channel. Anterograde movement from the ER in basolaterally directed transport vesicles (BE) would enable the toxin to move along a secretory pathway to the basolateral membrane. Fusion with the basolateral membrane would position the A1-fragment at a site near the Gs/adenylate cyclase where it could mediate activation and toxic action.

trimeric GTP-binding protein that activates adenylate cyclase. The concomitant elevation in cAMP levels leads to electrogenic Cl^- secretion – the primary ion transport event responsible for secretory diarrhoea in humans.

For Ctx and Etx to exhibit full toxicity, the A-subunits must undergo proteolytic cleavage or 'nicking' at Arg-192 to give separate A1- and A2- fragments (Lencer *et al.*, 1997). In the case of Ctx, extracellular proteases, such as HA protease produced by *V. cholerae*, can efficiently nick and activate the A-subunit (Booth *et al.*, 1984). By contrast, Etx from ETEC, as well as recombinant Ctx produced in *E. coli,* are normally isolated with their A-subunits intact; trypsin or other gut-associated proteases have been postulated to accomplish toxin activation in such cases (Hunt and Hardy, 1991). Recent studies on T84 cells have shown that a serine protease that efficiently nicks and activates both CtxA and EtxA is present either on the apical surface or in apically derived endosomes (Lencer *et al.*, 1997). In this respect, no difference was found in toxicity of commercial (nicked) preparations of Ctx and recombinant Ctx that was either unnicked or nicked with trypsin, implying that proteolytic activation is not the rate-determining step in toxin action.

A comparison of the relative toxicity of Ctx and Etx in polarized human epithelial (T84) cells revealed that Ctx was the more potent of the two toxins. The underlying structural basis for this difference in toxicity was recently assessed by engineering a set of mutant and hybrid Ctx/Etx toxins (Rodighiero *et al.*, 1999). This revealed that the differential toxicity of Ctx and Etx was not due to differences in the A-subunit's C-terminal KDEL targeting motif (which is RDEL in Etx), since a KDEL to RDEL substitution had no effect on Ctx activity. Moreover, it could not be attributed to either the enzymically active A1-fragment, since hybrid toxins in which the A1-fragment of Ctx was substituted for that of Etx had exactly the same potency as that of authentic Ctx, or the B-subunit, since the replacement of the B-subunit in Ctx for that of Etx caused no alteration in toxicity. Remarkably, the difference in toxicity could be mapped to the 10 amino acid segment of the A2-fragment that penetrates the central pore of the B-subunit pentamer. This region is responsible for maintaining A-/B-subunit interaction.

A comparison of the *in vitro* stability of two hybrid toxins, differing only in this 10 amino acid segment, revealed that the Ctx A2-segment conferred a greater stability to the interaction between the A- and B-subunits than the corresponding segment from Etx A2. This suggests that the reason for the relative potency of Ctx compared with Etx stems from the increased ability of the A2-fragment of Ctx to maintain holotoxin stability during uptake and transport into intestinal epithelia. These findings highlight the importance of a region in the toxin that has hitherto been generally overlooked, and provides a possible contributory explanation for the difference in severity of cholera and traveller's diarrhoeal disease.

TOXIN ACTION AND THE IMMUNE SYSTEM

In addition to their established roles as mediators of diarrhoeal disease, Ctx and Etx are now recognized as having remarkable immunological properties. The toxins trigger extremely potent anti-toxin antibody responses following systemic immunization, even when administered in the absence of a conventional adjuvant. They also elicit vigorous mucosal responses, which makes them almost unique among soluble proteins.

These findings are interesting in their own right, but studies have also shown that the toxins can act as adjuvants, stimulating either systemic or mucosal responses to other co-administered or conjugated antigens. However, concerns over the inherent toxicity of Ctx and Etx have led to attempts to separate their toxic and adjuvant properties either by creating non-toxic variants or by using components of the toxins devoid of diarrhoeagenic activity. Recombinant preparations of B-subunits, completely devoid of any toxicity, retain significant adjuvant activity. However, somewhat paradoxically, the B-subunits of Ctx and Etx also promote tolerance to autoantigens. These toxins can therefore be considered as potent immunomodulators, which function by interacting in a complex fashion with the immune system.

Why are cholera toxin and related enterotoxins such potent immunogens?

In contrast to most soluble proteins, Ctx and Etx are exceptionally potent immunogens. Normally, the parenteral immunization of a protein leads to low-level antibody responses unless it is administered in combination with a suitable adjuvant, such as Freund's adjuvant or Alum. However, Ctx and Etx induce high-level antibody responses even when injected in normal saline or water. In addition, if Ctx or Etx is administered to mucosal sites, it induces strong local IgA and systemic antibody production (Nashar *et al.*, 1993, 1996b), whereas conventional soluble antigens usually either fail to trigger an immune response or they induce specific tolerance.

Immunization of mice or rabbits with Ctx or Etx results in high-titre antibody responses directed against the B-subunit component of the toxin, suggesting that the B-subunit is immunodominant. When congenic strains of mice were orally immunized with recombinant preparations of the B-subunits of Ctx or Etx, it was found that the magnitude of serum anti-B-subunit responses was H-2 linked. For example, mice of the H-2^b haplotype (B10 or C57/BL/6) responded exceptionally well to CtxB, whereas H-2^d mice (B10.D2 or BALB/c) proved to be high responders to EtxB, but relatively low responders to CtxB (Nashar and Hirst, 1995). These observations are consistent with the recent identification of different immunodominant T-cell epitopes in CtxB (residues 89–100) and EtxB (residues 36–44), respectively (Cong et al., 1996; Takahashi et al., 1996).

Since the B-subunit's primary function is to bind to cell surfaces, site-directed mutagenesis was used to generate a mutant EtxB subunit containing a glycine to aspartate substitution at residue 33 so as to inhibit receptor recognition (Nashar et al., 1996). The resultant mutant, EtxB(G33D), formed indistinguishable pentameric complexes to wild-type B-subunit, as revealed by X-ray crystallography, and retained all of its physicochemical properties, except for the ability to bind to GM1 ganglioside (Nashar et al., 1996b; Merritt et al., 1997). A comparison of the antibody responses elicited by EtxB and EtxB(G33D) demonstrated that the potent immunogenicity of the B-subunit is fully dependent on receptor recognition (Nashar et al., 1996b). Mice immunized subcutaneously (s.c.) or orally with EtxB produced high-titre antibody responses to the B-subunit, whereas those given EtxB(G33D) produced only meagre responses following s.c. injection, and failed to elicit any response after oral immunization.

An analysis of in vitro lymph node responses from mice immunized with EtxB revealed that addition of wild-type EtxB strikingly altered the distribution of lymphocyte subsets compared with those found in the presence of EtxB(G33D). Notably, by day 4 of culture EtxB had caused an increase in the proportion of B-cells, many of which were activated, as evidenced by the upregulation of the low-affinity IL-2 receptor (CD25) (Nashar et al., 1996b). In addition, the presence of EtxB in such cultures triggered the complete depletion of CD8+ T-cells (Nashar et al., 1996b).

A similar pattern of responses was observed when EtxB was added to cultures of mesenteric lymph node cells responding to an unrelated antigen, namely ovalbumin (Nashar et al., 1996a). Subsequent studies revealed that addition of EtxB, but not EtxB(G33D), to lymphocyte cultures, triggered CD8+ T-cells to undergo apoptosis (Nashar et al., 1996a). This occurred over a period of 12–16 h and explains the observed depletion of this population of T-cells from lymph node cells cultured in the presence of EtxB.

A similar effect may also occur in vivo, since oral administration of CtxB causes a reduction in CD8+ T-cells from both the Peyer's patch and intraepithelial lymphocyte (IEL) compartments (Elson et al., 1995). It is, however, not yet clear what effects depletion of this subset of cells would have on the outcome of the immune response to EtxB or CtxB in vivo. CD8+ T-cells are associated with a regulatory role in the immune response, arising from production of specific cytokines such as interferon-γ (IFNγ). The loss of immunoregulation by this cytokine should favour induction of T-helper 2 (Th2) T-cell responses, characterized by the production of IL-4, IL-5 and IL-10, and induction of high levels of antibodies of the IgG1 and IgE classes, as well as suppression of cell-mediated immunity. Consistent with this is the observation that the IgG1 subclass predominates following immunization with EtxB, and that EtxB induces a Th2-dominated response to coadministered antigens (see below).

The effect of EtxB and CtxB on B-cells and CD4+ T-cells has also been investigated (Figure 6.7). Incubation of EtxB with lymphocyte cultures, depleted of adherent cells, caused the generalized activation of naïve B-cells with upregulation of surface markers such as major histocompatibility complex (MHC) class II, B7, CD25, CD40 and ICAM-1 (Nashar et al., 1997). This occurred in the absence of significant proliferation in cultures containing both B- and T-cells, indicating that EtxB does not have mitogenic activity.

Further investigations using highly purified B-cells revealed that EtxB caused upregulated expression of MHC class II and CD25, implying that the effect on other surface markers was likely to be the consequence of increased interaction with CD4+ T-cells present in the original cultures (Nashar, unpublished observations). CtxB elevates the expression of MHC class II on naïve B-cells (Francis et al., 1992). Preliminary studies have also revealed that EtxB-receptor binding triggers upregulation of several adhesion and co-stimulatory molecules on CD4+ T-cells, and may trigger secretion of Th2-associated cytokines (Nashar, Lang Williams and Hirst, unpublished observations). While the degree of surface expression of MHC molecules, co-stimulatory and adhesion proteins, and secreted cytokines may ultimately enhance or modulate the T-cell response to presented antigens, productive processing of the protein and loading of antigenic epitopes on to MHC class II molecules is essential for T-cell receptor (TcR) signalling.

Normally, antigen-presenting cells (APCs), such as dendritic cells, macrophages, B-cells and intestinal

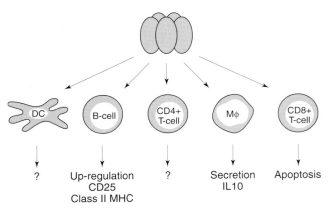

FIGURE 6.7 Immunomodulation of leucocytes by the B-subunits of Ctx or Etx. Direct effects include induction of CD25 and class II MHC on B-cells, apoptosis of CD8+ T-cells, activation of macrophages (MΦ) with release of IL-10 but not IL-12. Downstream events resulting from such alterations will lead to an alteration in the environment in which immune responses are generated, with a resultant bias towards the generation of Th2 helper T-cell responses. The direct effects of the B-subunits on CD4+ T-cells and on dendritic cells (DC) remain to be established.

epithelial cells, internalize foreign antigen by macropinocytosis, phagocytosis and receptor-mediated endocytosis (for a review, see Watts, 1997). This results in the antigen's being taken into endosomes and subsequently trafficked through intracellular compartments of distinct composition and pH. The early endosomes are progressively acidified and mature into late endosomes that fuse with lysosomes, where their contents are proteolytically digested. The activity of proteases at various stages in the pathway can generate peptides fragments containing antigenic epitopes. Newly synthesized MHC class II molecules, in association with invariant-chain protein (Ii), are transported via a secretory pathway that intersects with the endocytic pathway containing processed antigen. The peptide epitope is loaded on to the class II molecule, Ii is removed and the assembled complex is routed to the cell surface.

The uptake, processing and presentation of toxin-derived epitopes by APCs remain to be fully evaluated. It is, however, conceivable that the ability of the B-subunits to bind to GM1, trigger uptake and sequester endosomal trafficking pathways might influence the generation of toxin-derived epitopes and their loading on to class II molecules. This may be most relevant in relation to the APC function of B-cells, where levels of antigen uptake, other than specific uptake through surface membrane immunoglobulin, are normally low. Thus, the GM1-mediated uptake of B-subunits might allow non-toxin specific B-cells to function as APCs in the presentation of toxin-derived

epitopes. In addition, toxins might alter intracellular trafficking pathways, causing the re-routing of antigen to intracellular compartments containing different proteolytic enzymes, and thereby altering peptide epitope liberation and destruction, and possibly improving peptide loading on to MHC class II molecules.

Somewhat paradoxically, Harding and co-workers (1996) recently demonstrated that pretreatment of *Listeria*-elicited peritoneal macrophages with Ctx inhibited the processing and presentation of hen egg lysozyme (HEL) to a HEL epitope-specific T-cell hybridoma (Matousek *et al.*, 1996). Similar inhibition of antigen processing by *E. coli* enterotoxin was also found (Matousek *et al.*, 1997).

Studies using the macrophage cell line J774 have revealed that recombinant EtxB also inhibits the processing and presentation of heterologous antigens such as ovalbumin and HEL to T-cell specific hybridomas (Millar and Hirst, unpublished observations). Intriguingly, the J774 cells become highly vacuolated, suggesting that EtxB induces an alteration in the sorting and trafficking of intracellular membranes. This finding is reminiscent of the effect of a vacuolating toxin, VacA, from *Helicobacter pylori*, which decreases invariant chain-dependent antigen processing, MHC class II loading and presentation of tetanus toxin (TT) by B-cells (Molinari *et al.*, 1998). VacA induces acidic vacuoles containing late endosomal and lysosomal markers (Molinari *et al.*, 1997) and perhaps sequestering the essential antigen-processing machinery needed for efficient TT presentation.

How can inhibition of antigen processing be reconciled with the observed increase in immune responses generated by the enterotoxins? Sustained T-cell responses and memory responses may require antigen to persist, undegraded, for extended periods. Retention of antigen in mildly acidic compartments has been observed in immature dendritic cells, and it has been postulated that delayed processing may be important in allowing efficient antigen presentation to T-cells, following migration to the lymph node (Lutz *et al.*, 1997).

It thus seems likely that the potent immunogenicity of Ctx, Etx and their respective B-subunits is due to a multitude of effects that can be chiefly attributed to the B-subunits' ability to bind to cell-surface receptors, namely, (a) improved uptake into APCs, (b) alterations in membrane trafficking leading to persistence and sustained presentation of the antigen, (c) the potential recruitment of B-cells into the process of antigen presentation, (d) the induced expression of co-stimulatory and adhesion molecules leading to enhanced cognate interaction between APCs and CD4+ T-cells, and (e) an alteration in the balance of cytokine secretion resulting in the differentiation of CD4+ Th2 cells and their subsequent help in the expansion of toxin-specific B-cells secreting high levels of antibodies directed against the B-subunit.

Adjuvant properties

The discovery that Ctx could act as an adjuvant for antibody responses towards other antigens was first reported in 1972 (Northrup and Fauci, 1972). It was subsequently shown that mucosal immune responses to a wide variety of soluble antigens could be induced if they were administered in combination with small quantities of Ctx (Elson and Ealding, 1984). This provided a unique opportunity to investigate the mucosal immune system and to develop effective mucosally delivered vaccines.

The explanation of why Ctx and Etx act as adjuvants to other antigens has been confounded by apparently conflicting observations on the importance and significance of the individual A- and B-subunits. For example, the analysis of a mutant of *E. coli* heat-labile enterotoxin, with a Glu to Lys substitution in residue 112 of the A-subunit, revealed that it was inactive both as a toxin and as an adjuvant (Lycke *et al.*, 1992). This led to the conclusion that adjuvanticity and toxicity were inseparable, and that the adjuvanticity was dependent on the ADP-ribosylation function of the A-subunit. However, subsequent studies have provided definitive evidence that mutant toxins, inactive in a wide range of *in vitro* and *in vivo* ADP-ribosylation assays, still function as adjuvants (Douce *et al.*, 1995, 1997; de Haan *et al.*, 1996).

In addition, it was observed that recombinant EtxB acted as a poor adjuvant, while CtxB lacked adjuvanticity altogether (Douce *et al.*, 1995, 1997). This has led to the view that a combination of properties in the A-subunit (both enzymic and non-enzymic) and the delivery function of the B-subunits confer adjuvanticity to these toxins. The A-subunit contains important targeting signals such as a C-terminal–KDEL retention sequence and an ARF-binding site, which might alter vesicular traffic and thereby enhance the targeting of coadministered antigens to compartments where improved processing and presentation can occur. However, it should be noted that there has been a number of reports indicating that the B-subunits (devoid of any contaminating A-subunit) possess immunomodulatory activity (Francis *et al.*, 1992; Green *et al.*, 1996; Williams *et al.*, 1997). For example, an intranasal vaccine comprising a mixture of glycoproteins from herpes simplex virus (HSV), administered with recombinant EtxB as the adjuvant, afforded high-level protection to ocular challenge with virulent HSV (Richards, Hill, Aman, Hirst and Williams, unpublished observations).

Thus, the B-subunits of Etx and Ctx should not be considered solely as delivery vehicles for their respective A-subunits, but as potentiators of immune responses in their own right. This should not be surprising given the profound effects of the B-subunits on leucocyte populations and their inherent potent immunogenicity. Therefore, Ctx and Etx should be considered as composite molecules, in which B-subunit receptor interaction activates signalling pathways that activate APCs (and other immune cells), while the A-subunit further augments uptake and traffic of coadministered antigen into appropriate intracellular compartments and also elevates cAMP, which alters gene expression and cell activation. Not too surprisingly, therefore, the fully active holotoxin remains the best adjuvant, although, because of its inherent toxicity, it is unlikely to find use as a component of human vaccines. By contrast, the non-toxic derivatives of Etx, as well as EtxB, are emerging as safe, credible alternatives.

Toxin B-subunits as therapeutic agents

Recent studies have shown that CtxB and EtxB potentiate immunological tolerance, which is something of a paradox given their evident adjuvant properties. Czerkinsky, Holmgren and colleagues have coupled various antigens to CtxB, and shown that the oral administration of the conjugates induces peripheral tolerance to the attached antigen (Sun *et al.*, 1994, 1996; Czerkinsky *et al.*, 1996). Their initial studies, based on the coupling of sheep red blood cells and bovine gamma globulin to CtxB, have now been extended to myelin basic protein (MBP) and insulin. These latter conjugates have been shown effectively to suppress systemic T-cell reactivity in experimental allergic encephalomyelitis (EAE) in the rat and diabetes in the NOD mouse, respectively (Sun *et al.*, 1994, 1996; Bergerot *et al.*, 1997). In addition, Williams and colleagues have shown that EtxB [but not EtxB(G33D)] can completely prevent the induction of collagen-induced arthritis (CIA) in DBA/1 mice when given subcutaneously at the same time as injection with collagen (Williams *et al.*, 1997).

The fact that EtxB used in these experiments was administered via a non-mucosal route indicates that the toleragenic activity of the B-subunit is not due to induction of oral tolerance in a conventional sense, but rather to a modulation of the immune response to the autoantigen. In this respect, the protection from induction of arthritis by EtxB was found to coincide with a shift in the nature of the immune response to collagen, away from a Th1-dominated response to one favouring Th2 activation, characterized by high levels of IL-4 secretion and increased relative IgG1 production (Williams *et al.*, 1997). Thus, EtxB and CtxB can be considered as agents that modulate the nature of T-helper cell differentiation and/or activity by suppressing pro-inflammatory Th1 responses and stimulating Th2 responses. This also explains why the B-subunit can be both a putative adjuvant and a toleragen, since in both circumstances a shift towards a Th2 response will be beneficial.

In the case of vaccines, in which B-subunits are incorporated as adjuvants, a Th2 response should increase the level of antigen-specific secretory IgA and serum IgG production, thereby affording greater protection against infectious agents and virulence determinants. Also, an EtxB-mediated Th2 response to autoantigens should deviate the immune system away from inflammatory responses that normally lead to tissue damage and disease. A range of inflammatory disorders, including multiple sclerosis, type-I diabetes and autoimmune uveitis, is thought to be initiated by autoreactive Th1 cells (Rocken *et al.*, 1996) and evidence from relevant experimental models suggests that protection can be afforded by the induction of a Th2 response.

CONCLUDING REMARKS

Cholera toxin and related enterotoxins from *E. coli* continue to be a source of considerable fascination. Whilst much is now known about the biogenesis of these toxins, the precise molecular basis by which, for example, ToxR senses and alters Ctx expression or how the toxins are translocated across the bacterial outer membrane remain challenging areas for future investigation.

The recent advances in our understanding of the mechanisms of toxin activity, particularly on cells of the immune system, have clearly opened up new possibilities for their use as adjuvants and immunotherapeutic agents. Again, the precise mechanisms by which the toxin subunits switch on the immune system remain to be explored.

It is thus quite evident that these toxins will remain excellent models for investigating cellular processes in prokaryotic and eukaryotic systems. Although they will continue to be agents of disease, it is also clear that they have every prospect of making an enormous impact in vaccination against infectious diseases and on the prevention of human immune disorders.

ACKNOWLEDGEMENTS

I would like to thank my colleagues who have contributed to many of the achievements documented in this review, particularly Simon Hardy, Jan Holmgren, Toufic Nashar, Jun Yu and Neil Williams; and I thank Kate Davies for her careful and meticulous assistance in preparing this manuscript.

TRH is supported by grants from The Wellcome Trust, the Medical Research Council, the Biotechnology and Biological Sciences Research Council and the Arthritis and Rheumatism Council.

REFERENCES

Agbonlahor, D.E. and Odugbemi, T.O. (1982). Enteropathogenic, enterotoxigenic and enteroinvasive *Escherichia coli* isolated from acute gastroenteritis patients in Lagos, Nigeria. *Trans. R. Soc. Trop. Med. Hyg.* **76**, 265–7.

Albert, M.J., Ansaruzzaman, M., Bardhan, P.K., Faruque, A.S.G., Faruque, S.M., Islam, M.S., Mahalanabis, D., Sack, R.B., Salam, M.A., Siddique, A.K., Yunus, M.D. and Zaman, K. (1993). Large epidemic of cholera-like disease in Bangladesh caused by *Vibrio cholerae* 0139 synonym Bengal. *Lancet* **342**, 387–90.

Albright, L.M., Huala, E. and Ausubel, F.M. (1989). Prokaryotic signal transduction mediated by sensor and regulator protein pairs. *Annu. Rev. Genet.* **23**, 311–16.

Arnone, A., Bier, C.J., Cotton, F.A., Day, V.W., Hazen, E.E.J., Richardson, D.C., Richardson, J.S. and Yonath, A. (1971). A high resolution structure of an inhibitor complex of the extracellular nuclease of *Staphylococcus aureus*. *J. Biol. Chem.* **245**, 2302–16.

Backstrom, M., Shahabi, V., Johansson, S., Teneberg, S., Kjellberg, A., Miller-Podraza, H., Holmgren, J. and Lebens, M. (1997). Structural basis for differential receptor binding of cholera and *Escherichia coli* heat-labile toxins: influence of heterologous amino acid substitutions in the cholera B-subunit. *Mol. Microbiol.* **24**, 489–97.

Bardwell, J.C.A. (1994). Building bridges: disulfide bond formation in the cell. *Mol. Microbiol.* **14**, 199–205.

Bardwell, J.C.A., McGovern, K. and Beckwith, J. (1991). Identification of a protein required for disulfide bond formation *in vivo*. *Cell* **65**, 581–9.

Bastiaens, P.I.H., Majoul, I.V., Verveer, P.J., Soling, H.D. and Jovin, T.M. (1996). Imaging the intracellular trafficking and state of the AB$_5$ quaternary structure of cholera toxin. *EMBO J.* **15**, 4246–53.

Baudry, B., Fasano, A., Ketley, J. and Kaper, J.B. (1992). Cloning of a gene (*zot*) encoding a new toxin produced by *Vibrio cholerae*. *Infect. Immun.* **60**, 428–34.

Bergerot, I., Ploix, C., Petersen, J., Moulin, V., Rask, C., Fabien, N., Lindblad, M., Mayer, A., Czerkinsky, C., Holmgren, J. and Thivolet, C. (1997). A cholera toxoid–insulin conjugate as an oral vaccine against spontaneous autoimmune diabetes. *Proc. Natl Acad. Sci. USA* **94**, 4610–14.

Black, R.E. (1986). The epidemiology of cholera and enterotoxigenic *E. coli* diarrheal disease. In: *Development of Vaccines and Drugs against Diarrhoea* (eds A.L. and R.M.J. Holmgren), pp. 23–32. Studentlitteratur, Lund.

Black, R.E., Merson, M.H., Rahman, A.S.M.M., Yunus, M., Alim, A.R.M.A., Huq, I., Yolken, R.H. and Curlin, G.T. (1980). A two-year study of bacterial, viral and parasitic agents associated with diarrhoea in rural Bangladesh. *J. Infect. Dis.* **142**, 660–4.

Black, R.E., Merson, M.H., Huq, I., Alim, A.R.M. and Yunus, M.D. (1981). Incidence of rotavirus and *Escherichia coli* diarrhoea in rural Bangladesh. *Lancet* **vi**, 141–3.

Booth, B.A., Boesman-Finkelstein, M. and Finkelstein, R.A. (1984). *Vibrio cholerae* hemagglutinin protease nicks cholera enterotoxin. *Infect. Immun.* **45**, 558–60.

Brickman, T.J., Boesman-Finkelstein, M., Finkelstein, R.A. and McIntosh, M.A. (1990). Molecular cloning and nucleotide sequence analysis of cholera toxin genes of the ctxA⁻ *Vibrio cholerae* strain Texas Star-SR. *Infect. Immun.* **58**, 4142–4.

Champion, G.A., Neely, M.N., Brennan, M.A. and DiRita, V.J. (1997). A branch in the ToxR regulatory cascade of *Vibrio cholerae* revealed by characterization of toxT mutant strains. *Mol. Microbiol.* **23**, 323–31.

Clements, J.D. and Finkelstein, R.A. (1979). Isolation and characterisation of homogeneous heat-labile enterotoxins with high specific activity from *Escherichia coli* cultures. *Infect. Immun.* **24**, 760–9.

Clements, J.D., Yancy, R.J. and Finkelstein, R.A. (1980). Properties of homogeneous heat-labile enterotoxin from *Escherichia coli*. *Infect. Immun.* **29**, 91–7.

Cong, Y.Z., Bowdon, H.R. and Elson, C.O. (1996). Identification of immunodominant T-cell epitope on cholera toxin. *Eur. J. Immunol.* **26**, 2587–94.

Cuatrecasas, P. (1973). Gangliosides and membrane receptors for cholera toxin. *Biochemistry* **12**, 3558–66.

Czerkinsky, C., Sun, J.B., Lebens, M., Li, B.L., Rask, C., Lindblad, M. and Holmgren, J. (1996). Cholera toxin B-subunit as transmucosal carrier delivery and immunomodulating system for Induction of antiinfective and antipathological immunity. *Ann. NY Acad. Sci.* **778**, 185–93.

Dallas, W.S. (1983). Conformity between heat-labile toxin genes from human and procine enterotoxigenic *Escherichia coli*. *Infect. Immun.* **40**, 647–52.

Dallas, W.S. and Falkow, S. (1979). The molecular nature of heat-labile enterotoxin from *Escherichia coli*. *Nature (Lond.)* **277**, 406–7.

Dallas, W.S. and Falkow, S. (1980). Amino acid sequence homology between cholera toxin and *Escherichia coli* heat-labile toxin. *Nature (Lond.)* **288**, 499–501.

Dallas, W.S., Gill, D.M. and Falkow, S. (1979). Cistrons encoding *Escherichia coli* heat-labile enterotoxin (LT) of *Escherichia coli*. *J. Bacteriol.* **139**, 850–8.

Dams, E., De Wolf, M. and Dierick, W. (1991). Nucleotide sequence analysis of the CT operon of *Vibrio cholerae* classical strain 569B. *Biochim. Biophys. Acta* **1090**, 139–41.

Daniels, C.J., Bole, D.G., Quay, S.C. and Oxender, D.L. (1981). Role of membrane potential in the secretion of protein into the periplasm of *Escherichia coli*. *Proc. Natl Acad. Sci. USA* **78**, 5396–400.

Dartigalongue, C. and Raina, S. (1998). A new heat-shock gene, *ppiD*, encodes a peptidyl-prolyl isomerase required for folding of outer membrane proteins in *Escherichia coli*. *EMBO J.* **17**, 3968–80.

de Haan, L., Verweij, W.R., Feil, I.K., Lijnema, T.H., Hol, W.G.J., Agsteribbe, E. and Wilschut, J. (1996). Mutants of the *Escherichia coli* heat-labile enterotoxin with reduced ADP-ribosylation activity or no activity retain the immunogenic properties of the native holotoxin. *Infect. Immun.* **64**, 5413–16.

De, S.N., Bahattacharya, K. and Sarkar, J.K. (1956). A study of the pathogenicity of strains of *Bacterium coli* from acute and chronic enteritis. *J. Pathol. Bacteriol.* **71**, 201–9.

DeMol, P., Brasseur, D., Hemelhof, W., Kalala, T., Butzler, J.P. and Vis, H.L. (1983). Enteropathogenic agents in children with diarrhoea in rural Zaire. *Lancet* **i**, 516–18.

DiRita, V.J. (1992). Coordinate expression of virulence genes by ToxR in *Vibrio cholerae*. *Mol. Microbiol.* **6**, 451–8.

DiRita, V.J., Neely, M., Taylor, R.K. and Bruss, P.M. (1996). Differential expression of the ToxR regulon in classical and El Tor biotypes of *Vibrio cholerae* is due to biotype-specific control over toxT expression. *Proc. Natl Acad. Sci. USA* **93**, 7991–5.

Douce, G., Turcotte, C., Cropley, I., Roberts, M., Pizza, M., Domenghini, M., Rappuoli, R. and Dougan, G. (1995). Mutants of *Escherichia coli* heat-labile enterotoxin lacking ADP-ribosyltransferase activity act as nontoxic, mucosal adjuvants. *Proc. Natl Acad. Sci. USA* **92**, 1644–8.

Douce, G., Fontana, M., Pizza, M., Rappuoli, R. and Dougan, G. (1997). Intranasal immunogenicity and adjuvanticity of site-directed mutant derivatives of cholera toxin. *Infect. Immun.* **65**, 2821–8.

Driessen, A.J.M. (1992). Precursor protein translocation by the *Escherichia coli* translocase is directed by the proton motive force. *EMBO J.* **11**, 847–53.

Driessen, A.J.M. and Wicker, W. (1991). Proton transfer is rate-limiting for translocation of precursor proteins by the *Escherichia coli* translocase. *Proc. Natl Acad. Sci. USA* **88**, 2471–5.

Driessen, A.J.M., Fekkes, P. and van der Wolk, J.P.W. (1998). The Sec system. *Curr. Opin. Microbiol.* **1**, 216–22.

Dykes, C.W., Halliday, I.J., Hobden, A.N., Read, M.J. and Harford, S. (1985). A comparison of the nucleotide sequence of the A subunit of heat-labile enterotoxin and cholera toxin. *FEMS Microbiol. Lett.* **26**, 171–4.

Echeverria, P., Seriwarana, J., Taylor, D.N., Yanggratoke, S. and Tirapat, C. (1985). A comparative study of enterotoxigenic *Escherichia coil*, *Shigella*, *Aeromonas* and *Vibrio* as aetiologies of diarrhoea in Northeastern Thailand. *Am. J. Trop. Med. Hyg.* **35**, 547–54.

Elson, C.O. and Ealding, W. (1984). Generalized systemic and mucosal immunity in mice after mucosal stimulation with cholera toxin. *J. Immunol.* **132**, 2736–41.

Elson, C.O., Holland, S.P., Dertzbaugh, M.T., Cuff, C.F. and Anderson, A.O. (1995). Morphologic and functional alterations of mucosal T-cells by cholera toxin B-subunit. *J. Immunol.* **154**, 1032–40.

Evans, D.G., Olarte, J., DuPont, H.L., Evans, D.J.J., Galindo, E., Portnoy, B.L. and Conklin, R.H. (1977). Enteropathogens associated with paediatric diarrhoea in Mexico City. *J. Pediatr.* **91**, 65–8.

Fasano, A., Baudry, B., Pumplin, D.W., Wasserman, S.S., Tall, B.D., Ketley, J.M. and Kaper, J.B. (1991). *Vibrio cholerae* produces a second enterotoxin, which affects intestinal tight junctions. *Proc. Natl Acad. Sci. USA* **88**, 5242–6.

Finkelstein, R.A., Atthasampunu, P., Chulasmya, M. and Charunmethee, P. (1966). Pathogenesis of experimental cholera: biologic activities of purified procholeragen. *J. Immunol.* **96**, 440–9.

Finkelstein, R.A. and LoSpalluto, J.J. (1969). Pathogenesis of experimental cholera: preparation and isolation of choleragen and choleragenoid. *J. Exp. Med.* **130**, 185–202.

Finkelstein, R.A., Boseman, M., Neoh, S.H., LaRue, M.K. and Delaney, R. (1974a). Dissociation and recombination of the subunits of the cholera enterotoxin (choleragen). *J. Immunol.* **113**, 145–50.

Finkelstein, R.A., Vasil, M.L. and Holmes, R.K. (1974b). Studies on toxinogenesis in *Vibrio cholerae* I. Isolation of mutants with altered toxinogenicity. *J. Infect. Dis.* **129**, 117–23.

Fitts, R., Reuveny, Z., van Amsterdam, J., Mulholland, J. and Botstein, D. (1987). Substitution of tyrosine for either cysteine in β-lactamase prevents release from the membrane during secretion. *Proc. Natl Acad. Sci. USA* **84**, 8540–3.

Francetic, O. and Pugsley, A.P. (1996). The cryptic general secretory pathway (*gsp*) operon of *Escherichia coli* K-12 encodes functional proteins. *J. Bacteriol.* **178**, 3544–9.

Francis, M.L., Ryan, J., Jobling, M.G., Holmes, R.K., Moss, J. and Mond, J.J. (1992). Cyclic AMP-independent effects of cholera toxin on B-cell activation. 2. Binding of ganglioside-GM1 induces B-cell activation. *J. Immunol.* **148**, 1999–2005.

Geary, S.J., Marchlewicz, B.A. and Finkelstein, R.A. (1982). Comparison of heat-labile enterotoxins from porcine and human strains of *Escherichia coli*. *Infect. Immun.* **36**, 215–20.

Geller, B.L. (1991). Energy requirements for protein translocation across the *Escherichia coli* inner membrane. *Mol. Microbiol.* **5**, 2093–8.

Gennaro, M.L. and Greenaway, P.J. (1983). Nucleotide sequences within the cholera toxin operon. *Nucl. Acids Res* **11**, 3855–61.

Gennaro, M.L., Greenaway, P.J. and Broadbent, D.A. (1982). Expression of biologically active cholera toxin in *Escherichia coli*. *Nucl. Acids Res.* **10**, 4883–90.

Gill, D.M. (1976). The arrangement of subunits in cholera toxin. *Biochemistry* **15**, 1242–8.

Gill, D.M., Clements, J.D., Roberston, D.C. and Finkelstein, R.A. (1981). Subunit number and arrangement in *Escherichia coli* heat-labile entertoxin. *Infect. Immun.* **33**, 677–82.

Goldberg, I. and Mekalanos, J.J. (1986). Effect of recA mutation on cholera toxin gene amplification and deletion events. *J. Bacteriol.* **165**, 723–31.

Green, E.A., Botting, C., Webb, H.M., Hirst, T.R. and Randall, R.E. (1996). Construction, purification and immunogenicity of antigen-antibody-LTB complexes. *Vaccine* **14**, 949–58.

Gyles, C.L., Palchaudhuri, S. and Maas, W. (1977). Naturally occuring plasmid carrying genes for enterotoxin production and drug resistance. *Science* **198**, 198–9.

Hardy, S.J.S. and Hedges, P.A. (1996). Reduced B subunit of heat-labile enterotoxin associates with membranes *in vivo*. *Eur. J. Biochem.* **236**, 412–18.

Hardy, S.J.S., Holmgren, J., Johansson, S., Sanchez, J. and Hirst, T.R. (1988). Coordinated assembly of multisubunit proteins: Oligomerization of bacterial enterotoxins *in vivo* and *in vitro*. *Proc. Natl Acad. Sci. USA* **85**, 7109–13.

Hayano, T., Takahashi, N., Kato, S., Maki, N. and Suzuki, M. (1991). Two distinct forms of peptidylprolyl-*cis*–*trans*-isomerase are expressed separately in periplasmic and cytoplasmic compartments of *Escherichia coli* cells. *Biochemistry* **30**, 3041–8.

Hengge, R. and Boos, W. (1985). Defective secrtion of maltose- and ribose-binding proteins caused by a truncated periplasmic protein in *Escherichia coli*. *J. Bacteriol.* **162**, 972–8.

Herrington, D.A., Hall, R.H., Losonsky, G., Mekalanos, J.J., Taylor, R.K. and Levine, M.M. (1988). Toxin, toxin-coregulated pili and the *toxR* regulon are essential for *Vibrio cholerae* pathogenesis in humans. *J. Exp. Med.* **168**, 1487–92.

Hillary, J.B. (1998). Investigations into the molecular mechanism of toxin secretion from *Vibrio cholerae*. Ph.D. Thesis, University of Bristol, UK.

Hirst, T.R. (1991). Assembly and secretion of oligomeric toxins by Gram negative bacteria. In: *Sourcebook of Bacterial Protein Toxins* (eds J. Alouf and J. Freer), pp. 75–100. Academic Press, London.

Hirst, T.R. (1995). Biogenesis of cholera toxin and related oligomeric enterotoxins. In: *Bacterial Toxins and Virulence Factors in Disease* (eds J. Moss, B. Iglewski, M. Vaughan and A.T. Tu), pp. 123–84. Marcel Dekker, New York.

Hirst, T.R. and Holmgren, J. (1987a). Transient entry of enterotoxin subunits into the periplasm occurs during their secretion from *Vibrio cholera*. *J. Bacteriol.* **169**, 1037–45.

Hirst, T.R. and Holmgren, J. (1987b). Conformation of protein secreted across bacterial outer membranes: a study of enterotoxin translocation from *Vibrio cholerae*. *Proc. Natl Acad. Sci. USA* **84**, 7418–22.

Hirst, T.R. and Welch, R.A. (1988). Mechanisms for secretion of extracellular proteins by Gram-negative bacteria. *Trends Biochem. Sci.* **13**, 265–9.

Hirst, T.R., Randall, L.L. and Hardy, S.J.S. (1983). Assembly *in vivo* of enterotoxin from *Escherichia coli*: formation of the B subunit oligomer. *J. Bacteriol.* **153**, 21–6.

Hirst, T.R., Randall, L.L. and Hardy, S.J.S. (1984a). Cellular location of heat-labile enterotoxin in *Escherichia coli*. *J. Bacteriol.* **157**, 637–42.

Hirst, T.R., Sanchez, J., Kaper, J.B., Hardy, S.J.S. and Holmgren, J. (1984b). Mechanism of toxin secretion by *Vibrio cholerae* investigated in strains harboring plasmids that encode heat-labile enterotoxins of *Escherichia coli*. *Proc. Natl Acad. Sci. USA* **81**, 7752–6.

Ho, A.S.Y., Mietzner, T.A., Smith, A.J. and Schoolnik, G.K. (1990). The pili of *Aeromonas hydrophila*: identification of an environmentally regulated mini pilin. *J. Exp. Med.* **172**, 795–806.

Hofstra, H. and Witholt, B., (1984). Kinetics of synthesis, processing and membrane transport of heat-labile enterotoxin, a periplasmic protein in *Escherichia coli*. *J. Biol. Chem.* **259**, 5182–7.

Hofstra, H. and Witholt, B. (1985). Heat-labile enterotoxin in *Escherichia coli*: kinetics of association of subunits into periplasmic holotoxin. *J. Biol. Chem.* **260**, 16037–44.

Hol, W.G.J., Sixma, T.K. and Merritt, E.A. (1995). Structure and function of *E. coli* heat-labile enterotoxin and cholera toxin B pentamer. In: *Bacterial Toxins and Virulence Factors in Disease* (ed. A.T. Tu), pp. 185–223. Marcel Dekker, New York.

Holmes, R.K., Vasil, M.L. and Finkelstein, R.A. (1975). Studies on toxigenesis in *Vibrio cholerae* III. Characterization of nontoxinogenic mutants *in vitro* and in experimental animals. *J. Clin. Invest.* **55**, 551–60.

Holmgren, J. (1973). Comparison of the tissue receptors for *Vibrio cholerae* and *Escherichia coli* enterotoxins by means of gangliosides and natural cholera toxoid. *Infect. Immun.* **8**, 851–9.

Hunt, P.D. and Hardy, S.J.S. (1991). Heat-labile enterotoxin can be released from *Escherichia coli* cells by host intestinal factors. *Infect. Immun.* **59**, 168–71.

Inoue, T., Tsuji, T., Koto, M., Imamura, S. and Miyama, A. (1993). Amino acid sequence of heat-labile enterotoxin from chicken enterotoxigenic *Escherichia coli* is identical to that of human strain H 10407. *FEMS Microbiol. Lett.* **108**, 157–62.

Ito, K. and Beckwith, J.R. (1981). Role of the mature protein sequence of maltose binding protein in its secretion across the *E. coli* cytoplasmic membrane. *Cell* **25**, 143–50.

Jaenicke, R. (1987). Folding and association of proteins. *Prog. Biophys. Mol. Biol.* **49**, 117–237.

Jaenicke, R. (1991). Protein folding: local structures, domains, subunits and assemblies. *Biochemistry* **30**, 3147–61.

Jiang, B. and Howard, S.P. (1992). The *Aeromonas hydrophila exeE* gene, required both for protein secretion and normal outer membrane biogenesis, is a member of a general secretion pathway. *Mol. Microbiol.* **10**, 1351–61.

Jobling, M.G. and Holmes, R.K. (1991). Analysis of structure and function of the B subunit of cholera toxin by the use of site-directed mutagenesis. *Mol. Microbiol.* **5**, 1755–67.

Jobling, M.G. and Holmes, R.K. (1992). Fusion proteins containing the A2 domain of cholera toxin assemble with B-polypeptides of cholera toxin to form immunoreactive and functional holotoxin-like chimeras. *Infect. Immun.* **60**, 4915–24.

Kamitani, S., Akiyama, Y. and Ito, K. (1992). Identification and characterization of an *Escherichia coli* gene required for the formation of correctly folded alkaline phosphatase, a periplasmic enzyme. *EMBO J.* **11**, 57–62.

Karlsson, K.A., Teneberg, S., Angstrom, J., Kjellberg, A., Hirst, T.R., Berstrom, J. and Miller-Podraza, H. (1996). Unexpected carbohydrate cross-binding by *Escherichia coli* heat-labile enterotoxin. Recognition of human and rabbit target cell glycoconjugates in comparison with cholera toxin. *Bioorg. Med. Chem.* **11**, 1919–28.

Koch, R. (1884). An address on cholera and its bacillus. *Br. Med. J.* **2**, 403–7.

Koshland, D. and Botstein, D. (1982). Evidence for post-translational translocation of B-lactamase across the bacterial inner membrane. *Cell* **30**, 893–902.

Koster, M., Bitter, W., deCock, H., Allaoui, A., Cornelis, G.R. and Tommassen, J. (1997). The outer membrane component, YscC, of the Yop secretion machinery of *Yersinia enterocolitica* forms a ring-shaped multimeric complex. *Mol. Microbiol.* **26**, 789–97.

Kunkel, S.L. and Robertson, D.C. (1979). Purification and chemical characterization of the heat-labile enterotoxin produced by enterotoxigenic *Escherichia coli. Infect. Immun.* **25**, 586–96.

Kuziemko, G.M., Stroh, M. and Stevens, R.C. (1996). Cholera toxin binding affinity and specificity for gangliosides determined by surface plasmon resonance. *Biochemistry* **35**, 6375–84.

Lang, K., Schmid, F.X. and Fischer, G. (1987). Catalysis of protein folding by prolyl isomerase. *Nature* **329**, 268–70.

Lencer, W.I., Delp, C., Neutra, M.R. and Madera, J.L. (1992). Mechanism of action on a polarized human epithelial cell line: role of vesicular traffic. *J. Cell Biol.* **117**, 1197–209.

Lencer, W.I., Dealmeida, J.B., Moe, S., Stow, J.L., Ausiello, D.A. and Madara, J.L. (1993). Entry of cholera toxin into polarized human intestinal epithelial cells: identification of an early Brefeldin A sensitive event required for a (1) peptide generation. *J. Clin. Invest.* **92**, 2941–51.

Lencer, W.I., Constable, C., Moe, S., Jobling, M.G., Webb, H.M., Ruston, S., Madara, J.L., Hirst, T.R. and Holmes, R.K. (1995). Targeting of cholera toxin and *Escherichia coli* heat-labile toxin in polarized epithelia – role of COOH-terminal KDEL. *J. Cell Biol.* **131**, 951–62.

Lencer, W.I., Constable, C., Moe, S., Rufo, P.A., Wolf, A., Jobling, M.G., Ruston, S.P., Madara, J.L., Holmes, R.K. and Hirst, T.R. (1997). Proteolytic activation of cholera toxin and *Escherichia coli* labile toxin by entry into host epithelial cells: signal transduction by a protease-resistant toxin variant. *J. Biol. Chem.* **272**, 15562–8.

Leong, J., Vinal, A.C. and Dallas, W.S. (1985). Nucleotide sequence comparison between heat-labile toxin B-subunit cistrons from *Escherichia coli* of human and porcine origin. *Infect. Immun.* **48**, 73–7.

Liu, J. and Walsh, C.T. (1990). Peptidyl-prolyl *cis–trans*-isomerase a periplasmic homologue of cyclophilin that is not inhibited by cyclosporin A. *Proc. Natl Acad. Sci. USA* **87**, 4028–32.

Lockman, H. and Kaper, J.B. (1983). Nucleotide sequence analysis of the A2 and B subunits of *Vibrio cholerae* enterotoxin. *J. Biol. Chem.* **258**, 13722–6.

Lockman, H.A., Galen, J.E. and Kaper, J.B. (1984). *Vibrio cholerae* enterotoxin genes: nucleotide sequence of DNA encoding ADP-ribosyl transferase. *J. Bacteriol.* **159**, 1086–9.

Lu, H.M. and Lory, S. (1996). A specific targeting domain in mature exotoxin A is required for its extracellular secretion from *Pseudomonas aeruginosa. EMBO J.* **15**, 429–36.

Lutz, M.B., Rovere, P., Kleijmeer, M.L., Rescigno, M., Assman, C.U., Oorschot, V.M.J., Geuze, H.L., Trucy, J., Demandolx, D., Davoust, J. and Ricciadi-Castagnoli, P. (1997). Intracellular routes and selective retention of antigens in mildy acidic cathepsin D/lysosome-associated membrane protein-I/MHC class II-positive vesicles in immature dendritic cells. *J. Immunol.* **125**, 672–83.

Lycke, N., Tsuji, T. and Holmgren, J. (1992). The adjuvant effect of *Vibrio cholerae* and *Escherichia coli* enterotoxins is linked to their ADP-ribosyl transferase activity. *Eur. J. Immunol.* **22**, 2277–81.

Majoul, I.V., Bastiaens, P.I.H. and Soling, H.D. (1996). Transport of an external *lys asp glu leu* (KDEL) protein from the plasma membrane to the endoplasmic reticulum: studies with cholera toxin *in vero* cells. *J. Cell Biol.* **133**, 777–80.

Majoul, I., Ferrari, D. and Soling, H.D. (1997). Reduction of protein disulfide bonds in an oxidizing environment – the disulfide bridge of cholera toxin A-subunit is reduced in the endoplasmic reticulum. *FEBS Lett.* **401**, 104–8.

Marsh, J.W. and Taylor, R.K. (1998). Identification of the *Vibrio cholerae* type 4 prepilin peptidase required for cholera toxin secretion and pilus formation. *Mol. Microbiol.* **29**, 1481–92.

Matousek, M.P., Nedrud, J.G. and Harding, C.V. (1996). Distinct effects of recombinant cholera toxin B subunit and holotoxin on different stages of class II MHC antigen-processing and presentation by macrophages. *J. Immunol.* **156**, 4137–45.

Matousek, M.P., Nedrud, J.G., Cieplak, W. and Harding, C.V. (1997). E-coli heat labile enterotoxin and cholera toxin inhibit class II MHC antigen processing. *J. Allerg. Clin. Immunol.* **99**, 938–8.

Mekalanos, J.J. (1983). Duplication and amplification of toxin genes in *Vibrio cholerae. Cell* **35**, 253–63.

Mekalanos, J.J. (1992). Environmental signal controlling expression of virulence determinants. *J. Bacteriol.* **174**, 1–7.

Mekalanos, J.J., Swartz, D.J., Pearson, G.D.N., Harford, N., Groyne, F.and de Wilde, M. (1983). Cholera toxin genes: nucleotide sequence deletion analysis and vaccine development. *Nature (Lond.)* **306**, 551–7.

Merritt, E.A., Sarfaty, S., van den Akker, F., L'Hoir, C., Martial, J.A. and Hol, W.G.J. (1994). Crystal structure of cholera toxin B-pentamer bound to receptor GM1 pentasaccharide. *Prot. Sci.* **3**, 166–75.

Merritt, E.A., Sarfaty, S., Jobling, M.G., Chang, T., Holmes, R.K., Hirst, T.R. and Hol, W.G.J. (1997). Structural studies of receptor binding by cholera toxin mutants. *Prot. Sci.* **6**, 1516–28.

Merritt, E.A., Kuhn, P., Sarfaty, S., Erbe, J.L., Holmes, R.K. and Ho, W.G.J. (1998). The 1.25 angstrom resolution refinement of the cholera toxin B-pentamer: evidence of peptide backbone strain at the receptor-binding site. *J. Mol. Biol.* **282**, 1043–59.

Miller, V.L. and Mekalanos, J.J. (1984). Synthesis of cholera toxin is positively regulated at the transcriptional level by *toxR. Proc. Natl Acad. Sci. USA* **81**, 3471–5.

Miller, V.L. and Mekalanos, J.J. (1988). A novel suicide vector and its use in construction of insertion mutation: osmoregulation of outer membrane proteins and virulence determinants in *Vibrio cholerae* requires *toxR. J. Bacteriol.* **170**, 2575–83.

Miller, V.L., Taylor, R.K. and Mekalanos, J.J. (1987). Cholera toxin transcriptional activator ToxR is a transmembrane DNA binding protein. *Cell* **48**, 271–9.

Minsky, A., Summers, R.G. and Knowles, J.R. (1986). Secretion of β-lactamase into the periplasm of *Escherichia coli*: evidence for a distinct release step associated with a conformational change. *Proc. Natl Acad. Sci. USA* **83**, 4180–4.

Missiakas, D. and Raina, S. (1997). Protein folding in the bacterial periplasm. *J. Bacteriol.* **179**, 2465–71.

Molinari, M., Galli, C., Norais, N., Telford, J.L., Rappuoli, R., Luzio, J.P. and Montecucco, C. (1997). Vacuoles induced by *Helicobacter pylori* contain both late endosomal and lysosomal markers. *J. Biol. Chem.* **272**, 25339–44.

Molinari, M., Salio, M., Galli, C., Norais, N., Rappuoli, R., Lanavecchia, A. and Montecucco, C. (1998). Selective inhibition of Ii-dependent antigen presentation by *Helicobacter pylori* toxin VacA. *J. Exp. Med.* **187**, 135–40.

Moon, H.W. and Bunn, T.O. (1993). Vaccines for preventing enterotoxigenic *Escherichia coli* infections in farm animals. *Vaccine* **11**, 213–20.

Moss, J. and Vaughan, M. (1992). Activation of cholera toxin and E. coli heat-labile enterotoxins by ADP-ribosylation factors, 20kDa guanine nucleotide-binding proteins. In: *Bacterial Protein Toxins* (eds B. Witholt, J.E. Alouf, G.J. Boulnois, P. Cossart, B.W. Dijkstra, P. Falmague, F.J. Fehrenbach, J. Freer, H. Niemann, R. Rappuoli and T. Wadström), pp. 220–8. Gustav Fischer, Stuttgart.

Moss, J., Tsai, S.C. and Vaughan, M. (1994). Activation of cholera toxin by ADP-ribosylation factors. *Meth. Enzymol.* **235**, 640–7.

Murzin, A. (1993). OB (oligonucleotide/oligosaccharide binding)-fold: common structural and functional solution for nonhomologous sequences. *EMBO J.* **12**, 861–7.

Nakashima, K., Eguchi, Y. and Nakasone, N. (1995). Characterization of an enterotoxin produced by *Vibrio cholerae* O139. *Microbiol. Immunol.* **39**, 87–94.

Nambiar, M.P., Oda, T., Chen, C.H., Kuwazuru, Y. and Wu, H.C. (1993). Involvement of the Golgi region in the intracellular trafficking of cholera toxin. *J. Cell. Physiol.* **154**, 222–8.

Nashar, T.O. and Hirst, T.R. (1995). Immunoregulatory role of H-2 and intra-H-2 alleles on antibody responses to recombinant preparations of B-subunits of *Escherichia coli* heat-labile enterotoxin (rEtxB) and cholera toxin (rCtxB). *Vaccine* **13**, 803–10.

Nashar, T.O., Amin, T., Marcello, A. and Hirst, T.R. (1993). Current progress in the development of the B subunits of cholera toxin and *Escherichia coli* heat-labile enterotoxin as carriers for the oral delivery of heterologous antigens and epitopes. *Vaccine* **11**, 235–40.

Nashar, T.O., Williams, N.A. and Hirst, T.R. (1996a). Cross-linking of cell-surface ganglioside GM1 induces the selective apoptosis of mature CD8+ T-lymphocytes. *Int. Immunol.* **8**, 731–6.

Nashar, T.O., Webb, H.M., Eaglestone, S., Williams, N.A. and Hirst, T.R. (1996b). Potent immunogenicity of the B-subunits of *Escherichia coli* heat-labile enterotoxin: receptor binding is essential and induces differential modulation of lymphocyte subsets. *Proc. Natl Acad. Sci. USA* **93**, 226–30.

Nashar, T.O., Hirst, T.R. and Williams, N.A. (1997). Modulation of B-cell activation by the B-subunit of *Escherichia coli* heat-labile enterotoxin: receptor interaction upregulates MHC Class II, B7, CD40, CD25 and ICAM1. *Immunology* **91**, 572–8.

Nataro, J.P. and Kaper, J.B. (1998). Diarrheagenic *Escherichia coli*. *Clin. Microbiol. Rev.* **11**, 142–201.

Neill, R.J., Ivins, B.E. and Holmes, R.K. (1983). Synthesis and secretion of the plasmid-coded heat-labile enterotoxin of *Escherichia coli* in *Vibrio cholerae*. *Science* **221**, 289–91.

Northrup, R.S. and Fauci, A.S. (1972). Adjuvant effect of cholera toxin on the immune response of the mouse to sheep red blood cells. *J. Infect. Dis.* **125**, 672–3.

Orlandi, P.A. (1997). Protein disulfide isomerase mediated reduction of the a subunit of cholera toxin in a human intestinal cell line. *J. Biol. Chem.* **272**, 4591–9.

Orlandi, P.A. and Fishman, P.H. (1993). Orientation of cholera toxin bound to target cells. *J. Biol. Chem.* **268**, 17038–44.

Orlandi, P.A., Critchley, D.R. and Fishman, P.H. (1994). The heat-labile enterotoxin of *Escherichia coli* binds to polylactosaminoglycan-containing receptors in CACO-2 human intestinal epithelial cells. *Biochemistry* **33**, 12886–95.

Overbye, L.J., Sandkvist, M. and Bagdasarian, M. (1993). Genes required for extracellular secretion of enterotoxin are clustered in *Vibrio cholerae*. *Gene* **132**, 101–6.

Palva, E.T., Hirst, T.R., Hardy, S.J.S., Holmgren, J. and Randall, L.L. (1981). Synthesis of a precursor to the B subunit of heat-labile enterotoxin in *Escherichia coli*. *J. Bacteriol.* **146**, 325–30.

Pearson, G.D.N. and Mekalanos, J.J. (1982). Molecular cloning of *Vibrio cholerae* enterotoxin genes in *Escherichia coli* K-12. *Proc. Natl Acad. Sci. USA* **79**, 2976–80.

Pearson, G.D.N., Woods, A., Chiang, S.L. and Mekalanos, J.J. (1993). Ctx genetic element encodes a site-specific recombination system and an intestinal colonization factor. *Proc. Natl Acad. Sci. USA* **90**, 3750–4.

Peek, J.A. and Taylor, R.K. (1992). Characterization of a periplasmic thiol:disulfide interchange protein required for the functional maturation of secreted virulence factors of *Vibrio cholerae*. *Proc. Natl Acad. Sci. USA* **89**, 6210–14.

Pugsley, A.P. (1993). The complete general secretory pathway in gram-negative bacteria. *Microbiol. Rev.* **57**, 50–108.

Pugsley, A.P., d'Enfert, C., Reyss, I. and Kornacker, M.G. (1990). Genetics of extracellular protein secretion by Gram negative bacteria. *Annu. Rev. Genet.* **24**, 67–90.

Pugsley, A.P., Kornacker, M.G. and Poquet, I. (1991). The general protein-export pathway is directly required for extracellular pullulanase secretion in *Escherichia coli* K12. *Mol. Microbiol.* **5**, 343–52.

Pugsley, A.P., Francetic, O., Possot, O.M., Sauvonnet, N. and Hardie, K.R. (1997). Recent progress and future directions in studies of the main terminal branch of the general secretory pathway in Gram-negative bacteria – A review. *Gene* **192**, 13–19.

Randall, L.L. and Hardy, S.J.S. (1989). Unity in function in the absence of consensus in sequence: role of the leader peptides in export. *Science* **243**, 1156–9.

Rocken, M., Racke, M. and Shevach, E.M. (1996). IL-4-induced immune deviation as antigen-specific therapy for inflammatory autoimmune disease. *Immunol. Today* **17**, 225–31.

Rodighiero, C., Aman, A.T., Kenny, M.J., Moss, J., Lencer, W.I. and Hirst, T.R. (1999) Structural basis for the differential toxicity of cholera toxin and *E. coli* heat-labile enterotoxin. *J. Biol. Chem.* **274**, 3962–9.

Ruddock, L.W., Ruston, S.P., Kelly, S.M., Price, N.C., Freedman, R.B. and Hirst, T.R. (1995). Kinetics of acid-mediated disassembly of the B subunit pentamer of *Escherichia coli* heat-labile enterotoxin: molecular basis of pH stability. *J. Biol. Chem.* **270**, 29953–8.

Ruddock, L.W., Coen, J.J.F., Cheesman, C., Freedman, R.B. and Hirst, T.R. (1996). Assembly of the B subunit pentamer of *Escherichia coli* heat-labile enterotoxin: kinetics and molecular basis of rate-limiting steps *in vitro*. *J. Biol. Chem.* **271**, 19118–23.

Sack, R.B. (1996). Emergence of *Vibrio cholerae* O139. *Curr. Clin. Top. Infect. Dis.* **16**, 172–93.

Sanchez, J. and Holmgren, J. (1989). Recombinant system for overexpression of cholera toxin B subunit in *Vibrio cholerae* as a basis for vaccine development. *Proc. Natl Acad. Sci. USA* **86**, 481–5.

Sandkvist, M., Hirst, T.R. and Bagdasarian, M. (1987). Alterations at the carboxyl terminus change assembly and secretion properties of the B subunit of *Escherichia coli* heat-labile enterotoxin. *J. Bacteriol.* **169**, 4570–6.

Sandkvist, M., Hirst, T.R. and Bagdasarian, M. (1990). Minimal deletion of amino acids from the carboxyl terminus of the B subunit of heat-labile enterotoxin causes defects in its assembly and release from the cytoplasmic membrane of *Escherichia coli*. *J. Biol. Chem.* **265**, 15239–44.

Sandkvist, M., Morales, V. and Bagdasarian, M. (1993). A protein required for secretion of cholera toxin through the outer membrane of *Vibrio cholerae*. *Gene* **123**, 81–6.

Sandkvist, M., Michel, L.O., Hough, L.P., Morales, V.M., Bagdasarian, M., Koomey, M., DiRita, V.J. and Bagdasarian, M. (1997). General secretion pathway (*eps*) genes required for toxin secretion and outer membrane biogenesis in *Vibrio cholerae*. *J. Bacteriol.* **179**, 6994–7003.

Sandvig, K., Garred, O. and Van Deurs, B. (1996). Thapsigargin induced transport of cholera toxin to the endoplasmic reticulum. *Proc. Natl Acad. Sci. USA* **93**, 12339–43.

Schonberger, O., Hirst, T.R. and Pines, O. (1991). Targeting and assembly of an oligomeric bacterial enterotoxoid in the endoplasmic reticulum of *Saccharomyces cerevisiae*. *Mol. Microbiol.* **5**, 2663–71.

Simonen, M. and Palva, I. (1993). Protein secretion in *Bacillus* species. *Microbiol. Rev.* **57**, 109–37.

Sixma, T.K., Pronk, S.E., Kalk, K.H., Vanzanten, B.A.M., Berghuis, A.M. and Hol, W.G.J. (1992). Lactose binding to heat-labile enterotoxin revealed by X-ray crystallography. *Nature* **355**, 561–4.

Sixma, T.K., Kalk, K.H., Vanzanten, B.A.M., Dauter, Z., Kingma, J., Witholt, B. and Hol, W.G.J. (1993). Refined structure of *Escherichia coli* heat-labile enterotoxin, a close relative of cholera toxin. *J. Mol. Biol.* **230**, 890–918.

Smith, H.W. and Halls, S. (1967). Observations by the ligated intestinal segment and oral inoculation methods on *Escherichia coli* infections in pigs, calves, lambs and rabbits. *J. Pathol. Bacteriol.* **93**, 499.

Smith, H.W. and Linggood, M.A. (1971). The transmissible nature of enterotoxin production in a human enteropathogenic strain of *Escherichia coli. J. Med. Microbiol.* **4**, 301–5.

So, M., Dallas, W.S. and Falkow, S. (1978). Characterization of an *Escherichia coli* plasmid encoding for synthesis of heat-labile toxin: molecular cloning the the toxin determinant. *Infect. Immun.* **21**, 405–11.

Spicer, E.K. and Noble, J.A. (1982). *Escherichia coli* heat-labile enterotoxin. Nucleoside sequence of the A subunit gene. *J. Biol. Chem.* **257**, 5716–21.

Steffen, R. (1986). Epidemiologic studies of travellers' diarrhea, severe gastrointestinal infections and cholera. *Rev. Infect. Dis.* **8**, S122-30.

Streatfield, S.J., Sandkvist, M., Sixma, T.K., Bagdasarian, M., Hol, W.G.J. and Hirst, T.R. (1992). Intermolecular interactions between the A and B subunits of heat-labile enterotoxin from *Escherichia coli* promote holotoxin assembly and stability *in vivo. Proc. Natl Acad. Sci. USA* **89**, 12140–4.

Sun, J.B., Holmgren, J. and Czerkinsky, C. (1994). Cholera toxin B subunit – an efficient transmucosal carrier delivery system for induction of peripheral immunological tolerance. *Proc. Natl Acad. Sci. USA* **91**, 10795–9.

Sun, J.B., Rask, C., Olsson, T., Holmgren, J. and Czerkinsky, C. (1996). Treatment of experimental autoimmune encephalomyelitis by feeding myelin basic protein conjugated to cholera-toxin-subunit. *Proc. Natl Acad. Sci. USA* **93**, 7196–201.

Takahashi, I., Kiyono, H., Jackson, R.J., Fujihashi, K., Staats, H.F., Hamada, S., Clements, J.D., Bost, K.L. and McGhee, J.R. (1996). Epitope maps of the *Escherichia coli* heat-labile toxin B-subunit for development of synthetic oral vaccine. *Infect. Immun.* **64**, 1290–8.

Takao, T., Watanabe, H. and Shimonishi, Y. (1985). Facile identification of protein sequences by mass spectrometry – B-subunit of *Vibrio cholerae* classical biotype INABA 569B toxin. *Eur. J. Biochem.* **146**, 503–8.

Taylor, R.K., Miller, V.L., Furlong, D. and Mekalanos, M.M. (1987). The use of *phoA* gene fusions to identify a pilus colonization factor coordinately regulated with cholera toxin. *Proc. Natl Acad. Sci. USA* **84**, 2833–7.

Teneberg, S., Hirst, T.R., Angstrom, J. and Karlsson, K.A. (1994). Comparison of the glycolipid-binding specificities of cholera toxin and porcine *Escherichia coli* heat-labile enterotoxin – identification of a receptor-active non-ganglioside glycolipid for the heat-labile toxin in infant rabbit small intestine. *Glycon. J.* **11**, 533–40.

Tomb, J.F. (1992). A periplasmic protein disulphide oxidoreductase is required for transformation of *Haemophilus influenzae* RD. *Proc. Natl Acad. Sci. USA* **89**, 10252–6.

Trucksis, M., Galen, J.E., Michalski, J., Fasano, A. and Kaper, J.B. (1993). Accessory cholera enterotoxin (Ace), the third toxin of a *Vibrio cholerae* virulence cassette. *Proc. Natl Acad. Sci. USA* **90**, 5267–71.

Tsuji, T., Joya, J.E., Yoa, S., Honda, T. and Miwatani, T. (1988). Purification and characterisation of heat-labile enterotoxinisolated from chicken enterotoxigenic *Escherichia coli. FEMS Microbiol. Lett.* **52**.

van Heyningen, W.E., Carpenter, C.C.J., Pierce, N.F. and Greenough, B.I. (1971). Deactivation of cholera toxin by ganglioside. *J. Infect. Dis.* **124**, 415–18.

Waldman, R. (1998). Cholera vaccination in refugee settings. *JAMA* **279**, 552–3.

Waldor, M.K. and Mekalanos, J.J. (1996). Lysogenic conversion by a filamentous phage encoding cholera toxin. *Science* **272**, 1910–14.

Walker, J.E., Saraste, M., Runswick, M.J. and Gay, N.J. (1982). Distantly related sequences in the α- and β-subunits of ATP-synthase, myosin, kinases and other ATP-requiring enzymes and a common nucleotide binding fold. *EMBO J.* **1**, 945–51.

Watts, C. (1997). Capture and processing of exogenous antigens for presentation on MHC molecules. *Annu. Rev. Immunol.* **15**, 821–50.

Wilholt, B., Hofstra, H., Kingma, J., Proak, S.E., Hol, W.G.J. and Drenth, J. (1988). Studies on the synthesis and structure of the heat-labile enterotoxin (LT) of *Escherichia coli*. In: *Bacterial Protein Toxins* (eds J.E.A.F. Fehrenback, P. Falmagne, W. Goebel, J. Jeljaszewicz, D. Jurgen and R. Rappuoli), pp. 3–12. Gustav Fischer, Stuttgart.

Williams, N.A., Stasiuk, L.M., Nashar, T.O., Richards, C.M., Lang, A.K., Day, M.J. and Hirst, T.R. (1997). Prevention of autoimmune disease due to lymphocyte modulation by the B-subunit of *Escherichia coli* heat-labile enterotoxin. *Proc. Natl Acad. Sci. USA* **94**, 5290–5.

Wolf, A.A., Jobling, M.G., Wimer-Mackin, S., Ferguson-Maltzman, M., Madara, J.L., Holmes, R.K. and Lencer, W.I. (1998). Ganglioside structure dictates signal transduction by cholera toxin and association with caveolae-like membrane domains in polarized epithelia. *J. Cell Biol.* **141**, 917–27.

Xia, Q.-C., Chang, D., Blacher, R. and Lai, C.-Y. (1984). The primary structure of the COOH-terminal half of cholera toxin A1 containing the ADP-ribosylation site. *Arch. Biochem. Biophys.* **234**, 363–70.

Yamamoto, T. and Yokota, T. (1982). Release of heat-labile entertoxin subunits in *Escherichia coli. J. Bacteriol.* **150**, 1482–4.

Yamamoto, T. and Yokota, T. (1983). Sequence of heat-labile enterotoxin of *Escherichia coli* pathogenic for humans. *J. Bacteriol.* **155**, 728–33.

Yamamoto, T., Tamura, T. and Yokota, T. (1984). Primary structure of heat-labile enterotoxin produced by *Escherichia coli* pathogenic for humans. *J. Biol. Chem.* **259**, 5037–44.

Yamamoto, T., Gojobori, T. and Yokota, T. (1987). Evolutionary origin of pathogenic determinants in enterotoxingenic *Escherichia coli* and *Vibrio cholerae* 01. *J. Bacteriol.* **169**, 1352–7.

Yu, J. and Hirst, T.R. (1995). A pleiotropic secretion mutant of *Aeromonas hydrophila* is unable to secrete heterologously expressed *Escherichia coli* enterotoxin: implication for common mechanisms of protein secretion. *Biochem. Soc. Trans.* **23**, S34.

Yu, J., Webb, H. and Hirst, T.R. (1992). A homologue of the *Escherichia coli* DsbA protein involved in disulfide bond formation is required for enterotoxin biogenesis in *Vibrio cholerae. Mol. Microbiol.* **6**, 1949–58.

Zhang, R.G., Scott, D.L., Westbrook, M.L., Nance, S., Spangler, B.D., Shipley, G.G. and Westbrook, E.M. (1995a). The three-dimensional crystal structure of cholera toxin. *J. Mol. Biol.* **251**, 563–73.

Zhang, R.G., Westbrook, M.L., Westbrook, E.M., Scott, D.L., Otwinowski, Z., Maulik, P.R., Reed, R.A. and Shipley, G.G. (1995b). The 2.4 Angstrom crystal structure of cholera toxin B subunit pentamer: choleragenoid. *J. Mol. Biol.* **251**, 550–62.

7

Bordetella pertussis protein toxins

Camille Locht and Rudy Antoine

INTRODUCTION

Bordetella virulence factors: toxins and adhesins

The genus *Bordetella* contains several species able to cause respiratory tract infections. *Bordetella pertussis* is the aetiological agent of whooping cough in humans, and *Bordetella parapertussis* causes a milder whooping cough-like disease in humans. *Bordetella bronchiseptica* and *Bordetella avium* are animal pathogens, causing atrophic rhinitis in pigs and kennel cough in dogs, and rhinotracheitis in birds, respectively.

The molecular mechanisms of *B. pertussis* infections are increasingly well understood. Many virulence factors have been identified and characterized at the molecular level. They can be grouped into two major categories: adhesins and toxins. The major adhesins include filamentous haemagglutinin, fimbriae, a 69-kDa protein named pertactin, the tracheal colonization factor and the serum resistance protein Brk. There are probably others that have yet to be identified.

In addition to the adhesins, *B. pertussis* produces a number of toxins, such as pertussis toxin, adenylate cyclase toxin, dermonecrotic toxin (DNT), tracheal cytotoxin (TCT) and endotoxin or lipopolysaccharide (LPS). Some of these toxins are protein toxins, others are non-proteinaceous.

NON-PROTEIN BORDETELLA PERTUSSIS TOXINS

Bordetella pertussis produces at least two non-protein toxins, TCT and LPS. TCT is a low-molecular-mass glycopeptide (Cookson *et al.*, 1989a). Its primary structure has been determined by fast atom bombardment mass spectrometry as a 921-Da *N*-actetylglucosaminyl-1,6-anhydro-*N*-acetylmuramylalanyl-γ-glutamyl-diaminopimelylalanine, identical to the *Neisseria gonorrhoeae* ciliostatic anhydropeptidoglycan and to the slow-wave sleep-promoting factor (Cookson *et al.*, 1989b). It destroys ciliated epithelial cells of the respiratory tract, and causes an increase in the number of cells with sparse ciliation and extrusion of cells from the epithelial surface (Wilson *et al.*, 1991). *In vitro*, TCT inhibits DNA synthesis in tracheal epithelial cells and induces the production of intracellular interleukin-1 and nitric oxide, which are most probably the triggers of the TCT-mediated cytopathy (Heiss *et al.*, 1994). The alanyl-γ-glutamyl-diaminopimelate moiety appears to be the smallest derivative of TCT with biological activity. However, the alanyl group can be replaced by other amino acids or blocking groups. Within this active substructure, main-chain chirality and all functional groups appear to be essential for toxicity (Luker *et al.*, 1995). TCT is also produced by the other *Bordetella* species, suggesting that it may be of general importance in bordetellosis (Gentry-Weeks *et al.*, 1988).

The *B. pertussis* LPS can sometimes act synergistically

with other toxins, such as TCT or pertussis toxin. By itself, it expresses endotoxin activities similar to those of LPS from enteric bacteria. It is lethal in galactosamine-sensitized mice, and pyrogenic and mitogenic in spleen cell cultures. It activates macrophages and induces the production of tumour necrosis factor (Watanabe *et al.*, 1990). Although all *Bordetella* species produce LPS, their structures vary somewhat between the different species, which may reflect the differences in the strength of some biological activities among the *Bordetella* species. These structural differences have been exploited to develop monoclonal anti-LPS antibodies, able to discriminate among the various *Bordetella* species (Gustafsson *et al.*, 1988). In addition, the lipid A moieties may vary from strain to strain within the same *Bordetella* species (Zarrouk *et al.*, 1997). *Bordetella pertussis* contains two different LPS classes, termed LPSI and LPSII (Le Dur *et al.*, 1980), which differ in their polysaccharide moieties. Furthermore, the ratio of LPSI to LPSII may differ between variants of the same strain (Ray *et al.*, 1991).

ADENYLATE CYCLASE TOXIN

Adenylate cyclase (AC) toxin has initially attracted attention because of several unusual properties for a bacterial adenylate cyclase. It has been identified as an extracytoplasmic enzyme, based on the fact that it is found in the *B. pertussis* culture medium and that trypsin treatment of intact bacteria destroys the enzyme (Hewlett *et al.*, 1976). Since ATP, the substrate for adenylate cyclases, is usually found in intracytoplasmic compartments, the extracellular location of AC was surprising, and has therefore been linked directly to a possible contribution to virulence of *B. pertussis*. A second unusual feature is the fact that the enzymic activity of AC is increased up to 1000-fold by the eukaryotic intracellular protein calmodulin (Wolff *et al.*, 1980). This suggested that AC penetrates into eukaryotic cells, which was substantiated by the direct demonstration that AC can be internalized by cells and catalyses the conversion of intracellular ATP to cAMP (Confer and Eaton, 1982). The elevated levels of cAMP in these cells disrupt normal cellular functions, thereby qualifying AC as a bacterial toxin.

The enzyme mechanism of adenylate cyclase

Molecular cloning of the AC structural gene and the determination of its sequence unequivocally established that the *B. pertussis* AC is a bifunctional protein of 1706 amino acids (Glaser *et al.*, 1988a, b). The protein carries both adenylate cyclase and haemolytic activities. The cyclase domain comprises the 450 *N*-terminal residues, whereas the haemolytic domain is located in the *C*-terminal region composed of approximately 1200 residues that share sequence similarities with *Escherichia coli* α-haemolysin and *Pasteurella haemolytica* leucotoxin.

Moderate proteolytic cleavage of the full-length AC yields an approximately 45–50-kDa *N*-terminal polypeptide with full catalytic activity. This polypeptide contains both the calmodulin-binding and the ATP-binding sites (Ladant, 1988). Limited tryptic digestion of this catalytic domain results in two fragments of 25 and 18 kDa, the *N*-terminal and *C*-terminal fragments, respectively, of this domain. The 25-kDa peptide catalyses the formation of cAMP at a level of approximately 0.1% of the wild-type level in the presence of calmodulin. The addition of calmodulin to this fragment does not increase the activity. However, when it is combined with the 18-kDa fragment in the presence of calmodulin, enzymic activity is regained.

The catalytic domain of AC is homologous to the central part of the *Bacillus anthracis* calmodulin-sensitive adenylate cyclase (Escuyer *et al.*, 1988). Among the most conserved regions is a 24-residue stretch located in the putative ATP-binding site. Lys58 and Lys65 are critical residues for the expression of enzymic activity (Glaser *et al.*, 1989). Both are conserved in the *B. anthracis* cyclase. Replacement by either residue with glutamine results in drastically reduced enzyme activity, whereas calmodulin-binding is not severely affected. Other critical residues include Asp188, Asp190, His298 and Glu301 (Glaser *et al.*, 1991). When these residues are altered, the ability of AC to bind analogues of ATP is substantially diminished. These observations, together with molecular modelling based on known crystal structures of ATP-binding proteins, led to a model in which Asp188 and Asp190 interact with Mg^{2+}–ATP by accepting a hydrogen bond from a water ligand of Mg^{2+} and/or by directly co-ordinating the Mg^{2+} through the carboxylate group (Figure 7.1). In this model, Lys58 or Lys65 may then interact with the α-phosphate group of Mg^{2+}–ATP.

Although not directly located in the calmodulin-binding site, His298 and Glu301 may, nevertheless, participate in the activation by calmodulin.

His63 may be involved in the reaction mechanism as a general acid/base catalyst in a charge-relay system (Munier *et al.*, 1992), because substitutions of His63 decrease catalytic activity by two or more orders of magnitude and alter the kinetic properties of the enzyme. These effects vary with pH and the direction of the reaction. A substitution of His63 by arginine best catalyses the formation of cAMP from ATP, with a pH

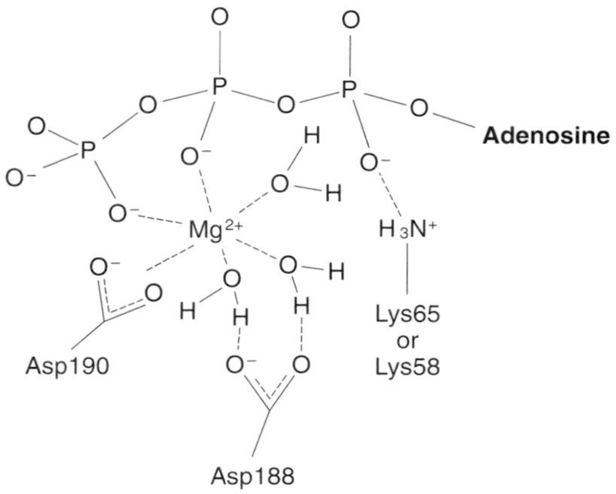

FIGURE 7.1 ATP-binding site of AC. Mg^{2+}-ATP is positioned in the active site by interaction which Asp188 and Asp190, either through direct co-ordination of Mg^{2+} by the carboxylate group of one the aspartate side-chains, or indirectly via a water ligand of Mg^{2+} hydrogen-bonded to one of the carboxylate groups. Lys58 or Lys65 interacts directly via its side-chain amino group with the α-phosphate of Mg^{2+}–ATP. (Adapted from Glaser *et al.*, 1991.)

optimum shifted towards the alkaline side. In contrast, replacement of His63 by glutamate best catalyses the reverse reaction, i.e. the formation of ATP from cAMP, with a pH optimum shifted towards the acidic side.

Activation by calmodulin

The enzymic activity of AC is increased by several orders of magnitude in the presence of calmodulin. This protein binds at a 1:1 stoichiometry to AC (Ladant, 1988) and stimulates its enzymic activity in both the presence and the absence of calcium (Greenlee *et al.*, 1982). However, at low concentrations of calmodulin, activation is calcium dependent, whereas at high concentrations it is not (Kilhoffer *et al.*, 1983). Both types of activation occur at the level of catalysis rather than at the level of ATP binding.

Calmodulin binding involves Trp242, located in the *C*-terminal portion of the catalytic domain. A substitution of this residue by aspartate or glycine strongly decreases the affinity of the enzyme for calmodulin. A substitution by valine has much less effect (Glaser *et al.*, 1989). Therefore, it has been proposed that the replacement of Trp242 by aspartate or glycine may introduce helix breaking residues in a predicted α-helix, and that the secondary structure rather than the primary structure is important for calmodulin binding. Nuclear magnetic resonance (NMR) and circular dichroic studies have indicated that a synthetic peptide of 20 amino acids surrounding Trp242 has a tendency to form a basic amphiphilic helix, although there was no evidence of a highly populated regular

conformation (Precheur *et al.*, 1991). An α-helix around Trp242 could correspond to a dipolar moment to orientate calmodulin correctly. Binding of calmodulin to AC changes the intrinsic fluorescence properties of Trp242 (Gilles *et al.*, 1990). Subsequently, it was confirmed that a 72-residue peptide containing Trp242 accounts for 90% of the binding energy of the AC–calmodulin complex (Bouhss *et al.*, 1993), and that it is essentially the hydrophobic side in the predicted helix that plays the major role in this interaction. Binding of calmodulin to this side may induce conformational changes that may affect catalysis.

Activation by calcium

Calcium influences several properties of AC, probably by direct interaction with the toxin. It changes the chromatographic behaviour of AC, its Stokes radius, as well as its electrophoretic mobility, and stabilizes it against heat inactivation (Masure *et al.*, 1988). It also influences the enzymic activity. AC activity is enhanced at lower concentrations, but inhibited at higher concentrations of calcium. At higher concentrations, calcium probably competes with magnesium for ATP binding, and Ca^{2+}–ATP is not a substrate for AC.

However, calcium has a much more pronounced effect on toxicity than on enzymic activity. Toxic activity, as well as haemolysis are substantially reduced in the absence of extracellular calcium (Hanski and Farfel, 1985; Masure *et al.*, 1988; Ehrmann *et al.*, 1991). The calcium dependency of toxicity can be related to a calcium-induced structural change in AC, which is necessary for binding and/or insertion of the toxin into the target cells. A calcium-induced conformational change of the toxin has been shown by a shift in intrinsic tryptophan fluorescence, an alteration in anti-AC monoclonal antibody binding and protection from mild tryptic digestion (Hewlett *et al.*, 1991).

Ultrastructural electron microscopy revealed that the structure of AC changes from a closed globular form at low calcium concentrations to an open semicircular conformation at high calcium concentrations. The calcium probably binds to a glycine-rich repeat region that contains 38 copies of the nonapeptide (L/I/F)-X-G-G-X-G-(N/D)-D-X located between positions 913 and 1612 (Glaser *et al.*, 1988a). AC can bind to erythrocytes in the absence of calcium, but the generation of cAMP requires a subsequent incubation step in the presence of calcium (Rogel *et al.*, 1991), suggesting that calcium plays a role in cell entry. Although the haemolytic activity of AC is also calcium dependent, this activity is substantially inhibited at the concentrations of maximal toxic activity (Ehrmann *et al.*, 1991), suggesting that the toxic effect can be mechanistically dissociated from haemolysis. Kinetic differences between these two activities have also been observed. Whereas the AC-mediated increase in cAMP concentrations in toxin-treated sheep erythrocytes is immediate, the appearance of haemolysis requires approximately 1 h (Rogel et al., 1991). Addition of exogenous calmodulin potentiates AC-mediated haemolysis, but blocks cAMP formation.

The haemolytic domain of adenylate cyclase

The purified catalytic domain is not, by itself, able to penetrate into eukaryotic cells. Penetration requires an additional polypeptide (Donovan et al., 1989), which corresponds to the C-terminal 1300 residues of the full-length protein. In-frame deletions in the 3' region of the AC gene abolish toxic activity, yet leave enzymic activity intact (Bellalou et al., 1990). The same mutations also severely reduce haemolytic activities, indicating that the haemolytic domain plays a critical role in the cell binding and/or entry of AC. Furthermore, recombinant AC molecules that lack the catalytic domain exhibit full haemolytic activity, establishing that the haemolytic domain, and hence the cell binding/entry domain, is located in a region independent of the catalytic domain (Sakamoto et al., 1992). This domain can associate with cells, insert into membranes and form pores. Although AC is a member of the RTX (repeats in toxins) toxin family, its specific haemolytic activity is rather low compared with that of the other members of the family (Bellalou et al., 1990), suggesting that the main role of the haemolytic domain is not to lyse red blood cells, but to deliver the catalytic domain into the target cell.

The complete haemolytic domain is required for the tight association of AC with membranes, as well as for its haemolytic and toxic activities, since a deletion of as few as 75 residues from the C-terminal end abolished all those activities (Iwaki et al., 1995). However, pairwise associations of different inactivated mutant AC with non-overlapping deletions are, in some instances, able to restore toxin activity, suggesting that the toxin may form dimers or higher oligomers. Since purified AC toxin is a monomer, as evidenced by gel permeation chromatography, multimerization probably occurs upon interaction with the target cell surface. Some pairs express both cytotoxic and haemolytic activities, others only restore cytotoxic activity and no haemolytic activity. In vitro complementation experiments also indicated that the last 217 residues form a domain required for toxicity, and that the region between residues 624 and 780 is required for the delivery of the catalytic domain into the target cells. This portion contains part of the most hydrophobic region of the molecule. However, these experiments indicate, overall, that the haemolytic domain of AC toxin has no dedicated translocation and target-cell binding regions.

Adenylate cyclase entry into target cells

Since AC has no apparent translocation and target-cell binding regions, it probably does not recognize specific receptors on the surface of the target cells. Accordingly, AC binds to a large variety of target cells, and the intracellular concentration of cAMP increases instantaneously upon incubation with AC (Gentile et al., 1988), suggesting that intoxication does not require an endocytic vesicle trafficking step. Furthermore, endocytosis inhibitors do not block the accumulation of intracellular cAMP upon AC treatment, and pretreatment of target cells with proteases or cycloheximide does not affect AC toxicity (Gordon et al., 1989), consistent with the notion that AC toxicity does not involve a protein receptor.

Instead, AC may interact directly with lipids. AC can bind and elicit marker release from liposomes. Polylysines and other polycations are able to inhibit AC penetration (Raptis et al., 1989), which suggests that charge–charge interactions are important for the binding and/or entry of AC. Binding of calmodulin to AC is not required for entry. In fact, in certain instances, calmodulin may inhibit AC entry (Shattuck and Storm, 1985). However, this inhibitory effect may depend on the target cells (Gentile et al., 1990). Toxin analogues with very low affinity for calmodulin are able to deliver their catalytic domain efficiently into the cytoplasm of target cells, demonstrating that high-affinity calmodulin binding is not crucial for entry (Heveker and Ladant, 1997). It is, however, important for efficient production of intracellular cAMP, which may explain why alterations of Trp242, within the calmodulin-binding site, result in a decrease in intracellular cAMP accumulation (Oldenburg et al., 1992).

At least three consecutive steps can be distinguished in the entry process of AC: membrane insertion, translocation and intracellular cleavage (Rogel and Hanski, 1992). Membrane insertion occurs even at low temperatures, whereas translocation only occurs at temperatures above 20°C. In addition, translocation requires higher calcium concentrations than membrane insertion. It is likely that the conformational change induced by high concentrations of calcium triggers the formation of a transmembrane channel, presumably composed of hydrophobic helices present in the central region of the molecule. Once internalized, the catalytic domain is cleaved, induces the formation of cAMP, and can then be rapidly inactivated or degraded, or remains preserved, depending on the cell type.

Palmitoylation of adenylate cyclase

Toxic activities and haemolysis, but not cell-association of AC require a post-translational modification of the toxin, which is catalysed by the bacterial enzyme CyaC (Figure 7.2). Recombinant AC produced in *E. coli* in the absence of CyaC has no toxic activity (Rogel *et al.*, 1989). The addition of CyaC is sufficient to confer full invasive and haemolytic activities to recombinant AC (Sebo *et al.*, 1991). CyaC is homologous to the *E. coli* HlyC, required for the activation of *E. coli* α-haemolysin (Barry *et al.*, 1991). Mutations in the *cyaC* gene result in the production and secretion of biologically inactive AC, although its enzymic activity, electrophoretic mobility, and ability to bind calcium and to undergo calcium-dependent conformational change remain intact (Hewlett *et al.*, 1993). In addition, AC from the mutant strain is able to bind to the surface of eukaryotic cells, where, in contrast to the AC from wild-type strains, it remains sensitive to trypsin. Injection of AC isolated from the mutant strain into cells results in toxic activities, which suggests that the lack of the CyaC-catalysed modification results in a defect in insertion and membrane translocation of the catalytic domain.

The use of mass spectrometry has identified the modification of AC by CyaC as an amide-linked palmitoylation on the ε-amino group of Lys983 (Hackett *et al.*, 1994). Expression of AC in *E. coli* in the presence of CyaC also results in acylation of Lys983 (Hackett *et al.*, 1995). However, in contrast to *B. pertussis*, where the Lys983 residue is exclusively palmitoylated, the Lys983 of the recombinant form is about 87% palmitoylated and 13% myristoylated. Furthermore, recombinant AC contains an additional palmitoylation on approximately 65% of the molecules on Lys860. Compared with natural AC, the recombinant protein has reduced haemolytic activity, but normal toxic activity, indicating that the nature of the fatty-acylation also influences the toxin activities.

Secretion of adenylate cyclase

Although palmitoylation occurs in the cytosol of the bacterial cell, it is not required for AC secretion from *B. pertussis*. Similarly to the other RTX toxins, secretion occurs via a specific pathway, which makes use of three distinct accessory proteins (Glaser *et al.*, 1988b). These three proteins, named CyaB, CyaD and CyaE, are encoded by genes located directly downstream of the AC structural gene, *cyaA* (Figure 7.2). CyaB and CyaD are homologous to the *E. coli* HlyB and HlyD involved in the transport of the *E. coli* α-haemolysin. CyaE is homologous to TolC, which is also required for haemolysin secretion in *E. coli*. As for other RTX toxins, AC secretion does not involve an *N*-terminal signal peptide. Instead, the secretion determinant is located in its *C*-terminal domain. AC can be secreted by the *E. coli* HlyBD/TolC translocator through both membranes of *E. coli* without a periplasmic intermediate (Sebo and Ladant, 1993). Deletion of the last 74 residues of AC reduces translocator-dependent secretion, without abolishing it, indicating the presence of additional secretion signals.

Biological and immunological activities of adenylate cyclase

The involvement of AC in virulence was first demonstrated by the use of transposon mutagenesis. AC-deficient Tn5 insertion mutants are drastically reduced in virulence. In the lethal challenge infant mouse model, the LD_{50} of these mutants is substantially higher than that of the wild-type parent strain upon intranasal infection (Weiss *et al.*, 1984). When sublethal doses of *B. pertussis* are used, wild-type strains rapidly proliferate in the lungs of infant mice and are cleared approximately 40 days after challenge. In contrast, AC mutants do not proliferate and are cleared after about 10 days, suggesting that AC plays a role in the initial phases of colonization (Goodwin and Weiss, 1990). Using more defined mutants, both the catalytic and the haemolytic domains were found to be required for virulence (Khelef *et al.*, 1992). A single amino acid substitution in the catalytic site results in a greater than 1000-fold reduction in the pathogenic effect to new-born mice (Gross *et al.*, 1992). In addition,

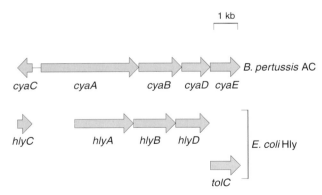

FIGURE 7.2 Molecular structure of AC gene locus. AC is encoded by the *cyaA* gene. The *cyaBDE* genes are involved in AC secretion through the bacterial envelope, whereas *cyaC* encodes the AC-modifying enzyme. The arrows represent the length of the different genes and their direction of transcription. The *cyaABCD* gene products are homologous to the *E. coli* *hlyABCD* gene products, and the *cyaE* gene product is homologous to TolC. In *E. coli*, the *hlyABCD* genes are grouped within a polycistronic operon, whereas *tolC* is located elsewhere in the chromosome.

calmodulin-binding plays a role in virulence, since a *B. pertussis* strain carrying an alteration of Trp242 in the calmodulin-binding region shows decreased virulence (Oldenburg *et al.*, 1993).

Bordetella bronchiseptica also produces AC toxin. The sequence of this toxin is very similar to that of the *B. pertussis* AC (Betsou *et al.*, 1995). Human isolates of *B. bronchiseptica* that do not produce AC are unable to induce lethality upon intranasal infection of mice (Gueirard *et al.*, 1995), suggesting that AC also plays a significant role in the pathogenesis of *B. bronchiseptica* infections.

At the cellular level, AC inhibits chemiluminescence, chemotaxis and superoxide production of human polymorphonuclear leucocytes without affecting phagocytic activities (Friedman *et al.*, 1987). In macrophages, it induces apoptosis (Khelef and Guiso, 1995). Both the enzymic and haemolytic activities are required for this effect. Such perturbances of the cellular immune functions may be important during the initial steps of infection and in bacterial survival within the host.

AC may also serve as a protective antigen. Whooping cough patients and children immunized with whole-cell pertussis vaccines produce antibodies to AC (Farfel *et al.*, 1990). These antibodies persist into adulthood, cross the placenta and disappear a few months after birth. Active immunization with AC protects mice against a *B. pertussis* respiratory challenge (Guiso *et al.*, 1991). Surprisingly, the *B. pertussis* AC does not protect against *B. parapertussis* infections, and the *B. parapertussis* AC does not protect against *B. pertussis* infection, although both toxins are very similar and protected against homologous challenge (Khelef *et al.*, 1993). Interestingly, recombinant AC purified from *E. coli* in the absence of CyaC is not protective. It acquires its protective potential upon activation by CyaC (Betsou *et al.*, 1993). However, even after activation by CyaC, the recombinant AC is still less protective than the authentic *B. pertussis* AC, indicating that the nature of the acyl group is also important for protection.

Applications of adenylate cyclase

Besides being a potent vaccinating antigen, other useful applications of AC have been proposed. The capacity of membrane translocation of the catalytic domain has been explored to introduce foreign peptides fused to AC into the cytoplasm of target cells. Insertion of a CD8[+] T-cell epitope from the nucleoprotein of lymphocytic choriomeningitis virus into the AC catalytic domain results in its presentation in a major histocompatibility complex (MHC) class-I restricted manner and elicits specific cytotoxic T-lymphocyte responses in mice

immunized with the recombinant protein. In addition, the immunized mice are protected against a lethal intracerebral challenge with virulent virus in a CD8[+] T-cell-dependent fashion (Saron *et al.*, 1997). Inactivation of the enzymic domain does not alter the CTL response. However, the intact haemolytic domain is required, indicating that genetically detoxified yet invasive recombinant AC, carrying protective CTL epitopes, can be used to induce protective immunity.

Another original application makes use of the ability of two non-overlapping AC fragments to reassociate in the presence of calmodulin and yield a fully active complex. In the absence of calmodulin, the two fragments do not associate. However, when fused to interacting proteins or protein domains, the two fragments can reassociate, even in the absence of calmodulin, and reconstitute enzymic activity (Karimova *et al.*, 1998). Reassociated AC activity can then easily be monitored in an AC-deficient *E. coli* strain, using indicator plates, since cAMP is essential for the activation of the genes involved in lactose or maltose catabolism in *E. coli*. This constitutes a powerful bacterial two-hybrid system for the selection or screening of functional interactions between proteins.

DERMONECROTIC TOXIN

In contrast to most *B. pertussis* toxins that are exotoxins, the localization of DNT is intracellular (Cowell *et al.*, 1979). Subjecting whole cells to osmotic shock, which releases periplasmic proteins, does not release DNT. Separation of soluble proteins from membrane-bound proteins revealed DNT in the soluble fraction, indicating its cytoplasmic localization. Consistent with this finding, trypsin treatment of whole cells does not affect DNT activity, whereas trypsin treatment of a cell lysate diminishes its activity by over 90%. Interestingly, DNT was one of the first *B. pertussis* virulence factors described and originally named endotoxin (Bordet and Gengou, 1909), already reflecting the intracellular nature of this toxin. It has subsequently been termed heat-labile toxin, because it is completely inactivated by heating at 56°C for 60 min (Livey and Wardlaw, 1984). The term DNT is due to the characteristic skin lesion that the toxin produces when injected into rabbits, mice or guinea pigs (Bordet and Gengou, 1909). Intravenous inoculation of even low doses of DNT is lethal for mice (Kume *et al.*, 1986).

Role of dermonecrotic toxin in pathogenesis

The role, if any, of DNT in whooping cough is not clear. The production of DNT is not required for lethal infection in the intranasal *B. pertussis* challenge model of infant mice (Weiss and Goodwin, 1989). In addition, different *Bordetella* strains isolated from humans may vary in their ability to produce DNT, and there is no correlation between the level of DNT production and infectivity in humans. In fact, most isolates did not produce DNT, suggesting that this toxin does not play an important role in virulence for humans (Gueirard and Guiso, 1993).

Although a role of DNT in human infections and in murine models has not yet been established, this toxin may be important in the pathogenesis of *B. bronchiseptica* infections in pigs. In an attempt to correlate certain phenotypic characteristics with the ability of *B. bronchiseptica* to cause respiratory pathology in piglets, several strains differing in the production of defined virulence factors have been compared in a neonatal swine infectious model (Roop *et al.*, 1987). In this model, AC appears to be important for colonization, but only strains that synthesize high levels of DNT are able to produce nasal and lung lesions in those animals. These lesions vary from moderate to severe, establishing a direct correlation between the ability of *B. bronchiseptica* to induce nasal and lung lesions in neonatal piglets and their level of DNT production.

Bordetella bronchiseptica strains lacking DNT have been derived from virulent strains by serial passages. Intranasal or intramuscular inoculation of these strains into guinea pigs results in no clinical manifestations but, instead, results in protection against subsequent challenge with the wild-type strain. Although the precise mutations have not yet been defined, the phenotype of the mutants appears to be stable after 50 subcultures *in vitro* or 20 passages in mice, and has been proposed for use as a live attenuated vaccine against atrophic rhinitis in pigs (Nagano *et al.*, 1988). It should be pointed out, however, that in addition to the lack of DNT, the mutants were deficient in haemolytic activity.

Molecular and cellular actions of dermonecrotic toxin

The pathology induced by DNT on swine respiratory tissues is characteristic for the lesions observed in atrophic rhinitis induced by *B. bronchiseptica* infections of piglets. The most evident histological manifestations are ultrastructural changes in the nasal turbinate bones of the pigs (Silvera *et al.*, 1982). The degenerative changes observed by electron microscopy are most severe in osteoblasts, suggesting that DNT may impair osteogenesis. The effect of DNT on osteoblastic cell lines, such as the MC3T3-E1 cells, has therefore been extensively studied. The toxin induces morphological

changes in osteoblasts, i.e. a transition from well-stretched, spindle-shaped cells to spherical or block forms (Horiguchi *et al.*, 1991). Upon DNT treatment, the cells lose their extensions and form small blebs on their surface. However, they are not lysed, remain viable and continue to proliferate in the presence of DNT. Alkaline phosphatase activity is reduced in treated cells compared with the control, as is the accumulation of type-I collagen. Both of these markers are closely linked to differentiation, which suggests that DNT impairs the ability of these cells to differentiate, thereby providing evidence that the toxin acts on osteogenesis.

In addition to inducing these morphological changes and inhibiting differentiation, DNT stimulates DNA and protein syntheses in these cells (Horiguchi *et al.*, 1993, 1994). Unlike that of control cells, confluent growth of DNT-treated MC3T3-E1 cells results in the accumulation of irregularly shaped polynucleated cells. Measurements of [methyl-^{3}H]thymidine incorporation indicated that this may correspond to DNT-stimulated DNA synthesis, since it can be inhibited by specific DNA replication inhibitors. However, the stimulatory effect of DNT is only seen in confluent cultures and not in actively proliferating cells. DNT does not induce cell proliferation, suggesting that the cells are blocked after mitosis at a subsequent cyto-kinesis step. In contrast to the stimulation of DNA synthesis, the induction of protein synthesis by DNT occurs even in actively proliferating cells (Horiguchi *et al.*, 1994) and has a shorter lag period.

Although DNA and protein syntheses are affected by treatment with DNT, the most dramatic effect observed is that of the morphological changes, which are accompanied by the assembly of actin stress fibres and focal adhesions (Horiguchi *et al.*, 1995). These ultrastructural rearrangements are regulated by small GTP-binding proteins of the Rho family, and treatment of MC3T3-E1 cells with DNT results in an electrophoretic mobility shift of cellular Rho proteins (Horiguchi *et al.* 1995). This mobility shift can be reproduced *in vitro* using purified recombinant RhoA, indicating that DNT modifies Rho proteins.

The analysis of the amino acid sequence of DNT-modified RhoA indicated that its Gln63 residue is deaminated to glutamate (Horiguchi *et al.*, 1997). Substitution of Gln63 by Glu through site-directed mutagenesis of the RhoA cDNA results in the same electrophoretic mobility as that observed for DNT-treated wild-type RhoA. Mutant RhoA or toxin-treated wild-type RhoA has the same affinity for GTP, but approximately 10-fold reduced GTP hydrolase activity compared with untreated wild-type RhoA. When the cDNA encoding mutant RhoA was transfected into C3H10T1/2 cells, the cells produced extensive actin stress fibres similar to those observed in cells treated with DNT. Therefore, it is likely that the DNT-induced stress fibre formation and morphological changes are due to the constitutive activation of RhoA by the inhibition of its GTPase activity via deamination of Gln63. DNT-catalysed deamination of Rho also appears to induce the proliferation of cytoplasmic membrane organelles, such as the Golgi apparatus, the endoplasmic reticulum (ER) and the mitochondria, as well as the formation of plasmalemmal calveaolae (Senda *et al.*, 1997). The focal adhesion phenotype may be related to the stimulation of tyrosine phosphory-lation of focal adhesion kinase and paxillin upon deamination of Rho (Lacerda *et al.*, 1997).

Molecular structure of dermonecrotic toxin

Molecular cloning of the DNT genes from *B. pertussis* and *B. bronchiseptica* showed approximately 99% sequence identity between the two homologues (Pullinger *et al.*, 1996). Southern blot analyses indicated that the *dnt* gene is also present in *B. parapertussis* (Walker and Weiss, 1994). Although *B. avium* produces a DNT (Gentry-Weeks *et al.*, 1988), the *B. pertussis dnt* DNA does not hybridize to *B. avium* DNA. Amino acid sequence comparisons revealed that the *C*-terminal portions of DNT and CNF share significant similarities (Walker and Weiss, 1994). The *N*-terminal domains are not homologous. Instead, the *N*-terminal domain of CNF shows sequence similarities to the *N*-terminal domain of the *Pasteurella multocida* dermonecrotic toxin, an organism also implicated in atrophic rhinitis. Functional analyses of CNF have indicated that the *N*-terminal region of this toxin contains the cell-bind-ing domain, whereas its *C*-terminal domain is respon-sible for the catalytic activity (Lemichez *et al.*, 1997). These observations strongly suggest that the DNT-mediated deamination of Rho is also catalysed by its *C*-terminal domain, and that the *N*-terminal portion of DNT contains the receptor-binding domain. In addition, these observations suggest that CNF and the *Pasteurella* toxin recognize analogous receptors on their target cells, which may be different from those that are recognized by DNT. Owing to the lack of significant sequence similarities in their *C*-terminal domains, it is also likely that the *Pasteurella* toxin expresses an enzymic activity that is different from that of DNT and CNF.

The *B. pertussis* and the *B. bronchiseptica* DNT contain 1451 amino acids and have a predicted molecular mass of approximately 159 kDa (Pullinger *et al.*, 1996; Walker and Weiss, 1994). Their calculated isoelectric points of 6.63 are in good agreement with the experi-mentally determined values of 6.3–6.7. Interestingly,

DNT contains a putative purine nucleotide-binding site, and a substitution of the highly conserved lysine by site-directed mutagenesis completely destroys toxic activity of recombinant DNT (Pullinger *et al.*, 1996). Although this site is located in the region homologous to CNF, the conserved lysine is not present in CNF, suggesting that if this is a nucleotide-binding site in DNT, possibly to provide energy at some stage, it may not be so in CNF.

PERTUSSIS TOXIN

Among the *B. pertussis* toxins, pertussis toxin (PTX) has the most complex structure, being composed of five dissimilar subunits, named S1–S5, according to their decreasing molecular masses. The M_r of these subunits range from 26 000 for the largest to 11 000 for the smallest subunit. PTX is a member of the A–B toxin family, in which the A moiety, composed of the S1-sub-unit, expresses enzymic activity, and the B moiety, composed of subunits S2–S5, is responsible for the binding of the toxin to the target-cell receptors. The B oligomer of PTX can be subdivided into two dimers, named D1 and D2, and composed of subunits S2–S4 and subunits S3–S4, respectively (Tamura *et al.*, 1982). The crystal structure of the toxin at 2.9 Å resolution (Stein *et al.*, 1994a) shows that PTX has the shape of a pyramid with a triangular base (Figure 7.3). The B oligomer constitutes the base of the pyramid in the order S5-S2-S4-S3-S4. Although S2 and S3 share about 70% amino acid identity, the wild-type holotoxin always incorporates the S2- and S3-subunits at their correct position. The S1-subunit is located on the tip of the pyramid. The structural arrangement of the B oligomer resembles that of symmetrical B pentamers of other toxins, such as cholera toxin and Shiga toxin.

The molecular action of PTX involves three distinct steps: binding of the toxin to its receptors via the B oligomer, membrane translocation of the S1-subunit and intracellular expression of the ADP-ribosyl-transferase activity catalysed by the S1-subunit.

All three steps are required for the full expression of most – albeit not all – biological activities of PTX. Depending on the target cell, the physiological effects of PTX may vary considerably. These effects are responsible for most of the systemic features of whooping cough (Pittman, 1984). The existence of PTX was therefore first predicted by the expression of its numerous biological activities detected after infection with *B. pertussis* or administration of whole-cell pertussis vaccines. These activities include histamine sensitization, islet activation, induction of leucocytosis, immunopotentiation and many others (reviewed by

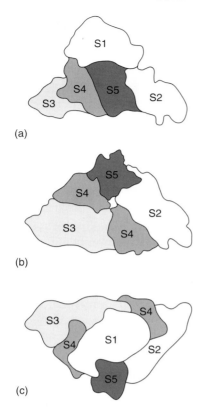

(a)

(b)

(c)

FIGURE 7.3 Schematic representation of the PTX structure. The lateral view (a) shows the side of the pyramidal structure of PTX with the S2–S5-subunits at the base and S1 on the top. The bottom view (b) shows the triangular base of the pyramidal structure with the B oligomer ring in the S5-S2-S4-S3-S4 ordering. The top view (c) shows the location of S1 on the top of the pyramid. This representation is drawn as an outline of PTX based on the space-filling model of the PTX crystal structure.

Munoz, 1985). Some of the pharmacological effects of whole-cell pertussis vaccines were recognized as early as the 1940s (Parfentjev and Goodline, 1948). However, considering the wide diversity of biological activities of pertussis vaccines, they were initially believed to be caused by distinct *B. pertussis* products.

Biogenesis of pertussis toxin

The PTX-subunits are individually produced as pre-proteins containing typical signal peptides at their N-terminal extremities (Locht and Keith, 1986; Nicosia *et al.*, 1986). Considering the similarities of the PTX signal peptides with standard signal peptides, it is likely that the individual subunits are translocated through the inner membrane by a Sec-dependent mechanism. DNA fragments encoding elements of the *B. pertussis* Sec apparatus have recently been identified in the course of systematic *B. pertussis* DNA sequenc-

ing. At the periplasmic side of the inner membrane the signal peptides are then removed, and the mature subunits can assemble into the holotoxin molecule.

In the absence of S1, this assembly produces a fully functional B oligomer. The association of the S1-subunit with the B oligomer requires the intramolecular disulfide bond of S1, as well as its C-terminal domain (Antoine and Locht, 1990). Secretion of the toxin through the outer membrane does not require the presence of S1, and the assembled B oligomer can be efficiently secreted even in the absence of S1. The S1-subunit alone cannot be secreted by *B. pertussis*, indicating that the secretion determinants are located in the B oligomer. However, even in the unsecreted form, the assembled toxin in the periplasm is fully active, indicating that the PTX secretion does not involve an activation step. Nevertheless, the expression of maximal virulence of *B. pertussis* in animal models requires not only the production of PTX, but also its secretion (Weiss and Goodwin, 1989).

The secretion of PTX through the outer membrane depends on the expression of accessory genes located directly downstream of the five structural genes (Weiss *et al.*, 1993). The products of these genes, named *ptl* genes (for pertussis toxin liberation), share high sequence similarities with the gene products of the *Agrobacterium tumefaciens virB* operon and the *E. coli* proteins involved in DNA transfer of broad-host range conjugative plasmids (Figure 7.4). Some of these proteins are also homologous to *Helicobacter pylori* virulence-associated gene products. The VirB proteins of *A. tumefaciens* are responsible for the transport of its T-DNA across the bacterial cell wall directly into the plant host cell (Ward *et al.*, 1991).

The expression of the *ptl* genes depends on the *ptx* promoter. A strain lacking the *ptx* promoter, but containing all other portions of the *ptx/ptl* locus, does not produce Ptl proteins (Kotob *et al.*, 1995). Polar disruption of the intergenic region between the *ptx* and the *ptl* genes in *Bordetella pertussis* abolishes PTX secretion (Antoine *et al.*, 1996). *B. bronchiseptica*, which generally does not produce PTX, but contains the *ptx* genes, also apparently contains intact *ptl* genes. Insertion of a functional promoter 5′ to the *ptx* genes results in the production of Ptl proteins and in the secretion of PTX. Together, these observations indicate that the *ptl* genes probably constitute with the *ptx* genes a single polycistronic operon, thus composed of the five *ptx* genes followed by nine *ptl* genes (Kotob *et al.*, 1995; Baker *et al.*, 1995; Antoine *et al.*, 1996; Farizo *et al.*, 1996).

Some of the Ptl proteins have been detected by immunoblot analysis, and their subcellular location has been determined (Johnson and Burns, 1994). PtlE, PtlF

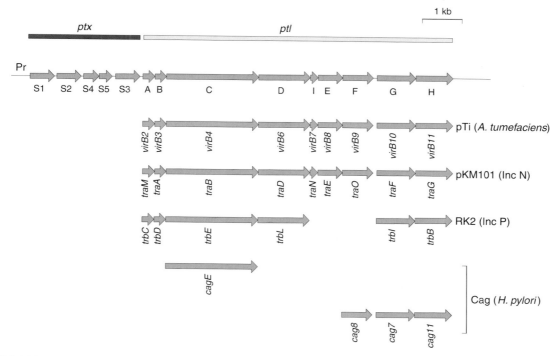

FIGURE 7.4 Molecular structure of the *ptx/ptl* locus. The genes encoding the PTX-subunits S1–S5, as well as the 9 *ptl* genes (*ptlABCDEFGHI*), are represented by the arrows that show the length and transcriptional direction of the genes (top lines). Pr designates the promoter region. The *ptl* gene products are homologous to the *virB* and *tra* gene products of *A. tumefaciens* and pKM101, respectively. Some of these gene products are also homologous to *trp* and *cag* gene products of RK2 and *H. pylori*, respectively.

and PtlG are associated with the membranes of *B. pertussis*, suggesting that they may be part of a gate or channel important for PTX transport. Similarly to its homologue in *A. tumefaciens*, PtlE is most probably an inner-membrane protein that may be associated with outer-membrane components. PtlF may be an outer-membrane protein or, like its VirB9 homologue, associated with both the outer and inner membranes. PtlF forms a complex with PtlI. In a non-polar *ptlI* mutant, PtlF is not detected. PtlI and PtlF coprecipitate with anti-PtlF antibodies, and under non-reducing conditions, PtlI and PtlF migrate as a complex in sodium dodecyl sulfate–polyacrylamide gel electrophoresis, indicating that PtlI binds to PtlF by disulfide bond formation (Farizo *et al.*, 1996). PtlG cofractionates with the membranes and is insoluble in Triton X-100, suggesting an inner-membrane location. However, similarly to VirB10, it may also tightly interact with components of the outer membrane. The precise architecture of this structure relevant to its function, however, is not yet known.

Interestingly, two of the Ptl proteins, PtlC and PtlH, contain putative ATP-binding sites in predicted cytoplasmic locations and might therefore provide the energy necessary for membrane translocation. Alterations

of the putative nucleotide-binding site of PtlH have recently been shown to inhibit PTX secretion (Kotob and Burns, 1997). Furthermore, overproduction of a mutant PtlH form in a *ptlH* wild-type *B. pertussis* background also inhibits PTX secretion, suggesting that PtlH either functions as a multimer or interacts with other components of the secretion apparatus.

Receptor binding and translocation

Once PTX reaches the extrabacterial environment, it can interact with specific receptors on host target cells. This binding occurs via the B oligomer. Virtually all cells that have been studied so far contain PTX receptors on their surface. However, a distinct universal receptor has not been identified. Rather, different cell types may express different PTX receptors. Nevertheless, a common feature of PTX receptors is that they are glycoconjugates, generally sialoglycoproteins (Armstrong *et al.*, 1988). These proteins may vary in size from 43 kDa to over 165 kDa (Brennan *et al.*, 1988; Clark and Armstrong, 1990; Rogers *et al.*, 1990). Several glycoproteins, such as fetuin, haptoglobin and transferrin, have been used as model systems to study the PTX–receptor interactions.

Chinese hamster ovary (CHO) cells are used in many laboratories as model cell lines to study the cellular action of PTX. These cells cluster in the presence of the toxin. They contain a 165-kDa glycoprotein at their surface that is recognized by PTX or purified B oligomer (Brennan *et al.*, 1988). The sialic acid moiety on the *N*-linked oligosaccharide of this protein is an important constituent of the receptor, since a CHO-cell line that specifically lacks the terminal oligosaccharide sequence on glycoproteins does not cluster in the presence of PTX, and treatment of CHO cells with sialidase also abolishes PTX-mediated clustering. Perhaps more relevant cellular models for the pathogenesis of whooping cough are T-lymphocytes. Chemical cross-linking and blotting techniques have demonstrated that PTX specifically binds to a 43-kDa membrane protein of human T-lymphocytes (Rogers *et al.*, 1990). In addition, PTX binds to a 70-kDa protein on human lymphocytes (Clark and Armstrong, 1990). This protein may be related to the 73-kDa LPS receptor. However, LPS and PTX bind to different domains of the 73-kDa protein (Lei and Morrison, 1993).

Although PTX interacts mainly with glycoprotein receptors, it can also bind to glycolipids. B oligomer can efficiently bind to lipid vesicles that contain the ganglioside GD1a. The binding is decreased upon treatment of the vesicles with neuraminidase, suggesting that the sialic acid residues on the GD1a are important for PTX binding (Hausman and Burns, 1993). In the absence of gangliosides, PTX can also bind to phospholipid vesicles, but this binding requires the presence of ATP and reducing agents. It occurs via the S1-subunit, and not via the B oligomer. Isolated S1 also binds to phospholipid vesicles, whereas isolated B oligomer does not. Binding of S1 to these membranes requires the *C*-terminal, hydrophobic portion of the subunit (Hausman and Burns, 1992).

The variety of different glycoproteins that function as receptors for PTX may perhaps be related to the complex structure of the B oligomer. The B oligomer contains at least two binding domains with distinct specificities. For example, D1 (S2 and S4) is able to bind to haptoglobin, whereas D2 (S3 and S4) binds to receptors on the surface of CHO cells. The crystal structure of PTX complexed with a soluble oligosaccharide from transferrin showed that the terminal sialic acid–galactose moiety binds to both the S2- and S3-subunits. The carbohydrate-binding domains include residues 101–105 in both subunits (Stein *et al.*, 1994b). Interestingly, deletion of Asn105 of S2 abolishes haptoglobin binding of PTX, and deletion of the analogous Lys105 in S3 abolishes interaction of PTX with CHO cells (Lobet *et al.*, 1993). However, both subunits bind to the sugar moieties in essentially the

same manner, and hydrogen-bond contacts with the sialic acid indicate amino acid residues that are common to S2 and S3 (Stein *et al.*, 1994b). This suggests that the specificities involve determinants that are outside the sialic acid sugar ring-binding site. Indeed, other regions, in particular the *N*-terminal domain, may also be involved in receptor recognition (Saukkonen *et al.*, 1992).

Target-cell entry

Unlike diphtheria toxin and *Pseudomonas aeruginosa* exotoxin A, PTX does not contain a well-identified toxin translocation domain in its B moiety. It is not clear whether target-cell membrane translocation of the S1-subunit requires the help of the B oligomer. In fact, the ability of S1 to bind directly to phospholipid bilayers (Hausman and Burns, 1992) suggests that the role of the B oligomer is simply to deliver the S1-subunit close enough to the cell membrane to allow it to interact directly with the membrane. Since the *C*-terminal, hydrophobic region of S1 is required for this interaction, it is conceivable that this portion of the molecule serves to translocate the subunit through the host cell membrane, even in the absence of the B oligomer. This assumption is supported by the observation that S1-mediated membrane interaction of holotoxin only occurs in the presence of ATP and reducing agents, conditions which dissociate the S1-subunit from the B oligomer.

The hydrophobic domain of S1 that is required for its interaction with phospholipids is buried within the holotoxin structure and can only be exposed upon dissociation of S1 from the B oligomer. This activation step may occur in specialized intracellular compartments. As suggested by observations *in vitro*, ATP may be a key factor for this activation step. If this also occurs *in vivo*, ATP may bind to the toxin in the ER, trigger the dissociation of S1 from the B-subunit and allow S1 to bind to and translocate through the ER membrane (Hazes *et al.*, 1996).

Several experimental data suggest that PTX follows the retrograde transport system. However, there is no clear evidence yet that it follows it all the way to the ER. Subcellular fractionation experiments showed that PTX travels to the Golgi complex upon receptor-mediated endocytosis (El Baya *et al.*, 1997). In addition, brefeldin A-treatment blocks ADP-ribosylation *in vivo* by externally added PTX (Xu and Barbieri, 1995), suggesting the involvement of Golgi-mediated intracellular trafficking beyond the endosome. CHO cells with a temperature-sensitive Golgi apparatus are not ADP-ribosylated *in vivo* at high temperatures. Moreover, NH_4Cl transiently inhibits ADP-ribosylation *in vivo*, suggesting that intoxication of CHO cells by PTX involves acidification of endosomes. Together, these observations are consistent with the model in which the toxin enters the cell by receptor-mediated endocytosis, travels through acidified vesicles and the Golgi apparatus, possibly all the way to the ER, where it may be activated by the dissociation of S1 from the B oligomer, and translocates the enzymically active subunit into the cytosol.

ADP-ribosyltransferase activity and enzyme mechanism

Upon arrival at the cytoplasmic side of the membrane, S1 expresses an ADP-ribosyltransferase activity using NAD^+ as the ADP-ribosyl donor and the α-subunits of G-proteins as acceptor substrate (Katada and Ui, 1982). The class of G-proteins that serve as substrates for PTX are the Gi/o proteins, which are involved in eukaryotic signal transduction. They are heterotrimeric and composed of α, β and γ-subunits. The precise ADP-ribosyl acceptor site is Cys-351, located near the C-terminus of the Gi/o α-subunits.

Only in their trimeric forms are the G-proteins efficient PTX substrates, whereas isolated $G\alpha$ are much less well ADP-ribosylated (Katada *et al.*, 1986). However, peptides composed of only the C-terminal residues of $G\alpha$ may also be efficiently ADP-ribosylated by PTX (Graf *et al.*, 1992), suggesting that the C-terminal region of the full-length $G\alpha$ may be relatively unavailable in the monomeric form. The C-proximal portion of S1 is involved in the interaction of the enzyme with the $G\alpha$ proteins. Deletion of this region results in a drastic decrease in ADP-ribosylation efficiency, whereas binding to the donor substrate NAD^+ and the catalytic rate are not affected (Locht *et al.*, 1990; Cortina *et al.*, 1991).

In the absence of the G-protein PTX catalyses the transfer of the ADP-ribose moiety of NAD^+ to water, a reaction known as NAD^+-glycohydrolysis (for review, see Locht and Antoine, 1995). This property is also shared with cholera toxin. In contrast, the NAD^+-glycohydrolase activities of diphtheria toxin and exotoxin A are several orders of magnitude less efficient. However, it is unlikely that the PTX-catalysed NAD^+-glycohydrolysis plays a significant role *per se* in the biological activities of PTX as they relate to pathogenesis. The affinity constant of PTX for NAD^+ is in the micromolar range, as evidenced by fluorescent quenching and K_m determinations. In contrast, the affinity of cholera toxin for NAD^+ is in the millimolar range. The higher affinity of PTX for NAD^+ may be in part attributed to the presence of a tryptophan residue, Trp26, located in the NAD^+-binding site. This residue is not conserved in cholera toxin, and alteration

of the PTX Trp26 results in a significant decrease in NAD^+-affinity of S1.

Two catalytic residues have been identified in S1, Glu129 (Antoine *et al.*, 1993) and His35 (Antoine and Locht, 1994). Glu129 is conserved in all known ADP-ribosylating toxins, whereas His35 is only conserved in cholera toxin and the mosquitocidal toxin produced by *Bacillus sphaericus*. It is not present in diphtheria toxin and exotoxin A. The absence of this catalytic residue in the latter two toxins may explain why their NAD^+-glycohydrolysis reaction is much less efficient than that of PTX and cholera toxin. The acceptor substrate of diphtheria toxin and exotoxin A is diphthamide, a modified histidine residue in EF2, which perhaps does not require activation by a catalytic His residue.

We have proposed a model for the PTX-catalysed ADP-ribosyltransferase activity (Locht and Antoine, 1995). The model was based on a number of different observations reported in the literature. These include the catalytic role of Glu129 in the NAD^+-glycohydrolase reaction, the difference in catalytic rates between NAD^+-glycohydrolysis and ADP-ribosylation, the differences in catalytic NAD^+-glycohydrolase properties of PTX and cholera toxin compared with diphtheria toxin and exotoxin A, the invariable conservation of the catalytic Glu residues among all ADP-ribosylating toxins, the less well conserved catalytic His residues, the differences in acceptor substrate specificities of the various toxins, the stereochemistry of the products of some of the reactions, active site geometry of the toxins with known structures, and the proposed mechanism of hydroxide-catalysed NAD^+-glycohydrolysis.

The model predicts that Glu129 weakens the *N*-glycosidic bond linking ADP-ribose to the nicotinamide ring by stabilizing or promoting the formation of an oxocarbenium-like intermediate. This may occur through ionization of the nicotinamide ribose diol, which would result in intramolecular electrostatic stabilization of the intermediate (Figure 7.5). The existence of an oxocarbenium-like ribose ring at the transition state has been demonstrated by isotope effect characterization of the transition state in the PTX-catalysed NAD^+-glycohydrolase reaction (Scheuring and Schramm, 1997a). However, at least during the ADP-ribosyltransferase reaction, this transition-state intermediate is probably not a stable oxocarbenium, since the existence of such an intermediate would result in an S_N1-type reaction, whereas ADP-ribosylating toxins probably catalyse their reaction by an S_N2-type mechanism.

His35 was proposed to act on the acceptor substrate by increasing its nucleophilicity via hydrogen bonding (Figure 7.5). The activated nucleophile can then attack

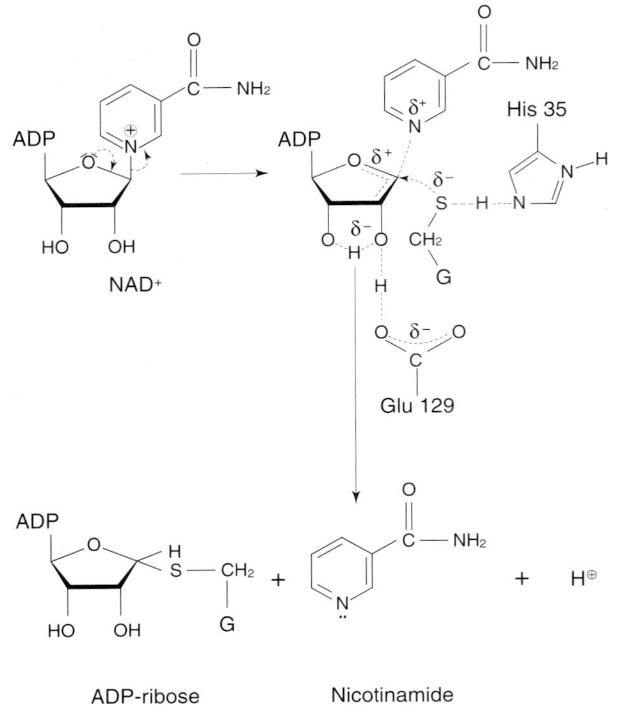

FIGURE 7.5 Model of the enzyme mechanism of PTX-catalysed ADP-ribosylation. The carboxylate group of the Glu129 side-chain interacts with one of the two hydroxyl groups of ribose, thereby favouring the development of an oxocarbenium-like transition state intermediate. The weakened *N*-glycosidic bond of this intermediate can then be attacked by the sulfur of the cysteine side-chain in the Giα protein. This sulfur acts as a nucleophile that can be activated by deprotonation induced by His35 of S1, via interaction with its ε nitrogen. (Adapted from Locht and Antoine, 1995.)

the weakened *N*-glycosidic bond, which results in the transfer of the ADP-ribose moiety of NAD$^+$ to the cysteine of Giα or to water through an S_N2-type mechanism. Observed D_2O kinetic isotope effects are consistent with the activation of the acceptor cysteine through deprotonation prior to the nucleophilic attack (Scheuring and Schramm, 1997b). The sulfur from the acceptor cysteine within the Giα-protein is a more active nucleophile than the oxygen of water. Consistently, His35 substitutions in PTX have a more drastic effect on the ADP-ribosyltransferase reaction than on the NAD$^+$-glycohydrolase reaction (Xu *et al.*, 1994). Whereas the formation of an oxocarbenium-like transition state intermediate stabilized by a catalytic glutamate residue may be shared by all ADP-ribosyl-transferases, the activation of the attacking nucleophile by deprotonation via a catalytic histidine may be more specific to PTX. However, given the conservation of the catalytic His residue in PTX, cholera toxin and mosquitocidal toxin, its role in increasing the nucleophilic-ity of the acceptor substrate may be conserved in this

subgroup of toxins. In diphtheria toxin and exotoxin A, this part of the reaction may perhaps be fulfilled direct-ly by their acceptor substrate diphthamide itself.

Biological activities and role of pertussis toxin in pathogenesis

Pertussis toxin-mediated ADP-ribosylation of the Giα-proteins results in uncoupling of the G-proteins from their cognate receptors. Initially, it was observed that the secretion of insulin by pancreatic B-cells is inhibit-ed via α$_2$-adrenergic receptors, and that this inhibition is lost upon ADP-ribosylation of Giα by PTX (Murayama and Ui, 1983). In the inactive state, the receptor is linked to its corresponding G-protein. The G-protein is then in its trimeric form, containing GDP bound to the α-subunit. Recognition of the receptor by an extracellular signal results in the dissociation of the trimeric G-protein from the signal–receptor complex and in the replacement of GDP by GTP. This activates the G-protein, which subsequently dissociates into Gα$_{GTP}$ and Gβγ. Gα$_{GTP}$ may then productively interact with the effector protein. The intrinsic GTP-hydrolase activity of the G-protein cleaves GTP into GDP and phosphate, thereby inactivating the Gα-protein again, promoting its dissociation from the effector and its reas-sociation with the β- and γ-subunits.

Because of the uncoupling of the Gi protein from its receptor by PTX-catalysed ADP-ribosylation, signals that normally activate the process are no longer able to transmit their trigger to the effector molecule. The first effector recognized in this signal transduction cascade was the membrane-bound adenylate cyclase. However, other effectors have now been identified and include ion channels, phosphodi-esterases and enzymes that produce diacylglycerol and inositol phosphates (Ui, 1988). The extracellular signals may also be very diverse and range from neuro-transmitters and hormones to light. The vast diversity of the biological activities of PTX can thus easily be accounted for by its enzymic ADP-ribosyltransferase activity. However, some biological activities of PTX, such as its mitogenicity and its ability to aggluti-nate erythrocytes, are independent of the enzymic activity.

Biological effects that have been related to the action of PTX in tissue cultures include suppression of cell proliferation, morphological changes, exocrine secre-tion, inhibition of histamine secretion, stimulation of lipolysis and many others. *In vivo* PTX induces islet activation, lymphocytosis, histamine sensitization and increase in vascular permeability (Munoz, 1985), some of which are the hallmarks of systemic pertussis in patients (Pittman, 1984).

In the infant mouse model, a PTX-deficient mutant is severely impaired in the ability to cause lethal infections. However, at lower doses, the lung-colonization profile of such mutants is not significantly different from that of wild-type *B. pertussis* in the initial phase of the infection. None the less, the mutant strain is cleared somewhat more rapidly than the wild-type strain, indicating that PTX is not required for the initial stages of infection, but may play a critical role later on (Goodwin and Weiss, 1990; Khelef et al., 1992).

Given its role in pathogenesis, it is not surprising that PTX is a major protective antigen in pertussis vaccines. It has also been included in all of the new, acellular vaccines, which are destined to replace current, whole-cell vaccines in many countries (Ad hoc group, 1988). Based on the knowledge of the molecular mechanisms of PTX action, genetically detoxified analogues have been developed (Pizza et al., 1989) that show promise as vaccine candidates in that they induce high levels of protection without causing harmful side-effects sometimes observed with whole-cell pertussis vaccines.

ACKNOWLEDGEMENTS

We gratefully acknowledge all the workers in the *Bordetella* toxin field for the many exciting experiments and the excellent work that has been done over the years. The work in our laboratory was supported by the Institut Pasteur de Lille, Région Nord-Pas-de-Calais, INSERM, Ministère de la Recherche.

REFERENCES

Ad hoc group for the study of pertussis vaccines (1988). Placebo-controlled trial of two acellular pertussis vaccines in Sweden – protective efficacy and adverse events. *Lancet* ii, 955–60.

Antoine, R. and Locht, C. (1990). Roles of the disulfide bond and the carboxy-terminal region of the S1 subunit in the assembly and biosynthesis of pertussis toxin. *Infect. Immun.* **58**, 1518–26.

Antoine, R. and Locht, C. (1994). The NAD-glycohydrolase activity of the pertussis toxin S1 subunit: involvement of the catalytic His-35 residue. *J. Biol. Chem.* **269**, 6450–7.

Antoine, R., Tallett, A., van Heyningen, S. and Locht, C. (1993). Evidence for a catalytic role of glutamic acid 129 in the NAD-glycohydrolase activity of the pertussis toxin S1 subunit. *J. Biol. Chem.* **268**, 24149–55.

Antoine, R., Raze, D. and Locht, C. (1996). Genetic analysis of the pertussis toxin locus. *Zbl. Bakteriol.* **28**, 44–5.

Armstrong, G.D., Howard, L.A. and Peppler, M.S. (1988). Use of glycosyltransferases to restore pertussis toxin receptor activity to asialogalactofetuin. *J. Biol. Chem.* **263**, 8677–84.

Baker, S.M., Masi, A., Liu, D.-F., Novitsky, B.K. and Deich, R.A. (1995). Pertussis toxin export genes are regulated by the *ptx* promoter and may be required for efficient translation of *ptx* mRNA in *Bordetella pertussis. Infect. Immun.* **63**, 3920–6.

Barry, E.M., Weiss, A.A., Ehrmann, I.E., Gray, M.C., Hewlett, E.L. and Goodwin, M.S. (1991). *Bordetella pertussis* adenylate cyclase toxin and hemolytic activities require a second gene, *cyaC*, for activation. *J. Bacteriol.* **173**, 720–6.

Bellalou, J., Sakamoto, H., Ladant, D., Geoffroy, C. and Ullmann, A. (1990). Deletions affecting hemolytic and toxin activities of *Bordetella pertussis* adenylate cyclase. *Infect. Immun.* **58**, 3242–7.

Betsou, F., Sebo, P. and Guiso, N. (1993). CyaC-mediated activation is important not only for toxic but also for protective activities of *Bordetella pertussis* adenylate cyclase-hemolysin. *Infect. Immun.* **61**, 3583–9.

Betsou, F., Sismeiro, O., Danchin, A. and Guiso, N. (1995). Cloning and sequence of the *Bordetella bronchiseptica* adenylate cyclase-hemolysin-encoding gene: comparison with the *Bordetella pertussis* gene. *Gene* **162**, 165–6.

Bordet, J. and Gengou, O. (1909). L'endotoxine coquelucheuse. *Ann. Inst. Pasteur (Paris)* **23**, 415–19.

Bouhss, A., Krin, E., Munier, H., Gilles, A.M., Danchin, A., Glaser, P. and Barzu, O. (1993). Cooperative phenomena in binding and activation of *Bordetella pertussis* adenylate cyclase by calmodulin. *J. Biol. Chem.* **268**, 1690–4.

Brennan, M.J., David, J.L., Kenimer, J.G. and Manclark, C.R. (1988). Binding of pertussis toxin to a 165-kilodalton Chinese hamster ovary cell glycoprotein. *J. Biol. Chem.* **263**, 4895–9.

Clark, C.G. and Armstrong, G.D. (1990). Lymphocyte receptors for pertussis toxin. *Infect. Immun.* **58**, 3840–6.

Confer, D.L. and Eaton, J.W. (1982). Phagocyte impotence caused by an invasive bacterial adenylate cyclase. *Science* **217**, 948–50.

Cookson, B.T., Cho, H.L., Herwaldt, L.A. and Goldman, W.E. (1989a). Biological activities and chemical composition of purified tracheal cytotoxin of *Bordetella pertussis. Infect. Immun.* **57**, 2223–9.

Cookson, B.T., Tyler, A.N. and Goldman, W.E. (1989b). Primary structure of the peptidoglycan-derived tracheal cytotoxin of *Bordetella pertussis. Biochemistry* **28**, 1744–9.

Cortina, G., Krueger, K.M. and Barbieri, J.T. (1991). The carboxyl terminus of the S1 subunit of pertussis toxin confers high affinity binding to transducin. *J. Biol. Chem.* **266**, 23810–14.

Cowell, J.L., Hewlett, E.L. and Manclark, C.R. (1979). Intracellular localization of the dermonecrotic toxin of *Bordetella pertussis. Infect. Immun.* **25**, 896–901.

Donovan, M.G., Masure, H.R. and Storm, D.R. (1989). Isolation of a protein fraction from *Bordetella pertussis* that facilitates entry of the calmodulin-sensitive adenylate cyclase into animal cells. *Biochemistry* **28**, 8124–9.

Ehrmann, I.E., Gray, M.C., Gordon, V.M., Gray, L.S. and Hewlett, E.L. (1991). Hemolytic activity of adenylate cyclase toxin from *Bordetella pertussis. FEBS Lett.* **278**, 79–83.

El Baya, A., Linnemann, R., von Olleschik-Elbheim, L., Robenek, H. and Schmidt, M.A. (1997). Endocytosis and retrograde transport of pertussis toxin to the Golgi complex as a prerequisite for cellular intoxication. *Eur. J. Cell Biol.* **73**, 40–8.

Escuyer, V., Duflot, E., Sezer, O., Danchin, A. and Mock, M. (1988). Structural homology between virulence-associated bacterial adenylate cyclases. *Gene* **71**, 293–8.

Farfel, Z., Konen, S., Wiertz, E., Klapmuts, R., Addy, P.A. and Hanski, E. (1990). Antibodies to *Bordetella pertussis* adenylate cyclase are produced in man during pertussis infection and after vaccination. *J. Med. Microbiol.* **32**, 173–7.

Farizo, K.M., Cafarella, T.G. and Burns, D.L. (1996). Evidence for a ninth gene, *ptlI*, in the locus encoding the pertussis toxin secretion system of *Bordetella pertussis* and formation of a PtlI-PtlF complex. *J. Biol. Chem.* **271**, 31643–9.

Friedman, R.L., Fiederlein, R.L., Glasser, L. and Galgiani, J.N. (1987). *Bordetella pertussis* adenylate cyclase: effects of affinity-purified adenylate cyclase on human polymorphonuclear leukocyte functions. *Infect. Immun.* **55**, 135–40.

Gentile, F., Raptis, A., Knipling, L.G. and Wolff, J. (1988). *Bordetella pertussis* adenylate cyclase. Penetration into host cells. *Eur. J. Biochem.* **175**, 447–53.

Gentile, F., Knipling, L.G., Sackett, D.L. and Wolff, J. (1990). Invasive adenylyl cyclase of *Bordetella pertussis*. Physical, catalytic, and toxic properties. *J. Biol. Chem.* **265**, 10686–92.

Gentry-Weeks, C.R., Cookson, B.T., Goldman, W.E., Rimler, R.B., Porter, S.B. and Curtiss, R. (1988). Dermonecrotic toxin and tracheal cytotoxin, putative virulence factors of *Bordetella avium*. *Infect. Immun.* **56**, 1698–707.

Gilles, A. M., Munier, H., Rose, T., Glaser, P., Krin, E., Danchin, A., Pellecuer, C. and Barzu, O. (1990). Intrinsic fluorescence of a truncated *Bordetella pertussis* adenylate cyclase expressed in *Escherichia coli*. *Biochemistry* **29**, 8126–30.

Glaser, P., Ladant, D., Sezer, O., Pichot, F., Ullmann, A. and Danchin, A. (1988a). The calmodulin-sensitive adenylate cyclase of *Bordetella pertussis*: cloning and expression in *Escherichia coli*. *Mol. Microbiol.* **2**, 19–30.

Glaser, P., Sakamoto, H., Bellalou, J., Ullmann, A. and Danchin, A. (1988b). Secretion of cyclolysin, the calmodulin-sensitive adenylate cyclase-haemolysin bifunctional protein of *Bordetella pertussis*. *EMBO J.* **7**, 3997–4004.

Glaser, P., Elmaoglou-Lazaridou, A., Krin, E., Ladant, D., Barzu, O. and Danchin, A. (1989). Identification of residues essential for catalysis and binding of calmodulin in *Bordetella pertussis* adenylate cyclase by site-directed mutagenesis. *EMBO J.* **8**, 967–72.

Glaser, P., Munier, H., Gilles, A.M., Krin, E., Porumb, T., Barzu, O., Sarfati, R., Pellecuer, C. and Danchin, A. (1991). Functional consequences of single amino acid substitutions in calmodulin-activated adenylate cyclase of *Bordetella pertussis*. *EMBO J.* **10**, 1683–8.

Goodwin, M.S. and Weiss, A.A. (1990). Adenylate cyclase toxin is critical for colonization and pertussis toxin is critical for lethal infection by *Bordetella pertussis* in infant mice. *Infect. Immun.* **58**, 3445–7.

Gordon, V.M., Young, W.W., Jr, Lechler, S.M., Gray, M.C., Leppla, S.H. and Hewlett, E.L. (1989). Adenylate cyclase toxins form *Bacillus anthracis* and *Bordetella pertussis*. Different processes for interaction with and entry into target cells. *J. Biol. Chem.* **264**, 14792–6.

Graf, R., Codina, J. and Birnbaumer, L. (1992). Peptide inhibitors of ADP-ribosylation by pertussis toxin are substrates with affinities comparable to those of the trimeric GTP-binding proteins. *Mol. Pharmacol.* **42**, 760–4.

Greenlee, D.V., Andreasen, T.J. and Storm, D.R. (1982). Calcium-independent stimulation of *Bordetella pertussis* adenylate cyclase by calmodulin. *Biochemistry* **21**, 2759–64.

Gross, M.K., Au, D.C., Smith, A.L. and Storm, D.R. (1992). Targeted mutations that ablate either the adenylate cyclase or hemolysin function of the bifunctional CyaA toxin of *Bordetella pertussis* abolish virulence. *Proc. Natl Acad. Sci. USA* **89**, 4898–902.

Gueirard, P. and Guiso, N. (1993). Virulence of *Bordetella bronchiseptica*: role of adenylate cyclase. *Infect. Immun.* **61**, 4072–8.

Gueirard, P., Weber, C., Le Coustumier, A. and Guiso, N. (1995). Human *Bordetella bronchiseptica* infection related to contact with infected animals: persistence of bacteria in host. *J. Clin. Microbiol.* **33**, 2002–6.

Guiso, N., Szatanik, M. and Rocancourt, M. (1991). Protective activity of *Bordetella* adenylate cyclase-hemolysin against bacterial colonization. *Microb. Pathog.* **11**, 423–31.

Gustafsson, B., Lindquist, U. and Andersson, M. (1988). Production and characterization of monoclonal antibodies directed against *Bordetella pertussis* lipopolysaccharide. *J. Clin. Microbiol.* **26**, 188–93.

Hackett, M., Guo, L., Shabanowitz, J., Hunt, D.F. and Hewlett, E.L. (1994). Internal lysine palmitoylation in adenylate cyclase toxin from *Bordetella pertussis*. *Science* **266**, 433–5.

Hackett, M., Walker, C.B., Guo, L., Gray, M.C., Van Cuyk, S., Ullmann, A., Shabanowitz, J., Hunt, D.F., Hewlett, E.L. and Sebo, P. (1995). Hemolytic, but not cell-invasive activity, of adenylate cyclase toxin is selectively affected by differential fatty-acylation in *Escherichia coli*. *J. Biol. Chem.* **270**, 20250–3.

Hanski, E. and Farfel, Z. (1985). *Bordetella pertussis* invasive adenylate cyclase. Partial resolution and properties of its cellular penetration. *J. Biol. Chem.* **260**, 5526–32.

Hausman, S.Z. and Burns, D.L. (1992). Interaction of pertussis toxin with cells and model membranes. *J. Biol. Chem.* **267**, 13735–9.

Hausman, S.Z. and Burns, D.L. (1993). Binding of pertussis toxin to lipid vesicles containing glycolipids. *Infect. Immun.* **61**, 335–7.

Hazes, B., Boodhoo, A., Cockle, S.A. and Read, R.J. (1996). Crystal structure of the pertussis toxin-ATP complex: a molecular sensor. *J. Mol. Biol.* **258**, 661–71.

Heiss, L.N., Lancaster, J.R., Jr, Corbett, J.A. and Goldman, W.E. (1994). Epithelial autotoxicity of nitric oxide: role in the respiratory cytopathology of pertussis. *Proc. Natl Acad. Sci. USA* **91**, 267–70.

Heveker, N. and Ladant, D. (1997). Characterization of mutant *Bordetella pertussis* adenylate cyclase toxins with reduced affinity for calmodulin. Implications for the mechanism of toxin entry into target cells. *Eur. J. Biochem.* **243**, 643–9.

Hewlett, E.L., Urban, M.A., Manclark, C.R. and Wolff, J. (1976). Extracytoplasmic adenylate cyclase of *Bordetella pertussis*. *Proc. Natl Acad. Sci. USA* **73**, 1926–30.

Hewlett, E.L., Gray, L., Allietta, M., Ehrmann, I., Gordon, V.M. and Gray, M.C. (1991). Adenylate cyclase toxin from *Bordetella pertussis*. Conformational change associated with toxin activity. *J. Biol. Chem.* **266**, 17503–8.

Hewlett, E.L., Gray, M.C., Ehrmann, I.E., Maloney, N.J., Otero, A.S., Gray, L., Allietta, M., Szabo, G., Weiss, A.A. and Barry, E.M. (1993). Characterization of adenylate cyclase toxin from a mutant of *Bordetella pertussis* defective in the activator gene, *cyaC*. *J. Biol. Chem.* **268**, 7842–8.

Horiguchi, Y., Nakai, T. and Kume, K. (1991). Effects of *Bordetella bronchiseptica* dermonecrotic toxin on the structure and function of osteoblastic clone MC3T3-E1 cells. *Infect. Immun.* **59**, 1112–16.

Horiguchi, Y., Sugimoto, N. and Matsuda, M. (1993). Stimulation of DNA synthesis in osteoblast-like MC3T3-E1 cells by *Bordetella bronchiseptica* dermonecrotic toxin. *Infect. Immun.* **61**, 3611–15.

Horiguchi, Y., Sugimoto, N. and Matsuda, M. (1994). *Bordetella bronchiseptica* dermonecrotizing toxin stimulates protein synthesis in an osteoblastic clone, MC3T3-E1 cells. *FEMS Microbiol. Lett.* **120**, 19–22.

Horiguchi, Y., Senda, T., Sugimoto, N., Katahira, J. and Matsuda, M. (1995). *Bordetella bronchiseptica* dermonecrotizing toxin stimulates assembly of actin stress fibers and focal adhesions by modifying the small GTP-binding protein Rho. *J. Cell Sci.* **108**, 3243–51.

Horiguchi, Y., Inoue, N., Masuda, M., Kashimoto, T., Katahira, J., Sugimoto, N. and Matsuda, M. (1997). *Bordetella bronchiseptica* dermonecrotizing toxin induces reorganization of actin stress fibers through deamination of Gln-63 of the GTP-binding protein Rho. *Proc. Natl Acad. Sci. USA* **94**, 11623–6.

Iwaki, M., Ullmann, A. and Sebo, P. (1995). Identification by *in vitro* complementation of regions required for cell-invasive activity of *Bordetella pertussis* adenylate cyclase toxin. *Mol. Microbiol.* **17**, 1015–24.

Johnson, F.D. and Burns, D.L. (1994). Detection and subcellular localization of three Ptl proteins involved in the secretion of pertussis toxin from *Bordetella pertussis*. *J. Bacteriol.* **176**, 5350–6.

Karimova, G., Pidoux, J., Ullmann, A. and Ladant, D. (1998). A bacterial two-hybrid system based on a reconstituted signal transduction pathway. *Proc. Natl Acad. Sci. USA* **95**, 5752–6.

Katada, T. and Ui, M. (1982). Direct modification of the membrane adenylate cyclase system by islet-activating protein due to ADP-ribosylation of a membrane protein. *Proc. Natl Acad. Sci. USA* **79**, 3129–33.

Katada, T., Oinuma, M. and Ui, M. (1986). Two guanine nucleotide-binding proteins in rat brain serving as the specific substrate of islet-activating protein, pertussis toxin. Interaction of the α-subunits with βγ-subunits in development of their biological activities. *J. Biol. Chem.* **261**, 8182–91.

Khelef, N. and Guiso, N. (1995). Induction of macrophage apoptosis by *Bordetella pertussis* adenylate cyclase-hemolysin. *FEMS Microbiol. Lett.* **134**, 27–32.

Khelef, N., Sakamoto, H. and Guiso, N. (1992). Both adenylate cyclase and hemolytic activities are required by *Bordetella pertussis* to initiate infection. *Microb. Pathog.* **12**, 227–35.

Khelef, N., Danve, B., Quentin-Millet, M.J. and Guiso, N. (1993). *Bordetella pertussis* and *Bordetella parapertussis*: two immunologically distinct species. *Infect. Immun.* **61**, 486–90.

Kilhoffer, M.C., Cook, G.H. and Wolff, J. (1983). Calcium-independent activation of adenylate cyclase by calmodulin. *Eur. J. Biochem.* **133**, 11–15.

Kotob, S.I. and Burns, D.L. (1997). Essential role of the consensus nucleotide-binding site of PtlH in secretion of pertussis toxin from *Bordetella pertussis*. *J. Bacteriol.* **179**, 7577–80.

Kotob, S.I., Hausman, S.Z. and Burns, D.L. (1995). Localization of the promoter for the *ptl* genes in *Bordetella pertussis*, which encode proteins essential for secretion of pertussis toxin. *Infect. Immun.* **63**, 3227–30.

Kume, K., Nakai, T., Samejima, Y. and Sugimoto, C. (1986). Properties of dermonecrotic toxin prepared from sonic extracts of *Bordetella bronchiseptica*. *Infect. Immun.* **52**, 370–7.

Lacerda, H.M., Pullinger, G.D., Lax, A.J. and Rozengurt, E. (1997). Cytotoxic necrotizing factor 1 from *Escherichia coli* and dermonecrotic toxin from *Bordetella bronchiseptica* induce p21(rho)-dependent tyrosine phosphorylation of focal adhesion kinase and paxillin in Swiss 3T3 cells. *J. Biol. Chem.* **272**, 9587–96.

Ladant, D. (1988). Interaction of *Bordetella pertussis* adenylate cyclase with calmodulin. Identification of two separated calmodulin-binding domains. *J. Biol. Chem.* **263**, 2612–18.

Le Dur, A., Chaby, R. and Szabo, L. (1980). Isolation of two protein-free and chemically different lipopolysaccharides from *Bordetella pertussis* phenol-extracted endotoxin. *J. Bacteriol.* **143**, 78–88.

Lei, M.G. and Morrison, D.C. (1993). Evidence that lipopolysaccharide and pertussis toxin bind to different domains on the same p73 receptor on murine splenocytes. *Infect. Immun.* **61**, 1359–64.

Lemichez, E., Flatau, G., Bruzzone, M., Boquet, P. and Gauthier, M. (1997). Molecular localization of the *Escherichia coli* cytotoxic necrotizing factor CNF1 cell-binding and catalytic domains. *Mol. Microbiol.* **24**, 1061–70.

Livey, I. and Wardlaw, A.C. (1984). Production and properties of *Bordetella pertussis* heat labile toxin. *J. Med. Microbiol.* **17**, 91–103.

Lobet, Y., Feron, C., Dequesne, G., Simoen, E., Hauser, P. and Locht, C. (1993). Site-specific alterations in the B-oligomer that affect receptor-binding activities and mitogenicity of pertussis toxin. *J. Exp. Med.* **177**, 79–87.

Locht, C. and Antoine, R. (1995). A proposed mechanism of ADP-ribosylation catalyzed by the pertussis toxin S1 subunit. *Biochimie* **77**, 333–40.

Locht, C. and Keith, J.M. (1986). Pertussis toxin gene: nucleotide sequence and genetic organization. *Science* **232**, 1258–64.

Locht, C., Lobet, Y., Feron, C., Cieplak, W. and Keith, J.M. (1990). The role of cysteine 41 in the enzymatic activities of the pertussis toxin S1 subunit as investigated by site-directed mutagenesis. *J. Biol. Chem.* **265**, 4552–9.

Luker, K.E., Tyler, A.N., Marshall, G.R. and Goldman, W.E. (1995). Tracheal cytotoxin structural requirements for respiratory epithelial damage in pertussis. *Mol. Microbiol.* **16**, 733–43.

Masure, H.R., Oldenburg, D.J., Donovan, M.G., Shattuck, R.L. and Storm, D.R. (1988). The interaction of Ca^{2+} with the calmodulin-sensitive adenylate cyclase from *Bordetella pertussis*. *J. Biol. Chem.* **263**, 6933–40.

Munier, H., Bouhss, A., Krin, E., Danchin, A., Gilles, A.M., Glaser, P. and Barzu, O. (1992). The role of histidine 63 in the catalytic mechanism of *Bordetella pertussis* adenylate cyclase. *J. Biol. Chem.* **267**, 9816–20.

Munoz, J.J. (1985). Biological activities of pertussigen (pertussis toxin). In: *Pertussis Toxin* (eds R.D. Sekura, J. Moss and M. Vaughn), pp. 1–18. Academic Press, Orlando, FL.

Murayama, T. and Ui, M. (1983). Loss of the inhibitory function of the guanine nucleotide regulatory component of adenylate cyclase due to its ADP-ribosylation by islet-activating protein, pertussis toxin, in adipocyte membrane. *J. Biol. Chem.* **258**, 381–90.

Nagano, H., Nakai, T., Horiguchi, Y. and Kume, K. (1988). Isolation and characterization of mutant strains of *Bordetella bronchiseptica* lacking dermonecrotic toxin-producing ability. *J. Clin. Microbiol.* **26**, 1983–7.

Nicosia, A., Perugini, M., Franzini, C., Casagli, M.C., Borri, M.G., Antoni, G., Almoni, M., Neri, P., Ratti, G. and Rappuoli, R. (1986). Cloning and sequencing of the pertussis toxin gene: operon structure and gene duplication. *Proc. Natl Acad. Sci. USA* **83**, 4631–5.

Oldenburg, D.J., Gross, M.K., Wong, C.S. and Storm, D.R. (1992). High-affinity calmodulin binding is required for the rapid entry of *Bordetella pertussis* adenylyl cyclase into neuroblastoma cells. *Biochemistry* **31**, 8884–91.

Oldenburg, D.J., Gross, M.K., Smith, A.L. and Storm, D.R. (1993). Virulence of a *Bordetella pertussis* stain expressing a mutant adenylyl cyclase with decreased calmodulin affinity. *Microb. Pathogen.* **14**, 489–93.

Parfentjev, I.A. and Goodline, M.A. (1948). Histamine shock in mice sensitized with *Hemophilus pertussis* vaccine. *J. Pharmacol. Exp. Ther.* **92**, 411–13.

Pittman, M. (1984). The concept of pertussis as a toxin mediated disease. *Pediatr. Infect. Dis.* **3**, 467–86.

Pizza, M., Covacci, A., Bartoloni, A., Perugini, M., Nencioni, L., De Magistris, T., Villa, L., Nucci, D., Manetti, R., Bugnoli, M., Giovannoni, F., Olivieri, R., Barbieri, J.T., Sato, H. and Rappuoli, R. (1989). Mutants of pertussis toxin suitable for vaccine development. *Science* **246**, 497–500.

Precheur, B., Siffert, O., Barzu, P. and Craescu, C.T. (1991). NMR and circular dichroic studies on the solution conformation of a synthetic peptide derived from the calmodulin-binding domain of *Bordetella pertussis* adenylate cyclase. *Eur. J. Biochem.* **196**, 67–72.

Pullinger, G.D., Adams, T.E., Mullan, P.B., Garrod, T.I. and Lax, A.J. (1996). Cloning, expression, and molecular characterization of the dermonecrotic toxin gene of *Bordetella* spp. *Infect. Immun.* **64**, 4163–71.

Raptis, A., Knipling, L.G., Gentile, F. and Wolff, J. (1989). Modulation of invasiveness and catalytic activity of *Bordetella pertussis* adenylate cyclase by polycations. *Infect. Immun.* **57**, 1066–71.

Ray, A., Redhead, K., Selkirk, S. and Poole, S. (1991). Variability in LPS composition, antigenicity and reactogenicity of phase variants of *Bordetella pertussis*. *FEMS Microbiol. Lett.* **63**, 211–17.

Rogel, A. and Hanski, E. (1992). Distinct steps in the penetration of adenylate cyclase toxin of *Bordetella pertussis* into sheep erythrocytes. Translocation of the toxin across the membrane. *J. Biol. Chem.* **267**, 22399–605.

Rogel, A., Schultz, J.E., Brownlie, R.M., Coote, J.G., Parton, R. and Hanski, E. (1989). *Bordetella pertussis* adenylate cyclase: purification and characterization of the toxic form of the enzyme. *EMBO J.* **8**, 2755–60.

Rogel, A., Meller, R. and Hanski, E. (1991). Adenylate cyclase toxin from *Bordetella pertussis*. The relationship between induction of cAMP and hemolysis. *J. Biol. Chem.* **266**, 3154–61.

Rogers, T.S., Corey, S.J. and Rosoff, P.M. (1990). Identification of a 43-kilodalton human T lymphocyte membrane protein as a receptor for pertussis toxin. *J. Immunol.* **145**, 678–83.

Roop, R.M.d., Veit, H.P., Sinsky, R.J., Veit, S.P., Hewlett, E.L. and Kornegay, E.T. (1987). Virulence factors of *Bordetella bronchiseptica* associated with the production of infectious atrophic rhinitis and pneumonia in experimentally infected neonatal swine. *Infect. Immun.* **55**, 217–22.

Sakamoto, H., Bellalou, J., Sebo, P. and Ladant, D. (1992). *Bordetella pertussis* adenylate cyclase toxin. Structural and functional independence of the catalytic and hemolytic activities. *J. Biol. Chem.* **267**, 13598–602.

Saron, M.F., Fayolle, C., Sebo, P., Ladant, D., Ullmann, A. and Leclerc, C. (1997). Anti-viral protection conferred by recombinant adenylate cyclase toxins from *Bordetella pertussis* carrying a CD8+ T cell epitope from lymphocytic choriomeningitis virus. *Proc. Natl Acad. Sci. USA* **94**, 3314–19.

Saukkonen, K., Burnette, W.N., Mar, V.L., Masure, H.R. and Tuomanen, E.I. (1992). Pertussis toxin has eukaryotic-like carbohydrate recognition domains. *Proc. Natl Acad. Sci. USA* **89**, 118–22.

Scheuring, J. and Schramm, V.L. (1997a). Kinetic isotope effect characterization of the transition state for oxidized nicotinamide adenine nucleotide hydrolysis by pertussis toxin. *Biochemistry* **36**, 4526–34.

Scheuring, J. and Schramm, V.L. (1997b). Pertussis toxin: transition state analysis for ADP-ribosylation of G-protein peptide alpha3C20. *Biochemistry* **36**, 8215–23.

Sebo, P. and Ladant, A. (1993). Repeat sequences in the *Bordetella pertussis* adenylate cyclase toxin can be recognized as alternative carboxy-proximal secretion signals by the *Escherichia coli* alpha-haemolysin translocator. *Mol. Microbiol.* **9**, 999–1009.

Sebo, P., Glaser, P., Sakamoto, H. and Ullmann, A. (1991). High-level synthesis of active adenylate cyclase toxin of *Bordetella pertussis* in a reconstructed *Escherichia coli* system. *Gene* **104**, 19–24.

Senda, T., Horiguchi, Y., Umemoto, M., Sugimoto, N. and Matsuda, M. (1997). *Bordetella bronchiseptica* dermonecrotizing toxin, which activates a small GTP-binding protein rho, induces membrane organelle proliferation and caveolae formation. *Exp. Cell Res.* **230**, 163–8.

Shattuck, R.L. and Storm, D.R. (1985). Calmodulin inhibits entry of *Bordetella pertussis* adenylate cyclase into animal cells. *Biochemistry* **24**, 6323–8.

Silvera, D., Edington, N. and Smith, I.M. (1982). Ultrastructural changes in the nasal turbinate bones of pigs in early infection with *Bordetella bronchiseptica*. *Res. Vet. Sci.* **33**, 37–42.

Stein, P.E., Boodhoo, A., Armstrong, G.D., Cockle, S.A., Klein, M.H. and Read, R. J. (1994a). The crystal structure of pertussis toxin. *Structure* **2**, 45–57.

Stein, P.E., Boodhoo, A., Armstrong, G.D., Heerze, L.D., Cockle, S.A., Klein, M.H. and Read, R.J. (1994b). Structure of a pertussis toxin–sugar complex as a model for receptor binding. *Nat. Struct. Biol.* **1**, 591–6.

Tamura, M., Nogimori, K., Murai, S., Yajima, M., Ito, K., Katada, T., Ui, M. and Ishii, S. (1982). Subunit structure of islet-activating protein, pertussis toxin, in conformity with the A-B model. *Biochemistry* **21**, 5516–22.

Ui, M. (1988). The multiple biological activities of pertussis toxin. In: *Pathogenesis and Immunity in Pertussis* (eds A.C. Wardlaw and R. Parton), pp. 121–45. John Wiley, Chichester.

Walker, K.E. and Weiss, A.A. (1994). Characterization of the dermonecrotic toxin in members of the genus *Bordetella*. *Infect. Immun.* **62**, 3817–28.

Ward, J.E., Dale, E.M. and Binns, A.N. (1991). Activity of the *Agrobacterium* T-DNA transfer machinery is affected by *virB* gene products. *Proc. Natl Acad. Sci. USA* **88**, 9350–4.

Watanabe, M., Takimoto, H., Kumazawa, Y. and Amano, K. (1990). Biological properties of lipopolysaccharides from *Bordetella* species. *J. Gen. Microbiol.* **136**, 489–93.

Weiss, A.A. and Goodwin, M.S. (1989). Lethal infection by *Bordetella pertussis* mutants in the infant mouse model. *Infect. Immun.* **57**, 3757–64.

Weiss, A.A., Hewlett, E.L., Myers, G.A. and Falkow, S. (1984). Pertussis toxin and extracytoplasmic adenylate cyclase as virulence factors of *Bordetella pertussis*. *J. Infect. Dis.* **150**, 219–22.

Weiss, A.A., Johnson, F.D. and Burns, D.L. (1993). Molecular characterization of an operon required for pertussis toxin secretion. *Proc. Natl Acad. Sci. USA* **90**, 2970–4.

Wilson, R., Read, R., Thomas, M., Rutman, A., Harrison, K., Lund, V., Cookson, B., Goldman, W., Lambert, H. and Cole, P. (1991). Effects of *Bordetella pertussis* infection on human respiratory epithelium *in vivo* and *in vitro*. *Infect. Immun.* **59**, 337–45.

Wolff, J., Cook, G.H., Goldhammer, A.R. and Berkowitz, S.A. (1980). Calmodulin activates prokaryotic adenylate cyclase. *Proc. Natl Acad. Sci. USA* **77**, 3841–4.

Xu, Y. and Barbieri, J.T. (1995). Pertussis toxin-mediated ADP-ribosylation of target proteins in Chinese hamster ovary cells involves a vesicle trafficking mechanism. *Infect. Immun.* **63**, 825–32.

Xu, Y., Brabançon-Finck, V. and Barbieri, J.T. (1994). Role of histidine 35 of the S1 subunit of pertussis toxin in the ADP-ribosylation of transducin. *J. Biol. Chem.* **269**, 9993–9.

Zarrouk, H., Karibian, D., Bodie, S., Perry, M.B., Richards, J.C. and Caroff, M. (1997). Structural characterization of the lipids A of three *Bordetella bronchiseptica* strains: variability of fatty acid substitution. *J. Bacteriol.* **179**, 3756–60.

8

Clostridial toxins acting on the cytoskeleton

Monica Thelestam, Esteban Chaves-Olarte, Michael Moos and Christoph von Eichel-Streiber

INTRODUCTION

Bacteria of the genus *Clostridium* cause dramatic diseases such as tetanus, botulism, gas gangrene and pseudomembranous colitis, all of which are typical toxin diseases. This chapter will consider those clostridial toxins that exert their primary cellular effects by attacking the cytoskeleton. In mammalian cells, the cytoskeleton is composed of three major systems: the microtubules (MT), the intermediate filaments (IF) and the microfilaments, which build the actin cytoskeleton (ACSK). The toxins dealt with here attack the ACSK, including its controlling small GTPases.

The toxins

Serotype C strains of *Clostridium botulinum*, besides the neurotoxin C1, produce two additional toxins (denoted C2 and C3), which affect the ACSK directly by ADP-ribosylation of actin (C2) or indirectly by ADP-ribosylation of the small GTPase Rho (C3). C2-like ADP-ribosyltransferases (ADPRTs) are also produced by *C. perfringens* and *C. spiroforme* (Table 8.1). In addition, an actin-specific ADPRT is produced by one clinical isolate of *C. difficile* (Perelle *et al.*, 1997), the major toxins of which, however, are the glucosyltransferases TcdA and TcdB. *Clostridium limosum* produces a C3-like ADPRT, which attacks Rho and has the same highly conserved catalytic domain as C3 (Table 8.1).

Toxins A and B (TcdA, TcdB) produced by *C. difficile* cause antibiotic-associated diarrhoea and pseudo-membranous colitis (Lyerly and Wilkins, 1995). These toxins are prototypes of the family of large clostridial cytotoxins (LCTs), which also comprises the haemorrhagic and lethal toxins (TcsH, TcsL) of *C. sordellii* and the α-toxin (Tcnα) of *C. novyi*. All LCTs possess glycosyltransferase activity. The ACSK collapse induced by the LCTs is caused by glycosylation of a variety of small GTPases, which control the ACSK through their effector proteins.

It should be mentioned that bacteria of several other genera also secrete toxins that attack the ACSK. Examples include the *Escherichia coli* cytotoxic necrotizing factor, the dermonecrotic toxin elaborated by *Bordetella bronchiseptica*, the *Vibrio cholerae* zonula occludens toxin, the *Bacteroides fragilis* enterotoxin, as well as the C3-like toxins from *Bacillus cereus*, *Legionella pneumophila* and *Staphylococcus aureus*. The majority of toxins that act on the ACSK cause its collapse. The opposite effect, a prominent cytoskeletal stabilization, is induced only by the cytotoxic necrotizing factors and the dermonecrotic toxin produced by *E. coli* and *B. bronchiseptica*, respectively (see Chapter 2).

The targets

Actin is one of the most abundant cellular proteins and occurs as polymeric (filamentous) F-actin or monomeric (globular) G-actin. Filaments are polymerized and

147

TABLE 8.1 Clostridial toxins acting on the actin cytoskeleton

Source	Toxin[a]	Cellular target(s)
ADP-ribosyltransferases[*]		
C. botulinum	C3 exoenzyme	Rho
C. limosum	C3-like toxin	Rho
C. botulinum	C2 toxin	Non-muscle G-actin
C. perfringens	Iota toxin	G-actin (also muscle)
C. spiroforme	Iota-like toxin	G-actin (also muscle)
C. difficile	CDT	Non-muscle G-actin
Glycosyltransferases[+]		
C. difficile	TcdA-10463	Rho, Rac, Cdc42, Rap[b]
C. difficile	TcdB-10463	Rho, Rac, Cdc42[c]
C. difficile	TcdB-1470	Rac, Rap, Ral[d]
C. sordellii	TcsL-9048	Rac, Cdc42, Rap, Ras[e]
C. sordellii	TcsL-82; TcsL-1522	Rac, Rap, Ral, Ras[f]
C. sordellii	TcsL-6018	Rac, Rap, Ral, Ras[e,g]
C. sordellii	TcsH-9048	Rho, Rac, Cdc42[h]
C. novyi	Tcnα–19402	Rho, Rac, Cdc42[i]

[*]Cofactor: NAD.

[+]Cofactor: UDP-Glc, except for Tcnα, which uses UDP-GlcNAc.

[a]**Nomenclature of LCTs:** The two toxins from *C. difficile*, originally designated toxin A and toxin B, appear in current reports under different abbreviations, e.g. ToxA and ToxB (Aktories 1997a), TxA and TxB (Souza *et al.*, 1997), or CdA and CdB (Fiorentini *et al.*, 1998). Since we knew that the number of ACSK-affecting clostridial toxins will grow, we introduced a systematic nomenclature in 1996 with four-letter abbreviations for the LCT family (Eichel-Streiber *et al.*, 1996). The initial T indicates <u>toxin</u> in contrast to the non-toxic enzymes, which are abundant in clostridia. The two middle letters are the initials of the producing organism. The fourth character alludes to the traditional toxin name and should be followed by a number specifying the strain. With this nomenclature, the *C. difficile* toxins are designated TcdA and TcdB, whereas TcsH and TcsL denote the *C. sordellii* haemorrhagic and lethal toxins, previously abbreviated HT and LT, respectively. The former 'alpha-novyi-toxin' from *C. novyi* is termed Tcnα. It can be seen from the table that different isoforms of the same toxin may have different substrate specificities, further underlining the need for this type of exact nomenclature. In this chapter the *C. difficile* toxins from VPI-10463 (reference strain used in laboratories world-wide) are denoted as TcdA and TcdB. The other LCTs are given with strain number when essential.

References to substrate specificities:

[b]Just *et al.* (1995a), Chaves-Olarte *et al.* (1997).

[c]Just *et al.* (1995b).

[d]Weidmann (1998).

[e]Hofmann *et al.* (1996).

[f]Popoff *et al.* (1996), Chaves-Olarte *et al.* (unpublished).

[g]Hofmann *et al.* (1996).

[h]Genth *et al.* (1996).

[i]Selzer *et al.* (1996).

depolymerized in a highly dynamic fashion, depending on cellular needs. A large number of actin-binding proteins operates at various levels to control the cytoskeleton. The various components and properties of the ACSK have been reviewed (Ayscough and Drubin, 1996; Molitoris, 1997; Molitoris *et al.*, 1997). The ACSK has structural functions, such as maintenance of the cell shape and motility. These functions depend on the controlled formation and breakdown of cellular stress fibres, lamellipodia, filopodia, contractile ring and focal adhesion plaques. The Rho family of small GTPases appears to be the most important regulators of these processes (Hall, 1998).

Small GTPases are molecular switches (Figure 8.1) which cycle between a GTP-loaded active state and a GDP-loaded inactive state. Accessory proteins control this cycle at certain key points. For instance, guanine nucleotide exchange factors (GEFs) activate small GTPases by promoting the GTP-loaded state, and GTPase activating proteins (GAPs) promote the inactive state by enhancing the intrinsic hydrolytic activity of the GTPase. In the case of Rho-family GTPases, the inactive state is maintained in the presence of GDP dissociation inhibitors (GDIs), which prevent the removal of GDP. Small GTPases co-ordinate intracellular signalling cascades initiated from the outside in response to metabolic and physiological needs of cells. This is accomplished by direct interaction of the active GTPases with downstream effectors, which usually initiate phosphorylation–dephosphorylation cascades (Hotchin and Hall, 1996; Van Aelst and

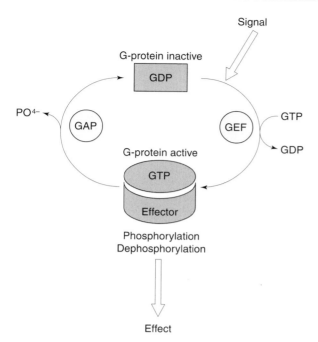

FIGURE 8.1 Schematic view of the GTPase cycle. See text for further explanation.

D'Souza-Schorey, 1997; Amano *et al.*, 1998; Hall, 1998; Ren and Schwartz, 1998, Scheffzek *et al.*, 1998; and see Chapter 2 for further details).

GENETIC ORGANIZATION OF THE TOXIN GENES

Binary actin-modifying ADP-ribosyltransferases

The two groups of toxins

The prototype of a binary toxin, *C. botulinum* C2, is composed of a light chain (designated component C2-I) and a heavy chain (component C2-II). C2 and the so-called 'iota-like' binary actin-modifying ADP-ribosyltransferases have been put into two separate classes of toxins. The reason is that polyclonal antisera directed against *C. perfringens* iota toxin (Ia, Ib) cross-detect the binary toxins of *C. spiroforme* (Sa, Sb) and *C. difficile* (CDTa, CDTb) but do not recognize the C2 toxin of *C. botulinum* (Popoff *et al.*, 1988). Moreover, sequence analysis of Ia and Ib (Perelle *et al.*, 1993, 1995) and CDTa and CDTb (Perelle *et al.*, 1997) has uncovered an 81% and 84% amino acid homology, respectively, between these components. The lower homology between iota-like toxins and C2 has been confirmed recently, after sequencing of the C2-I and C2-II genes (Fujii *et al.*, 1996; Kimura *et al.*, 1998).

The toxin genes

Component I of C2 encompasses 1293 bp, separated by 246 bp from the downstream 2163 bp component II (Fujii *et al.*, 1996; Kimura *et al.*, 1998). DNA-hybridizations, using C2 as the probe to identify homologous sequences in other clostridial species, identified such sequences only in *C. botulinum* serotypes C and D. None of the other *C. botulinum* serotypes, nor *C. butyricum*, *C. perfringens*, *C. hastiforme* or *C. difficile* contained genes homologous to C2. Hybridizations using the phages purified from serotype C and D strains were all negative, proving that C2 is not phage but chromosomally encoded (Fujii *et al.*, 1996; Kimura *et al.*, 1998). Comparative polymerase chain reaction (PCR)-amplifications with four different pairs of primers were performed and led to the differentiation of three groups of C2-II genes, indicating gene diversity between individual isolates. The relative molecular masses of purified C2-II protein from group 2 (95 000) and groups 1 and 3 (105 000) differ (Kimura *et al.*, 1998).

The genetic organization of the *C. perfringens* iota toxin, the *C. spiroforme* iota-like toxin and the binary C2 toxin is similar, with upstream catalytic domains separated by approximately 240 bp from their downstream binding domains (Perelle *et al.*, 1993; Gibert *et al.*, 1997). The two components are transcribed in the same direction. In the case of C2, promoters have been identified upstream of each component gene, so that transcription on polycistronic mRNA species is most likely (Kimura *et al.*, 1998). In contrast with this, long monocistronic mRNA species have been found after iota and *C. difficile* binary toxin transcription (Perelle *et al.*, 1997).

Description of the proteins

The C2-I and C2-II proteins exist independently in culture supernatant, and are not linked by either covalent or non-covalent bonds. C2-I was recognized as the mono-ADP-ribosyltransferase, and C2-II has binding activity targeted to cell surface receptors. The C2-I protein is composed of 431 amino acids with a relative molecular mass of 49 400 (Fujii *et al.*, 1996), and component C2-II of 721 amino acids forming a protein with a relative molecular mass of 80 500 (Kimura *et al.*, 1998).

In C2-I the NAD$^+$ binding domain EXXXXW typical of ADP-ribosyltransferases is positioned near the C-terminus, whereas in 'iota-like' toxins it is near the N-terminus (Perelle *et al.*, 1993). In iota toxin, Glu-380 and Glu-378 were identified as being part of the active site (Van Damme *et al.*, 1996). Similar investigations in C2-I have not yet been published. Sequence comparison of the binding domain of C2-II and the iota toxin component Ib revealed only 39% amino acid identity (Kimura *et al.*, 1998), whereas iota-toxin and iota-like toxins share at least 80% homology. The ATP/GTP binding motif GXXXXGK(T/S) is found in *C. perfringens* iota Ib (314-ASSDQGKT-321; Perelle *et al.*, 1993) and *C. spiroforme* Sb (318-ASTDQGKT-325, EMBL-Acc.No. X97969) but not in *C. botulinum* C2-II (Kimura *et al.*, 1998).

Evolutionary aspects

The fact that the amino acid homology between *C. botulinum* C2-I and *C. perfringens* iota Ia is only about 10% and between C2-II and Ib 39% suggests that C2 and the iota-like toxins developed independently over millions of years of evolution, although they belong to the same group of ADP-ribosylating enzymes (Fuji *et al.*, 1996; Kimura *et al.*, 1998). Considering the high homology of the coding sequences and the low similarity found in their flanking non-coding regions, evolution from a common ancestor gene has been discussed for *C. perfringens* iota toxin and the *C. difficile* binary toxin (Perelle *et al.*, 1997).

Rho-modifying ADP-ribosyltransferases

C3 or C3-like exoenzymes

Within this family of Rho-modifying ADP-ribosyltransferases the *C. botulinum* exoenzyme C3 is differentiated from C3-like exoenzymes. Using polyclonal antisera, *C. botulinum* C3 exoenzymes themselves can be divided roughly into two groups. Sequencing of the genes revealed conservation of approximately 90% within and not less than 60% between the two groups (Moriishi *et al.*, 1993). Both types of C3 enzymes occur in serotype C and D strains. The similarity between C3 and C3-like enzymes is lower; for instance, C3 displays a 33–35% homology to *S. aureus* Edin (Inoue *et al.*, 1991).

The toxin genes

In contrast to the binary toxins, the C3 exoenzyme of *C. botulinum* is encoded by phages that are found in serotype C and D strains. The nucleotide sequences of phages isolated from both serotype strains are highly homologous (Hauser *et al.*, 1993). Besides the C3 exoenzyme, the phages also carry the genes for neurotoxin production. After nitrosoguanidine treatment, a phage was generated that lacked the C3 sequence (Oguma *et al.*, 1975). Comparative sequence analysis revealed that a fragment of 21.5 kb is missing in the C3 cured phage (Hauser *et al.*, 1993). This distinct element displays the following similarities with the site-specific transposon family of Tn554: asymmetric ends, absence of inverted or direct terminal repeats and flanking by a 6-bp 'core motif' AAGGAG (Hauser *et al.*, 1995). The C3 gene can be said to be 'twice mobile', first

as part of a mobile phage and secondly as part of the 21.5-kb mobile element.

Description of the proteins

All C3 and C3-like exoenzymes are produced as single-chain precursor proteins of 244–250 amino acids. After cleavage of a signal peptide of approximately 40–50 amino acids, the resulting polypeptide of amino acid 210–215 is released (Nemoto *et al.*, 1991; Popoff *et al.*, 1991; Inoue *et al.*, 1991; Just *et al.*, 1995c). The molecular sizes of the mature proteins vary between 23 and 24 kDa. A glutamine residue (Aktories *et al.*, 1995; Saito *et al.*, 1995; Bohmer *et al.*, 1996) is highly conserved (around position 174) and is proposed as an active site involved in NAD$^+$ binding. Comparison of ADP-ribosylating enzymes, expressed by eukaryotes, bacteria and T-even bacteriophages, has allowed identification of the three conserved consensus sequences of their NAD-binding site (Domenighini and Rappuoli, 1996).

Evolutionary aspects

The similar molecular properties of C3 of *C. botulinum* and the C3-like enzymes from *C. limosum* (Just *et al.*, 1992a), *B. cereus* (Just *et al.*, 1992b), *S. aureus* (Inoue *et al.*, 1991; Sugai *et al.*, 1992) and *L. pneumophila* (Belyi *et al.*, 1991) suggest that C3-like enzymes were distributed by horizontal gene transfer. A common ancestral origin and a divergent evolution and dissemination of the C3 genes in these bacteria may have occurred (Hauser *et al.*, 1993). The C3 gene in *C. botulinum* is distributed between type C and D strains by phages that carry a 21.5-kb mobile element.

Comparison of the toxins of the large clostridial cytotoxin family

The genes of the toxins TcdA and TcdB of *C. difficile*, TcsL and TcsH of *C. sordellii* and Tcnα of *C. novyi* have been cloned and, except for *tcsH* of *C. sordellii*, their sequences have been determined (Dove *et al.*, 1990; Barroso *et al.*, 1990; Sauerborn and Eichel-Streiber, 1990; Eichel-Streiber *et al.*, 1990; Hofmann *et al.*, 1995; Green *et al.*, 1995). After the glycosyltransferase activity of LCTs was discovered (Just *et al.*, 1995) it became apparent that LCTs share DNA-sequence homologies and the effector mechanism, thus forming a novel family of toxins (Eichel-Streiber *et al.*, 1996). While sequencing other toxins of different *C. difficile* isolates, it was noticed that the toxins are polymorphic, showing different sequences in various isolates within one species (Rupnik *et al.*, 1998). The same picture emerges when observing target protein specificities of the TcsL preparations examined previously (Popoff *et al.*, 1996; Hofmann *et al.*, 1996).

Owing to the toxin polymorphisms within different as well as one and the same species, it is imperative to determine the exact origin of each toxin before its utilization. Only this makes the distinction between the biological effects of the different toxins possible and, in particular, reproducible (see comment in footnote to Table 8.1). As an example, the *C. difficile* toxins TcdB-10463 and TcdB-1470 (Eichel-Streiber *et al.*, 1995) and the *C. sordellii* toxins TcsL-82 (Popoff *et al.*,. 1996) and TcsL-9048 (Hofmann *et al.*, 1996) are discussed here. Looking at the sequences reveals TcdB-10463 and TcdB-1470 to be similar. Observing the pattern of modified GTPases as well as the morphology of the poisoned cells, however, TcdB is different from TcdB-1470, which induces the cytopathic effect (CPE) otherwise typical for *C. sordellii* TcsL-82. Comparing TcsL-82 and TcsL-6018, a similar CPE can be monitored, but the target GTPase spectrum is different (Popoff *et al.*, 1996; Hofmann *et al.*, 1996).

Genetic organization of the large clostridial cytotoxin genes

The currently published data on *C. difficile* and *C. sordellii* toxins indicate that they are encoded on chromosomal DNA. In contrast to this, a phage carries the *tcnα* gene (Schallehn *et al.*, 1980) and toxinogenic *C. novyi* strains can be cured of the phage. The derived atoxinogenic strain regains toxicity after reinfection with the isolated phage. During toxin preparation, this effect has to be taken into account, since toxinogenic strains may accidentally lose the phage. Tcnα has the lowest similarity to the other members of the LCT family (Hofmann *et al.*, 1995; Eichel-Streiber *et al.*, 1996). However, the organization of the genes seems to be similar for all LCTs. Upstream and downstream of the toxin genes, homologous open reading frames (ORFs) were discovered (Braun *et al.*, 1996, Green *et al.*, 1995; Hofmann *et al.*, 1995, and unpublished data). Because of the obvious high homology in the organization of the toxin genes, and since the analysis is most advanced for *C. difficile*, the genetic organization of the *C. difficile* pathogenicity locus (PaLoc) (see below and Figure 8.2) will be presented here as an example. For details about the genetic environment of the *C. difficile* PaLoc, the reader is referred to the publication of Braun *et al.* (1996).

Clostridium difficile *toxins are encoded by two separate genes*

Shortly after the discovery that *C. difficile* toxins cause diarrhoea in humans, both toxins were purified (Banno *et al.*, 1981, 1984; Taylor *et al.*, 1981; Sullivan *et al.*, 1982) and characterized separately. Cloning and sequencing of the toxin genes proved the existence of two separate toxin genes, *tcdA* and *tcdB*, that possess great homology (Eichel-Streiber *et al.*, 1992, and references therein). In earlier chemical studies, cross-reactive monoclonal antibodies were interpreted as evidence for such a homology. However, the contention that the larger toxin A was the progenitor of the smaller toxin B (Eichel-Streiber *et al.*, 1987) was contradicted by cloning experiments (Dove *et al.*, 1990; Barroso *et al.*, 1990; Eichel-Streiber and Sauerborn, 1990; Eichel-Streiber *et al.*, 1990).

The pathogenicity locus

Sequence comparison of toxinogenic and atoxinogenic isolates determined a fundamental difference between

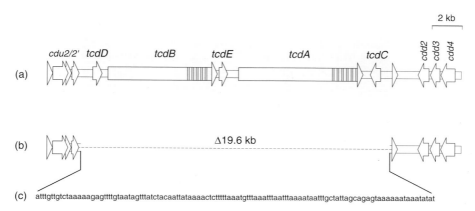

FIGURE 8.2 The pathogenicity locus (PaLoc) of *C. difficile*. (a) Open reading frames according to Braun *et al.* (1996), with the two toxin genes *tcdA* and *tcdB* and the three accessory genes *tcdC–E*. For *tcdD* a positive regulatory effect is proven (Moncrief *et al.*, 1997), *tcdC* is supposed to have negative regulatory influence on toxin transcription (and expression) (Hundsberger *et al.*, 1997). (b) Length of 19 600 bp of the PaLoc in toxinogenic *C. difficile* strains. (c) The 115 bp insertion replacing the PaLoc in atoxinogenic strains (Braun *et al.*, 1996).

such isolates (Hammond and Johnson, 1995; Braun *et al.*, 1996). So far, the examined strains showed one single integration site for the 19.6-kb sequence coding for the two toxin genes, *tcdA* and *tcdB*, as well as for three further toxin-associated genes, *tcdC, tcdD* and *tcdE* (Figure 8.2) The genetic element is designated the pathogenicity locus (PaLoc; Braun *et al.*, 1996) or toxic element (Hammond and Johnson, 1995). Integration of the PaLoc into its insertion site always occurs in one orientation. In atoxinogenic strains, the PaLoc sequence is replaced by a 115-bp sequence containing a direct repeat but otherwise no remarkable features (Braun *et al.*, 1996).

The accessory proteins within the pathogenicity locus

Owing to the high relative molecular mass of the toxins (578 000 = TcdA 308 000 + TcdB 270 000), it was expected that production of the toxins would involve regulation at the DNA and/or protein level. For the regulatory effect on toxin production the three accessory proteins TcdC-E were identified as likely candidates. There are no homologies of TcdC and TcdE to other known proteins in the database. These searches identified homologous proteins solely for TcdD. TcdD shows homologies with other DNA-binding regulatory proteins, their typical length and a helix-turn-helix motif typically inherent in DNA-binding regulatory proteins (Hundsberger *et al.*, 1997). Particularly noticeable is the homology to ORF22, a positive transcription regulator of botulinum toxins, and to a similar protein identified upstream of the tetanus toxin gene (see Chapter, 9 and 10 in this book).

Transcription analysis of the tcd genes

The latter claim was supported by extensive transcriptional analyses, using the reverse transcriptase (RT)-PCR technique (Hundsberger *et al.*, 1997) and reporter gene assays in the host *E. coli* (Moncrief *et al.*, 1997)

The direction of transcription of the *tcdC* gene is opposite to that of the other four *TCD* genes. Owing to the organization of the toxin genes within the PaLoc, transcription and translation of TcdB versus TcdA is clearly shifted toward the *tcdA* gene. This concurs with the findings that transcription of *tcdA* and *tcdB* occurs mono- and polycistronically and that both toxins are purified from culture supernatants at a ratio of 1:2–3 (Hundsberger *et al.*, 1997).

During the early growth phase of *C. difficile* the *tcdC* gene shows a high transcription rate, which is lowered during the late logarithmic and stationary growth (Hundsberger *et al.*, 1997). The *tcdC* gene and the other four genes *tcdA, tcdB, tcdD* and *tcdE* of the PaLoc are

inversely transcribed during bacterial growth. From these observations, a model of the regulation of toxin expression was developed (Hundsberger *et al.* 1997), wherein the positive regulation by TcdD is complemented by a negative regulation probably governed by TcdC (Figure 8.3).

The promoters of the toxin genes feature an exceptionally long spacing between the TATA-box and the starting point of translation, which is typical for clostridial toxins (Hundsberger *et al.*, 1997). Since some of *C. difficile* promoters were shown also to be active in the Gram-negative host *E. coli* (Dailey and Schloemer, 1988), Moncrief *et al.* (1997) have conducted reporter gene assays in the heterologous host *E. coli*. These showed a positive regulatory effect of TcdD on the transcription of both *ptcdA* and *ptcdB*, the promoters of the toxin genes.

Regulation of toxin formation

A continuous production of toxins would be strenuous for the organism and also a waste of resources. An additional puzzle is the simultaneous production of two such similar molecules that obviously have different functions. Thus, it seems logical that the production of the toxins is not continuous but enhanced during the late-logarithmic growth phase of cultures. Contrary to the enterotoxin of *C. perfringens*, the toxins are not essential components of the sporulation apparatus, since atoxinogenic *C. difficile* isolates are able to sporulate. Thus, it is not yet known which physiological tasks are assigned to the respective *C. difficile* toxins.

The above-mentioned reporter gene assays in the heterologous host *E. coli* gave a clear indication of the positive regulatory function of TcdD. In these assays

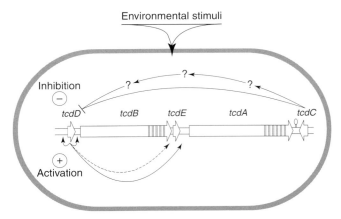

FIGURE 8.3 Putative regulatory circuit of the *C. difficile* toxins. See text for further explanations and Hundsberger *et al.* (1997) for experimental details.

the TcdA-repeat was cloned under the control of the *tcdA* or *tcdB* promoters. Expression of the TcdD protein *in trans* had a positive regulatory effect on the expression of the TcdA-repeat, which was detected with specific monoclonal antibodies (Moncrief *et al.*, 1997). Owing to the pattern of *tcdC* versus *tcdD* gene transcription and the molecular attributes of TcdC, it was assumed that TcdC, as an acidic molecule, interacts (directly or indirectly) with TcdD, thereby blocking its positive regulatory effect during early growth phases (Hundsberger *et al.*, 1997). As soon as more TcdD is produced, excessive transcription of all *tcd* genes except *tcdC* occurs. In contrast, *tcdC* transcription is actively blocked by the run-through transcription of *tcdA* into *tcdC* and the ensuing generation of anti-sense RNA (Soehn *et al.*, 1998).

The model in Figure 8.3 leaves open the position of the inducers of the regulatory cycle. In analogy to other bacterial systems, we firmly believe that there are environmental influences inducing the regulatory cycle and thereby toxin production. In spite of an extensive search, only a few triggers of toxin production have been noted. Stressful events such as increased oxygen tension, elevated temperature, limited nutrients, low concentrations of antibiotics and a high density of bacteria (in culture and in the gut) seem to be such triggers (Onderdonk *et al.*, 1979; Nakamura *et al.*, 1982; Moncrief *et al.*, 1997). However, Yamakawa *et al.* (1996) showed that, in minimal medium under limited supply of biotin, *C. difficile* grow to lower density, whereas toxins were produced in much higher amounts.

Domain structure of the toxins

Confirming an earlier hypothesis (Eichel-Streiber *et al.*, 1996), our three-domain model for the LCTs was substantiated in recent reports. Thus, LCTs contain a repetitive C-terminal ligand domain, a central translocation domain and an N-terminal catalytic domain. This structure is valid for all toxins of the family of LCTs (Figure 8.4). It reflects the way that the toxins enter the cell from the outside, are taken up by receptor-mediated endocytosis, translocated to the cytoplasm and release their enzymic activity within the cell. This process shows many similarities to the intoxication processes for other bacterial toxins described in this volume.

Ligand domain

A repetitive C-terminal domain of all LCTs can be discerned easily from the rest of the toxin molecule by its characteristic structure. For these proteins, combined repetitive oligopeptides (CROPs; Eichel-Streiber, 1993 and references therein) were defined, the central motif of which consists of a triad of aromatic amino acids

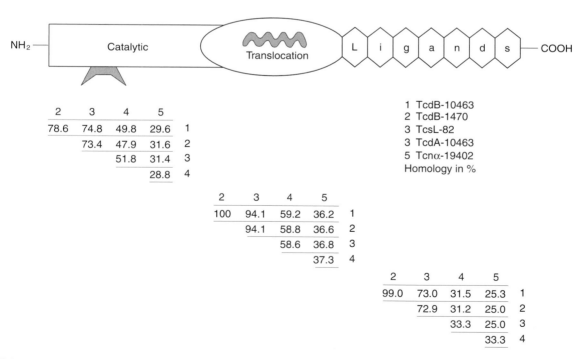

FIGURE 8.4 Three-domain model and homology between known LCTs. Three-domain model valid for all toxins of the LCT family as proposed by Eichel-Streiber *et al.* (1996). The numbers below the figure are the calculated homology (%) between the LCTs listed.

(mostly YYF), which are either 20–22 or 50 amino acids long. The repetitive character of this toxin fraction is more evident in TcdA. Its C-terminal domain is constructed in a repetitive manner, not only on the protein basis but also in the gene itself, so that for *tcdA*, short repetitive oligonucleotides (SRONs; Eichel-Streiber and Sauerborn, 1990) were defined. When they were first described, it was suggested that this part of the toxin was an area for modification by homologous recombination (Eichel-Streiber and Sauerborn, 1990). This assumption was corroborated by the finding that TcdA varies in length (Rupnik *et al.*, 1997). At the same time, this domain of the PaLoc is the only region where length variations occur more frequently (Rupnik *et al.*, 1998).

In one of the first publications on the DNA sequence of toxin A (*tcdA*), its homology to glucosyl-transferases (GTF) of *Streptococcus mutans* was also described (Eichel-Streiber and Sauerborn, 1990). The homologies were at first confined to the repetitive areas of the GTFs and to the repetitive part of the TcdA. This region of the LCTs participates in the interaction between the cellular receptors and the toxin (Price *et al.*, 1987; Krivan *et al.*, 1986; Sauerborn *et al.*, 1997). Taking the known findings about TcdA as an example, and also considering the homologies of all LCTs, an interaction with oligosaccharide side-chains of the respective receptors was postulated for the C-terminal area of all LCTs (Eichel-Streiber *et al.*, 1996). Such an interaction has clearly been proven for TcdA (Krivan *et al.*, 1986; Tucker and Wilkins, 1991). A direct fluorescent image of the binding of the repetitive domain to cells was accomplished by using monoclonal antibodies binding to different epitopes within TcdA (Sauerborn *et al.*, 1997). One particular monoclonal antibody that blocked binding of the TcdA by interaction with epitopes in the CROPs, thus neutralizing the effect of the toxin, was especially helpful in achieving this goal (Sauerborn *et al.*, 1997). That the binding can be competitively prevented by adding recombinant fragments of the C-terminal TcdA domain has been shown by several groups (Ryan *et al.*, 1997; Sauerborn *et al.*, 1997).

Translocation domain

A comparative analysis of the toxins indicated the existence of a second domain, marked by an accumulation of hydrophobic amino acids and containing all the transmembrane helices of the molecules predicted by computer analysis. It was therefore postulated that this central domain of the toxins is essential in the translocation of the toxins across the cell membrane (Eichel-Streiber *et al.*, 1996). Experimental evidence for this hypothesis has still to be gained.

Catalytic domain

By comparative sequencing of two TcdB variants with different cytotoxic effects, the N-terminal 879 amino acids were predicted as the domain carrying the enzymic activity (Eichel-Streiber *et al.*, 1995). By now, subclonings of the respective *tcdB* gene fragments have restricted the area of catalytic activity to the 546–557 N-terminal amino acids (Hofmann *et al.*, 1997; Wagenknecht-Wiesner *et al.*, 1997). The shortest active fragment comprises 467 amino acids (Wagenknecht-Wiesner *et al.*, 1997).

The enzymic reaction requires an interaction between toxin and target GTPases, the hydrolysis of activated sugars (most often uridine-diphospho-glucose, UDP-Glc) and the covalent attachment of the sugar to a threonine within the effector domain of the GTPase. In mannosyltransferases, site-directed mutations revealed that a motif DXD was important for transfer of the sugar to the target protein (Wiggins and Munro, 1998). Comparing all LCTs of different origins, one area from amino acid 261–295 is highly conserved within all toxin sequences. By mutagenesis of this motif in TscL-6018 and TcdB-10463, transferase activity can be diminished or even eliminated by substituting alanine for the aspartate in either or both positions (Busch *et al.*, 1998). The DXD motif seems to participate in co-ordination of the divalent cation shown to increase the hydrolase activity (Genth *et al.*, 1996; Ciesla and Bobak, 1998) and/or in the binding of UDP-Glc.

With these experiments, one of the three areas of the toxins involved in the catalytic process could be identified. According to current studies, the area related to the hydrolase activity seems restricted to amino acids 516–546 (Hofmann *et al.*, 1998). Chimeras between TcdB and TcsL identified the region that may be responsible for the protein substrate specificity of these LCTs. The exchange of amino acids 364–516 between TcsL-6018 and TcdB-10463 resulted in chimeric toxins with interchanged substrate patterns (Hofmann *et al.*, 1998). However, the identity of this region between TcsL-82 and TcdB-1470, which has a TcsL-like substrate pattern, is 75%, i.e. very similar to the identity in the same region (79%) between TcdB-1470 and TcdB-10463. Since the substrate pattern of TcdB-10463 is different from that of both TcsL-82 and TcdB-1470, additional determinants of specificity should exist in the LCTs.

Genetic polymorphism among large clostridial cytotoxins

Gross changes in the *C. difficile* PaLoc sequence were found, especially in the *tcdA* gene region, which exhibits different lengths at its highly repetitive 3'-end. At a second site at the 5'-end of the PaLoc, excessively frequent but less significant length variations are observed (Rupnik *et al.*, 1998). So far, only two isolates are known that contain additional DNA in the PaLoc, a 1.1-kb insertion between *tcdE* and *tcdA* (Soehn *et al.*, 1998) and an insertion of 2 kb within the toxin A gene (Mehlig and Eichel-Streiber, unpublished data). From that, the known size variation of the PaLoc (standard size 19.6 kb) between 15.2 and 21.6 kb can be inferred.

Sequencing data (Eichel-Streiber *et al.*, 1995; Rupnik *et al.*, 1997, 1998) and differences in enzymic activity show that clostridial toxins are polymorphic. These polymorphisms were first shown when studying *C. difficile* toxins producing a CPE typical for *C. sordellii* (Eichel-Streiber *et al.*, 1995). Defining the DNA sequence of TcdB-1470 of *C. difficile* isolate 1470 showed a surplus of 156 divergences in the 868 N-terminal amino acids. In contrast, the rest of the molecule shows eight deviations in 1498 amino acids. Not only the altered CPE, but also the polymorphisms in the restriction sites of the respective gene fragment are linked to these changes. The restriction fragment length polymorphisms thus developing were used to study systematically an extended collection of *C. difficile* strains (Rupnik *et al.*, 1998). Eleven different toxinotypes were defined and this typing may now become useful for the epidemiological differentiation of clinical isolates. More important in this context, however, is the fact that the various toxin types also show differences in their enzymic properties, also resulting in divergences of the GTPase modification spectrum (Rupnik and Eichel-Streiber, unpublished data).

As a résumé of the genetic analysis and application of the toxins in biological assays (see below), it should again be emphasized that LCT-toxins, even of the same species, have polymorph genes and differing enzymic properties. Thus, comparison and repetition of published data is only possible if the partners are exactly defined (as, for example, in Schmidt *et al.*, 1998).

MOLECULAR MODES OF ACTION OF TOXINS

The toxins dealt with here induce a collapse of the ACSK by either ADP-ribosylation or glycosylation of target proteins. The primary targets for these covalent reactions differ, but the modification in each case leads to either complete cell rounding with massive detachment of cells or retraction only of the cell body, leaving long protrusions, which keep the cells

attached to solid supports, at least during the first 24 h of intoxication. The target modifications are essentially irreversible, and newly synthesized GTPases most probably are immediately modified. Thus, the cellular effects also in general are irreversible; the rounded or retracted cells do not return to their original morphology. Neither are they able to undergo cytokinesis, because they fail to assemble the contractile ring. Thus, cell proliferation is blocked and the cells begin to deteriorate after hours to days, depending on the toxin. Below, we will treat the clostridial toxins as two large groups: the ADPRTs and the LCTs (Table 8.1). The greater emphasis on the LCTs reflects their more recent discovery. Furthermore, the molecular/structural features of ADPRTs are dealt with elsewhere in this volume (see Chapter 3).

ADP-ribosyltransferases

ADP-ribosyltransferases catalyse the hydrolysis of NAD^+ and the transfer of the resulting ADP-ribose on to various target proteins. The catalytic sites of the clostridial ADPRTs have been studied recently by photoaffinity labelling with [carbonyl-^{14}C]NAD$^+$ (reviewed in Aktories, 1997a). Glu-residues have pivotal roles in the active sites of ADPRTs. The C2 and C3 toxins are mono-ADPRTs, and ADP-ribosylation by these transferases is reversible only at high concentrations of nicotinamide (30–50 mM), i.e. not under physiological conditions.

C2 and C2-like toxins ADP-ribosylate G-actin

In general, GTP-binding proteins serve as targets for bacterial ADPRTs. The C2-like toxins represent an exception to this rule as they attack actin, which is an ATP-binding protein. ADP-ribosylation takes place on the Arg177 of monomeric G-actin, which is the target of all toxins in this family (Aktories *et al.*, 1992). However, the toxins differ with respect to the actin isoforms attacked. Six highly homologous mammalian isoforms are known: skeletal and cardiac muscle α-actins, smooth muscle α- and γ-actin, and cytoplasmic (non-muscle) β- and γ-actin. Although Arg177 is conserved in all of the isoforms, the *C. botulinum* C2 toxin specifically ADP-ribosylates cytoplasmic actins and smooth muscle γ-actin, but not the α-actin isoforms. In contrast, the *C. perfringens* iota-toxin ADP-ribosylates all actin isoforms studied (Aktories *et al.*, 1992).

ADP-ribosylation at Arg177 inhibits the intrinsic ATPase activity of actin (Reuner *et al.*, 1996a). However, the reason that these toxins effectively produce a depolymerization of cellular actin filaments is that ADP-ribosylated G-actin monomers bind to and 'cap' the growing (barbed) end of F-actin filaments. The 'capping' prevents the attachment of additional monomers to this faster growing end of F-actin. As the polymerization is less efficient at the other (pointed) end of F-actin, the critical actin concentration for polymerization is increased. Thus, ADP-ribosylated G-actin becomes trapped in a monomeric pool. ADP-ribosylation of G-actin also inhibits the nucleation activity of the gelsolin–actin complex, further adding to the net depolymerization of F-actin in cells treated with these toxins. The redistribution from F- to G-actin leads to a collapse of the ACSK and rounding up of the intoxicated cells. Microinjection of ADP-ribosylated G-actin induced the same cellular effect as the C2-toxin (Kiefer *et al.*, 1996; Reuner *et al.*, 1996b). In conclusion, the C2-toxins prevent actin filament polymerization and induce a depolymerization of already existing filaments with a concomitant increase in the G-actin content of the cells (see original references in Aktories, 1997c).

Exoenzyme C3 and C3-like transferases ADP-ribosylate Rho protein

The *C. botulinum* exoenzyme C3 and the C3-like enzymes from other species mono-ADP-ribosylate members of the Rho protein subfamily (RhoA, B and C) at Asn41, thereby inactivating this GTP-binding protein (for earlier references see Aktories and Koch, 1997; and Chapter 2). Recombinant or purified Rho proteins are more efficiently ADP-ribosylated in the GDP-loaded than in the GTP-loaded state. In contrast, ADP-ribosylation of Rho in cell lysates is enhanced by GTPγS or GTP, probably because they facilitate the dissociation of GDI from Rho. Accordingly, free Rho protein is a better substrate for C3 than GDI-associated Rho. ADP-ribosylation of cellular Rho induces retraction and rounding of the cell, with protrusions remaining around cell bodies (Aktories and Koch, 1997).

Glycosyltransferases (large clostridial cytotoxins)

Cytopathic effects of large clostridial cytotoxins

The CPE induced by LCTs on cells in culture is one of the most studied biological effects of these toxins. The CPE of TcdB consist of the collapse of the ACSK accompanied by cytoplasmic retraction and cell rounding (Figure 8.5). Since long protrusions remain radiating around the cell body this particular morphology was named actinomorphic, arborizing or neurite-like. TcdA and Tcnα elicit a similar CPE whereas, in contrast, cells treated with TcsL round up without leaving any protrusions (Figure 8.5). Differences in the cytotoxic potencies are also found within the LCT family (see below).

Enzymic mode of action of large clostridial cytotoxins

The enzymic activity of the LCTs was only recently discovered (Just *et al.*, 1995a,b), although their ACSK-collapsing effect was described much earlier (Thelestam and Brönnegård, 1980; Ottlinger and Lin, 1988; Oksche *et al.*, 1992). The clue to the primary GTPase target came from the observation that C3-induced ADP-ribosylation of Rho was prevented in lysates of cells treated with TcdB (Just *et al.*, 1994) or TcdA (Just *et al.*, 1995d). This, together with the fact that the TcdB/TcdA-induced CPEs greatly resembled the C3-induced CPE, pointed to Rho as a plausible target also for these toxins. Mass spectroscopy was the key method for defining the molecular mode of modification, which was the covalent attachment of a glucose moiety from UDP-Glc to a conserved threonine (Thr-35/37) in the effector domain of small GTPases. Shortly after this modification had first been described for TcdB (Just *et al.*, 1995b), the glycosyltransferase activities of TcdA (Just *et al.*, 1995a), TcsL (Popoff *et al.*, 1996; Hofmann *et al.*, 1996), TcsH (Genth *et al.*, 1996) and Tcnα (Selzer *et al.*, 1996) were reported.

This concept was confirmed by analysis of a mutant cell, resistant to the *C. difficile* toxins. Its resistance was due to a low cellular level of UDP-glucose (Chaves-Olarte *et al.*, 1996), caused by a point mutation that inactivates the UDP-glucose:pyrophosphorylase gene (Flores-Diáz *et al.*, 1997). Thus, the mutant cell is resistant to all LCTs using UDP-glucose as cofactor, but sensitive to Tcnα, which transfers *N*-acetyl-glucosamine from UDP-*N*-acetyl-glucosamine (UDP-GlcNAc) (Selzer *et al.*, 1996). The LCTs are capable of hydrolysing the sugar nucleotide alone (Just *et al.*, 1995a), in analogy with the *in vitro* cleavage of NAD^+ by the ADP-ribosyltransferases. We found TcdB to be considerably more active than TcdA, both as a glucosyltransferase (see below) and as glucohydrolase (Chaves-Olarte *et al.*, 1997). In agreement with this, the hydrolase activity of TcdB was recently reported to be approximately five times greater than that of TcdA, whereas the K_m for UDP-Glc was similar for both toxins (Ciesla and Bobak, 1998).

Large clostridial cytotoxins glycosylate several small GTPases

In contrast to C3, which modifies only Rho, every LCT modifies more than a single substrate. Once UDP-Glc and UDP-GlcNAc were identified as cofactors for LCTs, these nucleotides in radioactive form were used to identify the actual substrates of each toxin. Rho, Rac and Cdc42 were found to be modified by TcdB, TcdA (Just *et al.*, 1995a, b) and Tcnα (Selzer *et al.*, 1996). TcdA, in addition, labels Rap1 and Rap2 (Chaves-Olarte *et al.*, 1997), whereas TcsL modifies Rac, Ras, Rap and Ral (Popoff *et al.*, 1996; Hofmann *et al.*, 1996; and see Table 8.1). A correlation between the modified substrates and the different types of CPE induced by each toxin has been suggested. Inactivation of Rho alone may be responsible for the characteristic arborizing effect, as overexpression of RhoA in target cells counteracted the effect of TcdB (Giry *et al.*, 1995). The relevant target for induction of the rounding CPE by TcsL is not yet known.

The glucose moiety is transferred to a conserved threonine (35 in Ras, 37 in Rho), which is located within the effector region of the small GTPases. This region undergoes a conformational change, depending on whether the protein is loaded with GTP or with GDP. In the GTP-loaded state, the lateral chain of the conserved Thr35/37 is facing the internal core of the GTPase whereas, in the GDP-loaded state, it faces the external medium. This phenomenon explains why the small GTPases are more easily glycosylated in the GDP-loaded than in the GTP-loaded state (Aktories and Just, 1995).

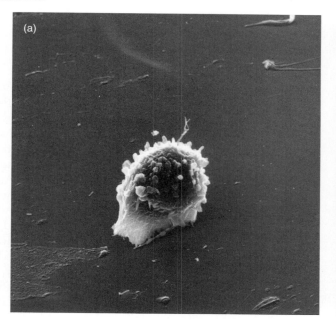

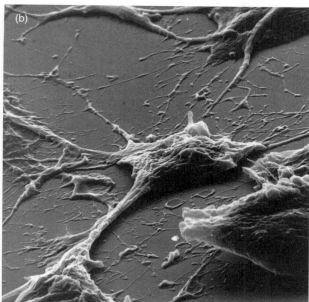

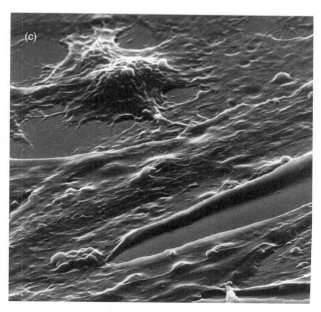

FIGURE 8.5 Cytopathic effect induced by LCTs. Chinese hamster Don fibroblasts were treated with TcsL-1522 (a) or TcdB-10463 (b) or left untreated (c). The toxin-induced morphology was analysed by scanning electron microscopy.

UPTAKE OF CLOSTRIDIAL
TOXINS IN CELLS

As the clostridial ADPRTs and LCTs have intracellularly located cofactors and target proteins, they need to be internalized to exert their cytotoxic actions. In general, intracellularly acting protein toxins enter cells by a multistep process: binding to specific receptors, uptake by endocytosis and transport in vesicles to their final destinations. Enzymic processing (nicking) of toxin molecules may be required and take place intravesicularly or in the cytosol after toxin translocation across the membrane of the terminal vesicle (for details of this cellular process, see Chapter 4).

Cellular internalization of ADP-ribosyltransferases

C2-toxin and the related ADPRTs have the enzymic and receptor-binding activities in two separate components, C2I and C2II. The C2II component has to be activated by proteolytic cleavage, after which it can bind to the target cell surface. The association of activated C2II with its receptor creates a binding site for the catalytic component C2I (Ohishi and Yanagimoto, 1992). Recently, the N-terminal portion of C2I (C2IN) was identified as the contact site for binding to C2II, whereas the catalytic activity of C2I was located in the C-terminal part (Barth et al., 1998). A C3–C2IN fusion protein was constructed, which, when applied together with C2II, efficiently intoxicated cells by Rho-specific ADP-ribosylation (Barth et al., 1998), confirming that the low cellular activity of C3 depends on its lack of a binding domain (see below).

The nature of the receptor for C2II is not known, but a C2-toxin resistant cell line (CHO-C2RK14) with a putative receptor deficiency has been described (Fritz et al., 1995) and it was also insensitive to the C3–C2IN fusion protein (Barth et al., 1998). After cell surface binding, the entry of C2 toxin into the cell appears to take place by receptor-mediated endocytosis (Simpson, 1989; Ohishi and Yanagimoto, 1992). Both C2I and C2II reach the same endosomal compartment (Ohishi and Yanagimoto, 1992). The C2II component is capable of inducing cation-selective channels in artificial membranes (Schmid et al., 1994). Analogous with postulates for other endocytosed toxins, the channel-forming ability of C2II may play a role in translocation of the catalytic component across the vesicular membrane, although no further details of this entry mechanism are known. The binding components of iota toxin and the C. spiroforme iota-like toxin (but not that of C2) are interchangeable. They can even translocate the C. difficile ADPRT into cells (Popoff et al., 1988).

The C3 and C3-like enzymes lack a typical binding subunit or domain and apparently do not bind to specific receptors on cell membranes. Indeed, when applied extracellularly, they have a considerably lower cytotoxic potency than when microinjected. Certain cultured cells do respond to C3, but only when exposed to high concentrations (>10 µg ml^{-1}) for a long time (usually at least 24 h). The cellular entry of C3 probably takes place non-specifically, e.g. via pinocytosis (Aktories, 1997b), but its uptake can be increased by permeabilizing cells with a suitable pore-forming agent.

Cellular internalization of large clostridial cytotoxins

Binding of large clostridial cytotoxins

TcdA binds to the trisaccharide Galα1–3Galβ1–4GlcNAc (Krivan et al., 1986), which is expressed on rabbit intestinal brush border membranes, but not in human tissues. However, TcdA also binds to the Galβ1–4GlcNAc structure found on certain human blood group antigens (Tucker and Wilkins, 1991). Moreover, it bound with similar affinity to GalNAcβ1–3Galβ1–4GlcNAc (Teneberg et al., 1996), which is also present in human tissues. Thus, the disaccharide Galβ1–4GlcNAc appears to be a minimum structure required for binding of TcdA. Recently, ^{3}H-TcdA was reported to bind specifically to human colonic epithelial cells isolated from endoscopic biopsies (Smith et al., 1997). Galactosidase treatment reduced the binding, consistent with an earlier suggestion (Tucker and Wilkins, 1991) that Lewis X, Y and I antigens might be involved in TcdA binding to the human intestine. A rabbit ileal receptor protein was demonstrated (Pothoulakis et al., 1991) and more recently identified as sucrase-isomaltase (Pothoulakis et al., 1996). However, this enzyme is absent from several cell types that are sensitive to TcdA, and it is not expressed in the colon of either rabbit or human (Pothoulakis et al., 1996). Thus, the human colonocyte membrane protein or lipid carrying the TcdA receptor structure remains to be identified.

Since no carbohydrate structure with specific affinity for TcdB has been identified, it was suggested that this toxin was associated non-specifically with membrane lipid components (Lyerly and Wilkins, 1995). However, we recently demonstrated the specific binding of TcdB to unknown receptors on the surface of fibroblasts (Chaves-Olarte et al., 1997). The variant toxin TcdB-1470 appears to bind to the same receptor as TcdB (Chaves-Olarte et al., 1999), which is reasonable in view of the 99% sequence identity between receptor-binding domains of these two toxins (Eichel-Streiber et al., 1995). The molecular properties and nature of the specific receptor used by TcdB have not been established.

However, it must be broadly distributed, since all mammalian cells tested so far (except for the above-mentioned UDP-Glc-deficient mutant) are sensitive to TcdB. The cellular receptors for other LCTs have not been studied.

Uptake of large clostridial cytotoxins

TcdA (Henriques *et al.*, 1987), TcdB (Florin and Thelestam, 1986) and TcsL (Popoff *et al.*, 1996) require passage via an acidic intracellular compartment(s) in order to intoxicate cells after extracellular addition. EM studies suggested that TcdA is taken up by receptor-mediated endocytosis (Eichel-Streiber *et al.*, 1991). There is no reason to believe that this general principle would not hold for the other LCTs. TcdB is apparently delivered to lysosomes before its release to the cytosol (Florin and Thelestam, 1986), whereas the routes of TcdA and TcsL are not clear. All LCTs tested so far are able to intoxicate cells upon microinjection, indicating that they can act intracellularly, without specific intravesicular processing (Müller *et al.*, 1992; Chaves-Olarte *et al.*, 1997). Recent studies on fragments of TcdA, TcdB and TcsL support the notion that these toxins may not need any vesicular activation since, upon microinjection, the catalytic N-terminal fragments had cytotoxic potency equal to that of the holotoxins (Hofmann *et al.*, 1997; Chaves-Olarte *et al.*, 1997). However, this does not exclude a need for intravesicular enzymic cleavage of the large holotoxin to enable release of the smaller catalytic N-terminal fragment into the cytosol. The nature of this putative processing and the molecular details of toxin membrane translocation are not known. The cellular uptake of other LCTs has not been studied.

Why do large clostridial cytotoxins have different cytotoxic potencies?

Although TcdA and TcdB have an identical enzymic mode of action and almost identical substrate patterns, they differ remarkably in cytotoxic potencies, TcdB being 10^3–10^4 times more potent than TcdA. This difference was thought previously to depend exclusively on receptor differences. However, recent work demonstrated that the enzymic potency differs considerably between the toxins (Chaves-Olarte *et al.*, 1997). TcdB labels its substrates in the test tube 100-fold more efficiently than TcdA. Moreover, a 100-fold difference in cytotoxic potencies remained between TcdB and TcdA, even when the toxins were microinjected into cells (Chaves-Olarte *et al.*, 1997).

TcdB also has a 10^3–10^4-fold higher cytotoxic potency than TcsL-1522. In this case, the difference in cytotoxicity may arise from the different substrate patterns

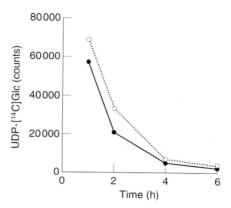

FIGURE 8.6 Glucohydrolase activity of LCTs. UDP-[^{14}C]glucose was incubated with 100 μg ml^{-1} purified TcdB-10463 (○) or TcsL-1522 (●). The reaction was stopped at the indicated times by heating at 95°C. Remaining non-hydrolysed UDP-[^{14}C]glucose was adsorbed on to activated charcoal and quantified by scintillation counting.

of the toxins. However, the variant toxin TcdB-1470, which has a TcsL-like substrate pattern but showing 99% identity with the C-terminal domain of TcdB, is equal in potency, in terms of cytotoxicity, to TcdB (Chaves-Olarte *et al.*, 1999). Thus, the difference in cytotoxicity between TcdB and TcsL depends either on a difference in enzymic potency, as in the case of TcdA, or on a difference in the binding and internalization into cells. The second explanation is more plausible, since TcdB and TcsL turn out to have very similar enzymic potencies *in vitro* (Figure 8.6) as well as cytotoxic titres when microinjected (Chaves-Olarte *et al.*, 1999). In conclusion, the cytotoxic potencies of LCTs depend on both the enzymic potencies and the efficiencies of receptor-binding, which differ for different LCTs.

MOLECULAR AND CELLULAR CONSEQUENCES OF INTOXICATION

Which are the molecular consequences of GTPase modifications?

The consequences of LCT-mediated glucosylation for the intrinsic properties of small GTPases, as well as for their interactions with modulator proteins, have been studied in some detail. Glucosylation of Ras weakens its affinity for GTP by accelerating the nucleotide dissociation rate, and glucosylated Ras had a four times slower intrinsic GTPase activity (Popoff *et al.*, 1996). If the GTPase activity is impaired, then an activation rather than an inactivation of the molecule would be expected, as was later shown to occur upon deamidation of Gln-63 in Rho by the *E. coli* cytotoxic

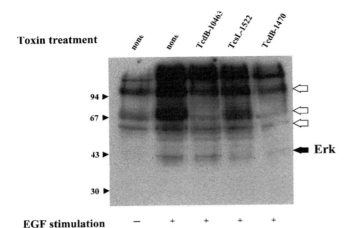

Toxin treatment

94 ▶
67 ▶
43 ▶ ◀ **Erk**
30 ▶

EGF stimulation — + + + +

FIGURE 8.7 Effect of LCTs on EGF-induced tyrosine phosphorylation. Serum-starved 3T3 cells were left untreated or intoxicated with TcdB-10463, TcsL-1522 or TcdB-1470. Then cells were stimulated with EGF for 5 min, lysed in Laemmli sample buffer and proteins resolved by 12.5% SDS–PAGE. Proteins were transferred to a nitrocellulose membrane and probed with anti-phosphotyrosine antibodies. Immune complexes were visualized with a chemiluminescence Western blotting kit using peroxidase-labelled secondary antibody.

necrotizing factor (Flatau *et al.*, 1997). Although the findings by Popoff *et al.* (1996) did not explain why the glucosylation on Thr-35 impairs the functional activity of the modified small GTPase, it was demonstrated that Ras signalling to the mitogen-activated protein (MAP) kinases ERK1 and ERK2 was prevented (Popoff *et al.*, 1996). In accordance with this, Herrmann *et al.* (1998) showed recently that Ras-glucosylation blocked the binding of the Ras effector Raf, thereby inhibiting the Ras-mediated signalling cascade.

A comparative study of the consequences of ADP-ribosylation and glucosylation of Rho, Rac and Cdc42 demonstrated interesting differences between these two types of modification (Sehr *et al.*, 1998). ADP-ribosylation decreased the rate of GTPγS release from Rho, whereas the release was increased by glucosylation. ADP-ribosylation of Rho increased its GTPase activity, whereas glucosylation reduced it. ADP-ribosylation did not affect the interaction of RhoA with the binding domain of protein kinase N. In contrast, Rho-glucosylation inhibited this interaction. The functional inactivation of Rho family GTPases by glucosylation therefore appeared to be mainly caused by an inhibition of the interaction between the GTPase and its effector protein(s) (Sehr *et al.*, 1998), in accordance with the above-mentioned findings on Ras (Herrmann *et al.*, 1998).

The blockade in Ras interaction with downstream effectors, in particular with Raf, explains the ability of TcsL to block the epidermal growth factor (EGF)-mediated phosphorylation of MAP kinases. This particular cascade is not inhibited by the other LCTs, which do not modify Ras. However, these LCTs inhibit other growth factor-induced phosphorylation events (Figure 8.7). The use of different LCTs may thus help to identify unknown downstream effectors in the signalling cascades mediated by small GTPases such as Rho, Rac and Cdc42. For further discussion of downstream effects, the reader is referred to the section dealing with the toxins as tools.

Cellular consequences of actin depolymerization

Studies on the cytotoxic effects of actin-ADP-ribosylating toxins up to 1997 have been reviewed and summarized as a model (Aktories, 1997c, and references therein). As discussed above, a collapse of the ACSK also results from glycosylation of small GTPases by the LCTs. The exact pathways leading to cell death after this attack have not been clarified in molecular detail for any of the LCTs. Studies on this aspect are complicated by the fact that (a) several GTPases are affected, (b) each GTPase may have several downstream signalling pathways, and (c) the GTPases also interact with each other in a complex network. Furthermore, (d) there are tissue-specific variations in the expression of small GTPases (Fritz *et al.*, 1994), so that the response to an LCT may differ from one cell type to another. Considering this complexity, it is currently not feasible to create a single picture of how cells are affected, but a number of different experimental systems is reviewed in the next section.

CLOSTRIDIAL TOXINS AS TOOLS IN CELL BIOLOGY

General considerations

The toxins, which are important determinants in diseases, have been used as 'tools' for prophylactic and diagnostic purposes (Torres *et al.*, 1995; Ryan *et al.*, 1997; Sauerborn *et al.*, 1997). In addition, clostridial toxins, like those of other bacteria, may be suitable for creating hybrid molecules for therapeutic use, e.g. as cytostatic immunotoxins for treatment of neoplastic tumours. The binding components of the binary C2-like toxins might potentially be used as protein carriers for delivery of fusion proteins into cells. For example, mammalian cells might be transfected with the C2II binding component for 'on-demand' expression, making the expressing cells selectively sensitive to C2I-chimeras with pharmacologically active molecules (Simpson, 1997). The diagnostic and therapeutic applications of toxins will not be considered further in this chapter (but see Kushnaryov *et al.*, 1992; Bartlett, 1996, 1997; Lyerly *et al.*, 1998; Rupnik *et al.*, 1998). Instead, we will discuss how the currently existing battery of clostridial toxins can be used in cell biology and pharmacology to clarify cellular processes in which the toxin target molecules have a function.

As tools in cell science, both the C2 and C3 toxins have the advantage of a strictly selective action on a single substrate molecule, i.e. monomeric G-actin and Rho, respectively. The LCTs, in contrast, have the advantages of being taken up very efficiently in target cells and of targeting a broad range of GTPases. Thus, at least two of the LCTs attacking GTPases of the Rho- and Ras-subfamilies, respectively (see Table 8.1), could be used first to clarify whether or not the investigated cellular process involves any of these ACSK-controlling molecules. If by this approach small GTPases appear to be required, the use of C3 alone will reveal whether Rho is essential. LCTs in various combinations with each other and with alternative approaches will help to clarify further which small GTPase(s) is required in the process investigated. Finally, C2 (and cytochalasin) can be applied to find out whether attacking actin alone would suffice for blocking the process.

In the following, we will mention some of the most recent cell biological studies in which the clostridial toxins acting on the cytoskeleton have been applied as tools. It is beyond the scope of this chapter to make the overview below exhaustive; rather the intention is to exemplify how these toxins have been utilized to answer cell biological questions of the most diverse character.

Applications

Detection of cellular Rho

When [^{32}P]NAD with a high specific activity is used, the ADP-ribosylation by C3 is an extremely sensitive method for detecting a small amount of Rho. A corresponding target detection is of course valid for every ADP-ribosylating toxin. However, it is particularly useful with C3, since Rho is not abundant in cells, and the sensitivity in the radioactive reaction is higher than in the antibody reaction with Rho. Moreover, C3 is completely specific for Rho in contrast to (at least) commercially available antibodies, which may cross-react with other Rho family proteins. Indeed, as evidenced above, the C3 transferase played an important role in identifying Rho as a major substrate for TcdB (Just *et al.*, 1994).

Regulation of actin structures

C3 and the B fragment of diphtheria toxin have been genetically fused to a chimera, DC3B, which inhibited the bacterially induced microfilament assembly in Vero cells infected by *Listeria monocytogenes* (Aullo *et al.*, 1993). This sophisticated use of C3 as a tool revealed that the bacterium utilizes cellular Rho to create its conspicuous 'actin comet tails'. C2 toxin, together with other tools, was used to show that cells autoregulate their synthesis and polymerization of actin, depending on the cytosolic concentration of G-actin (Reuner *et al.*, 1996b). Hall and collaborators, using C3 in conjunction with molecular genetic methods, have systematically clarified the roles played by different Rho family proteins (Rho, Rac and Cdc42) in growth factor-induced assembly of actin structures, such as stress fibres, focal adhesions, filopodia and the cortical actin network needed for membrane ruffling (for original references see Hall, 1998). TcdB was recently utilized in a basic study on GTPγS-induced actin polymerization, mediated by ATP- and phosphoinositide-independent signalling via Rho-family proteins (Katanaev and Wymann, 1998).

Cells of the immune system

Many investigations have centred on the role of the ACSK in ligand-evoked signal transduction and secretion in neutrophils as well as in lymphoid cell motility. For instance, a role for Rho in fMLP-dependent activation of phospholipase D was demonstrated with C3 in human neutrophils (Fensome *et al.*, 1998). C2 toxin (and cytochalasin) helped to uncover the existence of multiple cation entry pathways in neutrophils (Wenzel-Seifert *et al.*, 1997), and the same agents were used to identify the T-cell ASCK as a major motor for sustaining signal transduction (Valitutti *et al.*, 1995). Furthermore, C3 strongly inhibited the lymphocyte shape changes associated with extension and retraction of pseudopodia, as well as invasion of the cells through a monolayer of fibroblast-like cells. Thus, the invasion-bound motility of lymphocytes depends on a Rho-mediated signal transduction pathway (Verschueren *et al.*, 1997).

A recent report described inactivation of Rho function in the thymus by thymic targeting of a transgene encoding C3 (Henning *et al.*, 1997). This elegant approach established Rho as a critical integrator of proliferation and cell survival signals in early thymic development, since a lack of Rho function severely impaired the generation of normal numbers of thymocytes and mature peripheral T-cells (Henning *et al.*, 1997). Inactivation of Rho with C3 also induced multinucleation and apoptosis in murine T-lymphoma EL4 cells (Moorman *et al.*, 1996). Accordingly, both TcdA (Fiorentini *et al.*, 1992) and TcdB (Shoshan *et al.*, 1990) induce multinucleation in T- and B-cells, respectively.

Studies on the effects of C2 and C3 on degranulation of mast cells indicated that the responsiveness towards C2 depends largely on mast cell attachment (Wex *et al.*, 1997a) and suggested that Rho proteins are involved in activation of mast cells (Wex *et al.*, 1997b). Application to rat basophilic leukaemia cells of TcdB in combination with C3 and C2 showed that Rho subfamily proteins regulate the secretion of serotonin independently of the ACSK (Prepens *et al.*, 1996, 1997).

Endothelial and epithelial cells

The importance of the ACSK for endothelial integrity and barrier function has been demonstrated with C2 (Ermert *et al.*, 1996, 1997) and TcdB (Hippenstiel *et al.*, 1997). Furthermore, human umbilical vein endothelial cell migration and wound repair was inhibited by C3 and TcdA, suggesting that Rho family GTPases control the cytoskeletal reorganization in migrating endothelial cells (Aepfelbacher *et al.*, 1997). A recent study with TcdB and TcsL showed that Rho protein inhibition blocked membrane translocation and activation of protein kinase C (PKC) in endothelial and epithelial cells (Hippenstiel *et al.*, 1998).

The first demonstration that Rho regulates tight junctions and perijunctional actin organization in polarized intestinal epithelial cells was performed with C3 and TcdB (Nusrat *et al.*, 1995). More recently, Rho proteins in intestinal cells were shown to be essential in growth factor-induced cell migration to re-establish mucosal integrity after wounding (Santos *et al.*, 1997). In addition, TcdB-induced apoptosis in cultured intestinal cells suggested that Rho proteins may play a pivotal role in triggering apoptosis (Fiorentini *et al.*, 1998).

Neuronal cells

Thrombin-induced apoptosis in cultured neurons and astrocytes was reported to involve signalling from the thrombin receptor to the ACSK via Rho, since C3 in this system attenuated cell death (Donovan *et al.*, 1997). More recently, the same group (Donovan and Cunningham, 1998) observed that thrombin could protect astrocytes from hypoglycaemia, and since C3 attenuated this protection, apparently it also involved signalling via Rho. Based on kinetic comparisons, the authors concluded that the pathways for cell death and cell protection may share initial signalling proteins, but differences in the amplitude as well as the duration of the signal may result in different final pathways (Donovan and Cunningham, 1998). A study on astrocytoma cells revealed that thrombin could activate Rho-dependent pathways for both DNA synthesis and cell rounding, the latter response being mediated in part through myosin phosphorylation (Majumdar *et al.*,

1998). Several different kinds of pharmacologically active inhibitors, including C3, enabled these authors to discriminate between mitogenesis and cell rounding. C3 and a dominant-negative RhoA mutant were utilized to demonstrate in mouse neuroblastoma cells that lysophosphatidic acid (LPA)-induced ACSK contraction, but not stress fibre formation, required translocation of RhoA from the cytosol to the plasma membrane (Kranenburg *et al.*, 1997).

Muscle cells

In myocytes RhoA was suggested as the mediator for the angiotensin II-triggered formation of premyofibrils based on the inhibitory effect of C3 (Aoki *et al.*, 1998). By exposing guinea pig smooth muscle to TcdB, a role was attributed to Rho proteins in the carbachol-induced force and myosin light chain phosphorylation taking place in this tissue (Lucius *et al.*, 1998).

Diverse model systems

Several workers using clostridial toxins have demonstrated a role of Rho in stimulation of cellular phospholipases C and D (Schmidt *et al.*, 1996a, b; Ojio *et al.*, 1996, 1998; Kato *et al.*, 1997). For example, in human embryonic kidney (HEK) cells phospholipase D (PLD) is stimulated by carbachol via the muscarinic acetylcholine receptor (mAChR) or by PMA via PKC. Treatment of these cells with TcdB or C3 blocked the carbachol-mediated PLD stimulation, while the PMA-mediated stimulation was not affected (Schmidt *et al.*, 1996a, b). In contrast, TcsL or TcdB-1470, which inactivate Ral but not Rho, did not affect the carbachol-mediated stimulation, whereas the PMA-mediated stimulation of PLD was almost completely blocked (Schmidt *et al.*, 1998). To identify the small GTPase regulating the PMA-mediated activation, Ras, Rac and Ral were transfected into the HEK cells. Only Ral-transfection could overcome the blockage by TcsL and TcdB-1470. Overall, this elegant study indicated that Rho is essential for stimulation of PLD via the mAChR, while Ral is necessary for the PLD activation by PMA via PKC (Schmidt *et al.*, 1998).

The use of C3 and Rho mutants demonstrated that Rho in rat hepatoma cells stimulates the actomyosin system, leading to an increased invasiveness of these cells (Yoshioka *et al.*, 1998). Using a double subgenomic recombinant capable of expressing active C3 in intact L929 fibroblasts, inactivation of Rho was observed to disrupt cellular attachment and induce apoptosis (Bobak *et al.*, 1997). The combined utilization of TcsL, TcdA, TcdB and C3 demonstrated that the Rho subfamily proteins have differential roles in glucose- and calcium-induced secretion from pancreatic beta cells (Kowluru *et al.*, 1997).

The formation of actin pedestals upon infection of HeLa cells with enteropathogenic *E. coli* was independent of Rho, Rac and Cdc42, whereas these GTPases may be involved in the cellular uptake of the bacteria (Ben-Ami *et al.*, 1998; Ebel *et al.*, 1998). Indeed, active Rho was essential for *Shigella* invasion of epithelial cells (Watarai *et al.*, 1997).

Conclusions

High potency and selective attack on a specific target(s) make a toxin useful for studying the cellular functions of the particular target and associated signalling pathways. The overview above demonstrates that the ACSK interacting clostridial toxins, used together as a battery of tools, may enable quite sophisticated approaches to answer specific questions in current cell science. However, there are also limitations which should not be forgotten. A weak point with C3 is its low cytotoxicity owing to inefficient cellular uptake. This problem can be overcome by microinjection of C3 (Hall, 1998), by applying C3 as a fusion protein (Aullo *et al.*, 1993), or by expressing C3 directly in the eukaryotic target cell (e.g. Bobak *et al.*, 1997; Henning *et al.*, 1997). A potential problem with the LCTs is that each of them affects several different GTPase targets. Since each small GTPase has several functions, many of which are unknown, results have to be interpreted with great caution. Moreover, it is very important to include adequate controls which demonstrate that the small GTPase(s) in the particular cells studied have really been modified by the toxin used. A 'differential' labelling (glucosylation or ADP-ribosylation) of the GTPase target(s) in lysates from the toxin-treated cells is recommended.

In summary, the clostridial toxins are useful tools, particularly when combined with each other and preferably also with other approaches. For instance, dominant negative mutants of the small GTPases help to discriminate the roles of different GTPases (as in Santos *et al.*, 1997), and additional inhibitors, such as the cytochalasins, must be applied to discount pleiotropic effects that are merely secondary phenomena to actin depolymerization.

CLOSTRIDIAL TOXINS IN DISEASE

ADP-ribosyltransferases

The exoenzyme C3 appears to be non-toxic in animals since i.p. injection of as much as 100 μg C3 into mice had no obvious effect (Aktories and Koch, 1997). In contrast, the LD_{50} of the C2 toxin was approximately 5 and 50 ng per mouse following i.v. and i.p. administration, respectively. Administered in rats, the C2 toxin caused hypotension, haemorrhaging of the lungs and fluid accumulation around the lungs (for references, see Aktories *et al.*, 1992). These symptoms probably arise from an increase in vascular permeability, which has been indirectly evidenced in several later studies reporting on the action of C2 on endothelial cell systems (see 'Clostridial toxins as tools in cell biology', above). In a mouse intestinal loop assay, C2 also caused fluid accumulation, appearing within 1–2 h and increasing for at least 10 h. An acute inflammation and severe damage to the intestinal mucosa accompanied the fluid (Ohishi *et al.*, 1983). Speculatively, such effects of C2 could have some significance in infant botulism when caused by *C. botulinum* strains producing this toxin. However, no convincing evidence exists for any significant role of either C2 toxin or exoenzyme C3 in disease caused by this bacterium (Simpson, 1997). The iota and iota-like toxins might contribute to muscle necrosis in gas gangrene, since these toxins (in contrast to other C2 toxins) attack the type of G-actin present in muscle cells.

Glycosyltransferases (large clostridial cytotoxins)

In the pathogenesis of gas gangrene, the α-toxin of *C. perfringens* has a proven role (Awad *et al.*, 1995), whereas it is unknown whether the LCTs produced by *C. sordellii* and *C. novyi* have any significance. Conceivably, these LCTs could contribute to the prominent tissue necrosis in cases of gas gangrene where these bacteria happen to be involved.

Among the currently known LCTs (Table 8.1), the *C. difficile* toxins TcdA and TcdB are undisputed virulence factors and induction of antibiotic-associated diarrhoea and colitis is well documented. The many recent reviews describing *C. difficile* infection in terms of epidemiology, clinical features and diagnosis (Lyerly and Wilkins, 1995; Dodson and Borriello, 1996; Jumaa *et al.*, 1996; Pothoulakis, 1996; Bartlett, 1997; Linevsky and Kelly, 1997; Kelly and LaMont, 1998) are an indication of the ever-increasing importance of this disease. The following discussions concentrate on the current intensive efforts to understand how the symptoms in *C. difficile* disease are elicited by TcdA and TcdB.

Both TcdA and TcdB are involved in the pathogenesis of Clostridium difficile

Studies in the 1980s indicated that TcdA causes strong inflammation and fluid accumulation in animal intestinal loops, while TcdB alone was without effect in such assays (for earlier references, see Thelestam *et al.*, 1997). Thus, TcdA was described as an enterotoxin and regarded as the major virulence factor, while TcdB, the more potent cytotoxin, was believed to be unimportant in the pathogenesis of the *C. difficile* disease (for references to earlier work, see Lyerly and Wilkins, 1995). More recently, the barrier function of polarized intestinal epithelial T84 cells *in vitro* was shown to be adversely affected by both toxins, although TcdA had a 10-fold stronger effect than TcdB in this model (Hecht *et al.*, 1988, 1992). This may be due to the fact that TcdA binds more efficiently than TcdB to T84 cells (Chaves-Olarte *et al.*, 1997). In contrast, the observation of Riegler *et al.* (1995) that TcdB was 10 times more potent than TcdA in causing damage to mucosal strips of normal human colonic epithelium reflects the higher cytotoxic and enzymic potency of TcdB (Chaves-Olarte *et al.*, 1997) and indicates an essential role for TcdB in human disease. This conclusion is supported by a recent study on treatment of *C. difficile* infection in hamster with hen antibodies against recombinant TcdA and TcdB. Either antitoxin alone was not fully efficient, but the combined antibodies given to hamsters for 4 days, starting 8 h after *C. difficile* challenge, resulted in 100% protection (Kink and Williams, 1998). Furthermore, it is well known that TcdB is enterotoxic in intestinal loop assays, provided a trace amount of TcdA is present, suggesting that the toxins may act synergistically in the clinical situation (Lyerly and Wilkins, 1995). In conclusion, both toxins appear important for creating the intestinal symptoms in human *C. difficile* infection. This seems reasonable from a teleological point of view; since TcdB exists and is produced at a high cost of energy, there should be some reason other than just to please toxicologists and cell biologists.

Direct toxin effects on epithelial cells

As stated above, both toxins perturb cytoskeletal structures and tight junction permeability (Hecht *et al.*, 1988, 1992) and damage human colonic epithelium (Riegler *et al.*, 1995). More recently, Riegler *et al.* (1997) demonstrated that EGF could reduce these effects if added to the serosal side of colonic strips prior to toxin treatment from the mucosal side (Riegler *et al.*, 1997). EGF might stabilize the ACSK by triggering signalling cascades, which switch the small GTPases into the GTP-loaded active state in which they are poorer substrates for the toxins. Using unpolarized IEC-6 and polarized Caco-2 intestinal cells in an *in vitro* wound-induced migration assay, Santos *et al.* (1997) demonstrated the importance of Rho for cell migration in both cell lines. Thus, impairment of Rho function could be of significance in the disease mechanism, not only through the loss of epithelial barrier function

(Nusrat *et al.*, 1995), but also by preventing the cell migration which is essential for mucosal restitution (Santos *et al.*, 1997). All of these observations are consistent with cellular uptake of both TcdA and TcdB in intestinal epithelial cells, resulting in a direct ACSK-collapse due to glucosylation of small GTPases.

Activation of the immune system

Because of the strong inflammatory reaction in *C. difficile* colitis, several recent reports have focused on the responses of immune cells to the toxins. TcdB was found to be approximately 1000 times more potent than TcdA in stimulating tumour necrosis factor-α (TNFα)-release from cultured monocytes (Souza *et al.*, 1997). An intense neutrophil migration, mediated by macrophage-derived TNFα and lipoxygenase products, resulted. Both TcdA and TcdB at low concentrations activated monocytes *in vitro* to release factors, such as IL-8, that facilitate neutrophil extravasation and tissue infiltration (Linevsky *et al.*, 1997). In accordance with this, neutrophil migration evoked by TcdA in the peritoneal cavities of rats was partially dependent on macrophage-derived cytokines (Rocha *et al.*, 1997). Also, mast cells responded to TcdA by releasing TNFα, and their functions and survival were impaired by prolonged incubation with the toxin (Calderón *et al.*, 1998). These effects could hamper the capacity of mast cells to counteract the infection, thus contributing to the prolonged pathogenic effects of the *C. difficile* toxins. Consistent with these studies, IL-11 administered before TcdA to rat ileal loops reduced all the adverse intestinal effects observed in the controls without IL-11. The authors speculate that IL-11 might inhibit the release of inflammatory mediators (Castagliuolo *et al.*, 1997a).

How do the toxins reach the immune cells? A direct monocyte/macrophage stimulation by the *C. difficile* toxins in the colon requires that they gain access to the lamina propria. The barrier function of polarized intestinal epithelial cells was decreased after toxin treatment (see above). However, the monolayer permeability in those studies was measured as a transepithelial leakage of mannitol, which is a much smaller molecule than the toxins. The study by Riegler *et al.* (1995) showed that human colonic mucosa exposed to either TcdA or TcdB was damaged in a patchy pattern similar to classical findings *in vivo* of pseudomembranes in human *C. difficile* colitis. With this background, Linevsky *et al.* (1997) suggested that the toxin molecules could cross the epithelium through such damaged patchy areas. This would enable activation of tissue macrophages to produce the proinflammatory cytokines mentioned above. Thereby, the strong acute inflammatory cell infiltration and further mucosal injury would be initiated and ultimately result in focal pseudomembrane formation (Linevsky *et al.*, 1997).

Neuronal activation by TcdA

In addition to these (presumably) direct effects on immune cells, earlier experiments in animal models suggested an involvement of the neuronal system in the intestinal effects of TcdA (Castagliuolo *et al.*, 1994; Kelly *et al.*, 1994). Recently, the same group has substantiated these observations by showing increased substance P (SP) responses in dorsal root ganglia and intestinal macrophages during TcdA enteritis in rats (Castagliuolo *et al.*, 1997b). Injection of TcdA into rat ileum caused an increased SP content in lumbar dorsal root ganglia and mucosal scrapings 30–60 min after TcdA administration. Compared with control cells, the lamina propria macrophages obtained from TcdA-injected loops released greater amounts of TNFα and SP, and pretreatment of the rats with an SP antagonist inhibited the TNFα release (Castagliuolo *et al.*, 1997b). Interestingly, SP receptors are massively increased in small blood vessels and lymphoid aggregates in the human pseudomembranous colitis bowel (Mantyh *et al.*, 1996). Therefore, an additional (or alternative?) scenario in the pathogenesis of *C. difficile* disease would be that binding of TcdA to receptors on intestinal cells first elicits a transepithelial signal of unknown nature. The signal could induce a release of mediators on the basolateral side of the epithelium, which in turn activate neurons that trigger fluid secretion, as well as the inflammatory events detailed above. This process would open tight junctions enough to allow TcdB to move across the epithelium and reach the basolateral cell surfaces where receptors for this toxin may be more enriched (as they are in polarized T84 cells; Chaves-Olarte *et al.*, unpublished). TcdB could then exert its cell-damaging activity, which would aggravate the mucosal necrosis and inflammation.

The reasons for proposing this alternative model are that (a) TcdA has a very much lower enzymic potency than TcdB, (b) there is strong evidence from animal models for a neuronal activation by TcdA, and (c) the receptors enabling TcdB to exert its potent cell damaging activity appear to be expressed preferentially on the basolateral side. The putative transepithelial signalling capacity of TcdA, but not TcdB, might be connected with its larger *C*-terminal ligand domain, which appears to be exposed on the surface of the native toxin molecule (Sauerborn *et al.*, 1994). In contrast, TcdB is likely to have a different 3D configuration, exposing also *N*-terminal parts of the molecule, since its B-cell epitopes, in contrast to those of TcdA,

appear distributed over the entire sequence of the molecule (Sauerborn *et al.*, 1994). The proposed model also takes into account the strong effects in animal intestinal loops elicited by TcdA alone.

In conclusion, there are several mechanisms by which *C. difficile* toxins may elicit intestinal disease. As schematically depicted in Figure 8.8, these mechanisms include (from left to right) a transepithelial neuro-immune activation by TcdA leading to fluid secretion and an opening of tight junctions. This permits the migration of neutrophils and enables TcdB to intoxicate enterocytes from the basolateral side, leading to cell rounding and detachment. Taken together, these processes will cause gross pathological changes, leading ultimately to the generation of pseudo-membranes. It is not yet possible to assign every single process elicited by the toxins a definitive role in the pathogenesis. However, new reports highlighting these events are rapidly accumulating and will soon generate the basis for a clearer picture.

CONCLUDING REMARKS

How do the clostridia profit from their large arsenal of toxins?

Clostridia are ancient 'toxic bacteria' that have adapted to their environment over millions of years. They have a biphasic 'lifestyle', shuttling between an actively growing phase and the sporulation state. While other bacteria cannot take a rest, clostridia survive after sporulation even in hostile surroundings. Thus, they may even kill their host and still survive. Killing the host will transform it into a large anaerobic fermenter. The life cycle ends after extensive multiplication, giving rise to dissemination and spread of the bacterial spores. Toxins could be instrumental for bacterial spread.

How to be effective as a tiny bacterium within a complex host

Clostridia, in general, serve as brilliant examples of adaptation to the eukaryotic host. The neurotoxins effectively block the vesicular exocytosis machinery by proteolysis of several SNARE proteins. For the second largest group of clostridial toxins, the LCTs, again an enormously effective adaptive evolution has occurred. By attacking GTPases, the clostridial toxins target key proteins rather than one of several prominent components (e.g. actin) of the cytoskeleton. The mechanism is efficient and very clever in the process of pathogenesis.

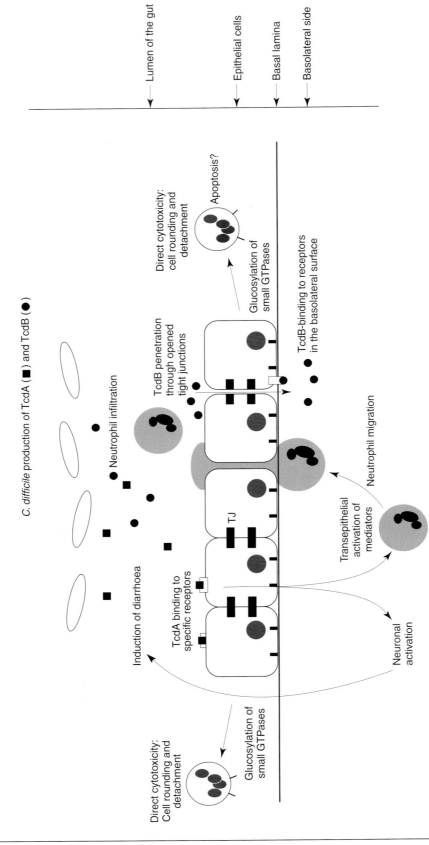

FIGURE 8.8 Summary of observed effects of the *C. difficile* toxins on the intestinal mucosa. See text for further explanation.

Glycosylation and ADP-ribosylation occur in bacteria and mammalian cells

The ADPRTs mimic a process that occurs endogenously in mammalian cells. Investigations into the catalytic sites of pro- and eukaryotic ADP-ribosyltransferases revealed conserved active amino acid residues (Domenighini and Rappuoli, 1996). With its O-N-acetylglucosaminidation activity, Tcnα mimics endogenous, post-translational modifications performed in eukaryotic cells. Evidence is accumulating that mammalian cells mono-O-glycosylate several important proteins, using UDP-GlcNAc as cofactor (Hayes and Hart, 1998; Snow and Hart, 1998). It is a matter of debate whether such mono-O-glycosylations on threonine and serine residues could be as important as phosphorylation in controlling protein–protein interactions in cells (Snow and Hart, 1998).

Since UDP-Glc is the most important activated glucose molecule in cells, it seems logical that the majority of identified LCTs utilize UDP-Glc as a cofactor. UDP-Glc is required in the quality control of proteins in the ER (Hammond and Helenius, 1995) and in the biosynthesis of glucosylceramide, a precursor for hundreds of cellular glycosphingolipids (Ichikawa and Hirabayashi, 1998). Nevertheless, mono-O-glucosylation of cellular proteins with UDP-Glc as the cofactor does not seem to take place normally in mammalian cells.

The occurrence of O-glycosylation in clostridia and mammalian cells indicates that this mechanism is a phylogenetically ancient process that underwent adaptation, specialization and optimization. Wiggins and Munro (1998) recently reported that, in yeast mannosyltransferases, a DXD motif was identified that is also present in LCTs. ADP-ribosylation and glycosylation is not exclusively reported in prokaryotes, possibly a reflection of exchange of successful genetic programs and information among microbes.

Actin as the prominent target of the cytoskeleton

Considering the known activities of bacterial toxins, one might get the impression that the ACSK is the most important part of the cytoskeleton. Several plants, in contrast, produce well-known non-protein toxins, such as colchicine, the vinca alkaloids and taxol, which directly attack the MTs (Thelestam and Gross, 1990), thus complementing nature's arsenal of cytoskeleton-attacking substances. Since only C2 of all the clostridial toxins interferes with the ACSK by attacking actin directly, this seems not to be the most effective strategy for cytoskeletal attack.

The involvement of Rho-subfamily GTPases in the dynamics of the ASCK is well proven. An interesting finding in mid-1998 was the discovery that Rho selectively stabilized a subset of MTs during the cell polarization induced by LPA (Cook et al., 1998). Thereby, Rho is the first signal transducing molecule identified to be capable of regulating the MT. This first report on the co-ordinated regulation of both the MT and the ACSK via Rho might be only the tip of an iceberg. The concept of a combined action of signalling GTPases and all cytoskeletal components may even be broadened, since IFs could be indirectly affected owing to their preferential coalignment with the subset of stabilized MTs controlled by Rho (Gurland and Gundersen, 1995).

Exploitation of clostridial toxins in science

Bacterial toxins have been used as tools ever since their mechanisms of action were resolved. The glycosyltransferase activity of LCTs was discovered at an appropriate time when their targets, the small GTP-binding proteins, were recognized as highly important in molecular and cellular biology research. By combining other cell science approaches with artificial intoxication using LCTs (or C2/C3-like toxins), the toxins discussed in this chapter turned into invaluable tools for learning more about the details of cell signalling. This field is expanding rapidly, and many new insights will be gained with these tools.

Learning from bacterial toxins – exploitation of functional principles

The information gained from clostridia, as well as from other bacteria, offers the opportunity to exploit functional domains for the development of new therapeutic approaches in the treatment of diseases. Ligand domains interact in a highly specific manner with their receptor molecules. As part of recombinant proteins, they will be useful in targeting pharmaceutical agents to defined cells, exploiting the best possible process in terms of specificity and delivery. Translocation domains of bacterial toxins will teach us how to circumvent lipid membranes and gain access to the hidden parts of the human body. These domains will become essential components of pharmacological targeting systems. After resolution of its mechanism of action, botulinum toxin is now a valuable therapeutic agent. Concerning LCTs, attenuated versions of these toxins have to be developed which could then be used for treating diseases in which a hyperfunction of Ras-superfamily GTPases must be counteracted. One example would be the use of TcsL for the treatment of Ras-related tumours.

Undiscovered toxins?

There will surely be other toxins targeting other Ras-superfamily GTPases (as well as other major processes of the cell), since such offensive strategies are very sophisticated and effective. A most recent example of toxins following the principle is the *E. coli* cytotoxic necrotizing factor, which also attacks Rho, but in contrast to LCTs acts by the deamidation of glutamine-63 (see Chapter 2).

Bacteria are players, and the toxins are their best cards. Desperate to win, they developed every possible trick, and we are sure that we will encounter many more surprises. 'Learning by doing' generates experience, and there has been plenty of time for bacteria to gain those experiences. The challenges posed by the little magical cellular machines and the bacterial weapons attacking them will continue to keep hordes of scientists in motion.

ACKNOWLEDGEMENTS

Work performed in the authors' laboratories was supported by the Swedish Medical Research Council (05969), the Swedish Cancer Society (3826-B96–01XAB), the Karolinska Institutet Research Funds, the Magnus Bergvall Foundation and the Deutsche Forschungsgemeinschaft (Ei2068–1).

REFERENCES

Aepfelbacher, M., Essler, M., Huber, E., Sugai, M. and Weber, S.C. (1997). Bacterial toxins block endothelial wound repair – evidence that Rho GTPases control cytoskeletal rearrangements in migrating endothelial cells. *Arterioscler. Thromb. Vasc. Biol.* **17**, 1623–9.

Aktories, K. (1997a). Identification of the catalytic site of clostridial ADP-ribosyltransferases. *Adv. Exp. Med. Biol.* **419**, 53–60.

Aktories, K. (1997b). Bacterial toxins that target Rho proteins. *J. Clin. Invest.* **99**, 827–9.

Aktories, K. (ed.) (1997c). *Bacterial Toxins – Tools in Cell Biology and Pharmacology*, pp. 1–308. Chapman & Hall, Weinheim.

Aktories, K. and Just, I. (1995). Monoglucosylation of low-molecular-mass GTP-binding proteins by clostridial cytotoxins. *Trends Cell Biol.* 5, 441–3.

Aktories, K. and Koch, G. (1997). *Clostridium botulinum* ADP-ribosyltransferase C3. In: *Bacterial Toxins – Tools in Cell Biology and Pharmacology* (ed. K. Aktories), pp. 61–9. Chapman & Hall, Weinheim.

Aktories, K., Wille, M. and Just, I. (1992). Clostridial actin ADP-ribosylating toxins. *Curr. Top. Microbiol. Immunol.* **175**, 97–113.

Aktories, K., Jung, M., Bohmer, J., Fritz, G., Vandekerckhove, J. and Just, I. (1995). Studies on the active-site structure of C3-like exoenzymes: involvement of glutamic acid in catalysis of ADP-ribosylation. *Biochimie* **77**, 326–32.

Amano, M., Fukata, Y. and Kaibuchi, K. (1998). Regulation of cytoskeleton and cell adhesisons by the small GTPase Rho and its targets. *Trends Cardiovasc. Med.* **8**, 162–8.

Aoki, H., Izumo, S. and Sadoshima, J. (1998). Angiotensin II activates RhoA in cardiac myocytes: a critical role of RhoA in angiotensin II-induced premyofibril formation. *Circ. Res.* **82**, 666–76.

Aullo, P., Giry, M., Olsnes, S., Popoff, M.R., Kocks, C. and Boquet, P. (1993). A chimeric toxin to study the role of the 21 kDa GTP binding protein rho in the control of actin filament assembly. *EMBO J.* **12**, 921–31.

Awad, M.M., Bryant, A.E., Stevens, D.L. and Rood, J. (1995). Virulence studies on chromosomal α-toxin and θ-toxin mutants constructed by allelic exchange provide genetic evidence for the essential role of α-toxin in *Clostridium perfringens*-mediated gas gangrene. *Mol. Microbiol.* **15**, 191–202.

Ayscough, K.R. and Drubin, D.G. (1996). ACTIN: general principles from studies in yeast. *Annu. Rev. Cell Dev. Biol.* **12**, 129–60.

Banno, T., Kobayashi, K., Watanabe, K., Ueno, K. and Nozawa, Y. (1981). Two toxins (D-1 and D-2) of *Clostridium difficile* causing antibiotic-associated colitis: purification and some characterization. *Biochem. Int.* **2**, 629–35.

Banno, Y., Kobayashi, T., Kono, H., Watanabe, K., Ueno, K. and Nozawa, Y. (1984). Biochemical characterization and biologic actions of two toxins (D-1 and D-2) from *Clostridium difficile*. *Rev. Infect. Dis.*, **6**, Suppl. 1, S11–20.

Barroso, L.A., Wang, S.Z., Phelps, C.J., Johnson, J.L. and Wilkins, T.D. (1990). Nucleotide sequence of *Clostridium difficile* toxin B gene. *Nucleic Acids Res.* **18**, 4004.

Barth, H., Hofmann, F., Olenik, C., Just, I. and Aktories, K. (1998). The N-terminal part of the enzyme component (C2I) of the binary *Clostridium botulinum* C2 toxin interacts with the binding component C2II and functions as a carrier system for a Rho ADP-ribosylating C3-like fusion toxin. *Infect. Immun.* **66**, 1364–9.

Bartlett, J.G. (1996). Management of *Clostridium difficile* infection and other antibiotic-associated diarrhoeas. *Eur. J. Gastroenterol. Hepatol.* **8**, 1054–61.

Bartlett, J.G. (1997). *Clostridium difficile* infection: pathophysiology and diagnosis. *Semin. Gastrointest. Dis.* **8**, 12–21.

Belyi, Yu.F., Tartakovskii, I.S., Vertiev, Yu.V. and Prosorovskii, S.V. (1991). ADP-ribosyltransferase activity of *Legionella pneumophila* is stimulated by the presence of macrophage lysates. *Biomed. Sci.* **2**, 94–6.

Ben-Ami, G., Ozeri, V., Hanski, E., Hofmann, F., Aktories, K., Hahn, K.M., Bokoch, G.M. and Rosenshine, I. (1998). Agents that inhibit Rho, Rac, and Cdc42 do not block formation of actin pedestals in HeLa cells infected with enteropathogenic *Escherichia coli*. *Infect. Immun.* **66**, 1755–8.

Bobak, D., Moorman, J., Guanzon, A., Gilmer, L. and Hahn, C. (1997). Inactivation of the small GTPase Rho disrupts cellular attachment and induces adhesion-dependent and adhesion-independent apoptosis. *Oncogene* **15**, 2179–89.

Bohmer, J., Jung, M., Sehr, P., Fritz, G., Popoff, M., Just, I. and Aktories, K. (1996). Active site mutation of the C3-like ADP-ribosyltransferase from *Clostridium limosum* – analysis of glutamic acid 174. *Biochemistry* **35**, 282–9.

Braun, V., Hundsberger, T., Leukel, P., Sauerborn, M. and von Eichel-Streiber, C. (1996). Definition of the single integration site of the pathogenicity locus in *Clostridium difficile*. *Gene* **181**, 29–38.

Busch, C., Hofmann, F., Selzer, J., Munro, S., Jeckel, D. and Aktories, K. (1998). A common motif of eukaryotic glycosyltransferases is essential for the enzyme activity of large clostridial cytotoxins. *J. Biol. Chem.* **273**, 19566–72.

Calderón, G.M., Torres-López, J., Lin, T.-J., Chaves, B., Hernández, M., Munoz, O., Befus, A.D. and Enciso, J.A. (1998). Effects of toxin A from *Clostridium difficile* on mast cell activation and survival. *Infect. Immun.* **66**, 2755–61.

Castagliuolo, I., LaMont, J.T., Letourneau, R., Kelly, C.P., O'Keane, J.C., Jaffer, A., Theoharides, T.C. and Pothoulakis, C. (1994). Neuronal involvement in the intestinal effects of *Clostridium difficile* toxin A and *Vibrio cholerae* enterotoxin in rat ileum. *Gastroenterology* **107**, 657–65.

Castagliuolo, I., Kelly, C.P., Qiu, B.S., Nikulasson, S.T., LaMont, J.T. and Pothoulakis, C. (1997a). Il-11 inhibits *Clostridium difficile* toxin A enterotoxicity in rat ileum. *Am. J. Physiol.* **273**, G333–43.

Castagliuolo, I., Keates, A.C., Qiu, B., Kelly, C.P., Nikulasson, S., Leeman, S.E. and Pothoulakis, C. (1997b). Increased substance P responses in dorsal root ganglia and intestinal macrophages during *Clostridium difficile* toxin A enteritis in rats. *Proc. Natl Acad. Sci. USA* **94**, 4788–93.

Chaves-Olarte, E., Florin, I., Boquet, P., Popoff, M., Eichel-Streiber, Cv. and Thelestam, M. (1996). UDP-Glucose deficiency in a mutant cell line protects against glucosyltransferase toxins from *Clostridium difficile* and *Clostridium sordellii*. *J. Biol. Chem.* **271**, 6925–32.

Chaves-Olarte, E., Weidmann, M., Eichel-Streiber, Cv. and Thelestam, M. (1997). Toxins A and B from *Clostridium difficile* differ with respect to enzymatic potencies, cellular substrate specificities, and surface binding to cultured cells. *J. Clin. Invest.* **100**, 1734–41.

Chaves-Olarte, E., Löw, P., Freer, E., Norlin, T., Weidmann, M., Eichel-Streiber, C.V. and Thelestam, M. (1999). A novel cytotoxin from *Clostridium difficile* serogroup F is a functional hybrid between two other large clostridial cytotoxins. *J. Biol. Chem.* **274**, 1–7 (in press).

Ciesla, W.P. and Bobak, D.A. (1998). *Clostridium difficile* toxins A and B are cation-dependent UDP-glucose hydrolases with differing catalytic activities. *J. Biol. Chem.* **273**, 16021–6.

Cook, T.A., Nagasaki, T. and Gundersen, G.G. (1998). Rho guanosine triphosphatase mediates the selective stabilisation of microtubules induced by lysophosphatidic acid. *J. Cell Biol.* **141**, 175–85.

Dailey, D.C. and Schloemer, R.H. (1988). Identification and characterization of *Clostridium difficile* promoter element that is functional in *Escherichia coli*. *Gene* **70**, 343–50.

Dodson, A.P. and Borriello, S.P. (1996). *Clostridium difficile* infection of the gut. *J. Clin. Pathol.* **49**, 529–32.

Domenighini, M. and Rappuoli, R. (1996). Three conserved consensus sequences identify the NAD-binding site of ADP-ribosylating enzymes, expressed by eukaryotes, bacteria and T-even bacteriophages. *Mol. Microbiol.* **21**, 667–74.

Donovan, F.M. and Cunningham, D.D. (1998). Signaling pathways involved in thrombin-induced cell protection. *J. Biol. Chem.* **273**, 12746–52.

Donovan, F.M., Pike, C.J., Cotman, C.W. and Cunningham, D.D. (1997). Thrombin induces apoptosis in cultured neurons and astrocytes via a pathway requiring tyrosine kinase and RhoA activities. *J. Neurosci.* **17**, 5316–26.

Dove, C.H., Wang, S.Z., Price, S.B., Phelps, C.J., Lyerly, D.M., Wilkins, T.D. and Johnson, J.L. (1990). Molecular characterization of the *Clostridium difficile* toxin A gene. *Infect. Immun.* **58**, 480–8.

Ebel, F., Eichel-Streiber, Cv., Rohde, M. and Chakraborty, T. (1998). Small GTP-binding proteins of the Rho- and Ras-subfamilies are not involved in the actin rearrangements induced by attaching and effacing *Escherichia coli*. *FEMS Microbiol. Lett.* **163**, 107–12.

Eichel-Streiber, Cv. (1993). Molecular biology of *Clostridium difficile* toxins. In: *Genetics and Molecular Biology of the Anaerobes* (ed. M. Sebald), pp. 264–89. Springer, New York.

Eichel-Streiber, Cv. and Sauerborn, M. (1990). *Clostridium difficile* toxin A carries a *C*-terminal repetitive structure homologous to the carbohydrate binding region of streptococcal glycosyltransferases. *Gene* **96**, 107–13.

Eichel-Streiber, Cv., Harperath, U., Bosse, D. and Hadding, U. (1987). Purification of two high molecular weight toxins of *Clostridium difficile* which are antigenically related. *Microb. Pathog.* **2**, 307–18.

Eichel-Streiber, Cv., Laufenberg Feldmann, R., Sartingen, S., Schulze, J. and Sauerborn, M. (1990). Cloning of *Clostridium difficile* toxin B gene and demonstration of high *N*-terminal homology between toxin A and B. *Med. Microbiol. Immunol. Berl.* **179**, 271–9.

Eichel-Streiber, Cv., Warfolomeow, I., Knautz, D., Sauerborn, M. and Hadding, U. (1991). Morphological changes in adherent cells induced by *Clostridium difficile* toxins. *Biochem. Soc. Trans.* **19**, 1154–60.

Eichel-Streiber, Cv., Laufenberg Feldmann, R., Sartingen, S., Schulze, J. and Sauerborn, M. (1992). Comparative sequence analysis of the *Clostridium difficile* toxins A and B. *Mol. Gen. Genet.*, **233**, 260–8.

Eichel-Streiber, Cv., Meyer zu Heringdorf, M., Habermann, E. and Hartingen, S. (1995). Closing in on the toxic domain through analysis of a variant *Clostridium difficile* cytotoxin B. *Mol. Microbiol.* **17**, 313–21.

Eichel-Streiber, Cv., Boquet, P., Sauerborn, M. and Thelestam, M. (1996). Large clostridial cytotoxins – a family of glycosyltransferases modifying small GTP-binding proteins. *Trends Microbiol.* **4**, 375–82.

Ermert, L., Rossig, R., Hansen, T., Schutte, H., Aktories, K. and Seeger, W. (1996). Differential role of actin in lung endothelial and epithelial barrier properties in perfused rabbit lungs. *Eur. Respir. J.* **9**, 93–9.

Ermert, L., Duncker, H.-R., Brückner, H., Grimminger, F., Hansen, T., Rössig, R., Aktories, K. and Seeger, W. (1997). Ultrastructural changes of lung capillary endothelium in response to botulinum C2 toxin. *J. Appl. Physiol.* **82**, 382–8.

Fensome, A., Whatmore, J., Morgan, C., Jones, D. and Cockcroft, S. (1998). ADP-ribosylation factor and Rho proteins mediate fMLP-dependent activation of phospholipase D in human neutrophils. *J. Biol. Chem.* **273**, 13157–64.

Fiorentini, C., Chow, S.C., Mastrantonio, P., Jeddi-Tehrani, M. and Thelestam, M. (1992). *Clostridium difficile* toxin A induces multinucleation in the human leukemic T cell line JURKAT. *Eur. J. Cell Biol.* **57**, 292–7.

Fiorentini, C., Fabbri, A., Falzano, L., Fattorossi, A., Matarrese, P., Rivabene, R. and Donelli, G. (1998). *Clostridium difficile* toxin B induces apoptosis in intestinal cultured cells. *Infect. Immun.* **66**, 2660–5.

Flatau, G., Lemichez, E., Gauthier, M., Chardin, P., Paris, S., Fiorentini, C. and Boquet, P. (1997). Toxin-induced activation of the G protein Rho by deamidation of glutamine. *Nature* **387**, 729–33.

Flores-Díaz, M., Alape-Girón, A., Persson, B., Pollesello, P., Moos, M., Eichel-Streiber, Cv., Thelestam, M. and Florin, I. (1997). Cellular UDP-glucose deficiency caused by a single point mutation in the UDP-glucose pyrophosphorylase gene. *J. Biol. Chem.* **272**, 23784–91.

Florin, I. and Thelestam, M. (1986). Lysosomal involvement in cellular intoxication with *Clostridium difficile* toxin B. *Microb. Pathog.* **1**, 373–85.

Fritz, G., Lang, P. and Just, I. (1994). Tissue-specific variations in the expression and regulation of the small GTP-binding protein Rho. *Biochim. Biophys. Acta* **1222**, 331–8.

Fritz, G., Schroeder, P. and Aktories, A. (1995). Isolation and characterization of a *Clostridium botulinum* C2 toxin-resistant cell line: evidence for possible involvement of the cellular C2II receptor in growth regulation. *Infect. Immun.* **63**, 2334–40.

Fujii, N., Kubota, T., Shirakawa, S., Kimura, K., Ohishi, I., Moriishi, K., Isogai, E. and Isogai, H. (1996). Characterization of component-I gene of botulinum C2 toxin and PCR detection of its gene in clostridial species. *Biochem. Biophys. Res. Commun.* **220**, 353–9.

Genth, H., Hofmann, F., Selzer, J., Rex, G., Aktories, K. and Just, I. (1996). Difference in protein substrate specificity between haemorrhagic toxin and lethal toxin from *Clostridium sordellii*. *Biochem. Biophys. Res. Commun.* **229**, 370–4.

Gibert, M., Perelle, S., Daube, G. and Popoff, M.R. (1997). *Clostridium spiroforme* toxin genes are related to *C. perfringens* iota toxin genes but have a different genomic localization. *Syst. Appl. Microbiol.* **20**, 337–47.

Giry, M., Popoff, M.R., Eichel-Streiber, Cv. and Boquet, P. (1995). Transient expression of RhoA, -B, and -C GTPases in HeLa cells potentiates resistance to *Clostridium difficile* toxins A and B but not to *Clostridium sordellii* lethal toxin. *Infect. Immun.* **63**, 4063–71.

Green, G.A., Schue, V. and Monteil, H. (1995). Cloning and characterization of the cytotoxin L-encoding gene of *Clostridium sordellii*: homology with *Clostridium difficile* cytotoxin B. *Gene* **161**, 57–61.

Gurland, G. and Gundersen, G.G. (1995). Stable, detyrosinated microtubules function to localize vimentin intermediate filaments in fibroblasts. *J. Cell Biol.* **131**, 1275–90.

Hall, A. (1998). Rho GTPases and the actin cytoskeleton. *Science* **279**, 509–14.

Hammond, C. and Helenius, A. (1995). Quality control in the secretory pathway. *Curr. Opin. Cell Biol.* **7**, 523–9.

Hammond, G.A. and Johnson, J.L. (1995). The toxigenic element of *Clostridium difficile* strain VPI 10463. *Microb. Pathog.* **19**, 203–13.

Hauser, D., Gibert, M., Eklund, M.W., Boquet, P. and Popoff, M.R. (1993). Comparative analysis of C3 and botulinal neurotoxin genes and their environment in *Clostridium botulinum* types C and D. *J. Bacteriol.* **175**, 7260–8.

Hauser, D., Gibert, M., Marvaud, J.C., Eklund, M.W. and Popoff, M.R. (1995). Botulinal neurotoxin C1 complex genes, clostridial neurotoxin homology and genetic transfer in *Clostridium botulinum*. *Toxicon* **33**, 515–26.

Hayes, B.K. and Hart, G.W. (1998). Protein O-GlcNAcylation: potential mechanisms for the regulation of protein function. *Adv. Exp. Med. Biol.* **435**, 85–94.

Hecht, G., Pothoulakis, C., LaMont, J.T. and Madara, J. (1988). *Clostridium difficile* toxin A perturbs cytoskeletal structure and tight junction permeability of cultured human intestinal epithelial monolayers. *J. Clin. Invest.* **82**, 1516–24.

Hecht, G., Koutsouris, A., Pothoulakis, C., LaMont, J.T. and Madara, J. (1992). *Clostridium difficile* toxin B disrupts the barrier function of T84 monolayers. *Gastroenterology* **102**, 416–23.

Henning, S.W., Galandrini, R., Hall, A. and Cantrell, D.A. (1997). The GTPase Rho has a critical regulatory role in thymus development. *EMBO J.* **16**, 2397–407.

Henriques, B., Florin, I. and Thelestam, M. (1987). Cellular internalisation of *Clostridium difficile* toxin A. *Microb. Pathogen.* **2**, 455–63.

Herrmann, C., Ahmadian, M.R., Hofmann, F. and Just, I. (1998). Functional consequences of monoglucosylation of Ha-Ras at effector domain amino acid threonine 35. *J. Biol. Chem.* **273**, 16134–9.

Hippenstiel, S., Tannert-Otto, S., Vollrath, N., Krull, M., Just, I., Aktories, K., Ecihel-Streiber, Cv. and Suttorp, N. (1997). Glucosylation of small GTP-binding proteins disrupts endothelial barrier function. *Am. J. Physiol.* **272**, L38–43.

Hippenstiel, S., Kratz, T., Krüll, M., Seybold, J., Eichel-Streiber, Cv. and Suttorp, N. (1998). Rho protein inhibition blocks protein kinase C translocation and activation. *Biochem. Biophys. Res. Commun.* **245**, 830–4.

Hofmann, F., Herrmann, A., Habermann, E. and von Eichel-Streiber, C. (1995). Sequencing and analysis of the gene encoding the alpha-toxin of *Clostridium novyi* proves its homology to toxins A and B of *Clostridium difficile*. *Mol. Gen. Genet.* **247**, 670–9.

Hofmann, F., Rex, G., Aktories, K. and Just, I. (1996). The Ras-related protein Ral is monoglucosylated by *Clostridium sordellii* lethal toxin. *Biochem. Biophys. Res. Commun.* **227**, 77–81.

Hofmann, F., Busch, C., Prepens, U., Just, I. and Aktories, K. (1997). Localization of the glucosyltransferase activity of *Clostridium difficile* toxin B to the *N*-terminal part of the holotoxin. *J. Biol. Chem.* **272**, 11074–8.

Hofmann, F., Busch, C. and Aktories, K. (1998). Chimeric clostridial cytotoxins: identification of the *N*-terminal region involved in protein substrate recognition. *Infect. Immun.* **66**, 1076–81.

Hotchin, N.A. and Hall, A. (1996). Regulation of the actin cytoskeleton, integrins and cell growth by the Rho family small GTPases. *Cancer Surv.* **27**, 311–22.

Hundsberger, T., Braun, V., Weidmann, M., Leukel, P., Sauerborn, M. and von Eichel-Streiber, C. (1997). Transcription analysis of the genes tcdA-E of the pathogenicity locus of *Clostridium difficile*. *Eur. J. Biochem.* **244**, 735–42.

Ichikawa, S. and Hirabayashi, Y. (1998). Glucosylceramide synthase and glycosphingolipid synthesis. *Trends Cell Biol.* **8**, 198–202.

Inoue, S., Sugai, M., Murooka, Y., Paik, S.Y., Hong, Y.M., Ohgai, H. and Suginaka, H. (1991). Molecular cloning and sequencing of the epidermal cell differentiation inhibitor gene from *Staphylococcus aureus*. *Biochem. Biophys. Res. Commun.* **174**, 459–64.

Jumaa, P., Wren, B. and Tabaqchali, S. (1996). Epidemiology and typing of *Clostridium difficile*. *Eur. J. Gastroenterol. Hepatol.* **8**, 1035–40.

Just, I., Mohr, C., Schallehn, G., Menard, L., Didsbury, J.R., Vandekerckhove, J., van Damme, J. and Aktories, K. (1992a). Purification and characterization of an ADP-ribosyltransferase produced by *Clostridium limosum*. *J. Biol. Chem.* **267**, 10274–80.

Just, I., Schallehn, G. and Aktories, K. (1992b). ADP-ribosylation of small GTP-binding proteins by *Bacillus cereus*. *Biochem. Biophys. Res. Commun.* **183**, 931–6.

Just, I., Fritz, G., Aktories, K., Giry, M., Popoff, M.R., Boquet, P., Hegenbarth, S. and Eichel-Streiber, Cv. (1994). *Clostridium difficile* toxin B acts on the GTP-binding protein Rho. *J. Biol. Chem.* **269**, 10706–12.

Just, I., Wilm, M., Selzer, J., Rex, G., Eichel-Streiber, Cv., Mann, M. and Aktories, K. (1995a). The enterotoxin from *Clostridium difficile* (ToxA) monoglucosylates the Rho proteins. *J. Biol. Chem.* **270**, 13932–6.

Just, I., Selzer, J., Wilm, M., Eichel-Streiber, Cv., Mann, M. and Aktories, K. (1995b). Glucosylation of Rho proteins by *Clostridium difficile* toxin B. *Nature* **375**, 500–3.

Just, I., Selzer, J., Jung, M., van Damme, J., Vandekerckhove, J. and Aktories, K. (1995c). Rho-ADP-ribosylating exoenzyme from *Bacillus cereus*. Purification, characterization, and identification of the NAD-binding site. *Biochemistry* **34**, 334–40.

Just, I., Selzer, J., Eichel-Streiber, Cv. and Aktories, K. (1995d). The low molecular mass GTP-binding protein Rho is affected by toxin A from *Clostridium difficile*. *J. Clin. Invest.* **95**, 1026–31.

Katanaev, V.L. and Wymann, M. (1998). GTPγS-induced actin polymerisation *in vitro*: ATP- and phosphoinositide-independent signalling via Rho-family proteins and a plasma membrane-associated guanine nucleotide exchange factor. *J. Cell Sci.* **111**, 1583–94.

Kato, Y., Banno, Y., Dohjima, T., Kato, N., Watanabe, K., Tatematsu, N. and Nozawa, Y. (1997). Involvement of Rho family proteins in prostaglandin F2 alpha (PGF2 alpha)-mediated phospholipase D (PLD) activation in the osteoblast-like cell line MC3T3-E1. *Prostaglandins* **54**, 475–92.

Kelly, C.P. and LaMont, J.T. (1998). *Clostridium difficile* infection. *Annu. Rev. Med.* **49**, 375–90.

Kelly, C.P., Pothoulakis, C. and LaMont, J.T. (1994). *Clostridium difficile* colitis. *N. Engl. J. Med.* **330**, 257–62.

Kiefer, G., Lerner, M., Sehr, P., Just, I. and Aktories, K. (1996). Cytotoxic effects by microinjection of ADP-ribosylated skeletal muscle G-actin in PtK2 cells in the absence of *Clostridium perfringens* iota toxin. *Med. Microbiol. Immunol.* **184**, 175–80.

Kimura, K., Kubota, T., Ohishi, I., Isogai, E., Isogai, H. and Fujii, N. (1998). The gene for component-II of botulinum C2 toxin. *Vet. Microbiol.* **62**, 27–34.

Kink, J.A. and Williams, J.A. (1998). Antibodies to recombinant *Clostridium difficile* toxins A and B are an effective treatment and prevent relapse of *C. difficile*-associated disease in a hamster model of infection. *Infect. Immun.* **66**, 2018–25.

Kowluru, A., Li, G., Rabaglia, M.E., Segu, V.B., Hofmann, F., Aktories, K. and Metz, S.A. (1997). Evidence for differential roles of the Rho subfamily of GTP-binding proteins in glucose- and calcium-induced insulin secretion from pancreatic beta cells. *Biochem. Pharmacol.* **54**, 1097–108.

Kranenburg, O., Poland, M., Gebbink, M., Oomen, L. and Moolenaar, W.H. (1997). Dissociation of LPA-induced cytoskeletal contraction from stress fibre formation by differential localization of RhoA. *J. Cell Sci.* **110**, 2417–27.

Krivan, H.C., Clark, G.F., Smith, D.F. and Wilkins, T.D. (1986). Cell surface binding site for *Clostridium difficile* enterotoxin: evidence for a glycoconjugate containing the sequence Galα1–3Galβ1–4GlcNAc. *Infect. Immun.* **53**, 573–81.

Kushnaryov, V.M., Redlich, P.N., Sedmak, J.J., Lyerly, D.M. and Wilkins, T.D. (1992). Cytotoxicity of *Clostridium difficile* toxin A for human colonic and pancreatic carcinoma cell lines. *Cancer Res.* **52**, 5096–9.

Linevsky, J.K. and Kelly, C.P. (1997). *Clostridium difficile* colitis. In: *Gastrointestinal Infections* (ed. J.T. LaMont), pp. 293–325. Marcel Dekker, New York.

Linevsky, J.K., Pothoulakis, C., Keates, S., Warny, M., Keates, A.C., LaMont, J.T. and Kelly, C.P. (1997). IL-8 release and neutrophil activation by *Clostridium difficile* toxin-exposed human monocytes. *Am. J. Physiol.* **273**, G1333–40.

Lucius, C., Arner, A., Steusloff, A., Troschka, M., Hofmann, F., Aktories, K. and Pfitzer, G. (1998). *Clostridium difficile* toxin B inhibits carbachol-induced force and myosin light chain phosphorylation in guinea pig smooth muscle: role of Rho proteins. *J. Physiol.* **506**, 83–93.

Lyerly, D.M. and Wilkins, T.D. (1995). *Clostridium difficile*. In: *Infections of the Gastrointestinal Tract* (eds M.J. Blaser *et al.*), pp. 867–91. Raven Press, New York.

Lyerly, D.M., Neville, L.M., Evans, D.T., Fill, J., Allen, S., Green, W., Sautter, R., Hnatuck, P., Torpey, D.J. and Schwalbe, R. (1998). Multicenter evaluation of the *Clostridium difficile* TOX A/B TEST. *J. Clin. Microbiol.* **36**, 184–90.

Majumdar, M., Seasholtz, T.M., Goldstein, D., de LAnerolle, P. and Heller Brown, J. (1998). Requirement for Rho-mediated myosin light chain phosphorylation in thrombin-stimulated cell rounding and its dissociation from mitogenesis. *J. Biol. Chem.* **273**, 10099–106.

Mantyh, C.R., Maggio, J.E., Mantyh, P.W., Vigna, S.R. and Pappas, T.N. (1996). Increased substance P receptor expression by blood vessels and lymphoid aggregates in *Clostridium difficile*-induced pseudomembranous colitis. *Dig. Dis. Sci.* **41**, 614–20.

Molitoris, B.A. (1997). Putting the actin cytoskeleton into perspective: pathophysiology of ischemic alterations. *Am. J. Physiol.* **272**, F430–3.

Molitoris, B.A., Leiser, J. and Wagner, M.C. (1997). Role of the actin cytoskeleton in ischemia-induced cell injury and repair. *Pediatr. Nephrol.* **11**, 761–7.

Moncrief, J.S., Barroso, L.A. and Wilkins, T.D. (1997). Positive regulation of *Clostridium difficile* toxins. *Infect. Immun.* **65**, 1105–8.

Moorman, J.P., Bobak, D.A. and Hahn, C.S (1996). Inactivation of the small GTP-binding protein Rho induces multinucleate cell formation and apoptosis in murine T lymphoma EL4. *J. Immunol.* **156**, 4146–53.

Moriishi, K., Syuto, B., Saito, M., Oguma, K., Fujii, N., Abe, N. and Naiki, M. (1993). Two different types of ADP-ribosyltransferase C3 from *Clostridium botulinum* type D lysogenized organisms. *Infect. Immun.* **61**, 5309–14.

Müller, H., Eichel-Streiber, Cv. and Habermann, E. (1992). Morphological changes of cultured endothelial cells after microinjection of toxins that act on the cytoskeleton. *Infect. Immun.* **60**, 3007–10.

Nakamura, S., Mikawa, M., Tanabe, N., Yamakawa, K. and Nishida, S. (1982). Effect of clindamycin on cytotoxin production by *Clostridium difficile*. *Microbiol. Immunol.* **26**, 985–92.

Nemoto, Y., Namba, T., Kozaki, S. and Narumiya, S. (1991). *Clostridium botulinum* C3 ADP-ribosyltransferase gene. Cloning, sequencing, and expression of a functional protein in *Escherichia coli*. *J. Biol. Chem.* **266**, 19312–19.

Nusrat, A., Giry, M., Turner, J.R., Colgan, S.P., Parkos, C.A., Carnes, D., Lemichez, E., Boquet, P. and Madara, J.L. (1995). Rho protein regulates tight junctions and perijunctional actin organization in polarized epithelia. *Proc. Natl Acad. Sci. USA* **92**, 10629–63.

Oguma, K., Iida, H. and Inoue, K. (1975). Observations on nonconverting phage, c-n71, obtained from a nontoxigenic strain of *Clostridium botulinum* type C. *Jpn. J. Microbiol.* **19**, 167–72.

Ohishi, I. (1983). Response of mouse intestinal loop to botulinum C2 toxin: enterotoxic activity induced by cooperation of nonlinked protein components. *Infect. Immun.* **40**, 691–5.

Ohishi, I. and Yanagimoto, A. (1992). Visualizations of binding and internalization of two nonlinked protein components of botulinum C2 toxin in tissue culture cells. *Infect. Immun.* **60**, 4648–55.

Ojio, K., Banno, Y., Nakashima, S., Kato, N., Watanabe, K., Lyerly, D.M., Miyata, H. and Nozawa, Y. (1996). Effect of *Clostridium difficile* toxin B on IgE receptor-mediated signal transduction in rat basophilic leukaemia cells: inhibition of phospholipase D activation. *Biochem. Biophys. Res. Commun.* **224**, 591–6.

Ojio, K., Banno, Y., Hayakawa, K., Ito, Y., Kato, N., Watanabe, K., Miyata, H. and Nozawa, Y. (1998). Role of Rho family GTP-binding proteins in IgE receptor-mediated phospholipase D activation in mast cells. *Biomed. Res.* **19**, 53–63.

Oksche, A., Nakov, R. and Habermann, E. (1992). Morphological and biochemical study of cytoskeletal changes in cultured cells after extracellular application of *Clostridium novyi* alpha-toxin. *Infect. Immun.* **60**, 3002–6.

Onderdonk, A.B., Lowe, B.R. and Bartlett, J.G. (1979). Effect of environmental stress on *Clostridium difficile* toxin levels during continuous cultivation. *Appl. Environ. Microbiol.* **38**, 637–41.

Ottlinger, M.E. and Lin, S. (1988). *Clostridium difficile* toxin B induces reorganization of actin, vinculin, and talin in cultured cells. *Exp. Cell Res.* **174**, 215–29.

Perelle, S., Gibert, M., Boquet, P. and Popoff, M.R. (1993). Characterization of *Clostridium perfringens* iota-toxin genes and expression in *Escherichia coli*. *Infect. Immun.* **61**, 5147–56.

Perelle, S., Gibert, M., Boquet, P. and Popoff, M.R. (1995). Characterization of *Clostridium perfringens* Iota-toxin genes and expression in *Escherichia coli*. *Infect. Immun.* **63**, 4967.

Perelle, S., Domenighini, M. and Popoff, M.R. (1996). Evidence that Arg-295, Glu-378, and Glu-380 are active-site residues of the ADP-ribosyltransferase activity of iota toxin. *FEBS Lett.* **395**, 191–4.

Perelle, S., Gibert, M., Bourlioux, P., Corthier, G. and Popoff, M. (1997). Production of a complete binary toxin (actin-specific ADP-ribosyltransferase) by *Clostridium difficile* CD196. *Infect. Immun.* **65**, 1402–7.

Popoff, M., Rubin, E.J., Gill, D.M. and Boquet, P. (1988). Actin-specific ADP-ribosyltransferase produced by a *Clostridium difficile* strain. *Infect. Immun.* **56**, 2299–306.

Popoff, M.R., Hauser, D., Boquet, P., Eklund, M.W. and Gill, D.M. (1991). Characterization of the C3 gene of *Clostridium botulinum* types C and D and its expression in *Escherichia coli. Infect. Immun.* **59**, 3673–9.

Popoff, M.R., Chaves-Olarte, E., Lemichez, E., Eichel-Streiber, Cv., Thelestam, M., Chardin, P., Cussac, D., Antonny, B., Chavrier, P., Flatau, G., Giry, M., Gunzburg, Jd. and Boquet, P. (1996). Ras, Rap, and Rac small GTP-binding proteins are targets for *Clostridium sordellii* lethal toxin glucosylation. *J. Biol. Chem.* **271**, 10217–24.

Pothoulakis, C. (1996). Pathogenesis of *Clostridium difficile*-associated diarrhoea. *Eur. J. Gastroenterol. Hepatol.* **8**, 1041–7.

Pothoulakis, C., LaMont, J.T., Eglow, R., Gao, N., Rubins, J.B., Theoharides, T.C. and Dickey, B.F. (1991). Characterization of rabbit ileal receptors for *Clostridium difficile* toxin A. Evidence for a receptor-coupled G protein. *J. Clin. Invest.* **88**, 119–25.

Pothoulakis, C., Gilbert, R.J., Cladaras, C., Semenza, G., Hitti, Y., Montcrief, J.S., Linevsky, J., Kelly, C.P., Nikulasson, S., Desai, H.P., Wilkins, T.D. and LaMont, J.T. (1996). Rabbit sucrase-isomaltase contains a functional intestinal receptor for *Clostridium difficile* toxin A. *J. Clin Invest.* **98**, 641–9.

Prepens, U., Just, I., Eichel-Streiber, Cv. and Aktories, K. (1996). Inhibition of Fc epsilon-RI-mediated activation of rat basophilic leukaemia cells by *Clostridium difficile* toxin B (monoglucosyltransferase). *J. Biol. Chem.* **271**, 7324–9.

Prepens, U., Just, I., Hofmann, F. and Aktories, K. (1997). ADP-ribosylating and glucosylating toxins as tools to study secretion in RBL cells. *Adv. Exp. Med. Biol.* **419**, 349–53.

Price, S.B., Phelps, C.J., Wilkins, T.D. and Johnson, J.L. (1987). Cloning of the carbohydrate-binding portion of the toxin A gene of *Clostridium difficile. Curr. Microbiol.* **16**, 55–60.

Ren, X.D. and Schwartz, M.A. (1998). Regulation of inositol lipid kinases by Rho and Rac. *Curr. Opin. Genet. Dev.* **8**, 63–7.

Reuner, K.H., Dunker, P., van der Does, A., Wiederhold, M., Just, I., Aktories, K. and Katz, N. (1996a). Regulation of actin synthesis in rat hepatocytes by cytoskeletal rearrangements. *Eur. J. Cell Biol.* **69**, 189–96.

Reuner, K.H., van der Does, A., Dunker, P., Just, I., Aktories, K. and Katz, N. (1996b). Microinjection of ADP-ribosylated actin inhibits actin synthesis in hepatocyte-hepatoma hybrid cells. *Biochem. J.* **319**, 843–9.

Riegler, M., Sedivy, R., Pothoulakis, C., Hamilton, G., Zacherl, J., Bischof, G., Cosentini, E., Feil, W., Schiessel, R., LaMont, J.T. and Wenzl, E. (1995). *Clostridium difficile* toxin B is more potent than toxin A in damaging human colonic epithelium *in vitro. J. Clin. Invest.* **95**, 2004–11.

Riegler, M., Sedivy, R., Sogukoglu, T., Castagliuolo, I., Pothoulakis, C., Cosentini, E., Bischof, G., Hamilton, G., Teleky, B., Feil, W., LaMont, J.T. and Wenzl, E. (1997). Epidermal growth factor attenuates *Clostridium difficile* toxin A- and B-induced damage of human colonic mucosa. *Am. J. Physiol.* **273**, G1014–22.

Rocha, M.F., Maia, M.E., Bezerra, L.R., Lyerly, D.M., Guerrant, R.L., Ribeiro, R.A. and Lima A.A. (1997). *Clostridium difficile* toxin A induces the release of neutrophil chemotactic factors from rat peri-toneal macrophages: role of interleukin-1b, tumor necrosis factor alpha, and leukotrienes. *Infect. Immun.* **65**, 2740–6.

Rupnik, M., Braun, V., Soehn, F., Janc, M., Hofstetter, M., Laufenberg Feldmann, R. and von Eichel-Streiber, C. (1997). Characterization of polymorphisms in the toxin A and B genes of *Clostridium difficile. FEMS Microbiol. Lett.* **148**, 197–202.

Rupnik, M., Avesani, V., Janc, M., Eichel-Streiber, Cv. and Delmée, M. (1998). A novel toxinotyping scheme and correlation of toxinotypes with serogroups of *Clostridium difficile* isolates. *J. Clin. Microbiol.* **36**, 2240–7.

Ryan, E.T., Butterton, J.R., Smith, R.N., Carroll, P.A., Crean, T.I. and Calderwood, S.B. (1997). Protective immunity against *Clostridium difficile* toxin A induced by oral immunization with a live, attenuated *Vibrio cholerae* vector strain. *Infect. Immun.* **65**, 2941–9.

Saito, Y., Nemoto, Y., Ishizaki, T., Watanabe, N., Morii, N. and Narumiya, S. (1995). Identification of Glu173 as the critical amino acid residue for the ADP-ribosyltransferase activity of *Clostridium botulinum* C3 exoenzyme. *FEBS Lett.* **371**, 105–9.

Santos, M.F., McCormack, S.A., Guo, Z., Okolicany, J., Zheng, Y., Johnson, L.R. and Tigyi, G. (1997). Rho proteins play a critical role in cell migration during the early phase of mucosal restitution. *J. Clin. Invest.* **100**, 216–25.

Sauerborn, M. and von Eichel-Streiber, C. (1990). Nucleotide sequence of *Clostridium difficile* toxin A. *Nucleic Acids Res.* **18**, 1629–30.

Sauerborn, M., Hegenbarth, S., Laufenberg-Feldmann, R., Leukel, P. and Eichel-Streiber, Cv. (1994). Monoclonal antibodies discriminating between *Clostridium difficile* toxins A and B. In: *Bacterial Protein Toxins* (eds J. Freer, R. Aitken, J.E. Alouf, G. Boulnois, P. Falmagne, F. Fehrenbach, C. Montecucco, Y. Piemont, R. Rappuoli, T. Wadströw and B. Witholt), pp. 510–11. Gustav Fischer, Stuttgart.

Sauerborn, M., Leukel, P. and Eichel-Streiber, Cv. (1997). The C-terminal ligand-binding domain of *Clostridium difficile* toxin A (TcdA). abrogates TcdA-specific binding to cells and prevents mouse lethality. *FEMS Microbiol. Lett.* **155**, 45–54.

Schallehn, G., Eklund, M.W. and Brandis, H. (1980). Phage conversion of *Clostridium novyi* type A. *Zentralbl. Bakteriol. A* **247**, 95–100.

Scheffzek, K., Ahmadian, M.R. and Wittinghofer, A. (1998). GTPase-activating proteins: helping hands to complement an active site. *Trends Biochem Sci*, **23**, 257–62.

Schmid, A., Benz, R., Just, I. and Aktories, K. (1994). Interaction of *Clostridium botulinum* C2 toxin with lipid bilayer membranes. Formation of cation-selective channels and inhibition of channel function by chloroquine. *J. Biol. Chem.* **269**, 16706–11.

Schmidt, M., Bienek, C., Rumenapp, U., Zhang, C., Lummen, G., Jakobs, K.H., Just, I., Aktories, K., Moos, M. and Eichel-Streiber, Cv. (1996a). A role for Rho in receptor- and G protein stimulated phospholipase C. Reduction in phosphatidylinositol 4,5-bisphosphate by *Clostridium difficile* toxin B. *Naunyn-Schmiedebergs Arch. Pharmacol.* **354**, 87–94.

Schmidt, M., Rumenapp, U., Bienek, C., Keller, J., Eichel-Streiber, Cv. and Jakobs, K.H. (1996b). Inhibition of receptor signaling to phospholipase D by *Clostridium difficile* toxin B. Role of Rho proteins. *J. Biol. Chem.* **271**, 2422–6.

Schmidt, M., Voss, M., Thiel, M., Bauer, B., Grannass, A., Tapp, E., Cool, R.H., Gunzburg, Jd., Eichel-Streiber, Cv. and Jakobs, K.H. (1998). Specific inhibition of phorbol ester-stimulated phospholipase D by *Clostridium sordellii* lethal toxin and *Clostridium difficile* toxin B-1470 in HEK-293 cells. *J. Biol. Chem.* **273**, 7413–22.

Sehr, P., Joseph, G., Genth, H., Just, I., Pick, E. and Aktories, K. (1998). Glucosylation and ADP-ribosylation of Rho proteins: effects on nucleotide binding, GTPase activity, and effector coupling. *Biochemistry* **37**, 5296–304.

Selzer, J., Hofmann, F., Rex. G., Wilm, M., Mann, M., Just, I. and Aktories, K. (1996). *Clostridium novyi* alpha-toxin-catalyzed incorporation of GlcNAc into Rho subfamily proteins. *J. Biol. Chem.* **271**, 25173–4.

Shoshan, M.C., Åman, P., Skoog, S., Florin, I. and Thelestam, M. (1990). Microfilament-disrupting *Clostridium difficile* toxin B causes multinucleation of transformed cells but does not block capping of membrane Ig. *Eur. J. Cell Biol.* **53**, 357–63.

Simpson, L.L. (1989). The binary toxin produced by *Clostridium botulinum* enters cells by receptor-mediated endocytosis to exert its pharmacologic effects. *J. Pharmacol. Exp. Ther.* **251**, 1223–8.

Simpson, L.L. (1997). The role of the *Clostridium botulinum* C2 toxin as a research tool to study eucaryotic cell biology. In: *Bacterial Toxins – Tools in Cell Biology and Pharmacology* (ed. K. Aktories), pp. 117–28. Chapman & Hall, Weinheim.

Smith, J.A., Cooke, D.L., Hyde, S., Borriello, S.P. and Long, R.G. (1997). *Clostridium difficile* toxin A binding to human intestinal epithelial cells. *J. Med. Microbiol.* **46**, 953–8.

Snow, D.M. and Hart, G.W (1998). Nuclear and cytoplasmic glycosylation. *Int. Rev. Cytol.* **181**, 43–74.

Soehn, F., Wagenknecht Wiesner, A., Leukel, P., Kohl, M., Weidmann, M., von Eichel-Streiber, C. and Braun, V. (1998). Genetic rearrangements in the pathogenicity locus of *Clostridium difficile* strain 8864 – implications for transcription, expression and enzymatic activity of toxins A and B. *Mol. Gen. Genet.* **258**, 222–32.

Souza, M.H., Melo-Filho, A.A., Rocha, M.F., Lyerly, D.M., Cunha, F.Q., Lima, A.A. and Ribeiro, R.A. (1997). The involvement of macrophage-derived tumour necrosis factor and lipoxygenase products on the neutrophil recruitment induced by *Clostridium difficile* toxin B. *Immunology* **91**, 281–8.

Sugai, M., Chen, C.H. and Wu, H.C. (1992). Bacterial ADP-ribosyltransferase with a substrate specificity of the rho protein disassembles the Golgi apparatus in Vero cells and mimics the action of brefeldin A. *Proc. Natl Acad. Sci. USA* **89**, 8903–7.

Sullivan, N.M., Pellet, S., Wilkins, T.D. (1982). Purification and characterization of toxins A and B of *Clostridium difficile*. *Infect. Immun.* **35**, 1032–40.

Taylor, N.S., Thorne, G.M. and Bartlett, J.G. (1981). Comparison of two toxins produced by *Clostridium difficile*. *Infect. Immun.* **34**, 1036–43.

Teneberg, S., Lönnroth, I., Torres Lopez, J.F., Galili, U., Halvarsson, M.O., Ångström, J. and Karlsson, K.A. (1996). Molecular mimicry in the recognition of glycosphingolipids by Galα3Galβ4GlcNAcβ-binding *Clostridium difficile* toxin A, a human natural anti α-galactosyl IgG and the monoclonal antibody Gal-13: characterization of a binding-active human glycosphingolipid, non-identical with the animal receptor. *Glycobiology* **6**, 599–609.

Thelestam, M. and Brönnegård, M. (1980). Interaction of cytopathogenic toxin from *Clostridium difficile* with cells in tissue culture. *Scand. J. Inf. Dis. Suppl.* **22**, 16–29.

Thelestam, M. and Gross, R. (1990). Toxins acting on the cytoskeleton. In: *Handbook of Toxinology* (eds W.T. Shier and D. Mebs), pp. 423–92. Marcel Dekker, New York.

Thelestam, M., Florin, I. and Chaves-Olarte, E. (1997). *Clostridium difficile* toxins. In: *Bacterial Toxins – Tools in Cell Biology and Pharmacology* (ed. K. Aktories), pp. 141–58. Chapman & Hall, Weinheim.

Torres, J.F., Lyerly, D.M., Hill, J.E. and Monath, T.P. (1995). Evaluation of formalin-inactivated *Clostridium difficile* vaccines administered by parenteral and mucosal routes of immunization in hamsters. *Infect. Immun.* **63**, 4619–27.

Tucker, K.D. and Wilkins, T.D. (1991). Toxin A of *Clostridium difficile* binds to the human carbohydrate antigens I, X and Y. *Infect. Immun.* **59**, 73–8.

Valitutti, S., Dessing, M., Aktories, K., Gallati, H. and Lanzavecchia, A. (1995). Sustained signaling leading to T cell activation results from prolonged T cell receptor occupancy. Role of T cell actin cytoskeleton. *J. Exp. Med.* **181**, 577–84.

Van Aelst, L. and D'Souza-Schorey, C. (1997). Rho GTPases and signaling networks. *Genes Dev.* **11**, 2295–322.

Van Damme, J., Jung, M., Hofmann, F., Just, I., Vandekerckhove, J. and Aktories, K. (1996). Analysis of the catalytic site of the actin ADP-ribosylating *Clostridium perfringens* iota toxin. *FEBS Lett.* **380**, 291–5.

Verschueren, H., De Baetselier, P., De Braekeleer, J., Dewit, J., Aktories, K. and Just, I. (1997). ADP-ribosylation of Rho-proteins with botulinum C3 exoenzyme inhibits invasion and shape changes of T-lymphoma cells. *Eur. J. Cell Biol.* **73**, 182–7.

Wagenknecht-Wiesner, A., Weidmann, M., Braun, V., Leukel, P., Moos, M. and Eichel-Streiber, Cv. (1997). Delineation of the catalytic domain of *Clostridium difficile* toxin B-10463 to an enzymatically active *N*-terminal 467 amino acid fragment. *FEMS Microbiol. Lett.* **152**, 109–16.

Watarai, M., Kamata, Y., Kozaki, S. and Sasakawa, C. (1997). Rho, a small GTP-binding protein, is essential for *Shigella* invasion of epithelial cells. *J. Exp. Med.* **185**, 281–92.

Weidmann, M. (1998). Vergleichende Untersuchungen zur Glukosyltransferase-Reaktion großer clostridialer Zytotoxine. Ph.D. thesis, Faculty of Biology, Johannes Gutenberg-Universität, Mainz, Germany.

Wenzel-Seifert, K., Lentzen, H., Aktories, K. and Seifert, R. (1997). Complex regulation of human neutrophil activation by actin filaments: dihydrocytochalasin B and botulinum C2 toxin uncover the existence of multiple cation entry pathways. *J. Leuk. Biol.* **61**, 703–11.

Wex, C.B., Koch, G. and Aktories, K. (1997a). Effects of *Clostridium botulinum* C2 toxin-induced depolymerisation of actin on degranulation of suspended and attached mast cells. *Naunyn-Schmiedebergs Arch. Pharmacol.* **355**, 319–27.

Wex, C.B., Koch, G. and Aktories, K. (1997b). Effects of *Clostridium difficile* toxin B on activation of rat peritoneal mast cells. *Naunyn-Schmiedebergs Arch. Pharmacol.* **355**, 328–34.

Wiggins, C.A.R. and Munro, S. (1998). Activity of the yeast MNN1 alpha-1,3-mannosyltransferase requires a motif conserved in many other families of glycosyltransferases. *Proc. Natl Acad. Sci. USA* **95**, 7945–50.

Yamakawa, K., Karasawa, T., Ikoma, S. and Nakamura, S. (1996). Enhancement of *Clostridium difficile* toxin production in biotin-limited conditions. *J. Med. Microbiol.* **44**, 111–14.

Yoshioka, K., Matsumura, F., Akedo, H. and Itoh, K. (1998). Small GTP-binding protein Rho stimulates the actomyosin system, leading to invasion of tumor cells. *J. Biol. Chem.* **273**, 5146–54.

Structural and genomic features of clostridial neurotoxins

Michel R. Popoff and Jean-Christophe Marvaud

INTRODUCTION

Tetanus toxin (TeTx) and botulinum neurotoxins (BoNTs) are produced by *Clostridium tetani* and *C. botulinum*, respectively, and are responsible for severe diseases in human and animals, which are characterized by specific neurological disorders. Tetanus is accompanied by spastic paralysis and botulism by flaccid paralysis. Despite the opposing clinical symptoms, these toxins display a similar structure and a similar mechanism of action at the cellular level. They inhibit the exocytosis of neurotransmitter. However, TeTx acts on inhibitory interneuron synapses in the central nervous system, whereas BoNTs block the acetylcholine release at the neuromuscular junctions of the motor neuron endings. Important insights into the mode of action have been gained in recent years (see Chapter 10). The molecular analysis of neurotoxins made a significant contribution to these studies.

The seven toxinotypes of BoNTs (A, B, C1, D, E, F and G) based on their antigenic properties and the unique type of TeTx are synthesized as a single protein (approximately 150 kDa) which is proteolytically activated into a light chain (L) (approximately 50 kDa) and a heavy chain (H) (approximately 100 kDa). Both chains remain linked by a disulfide bridge. In natural conditions or *in vitro* culture, BoNTs are associated with non-toxic proteins (ANTPs) by non-covalent bonds, and form botulinum complexes. Some of the ANTPs exhibit haemagglutinin activity. The size of botulinum complexes varies from 230 to about 900 kDa depending on toxinotypes and strains (Sakaguchi *et al.*, 1988). ANTPs dissociate from BoNT at high pH (pH 8 and above) and reassociate spontaneously when the pH is lowered. The main role of ANTPs is supposed to protect BoNTs from gastric acidity and digestive proteases, and to mediate the BoNT absorption through the intestinal mucosa (Hambleton, 1992). In contrast, TeTx does not form complexes.

The structural genes of the clostridial neurotoxins and the various ANTP genes have been characterized. Genetic analysis has provided information on gene organization, gene transfer and regulation of gene expression, and has facilitated the development of variant toxins for vaccination and therapeutic use, and tools for cell biology.

NEUROTOXIN-PRODUCING CLOSTRIDIA

TeTx-producing *Clostridia* display homogeneous bacteriological characteristics and belong to a uniform group forming the *C. tetani* species. Some *C. tetani* strains are non-toxigenic, and are indistinguishable from the toxigenic strains by their phenotypic characteristics and DNA/DNA homology, except for the production of toxin.

In contrast, the BoNT-producing *Clostridia* are heterogeneous in their physiological and genetic characteristics, and belong to several species and groups

174

(Table 9.1). The species *C. botulinum* was originally identified by production of BoNT, which typically induces flaccid paralysis in experimental animals. It was soon discovered that BoNT fell into seven different toxin types (A, B, C1, D, E, F and G) according to their antigenic properties. The *C. botulinum* species was divided into four physiological groups:

- group I: *C. botulinum* A, and proteolytic strains of *C. botulinum* B and F
- group II: *C. botulinum* E, and glucolytic strains of *C. botulinum* B and F
- group III: *C. botulinum* C and D
- group IV: *C. botulinum* G. This group, which also includes non-toxic strains previously identified as *C. subterminale* and *C. hastiforme*, is metabolically distinct from the other groups and has been assigned to a different species called *C. argentinense* (Suen *et al.*, 1988).

The taxonomic position of *C. botulinum* has become more ambiguous since the discovery that BoNT can be produced by *Clostridium* strains clearly distinct from already defined *C. botulinum,* and biochemically and genetically related to different species such as *C. butyricum* and *C. baratii* (Hall *et al.*, 1985; McCroskey *et al.*, 1986, 1991).

In each group all of the strains, regardless of toxin types, are closely related according to their phenotypic properties, DNA/DNA homology and 16S rRNA analysis. The atypical toxigenic *C. butyricum* and *C. baratii* strains are phenotypically and genetically related to the type strains of these species and not to the other BoNT-producing *Clostridia* (Popoff, 1995).

HABITAT OF NEUROTOXIN-PRODUCING CLOSTRIDIA

The main habitat of the neurotoxin-producing *Clostridia*, like other *Clostridium* species, is the environment. These bacteria are sporulating microorganisms, and the spores are resistant to extreme conditions (heat, dryness, radiation, chemicals, oxygen), which enables them to survive for very long periods. *Clostridia* are therefore ubiquitous and widely distributed in the environment. However, spore germination and cell division take place in anaerobic conditions and when nutritional requirements are available. This restricts the habitat of the *Clostridia* to anaerobic to low oxygen tension areas and those containing sufficient amounts of organic materials.

Clostridium tetani is a ubiquitous organism commonly found in soils throughout the world. However, it is more widespread in southern countries of the northern hemisphere, and the incidence of tetanus is higher in warm countries than in northern areas.

The various toxinotypes of *C. botulinum* are not equally distributed. The main habitat of toxinotypes A, B, E, F and G is soil, sea and freshwater sediments. Toxinotypes A and B occur more frequently in soil samples, but their geographical distribution differs. *Clostridium botulinum* A is predominant in the western United States, South America and China; *C. botulinum* B in the eastern United States and continental Europe; *C. botulinum* E is more predominant in sea or lake sediments and fish than in soil, mainly in northern areas of the northern hemisphere; *C. botulinum* C and D seem to be obligate parasites in animals and birds, and are mainly found in warm regions. Cadavers of animals dying of botulism are the main sources of these organisms.

Clostridium butyricum and *C. baratii* are very widespread bacteria in the environment including soil, sediments, intestinal contents of healthy animals and human faeces. The ecology of the neurotoxigenic strains from these species remains undefined (Popoff, 1995).

EPIDEMIOLOGY

Tetanus

Tetanus has always been associated with wound infections in humans and animals. Wounds associated with tissue necrosis and anaerobic conditions are favourable to *C. tetani* growth and toxin production. There are no reports of tetanus resulting from ingestion of preformed toxin in food or by absorption of TeTx produced by *C. tetani* in the intestine. The lack of association of TeTx with ANTP to form complexes could result in instability of TeTx in the digestive tract, and could explain why TeTx is not active by the oral route.

Human botulism

Four forms of human botulism are recognized according to the mode of acquisition.

Food-borne botulism

Food-borne botulism is due to ingestion of food in which *C. botulinum* has developed and produced sufficient amounts of BoNT. The incidence and type of botulism depend on the occurrence of *C. botulinum* in the environment and, subsequently, in food and cooking practices. Type A is predominant in America, type B in Europe and type E in the cold regions of the northern hemisphere.

In the majority of outbreaks, botulism is due to home-canned, home-fermented products or home-processed low-acid vegetables.

Infant botulism

Infant botulism results from ingestion of *C. botulinum*

spores that germinate, multiply and produce BoNT in the infant's gastrointestinal tract. *Clostridium botulinum* and BoNT are recovered from stools of affected infants and toxin is seldom found in the serum; in food-borne botulism, however, toxaemia is common.

Most of the *C. botulinum* strains involved in infant botulism belong to group I (*C. botulinum* A and proteolytic strains of *C. botulinum* B); *C. botulinum* C, *C. butyricum*-producing ButNT/E, *C. botulinum* F and *C. baratii* have also been implicated in certain cases.

Clostridium botulinum A strains are divided into two groups termed A1 and A2, according to the restriction fragment length polymorphism analysis of the polymerase chain reaction (PCR) products with *Rsa*I and *Sau*3A. The strains isolated from infant botulism in Japan fall into group A2, the strains from food-borne and infant botulism from the United States and the UK fall into groups A1 and A2 (Cordoba *et al.*, 1995).

Toxin infection in adults

Colonization by *C. botulinum* and BoNT production in the intestine have been recognized as the botulism origin in certain adults. Predisposing factors such as intestinal surgery, antimicrobial agents, chronic inflammation and necrotic lesions of the intestinal mucosa could support the growth of *C. botulinum*. *Clostridium botulinum* A and B from group I are generally involved.

Wound botulism

Wounds can be colonized by *C. botulinum*, resulting in the onset of botulism. However, the occurrence of wound botulism is much lower than that of tetanus. *Clostridium botulinum* involved in wound botulism belongs to the toxinotypes A and B from group I.

Animal botulism

Clostridium botulinum C and D are the most common sources of botulism in animals throughout the world. *Clostridium botulinum* C is mainly found in birds and mink, whereas *C. botulinum* D is associated with botulism in bovine and other mammals. Sporadic cases of animal botulism due to other toxinotypes have been reported. Animal botulism results from toxin ingestion or toxic infection.

GENETIC ORGANIZATION AND GENOMIC LOCALIZATION OF THE CLOSTRIDIAL NEUROTOXIN LOCUS

Genetic organization

The genes encoding the neurotoxins and ANTPs have been cloned and sequenced in representative clostridial strains of each toxinotype. The neurotoxin and ANTP genes are clustered in close proximity, and constitute the botulinum or tetanus neurotoxin locus. The organization of the botulinum locus is conserved in the 3' part but differs slightly in the 5' part in the different toxinotype of the BoNT-producing *Clostridia*.

The *bont* genes are localized in the 3' part of the locus and are preceded by the genes of the non-toxic non-haemagglutinin (NTNH) components. *ntnh* and *bont* genes are transcribed in the same orientation (Figure 9.1).

The haemagglutinin (HA) genes (*ha*33, *ha*17 and *ha*70) are upstream of the *ntnh–bont* genes and are transcribed in the opposite orientation. The *ha* genes are missing in the non-haemagglutinating toxinotypes A2, E and F. The *ha* genes of *C. botulinum* G only comprise *ha*17 and *ha*70. In the toxinotype A2, E and F, a gene (*p*47) encoding a 47-kDa protein is immediately upstream of the *ntnh* gene, and both genes are transcribed in the same orientation. In addition, a gene (*orf*X1) encoding a 18-kDa protein which is not related to *ha*17 lies upstream of *p*47.

A gene (*bot*R, previously called *orf*21 or *orf*22) encoding a 21–22-kDa protein which has features of a regulatory protein, is localized in the 5' part of the botulinum locus in *C. botulinum* C and D, whereas it is localized between *ntnh–bont* and *ha* genes in *C. botulinum* A1, B and G (Figure 9.1).

In *C. tetani*, one gene (*tet*R) equivalent to *bot*R was found upstream of the *tetx* gene (Figure 9.1).

Usually, one clostridial strain produces only one type of neurotoxin and the botulinum locus is present in only one copy on the genome, as suggested by Southern blotting of DNA fragments separated by pulse-field gel electrophoresis (Lin and Johnson, 1995; Hutson *et al.*, 1996). Rare strains synthesize two types of BoNT. Therefore, BoNT/A–BoNT/B-, BoNT/A–BoNT/F- and BoNT/B–BoNT/F-producing strains have been isolated (Hutson *et al.*, 1996; Henderson *et al.*, 1997). Investigations on a A–B strain show that this strain contains two *bont* genes related to those of *C. botulinum* A2 and proteolytic *C. botulinum* B, respectively (Fujinaga *et al.*, 1995). In the multiple toxin-producing strains, the neurotoxins are usually produced in different proportions. Thus, in B-a and B-f strains, 10 times more BoNT/B is produced than BoNT/A and BoNT/F (Henderson *et al.*, 1997).

Recently, it has been reported that clostridial strains contain silent neurotoxin genes. Several *C. botulinum* A strains isolated from food-borne and infant botulism contain a silent *bont*B gene. The characterization of strain NCTC2916 shows that this strain has two loci, A and B, which are chromosomally localized and are separated by approximately 40 kbp. The botulinum B locus

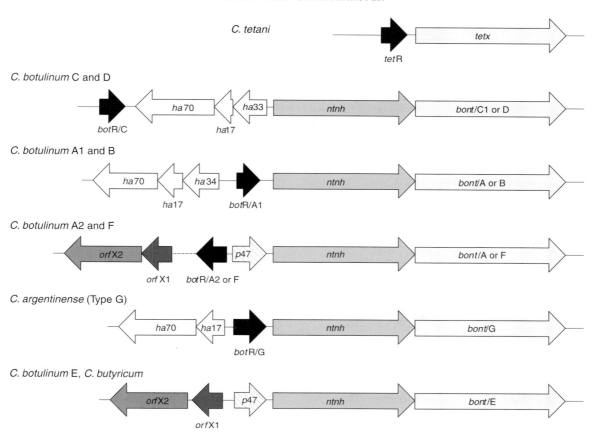

FIGURE 9.1 Genetic organization of the botulinum and tetanus locus in the different *C. botulinum* toxinotypes and *C. tetani*.

consists of *bont*/B, *ntnh*, *orf*21, *ha*33 and *ha*11 genes. The *bont*/A gene sequence is identical to that of *C. botulinum* A1 strains, but the organization of the botulinum A locus is similar to that of *C. botulinum* A2 and F strains. The *bont*/B nucleotide sequence is related to that of *C. botulinum* B strains (97% identity), but it has a stop mutation in position 128 and two base deletions (positions 2839 and 2944) resulting in reading frameshifts and multiple stop codons (Hutson *et al.*, 1996). Silent *bont*/B also exist in non-toxigenic *C. subterminale* strains (Franciosa *et al.*, 1994; Jovita *et al.*, 1998).

Genomic localization

In *C. tetani* and *C. argentinense*, the neurotoxin gene has been localized on a large plasmid (51 and 73 MDa, respectively) (Table 9.1).

Plasmids of various sizes and bacteriophages have been found in *C. botulinum* A, B, E and F, but toxigenicity has not been associated with the presence of these genetic elements (Strom *et al.*, 1984; Weickert *et al.*, 1986; Eklund, 1989). The neurotoxin genes of these toxin types have been cloned from chromosomal DNA, confirming the chromosomal localization of these genes (Table 9.1). In neurotoxigenic *C. butyricum*

strains, the *bont*/E gene has been found in a plasmid using PCR amplification (Hauser *et al.*, 1992), but by DNA/DNA hybridization this gene seems to be localized on the chromosomal DNA (Zhou *et al.*, 1993).

In *C. botulinum* C and D, it has been clearly shown that BoNT is encoded by bacteriophages. *Clostridium botulinum* C and D strains cured of their phages produce neither BoNT/C1 nor BoNT/D, respectively, but they do continue to produce C2 toxin. Such organisms could be converted into neurotoxigenic strains C or D by reinfection with phages obtained from toxigenic *C. botulinum* C or D strains. The BoNT/C1 and BoNT/D genes have been cloned and sequenced from purified phage DNA of *C. botulinum* C-468 and *C. botulinum* D-1873, respectively (reviewed in Popoff and Eklund, 1995).

NEUROTOXIN GENE TRANSFER

The similarity between BoNTs and TeTx, and the facts that different *Clostridium* species can produce BoNT and that some strains contain combinations of *bont* genes (A and B, A and silent B, A and F, B and F) (Franciosa *et al.*, 1997; Henderson *et al.*, 1997), strongly suggest that *bont* and *tetx* genes derive from a common ancestor and have

TABLE 9.1 Main properties of the neurotoxin-producing *Clostridium* and associated affections

Clostridium species	Group	Toxin	Complex form	Genomic localization of the neurotoxin genes	Proteolysis	Lipase	Disease
C. botulinum	I	A1 B A2, F	M, L, LL M, L M	Chromosome	+	+	Botulism in humans: food-borne botulism; infant botulism; wound botulism
C. botulinum	II	B E F	M, L M M	Chromosome	−	+	Food-borne botulism in humans; rare cases of animal botulism
C. botulinum	III	C D	M, L M, L	Phage	−	−	Animal botulism: C, birds, mink; D, mammals
C. argentinense	IV	G	L	Plasmid	+	−	No diseases reported
C. butyricum		E	M	Chromosome	−	−	Infant botulism
C. baratii		F	M	?	−	−	Human botulism
C. tetani		TeTx	no	Plasmid	−	−	Human and animal tetanus

been transferred between *Clostridium* strains. Non-toxigenic derivatives were shown in certain toxinotypes, such as in *C. botulinum* B (Yamakawa *et al.*, 1997), indicating instability of a DNA fragment harbouring the *bont* genes. Genetics in *Clostridum* is still poorly understood, but it can be assumed that gene transfer can be mediated by mobile elements.

Plasmids

As the *tetx* and *bont*/G genes have been localized on large plasmids in *C. tetani* and *C. argentinense*, respectively (Finn *et al.*, 1984; Eklund *et al.*, 1988), their transfer could be achieved by mobilization of the corresponding plasmids. Conjugation and mobilization of large plasmids in *Clostridium*, such as *C. perfringens*, have already been reported (Brefort *et al.*, 1977). However, up to now, only non-toxigenic *C. argentinense* and *C. tetani* variants free of plasmids have been obtained.

Bacteriophages

In *C. botulinum* from group III, it has been clearly shown that bacteriophages mediate the neurotoxin gene transfer. Bacteriophages carrying *bont*/C1 and *bont*/D genes can be transferred in different *Clostridium* host strains and determine the toxinotype of these strains. Moreover, in *C. novyi*, the alpha toxin gene is also localized on a bacteriophage. The *C. botulinum* C or D free of bacteriophage are indistinguishable from *C. novyi* cured of bacteriophage, and can be reinfected with *C. novyi* bacteriophage (Eklund *et al.*, 1971, 1974; Eklund and Poysky, 1974).

A pseudolysogenic relationship corresponding to the presence of bacteriophages free within the bacterial cytoplasm exists between these phages and hosts. Thus, variants free of bacteriophages can be obtained at high frequency using curing reagents such as acridine orange and UV. Under laboratory culture conditions, a proportion of the bacteria depending on the strain and growth conditions (temperature, salinity) is lysed and free bacteriophages are released. Other bacteria lose their bacteriophages and can be reinfected with free bacteriophages. Such lysogeny and reinfection cycles probably occur in the environment (soil, intestinal tract of birds and animals) and account for the isolation of non-toxigenic or low toxin producer variants (Eklund and Dowell, 1987).

In *C. botulinum* A and F, involvement of bacteriophage has been suggested on the basis of the identification of a gene (*lyc*) in the vicinity of the *bont* genes. *lyc*/A and *lyc*/F genes have been mapped approximately 1 kb downstream from the corresponding *bont* genes, and are partially related to bacteriophage genes encoding lytic enzymes in *Lactobacillus* and *Streptococcus pneumoniae* (Henderson *et al.*, 1997). As lytic enzymes are involved in the bacteriophage life cycle, the presence of the *lyc* gene in the vicinity of *bont*/A gene in NCTC2916 could indicate that the botulinum locus is part of an integrated prophage.

Transposon

Transposable elements have been demonstrated in *C. botulinum* C and D. The exoenzyme C3 gene, which encodes an ADP-ribosyltransferase specific for Rho protein, is also harboured by the same bacteriophages carrying *bont*/C1 and *bont*/D genes. Genetic analysis of a mutant phage (CN) showed a deletion of a 21.5-kb fragment containing the C3 gene. This fragment was found in several bacteriophages type C and type D, and is delineated by the 6-bp core motif AAGGAG. This motif is found in only one copy at the deletion junction in the CN-phage DNA. The DNA sequences from phages C and D which flank the 21.5-kb fragment are unrelated. However, the sequence on the left end remains homologous on a 61-bp stretch upstream of the core motif, whereas the sequences diverge immediately downstream from the core motif on the right end. The 21.5-kb fragment seems to be a mobile DNA element responsible for spreading the C3 gene in *C. botulinum* C and D, and has a similar feature to that of the site-specific transposon family of Tn1554 including:

(a) asymmetric ends
(b) the absence of either inverted or terminal repeats
(c) the presence of a 6-bp core motif sequence at both the insertion junctions and the insertion site itself (Hauser *et al.*, 1993).

Transposable elements, analogous to that encompassing the C3 gene, could account for the different localizations of the neurotoxin genes (chromosomal, plasmid, bacteriophage) and, subsequently, for the gene transfer between *Clostridium* strains. Such elements have not yet been clearly identified. However, nucleotide sequence analysis in *C. botulinum* A suggests their possible existence. A 97 nucleotide stretch downstream of the stop codon of the *bont*/A gene is identical in *C. botulinum* A strains 62 and NTCT2916, whereas the following nucleotides are totally unrelated between the two strains. The 97 nucleotide stretch could be part of a mobile DNA element encompassing the *bont*/A and *antp* genes. In addition, the different surrounding sequences indicate that *bont*/A gene has a different localization on the chromosome of strains 62 and NCTC2916. This is confirmed by a different restriction profile of the 3′ part of the *bont*/A gene in the two strains. An *Eco*RI site lies 1 kbp upstream of the stop codon of the *bont*/A gene and a probe was selected between the *Eco*RI site in the coding sequence and the stop codon. Southern blot showed a different pattern between strains 62 and NCTC 2916 (Figure 9.2). Three other strains showed an *Eco*RI profile similar to that of strain 62 (Figure 9.2). At least two sites of *bont*/A localization exist on the chromosome of *C. botulinum* A and reinforce the idea of the involvement of a mobile DNA element.

Neurotoxigenic *C. butyricum* strains probably originated from non-toxigenic *C. butyricum* strains by acquisition of the *bont*/E gene from *C. botulinum* E (Poulet, 1992). The *bont*/E gene and its flanking regions are absent in non-toxigenic *C. butyricum* strains, suggesting a possible gene mobilization by a mobile DNA element (Hauser *et al.*, 1992). This gene has been transferred from a neurotoxigenic *C. butyricum* strain to a non-toxigenic *C. botulinum* E strain by a protocol resembling transduction with a defective phage (Zhou *et al.*, 1993). The precise method of molecular transfer has not been elucidated as DNA/DNA hybridization studies suggest the *bont*/E gene is localized on chromosomal DNA and not on phage DNA in toxigenic *C. butyricum* (Zhou *et al.*, 1993).

In addition, rearrangement, probably by homologous recombination, seems to have occurred between genes from the botulinum locus of different toxinotypes. This is supported by atypical strains which show

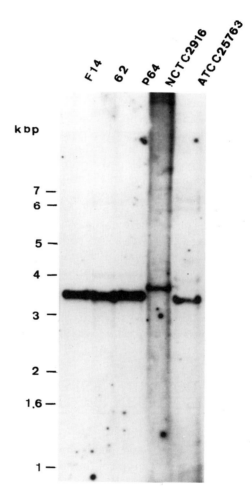

FIGURE 9.2 *Eco*RI restriction profile of the 3′ part of the *bont*/A gene in *C. botulinum* A strains F14, 62, P64, NCTC2916 and ATCC25763. Genomic DNA digested by *Eco*RI was hybridized with a probe localized between the *Eco*RI site, mapped on to the coding region 1 kbp upstream of the stop codon, and the stop codon. When compared to the other strains, a second *Eco*RI site mapped by hybridization lies in a different position downstream of the 97 conserved bp of the 3′ non-coding region in strain NCTC2916.

genes with hybrid sequences from classical strains: mosaic *bont* gene from *C. botulinum* C and D (Moriishi *et al.*, 1996), mosaic *ntnh* gene from *C. botulinum* A and C (Kubota *et al.*, 1996) or from *C. botulinum* A and B (Hutson *et al.*, 1996).

IDENTIFICATION OF NEUROTOXIN-PRODUCING *CLOSTRIDIUM* BY GENETIC METHODS

The routine identification of neurotoxin-producing *Clostridium* is hampered by the need for anaerobic conditions for growth and the variable bacteriological characteristics of the different groups. Moreover, the use of mouse bioassay for toxin typing is now more restrictive. The molecular characterization of *bont* genes offers the possibility of using genetic methods such as PCR and DNA/DNA hybridization. Several PCR-based detection tests for individual botulinum types have been developed (Campbell *et al.*, 1993b; Fach *et al.*, 1993; Szabo *et al.*, 1993; Franciosa *et al.*, 1994). To minimize the number of PCRs, a multiplex PCR, consisting of a mix of primers specific to the neurotoxin types (A, B, E and F) involved in human botulism, was designed (Henderson *et al.*, 1997). We have developed a primer pair, partially degenerated on their 3′ end, able to amplify a conserved region of *bont* genes from type A, B, E, F and G. Internal probes specific for each type allow the typing by hybridization. This method was used in contaminated food samples and showed a 95.6% correlation with the standard method (Fach *et al.*, 1995). A similar strategy was used by other authors (Campbell *et al.*, 1993b; Aranda *et al.*, 1997). A single internal probe was designed to recognize the amplification products, without distinction of type, in the work of Aranda *et al.* (1997). A double PCR procedure was also proposed for identification of *C. botulinum* C and D (Fach *et al.*, 1996).

NEUROTOXIN GENES AND NEUROTOXIN PROTEINS

The *tetx* and *bont* genes of the different toxinotypes and from various strains have been characterized. The genes are preceded by a typical ribosome binding site rich in G, six to seven nucleotides upstream of the ATG translational start codon. A hairpin loop structure forming a Rho-independent transcriptional termination is found downstream of the stop codon in the strains that have been sequenced for a sufficient distance in the 3′ *bont* region.

The deduced TeTx and BoNT proteins contain 1251–1315 residues. The longer sequence is that of TeTx (1315 residues). No signal peptide has been detected in the neurotoxin primary structure. It was assumed that the neurotoxins are not secreted and appeared late (1–4 days) in the culture supernatant by bacterial lysis. However, it was found that lysis is not required for BoNT release from the bacteria. BoNT is synthesized in the bacterial cytoplasm, probably during the exponential growth phase, translocated across the cytoplasmic membrane and then exported to the extracellular medium via cell-wall exfoliation (Call *et al.*, 1995).

The overall similarity between the available neurotoxin sequences ranges from 34% to 97% identity. The neurotoxin sequences from the same toxinotype and from the same physiological groups are almost identical, except BoNT/A1 and BoNT/A2 which are highly related (90% identity) but differ in 129 residues (Binz *et al.*, 1990a; Thompson, 1990; Willems *et al.*, 1993). BoNTs of the same toxinotype but produced by strains from different *C. botulinum* groups or *Clostridium* species show a high level of identity (70–97%). BoNT sequences from *C. botulinum* E and *C. butyricum* have 97% identity (Poulet *et al.*, 1992; Fujii *et al.*, 1993a). BoNT/B sequences from proteolytic and non-proteolytic *C. botulinum* B share 93% identity (Whelan *et al.*, 1992; Hutson *et al.*, 1994), but BoNT/F from *C. botulinum* F and *C. baratii* have only 70% identity (East *et al.*, 1992; Thompson *et al.*, 1993) (Figure 9.3). Both variants of BoNT/B seem to be equipotent. No differences in epidemiology and clinical manifestations have been reported in botulism type B due to proteolytic and non-proteolytic strains. The relatedness of the clostridial neurotoxins is also confirmed immunologically. Common monoclonal antibodies which recognize TeTx and the various types of BoNTs have been recognized (Tsuzuki *et al.*, 1988). However, neutralizing antibodies are specific to each toxinotype, except for toxinotypes C and D which share cross-reactions. BoNTs and TeTx have specific epitopes in addition to common epitopes.

The relatedness is much lower between TeTx (Eisel *et al.*, 1986; Fairweather and Lyness, 1986) and BoNT of different toxinotypes (34–64% identity). Sequence divergence analysis indicates different evolutionary branches of the clostridial neurotoxins (Figure 9.4). TeTx is distantly related to the BoNTs. Bont/B is the most closely related to TeTx (42% identity). BoNT/B and BoNT/G (Campbell *et al.*, 1993a) seem to evolve from the same phylogenetic branch. BoNT/A is the most potent BoNT and it is four times more active than TeTx. The phylogeny tree (Figure 9.4) shows that BoNT/A is localized on a separate branch from the other neurotoxins. The BoNT/C1 and BoNT/D share 53.8% identity, and are relatively distinct from the other BoNT

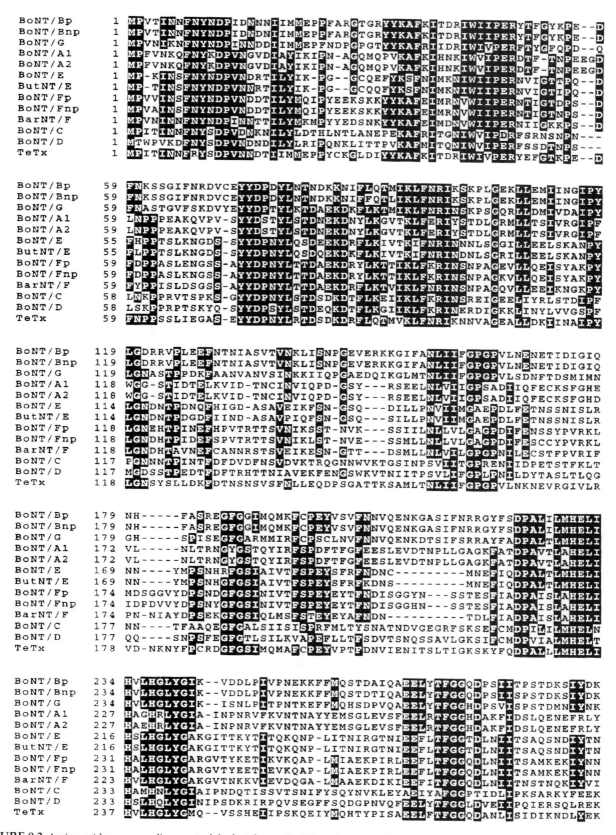

FIGURE 9.3 Amino acid sequence alignment of the botulinum (BoNT) and tetanus (TeTx) neurotoxins.

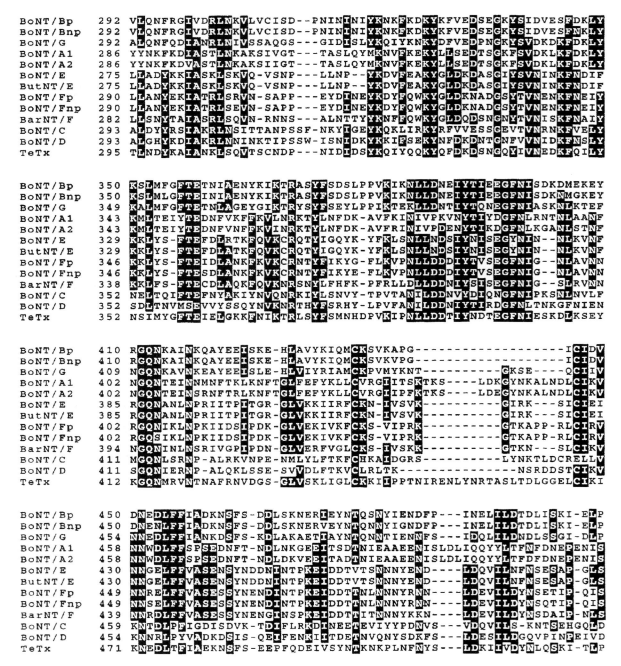

FIGURE 9.3 *Continued.*

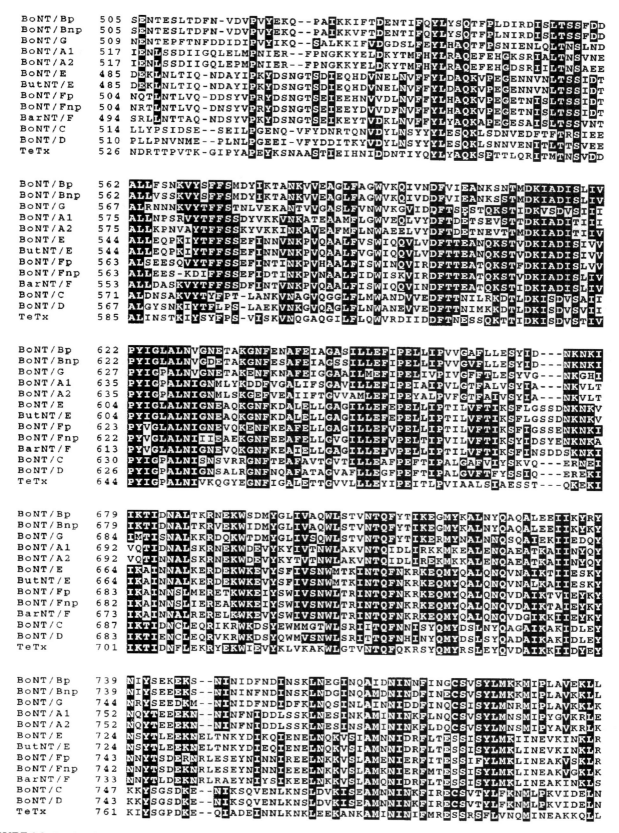

FIGURE 9.3 *Continued.*

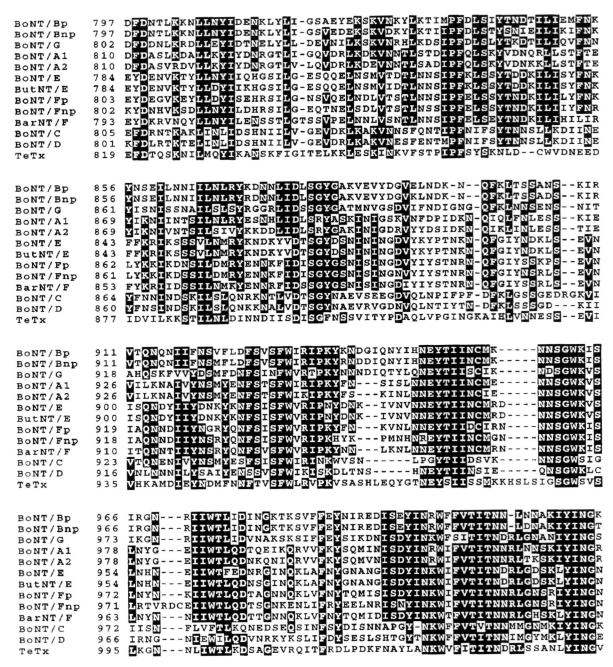

FIGURE 9.3 *Continued.*

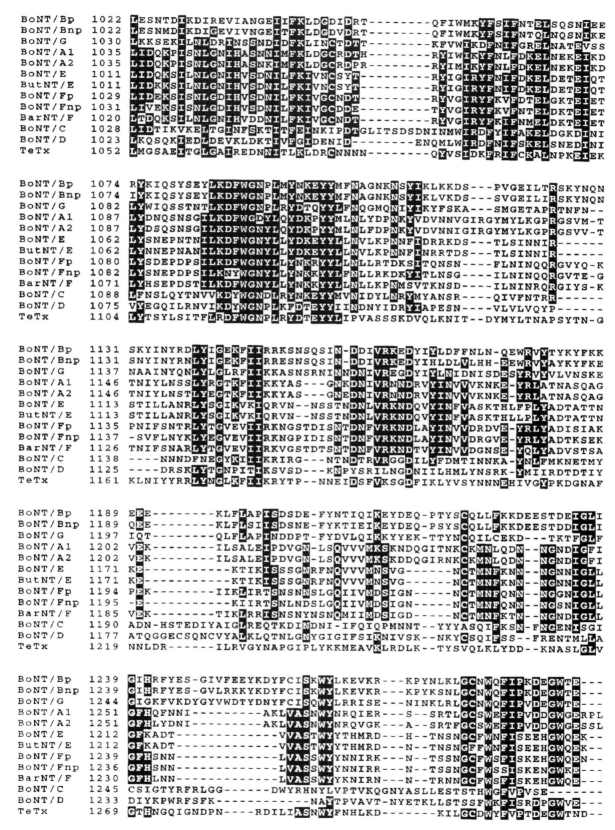

FIGURE 9.3 *Continued.*

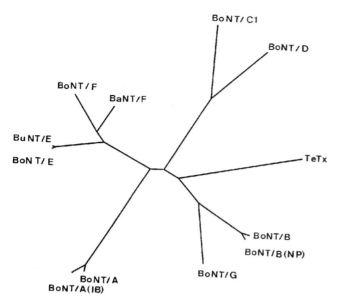

FIGURE 9.4 Phylogeny tree of the clostridial neurotoxins obtained by the Phylip program.

(Binz *et al.*, 1990b; Hauser *et al.*, 1990; Kimura *et al.*, 1990; Sunagawa *et al.*, 1992) (Figure 9.4). This phylogenetic relatedness is in agreement with the existence of a high degree of immunological cross-reaction between BoNT/C1 and BoNT/D (Kozaki *et al.*, 1989a). BoNT/E, BoNT/F and their derivatives in *C. butyricum* and *C. baratii* are localized on a phylogenetic branch separate from the other neurotoxins. This relationship is corroborated by immunological cross-reactions between BoNT/E and BoNT/F (Kozaki *et al.*, 1989a).

Analysis of the identity level suggests that the clostridial neurotoxins derive from a common ancestor. The homology could be interpreted as the time distances of the neurotoxin gene dissemination (Collins and East, 1997). Thus, *bont* genes could have been transferred from *C. botulinum* E to *C. butyricum* (97% identity), subsequent to the transfer of *bont*/B between proteolytic and non-proteolytic *C. botulinum* B strains (93% identity), and to the segregation between *bont*/C1 and *bont*/D (54% identity). The degree of relatedness varies according to the neurotoxin region which is considered. For example, the level of identity between the L-chains of BoNT/Bp and BoNT/Bnp is 97.7%, and between the H-chains is 90.2%; the BoNT/A1 and BoNT/A2 L-chains are 94.9% identical and the H-chains are 87.1%; and the BoNT/C1 and BoNT/D L-chains have a 46.5% identity level, whereas the H-chains have a 57.3% identity level (Henderson *et al.*, 1997). This indicates that the different domains could evolve separately.

The different evolution of the clostridial neurotoxin domains is supported by the recent evidence of mosaic toxins. It was previously found that some BoNTs are hybrids between types C and D, as they are recognized by specific antibodies anti-type C and anti-type D raised against typical strains. The molecular characterization of BoNT/D from strain Dsa shows that its *N*-terminal region (residues 1–522) is highly related to the corresponding region of BoNT/D (96% identity), and the *C*-terminal region (residues 945–1285) is related to the corresponding region of BoNT/C (74% identity). Inversely, BoNT/C produced by strain C6813 has a *N*-terminal region related to BoNT/C and the *C*-terminal region to BoNT/D. The central region (residues 523–944) is common (83–92% identity) to the hybrid and typical C and D strains (Moriishi *et al.*, 1996).

FUNCTIONAL DOMAINS OF THE CLOSTRIDIAL NEUROTOXINS

Three functional domains have been assigned to the neurotoxin molecules. The *C*-terminal half of the H-chain (H_c) is involved in the binding to a surface nerve-cell receptor, while the *N*-terminal half of the H-chain is involved in the translocation of the toxin across the cell membrane. The L-chain has been implicated in the intracellular blockade of neurotransmitter release (Figure 9.4). Electron crystallography of TeTx shows an asymmetric three-lobe structure compatible with the three functional domains model (Robinson *et al.*, 1988).

The multiple alignment carried out by the MACAW program shows several blocks of similarity between the BoNT and TeTx sequences (Figures 9.3 and 9.5) (Popoff and Eklund, 1995).

Enzymic domain

The L-chains display two main blocks of high similarity (Figure 9.5): the first 145 residues; and the central region from residue 220 to about 350 which contains the zinc-dependent metalloprotease active site (HExxH) (Schiavo *et al.*, 1993). Glu271 in TeTx, which is also probably involved in zinc co-ordination, is conserved in all BoNT sequences (Figure 9.6). It has been confirmed that the consensus motif HExxH is the functional site in TeTx. Mutational replacement of His233, which constitutes the metal binding site, by Cys or Val (Höhne-Zell *et al.*, 1993; Dayanithi *et al.*, 1994), or changing Glu234, which probably plays the catalytic role, to Ala (Li *et al.*, 1994) abolished the toxicity of the TeTx L-chain. Moreover, antibodies raised against synthetic peptides containing the zinc-binding motif efficiently neutralized TeTx and BoNT/A (Bartels *et al.*, 1994). However, the major part of the L-chain is required for the activity. Deletions of eight *N*-terminal residues of TeTx and BoNT/A did not affect toxicity, but the deletion of 10 *N*-terminal residues drastically reduces its toxicity. Derivatives of the TeTx and BoNT/A L-chains lacking 65 and 32 *C*-terminal residues, respectively, are still active, while deletions of 68 and 57 *C*-terminal residues, respectively, abolished toxicity (Kurazono *et al.*, 1992). The regions surrounding the active site are probably involved in the recognition and binding to the substrate, and maintain the right conformation of the molecule. The sequences of the L-chains which interact with the binding sites of the substrate have not yet been characterized. A common motif, consisting of three negatively charged and three hydrophobic residues predicted to form an α-helix (SNARES motif) in synaptobrevin (two copies), SNAP25 (four copies) and syntaxin (two copies), is considered as the specific neurotoxin recognition site (Pelizzari *et al.*, 1997; Washbourne *et al.*, 1997). A basic site is also required for the recognition of synaptobrevin by TeTx. The binding of TeTx to the acidic and basic motifs surrounding the cleavage site could induce a conformational change in the TeTx L-chain in the active form, according to an allosteric model of the enzyme (Cornille *et al.*, 1997).

The two Cys residues involved in the disulfide bridge between the L- and H-chains can also be aligned (positions 439 and 467 in the TeTx sequence) (Figure 9.6). The interchain disulfide bond and the participating Cys are required for internalization of the neurotoxins and not

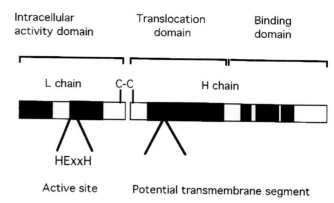

FIGURE 9.5 Schematic presentation of the functional domains of BoNT and TeTx. The black boxes represent conserved sequences.

for the enzymic activity. TeTx and BoNT/A L-chains lacking Cys439 and Cys430, respectively, are active only when applied intracellularly (Kurazono *et al.*, 1992; de Paiva *et al.*, 1993). It is not known whether the reductive cleavage, necessary for the release of the L-chain into the cytosol, occurs before or after the L-chain emerges from the endosome into the cytosol.

The proteolytic activation of the clostridial neurotoxins results in the cleavage of the precursor protein between the two conserved Cys, and removal of residues in the *N*-terminal part of the H-chain. Thus, the activation of BoNT/A by the endogeneous protease of *C. botulinum* consists of the cleavage of Lys438–Thr439 and the removal of 10 amino acids (Thr439–Lys448) (Krieglstein *et al.*, 1994). In Bont/E, the cleavage occurs

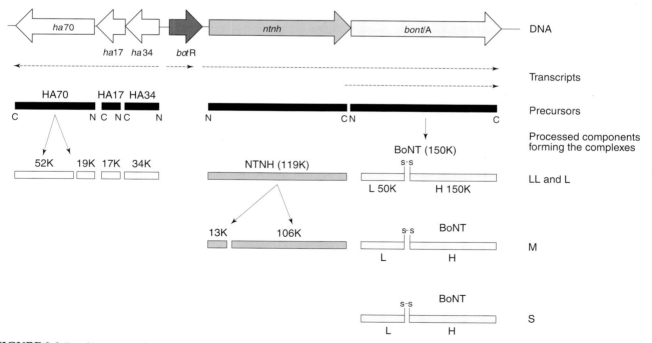

FIGURE 9.6 Botulinum complex genes, transcription, translation and protein processing of the complex components in *C. botulinum* A. N, *N*-terminal part; C, *C*-terminal part.

at Lys419–Gly420, and the residues Arg422–Lys423 and Gly420–Ile–Arg422 are excised (Antharavally and Das Gupta, 1997). Proteases produced by *C. tetani* cleave TeTx at Glu449 and Ala456. Other proteases cut at various residues within the loop formed by the interchain disulfide bond. All the dichain derivatives of TeTx are active, indicating that the exact cleavage site is not critical (Krieglstein *et al.*, 1991).

Translocation domain

A highly conserved region (residues 550–860) is part of the *N*-terminal half of the neurotoxin H-chains. Fragments of the *N*-terminal part of TeTx H-chain interact with lipid membranes at acidic pH (Roa and Boquet, 1985). The *N*-terminal parts of the neurotoxin H-chains form channels through lipid bilayers (Högy *et al.*, 1992). Channels in phospholipid vesicles induced by BoNT have been visualized and seem to be the result of the interaction of four whole neurotoxin molecules (Schmid *et al.*, 1993). Organization of BoNT monomers into a oligomeric structure could also be required for the formation of ion channels through the membrane (Ledoux *et al.*, 1994). The segments 669–691 of TeTx and 660–691 of BoNT/A are predicted to be a transmembrane domain and to form a plausible four α-helix bundle structure. The corresponding synthetic peptides form ion channels in planar lipid bilayers (Montal *et al.*, 1992). Beise *et al.* (1994) identified possible helical structures able to form ion channels between residues 700 and 850 of TeTx, and Lebeda and Olson (1995) reported four predicted transmembrane segments between positions 606 and 693 of theBoNT/A H-chain, which are conserved in the other BoNTs and TeTx H-chains. Analysis with the TopPred II program shows that the region between residues 669 and 689 of TeTx, 667 and 687 of BoNT/A and those in equivalent regions of the other BoNT H-chains are the most probable candidates for transmembrane segments. This domain is probably involved in the translocation of the L-chain across the vesicular membrane (Hauser *et al.*, 1995).

The crystal structure of BoNT/A has been recently solved. The *N*-terminal part of the H chain is characterized by a pair of unusually long α-helices corresponding to residues 685–827. The antiparallel and amphipathic helices are organized like a coiled-coil with shorter helices at either ends that wrap around the catalytic domain (Lacy *et al.*, 1998).

Receptor-binding domain

A second conserved region (residues 930–1150 from the TeTx sequence) is localized in the *C*-terminal half of the H-chains, termed H_c or *C*-fragment (F_c), which is involved in the binding of the neurotoxin to cell-surface receptor(s). The approximately 170 *C*-terminal residues of the H-chains are unrelated, and probably reflect the fact that each neurotoxin recognizes a specific receptor. Neurotoxins bind to gangliosides of the G_{1b} series, but the optimal binding is obtained in non-physiological conditions (low pH, low ionic strength). It has been proposed that the receptor may be a ganglioside–protein complex (Montecucco, 1986). Synaptotagmin has been proposed as the receptor of BoNT/B (Nishiki *et al.*, 1996).

The X-ray crystal structure of the TeTx *C*-fragment shows two interconnected domains containing predominantly β-strands. The *N*-terminal domain (residues 864–1107) is a jellyroll motif composed of 14 antiparallel β-strands and two α-helices, and the *C*-terminal domain (residues 1127–1315) has a six-strand β-barrel constructed from three independent, non-parallel β-sheets, upon which is a β-hairpin triplet (Umland *et al.*, 1997). A similar structure has been found in the *C*-terminal part of the BoNT/A H chain consisting of β-strands organized in two subdomains connected by an α-helix (Lacy *et al.*, 1998). The structure of *C*-fragment in β-sheets is analogous to that of the receptor-binding domains of toxins such as diphtheria toxin and *Bacillus anthracis* toxins (Choe *et al.*, 1992; Petosa *et al.*, 1997).

The Hc regions of TeTx (residues from 858 to 1315) and BoNT/A bind to gangliosides (Kozaki *et al.*, 1989b; Halpern *et al.*, 1990). Deletion of the 263 *N*-terminal and five *C*-terminal residues of Hc from TeTx did not affect the binding capacity. This indicates that the binding site of TeTx is localized mainly in the *C*-terminal domain from residue 1121 to residue 1310 (Halpern, 1995). It seems that the 34 *C*-terminal residues are sufficient to support binding to 1b series gangliosides, and that His1293 is a critical residue for binding (Shapiro *et al.*, 1997).

The *C*-terminal domains of the Hc regions of TeTx and BoNT display three conserved Trp (positions 1298, 1303 and 1312 in the TeTx sequence), and are terminated by charged residues. The third Trp residue is missing only in the BoNT/C1 sequence. The terminal charged residues and the third Trp residue do not seem to be important for binding as their deletion in TeTx did not alter the binding activity (Halpern, 1995). The role of the other conserved Trp residues has not been investigated.

A disulfide bond is localized in the Hc region of TeTx between Cys1077 and Cys1093, and also in BoNT/A (Cys1234–Cys1279), BoNT/E (Cys1196–Cys1237) and, probably, BoNT/B (Cys1220–Cys1257 or Cys1277). This disulfide bridge is probably not involved in binding as its deletion did not inhibit it (Halpern and Loftus, 1993).

The neurotoxins have several Cys residues (for example 10 in TeTx, nine in BoNT/A, 10 in BoNT/B and eight in BoNT/E). The role of the Cys residues, except for those involved in the disulfide bridge, is unknown. It appears that at least one Cys as free –SH is necessary for the full activity of the neurotoxins. The neurotoxin conformational changes which occur *in vivo* could result from –SH/–S–S– exchanges between the interchain disulfide bond and free –SH present in the neurotoxins (Antharavally *et al.*, 1998).

In contrast to BoNTs, which are delivered into the cytosol of the nerve extremities, TeTx is transported into a vesicular compartment retrogradely along the axon. TeTx could recognize a specific receptor of vesicles involved in the retrograde transport. The H$_c$ fragment controls the retrograde transport, and H$_c$ of TeTx, obtained by protease digestion, or recombinant H$_c$ was transported by neurons in a similar manner to native toxin without causing clinical symptoms. Moreover, H$_c$ chemically conjugated or genetically fused to various proteins was internalized by neurons and was able to migrate retrogradely within the axons (reviewed in Coen *et al.*, 1997). An important application is that H$_c$ could be use to deliver biologically active proteins into the central nervous system for therapeutic use.

NTNH GENES AND NTNH PROTEINS

The DNA flanking regions of the *bont* genes have been cloned and sequenced in various strains. A gene encoding a protein of 120–139 kDa has been identified immediately upstream of the *bont* genes in all of the

BoNT-producing *Clostridium* that have been characterized (Figure 9.1). The *ntnh* genes are closely linked to the *bont* genes. They are separated by only 11–48 nucleotides (84 in *C. argentinense*) and are transcribed in the same orientation. In contrast to *C. botulinum*, this gene is missing in *C. tetani*.

The NTNH proteins do not contain either a sequence predicted to be a signal peptide or Cys residues able to form an interchain disulfide bridge.

In *C. botulinum* D strain CB16, it has been shown that NTNH/D [138.7 kDa migrating in sodium dodecyl sulfate–polyacrylamide gel electrophoresis (SDS–PAGE) at 130 kDa] is cleaved specifically at the Thr140–Ser141 site yielding 115 and 15 kDa proteins. The cleavage is mediated by a protease produced by *C. botulinum*, which seems to be specific as the cleavage site Thr–Ser is unusual for the known families of proteases. Interestingly, the unnicked NTNH/D (130 kDa) was associated with the L forms (500 kDa) of botulinum complexes and the nicked form (115 and 15 kDa) with the M forms (300 kDa). The L forms result from the association of the HA proteins with the M forms which consist of the NTNH/D and BoNT/D proteins (Table 9.2). This suggests that the *N*-terminal region of NTNH in the vicinity of the cleavage site is probably involved in the binding of the HA proteins to NTNH/D (Tsuzuki *et al.*, 1992; Ohyama *et al.*, 1995). The Thr–Ser cleavage site is conserved in NTNH/C, and a similar feature is probably involved in *C. botulinum* C (Hauser *et al.*, 1994).

In *C. botulinum* A (Figure 9.6), the processing of NTNH/A occurs at the site Pro144–Phe145. Similar to NTNH/D, the uncleaved NTNH/A (120 kDa) is associated with the L and LL forms of botulinum A complexes, and the processed form (106 and 13 kDa)

TABLE 9.2 Composition of botulinum complexes type A and D according to Ohyama *et al.* (1995) and Inoue *et al.* (1996)

	Complex size		Haemagglutinating activity	*C. botulinum* A			*C. botulinum* D	
				Composition	Size (kDa)	Molar ratio	Composition	Size (kDa)
LL Dimer of 16s complex	19S	900 kDa	+	BoNT/A	150	1		
				NTNH/A uncleaved	119	1		
				HA70 cleaved	19–52	2–3		
				HA35	35	8		
				HA17	17	3		
L	16S	500 kDa	+	BoNT/A	150	1	BoNT/D	150
				NTNH/A uncleaved	119	1	NTNH/D uncleaved	130
				HA70 cleaved	19–52	3	HA70 cleaved	22–55
				HA35	35	4	HA33	33
				HA17	17	3	HA17	17
M	12S	300 kDa	–	BoNT/A	150	1	BoNT/D	150
				NTNH/A cleaved	13–106	1	NTNH/D cleaved	15–115
S	7S	150 kDa	–	BoNT/A	150	1	BoNT/D	150

is only associated with the M forms. Thus, it seems that the absence of cleavage of NTNH prevents the formation of M complexes (Inoue *et al.*, 1996).

NTNH/E and NTNH/F from *C. botulinum* E and F, respectively, which only produce non-haemagglutinating M complex, have a common deletion of 33 residues in the equivalent region of NTNH/A, NTNH/C and NTNH/D containing the cleavage site and probably the binding site to the HA components (Fujita *et al.*, 1995) (Figure 9.7). Thus, it seems that the inability of *C. botulinum* E and F to produce L form complexes could result from the absence of HA or other related protein binding to NTNH, and from the absence of a putative binding site in NTNH/E and NTNH/F.

The NTNH sequence from *C. butyricum* E is identical to that of *C. botulinum* E, whereas the BoNT sequences from the two *Clostridium* species differ in 39 amino acids (Poulet *et al.*, 1992; Fujii *et al.*, 1993b).

The NTNH proteins show an identity level from 76 to 83.5% (Figure 9.7). They are more highly conserved than the BoNTs. NTNHs and BoNTs have almost the same length but are weakly related (29–41% identity). The most conserved region between NTNHs and BoNTs resides in the *N*-terminal residues. Pairwise analysis of NTNH sequences revealed that the degree of identity varies along their length. Therefore, the *N*-terminal 600 residues of NTNH/A1 and NTNH/Bp are 99% identical, whereas the sequence downstream from position 600 shows a lesser identity level (66%) (East *et al.*, 1996). NTNH/A2 has a mosaic structure. The *C*-terminal region shows a high sequence identity with NTNH/A1 (93% from position 692 to the end), and the *N*-terminal part is highly related to that of NTNH/C (97% identity) (Kubota *et al.*, 1996). In *C. botulinum* A strains possessing a silent *bont*/B gene, the *ntnh* gene upstream from the defective *bont*/B gene has a particular feature. It encodes a chimeric protein with the *N*- and *C*- terminal parts very similar to the corresponding regions of NTNH/B, and with a 471 amino acid sequence in the central region which is identical to NTNH/A (Hutson *et al.*, 1996). This indicates that NTNH could represent a hot spot for recombination events within the locus of *botulinum* complex genes.

As NTNHs are highly conserved and their genes are located in the close vicinity of *bont* genes in all strains which have been characterized, they seem to have a central role in the structure and/or the function of the botulinum complexes. The NTNH processing is probably crucial for the formation of the different complex sizes. The unprocessed NTNH is only found in L and LL forms, and the processed form in the M complexes (Figure 9.6). Their function seems to be involved in the protection of the BoNTs against proteases and the acidic environment. It is intriguing that the *bont* genes have

been subjected to a higher divergence than the *ntnh* genes, although both groups of genes are in the same genetic environment. The apparently high frequency of recombination in *ntnh* genes could explain the overall high similarity of these genes in the different *C. botulinum* toxinotypes.

HA GENES AND HA PROTEINS

Haemagglutinating activity is only associated with the large forms (L and LL) of botulinum complexes (Table 9.2). In *C. botulinum* C, it was first found that *ha* is localized on the same phage DNA that harbours the *bont*/C gene. The *ha* gene expressing haemagglutining activity was cloned in *E. coli*, and was found to encode a 33-kDa protein (HA33). The recombinant protein in *E. coli* exhibited haemagglutining activity. Subsequently, a *ha33* gene was mapped immediately upstream of the *ntnh–bont* genes in *C. botulinum* C and in the opposite orientation (Tsuzuki *et al.*, 1990; Hauser *et al.*, 1994). The cloning and sequencing of the flanking regions showed that two other genes (*ha17* and *ha70*) are closely linked to *ha33* in *C. botulinum* C (Hauser *et al.*, 1994) (Figure 9.1).

In *C. botulinum* A and B from group I, it was reported that HA purified from the culture supernatant consists of proteins of 17, 21.5, 35 and 57 kDa (Somers and Das Gupta, 1991). The haemagglutinating activity could not be demonstrated in the individual proteins isolated from SDS–PAGE. Because haemagglutinin results in general from the assembly of several components, it is assumed that the other proteins are additional haemagglutinin components (Somers and Das Gupta, 1991). HA33 was purified from *C. botulinum* A and was found to be the major haemagglutinin component (Fu *et al.*, 1997). Then, the *ha* genes were identified and sequenced in the DNA region upstream of *ntnh–bont* genes in *C. botulinum* A. The *N*-terminal sequencing of the purified proteins from *C. botulinum* permitted the establishment of their correspondence with the identified genes. The 17- and 35-kDa proteins are encoded by *ha17* and *ha35* genes, respectively, and the 21.5- and 57-kDa proteins result from the cleavage of a precursor encoded by *ha70*. In addition, it was demonstrated that the HA components are only associated with L forms of botulinum A complexes (Fujita *et al.*, 1995). In *C. botulinum* A, the molar ratio of HA35 is 4 in the L complex and 8 in the LL complex. The molar ratio of the other components of the complex is the same in both complex forms. This indicates that the 19S (LL) is a dimer of the 16S (L) complex cross-linked by the HA35 components (Inoue *et al.*, 1996) (Figure 9.6). Electron crystallography of the type

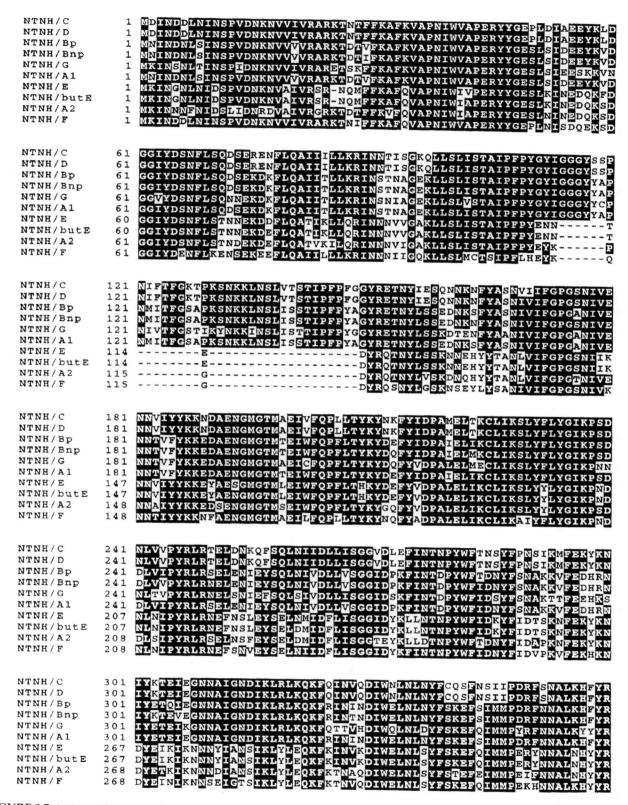

FIGURE 9.7 Amino acid sequence alignment of NTNHs. NTNH/A2, NTNH/E and NTNH/F show a deletion between positions 114 and 148 of the other NTNHs.

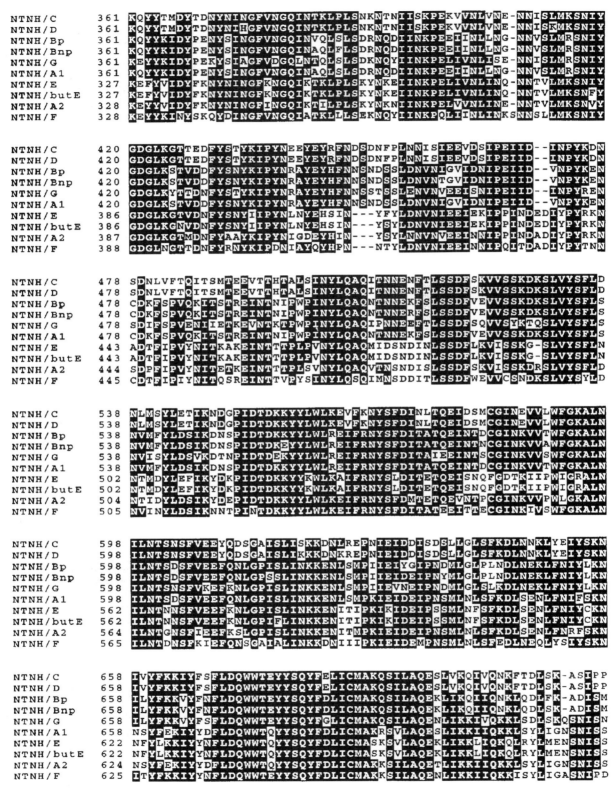

FIGURE 9.7 *Continued.*

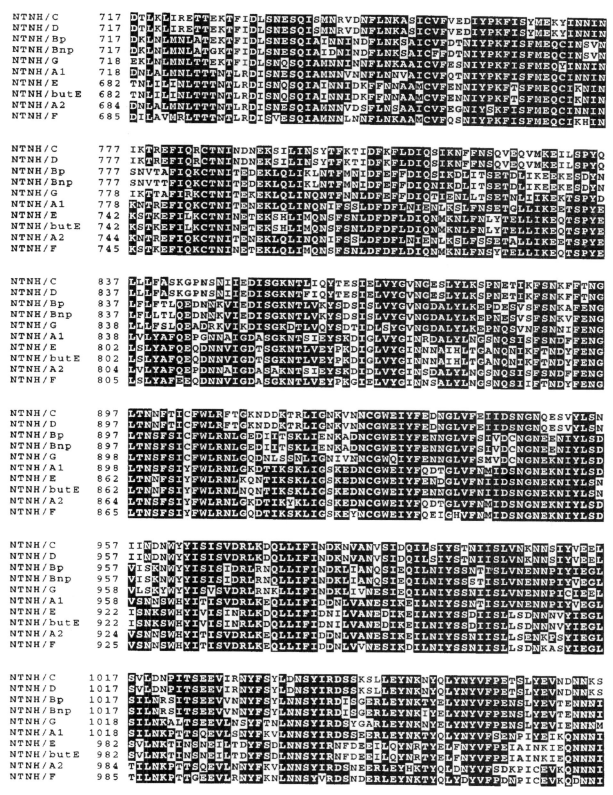

FIGURE 9.7 *Continued.*

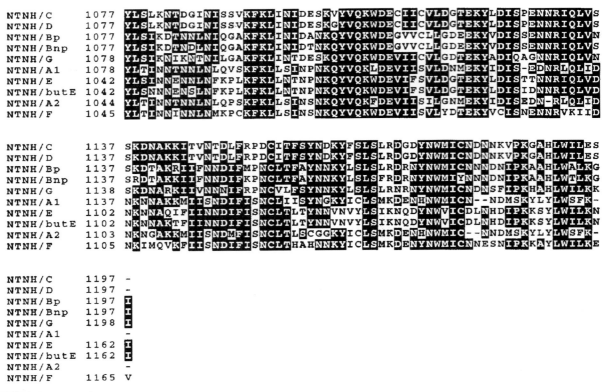

FIGURE 9.7 *Continued.*

A LL complex shows an organization in six cylinders approximately 40 Å in diameter by 100 Å high, arranged in a triangular pattern and hollow in the middle (Burkard *et al.*, 1997). Similar to *C. botulinum* A, the haemagglutinin components of *C. botulinum* D (HA17, HA22, HA33 and HA55) were found only in the L forms of botulinum complexes by SDS–PAGE (Ohyama *et al.*, 1995; Nakajima *et al.*, 1998). *Clostridium botulinum* B and G show an equivalent organization of the *ha* genes compared to that in *C. botulinum* A1, but in *C. argentinense*, *ha*33 is missing (East *et al.*, 1996; Bhandari *et al.*, 1997) (Figure 9.1).

The HA proteins do not contain a signal peptide. HA from *C. botulinum* A and from the proteolytic and non-proteolytic *C. botulinum* B are very similar (82–90% identity), but they are less related (36-40% identity) to HA33 from *C. botulinum* C and *C. botulinum* D which are identical. HA does not present significant homology with other known protein sequences.

The role of the HA components is not well understood. It had already been found that the orally ingested botulinum complex is absorbed from the upper small intestine into the lymphatic system, and that it then enters the bloodstream and reaches peripheral nerves. The function of HA components was investigated in the botulinum C complex. It was reported that the 16S complex, which contains HA components, bound intensively to the brush border of guinea pig intestinal cells as assayed by immunostaining, whereas 12S and 7S complexes, which do not encompass HA, did not. Moreover, the absorption through the intestinal barrier was more effective with the 16S complex (100 times more effective) than with the 12S and 7S complexes. The stability of the three complex forms in the guinea pig digestive juice was similar. It was concluded that HAs are required for the efficient absorption of the toxin and that they are not involved in the protection of the neurotoxin against the digestive proteolytic activity (Fujinaga *et al.*, 1997). However, whether only the neurotoxin or the whole complex can cross the intestinal barrier, and the mode of oral toxicity of the 12S complex and that of the non-haemagglutinating strains, remain to be determined. In contrast, HA33 from *C. botulinum* A was resistant to proteases and could play a role in the protection of BoNT/A (Fu *et al.*, 1997). In another study, it was found that the 19S complex (900 kDa) of type A is resistant to the gastrointestinal environment, and that BoNT/A and complex components are degraded separately. It seems that BoNT/A and complex components protect one another during exposure to pH extremes and gastrointestinal proteases (Chen *et al.*, 1998).

ha 17 is localized downstream of *ha*33, and has been identified in *C. botulinum* A, C, D and G (Hauser *et al.*, 1995; Henderson *et al.*, 1996; Bhandari *et al.*, 1997). Their level of identity is from 58% to 66%, and they are not related to other known proteins. Their function remains unknown.

HA70 has been found in the botulinum locus of *C. botulinum* A, B, C and D. The gene encodes a protein with a predicted size of 70 kDa which is cleaved into two proteins, one (19–22 kDa) corresponds to the N-terminal part and the other (52 kDa) corresponds to the C-terminal part (Somers and Das Gupta, 1991; Ohyama *et al.*, 1995; Inoue *et al.*, 1996). In *C. botulinum* D, the cleavage sites have been reported to be at Lys-6 and Lys-192 (Ohyama *et al.*, 1995) which could correspond to a trypsin-like protease probably produced by *C. botulinum*. In *C. botulinum* A and C, several proteins between 19 and 23 kDa coexist which correspond to different cleavages in the N-terminal part of the HA70 precursor (Inoue *et al.*, 1996). The 52-kDa protein shows a significant similarity with the enterotoxin from *C. perfringens* (CPE) (Hauser *et al.*, 1995). HA70/C shares an overall identity of 27.6% with CPE. The higher level of identity was found between the N-terminal parts of the 52-kDa protein and CPE (Hanna *et al.*, 1992). CPE is a protein which recognizes a cell-surface receptor and inserts into the membrane forming a large complex (160 kDa) with an additional membrane protein leading to the formation of pores across the cell membrane. Interestingly, the CPE receptor is related to a trans-membrane protein of blood–brain barrier endothelial cells (Chen *et al.*, 1998). However, it is not known whether HA70 shares common functional activity with CPE.

In *C. botulinum* A2, E and F, a gene encoding a 47–48-kDa protein was identified immediately upstream of the *ntnh* gene (Figure 9.1). This gene is separated by only 15 nucleotides from the *ntnh* gene, and both genes are transcribed in the same orientation. P47 from *C. botulinum* A2 and F are identical, and P48 from *C. botulinum* E has a 80% identity level with P47 from *C. botulinum* A2 and F. A gene encoding for a homologous protein to P47 and P48 was mapped and partially characterized in the locus of botulinum complex genes in toxigenic *C. butyricum* and *C. botulinum* A1 with a silent *bont*/B gene (Hutson *et al.*, 1996; Li *et al.*, 1998). The function of P47 and P48 remains unknown. However, it is interesting to note that *p47*–*p48* genes are found only in the locus of botulinum complex genes which lack the *ha* genes. In addition, *C. botulinum* A2, E and F, and toxigenic *C. butyricum*, show the presence of two genes (*orf*X1 and *orf*X2) in the 5′ part of the botulinum locus which are not related to complex genes from the other toxinotypes (Kubota *et al.*, 1998).

REGULATION OF NEUROTOXIN AND ANTP GENES

The synthesis of many virulence factors is subjected to regulation. Environmental factors can stimulate a sensor protein which activates a cascade of protein kinases. Finally, the virulence factor genes are regulated at the transcriptional level. Findings on toxin gene regulation in *C. botulinum* and *C. tetani* have emerged recently.

Transcriptional analysis

DNA sequencing of the botulinum locus showed that the *bont* and *antp* genes are in the vicinity of each other. Two clusters of genes can be distinguished: the cluster of *ntnh–bont* genes and the one of *ha* genes.

ntnh and *bont* genes are transcribed in the same orientation and are closely linked. This suggests that both genes are organized in an operon. In *C. botulinum* A2, E and F, a third gene (*p47*, *p48*) is localized immediately upstream of the *ntnh* gene and is transcribed in the same orientation (Figure 9.1).

The second cluster of genes encompasses the three genes encoding the HA components (*ha*33, *ha*17 and *ha*70). The three genes are in close proximity and are transcribed in the opposite direction to that of the *ntnh–bont* genes.

Transcriptional analysis has been investigated in *C. botulinum* A and C (Hauser *et al.*, 1995; Henderson *et al.*, 1996) (Figure 9.8). In *C. botulinum* C, a transcription start site was mapped 113 nucleotides upstream the ATG of *ntnh*/C, and another one 100 nucleotides upstream of the initial codon of *bont*/C1. Both genes are preceded by a consensus ribosome binding site (GGAGG). The analysis of mRNA by reverse transcriptase PCR (RT-PCR) showed that a mRNA overlapped *bont*/C1 and *ntnh*/C1 genes, possibly encoding both genes (Hauser *et al.*, 1995). Similar results were found in *C. botulinum* A (Henderson *et al.*, 1996). Two transcripts have been identified for the *bont*/A gene. The longer one (7.5 kb) encompasses the *ntnh–bont*/A genes, and the shorter one (4 kb) corresponds to *bont*/A alone. This indicates that *bont*/A and *bont*/C1 are transcribed as a mono- or a bicistronic messenger in association with the corresponding *ntnh* gene.

Using RT-PCR, it was found that at least one mRNA overlaps the three *ha* genes in *C. botulinum* C (Hauser *et al.*, 1995). By Northern blots in *C. botulinum* A, a single 3.2-kb transcript encompasses *ha*35, *ha*17 and *ha*70 genes (Henderson *et al.*, 1996). The *ha* genes seem to form a tricistronic operon. However, the HA35 or HA33 components in *C. botulinum* A and C, respectively, were produced in larger amount than the other HA components (Inoue *et al.*, 1996). This suggests that the relative levels of synthesis of the three HA components are controlled at the translational level as was found for other toxins, such as cholera toxin (Hirst, 1995).

The genetic organization in the other *C. botulinum* types shows that *bont* and *antp* genes are localized in two clusters which can be considered as polycistronic units by comparison with the findings reported in *C. botulinum* A and C.

botR and tetR genes

It was first reported that in *C. botulinum* C, a gene (*bot*R/C previously called *orf*22) encodes a 22-kDa protein having the features of a DNA-binding protein: basic pI (10.4) and the presence of a helix–turn–helix motif (Hauser *et al.*, 1995). *bot*R/C is localized upstream of the *ha* genes (Figure 9.1). A homologous gene (*bot*R/A, previously called *orf*21) was characterized in *C. botulinum* A (Henderson *et al.*, 1996). *bot*R/A has a location different to *bot*R/C, as it is inserted between the *ha* and *ntnh–bont* genes. This gene is also conserved in proteolytic and non-proteolytic *C. botulinum* B, *C. botulinum* D, F and G, and in *C. tetani* (East *et al.*, 1996; Henderson *et al.*, 1996; Bhandari *et al.*, 1997; Li *et al.*, 1998). However, it has not yet been detected in *C. botulinum* E. *tet*R is the only gene related to a gene from the botulinum locus which is conserved in *C. tetani*.

BotRs from the different toxinotypes and TetR have 51–97% identity, and are related to other known regulatory proteins such as Uvia (25–28% identity) which regulates the bacteriocin production in *C. perfringens*, Msmr protein (21–26%) which regulates the sugar transport in *Streptococcus mutans* and TxeR (20–24%) in

C. botulinum C

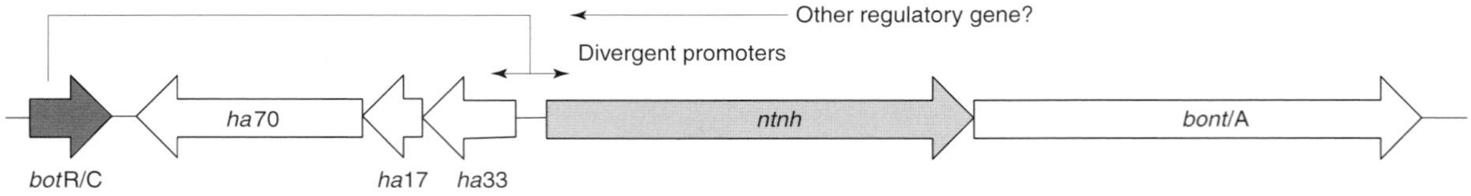


```
          -35                        *    -10
ATTTTAGGTTTACAAAAgacgatgaagaaATGTTATATATAAgtg....98 nt...-> ntnh-bont/C1
ATTTTAGGTTTACAAAAtttggcttaaatATGTTATATGTAAgtg....31 nt...-> ha33
```

C. botulinum A

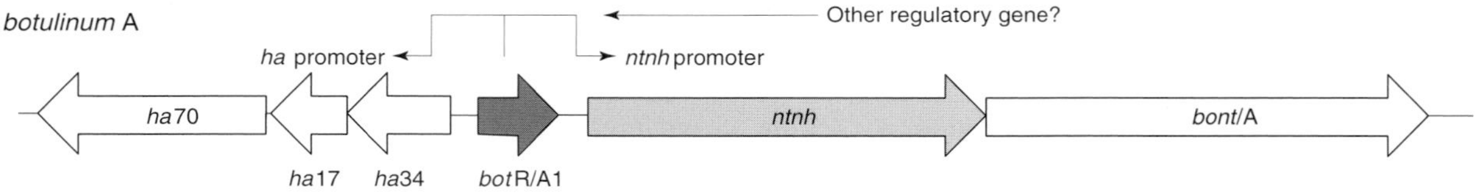


```
          -35                            -10       *
ATTTTAGGTTTACAAAAaatagtgtggctATGTTATATATAAatg....96 nt...-> ntnh-bont/A
                                                   *
ATTTTAGGTTTACAAAAaataacttgattATGTTATATGTTAtat....34 nt...-> ha34
```

Corresponding sequences in other *Clostridium botulinum*

```
C. botulinum B        ATTTTAGGTTTACAAAAaatagtgtggctATGTTATATATAA.......103 nt..-> ntnh-bont/A
                      ATTTTAGGTTTACAAAAaataatttgattATGTTATATGTTA........25 nt..-> ha33/B

C. botulinum D        ATTTTAGGTTTACAAAAgacgatgaagaaATGTTATATATAA........99 nt..-> ntnh-bont/D
                      ATTTTAGGTTTACAAAAtttggcttaaatATGTTATATGTAA........21 nt..-> ha33/D

C. argentinense G     GTTTTAGGTTTACAAAAaataatgtagtcATGTTATATATAA.......116 nt..-> ntnh-bont/G
                      AATTTAGGTTTACAAAAaataatataattATGTTATATATAA........80 nt..-> ha17/G

C.tetani              ATTTTCAGTTTACAAAAaataacctgattATGTTATATGTAA.........8 nt..-> tetx

C. botulinum A2       TTAATACCTTTACAAAAattaatttagatAAGTTATATATTT........32 nt..-> P47
C. botulinum F        TTAATAAGTTTACAAAAattcatttagatAAGTTATATATTT........32 nt..-> P47
```

FIGURE 9.8 Schematic representation of the regulation the *bont* and *antp* gene expression in *C. botulinum* C and A, and promoter conserved sequences in *C. botulinum* and *C. tetani* (summarized data from Bhandari *et al.*, 1997 and Henderson *et al.*, 1997). The * indicates transcriptional start sites which have been experimentally mapped.

C. difficile (Figure 9.9). Interestingly, *txe*R has a similar location to *bot*R and *tet*R, as it lies upstream of the locus of the toxin A and toxin B genes. It was shown that TxeR is a positive regulator by studying the expression of a reporter gene fused to *tox*A and *tox*B gene promoter in *E. coli* (Moncrief *et al.*, 1997). In addition, BotR and TetR show some similarity, but to a lower extent, with the −35 binding domain of sigma factors.

The function of *bot*R/A was analysed by over-expressing this gene in *C. botulinum* A. *bot*R/A was cloned under the control of its own promoter in a high copy number shuttle vector which was transferred in *C. botulinum* A by electroporation. A significant increase in the BoNT/A and ANTPs production and in the corresponding mRNA levels was detected in the recombinant strain. In contrast, partial inhibition of the *bot*R/A expression by antisense mRNA resulted in a lower production of BoNT/A and ANTPs. It was concluded

that BotR/A is a positive regulator at the transcriptional level of *bont* and *antp* genes. In the same way, it was shown that TetR is a positive regulator of the *tetx* gene expression, and that BotR/A and to a lesser extent BotR/C are functional in *C. tetani*. This indicates that the regulation of the *bont* and *tetx* genes is conserved in *C. botulinum* and *C. tetani*, and this constitutes further evidence that *bont* and *tetx* gene loci derive from a common ancestor (Marvaud *et al.*, 1998a, b).

Moreover, it was found that BotR/A interacts directly with the promoter region upstream of the *ntnh–bont* and *ha* genes (Marvaud *et al.*, 1998a). Interestingly, the −10 and −35 regions of the two operons are conserved in *C. botulinum* A and C (Henderson *et al.*, 1996), and also in *C. botulinum* A2, F, G and in *C. tetani* (Henderson *et al.*, 1996; Bhandari *et al.*, 1997). These regions could represent the target sites of BotR and TetR (Figure 9.8).

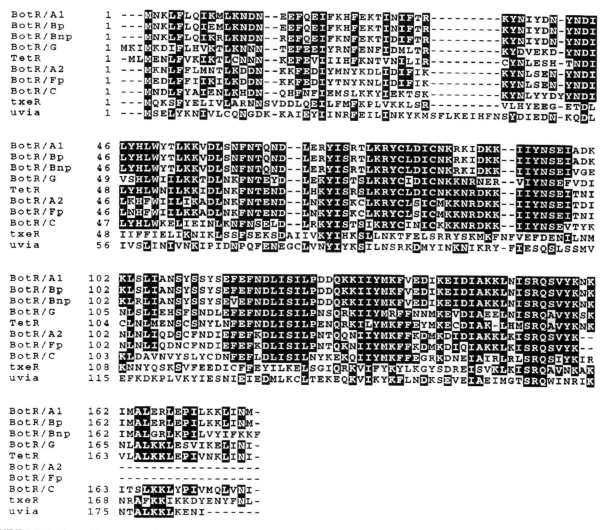

FIGURE 9.9 Amino acid sequence alignment of BotR, TetR, Uvia from *C. perfringens*, and TxeR from *C. difficile*.

It is not known whether BotR and TetR are involved in a regulation cascade and what environmental signals trigger the production of the neurotoxins and ANTPs. It was reported that the nitrogen source, such as small peptides, modulates the TeTx production (Porfirio *et al.*, 1997). However, the involvment of a two-component signal transduction remains to be proven. In *C. difficile*, catabolic repression of the expression of toxin A and toxin B genes by rapidly metabolizable sugars such as glucose has been demonstrated (Dupuy and Sonenshein, 1998). No such regulation has been identified in *C. botulinum* and *C. tetani*.

CONCLUDING REMARKS

The mechanism of action of BoNTs and TeTx is now well understood, although some points remain to be clarified, such as the specific neuronal receptor(s) and the support for the retrograde transport of TeTx. In contrast, the structure and function of the ANTPs are still unexplained: why do BoNTs form complexes and TeTx does not? Why does the same strain produce different *botulinum* complex sizes? Do NTNH and HA have the same function? What is the basis of the BoNT resistance to acidity and proteases? Are the ANTPs involved in the absorption of BoNTs through the intestinal mucosa?

Understanding of ANTPs is important for a better appreciation of the early steps of food-borne botulism, and to determine efficient measures of prevention. ANTPs could also be used to protect pharmaceutical drugs in the digestive tract and to mediate their absorption by the intestinal mucosa.

REFERENCES

Antharavally, B.S. and Das Gupta, B.R. (1997). Covalent structure of botulinum neurotoxin type E: location of sulfhydryl groups, and disulfide bridges and identification of *C*-termini of light and heavy chains. *J. Protein Chem.* **16**, 787–99.

Antharavally, B., Tepp, W. and Das Gupta, B.R. (1998). Status of Cys residues in the covalent structure of botulinum neurotoxin types A, B, and E. *J. Protein Chem.* **17**, 187–96.

Aranda, E., Rodriguez, M.M., Asensio, M.A. and Cordoba, J.J. (1997). Detection of *Clostridium botulinum* types A, B, E, and F in foods by PCR and DNA probe. *Lett. Appl. Microbiol.* **25**, 186–90.

Bartels, F., Bergel, H., Bigalke, H., Frevert, J., Halpern, J. and Middlebrook, J. (1994). Specific antibodies against the Zn^{2+}-binding domain of clostridial neurotoxins restore exocytosis in chromaffin cells treated with tetanus or Botulinum A neurotoxin. *J. Biol. Chem.* **269**, 8122–7.

Beise, J., Hahnen, J., Andersen-Beckh, B. and Dreyer, F. (1994). Pore formation by tetanus toxin, its chain and fragments in neuronal membranes and evaluation of the underlying motifs in the structure of the toxin molecule. *Arch. Pharmacol.* **349**, 66–73.

Bhandari, M., Campbell, K.D., Collins, M.D. and East, A.K. (1997). Molecular characterization of the clusters of genes encoding the botulinum neurotoxin complex in *Clostridium botulinum* (*Clostridium argentinense*) type G and nonproteolytic *Clostridium botulinum* type B. *Curr. Microbiol.* **35**, 207–14.

Binz, T., Kurazono, H., Wille, M., Frevert, J., Wernars, K. and Niemann, H. (1990a). The complete sequence of botulinum neurotoxin type A and comparison with other clostridial neurotoxins. *J. Biol. Chem.* **265**, 9153–8.

Binz, T., Kurazono, H., Popoff, M.R., Eklund, M.W., Sakaguchi, G., Kozaki, S., Krieglstein, K., Henschen, A., Gill, D.M. and Niemann, H. (1990b). Nucleotide sequence of the gene encoding *Clostridium botulinum* neurotoxin type D. *Nucleic Acid. Res.* **18**, 5556.

Brefort, G., Magot, M., Ionesco, H. and Sebald, M. (1977). Characterization and transferability of *Clostridium perfringens* plasmids. *Plasmid* **1**, 52–66.

Burkard, F., Chen, F., Kuziemko, G.M. and Stevens, R.C. (1997). Electron density projection map of the botulinum neurotoxin 900-kilodalton complex by electron miscrocopy. *J. Struct. Biol.* **120**, 78–84.

Call, J.E., Cooke, P.H. and Miller, A.J. (1995). *In situ* characterization of *Clostridium botulinum* neurotoxin synthesis and export. *J. Appl. Bacteriol.* **79**, 257–63.

Campbell, K., Collins, M.D. and East, A.K. (1993a). Nucleotide sequence of the gene coding for *Clostridium botulinum* (*Clostridium argentinense*) type G neurotoxin: genealogical comparison with other clostridial neurotoxins. *Biochim. Biophys. Acta* **1216**, 487–93.

Campbell, K.D., Collins, M.D. and East, A.K. (1993b). Gene probes for identification of the botulinal neurotoxin gene and specific identification of neurotoxin types B, E and F. *J. Clin. Microbiol.* **31**, 2255–62.

Chen, F., Kuziemko, G.M. and Stevens, R.C. (1998). Biophysical characterization of the stability of the 150-kilodalton botulinum toxin, the nontoxic component, and the 900-kilodalton botulinum toxin complex species. *Infect. Immun.* **66**, 2420–5.

Chen, Z., Zandonalti, M., Jakubowski, D. and Fox, H.S. (1998). Brain capillary endothelial cells express MBEC1, a protein that is related to the *Clostridium perfringens* enterotoxin receptors. *Lab. Investig.* **78**, 353–63.

Choe, S., Bennet, M.J., Fujii, G., Curni, P.M., Kantardjieff, K.A., Collier, R.J. and Eisenberg, D. (1992). The crystal structure of diphtheria toxin. *Nature* **357**, 216–22.

Coen, L., Osta, R., Maury, M. and Brulet, P. (1997). Construction of hybrid proteins that migrate retrogradely and transynaptically into the central nervous sytem. *Proc. Natl Acad. Sci. USA* **94**, 9400–5.

Collins, M.D. and East, A.K. (1997). Phylogeny and taxonomy of the food-borne pathogen *Clostridium botulinum* and its neurotoxins. *J. Appl. Microbiol.* **84**, 5–17.

Cordoba, J.J., Collins, M.D. and East, A.K. (1995). Studies on the genes encoding botulinum neurotoxin type A of *Clostridium botulinum* from a variety of sources. *System. Appl. Microbiol.* **18**, 13–22.

Cornille, F., Martin, L., Lenoir, C., Cussac, D., Doques, B.P. and Fournie-Zaluski, M.C. (1997). Cooperative exosite-dependent cleavage of synaptobrevin by tetanus toxin light chain. *J. Biol. Chem.* **272**, 3459–64.

Dayanithi, G., Stecher, B., Höhne-Zell, B., Yamasaki, S., Binz, T., Weller, U., Niemann, H. and Gratzl, M. (1994). Exploring the functional domain and the target of the tetanus toxin light chain in neurophysial terminals. *Neuroscience* **58**, 423–31.

de Paiva, A., Poulain, B., Lawrence, G.W., Shone, C.C., Tauc, L. and Dolly, J.O. (1993). A role for the interchain disulfide or its participating thiols in the internalization of botulinum neurotoxin A revealed by a toxin derivative that binds to ecto-acceptors and inhibits transmitter release intracellularly. *J. Biol. Chem.* **268**, 20838–44.

Dupuy, B. and Sonenshein, A.L. (1998). Regulated transcription of *Clostridium difficile* toxin genes. *Mol. Microbiol.* **27**, 107–20.

East, A.K., Richardson, P.T., Allaway, D., Collins, M.D., Roberts, T.A. and Thompson, D.E. (1992). Sequence of the gene encoding type F neurotoxin of *Clostridium botulinum. FEMS Microbiol. Lett.* **96**, 225–30.

East, A.K., Bhandari, M., Stacey, J.M., Campbell, K.D. and Collins, M.D. (1996). Organization and phylogenetic interrelationships of genes encoding components of the botulinum toxin complex in proteolytic *Clostridium botulinum* types A, B, and F: evidence of chimeric sequences in the gene encoding the nontoxic nonhemagglutinin component. *Int. J. Syst. Bacteriol.* **46**, 1105–12.

Eisel, U., Jarausch, W., Goretzki, K., Henschen, A., Engels, J., Weller, U., Hudel, M., Habermann, E. and Niemann, H. (1986). Tetanus toxin: primary structure, expression in *E. coli*, and homology with botulinum toxins. *EMBO J.* **5**, 2495–502.

Eklund, M.W. and Dowell, J. (1987). *Avian Botulism*. Charles C. Thomas, Springfield, IL.

Eklund, M.W. and Poysky, F.T. (1974). Interconversion of type C and D strains of *Clostridium botulinum* by specific bacteriophages. *Appl. Microbiol.* **27**, 251–8.

Eklund, M.W., Poysky, F.T., Reed, S.M. and Smith, C.A. (1971). Bacteriophage and the toxigenicity of *Clostridium botulinum* type C. *Science* **172**, 480–2.

Eklund, M.W., Poysky, F.T., Meyers, J.A. and Pelroy, G.A. (1974). Interspecies conversion of *Clostridium botulinum* type C to *Clostridium novyi* type A by bacteriophage. *Science* **172**, 480–2.

Eklund, M.W., Poysky, F.T., Mseitif, L.M. and Strom, M.S. (1988). Evidence for plasmid-mediated toxin and bacteriocin production in *Clostridium botulinum* type G. *Appl. Environ. Microbiol.* **54**, 1405–8.

Eklund, M.W., Poysky, F.T. and Habig, W.H. (1989). Bacteriophages and plasmids in *Clostridium botulinum* and *Clostridium tetani* and their relationship to production of toxins. In: *Botulinum Neurotoxin and Tetanus Toxin* (ed. L.L. Simpson), pp. 25–51. Academic Press, San Diego, CA.

Fach, P., Hauser, D., Guillou, J.P. and Popoff, M.R. (1993). Polymerase chain reaction for the rapid identification of *Clostridium botulinum* type A strains and detection in food samples. *J. Appl. Bacteriol.* **75**, 234–9.

Fach, P., Gibert, M., Grifais, R., Guillou, J.P. and Popoff, M.R. (1995). PCR and gene probe identification of botulinum neurotoxin A-, B-, E-, F-, and G-producing *Clostridium* spp. and evaluation in food samples. *Appl. Environ. Microbiol.* **61**, 389–92.

Fach, P., Gibert, M., Griffais, R. and Popoff, M.R. (1996). Investigation of animal botulism outbreaks by PCR and standard methods. *FEMS Immunol. Med. Microbiol.* **13**, 279–85.

Fairweather, N.F. and Lyness, V.A. (1986). The complete nucleotide sequence of tetanus toxin. *Nucleic Acids Res.* **14**, 7809–12.

Finn, C.W., Silver, R.P., Habig, W.H., Hardegree, M.C., Zon, G. and Garon, C.F. (1984). The structural gene for tetanus neurotoxin is on a plasmid. *Science* **224**, 881–4.

Franciosa, G., Ferreira, J.L. and Hatheway, C.L. (1994). Detection of type A, B, and E botulism neurotoxin genes in *Clostridium botulinum* and other *Clostridium* species by PCR: evidence of unexpressed type B toxin genes in type A toxigenic organisms. *J. Clin. Microbiol.* **32**, 1911–17.

Franciosa, G., Fenicia, L., Pourshaban, M. and Aureli, P. (1997). Recovery of a strain of *Clostridium botulinum* producing both neurotoxin A and neurotoxin B from canned macrobiotic food. *Appl. Environ. Microbiol.* **63**, 1148–50.

Fu, F.N., Sharma, S.K. and Singh, B.R. (1997). A protease-resistant novel hemagglutinin purified from type A *Clostridium botulinum*. *J. Protein Chem.* **17**, 53–60.

Fujii, N., Kimura, K., Yokosawa, N., Yashiki, N., Tsuzuki, K. and Oguma, K. (1993a). The complete nucleotide sequence of the gene encoding the nontoxic component of *Clostridium botulinum* type E progenitor toxin. *J. Gen. Microbiol.* **139**, 79–86.

Fujii, N., Kimura, K., Yokosawa, N., Oguma, K., Yashiki, T., Takeshi, K., Ohyama, T. and Isogai, H. (1993b). Similarity in nucleotide sequence of the gene encoding nontoxic component of botulinum toxin produced by toxigenic *Clostridium butyricum* strain BL6340 and *Clostridium botulinum* type E strain Mashike. *Microbiol. Immunol.* **37**, 395–8.

Fujinaga, Y., Takeshi, K., Inoue, K., Fujita, R., Ohyama, T., Moriishi, K. and Oguma, K. (1995). Type A and B neurotoxin genes in a *Clostridium botulinum* type AB strain. *Biochem. Biophys. Res. Commun.* **213**, 737–45.

Fujinaga, Y., Inoue, K., Watanabe, S., Yokota, K., Hirai, Y., Nagamachi, E. and Oguma, K. (1997). The haemagglutinin of *Clostridium botulinum* type C progenitor toxin plays an essential role in binding of toxin to the epithelial cells of guinea pig intestine, leading to the efficient absorption of the toxin. *Microbiology* **143**, 3841–7.

Fujita, R., Fujinaga, Y., Inoue, K., Nakajima, H., Kumon, H. and Oguma, K. (1995). Molecular characterization of two forms of nontoxic-non hemagglutinantinin components of *Clostridium botulinum* type A progenitor toxins. *FEBS Lett.* **376**, 41–4.

Hall, J.D., McCroskey, L.M., Pincomb, B.J. and Hatheway, C.L. (1985). Isolation of an organism resembling *Clostridium barati* which produces type F botulinal toxin from an infant with botulism. *J. Clin. Microbiol.* **21**, 654–5.

Halpern, J. (1995). Tetanus toxin. In: *Bacterial Toxins and Virulence Factors in Disease*, Vol. 8 (eds J. Moss, B. Iglewski, M. Vaughan and A. T. Tu), pp. 521–41. Marcel Dekker, New York.

Halpern, J.L. and Loftus, A. (1993). Characterization of the receptor-binding domain of tetanus toxin. *J. Biol. Chem.* **268**, 11188–92.

Halpern, J.L., Habig, W.H., Neale, E.A. and Stibitz, S. (1990). Cloning and expression of functional fragment C of tetanus toxin. *Infect. Immun.* **58**, 1004–9.

Hambleton, P. (1992). *Clostridium botulinum* toxins: a general review of involvement in disease, structure, mode of action and preparation for clinical use. *J. Neurol.* **239**, 16–20.

Hanna, P.C., Wieckoski, E.U., Mietzner, T.A. and McClane, B.A. (1992). Mapping of functional regions of *Clostridium perfringens* type A enterotoxin. *Infect. Immun.* **60**, 2110–14.

Hauser, D., Eklund, M.W., Kurazono, H., Binz, T., Niemann, H., Gill, D.M., Boquet, P. and Popoff, M.R. (1990). Nucleotide sequence of *Clostridium botulinum* C1 neurotoxin. *Nucleic Acid. Res.* **18**, 4924.

Hauser, D., Gibert, M., Boquet, P. and Popoff, M.R. (1992). Plasmid localization of a type E botulinal neurotoxin gene homologue in toxigenic *Clostridium butyricum* strains, and absence of this gene in non-toxigenic *C. butyricum* strains. *FEMS Microbiol. Lett.* **99**, 251–6.

Hauser, D., Gibert, M., Eklund, M.W., Boquet, P. and Popoff, M.R. (1993). Comparative analysis of C3 and botulinal neurotoxin genes and their environment in *Clostridium botulinum* C and D. *J. Bacteriol.* **175**, 7260–8.

Hauser, D., Eklund, M.W., Boquet, P. and Popoff, M.R. (1994). Organization of the botulinum neurotoxin C1 gene and its associated non-toxic protein genes in *Clostridium botulinum* C468. *Mol. Gen. Genet.* **243**, 631–40.

Hauser, D., Gibert, M., Marvaud, J.C., Eklund, M.W. and Popoff, M.R. (1995). Botulinal neurotoxin C1 complex, Clostridial neurotoxin homology and genetic transfer in *Clostridium botulinum*. *Toxicon* **33**, 515–26.

Henderson, I., Whelan, S.M., Davis, T.O. and Minton, N.P. (1996). Genetic characterization of the botulinum toxin complex of *Clostridium botulinum* strain NCTC2916. *FEMS Microbiol. Lett.* **140**, 151–8.

Henderson, I., Davis, T., Elmore, M. and Minton, N. (1997). The genetic basis of toxin production in *Clostridium botulinum* and *Clostridium tetani*. In: *The Clostridia: Molecular Biology and Pathogenesis* (ed. I. Rood), pp. 261–94. Academic Press, New York.

Hirst, T.R. (1995). Biogenesis of cholera toxin and related oligomeric toxins. In: *Bacterial Toxins and Virulence Factors in Disease* (eds J. Moss, B. Iglewsski, M. Vaughan and A.T. Tu), pp. 123–84. Marcel Dekker, New York.

Högy, B., Dauzenroth, M.E., Hudel, M., Weller, U. and Habermann, E. (1992). Increase of permeability of synaptosomes and liposomes by the heavy chains of tetanus toxin. *Toxicon* **30**, 63–76.

Höhne-Zell, B., Stecher, B. and Gratzl, M. (1993). Functional characterization of the catalytic site of the tetanus toxin light chain using permeabilized chromaffin cells. *FEBS Lett.* **336**, 175–80.

Hutson, R.A., Collins, M.D., East, A.K. and Thompson, D.E. (1994). Nucleotide sequence of the gene coding for non-proteolytic *Clostridium botulinum* type B neurotoxin: comparison with other Clostridial neurotoxins. *Curr. Microbiol.* **28**, 101–10.

Hutson, R.A., Zhou, Y., Collins, M.D., Johnson, E.A., Hatheway, C.L. and Sugiyama, H. (1996). Genetic characterization of *Clostridium botulinum* type A containing silent type B neurotoxin gene sequences. *J. Biol. Chem.* **271**, 10786–92.

Inoue, K., Fujinaga, Y., Watanabe, T., Ohyama, T., Takeshi, K., Moriishi, K., Nakajima, H., Inoue, K. and Oguma, K. (1996). Molecular composition of *Clostridium botulinum* type A progenitor toxins. *Infect. Immun.* **64**, 1589–94.

Jovita, M.R., Collins, M.D. and East, A.K. (1998). Gene organization and sequence determination of the two botulinum neurotoxin gene clusters in *Clostridium botulinum*. *Curr. Microbiol.* **36**, 226–31.

Kimura, K., Fujii, N., Tsuzuki, K., Murakami, T., Indoh, T., Yokosawa, N., Takeshi, K., Syuto, B. and Oguma, K. (1990). The complete nucleotide sequence of the gene coding for botulinum type C1 toxin in the C–St phage genome. *Biochem. Biophys. Res. Commun.* **171**, 1304–11.

Kozaki, S., Kamata, Y., Takahashi, M., Shimizu, T. and Sakaguchi, G. (1989a). Antibodies against Botulinum Neurotoxin. In *Botulinum Neurotoxin and Tetanus Toxin* (ed. L.L. Simpson), pp. 301–18. Academic Press, San Diego, CA.

Kozaki, S., Kamata, Y., Ogasawara, J. and Sakaguchi, G. (1989b). Immunological characterization of papain-induced fragments of *Clostridium botulinum* A neurotoxin and interaction of the fragments with brain synaptosomes. *Infect. Immun.* **57**, 2634–9.

Krieglstein, K.G., Henschenn, A.H., Weller, U. and Habermann, E. (1991). Limited proteolysis of tetanus toxin. Relation to activity and identification of cleavage sites. *Eur. J. Biochem.* **202**, 41–51.

Krieglstein, K.G., Das Gupta, B.R. and Henschen, A.H. (1994). Covalent structure of botulinum neurotoxin A: location of sulfhydryl bridges and identification of *C*-termini of light and heavy chains. *J. Protein Chem.* **13**, 49–57.

Kubota, T., Shirakawa, S., Kozaki, S., Isogai, E., Isogai, H., Kimura, K. and Fujii, N. (1996). Mosaic structure of the nontoxic-nonhemagglutinating component gene in *Clostridium botulinum* type A strain isolated from infant botulism in Japan. *Biochem. Biophys. Res. Commun.* **224**, 843–8.

Kubota, T., Yonekura, N., Hariya, Y., Isogai, E., Isogai, H., Amano, K. and Fujii, N. (1998). Gene arrangement in the upstream region of *Clostridium botulinum* type E and *Clostridium butyricum* BL6340 progenitor toxin genes is different from that of other types. *FEMS Microbiol. Lett.* **158**, 215–21.

Kurazono, H., Mochida, S., Binz, T., Eisel, U., Quanz, M., Grebenstein, O., Wernars, K., Poulain, B., Tauc, L. and Niemann, H. (1992). Minimal essential domains specifying toxicity of the light chains of Tetanus Toxin and Botulinum Neurotoxin type A. *J. Biol. Chem.* **2667**, 14721–9.

Lacy, D.B., Tepp, W., Cohen, A.C., Das Gupta, B.R. and Steven, R.C. (1998). Crystal structure of botulinum neurotoxin type A and implications for toxicity. *Nature Struct. Biol.* **5**, 898–902.

Lebeda, F.J. and Olson, M.A. (1995). Structural predictions of the channel-forming region of botulinum neurotoxin heavy chain. *Toxicon* **33**, 559–67.

Ledoux, D.N., Be, X.H. and Singh, B.R. (1994). Quaternary structure of Botulinum and Tetanus neurotoxins as probed by chemical cross-linking and native gel electrophoresis. *Toxicon* **32**, 1095–104.

Li, B., Qian, X., Sarkar, H.K. and Singh, B.R. (1998). Molecular characterization of type E *Clostridium botulinum* and comparison to other types of *Clostridium botulinum*. *Biochim. Biophys. Acta* **1395**, 21–7.

Li, Y., Foran, P., Fearweather, N.F., de Paiva, A., Weller, U., Dougan, G. and Dolly, O. (1994). A single mutation in the recombinant light chain of Tetanus Toxin abolishes its proteolytic activity and removes the toxicity seen after reconstruction with heavy chain. *Biochemistry* **33**, 7014–20.

Lin, W.J. and Johnson, E.A. (1995). Genome analysis of *Clostridium botulinum* type A by pulsed-field gel electrophoresis. *Appl. Environ. Microbiol.* **61**, 4441–7.

Marvaud, J.C., Gibert, M., Inoue, K., Fujinaga, V., Oguma, K. and Popoff, M.R. (1998a). *bot*R is a positive regulator of botulinum neurotoxin and associated non toxic protein genes in *Clostridium botulinum* A. *Mol. Microbiol.* **29**, 1009–18.

Marvaud, J.C., Eisel, U., Binz, T., Niemann, H. and Popoff, M.R. (1998b). TetR is a positive regulator of the tetanus Toxin gene in *Clostridium tetani* and is homologous to *bot*R. *Infect. Immun.* **66**, 5698–702.

McCroskey, L.M., Hatheway, C.L., Fenicia, L., Pasolini, B. and Aureli, P. (1986). Characterization of an organism that produces type E botulinal toxin but which resembles *Clostridium butyricum* from the feces of an infant with type E botulism. *J. Clin. Microbiol.* **23**, 201–4.

McCroskey, L.M., Hatheway, C.L., Woodruff, B.A., Greenberg, J.A. and Jurgenson, P. (1991). Type F botulism due to neurotoxigenic *Clostridium baratii* from an unknown source in an adult. *J. Clin. Microbiol.* **29**, 2618–20.

Moncrief, J.S., Barroso, L.A. and Wilkins, T.D. (1997). Positive regulation of *Clostridium difficile* toxins. *Infect. Immun.* **65**, 1105–8.

Montal, M.S., Blewitt, R., Tomich, J.M. and Montal, M. (1992). Identification of an ion channel-forming motif in the primary structure of tetanus and botulinum neurotoxins. *FEBS Lett.* **313**, 12–18.

Montecucco, C. (1986). How do tetanus and botulinum toxins bind to neuronal membranes? *Trends Biochem. Sci.* **11**, 314–17.

Moriishi, K., Koura, M., Aba, N., Fujii, N., Fujinaga, Y., Inoue, K. and Oguma, K. (1996). Mosaic structures of neurotoxins produced from *Clostridium botulinum* types C and D. *Biochim. Biophys. Acta* **1307**, 123–6.

Nakajima, H., Inove, K., Ikeda, T., Fujinaga, Y., Sunagawa, H., Takeshi, K., Ohyama, T., Watanabe, T., Inove, K. and Oguma, K. (1998). Molecular composition of the 16S toxinpzoduccol by a

Clostridium botulinum type D strain, 1873. *Microbiol. Immunol.* **42**, 599–605.

Nishiki, T., Tokuyama, Y., Kamata, Y., Nemoto, Y., Yoshida, A., Sato, K., Sekigichi, M., Taakahashi, M. and Kozaki, S. (1996). The high-affinity of *Clostridium botulinum* type B neurotoxin to synapto-tagmin II associated with gangliosides G_{T1B}/G_{D1a}. *FEBS Lett.* **378**, 253–7.

Ohyama, T., Watanabe, T., Fujinaga, Y., Inoue, K., Sunagawa, H., Fujii, N., Inoue, K. and Oguma, K. (1995). Characterization of nontoxic-nonhemagglutinin component of the two types of progenitor toxin (M and L) produced by *Clostridium botulinum* type C CB-16. *Microbiol. Immunol.* **39**, 457–65.

Pelizzari, R., Mason, S., Shone, C.C. and Montecucco, C. (1997). The interaction of synaptic vesicle-associated membrane protein/synaptobrevin with botulinum neurotoxins D and F. *FEBS Lett.* **409**, 339–42.

Petosa, C., Collier, J.R., Klimpel, K.R., Leppla, S.H. and Liddington, R.C. (1997). Crystal structure of the anthrax toxin protective antigen. *Nature (Lond.)* **385**, 833–8.

Popoff, M.R. (1995). Ecology of neurotoxigenic strains of Clostridia. In: *Clostridial Neurotoxins*, Vol. 195. *Current Topics in Microbiology and Immunology* (ed. C. Montecucco), pp. 1–29. Springer, Heidelberg.

Popoff, M.R. and Eklund, M.W. (1995). Tetanus and botulinum neurotoxins: genetics and molecular mode of action. In: *Molecular Approaches to Food Safety Issues Involving Toxic Microorganisms* (eds M.W. Eklund, J. Richards and K. Mise), pp. 481–511. Alaken, Fort Collins, CO.

Porfirio, Z., Prado, S.M., Vancetto, M.D.C., Fratelli, F., Alves, E.W., Raw, I., Fernandes, B.L., Camargo, A.C.M. and Lebrun, I. (1997). Specific peptides of casein pancreatic digestion enhance the production of tetanus toxin. *J. Appl. Microbiol.* **83**, 678–84.

Poulet, S., Hauser, D., Quanz, M., Niemann, H. and Popoff M.R. (1992). Sequences of the botulinal neurotoxin E derived from *Clostridium botulinum* type E (strain Beluga) and *Clostridium butyricum* (strains ATCC43181 and ATCC43755). *Biochem. Biophys. Res. Commun.* **183**, 107–13.

Roa, M. and Boquet, P. (1985). Interaction of tetanus toxin with lipid vesicles at low pH. *J. Biol. Chem.* **260**, 6827–35.

Robinson, J.P., Schmid, M.F., Morgan, D.G. and Chiu, W. (1988). Three-dimensional structure analysis of tetanus toxin by electron crystallography. *J. Mol. Biol.* **200**, 367–75.

Sakaguchi, G., Ohishi, I. and Kozaki,.S. (1988). Botulism – structure and chemistry of botulinum. In: *Handbook of Natural Toxins*, Vol. 4 (eds M.C. Hardegree and A.T. Tu), pp. 191–216. Marcel Dekker, New York.

Schiavo, G., Poulain, B., Benfenati, F., DasGupta, B.R. and Montecucco, C. (1993). Novel targets and catalytic activities of bacterial toxin. *Trends Microbiol. Sci.* **1**, 170–4.

Schmid, M.F., Robinson, J.P. and Das Gupta, B.R. (1993). Direct visualisation of botulinum neurotoxin-induced channels in phospholipid vesicles. *Nature (Lond.)* **364**, 827–30.

Shapiro, R.E., Spech, C.D., Collins, B.E., Woods, A.S., Cotter, R.J. and Schnaar, R.L. (1997). Identification of a ganglioside recognition domain of tetanus toxin using a novel ganglioside photoaffinity ligand. *J. Biol. Chem.* **272**, 30380–6.

Somers, E. and Das Gupta, B.R. (1991). *Clostridium botulinum* types A, B, C1, and E produce proteins with or without hemagglutinating activity: do they share common amino acid sequences and genes? *J. Protein Chem.* **10**, 415–25.

Strom, M.S., Eklund, M.W. and Poysky, F.T. (1984). Plasmids in *Clostridium botulinum* and related species. *Appl. Environ. Microbiol.* **48**, 956–63.

Suen, J.C., Hatheway, C.L., Steigerwalt, A.G. and Brenner, D.J. (1988). *Clostridium argentinense* sp. nov.: a genetically homogeneous group composed of all strains of *Clostridium botulinum* toxin type G and some nontoxigenic strains previously identified as *Clostridium subterminale* or *Clostridium hastiforme*. *Int. J. Syst. Bacteriol.* **38**, 375–81.

Sunagawa, H., Ohyama, T., Watanabe, T. and Inoue, K. (1992). The complete amino acid sequence of the *Clostridium botulinum* type D neurotoxin, deduced by nucleotide sequence analysis of the encoding phage d-16 phi genome. *J. Vet. Med. Sci.* **54**, 905–13.

Szabo, E.A., Pemberton, J.M. and Desmarchellier, P.M. (1993). Detection of the genes encoding botulinum neurotoxin types A to E by the polymerase chain reaction. *Appl. Environ. Microbiol.* **59**, 3011–20.

Thompson, D.E., Brehm, J.K., Oultram, J.D., Swinfield, T.J., Shone, C.C., Atkinson, T., Melling, J. and Minton, N.P. (1990). The complete amino acid sequence of the *Clostridium botulinum* type A neurotoxin, deduced by nucleotide sequence analysis of the encoding gene. *Eur. J. Biochem.* **189**, 73–81.

Thompson, D.E., Hutson, R.A., Easy, A.K., Allaway, D., Collins, M.D. and Richardson, P.T. (1993). Nucleotide sequence of the gene coding for *Clostridium barati* type F neurotoxin: comparison with other clostridial neurotoxins. *FEMS Microbiol. Lett.* **108**, 175–82.

Tsuzuki, K., Yokosawa, N., Syuto, B., Ohishi, I., Fujii, N., Kimura, K. and Oguma, K. (1988). Establishment of a monoclonal antibody recognizing an antigenic site common to *Clostridium botulinum* type B, C1, D, and E toxins and Tetanus Toxin. *Infect. Immun.* **56**, 898–902.

Tsuzuki, K., Kimura, K., Fujii, N., Yokosawa, N., Murakami, I.T. and Oguma, K. (1990). Cloning and complete nucleotide sequence of the gene for the main component of hemagglutinin produced by *Clostridium botulinum* type C. *Infect. Immun.* **58**, 3173–7.

Tsuzuki, K., Kimura, K., Fujii, N., Yokosawa, N. and Oguma, K. (1992). The complete nucleotide sequence of the gene coding for the non-toxic-nonhemagglutinin component of *Clostridium botulinum* T type C progenitor toxin. *Biochem. Biophys. Res. Commun.* **183**, 1273–9.

Umland, T.C., Wingert, L.M., Swaminathan, S., Furey, W.F., Schmidt, J.J. and Sax, M. (1997). The structure of the receptor binding fragment H_c of tetanus neurotoxin. *Nat. Struct. Biol.* **4**, 788–92.

Washbourne, P., Pellizzari, R., Baldini, G., Wilson, M.C. and Montecucco, C. (1997). Botulinum neurotoxin types A and E require the SNARE motif in SNAP-25 for proteolysis. *FEBS Lett.* **418**, 1–5.

Weickert, M.J., Chambliss, G.H. and Sugiyama, H. (1986). Production of toxin by *Clostridium botulinum* type A strains cured of plasmids. *Appl. Environ. Microbiol.* **51**, 52–6.

Whelan, S.M., Elmore, M.J., Bodsworth, N.J., Brhem, J.K., Atkinson, T. and Minton, N.P. (1992). Molecular cloning of the *Clostridium botulinum* structural gene encoding the type B neurotoxin and determination of its entire nucleotide sequence. *Appl. Environ. Microbiol.* **58**, 2345–54.

Willems, A., East, A.K., Lawson, P.A. and Collins, M.D. (1993). Sequence of the gene coding for the neurotoxin of *Clostridium botulinum* type A associated with infant botulism: comparison with other clostridial neurotoxins. *Res. Microbiol.* **144**, 547–56.

Yamakawa, K., Karasawa, T., Kakinuma, H., Maruyama, H., Takahashi, H. and Nakamura, S. (1997). Emergence of *Clostridium botulinum* typeB-like nontoxigenic organisms in a patient with type B infant botulism. *J. Clin. Microbiol.* **35**, 2163–4.

Zhou, Y., Sugiyama, H. and Johnson, E.A. (1993). Transfer of neuro-toxigenicity from *Clostridium butyricum* to a nontoxigenic *Clostridium botulinum* type E-like strain. *Appl. Environ. Microbiol.* **59**, 3825–31.

10

Pathophysiological properties of clostridial neurotoxins

Judit Herreros, Giovanna Lalli, Cesare Montecucco and Giampietro Schiavo

INTRODUCTION

In the 100 years following the isolation of the causative agents of tetanus and botulism, large amounts of data have been collected on these pathologies and the class of molecules responsible for their symptoms, the clostridial neurotoxins (CNTs). These findings encompass several disciplines, including microbiology, medicine, pharmacology, physiology and biochemistry. One could argue that, after such prolonged investigation, this subject may have exhausted its interest and novelty, leaving only the details to be uncovered. But, like the phoenix, which rose from its ashes, CNTs have the ability to renew their fascination for the scientific community and to fuel general interest.

Since the publication of the first edition of this book, featuring the seminal review of Niemann (1991), two major steps have enlightened research into CNTs. In fact, if the 1980s were the years for the cloning of CNTs' genes, the 1990s have seen the unravelling of their intracellular activity and the characterization of the synaptic targets of these neurotoxins. Furthermore, the structure of the entire toxin molecule has revealed important insights into a complex machinery able to recognize, enter and modify the physiology of neuronal cells.

These discoveries have led to the use of CNTs as molecular tools for the dissection of regulated secretion and intracellular trafficking in a variety of cells. Moreover, this last decade has witnessed the increasing use of bot-

ulinum neurotoxins (BoNTs) in human therapy and the metamorphosis from a terrible poison to a very useful and safe therapeutic agent. This chapter offers a summary of the biological properties and pathophysiology of tetanus and botulinum neurotoxins, and attempts to highlight the areas in which these molecules could reveal their vast potential.

TETANUS AND BOTULINUM NEUROTOXINS: THE ORIGINS AND THE DISEASES

Despite the recent fame attained by CNTs, the clinical effects of one of them have been known since the beginning of medical literature (Figure 10.1). It was Hippocrates who described for the first time the symptoms of a sailor affected by a syndrome characterized by hypercontraction of the skeletal muscles (Major, 1945). He termed such a spastic paralysis tetanus (from the Greek word τεταvος, tension).

Tetanus is often fatal and death occurs by respiratory or heart failure (Bleck, 1989). Thought to be of nervous origin, this disease was shown to be transmissible and caused by a bacterium (Carle and Rattone, 1884), which was isolated and named *Clostridium tetani* (Figure 10.1) (Kitasato, 1889). *Clostridium tetani* is a rod-shaped, strictly anaerobic bacterium, which frequently harbours a sub-terminal spore, thus resembling a drumstick (*clostridium* in Latin). It is widespread in nature in the

202

TeNT		BoNTs	
1884	Tetanus is a transmissible disease	1897	Clinical description of botulism
1889	Isolation of *C. tetani*	1897	Isolation of *C. botulinum*
1890	TeNT described	1923	Action at cholinergic synapse
1892	Neuroselective binding	1949	Neurotransmission blockade revealed
1903	Retrograde transport	1951	Wound botulism
1925	Toxoid and vaccination	1971	Binding to gangliosides
1955	Neurotransmission blockade revealed	1973	First clinical use of BoNT/A
1961	Binding to gangliosides	1976	Infant botulism [c]
1976	Transsynaptic transfer	1990/93	Complete DNA sequences [d]
1986	Complete DNA sequence	1992/93	Enzymatic action
1992	Enzymic action [a]	1992/93	Targets identified [e]
1992	Target identified [b]	1998	Crystal structure (BoNT/A)
1997	Crystal structure of H$_C$		

FIGURE 10.1 Landmarks in CNTs research. For TeNT: Carle and Rattone (1884), Kitasato (1889), Faber (1890), Bruschettini (1892), Meyer (1903), Ramon and Descombey (1925), Brooks *et al.* (1955), van Heyningen (1961), Schwab and Thoenen (1976), Eisel *et al.* (1986), Umland *et al.* (1997), [a]Schiavo *et al.* (1992a); [b]Schiavo *et al.* (1992c). For BoNTs: van Ermengem (1897), Edmunds (1923), Burgen *et al.* (1949), Davis *et al.* (1951), Simpson (1971), Scott *et al.* (1973), Schiavo *et al.* (1992b), Lacy *et al.* (1998), [c]Midura and Arnon (1976), Pickett *et al.* (1976); [d]Minton (1995); [e]Blasi *et al.* (1993a, b), Schiavo *et al.* (1992c), see also Table 10.1.

non-vegetative form (spores), which can germinate in the presence of low oxygen and an abundance of nutrients (Popoff, 1995). Such conditions may be present in anaerobic wounds (Figure 10.2), where toxigenic bacteria produce a cytoplasmic protein toxin that is released following autolysis. This toxin, named tetanus neurotoxin (TeNT), is responsible for all the symptoms of tetanus (Faber, 1890; Tizzoni and Cattani, 1890), which comprise risus sardonicus, trismus (lockjaw), opisthotonus and generalized persistent reflex spasms triggered by any kind of sensory or motor stimulation (Bleck, 1989).

Unlike tetanus, adult botulism (from the Latin word *botulus*, sausage) was first recognized at the beginning of the nineteenth century (Figure 10.1) (Kerner, 1817), and infant botulism was described only in the 1970s (Midura and Arnon, 1976; Pickett *et al.*, 1976; Arnon, 1995). A minor form of botulism, termed wound botulism, was reported in the 1950s (Smith and Sugiyama, 1988). This late recognition may be attributed to the much less dramatic clinical manifestations of botulism compared with those of tetanus.

Botulism is characterized by a generalized muscular weakness, which compromises ocular and throat muscles functions and later extends to all skeletal muscles (Sakaguchi, 1983; Smith and Sugiyama, 1988). A generalized flaccid paralysis accompanied by impaired respiratory and autonomic function develops in the more severe forms and death may result from respiratory failure (Smith and Sugiyama, 1988; Hatheway, 1995). Massive fatal outbreaks of botulism are not infrequent in animals, both in the wild and on farms (Smith and Sugiyama, 1988). Botulism is an intestinal toxaemia

caused by neurotoxins produced by toxigenic strains of *Clostridium botulinum* (van Ermengem, 1897; Arnon, 1995), *Clostridium barati* and *Clostridium butirycum* (Hall *et al.*, 1985; Aureli *et al.*, 1986; McCroskey *et al.*, 1986; Suen *et al.*, 1988) (Figure 10.2). Seven different BoNT are distinguished on the basis of a serological classification (serotypes A to G). Strict requirements for spore germination and vegetative growth may explain why, unlike tetanus, botulism very rarely follows wound infection with spores of *C. botulinum* (wound botulism) (Hatheway, 1995).

Adult botulism is caused by the ingestion of the neurotoxin present in foods contaminated with spores of *C. botulinum* and preserved under anaerobic conditions favouring germination, proliferation and neurotoxin production (Sakaguchi, 1983; Smith and Sugiyama, 1988; Hatheway, 1995). In contrast, infant botulism is an infection of the intestinal tract in which toxigenic *C. botulinum* spores germinate and the vegetative form of the bacteria colonizes the lumen of the large intestine, releasing the neurotoxin. Intestine colonization by *C. botulinum* is strictly dependent on the absence of a normal microflora, a condition present in children less than 6 months old or, less frequently, in adults treated with a broad-spectrum antibiotic or with an underlying alteration of normal intestinal anatomy as a result of surgery or inflammatory bowel disease (Arnon, 1995).

Unlike TeNT, BoNTs are produced as complexes with other non-toxic proteins (Minton, 1995; Inoue *et al.*, 1996; Popoff and Marvaud, 1999), which protect them from the denaturing and proteolytic conditions found in the stomach lumen. Once in the intestine, the slightly alkaline pH causes dissociation of the complexes and

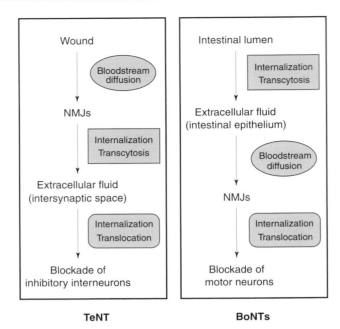

TeNT   **BoNTs**

FIGURE 10.2 Mechanism of action of TeNT and BoNTs. *Clostriduim tetani* spores enter an organism via wounds, where they can germinate if suitable anaerobic conditions exist. The vegetative form of the bacteria produces the neurotoxin, which is released following autolysis and can then diffuse through the bloodstream to reach the NMJs. Here, specific and high-affinity receptors are responsible for the binding and internalization of the toxin. TeNT is then retrogradely transported to the soma and dendrites of the motor neurons in the spinal cord, where it is released into the intersynaptic space. TeNT migrates across the synaptic cleft and is internalized by inhibitory interneurons. Within the inhibitory interneurons, TeNT follows a pathway that allows the release of the L-chain in the cytosol. Here, the zinc-endoprotease activity of TeNT causes the cleavage of VAMP and, as a consequence, the blockade of the release of inhibitory neurotransmitters. This event produces the clinical symptoms of tetanus.

The main gateway for BoNT entry into an organism is by the ingestion of food contaminated by *C. botulinum* spores, where BoNTs have been synthesized. BoNTs are synthesized as part of large complexes, which protect the neurotoxin moiety from the harsh conditions in the stomach. BoNTs are internalized by epithelial cells of the intestine and transcytosed towards the basolateral membrane. Following their release into the extracellular fluid, the neurotoxin diffuses through the bloodstream and reaches the NMJs. Here, serotype-specific receptors bind BoNTs, which are then internalized and translocated to the cytoplasm. This process allows the cleavage of the target proteins by the L-chains. All symptoms of botulism can be ascribed to the inhibition of neurotransmission at the cholinergic synapse caused by SNARE cleavage.

TeNT and BoNTs therefore display very similar strategies in their route of intoxication. The different clinical symptoms derive only from the different sites and order of execution of their transcytotic, diffusion and translocation programmes. For BoNTs, transcytosis occurs in the intestinal epithelium, whilst for TeNT this happens within motor neurons (retrograde transport). During this transcytosis, CNTs remain in the lumen of the transport vesicles, which protects the cell from SNARE cleavage. CNTs are translocated into the cytoplasm only during the second internalization process (NMJ for BoNT/A and spinal inhibitory interneurons for TeNT). At this point, SNARE cleavage, and therefore inhibition of neurotransmitter release, can occur.

the neurotoxin reaches the bloodstream by transcytosis from the apical to the basolateral side of intestinal epithelial cells (Maksymowych and Simpson, 1998).

Tetanus and botulism can be completely prevented by anti-toxin-specific antibodies, further underlining the notion that a single protein is responsible for all clinical symptoms (Galazka and Gasse, 1995; Middlebrook and Brown, 1995). Toxin-neutralizing antibodies can be acquired passively by injection of purified immunoglobulins or actively, following vaccination with the formaldehyde-inactivated toxin (toxoid) (Galazka and Gasse, 1995). Recently, experimental vaccines based on the carboxyl terminal third of the TeNT or BoNT molecules have been developed (Middlebrook and Brown, 1995). Whilst TeNT vaccination is compulsory in most countries, owing to the rarity of botulism the

vaccination with BoNT toxoid is performed only on people involved in manipulation of toxigenic clostridia and of large amounts of BoNTs. Vaccination does not appear necessary for scientists working with BoNTs: the avoidance of using needles and the availability in the laboratory of anti-BoNT serum seem to offer sufficient safety measures.

CLOSTRIDIAL NEUROTOXINS AND SYNAPTIC ACTIVITY

Clostridial neurotoxins are the most toxic substances known, with mouse LD_{50} ranging between 0.1 and 1 ng kg^{-1} of body weight. Different animals show a great range of sensitivity to TeNT and to the different

BoNTs, with mammals (except for rats) being among the most susceptible species, and rats, birds, reptiles and amphibians being resistant to different extents (Payling-Wright, 1955; Gill, 1982). The absolute neurospecificity of TeNT and BoNTs and their enzymic activity are at the basis of such high toxicity. The time of onset of paralysis after CNT intoxication is dependent on the species, dose and route of application. However, a lag phase ranging from several hours to days is always present between the CNTs' challenge and the manifestation of the earliest clinical symptoms.

After entering the general circulation, CNTs bind specifically to the presynaptic membrane of the neuromuscular junctions (NMJs) (Figure 10.2). CNTs associate rapidly and with high affinity to presynaptic receptor(s) (see below). This fact accounts for the limited spreading around the site of injection experienced in clinical treatments and for the extremely low LD_{50} values. After binding, BoNTs enter the neuronal cytosol and block the release of acetylcholine (ACh) at the NMJ, thus causing a flaccid paralysis (Burgen et al., 1949; Mellanby, 1984; Simpson, 1989). TeNT also binds to the presynaptic membrane of motor neurons, but it is then transported retrogradely up to the spinal cord, where it accumulates in the ventral horn of the grey matter (Bruschettini, 1892; Habermann and Dimpfel, 1973; Erdmann et al., 1975; Price et al., 1975; Stöckel et al., 1975). Neuromuscular stimulation enhances the extent of uptake of CNTs (Kryzhanovsky, 1958; Hughes and Whaler, 1962; Wellhöner et al., 1973; Habermann et al., 1980; Schmitt et al., 1981). Within the spinal cord, TeNT migrates across the synaptic cleft into coupled inhibitory interneurons (Schwab and Thoenen, 1976; Evinger and Erichsen, 1986), where it blocks the release of inhibitory neurotransmitters (Brooks et al., 1955; Curtis et al., 1976; Benecke et al., 1977; Bergey et al., 1983). The blockade impairs the balance between inhibitory and excitatory afferents on the motor neurons, which ensures co-ordinated muscle contraction. This causes the spastic paralysis characteristic of tetanus (Mellanby and Green, 1981; Simpson, 1989; Wellhöner, 1992).

Unlike BoNTs, TeNT can also bind to sensory and adrenergic neurons in which a similar retrograde transport pathway takes place (Erdmann et al., 1975; Price et al., 1975; Stöckel et al., 1975). None the less, excitatory synapses appear not to be compromised in the early stages (reviewed in Wellhöner, 1992), although they may be inhibited at later times of TeNT intoxication (Takano et al., 1983). This specificity for inhibitory versus excitatory synapses is maintained when TeNT is applied directly into the CNS (Mellanby and Thompson, 1977; Mellanby et al., 1977; Calabresi et al., 1989) and it may underlie the neurodegenerative and epileptogenic effects of TeNT, which result from unopposed

release of glutamate from excitatory central synapses (Bagetta et al., 1991; Jefferys and Whittington, 1996; Leist et al., 1997). The selective action of TeNT on inhibitory synapses within spinal cord may be enhanced by the anatomical organization of the tissue, since it is not preserved in spinal cord neurons in culture (Bergey et al., 1983; Williamson et al., 1992, 1996).

CNTs do not kill intoxicated neurons. Their extreme animal toxicity is due to the essential role of synaptic transmission in animal physiology and behaviour. Following BoNTs poisoning, NMJs become paralysed, but the motor neuron and the innervated muscle fibre remain alive. However, the muscle undergoes a transient atrophy (Duchen and Tonge, 1973; Angaut-Petit et al., 1990; Bambrick and Gordon, 1994), which progresses for 4–6 weeks. Unlike the denervation obtained by other means such as nerve ligation, anatomical contacts between the nerve and muscle are maintained in BoNT-treated animals and there is no apparent loss of motor axons.

For the above reasons, BoNTs are increasingly used to study NMJ degeneration and regeneration. Paralysed muscle fibres release a panel of trophic factors (Hassan et al., 1994; Sala et al., 1995), which induce an enlargement of the motor endplate and the emergence of sprouts from the endplate, the terminal part of the axon and the more distal nodes of Ranvier. Following axonal sprouting and the establishment of new functional NMJs, the muscle regains its normal size, reversing completely the muscle atrophy induced by BoNT (reviewed in Borodic et al., 1994). The opposite clinical symptoms of tetanus and botulism result from a difference in the intracellular trafficking of TeNT and BoNTs, rather than from a different mechanism of action (Figure 10.2). The remarkable specificity of TeNT and BoNTs for central and peripheral synapses, respectively, exists only at pharmacological doses (picomolar and lower). It is largely lost at the concentrations used in the laboratory where, in order to rapidly obtain relevant effects, a much higher dosage is frequently applied. Under such conditions, TeNT also inhibits peripheral synapses (Matsuda et al., 1982; Mellanby, 1984; Habermann and Dreyer, 1986; Wellhöner, 1992). Even at high concentrations, the action of CNTs is only at the presynaptic level, where they cause a persistent inhibition of the exocytosis of a variety of neurotransmitters (reviewed in Wellhöner, 1992).

Analysis of the synaptic action of clostridial neurotoxins

Morphological examination of synapses intoxicated with CNTs does not reveal major alterations in the overall structure and in the number and distribution of small synaptic vesicles (SSVs) and large dense core granules. The only change is an increase in the number of SSVs close to the cytosolic face of the presynaptic membrane (Kryzhanovsky *et al.*, 1971; Pozdynakov *et al.*, 1972; Duchen, 1973; Mellanby *et al.*, 1988; Neale *et al.*, 1989; Pecot-Dechavassine *et al.*, 1991; Hunt *et al.*, 1994; Osen Sand *et al.*, 1996) and the disappearance of the small membrane invaginations believed to represent SSV fusion events (Pumplin and Reese, 1977).

Despite the lack of detectable structural modifications, the analysis of synapses intoxicated with CNTs revealed profound alterations in the electrophysiological parameters. Following the seminal work of Burgen and co-workers (Burgen *et al.*, 1949), the consequences of CNT poisoning have been studied on different classes of synaptic terminals. However, a dataset large enough to compare the effects of TeNT and of BoNTs is only available for the NMJ (reviewed in Van der Kloot and Molgo, 1994; Montecucco and Schiavo, 1995; Poulain *et al.*, 1995). NMJ preparations also allow the study of the effect of pharmacological agents when neurotransmitter release is virtually blocked by CNTs, a condition in which other systems show their limitations (Poulain *et al.*, 1995; B. Poulain *et al.*, personal communication).

The conclusions emerging from this massive amount of data can be summarized as follows.

- At the NMJ, nerve stimulation triggers the synchronous release of hundreds of quanta which causes a post-synaptic response, the endplate potential (EPP). CNTs cause an extensive and persistent blockade of EPPs. This inhibition accounts for the impaired synaptic transmission and extends to a variety of synapses tested *in vitro*.
- CNTs reduce the frequency, but not the amplitude, of evoked and spontaneous miniature endplate potentials (MEPPs), events reflecting the release of a

single SSV. Hence, CNTs lower the number of SSVs capable of being released, without affecting their ACh content. Despite the strong reduction in frequency, MEPPs are not completely abolished at poisoned nerve terminals.

- CNTs do not interfere with the processes of neurotransmitter synthesis, uptake and storage (Gundersen, 1980).
- CNTs affect neither the propagation of the nerve impulse nor calcium homeostasis at the synaptic terminals (Gundersen *et al.*, 1982; Dreyer *et al.*, 1983; Mallart *et al.*, 1989).
- Sub-MEPPs and giant MEPPs, spontaneous events with an amplitude different from that of normal MEPPs, are not inhibited. Giant MEPPs may even increase at the intoxicated NMJ (Thesleff, 1986; Molgo *et al.*, 1990; Van der Kloot and Molgo, 1994; Herreros *et al.*, 1995; Sellin *et al.*, 1996). This result indicates that the exocytic machinery responsible for sub- and giant MEPPs is insensitive to the CNTs' action and therefore different from that of classical SSV exocytosis (Bauerfeind *et al.*, 1994).

Despite the similarities in their presynaptic inhibitory action, some important differences were revealed through the detailed study of the blockade of neurotransmitter release induced by CNTs (Van der Kloot and Molgo, 1994; Poulain *et al.*, 1995). As a result, CNTs can be divided into two groups on the basis of their electrophysiological behaviour (Table 10.1).

At the NMJ, BoNT/A, /E and /C inhibit quantal release without altering its synchrony. On the contrary, TeNT, BoNT/B, /D and /F cause a desynchronization of the quanta released after depolarization, although, in the case of BoNT/F, this effect is only partial (Harris and Miledi, 1971; Dreyer and Schmitt, 1983; Bevan and Wendon, 1984; Gansel *et al.*, 1987; Molgo *et al.*, 1989; J. Molgo, personal communication). Neurotransmitter release is enhanced by increasing calcium entry at the

TABLE 10.1 CNTs: effects on the Ach release at NMJ, their targets and cleavage sites[a]

Neurotoxin	Evoked quantal release	Rescue by AP or high Ca^{2+}	Rescue by α-LTX	Target	Cleavage site
BoNT/A	Synchronous	Yes	Yes	SNAP-25	ANQ–RAT
BoNT/E	Synchronous	Yes (partial)[b]	Yes/No[c]	SNAP-25	IDR–IME
BoNT/C	Synchronous[b]	Yes[b]	Yes[b]	Syntaxin	TKK–AVK
				SNAP-25	NQR–ATK
BoNT/B	Asynchronous	No	No	VAMP	ASQ–FET
BoNT/D	Asynchronous	No	No	VAMP	DQK–LSE
BoNT/F	Partially synchronous[b]	No	Not known	VAMP	RDQ–KLS
BoNT/G	Not known	Not known	Not known	VAMP	TSA–AKL
TeNT	Asynchronous	No	No	VAMP	ASQ–FET

[a]References in the text.
[b]B. Poulain and J. Molgo, personal communication.
[c]The rescue with α-LTX of NMJs intoxicated with BoNT/E differ between rat and frog NMJs. The recovery is total in frog NMJ, and null in rat NMJ.

nerve terminal and this experimental strategy was extensively used to monitor the residual level of presynaptic activity following CNT intoxication.

Aminopyridines (AP), by inhibiting potassium channels and thus indirectly increasing presynaptic calcium levels, are potent antagonists of the inhibitory action of BoNT/A and partial antagonists of BoNT/E (Table 10.1), providing the block of neurotransmission is not complete. A similar treatment of NMJs intoxicated with BoNT/B, /D, /F or TeNT also increases the probability of quantal release but, unlike BoNT/A and /E, the release remains largely asynchronous (Lundh *et al.*, 1977; Sellin, 1987; Molgo *et al.*, 1990; Wellhöner, 1992; Poulain *et al.*, 1995). In this case, the asynchrony of the release prevents the formation of EPPs and, consequently, the recovery of neurotransmission. The result obtained with AP is strengthened by other agents known to increase the intracellular calcium concentration, such as ionophores. In these conditions, the inhibition caused by BoNT/A and BoNT/C is partially reversed, whilst a very poor or no effect is noticed on BoNT/E- and TeNT-treated preparations, respectively (Cull-Candy *et al.*, 1976; Ashton *et al.*, 1993; Banerjee *et al.*, 1996; Capogna *et al.*, 1997).

A difference in the mechanism of action of these two groups of neurotoxins is also suggested by experiments of double poisoning of the NMJ with CNT, followed by α-latrotoxin (α-LTX). This neurotoxin is the active component of black widow spider venom and causes a massive release of SSVs in a variety of nerve terminals (Rosenthal and Meldolesi, 1989; Grishin, 1998). α-LTX counteracts only the action of BoNT/A, /C and /E, but it is not effective on the blockade caused by TeNT or BoNT/B (Gansel *et al.*, 1987) (Table 10.1).

Following these extensive electrophysiological studies, it is clear that CNTs act on components of presynaptic machinery playing essential roles in neurotransmitter release. CNTs fall into two groups with different targets within the presynaptic terminal: BoNT/A, /C and /E constitute the first functional group and TeNT, BoNT/B, /D and /F the second. This functional grouping is now fully supported by the identification of the molecular targets of each CNT (Table 10.1) and gives important insights into the function of these proteins in neurotransmitter release (see below).

STRUCTURE–FUNCTION RELATIONSHIPS

As expected from their similar inhibitory activity at the synapse, CNTs share a closely related structural organization. CNTs are produced as inactive polypeptide chains of 150 kDa, and they are released in the culture medium only after bacterial lysis, as expected by the absence of a leader sequence. No proteins are associated with TeNT, whereas BoNTs are released in the form of multimeric protein complexes with a relative molecular mass up to 900 000 (progenitor toxins) (Minton, 1995; Inoue *et al.*, 1996; Popoff and Marvaud, 1999).

Progenitor toxins shield the neurotoxin moiety from denaturing conditions (Sakaguchi, 1983), allowing BoNTs to survive the harsh conditions of the stomach and reach the intestine, where the slightly alkaline pH induces dissociation of the complex. The inactive single chain of CNTs is activated by endogenous or exogenous proteases (DasGupta, 1989; Weller *et al.*, 1989; Krieglstein *et al.*, 1991; DasGupta, 1994), which nick an exposed loop subtended by a highly conserved disulfide bridge (Figure 10.3). The heavy chain (H, 100 kDa) and the light chain (L, 50 kDa) thus generated, remain associated via non-covalent protein–protein interactions and the conserved interchain disulphide bond, the integrity of which is essential for neurotoxicity (Schiavo *et al.*, 1990a; de Paiva *et al.*, 1993a). Another site for preferential proteolysis is in the middle of the H-chain: treatment of CNTs with papain generates a C-terminal fragment, termed H_C, and a heterodimer composed of the L-chain and the N-terminal portion of the H-chain (Figure 10.3) (Bizzini *et al.*, 1977; Helting and Zwisler, 1977; Matsuda and Yoneda, 1977; Neubauer and Helting, 1981).

The existence of these three functional domains assembled together in a modular fashion was recently confirmed by the crystallization and structural determination of BoNT/A at 3.3 Å resolution (Stevens *et al.*, 1991; Lacy *et al.*, 1998). BoNT/A is an elongated molecule measuring approximately $45 \times 105 \times 130$ Å, presenting a linear arrangement of the three domains (Figure 10.3). These domains are structurally distinct with the exception of a large loop in the N-terminal part of the H-chain, which wraps around the perimeter of the L-chain.

The C-terminal portion is implicated in the neurospecific binding of CNTs (Bizzini *et al.*, 1977; Morris *et al.*, 1980; Weller *et al.*, 1986; Halpern and Neale, 1995) and is composed of two distinct subdomains, rich in β-structure and of almost identical size ($32 \times 37 \times 38$ Å). The N-terminal subdomain (H_{CN}) has two seven-stranded β-sheets coupled in a jelly-roll motif closely similar to that present in the carbohydrate binding moiety of plant lectins and in other oligosaccharide-binding proteins. The C-terminal portion (H_{CC}) of the H-chain adopts a modified β-trefoil fold with a six-stranded β-barrel, which is present in several proteins involved in recognition and binding functions such as trypsin inhibitors. A β-hairpin capping the base of the domain completes the fold. The entire H$_C$ domain remains completely isolated from the rest of the molecule, such that all the surface loops are accessible and therefore available for binding. The sequence of H$_{CN}$ is highly conserved among CNTs, thus suggesting that it may have a closely similar three-dimensional structure.

This hypothesis was recently confirmed by the crystallization and structure determination at 2.7 Å of the

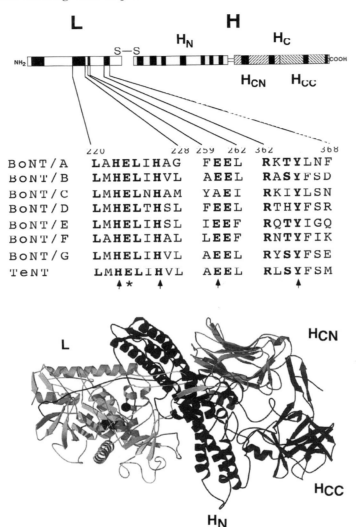

FIGURE 10.3 The three functional domain structure of CNTs. The upper panel shows the structure of activated di-chain CNT. The neurotoxin is composed of two polypeptide chains held together by a single disulfide bridge, which must be reduced intracellularly for full activity. The C-terminal portion of the heavy chain (H, 100 kDa) is responsible for neurospecific binding (domain H$_C$), whilst the N-terminus (H$_N$) is implicated in the translocation of the light chain in the cytosol and pore formation. Structurally, H$_C$ can be further subdivided into two portions of 25 kDa, H$_{CN}$ and H$_{CC}$. The light chain (L, 50 kDa) is a zinc-endopeptidase responsible for the intracellular activity of CNTs. The segments presenting high homology between different serotypes are in black. The highest homology is shown by a short segment in the central part of the L-chain, which contains the zinc-binding motif of metallo-endopeptidases (Jiang and Bond, 1992). The sequences of CNTs containing all the amino acid residues in close proximity to the zinc atom in the active site are shown in the middle panel. Two histidines and one glutamic acid directly co-ordinate the zinc atom in the crystallographic structure of BoNT/A (lower panel), whilst the distal tyrosine residue is supposed to contribute either to substrate recognition or to stabilization of the active site (arrows). The glutamic acid of the HExxH motif is likely to co-ordinate the water molecule responsible for the target hydrolysis (asterisk). The structure of BoNT/A (kindly provided by Dr R.C. Stevens) highlights the three domain structure, with the zinc atom (black) held in the centre of the active site of the L-chain. This domain is composed of a mixture of β-strands and α-helices and measures 55 × 55 × 62 Å. The domain H$_N$ (28 × 32 × 105 Å) is characterized by two 105-Å long α-helices and a long loop interacting with the L-chain. The helices are amphipathic and, despite the absence of a strict heptad repeat, they twist around each other in a coiled-coil-like fashion. The C-terminal portion is composed of two distinct subdomains, rich in β-structure and almost identical in size (32 × 37 × 38 Å). The N-terminal subdomain (H$_{CN}$) has two seven-stranded β-sheets, whilst the C-terminal portion (H$_{CC}$) of the H-chain adopts a modified β-trefoil fold with a β-hairpin capping the base of the domain.

isolated H_C domain of TeNT (Umland *et al.*, 1997). The major differences appear in the loops of H_{CC}, where sequence is poorly conserved among CNTs. The removal from H_C of its *N*-terminal domain does not reduce membrane binding, whereas the deletion of only 10 residues from the *C*-terminus abolishes its binding to spinal cord neurons (Halpern and Loftus, 1993).

The *N*-terminal part of H-chain (H_N) is implicated in the pH-dependent membrane penetration and translocation of the catalytic domain into the cytosol (see below). It is composed of a loop interacting with the L-chain and a central body of $28 \times 32 \times 105$ Å, the main structural units of which are two 105-Å long α-helices encompassing the entire length of the domain. These helices are amphipathic and, despite the absence of a strict heptad repeat, they twist around each other in a coiled-coil-like fashion. Similar long pairs of α-helices have recently been observed in colicin Ia, in the nucleotide exchange factor GrpE and in some viral proteins, but are absent from other proteins known to interact with the lipid bilayer as a function of pH, such as diphtheria and *Pseudomonas aeruginosa* toxins (Lacy *et al.*, 1998).

CNTs exhibit a pore-forming activity which is localized to the H_N domain (Hoch *et al.*, 1985; Donovan and Middlebrook, 1986; Blaustein *et al.*, 1987; Shone *et al.*, 1987; Gambale and Montal, 1988). Following this observation, Oblatt-Montal *et al.* (1995) found that the amphipathic segment 659–681 of BoNT/A, predicted to have a helical structure, increases the permeability of lipid bilayers, implicating it in the pore formation. In the structure, the sequence corresponding to the peptide adopts a strand-like conformation and lies against the two main α-helices. This may suggest an involvement of this portion of the molecule in a pH-dependent conformational change, causing the exposure of previously hidden hydrophobic surfaces.

The L-chain is responsible for the intracellular activity of CNTs (Penner *et al.*, 1986; Poulain *et al.*, 1988; Ahnert-Hilger *et al.*, 1989; Bittner *et al.*, 1989a, b; Mochida *et al.*, 1989; Weller *et al.*, 1991). In agreement with secondary structure predictions (Singh and DasGupta, 1989; Lebeda and Olson, 1994), this domain is composed of a mixture of β-strands and α-helices and measures $55 \times 55 \times 62$ Å. The catalytic site contains the zinc atom essential for CNT activity (dark circle in Figure 10.3). One atom of zinc is bound to the L-chain of TeNT, BoNT/A, /B and /F (Schiavo *et al.*, 1992a, b; Schiavo *et al.*, 1993a), whilst BoNT/C binds two atoms of zinc with different affinities (Schiavo *et al.*, 1995). In this regard, BoNT/C is similar to neutrophil collagenase, which has one exchangeable metal ion in the active site and a second strongly bound zinc atom thought to play

a structural role (Lovejoy *et al.*, 1994). Heavy metal chelators are effective to remove the bound zinc and generate inactive apo-neurotoxins (Bhattacharyya and Sugiyama, 1989; Schiavo *et al.*, 1992a; Simpson *et al.*, 1993; Höhne-Zell *et al.*, 1994; Adler *et al.*, 1997). Zinc removal appears to have limited effects on the secondary structure of L-chain of TeNT (De Filippis *et al.*, 1995), whilst more drastic changes are observed in BoNT/A (Fu *et al.*, 1998).

The catalytic site of BoNT/A has an overall negative surface charge and is buried in the structure of the protein, being accessible from the outside only via a large channel of $12 \times 15 \times 35$ Å, which accommodates the substrate. In the di-chain CNTs, entry to the channel is partially blocked by the large loop wrapping the L-chain and by the translocation domain itself (Lacy *et al.*, 1998).

Although the resolution of the structure presently available is not sufficient to establish the exact bond lengths and the position of water molecules in the active site, it gives precise information on the identity of the ligands co-ordinating the metal atom and confirms the large number of biochemical and biophysical studies aimed at unravelling the nature of the catalytic site. In fact, as previously suggested by chemical modification and mutagenesis experiments performed on the L chains of TeNT and BoNT/A, /B and /E (Schiavo *et al.*, 1992a, b; Li *et al.*, 1994; Yamasaki *et al.*, 1994a), the zinc atom is held in place by two histidine residues (H_{222} and H_{226} in BoNT/A), located in the central and most conserved portion of CNTs. These two residues are present in the sequence HexxH, which is the motif characterizing the catalytic site of zinc-endopeptidases (Jongeneel *et al.*, 1989; Vallee and Auld, 1990; Kurazono *et al.*, 1992; Schiavo *et al.*, 1992a, b; Wright *et al.*, 1992). The glutamic acid residue in the HExxH motif co-ordinates the water molecule necessary for the catalysis (third ligand), while another glutamic acid residue (E_{261} in BoNT/A) is the fourth zinc ligand.

Recent spectroscopic studies suggested the presence of a tyrosine in the vicinity of the metal atom (Morante *et al.*, 1996), as previously observed for the protease astacin and for the members of the metzincins superfamily (Hase and Finkelstein, 1993; Stocker and Bode, 1995). This result was confirmed by the presence in the structure of a tyrosine residue (Y_{365}) as a putative fifth zinc ligand. The relatively large distance from the zinc atom (5 Å) suggests that this residue could be involved in substrate binding or in the stabilization of the active site.

The catalytic site of CNT presents structural similarity with other metalloendopeptidases of the zincin and metzincin superfamilies (Jiang and Bond, 1992; Hase and Finkelstein, 1993; Stocker and Bode, 1995), but this similarity is limited only to the α-helix containing the HExxH motif. This structural observation is in agreement with the finding that the absorption spectrum of cobalt-substituted TeNT is different from both the analogous complex formed by thermolysin and astacin (Tonello et al., 1996), and supports the hypothesis that CNTs constitute a novel class of metalloproteases with a different mode of zinc co-ordination (Montecucco and Schiavo, 1995).

The structural organization of CNTs is functionally related to the mechanism adopted by these neurotoxins to intoxicate neurons, consisting of four steps: (a) neurospecific binding, (b) internalization, (c) membrane translocation, and (d) intracellular enzymic activity (Schmitt et al., 1981; Simpson, 1986; Montecucco et al., 1994).

NEUROSPECIFIC BINDING

Following diffusion from the site of entry in the body (Figure 10.2), CNTs bind to the presynaptic membrane of cholinergic nerve terminals (reviewed by Simpson, 1989; Wellhöner, 1992; Halpern and Neale, 1995). TeNT may also bind to sympathetic and adrenergic fibres. In vitro, CNTs are capable of binding to a variety of non-neuronal cells, but only at concentrations several orders of magnitude higher than those clinically relevant. Two classes of presynaptic binding sites with subnanomolar and nanomolar affinities have been described (Critchley et al., 1988; Halpern and Neale, 1995) and the available evidence suggests that the H_C domain plays a major role in the binding processes (Bizzini et al., 1977; Helting et al., 1977; Morris et al., 1980; Shone et al., 1985; Weller et al., 1986; Kozaki et al., 1989).

A large number of studies demonstrated the involvement of polysialogangliosides in CNT binding (for a complete series of references see Montecucco, 1986; Halpern and Neale, 1995). In fact, CNTs show maximal binding to members of the G1b series (GD1b, GT1b and GQ1b), which are able, upon preincubation, to protect the NMJ from BoNT-dependent inhibition of neurotransmitter release and partially abolish the retrograde transport of TeNT (Stöckel et al., 1977). Preincubation with exogenous polysialogangliosides also increases the sensitivity of cultured chromaffin cells to TeNT and BoNT/A (Marxen et al., 1989), and the removal of sialic acid residues from the membrane surface with neuraminidase decreases, but does not abolish, CNT binding.

It is unlikely, however, that the binding to polysialogangliosides totally accounts for the absolute neurospecificity of these neurotoxins (Mellanby and Green, 1981; Montecucco, 1986). Experiments carried out with cells in culture and brain homogenates have indicated that proteins may be involved in toxin binding (Pierce et al., 1986; Yavin and Nathan, 1986; Parton et al., 1988; Schiavo et al., 1991). The lectin-like and protein binding domains present in the H_C domain of TeNT and BoNT/A (Umland et al., 1997; Lacy et al., 1998) suggest that CNTs may bind strongly and specifically to the presynaptic membrane owing to multiple interactions with sugar and protein binding sites (Montecucco, 1986). Recent experiments have provided evidence in favour of such a double receptor model by showing that BoNT/B interacts with the intravesicular domain of the synaptic vesicle proteins synaptotagmin I and II in the presence of polysialogangliosides (Nishiki et al., 1994; Nishiki et al., 1996a, b; Kozaki et al., 1998). This result was very recently extended to BoNT/A and /E (Li and Singh, 1998), thus suggesting that synaptotagmin may act as receptor for all BoNTs. However, competition experiments clearly demonstrated that different BoNTs serotypes do not share the same receptor (Evans et al., 1986; Habermann and Dreyer, 1986), predicting a model in which each BoNT binds to a different synaptotagmin isoform (Schiavo et al., 1998). The sensitivity of a particular cell type for different BoNTs serotypes would then depend on the repertoire of expressed synaptotagmins.

Generally, receptors for pathogens and virulence factors, such as toxins and viruses, are surface molecules essential for cell physiology, and their study has led to important progress in cell biology and neuroscience. The identification of the receptors of the CNTs is particularly relevant for several theoretical and practical reasons. In fact, both TeNT and BoNTs bind to the presynaptic membranes of alpha-motor neurons, but then they follow different intracellular trafficking pathways. Electrophysiological studies have clearly shown that BoNTs block neuroexocytosis at peripheral terminals, whereas TeNT causes the same effect on inhibitory synapses of the spinal cord.

The different fate of TeNT and BoNTs must be determined by specific receptors, which target them to different intracellular routes. The identification of the receptor(s) for TeNT present on peripheral motor neurons will provide a route of entry to the gateway leading from the peripheral to the CNS, offering new insights for the delivery of biological and pharmacological agents into the spinal cord. In this regard, recombinant TeNT-superoxide dismutase proteins are internalized and retrogradely transported as efficiently as the H_C alone (Figueiredo et al., 1997). The isolation of BoNT receptors will help us to clarify important aspects of synaptic endocytosis and contribute to improving present therapeutic protocols for BoNT/A treatment.

To reach its final site of action, TeNT has to bind and enter two different neurons: a peripheral motor neuron and an inhibitory interneuron of the spinal cord (Figure 10.2). The comparison of the toxicity of TeNT and its fragments at peripheral and central level clearly indicates that TeNT binding to NMJs and to presynaptic terminals in the spinal cord is different (Shumaker *et al.*, 1939; Takano *et al.*, 1989). Niemann (1991) suggested a model in which polysialogangliosides act as peripheral receptors for TeNT, mediating its retrograde transport to the CNS, where it binds to a second different acceptor. Although very appealing, this model suffers from the drawback of the low affinity and low specificity of polysialogangliosides as the only TeNT receptor in the periphery, where a high-affinity interaction is required to account for the extremely low doses of TeNT causing clinical symptoms *in vivo*.

A possible scenario reconciling the data presently available envisages that both glycoprotein and glycolipids are involved in the high-affinity peripheral binding of CNTs. This double receptor complex (Montecucco, 1986) would be responsible for its inclusion in an endocytic vesicle undergoing retrograde transport along the axon, whereas the receptors for BoNTs would guide them inside vesicles that acidify within the NMJ. After sorting in the soma of the motor neuron, the vesicles carrying TeNT will then release the toxin into the intersynaptic space, where it will enter the inhibitory interneurons, possibly via synaptic vesicle endocytosis (Matteoli *et al.*, 1996).

INTERNALIZATION

Since the catalytic activity of the L-chains is directed towards intracellular targets, at least this toxin domain must reach the cytosol. *In vivo*, CNTs do not enter the cell directly via the plasma membrane, but rather are endocytosed inside acidic cellular compartments. Electron microscopic studies have shown that, after binding, CNTs enter the lumen of vesicular structures in a temperature- and energy-dependent process (Dolly *et al.*, 1984; Critchley *et al.*, 1985; Black and Dolly, 1986a, b; Staub *et al.*, 1986; Parton *et al.*, 1987). Parton *et al.* (1987) examined the binding and internalization of gold-labelled TeNT in dissociated spinal cord neurons. The toxin accumulated in coated pits on the cell surface, followed by internalization in a variety of vesicular structures, such as endosomes, multivesicular bodies, tubules and to a lesser extent SSVs. In contrast, Montesano *et al.* (1982) found that TeNT accumulates in uncoated pits and non-clathrin-coated vesicles in liver cells.

These discrepancies are difficult to reconcile in a unitary view and are probably a consequence of the different cell system and the high TeNT concentration used by one group (Montesano *et al.*, 1982). Further insights into the internalization pathway used by TeNT are provided by a recent study on hippocampal neurons (Matteoli *et al.*, 1996). In this system, TeNT co-localizes with SSV markers following membrane depolarization, suggesting its internalization via SSVs. It is well established that nerve stimulation facilitates intoxication (Ponomarev, 1928; Kryzhanovsky, 1958; Hughes and Whaler, 1962; Wellhöner *et al.*, 1973; Habermann *et al.*, 1980; Schmitt *et al.*, 1981). A high rate of neuroexocytosis correlates with a high rate of synaptic vesicle recycling via endocytosis, the two processes being tightly coupled (Schweizer *et al.*, 1995; Cremona and De Camilli, 1997).

A possible explanation for the shorter onset of paralysis induced by CNTs under nerve stimulation is that they enter the synaptic terminal inside the lumen of SSVs. The accessibility of the internal lumen of SSVs to extracellular agents during neurotransmitter release is demonstrated by the binding and uptake of antibodies specific for lumenal epitopes of a SSV protein (Matteoli *et al.*, 1992; Kraszewski *et al.*, 1995; Mundigl *et al.*, 1995). According to this hypothesis, CNTs use SSVs as Trojan horses, to gain entry to CNS neurons. Despite the attractiveness of this model for the endocytosis of CNTs in central neurons, it is unlikely that this mechanism is responsible for the uptake of TeNT at the NMJ.

In fact, three experimental findings contrast with this hypothesis.

First, high-frequency stimulation increases the rate of intoxication but not the binding of TeNT to the NMJ (Schmitt *et al.*, 1981). If the toxin receptor is exposed during neuroexocytosis to allow the toxin to bind and then be endocytosed, an increase in the stimulation rate should also increase the total number binding sites present at the NMJ. Second, TeNT is not active on a NMJ maintained at 18°C even in the presence of high-frequency stimulation and massive neurotransmitter release, while it is fully inhibitory at 25°C (Schmitt *et al.*, 1981). Third, the uptake and retrograde transport of TeNT is perfectly functional in NMJ intoxicated with BoNT/A, where neurotransmitter release is completely blocked (Habermann and Erdmann, 1978).

The latter fact is in complete agreement with the notion that retrograde transport of various substances, including horseradish peroxidase, is not impaired in silenced NMJ (Kristensson and Olsson, 1978; Kemplay and Cavanagh, 1983). Preliminary experiments on peripheral motor neurons indicate that there is very limited co-localization between TeNT and

SSV markers (G. Lalli, unpublished results), thus indicating that the pathway of internalization and intracellular trafficking of TeNT may be different in peripheral and central neurons.

As discussed above, TeNT and BoNTs have to enter different vesicles at the NMJ in order to account for their different fates. It is perfectly plausible, however, that BoNTs and TeNT share the same internalization mechanism in the neuron in which they block neurotransmitter release (NMJs for BoNT/A and inhibitory interneurons for TeNT) (Figure 10.2). Future experiments aiming towards the identification of the peripheral BoNT receptors and the central and peripheral TeNT receptors will be instrumental to clarify this important aspect of CNTs trafficking and will give us invaluable insights into mechanisms of endocytosis at the nerve terminal.

TRANSLOCATION INTO THE NEURONAL CYTOSOL

Once internalized, the L-chains must cross the vesicle membrane to reach the cytosol in order to display their activity. The different trafficking of TeNT and BoNT at the NMJ clearly indicates that internalization is not directly linked to membrane translocation. Therefore, internalization and membrane translocation are clearly distinct steps in the process of cell intoxication, as is the case for most bacterial toxins acting in the cytoplasm (Menestrina *et al.*, 1994). Compelling evidence indicates that di-chain CNTs have to be exposed to a low pH step for nerve intoxication to occur (Simpson, 1982, 1983; Adler *et al.*, 1994; Simpson *et al.*, 1994; Williamson and Neale, 1994; Matteoli *et al.*, 1996). The introduction of a non-acid-treated L-chain in the cytosol is sufficient to block exocytosis (Penner *et al.*, 1986; Poulain *et al.*, 1988; Ahnert-Hilger *et al.*, 1989; Bittner *et al.*, 1989a, b; Mochida *et al.*, 1989; Weller *et al.*, 1991). It is therefore likely that low pH is crucial for the process of membrane translocation of the L-chain from the internalized vesicle lumen to the cytosol, as previously demonstrated for several bacterial protein toxins (Montecucco *et al.*, 1994).

The membrane interaction of CNTs has been studied with model membranes (reviewed in Montecucco *et al.*, 1994) and, more rarely, in cell cultures (Beise *et al.*, 1994). These studies have shown that low pH triggers a structural change in CNTs from a hydrophilic neutral form to an acid conformation characterized by a higher degree of hydrophobicity. This transition enables the penetration of both the H- and L-chains into the hydrophobic core of the lipid bilayer (Boquet and Duflot, 1982; Cabiaux *et al.*, 1985; Montecucco *et al.*,

1986, 1989; Menestrina *et al.*, 1989; Schiavo *et al.*, 1990b). Following membrane insertion, CNTs form cation-selective ion channels in planar lipid bilayers with conductances of a few tens of picosiemens and permeable to molecules smaller than 700 Da. Experimental evidence indicates that CNT channels are formed by the oligomerization of the H_N domain (Donovan and Middlebrook, 1986; Shone *et al.*, 1987; Menestrina *et al.*, 1989; Schmid *et al.*, 1993) which, at neutral pH, consists of a pair of 105-Å long α-helices buried in the core domain. These helices are flanked by a short amphipathic segment, which has the ability to form channels with properties similar to those of the intact toxin molecule (Montal *et al.*, 1992; Oblatt-Montal *et al.*, 1995). On this basis, it was proposed that the channel is formed by a toxin tetramer that brings four of these amphipathic segments into strict proximity with their hydrophilic residues lining the lumen of the pore (Montal *et al.*, 1992; Oblatt-Montal *et al.*, 1995). This is in agreement with the three-dimensional image reconstruction of the channel formed by BoNT/B in phospholipid bilayers (Schmid *et al.*, 1993).

These toxin channels are likely to be related to the translocation of the L-domain across the endocytic vesicle membrane into the neuronal cytosol, but no general consensus exists on the precise mechanism of this process (Montecucco *et al.*, 1994). A model which explains all available experimental data proposes that the L-chain translocates across the vesicle membrane through a channel open laterally to lipids, rather than inside a proteinaceous pore (Montecucco *et al.*, 1994). The H- and L-polypeptide chains are supposed to change conformation at low pH in a concerted fashion, both of them entering into contact with the hydrophobic core of the lipid bilayer after the exposure of hydrophobic surfaces. The acidic toxin form may have the properties of a molten globule (Bychkova *et al.*, 1988; van der Goot *et al.*, 1991).

The H-chain forms a transmembrane hydrophilic cleft that nests the passage of the partially unfolded L-chain with its hydrophobic segments facing the lipids. The cytosolic neutral pH induces the L-chain to refold and to regain its water-soluble neutral conformation following reduction of the interchain disulfide bond. Cytosolic chaperones may be involved in facilitating the exit of the L-chain from the vesicle membrane and in promoting its cytosolic refolding. CNT refolding is further complicated by the absolute requirement of zinc for its catalytic activity. In fact, the protonation of the histidines co-ordinating the zinc ion at low pH is expected to release the metal atom, which has to be acquired again in the cytosol.

As the L-chain is released from the vesicle membrane, the transmembrane hydrophilic cleft of the H-chain tightens up to reduce the amount of hydrophilic protein surface exposed to the hydrophobic core of the membrane. However, this process leaves a peculiarly shaped channel across the membrane, with two rigid protein walls and a flexible lipid seal on one side, which is proposed to be the structure responsible for the ion-conducting properties of CNTs. Pore formation is therefore a consequence of membrane translocation, rather than its prerequisite.

INTRACELLULAR ZINC-ENDOPEPTIDASE ACTIVITY

The catalytic nature of these neurotoxins was discovered following the observation that all the sequences of the L-chain of CNTs contain the His-Glu-Xaa-Xaa-His zinc-binding motif of zinc-endopeptidases (Kurazono et al., 1992; Schiavo et al., 1992a, b; Wright et al., 1992). Following this observation, it was soon demonstrated that TeNT was blocking ACh release at synapses of the buccal ganglion of Aplysia californica via a zinc-dependent protease activity (Schiavo et al., 1992a).

The eight CNTs are remarkably specific proteases: among the many proteins and synthetic substrates assayed so far, only three targets, all members of the SNARE (SNAP-receptor) family (Söllner et al., 1993; Bock and Scheller, 1997), have been identified (Table 10.1). TeNT and BoNT/B, /D, /F and /G cleave vesicle-associated membrane protein (VAMP), but each at a different site (Schiavo et al., 1992c, 1993a, b, 1994; Yamasaki et al., 1994b, c); BoNT/A and /E cleave synap-topsomal-associated membrane protein of 25 kDa (SNAP-25) at two different sites and BoNT/C cleaves both syntaxin and SNAP-25 (Blasi et al., 1993a, b; Schiavo et al., 1993b, c, 1995; Binz et al., 1994; Foran et al., 1996; Osen Sand et al., 1996; Williamson et al., 1996; Vaidyanathan et al., 1999).

Clostridial neurotoxin targets

Synaptosomal-associated membrane protein

SNAP-25 was originally described as the major palmitoylated protein in the CNS (Oyler et al., 1989; Hess et al., 1992; Wilson et al., 1996). As schematically depicted in Figure 10.4, this protein lacks a classical transmembrane segment and its membrane binding is thought to be mediated by the palmitoylation of cysteines located in the middle of the polypeptide chain (Hess et al., 1992; Veit et al., 1996). SNAP-25 interacts with the other t-SNARE, syntaxin and VAMP to form a protein complex known as the synaptic SNARE complex, which constitutes the core of the neuroexocytic apparatus (Söllner et al., 1993). In addition, SNAP-25 forms a sto-

ichiometric complex with the putative calcium sensor synaptotagmin, and this interaction is believed to be important in a late step of the calcium-dependent phase of neurotransmitter release (Banerjee et al., 1996; Schiavo et al., 1997).

SNAP-25 is required for axonal growth during neuronal development and in nerve terminal plasticity in the mature nervous system (Geddes et al., 1990; Osen Sand et al., 1993). SNAP-25 is conserved from yeast to humans (Wilson et al., 1996). It is present in endocrine tissue (Jacobsson et al., 1994) and is expressed in the nervous system in two developmentally regulated isoforms (SNAP-25A and B) (Bark and Wilson, 1994). Recently, SNAP-25 analogues (termed SNAP-23 and SNAP-29) with a broader distribution were identified (Ravichandran et al., 1996; Inoue et al., 1997; Mollinedo and Lazo, 1997; Sadoul et al., 1997; Wang et al., 1997; Steegmaier et al., 1998). In mast cells, SNAP-23 relocates in response to stimulation from the plasma membrane to granule membranes. After relocation, SNAP-23 is required for exocytosis, implying a crucial role of this SNAP-25 isoform in promoting membrane fusion (Guo et al., 1998).

Syntaxin

Syntaxin is a typical type-II membrane protein, located mainly on the neuronal plasmalemma (Bennett et al., 1992; Inoue and Akagawa, 1992). The N-terminal

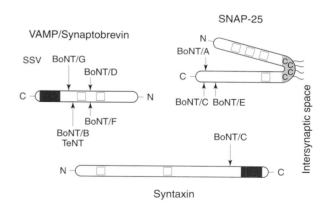

FIGURE 10.4 Schematic structure of SNARE proteins. SNAP-25 is bound to the presynaptic membrane by palmitoylated cysteine residues located in the middle of the polypeptide chain. Syntaxin is a typical type-II membrane protein, localized mainly on the neuronal plasmalemma. The N-terminal portion is exposed to the cytosol and is followed by a transmembrane domain and few extracellular residues. VAMP/synaptobrevin is also a type-II membrane protein with a short C-terminal tail protruding in the vesicle lumen and a transmembrane segment (dark grey box), followed by a 60-residue long cytosolic part. This portion is highly conserved among isoforms and species, whereas the N-terminal portion is poorly conserved and rich in prolines. The arrows indicate the sites of cleavage by CNTs, while the SNARE motifs required for CNT-substrate interaction are shown by small squares.

portion is exposed to the cytosol and is followed by a transmembrane domain and a few extracellular residues (Figure 10.4). Syntaxin is associated with calcium channels in the active zones (Stanley, 1997), but it is also present on most of the neuronal cell membrane (Garcia *et al.*, 1995).

Syntaxin is also localized on chromaffin granules (Tagaya *et al.*, 1995) and undergoes, together with SNAP-25, a recycling process in organelles indistinguishable from SSV (Walch-Solimena *et al.*, 1995). Syntaxin interacts in a calcium-dependent manner with some isoforms of the SSV protein synaptotagmin (Südhof, 1995) via a domain that is also responsible for the interaction with VAMP and α-SNAP (soluble NSF accessory protein) (Hayashi *et al.*, 1995; Kee *et al.*, 1995; Lin and Scheller, 1997).

Syntaxins constitute a large protein family with homologues in yeast and plants, and a vast syntaxin polymorphism exists within the nervous tissue (Bennett *et al.*, 1993; Bock and Scheller, 1997). Syntaxins are essential for neuronal development and survival, since BoNT/C, unlike the other CNTs, acts as a cytotoxic factor in neurons (Kurokawa *et al.*, 1987; Osen Sand *et al.*, 1996). Several isoforms undergo a complex pattern of alternative splicing and expression during long-term potentiation (LTP), thus suggesting that syntaxins are involved in synaptic plasticity (Hicks *et al.*, 1997; Rodger *et al.*, 1998). This differential expression could be important for a direct modulation of calcium entry via selective interaction with specific calcium channels, in addition to the formation of distinct SNARE complexes with different SNAP-25 and VAMP isoforms.

Vesicle-associated membrane protein

VAMP (also known as synaptobrevin) is a protein of 13 kDa specifically localized to SSVs, dense core granules and synaptic-like microvesicles, and is the prototype of vesicular SNAREs (v-SNARE) (Südhof, 1995). Several different isoforms have been identified, but only three have been extensively characterized: VAMP-1, VAMP-2 and cellubrevin (Trimble *et al.*, 1988; Baumert *et al.*, 1989; McMahon *et al.*, 1993). VAMP isoforms are present in all vertebrate tissues, their relative amount and distribution differing between tissues and cell-types (Trimble, 1993; Rossetto *et al.*, 1996).

Structurally, VAMP can be divided into four functional domains (Figure 10.4) (Trimble *et al.*, 1988; Baumert *et al.*, 1989). The *N*-terminal portion is proline rich and isoform specific. In contrast, the central portion of the cytoplasmic domain is very conserved through evolution and contains the coiled-coil regions responsible for SNARE complex formation (Hayashi *et al.*, 1994; Kee *et al.*, 1995; Lin and Scheller, 1997; Poirier *et al.*, 1998; Sutton *et al.*, 1998). The protein is anchored

to the SSV by a single transmembrane domain, which is followed by a short and poorly conserved intravesicular portion.

On the SSV, VAMP is associated with synaptophysin and with subunits of the V-ATPase (Calakos and Scheller, 1994; Edelmann *et al.*, 1995; Washbourne *et al.*, 1995; Galli *et al.*, 1996).

Are SNARE proteins the only targets of clostridial neurotoxin activity?

Strikingly, TeNT and BoNT/B cleave VAMP at the same peptide bond (Gln76–Phe77) (Schiavo *et al.*, 1992c), yet when injected into the animal they cause the opposite symptoms of tetanus and botulism. This observation has been particularly relevant, because it clearly demonstrates that the different symptoms derive from different sites of intoxication rather than from a different intracellular mechanism of action. Recombinant VAMP, SNAP-25 and syntaxin are cleaved at the same peptide bonds as the corresponding cellular proteins, thus indicating that no additional endogenous factors are necessary for the proteolytic activity of the CNTs.

These findings have been exploited to develop *in vitro* assays of the metalloprotease activity of CNTs (Ekong *et al.*, 1995, 1996; Schiavo and Montecucco, 1995; Hallis *et al.*, 1996; Soleilhac *et al.*, 1996). The proteolytic activity of the CNTs can be assayed in neuronal cells in culture, using anti-SNARE antibodies. A sensitive single cell assay can thus be performed by following the progressive loss of SNARE staining as its proteolysis progresses (Matteoli *et al.*, 1996; Osen Sand *et al.*, 1996; Williamson *et al.*, 1996). Several isoforms of SNARE proteins have been identified in different species and tissues and only some of them are susceptible to proteolysis by CNT. In general, a SNARE protein may be resistant to a neurotoxin because of mutations at the cleavage site or in other regions involved in neurotoxin binding.

Several experimental data support a direct correlation between CNT-induced proteolysis of SNAREs and the inhibition of neurotransmitter release.

- The intracellular activity of the CNTs is inhibited by specific inhibitors of zinc-endopeptidases such as phosphoramidon and captopril (Schiavo *et al.*, 1992a; de Paiva *et al.*, 1993b; Gaisano *et al.*, 1994; Höhne-Zell *et al.*, 1994; Deshpande *et al.*, 1995).
- VAMP, SNAP-25 or syntaxin is cleaved in synaptosomes, PC12 and neuronal cells intoxicated with CNTs with a corresponding inhibition of neuroexocytosis (Link *et al.*, 1992; Blasi *et al.*, 1993a, b; Adler *et al.*, 1994; Höhne-Zell *et al.*, 1994; Herreros *et al.*, 1995;

Papini *et al.*, 1995; Foran *et al.*, 1996; Matteoli *et al.*, 1996; Osen Sand *et al.*, 1996; Williamson *et al.*, 1996; Capogna *et al.*, 1997).

- Peptides spanning the cleavage site of VAMP inhibit TeNT and BoNT/B activity in *Aplysia*, in squid neurons or in chromaffin cells (Schiavo *et al.*, 1992a; Dayanithi *et al.*, 1994; Hunt *et al.*, 1994).
- VAMP-specific antibodies prevent the inhibition of neurotransmitter release in *Aplysia* neurons induced by TeNT and BoNT/B, but not that caused by BoNT/A (Poulain *et al.*, 1993).
- An antibody against the zinc-binding region inhibits the activity of TeNT and BoNT/A in chromaffin cells (Bartels *et al.*, 1994).
- Mutation of the two histidine residues in the active site of TeNT (H_{232} and H_{237}) which co-ordinates the zinc atom abolishes zinc binding and toxicity (Yamasaki *et al.*, 1994a).
- In TeNT, mutation of the glutamate residue co-ordinating the water molecule important in catalysis (E234Q) does not alter zinc and substrate binding, but blocks toxin action when assayed on NMJ preparations (Li *et al.*, 1994), catecholamine release in PC12 cells (Höhne-Zell *et al.*, 1993) and in *Drosophila* embryonic nervous system (Sweeney *et al.*, 1995).
- The effects of neuronal expression of L-chain of TeNT (Sweeney *et al.*, 1995) and of mutating neuronal VAMP (Deitcher *et al.*, 1998) in *Drosophila* are virtually identical and include the complete inhibition of evoked synaptic transmission and a reduction in MEPPs frequency.
- The blockade of Ca^{2+}-dependent insulin secretion in pancreatic cells by TeNT can be overcome by expressing VAMP or its homologue cellubrevin, which have been made resistant to TeNT proteolysis (Regazzi *et al.*, 1996). Hence, members of the VAMP family are the only targets of TeNT relevant for inhibition of exocytosis in this system. The same experimental approach has recently been repeated for SNAP-25 isoforms (Sadoul *et al.*, 1997; P. Washbourne, personal communication).
- In leech, VAMP is cleaved by TeNT, whereas SNAP-25 is resistant to BoNT/A (Bruns *et al.*, 1997). This correlates with an inhibition of synaptic transmission following injection of the L-chain of TeNT and the lack of effect of injected BoNT/A L-chain.
- Non-neuronal toxicity can often be accounted for by the susceptibility of homologues of the synaptic SNAREs to the toxins. For example, the vertebrate VAMP homologue, cellubrevin, is ubiquitously expressed, and is sensitive to TeNT (McMahon *et al.*, 1993); in contrast, the ubiquitously expressed *Drosophila* VAMP is not susceptible to TeNT (Sweeney *et al.*, 1995). This correlates with the toxic

effects of TeNT in mammalian non-neuronal cells (Eisel *et al.*, 1993) and the apparent lack of non-neuronal defects in *Drosophila* (Sweeney *et al.*, 1995).

Taken together, these findings provide direct evidence for the involvement of VAMP, SNAP-25 and syntaxin in regulated secretion, and they provide a molecular explanation for the pathogenesis of tetanus and botulism.

Full inhibition of neurotransmitter release is not accompanied by a parallel full proteolysis of the SNARE proteins present within a nerve terminal (Foran *et al.*, 1996; Osen Sand *et al.*, 1996; Williamson *et al.*, 1996; Bruns *et al.*, 1997). This result is best explained by the existence within the nerve terminal of different pools of SNARE proteins with different availability to binding and proteolysis by the CNTs. A large proportion of a given SNARE protein may be not accessible owing to specific protein–protein interactions or physical compartmentalization. The recently published structure of the synaptic SNARE complex (Sutton *et al.*, 1998) reveals that the soluble domains of SNAP-25, syntaxin 1 and VAMP 2 form a highly twisted four-helix bundle with a hydrophobic core, with the CNT cleavage sites engaged in the helical bundle. Remarkably, complex formation induces a major conformational change in SNAP-25 and VAMP, which switch from largely unordered to helical-rich conformations.

The resistance of the SNARE proteins engaged in the SNARE complex towards CNTs proteolysis (Hayashi *et al.*, 1994; Pellegrini *et al.*, 1994) can now be explained on the basis of both the transition from a protease-sensitive to a protease-insensitive structure of the individual SNARE components (Fasshauer *et al.*, 1997) and the masking of the CNT cleavage sites within the four-helix bundle of the SNARE complex. However, it appears that those SNARE molecules that are ready to enter the neuroexocytosis cascade are available for proteolysis by the CNTs, and it is their cleavage that results in the block of neurotransmitter release (Xu *et al.*, 1998).

In addition to the well-characterized zinc-endopeptidase activity, some experimental evidence suggests that CNTs could display other inhibitory activities at the synapse.

- A protease-deficient mutant of TeNT where the glutamic acid residue of the active site was substituted for an alanine (E234A) showed some residual inhibition of neurotransmission in *Aplysia* neurons (Binz *et al.*, 1992; Ashton *et al.*, 1995). The inactive CNTs may still bind their substrates, thus acting as competitive inhibitors of the normal interactions between SNAREs and other components of the neurosecretory machinery, especially when applied at concentrations well above the minimal levels required for toxicity.

- TeNT stimulates transglutaminase, a cytosolic enzyme catalysing the cross-linking of glutamine residues to primary amino groups (Facchiano and Luini, 1992). Synapsin, a phosphoprotein important for SSV mobilization (Valtorta et al., 1992), is a preferred substrate, at least in vitro (Facchiano et al., 1993a). It was suggested that toxin-activated transglutaminase could cross-link synapsin present on SSVs to the actin cytoskeleton, thus rendering them unavailable for exocytosis. This would cause a long-term inhibition of neurotransmitter release superimposed on the rapid inhibition due to VAMP proteolysis (Facchiano et al., 1993a, b). However, the in vivo evidence for a role of transglutaminase in TeNT intoxication is conflicting. While some effects of transglutaminase inhibitors have been found on TeNT action in Aplysia (Ashton et al., 1995) and synaptosomes (Ashton and Dolly, 1997), three other studies on neurons and insulin-secreting cells failed to find any effect on toxicity of either transglutaminase inhibitors or other conditions affecting transglutaminase activity (Coffield et al., 1994; Gobbi et al., 1996; Regazzi et al., 1996).
- Disruption of the actin or microtubule cytoskeleton reduces TeNT toxicity in mammalian synaptosomes (Ashton and Dolly, 1997). A possible explanation is that the toxin or its substrate affects the mobilization of SSVs from the cytoskeleton-linked reserve pool. However, the attenuation of TeNT effects could be due to a requirement of the cytoskeleton for toxin action, for example in the transport of TeNT-containing vesicles from their uptake sites.
- Several CNTs are substrates for src protein kinase in vitro and in vivo (Ferrer Montiel et al., 1996). This modification increases both catalytic activity and stability, suggesting that in neurons, a biologically relevant form of the toxin may be phosphorylated. This finding also raises the possibility that phosphorylated tyrosine residues of the toxin could bind cellular proteins containing SH2 domains, thus interfering with signal transduction pathways.
- BoNT/A intoxication causes a decrease in arachidonic acid (AA) release, which overlaps with the block of ACh secretion, suggesting that AA deprivation may affect neuroexocytosis (Ray et al., 1993). An increase in AA via exogenous phospholipase A_2, or its activators, such as mellitin, protects against the effects of BoNT/A.

Further experiments are necessary to confirm the contribution of the above activities to the intoxication process at clinically relevant CNT concentrations.

The enzymic nature of the intracellular activity of CNTs is at the basis of their potency. In fact, it can be estimated that, in Aplysia neurons, fewer than 10 molecules of TeNT L-chain are sufficient to cause a 50% inhibition of neuroexocytosis within 20 min at 20°C (B. Poulain, personal communication). Considering the higher body temperature of mammals and the long onset of tetanus and botulism in humans, it would be plausible if a single toxin molecule is sufficient to intoxicate one synapse.

Basis for clostridial neurotoxin recognition of SNARE proteins

An inspection of the sequences of the three synaptic SNAREs at the CNT cleavage sites (Table 10.1) reveals no conserved patterns that could account for the specificity of these zinc proteases for their intraneuronal targets. Hence, while the overall architecture of the active sites of these enzymes is expected to be similar, each CNT must differ in detailed spatial organization in order to accommodate and hydrolyse such different peptide bonds. Biochemical studies have uncovered several peculiarities of CNTs. Among them are the following.

- Short peptides encompassing the cleavage site are not cleaved, although they bind the toxin, whilst longer segments are cleaved (Shone et al., 1993; Cornille et al., 1994, 1995, 1997; Foran et al., 1994; Shone and Roberts, 1994; Yamasaki et al., 1994a).
- TeNT and BoNT/B hydrolyse the same peptide bond of VAMP, but they have different requirements in terms of the minimal length of the peptide acting as substrate. For example, the minimal VAMP segment cleaved by BoNT/B is 44–94 but 33–94 for TeNT (Foran et al., 1994).
- Some CNTs hydrolyse a peptide bond, leaving untouched peptide bond(s) with identical sequences located in other portions of the substrate. Other CNTs have some degree of flexibility on the peptide bond cleaved: BoNT/B cleaves the peptide bond Q–F, present in the natural substrate VAMP, but is also able to hydrolyse peptides in which the above bond was substituted by N–F, A–F or Q–Y.
- In general, CNTs are more effective at cleaving membrane-bound substrate than the recombinant soluble molecule. However, BoNT/C only cleaves membrane-bound syntaxin and SNAP-25 (Blasi et al., 1993b; Schiavo et al., 1995) and is ineffective on the isolated molecules.

These findings clearly indicate that the toxin–substrate interaction requires some other common structural element of the SNARE sequence or structure serving as recognition motifs for the CNTs.

Comparison of the sequence of synaptic SNARE proteins of different species shows the presence of a 10 residue-long motif, termed the SNARE motif (Rossetto *et al.*, 1994). This motif has the sequence xh– –xh–xhp (x, any amino acid; h, hydrophobic residue; –, acidic residue; p, polar residue) and forms an *α*-helix with a negatively charged face flanked by a hydrophobic face (Rossetto *et al.*, 1994; Pellizzari *et al.*, 1996, 1997; Washbourne *et al.*, 1997). Multiple copies of this motif are present in syntaxin, SNAP-25 and VAMP/synaptobrevin (Figure 10.4) and are proposed to act as CNT recognition sites. The majority of them are included in the four-helix bundle structure of the synaptic SNARE complex or, in one case, in the three-helix bundle at the *N*-terminus of syntaxin (Fernandez *et al.*, 1998; Sutton *et al.*, 1998). While direct structural evidence, for example, co-crystallization of CNTs L-chain with its substrate, is lacking, several biochemical and electrophysiological experiments support this proposal.

In fact, only peptides that include at least one SNARE motif are cleaved *in vitro* by CNTs (Shone *et al.*, 1993; Cornille *et al.*, 1994, 1997; Foran *et al.*, 1994). The SNARE motif is exposed on the protein surface on native non-assembled SNARE proteins, as shown by the binding of anti-SNARE motif antibodies. These antibodies cross-react among the three SNAREs and inhibit the proteolytic activity of the neurotoxins (Pellizzari *et al.*, 1996). CNTs also cross-inhibit each other (Pellizzari *et al.*, 1996), possibly by interaction with the SNARE motif. Extensive site-directed mutagenesis of VAMP and SNAP-25 demonstrates that different CNTs bind different SNARE motifs (Pellizzari *et al.*, 1996, 1997; Washbourne *et al.*, 1997). Resistance to CNTs *in vivo* is sometimes associated with mutations in SNARE motifs, or with deviations from the consensus. In fact, the *Drosophila* VAMP lacks one of the three acidic residues in each SNARE motif, and is not cleaved by TeNT (Sweeney *et al.*, 1995).

In conclusion, these studies suggest that recognition of the SNARE motif is a major determinant of the specificity of the CNTs for their targets. This binding is followed by interactions with other regions of the sequence that are different for each SNARE, including the segment containing the peptide bond to be cleaved. The relative contribution of these interactions for the specificity and strength of CNT binding to each SNARE remains to be determined. However, it can be predicted that hydrolysis of the substrate region bound to the CNT active site causes a decrease in binding affinity, which is expected to lead to the rapid release of the two fragments generated.

USE OF CLOSTRIDIAL NEUROTOXINS IN THE STUDY OF REGULATED SECRETION

The zinc-dependent protease activity of different CNTs, with the exception of BoNT/B and TeNT that cleave VAMP at the same site, is directed against distinct peptide bonds in their target SNAREs (Table 10.1). CNTs are therefore highly specific tools for the investigation of the role of their targets in different cellular processes.

The three SNAREs are not cleavable by the CNTs when they form a heterotrimer (Hayashi *et al.*, 1994). However, they can be cleaved by the CNTs when they are in a monomeric form or when SNAP-25 forms a heterodimer with syntaxin or synaptotagmin, or when VAMP is interacting with other SSV proteins (Calakos and Scheller, 1994; Galli *et al.*, 1994; Edelmann *et al.*, 1995; Washbourne *et al.*, 1995; Schiavo *et al.*, 1997). The cleaved fragments are able to enter a SNARE complex, although this is less stable than the one generated with the full-length proteins (Hayashi *et al.*, 1995; Otto *et al.*, 1995).

BoNT/A removes only nine residues from the SNAP-25 *C*-terminus, yet this is sufficient to impair neurotransmitter release and other forms of regulated secretion (Huang *et al.*, 1998). Recent experiments using flash photolysis of caged intracellular calcium under conditions of BoNT/A intoxication suggest that this nine-residue peptide may couple the calcium sensor to the final step of exocytosis (Xu *et al.*, 1998).

Several lines of evidence point to SNAREs as the major determinants of specificity in the interaction between vesicles and their target membranes (reviewed in Götte and von Mollard, 1998). The cleavage of VAMP and syntaxin by CNTs leads to the release of large portions of their cytosolic domains into the cytosol, suggesting that the interaction between SSVs and the presynaptic membrane should be impaired in CNT intoxicated synapses. However, poisoned, and therefore electrically silent, synapses contain increased numbers of SSV 'docked' at the presynaptic membrane, as shown by electron microscopy (Mellanby *et al.*, 1988; Neale *et al.*, 1989; Hunt *et al.*, 1994; Broadie *et al.*, 1995; O'Connor *et al.*, 1997).

The active zone of the synapse is a complex macromolecular structure, in which several components of the vesicle fusion machinery are assembled. While SNAREs could act as major determinants for the specificity of membrane trafficking in some cases, it appears that the functional docking of a SSV results from multiple interactions at the active zone (Schiavo *et al.*, 1998). After CNT treatment, these other interactions can give rise to SSV closely apposed to the presynaptic membrane, which appear 'docked' by electron microscopy criteria, but are unable to undergo fusion. SNAREs must therefore play a key role in the fusion process itself. Evidence that SNAREs are the minimal machinery required for membrane fusion has recently been provided by experiments demonstrating the CNT-sensitive heterotypic fusion of liposomes containing only VAMP, syntaxin and SNAP-25 (Weber *et al.*, 1998).

Given the general role that SNAREs play in vesicular trafficking, the use of CNTs can be extended to a variety of cell types and biochemical preparations. In the case of neuronal cells or synaptosomes, simply incubating them with CNTs is sufficient to cause SNARE cleavage. In contrast, non-neuronal cells have to be permeabilized or microinjected (reviewed in Mochida *et al.*, 1990; Ahnert-Hilger and Weller, 1997). Alternatively, cells can be transfected with the gene encoding the light chain (Eisel *et al.*, 1993; Sweeney *et al.*, 1995; Aguado *et al.*, 1997; Lang *et al.*, 1997a; Hackam *et al.*, 1998; Huang *et al.*, 1998). Detailed protocols for the use of CNTs in different cell-types and organelle preparations are now available (Blasi *et al.*, 1997; Lang *et al.*, 1997b).

Use of botulinum neurotoxins in human therapy

BoNTs are used for the treatment of several human diseases characterized by the hyperfunction of cholinergic terminals (Jankovic and Hallett, 1994). Injections of minute amounts of BoNT into the hyperactive muscle(s) lead to a suppression of symptoms lasting for a few months. BoNT/A is the most widely used serotype, but other BoNTs are currently under clinical trials (Eleopra *et al.*, 1997, 1998). BoNT is currently the best available treatment for dystonias and, in ophthalmology, for certain types of strabismus and blepharospasm.

Its use is now being extended to several other human pathologies, such as hemifacial spasm and achalasia (Jankovic and Hallett, 1994; Montecucco *et al.*, 1996) and also in topical applications for cosmetic use (Foster *et al.*, 1996; Carruthers and Carruthers, 1998). The treatment can be repeated several times without major side-effects such as the development of an immune response. If anti-neurotoxin antibodies are produced, treatment can be continued with another BoNT serotype.

In contrast to BoNTs, TeNT is not used in therapy but only in animal models to induce experimental epilepsy and neuronal degeneration (Brace *et al.*, 1985; Bagetta *et al.*, 1991). Recently, the binding domain of TeNT, which retains the neurospecificity and uptake properties of the entire toxin, has been used as a carrier of lysosomal hydrolase (Dobrenis *et al.*, 1992) and of superoxide dismutase inside cells (Francis *et al.*, 1995; Figueiredo *et al.*, 1997).

CONCLUDING REMARKS

The detailed study of CNTs has recently provided a large quantity of novel information that has revolutionized our understanding of the process of neuroexocytosis. At the same time, this knowledge has provided a platform for further discoveries and promoted their use as tools in neuroscience and cell biology. Is this the last burst of scientific attention for CNTs, or will the clostridial phoenix rise again? Several pieces of information suggest that the latter will indeed be the case.

A remarkable development has been the ever-growing utilization of BoNTs in the therapy of an expanding group of human syndromes and for cosmetic application. The absolute neurospecificity of CNTs and their ability to enter nerve cells makes them ideal carrier proteins. Even though the number of examples where this delivery technology has been applied is limited, the molecular definition of CNT receptors will certainly boost this field and provide important insights into the retrograde transport in motor neurons and into endocytic processes at the synapse.

Furthermore, the availability of the complete structure of BoNT/A provides the opportunity to design innovative zinc-endoproteases based on the structure of these neurotoxins, which are specific for CNT-resistant SNAREs or other proteins of biological interest. Protein modelling of the crystal structure of BoNT/A will also be essential for the design of novel, specific CNT inhibitors (Adler *et al.*, 1998; Martin *et al.*, 1998; Schmidt *et al.*, 1998) to be used not only in the restricted environment of the research laboratory, but also in the clinical treatment of tetanus and botulism.

ACKNOWLEDGEMENTS

We apologize to our colleagues whose work has been omitted because of space limitations. We are grateful to R.C. Stevens for the structure of BoNT/A shown in Figure 10.3, B. Poulain and J. Molgo for helpful discussion and for critical reading of the manuscript, and R. Eglesfield for excellent secretarial assistance. Work in the authors' laboratories is supported by the Human Frontier Science Program (J.H.), Telethon-Italia Grant 473 and MURST (C.M.) and by the Imperial Cancer Research Fund (G.L. and G.S.)

REFERENCES

Adler, M., Deshpande, S.S., Sheridan, R.E. and Lebeda, F. J. (1994). Evaluation of captopril and other potential therapeutic compounds in antagonizing botulinum toxin-induced muscle paralysis. In: *Therapy with Botulinum Toxin* (eds J. Jankovic and M. Hallett). Marcel Dekker, New York.

Adler, M., Dinterman, R.E. and Wannemacher, R.W. (1997). Protection by the heavy-metal chelator *N,N,N′,N′*-tetrakis (2-pyridyl-methyl)ethylenediamine (TPEN) against the lethal action of botulinum neurotoxin-A and neurotoxin-B. *Toxicon* **35**, 1089–100.

Adler, M., Nicholson, J.D. and Hackley, B.E. (1998). Efficacy of a novel metalloprotease inhibitor on botulinum neurotoxin B activity. *FEBS Lett.* **429**, 234–8.

Aguado, F., Gombau, L., Majo, G., Marsal, J., Blanco, J. and Blasi, J. (1997). Regulated secretion is impaired in ATT-20 endocrine-cells stably transfected with botulinum neurotoxin type-A light-chain. *J. Biol. Chem.* **272**, 26005–8.

Ahnert-Hilger, G. and Weller, U. (1997). Application of alpha-toxin and streptolysin O in cell biology. In: *Bacterial Toxins. Tools in Cell Biology and Pharmacology* (ed. K. Aktories), pp. 259–72. Chapman & Hall, Weinheim.

Ahnert-Hilger, G., Weller, U., Dauzenroth, M.E., Habermann, E. and Gratzl, M. (1989). The tetanus toxin light chain inhibits exocytosis. *FEBS Lett.* **242**, 245–8.

Angaut-Petit, D., Molgo, J., Comella, J.X., Faille, L. and Tabti, N. (1990). Terminal sprouting in mouse neuromuscular junctions poisoned with botulinum type A toxin: morphological and electrophysiological features. *Neuroscience* **37**, 799–808.

Arnon, S.S. (1995). Botulism as an intestinal toxaemia. In: *Infections of the Gastrointestinal Tract* (eds M.J. Blaser, P.D. Smith, J.I. Ravdin, H.B. Greenberg and R.L. Guerrant), pp. 257–71. Raven Press, New York.

Ashton, A.C. and Dolly, J.O. (1997). Microtubules and microfilaments participate in the inhibition of synaptosomal noradrenaline release by tetanus toxin. *J. Neurochem.* **68**, 649–58.

Ashton, A.C., dePaiva, A.M., Poulain, B., Tauc, L. and Dolly, J.O. (1993). Factors underlying the characteristic inhibition of the neuronal release of neurotransmitters by tetanus and various botulinum toxin. In: *Botulinum and Tetanus Toxin. Neurotransmission and Biomedical Aspects* (ed. B.R. Das Gupta), pp. 191–213. Plenum Press, New York.

Ashton, A.C., Li, Y., Doussau, F., Weller, U., Dougan, G., Poulain, B. and Dolly, J.O. (1995). Tetanus toxin inhibits neuroexocytosis even when its Zn^{2+}-dependent protease activity is removed. *J. Biol. Chem.* **270**, 31386–90.

Aureli, P., Fenicia, L., Pasolini, B., Gianfranceschi, M., McCroskey, L.M. and Hatheway, C.L. (1986). Two cases of type E infant botulism caused by neurotoxigenic *Clostridium butyricum* in Italy. *J. Infect. Dis.* **154**, 207–11.

Bagetta, G., Nistico, G. and Bowery, N.G. (1991). Characteristics of tetanus toxin and its exploitation in neurodegenerative studies. *Trends Pharmacol. Sci.* **12**, 285–9.

Bambrick, L. and Gordon, T. (1994). Neurotoxins in the study of neural regulation of membrane proteins in skeletal muscle. *J. Pharmacol. Toxicol. Methods* **32**, 129–38.

Banerjee, A., Kowalchyk, J.A., DasGupta, B.R. and Martin, T.F.J. (1996). SNAP-25 is required for a late postdocking step in Ca^{2+}-dependent exocytosis. *J. Biol. Chem.* **271**, 20227–30.

Bark, I.C. and Wilson, M.C. (1994). Human cDNA clones encoding two different isoforms of the nerve terminal protein SNAP-25. *Gene* **139**, 291–2.

Bartels, F., Bergel, H., Bigalke, H., Frevert, J., Halpern, J. and Middlebrook, J. (1994). Specific antibodies against the Zn^{2+}-binding domain of clostridial neurotoxins restore exocytosis in chromaffin cells treated with tetanus or botulinum A neurotoxin. *J. Biol. Chem.* **269**, 8122–7.

Bauerfeind, R., Hüttner, W.B., Almers, W. and Augustine, G.J. (1994). Quantal neurotransmitter release from early endosomes? *Trends Cell Biol.* **4**, 155–6.

Baumert, M., Maycox, P.R., Navone, F., De Camilli, P. and Jahn, R. (1989). Synaptobrevin: an integral membrane protein of 18,000 daltons present in small synaptic vesicles of rat brain. *EMBO J.* **8**, 379–84.

Beise, J., Hahnen, J., Andersen-Beckh, B. and Dreyer, F. (1994). Pore formation by tetanus toxin, its chain and fragments in neuronal membranes and evaluation of the underlying motifs in the structure of the toxin molecule. *Naunyn Schemiedebergs Arch. Pharmacol.* **349**, 66–73.

Benecke, R., Takano, K., Schmidt, J. and Henatsch, H.D. (1977). Tetanus toxin induced actions on spinal Renshaw cells and Ia-inhibitory interneurones during development of local tetanus in the cat. *Exp. Brain Res.* **27**, 271–86.

Bennett, M.K., Calakos, N. and Scheller, R.H. (1992). Syntaxin: a synaptic protein implicated in docking of synaptic vesicles at presynaptic active zones. *Science* **257**, 255–9.

Bennett, M.K., Garcia-Arraras, J.E., Elferink, L.A., Peterson, K., Fleming, A.M., Hazuka, C.D. and Scheller, R.H. (1993). The syntaxin family of vesicular transport receptors. *Cell* **74**, 863–73.

Bergey, G.K., MacDonald, R.L., Habig, W.H., Hardegree, M.C. and Nelson, P.G. (1983). Tetanus toxin: convulsant action on mouse spinal cord neurons in culture. *J. Neurosci.* **3**, 2310–23.

Bevan, S. and Wendon, L.M. (1984). A study of the action of tetanus toxin at rat soleus neuromuscular junctions. *J. Physiol. (Lond.)* **348**, 1–17.

Bhattacharyya, S.D. and Sugiyama, H. (1989). Inactivation of botulinum and tetanus toxins by chelators. *Infect. Immun.* **57**, 3053–7.

Binz, T., Grebenstein, O., Kurazono, H., Eisel, H., Wernars, K., Popoff, M., Mochida, S., Poulain, B., Tauc, L., Kozaki, S. and Niemann, H. (1992). Molecular biology of the L chains of clostridial neurotoxins. In: *Bacterial Protein Toxin* (eds B. Witholt *et al.*), pp. 56–65. Gustav Fisher, Stuttgart.

Binz, T., Blasi, J., Yamasaki, S., Baumeister, A., Link, E., Südhof, T.C., Jahn, R. and Niemann, H. (1994). Proteolysis of SNAP-25 by types E and A botulinal neurotoxins. *J. Biol. Chem.* **269**, 1617–20.

Bittner, M.A., DasGupta, B.R. and Holz, R.W. (1989a). Isolated light chains of botulinum neurotoxins inhibit exocytosis. Studies in digitonin-permeabilized chromaffin cells. *J. Biol. Chem.* **264**, 10354–60.

Bittner, M.A., Habig, W.H. and Holz, R.W. (1989b). Isolated light chain of tetanus toxin inhibits exocytosis: studies in digitonin-permeabilized cells. *J. Neurochem.* **53**, 966–8.

Bizzini, B., Stoeckel, K. and Schwab, M. (1977). An antigenic polypeptide fragment isolated from tetanus toxin: chemical characterization, binding to gangliosides and retrograde axonal transport in various neuron systems. *J. Neurochem.* **28**, 529–42.

Black, J.D. and Dolly, J.O. (1986a). Interaction of ^{125}I-labeled botulinum neurotoxins with nerve terminals. I. Ultrastructural autoradiographic localization and quantitation of distinct membrane acceptors for types A and B on motor nerves. *J. Cell Biol.* **103**, 521–34.

Black, J.D. and Dolly, J.O. (1986b). Interaction of ^{125}I-labeled botulinum neurotoxins with nerve terminals. II. Autoradiographic evidence for its uptake into motor nerves by acceptor-mediated endocytosis. *J. Cell Biol.* **103**, 535–44.

Blasi, J., Chapman, E.R., Link, E., Binz, T., Yamasaki, S., De Camilli, P., Südhof, T.C., Niemann, H. and Jahn, R. (1993a). Botulinum neurotoxin A selectively cleaves the synaptic protein SNAP-25. *Nature* 365, 160–3.

Blasi, J., Chapman, E.R., Yamasaki, S., Binz, T., Niemann, H. and Jahn, R. (1993b). Botulinum neurotoxin C1 blocks neurotransmitter release by means of cleaving HPC-1/syntaxin. *EMBO J.* 12, 4821–8.

Blasi, J., Link, E. and Jahn, R. (1997). Isolated nerve terminals as a model system for the study of botulinum and tetanus toxin. In: *Bacterial Toxins. Tools in Cell Biology and Pharmacology* (ed. K. Aktories), pp. 193–211. Chapman & Hall, London.

Blaustein, R.O., Germann, W.J., Finkelstein, A. and DasGupta, B.R. (1987). The N-terminal half of the heavy chain of botulinum type A neurotoxin forms channels in planar phospholipid bilayers. *FEBS Lett.* 226, 115–20.

Bleck, T.P. (1989). Clinical aspects of tetanus. In: *Botulinum Neurotoxins and Tetanus Toxin* (ed. L.L. Simpson), pp. 379–98. Academic Press, San Diego, CA.

Bock, J.B. and Scheller, R.H. (1997). A fusion of new ideas. *Nature* 387, 133–5.

Boquet, P. and Duflot, E. (1982). Tetanus toxin fragment forms channels in lipid vesicles at low pH. *Proc. Natl Acad. Sci. USA* 79, 7614–18.

Borodic, G.E., Ferrante, R.J., Perace, L.B. and Alderson, K.(1994). Pharmacology and histology of the therapeutic application of botulinum toxin. In: *Therapy with Botulinum Toxin* (eds J. Jankovic and M. Hallett), pp. 119–57. Marcel Dekker, New York.

Brace, H.M., Jefferys, J.G. and Mellanby, J. (1985). Long-term changes in hippocampal physiology and learning ability of rats after intrahippocampal tetanus toxin. *J. Physiol. (Lond.)* 368, 343–57.

Broadie, K., Prokop, A., Bellen, H.J., O'Kane, C.J., Schulze, K.L. and Sweeney, S.T. (1995). Syntaxin and synaptobrevin function downstream of vesicle docking in *Drosophila. Neuron* 15, 663–73.

Brooks, V.B., Curtis, D.R. and Eccles, J.C. (1955). Mode of action of tetanus toxin. *Nature* 175, 120–1.

Bruns, D., Engers, S., Yang, C., Ossig, R., Jeromin, A. and Jahn, R. (1997). Inhibition of transmitter release correlates with the proteolytic activity of tetanus toxin and botulinus toxin A in individual cultured synapses of *Hirudo medicinalis. J. Neurosci.* 17, 1898–910.

Bruschettini, A. (1892). Sulla diffusione del veleno del tetano nell'organismo. *Rif Med.* 8, 270–3.

Burgen, A.S.V., Dickens, F. and Zatman, L.J. (1949). The action of botulinum toxin on the neuro-muscular junction. *J. Physiol. (Lond.)* 109, 10–24.

Bychkova, V.E., Pain, R.H. and Ptitsyn, O.B. (1988). The 'molten globule' state is involved in the translocation of proteins across membranes? *FEBS Lett.* 238, 231–4.

Cabiaux, V., Lorge, P., Vandenbranden, M., Falmagne, P. and Ruysschaert, J.M. (1985). Tetanus toxin induces fusion and aggregation of lipid vesicles containing phosphatidylinositol at low pH. *Biochem. Biophys. Res. Commun.* 128, 840–9.

Calabresi, P., Benedetti, M., Mercuri, N.B. and Bernardi, G. (1989). Selective depression of synaptic transmission by tetanus toxin: a comparative study on hippocampal and neostriatal slices. *Neuroscience* 30, 663–70.

Calakos, N. and Scheller, R.H. (1994). Vesicle-associated membrane protein and synaptophysin are associated on the synaptic vesicle. *J. Biol. Chem.* 269, 24534–7.

Capogna, M., McKinney, R.A., O'Connor, V., Gahwiler, B.H. and Thompson, S.M. (1997). Ca^{2+} or Sr^{2+} partially rescues synaptic transmission in hippocampal cultures treated with botulinum toxin-A and toxin-C, but not tetanus toxin. *J. Neurosci.* 17, 7190–202.

Carle, A. and Rattone, G. (1884). Studio eperimentale sull'eziologia del tetano. *Giorn. Accad. Med. Torino* 32, 174–9.

Carruthers, A. and Carruthers, J. (1998). Clinical indications and injection technique for the cosmetic use of botulinum A exotoxin. *Dermatol. Surg.* 24, 1189–94.

Coffield, J.A., Considine, R.V., Jeyapaul, J., Maksymowych, A.B., Zhang, R.D. and Simpson, L.L. (1994). The role of transglutaminase in the mechanism of action of tetanus toxin. *J. Biol. Chem.* 269, 24454–8.

Cornille, F., Goudreau, N., Ficheux, D., Niemann, H. and Roques, B.P. (1994). Solid-phase synthesis, conformational analysis and *in vitro* cleavage of synthetic human synaptobrevin II 1–93 by tetanus toxin L chain. *Eur. J. Biochem.* 222, 173–81.

Cornille, F., Deloye, F., Fournie-Zaluski, M.C., Roques, B.P. and Poulain, B. (1995). Inhibition of neurotransmitter release by synthetic proline-rich peptides shows that the N-terminal domain of vesicle-associated membrane protein/synaptobrevin is critical for neuro-exocytosis. *J. Biol. Chem.* 270, 16826–32.

Cornille, F., Martin, L., Lenoir, C., Cussac, D., Roques, B.P. and Fourniezaluski, M.C. (1997). Cooperative exocytosis-dependent cleavage of synaptobrevin by tetanus toxin light-chain. *J. Biol. Chem.* 272, 3459–64.

Cremona, O. and De Camilli, P. (1997). Synaptic vesicle endocytosis. *Curr. Opin. Neurol.* 7, 323–30.

Critchley, D.R., Nelson, P.G., Habig, W.H. and Fishman, P.H. (1985). Fate of tetanus toxin bound to the surface of primary neurons in culture: evidence for rapid internalization. *J. Cell Biol.* 100, 1499–507.

Critchley, D.R., Parton, R.G., Davidson, M.D. and E.J. (1988). Characterisation of tetanus toxin binding and internalisation by neuronal tissue. In: *Neurotoxins in Neurochemistry* (eds J.O. Dolly and E. Horwood), pp. 110–22. Halstead Press, London.

Cull-Candy, S.G., Lundh, H. and Thesleff, S. (1976). Effects of botulinum toxin on neuromuscular transmission in the rat. *J. Physiol. (Lond.)* 260, 177–203.

Curtis, D.R., Game, C.J., Lodge, D. and McCulloch, R.M. (1976). A pharmacological study of Renshaw cell inhibition. *J. Physiol. (Lond.)* 258, 227–42.

DasGupta, B.R. (1989). The structure of botulinum neurotoxin. In: *Botulinum Neurotoxin and Tetanus Toxin* (ed. L.L. Simpson), pp. 53–67. Academic Press, San Diego.

DasGupta, B.R. (1994). Structures of botulinum neurotoxin, its functional domains and perspectives on the crystalline type A toxin. In: *Therapy with Botulinum Toxin* (eds J. Jankovic and M. Hallett), pp. 15–39. Marcel Dekker, New York.

Davis, J.B., Mattman, L.H. and Wiley, M. (1951). *Clostridium botulinum* in a fatal wound infection. *JAMA* 146, 646–8.

Dayanithi, G., Stecher, B., Höhne-Zell, B., Yamasaki, S., Binz, T., Weller, U., Niemann, H. and Gratzl, M. (1994). Exploring the functional domain and the target of the tetanus toxin light chain in neurohypophysial terminals. *Neuroscience* 58, 423–31.

De Filippis, V., Vangelista, L., Schiavo, G., Tonello, F. and Montecucco, C. (1995). Structural studies on the zinc-endopeptidase light chain of tetanus neurotoxin. *Eur. J. Biochem.* 229, 61–9.

de Paiva, A., Poulain, B., Lawrence, G.W., Shone, C.C., Tauc, L. and Dolly, J.O. (1993a). A role for the interchain disulfide or its participating thiols in the internalization of botulinum neurotoxin A revealed by a toxin derivative that binds to ecto-acceptors and inhibits transmitter release intracellularly. *J. Biol. Chem.* 268, 20838–44.

de Paiva, A., Ashton, A.C., Foran, P., Schiavo, G., Montecucco, C. and Dolly, J.O. (1993b). Botulinum A like type B and tetanus toxins fulfils criteria for being a zinc-dependent protease. *J. Neurochem.* 61, 2338–41.

Deitcher, D.L., Ueda, A., Stewart, B.A., Burgess, R.W., Kidokoro, Y. and Schwarz, T.L. (1998). Distinct requirements for evoked and spontaneous release of neurotransmitter are revealed by mutations in the *Drosophila* gene neuronal-synaptobrevin. *J. Neurosci.* **18**, 2028–39.

Deshpande, S.S., Sheridan, R.E. and Adler, M. (1995). A study of zinc-dependent metalloendopeptidase inhibitors as pharmacological antagonists in botulinum neurotoxin poisoning. *Toxicon* **33**, 551–7.

Dobrenis, K., Joseph, A. and Rattazzi, M.C. (1992). Neuronal lysosomal enzyme replacement using fragment C of tetanus toxin. *Proc. Natl Acad. Sci. USA* **89**, 2297–301.

Dolly, J.O., Black, J., Williams, R.S. and Melling, J. (1984). Acceptors for botulinum neurotoxin reside on motor nerve terminals and mediate its internalization. *Nature* **307**, 457–60.

Donovan, J.J. and Middlebrook, J.L. (1986). Ion-conducting channels produced by botulinum toxin in planar lipid membranes. *Biochemistry* **25**, 2872–6.

Dreyer, F. and Schmitt, A. (1983). Transmitter release in tetanus and botulinum A toxin-poisoned mammalian motor endplates and its dependence on nerve stimulation and temperature. *Pflugers Arch.* **399**, 228–34.

Dreyer, F., Mallart, A. and Brigant, J.L. (1983). Botulinum A toxin and tetanus toxin do not affect presynaptic membrane currents in mammalian motor nerve endings. *Brain Res.* **270**, 373–5.

Duchen, L.W. (1973). The effects of tetanus toxin on the motor endplates of the mouse. An electron microscopic study. *J. Neurol. Sci.* **19**, 153–67.

Duchen, L.W. and Tonge, D.A. (1973). The effects of tetanus toxin on neuromuscular transmission and on the morphology of motor end-plates in slow and fast skeletal muscle of the mouse. *J. Physiol. (Lond.)* **228**, 157–72.

Edelmann, L., Hanson, P.I., Chapman, E.R. and Jahn, R. (1995). Synaptobrevin binding to synaptophysin: a potential mechanism for controlling the exocytotic fusion machine. *EMBO J.* **14**, 224–31.

Edmunds, C. and Long, P.H. (1923). Contribution to the pathologic physiology of botulism. *JAMA* **81**, 542–7.

Eisel, U., Jarausch, W., Goretzki, K., Henschen, A., Engels, J., Weller, U., Hudel, M., Habermann, E. and Niemann, H. (1986). Tetanus toxin: primary structure, expression in *E. coli* and homology with botulinum toxins. *EMBO J.* **5**, 2495–502.

Eisel, U., Reynolds, K., Riddick, M., Zimmer, A., Niemann, H. and Zimmer, A. (1993). Tetanus toxin light chain expression in Sertoli cells of transgenic mice causes alterations of the actin cytoskeleton and disrupts spermatogenesis. *EMBO J.* **12**, 3365–72.

Ekong, T.A., McLellan, K. and Sesardic, D. (1995). Immunological detection of *Clostridium botulinum* toxin type A in therapeutic preparations. *J. Immunol. Methods* **180**, 181–91.

Ekong, T.A., McLellan, K. and Sesardic, D. (1996). Recombinant SNAP-25 is an effective substrate for *Clostridium botulinum* type A toxin endopeptidase activity *in vitro*. *Microbiology* **143**, 3337–47.

Eleopra, R., Tugnoli, V., Rossetto, O., Montecucco, C. and De Grandis, D. (1997). Botulinum neurotoxin serotype C: a novel effective botulinum toxin therapy in human. *Neurosci. Lett.* **224**, 91–4.

Eleopra, R., Tugnoli, V., Rossetto, O., De Grandis, D. and Montecucco, C. (1998). Botulinum neurotoxin serotype A and E in human: evidence of a different temporal profile in the neuromuscular block induced. *Neurosci. Lett.* **224**, 91–4.

Erdmann, G., Wiegand, H. and Wellhoner, H.H. (1975). Intraaxonal and extraaxonal transport of ^{125}I-tetanus toxin in early local tetanus. *Naunyn Schemiedebergs Arch. Pharmacol.* **290**, 357–73.

Evans, D.M., Williams, R.S., Shone, C.C., Hambleton, P., Melling, J. and Dolly, J.O. (1986). Botulinum neurotoxin type B. Its purification, radioiodination and interaction with rat-brain synaptosomal membranes. *Eur. J. Biochem.* **154**, 409–16.

Evinger, C. and Erichsen, J.T. (1986). Transsynaptic retrograde transport of fragment C of tetanus toxin demonstrated by immunohistochemical localization. *Brain Res.* **380**, 383–8.

Faber, K. (1890). Die Pathogenie des Tetanus. *Berl. Klin. Wochenschr.* **27**, 717–20.

Facchiano, F. and Luini, A. (1992). Tetanus toxin potently stimulates tissue transglutaminase. A possible mechanism of neurotoxicity. *J. Biol. Chem.* **267**, 13267–71.

Facchiano, F., Benfenati, F., Valtorta, F. and Luini, A. (1993a). Covalent modification of synapsin I by a tetanus toxin-activated transglutaminase. *J. Biol. Chem.* **268**, 4588–91.

Facchiano, F., Valtorta, F., Benfenati, F. and Luini, A. (1993b). The transglutaminase hypothesis for the action of tetanus toxin. *Trends Biochem. Sci.* **18**, 327–9.

Fasshauer, D., Otto, H., Eliason, W.K., Jahn, R. and Brunger, A.T. (1997). Structural-changes are associated with soluble *N*-ethylmaleimide-sensitive fusion protein attachment protein-receptor complex-formation. *J. Biol. Chem.* **272**, 28036–41.

Fernandez, I., Ubach, J., Dulubova, I., Zhang, X.Y., Südhof, T.C. and Rizo, J. (1998). 3-Dimensional structure of an evolutionarily conserved *N*-terminal domain of syntaxin 1A. *Cell* **94**, 841–9.

Ferrer Montiel, A.V., Canaves, J.M., DasGupta, B.R., Wilson, M.C. and Montal, M. (1996). Tyrosine phosphorylation modulates the activity of clostridial neurotoxins. *J. Biol. Chem.* **271**, 18322–5.

Figueiredo, D.M., Hallewell, R.A., Chen, L.L., Fairweather, N.F., Dougan, G., Savitt, J.M., Parks, D.A. and Fishman, P.S. (1997). Delivery of recombinant tetanus-superoxide dismutase proteins to central-nervous-system neurons by retrograde axonal-transport. *Exp. Neurol.* **145**, 546–54.

Foran, P., Shone, C.C. and Dolly, J.O. (1994). Differences in the protease activities of tetanus and botulinum B toxins revealed by the cleavage of vesicle-associated membrane protein and various sized fragments. *Biochemistry* **33**, 15365–74.

Foran, P., Lawrence, G.W., Shone, C.C., Foster, K.A. and Dolly, J.O. (1996). Botulinum neurotoxin C1 cleaves both syntaxin and SNAP-25 in intact and permeabilized chromaffin cells: correlation with its blockade of catecholamine release. *Biochemistry* **35**, 2630–6.

Foster, J.A., Barnhorst, D., Papay, F., Oh, P.M. and Wulc, A.E. (1996). The use of botulinum A toxin to ameliorate facial kinetic frown lines. *Ophthalmology* **103**, 618–22.

Francis, J.W., Hosler, B.A., Brown, R.H., Jr and Fishman, P.S. (1995). CuZn superoxide dismutase (SOD-1):tetanus toxin fragment C hybrid protein for targeted delivery of SOD-1 to neuronal cells. *J. Biol. Chem.* **270**, 15434–42.

Fu, F.N., Lomneth, R.B., Cai, S.W. and Singh, B.R. (1998). Role of zinc in the structure and toxic activity of botulinum neurotoxin. *Biochemistry* **37**, 5267–78.

Gaisano, H.Y., Sheu, L., Foskett, J.K. and Trimble, W.S. (1994). Tetanus toxin light chain cleaves a vesicle-associated membrane protein (VAMP) isoform 2 in rat pancreatic zymogen granules and inhibits enzyme secretion. *J. Biol. Chem.* **269**, 17062–6.

Galazka, A. and Gasse, F. (1995). The present status of tetanus and tetanus vaccination. *Curr. Top. Microbiol. Immunol.* **195**, 31–53.

Galli, T., Chilcote, T., Mundigl, O., Binz, T., Niemann, H. and De Camilli, P. (1994). Tetanus toxin-mediated cleavage of cellubrevin impairs exocytosis of transferrin receptor-containing vesicles in CHO cells. *J. Cell Biol.* **125**, 1015–24.

Galli, T., McPherson, P.S. and De Camilli, P. (1996). The V_0 sector of the V-ATPase, synaptobrevin and synaptophysin are associated on synaptic vesicles in a Triton X-100-resistant, freeze–thawing sensitive, complex. *J. Biol. Chem.* **271**, 2193–8.

Gambale, F. and Montal, M. (1988). Characterization of the channel properties of tetanus toxin in planar lipid bilayers. *Biophys. J.* **53**, 771–83.

Gansel, M., Penner, R. and Dreyer, F. (1987). Distinct sites of action of clostridial neurotoxins revealed by double-poisoning of mouse motor nerve terminals. *Pflugers Arch.* **409**, 533–9.

Garcia, E.P., McPherson, P.S., Chilcote, T.J., Takei, K. and De Camilli, P. (1995). rbSec1A and B colocalize with syntaxin 1 and SNAP-25 throughout the axon, but are not in a stable complex with syntaxin. *J. Cell Biol.* **129**, 105–20.

Geddes, J.W., Hess, E.J., Hart, R.A., Kesslak, J.P., Cotman, C.W. and Wilson, M.C. (1990). Lesions of hippocampal circuitry define synaptosomal-associated protein-25 (SNAP-25) as a novel presynaptic marker. *Neuroscience* **38**, 515–25.

Gill, D.M. (1982). Bacterial toxins: a table of lethal amounts. *FEMS Microbiol. Rev.* **46**, 86–94.

Gobbi, M., Frittoli, E. and Mennini, T. (1996). Role of transglutaminase in [^{3}H]5-HT release from synaptosomes and in the inhibitory effect of tetanus toxin. *Neurochem. Int.* **29**, 129–34.

Götte, M. and von Mollard, G.F. (1998). A new beat for the SNARE drum. *Trends Cell Biol.* **8**, 215–18.

Grishin, E. V. (1998). Black-widow spider toxins – the present and the future. *Toxicon* **36**, 1693–701.

Gundersen, C.B. (1980). The effects of botulinum toxin on the synthesis, storage and release of acetylcholine. *Prog. Neurobiol.* **14**, 99–119.

Gundersen, C.B., Katz, B. and Miledi, R. (1982). The antagonism between botulinum toxin and calcium in motor nerve terminals. *Proc. R. Soc. Lond. B Biol. Sci.* **216**, 369–76.

Guo, Z.H., Turner, C. and Castle, D. (1998). Relocation of the t-SNARE SNAP-23 from lamellipodia-like cell-surface projections regulates compound exocytosis in mast-cells. *Cell* **94**, 537–48.

Habermann, E. and Dimpfel, W. (1973). Distribution of ^{125}I-tetanus toxin and ^{125}I-toxoid in rats with generalized tetanus, as influenced by antitoxin. *Naunyn Schemiedebergs Arch. Pharmacol.* **276**, 327–40.

Habermann, E. and Dreyer, F. (1986). Clostridial neurotoxins: handling and action at the cellular and molecular level. *Curr. Top. Microbiol. Immunol.* **129**, 93–179.

Habermann, E. and Erdmann, G. (1978). Pharmacokinetic and histoautoradiographic evidence for the intraaxonal movement of toxin in the pathogenesis of tetanus. *Toxicon* **16**, 611–23.

Habermann, E., Dreyer, F. and Bigalke, H. (1980). Tetanus toxin blocks the neuromuscular transmission *in vitro* like botulinum A toxin. *Naunyn Schemiedebergs Arch. Pharmacol.* **311**, 33–40.

Hackam, D.J., Rotstein, O.D., Sjolin, C., Schreiber, A.D., Trimble, W.S. and Grinstein, S. (1998). v-SNARE dependent secretion is required for phagocytosis. *Proc. Natl Acad. Sci. USA* **95**, 11691–6.

Hall, J.D., McCroskey, L.M., Pincomb, B.J. and Hatheway, C.L. (1985). Isolation of an organism resembling *Clostridium barati* which produces type F botulinal toxin from an infant with botulism. *J. Clin. Microbiol.* **21**, 654–5.

Hallis, B., James, B.A. and Shone, C.C. (1996). Development of novel assays for botulinum type A and B neurotoxins based on their endopeptidase activities. *J. Clin. Microbiol.* **34**, 1934–8.

Halpern, J.L. and Loftus, A.(1993). Characterization of the receptor-binding domain of tetanus toxin. *J. Biol. Chem.* **268**, 11188–92.

Halpern, J.L. and Neale, E.A. (1995). Neurospecific binding, internalization and retrograde axonal transport. *Curr. Top. Microbiol. Immunol.* **195**, 221–41.

Harris, A.J. and Miledi, R. (1971). The effect of type D botulinum toxin on frog neuromuscular junctions. *J. Physiol. (Lond.)* **217**, 497–515.

Hase, C.C. and Finkelstein, R.A. (1993). Bacterial extracellular zinc-containing metalloproteases. *FEMS Microbiol. Rev.* **57**, 823–37.

Hassan, S.M., Jennekens, F.G., Wieneke, G. and Veldman, H. (1994). Calcitonin gene-related peptide-like immunoreactivity, in botulinum toxin-paralysed rat muscles. *Neuromusc. Disord.* **4**, 489–96.

Hatheway, C.L. (1995). Botulism: the present status of the disease. *Curr. Top. Microbiol. Immunol.* **195**, 55–75.

Hayashi, T., McMahon, H., Yamasaki, S., Binz, T., Hata, Y., Südhof, T.C. and Niemann, H. (1994). Synaptic vesicle membrane fusion complex: action of clostridial neurotoxins on assembly. *EMBO J.* **13**, 5051–61.

Hayashi, T., Yamasaki, S., Nauenburg, S., Binz, T. and Niemann, H. (1995). Disassembly of the reconstituted synaptic vesicle membrane fusion complex *in vitro*. *EMBO J.* **14**, 2317–25.

Helting, T.B. and Zwisler, O. (1977). Structure of tetanus toxin. I. Breakdown of the toxin molecule and discrimination between polypeptide fragments. *J. Biol. Chem.* **252**, 187–93.

Helting, T.B., Zwisler, O. and Wiegandt, H. (1977). Structure of tetanus toxin. II. Toxin binding to ganglioside. *J. Biol. Chem.* **252**, 194–8.

Herreros, J., Miralles, F.X., Solsona, C., Bizzini, B., Blasi, J. and Marsal, J. (1995). Tetanus toxin inhibits spontaneous quantal release and cleaves VAMP/synaptobrevin. *Brain Res.* **699**, 165–70.

Hess, D.T., Slater, T.M., Wilson, M.C. and Skene, J.H. (1992). The 25 kDa synaptosomal-associated protein SNAP-25 is the major methionine-rich polypeptide in rapid axonal transport and a major substrate for palmitoylation in adult CNS. *J. Neurosci.* **12**, 4634–41.

Hicks, A., Davis, S., Rodger, J., Helme-Guizon, A., Laroche, S. and Mallet, J. (1997). Synapsin I and syntaxin 1B: key elements in the control of neurotransmitter release are regulated by neuronal activation and long-term potentiation *in vivo*. *Neuroscience* **79**, 329–40.

Hoch, D.H., Romero-Mira, M., Ehrlich, B.E., Finkelstein, A., DasGupta, B.R. and Simpson, L.L. (1985). Channels formed by botulinum, tetanus and diphtheria toxins in planar lipid bilayers: relevance to translocation of proteins across membranes. *Proc. Natl Acad. Sci. USA* **82**, 1692–6.

Höhne-Zell, B., Stecher, B. and Gratzl, M. (1993). Functional characterization of the catalytic site of the tetanus toxin light chain using permeabilized adrenal chromaffin cells. *FEBS Lett.* **336**, 175–80.

Höhne-Zell, B., Ecker, A., Weller, U. and Gratzl, M. (1994). Synaptobrevin cleavage by the tetanus toxin light chain is linked to the inhibition of exocytosis in chromaffin cells. *FEBS Lett.* **355**, 131–4.

Huang, X.H., Wheeler, M.B., Kang, Y.H., Sheu, L., Lukacs, G.L., Trimble, W.S. and Gaisano, H.Y. (1998). Truncated SNAP-25 (1–197), like botulinum neurotoxin A, can inhibit insulin secretion from HIT-T15 insulinoma cells. *Mol. Endocrinol.* **12**, 1060–70.

Hughes, R. and Whaler, B. C. (1962). Influence of nerve-endings activity and of drugs on the rate of paralysis of rat diaphragm preparations by *Clostridium botulinum* type A toxin. *J. Physiol. (Lond.)* **160**, 221–33.

Hunt, J.M., Bommert, K., Charlton, M.P., Kistner, A., Habermann, E., Augustine, G.J. and Betz, H. (1994). A post-docking role for synaptobrevin in synaptic vesicle fusion. *Neuron* **12**, 1269–79.

Inoue, A. and Akagawa, K. (1992). Neuron-specific antigen HPC-1 from bovine brain reveals strong homology to epimorphin, an essential factor involved in epithelial morphogenesis: identification of a novel protein family. *Biochem. Biophys. Res. Commun.* **187**, 1144–50.

Inoue, K., Fujinaga, Y., Watanabe, T., Ohyama, T., Takeshi, K., Moriishi, K., Nakajima, H., Inoue, K. and Oguma, K. (1996). Molecular composition of *Clostridium botulinum* type A progenitor toxins. *Infect. Immun.* **64**, 1589–94.

Inoue, T., Mandon, B., Nielsen, S. and Knepper, M.A. (1997). Expression of snap23, the missing snare, in rat collecting duct principal cells. *J. Am. Soc. Nephrol.* **8**, A0296.

Jacobsson, G., Bean, A.J., Scheller, R.H., Juntti-Berggren, L., Deeney, J.T., Berggren, P.O. and Meister, B. (1994). Identification of synaptic proteins and their isoform mRNAs in compartments of pancreatic endocrine cells. *Proc. Natl Acad. Sci. USA* **91**, 12487–91.

Jankovic, J. and Hallett, M. (eds) (1994). *Therapy with Botulinum Toxin.* Marcel Dekker, New York.

Jefferys, J.G. and Whittington, M.A. (1996). Review of the role of inhibitory neurons in chronic epileptic foci induced by intracerebral tetanus toxin. *Epilepsy Res.* **26**, 59–66.

Jiang, W. and Bond, J.S. (1992). Families of metalloendopeptidases and their relationships. *FEBS Lett.* **312**, 110–4.

Jongeneel, C.V., Bouvier, J. and Bairoch, A. (1989). A unique signature identifies a family of zinc-dependent metallopeptidases. *FEBS Lett.* **242**, 211–14.

Kee, Y., Lin, R.C., Hsu, S.C. and Scheller, R.H. (1995). Distinct domains of syntaxin are required for synaptic vesicle fusion complex formation and dissociation. *Neuron* **14**, 991–8.

Kemplay, S. and Cavanagh, J.B. (1983). Effects of acrylamide and botulinum toxin on horseradish peroxidase labelling of trigeminal motor neurons in the rat. *J. Anat.* **137**, 477–82.

Kerner, J. (1817). Medizinische Polizen. Vergiftung durch verborbene Würste. *Tübinger Blatter* **3**, 1–25.

Kitasato, S. (1889). Über den Tetanusbazillus. *Z. Hyg. Infektkr.* **7**, 225–30.

Kozaki, S., Miki, A., Kamata, Y., Ogasawara, J. and Sakaguchi, G. (1989). Immunological characterization of papain-induced fragments of *Clostridium botulinum* type A neurotoxin and interaction of the fragments with brain synaptosomes. *Infect. Immun.* **57**, 2634–9.

Kozaki, S., Kamata, Y., Watarai, S., Nishiki, T. and Mochida, S. (1998). Ganglioside GT1b as a complementary receptor component for *Clostridium botulinum* neurotoxins. *Microb. Pathog.* **25**, 91–9.

Kraszewski, K., Mundigl, O., Daniell, L., Verderio, C., Matteoli, M. and De Camilli, P. (1995). Synaptic vesicle dynamics in living cultured hippocampal neurons visualized with CY3-conjugated antibodies directed against the lumenal domain of synaptotagmin. *J. Neurosci.* **15**, 4328–42.

Krieglstein, K.G., Henschen, A.H., Weller, U. and Habermann, E. (1991). Limited proteolysis of tetanus toxin. Relation to activity and identification of cleavage sites. *Eur. J. Biochem.* **202**, 41–51.

Kristensson, K. and Olsson, T. (1978). Uptake and retrograde axonal transport of horseradish peroxidase in botulinum-intoxicated mice. *Brain Res.* **155**, 118–23.

Kryzhanovsky, G.N. (1958). Central nervous changes in experimental tetanus and the mode of action of the tetanus toxin. Communication I. Irradiation of the excitation on stimulating the tetanized limb. *Bull. Exp. Biol. Med.* **44**, 1456–64.

Kryzhanovsky, G.N., Pozdynakov, O.M., D'yakonova, M.V., Polgar, A.A. and Smirnova, V.S. (1971). Disturbance of neurosecretion in myoneural junctions of muscle poisoned with tetanus toxin. *Bull. Exp. Biol. Med.* **72**, 1387–91.

Kurazono, H., Mochida, S., Binz, T., Eisel, U., Quanz, M., Grebenstein, O., Wernars, K., Poulain, B., Tauc, L. and Niemann, H. (1992). Minimal essential domains specifying toxicity of the light chains of tetanus toxin and botulinum neurotoxin type A. *J. Biol. Chem.* **267**, 14721–9.

Kurokawa, Y., Oguma, K., Yokosawa, N., Syuto, B., Fukatsu, R. and Yamashita, I. (1987). Binding and cytotoxic effects of *Clostridium botulinum* type A, C1 and E toxins in primary neuron cultures from foetal mouse brains. *J. Gen. Microbiol.* **133**, 2647–57.

Lacy, D.B., Tepp, W., Cohen, A.C., DasGupta, B.R. and Stevens, R.C. (1998). Crystal structure of botulinum neurotoxin type A and implications for toxicity. *Nature Struct. Biol.* **5**, 898–902.

Lang, J.C., Zhang, H., Vaidyanathan, V.V.A., Sadoul, K., Niemann, H. and Wollheim, C.B. (1997a). Transient expression of botulinum neurotoxin C light-chain differentially inhibits calcium and glucose-induced insulin-secretion in clonal beta-cells. *FEBS Lett.* **419**, 13–17.

Lang, J., Regazzi, R. and Wollheim, C.B. (1997b). Clostridial toxins and endocrine secretion: their use in insuline secreting cells. In: *Bacterial Toxins. Tools in Cell Biology and Pharmacology* (ed. K. Aktories), pp. 217–37. Chapman & Hall, London.

Lebeda, F.J. and Olson, M.A. (1994). Secondary structure prediction for the clostridial neurotoxins. *Proteins: Struct. Funct. Genet.* **20**, 293–300.

Leist, M., Fava, E., Montecucco, C. and Nicotera, P. (1997). Peroxynitrite and nitric oxide donors induce neuronal apoptosis by eliciting autocrine excitotoxicity. *Eur. J. Neurosci.* **9**, 1488–98.

Li, L. and Singh, B.R. (1989). Isolation of synaptotagmin as a receptor for types A and *E. botulinum* neurotoxin and analysis of their comparative binding using a new microtiter plate assay. *J. Nat. Toxins* **7**, 215–26.

Li, Y., Foran, P., Fairweather, N.F., de Paiva, A., Weller, U., Dougan, G. and Dolly, J.O. (1994). A single mutation in the recombinant light chain of tetanus toxin abolishes its proteolytic activity and removes the toxicity seen after reconstitution with native heavy chain. *Biochemistry* **33**, 7014–20.

Lin, R.C. and Scheller, R.H. (1997). Structural organization of the synaptic exocytosis core complex. *Neuron* **19**, 1087–94.

Link, E., Edelmann, L., Chou, J.H., Binz, T., Yamasaki, S., Eisel, U., Baumert, M., Sudhof, T.C., Niemann, H. and Jahn, R. (1992). Tetanus toxin action: inhibition of neurotransmitter release linked to synaptobrevin proteolysis. *Biochem. Biophys. Res. Commun.* **189**, 1017–23.

Lovejoy, B., Cleasby, A., Hassell, A.M., Longley, K., Luther, M.A., Weigl, D., McGeehan, G., McElroy, A.B., Drewry, D., Lambert, M.H. and Jordan, S.R. (1994). Structure of the catalytic domain of fibroblast collagenase complexed with an inhibitor. *Science* **263**, 2375–7.

Lundh, H., Leander, S. and Thesleff, S. (1977). Antagonism of the paralysis produced by botulinum toxin in the rat. *J. Neurol. Sci.* **32**, 29–43.

Major, R.H. (1945). *Classic Descriptions of Disease*, 3rd edn. Charles C. Thomas, Springfield, IL.

Maksymowych, A.B. and Simpson, L.L. (1998). Binding and transcytosis of botulinum neurotoxin by polarized human colon-carcinoma cells. *J. Biol. Chem.* **273**, 21950–7.

Mallart, A., Molgo, J., Angaut-Petit, D. and Thesleff, S. (1989). Is the internal calcium regulation altered in type A botulinum toxin-poisoned motor endings? *Brain Res.* **479**, 167–71.

Martin, L., Cornille, F., Coric, P., Roques, B.P. and Fourniezaluski, M.C. (1998). Beta-amino thiols inhibit the zinc metallopeptidase activity of tetanus toxin light chain. *J. Med. Chem.* **41**, 3450–60.

Marxen, P., Fuhrmann, U. and Bigalke, H. (1989). Gangliosides mediate inhibitory effects of tetanus and botulinum A neurotoxins on exocytosis in chromaffin cells. *Toxicon* **27**, 849–59.

Matsuda, M. and Yoneda, M. (1977). Antigenic substructure of tetanus neurotoxin. *Biochem. Biophys. Res. Commun.* **77**, 268–74.

Matsuda, M., Sugimoto, N., Ozutsumi, K. and Hirai, T. (1982). Acute botulinum-like intoxication by tetanus neurotoxin in mice. *Biochem. Biophys. Res. Commun.* **104**, 799–805.

Matteoli, M., Takei, K., Perin, M.S., Südhof, T.C. and De Camilli, P. (1992). Exo-endocytotic recycling of synaptic vesicles in developing processes of cultured hippocampal neurons. *J. Cell Biol.* **117**, 849–61.

Matteoli, M., Verderio, C., Rossetto, O., Iezzi, N., Coco, S., Schiavo, G. and Montecucco, C. (1996). Synaptic vesicle endocytosis mediates the entry of tetanus neurotoxin into hippocampal neurons. *Proc. Natl Acad. Sci. USA* **93**, 13310–15.

McCroskey, L.M., Hatheway, C.L., Fenicia, L., Pasolini, B. and Aureli, P. (1986). Characterization of an organism that produces type E botulinal toxin but which resembles *Clostridium butyricum* from the feces of an infant with type E botulism. *J. Clin. Microbiol.* **23**, 201–2.

McMahon, H.T., Ushkaryov, Y.A., Edelmann, L., Link, E., Binz, T., Niemann, H., Jahn, R. and Sudhof, T.C. (1993). Cellubrevin is a ubiquitous tetanus-toxin substrate homologous to a putative synaptic vesicle fusion protein. *Nature* **364**, 346–9.

Mellanby, J. (1984). Comparative activities of tetanus and botulinum toxins. *Neuroscience* **11**, 29–34.

Mellanby, J. and Green, J. (1981). How does tetanus toxin act? *Neuroscience* **6**, 281–300.

Mellanby, J. and Thompson, P.A. (1977). Tetanus toxin in the rat hippocampus. *J. Physiol. (Lond.)* **269**, 44–5P.

Mellanby, J., George, G., Robinson, A. and Thompson, P. (1977). Epileptiform syndrome in rats produced by injecting tetanus toxin into the hippocampus. *J. Neurol. Neurosurg. Psychiatry* **40**, 404–14.

Mellanby, J., Beaumont, M.A. and Thompson, P.A. (1988). The effect of lanthanum on nerve terminals in goldfish muscle after paralysis with tetanus toxin. *Neuroscience* **25**, 1095–106.

Menestrina, G., Forti, S. and Gambale, F. (1989). Interaction of tetanus toxin with lipid vesicles. Effects of pH, surface charge and transmembrane potential on the kinetics of channel formation. *Biophys. J.* **55**, 393–405.

Menestrina, G., Schiavo, G. and Montecucco, C. (1994). Molecular mechanisms of action of bacterial protein toxins. *Mol. Aspects Med.* **15**, 79–193.

Meyer, H. and Ransom, F. (1903). Untersuchungen über den Tetanus. *Arch. Exp. Pathol. Pharmakol.* **49**, 369–416.

Middlebrook, J.L. and Brown, J.E. (1995). Immunodiagnosis and immunotherapy of tetanus and botulinum neurotoxins. *Curr. Top. Microbiol. Immunol.* **195**, 89–122.

Midura, T.F. and Arnon, S.S. (1976). Infant botulism. Identification of *Clostridium botulinum* and its toxins in faeces. *Lancet* **ii**, 934–6.

Minton, N.P. (1995). Molecular genetics of clostridial neurotoxins. *Curr. Top. Microbiol. Immunol.* **195**, 161–94.

Mochida, S., Poulain, B., Weller, U., Habermann, E. and Tauc, L. (1989). Light chain of tetanus toxin intracellularly inhibits acetylcholine release at neuro-neuronal synapses and its internalization is mediated by heavy chain. *FEBS Lett.* **253**, 47–51.

Mochida, S., Poulain, B., Eisel, U., Binz, T., Kurazono, H., Niemann, H. and Tauc, L. (1990). Molecular biology of Clostridial toxins: expression of mRNAs encoding tetanus and botulinum neurotoxins in *Aplysia* neurons. *J. Physiol. Paris* **84**, 278–84.

Molgo, J., Siegel, L.S., Tabti, N. and Thesleff, S. (1989). A study of synchronization of quantal transmitter release from mammalian motor endings by the use of botulinal toxins type A and D. *J. Physiol. (Lond.)* **411**, 195–205.

Molgo, J., Comella, J.X., Angaut-Petit, D., Pecot-Dechavassine, M., Tabti, N., Faille, L., Mallart, A. and Thesleff, S. (1990). Presynaptic actions of botulinal neurotoxins at vertebrate neuromuscular junctions. *J. Physiol. Paris* **84**, 152–66.

Mollinedo, F. and Lazo, P.A. (1997). Identification of two isoforms of the vesicle-membrane fusion protein SNAP-23 in human neutrophils and HL-60 cells. *Biochem. Biophys. Res. Commun.* **231**, 808–12.

Montal, M.S., Blewitt, R., Tomich, J.M. and Montal, M. (1992). Identification of an ion channel-forming motif in the primary structure of tetanus and botulinum neurotoxins. *FEBS Lett.* **313**, 12–18.

Montecucco, C. (1986). How do tetanus and botulinum toxins bind to neuronal membranes? *Trends Biochem. Sci.* **11**, 315–17.

Montecucco, C. and Schiavo, G. (1995). Structure and function of tetanus and botulinum neurotoxins. *Q. Rev. Biophys.* **28**, 423–72.

Montecucco, C., Schiavo, G., Brunner, J., Duflot, E., Boquet, P. and Roa, M. (1986). Tetanus toxin is labeled with photoactivatable phospholipids at low pH. *Biochemistry* **25**, 919–24.

Montecucco, C., Schiavo, G. and Dasgupta, B.R. (1989). Effect of pH on the interaction of botulinum neurotoxins A, B and E with liposomes. *Biochem. J.* **259**, 47–53.

Montecucco, C., Papini, E. and Schiavo, G. (1994). Bacterial protein toxins penetrate cells via a four-step mechanism. *FEBS Lett.* **346**, 92–8.

Montecucco, C., Schiavo, G., Tugnoli, V. and de Grandis, D. (1996). Botulinum neurotoxins: mechanism of action and therapeutic applications. *Mol. Med. Today* **2**, 418–24.

Montesano, R., Roth, J., Robert, A. and Orci, L. (1982). Non-coated membrane invaginations are involved in binding and internalization of cholera and tetanus toxin. *Nature* **296**, 651–3.

Morante, S., Furenlid, L., Schiavo, G., Tonello, F., Zwilling, R. and Montecucco, C. (1996). X-ray absorption spectroscopy study of zinc coordination in tetanus neurotoxin, astacin, alkaline protease and thermolysin. *Eur. J. Biochem.* **235**, 606–12.

Morris, N.P., Consiglio, E., Kohn, L.D., Habig, W.H., Hardegree, M.C. and Helting, T.B. (1980). Interaction of fragment B and C of tetanus toxin with neural and thyroid membranes and with gangliosides. *J. Biol. Chem.* **255**, 6071–6.

Mundigl, O., Verderio, C., Krazewski, K., De Camilli, P. and Matteoli, M. (1995). A radioimmunoassay to monitor synaptic activity in hippocampal neurons *in vitro*. *Eur. J. Cell Biol.* **66**, 246–56.

Neale, E.A., Habig, W.H., Schrier, B.K., Bergey, G.K., Bowers, L.M. and Koh, J. (1989). Application of tetanus toxin for structure–function studies in neuronal cell cultures. In: *Eighth International Conference on Tetanus* (eds G. Nistico', B. Bizzini, B. Bytchencko and R. Triau), pp. 66–70. Pythagora Press, Rome.

Neubauer, V. and Helting, T.B. (1981). Structure of tetanus toxin: the arrangement of papain digestion products within the heavy chain-light chain framework of extracellular toxin. *Biochim. Biophys. Acta* **668**, 141–8.

Niemann, H. (1991). Molecular biology of clostridial neurotoxins. In: *A Sourcebook of Bacterial Protein Toxins* (eds J.E. Alouf and J.H. Freer), pp. 303–48. Academic Press, London.

Nishiki, T., Kamata, Y., Nemoto, Y., Omori, A., Ito, T., Takahashi, M. and Kozaki, S. (1994). Identification of protein receptor for *Clostridium botulinum* type B neurotoxin in rat brain synaptosomes. *J. Biol. Chem.* **269**, 10498–503.

Nishiki, T., Tokuyama, Y., Kamata, Y., Nemoto, Y., Yoshida, A., Sato, K., Sekiguchi, M., Takahashi, M. and Kozaki, S. (1996a). The high-affinity binding of *Clostridium botulinum* type B neurotoxin to synaptotagmin II associated with gangliosides GT1b/GD1a. *FEBS Lett.* **378**, 253–7.

Nishiki, T., Tokuyama, Y., Kamata, Y., Nemoto, Y., Yoshida, A., Sekiguchi, M., Takahashi, M. and Kozaki, S. (1996b). Binding of botulinum type B neurotoxin to Chinese hamster ovary cells transfected with rat synaptotagmin II cDNA. *Neurosci. Lett.* **208**, 105–8.

O'Connor, V., Heuss, C., Debello, W.M., Dresbach, T., Charlton, M.P., Hunt, J.H., Pellegrini, L.L., Hodel, A., Burger, M.M., Betz, H., Augustine, G.J. and Schafer, T. (1997). Disruption of syntaxin-mediated protein interactions blocks neurotransmitter secretion. *Proc. Natl Acad. Sci. USA* **94**, 12186–91.

Oblatt-Montal, M., Yamazaki, M., Nelson, R. and Montal, M. (1995). Formation of ion channels in lipid bilayers by a peptide with the predicted transmembrane sequence of botulinum neurotoxin A. *Protein Sci.* **4**, 1490–7.

Osen Sand, A., Catsicas, M., Staple, J.K., Jones, K.A., Ayala, G., Knowles, J., Grenningloh, G. and Catsicas, S. (1993). Inhibition of axonal growth by SNAP-25 antisense oligonucleotides *in vitro* and *in vivo*. *Nature* **364**, 445–8.

Osen Sand, A., Staple, J.K., Naldi, E., Schiavo, G., Rossetto, O., Petitpierre, S., Malgaroli, A., Montecucco, C. and Catsicas, S. (1996). Common and distinct fusion proteins in axonal growth and transmitter release. *J. Comp. Neurol.* **367**, 222–34.

Otto, H., Hanson, P.I., Chapman, E.R., Blasi, J. and Jahn, R. (1995). Poisoning by botulinum neurotoxin A does not inhibit formation or disassembly of the synaptosomal fusion complex. *Biochem. Biophys. Res. Commun.* **212**, 945–52.

Oyler, G.A., Higgins, G.A., Hart, R.A., Battenberg, E., Billingsley, M., Bloom, F.E. and Wilson, M.C. (1989). The identification of a novel synaptosomal-associated protein, SNAP-25, differentially expressed by neuronal subpopulations. *J. Cell Biol.* **109**, 3039–52.

Papini, E., Rossetto, O. and Cutler, D.F. (1995). Vesicle-associated membrane protein (VAMP)/synaptobrevin-2 is associated with dense core secretory granules in PC12 neuroendocrine cells. *J. Biol. Chem.* **270**, 1332–6.

Parton, R.G., Ockleford, C.D. and Critchley, D.R. (1987). A study of the mechanism of internalisation of tetanus toxin by primary mouse spinal cord cultures. *J. Neurochem.* **49**, 1057–68.

Parton, R.G., Ockleford, C.D. and Critchley, D.R. (1988). Tetanus toxin binding to mouse spinal cord cells: an evaluation of the role of gangliosides in toxin internalization. *Brain Res.* **475**, 118–27.

Payling-Wright, G. (1955). The neurotoxins of *Clostridium botulinum* and *Clostridium tetani*. *Pharmacol. Rev.* **7**, 413–65.

Pecot-Dechavassine, M., Molgo, J. and Thesleff, S. (1991). Ultrastructure of botulinum type-A poisoned frog motor nerve terminals after enhanced quantal transmitter release caused by carbonyl cyanide m-chlorophenylhydrazone. *Neurosci. Lett.* **130**, 5–8.

Pellegrini, L.L., O'Connor, V. and Betz, H. (1994). Fusion complex formation protects synaptobrevin against proteolysis by tetanus toxin light chain. *FEBS Lett.* **353**, 319–23.

Pellizzari, R., Rossetto, O., Lozzi, L., Giovedi, S., Johnson, E., Shone, C.C. and Montecucco, C. (1996). Structural determinants of the specificity for synaptic vesicle-associated membrane protein/synaptobrevin of tetanus and botulinum type B and G neurotoxins. *J. Biol. Chem.* **271**, 20353–8.

Pellizzari, R., Mason, S., Shone, C.C. and Montecucco, C. (1997). The interaction of synaptic vesicle-associated membrane protein/synaptobrevin with botulinum neurotoxins D and F. *FEBS Lett.* **409**, 339–42.

Penner, R., Neher, E. and Dreyer, F. (1986). Intracellularly injected tetanus toxin inhibits exocytosis in bovine adrenal chromaffin cells. *Nature* **324**, 76–8.

Pickett, J., Berg, B., Chaplin, E. and Brunstetter-Shafer, M.A. (1976). Syndrome of botulism in infancy: clinical and electrophysiologic study. *N. Engl. J. Med.* **295**, 770–2.

Pierce, E.J., Davison, M.D., Parton, R.G., Habig, W.H. and Critchley, D.R. (1986). Characterization of tetanus toxin binding to rat brain membranes. Evidence for a high-affinity proteinase-sensitive receptor. *Biochem. J.* **236**, 845–52.

Ponomarev, A.W. (1928). Zur Frage der Pathogenese des Tetanus und des Fortbewegungsmechanismus des Tetanustoxin entlang der Nerven. *Z. Ges. Exp. Med.* **61**, 93–106.

Popoff, M.R. (1995). Ecology of neurotoxigenic strains of clostridia. *Curr. Top. Microbiol. Immunol.* **195**, 1–29.

Popoff, M.R. and Marvaud, J.-C. (1999). Structural and genomic features of clostridial neurotoxins. In: *Bacterial Toxins: a Comprehensive Sourcebook* (eds J. Alouf and J. Freer), pp. 174–201. Academic Press, London.

Poulain, B., Tauc, L., Maisey, E.A., Wadsworth, J.D., Mohan, P.M. and Dolly, J.O. (1988). Neurotransmitter release is blocked intracellularly by botulinum neurotoxin and this requires uptake of both toxin polypeptides by a process mediated by the larger chain. *Proc. Natl Acad. Sci. USA* **85**, 4090–4.

Poulain, B., Rossetto, O., Deloye, F., Schiavo, G., Tauc, L. and Montecucco, C. (1993). Antibodies against rat brain vesicle-associated membrane protein (synaptobrevin) prevent inhibition of acetylcholine release by tetanus toxin or botulinum neurotoxin type B. *J. Neurochem.* **61**, 1175–8.

Poulain, B., Molgo, J. and Thesleff, S. (1995). Quantal neurotransmitter release and the clostridial neurotoxins' targets. *Curr. Top. Microbiol. Immunol.* **195**, 243–55.

Pozdynakov, O.M., Polgar, A.A., Smirnova, V.S. and Kryzhanovskii, G.N. (1972). Changes in the ultrastructure of the neuromuscular synapse produced by tetanus toxin. *Bull. Exp. Biol. Med.* **74**, 852–5.

Price, D.L., Griffin, J., Young, A., Peck, K. and Stocks, A. (1975). Tetanus toxin: direct evidence for retrograde intraaxonal transport. *Science* **188**, 945–7.

Pumplin, D.W. and Reese, T.S. (1977). Action of brown widow spider venom and botulinum toxin on the frog neuromuscular junction examined with the freeze-fracture technique. *J. Physiol. (Lond.)* **273**, 443–57.

Ramon, G. and Descombey, P.A. (1925). Sur l'immunization antitetanique et sur la production de l'antitoxine tetanique. *C. R. Soc. Biol. Fil.* **93**, 508–98.

Ravichandran, V., Chawla, A. and Roche, P.A. (1996). Identification of a novel syntaxin- and synaptobrevin/VAMP-binding protein, SNAP-23, expressed in non-neuronal tissues. *J. Biol. Chem.* **271**, 13300–3.

Ray, P., Berman, J.D., Middleton, W. and Brendle, J. (1993). Botulinum toxin inhibits arachidonic acid release associated with acetylcholine release from PC12 cells. *J. Biol. Chem.* **268**, 11057–64.

Regazzi, R., Sadoul, K., Meda, P., Kelly, R.B., Halban, P.A. and Wollheim, C.B. (1996). Mutational analysis of VAMP domains implicated in Ca^{2+}-induced insulin exocytosis. *EMBO J.* **15**, 6951–9.

Rodger, J., Davis, S., Laroche, S., Mallet, J. and Hicks, A. (1998). Induction of long-term potentiation *in vivo* regulates alternate splicing to alter syntaxin 3 isoform expression in rat dentate gyrus. *J. Neurochem.* **71**, 666–75.

Rosenthal, L. and Meldolesi, J. (1989). Alpha-latrotoxin and related toxins. *Pharmacol. Ther.* **42**, 115–34.

Rossetto, O., Schiavo, G., Montecucco, C., Poulain, B., Deloye, F., Lozzi, L. and Shone, C.C. (1994). SNARE motif and neurotoxins. *Nature* **372**, 415–6.

Rossetto, O., Gorza, L., Schiavo, G., Schiavo, N., Scheller, R.H. and Montecucco, C. (1996). VAMP/synaptobrevin isoforms 1 and 2 are widely and differentially expressed in nonneuronal tissues. *J. Cell Biol.* **132**, 167–79.

Sadoul, K., Berger, A., Niemann, H., Weller, U., Roche, P.A., Klip, A., Trimble, W.S., Regazzi, R., Catsicas, S. and Halban, P.A. (1997). SNAP-23 is not cleaved by botulinum neurotoxin-E and can replace SNAP-25 in the process of insulin-secretion. *J. Biol. Chem.* **272**, 33023–7.

Sakaguchi, G. (1983). *Clostridium botulinum* toxins. *Pharmacol. Ther.* **19**, 165–94.

Sala, C., Andreose, J.S., Fumagalli, G. and Lomo, T. (1995). Calcitonin gene-related peptide: possible role in formation and maintenance of neuromuscular junctions. *J. Neurosci.* **15**, 520–8.

Schiavo, G. and Montecucco, C. (1995). Tetanus and botulism neurotoxins: isolation and assay. *Methods Enzymol.* **248**, 643–52.

Schiavo, G., Papini, E., Genna, G. and Montecucco, C. (1990a). An intact interchain disulfide bond is required for the neurotoxicity of tetanus toxin. *Infect. Immun* **58**, 4136–41.

Schiavo, G., Boquet, P., Dasgupta, B.R. and Montecucco, C. (1990b). Membrane interactions of tetanus and botulinum neurotoxins: a photolabelling study with photoactivatable phospholipids. *J. Physiol. Paris* **84**, 180–7.

Schiavo, G., Ferrari, G., Rossetto, O. and Montecucco, C. (1991). Specific cross-linking of tetanus toxin to a protein of NGF-differentiated PC12 cells. *FEBS Lett.* **290**, 227–30.

Schiavo, G., Poulain, B., Rossetto, O., Benfenati, F., Tauc, L. and Montecucco, C. (1992a). Tetanus toxin is a zinc protein and its inhibition of neurotransmitter release and protease activity depend on zinc. *EMBO J.* **11**, 3577–83.

Schiavo, G., Rossetto, O., Santucci, A., DasGupta, B.R. and Montecucco, C. (1992b). Botulinum neurotoxins are zinc proteins. *J. Biol. Chem.* **267**, 23479–83.

Schiavo, G., Benfenati, F., Poulain, B., Rossetto, O., Polverino de Laureto, P., DasGupta, B.R. and Montecucco, C. (1992c). Tetanus and botulinum-B neurotoxins block neurotransmitter release by proteolytic cleavage of synaptobrevin. *Nature* **359**, 832–5.

Schiavo, G., Shone, C.C., Rossetto, O., Alexander, F.C. and Montecucco, C. (1993a). Botulinum neurotoxin serotype F is a zinc endopeptidase specific for VAMP/synaptobrevin. *J. Biol. Chem.* **268**, 11516–9.

Schiavo, G., Rossetto, O., Catsicas, S., Polverino de Laureto, P., DasGupta, B.R., Benfenati, F. and Montecucco, C. (1993b). Identification of the nerve terminal targets of botulinum neurotoxin serotypes A, D and E. *J. Biol. Chem.* **268**, 23784–7.

Schiavo, G., Santucci, A., Dasgupta, B.R., Mehta, P.P., Jontes, J., Benfenati, F., Wilson, M.C. and Montecucco, C. (1993c). Botulinum neurotoxins serotypes A and E cleave SNAP-25 at distinct COOH-terminal peptide bonds. *FEBS Lett.* **335**, 99–103.

Schiavo, G., Malizio, C., Trimble, W.S., Polverino de Laureto, P., Milan, G., Sugiyama, H., Johnson, E.A. and Montecucco, C. (1994). Botulinum G neurotoxin cleaves VAMP/synaptobrevin at a single Ala-Ala peptide bond. *J. Biol. Chem.* **269**, 20213–6.

Schiavo, G., Shone, C.C., Bennett, M.K., Scheller, R.H. and Montecucco, C. (1995). Botulinum neurotoxin type C cleaves a single Lys-Ala bond within the carboxyl-terminal region of syntaxins. *J. Biol. Chem.* **270**, 10566–70.

Schiavo, G., Stenbeck, G., Rothman, J.E. and Sollner, T.H. (1997). Binding of the synaptic vesicle v-SNARE, synaptotagmin, to the plasma-membrane t-SNARE, SNAP-25, can explain docked vesicles at neurotoxin-treated synapses. *Proc. Natl Acad. Sci. USA* **94**, 997–1001.

Schiavo, G., Osborne, S.L. and Sgouros, J.G. (1998). Synaptotagmins – more isoforms than functions. *Biochem. Biophys. Res. Commun.* **248**, 1–8.

Schmid, M.F., Robinson, J.P. and DasGupta, B.R. (1993). Direct visualization of botulinum neurotoxin-induced channels in phospholipid vesicles. *Nature* **364**, 827–30.

Schmitt, A., Dreyer, F. and John, C. (1981). At least three sequential steps are involved in the tetanus toxin-induced block of neuromuscular transmission. *Naunyn Schemiedebergs Arch. Pharmacol.* **317**, 326–30.

Schwab, M.E. and Thoenen, H. (1976). Electron microscopic evidence for a transsynaptic migration of tetanus toxin in spinal cord motoneurons: an autoradiographic and morphometric study. *Brain Res.* **105**, 213–27.

Schweizer, F.E., Betz, H. and Augustine, G.J. (1995). From vesicle docking to endocytosis: intermediate reactions of exocytosis. *Neuron* **14**, 689–96.

Scott, A.B., Rosenbaum, A. and Collins, C.C. (1973). Pharmacologic weakening of extraocular muscles. *Invest. Opthalmol.* **92**, 924–7.

Sellin, L.C. (1987). Botulinum toxin and the blockade of neurotransmitter release. *Asia Pac. J. Pharmacol.* **2**, 203–22.

Sellin, L.C., Molgo, J., Tornquist, K., Hansson, B. and Thesleff, S. (1996). On the possible origin of giant or slow-rising miniature end-plate potentials at the neuromuscular junction. *Pflugers Arch.* **431**, 325–34.

Shone, C.C. and Roberts, A.K. (1994). Peptide substrate specificity and properties of the zinc-endopeptidase activity of botulinum type B neurotoxin. *Eur. J. Biochem.* **225**, 263–70.

Shone, C.C., Hambleton, P. and Melling, J. (1985). Inactivation of *Clostridium botulinum* type A neurotoxin by trypsin and purification of two tryptic fragments. *Eur. J. Biochem.* **151**, 75–82.

Shone, C.C., Hambleton, P. and Melling, J. (1987). A 50-kDa fragment from the NH₂-terminus of the heavy subunit of *Clostridium botulinum* type A neurotoxin forms channels in lipid vesicles. *Eur. J. Biochem.* **167**, 175–80.

Shone, C.C., Quinn, C.P., Wait, R., Hallis, B., Fooks, S.G. and Hambleton, P. (1993). Proteolytic cleavage of synthetic fragments of vesicle-associated membrane protein, isoform-2 by botulinum type B neurotoxin. *Eur. J. Biochem.* **217**, 965–71.

Shumaker, H.B., Lamont, A. and Firor, W.M. (1939). The reaction of 'tetanus sensitive' and 'tetanus resistant' animals to the injection of tetanal toxin into the spinal cord. *J. Immunol.* **37**, 425–33.

Simpson, L.L. (1971). Ganglioside inactivation of botulinum toxin. *J. Neurochem.* **18**, 1341–3.

Simpson, L.L. (1982). The interaction between aminoquinolines and presynaptically acting neurotoxins. *J. Pharmacol. Exp. Ther.* **222**, 43–8.

Simpson, L.L. (1983). Ammonium chloride and methylamine hydrochloride antagonize clostridial neurotoxins. *J. Pharmacol. Exp. Ther.* **225**, 546–52.

Simpson, L.L. (1986). Molecular pharmacology of botulinum toxin and tetanus toxin. *Annu. Rev. Pharmacol.* **26**, 427–53.

Simpson, L.L. (ed.) (1989). *Botulinum Neurotoxin and Tetanus Toxin.* Academic Press, San Diego.

Simpson, L.L., Coffield, J.A. and Bakry, N. (1993). Chelation of zinc antagonizes the neuromuscular blocking properties of the seven serotypes of botulinum neurotoxin as well as tetanus toxin. *J. Pharmacol. Exp. Ther.* **267**, 720–7.

Simpson, L.L., Coffield, J.L. and Bakry, N. (1994). Inhibition of vacuolar adenosine triphosphatase antagonizes the effects of clostridial neurotoxins but not phospholipase A2 neurotoxins. *J. Pharmacol. Exp. Ther.* **269**, 256–69.

Singh, B.R. and DasGupta, B.R. (1989). Structure of heavy and light chain subunits of type A botulinum neurotoxin analyzed by circular dichroism and fluorescence measurements. *Mol. Cell Biochem.* **85**, 67–73.

Smith, L.D. and Sugiyama, H. (1988). *Botulism: the Organism, its Toxins, the Disease.* C.C. Thomas, Springfield, IL.

Soleilhac, J.M., Cornille, F., Martin, L., Lenoir, C., Fournie-Zaluski, M.C. and Roques, B.P. (1996). A sensitive and rapid fluorescence-based assay for determination of tetanus toxin peptidase activity. *Anal. Biochem.* **241**, 120–7.

Söllner, T., Whiteheart, S.W., Brunner, M., Erdjument-Bromage, H., Geromanos, S., Tempst, P. and Rothman, J.E. (1993). SNAP receptors implicated in vesicle targeting and fusion. *Nature* **362**, 318–24.

Stanley, E.F. (1997). The calcium-channel and the organization of the presynaptic transmitter release face. *Trends Neurosci.* **20**, 404–9.

Staub, G.C., Walton, K.M., Schnaar, R.L., Nichols, T., Baichwal, R., Sandberg, K. and Rogers, T.B. (1986). Characterization of the binding and internalization of tetanus toxin in a neuroblastoma hybrid cell line. *J. Neurosci.* **6**, 1443–51.

Steegmaier, M., Yang, B., Yoo, J.S., Huang, B., Shen, M., Yu, S., Luo, Y. and Scheller, R.H. (1998). 3 novel proteins of the syntaxin/SNAP-25 family. *J. Biol. Chem.* **273**, 34171–9.

Stevens, R.C., Evenson, M.L., Tepp, W. and DasGupta, B.R. (1991). Crystallization and preliminary X-ray analysis of botulinum neurotoxin type A. *J. Mol. Biol.* **222**, 877–80.

Stöckel, K., Schwab, M. and Thoenen, H. (1975). Comparison between the retrograde axonal transport of nerve growth factor and tetanus toxin in motor, sensory and adrenergic neurons. *Brain Res.* **99**, 1–16.

Stöckel, K., Schwab, M. and Thoenen, H. (1977). Role of gangliosides in the uptake and retrograde axonal transport of cholera and tetanus toxin as compared to nerve growth factor and wheat germ agglutinin. *Brain Res.* **132**, 273–85.

Stocker, W. and Bode, W. (1995). Structural features of a superfamily of zinc-endopeptidases: the metzincins. *Curr. Opin. Struct. Biol.* **5**, 383–90.

Südhof, T.C. (1995). The synaptic vesicle cycle: a cascade of protein–protein interactions. *Nature* **375**, 645–53.

Suen, J.C., Hatheway, C.L., Steigerwalt, A.G. and Brenner, D.J. (1988). Genetic confirmation of identities of neurotoxigenic *Clostridium baratii* and *Clostridium butyricum* implicated as agents of infant botulism. *J. Clin. Microbiol.* **26**, 2191–2.

Sutton, R.B., Fasshauer, D., Jahn, R. and Brunger, A.T. (1998). Crystalstructure of a SNARE complex involved in synaptic exocytosis at 2.4 Angstrom resolution. *Nature* **395**, 347–53.

Sweeney, S.T., Broadie, K., Keane, J., Niemann, H. and O'Kane, C.J. (1995). Targeted expression of tetanus toxin light chain in *Drosophila* specifically eliminates synaptic transmission and causes behavioral defects. *Neuron* **14**, 341–51.

Tagaya, M., Toyonaga, S., Takahashi, M., Yamamoto, A., Fujiwara, T., Akagawa, K., Moriyama, Y. and Mizushima, S. (1995). Syntaxin 1 (HPC-1) is associated with chromaffin granules. *J. Biol. Chem.* **270**, 15930–3.

Takano, K., Kirchner, F., Terhaar, P. and Tiebert, B. (1983). Effect of tetanus toxin on the monosynaptic reflex. *Naunyn Schemiedebergs Arch. Pharmacol.* **323**, 217–20.

Takano, K., Kirchner, F., Gremmelt, A., Matsuda, M., Ozutsumi, N. and Sugimoto, N. (1989). Blocking effects of tetanus toxin and its fragment [A-B] on the excitatory and inhibitory synapses of the spinal motoneurone of the cat. *Toxicon* **27**, 385–92.

Thesleff, S. (1986). Different kinds of acetylcholine release from the motor nerve. *Int. Rev. Neurobiol.* **28**, 59–88.

Tizzoni, G. and Cattani, G. (1890). Uber das Tetanusgift. *Zentralbl. Bakt.* **8**, 69–73.

Tonello, F., Morante, S., Rossetto, O., Schiavo, G. and Montecucco, C. (1996). Tetanus and botulism neurotoxins: a novel group of zinc-endopeptidases. *Adv. Exp. Med. Biol.* **389**, 251–60.

Trimble, W.S. (1993). Analysis of the structure and expression of the VAMP family of synaptic vesicle proteins. *J. Physiol. Paris* **87**, 107–15.

Trimble, W.S., Cowan, D.M. and Scheller, R.H. (1988). VAMP-1: a synaptic vesicle-associated integral membrane protein. *Proc. Natl Acad. Sci. USA* **85**, 4538–42.

Umland, T.C., Wingert, L.M., Swaminathan, S., Furey, W.F., Schmidt, J.J. and Sax, M. (1997). Structure of the receptor binding fragment Hc of tetanus toxin. *Nature Struct. Biol.* **4**, 788–92.

Vaidyanathan, V.V., Yoshino, K., Jahnz, M., Dorries, C., Bade, S., Nauenburg, S., Niemann, H. and Binz, T. (1999). Proteolysis of SNAP-25 isoforms by botulinum neurotoxin types A, C and E:

Domains and amino acid residues controlling the formation of enzyme–substrate complexes and cleavage. *J. Neurochem.* **72**, 327–37.

Vallee, B.L. and Auld, D.S. (1990). Zinc coordination, function and structure of zinc enzymes and other proteins. *Biochemistry* **29**, 5647–59.

Valtorta, F., Greengard, P., Fesce, R., Chieregatti, E. and Benfenati, F. (1992). Effects of the neuronal phosphoprotein synapsin I on actin polymerization. I. Evidence for a phosphorylation-dependent nucleating effect. *J. Biol. Chem.* **267**, 11281–8.

van der Goot, F.G., Gonzalez-Manas, J.M., Lakey, J.H. and Pattus, F. (1991). A 'molten-globule' membrane-insertion intermediate of the pore-forming domain of colicin A. *Nature* **354**, 408–10.

Van der Kloot, W. and Molgo, J. (1994). Quantal acetylcholine release at the vertebrate neuromuscular junction. *Physiol. Rev.* **74**, 899–991.

van Ermengem, E. (1897). Uber ein neuenanaeroben Bacillus und seine Beziehungen zum Botulismus. *Z. Hyg. Infektkr.* **26**, 1–56.

van Heyningen, W.E.a.M., P.A. (1961). The fixation of tetanus toxin by ganglioside. *J. Gen. Microbiol.* **24**, 107–19.

Veit, M., Söllner, T. and Rothman, J.E. (1996). Multiple palmitoylation of synaptotagmin and the tSNARE SNAP-25. *FEBS Lett.* **385**, 119–23.

Walch-Solimena, C., Blasi, J., Edelmann, L., Chapman, E.R., von Mollard, G.F. and Jahn, R. (1995). The t-SNAREs syntaxin 1 and SNAP-25 are present on organelles that participate in synaptic vesicle recycling. *J. Cell Biol.* **128**, 637–45.

Wang, G., Witkin, J.W., Hao, G., Bankaitis, V.A., Scherer, P.E. and Baldini, G.(1997). Syndet is a novel SNAP-25 related protein expressed in many tissues. *J. Cell Sci.* **110**, 505–13.

Washbourne, P., Schiavo, G. and Montecucco, C. (1995). Vesicle-associated membrane protein-2 (synaptobrevin-2) forms a complex with synaptophysin. *Biochem. J.* **305**, 721–4.

Washbourne, P., Pellizzari, R., Baldini, G., Wilson, M.C. and Montecucco, C. (1997). Botulinum neurotoxin type-A and type-E require the SNARE motif in SNAP-25 for proteolysis. *FEBS Lett.* **418**, 1–5.

Weber, T., Zemelman, B.V., McNew, J.A., Westermann, B., Gmachl, M.J., Parlati, F., Söllner, T.H. and Rothman, J.E. (1998). SNAREpins: minimal machinery for membrane fusion. *Cell* **92**, 759–72.

Weller, U., Taylor, C.F. and Habermann, E. (1986). Quantitative comparison between tetanus toxin, some fragments and toxoid for binding and axonal transport in the rat. *Toxicon* **24**, 1055–63.

Weller, U., Dauzenroth, M.E., Meyer zu Heringdorf, D. and Habermann, E. (1989). Chains and fragments of tetanus toxin. Separation, reassociation and pharmacological properties. *Eur. J. Biochem.* **182**, 649–56.

Weller, U., Dauzenroth, M.E., Gansel, M. and Dreyer, F. (1991). Cooperative action of the light chain of tetanus toxin and the heavy chain of botulinum toxin type A on the transmitter release of mammalian motor endplates. *Neurosci. Lett.* **122**, 132–4.

Wellhöner, H.H. (1992). Tetanus and botulinum neurotoxins. In: *Handbook of Experimental Pharmacology* (eds H. Herchen and F. Hucho), pp. 357–417. Springer, Berlin.

Wellhöner, H.H., Hensel, B. and Seib, U.C. (1973). Local tetanus in cats: neuropharmacokinetics of ^{125}I-tetanus toxin. *Naunyn Schemiedebergs Arch. Pharmacol.* **276**, 375–86.

Williamson, L.C. and Neale, E.A. (1994). Bafilomycin A1 inhibits the action of tetanus toxin in spinal cord neurons in cell culture. *J. Neurochem.* **63**, 2342–5.

Williamson, L.C., Fitzgerald, S.C. and Neale, E.A. (1992). Differential effects of tetanus toxin on inhibitory and excitatory neurotransmitter release from mammalian spinal cord cells in culture. *J. Neurochem.* **59**, 2148–57.

Williamson, L.C., Halpern, J.L., Montecucco, C., Brown, J.E. and Neale, E.A. (1996). Clostridial neurotoxins and substrate proteolysis in intact neurons: botulinum neurotoxin C acts on synaptosomal-associated protein of 25 kDa. *J. Biol. Chem.* **271**, 7694–9.

Wilson, M.C., Mehta, P.P. and Hess, E.J. (1996). SNAP-25, enSNAREd in neurotransmission and regulation of behaviour. *Biochem. Soc. Trans.* **24**, 670–6.

Wright, J.F., Pernollet, M., Reboul, A., Aude, C. and Colomb, M.G. (1992). Identification and partial characterization of a low affinity metal-binding site in the light chain of tetanus toxin. *J. Biol. Chem.* **267**, 9053–8.

Xu, T., Binz, T., Niemann, H. and Neher, E. (1998). Multiple kinetic components of exocytosis distinguished by neurotoxin sensitivity. *Nature Neurosci.* **1**, 192–200.

Yamasaki, S., Baumeister, A., Binz, T., Blasi, J., Link, E., Cornille, F., Roques, B., Fykse, E.M., Südhof, T.C., Jahn, R. and Niemann, H. (1994a). Cleavage of members of the synaptobrevin/VAMP family by types D and F botulinal neurotoxins and tetanus toxin. *J. Biol. Chem.* **269**, 12764–72.

Yamasaki, S., Binz, T., Hayashi, T., Szabo, E., Yamasaki, N., Eklund, M., Jahn, R. and Niemann, H. (1994b). Botulinum neurotoxin type G proteolyses the Ala81-Ala82 bond of rat synaptobrevin 2. *Biochem. Biophys. Res. Commun.* **200**, 829–35.

Yamasaki, S., Hu, Y., Binz, T., Kalkuhl, A., Kurazono, H., Tamura, T., Jahn, R., Kandel, E. and Niemann, H. (1994c). Synaptobrevin/vesicle-associated membrane protein (VAMP) of *Aplysia californica*: structure and proteolysis by tetanus toxin and botulinal neurotoxins type D and F. *Proc. Natl Acad. Sci. USA* **91**, 4688–92.

Yavin, E. and Nathan, A. (1986). Tetanus toxin receptors on nerve cells contain a trypsin-sensitive component. *Eur. J. Biochem.* **154**, 403–7.

11

The family of Shiga toxins

David W.K. Acheson and Gerald T. Keusch

INTRODUCTION

The Shiga family of toxins is comprised of a group of genetically and functionally related molecules whose original family member was described 100 years ago. Up until the early 1980s this group of toxins was little more than a scientific curiosity without a clear role in disease pathogenesis. However, since the discovery of these toxins in *Escherichia coli* and other Enterobacteriaciae, and their association with significant clinical diseases such as haemorrhagic colitis and haemolytic uraemic syndrome, their importance in disease processes has been brought to the fore. During the course of this chapter we will discuss some of the historical aspects of the Shiga toxins, along with their basic genetics and biology. We will outline how to purify and assay for the toxins, and discuss their role in disease pathogenesis as well as offer a brief outline of the clinical relevance of the Shiga toxins.

HISTORY

The first complete description of Shiga's bacillus (the prototype organism of the genus *Shigella*), now termed *Shigella dysenteriae*, was made by the Japanese microbiologist Kiyoshi Shiga in 1898, following an extensive epidemic of lethal dysentery in Japan (Shiga, 1898). Simon Flexner made the initial observations suggesting the existence of Shiga toxin, although much of the effect he observed was due to endotoxin (Flexner, 1900). In 1903, Neisser and Shiga reported on the lethal effects of bacterial extracts of Shiga's bacillus given to rabbits (Neisser and Shiga, 1903); however, the credit for the discovery of Shiga toxin is generally accorded to Conradi (1903), who described many of its properties in the same year. Among the most prominent effects of crude toxin was its ability to cause limb paralysis followed by death when given parenterally to certain experimental animals. These neurological manifestations led to the designation of the factor as Shiga 'neurotoxin', a term which remained in use for nearly 70 years. Initial purification of toxin was fraught with problems related to contamination with lipopolysaccharide (LPS). Definitive evidence that the 'neurotoxin' was separable from LPS was not obtained until 34 years after the original report of Shiga toxin, when chemical fractionation was employed (Boivin and Mesrobeanu, 1937).

From the time of its description until today, the role of Shiga toxin in the pathogenesis of Shigellosis has been the subject of debate (Keusch *et al.*, 1986). For the first five decades, ability to interpret studies of Shiga toxin was clouded initially by the inevitable LPS contamination, and later by the lack of model systems suitable to study pathogenesis of intestinal manifestations of the diseases. Thus, most experimental animal studies employed parenteral inoculation of what turned later to be impure material and focused on the paralysis caused by the protein Shiga 'neurotoxin'. In

1953 van Heyningen and colleagues achieved a significant increase in purification and yield of Shiga toxin (van Heyningen and Gladstone, 1953), although later studies were to demonstrate that this preparation had at least 20 protein bands and was no more than 5% toxin (Keusch and Jacewicz, 1975). This material was used to study the pathogenesis of the neurotoxin effect in animals (Bridgewater *et al.*, 1955; Howard, 1955). The studies suggested that the underlying lesion caused by Shiga toxin was focal endothelial cell damage, resulting in focal bleeding in the spinal cord and secondary neurological manifestations. In 1960, Vicari *et al.* found that van Heyningen's preparation of toxin was cytotoxic to certain epithelial cells in culture (Vicari *et al.*, 1960).

In 1969 there was an extensive outbreak of epidemic *S. dysenteriae* type 1 infection in Central America and Mexico, with significant morbidity and mortality (Gangarosa *et al.*, 1970). This led to renewed interest in Shiga toxin as a virulence factor of Shigella. By 1972 it was reported that cell-free supernatants of *S. dysenteriae* 1 cultures could reproduce the major features of human Shigellosis in a rabbit ligated ileal loop model, including intestinal fluid secretion and inflammatory enteritis (Keusch *et al.*, 1972a, b). Since then great progress has been made in the biochemistry, physiology and genetics of Shiga toxin, as discussed in the later sections of this chapter and in recent reviews (Nataro and Kaper, 1998; Paton and Paton, 1998).

In 1978 *S. dysenteriae* was associated with the clinical condition known as haemolytic uraemic syndrome (HUS), a triad of renal failure, thrombocytopenia and haemolytic anaemia (Koster *et al.*, 1978). It was not until the recognition of Shiga toxins in *E. coli* and their association with HUS in the early 1980s that the role of Shiga toxin from *S. dysenteriae* in HUS could be truly appreciated. Over the last 15 years we have come to realize that the Shiga family of toxins is in fact a major cause of disease in many developed countries.

In the latter half of the 1970s Konowalchuk *et al.* (1977) reported that culture filtrates of several different strains of *E. coli* were cytotoxic for Vero cells. This cytolethal activity was heat labile, and not neutralized by antiserum to the classical cholera-like *E. coli* heat-labile enterotoxin. Konowalchuk subsequently investigated 136 *E. coli* strains and found only 10 to be capable of producing this Vero cell cytotoxin. It soon became apparent that the cytotoxin was probably not due to a single protein as careful investigation of one strain (*E. coli* O26:H30) revealed two variant toxin activities with a p*I* of 7.2 and 6.8 (Konowalchuk *et al.*, 1978). Following these observations, Wade *et al.* (1979) in the UK noted the presence of cytotoxin-producing *E. coli* O26 strains in association with bloody diarrhoea. This finding was supported by reports from Scotland *et al.* (1979) in the UK and Wilson and Bettleheim (1980) in New Zealand.

In 1983 Riley *et al.* (1983) reported on the association between a rare serotype of *E. coli* (O157:H7) and the development of haemorrhagic colitis. This strain of *E. coli* was later shown to carry two bacteriophages, each of which encoded a different type of Shiga toxin, thus confirming the earlier suggestions of multiple cytotoxins made by these studies. At around the same time Karmali *et al.* (1983) made the association between the presence of Shiga toxins produced by *E. coli* and the development of HUS. This heralded the onset of a new era in the Shiga toxin field. Currently, over 200 different types of *E. coli* make Shiga toxins, many of which have been associated with human disease (Acheson and Keusch, 1996). Shiga toxin-producing *E. coli* (STEC) are the most common cause of acute renal failure in the USA, and result in an estimated 250–500 deaths annually in the USA alone. Much has been learned about the biochemistry and genetics of these toxins in recent years, and the following sections will review the latest aspects of the Shiga toxin family.

NOMENCLATURE

The nomenclature for the Shiga toxin family has become confusing. In 1972, a toxin causing fluid secretion by rabbit small bowel was identified in *S. dysenteriae* type 1 and named Shigella enterotoxin. This was later shown to be identical to Shiga neurotoxin. Following the original description of the cytotoxic effects of these toxins on Vero cells they were given the name Verotoxin. This name is still used by many workers in the field, and the Verotoxin-producing *E. coli* are known as VTEC. However, it became apparent in the mid-1980s that these newly described *E. coli* toxins were very similar to Shiga toxin and were neutralized by antisera to Shiga toxin, and so they were given the name of Shiga-like toxin. By 1996, when it was known that these toxins showed a common mechanism of action and cellular binding site, an international group of investigators decided simply to call this group of biologically homogenous toxins, irrespective of their bacterial origin, the Shiga toxins (Stx) after the original description nearly 100 years ago (Calderwood *et al.*, 1996). The gene designation (*stx*) was already well established and the new nomenclature has maintained the *stx* gene designation. Currently, there are six members of this family, as shown in Table 11.1. The toxins are divided into two main groups: Shiga toxin from *S. dysenteriae* type 1 and Shiga toxin 1 form one group; and the Shiga toxin 2 family forms the other group. As

TABLE 11.1 The Shiga toxin family

Name	Gene	Protein	Comments
Shiga toxin	*stx*	Stx	From *S. dysenteriae*
Shiga toxin 1	*stx*$_1$	Stx1	
Shiga toxin 2	*stx*$_2$	Stx2	
Shiga toxin 2c	*stx*$_{2c}$	Stx2c	
Shiga toxin 2d	*stx*$_{2d}$	Stx2d	Mucus activatable
Shiga toxin 2e	*stx*$_{2e}$	Stx2e	Associated with porcine oedema disease

will be discussed later, Shiga toxin from *S. dysenteriae* and Stx1 from *E. coli* are essentially the same molecule. Shiga toxin 2 is significantly different and is made up of a number of subfamilies, as outlined in Table 11.1. The characteristics of these protein strains will be discussed below.

SHIGA TOXIN PURIFICATION AND CHARACTERIZATION

Purification of Shiga toxin remained a problem until the early 1980s when small amounts of highly purified toxin were obtained by Olsnes and Eiklid (1980) and O'Brien *et al.* (1980) by ion-exchange column and antibody affinity chromatography, respectively. The first successful large-scale purification was reported by Donohue-Rolfe *et al.* (1984) using chromatofocusing as the principal technique. Their method resulted in a yield of nearly 1 mg of pure toxin from 3 litres of culture. The purified toxin was isoelectric at pH 7.2, and consisted of two peptide bands of approximately 7 and 32 kDa. Toxin production was enhanced by low iron concentration. Although O'Brien *et al.* (1980) used chelex-treated medium to reduce iron content, Donohue-Rolfe *et al.* (1984) found that the iron content of Syncase medium was sufficiently low to support maximum toxin production without the need for further manipulation. In 1989, Donohue-Rolfe *et al.* (1989a) reported the use of a one-step affinity purification system that was suitable for the purification of all Shiga toxin family members. This involved the coupling of a glycoprotein toxin receptor analogue present in hydatid cyst fluid, P1-blood group active glycoprotein (P1gp), to Sepharose 4B and using this as an affinity chromatography matrix. Toxin bound to immobilized P1gp and could be eluted with 4.5 M MgCl$_2$. Renatured toxin was fully biologically active, and as much as 10 mg of pure toxin could be obtained from a 20-litre fermenter culture. When the toxin expressing *E. coli* C600 933W (for Stx2) and *E. coli* HB101 H-19B (for Stx1) lysogens

are used the levels of toxin in the cultures can be increased significantly. Additionally, the bacteriophage-inducing drug mitomycin C can be used to increase the levels of Stx2 expression up to 200 mg per 20-litre fermenter culture. The P1gp purification method has been used to purify successfully Stx1, Stx2, Stx2c, Stx2d and Stx2e. Others have published alternative purification systems for Shiga toxins that include the use of standard and high-performance liquid chromatography (Noda *et al.*, 1987), and receptor–ligand-based chromatography (Ryd *et al.*, 1989) that works from the same principle as the P1gp system discussed above.

Purified Shiga toxin of any type (Stx1 or Stx2) is approximately 70 kDa and separates into two peptide bands in sodium dodecyl sulfate (SDS)–polyacrylamide gels under reducing conditions (Donohue-Rolfe *et al.*, 1984). The larger A subunit (31–32 kDa) has been isolated and shown to mediate the inhibitory effect of Shiga toxin on protein synthesis in cell-free systems (Reisbig *et al.*, 1981). The smaller multimeric B subunit is responsible for the toxins' binding to cell-surface receptors (see below), and has been shown to bind to receptor-positive cells and competitively inhibit both the binding and cytotoxicity of holotoxin (Mobassaleh *et al.*, 1988; Donohue-Rolfe *et al.*, 1989b). Biochemical cross-linking studies have demonstrated that the B subunit is pentameric for both Shiga toxin 1 and 2, and linked to a single A subunit. X-ray crystallographic analysis of Shiga toxin and Stx1 B subunits have confirmed this structure, revealing a pentameric ring of five B subunit monomers which surrounds a helix at the C-terminus of the A subunit (Stein *et al.*, 1992; Fraser *et al.*, 1994). When the A subunit is nicked with trypsin and reduced an A1 portion of approximately 28 kDa and an A2 peptide of approximately 4 kDa are separated. The A1 fragment contains the enzymically active portion of the toxin molecule, and the A2 component is required to bind non-covalently the whole A subunit to the B pentamer (Austin *et al.*, 1994). The A2 fragment may be important in assembly of the holotoxin (Austin *et al.*, 1994), and the presence of a disulfide bridge appears necessary for pentamer formation (Jackson *et al.*, 1990a). The B subunit monomer consists of two, three-stranded, antiparallel β-sheets and an α-helix and, upon pentamer formation, the resulting central pore is formed by three long helices in which the exposed residues are Asn35, Ser38, Ser43, Ile45 and Thr46 (Sixma *et al.*, 1993), with the A2 fragment sitting in the central pore. The α-helix of the 293-amino acid StxA subunit contains nine residues (Ser279–Met287) which penetrate the non-polar pore of the B subunit pentamer. Jemal *et al.* (1995) showed by site-directed mutagenesis the role of two residues bordering this α-helix, Asp278 and Arg288, in coupling the C-terminus of StxA to the B pentamer.

There are several reports documenting the critical amino acid residues relating to the structure and function of Stx. Stx1, Stx2 and Stx2e and the plant toxin ricin, all of which inhibit protein synthesis by the same mechanism of action, share two areas of homology in their A subunits (Figure 11.1) (Jackson et al., 1990b). Hovde et al. (1988) showed that a glutamic acid residue at position 167 (area 1) was critical for biological activity of Shiga toxin. A conservative substitution of Glu167 to aspartic acid resulted in a 1000-fold drop in biological activity, although some activity still remained. Jackson et al. (1990b) undertook similar experiments in Stx2 and found that the same substitution of an aspartic acid for glutamic acid 167 of the Stx2 A subunit decreased the capacity of the polypeptides to inhibit protein synthesis by at least 100-fold in a cell-free system, and resulted in a 1000-fold reduction in toxicity to Vero cells. When amino acids 202–213 (area 2) of Stx2 A were deleted, the resulting A subunit was able to assemble into holotoxin, but there was no demonstrable cytotoxicity.

Further structure–function relationships of Stx2, reported by Perera et al. (1991), revealed that proteins lacking the last four amino acids of the C-terminus of the B subunit were not cytotoxic. However, it is not clear whether this was due to lack of B subunit binding to the receptor, or failure of the mutated B subunit to form pentamers with an A subunit. Deletion of the amino-terminal amino acids (3–18) of the Stx2 A subunit also resulted in abolition of cytotoxic activity, presumably for the same reasons.

Jackson et al. (1990a) found that mutagenesis of the hydrophilic region of the Stx1 B subunit rendered the molecule non-toxic without affecting holotoxin assembly. Double mutations of D16H and D17H resulted in a loss of Stx1B subunit binding but did not prevent holotoxin assembly, as determined by immunoprecipitation studies. Tyrrell et al. (1992) showed that specific mutations in Stx B subunits could either affect the receptor specificity (Gb3 vs Gb4) or result in loss of binding altogether.

While the StxA subunit is capable of inhibiting protein synthesis and causing cell death, the Stx B subunits

are not considered to be enzymatically active, rather serving as the binding moiety for Stx holotoxin. This view may need to be treated with caution in view of work by Mangeny et al. (1991) and others, who found that the Stx1 B subunit alone was capable of inducing apoptosis in CD77-positive Burkitt's lymphoma cells. CD77 has been characterized as globotriaosyl-ceramide (Gb3), which is the same as the Stx1 and Stx2 receptor.

The Shiga toxin 1 subfamily and the Shiga toxin 2 subfamily are separable by the fact that they are not neutralized by heterologous polyclonal antisera, although at least one monoclonal antibody has been described that will neutralize both Stx1 and Stx2 (Donohue-Rolfe et al., 1989a). O'Brien and LaVeck (1983) purified Stx1 from E. coli H30 using a multistep procedure and found the mobility of the two subunits to be identical with Shiga toxin from S. dysenteriae type 1 on SDS–polyacrylamide gel electrophoresis (PAGE). We now know that Shiga toxin from S. dysenteriae and Stx1 are essentially identical molecules with only three nucleotide differences that result in a single conservative amino acid change in the A subunit (Strockbine et al., 1988). Stx1 proteins form a relatively homogeneous family and little significant variation has been reported within that family. In contrast, the Stx2 family is diverse and contains members with a variety of properties. The B subunit is responsible for the majority of the significant functional differences within the Stx2 subfamily (Schmitt et al., 1991; Lindgren et al., 1994). Different members of the Stx2 subfamily also have variable biological effects on tissue culture cells. The specific activity of Stx2c and Stx2d has been reported to be lower for Vero cells than that of Stx2 (Melton-Celsa and O'Brien, 1998). This difference is thought to be due to a single amino acid change in the B subunit (Lindgren et al., 1994). Despite this difference in specific activities in vitro, the 50% lethal doses (LD_{50}) for mice injected with either Stx2 or Stx2d are equivalent. Stx2e behaves in the opposite way in that the LD_{50} is about 100-fold less in mice compared with Stx2, but their specific activity on Vero cells is similar (Melton-Celsa and O'Brien, 1998).

Stx2d is activated by intestinal mucus, and this is a characteristic that makes Stx2d different from other members of the Stx2 subfamily. Stx2d and Stx2c have the same mature B subunit amino acid sequence and therefore the A subunit is considered to be the locus of this biological difference. The A subunits of these two molecules have two differences in the amino acid sequence of the A2 fragment, and these differences appear to be responsible for the mucus activatable nature of Stx2d (Melton-Celsa and O'Brien, 1998).

Area 1					Area 2				
Stx1	E167	A	L	R	F171	Stx1	W203	G	R	L	S207
Stx2	E167	A	L	R	F171	Stx2	W202	G	R	L	S206
Stx2e	E167	A	L	R	F171	Stx2e	W202	G	R	L	S206
Ricin	E177	A	L	R	F181	Ricin	W211	G	R	L	S215

FIGURE 11.1 Areas of sequence homology in Stx1, Stx2, Stx2e and ricin.

GENETICS

Shiga toxins are either phage encoded (Stx1 and Stx2) or chromosmally encoded (Stx, Stx2e) (Newland *et al.*, 1985; Weinstein *et al.*, 1988; O'Brien *et al.*, 1989). The finding that Shiga toxins are bacteriophage encoded goes back to the early 1970s when Smith and Linggood (1971) reported that lysates of H-19B, an *E. coli* O26:H11 strain isolated from an outbreak of infantile diarrhoea, could transfer enterotoxinogenicity to *E. coli* K12 *in vitro*. This phage, known as H-19B, was subsequently shown to encode Stx1 and to have DNA sequence homology with phage lambda (λ) (Huang *et al.*, 1986, 1987). A second lambdoid phage, designated 933W, was isolated from a clinical O157:H7 isolate that was responsible for an outbreak of haemorrhagic colitis in 1982 (Riley *et al.*, 1983; O'Brien *et al.*, 1984; Newland *et al.*, 1985). Stx-encoding phages are lambda-like and the regulatory components relating to induction and phage gene control appear to be similar to those in λ. Clinical STEC isolates may contain either one type of Stx or more than one. Many clinical O157:H7 isolates harbour distinct Stx1 and Stx2 bacteriophages. The nucleotide sequences of the Stx genes have been published by various groups (Calderwood *et al.*, 1987; De Grandis *et al.*, 1987; Jackson *et al.*, 1987a, b; Kozlov *et al.*, 1988; Strockbine *et al.*, 1988). The different Stx operons have a similar structure and are composed of a single transcriptional unit which consists of one copy of the A subunit gene followed by the B subunit gene. Habib and Jackson (1992) initially suggested that the B subunit has its own promoter in the Shiga toxin operon. They later reported that B subunit translation was augmented because it was considered to have a stronger ribosomal binding site compared with the A subunit. The results of this would be more B subunit translation, therefore providing more B subunit for the A1–B5 structure of the Stx holotoxin (Habib and Jackson, 1993).

Shiga toxin from *S. dysenteriae* and Stx1 from *E. coli* are highly conserved, and there are only three nucleotide differences in three codons of the A subunit of Stx1 which results in a single amino acid change (Thr45 for Ser45). The calculated masses for the processed A and B subunits are 32 225 and 7691, respectively. Signal peptides of 22 and 20 residues were present for the A and B subunits. The production of both Stx and Stx1 is regulated by the concentration of iron in the growth medium.

Until recently, very little was known about the regulation of Stx production *in vitro* and essentially nothing about Stx regulation *in vivo*. The production of both Stx from *S. dysenteriae* and Stx1 *in vitro* is regulated by *fur* via the concentration of iron in the growth medium. The iron regulation involves the *fur* gene, whose protein product acts as a transcriptional repressor. In the promoter regions of both Stx and Stx1 there is a 21 base pair dyad repeat that appears to be a binding site for the Fur protein (Calderwood and Mekalanos, 1987; Strockbine *et al.*, 1988). The Stx2 operon does not have an upstream Fur binding site and is not iron regulated *in vitro*. Work from our own group (Muhldorfer *et al.*, 1996) using a Stx2A–phoA fusion led to the conclusion that the location of the Stx2AB operon in the genome of a bacteriophage exerts an important influence on the expression of this toxin. Two mechanisms in which the bacteriophage influenced toxin production were proposed. The first was through an increase in the number of toxin gene copies brought about by phage replication. The second mechanism we proposed was the existence of a phage-encoded regulatory molecule whose activity was dependent on phage induction. Subsequent studies by Neely and Friedman (1998) using the H19-B Stx1-encoding bacteriophage added to this picture of toxin regulation. They compared the regions adjacent to the Stx1 and Stx2 genes in H-19B and 933W and phage λ, and found significant areas of homology.

The Q gene is situated just upstream of the *stx* gene and the Q product functions as a transcription antiterminator that regulates expression of late phage genes by modifying transcription complexes initiated at the late promoter $P_{R'}$. Neely and Friedman (1998) examined the Q-mediated read-through in H-19B and found that there was a greater than 100-fold read-through at $t_{R'}$, a terminator situated just downstream of Q. Their data indicate that the Q protein is acting as a positive regulator on the expression of Stx1 and Stx2, and most probably Q represents the phage-encoded positive regulator that we originally described. Neely and Friedman (1998) propose that with phage induction, Q binds to the *qut* site within $P_{R'}$ and modifies RNA-polymerase, resulting in read-through of terminators ($t_{R'}$) and therefore enhanced expression of toxin and other downstream genes (Neely and Friedman, 1998).

The gene for Stx2 was originally cloned from *E. coli* 933W and when compared with Stx1 was found to be organized in a similar way (Newland *et al.*, 1987). Comparison of the nucleotide sequence of the A and B subunits of Stx1 and 2 showed 57% and 60% homology, respectively, with 55% and 57% amino acid homology (Jackson *et al.*, 1987b).

Despite this degree of homology, Stx1 and Stx2 are immunologically distinct, and neither is cross-neutralized by polyclonal antibody raised to the other toxin. Although some recent data published in abstract form suggest that rabbits immunized with Stx2 toxoid are protected against challenge by Stx1 (Ludwig *et al.*, 1997), the reverse has not been demonstrated. Stx2c is very similar to Stx2 with identical A subunits and 97% amino acid homology in the B subunits. Stx2d has the same B subunit as Stx2c, although the A subunit is slightly different (99% homology) (Melton-Celsa and O'Brien, 1998). The Stx2e gene found in porcine strains has been compared with Stx2 and found to have a 94% homology of nucleotide sequence between the A subunits (93% deduced amino acid sequence homology) and 79% nucleotide sequence homology between the B subunits (84% deduced amino acid sequence homology). The degree of homology between Stx2e and Stx1 was 55–60%, which is similar to the degree of homology between Stx1 and Stx2 (Weinstein et al., 1988).

MECHANISM OF ACTION

Reisbig et al. (1981) first reported that Shiga toxin could irreversibly inhibit protein synthesis by a highly specific action on the 60S mammalian ribosomal subunit. Brown et al. (1980) and others (O'Brien et al., 1980; Olsnes et al., 1981) found that the A subunit was activated by proteolysis and reduction, leading to the removal of a small A2 peptide from the biologically active A1 subunit. In these properties, Shiga toxin resembled the plant toxin, ricin. The possible structural relevance of this became apparent in 1987 when Calderwood et al. (1987) cloned and sequenced Stx1 and found significant homology between the A subunits of Stx1 and the previously sequenced ricin (see Figure 11.1). In the 73-amino acid segment from residues 138 to 210 of Stx there were 32% identical and 53% chemically conserved residues. The same year Endo and Tsurugi (1987) determined the enzymic activity of ricin as an N-glycosidase that hydrolysed adenine 4324 of the 28S ribosomal RNA of the 60S ribosomal subunit. Igarashi et al. (1987) found that Stx1 from E. coli O157:H7 inactivated the 60S ribosomal subunits of rabbit reticulocytes and blocked elongation factor-1-dependent binding of aminoacyl-tRNA to ribosomes. Shortly thereafter, Endo et al. (1988) reported that Shiga toxin had identical enzymic specificity to ricin. All Stx proteins share this property; hence this is one of the criteria used to define the Shiga family of toxins. Skinner and Jackson (1997), using site-directed mutagenesis coupled with N-bromosuccinimide modification, showed that the sole tryptophan residue of StxA1 is required for binding of the A subunit to the 28S rRNA backbone.

A number of laboratories has cloned and expressed the Stx1 B subunit (Calderwood et al., 1990; Ramotar et al., 1990; Acheson et al., 1993a). The purified Stx1 B subunit has been expressed from both E. coli and V. cholerae strains, and appears to form natural functional pentamers when expressed alone. In contrast, expression of Stx2 B subunits has been much more difficult. Acheson et al. (1995) cloned and expressed the Stx2 B subunit in a variety of vectors using different promoters and, while it was possible to obtain modest levels of B subunit expression, the multimeric forms of the B subunit appeared to be unstable.

The enzymically active A subunit has a trypsin-sensitive region, and after cleavage results in two disulfide-bonded fragments (A1 and A2). The role of this disulfide bond was studied by Garred et al. (1997). They eliminated it by mutating C242 to serine. In T47D cells the mutated toxin was more toxic than wild-type toxin after a short incubation, whereas after longer incubation times wild-type toxin was more toxic. After prebinding of Stx C242S to wells coated with the Stx receptor glycolipid Gb3, trypsin treatment induced dissociation of A1 from the toxin–receptor complex, demonstrating that in addition to stabilizing the A-chain the disulfide bond prevents dissociation of the A1 fragment from the toxin–receptor complex. Cleavage of the Stx A fragment at the trypsin-sensitive site will increase its enzymic activity. As noted above, the two A subunit fragments are normally linked via a disulfide bond. The disulfide loop contains the sequence Arg–X–X–Arg, which is a consensus motif for cleavage by the membrane-anchored protease furin. Garred et al. (1995a) found that a soluble form of furin cleaves intact A-chain producing A1 and A2 fragments. LoVo cells, which normally fail to produce functional furin, cleave the toxin very slowly, yet cleave the A subunit very effectively when transfected with a furin-encoding cosmid. These results indicate that cleavage of Shiga toxin is important for intoxication of cells, and that furin can cleave and thereby activate Stx in vivo in intact cells (Garred et al., 1995a).

TOXIN RECEPTORS

The search for Shiga toxin receptors on mammalian cells began in 1977, when Keusch and Jacewicz (1977) reported that toxin-sensitive cells in tissue culture removed toxin bioactivity from the medium, whereas toxin-resistant cells did not. These studies also suggested that the receptor was carbohydrate in nature, and that the toxin was a sugar-binding protein or lectin. Jacewicz et al. (1986) also extracted a toxin-binding constituent from toxin-sensitive HeLa cells and from rabbit jejunal microvillus membranes (MVMs). The MVM binding site was shown to be Gb3 by thin-layer chromatography (TLC) methods (Jacewicz et al., 1986) and later confirmed by high-performance liquid chromatography of derivatized glycolipids, which also demonstrated it to be the hydroxylated fatty acid variety of Gb3 (Mobassaleh *et al.*, 1989). Based on *in vitro* solid-phase binding studies using isolated glycolipids, Lindberg *et al.* (1987) reported that toxin bound to P-blood group active glycolipid, Gb3, which consists of a trisaccharide of galactose α1–4-galactose-β1–4-glucose linked to ceramide, and that Gb3 inhibited biological activity of Shiga toxin in cell culture systems.

The function of Gb3 as a receptor has been shown in several ways. Most relevant to the intestinal effects of Shiga toxin is the work of Mobassaleh *et al.* (1988). They reported that infant rabbits are not susceptible to the fluid secretory effects of Shiga toxin before 16 days of age, and that the age-related sensitivity correlated with developmentally regulated Gb3 levels in rabbit intestinal MVMs. Additional evidence suggesting that Gb3 is the receptor mediating the fluid secretory response of rabbit intestine has been reported by Kandel *et al.* (1989). These investigators found that Shiga toxin leads to a significant decrease in neutral sodium absorption in rabbit jejunum, with no alteration in substrate-coupled sodium absorption or active chloride secretion, suggesting that toxin acted on the absorptive villus cell and not on the secretory crypt cell. When villus and crypt cells were separately isolated from rabbit jejunum it was found that the former alone expressed Gb3, bound toxin and was susceptible to its effect on protein synthesis. The interpretation of these data is that Shiga toxin targets the villus cell because it expresses the Gb3 receptor, and that inhibition of villus cell protein synthesis results in diminished sodium absorptive capacity. This targeting to villus but not crypt cells was subsequently confirmed using human-derived villus-like or crypt-like cell lines derived from colon cancers (Jacewicz *et al.*, 1995).

One of the main differentiating features of Stx2 and Stx2e is their differential cytotoxic activities on HeLa and Vero cells. The latter is substantially more active on Vero cells, reportedly owing to differences in receptor specificity of the toxin, mediated in turn by important differences in the B subunits (Weinstein *et al.*, 1989). De Grandis *et al.* (1989) showed that Stx2e from a pig isolate bound less well to Gb3 separated on TLC plates compared with Stx1 or Stx2. Instead, Stx2e bound preferentially to globotetraosylceramide (Gb4), another neutral glycolipid, which has a subterminal Gal-α1–4-Gal disaccharide. Samuel *et al.* (1990) compared the binding specificity of various Stx2 family members. By preparing *E. coli* strains carrying various fusions between Stx2 and 2e genes for the A and B subunits, a number of hybrids was obtained. These studies confirmed that the B subunits conferred cell specificity for

the cytotoxic effects. Stx2e bound preferentially to internal Gal-α1–4-Gal sequences in globotetraosylceramide and galactosylglobotetraosylceramide. These specificities were important *in vivo* as well, as there was a 400-fold difference in mouse lethality between Stx2 and Stx2e.

The sensitivity of cells to Stx toxins is, in many cases, related to the number of toxin receptors. Measures that increase or decrease toxin receptor expression directly alter responses to these potent molecules (Jacewicz *et al.*, 1994). In addition, toxin activity is modulated by the composition of the fatty acids of the lipid ceramide moiety (Kiarash *et al.*, 1994), in particular fatty acid carbon chain length, which alters the intracellular uptake pathway of toxin and its biological activity. This may explain why some cells can express Gb3 but fail to respond to toxin (Sandvig *et al.*, 1996). Arab and Lingwood (1996) demonstrated the importance of the surrounding lipid environment on the availability of glycolipid carbohydrate for Stx binding. The lipid heterogeneity of Gb3 appears to be important in Stx binding and may define a growth-related signal transduction pathway used by Stx (Lingwood, 1996).

Sensitivity of cells to Stx can also be variably affected by regulation of receptor expression. For example, rabbit intestinal brush border membrane Gb3 is both developmentally and maturationally regulated via the biosynthetic Gb3-galtosyltransferase and degradative α-galactosidase enzymes (Mobassaleh *et al.*, 1994). Gb3 is also maturationally regulated in cultured human intestinal epithelial cells and is induced by exposure of villus-like CaCo2A cells but not crypt-like T84 cells to known regulators of gene transcription (e.g. sodium butyrate) (Jacewicz *et al.*, 1995). Expression of Gb3 in CaCo2A intestinal epithelial cells coincides with expression of villus cell differentiation markers such as alkaline phosphatase, lactase and sucrase. The effect of butyrate may be pertinent in the human colon, the site of STEC infection, because butyrate is normally produced by the enteric flora in concentrations high enough to induce Gb3 *in vitro*.

After binding, toxin is internalized by receptor-mediated endocytosis at clathrin-coated pits. Brefeldin A, which interrupts the intracellular movement of vesicles at the Golgi stack, blocks Stx binding, uptake and proteolysis (Garred *et al.*, 1995b). This suggests that transport of toxin through the Golgi is necessary to reach the ribosomal target (Sandvig *et al.*, 1991; Sandvig and van Deurs, 1996). The requirement of transport of toxin beyond the Golgi to the cytoplasm has been confirmed by the use of ilimaquinone, a sea sponge metabolite that causes the breakdown of Golgi membranes and inhibits the retrograde transport of proteins to the endoplasmic reticulum (Nambiar and Wu, 1995).

PATHOGENESIS

Shiga toxin-producing bacteria

Shiga toxin-producing bacteria fall into three major groups. The prototype is the mucosa-invasive *Shigella dysenteriae* type 1, the cause of severe bacillary dysentery in many developing countries in different parts of the world. The second major group of Stx-producing bacteria is non-invasive *E. coli*. Currently, over 200 different *E. coli* types have been found to produce Stx family toxins. Of these, at least 60 serotypes have been associated with disease in humans. *Escherichia coli* O157:H7 was the first to gain significant notoriety following two outbreaks of fast-food hamburger-associated bloody diarrhoea in 1982 (Riley *et al.*, 1983) and a large outbreak on the west coast of the USA in 1993 in which over 750 people became sick, 55 developed haemolytic uraemic syndrome and four died. Since then many O157 outbreaks have been reported in the USA and other parts of the world, including a series of outbreaks in Japan involving close to 10 000 patients in 1996; O157:H7 is generally considered to be the major STEC pathogen causing human disease. However, the relative ease by which O157:H7 is isolated in the clinical microbiology laboratory, because of its slow sorbitol fermentation compared with other *E. coli*, may have resulted in an overestimate of the importance of this serotype compared with non-O157:H7 STEC. A number of other serotypes, including O111, O26, O103, O121, and O145 and O157:H–, has been isolated with increasing frequency from patients with bloody diarrhoea and HUS in many different parts of the world (Paton and Paton, 1998). The third group of Stx-producing bacteria is other members of the family Enterobacteriaceae which have been implicated in cases of HUS. Stx-producing *Citrobacter freundii* were associated with a sizeable outbreak of HUS in Europe (Tschape *et al.*, 1995) and a Stx2-producing *Enterobacter cloaceae* was associated with a case of HUS in Australia (Paton and Paton, 1996). Stx1 production has also been described from strains of *Aeromonas hydrophila* and *A. caviae* (Haque *et al.*, 1996).

Effects of Shiga toxins on different cell types

As already discussed, different cells have different susceptibility to the various members of the Stx family of toxins. Initial studies using Shiga toxin employed cell types that were not directly relevant to the Stx-mediated disease in humans. These initial data showed that receptor number determines sensitivity, with a suggestion of cooperativity as receptor density increased and toxicity increased even more sharply (Jacewicz *et al.*, 1989). More recently, a variety of investigators has examined more relevant cell types in a systemic manner according to the assumption that toxin is produced first in the intestinal lumen, crosses the intestinal mucosa where it encounters various leucocyte populations, and gains access to the circulation where it may affect other circulating blood cells or attack the endothelial cell lining of blood vessels in toxin-sensitive organs.

In the initial studies of the 'enterotoxin' (fluid secretory) activity of Stx in ligated rabbit small bowel loops, epithelial cell damage and apoptosis of villus tip cells were noted (Keusch *et al.*, 1972a, b). Subsequent *in vivo* electrolyte transport and unidirectional flux measurements using rabbit intestinal mucosal sheets mounted in an Ussing chamber revealed that villus cell sodium absorption was diminished by Stx, with no alteration in substrate-coupled sodium absorption or active chloride secretion. This suggested that toxin acted on the absorptive villus cell and not on the secretory crypt cell (Donowitz *et al.*, 1975). Subsequent studies, summarized above, support this conclusion, showing that Shiga toxin targets the villus cell because they, and not crypt cells, express the Gb3 receptor and are susceptible to inhibition of villus cell protein synthesis, resulting in diminished sodium absorptive capacity (Kandel *et al.*, 1989).

Obrig and co-workers (Obrig *et al.*, 1988; Louise and Obrig, 1991) were among the first to recognize that HUS might be due to toxin effects on endothelial cells. They reported that preincubation of human umbilical vein endothelial cells (HUVEC) with LPS or LPS-induced cytokines (IL-1) or tumour necrosis factor (TNF) induced Gb3 and converted the relatively Stx-resistant HUVEC into relatively responsive cells. These investigators later reported that human glomerular endothelial cells (GEC) constitutively produced Gb3 and were not further induced by cytokines, suggesting a reason why these cells may be a preferred target for Stx (Obrig *et al.*, 1993). The significance of this is now uncertain as Monnens *et al.* (1998) have reported conflicting data demonstrating that GECs required pre-exposure to TNFα in order to become sensitive to Stx. Inter-individual variation may be the explanation for these different observations. The effects of Stx1 on cerebral endothelial cells (CEC) was examined by Hutchinson *et al.* (1998), who found that CECs were sensitive to Stx1 but that this sensitivity could be enhanced significantly with exposure to IL-1β and TNFα. In relation to the intestine, recent data from our laboratory (Jacewicz *et al.*, unpublished observations) using human intestinal microvascular endothelial cells showed that these cells are very sensitive to both Stx1 and Stx2, and were not made more sensitive by exposure to proinflammatory cytokines or LPS.

Shiga toxin has also been shown to interact with polymorphonuclear leucocytes and macrophages. King *et al.* (1998) have demonstrated that Stx1 induces release of reactive intermediates from PMNs and causes a reduction in their phagocytosis without inducing apoptosis or necrosis. Barrett *et al.* (1990) found that when murine peritoneal macrophages were exposed to Stx2 there was an increase in TNF bioactivity. Tesh *et al.* (1994) showed that murine peritoneal macrophages, human peripheral blood monocytes and human monocytic cell lines (Ramegowda and Tesh, 1996) were relatively refractory to the cytotoxic action of Stxs but responded by secreting the proinflammatory cytokines TNFα and IL-1 in a dose-dependent manner. Human peripheral blood monocytes have also been reported (Van Setten *et al.*, 1996) to make TNFα, IL-1β, IL-6 and IL-8 in response to Stx1.

STEC in relation to the gastrointestinal tract

STEC are ingested by mouth, colonize portions of the lower intestine and then produce Shiga toxins. Two recent reviews (Paton and Paton, 1998; Nataro and Kaper, 1998) discuss many aspects of the pathogenesis of STEC and their diagnosis, and the reader is referred to these papers for a detailed discussion. However, there are several important issues that warrant emphasis. First, although it is known that STEC are present in the intestine and that they produce Stx, nothing is known about the regulation of Stx expression *in vivo*. It is an axiom of pathogenic microbiology that virulence factors are regulated *in vivo* in a manner that may or may not be replicated *in vitro*, and that some virulence factors are produced only in the former milieu. As mentioned previously, Stx and Stx1, but not Stx2, are regulated *in vitro* by the Fur system. Whether or not this operates *in vivo* is unknown. Other possible factors affecting toxin concentrations in intestinal secretions *in vivo* include the use of antibiotics. Subinhibitory concentrations of certain antibiotics increase the production and release of Stx *in vitro* (Karch *et al.*, 1985, 1986; Walterspiel *et al.*, 1992). Recent data from our laboratory suggest that this may be due to the induction of the Stx-encoding bacteriophages, and that this may have important *in vivo* relevance (Acheson *et al.*, unpublished observations).

The mechanisms by which Stx crosses the intestinal barrier are poorly understood. We have used a simplified *in vitro* model of intestinal mucosa to address this question using cultured intestinal epithelial cells grown on permeable polycarbonate filters (Acheson *et al.*, 1996a). These monolayers develop functional tight junctions, with high transcellular electrical resistance. Biologically active Stx translocates across these epithelial cell barriers in an apical to basolateral direction without disrupting the tight junctions, apparently through the cells (transcellular) rather than between the cells (paracellular). This pathway appears to be energy dependent, saturable and directional. Quantitative measurements of toxin transfer in this artificial system indicate that if the same events were to occur to the same extent *in vivo*, sufficient toxin molecules would cross the intestinal epithelial cell barrier to initiate the endothelial cell pathology associated with the microangiopathy seen in STEC-related disease.

DETECTION

There is a variety of ways in which Shiga toxins can be detected. The use of tissue culture cytotoxicity assays is one of the classic methods and has the advantage of sensitivity. The disadvantage of cytotoxicity assays is that they require tissue culture, are relatively expensive, time-consuming and require confirmation of specificity using anti-Stx antibodies. Vero cells and HeLa cells are the two typical cell lines that have been used in the majority of laboratories for these assays. When cells are exposed to the toxin there are morphological changes, and detachment of the monolayer when the cells are killed. This can be assessed subjectively using microscopy or quantitated using cell stains such as crystal violet (Gentry and Dalrymple, 1980). Measuring the levels of radiolabelled amino acid incorporation is another approach, which directly measures the inhibition of protein synthesis and not cell death. This has the advantage of being faster, in that the time of toxin exposure can be less (as little as 3 h in some cases), but the disadvantages of being somewhat cumbersome and costly (Keusch *et al.*, 1988).

A variety of enzyme immunoassays is also available for the detection and quantitation of Shiga toxins, although purified toxin is needed for quantitative assays. Various capture systems have been described, including the use of monoclonal or polyclonal antibodies, hydatid cyst material containing the P1 glycoprotein (Acheson *et al.*, 1990), and the glycolipids Gb3 and Gb4 (Ashkenazi and Cleary, 1989, 1990; Acheson *et al.*, 1993b). Currently, there are two commercially available enzyme immunoassays available in kit form, Premier EHEC (Meridian Diagnostics Inc., Cincinnati, OH, USA) and ProSpect T shiga toxin *E. coli* (Alexon-Trend Inc., San Jose, CA, USA). Both are simple to use, and both detect Stx1 and Stx2.

The use of genetic methods to detect Stx or STEC includes standard polymerase chain reaction (PCR) or gene-probe methodologies and are discussed extensively in a recent review by Paton and Paton (1998).

CLINICAL ASSOCIATIONS

The evidence implicating Shiga toxin in the pathogenesis of Shigellosis is not conclusive, and is based on animal models and *in vitro* studies in cell culture. None of the animal models used, including oral infection in primates, truly mimics human infection. Although Shiga toxin causes fluid secretion in rabbits (Keusch *et al.*, 1972a) and results in inflammatory enteritis in the rabbit model (Keusch *et al.*, 1972b), and is cytotoxic to human colonic epithelial cells (Moyer *et al.*, 1987) and thus can cause manifestations of clinical Shigellosis, the interpretation is complicated because *Shigella* are invasive and multiply within epithelial cells. Fontaine *et al.* (1988) created a Shiga toxin deletion mutant strain of *S. dysenteriae* type 1, and compared its clinical effects in the monkey with a wild-type strain. The toxin-negative mutant caused disease when fed to monkeys, but it was much less haemorrhagic than the wild-type, suggesting that the toxin plays a role in the haemorrhagic component of the dysenteric phase of Shigellosis, but is not a necessary factor for the initiation of clinical disease. Thus, the presence of Stx may add to the level of intestinal haemorrhage *in vivo* in patients with Shigellosis. There is strong epidemiological evidence to link the one Stx-producing *Shigella* species (*S. dysenteriae*) with HUS in many parts of the world, particularly South Africa and Bangladesh, suggesting that once systemic exposure to the toxin occurs there is an increased risk of HUS.

Evidence that the Shiga toxins from *E. coli* are involved in disease pathogenesis is much stronger. However, the evidence is predominantly epidemiological because it is unethical to challenge human subjects with STEC strains. There is strong epidemiological evidence for the association of Stx1- and/or Stx2-producing STEC with both outbreaks and sporadic disease. Typically, the disease begins with watery diarrhoea, abdominal pain, some nausea and vomiting but little fever. This may or may not progress to bloody diarrhoea, which in its most profound form causes haemorrhagic colitis and haemolytic uraemic syndrome. Exposure to STEC has also been associated with the development of thrombotic thrombocytopenic purpura (TTP) (Keusch and Acheson, 1997). Of all the clinical associations with STEC, the development of haemolytic uraemic syndrome (HUS) is the most feared complication. Karmali *et al.* (1985) first noted a link between Stx-producing *E. coli* and HUS, a finding that has subsequently been corroborated by many investigators (Karmali, 1989). It is not clear what predisposes an individual to develop HUS. In various large outbreaks of STEC and in dysentery due to *S. dysenteriae* type 1 it appears that between 5% and 10% of patients who become infected will go on to develop HUS, and of those that get HUS around 5% will die during the acute part of the disease.

Previously it was considered that STEC, especially O157:H7, were mainly related to outbreaks. As surveillance for O157:H7 and other STEC has increased it has become clear that they are linked with considerable sporadic disease (Acheson and Keusch, 1996). For instance, a recent report from Virginia in the USA isolated 11 STEC of 270 diarrhoeal samples examined for bacterial enteric pathogens. Only six were O157:H7 and five were non-O157:H7, including one O111 and one O103, both of which were from patients with bloody diarrhoea (Park *et al.*, 1996). In this study STEC were almost as common as *Salmonella* ($n = 13$) and exceeded both *Campylobacter* ($n = 7$) and *Shigella* ($n = 4$). Other, larger, studies have confirmed the prevalence of STEC in the USA, which is found in 0.5–1% of stools submitted to clinical laboratories for testing. Of those from which isolates are obtained 30–50% are non-O157:H7 STEC (Park *et al.*, 1996; Acheson *et al.*, 1998).

SHIGA TOXIN VACCINES

That Shiga toxins play an important role in the development of HUS and haemorrhagic colitis is now indisputable. Consequently, a number of investigators has considered the possibility of developing vaccines directed toward Shiga toxins in an attempt to prevent or ameliorate these complications. One option that has been examined is the use of the non-toxic B subunit as a vaccine antigen. Purified Stx1 B induces a protective immune response when given parenterally to rabbits, and may be a suitable antigen for use as a human vaccine (Mullett *et al.*, 1994). A second option that our laboratory has explored is the enteral delivery of the Stx1 B subunit. This has been undertaken by cloning the Stx1 B subunit gene into the *Vibrio cholerae* vaccine strain CVD103-HgR (Acheson *et al.*, 1996b). When this strain was fed to rabbits, the animals developed serum antibodies that were capable of neutralizing Stx1 holotoxin in a tissue culture assay. An alternative strategy is to undertake site-directed mutagenesis of the *stx* genes to produce biologically inactive products that retain appropriate immunogenicity and induce neutralizing antibodies.

A third strategy, adopted by a number of groups, is to produce mouse monoclonal antibodies directed towards Stx1 and Stx2. Such antibodies can then be humanized for use in humans. Two murine antibodies, 13C4 and 11E10, directed towards Stx1 and Stx2, respectively, have been humanized (Edwards *et al.*, 1997). Studies to examine the potential of these antibodies for passive immunotherapy are underway. A fourth strategy is to use an inactivated holotoxin vaccine. This is a direct, simple and rapid approach that is analogous to the use of toxoid vaccines in the therapy of tetanus and diphtheria. This approach requires the production of large amounts of purified Stx that is then chemically inactivated and used to raise polyclonal hyperimmune antiserum in human volunteers (Keusch *et al.*, 1998). Such antiserum can then be used as a passsive immunotherapeutic agent in patients infected with STEC.

SUMMARY

Shiga toxins are now well characterized genetically and biochemically, and much is known about their biological actions. The precise role of Stx in *S. dysenteriae*-related disease is still unclear, other than the strong association of toxin with the development of HUS. The role of Stx1 and Stx2 in STEC-related disease is much more clear-cut, although many questions remain concerning the pathways by which toxin actually moves from the intestinal lumen to its final site of action at distal sites such as the kidney and the brain. We believe that the microvascular endothelial cell is the primary target of the toxin; however, exactly how the toxin is affecting these cells remains unanswered. For example, is it simply a cytotoxic effect or are there more subtle perturbations of the endothelial cell physiology? Are these effects due to toxin alone or related to the action of multiple factors induced either locally or systemically? And, finally, how are these events translated into clinical illness? Further study will increase our understanding of the role of these toxins in human and animal disease and will lead to better diagnostics, and to a vaccine or therapy to prevent the severe consequences of STEC and, possibly, *S. dysenteriae* type 1 infections.

REFERENCES

Acheson, D.W.K. and Keusch, G.T. (1996). Which Shiga toxin-producing types of *E. coli* are important? *ASM News* **62**, 302–6.

Acheson, D.W.K., Keusch, G.T., Lightowers, M. and Donohue-Rolfe, A. (1990). Enzyme linked imunosorbent assay for Shiga toxin and Shiga-like toxin II using P1 glycoprotein from hydatid cysts. *J. Infect. Dis.* **161**, 134–7.

Acheson, D.W.K., Calderwood, S.B., Boyko, S.A., Lincicome, L.L., Kane, A.V., Donohue-Rolfe, A. and Keusch, G.T. (1993a). A comparison of Shiga-like toxin I B subunit expression and localization in *Escherichia coli* and *Vibrio cholerae*, using *trc* or iron-regulated promoter systems. *Infect. Immun.* **61**, 1098–104.

Acheson, D.W.K., Jacewicz, M., Kane, A.V., Donohue-Rolfe, A. and Keusch, G.T. (1993b). One step high yield purification of Shiga-like toxin II variants and quantitation using enzyme linked immunosorbent assays. *Microb. Pathogen.* **14**, 57–66.

Acheson, D.W.K., DeBreucker, S.A., Jacewicz, M., Lincicome, L.L., Donhue-Rolfe, A., Kane, A.V. and Keusch, G.T. (1995). Expression and purification of Shiga-like toxin II B subunits. *Infect. Immun.* **63**, 301–8.

Acheson, D.W.K., Moore, R., DeBreuker S. *et al.* (1996a). Translocation of Shiga-like toxins across polarized intestinal cells in tissue culture. *Infect. Immun.* **64**, 3294–300.

Acheson, D.W.K., Levine, M.M., Kaper, J.B. and Keusch, G.T. (1996b). Protective immunity to Shiga-like toxin 1 following oral immunization with Shiga-like toxin 1B producing *Vibrio cholerae* CVD103-HgR. *Infect. Immun.* **64**. 355–7.

Acheson, D.W.K., Frankson, K. and Willis, D. (1998). Multicenter prevalence study of Shiga toxin-producing *Escherichia coli*. In: *Annual Meeting of the American Society for Microbiology*, Atlanta, 1998 C-205.

Arab, S. and Lingwood, C.A. (1996). Influence of phospholipid chain length on verotoxin/globotriaosyl ceramide binding in model membranes: comparison of a supported bilayer film and liposomes. *Glycoconj. J.* **13**, 159–66.

Ashkenazi, S. and Cleary, T.G. (1989). Rapid method to detect Shiga toxin and Shiga-like toxin I based on binding to globotriaosylceramide (Gb3), their natural receptor. *J. Clin. Microbiol.* **27**, 1145–50.

Ashkenazi, S. and Cleary, T.G. (1990). A method for detecting Shiga toxin and Shiga-like toxin I in pure and mixed culture. *J. Med. Microbiol.* **32**, 255–61.

Austin, P.R., Jablonski, P.E., Bohach, G.A., Dunker, A.K. and Hovde, C.J. (1994). Evidence that the A2 fragment of Shiga-like toxin type I is required for holotoxin integrity. *Infect. Immun.* **62**, 1768–75.

Barrett, T.J., Potter, M.E. and Strockbine, N.A. (1990). Evidence for participation of the macropage in Shiga-like toxin II-induced lethality in mice. *Microb. Pathog.* **9**, 95–103.

Boivin, A. and Mesrobeanu, L. (1937). Recherches sur les toxines des bacilles dysentériques, sur l'identité entre la toxine thermobile de Shiga et l'exotoxine présente dans les filtraté des cultures sur bouillon de la même bactérie. *C. R. Soc. Biol.* **126**, 323–5.

Bridgewater, F.A.J., Morgan, R.S., Rowson, K.E.K. and Payling-Wright, G. (1955). The neurotoxin of *Shigella shigae*. Mophological and functional lesions produced in the central nervous system of rabbits. *Br. J. Exp. Pathol.* **36**, 447–53.

Brown, J.E., Ussery, M.A., Leppla, S.H. and Rothman, S.W. (1980). Inhibition of protein synthesis by Shiga toxin. Activation of the toxin and inhibition of peptide elongation. *FEBS. Lett.* **117**, 84–8.

Calderwood, S.B. and Mekalanos, J.J. (1987). Iron regulation of Shiga-like toxin expression in *Escherichia coli* is mediated by the *fur* locus. *J. Bacteriol.* **169**, 4759–64.

Calderwood, S.B., Auclair, F., Donohue-Rolfe, A., Keusch, G.T. and Mekalanos, J.J. (1987). Nucleotide sequence of the Shiga-like toxin genes of *Escherichia coli*. *Proc. Natl Acad. Sci. USA* **84**, 4364–8.

Calderwood, S.B., Acheson, D.W.K., Goldberg, M.B., Boyko, S.A. and Donohue-Rolfe, A. (1990). A system for production and rapid purification of large amounts of Shiga toxin/Shiga-like toxin I B subunit. *Infect. Immun.* **58**, 2977–82.

Calderwood, S.B., Acheson, D.W.K., Keusch, G.T., Barratt, T.J., Griffin, P.M., Swarminathan, B., Kaper, J.B., Levine, B.S., Karch, H., O'Brien, A.D., Obrig, T.G., Takeda, Y., Tarr, P.I. and Wachsmith, I.K. (1996). Proposed new nomenclature for SLT (VT) family. *ASM News* **62**, 118-19.

Conradi, H. (1903). Über lösliche, durch aseptische Autolyste erhatlene Giftstoffe von Ruhr- und typhus-bazillen. *Dtsch. Med. Wochenschr.* **20**, 26–8.

De Grandis. S., Ginsberg, J., Toone, M., Climie, S., Friesen, J. and Brunton, J. (1987). Nucleotide sequence and promoter mapping of the *Escherichia coli* Shiga-like toxin operon of bacteriophage H-19B. *J. Bacteriol.* **169**, 4313–19.

De Grandis, S., Law, H., Brunton, J., Gyles, C. and Lingwood, C.A. (1989). Globotetrosylceramide is recognized by the pig edema disease toxin. *J. Biol. Chem.* **264**, 12520–5.

Donohue-Rolfe, A., Keusch, G.T., Edson, C., Thorley-Lawson, D. and Jacewicz, M. (1984). Pathogenesis of *Shigella* diarrhea. IX. Simplified high yield purification of Shigella toxin and characterization of subunit composition and function by the use of subunit specific monclonal and polyclonal antibodies. *J. Exp. Med.* **160**, 1767–81.

Donohue-Rolfe, A., Acheson, D.W.K., Kane, A.V. and Keusch, G.T. (1989a). Purification of Shiga toxin and Shiga-like toxin I and II by receptor analogue affinity chromatography with immobilized P1 glycoprotein and the production of cross-reactive monoclonal antibodies. *Infect. Immun.* **57**, 3888–93.

Donohue-Rolfe, A., Jacewicz, M. and Keusch, G.T. (1989b). Isolation and characterization of functional Shiga toxin subunits and renatured holotoxin. *Mol. Microbiol.* **3**, 1231–6.

Donowitz, M., Keusch, G.T. and Binder, H.J. (1975). Effect of *Shigella* enterotoxin on electrolyte transport in rabbit ileum. *Gastroenterology* **69**, 1230–7.

Edwards, A., Arbuthnott, K., Stinson, J.R., Wong, C.H., Schmitt, C. and O'Brien, A.D. (1997). Humanization of monoclonal antobodies against *Escherichia coli* toxins Stx1 and Stx2. In: *VTEC '97: 3rd International Symposium and Workshop on Shiga Toxin (Verocytotoxin)-producing Escherichia coli Infections*, Abstract V110/V11, p. 113.

Endo, Y. and Tsurugi, K. (1987). RNA N-glycosidase activity of ricin A-chain. Mechanism of actin of the toxic lectin ricin on eukaryotic ribosomes. *J. Biol. Chem.* **262**, 8128–30.

Endo, Y., Tsurgi, K., Yutsudo, T., Takeda, Y, Igasawara, T. and Igarashi, E. (1988). Site of action of A Vero toxin (VT2) from *Escherichia coli* O157:H7 and of Shiga toxin on eukaryotic ribosomes. *Eur. J. Biochem.* **171**, 45–50.

Flexner, S. (1900). On the etiology of tropical dysentery. *Bull. Johns Hopkins Hosp.* **11**, 231–42.

Fontaine, A., Arondel, J. and Sansonetti, P.J. (1988). Role of Shiga toxin in the pathogenesis of bacillary dysentery, studied by using a Tox-mutant of *Shigella dysenteriae* 1. *Infect. Immun.* **56**, 3099–109.

Fraser, M.E., Chernaia, M.M., Kozlov, Y.V. and James, M.N. (1994). Crystal structure of the holotoxin from *Shigella dysenteriae* at 2.5 Å resolution. *Nat. Struct. Biol.* **1**, 59–64.

Gangarosa, E.J., Perera, D.R., Mata, L.J., Mendiazabal-Morris, Guzman, C. and Reller, L.B. (1970). Epidemic shiga bacillus dysentery in central America. II. Epidemiologic studies in 1969. *J. Infect. Dis.* **122**, 181–90.

Garred, O., van Deurs, B. and Sandvig, K. (1995a). Furin-induced cleavage and activation of Shiga toxin. *J. Biol. Chem.* **270**, 10817–21.

Garred, O., Dubinina, E. and Holm, P.K., Olnes, S., van Deurs, B., Kozlov, J.V. and Sandvig, K. (1995b). Role of processing and intracellular transport for optimal toxicity of Shiga toxin and toxin mutants. *Exp. Cell Re*s. **218**, 39–49.

Garred, O., Dubinina, E., Polessakaya, A., Olsnes, S., Kozlov, J. and Sandvig, K. (1997). Role of the disulfide bond in Shiga toxin A-chain for toxin entry into cells. *J. Biol. Chem.* **272**, 11414–19.

Gentry, M.K. and Dalrymple, J.M. (1980). Quantitative microtitre cytotoxicity assay for *Shigella* toxin. *J. Clin. Microbiol.* **12**, 361–6.

Habib, N.F. and Jackson, M.P. (1992). Identification of a B subunit gene promoter in the Shiga toxin operon of *Shigella dysenteriae* 1. *J. Bacteriol.* **174**, 6498–507.

Habib, N.F. and Jackson, M.P. (1993). Secondary structure in differential expression of Shiga toxin genes. *J. Bacteriol.* **175**, 597–603.

Haque, Q.M., Sugiyama, A., Iwade, Y., Midorikawa, Y. and Yamauchi, T. (1996). Diarrheal and environmental isolates of *Aeromonas* spp. produce a toxin similar to Shiga-like toxin I. *Curr. Microbiol.* **32**, 239–45.

Hovde, C.J., Calderwood, S.B., Mekalanos, J.J. and Collier, R.J. (1988). Evidence that glutamic acid 167 is an active-site residue of Shiga-like toxin I. *Proc. Natl Acad. Sci. USA* **85**, 2568–72.

Howard, J.G. (1955). Observations on the intoxication produced in mice and rabbits by the neurotoxin of *Shigella shigae*. *Br. J. Exp. Pathol.* **36**, 439–46.

Huang, A., De Grandis, S., Friesen, J., Karmali, M., Petric, M., Congi, R. and Brunton, J.L. (1986). Cloning and expression of the gene specifying Shiga-like toxin production in *Escherichia coli*. H19. *J. Bacteriol.* **166**, 375–9.

Huang, A., Friesen, J. and Brunton, J.L. (1987). Characterization of a bacteriophage that carries the genes for production of Shiga-like toxin I in *Escherichia coli*. *J. Bacteriol.* **169**, 4308–12.

Hutchinson, J.S., Stanimirovic, D., Shapiro, A. and Armstrong G.D. (1998). Shiga toxin toxicity in human cerebral endothelial cells. In: *Escherichia coli O157:H7 and Other Shiga Toxin Producing* E. coli *Strains* (eds J.B. Kaper and A.D. O'Brien), pp. 323–8. American Society for Microbiology, Washington, DC.

Igarashi, K., Ogasswara, T., Ito, K., Yutsudo, T. and Takada, Y. (1987). Inhibition of elongation factor 1-dependent amionacyl-tRNA binding to ribosomes by Shiga-like toxin I (VT1) from *Escherichia coli* O157:H7 and by Shiga toxin. *FEMS Microbiol. Lett.* **44**, 91–4.

Jacewicz, M., Clausen, H., Nudelman, E., Donohue-Rolfe, A. and Keusch, G.T. (1986). Pathogenesis of *Shigella* diarrhea. XI Isolation of a shigella toxin-binding glycolipid from rabbit jejunum and HeLa cells and its identification as globotriosylceramide. *J. Exp. Med.* **163**, 1391–404.

Jacewicz, M., Feldman, H.A., Donhoue-Rolfe, A., Balasubramanian, K.A. and Keusch, G.T. (1989). Pathogenesis of *Shigella* diarrhea XIV. Analysis of Shiga toxin receptors on cloned HeLa cells. *J. Infect. Dis.* **159**, 881–9.

Jacewicz, M.S., Mobassaleh, M., Gross, S.K., Balasubramanion, K.A., David, P.F., Raghavan, S., McCluer, R.H. and Keusch, G.T. (1994). Pathogenesis of *Shigella* diarrhea. XVII. A mammalian cell membrane glycolipid, Gb3, is required but not sufficient to confer sensitivity to Shiga toxin. *J. Infect. Dis.* **169**, 538–46.

Jacewicz, M.S., Mobassaleh, M., Acheson, D.W.K., Mobassaleh, M., Donahue-Rolfe, A., Balasubramanion, K.A. and Keusch, G.T. (1995). Maturational regulation of globotriaosylceramide, the Shiga-like toxin I receptor, by butyrate in intestinal epithelial lines. *J. Clin. Invest.* **96**, 1328–35.

Jackson, M.P., Neill, R.J., O'Brien, A.D., Holmes, R.K. and Newland, J.W. (1987a). Nucleotide sequence analysis and comparison of the structural genes for Shiga-like toxin I and Shiga-like toxin II encoded by bacteriophages from *Escherichia coli* 933. *FEMS Microbiol. Lett.* **44**, 109–14.

Jackson, M.P., Newland, J.W., Holmes, R.K. and O'Brien, A.D. (1987b). Nucleotide sequence analysis of the structural genes for Shiga-like toxin I encoded by bacteriophage 933J from *Escherichia coli*. *Microb. Pathogen.* **2**, 147–53.

Jackson, M.P., Wadolkowski, E.A. Weinstein, D.L., Holmes, R.K. and O'Brien, A.D. (1990a). Functional analysis of the Shiga toxin and Shiga-like toxin type II variant binding subunit by using site-directed mutagenesis. *J. Bacteriol.* **172**, 653–8.

Jackson, M.P., Deresiewicz, R.L. and Calderwood, S.B. (1990b). Mutational analysis of the Shiga toxin and Shiga-like toxin II enzymatic subunits. *J. Bacteriol.* **172**, 3346–50.

Jemal, C., Haddad, J.E., Begum, D. and Jackson, M.P. (1995). Analysis of Shiga toxin subunit association by using hybrid A polypeptides and site-specific mutagenesis. *J. Bacteriol.* **177**, 3128–32.

Kandel, G., Donohue-Rolfe, A., Donowitz, M. and Keusch, G.T. (1989). Pathogenesis of *Shigella* diarrhea XVI. Selective targeting of Shiga toxin to villus cells of rabbit jejunum explains the effect of the toxin on intestinal electrolyte transport. *J. Clin. Invest.* **84**, 1509–17.

Karch, H., Goroncy-Bermes, P., Opferkuch, W., Droll, H.P. and O'Brien, A.D. (1985). Subinhibitory concentrations of antibiotics modulate amount of Shiga-like toxin produced by *Escherichia coli*. In: *The Influence of Antibiotics on the Host–Parasite Relationship* (eds D. Adam, H. Hahr and W. Opferkuch), pp. 239–45. Springer, Berlin.

Karch, H., Strockbine, N.A. and O'Brien, A.D. (1986). Growth of *Escherichia coli* in the presence of trimethoprim-sulfamethoxazole facilitates detection of Shiga-like toxin producing strains by colony blot assay. *FEMS Microbiol. Lett.* **35**, 141–5.

Karmali, M.A. (1989). Infection by verocytotoxin-producing *Escherichia coli*. *Microbiol. Rev.* **2**, 15–38.

Karmali, M.A., Steele, B.T., Petric, M. and Lim, C. (1983). Sporadic cases of hemolytic-ureamic syndrome associated with faecal cytotoxin and cytotoxin-producing *Escherichia coli* in stools. *Lancet* **ii**, 619–20.

Karmali, M.A., Petric, M., Lim, C., Fleming, P.C., Arbus, G.A. and Lior, H. (1985). The association between idiopathic hemolytic uremic syndrome and infection by verotoxin producing *Escherichia coli*. *J. Infect. Dis.* **151**, 775–82.

Keusch, G.T. and Acheson, D.W.K. (1997). Thrombotic thrombocytopenic purpura associated with Shiga toxins. *Semin. Hematol.* **34**, 106–16.

Keusch, G.T. and Jacewicz, M. (1975). Pathogenesis of *Shigella* diarrhea.V. Relationship of Shiga enterotoxin, neurotoxin, and cytotoxin. *J. Infect. Dis.* **131**, S33–9.

Keusch, G.T. and Jacewicz, M. (1977). Pathogenesis of *Shigella* diarrhea. VII. Evidence for a cell membrane toxin receptor involving β1–4 linked *N*-acetyl-D-glucosamine oligomers. *J. Exp. Med.* **146**, 535–46.

Keusch, G.T., Grady, G.F., Mata, L.J. and McIver, J. (1972a). The pathogenesis of *Shigella* diarrhea. I. Enterotoxin production by *Shigella dysenteriae* 1. *J. Clin. Invest.* **51**, 1212–18.

Keusch, G.T., Grady, G.F., Takeuchi, A. and Sprinz, H. (1972b). The pathogenesis of *Shigella* diarrhea II. Enterotoxin induced acute enteritis in the rabbit ileum. *J. Infect. Dis.* **126**, 92–5.

Keusch, G.T., Donohue-Rolfe, A. and Jacewicz, M. (1986). *Shigella* toxins: description and role in diarrhea and dysentery. In: *Pharmacology of Bacterial Toxins* (eds F. Dorner and J. Drews), pp. 235–70. Pergamon Press, Oxford.

Keusch, G.T., Donohue-Rolfe, A., Jacewicz, M. and Kane, A.V. (1988). Shiga toxin: production and purification. *Methods Enzymol.* **165**, 152–62.

Keusch, G.T., Acheson, D.W.K., Marchant, C. and Mciver, J. (1998). Toxoid-based active and passive immunization to prevent and/or modulate hemolytic-uremic syndrome due to Shiga toxin-producing *Escherichia coli*. In: E. coli *0157: H7 and Other Shigi Toxin-producing* E. coli *Strains* (eds A.D. O'Brien and J.B. Kapar), pp. 409–18. American Society for Microbiology.

Kiarash, A., Boyd, G. and Lingwood, C.A. (1994). Glycosphingolipid receptor function is modified by fatty acid content. Verotoxin 1 and verotoxin 2c preferentially recognize different globotriaosyl ceramide fatty acid homologues. *J. Biol. Chem.* **269**, 1139–46.

King, A.J., Sundaram, S., Cendoroglo, M., Acheson, D.W.K. and Keusch, G.T. (1998). Shiga toxin induces superoxide production in polymorphonuclear cells with subsequent impairment of phagocytosis and responsiveness to phorbol esters. *J. Infect. Dis.* **179**, 503–7.

Konowalchuk, J., Speirs, J.I. and Stavric, S. (1977). Vero response to a cytotoxin of *Escherichia coli*. *Infect. Immun.* **18**, 775–9.

Konowalchuk, J., Dickie, N., Stavric, S. and Speirs, J.I. (1978). Properties of an *Escherichia coli* cytotoxin. *Infect. Immun.* **20**, 575–7.

Koster. F., Levin. J. and Walker, L. Tung, K.S., Gilman, R.H., Rahaman, M.M., Majid, M.A., Islam, S. and Williams, R.C. Jr. (1978). Hemolytic-uremic syndrome after shigellosis. *N. Engl. J. Med.* **298**, 927–33.

Kozlov, Y.V., Kabishev, A.A., Lukyanov, E.V. and Bayev, A.A. (1988). The primary structure of the operons coding for *Shigella dysenteriae* toxin and temperate phage H30 Shiga-like toxin. *Gene* **67**, 213–21.

Lindberg, A.A., Brown, J.E., Stromberg, N., Westling-Ryd, M., Schultz, J.E. and Karlson, K. (1987). Identification of the carbohydrate receptors for Shiga toxin produced by *Shigella dysenteriae* type 1. *J. Biol. Chem.* **262**, 1779–85.

Lindgren, S.W., Samuel, J.E., Schmitt, C.K. and O'Brien, A.D. (1994). The specific activities of Shiga-like toxin type II (SLT-II) and SLT-II-related toxins of enterohemorrhagic *Escherichia coli* differ when measured by Vero cell cytotoxicity but not by mouse lethality. *Infect. Immun.* **62**, 623–31.

Lingwood, C.A. (1996). Role of verotoxin receptors in pathogenesis. *Trends Microbiol.* **4**, 147–53.

Louise, C.B. and Obrig, T.G. (1991). Shiga toxin associated hemolytic uremic syndrome: combined cytotoxic effects of Shiga toxin IL-1, and tumor necrosis factor alpha on human vascular endothelial cells *in vitro*. *Infect. Immun.* **59**, 4173–9.

Ludwig, K., Karmali, M.A., Winkler, M. and Petric, M. (1997). Rabbits immunized with Verocytotoxin 2 (VT2) toxoid are cross-protected against challenge by intravenous VT1. In: *3rd International Symposium on Shiga Toxin-producing* Escherichia coli *Infections*.

Mangeny, M., Richard, Y., Coulard, D., Tursz, T. and Wiels, J. (1991). CD77: an antigen of germinal center B cells entering apoptosis. *Eur. J. Immunol.* **21**, 1131.

Melton-Celsa, A.R. and O'Brien, A.D. (1998). Structure, biology, and relative toxicity of Shiga toxin family members for cells and animals. In: Escherichia coli *O157:H7 and Other Shiga Toxin Producing* E. coli *Strains* (eds J.B. Kaper and A.D. O'Brien), pp. 121–8. American Society for Microbiology, Washington, DC.

Mobassaleh, M., Donohue-Rolfe, A., Jacewicz, M., Grand, R.J. and Keusch, G.T. (1988). Pathogenesis of Shigella diarrhea: evidence for a developmentally regulated glycolipid receptor for Shigella toxin involved in the fluid secretory response of rabbit small intestine. *J. Infect. Dis.* **157**, 1023–31.

Mobassaleh, M., Gross, S.K., McCluer, R.H., Donohue-Rolfe, A. and Keusch, G.T. (1989). Quantitation of the rabbit intestinal glycolipid receptor for Shiga toxin. Further evidence for the developmental regulation of globotriaosylceramide in microvillus membranes. *Gastroenterology* **97**, 384–91.

Mobassaleh, M., Koul, O., Mishra, K., McCluer, R.H. and Keusch, G.J. (1994). Developmental regulation of intestinal Gb3 galactosyltransferase and α-galactosidase controls Shiga toxin receptors. *Am. J. Physiol.* **267**, G618–24.

Monnens, L., Savage, C.O. and Taylor, C.M. (1998). Pathophysiology of hemolytic-uremic syndrome. In: Escherichia coli *O157:H7 and Other Shiga Toxin Producing* E. coli *Strains* (eds J.B. Kaper and A.D. O'Brien), pp. 287–92. American Society for Microbiology, Washington, DC.

Moyer, M.P., Dixon, P.S., Rothman, S.W. and Brown, J.E. (1987). Cytotoxicity of Shiga toxin for primary cultures of human colonic and ileal epithelial cells. *Infect. Immun.* **55**, 1533–5.

Muhldorfer, I., Hacker, J., Keusch, G.T., Acheson, D.W.K., Tschape, H., Kane, A.V., Ritter, A., Olschlager, T. and Donohue-Rolfe, A. (1996). Regulation of the Shiga-like toxin II operon in *Escherichia coli*. *Infect. Immun.* **64**, 495–502.

Mullett, C., Acheson, D.W.K., Tseng, L.Y. and Boedecker, E. (1994). Parenteral administration with recombinant SLT-IB subunit protects against disease induced by *E. coli* strain RDEC-H19A. In: *2nd International Symposium and Workshop of Verocytotoxin (Shiga-like Toxin)-producing* Escherichia coli *Infections*, Abstract O3.2.

Nambiar, M.P. and Wu, H.C. (1995). Ilimaquinone inhibits the cytotoxicities of ricin, diphtheria toxin, and other proten toxins in Vero cells. *Exp. Cell Res.* **219**, 671–8.

Nataro, J.P. and Kaper, J.B. (1998). Diarrheagenic *Escherichia coli*. *Clin. Microbiol. Rev.* **11**, 142–201.

Neely, M.N. and Friedman, D.I. (1998). Function and genetic analysis of regulatory regions of coliphage H-19B: location of shiga-like toxin and lysis genes suggest a role for phage functions in toxin release. *Mol. Microbiol.* **28**, 1255–67.

Neisser, M. and Shiga, K. (1903). Ueber freie receptoren von typhus-und dysenterie-bazillen und über das dysenterie-toxin. *Dtsch. Med. Wochenschr.* **29**, 61–2.

Newland, J.W., Strockbine, N.A., Miller, S.F., O'Brien, A.D. and Holmes, R.K. (1985). Cloning of Shiga-like toxin structural genes from a toxin converting phage of *Escherichia coli*. *Science* **230**, 179–81.

Newland, J.W., Strockbine, N.A. and Neill, R.J. (1987). Cloning of genes for production of *Escherichia coli* Shiga-like toxin II. *Infect. Immun.* **55**, 2675–80.

Noda, M., Yutsudo, T., Nakabayashi, N., Hirayama, T. and Takeda, Y. (1987). Purification and some properties of Shiga-like toxin from *Escherichia coli* O157:H7 that is immunologically identical to Shiga toxin. *Microb. Pathogen.* **2**, 339–49.

O'Brien, A.D. and LaVeck, G.D. (1983). Purification and characterization of *Shigella dysenteriae*1-like toxin produced by *Escherichia coli*. *Infect. Immun.* **40**, 675–83.

O'Brien, A.D., LaVeck, G.D., Griffin, D.E. and Thompson, M.R. (1980). Characterization of *Shigella dysenteriae* 1 (Shiga) toxin purified by anti-Shiga toxin affinity chromatography. *Infect. Immun.* **30**, 170–9.

O'Brien, A.D., Newland, J.W., Miller, S.F., Holmes, R.K., Williams Smith, H. and Formal, S.B. (1984). Shiga-like toxin-converting phages from *Escherichia coli* strains that cause hemorrhagic colitis or infantile diarrhea. *Science* **226**, 694–6.

O'Brien, A.D., Marques, L.R.M., Kerry, C.F., Newland, J.W. and Holmes, R.K. (1989). Shiga-like toxin converting phage of enterohemorrhagic *Escherichia coli* strain 933. *Microb. Pathogen.* **6**, 381–90.

Obrig, T.G., Del Vecchio, P.J., Brown, J.E., Moran, T.P., Rowland, B.M., Judge, T.K. and Rothman, S.W. (1988). Direct cytotoxic action of Shiga toxin on human vascular endothelial cells. *Infect. Immun.* **56**, 2372–8.

Obrig, T.G., Louise, C.B., Lingwood, C.A., Boyd, B., Barley-Maloney, L. and Daniel, T.O. (1993). Endothelial heterogeneity in Shiga toxin receptors and responses. *J. Biol. Chem.* **268**, 15484–8.

Olsnes, S. and Eiklid, K. (1980). Isolation and characterization of *Shigella shigae* cytotoxin. *J. Biol. Chem.* **255**, 284–9.

Olsnes, S., Reisbig, R. and Eiklid K (1981). Subunit structure of Shigella cytotoxin. *J. Biol. Chem.* **256**, 8732–88.

Park, C.H., Gates, K.M. and Hixon, D.L. (1996). Isolation of Shiga-like toxin producing *Escherichia coli* (O157 and non-O157) in a community hospital. *Diagn. Microbiol. Infect. Dis.* **26**, 69–72.

Paton, A.W. and Paton, J.C. (1996). *Enterobacter cloacae* producing a Shiga-like toxin II-related cytotoxin associated with a case of hemolytic-uremic syndrome. *J .Clin. Microbiol.* **34**, 463–5.

Paton, J.C. and Paton, A.W. (1998). Pathogenesis and diagnosis of Shiga toxin-producing *Escherichia coli* infections. *Clin. Microbiol. Rev.* **11**, 450–79.

Perera, L.P., Samuel, J.E., Holmes, R.K. and O'Brien, A.D. (1991). Mapping the minimal contiguous gene segment that encodes functionally active Shiga-like toxin II. *Infect. Immun.* **59**, 829–35.

Ramegowda, B. and Tesh, V.L., (1996). Differentiation-associated toxin receptor modulation, cytokine production, and sensitivity to Shiga-like toxins in human monocytes and monocytic cell lines. *Infect. Immun.* **64**, 1173–80.

Ramotar, K., Boyd, B., Tyrrell, G., Gariepe, J., Lingwood, C. and Brunton, J. (1990). Characterization of Shiga-like toxin I B subunit purified from over-producing clones of the SLT-I B cistron. *Biochemistry* **272**, 805–11.

Reisbig, R., Olsnes, S. and Eiklid, K. (1981). The cytotoxic activity of *Shigella* toxin. Evidence for catalytic inactivation of the 60S ribosomal subunit. *J. Biol. Chem.* **256**, 8739–44.

Riley, L.W., Temis, R.S., Helgerson, S.D., McGee, H.B., Wells, J.G., Davis, B.R., Herbert, R.J., Olcott, E.S., Johnson, L.M., Hargrett, N.T., Blake, P.A. and Cohen, M.L. (1983). Hemorrhagic colitis associated with a rare *Escherichia coli* serotype. *N. Engl. J. Med.* **308**, 681–5.

Ryd, M., Alfredsson, H., Blomberg, L,. Andersson, A. and Lindberg, A.A. (1989). Purification of Shiga toxin by alpha-D-galactose-(1–4)-beta-D-galactose-(1–4)-beta-D-glucose–receptor ligand-based chromatography. *FEBS Lett.* **258**, 320–2.

Samuel, J.E., Perera, L.P., Ward, S., O'Brien, A.D., Ginsburg, V. and Krivan, H.C. (1990). Comparison of the glycolipid receptor specificities of Shiga-like toxin type II and Shiga-like toxin type II variants. *Infect. Immun.* **58**, 611–18.

Sandvig, K. and van Deurs, B. (1996). Endocytosis, intracellular transport, and cytotoxic action of Shiga toxin and ricin. *Physiol. Rev.* **76**, 949–66.

Sandvig, K., Prydzk, K., Ryd, M. and van Deurs, B. (1991). Endocytosis and intracellular transport of the glycolipid-binding ligand Shiga toxin in polarized MDCK cells. *J. Cell Biol.* **113**, 553–62.

Sandvig, K., Garred, O., van Helvoort, A., van Meer, G. and van Deurs, B. (1996). Importance of glycolipid synthesis for butyric acid-induced sensitisation to Shiga toxin and intracellular sorting of toxin in A431 cells. *Mol. Biol. Cell.* **7**, 1391–404.

Schmitt, C.K., Mckee, M.L. and O'Brien, A.D. (1991). Two copies of Shiga-like toxin II-related genes common in enterohemorrhagic *Escherichia coli* strains are responsible for the antigenic heterogeneity of the O157:H-strain E32511. *Infect. Immun.* **59**, 1065–73.

Scotland, S.M., Day, N.P. and Rowe, B. (1979). Production by strains of *Escherichia coli* of a cytotoxin (VT) affecting Vero cells. *Soc. Gen. Microbiol. Qt.* **6**, 156–7.

Shiga, K. (1898). Ueber den Dysenteriebacillus (*Bacillus dysenteriae*). *Zentralbl. Bakt. Parasit. Abt. 1 Orig.* **24**, 817–24.

Sixma, T.K., Stein, P.E., Hol, W.G.J. and Read, R.J. (1993). Comparison of the B pentamers of heat-labile enterotoxin and Verotoxin-1: two structures with remarkable similarity and dissimilarity. *Biochem. J.* **32**, 191–8.

Skinner, L.M. and Jackson, M.P. (1997). Investigation of ribosome binding by the Shiga toxin A1 subunit, using competition and site-directed mutagenesis. *J. Bacteriol.* **179**, 1368–74.

Smith, H.W. and Linggood, M.A. (1971). The transmissible nature of enterotoxin production in a human enteropathogenic strain of *Escherichia coli*. *J. Med. Microbiol.* **4**, 301–5.

Stein, P.E., Boodhoo, A., Tyrrell, G.J., Brunton, J.L. and Read, R.J. (1992). Crystal structure of the cell-binding B oligomer of verotoxin-1 from *E. coli*. *Nature* **355**, 748–50.

Strockbine, N.A., Jackson, M.P., Sung, L.M., Holmes, R.K. and O'Brien, A.D. (1988). Cloning and sequencing of the genes for Shiga toxin from *Shigella dysenteriae* type 1. *J. Bacteriol.* **170**, 1116–22.

Tesh, V.L., Ramegowda, B. and Samuel, J.E. (1994). Purified Shiga-like toxins induce expression of proinflammatory cytokines from murine peritoneal macrophages. *Infect. Immun.* **62**, 5085–94.

Tschape, H., Prager, R., Steckel, W., Fruth, A., Tietze, E. and Bohme, G. (1995). Verotoxinogenic *Citrobacter freundii* associated with severe gastroenteritis and cases of haemolytic uraemic syndrome in a nursery school. Green butter as the infection source. *Epidemol. Infect.* **114**, 441–50.

Tyrrell, G.J., Ramotar, K., Toye, B., Body, B., Lingwood, C.A. and Brunton, J.L. (1992). Alteration of the carbohydrate binding specificity of verotoxin from Galα1–4Gal to Gal NAcβ1–3Gal α1–4Gal and vice versa by site-directed mutagenesis of the binding subunit. *Proc. Natl Acad. Sci. USA* **89**, 524–8.

van Heyningen, W.E. and Gladstone, G.P. (1953). The neurotoxin of *Shigella dysenteriae* 1. Production, purification and properties of the toxin. *Br. J. Exp. Pathol.* **34**, 202–16.

van Setten, P.A., Monnens, L.A.H., Verstraten, G., van den Heuvel, L.P. and van Hinsbergh, V.W. (1996). Effects of verocytotoxin-1 on nonadherent human monocytes: binding characteristics, protein synthesis and induction of cytokine release. *Blood* **88**, 174–83.

Vicari, G., Olitzki, A.L. and Olitzki, Z. (1960). The action of the thermolabile toxin of *Shigella dysenteriae* on cells cultivated *in vitro*. *Br. J. Exp. Pathol.* **41**, 179–89.

Wade, W.G., Thom, B.T. and Evens, N. (1979). Cytotoxic enteropathogenic *Escherichia coli*. *Lancet* **ii**, 1235–36.

Walterspiel, J.N., Ashkenazi, S., Morrow, A.L. and Cleary, T.G. (1992). Effect of subinhibitory concentrations of antibiotics on extracellular Shiga-like toxin 1. *Infection* **20**, 25–9.

Weinstein, D.L., Jackson, M.P., Samuel, J.E., Holmes, R.K. and O'Brien, A.D. (1988). Cloning and sequencing of a Shiga-like toxin type II variant from an *Escherichia coli* strain responsible for edema disease of swine. *J. Bacteriol.* **170**, 4223–30.

Weinstein, D.L., Jackson, M.P., Perera, L.P., Holmes, R.K. and O'Brien, A.D. (1989). *In vivo* formation of hybrid toxins comprising Shiga toxin and the Shiga-like toxins and a role of the B subunit in localization and cytotoxic activity. *Infect. Immun.* **57**, 3743–50.

Wilson, M.W. and Bettelheim, K.A. (1980). Cytotoxic *Escherichia coli* serotypes. *Lancet* **ii**, 201.

12

The bifactorial *Bacillus anthracis* lethal and oedema toxins

Stephen A. Leppla

INTRODUCTION

Anthrax was recognized for many centuries as a serious disease of animals and humans, one that inflicted great losses in agricultural economies and caused significant disease in humans. Thus, anthrax was a major concern to the pioneers of microbiology, and its study by Pasteur, Koch, Mechnikoff and others established many of the basic principles of infectious diseases.

In domestic livestock and wild animals, symptoms of infection by *Bacillus anthracis* are rarely evident until the animal becomes lethargic several hours before death. Necropsy shows extensive oedema in many tissues and concentrations of bacteria in blood that may exceed 10^8 per millilitre (Turnbull, 1990a). Most human cases arise from contact with infected animals or spores present in animal products (wool, leather, bone meal), and begin as cutaneous infection. This is easily treated with antibiotics if correctly diagnosed. The less frequent but more dangerous gastrointestinal and respiratory forms of anthrax show a rapid progression and must be recognized early to be treated successfully (Franz *et al.*, 1997; Friedlander and Brachman, 1998).

The virulence of *B. anthracis* for animals and humans depends on the production of two recognized virulence factors, the gamma-linked poly-D-glutamic acid capsule, and the three-component protein exotoxin that is termed anthrax toxin (Smith *et al.*, 1955; Keppie *et al.*, 1963). The capsule appears to protect bacteria from phagocytosis, and therefore plays an essential role during establishment of an infection. The protein exotoxin may also help to establish an infection by incapacitating phagocytes (Keppie *et al.*, 1963; O'Brien *et al.*, 1985; Wade *et al.*, 1985), but its more obvious role is to cause the extensive tissue oedema that appears to be a principal cause of death. It is generally accepted that the pathological effects causing death in infected animals are due to the action of the toxin (Smith and Stoner, 1967). A number of reviews has discussed the toxin and its role in pathogenesis and immunity (Hambleton *et al.*, 1984; Stephen, 1986; Hambleton and Turnbull, 1990; Leppla, 1991a, 1995). The published proceedings of international workshops on anthrax held in 1989 and 1995 contain many short papers, of which a significant number discusses some aspect of the toxin (Turnbull, 1990b, 1996).

Virulent strains of *B. anthracis* contain two large plasmids, pXO1 and pXO2. The genes coding for toxin are contained on pXO1 (Mikesell *et al.*, 1983; Thorne, 1985), and the genes for capsule are on pXO2 (Green *et al.*, 1985; Uchida *et al.*, 1985). Virulence requires the presence of both plasmids, as is evident from comparison of strains that have lost either of the two plasmids. Strains lacking plasmid pXO1 do not produce toxin and are essentially avirulent (Ivins *et al.*, 1986; Uchida *et al.*, 1986). Strains lacking pXO2 are at least 10^5-fold less virulent than wild-type (Ivins *et al.*, 1986; Welkos and Friedlander, 1988). Although it is possible that these plasmids code for other materials that contribute to virulence, no such materials have yet been identified. *Bacillus anthracis* produces several other secreted and cytosolic materials that are potentially harmful to animal hosts; these include phospholipases, proteases and a thiol-activated haemolysin (unpublished studies of SHL). However, none of these materials has been shown to contribute to pathogenesis or to be plasmid encoded.

The anthrax toxin complex, the principal subject of this chapter, merits study both because it is a principal virulence factor of *B. anthracis*, and because it provides an attractive model for studying interactions of protein ligands with eukaryotic target cells. Foremost among these is the fact that this toxin complex contains three proteins that are individually non-toxic. These proteins are designated protective antigen (PA), lethal factor (LF) and oedema factor (EF). Toxic activity is obtained only when the proteins are administered in pairwise combinations. Two different toxic activities are produced. The combination of PA with LF, which causes rapid death of certain animal species when injected intravenously, is designated 'lethal toxin'. The combination of PA with EF, which causes oedema when injected intradermally, is designated 'oedema toxin'.

Work in recent years, to be detailed in this chapter, demonstrated that PA binds to receptors on eukaryotic cells and mediates the internalization of LF and EF to the cytosol. EF is an adenylate cyclase; it converts ATP to unphysiologically high concentrations of cAMP that cause metabolic perturbations (Leppla, 1982, 1984). LF was recently proven to be a metalloprotease (Klimpel *et al.*, 1994; Hammond and Hanna, 1998), and it was shown to cleave the mitogen activated protein kinase kinases 1 and 2 (MAPKK1 and 2, or MEK1 and MEK2) (Duesbery *et al.*, 1998).

The anthrax toxins can be viewed as fitting the A/B model described by Gill (1978), where the A moiety is a catalytic polypeptide (i.e. enzyme) and the B moiety is the receptor binding region. Anthrax toxin fits this pattern. The PA protein binds to receptors, and can therefore be considered the B moiety. LF and EF are each alternate A moieties. Anthrax toxin is unusual in that the A and B moieties are separate proteins, whereas in most toxins these functions are performed by separate domains on a single polypeptide. Several other toxins are now known that have separate protein components. Most similar to anthrax toxins is a group of related clostridial toxins. In these toxins, the receptor recognition protein that resembles PA is activated by proteolysis and then binds and promotes the internalization of an actin ADP-ribosylating protein (Aktories and Wegner, 1992).

The anthrax oedema toxin is one of two known bacterial 'invasive' adenylate cyclases that were recognized at approximately the same time; the other is the *Bordetella pertussis* cyclase, which is discussed in another chapter of this book (Chapter 7). Both adenylate cyclases require a eukaryotic protein, calmodulin, as a cofactor for enzymic activity. The structural and functional similarities of the two cyclases are discussed in a later section.

Some confusion may arise from use of the designation PA for a component of a toxin. Before discovery of the toxin, it was shown that culture supernatants of *B. anthracis* could immunize animals against infection (Gladstone, 1948). Only after the toxin was discovered did it become clear that the 'protective antigen' in the culture supernates was a component of the toxin. PA remains the principal and essential immunogen in both killed and live anthrax vaccines (Hambleton *et al.*, 1984; Ivins and Welkos, 1988). Therefore, study of anthrax toxin has direct application to development of improved vaccines. This review will not attempt to cover in depth the issues of anthrax immunity and vaccines.

THE GENETICS OF TOXIN AND VIRULENCE

Plasmids pXO1 and pXO2

The possibility that the genes for *B. anthracis* virulence factors might be extrachromosomal was suggested by the classical studies of Louis Pasteur and Max Sterne, who each isolated variants of *B. anthracis* that had reduced virulence. The ease with which these variants were obtained suggested that the genes for toxin and capsule might be extrachromosomal. After some initial difficulties, plasmids associated with toxin and capsule were discovered (Mikesell *et al.*, 1983; Green *et al.*, 1985; Uchida *et al.*, 1985). Recognition and characterization of the plasmids was delayed because these plasmids are very large, and are destroyed by shear unless special precautions are employed.

The *B. anthracis* plasmids can each be selectively cured, pXO1 by repeated passage at 42°C, and pXO2 by growth in novobiocin (Figure 12.1). Comparison of plasmid cured variants proved that pXO1 is needed for production of toxin, and pXO2 for production of capsule (Thorne, 1985). Conjugal transfer of pXO1 to plasmid-cured strains of *B. anthracis* confirmed that all the genes necessary for toxin production are contained on the plasmid (Thorne, 1985; Heemskerk and Thorne, 1990). These results explain the properties and the efficacy of the anthrax vaccines developed by Louis Pasteur and Max Sterne. It is now well established that *B. anthracis* strains must produce PA in order to induce protective immunity (Ivins and Welkos, 1988).

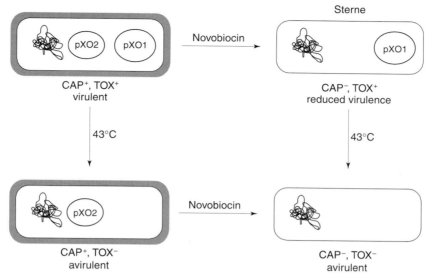

FIGURE 12.1 The four variants of *B. anthracis*. In the bacterium represented at the upper left, the random pattern denotes the chromosomal DNA, labelled ellipses denote plasmids, and the shaded zone at the outer edge denotes the polyglutamate capsule. Labelled arrows identify treatments that lead to loss of each plasmid. Cap⁺ or Cap⁻ denotes capsule phenotype, and Tox⁺ or Tox⁻ denotes toxin production phenotype.

Elimination of the pXO1 plasmid yields an avirulent strain, but one that does not induce immunity. In retrospect, it is evident that Louis Pasteur's attenuation of the virulence of *B. anthracis* cultures by growth at 42°C was due to partial curing of plasmid pXO1. The cultures he used successfully to immunize sheep probably contained a small number of virulent (pXO1⁺, pXO2⁺) bacteria with a larger number of avirulent (pXO1⁻, pXO2⁺) bacteria. It appears that the efficacy of Pasteur's vaccine depended on the presence of the small number of virulent bacteria, which would induce antibodies to PA. While effective, these vaccines had to be carefully prepared, because they could cause infection if the fraction of virulent organisms was too high.

Max Sterne's important contribution was carefully to analyse the rare, spontaneous non-capsulated variants appearing on agar plates, and to show that they were greatly reduced in virulence (Sterne, 1937). These variants, now known to have lost the pXO2 plasmid, were effective animal vaccines, and did not revert to virulence. The (pXO1⁺, pXO2⁻) 'Sterne' strain was immediately adopted and continues in use today as the preferred live animal vaccine.

Both of these large plasmids have been physically characterized (Kaspar and Robertson, 1987), restriction maps obtained, and the origins of replication cloned (Uchida *et al.*, 1987; Robertson *et al.*, 1990). Plasmid pXO1 is 170–185 kbp (Kaspar and Robertson, 1987; Heemskerk and Thorne, 1990; Robertson *et al.*, 1990) and pXO2 is 90–95 kbp (Uchida *et al.*, 1985; Robertson

et al., 1990). The three toxin genes are separated by about 3 kbp, and the EF gene is located approximately 20 kbp from the PA gene. The location on pXO2 and the DNA sequences of a cluster of genes involved in both synthesis and degradation of the capsule have also been determined (Makino *et al.*, 1989; Uchida *et al.*, 1993b). The DNA sequences of the toxin and *atxA* genes are discussed in a later section.

Genes specifying toxin structure and expression

The recognition that anthrax toxin was encoded on pXO1 facilitated the cloning of the genes encoding PA (*pag*), LF (*lef*) and EF (*cya*). Libraries of pXO1 DNA were screened by immunoblotting to identify clones containing *pag* and *lef* (Vodkin and Leppla, 1983; Robertson and Leppla, 1986) and *cya* (Tippetts and Robertson, 1988). Cloning of *cya* was also achieved independently, using a genetic selection for complementation of an adenylate cyclase mutant of *Escherichi coli* (Mock *et al.*, 1988).

The nucleotide sequences were determined for PA (Welkos *et al.*, 1988), LF (Bragg and Robertson, 1989) and EF (Escuyer *et al.*, 1988; Robertson *et al.*, 1988). Each of the genes has a G + C content of about 30% (Table 12.1), similar to that of the *B. anthracis* genomic DNA (35% G + C). Upstream of the ATG start codons in each of the genes is an appropriately located ribosome binding site, AAAGGAG for the PA and LF genes, and AAAGGAGGT for the EF gene. Each of the three genes shows a typical bacillus signal sequence of 29–33 amino acids (aa), with cleavage occurring after an Ala or Gly. The putative open reading frames of all three genes end at TAA codons. Following the end of the PA gene is an inverted repeat that may act as a transcriptional stop; no similar structures are present in the LF or EF regions.

Production of both capsule (Meynell and Meynell, 1964) and PA (Gladstone, 1948) by *B. anthracis* is dependent on addition of bicarbonate or CO_2. Early studies showed that stimulation of PA synthesis by bicarbonate requires the presence of a gene located on pXO1 (Bartkus and Leppla, 1989). This gene, *atxA*, was mapped by transposon mutagenesis (Hornung and Thorne, 1991) and cloned and sequenced (Uchida *et al.*, 1993a; Koehler *et al.*, 1994). It is located between *cya* and *pag* but is transcribed in the opposite direction. The product of the *atxA* gene, a protein of 56 kDa, increases by at least 10-fold the transcription from an otherwise silent start site, P1, located at bp −58 relative to the start codon. Constitutive, low-level transcription initiates at another site, P2, at bp −26. Both of these start sites, as well as the −35 and −10 regions corresponding to the P2 site, are located in a potential 58-bp stem–loop structure (Welkos *et al.*, 1988). Disruption of the *atxA* gene and complementation with an *atxA*-containing plasmid demonstrated that this positive regulator is needed for transcription of all three toxin genes, and that *B. anthracis* strains lacking *atxA* are less virulent for mice (Uchida *et al.*, 1993a; Dai *et al.*, 1995).

Regulation of toxin gene expression does not occur through variations in *atxA* transcription or protein expression, because these are regulated by temperature but not by CO_2 concentrations (Dai and Koehler, 1997). The AtxA protein has no sequence similarity to transcriptional activators that bind to DNA. Furthermore, no common sequences can be identified in the regions upstream of *pag*, *cya* and *lef*. Therefore, it was interesting to examine how the *atxA* gene product might be functioning. Reporter plasmids containing varying portions of the region upstream of the *pag* coding sequence showed that a 111-bp region upstream of the ATG is necessary for the *atxA* gene product to activate transcription (Dai *et al.*, 1995). Although it is suspected that the *atxA* product binds to this region, DNA binding has not yet been proven. Further studies on *atxA* suggest that it may be a global regulator of gene expression in *B. anthracis*, because transcription of at least 10 (non-toxin) genes on pXO1 requires *atxA* (Hoffmaster and Koehler, 1997). An analogous regulatory gene involved in the CO_2-dependent control of capsule synthesis was found on pXO2 and the gene was sequenced (Vietri *et al.*, 1995). Cross-talk between the regulatory genes on pXO1 and pXO2 was demonstrated (Guignot *et al.*, 1997; Uchida *et al.*, 1997).

Toxin gene sequence homologies and variations

Subsequent to the sequence determination of PA (Welkos *et al.*, 1988), other genes have been found that are closely related. All appear to encode two-compo-

TABLE 12.1 Properties of the anthrax toxin proteins

	PA[a]	LF[b]	EF[c]
%G + C in gene	31%	30%	29%
AA residues in mature protein	735	776	767
AA residues in signal sequence	29	33	33
Sequence at signal peptide cleavage site[d]	IQA*E	VQG*A	VNA*M
$M_r \times 10^3$			
calculated from sequence	82.7	90.2	88.8
by SDS electrophoresis[e]	85	83	89
by SDS electrophoresis[f]	85	87	86
Isoelectric point			
calculated from sequence	5.6	6.1	6.8
measured[f]	5.5	5.8	5.9
measured[g]	5.5	5.8	6.4
Database accession numbers	M22589	M29081	M23179
		M30210	M24074

[a]Except as noted, data are from Welkos *et al.* (1988)
[b]Except as noted, data are from Bragg and Robertson (1989)
[c]Except as noted, data are from Robertson *et al.* (1988) and Escuyer *et al.* (1988).
[d]Asterisk shows site of signal peptide cleavage.
[e]From Leppla (1988).
[f]From Quinn *et al.* (1988).
[g]Unpublished studies by S.F. Little.

TABLE 12.2 Protein neighbours of PA

Protein	Accession number	Reference
Clostridium perfringens iota toxin component Ib	EMBL X73562	Perelle *et al.* (1993)
Clostridium spiroforme toxin component Sb	EMBL X97969	Gibert *et al.* (1997)
Clostridium difficile binary toxin component CDTb	GENBANK L76081	Perelle *et al.* (1997)
Clostridium botulinum C2 toxin component II	Reserved	Kimura *et al.* (1998)
Bacillus cereus vegetative insecticidal protein	Reserved	Warren *et al.* (1996)

nent toxins, such as the anthrax toxins, but the second, catalytic component of each of these has ADP-ribosylation activity. The PA homologues are listed in Table 12.2, and the properties of the proteins are discussed in a later section.

The EF gene, *cya* (Escuyer *et al.*, 1988; Robertson *et al.*, 1988), has homology to another 'invasive' adenylate cyclase, that of *Bordetella pertussis* (Glaser *et al.*, 1988; Hanski and Coote, 1991); see also Chapter 7 in this book. EF has homology to the *B. pertussis* cyclase in three regions. From studies of the pertussis cyclase, the area of homology to EF is known to be the catalytic domain. Details of the amino acid sequence homologies of the two cyclases are discussed later. Searching the newly available database of incomplete bacterial genome sequences (www.ncbi.nlm.nih.gov/BLAST/unfinishedgenome.html) identifies two sequences with homology to EF. These are in *Yersinia pestis* and *Pseudomonas aeruginosa*. This suggests that these and probably other bacterial pathogens also utilize 'invasive' adenylate cyclases as virulence factors. The EF gene also has strong homo-

logy to the LF gene, *lef* (Bragg and Robertson, 1989). The amino-terminal 250 residues of the two proteins are very similar. Beyond residues 250, LF does not have significant sequence homology to EF or to other genes currently in the accessible databases, with the exception of very limited homology to metalloproteases. The value of the latter fact in identifying the mechanism of LF action is discussed later.

To date, little evidence has been obtained for sequence variations between *B. anthracis* toxin genes of different isolates. The two *cya* genes that were independently cloned and sequenced appeared to have a few differences, but this was later recognized to be due to a sequencing error, and both are now considered identical (see annotations to GenBank Accession M24074).

Role of the toxin in virulence

Recent years have seen the development of a number of tools for achieving genetic transfers of DNA into *B. anthracis*. These include transduction and conjugation (Battisti *et al.*, 1985; Heemskerk and Thorne, 1990) and electroporation (Bartkus and Leppla, 1989). Transposon Tn*916* has been used to produce aromatic amino acid requiring mutants (Ivins *et al.*, 1988), and the elegant methods developed for use of Tn*917* in *Bacillus subtilis* have been adapted to *B. anthracis* (Heemskerk and Thorne, 1990). A conjugational transfer system was used to transfer a mutated PA gene into the Sterne strain, replacing the resident PA gene (Cataldi *et al.*, 1990). The latter method was extended to produce strains producing every combination of the three components (Table 12.3). The only strain that retained any virulence was RP9, which makes PA and LF (Pezard *et al.*, 1991, 1993). This clearly proves that LF is the more important virulence factor, whereas EF makes a smaller contribution.

TABLE 12.3 Virulence properties of the modified *B. anthracis* strains

Strain name	Parental strain	Plasmid content	Proteins produced	LD_{50} for mice	Characteristics
7702	–	pXO1	PA LF EF	10^6	Sterne type
7700	7702	– (cured)		$>10^9$	Plasmid free
RP8	7702	pXO1 pag 322	LF EF	$>10^9$	
RP9	7702	pXO1 cya 303	PA LF	10^7	
RP10	7702	pXO1 lef 238	PA EF	$>10^9$	Causes oedema
RP4	RP9	pXO1 cya 303 pag 652	LF	$>10^9$	
RP31	RP10	pXO1 lef 238 pag 652	EF	$>10^9$	
RP42	RP10	pXO1 lef 238 cya 303	PA	$>10^9$	

Toxin genes were inactivated by insertion of erythromycin or kanamycin resistance genes (Cataldi *et al.*, 1990; Pezard *et al.*, 1991, 1993). Swiss mice were injected subcutaneously with spore suspensions and lethality was monitored.

THE PROTEINS

Production of toxin from *Bacillus anthracis* and *Bacillus subtilis*

Study of the anthrax toxin proteins has been greatly facilitated by the relative ease with which they can be prepared in milligram amounts. Extensive work in the period 1940–1965 led to development of synthetic media that support good production of PA (Puziss *et al.*, 1963; Haines *et al.*, 1965). Key ingredients of synthetic media are bicarbonate (Gladstone, 1946) to activate toxin gene transcription (Bartkus and Leppla, 1989), and a buffering system that maintains the pH above 7.0 (Strange and Thorne, 1958). Particular amino acid combinations promote toxin production, but this effect is complex because systematic trials failed to show that individual amino acids controlled toxin synthesis (S.H. Leppla, unpublished studies). Later work led to development of a completely synthetic medium, designated R, which was claimed to increase yields further (Ristroph and Ivins, 1983). Subsequently, slight medium modifications (RM and RMM) were made, and effective methods were developed for isolation of the toxin proteins from the cell-free supernatant of 50-l fermenter cultures (Leppla, 1988, 1991b). With these methods, a 50-l culture typically yields 500 mg PA, 100 mg LF and 40 mg EF.

Much effort has been put into producing PA in the absence of LF and EF, because PA is the major component of all anthrax vaccines. For this purpose, the PA gene and upstream sequences were cloned into the staphylococcal vector pUB110, which is a high copy number plasmid in bacilli. The resulting vectors, pPA101 and pPA102, expressed PA at 20–40 mg l^{-1} from *B. subtilis*, but the PA was rapidly destroyed by extracellular proteases (Ivins and Welkos, 1986). Expression from pPA101 in *B. subtilis* WB600, a strain in which six extracellular proteases are inactivated, improved yields of intact PA to about 5 mg l^{-1} (Miller *et al.*, 1998). However, even with the elimination of six proteases, degradation of PA in the WB600 cultures is still a major problem compared with the situation in *B. anthracis* culture supernatants.

An effective alternative for production of PA and LF in *B. anthracis* is the shuttle vector pYS5, constructed from pPA102 and a pBR322-derived plasmid (Singh *et al.*, 1989b). In the original pPA102 and the derivative pYS5 vectors, the *pag* structural gene and about 150 bp of upstream sequence follow a truncated bleomycin resistance gene originating from pUB110 (S.H. Leppla, unpublished results). Because this vector lacks the *atxA* regulator gene, transcription from the PA promoter is probably poor, explaining the failure to obtain significant PA expression in RM medium.

However, in a rich medium (FA) under optimal conditions, PA is produced at 50 mg l^{-1}. In this case, transcription probably originates from the bleomycin promoter. The pYS5 vector has been used successfully by the author's colleagues to produce PA and many PA mutants. Similarly, the original pPA102 plasmid has been used in a sporulation deficient *B. anthracis* strain grown in a rich medium such as FA under optimized fermenter conditions to obtain PA at 20–30 mg l^{-1} (Farchaus *et al.*, 1998). Efficient production of LF has been achieved in a modified pYS5 vector by replacing the region encoding PA aa 167–735 with a Factor Xa cleavage site and the LF gene. The resulting fusion protein, PA20-LF, is produced in levels similar to those of PA, and the LF protein is obtained by cleavage with Factor Xa (Klimpel *et al.*, 1994). In general, these findings emphasize the advantages of producing a protein in its original host, where there has been natural selection to adapt the toxin to the host's secretion apparatus and to resist the host's proteases.

Production of toxin components and fusion proteins in *Escherichia coli*

For production of mutant PA proteins, *E. coli* systems offer obvious advantages. T7-based vectors that include signal peptides were found successfully to secrete PA and PA mutant proteins to the *E. coli* periplasm with yields after purification of 0.5 mg l^{-1} (Sharma *et al.*, 1996; Benson *et al.*, 1998). A T5-based vector was used to produce LF with an *N*-terminal, 6-His sequence, which enabled convenient affinity purification at yields of 1.5 mg l^{-1} (Gupta *et al.*, 1998). *Escherichia. coli* has also been used to produce a number of modified LF and EF proteins. Fusion proteins in which LF residues 1–254 are attached to other polypeptides have been made in several standard expression systems (Arora and Leppla, 1993; Ballard *et al.*, 1996).

Toxin purification

The three toxin proteins together constitute more than 50% of the protein present in *B. anthracis* Sterne culture supernatants grown in R medium. This makes purification of the proteins relatively easy once the proteins have been protected from proteases and concentrated. Recovery from culture supernatants has been done by hydrophobic 'salting out' on to agarose resins (Leppla, 1991b), but more recently is done by ultrafiltration. Effective purification steps include chromatography on anion-exchange resins (Wilkie and Ward, 1967) or hydroxyapatite (Leppla, 1988). Detailed protocols for purification are available (Leppla, 1988, 1991b; Quinn *et al.*, 1988; Farchaus *et al.*, 1998). Immunoadsorbants prepared from monoclonal antibodies are useful for purifying less stable PA mutant proteins that need to be grown in culture medium supplemented with horse serum to inhibit degradation (Singh *et al.*, 1989b, 1991).

Relatively few procedures for purification of native LF and EF have been reported, and this is an area requiring additional effort. An EF protein truncated at the *N*-terminus by 261 aa (Cya 62) was expressed at 2% of total protein in *E. coli* and purified to homogeneity (Labruyere *et al.*, 1990). This protein retains full adenylate cyclase activity and has been used to characterize the catalytic properties of EF.

Structural features common to the three toxin components

All three of the anthrax toxin proteins are similar in size and charge (Table 12.1). The masses calculated from the DNA sequences agree reasonably well with those estimated in several different laboratories by sodium dodecyl sulfate (SDS) electrophoresis. It was noted some years ago that extracellular bacterial proteins generally have a low cysteine content (Pollock and Richmond, 1962). This generalization appears to hold for a number of other secreted bacterial toxins. Perhaps most striking is the absence of cysteines in the *Bordetella pertussis* adenylate cyclase, a protein of 1706 residues (Glaser *et al.*, 1988) and in several other RTX toxins (Welch, 1991).

Overview of toxin binding and internalization by cells

The data on toxin structure discussed above combined with studies on interaction with cells to be discussed

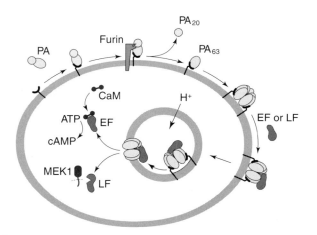

FIGURE 12.2 Steps in anthrax toxin action. Toxin components and cellular proteins that interact during LF and EF internalization are as those described in the text. CaM is calmodulin, and MEK1 is mitogen-activated protein kinase kinase 1 (MAPKK1 or MEK1).

below led to a model of toxin uptake depicted in Figure 12.2 (Petosa *et al.*, 1997). PA binds to cell surface receptors and is cleaved by a cell surface protease, principally furin, with release of the 20-kDa *N*-terminal fragment. PA63 then oligomerizes and also binds LF or EF. The complex is internalized by endocytosis, and acidification of the vesicle causes insertion of the PA63 heptamer into the endosomal membrane to produce a channel through which LF or EF translocate to the cytosol. Details of the individual steps are discussed in the following sections.

Protective antigen structure and function

The crystal structure from X-ray diffraction

The structure of PA was solved by X-ray diffraction (Petosa *et al.*, 1997). PA is a long, flat protein that is rich in β-sheet structure (Figure 12.3, in the colour plate section). Four domains are distinguished, which correspond in general to functional regions previously defined by analysis of large fragments produced by trypsin and chymotrypsin. Domain 1 (aa 1–258) contains two tightly bound calcium ions, and a large flexible loop (aa 162–175) that includes the sequence $_{164}$RKKR$_{167}$, which is cleaved during proteolytic activation. Domain 2 (aa 259–487) contains several very long β-strands and forms the core of the membrane-inserted channel. It also has a large flexible loop (aa 303–319) implicated in membrane insertion. Domain 3 (aa 488–595) has no known function. Domain 4 is loosely associated with the other three domains and is involved in receptor binding.

Proteolytic activation, nicking, oligomerization and binding of lethal factor and oedema factor

The two flexible loops mentioned above contain sites uniquely sensitive to proteolytic cleavage, as was recognized even before the aa sequence was known. The sequence $_{164}$RKKR$_{167}$ is extremely sensitive to cleavage by trypsin, clostripain and other proteases that recognize basic residues. PA in solution at 1 mg ml^{-1} is completely cleaved in 30 min when treated with 0.1 µg ml^{-1} of trypsin (Leppla *et al.*, 1988; Singh *et al.*, 1989a). The fragments of 20 and 63 kDa (designated PA20 and PA63, respectively) do not easily dissociate. However, at pH 7.5, concentrated solutions of the nicked PA, incubated for several hours, form a precipitate, which can be shown by SDS gel electrophoresis to contain PA63. Chromatography of trypsin-treated PA on the MonoQ anion exchange resin (Pharmacia) using a pH 9.0 aminoethanol buffer system and a NaCl gradient yields an early peak of PA20 and a later peak containing PA63. *N*-terminal sequencing shows that PA20 begins with residue 8, indicating that a second cleavage had occurred at Arg7, and PA63 begins with residue 168, showing that cleavage had occurred following the cluster of four basic residues.

PA63 purified on MonoQ resin remains soluble indefinitely if kept at pH 9.0. Analysis by non-denaturing gel electrophoresis and gel filtration chromatography showed this material to be large, at least 300 kDA, suggesting it was a pentamer or hexamer of the 63-kDa peptide (Leppla *et al.*, 1988; Singh *et al.*, 1994). This oligomer is extremely stable. By transmission electron microscopy after negative staining, the oligomer was shown to be a heptamer (Milne *et al.*, 1994). This species was crystallized and a structure determined by X-ray diffraction, which confirmed the heptameric nature of the oligomer (Petosa *et al.*, 1997).

The functional importance of the trypsin-sensitive site at residues 164–167 became evident when it was noted that PA incubated with cells becomes nicked at this site (Leppla *et al.*, 1988). Only PA63 remains bound to cells; PA20 can be detected in the supernate if this is concentrated before analysis. Cleavage occurs at 4°C and also on cells that have been fixed chemically, situations where PA would be restricted to the cell surface. A PA mutant in which the $_{164}$RKKR$_{167}$ sequence is deleted is not cleaved by trypsin or on the surface of cells and is non-toxic (Singh *et al.*, 1989b). These experiments strongly suggest that PA63 is the active species and is sufficient for delivery of LF or EF. This was proven directly by showing that purified, heptameric PA63 is toxic to macrophages when combined with LF. Finally, PA incubated with cells at 37°C and allowed to internalize forms oligomers that are stable to heating in SDS (Milne *et al.*, 1994). Presumably, acidification of early endosomes causes the oligomer to insert into membranes to form a second, stable type of heptamer (to be discussed later).

A detailed analysis of PA residues 164–167 showed that cleavage by cellular proteases requires the minimum sequence RxxR (Klimpel *et al.*, 1992). Semi-random, codon-based cassette mutagenesis produced more than 30 PA proteins altered in this region. After noting that all toxic mutants contained Arg at positions 164 and 167, additional mutants were made with defined sequences. This result, combined with inhibitor studies and rapidly advancing knowledge about cellular processing proteases, proved that the cellular protease that most rapidly activates PA is furin, also called PACE. This finding was extended by showing that the proteolytic activation of a number of bacterial toxins depends on furin (Gordon *et al.*, 1995, 1997).

The other site that is uniquely sensitive to protease is the large loop in domain 2, aa 303–319. Cleavage in this region to produce fragments of 47 and 37 kDa occurs when purified PA is incubated with *B. anthracis* culture supernates, probably due to the action of a metalloprotease of the thermolysin type. Cleavage at the same site is obtained with approximately 1 μg ml^{-1} of chymotrypsin or thermolysin, or higher concentrations of several bacillus metalloproteases (e.g. *Bacillus polymyxa* protease). The fragments produced by cleavage can be separated by chromatography on MonoS cation-exchange columns (Pharmacia) at pH 5.0 or 6.5 in 7 M urea. *N*-terminal sequencing showed that cleavage by chymotrypsin occurs *C*-terminal to the paired Phe residues, aa 313–314 in the sequence SFFDI, while cleavage by thermolysin occurs *N*-terminal to these two residues (S.H. Leppla, unpublished work; Novak *et al.*, 1992). The great sensitivity of residues 313–314 must be due to the presence of paired hydrophobic residues and their exposure on the protein's surface. Mutagenesis of this region proved that it also is functionally important. Deletion of the pair of Phe, aa 313–314, completely inactivates PA (Singh *et al.*, 1994). The altered protein is resistant to chymotrypsin and thermolysin, as expected. The mutant PA binds to cells, becomes nicked, and internalizes LF to endosomes, but fails to translocate LF to the cytosol (Novak *et al.*, 1992; Singh *et al.*, 1994).

Cleavage of PA with both trypsin and chymotrypsin generates three fragments. Analysis of these shows that residues 168–312 comprise part of the binding site for LF and EF, while residues 315–735 contain the cell receptor binding site. For example, when receptor binding studies were done with radiolabelled PA that was partially nicked at residues 313–314, the cell-bound material contained both 63- and 47-kDa fragments. This result showed that the cell recognition domain is entirely contained in the 47-kDa fragment, residues 315–735.

The PA63 heptamer binds tightly to LF, as can be demonstrated by several methods, including sedimentation equilibrium (Singh *et al.*, unpublished data) and gel electrophoresis. On non-denaturing 5% polyacrylamide gels run at pH 8.5 in the presence of detergents, LF has high electrophoretic mobility, followed by PA, and then the PA63 oligomer. The PA63 moves as a sharp band, much more slowly than PA. In mixtures of PA63 and LF, several very closely spaced bands are seen, migrating even more slowly than PA63. These contain oligomeric PA63 with increasing numbers of bound LF molecules. The fact that distinct species are seen suggests that the rate of dissociation of LF from the complex is very low. Titration with increasing amounts of LF shows that the heptameric PA63 can bind a maximum of seven LF molecules (Singh *et al.*, 1999).

Functional sites of protective antigen defined by site-specific mutagenesis

The C-terminal domain 4 of PA was initially implicated in receptor binding by analysis of mutants truncated to varying extents (Singh et al., 1991). Truncation by 3, 5 or 7 aa causes a 10-fold decrease in toxicity and affinity for the receptor. Truncation by 12 or more aa leads to complete loss of toxicity and destabilization of the protein. Although this suggests that the terminal residues interact with the cellular receptor, later studies suggest that the effects of these truncations may be indirect, by causing changes to the folding of domain 4. Thus, in work directed towards the use of anthrax toxin fusion proteins as therapeutic agents, site-specific mutagenesis was used to identify residues in domain 4 that interact directly with the cellular receptor (Varughese et al., 1999). Two large flexible loops, aa 679–693 and aa 704–723, were mutated in a semi-random approach and more than 50 mutant PA proteins were purified. Surprisingly, the results proved that these two loops are not involved in receptor binding. Instead, a region on the opposite side of domain 4 was implicated in receptor binding because fortuitous mutations there are non-toxic. In particular, substitution of N657 or N682 inactivates PA by preventing its binding to the receptor. These two residues are adjacent on the surface of PA and appear to identify a critical area involved in receptor interaction (Varughese et al., 1999). Other evidence that this region constitutes the receptor binding domain is that a CNBr fragment containing PA residues 663–735 competes with PA for binding to cells (Noskov et al., 1996a).

Site-specific mutagenesis was instrumental in proving the functional importance of trypsin- and chymotrypsin-sensitive sites in PA, as discussed above. Mutagenesis of other regions was carried out to facilitate use of PA as a reagent in cell biology or therapeutics. An example was the creation of a set of mutants in which Ser or Thr residues distributed throughout the PA sequence were substituted with Cys (Klimpel and Leppla, 1996). The introduced Cys residue in such a protein provides a unique site for chemical conjugation to small molecules or other proteins. Although the Ser/Thr to Cys substitutions were chosen to be conservative, four of the 12 mutant proteins obtained are non-toxic. In two of these mutants, T357C and S429C, the mutation is in a region highly conserved in the proteins of the PA family. For example, S429 is conserved in five of the six proteins. Several of the Cys substitution mutants were also useful in the X-ray diffraction determination of the PA structure (Petosa et al., 1997). Extending these analyses will be helpful in obtaining a detailed understanding of PA function.

Functional sites of protective antigens defined by monoclonal antibodies

Monoclonal antibodies to PA have been assigned by competition to at least 23 different epitopes (Little et al., 1988). However, very few of these neutralize the toxin. Three antibodies, including 3B6, react with a region that includes or overlaps with the receptor-binding domain. This follows from the ability of 3B6 to block binding of radiolabelled PA to cells and to neutralize toxicity, but only if preincubated with PA prior to addition to cells. Antibody 3B6 reacts with PA63 and the 47-kDa chymotrypsin fragment, confirming the assignment of the receptor binding region to this C-terminal region (Little and Lowe, 1991; Little et al., 1996). Further mapping showed that the epitope includes a region between aa Asp671 and Ile721.

The model for interaction of PA63 with LF predicts that neutralizing monoclonal antibodies may also be found that block LF and EF binding to PA63. One antibody obtained by immunization with PA63 does have this property. Antibody 1G3 is unique because it reacts with PA63 but not with PA, it blocks LF binding to PA63, and it neutralizes toxin even when added at amounts far below those of PA (Little et al., 1996). The latter finding can be explained by considering that the antibody will react only with the PA63 generated by nicking of the small fraction of PA that binds to receptor, and will not be 'wasted' reacting with the native PA that is in solution. The 1G3 antibody is also the only one of those tested that was able to delay the time to death of guinea pigs challenged with virulent B. anthracis (Little et al., 1997).

Surprisingly, none of the more than 35 monoclonal antibodies obtained by immunization with PA is reactive with PA20. Perhaps cleavage of PA on the surface of cells in the immunized animal releases PA20 as a peptide that is too small to induce antibody efficiently. In contrast, the cell-bound PA63 may be efficiently presented to antibody-producing cells. Monoclonal antibodies to PA20 were later obtained by immunizing mice with purified PA20 (S. Little and S.H. Leppla, unpublished work).

Protein neighbours of protective antigen

The five other proteins with homology to PA are listed in Table 12.2. The high degree of similarity of these proteins argues that all are derived from a common ancestor (Figure 12.4). It is highly probable that all of these proteins will share the same overall structural design as PA. All six proteins have perfect conservation of the five Asp and Glu residues in PA whose side-chain carboxyl groups chelate the two calcium ions. Although calcium has not been identified in the other toxins, it would be surprising if it were not present.

```
                     *  *  * ***   *  ***
PA        ELKQKSSNSRKKRSTS-AGPTVPDRDNDGIPDSLEVEGYTVD
VIP1      LFTQK-----MKR----EIDEDTDTDGDSIPDLWEENGYTIH
Iota Ib   FFDVR-FFS-----AA-WEDEDLDTDNDNIPDAYEKNGYTIK
Spirof    FFDLK-LKSRSARLASGWGDEDLDTDNDNIPDAYEKNGYTIK
CDTb      FFDPK-LMS-------DWEDEDLDTDNDNIPDSYERNGYTIK
C2 II     LFSNA-----KLK-----ANANRDTDRDGIPDEWEINGYTVM
                              ^  ^ ^        ^
```

* fully conserved residue
^ fully conserved residue having side chain that chelates calcium ion

FIGURE 12.4 Sequences of PA family members in the region containing the protease activation and calcium binding sites. Protein designations at the left correspond to those in Table 12.2. VIP1 is vegetative insecticidal protein 1.

While several of the proteins are known, like PA, to require proteolytic activation, none of the other proteins has a fully furin-susceptible cleavage site. The *C. spiroforme* toxin has a minimum furin site, RSAR, and the C2 and VIP1 proteins have pairs of two basic amino acids in the putative region. Perhaps the four clostridial toxins, because they are produced by anaerobic pathogens that grow in restricted spaces, have evolved to depend on activation by proteases of the producing bacteria.

Sequence similarity of the proteins decreases greatly in the regions corresponding to PA domain 4, suggesting that the toxins have either evolved from a common ancestor to bind effectively to cellular receptors in different host organisms, or undergone recombination to change receptor specificity completely. Receptors have not been identified for any of these toxins, and progress in understanding the toxin–receptor interaction will be limited until this is achieved.

Oedema factor structure and function

Oedema factor domain structure

Comparison of their sequences showed that the *N*-terminal 250 amino acids of LF and EF have substantial homology, consistent with their common ability to bind to PA63. The region beyond residue 250 has homology to the *B. pertussis* cyclase and is an active enzyme when expressed as a recombinant protein (Labruyere *et al.*, 1991). Because relatively little experimental work has been done on EF, our knowledge about the structural domains of EF is largely derived by comparison with the pertussis cyclase. The amino acid sequences of EF and of *B. pertussis* adenylate cyclase show several regions with substantial homology. It was expected that the regions of homologies would identify essential parts of the catalytic domain. The first of these regions contains the sequences GVATKGxxxxGKS (residues 309–321) in EF and GVATKGxxxxAKS (residues 54–66) in the *B. pertussis* cyclase. The EF sequence exactly

matches the consensus sequence, GxxxxGKS, recognized in a large number of nucleoside triphosphate binding proteins (Higgins *et al.*, 1986). Mutagenesis in this region to replace Lys320 with Met in EF (Xia and Storm, 1990) and to replace Lys65 with Gln in the pertussis cyclase (Glaser *et al.*, 1989) caused more than a 10-fold loss of catalytic activity. Substitution of Lys346 with Gln also destroyed catalytic activity, demonstrating the importance of this site (Labruyere *et al.*, 1991). Other regions of homology to the pertussis cyclase can be assumed to be involved in binding the activator of these cyclases, calmodulin. However, prediction of calmodulin binding sites from amino acid sequence has been difficult. From a number of studies searching for the calmodulin binding site of the pertussis cyclase, the finding most relevant to EF was that a peptide corresponding to EF residues 499–532 binds tightly to calmodulin (Munier *et al.*, 1993). As might be predicted from the amino acid sequence homology of the two cyclases, there is also some serological cross-reactivity (Goyard *et al.*, 1989).

Information about the structure of EF was also deduced from a set of monoclonal antibodies (Little *et al.*, 1994). Cleavage in formic acid produced the three fragments of 18, 53 and 17 kDa expected from cleavage at Asp–Pro bonds, and these were used to map the antibodies. The two antibodies which blocked binding of EF to PA on cells reacted with the *N*-terminal 18-kDa fragment, consistent with its role in binding to PA. Several other antibodies inhibited adenylate cyclase activity.

Oedema factor catalytic activity

The enzymic properties of the *B. anthracis* adenylate cyclase have received only limited study (Leppla, 1982, 1984; Labruyere *et al.*, 1990). This is surprising, because the enzyme can be obtained more easily than most other bacterial and eukaryotic adenylate cyclases (Leppla, 1991b; Labruyere *et al.*, 1991), and because it has high catalytic activity and considerable potential as a pharmacological tool for transiently increasing intracellular cAMP concentrations in eukaryotic cells. EF has high catalytic activity, with a $V_{max} = 1.2$ mmol cAMP min^{-1} mg^{-1} protein (see Figure 7 of Leppla, 1984), similar to the values of 1.6 mmol cAMP min^{-1} mg^{-1} protein reported for the *B. pertussis* cyclase (Ladant *et al.*, 1986; Rogel *et al.*, 1988). The K_m for ATP in the presence of Mg^{2+} is 0.16 mM. The enzyme activity is very sensitive to Ca^{2+}, showing optimum activity at 0.2 mM and inhibition at higher concentrations. Apparently, a low concentration of Ca^{2+} is needed to activate calmodulin, while a higher concentration directly inhibits the catalytic centre.

Mn^{2+} can substitute for Ca^{2+} in activation of calmodulin and does not cause inhibition at the catalytic centre, so its use as the only added divalent ion makes the enzyme activity insensitive to small variations in free Ca^{2+}.

The enzyme activity of EF has an absolute requirement for calmodulin. In the presence of 50 μM Ca^{2+}, the concentration of calmodulin giving half-maximal activity is 2.0 nM. When the Ca^{2+} is chelated by excess EGTA, calmodulin still can activate EF, but 5 μM is needed to obtain equivalent activity. Thus, calmodulin can activate EF even when it contains no bound Ca^{2+}. The *B. anthracis* cyclase differs from the pertussis enzyme in that no degraded protein species can be generated that are active in the absence of calmodulin. This was consistent with the observation that *E. coli* strains containing the EF gene, but not calmodulin, produced no detectable cAMP and did not have any adenylate cyclase activity (Mock *et al.*, 1988), in spite of the probable presence of breakdown fragments of the EF protein.

Lethal factor structure and function

Lethal factor domain structure

The *N*-terminal 250 aa of LF and EF have substantial sequence similarity. This was the first evidence that this region constitutes the PA-binding domain. Many types of evidence now support this conclusion. Most direct is the construction of fusion proteins in which LF aa 1–254 are fused to catalytic domains of other toxins (Arora *et al.*, 1992; Arora and Leppla, 1993, 1994). In each case, these fusion proteins have toxicities consistent with those expected from delivery of the fused catalytic domain. Residue 1–254 of LF (LFn), expressed and purified from *E. coli*, competes with LF and thereby blocks the uptake of LF and LF fusion proteins into cells (Milne *et al.*, 1995; Zhao *et al.*, 1995). A smaller fragment generated by CNBr cleavage, aa 41–244, also competes with LF (Noskov *et al.*, 1996b).

Distal to the *N*-terminal approximately 250 aa is a unique structural feature of LF – a sequence of 19 amino acids that is repeated five times within aa 282–383 (Bragg and Robertson, 1989). The repeats are imperfect, having only about 60% homology. It is unlikely that this region serves a catalytic function. Comparison of the repeats suggests that repeat 1 was the ancestral sequence, which was duplicated to make repeat 2, repeat 2 was duplicated to make repeat 3, and then the pair 2 + 3 was duplicated to make 4 + 5. This suggests that there was selection to increase the length of this region. Perhaps the repeats serve to increase the physical separation of the two terminal domains.

Direct evidence for the domain assignments listed above was obtained by linker insertion mutagenesis (Quinn *et al.*, 1991). Insertions of two or four amino acids were obtained at 17 random sites throughout the protein. Most insertions in the *N*-terminal region eliminate toxicity and the ability to bind to PA, whereas insertions in the *C*-terminal region destroy toxicity without decreasing binding to PA, consistent with this being the catalytic domain.

It was widely accepted in the older anthrax toxin literature that LF and EF are serologically distinct; no cross-reactivity was detected with polyclonal antibodies in immunodiffusion, and this was confirmed with purified components (Quinn *et al.*, 1988). However, once monoclonal antibodies to LF and EF were obtained, several were found that cross-react. In a set of 61 monoclonal antibodies to LF, three cross-react with EF (Little *et al.*, 1990). Eight of the LF monoclones could neutralize lethal toxin in either *in vivo* or *in vitro* tests, and six of these were shown to do so by preventing LF binding to PA63. Surprisingly, only two of the six that blocked LF binding cross-reacted with EF. This suggests that the sequences shared by LF and EF and presumably involved in binding to PA63 are not strongly immunogenic.

Lethal factor metalloprotease activity

The recognition that LF contains a site characteristic of zinc metalloproteases (Klimpel *et al.*, 1994) started a process that eventually led to the identification of its catalytic activity. Substitution of Ala for H686, E687 or H690 in the sequence $_{686}$-HEFGH-$_{690}$ destroys the toxicity of LF for macrophages and decreases its zinc binding ability. This result suggested that LF is a zinc metalloprotease. This hypothesis was supported by the discovery that inhibitors of zinc-dependent aminopeptidases (e.g. bestatin, aromatic amino acid amides and hydroxamates) protect macrophages from LF (Klimpel *et al.*, 1994; Menard *et al.*, 1996a). However, searches for substrates were unsuccessful. The example of the clostridial neurotoxins suggested that LF might be a highly specific protease, able to cleave only a few cellular proteins.

Discovery of a substrate for LF by the author's collaborators grew out of an effort to identify additional cells sensitive to the lethal toxin. The cytotoxicity of lethal toxin was tested on 60 human tumour cell lines that constitute a panel used to screen for anti-neoplastic agents (Weinstein *et al.*, 1997). The growth of some cell lines was inhibited, but no cells were lysed in the dramatic manner of mouse macrophages. Several years later, the database was queried to find drugs with the same action spectrum as the chemical PD98059, which has come into use as a relatively specific inhibitor of the

```
MAPPK1      PKKKPTP-IQL
MAPKK2      LARRKPVLP-ALT
```

FIGURE 12.5 Sequences cleaved by LF. The *N*-terminal sequences of mitogen-activated protein kinase kinases 1 and 2 are shown. The dash indicates the bond cleaved by LF.

mitogen activated protein kinase (MAPK) pathway, and specifically of the enzyme MAPKK1, also designated MEK1 (Alessi *et al.*, 1995). Of the approximately 60 000 drugs that had been tested against this panel of cells, LF was the one most similar in action to PD98059. The implication that LF acted on the MAPK pathway led to a series of experiments, resulting in the discovery that LF blocks the MAPK pathway and does so by cleaving seven amino acids from the *N*-terminus of MAPKK1 (Duesbery *et al.*, 1998). It was then found that MAPKK2 is also cleaved (Figure 12.5). Comparison of these two proteins suggests that LF recognizes a motif containing several basic residues and two prolines.

Lethal factor fusion proteins

Extensive work has been done to exploit the ability of the anthrax toxin to translocate polypeptides to the cytosol of eukaryotic cells. As noted above, fusion proteins have been made in which the catalytic domain of LF is replaced by other polypeptides. Fusion of sequences containing epitopes known to be presented on MHC class I molecule are able to sensitize antigen presenting cells to lysis by specfic cytotoxic T-cells (Goletz *et al.*, 1997a) and to induce cytotoxic T-cell responses in animals (Ballard *et al.*, 1996, 1998). In the case of an epitope of listeriolysin that is known to induce protective immunity, immunization with only 300 fmol (10 ng) of the LF fusion protein induced protective immunity in mice. Several types of controls showed that successful presentation of the epitope depended on active PA. When the LF fusion protein contained HIV gp120, proteasome activity was needed to process the peptide for presentation (Goletz *et al.*, 1997b). Of several bacterial toxins that have been tested for presentation of CTL epitopes, the anthrax toxin system appears to offer advantages (Goletz *et al.*, 1997b), in particular its efficiency and the flexibility in design of LF fusions and their ease of preparation.

CELLULAR MECHANISMS OF ACTION

The anthrax toxins interact to deliver the catalytic components, LF and EF, to the cytosol of target cells. The role of PA is to cause binding of LF and EF to the cell surface, so that they will be internalized by endocyto-

sis, and to provide a membrane channel for their translocation from endosomes to the cytosol. The effect of EF delivery to the cytosol is predictable from its adenylate cyclase activity and the well-known effects of cAMP and of other toxins that elevate cytosolic cAMP concentrations. LF is now known to act as a metalloprotease, which cleaves MAPKK1 and possibly other important cytosolic proteins. This initiates unknown processes, which lead to the lysis of macrophages and to the death of animals.

Cell specificity and receptors

Nearly all types of eukaryotic cells possess receptors for PA. Binding isotherms are easily performed and interpreted because non-specific binding of radioiodinated PA is consistently less than 20% of the total binding. Analysis of the binding data by the LIGAND program (Munson and Rodbard, 1980) shows that cells possess a single class of high-affinity receptors with association constants of approximately 10^{-9} M (Singh *et al.*, 1990; Escuyer and Collier, 1991). Most types of cells have between 5000 and 50 000 receptors. Cells with about 50 000 receptors include L-6 rat myoblast and the human melanoma cell line LOX IMVI. The only two cell types known to lack receptors are GG2EE, a macrophage line derived from C3H/HeJ mice (Jin *et al.*, 1998), and Raji, a human lymphoma. Cross-linking studies suggested a mass of 90 000 for the receptor (Escuyer and Collier, 1991). Removal of Ca^{2+} decreases binding of PA by about 50% (Bhatnagar *et al.*, 1989). The modest number of receptors, their sensitivity to trypsin, the low non-specific binding, and the linear binding curves suggest that the receptor is a single-cell surface protein. If this is true, then it should prove possible to identify and isolate the receptor using standard techniques.

The author and colleagues have selected CHO cell mutants lacking functional PA receptors (unpublished work). Mutagenized cells were treated with a fusion protein containing LF residues 1–254 and the ADP-ribosylation domain of *Pseudomonas* exotoxin A (PE). Ten independent mutants lacking functional receptor were obtained from several different experiments. The mutant cells grow more slowly than parental cells, suggesting that the PA receptor plays an important role in essential cellular processes. No cross-resistance was found to other toxins, showing that the PA receptor is distinct from other known toxin receptors. All mutants map to the same complementation group, indicating that a single locus codes for the protein that acts as PA receptor. These receptor-deficient CHO cells are being used in ongoing attempts to identify or clone the receptor.

Proteolytic activation of protective antigen by furin and other proteases

PA bound to cellular receptors must be activated by proteolytic cleavage at aa 167 to acquire the ability to bind LF and EF and to oligomerize (discussed above). The cellular enzyme that most rapidly activates PA is furin (Klimpel *et al.*, 1992), because furin-deficient CHO and LoVo cells are much less sensitive than parental cells to the combination of PA and LFn-PEIII (Gordon *et al.*, 1995). However, the cleavage site, RKKR, has more basic residues than the minimum furin recognition sequence RxxR, and appears to be susceptible to other cellular proteases. Thus, furin-deficient cells retain some sensitivity to wild-type PA but are totally resistant to PA mutants with a RAAR sequence. It has been found for several other toxins and viruses that furin is the dominant activating enzyme, but that other cellular proteases can also perform the activation, although less efficiently.

Toxin internalization and translocation

Proteolytically activated PA can bind LF and EF with high affinity. This can be demonstrated through biophysical methods such as sedimentation equilibrium, gel filtration and gel electrophoresis (discussed above). Although an association constant of 0.01 nM was reported earlier from an immunochemical technique (Leppla, 1991a), later studies suggest that the affinity may be lower, e.g. 0.24 nM (Novak *et al.*, 1992). However, the gel electrophoresis data discussed above suggest that LF binding to the PA63 heptamer is multivalent and potentially co-operative, so affinities measured may vary, depending on the degree to which physical methods mimic the multivalent conditions likely to exist on cells.

The anthrax toxins enter cells by receptor-mediated endocytosis and pass through acidic vesicles, because the effects of the toxins are blocked by pharmacological agents that block this process. Thus, the effects of oedema toxin are blocked by cytochalasin D and by amines (Gordon *et al.*, 1988, 1989), and the effects of lethal toxin are blocked by amines (Friedlander, 1986) and by specific inhibitors of the endosomal proton pump (Menard *et al.*, 1996b). The latter report suggested that the ability of bafilomycin to inhibit even when added 30 min after LF meant that the toxin translocated from late endosomes. However, the results may also be viewed as evidence that the toxin is activated and internalized slowly.

PA63 has the ability to insert into the plasma membrane or artificial lipid membranes, events presumed to reflect the normal acid-dependent insertion into endosomal membranes. This process requires proteolytic activation, because native PA is unable to insert. Insertion in the plasma membrane to form an ion-conductive channel, as measured by Rb^+ or Na^+ fluxes, occurs when cells containing surface-bound, nicked PA are briefly treated with pH 5.0 buffer (Milne and Collier, 1993; Singh *et al.*, 1994). If LF is bound to these cells, then acidification leads to toxicity, in a process that no longer requires acidification of cellular compartments (Friedlander, 1986; Gordon *et al.*, 1988). PA63, but not PA, forms defined ion-conductive channels in artificial lipid membranes containing phospholipids (Blaustein *et al.*, 1989; Finkelstein, 1994). Insertion requires lowering the pH to 6.5. The electrical properties of the channels are pH and voltage dependent, and the channel can be blocked by large, quaternary ammonium ions. The passage of smaller quaternary ammonium ions leads to an estimate that the lumen of the channel is approximately 11 Å at its most narrow point.

The structure determined for the heptameric form of PA63 (Petosa *et al.*, 1997) provides a satisfying explanation of the previously measured biochemical and cellular properties of the toxin. Furthermore, comparison with other heptameric channel-forming proteins led to the recognition that the flexible loop containing the chymotrypsin-sensitive site has the properties of an amphipathic beta hairpin, with alternating hydrophobic and hydrophilic residues (Figure 12.6) (Petosa *et al.*, 1997). In staphylococcal alpha toxin, a similar beta hairpin assembles into a 14-stranded beta barrel where the hydrophobic residues face the lipid and the hydrophilic residues face the lumen of the channel (Song *et al.*, 1996) (Figure 12.7). It is reasonable to assume that PA forms a very similar structure, and a model has been presented based on the similarity to the alpha toxin.

All members of the PA family have a similar amphipathic beta hairpin that is likely to function in the same way. The pair of Phe residues, Phe313–Phe314, previously shown to be essential for PA activity, are at the turn of the loop. It is interesting that all members of the family except for PA have a His-Ser or His-Thr sequence following the beta hairpin. Protonation of this or other His residues may be involved in the acid dependence of membrane insertion (Petosa *et al.*, 1997). Direct evidence that the channel has the proposed structure comes from study of Cys substitution mutants in each residue of aa 302–325 (Benson *et al.*, 1998). Channels generated by these proteins in artificial lipid membranes are blocked by a sulfhydryl-reactive quaternary amine reagent, MTS-ET, for those mutants in which the Cys replaces a hydrophilic, lumen-facing residue, but not for the others, where the Cys is predicted to face the lipid bilayer.

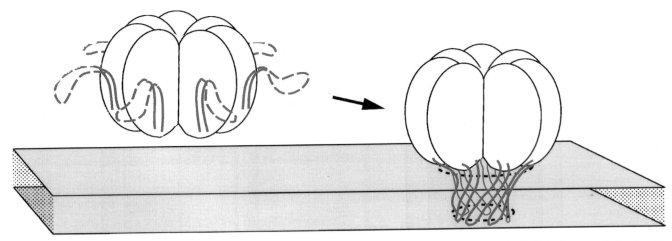

FIGURE 12.6 Model for insertion of amphipathic hairpins to produce β-barrel channel. The heptameric PA63 is shown with the flexible loop containing residues 300–325 shown as a dotted line.

It is not established whether PA63 must oligomerize before inserting in membranes, nor is it established whether oligomerization necessarily precedes LF binding. Analysis *in vitro* shows that monomeric PA63 can bind LF (Singh *et al.*, 1999) and that PA63 can oligomerize in the absence of LF. This suggests that oligomerization and LF binding can occur in parallel rather than sequentially. The PA63 heptamer that assembles on the receptor is viewed as a pre-pore. Experimentally, it is characterized as being quite stable, presumably being equivalent to the PA63 oligomer that is produced *in vitro* by trypsin treatment and MonoQ chromatography, as described above. This form is stable to gel filtration, non-denaturing gel electrophoresis,

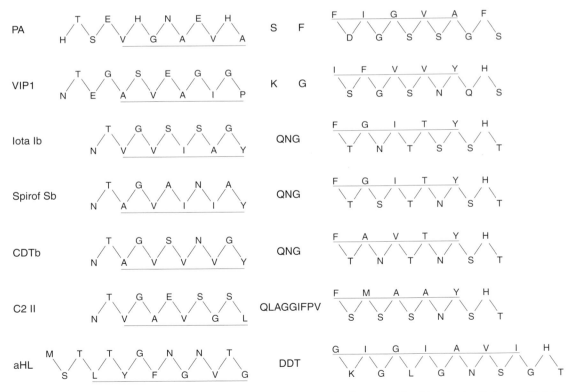

FIGURE 12.7 Sequences from PA family members that are known or proposed to form amphipathic hairpins that assemble into membrane-spanning barrel structures. Hydrophobic residues (underlined) will face the lipid bilayer and the alternating hydrophilic residues will face the lumen of the channel. Residues in the central region (SF in the case of PA) form the bend of the hairpin and will end up on the cytosolic side of the membrane.

ultracentrifugation and crystallization, provided the pH is maintained at >8.5. However, this form does not survive heating in SDS, as is done in preparing samples for gel electrophoresis. Transition to an active pore, involving insertion of the chymotrypsin-sensitive loop into the membrane, is an acid-dependent step. The inserted pore becomes even more stable, as demonstrated by the appearance of SDS and heat-stable oligomers when PA-treated cells are examined by SDS gel electrophoresis (Milne *et al.*, 1994).

It is assumed that LF and EF translocate through the lumen of the PA63 heptamer, passing as an unfolded polypeptide, but little is known of how this process occurs. It was originally suggested that translocation began at the *N*-terminus of LF, but this can now be discounted because fusion proteins with polypeptides attached at either end of LF residues 1–254 are equally potent (Arora and Leppla, 1994; Milne *et al.*, 1995; Ballard *et al.*, 1996). It follows that LF must insert an internal sequence, presumably a hairpin loop, suggesting parallels with signal sequence dependent mechanisms of protein secretion. Another poorly understood question is what drives translocation and what types of sequences or structures are compatible with translocation. Studies of diphtheria toxin fusion proteins show that tightly folded protein domains are not translocated (Klingenberg and Olsnes, 1996), and it can be predicted that similar constraints will apply to the anthrax toxin system.

Cellular mechanisms of action

Oedema factor action on cells

Oedema toxin treatment of nearly all cell types causes an increase in cAMP. This provides additional evidence that most cell types possess receptors and a protease able to activate PA. The maximum levels of cAMP reached in treated cells differ greatly, from levels that barely exceed the basal concentration of about 2 µmol cAMP per mg cell protein to levels of 2000 µmol cAMP per mg protein (Leppla, 1984; Gordon *et al.*, 1988, 1989). At the higher cAMP concentrations, 20–50% of the cellular ATP has been converted to cAMP. Full activation of cAMP-dependent protein kinase requires only a modest elevation of cAMP concentration above basal levels, so the much higher cAMP concentrations generated in some cell types may have no additional physiological effect, except possibly through depletion of ATP. Very high cAMP concentrations are not lethal to cultured cells and toxin-treated cells can recover after removal of toxin. Unlike cholera toxin, the effects of oedema toxin are rapidly reversed, apparently because EF is unstable in the cytosol (Leppla, 1982), with a half-life of less than 2 h.

Lethal factor action on cells

In contrast to the broad range of cells sensitive to oedema toxin, only one cell type responds in an acute manner to lethal toxin. Treatment of mouse and rat macrophages with lethal toxin causes lysis in 90–120 min (Friedlander, 1986). Both resident and elicited primary macrophages are susceptible, as are certain macrophage-like cells lines, including J774A.1 and RAW 264.7. Certain other types of cultured cells are inhibited in growth if plated at low cell densities and treated with lethal toxin for 3–7 days. For example, BHK cells show a 50% inhibition of growth after 3 days (S.H. Leppla, unpublished). One mouse macrophage that is resistant to lethal toxin, IC-21, also resists the toxic action of LF that is internalized artificially, suggesting that resistance lies at a step subsequent to the enzymic action of LF (Singh *et al.*, 1989a).

The earliest measured changes in lethal toxin-treated macrophages are increases in K^+ and Rb^+ ion fluxes at 45 min, decreases in ATP concentrations and release of superoxide at 60 min, and changes in morphology at 75 min, which progress to lysis at 90 min (Bhatnagar *et al.*, 1989; Hanna *et al.*, 1992, 1995; Hanna, 1998). The changes occurring after 60 min can be blocked by removal of extracellular calcium and by osmotic stabilizers, suggesting that membrane integrity is compromised. Reducing agents also protect mouse macrophages from lethal toxin (Hanna *et al.*, 1995), indicating that the oxidative burst is required for cell lysis. The resistance of the IC-21 macrophage can be attributed to its inability to mount an oxidative burst. Sublytic concentrations of lethal toxin as low as 10^{-9} µg ml^{-1} induce RAW 264.7 macrophages to synthesize and release TNF and IL-1, with the TNF response being maximal at 120 min (Hanna *et al.*, 1993).

The recent demonstration that LF acts as a metallo-protease in the cytosol of cells, cleaving MAPKK1 and MAPKK2 (Duesbery *et al.*, 1998), has opened the way to explaining the molecular events that lead to the cellular responses detailed above. However, it has been difficult to identify known pathways by which MAPKK cleavage could lead to the dramatic cytotoxic events caused by lethal toxin. MAPKK is essential to the control of transcription and development, but is not known to act directly on pathways required for short-term homeostasis.

The difficulty of connecting MAPKK cleavage to rapid macrophage lysis suggests that there may be other cytosolic proteins that are cleaved by LF. Studies can be anticipated to identify the optimum amino acid sequence recognized by LF and to identify other substrates directly.

In vivo actions

Anthrax toxin was first recognized by noting the oedema caused by intradermal injection of culture supernatants or plasma of infected animals (Smith *et al.*, 1955). Fractionation showed that this effect was due to the mixture of components now termed oedema toxin. Oedema toxin is believed to play an important role early in an anthrax infection because elevated cAMP concentrations incapacitate phagocytic cells (Confer and Eaton, 1982). Thus, human PMNs treated with purified oedema toxin are unable to phagocytose opsonized *B. anthracis* (O'Brien *et al.*, 1985). Oedema toxin also has profound effects on chemotaxis and LPS priming of PMNs (Wade *et al.*, 1985; Wright *et al.*, 1988). Human monocytes treated with oedema toxin produced more interleukin-1 (IL-6) and were blocked in their lipopolysaccharide-induced tumour necrosis factor (TNF) production (Hoover *et al.*, 1994).

A critical role of phagocytes early in infection was also suggested by studies showing that inbred mouse strains with a lower natural resistance to *B. anthracis* infection are slower to mobilize phagocytes to the site of infection (Welkos *et al.*, 1990). This argues that the early mobilization and activation of phagocytes is critical for clearance of *B. anthracis*, and supports the view that the oedema toxin may play an essential role at this stage in infection.

The extensive studies done prior to 1965 on the pathophysiological effects of toxin and infection in animals have been critically reviewed (Stephen, 1986). Unfortunately, those early studies employed toxin preparations of uncertain purity. Furthermore, those early studies were mostly descriptive and, on re-examination, provide few insights into the cellular systems that are damaged by toxin *in vivo*. However, it is accepted that the effects of lethal toxin administration closely match those observed in animals dying from infection with virulent *B. anthracis*.

The best-characterized *in vivo* action of the lethal toxin is its rapid toxicity in rats, where it causes overwhelming pulmonary oedema and death in as little as 38 min (Haines *et al.*, 1965; Ezzell *et al.*, 1984). The Fischer 344 rat is particularly sensitive to the toxin. The oedematous fluid in the lungs of the rats has the protein composition of serum, implying a cytotoxic effect that leads to failure of the endothelial barrier, rather than stimulation of fluid secretion. However, when the lethal toxin was tested on cultured endothelial cells, no cytotoxic effects were seen. The lethality of anthrax lethal toxin for mice requires macrophages, because macrophage depletion of mice by silica injection confers resistance, and reconstitution with RAW264.7 macrophages restores sensitivity

(Hanna *et al.*, 1993). Furthermore, antibodies to TNF and especially IL-1 protect mice from lethal toxin, arguing that it is the release of these cytokines that causes the shock-like response and death.

The studies *in vitro* and *in vivo* on macrophages point to a hypothesis for the way in which anthrax lethal toxin kills animals. This hypothesis, detailed in a recent review (Hanna, 1998), is that macrophages are induced to synthesize and then release unphysiologically high concentrations of cytokines, in particular IL-1 and TNF, which would cause a shock-like response. Low concentrations of lethal toxin induce synthesis of these cytokines, of which TNF is largely secreted and IL-1 retained in the cell (Hanna *et al.*, 1993). At a later stage in an infection, higher concentrations of toxin would lyse the macrophages, releasing a burst of IL-1 and other cytokines. There remains the intriguing question of why the especially sensitive Fischer 344 rat dies in as little as 38 min. Perhaps this rat strain contains macrophages that are highly sensitive, or additional cell types that respond to lethal toxin.

CONCLUSIONS AND SUMMARY

It is clear from the data presented above that the anthrax toxin is the principal protein virulence factor of *B. anthracis*. Strains unable to produce the toxin are avirulent, and animals are protected against infection only if they possess antibodies to the toxin. Other bacterial pathogens depend on protein exotoxins for their virulence, but in most cases the toxins have a less complex design. A unique feature of the anthrax toxins is the segregation of the receptor binding and translocation functions on a protein separate from the effector function. This binary design is clearly an effective one, because it is utilized by at least five other toxins produced by Gram-positive pathogens.

Rapid progress has occurred in recent years in understanding the interaction of the anthrax toxin proteins with cells. Much descriptive information about the toxin receptor has been acquired, although its identity remains to be determined. The requirement for and mechanism of proteolytic activation of PA has been resolved. The structure of the protein-conducting channel produced by PA is now well understood. The long-standing question of the catalytic activity of LF has been answered with the demonstration that it acts as a protease. We can now look forward to obtaining a detailed understanding of the catalytic properties of LF and, eventually, to solving its complete crystal structure by X-ray diffraction.

There is renewed interest in the development of improved anthrax vaccines. The new knowledge about toxin structure and function will be useful in the design of candidate vaccines. However, much needs to be learned about the mechanisms of immunity to anthrax before the knowledge about toxin structure can be used in ration design of improved immunogens.

There remain important questions about how the oedema and lethal toxins act in cells and animals to cause the pathology associated with anthrax infection. Important progress can be anticipated in connecting the initial catalytic steps, in particular the proteolytic action of LF, to the subsequent pathological effects in cells and tissues. The fact that these dominant virulence factors, LF and EF, act by enzymic mechanisms offers the possibility that small molecule inhibitors might be developed to block these effects.

REFERENCES

Aktories, K. and Wegner, A. (1992). Mechanisms of the cytopathic action of actin-ADP-ribosylating toxins. *Mol. Microbiol.* **6**, 2905–8.

Alessi, D.R., Cuenda, A., Cohen, P., Dudley, D.T. and Saltiel, A.R. (1995). PD 098059 is a specific inhibitor of the activation of mitogen-activated protein kinase kinase *in vitro* and *in vivo*. *J. Biol. Chem.* **270**, 27489–94.

Arora, N. and Leppla, S.H. (1993). Residues 1–254 of anthrax toxin lethal factor are sufficient to cause cellular uptake of fused polypeptides. *J. Biol. Chem.* **268**, 3334–41.

Arora, N. and Leppla, S.H. (1994). Fusions of anthrax toxin lethal factor with shiga toxin and diphtheria toxin enzymatic domains are toxic to mammalian cells. *Infect. Immun.* **62**, 4955–61.

Arora, N., Klimpel, K.R., Singh, Y. and Leppla, S.H. (1992). Fusions of anthrax toxin lethal factor to the ADP-ribosylation domain of *Pseudomonas* exotoxin A are potent cytotoxins which are translocated to the cytosol of mammalian cells. *J. Biol. Chem.* **267**, 15542–8.

Ballard, J.D., Collier, R.J. and Starnbach, M.N. (1996). Anthrax toxin-mediated delivery of a cytotoxic T-cell epitope *in vivo*. *Proc. Natl Acad. Sci. USA* **93**, 12531–4.

Ballard, J.D., Doling, A.M., Beauregard, K., Collier, R.J. and Starnbach, M.N. (1998). Anthrax toxin-mediated delivery *in vivo* and *in vitro* of a cytotoxic T-lymphocyte epitope from ovalbumin. *Infect. Immun.* **66**, 615–19.

Bartkus, J.M. and Leppla, S.H. (1989). Transcriptional regulation of the protective antigen gene of *Bacillus anthracis*. *Infect. Immun.* **57**, 2295–300.

Battisti, L., Green, B.D. and Thorne, C.B. (1985). Mating system for transfer of plasmids among *Bacillus anthracis*, *Bacillus cereus*, and *Bacillus thuringiensis*. *J. Bacteriol.* **162**, 543–50.

Benson, E.L., Huynh, P.D., Finkelstein, A. and Collier, R.J. (1998). Identification of residues lining the anthrax protective antigen channel. *Biochemistry* **37**, 3941–8.

Bhatnagar, R., Singh, Y., Leppla, S.H. and Friedlander, A.M. (1989). Calcium is required for the expression of anthrax lethal toxin activity in the macrophage-like cell line J774A.1. *Infect. Immun.* **57**, 2107–14.

Blaustein, R.O., Koehler, T.M., Collier, R.J. and Finkelstein, A. (1989). Anthrax toxin: channel-forming activity of protective antigen in planar phospholipid bilayers. *Proc. Natl Acad. Sci. USA* **86**, 2209–13.

Bragg, T.S. and Robertson, D.L. (1989). Nucleotide sequence and analysis of the lethal factor gene (lef) from *Bacillus anthracis*. *Gene* **81**, 45–54.

Cataldi, A., Labruyere, E. and Mock, M. (1990). Construction and characterization of a protective antigen-deficient *Bacillus anthracis* strain. *Mol. Microbiol.* **4**, 1111–17.

Confer, D.L. and Eaton, J.W. (1982). Phagocyte impotence caused by an invasive bacterial adenylate cyclase. *Science* **217**, 948–50.

Dai, Z. and Koehler, T.M. (1997). Regulation of anthrax toxin activator gene (*atxA*) expression in *Bacillus anthracis*: temperature, not CO2/bicarbonate, affects *AtxA* synthesis. *Infect. Immun.* **65**, 2576–82.

Dai, Z., Sirard, J.C., Mock, M. and Koehler, T.M. (1995). The *atxA* gene product activates transcription of the anthrax toxin genes and is essential for virulence. *Mol. Microbiol.* **16**, 1171–81.

Duesbery, N.S., Webb, C.P., Leppla, S.H., Gordon, V.M., Klimpel, K.R., Copeland, T.D., Ahn, N.G., Oskarsson, M.K., Fukasawa, K., Paull, K.D. and Vande, W.G. (1998). Proteolytic inactivation of MAP-kinase-kinase by anthrax lethal factor. *Science* **280**, 734–7.

Escuyer, V. and Collier, R.J. (1991). Anthrax protective antigen interacts with a specific receptor on the surface of CHO-K1 cells. *Infect. Immun.* **59**, 3381–6.

Escuyer, V., Duflot, E., Sezer, O., Danchin, A. and Mock, M. (1988). Structural homology between virulence-associated bacterial adenylate cyclases. *Gene* **71**, 293–8.

Ezzell, J.W., Ivins, B.E. and Leppla, S.H. (1984). Immunoelectrophoretic analysis, toxicity and kinetics of *in vitro* production of the protective antigen and lethal factor components of *Bacillus anthracis* toxin. *Infect. Immun.* **45**, 761–7.

Farchaus, J.W., Ribot, W.J., Jendrek, S. and Little, S.F. (1998). Formation, purification and characterization of protective antigen from a recombinant, avirulent strain of *Bacillus anthracis*. *Appl. Environ. Microbiol.* **64**, 982–91.

Finkelstein, A. (1994). The channel formed in planar lipid bilayers by the protective antigen component of anthrax toxin. *Toxicology* **87**, 29–41.

Franz, D.R., Jahrling, P.B., Friedlander, A.M., McClain, D.J., Hoover, D.L., Bryne, W.R., Pavlin, J.A., Christopher, C.W. and Eitzen, E.M. (1997). Clinical recognition and management of patients exposed to biological warfare agents. *JAMA* **278**, 399–411.

Friedlander, A.M. (1986). Macrophages are sensitive to anthrax lethal toxin through an acid-dependent process. *J. Biol. Chem.* **261**, 7123–6.

Friedlander, A.M. and Brachman, P.S. (1998). Anthrax. In: *Vaccines* (eds S.A. Plotkin and E.A. Mortimer), pp. 729–39. W.B. Saunders, Philadelphia, PA.

Gibert, M., Perelle, S., Daube, G. and Popoff, M.R. (1997). *Clostridium spiroforme* toxin genes are related to *C. perfringens* iota toxin genes but have a different genomic localization. *Syst. Appl. Microb.* **20**, 337–47.

Gill, D.M. (1978). Seven toxin peptides that cross cell membranes. In: *Bacterial Toxins and Cell Membranes* (eds J. Jeljaszewicz and T. Wadstrom), pp. 291–332. Academic Press, New York.

Gladstone, G.P. (1946). Immunity to anthrax. Protective antigen present in cell-free culture filtrates. *Br. J. Exp. Pathol.* **27**, 349–418.

Gladstone, G.P. (1948). Immunity to anthrax. Production of the cell-free protein antigen in cellophane sacs. *Br. J. Exp. Pathol.* **29**, 379.

Glaser, P., Ladant, D., Sezer, O., Pichot, F., Ullmann, A. and Danchin, A. (1988). The calmodulin-sensitive adenylate cyclase of *Bordetella pertussis*: cloning and expression in *Escherichia coli*. *Mol. Microbiol.* **2**, 19–30.

Glaser, P., Elmaoglou-Lazaridou, A., Krin, E., Ladant, D., Barzu, O. and Danchin, A. (1989). Identification of residues essential for catalysis and binding of calmodulin in *Bordetella pertussis* adenylate cyclase by site-directed mutagenesis. *EMBO J.* **8**, 967–72.

Goletz, T.J., Klimpel, K.R., Arora, N., Leppla, S.H., Keith, J.M. and Berzofsky, J.A. (1997a). Targeting HIV proteins to the major histocompatibility complex class I processing pathway with a novel gp120-anthrax toxin fusion protein. *Proc. Natl Acad. Sci. USA* **94**, 12059–64.

Goletz, T.J., Klimpel, K.R., Leppla, S.H., Keith, J.M. and Berzofsky, J.A. (1997b). Delivery of antigens to the MHC class I pathway using bacterial toxins. *Hum. Immunol.* **54**, 129–36.

Gordon, V.M., Leppla, S.H. and Hewlett, E.L. (1988). Inhibitors of receptor-mediated endocytosis block the entry of *Bacillus anthracis* adenylate cyclase toxin but not that of *Bordetella pertussis* adenylate cyclase toxin. *Infect. Immun.* **56**, 1066–9.

Gordon, V.M., Young, W.W., Jr, Lechler, S.M., Gray, M.C., Leppla, S.H. and Hewlett, E.L. (1989). Adenylate cyclase toxins from *Bacillus anthracis* and *Bordetella pertussis*. Different processes for interaction with and entry into target cells. *J. Biol. Chem.* **264**, 14792–6.

Gordon, V.M., Klimpel, K.R., Arora, N., Henderson, M.A. and Leppla, S.H. (1995). Proteolytic activation of bacterial toxins by eukaryotic cells is performed by furin and by additional cellular proteases. *Infect. Immun.* **63**, 82–7.

Gordon, V.M., Benz, R., Fujii, K., Leppla, S.H. and Tweten, R.K. (1997). *Clostridium septicum* alpha-toxin is proteolytically activated by furin. *Infect. Immun.* **65**, 4130–4.

Goyard, S., Orlando, C., Sabatier, J.M., Labruyere, E., d'Alayer, J., Fontan, G., van Rietschoten, J., Mock, M., Danchin, A., Ullmann, A. and Monneron, A. (1989). Identification of a common domain in calmodulin-activated eukaryotic and bacterial adenylate cyclases. *Biochemistry* **28**, 1964–7.

Green, B.D., Battisti, L., Koehler, T.M., Thorne, C.B. and Ivins, B.E. (1985). Demonstration of a capsule plasmid in *Bacillus anthracis*. *Infect. Immun.* **49**, 291–7.

Guignot, J., Mock, M. and Fouet, A. (1997). AtxA activates the transcription of genes harbored by both *Bacillus anthracis* virulence plasmids. *FEMS Microbiol. Lett.* **147**, 203–7.

Gupta, P., Batra, S., Chopra, A.P., Singh, Y. and Bhatnagar, R. (1998). Expression and purification of the recombinant lethal factor of *Bacillus anthracis*. *Infect. Immun.* **66**, 862–5.

Haines, B.W., Klein, F. and Lincoln, R.E. (1965). Quantitative assay for crude anthrax toxins. *J. Bacteriol.* **89**, 74–83.

Hambleton, P. and Turnbull, P.C. (1990). Anthrax vaccine development: a continuing story. *Adv. Biotechnol. Processes* **13**, 105–22.

Hambleton, P., Carman, J.A. and Melling, J. (1984). Anthrax: the disease in relation to vaccines. *Vaccine* **2**, 125–32.

Hammond, S.E. and Hanna, P.C. (1998). Lethal factor active-site mutations affect catalytic activity *in vitro*. *Infect. Immun.* **66**, 2374–8.

Hanna, P. (1998). Anthrax pathogenesis and host response. *Curr. Top. Microbiol. Immunol.* **225**, 13–35.

Hanna, P.C., Kouchi, S. and Collier, R.J. (1992). Biochemical and physiological changes induced by anthrax lethal toxin in J774 macrophage-like cells. *Mol. Biol. Cell* **3**, 1269–77.

Hanna, P.C., Acosta, D. and Collier, R.J. (1993). On the role of macrophages in anthrax. *Proc. Natl Acad. Sci. USA* **90**, 10198–201.

Hanna, P.C., Kruskal, B.A., Ezekowitz, R.A., Bloom, B.R. and Collier, R.J. (1995). Role of macrophage oxidative burst in the action of anthrax toxin lethal toxin. *Mol. Med.* **1**, 7–18.

Hanski, E. and Coote, J.G. (1991). *Bordetella pertussis* adenylate cyclase toxin. In: *Sourcebook of Bacterial Protein Toxins* (eds J.E. Alouf and J.H. Freer), pp. 349–66. Academic Press, London.

Heemskerk, D.D. and Thorne, C.B. (1990). Genetic exchange and transposon mutagenesis in *Bacillus anthracis*. *Salisbury Med. Bull.* **68**, Special Suppl., 63–7.

Higgins, C.F., Hiles, I.D., Salmond, G.P.C., Gill, D.R., Downie, J.A., Evans, I.J., Holland, I.B., Gray, L., Buckel, S.D., Bell, A.W. and Hermodson, M.A. (1986). A family of related ATP-binding subunits

coupled to many distinct biological processes in bacteria. *Nature* **323**, 448–50.

Hoffmaster, A.R. and Koehler, T.M. (1997). The anthrax toxin activator gene *atxA* is associated with CO_2-enhanced non-toxin gene expression in *Bacillus anthracis*. *Infect. Immun.* **65**, 3091–9.

Hoover, D.L., Friedlander, A.M., Rogers, L.C., Yoon, I.K., Warren, R.L. and Cross, A.S. (1994). Anthrax edema toxin differentially regulates lipopolysaccharide-induced monocyte production of tumor necrosis factor alpha and interleukin-6 by increasing intracellular cyclic AMP. *Infect. Immun.* **62**, 4432–9.

Hornung, J.M. and Thorne, C.B. (1991). Insertion mutations affecting pXO1-associated toxin production in *Bacillus anthracis*. *91st Annu. Meet. Am. Soc. Microbiol.* **98**, Abst. D-121 (Abstract).

Ivins, B.E. and Welkos, S.L. (1986). Cloning and expression of the *Bacillus anthracis* protective antigen gene in *Bacillus subtilis*. *Infect. Immun.* **54**, 537–42.

Ivins, B.E. and Welkos, S.L. (1988). Recent advances in the development of an improved, human anthrax vaccine. *Eur. J. Epidemiol.* **4**, 12–19.

Ivins, B.E., Ezzell, J.W., Jr, Jemski, J., Hedlund, K.W., Ristroph, J.D. and Leppla, S.H. (1986). Immunization studies with attenuated strains of *Bacillus anthracis*. *Infect. Immun.* **52**, 454–8.

Ivins, B.E., Welkos, S.L., Knudson, G.B. and Leblanc, D.J. (1988). Transposon Tn*916* mutagenesis in *Bacillus anthracis*. *Infect. Immun.* **56**, 176–81.

Jin, F., Nathan, C.F., Radzioch, D. and Ding, A. (1998). Lipopolysaccharide-related stimuli induce expression of the secretory leukocyte protease inhibitor, a macrophage-derived lipopolysaccharide inhibitor. *Infect. Immun.* **66**, 2447–52.

Kaspar, R.L. and Robertson, D.L. (1987). Purification and physical analysis of *Bacillus anthracis* plasmids pXO1 and pXO2. *Biochem. Biophys. Res. Commun.* **149**, 362–8.

Keppie, J., Harris-Smith, P.W. and Smith, H. (1963). The chemical basis of the virulence of *Bacillus anthracis*. IX. Its aggressins and their mode of action. *Br. J. Exp. Pathol.* **44**, 446–53.

Kimura, K., Kubota, T., Ohishi, I., Isogai, E., Isogai, H. and Fujii, N. (1998). The gene for component-II of botulinum C2 toxin. *Vet. Microbiol.* **62**, 27–34.

Klimpel, K.R. and Leppla, S.H. (1996). Cysteine mutants of anthrax toxin protective antigen as tools to probe structure and function. *Salisbury Med. Bull.*, **87**, Special Suppl., 93–4.

Klimpel, K.R., Molloy, S.S., Thomas, G. and Leppla, S.H. (1992). Anthrax toxin protective antigen is activated by a cell-surface protease with the sequence specificity and catalytic properties of furin. *Proc. Natl Acad. Sci. USA* **89**, 10277–81.

Klimpel, K.R., Arora, N. and Leppla, S.H. (1994). Anthrax toxin lethal factor contains a zinc metalloprotease consensus sequence which is required for lethal toxin activity. *Mol. Microbiol.* **13**, 1093–100.

Klingenberg, O. and Olsnes, S. (1996). Ability of methotrexate to inhibit translocation to the cytosol of dihydrofolate reductase fused to diphtheria toxin. *Biochem. J.* **313**, 647–53.

Koehler, T.M., Dai, Z. and Kaufman-Yarbray, M. (1994). Regulation of the *Bacillus anthracis* protective antigen gene: CO_2 and a trans-acting element activate transcription from one of two promoters. *J. Bacteriol.* **176**, 586–95.

Labruyere, E., Mock, M., Ladant, D., Michelson, S., Gilles, A.M., Laoide, B. and Barzu, O. (1990). Characterization of ATP and calmodulin-binding properties of a truncated form of *Bacillus anthracis* adenylate cyclase. *Biochemistry* **29**, 4922–8.

Labruyere, E., Mock, M., Surewicz, W.K., Mantsch, H.H., Rose, T., Munier, H., Sarfati, R.S. and Barzu, O. (1991). Structural and ligand-binding properties of a truncated form of *Bacillus anthracis* adenylate cyclase and of a catalytically inactive variant in which glutamine substitutes for lysine-346. *Biochemistry* **30**, 2619–24.

Ladant, D., Brezin, C., Alonso, J.M., Crenon, I. and Guiso, N. (1986). *Bordetella pertussis* adenylate cyclase. Purification, characterization, and radioimmunoassay. *J. Biol. Chem.* **261**, 16264–9.

Leppla, S.H. (1982). Anthrax toxin edema factor: a bacterial adenylate cyclase that increases cyclic AMP concentrations of eukaryotic cells. *Proc. Natl Acad. Sci. USA* **79**, 3162–6.

Leppla, S.H. (1984). *Bacillus anthracis* calmodulin-dependent adenylate cyclase: chemical and enzymatic properties and interactions with eucaryotic cells. In: *Advances in Cyclic Nucleotide and Protein Phosphorylation Research*, Vol. 17 (ed. P. Greengard), pp. 189–98. Raven Press, New York.

Leppla, S.H. (1988). Production and purification of anthrax toxin. In: *Methods in Enzymology*, Vol. 165 (ed. S. Harshman), pp. 103–16. Academic Press, Orlando, FL.

Leppla, S.H. (1991a). The anthrax toxin complex. In: *Sourcebook of Bacterial Protein Toxins* (eds J.E. Alouf and J.H. Freer), pp. 277–302. Academic Press, London.

Leppla, S.H. (1991b). Purification and characterization of adenylyl cyclase from *Bacillus anthracis*. In: *Methods in Enzymology*, Vol. 195 (eds R.A. Johnson and J.D. Corbin), pp. 153–68. Academic Press, San Diego, CA.

Leppla, S.H. (1995). Anthrax toxins. In: *Bacterial Toxins and Virulence Factors in Disease. Handbook of Natural Toxins*, Vol. 8 (eds J. Moss, B. Iglewski, M. Vaughan and A. Tu), pp. 543–72. Marcel Dekker, New York.

Leppla, S.H., Friedlander, A.M. and Cora, E. (1988). Proteolytic activation of anthrax toxin bound to cellular receptors. In: *Bacterial Protein Toxins* (eds F. Fehrenbach, J.E. Alouf, P. Falmagne, W. Goebel, J. Jeljaszewicz, D. Jurgen and R. Rappuoli), pp. 111–12. Gustav Fischer, New York.

Little, S.F. and Lowe, J.R. (1991). Location of receptor-binding region of protective antigen from *Bacillus anthracis*. *Biochem. Biophys. Res. Commun.* **180**, 531–7.

Little, S.F., Leppla, S.H. and Cora, E. (1988). Production and characterization of monoclonal antibodies to the protective antigen component of *Bacillus anthracis* toxin. *Infect. Immun.* **56**, 1807–13.

Little, S.F., Leppla, S.H. and Friedlander, A.M. (1990). Production and characterization of monoclonal antibodies against the lethal factor component of *Bacillus anthracis* lethal toxin. *Infect. Immun.* **58**, 1606–13.

Little, S.F., Leppla, S.H., Burnett, J.W. and Friedlander, A.M. (1994). Structure–function analysis of *Bacillus anthracis* edema factor by using monoclonal antibodies. *Biochem. Biophys. Res. Commun.* **199**, 676–82.

Little, S.F., Novak, J.M., Lowe, J.R., Leppla, S.H., Singh, Y., Klimpel, K.R., Lidgerding, B.C. and Friedlander, A.M. (1996). Characterization of lethal factor binding and cell receptor binding domains of protective antigen of *Bacillus anthracis* using monoclonal antibodies. *Microbiology* **142**, 707–15.

Little, S.F., Ivins, B.E., Fellows, P.F. and Friedlander, A.M. (1997). Passive protection by polyclonal antibodies against *Bacillus anthracis* infection in guinea pigs. *Infect. Immun.* **65**, 5171–5.

Makino, S., Uchida, I., Terakado, N., Sasakawa, C. and Yoshikawa, M. (1989). Molecular characterization and protein analysis of the cap region, which is essential for encapsulation in *Bacillus anthracis*. *J. Bacteriol.* **171**, 722–30.

Menard, A., Papini, E., Mock, M. and Montecucco, C. (1996a). The cytotoxic activity of *Bacillus anthracis* lethal factor is inhibited by leukotriene A4 hydrolase and metallopeptidase inhibitors. *Biochem. J.* **320**, 687–91.

Menard, A., Altendorf, K., Breves, D., Mock, M. and Montecucco, C. (1996b). The vacuolar ATPase proton pump is required for the cytotoxicity of *Bacillus anthracis* lethal toxin. *FEBS Lett.* **386**, 161–4.

Meynell, E. and Meynell, G.G. (1964). The roles of serum and carbon dioxide in capsule formation by *Bacillus anthracis*. *J. Gen. Microbiol.* **34**, 153–64.

Mikesell, P., Ivins, B.E., Ristroph, J.D. and Dreier, T.M. (1983). Evidence for plasmid-mediated toxin production in *Bacillus anthracis*. *Infect. Immun.* **39**, 371–6.

Miller, J., McBride, B.W., Manchee, R.J., Moore, P. and Baillie, L.W.J. (1998). Production and purification of recombinant protective antigen and protective efficacy against *Bacillus anthracis*. *Lett. Appl. Microbiol.* **26**, 56–60.

Milne, J.C. and Collier, R.J. (1993). pH-dependent permeabilization of the plasma membrane of mammalian cells by anthrax protective antigen. *Mol. Microbiol.* **10**, 647–53.

Milne, J.C., Furlong, D., Hanna, P.C., Wall, J.S. and Collier, R.J. (1994). Anthrax protective antigen forms oligomers during intoxication of mammalian cells. *J. Biol. Chem.* **269**, 20607–12.

Milne, J.C., Blanke, S.R., Hanna, P.C. and Collier, R.J. (1995). Protective antigen-binding domain of anthrax lethal factor mediates translocation of a heterologous protein fused to its amino- or carboxy-terminus. *Mol. Microbiol.* **15**, 661–6.

Mock, M., Labruyere, E., Glaser, P., Danchin, A. and Ullmann, A. (1988). Cloning and expression of the calmodulin-sensitive *Bacillus anthracis* adenylate cyclase in *Escherichia coli*. *Gene* **64**, 277–84.

Munier, H., Blanco, F.J., Precheur, B., Diesis, E., Nieto, J.L., Craescu, C.T. and Barzu, O. (1993). Characterization of a synthetic calmodulin-binding peptide derived from *Bacillus anthracis* adenylate cyclase. *J. Biol. Chem.* **268**, 1695–701.

Munson, P.J. and Rodbard, D. (1980). Ligand: a versatile computerized approach for characterization of ligand-binding systems. *Anal. Biochem.* **107**, 220–39.

Noskov, A.N., Kravchenko, T.B. and Noskova, V.P. (1996a). Vyiavlenie funktsional'no aktivnykh domenov v molekule protektivnogo antigena sibireiazvennogo ekzotoksina. [Detection of the functionally active domains in the molecule of protective antigen of the anthrax exotoxin.] *Mol. Gen. Mikrobiol. Virusol.* 16–20.

Noskov, A.N., Kravchenko, T.B. and Noskova, V.P. (1996b). Vyiavlenie funktsional'no aktivnykh domenov v molekule letal'nogo faktora sibireiazvennogo ekzotoksina. [Detection of functionally active domains in the molecule of the lethal factor of the anthrax exotoxin.] *Mol. Gen. Mikrobiol. Virusol.* 20–2.

Novak, J.M., Stein, M.P., Little, S.F., Leppla, S.H. and Friedlander, A.M. (1992). Functional characterization of protease-treated *Bacillus anthracis* protective antigen. *J. Biol. Chem.* **267**, 17186–93.

O'Brien, J., Friedlander, A., Dreier, T., Ezzell, J. and Leppla, S. (1985). Effects of anthrax toxin components on human neutrophils. *Infect. Immun.* **47**, 306–10.

Perelle, S., Gibert, M., Boquet, P. and Popoff, M.R. (1993). Characterization of *Clostridium perfringens* iota-toxin genes and expression in *Escherichia coli*. *Infect. Immun.* **61**, 5147–56.

Perelle, S., Gibert, M., Bourlioux, P., Corthier, G. and Popoff, M.R. (1997). Production of a complete binary toxin (actin-specific ADP-ribosyltransferase) by *Clostridium difficile* CD196. *Infect. Immun.* **65**, 1402–7.

Petosa, C., Collier, R.J., Klimpel, K.R., Leppla, S.H. and Liddington, R.C. (1997). Crystal structure of the anthrax toxin protective antigen. *Nature* **385**, 833–8.

Pezard, C., Berche, P. and Mock, M. (1991). Contribution of individual toxin components to virulence of *Bacillus anthracis*. *Infect. Immun.* **59**, 3472–7.

Pezard, C., Duflot, E. and Mock, M. (1993). Construction of *Bacillus anthracis* mutant strains producing a single toxin component. *J. Gen. Microbiol.* **139**, 2459–63.

Pollock, M.R. and Richmond, M.H. (1962). Low cysteine content of bacterial extracellular proteins: its possible physiological significance. *Nature* **194**, 446–9.

Puziss, M., Manning, L.C., Lynch, J.W., Barclay, E., Abelow, I. and Wright, G.G. (1963). Large-scale production of protective antigen of *Bacillus anthracis* in anaerobic cultures. *Appl. Microbiol.* **11**, 330–4.

Quinn, C.P., Shone, C.C., Turnbull, P.C. and Melling, J. (1988). Purification of anthrax-toxin components by high-performance anion-exchange, gel-filtration and hydrophobic-interaction chromatography. *Biochem. J.* **252**, 753–8.

Quinn, C.P., Singh, Y., Klimpel, K.R. and Leppla, S.H. (1991). Functional mapping of anthrax toxin lethal factor by in-frame insertion mutagenesis. *J. Biol. Chem.* **266**, 20124–30.

Ristroph, J.D. and Ivins, B.E. (1983). Elaboration of *Bacillus anthracis* antigens in a new, defined culture medium. *Infect. Immun.* **39**, 483–6.

Robertson, D.L. and Leppla, S.H. (1986). Molecular cloning and expression in *Escherichia coli* of the lethal factor gene of *Bacillus anthracis*. *Gene* **44**, 71–8.

Robertson, D.L., Tippetts, M.T. and Leppla, S.H. (1988). Nucleotide sequence of the *Bacillus anthracis* edema factor gene (cya): a calmodulin-dependent adenylate cyclase. *Gene* **73**, 363–71.

Robertson, D.L., Bragg, T.S., Simpson, S., Kaspar, R., Xie, W. and Tippetts, M.T. (1990). Mapping and characterization of *Bacillus anthracis* plasmids pXO1 and pXO2. *Salisbury Med. Bull.* **68**, Special Suppl., 55–8.

Rogel, A., Farfel, Z., Goldschmidt, S., Shiloach, J. and Hanski, E. (1988). *Bordetella pertussis* adenylate cyclase: identification of multiple forms of the enzyme by antibodies. *J. Biol. Chem.* **263**, 13310–16.

Sharma, M., Swain, P.K., Chopra, A.P., Chaudhary, V.K. and Singh, Y. (1996). Expression and purification of anthrax toxin protective antigen from *Escherichia coli*. *Protein Exp. Purif.* **7**, 33–8.

Singh, Y., Leppla, S.H., Bhatnagar, R. and Friedlander, A.M. (1989a). Internalization and processing of *Bacillus anthracis* lethal toxin by toxin-sensitive and -resistant cells. *J. Biol. Chem.* **264**, 11099–102.

Singh, Y., Chaudhary, V.K. and Leppla, S.H. (1989b). A deleted variant of *Bacillus anthracis* protective antigen is non-toxic and blocks anthrax toxin action *in vivo*. *J. Biol. Chem.* **264**, 19103–7.

Singh, Y., Leppla, S.H., Bhatnagar, R. and Friedlander, A.M. (1990). Basis of cellular sensitivity and resistance to anthrax lethal toxin. *Salisbury Med. Bull.* **68**, 46–8.

Singh, Y., Klimpel, K.R., Quinn, C.P., Chaudhary, V.K. and Leppla, S.H. (1991). The carboxyl-terminal end of protective antigen is required for receptor binding and anthrax toxin activity. *J. Biol. Chem.* **266**, 15493–7.

Singh, Y., Klimpel, K.R., Arora, N., Sharma, M. and Leppla, S.H. (1994). The chymotrypsin-sensitive site, FFD315, in anthrax toxin protective antigen is required for translocation of lethal factor. *J. Biol. Chem.* **269**, 29039–46.

Singh, Y., Klimpel, K.R., Goel, S., Swaim, P.K. and Leppla, S.H. (1999). Oligomerization of anthrax toxin protective antigen and binding of lethal factor during endocytic uptake into mammalian cells. *Infect. Immun.* **67**, 1853–9.

Smith, H. and Stoner, H.B. (1967). Anthrax toxic complex. *Fed. Proc.* **26**, 1554–7.

Smith, H., Keppie, J. and Stanley, J.L. (1955). The chemical basis of the virulence of *Bacillus anthracis*. V. The specific toxin produced by *B. anthracis in vivo*. *Br. J. Exp. Pathol.* **36**, 460–72.

Song, L., Hobaugh, M.R., Shustak, C., Cheley, S., Bayley, H. and Gouaux, J.E. (1996). Structure of staphylococcal alpha hemolysin, a heptameric transmembrane pore. *Science* **274**, 1859–66.

Stephen, J. (1986). Anthrax toxin. In: *Pharmacology of Bacterial Toxins* (eds F. Dorner and J. Drews), pp. 381–95. Pergamon Press, Oxford.

Sterne, M. (1937). Variation in *Bacillus anthracis*. *Onderstepoort J. Vet. Sci. Anim. Ind.* **8**, 271–349.

Strange, R.E. and Thorne, C.B. (1958). Further purification of the protective antigen of *Bacillus anthracis* produced *in vitro*. *J. Bacteriol.* **76**, 192–201.

Thorne, C.B. (1985). Genetics of *Bacillus anthracis*. In: *Microbiology–85* (eds L. Lieve, P.F. Bonventre, J.A. Morello, S. Schlessinger, S.D. Silver and H.C. Wu), pp. 56–62. American Society for Microbiology, Washington, DC.

Tippetts, M.T. and Robertson, D.L. (1988). Molecular cloning and expression of the *Bacillus anthracis* edema factor toxin gene: a calmodulin-dependent adenylate cyclase. *J. Bacteriol.* **170**, 2263–6.

Turnbull, P.C.B. (1990a). *Salisbury Med. Bull.* **68**, Special Suppl., 1–105.

Turnbull, P.C.B. (1990b). Terminal bacterial and toxin levels in the blood of guinea pigs dying of anthrax. *Salisbury Med. Bull.* **68**, 53–5.

Turnbull, P.C.B. (1996). *Salisbury Med. Bull.* **87**, Special Suppl., 1–139.

Uchida, I., Sekizaki, T., Hashimoto, K. and Terakado, N. (1985). Association of the encapsulation of *Bacillus anthracis* with a 60 megadalton plasmid. *J. Gen. Microbiol.* **131**, 363–7.

Uchida, I., Hashimoto, K. and Terakado, N. (1986). Virulence and immunogenicity in experimental animals of *Bacillus anthracis* strains harbouring or lacking 110 MDa and 60 MDa plasmids. *J. Gen. Microbiol.* **132**, 557–9.

Uchida, I., Hashimoto, K., Makino, S., Sasakawa, C., Yoshikawa, M. and Terakado, N. (1987). Restriction map of a capsule plasmid of *Bacillus anthracis*. *Plasmid* **18**, 178–81.

Uchida, I., Hornung, J.M., Thorne, C.B., Klimpel, K.R. and Leppla, S.H. (1993a). Cloning and characerization of a gene whose product is a *trans*-activator of anthrax toxin synthesis. *J. Bacteriol.* **175**, 5329–38.

Uchida, I., Makino, S., Sasakawa, C., Yoshikawa, M., Sugimoto, C. and Terakado, N. (1993b). Identification of a novel gene, *dep*, associated with depolymerization of the capsular polymer in *Bacillus anthracis*. *Mol. Microbiol.* **9**, 487–96.

Uchida, I., Makino, S., Sekizaki, T. and Terakado, N. (1997). Cross-talk to the genes for *Bacillus anthracis* capsule synthesis by *atxA*, the gene encoding the trans-activator of anthrax toxin synthesis. *Mol. Microbiol.* **23**, 1229–40.

Varughese, M., Teixeira, A.V., Liu, S. and Leppla, S.H. (1999). Indentification of a receptor-binding region within domain 4 of the protective antigen component of anthrax toxin. *Infect. Immun.* **67**, 1860–5.

Vietri, N.J., Marrero, R., Hoover, T.A. and Welkos, S.L. (1995). Identification and characterization of a trans-activator involved in the regulation of encapsulation by *Bacillus anthracis*. *Gene* **152**, 1–9.

Vodkin, M.H. and Leppla, S.H. (1983). Cloning of the protective antigen gene of *Bacillus anthracis*. *Cell* **34**, 693–7.

Wade, B.H., Wright, G.G., Hewlett, E.L., Leppla, S.H. and Mandell, G.L. (1985). Anthrax toxin components stimulate chemotaxis of human polymorphonuclear neutrophils. *Proc. Soc. Exp. Biol. Med.* **179**, 159–62.

Warren, G.W., Koziel, M.G., Mullins, M.A., Nye, G.J., Carr, B., Desai, N., Kostischka, K., Duck, N.B. and Estruch, J.J. (1996). Novel pesticidal proteins and strains. Patent application WO 96/10083 (1996). World Intellectual Patent Organization (Abstract).

Weinstein, J.N., Myers, T.G., O'Connor, P.M., Friend, S.H., Fornace, A.J.J., Kohn, K.W., Fojo, T., Bates, S.E., Rubinstein, L.V., Anderson, N.L., Buolamwini, J.K., van, O.W., Monks, A.P., Scudiero, D.A., Sausville, E.A., Zaharevitz, D.W., Bunow, B., Viswanadhan, V.N., Johnson, G.S., Wittes, R.E. and Paull, K.D. (1997). An information-intensive approach to the molecular pharmacology of cancer. *Science* **275**, 343–9.

Welch, R.A. (1991). Pore-forming cytolysins of Gram-negative bacteria. *Mol. Microbiol.* **5**, 521–8.

Welkos, S.L. and Friedlander, A.M. (1988). Comparative safety and efficacy against *Bacillus anthracis* of protective antigen and live vaccines in mice. *Microb. Pathog.* **5**, 127–39.

Welkos, S.L., Lowe, J.R., Eden-McCutchan, F., Vodkin, M., Leppla, S.H. and Schmidt, J.J. (1988). Sequence and analysis of the DNA encoding protective antigen of *Bacillus anthracis*. *Gene* **69**, 287–300.

Welkos, S., Becker, D., Friedlander, A. and Trotter, R. (1990). Pathogenesis and host resistance to *Bacillus anthracis*: a mouse model. *Salisbury Med. Bull.* **68**, Special Suppl., 49–52.

Wilkie, M.H. and Ward, M.K. (1967). Characterization of anthrax toxin. *Fed. Proc.* **26**, 1527–31.

Wright, G.G., Read, P.W. and Mandell, G.L. (1988). Lipopolysaccharide releases a priming substance from platelets that augments the oxidative response of polymorphonuclear neutrophils to chemotactic peptide. *J. Infect. Dis.* **157**, 690–6.

Xia, Z.G. and Storm, D.R. (1990). A-type ATP binding consensus sequences are critical for the catalytic activity of the calmodulin-sensitive adenylyl cyclase from *Bacillus anthracis*. *J. Biol. Chem.* **265**, 6517–20.

Zhao, J., Milne, J.C. and Collier, R.J. (1995). Effect of anthrax toxin's lethal factor on ion channels formed by the protective antigen. *J. Biol. Chem.* **270**, 18626–30.

13

Helicobacter pylori vacuolating cytotoxin and associated pathogenic factors

Cesare Montecucco, Emanuele Papini, Marina de Bernard, John L. Telford and Rino Rappuoli

INTRODUCTION

Heliobacter pylori was cultivated in pure culture and shown to be associated with severe gastroduodenal diseases only in the 1980s (Warren and Marshall, 1983; Marshall *et al.*, 1985), despite the fact that it has probably infected humans since their origin (Blaser, 1997). This makes one think once more how little we know of the microbial world.

After an initial lag, due to the scepticism of the scientific community, an increasing amount of research has focused on *H. pylori*, second in the area of medical microbiology only to HIV. This has led to the sequencing of the *H. pylori* pathogenicity island (Censini *et al.*, 1996). Shortly afterwards, the sequence of the entire circular genome of *H. pylori* strain 26695, which consists of 1 667 867 base pairs and is predicted to code for 1590 proteins, was reported by Tomb *et al.* (1997). We now know several features of *H. pylori* infection and of the host reaction, and we are beginning to uncover the molecular mechanisms of the action of *H. pylori* virulence factors (Telford *et al.*, 1997). *Helicobacter pylori* infection affects billions of people in the world, with peaks of 90% of the population infected in countries with poor sanitary conditions and a low socio-economical level.

Helicobacter pylori is a specific pathogen of humans, and available evidence indicates that *H. pylori* infection is acquired mainly via the oro-faecal route and much less frequently via the oro-oral route (Figura, 1996; McGuigan, 1996; Covacci and Rappuoli, 1998).

Helicobacter pylori is a spiral-shaped Gram-negative bacterium endowed with polar flagella and with very powerful urease activity (Blaser, 1993). These features allow it to colonize and to survive in the stomach by buffering its surface within the stomach lumen and to penetrate across the mucus layer that covers and protects the stomach epithelial cells. *Helicobacter pylori* resides within the mucus and on the apical surface of epithelial cells, where it attaches firmly via adhesin molecules and modifications of cell membrane proteins and of cytoskeletal proteins (Dytoc *et al.*, 1993; Smoot *et al.*, 1993; Segal *et al.*, 1996). The mucus layer and the apical domain of epithelial cells of the stomach is the ecological niche 'chosen' by *H. pylori*, an environment which requires special features to survive, and where little competition from other bacterial species may be experienced.

Helicobacter pylori infection lasts for decades, and some 20% of infected patients develop severe gastroduodenal diseases, including ulceration of the stomach or duodenum mucosa and, in some, gastric adenocarcinomas or lymphomas (Isaacson, 1994; Parsonnet *et al.*, 1994; Correa, 1995; Goodwin, 1997). Several factors contribute to the pathogenesis and severity of *H. pylori*-associated disease, including diet, genetic factors, acid hypersecretion and stress, but the most important one appears to be the type of *H. pylori* strain (Telford *et al.*, 1997). In fact, the *H. pylori* species includes a large number of different strains with different genetic and phenotypic structure, among which extensive horizontal DNA transfer takes place (Logan and Berg, 1996;

Covacci and Rappuoli, 1998; Kuipers *et al.*, 1998). With respect to pathogenicity, one major feature is the presence of a 40-kb pathogenicity island, marked by the presence of the cytotoxin associated gene A (*cagA* gene) (Censini *et al.*, 1996; Parsonnet *et al.*, 1997).

Helicobacter pylori strains can be divided into two broad families: type I, which includes cagA$^+$ strains, and type II, which groups cagA$^-$ strains (Xiang *et al.*, 1995; Censini *et al.*, 1996; Covacci and Rappuoli, 1998), and frequently the two type of strains coexist within the same patient (Figura *et al.*, 1998). All strains produce urease, without which *H. pylori* would die soon after entry into the stomach lumen. Nearly all type-I strains produce vacuolating cytotoxin (VacA) and almost invariably CagA protein (Covacci and Rappuoli, 1998). Type-I *H. pylori* strains are isolated from most human biopsies of patients affected by gastric or duodenal ulcers, suggesting a correlation between the infection with toxigenic strains and the pathogenesis of gastro-duodenal lesions (Cover *et al.*, 1993b; Xiang *et al.*, 1995). Moreover, sonicated extracts of type-I, but not type-II, strains orally administered to mice induce pathological lesions similar to those observed in human biopsies (Marchetti *et al.*, 1995).

These findings have prompted several laboratories to investigate closely the activity of VacA and the associated virulence factors. This review will discuss the present status of knowledge on the cellular lesions caused by protein virulence factors produced by *H. pylori*.

VACUOLATING CYTOTOXIN (VACA)

Cell vacuolization

Leunk *et al.* (1988) were the first to report that the supernatant of cultures of about half of the *H. pylori* isolates were capable of inducing eukaryotic cells in culture to form large cytoplasmic vacuoles. They attributed this effect to the presence of a protein that was effective on cells derived from different animals. Hence, contrary to *H. pylori*, which is a strictly human pathogen, its vacuolating activity is not species specific, although different cell lines vary in their susceptibility to vacuolate (Yahiro *et al.*, 1997; Leunk *et al.*, 1988; de Bernard *et al.*, 1998a), with HeLa cells being the more prone to vacuolation.

A peculiar behaviour is that of the HL60 cells, which are toxin resistant (Yahiro *et al.*, 1997), but do vacuolate if they are pretreated with 12-0-tetrolecanoylphorbol-13-0-acetate (TPA) (de Bernard *et al.*, 1998b). Vacuolation is particularly evident in sparse cells in culture, is potentiated by membrane permeant weak bases, such as ammonia and amines, and depends on cell density. Confluent cells and polarized epithelial cell monolayers in culture show little, if any, vacuolization (de Bernard *et al.*, 1998a; Papini *et al.*, 1998). *In vivo*, vacuoles can be seen in cells of the stomach mucosa of mice (Marchetti *et al.*, 1995) or beagle dogs (Rossi *et al.*, 1999) fed with supernatant of pathogenic strains of *H.*

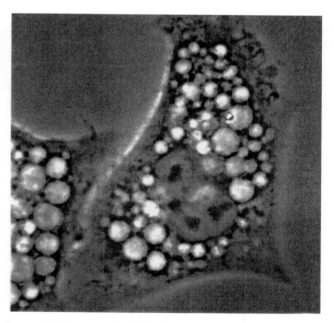

FIGURE 13.1 Vacuole formation and growth in HeLa cells exposed to the vacuolating cytotoxin of *H. pylori*. The picture shows two frames taken from a videomicroscope investigation of HeLa cells after 30 min (left panel) and 180 min (right panel) from the addition of 200 nM VacA in the presence of 5 mM ammonium chloride.

pylori, as well as in human biopsies (Tricottet *et al.*, 1986; Fiocca *et al.*, 1992).

In cultured HeLa cells, vacuoles arise in the perinuclear cell area and then increase in number and size to fill the entire cytoplasm (Cover *et al.*, 1992; Montecucco *et al.*, unpublished observations), as shown in Figure 13.1. This is believed to cause a state of cell sufferance that eventually leads to death by necrosis (Figura *et al.*, 1989).

Vacuoles actively accumulate weak base dyes, including neutral red, which is membrane permeant and becomes trapped inside acidic intracellular compartments following protonation (Cover *et al.*, 1992). This important finding revealed that the vacuolar lumen is acidic and, at the same time, it provided the basis for a quantitative assay of the total volume occupied by vacuoles in a cell culture.

The factor responsible for cytotoxicity was isolated via several steps of chromatography and was identified as a protein of approximately 90 kDa, with an amino-terminal sequence of overall hydrophobic nature (Cover and Blaser, 1992). Alternative purification protocols have been developed subsequently (Manetti *et al.*, 1995; Icatlo *et al.*, 1998; Reyrat *et al.*, 1998). Antibodies raised against the purified protein prevented the cell vacuolation induced by *H. pylori* supernatants, as did some antisera derived from *H. pylori*-infected patients (Cover and Blaser, 1992; Manetti *et al.*, 1995). The toxin was termed VacA because it was the first protein factor identified in *H. pylori* supernatant endowed with cell vacuolating activity.

Gene structure and biosynthesis

The partial amino acid sequence of VacA allowed its cloning with degenerate primers (Cover *et al.*, 1994; Phadnis *et al.*, 1994; Schmitt and Haas, 1994; Telford *et al.*, 1994) and, at present, the gene structure of VacA in several toxigenic strains is known. Also, the *vacA* gene conforms to the characteristic large genetic variability of this bacterial genus, presenting different gene sequences in different strains. The analysis of the deduced amino acid sequence shows no overall similarity to any known protein. However, four different domains can be identified by secondary structure prediction methods, as depicted in Figure 13.2. Beginning at the amino-terminus, there is an amino-terminal signal sequence (33 residues long), which determines the export of the toxin from the cytosol to the periplasm. This is followed by a region predicted to consist of α-helices and of β-pleated segments, which corresponds to a protein domain of 37 kDa. This region begins with a 32-residue-long hydrophobic segment and ends with the sequence AKNDKNESAKND-

KQES, which contains a hydrophilic double repeat devoid of secondary structure, likely to be exposed on the protein surface.

These features are common to protease-sensitive loops and, indeed, part of the VacA toxin released in the medium is nicked at this point (Telford *et al.*, 1994). The following domain corresponds to a segment of 58 kDa predicted to consist of a first region rich in β-pleated sheets and a second one with both α and β secondary structure elements, which strongly indicates the existence of two subdomains, possibly involved in different aspects of cell intoxication (Figure 13.2). This suggestion is further supported by the fact that the second subdomain exhibits considerable genetic diversity, whereas the first subdomain does not. Moreover, a 23 amino acid insertion is frequently present between the two subdomains (Atherton *et al.*, 1995).

The fourth part has three distinctive features: a pair of cysteines separated by 10 residues, followed by a 35-kDa region rich in amphipathic β-pleated segments, ending with a motif consisting of alternating hydrophobic residues with a C-terminal phenylalanine. These structural elements characterize a domain that

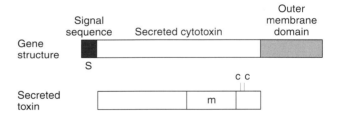

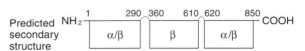

FIGURE 13.2 Schematic structure of the *vacA* gene and the VacA protein toxin of *H. pylori*. The VacA toxin is made in the bacterial cytosol as a 140-kDa protein, which is recognized by an inner membrane protein transport apparatus via the amino-terminal signal sequence (black), which is then removed by periplasmic proteases. The carboxyl-terminal 35-kDa domain (grey) is in-built protein translocating machinery, which mediates the movement of the secreted part of the toxin across the membrane on the cell surface. Here, the toxin is released from the bacterium upon proteolysis at an undefined place(s) located after the cysteine doublet indicated in the figure (middle panel). The lower panel shows that secondary structure prediction methods indicate the existence of three domains with different folding patterns. Based on the effect of variations in the m-region on cell-binding properties of VacA and on the deleterious effects of N-terminal deletions, it is tempting to suggest that the N-terminal α/β-region is an enzyme, the middle β-region is involved in membrane translocation and the C-terminal α/β-region is responsible for toxin binding to the cell surface.

is capable of translocating the portion of the polypeptide chain present at its amino-terminus across the outer membrane of Gram-negatives (Wandersman, 1992). Such a membrane translocating unit is present in several proteins of Gram-negatives and was first characterized in the IgA protease from *Neisseriae gonorrhoeae* (Pohlner *et al.*, 1987).

After translocation, part of the 95-kDa protein is released from the outer bacterial membrane, following proteolysis by yet unidentified proteases. Thus, in a culture of *H. pylori*, part of VacA remains associated with the bacterium and the remaining part is released in the medium, where it can be additionally cleaved within the repeat connecting the p37 and p58 domains, which remain associated via non-covalent forces (Telford *et al.*, 1994).

Genetic variability in the *vacA* gene

The characterization of many clinical isolates has shown that there is a considerable quantitative and qualitative variation in the VacA released by the various *H. pylori* strains (Atherton *et al.*, 1995; Cover, 1996). Signal sequences can be grouped in at least three different types (s1a, s1b and s2) and another highly divergent segment, designated as the m-region, is present within p58. The various m-sequences were grouped into two main groups, m1 and m2 (Atherton *et al.*, 1995).

Strains with s1/m1 produce generally high levels of VacA protein and are highly toxigenic in the standard HeLa cell vacuolation assay. Most strains that have a *vacA* gene with the s2 signal peptide fail to release the toxin from the bacteria, and it is likely that this signal peptide does not function properly (Atherton *et al.*, 1995). Strains expressing an s1/m2 toxin produce significant quantities of toxin, which assemble into the correct structure but which have no activity on HeLa cells. This raises the questions as to why these genes have been conserved and why there is no significant difference in disease association between m1 and m2 strains (Go *et al.*, 1998).

Recent work provides an explanation for this apparent paradox. The definition of tox⁻ is strictly linked to the vacuolating assay that is almost invariably performed on HeLa cells in sparse culture, and m2-type toxins have no activity in this assay and have been designated tox⁻. It has recently been reported, however, that m2-type toxins are strongly toxic to a rabbit epithelial cell line, RK13, or to primary cultured cells from human gastric biopsies (Pagliaccia *et al.*, 1998). Toxicity correlates with the ability to bind to the surface of the cell: m1-type toxins bind and intoxicate both HeLa and RK13 cells, whereas m2-type toxins bind and intoxicate only RK13 cells. Hence, it appears that the m-

region is involved in target cell interaction, and the two alleles may have evolved to bind different or polymorphic receptors.

Also relevant is the finding that *H. pylori* strain 95–54, which produces an m2-type toxin, is as active as the standard toxigenic CCUG strain in lowering the transepithelial resistance of a polarized cell monolayer, when bound to the apical domain (Pelicic *et al.*, submitted). This assay mimics the situation *in vivo* of the bacteria adherent to the apical surface of polarized epithelial cells, thus revealing the biological properties of the toxin molecules, which are displayed by VacA i*n vivo* (see also the section on 'Cell binding and entry').

Strains containing *vacA* genes with the s2 signal peptide are relatively rare, and it is not clear where these alleles have arisen. In marked contrast, m2 strains are found frequently and reflect functional polymorphism. Interestingly, whereas approximately 80% of strains found in Western populations are s1/m1 and 20% are s1/m2, in Chinese populations this figure is reversed (Pan *et al.*, 1998; Ji *et al.*, submitted). The high incidence of s1/m2 strains in China may indicate genetic bottlenecking and expansion. However, given the difference in cell binding specificities between m1 and m2 alleles, there may have been selection for such a VacA form in the Chinese population.

Analysis of the *vacA* gene sequences from a number of Chinese strains has revealed that there is significant variation and that Chinese strains form a distinct evolutionary group. Chinese strains are significantly more different from Western strains than they are from each other (Pan *et al.*, 1998; van der Ende *et al.*, 1998; Ji *et al.*, submitted). This difference is found throughout the gene, including the m-region of m1 strains. In m2 strains, the situation is quite different. In these strains, the regions coding for the 37-kDa subunit and the outer membrane exporter show the same geographical variation found in m1 strains. However, the m-region is highly conserved between Western and Chinese strains (Ji *et al.*, submitted).

It is generally considered that *H. pylori* has been a parasite of humans since their appearance on earth, and it is thus likely that the difference in sequence of the *vacA* gene in Western and Chinese populations reflects the separation of these populations during the first human migrations. The most likely explanation for the similarity in sequence of the m2 allele between East and West is that this region has been acquired more recently by horizontal transfer of DNA, perhaps from a related *Helicobacter* species, and has subsequently spread throughout the world.

Other parts of the molecule show a higher degree of similarity, although the effect of minor substitution cannot be appreciated because of the present lack of knowledge about the protein regions directly involved in cell binding and entry and about the intracellular activity of the toxin. The 35-kDa *C*-terminal domain, which acts as protein translocating machinery, is highly conserved among strains.

Based on the identification of multiple combinations of the divergent s- and m-regions, it has been suggested that horizontal DNA transfer from other species is at the origin of the mosaic organization of the vacA gene, in a similar way to that observed for the IgA protease locus of *N. gonorrhoeae* (Halter *et al.*, 1989).

Structure

The toxin released from *H. pylori* is a 95-kDa protein with a strong tendency to oligomerize (Cover and Blaser, 1992), the structure of which has been investigated by advanced electron microscopic techniques (Lupetti *et al.*, 1996; Cover *et al.*, 1997; Lanzavecchia *et al.*, 1998). The oligomeric toxin has a flower-shaped structure consisting of six or seven monomers, as shown in Figure 13.3. Each monomer in the oligomer is clearly structured in two subunits, which represent the 37- and 58-kDa subunits described above. The oligomer composition is influenced by the length of the hydrophilic loop between the two

subunits: more than 70% of the toxin produced by strain CCUG17784 is in heptameric form, but the same strain engineered to remove 46 amino acids from the loop region produces predominantly hexamers (Burroni *et al.*, 1998). Hence, it is likely that the oligomeric structure is maintained by interactions involving both the 37- and 58-kDa subunits.

Accordingly, the 58-kDa subunit, expressed in the absence of the 37-kDa subunit, is capable of autonomous folding into a soluble structure (Reyrat *et al.*, submitted) but does not form hexamers or heptamers. Images of the free 58-kDa subunit obtained by computer analysis of electron micrographs are as the peripheral petals of the oligomer. Hence, the toxin appears to be organized as a ring of 58-kDa subunits with the 37-kDa subunits arranged in a smaller ring above and to the centre of the oligomer. We have proposed a model in which each 37-kDa subunit sits on top of the 58-kDa subunit of the adjacent monomer, such that the oligomer is held together by the intercalation of adjacent monomers (see Figure 13.3).

This oligomeric form of VacA is weakly active. The toxin is strongly activated by transient exposure to pH values <5.5, and it is resistant to acid solutions at pH 1.5 (de Bernard *et al.*, 1995). VacA also shows a remarkable resistance to temperature (Leunk *et al.*, 1988; Yahiro *et al.*, 1997). Although well correlated to the physiological environment where the toxin operates, the structural basis of the remarkable acid resistance of VacA is not known, but it is clear that VacA changes structure as the pH is lowered, and that this structural transition takes place in a narrow range centred around pH 5.2 (de Bernard *et al.*, 1995; Molinari *et al.*, 1998a). The neutral and acid forms of VacA have different spectroscopic properties with different secondary structures. Upon neutralization, VacA does not regain its neutral structure within hours, as shown by spectroscopy and by limited proteolysis (de Bernard *et al.*, 1995), but it may do so with very slow kinetics. On the basis of these findings, as well as of additional data discussed below (see the section on 'Effect of VacA on the trans-epithelial resistance and permeability of polarized epithelial monolayers'), we propose that VacA released from bacteria can exist in different forms endowed with different biological activity, as depicted in Figure 13.4.

Cover *et al.* (1997) have shown that acidification induces a dissociation of oligomers into monomers, which can be separated by ultracentrifugation on glycerol gradients. The monomers expose hydrophobic segments on their protein surface, which avidly bind hydrophobic fluorescent dyes such as ANS, and mediate the interaction of both p37 and p58 with the hydrophobic interior of the phospholipid bilayer

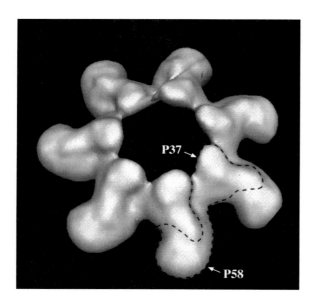

FIGURE 13.3 Three-dimensional reconstruction of the heptameric form of VacA from electron micrographs of quick-frozen deep-etched preparations (Lanzavecchia *et al.*, 1998). The ring formed by the 37-kDa subunit and the petals formed by the 58-kDa subunits are clearly visible. The dashed line represents the possible position of a 95-kDa monomer in a model in which each 37-kDa subunit sits on top of the 58-kDa subunit of the adjacent monomer such that the oligomer is held together by the intercalation of adjacent monomers.

(Molinari *et al.*, 1998a). The pH dependence of the low pH-induced toxin activation is identical to that of ANS binding, suggesting the possibility that the rate-determining step of VacA cell intoxication is its membrane penetration (see the section on 'Cell binding and entry' for a more complete discussion).

Little is known of the structure of VacA on the outer membrane of *H. pylori*, the presence of which can be easily demonstrated by immunofluorescent staining with anti-toxin antibodies. In fact, although in this initial phase research on VacA has focused on the released toxin, the bacterial-associated toxin *in vivo* may be as important as the released form, if not the predominant one. Recent work on monolayers of epithelial cells shows that bacterial-associated toxin does not require acid exposure to be active (Pelicic *et al.*, submitted). This is an indication that VacA bound to the external surface of the outer membrane may be predominantly monomeric. It is likely that the presence of the COOH-terminal domain and the interactions with the lipids of the outer membrane influence the structure of bacterial-bound VacA and, hence, that it is to be considered as an additional structurally different form of toxin (Figure 13.4).

Vacuole biogenesis

Under optimal conditions, that is using sparse HeLa cells in culture in the presence of ammonium ions, the addition of 100 nM VacA causes the appearance of small translucid vacuoles in the perinuclear area within half an hour (Figure 13.1). Videomicroscopy shows that such vacuoles are highly mobile, but remain in the area of formation, as if they were linked to cytoskeletal elements (Montecucco *et al.*, unpublished results). Indeed, microtubule depolymerizing agents, such as nocodazole, inhibit vacuolar growth (Papini *et al.*, 1994). Vacuoles are round and grow in size to fill the entire cytosol within a few hours, as shown in the right-hand panel of Figure 13.1.

In the absence of membrane-permeable amines, vacuoles develop similarly, although with a slower kinetics. VacA-induced vacuoles are larger and less numerous than those induced by permeant amines (Yamashiro and Maxfield, 1987; DeCourey and Storrie, 1991; Cover *et al.*, 1992; Montecucco *et al.*, unpublished observations). Vacuoles appear to be dynamic membrane-bound endocytic structures because they can incorporate fluid phase markers, i.e. dyes present

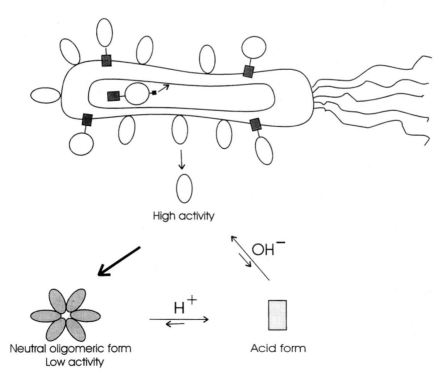

High activity

Neutral oligomeric form
Low activity

OH⁻

H⁺

Acid form

FIGURE 13.4 The VacA protein can exist in different structures endowed with different biological activity. The toxin bound on the external surface of the outer membrane of *H. pylori* is active and can induce the drop of the trans-epithelial resistance of polarized epithelial monolayers. VacA is released by selective proteolysis and, at neutral pH, it aggregates in the medium to form inactive high molecular weight hexamers/heptamers. Acid exposure induces oligomer dissociation into monomers and ensuing neutralization leads to yet another toxin form, which is monomeric and capable of inducing a decrease in the epithelial resistance as well as vacuolation of cultured cells. This toxin may slowly reform inactive oligomers.

in the extracellular medium (Cover *et al.*, 1992; Papini *et al.*, 1994), as well as BSA-gold localized inside lysosomes (Montecucco *et al.*, unpublished observations), and because vacuoles disappear within hours of the replacement of the toxin-containing medium (Cover *et al.*, 1992). Vacuoles are acidic because their limiting membrane contains an active vacuolar ATPase proton pump (v-ATPase) (Papini *et al.*, 1996), whose inhibition with bafilomycins prevents vacuole formation (Cover *et al.*, 1993a; Papini *et al.*, 1993). Moreover, a monoclonal antibody specific for the 116-kDa component of human vacuolar ATPase, which inhibits proton pumping activity (Sato and Toyama, 1994), also inhibits vacuolation (Papini *et al.*, 1996). Bafilomycins not only prevent vacuolation, but also cause vacuoles to disappear rapidly when added to vacuolated cells, as detected by optical microscopy and neutral red uptake (Papini *et al.*, 1993). This indicates that vacuoles remain swollen and morphologically evident as long as the proton pump operates, pinpointing the role in swelling of protonated bases within the vacuolar lumen, and are osmotically active.

Several intracellular compartments are acidic and marked by the presence of the v-ATPase. As shown in Figure 13.5, vacuoles contain membrane protein markers of late endosomes and of lysosomes (Papini *et al.*, 1994; Molinari *et al.*, 1997) and are capable of incorporating fluid-phase markers of the extracellular medium as well as non-digested material present inside lysosomes (Catrenich and Chestnut, 1992; Cover *et al.*, 1992; Papini *et al.*, 1994; our unpublished results). Electron microscopy reveals that vacuoles in biopsies of stomach mucosa of *H. pylori*-infected patients and of HeLa cells in culture contain some electron-dense material, but are mostly devoid of the large array of multivesicular bodies characteristic of late endosomes (LE) and lysosomes (LY) (Leunk *et al.*, 1988; Cover *et al.*, 1992; Ricci *et al.*, 1997; our unpublished results). Hence, VacA induces a large rearrangement of the organization of LE and LY, with extensive membrane fusion, which can in principle be of at least five different types: (a) homotypic fusion between LE, (b) homotypic fusion between LY, (c) heterotypic fusion between LE and LY, (d) homotypic fusion between internal membranes of LE and of LY, and (e) heterotypic fusion of internal membrane with the limiting organelle membrane.

Alterations of intracellular trafficking in vacuolated cells

LE are a cross-roads of intense protein and membrane trafficking, where the endocytic route crosses the biosynthetic and exocytic route at the level of the trans-Golgi network (TGN) (Figure 13.5). Molecules to be degraded meet within LE and LY with acidic hydrolases coming from the TGN, after synthesis and glycosylation in the endoplasmic reticulum and Golgi. VacA causes a sort of 'traffic jam' at the LE level, with two main consequences, which are evident early after application of VacA, before any vacuolation is manifest: (a) decreased rate of lysosomal protein degradation, and (b) extracellular release of acid hydrolases. Both phenomena are likely to be directly relevant for infection and disease pathogenesis. In fact, protein degradation is an essential function of cell life, which allows for the removal of non-functional cell membrane proteins and of extracellular ligands and re-utilization of amino acids. Moreover, the processing of protein antigens by antigen processing cells (APC) is essentially a specialized type of degradation that takes place in minor proportion in early endosomal compartments and in major proportion inside the antigen processing compartment, which is a specialized form of LE/LY compartment, capable of fusion with the plasma membrane (Watts, 1997).

VacA was recently shown to inhibit the degradation of tetanus toxoid epitopes in the LE processing compartment of tetanus toxoid-specific APC (Molinari *et al.*, 1998b), which were generated by transformation and immortalization of B-cells with Epstein–Barr virus (Lanzavecchia, 1985). Consequently, the stimulation of T-cell clones specific for epitopes generated in the APC compartment was strongly inhibited by VacA, while that of T-cell clones specific for epitopes generated in the early compartment was unaffected (Molinari *et al.*, 1998b). Thus, VacA can be used to distinguish between epitopes generated in early and late endocytic compartments, because the former is unaffected, while the latter is inhibited. The generally modest human immune response to *H. pylori* antigens has been discussed before (Blaser, 1993, 1997), and one report had described that *H. pylori* cultures possess a heat-labile and trypsin-sensitive monocyte suppressing activity (Knipp *et al.*, 1993). Thus, the depression of antigen processing and presentation by VacA could be part of a strategy of survival for *H. pylori*, since depression of antigen processing by APC of the mucosa could contribute significantly to the long-lasting infection that *H. pylori* establishes in the human stomach.

A second consequence of VacA-induced alteration of the LE/LY compartments concerns the level of cellular trafficking of lysosomal acid hydrolases, which are made in the ER as pre-pro-enzymes and are tagged in the Golgi in such a way as to be recognized by tag-specific receptors, which drive them from the TGN to LE.

In the acidic endosomal lumen, the precursor is converted to the active enzyme: the pre sequence is removed in LE and the pro segment is released inside LY. A minor part of pre-pro-enzymes escapes capture by the receptor in the TGN and is released to the medium, from where they are recaptured by a plasma membrane receptor and brought into the endocytic route to reach their final LY destination (Kornfeld and Mellman, 1989). In VacA exposed cells, a very minor amount of cathepsin D reaches LY, while about half of pre-pro-cathepsin D is diverted to the extracellular medium, in such a quantity as to exceed the re-capture mediated by the plasma membrane receptor. If pre-pro-

acid hydrolases are released on the acidic apical domain of the stomach epithelial cells, they can be converted into the active form and thus degrade the protective mucus layer. This degradation activity would add to the production by *H. pylori* of mucin degrading enzymes (Slomiany and Slomiany, 1991). The thickness of this layer is reduced in *H. pylori* patients (Frieri *et al.*, 1995), and it is believed that this contributes greatly to the progression of the disease, because the mucus layer is a semipermeable barrier, very permeable to protons in the direction from the

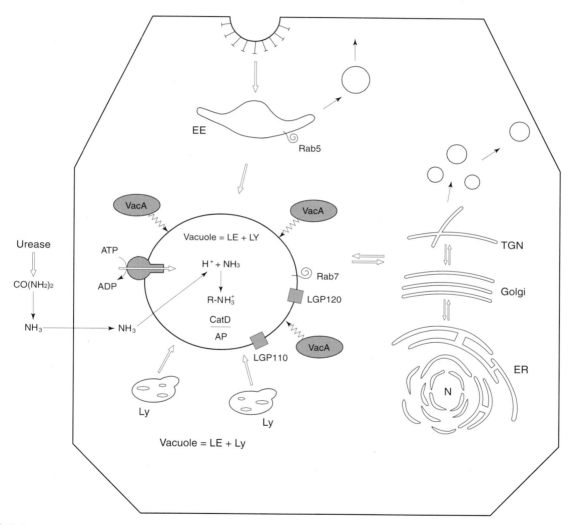

FIGURE 13.5 VacA-induced vacuoles contain late endosomal and lysosomal markers. VacA-induced vacuoles derive from membrane fusion events taking place within and between late endosomes and lysosomes. Accordingly, they contain LGP 110 and 120, which are lysosomal membrane proteins, cathepsin D and acid phosphatases, which derive from the lumen of these acidic compartments, and rab7, which is a small GTP-binding protein associated with late endosomes. The vacuoles are acidic following the activity of a vacuolar ATPase proton pump, which causes the accumulation of membrane permeant weak bases, including the ammonia released by urea hydrolysis catalysed by the *H. pylori* urease. Such massive membrane fusion and swelling of late endocytic compartments causes a 'jamming' of membrane and protein traffic taking place at late endocytic stages, with several consequences, which are discussed in the section on 'Effect of VacA on the trans-epithelial resistance and permeability of polarized epithelial monolayers'.

epithelial cells to the stomach lumen and very permeable to bicarbonate anions in the opposite direction (Allen and Garner, 1980; Williams and Turnberg, 1981; Takeuchi *et al.*, 1983; Bhaskar *et al.*, 1992). As a result, the pH below the mucus layer above the epithelial cells is much higher than in the lumen, but becomes more acidic when much of the layer is eroded. The remaining cell-retained pre-pro-Cathepsin D is normally targeted to its final destination, the lysosomes, but is not processed into the mature enzymic form, in agreement with the fact that VacA also impairs acid hydrolysis in these organelles.

The dysfunction of endo/lysosomal compartments described above may just reflect the initial phases of the complex membrane reorganization resulting in vacuolation, but may also result from partial neutralization of the lumen of such compartments. In fact, both the intrinsic activity of acidic hydrolases and their affinity to mannose-6-P receptor, responsible for correct lysosomal delivery, are impaired by an increase in the lumenal pH of LE/LY compartments. Actually, evaluation of the internal pH of these compartments obtained with the use of endocytosed FITC-dextran shows that VacA causes the endo/lysosomal pH to increase about half a unit above its normal value (Satin *et al.*, 1997). Such a pH increase may result from VacA-associated channel activity at the level of the membrane of the endocytic vesicles, disturbing the homeostasis of intralumenal acidification (see the section on 'Cell binding and entry').

Effect of VacA on the trans-epithelial resistance and permeability of polarized epithelial monolayers

Helicobacter pylori adheres strongly to the apical surface of epithelial cells via interactions with glycoproteins, and binding is followed by rearrangements of the plasma membrane and of the underlying actin meshwork, leading to an almost irreversible binding (Dytoc *et al.*, 1993; Smoot *et al.*, 1993; Boren *et al.*, 1994; Segal *et al.*, 1996). A cellular system *in vitro* that better approaches the situation *in vivo* is that of polarized epithelial cell monolayers grown on filters. Upon reaching confluence, these cells develop tight junctions as well as other types of intercellular junctions, which ensure insulation between an apical domain and a basolateral domain. The trans-epithelial resistance (TER) measures the degree of sealing, and different TER values are attained with different cell lines (Kraehenbuhl and Neutra, 1992; Eaton and Simons, 1995).

Low pH-activated VacA added apically causes the rapid drop of TER from the initial value to 1000–1500 ohm cm^2, which is then maintained for days

(Papini *et al.*, 1998). This effect is much weaker when induced by non-activated VacA and manifests solely with cell monolayers of TER >1500 ohm cm^2; hence, it is not visible with CACO2 cell monolayers, which have a typical TER of 600–800 ohm cm^2. In parallel, the paracellular, but not the transcellular, route of permeability to small organic molecules and to ions such as Fe^{3+} and Ni^{2+}, which are essential for *H. pylori* growth, is increased. This decrease in epithelial resistance takes place with virtually no vacuolation and inhibition of lysosomal degradation capability of epithelial cells. Together with the lack of effect of V-ATPase inhibitors, this clearly indicates that epithelia permeabilization is not the consequence of cell vacuolation. At the same time, no morphological alterations of the tight junctions, the intercellular structures mainly responsible for paracellular permeability, were apparent (Papini *et al.*, 1998). These data indicate that VacA causes a subtle change in the structure of tight junctions, which maintains the polarized monolayer organization while permitting the passage of molecules essential for bacterial growth. The permeability of the mucus layer to Fe^{3+} and Ni^{2+} and to organic anions is not known but, given its selectivity, it is expected to be very low. Hence, it is possible that the major role of VacA is that of allowing *H. pylori* to drain essential nutrients from the mucosa. This function has been proposed by Blaser (1993) to be part of the inflammatory reaction following damage of the epithelial monolayer integrity, but VacA alone is capable of reaching the same goal with a more subtle and less damaging alteration. This function is strictly linked to VacA, with no participation of other *H. pylori* molecules, because toxigenic strains of *H. pylori* grown on the apical domain of epithelial monolayers induce the same TER drop, whereas isogenic VacA-mutants of *H. pylori* do not (Pelicic *et al.*, submitted).

It is interesting to note that toxigenic strains of *H. pylori* do not require low pH exposure to induce the TER drop, indicating that VacA is already in an active form when bound to the extracellular surface of the outer membrane of *H. pylori* (Pelicic *et al.*, submitted). This is strongly indicative of the fact that bacteria-associated VacA is maintained in a monomeric form, ready to act upon the cell, as shown in Figure 13.4. Even more relevant is the finding that *H. pylori* 95–54 strain, which produces an m2-type toxin, devoid of vacuolating activity on HeLa or MDCK cells, is as active as the standard toxigenic CCUG strain in lowering the TER of polarized MDCK cell monolayers (Pelicic *et al.*, submitted). This assay mimics the situation *in vivo* of the bacteria adherent to the apical surface of polarized epithelial cells, thus revealing biological properties of the toxin molecules that are displayed by VacA *in vivo*.

Cell binding and entry

The first step in the mechanism of cell intoxication by bacterial protein toxins is binding to the cell surface. Receptors of many toxins have been identified, and they can be of any chemical nature, including glycolipids and glycoproteins. It is not infrequent that a toxin has both high- and low-affinity binding sites in such a way that different cellular structures may mediate binding and action, depending on the toxin concentration. Specific binding of VacA to target cells has been described (Massari et al., 1998), and a recent report describes the involvement of a 140-kDa glycoprotein in the binding of VacA to human gastric cancer cell lines (Yahiro et al.,1997). It is possible that more than one type of cell surface molecule can act as a VacA binding site, since exposure of cells to neuraminidase decreases, but does not abolish toxicity, while trypsin is ineffective (Molinari et al., unpublished results). Whatever the nature of VacA binding sites, it is very likely that bacteria-associated VacA may contribute substantially to the binding of H. pylori to cells.

Binding of VacA is followed by internalization in HeLa cells, as detected by immunofluorescence and ultrastructural immunocytochemistry with antibodies specific for the toxin (Garner and Cover, 1996; Ricci et al., 1997), as occurs with all bacterial protein toxins with intracellular targets (Montecucco et al., 1994). However, there is no conclusive evidence that endocytosis is a prerequisite for VacA intracellular action as there is for diphtheria toxin or the clostridial neurotoxins (Menestrina et al., 1994). At the moment one cannot exclude the possibility that VacA is an intracellular enzyme and that, as such, one or a few molecules are sufficient to cause vacuolation. If these few molecules enter the cytosol from the plasma membrane, they are below detection by presently available methods, while the bulk of toxin molecules cleared from the plasma membrane following the continuous process of endocytosis is clearly visible.

The biochemical activity of VacA is not known, but there is evidence that it can induce vacuole formation from the cytosol. Bacterial protein toxins acting on intracellular targets consist of two protomers and are therefore termed A–B-type toxins. A is the active subunit, which displays its catalytic activity in the cytosol, whereas B is the part of the molecule responsible for both cell surface binding and membrane translocation of protomer A in the cytosol. A–B-type toxins follow two different routes of entry into cells, which differ in their requirement for a passage through an acidic compartment (Montecucco et al., 1994). A first group of toxins binds to a receptor, and the toxin–receptor complex is endocytosed and reach-

es acidic compartments endowed with a vacuolar ATPase proton pump. Low pH triggers the conformational rearrangement of the toxin to an acidic structure, characterized by the exposure of hydrophobic patches on the protein surface. The toxin becomes hydrophobic, and thus capable of inserting into the lipid bilayer of the limiting membrane of the intracellular organelle, and the catalytic subunit translocates into the cytosol.

At variance, another group of A–B bacterial toxins binding to sugar-containing receptors (glycolipids or glycoproteins) via an oligomeric B protomer, is endocytosed, but then takes the route leading from LE to TGN. They then move retrogradely through the Golgi to the ER, from where they escape into the cytosol. The same route of entry is used by plant ribosome inactivating toxins, which include ricin, abrin, modeccin, viscumin and many other toxins all characterized by their ability to bind sugars on the cell surface (Sandvig and van Deurs, 1996; Montecucco, 1998; Chapter 4, this volume). Characteristically, the cellular toxicity of all these toxins is strongly inhibited by brefeldin A, a powerful Golgi disrupting agent.

The two pathways of toxin entry are differently affected by agents which buffer the intralumenal pH of acidic compartments. The cellular toxicity exerted by toxins taking the former route is largely prevented by ammonia and related compounds, whereas toxins using the latter route are poorly or not affected. As mentioned above, VacA-induced vacuolation is potentiated by the presence of millimolar ammonium in the cell culture medium, clearly indicating that VacA does not require the passage through an acidic compartment to translocate into the cytosol. However, VacA is strongly potentiated by short exposure to low pH and thus this toxin could acquire the low pH-induced membrane translocating capability before cell binding, at variance with diphtheria and related toxins, which change conformation inside acidic intracellular compartments.

Toxins that translocate their catalytic subunit following a low pH-induced conformational change exhibit a common set of properties: (a) they consist of three domains endowed with different functions: binding, membrane translocation and catalysis; (b) their cell toxicity is strongly inhibited by agents that neutralize acidic compartments and is unaffected by brefeldin A; (c) they show a marked increase in the binding of hydrophobic or amphipathic molecules, such as detergents and fluorescent hydrophobic probes (such as ANS) when going from neutral pH to low pH (Boquet et al., 1984; Defrise-Quertain et al., 1989); (d) at neutral pH they interact with the surface of lipid bilayers, generally with a strong preference for negatively charged lipids, but at low pH they penetrate the hydrophobic core of the bilayer and interact with the hydrocarbon chains of lipids (Montecucco *et al.*, 1991); (e) at low pH the B protomer of these toxins forms transmembrane ion channels, a property which is generally assigned to the intermediate domain.

VacA shares all of these properties but one. (a) Its cellular toxicity is not affected by brefeldin A (our unpublished observations), and it is also unaffected by weak amines, which, on the contrary, potentiate its vacuolating activity. (b) VacA binds ANS at low pH and, remarkably, the pH dependence of this phenomenon is identical to that of the low pH toxin activation. This result is highly suggestive of the possibility that the rate-determining step of the entire mechanism of cell intoxication of VacA is membrane insertion and translocation (Molinari *et al.*, 1998a). (c) VacA is capable of penetrating the lipid bilayer, as demonstrated by the labelling of both p37 and p58 with membrane photoactive probes, indicating that both toxin fragments expose hydrophobic segments on their surface at low pH (Moll *et al.*, 1996; Molinari *et al.*, 1998a). (d) VacA forms ion channels across lipid bilayers. A crude assay of this property consists of the incubation of the toxin with liposomes filled with potassium, whose toxin-mediated release can be monitored with a potassium electrode (Boquet and Duflot, 1982). A rigorous approach is the electrophysiological one, which can be performed with planar lipid bilayers or with patch-clamped cells (see Chapter 14). In artificial planar membranes, VacA forms anion-selective and voltage-dependent channels only at low pH or after low-pH pre-activation (Tombola *et al.*, 1998).

These results show that VacA conforms to diphtheria toxin and related three-domain toxins with respect to the cell entry process. None the less, VacA cell intoxication is clearly non-inhibited by ammonium ions. A possible explanation is that VacA has already been activated by low pH and hence ammonium ions can no longer inhibit such a process, which leaves in evidence the potentiation effect due to the vacuolar accumulation mechanism discussed above. This opens up the possibility that VacA can enter the cytosol from the plasma membrane. If this is the case, VacA should form ion channels on the plasma membrane of cells in culture after low pH-activation and not in the neutral form. Indeed, by using patch-clamp techniques, low pH pre-activated VacA induces, at very low doses and within minutes, anion-selective channels in the plasma membrane of HeLa cells. The induction of this current leads, as tested by voltage-dependent probes, to a straight depolarization of the plasma membrane, which is consistent with an

efflux of Cl^- and HCO_3^- from the cell cytosol into the extracellular medium (Brutsche *et al.*, in preparation). Such increased anionic permeability of the plasma membrane may favour, after endocytosis and targeting to the membrane of late endosomes of VacA channels, the turnover of the electrogenic V-ATPase, and contribute strongly to vacuolation by generating an osmotically driven water uptake in the endosomal vesicle (Tombola *et al.*, 1998).

The ion channels formed by VacA in pure lipid bilayers and in the cell plasmalemma have similar low conductance (10–30 pS in 2 M KCl), suggesting a size incompatible with the passage of a polypeptide chain. Protein translocating channels of the ER and of mitochondria have much larger conductances (Simon and Blobel, 1991; Hill *et al.*, 1998; Kunkele *et al.*, 1998). The ER channel is open laterally to lipids, a feature which distinguishes it from ion channels, the walls of which consist of protein only, and this difference must be related to the different nature of the transported material (Martoglio *et al.*, 1995; Martoglio and Dobberstein, 1995). Indeed, proteins, apart from being much larger than single ions, are composed of both hydrophilic and hydrophobic amino acids with lateral chains of different shape and volumes. The possibility of accommodating hydrophobic and bulky protein segments is greatly facilitated by their insertion among the flexible hydrocarbon chain of lipids, thus reducing effectively the high energetic cost of protein membrane translocation.

Although highly debated, the chemical nature of the toxin channel that mediates the translocation of the active subunit in the cytosol is not yet defined. Two models have been advanced (for more complete discussions see Montecucco *et al.*, 1991, 1992).

In the tunnel model, on the acidic side of the membrane the toxin changes conformation in such a way that the B protomer penetrates into the lipid bilayer, forming a protein tunnel inside which the unfolded A chain transverses the membrane to refold into its active conformation on the neutral cytosolic side (Boquet and Duflot, 1982; Finkelstein, 1990). In this 'tunnel' model, the formation of a transmembrane ion-conducting pore is a prerequisite for translocation. This model leaves unexplained two main experimental findings: (a) the membrane-inserted A chain is in contact with the fatty acid chains of phospholipids, i.e. it is not shielded from lipids inside the B protomer tunnel; (b) although there is no direct relationship between channel size and conductance, values of the order of units – tens of picosiemens – do not fit with the dimensions expected for a protein channel chain with its lateral groups. In another view, the translocation of the catalytic domain is proposed to take place at the lipid–protein boundary, rather than inside a proteinaceous pore (Bisson and Montecucco, 1987; Montecucco *et al.*, 1991, 1994). This cleft model proposes that the two toxin subunits change conformation at low pH in a concerted fashion in such a way that both of them expose hydrophobic surfaces and enter in contact with the hydrophobic core of the lipid bilayer. The 'acid' form of the toxin may have properties of a molten globule (Bychkova *et al.*, 1988; van der Goot *et al.*, 1991). In its 'acid' conformation, the B protomer may form a transmembrane cleft which facilitates the passage of the partially unfolded A chain, which remains in contact with lipids. The cytosolic neutral pH induces refolding of the catalytic subunit into its enzymically active conformation. While the A chain is leaving the membrane, the transmembrane hydrophilic cleft of the B protomer tightens up to reduce the amount of hydrophilic protein surface exposed to the hydrocarbon chains of lipid. This leaves across the membrane a peculiarly shaped channel which has two rigid protein walls and a small mobile lipid seal on one side, which is proposed to be the structure responsible for the ion-conducting properties of VacA and of the other A–B toxins.

In the 'cleft' model, the ion channel is a consequence of membrane translocation, rather than a prerequisite. The ion transport is performed by a transmembrane structure, which derives from the one involved in the transport of the A chain, but it is not the same molecular entity, as proposed by the 'tunnel' model. In general, there is no evidence that the B channel contributes significantly to toxicity, but this may not be the case for VacA.

Intracellular activity

Neither the target nor the biochemical/biophysical activities of VacA responsible for the induction of intracellular vacuoles and for the decrease of the trans-epithelial resistance of polarized epithelial cell monolayers are known. As briefly mentioned above, VacA is capable of inducing vacuoles from the cytosol.

VacA was introduced into the cytosol of HeLa cells by microinjection or by transfection and expression, and in both cases it induced vacuoles that are in all respects similar to those caused by the external addition of the toxin (de Bernard *et al.*, 1997; Cover, 1998). To locate the portion of the toxin molecule responsible for this cytosolic activity, the VacA gene encoding for the 95-kDa protein was progressively deleted at the *C*-terminus and at the *N*-terminus, transfected into HeLa cells and assayed for vacuolization (de Bernard *et al.*, 1998c). The toxin does not tolerate *N*-terminal deletions, which shorten the p37

part, while almost the entire p58 can be dispensed without loss of function, since fragment 1–511 is active. The residual portion of p58 which is required for activity is probably needed for the folding of p37. These results are in keeping with the possibility that VacA is an A–B-type toxin with p37 corresponding to the A subunit and p58 corresponding to protomer B.

Factors that remain to be determined include (a) to what extent the ion channel activity of VacA contributes to cell intoxication, and (b) what sort of activity p37 displays in the cytosol. As discussed above, the toxin anion channel is expected to play a role in vacuolar swelling and in the decrease in the acidity of vacuoles with respect to the LE/LY compartments. However, it appears that several events of membrane fusion take place during vacuolization, and the link between operation of the toxin channel and membrane fusion is not established.

Alternatively, the *N*-terminal p37 fragment of VacA could be an enzyme that modifies a cytosolic, or a cytosol-exposed, target. No evident motifs are present in the primary structure of p37 that could give a clue about its possible function, as was the case for the clostridial neurotoxins and for the lethal factor of *B. anthracis* (Schiavo *et al.* 1992; see Chapters 10 and 12, this volume). Cover (1996) has noted weak similarities with serine proteases, but, despite attempts with hundreds of small peptides of different sequence, a precise proteolytic activity has not yet been demonstrated (our unpublished observations). The toxin does not promote homotypic fusion of purified LE *in vitro*, and vacuolation depends strictly on the presence of a functional rab7 molecule (Papini *et al.*, 1997). It is tempting to suggest that VacA acts on a rab7-interacting protein, which is involved in the regulation of membrane trafficking at the late endocytic stages as well as of the organization of tight junctions. Alternatively, VacA could act on two cytosolic targets.

While proteins interacting with rab5 are being characterized (Stenmark *et al.*, 1995; Simonsen *et al.*, 1998), research on partners of rab7 is not comparably advanced. The identification of the biochemical activity of VacA is expected not only to unravel its mechanism of action, but also to contribute significantly to the understanding of how cells regulate the membrane dynamics at late endosomes and lysosomes.

NEUTROPHIL ACTIVATING PROTEIN (HPNAP)

A very common histological feature of *H. pylori* gastritis is the infiltration of the infected stomach mucosa by neutrophils and mononuclear inflammatory cells

(Warren and Marshall, 1983; Marshall et al., 1985; Goodwin et al., 1986; Bayerdorffer et al., 1992; Fiocca et al., 1992).

There is good correlation between the degree of mucosal damage and neutrophil infiltration (Warren and Marshall, 1983; Davies et al., 1992, 1994; Fiocca et al., 1994). Several studies have shown the presence in H. pylori extracts of protein component(s) that attract and activate neutrophils and other inflammatory cells (Karttunen et al., 1990; Mai et al., 1991, 1992; Craig et al., 1992; Nielsen and Andersen, 1992a, b; Kozol et al., 1993; Reymunde et al., 1993; Evans et al., 1995a; Marchetti et al., 1995). Helicobacter pylori strains capable of neutrophil activation were found more frequently in patients affected by peptic ulcer disease than in those with active chronic gastritis only (Rautelin et al., 1993). Yoshida et al. (1993) identified in H. pylori water extracts a protein component capable of promoting neutrophil adhesion to endothelial cells. This protein was purified and found to be a 150-kDa oligomer composed of identical 15-kDa subunits, and was termed HPNAP because of its capacity to induce neutrophils to produce reactive oxygen radicals (Evans et al., 1995a). There is considerable variation in the level of neutrophil adhesion-promoting activity among different H. pylori strains, which suggests a variable level of expression of the protein, similar to that found for VacA (Evans et al., 1995a). The amino acid sequence of HPNAP shows significant similarity with a family of bacterial ferritins, cytosolic proteins capable of binding iron (Evans et al., 1995b).

However, H. pylori produces another such protein, the iron-binding properties of which have been demonstrated (Frazier et al., 1993), while iron binding to HPNAP remains to be determined. This protein was recently reported to bind to mucin, a highly glycosylated protein which is the main component of the mucus layer (Namavar et al., 1998), and to neutrophil glycosphingolipids (Teneberg et al., 1997). The bacterial ferritin family of proteins includes members such as the E. coli protein Dps, which binds DNA with no apparent sequence specificity, and this is believed to protect DNA from oxidative damage (Almiron et al., 1992; Martinez and Kolter, 1997). The structures of the bacterial ferritin Bfr (Frolow et al., 1994) and of Dps (Grant et al., 1998) have been determined. The monomeric proteins have an almost identical fold characterized by a four-helix bundle. The monomers assemble in trimers, and four trimers form the dodecameric Dps molecule, which has a diameter of about 9.0 nm with a hollow core of about 4.5 nm. The decameric composition of the HPNAP oligomer has been deduced from chromatographic sieving, a technique with a level of uncertainty such that the data may

well be compatible with 12 tightly packed monomers.

Purified HPNAP not only induces adhesion of neutrophils to endothelial cells, an activity inhibited by anti-HPNAP-specific antibodies, but also strongly stimulates the production of reactive oxygen radicals, as demonstrated by the reduction of NBT (Evans et al., 1995a), and by the quantitative and sensitive assay performed with acid peroxidase-mediated oxidation of homovanillic acid (Menegazzi et al., 1991; Satin et al., in preparation). HPNAP acts via a cascade of intracellular activation events, including an increase in cytosolic calcium ion concentration and phosphorylation of proteins, leading to assembly of functional NADPH oxidase on the neutrophil plasma membrane (Satin et al., in preparation). HPNAP as a neutrophil activator is, however, less powerful than PMA or fMLP. This is in keeping with the hypothesis of Blaser (1993) that H. pylori induces a moderate inflammatory reaction leading to alteration of the epithelial tight junctions and basal membranes, such as to promote the release of nutrients from the mucosa to support the growth of H. pylori residing on the apical region and within the mucus layer. Alternatively, it is possible that this is an unwanted side activity of HPNAP, secondary to its main biological function, which remains to be determined.

UREASE

An essential aspect of H. pylori physiology is the production of a powerful urease, the enzymic characteristics of which are well suited to operate in the H. pylori environment, i.e. acidity and low urea concentrations (Marshall and Langton, 1986; Mobley et al., 1988, 1995; Dunn et al., 1990; Cesareo and Langton, 1992; Labigne et al., 1996; Mobley, 1996). Urease is an essential virulence factor of H. pylori, since isogenic urease-lacking strains are unable to colonize the stomach (Eaton et al., 1991). It is believed that the urease allows H. pylori to survive short exposure to very acidic conditions of the stomach lumen before H. pylori swims into the mucus layer (Marshall et al., 1990; Clyne et al., 1995; Bauerfeind et al., 1997). Its role in the survival of H. pylori within the mucus and on the apical domain of the epithelia is less clear and depends on the degree of acidity of these environments (Allen and Garner, 1980; Williams and Turnberg, 1981; Takeuchi et al., 1983; Bhaskar et al., 1992). There is evidence that the urease activity prevents growth of H. pylori at neutral pH in the presence of urea because H. pylori does not tolerate alkaline environments (Perez-Perez et al., 1992; Segal et al., 1992; Clyne et al., 1995).

The biosynthesis of urease is governed by a seven-gene cluster, which includes the genes encoding for

UreA (26.5 kDa) and UreB (60.3 kDa) subunits of the urease and for five accessory proteins, which are responsible for Ni^{2+} uptake and insertion into the active site of the apo-enzyme (Labigne and de Reuse, 1991). Urease is a dodecameric protein composed of six UreA and six UreB subunits arranged as a double ring 13 nm in diameter (Austin *et al.*, 1991). The atomic structure of the related urease from *Klebsiella aerogenes* has been solved (Jabri *et al.*, 1995) and all of the residues involved in the co-ordination of the two active site Ni^{2+} ions are strictly conserved in *H. pylori*.

The amount of urease produced by the bacterium varies with culture conditions and it can account for as much as 10% of total bacterial protein (Bauerfeind *et al.*, 1997). It is synthesized without a leader sequence and accumulates in the cytosol together with the heat shock protein HspB. However, the two proteins are released upon bacterial autolysis and adsorb efficiently on the extracellular surface of the outer membrane of viable bacteria (Bode *et al.*, 1989; Phadnis *et al.*, 1996). This mechanism of surface 'expression' appears to operate also *in vivo*, since urease and HspB are found on the surface of *H. pylori* present within human biopsies (Dunn *et al.*, 1997).

The role of urease in the pathogenesis of *H. pylori*-associated diseases is not limited to colonization. Urease is a major antigen in the human immune response to *H. pylori* (Michetti *et al.*, 1996). It also participates in the recruitment of neutrophils and monocytes in the inflamed mucosa (Mai *et al.*, 1992) and in the activation of mononuclear phagocytes, which are induced to produce proinflammatory cytokines (Harris *et al.*, 1996).

Urease should be considered in all respects as a toxin acting outside cells because of its enzymic activity, which produces ammonia. Many studies have considered the biological effects of ammonia as a toxic agent, by itself or in conjunction with neutrophil metabolites, both on cultured cells and on stomach tissue preparations (Xu *et al.*, 1990; Cover *et al.*, 1991; Megraud *et al.*, 1992; Suzuki *et al.*, 1992; Ricci *et al.*, 1993, 1997; Tsuji *et al.*, 1996; Brzozowski *et al.*, 1996; Sommi *et al.*, 1996; Mayo *et al.*, 1997). Ammonia and membrane-permeant weak bases have long been known to be toxic to cells, where they induce a variety of toxic effects including swelling of intracellular acidic compartments, alteration of vesicular membrane trafficking, depression of protein synthesis and of ATP production, and arrest of the cell cycle (de Duve *et al.*, 1974). In general, toxic doses are rather high, and the dose dependence of the toxic effects strongly depends on cell type and culture conditions. In several studies, very large ammonia concentrations have been used (from 10 to 250 mM), and it is not clear how they can be compared with the

in vivo quantities produced by *H. pylori*. Moreover, circulation of extracellular fluids should lead to a rapid dilution of the ammonia produced, with a marked gradient of concentration from the bacterial surface to the medium surrounding the cells of the host.

Ammonia can also cause damage in association with other elements involved in *H. pylori*-dependent diseases. It has been proposed that neutrophils and monocytes contribute the *H. pylori*-induced carcinogenesis via production of oxygen radicals, which alter the genome of epithelial cells in such a way as to induce their transformation to cancer cells. It has also been proposed that ammonia can react with intermediates produced by the activity of neutrophil myeloperoxidase to form carcinogenic agents, but again it is not clear to what extent these reactions can take place *in vivo*, given the infrequent contact between the bacterium and neutrophils (Suzuki *et al.*, 1992; Mobley, 1996). The picture is more defined in systems *in vitro*. In the presence of ammonium chloride, VacA-induced vacuolization proceeds more rapidly and vacuoles are larger and more swollen when compared with the vacuoles generated by the toxin alone and with those induced by ammonia (Cover *et al.*, 1992; Ricci *et al.*, 1997; our unpublished observations). Although there are differences among cells lines and in the purity of the VacA preparations used in different studies, it appears that VacA induces a reorganization of LE/LY with extensive membrane fusion and formation of tubulo-vesicular compartments, which are moderately swollen in the absence of membrane-permeant bases. At early stages, such vacuoles are not visible under the optical microscope, while the lesions caused to membrane trafficking and endocytic proteolysis are fully developed. The presence of bases together with the toxin, or their addition after incubation with VacA and subsequent washings, causes swelling, with the appearance of macroscopic translucent vacuoles. Hence, urease and VacA substantially co-operate in damaging cells (Ricci *et al.*, 1997).

CYTOTOXIN ASSOCIATED GENE A (CAGA)

The *cagA* gene encodes for a 128-kDa protein, an immunodominant surface antigen nearly always present in *H. pylori* strains associated with the more severe forms of disease, and therefore frequently associated with the presence of VacA (Covacci *et al.*, 1993; Xiang *et al.*, 1995; Weel *et al.*, 1996). The relation between CagA and VacA remains undetermined, particularly in the light of the finding that knock-out of the *cagA* gene does not affect the production of VacA, nor does it affect the ability of *H. pylori* to induce the release of IL-8

(Tummuru *et al.*, 1994; Crabtree *et al.*, 1995). Although there is no known function for CagA, the *cagA* gene is part of a large pathogenicity island, which has been acquired by horizontal transfer of DNA (Covacci and Rappuoli, 1998). The origin of the cag pathogenicity island is not known, but difference in GC content from the rest of the *H. pylori* genome indicates that it has been acquired from another species or even genus. The pathogenicity island marked by the *cagA* gene codes for at least 40 genes involved in different pathogenic processes, including the induction of inflammatory mediators in gastric epithelial cells, induction of pedestal formation, and modification of signal transduction in target cells. Similarities of some of these genes with genes of known function in other bacteria indicate that the pathogenicity island codes for a complex secretion system involved in the release of macromolecules from the bacteria. It is believed that different molecules, secretion of which is controlled by these genes, are directly responsible for the observed pathogenic phenomena (Censini *et al.*, 1996; Covacci and Rappuoli, 1998).

ACKNOWLEDGEMENTS

Work carried out in the authors' laboratories is supported by European Community grants (TMR FMRX CT96 0004 and Biomed BMH4 CT97 2410), the Armenise-Harvard Medical School Foundation, the Progetto Finalizzato CNR Biotecnologie and the MURST 40% Project on Inflammation.

REFERENCES

Allen, A. and Garner, A. (1980). Mucus and bicarbonate secretion in the stomach and their possible role in mucosal protection. *Gut* **21**, 249–62.

Allen, A., Flemstrom, G., Garner, A. and Kivilaakso, E. (1993). Gastroduodenal mucosal protection. *Physiol. Rev.* **73**, 823–57.

Almiron, M., Link, A.J., Furlong, D. and Kolter, R. (1992). A novel DNA-binding protein with regulatory and protective roles in starved *Escherichia coli*. *Genes* **6**, 2646–54.

Atherton, J.C., Cao, P., Peek, R.M., Jr, Tummuru, M.K., Blaser, M.J. and Cover, T.L. (1995). Mosaicism in vacuolating cytotoxin alleles of *Helicobacter pylori*. Association of specific VacA types with cytotoxin production and peptic ulceration. *J. Biol. Chem.* **270**, 17771–7.

Austin, J.W., Doig, P., Stewart, M. and Trust, T.J. (1991). Macromolecular structure and aggregation states of *Helicobacter pylori* urease. *J. Bacteriol.* **173**, 5663–7.

Bauerfeind, P., Garner, R., Dunn, B.E. and Mobley, H.L. (1997). Synthesis and activity of *Helicobacter pylori* urease and catalase at low pH. *Gut* **40**, 25–30.

Bayerdorffer, E., Lehn, N., Hatz, R., Mannes, G.A., Oertel, H., Sauerbruch, T. and Stolte, M. (1992). Difference in expression of *Helicobacter pylori* gastritis in antrum and body. *Gastroenterology* **102**, 1575–82.

Bhaskar, K.R., Garik, P., Turner, B.S., Bradley, J.D., Bansil, R., Stanley, H.E. and LaMont, J.T. (1992). Viscous fingering of HCl through gastric mucin. *Nature* **360**, 458–61.

Bisson, R. and Montecucco, C. (1987). Diphtheria toxin membrane translocation: an open question. *Trends Biochem. Sci.* **12**, 181–2.

Blaser, M.J. (1993). *Helicobacter pylori*: microbiology of a 'slow' bacterial infection. *Trends Microbiol.* **1**, 255–9.

Blaser M.J. (1997). Not all *Helicobacter pylori* strains are created equal: should all be eliminated? *Lancet* **349**, 1020–2.

Bode, G., Malfertheiner,, P., Nilius, M., Lehnhardt, G. and Ditschuneit, H. (1989). Ultrastructural localisation of urease in outer membrane and periplasm of *Campylobacter pylori*. *J. Clin. Pathol.* **42**, 778–9.

Boquet, P. and Duflot, E. (1982). Tetanus toxin fragment forms channels in lipid vesicles at low pH. *Proc. Natl Acad. Sci. USA* **79**, 7614–18.

Boquet, P., Duflot, E. and Hauttecoeur, B. (1984). Low pH induces a hydrophobic domain in the tetanus toxin molecule. *Eur. J. Biochem.* **144**, 339–44.

Boren, T., Normark, S. and Falk, P. (1994). *Helicobacter pylori*: molecular basis for host recognition and bacterial adherence. *Trends Microbiol.* **2**, 221–8.

Brzozowski, T., Konturek, P.C., Konturek, S.J., Ernst, H., Sliwowski, K. and Hahn, E.G. (1996). Mucosal irritation, adaptive cytoprotection, and adaptation to topical ammonia in the rat stomach. *Scand. J. Gastroenterol.* **31**, 837–46.

Burroni, D., Lupetti, P., Pagliaccia, C., Reyrat, J.M., Dallai, R., Rappuoli, R. and Telford, J.L. (1998). Deletion of the major proteolytic site of the *Helicobacter pylori* cytotoxin does not influence toxin activity but favors assembly of the toxin into hexameric structures. *Infect. Immun.* **66**, 5547–50.

Bychkova, V.E., Pain, R.H. and Ptitsyn, O.B. (1988). The 'molten globule' state is involved in the translocation of proteins across membranes? *FEBS Lett.* **238**, 231–4.

Catrenich, C.E. and Chestnut, M.H. (1992). Character and origin of vacuoles induced in mammalian cells by the cytotoxin of *Helicobacter pylori*. *J. Med. Microbiol.* **37**, 389–95.

Censini, S., Lange, C., Xiang, Z., Crabtree, J.E., Ghiara, P., Borodovsky, M., Rappuoli, R. and Covacci, A. (1996). Cag, a pathogenicity island of *Helicobacter pylori*, encodes type I-specific and disease-associated virulence factors. *Proc. Natl Acad. Sci. USA* **93**, 14648–53.

Cesareo, S.D. and Langton, S.R. (1992). Kinetic properties of *Helicobacter pylori* urease compared with jack bean urease. *FEMS Microbiol. Lett.* **78**, 15–21.

Clyne, M., Labigne, A. and Drumm B. (1995). *Helicobacter pylori* requires an acidic environment to survive in the presence of urea. *Infect. Immun.* **63**, 1669–73.

Correa, P. (1995). *Helicobacter pylori* and gastric carcinogenesis. *Am. J. Surg. Pathol.* **19**, 37–43.

Covacci, A. and Rappuoli, R. (1998). *Helicobacter pylori*: molecular evolution of a bacterial quasi-species. *Curr. Opin. Microbiol.* **1**, 96–102.

Covacci, A., Censini, S., Bugnoli, M., Petracca, R., Burroni, D., Macchia, G., Massone, A., Papini, E., Xiang, Z., Figura, N. and Rappuoli, R. (1993). Molecular characterization of the 128-kDa immunodominant antigen of *Helicobacter pylori* associated with cytotoxicity and duodenal ulcer. *Proc. Natl Acad. Sci. USA* **90**, 5791–5.

Cover, T.L. (1996). The vacuolating cytotoxin of *Helicobacter pylori*. *Mol. Microbiol.* **20**, 241–6.

Cover, T.L. (1998). An intracellular target for *Helicobacter pylori* vacuolating toxin. *Trends Microbiol.* **6**, 127–8.

Cover, T.L. and Blaser, M.J. (1992). Purification and characterization of the vacuolating toxin from *Helicobacter pylori*. *J. Biol. Chem.* **267**, 10570–5.

Cover, T.L., Puryear, W., Perez-Perez, G.I. and Blaser, M.J. (1991). Effect of urease on HeLa cell vacuolation induced by *Helicobacter pylori* cytotoxin. *Infect. Immun.* **59**, 1264–70.

Cover, T.L., Susan, A.H. and Blaser, M.J. (1992). Characterization of HeLa cell vacuoles induced by *Helicobacter pylori* broth culture supernatant. *Human Pathol.* **23**, 1004–10.

Cover, T.L., Reddy, L.Y. and Blaser, M.J. (1993a). Effects of ATPase inhibitors on the response of HeLa cells to *Helicobacter pylori* vacuolating toxin. *Infect. Immun.* **61**, 1427–31.

Cover, T.L., Cao, P., Lind, C.D., Tham, K.T. and Blaser, M.J. (1993b). Correlation between vacuolating cytotoxin production by *Helicobacter pylori* isolates *in vitro* and *in vivo*. *Infect. Immun.* **61**, 5008–12.

Cover, T.L., Tummuru, M.K.R., Cao, P., Thompson, S.A. and Blaser, M.J. (1994). Divergence of genetic sequences for the vacuolating cytotoxin among *Helicobacter pylori* strains. *J. Biol. Chem.* **269**, 10566–73.

Cover, T. L., Hanson, P.I. and. Heuser, J.E. (1997). Acid-induced dissociation of VacA, the *Helicobacter pylori* vacuolating toxin, reveals its pattern of assembly. *J. Cell Biol.* **138**, 759–69.

Crabtree, J.E., Xiang, Z., Lindley, I.J., Tompkins, D.S., Rappuoli, R. and Covacci, A. (1995). Induction of interleukin-8 secretion from gastric epithelial cells by a cagA negative isogenic mutant of *Helicobacter pylori*. *J. Clin. Pathol.* **48**, 967–9.

Craig, P.M., Territo, M.C., Karnes, W.E. and Walsh, J.H. (1992). *Helicobacter pylori* secretes a chemotactic factor for monocytes and neutrophils. *Gut* **33**, 1020–3.

Davies, G.R., Simmonds, N.J., Stevens, T.R., Grandison, A., Blake, D.R. and Rampton, D.S. (1992). Mucosal reactive oxygen metabolite production in duodenal ulcer disease. *Gut* **33**, 1467–72.

de Bernard, M., Papini, E., de Filippis, V., Gottardi, E., Telford., J., Manetti, R., Fontana, A., Rappuoli, R. and Montecucco, C. (1995). Low pH activates the vacuolating toxin of *Helicobacter pylori*, which becomes acid and pepsin resistant. *J. Biol. Chem.* **270**, 23937–40.

de Bernard, M., Aricò, B., Papini, E., Rizzuto, R., Grandi, G., Rappuoli, R. and Montecucco, C. (1997). *Helicobacter pylori* toxin VacA induces vacuole formation by acting in the cell cytosol. *Mol. Microbiol.* **26**, 665–74.

de Bernard, M., Moschioni, M., Papini, E., Telford, J.L., Rappuoli, R. and Montecucco, C. (1998a) Cell vacuolization induced by *Helicobacter pylori* VacA toxin: cell line sensitivity and quantitative estimation. *Toxicol. Lett.* **99**, 109–15.

de Bernard, M., Moschioni, M., Papini, E., Telford, J.L., Rappuoli, R. and Montecucco, C. (1998b). TPA and butyrate increase cell sensitivity to the vacuolating toxin of *Helicobacter pylori*. *FEBS Lett.* **436**, 218–22.

de Bernard, M., Burroni, D., Papini, E., Rappuoli, R., Telford, J.L. and Montecucco, C. (1998c). Identification of the *Helicobacter pylori* VacA toxin domain active in the cell cytosol. *Infect. Immun.* **66**, 6014–16.

DeCourey, K. and Storrie, B. (1991). Osmotic swelling of endocytic compartments induced by internalized sucrose is restricted to mature lysosomes in cultured mammalian cells. *Exp. Cell. Res.* **192**, 52–60.

De Duve, C., de Barsy, T., Poole, B., Trouet, A., Tulkens, P. and Van Hoof, F. (1974). Lysosomotropic agents. *Biochem. Pharmacol.* **23**, 2495–531.

Defrise-Quertain, F., Cabiaux, V., Vandenbranden, M., Wattiez, R., Falmagne, P. and Ruysschaert, J.M. (1989). pH-dependent bilayer destabilization and fusion of phospholipidic large unilamellar vesicles induced by diphtheria toxin and its fragments A and B. *Biochemistry* **28**, 3406–13.

Dunn, B.E., Campbell, G.P., Perez-Perez, G.I. and Blaser, M.J. (1990). Purification and characterization of urease from *Helicobacter pylori*. *J. Biol. Chem.* **265**, 9464–9.

Dunn, B.E., Vakil, N.B., Schneider, B.G., Miller, M.M., Zitzer, J.B., Peutz, T. and Phadnis, S.H. (1997). Localization of *Helicobacter pylori* urease and heat shock protein in human gastric biopsies. *Infect. Immun.* **65**, 1181–8.

Dytoc, M., Gold, B., Louie, M., Huesca, M., Fedorko, L., Crowe, S., Lingwood, C., Brunton, J. and Sherman, P. (1993). Comparison of *Helicobacter pylori* and attaching-effacing *Escherichia coli* adhesion to eukaryotic cells. *Infect. Immun.* **61**, 448–56.

Eaton, K.A., Brooks, C.L., Morgan, D.R. and Krakowka, S. (1991). Essential role of urease in pathogenesis of gastritis induced by *Helicobacter pylori* in gnotobiotic piglets. *Infect. Immun.* **59**, 2470–5.

Eaton, S. and Simons, K. (1995). Apical, basal, and lateral cues for epithelial polarization. *Cell* **82**, 5–8.

Evans, D.J., Jr, Evans, D.G., Takemura, T., Nakano, H., Lampert, H.C., Graham, D.Y., Granger, D.N. and Kvietys, P.R. (1995a). Characterization of a *Helicobacter pylori* neutrophil-activating protein. *Infect. Immun.* **63**, 2213–20.

Evans, D.J., Jr, Evans, D.G., Lampert, H.C. and Nakano, H. (1995b). Identification of four new prokaryotic bacterioferritins, from *Helicobacter pylori*, *Anabaena variabilis*, *Bacillus subtilis* and *Treponema pallidum*, by analysis of gene sequences. *Gene* **153**, 123–7.

Figura, N. (1996). Mouth-to-mouth resuscitation and *Helicobacter pylori* infection. *Lancet* **347**, 1342.

Figura, N., Guglielmetti, P., Rossolini, A., Barberi, A., Cusi, G., Musmanno, R.A., Russi, M. and Quaranta, S. (1989). Cytotoxin production by *Campylobacter pylori* strains isolated from patients with peptic ulcers and from patients with chronic gastritis only. *J. Clin. Microbiol.* **27**, 225–6.

Figura, N., Vindigni, C., Covacci, A., Presenti, L., Burroni, D., Vernillo, R., Banducci, T., Roviello, F., Marrelli, D., Biscontri, M., Kristodhullu, S., Gennari, C. and Vaira, D. (1998). CagA positive and negative *Helicobacter pylori* strains are simultaneously present in the stomach of most patients with non-ulcer dyspepsia: relevance to histological damage. *Gut* **42**, 772–8.

Finkelstein, A. (1990). Channels formed in phospholipid bilayer membranes by diphtheria, tetanus, botulinum and anthrax toxin. *J. Physiol. Paris* **84**, 188–90.

Fiocca, R., Villani, L., Luinetti, O., Gianotti, A., Perego, M., Alvisi, C., Turpini, F. and Solcia, E. (1992). *Helicobacter* colonization and histopathological profile of chronic gastritis in patients with or without dyspepsia, mucosal erosion and peptic ulcer: a morphological approach to the study of ulcerogenesis in man. *Virchows Arch. Pathol. Anat. Histol.* **420**, 489–92.

Fiocca, R., Luinetti, O., Villani, L., Chiaravalli, A.M., Capella, C. and Solcia, E. (1994). Epithelial cytotoxicity, immune responses, and inflammatory components of *Helicobacter pylori* gastritis. *Scand. J. Gastroenterol.* **205**, 11–21.

Frazier, B.A., Pfeifer, J.D., Russell, D.G., Falk, P., Olsen, A.N., Hammar, M., Westblom, T.U. and Normark, S.J. (1993). Paracrystalline inclusions of a novel ferritin containing nonheme iron, produced by the human gastric pathogen *Helicobacter pylori*: evidence for a third class of ferritins. *J. Bacteriol.* **175**, 966–72.

Frieri,G., De Petris, G., Aggio, A., Santarelli, D., Ligas, E., Rosoni, R. and Caprilli, R. (1995). Gastric and duodenal juxtamucosal pH and *Helicobacter pylori*. *Digestion* **56**, 107–10.

Frolow, F., Kalb, A.J. and Yariv, J. (1994). Structure of a unique twofold symmetric haem-binding site. *Nat. Struct. Biol.* **1**, 453–60.

Garner, J.A. and Cover, T.L. (1996). Binding and internalization of the *Helicobacter pylori* vacuolating cytotoxin by epithelial cells. *Infect. Immun.* **64**, 4197–203.

Go, M.F., Cissell, L. and Graham, D.Y. (1998). Failure to confirm association of vac A gene mosaicism with duodenal ulcer disease. *Scand. J. Gastroenterol.* **33**, 132–6.

Goodwin, C.S., Armstrong, J.A. and Marshall, B.J. (1986). *Campylobacter pyloridis*, gastritis and peptic ulceration. *J. Clin. Pathol.* **39**, 353–65.

Goodwin, C.S. (1997). *Helicobacter pylori* gastritis, peptic ulcer, and gastric cancer: clinical and molecular aspects. *Clin. Infect. Dis.* **25**, 1017–19.

Grant, R.A., Filman, D.J., Finkel, S.E., Kolter R, Hogle, J.M. (1998). The crystal structure of Dps, a ferritin homolog that binds and protects DNA. *Nat. Struct. Biol.* **5**, 294–303.

Halter, R., Pohlner, J. and Meyer, T.F. (1989). Mosaic-like organization of IgA protease genes in *Neisseria gonorrhoeae* generated by horizontal genetic exchange *in vivo*. *EMBO J.* **8**, 2737–44.

Harris, P.R., Mobley, H.L., Perez-Perez, G.I., Blaser, M.J. and Smith, P.D. (1996). *Helicobacter pylori* urease is a potent stimulus of mononuclear phagocyte activation and inflammatory cytokine production. *Gastroenterology* **111**, 419–25.

Hill, K., Model, K., Ryan, M.T., Dietmeier, K., Martin, F., Wagner, R. and Pfanner, N. (1998) Tom40 forms the hydrophilic channel of the mitochondrial import pore for preproteins. *Nature* **395**, 516–21.

Icatlo, F.C., Kuroki, M., Kobayashi, C., Yokoyama, H., Ikemori, Y., Hashi, T. and Kodama, Y. (1998). Affinity purification of *Helicobacter pylori* urease. Relevance to gastric mucin adherence by urease protein. *J. Biol. Chem.* **273**, 18130–8.

Isaacson, P.G. (1994). Gastric lymphoma and *Helicobacter pylori*. *N. Engl. J. Med.* **330**, 1310–11.

Jabri, E., Carr, M.B., Hausinger, R.P. and Karplus, P.A. (1995). The crystal structure of urease from *Klebsiella aerogenes*. *Science* **268**, 998–1004.

Ji, X., Burroni, D., Pagliaccia, C., Xu, G., Rappuoli, R., Reyrat, J.M. and Telford, J.L. Evolution and geographic variation in the conserved and functionally polymorphic regions of the *Helicobacter pylori* cytotoxin gene. Submitted.

Karttunen, R., Andersson, G., Poikonen, K., Kosunen, T.U., Karttunen, T., Juutinen, K. and Niemela, S. (1990). *Helicobacter pylori* induces lymphocyte activation in peripheral blood cultures. *Clin. Exp. Immunol.* **82**, 485–8.

Knipp, U., Birkholz, S., Kaup, W. and Opferkuch, W. (1993) Immune suppressive effects of *Helicobacter pylori* on human peripheral blood mononuclear cells. *Med. Microbiol. Immunol.* **182**, 63–76.

Kornfeld, S. and Mellman, I. (1989). The biogenesis of lysosomes. *Annu. Rev. Cell Biol.* **5**, 483–525.

Kozol, R., McCurdy, B. and Czanko, R. (1993). A neutrophil chemotactic factor present in *H. pylori* but absent in *H. mustelae*. *Dig. Dis. Sci.* **38**, 137–41.

Kraehenbuhl, J.P. and Neutra, M.R. (1992). Molecular and cellular basis of immune protection of mucosal surfaces. *Physiol. Rev.* **72**, 853–79.

Kuipers, E.J., Israel, D.A., Kusters, J.G. and Blaser, M.J. (1998). Evidence for a conjugation-like mechanism of DNA transfer in *Helicobacter pylori*. *J. Bacteriol.* **180**, 2901–5.

Kunkele, K.P., Heins, S., Dembowski, M., Nargang, F.E., Benz, R., Thieffry, M., Walz, J., Lill, R., Nussberger, S. and Neupert, W. (1998). The preprotein translocation channel of the outer membrane of mitochondria. *Cell* **93**, 1009–19.

Labigne, A. and de Reuse, H. (1996). Determinants of *Helicobacter pylori* pathogenicity. *Infect. Agents Dis.* **5**, 191–202.

Labigne A, Cussac, V. and Courcoux, P. (1991). Shuttle cloning and nucleotide sequences of *Helicobacter pylori* genes responsible for urease activity. *J. Bacteriol.* **173**, 1920–31.

Lanzavecchia, A. (1985). Antigen-specific interaction between T and B cells. *Nature* **314**, 537–9.

Lanzavecchia, S., Bellon, P.L., Lupetti, P., Dallai, R., Rappuoli, R. and Telford, J.L. (1998). Three-dimensional reconstruction of metal replicas of the *Helicobacter pylori* vacuolating cytotoxin. *J. Struct. Biol.* **121**, 9–18.

Leunk, R.D., Johnson, P.T., David, B.C., Kraft, W.G. and Morgan, D.R. (1988). Cytotoxin activity in broth-culture filtrates of *Campylobacter pylori*. *J. Med. Microbiol.* **26**, 93–9.

Logan, R.P. and Berg, D.E. (1996). Genetic diversity of *Helicobacter pylori*. *Lancet* **348**, 1462–3.

Lupetti, P., Heuser, J.E., Manetti, R., Lanzavecchia, S., Bellon, P.L., Dallai, R., Rappuoli, R. and Telford, J.L. (1996). Oligomeric and subunit structure of the *Helicobacter pylori* vacuolating cytotoxin. *J. Cell Biol.* **133**, 801–7.

Mai, U.E., Perez-Perez, G.I., Wahl, L.M., Wahl, S.M., Blaser, M.J. and Smith, P.D. (1991). Soluble surface proteins from *Helicobacter pylori* activate monocytes/macrophages by lipopolysaccharide-independent mechanism. *J. Clin. Invest.* **87**, 894–900.

Mai, U.E., Perez-Perez, G.I., Allen, J.B., Wahl, S.M., Blaser, M.J. and Smith, P.D. (1992). Surface proteins from *Helicobacter pylori* exhibit chemotactic activity for human leukocytes and are present in gastric mucosa. *J. Exp. Med.* **175**, 517–25.

Manetti, R., Massari, P., Burroni, D., de Bernard, M., Marchini, A., Olivieri, R., Papini, E., Montecucco, C., Rappuoli, R. and Telford, J.L. (1995). *Helicobacter pylori* cytotoxin: importance of native conformation for induction of neutralizing antibodies. *Infect. Immun.* **63**, 4476–80.

Marchetti, M., Aricò, B., Burroni, D., Figura, N., Rappuoli, R. and Ghiara, P. (1995). Development of a mouse model of *Helicobacter pylori* infection that mimics human disease. *Science* **265**, 1656–8.

Marshall, B.J. and Langton, S.R. (1986). Urea hydrolysis in patients with *Campylobacter pyloridis* infection. *Lancet* **i**, 965–96.

Marshall, B.J., Armstrong, J.A., McGeche, D.B. and Glancy, R.J. (1985). Attempt to fulfil Koch's postulates for pyloric *Campylobacter*. *Med. J. Austr.* **142**, 436–9.

Marshall, B.J., Barrett, L.J., Prakash, C., McCallum, R.W. and Guerrant, R.L. (1990). Urea protects *Helicobacter* (*Campylobacter*) *pylori* from the bactericidal effect of acid. *Gastroenterology* **99**, 697–702.

Martinez, A. and Kolter, R. (1997). Protection of DNA during oxidative stress by the nonspecific DNA-binding protein. *J. Bacteriol.* **179**, 5188–94.

Martoglio, B. and Dobberstein, B. (1998). Signal sequences: more than just greasy peptides.*Trends Cell Biol.* **8**, 410–15.

Martoglio, B., Hofmann, M.W., Brunner, J. and Dobberstein, B. (1995). The protein-conducting channel in the membrane of the endoplasmic reticulum is open laterally toward the lipid bilayer. *Cell* **81**, 207–14.

Massari, P., Manetti, R., Burroni, D., Nuti, S., Norais, N., Rappuoli, R. and Telford, J.L. (1998). Binding of the *Helicobacter pylori* vacuolating cytotoxin to target cells. *Infect. Immun.* **66**, 3981–4.

Mayo, K., Held, M., Wadstrom, T. and Megraud, F. (1997). *Helicobacter pylori* – human polymorphonuclear leucocyte interaction in the presence of ammonia. *Eur. J. Gastroenterol. Hepatol.* **9**, 457–61.

McGuigan, J.E. (1996). *Helicobacter pylori*: the versatile pathogen. *Dig. Dis.* **14**, 289–303.

Megraud, F., Neman-Simha, V. and Brugman, D. (1992) Further evidence of the toxic effect of ammonia produced by *Helicobacter pylori* urease on human epithelial cells. *Infect. Immun.* **60**, 1858–63.

Menegazzi, R., Zabucchi, G., Zuccato, P., Cramer, R., Piccinini, C. and Patriarca, P. (1991). Oxidation of homovanillic acid as a selective assay for eosinophil peroxidase in eosinophil peroxidase-myeloperoxidase mixtures and its use in the detection of human eosinophil peroxidase deficiency. *J. Immunol. Methods* **137**, 55–63.

Menestrina, G., Schiavo, G. and Montecucco, C. (1994). Molecular mechanisms of action of bacterial protein toxins. *Mol. Aspects Med.* **15**, 79–193.

Michetti, P., Wadstrom, T., Kraehenbuhl, J.P., Lee, A., Kreiss, C. and Blum, A.L. (1996). Frontiers in *Helicobacter pylori* research: pathogenesis, host response, vaccine development and new therapeutic approaches. *Eur. J. Gastroenterol. Hepatol.* **8**, 712–22.

Mobley, H.L. (1996). The role of *Helicobacter pylori* urease in the pathogenesis of gastritis and peptic ulceration. *Aliment. Pharmacol. Ther.* **10**, 57–64.

Mobley, H.L., Cortesia, M.J., Rosenthal, L.E. and Jones, B.D. (1988). Characterization of urease from *Campylobacter pylori*. *J. Clin. Microbiol.* **26**, 831–6.

Mobley, H.L., Island, M.D. and Hausinger, R.P. (1995). Molecular biology of microbial ureases. *Microbiol. Rev.* **59**, 451–80.

Molinari, M., Galli, C., Norais, N., Telford, J.L., Rappuoli, R., Luzio, J.P. and. Montecucco, C. (1997). Vacuoles induced by *Helicobacter pylori* toxin contain both late endosomal and lysosomal markers. *J. Biol. Chem.* **272**, 25339–44.

Molinari, M., Galli, C., de Bernard, M., Norais, N., Ruysschaert, J.M., Rappuoli, R. and Montecucco, C. (1998a). The acid activation of *Helicobacter pylori* toxin VacA: structural and membrane binding studies. *Biochem. Biophys. Res. Commun.* **248**, 334–40.

Molinari, M., Salio, M., Galli, C., Norais, N., Rappuoli, R., Lanzavecchia, A. and Montecucco, C. (1998b). Selective inhibition of Li-dependent antigen presentation by *Helicobacter pylori* toxin VacA. *J. Exp. Med.* **187**, 135–40.

Moll, G., Papini, E., Colonna, R., Burroni, D., Telford, J.L., Rappuoli, R. and Montecucco, C. (1995). Lipid interaction of the 37-kDa and 58-kDa fragments of the *Helicobacter pylori* cytotoxin. *Eur. J. Biochem.* **234**, 947–52.

Montecucco, C. (1998). Protein toxins and membrane transport. *Curr. Opin. Cell Biol.* **10**, 530–6.

Montecucco, C., Papini, E. and Schiavo, G. (1991). Molecular models of toxin membrane translocation. In: *Sourcebook of Bacterial Protein Toxins* (eds J.E. Alouf and J.H. Freer), pp. 45–56. Academic Press, London.

Montecucco, C., Papini, E., Schiavo, G., Padovan, E. and Rossetto, O. (1992). Ion channel and membrane translocation of diphtheria toxin. *FEMS Microbiol. Immun.* **5**, 101–11.

Montecucco, C., Papini, E. and Schiavo, G. (1994) Bacterial protein toxins penetrate cells via a four-step mechanism. *FEBS Lett.* **346**, 92–8.

Namavar, F., Sparrius, M., Veerman, E.C., Appelmelk, B.J. and Vandenbroucke-Grauls, C.M. (1998). Neutrophil-activating protein mediates adhesion of *Helicobacter pylori* to sulfated carbohydrates on high-molecular-weight salivary mucin. *Infect. Immun.* **66**, 444–7.

Nielsen, H. and Andersen, L.P. (1992a). Activation of human phagocyte oxidative metabolism by *Helicobacter pylori*. *Gastroenterology* **103**, 1747–53.

Nielsen, H. and Andersen, L.P. (1992b). Chemotactic activity of *Helicobacter pylori* sonicate for human polymorphonuclear leucocytes and monocytes. *Gut* **33**, 738–42.

Pagliaccia, C., de Bernard, M., Lupetti, P., Ji, X., Burroni, D., Cover, T.L., Papini, E., Rappuoli, R., Telford, J.L. and Reyrat, J.M. (1998). The m2 form of the *Helicobacter pylori* cytotoxin has cell type-specific vacuolating activity. *Proc. Natl Acad. Sci. USA* **95**, 10212–27.

Pan, Z.J., Berg, D.E., van der Hulst, R.W., Su, W.W., Raudonikiene, A., Xiao, S.D., Dankert, J., Tytgat, G.N. and van der Ende, A. (1998). Prevalence of vacuolating cytotoxin production and distribution of distinct vacA alleles in *Helicobacter pylori* from China. *J. Infect. Dis.* **178**, 220–6.

Papini, E. Bugnoli, M., de Bernard, M., Figura, N., Rappuoli, R. and Montecucco, C. (1993). Bafomycin A1 inhibits *Helicobacter pylori*-induced vacuolization of HeLa cells. *Mol. Microbiol.* **7**, 323–7.

Papini, E., de Bernard, M., Milia, E., Zerial, M., Rappuoli, R. and Montecucco, C. (1994). Cellular vacuoles induced by *Helicobacter pylori* originate from late endosomal compartments. *Proc. Natl Acad. Sci. USA* **91**, 9720–4.

Papini, E., Gottardi, E., Satin, B., de Bernard, M., Telford, J., Massari, P., Rappuoli, R., Sato S.B. and Montecucco, C. (1996). The vacuolar ATPase proton pump on intracellular vacuoles induced by *Helicobacter pylori*. *J. Med. Microbiol.* **44**, 1–6.

Papini, E., Satin, B., Bucci, C., de Bernard, M., Telford, J.L., Manetti, R., Rappuoli, Zerial, M. and Montecucco, C. (1997). The small GTP binding protein rab7 is essential for cellular vacuolation induced by *Helicobacter pylori* cytotoxin. *EMBO J.* **16**, 15–24.

Papini, E., Satin, B., Norais, N., de Bernard, M., Telford, J.L., Rappuoli, R. and Montecucco, C. (1998). Selective increase of the permeability of polarized epithelial cell monolayers by *Helicobacter pylori* vacuolating toxin. *J. Clin. Invest.* **102**, 813–20.

Parsonnet, J., Hansen, S., Rodriguez, L., Gelb, A., Warnke, A., Jellum, E., Orentreich, N., Vogelman, J. and Friedman, G. (1994). *Helicobacter pylori* infection and gastric lymphoma. *N. Engl. J. Med.* **330**, 1267–71.

Parsonnet, J., Replogle, M., Yang, S. and Hiatt, R. (1997). Seroprevalence of CagA-positive strains among *Helicobacter pylori*-infected, healthy young adults. *J. Infect. Dis.* **175**, 1240–2.

Pelicic, V., Reyrat, J.M., Sartori, L., Pagliaccia, C., Rappuoli, R., Telford, J.L., Montecucco, C. and Papini, E. Effect of *Helicobacter pylori* VacA cytotoxin on epithelial permeability: a genetic analysis. Submitted.

Perez-Perez, G.I., Olivares, A.Z., Cover, T.L. and Blaser, M.J. (1992). Characteristics of *Helicobacter pylori* variants selected for urease deficiency. *Infect. Immun.* **60**, 3658–63.

Phadnis, S.H., Ilver, D., Janzon, L., Normark, S. and Westblom, T.U. (1994). Pathological significance and molecular characterization of the vacuolating toxin gene of *Helicobacter pylori*. *Infect. Immun.* **62**, 1557–65.

Phadnis, S.H., Parlow, M.H., Levy, M., Ilver, D., Caulkins, C.M., Connors, J.B. and Dunn, B.E. (1996). Surface localization of *Helicobacter pylori* urease and a heat shock protein homolog requires bacterial autolysis. *Infect. Immun.* **64**, 905–12.

Pohlner, J., Halter, R., Beyreuther, K. and Meyer, T.F. (1987). Gene structure and extracellular secretion of *Neisseria Gonorrhoeae* IgA protease. *Nature* **325**, 458–62.

Rautelin, H., Blomberg, B., Fredlund, H., Jarnerot, G. and Danielsson, D. (1993). Incidence of *Helicobacter pylori* strains activating neutrophils in patients with peptic ulcer disease. *Gut* **34**, 599–603.

Reymunde, A., Deren, J., Nachamkin, I., Oppenheim, D. and Weinbaum, G. (1993). Production of chemoattractant by *Helicobacter pylori*. *Dig. Dis. Sci.* **38**, 1697–701.

Reyrat, J.M., Charrel, M., Pagliaccia, C., Burroni, D., Lupetti, P., de Bernard, M., Xi, J, Norais, N., Papini, E., Dallai, R., Rappuoli, R. and Telford, J.L. (1998). Characterization of a monoclonal antibody and its use to purify the cytotoxin of *Helicobacter pylori*. *FEMS Lett.* **165**, 79–84.

Reyrat, J.M., Lanzavecchia, S., Lupetti, P., de Bernard, M., Pagliaccia, C., Pelicic, V., Charrel, M., Ulivieri, C., Norais, N., Ji, X., Cabiaux, V., Papini, E., Rappuoli, R. and Telford, J.L. Cell interaction, and 3D imaging of the 58 kDa subunit of VacA, the vacuolating cytotoxin of *Helicobacter pylori*. Submitted.

Ricci, V., Sommi, P., Cova, E, Fiocca, R., Romano, M., Ivey, K.J., Solcia, E. and Ventura, U. (1993). Na$^+$,K($^+$)-ATPase of gastric cells. A target of *Helicobacter pylori* cytotoxic activity. *FEBS Lett.* **334**, 158–60.

Ricci, V., Sommi, P., Fiocca, R., Romano, M., Solcia, E. and Ventura, U. (1997). *Helicobacter pylori* vacuolating toxin accumulates within the endosomal-vacuolar compartment of cultured gastric cells and potentiates the vacuolating activity of ammonia. *J. Pathol.* **183**, 453–9.

Rossi, G., Rossi, M., Vitali, C.G., Fortuna, D., Burroni, D., Pancotto, L., Capecchi, S., Sozzi, S., Renzoni, G., Braca, G., Del Giudice, G., Rappuoli, R., Ghiara, P. and Taccini, E. (1999). A symptomatic xenobiotic Beagle dog model for acute and chronic infection with *Helicobacter pylori*. *Infect. Immun.* (in press).

Sandvig, K. and van Deurs, B. (1996). Endocytosis, intracellular transport, and cytotoxic action of Shiga toxin and ricin. *Physiol. Rev.* **76**, 949–66.

Satin, B., Norais, N., Telford, J.L., Rappuoli, R., Murgia, M., Montecucco, C. and Papini, E. (1997). Vacuolating toxin of *Helicobacter pylori* inhibits maturation of procathepsin D and degradation of Epidermal Growth Factor in HeLa cells through a partial neutralization of acidic intracellular compartments. *J. Biol. Chem.* **272**, 25022–8.

Sato, S.B. and Toyama, S. (1994). Interference with the endosomal acidification by a monoclonal antibody directed toward the 116 (100)-kDa subunit of the vacuolar type proton pump. *J. Cell Biol.* **127**, 39–53.

Schiavo, G., Poulain, B., Rossetto, O., Benfenati, F., Tauc, L. and Montecucco, C. (1992). Tetanus toxin is a zinc protein and its inhibition of neurotransmitter release and protease activity depend on zinc. *EMBO J.* **11**, 3577–83.

Schmitt, W. and Haas, R. (1994). Genetic analysis of the *Helicobacter pylori* vacuolating cytotoxin: structural similarities with the IgA protease type of exported protein. *Mol. Microbiol.* **12**, 307–19.

Segal, E.D., Shon, J. and Tompkins, L.S. (1992). Characterization of *Helicobacter pylori* urease mutants. *Infect. Immun.* **60**, 1883–9.

Segal, E.D., Falkow, S. and Tompkins, L.S. (1996). *Helicobacter pylori* attachment to gastric cells induces cytoskeletal rearrangements and tyrosine phosphorylation of host cell proteins. *Proc. Natl Acad. Sci. USA* **93**, 1259–64.

Simon, S.M. and Blobel, G. (1991). A protein-conducting channel in the endoplasmic reticulum. *Cell* **65**, 371–80.

Simonsen, A., Lippe, R., Christoforidis, S., Gaullier, J.M., Brech, A., Callaghan, J., Toh, B.H., Murphy, C., Zerial, M. and Stenmark, H. (1998). EEA1 links PI(3)K function to Rab5 regulation of endosome fusion. *Nature* **394**, 494–8.

Slomiany, B.L. and Slomiany, A. (1991). Role of mucus in gastric mucosal protection. *J. Physiol. Pharmacol.* **42**, 147–61.

Smoot, D.T., Resau, J.H., Naab, T., Desbordes, B.C., Gilliam, T., Bull-Henry, K., Curry, S.B., Nidiry, J., Sewchand, J., Mills-Robertson, K., Frontin, K., Abebe, E., Dillon, M., Chippendale, G.R., Phelps, P.C., Scott, V.F. and Mobley, H.L.T. (1993). Adherence of *Helicobacter pylori* to cultured human gastric epithelial cells. *Infect. Immun.* **61**, 350–5.

Sommi, P., Ricci, V., Fiocca, R., Romano, M., Ivey, K.Y., Cova, E., Solcia, E. and Ventura, U. (1996). Significance of ammonia in the genesis of gastric epithelial lesions induced by *Helicobacter pylori*: an *in vitro* study with different bacterial strains and urea concentrations. *Digestion* **57**, 299–304.

Stenmark, H., Vitale, G., Ullrich, O. and Zerial, M. (1995). Rabaptin-5 is a direct effector of the small GTPase Rab5 in endocytic membrane fusion. *Cell* **83**, 423–32.

Suzuki, M., Miura, S., Suematsu, M., Fukumura, D., Kurose, I., Suzuki, H., Kai, A., Kudoh, Y., Ohashi, M. and Tsuchiya, M. (1992). *Helicobacter pylori*-associated ammonia production enhances neutrophil-dependent gastric mucosal cell injury. *Am. J. Physiol.* **263**, G719-25.

Takeuchi, K., Magee, D., Critchlow, J., Matthews, J. and Silen, W. (1983). Studies of the pH gradient and thickness of frog gastric mucus gel. *Gastroenterology* **84**, 331–40.

Telford, J.L., Ghiara, P., Dell'Orco, M., Comanducci, M., Burroni, D., Bugnoli, M., Tecce, M.F., Censini, S., Covacci, A., Xiang, Z., Papini, E., Montecucco, C., Parente, L. and Rappuoli, R. (1994). Purification and characterization of the vacuolating toxin from *Helicobacter pylori*. *J. Exp. Med.* **179**, 1653–8.

Telford, J.L., Covacci, A., Rappuoli, R. and Ghiara, P. (1997). Immunobiology of *Helicobacter pylori* infection. *Curr. Opin. Immunol.* **9**, 498–503.

Teneberg, S., Miller-Podraza, H., Lampert, H.C., Evans, D.J., Jr, Evans, D.G., Danielsson, D. and Karlsson, K.A. (1997). Carbohydrate binding specificity of the neutrophil-activating protein of *Helicobacter pylori*. *J. Biol. Chem.* **272**, 19067–71.

Tomb, J.F., White, O., Kerlavage, A.R., Clayton, R.A., Sutton, G.G., Fleischmann, R.D., Ketchum, K.A., Klenk, H.P., Gill, S., Dougherty, B.A., Nelson, K., Quackenbush, J., Zhou, L., Kirkness, E.F., Peterson, S., Loftus, B., Richardson, D., Dodson, R. Khalak, H.G., Glodek, A., McKenney, K., Fitzegerald, L.M., Lee, N., Adams, M.D., Hickey, E.K., Berg, D.E., Gocayne, J.D., Utterback, T.R., Peterson, J.D., Kelley, J.M., Cotton, M.D., Weidman, J.M., Fujli, G., Bowman, C., Watthey, L., Wallin, E., Hayes, W.S., Borodorsky, M., Karp, P.D., Smith, H.O., Fraser, C.H. and Venter, J.C. (1997). The complete genome sequence of the gastric pathogen *Helicobacter pylori*. *Nature* **388**, 539–47.

Tombola, F., Carlesso, C., Szabò, I., de Bernard, M., Reyrat, J.M., Telford, J.L., Rappuoli, R., Montecucco, C., Papini, E. and Zoratti, M. (1998). *Helicobacter pylori* vacuolating toxin forms anion-selective channels in planar lipid bilayers: possible implications for the mechanism of cellular vacuolation. *Biophys. J.* **96**, 1401–9.

Tricottet, V., Bruneval, P., Vire, O., Camilleri, J.P., Bloch, F., Bonte, N and Roge, J. (1986). *Campylobacter*-like organisms and surface epithelium abnormalities in active, chronic gastritis in humans: and ultrastructural study. *Ultrastruct. Pathol.* **10**, 113–22.

Tsuji, S., Kawano, S., Tsujii, M., Takei, Y., Tanaka, M., Sawaoka, H., Nagano, K., Fusamoto, H. and Kamada, T. (1996). *Helicobacter pylori* extract stimulates inflammatory nitric oxide production. *Cancer Lett.* **108**, 195–200.

Tummuru, M.K., Cover, T.L. and Blaser, M.J. (1994). Mutation of the cytotoxin-associated cagA gene does not affect the vacuolating cytotoxin activity of *Helicobacter pylori*. *Infect. Immun.* **62**, 2609–13.

van der Ende, A., Pan, Z.J., Bart, A., van der Hulst, R.W., Feller, M., Xiao, S.D., Tytgat, G.N. and Dankert, J. (1998). CagA-positive *Helicobacter pylori* populations in China and the Netherlands are distinct. *Infect. Immun.* **66**, 1822–6.

van der Goot, F.G., Gonzalez-Manas, J.M., Lakey, J.H. and Pattus, F. (1991). A 'molten-globule' membrane-insertion intermediate of the pore-forming domain of colicin A. *Nature* **354**, 408–10.

Wandersman, C. (1992). Secretion across the bacterial outer membrane. *Trends Genet.* **8**, 317–22.

Warren, J.R. and Marshall, B.J. (1983). Unidentified curved bacilli on gastric epithelium in active chronic gastritis. *Lancet* **i**, 1273–5.

Watts, C. (1997). Capture and processing of exogenus antigen for presentation on MHC molecules. *Annu. Rev. Immunol.* **15**, 821–50.

Weel, J.F., van der Hulst, R.W., Gerrits, Y., Roorda, P., Feller, M., Dankert, J., Tytgat, G.N. and van der Ende, A. (1996) The interrelationship between cytotoxin-associated gene A, vacuolating cytotoxin, and *Helicobacter pylori*-related diseases. *J. Infect. Dis.* **173**, 1171–5.

Williams, S.E. and Turnberg, L.A. (1981). Demonstration of a pH gradient across mucus adherent to rabbit gastric mucosa: evidence for a 'mucus-bicarbonate' barrier. *Gut* **22**, 94–6.

Xiang, Z., Censini, S., Bayeli, P.F., Telford, J.L., Figura, N., Rappuoli, R. and Covacci, A. (1995). Analysis of expression of CagA and VacA virulence factors in 43 strains of *Helicobacter pylori* reveals that clinical isolates can be divided into two major types and that CagA is not necessary for expression of the vacuolating cytotoxin. *Infect. Immun.* **63**, 94–8.

Xu, J., Goodwin, C.S., Cooper, M. and Robinson, J. (1990). Intracellular vacuolization caused by the urease of *Helicobacter pylori*. *J. Infect. Dis.* **161**, 1302–4.

Yahiro, K., Niidome, T., Hatakeyama, T., Aoyagi, H., Kurazono, H., Padilla, P.I., Wada, A. and Hirayama, T. (1997). *Helicobacter pylori* vacuolating cytotoxin binds to the 140-kDa protein in human gastric cancer cell lines, AZ-521 and AGS. *Biochem. Biophys. Res. Commun.* **238**, 629–32.

Yamashiro, D.J. and Maxfield, F.R. (1987). Acidification of morphologically distinct endosomes in mutant and wild-type Chinese hamster ovary cells. *J. Cell Biol.* **105**, 2723–33.

Yoshida, N., Granger, D.N., Evans, D.J., Jr, Evans, D.G., Graham, D.Y., Anderson, D.C., Wolf, R.E. and Kvietys, P.R. (1993). Mechanisms involved in *Helicobacter pylori*-induced inflammation. *Gastroenterology* **105**, 1431–40.

SECTION II

MEMBRANE-DAMAGING TOXINS

14

Biophysical methods and model membranes for the study of bacterial pore-forming toxins

Gianfranco Menestrina and Beatrix Vécsey Semjén

INTRODUCTION

A major aspect of toxin–cell interaction is represented by surmounting the permeability barrier of an intact cell membrane. There are at least two major routes along which a bacterial protein toxin can interfere with the integrity of this barrier: (a) by catalysing the translocation of a toxic component into the cell; or (b) by increasing the permeability of the membrane to a number of small molecules. Interestingly, both lead to the formation of new exogenous toxin-formed channels in the cell membrane. Toxins that follow the first route are those of the A–B supergroup, while toxins of the second kind belong to the membrane-damaging/pore-forming supergroup. Only because of these functional similarities will we refer to both of them by the general term of pore-forming toxin (PFT), although this is normally reserved for members of the second group. In the first part of this chapter some of the biophysical models and techniques used to characterize the formation and properties of these exogenous pores are reviewed. A summary of the most recent data on bacterial PFT that were obtained with such methods is given in the second part.

BIOPHYSICAL MODELS AND METHODS FOR STUDYING PORE-FORMING TOXINS

A number of biophysical approaches has been developed to study the structure–function correlation of PFT

on synthetic and biological model membranes. They derive from the efforts to find simplified systems mimicking the cell membrane, where PFT pores could be studied under fully controlled experimental conditions. This presentation is by no means exhaustive and deals only with some of the methods with which the authors are more familiar.

Lipid monolayers

When strongly amphiphilic molecules such as phospholipids are deposited at an air–water interface they spontaneously form a monomolecular film. Seen from the water phase it appears as a continuous layer of polar headgroups and hence represents perhaps the simplest physical model of a cell membrane. A set-up for monolayer experiments is shown in Figure 14.1a. It utilizes a Teflon trough to prepare the film, one or two computer-controlled movable barriers that restrict the film to the desired area and a microbalance which measures the surface pressure of the interface, for example, by the Wilhelmy method, using a platinum plate as a sensor. To avoid the presence of contaminant surfactants, special care must be used in preparing the solutions, cleaning the apparatus and repeatedly wiping the water surface by careful suction of the top layer. Since phospholipids are virtually insoluble all applied molecules remain at the interface. Accordingly, the area occupied by each molecule is easily determined from the area of the film and the

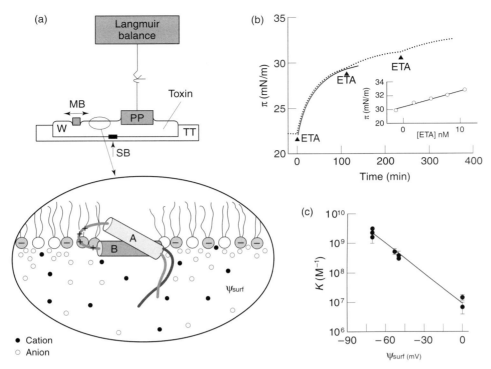

FIGURE 14.1 Surface active toxins in lipid monolayers. (a) Schematic representation of the apparatus for lipid monolayers. In a Teflon trough (TT) a water phase (W) stirred with a magnetic spin-bar (SB) is covered with a lipid film. The lateral pressure of the film is measured with a Langmuir balance equipped with a platinum plate (PP) and is controlled by a movable barrier (MB). A toxin is injected in the subphase. In the balloon there is an exploded view of part of the toxin interacting with the layer. An electrostatic interaction occurs between charged residues on the protein and the surface potential ψ_{surf}, generated by negatively charged lipids. (b) Surface pressure increase induced by successive additions of *P. aeruginosa* exotoxin A (ETA). The time course is first order, as indicated by the fit with a single exponential. In the inset a plot of π vs ln C that allows estimating the molecular area occupied by ETA in the film. (c) Dependence of the partition constant K on the surface potential of the monolayer ψ_{surf}.

number of molecules applied. At a surface pressure of 22 mN m^{-1}, which corresponds to a lateral packing of the lipid chains similar to that observed in cell membranes, the molecular area occupied by a phospholipid is about 0.7 nm^2, and that occupied by a sterol 0.55 nm^2.

This model can provide useful information on the tendency of membrane-active toxins to insert into the lipid phase. In fact, when injected into the sub-phase, they migrate to the water–air interface, where they partition into the monolayer, increasing its surface pressure (Figure 14.1b). Since these proteins have a stable soluble form, the number of molecules that go into the lipid layer is variable, depending on a partition coefficient, and cannot be easily determined. Therefore the estimate of the average molecular area A_m occupied by each toxin molecule is less straightforward. In some cases it is possible to derive A_m by application of the Gibb's equation (Bougis *et al.*, 1981; Nordera *et al.*, 1997). This states that A_m is proportional to the slope of a plot of the increase in the surface

pressure (π) induced by the toxin, versus the logarithm of its bulk concentration (C):

$$A_m = RT/N_A \cdot (d\pi/d \ln C)^{-1} \qquad (14.1)$$

where N_A is Avogadro's number, R the gas constant and T the absolute temperature (Figure 14.1b inset). Molecular areas for several bacterial protein and peptide toxins have been determined in this way. They range from 0.6 to 1.7 nm^2 for the lipocyclopeptides of *Pseudomonas syringae* and *Bacillus* (Maget-Dana and Peypoux, 1994; Dalla Serra *et al.*, 1999a) to the much larger value of 21 nm^2 estimated for diphtheria toxin by a complementary technique (Demel *et al.*, 1991). In the latter case, the toxin was labelled with a β-emitter and its monolayer concentration was directly determined from the surface radioactivity. When these two techniques have been directly compared using the same toxin and the same experimental conditions they have given consistent results (Bougis *et al.*, 1981).

Lipid monolayers have been extensively used in the study of single chain A–B toxins (Demel *et al.*, 1991; Schiavo *et al.*, 1991; Nordera *et al.*, 1997). These toxins interact with lipid membranes via a common mechanism that requires: (a) a pH-induced conformational transition exposing new hydrophobic regions; (b) a lipid-dependent binding to the membrane surface; (c) the insertion of one or more hydrophobic stretches of the polypeptide into the lipid phase; and (d) the aggregation of several polypeptide entities in the lipid bilayer with the formation of structures enabling the passage of otherwise impermeable molecules. At least the first three steps can be conveniently studied with lipid monolayers. With *Pseudomonas aeruginosa* exotoxin A (ETA) it was observed that after injection into the subphase it increased the surface pressure of lipid monolayers, inserting at least a portion of the molecule in the lipid film (Nordera *et al.*, 1997). The time course of this interaction was well described by a single exponential relaxation (Figure 14.1b), the time constant of which was inversely related to ETA concentration. The whole absorption process was regulated by a simple set of equations which apply to a Langmuir absorption isotherm:

$$\pi(t) = \pi_{ss}(1 - \exp(-t/\tau)) \quad (14.2)$$

$$\pi_{ss} = KT_0/(1 + KT_0) \quad (14.3)$$

$$\tau = 1/(k_a T_0 + k_d) \quad (14.4)$$

$$K = k_a/k_d \quad (14.5)$$

where $\pi(t)$ and π_{ss} are the lateral pressure of the film at time t and at steady-state; τ is the time constant; T_0 the free toxin concentration; K the equilibrium constant for the adsorption and k_a, k_d the rate constants for adsorption and desorption. In addition to the pH of the subphase, the surface pressure increase induced by ETA (and other A–B toxins) depends on the lipid composition of the monolayer. In particular, acidic lipids such as PG, PS, PI or PA promote the interaction. It can be demonstrated that this does not depend on the particular nature of the acidic lipid, but rather on the negative potential generated at the surface of the lipid monolayer. In fact, similar effects were observed by changing the negative surface potential in different ways, for example, with different acidic lipid species, different percentages of the negatively charged component or different ionic strengths. In all cases, increasing the negative surface potential promoted the extent and the rate of insertion of ETA. A direct electrostatic interaction between a positive charge on ETA and the negative surface potential (ψ) of the

membrane was implicated. All constants could be predicted from usual reaction rate theory; in particular, the partition constant K was:

$$K = K^0 \exp(-Q\psi) \quad (14.6)$$

where Q is the positive charge on ETA (in electronic units); ψ is the reduced potential (i.e. $e\psi/kT$, roughly equal to $\psi/25$ mV); $K^0 = k_a^0/k_d^0$ is the intrinsic equilibrium constant (independent of the surface potential). Besides the intrinsic constants, one can estimate the protein charge $Q = +1.9 \pm 0.3$ (Figure 14.1c). This example shows how intrinsic molecular data on the toxin membrane interaction can be obtained from such simple models.

Planar lipid bilayers

Planar lipid membranes (PLM) are thermodynamically stable portions of a lipid bilayer prepared on a hole in a Teflon septum which separates two aqueous solutions (Figure 14.2a). They are made of purified lipids and are quite useful in the study of protein channels and particularly PFT in a simple membrane environment (Hanke and Schlue, 1993; Kagan and Sokolov, 1997). Among the advantages offered are (a) easy control of all the physicochemical parameters on each side of the bilayer, for example, lipid composition of the membrane, chemical composition and temperature of the water phase, applied voltage; (b) current resolution that allows the detection of single channel events at a typical bandwidth of 1 kHz. Among the drawbacks are (a) sensitivity to small amounts of impurities; (b) necessity that the channel self-incorporates to the lipid film from the water phase. With the majority of PFT the latter is usually not a problem because an inherent property of such proteins is their ability to cross a water phase before reaching their final membrane target. Furthermore, interesting methods have been developed to circumvent this problem when present. For instance, the reconstitution of toxin receptors into the PLM has been achieved either by folding bilayers from monolayers, produced by addition of native cell membrane vesicles in the subphase, as with aerolysin (Gruber *et al.*, 1994), or by fusing receptor-containing vesicles to preformed PLM by osmotic gradients, as with δ-endotoxins of *Bacillus thuringiensis* (Lorence *et al.*, 1995; Schwartz *et al.*, 1997). PLM with improved stability and current resolution have been prepared on the tip of glass micropipettes similar to those used in the patch-clamp technique, either by dipping the pipette through a lipid monolayer [used with the enterotoxin of *Clostridium perfringens* type A (Sugimoto *et al.*, 1988) and with staphylococcal δ-toxin

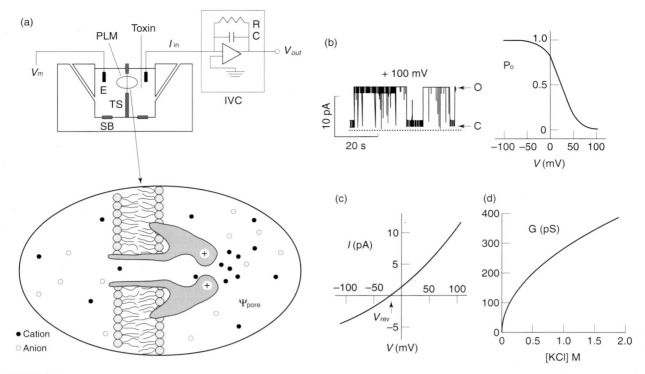

FIGURE 14.2 Pore-forming toxins in planar lipid membranes. (a) Schematic representation of a cell for the preparation of PLM. A supported bilayer is prepared by apposing two lipid monolayers on to a small (0.1–0.2 mm) hole, punched in a thin (12 μm) Teflon foil separating two water-filled compartments. Two electrodes (E) are used, one to apply the voltage, V_m, the other to drive the current (I_{in}) into the current/voltage converter (IVC), a virtual grounded operational amplifier with feedback resistance (R), usually 10^9 Ω, and capacitance (C) of 1 pF (time resolution 1 ms). Buffer solutions are stirred with magnetic spin-bars (SB) and the toxin is applied in the *cis* compartment. In the balloon an exploded view of the pore inserted into the bilayer. An electrostatic filter at its entrance, formed by positively charged residues, generates the entrance potential ψ_{pore} that attracts anions and repels cations. (b) Current fluctuations due to a single pore that is either open (o) or closed (c). The probability that the pore is open, P_o, is voltage dependent. (c) Current voltage curve in the presence of a concentration gradient (*cis* side more concentrated). Owing to the electrostatic filter the pore is anion selective, as indicated by the negative reversal voltage, V_{rev}, that has to be applied to stop the anion current. For the same reason the I/V is non-linear. (d) Dependence of the single channel conductance on the salt concentration. A typical effect of the fixed charges is a square root dependence.

(Mellor *et al.*, 1988)] or by patching giant vesicles [as for the haemolysin of *Gardnerella vaginalis* (Moran *et al.*, 1991)].

Measurement of molecular parameters

Pore size

The background current of an unmodified PLM is very low. Upon addition of small amounts of a PFT, uniform step increases in the current are observed representing single channel events (Figure 14.2b). From this the conductance G of the pore is derived. Since pores formed by toxin are usually relatively large and filled with water it is expected that their conductance depends linearly on the conductivity of the solution according to

$$G = \sigma \pi r^2 / l \qquad (14.7)$$

where σ is the solution conductivity; r and l are radius and length of the pore, respectively. This assumes that

the pore is simply a cylindrical hole filled with water, where the mobility of ions is similar to that in the bulk aqueous solution. In those cases in which such linear relationship is observed Equation (14.7) allows an estimate of pore radius provided a guess is made on its length. Despite its evident roughness, this relation has been used with some success in a number of cases. An improved method was recently introduced (Krasilnikov *et al.*, 1992a; Parsegian *et al.*, 1995), based on measuring G in the presence of sugars of different size. If the sugars are able to enter the pore, G is decreased, if they are too big, G remains unchanged. It is therefore easy to determine the size of the biggest molecule allowed inside the pore. When compared to the previous method it may give somewhat different results. In fact, if the shape of the pore is not that of a regular cylinder the first method provides an average radius, whereas the second gives essentially the radius of the entrance of the pore.

Effects of pore charges

Very often the conductance tends to increase sublinearly with the ionic concentration of the solution, an effect usually called saturation (Figure 14.2d). This may derive from the presence of an excess of charges of one sign at the entrance of the pore. In fact, the presence of fixed charges at the pore entrance generates a local field that attracts and concentrates the counter-ions and eventually originates a larger then expected current. Owing to the charge screening effect, this attraction is maximum at low ionic strength, but decreases when the salt concentration is increased, thus generating the observed saturation. A good indication of this effect is the observation of an approximately square-root dependence of the conductance on the salt concentration. In fact, a more realistic expression of the conductance of the ion species, i, through the pore is:

$$G_i = (\pi r^2/l)u_i c_{i0} \exp(z_i \psi'_{pore}) \qquad (14.8)$$

where u_i, c_{i0} and z_i are the mobility, bulk concentration and valence of the ion i, and ψ'_{pore} is the reduced potential at the pore entrance. This may be approximated (Menestrina and Antolini, 1981) by

$$G_i = G^*_{i0}\kappa = (G_{i0}c_{i0})^{1/2} \qquad (14.9)$$

where κ is the Debye–Hückel coefficient.

Interestingly, the fixed charges, which are provided in general by charged residues, can give rise to two other phenomena often observed with toxin pores: cation/anion selectivity and non-linearity of the current/voltage (I/V) curve. The selectivity originates from the fact that the presence at the pore entrance of an uncompensated charge, for instance positive, would attract anions and repel cations, thus giving rise to a larger anion flux, i.e. anion selectivity. The opposite would be generated by a negative charge. As a rule of thumb, we can write (Cescatti et al., 1991):

$$P_-/P_+ = \exp\psi'_{pore}u_-/u_+ \qquad (14.10)$$

where P_-, P_+, u_- and u_+ are permeability and mobility of anion and cation, respectively. As an example, an entrance potential of 25 mV would imply that anions are approximately three times more permeant than cations. It should be emphasized that such selectivity, which we will call elecrostatic filtering, can discriminate ions only on the basis of their charge. This means it is is rather poor if compared, for example, to the selectivity of endogenous ion channels found in animal or plant cells, which discriminate ions also according to their crystal radius. However, this is only natural for toxin pores that are supposed to inflict a rather unse-

lective damage to target cells. The recently determined structure of the eukaryotic potassium channel (Doyle et al., 1998) has shown that it also exploits the electrostatic filter provided by the attraction of a cluster of negative charges located around its entrances to repel anions and attract cations, before filing them to a very narrow passage in the lumen. This provides an exquisite selectivity for K^+ (the real selectivity filter).

A non-linear I/V characteristic, for example a pore conductance larger at negative potentials than at positive (or vice versa), may result from an asymmetric distribution of the charges on the two pore mouths (see the model in Figure 14.2). In fact, in this case the accumulation effect would be different at the two pore entrances and this would generate different currents when, for example, cations are pushed from the cis to the trans entrance or the other way round by opposite applied voltages. Again we can approximate to:

$$I(+V)/I(-V) = \exp\psi'_{pore} \qquad (14.11)$$

These concepts can often be applied to PFT pores (Ropele and Menestrina, 1989; Benz et al., 1994a; Maier et al., 1996). For example, they have been used in a study of the effects of modification of charged lysine residues on the *Staphylococcus aureus* α-toxin pore (Cescatti et al., 1991), arriving at a model of the distribution of charges at the pore's mouths that is consistent with the recently determined 3D structure (Song et al., 1996).

Channel gating

Very often a toxin pore does not stay open all the time but rather fluctuates between a high conductance state, open, and a low conductance state, closed (Figure 14.2b). This behaviour, which is typical of an endogenous channel, is generally referred to as 'gating' of the pore. It might be closely controlled by the voltage applied as with α-haemolysin of *Escherichia coli* (Menestrina and Ropele, 1989). In general, it does not imply that the pore is removed from the membrane, but rather that it is in a different state. Gating can be due either to a conformational change of a part of the pore, for example a domain of the protein that folds back to block its mouth, as found in the rectifying K^+ channel (ball and chain model; Hoshi et al., 1990), or even to a change in the ionization state of the pore lumen, as proposed for the *S. aureus* α-toxin pore (Korchev et al., 1997).

Comparison of macroscopic currents and single events

PLM provide a model for detecting the contribution to the membrane current of single pores. However, they are stable enough to allow the characterization of the currents deriving from the simultaneous presence of several thousand of these pores in a so-called multi-channel experiment. The information that can be drawn from such experiments is very often the same as from the single channel, with the advantage that they provide values that are averages of many individual contributions. In general, a good rule which may be helpful in avoiding artefacts is that single channel and multi-channel experiments give the same result when determining the same property.

Noise measurements

As an example the pore conductance and lifetime can be obtained from noise analysis of multi-channel traces. In fact, since pores are discrete molecules which open and close stochastically, the number of open channels in a voltage-clamped membrane fluctuates even at equilibrium, generating an extra noise in the current traces. These can be readily distinguished from common sources of electrical noise such as thermal, shot or $1/f$ noise, by a spectral (Fourier) analysis. Assuming that the channels are constant in number, all equal and independent and have only two possible states (open and closed), their conductance and lifetime can be estimated. This technique was successfully applied, for example, to *P. aeruginosa* ETA (Gambale *et al.*, 1992). A similar approach has shown that current noise through one single open *S. aureus* α-toxin channel can also be analysed to give information on the ionization state of the pore lumen (Bezrukov and Kasianowicz, 1993) or on its occupancy by large molecules (Bezrukov *et al.*, 1996).

Patch clamp

This technique allows the detection of single channels on the membrane of intact cells. It is in some respects similar to PLM but has a higher resolution, allowing current measurements with less than 0.1 pA noise at 10 kHz bandwidth. Although developed to study endogenous channels, this powerful technique offers a unique means of investigating the mechanism of cell attack by PFT at a molecular level and directly on target cells. Two typical configurations have been used in these experiments.

Whole cell recording

After obtaining a giga-seal between the patch pipette and the cell membrane, the patch is broken and a low-resistance access to the interior of the cell is gained. At this point, one can measure the current flowing through the whole cell membrane from the cytoplasm (held at a clamped potential) to the external solution

(the reference). Despite the problem that the cell interior is rapidly perfused by the patch pipette solution (with obvious loss of many cytoplasmic components), and that endogenous channels usually have to be blocked, the method has been used successfully to study bacterial PFT (Beise *et al.*, 1994; Menestrina *et al.*, 1996; Sheridan, 1998).

Outside-out excised patch

Pulling away the pipette from the cell after the whole cell configuration is achieved causes the resealing of the pipette with a piece of cell membrane which preserves the correct orientation of all membrane components (outside-out excised patch). This smaller portion of the original cell envelope has the advantage that it permits current measurements with a higher sensitivity and lower background contribution of the endogenous channels than the whole cell. Furthermore, the use of secondary perfusion pipettes allows abrupt changes in the external medium (within milliseconds), which are useful for applying the toxin as well as studying the chemical or pharmacological control of these channels. Most patch-clamp studies of PFT include this approach.

Although not yet fully exploited in this direction, the patch technique can also help in understanding the mode of action of PFT and other toxins that specifically interact with the endogenous cell channels. Examples are the crystal toxins of *B. thuringiensis*, which block K^+ channels (Liebig *et al.*, 1995), and *S. aureus* leucotoxins (Staali *et al.*, 1998), which interact with Ca^{2+} channels.

Lipid vesicles

Unilamellar lipid vesicles are quite realistic models for cell envelopes, being completely sealed lipid bilayers. Several methods are used to produce and characterize vesicles with diameters ranging from 30 nm to several micrometres (New, 1990). One of the advantages offered by vesicles is that their lipid composition can be varied almost at will, without compromising their stability. This provides a very useful system for investigating the lipid requirements of the PFT. Another advantage is that the information they provide is averaged from a large population of equivalent objects (typically 10^{10}–10^{12}) and thus it is less prone to the artefacts that may occur when just a few copies of the pores are studied in a PLM or in a cell patch. The incorporation of bacterial PFT into a vesicle is normally very simple; soluble toxins are mixed with vesicles in the water phase and in the presence of the lipid they spontaneously undergo a transition from hydrophilic to amphiphiphilic, which allows them to insert into the membrane. A number of techniques can be applied to demonstrate the incorporation of new channels into the membrane of the vesicle and to measure the flux of ions, metabolites, neutral solutes or even proteins across them. Many of these methods employ radioactive molecules, but fluorescent dyes are becoming most popular. To review all such approaches is outside the scope of this chapter; rather, we will describe a few spectroscopic techniques, based on the emission or absorption of light, that have been fruitfully and extensively used in the study of bacterial PFT.

Detecting toxin binding

The first information that can be derived from the study of PFT–vesicle interaction is the rate of toxin binding. For this it is necessary to have a signal that affects whether the toxin is in the lipid or in the water phase. This has been accomplished using both intrinsic and extrinsic chromophores. The most used intrinsic chromophore is tryptophan, whose fluorescence is environment dependent. Because the quantum yield and frequency changes due to polarity variation are not very strong, in some cases suitable quenchers have been used that enhance the differences. If only one Trp residue is present in the toxin, the use of quenchers with a precise location in the lipid bilayer, for example brominated lipids with bromine ions at different positions, can provide information on the membrane topology of the fluorophore (Abrams and London, 1992). Alternatively, the toxin may be labelled with an environment-sensitive fluorophore at precise positions, for example by modification of single cysteins with acrylamide, and the fate of the label during interaction is followed. Coupled with the cystein scanning technique, this has allowed the precise determination of which parts of the PFT molecule interact with the membrane (Palmer et al., 1996).

Release of fluorescent molecules

Vesicles can be loaded with fluorescent molecules that have a different quantum yield, whether they are inside or outside the vesicle. Examples are concentration-dependent dyes, such as calcein and carboxyfluorescein, that undergo self-quenching in the inner compartment, or ion couples, such as SPQ-Cl and ANTS-DPX, which will emit fluorescence only when dissociated, usually in the outer solution (Ropele and Menestrina, 1989; Vécsey-Semjén et al., 1996). Loading is normally achieved by preparing the vesicles in the presence of the quenched dye and then removing the untrapped material by washing on a column of desalting gel. The vesicles are then exposed to the PFT, and the time course of the marker release is measured with a fluorometer. This reflects the kinetics of the formation of pores in the vesicle, since the efflux of the marker from one single vesicle is normally too fast to be detected. This approach has a number of advantages. It is easy, straightforward and, in the case of concentration-dependent dyes, it allows discrimination between two mechanisms of release, all-or-nothing or graded, just by determining the degree of self-quenching of the dye retained in the vesicles at steady-state. All-or-nothing release indicates the formation of stable pores that are open long enough for the vesicle to release all of its content; graded release indicates transient pores. Particularly noteworthy is also the possibility of performing experiments with a microplate reader instead of a conventional fluorometer, which allows the simultaneous determination of the reaction kinetics from a large number of small samples (Dalla Serra et al., 1999a).

Influx of neutral molecules (osmotic swelling)

If one wants to study the properties of channels that are already inserted into the vesicles, the osmotic swelling assay can be used. This is based on the fact that vesicles are good osmometers and have a high intrinsic permeability to water. When they are placed in a hyperosmotic medium they first rapidly shrink, as water flows out along the osmotic gradient, then reswell as the permeable osmolite flows in, followed by water. The ensuing volume changes can be observed using 90° light scattering, which increases as vesicles shrink and decreases when they swell. Since the intrinsic permeability of the lipid membrane to most solutes is quite low, the rate of reswelling reports the permeation of the osmolite through the channels. If the osmotic gradient is generated with a dissociated salt, it is rate limited by the least permeable ion species. Because shrinking and swelling are quite fast (shrinking is normally finished in around 200 ms and swelling takes from a few to some seconds depending on the osmolite), a stopped-flow apparatus is required which ensures a rapid and complete mixing and a good time resolution. By choosing osmolites with different sizes it is possible to evaluate the cut-off of the pores as well as their selectivity (but limited to the least permeable species). This technique can also be applied to vesicles prepared in minute amounts from natural tissues, and has been widely used for δ-endotoxins of B. thuringiensis (Carroll and Ellar, 1993).

Measurements of membrane potential

The creation and/or dissipation of a transmembrane potential through the vesicle can be used to detect the effects of PFT. The voltage is measured with a fluorescent lipophilic ion whose partitioning between the water and lipid phase depends on the membrane potential and whose quantum yield is different in the two phases (Plasek and Sigler, 1996). Since lipid vesicles have no internal potential, this is artificially generated, for example by creating a K^+ gradient across the membrane and applying valinomycin, a specific K^+ carrier. A Nernst potential is thus generated through the membrane, which prevents further movement of the ions and dissipation of the gradient. Insertion of unselective (or poorly selective) toxin channels will dissipate such a potential, leading to a detectable fluorescence change. Intriguingly, the dependence of toxin insertion on the transmembrane voltage can be studied with this technique simply by creating different voltages with different K^+ gradients [as shown with diphtheria toxin (Shiver and Donovan, 1987) and tetanus toxin (Menestrina *et al.*, 1989)]. Alternatively, if the toxin channels are selective enough, they may be directly used to establish the transmembrane potential using suitable gradients. From the magnitude and the sign of the established voltage it is possible to determine the extent and the quality of the channel selectivity. In this way, the cation selectivity of *E. coli* α-haemolysin pores observed in PLM and cells was confirmed in vesicles (Menestrina *et al.*, 1996).

Demonstrating the formation of toxin oligomers

As we will see in the following section, most (if not all) PFT form oligomeric channels. Some general questions arise: if and where oligomerization takes place, what is the oligomer size and is it fixed or not? In some cases the oligomers are stable enough to allow biochemical isolation, but this is not always true. A very powerful approach exploits the resonant energy transfer effect (RET). This occurs between two fluorophores with overlapping excitation and emission spectrum, provided they are located within a short distance, typically in the range of 1–3 nm (Yguerabide, 1994). Mixing a population of toxin monomers labelled with the two fluorophores leads to RET only if the monomers come in close contact, for example, by oligomerization. More generally, information on the occurrence and extent of aggregation is embodied in the kinetics of pore formation that is easily assessed from release experiments. It is therefore very useful to have models for the interpretation of such these kinetics that allow the extrapolation of some basic notions such as equilibrium and rate constants for aggregation and size of the conducting oligomer. A very useful treatment of such a general problem has been elaborated by Schwarz and co-workers (Rex and Schwarz, 1998) and applied to several pore-forming peptides of animal origin. It offers analytical solutions but requires precise modelling of all the steps involved, which should be kept to a minimum number. Alternatively, a more statistical approach has

been presented (Parente *et al.*, 1990; Rapaport *et al.*, 1996). In this Parente–Rapaport model the process of permeabilization is divided into two steps: partitioning of toxin monomers into the lipid bilayer (which is assumed to be very fast), and aggregation of membrane-inserted monomers (assumed to be rate limiting). When an aggregate has reached a critical size, comprising M monomers, a conducting unit is formed and release occurs. The incorporation process is governed by a partition coefficient, Γ:

$$\Gamma = T_0/(T_0 - T_b)L \qquad (14.12)$$

where T_0 and L are the total concentration of toxin and lipid and T_b is that of bound toxin. The aggregation process is characterized by c and d, the forward and backward rate constants. For simplicity they are assumed to be the same at each step of the aggregation. The percentage of release is then given by:

$$P(\%) = 100 \sum_{i=M}^{N} A_i \cdot Z(M,t,i,c,d) \qquad (14.13)$$

where A_i is the fraction of vesicles with i monomers bound, which is time independent resulting from the fast partitioning step, N is the maximum number of toxin molecules that can be bound to one vesicle, and $Z(M,t,i,c,d)$ is the probability that a vesicle containing i bound monomers will also have an aggregate of order not smaller than M at time t. This model has been used successfully with several toxins including *S. aureus* leucotoxins (see Ferreras *et al.*, 1998) which reports a complete expression of A_i and the probability factor Z.

BACTERIAL TOXINS THAT FORM PORES IN MEMBRANES

The following section is a short review on bacterial toxins that produce well-defined lesions in cell membranes thereby increasing cell permeability to ions and small molecules (membrane-damaging toxins), or even other polypeptides (A–B toxins) (see Table 14.1). The permeabilization may overcome the capacity of the cell to maintain its 'milieu interieur' by active transport, and often leads to apoptosis. The number of toxins belonging to this group is large and steadily growing. They have been further divided into subgroups according to structural, functional and/or genetic characteristics. Because most of them are treated in detail in other chapters of this book, we will focus here on the information on their pore properties that was obtained by the previously described techniques.

TABLE 14.1 Properties of bacterial protein toxins forming channels in lipid membranes

Organism	Toxin	Acronym	Molecular weight[a] (kDa)	Lipid preference	Selectivity	Conductance (pS)	Buffer[b] (mM)	Radius (nm)	Reference
RTX toxins									
Escherichia coli (chromosome encoded)	α-Haemolysin	HlyA	110	Acidic phospholipids	Cation	200	100 NaCl	0.6	1
Escherichia coli (plasmid encoded)	α-Haemolysin	HlyA	110		Cation	500	150 KCl	1	2
Escherichia coli 0157:H7	EHEC-haemolysin	Ehx	110		Cation	550	150 KCl	1.3	3
Proteus vulgaris	Haemolysin	PvxA	110		Cation	500	150 KCl	1	4
Morganella morganii	Haemolysin	MmxA	110		Cation	500	150 KCl	1	4
Actinobacillus pleuropneumoniae	RTX-toxin I	ApxIA	107		Cation	540	150 KCl	1.2	5
Actinobacillus pleuropneumoniae	RTX-toxin II	ApxIIA	103		Cation	620	150 KCl	1.2	5
Actinobacillus pleuropneumoniae	RTX-toxin III	ApxIIIA	120		Cation	95	150 KCl	0.9	5
Actinobacillus actinomycetemcomitans	Leucotoxin	AaltA	116	Acidic phospholipids	Cation	400	140 NaCl	0.8	6, 7
Bordetella pertussis	Adenylate-cyclase/ haemolysin	CyaA	177	Acidic phospholipids	Cation	27	1 M KCl	0.3	8, 9
Aeromonas and Vibrio haemolysins									
Aeromonas hydrophila	Aerolysin	AerA	52		Anion	145	300 NaCl	1.9–2.3	10, 11
Aeromonas sobria (strain AB3)	Aerolysin		54		Anion	70	100 KCl	0.4	12
Vibrio cholerae El Tor	Labile haemolysin	HlyA	65		Anion	22	100 KCl	0.6–0.8	13, 14
Vibrio cholerae non-O1	Labile haemolysin		65		Anion	25	150 NaCl	0.9–1.0	15
Other Gram-negative PFT									
Pseudomonas aeruginosa	Leucotoxin	CTX	29	Cholesterol	Non-selective	120	140 K-gluconate	1.2	16, 17
Legionella pneumophila	Legiolysin				Non-selective			1.5	18
Serratia marcescens	Cytolysin	ShlA	165		Weak anion	150	100 KCl	0.5	19
Staphylococcal leucotoxins									
Staphylococcus aureus	α-Toxin	Hla	33	PC + cholesterol or PS	Anion	90	100 KCl	0.6	20, 21
Staphylococcus aureus	γ-Haemolysin A+B	HlgA + HlgB	32+34	PC+cholesterol	Cation	110	100 NaCl	0.6	22, 23
Staphylococcus aureus	γ-Haemolysin C+B	HlgC + HlgB	32+34	PC+cholesterol	Cation	185	100 NaCl	0.7	22, 23
Cholesterol-binding cytolysins									
Clostridium tetani	Tetanolysin	TTL	48–53	Cholesterol	Cation	10–400[c]	145 NaCl		24
Clostridium perfringens	Perfringolysin O (θ-toxin)	PFO	53	Cholesterol	2–30 nS[c,d]	<100	100 KCl		25
Streptococcus pneumoniae	Pneumolysin	PLY	53	Ergosterol	Cation Non-selective	≈ 35, ≈ 300 ≈ 12 nS[c,e]	100 KCl		26

(Continued overleaf.)

TABLE 14.1 *Continued*

Organism	Toxin	Acronym	Molecular weight[a] (kDa)	Lipid preference	Selectivity	Conductance (pS)	Buffer[b] (mM)	Radius (nm)	Reference
Bacillus thuringiensis	δ-Endotoxin	CryIAc	66–67		High cation	200–4000[c]	300 KCl; pH 9.7	1.2–1.3	28, 29
Bacillus thuringiensis	δ-Endotoxin	CryIIA	66–67		Non-selective	25	300 KCl; pH 9.7		30
Bacillus thuringiensis	δ-Endotoxin	CryIIIA	66–67		High cation	4000	300 KCl; pH 9.7		28
Bacillus thuringiensis	δ-Endotoxin	CryIC	66–67		Cation	100–200	150 KCl; pH 9.5		31, 32
Bacillus thuringiensis	δ-Endotoxin	CytA	23–25		High cation	40	300 KCl; pH 9.5		33
Other Gram-positive PFT									
Gardnerella vaginalis	Haemolysin	Gvh	59	Cholesterol + acidic lipids	Cation	125	150 KCl	2.3	34
Clostridium septicum	α-Toxin	AT	41			70	100 KCl	0.6–0.8	35, 36
Clostridium perfringens type A	Enterotoxin	CPE	35			40–450[c]	130 NaCl	37	
Small peptides									
Pseudomonas syringae	Syringomycin E	SRE	1.5	Various sterols	Anion	7, ~50[d]	100 NaCl	0.7–1.7	38
Pseudomonas syringae	Syringopeptin 25	SP25	2.9		Anion	7, 40[d]	100 NaCl	0.9	39
Bacillus subtilis	Iturin A		1	Cholesterol	Weak anion	10–900[c]	1 M KCl		40
Bacillus subtilis	Mycosubtilin		1	Cholesterol	Weak anion	10–300[c]	1 M KCl		41
Bacillus subtilis	Surfactin		1		Cation	130	1 M KCl		42
Staphylococcus aureus	δ-Toxin		2.6		Cation	70–100, 450[c,d]	500 KCl	0.4	43
AB toxins									
Vibrio cholerae	Cholera toxin	CT	86	Ganglioside GM₁	Anion	55	100 KCl; pH 4.5	1.0	44
Vibrio cholerae	Choleragen	CT-B	55	Ganglioside GM₁	Anion	125	100 KCl; pH 4.5	1.1	44
Bacillus anthracis	Anthrax toxin Protective antigen	PA	63		High cation	165	1 M KCl; pH 5.5	0.6	45, 46
Corynebacterium diphtheria	Diphtheria toxin	DT	62	Acidic phospholipids	Anion[f]/cation[g]	20	1 M NaCl; pH 4.9	>0.9	47, 48
Pseudomonas aeruginosa	Exotoxin A	ETA	62	Acidic phospholipids		30	100 KCl; pH 5.5		49
Clostridium tetani	Tetanus toxin	TeNT	150	Acidic phospholipids	Cation	30	0.5 M KCl; pH 4.8		48, 50
Clostridium botulinum	Botulinum toxin	BoNT	150		Cation	12	100 NaCl; pH 6.6		48, 51

Notes: [a] active form; if not otherwise specified; [b] neutral pH if not otherwise specified; [c] broad range; [d] two forms; [e] three forms; [f] below pH 4; [g] above pH 4. References: 1: Menestrina *et al.* (1987); 2: Benz *et al.* (1989); 3: Schmidt *et al.* (1996); 4: Benz *et al.* (1994b); 5: Maier *et al.* (1996); 6: Lear *et al.* (1995); 7: Menestrina *et al.* (1994); 8: Benz *et al.* (1994a); 9: Ehrmann *et al.* (1991); 10: Wilmsen *et al.* (1990); 11: Tschödrich-Rotter *et al.* (1996); 12: Chakraborty *et al.* (1990); 13: Ikigai *et al.* (1997); 14: Menzl *et al.* (1996); 15: Krasilnikov *et al.* (1992b); 16: Weiner *et al.* (1985); 17: Dreyer *et al.* (1990); 18: Kirby *et al.* (1998); 19: Schönherr *et al.* (1994); 20: Menestrina (1986); 21: Korchev *et al.* (1995b); 22: Ferreras *et al.* (1996); 23: Ferreras *et al.* (1998); 24: Blumenthal and Habig (1984); 25: Menestrina *et al.* (1990); 26: Korchev *et al.* (1998); 27: Grochulski *et al.* (1995); 28: Slatin *et al.* (1990); 29: Schwartz *et al.* (1993); 30: English *et al.* (1994); 31: Schwartz *et al.* (1993); 32: Lorence *et al.* (1995); 33: Knowles *et al.* (1989); 34: Moran *et al.* (1991); 35: Sellman and Tweten (1997); 36: Ballard *et al.* (1993); 37: Sugimoto *et al.* (1988); 38: Feigin *et al.* (1996); 39: Dalla Serra *et al.* (1999b); 40: Maget-Dana and Peypoux (1994); 41: Maget-Dana and Ptak (1990); 42: Sheppard *et al.* (1991); 43: Mellor *et al.* (1988); 44: Krasilnikov *et al.* (1991); 45: Blaustein *et al.* (1989); 46: Blaustein and Finkelstein (1990); 47: Donovan *et al.* (1981); 48: Hoch *et al.* (1985); 49: Gambale *et al.* (1992); 50: Gambale and Montal (1988); 51: Donovan and Middlebrook (1986).

Pore-forming protein toxin from Gram-negative bacteria

RTX toxins

RTX toxins (repeat in toxins) are a group of structurally and functionally related toxins produced by Gram-negative bacteria (Coote, 1992). They all carry several copies of a characteristic glycine and aspartate-rich nonapeptide involved in Ca^{2+}-dependent activity (Boehm *et al.*, 1990). The first was identified in haemolytic strains of *E. coli* infecting extra-intestinal sites, but many more were found in other Gram-negative species (for example, *Pasteurella*, *Actinobacillus*, *Bordetella*). Although most of them are haemolytic, RTX toxins are targeted to mammalian cells of the immune system and are more properly defined as leucotoxins. Some of them, for example the leucotoxins of *Actinobacillus actinomycetemcomitans* (AaltA) and *Pasteurella haemolytica* (LktA), are unable to lyse red blood cells (RBC). Although a specific receptor for AaltA was identified, RTX toxins are post-translationally acylated with a fatty acyl chain (Hardie *et al.*, 1991; Issartel *et al.*, 1991) and probably use lipids as low-affinity acceptors. They share several common functional regions. The repeat domain is folded in an unusual spring-like structure stabilized by Ca^{2+}, called a parallel β-barrel or β-superhelix (Baumann *et al.*, 1993). They induce haemolysis via a colloid osmotic shock following the formation of hydrophilic pores. The pore-forming ability of many RTX toxins has been demonstrated in model membranes, such as vesicles and PLM.

Escherichia coli haemolysin (HlyA) is perhaps the best studied RTX toxin. It binds to cell and model membranes in a Ca^{2+}-dependent way, the minimum requirement being that bacteria are grown on a Ca^{2+}-containing medium (Dobereiner *et al.*, 1996). Membrane binding involves both electrostatic and hydrophobic interactions (Ostolaza *et al.*, 1997). Membrane-bound and membrane-inserted forms adopt different conformations, although the secondary structure of the protein remains largely unchanged (Bakas *et al.*, 1998). The requirement of aggregation for the formation of pores has been proposed (Ostolaza *et al.*, 1993), but the size of the aggregate has not been established. In RBC, heterogeneous pore/membrane lesions, which increased in size over time, were also reported (Moayeri and Welch, 1994). The channels formed in PLM are cation selective with a pH-dependent conductance of around 300 pS in 0.1 M KCl at pH 7.0. This increases with approximately the square root of the salt concentration (Menestrina *et al.*, 1987; Benz *et al.*, 1989; Ropele and Menestrina, 1989), suggesting the presence of an electrostatic filter formed by negative charges. By this analysis, HlyA pore

radius was estimated to be 0.6 nm and the number of fixed charges at the entrance of the pore was approximately 1.3. Channel opening is strictly voltage dependent (Menestrina and Ropele, 1989). HlyA pores on human macrophages were observed by the patch-clamp technique (Menestrina *et al.*, 1996) and gave very similar results to those seen in PLM. The RTX haemolysins of *P. mirabilis*, *M. morganii* (Benz *et al.*, 1994b) and enterohaemorragic *E. coli* O157:H7 formed quite similar channels in PLM (Schmidt *et al.*, 1996).

Actinobacillus pleuropneumoniae produces three distinct RTX toxins, named ApxI, II, and III. The first two are haemolytic whereas ApxIII (also called pleurotoxin) is only leucotoxic. All three Apx toxins form channels in PLM (Maier *et al.*, 1996), although the activity of ApxIII and II is smaller than that of ApxI. The observed dose–response curve had a slope of two to three in a double logarithmic plot, suggesting an association–dissociation mechanism. Different conductances were observed, with the main level at 540, 620 and 95 pS for ApxI, II and III (in 0.15 M KCl). All pores were slightly cation selective and displayed some saturation compatible with an electrostatic filter effect, as seen with *E. coli* HlyA. Estimated pore radii were 1.2, 1.2 and 0.9 nm for ApxI, II and III, consistent with osmotic protection data (Lalonde *et al.*, 1989). Fixed charges were approximately 2.3 for ApxI and II and slightly less for ApxIII. *Actinobacillus actinomycetemcomitans* AaltA, despite having a specific protein receptor, β2 integrin, on target cells (Lally *et al.*, 1997), can also form channels in synthetic PE/PS planar bilayers (Lear *et al.*, 1995). The pores have multiple conductance levels with an upper value of around 400 pS in physiological salt. As for *E. coli* HlyA, they are voltage gated.

The 177-kDa adenylate cyclase-haemolysin, CyaA, produced by *Bordetella pertussis* (Hanski, 1989) and *B. bronchiseptica* (Gueirard and Guiso, 1993), is a bifunctional protein comprising an enzymic adenylate cyclase domain at the N-terminus and an RTX haemolysin at the C-terminus, both important for virulence. Apparently, the enzyme bypasses the receptor-mediated entry pathway and the delivery of the catalytic domain across the cytoplasmic membrane is mediated by the functional RTX domain (Hewlett *et al.*, 1993). In the cytoplasm, the adenylate cyclase elevates cAMP levels resulting in altered cellular functions including activation of Ca^{2+} channels, which was evidenced by whole cell recording. Ca^{2+} is required for the delivery of the catalytic domain, for the haemolytic activity and for improving the formation of trimeric cation-selective (Ca^{2+} permeable) channels in PLM (Benz *et al.*, 1994a; Szabo *et al.*, 1994). The conductance of the pore was small, 27 pS in 1 M KCl, consistent with the fact that its radius, measured in an osmotic protection assay on RBC, was about 0.3 nm (Ehrmann *et al.*, 1991). Selectivity and saturation were, also in this case, compatible with the existence of an electrostatic filter of negative charges. This is a unique example of a toxin which has conjugated in the same molecule the main features of the two groups of toxins discussed here: A–B and membrane-damaging.

Pseudomonas aeruginosa *cytotoxin*

Pseudomonas aeruginosa secretes a 33-kDa protoxin which is proteolytically processed at the *C*-terminus to produce two active forms (26.5 and 28.5 kDa) of a cytotoxin (CTX). CTX, injected in rats, causes extensive damage to kidney tubules; *in vitro*, it is active against many eukaryotic cells, especially white cells, for example polymophonuclear leucocytes (PMN) and macrophages. Cell cytotoxicity requires binding to a receptor, insertion, oligomerization and formation of pores with a functional radius of about 1 nm (Suttorp *et al.*, 1985; Lutz *et al.*, 1987). In rabbit RBC the receptor is an oligomer of a 28-kDa membrane protein recognized as CHIP28 (Lutz *et al.*, 1993), a member of the ubiquitous water-channel family. The receptor does not participate in forming the pore, rather the toxin alone makes a pentameric oligomer both on RBC membranes and on cholesterol-containing vesicles (Ohnishi *et al.*, 1994). CTX permeabilizes phosphatidylcholine/cholesterol PLM (Weiner *et al.*, 1985), but no defined conductance could be measured. Single-channel events were detected instead by the patch-clamp technique in excised outside-out patches of bovine medullary chromaffin cells (Dreyer *et al.*, 1990). These pores were non-selective with a conductance of 120 pS in 0.14 M K-gluconate, and voltage-gated, as confirmed by whole-cell experiments.

Serratia *Ca^{2+}-independent haemolysins*

All strains of *Serratia marcescens* produce a cell-associated 165 kDa Ca^{2+}-independent haemolysin, ShlA, which is activated and released with the aid of the product of a second gene, ShlB (Ondraczek *et al.*, 1992). ShlA was thought to form small pores with a cut-off of around 1000 Da in the RBC membrane (Schiebel and Braun, 1989), but lately the pores were shown in osmotic protection experiments to vary in size with temperature; the radius was 0.5–0.8 nm at 0°C but 1.2–1.5 nm at 20°C. In PLM, weakly anion-selective channels were observed with a broad distribution of conductances, confirming a variable size of the pore (Schönherr *et al.*, 1994). The average value in 0.1 M KCl was 150 pS, suggesting a radius of 0.5 nm. Apparently, ShlA monomers can form transmembrane pores, but can also associate into multimers, for example tetramers and larger oligomers, thus increasing pore radius.

Aeromonas hydrophila *aerolysin*

Aerolysin appears to be the main virulence factor of infections caused by *Aeromonas hydrophila* and *A. sobria* (Parker *et al.*, 1996). The secreted protoxin dimer is proteolytically activated and binds to a glycosylphosphatidylinositol (GPI)-anchored glycoprotein receptor on target cells (Abrami *et al.*, 1998). Aerolysin dimer consists mainly of β-structure and reorganizes into a mushroom-shaped heptameric channel in membranes (Wilmsen *et al.*, 1992; Parker *et al.*, 1994). On insertion this forms a voltage-dependent, slightly anion-selective pore in PLM (Wilmsen *et al.*, 1990). The channels have uniform size, 145 pS in 0.3 M NaCl, and rapidly close at voltages larger than 70 mV. Zn^{2+} induces their reversible block from the *cis* and the *trans* side because two to three Zn^{2+} ions can bind in the pore lumen. Analysing the transport of ions through the aerolysin pore in erythrocyte membranes by scanning microphotolysis (Tschödrich-Rotter *et al.*, 1996), the transport rates for two anions (lucifer yellow, $1.8 \pm 0.4 \times 10^{-3}$ s^{-1} and carboxyfluorescein, $0.3 \pm 0.1 \times 10^{-3}$ s^{-1}) and for Ca^{2+} ($8.2 \pm 2.3 \times 10^{-3}$ s^{-1}) were determined. Taking steric hindrance and viscous drag into account, the radius of the channel was estimated as 1.9–2.3 nm. This was consistent with marker-release experiments from resealed erythrocyte ghosts (Howard and Buckley, 1982) indicating a pore cut-off for molecules of 3500 Da (radius 1.8 nm). Similar pores were observed with an aerolysin from *A. sobria* (Chakraborty *et al.*, 1990).

Vibrio *haemolysins*

Vibrio cholerae O1 El Tor thermolabile haemolysin (HlyA) is produced as a preprotoxin of 82 kDa (Yamamoto *et al.*, 1990). The signal sequence is cleaved cotranslationally and the inactive protoxin of 79 kDa is further processed to a 60–65-kDa active haemolysin by a bacterial protease. The monomer has to oligomerize to become active. In phosphatidyl choline/cholesterol PLM the oligomer formed anion-selective, voltage-dependent channels with a long lifetime. The single channel conductance was 22 pS in 0.1 M KCl and 350 pS in 1 M KCl (Menzl *et al.*, 1996; Ikigai *et al.*, 1997). The radius of the HlyA pores in erythrocyte membranes was sized to 0.6–0.8 nm in osmotic protection assays (Zitzer *et al.*, 1995; Ikigai *et al.*, 1996). The oligomer is sodium dodecyl sulfate (SDS) stable and appears as a 200–220-kDa band, although several higher molecular mass bands were also observed, suggesting oligomers of varying numbers of subunits. Cytolysins with similar properties are produced by *V. cholerae* non-O1 (Krasilnikov *et al.*, 1992b; Zitzer *et al.*, 1997) and *V. vulnificus* (Yamamoto *et al.*, 1990).

Legiolysin from **Legionella pneumophila**

Invasive Gram-negative bacteria usually are non-haemolytic, or very poorly haemolytic. However, evidence for the production of a weak haemolysin by *L. pneumophila* suggested that they might be involved in the contact lysis of the host membrane enclosing the intracellular bacterium and have a role in invasiveness. On contact *L. pneumophila* haemolysin forms pores of 1.5 nm radius in cell membranes, permeable to large anions and cations (Kirby *et al.*, 1998).

Pore-forming protein toxins from Gram-positive bacteria

Staphylococcus aureus α-toxin

Staphylococcus aureus α-toxin is probably the best understood PFT. It is secreted by most pathogenic strains as a water-soluble polypeptide of 33 kDa. It binds specifically and non-specifically to various cells (Bhakdi and Tranum-Jensen, 1991), oligomerizes and forms transmembrane channels which appear in electron microscopy as hollow cylinders protruding from the membrane. The three-dimensional structure of oligomers formed in deoxycholate (DOC) revealed a mushroom-shaped, hollow heptamer composed of a β-structure (Song *et al.*, 1996). In artificial lipid layers hexamers and incomplete ring structures were also observed (Czajkowsky *et al.*, 1998). In PLM of purified phospholipids the toxin forms voltage-dependent, anion-selective channels of uniform conductance, around 90 pS in 0.1 M KCl, with a nearly linear relationship with salt concentration (Menestrina, 1986). This allowed, using Equation (14.7), an estimation of the radius of the pore of 0.6 nm. At variance with this, the method of the lowest impermeable sugar (Krasilnikov *et al.*, 1992a) gave a radius of 1.4 nm. As explained before, the reason for this discrepancy lies in the fact that the lumen of the pore is not regularly shaped: the transmembrane part has a radius of 0.7 nm, the central part of 0.5 nm and the entrance of 1.5 nm (Song *et al.*, 1996). Hence, the conductance method finds an average value, whereas the sugar method sizes the entrance. In addition, the linear polymeric non-electrolytes (PEGs) used have a favourable interaction within the pore, which accepts molecules apparently larger than its size (probably unfolded) because its interior is partially lined by hydrophobic residues (Bezrukov *et al.*, 1996). Protons and divalent cations, particularly Zn^{2+}, can close preformed channels. This gating is not due to a conformational change squeezing the pore lumen, since this change is less than twofold (Korchev *et al.*, 1995a). Rather, it results from a direct binding of these ions to titratable protein groups (Menestrina, 1986; Korchev *et al.*, 1997).

Channel protonation increases the anion selectivity and the conductance of the pore, whereas neutralization of lysine residues reduces both, suggesting the presence of an electrostatic filtering effect. The number of ionizable residues in the channel interior was determined to be four (Bezrukov and Kasianowicz, 1993), whereas that of the lysine residues at its entrance was three (Cescatti *et al.*, 1991). Whole cell recordings of Lettrè-cells exposed to α-toxin indicated the formation of single channels of comparable properties (Korchev *et al.*, 1995b). In a Langmuir trough the surface pressure increase induced by α-toxin was 0.1 mN m^{-1} with a dimyristoyl phosphatidyl choline (DMPC) monolayer and 2.7 mN m^{-1} with a dimyristoyl phosphatidyl serine (DMPS), one showing a higher insertion rate in the presence of negatively charged lipids (Ellis *et al.*, 1997) which was seen also in vesicles (Vécsey-Semjén *et al.*, 1996).

Staphylococcus aureus γ-haemolysins and leucotoxins

These bi-component toxins are targeted towards cells of the immune system including monocytes, macrophages and PMNs (Prévost *et al.*, 1995). They include proteins belonging to either an S or F subtype (32 and 34 kDa, respectively). Combination of an S and F component acts synergically to form a membrane-damaging toxin. The S component can bind to a specific receptor linked to Ca^{2+} channels on the surface of PMNs (Colin *et al.*, 1994). Binding triggers the specific flow of divalent cations, and is followed by the formation of a non-selective pore which allows the flux of monovalent ions and ethidium bromide. The non-specific pores can be inhibited by Zn^{2+} while the specific pores remain open (Staali *et al.*, 1998). The γ-haemolysins (HlgA, HlgB and HlgC) are active also against RBC and model membranes like vesicles (Ferreras *et al.*, 1998) and PLM (Ferreras *et al.*, 1996). The ion channels in PLM are slightly cation selective and have a conductance of 110 and 180 pS in 0.1 M NaCl for HlgA + HlgB and HlgC + HlgB, respectively. Formation of hetero-oligomers (hexamers and higher molecular mass oligomers), followed by the opening of pores, seems to be the mechanism underlying membrane damage, as implicated also by the appearance of ring-shaped complexes on RBC membranes seen by electron microscopy (Sugawara *et al.*, 1997). The radius of the effective pores was estimated to be 1.0–1.2 nm using PEGs as osmotic protectants (Ferreras *et al.*, 1998). The initial velocity of calcein release caused by HlgA + HlgB was 10 times higher than that caused by HlgC + HlgB, owing to a faster oligomerization of the first couple on pure lipid membranes. These toxins share sequence homology, as well as functional similarity, with α-toxin, suggesting that they form a unique protein family.

Gardnerella vaginalis *haemolysin*

Gardnerella vaginalis, the main aetiological agent of 'non-specific' vaginosis, produces a haemolysin (Gvh) which binds to membranes in a cholesterol-dependent way (Kretzschmar *et al.*, 1991; Cauci *et al.*, 1993a). After binding the toxin forms large pores (approximate radius 2.3 nm) in target membranes, such as human erythrocytes, neutrophils and endothelial cells. Its activity is enhanced by negatively charged phospholipids in the target lipid bilayer or by acidification of the medium and is decreased by divalent cations and low ionic strength, despite normal binding (Cauci *et al.*, 1993b). Pore formation occurs by aggregation of monomers into high molecular weight oligomers, appearing on Western blot as a trimeric 200-kDa single band (Shubair *et al.*, 1993). In PLM it forms cation-selective channels with conductance of 130 pS in 0.15 M KCl and a higher open probability at positive voltages (Moran *et al.*, 1991). Patch-clamp of giant proteoliposomes gave similar results.

Bacillus thuringiensis *crystal endotoxins*

Bacillus thuringiensis synthesizes insecticidal proteins, called δ-endotoxins, which crystallize and form parasporal inclusions (Gill *et al.*, 1992). They are divided into two families, crystal-forming (Cry) and cytolytic (Cyt), different in size and cell recognition. Cry δ-endotoxins are the largest group, with 13 subclasses. Produced as protoxins, they are activated by insect gut proteases to a 66–67-kDa PFT that binds to specific mid-gut epithelial receptors expressed in insects and nematodes and hence are of primary interest as biopesticides (Cannon, 1996). The 120-kDa aminopeptidase N (APN) was recognized as receptor for CryIAc. After receptor binding they form leaky channels in gut epithelial membranes, killing the insect by starvation or septicaemia. The X-ray crystallographic structure of CryIIIA and CryIAa has revealed the same three domain structure, likely to be common to all Cry endotoxins (Li *et al.*, 1991; Grochulski *et al.*, 1995). The effects of Cry toxins on model membranes prepared from sensitive insect tissue, i.e. brush border membrane vesicles (BBMV), were investigated by osmotic swelling experiments (Carroll and Ellar, 1993). Addition of CryIA led to an increased membrane permeability for ions and neutral solutes, while addition of CryIB caused no detectable changes. CryIAc also formed pores in mid-gut BBMV at alkaline pH. Using non-electrolyte osmotic protectants, the radius of the pore was estimated to 1.2–1.3 nm (Carroll and Ellar, 1997). While Cry toxins form channels in lipid bilayers even in the absence of specific receptors (Slatin *et al.*, 1990; Schwartz *et al.*, 1993), reconstitution of the APN receptor into the PLM increased enormously the pace of channel formation (Lorence *et al.*, 1997;

Schwartz *et al.*, 1997). The channels may have different properties depending on the particular toxin, the method of activation and the presence of the receptor. In general, they are cation selective and pH dependent (an alkaline pH as in the insect gut is required), and exhibit several conductance states with a broad range. For example, the conductance of CryIAa channels in 0.15 M KCl at pH 9.0 was 450 pS with multiple substrates (Grochulineski *et al.*, 1995). CryIC formed channels which were cation selective with 100–200 pS conductance at pH 9.5 (but anion selective with 25–35 pS conductance at pH 6.0) in pure lipid PLM (Schwartz *et al.*, 1993), whereas in PLM fused with BBMV at pH 9.0 three conductance levels were observed of 50, 106 and 360 pS (Lorence *et al.*, 1995). In contrast, CryIIA endotoxin formed small, voltage-dependent channels with 25 pS conductance and no ion selectivity (English *et al.*, 1994). Voltage-clamp experiments on isolated insect mid-gut showed the inhibition of short-circuit current by CryIAa (and to a much lower extent by CryIAc), implying the block of pre-existing K^+ conductance pathways (Liebig *et al.*, 1995).

Cyt δ-endotoxins are smaller, one-domain proteins, with an entirely different fold (Li *et al.*, 1996). The 28–29-kDa protoxin is processed, resulting in a 23–25-kDa active toxin specific for dipteran larvae. It probably binds to phospholipids with unsaturated fatty acyl chain in the *syn-2* position and spontaneously inserts into the bilayer. Although the transmembrane pore is probably a β-sheet formed by toxin multimers, two helices (A and C) are major structural elements expected to be involved in membrane interactions (Gazit *et al.*, 1997). CytA forms highly cation-selective channels in phosphatidyl ethanolamine (PE) planar lipid bilayers at pH 9.5. The single channel conductance is 40 pS in 300 mM KCl, and 5 mM Ca^{2+} greatly reduced the rate and number of channel openings (Knowles *et al.*, 1989).

Cholesterol-binding toxins

Perhaps the most numerous family of Gram-positive PFT, with more than 20 members, is the cholesterol-binding (thiol-activated) cytolysins (Alouf and Geoffroy, 1991). The three-dimensional X-ray structure of one of them, perfringolysin O (PFO), in the water-soluble, monomeric form was recently solved (Rossjohn *et al.*, 1997). Electron microscopy showed that cholesterol-binding toxins (CBT) in contact with cell membranes containing cholesterol undergo a conformational change which initiates the formation of membrane-bound oligomers which may appear either as ring-shaped structures or as protein arcs faced by a free edge of lipid. The large rings contain 50–80 subunits and form pores of variable sizes, letting through even haemoglobin (molecular radius 3.5 nm). The arc-shaped oligomers are also able to form functional trans-membrane pores, as shown by a mutant of streptolysin O (SLO) defective in oligomerization. This mutant was still able to form hybrid oligomers with native SLO, albeit at a lower rate. The unfinished pores were permeable to salt and calcein, but in contrast to pores formed by native SLO, were not permeable to large dextrans (Palmer *et al.*, 1998). CBT readily inter-act with model membranes provided they contain enough cholesterol. The role of the cholesterol 3β-OH group in membrane insertion was demonstrated with Langmuir films (Alouf *et al.*, 1984). The formation of CBT pores in lipid vesicles has been studied by a num-ber of authors and techniques. Pore formation by tetanolysin in *Mycoplasma gallisepticum* was shown by the osmotic swelling technique (Rottem *et al.*, 1990). In contrast, only a few reports of activity on PLM exist. PFO (Menestrina *et al.*, 1990) and pneumolysin (Korchev *et al.*, 1992) were reported to induce channels with a wide spectrum of conductance values, ranging from very small (30–50 pS) to very large (12–25 nS). Apparently the largest pores, albeit few, were quite sta-ble. It is possible that they correspond to the ring lesions, whereas intermediate values could derive from the arc structures. Some divalent cations, in particular Zn^{2+}, can close pre-formed pores and protect cells (Menestrina *et al.*, 1990). Listeriolysin O and legiolysin perforate human macrophage vacuoles at acidic pH (Beauregard *et al.*, 1997), thus possibly contributing to bacterial invasiveness.

Clostridium perfringens *type A enterotoxin*

Clostridium perfringens type A is the most common cause of food poisoning (Kokai-Kun and McClane, 1997), causing a diarrhoeal-type disease, and has been implicated also in non-foodborne diarrhoea and SID (sudden infant death). The 35-kDa enterotoxin [*Clostridium perfringens* enterotoxin (CPE)] binds to epithelial cells of the small intestine via a hydrophobic, 22-kDa receptor (which shows similarities to the rat androgen withdrawal apoptosis protein), it inserts into their membranes and finally forms a large hetero-com-plex (Katahira *et al.*, 1997). The complex increases the permeability of the cell membrane, forming a hole for ions and small molecules (up to 200 Da) which, in the absence of osmotic protectants, evolves to a larger one (letting through molecules up to 3 kDa). If inserted in the lipid bilayer by sonication, the toxin forms discrete ion channels in asolectin PLM with multiple conduc-tance levels ranging from 40 to 450 pS (Sugimoto *et al.*, 1988), which can explain at least in part its physiolog-ical effects on cells.

Clostridium septicum α-toxin

Similarly to aerolysin, *Cl. septicum* α-toxin (AT) is secreted as an inactive protoxin which requires prote-olytic activation by furin after membrane binding (Gordon *et al.*, 1997). After removal of a 4-kDa pro-peptide at the *C*-terminus the 41-kDa active toxin oligomerizes and forms a pre-pore complex of >200 kDa. At low temperatures AT^{oligo} is trapped in the pre-pore complex. When the temperature is raised to over 15°C the inactive AT^{oligo} inserts into the membrane and forms a transmembrane channel (Sellman and Tweten, 1997). In azolectin bilayers volt-age-dependent channels with conductance of 70 pS in 0.1 M KCl and an effective radius of at least 0.6–0.8 nm were observed (Ballard *et al.*, 1993).

Pore-forming bacterial peptides

Pseudomonads cyclic *lipodepsipeptides*

Pseudomonas syringae is a phytopathogenic bacterium with antifungal activity. Its virulence is related to the production of a group of toxins composed of a long unbranched 3-hydroxy fatty acid chain linked to a cyclic cationic peptide, which is either a nonapeptide (syringomycin, syringostatin and syringotchoxin) or a longer peptide of 22 or 25 residues (syringopeptins). All these cyclic lipopeptides (CLP) induce RBC haemolysis (Dalla Serra *et al.*, 1999a) via a colloid-osmotic shock caused by the formation of pores. The functional radius of these pores was estimated to be about 0.9 nm for syringopeptins, but was variable, ranging from 0.7 to 1.7 nm in a dose-dependent way for syringomycin E (SRE). They partition into preformed lipid monolayers (with a preference for films containing sterols) occupying molecular areas from 0.6 to 1.7 nm^2, depending on the peptide and the film. They form oligomeric pores comprising five to six monomers in lipid vesicles. In PLM syringopeptin opens an anion-selective channel (Dalla Serra *et al.*, 1999b), with a conductance of 40 pS in 0.1 M NaCl. Channel opening is strongly voltage dependent, requiring a negative potential on the same side as the toxin. An additional pore state of around 7 pS conductance and 30 ms lifetime is present. SRE makes similar channels (Feigin *et al.*, 1996).

Bacillus subtilis *cyclolipopeptides*

Gram-positive *Bacillus* species produce similar lipo-peptides, the difference being that the fatty acid chain is normally branched. The cyclic peptide always con-tains seven residues and the whole molecule is either neutral or anionic (Maget-Dana and Peypoux, 1994). Examples are iturins, mycosubtilins, bacillomycins and surfactins. They are haemolytic, surface active (occupying molecular areas of 0.8–1.7 nm^2), sterol seeking and able to form ion-conducting oligomeric pores in PLM. However, such pores are quite different from those of pseudomonads CLP; their opening is voltage independent, and the conductances are larger and can increase up to 50 times with time and CLP dose. Pores formed by iturin A and mycosubtilin, two uncharged molecules, are poorly anion selective (Maget-Dana and Ptak, 1990), whereas those formed by surfactin, which bears two negative charges, are cation selective (Sheppard et al., 1991), confirming the validity of the electrostatic attraction hypothesis.

Staphylococcus aureus δ-lysin

Staphylococcus aureus δ-lysin is a 26-residue polypeptide that lyses different types of cells and artificial membranes and has some structural analogy to melittin (Lee et al., 1987; Thiaudière et al., 1991). Two variants have been identified that differ by as much as nine amino acid residues but preserve the amphipathic profile. The monomeric peptide is a poorly structured molecule in aqueous environment, but an α-helix in organic solutions. It self-aggregates extensively, the minimum oligomer size being a tetramer. On lipid binding it adopts an extended helical form. By lateral diffusion six to eight monomers form a discrete membrane channel which in PLM is weakly cation selective (Mellor et al., 1988). Two different types, small and large, have been observed with conductances of 70–100 pS and 450 pS, respectively.

Toxins with intracellular targets (A–B type)

These toxins have two domains, A and B, with different physiological roles. The domains are either different components, different subunits of the same protein or separable parts of a unique polypeptide chain. They usually bind to specific cell surface receptors, via the B part. They are then cleaved and internalized by receptor-mediated endocytosis. Along the endocytic pathway, as the luminal pH decreases, they undergo partial unfolding that triggers membrane insertion and facilitates translocation of the unfolded A component. After separation from B, the A fragment refolds into the cytosol and exerts its catalytic activity. Almost all of these toxins have been shown to open ion channels into cells and model membranes via the B part, although it is unknown how this property may relate to the translocation of A.

Toxins increasing intracellular cAMP

Cholera toxin (CT), and the closely related heat-labile enterotoxins of *Escherichia coli*, *Vibrio mimicus* and *Aeromonas hydrophila*, consist of one A1 subunit linked via an A2 subunit to five B subunits (AB$_5$ topology). The B pentamer binds to ganglioside GM$_1$ on the cell membrane and is internalized through non-clathrin-coated vesicles. CT-A contains the C-terminal KDEL sequence coding for retrograde transport to the endoplasmic reticulum (ER) where the disulfide bond between A1–A2 is reduced and A1 exerts its catalytic activity on the cytosolic G$_s$ proteins (Majoul et al., 1996). CT and the CT-B pentamer are able to permeabilize GM$_1$-containing liposomes (Moss et al., 1976) and to open pores in PLM (Krasilnikov et al., 1991). CT channels appear only at acidic pH (pH 4.5), they are anion selective and have a conductance of 55 pS in 0.1 M KCl. Channels formed by CT-B alone appear under similar conditions but are slightly larger, 125 pS conductance and 1.1 nm radius. It was suggested that translocation of the A1 subunit could occur through this pore (tunnel model), but the three-dimensional structure would exclude it and channel formation might only be a side-effect.

Anthrax toxin, secreted by *Bacillus anthracis*, is a three-component toxin which acts in binary combinations to generate two different toxic responses. Protective antigen (PA, 83 kDa) binds to high-affinity cell-surface receptors and a 20-kDa N-terminal fragment is then cleaved by cell-surface proteases. Its removal from the remaining PA$_{63}$ uncovers a site that can bind either oedema factor (EF) or lethal factor (LF). The receptor-bound PA-LF or PA-EF complex is internalized and in the endosomal compartment PA$_{63}$ undergoes oligomerization and facilitates the translocation of LF and/or EF (Singh et al., 1989). Once in the cytoplasm, EF increases intracellular cAMP levels, whereas LF cleaves and inactivates the mitogen-activated protein kinase kinase (MAPKK1 + 2) leading to disruption of this signal transduction pathway (Duesbery et al., 1998). At acidic pH, PA$_{63}$, but not PA, forms heptamers, inserts irreversibly in the plasma membrane of mammalian cells and forms well-defined ion channels which release monovalent cations (Milne and Collier, 1993). Under similar conditions K$^+$ is released from asolectin vesicles while calcein (an anion of 0.6 nm radius) is retained (Koehler and Collier, 1991), suggesting that the channels are either too small or too selective for its release. In fact, in PLM experiments the PA$_{63}$ channel is strongly voltage dependent and almost ideally selective for univalent cations. It has a conductance of 165 pS in 1 M KCl and an estimated radius >0.5 nm (Blaustein et al., 1989). The fact that both LF and EF are able to block these channels (Finkelinestein, 1994; Zhao et al., 1995) lends credence to the model where EF and LF pass through PA$_{63}$ pores to reach the cytoplasm.

Toxins suppressing protein synthesis

Diphtheria toxin (DT) is secreted by *Corynebacterium diphtheria* as a monomeric protein that binds to the heparin-binding EGF-like growth factor precursor on the cell surface and is cleaved by furin into two disulfide linked fragments, A and B. After reduction of the disulfide bond, the A fragment is internalized and ADP ribosylates elongation factor 2 (EF-2), thereby inhibiting protein synthesis. The B fragment consists of two domains: a receptor-binding domain and a membrane translocation domain (T) composed of 10 α-helices, two of which (TH8 and TH9) might insert into the membrane in an acidic environment. DT inserts into lipid monolayers occupying an area of 21 nm^2, which approximates the cross-section of the whole toxin (Demel *et al.*, 1991). It permeabilizes vesicles by forming pores of 1.0 nm radius (Kagan *et al.*, 1981; Shiver and Donovan, 1987) and opens an ion channel in PLM of 20 pS in 1 M NaCl (Donovan *et al.*, 1981). Membrane insertion and channel formation depend on the presence of acidic pH, negatively charged phospholipids and a *trans*-negative membrane potential. DT forms 30 pS cation channels in the membrane of Vero cells, in association with the translocation of the A fragment, as shown by patch-clamp (Eriksen *et al.*, 1994). The T domain, a 13 kDa cyanogen-bromide fragment harbouring the middle region of B fragment (Deleers *et al.*, 1983), and even a 61 amino acid stretch containing only the two hydrophobic helices, TH8 and TH9 (Huynh *et al.*, 1997), form ion channels in PLM with properties similar to the full-length toxin. Mutants of the TH8–9 fragment, scanned by single Cys insertions, formed channels showing only one transition after reaction with sulfydryl-specific reagents, suggesting that they were monomeric. Moreover, the pattern of accessibility was incompatible with either α-helical or β-structure, implying a flexible organization of this loop in the membrane. The idea that the channel formed by the TH8–9 fragment is used for translocation of the A fragment is further complicated by the fact that the A domain alone forms a pore of 1.5 nm radius in asolectin vesicles at pH 4.0, which might mediate self-translocation (Jiang *et al.*, 1989).

Exotoxin A from *P. aeruginosa* (ETA) is secreted as a single-chain, three-domain protein (Allured *et al.*, 1986). Domain Ia binds to the α_2-macroglobulin receptor. Domain II, composed of six α-helices (A to F), is the translocation part and domain III the catalytic one with a REDLK sequence at the *C*-terminal targeting the toxin to the ER. Eventually, domain II is cleaved between helix A and B and its *C*-terminal part is translocated to the cytosol together with domain III, which ADP ribosylates EF-2. In lipid vesicles ETA induces both aggregation and permeabilization (Zalman and Wisnieski,

1985; Menestrina *et al.*, 1991). Maximal effect requires low pH and the presence of negatively charged lipids. The role of the negative surface potential was demonstrated both on vesicles (Rasper and Merrill, 1994) and on lipid monolayers (Nordera *et al.*, 1997). At pH 5.5 ETA forms strongly voltage-dependent transmembrane channels in PLM with 50–60 pS conductance (Gambale *et al.*, 1992). The channels are open at negative transmembrane potentials. ETA occupies an average area of 4.6 nm^2 in lipid monolayers, much smaller than DT. This suggests that only part of the molecule is inserted, most probably the two amphiphilic helices A and B of domain II, as indicated by the fact that lipid binding strongly facilitates proteolytic cleavage at the Arg loop between them (Nordera and Menestrina, 1998).

Toxins inhibiting neurotransmitter release

Clostridium tetani tetanus toxin (TeNT) is released as a 150-kDa polypeptide chain which is enzymatically processed into two fragments, A and B (light and heavy chain), linked by a disulfide bridge. The *C*-terminal part of the heavy chain is responsible for receptor binding, whereas the *N*-terminal part may allow cell entry of the light chain (Johnstone *et al.*, 1990). The light chain is a Zn^{2+} endopeptidase, which hydrolyses VAMP/synaptobrevin II (Schiavo *et al.*, 1992), thus blocking exocytosis. TeNT penetrates lipid monolayers (Schiavo *et al.*, 1991) and vesicles (Montecucco *et al.*, 1988; Menestrina *et al.*, 1989) and the interaction is dependent, as for DT and ETA, on acidic pH, *trans*-negative voltage and the presence of acidic phospholipids (for example, phosphatidyl inositol or phosphatidyl serine), whereas gangliosides had only a minor effect. Under similar conditions TeNT forms voltage-dependent, cation-selective channels in PLM (Borochov-Neori *et al.*, 1984; Hoch *et al.*, 1985). Channel-forming activity is expressed also by the heavy chain alone and in particular by its *N*-terminal part. These channels flicker continuously, generating bursts of openings (lifetime $\approx$ 1 ms) separated by long closing periods (Gambale and Montal, 1988; Rauch *et al.*, 1990). Acidic pH promotes channel opening, but lowers the single channel conductance, 30 pS in 0.5 M KCl. Acidic phospholipids instead increase both. Pores were detected on outside-out patches of spinal cord neuron membrane at pH 5.0 but not at pH 7.4 (Beise *et al.*, 1994). The channels showed voltage dependence and 38 pS conductance. The heavy chain alone formed pores at both pH 5 and 7 with increasing conductance at neutral pH.

Clostridium botulinum neurotoxins (BoNT) include seven different serotypes. BoNTs, like TeNT, are 150-kDa polypeptides proteolysed into two fragments, the

50-kDa L-chain and the 100-kDa H-chain, linked by a disulfide bond. Intoxication involves membrane attachment of the C-terminal half of the H-chain, endocytosis and membrane translocation of the L-chain. BoNT acts on the presynaptic terminal of peripheral motor neurons. Types B, D, F and G cleave VAMP/synaptobrevin (as TeNT), type C cleaves syntaxin, and types A and E cleave SNAP-25, thus blocking acetylcholine release. BoNT type A, B and E interact with lipid vesicles with the same requirements as TeNT (Montecucco *et al.*, 1989). Both the H- and L-chains are involved in the interaction, but the permeabilization ability is limited to the N-terminal region of the H-chain (Shone *et al.*, 1987). Full-length BoNT type A, B and C form channels in PLM (Hoch *et al.*, 1985; Donovan and Middlebrook, 1986; Blaustein *et al.*, 1987) stimulated by low pH and negative *trans* voltages, similar to TeNT. The channels are cation selective and have a small conductance, typically 12 pS in 0.1 M NaCl. Cation channels with bursting kinetics and conductance of 27 pS in 0.2 M KCl were also observed in excised outside-out patches from BoNT-treated PC12 cells (Sheridan, 1998). Larger channels with conductance five or more times the primitive value were also observed, suggesting that the major problem of the tunnel model, i.e. that the pore is too small to let the A fragment through, may be circumvented by aggregation of several units.

REFERENCES

Abrami, L., Fivaz, M., Glauser, P.E., Parton, R.G. and vanderGoot, F.G. (1998). A pore-forming toxin interacts with a GPI-anchored protein and causes vacuolation of the endoplasmic reticulum. *J. Cell Biol.* **140**, 525–40.

Abrams, F.S. and London, E. (1992). Calibration of the parallax fluorescence quenching method for determination of membrane penetration depth: refinement and comparison of quenching by spin-labelled and brominated lipids. *Biochemistry* **31**, 5312–22.

Allured, V.S., Collier, R.J., Carroll, S.F. and McKay, D.B. (1986). Structure of exotoxin A of *Pseudomonas aeruginosa* at 3.0-Ångstrom resolution. *Proc. Natl Acad. Sci. USA* **83**, 1320–4.

Alouf, J.E. and Geoffroy, C. (1991). The family of the antigenically-related, cholesterol-binding ('sulphydryl-activated') cytolytic toxins. In: *Sourcebook of Bacterial Protein Toxins* (eds J.E. Alouf and J.H. Freer), pp. 147–86. Academic Press, London.

Alouf, J.E., Geoffroy, C., Pattus, F. and Verger, R. (1984). Surface properties of bacterial sulphydryl-activated cytolytic toxins. *Eur. J. Biochem.* **141**, 205–10.

Bakas, L., Veiga, M.P., Soloaga, A., Ostolaza, H. and Goni, F.M. (1998). Calcium-dependent conformation of E. coli alpha-haemolysin. Implications for the mechanism of membrane insertion and lysis. *Biochim. Biophys. Acta* **1368**, 225–34.

Ballard, J., Sokolov, Y., Yuan, W.L., Kagan, B.L. and Tweten, R.K. (1993). Activation and mechanism of *Clostridium septicum* alpha toxin. *Mol. Microbiol.* **10**, 627–34.

Baumann, U., Wu, S., Flaherty, K.M. and McKay, D.B. (1993). Three-dimensional structure of the alkaline protease of *Pseudomonas aeruginosa*: a two-domain protein with a calcium-binding beta-roll motif. *EMBO J.* **12**, 3357–64.

Beauregard, K.E., Lee, K.D., Collier, R.J. and Swanson, J.A. (1997). pH-dependent perforation of macrophage phagosomes by listeriolysin O from *Listeria monocytogenes*. *J. Exp. Med.* **186**, 1159–63.

Beise, J., Hahnen, J., Andersen-Beckh, B. and Dreyer, F. (1994). Pore formation by tetanus toxin, its chain and fragments in neuronal membranes and evaluation of the underlying motifs in the structure of the toxin molecule. *Naunyn-Schmiedeberg's Arch. Pharmacol.* **349**, 66–73.

Benz, R., Schmid, A., Wagner, W. and Goebel, W. (1989). Pore formation by the *Escherichia coli* haemolysin: evidence for an association–dissociation equilibrium of the pore-forming aggregates. *Infect. Immun.* **57**, 887–95.

Benz, R., Maier, E., Ladant, D., Ullmann, A. and Sebo, P. (1994a). Adenylate cyclase toxin (CyaA) of *Bordetella pertussis*. Evidence for the formation of small ion-permeable channels and comparison with HlyA of *Escherichia coli*. *J. Biol. Chem.* **269**, 27231–9.

Benz, R., Hardie, K.R. and Hughes, C. (1994b). Pore formation in artificial membranes by the secreted haemolysins of *Proteus vulgaris* and *Morganella morganii*. *Eur. J. Biochem.* **220**, 339–47.

Bezrukov, S.M. and Kasianowicz, J.J. (1993). Current noise reveals protonation kinetics and number of ionizable sites in an open protein ion channel. *Phys. Rev. Lett.* **70**, 2352–5.

Bezrukov, S.M., Vodyanoy, I., Brutyan, R.A. and Kasianowicz, J.J. (1996). Dynamics and free energy of polymers partitioning into a nanoscale pore. *Macromolecules* **29**, 8517–22.

Bhakdi, S. and Tranum-Jensen, J. (1991). Alpha-toxin of *Staphylococcus aureus*. *Microbiol. Rev.* **55**, 733–51.

Blaustein, R.O. and Finkelstein, A. (1990). Voltage-dependent block of anthrax toxin channels in planar phospholipid bilayer membranes by symmetric tetraalkylammonium ions – effects on macroscopic conductance. *J. Gen. Physiol.* **96**, 905–19.

Blaustein, R.O., Germann, W.J., Finkelstein, A. and DasGupta, B.R. (1987). The N-terminal half of the heavy chain of botulinum type A neurotoxin forms channels in planar phospholipid bilayer. *FEBS Lett.* **226**, 115–20.

Blaustein, R.O., Koehler, T.M., Collier, R.J. and Finkelstein, A. (1989). Anthrax toxin: channel-forming activity of protective antigen in planar phospholipid bilayers. *Proc. Natl Acad. Sci. USA* **86**, 2209–13.

Blumenthal, R. and Habig, W.H. (1984). Mechanism of tetanolysin-induced membrane damage: studies with black lipid membranes. *J. Bacteriol.* **151**, 321–3.

Boehm, D.F., Welch, R.A. and Snyder, I.S. (1990). Domains of *Escherichia coli* haemolysin (HlyA) involved in binding of calcium and erythrocyte membranes. *Infect. Immun.* **58**, 1959–64.

Borochov-Neori, H., Yavin, E. and Montal, M. (1984). Tetanus toxin forms channels in planar lipid bilayers containing gangliosides. *Biophys. J.* **45**, 83–5.

Bougis, P., Rochat, H., Pieroni, G. and Verger, R. (1981). Penetration of phospholipid monolayers by cardiotoxins. *Biochemistry* **20**, 4915–20.

Cannon, R.J.C. (1996). *Bacillus thuringiensis* use in agriculture: a molecular perspective. *Biol. Rev.* **71**, 561–636.

Carroll, J. and Ellar, D.J. (1993). An analysis of *Bacillus thuringiensis* delta-endotoxin action on insect-mid-gut-membrane permeability using a light-scattering assay. *Eur. J. Biochem.* **214**, 771–8.

Carroll, J. and Ellar, D.J. (1997). Analysis of the large aqueous pores produced by a *Bacillus thuringiensis* protein insecticide in *Manduca sexta* mid-gut-brush-border-membrane vesicles. *Eur. J. Biochem.* **245**, 797–804.

Cauci, S., Monte, R., Ropele, M., Missero, C., Not, T., Quadrifoglio, F. and Menestrina, G. (1993a). Pore-forming and haemolytic properties of the *Gardnerella vaginalis* cytolysin. *Mol. Microbiol.* **9**, 1143–55.

Cauci, S., Monte, R., Quadrifoglio, F., Ropele, M. and Menestrina, G. (1993b). Ionic factors regulating the interaction of *Gardnerella vaginalis* haemolysin with red blood cells. *Biochim. Biophys. Acta* **1153**, 53–8.

Cescatti, L., Pederzolli, C. and Menestrina, G. (1991). Modification of lysine residues of *S. aureus* α-toxin: effects on its channel forming properties. *J. Memb. Biol.* **119**, 53–64.

Chakraborty, T., Schamid, A., Notermans, S. and Benz, R. (1990). Aerolysin from *Aeromonas sobria*: evidence for formation of ion-permeable channels and comparison with alpha-toxin of *Staphylococcus aureus*. *Infect. Immun.* **58**, 2127–32.

Colin, D.A., Mazurier, I., Sire, S. and Finck-Barbançon, V. (1994). Interaction of the two components of leukocidin from *Staphylococcus aureus* with human polymorphonuclear leukocyte membranes: sequential binding and subsequent activation. *Infect. Immun.* **62**, 3184–8.

Coote, J.G. (1992). Structural and functional relationships among the RTX toxin determinants of Gram-negative bacteria. *FEMS Microbiol. Rev.* **88**, 137–62.

Czajkowsky, D.M., Sheng, S.T. and Shao, Z.F. (1998). Staphylococcal alpha-haemolysin can form hexamers in phospholipid bilayers. *J. Mol. Biol.* **276**, 325–30.

Dalla Serra, M., Fagiuoli, G., Nordera, P., Bernhart, I., Della Volpe, C., Di Giorgio, D., Ballio, A. and Menestrina, G. (1999a). The interaction of antifungal lipodepsipeptides from *Pseudomonas syringae* with biological and model membranes. *Mol. Plant-Miorobe Interact.* **12** (in press).

Dalla Serra, M., Nordera, P., Beruhart, I., Di Giorgio, D., Ballio, A. and Menestrina, G. (1999b). Conductive properties and gating of channels formed by syringopeptin 25, an antifungal lipodepsipeptide from *Pseudomonas syringae*, in planar lipid membranes. *Mol. Plant-Miorobe Interact.* **12** (in press).

Deleers, M., Beugnier, N., Falmagne, P., Cabiaux, V. and Ruysschaert, J.-M. (1983). Localization in diphtheria toxin fragment B of a region that induces pore formation in planar lipid bilayers at low pH. *FEBS Lett.* **160**, 82–6.

Demel, R.A., Schiavo, G., de Kruijff, B. and Montecucco, C. (1991). Lipid interaction of diphtheria toxin and mutants. A study with phospholipids and protein monolayers. *Eur. J. Biochem.* **197**, 481–6.

Dobereiner, A., Schmid, A., Ludwig, A., Goebel, W. and Benz, R. (1996). The effects of calcium and other polyvalent cations on channel formation by *Escherichia coli* alpha-haemolysin in red blood cells and lipid bilayer membranes. *Eur. J. Biochem.* **240**, 454–60.

Donovan, J.J. and Middlebrook, J.L. (1986). Ion-conducting channels produced by botulinum toxin in planar lipid membranes. *Biochemistry* **25**, 2872–6.

Donovan, J.J., Simon, M.I., Draper, R.K. and Montal, M. (1981). Diphtheria toxin forms transmembrane channels in planar lipid bilayers. *Proc. Natl Acad. Sci. USA* **78**, 172–6.

Doyle, D.A., Morais Cabral, J., Pfuetzner, R.A., Kuo, A., Gulbis, J.M., Cohen, S.L., Chait, B.T. and MacKinnon, R. (1998). The structure of the potassium channel: molecular basis of K^+ conduction and selectivity. *Science* **280**, 69–77.

Dreyer, F., Maric, K. and Lutz, F. (1990). The use of patch-clamp technique for the study of the mode of action of bacterial toxins. In: *Bacterial Protein Toxins (Zbl. Bakt. Suppl.)* (eds R. Rappuoli, J.E. Alouf, P. Falmagne, F.J. Fehrenbach, J. Freer, R. Gross, J. Jeljasewicz, C. Montecucco, M. Tomasi, B. Wadstrom and T. Witholt), pp. 227–34. Gustav Fisher, Stuttgart.

Duesbery, N.S., Webb, C.P., Leppla, S.H., Gordon, V.M., Klimpel, K.R., Copeland, T.D., Ahn, N.G., Oskarsson, M.K., Fukasawa, K., Paull, K.D. and Vande Woude, G.F. (1998). Proteolytic inactivation of MAP-kinase-kinase by anthrax lethal factor. *Science* **280**, 734–7.

Ehrmann, I.E., Gray, M.C., Gordon, V.M., Gray, L.S. and Hewlett, E.L. (1991). Haemolytic activity of adenylate cyclase toxin from *Bordetella pertussis*. *FEBS Lett.* **278**, 79–83.

Ellis, M.J., Hebert, H. and Thelestam, M. (1997). *Staphylococcus aureus* alpha-toxin: characterization of protein/lipid interactions, 2D crystallization on lipid monolayers, and 3-D structure. *J. Struct. Biol.* **118**, 178–88.

English, L., Robbins, H.L. and Von, M.A. (1994). Mode of action of CryIIA: a *Bacillus thuringiensis* delta-endotoxin. *Insect Biochem. Mol. Biol.* **24**, 1025–35.

Eriksen, S., Olsnes, S., Sandvig, K. and Sand, O. (1994). Diphtheria toxin at low pH depolarizes the membrane, increases the membrane conductance and induces a new type of ion channel in Vero cells. *EMBO J.* **13**, 4433–9.

Feigin, A.M., Takemoto, J.Y., Wangspa, R., Teeter, J.H. and Brand, J.G. (1996). Properties of voltage-gated ion channels formed by syringomycin E in planar lipid bilayers. *J. Memb. Biol.* **149**, 41–7.

Ferreras, M., Menestrina, G., Foster, T.J., Colin, D.A., Prévost, G. and Piémont, Y. (1996). Permeabilization of lipid bilayers by *Staphylococcus aureus* gamma-toxins. In: *Bacterial Protein Toxins* (eds P.L. Frandsen, J.E. Alouf, P. Falmagne, F.J. Fehrenbach, J.H. Freer, C. Montecucco, S. Olsnes, R. Rappuoli and T. Wadstrom), pp. 105–6. Gustav-Fischer, Stuttgart.

Ferreras, M., Höper, F., Dalla Serra, M., Colin, D.A., Prévost, G. and Menestrina, G. (1998). The interaction of *Staphylococcus aureus* bicomponent gamma haemolysins and leucocidins with cells and model membranes. *Biochim. Biophys. Acta* **1414**, 108–26.

Finkelstein, A. (1994). The channel formed in planar lipid bilayers by the protective antigen component of anthrax toxin. *Toxicology* **87**, 29–41.

Gambale, F. and Montal, M. (1988). Characterization of the channel properties of tetanus toxin in planar lipid bilayers. *Biophys. J.* **53**, 771–83.

Gambale, F., Rauch, G., Belmonte, G. and Menestrina, G. (1992). Properties of *Pseudomonas aeruginosa* exotoxin A ionic channel incorporated in planar lipid bilayers. *FEBS Lett.* **306**, 41–5.

Gazit, E., Burshtein, N., Ellar, D.J., Sawyer, T. and Shai, Y. (1997). *Bacillus thuringiensis* cytolytic toxin associates specifically with its synthetic helices A and C in the membrane-bound state. Implications for the assembly of oligomeric transmembrane pores. *Biochemistry* **36**, 15546–54.

Gill, S.S., Cowles, E.A. and Pietrantonio, P. (1992). The mode of action of *Bacillus thuringiensis* endotoxins. *Annu. Rev. Entomol.* **37**, 615–36.

Gordon, V.M., Benz, R., Fujii, K., Leppla, S.H. and Tweten, R.K. (1997). *Clostridium septicum* alpha-toxin is proteolytically activated by furin. *Infect. Immun.* **65**, 4130–4.

Grochulski, P., Masson, L., Borisova, S., Pusztai-Carey, M., Schwartz, J.L., Brousseau, R. and Cygler, M. (1995). *Bacillus thuringiensis* CryIA(a) insecticidal toxin: crystal structure and channel formation. *J. Mol. Biol.* **254**, 447–64.

Gruber, H.J., Wilmsen, H.U., Cowell, S., Schindler, H. and Buckley, J.T. (1994). Partial purification of the rat erythrocyte receptor for the channel-forming toxin aerolysin and reconstitution into planar lipid bilayers. *Mol. Microbiol.* **14**, 1093–101.

Gueirard, P. and Guiso, N. (1993). Virulence of *Bordetella bronchiseptica*: role of adenylate cyclase-haemolysin. *Infect. Immun.* **61**, 4072–8.

Hanke, W. and Schlue, W.-R. (1993). *Planar Lipid Bilayers. Methods and Applications.* Academic Press, London.

Hanski, E. (1989). Invasive adenylate cyclase toxin of *Bordetella pertussis. Trends Biochem. Sci.* **14**, 459–63.

Hardie, K.R., Issartel, J.P., Koronakis, E., Hughes, C. and Koronakis, V. (1991). *In vitro* activation of *Escherichia coli* prohaemolysin to the mature membrane-targeted toxin requires HlyC and a low molecular-weight cytosolic polypeptide. *Mol. Microbiol.* **5**, 1669–79.

Hewlett, E.L., Gray, M.C., Ehrmann, I.E., Maloney, N.J., Otero, A.S., Gray, L., Allietta, M., Szabo, G., Weiss, A.A. and Barry, E.M. (1993). Characterization of adenylate cyclase toxin from a mutant of *Bordetella pertussis* defective in the activator gene, cyaC*. *J. Biol. Chem.* **268**, 7842–8.

Hoch, D.H., Romero-Mira, M., Ehrlich, B.E., Finkelstein, A., DasGupta, B.R. and Simpson, L.L. (1985). Channels formed by botulinum, tetanus, and diphtheria toxins in planar lipid bilayers: relevance to translocation of proteins across membranes. *Proc. Natl Acad. Sci. USA* **82**, 1692–6.

Hoshi, T., Zagotta, W.N. and Aldrich, R.W. (1990). Biophysical and molecular mechanism of *Shaker* potassium channel inactivation. *Science* **250**, 533–8.

Howard, S.P. and Buckley, J.T. (1982). Membrane glycoprotein receptor and hole-forming properties of a cytolytic protein toxin. *Biochemistry* **21**, 1662–7.

Huynh, P.D., Cui, C., Zhan, H.J., Oh, K.J., Collier, R.J. and Finkelstein, A. (1997). Probing the structure of the diphtheria toxin channel – reactivity in planar lipid bilayer membranes of cysteine-substituted mutant channels with methanethiosulphonate derivatives. *J. Gen. Physiol.* **110**, 229–42.

Ikigai, H., Akatsuka, A., Tsujiyama, H., Nakae, T. and Shimamura, T. (1996). Mechanism of membrane damage by El Thor haemolysin of *Vibrio cholerae* 01. *Infect. Immun.* **64**, 2968–73.

Ikigai, H., Ono, T., Iwata, M., Nakae, T. and Shimamura, T. (1997). El Thor haemolysin of *Vibrio cholerae* O1 forms channels in planar lipid bilayer membranes. *FEMS Microbiol. Lett.* **150**, 249–54.

Issartel, J-P., Koronakis, V. and Hughes, C. (1991). Activation of *Escherichia coli* prohaemolysin to the mature toxin by acyl carrier protein-dependent fatty acylation. *Nature* **351**, 759–61.

Jiang, G-S., Solow, R. and Hu, V.W. (1989). Fragment A of diphtheria toxin causes pH-dependent lesions in model membranes. *J. Biol. Chem.* **264**, 17170–3.

Johnstone, S.R., Morrice, L.M. and van Heyningen, S. (1990). The heavy chain of tetanus toxin can mediate the entry of cytotoxic gelonin into intact cells. *FEBS Lett.* **265**, 101–3.

Kagan, B.L. and Sokolov, Y. (1997). Use of lipid bilayer membranes to detect pore formation by toxins. In: *Bacterial Pathogenesis, Selected Methods in Enzymology*, pp. 395–409. Academic Press, San Diego.

Kagan, B.L., Finkelstein, A. and Colombini, M. (1981). Diphtheria toxin fragment forms large pores in phospholipid bilayer membranes. *Proc. Natl Acad. Sci. USA* **78**, 4950–4.

Katahira, J., Sugiyama, H., Inoue, N., Horiguchi, Y., Matsuda, M. and Sugimoto, N. (1997). *Clostridium perfringens* enterotoxin utilizes two structurally related membrane proteins as functional receptors *in vivo. J. Biol. Chem.* **272**, 26652–8.

Kirby, J.E., Vogel, J.P., Andrews, H.L. and Isberg, R.R. (1998). Evidence for pore-forming ability by *Legionella pneumophila. Mol. Microbiol.* **27**, 323–36.

Knowles, B.H., Blatt, M.R., Tester, M., Horsnell, J.M., Carroll, J., Menestrina, G. and Ellar, D.J. (1989). A cytolytic delta-endotoxin from *Bacillus thuringiensis* var. *israelensis* forms cation selective channels in planar lipid bilayers. *FEBS Lett.* **244**, 259–62.

Koehler, T.M. and Collier, R.J. (1991). Anthrax toxin protective antigen: low-pH-induced hydrophobicity and channel formation in liposomes. *Mol. Microbiol.* **5**, 1501–6.

Kokai-Kun, J.F. and McClane, B.A. (1997). The *Clostridium perfringens* enterotoxin. In: *The Clostridia: Molecular Biology and Pathogenesis* (eds J.I. Rood, B.A. McClane, J.G. Songer and R.W. Titball), pp. 325–57. Academic Press, London.

Korchev, Y.E., Bashford, C.L. and Pasternak, C.A. (1992). Differential sensitivity of Pneumolysin-induced channels to gating by divalent cations. *J. Memb. Biol.* **127**, 195–203.

Korchev, Y.E., Bashford, C.L., Alder, G.M., Kasianowicz, J.J. and Pasternak, C.A. (1995a). Low conductance states of a single ion channel are not 'closed'. *J. Memb. Biol.* **147**, 233–9.

Korchev, Y.E., Alder, G.M., Bakhramov, A., Bashford, C.L., Joomun, B.S., Sviderskaya, E.V., Usherwood, P.N.R. and Pasternak, C.A. (1995b). *Staphylococcus aureus* alpha-toxin-induced pores: channel-like behaviour in lipid bilayers and clamped cells. *J. Memb. Biol.* **143**, 143–51.

Korchev, Y.E., Bashford, C.L., Alder, G.M., Apel, P.Y., Edmonds, D.T., Lev, A.A., Nandi, K., Zima, A.V. and Pasternak, C.A. (1997). A novel explanation for fluctuations of ion current through narrow pores. *FASEB J.* **11**, 600–8.

Korchev, Y.E., Bashford, C.L., Pederzolli, C., Pasternak, C.A., Morgan, P.J., Andrew, P.W. and Mitchell, T.J. (1998). A conserved tryptophan in pneumolysin is a determinant of the characteristics of channels formed by pneumolysin in cells and planar lipid bilayers. *Biochem. J.* **329**, 571–7.

Krasilnikov, O.V., Muratkhodjaev, J.N., Voronov, S.E. and Yezepchuk, Y.V. (1991). The ionic channels formed by cholera toxin in planar bilayer lipid membranes are entirely attributable to its B-subunit. *Biochim. Biophys. Acta* **1067**, 166–70.

Krasilnikov, O.V., Sabirov, R.Z., Ternovsky, O.V., Merzlyak, P.G. and Muratkhodjaev, J.N. (1992a). A simple method for the determination of the pore radius of channels in planar lipid bilayer membranes. *FEMS Microbiol. Immunol.* **105**, 93–100.

Krasilnikov, O.V., Muratkhodjaev, J.N. and Zitzer, A.O. (1992b). The mode of action of *Vibrio cholerae* cytolysin.The influences on both erythrocytes and planar lipid bilayers. *Biochim. Biophys. Acta* **1111**, 7–16.

Kretzschmar, U.M., Hamman, R. and Kutzner, H.J. (1991). Purification and characterization of *Gardnerella vaginalis* haemolysin. *Curr. Microbiol.* **23**, 7–13.

Lally, E.T., Kieba, I.R., Sato, A., Green, C.L., Rosenbloom, J., Korostoff, J., Wang, J.F., Shenker, B.J., Ortlepp, S., Robinson, M.K. and Billings, P.C. (1997). RTX toxins recognize a beta 2 integrin on the surface of human target cells. *J. Biol. Chem.* **272**, 30463–9.

Lalonde, G., McDonald, T.V., Gardner, P. and O'Hanley, P.D. (1989). Identification of a haemolysin from *Actinobacillus pleuropneumoniae* and characterization of its channel properties in planar phospholipid bilayers. *J. Biol. Chem.* **264**, 13559–64.

Lear, J.D., Furblur, U.G., Lally, E.T. and Tanaka, J.C. (1995). *Actinobacillus actinomycetemcomitans* leukotoxin forms large conductance, voltage-gated ion channels when incorporated into planar lipid bilayers. *Biochim. Biophys. Acta* **1238**, 34–41.

Lee, K.H., Fitton, J.E. and Würrich, K. (1987). Nuclear magnetic resonance investigation of the conformation of δ-haemolysin bound to dodecylphosphocoline micelles. *Biochim. Biophys. Acta* **911**, 144–53.

Li, J., Carroll, J. and Ellar, D.J. (1991). Crystal structure of insecticidal δ-endotoxin from *Bacillus thuringiensis* at 2.5 Å resolution. *Nature* **353**, 815–21.

Li, J., Koni, P.A. and Ellar, D.J. (1996). Structure of the mosquitocidal delta-endotoxin CytB from *Bacillus thuringiensis* sp. *kyushuensis* and implications for membrane pore formation. *J. Mol. Biol.* **257**, 129–52.

Liebig, B., Stetson, D.L. and Dean, D.H. (1995). Quantification of the effect of *Bacillus thuringiensis* toxins on short-circuit current in the mid-gut of *Bombyx mori*. *J. Insect Physiol.* **41**, 17–22.

Lorence, A., Darszon, A., Diaz, C., Liévano, A., Quintero, R. and Bravo, A. (1995). Delta-endotoxins induce cation channels in *Spodoptera frugiperda* brush border membranes in suspension and in planar lipid bilayers. *FEBS Lett.* **360**, 217–22.

Lorence, A., Darszon, A. and Bravo, A. (1997). Aminopeptidase dependent pore formation of *Bacillus thuringiensis* Cry1Ac toxin on *Trichoplusia ni* membranes. *FEBS Lett.* **414**, 303–7.

Lutz, F., Maurer, M. and Failing, K. (1987). Cytotoxic protein from *Pseudomonas aeruginosa*: formation of hydrophilic pores in Erlicha ascites tumour cells and effect on cell viability. *Toxicon* **25**, 293–305.

Lutz, F., Mohr, M., Grimmig, M., Leidolf, R. and Linder, R. (1993). *Pseudomonas aeruginosa* cytotoxin-binding protein in rabbit erythrocyte membranes. An oligomer of 28 kDaa with similarity to transmembrane channel proteins. *Eur. J. Biochem.* **217**, 1123–8.

Maget-Dana, R. and Peypoux, F. (1994). Iturins, a special class of pore-forming lipopeptides: biological and physicochemical properties. *Toxicology* **87**, 151–74.

Maget-Dana, R. and Ptak, M. (1990). Iturin lipopeptides: interaction of mycosubtilin with lipids in planar membranes and mixed monolayers. *Biochim. Biophys. Acta* **1023**, 34–40.

Maier, E., Reinhard, N., Benz, R. and Frey, J. (1996). Channel-forming activity and channel size of the RTX toxins ApxI, ApxII, and ApxIII of *Actinobacillus pleuropneumoniae*. *Infect. Immun.* **64**, 4415–23.

Majoul, I.V., Bastiaens, L.H. and Söling, H-D. (1996). Transport of an external Lys-Asp-Glu-Leu (KDEL) protein from the plasma membrane to the endoplasmic reticulum: studies with cholera toxin in Vero cells. *J. Cell Biol.* **133**, 777–89.

Mellor, I.R., Thomas, D.H. and Sansom, M.S.P. (1988). Properties of ion channels formed by *Staphylococcus aureus* delta-toxin. *Biochim. Biophys. Acta* **942**, 280–94.

Menestrina, G. (1986). Ionic channels formed by *Staphylococcus aureus* alpha-toxin: voltage dependent inhibition by di- and trivalent cations. *J. Memb. Biol.* **90**, 177–90.

Menestrina, G. and Antolini, R. (1981). Ion transport through haemocyanin channels in oxidized cholesterol artificial bilayer membranes. *Biochim. Biophys. Acta* **643**, 616–25.

Menestrina, G. and Ropele, M. (1989). Voltage-dependent gating properties of the channel formed by *E. coli* haemolysin in planar lipid membranes. *Biosci. Rep.* **9**, 465–73.

Menestrina, G., Mackman, N., Holland, I.B. and Bhakdi, S. (1987). *Escherichia coli* haemolysin forms voltage-dependent channels in lipid membranes. *Biochim. Biophys. Acta* **905**, 109–17.

Menestrina, G., Forti, S. and Gambale, F. (1989). Interaction of tetanus toxin with lipid vesicles. Effects of pH, surface charge and transmembrane potential on the kinetics of channel formation. *Biophys. J.* **55**, 393–405.

Menestrina, G., Bashford, C.L. and Pasternak, C.A. (1990). Pore-forming toxins: experiments with *S. aureus* α-toxin, *C. perfringens* θ-toxin and *E. coli* haemolysin in lipid bilayers, liposomes and intact cells. *Toxicon* **28**, 477–91.

Menestrina, G., Pederzolli, C., Forti, S. and Gambale, F. (1991). Lipid interaction of *Pseudomonas aeruginosa* exotoxin A: acid triggered aggregation and permeabilization of lipid vesicles. *Biophys. J.* **60**, 1388–400.

Menestrina, G., Moser, C., Pellett, S. and Welch, R.A. (1994). Pore-formation by *Escherichia coli* haemolysin (HlyA) and other members of the RTX toxins family. *Toxicology* **87**, 249–67.

Menestrina, G., Pederzolli, C., Dalla Serra, M., Bregante, M. and Gambale, F. (1996). Permeability increase induced by *Escherichia coli* haemolysin A in human macrophages is due to the formation of ionic pores: a patch clamp characterization. *J. Memb. Biol.* **149**, 113–21.

Menzl, K., Maier, E., Chakraborty, T. and Benz, R. (1996). HlyA haemolysin of *Vibrio cholerae* 01 biotype El Thor. *Eur. J. Biochem.* **240**, 646–54.

Milne, J.C. and Collier, R.J. (1993). pH-dependent permeabilization of the plasma membrane of mammalian cells by anthrax protective antigen. *Mol. Microbiol.* **10**, 647–53.

Moayeri, M. and Welch, R.A. (1994). Effects of temperature, time, and toxin concentration on lesion formation by the *Escherichia coli* haemolysin. *Infect. Immun.* **62**, 4124–34.

Montecucco, C., Schiavo, G., Gao, Z., Bauerlein, E., Boquet, P. and DasGupta, B.R. (1988). Interaction of botulinum and tetanus toxins with the lipid bilayer surface. *Biochem. J.* **251**, 379–83.

Montecucco, C., Schiavo, G. and DasGupta, B.R. (1989). Effect of pH on the interaction of botulinum neurotoxins A, B and E with liposomes. *Biochem. J.* **259**, 47–53.

Moran, O., Zegarra-Moran, O., Virginio, C. and Rottini, G. (1991). Voltage dependent cationic channels formed by a cytolytic toxin produced by *Gardnerella vaginalis*. *FEBS Lett.* **283**, 317–20.

Moss, J., Richards, R.L., Alving, C.R. and Fishman, P.H. (1976). Effect of the A and B protomers of choleragen on release of trapped glucose from liposomes containing or lacking gangliosides GM1. *J. Biol. Chem.* **252**, 797–8.

New, R.R.C. (1990). *Liposomes: A Practical Approach*. IRL Press, Oxford.

Nordera, P. and Menestrina, G. (1998). Proteolytic cleavage of *Pseudomonas aeruginosa* exotoxin A in the presence of lipid bilayers of different composition. *FEBS Lett.* **421**, 268–72.

Nordera, P., Dalla Serra, M. and Menestrina, G. (1997). The adsorption of *Pseudomonas aeruginosa* exotoxin A to phospholipid monolayers is controlled by pH and surface potential. *Biophys. J.* **73**, 1468–78.

Ohnishi, M., Hayashi, T., Tomita, T. and Terawaki, Y. (1994). Mechanism of the cytolytic action of *Pseudomonas aeruginosa* cytotoxin: oligomerization of the cytotoxin on target membranes. *FEBS Lett.* **356**, 357–60.

Ondraczek, R., Hobbie, S. and Braun, V. (1992). *In vitro* activation of *Serratia marcescens* haemolysin through modification and complementation. *J. Bacteriol.* **174**, 5086–94.

Ostolaza, H., Bartolomé, B., de Zárate, I.O., de la Cruz, F. and Goñi, F.M. (1993). Release of lipid vesicle contents by the bacterial protein toxin α-haemolysin. *Biochim. Biophys. Acta* **1147**, 81–8.

Ostolaza, H., Bakas, L. and Goñi, F.M. (1997). Balance of electrostatic and hydrophobic interactions in the lysis of model membranes by *E. coli* alpha-haemolysin. *J. Memb. Biol.* **158**, 137–45.

Palmer, M., Saweljew, P., Vulicevic, I., Valeva, A., Kehoe, M. and Bhakdi, S. (1996). Membrane-penetrating domain of Streptolysin O identified by cysteine scanning mutagenesis. *J. Biol. Chem.* **271**, 26664–7.

Palmer, M., Harris, R., Freytag, C., Kehoe, M., Tranum-Jensen, J. and Bhakdi, S. (1998). Assembly mechanism of the oligomeric streptolysin O pore: the early membrane lesion is lined by a free edge of the lipid membrane and is extended gradually during oligomerization. *EMBO J.* **17**, 1598–605.

Parente, R.A., Nir, S. and Szoka, F.C., Jr (1990). Mechanism of leakage of phospholipid vesicle contents induced by the peptide GALA. *Biochemistry* **29**, 8720–8.

Parker, M.W., Buckley, J.T., Postma, J.P.M., Tucker, A.D., Leonard, K., Pattus, F. and Tsernoglou, D. (1994). Structure of the *Aeromonas* toxin proaerolysin in its water-soluble and membrane-channel state. *Nature* **367**, 292–5.

Parker, M.W., Buckley, J.T., van der Goot, F.G. and Tsernoglou, D. (1996). Structure and assembly of the channel-forming *Aeromonas* toxin aerolysin. In: *Protein Toxin Structure* (ed. M.W. Parker), pp. 79–95. R.G. Landes, Georgetown.

Parsegian, V.A., Bezrukov, S.M. and Vodyanoy, I. (1995). Watching small molecules move: interrogating ion channels using neutral solutes. *Biosci. Rep.* **15**, 503–14.

Plasek, J. and Sigler, K. (1996). Slow fluorescent indicators of membrane potential: a survey of different approaches to probe response analysis. *J. Photochem. Photobiol. B – Biology* **33**, 101–24.

Prévost, G., Cribier, B., Couppié, P., Petiau, P., Supersac, G., Finck-Barbançon, V., Monteil, H. and Piémont, Y. (1995). Panton-Valentine leucocidin and gamma-haemolysin from *S. aureus* ATCC 49775: distinct genetic loci and different biological activities of the six toxic couples generated. *Infect. Immun.* **63**, 4121–9.

Rapaport, D., Peled, R., Nir, S. and Shai, Y. (1996). Reversible surface aggregation in pore formation by pardaxin. *Biophys. J.* **70**, 2502–12.

Rasper, D.M. and Merrill, A.R. (1994). Evidence for the modulation of *Pseudomonas aeruginosa* exotoxin A-induced pore formation by membrane surface charge density. *Biochemistry* **33**, 12981–9.

Rauch, G., Gambale, F. and Montal, M. (1990). Tetanus toxin channel in phosphatidylserine planar bilayers: conductance states and pH dependence. *Eur. Biophys. J.* **18**, 79–83.

Rex, S. and Schwarz, G. (1998). Quantitative studies on the melittin-induced leakage mechanism of lipid vesicles. *Biochemistry* **37**, 2336–45.

Ropele, M. and Menestrina, G. (1989). Electrical properties and molecular architecture of the channel formed by *E. coli* haemolysin in planar lipid membranes. *Biochim. Biophys. Acta* **985**, 9–18.

Rossjohn, J., Feil, S.C., McKinstry, W.J., Tweten, R.K. and Parker, M.W. (1997). Structure of a cholesterol-binding, thiol-activated cytolysin and a model of its membrane form. *Cell* **89**, 685–92.

Rottem, S., Groover, K., Habig, W.H., Barile, M.F. and Hardegree, M.C. (1990). Transmembrane diffusion channels in *Mycoplasma gallisepticum* induced by tetanolysin. *Infect. Immun.* **58**, 598–602.

Schiavo, G., Demel, R.A. and Montecucco, C. (1991). On the role of polysialoglycosphingolipids as tetanus toxin receptors. *Eur. J. Biochem.* **199**, 705–11.

Schiavo, G., Benfenati, F., Poulain, B., Rossetto, O., Polverino de Laureto, P., DasGupta, B.R. and Montecucco, C. (1992). Tetanus and botulinum-B neurotoxins block neurotransmitter release by proteolytic cleavage of synaptobrevin. *Nature* **359**, 832–5.

Schiebel, E. and Braun, V. (1989). Integration of the *Serratia marcescens* haemolysin into human erythrocyte membranes. *Mol. Microbiol.* **3**, 445–53.

Schmidt, H., Maier, E., Karch, H. and Benz, R. (1996). Pore-forming properties of the plasmid-encoded haemolysin of enterohaemorrhagic *Escherichia coli* O157:H7. *Eur. J. Biochem.* **241**, 594–601.

Schönherr, R., Hilger, M., Broer, S., Benz, R. and Braun, V. (1994). Interaction of *Serratia marcescens* haemolysin (ShlA) with artificial and erythrocyte membranes. *Eur. J. Biochem.* **223**, 655–63.

Schwartz, J.-L., Garneau, L., Savaria, D., Masson, L., Brousseau, R. and Rousseau, E. (1993). Lepidopteran-specific crystal toxins from *Bacillus thuringiensis* form cation- and anion-selective channels in planar lipid membranes. *J. Memb. Biol.* **132**, 53–62.

Schwartz, J.-L., Lu, Y.J., Sohnlein, P., Brousseau, R., Laprade, R., Masson, L. and Adang, M.J. (1997). Ion channels formed in planar lipid bilayers by *Bacillus thuringiensis* toxins in the presence of *Manduca sexta* midgut receptors. *FEBS Lett.* **412**, 270–6.

Sellman, B.R. and Tweten, R.K. (1997). The propeptide of *Clostridium septicum* alpha toxin functions as an intramolecular chaperone and is a potent inhibitor of alpha toxin-dependent cytolysis. *Mol. Microbiol.* **25**, 429–40.

Sheppard, J.D., Jumarie, C., Cooper, D.G. and Laprade, R. (1991). Ionic channels induced by surfactin in planar lipid bilayer membranes. *Biochim. Biophys. Acta* **1064**, 13–23.

Sheridan, R.E. (1998). Gating and permeability of ion channels produced by botulinum toxins types A and E in PC12 cell membranes. *Toxicon* **36**, 703–17.

Shiver, J.W. and Donovan, J.J. (1987). Interaction of diphtheria toxin with lipid vesicles: determinants of ion channel formation. *Biochim. Biophys. Acta* **903**, 48–55.

Shone, C.C., Hambleton, P. and Melling, J. (1987). A 50-kDa fragment from the NH_2-terminus of the heavy subunit of *Clostridium botulinum* type A neurotoxin forms channels in lipid vesicles. *Eur. J. Biochem.* **167**, 175–80.

Shubair, M., Snyder, I.S. and Larsen, B. (1993). *Gardnerella vaginalis* haemolysin. I. Production and purification. *Immunol. Infect. Dis.* **3**, 135–42.

Singh, Y., Leppla, S.H., Bhatnagar, R. and Friedlander, A.M. (1989). Internalization and processing of *Bacillus anthracis* lethal toxin by toxin-sensitive and -resistant cells. *J. Biol. Chem.* **264**, 11099–102.

Slatin, S.L., Abrams, C.K. and English, L. (1990). Delta-endotoxins form cation-selective channels in planar lipid bilayers. *Biochem. Biophys. Res. Commun.* **169**, 765–72.

Song, L., Hobaugh, M.R., Shaustak, C., Cheley, S., Bayley, H. and Gonaux, J.E. (1996). Structure of staphylococcal alpha-haemolysin, a heptameric transmembrane pore. *Science* **274**, 1859–66.

Staali, L., Monteil, H. and Colin, D.A. (1998). The staphylococcal pore-forming leukotoxins open Ca^{2+} channels in the membrane of human polymorphonuclear neutrophils. *J. Memb. Biol.* **162**, 209–16.

Sugawara, N., Tomita, T. and Kamio, Y. (1997). Assembly of *Staphylococcus aureus* gamma-haemolysin into a pore-forming ring-shaped complex on the surface of human erythrocytes. *FEBS Lett.* **410**, 333–7.

Sugimoto, N., Takagi, M., Ozutsumi, K., Harada, S. and Matsuda, M. (1988). Enterotoxin of *Clostridium perfringens* type A forms ion-permeable channels in a lipid bilayer membrane. *Biochem. Biophys. Res. Commun.* **156**, 551–6.

Suttorp, N., Seeger, W., Uhl, J., Lutz, F. and Roka, L. (1985). *Pseudomonas aeruginosa* cytotoxin stimulates prostacyclin production in cultured pulmonary artery endothelial cells: membrane attack and calcium influx. *J. Cell. Physiol.* **123**, 64–72.

Szabo, G., Gray, M.C. and Hewlett, E.L. (1994). Adenylate cyclase toxin from *Bordetella pertussis* produces ion conductance across artificial lipid bilayers in a calcium- and polarity-dependent manner. *J. Biol. Chem.* **269**, 22496–9.

Thiaudière, E., Siffert, O., Talbot, J.-C., Bolard, J., Alouf, J.E. and Dufourcq, J. (1991). The amphiphilic α-helix concept. Consequences on the structure of staphylococcal δ-toxin in solution and bound to lipid bilayers. *Eur. J. Biochem.* **195**, 203–13.

Tschödrich-Rotter, M., Kubitscheck, U., Ugochukwu, G., Buckley, J.T. and Peters, R. (1996). Optical single-channel analysis of the aerolysin pore in erythrocyte membranes. *Biophys. J.* **70**, 723–32.

Vécsey-Semjén, B., Möllby, R. and van der Goot, F.G. (1996). Partial C-terminal unfolding is required for channel formation by staphylococcal alpha-toxin. *J. Biol. Chem.* **271**, 8655–60.

Weiner, R.N., Schneider, E., Haest, C.W.M., Deuticke, B., Benz, R. and Frimmer, M. (1985). Properties of the leak permeability induced by a cytotoxic protein of *Pseudomonas aeruginosa* (PACT) in rat erythrocytes and black lipid membranes. *Biochim. Biophys. Acta* **820**, 173–82.

Wilmsen, H.U., Pattus, F. and Buckley, J.T. (1990). Aerolysin, a haemolysin from *Aeromonas hydrophila*, forms voltage gated channels in planar lipid bilayers. *J. Membrane Biol.* **29115**, 71–81.

Wilmsen, H.U., Leonard, K.R., Tichelaar, W., Buckley, J.T. and Pattus, F. (1992). The aerolysin membrane channel is formed by heptamerization of the monomer. *EMBO J.* **11**, 2457–63.

Yamamoto, K., Wright, A.C., Kaper, J.B. and Morris, J.G.J. (1990). The cytolysin gene of *Vibrio vulnificus:* sequence and relationship to the *Vibrio cholerae* el Thor haemolysin gene. *Infect. Immun.* **58**, 2706–9.

Yguerabide, J. (1994). Theory for establishing proximity relations in biological membranes by excitation energy transfer measurements. *Biophys. J.* **66**, 683–93.

Zalman, L.S. and Wisnieski, B.J. (1985). Characterization of the insertion of *Pseudomonas* exotoxin A into membranes. *Infect. Immun.* **50**, 630–5.

Zhao, J., Milne, J.C. and Collier, R.J. (1995). Effects of anthrax toxin's lethal factor on ion channels formed by the protective antigen. *J. Biol. Chem.* **270**, 18626–30.

Zitzer, A., Walev, I., Palmer, M. and Bhakdi, S. (1995). Characterization of *Vibrio cholerae* El Thor cytolysin as an oligomerizing pore-forming toxin. *Med. Microb. Immunol.* **184**, 37–44.

Zitzer, A., Palmer, M., Weller, U., Wassenaar, T., Biermann, C., Tranum Jensen, J. and Bhakdi, S. (1997). Mode of primary binding to target membranes and pore formation induced by *Vibrio cholerae* cytolysin (haemolysin). *Eur. J. Biochem.* **247**, 209–16.

15

Membrane-damaging and cytotoxic phospholipases

Richard W. Titball

SUBSTRATES FOR PHOSPHOLIPASES

Phospholipids are ubiquitous in biological systems. They are key components of biological membranes, and derivatives of phospholipids such as diacylglycerol and inositol triphosphates serve as messengers within cells (Berridge, 1987; Exton, 1990). The generalized structure of a glycerolphospholipid (Figure 15.1) reveals a polar head group linked via a phosphate group and a glycerol backbone to non-polar hydrocarbon (fatty acyl) tails. The nature of the head group can be used to categorize different types of phospholipid. For example, the choline group is present in phosphatidylcholine, whilst phosphatidylinositol has an inositol sugar head group.

Sphingolipids (such as sphingomyelin) contain a modified glycerol backbone linked to fatty acyl tails, and this entire moiety is called a ceramide. Sphingomyelin usually contains a choline head group.

Within these broad groups, the hydrocarbon tails vary with respect to both their length and degree of saturation. The non-polar tail groups mean that phospholipids are sparingly soluble in water and the majority of phospholipid in eukaryotes is found in cell membranes, where it can adopt a bilayer configuration, with the tail groups embedded in the membrane and the polar head groups exposed on the surface.

Phospholipases hydrolyse phospholipids, and the location of the hydrolysed bond is used to characterize phospholipases as types A_1, A_2, C or D (Figure 15.1: PLA_1, PLA_2, PLC or PLD; Möllby, 1978). A wide variety of assays can be used to measure the activity of these enzymes. Colorimetric assays have been described that use water-soluble phospholipid derivatives such as p-nitrophenolphosphorylcholine (pNPPC), in which the phosphorylcholine head group is linked to p-nitrophenol. (Kurioka and Matsuda, 1976). Since these substrates lack the glycerol backbone and the fatty acyl tails, they can only serve as substrates for PLCs and PLDs. Other solution phase assays use phospholipid that has been emulsified with detergents to generate micelles (Waite, 1987) or use lipoproteins such as egg-yolk lipoprotein (Waite, 1987).

All of these assay systems are useful for measuring phospholipase activity, but they reveal little about the ways in which phospholipases interact with membrane phospholipids. Artificial membranes, such as monomolecular films, or liposomes, allow studies where the form of the phospholipid more closely approximates the cell membrane. However, it is clear that even these systems do not faithfully mimic the cell membrane, since phospholipases that are active in these systems may be inactive towards eukaryotic cells. This might be for several reasons; living cell membranes are complex mixtures of different phospholipids and proteins, and are able to repair limited damage. In addition, there is evidence that cellular metabolism is modulated in phospholipase-treated cells and that gross effects on cell membranes (e.g. cell lysis) are only

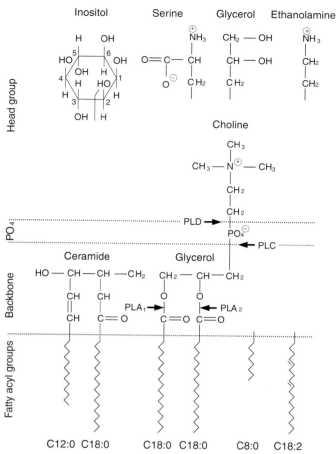

FIGURE 15.1 Generalized structure of phospholipids. The entire phosphatidylcholine molecule is shown, along with examples of the differences that are seen in other types of phospholipids. Inositol, serine, glycerol or ethanolamine head groups are found in other types of phospholipid. In sphingomyelins, the backbone includes a $(CH_2)_{12}$, but the CH_3 fatty acyl chain is termed a ceramide, and the head group is usually choline. In all phospholipids, the fatty acyl tails are typically 16–18 carbon chain lengths, but may be shorter or longer and may be unsaturated. The sites of phospholipid cleavage by PLA_1, PLA_2, PLC or PLD enzymes are shown arrowed.

one manifestation of the effect of the enzyme on the cell. The eukaryotic cell type most frequently used to measure the activity of phospholipases towards membranes is the erythrocyte (Möllby, 1978; Titball, 1993), because of its availability and the ease with which cell lysis can be measured.

These different assay systems are of more than incidental significance: they can reveal the mechanisms by which different enzymes recognize and hydrolyse different forms of phospholipids.

PHOSPHOLIPASES PRODUCED BY BACTERIA

The production of phospholipases by a variety of bacterial species has been reported and many of these bacteria are pathogens of humans or of animals. These bacterial enzymes can be grouped not only according to the site of cleavage of phospholipids (i.e. as PLA, PLC or PLD enzymes), but also according to the preferred substrate (Table 15.1).

Perhaps not surprisingly, for those proteins whose deduced amino acids sequences are known, this classification also leads to the grouping of enzymes according to amino acid sequence homology. Some of the bacterial phospholipases are lethal toxins, and this might actually be a consequence of their roles in scavenging PO_4^{2-}, which is only available in limited free amounts in the body. But the fine distinction between toxic and non-toxic phospholipases is not discussed here – activity towards membrane phospholipids might be a better indicator of the potential involvement of a phospholipase in disease. This chapter aims to consider membrane-active bacterial phospholipases, the roles that these enzymes play in the pathogenesis of disease, and the molecular basis of membrane interactions.

TABLE 15.1 Membrane-active bacterial phospholipases: categorized according to site of substrate cleavage, substrate specificity and amino acid sequence homology

PHOSPHOLIPASES A
Vibrionacae phospholipases A$_2$

Vibrio parahaemolyticus thermolabile haemolysin
Vibrio cholera phospholipase[a]
Aeromonas hydrophila GCAT
Aeromonas salmonicida GCAT

Other phospholipases A$_2$

Helicobacter pylori[b]

PHOSPHOLIPASES C
Phosphatidylcholine preferring
zincmetallophospholipases

Clostridium perfringens α-toxin
Clostridium novyi γ-toxin
Clostridium bifermentans PLC
Clostridium absonum PLC[b]
Clostridium barati PLC[b]
Bacillus cereus PC-PLC
Listeria monocytogenes PLC B
Pseudomonas fluorescens PLC

Gram-negative PLCs

Pseudomonas aeruginosa PLC-N
Pseudomonas aeruginosa PLC-H
Mycobacterium tuberculosis MpcA
Mycobacterium tuberculosis MpcB
Mycobacterium bovis PLC
Mycobacterium marinum PLC
Burkholderia cepacia PLC[a]
Burkholderia mallei PLC[a]
Burkholderia pseudomallei PLC[a]

Other PLCs

Legionella pneumophila PLC[b]
Ureaplasma urealyticum PLC[b]
Helicobacter pylori PLC[b]
Clostridium novyi β-toxin[b]

Sphingomyelinases
Gram-positive sphingomyelinases

Bacillus cereus Smase
Staphylococcus aureus β-toxin
Leptospira interrogans haemolysin

Phosphatidylinositol hydrolysing
Gram-positive enzymes

Bacillus cereus PI-PLC
Bacillus thuringiensis PI-PLC
Clostridium novyi PI-PLC[b]
Listeria monocytogenes PLC A
Staphylococcus aureus PI-PLC

PHOSPHOLIPASES D
Brown recluse spider-related enzymes

Corynebacterium pseudotuberculosis
Corynebacterium ulcerans
Arcanobacterium haemolyticum

Other phospholipases D

Mycobacterium tuberculosis[b]
Mycobacterium smegmatis[b]
Haemophilus parainfluenzae[b]
Escherichia coli[b]
Vibrio damsela haemolysin

[a]Assigned to this group on the basis of deduced amino acid sequence homology. Substrate specificity not fully determined.
[b]Assigned to this group of the basis of DNA hybridization studies and/or studies with purified phospholipase. Precise substrate specificity has yet to be determined.

Phospholipases A

The production of PLAs by a wide range of bacteria has been reported, but many of these enzymes are intracellular or membrane bound and are required for modifications to bacterial membranes in response to different cultural conditions (Möllby, 1978; Waite, 1987). These enzymes are not considered further in this chapter.

Extracellular PLAs, which in some cases have been shown to be haemolytic, are produced by members of the *Vibrionacae* (Fiore *et al.*, 1997), by *Salmonella newport* (Asnani and Asnani, 1991) and by *Helicobacter pylori* (Ottlecz *et al.*, 1993). Depending on whether these enzymes show specificity for the 1- or 2-acyl position, they are classified as phospholipases PLA$_1$ or PLA$_2$. The *Vibrio parahaemolyticus* thermolabile haemolysin is reported to be a PLA$_2$ (Shinoda *et al.*,1991) and shows 77% similarity to the phospholipase produced by *Vibrio cholera* (Fiore *et al.*, 1997). These enzymes are also related to the glycerophospholipid-cholesterol acyltransferases (GCAT) produced by *Aeromonas salmonicida* (28% identity) and *Aeromonas hydrophila* (26% identity) (Fiore *et al.*, 1997), which can act as PLA$_2$s in the absence of acyl receptors (Thornton *et al.*, 1988).

Phospholipases C

Gram-positive zincmetallophospholipases C

All of these enzymes share the properties of being inactivated by treatment with compounds capable of chelating divalent cations (such as EDTA) and reactivated by zinc ions (Krug and Kent 1984; Gerasimene *et al.*, 1985; Geoffroy *et al.*, 1991). The zincmetallophospholipases C have molecular sizes in the range 29–43 kDa, and a comparison of the deduced amino acid sequences of the enzymes produced by the *Clostridium perfringens* (α-toxin; Titball *et al.*, 1989), *Clostridium novyi* (γ-toxin; Tsutsui *et al.*, 1995), *Clostridium bifermentans* (PLC; Tso and Siebel, 1989), *Listeria monocytogenes* (PLC-B or PC-PLC; Vazquez-Boland *et al.*, 1992) and *Bacillus cereus* (PC-PLC; Johansen *et al.*, 1988) reveals significant homology (29–58%; Titball, 1993; Tsutsui *et al.*, 1995). The *Clostridium absonum* and *Clostridium barati* are also thought to be zincmetallophospholipases C (Nakamura *et al.*, 1973; Titball, 1997). The zincmetallophospholipases C are all active towards phosphatidyl choline but, apart from this, the individual enzymes have different activities towards other phospholipids. For example, only some of these enzymes are active towards sphingomyelin (Titball, 1993) and the enzymes also show variations in their specific activities, inhibition by halides and activation by salts (Portnoy *et al.*, 1994). These differences provide a unique opportunity to dissect structure–function relationships.

On the basis of an amino acid sequence alignment, the zincmetallophospholipases C can be divided into two groups; single-domain proteins and two-domain proteins. The single-domain proteins (*Listeria monocytogenes* PLC-B and *Bacillus cereus* PC-PLC) consist of approximately 250 amino acids, and the crystal structure of the *B. cereus* PC-PLC (Hough *et al.*, 1989) revealed a seven-helix protein with the putative active site located in a cleft which was also occupied by three zinc ions (Figure 15.2, see colour plate section).

The two-domain proteins are typically composed from 370 amino acids. These proteins are typified by *C. perfringens* α-toxin, which has one domain that is structurally similar to *B. cereus* PC-PLC and a second domain (carboxy-terminal domain) composed from beta-sheet arranged in a Greek key fold and forming a so-called 'C2-fold domain' (Naylor *et al.*, 1998). When the first domain (amino-terminal domain) of α-toxin was produced in *E. coli* and purified, it was found to have PLC activity towards egg yolk lipoprotein or detergent solubilized phospholipids, but it was devoid of haemolytic activity. Therefore, it appears that the amino-terminal domain of α-toxin is a structural and functional homologue of the entire *B. cereus* PC-PLC.

The crystal structures of *B. cereus* PC-PLC and *C. perfringens* α-toxin have provided insight into the locations and roles of the zinc ions. In both proteins, three zinc ions are located in the putative active site cleft. One of the zinc ions is less firmly bound than the others, and it is likely that this is the zinc ion removed by EDTA treatment (Hough *et al.*, 1989; Titball and Rubidge, 1990; Nagahama *et al.*, 1995). Crystallographic analysis of the *B. cereus* PC-PLC with substrate or substrate derivative analogues (tris- and 3,4-dihexanoyloxybutylphosphorylcholine; Hansen *et al.*, 1993a, b) suggests that the zinc ions form part of the substrate binding pocket. These zinc ions also play a structural role by bridging helices in these proteins, and this might explain, in part, the high thermal stability of these proteins. The zinc-co-ordinating residues, identified in the crystal structures of *B. cereus* PC-PLC (Hough *et al.*, 1989) and *C. perfringens* α-toxin (Naylor *et al.*, 1998), are conserved in all of the other zincmetallophospholipases for which deduced amino acid sequences are available (Titball, 1993), suggesting that these enzymes all have similar active site architectures. Site-directed mutagenesis of these residues in *C. perfringens* α-toxin results in proteins devoid of PLC activity. Significantly, these mutant proteins were also non-haemolytic and non-lethal (Guillouard *et al.*, 1996; Nagahama *et al.*, 1995, 1997), confirming that the phospholipase activity is essential for all biological activities of α-toxin.

Until recently it was thought that zincmetallophospholipases C were produced only by Gram-positive bacteria. The *Pseudomonas fluoresecens* PLC is inactivated by treatment with a zinc-chelating compound (*o*-phenanthroline), which can be reversed on incubation with an excess of zinc ions (Crevel *et al.*, 1994; Ivanov *et al.*, 1996). The N-terminal amino acid sequence of the protein also suggests that the enzyme is related to the Gram-positive zincmetallophospholipases C (Crevel *et al.*, 1994).

These features, along with the apparent molecular mass of the protein (39.5 kDa), suggest that the enzyme is a zincmetallophospholipase C and indicate that the *P. fluorescens* enzyme is not related to the *Pseudomonas aeruginosa* PLCs (see below). It is possible that the enzymes produced by *Pseudomonas schuylkilliensis* (Arai *et al.*, 1974) and *Pseudomonas aureofaciens* (Sonoki and Ikezawa, 1975), which are also inhibited by treatment with EDTA, and have apparent molecular sizes of 23 kDa and 35 kDa, respectively, are also zincmetallophospholipases, but these enzymes have received little attention recently.

Gram-negative phospholipases C

There is significant amino acid sequence homology between the two PLCs produced by *P. aeruginosa* (PlcH encoded by *plcH* (previously known as *plcS*) and PlcN encoded by *plcN*; Ostroff *et al.*, 1990) and the PLCs produced by *Burkholderia pseudomallei* and *Burkholderia mallei* (Mack *et al.*, unpublished). On the basis of Southern blotting, the PLC produced by *Burkholderia cepacia* also appears to be a member of this group of enzymes (Vasil *et al.*, 1990). Although PlcH and PlcN show a high degree of overall homology (59%; Ostroff *et al.*, 1990), these PLCs differ in their properties. PlcH is haemolytic and able to hydrolyse phosphatidylcholine and sphingomyelin, whereas PlcN is non-haemolytic and able to hydrolyse phosphatidylcholine and phosphatidylserine. The proteins also have markedly different isoelectric points (p*I* 8.8 for PlcN and p*I* 5.5 for PlcH). The greatest homology between these PLCs occurs in the N-terminal regions, suggesting that the C-terminal regions are responsible for the different properties of these proteins (Ostroff *et al.*, 1990).

Expression of *P. aeruginosa* PlcH (but not PlcN) appears to be dependent on the co-expression of a protein (PlcR) that is encoded downstream of *plcH* and forms part of an operon, termed *plcHR* (Shen *et al.*, 1987). The function of *plcR* is not fully elucidated, but it is thought either to assist in the export of PlcH from the bacterial cell, or to modify post-translationally the structure of PlcH (Cota-Gomez *et al.*, 1997). Intriguingly, the deduced amino acid sequence of PlcR

shows significant homology with calmodulin-type calcium binding proteins (Cota-Gomez *et al.*, 1997) and Ca^{2+} ions appear to be required for activity of PlcH. In the case of *C. perfringens* α-toxin, Ca^{2+} ions are required for recognition of membrane phospholipids by the protein, and it is possible that the PlcR/Ca^{2+} complex performs a similar function for PlcH.

Until recently it was thought that two PLCs were produced by *Mycobacterium tuberculosis* (HL-A or MpcA and HL-B or MpcB (Leao *et al.*, 1995; Johansen *et al.*, 1996) with molecular masses of approximately 56 kDa and 30–40% amino acid sequence identity with the *P. aeruginosa* PLCs. The genome sequence of *M. tuberculosis* (Cole *et al.*, 1998) reveals that *mpcA* and *mpcB* are tandemly located and that upstream is a third open reading frame, which could encode another PLC (*mpcC*).

Interestingly, part of a PLC-encoding gene (*mpcD*) is located elsewhere on the genome. Only MpcA and MpcB have been studied in any detail and the smaller sizes of these proteins, in comparison with the *P. aeruginosa* enzymes, are due to several deletions within the open reading frame and the absence of the C-terminal polypeptide. MpcA expressed as a fusion with glutathione-S-transferase in *E. coli* possessed haemolytic activity (Leao *et al.*, 1995), but in *M. tuberculosis* the enzyme is thought to be surface bound (Johansen *et al.*, 1996), and this might explain the contact-dependent haemolytic activity associated with the most virulent strains of *M. tuberculosis* (Leao *et al.*, 1995).

Other phospholipases C

The relationship of other Gram-negative PLCs to the above enzymes is not known. Many of these enzymes are cell associated and are produced by intracellular pathogens, although the significance of this relationship awaits clarification. The *Legionella pneumophila* enzyme (M_r 50 000–54 000) requires divalent cations, and treatment with EDTA abolishes activity (Baine, 1988). However, the enzyme is not reactivated by treatment with Zn^{2+} ions, suggesting that it is not a zincmetallophospholipase C. The activity of the *Ureaplasma urealyticum* (De Silva and Quinn, 1987), the oral spirochaetes such as *Treponema denticola* (Siboo *et al.*, 1989) and *Helicobacter pylori* (Weitkamp *et al.*, 1993) enzymes has only been demonstrated using ρNPPC, so it is not possible to conclude whether these PLCs are phosphatidylcholine-preferring enzymes or are sphingomyelinases.

Sphingomyelinases C

The head groups of phosphatidylcholine and sphingomyelin are usually identical, the difference residing in the glycerol and fatty acyl tail groups of the phospholipid. Some bacterial phospholipases C show a remarkable specificity for sphingomyelin, which must reflect their ability to recognize these structural differences. *Bacillus cereus* sphingomyelinase (Yamada *et al.*, 1988; Gilmore *et al.*, 1989) and *Staphylococcus aureus* sphingomyelinase (β-toxin; Projan *et al.*, 1989) show 56% amino acid sequence identity over 200 amino acids and both have molecular masses of approximately 39 kDa. Both enzymes require Mg^{2+} or Co^{2+} ions for activity (Gerasimene *et al.*, 1985; Ikezawa *et al.*, 1986). The sphingomyelinase produced by *Leptospira interrogans* is related to the *B. cereus* and *S. aureus* enzymes, but appears to be produced as a high molecular mass form (63 kDa) that is processed at the C-terminus to yield a 39-kDa mature protein (Segers *et al.*, 1990)

Phosphatidylinositol-hydrolysing enzymes

Several bacterial enzymes ($M_r \sim 34\,000$) that are specific for phosphatidylinositol have been described. These enzymes (*B. cereus* PI-PLC; Kuppe *et al.*, 1989; *B. thuringiensis* PI-PLC; Henner *et al.*, 1988; *L. monocytogenes* PLC-A; Leimeister-Wächter *et al.*, 1991; Mengaud *et al.*, 1991) are all related at the amino acid sequence level (Mengaud *et al.*, 1991). They are all type-II PI-PLCs, that is they are soluble and are able to hydrolyse phosphatidylinositol and glucosyl phosphatidylinositol (GPI). It is possible that the *S. aureus* PI-PLC (apparent M_r 20 000–33 000: Ikezawa, 1986) and the *C. novyi* PI-PLC (apparent M_r 30 000: Ikezawa, 1986) are also structurally related to the *Bacillus* and *Listeria* enzymes.

Unlike the mammalian enzymes (type-1 enzymes), none of the bacterial PI-PLCs reported to date is a metalloenzyme – in fact Zn^{2+}, Mg^{2+} or Ca^{2+} ions inhibit the activity of the *B. cereus* and *S. aureus* enzymes (Ikezawa, 1991; Heinz *et al.*, 1998). Also unlike mammalian PI-PLCs, the bacterial enzymes are not able to hydrolyse phosphatidylinositol phosphates to generate inositol triphosphate secondary messengers. The bacterial PI-PLCs are also distinct from the GPI-PLC produced by *Trypanosoma brucei*, which behaves as an integral membrane-bound enzyme (Hereld *et al.*, 1988).

Significant insight into the relationship between the structures and functions of the bacterial (*B. cereus* PI-PLC) and mammalian enzymes (rat PI-PLC-δ1) has been obtained by comparing the crystal structures of these proteins (Figure 15.3, in colour plate section: Heinz *et al.*, 1998). The bacterial enzyme is a single domain, which folds as a $\beta\alpha_8$ barrel, with the active site located at the C-terminal end in an open cleft (Heinz *et al.*, 1998). The crystal structure of the *L. monocytogenes* PLC-A (Moser *et al.*, 1997) also reveals a single domain with very similar architecture to the *B. cereus* enzyme – even though these proteins show only 24% sequence identity.

The mammalian enzyme also has a $\beta\alpha_8$ barrel domain, which contains the active site, and also has a C2-fold domain, an EF-hand domain and a PH domain (Figure 15.3; Heinz *et al.*, 1998). Although the active site domain of the mammalian enzyme shows a related topology to the entire bacterial PI-PLC, significant amino acid sequence identity (26%) is found only between the N-terminal halves of the catalytic domains. On the basis of amino acid sequence alignments, it seems likely that the *T. brucei* enzyme is a single-domain protein, which is folded like the bacterial enzymes (Heinz *et al.*, 1998). The crystal structures of these proteins also confirm that only the mammalian enzymes bind a single catalytically essential Ca^{2+} ion within the active site cleft, and also show that the C2-fold domain of the mammalian enzymes co-ordinates Ca^{2+} ions (Heinz *et al.*, 1998).

Phospholipases D

The best-studied group of bacterial phospholipases D (PLDs) includes the enzymes produced by *Corynebacterium pseudotuberculosis*, *Corynebacterium ulcerous* and *Arcanobacterium haemolyticum*, which show 64–97% identity at the amino acid sequence level (Songer, 1997) and are related to the PLD in the venom of the brown recluse spider (Truett and King, 1993; Songer, 1997). These PLDs are haemolytic only in the presence of the cholesterol oxidase of *Rhodococcus equi*, and pretreatment of erythrocytes with *C. pseudotuberculosis* PLD actually renders the cells resistant to lysis with *C. perfringens* α-toxin or *S. aureus* β-toxin (Soucek *et al.*, 1971).

PLDs have also been reported to be produced by a variety of other bacteria including *Haemophilus influenzae*, *Salmonella typhimurium*, *Vibrio damsela* (Waite, 1987) and *M. tuberculosis* (Johansen *et al.*, 1996). These enzymes have generally not been well studied, although the *V. damsela* PLD is a haemolysin, which is active preferentially towards sphingomyelin, and degrades phospholipids in Madin-Darby canine kidney cells (Kothary and Kreger, 1985). As in the *C. pseudotuberculosis* enzyme, treatment of erythrocytes with the *V. damsela* PLD renders them more resistant to lysis with *S. aureus* β-toxin (Kreger *et al.*, 1987).

THE MOLECULAR BASIS OF SUBSTRATE SPECIFICITY

One of the striking features of the phospholipases is the wide variation in substrate specificity of proteins that are structurally related. For example, although phosphatidylcholine is the preferred substrate for the zincmetallophospholipases, individual enzymes show differences in their abilities to hydrolyse other phospholipids. The molecular basis for these differences is becoming clearer as crystal structures of these proteins, in some cases complexed with substrate analogues, are becoming available.

One of the key mechanisms by which phospholipases recognize (and discriminate between) their substrates involves the interaction of the head group with the active site. In part, the specificity of the enzyme for the substrate must reflect the charge on the phospholipid head group and the corresponding charge distribution in the active site cleft. Phosphatidylserine is the only phospholipid to carry a net negative charge, and it may be significant that *P. aeruginosa* PlcN (p*I* 8.8), which is a basic protein, is able to hydrolyse this phospholipid, whereas PlcH is acidic (p*I* 5.5) and cannot.

The molecular mechanisms by which phospholipases bind to the head group have been investigated for *B. cereus* PC-PLC (Hanssen *et al.*, 1993a, b) and *B. cereus* PI-PLC (Heinz *et al.*, 1998). *Bacillus cereus* PC-PLC appears to recognize, primarily, the phosphate moiety that displaces water molecules and becomes co-ordinated to the three zinc ions in the active site cleft (Hansen *et al.*, 1993b). Additional binding occurs via hydrogen bonds between various amino acid sidechains and, especially, the choline head group and the carbonyl group (El-Sayed *et al.*, 1985; Hansen *et al.*, 1993b). The active site of *B. cereus* PI-PLC is relatively wide, to allow entry of the bulky inositol head group, which is recognized primarily via the 4- and 5-hydroxyl groups and various polar and charged side-chains of amino acids. In fact, the close interactions between these hydroxyl groups and the active site explain why the enzyme is not able to cleave inositol bisphosphate, since the additional phosphate groups cannot be accommodated within the active site cleft (Heinz *et al.*, 1998).

The importance of the recognition of the fatty acyl tails of phospholipids is more variable between different bacterial phospholipases. The recognition of substrate by the bacterial PI-PLCs does not appear to be dependent on the fatty acyl tail groups, and these enzymes show no stereospecificity towards the diacylglycerol moiety (Bruznik *et al.*, 1992). In contrast, the fatty acyl tails are recognized by zincmetallophopholipases such as *B. cereus* PC-PLC and *C. perfringens* α-toxin. The fatty acyl tails must be at least C_6 in length for recognition by *B. cereus* PC-PLC (El-Sayed *et al.*, 1985), and the recognition of phospholipids incorporated into liposomes by *C. perfringens* α-toxin is dependent on the presence of unsaturated fatty acyl tail groups that are C_{14} or less in length (Nagahama *et al.*, 1996).

Some of these apparent substrate recognition events might reflect changes in membrane fluidity, especially when liposome assay systems are used. However, it is clear that, for some enzymes, recognition of this region of phospholipids is a key event; since the head groups of sphingomyelin and phosphatidylcholine are usually identical (i.e. phosphorylcholine), the discrimination of these molecules must be dependent on recognition of the glycerol backbone and fatty acyl tail.

ROLES OF PHOSPHOLIPASES IN DISEASE

In 1941, pioneering work by MacFarlane and Knight showed that a bacterial toxin produced by *C. perfringens* was a PLC (MacFarlane and Knight, 1941). This finding stimulated interest in the roles of PLCs produced by other pathogenic bacteria, and we are still discovering the diverse roles of these enzymes in the pathogenesis of diseases caused by bacteria.

Colonization and invasion of host tissues

Whilst there is no clear evidence that PLCs are essential for the colonization of host tissues, there are several reports that suggest a role for these enzymes in this process.

One barrier to colonization and invasion of the respiratory tract is the lung surfactant, which is rich in phosphatidylcholine. The degradation of lung surfactant has been demonstrated *in vitro* using the α-toxin from *C. perfringens* (Holm *et al.*, 1991). *Clostridium perfringens* is not a pathogen of the respiratory tract, but there is no reason to suppose that phospholipases produced by respiratory tract pathogens would not be equally effective in degrading the lung surfactant and thereby enhancing the colonization of underlying tissues. It is known that the *P. aeruginosa* PLCs are produced *in vivo*, and cystic fibrosis patients develop antibody responses to these proteins (Granström *et al.*, 1984). The finding that a Δ*plcS*–Δ*plcN* mutant of *P. aeruginosa* was less able than the wild-type to induce injury to alveolar epithelial cells (Saiman *et al.*, 1992) certainly suggests that the phospholipases do play a role in disease of the respiratory tract. The *P. aeruginosa* enzymes might play a second role in respiratory tract infections by converting phosphatidylcholine to choline-betaine, which then accumulates within the bacterial cell and acts as a protectant against the high osmotic strength environment in the lung (Shortridge *et al.*, 1992). The phospholipases of other pathogens of the respiratory tract might play similar roles.

The suggestion that phospholipases produced by respiratory tract pathogens might enhance colonization has led to the proposal that these enzymes might also play similar roles in the case of pathogens that colonize the stomach. For example, it has been suggested that the PLA$_2$ produced by *H. pylori* plays a role in the degradation of the phosphatidylcholine-rich stomach lining and thereby allows access to underlying tissues (Langton and Cesareo, 1992). This suggestion is supported by the finding that bismuth salts, which are often used to treat peptic ulcers, are potent inhibitors of the *H. pylori* PLA$_2$ activity (Ottlecz *et al.*, 1993). The phospholipases produced by oral spirochaetes might perform similar functions (Siboo *et al.*, 1989)

A second possible role for phospholipases in the invasion of host tissues arises from a recent study with *B. cereus* PC-PLC, which has shown increased matrix metalloprotease production and enhanced permeability of monolayers after treatment of epithelial cell cultures with this enzyme (Firth *et al.*, 1997). Whether this mechanism also operates *in vivo*, and with PLCs produced by pathogenic bacteria, awaits investigation. However, if demonstrated, this process could allow the release of nutrients on to the epithelial cell surface or allow the bacteria to gain access into the body.

Growth and spread of infection in the host

The finding that some phospholipases are produced by bacteria that are not pathogenic for humans but are saprophytes has prompted suggestions that these enzymes form part of an enzyme cascade involved in phosphate scavenging. The production of the *B. cereus* PLC is induced under low P$_i$ conditions, lending weight to this suggestion (Guddal *et al.*, 1989). It is worth noting that the PI-PLCs release GPI-anchored proteins, such as alkaline phosphatase and alkaline phosphodiesterase, from the surfaces of eukaryotic cells (Ikezawa, 1986), and it is conceivable that these released enzymes contribute to the phosphate-scavenging process. In addition to their possible phosphate-scavenging functions, some phospholipases appear to play a direct role in the spread of infection in the host. Perhaps the most dramatic examples of this are seen with gas gangrene infections in humans caused by *C. perfringens* and lymphadenitis or lymphangitis in ruminants or horses caused by *C. pseudotuberculosis*.

The development of gas gangrene is associated with two events. Firstly, traumatic injury to soft tissues, such as those sustained during road traffic accidents or during warfare, allows entry of the bacterium (which is present in soil and decaying organic matter) into the host (Willis, 1969). Secondly, disruption to the blood supply, which might occur following the severing of major blood vessels, results in anoxic conditions in tissues necessary for the growth of *C. perfringens* – an anaerobe. In the mouse model of disease, gas gangrene can be caused by the inoculation of a large challenge dose of bacteria into the hind limb (Williamson and Titball, 1993; Awad *et al.*, 1995). The disease progresses rapidly, with spread of the infection from the initial focus followed by extensive muscle necrosis and swelling of the affected limb and is fatal in the majority of animals within 18 h (Awad *et al.*, 1995).

For many decades, it was suggested that α-toxin was the major virulence determinant associated with the spread of the infection from the initial focus into adjacent tissues (McNee and Dunn, 1917; MacFarlane and Knight, 1941). Evidence that α-toxin played a key role in this process was provided when a *plc*-mutant of *C. perfringens* was tested in the murine model with little evidence of infection, minimal muscle necrosis and no deaths of experimental animals (Awad *et al.*, 1995). The *plc*-mutant containing the cloned α-toxin gene was as virulent as the wild-type strain (Awad *et al.*, 1995).

The mechanisms by which α-toxin allows the spread of the bacterium into previously healthy tissues is not fully elucidated. However, the toxin is thought to diffuse away from the site of infection into adjacent healthy tissues. Here α-toxin might cause changes to the microcirculatory system in the limb, which results in the anoxic conditions required for growth of the bacterium. Certainly, α-toxin induces platelet aggregation (Ohsaka *et al.*, 1978) and muscle contraction (Fujii and Sakurai, 1989), both of which would reduce the blood supply to tissues. It seems likely that these effects are a result of activation of the arachidonic acid cascade and protein kinase C. The toxin might also damage healthy cells, allowing the release of nutrients, and modulate the immune response, thereby allowing the spread of the infection. These effects are discussed in more detail below.

The PLD produced by *C. pseudotuberculosis* also appears to play a key role in the growth and colonization of tissues by the bacterium. Studies in sheep (Hodgson *et al.*, 1992) or in goats (McNamara *et al.*, 1994) have shown that subcutaneous abscesses, which are characteristic of infection with a wild-type strain, were not apparent in most of the animals challenged with a *pld*-mutant. In contrast to the *C. perfringens* α-toxin, which appears to play a role in reducing the blood supply to infected tissues, the PLD produced by *C. pseudotuberculosis* causes increased vascular permeability, and this might allow the bacterium to spread from the primary sites of infection to the regional lymph nodes, where chronic abscesses become established (Muckle and Gyles, 1983; McNamara *et al.*, 1994). Apart from the haemolytic activity of the enzyme, the effects of the *C. pseudotuberculosis* PLD on eukaryotic cells are not known. However, since one effect of the leukotrienes is to increase vascular permeability, it is tempting to speculate that the ceramide phosphate product plays a role in activation of the arachidonic acid cascade. In this context, it may be significant that some PLCs (and perhaps eukaryotic membrane PLCs) are able to hydrolyse ceramide phosphate to generate diacylglycerol (Waite, 1987).

The PLCs produced by intracellular pathogens also play key roles in the spread of the bacterium. *Listeria monocytogenes* produces two PLCs; PLC-A is active against phosphatidylinositol, whilst PLC-B preferentially hydrolyses phosphatidylcholine (Mengaud *et al.*, 1991; Leimeister-Wächter *et al.*, 1991; Vazquez-Boland *et al.*, 1992). Initial studies with single Δ*plcA* (Camilli *et al.*, 1991) or Δ*plcB* (Vazquez-Boland *et al.*, 1992) mutants suggested that these enzymes played a minor role, respectively, in the escape of the bacterium from the phagosome and the spread of the bacterium from cell to cell (Figure 15.4).

More recent studies have suggested that these enzymes have overlapping functions, and a Δ*plcA* Δ*plcB* double mutant was 500-fold attenuated in a mouse model of disease (Smith *et al.*, 1995). In cell cultures, the double mutant was markedly less effective in escaping from the phagosome and in spreading from cell to cell (Smith *et al.*, 1995). These findings suggest that the enzymes contribute to degradation of the phagosome membrane and degradation of the double membrane

vacuole that forms when bacteria spread from cell to cell. It is also worth noting that these effects are further potentiated by listeriolysin O (Goldfine *et al.*, 1995), a pore-forming toxin, which is produced by *L. monocytogenes*.

Whilst most of this chapter is devoted to the consideration of the roles of phospholipases in the pathogenesis of diseases of mammalian hosts, it is also worth highlighting the possibility that these enzymes play a role in diseases of non-mammalian hosts. Perhaps the most dramatic example of this is seen in the case of the PLCs of *P. aeruginosa*, some strains of which are able to infect both mammalian and plant hosts (Rahme *et al.*, 1995). Whilst PlcH is known to play an important role in disease of mammalian hosts, for example in burns infections (Ostroff *et al.*, 1990; Rahme *et al.*, 1995) and in respiratory tract infections (Saiman *et al.*, 1992), the enzyme also appears to play a key role in growth in plant (*Arabidopsis thaliana*) tissues (Rahme *et al.*, 1995). Whether the enzyme exerts similar effects in these markedly different host species awaits investigation.

Other PLCs appear to play less important roles in the infection process. The pathological effects of wild-type and β-toxin (Hlb) negative mutants of *S. aureus* were almost identical in the mouse model of mastitis and, suprisingly, the level of tissue colonization was greater with the Hlb⁻ mutant (Bramley *et al.*, 1989). Whilst this observation suggests that β-toxin does not play a role in disease, it should be borne in mind that *S. aureus* causes a wide variety of different diseases in humans and animals. It seems unlikely that the bacterium would retain a gene encoding a protein that was not required for survival or growth of the bacterium in any niche. Perhaps the enhanced colonization levels of the Hlb⁻ mutant of *S. aureus* in the mastitis model indicate the role of Hlb in the development of a stable host–pathogen relationship.

Avoidance of host defence mechanisms

The cytolytic properties of some PLCs extend to cells other than erythrocytes, and cytolytic activity towards host phagocytic cells cultured *in vitro* has been reported (Titball *et al.*, 1993). There is no evidence that lysis of phagocytic cells occurs on a large scale *in vivo*, but there is good evidence that PLCs can modulate the host response to infection in more subtle ways, thereby allowing the growth of bacteria. The usual host response to infection is to mount a dramatic inflammatory response with the migration of phagocytes into the infected tissues. In contrast, it has been known for many years that tissue samples taken from gas

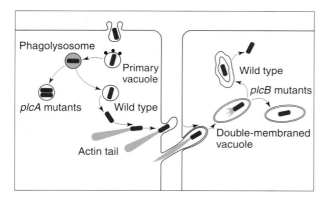

FIGURE 15.4 Roles of the *L. monocytogenes* PLC-A and PLC-B in escape from the primary phagosome and spread from cell to cell. *plcA* mutants are not able to escape from the phagosome and therefore accumulate within the phagosome. *plcB* mutants are not able to escape from the double-membrane vacuole that forms when the bacteria spread from cell to cell. Reproduced from Songer (1997), with permission from Elsevier Science Publishers Ltd.

gangrenous limbs were devoid of phagocytic cells (McNee and Dunn, 1917). This pattern is also seen in tissues taken from mice experimentally infected with wild-type *C. perfringens*, but not in tissues taken from mice experimentally infected with a Δ*plc* mutant, where there was the expected influx of phagocytes into infected tissues (Stevens *et al.*, 1997).

In an elegant study, it was shown that α-toxin caused the upregulation of cell adherence markers (ICAM, ELAM and p-selectin) on the surface of endothelial cells lining blood vessels (Bryant and Stevens, 1996; Bunting *et al.*, 1997). In tissues taken from mice infected with wild-type strains, phagocytic cells bound to these markers and accumulated within blood vessels serving the site of infection. Therefore, it appears that α-toxin causes mis-trafficking of host phagocytes, allowing the bacterium to grow in host tissues. Whether other bacterial phospholipases are also able to cause host phagocyte mis-trafficking awaits investigation.

Damage to the host

Early studies with PLCs focused on their cytolytic properties, and there was a general assumption that these enzymes caused widespread cell lysis in the host. Although lysis of erythrocytes does occur in experimental gas gangrene (evidenced as haematuria), there is little evidence that extensive haemolysis occurs during natural cases of gas gangrene or any other diseases involving phospholipase-producing bacteria. Rather, it appears that these enzymes damage the host by perturbing host cell metabolism.

Many derivatives of phospholipids act as secondary messengers within eukaryotic cells (Exton, 1990). Diacylglycerol, which may be generated after cleavage of glycerophospholipids by PLCs, has profound effects on cellular metabolism (Exton, 1990). Diacylglcerol lipase is able to convert diacylglycerol into arachidonic acid, which is then able to enter the arachidonic acid cascade (Samuelsson, 1983), with the resultant generation of prostaglandins, thromboxanes and leukotrienes (Figure 15.5). Treatment of cultured cells or isolated tissues with a variety of phospholipases including *C. perfringens* α-toxin (Fujii and Sakurai, 1989; Bunting *et al.*, 1997) or *P. aeruginosa* Plc-H (König *et al.*, 1997) result in the generation of these molecules. The diacylglycerol might also activate endogenous protein kinase C, which is essential for both short- and long-term effects on cellular metabolism (Nishizuka, 1992; Bunting *et al.*, 1997). The elevated production of matrix metalloproteases by cells treated with *B. cereus* PC-PLC is also thought to be due to activated protein kinase C (Firth *et al.*, 1997).

The effects of bacterial phospholipases on mammalian cells might be further potentiated as a result of activation of endogenous membrane phospholipases and especially phospholipases A_2, C and D (Gustafson and Tagesson, 1990; Sakurai *et al.*, 1993; Sakurai *et al.*, 1994; Ochi *et al.*, 1996). In studies with rabbit erythrocyte membranes, it has been shown that endogenous PLC is rapidly activated, followed later by the activation of PLD. These activated mammalian enzymes might themselves generate further substrates for the arachidonic acid cascade. The mechanism by which the mammalian phospholipases are activated has been the subject of speculation. Some workers have suggested that activated protein kinase C is able to activate the mammalian phospholipases, either directly or indirectly (Waite, 1987; Exton, 1990; Nishizuka, 1992). An alternative explanation is that the α-toxin mediates the effect via activation of GTP-binding proteins, which in turn activate mammalian phospholipases (Sakurai *et al.*, 1994).

In addition to the further activation of the arachidonic acid cascade, mammalian PLCs hydrolyse phosphatidylinositol diphosphate (PIP_2) to generate the secondary messenger inositol triphosphate (IP_3; Sakurai *et al.*, 1993). The IP_3 would have a variety of effects on cells. For example, the release of calcium from the endoplasmic reticulum is triggered by IP_3, causing the opening of calcium gates in the cell membrane (Fujii *et al.*, 1986). Boethius *et al.* (1973) have shown that muscle cells treated with α-toxin become inexcitable, and this might explain the cardiotoxic effects of the α-toxin (Asmuth *et al.*, 1995). The elevated intracellular calcium levels would also contribute to the activation of endogenous membrane phospholipases described in the previous paragraph.

The effects of α-toxin on eukaryotic cells might be potentiated by the metabolic status of the cells. A mutation in the UDP-glucose pyrophosphorylase gene in Chinese hamster fibroblasts, leading to UDP-glucose deficiency, rendered cells 105 times more sensitive to *C. perfringens* α-toxin (Flores-Díaz *et al.*, 1997). UDP-glucose deficiency is also a consequence in ischaemic cells. Ischaemia would be expected in tissues infected with *C. perfringens*, leading to the suggestion that these tissues would be more susceptible to the effects of α-toxin. The molecular basis for this enhanced susceptibility has not been fully elucidated. However, UDP-glucose-deficient cells contained normal levels of phosphatidylcholine, suggesting that alterations to the outer membrane leaflet were not responsible for the enhanced sensitivity. Interestingly, these cells were not more susceptible to *B. cereus* PC-PLC, and the effects of α-toxin on the cells could not be blocked

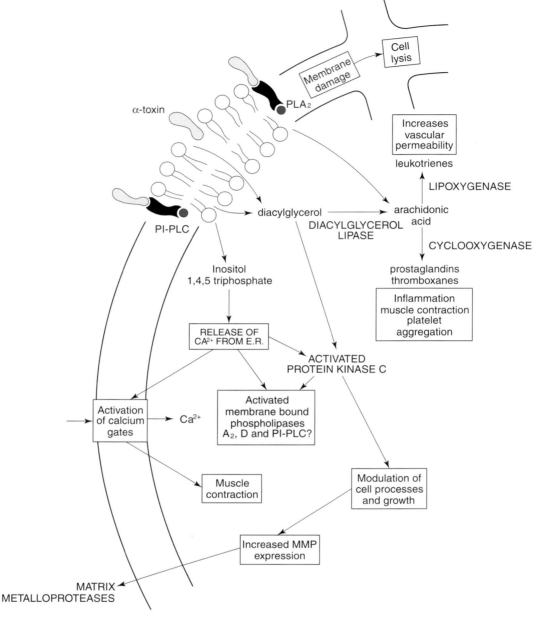

FIGURE 15.5 Effects of *C. perfringens* α-toxin on mammalian cell metabolism. Adapted from Titball (1993), with permission from the American Society for Microbiology.

using inhibitors of the arachidonic acid cascade or using inhibitors of eukaryotic phospholipases (Flores-Díaz *et al.*, 1998).

Whilst the above paragraphs discuss the effects of glycerophospholipid hydrolysis on cellular metabolism, it is becoming apparent that the products of sphingolipid hydrolysis (i.e. ceramide) by sphingomyelinases might also modulate cellular metabolism. Although the ability of bacterial enzymes to activate these pathways has not been studied in detail, it is per-

haps worth pointing out that ceramide is thought to play an important role in the induction of apoptotic responses (Spiegel *et al.*, 1996).

When viewed together, these elegant studies tend to support the hypothesis that the bacterial enzymes are mimicking the effects of the mammalian enzymes and thereby modulating cellular metabolism to the detriment of the host. The effects described above are all relatively short term. It is also worth bearing in mind that long-term perturbation of cellular messengers such as protein kinase C has been linked with the transformation of cultured cells. Such effects might suggest a link between the development of cancers and the long-term exposure to bacterial phospholipases (Parkinson, 1987).

INTERACTION OF PHOSPHOLIPASES WITH MEMBRANE PHOSPHOLIPIDS

A central feature of phospholipases that play roles in the pathogenesis of disease is their ability to interact with membranes, whether from outside the cell or from inside phagosomes. Bacterial phospholipases vary widely in their abilities to interact with membrane phospholipids, and this has resulted in the suggestion that those enzymes that play important roles in the pathogenesis of disease can be identified on the basis of their haemolytic activity. To some extent this proposal is supported by experimental evidence. Of the two related PLCs produced by *P. aeruginosa* PlcH is haemolytic and able to induce the release of inflammatory mediators from granulocytes, whereas PlcN is devoid of both activities (König *et al.*, 1997). A similar pattern is observed when the properties of the related *C. perfringens* α-toxin and *B. cereus* PC-PLC are compared; only the *C. perfringens* α-toxin is haemolytic, able to cause release of inflammatory mediators from granulocytes (König *et al.*, 1997) and lethal when administered to animals (Titball *et al.*, 1991).

The mechanisms that allow some enzymes to interact with membrane phospholipids have been the subject of various studies and various theories have been developed to explain the mode of action of these enzymes. One theory proposes that membrane-active enzymes have structural features that allow the insertion into membranes. A second theory suggests that membrane-active enzymes are able to hydrolyse both phosphatidylcholine and sphingomyelin. These theories are discussed in more detail below.

Insertion into the membrane bilayer

It is generally assumed that the surface of the phospholipid bilayer consists of tightly packed head groups, which mask the hydrophobic tail groups. This model would suggest that the bond susceptible to cleavage by the phospholipase is not exposed to the aqueous phase. In reality, the phospholipids adopt a 'hockey stick' conformation in the bilayer, and the susceptible bond might be partially accessible to an enzyme in the aqueous phase (Figure 15.6). However, even allowing for the 'hockey stick' conformation of

phospholipids, it is apparent that the active site of the enzyme could not be docked with the substrate without the phospholipase's either becoming inserted into the membrane or having a mechanism that allowed retraction of the phospholipid out of the membrane.

Evidence supporting the 'insertion' hypothesis was originally reported by van Deenen *et al.* (1976), who measured the activity of phospholipases towards phospholipid monolayers maintained at increasing lateral pressures. This study (Table 15.2) showed that haemolytic phospholipases (such as *C. perfringens* α-toxin) were able to insert into monolayers maintained at lateral pressures above those found in erythrocytes. In contrast, non-haemolytic enzymes (such as *B. cereus* PC-PLC) were able to hydrolyse phospholipids only in monolayers maintained at low lateral pressures.

Surface-exposed hydrophobic amino acids might become buried in membranes

What structural features of phospholipases might allow insertion into the membrane? On comparing the structures of *B. cereus* PC-PLC (Hough *et al.*, 1989) and *C. perfringens* α-toxin (Naylor *et al.*, 1998), it is apparent that both enzymes possess structurally similar catalytic domains, but that only α-toxin has a second domain (carboxy-terminal domain). If the carboxy-terminal domain is removed from α-toxin, the PLC activity is

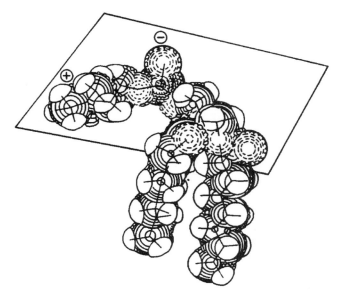

FIGURE 15.6 Space-filling and energy minimized model of dioctanoyl phosphatidylcholine. The regions of the phospholipid that are recognized by the *B. cereus* PC-PLC are shown boxed. The charged regions on the phosphorylcholine head group are also shown. Adapted from El-Sayed *et al.* (1985), with permission from Elsevier Science Publishers Ltd.

TABLE 15.2 Effect of lateral pressure on the hydrolysis of phospholipids in monolayers by different phospholipases

Source of enzyme	Type	Haemolysis of human erythrocytes	Maximum lateral pressure at which hydrolysis ceases (dyne cm^{-1})
Pig pancreas	A2	−	16.5
Cabbage	D	−	20.5
Crotalus adamanteus venom	A2	−	23
B. cereus PC-PLC	C	−	31
Naja naja venom	A2	+	34.8
Bee venom	A2	+	35.3
S. aureus	C	+	>40
C. perfringens	C	+	>40

Modified from Möllby (1978), with the permission from Academic Press Ltd.

retained but haemolytic activity is lost (Titball *et al.*, 1991). The possibility that the carboxy-terminal domain contains a second active site, responsible for membrane phospholipid hydrolysis, appears unlikely, since the isolated carboxy-terminal domain does not possess haemolytic (or indeed any detectable enzymic) activity (Titball *et al.*, 1993). When the two domains of α-toxin are mixed together in solution, haemolytic activity is restored, confirming that both the catalytic domain and the carboxy-terminal domain are required for membrane activity (Titball *et al.*, 1993).

Several regions on the surface of α-toxin have been identified as having possible membrane binding functions (Naylor *et al.*, 1998). Trp$_{214}$ (*N*-terminal domain) Tyr$_{331}$ and Phe$_{334}$ (*C*-terminal domain) are surface exposed and are positioned such they could interact with the phospholipid tail groups on membrane binding (Figure 15.7, in colour plate section). In the *B. cereus* PC-PLC, Trp$_{214}$ is replaced with a pair of antiparallel helices and Tyr$_{331}$ and Phe$_{334}$ are not present, since there is no carboxy-domain in this enzyme. Other bacterial phospholipases might exploit similar mechanisms to allow insertion into the membrane; the *B. cereus* PI-PLC has a 'ridge' of hydrophobic amino acids, which are thought to become inserted into the membrane (Heinz *et al.*, 1998).

C2 domains might be involved in phospholipid recognition

A second mechanism that allows binding of α-toxin to membranes involves the co-ordination of calcium ions by sites formed partially by phospholipid phosphate groups and partially by side-chains in the carboxy-terminal domain (Guillouard *et al.*, 1997).

For many years it has been known that calcium ions are essential for the haemolytic activity of α-toxin (MacFarlane and Knight, 1941) and for the hydrolysis of monomolecular films of phosphatidylcholine (Bangham and Dawson, 1962; Moreau *et al.*, 1988). However, analysis of purified α-toxin has not revealed the presence of bound calcium ions (Guillouard *et al.*, 1997), suggesting that the calcium binding sites were of a low affinity – which would be expected if the Ca^{2+} ions were partially co-ordinated by membrane phospholipid. Although the location of calcium ions in the molecule has not been determined, protein crystallized in the presence of cadmium ions did reveal potential calcium-binding sites in the carboxy-terminal domain of α-toxin (calcium and cadmium have similar charges and ionic radii; Naylor *et al.*, 1998). The fold of the carboxy-terminal domain, and the ability of this region to recognize phospholipids via Ca^{2+} co-ordination, confirms that this domain is a 'C2-fold domain'. C2-fold domains have not previously been found in prokaryotes, and are typically found in proteins that are involved in lipid metabolism in higher eukaryotes, such as synaptotagmin and phosphatidylinositol-specific PLC (Naylor *et al.*, 1998).

In this context, it is interesting to note that the mammalian PI-PLC-δ1 has a C2-fold domain (Figure 15.2), and models suggest that this domain may be involved in Ca^{2+}-mediated interaction with the membrane surface (Heinz *et al.*, 1998). The crystal structure of mammalian lipoxygenase also reveals an *N*-terminal C2-like domain (Gillmor *et al.*, 1997), and this is reflected in the earlier observation that the deduced amino acid sequences of the *C*-terminal domain of α-toxin and the *N*-terminal region of arachidonate-5-lipoxygenase showed significant homology (Titball *et al.*, 1991).

Although these models can explain how phospholipases become partially inserted into the membrane, they do not fully explain how the active site of the enzymes is brought into juxtaposition with the susceptible bond in the phospholipid. It is possible that the enzymes partially retract the phospholipid from the membrane; alternatively, the phospholipase may undergo a conformational change, which inserts the active site region into the membrane (Blow, 1991).

The properties of C2 domains vary

All of the zincmetallophospholipases C show a one- or two-domain organization and, although the two-domain enzymes are all haemolytic, the specific activities of these enzymes are different. This might be because the carboxy-terminal domains of different enzymes are not structurally identical and have different properties. The carboxy-terminal domain of the *C. bifermentans* enzyme shows 50% identity with the carboxy-terminal domain of *C. perfringens* α-toxin and it is therefore likely to be a C2-fold domain. However, the *C. bifermentans* enzyme is only weakly haemolytic, even though the PLC activity is close to that of α-toxin. Fusion of the α-toxin carboxy-terminal domain to the *C. bifermentans* amino-terminal domain significantly enhanced the haemolytic activity of the enzyme (submitted). Interestingly, the residues identified from the model of the α-toxin as playing a key role in membrane binding are replaced with a functionally different type of amino acid in the *C. bifermentans* C-terminal domain (Naylor *et al.*, 1998).

Substrate specificity and membrane phospholipid hydrolysis

For some years, it was suggested that haemolytic PLCs were able to hydrolyse both phosphatidylcholine and sphingomyelin, which are major components of the outer leaflet of mammalian cell membranes, whereas non-haemolytic enzymes were able to hydrolyse only one type of phospholipid (Ostroff *et al.*, 1990; Titball, 1993). For example, the haemolytic *P. aeruginosa* PlcH enzyme can hydrolyse both phosphatidylcholine and sphingomyelin, whilst the related *P. aeruginosa* PlcN enzyme is able to hydrolyse only phosphatidylcholine and is non-haemolytic (Ostroff *et al.*, 1990).

One key difference between *B. cereus* PC-PLC and *C. perfringens* α-toxin is the ability of the latter to hydrolyse sphingomyelin and phosphatidylcholine (Krug and Kent, 1984) whereas the *B. cereus* enzyme is able to hydrolyse only phosphatidylcholine (Otnaess *et al.*, 1977). Significantly, a mixture of *B. cereus* PC-PLC and *B. cereus* sphingomyelinase does cause haemolysis, and this mixture is termed cereolysin A–B (Gilmore *et al.*, 1989).

How is *B. cereus* PC-PLC active against erythrocyte membranes when combined with a sphingomyelinase, when it is apparently not able to insert into membrane bilayers? One possibility is that the sphingomyelinase enzyme first hydrolyses sphingomyelin, and that the resultant changes in membrane organization (and lateral pressure) then allow the phosphatidylcholine hydrolysing enzyme to act on membrane phospholipids. Experimental evidence does lend some support to this suggestion; sphingomyelin plays a key structural role in the outer leaflet of membranes and erythrocytes incubated with *B. cereus* (or *S. aureus*) sphingomyelinase are not lysed unless they are subsequently exposed to an additional stress such as cooling below the phase transition temperature (termed hot–cold lysis) or treatment with divalent cation chelating agents (Bernheimer *et al.*, 1974; Ikezawa *et al.*, 1986). Viewed together, these findings tend to confirm that when sphingomyelin and phosphatidylcholine are hydrolysed, the cell membrane is sufficiently weakened, which results in lysis.

Membrane-active enzymes may be non-haemolytic

The above findings suggest that haemolytic activity is indicative of the hydrolysis of both sphingomyelin and phosphatidylcholine, and that haemolytic enzymes (or mixtures of enzymes) are able to penetrate membrane bilayers. Haemolysis is certainly an extreme measure of cell membrane damage and remains a useful indicator of the interaction of an enzyme with mammalian cells. However, it should not be viewed as the sole indicator of membrane phospholipid hydrolysis. As noted above, bacterial sphingomyelinases are able to hydrolyse membrane sphingomyelin without cell lysis. In other cases, non-haemolytic enzymes might act synergistically with other bacterial products to cause haemolysis. One example of this synergy is seen with the PLCs produced by *L. monocytogenes*. Although PLC-A and PLC-B are considered to be non-haemolytic (or at least very weakly haemolytic; Geoffroy *et al.*, 1991), they play key roles in disruption of the phagosome and the double-membrane vacuole formed during cell-to-cell spread (Smith *et al.*, 1995). It is significant that these phospholipases appear to act together, and with the pore-forming toxin listeriolysin O, to achieve these effects (Goldfine *et al.*, 1995; Smith *et al.*, 1995).

VACCINES AND THERAPEUTIC APPLICATIONS OF PHOSPHOLIPASES

Membrane probes

For many years phospholipases have been used as membrane probes – either specifically to hydrolyse different classes of phospholipids in membranes or to mimic the effects of eukaryotic PLCs. In the past, the use of PLCs in this way has been fraught with many problems; the enzymes were isolated from bacteria, which often produced other membrane-active toxins, and the difficulties of isolating 'pure' PLCs meant that it was difficult to be certain that the effects observed were due to the bacterial PLCs. The availability of cloned gene products means that better-defined preparations of phospholipases are now available to support these studies.

Vaccines

Given the key roles that some phospholipases play in the pathogenesis of disease, it is not surprising that inactivated forms of the enzymes have been considered as components of vaccines. In this respect, *C. perfringens* α-toxin has attracted most attention, but the toxic nature of this enzyme meant that early vaccination studies used formaldehyde toxoids, prepared from α-toxin that was isolated from the supernatant fluid of cultures of *C. perfringens* (Evans *et al.*, 1945; Boyd *et al.*, 1972; Kameyama *et al.*, 1975). Whilst these toxoids did induce protection against gas gangrene, it was difficult to ensure that the vaccine did not contain other toxoided *C. perfringens* proteins, and therefore to be sure that the protective component in the vaccine was α-toxoid.

Several approaches for the generation of genetically engineered toxoids have been reported. Site-directed mutant forms of the α-toxin with reduced enzymic and cytolytic activity have been reported (Nagahama *et al.*, 1995; Guillouard *et al.*, 1996; Nagahama *et al.*, 1997). These mutated proteins could, in principle, be developed as vaccines if multiple mutations were incorporated to ensure that reversion was an unlikely event, and to reduce the activity of the enzyme to a negligible level. Alternatively, the carboxy-terminal domain of α-toxin, produced using genetic engineering of *E. coli*, could be used as a non-toxic vaccine. Immunization of mice with this vaccine induced high levels of antibody, which protected mice against experimental gas gangrene cause by *C. perfringens* (Williamson and Titball, 1993). This vaccine might be used to protect individuals at risk from developing gas gangrene, such as the elderly or individuals due to undergo surgical procedures involving the lower gastrointestinal tract. The vaccine has also been proposed for the prevention of enterotoxigenic diseases of domesticated livestock, including necrotic enteritis in chickens and sudden death disease in calves and piglets (Ginter *et al.*, 1996).

The finding that the *C. pseudotuberculosis* PLD plays a key role in the pathogenesis of caseous lymphadenitis in sheep and goats has prompted suggestions that either a PLD mutant might itself be used as a vaccine (Hodgson *et al.*, 1992) or an enzymically inactive form of the PLD might be used as a component of a subunit vaccine (Haynes *et al.*, 1992).

The possibility that phospholipases produced by other bacteria might be used as components of vaccines has yet to be investigated. However, it should be borne in mind that, in the case of intracellular pathogens, it may be difficult to induce an appropriate protective immune response. Furthermore, many of the enzymes produced by extracellular pathogens do not play the same overriding roles in the pathogenesis of disease as the α-toxin of *C. perfringens* or the PLD of *C. pseudotuberculosis*.

Phospholipase C conjugates

The ability of phospholipases to damage host cells has attracted attention as the basis of immunotoxins for treating some tumours. The first studies used *C. perfringens* α-toxin linked to an antibody against a eukaryotic cell surface marker (Chovnick *et al.*, 1991). Whilst these immunotoxins showed retention of both PLC and antibody-binding activity, the systemic toxicity effects were considered to be too great for use of the immunotoxin in humans. A variety of approaches has been suggested to resolve this problem. One uses the targeting of low doses of the immunotoxin to the cancer cell surface, where the immunotoxin is then able to cause release of liposome-encapsulated drugs. Studies *in vitro* and *in vivo* have confirmed that this approach can be used to enhance the activity of liposome-encapsulated anti-cancer drugs (Carter *et al.*, 1998)

Design of therapeutics

When bacterial phospholipases play key roles in the pathogenesis of disease, it is feasible that enzyme inhibitors could be used to treat these diseases. For example, bismuth salts, which can be used to treat gastric and duodenal ulcers, are thought to inactivate the *H. pylori* PLA_2, and it may be possible to design more effective enzyme inhibitors. One limitation of this approach is that such inhibitors might also be active against eukaryotic PLCs and one report describes immunological cross-reactivity between *B. cereus* PC-PLC and mammalian phosphatidylcholine-preferring PLCs (Clark *et al.*, 1986). The possibility that prokaryotic and eukaryotic phospholipases have similar structures or molecular modes of action suggests that the bacterial enzymes might be used to design and test drugs for use against eukaryotic PLCs or lipid binding enzymes.

Such drugs might be used to treat a variety of inflammatory diseases in humans. In this respect it is interesting to note that the *C*-terminal domain of the *C. perfringens* α-toxin is folded as a C2-domain that is found frequently in eukaryotic enzymes such as synaptotagmin, phosphatidylinositol PLC and arachidonate-5-lipoxygenase (Naylor *et al.*, 1998).

FUTURE DIRECTIONS FOR RESEARCH

For many years, studies with bacterial phospholipases were limited to the enzymes produced by *C. perfringens* and *B. cereus*. More recently, studies have been carried out with phospholipases produced by other pathogens. However, there are still many phospholipases produced by pathogens that are poorly characterized and for which we have not determined their roles in the pathogenesis of disease. Why are some enzymes membrane bound and others exported from the cell and, more generally, are the enzymes produced by intracellular pathogens specially adapted for this lifestyle?

In general, the interaction of phospholipases with membrane phospholipids is thought to be a key event, but we have limited knowledge of how this occurs. Haemolysis is an indicator of membrane interaction, but the hydrolysis of membrane phospholipids can occur without haemolysis. Why are some enzymes active towards liposomes but inactive towards erythrocytes? Why are mixtures of phosphatidylcholine-hydrolysing enzymes and sphingomyelinases haemolytic when the individual enzymes are not? Clearly, substrate specificity is one of the factors that determines the activity of enzymes – but how is this specificity achieved at a molecular level? In spite of research over the past 10 years, we do not fully understand how phospholipases exert their many effects on eukaryotic cells. What role do eukaryotic enzymes play in these processes? How do different enzymes exert different effects on different cell types?

The information that will be derived from these studies over the next decade is of more than philosophical significance. It will allow a variety of new approaches to the control of diseases. In some cases, vaccines will be devised and enzyme inhibitors could be used to treat bacterial diseases. These opportunities provide exciting new challenges to investigators in a rapidly expanding field.

ACKNOWLEDGEMENTS

I am grateful to Dr Claire Naylor for kindly providing the picture of the crystal structures of *B. cereus* PC-PLC and *C. perfringens* α-toxin, Dr Dirk Heinz for providing the picture of the crystal structures of *B. cereus* PI-PLC and rat PI-PLC-δ1 and Dr Lesley Harrington for proofreading this manuscript.

REFERENCES

Arai, M., Matsunaga, K. and Murao, S. (1974). Purification and some properties of phospholipase C of *Pseudomonas schuylkilliensis*. *J. Agric. Chem. Soc. (Japan)* **7**, 409–15.

Asmuth, D.M., Olson, R.D., Hackett, S.P., Bryant, A.E., Tweten, R.K., Tso, J.Y., Zollerman, T. and Stevens, D.L. (1995). Effects of *Clostridium perfringens* recombinant and crude phospholipase C and θ-toxin on rabbit hemodynamic parameters. *J. Infect. Dis.* **172**, 1317–23.

Asnani, N. and Asnani, P.J. (1991). Structural and functional changes in rabbit ileum by purified extracellular phospholipase A of *Salmonella newport*. *Folia Microbiol.* **36**, 572–7.

Awad, M.M., Bryant, A.E., Stevens, D.L. and Rood, J.I. (1995). Virulence studies on chromosomal α-toxin and θ-toxin mutants constructed by allelic exchange provide genetic evidence for the essential role of α-toxin in *Clostridium perfringens*-mediated gas gangrene. *Mol. Microbiol.* **15**, 191–202.

Baine, W.B. (1988). A phospholipase C from the Dallas 1E strain of *Legionella pneumophila* serogroup 5: purification and characterisation of conditions for optimal activity with an artificial substrate. *J. Gen. Microbiol.* **134**, 489–98.

Bangham, A.D. and Dawson, R.M.C. (1962). Electrokinetic requirements for the reaction between *Cl. perfringens* α-toxin (phospholipase C) and phospholipid substrates. *Biochim. Biophys. Acta* **59**, 103–15.

Bernheimer, A.W., Avigad, L.S. and Kim, K.S. (1974). Staphylococcal sphingomyelinase (β-hemolysin). *Ann. NY Acad. Sci.* **236**, 292–305.

Berridge, M.J. (1987). Inositol triphosphate and diacylglycerol: two interacting secondary messengers. *Annu. Rev. Biochem.* **56**, 159–93.

Blow, D. (1991). Lipases reach the surface. *Nature* **351**, 444–5.

Boethius, J., Rydqvist, B., Möllby, R. and Wadström, T. (1973). Effect of a highly purified phospholipase C on some electrophysiological properties of the frog muscle fibre membrane. *Life Sci.* **13**, 171–6.

Boyd, N.A., Thomson, R.O. and Walker, P.D. (1972). The prevention of experimental *Clostridium novyi* and *Cl. perfringens* gas gangrene in high-velocity missile wounds by active immunisation. *J. Med. Microbiol.* **5**, 467–72.

Bramley, A.J., Patel, A.H., O'Reilly, M., Foster, R. and Foster, T.J. (1989). Roles of alpha-toxin and beta-toxin in virulence of *Staphylococcus aureus* for the mouse mammary gland. *Infect. Immun.* **57**, 2489–94.

Bruznik, K.S., Morocho, A.M., Jhon, D.-Y., Rhee, S.G., and Tsai, M.-D. (1992) Phospholipids chiral at phosphorous. Stereochemical mechanism for the formation of inositol 1-phosphate catalysed by phosphatidylinositol-specific phospholipase C. *Biochemistry* **31**, 5183–93.

Bryant, A.E. and Stevens, D.L. (1996). Phospholipase C and perfringolysin O from *Clostridium perfringens* upregulate endothelial cell-leukocyte adherence molecule 1 and intercellular leukocyte adherence molecule 1 expression and induce interleukin-8 synthesis in cultured human umbilical vein endothelial cells. *Infect. Immun.* **64**, 358–62.

Bunting, M., Lorant, D.E., Bryant, A.E., Zimmerman, G.A., McIntyre, T.M., Stevens, D.L. and Prescott, S.M. (1997).

Alpha-toxin from *Clostridium perfringens* induces proinflammatory changes in endothelial cells. *J. Clin. Invest.* **100**, 565–74.

Camilli, A., Goldfine, H. and Portnoy, D.A. (1991). *Listeria monocytogenes* mutants lacking phosphatidylinositol-specific phospholipase C are avirulent. *J. Exp. Med.* **173**, 751–4.

Carter, G., White, P., Fernie, P., King, S., Hamilton, A., McLean, G., Titball, R.W. and Carr, F.J. (1998). Enhanced antitumour effect of liposomal daunorubicin using antibody-phospholipase C conjugates or fusion protein. *Int. J. Oncol.* **13**, 819–25.

Chovnick, A., Schneider, W.P. Tso, J.Y., Queen, C. and Chang, C.N. (1991). A recombinant membrane-acting immunotoxin. *Cancer Res.* **51**, 465–7.

Clark, M.A., Shorr, R.G.L. and Bomalaski, J.S. (1986). Antibodies prepared to *Bacillus cereus* phospholipase C cross-react with a phosphatidylcholine preferring phospholipase C in mammalian cells. *Biochem. Biophys. Res. Commun.* **140**, 114–19.

Cole, S.T., Brosch, R., Parkhill, J., Garnier, T., Churcher, C., Harris, D., Gordon, S.V., Eiglmeier, K., Gas, S., Barry III, C.E., Tekaia, F., Badcock, K., Basham, D., Brown, D., Chillingworth, T., Connor, R., Davies, R., Devlin, K., Feltwell, T., Gentles, S., Hamlin, N., Holroyd, S., Hornsby, T., Jagels, K., Krogh, A., McLean, J., Moule, S., Murphy, L., Oliver, K., Osborne, J., Quail, M.A., Rajandream, M.-A., Rogers, J., Rutter, S., Seeger, K., Skelton, J., Squares, R., Squares, S., Sulston, J.E., Taylor, K., Whitehead, S. and Barrell, B. (1998). Deciphering the biology of *Mycobacterium tuberculosis* from the complete genome sequence. *Nature* **393**, 537–44.

Cota-Gomez, A., Vasil, A.I., Kadurugamuwa, J., Beveride, T.J., Schweizer, H.P. and Vasil, M.L. (1997). PlcR1 and PlcR2 are putative calcium-binding proteins required for secretion of the hemolytic phospholipase C of *Pseudomonas aeruginosa*. *Infect. Immun.* **65**, 2904–13.

Crevel, I., Sally, U., Carne, A. and Katan, M. (1994). Purification and properties of zinc-metallophospholipase C from *Pseudomonas fluorescens*. *Eur. J. Biochem.* **224**, 845–52.

De Silva, N.S. and Quinn, P.A. (1987). Rapid screening for phospholipase C activity in mycoplasmas. *J. Clin. Microbiol.* **25**, 729–31.

El-Sayed, M.Y., DeBose, C.D., Coury, L.A. and Roberts, M.F. (1985). Sensitivity of phospholipase C (*Bacillus cereus*) activity to phosphatidylcholine structural modifications. *Biochim. Biophys. Acta* **837**, 325–35.

Evans, D.G. (1945). The *in-vitro* production of α-toxin, θ-haemolysin and hyaluronidase by strains of *Cl. welchii* type A, and the relationship of *in-vitro* properties to virulence for guinea-pigs. *J. Path. Bact.* **57**, 77–85.

Exton, J.H. (1990). Signalling through phosphatidylcholine breakdown. *J. Biol. Chem.* **265**, 1–4.

Fiore, A.E., Michalski, J.M., Russell, R.G., Sears, C.L. and Kaper, J.B. (1997). Cloning, characterisation, and chromosomal mapping of a phospholipase (lecithinase) produced by *Vibrio cholerae*. *Infect. Immun.* **65**, 3112–17.

Firth, J.D., Putnins, E.E., Larjava, H. and Uitto, V.-J. (1997). Bacterial phospholipase C upregulates matrix metalloproteinase expression by cultured epithelial cells. *Infect. Immun.* **65**, 4931–6.

Flores-Díaz, M., Alape-Girón, Persson, B., Pollosello, P., Moos, M., von Eichel-Streiber, C., Thelestam, M. and Florin, I. (1997). Cellular UDP-glucose deficiency caused by a single point mutation in the UDP-glucose pyrophosphorylase gene. *J. Biol. Chem.* **272**, 23784–91.

Flores-Díaz, M., Alape-Girón, Titball, R.W., Moos, M., Guillouard, I., Cole, S., Howells, A.M., von Eichel-Streiber, C. and Thelestam, M. (1998). UDP-glucose deficiency causes hypersensitivity to the cytotoxic effect of *Clostridium perfringens* phospholipase C. *J. Biol. Chem.* **273**, 24433–8.

Fujii, Y. and Sakurai, J. (1989). Contraction of the rat isolated aorta caused by *Clostridium perfringens* alpha-toxin (phospholipase C); evidence for the involvement of arachidonic acid metabolism. *Br. J. Pharmacol.* **97**, 119–24.

Fujii, Y., Nomura, S., Oshita, Y. and Sakurai, J. (1986). Excitatory effect of *Clostridium perfringens* alpha toxin on the rat isolated aorta. *Br. J. Pharmacol.* **88**, 531–9.

Geoffroy, C., Raveneau, J., Beretti, J.-L., Lechroisey, A., Vaquez-Boland, J.-A., Alouf, J.E. and Berche, P. (1991). Purification and characterisation of an extracellular 29-kilodalton phospholipase C from *Listeria monocytogenes*. *Infect. Immun.* **59**, 2382–8.

Gerasimene, G.B., Makaryunaite, Y.P., Kulene, V.V., Glemzha, A.A. and Yanulaitene, K.K. (1985). Some properties of phospholipases C from *Bacillus cereus*. *Prikl. Biokhimi Mikrobiol.* **21**, 184–9.

Gillmor, S.A., Villaseñor, A. Fletterick, R., Sigal, E. and Browner, M.F. (1997). The structure of mammalian 15-lipoxygenase reveals similarity to the lipases and the determinants of substrate specificity. *Nature Struct. Biol.* **4**, 1003–9.

Gilmore, M.S., Cruz-Rodz, A.L., Leimeister-Wachter, M., Kreft, J. and Goebbel, W. (1989). A *Bacillus cereus* cytolytic determinant, cereolysin AB, which comprises the phospholipase C and sphingomyelinase genes; nucleotide sequence and genetic linkage. *J. Bacteriol.* **171**, 744–53.

Ginter, A., Williamson, E.D., Dessy, F., Coppe, Ph., Fearn, A. and Titball, R.W. (1996). Molecular variation between the α-toxins from the type strain (NCTC8237) and clinical isolates of *Clostridium perfringens* associated with disease in man and animals. *Microbiology* **142**, 191–8.

Goldfine, H., Knob, C., Alford, D. and Bentz, J. (1995). Membrane permeabilization by *Listeria monocytogenes* phosphatidylinositol-specific phospholipase C is independent of phospholipid hydrolysis and cooperative with listeriolysin O. *Proc. Natl Acad. Sci. USA* **92**, 2979–83.

Granström, M., Erickson, A., Strandvik, B., Wretlind, B., Pavlovskis, O.R., Berka, R. and Vasil, M. (1984). Relationship between antibody response to *Pseudomonas aeruginosa* exoproteins and colonisation/infection in patients with cystic fibrosis. *Acta. Paediatr. Scand.* **73**, 772–7.

Guddal, P.H., Johansen, T., Schulstad, K. and Little, C. (1989). Apparent phosphate retrieval system in *Bacillus cereus*. *J. Bacteriol.* **17**, 5702–6.

Guillouard, I., Garnier, T. and Cole, S.T. (1996). Use of site-directed mutagenesis to probe structure–function relationships of alpha-toxin from *Clostridium perfringens*. *Infect. Immun.* **64**, 2440–4.

Guillouard, I., Alzari, P.M., Saliou, B. and Cole, S.T. (1997). The carboxy-terminal C2-like domain of the α-toxin from *Clostridium perfringens* mediates calcium-dependent membrane recognition. *Mol. Microbiol.* **26**, 867–76.

Gustafson, C. and Tagesson, C. (1990). Phospholipase C from *Clostridium perfringens* stimulates phospholipase A2 -mediated arachidonic acid release in cultured intestinal epithelial cells (INT 407). *Scand. J. Gastroenterol.* **25**, 363–71.

Hansen, S., Hansen, L.K. and Hough, E. (1993a). The crystal structure of tris-inhibited phospholipase C from *Bacillus cereus* at 1.9A resolution. *J. Mol. Biol.* **231**, 870–6.

Hansen, S., Hough, E., Svensson, L.A., Wong, Y.-L. and Martin, S.F. (1993b). Crystal structure of phospholipase C from *Bacillus cereus* complexed with a substrate analog. *J. Mol. Biol.* **234**, 179–87.

Haynes, J.A., Tkalcevic, J. and Nisbet, I.T. (1992). Production of an enzymatically inactive analog of phospholipase D from *Corynebacterium pseudotuberculosis*. *Gene* **119**, 119–21.

Heinz, D.W., Essen, L.-O. And Williams, R.L. (1998). Structural and mechanistic comparison of prokaryotic and eukaryotic phosphoinositide-specific phospholipases C. *J. Mol. Biol.* **275**, 635–50.

Henner, D.J., Yang, M., Chen, E., Hellmiss, R., Rodriguez, H. and Low, M.G. (1988). Sequence of the *Bacillus thuringiensis* phosphatidylinositol specific phospholipase C. *Nucleic Acids Res.* **16**, 10383.

Hereld, D., Hart, G.W. and Englund, P.T. (1988). cDNA encoding the glycosyl-phosphatidylinositol-specific phospholipase C of *Trypanosoma brucei*. *Proc. Natl Acad. Sci. USA* **85**, 8914–18.

Hodgson, A.L., Krywult, J., Corner, L.A., Rothel, J.S. and Radford, A.J. (1992). Rational attenuation of *Corynebacterium pseudotuberculosis*: potential cheesy gland vaccine and live delivery vehicle. *Infect. Immun.* **60**, 2900–5.

Holm, B.A., Keicher, L., Liu, M., Sokolowski, J. and Enhorning, G. (1991). Inhibition of pulmonary surfactant function by phospholipases. *J. Appl. Physiol.* **71**, 317–21.

Hough, E., Hansen, L.K., Birkness, B., Jynge, K., Hansen, S., Hordik, A., Little, C., Dodson, E. and Derewenda, Z. (1989). High resolution (1.5A) crystal structure of phospholipase C from *Bacillus cereus*. *Nature* **338**, 357–60.

Ikezawa, H. (1986). The physiological action of bacterial phosphatidylinositol-specific phospholipase C. The release of ectoenzymes and other effects. *J. Toxicol. Toxin Rev.* **5**, 1–24.

Ikezawa, H. (1991). Bacterial PIPLCs-unique properties and usefulness in studies on GPI anchors. *Cell Biol. Int. Rep.* **15**, 1115–31.

Ikezawa, H., Matsushita, M., Tomita, M. and Taguchi, R. (1986). Effects of metal ions on sphingomyelinase activity of *Bacillus cereus*. *Arch. Biochem. Biophys.* **249**, 588–95.

Ivanov, I., Titball, R.W. and Kostadinova, S. (1996). Characterisation of phospholipase C of *Pseudomonas fluorescens*. *Microbiologica* **19**, 113–21.

Johansen, K.A., Gill, R.E. and Vasil, M.L. (1996). Biochemical and molecular analysis of phospholipase C and phospholipase D activity in Mycobacteria. *Infect. Immun.* **64**, 3259–66.

Johansen, T., Holm, T., Guddal, P.H., Sletten, K., Haugli, F.B. and Little, C. (1988). Cloning and sequencing of the gene encoding the phosphatidylcholine-preferring phospholipase C of *Bacillus cereus*. *Gene* **65**, 293–304.

Kameyama, S., Sato, H. and Murata, R. (1975). The role of α-toxin of *Clostridium perfringens* in experimental gas gangrene in guinea pigs. *Jpn. J. Med. Sci. Biol.* **25**, 200.

König, B., Vasil, M.L. and König, W. (1997). Role of hemolytic and non-hemolytic phospholipase C from *Pseudomonas aeruginosa* for inflammatory mediator release from human granulocytes. *Int. Arch. Allergy Immunol.* **112**, 115–24.

Kothary, M.H. and Kreger, A.S. (1985). Purification and characterisation of an extracellular cytolysin produced by *Vibrio damsela*. *Infect. Immun.* **49**, 25–31.

Kreger, A.S., Bernheimer, A.W., Etkin, L.A. and Daniel, L.W. (1987). Phospholipase D activity of *Vibrio damsela* cytolysin and its interaction with sheep erythrocytes. *Infect. Immun.* **55**, 3209–12.

Krug, E.L. and Kent, C. (1984). Phospholipase C from *Clostridium perfringens*: preparation and characterisation of homogenous enzyme. *Arch. Biochem. Biophys.* **231**, 400–10.

Kuppe, A., Evans, L.M., McMillen, D.A. and Griffith, O.H. (1989). Phosphatidylinositol-specific phospholipase C of *Bacillus cereus*: cloning, sequencing and relationship to other phospholipases. *J. Bacteriol.* **171**, 6077–83.

Kurioka, S. and Matsuda, M. (1976). Phospholipase C assay using *p*-nitrophenylphosphorylcholine and its application to studying the metal and detergent requirement of the enzyme. *Anal. Biochem.* **75**, 281–9.

Langton, S.R. and Cesareo, S.D. (1992) *Helicobacter pylori* associated phospholipase A2 activity: a factor in peptic ulcer production. *J. Clin Pathol.* **45**, 221–4.

Leao, S.C., Rocha, C.L., Murillo, L.A., Parra, C.A. and Patarroyo, M.E. (1995). A species-specific nucleotide sequence of *Mycobacterium tuberculosis* encodes a protein that exhibits hemolytic activity when expressed in *Escherichia coli*. *Infect. Immun.* **63**, 4301–6.

Leimeister-Wächter, M., Domann, E. and Chakraborty, T. (1991). Detection of a gene encoding a phosphatidylinositol-specific phospholipase C that is co-ordinately expressed with listeriolysin in *Listeria monocytogenes*. *Mol. Microbiol.* **5**, 361–6.

MacFarlane, M.G. and Knight, B.C.J.G. (1941). The biochemistry of bacterial toxins. I. Lecithinase activity of *Cl. welchii* toxins. *Biochem. J.* **35**, 884–902.

McNamara, P.J., Bradley, G.A. and Songer, J.G. (1994). Targeted mutagenesis of the phospholipase D gene results in decreased virulence of *Corynebacterium pseudotuberculosis*. *Mol. Microbiol.* **12**, 921–30.

McNee, J.W. and Dunn, J.S. (1917). The method of spread of gas gangrene into living muscle. *Br. Med. J.* **i**, 727–9.

Mengaud, J., Barun-Breton, C. and Cossart, P. (1991). Identification of a phosphatidylinositol-specific phospholipase C in *Listeria monocytogenes*: a novel type of virulence factor? *Mol. Microbiol.* **5**, 367–72.

Möllby, R. (1978). Bacterial phospholipases. In: *Bacterial Toxins and Cell Membranes* (eds J. Jeljaszewicz and T. Wadström), pp. 367–424. Academic Press, London.

Moreau, H., Pieroni, G., Jolivet-Raynaud, C., Alouf, J.E. and Verger, R. (1988). A new kinetic approach for studying phospholipase C (*Clostridium perfringens* α-toxin) activity on phospholipid monolayers. *Biochemistry* **27**, 2319–23.

Moser, J., Gerstel, B., Meyer, J.E.W., Chakraborty, T., Wehland, J. and Heinz, D.W. (1997). Crystal structure of the phosphatidylinositol-specific phospholipase C from the human pathogen *Listeria monocytogenes*. *J. Mol. Biol.* **273**, 269–82.

Muckle, C.A. and Gyles, C.L. (1983). Relation of lipid content and exotoxin production to virulence of *Corynebacterium pseudotuberculosis* in mice. *Curr. Microbiol.* **13**, 57–60.

Nagahama, M., Okagawa, Y., Nakayama, T., Nishioka, E. and Sakurai, J. (1995). Site-directed mutagenesis of histidine residues in *Clostridium perfringens* alpha-toxin. *J. Bacteriol.* **177**, 1179–85.

Nagahama, M., Michiue, K. and Sakurai, J. (1996). Membrane-damaging action of *Clostridium perfringens* alpha-toxin on phospholipid liposomes. *Biochim. Biophys. Acta* **1280**, 120–6.

Nagahama, M., Nakayama, T., Michiue, K. and Sakurai, J. (1997). Site-specific mutagenesis of *Clostridium perfringens* alpha-toxin: replacement of Asp-56, Asp-130 or Glu-152 causes loss of enzymatic and hemolytic activities. *Infect. Immun.* **65**, 3489–92.

Nakamura, S., Shimamura, T., Hayase, M. and Nishida, S. (1973). Numerical taxonomy of saccharolytic clostridia, particularly *Clostridium perfringens*-like strains: descriptions of *Clostridium absonum* sp.n. and *Clostridium paraperfringens*. *Int. J. Syst. Bacteriol.* **23**, 419–29.

Naylor, C.E., Eaton, J.T., Howells, A., Justin, N., Moss, D.S., Titball, R.W. and Basak, A.K. (1998). The structure of the key toxin in gas gangrene has a prokaryotic calcium-binding C2 domain. *Nature Struct. Biol.* **5**, 738–46.

Nishizuka, Y. (1992). Intracellular signalling by hydrolysis of phospholipids and activation of protein kinase C. *Science* **258**, 607–14.

Ochi, S., Hashimoto, K., Nagahama, M. and Sakurai, J. (1996). Phospholipid metabolism induced by *Clostridium perfringens* alpha-toxin elicits a hot-cold type of hemolysis in rabbit erythrocytes. *Infect. Immun.* **64**, 3930–3.

Ohsaka, A., Tsuchiya, M., Oshio, C., Miyaira, M., Suzuki, K. and Yamakawa, Y. (1978). Aggregation of platelets in the microcirculation of the rat induced by the α-toxin (phospholipase C) of *Clostridium perfringens*. *Toxicon* **16**, 333–41.

Ostroff, R.M., Vasil, A.I. and Vasil, M.L. (1990). Molecular comparison of a nonhaemolytic and a haemolytic phospholipase C from *Pseudomonas aeruginosa*. *J. Bacteriol.* **172**, 5915–23.

Otnaess, A-B., Little, C., Sletten, K., Wallin, R., Johnsen, S., Flensgrud, R. and Prydz, H. (1977). Some characteristics of phospholipase C from *Bacillus cereus*. *Eur. J. Biochem.* **79**, 459–68.

Ottlecz, A., Romero, J.J., Hazell, S.L., Graham, D.Y., and Lichtenberger, L.M. (1993). Phospholipase activity of *Helicobacter pylori* and its inhibition by bismuth salts. *Dig. Dis. Sci.* **38**, 2071–80.

Parkinson, E.K. (1987). Phospholipase C mimics the differential effects of phorbol-12-myristate-13-acetate on the colony formation and cornification of cultured normal and transformed human keratinocytes. *Carcinogenesis* **8**, 857–60.

Portnoy, D.A., Smith, G.A. and Goldfine, H. (1994). Phospholipases C and the pathogenesis of Listeria. *Braz. J. Med. Biol. Res.* **27**, 357–61.

Projan, S.J., Kornblum, J., Kreiswirth, B., Moghazeh, S.L., Eiser, W. and Novick, R.P. (1989). Nucleotide sequence of the β-hemolysin gene of *Staphylococcus aureus*. *Nucleic Acids Res.* **17**, 3305.

Rahme L.G., Stevens E.J., Wolfort S.F., Shao J., Tompkins R.G., Ausubel F.M. (1995). Common virulence factors for bacterial pathogenicity in plants and animals. *Science* **268**, 1899–902.

Saiman, L., Cacalano, G., Gruenert, D. and Prince, A. (1992). Comparison of adherence of *Pseudomonas aeruginosa* to respiratory epithelial cells from cystic fibrosis patients and healthy subjects. *Infect. Immun.* **60**, 2808–14.

Sakurai, J., Ochi, S. and Tanaka, H. (1993). Evidence for coupling of *Clostridium perfringens* alpha-toxin-induced hemolysis to stimulated phosphatidic acid formation in rabbit erythrocytes. *Infect. Immun.* **61**, 3711–18.

Sakurai, J., Ochi, S. and Tanaka, H. (1994). Regulation of *Clostridium perfringens* alpha-toxin-activated phospholipase C in rabbit erythrocyte membranes. *Infect. Immun.* **62**, 717–21.

Samuelsson, B. (1983). Leukotrienes: mediators of immediate hypersensitivity reactions and inflammation. *Science* **220**, 568–75.

Segers, R.P.A.M., van der Drift, A., de Njis, A., Corcione, P., van der Zeijst, B.A.M. and Gaastra, W. (1990). Molecular analysis of a sphingomyelinase C gene from *Leptospira interrogans* serovar hardjo. *Infect. Immun.* **58**, 2177–85.

Shen, B.-F., Tai, P.C., Pritchard, A.E. and Vasil, M.L. (1987). Nucleotide sequence and expression in *Escherichia coli* of the inphase overlapping *Pseudomonas aeruginosa* plcR genes. *J. Bacteriol.* **169**, 4602–7.

Shinoda, S., Matsuoaka, H., Tsuchie, T., Miyoshi, S.-I., Yamamoto, S., Taniguchi, H. and Miuguchi, Y. (1991). Purification and characterisation of a lecithin-dependent haemolysin from *Escherichia coli* transformed by a *Vibrio parahaemolyticus* gene. *J. Gen. Microbiol.* **137**, 2705–11.

Shortridge, V.D., Lazdunski, A. and Vasil, M.L. (1992). Osmoprotectants and phosphate regulate expression of phospholipase C in *Pseudomonas aeruginosa*. *Mol. Microbiol.* **6**, 863–71.

Siboo R., al-Joburi W., Gornitsky M. and Chan E.C. (1989) Synthesis and secretion of phospholipase C by oral spirochetes. *J. Clin. Microbiol.* **27**, 568–70.

Smith, G.A., Marquis, H., Jones, S., Johnston, N.C., Portnoy, D.A. and Goldfine, H. (1995). The two distinct phospholipase C of *Listeria monocytogenes* have overlapping roles in escape from a vacuole and cell-to-cell spread. *Infect. Immun.* **63**, 4231–7.

Songer, J.G. (1997). Bacterial phospholipases and their role in virulence. *Trends Microbiol.* **5**, 156–61.

Sonoki, S. and Ikezawa, H. (1975). Studies on phospholipase C from *Pseudomonas aureofaciens*. I. Purification and some properties of phospholipase C. *Biochim. Biophys. Acta* **403**, 412–24.

Soucek, A., Michalec, C. and Souckova, A. (1971). Identification and characterisation of a new enzyme of the group 'phospholipase D' isolated from *Corynebacterium ovis*. *Biochim. Biophys. Acta* **227**, 116–28.

Spiegel, S., Foster, D. and Kolesnick, R. (1996). Signal transduction through lipid second messengers. *Curr. Opin. Cell. Biol.* **8**, 159–67.

Stevens D.L., Tweten R.K., Awad M.M., Rood J.I. and Bryant A.E. (1997). Clostridial gas gangrene: evidence that alpha and theta toxins differentially modulate the immune response and induce acute tissue necrosis. *J. Infect. Dis.* **176**, 189–95.

Thornton, J., Howard, S.P. and Buckley, J.T. (1988). Molecular cloning of a phospholipid-cholesterol acyltransferase from *Aeromonas hydrophila*. Sequence homologies with lecithin-cholesterol acyltransferases and other lipases. *Biochim. Biophys. Acta* **959**, 153–9.

Titball, R.W. (1993). Bacterial phospholipases C. *Microbiol. Rev.* **57**, 347–66.

Titball, R.W. (1997). Clostridial phospholipases. In: *The Clostridia: Molecular Biology and Pathogenesis* (eds J.I. Rood, B.A. McClane, J.G. Songer and R.W. Titball), pp. 223–42. Academic Press, London.

Titball, R.W. and Rubidge, T. (1990). The role of histidine residues in the alpha-toxin of *Clostridium perfringens*. *FEMS Microbiol. Lett.* **68**, 261–6.

Titball, R.W., Hunter, S.E.C., Martin, K.L., Morris, B.C., Shuttleworth, A.D., Rubidge, T., Anderson, D.W. and Kelly, D.C. (1989). Molecular cloning and nucleotide sequence of the alpha-toxin (phospholipase C) of *Clostridium perfringens*. *Infect. Immun.* **57**, 367–76.

Titball, R.W., Leslie, D.L., Harvey, S. and Kelly, D.C. (1991). Haemolytic and sphingomyelinase activities of *Clostridium perfringens* alpha-toxin are dependent on a domain homologous to that of an enzyme from the human arachidonic acid pathway. *Infect. Immun.* **59**, 1872–4.

Titball, R.W., Fearn, A.M. and Williamson, E.D. (1993). Biochemical and immunological properties of the C-terminal domain of the alpha-toxin of *Clostridium perfringens*. *FEMS Microbiol. Lett.* **110**, 45–50.

Truett, A.P., III and King, L.E., Jr (1993). Sphingomyelinase D; a pathogenic agent produced by bacteria and arthropods. *Adv. Lipid Res.* **26**, 275–91.

Tso, J.Y. and Siebel, C. (1989). Cloning and expression of the phospholipase C gene from *Clostridium perfringens* and *Clostridium bifermentans*. *Infect. Immun.* **57**, 468–76.

Tsutsui, K., Minami, J., Matsushita, O., Katayama, S.-I., Taniguchi, Y., Nakamura, S., Nishioka, M. and Okabe, A. (1995). Phylogenetic analysis of phospholipase C genes from *Clostridium perfringens* types A to E and *Clostridium novyi*. *J. Bacteriol.* **177**, 7164–70.

van Deenen, L.L.M., Demel, R.A., Guerts van Kessel, W.S.M., Kamp, H.H., Roelofsen, B., Verkleij, A.J., Wirtz, K.W.A. and Zwaal, R.F.A. (1976). Phospholipases and monolayers as tools in studies on membrane structure. In *The Structural Basis of Membrane Function* (eds Y. Hatefi and L. Djavadi-Ohaniance), pp. 21–38. Academic Press, New York.

Vasil, M.L., Kreig, D.P., Kuhns, J.S., Ogle, J.W., Shortridge, V.D., Ostroff, R.M. and Vasil, A.I. (1990). Molecular analysis of haemolytic and phospholipase C activities of *Pseudomonas cepacia*. *Infect. Immun.* **58**, 4020–9.

Vazquez-Boland, J.-A., Kocks, C., Dramsi, S., Ohayon, H., Geoffroy, C., Mengaud, J. and Cossart, P. (1992). Nucleotide sequence of the lecithinase operon of *Listeria monocytogenes* and possible role of lecithinase in cell-cell spread. *Infect. Immun.* **60**, 219–30.

Waite, M. (1987). *Handbook of Lipid Research*, Vol. 5, *The Phospholipases*. Plenum Press, New York.

Weitkamp, J.-H., Pérez-Pérez, G.I., Bode, G., Malfertheiner, P. and Blaser, M.J. (1993). Identification and characterisation of *Helicobacter pylori* phospholipase C activity. *Zentrlbl. Bakt.* **280**, 11–27.

Willis, T.A. (1969). *Clostridia of Wound Infection*. Butterworth, London.

Williamson, E.D. and Titball, R.W. (1993). A genetically engineered vaccine against the alpha-toxin of *Clostridium perfringens* also protects mice against experimental gas gangrene. *Vaccine* **11**, 1253–8.

Yamada, A., Tsukagoshi, N., Ukada, S., Sasaki, T., Makino, S., Nakamura, S., Little, C., Tomita, M. and Ikezawa, H. (1988). Nucleotide sequence and expression in *Escherichia coli* of the gene coding for sphingomyelinase of *Bacillus cereus*. *Eur. J. Biochem.* **175**, 213–20.

16

The family of the multigenic encoded RTX toxins

Albrecht Ludwig and Werner Goebel

INTRODUCTION

Cytolysins represent important virulence factors of many pathogenic bacteria. These toxins disrupt animal and/or human cells by damaging their plasma membrane. Several types of cytolysins can be distinguished, based on the mechanism of action against target cell membranes: (a) pore-forming cytolysins; (b) enzymically active cytolysins, which degrade membrane lipids; and (c) surfactant cytolysins, which solubilize cell membranes by a detergent-like action. By far the most well-known bacterial cytolysins are pore-forming proteins. The generation of transmembrane pores is cytotoxic and may cause osmotic cell lysis. Sublytic concentrations of pore-forming cytolysins may also induce secondary reactions in target cells and thereby influence the physiological functions of these cells.

The bacterial pore-forming cytolysins represent a heterogeneous group of exotoxins. Several families of structurally and functionally related pore-forming cytolysins have been identified among both the Gram-positive and Gram-negative bacteria. Pore-forming cytolysins of Gram-positive bacteria are generally synthesized with an *N*-terminal signal peptide that is cleaved during Sec-dependent transport across the cytoplasmic membrane, but they are usually active *per se* and do not require any activation. In contrast, those of Gram-negative bacteria are commonly synthesized as inactive protoxins, which are subsequently converted into the active toxins by either modification or proteolytic processing. The secretion of these toxins across the inner and outer membranes of the Gram-negative

bacteria is mediated by complex secretion systems that are either dependent or independent of the *sec* genes.

The RTX toxins represent the largest family of bacterial pore-forming cytolysins; they are widespread among Gram-negative pathogens and will be discussed in detail in this chapter.

GENERAL CHARACTERISTICS OF RTX TOXINS

The RTX toxin family is distinguished from other groups of pore-forming cytolysins by a number of common traits.

1. RTX toxins are synthesized as inactive proteins with molecular masses typically around 100 to 120 kDa. Two more distantly related members of this toxin family, adenylate cyclase toxin from *Bordetella pertussis* and RtxA from *Vibrio cholerae*, have, however, predicted sizes of 177 kDa and 500 kDa, respectively. The C-terminal half of RTX toxin proteins includes a tandem array of glycine- and aspartate-rich nonameric repeats with the consensus sequence UXGGXG(N/D)DX, where U is a large hydrophobic amino acid and X is an arbitrary amino acid. The designation 'RTX toxins' refers to these characteristic repeats (RTX stands for repeats in toxin). The number of repeats varies among the RTX toxins between about 10 and 40.

2. RTX toxins are post-translationally activated by a modification that is mediated by an accessory cytoplasmic protein. So far, two RTX toxins, *Escherichia coli* α-haemolysin and adenylate cyclase toxin from

Bordetella pertussis, have been shown to be activated by covalent acylation of specific internal lysine residues; it is likely that the other members of this toxin family are activated by a similar mechanism.

3. The RTX toxins lack a cleavable *N*-terminal signal peptide and their secretion is not *sec* dependent; they are thus not exported via the general secretory pathway (also designated type-II secretion pathway). The extracellular secretion of the RTX toxins rather proceeds via the type-I secretion pathway, which allows direct translocation of the toxins across both the inner and the outer membrane in one step, without any detectable periplasmic intermediate. Secretion of the RTX toxins thereby depends on specific, highly conserved export systems, composed of three envelope proteins. One is an inner membrane ATPase belonging to the superfamily of ATP-binding cassette (ABC) transporters, which include prokaryotic and eukaryotic proteins involved in the import and export of a variety of hydrophilic substrates across cell membranes (Higgins, 1992). The ABC protein provides energy for the secretory process through hydrolysis of ATP. The second protein is also anchored in the inner membrane and belongs to the family of membrane fusion proteins (MFP). The MFPs are transport accessory proteins found mostly in Gram-negative bacteria, where they function in conjunction with inner membrane transporters such as ABC proteins (Dinh *et al.*, 1994). They are assumed to be involved in localized fusions between the inner and outer membrane, since they have a periplasmic domain, which may interact with the outer membrane. The third component of the exporter is an outer membrane protein with a typical *N*-terminal signal sequence. The entire secretion apparatus comprising the three proteins is usually designated as an ABC exporter. The transport signal that targets RTX toxins for export is a structure located within the *C*-terminal ~60 amino acids of the toxin protein. This secretion signal is not processed during secretion.

4. The activity of the RTX toxins is Ca^{2+} dependent. The calcium ions bind in an unknown stoichiometry to the signature repeat domain, which thereby presumably forms a very stable β-roll structure. Ca^{2+} binding most likely occurs after secretion of the RTX toxins.

5. RTX toxins form transient, cation-selective pores of different sizes in lipid membranes.

6. The genes specifically required for synthesis, activation and secretion of a particular RTX toxin are clustered either on the bacterial chromosome or (less frequently) on a plasmid and usually represent a single operon. This operon typically contains four contiguous genes in the order C-A-B-D. Gene A is the structural gene of the toxin protein, gene C encodes the activator protein and the genes B and D encode the ABC protein and the MFP component, respectively, of the ABC exporter. The gene encoding the outer membrane component of the secretion apparatus is in most cases not linked to this gene cluster but located elsewhere on the chromosome.

The RTX toxin family includes both cytolysins active against a broad spectrum of cell types from a variety of species (e.g. *E. coli* α-haemolysin and ApxI from *Actinobacillus pleuropneumoniae*) and cytolytic toxins with narrow target cell and host species reactivities (e.g. leucotoxins from *Pasteurella haemolytica* and *Actinobacillus actinomycetemcomitans*). The broad-range cytolysins are commonly referred to as haemolysins, owing to their ability to lyse erythrocytes.

MEMBERS OF THE RTX TOXIN FAMILY

Escherichia coli α-haemolysin

α-Haemolysin from *E. coli* is one of the best-characterized members of the RTX toxin family. It is frequently produced by *E. coli* strains causing urinary tract and other extra-intestinal infections and contributes significantly to the virulence of these strains, as shown in several animal models (Welch *et al.*, 1981; Hacker *et al.*, 1983; Smith and Huggins, 1985). α-Haemolysin is less frequently associated with *E. coli* strains causing diarrhoeal diseases, but a recent study suggests that it might also play a role in enteropathy (Elliott *et al.*, 1998).

Biological activity of *Escherichia coli* α-haemolysin

The pathophysiological function of α-haemolysin in extra-intestinal *E. coli* infections is probably multifactorial, owing to the action of this toxin on a wide range of cell types. α-Haemolysin lyses erythrocytes from many species, but also exhibits strong cytotoxic and cytolytic activity against a variety of nucleated cells. In particular, it potently kills immune cells involved in first-line defence mechanisms, including polymorphonuclear leucocytes (PMN) and monocytes (Cavalieri and Snyder, 1982; Gadeberg and Orskov, 1984; Bhakdi *et al.*, 1989, 1990), it is cytotoxic for T-lymphocytes (Jonas *et al.*, 1993) and it can also cause local tissue damage by destroying tissue cells (Keane *et al.*, 1987; Mobley *et al.*, 1990; O'Hanley *et al.*, 1991). The toxin thus directly promotes bacterial infection. The lytic action of α-haemolysin on red blood cells may further promote bacterial growth in the host as it increases the concentration of available iron.

Very low, sublytic doses of α-haemolysin provoke a broad spectrum of secondary reactions on target cells, which in turn cause short- and long-range effects in the host; this contributes also to the pathogenesis of infection. α-Haemolysin induces, for example, the production and release of inflammatory lipid mediators (e.g. leukotrienes and hydroxyeicosatetraenoic acids) in leucocytes and platelets (König et al., 1986, 1990; Grimminger et al., 1990, 1991a), it stimulates the secretion of interleukin-1 (IL-1) from monocytes and other cells (Bhakdi et al., 1990; May et al., 1996) and provokes an oxidative burst and degranulation in PMN (Bhakdi et al., 1989; Bhakdi and Martin, 1991; Grimminger et al., 1991b). Endothelial cells are also highly susceptible to α-haemolysin. Sublytic concentrations of the toxin stimulate, for example, the arachidonate metabolism and production of vasodilatory agents in these cells, and increase the permeability of endothelial cell monolayers (Suttorp et al., 1990; Grimminger et al., 1997). α-Haemolysin apparently elicits several cellular reactions via activation of the phosphatidylinositol hydrolysis-related signal transduction pathway (Grimminger et al., 1991b, 1997; König and König, 1993), but secondary reactions may also be triggered by the passive influx of Ca^{2+} into the target cells through the pores that are formed by the toxin (Suttorp et al., 1990).

Synthesis and secretion of α-haemolysin

Genetic determinants of α-haemolysin are generally found on large, transmissible plasmids in animal isolates of E. coli or on the chromosome of E. coli strains causing urinary tract infections in humans (de la Cruz et al., 1980; Welch et al., 1981; Müller et al., 1983). Several uropathogenic E. coli isolates carry two α-haemolysin determinants within unique chromosomal inserts, called pathogenicity islands, which are absent in the non-pathogenic E. coli laboratory strain K-12 (Blum et al., 1994; Swenson et al., 1996; Hacker et al., 1997).

The E. coli α-haemolysin determinant was the first RTX toxin determinant that was cloned and characterized (Welch et al., 1981; Goebel and Hedgpeth, 1982; Felmlee et al., 1985; Mackman et al., 1985; Hess et al., 1986). It represents a classical RTX toxin operon and comprises four genes arranged in the order hlyC, hlyA, hlyB and hlyD, which encode proteins of 20 kDa, 110 kDa, 80 kDa and 55 kDa, respectively. The hlyA gene product (1024 residues) is the protein component of α-haemolysin, which represents an inactive (non-haemolytic) protoxin (pro-HlyA), HlyC is required for the post-translational activation of pro-HlyA to the mature α-haemolysin (HlyA), and HlyB and HlyD are the two inner membrane components of the α-haemolysin secretion apparatus. Transcription of the hlyCABD operon is strongly polar, owing to the presence of a rho-independent terminator

in the hlyA-hlyB intergenic region (Welch and Pellett, 1988). Wild-type expression of extracellular α-haemolysin depends on the E. coli RfaH protein (18.3 kDa), which enhances the transcriptional elongation through the hly operon, most likely by acting as a transcriptional antiterminator (Bailey et al., 1996; Leeds and Welch, 1996). In addition to RfaH, a short cis-acting sequence termed JUMPstart element or ops element (operon polarity suppressor) has recently been shown to be essential for the suppression of transcription polarity in the E. coli α-haemolysin operon (Nieto et al., 1996; Leeds and Welch, 1997); it is located 3' of the main promoters in the non-translated leader sequences of plasmid-borne and chromosomal α-haemolysin determinants. Similar JUMPstart elements have previously been identified in the leader regions of other E. coli operons whose transcription is also enhanced by RfaH, including those directing synthesis of the lipopolysaccharide (LPS) core (rfaQ-K), production of the F-factor sex pilus (traY-Z) and production of group II polysaccharide capsules (kps). RfaH and the JUMPstart (ops) element apparently contribute to the same mechanism of transcriptional elongation/antitermination (Bailey et al., 1996; Leeds and Welch, 1997).

Pro-HlyA is activated intracellularly by covalent, quantitative acylation of two internal lysine residues, Lys564 and Lys690 (Issartel et al., 1991; Stanley et al., 1994, 1998; Ludwig et al., 1996). This activation is mediated by the co-synthesized cytoplasmic protein HlyC, which acts as an acyltransferase using acylated acyl carrier protein (acyl-ACP) as the fatty acid donor (Issartel et al., 1991; Trent et al., 1998). Interestingly, the amino acid sequence of HlyC shows no similarity to other known acyltransferases. The acylation of pro-HlyA by HlyC requires short amino acid sequences spanning the two acylation sites. These sequences probably serve as independent HlyC recognition domains (Ludwig et al., 1996; Stanley et al., 1996). Under in vitro conditions, HlyC can acylate pro-HlyA with a range of fatty acids (Issartel et al., 1991), but it remains to be determined which acyl groups are bound to pro-HlyA in vivo. The acylation of the haemolysin protein is not required for the subsequent secretion of the toxin (Ludwig et al., 1987).

Escherichia coli α-haemolysin was the first protein shown to be exported via a type-I secretion system. In particular, it has been demonstrated that the secretion of HlyA depends on a specific transport apparatus comprising HlyB, HlyD and the E. coli outer membrane protein TolC, which is not genetically linked to the hly gene cluster (Wagner et al., 1983; Wandersman and Delepelaire, 1990). In addition, the secretion of α-haemolysin has been shown to be sec independent (Gentschev et al., 1990; Blight and Holland, 1994) and to proceed across the inner and outer membrane of E.

coli in a single step without accumulation of the toxin in the periplasmic space (Gray *et al.*, 1986; Felmlee and Welch, 1988; Koronakis *et al.*, 1989).

The secretion signal of α-haemolysin has been localized within the *C*-terminal 50–60 amino acids of the haemolysin protein (Gray *et al.*, 1986; Koronakis *et al.*, 1989; Hess *et al.*, 1990; Kenny *et al.*, 1992; Jarchau *et al.*, 1994), but the precise nature of this signal is poorly understood. The *C*-terminal targeting signals of RTX toxins share little sequence similarity, although efficient heterologous complementation between the exporters of various RTX toxins has been observed (Koronakis *et al.*, 1987; Chang *et al.*, 1989a; Gygi *et al.*, 1990; Highlander *et al.*, 1990; Masure *et al.*, 1990). It has been suggested that the transport signals of HlyA and other RTX toxins may at least partially be defined by higher order structures (Hess *et al.*, 1990; Stanley *et al.*, 1991; Zhang *et al.*, 1993a, 1995). However, it has also been proposed that a small number of individual residues dispersed throughout the *C*-terminal region of HlyA may represent critical contact sites required for the interaction with the transport apparatus, irrespective of a specific secondary structure (Kenny *et al.*, 1992, 1994; Chervaux and Holland, 1996).

The *E. coli* α-haemolysin transport apparatus represents the prototype of an ABC exporter. HlyB, the ABC protein of this export system, consists of an *N*-terminal integral membrane domain, predicted to contain six to eight transmembrane segments, and a large *C*-terminal cytoplasmic domain carrying an ATP-binding cassette (Wang *et al.*, 1991; Gentschev and Goebel, 1992; Koronakis *et al.*, 1993). HlyD, the MFP component of the α-haemolysin exporter, is proposed to have a single transmembrane segment with the *N*-terminal end in the cytoplasm and a large *C*-terminal domain in the periplasm (Wang *et al.*, 1991; Schülein *et al.*, 1992). The *C*-terminal domain of HlyD is believed to interact with TolC in the outer membrane, thereby bridging the periplasmic space (Schlör *et al.*, 1997; Thanabalu *et al.*, 1998). Electron microscopy of two-dimensional TolC crystals revealed a trimeric, porin-like structure of TolC (Koronakis *et al.*, 1997). Each 51.5-kDa TolC monomer comprises a membrane domain, predicted to form a β-barrel, and a *C*-terminal periplasmic domain, which may form part of the bridge to the inner membrane components of the transport complex (Koronakis *et al.*, 1997). Lipid bilayer experiments demonstrated that TolC forms transmembrane channels (Benz *et al.*, 1993). Thus, these data suggest that HlyA crosses the outer membrane through a pore formed by TolC.

Recently, Thanabalu *et al.* (1998) demonstrated that HlyB and HlyD form a stable inner membrane (IM) complex, independent of TolC and in the absence of HlyA. The data of these authors further suggest that

binding of HlyA to the preformed HlyB/HlyD complex induces this complex to contact the TolC exit pore in the outer membrane (OM) via HlyD, resulting in the assembly of a trans-periplasmic export channel. HlyD, the IM bridging component of the α-haemolysin transport apparatus, apparently oligomerizes to IM trimers, which corresponds to the TolC trimers in the outer membrane (Thanabalu *et al.*, 1998). The stoichiometry of HlyB in the HlyA secretion apparatus remains to be determined. The IM–OM bridging seems to be transient, i.e. the components of the exporter revert to the IM and OM states after translocation of α-haemolysin (Thanabalu *et al.*, 1998).

It is assumed that HlyB recognizes the secretion signal of HlyA (Oropeza-Wekerle *et al.*, 1990; Letoffe *et al.*, 1996), and this is lent support by the finding that mutations in the secretion signal are partially compensated by suppressor mutations in HlyB (Zhang *et al.*, 1993b; Sheps *et al.*, 1995). Nevertheless, data obtained by Thanabalu *et al.* (1998) suggest that both HlyB and HlyD contribute to the recruitment of HlyA. Binding and hydrolysis of ATP by the *C*-terminal domain of HlyB is directly coupled to the export of HlyA (Koronakis *et al.*, 1995). In addition to ATP hydrolysis, total proton motive force is required at an early stage of HlyA secretion (Koronakis *et al.*, 1991)

The haemolytic activity of α-haemolysin depends not only on the acylation of the haemolysin protein, but also on the binding of an unknown number of Ca^{2+} ions to the signature nonapeptide repeats (Ludwig *et al.*, 1988; Boehm *et al.*, 1990a, b). HlyA contains a series of 13 glycine-rich nonameric repeats in the region between the distal acylation site and the secretion signal. Two additional repeats are present in the sequence separating the two acylation sites. The binding of Ca^{2+} probably induces a particular toxin conformation that is essential for the lytic activity. The repeats presumably bind Ca^{2+} in a parallel β-roll structure (Baumann *et al.*, 1993). Binding of Ca^{2+} to HlyA most likely occurs outside the *E. coli* cell, i.e. after secretion, because the cytoplasmic Ca^{2+} concentration of *E. coli* is tightly regulated to a very low, constant level of about 0.1 μM (Gangola and Rosen, 1987). HlyA, however, requires Ca^{2+} concentrations of >100 μM for full haemolytic activity (Ostolaza *et al.*, 1995; Döbereiner *et al.*, 1996). The repeat domain of intracellular HlyA is therefore assumed to be flexible, allowing translocation of the toxin across the *E. coli* envelope in an unfolded state.

The analysis of active α-haemolysin purified from *E. coli* culture supernatants revealed that the extracellular HlyA is complexed with LPS (Bohach and Snyder, 1985; Ostolaza *et al.*, 1991). The physical interaction between HlyA and LPS appears to be functionally important. Stanley *et al.* (1993) and Bauer and Welch (1997) demonstrated that extracellular α-haemolysin from *E. coli* strains carrying mutations in the genes *rfaC* or *rfaP*, which are involved in the synthesis of the inner core of LPS, has greatly reduced haemolytic activity as compared with α-haemolysin secreted by wild-type *E. coli*. The data further suggest that LPS with an intact inner core is important for stabilizing the active conformation of HlyA and for protecting HlyA from aggregation or degradation (Stanley *et al.*, 1993; Bauer and Welch, 1997).

Interaction of α-haemolysin with target membranes

The interaction of α-haemolysin with eukaryotic cell membranes occurs in distinct steps. In particular, binding of the toxin to the membrane surface and membrane insertion are separable events (Eberspächer *et al.*, 1989; Bakas *et al.*, 1996; Bauer and Welch, 1996a). Transition from the membrane-bound form to the inserted form is associated with a conformational change in the haemolysin protein (Moayeri and Welch, 1997). Whether α-haemolysin binds to a specific cell surface receptor has not been unequivocally resolved. Studies performed by Bauer and Welch (1996a) suggested that the number of HlyA binding sites on sheep erythrocytes is limited to about 4000. This would argue in favour of an interaction with a specific cell surface component. However, Eberspächer *et al.* (1989) did not observe saturability of the binding of HlyA to erythrocytes. Lally *et al.* (1997) suggested that a member of the β2 integrin family, lymphocyte function-associated antigen 1 (LFA-1), serves as a cell surface receptor for *E. coli* α-haemolysin and for the related *Actinobacillus actinomycetemcomitans* leucotoxin (LtxA) on human leucocytes. Direct binding to the heterodimeric LFA-1 was, however, only shown for LtxA.

Insertion of α-haemolysin into the membrane leads to the formation of transmembrane pores, which may cause the osmotic lysis of target cells (Bhakdi *et al.*, 1986). The physical properties of these pores have been analysed using artificial lipid bilayers. These studies demonstrated that HlyA forms uniform, transient pores, which are hydrophilic and cation selective (Menestrina *et al.*, 1987; Benz *et al.*, 1989). Based on the single-channel conductance, the effective pore diameter was estimated to be at least 1 nm (Benz *et al.*, 1989). The pores generated by HlyA in the plasma membrane of macrophages have almost identical characteristics to those formed in planar lipid bilayers, as shown in patch-clamp experiments (Menestrina *et al.*, 1996).

A hydrophobic domain located in the *N*-terminal half of the otherwise hydrophilic HlyA protein is essential for pore formation and for lysis of erythrocytes (Ludwig *et al.*, 1987, 1991). This domain is predicted to contain four hydrophobic, membrane-spanning α-helices, flanked by sequences with amphipathic properties. The amino acid

sequence spanning the hydrophobic domain may thus directly be involved in the generation of the pore structure. The hydrophobic domain is highly conserved in other RTX toxins and its importance for the cytolytic activity has been confirmed in several cases (Glaser *et al.*, 1988; Cruz *et al.*, 1990).

Pore formation by HlyA in artificial lipid bilayers is independent of Ca^{2+} and not affected by deletion of the repeats (Ludwig *et al.*, 1988; Döbereiner *et al.*, 1996). In addition, pro-HlyA forms pores in lipid bilayers, which have similar properties to those generated by the acylated toxin, although the pore-forming activity of pro-HlyA is considerably reduced compared with that of HlyA (Ludwig *et al.*, 1996). Ca^{2+} binding and acylation thus appear to be critical for the interaction between HlyA and eukaryotic cell membranes at some stage preceding pore formation; their precise function is, however, not clear. A role for Ca^{2+} binding and acylation in binding of α-haemolysin to target cell membranes has been suggested by several research groups (Ludwig *et al.*, 1988, 1996; Boehm *et al.*, 1990a, b; Issartel *et al.*, 1991), but this has also been challenged in a number of recent studies (Ostolaza and Goni, 1995; Bauer and Welch, 1996a; Soloaga *et al.*, 1996; Bakas *et al.*, 1998).

Whether α-haemolysin pores are formed by HlyA monomers or oligomers is controversial. Erythrocyte membranes lysed by high doses of α-haemolysin do not exhibit recognizable pores when examined by electron microscopy. In addition, HlyA has been recovered exclusively in monomeric form from deoxycholate-solubilized erythrocyte membranes (Bhakdi *et al.*, 1986), suggesting either that the pores are formed by HlyA monomers or that HlyA oligomers are unstable. The biophysical characteristics of the pores formed by *E. coli* α-haemolysin in planar lipid bilayers are indeed compatible with the hypothesis that there is an association–dissociation equilibrium between non-conducting HlyA monomers and conducting (pore-forming) oligomers in the membrane (Benz *et al.*, 1989). Complementation between certain mutant HlyA variants to produce haemolytic activity further argues in favour of an oligomerization of HlyA (Ludwig *et al.*, 1993). In view of the absence of a structurally visible α-haemolysin pore, it has also been suggested that the cytolytic activity of this toxin may be due to a mechanism different from the generation of discrete-sized transmembrane pores (Ostolaza *et al.*, 1993; Moayeri and Welch, 1994).

Other RTX toxins of Enterobacteriaceae

A haemolysin related to *E. coli* α-haemolysin was recently identified in enterohaemorrhagic *E. coli* (EHEC) strains of serotype O157:H7. These strains are the predominant cause of haemorrhagic colitis and haemolytic uraemic syndrome in humans. The genetic determinant of the EHEC haemolysin is located on a 90-kb plasmid, pO157, which is present in almost all clinical *E. coli* O157 isolates, and represents a typical RTX toxin operon (EHEC-*hlyCABD*) (Schmidt *et al.*, 1995, 1996a). The amino acid sequence of EHEC-HlyA is 61% identical to that of HlyA.

EHEC-HlyA (also designated EhxA) has haemolytic and leucotoxic activity, but the available data suggest that it exhibits a stronger target cell specificity than α-haemolysin (Bauer and Welch, 1996b). The pore-forming characteristics of EHEC haemolysin are very similar to those of *E. coli* α-haemolysin (Schmidt *et al.*, 1996b). The role of EHEC haemolysin in the virulence of *E. coli* O157:H7 has yet to be established.

An exotoxin antigenically related to α-haemolysin has been identified in enteroaggregative *E. coli* (EAggEC) strains (Baldwin *et al.*, 1992). Haemolysins that are structurally, functionally and genetically highly related to *E. coli* α-haemolysin have also been detected in other species belonging to the *Enterobacteriaceae*, particularly in *Proteus vulgaris* and *Morganella morganii* (Koronakis *et al.*, 1987; Welch, 1987; Eberspächer *et al.*, 1990; Benz *et al.*, 1994a).

Leucotoxin of *Pasteurella haemolytica*

Pasteurella haemolytica biotype A, serotype 1, the causative agent of bovine pneumonic pasteurellosis, produces a leucotoxin, LktA, which belongs to the RTX toxin family. This toxin differs from the other RTX toxins in that it specifically kills ruminant leucocytes. Leucocytes from other species are not affected by LktA (Shewen and Wilkie, 1982; Brown *et al.*, 1997). LktA exhibits only weak haemolytic activity against ruminant and certain non-ruminant erythrocytes (Forestier and Welch, 1990; Highlander *et al.*, 1990; Murphy *et al.*, 1995).

Results from *in vivo* studies performed with LktA-negative mutants of *P. haemolytica* A1 suggest that the leucotoxin plays a central role in the pathogenesis of bovine pneumonic pasteurellosis (Petras *et al.*, 1995; Tatum *et al.*, 1998). High concentrations of LktA are cytocidal to both ruminant alveolar macrophages and neutrophils, which suggests that the toxin helps the bacteria evade phagocytic killing in the lung. Thus, LktA probably promotes bacterial proliferation and survival at the site of infection. Low doses of LktA have pleiotropic effects on bovine leucocytes. In particular, LktA stimulates bovine neutrophils to produce reactive oxygen intermediates, to degranulate, and to produce and release eicosanoids such as leukotriene B$_4$ and 5-hydroxyeicosatetraenoic acid (Czuprynski *et al.*, 1991; Henricks *et al.*, 1992; Maheswaran *et al.*, 1993; Wang *et al.*, 1998). The leucotoxin further induces the expression

and secretion of inflammatory cytokines (TNF-α, interleukin-1β) from bovine alveolar macrophages (Yoo *et al.*, 1995). These cellular reactions may contribute to the severe inflammation that characterizes acute pulmonary pasteurellosis ('shipping fever') in cattle. Low concentrations of LktA also impair the proliferation of bovine peripheral blood monocytes and inhibit the expression of major histocompatibility complex class-II antigens in these cells (Czuprynski and Ortiz-Carranza, 1992; Hughes *et al.*, 1994). In addition, the leucotoxin induces bovine leucocytes to undergo apoptosis (Stevens and Czuprynski, 1996).

The chromosomal *lktCABD* operon required for synthesis and secretion of LktA resembles the *E. coli* α-haemolysin determinant with respect to gene organization and transcriptional polarity (Lo *et al.*, 1987; Highlander *et al.*, 1989, 1990; Strathdee and Lo, 1989a, b). The proteins encoded by the *lkt* genes, LktC (20 kDa), LktA (102 kDa), LktB (80 kDa) and LktD (55 kDa), are structurally and functionally homologous to HlyC, HlyA, HlyB and HlyD, respectively, of *E. coli* (Strathdee and Lo, 1987, 1989a; Chang *et al.*, 1989a; Clinkenbeard *et al.*, 1989; Forestier and Welch, 1990; Highlander *et al.*, 1990). The amino acid sequence of LktA is 36.4% identical to that of *E. coli* HlyA. The most pronounced structural difference between these two toxins is that LktA has a shorter repeat domain, consisting of only eight instead of 13 consecutive glycine-rich nonapeptide repeats (Strathdee and Lo, 1987). In addition, one of the two acylation sites identified in HlyA, Lys690, is not conserved in LktA. The latter finding suggests that the acylation patterns of HlyA and LktA are different, although the acylation process *per se* is presumed to be mechanistically similar for both toxins (Pellett and Welch, 1996).

Burrows *et al.* (1993) reported that all 16 serotypes of *P. haemolytica* produce leucotoxins that are immunologically, functionally and genetically highly related but not identical to LktA of *P. haemolytica* serotype 1. A leucotoxin related to LktA but exhibiting a lower degree of species specificity has also been identified in a *P. haemolytica*-like bacterium which was isolated from pigs with enteritis (Chang *et al.*, 1993a).

RTX toxins of *Actinobacillus pleuropneumoniae*

Actinobacillus pleuropneumoniae is the aetiological agent of porcine pleuropneumonia, a contagious pulmonary disease of pigs causing important economic losses in industrialized pig production. Three different exotoxins belonging to the RTX toxin family have been identified in *A. pleuropneumoniae*, ApxI, ApxII and ApxIII. ApxI (previously named HlyI or ClyI) is strongly haemolytic and strongly cytotoxic for phagocytic cells, ApxII (originally designated HlyII, ClyII, App or Cyt) is weakly haemolytic and moderately cytotoxic and ApxIII (formerly named Ptx, ClyIII or Mat) is strongly cytotoxic against alveolar macrophages and neutrophils but has no haemolytic activity (Frey *et al.*, 1993b; Frey, 1995). All three Apx toxins form cation-selective pores in lipid bilayer membranes (Maier *et al.*, 1996). It is remarkable that the ApxIII pores are considerably smaller than those generated by ApxI and ApxII.

The strains of the 12 serotypes of *A. pleuropneumoniae* biotype 1 generally produce either one or two of the Apx toxins: serotypes 1, 5, 9 and 11 produce ApxI and ApxII; serotypes 2, 3, 4, 6 and 8 produce ApxII and ApxIII; serotypes 7 and 12 synthesize only ApxII and serotype 10 produces only ApxI (Kamp *et al.*, 1991, 1994; Frey *et al.*, 1992, 1993a; Macdonald and Rycroft, 1992; Jansen *et al.*, 1993b, 1994; Beck *et al.*, 1994). In the tested *A. pleuropneumoniae* strains belonging to the uncommon biotype 2 only ApxII has been found (Kamp *et al.*, 1994).

The three Apx toxins are major virulence factors of *A. pleuropneumoniae* but their precise functions remain to be determined (Cullen and Rycroft, 1994; Tascon *et al.*, 1994; Jansen *et al.*, 1995; Reimer *et al.*, 1995). Interestingly, the virulence of the *A. pleuropneumoniae* serotypes depends on which Apx toxins they synthesize (Frey, 1995): Serotypes secreting ApxI are exceptionally virulent and cause acute pleuropneumonia with high mortality. The serotypes secreting only ApxII or ApxII and ApxIII, however, are usually moderately virulent, and often cause chronic infections with low mortality. Furthermore, serotypes that produce two different Apx toxins are generally more virulent than those producing only one of these toxins, which suggests that the Apx toxins synergize.

ApxI (110 kDa) is highly related to *E. coli* α-haemolysin (56% amino acid identity) and exhibits weaker similarity to LktA of *P. haemolytica* (41% identity) (Frey *et al.*, 1991; Jansen *et al.*, 1993b). The *A. pleuropneumoniae* serotypes that produce ApxI generally contain a complete ApxI determinant comprising the four genes *apxIC*, *apxIA*, *apxIB* and *apxID*, which are organized in a typical RTX toxin operon (Frey *et al.*, 1993a, 1994; Jansen *et al.*, 1993b, 1994). Transcription of the *apxICABD* operon appears to be positively controlled by high concentrations of Ca^{2+} (Gygi *et al.*, 1992; Frey *et al.*, 1994). A regulation by Ca^{2+} has not been reported for other RTX toxin determinants. All serotypes of *A. pleuropneumoniae* biotype 1, which do not produce ApxI, contain a truncated *apxI* operon (Frey *et al.*, 1993a; Jansen *et al.*, 1993b, 1994). Particularly, in serotypes 2, 4, 6, 7, 8 and 12 *apxIC* and most of the ApxI structural gene *apxIA* is deleted so that the distal half of the operon, including the 3'-terminal

region of *apxIA* and the intact *apxIB* and *apxID* genes, is directly fused to the *apxI* promoter region. These serotypes are thus still able to express a functional ApxI secretion apparatus. The strains of serotype 3 have lost all four structural *apxI* genes and retain only sequences of the *apxI* promoter region.

ApxII strongly resembles LktA of *P. haemolytica* (67% identity) and exhibits a lower degree of similarity to ApxI (46% identity) and *E. coli* HlyA (47% identity). Like LktA, ApxII contains only eight consecutive glycine-rich nonapeptide repeats, while a series of 13 repeats has been identified in ApxI (Chang *et al.*, 1989b; Frey *et al.*, 1991; Smits *et al.*, 1991). The ApxII-producing strains generally harbour a truncated ApxII determinant, comprising only the activator gene *apxIIC*, the gene *apxIIA* encoding the 102.5-kDa toxin protein and the 5' end of a B-like gene (Smits *et al.*, 1991; Jansen *et al.*, 1992, 1994; Frey *et al.*, 1993a). Most of the B-like gene and the complete D-like gene were probably lost during evolution by deletion. However, ApxII can be secreted via the ApxI exporter, because ApxIB and ApxID are encoded in all serotypes of *Actinobacillus pleuropneumoniae* biotype 1, except in serotype 3, by a complete or truncated *apxI* operon (Smits *et al.*, 1991; Jansen *et al.*, 1992, 1993b, 1995; Frey *et al.*, 1993a; Reimer *et al.*, 1995).

ApxIII has an apparent molecular mass of about 120 kDa, but the toxin gene *apxIIIA* encodes a protein of only 113 kDa containing an array of 13 glycine-rich repeats (Macdonald and Rycroft, 1992; Chang *et al.*, 1993b; Jansen *et al.*, 1993a). The amino acid sequence of ApxIII is 50% identical to that of ApxI and *E. coli* HlyA and 41% identical to that of ApxII (Chang *et al.*, 1993b; Jansen *et al.*, 1993a). *Actinobacillus pleuropneumoniae* serotypes producing ApxIII contain a complete *apxIIICABD* operon (Chang *et al.*, 1993b; Jansen *et al.*, 1993a, 1994). The ApxIII transport proteins ApxIIIB and ApxIIID may, like ApxIB/ApxID, also be involved in the secretion of ApxII (Macdonald and Rycroft, 1993).

Actinobacillus suis, an opportunistic pathogen that causes septicaemia in young pigs, produces a haemolysin identical at the amino acid level to ApxII of *A. pleuropneumoniae* (Burrows and Lo, 1992). Furthermore, the determinant encoding this toxin resembles the truncated ApxII determinant of *A. pleuropneumoniae*, as it contains only a C-like and an A-like gene (designated *ashC* and *ashA*, respectively). The B- and D-like proteins required for the secretion of the *A. suis* haemolysin are encoded on a separate region of the chromosome, probably as part of another RTX toxin operon (Burrows and Lo, 1992). Indeed, Kamp *et al.* (1994) found a toxin similar to ApxI in all tested *A. suis* strains.

RTX toxins antigenically related to ApxII have also been detected in *Actinobacillus equuli* and *Actinobacillus lignieresii* (Burrows and Lo, 1992).

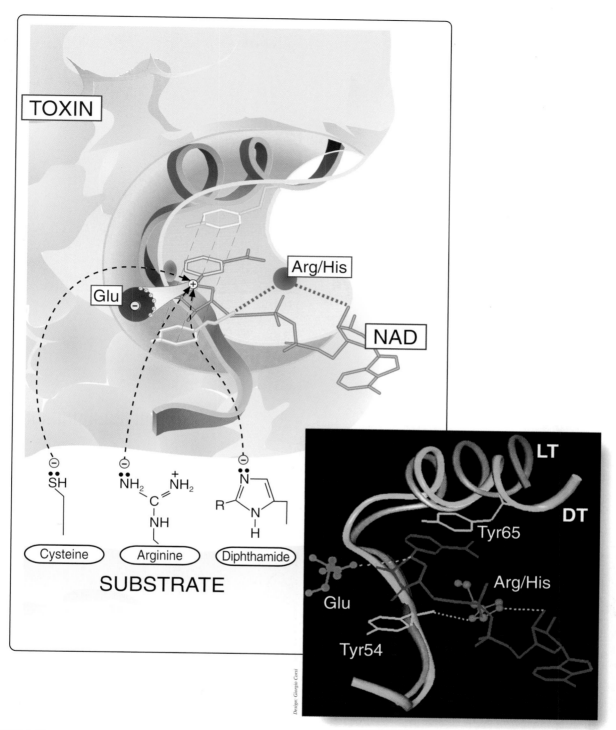

FIGURE 3.11 Catalytic site of ADP-ribosylating toxins. Upper panel: a schematic representation of a possible common mechanism of catalysis is illustrated: the NAD molecule (red) is docked inside the cavity by means of stacking interactions provided by the two aromatic rings (yellow) that protrude from the scaffold of the β–α structure. The catalytic glutamic acid (purple) and its possible interactions with the acceptor residues of the various substrates are also reported. The Arg–His residue (green) provides stabilizing interactions with the backbone of the cavity and seems also to be responsible for the correct positioning of NAD inside the pocket. Lower panel: superimposition of the backbones of the three-dimensional structures of the β–α motif of DT (yellow) and LT (blue). The NAD molecule has been introduced inside the cavity as it is folded in the complex with the DT structure (Bell and Eisenberg, 1997). The essential amino acids together with their main interactions are again reported.

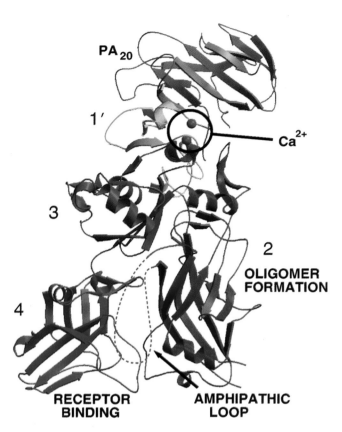

FIGURE 12.3 PA crystal structure. The functions associated with each region of PA are listed. For further details on the structure, see the original publication (Petosa *et al.*, 1997).

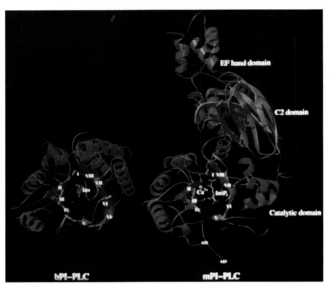

FIGURE 15.3 Comparison of the crystal structures of the *B. cereus* PI-PLC (bPI-PLC) bound to inositol and rat PI-PLCδ1 bound to inositol triphosphate (mPI-PLC). α-Helices are shown in red, β-strands in blue and loops in green. The Ca²⁺ in the active site of the mammalian enzyme is shown as a yellow sphere. Reproduced from Heinz *et al.* (1998), with permission from Academic Press Ltd.

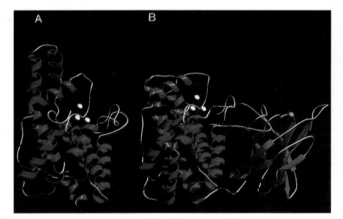

FIGURE 15.2 Comparison of the crystal structures of the *B. cereus* PC-PLC A; Hough *et al.*, 1989) with the *C. perfringens* α-toxin B; Naylor *et al.*, 1998). α-Helices are shown in red, β-strands in blue and loops in green. Zinc residues found in the active site of the enzymes are shown as grey spheres. The location of the bound cadmium ion in the *C. perfringens* α-toxin is shown as a blue sphere.

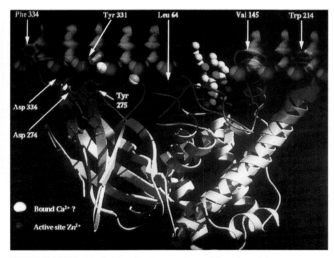

FIGURE 15.7 Model for the interaction of the *C. perfringens* α-toxin with membrane phospholipids. Reproduced from Naylor *et al.* (1998), with permission from Macmillan Magazines Ltd.

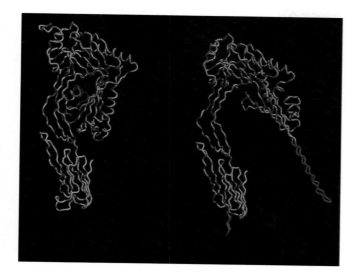

FIGURE 27.4 PFO membrane insertion (see text). On the left-hand side is a 'worm' picture of PFO based on the crystal structure. The wound-up helices in domain 3 are shown in red. On the right-hand side the wound-up helices are shown to extend to form a β-hairpin (in red) and the Trp-rich loop is shown extended (in blue). This picture was generated with the program MidasPlus (Ferrin *et al.*, 1988).

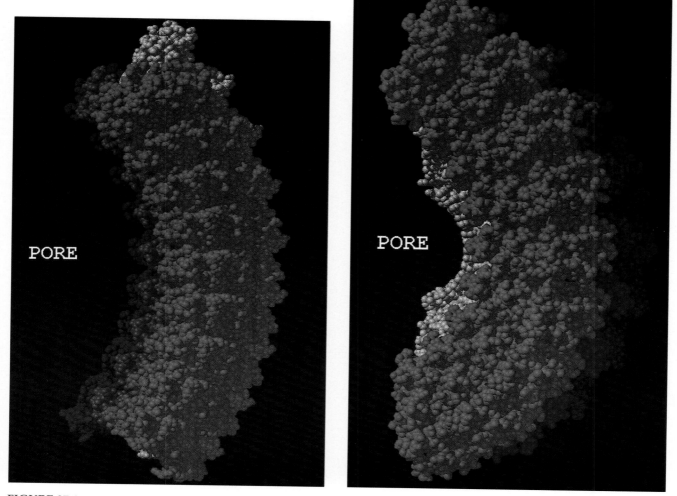

FIGURE 27.6 Alternate models for PFO oligomers. Top views of alternate models of a partial oligomer are shown. The colour coding is as follows: domain 1 is in red, domain 2 is in green, domain 3 is in yellow and domain 4 is in blue. (a) Domain 4 insertion model. (b) Domain 3 insertion model. These pictures were generated with the program MidasPlus (Ferrin *et al.*, 1988).

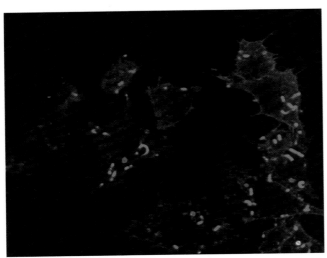

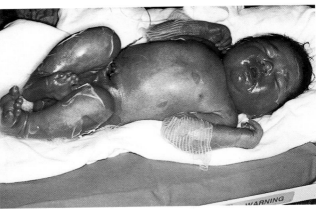

FIGURE 37.1 Generalized SSSS in a newborn caused by an ETA-producing strain of *S. aureus*. Reproduced from Haddad *et al.* (1991) with permission from the publisher of *Arch. Fr. Pediatr.*

FIGURE 28.2 The macrophage-like cell line J774 was infected with *L. monocytogenes* expressing HA-tagged listeriolysin. Cells were stained with a monoclonal anti-HA antibody followed by an Cy-3 labelled secondary antibody (red). Cytosolic listeria recruiting actin were stained with FITC-Phalloidin (green).

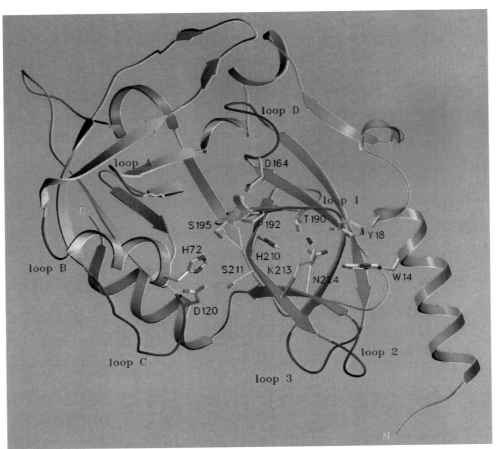

FIGURE 37.4 Schematic drawing of the three-dimensional structure of ETA. The two domains, built around a six-stranded antiparallel β-sheet characteristic of the trypsin-like serine protease fold, are shown in green for domain 1 and in red for domain 2. Reproduced from Cavarelli *et al.* (1997) with permission from the publisher of *Structure*.

Leucotoxin of *Actinobacillus* actinomycetemcomitans

Actinobacillus actinomycetemcomitans is considered to be a primary pathogenic agent in various periodontal diseases, especially in localized juvenile periodontitis. The best-characterized potential virulence factor of this bacterium is the 116-kDa leucotoxin, LtxA, which belongs to the RTX toxin family. This leucotoxin is unique among the RTX toxins in that it specifically disrupts human and primate polymorphonuclear leucocytes, monocytes and T-lymphocytes (Taichman *et al.*, 1980; Tsai *et al.*, 1984; Mangan *et al.*, 1991). Other human cells or leucocytes of other species are not killed by this toxin. LtxA is produced by most strains of *A. actinomycetemcomitans* isolated from patients with periodontitis, but it is usually not synthesized by the few strains that are cultured from healthy individuals (Zambon *et al.*, 1983), which suggests that it plays a significant role in virulence. Like other RTX toxins, LtxA is presumed to help *A. actinomycetemcomitans* to evade host defence cells at the site of infection. Studies performed with planar lipid bilayer membranes and intact target cells have shown that the leucotoxin of *A. actinomycetemcomitans* has pore-forming activity (Lear *et al.*, 1995; Karakelian *et al.*, 1998).

The chromosomal determinant of *A. actinomycetemcomitans* leucotoxin represents a typical RTX toxin operon; it contains the four genes *ltxC*, *ltxA*, *ltxB* and *ltxD* (the genes are alternatively designated *lktC*, *A*, *B*, *D* or *AaLtC*, *A*, *B*, *D*), which are required for the synthesis of the toxin and for its transport to the bacterial cell surface (Lally *et al.*, 1989, 1991; Kraig *et al.*, 1990; Spitznagel *et al.*, 1991). However, in contrast to the other known RTX toxins, LtxA is not secreted extracellularly but remains associated with the bacterium (Tsai *et al.*, 1984; Lally *et al.*, 1989). Ohta *et al.* (1993) reported that LtxA binds ionically to nucleic acids located on the cell surface of *A. actinomycetemcomitans*. Interestingly, LtxA is more related to *E. coli* α-haemolysin (51% identity) than to the leucotoxin of *P. haemolytica* (43% identity) (Kraig *et al.*, 1990).

The lack of leucotoxin production in some strains of *A. actinomycetemcomitans* seems to be due to poor transcription of the *ltx* operon. Indeed, the complete *ltx* gene cluster appears to be present in all *A. actinomycetemcomitans* isolates examined so far, but the level of leucotoxin expression varies considerably among different strains (Spitznagel *et al.*, 1991; Poulsen *et al.*, 1994; Hritz *et al.*, 1996; Kolodrubetz *et al.*, 1996). Highly leucotoxic strains of *A. actinomycetemcomitans* lack a sequence of about 530 nucleotides, which is present in minimally and moderately leucotoxic strains in the 5'-flanking region of the *ltx* gene cluster (Brogan *et al.*, 1994; Hritz *et al.*, 1996; Kolodrubetz *et al.*, 1996). This sequence seems to influence the transcription of the *ltx* operon.

RTX toxin of *Vibrio cholerae*

Recently, an RTX toxin gene cluster was identified in *Vibrio cholerae*; it is physically linked to the cholera toxin (CTX) element on the *V. cholerae* genome and contains four genes in the order *rtxA*, *rtxC*, *rtxB* and *rtxD*, which are related to the A, C, B and D genes, respectively, of other RTX toxin determinants (Lin *et al.*, 1999). The operon structure of the *V. cholerae* RTX toxin gene cluster is unique because the *rtxC* and *rtxA* genes are inverted and consequently transcribed divergently from *rtxB* and *rtxD*. The toxin, RtxA, resembles other members of the RTX family of haemolysins/leucotoxins in that it contains the glycine- and aspartate-rich repeated motif in the C-terminal region. RtxA is, however, 4546 amino acids in size and has a predicted molecular mass of about 500 kDa, i.e. it is more than four times the size of *E. coli* HlyA. Large regions of RtxA do not share sequence similarity to proteins in the database. RtxA has cytotoxic activity but it does not appear to exhibit haemolytic activity (Lin *et al.*, 1999).

The intact *rtx* gene cluster was found in *V. cholerae* serotype O1 strains belonging to the biotype El Tor and in *V. cholerae* O139, which emerged as aetiological agents of epidemic cholera in 1961 and 1992, respectively. In addition, the complete *rtxACBD* gene cluster was detected in several environmental *V. cholerae* isolates. However, *V. cholerae* O1 strains of the classical biotype, the prevalent agents of cholera before 1961, contain a deletion within the *rtx* gene cluster. This deletion removes the 5' end of *rtxA*, all of *rtxC*, the 5' end of *rtxB*, and any regulatory element that lies between these genes. Classical strains of *V. cholerae* are therefore defective in the production of the cytotoxin RtxA (Lin *et al.*, 1999).

Adenylate cyclase toxin of *Bordetella pertussis*

Bordetella pertussis, the causative agent of whooping cough, produces another unusually large RTX toxin known as adenylate cyclase toxin (AC toxin, CyaA). The CyaA protein consists of 1706 amino acid residues and has a deduced molecular mass of 177 kDa (apparent molecular mass, 200–220 kDa) (Glaser *et al.*, 1988; Hewlett *et al.*, 1989; Rogel *et al.*, 1989; Bellalou *et al.*, 1990a). AC toxin is unique among the RTX toxins because it is a bifunctional toxin endowed with both cell-invasive adenylate cyclase (cytotoxic) activity and haemolytic (pore-forming) activity. The enzymic and the haemolytic activity are associated with different regions of the CyaA protein. In particular, the amino-terminal ~400 amino acids of CyaA constitute the adenylate cyclase domain, while the carboxy-terminal ~1300 residues represent the haemolysin domain (Bellalou *et al.*, 1990b; Ehrmann *et al.*, 1992; Sakamoto *et al.*, 1992). Only the haemolysin moiety of CyaA exhibits sequence similarity to the other members of the RTX toxin family (Glaser *et al.*, 1988), suggesting that the *cyaA* gene arose by fusion of an RTX toxin gene with an adenylate cyclase gene. The haemolysin domain of CyaA diverges significantly from the other RTX toxins (the amino acid sequence is only 25% and 22% identical to that of *E. coli* HlyA and *P. haemolytica* LktA, respectively). CyaA particularly contains a very large repeat region comprising about 40 glycine-rich nonameric repeats (Glaser *et al.*, 1988).

AC toxin forms very small cation-selective pores in lipid bilayers (Benz *et al.*, 1994b; Szabo *et al.*, 1994) and it has only weak haemolytic activity (Bellalou *et al.*, 1990b; Rogel *et al.*, 1991). Indeed, the haemolytic domain of CyaA appears to be primarily required for the binding of AC toxin to target cells such as leucocytes, epithelial cells and erythrocytes, and for the subsequent translocation of the AC domain through the cell membrane into the cytoplasm of the target cell. After activation by calmodulin, the internalized AC domain catalyses the uncontrolled formation of supraphysiological levels of cyclic AMP from endogenous ATP, thereby disrupting normal cellular functions (Wolff *et al.*, 1980; Confer and Eaton, 1982; Bellalou *et al.*, 1990b; Rogel *et al.*, 1991; Iwaki *et al.*, 1995).

The molecular mechanism by which the AC domain of CyaA is delivered into the target cell has not been completely determined. However, the penetration through the cell membrane depends on a conformational change of CyaA, which is induced by Ca^{2+} concentrations in the millimolar range. The membrane binding capability and the haemolytic activity of AC toxin require significantly lower Ca^{2+} concentrations (Hewlett *et al.*, 1991; Rogel *et al.*, 1991; Rogel and Hanski, 1992). Rose *et al.* (1995) suggested that binding of Ca^{2+} to a small number (three to five) of high-affinity calcium binding sites present in CyaA might be necessary for the binding of the toxin to cell membranes and for the lysis of erythrocytes, while binding of Ca^{2+} to about 40–45 low-affinity binding sites located in the repeat region of CyaA might promote entry of the AC domain into the target cell by inducing a large conformational change in the toxin. Translocation of the AC domain into the cell probably does not proceed through the small transmembrane pore that is formed by the haemolysin moiety of CyaA. Invasion of cells by AC toxin apparently depends on the electrical potential across the plasma membrane (Otero *et al.*, 1995).

AC toxin is an essential virulence factor of *B. pertussis*; it is particularly involved in the early steps of respiratory tract colonization (Goodwin and Weiss, 1990; Gross *et al.*, 1992; Khelef *et al.*, 1994). Based on its cytotoxic activity, CyaA induces apoptosis of alveolar macrophages (Khelef and Guiso, 1995; Gueirard *et al.*, 1998). The cytotoxic activity of CyaA is also specifically required for the creation of pulmonary lesions and for lethality (Khelef *et al.*, 1994).

The chromosomal AC toxin determinant of *B. pertussis* exhibits several interesting features. In contrast to the other RTX toxin determinants, it consists of five closely linked genes arranged in the order *cyaC*, *cyaA*, *cyaB*, *cyaD* and *cyaE* (Glaser *et al.*, 1988; Barry *et al.*, 1991). The toxin structural gene, *cyaA*, and the three genes downstream of *cyaA* form an operon, which is transcribed from a strong promoter located in the *cyaC*–*cyaA* intergenic region (Laoide and Ullmann, 1990). Transcription initiation at this promoter is regulated in response to environmental signals by the BvgA/BvgS two-component signal transduction system (Laoide and Ullmann, 1990; Karimova *et al.*, 1996; Steffen *et al.*, 1996). The BvgA/BvgS system is also involved in the regulation of other virulence factors of *B. pertussis*. The three genes *cyaB*, *cyaD* and *cyaE* encode the components of the AC toxin secretion apparatus. CyaB and CyaD share striking homology with the B and D proteins encoded by other RTX toxin determinants, and CyaE is the outer membrane component of the AC toxin exporter (Glaser *et al.*, 1988). Thus, in the case of AC toxin, the genes of all three transport proteins are linked to the toxin gene. *cyaC* is homologous to the C genes of the other RTX toxin determinants and required for the post-translational activation of CyaA. However, *cyaC* is orientated oppositely as compared with the *cyaABDE* operon and transcribed from its own promoter (Barry *et al.*, 1991). CyaA is activated in *B. pertussis* by palmitoylation of a single lysine residue, Lys983, which corresponds to Lys690 of *E. coli* HlyA (Hackett *et al.*, 1994). Non-activated adenylate cyclase toxin possesses AC activity but is devoid of invasive (cytotoxic) and haemolytic activities (Rogel *et al.*, 1989; Barry *et al.*, 1991; Hewlett *et al.*, 1993).

Adenylate cyclase toxin is also produced by *Bordetella parapertussis* and *Bordetella bronchiseptica* (Bellalou *et al.*, 1990a; Betsou *et al.*, 1995).

OTHER PROTEINS OF GRAM-NEGATIVE BACTERIA SECRETED BY ABC EXPORTERS

In addition to the RTX toxins, a range of other proteins from various, mostly pathogenic, Gram-negative bacteria is secreted by type-I secretion systems (i.e. by ABC exporters). These proteins include several protein families that differ significantly in their activities or functions. There is no significant overall sequence similarity between exoproteins employing the type-I pathway for secretion but exhibiting different functions. However, most proteins secreted by ABC exporters contain a series of glycine- and aspartate-rich Ca^{2+}-binding nonapeptide repeats that resemble those of the RTX toxins, and in addition most of them possess an uncleaved transport signal at the C-terminal end. Proteins exhibiting these features are grouped together with the RTX toxins in the large family of RTX exoproteins (Table 16.1).

A prominent subgroup among the RTX exoproteins is the serralysin family of bacterial metalloproteases, which includes the four proteases PrtA, PrtB, PrtC and PrtG of *Erwinia chrysanthemi* (Delepelaire and Wandersman, 1989, 1990, 1991; Letoffe *et al.*, 1990; Ghigo and Wandersman, 1992, 1994), the protease SM (PrtSM, PrtA) of *Serratia marcescens* (Nakahama *et al.*, 1986; Letoffe *et al.*, 1991; Akatsuka *et al.*, 1997), the alkaline protease (AprA) of *Pseudomonas aeruginosa* (Guzzo *et al.*, 1991; Duong *et al.*, 1992, 1996), and the protease ZapA of *Proteus mirabilis* (Wassif *et al.*, 1995). These metalloproteases have molecular masses between 50 and 55 kDa and are highly related to each other (50–60% identity).

The RTX exoprotein family further includes the homologous lipases from *S. marcescens* (LipA, 65 kDa) and *Pseudomonas fluorescens* (Lip, 50 kDa) (Akatsuka *et al.*, 1994, 1995; Duong *et al.*, 1994; Li *et al.*, 1995), the homologous S-layer proteins SlaA (100 kDa) and RsaA (98 kDa) from *S. marcescens* and *Caulobacter crescentus*, respectively (Awram and Smit, 1998; Kawai *et al.*, 1998), the related Fe-regulated proteins FrpA (122 kDa) and FrpC (198 kDa) of *Neisseria meningitidis* (Thompson *et al.*, 1993a, b), and the nodulation-signalling protein NodO (30 kDa) from *Rhizobium leguminosarum* (Economou *et al.*, 1990; Scheu *et al.*, 1992; Finnie *et al.*, 1997).

The *E. coli* bacteriocin colicin V (CvaC, 9 kDa) and the 19-kDa haem-binding protein HasA from *S. marcescens* are also secreted by type-I secretion systems, but these two proteins differ from the other proteins specified above in that they lack the nonapeptide repeats (Gilson *et al.*, 1990; Letoffe *et al.*, 1994a, b; Binet and Wandersman, 1996). Colicin V further lacks the C-terminal secretion signal but contains instead an N-terminal transport signal of the double-glycine type, which is cleaved concomitant with secretion (Gilson *et*

TABLE 16.1 Members of the RTX exoprotein family

RTX toxins	
Escherichia coli	α-Haemolysin (HlyA)
	EHEC-Haemolysin (EHEC-HlyA, EhxA)
	EAggEC exotoxin
Proteus vulgaris	Haemolysin
Morganella morganii	Haemolysin
Pasteurella haemolytica	Leucotoxin (LktA)
Pasteurella haemolytica-like	Leucotoxin (PllktA)
Actinobacillus pleuropneumoniae	ApxI (Haemolysin)
	ApxII (Haemolysin)
	ApxIII (Cytotoxin)
Actinobacillus suis	Haemolysin (AshA)
Actinobacillus actinomycetemcomitans	Leucotoxin (LtxA)
Vibrio cholerae	Cytotoxin (RtxA)
Bordetella pertussis	Adenylate cyclase toxin (CyaA)
Moraxella bovis	Haemolysin
Enterobacter cloacae	Haemolysin
Metalloproteases	
Erwinia chrysanthemi	Proteases PrtA, PrtB, PrtC, PrtG
Serratia marcescens	Protease SM (PrtSM, PrtA)
Pseudomonas aeruginosa	Alkaline protease (AprA)
Proteus mirabilis	Protease ZapA
Lipases	
Serratia marcescens	Lipase (LipA)
Pseudomonas fluorescens	Lipase (Lip)
S-layer proteins	
Serratia marcescens	SlaA
Caulobacter crescentus	RsaA
Nodulation-signalling protein	
Rhizobium leguminosarum	NodO
Protein of unknown function	
Neisseria meningitidis	FrpA, FrpC

al., 1990; Havarstein et al., 1994; van Belkum et al., 1997). The transport signal of colicin V is, however, not related to the N-terminal signal peptides that direct proteins across the cytoplasmic membrane via the Sec pathway.

The various ABC exporters of Gram-negative bacteria are generally composed of an inner membrane ABC protein, an accessory inner membrane protein belonging to the MFP family and an outer membrane protein (Binet et al., 1997). Corresponding components of different ABC exporters exhibit striking homology (between about 20 and 90% amino acid sequence identity). Each ABC exporter is usually dedicated to the secretion of a particular protein or a family of closely related proteins. Heterologous complementation between ABC exporters belonging to related exoproteins generally occurs with high efficiency, but only poor cross-complementation is usually observed between ABC exporters of unrelated exoproteins (Delepelaire and Wandersman, 1990; Fath et al., 1991; Guzzo et al., 1991; Letoffe et al., 1991; Scheu et al., 1992; Suh and Benedik, 1992; Thompson and Sparling, 1993; Duong et al., 1994; Akatsuka et al., 1997). The ABC protein is assumed to be critical for the substrate specificity of ABC exporters. Binding of the secretory protein to the ABC protein (or to the preformed ABC protein/MFP complex) apparently triggers the assembly of the inner and outer membrane components of an ABC exporter into a functional transport complex (Binet and Wandersman, 1995; Duong et al., 1996; Letoffe et al., 1996; Akatsuka et al., 1997; Thanabalu et al., 1998).

ABC exporter genes are usually adjacent to the structural gene(s) of the secreted protein(s). In several cases, the genes of all three transport proteins are linked to the exoprotein gene(s). For example, the four metalloproteases of E. chrysanthemi (PrtA, B, C, G) are encoded by contiguous genes that are clustered with the three genes, prtD, prtE and prtF, encoding the ABC protein, the MFP and the outer membrane component, respectively, of their common secretion apparatus (Letoffe et al., 1990; Ghigo and Wandersman, 1992). Similarly, the alkaline protease structural gene of P. aeruginosa, aprA, is clustered with the corresponding exporter genes, aprD, aprE and aprF (Duong et al., 1992). This type of exporter gene organization has also been found in the AC toxin determinant of B. pertussis (see above). However, the hasA gene of S. marcescens, the rsaA gene encoding the S-layer protein of C. crescentus, the E. coli colicin V gene (cvaC) and most RTX toxin genes are clustered only with the genes encoding the two inner membrane components of the respective transport apparatus, while the outer membrane component is encoded by an unlinked gene (Gilson et al., 1990; Binet and Wandersman, 1996; Awram and Smit, 1998). Interestingly, the ABC transport systems for E. coli α-haemolysin and colicin V employ the same outer membrane component, the

TolC protein (Gilson et al., 1990; Wandersman and Delepelaire, 1990; Hwang et al., 1997).

Some RTX exoprotein genes are not clustered with corresponding exporter genes. This is particularly the case when the ABC exporter is involved in the secretion of several unrelated proteins. For example, in S. marcescens three quite different proteins, the metalloprotease (PrtSM, PrtA), the lipase (LipA) and the S-layer protein (SlaA), are secreted by a single ABC exporter, the Lip exporter, which is encoded by the lipBCD operon (Akatsuka et al., 1995, 1997; Kawai et al., 1998). Only the slaA gene is clustered with the lipBCD genes, while lipA and the metalloprotease gene are unlinked (Kawai et al., 1998). Furthermore, the nodO gene of R. leguminosarum is not linked to the prsDE genes, which encode the ABC protein and the MFP component of a type-I secretion system required for the secretion of NodO and several other unrelated proteins (Scheu et al., 1992; Finnie et al., 1997). Remarkably, NodO forms cation-selective channels in lipid bilayers, suggesting that it facilitates nodulation-signalling by forming channels in plant cell plasma membranes (Sutton et al., 1994).

The functional significance of the calcium-binding repeats in the RTX exoproteins is unclear. However, it is possible that the binding of Ca^{2+} generally facilitates the folding of these proteins in the external medium, following secretion in a presumably extended conformation. Determination of the crystal structure of the alkaline protease (AprA) from P. aeruginosa and of the metalloprotease PrtSM from S. marcescens revealed that the glycine-rich nonameric repeats form a parallel β-roll structure upon binding of Ca^{2+} ions, in which successive β-strands are wound in a right-handed spiral (Baumann et al., 1993; Baumann, 1994). The GGXGXD motif of each repeat forms two half sites for Ca^{2+} binding, while the remaining three residues of the repeats (XUX) form a short β-strand. Each Ca^{2+} ion is bound in a six-co-ordinate, approximately octahedral site between a pair of loops in the β-roll. It is presumed that the repeats of other RTX exoproteins also bind Ca^{2+} in a parallel β-roll structure.

In the absence of Ca^{2+}, the parallel β-roll structure is probably unstable, which could facilitate extracellular secretion of the RTX exoproteins in an unfolded form. After membrane translocation, the calcium ions present in the extracellular medium may allow the rapid formation of a very stable β-roll structure and hence induce the protein to adopt a well-defined tertiary structure (Baumann et al., 1993). Remarkably, both HasA and colicin V, which lack the nonapeptide repeats, are small proteins. They may therefore be able to fold rapidly without requiring stabilization of a specific tertiary structure by co-ordination of calcium ions (Finnie et al., 1997).

Recently, several relatively large proteins lacking the glycine-rich nonameric repeats but containing other repeated sequence motifs have been shown to be secreted via the PrsDE type-I secretion system of *R. leguminosarum* (Finnie *et al.*, 1997, 1998). The function of the novel repeats in these proteins remains to be determined.

EVOLUTION OF THE RTX TOXINS

It is tempting to presume that the various RTX exoproteins have a common origin. A progenitor of the RTX exoprotein determinants possibly arose by combination of a set of genes encoding the components of an ancient ABC exporter with a gene encoding a progenitor of the RTX exoproteins, containing Ca^{2+}-binding nonameric repeats and a *C*-terminal secretion signal. The exoprotein then possibly diverged during evolution to acquire haemolytic, leucotoxic, proteolytic or other activities, while the transport proteins were more conserved owing to greater functional constraints (Welch, 1991).

Spreading of the RTX toxin determinants within the Gram-negative bacteria was probably promoted by horizontal gene transfer. The localization of *E. coli* α-haemolysin determinants within chromosomal pathogenicity islands or on transmissible plasmids clearly supports this hypothesis. Several plasmid-encoded α-haemolysin determinants have been shown to be flanked by insertion sequence (IS)-like elements (Zabala *et al.*, 1984; Knapp *et al.*, 1985), suggesting that IS-mediated gene transfer mechanisms played a role in the dissemination of RTX toxin determinants.

The G + C content of *E. coli* α-haemolysin determinants is about 40%, which is 10% lower than the average for the *E. coli* genome (Felmlee *et al.*, 1985; Hess *et al.*, 1986). This further suggests that the *hly* genes have been acquired by *E. coli* from another organism. In *Actinobacillus* species, *Proteus vulgaris* and *Pasteurella haemolytica* the overall base composition of RTX toxin determinants is very similar to that of the bacterial genomes (about 40% G + C in each case), but it is not known whether the *E. coli* α-haemolysin determinants originate from one of these bacteria.

Adenylate cyclase toxin (CyaA) from *B. pertussis* and RtxA from *V. cholerae* are the most distantly related members of the RTX toxin family, as indicated by the low degree of sequence similarity between these toxins and the other RTX toxins, and by the unique operon structures of the *cya* and *rtx* gene clusters. In the case of adenylate cyclase toxin, this is also evident from several other findings, including (a) the fusion of the original RTX toxin gene with an adenylate cyclase gene, possibly of eukaryotic origin; (b) the unusually high G + C content of the *cya* genes (66 %), which is similar to that of other genes of *B. pertussis*; and (c) the regulation of the *cyaABDE* operon by a two-component signal transduction system.

Interestingly, the phylogenetic tree of the RTX toxins obtained by alignment of the corresponding amino acid sequences does not provide clues about how host species and target cell specificities arose during evolution of this toxin family, because the RTX leucotoxins and haemolysins are not clustered in distinct branches (Welch, 1995). The structural basis of the cell type and species specificity of RTX toxins is currently unresolved, but the construction of hybrid toxins by exchange of portions between different RTX toxins is a useful approach to identify domains important for target cell specificity (Forestier and Welch, 1991; McWhinney *et al.*, 1992; Lally *et al.*, 1994; Westrop *et al.*, 1997).

CONCLUDING REMARKS

During the 1990s, cytolysins from a variety of Gram-negative bacteria have been recognized as members of the RTX toxin family. It is likely that additional RTX toxins will be discovered and characterized in the future. Preliminary data suggest, for example, that RTX haemolysins are also produced by *Moraxella bovis* and *Enterobacter cloacae* (Prada and Beutin, 1991; Gray *et al.*, 1995). Several RTX toxins have been shown to represent important virulence factors, but in most cases their precise pathophysiological functions are still poorly defined. Many questions regarding the genetics, regulation, activation, secretion and mode of action of RTX toxins remain to be answered. Clearly, further work is required to elucidate the structural and functional properties of the RTX toxins and to determine their roles in the pathogenesis of bacterial infections.

REFERENCES

Akatsuka, H., Kawai, E., Omori, K., Komatsubara, S., Shibatani, T. and Tosa, T. (1994). The *lipA* gene of *Serratia marcescens* which encodes an extracellular lipase having no N-terminal signal peptide. *J. Bacteriol.* **176**, 1949–56.

Akatsuka, H., Kawai, E., Omori, K. and Shibatani, T. (1995). The three genes *lipB*, *lipC*, and *lipD* involved in the extracellular secretion of the *Serratia marcescens* lipase which lacks an N-terminal signal peptide. *J. Bacteriol.* **177**, 6381–9.

Akatsuka, H., Binet, R., Kawai, E., Wandersman, C. and Omori, K. (1997). Lipase secretion by bacterial hybrid ATP-binding cassette exporters: molecular recognition of the LipBCD, PrtDEF, and HasDEF exporters. *J. Bacteriol.* **179**, 4754–60.

Awram, P. and Smit, J. (1998). The *Caulobacter crescentus* paracrystalline S-layer protein is secreted by an ABC transporter (type I) secretion apparatus. *J. Bacteriol.* **180**, 3062–9.

Bailey, M.J.A., Hughes, C. and Koronakis, V. (1996). Increased distal gene transcription by the elongation factor RfaH, a specialized homologue of NusG. *Mol. Microbiol.* **22**, 729–37.

Bakas, L., Ostolaza, H., Vaz, W.L.C. and Goni, F.M. (1996). Reversible adsorption and nonreversible insertion of *Escherichia coli* α-haemolysin into lipid bilayers. *Biophys. J.* **71**, 1869–76.

Bakas, L., Veiga, M.P., Soloaga, A., Ostolaza, H. and Goni, F.M. (1998). Calcium-dependent conformation of *E. coli* α-haemolysin. Implications for the mechanism of membrane insertion and lysis. *Biochim. Biophys. Acta* **1368**, 225–34.

Baldwin, T.J., Knutton, S., Sellers, L., Manjarrez Hernandez, H.A., Aitken, A. and Williams, P.H. (1992). Enteroaggregative *Escherichia coli* strains secrete a heat-labile toxin antigenically related to *E. coli* haemolysin. *Infect. Immun.* **60**, 2092–5.

Barry, E.M., Weiss, A.A., Ehrmann, I.E., Gray, M.C., Hewlett, E.L. and Goodwin, M.St.M. (1991). *Bordetella pertussis* adenylate cyclase toxin and haemolytic activities require a second gene, *cyaC*, for activation. *J. Bacteriol.* **173**, 720–6.

Bauer, M.E. and Welch, R.A. (1996a). Association of RTX toxins with erythrocytes. *Infect. Immun.* **64**, 4665–72.

Bauer, M.E. and Welch, R.A. (1996b). Characterization of an RTX toxin from enterohemorrhagic *Escherichia coli* O157:H7. *Infect. Immun.* **64**, 167–75.

Bauer, M.E. and Welch, R.A. (1997). Pleiotropic effects of a mutation in *rfaC* on *Escherichia coli* haemolysin. *Infect. Immun.* **65**, 2218–24.

Baumann, U. (1994). Crystal structure of the 50 kDa metallo protease from *Serratia marcescens*. *J. Mol. Biol.* **242**, 244–51.

Baumann, U., Wu, S., Flaherty, K.M. and McKay, D.B. (1993). Three-dimensional structure of the alkaline protease of *Pseudomonas aeruginosa*: a two-domain protein with a calcium binding parallel beta roll motif. *EMBO J.* **12**, 3357–64.

Beck, M., van den Bosch, J.F., Jongenelen, I.M., Loeffen, P.L., Nielsen, R., Nicolet, J. and Frey, J. (1994). RTX toxin genotypes and phenotypes in *Actinobacillus pleuropneumoniae* field strains. *J. Clin. Microbiol.* **32**, 2749–54.

Bellalou, J., Ladant, D. and Sakamoto, H. (1990a). Synthesis and secretion of *Bordetella pertussis* adenylate cyclase as a 200-kilodalton protein. *Infect. Immun.* **58**, 1195–200.

Bellalou, J., Sakamoto, H., Ladant, D., Geoffroy, C. and Ullmann, A. (1990b). Deletions affecting haemolytic and toxin activities of *Bordetella pertussis* adenylate cyclase. *Infect. Immun.* **58**, 3242–7.

Benz, R., Schmid, A., Wagner, W. and Goebel, W. (1989). Pore formation by the *Escherichia coli* haemolysin: evidence for an association-dissociation equilibrium of the pore-forming aggregates. *Infect. Immun.* **57**, 887–95.

Benz, R., Maier, E. and Gentschev, I. (1993). TolC of *Escherichia coli* functions as an outer membrane channel. *Zbl. Bakt.* **278**, 187–96.

Benz, R., Hardie, K.R. and Hughes, C. (1994a). Pore formation in artificial membranes by the secreted haemolysins of *Proteus vulgaris* and *Morganella morganii*. *Eur. J. Biochem.* **220**, 339–47.

Benz, R., Maier, E., Ladant, D., Ullmann, A. and Sebo, P. (1994b). Adenylate cyclase toxin (CyaA) of *Bordetella pertussis*: evidence for the formation of small ion-permeable channels and comparison with HlyA of *Escherichia coli*. *J. Biol. Chem.* **269**, 27231–9.

Betsou, F., Sismeiro, O., Danchin, A. and Guiso, N. (1995). Cloning and sequence of the *Bordetella bronchiseptica* adenylate cyclase-haemolysin-encoding gene: comparison with the *Bordetella pertussis* gene. *Gene* **162**, 165–6.

Bhakdi, S. and Martin, E. (1991). Superoxide generation by human neutrophils induced by low doses of *Escherichia coli* haemolysin. *Infect. Immun.* **59**, 2955–62.

Bhakdi, S., Mackman, N., Nicaud, J.-M. and Holland, I.B. (1986). *Escherichia coli* haemolysin may damage target cell membranes by generating transmembrane pores. *Infect. Immun.* **52**, 63–9.

Bhakdi, S., Greulich, S., Muhly, M., Eberspächer, B., Becker, H., Thiele, A. and Hugo, F. (1989). Potent leukocidal action of *Escherichia coli* haemolysin mediated by permeabilization of target cell membranes. *J. Exp. Med.* **169**, 737–54.

Bhakdi, S., Muhly, M., Korom, S. and Schmidt, G. (1990). Effects of *Escherichia coli* haemolysin on human monocytes. Cytocidal action and stimulation of interleukin 1 release. *J. Clin. Invest.* **85**, 1746–53.

Binet, R. and Wandersman, C. (1995). Protein secretion by hybrid bacterial ABC-transporters: specific functions of the membrane ATPase and the membrane fusion protein. *EMBO J.* **14**, 2298–306.

Binet, R. and Wandersman, C. (1996). Cloning of the *Serratia marcescens hasF* gene encoding the Has ABC exporter outer membrane component: a TolC analogue. *Mol. Microbiol.* **22**, 265–73.

Binet, R., Letoffe, S., Ghigo, J.M., Delepelaire, P. and Wandersman, C. (1997). Protein secretion by Gram-negative bacterial ABC exporters – a review. *Gene* **192**, 7–11.

Blight, M.A. and Holland, I.B. (1994). Heterologous protein secretion and the versatile *Escherichia coli* haemolysin translocator. *Tibtech* **12**, 450–5.

Blum, G., Ott, M., Lischewski, A., Ritter, A., Imrich, H., Tschäpe, H. and Hacker, J. (1994). Excision of large DNA regions termed pathogenicity islands from tRNA-specific loci in the chromosome of an *Escherichia coli* wild-type pathogen. *Infect. Immun.* **62**, 606–14.

Boehm, D.F., Welch, R.A. and Snyder, I.S. (1990a). Calcium is required for binding of *Escherichia coli* haemolysin (HlyA) to erythrocyte membranes. *Infect. Immun.* **58**, 1951–8.

Boehm, D.F., Welch, R.A. and Snyder, I.S. (1990b). Domains of *Escherichia coli* haemolysin (HlyA) involved in binding of calcium and erythrocyte membranes. *Infect. Immun.* **58**, 1959–64.

Bohach, G.A. and Snyder, I.S. (1985). Chemical and immunological analysis of the complex structure of *Escherichia coli* alpha-haemolysin. *J. Bacteriol.* **164**, 1071–80.

Brogan, J.M., Lally, E.T., Poulsen, K., Kilian, M. and Demuth, D.R. (1994). Regulation of *Actinobacillus actinomycetemcomitans* leukotoxin expression: analysis of the promoter regions of leukotoxic and minimally leukotoxic strains. *Infect. Immun.* **62**, 501–8.

Brown, J.F., Leite, F. and Czuprynski, C.J. (1997). Binding of *Pasteurella haemolytica* leukotoxin to bovine leukocytes. *Infect. Immun.* **65**, 3719–24.

Burrows, L.L. and Lo, R.Y.C. (1992). Molecular characterization of an RTX toxin determinant from *Actinobacillus suis*. *Infect. Immun.* **60**, 2166–73.

Burrows, L.L., Olah-Winfield, E. and Lo, R.Y.C. (1993). Molecular analysis of the leukotoxin determinants from *Pasteurella haemolytica* serotypes 1 to 16. *Infect. Immun.* **61**, 5001–7.

Cavalieri, S.J. and Snyder, I.S. (1982). Effect of *Escherichia coli* alpha-haemolysin on human peripheral leukocyte viability *in vitro*. *Infect. Immun.* **36**, 455–61.

Chang, Y.-F., Young, R., Moulds, T.L. and Struck, D.K. (1989a). Secretion of the *Pasteurella* leukotoxin by *Escherichia coli*. *FEMS Microbiol. Lett.* **60**, 169–74.

Chang, Y.-F., Young, R. and Struck, D.K. (1989b). Cloning and characterization of a haemolysin gene from *Actinobacillus (Haemophilus) pleuropneumoniae*. *DNA* **8**, 635–47.

Chang, Y.-F., Ma, D.-P., Shi, J. and Chengappa, M.M. (1993a). Molecular characterization of a leukotoxin gene from a *Pasteurella haemolytica*-like organism, encoding a new member of the RTX toxin family. *Infect Immun.* **61**, 2089–95.

Chang, Y.-F., Shi, J., Ma, D.-P., Shin, S.J. and Lein, D.H. (1993b). Molecular analysis of the *Actinobacillus pleuropneumoniae* RTX toxin-III gene cluster. *DNA Cell Biol.* **12**, 351–62.

Chervaux, C. and Holland, I.B. (1996). Random and directed mutagenesis to elucidate the functional importance of helix II and F-989 in the C-terminal secretion signal of *Escherichia coli* haemolysin. *J. Bacteriol.* **178**, 1232–6.

Clinkenbeard, K.D., Mosier, D.A. and Confer, A.W. (1989). Transmembrane pore size and role of cell swelling in cytotoxicity caused by *Pasteurella haemolytica* leukotoxin. *Infect. Immun.* **57**, 420–5.

Confer, D.L. and Eaton, J.W. (1982). Phagocyte impotence caused by an invasive bacterial adenylate cyclase. *Science* **217**, 948–50.

Cruz, W.T., Young, R., Chang, Y.F. and Struck, D.K. (1990). Deletion analysis resolves cell-binding and lytic domains of the *Pasteurella* leukotoxin. *Mol. Mocrobiol.* **4**, 1933–9.

Cullen, J.M. and Rycroft, A.N. (1994). Phagocytosis by pig alveolar macrophages of *Actinobacillus pleuropneumoniae* serotype 2 mutant strains defective in haemolysin II (ApxII) and pleurotoxin (ApxIII). *Microbiol.* **140**, 237–44.

Czuprynski, C.J. and Ortiz-Carranza, O. (1992). *Pasteurella haemolytica* leukotoxin inhibits mitogen-induced bovine peripheral blood mononuclear cell proliferation *in vitro*. *Microb. Pathog.* **12**, 459–63.

Czuprynski, C.J., Noel, E.J., Ortiz-Carranza, O. and Srikumaran, S. (1991). Activation of bovine neutrophils by partially purified *Pasteurella haemolytica* leukotoxin. *Infect. Immun.* **59**, 3126–33.

de la Cruz, F., Müller, D., Ortiz, J.M. and Goebel, W. (1980). Hemolysis determinant common to *Escherichia coli* haemolytic plasmids of different incompatibility groups. *J. Bacteriol.* **143**, 825–33.

Delepelaire, P. and Wandersman, C. (1989). Protease secretion by *Erwinia chrysanthemi*. Proteases B and C are synthesized and secreted as zymogens without a signal peptide. *J. Biol. Chem.* **264**, 9083–9.

Delepelaire, P. and Wandersman, C. (1990). Protein secretion in Gram-negative bacteria. The extracellular metalloprotease B from *Erwinia chrysanthemi* contains a C-terminal secretion signal analogous to that of *Escherichia coli* α-haemolysin. *J. Biol. Chem.* **265**, 17118–25.

Delepelaire, P. and Wandersman, C. (1991). Characterization, localization and transmembrane organization of the three proteins PrtD, PrtE and PrtF necessary for protease secretion by the Gram-negative bacterium *Erwinia chrysanthemi*. *Mol. Microbiol.* **5**, 2427–34.

Dinh, T., Paulsen, I.T. and Saier, M.H. (1994). A family of extracytoplasmic proteins that allow transport of large molecules across the outer membranes of Gram-negative bacteria. *J. Bacteriol.* **176**, 3825–31.

Döbereiner, A., Schmid, A., Ludwig, A., Goebel, W. and Benz, R. (1996). The effects of calcium and other polyvalent cations on channel formation by *Escherichia coli* α-haemolysin in red blood cells and lipid bilayer membranes. *Eur. J. Biochem.* **240**, 454–60.

Duong, F., Lazdunski, A., Cami, B. and Murgier, M. (1992). Sequence of a cluster of genes controlling synthesis and secretion of alkaline protease in *Pseudomonas aeruginosa*: relationships to other secretory pathways. *Gene* **121**, 47–54.

Duong, F., Soscia, C., Lazdunski, A. and Murgier, M. (1994). The *Pseudomonas fluorescens* lipase has a C-terminal secretion signal and is secreted by a three-component bacterial ABC-exporter system. *Mol. Microbiol.* **11**, 1117–26.

Duong, F., Lazdunski, A. and Murgier, M. (1996). Protein secretion by heterologous bacterial ABC-transporters: the C-terminus secretion signal of the secreted protein confers high recognition specificity. *Mol. Microbiol.* **21**, 459–70.

Eberspächer, B., Hugo, F. and Bhakdi, S. (1989). Quantitative study of the binding and haemolytic efficiency of *Escherichia coli* haemolysin. *Infect. Immun.* **57**, 983–8.

Eberspächer, B., Hugo, F., Pohl, M. and Bhakdi, S. (1990). Functional similarity between the haemolysins of *Escherichia coli* and *Morganella morganii*. *J. Med. Microbiol.* **33**, 165–70.

Economou, A., Hamilton, W.D.O., Johnston, A.W.B. and Downie, J.A. (1990). The *Rhizobium* nodulation gene *nodO* encodes a Ca^{2+}-binding protein that is exported without N-terminal cleavage and is homologous to haemolysin and related proteins. *EMBO J.* **9**, 349–54.

Ehrmann, I.E., Weiss, A.A., Goodwin, M.S., Gray, M.C., Barry, E. and Hewlett, E.L. (1992). Enzymatic activity of adenylate cyclase toxin from *Bordetella pertussis* is not required for hemolysis. *FEBS Lett.* **304**, 51–6.

Elliott, S.J., Srinivas, S., Albert, M.J., Alam, K., Robins-Browne, R.M., Gunzburg, S.T., Mee, B.J. and Chang, B.J. (1998). Characterization of the roles of haemolysin and other toxins in enteropathy caused by alpha-haemolytic *Escherichia coli* linked to human diarrhea. *Infect. Immun.* **66**, 2040–51.

Fath, M.J., Skvirsky, R.C. and Kolter, R. (1991). Functional complementation between bacterial MDR-like export systems: colicin V, alpha-haemolysin, and *Erwinia* protease. *J. Bacteriol.* **173**, 7549–56.

Felmlee, T. and Welch, R.A. (1988). Alterations of amino acid repeats in the *Escherichia coli* haemolysin affect cytolytic activity and secretion. *Proc. Natl Acad. Sci. USA* **85**, 5269–73.

Felmlee, T., Pellett, S. and Welch, R.A. (1985). Nucleotide sequence of an *Escherichia coli* chromosomal haemolysin. *J. Bacteriol.* **163**, 94–105.

Finnie, C., Hartley, N.M., Findlay, K.C. and Downie, J.A. (1997). The *Rhizobium leguminosarum prsDE* genes are required for secretion of several proteins, some of which influence nodulation, symbiotic nitrogen fixation and exopolysaccharide modification. *Mol. Microbiol.* **25**, 135–46.

Finnie, C., Zorreguieta, A., Hartley, N.M. and Downie, J.A. (1998). Characterization of *Rhizobium leguminosarum* exopolysaccharide glycanases that are secreted via a type I exporter and have a novel heptapeptide repeat motif. *J. Bacteriol.* **180**, 1691–9.

Forestier, C. and Welch, R.A. (1990). Nonreciprocal complementation of the *hlyC* and *lktC* genes of the *Escherichia coli* haemolysin and *Pasteurella haemolytica* leukotoxin determinants. *Infect. Immun.* **58**, 828–32.

Forestier, C. and Welch, R.A. (1991). Identification of RTX toxin target cell specificity domains by use of hybrid genes. *Infect. Immun.* **59**, 4212–20.

Frey, J. (1995). Virulence in *Actinobacillus pleuropneumoniae* and RTX toxins. *Trends Microbiol.* **3**, 257–61.

Frey, J., Meier, R., Gygi, D. and Nicolet, J. (1991). Nucleotide sequence of the haemolysin I gene from *Actinobacillus pleuropneumoniae*. *Infect. Immun.* **59**, 3026–32.

Frey, J., van den Bosch, H., Segers, R. and Nicolet, J. (1992). Identification of a second haemolysin (HlyII) in *Actinobacillus pleuropneumoniae* serotype 1 and expression of the gene in *Escherichia coli*. *Infect. Immun.* **60**, 1671–6.

Frey, J., Beck, M., Stucki, U. and Nicolet, J. (1993a). Analysis of haemolysin operons in *Actinobacillus pleuropneumoniae*. *Gene* **123**, 51–8.

Frey, J., Bosse, J.T., Chang, Y.-F., Cullen, J.M., Fenwick, B., Gerlach, G.F., Gygi, D., Haesebrouck, F., Inzana, T.J., Jansen, R., Kamp, E.M., Macdonald, J., MacInnes, J.I., Mittal, K.R., Nicolet, J., Rycroft, A.N., Segers, R.P.A.M., Smits, M.A., Stenbaek, E., Struck, D.K., van den Bosch, J.F., Willson, P.J. and Young, R. (1993b). *Actinobacillus pleuropneumoniae* RTX-toxins: uniform designation of haemolysins, cytolysins, pleurotoxin and their genes. *J. Gen. Microbiol.* **139**, 1723–8.

Frey, J., Haldimann, A., Nicolet, J., Boffini, A. and Prentki, P. (1994). Sequence analysis and transcription of the *apxI* operon (haemolysin I) from *Actinobacillus pleuropneumoniae*. *Gene* **142**, 97–102.

Gadeberg, O.V. and Orskov, I. (1984). *In vitro* cytotoxic effect of α-haemolytic *Escherichia coli* on human blood granulocytes. *Infect. Immun.* **45**, 255–60.

Gangola, P. and Rosen, B.P. (1987). Maintenance of intracellular calcium in *Escherichia coli*. *J. Biol. Chem.* **262**, 12570–4.

Gentschev, I. and Goebel, W. (1992). Topological and functional studies on HlyB of *Escherichia coli*. *Mol. Gen. Genet.* **232**, 40–8.

Gentschev, I., Hess, J. and Goebel, W. (1990). Change in the cellular localization of alkaline phosphatase by alteration of its carboxy-terminal sequence. *Mol. Gen. Genet.* **222**, 211–16.

Ghigo, J.-M. and Wandersman, C. (1992). Cloning, nucleotide sequence and characterization of the gene encoding the *Erwinia chrysanthemi* B374 PrtA metalloprotease: a third metalloprotease secreted via a C-terminal secretion signal. *Mol. Gen. Genet.* **236**, 135–44.

Ghigo, J.-M. and Wandersman, C. (1994). A carboxyl-terminal four-amino acid motif is required for secretion of the metalloprotease PrtG through the *Erwinia chrysanthemi* protease secretion pathway. *J. Biol. Chem.* **269**, 8979–85.

Gilson, L., Mahanty, H.K. and Kolter, R. (1990). Genetic analysis of an MDR-like export system: the secretion of colicin V. *EMBO J.* **9**, 3875–84.

Glaser, P., Sakamoto, H., Bellalou, J., Ullmann, A. and Danchin, A. (1988). Secretion of cyclolysin, the calmodulin-sensitive adenylate cyclase-haemolysin bifunctional protein of *Bordetella pertussis*. *EMBO J.* **7**, 3997–4004.

Goebel, W. and Hedgpeth, J. (1982). Cloning and functional characterization of the plasmid-encoded haemolysin determinant of *Escherichia coli*. *J. Bacteriol.* **151**, 1290–8.

Goodwin, M.St.M. and Weiss, A.A. (1990). Adenylate cyclase toxin is critical for colonization and pertussis toxin is critical for lethal infection by *Bordetella pertussis* in infant mice. *Infect. Immun.* **58**, 3445–7.

Gray, J.T., Fedorka-Cray, P.J. and Rogers, D.G. (1995). Partial characterization of a *Moraxella bovis* cytolysin. *Vet. Microbiol.* **43**, 183–96.

Gray, L., Mackman, N., Nicaud, J.-M. and Holland, I.B. (1986). The carboxy-terminal region of haemolysin 2001 is required for secretion of the toxin from *Escherichia coli*. *Mol. Gen. Genet.* **205**, 127–33.

Grimminger, F., Walmrath, D., Birkemeyer, R.G., Bhakdi, S. and Seeger, W. (1990). Leukotriene and hydroxyeicosatetraenoic acid generation elicited by low doses of *Escherichia coli* haemolysin in rabbit lungs. *Infect. Immun.* **58**, 2659–63.

Grimminger, F., Scholz, C., Bhakdi, S. and Seeger, W. (1991a). Subhemolytic doses of *Escherichia coli* haemolysin evoke large quantities of lipoxygenase products in human neutrophils. *J. Biol. Chem.* **266**, 14262–9.

Grimminger, F., Sibelius, U., Bhakdi, S., Suttorp, N. and Seeger, W. (1991b). *Escherichia coli* haemolysin is a potent inductor of phosphoinositide hydrolysis and related metabolic responses in human neutrophils. *J. Clin. Invest.* **88**, 1531–9.

Grimminger, F., Rose, F., Sibelius, U., Meinhardt, M., Pötzsch, B., Spriestersbach, R., Bhakdi, S., Suttorp, N. and Seeger, W. (1997). Human endothelial cell activation and mediator release in response to the bacterial exotoxins *Escherichia coli* haemolysin and Staphylococcal α-toxin. *J. Immunol.* **159**, 1909–16.

Gross, M.K., Au, D.C., Smith, A.L. and Storm, D.R. (1992). Targeted mutations that ablate either the adenylate cyclase or haemolysin function of the bifunctional cyaA toxin of *Bordetella pertussis* abolish virulence. *Proc. Natl Acad. Sci. USA* **89**, 4898–902.

Gueirard, P., Druilhe, A., Pretolani, M. and Guiso, N. (1998). Role of adenylate cyclase-haemolysin in alveolar macrophage apoptosis during *Bordetella pertussis* infection *in vivo*. *Infect. Immun.* **66**, 1718–25.

Guzzo, J., Duong, F., Wandersman, C., Murgier, M. and Lazdunski, A. (1991). The secretion genes of *Pseudomonas aeruginosa* alkaline protease are functionally related to those of *Erwinia chrysanthemi* proteases and *Escherichia coli* α-haemolysin. *Mol. Microbiol.* **5**, 447–53.

Gygi, D., Nicolet, J., Frey, J., Cross, M., Koronakis, V. and Hughes, C. (1990). Isolation of the *Actinobacillus pleuropneumoniae* haemolysin gene and the activation and secretion of the prohaemolysin by the HlyC, HlyB and HlyD proteins of *Escherichia coli*. *Mol. Microbiol.* **4**, 123–8.

Gygi, D., Nicolet, J., Hughes, C. and Frey, J. (1992). Functional analysis of the Ca²⁺-regulated haemolysin I operon of *Actinobacillus pleuropneumoniae* serotype 1. *Infect. Immun.* **60**, 3059–64.

Hacker, J., Hughes, C., Hof, H. and Goebel, W. (1983). Cloned haemolysin genes from *Escherichia coli* that cause urinary tract infection determine different levels of toxicity in mice. *Infect. Immun.* **42**, 57–63.

Hacker, J., Blum-Oehler, G., Mühldorfer, I. and Tschäpe, H. (1997). Pathogenicity islands of virulent bacteria: structure, function and impact on microbial evolution. *Mol. Microbiol.* **23**, 1089–97.

Hackett, M., Guo, L., Shabanowitz, J., Hunt, D.F. and Hewlett, E.L. (1994). Internal lysine palmitoylation in adenylate cyclase toxin from *Bordetella pertussis*. *Science* **266**, 433–5.

Havarstein, L.S., Holo, H. and Nes, I.F. (1994). The leader peptide of colicin V shares consensus sequences with leader peptides that are common among peptide bacteriocins produced by Gram-positive bacteria. *Microbiol.* **140**, 2383–9.

Henricks, P.A.J., Binkhorst, G.J., Drijver, A.A. and Nijkamp, F.P. (1992). *Pasteurella haemolytica* leukotoxin enhances production of leukotriene B₄ and 5-hydroxyeicosatetraenoic acid by bovine polymorphonuclear leukocytes. *Infect. Immun.* **60**, 3238–43.

Hess, J., Wels, W., Vogel, M. and Goebel, W. (1986). Nucleotide sequence of a plasmid-encoded haemolysin determinant and its comparison with a corresponding chromosomal haemolysin sequence. *FEMS Microbiol. Lett.* **34**, 1–11.

Hess, J., Gentschev, I., Goebel, W. and Jarchau, T. (1990). Analysis of the haemolysin secretion system by PhoA-HlyA fusion proteins. *Mol. Gen. Genet.* **224**, 201–8.

Hewlett, E.L., Gordon, V.M., McCaffery, J.D., Sutherland, W.M. and Gray, M.C. (1989). Adenylate cyclase toxin from *Bordetella pertussis*: identification and purification of the holotoxin molecule. *J. Biol. Chem.* **264**, 19379–84.

Hewlett, E.L., Gray, L., Allietta, M., Ehrmann, I., Gordon, V.M. and Gray, M.C. (1991). Adenylate cyclase toxin from *Bordetella pertussis*: conformational change associated with toxin activity. *J.Biol. Chem.* **266**, 17503–8.

Hewlett, E.L., Gray, M.C., Ehrmann, I.E., Maloney, N.J., Otero, A.S., Gray, L., Allietta, M., Szabo, G., Weiss, A.A. and Barry, E.M. (1993). Characterization of adenylate cyclase toxin from a mutant of *Bordetella pertussis* defective in the activator gene, cyaC. *J. Biol. Chem.* **268**, 7842–8.

Higgins, C.F. (1992). ABC transporters – from microorganisms to man. *Annu. Rev. Cell. Biol.* **8**, 67–113.

Highlander, S.K., Chidambaram, M., Engler, M.J. and Weinstock, G.M. (1989). DNA sequence of the *Pasteurella haemolytica* leukotoxin gene cluster. *DNA* **8**, 15–28.

Highlander, S.K., Engler, M.J. and Weinstock, G.M. (1990). Secretion and expression of the *Pasteurella haemolytica* leukotoxin. *J. Bacteriol.* **172**, 2343–50.

Hritz, M., Fisher, E. and Demuth, D.R. (1996). Differential regulation of the leukotoxin operon in highly leukotoxic and minimally leukotoxic strains *of Actinobacillus actinomycetemcomitans*. *Infect. Immun.* **64**, 2724–9.

Hughes, H.P.A., Campos, M., McDougall, L., Beskorwayne, T.K., Potter, A.A. and Babiuk, L.A. (1994). Regulation of major histocompatibility complex class II expression by *Pasteurella haemolytica* leukotoxin. *Infect. Immun.* **62**, 1609–15.

Hwang, J., Zhong, X. and Tai, P.C. (1997). Interactions of dedicated export membrane proteins of the colicin V secretion system: CvaA, a member of the membrane fusion protein family, interacts with CvaB and TolC. *J. Bacteriol.* **179**, 6264–70.

Issartel, J.-P., Koronakis, V. and Hughes, C. (1991). Activation of *Escherichia coli* prohaemolysin to the mature toxin by acyl carrier protein-dependent fatty acylation. *Nature* **351**, 759–61.

Iwaki, M., Ullmann, A. and Sebo, P. (1995). Identification by *in vitro* complementation of regions required for cell-invasive activity of *Bordetella pertussis* adenylate cyclase toxin. *Mol. Microbiol.* **17**, 1015–24.

Jansen, R., Briaire, J., Kamp, E.M. and Smits, M.A. (1992). Comparison of the cytolysin II genetic determinants of *Actinobacillus pleuropneumoniae* serotypes. *Infect. Immun.* **60**, 630–6.

Jansen, R., Briaire, J., Kamp, E.M., Gielkens, A.L.J. and Smits, M.A. (1993a). Cloning and characterization of the *Actinobacillus pleuropneumoniae*-RTX-toxin III (ApxIII) gene. *Infect. Immun.* **61**, 947–54.

Jansen, R., Briaire, J., Kamp, E.M., Gielkens, A.L.J. and Smits, M.A. (1993b). Structural analysis of the *Actinobacillus pleuropneumoniae*-RTX-toxin I (ApxI) operon. *Infect. Immun.* **61**, 3688–95.

Jansen, R., Briaire, J., van Geel, A.B.M., Kamp, E.M., Gielkens, A.L.J. and Smits, M.A. (1994). Genetic map of the *Actinobacillus pleuropneumoniae* RTX-toxin (Apx) operons: characterization of the ApxIII operons. *Infect. Immun.* **62**, 4411–18.

Jansen, R., Briaire, J., Smith, H.E., Dom, P., Haesebrouck, F., Kamp, E.M., Gielkens, A.L.J. and Smits, M.A. (1995). Knockout mutants of *Actinobacillus pleuropneumoniae* serotype 1 that are devoid of

RTX toxins do not activate or kill porcine neutrophils. *Infect. Immun.* **63**, 27–37.

Jarchau, T., Chakraborty, T., Garcia, F. and Goebel, W. (1994). Selection for transport competence of C-terminal polypeptides derived from *Escherichia coli* haemolysin: the shortest peptide capable of autonomous HlyB/HlyD-dependent secretion comprises the C-terminal 62 amino acids of HlyA. *Mol. Gen. Genet.* **245**, 53–60.

Jonas, D., Schultheis, B., Klas, C., Krammer, P.H. and Bhakdi, S. (1993). Cytocidal effects of *Escherichia coli* haemolysin on human T lymphocytes. *Infect. Immun.* **61**, 1715–21.

Kamp, E.M., Popma, J.K., Anakotta, J. and Smits, M.A. (1991). Identification of haemolytic and cytotoxic proteins of *Actinobacillus pleuropneumoniae* by use of monoclonal antibodies. *Infect. Immun.* **59**, 3079–85.

Kamp, E.M., Vermeulen, T.M.M., Smits, M.A. and Haagsma, J. (1994). Production of Apx toxins by field strains of *Actinobacillus pleuropneumoniae* and *Actinobacillus suis*. *Infect. Immun.* **62**, 4063–5.

Karakelian, D., Lear, J.D., Lally, E.T. and Tanaka, J.C. (1998). Characterization of *Actinobacillus actinomycetemcomitans* leukotoxin pore formation in HL60 cells. *Biochim. Biophys. Acta* **1406**, 175–87.

Karimova, G., Bellalou, J. and Ullmann, A. (1996). Phosphorylation-dependent binding of BvgA to the upstream region of the *cyaA* gene of *Bordetella pertussis*. *Mol. Microbiol.* **20**, 489–96.

Kawai, E., Akatsuka, H., Idei, A., Shibatani, T. and Omori, K. (1998). *Serratia marcescens* S-layer protein is secreted extracellularly via an ATP-binding cassette exporter, the Lip system. *Mol. Microbiol.* **27**, 941–52.

Keane, W.F., Welch, R., Gekker, G. and Peterson, P.K. (1987). Mechanism of *Escherichia coli* alpha-haemolysin-induced injury to isolated renal tubular cells. *Am. J. Pathol.* **126**, 350–7.

Kenny, B., Taylor, S. and Holland, I.B. (1992). Identification of individual amino acids required for secretion within the haemolysin (HlyA) C-terminal targeting region. *Mol. Microbiol.* **6**, 1477–89.

Kenny, B., Chervaux, C. and Holland, I.B. (1994). Evidence that residues –15 to –46 of the haemolysin secretion signal are involved in early steps in secretion, leading to recognition of the translocator. *Mol. Microbiol.* **11**, 99–109.

Khelef, N. and Guiso, N. (1995). Induction of macrophage apoptosis by *Bordetella pertussis* adenylate cyclase-haemolysin. *FEMS Microbiol. Lett.* **134**, 27–32.

Khelef, N., Bachelet, C.-M., Vargaftig, B.B. and Guiso, N. (1994). Characterization of murine lung inflammation after infection with parental *Bordetella pertussis* and mutants deficient in adhesins or toxins. *Infect. Immun.* **62**, 2893–900.

Knapp, S., Then, I., Wels, W., Michel, G., Tschäpe, H., Hacker, J. and Goebel, W. (1985). Analysis of the flanking regions from different haemolysin determinants of *Escherichia coli*. *Mol. Gen. Genet.* **200**, 385–92.

Kolodrubetz, D., Spitznagel, J., Wang, B., Phillips, L.H., Jacobs, C. and Kraig, E. (1996). *cis* elements and *trans* factors are both important in strain-specific regulation of the leukotoxin gene in *Actinobacillus actinomycetemcomitans*. *Infect. Immun.* **64**, 3451–60.

König, B. and König, W. (1993). The role of the phosphatidylinositol turnover in 12-hydroxyeicosatetraenoic acid generation from human platelets by *Escherichia coli* α-haemolysin, thrombin and fluoride. *Immunology* **80**, 633–9.

König, B., König, W., Scheffer, J., Hacker, J. and Goebel, W. (1986). Role of *Escherichia coli* alpha-haemolysin and bacterial adherence in infection: requirement for release of inflammatory mediators from granulocytes and mast cells. *Infect. Immun.* **54**, 886–92.

König, B., Schönfeld, W., Scheffer, J. and König, W. (1990). Signal transduction in human platelets and inflammatory mediator release induced by genetically cloned haemolysin-positive and -negative *Escherichia coli* strains. *Infect. Immun.* **58**, 1591–9.

Koronakis, E., Hughes, C., Milisav, I. and Koronakis, V. (1995). Protein exporter function and *in vitro* ATPase activity are correlated in ABC-domain mutants of HlyB. *Mol. Microbiol.* **16**, 87–96.

Koronakis, V., Cross, M., Senior, B., Koronakis, E. and Hughes, C. (1987). The secreted haemolysins of *Proteus mirabilis*, *Proteus vulgaris*, and *Morganella morganii* are genetically related to each other and to the alpha-haemolysin of *Escherichia coli*. *J. Bacteriol.* **169**, 1509–15.

Koronakis, V., Koronakis, E. and Hughes, C. (1989). Isolation and analysis of the C-terminal signal directing export of *Escherichia coli* haemolysin protein across both bacterial membranes. *EMBO J.* **8**, 595–605.

Koronakis, V., Hughes, C. and Koronakis, E. (1991). Energetically distinct early and late stages of HlyB/HlyD-dependent secretion across both *Escherichia coli* membranes. *EMBO J.* **10**, 3263–72.

Koronakis, V., Hughes, C. and Koronakis, E. (1993). ATPase activity and ATP/ADP-induced conformational change in the soluble domain of the bacterial protein translocator HlyB. *Mol. Microbiol.* **8**, 1163–75.

Koronakis, V., Li, J., Koronakis, E. and Stauffer, K. (1997). Structure of TolC, the outer membrane component of the bacterial type I efflux system, derived from two-dimensional crystals. *Mol. Microbiol.* **23**, 617–26.

Kraig, E., Dailey, T. and Kolodrubetz, D. (1990). Nucleotide sequence of the leukotoxin gene from *Actinobacillus actinomycetemcomitans*: homology to the alpha-haemolysin/leukotoxin gene family. *Infect. Immun.* **58**, 920–9.

Lally, E.T., Golub, E.E., Kieba, I.R., Taichman, N.S., Rosenbloom, J., Rosenbloom, J.C., Gibson, C.W. and Demuth, D.R. (1989). Analysis of the *Actinobacillus actinomycetemcomitans* leukotoxin gene. Delineation of unique features and comparison to homologous toxins. *J. Biol. Chem.* **264**, 15451–6.

Lally, E.T., Golub, E.E., Kieba, I.R., Taichman, N.S., Decker, S., Berthold, P., Gibson, C.W., Demuth, D.R. and Rosenbloom, J. (1991). Structure and function of the B and D genes of the *Actinobacillus actinomycetemcomitans* leukotoxin complex. *Microb. Pathog.* **11**, 111–21.

Lally, E.T., Golub, E.E. and Kieba, I.R. (1994). Identification and immunological characterization of the domain of *Actinobacillus actinomycetemcomitans* leukotoxin that determines its specificity for human target cells. *J. Biol. Chem.* **269**, 31289–95.

Lally, E.T., Kieba, I.R., Sato, A., Green, C.L., Rosenbloom, J., Korostoff, J., Wang, J.F., Shenker, B.J., Ortlepp, S., Robinson, M.K. and Billings, P.C. (1997). RTX toxins recognize a β2 integrin on the surface of human target cells. *J. Biol. Chem.* **272**, 30463–9.

Laoide, B.M. and Ullmann, A. (1990). Virulence dependent and independent regulation of the *Bordetella pertussis cya* operon. *EMBO J.* **9**, 999–1005.

Lear, J.D., Furblur, U.G., Lally, E.T. and Tanaka, J.C. (1995). *Actinobacillus actinomycetemcomitans* leukotoxin forms large conductance, voltage-gated ion channels when incorporated into planar lipid bilayers. *Biochim. Biophys. Acta* **1238**, 34–41.

Leeds, J.A. and Welch, R.A. (1996). RfaH enhances elongation of *Escherichia coli hlyCABD* mRNA. *J. Bacteriol.* **178**, 1850–7.

Leeds, J.A. and Welch, R.A. (1997). Enhancing transcription through the *Escherichia coli* haemolysin operon, *hlyCABD*: RfaH and upstream JUMPStart DNA sequences function together via a postinitiation mechanism. *J. Bacteriol.* **179**, 3519–27.

Letoffe, S., Delepelaire, P. and Wandersman, C. (1990). Protease secretion by *Erwinia chrysanthemi*: the specific secretion functions are analogous to those of *Escherichia coli* α-haemolysin. *EMBO J.* **9**, 1375–82.

Letoffe, S., Delepelaire, P. and Wandersman, C. (1991). Cloning and expression in *Escherichia coli* of the *Serratia marcescens* metalloprotease gene: secretion of the protease from *E. coli* in the presence of the *Erwinia chrysanthemi* protease secretion functions. *J. Bacteriol.* **173**, 2160–6.

Letoffe, S., Ghigo, J.M. and Wandersman, C. (1994a). Secretion of the *Serratia marcescens* HasA protein by an ABC transporter. *J. Bacteriol.* **176**, 5372–7.

Letoffe, S., Ghigo, J.M. and Wandersman, C. (1994b). Iron acquisition from heme and hemoglobin by a *Serratia marcescens* extracellular protein. *Proc. Natl Acad. Sci. USA* **91**, 9876–80.

Letoffe, S., Delepelaire, P. and Wandersman, C. (1996). Protein secretion in Gram-negative bacteria: assembly of the three components of ABC protein-mediated exporters is ordered and promoted by substrate binding. *EMBO J.* **15**, 5804–11.

Li, X., Tetling, S., Winkler, U.K., Jaeger, K.-E. and Benedik, M.J. (1995). Gene cloning, sequence analysis, purification, and secretion by *Escherichia coli* of an extracellular lipase from *Serratia marcescens*. *Appl. Environ. Microbiol.* **61**, 2674–80.

Lin, W., Fullner, K.J., Clayton, R., Sexton, J.A., Rogers, M.B., Calia, K.E., Calderwood, S.B., Fraser, C. and Mekalanos, J.J. (1999). Identification of a *Vibrio cholerae* RTX toxin gene cluster that is tightly linked to the cholera toxin prophage. *Proc. Natl Acad. Sci. USA* **96**, 1071-6.

Lo, R.Y.C., Strathdee, C.A. and Shewen, P.E. (1987). Nucleotide sequence of the leukotoxin genes of *Pasteurella haemolytica* A1. *Infect. Immun.* **55**, 1987–96.

Ludwig, A., Vogel, M. and Goebel, W. (1987). Mutations affecting activity and transport of haemolysin in *Escherichia coli*. *Mol. Gen. Genet.* **206**, 238–45.

Ludwig, A., Jarchau, T., Benz, R. and Goebel, W. (1988). The repeat domain of *Escherichia coli* haemolysin (HlyA) is responsible for its Ca^{2+}-dependent binding to erythrocytes. *Mol. Gen. Genet.* **214**, 553–61.

Ludwig, A., Schmid, A., Benz, R. and Goebel, W. (1991). Mutations affecting pore formation by haemolysin from *Escherichia coli*. *Mol. Gen. Genet.* **226**, 198–208.

Ludwig, A., Benz, R. and Goebel, W. (1993). Oligomerization of *Escherichia coli* haemolysin (HlyA) is involved in pore formation. *Mol. Gen. Genet.* **241**, 89–96.

Ludwig, A., Garcia, F., Bauer, S., Jarchau, T., Benz, R., Hoppe, J. and Goebel, W. (1996). Analysis of the *in vivo* activation of haemolysin (HlyA) from *Escherichia coli*. *J. Bacteriol.* **178**, 5422–30.

Macdonald, J. and Rycroft, A.N. (1992). Molecular cloning and expression of *ptxA*, the gene encoding the 120-kilodalton cytotoxin of *Actinobacillus pleuropneumoniae* serotype 2. *Infect. Immun.* **60**, 2726–32.

Macdonald, J. and Rycroft, A.N. (1993). *Actinobacillus pleuropneumoniae* haemolysin II is secreted from *Escherichia coli* by *A. pleuropneumoniae* pleurotoxin secretion gene products. *FEMS Microbiol. Lett.* **109**, 317–22.

Mackman, N., Nicaud, J.-M., Gray, L. and Holland, I.B. (1985). Genetical and functional organisation of the *Escherichia coli* haemolysin determinant 2001. *Mol. Gen. Genet.* **201**, 282–8.

Maheswaran, S.K., Kannan, M.S., Weiss, D.J., Reddy, K.R., Townsend, E.L., Yoo, H.S., Lee, B.W. and Whiteley, L.O. (1993). Enhancement of neutrophil-mediated injury to bovine pulmonary endothelial cells by *Pasteurella haemolytica* leukotoxin. *Infect. Immun.* **61**, 2618–25.

Maier, E., Reinhard, N., Benz, R. and Frey, J. (1996). Channel-forming activity and channel size of the RTX toxins ApxI, ApxII, and ApxIII of *Actinobacillus pleuropneumoniae*. *Infect. Immun.* **64**, 4415–23.

Mangan, D.F., Taichman, N.S., Lally, E.T. and Wahl, S.M. (1991). Lethal effects of *Actinobacillus actinomycetemcomitans* leukotoxin on human T lymphocytes. *Infect. Immun.* **59**, 3267–72.

Masure, H.R., Au, D.C., Gross, M.K., Donovan, M.G. and Storm, D.R. (1990). Secretion of the *Bordetella pertussis* adenylate cyclase from *Escherichia coli* containing the haemolysin operon. *Biochem.* **29**, 140–5.

May, A.K., Sawyer, R.G., Gleason, T., Whitworth, A. and Pruett, T.L. (1996). *In vivo* cytokine response to *Escherichia coli* alpha-haemolysin determined with genetically engineered haemolytic and nonhemolytic *E. coli* variants. *Infect. Immun.* **64**, 2167–71.

McWhinney, D.R., Chang, Y.-F., Young, R. and Struck, D.K. (1992). Separable domains define target cell specificities of an RTX

haemolysin from *Actinobacillus pleuropneumoniae*. *J. Bacteriol.* **174**, 291–7.

Menestrina, G., Mackman, N., Holland, I.B. and Bhakdi, S. (1987). *Escherichia coli* haemolysin forms voltage-dependent ion channels in lipid membranes. *Biochim. Biophys. Acta* **905**, 109–17.

Menestrina, G., Pederzolli, C., Dalla Serra, M., Bregante, M. and Gambale, F. (1996). Permeability increase induced by *Escherichia coli* haemolysin A in human macrophages is due to the formation of ionic pores: a patch clamp characterization. *J. Membrane Biol.* **149**, 113–21.

Moayeri, M. and Welch, R.A. (1994). Effects of temperature, time, and toxin concentration on lesion formation by the *Escherichia coli* haemolysin. *Infect. Immun.* **62**, 4124–34.

Moayeri, M. and Welch, R.A. (1997). Prelytic and lytic conformations of erythrocyte-associated *Escherichia coli* haemolysin. *Infect. Immun.* **65**, 2233–9.

Mobley, H.L.T., Green, D.M., Trifillis, A.L., Johnson, D.E., Chippendale, G.R., Lockatell, C.V., Jones, B.D. and Warren, J.W. (1990). Pyelonephritogenic *Escherichia coli* and killing of cultured human renal proximal tubular epithelial cells: role of haemolysin in some strains. *Infect. Immun.* **58**, 1281–9.

Müller, D., Hughes, C. and Goebel, W. (1983). Relationship between plasmid and chromosomal haemolysin determinants of *Escherichia coli*. *J. Bacteriol.* **153**, 846–51.

Murphy, G.L., Whitworth, L.C., Clinkenbeard, K.D. and Clinkenbeard, P.A. (1995). Haemolytic activity of the *Pasteurella haemolytica* leukotoxin. *Infect. Immun.* **63**, 3209–12.

Nakahama, K., Yoshimura, K., Marumoto, R., Kikuchi, M., Lee, I.S., Hase, T. and Matsubara, H. (1986). Cloning and sequencing of *Serratia* protease gene. *Nucleic Acids Res.* **14**, 5843–55.

Nieto, J.M., Bailey, M.J.A., Hughes, C. and Koronakis, V. (1996). Suppression of transcription polarity in the *Escherichia coli* haemolysin operon by a short upstream element shared by polysaccharide and DNA transfer determinants. *Mol. Microbiol.*, **19**, 705–13.

O'Hanley, P., Lalonde, G. and Ji, G. (1991). Alpha-haemolysin contributes to the pathogenicity of piliated digalactoside-binding *Escherichia coli* in the kidney: efficacy of an alpha-haemolysin vaccine in preventing renal injury in the BALB/c mouse model of pyelonephritis. *Infect. Immun.* **59**, 1153–61.

Ohta, H., Hara, H., Fukui, K., Kurihara, H., Murayama, Y. and Kato, K. (1993). Association of *Actinobacillus actinomycetemcomitans* leukotoxin with nucleic acids on the bacterial cell surface. *Infect. Immun.* **61**, 4878–84.

Oropeza-Wekerle, R.-L., Speth, W., Imhof, B., Gentschev, I. and Goebel, W. (1990). Translocation and compartmentalization of *Escherichia coli* haemolysin (HlyA). *J. Bacteriol.* **172**, 3711–17.

Ostolaza, H. and Goni, F.M. (1995). Interaction of the bacterial protein toxin α-haemolysin with model membranes: protein binding does not always lead to lytic activity. *FEBS Lett.* **371**, 303–6.

Ostolaza, H., Bartolome, B., Serra, J.L., de la Cruz, F. and Goni, F.M. (1991). α-Haemolysin from *E. coli*. Purification and self-aggregation properties. *FEBS Lett.* **280**, 195–8.

Ostolaza, H., Bartolome, B., Ortiz de Zarate, I., de la Cruz, F. and Goni, F.M. (1993). Release of lipid vesicle contents by the bacterial protein toxin α-haemolysin. *Biochim. Biophys. Acta* **1147**, 81–8.

Ostolaza, H., Soloaga, A. and Goni, F.M. (1995). The binding of divalent cations to *Escherichia coli* α-haemolysin. *Eur. J. Biochem.* **228**, 39–44.

Otero, A.S., Yi, X.B., Gray, M.C., Szabo, G. and Hewlett, E.L. (1995). Membrane depolarization prevents cell invasion by *Bordetella pertussis* adenylate cyclase toxin. *J. Biol. Chem.* **270**, 9695–7.

Pellett, S. and Welch, R.A. (1996). *Escherichia coli* haemolysin mutants with altered target cell specificity. *Infect. Immun.* **64**, 3081–7.

Petras, S.F., Chidambaram, M., Illyes, E.F., Froshauer, S., Weinstock, G.M. and Reese, C.P. (1995). Antigenic and virulence properties of *Pasteurella haemolytica* leukotoxin mutants. *Infect. Immun.* **63**, 1033–9.

Poulsen, K., Theilade, E., Lally, E.T., Demuth, D.R. and Kilian, M. (1994). Population structure of *Actinobacillus actinomycetemcomitans*: a framework for studies of disease-associated properties. *Microbiol.* **140**, 2049–60.

Prada, J. and Beutin, L. (1991). Detection of *Escherichia coli* α-haemolysin genes and their expression in a human faecal strain of *Enterobacter cloacae*. *FEMS Microbiol. Lett.* **79**, 111–14.

Reimer, D., Frey, J., Jansen, R., Veit, H.P. and Inzana, T.J. (1995). Molecular investigation of the role of ApxI and ApxII in the virulence of *Actinobacillus pleuropneumoniae* serotype 5. *Microb. Pathog.* **18**, 197–209.

Rogel, A. and Hanski, E. (1992). Distinct steps in the penetration of adenylate cyclase toxin of *Bordetella pertussis* into sheep erythrocytes: translocation of the toxin across the membrane. *J. Biol. Chem.* **267**, 22599–605.

Rogel, A., Schultz, J.E., Brownlie, R.M., Coote, J.G., Parton, R. and Hanski, E. (1989). *Bordetella pertussis* adenylate cyclase: purification and characterization of the toxic form of the enzyme. *EMBO J.* **8**, 2755–60.

Rogel, A., Meller, R. and Hanski, E. (1991). Adenylate cyclase toxin from *Bordetella pertussis*: the relationship between induction of cAMP and hemolysis. *J. Biol. Chem.* **266**, 3154–61.

Rose, T., Sebo, P., Bellalou, J. and Ladant, D. (1995). Interaction of calcium with *Bordetella pertussis* adenylate cyclase toxin. *J. Biol. Chem.* **270**, 26370–6.

Sakamoto, H., Bellalou, J., Sebo, P. and Ladant, D. (1992). *Bordetella pertussis* adenylate cyclase toxin: structural and functional independence of the catalytic and haemolytic activities. *J. Biol. Chem.* **267**, 13598–602.

Scheu, A.K., Economou, A., Hong, G.F., Ghelani, S., Johnston, A.W.B. and Downie, J.A. (1992). Secretion of the *Rhizobium leguminosarum* nodulation protein NodO by haemolysin-type systems. *Mol. Microbiol.* **6**, 231–8.

Schlör, S., Schmidt, A., Maier, E., Benz, R., Goebel, W. and Gentschev, I. (1997). *In vivo* and *in vitro* studies on interactions between the components of the haemolysin (HlyA) secretion machinery of *Escherichia coli*. *Mol. Gen. Genet.* **256**, 306–19.

Schmidt, H., Beutin, L. and Karch, H. (1995). Molecular analysis of the plasmid-encoded haemolysin of *Escherichia coli* O157:H7 strain EDL 933. *Infect. Immun.* **63**, 1055–61.

Schmidt, H., Kernbach, C. and Karch, H. (1996a). Analysis of the EHEC *hly* operon and its location in the physical map of the large plasmid of enterohaemorrhagic *Escherichia coli* O157:H7. *Microbiol.* **142**, 907–14.

Schmidt, H., Maier, E., Karch, H. and Benz, R. (1996b). Pore-forming properties of the plasmid-encoded haemolysin of enterohemorrhagic *Escherichia coli* O157:H7. *Eur. J. Biochem.* **241**, 594–601.

Schülein, R., Gentschev, I., Mollenkopf, H.-J. and Goebel, W. (1992). A topological model for the haemolysin translocator protein HlyD. *Mol. Gen. Genet.* **234**, 155–63.

Sheps, J.A., Cheung, I. and Ling, V. (1995). Haemolysin transport in *Escherichia coli*. Point mutants in HlyB compensate for a deletion in the predicted amphiphilic helix region of the HlyA signal. *J. Biol. Chem.* **270**, 14829–34.

Shewen, P.E. and Wilkie, B.N. (1982). Cytotoxin of *Pasteurella haemolytica* acting on bovine leukocytes. *Infect. Immun.* **35**, 91–4.

Smith, H.W. and Huggins, M.B. (1985). The toxic role of alpha-haemolysin in the pathogenesis of experimental *Escherichia coli* infection in mice. *J. Gen. Microbiol.* **131**, 395–403.

Smits, M.A., Briaire, J., Jansen, R., Smith, H.E., Kamp, E.M. and Gielkens, A.L.J. (1991). Cytolysins of *Actinobacillus pleuropneumoniae* serotype 9. *Infect. Immun.* **59**, 4497–504.

Soloaga, A., Ostolaza, H., Goni, F.M. and de la Cruz, F. (1996). Purification of *Escherichia coli* pro-haemolysin, and a comparison with the properties of mature α-haemolysin. *Eur. J. Biochem.* **238**, 418–22.

Spitznagel, J., Kraig, E. and Kolodrubetz, D. (1991). Regulation of leukotoxin in leukotoxic and nonleukotoxic strains of *Actinobacillus actinomycetemcomitans*. *Infect. Immun.* **59**, 1394–401.

Stanley, P., Koronakis, V. and Hughes, C. (1991). Mutational analysis supports a role for multiple structural features in the C-terminal secretion signal of *Escherichia coli* haemolysin. *Mol. Microbiol.* **5**, 2391–403.

Stanley, P.L.D., Diaz, P., Bailey, M.J.A., Gygi, D., Juarez, A. and Hughes, C. (1993). Loss of activity in the secreted form of *Escherichia coli* haemolysin caused by an *rfaP* lesion in core lipopolysaccharide assembly. *Mol. Microbiol.* **10**, 781–7.

Stanley, P., Packman, L.C., Koronakis, V. and Hughes, C. (1994). Fatty acylation of two internal lysine residues required for the toxic activity of *Escherichia coli* haemolysin. *Science* **266**, 1992–6.

Stanley, P., Koronakis, V., Hardie, K. and Hughes, C. (1996). Independent interaction of the acyltransferase HlyC with two maturation domains of the *Escherichia coli* toxin HlyA. *Mol. Microbiol.* **20**, 813–22.

Stanley, P., Koronakis, V. and Hughes, C. (1998). Acylation of *Escherichia coli* haemolysin: a unique protein lipidation mechanism underlying toxin function. *Microbiol. Mol. Biol. Rev.* **62**, 309–33.

Steffen, P., Goyard, S. and Ullmann, A. (1996). Phosphorylated BvgA is sufficient for transcriptional activation of virulence-regulated genes in *Bordetella pertussis*. *EMBO J.* **15**, 102–9.

Stevens, P.K. and Czuprynski, C.J. (1996). *Pasteurella haemolytica* leukotoxin induces bovine leukocytes to undergo morphologic changes consistent with apoptosis *in vitro*. *Infect. Immun.* **64**, 2687–94.

Strathdee, C.A. and Lo, R.Y.C. (1987). Extensive homology between the leukotoxin of *Pasteurella haemolytica* A1 and the alpha-haemolysin of *Escherichia coli*. *Infect. Immun.* **55**, 3233–6.

Strathdee, C.A. and Lo, R.Y.C. (1989a). Cloning, nucleotide sequence, and characterization of genes encoding the secretion function of the *Pasteurella haemolytica* leukotoxin determinant. *J. Bacteriol.* **171**, 916–28.

Strathdee, C.A. and Lo, R.Y.C. (1989b). Regulation of expression of the *Pasteurella haemolytica* leukotoxin determinant. *J. Bacteriol.* **171**, 5955–62.

Suh, Y. and Benedik, M.J. (1992). Production of active *Serratia marcescens* metalloprotease from *Escherichia coli* by α-haemolysin HlyB and HlyD. *J. Bacteriol.* **174**, 2361–6.

Sutton, J.M., Lea, E.J.A. and Downie, J.A. (1994). The nodulation-signaling protein NodO from *Rhizobium leguminosarum* biovar *viciae* forms ion channels in membranes. *Proc. Natl Acad. Sci. USA* **91**, 9990–4.

Suttorp, N., Flöer, B., Schnittler, H., Seeger, W. and Bhakdi, S. (1990). Effects of *Escherichia coli* haemolysin on endothelial cell function. *Infect. Immun.* **58**, 3796–801.

Swenson, D.L., Bukanov, N.O., Berg, D.E. and Welch, R.A. (1996). Two pathogenicity islands in uropathogenic *Escherichia coli* J96: cosmid cloning and sample sequencing. *Infect. Immun.* **64**, 3736–43.

Szabo, G., Gray, M.C. and Hewlett, E.L. (1994). Adenylate cyclase toxin from *Bordetella pertussis* produces ion conductance across artificial lipid bilayers in a calcium- and polarity-dependent manner. *J. Biol. Chem.* **269**, 22496–9.

Taichman, N.S., Dean, R.T. and Sanderson, C.J. (1980). Biochemical and morphological characterization of the killing of human monocytes by a leukotoxin derived from *Actinobacillus actinomycetemcomitans*. *Infect. Immun.* **28**, 258–68.

Tascon, R.I., Vazquez-Boland, J.A., Gutierrez-Martin, C.B., Rodriguez-Barbosa, I. and Rodriguez-Ferri, E.F. (1994). The RTX haemolysins ApxI and ApxII are major virulence factors of the swine pathogen *Actinobacillus pleuropneumoniae*: evidence from mutational analysis. *Mol. Microbiol.* **14**, 207–16.

Tatum, F.M., Briggs, R.E., Sreevatsan, S.S., Zehr, E.S., Hsuan, S.L., Whiteley, L.O., Ames, T.R. and Maheswaran, S.K. (1998). Construction of an isogenic leukotoxin deletion mutant of *Pasteurella haemolytica* serotype 1: characterization and virulence. *Microb. Pathog.* **24**, 37–46.

Thanabalu, T., Koronakis, E., Hughes, C. and Koronakis, V. (1998). Substrate-induced assembly of a contiguous channel for protein export from *E. coli*: reversible bridging of an inner-membrane translocase to an outer membrane exit pore. *EMBO J.* **17**, 6487–96.

Thompson, S.A. and Sparling, P.F. (1993). The RTX cytotoxin-related FrpA protein of *Neisseria meningitidis* is secreted extracellularly by Meningococci and by HlyBD$^+$ *Escherichia coli*. *Infect. Immun.* **61**, 2906–11.

Thompson, S.A., Wang, L.L. and Sparling, P.F. (1993a). Cloning and nucleotide sequence of *frpC*, a second gene from *Neisseria meningitidis* encoding a protein similar to RTX cytotoxins. *Mol. Microbiol.* **9**, 85–96.

Thompson, S.A., Wang, L.L., West, A. and Sparling, P.F. (1993b). *Neisseria meningitidis* produces iron-regulated proteins related to the RTX family of exoproteins. *J. Bacteriol.* **175**, 811–18.

Trent, M.S., Worsham, L.M.S. and Ernst-Fonberg, M.L. (1998). The biochemistry of haemolysin toxin activation: characterization of HlyC, an internal protein acyltransferase. *Biochem.* **37**, 4644–52.

Tsai, C.-C., Shenker, B.J., DiRienzo, J.M., Malamud, D. and Taichman, N.S. (1984). Extraction and isolation of a leukotoxin from *Actinobacillus actinomycetemcomitans* with polymyxin B. *Infect. Immun.* **43**, 700–5.

van Belkum, M.J., Worobo, R.W. and Stiles, M.E. (1997). Double-glycine-type leader peptides direct secretion of bacteriocins by ABC transporters: colicinV secretion in *Lactococcus lactis*. *Mol. Microbiol.* **23**, 1293–301.

Wagner, W., Vogel, M. and Goebel, W. (1983). Transport of haemolysin across the outer membrane of *Escherichia coli* requires two functions. *J. Bacteriol.* **154**, 200–10.

Wandersman, C. and Delepelaire, P. (1990). TolC, an *Escherichia coli* outer membrane protein required for haemolysin secretion. *Proc. Natl Acad. Sci. USA* **87**, 4776–80.

Wang, R., Seror, S.J., Blight, M., Pratt, J.M., Broome-Smith, J.K. and Holland, I.B. (1991). Analysis of the membrane organization of an *Escherichia coli* protein translocator, HlyB, a member of a large family of prokaryote and eukaryote surface transport proteins. *J. Mol. Biol.* **217**, 441–54.

Wang, Z., Clarke, C. and Clinkenbeard, K. (1998). *Pasteurella haemolytica* leukotoxin-induced increase in phospholipase A$_2$ activity in bovine neutrophils. *Infect. Immun.* **66**, 1885–90.

Wassif, C., Cheek, D. and Belas, R. (1995). Molecular analysis of a metalloprotease from *Proteus mirabilis*. *J. Bacteriol.* **177**, 5790–8.

Welch, R.A. (1987). Identification of two different haemolysin determinants in uropathogenic *Proteus* isolates. *Infect. Immun.* **55**, 2183–90.

Welch, R.A. (1991). Pore-forming cytolysins of Gram-negative bacteria. *Mol. Microbiol.* **5**, 521–8.

Welch, R.A. (1995). Phylogenetic analyses of the RTX toxin family. In: *Virulence Mechanisms of Bacterial Pathogens* (eds J.A. Roth, C.A. Bolin, K.A. Brogden, F.C. Minion and M.J. Wannenmuehler), pp. 195–206. American Society for Microbiology, Washington, DC.

Welch, R.A. and Pellett, S. (1988). Transcriptional organization of the *Escherichia coli* haemolysin genes. *J. Bacteriol.* **170**, 1622–30.

Welch, R.A., Dellinger, E.P., Minshew, B. and Falkow, S. (1981). Haemolysin contributes to virulence of extra-intestinal *E. coli* infections. *Nature* **294**, 665–7.

Westrop, G., Hormozi, K., da Costa, N., Parton, R. and Coote, J. (1997). Structure-function studies of the adenylate cyclase toxin of *Bordetella pertussis* and the leukotoxin of *Pasteurella haemolytica* by heterologous C protein activation and construction of hybrid proteins. *J. Bacteriol.* **179**, 871–9.

Wolff, J., Cook, G.H., Goldhammer, A.R. and Berkowitz, S.A. (1980). Calmodulin activates prokaryotic adenylate cyclase. *Proc. Natl Acad. Sci. USA* **77**, 3841–4.

Yoo, H.S., Rajagopal, B.S., Maheswaran, S.K. and Ames, T.R. (1995). Purified *Pasteurella haemolytica* leukotoxin induces expression of inflammatory cytokines from bovine alveolar macrophages. *Microb. Pathog.* **18**, 237–52.

Zabala, J.C., Garcia-Lobo, J.M., Diaz-Aroca, E., de la Cruz, F. and Ortiz, J.M. (1984). *Escherichia coli* alpha-haemolysin synthesis and export genes are flanked by a direct repetition of IS91-like elements. *Mol. Gen. Genet.* **197**, 90–7.

Zambon, J.J., DeLuca, C., Slots, J. and Genco, R.J. (1983). Studies of leukotoxin from *Actinobacillus actinomycetemcomitans* using the promyelocytic HL-60 cell line. *Infect. Immun.* **40**, 205–12.

Zhang, F., Greig, D.I. and Ling, V. (1993a). Functional replacement of the haemolysin A transport signal by a different primary sequence. *Proc. Natl Acad. Sci. USA* **90**, 4211–15.

Zhang, F., Sheps, J.A. and Ling, V. (1993b). Complementation of transport-deficient mutants of *Escherichia coli* α-haemolysin by second-site mutations in the transporter haemolysin B. *J. Biol. Chem.* **268**, 19889–95.

Zhang, F., Yin, Y., Arrowsmith, C.H. and Ling, V. (1995). Secretion and circular dichroism analysis of the C-terminal signal peptides of HlyA and LktA. *Biochemistry* **34**, 4193–201.

17

The family of *Serratia* and *Proteus* cytolysins

Volkmar Braun and Ralf Hertle

INTRODUCTION

When the characterization of the cytolysin (haemolysin) of *Serratia marcescens* began (Braun *et al.*, 1985), it was known that *S. marcescens* secretes a number of exoenzymes, proteases, chitinases, a lipase and a nuclease, in contrast to the *Escherichia coli* K-12 laboratory strain, which does not secrete proteins. Both organisms belong to the Enterobacteriaceae family, and it was questioned which secretory activities of *S. marcescens* were lacking in *E. coli* that made *S. marcescens* secretion competent. The authors biochemically characterized an exoprotease of *S. marcescens* (Schmitz and Braun, 1985), and examined whether the unspecific exoprotease lyses erythrocytes in order to obtain a convenient and sensitive assay for the protease activity.

All *S. marcescens* strains tested lysed erythrocytes – not with the exoprotease, but with an unknown activity. A haemolytic activity on blood agar plates had been mentioned occasionally in the literature, but no data on a haemolysin were available. A book on the genus *Serratia* for clinical microbiologists, infectious disease physicians and epidemiologists (von Graevenitz and Rubin, 1980) does not contain the terms haemolysin and cytolysin, which shows that haemolysis was not considered to be a trait of *Serratia*. The haemolysin was almost neglected then because of the very small size of the zones of lysis around colonies on standard blood agar plates, which, as we now know, is caused by (a)

low diffusion due to the high molecular weight of the haemolysin (Braun *et al.*, 1987), (b) rapid aggregation of the haemolysin upon release from the cells (Schiebel *et al.*, 1989), (c) degradation of the haemolysin by exoproteases, and (d) synthesis of the haemolysin only under iron-limiting growth conditions (Poole and Braun, 1988).

On blood agar plates, *S. marcescens* can easily meet its iron requirement by active transport through ferric siderophores, haem (Angerer *et al.*, 1992) and haemoglobin (Ghigo *et al.*, 1997), resulting in repression of haemolysin synthesis. The authors have tested 11 *S. marcescens* strains and two *Serratia liquefaciens* strains, all of which are haemolytic (Ruan and Braun, 1990), and a subsequent survey of nosocomial *S. marcescens* strains identified 58 haemolytic strains among 60 strains tested (Carbonell and Vidotto, 1992). DNA probes of the *S. marcescens* haemolysin genes do not hybridize to DNA of *E. coli*, *Salmonella typhimurium*, *Proteus mirabilis*, *Proteus vulgaris*, *Citrobacter freundii*, *Enterobacter cloacae*, *Klebsiella aerogenes*, *Klebsiella pneumoniae*, *Shigella dysenteriae*, *Yersinia enterocolitica*, *Yersinia pseudotuberculosis*, *Listeria* sp., *Aeromonas* sp., *Legionella* sp. and *Meningococcus* strains (Ruan and Braun, 1990), and polyclonal antibodies against the *S. marcescens* haemolysin do not react with the rather similar haemolysin of *P. mirabilis*. Lack of DNA cross-hybridization and of antibody cross-reaction make the haemolysin a suitable marker for the diagnosis of *S. marcescens* and *S. liquefaciens*, which

---

349

are the most frequent *Serratia* strains among clinical isolates.

The *S. marcescens* haemolysin represents a new type of haemolysin and has been studied in great molecular detail with regard to structure, activation and secretion. It has nothing in common with the haemolysins (RTX toxins) of the *E. coli* type (Braun and Focareta, 1991; Braun *et al.*, 1993). Haemolysins homologous to the *S. marcescens* haemolysin have been found in *P. mirabilis* and *P. vulgaris* (Uphoff and Welch, 1990), *Haemophilus ducreyi* (Palmer and Munson, 1995) and *Edwardsiella tarda* (Hirono *et al.*, 1997). Furthermore, sequence motifs shown to be important for activity and secretion of the *S. marcescens* haemolysin (Schönherr *et al.*, 1993) are also contained in the filamentous haemagglutinin FHA of *Bordetella pertussis* (Willems *et al.*, 1994), the haemopexin HxuA/HxuB of *Haemophilus influenzae* (Cope *et al.*, 1995) and in the adhesins HMW2A/HMW2B of *H. influenzae* (Barenkamp and Geme, 1994). Similarly to the haemolysins of *S. marcescens* and the related haemolysins, the latter proteins also require a protein for secretion across the outer membrane; this protein displays sequence similarity to the *S. marcescens* secretory protein. Thus, the *S. marcescens* haemolysin forms the prototype of a new class of haemolysins and of a new secretory mechanism.

Knowledge of the structure, activation, mode of action and secretion of the *S. marcescens* haemolysin has implications for these proteins and will be discussed here after the *S. marcescens* haemolysin has been described.

TWO PROTEINS DETERMINE THE *SERRATIA MARCESCENS* HAEMOLYSIN

Bacterial protein toxins are necessarily chimeric proteins, because they must be hydrophilic to be soluble when released from the bacteria, and they also have to be hydrophobic to enter into or pass through the plasma membrane of eukaryotic cells. These properties frequently result in aggregation of the toxins in aqueous solution, and the *S. marcescens* haemolysin is no exception, which makes its biochemical characterization difficult (Braun *et al.*, 1985).

The breakthrough in the precise description of many bacterial toxins came with recombinant DNA techniques, which yielded accurate data on the number of proteins involved, their relative molecular masses, transcriptional regulation of protein synthesis and production of sufficient quantities of the proteins for functional studies. However, in the case of the *S.*

marcescens haemolysin (ShlA), biochemical experiments could only be performed after it was discovered that ShlA can be kept in soluble form in 6 M urea without loss of activity, and that precipitated ShlA aggregates can be solubilized in 6 M urea in an active form. ShlA even withstands irreversible denaturation by 10% trichloroacetic acid, since solubilization of the precipitate in 6 M urea results in a haemolytic solution. In contrast, these treatments denature ShlB, the outer membrane protein required for secretion of ShlA; however, ShlB, in contrast to ShlA, can be solubilized in active form in mild detergents such as octylglucoside.

The haemolysin was characterized in transformants of *E. coli* K-12 that carried both the *shlA shlB* genes, *shlA* or *shlB*, or mutated *shlA* and *shlB* on plasmids. In the cases examined, the properties of the *E. coli* transformants agree with the properties of wild-type *S. marcescens*.

Genetically, the haemolysin trait of *S. marcescens* is very stable. The authors have never observed spontaneous non-haemolytic mutants. A 7.5-kb chromosomal DNA fragment from *S. marcescens* renders *E. coli* K-12 transformants haemolytic. The analysis of the nucleotide sequence revealed two open reading frames, named *shlA* and *shlB*, which are transcribed from *shlB* to *shlA* (Figure 17.1). *shlA* encodes the haemolysin and *shlB* encodes an outer membrane protein that is required for the secretion of the haemolytic ShlA protein across the outer membrane into the culture medium. Mature ShlA is composed of 1578 amino acids and mature ShlB contains 539 amino acids. Both proteins are synthesized as precursors that contain signal peptides of 30 and 18 amino acids, respectively, which are cleaved off during export of the polypeptides across the cytoplasmic membrane (Poole *et al.*, 1988).

In *S. marcescens*, synthesis of the haemolysin is repressed by iron. In rich media, iron limitation by the iron chelator 2,2'-dipyridyl (0.3 mM) strongly increases haemolysin synthesis (Poole and Braun, 1988; Schiebel *et al.*, 1989), despite the numerous iron supply systems that exist in this organism (Angerer *et al.*, 1992). In *E. coli*, transcription of the *shlB shlA* genes is regulated by the Fur protein (Schäffer *et al.*, 1985) which, when loaded with Fe^{2+}, functions as a transcriptional repressor of iron-controlled genes. Fur–Fe^{2+} binds to the Fur box, a DNA consensus sequence composed of 19 nucleotides (Figure 17.2). A sequence similar to the Fur box is located in the −30 region of the promoter upstream of *shlB*. A *fur* deletion mutant of *E. coli* transformed with the *shlA shlB* genes produces 10 times more haemolysin than a *fur* wild-type strain grown in a medium that contains sufficient iron (Poole and Braun, 1988).

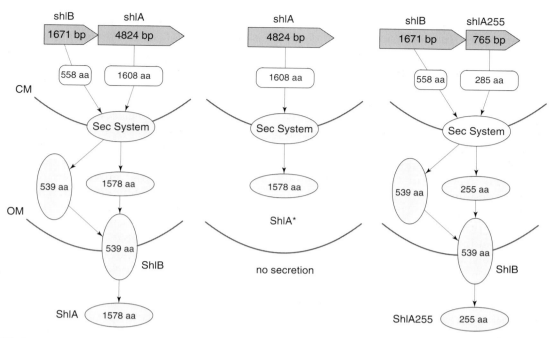

FIGURE 17.1 Arrangement and transcriptional polarity of the *shlA* and *shlB* genes. ShlB inserts into the outer membrane (OM), and activates and secretes ShlA (left), and the *C*-terminally truncated ShlA-255 (right). In the absence of ShlB, non-haemolytic ShlA* remains in the periplasm (centre).

The dual function of ShlB in secretion and lipid activation of ShlA

In an experiment to localize the haemolysin formed by an *E. coli shlA shlB* transformant, 480 haemolytic units were determined in the spent medium, 205 units associated with the cell surface, and 250 units within the cells. Of radiolabelled ShlA, 59% was found in the spent medium and 41% associated with the cells. In the absence of ShlB, inactive ShlA, termed ShlA*, remained completely within cells. Immunogold electron microscopy with anti-ShlA antibodies identified most of ShlA* in the periplasm and some in the cytoplasm. The haemolytic activity found in ShlA*-producing cells is 0.1% of the total activity of ShlA-producing cells (Schiebel *et al.*, 1989). These experiments demonstrate that ShlB is required for activation of ShlA* and for secretion of ShlA.

If activation of ShlA* is a catalytic reaction of ShlB, and ShlB forms pores for secretion of ShlA, much less ShlB than ShlA should fulfil these two functions. This is clearly not the case because, in crude cell extracts that contained much more ShlA* than ShlB, ShlA* was not activated by ShlB. In addition, the ShlB clone used synthesizes an *N*-terminal ShlA fragment, which we know is activated by ShlB; this presumably consumes ShlB and further reduces the amount of ShlB available to activate ShlA*. To demonstrate *in vitro* activation of ShlA* by ShlB uncoupled from secretion, both proteins

have to be overexpressed and synthesized in similar amounts (Ondraczek *et al.*, 1992); this reflects the *in vivo* situation where *shlB* is transcribed prior to *shlA* and at least as much ShlB as ShlA is formed. When both ShlA* and ShlB are highly purified by ion-exchange column chromatography, no active ShlA is formed.

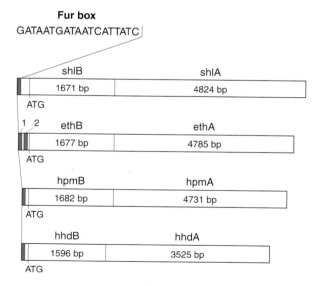

FIGURE 17.2 Demonstration of the Fur box at the promoter of the *shlB* gene. The Fe^{2+}-loaded Fur protein binds to the Fur box and inhibits transcription. Per cent identity of the sequence to the Fur box in *shlB* (52.6%), *ethB* (57.9%), *hpmB* (63.1%) and *hhdB* (50.0%).

The missing component in the assay is phosphatidylethanolamine (PE), which is the major phospholipid (90%) of the *E. coli* outer membrane. Phosphatidylserine, the biosynthetic precursor of PE, is similarly active but does not occur in the outer membrane, and phosphatidylglycerol shows 10% of the PE activity; phosphatidylcholine, cardiolipin, lyso-PE, phosphatidic acid, lipopolysaccharide and various detergents have no effect (Hertle *et al.*, 1997). Approximately four PE molecules bind so tightly to ShlA* without the help of ShlB that they remain bound to ShlA* during sodium dodecyl sulfate–polyacrylamide gel electrophoresis (SDS–PAGE), and are removed only by thin-layer chromatography with organic solvents. Binding of PE to ShlA* does not convert ShlA* to haemolytic ShlA. ShlB has to be added for ShlA*-PE activation.

Removal of the fatty acid at the C2 position of PE by phospholipase A_2 inactivates ShlA, and the resulting lyso-PE dissociates from ShlA. Evidence for the *in vivo* relevance of these *in vitro* results comes from experiments with an *E. coli* mutant devoid of PE owing to a mutation in the *pss* gene, which encodes phosphatidylserine synthase (DeChavigny *et al.*, 1991). The haemolytic activity of the *pss* mutant transformed with a *shlA shlB* plasmid is 9% of that of a *pss*+ strain, and the activity observed probably arises from the highly elevated phosphatidylglycerol content (46%). Inactive ShlA, termed ShlA° to differentiate it from the periplasmic ShlA*, is contained in the culture medium of the *pss* mutant and amounts to 16% (measured by complementation, see below) of the total haemolytic activity of a PE-synthesizing strain (Hertle *et al.*, 1997).

Domains in ShlA responsible for secretion, binding to eukaryotic membranes and pore formation

The 1608-amino-acid polypeptide encoded by the *shlA* gene can be divided into four functional regions. The *N*-terminal signal sequence serves for the Sec-dependent export of the ShlA protein (Braun and Focareta, 1991). The functional regions of the mature form have been unravelled by progressive *C*-terminal truncation of the ShlA protein through genetic means. Reduction in length by 25% results in a haemolysin that displays 20% of the wild-type haemolysis rate. Shorter ShlA fragments cause only residual or no haemolysis (Poole *et al.*, 1988). A 72-kDa fragment (the wild-type mature protein is 160 kDa) still binds to but does not lyse erythrocytes. It can be degraded by trypsin, whereas wild-type haemolysin becomes trypsin resistant upon integration into the erythrocyte membrane (Schiebel and Braun, 1989).

An *N*-terminal fragment of 238 amino acids (ShlA-238) is the shortest fragment that is still secreted. ShlA-238 also converts inactive ShlA*, isolated from the periplasm of an ShlB-negative strain, to active haemolysin. This kind of activation, which we term complementation because of its similarity to the α-complementation of β-galactosidase, was discovered when an *E. coli shlB shlA'* transformant that secretes a 269-residue ShlA fragment (ShlA-269) and an *E. coli shlA* transformant were streaked side by side on a blood agar plate. Neither of the individual transformants was haemolytic, since the ShlA-269 secreted by the ShlB protein is non-haemolytic, and ShlA* remaining in the periplasm in the absence of ShlB is also non-haemolytic. After incubation for several days, a zone of haemolysis appears around the ShlA*-producing cells, mainly on the side where the ShlA-269-producing cells are located. Apparently, ShlA-269 diffuses to the ShlA*-producing cells and gains access to ShlA*, a fraction of which is unspecifically released by partial lysis of some of the cells during the extended incubation period.

This experiment was then repeated with an ShlA-269-containing spent medium, which rendered ShlA* in a crude cell extract haemolytic. Activation of ShlA* by ShlA-269 was reversible, since chromatographic separation of ShlA* and ShlA-269 inactivated ShlA* (Ondraczek *et al.*, 1992). Complementation was also achieved with the smaller ShlA-238 and even with a trypsin degradation fragment of ShlA-269 consisting of only 149 *N*-terminal residues of ShlA (Schönherr *et al.*, 1993).

These data demonstrate that ShlA-238 contains all the information necessary for secretion and activation of ShlA*. For complementation, secretion of ShlA-238 by ShlB is required, since periplasmic ShlA-238 synthesized in the absence of ShlB does not activate ShlA*. Conversion of ShlA* to ShlA by the trypsin fragment also occurs only when the trypsin fragment is isolated from a secreted ShlA polypeptide. Phospholipase A_2 inactivates ShlA-255, which no longer complements ShlA* to ShlA; this indicates that PE binds to the *N*-terminus of ShlA, which is required for activation by ShlB and for the haemolytic activity (Hertle *et al.*, 1997).

ShlA contains the sequence ANPN twice; this sequence rarely occurs in proteins. Each of the first asparagines of the tetrapeptides, N-69 and N-109, was replaced individually by isoleucine, and N-69 was also replaced by lysine. None of the three mutant proteins is secreted and none is haemolytic in whole cells and after extraction with 6 M urea. The high specificity of these mutations is demonstrated by a mutant with a substitution of N-111 by isoleucine in which secretion and haemolytic activity are fully retained. *shlA* mutants with deletions covering N-69 (ShlAΔ68-97) and N-109 (ShlAΔ99-117) are not secreted and are non-haemolytic. The ShlA derivatives carrying point mutations and deletions gain activity by *in vitro* complementation with ShlA-269 (Schönherr *et al.*, 1993). These results clearly indicate that the N-terminal region of ShlA carries the information for secretion and activation by ShlB.

Superhaemolytic ShlA mutants have been isolated by treatment of plasmid-encoded *shlA* with hydroxylamine (Hilger and Braun, 1995). Three mutants with haemolysis rates 7–20-fold higher than that of cells producing wild-type ShlA were studied in some detail. Two mutants carry single amino acid replacements, glycine to aspartate at position 326 (G326D) and S386N, and the third mutant contains two mutations (G326D and N236D). The higher activity of the mutant ShlA proteins is mainly due to a greatly reduced aggregation. The half-life of wild-type ShlA activity in the spent medium is 2.5 min and those of the mutant ShlA proteins are 10, 30 and 40 min. The superhaemolytic mutants differ most strongly from cells producing wild-type ShlA by the failure of the mutant ShlA proteins to cause haemolysis at 0°C when they are dissolved in 6 M urea or 6 M guanidinium chloride to prevent spontaneous precipitation. At 0°C, adsorption of the mutant ShlA proteins to erythrocytes is much reduced. At 20°C, the mutant ShlA proteins in 6 M urea and 6 M guanidinium chloride lyse erythrocytes with rates similar to that of wild-type ShlA. Residues G-326 and S-386, and perhaps N-236, define important sites of ShlA activity, because they contribute to the aggregation of ShlA. The strong effects caused by these mutations is surprising if one considers the conservative nature of the amino acid replacements, and in addition the large size of ShlA.

ShlA forms pores in eukaryotic plasma membranes, but not in prokaryotic plasma membranes

ShlA causes haemolysis by forming pores in erythrocyte membranes (Braun *et al.*, 1987). The onset of haemolysis is progressively retarded and the haemolysis rate diminishes when oligosaccharides of increasing M_r are added to the assay at a concentration of 30 mM, which corresponds to the internal osmotic pressure of erythrocytes. Maltoheptaose (M_r 1152) is highly protective, and dextran 4 (M_r 4000) prevents haemolysis. Removal of dextran 4 and the surplus of haemolysin immediately results in haemolysis by the ShlA proteins that were inserted into the erythrocyte membrane in the presence of dextran 4. The prevention of haemolysis by oligosaccharides demonstrates

the formation of ShlA pores of a limited size range, through which water and smaller oligosaccharides flow into the erythrocytes and cause osmolysis. The larger oligosaccharides prevent osmolysis because they are too large to enter the erythrocytes through the ShlA pores and thus counterbalance the internal osmotic pressure of the erythrocytes. Comparison of the ShlA pores with the structurally defined pores (heptamers) of the *Staphylococcus aureus* α-toxin reveals that ShlA pores are smaller and vary in size, and are larger at 28°C than at 0°C (Schönherr *et al.*, 1994).

Protease digestion experiments have been used to determine how ShlA is inserted into the erythrocyte membrane. The haemolysin integrates into the erythrocytes such that it is not cleaved by trypsin added to erythrocytes and sealed right-side-out erythrocyte ghosts (Schiebel and Braun, 1989). In unsealed ghosts and inside-out vesicles, ShlA is cleaved by trypsin; this demonstrates the accessibility of ShlA to trypsin from the inside of the erythrocytes. The most sensitive cleavage site results in a 143-kDa and a 19.5-kDa fragment, both of which remain in the erythrocyte membrane. This site is close to the C-terminus of ShlA. Upon longer incubations, trypsin yields fragments of 138, 89 and 58 kDa. Only a few ShlA sites in the C-terminal half of ShlA, all of which are exposed at the inside of the erythrocytes, are amenable by trypsin. A genetically engineered N-terminal 72-kDa ShlA fragment can be completely degraded by trypsin, which shows that it adsorbs to, but does not integrate into, the erythrocyte membrane. Adsorption of the 72-kDa ShlA depends on co-synthesis with ShlB, which illustrates that the modification of ShlA by ShlB is required not only for insertion of ShlA into erythrocyte membranes, but also for adsorption to erythrocytes.

Examination of erythrocytes from various animal sources reveals no specificity for pore formation by ShlA. This raises the question of why ShlA does not kill bacterial producer cells. Activation during secretion across the outer membrane by ShlB could be a means of protecting the bacterial producer cells. To test this hypothesis, stable protoplast-type L-forms of *Proteus mirabilis* that do not contain an outer membrane were transformed with a plasmid carrying the *shlA shlB* genes (Sieben *et al.*, 1998). Inactive ShlA* is secreted by the L-form cells, and ShlB is associated with the cytoplasmic membrane. Addition of haemolytic ShlA to the L-form cells has no effect, which suggests that the prokaryotic cytoplasmic membrane is resistant to ShlA.

Pore formation by ShlA has also been examined in artificial lipid bilayers, which revealed water-filled channels with an inner diameter of 1–3 nm, depending on the conditions under which ShlA was prepared

(Schönherr *et al.*, 1994). The data suggest that preformed ShlA monomers and dimers can insert into erythrocyte membranes and that larger oligomers may form in the erythrocyte membranes at 28°C. However, oligomerization within the erythrocyte membrane is not required for pore formation, which occurs more rapidly at 0°C than at 28°C (Schiebel and Braun, 1989). At 0°C, the lateral mobility of integral membrane proteins is greatly reduced so that oligomerization cannot contribute much to pore formation.

ShlB has the potential to form membrane pores through which ShlA might be secreted

To examine the question of whether ShlB forms pores through which ShlA is secreted across the outer membrane, ShlB and deletion derivatives of ShlB were added to artificial lipid bilayer membranes, and the increase in the membrane conductance was measured. This procedure has been used successfully with FhuA, an outer membrane receptor for various bacterial viruses, bacterial toxins and the iron carrier ferrichrome. FhuA does not increase the conductance unless a surface-exposed loop is deleted, which converts the FhuA closed channel into a permanently open channel (Killmann *et al.*, 1993). Various fragments of ShlB were excised by genetic engineering, and the resulting proteins were extracted from the outer membrane by octylglucoside and then highly purified by two ion-exchange column chromatographies to remove completely the porins, which have a high propensity to form channels in lipid bilayer membranes.

Only three deletion derivatives could be solubilized and purified. Wild-type ShlB causes, with an irregular frequency, an increase in the conductance of 1 nS, which after a few milliseconds decreases to zero level; this indicates that ShlB has the potential to form a channel (Könninger *et al.*, 1999). ShlBΔ87–153 and ShlBΔ65–168 increase the membrane conductance stepwise by 1.2 nS, and ShlB(Δ126–200) by 1.5 nS, which demonstrates formation of rather stable single channels. Frequently, two or three channel opening events are observed, followed by one or two closing steps, and the open periods last longer than the closed periods. The results obtained with the deletion derivatives support the channel-forming properties of ShlB. The authors propose that ShlB forms a closed channel that can be opened when ShlA is secreted.

A topology model of ShlB predicts 20 transmembrane regions that are interconnected by loops at the cell surface and short turns in the periplasm (Könninger *et*

al., 1999). This model is based on the reaction of an antigenic epitope inserted at 22 sites along the ShlB polypeptide. Intact cells of 16 epitope mutants react with the monoclonal antibody, which demonstrates accessibility of the epitope at the cell surface. In six mutants, the antibody reacts only with the isolated outer membrane, which indicates a periplasmic or transmembrane location of the epitope. A computer-assisted program (Schirmer and Cowan, 1993) for the prediction of membrane-spanning β-strands of outer membrane proteins was used to localize the six latter epitope sites within the membrane half orientated toward the periplasm. According to this model, the deletions in the two ShlB mutants that form rather stable channels in artificial lipid bilayer membranes comprise portions of the two largest ShlB loops at the cell surface (extending from residues 60 to 97 and 120 to 213), two transmembrane segments and one periplasmic turn.

HPMA AND HPMB OF *PROTEUS MIRABILIS* AND *PROTEUS VULGARIS*

In *P. mirabilis* and *P. vulgaris*, haemolysins of the *S. marcescens* type (HpmA) and of the *E. coli* type (HlyA) (Koronakis *et al.*, 1987; Senior and Hughes, 1988) are found. The HpmA type is more common and is, for example, found in all 63 tested *P. mirabilis* strains and in 23 of 24 tested *P. vulgaris* strains isolated from various infections and normal faeces (Swihart and Welch, 1990). The *P. mirabilis* haemolysin, like the *S. marcescens* haemolysin, is determined by two genes, *hpmA* and *hpmB*, which display 52.1% nucleotide identity to

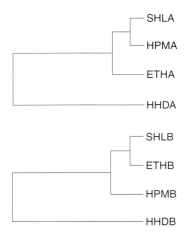

FIGURE 17.3 Dendrogram of the haemolysin proteins A and B from *S. marcescens* (ShlB/A), *E. tarda* (EthB/A), *P. mirabilis* (HpmB/A) and *H. ducreyi* (HhdB/A) as determined by the CLUSTAL program.

the *shlAB* genes (Figure 17.3) despite the 38 mol% G + C content in the *hpmAB* genes and 65 mol% in the *shlAB* genes (Uphoff and Welch, 1990). Two cysteine residues are contained in a highly conserved region between HpmA and ShlA, but substitution of single cysteines or both cysteines by serine in ShlA did not reduce ShlA activity (Schönherr *et al.*, 1993).

HpmB is required for the formation and secretion of haemolytic HpmA. The strong structure/function similarity of the *P. mirabilis* and *S. marcescens* haemolysins is demonstrated by the mutual functional replacement of the HpmAB and ShlAB proteins. HpmB can replace ShlB for the activation and secretion of ShlA, and ShlB can activate and secrete HpmA. In addition, ShlA-269 complements HpmA* synthesized without HpmB, and a non-haemolytic C-terminally truncated HpmA' fragment of approximately the size of ShlA-269 complements ShlA* (Ondraczek *et al.*, 1992).

HHDA AND HHDB OF *HAEMOPHILUS DUCREYI*

A 13-kb DNA fragment of *H. ducreyi* confers haemolytic activity to an *E. coli* transformant and complements a non-haemolytic transposon insertion mutant of *H. ducreyi* (Palmer and Munson, 1995). The haemolytic activity is determined by the *hhdA* and *hhdB* genes. The derived HhdA amino acid sequence is 28% similar and 48% identical to ShlA, and 47% similar and 26% identical to HpmA. HhdB displays 30% similarity and 50% identity to ShlB, and 53% similarity and 29% identity to HpmB (Figures 17.2 and 17.3). Chromosomal DNA of eight *H. ducreyi* strains hybridizes with an *hhdB hhdA* probe, which indicates that the haemolysin is highly prevalent in *H. ducreyi*.

ETHA AND ETHB OF *EDWARDSIELLA TARDA*

The fish pathogen *E. tarda* produces a haemolysin of the *S. marcescens* type. The EthA protein sequence deduced from the *ethA* nucleotide sequence displays 47, 37 and 23% sequence identity to ShlA, HpmA and HhdA, respectively, and EthB has 58, 51 and 26% identity to ShlB, HpmB and HhdB (Figures 17.2 and 17.3) (Hirono *et al.*, 1997). Formation and secretion of haemolytic EthA depends on EthB, but EthA agglutinates erythrocytes also when it is not co-synthesized with EthB. Haemagglutination is eightfold stronger after sonication of the cells than released EthA, which shows that mainly cell-associated EthA agglutinates erythrocytes. However, the strength of haemag-

glutination was not compared with that of EthA co-synthesized with EthB, since erythrocytes lysed. Lysis can be prevented by osmoprotection, as has been described for ShlA (Poole *et al.*, 1988; Schönherr *et al.*, 1994). Therefore, EthA-mediated haemagglutination could be a residual activity that amounts to only a few per cent of the activity of EthA co-synthesized with EthB.

Two putative Fur boxes are located upstream of *ethB* and both genes, *ethB* and the downstream *ethA*, are strongly transcribed only under iron-limiting conditions, as shown by reverse-transcription polymerase chain reaction (Hirono *et al.*, 1997).

SECRETION OF NON-HAEMOLYTIC ADHESINS BY A MECHANISM REMINISCENT OF SHLA SECRETION BY SHLB

Filamentous haemagglutinin FHA of *Bordetella pertussis*, the HMW1 and HMW2 adhesins of *Haemophilus influenzae*, and the HxuA haem–haemopexin receptor of *H. influenzae* type b are surface-exposed proteins, and portions are released into the culture medium. Like ShlA, these proteins contain the ANPNL motif once or twice (Table 17.1); when the motif is mutated, ShlA is inactivated (Schönherr *et al.*, 1993). Secretion requires the outer membrane proteins FhaC, HMWB1 and HMWB2, and HxuB, respectively.

Replacement of Asn137 by isoleucine totally abolishes the secretion of an *N*-terminal fragment (Fha44, 947 amino acids) of the *B. pertussis* haemagglutinin, whereas the replacement of Asn176 by isoleucine reduces Fha44 secretion by 80–90%. FhaC, in contrast to ShlB, does not secrete and activate HpmA of *P. mirabilis* (Jacob-Dubuisson *et al.*, 1997), and Fha44 does not complement ShlA* (Yang and Braun, unpublished). The analogy between secretion of the haemagglutinin and the *S. marcescens* type haemolysins is limited since, for efficient secretion, the C-terminal 150-kDa region of the 367-kDa haemagglutinin is important (Renauld-Mongénie *et al.*, 1996), whereas the C-terminus plays no role in ShlA secretion. Furthermore, an additional protein, designated HMWC, is required for secretion of HMW (Geme and Grass, 1998).

Replacement of Asp150 in the NPNGI motif of HMW1 does not affect secretion and processing of HMW1 (cleavage of a 441-residue *N*-terminal fragment), but conversion to IAIGI eliminates release from the cell surface and processing, while the mutant protein is still exposed at the cell surface (Geme and Grass, 1998).

TABLE 17.1 Common sequence motifs related to functionally essential regions of the *Serratia marcescens* haemolysin[a]

Serratia marcescens haemolysin ShlA	68-ANPNL
Edwardsiella tarda haemolysin EthA	68-ANPNL
Proteus mirabilis haemolysin HpmA	68-ANSNL
Haemophilus ducreyi haemolysin HhdA	66-ANPHL
Bordetella pertussis filamentous haemagglutinin FHA	65-KNPNL
Serratia marcescens haemolysin ShlA	109-NPNGIS
Edwardsiella tarda haemolysin EthA	108-NPNGIT
Proteus mirabilis haemolysin HpmA	109-NPNGIT
Haemophilus ducreyi haemolysin HhdA	107-NPNGMS
Bordetella pertussis filamentous haemagglutinin FHA	94-NPNGIS
Haemophilus influenzae Type b heme:hemopexin HxuA	86-NPNGVI
Haemophilus influenzae surface protein HMW	150-NPNGIT

[a]Numbering corresponds to the proteins after cleavage of the signal peptide, which is unusually long for FHA (71 residues) (Lambert-Buisine *et al.*, 1998) and not known for HMW (Barenkamp and Leininger, 1992).

Haemophilus influenzae requires haem, which it cannot synthesize, in the growth medium. Haem is taken up bound to the serum glycoprotein haemopexin by *H. influenzae* type b via the outer membrane protein HxuA. A second outer membrane protein, HxuB, is functionally equivalent to ShlB (Cope *et al.*, 1995). Non-typeable *H. influenzae* strains contain the same haem acquisition system (Cope *et al.*, 1995). HxuA of both *H. influenzae* types contains an NPNG motif which may be important for recognition by HxuB for secretion of HxuA to the cell surface and subsequently into the culture medium.

PATHOGENICITY OF *SERRATIA MARCESCENS* AND *PROTEUS MIRABILIS*

Serratia marcescens is an important opportunistic pathogen, which causes respiratory and urinary tract infections, bacteraemia, endocarditis, keratitis, arthritis and meningitis (Lyerly and Kreger, 1983; Maki *et al.*, 1973). *Serratia marcescens* can be detected on blood agar by its haemolytic activity. This activity has been ascribed to haemolysin ShlA, which is also a cytolysin and damages tissues, and causes the release of the inflammatory mediators leukotrienes (LTB4 and LTC4) from leucocytes and the release of histamine from rat mast cells (König *et al.*, 1987; Scheffer *et al.*, 1988). ShlA also contributes to the uropathogenicity of the pathogenic *E. coli* 536/21 after transformation with the *S. marcescens shlA shlB* genes (Marré *et al.*, 1989). *Proteus mirabilis* induces acute pyelonephritis by stimulating the host cells to produce interleukins, tumour necrosis factor-α (TNF-α) or prostaglandins (Peerbooms *et al.*, 1984). The major factors in this observed pathogenicity are the haemolysins/cytolysins ShlA and HpmA. However, the pathogenicity of *Serratia* and *Proteus* strains is a multifactorial process, and includes the haemolysin, urease, fimbriae, proteases, lipase and undefined determinants that facilitate invasion. These factors act in concert, and the resulting effects are adherence, host cell invasion, cytotoxic effects and final cytolysis. Because the haemolytic activity is mainly cell associated, its effect arises predominantly after adherence of the bacteria to the host cell tissue.

Adherence of *Serratia marcescens*

Colonization of epithelial tissue is only possible with efficient adherence of bacteria to the target cells. *Serratia marcescens* strains produce type-1 fimbriae (Yamamoto *et al.*, 1985; Leranoz *et al.*, 1997) or US5 pili (Kono *et al.*, 1984), which are involved in adherence on epithelial cells. Non-fimbriated mutants are markedly decreased in adherence. Fimbriae of *S. marcescens* also contribute to superoxide production in neutrophils (König *et al.*, 1987; Scheffer *et al.*, 1988) and phagocytosis (Mizunoe *et al.*, 1995). It has been shown that radiolabelled bacteria adhere to purified granulocytes and are subsequently phagocytosed. The superoxide production, determined by chemiluminescence response of bacteria, is dependent on fimbriae and also on haemolysin production (König *et al.*, 1987). However, *S. marcescens* secretes not only a haemolysin, but also exoproteases and lipases, which also induce chemiluminescence and contribute to the response to the haemolysin. The mutant strain W1436 lacks exoprotease and lipase activity and is still active in chemiluminescence, although less than the wild-type strain 5817, but is more active in histamine release (König *et al.*, 1987).

With purified ShlA, no typical oxidative burst is detectable; instead, granulocytes continuously increase chemiluminescence (Hertle, 1998). These data indicate that the oxidative burst detected with viable haemolytic *S. marcescens* strains is the sum of cytotoxic effects of various products secreted by the bacteria and not only of the cytolysin. The *S. marcescens* haemolysin contributes to colonization of the urinary tract epithelium in an experimental rat model (Marré *et al.*, 1989). An *E. coli* 536/21 transformant producing ShlA is five times more efficient in colonization than the ShlA-negative recipient strain, but it is not clear whether ShlA acts as a cell-bound adhesin or facilitates the colonization by its cytolytic activity. On epithelial cell cultures (HEp-2, HeLa), ShlA-producing *S. marcescens* strains W1128 or CDC04:H4, for example, are adherent (Hertle *et al.*, 1998a). A non-adherent *E. coli* strain BL21 transformed with the *shlAB* genes that produces 30-fold more haemolysin than the *Serratia* wild-type is less cytotoxic and not adherent. Therefore, it is clear that adherence is not primarily mediated by the haemolysin.

Invasion of *Serratia marcescens* and *Proteus mirabilis*

Serratia marcescens invades tissue cells (Hertle *et al.*, unpublished). With the commonly used gentamycin protection assay, intracellular *S. marcescens* wild-type strain W1126 and CDC04:H4 cells have been isolated from HEp-2 and HeLa cells. The cytotoxicity of these strains (due to ShlA) is a confounding problem and reduces the amount of viable bacteria protected against gentamycin, because the bacteria lyse their host cells and become amenable to gentamycin. To avoid this problem, an isogenic haemolytic-negative *S. marcescens* W1128 strain was constructed by site-directed mutagenesis (Hertle *et al.*, unpublished results). This mutant strain shows a lower invasiveness than the wild-type and is less cytotoxic in the LDH release assay.

The ability of *P. mirabilis* to adhere to and invade urothelial cells is closely coupled to swarming (differentiation into hyperflagellated, filamentous swarm cells). Invasion is predominantly determined by flagellin production, and the haemolysin plays a minor role in this process (Allison *et al.*, 1992). Invasion by swarm cells occurs within 30 min and is about 15-fold greater (*ca.* 0.18% entry) after 2 h than is invasion by vegetative cells (*ca.* 0.012% entry), which are internalized more slowly. A haemolysin-negative strain of WPM111 does not significantly alter virulence, but flagella contribute to the ability of *P. mirabilis* to colonize the urinary tract and cause acute pyelonephritis

in an experimental model of ascending urinary tract infection (Mobley *et al.*, 1996).

In addition, vegetative, non-swarming *P. mirabilis* cells invade epithelial cells efficiently and are found in endosomes and free in the cytoplasm (Oelschläger and Tall, 1996). Inhibition of eukaryotic protein synthesis by cycloheximide does not reduce bacterial uptake. In addition, Chippendale *et al.* (1994) have found that internalization of *P. mirabilis* into human renal proximal tubular epithelial cells (HRPTEC) is decreased by the HpmA haemolysin owing to cytolysis. Haemolysin-negative mutant strain WPM111 cells are recovered over the course of the experiment in progressively higher numbers (10–100-fold higher) than the haemolytic parent strain BA6163.

From these data it is concluded that adherence mediated by fimbriae or pili of the bacteria directs the haemolysin to the target membranes, and the pore-forming toxin becomes accessible to the membrane, but the haemolysin does not mediate adherence by itself. It is more likely that pore formation may play a role in invasion, but renders invasion studies difficult owing to its cytotoxicity.

Cytotoxic effects and cytolysis by the *Serratia marcescens* haemolysin

ShlA shows target cell specificity that differs from other bacterial pore-forming toxins, such as staphylococcal α-toxin or *E. coli* haemolysin, both of which induce ATP depletion and potassium efflux in nucleated cells at low toxin doses (Walev *et al.*, 1993, 1995).

ShlA is inactive on keratinocytes, endothelial cells and monocytes, all of which are specific target cells for α-toxin (Bhakdi *et al.*, 1991). ShlA induces ATP depletion and potassium efflux in epithelial cells and in fibroblasts (Hertle *et al.*, 1999). The depletion of the cellular ATP level is dramatic with sublytic doses of ShlA. Treatment of HEp-2 or HeLa cells with 1 µg ShlA ml^{-1} decreases the initial ATP level to 10% within 30 min. In parallel, intracellular potassium is released into the supernatant. Pore formation by ShlA in the plasma membrane leads to leakage of potassium. This is the signal for the Na/K-ATPase in the membrane to transport potassium into the cell, but the leakage through the ShlA pores is greater and the resulting effect is an extended ATP consumption. This effect is also observed with inactive ShlA* complemented with ShlA255, and underlines the efficiency of the complementation even on nucleated eukaryotic cells. Pores formed by ShlA in the eukaryotic cytoplasmic membrane are smaller than the 1–3 nm estimated for ShlA pores in artificial black lipid membranes or in erythrocytes. The pores permit the influx of propidium iodide and trypan blue. Propidium iodide (M_r 668) uptake of ShlA-treated T-cells was measured by flow cytometry. Cells were incubated with various toxin concentrations for 1 h at 37°C. T-cells, depleted in the ATP level to 10%, were used for fluorescence-assisted cell sorting (FACS) analysis. ShlA-treated cells show no propidium iodide influx. Using osmoprotective conditions with various oligosaccharides as described above, the ATP depletion was unaffected (Hertle *et al.*, 1999).

The observed ATP depletion caused by pore formation is reversible up to 80% of the initial ATP level, and depends on the initial drop of the ATP level. Depletion of the initial ATP level to lower than 20% decreases the recuperation significantly in HEp-2 cells (40% restored) because the cells begin to lyse. Prolonged toxin treatment leads to an increased ATP depletion of less than 5% and to subsequent cell lysis. However, the ATP depletion is reversible, and it is thought that this restoration is due to a repair or closure of the ShlA-produced pores in the cytoplasmic membrane. Interestingly, this repair is affected by cycloheximide, an inhibitor of protein synthesis, indicating a role of protein synthesis in this process. Cycloheximide treatment (5 µg ml⁻¹) of HEp-2 cells before or after addition of ShlA decreases the recuperative capacity of the nucleated cells significantly (Hertle *et al.*, 1999).

ShlA-mediated vacuolation

The osmotic imbalance, indicated by potassium efflux, ATP depletion and cell swelling, yields vacuolation (Hertle *et al.*, 1999). Vacuolation mediated by ShlA is seen with epithelial cells, but not with fibroblasts. Vacuolation is observed after ATP depletion to less than 5%. The affected cells begin to show small vacuoles all over the cytoplasm. With prolonged incubation, the small vacuoles fuse and yield large vacuoles with undefined shape that fill the entire cytoplasm (see Figure 17.4). The ShlA-treated cells retain a tight cytoplasmic membrane, which prevents the influx of propidium iodide and trypan blue. No cytotoxicity, as determined by the LDH assay, can be seen during vacuolation.

Vacuolation is seen not only with isolated ShlA, but also with haemolytic *S. marcescens* strains. Vacuolation is thought to be the result of an osmotically driven influx of water. Osmoprotection with oligosaccharides can suppress vacuolation mediated by ShlA (Hertle *et al.*, 1999). Oligosaccharides with a molecular mass up to 1152 Da (maltoheptaose) strongly reduce vacuolation and cytotoxicity, and oligosaccharides larger than 1400 Da (dextrin 15) completely inhibit vacuolation. Oligosaccharides with a molecular mass below 600–700 Da cannot stop or alter vacuolation. These data

demonstrate formation of pores by ShlA and a pore size in nucleated cells smaller than that in erythrocytes. Diluting out the oligosaccharides results in vacuolation and cytolysis of ShlA-pretreated cells. Unexpectedly, bacteria found in the large vacuoles formed in the late stage of infection are highly mobile. The reason for this is unknown, but the bacteria may undergo a transition to a highly mobile swarming state during the infection process.

Vacuolation has been observed with *Helicobacter pylori* cytotoxin VacA (Cover and Blaser, 1992). Cells exposed to VacA develop large vacuoles that originate from massive swelling of membranous compartments of the late stages of the endocytic pathway. These vacuoles are acidic (stainable with neutral red) and their membranes contain the vacuolar ATPase proton pump and the small GTP-binding protein rab7 (Cover *et al.*, 1993; Papini *et al.*, 1997). The vacuolar ATPase is inhibited by bafilomycin and vacuolation stops and reverts (Papini *et al.*, 1993). It is clear that VacA displays its toxic activity in the cell cytosol (Montecucco *et al.*, 1996; de Bernard *et al.*, 1997). VacA- and ShlA-induced vacuolation are different in many respects. Vacuoles induced by ShlA are not acidified, vacuolation is not inhibited by bafilomycin, and vacuolation cannot be reversed (Hertle *et al.*, 1999).

The *P. mirabilis* cytolysin HpmA acts as a potent cytotoxin against HRPTEC (Mobley *et al.*, 1991), Daudi, Raji (human β-cell lymphoma), U973 (human monocytes) and Vero cells (African green monkey kidney cells) (Swihart and Welch, 1990). Cytotoxicity was determined as release of lactate dehydrogenase (LDH), but the cellular effects were not studied in detail. ShlA is also cytotoxic in the LDH assay at very low doses (0.2 haemolytic units ml⁻¹ within 1 h of incubation) (Hertle *et al.*, 1999). As is the case with ShlA, cytolysis of target cells is very rapid and is due to pore formation by HpmA.

Pathogenicity of *Edwardsiella tarda* and *Haemophilus ducreyi*

Edwardsiella tarda synthesizes a haemolysin, EthA, which is homologous to ShlA (Hirono *et al.*, 1997). *Edwardsiella tarda* can penetrate and replicate in HEp-2 cells (Janda *et al.*, 1991), and invasion depends on the presence of EthA (Strauss *et al.*, 1997). In contrast to other bacteria that replicate in the cytoplasm, such as *Listeria monocytogenes* and *Shigella flexneri* (Theriot, 1995), *E. tarda* does not appear to harness actin or engage in direct cell-to-cell spread. Haemolysin production by *E. tarda* enhances HEp-2 cell invasion significantly, and a haemolysin mutant strain (Tn5 insertion into *ethB*) enters HEp-2 cells two or three

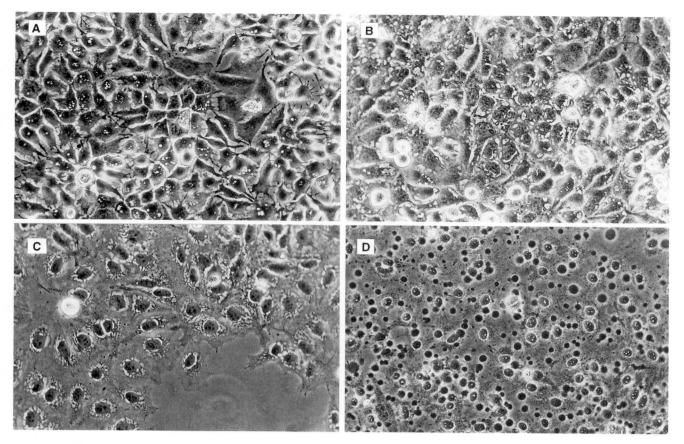

FIGURE 17.4 Photomicrograph of ShlA-induced vacuolation of HEp-2 cells after 10 min (A), 20 min (B), 30 min (C) and 45 min (D). Cells were cultured in a 24-well plate until they reached subconfluence. Medium was supplemented with 100 μl of a bacterial culture supernatant containing ShlA (30 HU ml^{-1}). After incubation at 37°C under an atmosphere of 5% CO_2, cells were examined by phase-contrast microscopy at ×320 magnification.

orders of magnitude less efficiently than the parental strain. This phenotype of the non-haemolytic strain raises the possibility that the haemolysin itself functions as an invasin. The role and the cellular effects of the isolated haemolysin have not been studied in detail, but may be very similar to those of ShlA and HpmA.

Haemophilus ducreyi, a bacterium that induces the sexually transmitted disease chancroid, also shows cytotoxic effects on HEp-2 or HeLa cells. Ulcer formation is dependent on iron deficiency (Sturm, 1997). This bacterium is also capable of adhering to epithelial cells and penetrating them (Lagergard *et al.*, 1993). Adherence is manifested after 15 min and reaches a maximum after 2–3 h. The cytotoxic agent is described as a cell-associated cytotoxin (Alfa, 1992) and characterized as haemolysin HhdA with sequence similarity to ShlA (Palmer and Munson, 1995). Attachment to epithelial cells is mediated by more than one mechanism. Proteinase K treatment, but not trypsinization of *H. ducreyi*, significantly reduces attachment, which suggests protein involvement. In addition, purified lipo-oligosaccharide (LOS) is able to inhibit attachment in a dose-dependent manner. It appears that the organism binds to fibronectin in the extracellular matrix of HFF cells (human foreskin fibroblasts), since competition studies using fibronectin have shown that it is able to reduce attachment significantly in a dose-dependent manner, whereas collagen does not.

One could hypothesize that the attachment of *H. ducreyi* involves both a protein mediator of attachment (likely pili) as well as LOS and that one or both of these bacterial components interacts with fibronectin in the extracellular matrix to mediate attachment to HFF cells (Alfa and De Gagne, 1997). It is thought that the first step in the pathogenesis of chancroid is the adherence of bacteria to epithelial cells, followed by the action of cytotoxin and further bacterial proliferation. It is suggested that this sequence of events results in the production of genital ulcers by *H. ducreyi*.

CONCLUSION

It has become obvious that the *Serratia* type of pore-forming cytotoxins are important virulence factors. They serve multiple purposes in the complicated interaction between parasite and host, and may in the future provide a more detailed view into the cell biology of infectious diseases.

REFERENCES

Alfa, M.J. (1992). Cytopathic effect of *Haemophilus ducreyi* for human foreskin cell culture. *J. Med. Microbiol.* **37**, 43–50.

Alfa, M.J. and DeGagne, P. (1997). Attachment of *Haemophilus ducreyi* to human foreskin fibroblasts involves LOS and fibronectin. *Microb. Pathog.* **22**, 39–46.

Allison, C., Coleman, N., Jones, P.L. and Hughes, C. (1992). Ability of *Proteus mirabilis* to invade human urothelial cells is coupled to motility and swarming differentiation. *Infect. Immun.* **60**, 4740–6.

Angerer, A., Klupp, B. and Braun, V. (1992). Iron transport systems of *Serratia marcescens*. *J. Bacteriol.* **174**, 1378–87.

Barenkamp, S.J. and Geme, J.W.S. (1994). Genes encoding high-molecular-weight adhesion proteins of nontypeable *Haemophilus influenzae* are part of gene clusters. *Infect. Immun.* **62**, 3320–8.

Barenkamp, S.J. and Leininger, E. (1992). Cloning, expression, and DNA sequence analysis of genes encoding nontyptable *Haemophilus influenzae* high-molecular weight surface-exposed proteins related to filamentous haemagglutinin of *Bordetella pertussis*. *Infect. Immun.* **60**, 1302–13.

Bhakdi, S. and Martin, E. (1991). Superoxide generation by human neutrophils induced by low doses of *Escherichia coli* haemolysin. *Infect. Immun.* **59**, 2955–62.

Braun, V. and Focareta, T. (1991). Pore-forming bacterial protein haemolysins (cytolysins). *Crit. Rev. Microbiol.* **18**, 115–58.

Braun, V., Günther, H., Neuss, B. and Tautz, C. (1985). Haemolytic activity of *Serratia marcescens*. *Arch. Microbiol.* **141**, 371–6.

Braun, V., Neuss, B., Ruan, Y., Schiebel, E., Schöffler, H. and Jander, G. (1987). Identification of the *Serratia marcescens* haemolysin determinant by cloning into *Escherichia coli*. *J. Bacteriol.* **169**, 2113–20.

Braun, V., Schönherr, R. and Hobbie, S. (1993). Enterobacterial haemolysins: activation, secretion and pore formation. *Trends Microbiol.* **1**, 211–16.

Carbonell, G.V. and Vidotto, M.C. (1992). Virulence factors in *Serratia marcescens*: cell-bound haemolysin and aerobactin. *J. Med. Biol. Res.* **25**, 1–8.

Chippendale, G.R., Warren, J.W., Tritillis, A.L. and Mobley, H.L.T. (1994). Internalization of *Proteus mirabilis* by human renal epithelial cells. *Infect. Immun.* **62**, 3114–21.

Cope, L., Yongev, R., Müller-Eberhard, U., and Hansen, E.C. (1995) A gene cluster involved in the utilization of both free heme and heme:hemopexin by *Haemophilus influenzae* type b. *J. Bacteriol.* **177**, 2644–53.

Cover, T. L. and Blaser, M.J. (1992). Purification and characterization of the vacuolating toxin from *Helicobacter pylori*. *J. Biol. Chem.* **267**, 10570–5.

Cover, T.L., Reddy, L.Y. and Blaser, M.J. (1993). Effects of ATPase inhibitors on the response of HeLa cells to *Helicobacter pylori* vacuolating toxin. *Infect. Immun.* **61**, 1427–31.

de Bernard, M., Arico, B., Papini, E., Rizzuto, R., Grandi, G., Rappuoli, R. and Montecucco, C. (1997). *Helicobacter pylori* toxin VacA induces vacuole formation by acting in the cell cytosol. *Mol. Microbiol.* **26**, 665–74.

DeChavigny, A., Heacock, P.N. and Downham, W. (1991) Sequence and inactivation of the *pss* gene of *Escherichia coli*. *J. Biol. Chem.* **266**, 5323–32.

Geme, J.W.S. and Grass, S. (1998). Secretion of the *Haemophilus influenzae* HMW1 and HMW2 adhesins involves a periplasmic intermediate and requires the HMWB and HMWC proteins. *Mol. Microbiol.* **27**, 617–30.

Ghigo, J.M., Letoffe, S. and Wandersman, C. (1997). A new type of hemophore-dependent haeme acquisition system of *Serratia marcescens* reconstituted in *Escherichia coli*. *J. Bacteriol.* **179**, 3572–9.

Hertle, R., Brutsche, S., Groeger, W., Hobbie, S., Koch, W., Könninger, U. and Braun, V. (1997). Specific phosphatidylethanolamine dependence of *Serratia marcescens* cytotoxin activity. *Mol. Microbiol.* **26**, 853–65.

Hertle, R., Weingardt-Kocher, S. and Schwarz, H. (1998). *Serratia marcescens* strain W1128 is invasive in epithelial cells in culture. Unpublished results.

Hertle, R., Hilger, M., Weingardt-Kocher, S. and Walev, I. (1999). Cytotoxic action of *Serratia marcescens* haemolysin on human epithelial cells. *Infect. Immun.* **67**, 817–25.

Hilger, M. and Braun, V. (1995). Superlytic haemolysin mutants of *Serratia marcescens*. *J. Bacteriol.* **177**, 7202–9.

Hirono, I., Tange, N. and Aoki, T. (1997). Iron-regulated haemolysin gene from *Edwardsiella tarda*. *Mol. Microbiol.* **24**, 851–6.

Jacob-Dubuisson, F., Buisine, C., Willery, E., Renauld-Mongénie, G. and Locht, C. (1997). Lack of functional complementation between *Bordetella pertussis* filamentous haemagglutinin and *Proteus mirabilis* HpmA haemolysin secretion machineries. *J. Bacteriol.* **179**, 775–83.

Janda, J.M., Abbott, S.L. and Oshiro, L.S. (1991). Penetration and replication of *Edwardsiella* spp. in HEp-2 cells. *Infect. Immun.* **59**, 154–61.

Killmann, H., Benz, R. and Braun, V. (1993). Conversion of the FhuA transport protein into a diffusion channel through the outer membrane of *Escherichia coli*. *EMBO J.* **12**, 3007–16.

Kono, K., Yamamoto, T., Kuroiwa, A. and Amoko, K. (1984). Purification and characterization of *Serratia marcescens* US5 pili. *Infect. Immun.* **46**, 295–300.

König, W., Faltin, Y., Scheffer, J., Schöffler, H. and Braun, V. (1987). Role of cell-bound haemolysin as a pathogenicity factor for *Serratia* infections. *Infect. Immun.* **55**, 2554–61.

Könninger, U.W., Hobbie, S., Benz, R. and Braun, V. (1999). The hemolysin-secreting ShlB protein of the outer membrane of *Serratia marcescens*: determination of surface-exposed residues and formation of ion-permeable pores by ShlB mutants in artificial bilayer membranes. *Mol. Microbiol.* (in press).

Koronakis, V., Cross, M., Senior, B., Koronakis, E. and Hughes, C. (1987). The secreted haemolysins of *Proteus mirabilis*, *Proteus vulgaris*, and *Morganella morganii* are genetically related to each other and to the alpha-haemolysin of *Escherichia coli*. *J. Bacteriol.* **169**, 1509–15.

Lagergard, T., Purven, M. and Frisk, A. (1993). Evidence of *Haemophilus ducreyi* adherence to and cytotoxin destruction of human epithelial cells. *Microb. Pathog.* **14**, 417–31.

Lambert-Buisine, C., Willery, E., Locht, C. and Jacob-Dubuisson, F. (1998). *N*-terminal characterization of the *Bordetella pertussis* filamentous haemagglutinin. *Mol. Microbiol.* **28**, 1283–93.

Leranoz, S., Orus, P., Berlanga, M., Dalet, F. and Vinas, M. (1997). New fimbrial adhesins of *Serratia marcescens* isolated from urinary tract infections: description and properties. *J. Urol.* **157**, 694–8.

Lyerly, D.M. and Kreger, A.S. (1983). Importance of *Serratia* protease in the pathogenesis of experimental *Serratia marcescens* pneumonia. *Infect. Immun.* **40**, 113–19.

Maki, D.G., Hennekens, C.G., Philips, C.W., Shaw, W.V. and Bennet, J.V. (1973). Nosocomial urinary tract infection with *Serratia marcescens*. *J. Infect. Dis.* **128**, 579–87.

Marré, R., Hacker, J. and Braun, V. (1989). The cell-bound haemolysin of *Serratia marcescens* contributes to uropathogenicity. *Microb. Pathog.* **7**, 153–6.

Mizunoe, Y., Matsumoto, T., Haraoka, M., Sakumoto, M., Kubo, S. and Kumazawa, J. (1995). Effect of pili of *Serratia marcescens* on superoxide production and phagocytosis of human polymorphonuclear leukocytes. *J. Urol.* **154**, 1227–30.

Mobley, H.T.L., Chippendale, G.R., Swihart, K.G. and Welch, R.A.. (1991). Cytotoxicity of the HpmA haemolysin and urease of *Proteus mirabilis* and *Proteus vulgaris* against cultured human renal proximal tubular epithelial cells. *Infect. Immun.* **59**, 2036–42.

Mobley, H.L., Belas, R., Lockatell, V., Chippendale, G., Trifillis, A.L., Johnson, D.E. and Warren, J.W. (1996). Construction of a flagellum-negative mutant of *Proteus mirabilis*: effect on internalization by human renal epithelial cells and virulence in a mouse model of ascending urinary tract infection. *Infect. Immun.* **64**, 5332–40.

Montecucco, C., Papini, E. and Schiavo, G. (1996). Bacterial protein toxins and cell vesicle trafficking. *Experientia* **52**, 1026–32.

Oelschläger, T.A. and Tall, B.D. (1996). Uptake pathways of clinical isolates of *Proteus mirabilis* into human epithelial cell lines. *Microb. Pathog.* **21**, 1–16.

Ondraczek, R., Hobbie, S. and Braun, V. (1992). *In vitro* activation of the *Serratia marcescens* haemolysin through modification and complementation. *J. Bacteriol.* **174**, 5086–94.

Palmer, K.L. and Munson, R.S., Jr (1995). Cloning and characterization of the genes encoding the haemolysin of *Haemophilus ducreyi*. *Mol. Microbiol.* **18**, 821–30.

Papini, E., Bugnoli, M., De Bernard, M., Figura, N., Rappuoli, R. and Montecucco, C. (1993). Bafilomycin A1 inhibits *Helicobacter pylori*-induced vacuolization of HeLa cells. *Mol. Microbiol.* **7**, 323–7.

Papini, E., Satin, B., Bucci, C., de Bernard, M., Telford, J.L., Manetti, R., Rappuoli, R., Zerial, M. and Montecucco, C. (1997). The small GTP binding protein rab7 is essential for cellular vacuolation induced by *Helicobacter pylori* cytotoxin. *EMBO J.* **16**, 15–24.

Peerbooms, P.G.H., Verweij, A.M.J. and MacLaren, D.M. (1984). Vero cell invasiveness of *Proteus mirabilis*. *Infect. Immun.* **43**, 1068–71.

Poole, K. and Braun, V. (1988). Influence of growth temperature and lipopolysaccharide on haemolytic activity of *Serratia marcescens*. *J. Bacteriol.* **170**, 5146–52.

Poole, K., Schiebel, E. and Braun, V. (1988). Molecular characterization of the haemolysin determinant of *Serratia marcescens*. *J. Bacteriol.* **170**, 3177–88.

Renauld-Mongénie, G., Cornette, J., Mielcarek, N., Menozzi, F.D. and Locht, C. (1996). Distinct roles of the *N*-terminal and *C*-terminal precursor domains in the biogenesis of the *Bordetella pertussis* filamentous haemagglutinin. *J. Bacteriol.* **178**, 1053–60.

Ruan, Y. and Braun, V. (1990). Haemolysin as a marker for *Serratia*. *Arch. Microbiol.* **154**, 221–5.

Schäffer, S. Hantke, K, and Braun, V. (1985). Nucleotide sequence of the iron regulatory gene *fur*. *Mol. Gen. Genet.* **201**, 204–12.

Scheffer, J., König, W., Braun, V. and Goebel, W. (1988). Comparison of four haemolysin-producing organisms (*Escherichia coli*, *Serratia marcescens*, *Aeromonas hydrophila*, and *Listeria monocytogenes*) for release of inflammatory mediators from various cells. *J. Clin. Microbiol.* **26**, 544–51.

Schiebel, E. and Braun, V. (1989). Integration of the *Serratia marcescens* haemolysin into human erythrocyte membranes. *Mol. Microbiol.* **3**, 445–53.

Schiebel, E., Schwarz, H. and Braun, V. (1989). Subcellular location and unique secretion of the haemolysin of *Serratia marcescens*. *J. Biol. Chem.* **264**, 16311–20.

Schönherr, R., Tsolis, R., Focareta, T. and Braun, V. (1993). Amino acid replacements in the *Serratia marcescens* haemolysin ShlA define sites involved in activation and secretion. *Mol. Microbiol.* **9**, 1229–37.

Schönherr, R., Hilger, M., Broer, S., Benz, R. and Braun, V. (1994). Interaction of *Serratia marcescens* haemolysin (ShlA) with artificial and erythrocyte membranes: demonstration of the formation of aqueous multistate channels. *Eur. J. Biochem.* **223**, 655–63.

Schmitz, G. and Braun, V. (1985). Cell-bound and secreted proteases of *Serratia marcescens*. *J. Bacteriol.* **161**, 1002–9.

Senior, B.W. and Hughes, C. (1988). Production and properties of haemolysins from clinical isolates of the Proteae. *J. Med. Microbiol.* **25**, 17–25.

Sieben, S., Hertle, R., Gumpert, J. and Braun, V. (1998). The *Serratia marcescens* haemolysin is secreted but not activated by stable protoplast-type L-forms of *Proteus mirabilis*. *Arch. Microbiol.* **170**, 236–42.

Sturm, A.W. (1997). Iron and virulence of *Haemophilus ducreyi* in a primate model. *Sex. Transm. Dis.* **24**, 64–8.

Strauss, E.J., Ghori, N. and Falkow, S. (1997). An *Edwardsiella tarda* strain containing a mutation in a gene with homology to *shlB* and *hpmB* is defective for entry into epithelial cells in culture. *Infect. Immun.* **65**, 3924–32.

Swihart, K.G. and Welch, R.A. (1990). The HpmA haemolysin is more common than HlyA among *Proteus* isolates. *Infect. Immun.* **58**, 1853–860.

Uphoff, T.S. and Welch, R.A. (1990). Nucleotide sequencing of the *Proteus mirabilis* calcium-independent haemolysin genes (*hpmA* and *hpmB*) reveals sequence similarity with the *Serratia marcescens* haemolysin genes (*shlA* and *shlB*). *J. Bacteriol.* **172**, 1206–16.

von Graevnitz, A. and Rubin, S.J. (1980). *The Genus* Serratia. CRC Press, Boca Raton, FL.

Walev, I., Martin, E., Jonas, D., Mohamadzadeh, M., Müller-Klieser, W., Kunz, L. and Bhakdi, S. (1993). Staphylococcal alpha-toxin kills human keratinocytes by permeabilizing the plasma membrane for monovalent ions. *Infect. Immun.* **61**, 4972–9.

Walev, I., Reske, K., Palmer, M., Valeva, A. and Bhakdi, S. (1995). Potassium-inhibited processing of IL-1*b* in human monocytes. *EMBO J.* **14**, 1607–14.

Willems, R.J., Geuijen, C., van der Heide, H.G.J., Renauld, G., Bertin, P., van den Akker, W.M.R., Locht, C. and Mooi, F.R. (1994). Mutational analysis of the *Bordetella pertussis fim/fha* gene cluster: Identification of a gene with sequence similarities to haemolysin accessory genes involved in export of FHA. *Mol. Microbiol.* **11**, 337–47.

Yamamoto, T., Ariyoshi, A. and Amako, K. (1985). Fimbriae-mediated adherence of *Serratia marcescens* strain US5 to human urinary bladder surface. *Microbiol. Immunol.* **29**, 677–81.

18

The channel-forming toxin aerolysin

J. Thomas Buckley

INTRODUCTION

Many species produce water-soluble proteins that are capable of forming channels in cell membranes. Examples range from a myriad of bacterial toxins to eukaryotic channel formers, such as perforin and complement. The great majority of these proteins can be grouped together because they employ the same strategy to produce an insertion-competent state; they oligomerize to generate amphipathic β-barrels that can insert into the bilayer and form channels. The toxin aerolysin is among the first of these proteins to be extensively characterized and it can be considered a useful prototype of the group. Aerolysin is secreted as a soluble protein by bacteria in the genus *Aeromonas*. The toxin is capable of existing in solution at concentrations above 50 mg ml^{-1}; yet under the right conditions, it is transformed into a highly stable transmembrane channel. Most of the steps leading to channel formation by aerolysin are now relatively well understood although, as we will see, the final and perhaps the most critical step remains largely a subject for speculation.

Aeromonads are ubiquitous Gram-negative bacteria that have been isolated from water supplies throughout the world. Members of the genus are pathogenic to many aquatic poikilotherms, including amphibians, fish and reptiles, and they are an important cause of losses in aquaculture. What is more, *Aeromonas* has been increasingly associated with human gastrointestinal

disease (see Austin *et al.*, 1996, for a comprehensive discussion of the genus). Like many other Gram-negative bacteria, and in contrast to *Escherichia coli* and *Salmonella typhimurium*, the aeromonads secrete many proteins into their environment. One of these proteins is aerolysin, which has been identified in all species in the genus. This toxin is known to contribute to the pathogenesis of the human and animal pathogen *A. hydrophila* (Chakraborty *et al.*, 1987; Wong *et al.*, 1998), which is certainly the most studied member of the genus, if not the most common.

Aerolysin from *A. hydrophila* was discovered by Bernheimer and Avigad, who partially purified the toxin and described some of its properties (Bernheimer and Avigad, 1974; Bernheimer *et al.*, 1975). It was later purified to homogeneity by Buckley *et al.* (1981) and the gene was cloned and sequenced by Howard and Buckley (1986, 1987). These authors noted that the sequence of the translated protein contained no hydrophobic stretches and they predicted that the protein would have extensive β structure, suggesting that the toxin could be analogous to porins, the channel-forming proteins found in the outer membranes of bacteria. Later, the cloned *A. hydrophila* gene for aerolysin was expressed in *A. salmonicida*, leading to the secretion of large amounts of the precursor form of the protein, proaerolysin, greatly facilitating its purification (Buckley, 1990).

Proaerolysin is comprised of a single chain of 470 amino acids and it has a molecular mass of approximately 52 000 kDa. The amino acid composition of the protein is not unusual, except for the very high tryptophan content (there are 19 tryptophans in the molecule). The protoxin has little or no tendency to aggregate or denature and it will withstand long-term storage at −20°C and repeated freezing and thawing with no loss of activity. The availability of large amounts of the protoxin in a stable form has certainly contributed to the rate at which we have increased our understanding of its mechanism of action.

STRUCTURAL FEATURES OF THE TOXIN

Proaerolysin from *A. hydrophila*, cloned and expressed in *A. salmonicida* and purified from the culture supernatant, was the first member of the group of hydrophilic channel-forming toxins to have its structure solved (Parker *et al.*, 1994). The structure provided the most important clue to the way in which these proteins are transformed from a water-soluble to an insertion-competent state. Proaerolysin consists of a small compact lobe, here called domain 1, and a much larger elongated lobe, which the author has divided into three additional domains based on structure–function studies (Figure 18.1). There is a disulfide bridge between the two cysteines in the small lobe that appears to be essential for toxin activity, perhaps because it has a role in folding (unpublished). Another bridge, which is not essential, joins the two cysteines at the top of the large

lobe in domain 2. The structure of the protein bears no resemblance to the structures of the colicins and toxins such as diphtheria toxin. These channel-forming proteins contain a hydrophobic α-helix that is in the interior of the soluble form of the protein, protected from the aqueous environment, but that can become exposed in a pH-dependent step that generates an insertion-competent state. There are no analogous hydrophobic helices in proaerolysin, indeed there are no hydrophobic regions in the protein that are long enough to span a lipid bilayer. This is one indication that aerolysin must use a different strategy to generate an insertion-competent state. Since domains 3 and 4 of the *Aeromonas* protein largely consist of β-sheet, it was proposed that the toxin becomes insertion competent in the same way as bacterial porins, which oligomerize to form amphipathic β-barrels.

Aerolysin is a dimer in the crystal, stabilized by extensive interactions between the small lobes. The toxin is a dimer in solution (van der Goot *et al.*, 1993a), and recently it was found that the large lobe by itself, that is proaerolysin without the small lobe, is monomeric, pointing to the importance of the small lobe in dimer stabilization (Diep *et al.*, personal communication). Presumably, dimerization provides an advantage to the protein, perhaps increasing its solubility or reducing its sensitivity to destruction by proteases, and some evidence has been presented for the latter (van der Goot *et al.*, 1993a). It may also be true that dimerization is necessary to generate a structure that is efficiently recognized by the secretion machinery

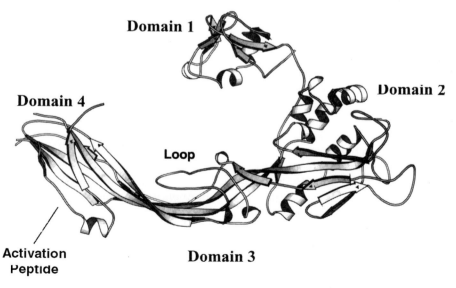

FIGURE 18.1 Crystal structure of the proaerolysin monomer, illustrating the location of the four functional domains as well as the activation peptide and the loop in domain 3.

responsible for transport across the outer membrane. In support of this, it was found that when the large lobe is expressed by itself it is much more poorly secreted (Diep *et al.*, personal communication).

SECRETION OF PROAEROLYSIN

Like most Gram-negative bacteria, and in contrast to *E. coli*, *Aeromonas* spp. secrete a number of proteins into their environment. In order for this to occur, the cytoplasmic membrane and the outer membrane of the bacteria must both be crossed. Bacteria have evolved several different mechanisms to accomplish this, but the most common one is a two-step pathway called the main terminal branch of the general secretory pathway (GSP). This pathway was first discovered by Pugsley during his studies of the secretion of pullulanase by *Klebsiella oxytoca* and it has recently been reviewed by him (Pugsley *et al.*, 1997). It is the GSP that aerolysin follows out of the cell; indeed, the study of aerolysin secretion has added significantly to our knowledge of the pathway. In the first step, which is cotranslational, the protein crosses the inner membrane in an unfolded state (Howard and Buckley, 1985a). The *sec* system is thought to be used for this step, based on the evidence that Pugsley has obtained in his studies of the secretion of pullulanase. The proaerolysin precursor expressed by the bacteria contains a typical *N*-terminal signal sequence that is required to direct its transport through the *sec* system. This signal is removed as the protein crosses the inner membrane and the resulting proaerolysin is released into the periplasmic space (Howard and Buckley, 1985a). The second step in secretion, by which secreted proteins such as proaerolysin cross the outer membrane, is much more poorly understood, although the process is clearly very different from inner membrane transport. One difference is in the state in which the protein crosses the membrane. Evidence obtained with proaerolysin and other secreted proteins has led to the surprising conclusion that these proteins fold into their native conformations in the periplasm before they leave the cell. Indeed, it has been shown that proaerolysin not only folds but also dimerizes before it crosses the outer membrane (Hardie *et al.*, 1995). Outer membrane transit appears to be a complex process based on the fact that a large number of auxiliary proteins is required [products of the *exe* genes in the case of *A. hydrophila* (Howard *et al.*, 1993; Jahagirdar and Howard, 1994)]. Details of the functions of these proteins have remained elusive; however, two of them appear to have adenosine triphosphate (ATP) binding sites, consistent with the observations that aerolysin

secretion from the periplasm is energy dependent, requiring both ATP and the electromotive force (Wong and Buckley, 1989; Letellier *et al.*, 1997). How energy is coupled to the secretory process has not been established.

Another of the proteins required for secretion, called ExeD in *A. hydrophila*, is thought to be located in the outer membrane (Howard *et al.*, 1996). The studies of analogous proteins by other groups have led to the suggestion that this protein forms a pore that somehow allows the passage of secreted proteins through the outer membrane (Daefler *et al.*, 1997; Bitter *et al.*, 1998). The secretory system must be able to allow passage of proteins that are destined for secretion, while blocking the exit of periplasmic and outer membrane proteins. This implies that secreted proteins have some signal that allows them to pass through the export machinery. No specific linear signalling information has been identified in any secreted protein, including aerolysin, although regions in exotoxin A and pullulanase have been reported to enable secretion of other proteins (Lu and Lory, 1996; Sauvonnet and Pugsley, 1996). It seems almost certain that if there is a general signal for secretion, it must be three-dimensional, generated as the molecule folds in the periplasm, rather than a simple linear signal sequence such as that used for inner membrane transit.

THE MECHANISM OF CHANNEL FORMATION

The process by which proaerolysin is converted from an inactive water-soluble protein to a transmembrane channel proceeds in several steps. These include activation of the protoxin, binding to specific receptors, oligomerization, and insertion into the membrane bilayer. Each of these steps is described in more detail below. Regions of the protein and individual amino acids that appear to be important in one or more steps are identified in Table 18.1.

Activation of the protoxin

Although aerolysin is far more active against cells that contain specific receptors, it is capable of forming channels in any lipid bilayer. This means that the toxin would represent a hazard to the bacteria that produce it and it is presumably for this reason, in order to avoid killing itself, that *Aeromonas* spp. secrete a totally inactive proform of the protein, which is called proaerolysin. Howard and Buckley observed some time ago that proaerolysin is larger than aerolysin by about 4.5 kDa and they showed that this is due to the

TABLE 18.1 Relationship between proaerolysin structure and function

Region of the toxin	Function	Residues identified
Small lobe (domain 1)	Dimer stabilization	Many residues
	Receptor binding	W45, I47, Y61, K66
Large lobe		
Domain 2	Receptor binding	W324, Y162
	Oligomerization	H107, H132, W371, W373
Domain 3	Oligomerization	Many residues comprising the β-strands or loop?
Domain 4	Activation peptide	Residues 425–470 (approx.)
	Insertion	G202

presence of an additional 40–50 amino acids at the C-terminus (Howard and Buckley, 1985b). Activation is accomplished by proteolytic nicking of proaerolysin in a highly mobile region that is not resolved in the crystal structure, but that is at the tip of the large lobe in domain 4 (Figure 18.1). This region contains cut sites for a variety of proteases, including trypsin, chymotrypsin and thermolysin (Garland and Buckley, 1988; van der Goot et al., 1992). There is also a cut site for furin and related proteases that are associated with mammalian cells, and cells that lack these enzymes appear to be less sensitive to aerolysin (van der Goot et al., personal communication). The rest of the molecule is quite resistant to proteolytic attack by many proteases, including trypsin.

It is easy to speculate that the way in which proaerolysin is activated in situ might depend on the location of the Aeromonas infection in the animal. Trypsin and chymotrypsin are obvious candidates for the activation of the toxin that is produced in the gut, and furin will activate any proaerolysin that reaches the target cell surface. Perhaps as an extra degree of insurance, the bacteria also secrete at least one protease that is capable of activating aerolysin (Garland and Buckley, 1988).

Once it is formed by proteolytic nicking, the distal C-terminal peptide dissociates from the rest of the toxin molecule spontaneously and it plays no role in the remaining steps of channel formation (van der Goot et al., 1994). In fact, if the peptide is prevented from leaving, the toxin is inactive. Interestingly, the situation appears to be somewhat different to the α-toxin from Clostridium septicum, which is an aerolysin homologue (see below). Rather than dissociating, the corresponding activation peptide appears to remain associated with the rest of the protein and to act as an intramolecular chaperone (Sellman and Tweten, 1997).

How the proteolytic nicking of aerolysin and dissociation of the peptide lead to activation has not been established. It is possible that activation lowers the stability of the dimer, promoting separation of the monomers, a step that must precede oligomerization. Alternatively, the presence of the peptide may physically prevent the monomers from dissociating, or its removal may lead to a structural change in the rest of the molecule that is essential for oligomerization to occur. In support of the latter possibility, using spectroscopic methods, the author's group has measured a small change in protein conformation accompanying activation (van der Goot et al., 1994).

Binding to target cells

Bernheimer and Avigad were the first to observe that different cells vary greatly in their sensitivity to aerolysin (Bernheimer et al., 1975) and Howard and Buckley (1982) were able to establish that sensitive mammalian erythrocytes contain protein receptors that bind the toxin with high affinity. Later, Gruber et al. (1994) provided evidence that incorporation of erythrocyte membrane components into planar lipid bilayers made them much more sensitive to aerolysin. This led to a method for the detection and successful purification and partial characterization of the erythrocyte aerolysin receptor (EAR) (Cowell et al., 1997). This surface molecule binds aerolysin and proaerolysin with a K_d lower than 10^{-9} M (unpublished observation). It is a 47-kDa glycoprotein, which appears to be a member of a small family of proteins that are involved in adenosine diphosphate (ADP)-ribosylation reactions, based on the homology of its N-terminal amino acid sequence with these proteins. However, its function as an enzyme appears to have nothing to do with its role in aerolysin binding. Membrane proteins from a number of other cells that bind aerolysin with similar affinity have since been identified (Nelson et al., 1997; Diep et al., 1998) and several of them have been further characterized. These include Thy-1, the major surface glycoprotein of rodent T-lymphocytes and mammalian neurons, contactin, which is found in the brain, and the variant surface glycoprotein (VSG) of trypanosomes. Remarkably, all of the aerolysin-binding proteins identified so far have one other common property: they are all anchored to the cytoplasmic membrane by glycosylphosphatidylinositol (GPI) anchors. These anchors consist of a core of several mannose residues attached to the C-terminus of the protein via an ethanolamine phosphate. The core may contain additional sugar residues, depending on the species and the cell type. The other end of the short sugar chain is attached to a phosphatidylinositol molecule which serves to anchor the entire assembly in the membrane (Englund, 1993). Proteins destined to be attached in this way are transcribed with a C-terminal signal sequence, which is removed in the endoplasmic reticulum and replaced with the GPI moiety. Although proteins that contain GPI anchors are generally poorly characterized, they have attracted increasing attention as it has become apparent that many of them are involved in cell–cell communication. More than 100 have been identified to date. How the GPI anchor affects their functions in the cell, if at all, has not been established.

The evidence that GPI-anchored proteins are aerolysin receptors *in situ* is quite compelling. Cells that have been treated with phosphatidylinositol-specific phospholipase C, which removes these proteins from the cell surface, are less sensitive to aerolysin, as are cells that have lost the ability to display GPI-anchored proteins on their surfaces (Nelson *et al.*, 1997). For example, the sensitivity of mouse T-lymphoma EL4 cells, which are killed by 1-h exposure to 10^{-11} M aerolysin, is four orders of magnitude higher than the sensitivity of the same cell line with a mutation in one of the genes required for GPI-anchor synthesis. Similarly, liposomes containing incorporated Thy-1 or EAR are three to four orders of magnitude more sensitive to aerolysin than liposomes that lack receptors (Diep *et al.*, 1998).

The proteins that have so far been shown to bind aerolysin appear to be structurally and functionally unrelated to each other. Aside from their ability to bind the toxin, the only other property that they share is their GPI anchors and this led us to suppose that the anchor itself is an important determinant of binding. This was confirmed by the results of an experiment in which deoxyribonucleic acid (DNA) for a GPI signal sequence was fused to the gene encoding a protein that normally does not bind aerolysin. When the construct was expressed in Chinese hamster ovary (CHO) cells, the protein appeared on the surface of the cells attached by a GPI anchor. It was shown that this form bound aerolysin, indicating that the anchor provided the necessary recognition signal for the toxin (Diep *et al.*, 1998).

Recently, it was shown that an aerolysin homologue, the α-toxin of *Clostridium septicum*, also binds GPI-anchored proteins (Gordon *et al.*, in preparation). Since the *C. septicum* protein does not have a small lobe (see below), this is a clue that the receptor-binding site in aerolysin is likely to be in the large lobe of the toxin. Interestingly, the two toxins do not seem to have the same relative affinities for different GPI-anchored proteins, indicating that their binding sites are not identical. Thus, although both toxins bind to contactin, only the *C. septicum* toxin appears to bind to the GPI-anchored folate receptor, and aerolysin binds far more strongly to Thy-1 (Gordon *et al.*, in preparation).

Perhaps the most important function of the toxin receptor is to bring aerolysin to the cell surface, thereby greatly increasing its effective concentration and promoting oligomerization. However, the special properties of GPI-anchored proteins may make them particularly well suited to be receptors for oligomeric channel-forming toxins for other reasons as well. Not only are they situated on the surfaces of all mammalian cells, but it seems likely that they have virtually unrestricted lateral mobility in the plane of the membrane. This is because they have no internal domains and presumably cannot be anchored to the cytoskeleton. This is an important property as aerolysin and the other members of this group of toxins must oligomerize by lateral movement before they insert into the membrane. Presumably, once it has bound the toxin diffuses in the plane of the membrane with the receptor, recruiting other aerolysin molecules to form the oligomer. The proximity of the anchor to the lipid bilayer may also be a feature that promotes insertion, although this will have to be established. Finally, it is even conceivable that the receptor somehow participates directly in the insertion process, perhaps by modifying the properties of the bilayer.

Oligomerization

Proaerolysin can bind to GPI-anchored proteins as well as aerolysin (to be published elsewhere), but activation of the protoxin is absolutely essential for the next step in channel formation, which is oligomerization. This is a concentration-dependent process that can even occur in solution. However, it is only appreciable at toxin concentrations higher than approx. 10 μM and most of the oligomers that are formed in solution rapidly aggregate and precipitate. In the absence of a receptor, only those oligomers that happen to be situated close to a bilayer when they are formed will insert and produce channels. In contrast, when the toxin can bind to a receptor like EAR or Thy-1 on a lipid bilayer, oligomerization and channel formation will occur at far lower toxin concentrations (lower than 0.1 nM).

The aerolysin oligomer is an extremely stable structure, easily visible after sodium dodecyl sulfate–polyacrylamide gel electrophoresis (SDS–PAGE) and not dissociated by weak acid or base, urea, detergents or elevated temperature. The complex is nevertheless noncovalent, as it can be dissociated by treatment with formic acid (unpublished results). The strength of the association is evidence of a co-operative interaction between a large number of amino acids in the oligomer. In support of this, image analysis of two-dimensional crystals of the oligomer has indicated that it is formed by seven aerolysin monomers (Wilmsen *et al.*, 1992). The heptameric symmetry has been confirmed by a direct measurement of the mass of the oligomer by MALDI-TOF mass spectroscopy, another testament to its stability (Moniatte *et al.*, 1996). Fitting the three-dimensional structure of the monomer into the images obtained from the two-dimensional crystals to the suggestion that it is the *β*-sheets in domain 3 of the monomer that interact to form the oligomer and that they generate a large amphipathic barrel consisting of an external hydrophobic surface and an interior which is hydrophilic (Wilmsen *et al.*, 1992). Formation of the barrel transforms the protein from the water-soluble state because the barrel can somehow insert into the membrane, forming a channel of fixed size. The author's group has not been able to produce working crystals of the aerolysin oligomer, so its three-dimensional structure is not available. However, the structure of the oligomer of *Staphylococcus aureus* α-toxin, which is a very similar protein, has been solved and found to be a heptamer containing an amphipathic barrel (see below).

Structure–function studies have provided some information about the mechanism of channel formation, although a great many details are still missing. Since the protein is a dimer in solution and the oligomer is a heptamer, it is difficult to imagine a model for oligomerization that does not require the monomers in the dimer to separate. What is more, it has been shown that if the monomers are locked together by a genetically engineered disulfide bridge, oligomerization is blocked, implying that dimer dissociation is necessary (Hardie *et al.*, 1995). Nevertheless, the author has not been able to detect the free monomer under any conditions, nor to measure dissociation of the dimer during oligomerization using any of the methods tried.

Using site-directed mutagenesis, it was found that at least two regions in domain 2 of the protein also play an important role in oligomerization (Table 18.1). One of these regions contains histidine 132, which must be protonated for oligomerization to occur (Buckley *et al.*, 1995), and the other is a tryptophan-rich region, which has a profound influence on the rate of oligomerization (van der Goot *et al.*, 1993b). For example, changing tryptophan 371 to leucine causes oligomerization to occur at much lower toxin concentrations. How domain 2 is involved in oligomerization is not obvious; however, it is possible that it affects the stability of the dimer, or that it is involved in any movement of the small lobe relative to the large lobe that might be necessary for the formation of stable oligomers.

The problem of insertion

The generation of an amphipathic barrel explains how aerolysin can be transformed from a soluble protein to an oligomer capable of spanning a lipid bilayer. However, it does not explain how the process of insertion actually occurs and the penetration of the bilayer by the oligomer is by far the most poorly understood step in channel formation. There are at least two intriguing puzzles associated with insertion. The first has to do with the orientation of the oligomer once it has formed on the surface of the cell. Based on our modelling of the structure of the oligomer and the evidence, the binding site for the receptor is located in domains 1 and 2, the amphipathic barrel is likely to be facing away from the lipid bilayer, opposite to the direction one would expect would be required for insertion. If this is the case, then somehow during the insertion process the orientation must be reversed, implying that the oligomer must be released from the receptor. However, measurements of oligomer binding to Thy-1 using surface plasmon resonance (Mackenzie and Buckley, unpublished) indicate that the oligomer binds to the receptor essentially irreversibly. This implies that some structural change that lowers the oligomer's affinity must precede or accompany insertion.

The second puzzle is one that applies to all proteins like aerolysin. Why do they insert into the membrane? What are the details of the steps that enable a (very large) amphipathic barrel to enter the lipid bilayer? The outside of the barrel is hydrophobic, yet somehow it manages to get through the hydrophilic outside surface of the membrane which is formed by the phospholipid polar head groups and water of hydration in order to bury itself in the hydrophobic interior. Presumably, some property of the oligomer must locally destabilize the bilayer, perhaps by causing the formation of a non-bilayer structure; however, there is as yet no evidence that this is what occurs.

Properties of the aerolysin channel

Aerolysin forms extraordinarily stable, homogeneous channels in planar lipid bilayers that display slight anion selectivity. The channels remain open between −70 and +70 mV, but beyond this range they undergo voltage-dependent closing (Wilmsen *et al.*, 1990).

Estimates of channel diameter vary somewhat, depending on the measuring procedure used. Image analysis of the two-dimensional crystalline arrays formed by the oligomer produced a value of 1.7 nm (Wilmsen *et al.*, 1992), while measurements of the release of small molecules from channels formed in erythrocytes yielded approx. 3 nm (Howard and Buckley, 1982) and scanning microphotolysis analysis of ion transport through aerolysin channels in red cells gave a value nearer to 4 nm (Tschodrich-Rotter *et al.*, 1996).

Consequences of binding and channel formation

When aerolysin forms channels in mammalian erythrocytes, osmotic swelling and cell lysis are the result, as the cells do not have the capacity to repair the damage to the membrane. However, this is not always the case with nucleated cells. Lymphocytes react differently to aerolysin depending on toxin concentration. At high concentrations, the cells die very quickly, presumably because the number of channels formed causes irreparable membrane damage or unacceptable changes in intracellular concentrations of critical small molecules and ions. However, at low concentrations of toxin (in the range of 10^{-10} M) programmed cell death or apoptosis results (Nelson *et al.*, in preparation). Similar observations have been made for several other channel-forming proteins (Mangan *et al.*, 1991; Jonas *et al.*, 1994). Why the cells respond in this way is not understood. In the case of aerolysin, one possibility is that the apoptotic signal is transmitted by aerolysin binding to GPI-anchored proteins such as Thy-1. Clustering of Thy-1, which would be caused by aerolysin oligomerization, has been shown to trigger apoptosis in T-lymphocytes (Hueber *et al.*, 1994). Recently, some experiments have been carried out that appear to rule out this possibility (Nelson *et al.*, in preparation). Instead, it seems likely that changes in cellular potassium or calcium due to the formation of a small number of channels at low toxin concentrations lead to the programmed cell death response (Duke *et al.*, 1994).

HOMOLOGUES AND ANALOGUES OF AEROLYSIN

The α-toxin of *Clostridium septicum*

It is generally true that the channel-forming toxins are a diverse group of proteins. There is only one subgroup which is large, consisting of the closely related 'oxygen-labile' toxins produced by a number of Gram-

positive bacteria. Among the many other members of the group, sequence similarities are commonly observed only between toxins of the same bacterial families. It is therefore remarkable that aerolysin has two relatives from species that are widely separated phylogenetically. One of these is the α-cytolysin of the Gram-positive bacterium *Clostridium septicum*, which is described elsewhere in this volume. Not only is this toxin homologous to aerolysin throughout its sequence (Ballard *et al.*, 1995), but functionally it is very similar to the *Aeromonas* protein. Like aerolysin it must be activated by proteolytic nicking near its C-terminus, and this is followed by the formation of extremely stable oligomers (Ballard *et al.*, 1993) which also appear to be heptameric, although this has not been confirmed. Many of the amino acid residues that are known to be involved in various steps of channel formation by aerolysin are conserved in this α-toxin (Ballard *et al.*, 1995). More remarkably, the *C. septicum* α-toxin also binds to GPI-anchored proteins on sensitive cells (see above).

Although aerolysin and α-toxin are obviously related proteins, there are several differences between the toxins, including the difference in the role of the activation peptide, which was pointed out above. For example, although the sequences of the two proteins are homologous, and although many structurally or functionally important residues are shared, only about 40% of the amino acids are identical. This is probably an indication that they diverged many years ago. A more important difference is that proaerolysin is nearly 100 amino acids longer than α-toxin. The extra amino acids are at the N-terminus of the *Aeromonas* toxin and they are responsible for the formation of domain 1 or the small lobe of the toxin. Thus, although the structure of *C. septicum* toxin has not yet been solved, it appears to be comprised of a single domain corresponding to the large lobe of the *Aeromonas* toxin. Several differences which have been observed between α-toxin and aerolysin may be due to the fact that only the latter has the small lobe. For example, proaerolysin is a dimer in solution (van der Goot *et al.*, 1993a), whereas α-toxin appears to be monomeric. The ability of the *Aeromonas* protein to dimerize may be conferred by domain 1, which forms extensive interactions with the other domain 1 in the dimer (Parker *et al.*, 1994). There is evidence that the small lobe may also contribute to the activity of aerolysin, either by increasing its ability to bind to receptors or by improving its ability to form channels. Recently, the author's group has expressed and purified a hybrid toxin containing the small lobe of aerolysin fused to α-toxin (Diep *et al.*, personal communication). It is correctly activated by proteolytic nicking, and it forms oligomers in the same way as both parents. Strikingly, it is more than 100 times more active than α-toxin against human erythrocytes; in fact, it is slightly more active than aerolysin against these cells.

Enterolobin

The existence of an aerolysin homologue in another bacterium is not astonishing, in spite of the fact that one species is a Gram-positive and the other a Gram-negative bacterium. However, aerolysin has another homologue from a family that is even more remote in evolution. This is the plant protein enterolobin that is found in the seeds of trees in the family *Enterolobium*. This cytolytic protein is an insecticide and may serve a defensive role in the plant. Recently, the entire protein has been sequenced by Sousa's group. The protein is about the same size as aerolysin and it shares 40% sequence identity with the bacterial toxin throughout its length (Fontes *et al.*, 1997). Several amino acids known to be important in aerolysin are found in similar positions in enterolobin, implying that they have a similar function. In contrast to proaerolysin, enterolobin does not seem to require activation, but like aerolysin itself, the purified form of the plant protein appears to form stable oligomeric structures that may be responsible for channel formation (Sousa, personal communication).

Pertussis toxin

There is a yet another toxin that shares some structural similarity with aerolysin. It was recently discovered that the fold of the small lobe of aerolysin is almost identical to a fold in the S2 and S3 subunits of pertussis toxin, which is produced by *Bordetella pertussis* (Rossjohn *et al.*, 1997). The common fold, which we have termed the APT domain, consists of two three-stranded anti-parallel β-sheets that encircle a pair of helices. This fold bears some similarity to the C-type lectins and Link modules that are involved in carbohydrate binding, implying a similar function in both aerolysin and pertussis toxin. In support of this, it was found that replacing any of the conserved surface residues in aerolysin (Table 18.1) decreases apparent binding to receptors such as Thy-1 and dramatically lowers the activity of the toxin. Indeed, the evidence obtained with both the large and small lobes of aerolysin, expressed and purified separately, indicates that both can bind to GPI-anchored proteins (Diep *et al.*, personal communication). This seems like an extraordinary coincidence, but it appears to account at least partially for the fact that aerolysin is far more active against erythrocytes and other cells than the *C. septicum* α-toxin, which does not contain a small lobe.

Staphylococcus aureus α-toxin

Although the *S. aureus* α-toxin bears little obvious structural or sequence similarity to aerolysin, this well-studied protein forms channels in a way that is completely analogous to the *Aeromonas* toxin. The structure of crystalline α-toxin oligomer has recently been solved (Song *et al.*, 1996). Like the aerolysin oligomer, the α-toxin oligomer is heptameric; however, there is an important difference between its structure and the structure predicted for the aerolysin oligomer, which was based on the structure of the monomer and image analysis of two-dimensional crystals of the oligomer. The barrel of the α-toxin oligomer is formed by a single loop contributed by each of the seven monomers, whereas it was suggested that the aerolysin heptamer is formed by oligomerization of the multi-stranded sheet in domain 3 of the protein (Parker *et al.*, 1994). Interestingly, aerolysin also contains a loop, formed by residues approximately 230–270 (Figure 18.1), which bears some structural similarity to the barrel-forming loop of α-toxin. We have shown that this loop must move for aerolysin oligomerization to occur (Rossjohn *et al.*, 1998), but we have not yet been able to determine whether it is the loop or the sheet in domain 3 that actually forms the amphipathic barrel. Additional work is underway to help to distinguish between these two options.

APPLICATIONS OF AEROLYSIN

Detection and study of GPI-anchored proteins

Aerolysin represents a new tool that can be used to detect GPI-anchored proteins and to study their function. One example is the first aerolysin-binding protein to be identified, EAR from mammalian erythrocytes, a previously unknown protein that was identified by a sandwich Western blotting procedure. This method can be very sensitive; for example, less than 1 ng of Thy-1 can be detected after SDS–PAGE, far less than can be measured by any other method. However, some GPI-anchored proteins cannot be detected in this way. Whether this is because some folding information necessary for binding is irreversibly lost upon SDS–PAGE or because they do not bind the toxin under any conditions has not been established. The author is currently taking another approach, using immobilized aerolysin, to determine whether one can 'fish' for novel GPI-anchored proteins.

Diagnosis of paroxysmal nocturnal haemoglobinuria

Paroxysmal nocturnal haemoglobinuria (PNH) is a stem-cell disorder in which circulating cells lose their ability to display GPI-anchored proteins on their surfaces (Rosse, 1997). Usually this is due to a spontaneous mutation in the PIG-A gene, which encodes one of the enzymes in the synthetic pathway for the GPI anchor (Takeda *et al.*, 1993). Although the disease is not common, it is scientifically of great interest and it has attracted a great deal of attention in the past few years. Clinically, it must be tested for and ruled out as part of the diagnosis of a number of other much more common blood disorders. The best diagnostic test for PNH has been the measurement of CD59, a GPI-anchored protein, using flow cytometry. However, this test is not sensitive and the equipment required is not widely available. Simpler tests, which rely on the fact that affected cells are more sensitive than normal cells to homologous complement, are even less sensitive and less specific than flow cytometry (Hall and Rosse, 1996). Because PNH cells do not display GPI-anchored proteins, they should be much less sensitive to aerolysin than normal cells. In the past few months it has been shown that this is the case for erythrocytes, T- and B-lymphocytes, and granulocytes and several assays have been developed which are more sensitive, more specific and less expensive than the diagnostic tests that have been available (Brodsky *et al.*, personal communication). The simplest of these assays is a simple haemolytic titration in which the rate of haemolysis of blood from patients with PNH is compared to the rate of normal haemolysis using a plate reader. This assay will reliably detect samples from patients who have fewer than 10% of their cells affected.

Isolation of intracellular parasites

It goes without saying that in order to form channels in cell membranes aerolysin must bind to the cell surface, form an amphipathic oligomer and insert into the bilayer. Intracellular parasites such as Leishmania, Theileria and trypanosomes are covered with a dense glycoprotein coating or glycocalyx which effectively shields their membranes from the toxin, because the protein either cannot bind, cannot oligomerize or cannot insert. As a result, these organisms are completely resistant to the toxin and this forms the basis of a method to purify them from tissues or from blood. In the first step, aerolysin is added to samples to a concentration that disrupts the host cells. The parasites are then separated from the cell fragments by a simple differential or gradient centrifugation procedure (Pearson *et al.*, 1982).

Biotechnological applications

Biotechnological applications of the α-toxin from *S. aureus* as a component of biosensors, or in drug delivery, have been explored by Hagan Bayley and his colleagues (Bayley, 1994, 1997a, 1997b; Chang *et al.*, 1995; Braha *et al.*, 1997). Aerolysin can be used for similar applications, and the *Aeromonas* has at least three advantages over the *S. aureus* toxin. Perhaps most importantly, aerolysin can be obtained as a proform which has no tendency to aggregate at any concentration. When needed, proaerolysin can be quickly and inexpensively converted to the active form of the toxin by treatment with trypsin or other proteases. The best method is to use immobilized trypsin, which can be rapidly and easily removed after treatment. Another advantage of aerolysin is that its receptor is known. This raises the possibility that the toxin can be targeted to specific cells or specific membrane surfaces. A third advantage that aerolysin has over *S. aureus* α-toxin as a biotechnological tool is the unusual stability of the oligomeric form of the protein (see above), which might be important in designing stable biosensors.

ACKNOWLEDGEMENTS

I would like to acknowledge the contributions of my many collaborators to the work described in this review. I am especially indebted to Franc Pattus, Mike Parker and Gisou van der Goot. Many students, postdoctoral fellows and research assistants have also been involved in these studies, including Dzung Diep, Tracy Lawrence and Kim Nelson, who are working on current projects. This work has been supported by the Natural Sciences and Engineering Research Council of Canada, and by the British Columbia Health Care Research Foundation.

REFERENCES

Austin, B., Altweg, M., Gosling, P.J. and Joseph, S.W. (eds) (1996). *The Genus Aeromonas*. John Wiley, Chichester.

Ballard, J., Sokolov, Y., Yuan, W.L., Kagan, B.L. and Tweten, R.K. (1993). Activation and mechanism of *Clostridium septicum* α-toxin. *Mol. Microbiol.* **10**, 627–34.

Ballard, J., Crabtree, J., Roe, B.A. and Tweten, R.K. (1995). The primary structure of *Clostridium septicum* α-toxin exhibits similarity with that of *Aeromonas hydrophila* aerolysin. *Infect. Immun.* **63**, 340–4.

Bayley, H. (1994). Triggers and switches in a self-assembling pore-forming protein. *J Cell. Biochem.* **56**, 177–82.

Bayley, H. (1997a). Building doors into cells. *Sci. Am.* **277**, 62–77.

Bayley, H. (1997b). Toxin structure: part of a hole? *Curr. Biol.* **7**, R763–7.

Bernheimer, A.W. and Avigad, L.S. (1974). Partial characterization of aerolysin, a lytic exotoxin from *Aeromonas hydrophila*. *Infect. Immun.* **9**, 1016–21.

Bernheimer, A.W., Avigad, L.S. and Avigad, G. (1975). Interactions between aerolysin, erythrocytes, and erythrocyte membranes. *Infect. Immun.* **11**, 1312–19.

Bitter, W., Koster, M., Latijnhouwers, M., de Cock, H. and Tommassen, J. (1998). Formation of oligomeric rings by XcpQ and PilQ, which are involved in protein transport across the outer membrane of *Pseudomonas aeruginosa*. *Mol. Microbiol.* **27**, 209–19.

Braha, O., Walker, B., Cheley, S., Kasianowicz, J.J., Song, L., Gouaux, J.E. and Bayley, H. (1997). Designed protein pores as components for biosensors. *Chem. Biol.* **4**, 497–505.

Buckley, J.T. (1990). Purification of cloned proaerolysin released by a low protease mutant of *Aeromonas salmonicida*. *Biochem. Cell Biol.* **68**, 221–4.

Buckley, J.T., Halasa, L.N., Lund, K.D. and MacIntyre, S. (1981). Purification and some properties of the haemolytic toxin aerolysin. *Can. J. Biochem.* **59**, 430–5.

Buckley, J.T., Wilmsen, H.U., Lesieur, C., Schulze, A., Pattus, F., Parker, M.W. and van der Goot, F.G. (1995). Protonation of histidine-132 promotes oligomerization of the channel-forming toxin aerolysin. *Biochemistry* **34**, 16450–5.

Chakraborty, T., Huhle, B., Hof, H., Bergbauer, H. and Goebel, W. (1987). Marker exchange mutagenesis of the aerolysin determinant in *Aeromonas hydrophila* demonstrates the role of aerolysin in *A. hydrophila*- associated systemic infections. *Infect. Immun.* **55**, 2274–80.

Chang, C.Y., Niblack, B., Walker, B. and Bayley, H. (1995). A photogenerated pore-forming protein. *Chem. Biol.* **2**, 391–400.

Cowell, S., Aschauer, W., Gruber, H.J., Nelson, K.L. and Buckley, J.T. (1997). The erythrocyte receptor for the channel-forming toxin aerolysin is a novel glycosylphosphatidylinositol-anchored protein. *Mol. Microbiol.* **25**, 343–50.

Daefler, S., Guilvout, I., Hardie, K.R., Pugsley, A.P. and Russel, M. (1997). The *C*-terminal domain of the secretin PulD contains the binding site for its cognate chaperone, PulS, and confers PulS dependence on pIVfl function. *Mol. Microbiol.* **24**, 465–75.

Diep, D.B., Nelson, K.L., Raja, S.M., Pleshak, E.N. and Buckley, J.T. (1998). Glycosylphosphatidylinositol anchors of membrane glycoproteins are binding determinants for the channel-forming toxin aerolysin. *J. Biol. Chem.* **273**, 2355–60.

Duke, R.C., Witter, R.Z., Nash, P.B., Young, J.D. and Ojcius, D.M. (1994). Cytolysis mediated by ionophores and pore-forming agents: role of intracellular calcium in apoptosis. *FASEB J.* **8**, 237–46.

Englund, P.T. (1993). The structure and biosynthesis of glycosyl phosphatidylinositol protein anchors. *Annu. Rev. Biochem.* **62**, 121–38.

Fontes, W., Sousa, M.V., Aragao, J.B. and Morhy, L. (1997). Determination of the amino acid sequence of the plant cytolysin enterolobin. *Arch. Biochem. Biophys.* **347**, 201–7.

Garland, W.J. and Buckley, J.T. (1988). The cytolytic toxin aerolysin must aggregate to disrupt erythrocytes, and aggregation is stimulated by human glycophorin. *Infect. Immun.* **56**, 1249–53.

Gruber, H.J., Wilmsen, H.U., Cowell, S., Schindler, H. and Buckley, J.T. (1994). Partial purification of the rat erythrocyte receptor for the channel-forming toxin aerolysin and reconstitution into planar lipid bilayers. *Mol. Microbiol.* **14**, 1093–101.

Hall, S.E. and Rosse, W.F. (1996). The use of monoclonal antibodies and flow cytometry in the diagnosis of paroxysmal nocturnal haemoglobinuria. *Blood* **87**, 5332–40.

Hardie, K.R., Schulze, A., Parker, M.W. and Buckley, J.T. (1995). *Vibrio* spp. secrete proaerolysin as a folded dimer without the need for disulphide bond formation. *Mol. Microbiol.* **17**, 1035–44.

Howard, S.P. and Buckley, J.T. (1982). Membrane glycoprotein receptor and hole-forming properties of a cytolytic protein toxin. *Biochemistry* **21**, 1662–7.

Howard, S.P. and Buckley, J.T. (1985a). Protein export by a Gram-negative bacterium: production of aerolysin by *Aeromonas hydrophila*. *J. Bacteriol.* **161**, 1118–24.

Howard, S.P. and Buckley, J.T. (1985b). Activation of the hole-forming toxin aerolysin by extracellular processing. *J. Bacteriol.* **163**, 336–40.

Howard, S.P. and Buckley, J.T. (1986). Molecular cloning and expression in *Escherichia coli* of the structural gene for the haemolytic toxin aerolysin from *Aeromonas hydrophila*. *Mol. Gen. Genet.* **204**, 289–95.

Howard, S.P. and Buckley, J.T. (1987). Nucleotide sequence of the gene for the haemolytic toxin aerolysin. *J. Bacteriol.* **169**, 2869–71.

Howard, S. P., Critch, J. and Bedi, A. (1993). Isolation and analysis of eight *exe* genes and their involvement in extracellular protein secretion and outer membrane assembly in *Aeromonas hydrophila*. *J. Bacteriol.* **175**, 6695–703.

Howard, S.P., Meiklejohn, H.G., Shivak, D. and Jahagirdar, R. (1996). A TonB-like protein and a novel membrane protein containing an ATP- binding cassette function together in exotoxin secretion. *Mol. Microbiol.* **22**, 595–604.

Hueber, A.O., Raposo, G., Pierres, M. and He, H.T. (1994). Thy-1 triggers mouse thymocyte apoptosis through a bcl-2-resistant mechanism. *J Exp. Med.* **179**, 785–96.

Jahagirdar, R. and Howard, S.P. (1994). Isolation and characterization of a second *exe* operon required for extracellular protein secretion in *Aeromonas hydrophila*. *J. Bacteriol.* **176**, 6819–26.

Jonas, D., Walev, I., Berger, T., Liebetrau, M., Palmer, M. and Bhakdi, S. (1994). Novel path to apoptosis: small transmembrane pores created by staphylococcal α-toxin in T-lymphocytes evoke internucleosomal DNA degradation. *Infect. Immun.* **62**, 1304–12.

Letellier, L., Howard, S.P. and Buckley, J.T. (1997). Studies on the energetics of proaerolysin secretion across the outer membrane of *Aeromonas* species. Evidence for a requirement for both the protonmotive force and ATP. *J. Biol. Chem.* **272**, 11109–13.

Lu, H.M. and Lory, S. (1996). A specific targeting domain in mature exotoxin A is required for its extracellular secretion from *Pseudomonas aeruginosa*. *EMBO J.* **15**, 429–36.

Mangan, D.F., Taichman, N.S., Lally, E.T. and Wahl, S.M. (1991). Lethal effects of *Actinobacillus actinomycetemcomitans* leukotoxin on human T-lymphocytes. *Infect. Immun.* **59**, 3267–72.

Moniatte, M., van der Goot, F.G., Buckley, J.T., Pattus, F. and van Dorsselaer, A. (1996). Characterisation of the heptameric pore-forming complex of the *Aeromonas* toxin aerolysin using MALDI-TOF mass spectrometry. *FEBS Lett.* **384**, 269–72.

Nelson, K.L., Raja, S.M. and Buckley, J.T. (1997) The glycosylphosphatidylinositol-anchored surface glycoprotein Thy-1 is a receptor for the channel-forming toxin aerolysin. *J. Biol. Chem.* **272**, 12170–4.

Parker, M.W., Buckley, J.T., Postma, J.P., Tucker, A.D., Leonard, K., Pattus, F. and Tsernoglou, D. (1994). Structure of the *Aeromonas* toxin proaerolysin in its water-soluble and membrane-channel states. *Nature* **367**, 292–5.

Pearson, T.W., Saya, L.E., Howard, S.P. and Buckley, J.T. (1982). The use of aerolysin toxin as an aid for visualization of low numbers of African trypanosomes in whole blood. *Acta Trop.* **39**, 73–7.

Pugsley, A.P., Francetic, O., Possot, O.M., Sauvonnet, N. and Hardie, K.R. (1997). Recent progress and future directions in studies of the main terminal branch of the general secretory pathway in Gram-negative bacteria – a review. *Gene* **192**, 13–19.

Rosse, W.F. (1997). Paroxysmal nocturnal haemoglobinuria as a molecular disease. *Medicine* (*Baltimore*) **76**, 63–93.

Rossjohn, J., Buckley, J.T., Hazes, B., Murzin, A.G., Read, R.J. and Parker, M.W. (1997). Aerolysin and pertussis toxin share a common receptor-binding domain. *EMBO J.* **16**, 3426–34.

Rossjohn, J., Raja, S.M., Nelson, K.L., Feil, S.C., van der Goot, F.G., Parker, M.W. and Buckley, J.T. (1998). Movement of a loop in domain 3 of aerolysin is required for channel formation. *Biochemistry* **37**, 741–6.

Sauvonnet, N. and Pugsley, A.P. (1996). Identification of two regions of *Klebsiella oxytoca* pullulanase that together are capable of promoting β-lactamase secretion by the general secretory pathway. *Mol. Microbiol.* **22**, 1–7.

Seliman, B.R. and Tweten, R.K. (1997). The propeptide of *Clostridium septicum* α-toxin functions as an intramolecular chaperone and is a potent inhibitor of α-toxin-dependent cytolysis. *Mol. Microbiol.* **25**, 429–40.

Song, L., Hobaugh, M.R., Shustak, C., Cheley, S., Bayley, H. and Gouaux, J.E. (1996). Structure of staphylococcal α-haemolysin, a heptameric transmembrane pore. *Science* **274**, 1859–66.

Takeda, J., Miyata, T., Kawagoe, K., Iida, Y., Endo, Y., Fujita, T., Takahashi, M., Kitani, T. and Kinoshita, T. (1993). Deficiency of the GPI anchor caused by a somatic mutation of the PIG-A gene in paroxysmal nocturnal haemoglobinuria. *Cell* **73**, 703–11.

Tschodrich-Rotter, M., Kubitscheck, U., Ugochukwu, G., Buckley, J.T. and Peters, R. (1996). Optical single-channel analysis of the aerolysin pore in erythrocyte membranes. *Biophys. J.* **70**, 723–32.

van der Goot, F.G., Lakey, J., Pattus, F., Kay, C.M., Sorokine, O., Van Dorsselaer, A. and Buckley, J.T. (1992). Spectroscopic study of the activation and oligomerization of the channel-forming toxin aerolysin: identification of the site of proteolytic processing. *Biochemistry* **31**, 8566–70.

van der Goot, F.G., Ausio, J., Wong, K.R., Pattus, F. and Buckley, J.T. (1993a) Dimerization stabilizes the pore-forming toxin aerolysin in solution. *J. Biol. Chem.* **268**, 18272–9.

van der Goot, F.G., Pattus, F., Wong, K.R. and Buckley, J.T. (1993b). Oligomerization of the channel-forming toxin aerolysin precedes insertion into lipid bilayers. *Biochemistry* **32**, 2636–42.

van der Goot, F.G., Hardie, K.R., Parker, M.W. and Buckley, J.T. (1994). The *C*-terminal peptide produced upon proteolytic activation of the cytolytic toxin aerolysin is not involved in channel formation. *J. Biol. Chem.* **269**, 30496–501.

Wilmsen, H.U., Pattus, F. and Buckley, J.T. (1990). Aerolysin, a haemolysin from *Aeromonas hydrophila*, forms voltage-gated channels in planar lipid bilayers. *J. Membr. Biol.* **115**, 71–81.

Wilmsen, H.U., Pattus, F., Tichelar, W., Buckley, J.T. and Leonard, K. (1992). The aerolysin membrane channel is formed by hepatmerization of the monomer. *EMBO J.* **11**, 2457–63.

Wong, C.Y., Heuzenroeder, M.W. and Flower, R.L. (1998). Inactivation of two haemolytic toxin genes in *Aeromonas hydrophila* attenuates virulence in a suckling mouse model. *Microbiology* **144**, 291–8.

Wong, K.R. and Buckley, J.T. (1989). Proton motive force involved in protein transport across the outer membrane of *Aeromonas salmonicida*. *Science* **246**, 654–6.

19

Haemolysins of *Vibrio cholerae* and other *Vibrio* species

Sumio Shinoda

INTRODUCTION

Bacteria of the genus *Vibrio* are normal habitants of the aquatic environment, but the 11 species listed in Table 19.1 are believed to be human pathogens (Chakraborty *et al.*, 1977; Janda *et al.*, 1988). These species can be classified into two groups according to the types of diseases they cause: one group causes gastrointestinal infections and the other extraintestinal infections. The pathogenic species produce various virulence factors including enterotoxin, haemolysin, cytotoxin, protease, siderophore, adhesive factor and haemagglutinin, with haemolysin being the one most widely distributed in the pathogenic vibrios (Table 19.2). Although the most important virulence factor of *Vibrio cholerae* O1 is cholera enterotoxin (CT), a haemolysin (El Tor haemolysin) produced by biovar *eltor* of the O1

serogroup is also reported to cause diarrhoea, and a similar haemolysin is a major factor in some pathogenic strains of non-O1 serogroups. A thermostable direct haemolysin (TDH) produced by *V. parahaemolyticus* is also believed to be a major virulence component of the species. This chapter offers a review of recent information on haemolysins of pathogenic vibrios with special emphasis on the above two representatives.

VIBRIO CHOLERAE HAEMOLYSINS

Of the currently recognized pathogenic vibrios, the species *V. cholerae* is the most extensively studied because it is the causative agent of the often lethal disease, cholera. Cholera is characterized by copious uncontrolled purging of rice water stools leading to severe electrolyte depletion, dehydration, acidosis, shock and, if left untreated, to death. Characteristic signs and symptoms of severely dehydrated patients include an increase in pulse rate and a decrease in pulse volume, hypotension, an increase in respiratory rate accompanied by deep respiration, sunken eyes and cheeks, dry mucous membranes, decreased skin turgor, abdominal pain, a decrease in urine output and thirst (Harvey *et al.*, 1966).

Vibrio cholerae is serologically classified by the major surface O antigen. Until the emergence of the O139 serogroup, only the O1 serogroup was thought to be

TABLE 19.1 Pathogenic species of the genus *Vibrio*

Gastrointestinal diseases	Extraintestinal diseases
V. cholerae O1	*V. vulnificus*
V. cholerae O139	*V. damsela*
V. cholerae non-O1	*V. alginolyticus*
V. mimicus	*V. cincinnatiensis*
V. parahaemolyticus	*V. metschnikovii*
V. fluvialis	
V. furnissii	
V. hollisae	
V. metschnikovii	

TABLE 19.2 Haemolysins produced by vibrios

Producing vibrio	Name or abbreviation	Remarks
1. Haemolysins related to El Tor haemolysin		
V. cholerae O1 biotype eltor	El Tor haemolysin	65 kDa (64 864 Da)[a], two-step processing for mature protein, colloid osmotic haemolysis, RIL[b] positive
V. cholerae non-O1	NAG haemolysin	Indistinguishable from El Tor haemolysin (biologically, physicochemically, immunologically)
V. mimicus	VMH	63 kDa (65 972 Da), two-step processing for mature protein, colloid osmotic haemolysis, RIL positive, 76% homology with El Tor haemolysin gene
V. fluvialis		84 260 Da (calculated from amino acid sequence deduced from open reading frame including signal peptide), 57% homology
V. furnissii		with El Tor haemolysin gene
V. anguillarum		
2. Haemolysins related to Vp-TDH		
V. parahaemolyticus		
KP$^+$	Vp-TDH	Dimer of 21 kDa (18 496 Da) subunit, thermostable (heating at 100°C), colloid osmotic cell lysis, RIL positive, cardiotoxicity
KP$^-$ but pathogenic	Vp-TRH	Closely related to Vp-TDH but thermolabile, virulence factor in diarrhoeal cases due to KP$^-$ strains
V. hollisae	Vh-rTDH	Thermolabile
V. cholerae non-O1	NAG-rTDH	Thermostable
V. mimicus	Vm-rTDH	Thermostable
3. Other haemolysins		
V. vulnificus	VVH	50 kDa (50 851 Da), colloid osmotic cell lysis, temperature-independent binding to cholesterol (suspected binding site)
V. damsela	Damselysin	69 kDa, phospholipase D
V. metschnikovii		50 kDa, colloid osmotic cell lysis
V. parahaemolyticus	LDH	43 kDa (41 453 Da) and 45 kDa (42 794 Da), lecithin-dependent indirect haemolysis (phospholipase A$_2$/lisophospholipase)
	δ-VPH	Produced by transformed *E. coli* cell, but not found in *V. parahaemolyticus* culture
V. cholerae		10 451 Da, a product of *hlx* gene

[a]Rabbit ileal loop test (a test for eneterotoxic activity).
[b]Molecular weight deduced from SDS–PAGE and amino acid sequence in parentheses.

the aetiological agent of cholera. All strains that were identified as *V. cholerae* on the basis of biochemical tests but were negative for agglutination with O1 antiserum were referred to as non-O1 *V. cholerae*. The emergence of *V. cholerae* O139 Bengal as the second aetiological agent of cholera in October 1992 in the southern Indian coastal city of Madras (Albert *et al.*, 1993; Ramamurthy *et al.*, 1993) necessitated the redefinition of non-O1 *V. cholerae* to non-O1, non-O139 *V. cholerae*. The majority of non-O1, non-139 strains do not produce cholera enterotoxin (CT) and are not associated with epidemic diarrhoea. However, non-O1, non-O139 strains are occasionally isolated from cases of diarrhoea (usually associated with consumption of shellfish) and have been isolated from a variety of extraintestinal infections, including wounds, ear, sputum, urine and cerebrospinal fluid (Morris, 1990). The non-O1, non-139 serogroups are regularly found in estuarine environments and infections due to these strains commonly originate from this type of environment.

CT is the major toxin causing the severe watery diarrhoea produced by *V. cholerae* O1 and O139 strains (see Chapter 6). Zonula occludens toxin (Zot) (Fasano *et al.*, 1991) and accessory cholera toxin (Ace) (Trucksis *et al.*, 1993) are also enterotoxic factors of these two serogroups. *Vibrio cholerae* non-O1, non-O139 strains produce a 17 amino acid heat-stable enterotoxin (NAG-ST) (Takeda *et al.*, 1991), as well as haemolysins.

El Tor strains of *V. cholerae* produce a haemolysin (El Tor haemolysin) which has a lytic effect on Vero and other mammalian cells in culture (Honda and Finkelstein, 1979). Some strains of *V. cholerae* non-O1 produce a haemolysin (NAG haemolysin) indistinguishable biologically, physicochemically and antigenically from El Tor haemolysin (Yamamoto *et al.*, 1986). Haemolysins were purified from culture filtrates of *V. cholerae* O1 El Tor strain and a non-O1 strain and their properties were investigated. Their molecular sizes were about 60 kDa, and the amino acid compositions were very similar. Both haemolysins were neutralized to the same extent with antiserum against the homologous and heterologous haemolysins, and Ouchterlony double immunodiffusion tests gave a common precipitin line. El Tor haemolysin and NAG haemolysin can cause bloody fluid accumulation in ligated rabbit ileal loops, in contrast to the watery fluid accumulation induced by CT (Ichinose *et al.*, 1987). Thus, the haemolysin may play a role in the pathogenesis of gastroenteritis caused by *V. cholerae* strains.

Although the apparent molecular size of El Tor haemolysin is 65 kDa when secreted and detected in the culture supernatant of *V. cholerae* strains, the structural gene (*hlyA*) of the haemolysin encodes an 82-kDa polypeptide. Yamamoto *et al.* (1990a) demonstrated that the El Tor haemolysin is synthesized as an 82-kDa precursor form (prepro-HlyA) consisting of a signal peptide (25 amino acid residues) at the *N*-terminus, a pro-region (132 residues) and a mature region (584 residues) at the *C*-terminus. The mature form of the haemolysin is produced by a two-step process: during passage through the inner membrane the signal peptide is cleaved off, followed immediately by extracellular secretion in which the pro-region is lost. This two-step process is necessary to activate the haemolysin (Hall and Drasar, 1990; Yamamoto *et al.*, 1990a; Nagamune *et al.*, 1996). Recently, the pro-region of HlyA was reported by Nagamune *et al.* (1997) to function as an intramolecular chaperone, which suggests that it plays a role in the correct folding of mature El Tor haemolysin. To analyse the role of the pro-region, Nagamune *et al.* (1997) substituted the native *hlyA* gene with the pro-region-deleted *hlyA* gene (*hlyA*Δ*pro*). The haemolytic activity of the mutant organism was markedly decreased and the product of the *hlyA*Δ*pro* gene, secreted in the periplasm, was degraded. The sequences of the pro-region and a molecular chaperone, Hsp90, were similar and the purified pro-region peptide also facilitated rematuration of the denatured mature HlyA. These results support the suggested role of intramolecular chaperone for the pro-region.

El Tor haemolysin is thought to act as a pore-forming toxin through oligomer formation. The estimated size of the pore is 1.2–1.6 nm (Ikigai *et al.*, 1996). Cholesterol in the target cell membrane was reported to play an important role in the assembly of the toxin oligomers needed for pore formation (Ikigai *et al.*, 1996). Formation of channels in planar lipid phosophatidylcholine–cholesterol-bilayer membranes was also demonstrated (Ikigai *et al.*, 1997). Saha and Banerjee (1997) showed carbohydrate-mediated regulation of interaction of El Tor haemolysin with erythrocyte and phospholipid vesicles. They concluded that (a) El Tor haemolysin is a monomer with distinct

domains associated with specific binding to carbohydrates and interaction with lipids; (b) the pore-forming property depends solely on the protein–lipid interaction with no evidence of the involvement of sugars; and (c) specific sugars can down-regulate the ability of the haemolysin to form pores in lipid bilayers. Menzl *et al.* (1996) showed some preferential movement of anions over cations through the pores at neutral pH by measuring zero-current membrane potentials and suggested that the El Tor haemolysin channel is moderately anion selective. Zitzer *et al.* (1997) studied the mode of action of El Tor haemolysin against rabbit and human erythrocytes and demonstrated a similar mode with classical pore-forming toxins such as staphylococcal α-toxin and the aerolysin of *Aeromonas hydrophila*.

The *tdh* gene encoding TDH of *V. cholerae* non-O1 (NAG-TDH) has 98.6% homology with *tdh2* which is thought to be the actual gene producing TDH of *V. parahaemolyticus* (Vp-TDH), so the properties of NAG-TDH are quite similar to Vp-TDH (Baba *et al.*, 1991).

A haemolysin gene (*hlx*) which is different from both El Tor haemolysin and NAG-TDH has been cloned from *V. cholerae* O1 (Nagamune *et al.*, 1995). The gene encodes a polypeptide of 92 amino acid residues with a calculated molecular mass of 10 451 Da which shows haemolytic activity when expressed in *Escherichia coli*. The gene was detected in most classical- and El Tor-biovar *V. cholerae* O1 and *V. mimicus*, but not in *V. parahaemolyticus*. Although the *E. coli* transformant having the *hlx* gene showed haemolytic activity against erythrocytes from various animal sources, the contribution of this gene to pathogenic effects of the organisms is unclear.

A haemolysin having phospholipase C activity was recently purified from *V. cholerae* O139 strain and characterized (Pal *et al.*, 1997).

VIBRIO PARAHAEMOLYTICUS HAEMOLYSINS

Vibrio parahaemolyticus inhabits coastal and estuarine waters and is recognized as a causative agent of gastroenteritis following consumption of seafood. This vibrio was discovered by Fujino *et al.* (1953) when an outbreak of food poisoning occurred in the southern suburbs of Osaka, Japan, in October 1950. The outbreak was due to a small, half-dried sardine, *Engraulis japonica* Hottuyn, called 'shirasu' in Japanese.

The outstanding features of *V. parahaemolyticus* infections are severe abdominal pain, diarrhoea, nausea, vomiting, mild fever and headache; diarrhoea (frequently bloody stools) and abdominal pain being the main symptoms (Miwatani and Takeda, 1976). The mean incubation period is 6–12 h and diarrhoea or soft stools persist for 4–7 days. The mortality rate is very low, although the first outbreak reported by Fujino *et al.* (1953) resulted in high mortality (20 victims out of 272 patients).

Although *V. parahaemolyticus* is a natural inhabitant of sea water, as shown above, only some isolates are pathogenic to humans. Miyamoto *et al.* (1969) reported that the haemolytic characteristics of this vibrio on a special agar medium (Wagatsuma's medium; Wagatsuma, 1968) correlated well with human pathogenicity. The haemolytic property on this medium was referred to as the Kanagawa phenomenon (KP) because its discoverers belonged to the Kanagawa Prefectural Public Health Laboratory. Sakazaki *et al.* (1968) examined the KP of the isolates from human patients and from ocean fish and sea water and found a good correlation between KP and pathogenicity.

Vibrio parahaemolyticus is the major causative agent of food poisoning in Japan, and it is thought that the Japanese custom of eating raw fish, such as 'sashimi' or 'sushi', is the main reason for this high frequency. However, fish which is to be eaten is handled very carefully in Japan, and is well cleaned and kept at a low temperature. Therefore, a secondary source of contamination of food is suspected: the utensils used to prepare seafood. Cooked foods other than seafoods are thought to be another cause of *V. parahaemolyticus* infection.

As stated, the KP is haemolysis on a special agar medium, Wagatsuma's agar, and is an index of pathogenicity of the vibrio. By the late 1970s, purification of the haemolysin causing KP (*V. parahaemolyticus* thermostable direct haemolysin: Vp-TDH; TDH referring to its stability at 100°C) was achieved, and Vp-TDH was shown to be a protein toxin without a lipid or carbohydrate moiety. The relative molecular mass of the haemolysin was determined to be approximately 42 000 by gel filtration and 21 000 by sodium dodecyl sulfate–polyacrylamide gel electrophoresis (SDS–PAGE). Thus, this haemolysin has a dimer structure that is composed of two identical subunits having a relative molecular mass of about 21 000. The amino acid sequence of the purified Vp-TDH has been determined (Tsunasawa *et al.*, 1987), and the subunit consists of 165 residues with one disulfide bond near the carboxy terminus. This result is in good agreement with that obtained from the nucleotide sequence of the gene-encoding TDH (*tdh*).

Purified haemolysin was not inactivated by heating at 60–100°C for 10 min. However, the haemolytic activity of the crude haemolysin preparation of the vibrio was partially inactivated by heating at around 60°C for 10 min, although it was not inactivated by heating at 80–90°C for 10 min. Miwatani *et al.* (1972) suggested that this phenomenon was similar to the Arrhenius effect of staphylococcal α-haemolysin reported by Arbuthnott (1970). Takeda *et al.* (1975a) attempted to elucidate the mechanism of the Arrhenius effect of crude haemolysin preparation of *V. parahaemolyticus*, and found a factor which inactivated TDH in 50–60°C incubation but not in 80–90°C incubation.

Vp-TDH lyses red blood cells from various animal sources, one exception being the horse (Honda and Iida, 1993). The toxin is also cytotoxic to various cultured cells such as HeLa cells, FL cells and foetal mouse heart cells (Honda and Iida, 1993). Tests have been made on the sensitivity of various cell lines to TDH and potential cell lines resistant to TDH in identification of a TDH receptor were studied (Tang *et al.*, 1997a). In the rabbit ileal loop test challenge with purified Vp-TDH at 100–250 µg loop^{-1} induced fluid accumulation (Miyamoto *et al.*, 1980; Honda and Iida, 1993), which suggests that Vp-TDH is enterotoxic. The amount inducing fluid accumulation is higher than that of enterotoxins from other diarrhoeagenic bacteria; for example, 0.2 µg of CT is enough to induce fluid accumulation in this assay. The contribution of Vp-TDH as the diarrhoeagenic factor has therefore been questioned. Other enterotoxic factors of *V. parahaemolyticus* such as one causing morphological change of Chinese hamster ovary (CHO) cells or cholera enterotoxin-like toxin immunologically cross-reactive in GM$_1$-ganglioside ELISA have also been suggested, but conclusive evidence of the contribution of these factors has not been demonstrated. Extensive study of Vp-TDH-producing or non-producing *V. parahaemolyticus* strains by rabbit ileal loop test suggested that only the former can induce fluid accumulation. Furthermore, intragastric administration of 6 µg of Vp-TDH to suckling mice caused diarrhoea and death (Miyamoto *et al.*, 1980), and Vp-TDH increased vascular permeability, which is one of the characteristics of various enterotoxins (Honda *et al.*, 1990). To obtain evidence for the role of Vp-TDH in the enteropathogenicity of *V. parahaemolyticus*, Nishibuchi *et al.* (1992) compared a KP-positive strain and its isogenic TDH-negative mutant, in which both the *tdh1* and *tdh2* genes were specifically inactivated by successive allelic exchange procedures. Whole-culture preparations of the parent strain gave positive fluid accumulation results in the rabbit ileal loop test, while the TDH-negative mutant induced no fluid accumulation in the model. Additional evidence for the enterotoxic activity of Vp-TDH was obtained with rabbit ileal tissue mounted in Ussing chambers, a sensitive technique for studying intestinal ion transport. Culture filtrates of the TDH-positive strain induced an increase in short circuit current, whereas no such increase was observed with the isogenic TDH-negative strain (Nishibuchi *et al.*, 1992). Returning the cloned *tdh* gene to the isogenic mutant complemented the mutation and restored the ability to increase short circuit current. Thus, Vp-TDH is believed to be an important virulence factor in *V. parahaemolyticus* gastroenteritis, although the mechanism inducing diarrhoea has not been identified.

The mode of action of Vp-TDH has not yet been completely elucidated, although the haemolysis is thought to proceed in a colloid osmotic manner which is shown in the section on *Vibrio vulnificus* haemolysin (VVH). Vp-TDH is believed to damage the erythrocyte membrane by acting as a pore-forming toxin (Honda *et al.*, 1992). The functional pore size made by the haemolysin has been estimated to be 2 nm. One of the essential mechanisms in the damage that Vp-TDH inflicts during the haemolytic process is apparently the phosphorylation of a 25-kDa protein on the erythrocyte membrane (Yoh *et al.*, 1996). They found that TDH induced the phosphorylation of two proteins of 22.5 and 25 kDa on membranes of human erythrocytes that are sensitive to TDH, but only 22.5-kDa protein was phosphorylated on the membranes of horse erythrocytes that are insensitive. Furthermore, a mutant which retained binding ability but had lost haemolytic activity phosphorylated only the 22.5-kDa protein on human erythrocyte membranes.

Vp-TDH causes haemolysis and lyses cultured cells in a Ca^{2+}-independent manner (Tang *et al.*, 1994, 1995). The receptor for Vp-TDH was previously reported as gangliosides GT_1 and GD_{1a} (Takeda *et al.*, 1975b; Takeda, 1983). However, conflicting results have been reported, although a putative receptor does seem to be present (Tang *et al.*, 1997a). Thus it remains to be clarified as to what entity acts as the receptor for Vp-TDH. A number of reports has described the structure–function relationship of Vp-TDH (Honda *et al.*, 1989; Baba *et al.*, 1991; Toda *et al.*, 1991; Tang *et al.*, 1994; Iida *et al.*, 1995). Tang *et al.* (1994) reported the isolation of an interesting mutant toxin of Vp-TDH in which the haemolytic activity was almost completely diminished but which retained the ability to bind to erythrocytes. The mutant toxin, R7, possessed an amino acid substitution at Gly62 to Ser62, suggesting that the site is functionally important for Vp-TDH. R7 is being used as a probe to identify the receptor for Vp-TDH (Tang *et al.*, 1997a). An attempt is being made to identify the receptor for Vp-TDH by isolating a TDH-resistant mutant cell line from a TDH-sensitive fibroblast cell line, Rat-1,

through mutagen treatment (Tang *et al.*, 1997a). Tang *et al.* (1997b) analysed the functional domains of TDH and demonstrated that the *N*-terminal region may be involved in the binding process while the region near the *C*-terminus may be involved in some postbinding process.

In addition to enterotoxicity, Vp-TDH possesses lethal toxicity, cytolytic activity (Sakurai *et al.*, 1976) and cardiotoxicity (Honda *et al.*, 1976a). *Vibrio parahaemolyticus* is lethal for various experimental animals. Honda *et al.* (1976b) purified the lethal toxin from culture filtrates of *V. parahaemolyticus* and found that it was identical to Vp-TDH. Intravenous injection of 5 μg of purified Vp-TDH killed mice within 1 min and even 1 μg of the toxin killed mice within 20 min (Honda *et al.*, 1976b). The toxin was lethal by intraperitoneal injection in mice, although less potent than by intravenous injection. After Vp-TDH injection, mice became motionless and sometimes developed cramps. Honda *et al.* (1976b) showed that the lethal activity was due to cardiotoxic activity. An electrocardiogram of rats injected with TDH showed wider and higher P waves, suggesting changes in conduction of the intra-atrial impulse, and an increase in voltage of QRS, suggesting changes in intraventricular impulses of electrical activation. Thereafter, the PQ intervals became longer, suggesting inhibition of atrio-ventricular conduction. When 7.5 μg of Vp-TDH was injected intravenously into rats weighing 448 g, ventricular flutter developed after about 50 s and the heart stopped after 148 s. The cardiotoxicity of Vp-TDH was also demonstrated using cultured mouse heart cells (Honda *et al.*, 1976a).

KP-negative strains are occasionally suspected as the causative agent of food poisoning. Some of these strains produce a TDH-related haemolysin called TRH, which is not detected by Wagatsuma's agar test (Honda *et al.*, 1988, 1989, 1990). TRH shows enterotoxicity and has common antigenicity with TDH, but is thermolabile. Kelly and Stroh (1988) reported that clinical isolates obtained from patients with locally acquired gastroenteritis in Canada all hydrolysed urea, but none of the isolates was KP-positive, suggesting that the urease-positive strains are the predominant biotype of *V. parahaemolyticus* associated with gastroenteritis in the Pacific northwest. However, Oosawa *et al.* (1996) reported evidence suggesting that urea hydrolysis is not a reliable marker for identifying *tdh*-carrying *V. parahaemolyticus* strains in Japan but may be a marker for *trh*-carrying strains. Okuda *et al.* (1997) analysed the *tdh* gene and *trh* genes in urease-positive (Ure^+) strains of *V. parahaemolyticus* isolated on the west coast of the United States. They reported a very strong correlation between the Ure^+ phenotype and the *trh* gene, as found in strains isolated in Asia (Suthienkul *et al.*, 1995).

A TDH gene (*tdh*) of 567 bp was cloned and the DNA sequence determined (Kaper *et al.*, 1984; Taniguchi *et al.*, 1985). Nine *tdh* genes have been demonstrated (Table 19.3), and *tdh2* is thought to be expressed in the KP-positive strain (Nishibuchi and Kaper, 1990, 1995). The *tdh* genes are usually, but not exclusively, located in the chromosome. All of the cloned *tdh* genes encode predicted protein products composed of 189 amino acid residues (including signal peptides) which have haemolytic and other biological activities. The nucleotide sequences of the various *tdh* genes are well conserved (>97% identity) and the protein products are immunologically indistinguishable. Similarity between *tdh2* and *trh1* is 68.6%. Genes similar to *tdh* are found in *V. cholerae* non-O1, *V. mimicus* and *V. hollisae* (Nishibuchi and Kaper, 1990). These vibrios produce haemolysin similar to Vp-TDH; the haemolysins from the former two are thermostable, but the haemolysin from the latter is thermolabile (Table 19.2).

In addition to TDH and TRH, production of two haemolysins, LDH (lecithin-dependent haemolysin) and δ-VPH, by *V. parahaemolyticus* was reported. Taniguchi *et al.* (1985) obtained a transformant producing LDH, which was designated as a thermolabile haemolysin. Shinoda *et al.* (1991) purified LDH, which was localized in the periplasmic space of the transformant cells. Although LDH of *V. parahaemolyticus* was reported as a phospholipase A$_2$, Shinoda *et al.* (1991) demonstrated that it should be classified as a phospholipase B or an atypical phospholipase to be

designated as phospholipase A$_2$/lysophospholipase. Taniguchi *et al.* (1985) cloned a gene encoding another haemolysin, δ-VPH. The production of δ-VPH was observed in *E. coli* transformed by the cloned gene, but not in *V. parahaemolyticus* culture, although all strains of the vibrio possessed the δ-VPH gene. Whether or not these two haemolysins contribute towards pathogenicity has not been determined.

VIBRIO VULNIFICUS HAEMOLYSIN

The first isolation of *V. vulnificus* was from a leg ulcer which was reported as a *V. parahaemolyticus* infection (Roland, 1970). Although the bacterium showed similar characteristics to *V. parahaemolyticus*, including having a slightly halophilic property and being negative for sucrose fermentation, it was different in that it fermented lactose (Hollis *et al.*, 1976). The bacterium was therefore called lactose-positive vibrio (Hollis *et al.*, 1976), and was subsequently termed *V. vulnificus* (Farmer, 1979). Isolates can be grouped according to the type of illness they cause, the primary septicaemia group or the wound-infection group (Blake *et al.*, 1980). The former group is remarkable for its high fatality rate in infections. *Vibrio vulnificus* is now recognized as being among the most rapidly fatal of human pathogens. In the majority of cases, the primary septicaemia is associated with the consumption of raw seafood, especially shellfish such as oyster, contaminated with the vibrio. In the United States, 95% of all

TABLE 19.3 Characteristics of *tdh* and *trh* genes with known nucleotide sequences (Nishibuchi and Kaper, 1995)

| | Origin | | | | |
| | Organism | | | | |
Designation	Species	Strain	Source	Location	% Similarity to the *tdh* 2 gene[a]
tdh 1	*V. parahaemolyticus*	WP-1	Clinical	Chromosome	97.0
tdh 2	*V. parahaemolyticus*	WP-1	Clinical	Chromosome	100
tdh 3	*V. parahaemolyticus*	AQ3376	Clinical	Chromosome	98.6
tdh 4	*V. parahaemolyticus*	AQ3376	Clinical	Plasmid	98.6
tdh 5	*V. parahaemolyticus*	AQ3860	Clinical	Chromosome	98.9
tdh A	*V. parahaemolyticus*	T4750	Clinical	Chromosome	99.6
tdh S	*V. parahaemolyticus*	T4750	Clinical	Chromosome	97.2
tdh X	*V. parahaemolyticus*	TH3766	Clinical	Chromosome	98.4
tdh /I	*V. parahaemolyticus*	TH012	Clinical	Chromosome	98.4
NAG-*tdh*	*V. cholerae* non-O1	91	Clinical	Plasmid	98.6
Vm-*tdh*	*V. mimicus*	6	Clinical	Chromosome	97.0
Vh-*tdh*	*V. hollisae*	9041	Clinical	Chromosome	93.3
trh 1	*V. parahaemolyticus*	AQ4037	Clinical	Chromosome	68.6
trh X	*V. parahaemolyticus*	TH3766	Clinical	Chromosome	68.1
trh 2	*V. parahaemolyticus*	AT4	Environmental	Chromosome	68.8

[a]Percentage similarity in the 567 bp coding region.

seafood-related deaths are due to *V. vulnificus*, most commonly from the consumption of raw oyster (Klontz *et al.*, 1988; Oliver, 1995). The primary septicaemia due to *V. vulnificus* is an opportunistic infection, although most patients with septicaemia have an underlying disease such as liver dysfunction, alcoholic cirrhosis or haemochromatosis which leads to an increased plasma iron level and decreased host defense system. Infections in patients having malignant tumour or acquired immune deficiency syndrome (AIDS) have also been reported. In two-thirds of patients, secondary skin lesions appear on the extremities and the trunk. Symptoms of the digestive tract such as diarrhoea or vomiting are very rare. Wound infection is characterized by the development of oedema, erythema or necrosis around a new wound exposed to sea water. This type of infection can occur in healthy persons as well as in compromised hosts, and may occasionally progress to septicaemia.

Vibrio vulnificus produces various extracellular toxic factors such as haemolysin or protease (Miyoshi *et al.*, 1993; Miyoshi and Shinoda, 1997). The protease (VVP) is a metalloprotease and degrades a number of biologically important proteins including elastin, fibrinogen and plasma protease inhibitors of complement components (Miyoshi *et al.*, 1995). It also enhances vascular permeability through activation of the Hageman factor-plasma kallikrein-kinin cascade (Miyoshi *et al.*, 1993) and/or exocytotic histamine release from mast cells (Miyoshi *et al.*, 1993), forming a haemorrhagic lesion which finally provokes severe dermonecrosis. Thus, VVP is the most probable inducer of oedema and bacterial invasion during an infection. It is also thought to supply iron, an essential element for bacterial growth, by co-operation with haemolysin; the haemolysin releases haemoglobin from erythrocytes and the protease liberates iron (protohaem) from the haemoglobin (Nishina *et al.*, 1992).

Gray and Kreger (1985) purified a heat-labile haemolysin from a clinical isolate of *V. vulnificus* by ammonium sulfate precipitation, gel filtration with Sephadex G-75, hydrophobic interaction chromatography with phenyl-Sepharose CL-4B and isoelectric focusing. This haemolysin has an isoelectric point of 7.1 and is a hydrophobic 56-kDa single-chain polypeptide. Oh *et al.* (1993) found that CHAPS [(3-cholamidopropyl) dimethylammonio-1-panesulfinate] treatment of a preparation obtained by ammonium sulfate precipitation was more efficient for the purification of the haemolysin; it had a relative molecular mass of 50 000 and they designated it VVH (*V. vulnificus* haemolysin). Western blot analysis with antibody against VVH showed the existence of a 50-kDa

haemolysin in all of nine *V. vulnificus* isolates tested (Oh *et al.*, 1994)

Okada *et al.* (1987) established murine hybridoma cell lines which produce monoclonal antibodies against haemolysin(s) of *V. vulnificus*. One monoclonal antibody among them recognized a 36-kDa protein and neutralized the haemolytic activity of crude toxin preparations. However, it is not clear whether this 36-kDa protein is a fragment of or a polypeptide associated with VVH.

The DNA fragment encoding the VVH gene was cloned and sequenced (Yamamoto *et al.*, 1990b); the sequence contained two open reading frames, *vvhA* and *vvhB*. *VvhA*, the structure gene for VVH, encodes a 50 851-Da polypeptide preceded by a 20 amino acid signal peptide. Part of this gene has considerable homology with the haemolysin gene of *V. cholerae*. The *vvhB* gene encodes an 18 082-Da polypeptide but the function of this polypeptide has not been determined. The gene encoding VVH has been used as a DNA probe for identification of the vibrio, because this fragment showed 100% specificity and sensitivity for *V. vulnificus* strains isolated from both clinical and environmental sources. Furthermore, Brauns *et al.* (1991) succeeded in detecting viable but non-culturable *V. vulnificus* cells using the polymerase chain reaction based on amplification of the VVH gene.

Purified VVH is active against erythrocytes from many animal species and against CHO and HeLa cells (Gray and Kreger, 1985). Mouse, sheep, pig, monkey, burro, cat and pigeon erythrocytes are more sensitive to the haemolysin, while those from rabbit, human and chicken are less sensitive (Yamanaka *et al.*, 1989).

The haemolytic process has been studied using sheep or mouse erythrocytes as target cells (Shinoda *et al.*, 1985; Yamanaka *et al.*, 1987b). Haemolysis induced by VVH is temperature dependent and is optimal between 30 and 37°C. At 4°C, although no erythrocytes are disrupted, VVH binds to them irreversibly. Therefore, haemolysis by VVH is believed to be a two-step process consisting of a temperature-independent toxin-binding step followed by a temperature-dependent cell-disruption step.

Most haemolysins are inactivated by the addition of lipids, such as phospholipids (Stollerman *et al.*, 1952), gangliosides (Takeda *et al.*, 1975b) or cholesterol (Hase *et al.*, 1976; Pridgent and Alouf, 1976), which is thought to constitute the toxin-binding site on the erythrocyte membrane. Shinoda *et al.* (1985) reported that a small amount of cholesterol abolished the haemolytic ability of VVH, whereas other lipids had no effect. No inhibition by cholesterol was evident when it was added after the toxin had bound to erythrocytes. VVH is active against cholesterol-phosphatidylcholine liposomes, but liposomes containing a negligible amount of cholesterol are less sensitive to VVH action (Yamanaka *et al.*, 1987a). These findings strongly suggest that cholesterol is the binding site for VVH.

Binding to cholesterol is a common and remarkable property of a group of haemolysins called thiol-activated haemolysins (TAHs) (see Chapter 24), including streptolysin O from group A streptococci and θ-toxin of *Clostridium perfringens* (Hase *et al.*, 1976; Pridgent and Alouf, 1976; Alouf, 1980). In addition to the cholesterol-binding characteristic, TAH loses activity when stored in air due to oxidation of cysteine residues, and is reactivated by sulfhydryl compounds, including dithiothreitol and glutathione (Cowell *et al.*, 1976; Hase *et al.*, 1976; Alouf, 1980). In contrast to TAHs, VVH is not inactivated by storage at 4°C for more than one month (Shinoda *et al.*, 1985) and is resistant to the sulfhydryl-blocking agents such as 4,4′-dithiopyridine and 5,5′-dithio-*bis*-(2-nitrobenzene) (Miyoshi *et al.*, 1985), which elicits inactivation of TAHs by modification of the cysteine residue (Cowell *et al.*, 1976; Hase *et al.*, 1976; Alouf, 1980). Thus, it appears that VVH is a new type of cholesterol-binding haemolysin.

At 37°C, VVH treatment of erythrocytes induced rapid K$^+$ efflux which was followed by the release of haemoglobin (Yamanaka *et al.*, 1987b). This phenomenon indicates that the temperature-dependent cell disruption step can be further divided into two stages: a small pore-forming stage causing K$^+$ efflux, and a membrane-bursting stage eliciting the release of haemoglobin (i.e. haemolysis). Intracellular K$^+$ initially passes through the small transmembrane-pore formed by VVH, then extracellular low-molecular-mass substances and water enter the erythrocyte via the pore, resulting in the physical explosion of erythrocyte membranes owing to the increased intracellular osmotic pressure. The initial pore-forming stage is temperature dependent while the later one is temperature independent. Haemoglobin release, not K$^+$ efflux, from VVH-treated erythrocytes was inhibited completely by addition of 30-mM (equivalent to intracellular haemoglobin) dextran 4 (molecular diameter of 3.5 nm) and partially by inulin (2.8 nm), although raffinose (1.2 nm) had no effect, indicating that the diameter of the transmembrane pore formed by VVH may be around 3 nm (Yamanaka *et al.*, 1987b).

Gray and Kreger (1985) and Shinoda *et al.* (1985) reported that more than one VVH molecule might be required to lyse a single erythrocyte. This suggestion was subsequently supported by experiments using liposomes (Yamanaka *et al.*, 1987a). When VVH was

allowed to act on cholesterol-phosphatidylcholine liposomes entrapping K$^+$, the leakage of K$^+$ accompanied by the formation of VVH oligomer with a relative molecular mass of *ca.* 200 000 was observed, suggesting that some VVH molecules assembled and formed a single pore on a target cell membrane.

The haemolytic reaction by VVH, like other bacterial cytolysins (Oberley and Duncan, 1971; Miyake *et al.*, 1989), is inhibited by the addition of a divalent cation (Mg^{2+}, Ca^{2+} or Mn^{2+}) (Shinoda *et al.*, 1985). Erythrocytes and VVH were incubated at 4°C in the presence of the cation, then erythrocytes were collected and rinsed with divalent cation-free saline. The rinsed erythrocytes were post-incubated at 37°C in a VVH- and divalent cation-free saline, after which marked haemolysis was observed (Shinoda *et al.*, 1985). Thus, the divalent cation does not prevent the binding of VVH to erythrocyte membranes. When VVH was allowed to act on erythrocytes in the presence of the cation and dextran 4 at 37°C, there was leakage of K$^+$ demonstrating the formation of transmembrane pores (Yamanaka *et al.*, 1987b). These findings indicate that divalent cations inhibit the final membrane burst stage of the haemolytic process. However, the inhibitory mechanism is unknown.

VVH is a hydrophobic protein which is gradually inactivated by autoaggregation. Miyoshi *et al.* (1997a) showed that VVH was modified to a hydrophilic and stable form by proteolysis with *V. vulnificus* protease.

VVH possesses vascular permeability-enhancing activity which is abolished by the simultaneous administration of an antihistaminic agent. Yamanaka *et al.* (1990) reported that VVH elicited the liberation of histamine from isolated mast cells accompanied by leakage of lactate dehydrogenase, indicating lysis of mast cells. Thus, VVH inoculated into dorsal skin may act directly on and disrupt mast cells, resulting in the leakage of histamine and enhancement of vascular permeability. Gray and Kreger (1987) stated that VVH injected subcutaneously into mice caused severe structural alteration of the skin, and that tissue damage induced by this haemolysin was very similar to that shown in *V. vulnificus* wound infection. In addition, the enzyme-linked immunosorbent assay using polyclonal or monoclonal antibody against purified VVH demonstrated toxin production *in vivo* (Gray and Kreger, 1989).

Gray and Kreger (1985) have reported that purified VVH administered by the intravenous route was lethal for mice (the 50% lethal dose was *ca.* 3.0 µg kg^{-1}), suggesting that VVH might contribute to the development of systemic *V. vulnificus* infection. However, Morris *et al.* (1987) demonstrated that the virulent potential of individual strains in mice did not correlate with *in vitro* ability to elaborate the haemolysin. Furthermore, Massad *et al.* (1988) and Wright and Morris (1991) reported that cytolysin-negative mutants were as virulent as wild-type strains in mouse models. Thus, VVH may be less important in the development of systemic *V. vulnificus* infection, although it is clear that VVH plays some role in the pathogenesis.

Testa *et al.* (1984) found that *V. vulnificus* produced phospholipase(s). The partially purified preparation could hydrolyse both acyl ester bonds of all classes of phospholipids except for sphingomyelin, demonstrating that the phospholipase preparation possesses both phospholipase A_2 and lysophospholipase activities. However, wide distribution of the phospholipase in *V. vulnificus* was not demonstrated in a survey of many clinical and environmental isolates (Miyoshi *et al.*, unpublished observation).

Vibrio mimicus haemolysins

Vibrio mimicus, a species closely related to *V. cholerae* (Davis *et al.*, 1981), is a causative agent of human gastroenteritis. Infection with pathogenic strains of *V. mimicus* induces various clinical symptoms, from watery to dysentery-like diarrhoea (Hoge *et al.*, 1989), suggesting that this pathogen produces many kinds of virulent factors. Enterotoxins similar to CT (Spira and Fedorka-Cray, 1984; Chowdhury *et al.*, 1987, 1991) and heat-stable enterotoxin (Gyobu *et al.*, 1988) have been found in some clinical strains to be causative factors of watery diarrhoea. However, some virulent strains lack the ability to produce any of these enterotoxins and, in these, haemolysin is thought to be a candidate for the enteropathogenic factor. For example, the good correlation between enteropathogenicity and haemolysin production in *V. parahaemolyticus* infection is well known (Nishibuchi *et al.*, 1992; Lin *et al.*, 1993), as shown above. Furthermore, in *V. cholerae*, the production of an enterotoxic haemolysin in addition to CT has been documented (Ichinose *et al.*, 1987). Honda *et al.* (1987) reported production of two types of haemolysin by *V. mimicus*: one was heat labile and immunologically similar to El Tor haemolysin, while another was heat stable and closely related to thermostable direct haemolysin produced by *V. parahaemolyticus*. These findings suggest that the haemolysins produced by *V. mimicus* are also virulent determinants, especially in dysentery-like diarrhoea.

The heat-stable *V. mimicus* haemolysin (Vm-TDH) was purified and characterized by Yoshida *et al.* (1991). Shinoda *et al.* (1993) studied the haemolytic mechanism of the heat-labile haemolysin, *V. mimicus*

haemolysin (VMH). The VMH was designated as a member of the pore-forming haemolysins and thought to form a transmembrane pore with a diameter of *ca*. 3 nm. The pore formed by VMH was permeable to water and monovalent ions, such as Na^+ and K^+, but impermeable by haemoglobin. This suggests that VMH causes haemolysis in a colloid osmotic manner.

Miyoshi *et al.* (1997b) purified the VMH with a relative molecular mass of 63 000 by ammonium sulphate precipitation, column chromatography on Phenyl Sepharose HP and Superose 6 HR. The haemolytic reaction induced by VMH continued up to disruption of all erythrocytes in the assay system. Moreover, VMH bound to erythrocyte ghosts showed sufficient ability to attack intact erythrocytes. These results suggest reversible binding of the toxin molecule to the membrane. The final cell-disrupting stage was effectively inhibited by various divalent cations, and some, such as Zn^{2+} and Cu^{2+}, blocked the pore-forming stage at higher concentration. Although VMH could disrupt various mammalian erythrocytes, including bovine, rabbit, sheep, human and mouse, those from horse were most sensitive to the haemolysin. Horse erythrocytes had the most toxin-binding sites and were haemolysed by the fewest membrane-bound toxin molecules, suggesting that toxin binding to, and pore formation on, horse erythrocytes take place most effectively.

The purified VMH induced fluid accumulation in ligated rabbit ileal loops in a dose-dependent manner (Miyoshi *et al.*, 1997b), and the antibody against the haemolysin obviously reduced enteropathogenicity of living *V. mimicus* cells. These findings clearly demonstrate that VMH relates to virulence of this human pathogen.

The structural gene of VMH has been cloned and the nucleotide sequence determined (Rahman *et al.*, 1997). A 2232-bp open reading frame codes a peptide of 744 amino acid residues, with a calculated relative molecular mass of 83 903. The sequence of the gene was closely related to the El Tor haemolysin gene shown above, with 76% homology. The 13 amino acid residue of the amino terminus of the 63-kDa mature VMH is identical from S-152 to T-164, as predicted from the nucleotide sequence. So, it seems that the mature VMH is processed by deletion of the first 151 amino acids, and the relative molecular mass is 65 972. Analysis of the deduced amino acid sequence showed the existence of a potential signal sequence of 24 amino acids at the amino terminus, suggesting that, like El Tor haemolysin, two-step processing also exists in VMH maturation. Kim *et al.* (1997) reported the nucleotide sequence of the *vmhA* gene encoding VMH.

Other vibrio haemolysins

In addition to the above, some vibrio species are pathogenic to humans or fish. *Vibrio fluvialis*, *V. furnissii* and *V. hollisae* cause gastroenteritis in humans, whereas *V. damsella* is a pathogen of extraintestinal infection. Extraintestinal infections by *V. alginolyticus* and *V. metscnikovii* have also been reported occasionally (Blake *et al.*, 1980). These vibrios pathogenic to humans inhabit brackish or sea-water environments; *V. anguillarum* is found in fresh or brackish water and is a fish pathogen.

These vibrios also produce haemolysins. *Vibrio fluvialis* has been implicated in causes of human food poisoning. Infection is usually associated with the consumption of seafood. Production of haemolysin(s) has been reported (Yamada *et al.*, 1988; Rahim and Aziz, 1996), but no detailed information on the characteristics is available.

Vibrio furnissii is a new species distinguished from *V. fluvialis* by its property of gas production, and is thought to cause diarrhoea in humans, although this has not been well established owing to the small number of clinical cases. It produces a haemolysin that is immunologically indistinguishable from that of *V. fluvialis* (Yamada *et al.*, 1988).

Vibrio hollisae sometimes causes gastroenteritis with symptoms including diarrhoea, abdominal pain and fever; in rare cases, it causes bacteraemia. A haemolysin (Vh-rTDH) is thought to be the diarrhoeagenic factor. Vh-rTDH resembles Vp-TDH, is 86% homologous in amino acid sequences but is heat labile (Yoh *et al.*, 1989).

A haemolysin gene of *V. anguillarum* was cloned (Hirono *et al.*, 1996). The open reading frame of the gene was 2253 bp and corresponded to a protein of 751 amino acid residues. The deduced amino acid sequence showed a significant degree of homology with that of El Tor haemolysin, VVH, *Aeromonas hydrophila* AHH1 haemolysin and *A. salmonicida* ASH1 haemolysin, and the overall amino acid identities were 57.3%, 25.5%, 46.2% and 43.7%, respectively. Contribution to pathogenicity of the species has not been documented.

Although *V. metschnikovii* was recognized more than 100 years ago, there have been few reports of isolation of this organism as a human pathogen. It has occasionally been implicated in bacteraemia, cholecystitis, diarrhoea and urinary tract infections. A haemolysin produced by this bacterium has been purified and characterized as a pore-forming toxin (Miyake *et al.*, 1988, 1989).

Vibrio damsela has been reported to cause wound infection. It may cause localized or wide-ranging cellulitis and sometimes more severe complications, such as disseminated intravascular coagulation. A haemolysin (damselysin) was purified and its phospholipase activity was reported (Kreger *et al.*, 1987). The gene encoding damselysin was cloned and sequenced but no homology was observed with the sequence of other vibrio haemolysins (Cutter and Kreger, 1990).

CONCLUSION

Haemolysins produced by pathogenic vibrio are classified into three groups, namely the El Tor haemolysin group, TDH group and others. This review focused on four haemolysins: El Tor haemolysin (*V. cholerae*), Vp-TDH (*V. parahaemolyticus*), VVH (*V. vulnificus*) and VMH (*V. mimicus*), which have been extensively studied. All four are pore-forming toxins, but their precise mechanism of action remains to be investigated. El Tor haemolysin, Vp-TDH and VMH cause diarrhoea, the major symptom of infection of these vibrios, although El Tor haemolysin is no more than a supplemental factor of CT, and VMH is only one of the many enterotoxic factors. Vp-TDH is thought to be the major pathogenic factor of *V. parahaemolyticus*, which is the most important food poisoning bacterium in Japan and other eastern and south-east Asian countries; the diarrhoeagenic mechanism, however, has not been identified. Furthermore, it is unclear whether a similar mechanism to the pore-forming action on erythrocytes contributes to cause diarrhoea. Molecular biological studies of the toxins, such as nucleotide and amino sequence, functional domain analysis, or mode of secretion from cell and processing have been well developed in the past decade. It is necessary for us in the next decade to link the progress in these molecular biological studies with elucidation of the pathogenic mechanism to facilitate prevention of the disease.

REFERENCES

Albert, M.J., Ansaruzzaman, M., Bardhan, P.K., Faruque, A.S.G., Faruque, S.M., Islam, M.S., Mahalanabis, D., Sack, R.B., Salam, M.A., Siddique, A.K., Yunus, M.D. and Zaman, K. (1993). Large epidemic of cholera-like disease in Bangladesh caused by *Vibrio cholerae* O139 synonym Bengal. *Lancet* **342**, 387–90.

Alouf, J.E. (1980). Streptococcal toxins (streptolysin O, strpetolysin S, erythrogenic toxin). *Pharmac. Ther.* **11**, 661–717.

Arbuthnott, J.P. (1970). Staphylococcal α-toxin. In *Microbial Toxins* (eds T.C. Montie, S. Kadis and S.J. Ajl), Vol. 3, pp. 189–236. Academic Press, New York.

Baba, K., Shirai, H., Terai, A., Kumagai, K., Takeda, Y. and Nishibuchi, M. (1991). Similarity of the *tdh* gene-bearing plasmids of *Vibrio cholerae* non-O1 and *Vibrio parahaemolyticus*. *Microb. Pathog.* **10**, 61–70.

Blake, P.A., Weaver, R.E. and Hollis, D.G. (1980). Disease of humans (other than cholera) caused by vibrios. *Annu. Rev. Microbiol.* **34**, 341–67.

Brauns, L.A., Hudson, M.C. and Oliver, J.D. (1991). Use of polymerase chain reaction in detection of culturable and nonculturable *Vibrio vulnificus* cells. *Appl. Env. Microbiol.* **57**, 2651–5.

Chakraborty, S., Nair, G.B. and Shinoda, S. (1997). Pathogenic vibrios in the natural aquatic environment. *Rev. Environ. Health* **12**, 63–80.

Chowdhury, M.A.R., Aziz, K.M.S., Kay, B.A. and Rahim, Z. (1987). Toxin production by *Vibrio mimicus* strains isolated from human and environmental sources in Bangladesh. *J. Clin. Microbiol.* **25**, 2200–3.

Chowdhury, M.A.R., Miyoshi, S. and Shinoda, S. (1991). Application of a direct culture method GM₁-enzyme-linked immnosorbent assay for detection of toxigenic *Vibrio mimicus*. *Biomedical Lett.* **44**, 31–4.

Cowell, J.L., Grushoff-Kosyk, P.S. and Bernheimer, A.W. (1976). Purification of cereolysin and the electrophoretic separation of the active (reduced) and inactive (oxidized) forms of the purified toxin. *Infect. Immun.* **14**, 144–54.

Cutter, D.L. and Kreger, A.S. (1990). Cloning and expression of the damselysin gene from *Vibrio damsela*. *Infect. Immun.* **58**, 266–8.

Davis, B.R., Fanning, G.R., Madden, J.M., Steigerwalt, A.G., Bradford, H.B., Jr, Smith, H.L. and Brenner, D.J. (1981). Characterization of biochemically atypical *Vibrio cholerae* strains and designation of a new pathogenic species, *Vibrio mimicus*. *J. Clin. Microbiol.* **14**, 631–9.

Farmer, J.J. (1979). *Vibrio (Beneckea) vulnificus*: the bacterium associated with sepsis, septicaemia, and the sea. *Lancet* **ii**, 903.

Fasano, A., Baudry, B., Pumplin, D.W., Wasserman, S.S., Tall, B.D., Kelly, J.M. and Kaper, J.B. (1991). *Vibrio cholerae* produces a second enterotoxin which affects intestinal tight junctions. *Proc. Natl Acad. Sci. USA.* **86**, 5242–6.

Fujino, T., Okuno, Y., Nakada, D., Aoyama, A., Fukai, K., Mukai, T. and Ueho, T. (1953). On the bacteriological examination of shirasu-food poisoning. *Med. J. Osaka Univ.* **4**, 299–304.

Gray, L.D. and Kreger, A.S. (1985). Purification and characterization of an extracellular cytolysin produced by *Vibrio vulnificus*. *Infect. Immun.* **48**, 62–72.

Gray, L.D. and Kreger, A.S. (1987). Mouse skin damage caused by cytolysin from *Vibrio vulnificus* and by *V. vulnificus* infection. *J. Infect. Dis.* **155**, 236–41.

Gray, L.D. and Kreger, A.S. (1989). Detection of *Vibrio vulnificus* cytolysin in *V. vulnificus*-infected mice. *Toxicon* **27**, 459–64.

Gyobu, Y., Kodama, H. and Uetake, H. (1988). Production and partial purification of a fluid-accumulating factor of non-O1 *Vibrio cholerae*. *Microbiol. Immunol.* **32**, 565–77.

Hall, R.H. and Drasar, B.S. (1990). *Vibrio cholerae* HlyA haemolysin is processed by proteolysis. *Infect. Immun.* **58**, 3375–9.

Harvey, R.M., Enson, Y., Lewis, M.L., Greenough, W.B., Ally, K.M. and Panno, R.A. (1966). Haemodynamic effects of dehydration and metabolic acidosis in Asiatic cholera. *Trans. Assoc. Am. Physiol.* **29**, 177–86.

Hase, J., Mitsui, K. and Shonaka, E. (1976). *Clostridium perfringens* exotoxins IV. Inhibition of the θ-toxin induced haemolysis by steroids and related compounds. *Jpn. J. Exp. Med.* **46**, 45–60.

Hirono, I., Masuda, T. and Aoki, T. (1996). Cloning and detection of the haemolysin gene of *Vibrio anguillarum*. *Microb. Pathog.* **21**, 173–82.

Hoge, C.W., Watsky, D., Pealer, R.N., Libonati, J.P., Israel, E. and Morris, J.G., Jr (1989). Epidemiology and spectrum of *Vibrio* infections in a Chesapeake Bay community. *J. Infect. Dis.* **160**, 985–93.

Hollis, D.G., Weaver, R.E., Baker, C.N. and Thronsberry, C. (1976). Halophilic *Vibrio* species isolated from blood cultures. *J. Clin. Microbiol.* **3**, 425–31.

Honda, T. and Finkelstein, R.A. (1979). Purification and characterization of a haemolysin produced by *Vibrio cholerae* biotype El Tor: another toxic substance produced by cholera vibrios. *Infect. Immun.* **26**, 1020–7.

Honda, T. and Iida, T. (1993) The pathogenicity of *Vibrio parahaemolyticus* and the role of the thermostable direct haemolysin and related haemolysins. *Rev. Med. Microbiol.* **4**, 106–13.

Honda, T., Goshima, K., Takeda, Y., Sugino, Y. and Miwatani, T. (1976a). Demonstration of cardiotoxic activity of thermostable direct haemolysin (lethal toxin) produced by *Vibrio parahaemolyticus*. *Infect. Immun.* **13**, 163–71.

Honda, T., Taga, S., Takeda, T., Hasibuan, M.A., Takeda, Y. and Miwatani, T. (1976b). Identification of lethal toxin with thermostable direct haemolysin produced by *Vibrio parahaemolyticus* and some physico-chemical properties of the purified toxin. *Infect. Immun.* **13**, 133–9.

Honda, T., Narita, I., Yoh, M. and Miwatani, T. (1987). Purification and properties of two haemolysins produced by *Vibrio mimicus*. *Jpn. J. Bacteriol.* **42**, 201.

Honda, T., Ni, Y. and Miwatani, T. (1988). Characterization of a haemolysin produced by a clinical isolate of Kanagawa phenomenon-negative *Vibrio parahaemolyticus* and related to the thermostable direct haemolysin. *Infect. Immun.* **56**, 961–5.

Honda, T., Ni, Y., Hori, S., Takakura, H., Tsunasawa, S., Sakiyama, F. and Miwatani, T. (1989). A mutant haemolysin with lower biological activity produced by a mutant *Vibrio parahaemolyticus*. *FEMS Microbiol. Lett.* **61**, 95–100.

Honda, T., Ni, Y., Hata, A., Yoh, M., Miwatani, T., Okamoto, T., Goshima, K., Takakura, H., Tsunasawa, S. and Sakiyama, F. (1990). Properties of a haemolysin related to the thermostable direct haemolysin produced by a Kanagawa phenomenon negative, clinical isolate of *Vibrio parahaemolyticus*. *Can. J. Microbiol.* **36**, 395–9.

Honda, T., Ni, Y., Miwatani, T., Adachi, T. and Kim, J. (1992). The thermostable direct haemolysin of *Vibrio parahaemolyticus* is a pore-forming toxin. *Can. J. Microbiol.* **38**, 1175.

Ichinose, Y., Yamamoto, K., Nakasone, N., Tanabe, M., Takeda, T., Miwatani, T. and Iwanaga, M. (1987). Enterotoxicity of El Tor-like haemolysin of non-O1 *Vibrio cholerae*. *Infect. Immun.* **55**, 1090–3.

Iida, T., Tang, G.-Q., Suttikulpitug, S., Yamamoto, K., Miwatani, T. and Honda, T. (1995). Isolation of mutant toxins of *Vibrio parahaemolyticus* haemolysin by *in vitro* mutagenesis. *Toxicon* **33**, 209–16.

Ikigai, H., Akatsuka, A., Tsujiyama, H., Nakae, T. and Shimamura, T. (1996). Mechanism of membrane damage by El Tor haemolysin of *Vibrio cholerae* O1. *Infect. Immun.* **64**, 2968–73.

Ikigai, H., Ono, T., Iwata, M., Nakae, T. and Shimamura, T. (1997). El Tor haemolysin of *Vibrio cholerae* O1 forms channels in planar lipid bilayer membranes. *FEMS Microbiol. Lett.* **150**, 249–54.

Janda, J.M., Powers, C., Bryant, R.G. and Abbott, S. (1988). Current perspectives on the epidemiology and pathogenesis of clinically significant *Vibrio* spp. *Clin. Microbiol. Rev.* **1**, 245–67.

Kaper, J.B., Campen, R.K., Seidler, R.J., Baldini, N.M. and Falkow, S. (1984) Cloning of the thermostable direct or Kanagawa phenomenon-associated haemolysin of *Vibrio parahaemolyticus*. *Infect. Immun.* **45**, 290–2.

Kelly, M.T. and Stroh, E.M. (1988). Temporal relationship of *Vibrio parahaemolyticus* in patients and the environment. *J. Clin. Microbiol.* **26**, 1754–6.

Kim, G.-T., Lee, J.-Y, Hu, S.-H, Yu, J.-H. and Kong, I.-S. (1997). Nucleotide sequence of the *vmhA* gene encoding haemolysin from *Vibrio mimicus*. *Biochim. Biophys. Acta* **1360**, 102–4.

Klontz, K.C., Lieb, S., Schreiber, M., Janowski, H.T., Baldy, L.M. and Gunn, R.A. (1988). Syndromes of *Vibrio vulnificus* infections: clinical and epidemiological features in Florida cases, 1981–1987. *Ann. Intern. Med.* **109**, 318–23.

Kreger, A.S., Bernheimer, A.W., Etkin, L.A. and Daniel, L.W. (1987). Phospholipase D activity of *Vibrio damsela* cytolysin and its interaction with sheep erythrocytes. *Infect. Immun.* **55**, 3209–12.

Lin, Z., Kumagai, K., Mekalanos, J.J. and Nishibuchi, M. (1993). *Vibrio parahaemolyticus* has a homolog of the *Vibrio cholerae toxRS* operon that mediates environmentally induced regulation of the thermostable direct haemolysin gene. *J. Bacteriol.* **175**, 3844–55.

Massad, G., Simpson, L.M. and Oliver, J.D. (1988). Isolation and characterization of haemolysin mutants of *Vibrio vulnificus*. *FEMS Microbiol. Lett.* **56**, 295–300.

Menzl, K., Maier, E., Chakraborty, T. and Benz, R. (1996). HlyA haemolysin of *Vibrio cholerae* O1 biotype El Tor. Identification of the haemolytic complex and evidence for the formation of anion-selective ion-permeable channels. *Eur. J. Biochem.* **240**, 646–54.

Miwatani, T. and Takeda, Y. (1976). *Vibrio parahaemolyticus. A Causative Bacterium of Food Poisoning.* Saikon Publishing Co., Tokyo.

Miwatani, T., Takeda, Y., Sakurai, J., Yoshihara, A. and Taga, S. (1972). Effect of heat (Arrhenius effect) on crude haemolysin of *Vibrio parahaemolyticus*. *Infect. Immun.* **6**, 1031–3.

Miyake, M., Honda, T. and Miwatani, T. (1988). Purification and characterization of *Vibrio metschnikovii* cytolysin. *Infect. Immun.* **56**, 954–60.

Miyake, M., Honda, T. and Miwatani, T. (1989). Effects of divalent cations and saccharides on *Vibrio metschnikovii* cytolysin-induced haemolysis of rabbit erythrocytes. *Infect. Immun.* **57**, 158–63.

Miyamoto, Y., Kato, T., Obara, Y., Akiyama, S., Takizawa, K. and Yamai, S. (1969). *In vitro* haemolytic characteristic *Vibrio parahaemolyticus*: its close relation with human pathogenicity. *J. Bacteriol.* **100**, 1147–9.

Miyamoto, Y., Obara, Y., Nikkawa, T., Yamai, S., Kato, T., Yamada, Y. and Ohashi, M. (1980). Simplified purification and biophysicochemical characteristics of Kanagawa phenomenon-associated haemolysin of *Vibrio parahaemolyticus*. *Infect. Immun.* **28**, 567–76.

Miyoshi, S. and Shinoda, S. (1997). Bacterial metalloprotease as the toxic factor in infection. *J. Toxicol. Toxin Rev.* **16**, 177–94.

Miyoshi, S., Yamanaka, H., Miyoshi, N. and Shinoda, S. (1985). Non-thiol-activated property of a cholesterol-binding haemolysin produced by *Vibrio vulnificus*. *FEMS Microbiol. Lett.* **30**, 213–16.

Miyoshi, S., Oh, E.-G., Hirata, K. and Shinoda, S. (1993). Exocellular toxic factors produced by *Vibrio vulnificus*. *J. Toxicol. Toxin Rev.* **12**, 253–88.

Miyoshi, S. Narukawa, H., Tomochika, K. and Shinoda, S. (1995). Actions of *Vibrio vulnificus* metalloprotease on human plasma proteinase-proteinase inhibitor systems: a comparative study of native protease with its derivative modified by polyethylene glycol. *Micobiol. Immunol.* **39**, 959–66.

Miyoshi, S., Fujii, S., Tomochika, K. and Shinoda, S. (1997a). Some properties of nicked *Vibrio vulnificus* haemolysin. *Microbial Pathog.* **23**, 235–9.

Miyoshi, S., Sasahara, K., Akamatsu, S., Rahman, M. M., Katsu, T., Tomochika, K. and Shinoda, S. (1997b). Purification and characterization of a haemolysin produced by *Vibrio mimicus*. *Infect. Immun.* **65**, 1830–5.

Morris, J.G. (1990). Non-O group 1 *Vibrio cholerae*: a look at the epidemiology of an occasional pathogen. *Epidemiol. Rev.* **12**, 179–91.

Morris, J.G., Jr, Wright, A.C., Simpson, L.M., Wood, P.K., Johnson, D.E. and Oliver, J.D. (1987). Virulence of *Vibrio vulnificus*: association with utilization of transferrin-bound iron, and lack of correlation with levels of cytotoxin or protease production. *FEMS Microbiol. Lett.* **40**, 55–9.

Nagamune, K., Yamamoto, K. and Honda, T. (1995). Cloning and sequencing of a novel haemolysis gene of *Vibrio cholerae*. *FEMS Microbiol. Lett.* **128**, 265–9.

Nagamune, K., Yamamoto, K., Naka, A., Matsuyama, J., Miwatani, T. and Honda, T. (1996). *In vitro* proteolytic processing and activation of the recombinant precursor of El Tor cytolysin/haemolysin (pro-HlyA) of *Vibrio cholerae* by soluble haemagglutinin/protease of *V. cholerae*, trypsin, and other proteases. *Infect. Immun.* **64**, 4655–8.

Nagamune, K., Yamamoto, K. and Honda, T. (1997). Intramolecular chaperone activity of the pro-region of *Vibrio cholerae* El Tor cytolysin. *J. Biol. Chem.* **272**, 1338–43.

Nishibuchi, M. and Kaper, J.B. (1990). Duplication of the thermostable direct haemolysin (*tdh*) gene in *Vibrio parahaemolyticus*. *Mol. Microbiol.* **4**, 87–99.

Nishibuchi, M. and Kaper, J.B. (1995). Thermostable direct haemolysin gene of *Vibrio parahaemolyticus*: a virulence gene acquired by a marine bacterium. *Infect. Immun.* **63**, 2093–9.

Nishibuchi, M., Fasano, A., Russel, R.G. and Kaper, J.B. (1992). Enterotoxigenicity of *Vibrio parahaemolyticus* with and without genes encoding thermostable direct haemolysin. *Infect. Immun.* **60**, 3539–45.

Nishina, Y., Miyoshi, S., Nagase, A. and Shinoda, S. (1992). Significant role of an exocellular protease in utilization of heme by *Vibrio vulnificus*. *Infect. Immun.* **60**, 2128–32.

Oberley, T.D. and Duncan, J.L. (1971). Characteristics of streptolysin O action. *Infect Immun.* **4**, 683–7.

Oh, E.-G., Tamanoi, Y., Toyoda, A., Usui, K., Miyoshi, S., Chang, D.-S. and Shinoda, S. (1993). Simple purification method for a *Vibrio vulnificus* haemolysin by hydrophobic column chromatography. *Microbiol. Immunol.* **37**, 975–8.

Oh, E.-G., Yamanaka, H., Ohmae, K., Kim, Y.-M., Miyoshi, S. and Shinoda, S. (1994). Homogeneity of haemolysin produced by *Vibrio vulnificus* isolates. *J. Natural Toxins* **3**, 117–23.

Okada, K., Miake, S., Moriya, T., Matsuyama, M. and Amako, K. (1987). Variability of haemolysin(s) produced by *Vibrio vulnificus*. *J. Gen. Microbiol.* **133**, 2853–7.

Okuda, J., Ishibashi, M., Abbot, S.L., Janda, J.M. and Nishibuchi, M. (1997). Analysis of the thermostable direct haemolysin (*tdh*) gene and the *tdh*-related haemolysin (*trh*) genes in urease-positive strains of *Vibrio parahaemolyticus* isolated on the West Coast of the United States. *J. Clin. Microbiol.* **35**, 1965–71.

Oliver, J.D. (1995). The viable but non-culturable state in the human pathogen *Vibrio vulnificus*. *FEMS Microbiol. Lett.* **133**, 203–8.

Oosawa, R., Okitsu, T., Morozumi, H. and Yamai, S. (1996). Occurrence of urease-positive *Vibrio parahaemolyticus* in Kanagawa, Japan, with specific reference to presence of thermostable direct haemolysin (TDH) and the TDH-related-haemolysin genes. *Appl. Environ. Microbiol.* **62**, 725–7.

Pal, S., Guhathakurta, B., Samal, D., Mallick, R. and Datta, A. (1997). Purification and characterization of a haemolysin with phospholipase C activity from *Vibrio cholerae* O139. *FEMS Microbiol. Lett.* **147**, 115–20.

Pridgent, D. and Alouf, J.E. (1976). Interaction of streptolysin O with sterols. *Biochim. Biophys. Acta* **443**, 288–300.

Rahim, Z. and Aziz, K.M. (1996). Factors affecting production of haemolysin by strains of *Vibrio fluvialis*. *J. Diarrhoeal Dis. Res.* **14**, 113–16.

Rahman, M.M., Miyoshi, S., Tomochika, K., Wakae, H. and Shinoda, S. (1997). Analysis of the structural gene encoding a haemolysin in *Vibrio mimicus*. *Microbiol. Immunol.* **41**, 169–73.

Ramamurthy, T., Garg, S., Sharma, R., Bhattacharya, S.K., Nair, G.B., Shimada, T., Takeda, T., Kurasawa, T., Kurazono, H., Pal, A. and Takeda, Y. (1993). Emergence of a novel strain of *Vibrio cholerae* with epidemic potential in southern and eastern India. *Lancet* **314**, 703–4.

Roland, F.P. (1970). Leg gangrene and endotoxin shock due to *Vibrio parahaemolyticus*. An infection acquired in New England coastal water. *New Engl. J. Med.* **282**, 1306.

Saha, N. and Banerjee, K.K. (1997). Carbohydrate-mediated regulation of interaction of *Vibrio cholerae* haemolysin with erythrocyte and phospholipid vesicle. *J. Biol. Chem.* **272**, 162–7.

Sakazaki, R., Iwanami, S. and Fukumi, H. (1968). Studies on the enteropathogenic, facultative halophilic bacteria, *Vibrio parahaemolyticus*. II. Serological characteristics. *Jpn. J. Med. Sci. Biol.* **21**, 313–24.

Sakurai, J., Honda, T., Jinguji, Y., Arita, M. and Miwatani, T. (1976). Cytotoxic effect of the thermostable direct haemolysin produced by *Vibrio parahaemolyticus* on FL cells. *Infect. Immun.* **13**, 876–83.

Shinoda, S., Miyoshi, S., Yamanaka, H. and Miyoshi-Nakahara, N. (1985). Some properties of *Vibrio vulnificus* haemolysin. *Microbiol. Immunol.* **29**, 583–90.

Shinoda, S., Matsuoka, H., Tsuchie, T., Miyoshi, S., Yamamoto, S., Taniguchi, H. and Mizuguchi, Y. (1991). Purification and characterization of a lecithin-dependent haemolysin from *Escherichia coli* transformed by a *Vibrio parahaemolyticus* gene. *J. Gen. Microbiol.* **137**, 1737–42.

Shinoda, S., Ishida, K., Oh, E.-G., Sasahara, K., Miyoshi, S., Chowdhury, M.A.R. and Yasuda, T. (1993). Studies on haemolytic action of a haemolysin produced by *Vibrio mimicus*. *Microbiol. Immunol.* **37**, 405–9.

Spira, W.M. and Fedorka-Cray, P.J. (1984). Purification of enterotoxin from *Vibrio mimicus* that appears to be identical to cholera toxin. *Infect. Immun.* **45**, 679–84.

Stollerman, G.H., Brodie, B.B. and Stelle, J.M. (1952). The relationship of streptolysin S inhibitor to phospholipids in the serum of human-beings and experimental animals. *J. Clin. Invest.* **31**, 180–7.

Suthienkul, O., Ishibashi, M., Iida, T., Nettip, N., Supavej, S., Eampokalap, B., Makino, M. and Honda, T. (1995). Urease production correlates with possession of the *trh* gene in *Vibrio parahaemolyticus* strains isolated in Thailand. *J. Infect. Dis.* **172**, 1405–8.

Takeda, T., Peina Y., Ogawa, A., Dohi, S., Abe, H., Nair, G.B. and Pal, S.C. (1991). Detection of heat-stable enterotoxin in a cholera toxin gene-positive strain of *Vibrio cholerae* O1. *FEMS Microbiol. Lett.* **80**, 23–8.

Takeda, Y. (1983). Thermostable direct haemolysin of *Vibrio parahaemolyticus*. *Pharmacol. Ther.* **19**, 123–46.

Takeda, Y., Hori, Y., Taga, S., Sakurai, J. and Miwatani, T. (1975a). Characterization of the temperature-dependent inactivating factor of the thermostable direct haemolysin in *Vibrio parahaemolyticus*. *Infect. Immun.* **12**, 449–54.

Takeda, Y., Takeda, T., Honda, T., Sakurai, J., Ohtomo, N. and Miwatani, T. (1975b). Inhibition of haemolytic activity of thermostable direct haemolysin of *Vibrio parahaemolyticus* by ganglioside. *Infect. Immun.* **12**, 931–3.

Tang, G.-Q. Iida, T., Yamamoto, K. and Honda, T. (1994). A mutant toxin of *Vibrio parahaemolyticus* thermostable direct haemolysin which has lost haemolytic activity but retains ability to bind to erythrocytes. *Infect. Immun.* **62**, 3299–304.

Tang, G.-Q., Iida, T., Yamamoto, K. and Honda, T. (1995). Ca^{2+}-independent cytotoxicity of *Vibrio parahaemolyticus* thermostable direct haemolysin (TDH) on Intestine 407, a cell line derived from human embryonic intestine. *FEMS Microbiol. Lett.* **134**, 233–8.

Tang, G.-Q., Iida, T., Inoue, H., Yutsudo, M., Yamamoto, K. and Honda, T. (1997a). A mutant cell line resistant to *Vibrio parahaemolyticus* thermostable direct haemolysin (TDH): its potential in identification of putative receptor for TDH. *Biochim. Biophys. Acta* **1360**, 277–82.

Tang, G.-Q., Iida, T., Yamamoto, K. and Honda, T. (1997b). Analysis of functional domains of *Vibrio parahaemolyticus* thermostable direct haemolysin using monoclonal antibodies. *FEMS Microbiol. Lett.* **150**, 289–96.

Taniguchi, H., Ohta, H., Ogawa, M. and Mizuguchi, Y. (1985). Cloning and expression in *Escherichia coli* of *Vibrio parahaemolyticus* thermostable direct haemolysin and thermolabile haemolysin genes. *J. Bacteriol.* **162**, 510–15.

Testa, J., Daniel, L.W. and Kreger, A.S. (1984). Extracellular phospholipase A$_2$ and lysophospholipase produced by *Vibrio vulnificus*. *Infect. Immun.* **45**, 458–63.

Toda, H., Sakiyama, F., Yoh, M., Honda, T. and Miwatani, T. (1991). Tryptophan 65 is essential for haemolytic activity of the thermostable direct haemolysin from *Vibrio parahaemolyticus*. *Toxicon* **29**, 837–44.

Trucksis, M., Galen, J.E., Michalski, J., Fasano, A. and Kaper, J.B. (1993). Accessory cholera enterotoxin (Ace), the third toxin of a *Vibrio cholerae* virulence cassette. *Proc. Natl Acad. Sci. USA* **90**, 5267–71.

Tsunasawa, S., Sugihara, A., Masaki, T., Sakiyama, F., Takeda, Y., Miwatani, T. and Narita, K. (1987). Amino acid sequence of thermostable direct haemolysin produced by *Vibrio parahaemolyticus*. *J. Biochem.* **101**, 111–21.

Wagatsuma, S. (1968). On a medium for hemolytic reaction. *Media Circle* **13**, 159–62.

Wright, A.C. and Morris, J.G. Jr (1991). The extracellular cytolysin of *Vibrio vulnificus*: inactivation and relationship to virulence in mice. *Infect. Immun.* **59**, 192–7.

Yamada, S., Matsushita, S., Kudoh, Y. and Ohashi, M. (1988). Purification and characterization of haemolysin produced by *Vibrio fluvialis*. *Adv. Res. Cholera Relat. Diarrhoeas* **4**, 111.

Yamamoto, K., Ichinose, Y., Nakasone, N., Tanabe, M., Nagahama, M. Sakurai, J. and Iwanaga, M. (1986). Identification of haemolysins produced by *Vibrio cholerae* non-O1 and *Vibrio cholerae* O1, biotype eltor. *Infect. Immun.* **51**, 927–31.

Yamamoto, K., Ichinose, Y., Shinagawa, H., Makino, K., Nakata, A., Iwanaga, M., Honda, T. and Miwatani, T. (1990a). Two-step processing for activation of the cytolysin/haemolysin of *Vibrio cholerae* O1 biotype El Tor: nucleotide sequence of the structural gene (*hly A*) and characterization of the processed products. *Infect. Immun.* **58**, 4106–16.

Yamamoto, K., Wright, A.C., Kaper, J.B. and Morris, J.G., Jr (1990b). The cytolysin gene of *Vibrio vulnificus*: sequence and relationship to the *Vibrio cholerae* El Tor haemolysin gene. *Infect. Immun.* **58**, 2706–9.

Yamanaka, H., Katsu, T., Satoh, T. and Shinoda, S. (1987a). Effect of *Vibrio vulnificus* haemolysin on liposome membranes. *FEMS Microbiol. Lett.* **44**, 253–8.

Yamanaka, H., Satoh, T., Katsu, T. and Shinoda, S. (1987b). Mechanism of haemolysis by *Vibrio vulnificus* haemolysin. *J. Gen. Microbiol.* **133**, 2859–64.

Yamanaka, H., Shimatani, S., Tanaka, M., Katsu, T., Ono, B. and Shinoda, S. (1989). Susceptibility of erythrocytes from several animal species to *Vibrio vulnificus* haemolysin. *FEMS Microbiol. Lett.* **61**, 251–6.

Yamanaka, H., Sugiyama, K., Furuta, H., Miyoshi, S. and Shinoda, S. (1990). Cytolytic action of *Vibrio vulnificus* haemolysin on mast cells from rat peritoneal cavity. *J. Med. Microbiol.* **32**, 39–43.

Yoh, M., Honda, T., Miwatani, T., Tsunasawa, S. and Sakiyama, F. (1989). Comparative amino acid sequence analysis of hemolysins produced by *Vibrio hollisae* and *Vibrio parahaemolyticus*. *J. Bacteriol.* **171**, 6859–61.

Yoh, M., Tang, G.-Q, Iida, T., Morinaga, N., Noda, M. and Honda, T. (1996). Phosphorylation of a 25-kDa protein is induced by thermostable direct haemolysin of *Vibrio parahaemolyticus*. *Int. J. Biochem. Cell Biol.* **28**, 1365–9.

Yoshida, H., Honda, T. and Miwatani, T. (1991). Purification and characterization of a haemolysin of *Vibrio mimicus* that relates to thermostable direct haemolysin of *Vibrio parahaemolyticus*. *FEMS Microbiol. Lett.* **84**, 249–54.

Zitzer, A., Palmer, M., Weller, U., Wassennaar, T., Biermann, C., Tranum-Jensen, J. and Bhakdi, S. (1997). Mode of primary binding to target membranes and pore formation induced by *Vibrio cholerae* cytolysin (haemolysin). *Eur. J. Biochem.* **247**, 209–16.

20

δ-Toxin, related haemolytic toxins and peptidic analogues

Jean Dufourcq, Sabine Castano and Jean-Claude Talbot

INTRODUCTION

δ-Toxin from *Staphylococcus aureus*, also called δ-haemolysin or δ-lysin, is a 26-residue long peptide encoded by the *hlg* gene (Janzon *et al.*, 1989) and secreted by almost all pathogenic strains (Freer and Arbuthnott, 1983). It was first identified owing to synergistic action with other toxins secreted by *S. aureus*, and is better characterized as a haemolytic factor different from the more well-known α- and β-haemolysins (Linder, 1984). Its properties were extensively reviewed in the 1970s and 1980s (McCartney and Arbuthnott, 1978; Freer and Arbuthnott, 1983; Freer *et al.*, 1984; Freer, 1986).

The peptide belongs to the family of amphipathic α-helical peptides which contains the vast majority of direct lytic factors (Cornut *et al.*, 1993; Saberwal and Nagaraj, 1994). They first act on membranes through direct contact with lipids. This plasma-membrane invasion results in drastic changes in the permeability barrier. The behaviour of δ-toxin will be discussed within the framework of our current knowledge of such a class of non-specific or weakly specific cytotoxins generally able to act as pleiotropic agents on most living cells. This class of haemolytic compounds is characterized by their wide variability in amino acid sequences. The behaviour of δ-toxin will be compared with that of other compounds with no sequence homology, but which share some or most of the properties of δ-toxin, the main one being haemolysis.

During the past few years, numerous reviews on haemolytic amphipathic peptides have been published (Dempsey, 1990; Cornut *et al.*, 1993; Saberwal and Nagaraj, 1994; Bechinger, 1997; Lohner and Epand, 1997) but δ-toxin is only marginally mentioned, probably because it is not antibacterial and its behaviour is more complex than that of other more soluble toxins.

This chapter will review extensively the recent literature, focusing on the haemolytic activity and mechanisms proposed today; it will also summarize what has been learned from studies on synthetic analogues. The other properties of δ-toxin will be discussed in parallel with toxins sharing its peculiar α-helical amphipathic structure. After recalling the basic knowledge on the isolation, sequence and structure of the toxin, its effects on lipids will be discussed, the properties of analogues summarized and their activities will be described with emphasis on haemolysis. This will facilitate discussion on structure–function relationships and on the mechanisms of action of δ-toxin.

AMINO-ACID COMPOSITIONS AND SEQUENCES OF δ-TOXIN, ANALOGUES AND RELATED TOXINS

δ-Toxin is a low M_r compound secreted by *S. aureus* with peak production at the end of the exponential growth phase and can be purified by different procedures

386

(Birkbeck and Freer, 1988). It is now obtained by reversed-phase high-performance liquid chromatography (RP-HPLC) as a single peak eluted in water-CH_3CN-rich medium indicative of strong interactions with the C_{18} chains of the stationary phase (Tappin *et al.*, 1988). The high solubility of the peptide in polar organic solvents such as methanol also reflects the fact the toxin is quite hydrophobic, as first documented by its rather poor water solubility and the discrepancies in the first defined relative molecular mass, due to aggregation (see below). It can be heated for 1 h at 90°C without change in activity, as opposed to most other *S. aureus* toxins. The monomeric toxin is a 26-residue long peptide with a relative molecular mass of 3003. It lacks Cys, Tyr, His, Arg and Pro. It has a single Trp which allows quantitative estimation of toxin concentration, using a molar extinction coefficient ε_{280} ~5600 M^{-1} cm^{-1}. About 46% of the residues are hydrophobic.

The sequence of the toxin determined by Fitton *et al.* (1980) is quite original and shows that the apolar residues are distributed all along the peptide (Figure 20.1). There is no apolar domain in the sequence but a clear alternation of polar and apolar residues, with a periodicity compatible with the amphipathic one of an α-helix, is well illustrated using the simplified binary reading (Figure 20.1). The charged residues are distributed with three acidic ones, namely $Asp_{4/11/18}$, more concentrated in the N-terminal part, while the four basic residues, $Lys_{14/22/25/26}$, are mainly clustered at the C-terminus. The N-terminus is formylated and the doublet $Lys_{25/26}$ over-compensates for the negative charge of the free C-terminal group. As a result, at pH 7.4, δ-toxin has no net charge, which makes it atypical in the cytotoxic field where most of the active compounds are basic.

As first proposed by Freer and Birkbeck (1982), assuming a folding in an α-helix, the helical wheel projection shows the characteristic amphipathic pattern of the toxin (Figure 20.2). This reveals an almost perfect segregation of all the apolar residues on one face while the polar and charged ones form another face of 180°. The hydrophobic residue Trp_{15} is the only odd residue located in the polar area. The perfect distribution of all the residues within such a characteristic periodic structure makes δ-toxin a classical member of the family of haemolytic amphipathic peptides, like melittin, the first well-characterized member. In this latter case, despite having the same total length, the regular amphipathic helical pattern only extends from residues 1–21 and it is more hydrophobic with a 120° wide polar sector (Figure 20.2). At first glance, the melittin sequence looks like a detergent with its hydrophobic residues in the 1–20 segment and the totally polar and basic residues in the 21–26 C-terminal peptide. However, the very recent characterization of a shorter fully active melittin-like peptide (Simmaco *et al.*, 1996), which lacks the C-terminal segment, containing only a single Lys_{21} and Gln_{22}, strongly emphasizes that it is the N-terminal segment with its typical secondary amphipathic structure (Figure 20.2) which endows the activity. The sequence of the shorter 22-residue long magainin isolated from skin secretions of different frogs has a more similar amphipathic profile to δ-toxin. Again, the amphipathy of the helix develops all along the sequence from the N- to the C-terminal. However, magainin differs significantly; it has a single acidic residue, Glu_{19}, and, like almost all the other antibacterial peptides, it is strongly basic owing to the presence of 5 Lys and 1 His. Pardaxin, secreted by sole, is the only other haemolytic peptide which has no net charge due to similar amounts of Asp and Lys in its 33-residue long sequence (Shai *et al.*, 1988). Negatively charged alamethicins and analogues can be active but have different hydrophobicity patterns (Cornut *et al.*, 1993; Bechinger, 1997). As soon as the amphipathy of such putative α-helical structures is recognized, it is possible to compare quantitatively their hydrophobic

	1				5					10					15					20						26
δ toxin																										
S. aureus human (Fitton *et al.*, 1980)	fM	A	Q	D	I	I	S	T	I	G	D	L	V	K	W	I	I	D	T	V	N	K	F	T	K	KC_{OO}
S. aureus canine (Fitton *et al.*, 1984)		A							V	E	F				L		A	E			E					
S. epidermis (McKevitt *et al.*, 1990)		A																						I	K	KC_{OO}
Binary reading	o	o	o/+	+	o	o	+	+	o	~	+	o	o	+	o	o	o	+	+	o	+	+	o	o/+	+	+
Analogue F (Alouf *et al.*, 1989)		L			L	L		S	L						L		S	W	L							
Analogue M		L			L	L		S	L	L					L		S	W	L							
Analogue M+ (Thiaudiere, 1990)		L		N	L	L		S	L	L	N				L		S	W	L	N				L	L	
5-20 analogue (Dhople and Nagaraj, 1993)					I	I	S	T	I	G	K	L	V	K	W	I	I	K	T	V						

FIGURE 20.1 Sequence of different natural δ-toxins and synthetic analogues. Binary reading corresponds to the code: hydrophobic residue o; hydrophilic residue +; indifferent ~. Other analogues with individual substitutions from these leading sequences are discussed in the text. In analogues, only changes with regard to *S. aureus* δ-toxin are reported. The C-terminus fragments of peptides M and M+ with progressive shortening are named L, corresponding to the sequence 4–26, K for the segment 8–26, J for the segment 11–26.

moments, μ_H (Eisenberg *et al.*, 1982; Cornut *et al.*, 1993). Owing to its quasi-ideal and long periodic structure, the μ_H value for δ-toxin (9.87) and the mean μ_H per residue, $\langle\mu_H\rangle$, (0.38) are about the highest values in the family of such toxins (Figure 20.3).

Other *S. aureus* strains and different staphylococcal species secrete peptides sharing the properties of δ-toxins. First, a new sequence of a 26-residue long haemolytic factor from *S. aureus* infecting dogs was described (Figure 20.1). It differs quite significantly by

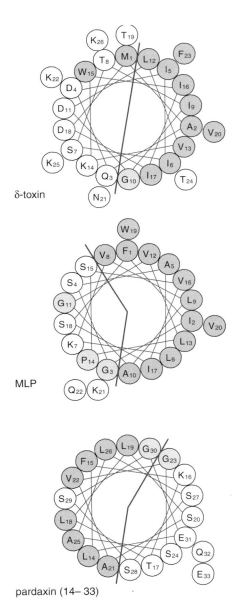

δ-toxin

MLP

pardaxin (14– 33)

FIGURE 20.2 Helical wheel projections of amphipathic cytotoxic peptides. Top: δ-toxin; centre: melittin-like peptide (MLP) isolated from frog *Rana temporaria* (Simmaco *et al.*, 1996); bottom: pardaxin from *Pardachirus marmoratus* (Shai *et al.*, 1988); only the segment 14–33 is represented in the wheel.

35% of the residues (Fitton *et al.*, 1984). All mutations occur without any drastic change to the hydropathic pattern of δ-toxin, as indicated by a 92% analogy in terms of hydrophilic to hydrophobic balance. The most significant changes are the systematic substitutions of Asp into Glu. The Glu3Ala, Thr$_{24}$Ile mutations result in an increased hydrophobicity and an enlarged hydrophobic face. The change of Trp$_{15}$ into Leu shows that this residue is not mandatory, contrasting with melittin, where such a Trp–Leu substitution drastically reduces the biological activity (Blondelle and Houghten, 1991). In comparison *S. epidermidis* secretes a much more closely related δ-toxin, since it only differs from the Gln3Ala substitution and the single deletion of Thr$_{24}$, which results in a 25-residue long peptide (McKevitt *et al.*, 1990) (Figure 20.1).

Some other staphylococcal species secrete δ-toxin-like compounds, i.e. haemolytic toxins which in their first assays have similar physical and/or biological behaviour to δ-toxins. However, when isolated and fully characterized, the corresponding peptides are different. A peptide known as gonococcal growth inhibiting factor was first isolated from *S. haemolyticus* (Frenette *et al.*, 1984). Like δ-toxin it shares a similar spectrum of haemolytic activity on erythrocytes of different origins and is inhibited by lecithins. However, it is antigenically totally different and further purification reveals that three closely related peptides with 75% sequence homology account for such activity. These peptides, about 44 residues long, have a net acidic character. Their sequences (Watson *et al.*, 1988) show no homology with that of δ-toxin. However, from these data one can anticipate that such peptides are strongly amphipathic and can be classified in the same family of haemolytic compounds. The putative folding of all the peptide in a single α-helix indicates a low amphipathy (Figure 20.3) with $\mu_H \sim 5.25$, i.e. $\langle\mu_H\rangle = 0.12$, but closer inspection suggests folding in two helical domains. The first putative helix from the *N*-terminus to Gly$_{15}$ or Lys$_{17}$ shows a clear segregation of the polar residues with an angle of 160° and $\mu_H \sim 4.5$, i.e. $\langle\mu_H\rangle = 0.27$. All of the mutations within the series of compounds which occur in positions 9,10,13,17 of the polar face fit with the same topology. Another very typical amphipathic helix runs to the *C*-terminus of the peptide. It can be up to 26 residues long and it has no net charge, with three Lys groups at the centre of the polar face compensated by Glu$_{19,32}$ and the *C*-terminal group located at the border of the apolar face (Figure 20.4). Again, within the three different peptides all mutations occur in the polar face, they are conservative and maintain the same segregation of polar and apolar residues with a high $\mu_H = 7.6$ value (Figure 20.3), i.e. $\langle\mu_H\rangle = 0.29$. Although this polar face covers about 180°, like that of δ-toxin, the charge

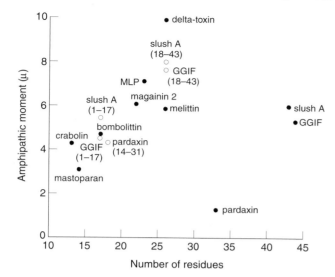

FIGURE 20.3 Amphipathic moment of the cytotoxic peptides when assuming ideal α-helix folding. Total hydrophobic moments were calculated using the Eisenberg concensus hydrophobic scale (Eisenberg *et al.*, 1982). GGIF: gonococcal growth inhibiting factor (Watson *et al.*, 1988); SLUSH: *Staphylococcus lugdunensis* synergistic haemolytic factor (Donvito *et al.*, 1997); MLP: melittin-like peptide (Simmaco *et al.*, 1996).

(1989) first developed a series of peptides to look at important residues in the sequence. Analogues were synthesized with a systematic change of apolar residues into Leu to obtain an ideal amphipathic helix. Therefore, Trp$_{15}$ was moved in position 16 and substituted by a polar Ser (analogue F, Figure 20.1). In another series, the Gly$_{10}$, implicated in putative conformational flexibility of the peptide, was also

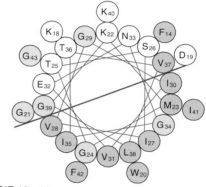

GGIF 18– 44

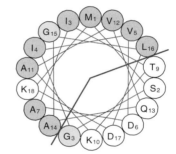

slush A 1–18

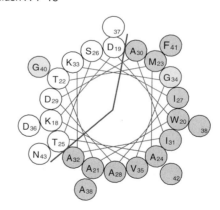

slush A 18– 43

topology differs. Finally, one should mention that this toxin has up to six Gly residues, at variance from the single Gly$_{10}$ of δ-toxin. This enriched amount of Gly is a rather general feature of the longer α-helical cytotoxins, often active as antibacterial compounds, as already seen in the magainin sequence with five Gly in a peptide of only 23 amino acids (Saberwal and Nagaraj, 1994).

A δ-toxin-like activity was also detected on some strains of *S. lugdunensis* (Vandenesch *et al.*, 1991) but these haemolytic peptides are encoded on a locus other than the hlg-related one (Vandenesch *et al.*, 1993). The recent isolation of the components responsible for such a haemolytic activity phenotypically similar to that of δ-toxin proved that at least three different but quite similar peptides, named SLUSH A, B, C, are involved. Their amphipathic pattern is identical to that described above for *S. haemolyticus*, with two putative helices with high μ$_H$ values. The first helix going from position 1→17 has a 100° polar face and μ$_H$ = 5.45. That running from Lys$_{18}$ to the C-terminus has a 140° hydrophilic face and μ$_H$ ~7.6, lying between those for melittin and δ-toxin and somewhat higher than those of pardaxin or magainin (Figure 20.3). It is quite similar to δ-toxin in its topology and length, eight residues are identical and all polar/apolar residues are in register with δ-toxin (Figure 20.4).

Owing to their relatively short lengths synthetic δ-toxins and analogues can be synthesized. Alouf *et al.*

FIGURE 20.4 Helical wheel representations of gonococcal growth inhibitory factor (GGIF) (Watson *et al.*, 1988) and *Staphylococcus lugdunensis* peptide SLUSH-A folded in two putative α-helical segments (Donvito *et al.*, 1997); top: C-terminus helix fragment 18–44 of GGIF; centre: the N-terminus helix fragment 1–18 of SLUSH; bottom: its C-terminus fragment 18–43.

replaced by Leu (analogue M, Figure 20.1) resulting in a perfect ideal helix able to run all along the peptide. Owing to its peculiarity of having no net charge, δ-toxin was also modified to carry a net positive charge, like the other members of the lytic peptide family. This results in a series of peptides called M+ (Figure 20.1) characterized by the systematic Asp→Asn substitution (Cornut *et al.*, personal communication). The more drastic Asp→Lys change produced strongly basic δ-toxin analogues (Dhople and Nagaraj, 1993).

STRUCTURE OF δ-TOXIN IN SOLUTION

In order to achieve a thorough understanding of the complex behaviour of these peptides towards bilayer binding, haemolysis and other biological activities, a knowledge of their structural features in solution is necessary.

Secondary structure

The putative α-helix of δ-toxin has been confirmed by several authors but they found very different amounts of α-helix in aqueous solution, from 40% to 87% (Colaccico *et al.*, 1977; Fitton, 1981; Lee *et al.*, 1987; Garone *et al.*, 1988). In fact, there is a strong dependence of the helicity on concentration, from 31% at 0.2 μM to 78% above 4 μM (Thiaudière *et al.*, 1991). Such behaviour is reminiscent of that observed for other cytotoxic peptides such as melittin, but such a structural transition occurs at concentrations two orders of magnitude higher (Cornut, 1993). δ-Toxin fragments and analogues are also mainly α-helical (Tappin *et al.*, 1988; Alouf *et al.*, 1989; Thiaudière, 1990; Thiaudière *et al.*, 1991; Dhople and Nagaraj, 1995), except for the shortest 1–11 and 11–26 peptides (Thiaudière, 1990; Thiaudière *et al.*, 1991). As shown by circular dichroism (CD) experiments, from 0.2 to 100 μM, peptide F, a 26-residue long peptide analogous to δ-toxin in section 21–26 but with Leu and Ser replacements in section 1–20, Trp_{16} in the apolar face and Trp15Ser change in the polar face (Figure 20.1), undergoes a structural transition very similar to that of δ-toxin. In contrast, deformylated peptide F with Gly10Leu change (peptide M, Figure 20.1) and the 23-residue C-terminal fragment of δ-toxin, are always predominantly α-helical in this concentration range (Thiaudière, 1990). This structure may be stabilized by the replacement of Gly_{10}, known as an α-helix breaker (Creighton, 1996). The presence of Asp_4 in these two analogues participates in α-helix stability since the shorter analogue of this series (19-residue long which lacks Asp_4) undergoes a

smooth concentration-dependent coil/α-helix transition. The shortest 16-residue long natural fragment 11–26 and analogue remain randomly coiled. When folded, all of these peptides from 19–26 residue long have 16–18 residues embedded in the α-helix, which should indicate that this helix might extend from Ser_7 to Phe_{23} in aqueous solution. In acidic methanol, nuclear magnetic resonance (NMR) measurements indicate an α-helix from residue 2–20 (Tappin *et al.*, 1988). This result has to be compared with the minimized energy calculation *in vacuo* which predicts an α-helix from residue 7–20 (Raghunathan *et al.*, 1990). Peptide M and shorter analogues, with Asp to Asn and apolar/uncharged polar replacements extended to the whole sequence, have been synthesized (see peptide M+, Figure 20.1). These analogues, unable to form intracatenary salt bridges, have a net charge up to +4 and can give rise to an amphipathic helix extending over the entire sequence owing to a Thr24Leu change (Cornut, 1993). This was actually observed for the longest peptides (23 and 26 residues). On the contrary, the shortest ones are less helical on increasing their charge from +2 to +4. This shows that, when the hydrophobic effect is less, due to a shorter amphipathic helix, the inability to form intrachain salt bridges inhibits helix folding but, when the hydrophobic effect is dominant, as for longer peptides, the charge has no effect on helix formation.

Oligomerization

The α-helix folding creates an apolar area whose interactions with the aqueous medium are energetically unfavourable. Therefore, self-association of these amphiphilic helices via their apolar faces should occur. The shift of Trp from position 15 (polar face) to 16 (apolar face) should result in drastic fluorescence changes when aggregation occurs. Whereas the emission of Trp_{15} random-coiled 11–26 fragment always remains at 351 nm (characteristic of water-exposed Trp) (Thiaudière, 1990), the fluorescence of Trp_{16} peptide F and Trp_{15} δ-toxin is very sensitive to increasing concentrations. Their emission is shifted from 344 to 333 nm and 351 to 331 nm, respectively, values characteristic of Trp in an apolar environment, in the same concentration range as helix folding (Thiaudière *et al.*, 1991). Nevertheless, while the transconformation develops, Trp_{16} remains more buried and immobilized than Trp_{15}, as proved by its lower emission wavelength and its higher fluorescence polarization. From anisotropy measurements of peptide F at low concentrations, it can be inferred that, when helical, these peptides associate in tetramers, burying the apolar faces of helices and exposing their polar ones to the aqueous solvent.

Multimerization

Beyond the region of coil to helix folding, the fluorescence wavelength and quantum yield of Trp_{16} of peptide F do not change, whereas the emission intensity of Trp_{15} δ-toxin is still increasing. The immobilization of both the fluorescent probes is still increasing, much more for Trp_{15} which is totally motionless within the timescale of fluorescence lifetimes. Time-resolved fluorescence anisotropy (Talbot *et al.*, personal communication) confirms that 26-residue long δ-toxin and peptide M at 20 µM reach high degrees of association (infinite rotational correlation times) but peptide F has a weaker tendency to self-associate, reflecting the α-helix destabilizing effect of Gly_{10}. One can anticipate the role of the numerous Gly in the related peptides such as gonococcal growth inhibitory factor and SLUSH (Figure 20.4). First, they should increase the conformational freedom and therefore decrease this tendency to self-associate in buffers. The shorter fragment (19 residues) of peptide M shows the same behaviour as δ-toxin, peptides F and M, but the shortening of the helix induces a shift of the self-association process to higher concentrations, with a saturation effect, around eight associated tetramers reached at 20 µM (Thiaudière *et al.*, personal communication). High degrees of aggregation of δ-toxin, from 21- to 100mer, have also been characterized by analytical centrifugation (Kreger *et al.*, 1971; Kantor *et al.*, 1972) and gel filtration (Fitton, 1981). Sedimentation equilibrium analysis (Thiaudière *et al.*, 1991) shows that δ-toxin undergoes an isodesmic self-association equilibrium of tetramers with $K = 1.5 \times 10^6 \ M^{-1}$. From hydrodynamic properties (sedimentation coefficient and quasi-elastic light scattering (QELS) measurements), δ-toxin is described as a rod-like model made of tetramers of four parallel or antiparallel helices with their axes perpendicular to the long axis of the rod (Thiaudière *et al.*, 1991) (Figure 20.5). The experimental results cannot be accounted for by another type of rod-like assembly or by a raft model (oblate ellipsoid or disc) as proposed by Raghunathan *et al.* (1990) from crystallographic data and energy calculation. The multimeric linear association of tetramers of δ-toxin involves interactions between the polar faces of α-helices of tetramers. This is documented by: (a) the shielding of Trp_{15} from the solvent (Thiaudière *et al.*, 1991); (b) its exposure at pH values under 4 and above 9 where δ-toxin is only tetrameric (Fitton, 1981); (c) the solubilization of poorly soluble peptide M from pH 7 to 2 (Thiaudière *et al.*, personal communication); (d) the decrease of scattered light by peptide M on increasing ionic strength (Thiaudière *et al.*, personal communication). These effects suggest intercatenary electrostatic interactions

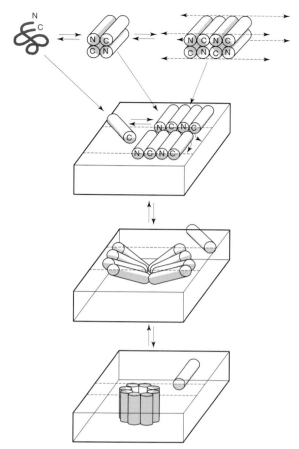

FIGURE 20.5 Various states of δ-toxin when alone in solution and bound to lipid bilayers. Schematic representation synthesizing information (adapted from Rhagunathan *et al.*, 1990; Thaudière *et al.*, 1991; Shai, 1995; Kerr *et al.*, 1996). Box represents the hydrophobic core of a lipid bilayer. The apolar faces are shaded.

between acidic and basic residues which should be favoured in an antiparallel orientation of helices, as predicted by Raghunathan *et al.* (1990).

In conclusion, δ-toxin behaviour in aqueous solution is now well understood. The natural toxin and analogues undergo a transition from disordered to α-helical secondary structure on increasing concentration, provided that the peptide length is sufficient (19 residues). This process is favoured when intracatenary salt bridges occur. These helices are amphipathic, with one polar face and one apolar face. In aqueous medium, they associate to form tetramers maintained by hydrophobic interactions. These tetramers associate in rod-like particles via electrostatic interactions, with antiparallel helices perpendicular to the long axis of the rod (Figure 20.5). The propensity to reach excessive degrees of polytetrameric association inducing precipitation is limited by Gly_{10} and an amphiphilic character not extended to the full peptide chain (Thr_{24} in the apolar side and no more than 18 residues in the α-helix). In the analogues, the longer the amphipathic helix and the more it is ideal, the higher the degree of self-association, with a tendency to precipitate for the most associated peptides.

INTERACTIONS WITH BILAYERS AND MEMBRANES – STRUCTURES OF BOUND δ-TOXIN

As soon as their amphipathy is well recognized, all such peptides will be classified as surface-seeking compounds.

Secondary structure in the bound state

NMR studies by Lee *et al.* (1987) showed that, in dodecylphosphocholine micelles, δ-toxin is α-helical from residue 5–23. This is confirmed by CD measurements of the toxin in detergents (Dhople and Nagaraj, 1995) or lysophosphatidylcholine micelles (Alouf *et al.*, 1989) and when bound to bilayers consisting of fluid dimyristoylphosphatidyl choline (DMPC) or of lecithin/phosphatidyl serine (PC/PS) (Thiaudière *et al.*, 1991; Cornut, 1993). The lytic peptides in association with lipids are always 70–80% α-helical, the only exception being the shortest fragment, 11–26, which is less structured (55%) (Thiaudière *et al.*, 1991).

Surface activity of δ-toxin

This develops as soon as its bulk concentration is higher than 10^{-7} M, i.e. when self-association occurs. The maximal drop in surface tension at the air–water interface, $\pi \sim 25.5$ mN m^{-1}, is very similar to that obtained for melittin (Bhakoo *et al.*, 1982). The natural δ-toxin penetrates phospholipid monolayers, whatever their polar head groups, up to critical pressures higher than 32 mN m^{-1}.

Direct interactions with lipid vesicles

The interactions of δ-toxin with lipid vesicles of different compositions and physical states were assessed from spectroscopic studies. CD or the single Trp fluorescence allows the quantitation of the affinity of the toxin, its fragments and analogues. A partition coefficient $\gamma \sim (5 \pm 3) \times 10^5$ was estimated for analogue F in the presence of egg lecithin vesicles (Thiaudière *et al.*, 1991). From CD changes, it was shown that the 11–26 fragment of the toxin has a strong affinity for lyso-lipid micelles only, a weaker affinity for fluid bilayers and a

decreased one for gel phase dipalmitoylphosphatidyl choline (DPPC). The affinity is modulated by the physical state of the lipids. The more they are packed and ordered, the lower the amount of bound toxin; about a 10-fold decrease is observed between highly fluctuating micelles and gel-phase bilayers (Thiaudière *et al.*, 1991). Analogues with decreasing lengths indicate that shortening the N-terminus results in a progressive decrease in affinity for lipids, but the 19-residue analogue K still has a strong partition coefficient ($\gamma \sim 1.5 \times 10^5$) for egg lecithin while the shorter 16-residue analogue does not significantly partition into zwitterionnic lipids. This analogue only adsorbs to any significant extent in a reversible way at the negatively charged interface of PS vesicles.

Binding to natural membranes was documented using labelled toxins. Radioactive δ-toxin binds to erythrocyte membranes with high affinity and ghosts show a higher amount of bound toxin (Nolte and Kapral, 1981). Rather similar findings were observed when using fluorescein isothiocyanate (FITC)-labelled toxin in the presence of leucocytes (Schmitz *et al.*, 1997) or fibroblasts (Durkin and Shier, 1980).

Location in membranes

In orientated bilayers prepared after co-solubilization of δ-toxin and DPPC in organic solvent, α-helical δ-toxin has no preferred orientation (Brauner *et al.*, 1987), at variance with melittin, which is found either flat at interface (Cornut *et al.*, 1996) or perpendicular to it (Frey and Tamm, 1991). The depth of δ-toxin in the bilayer depends on the lipid physical state (gel or fluid). Below the transition temperature, δ-toxin disturbs only the phosphate of phospholipids, as shown by ^{31}P-NMR, whereas in the fluid state, methylenes of acyl chains (^{2}H-NMR) are also perturbed (Dufourc *et al.*, 1990). Whatever the state of self-association in the aqueous phase, the fluorescence emission of Trp$_{15}$ of δ-toxin (polar face of the helix) is shifted to 340 ± 2 nm, depending on the lipid to which it is bound (Fitton, 1981; Yianni *et al.*, 1986; Thiaudière, 1990; Thiaudière *et al.*, 1991). This wavelength is characteristic of the polarity at the interface between the glycercol and the first methylenes of acyl chains of phospholipids. When bound, analogues with Trp$_{16\,or\,5}$ (apolar face of the soluble self-associated α-helices) emit light at 333 ± 1 nm, a value characteristic of the hydrophobic core of bilayers and, therefore, a deeper location for this side of the helix. The shortest fragments 11–26 of δ-toxin (Trp$_{15}$) (Thiaudière *et al.*, 1991), and of peptides M (Trp$_{16}$) (Thiaudière, 1990) and M+ (basic analogue) (Cornut, 1993) are always in a more polar environment than longer analogues. In the presence of negatively charged bilayers (PS), this difference vanishes, which indicates a stronger insertion. More efficiently than increases of ionic strength, polylysine can displace more easily the shorter peptides (16 and 19 residues) when they are less charged but it is always devoid of any efficiency against longer ones (23 and 26 residues) (Cornut et al., 1993) (Thiaudière et al., personal communication). A deeper location of longer helices can be explained by the prevalence of hydrophobic interactions over electrostatic ones. Time-resolved fluorescence anisotropy decays (Talbot et al., personal communication) of δ-toxin and 26-residue long analogues bound to PC:PS (4:1) vesicles indicate short rotational correlation times attributed to local motions (<1 ns for the indole ring and 2–3 ns for peptide backbone around the Trp) and a much longer one dependent of the state of self-association of peptides. At high surface densities of peptides, an infinite correlation time is observed, characteristic of bound self-associated toxins. On dilution of peptides in the bilayer, this long correlation time diminishes to values of about 30 ns, attributed to the motion of monomers in this viscous medium. The comparison of fluorescence quenching of Trp_{16} and Trp_5 analogues by water-soluble compounds and phospholipids nitroxylated at different positions of acyl chains (Talbot et al., personal communication) indicates that the orientation of these long peptides also changes according to their surface density. They could lie parallel to the bilayer plane at high dilution and be tilted or span the membrane at high concentration.

Channels of δ-toxin

Such a spanning cannot happen without masking the hydrophilic sector of the helix from the bilayer core; then, aggregates of helices exposing their apolar faces to lipids and forming a hydrophilic pore in the centre are expected. As already mentioned, a partition between water and lipid phases coupled with a twofold dissociation factor of soluble tetrameric (or more associated species) (Thiaudière et al., 1991; Cornut et al., 1993) is the best and simplest description of binding experiments. This has to be connected with the model of soluble rafts of δ-toxin interacting with membranes (Raghunathan et al., 1990). These soluble rafts do not exist but a similar model (Figure 20.5) could be applied to the long rods described above. On binding to the bilayer, they could be cut lengthwise into two parts. These long rafts made of antiparallel helices might lie (and dissociate) in the bilayer interface at low surface density or, on colliding with the long side of another raft at high surface density, tip down the bilayer and form spanning aggregates inducing channels or

pores according to the number of associated helices (Figure 20.5). Energy minimization favours channels made of six to eight antiparallel helices (Raghunathan et al., 1990), consistent with the relative molecular mass of δ-toxin complexes in Tween 80 (Freer and Birkbeck, 1982). The analysis of the experimental dependence of conductance with concentration and applied voltage through a planar bilayer (BLM) at the tip of an electrode is consistent with channels made of a bundle of seven helices that can form without any transmembrane potential (Mellor et al., 1988). There is also a voltage component to gating, as already observed with alamethicin (Menestrina, 1988) and melittin (Tosteson and Tosteson, 1984). Energy minimization followed by molecular dynamics calculations show that in the presence of a transbilayer voltage, an isolated helix pre-inserted in a trans or in a flat position remains in its position after 500 ps, but a trans-helix lies on the surface after 200 ps in the absence of voltage. The channels should derive from pre-associated helices lying at the bilayer interface but are not formed from pre-existing isolated transmembrane helices (Biggin and Sansom, 1996). In contrast with melittin, although δ-toxin was added to one side of the bilayer and in spite of its asymmetrical distribution of charges along the sequence, the current–voltage relationships are symmetrical around 0 mV (Mellor et al., 1988). This implies that spanning helices are directed in opposite ways just as in the energetically favoured antiparallel model of Raghunathan (Raghunathan et al., 1990). Modelling of such a channel consistent with the antiparallel assembly of the soluble multimer should be of great interest. Nevertheless only a barrel-stave made of six parallel helices has been described. In spite of the approximations made in molecular modeling the calculated improved organization of this bundle (Kerr and Sansom, 1993; Kerr et al., 1996) is a fascinating insight into a transmembrane channel. Schiffer Edmundson wheels and the direction of the hydrophobic moment of Eisenberg give a static view of apolar and polar faces of amphipathic helices, while consideration of side-chain mobility gives a dynamic view. The centre of the polar face can shift by 10° around the C_α of Asp_{11}. The most flexible side-chains are Gln_3, $Asp_{11 \text{ and } 18}$, $Lys_{14 \text{ and } 22}$, with 2 Å fluctuations (Kerr and Sansom, 1993). Among the analysed models, the best structure (Kerr et al., 1996) in terms of energy is a symmetrical hexamer with 16 ± 8°-tilted helices forming a left-handed supercoil. Intrahelical electrostatic interactions between Asp_{11} and Lys_{14}, Asp_{18} and Lys_{22}, hydrogen bonds between H in residue i and C=O in residue i-4 stabilize the α-helix conformation, whereas the tilted bundle is stabilized by interactions between Gln_3 and Gln_3 and Asp_4 of the two adjacent helices, between

Asn$_{21}$ and Asn$_{21}$ and Lys$_{22}$, with potential hydrogen bonds between Ser$_7$ and Thr$_8$ of one helix and Thr$_8$ and Ser$_7$ of neighbouring ones. The aqueous channel is 10 Å in diameter at the N- and C-terminal with three constrictions at Asp$_4$ (7 ± 2 Å), Lys$_{14}$ (5.6 ± 1 Å), and Lys$_{25}$ (5.7 ± 2.6 Å).   The solvation of the polar faces of δ-toxin helices results in a broadening to 6.6 Å at the level of Lys$_{14}$ and a decrease in the diameter to 5 Å at Lys$_{25}$. The fluctuations of the channel are restricted, but it appears that this molecular dynamics simulation over 100 ps does not enable the molecule to reach a stationary rate.

To sum up, the bound state of δ-toxin consists of a 19-residue long α-helix. Helices can lie parallel to the bilayer at the interface between polar heads and acyl chains of phospholipids, in monomeric or oligomeric forms. At higher surface densities, δ-toxin can associate in long rafts of antiparallel helices which can tip down into the bilayer, forming bundles stabilized by transmembrane voltage, with six or more helices, most probably antiparallel, surrounding a hydrophilic channel. Nevertheless, despite this description, it must be kept in mind that these interpretations in terms of orientation and self-association suffer from a great weakness because they always result from molecular modelling, with numerous oversimplifications, or from indirect interpretations of few experimental data.

PERTURBATIONS OF LIPIDS AND PERMEABILITY INDUCED BY δ-TOXIN

When an amphipathic peptide inserts into lipid bilayers, one expects several perturbations on the packing, thermotropic behaviour and stability of the lipid structures, as is well documented in the case of melittin and other toxins (Dempsey, 1990; Cornut et al., 1993; Bechinger, 1997). For δ-toxin, several physical methods were used to characterize such changes, from the very local ones in the acyl chain order and dynamics up to the more general stability of large unilamellar vesicles.

Sub-molecular changes in lipid order

Solid-state NMR of deuterated DMPC first documented very significant perturbations on both the head group and the acyl chain orders, which increase according to the amount of δ-toxin bound and are modulated by the physical state of lipids. For saturated lipids in the fluid phase at high temperatures, $T \gg T_m$, the addition of δ-toxin leads to a general decrease in the order parameters of all methylene groups of the acyl chains. When the lipid to peptide ratio R decreases from

$R = 50$ to $R = 10$, the higher the peptide content, the stronger the perturbation (Dufourc et al., 1990). This totally parallels the effects induced by melittin (Dufourc et al., 1986). When looking at selectively deuterated fluid palmitoyl-oleoylphosphatidyl choline, the α-CH$_2$ groups of the choline become somewhat less ordered at $R \leq 60$ (Rydall and MacDonald, 1992). When approaching the transition temperature towards the gel phase, T_m, an opposite effect is observed with an increase in the order of all the methylene groups of the acyl chains; this totally differs from what is observed for melittin (Dufourc et al., 1986). It can be accounted for either by a loss of the co-operativity of the thermotropic transition at such a high peptide content, $R \sim 10$–20, and/or by a change in the location of the toxin in relation to the lipids.

Morphological changes

When very high amounts of δ-toxin are bound, typically $R < 12$, quasi-isotropic lines are detected in NMR. At $R \sim 5$ all lipids are in a new structure able to perform very fast isotropic tumbling (Dufourc et al., 1990; Rydall and MacDonald, 1992). Electron microscopy images indicate that in such conditions very small mixed lipid–toxin particles can be formed (Freer et al., 1984). This totally fits with the same and better-documented behaviour observed for melittin (Dufourc et al., 1986; Dufourcq et al., 1986; Faucon et al., 1995). In such a case, melittin induces a more efficient fragmentation of the lecithin bilayers into co-micelles or disc-like particles whatever the physical state, fluid or gel, even at $R \sim 50$ (Dufourcq et al., 1986; Faucon et al., 1995). Although very drastic changes are observed at high peptide content, the collective properties of lecithins can be changed by very small amounts of δ-toxin. Despite no significant change being observed in the transition temperature, an important decrease in the lipid volume at the transition can be detected at one toxin bound for 10^3–10^4 lipid molecules (Neitchev et al., 1992). Such changes differ when comparing melittin and δ-toxin (Laggner et al., 1998).

The full titration of DMPC by δ-toxin was followed by looking at changes in the transition parameter by calorimetry. At very low amounts, $R \geq 600$, δ-toxin severely decreases and finally abolishes the pre-transition. On increasing the δ-toxin content, while the main temperature transition does not shift, its enthalpy is decreased to 80% of the value for peptide–DMPC mixtures at $R = 15$. Concomitantly, for $R < 100$, the highly co-operative transition can be decomposed from a sharp one, whose contribution decreases with the appearance of a broad one with a low enthalpy ~ 1 kcal mol^{-1}. This increases and develops at about 8°C above the main transition temperature. This new phase transition seen at $R \leq 60$ clearly indicates a poorly co-operative transition between lipids with rather similar order but also proves domain formation with peptide-rich and peptide-poor domains in the gel phase (Bhakoo et al., 1985). This agrees with dilatometric and X-ray data (Lohner et al., 1986; Laggner et al., 1998).

As soon as molecular and long-range lipid structures are perturbed by δ-toxin, the stability of vesicles can be questioned. The aggregation and fusion processes were documented quite early and compared with other toxins. On gel-phase lecithin vesicles, in contrast to melittin, δ-toxin fails to induce even aggregation, while in the fluid phase a slow increase in turbidity is observed with both peptides (Yianni et al., 1986). When looking at natural lecithins by light scattering, a drastic increase in the hydrodynamic radii of small unilamellar vesicles (SUVs) from 300 to about 3000 Å develops as soon as the amount of toxin is increased up to $R = 200$. Such an effect is also observed for the 11–26 fragment but it requires a much higher amount of peptide ($R \sim 3$) in order to obtain large vesicles (Thiaudière, 1990; Cornut et al., 1993). A more definitive proof of fusion of the vesicles or the membranes was obtained by Morgan et al. (1986). When mixing two vesicle populations of saturated lecithins with different chain lengths, in their gel phase, the presence of δ-toxin has no effect on lipid mixing. However, as soon as both lipids are in their fluid state, a fast and total mixing is observed, indicative of a stabilization of a new homogeneous bilayer resulting from the fusion of the two different ones.

Permeability increases induced by δ-toxin on lipid vesicles

δ-Toxin induces leakage of calcein when it binds to DPPC large unilamellar vesicles (LUVs). In the gel phase a lag period is always observed despite the fact that the binding is monotonous and fast. These kinetics differ from those observed for melittin. Moreover, the permeability increase induced by δ-toxin can be totally inhibited by high phosphate ionic strength, again contrasting with what is observed for melittin (Yianni et al., 1986). Melittin is therefore more efficient and induces leakage in all conditions. The ability of δ-toxin to induce leakage of entrapped small solutes was also studied by varying the physical state and the thickness of the bilayer by lengthening of the lipid acyl chains (Bhakoo et al., 1985). The efficiency of δ-toxin falls drastically for acyl chains longer than palmitoyl, while in contrast, melittin becomes more efficient up to stearyl chains. A more systematic study of the effects of ionic concentration on the leakage induced by different toxins, namely melittin, alamethicin and gramicidin S, led to the conclusion that each

toxin has a peculiar sensitivity to different ions (Portlock et al., 1990). Such effects are different when assayed on leakage of calcein on LUVs and on haemolysis. Similarly, polylysine has different promoting or inhibiting effects on lipid vesicles and erythrocytes, respectively. Therefore, the ability to induce leakage strongly depends not only on the very fine structure of the lipids but also on all the other interactions via small solutes in solution.

BIOLOGICAL ACTIVITIES OF δ-TOXIN

Haemolysis

As early as 1976, δ-toxin was shown to induce fast complete haemolysis in less than 10 min, without any lag time (Kapral, 1976). The amount of lysis increases almost linearly with toxin concentration with $LD_{50} \sim 1.3$ μM for human erythrocytes. It was further proved that K^+ leakage occurs more quickly than haemoglobin. Haemolysis can be totally inhibited by egg lecithin for a lipid to δ-toxin molar ratio ≈ 1, while conversely, palmitic acid increases it. Such features are similar to those found for melittin and related direct haemolytic amphipathic peptides acting first via lipids on plasma membrane. The haemolytic activity is quite sensitive to the oligomerization state of the toxin in the buffer, which depends on several physicochemical parameters (see above structure). In very dilute suspensions, i.e. at very low haematocrit, δ-toxin is as efficient as melittin in inducing haemolysis, while its efficiency is five to 10 times lower at higher haematocrit (4–5×10^7 cells ml^{-1}) (Cornut et al., 1993) when δ-toxin is self-associated in solution.

Functional changes induced on other blood cells

Recently, the binding of synthetic FITC-labelled δ-toxin to cells in whole blood showed that, at 1 μM, the toxin has a high affinity for neutrophils, a lower one for monocytes and lymphocytes, and the weakest one for erythrocytes (Schmitz et al., 1997). This hierarchy with seven times larger amount of toxin bound per neutrophil compared to monocytes, 30 times more than lymphocytes and up to ~200 times more than that of erythrocytes suggested more specific interactions with the leucocytes. This strong binding results in the activation of monocytes in a dose-dependent way, with a significant up-regulation of the expression of the complement receptor three at very low toxin concentrations (0.03 μM). Although δ-toxin at concentrations up to 10 μM does not induce any drastic metabolic burst, it enhances the lipopolysaccharide and α tumour necrosis factor priming in the micromolar range. Hence, selective activation, co-stimulatory effects on neutrophil priming and oxidative burst can occur without inducing significant cell lysis. Direct interaction of δ-toxin shows that monocytes and neutrophils are still viable after 1-h incubation, even at high concentrations up to 30 μM (Raulf *et al.*, 1990; Schmitz *et al.*, 1997).

δ-Toxin activity/action on other eukaryotic cells

Despite the fact that haemolysis and action on blood cells are the most studied effects, the amphipathic class of cytotoxins, to which δ-toxin is related, has generally a low specificity, being able to kill or strongly perturb almost any living cell. Thelestam and Möllby (1979) compared a wide set of cytotoxins, including most of those produced by *S. aureus*, on their ability to kill human diploid lung fibroblasts grown at confluence in monolayers. Toxins were classified according to their relative ability to induce leakage of amino acids, nucleotides (800 Da) and larger macromolecules such as ribonucleic acid (RNA). At a similar LD_{50} detergents induced concomitant leakage whatever the M_r of the permeants, while melittin induced such releases with a significant increase in LD_{50} according to the size of the permeant. δ-Toxin falls into a different group, being evermore selective than melittin, since a 70-fold higher LD_{50} is required for release for macromolecular compounds compared with that for small solutes. δ-Toxin shares the same selectivity as many other bacterial toxins such as aerolysin or streptolysin O. However, in this assay, δ-toxin is the least specific toxin from *S. aureus*. This is because α-toxin is unable to induce the leakage of high molecular weight compounds and β- and γ-toxins only are able to cause amino acid leakage (Thelestam and Möllby, 1979).

Further studies on fibroblasts grown synchronously proved that sensitivity differs according to the stage in the cell cycle. While cells in the mitotic phase are quite resistant up to 30 μM δ-toxin, cells in interphase, whether attached or trypsin solubilized, are sensitive to δ-toxin in the micromolar concentration range. After a fast increase in permeability to small solutes, severe morphological changes with bleb formation are detected after a 10-min incubation, before cells become turgid and are fragmented after 30 min. Labelled δ-toxin binds to the plasma membrane of fibroblasts, and only after the blebbing process can it enter the cell, remaining localized on both the plasmic and perinuclear membranes (Durkin and Shier, 1980).

Antibacterial activity

Many amphipathic peptides also have broad antibacterial activity (Saberwal and Nagaraj, 1994; Bechinger, 1997; Lohner and Epand, 1997). In this respect, δ-toxin differs strikingly from the very active melittin. Using a synthetic analogue which lacks $fMet_1$ but remains haemolytic (LD_{50} ~ 5 μM on guinea pig erythrocytes), no inhibition of growth of Gram-positive or Gram-negative bacteria was detected up to 50 μM (Dhople and Nagaraj, 1993). Even the permeability of *E. coli* spheroplasts to small solutes was not affected, at variance with the first studies using natural δ-toxin (Freer and Arbuthnott, 1983).

The amphipathic cytotoxic peptides were recently shown to inhibit the growth of *Mycoplasma* (Béven and Wroblewski, 1997). δ-Toxin analogues are also active but somewhat less than melittin, with minimal inhibitory concentration (MIC) values generally in the range of 15–25 μM, for those bearing a net positive charge, compared to an MIC value ~1 μM for melittin. This totally agrees with the notion that the overall positive charge of the peptide is a requirement for antibacterial activity, as clearly demonstrated for magainins (Matsuzaki *et al.*, 1997).

Other targets

Both membrane proteins and intracellular soluble proteins have to be taken into account, amphipathic helices being generally able to interact between themselves (see self-association) and with any other amphipathic structure, compensating for the hydrophobic effect. This is well documented by tight and stoichiometric binding of δ-toxin to calmodulin in the micromolar range, the affinity increasing on dilution, i.e. when the unstructured monomeric form of the toxin is favoured (Cox *et al.*, 1985). These 1:1 complexes were more accurately characterized and shown to occur when δ-toxin is in the presence of other calcium-binding proteins such as troponin C (Garone *et al.*, 1988). Such behaviour is shared by melittin (Maulet and Cox, 1983) and other haemolytic and ideally amphipathic Leu, Lys peptides (LK peptides) (Cox *et al.*, 1985; O'Niel and De Grado, 1990; Cornut, 1993) which also form high-affinity 1:1 complexes. It has been proposed that such complexes play a role in the inhibition of calmodulin activities, the most relevant in this discussion being its role in cell proliferation (Hait *et al.*, 1985) and in Ca^{2+}-dependent protein kinases (Katoh *et al.*, 1982; Pendel *et al.*, 1993).

Despite the fact that direct toxin–protein interactions are not mandatory, the activation of phospholipase A_2 and many other membrane enzymes should be considered as potentially important effects induced by δ-toxin (Durkin and Shier, 1980; Kasimir *et al.*, 1990).

STRUCTURE–ACTIVITY RELATIONSHIPS

What did we learn from studies of δ-toxin analogues? The relevant features of these amphipathic peptides in lytic properties are the importance of their hydrophobicity and charge.

Optimal hydrophobicity is required

The hydrophobic effect is mandatory for peptide insertion into membranes and it governs the binding process as well as the stationary position of peptides within the membrane. It was varied experimentally in two ways, first by a massive Leu substitution of residues on the apolar face and second by a progressive shortening of the peptides.

The 26-residue long analogues, despite being more ideally amphipathic, are less haemolytic than the natural toxin, as seen in Figure 20.6. This is interpreted as being due to the severe self-association of analogues in buffer, especially analogue M, which is the most hydrophobic one. Despite the large number of substituted residues, these analogues have an antigenic pattern identical to that of the natural toxin. This indicates that antibodies are essentially directed against epitopes on the polar face and can bind even when the toxin is embedded into lipids (Alouf *et al.*, 1989).

These conclusions also agree with the fact that shortening the analogues down to 19 residues increases the activity (Figure 20.6) owing to better solubility. Further shortening down to 16 residues abolishes the haemolytic activity whatever the composition of the peptide, i.e. 11–26 fragment of the toxin (Alouf *et al.*, 1989), the shortest analogue J (Thiaudière, 1990) and the fragment 5–20 of δ-toxin (Dhople and Nagaraj, 1995). It seems therefore that an optimal hydrophobicity for induction of haemolytic activity requires about eight Leu. This corresponds to about 8.5 kcal mol⁻¹ free energy of transfer from water to the membrane. One should emphasize that, as in the case of studies on several other toxins, early speculations about the need for helix breaker residue can now be interpreted differently. The Gly10Leu substitution decreases the activity for full-length peptides because it favours good packing. It then stabilizes aggregates in solution, and conversely decreases its binding to lipids. When decreasing the length and/or increasing the net peptide charge, which avoids severe self-association, Gly analogues are more active (Cornut, 1993). Similarly, in melittin, Pro was not mandatory for activity (Dempsey, 1990).

A special role was also anticipated for Trp_{15}. It could be seen as a limiting bulky group in the putative channel structure; however, its shift to position 16, i.e. on the opposite apolar face, does not severely change either the haemolytic or the channel-forming activity (Cornut, 1993; Kerr *et al.*, 1995).

Requirements for charged residues

As opposed to the apolar face, where weak selectivity of the residues was expected, the changes in the polar residues should lead to more drastic changes in the properties of the toxin. δ-Toxin has no net charge, which facilitates self-association (see above). What will happen when basic properties are conferred on the toxin? This is done either by altering the protonation state of Asp residues, which gives a peptide with +4 net charge at pH 5, or by selective systematic substitution of Asp into Asn (Figure 20.1). These changes result in about a 10-fold increase in the haemolytic activity, which occurs even for shorter analogues of 19 and 23 residues (Cornut, 1993) (Figure 20.6). A more drastic change in the sequence by Asp→Lys substitution changes the inactive 5–20 fragment into a peptide endowed with both haemolytic and antibacterial

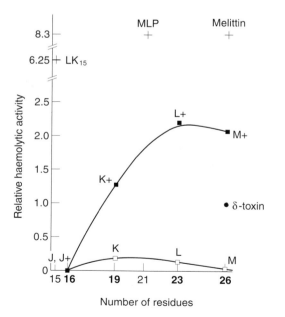

FIGURE 20.6 Relative haemolytic activity towards human erythrocytes, of δ-toxin, its analogues, melittin (Cornut, 1993), MLP (Simmaco *et al.*, 1996) and LK_{15} when varying the length of the peptides and their net charge (Cornut *et al.*, 1994).

activities (Dhople and Nagaraj, 1995). Rather similar changes have been made with melittin, by decreasing its +6 net charge. Its acetylation, which leaves only the charged Arg group with a +2 net charge, results in retention of full activity on both erythrocytes and lipid vesicles (Portlock *et al.*, 1990). Similarly, the newly isolated analogue melittin-like peptide (MLP) (Figure 20.2) has only two charged Lys residues and is fully active (Simmaco *et al.*, 1996).

It can therefore be concluded that the net positive charge of δ-toxin analogues very significantly decreas-

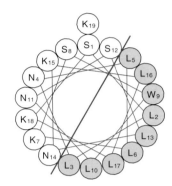

K+

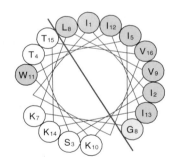

5– 20 analogue

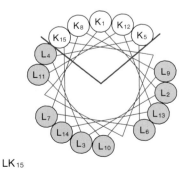

LK₁₅

FIGURE 20.7 Helical wheel representations of δ-toxin analogues with the most simplified and shortened sequences retaining full lytic activity. Top: peptide K⁺; centre: analogue 5–20 (Dhople and Nagaraj, 1993); bottom: LK₁₅ (Cornut *et al.*, 1994).

es self-association in buffer and perhaps also in the bound state, which improves the binding to lipids and membranes. The need for a charged peptide has recently been shown for antibacterial activities (Matsuzaki *et al.*, 1997), but short peptides down to 12 residues with +3 or +4 net charge and only constituted from Leu and Lys residues have proved to be stabilized in the α-helix configuration and are highly haemolytic (Castano *et al.*, 1997). Such rationalized peptide sequences are shown in Figure 20.7; indeed, quite severe changes in the δ-toxin sequence can still give highly active compounds.

The peculiar almost zwitterionic structure of δ-toxin, unusual for the lytic peptides, leads to the question of whether the toxin is only a haemolytic weapon for *S. aureus*. In this respect, it is somewhat less efficient than melittin and, despite a very sophisticated sequence, it is also less active than the minimalistic developed amphipathic basic peptides. The 1:1 Leu, Lys poly-peptides of about 20 residues with a similar ideal amphipathic pattern to δ-toxin, i.e. 180° apolar face, are very active haemolytic and antibacterial molecules (Blondelle and Houghten, 1992). Moreover, 2:1 molar ratio Leu, Lys peptides, which fold in an ideal amphi-pathic α-helix with a 260° apolar face, are even more active with only 15 residues (Figure 20.7) (Cornut *et al.*, 1994; Castano *et al.*, 1999a, b). Thus, one can speculate that δ-toxin can be much more efficient when it can acquire a strong basic character in specific conditions. This is what happens on decreasing the pH value for haemolysis (see above), but this could also happen when, being internalized in eukaryotic cells, δ-toxin is located in lysosomes, where the lowered local pH will increase its efficiency as a lytic compound. When staphylococci are in the terminal growing phase, pH decreases, then δ-toxin could acquire some anti-bacterial activity against other organisms competing for the same ecological niche. Then, considering cyto-toxicity alone, δ-toxin structure could allow some fine tuning of the activity according to the growth conditions. Indeed, this does not rule out the possibility of other pleiotropic actions which may relate to its mode of action, both on membranes and on other cell components.

Current models for the haemolytic activity

Owing to its amphipathic character, δ-toxin binds quite aspecifically to cell interfaces, burying its apolar moiety in the less polar medium. The membrane invasion should result in the peptide flat on the outer leaflet of bilayers and plasma membranes. This is consistant with several experimental data. First, on lipid vesicles, the asymmetric binding of δ-toxin proved to perturb the lipid head groups on the outer leaflet without any effect on the inner ones, over several hours (Marassi *et al.*, 1993). Second, on erythrocytes, the binding before lysis corresponds to a much smaller amount of δ-toxin bound compared with that found when using ghosts, where both membrane leaflets are accessible (Nolte and Kapral, 1981). Finally, in a rather crude experiment, δ-toxin linked to Sepharose beads retains activity (Lee and Haque, 1976). All such data favour as a first step the asymmetrical and flat location as seen in Figure 20.5.

Increasing the amount of such asymmetrically located peptide results in a severe tension in the membrane. Such constraints can be relaxed by the peptide translocation, which implies that at least a few transient peptide aggregates can form unstable pores or channels and then flip on to the other side. The degree of order and peptide–peptide contacts in the bound state are still poorly documented, which explains controversies and different models. Fluorescence data favour self-association in the bound state, but are insufficient to discriminate between channel structures and rafts or carpet models (Shai, 1995). It has, however, to be mentioned that the study of the ability of δ-toxin analogues to induce channels in black lipid membrane (BLM) does not correlate with the changes in their haemolytic activity (Kerr *et al.*, 1995). Therefore, despite the fact that channels are very probably formed and allow peptides to enter the cell, these steps could be events occurring concomitantly with lysis, but not causing it.

ACKNOWLEDGEMENT

We are grateful to Dr Tanja Pott for very kindly and cleverly drawing Figure 20.5.

REFERENCES

Alouf, J.E., Dufourcq, J., Siffert, O., Thiaudière, E. and Geoffroy, C. (1989). Interaction of staphylococcal δ-toxin and synthetic analogues with erythrocytes and phospholipid vesicles. *Eur. J. Biochem.* **183**, 381–90.

Bechinger, B. (1997). Structure and function of channel-forming peptides: magainins, melittin, cecropins and alamethicin. *J. Membr. Biol.* **156**, 197–211.

Béven, L. and Wroblewski, H. (1997). Effect of natural amphipathic peptides on viability, membrane potential, cell shape and mobility of mollicutes. *Res. Microbiol.* **148**, 163–75.

Bhakoo, M., Birkbeck, T.H. and Freer, J.H. (1982). Interaction of δ-haemolysin with phospholipid monolayers. *Biochemistry* **21**, 6883–9.

Bhakoo, M., Birkbeck, T.H. and Freer, J.H. (1985). Phospholipid-dependent changes in membrane permeability induced by staphylococcal δ-lysin and bee venom melittin. *Can. J. Biochem.* **63**, 1–6.

Biggin, P.C. and Sansom, M.S.P. (1996). Simulation of voltage-dependent interactions of α-helical peptides with lipid bilayers. *Biophys. Chem.* **60**, 99–110.

Birkbeck, T.H. and Freer, J.H. (1988). Purification and assay of staphylococcal δ-lysin. *Methods in Enzymol.* **165**, 16–23.

Blondelle, S.E. and Houghten, R.A. (1991). Probing the relationships between the structure and haemolytic activity of melittin with a complete set of Leu substitution analogues. *Peptide Res.* **4**, 12–18.

Blondelle, S.E. and Houghten, R.A. (1992). Design of model amphipathic peptides having pattern antimicrobial activities. *Biochemistry* **31**, 12688–94.

Brauner, J.W., Mendelsohn, R. and Prendergast, F.G. (1987). Attenuated total reflectance Fourier transform infrared studies of the interaction of melittin, two fragments of melittin, and δ-haemolysin with phosphatidylcholines. *Biochemistry.* **26**, 8151–8.

Castano, S., Cornut, I., Büttner, K., Dasseux, J.L. and Dufourcq, J. (1999a). The amphipathic helix concept: length effects on ideally amphiphathic Likj (with i = 2j) hemolytic peptides. *Biochim. Biophys. Acta*, **1416**, 161–75.

Castano, S., Desbat, B., Laguerre, M. and Dufourcq, J. (1999b). Structure, orientation and affinity for lipids and interfaces of ideally amphipathic lytic Likj peptides (i = 2j). *Biochim. Biophys. Acta* **1416**, 176–94.

Colaccico, G., Basu, M.K., Buchelew, A.R. and Berheimer, A.W.J. (1977). Surface properties of membrane systems. Transport of staphylococcal δ-toxin from aqueous to membrane phase. *Biochim. Biophys. Acta* **465**, 378–90.

Cornut, I. (1993). *Etude spectroscopie des interactions de cytotoxines amphipathiques avec des membranes phospholipidiques modèles et leurs activitiés sur des erythrocytes.* PhD thesis, University Bordeaux I, Bordeaux.

Cornut, I., Thiaudière, E. and Dufourcq, J. (1993). The amphipathic helix in cytotoxic peptides. In: *The Amphipathic Helix* (ed. R.M. Epand), pp. 173–219. CRC Press, Boca Raton, FL.

Cornut, I., Büttner, K., Dasseux, J.L. and Dufourcq, J. (1994). The amphipathic α-helix concept. Application to the *de novo* design of ideally amphipathic Leu, Lys peptides with hemolytic activity higher than that of melittin. *FEBS Lett.* **349**, 29–33.

Cornut, I., Desbat, B., Turlet, J.M. and Dufourcq, J. (1996). *In situ* study by polarization modulated Fourier transform infrared spectroscopy of the structure and orientation of lipids and amphipathic peptides at the air/water interface. *Biophys. J.* **70**, 305–12.

Cox, J.A., Comte, M., Fitton, J.E. and De Grado, W.F. (1985). The interaction of calmodulin with amphiphilic peptides. *J. Biol. Chem.* **260**, 2527–34.

Creighton, T.E. (1996). *Proteins, Structure and Molecular Properties.* W.H. Freeman, New York.

Dempsey, C.E. (1990). The actions of melittin on membranes. *Biochim. Biophys. Acta* **1031**, 143–61.

Dhople, V.H. and Nagaraj, R. (1993). δ-Toxin, unlike melittin, has only haemolytic activity and no antimicrobial activity: rationalization of this specific biological activity. *Biosci. Rep.* **13**, 245–50.

Dhople, V.H. and Nagaraj, R. (1995). Generation of analogues having potent antimicrobial and haemolytic activities with minimal changes from an inactive 16-residue peptide corresponding to the helical region of *Staphylococcus aureus* δ-toxin. *Prot. Eng.* **8**, 315–18.

Donvito, B., Etienne, J., Denoroy, L., Greenland, T., Benito, Y. and Vandenesch, F. (1997). Synergistic haemolytic activity of *Staphylococcus lugdunensis* is mediated by three peptides encoded by a *non-Agr* genetic locus. *Infect. Immun.* **65**, 95–100.

Dufourc, E.J., Smith, I.C.P. and Dufourcq, J. (1986). Molecular details of melittin-induced lysis of phospholipid membranes as revealed by deuterium and phosphorus NMR. *Biochemistry* **25**, 6448–55.

Dufourc, E.J., Dufourcq, J., Birkbeck, T.H. and Freer, J.H. (1990). δ-Haemolysin from *Staphylococcus aureus* and model membranes. A solid state ²H-NMR and ³¹P-NMR study. *Eur. J. Biochem.* **187**, 581–7.

Dufourcq, J., Faucon, J.F., Fourche, G., Dasseux, J.L., Le Maire, M. and Gulik-Krzywicki, T. (1986). Morphological changes of phosphatidylcholine bilayers induced by melittin: vesicularization, fusion, discoidal particles. *Biochim. Biophys. Acta* **859**, 33–48.

Durkin, J.P. and Shier, J.P. (1980). Cell cycle dependent resistance to staphylococcal delta-toxin induced lysis of cultured cells. *Biochem. Biophys. Res. Commun.* **94**, 980–7.

Eisenberg, D., Weiss, R.M. and Terwilliger, T.C. (1982). The hydrophobic moment: a measure of the amphiphilicity of α-helix. *Nature* **299**, 371–4.

Faucon, J.F., Bonmatin, J.M., Dufourcq, J. and Dufourc, E.J. (1995). Acyl chain length dependence in the stability of melittin-lecithin complexes. A light scattering and ³¹P-NMR study. *Biochim. Biophys. Acta* **1234**, 235–43.

Fitton, J.E. (1981). Physicochemical studies on δ-haemolysin, a staphylococcal cytolytic polypeptide. *FEBS Lett.* **130**, 257–60.

Fitton, J.E., Dell, A. and Shaw, W.V. (1980). The amino-acid sequence of δ-haemolysin of *Staphylococcus aureus*. *FEBS Lett.* **115**, 209–12.

Fitton, J.E., Hunt, D.F., Shabanovitz, J., Winston, S. and Dell, A. (1984). The amino acid sequence of delta-toxin purified from a canine isolate of *S. aureus*. *FEBS Letters.* **169**, 25–9.

Freer, J.H. (1986). Membrane damage caused by bacterial toxins: recent advances and new challenges. In: *Natural Toxins* (ed. J.B. Harris), pp. 189–211. Clarendon Press, Oxford.

Freer, J.H. and Birkbeck, T.H. (1982). Possible conformation of δ-lysin, a membrane-damaging peptide of *S. aureus*. *J. Theor. Biol.* **94**, 535–40.

Freer, J.H. and Arbuthnott, J.P. (1983). Toxins of *Staphylococcus aureus*. *Pharmacol. Ther.* **19**, 55–106.

Freer, J.H., Birkbeck, T.H. and Bhakoo, M. (1984). Interaction of staphylococcal delta-lysin with phospholipid mono and bilayers – a short review. In: *Bacterial Protein Toxins* (eds J.E. Alouf, J.H. Freer, R. Fehrenbach and Jeljaszewicz), pp. 181–98. Academic Press, London.

Frenette, M., Beaudet, R., Bisaillon, J.G., Sylvestre, M. and Portelance, V. (1984). Chemical and biological characterization of a gonococcal growth inhibitor produced by *S. haemolyticus* isolated from urogenital flora. *Infect. Immun.* **46**, 340–5.

Frey, S. and Tamm, L.K. (1991). Orientation of melittin in phospholipid bilayers. *Biophys. J.* **60**, 922–30.

Garone, L., Fitton, J.E. and Steiner, R.F. (1988). The interaction of δ-haemolysin with calmodulin. *Biophys. Chem.* **31**, 231–45.

Hait, W.N., Grais, C., Benz, C. and Cadman, E.C. (1985). Inhibition of growth of leukaemic cells by inhibitors of calmodulin: phenothiazins and melittin. *Cancer Chemother. Pharmacol.* **14**, 202–5.

Janzon, L., Löfdahk, S. and Arvidson, S. (1989). Identification and nucleotide sequence of the delta-lysin gene, adjacent to the accessory gene regulator, agr, of *Staphylococcus aureus*. *Mol. Gen. Genet.* **219**, 480–5.

Kantor, H.V., Temples, B. and Shaw, W.V. (1972). Delta-haemolysin: purification and characterization. *Arch. Biochem. Biophys.* **151**, 142–56.

Kapral, F.A. (1976). Effect of fatty acids on *Staphylococcus aureus* delta-toxin haemolytic activity. *Infect. Immun.* **13**, 114–19.

Kasimir, S., Schönfeld, W., Alouf, J.E. and König, W. (1990). Effect of *Staphylococcus aureus* δ-toxin on human granulocyte functions and platelet-activating factor metabolism. *Infect. Immun.* **58**, 1653–9.

Katoh, N., Raynor, R.L., Wise, B.R., Schatzman, R.C., Turner, R.C., Helfman, D.M., Fain, J.N. and Kuo, J.F. (1982). Inhibition by melittin of phospholipid-sensitive and calmodulin-sensitive calcium-dependent protein kinases. *Biochem. J.* **202**, 217–24.

Kerr, I.D. and Sansom, M.S.P. (1993). Hydrophilic surface maps of channel-forming amphipathic helices. *Eur. Biophys. J.* **22**, 269–77.

Kerr, I.D., Dufourcq, J., Rice, J.A., Fredkin, D.R. and Sansom, M.S.P. (1995). Ion-channel formation by synthetic analogues of staphylococcal δ-toxin. *Biochim. Biophys. Acta* **1236**, 219–27.

Kerr, I.D., Doak, D.G., Sankararamakhrishnan, R., Breed, J. and Sansom, M.S.P. (1996). Molecular modelling of staphylococcal delta-toxin ion channels by restrained molecular dynamics. *Prot. Eng.* **9**, 161–71.

Kreger, A.S., Kim, K.S., Zaboretzky, F. and Bernheimer, A.W. (1971). Purification and properties of δ-haemolysin. *Infect. Immun.* **3**, 449.

Laggner, P., Latal, A., Staudegger, E., Degovics, G. and Lohner, K. (1998). Modulation of lipid phase structure by lytic peptides. *Biophys. J.* **74**, A122.

Lee, S.H. and Haque, R. (1976). Lysis of erythrocytes by sepharose-δ-toxin complex. *Biochem. Biophys. Res. Commun.* **68**, 1116–18.

Lee, K.H., Fitton, J.E. and Wuthrich, K. (1987). Nuclear magnetic resonance investigation of the conformation of δ-haemolysin bound to dodecylphosphocholine micelles. *Biochim. Biophys. Acta* **911**, 144–53.

Linder, R. (1984). Alteration of mamallian membranes by the cooperative and antagonistic actions of toxins. *Biochim. Biophys. Acta* **779**, 423–45.

Lohner, K. and Epand, R.M. (1997). Membrane interactions of haemolytic and antibacterial peptides. *Adv. Biophys. Chem.* **6**, 53–66.

Lohner, K., Laggner, P. and Freer, J.H. (1986). Dilatometric and calorimetric studies of the effect of *S. aureus* delta-lysin on the phopholipid phase transition. *J. Solution Chem.* **15**, 189–98.

Marassi, F.M., Shivers, R.R. and MacDonald, P.M. (1993). Resolving the two monolayers of a lipid bilayer in giant unilamellar vesicles using deuterium nuclear magnetic resonance. *Biochemistry* **32**, 9936–43.

Matsuzaki, K., Sugishita, K., Mitsunori, H., Fujii, N. and Miyajima, K. (1997). Interactions of an antimicrobial magainin 2, with outer and inner membranes of Gram-bacteria. *Biochim. Biophys. Acta* **1327**, 119–30.

Maulet, Y. and Cox, J.A. (1983). Structural changes of melittin and calmodulin upon complex formation. *Biochemistry* **22**, 5680–6.

McCartney, A.C. and Arbuthnott, J.P. (1978). Mode of action of membrane damaging toxins produced by staphylococci. In: *Bacterial Toxins and Cell Membranes* (eds J. Jeljaszewicz and T. Wadstrom), pp. 89–127. Academic Press, London.

McKevitt, A., Bjorson, G.L., Mauracher, C.A. and Scheifele, D. (1990). Amino acid sequence of δ-like toxin from *Staphylococcus epidermidis*. *Infect. Immun.* **58**, 1473–5.

Mellor, I.R., Thomas, D.H. and Sansom, M.S.P. (1988). Properties of ion-channels formed by *Staphylococcus aureus* δ-toxin. *Biochim. Biophys. Acta* **942**, 280–94.

Menestrina, G. (1988). *E. coli* haemolysin permeabilizes unilamellar vesicles loaded with calcein by a single-hit mechanism. *FEBS Lett.* **232**, 217–20.

Morgan, C.G., Fitton, J.E. and Yianni, Y.P. (1986). Fusogenic activity of δ-haemolysin from *Staphylococcus aureus* in phospholipid vesicles in the liquid-crystalline state. *Biochim. Biophys. Acta* **863**, 129–38.

Neitchev, V., Löhner, K., Colotto, A. and Laggner, P. (1992). Effect of δ-lysin on the phase transitions of lipids. *Mol. Biol. Rep.* **16**, 249–53.

Nolete, F.S. and Kapral, F.A. (1981). Binding of radiolabeled *S. aureus* δ-toxin to human erythrocytes. *Infect. Immun.* **31**, 1086–93.

O'Niel, K. and De Grado, W. (1990). How Calmodulin binds its target: sequence independent recognition of amphipathic α-helices. *Trends Biochem. Sci.* **15**, 59–64.

Pendel, H.R., Xu, Y.H., Jarrett, H.W. and Carlson, G.M. (1993). The model calmodulin-binding peptide melittin inhibits phosphorylase kinase by interaction with its catalytic centre. *Biochemistry* **32**, 11865–72.

Portlock, S.H., Clague, M.J. and Cherry, R.J. (1990). Leakage of internal markers from erythrocytes and lipid vesicles induced by melittin, gramicidin S and alamethicin. *Biochim. Biophys. Acta* **1030**, 1–10.

Raghunathan, G., Seetharamulu, P., Brooks, B.R. and Guy, H.R. (1990). Models of δ-haemolysin membrane channels and crystal structures. *Proteins* **8**, 213–25.

Raulf, M., Alouf, J.E. and König, W. (1990). Effect of staphylococcal δ-toxin and melittin on leukotriene induction and metabolism of human polymorphonuclear granulocytes. *Infect. Immun.* **58**, 2678–82.

Rydall, J.R. and MacDonald, P.M. (1992). Influence of staphylococcal δ-toxin on the phosphatidylcholine headgroup as observed using ^{2}H-NMR. *Biochim. Biophys. Acta* **1111**, 211–20.

Saberwal, G. and Nagaraj, R. (1994). Cell-lytic and antibacterial peptides that act by perturbing the barrier function of membranes facets of their conformational features, structure–function correlations and membrane-perturbing abilities. *Biochim. Biophys. Acta* **1197**, 109–31.

Schmitz, F.J., Veldkamp, K.E., Van Kessel, K.P.M., Verhoef, J. and Van Strijp, J.A.G. (1997). δ-Toxin from *S. aureus* as a costimulator of human neutrophil oxidative burst. *J. Infect. Dis.* **176**, 1531–7.

Shai, Y. (1995). Membrane spanning polypeptides. *Trends Biochem. Sci.* **20**, 460–4.

Shai, Y., Fox, J., Carasch, C., Shih, Y.L., Edwards, C. and Lazarovici, P. (1988). Sequencing and synthesis of pardaxin, a polypeptide from the Red Sea Moses sole with ionophore activity. *FEBS Lett.* **242**, 161–6.

Simmaco, M., Mignogna, G., Canofeni, S., Miele, R., Mangoni, M.L. and Barra, D. (1996). Temporins, antimicrobial peptides from the European red frog *Rana temporaria*. *Eur. J. Biochem.* **242**, 788–92.

Tappin, M.J., Pastore, A., Norton, R.S., Freer, J.H. and Campbell, I.D. (1988). High-resolution H-NMR study of the solution structure of δ-haemolysin. *Biochemistry* **27**, 1643–7.

Thelestam, M. and Möllby, R. (1979). Classification of microbial, plant and animal cytolysins based on their membrane-damaging effects on human fibroblasts. *Biochim. Biophys. Acta* **557**, 156–69.

Thiaudière, E. (1990). *Relations structure-activité de peptides amphiphiles cytolytiques. Etude de la toxinelode straplylococcus aureus et d'analogus synthétiques, en solution et lies aux lipides.* PhD thesis. University of Bordeaux I, Bordeaux.

Thiaudière, E., Siffert, O., Talbot, J.C., Bolard, J., Alouf, J. and Dufourcq, J. (1991). The amphipathic helix concept; consequences for the structure of staphylococcal delta-toxin in solution and bound to lipids. *Eur. J. Biochem.* 203–13.

Tosteson, M.T. and Tosteson, D.C. (1984). Activation and inactivation of melittin channels. *Biophys. J.* **45**, 112–14.

Vandenesch, F., Storrs, M.J., Poitevin-Later, F., Etienne, J., Courvalin, P. and Fleurette, J. (1991). δ-Like haemolysin produced by *S. lugdunensis*. *FEMS Microbiol. Lett.* **78**, 65–8.

Vandenesch, F., Projean, S.J., Kreiswirth, B., Etienne, J. and Novick, R.P. (1993). Agr-related sequences in *S. lugdunensis*. *FEMS Microbiol. Lett.* **111**, 115–22.

Watson, D.C., Yaguchi, M., Bisaillon, J.G. and Beaudet, R. (1988). The amino acid sequence of a gonococcal growth inhibitor from *S. haemolyticus*. *Biochemistry* **252**, 87–93.

Yianni, Y.P., Fitton, J.E. and Morgan, C. (1986). Lytic effects of melittin and δ-haemolysin from *Staphylococcus aureus* on vesicles of dipalmitoylphosphatidylcholine. *Biochim. Biophys. Acta* **856**, 91–100.

The bi-component staphylococcal leucocidins and γ-haemolysins (toxins)

Gilles Prévost

INTRODUCTION

The terms 'leucocidins' and 'γ-haemolysins' represent toxins that were previously named according to some of their properties which were observed in the culture supernatant of strains of *Staphylococcus aureus*. In fact, as the most common target cells of these toxins are the polymorphonuclear cells, monocytes and macrophages, they should be called leucotoxins or, more accurately, staphylococcal bi-component leucotoxins/pore-forming leucotoxins. These toxins interact with the cell membrane and the synergy of the two distinct secreted proteins from class S and class F results in membrane damage. With the recent characterization of several related leucotoxins, they now constitute a true toxin family. Furthermore, a given strain of *S. aureus* may harbour more than one locus encoding these toxins, and each class S protein may create a leucotoxin with specific properties by combining with a class F protein. Hence, one question which may drive the following chapter is: why have various strains of *S. aureus* with stable genetics developed such a complex system? What is the eventual benefit for pathogenesis and, in terms of the virulence of this bacterium, which is one of the most well-adapted commensal and/or pathogenic organisms for humans, why is it one of the most commonly encountered bacterium in clinical laboratories, and one of the most common pathogens responsible for nosocomial infections?

Therefore, it is important to detail the diversity of these leucotoxins, their epidemiological distribution and clinical associations, and their biological implications in infections.

HISTORY OF THE STAPHYLOCOCCAL LEUCOTOXINS: 1900–1990s

The leucocidal effect of infection caused by *S. aureus* was first reported by Van de Velde in 1894 after its injection in the pleural cavity. Titration of this leucocidal activity was determined by the reduction of methylene blue by still viable leucocytes (Neisser and Wechsberg, 1901). The haemolytic and leucocytolytic activities were recognized as independent effects, since culture filtrates of different strains did not harbour comparable activities (Julianelle, 1922). This leucocytotoxic ability of certain strains was associated with the ability to cause furuncles (Panton and Valentine, 1932), and was active against human and rabbit granulocytes and macrophages, without being haemolytic. Finally, this product was named the Panton–Valentine leucocidin (PVL) (Wright, 1936), which was differentiated from the other staphylococcal toxins by having leucotoxic activity. Concurrently, Smith and Price (1938) recognized four distinct haemolysins: α, β, γ, δ, according to their activity on human, rabbit, sheep and horse erythrocytes.

Attempts were then made to determine whether all of these toxins were secreted by different strains or whether culture conditions influenced their production. Culture media and the 'dialysis sac-culture' method were developed that allowed the PVL to be produced in high amounts (Gladstone and Fildes, 1940; Gladstone and Glencross, 1960). These authors reported that the PVL was especially cytolytic against polymorphonuclear (PMN) cells, monocytes and macrophages from humans and rabbits. Several staining methods were developed to visualize the intoxicated cells that used compounds such as Trypan blue (Panton and Valentine, 1932), methylene blue (Neisser and Wechsberg, 1901), Congo red and Nil blue (Johanovsky, 1959), ethidium bromide and acridine orange (Pfanneberg et al., 1975).

In fact, Woodin (1960) discovered that the leucotoxic action previously reported in the supernatant fluids of strain V8 was a consequence of the synergistic interaction of two non-associated proteins called S (slow eluted = 38 kDa) and F (fast eluted = 32 kDa), according to their behaviour in carboxymethyl cellulose chromatography. These observations led to a reconsideration of previous data and development of diagnostic tests based on the use of specific antibodies (Gladstone and Van Heyningen, 1957; Towers and Gladstone, 1958).

Woodin was the first to undertake a study of the effect of Panton–Valentine leucocidin on target cells. He rapidly understood that this toxin modified the ionic exchanges and the metabolism of the cells and that it could be a useful tool for analysing different metabolic pathways in the cell (Woodin, 1972a). He reported that, according to the doses of PVL, proteins were extruded from cells (Woodin, 1962), the Na^+/K^+ pump was activated (Woodin and Wieneke, 1968), and that intracellular calcium accumulated (Woodin and Wieneke, 1963a). Also, adenylate cyclase activity was enhanced (Woodin, 1972b) and an increase in cell metabolism was observed (Woodin and Wieneke, 1963b, 1964). Finally, PVL induced increased turnover of membrane phospholipids, especially phosphoinositides (Woodin and Wieneke, 1967) as well as the activity of cation-sensitive phosphatases (Woodin and Wieneke, 1968; Woodin, 1972a). This line of research ceased on the death of the principal investigator. Of course, the chronology and the inter-relationships of the molecular events resulting in cell intoxication were not assessed at this time. However, the primary binding of the small protein to the membranes was reported (Woodin, 1972a).

Concurrently with the observations on the nature of PVL, the γ-haemolysin was finally shown to be a two-component toxin (Guyonnet and Plommet, 1970; Taylor and Bernheimer, 1974). However, this toxin was not compared with PVL, nor was the frequency of S.

aureus strains producing these leucotoxins examined. The absence of such a comparison has been responsible for the confusion between the leucotoxins. Furthermore, the γ-haemolysin is composed of three proteins constituting two distinct leucotoxic entities, which raised doubts about much of the earlier work prior to the reported sequences of the purified fractions (Cooney et al., 1993). Nevertheless, the purified γ-haemolysin was confirmed to be haemolytic on both human and rabbit erythrocytes (Guyonnet and Plommet, 1970; Möllby, 1989).

Another leucocidin was purified and characterized from a S. aureus strain isolated from a case of bovine mastitis (Soboll et al., 1973). It was cytolytic for bovine PMNs, but was not further studied until the leucotoxin genes for this P83 strain were cloned (Supersac et al., 1993; Choorit et al., 1995).

Other studies, sometimes with incompletely purified PVL, involved the examination and titration of anti-PVL antibodies as reporters of staphylococcal infection (Bänffer, 1962; Gladstone et al., 1962). The toxin was immunogenic in humans (Mudd et al., 1965; Bänffer and Franken, 1967) and immunization was recommended for a certain period in cases of chronic osteomyelitis. Further research revealed platelets as new target cells for PVL (Jeljaszewicz et al., 1976).

The impact of this leucotoxin on haemostasis and on the inflammatory response was also investigated in experimental animal models. The intravenous injection of PVL in rabbits or in mice was not lethal, but decreased the PMN counts in blood, followed by a reactive increase persisting for a few hours (Szmigielski et al., 1966, 1968). The observed increase was comparable with the effect of cytostatic agents (Szmigielski and Jeljaszewicz, 1976). The distribution of the toxin in mouse organs was studied by Grojec (1979). However, such model studies did not include consideration of a possible co-operativity of different virulence factors in vivo, when the bacteria were spreading through septicaemia (Grojec and Jeljaszewicz, 1985).

Noda et al. in the 1980s studied a leucocidin whose molecular masses of constitutive proteins were reported to be 32 and 31 kDa for the F and S components, respectively (Noda et al., 1980a). Despite the title of the report mentioning crystallization, the proteins were demonstrated to precipitate with high concentrations of $(NH_4)_2SO_4$, and were then resoluble. This leucocidin, which was not compared with the PVL, was cytolytic against human granulocytes, and also precipitable by antibodies. The binding of S prior to that of F at the surface of cell membranes (Noda et al., 1981) was confirmed and the number of binding sites per human granulocytes was determined as 5000 and 3000 for the S and the F components, respectively. The S component was reported to interact with GM1, which inhibited its binding to the membrane and caused its precipitation in agarose gels (Noda *et al.*, 1980b). However, the present author was unable to repeat these observations in the laboratory, although the B subunit of cholera toxin precipitates in the presence of GM1. The leucocidin induces an increase in the intracellular Ca^{2+} concentration (Noda *et al.* 1982), but in the absence of extracellular Ca^{2+}, the toxin remained cytotoxic and the remaining K^+ exchanges could be blocked by tetraethyl ammonium chloride, suggesting the blocking of the challenged pores (Morinaga *et al.*, 1987). The cellular phospholipase A_2 was affected by the action of the leucotoxin, as well as of the metabolism of the arachidonic acid (Noda *et al.*, 1982). Subsequent to the cell stimulation by the leucocidin, the metabolism of phosphoinositides was affected (Wang *et al.*, 1990), both mechanisms probably induced by a GTP-dependent signal transduction. ADP-ribosylation of rabbit erythrocyte membrane proteins by both α-toxin and leucocidin was reported (Kato and Noda, 1989), but was obtained *in vitro* with amounts of toxin 1000-fold higher than those including cytotoxic activity. These experiments remain to be confirmed. Finally, *in vitro* differentiated HL-60 promyelocytic cells became susceptible to this leucocidin (Morinaga *et al.*, 1993). Thus, the results obtained before 1990 with the staphylococcal leucocidins were often confusing, lacked comparisons between different molecular species and thereby stimulated further investigations.

STAPHYLOCOCCAL BI-COMPONENT LEUCOTOXINS: A FAMILY OF TOXINS

The different members of the family

It was observed that V8 strains collected from different international collections did not secrete comparable levels of leucocytoxicity after being identically cultivated. This led to the registration of a new certified V8 strain able to produce PVL and V8 protease activity (ATCC 49775). Such a feature might also be observed with the P83 strain producing Luk M/Luk F'-PV.

The staphylococcal bi-component leucotoxins (Table 21.1) can be obtained by using two kinds of culture media. PVL, LukE/LukD, LukM/LukF'-PV and LukS-I/LukF-I are significantly present in culture supernates after 15 h of culture in the defined YCP medium [Yeast extract (Oxoid), casaminoacids, sodium pyruvate, pH 7.5]. HlgA, HlgC and HlgB are rather more abundant when the strains are cultivated in heart infusion broth.

From the 1990s, two groups continued to study the leucotoxins. Kamio's group (Nariya *et al.*, 1993a) principally used hydroxyapatite and cation-exchange high-performance liquid chromatographies for purification, whereas the other group used cation-exchange chromatography followed by hydrophobic fast-performance liquid chromatography (Finck-Barbançon *et al.*, 1991; Prévost *et al.*, 1995a, b).

All of the relevant toxic proteins (Table 21.1) are clearly partitioned into two classes: class S proteins with a calculated molecular mass of approximately 32 kDa, and class F proteins with a calculated molecular mass around 35.5 kDa (Prévost *et al.*, 1994). All of the coresponding genes have been cloned using oligonucleotide probes whose sequence was deduced from *N*-terminal sequencing or from trypsin-generated peptides of the purified proteins.

These proteins become insoluble at pH <5.5 or at concentrations >7.5 mg ml^{-1} for the class S proteins. They are thermolabile and are more or less irreversibly denatured after 30 min at 60°C. They remain stable for several months at +4°C at a concentration of 0.5–1 mg ml^{-1} in 50 mM Na/phosphate, 200 mM NaCl, pH 7.5 buffer. They all promote neutralizing antibodies in immunized

TABLE 21.1 Characteristics of the secreted bi-component staphylococcal leucotoxins

Bacterial toxins	Protein designations	No. aa/molecular weight[a]/p*I*	Accession number (EMBL)
Panton–Valentine	LukS-PV	284/32 317/8.94	X72700
Leucocidin	LukF-PV	301/34 386/9.01	
γ-Haemolysin	HlgA, HgII, HγI	280/31 924/9.57	L01055, X64389,
	HlgC, LukS, LukS-R	286/30 688/9.22	X81586, S65052
	HlgB, Hγ, LukF-R	300/34 180/9.21	
LukM/LukF'-PV	LukM	277/31 694/9.21	D83951
	LukF'-PV	296/33 727/9.04	
LukE/LukD	LukE	286/32 239/9.88	Y13225
	LukD	301/34 302/9.02	
LukS-I/LukF-I	LukS-I	281/32 055/9.34	X79188
	LukF-I	300/33 802/7.54	

[a]Calculated molecular mass (Da).

rabbits with the formaldehyde-generated toxoid.

The first of the staphylococcal leucotoxins to be cloned and sequenced (Table 21.1) was the γ-haemolysin, both from a methicillin-resistant *S. aureus* strain (Rahman *et al.*, 1991, 1992; Kamio *et al.*, 1993) and from the Smith 5R strain (Cooney *et al.*, 1993), while the cloning of the presumed leucocidin R from P83 strain (Supersac *et al.*, 1993) revealed a 98% sequence identity, and a similar organization to the γ-haemolysin produced by the V8 strain (Prévost *et al.*, 1995a). The γ-haemolysin is in fact composed of two class S proteins, HlgA and HlgC, (Table 21.1) and one class F protein and, therefore, constitutes two leuco-toxins: HlgA/HlgB and HlgC/HlgB. The corresponding genes are organized in a locus probably located near the biotin chromosomal operon where *hlgA* was found 400 pb upstream to the contiguous and cotran-scribed *hlgC/hlgB* (Supersac *et al.*, 1993).

The genes encoding PVL were characterized in the V8 strain (ATCC 49775) by using an oligonucleotide probe based on the sequence of a tryptic peptide because of the similarities between the *N*-terminals of proteins constituting PVL and γ-haemolysin (Prévost *et al.*, 1995a). The genes *lukS-PV/lukF-PV*, as for *hlgC/hlgB*, are tandemly arranged (Figure 21.1), encod-ing the class S component preceding that encoding the class F components by one T. Genes are A-T rich (around 60%) and their homology is similar to the scores observed for the corresponding proteins. It was previously reported that PVL expression was brought about by the lysogenic conversion of a bacteriophage (Van der Vijver *et al.*, 1972). This claim was strengthened by the recent observation (Kaneko *et al.*, 1997a) that 3' sequences downstream of *lukF-PV* harbour repeats homologous with the recombinase-binding sites of several bacteriophages and a sequence similar to integrases.

A third locus has been characterized from the P83 strain in two steps (Choorit *et al.*, 1995; Kaneko *et al.*, 1997b) where LukM was first reported with a sequence similar to a class S protein, but with the function of a class F protein. In the second work, LukM was considered as a true class S protein together with the characterization of LukF-PV-like, or LukF'-PV in this text. LukF'-PV harboured 79% of sequence identity with LukF-PV. The proteins were characterized by the analysis of chromatographic eluates and by cross-immunoreactive properties with other proteins.

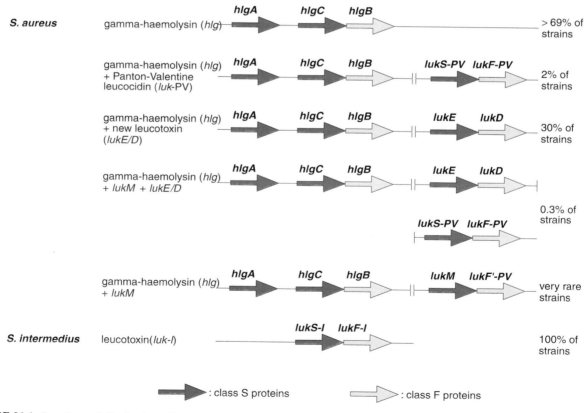

FIGURE 21.1 Genetics and distribution of leucotoxin-encoding genes among staphylococci.

Recently, two different cross-immunoreactive products with the previously purified proteins were characterized in the *S. aureus* Newman strain (Gravet *et al.*, 1998a, b). LukE was recognized as the class S protein and LukD as the class F protein. The reason why these two latter proteins (and LukM/LukF'-PV) were not recognized before was that they did not bear any strong cytotoxic activity against the usual target cells when tested (for example, erythrocytes, lymphocytes, polymorphonuclear cells, monocytes and macrophages), and they had similar apparent molecular masses to the two class S and F proteins. Furthermore, the differences in nucleotide sequence precluded the presence of large conserved sequences.

Another bi-component leucotoxin is expressed by *Staphylococcus intermedius* (Prévost *et al.*, 1997), which is neither a human commensal nor a human pathogen, contrasting with *S. aureus* (Mahoudeau *et al.*, 1997). However, this bacterium is responsible for pyodermitis in small carnivores such as dogs and cats. The genes *lukS-I* and *lukF-I* are tandemly arranged and cotranscribed. They were isolated after the blind application of the main purification procedure to culture supernates of strain ATCC 51874; testing for protein fractions having a leucocytotoxic activity on human PMNs only when they were mixed in pairs.

Finally, all genes encoding class F proteins are located dowstream to those encoding class S protein and are separated by one T. For all genes encoding a class F protein, a putative ribosome-binding site may be observed (LukF-PV, HlgB, LukD, LukF'-PV, LukF-I). This may help the translation efficacy by reducing the pausing of ribosomes. Although all of the tandemly arranged genes are cotranscribed (Figure 21.1), they may originate from distinct genetic elements. Promoter sequences are generally different, except for a conserved sequence, GNA(T)TAAAA, located from 31 to 35 bases upstream of the ATG codons. As the expression of these genes varies according to the culture media, fine differences in their regulation may be expected.

It appears that the calculated molecular masses of class S proteins are close to 32 kDa, with lengths varying from 277 to 286 amino acids (Table 21.1). These proteins are positively charged, with p*I* ranging from 8.9 to 9.9. The calculated molecular masses of the class F proteins are close to 34 kDa, with lengths ranging from 294 to 301 amino acids. They are also positively charged, with p*I* near 9.0, except for LukF-I, with a calculated p*I* equal to 7.5. The amino acid sequences of class S proteins show 59–79% identity, but only 20–27% with class F proteins (Figure 21.2). Class F proteins have 71–80% sequence identity. The amino acid sequence of α-toxin (Gray and Kehoe, 1984) shows 12–18% sequence identity with class S proteins, and

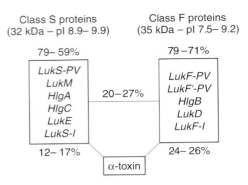

FIGURE 21.2 Identity scores between peptide sequences of class S and class F proteins of staphylococcal bi-component leucotoxins and staphylococcal α-toxin.

24–26% with class F proteins. Considering the partition of the class S and the class F proteins and the fact they may have originated from a common ancestor, α–toxin may also originate from an older ancestor. Moreover, when combined with either one class S or one class F protein, no further biological activity of the α-toxin was observed on human PMNs.

Distribution and epidemiology of leucotoxin-producing strains

As soon as the purified proteins constituting leucotoxins were obtained, toxoids were generated, following a formaldehyde denaturation step, in order to immunize rabbits and obtain polyclonal affinity-purified antibodies for use in an immunoprecipitation test (Ouchterlony, 1962; Finck-Barbançon *et al.*, 1991). These affinity-purified antibodies offered the opportunity to detect the homologous antigen by the formation of continuous immunoprecipitation lines with the tested antigens (Prévost *et al.*, 1995c; Gravet *et al.*, 1998). Such tests were validated by hybridization experiments for most of the toxins. It was demonstrated (Prévost *et al.*, 1995c) that more than 99% of *S. aureus* strains isolated in a routine hospital laboratory produced γ-haemolysin. Therefore, no specific clinical association was suggested for the leucotoxins. A retrospective, then prospective study (Finck-Barbançon *et al.*, 1991; Couppié *et al.*, 1994; Prévost *et al.*, 1995c) confirmed that PVL-producing strains accounted in Europe for 2% of the routine isolations of *S. aureus*. Furthermore, these PVL-producing strains were clearly associated with primary and necrotizing cutaneous infections, particularly with furuncles (Cribier *et al.*, 1992a–c; Couppié *et al.*, 1994; Prévost *et al.*, 1995c). It was observed that 90% of the PVL-producing strains originated from furuncles. Conversely, in 96% of furuncles, such a strain was isolated. Additionally, it was observed that in patients having recurrent furuncles, PVL-producing *S. aureus* strains were present in the anterior nares (Prévost *et al.*, 1995c). The *Sma*I-restricted total deoxyribonucleic acid (DNA) analysed by pulsed field gel electrophoresis showed identical patterns for isolates originating from both furuncles and from anterior nares of particular patients. This justified the decontamination of anterior nares of the patients concerned, in order to avoid chronic episodes. PVL-producing strains originating from furuncles were, in some cases, responsible for septicaemia with dissemination into pulmonary abscesses (Couppié *et al.*, 1997). This association with furuncles was also observed in Guyana. However, the situation seemed different in western Africa (Prévost *et al.*, 1998). In a cohort of 392 *S. aureus* isolates, 22.4% were PVL producers. This result raises questions about (a) the environmental influence on the bacterial flora and the induced risk of infections, and (b) the specific risks of bad sanitary conditions.

Concerning LukE/LukD, a distribution of the producing isolates of 30%, differing from those producing PVL or γ-haemolysin ($P < 0.01$), was observed (Gravet *et al.*, 1998b). In fact, all the samples were affected by such strains, but when *S. aureus* was isolated as the predominant bacterium or in pure culture from stools, most of the isolates (85%) were LukE/LukD- and enterotoxinA-producers (Gravet *et al.*, 1998a). These isolates were all obtained from quite old patients (mean age: 63 years) suffering from post-antibiotic diarrhoea and *S. aureus* isolates were all methicillin resistant. It was also observed that most of epidermolysin A and B-producing strains isolated in cases of bullous impetigo or staphylococcal scalded skin syndrome (SSSS) were also LukE/LukD-producers (Gravet *et al.*, 1998a). As enterotoxin A, epidermolysin A and B, and LukE/LukD production are all independent genetic features, the two new clinical associations of LukE/LukD in post-antibiotic diarrhoea and SSSS are noteworthy. Despite the fact that many *S. aureus* strains produce several virulence factors, co-operativity between them remains to be investigated.

The distribution of LukM/LukF'-PV among clinical strains was also investigated by the same immuno-precipitation test. Although more than 400 *S. aureus* isolates originating from human samples were tested, none of them produced this bi-component leucotoxin. Finally, only two out of the 21 specimens obtained from bovine mastitis produced this toxin (Gravet *et al.* 1998a). Finally, LukS-I/LukF-I was secreted by each of the 40 *S. intermedius* strains that were tested (Prévost *et al.*, 1995b).

MODE OF ACTION OF BI-COMPONENT LEUCOTOXINS

Target cells of the staphylococcal leucotoxins

The topic of the staphylococcal leucotoxins becomes more complex when considering the different target cells of these toxins. Up to now, the target cell specificity and cell susceptibility may occur not only on different cells of blood from humans, but also from animals (for example, rabbit, bovine, sheep, dog, rat) or biomembranes according to the leucotoxins. As reported in Table 21.2, the cell spectrum of the different major combinations of toxic components may be wide. This is shown in HlgA/HlgB which lyses PMNs from humans, rabbit, bovine, also erythrocytes from humans and rabbit (Supersac *et al.*, 1993; Prévost *et al.* 1995a), some T-lymphocytes, Jurkat cells (Meunier et al., 1996; Colin *et al.*, 1997) and phosphatidylcholine-cholesterol (1:1) biomembranes (Ferreras *et al.*, 1996; Meunier *et al.*, 1997). The results obtained recently on PMNs from other animals confirmed this ability (Szmigielski *et al.*, 1998) by measuring both the cell viability and the oxidative burst.

In particular, the specific activity of HlgA/HlgB, in an equimolar ratio on rabbit erythrocytes was 2.8×10^5 U mg^{-1} of each component, whereas that of our purified α-toxin (Bhakdi *et al.*, 1996) was 1.5×10^5 U mg^{-1}. It was 5×10^4 U mg^{-1} for HlgA/HlgB

TABLE 21.2 Recognized target cells to some of the major staphylococcal bi-component leucotoxins

	H-PMN[a]	H-T Lym	HRBC[b]	R-PMNs[c]	RRBC	B-PMNs[d]	SUV[e]
HlgA/HlgB	+++	++ (45%)	+++	+++	++++	+++	++
HlgC/HlgB	+++	−	−	+++	++	+	±
LPV	++++	−	−	++++	−	−	−
LukE/LukD	+ (Ca^{2+})[f]	−	−	+	−	nd[g]	−
LukM/LukF'-PV	+	−	−	+	−	nd	−
LukS-I/LukF-I	++++	−	−	++++	−	++	−

[a]Human polymorphonuclear leucocytes; [b]human red blood cells; [c]rabbit polymorphonuclear leucocytes; [d]bovine polymorphonuclear leucocytes; [e]small unilamellar vesicles; [f]cells from some donors bear susceptible Ca^{2+} channels; [g]not determined.

on human erythrocytes and that of α-toxin was significantly lower, with only $5 \times 10^2\,U\,mg^{-1}$ on the same cells.

The cell specificity of the PVL was much narrower since it is highly active (around $3 \times 10^7\,U\,mg^{-1}$) on human and rabbit PMNs only (Prévost et al., 1995a). In fact, myeloblastoid cells are sensitive to the PVL from the differentiation step of the metamyelocyte (Finck-Barbançon et al., 1993). Such cells as the HL60 lineage became susceptible only after their differentiation into myeloid cells by dimethyl sulfoxide (Meunier et al., 1995). LukS-I/LukF-I shared biological activities similar to that of PVL, except that they remained leucocytolytic $(2 \times 10^4\,U\,mg^{-1})$ against bovine PMNs (Prévost et al., 1995a).

HlgC/HlgB constitutes the other component couple of the γ-haemolysin (Cooney et al., 1993). Despite the fact that Kamio et al. (1993) named this couple leucocidin, because it was not haemolytic against human erythrocytes but exhibited a leucotoxic activity on human PMNs, it is haemolytic $(8 \times 10^4\,U\,mg^{-1})$ against rabbit red blood cells (Prévost et al., 1995a; Tomita and Kamio, 1997; Nariya et al., 1997b).

The recently characterized bi-component leuco toxin LukE/LukD brought to light another phenomenon (Gravet et al., 1998b). When tested against PMNs from different donors, it appeared that it was unable to permeabilize human PMNs, but LukE/HlgB was only able to for some of them. Such an observation suggests that complex and specific receptors are used by leucotoxins at the primary events of their interaction with cell membranes, and may have a critical role for the insertion of these toxins. LukE/LukD had no haemolytic activity. LukM/LukF'-PV had only a slight leucocytotoxicity upon human and rabbit PMNs among the cells tested.

As some S. aureus strains may produce one or two bi-component leucotoxins in addition to γ-haemolysin, the potential for heterologous combinations of one class S plus one class F component must be considered. From recent data, it appeared that all combinations with HlgA were more or less active on human PMNs (Gravet et al., 1986). Similarly, all combinations with HlgB were also active except for the pore formation with LukD and LukM. All of these observations gathered together support the variability of specific cell ligands according to the bi-component leucotoxins. Such ligands would be different from those allowing the binding of the α-toxin. The latter toxin was reported not to have significant activity on human PMNs, but rather on human platelets, T-lymphocytes, keratinocytes and fibroblasts (Walev et al., 1993; Jonas et al., 1994; Bhakdi et al., 1996). Considering the clinical associations of the bi-component leucotoxins such as PVL or LukE/LukD with tissue-necrotizing syndromes, it may be hypothesized

that some resident cells, such as dendrocytes, Langerhans cells, professional macrophages or keratinocytes, may constitute privileged target cells of these toxins. For example, keratinocytes represent an important mass of cells that express CD40 and secrete some inflammatory molecules such as granulocyte-macrophage–colony-stimulating factor (GM–CSF), interleukin-8 (IL-8) and tumour necrosis factor-α (TNF-α) (Péguet-Navarro et al., 1997).

Questioning about the cell receptor of leucotoxins

Noda et al. (1980b) were the first to report that GM1 in solution probably inactivated the leucotoxic activity of HlgA/HlgB. Other authors (Ozawa et al., 1994) also reported the binding of both HlgA and HlgC selectively to GM1, compared with other gangliosides. This binding modified the fluorescence properties of HlgA when bound to the membrane. However, the kinetic constant of this binding remains to be determined and binding was not confirmed by sodium dodecyl sulfate–polyacrylamide gel electrophoresis (SDS–PAGE) analysis. Moreover, it is surprising that GM1 constitutes the receptor of HlgA and HlgC, while HlgC/HlgB is biologically active on rabbit red blood cells (RBC) but not on human RBC. HlgA/HlgB lyses both cell types. GM1 is a minor component of blood-cell membranes and is much more prevalent in nerve cells. The author has never succeeded in confirming the results cited above and the previous N-neuraminidase treatment of human PMNs does not prevent biological activity of the leucotoxins. In addition, HlgA/HlgB was able to permeabilize small unilamellar vesicles lacking gangliosides (phosphatidyl-choline:cholesterol, 1:1) at quite a high dose of 95 nM (Ferrerras et al., 1996). Other experiments consisting of treatment of human PMNs with phosphatidylinositol-specific phospholipase C did not change the leucocytotoxic activity of the leucotoxins.

As HlgA/HlgB was biologically active on T-lymphocytes, surface antigen-directed monoclonal antibodies were used in order to screen the sensitive populations (Meunier et al., 1996; Colin et al., 1997). Despite using fluorescent monoclonal antibodies specific for CD3, 4, 8, 11, 16, 18, 29 and 45RA, no population appeared completely sensitive to HlgA/HlgB (Colin et al., 1997). By the same token, the same panel of monoclonal antibodies was unable to prevent binding and biological activity of the toxins. The binding ligands of the bi-component leucotoxins might be distributed only on some cells of the blood. It is not known whether these ligands belong to a single family of molecules, but it was observed that the binding of some class S proteins such as HlgA and LukS-PV may compete more or less at the surface of the membrane. Up to now, no argument can exclude firmly that these receptors are probably involved in one activation pathway of the cells (see below).

Sequential binding of class S, class F proteins

As previously reported (Woodin and Wieneke, 1968; Noda et al., 1981; Colin et al., 1994a), biological activity, for example release of hexosaminidase or lysozyme activities, was expressed only after the prior interaction of the small protein LukS-PV with the membranes of human PMNs. Application of LukF-PV prior to LukS-PV never led to any damage of the cell, as evidenced by fluorescence from probes such as Fluo-3 and Fura-2, which are sensitive to divalent cations, or ethidium entering by large pores and combining with nucleic acids (Meunier et al., 1995).

A troubling result was found by Ozawa et al. (1995), who reported binding of LukF (or HlgB) to human erythrocytes prior to that of HγI (HlgA), whereas the opposite was mentioned with rabbit erythrocytes (Noda et al., 1982). Whatever the cells tested, in two different laboratories, the binding of class S proteins was always recorded as a prerequisite before the binding of class F proteins (Colin et al., 1994a; Ferreras et al., 1996). More recently, the primary binding of class S proteins was directly and functionally shown by the study of site-directed mutants of HlgA and HlgC with human PMNs, rabbit erythrocytes and small unilamellar vesicles (SUV) (Meunier et al., 1997). Effectively, the T28D or T30D substitution of HlgA and HlgC, respectively, did not modify the binding properties of the recombinant and mutated proteins, compared with the wild-type proteins. However, these mutated proteins did not allow any secondary interaction of the class F protein (HlgB) with membranes and led to the loss of activity. Finally, it is surprising that the necessity of initial binding of class S proteins to be observed for different membranes, by different methods and by different authors should differ only on human erythrocytes, whereas it was conserved on basic SUV (Meunier et al., 1997).

The primary binding of class S proteins confers a specific function to the latter. It underlines that class S and class F proteins are not interchangeable and indicates that several adaptation steps are necessary before the soluble proteins are able to insert into membranes. The binding of S proteins probably induces conformational modifications of S themselves and/or of molecules at the surface of the membrane, allowing the secondary interaction of F proteins. This is fundamentally different to the binding of monomers of α-toxin (Song et al., 1996), where monomers are presumed equivalent.

Neutralizing antibodies inhibited the induced entry of Ca^{2+} by PVL on human PMNs, when they were applied prior to or simultaneously to the toxin (Finck-Barbançon et al., 1993). When applied between the application of the S and F component, no biological activity was observed. Finally, when applied during the permeabilizing process, only a partial inhibition of the entry of Ca^{2+} was recorded.

Pore formation by the leucotoxins

Spectrofluorimetry and flow cytometry with the use of molecular probes specifically sensitive to ions allowed the characterization of the early events of the cell intoxication. It was shown that expression of the biological activity of the bi-component leucotoxins consisted of a rapid osmotic shock, since the surface area of the human PMNs increased significantly within 10 min of intoxication (Meunier et al., 1995). At the same time as the nuclei were swelling, pseudopods were observed as well as lysis of the cells at high concentrations of toxin.

This osmotic shock was manifested by entry of Ca^{2+} and Mn^{2+} shown by the Fura-2 fluorescent probe, and Mg^{2+} by using Mag Fura-2 (Finck-Barbançon et al., 1993). These entries were time and dose dependent, but were stopped in the presence of 3 mM EGTA. The biological activity of PVL was dependent on the concentration of Zn^{2+}, since 0.2 mM Zn^{2+} was a permissive concentration for the entry of Ca^{2+}, whereas 2 mM Zn^{2+} inhibited this entry. Moreover, ethidium entered the cells as a result of leucotoxin activity. Such an event was significant in the absence or in the presence of 0.1 mM Ca^{2+} but was inhibited in the presence of 1 mM Ca^{2+} or 0.2 mM Zn^{2+}. With concentrations of LukS-PV = 2.3 nM and LukF-PV = 0.57 nM, it was shown that PVL was able to lyse human PMNs at a concentration of 3×10^6 cells ml^{-1}.

Using iodinated LukS-PV and LukF-PV, apparent dissociation constants were determined to be around 6 nM and 2 nM, respectively (Colin et al., 1994b). However, it seemed that iodination resulted in some loss of toxin activity because, when using a fluorescein-labelled protein (LukS-PV G10C-Fluorescein, HlgB-fluorescein), these constants were closer to 0.1 nM, as observed in our laboratory. Using radiolabelled molecules in binding studies, around 40 000 of each kind of protein bound at the surface of the membrane with a Hill coefficient close to 1, indicating the presence of one class of binding site for PVL. This situation seemed different when studing HlgA, because human PMNs appeared not to be easily saturable (Meunier et al., 1997). This may reflect an intrinsic affinity of HlgA for hydrophobic bilayers (Ferreras et al., 1996, 1998a).

Recently, Ca^{2+} influx and entry of ethidium generated by both PVL, HlgA/HlgB, HlgC/HlgB with human PMNs were shown to be independent phenomena (Staali et al., 1998). Effectively, by considering the time-course of the penetration of Mn^{2+} and ethidium, the authors recorded that the entry of ethidium was delayed by around 80–100 s compared with that for Mn^{2+}. The entry of both Ca^{2+} and Mn^{2+} was inhibited by the presence of either adenosine or verapamil (D 600), or econazole; the latter competely inhibiting such penetrations (Colin et al., 1996b). Concurrently to these inhibitions, the penetration of ethidium was maintained. Nifedipine, a dihydropyridine inhibitor, had no significant action on the three couples tested. These calcium channels are activated within 100 s following leucotoxin application. Moreover, by using Na-Green and PBFI fluorescent probes specific for Na^+ influx and K^+ efflux, respectively, the leucotoxins provoked their exchange according to the gradient of concentration. These fluxes were not sensitive to the application of the drugs cited above. However, in the presence of 0.2 mM Zn^{2+} + 0.1 mM Ca^{2+} it appeared that ethidium, Na^+ and K^+ exchanges remained inhibited, whereas divalent cation influxes were indicated by the increase of Fura-2 fluorescence. Furthermore, PMNs from some donors remained defective for any leucotoxin-mediated Ca^{2+} entry, whereas they were lysed and ethidium, Na^+, K^+ were exchanged. Therefore, it is proposed that leucotoxins, after interacting with cell membranes via an unknown receptor, are able to activate receptor-mediated and store-regulated Ca^{2+} channels and are probably able to form pores themselves that are specific to various monovalent cations. These phenomena remain undescribed for other bacterial pore-forming toxins. The cell-activated Ca^{2+} channels are likely to be characterized by particular agonists and inhibitors soon.

During these assays, it was observed that physiological concentrations of Ca^{2+} lowered dramatically the entry of ethidium, Na^+ and K^+. The effect of the bi-component leucotoxins might occur first or essentially by the activation of target cells where Ca^{2+} metabolism plays a critical role.

Structure–function relationships of bi-component leucotoxins

Current data should be considered at two different levels: those reporting that modified proteins have lost cytolytic functions, and those reporting that modified proteins failed in at least one of the presumed events leading to the pore formation.

The first approach consisted of characterization of leucotoxin-constituting proteins that lost biological activity. Nariya et al. (1993b) mentioned the purification of LS2, a C-terminal truncated LukS (HlgC), that did not harbour any leucotoxic activity. However, no toxin concentration was reported. Although the absence of binding to the membrane was not demonstrated, no binding to GM1 was recorded. The truncated protein, determined by C-terminal sequencing, ended 17 amino acid residues before that of LukS. Another amino acid substitution of HγII (HlgA):Lys^{217}Arg reduced the haemolytic activity on human erythrocytes to around 40% of the wild-type protein (Sudo et al., 1995). Variations are within the limit of significance and would have been presented with standard deviation data. Predictions of secondary structures (Chou and Fasman, 1974) indicated that the substitution changed the vicinity of Lys^{217} from a random coil to an α-helix structure.

By using conserved restriction sites within the genes encoding LukS (HlgC) and HγII (HlgA), different full-length chimeric proteins derived from the two latter were constructed and purified (Nariya and Kamio, 1995). The chimeric proteins were assayed with HlgB for the determination of their residual leucotoxic and haemolytic activities. Despite the fact that the source of cells was not indicated, and HlgA/HlgB was classed as non-leucocytolytic, the exchange of the 123 C-terminal residues of HlgC by those of HlgA restored haemolytic properties (assumed to be human RBC). In contrast, the substitution of the C-terminal 24 residues of HlgC by that of HlgA abolished leucocytotoxicity.

Recently, LukS (HlgC) was finally recognized as being haemolytic on rabbit RBC when combined with LukF (HlgB) (Nariya et al., 1997a). In this work, chimeric proteins derived from HlgC and LukS-PV were engineered by the polymerase chain reaction. The haemolytic activity of the proteins was tested on rabbit RBC where HlgC/HlgB was active but LukS-PV/HlgB was not. Interestingly, it was found that addition of the two residues Asp-Ile of HlgC in the position 12–13 of LukS-PV conferred haemolytic activity to the new protein when combined with HlgB. Conversely, when the two amino acids were deleted from HlgC, only 12% of the haemolytic activity remained. Such a result is in keeping with several possible mechanisms of the pore formation in erythrocytes and raises several important questions. (a) Does LukS-PV bind to the erythrocyte membrane? (b) Do the Asp^{12}Ile^{13} amino acids favour a more efficient interaction of the S component with HlgB or with the phospholipid membrane? (c) Do the Asp^{12}Ile^{13} amino acids modify the conformation of proteins that allow the orientation of new functional domains? Finally, the substitution of the C-terminus of HlgC by that of LukS-PV did not modify significantly the haemolytic activity. Another report showed

(Nariya *et al.*, 1997b) that the addition of the LukS (HlgC) [242]IleLysSerThr[245] peptide sequence in the corresponding location of Hlg2 (HlgA) restored leucotoxic activity of the protein when combined with HlgB. LukS or Hlg2 (HlgA-) [242]-IleLysArgSerSer mutant proteins, both inactivated by boiling, were phosphorylated *in vitro* by protein kinase A. These authors concluded that this sequence was a minimal segment for the leucotoxic function, and that the phosphorylation of LukS was a prerequisite for the function of LukS. These observations conflict with our reports where HlgA/HlgB was leucotoxic and haemolytic (Prévost *et al.*, 1995a; Ferreras *et al.*, 1996; Meunier *et al.*, 1997; Staali *et al.*, 1998). The sequences of HlgA (Prévost *et al.*, 1995a) and Hlg2 (Kamio *et al.*, 1993) exhibit two amino acid differences, Arg[258]Lys and His[248]Pro, but both were lacking for the IleLysArgSerThr sequence. Therefore, if Hlg2 is a natural mutant, the effects of these substitutions should be analysed on the impairment of pore formation. The phosphorylation of the boiled LukS essentially interrogates the specificity of this test of phosphorylation.

By both attenuated total reflectance–Fourier transformed infrared (ATR–FTIR) spectroscopy and secondary structures prediction, it can be seen that leucotoxins and particularly γ-haemolysin are composed predominantly (60–65%) of β-sheet structures (Meunier *et al.*, 1997). This feature spans the bi-component leucotoxins in the supergroup of pore-forming toxins with predominant β-sheet structures that include α-toxin, aerolysin and the protective antigen (PA) of anthrax toxin, and differs from colicins or diphtheria toxin by both having α-helix structures (Lesieur *et al.*, 1997). From the secondary structure predictions, it appeared that a 13-residue β-sheet was conserved at the *N*-terminal extremity of class S proteins (Figure 21.3). Structural predictions were used to generate mutations that would destroy the predicted β-sheet. First, these mutations focused on Thr[28] and Thr[30] of HlgA and HlgC, respectively, and substitutions by Asp were analysed (Colin *et al.*, 1996a; Meunier *et al.*, 1997). The combinations of both HlgAT28D/HlgB and HlgCT30D/HlgB resulted in toxins unable to induce calcein leakage from small unilamellar vesicles (phosphatidylcholine:cholesterol 1:1), or to increase the permeability of rabbit (and human for HlgAT28D) erythrocytes. However, competition experiments indicated that the mutated proteins could reduce the biological activities of the wild-type proteins suggesting that they bound efficiently to the membranes at the same sites. Non-specific binding to human PMNs was observed for HlgA. Moreover, the functional fluorescein-labelled HlgB did not bind to the membranes even when HlgAT28D or HlgCT30D were previously

bound. The lack of binding of HlgB was observed with HlgA T28D when it was previously applied and bound, presumably to aspecific sites. Human PMN alone did not retain the fluorescein-labelled HlgB. Further studies may demonstrate a critical role for Thr[28] and possibly other amino acids in the secondary interaction of the class F component. This amino acid can be aligned with His[35] of the α-toxin (Figure 21.3) which is involved in the interaction between monomers (Jursch *et al.*, 1994; Menzies and Kernodle, 1994; Walker *et al.*, 1995). These observations suggest that the monomers of the pore-forming leucotoxins interact by different amino acid interactions compared with those of α-toxin (Song *et al.*, 1996). From a comparison of the primary structures of α-toxin and the leucotoxins and the three-dimensional structure of α-toxin, it has been proposed recently that these toxins converge in some of their functional properties (Gouaux *et al.*, 1998). It was noticed that the particular structure of the stem domain of α-toxin was similarly conserved in the bi-component leucotoxins (Figure 21.4), indicating comparable transmembrane structures, although that of leucotoxins might be shorter. Understanding of the conformational changes between the water-soluble form and the lipid-stable one of the stem when pores are created is of fundamental importance. However, specificity of the pores formed by leucotoxins might differ slightly from those of α-toxin since many monovalent cations can go through (Na[+], K[+], ethidium), whereas propidium does not permeate α-toxin-created pores (Jonas *et al.*, 1994; Walev *et al.*, 1993; Abrami *et al.*, 1998). Such flexibility in the pores created by leucotoxins remains to be understood. Of the 28 conserved residues in these proteins, many are clustered in three-dimensional space proximal to the rim domain assumed to stabilize the pore-forming toxin on to/within the membrane. Most of the conserved residues are located inside the folding of the structure of the α-toxin, indicating that the concerned toxins, including aerolysin (Lesieur *et al.*, 1997), certainly proceed by similar steps which lead to the formation of pores. However, these toxins diverge

HlgA	26	AIT**Q**NIQFDFVKDK
HlgC	28	GVT**Q**NIQFDFVLDK
LukS-PV	26	GVT**Q**NIQFDFVKDK
LukM	28	GVT**Q**NVQFDFVKDK
LukE	26	GVT**Q**NVQFDFVKDK
LukS-I	28	GVT**Q**NIQFDFVKDP
α-Toxin	29	DKENGM**H**KKVFYSFIDDK
		* * * ** **

FIGURE 21.3 Sequence alignment of the predicted β-sheets of the class S proteins from staphylococcal bi-component leucotoxins with the corresponding one from the staphylococcal α-toxin. Asterisks indicate the most conserved residues.

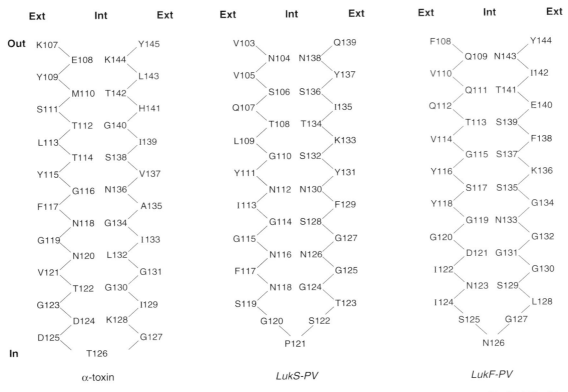

FIGURE 21.4 Schematic representation of the stem regions (double β-sheet structures) of α-toxin, LukS-PV and LukF-PV with mention of the exterior and interior of the pores and the insides and outsides of the membranes. Numbering of amino acids corresponds to their position in the mature proteins.

by their specific binding to target cells, the specificity/selectivity of the pores and possibly in the stoechiometry of the proteins in the pores and, of course, in their association with clinical syndromes.

Sugarawa and co-workers (Sugarawa *et al.*, 1997) published photomicrographs of hexameric ring-shaped structures from HlgA/HlgB-treated human RBC, and electrophoretic resolution of protein complexes of around 200 kDa was obtained. These authors determined the radius of the pores to be 1.2 Å by using non-ionic polyethylene glycols, a value similar to that found in the author's laboratory (Colin *et al.*, 1997) and similar to that of the α-toxin (Song *et al.*, 1996). These values remain to explain the specific monovalent cations of these pores. A recent work (Ferreras *et al.*, 1998b) also proposed a hexameric structure of the leucotoxin pore with an equivalent ratio between class S and class F proteins.

Secondary events following the action of bi-component leucotoxins

The membrane-mediated ionic exchanges induced by the bi-component leucotoxins have a profound impact on cell metabolism and may trigger cell-sig-

nalling pathways (Figure 21.5). After being primed by TNF-α, PVL-treated human PMNs expressed the heat-shock protein 72, but were more resistant to the cytolytic effect of the toxin (Köller *et al.*, 1993). In these experiments, the stimulation of LTB$_4$ secretion was observed, and further analysed (Hensler *et al.*, 1994b). This secretion of LTB$_4$ was dependent on the dose of PVL (at least 100 ng of each component/10^7 cells for >2 min) and the exposure time. Moreover, the action of PVL seemed to increase the half-life of the LTB$_4$ and inhibited its ω-oxidization by fourfold, irrespective of a previous treatment of cells with either GM-CSF, G-CSF or IL-3. The pretreatment of cells with GM-CSF or G-CSF stimulated the secretion of the LTB$_4$.

This chemotactic agent is not the only one to be induced. PVL also generated secretion of IL-8, but not its synthesis (König *et al.*, 1994). Moreover, protein kinases and calcium-channel inhibitors such as Tyrphostin 25 and Lavendustin A inhibited both the synthesis of IL-8 and its secretion, in the presence of PVL. This observation, together with a recent one concerning the activation of Ca^{2+} channels (Staali *et al.*, 1998), suggests that a common pathway triggers both the activation of these channels and the secretion of IL-8. It was also observed that the secretion of IL-8

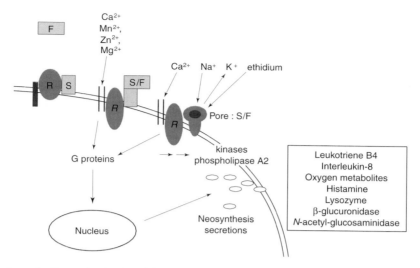

FIGURE 21.5 Proposed scheme for the mode of action of the staphylococcal bi-component leucotoxins, and the induced cell responses (human PMNs). S: class S proteins, F: class F proteins, R: cell receptor(s), R: activated receptor(s).

generated by streptococcal erythrogenic toxin A was not inhibited by the above protein kinase inhibitors, suggesting the existence of several pathways in human PMNs for release of this chemotactic and pain agent.

Investigations on G-protein pathways revealed that the prior ADP-ribosylation of heterotrimeric G-proteins by cholera and pertussis toxins decreased significantly PVL-stimulated LTB$_4$-generation (Hensler *et al.*, 1994a), although PVL still bound to the membrane and promoted cell lysis in the absence of calcium. The action of the PVL increased some GTP-binding protein functions, lowering translocation into the membrane of *ras*GAP and lipooxygenase. Despite the fact that PVL has never been observed to be translocated into cells, ADP-ribosylation of three proteins of 24, 40 and 45 kDa was reported, probably following activation of signal transduction pathways. In other experiments, the inflammatory potential of PVL (Luk-PV) and HlgC/HlgB (LukR) was compared in human PMNs (König *et al.*, 1995). PVL was a better inducer of release of LTB$_4$, IL-8, histamine, β-glucuronidase and lysozyme. The treated human PMN became twice as chemiluminescent over more time. As PVL induced the secretion of the vasodilatator histamine, together with IL-8 and LTB$_4$, the chemotactic invasion of human PMNs in tissues would be expected. As PVL-producing strains may produce PVL and γ-haemolysin, both possible couples with one class S and a class F component were compared for their ability to promote mediator release (König *et al.*, 1997). The results indicated that class S and class F components are interchangeable. Again, PVL was reported to be the most potent couple for the release of β-glucuronidase,

histamine and IL-8, but closer activity levels were observed with LukS-PV/HlgB, while HlgA/HlgB and HlgA/LukF-PV seemed to be less effective. From the results of the class S components studied, the following potency was proposed: LukS-PV > HlgC > HlgA.

ANIMAL MODELS

As PVL-producing strains were associated with furuncles, intradermal injections of PVL were assayed in rabbit skin (Ward and Turner, 1980; Cribier *et al.*, 1992). Increasing doses of PVL induced inflammatory reactions which increased with time over 48 h. Induction of oedema and erythema was observed by Cribier *et al.* with only 30 ng of each component and lesions becoming larger with 300 ng of the proteins. Slough formation was observed with 3000 ng, while no significant reaction was recorded with each of the proteins alone. Necrosis was associated with vasodilatation of capillaries, leucocytoclasis and PMN infiltration into surrounding tissues 24 h after the injection of the toxin. These observations were in accord with the release *in vitro* of inflammatory mediators (König *et al.*, 1995). Previous immunization of rabbits with formaldehyde-generated toxoid prevented the necrotizing effect of PVL (Cribier *et al.*, 1992a). Only slight PMN infiltration, without leucocytoclasis was observed. Comparison of dermonecrotizing properties between bi-component leucotoxins indicated that HlgA/LukF-PV was less dermonecrotizing than PVL, whereas HlgA/HlgB induced no comparable effect (Prévost *et al.*, 1995a). The recently characterized LukE/LukD and LukM/LukF'-PV bi-component leucotoxins were also highly dermonecrotizing (100–300 ng) for rabbit skin (Gravet *et al.*, 1998b). Despite promising results, *S. aureus* ATCC 49775-producing PVL always failed to develop a necrotizing lesion when intradermally injected in the experimental model. The limits of this model were paralleled by the difficulty to quantify the bacterial growth and to assess the expression of these genes encoding the toxins.

Another model was developed which was chosen on the basis of criteria including the existence of a parallel in human pathology, pre-existence of assays with staphylococci in this model, ease of sampling and processing of samples, slow clearance of the injected products, and the sensitivity of some animal cells to the toxins studied. Injections into the vitreous humour of the rabbit eye of bi-component leucotoxins constituted from PVL and γ-haemolysin were done, recording the clinical reactions in the anterior and posterior chambers of eye, their histological aspects and the biochemical assay of *N*-acetyl β-D-glucosaminidase in the vitreous humour (Siqueira *et al.*, 1997). In this model, as little as 30 ng of HlgA/LukF-PV provoked a significant inflammatory reaction within 48 h. Examination of the posterior chamber just allowed a view of the optic nerve and the infiltration of blood cells at around 200–300 cells per microscopic field was recorded. No reaction was evident when toxins were injected on to the conjunctiva. When injected under the conjunctiva, a local reaction was initiated. Finally, levels of glucosaminidase activity assay were relevant to the clinical and histological effects. In a second experiment, a γ-haemolysin-deficient mutant was created by a double crossing-over via a suicide vector in the Newman strain (Supersac *et al.*, 1998), and assayed in the animal model of experimental endophthalmitis. Compared with the wild-type strain, the infection caused by the injection of the mutant strain showed a strong inflammatory reaction in both posterior and anterior chambers, but a reduced reaction in the eyelid. The growth of the latter strain was simultaneously reduced by 100-fold. While the Newman strain did not produce high yields of α-toxin, the remaining inflammation generated by infection by the mutant is noteworthy. This strain produces enterotoxin A and LukE/LukD. The results may indicate how the bi-component leucotoxins such as γ-haemolysin may promote infection by *S. aureus* in some tissues.

CONCLUSION

The family of the staphylococcal bi-component leucotoxins consists of six class S proteins and five class F proteins with 30 possible binary combinations generating leucotoxins with specific biological activities. Several groups have studied these toxins over a period of 10 years and their independent investigations have often resulted in different designations for the same toxin, especially for the two most studied ones, γ-haemolysin and PVL. Some of the confusion originated from the diversity within some 'single' reference strains such as V8 (ATCC 27733 then ATCC 49775) that produced the PVL, γ-haemolysin and V8 protease, and probably strain P83. Also, further confusion originated from the absence of adequate comparison of purified toxins prior to the 1990s.

In an attempt to create a generic word that encompassed these toxins, they were called synergohymenotropic toxins (Supersac *et al.*, 1993). This designation has not been widely accepted since many bacterial toxins, bi- or multi-component toxins manifest their toxic activity by interaction with membranes, but cannot be equalized with leucotoxins; examples include diphtheria, pertussis and anthrax toxins, and some A/B/(C) bacteriocins (Donvito *et al.*, 1997; Cintas *et al.*, 1998) that possess α-helix-rich structures. Because they are both cell activators and permeabilizing agents, a unique pharmacological designation is not easy. However, a unifying feature of these toxins is their activity against PMN, macrophages and monocytes. Therefore, the designation of staphylococcal bi-component leucotoxins appears to be a realistic one.

Several of the staphylococcal bi-component leucotoxins are significantly associated, alone or with other virulence factors, with clinical syndromes such as furuncles, post-antibiotic diarrhoea due to predominant *S. aureus* and staphylococcal impetigo. Yet, the mechanistic links between such toxins and these syndromes remains obscure. Bi-component leucotoxins may in part account for the diversity of infections caused by *S. aureus*. This diversity may impair the immune defence. These toxins offer multiple challenges. First is the characterization of the different cell ligands that may suggest pathways of the evolutionary adaptation of such leucotoxins to human cell(s) antigens. Second is the understanding of the events following interaction between the toxic proteins and cell surface leading to both cell activation and pore formation. Resolution of the three dimensions will be of considerable importance in this field. Third is the characterization of the chronological sequence of mechanisms of intoxication and biological inter-relationships responsible for the cell response. Fourth is the documentation of their role in *S. aureus* infections. Finally, potential applications of bi-component leucotoxins in microbiology and other fields in medicine are likely to be of interest.

ACKNOWLEDGEMENTS

The author thanks Professor H. Monteil for his critical reading and Mrs D. Prévost for typing part of this manuscript. This work was supported by grant EA1318 from the Direction de la Recherche et des Etudes Doctorales (DRED).

REFERENCES

Abrami, L., Fivaz, M., Glauser, P.-E., Parton, R.G. and van der Goot, F.G. (1998). A pore-forming toxin interacts with a GPI-anchored protein and causes vacuolation of the endoplasmic reticulum. *J. Cell Biol.* **140**, 525–40.

Bänffer, J.R.F. (1962). Anti-leucocidin and mastitis puerperalis. *Br. Med. J.* **2**, 1224–31.

Bänffer, J.R.F. and Franken, J.F. (1967). Immunisation with leucocidin toxoid against staphylococcal infection. *Path. Microbiol.* **30**, 166–71.

Bhakdi, S., Bayley, H., Valeva, A., Walev, I., Walker, B., Weller, U., Kehoe, M. and Palmer, M. (1996). Staphylococcal alpha-toxin, streptolysin-O, and *Escherichia coli* haemolysin: prototypes of pore-forming bacterial cytolysins. *Arch. Microbiol.* **165**, 73–9.

Blobel, H., Wenk, K. and Kanoe, M. (1971). Effects of Panton-Valentine leukocidin of *Staphylococcus aureus* on leukocytes from patients with leukemia. *Infect. Immun.* **3**, 507–9.

Cintas, L.M., Casaus, P., Holo, H., Hernandez, P.E., Nes, I.F. and Havarstein, L.S. (1998). Enterocins L50A and L50B, two novel bacteriocins from *Enterococcus faecium* L50, are related to staphylococcal haemolysins. *J. Bacteriol.* **180**, 1988–94.

Choorit, W., Kaneko, J., Muramoto, K. and Kamio, Y. (1995). Existence of a new protein component with the same function as the LukF component of leukocidin or γ-haemolysin and its gene in *Staphylococcus aureus* P83. *FEBS Lett.* **357**, 260–4.

Chou, P.Y. and Fasman, G.D. (1974). Prediction of protein conformation. *Biochemistry* **13**, 222–45.

Colin, D.A., Mazurier, I., Sire, S. and Finck-Barbançon, V. (1994a). Interaction of the two components of leukocidin from *Staphylococcus aureus* with human polymorphonuclear leukocyte membranes: sequential binding and subsequent activation. *Infect. Immun.* **62**, 3184–8.

Colin, D.A., Prévost, G., Meunier, O., Sire, S., Mazurier, I. and. Supersac, G. (1994b). The leucocidin from *S. aureus*: physiopathology and sequence. In: *Bacterial Protein Toxins* (eds J. Freer *et al.*), pp. 347–8. Zentralblatt für Bakteriologie, Suppl. 24. Gustav Fischer, Stuttgart.

Colin, D.A., Meunier, O., Baba-Moussa, L., Ferreras, M., Supersac, G., Monteil, H., Menestrina, G. and Prévost, G. (1996a). A predicted β-sheet of class S components from staphylococcal SHT toxins is involved in their binding to membranes of neutrophils. *Proceedings of the 96th American Society of Microbiological General Congress*, p. 264, New Orleans.

Colin, D.A., Meunier, O., Staali, L., Monteil, H. and Prévost, G. (1996b). Action mode of two components pore-forming leucotoxins from *S. aureus*. *Med. Microbiol. Immunol.* **185**, 107.

Colin, D.A, Meunier, O., Staali, L., Prévost, G. and Monteil, H. (1997). Bi-component leukotoxins from *Staphylococcus aureus*. In: *Microbial Pathogenesis and Host Response* (eds Maloy *et al.*), Proceedings of the Cold Spring Harbor Laboratory on Microbial Pathogenesis and Host Response, p. 150. Cold Spring Harbor, New York.

Cooney, J., Kienle, Z., Foster, T.J. and O'Toole, P.W. (1993). The γ haemolysin locus of *Staphylococcus aureus* comprises three linked genes, two of which are identical to the genes for F and S components of leucocidin. *Infect. Immun.* **61**, 768–71.

Couppié, P., Cribier, B., Prévost, G., Grosshans, E. and Piémont, Y. (1994). Leucocidin from *Staphylococcus aureus* and cutaneous infections: an epidemiological study. *Arch. Dermatol.* **130**, 1208–9.

Couppié, P., Hommel, D., Prévost, G., Godart, M.C., Moreau, B., Sainte-Marie, D., Peneau, C., Hulin, A., Monteil, H. and Pradinaud, R. (1997). Septicémie à *Staphylococcus aureus*, furoncle et leucocidine de Panton et Valentine: 3 observations. *Ann. Dermatol. Vénéréol.* **124**, 684–6.

Cribier, B., Prévost, G., Couppié, P., Finck-Barbançon, V., Grosshans, E. and Piémont, Y. (1992a). *Staphylococcus aureus* leukocidin: a new virulence factor in cutaneous infections. *Dermatology* **185**, 175–80.

Cribier, B., Couppié, P., Finck-Barbançon, V., Prévost, G., Piémont, Y. and Grosshans, E. (1992b). La leucocidine: un nouveau facteur de virulence dans les infections cutanées. *Ann. Dermatol. Vénéréol.* **119**, 449.

Cribier, B., Finck-Barbançon, V., Prévost, G., Jaulhac, B. and Piémont, Y. (1992c). *Staphylococcus aureus* cutaneous infections: role of leucocidin in an experimental animal model. *J. Invest. Dermatol.* **98**, 835.

Donvito, B., Etienne, J., Denoroy, L., Greenland, T., Benito, Y. and Vandenesh, F. (1997). Synergistic haemolytic activity of *Staphylococcus lugdunensis* is mediated by three peptides encoded by a non-agr genetic locus. *Infect. Immun.* **65**, 95–100.

Ferreras, M., Menestrina, G., Foster, T., Colin, D.A., Prévost, G. and Piémont, Y. (1996). Permeabilization of lipid bilayers by *Staphylococcus aureus* γ-toxins. In: *Bacterial Protein Toxins* (eds P.L. Frandsen *et al.*), pp. 105–6. Zentralblatt für Bakteriology, Suppl. 28. Gustav Fischer, Stuttgart.

Ferreras, M., Colin, D.A., Prévost, G. and Menestrina, G. (1998a). Interaction of *Staphylococcus aureus* γ-toxins and leukocidins with lipid bilayers. In: *Bacterial Protein Toxins* (eds J. Hacker *et al.*), pp. 297–8. Zentralblatt für Bakteriologie, Suppl. 29. Gustav Fischer, Stuttgart.

Ferreras, M., Höper, F., Dalla Serra, M., Colin, D., Prévost, G. and Menestrina, G. (1998b). The interaction of *Staphylococcus aureus* bi-component γ-haemolysins and leucocidins with cells and lipid membranes. *Biochim. Biophys. Acta* **1414**, 108–26.

Finck-Barbançon, V., Prévost, G. and Piémont, Y. (1991). Improved purification of leucocidin from *Staphylococcus aureus* and toxin distribution among hospital strains. *Res. Microbiol.* **142**, 75–85.

Finck-Barbançon, V., Duportail, G., Meunier, O. and Colin, D.A. (1993). Pore formation by a two-component leukocidin from *Staphylococcus aureus* within the membrane of human polymorphonuclear leukocytes. *Biochim. Biophys. Acta* **1182**, 275–82.

Gladstone, G.P. and Fildes, P. (1940). A simple culture medium for general use without meat extract or peptone. *Br. J. Exp. Pathol.* **21**, 161–73.

Gladstone, G.P. and Van Heyningen, W.E. (1957). Staphylococcal leucocidin. *Br. J. Exp. Pathol.* **38**, 125–37.

Gladstone, G.P. and Glencross, E.J.G. (1960). Growth and toxin production of staphylococci in cellophane sacs *in vivo*. *Br. J. Exp. Pathol.* **41**, 313–33.

Gladstone, G.P., Mudd, S., Hoschstein, H.D. and Lenhardt, N.A. (1962). The assay of anti-staphylococcal leucocidal components (F and S) in human serum. *Br. J. Exp. Pathol.* **43**, 295–312.

Gouaux, J.E., Hobaugh, M. and Song, L. (1998). α-Haemolysin, γ haemolysin and leukocidin from *Staphylococcal aureus*: distant in sequence but similar in structure. *Protein Sci.* **6**, 2631–5.

Gravet, A., Baba Moussa, L., Werner, S., Staali, L., Meunier, O., Keller, D., Colin, D.A., Sanni, A., Monteil, H. and Prévost, G. (1998a). Characterization of a novel 30%-distributed staphylococcal leucotoxin and structure comparison with the other members of the toxin family. In: *Bacterial Protein Toxins* (eds J. Hacker *et al.*), pp. 401–2. Zentralblatt für Bakteriologie, Suppl. 29. A Gustav Fischer, Stuttgart.

Gravet, A., Colin, D., Keller, D., Girardot, R., Monteil, H. and Prévost, G. (1998b). Characterization of a novel structural member, LukE-LukD, of the bi-component staphylococcal leucotoxins family. *FEBS Lett.* **436**, 202–8.

Gray, G.S. and Kehoe, M. (1984). Primary sequence of the α-toxin gene from *Staphylococcus aureus* Wood 46. *Infect. Immun.* **46**, 615–18.

Grojec, P. (1979). Distribution of ^{131}I-labelled staphylococcal leucocidin in mouse organs. *Med. Dosw. Mikrobiol.* **31**, 209–16.

Grojec, P.L. and Jeljaszewicz, J. (1985). Staphylococcal leucocidin, Panton–Valentine type. *J. Toxicol. Toxin Rev.* **4**, 133–89.

Guyonnet, F. and Plommet, M. (1970). Hémolysine γ de *Staphylococcus aureus*: purification et propriétés. *Ann. Inst. Pasteur* **118**, 19–33.

Hensler, T., Köller, M., Prévost, G., Piémont, Y. and König, W. (1994a). GTP-binding proteins are involved in the modulated activity of human neutrophils treated by the Panton–Valentine leucocidin from *Staphylococcus aureus*. *Infect. Immun.* **62**, 5281–9.

Hensler, T., König, B., Prévost, G., Piémont, Y., Köller, M. and König, W. (1994b). LTB4- and DNA fragmentation induced by leukocidin from *Staphylococcus aureus*. The protective role of GM-CSF and G-CSF on human neutrophils. *Infect. Immun.* **62**, 2529–35.

Jeljaszewicz, J., Szmigielski, S. and Grojec, P. (1976). Staphylococcal leucocidin: stimulatory effect on granulopoiesis disturbed by cytostatic agents and review of the literature. In: *Staphylococci and Staphylococcal Diseases* (ed. J. Jeljaszewicz), pp. 639–59. Gustav Fischer, Stuttgart.

Johanowski, J. (1959). Die bedeutung des antileucozidins und antitoxins bei der immunität gegen staphylokokken. *Z. Immun. Forsch. Exp. Therapie* **116**, 318–28.

Jonas, D., Walev, I., Berger, T., Liebetrau, M., Palmer, M. and Bhakdi, S. (1994). Novel path to apoptosis: small transmembrane pores created by staphylococcal α-toxin in T-lymphocytes evoke internucleosomal DNA degradation. *Infect. Immun.* **62**, 1304–12.

Julianelle, L.A. (1922). Studies of haemolytic staphylococci. Haemolytic activity – biochemical reactions – serologic reactions. *J. Infect. Dis.* **31**, 256–84.

Jursch, R., Hildebrand, A., Hobom, G., Tranum-Jensen, J., Ward, R., Kehoe, M. and Bhakdi, S. (1994). Histidine residues near the *N*-terminus of *Staphylococcus* alpha-toxin as reporters of regions that are critical for oligomerization of pore-formation. *Infect. Immun.* **62**, 2249–56.

Kamio, Y., Rahman, A., Nariya, H., Ozawa, T. and Izaki, K. (1993). The two staphyloccocal bi-component toxins, leukocidin and γ-haemolysin, share one component in common. *FEBS Lett.* **321**, 15–18.

Kaneko, J., Kimura, T., Kawakami, Y., Tomita, T. and Kamio, Y. (1997a). Panton–Valentine genes in a phage-like particle isolated from mytomycin C-treated *Staphylococcus aureus* V8 (ATCC 49775). *Biosci. Biotech. Biochem.* **61**, 1960–2.

Kaneko, J., Muramoto, K. and Kamio, Y. (1997b). Gene of LukF-PV-like component of Panton–Valentine leucocidin in *Staphylococcus aureus* P83 is linked with *lukM*. *Biosci. Biotech. Biochem.* **61**, 541–4.

Kato, I. and Noda, M. (1989). ADP-ribosylation of cell membrane proteins by staphylococcal α-toxin and leucocidin in rabbits erythrocytes and polymorphonuclear leucocytes. *FEBS Lett.* **255**, 59–62.

Köller, M., Hensler, T., König, B., Prévost, G., Alouf, J. and König, W. (1993). Induction of heat-shock proteins by bacterial toxins, lipid mediators and cytokines in human leucocytes. *Zbl. Bakt.* **278**, 365–76.

König, B., Köller, M., Prévost, G., Piémont, Y., Alouf, J.E., Schreiner, A. and König, W. (1994). Activation of human effector cells by different bacterial toxins (leukocidin, alveolysin, erythrogenic toxin A): generation of interleukin-8. *Infect. Immun.* **62**, 4831–7.

König, B., Prévost, G., Piémont, Y. and König, W. (1995). Effects of *Staphylococcus aureus* leucocidins inflammatory mediator release from human granulocytes. *J. Infect. Dis.* **171**, 607–13.

König, B., Prévost, G. and König, W. (1997). Composition of staphylococcal bi-component toxins determines pathophysiological reactions. *J. Med. Microbiol.* **46**, 479–85.

Lesieur, C., Vécsey-Semjen, B., Abrami, L., Firaz, M. and van der Goot, G. (1997). Membrane insertion: the strategies of toxins. *Molecular Membrane Biology*, **14**, 45–64.

Mahoudeau, I., Delabranche, X., Prévost, G., Monteil, H. and Piémont, Y. (1997). Frequency of *Staphylococcus intermedius* in human isolates. *J. Clin. Microbiol.* **35**, 2153–4.

Menzies, B.E. and Kernodle, D.S. (1994). Site-directed mutagenesis of the alpha-toxin gene of *Staphylococcus aureus*: role of histidines in toxin activity *in vitro* and in a murine model. *Infect. Immun.* **62**, 1843–7.

Meunier, O., Falkenrodt, A., Monteil, H. and Colin, D.A. (1995). Application of flow cytometry in toxinology: pathophysiology of human polymorphonuclear leucocytes damaged by a pore-forming toxin from *Staphylococcus aureus*. *Cytometry* **21**, 241–7.

Meunier, O., Staali, L., Monteil, H. and Colin, D.A. (1996). Staphylococcal 'pore-forming toxins': analysis by flow cytometry. *Procedings of the 8th International Symposium on Staphylococci and Staphylococcal Infections*, p. 220, Aix les Bains.

Meunier, O., Ferreras, M., Supersac, G., Hoeper, F., Baba Moussa, L., Monteil, H., Colin, D.A., Menestrina, G. and Prévost, G. (1997). A predicted β-sheet from class S components of staphylococcal γ-haemolysin is essential for the secondary interaction of the class F component. *Biochim. Biophys. Acta* **1326**, 275–89.

Möllby, R. (1989). Isolation and properties of membrane damaging toxins. In: *Staphylococci and Staphylococcal Infections* (eds C.S.F. Easmon and C. Adlam), pp. 619–69. Academic Press, London.

Morinaga, N., Nagamori, M. and Kato, I. (1987). Suppressive effect of calcium on the cytotoxicity of staphylococcal leucocidin for HL60 cells. *FEMS. Lett.* **42**, 259–64.

Morinaga, N., Kato, I. and Noda, M. (1993). Changes in the susceptibility of 12-O-tetradecanoyl-phorbol 13-acetate (TPA)-treated HL60 cells to staphylococcal leucocidin. *Microbiol. Immunol.* 37, 537–41.

Mudd, S., Gladstone, G.P. and Lenhardt, N.A. (1965). The antigenicity in man of staphylococcal leucocidin toxoid, with notes on therapeutic immunization in chronic osteomyelitis. *Br. J. Exp. Pathol.* **46**, 455–72.

Nariya, H. and Kamio, Y. (1995). Identification of the essential region for LukS- and HgII-specific functions of staphylococcal leucocidin and γ-haemolysin. *Biosci. Biotech. Biochem.* **59**, 1603–4.

Nariya, H., Asami, I., Ozawa, T., Beppu, Y., Izaki, K. and Kamio, Y. (1993a). Improved method for purification of leukocidin and γ-haemolysin components from *Staphylococcus aureus*. *Biosci. Biotech. Biochem.* **57**, 2198–9.

Nariya, H., Izaki, K. and Kamio, Y. (1993b). The C-terminal region of the S component of staphylococcal leukocidin is essential for the biological activity of the toxin. *FEBS Lett.* **229**, 219–22.

Nariya, H., Shimatani, A., Tomita, T. and Kamio, Y. (1997a). Identification of the essential amino acids residues in LukS for the haemolytic activity of staphylococcal leukocidin towards rabbit erythrocytes. *Biosci. Biotech. Biochem.* **61**, 2095–9.

Nariya, H., Nishiyama, A. and Kamio, Y. (1997b). Identification of the minimal segment in which the threonine 246 residue is a potential phosphorylated site by protein kinase A for the LukS-specific function of staphylococcal leukocidin. *FEBS Lett.* **415**, 96–100.

Neisser, M. and Wechsberg, F. (1901). Ueber das Staphylotoxin. *Z. Hyg. Infecktkrankh.* **36**, 299–349.

Noda, M., Hyrayama, T., Kato, I. and Matsuda, F. (1980a). Crystallization and properties of staphylococcal leukocidin. *Biochim. Biophys. Acta* **633**, 33–44.

Noda, M., Kato, I., Matsuda, F. and Hyrayama, T. (1980b). Fixation and inactivation of staphylococcal leukocidin by phosphatidylcholine and ganglioside GM1 in rabbit polymorphonuclear leucocytes. *Infect. Immun.* **29**, 678–84.

Noda, M., Kato, I., Matsuda, F. and Hirayama, T. (1981). Mode of action of staphylococcal leucocidin: relationship between binding of ^{125}I-labeled S and F components of leucocidin to rabbit polymorphonuclear leukocytes and leucocidin activity. *Infect. Immun.* **34**, 362–367.

Noda, M., Kato, I., Hyrayama, T. and Matsuda, F. (1982). Mode of action of staphylococcal leucocidin: effects of the S and F components on the activities of membrane-associated enzymes of rabbit polymorphonuclear leukocytes. *Infect. Immun.* **35**, 38–45.

Ouchterlony, Ö. (1962). Diffusion in gel methods for immunological analysis. *Prog. Allergy* **6**, 30–154.

Ozawa, T., Kaneko, J., Nariya, H., Izaki, K. and Kamio, Y. (1994). Inactivation of γ-haemolysin HγII component by addition of monosialoganglioside GM1 to human erythrocyte. *Biosci. Biotech. Biochem.* **58**, 602–5.

Ozawa, T., Kaneko, J. and Kamio, Y. (1995). Essential binding of LukF of staphylococcal γ-haemolysin followed by the binding of HγII for the haemolysis of human erythrocytes. *Biosci. Biotech. Biochem.* **59**, 1181–3.

Panton, P.N. and Valentine, F.C.O. (1932). Staphylococcal toxin. *Lancet* **222**, 506–8.

Péguet-Navarro, M., Dalbiez-Gauthier, C., Moulon, C., Berthier, O., Réano, A., Gaucherand, M., Banchereau, J., Rousset, F. and Schmitt, D. (1997). CD40 ligation of human keratinocytes inhibits their proliferation and induces their differentiation. *J. Immunol.* **158**, 144–152.

Pfanneberg, T., Blobel, H. and Schaeg, W. (1975). Cytotoxic effects of a leukocidin from *Staphylococcus aureus*. *Zbl. Bakt. Hyg. I. Abt. Orig. A* **223**, 147–52.

Prévost, G., Supersac, G., Colin, D.A., Couppié, P., Sire, S., Hensler, T., Petiau, P., Meunier, O., Cribier, B., König, W. and Piémont, Y. (1994). The new family of leucotoxins from *S. aureus*: structural and biological properties. In: *Bacterial Protein Toxins* (eds J. Freer *et al.*), pp. 284–93. Zentralblatt für Bakteriologie, Suppl. 24. Gustav Fischer, Stuttgart.

Prévost, G., Cribier, B., Couppié, P., Petiau, P., Supersac, G., Finck-Barbançon, V., Monteil, H. and Piémont, Y. (1995a). Panton–Valentine leucocidin and γ-haemolysin from *Staphylococcus aureus* ATCC 49775 are encoded by distinct genetic loci and have different biological activities. *Infect. Immun.* **63**, 4121–9.

Prévost, G., Bouakham, T., Piémont, Y. and Monteil, H. (1995b). Characterization of a synergohymenotropic toxin from *Staphylococcus intermedius*. *FEBS Lett.* **376**, 135–40.

Prévost, G., Couppié, P., Prévost, P., Gayet, S., Petiau, P., Cribier, B., Monteil, H. and Piémont, Y. (1995c). Epidemiological data on *Staphylococcus aureus* strains producing synergohymenotropic toxins. *J. Med. Microbiol.* **42**, 237–45.

Prévost, G., Colin, D.A., Staali, L., Baba Moussa, L., Gravet, A., Werner, S., Sanni, A., Meunier, O. and Monteil, H. (1998). Les leucotoxines formant des pores de *Staphylococcus aureus*: variabilité des cellules-cible et deux processus pharmacologiques. *Pathol. Biol.* in press.

Rahman, A., Isaki, K., Kato, I. and Kamio, Y. (1991). Nucleotide sequence of leucocidin S-component gene (*lukS*) from methicillin-resistant *Staphylococcus aureus*. *Biochem. Biophys. Res. Commun.* **184**, 138–44.

Rahman, A, Nariya, H., Isaki, I., Kato, I. and Kamio, Y. (1992). Molecular cloning and nucleotide sequence of leucocidin F-component gene (*lukF*) from methicillin-resistant *Staphylococcus aureus*. *Biochem. Biophys. Res. Commun.* **184**, 640–6.

Siqueira, J.A., Speeg-Schatz, C., Freitas, F.I.S., Sahel, J., Monteil, H. and Prévost, G. (1997). Channel-forming leucotoxins from *Staphylococcus aureus* cause severe inflammatory reactions in a rabbit eye model. *J. Med. Microbiol.* **46**, 486–94.

Smith, M.L. and Price, S.A. (1938). *Staphylococcus* γ-haemolysin. *J. Pathol. Bacteriol.* **47**, 379–93.

Soboll, H., Ito, A., Schaeg, W. and Blobel, H. (1973). Leukocidin of staphylococci of different origins. *Zentralbl. Bakteriol. Hyg. 1 Abt. Orig. [A]* **224**, 184–93.

Song, L., Hobaugh, M.R., Shustak, C., Cheley, S., Bayley, H. and Gouaux, J.E. (1996). Structure of staphylococcal α-haemolysin, a heptameric transmembrane pore. *Science* **274**, 1859–66.

Staali, L., Monteil, H. and Colin, D.A. (1998). The pore-forming leukotoxins from *Staphylococcus aureus* open Ca^{2+} channels in human polymorphonuclear neutrophils. *J. Memb. Biol.* **162**, 209–16.

Sudo, K., Choorit, W., Asami, I., Kaneko, J., Muramoto, K. and Kamio, Y. (1995). Substitution of lysin for arginine in the N-terminal 217th amino acid residue of the HγII of staphylococcal γ-haemolysin lowers activity of the toxin. *Biosci. Biotech. Biochem.* **59**, 1786–9.

Sugarawa, N., Tomita, T. and Kamio, Y. (1997). Assembly of *Staphylococcus aureus* γ-haemolysin into a pore-forming ring-shaped complex on the surface of human erythrocytes. *FEBS Lett.* **410**, 333–7.

Supersac, G., Prévost, G. and Piémont, Y. (1993). Sequencing of leucocidin R from *Staphylococcus aureus* P83 suggests that staphylococcal leucocidins and γ-haemolysin are members of a single, two-component family of toxins. *Infect Immun.* **61**, 580–7.

Supersac, G., Piémont, Y., Kubina, M., Prévost, G. and Foster, T.J. (1998). Assessment of the role of γ-toxin in experimental endophthalmitis using a Δ*hlg* deficient mutant of *Staphylococcus aureus*. *Microb. Pathog.* **24**, 241–51.

Szmigielski, S. and Jeljaszewicz, J. (1976). Stimulatory effect of staphylococcal leukocidin on granulopoiesis disturbed by cytostatic agents. *Cancer Lett.* **1**, 299–303.

Szmigielski, S., Jeljaszewicz, J., Wiszinski, J. and Korbecki, M. (1966). Reaction of rabbit leucocytes to staphylococcal (Panton-Valentine) leukocidin *in vivo*. *J. Pathol. Bacteriol.* **84**, 599–604.

Szmigielski, S., Jeljaszewicz, J. and Zak, C. (1968). Leucocyte system of rabbits receiving repeated doses of staphylococcal leukocidin. *Pathol. Microbiol.* **31**, 328–36.

Szmigielski, S., Sobiczewska, E., Prévost, G., Monteil, H., Colin, D.A. and Jeljaszewicz, J. (1998). Effects of purified staphylococcal leucocidal toxins on isolated blood polymorphonuclear leukocytes and peritoneal macrophages *in vitro*. *Zbl. Bakt.* **288**, 383–94

Taylor, A.G. and Bernheimer, A.W. (1974). Further characterization of staphylococcal γ-haemolysin. *Infect. Immun.* **10**, 54–9.

Towers, A.G. and Gladstone, G.P. (1958). Two serological tests for staphylococcal infection. *Lancet* **275**, 1192–5.

Tomita, T. and Kamio, Y. (1997). Molecular biology of pore-forming cytolysins from *Staphylococcus aureus* α- and γ-haemolysins and leukocidin. *Biosci. Biotech. Biochem.* **61**, 565–72.

Van der Velde, H. (1894). Etude sur le mécanisme de la virulence du staphylocoque pyogène. *La Cellule* **10**, 401–60.

Van der Vijver, J.C.M., van Es-Boon, M. and Michel, M.F. (1972). Lysogenic conversion in *Staphylococcus aureus* to leucocidin production. *J. Virol.* **10**, 318–19.

Walev, I., Martin, E., Jonas, D., Mohamadzadeh, M., Müller-Klieser, W., Kunz, L. and Bhakdi, S. (1993). Staphylococcal alpha-toxin kills human keratinocytes by permeabilizing the plasma membrane for monovalent ions. *Infect. Immun.* **61**, 4972–9.

Walker, B., Braha, O., Cheley, S. and Bayley, H. (1995). An intermediate in the assembly of a pore-forming protein trapped with a genetically engineered switch. *Chem. Biol.* **2**, 99–105.

Wang, X., Noda, M. and Kato, I. (1990). Stimulatory effect of staphylococcal leucocidin on phosphoinositide metabolism in rabbit polymorphonuclear leucocytes. *Infect. Immun.* **58**, 2745–9.

Ward, P.D. and Turner, W.H. (1980). Identification of Panton–Valentine leukocidin as a potent dermonecrotic toxin. *Infect. Immun.* **28**, 393–7.

Woodin, A.M. (1960). Purification of the two components of leucocidin from *Staphylococcus aureus*. *Biochem. J.* **75**, 158–65.

Woodin, A.M. (1962). The extrusion of protein from the rabbit polymorphonuclear leukocyte treated with staphylococcal leucocidin. *Biochem. J.* **82**, 9–15.

Woodin, A.M. (1972a). The staphylococcal leukocidin. In: *The Staphylococci* (ed. J.O. Cohen), pp. 281–9. Wiley Interscience, New York.

Woodin, A.M. (1972b). Adenylate cyclase and the function of cyclic adenosine 3':5'-monophosphate in the leucocidin-treated leucocyte. *Biochim. Biophys. Acta* **286**, 406–15.

Woodin, A.M. and Wieneke, A.A. (1963a). The accumulation of calcium by the polymorphonuclear leucocyte treated with staphylococcal leucocidin and its significance in the extrusion of protein. *Biochem. J.* **87**, 487–95.

Woodin, A.M. and Wieneke, A.A. (1963b). The incorporation of radioactive phosphorus in the leucocyte during the extrusion of protein induced by staphylococcal leucocidin. *Biochem. J.* **87**, 480–7.

Woodin, A.M. and Wieneke, A.A. (1964). The participation of calcium, adenosine triphosphate and adenosine triphosphatase in the extrusion of granule proteins from polymorphonuclear leucocyte. *Biochem. J.* **90**, 498–509.

Woodin, A.M. and Wieneke, A.A. (1967). The participation of phospholipids in the interaction of leukocidin and the cell membrane of the polymorphonuclear leucocyte. *Biochem. J.* **105**, 1029–38.

Woodin, A.M. and Wieneke, A.A. (1968). The cation-sensitive phosphatases of the leucocyte cell membrane. *Biochem. Biophys. Res. Commun.* **33**, 558–52.

Wright, J. (1936). Staphylococcal leukocidin (Neisser–Weschberg type) and antileucocidin. *Lancet* **230**, 1002–4.

22

Enterococcus faecalis cytolysin and *Bacillus cereus* bi- and tri-component haemolysins

Michael S. Gilmore, Michelle C. Callegan and Bradley D. Jett

ENTEROCOCCUS FAECALIS CYTOLYSIN

Introduction

Over the past two decades enterococci have emerged as leading causes of nosocomial infection, and are now among the most commonly isolated micro-organisms from bacteraemia, surgical wound infections and urinary tract infections (Emori and Gaynes, 1993; Jarvis *et al.*, 1998). Moreover, enterococcal strains have been isolated that are resistant to all currently approved antibiotics (Handwerger *et al.*, 1992), highlighting the urgency to understand enterococcal host–parasite interactions, with the goal of identifying new therapeutic opportunities.

In an extensive study spanning the years 1995–97, *Enterococcus faecalis* caused 79% (11 958/15 203) of human enterococcal infections, with *E. faecium* the cause of most of the remaining 21% (Huycke *et al.*, 1998). While *E. faecium* was more likely to express antibiotic resistance, including vancomycin and ampicillin resistance, *E. faecalis* was observed to express a larger number of traits related to virulence. The subjects of enterococcal infection, clinical management, antibiotic resistance and pathogenesis have been reviewed recently (Eliopoulos and Eliopoulos, 1990; Herman and Gerding, 1991; Moellering, 1992; Jett *et al.*, 1994; Huycke *et al.*, 1998; Murray, 1998). This section critical-ly reviews the literature on the cytolysin of *E. faecalis*, a novel bacterial toxin, and its contribution to disease.

Early observations

The first systematic investigation of the *E. faecalis* cytolysin (previously termed 'haemolysin' or 'haemolysin/bacteriocin') was by E.W. Todd (1934), who observed that, while some group-D streptococcal (*E. faecalis*) strains produced noticeable haemolytic zones on blood agar, they failed to produce haemolytic activity when grown in broth. He therefore classified these β-haemolytic strains as 'pseudo-haemolytic', and was able to detect haemolytic activity in broth by manipulating the media formulation. The cytolysin was found to be acid and heat labile, but stable to oxygen and non-antigenic. The basis for medium-dependent differences in cytolysin activity may relate to pH effects on stability, differential induction or other factors. Because common liquid laboratory media do not support the production of detectable levels of cytolysin by *E. faecalis*, the results of studies *in vitro*, testing the contribution of cytolysin to phagocytic killing but lacking a control for the production of cytolysin in the liquid media used (e.g. Arduino *et al.*, 1993), should be examined critically.

Kobayashi (1940) and, later, Basinger and Jackson (1968) demonstrated that erythrocytes from various species were differentially susceptible to the lytic effects of the *E. faecalis* cytolysin. Rabbit, human, horse and dog erythrocytes are most susceptible to lysis, whereas sheep erythrocytes are largely refractory, and goose erythrocytes are completely resistant to lysis. Because sheep erythrocytes – the species commonly incorporated into blood agar plates – are essentially unaffected by the *E. faecalis* cytolysin, the variable haemolytic phenotype of this species has largely gone unnoticed.

Haemolytic strains of *E. faecalis* were observed to inhibit the growth of several Gram-positive species, including the species *Staphylococcus*, *Streptococcus* and *Clostridium* (Stark, 1960). This activity was ascribed unambiguously to the haemolysin by observing that haemolysin and bacteriocin activities were lost simultaneously upon UV irradiation, and simultaneously restored upon a second round of mutagenesis. Moreover, both activities were lost upon exposure to chloroform vapours (Brock and Davie, 1963). The spectrum of bacteriocin activity was observed to extend to members of the species *Leuconostoc*, *Lactobacillus*, *Streptococcus*, *Bacillus*, *Clostridium*, *Staphylococcus*, *Micrococcus* and *Corynebacterium* (Brock *et al.*, 1963).

With a view to exploring potential applications of the bacteriocin activity in ecological control, oral streptococci were observed to be uniformly susceptible to the *E. faecalis* cytolysin (Jett and Gilmore, 1990). Interestingly, *Bacillus polymyxa* was observed to be resistant, *Bacillus subtilis* was observed to be weakly susceptible, and the Gram-negative species *Escherichia coli* and *Proteus vulgaris* were completely resistant (Brock *et al.*, 1963). L-forms, however, were susceptible (Kalmanson *et al.*, 1970), suggesting that resistance derives from cell wall layers external to the cytoplasmic membrane, such as the outer membrane of Gram-negative bacteria.

In general, bacterial membranes are about three orders of magnitude more effective in competitively inhibiting target cell lysis by the *E. faecalis* cytolysin than membranes derived from eukaryotic cells (Basinger and Jackson, 1968). However, even among bacterial membranes, there appears to be substantial variation in inhibitory activity, with *E. coli* membranes being the least effective, followed by those of *Bacillus megaterium*, *E. faecalis* and *Micrococcus lysodeikticus* in increasing order of effectiveness (Basinger and Jackson, 1968). Lecithin was observed to be inhibitory to both bacteriocin and haemolytic activities (Brock and Davie, 1963). Together, this evidence is consistent in indicating that the target of the cytolysin is the cell membrane, prokaryotic or eukaryotic, and is otherwise receptor independent. Moreover, a gradient of susceptibility exists that may relate to lipid composition, bilayer surface charge, or both. Because the spectrum of activity of the *E. faecalis* lysin extends to a wide range of cell types, prokaryotic and eukaryotic, the term cytolysin was adopted in favour of haemolysin, or haemolysin/bacteriocin (Gilmore, 1991).

Cytolysin as a variable trait of *Enterococcus faecalis*

The β-haemolytic phenotype of cytolytic strains was used historically as a distinguishing feature in classifying a subset of group-D streptococci as the separate species *Streptococcus zymogenes* (Breed *et al.*, 1948). Plasmid curing and conjugal mating experiments, however, demonstrated that the *E. faecalis* cytolysin is a plasmid encoded variable trait of the species (Dunny and Clewell, 1975; Jacob *et al.*, 1975). Moreover, plasmids encoding the cytolysin, such as the archetype pAD1, encode a pheromone-responsive conjugation pathway capable of transferring between *E. faecalis* strains at high frequencies *in vitro* (Clewell *et al.*, 1982; Ike and Clewell, 1984; Clewell, 1993; Dunny and Leonard, 1997; Fujimoto *et al.*, 1997, 1998) and *in vivo* (Huycke *et al.*, 1992).

Cytolysin determinants on plasmids from diverse geographical sources were observed to be closely related (LeBlanc *et al.*, 1983) and belong to the same incompatibility group, IncHly (Colmar and Horaud, 1987). More recent studies, however, have identified cytolysin determinants on plasmids from other incompatibility groups as well as on the *E. faecalis* chromosome (Ike and Clewell, 1992). Studies currently in progress indicate that, in strains associated with a hospital ward bacteraemia outbreak (Huycke *et al.*, 1991), the cytolysin determinant exists on the chromosome in association with other disease-associated genes, including the surface protein, *esp*, in a structure resembling a pathogenicity island (Shankar and Gilmore, personal communication).

The *Enterococcus faecalis* cytolysin: a novel multicomponent bacterial toxin

Granato and Jackson (1969) recognized that cytolytic activity resulted from the interaction of multiple components. Complementation analysis, performed by observing cytolytic activity between neighbouring colonies on blood agar, identified two classes of non-cytolytic mutants from nitrosoguanidine mutagenesis. Cytolytic activity was observed to result from the independent secretion and extracellular interaction of at least two substances. The extracellular substances were termed 'component A' and 'component L', since A appeared to play an activator role, and L, when activated, possessed the lytic activity (Granato and

Jackson, 1971a, b). In later studies, transposon Tn917 insertional mutagenesis localized the cytolysin determinant to an 8-kb region on pAD1 (Ike *et al.*, 1990). These transposon mapping studies permitted direct cloning of restriction fragments related to cytolysin expression and ultimately reassembly of the functional cytolysin determinant in *E. coli* (Ike *et al.*, 1990).

Nucleotide sequence determination for the *E. faecalis* cytolysin determinant revealed a complex operon encoding six gene products (Figure 22.1). The first gene product of the cytolysin operon to be analysed at the molecular level was CylA (formerly 'component A'). CylA was shown to share amino acid similarity, as well as biochemical similarity, with serine proteases of the subtilisin class (Segarra *et al.*, 1991). Complementation studies further indicated that CylA is secreted independently from other components of the cytolysin gene cluster. Western immunoblot and mass spectrographic analysis suggests that the 34 811-Da CylA, like most secreted proteases of Gram-positive bacteria, is expressed as a pre-proenzyme (Booth *et al.*, 1996). Extracellular complementation analyses, as well as direct functional studies utilizing purified CylA (discussed below), support a role for CylA in activation of the cytolysin precursors via proteolytic cleavage (Segarra *et al.*, 1991; Booth *et al.*, 1996).

Immediately 5' to *cylA* is an open reading frame termed *cylB*. CylB is a member of the HlyB family of ATP-binding cassette (ABC) transporters, and was the first of this class to have been identified in a Gram-positive bacterial operon (Gilmore *et al.*, 1990; Fath and Kolter, 1993). Site-specific mutagenesis and complementation analysis demonstrated that functional CylB was required for externalization of both cytolysin precursors, $CylL_L$ and $CylL_S$ (Gilmore *et al.*, 1994). Interestingly, the apparent ATP binding domain of CylB, which putatively supplies the energetics for transfer, is required only for externalization of the larger cytolysin precursor, $CylL_L$, and not $CylL_S$. Although CylB is closely related to HlyB from the *E. coli* α-haemolysin operon, CylB activity could not be replaced functionally by HlyB, and no HlyD counterpart was found in the *E. faecalis* cytolysin operon (Gilmore *et al.*, 1990). Mutations in *cylB* did not affect expression and secretion of active CylA, demonstrating that the CylA protease is secreted independently of the cytolysin precursors, presumably by a conventional *secA–secY* type process (Segarra *et al.*, 1991).

A comparison of related secretion systems predicted that CylB possessed a functional cysteine protease domain at its amino terminus, which was proposed to function in removal of the leader peptides from the lysin precursors $CylL_L$ and $CylL_S$ during translocation across the cytoplasmic membrane (Havarstein *et al.*,

1995). As discussed below, amino acid sequence and mass spectrometric analysis of purified *E. faecalis* cytolysin precursors support this role for CylB in secretion-associated processing of cytolysin subunits (Booth *et al.*, 1996).

Upstream (5') to *cylB* is a reading frame termed *cylM* (Gilmore *et al.*, 1994). *cylM* encodes a large protein of 993 amino acids, and insertional mutations in the *cylM* open reading frame demonstrated an absolute requirement for expression of active cytolysin subunits (Gilmore *et al.*, 1994). CylM homologues have been described in a number of operons for biosynthesis of lantibiotics (Sahl *et al.*, 1995), and CylM-like proteins from lantibiotic systems demonstrate no similarity to any other class of proteins. While the enzymology of the CylM class of proteins in the maturation process of lantibiotics remains obscure, it is hypothesized that these proteins are directly or indirectly involved in the post-translational modification of lysin precursors. This post-translational modification involves dehydration of serine and threonine residues, and frequently thioether bond formation with adjacent cysteine residues to form lanthionine and methyl-lanthionine, respectively (Klein *et al.*, 1992; Schnell *et al.*, 1992).

The *E. faecalis* cytolysin structural components were recently shown to be novel divergent members of the lantibiotic family of small antibacterial peptides (Booth *et al.*, 1996). Early attempts to purify the *E. faecalis* cytolysin identified a lytic precursor that behaved as a single protein of 11 kDa apparent molecular mass (Granato and Jackson, 1971b). However, nucleotide sequence determination and complementation analysis revealed two small open reading frames 5' to *cylM*, which encoded separate subunits, $CylL_L$ (large) and $CylL_S$ (small), each of which was required for cytolytic activity (Gilmore *et al.*, 1994). These two reading frames bear little resemblance to each other, except for a run of 26 residues, of which 25 are identical. Interestingly, this highly conserved core sequence is located near the amino terminus of $CylL_L$, but more centrally within $CylL_S$. We speculate that this sequence targets the subunits for post-translational processing and/or secretion. All proteolytic processing of the subunits that occurs during maturation, occurs within this highly conserved core sequence.

Using *E. faecalis* strains defective in expression of one or the other component, and the cytolysin activator, both cytolysin subunits were purified in precursor form to homogeneity (Booth *et al.*, 1996). The secreted forms of the preactivated cytolysin precursors, $CylL_L'$ and $CylL_S'$, were demonstrated by mass spectrometry to be 4037.7 and 2630.9 Da, respectively (Booth *et al.*, 1996). The discrepancy between early and recent estimations of molecular mass of cytolysin precursors is probably

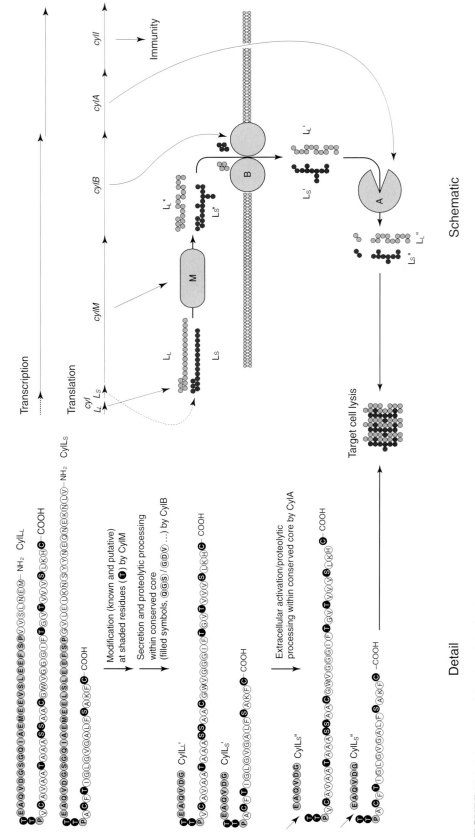

FIGURE 22.1 Schematic and detail models of *E. faecalis* cytolysin expression and maturation.

due to the formation of cytolysin aggregates in early purification attempts (Booth *et al.*, 1996).

Amino acid composition analysis and derivatization were used to demonstrate unambiguously that both cytolysin components, CylL$_L$ and CylL$_S$, possessed the lanthionine modifications that are the hallmark of the lantibiotic family. Moreover, Edman degradation and mass spectrometry demonstrated that the final activation of CylL$_L$ and CylL$_S$ involved the proteolytic removal of six amino-terminal amino acids by CylA cleavage within the highly conserved core sequence of each subunit (Booth *et al.*, 1996).

Global similarities exist between the organization of genes within the *cyl* operon and those within operons specifying lantibiotic biosynthesis by other Grampositive species. These similarities include: (a) inferred amino acid sequence of the cytolysin precursors, which possess hydrophilic amino-terminal and hydrophobic carboxy-terminal ends; (b) the small size of the Cyl precursors; (c) the cysteine-rich nature of the inferred products, with cysteine residues flanked by hydroxylated amino acids, which in lantibiotics are combined to form lanthionine and lanthionine derivatives; (d) the occurrence of similar auxiliary genes putatively encoding products that contribute to posttranslational modification; (e) the presence of dedicated secretion proteins related to the HlyB family; and (f) the presence of a subtilisin-class serine protease required for activation or processing (Gilmore *et al.*, 1994). Based on these similarities, the *E. faecalis* cytolysin was recently classified as a Type-A, lactocin S-subtype, pore-forming lantibiotic (Sahl *et al.*, 1995).

A common feature of lantibiotic operons is the presence of gene products conferring immunity, or selfprotection, from the bactericidal effect of the lantibiotic (Sahl *et al.*, 1995). Previous investigations involving transposon insertional mutagenesis of the *E. faecalis* cytolysin operon suggested that CylA, the activator protease, was related to immunity (Ike *et al.*, 1990; Segarra *et al.*, 1991). However, an analysis of enzyme–substrate interactions using purified CylA, CylL$_L$ and CylL$_S$ did not detect an activity consistent with immunity (Booth *et al.*, 1996). Subsequent genetic analysis of the region immediately 3′ to *cylA* revealed the presence of a previously unidentified open reading frame (Coburn *et al.*, 1999). Deletion analysis and sitedirected mutagenesis demonstrated that this open reading frame, which is dependent on transcription of *cylA*, confers immunity to the *E. faecalis* cytolysin. The immunity gene, inferred to encode a 327 amino acid polypeptide, represents to our knowledge the first lantibiotic immunity mechanism ascribable to a single gene product. The *E. faecalis* cytolysin immunity gene has been designated *cylI* (Coburn *et al.*, 1999).

Summarizing what is currently known about the *E. faecalis* cytolysin, the following model has been developed (Figure 22.1).

1. Cytolysin precursors are initially translated as unmodified peptides of 68 (CylL$_L$) and 63 (CylL$_S$) amino acids in length. Genes related to intracellular maturation (*cylM*) and secretion (*cylB*) of the cytolysin precursors (CylL$_L$ and CylL$_S$) are transcribed as a unit, perhaps including the cytolysin precursor genes as well (Gilmore *et al.*, 1994). The *cylA* and *cylI* genes are capable of being transcribed as a separate unit (Coburn *et al.*, 1999), although some read through of the *cylB/cylM* message appears likely.

2. CylM appears to modify the cytolysin precursors CylL$_L$ and CylL$_S$ intracellularly to CylL$_L$* and CylL$_S$*, by directly or indirectly catalysing the dehydration of hydroxyl amino acids and formation of lanthionine and β-methyllanthionine residues (Booth *et al.*, 1996).

3. Following intracellular modification, the cytolysin precursors are translocated across the cytoplasmic membrane by CylB (Gilmore *et al.*, 1990). During translocation, CylL$_L$* and CylL$_S$* undergo proteolytic processing to CylL$_L$′ and CylLs′, which involves cleavage within the highly conserved core region of each subunit, and the removal of 24 and 36 amino acid leader peptides from CylL$_L$* and CylL$_S$*, respectively (Booth *et al.*, 1996). This proteolytic step is predicted to be mediated by CylB itself (Havarstein *et al.*, 1995).

4. Once outside the cell, post-translationally modified and trimmed CylL$_L$′ and CylL$_S$′ are further processed and activated by CylA. This final step in maturation involves proteolytic trimming at the amino-termini, yielding active lytic subunits 38 and 21 amino acids in length, respectively. Target cell lysis requires interaction of both activated cytolysin subunits, since neither subunit alone possesses significant lytic activity (Booth *et al.*, 1996).

5. Immunity of the cytolysin-producing cell is provided by the gene product of *cylI* (Coburn *et al.*, 1999). The mechanism by which CylI prevents selflysis is currently under investigation (Figure 22.1).

While the *E. faecalis* cytolysin operon demonstrates many similarities at the genetic level with those of other lantibiotics, it is clearly unique among Gram-positive antibiotic peptides in two important respects. First, the *E. faecalis* cytolysin consists of two dissimilar subunits, both of which are post-translationally modified and required for target cell lysis. Secondly, the spectrum of activity of the *E. faecalis* cytolysin is not limited to prokaryotic cells. In fact, the *E. faecalis* cytolysin (first classified as a haemolysin) makes important contributions to virulence in a number of systems (Gilmore *et al.*, 1996; Huycke *et al.* 1998).

Role of cytolysin in *Enterococcus faecalis* virulence

Based on the broad activity spectrum of the *E. faecalis* cytolysin, it can be hypothesized to contribute to the virulence of the organism, or pathogenesis of infection, at any of several levels. As a potentially toxic cell lytic agent, the cytolysin could contribute directly to tissue destruction or evasion of clearance by potentially susceptible phagocytic cells (Miyazaki *et al.* 1993). The former activity would be detectable as an enhancement in the toxicity of infections caused by cytolytic strains, the latter activity would be detectable as a contribution to disease severity and/or disease incidence. The cytolytic activity could also compromise normal anatomical barriers and facilitate invasion of the organism into host tissues, thereby contributing to disease incidence, and potentially also severity. A third hypothetical role could be proposed based on the bacteriocin activity of the cytolysin, which could provide cytolytic strains with a competitive advantage and promote colonization. This activity would be reflected in an increased incidence of disease caused by cytolytic enterococci.

Epidemiological studies and controlled tests in animal models support a role for the *E. faecalis* cytolysin in disease. Few studies have been conducted specifically to determine the relative incidence of cytolytic versus non-cytolytic enterococci in the primary reservoir in healthy, hospital non-associated subjects – the gastrointestinal (GI) tract. Studies where β-haemolytic *S. zymogenes* (now known to be cytolytic strains of *E. faecalis*) were differentiated from non-haemolytic *S. faecalis* (*E. faecalis*) or *S. liquefaciens* (gelatinase-producing strains of *E. faecalis*) can provide insight into historical carriage rates. In a study of saliva carriage rates in 206 dental patients aged 4–34 years, 21.8% were observed to be colonized by enterococci. Of the 45 positive patients, *S. zymogenes* (cytolytic *E. faecalis*) was isolated from three (6.6%), indicating a low oral carriage rate of cytolytic *E. faecalis* (Williams *et al.*, 1950). More recently, Ike *et al.* (1987) observed that 17% of *E. faecalis* isolates from stools of healthy medical students exhibited the cytolytic phenotype. In contrast, 60% of clinical *E. faecalis* isolates expressed cytolysin, constituting a statistically significant enrichment (Ike *et al.*, 1987). This study has been cited for not examining the clonal relatedness of the infection derived isolates studied (Coque *et al.*, 1995); but in a study designed to ask the epidemiological question, 'Strains of which phenotype

cause the most disease?', clonality is irrelevant. (Clonality, or the emergence of pathogenic lineages, is a common motif in bacterial pathogenesis; Armstrong *et al.*, 1996.)

In a study comparing the incidence of cytolytic *E. faecalis* strains among stool isolates from healthy sources unrelated to health-care facilities with the incidence in bacteraemia or endocarditis clinical isolates, the following observations were made: 0/14 (0%) of faecal *E. faecalis* isolates from healthy subjects possessed the cytolysin determinant, 50% (34/68) of bacteraemia isolates were genotypically and phenotypically positive for cytolysin, and 11% (4/35) of a historical collection derived from endocarditis patients were cytolytic (Huycke and Gilmore, 1995). In a related study (Huycke *et al.*, 1991), 45% (85/190) of bacteraemia isolates were cytolysin positive. The enrichment in the cytolytic phenotype among bacteraemia isolates derived from blood cultures collected between 1985 and 1987 was observed to be significant at $P < 0.001$ (Huycke and Gilmore, 1995). Moreover, most of the cytolytic strains derived from bacteraemia patients (82%, 28/34) represented a highly pathogenic lineage that was capable of nosocomial transmission. A similar rate of cytolytic strains was observed for a smaller cohort of isolates (40%, 6/15) reported at about the same time (Libertin *et al.*, 1992).

In addition to a numerical enrichment in the cytolytic phenotype, bacteraemia patients infected with the cytolytic strain were observed to be at a five-fold increased risk of acutely terminal outcome (Huycke *et al.*, 1991). This indicates that the cytolytic phenotype may contribute both to an increased representation among disease isolates and to an increase in toxicity of infection. Cytolytic isolates were found to be more likely to possess multiple antibiotic resistances (Ike *et al.*, 1987; Huycke *et al.*, 1991); however, treatment modality was not associated with acutely terminal outcome, and all strains retained susceptibility to one or more antibiotic (Huycke *et al.*, 1991), discounting the contribution of resistance to this statistic.

The emergence of pathogenic, highly resistant cytolytic lineages among urinary tract isolates has also been observed (Hall *et al.*, 1992). Among a historical cohort, cytolytic enterococci were isolated in 28% of *Enterococcus*-positive urine cultures (Rantz and Kirby, 1943). Among a collection of isolates consisting of all group-D streptococci submitted to Centers for Disease Control (CDC) between 1968 and 1971, 29% (11/38) of *E. faecalis* isolates from urine cultures were observed to express the cytolytic phenotype (Facklam, 1972). With surprising consistency, 25% (6/24) of urine isolates in a more recent report expressed the cytolytic phenotype (Libertin *et al.*, 1992).

Because many infections were caused by pathogenic lineages of highly resistant, cytolytic isolates of *E. faecalis*, an analysis was undertaken that excluded isolates sharing pulse field gel electrophoresis patterns (Coque *et al.*, 1995). This adjustment in incidence resulted in the exclusion of 31% (86/278) of clinical isolates from the analysis, ignoring by design the contribution of any unusually pathogenic lineage, or any isolates capable of bed-to-bed transmission, to disease incidence. Rather than asking 'Which isolates possess an enhanced level of virulence (and therefore are disproportionately represented among disease isolates)?', such a study design asks 'Which different genetic lineages of *E. faecalis* are capable of causing disease?' Not unexpectedly, a variety of phenotypes, cytolytic as well as non-cytolytic, was observed in this study to be capable of causing infection, indicating that *E. faecalis* pathogenesis is multifactorial and not entirely dependent on any invariant trait (Coque *et al.*, 1995). Extrapolations beyond that should be considered in the context of the study design.

In summary, epidemiological studies consistently indicate that pathogenic clonal lineages of *E. faecalis* have emerged that possess the cytolytic phenotype and are enriched for antibiotic resistance, particularly among recent bloodstream isolates (Ike *et al.*, 1987; Huycke *et al.*, 1991; Huycke and Gilmore, 1995). Epidemiological evidence also indicates that infections caused by these nosocomially transmitted cytolytic strains are more severe (Huycke *et al.*, 1991). Less of an enrichment in the cytolytic phenotype appears among urinary tract isolates (Rantz and Kirby, 1943; Facklam, 1972; Libertin et al., 1992); and no evidence indicates that endocarditis or blood isolates collected primarily in the 1960s and 1970s are enriched for the cytolytic phenotype (Facklam, 1972; Huycke and Gilmore, 1995; Coque *et al.*, 1995). Both the clonality and restriction to studies of bloodstream isolates from the 1980s and 1990s suggest that the emergence of pathogenic lineages of cytolytic antibiotic-resistant *E. faecalis* is a recent phenomenon, which coincides with the emergence of enterococci as leading nosocomial pathogens (Huycke *et al.*, 1998). Nevertheless, as recognized by Koch, the utility of epidemiological studies is limited to drawing associations, whereas carefully controlled experimentation is required to establish cause and effect.

Direct evidence for a role in virulence of the *E. faecalis* cytolysin comes from animal studies using isogenic mutant strains. Ike *et al.* (1984) observed that cytolysin-producing enterococcal strains were significantly more lethal to mice when injected intraperitoneally, compared with isogenic cytolysin-deficient strains. These investigators also noted that the organism could be made even more virulent by a copy number mutation that resulted in a hyperhaemolytic phenotype (Ike *et al.*, 1984), establishing a dose–response relationship.

While lethality provided an unequivocal measure of the contribution of the cytolysin to toxicity of bolus injection of *E. faecalis*, it provided relatively little information on pathogenesis. In order to study the pathogenesis of *E. faecalis* infection, the present authors established several criteria for development of an infection model.

1. An infectious dose of relatively few organisms should establish infection to permit adaptation of the bacterium to conditions *in vivo* for appropriate expression of products in response to mammalian environmental cues.
2. The course of disease should develop at a measurable rate and be objectively quantifiable.
3. The evolution of disease should be observable in a non-destructive manner to permit multiple measures on a single animal, minimizing animal-to-animal variation.
4. The infection should be localized to enable histological assessment of its effect on mammalian tissues. Initial attempts to develop such a model in an immunologically intact animal were unsuccessful, as low numbers of enterococci are rapidly cleared.

For simplicity, rather than redirecting the effort towards chemically or genetically immunocompromised animal models, a naturally compromised tissue in an immunologically intact animal was used. The eye is lined with cells bearing CD-95 (Fas-ligand), which induces apoptosis in infiltrating immune cells, limiting the ability of the organ to mount a complete immune response (Griffith *et al.*, 1995). The intraocular infection, or endophthalmitis, model required an extremely low infection dose of fewer than 100 cfu of *E. faecalis* (Jett *et al.*, 1992, 1995; Stevens *et al.*, 1992). Enterococcal endophthalmitis was observed to evolve over the course of 3–5 days, and was quantifiable by non-destructive slit lamp biomicroscopic examination as well as electroretinography. Further, histological examination of tissue sections documented the progression of infection and inflammatory sequelae through various layers of organized tissues. By comparing isogenic strains variously defective in cytolysin expression, it was observed that cytolysin was expressed *in vivo*, and that cytolysin production led to rapid and complete destruction of the retinal architecture (Jett *et al.*, 1992; Stevens *et al.*, 1992; Jett *et al.*, 1995). Furthermore, while experimental endophthalmitis caused by non-cytolytic strains was resolved with intraocular antibiotics and anti-inflammatory agents, similar infections caused by isogenic cytolytic strains were completely refractory to treatment (Jett *et al.*, 1995), highlighting the toxicity of cytolytic *E. faecalis* infection.

The cytolysin contributes to acute toxicity in a rabbit model of enterococcal endocarditis (Chow *et al.*, 1993). In conjunction with expression of aggregation substance on the surface of *E. faecalis*, a cytolytic strain produced heart valve vegetations that were significantly more lethal (55%, 6/11, lethality) than those of an isogenic, non-cytolytic strain (15%, 2/13, P < 0.01). Finally, whereas the contribution of the cytolysin to toxicity could be reproduced faithfully in a murine peritonitis model, resulting in a 90% decrease in LD_{50}, similar effects could not be shown in a rat model demonstrating at least some degree of species specificity (DuPont *et al.*, 1998). The robust physiology of the rat may render it less susceptible to subtle variations in the virulence of nosocomial pathogens such as enterococci, as is known for its relative lack of susceptibility to LPS.

In summary, the cytolysin contributes to the toxicity of infection in murine peritonitis, and lupine endocarditis and endophthalmitis models, findings coherent with the observed association between cytolysin phenotype and acutely terminal outcome in human infection. A rat model proved to be uniquely refractory to the cytolysin activity over the range of conditions tested (Dupont *et al.*, 1998), and the basis for this exception remains to be determined. Whereas the cytolysin contributes to tissue destruction and lethality in most infection models, no role has yet been ascribed for its bacteriocin activity in promoting colonization in an animal model (Huycke *et al.*, 1995).

BACILLUS CEREUS MULTI-COMPONENT TOXINS

Introduction

Bacillus cereus is a Gram-positive spore-forming aerobic motile organism that has, until recently, attracted little attention in the clinical literature because of its low degree of pathogenicity and isolation as a common contaminant. However, this saprophytic soil inhabitant has become recognized as an opportunistic pathogen of increasing importance, primarily owing to the increasing incidence of non-food-related infections. Drobniewski (1993) reviewed and categorized the clinical infectious syndromes caused by *B. cereus*, which include localized traumatic or postoperative infections, bacteraemia and septicaemia, central nervous system infections, respiratory infections, cardiac infections and toxin-induced food poisoning. Among those groups at significant risk of

infection with *B. cereus* are neonates, intravenous drug misusers and the immunologically compromised (Drobniewski, 1993).

Bacillus cereus elaborates several toxins putatively involved in the pathogenesis of disease. Two of these toxin complexes include haemolysin BL, a tri-component enterotoxic and haemolytic complex, and cereolysin AB, the haemolytic activity of which is due to the activities of its two enzymic components, phospholipase C and sphingomyelinase. The properties of these two multi-component toxins, their modes of action and contributions to disease are the focus of this section.

Haemolysin BL

General characteristics

Bacillus cereus is a common aetiological agent of two clinical food poisoning syndromes – diarrhoeal and emetic (Granum and Lund, 1997). The diarrhoeal form of *B. cereus* food poisoning involves a longer incubation period than the emetic syndrome, with diarrhoea and cramping 8–16 h after ingestion of food containing the organism.

Initial interest in *B. cereus* as an agent of diarrhoeal disease brought about the characterization of an enterotoxin complex of two to three protein components, which caused pathological effects ranging from fluid accumulation in rabbit and mouse ileal loops, vascular permeability and necrosis following intradermal injection, to death of mice following intravenous injection, and cytotoxicity and haemolysis *in vitro* (Turnbull, 1981). Attempts by several different research groups to identify the individual proteins causing these effects led to the compilation of a somewhat confounding list of potential factors (oedema factor, diarrhoeagenic factor, vascular permeability factor, necrotic factor, mouse lethal factor), which have at some time been suggested as part of the same enterotoxin complex (Turnbull, 1981; Thompson *et al.*, 1984; Shinagawa *et al.*, 1991). The exact nature of the *B. cereus* enterotoxin complex has been a subject of controversy for several years.

Molecular characterization

Efforts to segregate and characterize the enterotoxin complex led to the isolation of a fraction of *B. cereus* culture filtrate possessing vascular permeability, ileal loop fluid accumulation and haemolytic activities (Thompson *et al.*, 1984). Beecher and MacMillan (1990) also determined that the haemolytic activity of a *B. cereus* enterotoxin complex was caused by a combination of at least two proteins, termed B and L to indicate their possible binding and lytic activities during haemolysis, respectively. These components (termed haemolysin BL) were distinguished immunologically from the other known *B. cereus* haemolytic components, cereolysin O and cereolysin AB, but it was not known whether the enterotoxin complexes described by Thompson *et al.* (1984) and Beecher and MacMillan (1990) were the same entity.

Further purification of the haemolysin BL components by anion-exchange chromatography and preparative polyacrylamide gel electrophoresis led to the identification of the proteins responsible for binding and lytic activities (Beecher and MacMillan, 1990). Lytic activity was shown to occur following the binding of component B (35 kDa) to erythrocytes, and only upon the addition of fractions containing two proteins of 36 kDa (L_1) and 45 kDa (L_2). Combinations of the three components were tested in a sheep erythrocyte gel diffusion assay to determine each component's contribution to the activity of haemolysin BL. Maximal haemolysis occurred when all three components were combined, and to a lesser extent when B and L_1 were combined. Limited haemolysis was observed when B and L_2 were combined, but no haemolysis was observed with individual components alone. In addition, an unusual haemolytic pattern was observed during maximal haemolysis in the gel diffusion assay. Erythrocytes adjacent to wells were not lysed. With time, however, lysis occurred at a specific distance from each well to form a discrete ring of haemolysis. Subsequent spectrophotometric analysis

and immunofluorescent staining of erythrocytes confirmed that component B binding was necessary for lysis, and that L_1 and L_2 independently recognized component B prior to lytic attack (Beecher and MacMillan, 1991).

In these particular studies, an alternative to the ileal loop fluid accumulation method of testing the enterotoxigenicity of *B. cereus* proteins was implemented to test the activity of haemolysin BL. The vascular permeability assay used was based on dermal microvasculature compromise caused by an oedematous reaction following injection of toxin (Glatz *et al.*, 1974; Turnbull, 1981). Tissue oedema surrounding the injection site was quantified by intravenous injection of Evans' blue dye, which produced an area of bluing surrounding the injection site. Testing of combinations of the three components in the vascular permeability assay (which correlated with *B. cereus* enterotoxicity) further suggested that these components composed or were similar to the enterotoxin complex described by Thompson *et al.* (1984), and that all three components were necessary for maximal activity.

Because of difficulties in producing sufficient quantities necessary for further characterization of the haemolysin BL components, efforts focused on cloning the genes encoding the haemolysin BL components and improving purification procedures to confirm that the enterotoxin complex and haemolysin BL were identical.

Using a combination of anion-exchange and hydroxylapatite chromatography, Beecher and Wong (1994)

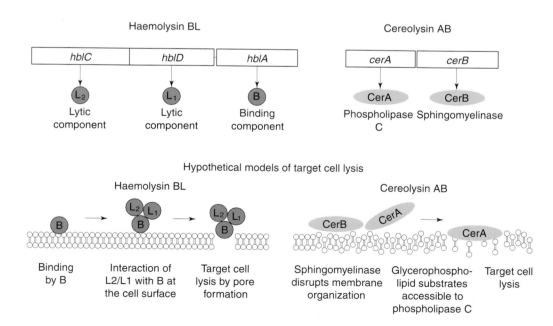

FIGURE 22.2 Haemolysin BL and cereolysin AB genetic maps and hypothetical models for target cell lysis. Genetic maps adapted from Gilmore *et al.* (1989) and Ryan *et al.* (1997).

were able to purify quantities of the components of haemolysin BL necessary to characterize quantitatively vascular permeability and haemolytic activities. Sodium dodecyl sulfate–polyacrylamide gel electrophoresis (SDS–PAGE) and isoelectric focusing of purified preparations of each component revealed the following physical characteristics: component B, 37.8 kDa, pI 5.34; component L_1, 38.5 kDa, pI 5.33; and component L_2, 43.2 kDa, pI 5.33. These values were similar to those of the enterotoxin complex of Thompson *et al.* (1984) and the DL toxin of Bitsaev and Ezepchuk (1987). The secondary structures of the three components of haemolysin BL have yet to be determined.

Combinations of the purified haemolysin BL components were tested for vascular permeability activity in rabbit skin. Dose-dependent dermal reactions were observed with combinations of all three components, with a slight oedematous response observed with the combination of B and L_1. Since the vascular permeability activity was shown to correlate with enterotoxigenic activity (Glatz *et al.*, 1974), the authors suggested that haemolysin BL was the vascular permeability factor described in several previous studies (Glatz *et al.*, 1974; Turnbull *et al.*, 1979a, b; Thompson *et al.*, 1984; Bitsaev and Ezepchuk, 1987; Turnbull and Kramer, 1983). Haemolysin BL and the enterotoxin complex of Thompson *et al.* (1984) were later confirmed immunologically as identical (Beecher *et al.*, 1995a).

Further analysis of purified haemolysin BL components in the ligated rabbit ileal loop model confirmed that all haemolysin BL components were necessary for maximal fluid accumulation. As little as 5 µg of each component elicited an enterotoxic response, a concentration comparable with the enterotoxic potency exhibited by cholera toxin (Finkelstein and LoSpalluto, 1972), thus positioning haemolysin BL as an important candidate for the diarrhoeagenic virulence factor of *B. cereus*.

Heinrichs *et al.* (1993) first reported the cloning and sequencing of the *hblA* gene encoding the B component of haemolysin BL, and the expression of the *hblA* gene product in *E. coli*. The 1.1-kb open reading frame of *hblA* encoded a peptide with a predicted molecular mass of 41 kDa, preceded by a potential ribosome binding site and 31-aa signal peptide. Northern blot analysis of early and late logarithmic-phase cultures also suggested that *hblA* may be located within a large 5.1-kb operon containing additional components of haemolysin BL. When combined with purified L_1 and L_2 components, recombinant *E. coli* lysates exhibited the ring-shaped haemolytic pattern characteristic of haemolysin BL in erythrocyte gel diffusion assays.

Ryan *et al.* (1997) later reported the cloning and sequencing of the genes encoding the L_2 and L_1 com-

ponents (*hblC* and *hblD*, respectively), and the characterization of the recombinant products expressed in *E. coli*. The nucleotide sequences of a cloned 5.5-kb *Eco*RI fragment revealed two open reading frames. The 1.3-kb open reading frame of *hblC* encoded a protein with a predicted molecular mass of 50 kDa, while the 1.2-kb open reading frame of *hblD* encoded a protein with a predicted mass of 41 kDa. Each sequence was preceded by putative ribosome binding sites, signal sequences and signal peptidase cleavage sites. Northern blot analysis of early logarithmic-phase cultures identified a single 5.5-kb transcript, which hybridized to probes generated from sequences of each of the three haemolysin BL components, suggesting that B, L_1 and L_2 are transcribed as part of a polycistronic message. The genetic arrangement of the haemolysin BL operon is shown in Figure 22.2 (adapted from Ryan *et al.*, 1997).

The results of dose–response turbidometric haemolytic analysis of the three haemolysin BL components also provided a hypothetical model for the unique ring-shaped (or discontinuous) haemolytic pattern observed in the gel diffusion assays (Beecher and Wong, 1994). The authors proposed that the B component primed erythrocytes to the lytic action of either or both L components, but inhibited lysis at high concentrations regardless of the concentrations of L_1 or L_2. The recent development of high-yield purification methods has furthered understanding of the interactions of the haemolysin BL components and the mechanisms of action of haemolysin BL.

Beecher and Wong (1997) used a novel haemolysis assay system to assess the priming activity of component B versus the lytic activity of the L_1 and L_2 components. The results showed that component B-induced priming of erythrocytes for the lytic action of the two L components (L_{1+2}) was temperature sensitive, and that the binding events of haemolysin BL components to erythrocytes were independent of one another. In addition, the discontinuous haemolytic zone phenomenon was caused by the inhibition of B by excess L_1 (and vice versa). Erythrocyte haemolysis was neutralized by inhibition of L components with L-specific antisera and by osmotic protection with carbohydrates. The results suggested that the haemolytic activity of haemolysin BL was due to (a) the interaction of at least the L components with the erythrocyte membrane, and (b) the formation of transmembrane pores, not enzymic lysis (Beecher and Wong, 1997).

Role in disease

Haemolysin BL has been proposed as an important virulence factor in diarrhoeal food poisoning syndromes (Beecher *et al.*, 1995a); however, the exact role of haemolysin BL in diarrhoeal disease has yet to be defined. The likelihood exists that the enterotoxigenicity of *B. cereus* is the result of one or more of the toxins or toxin complexes described over the past few decades. A recent report of a rapid polymerase chain reaction (PCR)-based assay for identification of enterotoxic *B. cereus* demonstrated that *hblA* was identified in only 45–50% of enterotoxic clinical isolates and was also identified in non-enterotoxic isolates (Mantynen and Lindstrom, 1998). Another report identified a non-haemolytic enterotoxin complex, distinct from haemolysin BL, in an enterotoxic strain of *B. cereus* (Lund and Granum, 1996, 1997). These findings highlight the likelihood that the enterotoxigenicity of *B. cereus* is multifactorial.

Haemolysin BL has also been proposed as a major virulence factor in necrotizing endophthalmitis, a highly virulent intraocular infection resulting from the introduction of *B. cereus* into the posterior segment of the eye. *Bacillus cereus* endophthalmitis is uniquely fulminant, and despite aggressive antibiotic and anti-inflammatory treatment, commonly results in loss of the eye within 48 h. Numerous reports have attributed the unique fulminance of *B. cereus* endophthalmitis to toxin production (O'Day et al., 1981; Turnbull and Kramer, 1983; Cowan et al., 1987; Davey and Tauber, 1987); however, surprisingly few studies have addressed the issue. Since haemolysin BL has been shown to exhibit haemolytic, dermonecrotic and emetic activities on a wide variety of cells and tissues, there exists the potential to exert multiple toxic effects on ocular tissues during fulminant endophthalmitis.

Beecher et al. (1995b) extended their research on haemolysin BL to include an analysis of the intraocular toxicity of haemolysin BL and other *B. cereus* toxins. Using a retinal button toxicity assay *in vitro*, the authors demonstrated that purified haemolysin BL and crude exotoxin preparations are toxic to retinal cells. In addition, purified haemolysin BL and crude exotoxin preparations caused intraocular inflammation resembling that of *B. cereus* endophthalmitis. The authors report, however, that haemolysin BL accounted for only half of the retinal toxicity of *B. cereus* supernatants, suggesting that full virulence is probably a multifactorial process.

Cereolysin AB

General characteristics

Bacillus cereus elaborates several membrane-degrading enzymes and exotoxins that are capable of exerting multiple effects on a variety of cells and tissues. These membranolytic toxins include phospholipases C with specificity for phosphatidylcholine and phosphatidylinositol, sphingomyelinase, a streptolysin O-like thiol-activated toxin (cereolysin O), and two other haemolysins termed II and III. The phosphatidylcholine-specific phospholipase C and sphingomyelinase of *B. cereus* have been used extensively in analyses of cell membrane structure (Coolbaugh and Williams, 1978; Little and Rumsby, 1980; Rao and Sundaram, 1993; Daniele et al., 1996), mammalian cell regulation and signal transduction (Levine et al., 1987; Diaz-Laviada et al., 1990; Diaz-Meco et al., 1992; Isakov, 1993; Jarvis et al., 1994; Johansen et al., 1994; Zhang et al., 1997), and for comparative studies of functionally related toxins secreted by other bacterial pathogens (*Clostridium perfringens*: Leslie et al., 1989; Saint-Joanis et al., 1989; Titball et al., 1989, 1993; *Staphylococcus aureus*: Verkleij et al., 1973; *Pseudomonas aeruginosa*: Crowell and Lutz, 1989; Konig et al., 1997). The two enzymes have been investigated in great detail separately, but also function together as a cytolytic unit, which will be described below.

Phospholipase C

The phosphatidylcholine-specific phospholipase C of *B. cereus* is a monomeric metalloenzyme containing three Zn^{2+} in its active site (Johansen et al., 1988; Hough et al., 1989; Hansen et al., 1993). Johansen et al. (1988) first reported the cloning and sequencing of the *plc* gene encoding phospholipase C and the expression of the *plc* gene product in *E. coli*. Sequencing of a 2.0-kb restriction fragment containing *plc* revealed a single, 0.7-kb gene encoding the secreted form of the peptide, which, upon SDS–PAGE analysis, migrated as a peptide of approximately 28 kDa. Further analysis determined that the *plc* gene product was probably translated as a 283-amino acid (aa) precursor with a 24-aa signal peptide and a 14-aa propeptide, and is secreted as a 245-aa monomer (Johansen et al., 1988). Structurally, phospholipase C is composed of 10 parallel and antiparallel α-helices that are folded into a single domain, with the three Zn^{2+} positioned at the bottom of a wide enzymic cleft located on the protein's surface (Hough et al., 1989).

The phospholipase C of *B. cereus* is of particular interest because of its potential for eliciting second messenger activity in mammalian cells. Incubation with phospholipase C resulted in activation of the arachidonic acid pathway via diacylglycerol in several different mammalian cell types (Levine et al., 1987; Isakov, 1993), and interference with protein kinase C regulation in oncogene-transformed fibroblasts (Diaz-Laviada et al., 1990; Diaz-Meco et al., 1992). The phospholipase C also causes various effects on different mammalian cell types, including activation and inactivation of neutrophils (Styrt et al., 1989; Wazny et al., 1990), and matrix metalloprotease upregulation in epithelial cells (Firth *et al.*, 1997), each of which could be involved in mediating tissue damage during infection.

Sphingomyelinase

The sphingomyelinase of *B. cereus* has also been cloned and sequenced (Johansen *et al.*, 1988; Yamada *et al.*, 1988) and its secondary structure postulated (Tomita *et al.*, 1990). The molecular mass of sphingomyelinase ranges from 23 to 41 kDa, depending upon the analytical method (Ikezawa *et al.*, 1978; Tomita *et al.*, 1982). Sequencing of a 7.5-kb restriction fragment containing the sphingomyelinase gene revealed a 1.0-kb open reading frame encoding a pre-peptide of 333 aa containing a signal peptide of 27 aa, with predicted molecular mass of 36.8 kDa. The secreted form of sphingomyelinase consisted of 306 aa with a predicted molecular mass of 34.2 kDa. Hydropathy profiles demonstrated that the *B. cereus* sphingomyelinase possessed several hydrophobic regions at the N-terminus that could be responsible for membrane binding (Tomita *et al.*, 1990). Furthermore, owing to the low α-helix content and the presence of a disulfide bond, the authors suggested that the secondary structure could be composed primarily of a loop or turn structure, allowing for the flexibility necessary to access the erythrocyte membrane (Tomita *et al.*, 1990, 1993).

Functional studies have determined that the ability of sphingomyelinase to adsorb on to erythrocyte membranes prior to haemolysis was stimulated by Mg^{2+} and inhibited by Ca^{2+} (Tomita *et al.*, 1983; Ikezawa *et al.*, 1986). A loosely associated Mg^{2+} located within the enzymic site has also been postulated (Tomita *et al.*, 1983; Yamada *et al.*, 1988). Guddal *et al.* (1989) reported that secretion of sphingomyelinase was repressed by inorganic phosphate (P_i) in the growth medium. The authors speculated that since *B. cereus* commonly inhabits a soil environment where P_i is a significant growth-limiting factor, sphingomyelinase (and phospholipase C) could be part of an efficient P_i retrieval-scavenging system. What contribution this system plays in virulence is not known.

Recently, interest has focused on bacterial sphingomyelinases as inducers of ceramide-induced apoptotic death of mammalian cells. Generation of intracellular ceramide through the sphingomyelin pathway has been shown to induce apoptosis in several mammalian cell lines (Hannun and Obeid, 1997). Exogenous treatment of human leukaemia cells with *B. cereus* sphingomyelinase resulted in significant increases in intracellular ceramide levels that, in other systems (Jarvis *et al.*, 1994), would ordinarily induce apoptosis (Zhang *et al.*, 1997). However, transfection of the *B. cereus* sphingomyelinase into human leukaemia cells was necessary for endogenous sphingomyelinase production, induction of intracellular ceramide and apoptotic cell death (Zhang *et al.*, 1997). A growing body of evidence has recently emerged, speculating on a role for bacteria-induced apoptosis in inflammation; however, whether a link exists between *B. cereus* sphingomyelinase and the induction of the apoptotic process during infection is not yet known.

Cereolysin AB

The sequence information provided by Yamada *et al.* (1988) revealed the close proximity of the sphingomyelinase gene to sequences homologous to the gene encoding phospholipase C. Gilmore *et al.* (1989) first reported the molecular characterization of cereolysin AB as a bi-component cytolytic unit comprising phosphatidylcholine-specific phospholipase C and sphingomyelinase. A cloned cytolytic determinant from *B. cereus*, originally reported to confer haemolytic activity to *E. coli* and *B. subtilis* (Kreft *et al.*, 1983), comprised tandem genes for phospholipase C (*cerA*, 5' open reading frame) and sphingomyelinase (*cerB*, 3' open reading frame). *cerA* showed amino acid sequence identity with the N-terminus of *B. cereus* phospholipase C (Otnaess *et al.*, 1977) and some nucleotide sequence identity with the *plc* gene of Johansen *et al.* (1988). In addition, *cerA* conferred a lecithinase-positive phenotype to *B. subtilis*, establishing the identity of the *cerA* product as phospholipase C. A comparison of the amino acid sequence of *cerB* to sequences in GenBank at the time revealed no known homologues; however, *cerB* conferred sphingomyelinase activity to a heterologous *B. subtilis* host, demonstrating heterologous expression and identifying the *cerB* product as a sphingomyelinase. The genetic arrangement of the cereolysin AB operon is shown in Figure 22.2 (adapted from Gilmore *et al.*, 1989).

Functionally, lysis of erythrocytes by CerAB was demonstrated to occur via sequential action on membranes by independent activities of sphingomyelinase and phospholipase C. The association of functional synergism between the two cereolysin AB components and the genetic linkage of the *cerA* and *cerB* genes suggested that the natural function of cereolysin AB is as a cytolytic unit. Cereolysin AB has been proposed as an evolutionary precursor of *Clostridium perfringens* α-toxin, a noted virulence factor with phospholipase C and sphingomyelinase activities (Leslie *et al.*, 1989). Although phospholipases C and sphingomyelinases from a variety of organisms have been suggested as key virulence factors in infectious diseases, no specific role for cereolysin AB in the virulence of *B. cereus* has been established.

It is possible that the components of cereolysin AB, either individually or as a bi-component cytolysin, could function to damage or lyse cells directly or induce the production of inflammatory mediators that could contribute to tissue damage during infection.

Other multi-component toxins of Bacillus cereus

In addition to the two multi-component haemolytic toxins addressed in this review, *B. cereus* is capable of producing over 20 additional exotoxins. Recently, a non-haemolytic enterotoxin (NHE) complex containing 39-, 45- and 105-kDa components has been described (Lund and Granum, 1997). Similarities between the 39-kDa NHE component and the L_1 component of haemolysin BL have been reported; however, the molecular nature of the NHE complex has yet to be elucidated. Other comprehensive reviews (Turnbull, 1981; Drobniewski, 1993; Granum, 1994) address the other *B. cereus* toxins putatively involved in virulence, which are beyond the scope of this review.

CONCLUSION

Studies of the pathogenesis of bacterial infections have focused on the contributions of specific toxins or other proteins to disease. The application of this approach to *B. cereus* is challenging because of the large number of proteins, some with multiple components, that could contribute to virulence during infection. The most plausible evidence that a bacterial gene product contributes to a disease is validated by generating isogenic mutants specifically deficient only in the virulence factor under investigation. This approach is presently being pursued to assess the contributions of haemolysin BL, cereolysin AB and other *B. cereus* toxins to ocular infection and other types of disease.

ACKNOWLEDGEMENTS

For their helpful discussions and assistance in preparation of this manuscript, we gratefully acknowledge Lynn E. Hancock, Brett D. Shepard, Phillip S. Coburn, Mary C. Booth Ph.D. and Viswanathan Shankar Ph.D. Portions of the work described in this review have been supported by US Public Health Service grants EY08289, EY06813 and AI41108, the US Department of Agriculture, the Oklahoma Center for the Advancement of Science and Technology, and Research to Prevent Blindness Inc.

REFERENCES

Arduino, R.C., Murray, B.E. and Rakita R.M. (1993). Role of antibodies and complement in the phagocytic killing of enterococci. *Infect. Immun.* **62**, 987–93.

Armstrong, G.L., Hollingsworth, J., Morris, J.G., Jr (1996). Emerging foodborne pathogens: *Escherichia coli* O157:H7 as a model of entry of a new pathogen into the food supply of the developed world. *Epidemiol. Rev.* **18**, 29–51.

Basinger, S.F. and Jackson, R.W. (1968). Bacteriocin (hemolysin) of *Streptococcus zymogenes*. *J. Bacteriol.* **96**, 1895–902.

Beecher, D.J. and MacMillan, J.D. (1990). A novel bicomponent hemolysin from *Bacillus cereus*. *Infect. Immun.* **58**, 2220–7.

Beecher, D.J. and MacMillan, J.D. (1991). Characterization of the components of hemolysin BL from *Bacillus cereus*. *Infect. Immun.* **59**, 1778–84.

Beecher, D.J. and Wong, A.C. (1994). Improved purification and characterization of hemolysin BL, a hemolytic dermonecrotic vascular permeability factor from *Bacillus cereus*. *Infect. Immun.* **62**, 980–6.

Beecher, D.J. and Wong, A.C. (1997). Tripartite hemolysin BL from *Bacillus cereus*. Hemolytic analysis of component interactions and a model for its characteristic paradoxical zone phenomenon. *J. Biol. Chem.* **272**, 233–9.

Beecher, D.J., Schoeni, J.L. and Wong, A.C. (1995a). Enterotoxic activity of hemolysin BL from *Bacillus cereus*. *Infect. Immun.* **63**, 4423–8.

Beecher, D.J., Pulido, J.S., Barney, N.P. and Wong, A.C. (1995b). Extracellular virulence factors in *Bacillus cereus* endophthalmitis: methods and implication of involvement of hemolysin BL. *Infect. Immun.* **63**, 632–9.

Bitsaev, A.R. and Ezepchuk, I.V. (1987). Molecular nature of the pathogenic effect induced by *B. cereus*. *Mol. Gen. Mikrobiol. Virusol.* **7**, 18–23.

Booth, M.C., Bogie, C.P., Sahl, H., Siezen, R.J., Hatter, K.L. and Gilmore, M.S. (1996). Structural analysis and proteolytic activation of *Enterococcus faecalis* cytolysin, a novel lantibiotic. *Mol. Microbiol.* **21**, 1175–84.

Breed, R.S., Murray, E.G.D. and Hitchens, A.P. (1948). *Bergey's Manual of Determinative Bacteriology*, 6th edn, p. 327. Williams & Wilkins, Baltimore, MD.

Brock, T.D. and Davie, J.M. (1963). Probable identity of group D hemolysin with bacteriocin. *J. Bacteriol.* **86**, 708–12.

Brock, T.D., Peacher, B.and Pierson D. (1963). Survey of the bacteriocins of enterococci. *J. Bacteriol.* **86**, 702–7.

Chow, J.W., Thal, L.A., Perri, M.B., Vazquez, J.A., Donabedian, S.M., Clewell, D.B. and Zervos, M.J. (1993). Plasmid-associated hemolysin and aggregation substance production contribute to virulence in experimental enterococcal endocarditis. *Antimicrob. Agents Chemother.* **37**, 2474–7.

Clewell, D.B. (1993). Bacterial sex pheromone-induced plasmid transfer. *Cell* **73**, 9–12.

Clewell, D.B., Tomich, P.K., Gawron-Burke, M.C., Franke, A.E., Yagi, Y. and An, F.Y. (1982). Mapping of *Streptococcus faecalis* plasmids pAD1 and pAD2 and studies relating to transposition of Tn*917*. *J. Bacteriol.* **152**, 1220–30.

Coburn, P.S., Hancock, L.E., Booth, M.C. and Gilmore, M.S. (1998). Identification and characterization of the immunity region of the *Enterococcus faecalis* cytolysin operon. In: *Third International Workshop on Lantibiotics and Related Modified Antibiotic Peptides*. Blaubeuren, Germany.

Coburn, P.S., Hancock, L.E., Booth, M.C. and Gilmore, M.S. (1999). A novel means of self-protection, unrelated to toxin activation, confers immunity to the bactericidal effects of the *Enterococcus faecalis* cytolysm. *Infect. Immun.* (in press).

Colmar, I. and Horaud, T. (1987). *Enterococcus faecalis* hemolysin-bacteriocin plasmids belong to the same incompatibility group. *Appl. Environ. Microbiol.* **53**, 567–70.

Coolbaugh, J.C. and Williams, R.P. (1978). Production and characterization of two hemolysins of *Bacillus cereus*. *Can. J. Microbiol.* **24**, 1289–95.

Coque, T.M., Patterson, J.E., Steckelberg, J.M. and Murray, B.E. (1995). Incidence of hemolysin, gelatinase, and aggregation substance among enterococci isolated from patients with endocarditis and other infections and from feces of hospitalized and community-based persons. *J. Infect. Dis.* **171**, 1223–9.

Cowan, D.L., Madden, W.M., Hatem, G.F. and Merritt, J.C. (1987). Endogenous *Bacillus cereus* panophthalmitis. *Ann. Ophthalmol.* **19**, 65–8.

Crowell, K.M. and Lutz, F. (1989). *Pseudomonas aeruginosa* cytotoxin: the influence of sphingomyelin on binding and cation permeability increase in mammalian erythrocytes. *Toxicon* **27**, 531–40.

Daniele, J.J., Maggio, B., Bianco, I.D., Goni, F.M., Alonso, A.and Fidelio, G.D. (1996). Inhibition by gangliosides of *Bacillus cereus* phospholipase C activity against monolayers, micelles and bilayer vesicles. *Eur. J. Biochem.* **239**, 105–10.

Davey, R.T. and Tauber, W.B. (1987). Posttraumatic endophthalmitis: the emerging role of *Bacillus cereus* infection. *Rev. Infect. Dis.* **9**, 110–23.

Diaz-Laviada, I., Larrodera, P., Diaz-Meco, M.T., Cornet, M.E., Guddal, P.H., Johansen, T. and Moscat, J. (1990). Evidence for a role of phosphatidylcholine-hydrolysing phospholipase C in the regulation of protein kinase C by *ras* and *src* oncogenes. *EMBO J.* **9**, 3907–12.

Diaz-Meco, M.T., Dominguez, I., Sanz, L., Municio, M.M., Berra, E., Cornet, M.E., Garcia de Herreros, A., Johansen, T. and Moscat, J. (1992). Phospholipase C-mediated hydrolysis of phosphatidylcholine is a target of transforming growth factor beta 1 inhibitory signals. *Mol. Cell Biol.* **12**, 302–8.

Drobniewski, F.A. (1993). *Bacillus cereus* and related species. *Clin. Microbiol. Rev.* **6**, 324–38.

Dunny, G.M. and Clewell, D.B. (1975). Transmissible toxin (hemolysin) plasmid in *Streptococcus faecalis* and its mobilization of a noninfectious drug resistance plasmid. *J. Bacteriol.* **124**, 784–90.

Dunny, G.M. and Leonard, B.A. (1997). Cell–cell communication in Gram-positive bacteria. *Annu. Rev. Microbiol.* **51**, 527–64.

DuPont, H., Montravers, P., Mohler, J. and Carbon, C. (1998). Disparate findings on the role of virulence factors of *Enterococcus faecalis* in mouse and rat models of peritonitis. *Infect. Immun.* **66**, 2570–5.

Eliopoulos, G.M. and Eliopoulos, C.T. (1990). Therapy of enterococcal infections. *Eur. J. Clin. Microbiol. Infect. Dis.* **9**, 118–26.

Emori, T.G. and Gaynes, R.P. (1993). An overview of nosocomial infections, including the role of the microbiology laboratory. *Clin. Microbiol. Rev.* **6**, 428–42.

Facklam, R.R. (1972). Recognition of group D streptococcal species of human origin by biochemical and physiological tests. *Appl. Microbiol.* **23**, 1131–9.

Fath, M.J. and Kolter, R. (1993). ABC transporters: bacterial exporters. *Microbiol. Rev.* **57**, 995–1017.

Finkelstein, R.A. and LoSpalluto, J.J. (1972). Crystalline cholera toxin and toxoid. *Science* **175**, 529–30.

Firth, J.D., Putnins, E. E., Larjava, H. and Uitto, V.J. (1997). Bacterial phospholipase C upregulates matrix metalloproteinase expression by cultured epithelial cells. *Infect. Immun.* **65**, 4931–6.

Fujimoto, S. and Clewell, D.B. (1998). Regulation of the pAD1 sex pheromone response of *Enterococcus faecalis* by direct interaction between the cAD1 peptide mating signal and the negatively regulating, DNA-binding TraA protein. *Proc. Natl Acad. Sci. USA.* **95**, 6430–5.

Fujimoto, S., Bastos, M., Tanimoto, K., An, F., Wu, K. and Clewell, D.B. (1997). The pAD1 sex pheromone response in *Enterococcus faecalis*. *Adv. Exp. Med. Biol.* **418**, 1037–40.

Gilmore, M.S. (1991). *Enterococcus faecalis* hemolysin/bacteriocin. In: *Genetics and Molecular Biology of Streptococci, Lactococci, and Enterococci* (eds G.M. Dunny, P.P. Cleary and L.L. McKay), pp. 206–13. American Society for Microbiology, Washington, DC.

Gilmore, M.S., Cruz-Rodz, A.L., Leimeister-Wachter, M., Kreft, J. and Goebel, W. (1989). A *Bacillus cereus* cytolytic determinant, cereolysin AB, which comprises the phospholipase C and sphingomyelinase genes: nucleotide sequence and genetic linkage. *J. Bacteriol.* **171**, 744–53.

Gilmore, M.S., Segarra, R.A. and Booth, M.C. (1990) An HylB-type function is required for expression of the *Enterococcus faecalis* hemolysin/bacteriocin. *Infect. Immun.* **58**, 3914–23.

Gilmore, M.S., Segarra, R.A., Booth, M.C., Bogie, C.P., Hall, L.R. and Clewell, D.B. (1994). Genetic structure of the *Enterococcus faecalis* plasmid pAD1-encoded cytolytic toxin system and its relationship to lantibiotic determinants. *J. Bacteriol.* **176**, 7335–44.

Gilmore, M.S., Skaugen, M. and Nes, I. (1996). *Enterococcus faecalis* cytolysin and lactocin S of *Lactobacillus sake*. *Antonie van Leeuwenhoek*. **69**, 129–38.

Glatz, B.A., Spira, W.M. and Goepfert, J.M. (1974). Alteration of vascular permeability in rabbits by culture filtrates of *Bacillus cereus* and related species. *Infect. Immun.* **10**, 299–303.

Granato, P.A. and Jackson, R.W. (1969). Bicomponent nature of lysin from *Streptococcus zymogenes*. *J. Bacteriol.* **100**, 865–8.

Granato, P.A. and Jackson, R.W. (1971a). Characterization of the A component of *Streptococcus zymogenes* lysin. *J. Bacteriol.* **107**, 551–6.

Granato, P.A. and Jackson, R.W. (1971b). Purification and characterization of the L component of *Streptococcus zymogenes* lysin. *J. Bacteriol.* **108**, 804–8.

Granum, P.E. (1994). *Bacillus cereus* and its toxins. *J. Appl. Bacteriol. Symp. Suppl.* **76**, 61-6S.

Granum, P.E. and Lund, T. (1997). *Bacillus cereus* and its food poisoning toxins. *FEMS Microbiol. Lett.* **157**, 223–8.

Griffith, T.S., Brunner, T., Fletcher, S.M., Green, D.R. and Ferguson, T.A. (1995). Fas ligand-induced apoptosis as a mechanism of immune privilege. *Science* **270**, 1189–92.

Guddal, P.H., Johansen, T., Schulstad, K. and Little, C. (1989). Apparent phosphate retrieval system in *Bacillus cereus*. *J. Bacteriol.* **171**, 5702–6.

Hall, L.M.C., Duke, B., Urwin, G. and Guiney, M. (1992). Epidemiology of *Enterococcus faecalis* urinary tract infection in a teaching hospital in London, United Kingdom. *J. Clin. Microbiol.* **30**, 1953–7.

Handwerger, S., Perlman, D.C., Altarac, D. and McAuliffe, V. (1992). Concomitant high-level vancomycin and penicillin resistance in clinical isolates of enterococci. *Clin. Infect. Dis.* **14**, 655–61.

Hannun, Y.A. and Obeid, L.M. (1997). Mechanisms of ceramide-mediated apoptosis. *Adv. Exp. Med. Biol.* **407**, 145–9.

Hansen, S., Hough, E., Svensson, L.A., Wong, Y.L. and Martin, S.F. (1993) Crystal structure of phospholipase C from *Bacillus cereus* complexed with a substrate analog. *J. Mol. Biol.* **234**, 179–87.

Havarstein, L.S., Diep, D.B. and Nes, I.F. (1995). A family of bacteriocin ABC transporters carry out proteolytic processing of their substrate concomitant with export. *Mol. Microbiol.* **16**, 229–40.

Heinrichs, J.H., Beecher, D.J., MacMillan, J.D. and Zilinskas, B.A. (1993). Molecular cloning and characterization of the *hblA* gene encoding the B component of hemolysin BL from *Bacillus cereus*. *J. Bacteriol.* **175**, 6760–6.

Herman, D.J. and Gerding, D.N. (1991). Antimicrobial resistance among enterococci. *Antimicrob. Agents Chemother.* **35**, 1–4.

Hough, E., Hansen, L.K., Birknes, B., Jynge, K., Hansen, S., Hordvik, A., Little, C., Dodson, E. and Derewenda, Z. (1989). High-resolution (1.5 Å) crystal structure of phospholipase C from *Bacillus cereus*. *Nature* **338**, 357–60.

Huycke, M.M. and Gilmore, M.S. (1995). Frequency of aggregation substance and cytolysin genes among enterococcal endocarditis isolates. *Plasmid* **34**, 152–6.

Huycke, M.M., Spiegel, C.A. and Gilmore, M.S. (1991). Bacteremia caused by hemolytic, high-level gentamicin-resistant *Enterococcus faecalis*. *Antimicrob. Agents Chemother.* **35**, 1626–34.

Huycke, M.M., Gilmore, M.S., Jett, B.D. and Booth, J.L. (1992). Transfer of pheromone-inducible plasmids between *Enterococcus faecalis* in the Syrian hamster gastrointestinal tract. *J. Infect. Dis.* **166**, 1188–91.

Huycke, M.M., Joyce, W.A. and Gilmore, M.S. (1995) *Enterococcus faecalis* cytolysin without effect on the intestinal growth of susceptible enterococci in mice. *J. Infect. Dis.* **172**, 273–6.

Huycke, M.M., Sahm, D.F. and Gilmore, M.S. (1998). Multiple-drug resistant enterococci: the nature of the problem and an agenda for the future. *Emerging Infect. Dis.* **4**, 239–49.

Ike, Y. and Clewell, D.B. (1984). Genetic analysis of pAD1 pheromone response in *Streptococcus faecalis* using transposon Tn*917* as an insertional mutagen. *J. Bacteriol.* **158**, 777–83.

Ike, Y. and Clewell, D.B. (1992). Evidence that the hemolysin/bacteriocin phenotype of *Enterococcus faecalis* subsp. *zymogenes* can be determined by plasmids in different incompatibility groups as well as by the chromosome. *J. Bacteriol.* **174**, 8172–7.

Ike, Y., Hashimoto, H. and Clewell, D.B. (1984). Hemolysin of *Streptococcus faecalis* subspecies *zymogenes* contributes to virulence in mice. *Infect. Immun.* **45**, 528–30.

Ike, Y., Hashimoto, H. and Clewell, D.B. (1987). High incidence of hemolysin production by *Enterococcus (Streptococcus) faecalis* strains associated with human parenteral infections. *J. Clin. Microbiol.* **25**, 1524–8.

Ike, Y., Clewell, D.B., Segarra, R.A. and Gilmore, M.S. (1990). Genetic analysis of the pAD1 hemolysin/bacteriocin determinant in *Enterococcus faecalis*: Tn*917* insertional mutagenesis and cloning. *J. Bacteriol.* **172**, 155–63.

Ikezawa, H., Mori, M., Ohyabu, T. and Taguchi, R. (1978). Studies on sphingomyelinase of *Bacillus cereus*. I. Purification and properties. *Biochim. Biophys. Acta* **528**, 247–56.

Ikezawa, H., Matsushita, M., Tomita, M. and Taguchi, R. (1986). Effects of metal ions on sphingomyelinase activity of *Bacillus cereus*. *Arch. Biochem. Biophys.* **249**, 588–95.

Isakov, N. (1993). Activation of murine lymphocytes by exogenous phosphatidylethanolamine- and phosphatidylcholine-specific phospholipase C. *Cell Immunol.* **152**, 72–81.

Jacob, A.E., Douglas, G.I. and Hobbs, S.J. (1975). Self-transferable plasmids determining the hemolysin and bacteriocin of *Streptococcus faecalis* var. *zymogenes*. *J. Bacteriol.* **121**, 863–72.

Jarvis, W.D., Fornari, F.A., Browning, J.L., Gewirtz, D.A., Kolesnick, R.N. and Grant, S. (1994). Attenuation of ceramide-induced apoptosis by diglyceride in human myeloid leukemia cells. *J. Biol. Chem.* **269**, 31685–92.

Jarvis, W.R., Gaynes, R.P., Horan, T.C., Alonso-Echanove, J., Emori, T.G., Fridkin, S.K., Lawton, R.M., Richards, M.J. and Wright, G.C. (1998). *National Nosocomial Infections Surveillance (NNIS) Report. Data from October 1986–April 1998, issued June 1998*, pp. 1–25. Centers for Disease Control and Prevention, 0Atlanta, GA.

Jett, B.D. and Gilmore, M.S. (1990). The growth inhibitory effect of the *Enterococcus faecalis* bacteriocin encoded by pAD1 extends to the oral streptococci. *J. Dent. Res.* **69**, 1640–5.

Jett, B.D., Jensen, H.G., Nordquist, R.E. and Gilmore, M.S. (1992). Contribution of the pAD1-encoded cytolysin to the severity of experimental *Enterococcus faecalis* endophthalmitis. *Infect. Immun.* **60**, 2445–52.

Jett, B. D., Huycke, M.M. and Gilmore, M.S. (1994). Virulence of enterococci. *Clin. Microbiol. Rev.* **7**, 462–78.

Jett, B.D., Jensen, H.G., Atkuri, R.V. and Gilmore, M.S. (1995). Evaluation of therapeutic measures for treating endophthalmitis caused by isogenic toxin producing and toxin non-producing *Enterococcus faecalis* strains. *Invest. Ophthalmol. Vis. Sci.* **36**, 9–15.

Johansen, T., Holm, T., Guddal, P. H., Sletten, K., Haugli, F.B. and Little, C. (1988). Cloning and sequencing of the gene encoding the phosphatidylcholine-preferring phospholipase C of *Bacillus cereus*. *Gene* **65**, 293–304.

Johansen, T., Bjorkoy, G., Overvatn, A., Diaz-Meco, M.T., Traavik, T., Moscat, J. (1994). NIH 3T3 cells stably transfected with the gene encoding phosphatidylcholine-hydrolyzing phospholipase C from *Bacillus cereus* acquire a transformed phenotype. *Mol. Cell. Biol.* **14**, 646–54.

Kalmanson, G.M., Hubert, E.G. and Guze, L.B. (1970). Effect of bacteriocin from *Streptococcus faecalis* on microbial L-forms. *J. Infec. Dis.* **121**, 311–15.

Klein, C., Kaletta, C., Schnell, N. and Entian, K.D. (1992). Analysis of genes involved in biosynthesis of the lantibiotic subtilin. *Appl. Environ. Microbiol.* **58**, 132–42.

Kobayashi, R. (1940). Studies concerning hemolytic streptococci: typing of human hemolytic streptococci and their relation to diseases and their distribution on mucous membranes. *Kitasato Arch. Exp. Med.* **17**, 218–41.

Konig, B., Vasil, M.L. and Konig, W. (1997). Role of hemolytic and non-hemolytic phospholipase C from *Pseudomonas aeruginosa* for inflammatory mediator release from human granulocytes. *Int. Arch. Allergy Immunol.* **112**, 115–24.

Kreft, J., Berger, H., Hartlein, M., Muller, B., Weidinger, G. and Goebel W. (1983). Cloning and expression in *Escherichia coli* and *Bacillus subtilis* of the hemolysin (cereolysin) determinant from *Bacillus cereus*. *J. Bacteriol.* **155**, 681–9.

LeBlanc, D.J., Lee, L.N., Clewell, D.B. and Behnke, D. (1983). Broad geographical distribution of a cytotoxin gene mediating beta-hemolysis and bacteriocin activity among *Streptococcus faecalis* strains. *Infect. Immun.* **40**, 1015–22.

Leslie, D., Fairweather, N., Pickard, D., Dougan, G. and Kehoe, M. (1989). Phospholipase C and haemolytic activities of *Clostridium perfringens* alpha-toxin cloned in *Escherichia coli*: sequence and homology with a *Bacillus cereus* phospholipase C. *Mol. Microbiol.* **3**, 383–92.

Levine, L., Xiao, D.M. and Little, C. (1987). Increased arachidonic acid metabolites from cells in culture after treatment with the phosphatidylcholine-hydrolyzing phospholipase C from *Bacillus cereus*. *Prostaglandins* **34**, 633–42.

Libertin, C.R., Dumitru, R. and Stein, D.S. (1992). The hemolysin/bacteriocin produced by enterococci is a marker of pathogenicity. *Diagn. Microbiol. Infect. Dis.* **15**, 115–20.

Little, C. and Rumsby, M.G. (1980). Lysis of erythrocytes from stored human blood by phospholipase C (*Bacillus cereus*). *Biochem. J.* **188**, 39–46.

Lund, T. and Granum, P.E. (1996). Characterisation of a non-haemolytic enterotoxin complex from *Bacillus cereus* isolated after a foodborne outbreak. *FEMS Microbiol. Lett.* **141**, 151–6.

Lund, T. and Granum, P.E. (1997). Comparison of biological effect of the two different enterotoxin complexes isolated from three different strains of *Bacillus cereus*. *Microbiology* **143**, 3329–36.

Mantynen, V. and Lindstrom, K. (1998). A rapid PCR-based DNA test for *enterotoxic Bacillus cereus*. *Appl. Environ. Microbiol.* **64**, 1634–9.

Miyazaki, S., Ohno, A., Kobayashi, I., Uji, T., Yamaguchi, K. and Goto, S. (1993). Cytotoxic effect of hemolytic culture supernatant from *Enterococcus faecalis* on mouse polymorphonuclear neutrophils and macrophages. *Microbiol. Immunol.* **37**, 265–70.

Moellering, R.C., Jr (1992). Emergence of *Enterococcus* as a significant pathogen. *Clin. Infect. Dis.* **14**, 1173–8.

Murray, B.E. (1998). Diversity among multidrug-resistant enterococci. *Emerging Infect. Dis.* **4**, 37–47.

O'Day, D.M., Smith, R.S., Gregg, C.R., Turnbull, P.C.B., Head, W.S., Ives, J.A. and Ho, P.C. (1981). The problem of *Bacillus* species infection with special emphasis on the virulence of *Bacillus cereus*. *Ophthalmology* **88**, 833–8.

Otnaess, A.B., Little, C., Slettin, K., Wallin, R., Johnsen, W., Flengsrud, R. and Prydz, H. (1977). Some characteristics of phospholipase C from *Bacillus cereus*. *Eur. J. Biochem.* **79**, 459–68.

Rantz, L.A. and Kirby, W.M.M. (1943). Enterococcic infections: an evaluation of the importance of fecal streptococci and related organisms in the causation of human disease. *Arch. Intern. Med.* **71**, 516–28.

Rao, N.M. and Sundaram, C.S. (1993). Sensitivity of phospholipase C (*Bacillus cereus*) activity to lipid packing in sonicated lipid mixtures. *Biochemistry* **32**, 8547–52.

Ryan, P.A., MacMillan, J.D. and Zilinskas, B.A. (1997). Molecular cloning and characterization of the genes encoding the L1 and L2 components of hemolysin BL from *Bacillus cereus*. *J. Bacteriol.* **179**, 2551–6.

Sahl, H., Jack, R.W. and Bierbaum, G. (1995). Biosynthesis and biological activities of lantibiotics with unique post-translational modifications. *Eur. J. Biochem.* **230**, 827–53.

Saint-Joanis, B., Garnier, T. and Cole, S.T. (1989). Gene cloning shows the alpha-toxin of *Clostridium perfringens* to contain both sphingomyelinase and lecithinase activities. *Mol. Gen. Genet.* **219**, 453–60.

Schnell, N., Engelke, G., Augustin, J., Rosenstein, R., Ungermann, V., Gotz, F. and Entian, K.D. (1992). Analysis of genes involved in biosynthesis of the lantibiotic epidermin. *Eur. J. Biochem.* **204**, 57–68.

Segarra, R.A., Booth, M.C., Morales, D.A., Huycke, M.M. and Gilmore, M.S. (1991). Molecular characterization of the *Enterococcus faecalis* cytolysin activator. *Infect. Immun.* **59**, 1239–46.

Shinagawa, K., Ueno, S., Konuma, H., Matsusaka, N. and Sugii, S. (1991). Purification and characterization of the vascular permeability factor produced by *Bacillus cereus*. *J. Vet. Med. Sci.* **53**, 281–6.

Stark, J.M. (1960). Antibiotic activity of haemolytic enterococci. *Lancet* **i**, 733–4.

Stevens, S.X., Jensen, H.G., Jett, B.D. and Gilmore, M.S. (1992). A hemolysin-encoding plasmid contributes to bacterial virulence in experimental *Enterococcus faecalis* endophthalmitis. *Invest. Ophthalmol. Vis. Sci.* **33**, 1650–6.

Styrt, B., Walker, R.D. and White, J.C. (1989). Neutrophil oxidative metabolism after exposure to bacterial phospholipase C. *J. Lab. Clin. Med.* **114**, 51–7.

Thompson, N.E., Ketterhagen, M.J., Bergdoll, M.S. and Schantz, E.J. (1984). Isolation and some properties of an enterotoxin produced by *Bacillus cereus*. *Infect. Immun.* **43**, 887–94.

Titball, R.W., Hunter, S.E., Martin, K.L., Morris, B.C., Shuttleworth, A.D., Rubidge, T., Anderson, D.W. and Kelly, D.C. (1989). Molecular cloning and nucleotide sequence of the alpha-toxin (phospholipase C) of *Clostridium perfringens*. *Infect. Immun.* **57**, 367–76.

Titball, R.W., Fearn, A.M. and Williamson, E.D. (1993) Biochemical and immunological properties of the C-terminal domain of the alpha-toxin of *Clostridium perfringens*. *FEMS Microbiol. Lett.* **110**, 45–50.

Todd, E.W. (1934). A comparative serological study of streptolysins derived from human and from animal infections, with notes on pneumococcal haemolysin, tetanolysin and staphylococcus toxin. *J. Pathol. Bacteriol.* **39**, 299–321.

Tomita, M., Taguchi, R. and Ikezawa, H. (1982). Molecular properties and kinetic studies on sphingomyelinase of *Bacillus cereus*. *Biochim. Biophys. Acta* **704**, 90–9.

Tomita, M., Taguchi, R. and Ikezawa, H. (1983). Adsorption of sphingomyelinase of *Bacillus cereus* onto erythrocyte membranes. *Arch. Biochem. Biophys.* **223**, 202–12.

Tomita, M., Nakai, K., Yamada, A., Taguchi, R. and Ikezawa, H. (1990). Secondary structure of sphingomyelinase from *Bacillus cereus*. *J. Biochem. (Tokyo)* **108**, 811–15.

Tomita, M., Ueda, Y., Tamura, H., Taguchi, R. and Ikezawa, H. (1993). The role of acidic amino-acid residues in catalytic and adsorptive sites of *Bacillus cereus* sphingomyelinase. *Biochim. Biophys. Acta* **1203**, 85–92.

Turnbull, P.C. (1981). *Bacillus cereus* toxins. *Pharmacol. Ther.* **13**, 453–505.

Turnbull, P.C. and Kramer, J.M. (1983). Non-gastrointestinal *Bacillus cereus* infections: an analysis of exotoxin production by strains isolated over a two-year period. *J. Clin. Pathol.* **10**, 1091–6.

Turnbull, P.C., Jorgensen, K., Kramer, J.M., Gilbert, R.J. and Parry, J.M. (1979a). Severe clinical conditions associated with *Bacillus cereus* and the apparent involvement of exotoxins. *J. Clin. Pathol.* **32**, 289–93.

Turnbull, P.C., Kramer, J.M., Jorgensen, K., Gilbert, R.J. and Melling, J. (1979b). Properties and production characteristics of vomiting, diarrheal, and necrotizing toxins of *Bacillus cereus*. *Am. J. Clin. Nutr.* **32**, 219–28.

Verkleij, A.J., Zwaal, R.F., Roelofsen, B., Comfurius, P., Kastelijn, D. and van Deenen, L.L. (1973). The asymmetric distribution of phospholipids in the human red cell membrane. A combined study using phospholipases and freeze-etch electron microscopy. *Biochim. Biophys. Acta* **323**, 178–93.

Wazny, T.K., Mummaw, N. and Styrt, B. (1990). Degranulation of human neutrophils after exposure to bacterial phospholipase C. *Eur. J. Clin. Microbiol. Infect. Dis.* **9**, 830–2.

Williams, N.B., Forbes, M.A., Blau, E. and Eickenberg, C.F. (1950). A study of the simultaneous occurrence of enterococci, lactobacilli, and yeasts in saliva from human beings. *J. Dent. Res.* **29**, 563–70.

Yamada, A., Tsukagoshi, N., Udaka, S., Sasaki, T., Makino, S., Nakamura, S., Little, C., Tomita, M. and Ikezawa, H. (1988) Nucleotide sequence and expression in *Escherichia coli* of the gene coding for sphingomyelinase of *Bacillus cereus*. *Eur. J. Biochem.* **175**, 213–20.

Zhang, P., Liu, B., Jenkins, G.M., Hannun, Y.A. and Obeid, L.M. (1997). Expression of neutral sphingomyelinase identifies a distinct pool of sphingomyelin involved in apoptosis. *J. Biol. Chem.* **272**, 9609–12.

23

Clostridium septicum pore-forming and lethal α-toxin

Rodney K. Tweten and Brett R. Sellman

INTRODUCTION

Clostridium septicum was the first pathogenic anaerobic micro-organism to be discovered. It was isolated in 1877 by Pasteur and Joubert and termed *Vibrion septique* (Pasteur and Joubert, 1877), and was later classified as a member of the genus *Clostridium*. Several of the pathogenic species of the clostridia cause gas gangrene and are commonly referred to as the histotoxic clostridia. *Clostridium septicum* causes a form of gas gangrene in the absence of any external trauma, called atraumatic or non-traumatic gas gangrene, or non-traumatic distal myonecrosis (Jendrzejewski *et al.*, 1978; Kizer and Ogle, 1981; Stevens *et al.*, 1990). This infection differs from the traumatic gas gangrene in that it occurs in individuals in the absence of any external wound or trauma. This form of myonecrosis is a fulminant and rapidly fatal infection with a reported mortality rate ranging from 50% to 80% (Arenas *et al.*, 1988; Stevens *et al.*, 1990; Corey, 1991; Cheng *et al.*, 1997) even with aggressive medical intervention. *Clostridium septicum* non-traumatic distal myonecrosis usually occurs in individuals with predisposing conditions such as an intestinal malignancy, neutropenia, leukaemia or diabetes (Alpern and Dowell, 1969; Kornbluth *et al.*, 1989; Stevens *et al.*, 1990; Bar-Joseph *et al.*, 1997). In all cases, the organism is believed to gain entry to the bloodstream by way of an intestinal lesion.

Clostridium septicum α-toxin was originally described by Bernheimer in 1944 (Bernheimer, 1944); however, it was only recently purified to homogeneity and characterized by Ballard *et al.* (1992). It is the only lethal factor produced by *C. septicum* and has a reported LD_{50} of approximately 10 µg kg^{-1} in mice (Bernheimer, 1944; Smith, 1975; Ballard *et al.*, 1992). It is likely that α-toxin is the major cause of the toxaemia and subsequent shock observed in non-traumatic gas gangrene in patients since purified toxin can cause shock and death in mice (Ballard *et al.*, 1992). To date, no other clostridial species has been found to produce a toxin similar to *C. septicum* α-toxin. It was previously suspected that both *C. chauvoei* and *C. histolyticum* produced toxins similar to *C. septicum* α-toxin based on serological data, although neither was found to produce toxins *in vitro* that cross-reacted with antibody to *C. septicum* α-toxin (Ballard *et al.*, 1992).

CLASSIFICATION OF CLOSTRIDIUM SEPTICUM α-TOXIN

α-Toxin belongs to a family of toxins termed pore-forming toxins which disrupt the membrane permeability by the formation of unregulated channels in the membrane. Although there are several classes of pore-forming toxins, *C. septicum* α-toxin most closely resembles those toxins which are typified by the *Staphylococcus aureus* α-haemolysin. This family of toxins is characterized by the fact that they form

435

homo-oligomeric complexes on the membrane consisting of five to seven toxin monomers. Some examples include *S. aureus* α-haemolysin (Gouaux *et al.*, 1997), the cytotoxin (CTX) produced by *Pseudomonas aeruginosa* (Xiong *et al.*, 1994), *C. septicum* α-toxin (Ballard *et al.*, 1992, 1995; Sellman *et al.*, 1997), *Aeromonas hydrophila* aerolysin (Howard and Buckley, 1985) and possibly the plant protein enterolobin from *Enterolobium contortisiliquum* which exhibits some primary structural similarities with the aerolysins (Fontes *et al.*, 1997). Another group of toxins that at this time cannot be considered to form homo-oligomeric membrane complexes, but may also be related to this group of toxins by their mechanism of pore formation, are the two-component cytolytic toxins exemplified by the leucocidins and γ-haemolysin of *S. aureus* (Prevost *et al.*, 1995). *Staphylococcus aureus* α-haemolysin exhibits weak similarity in its primary structure with the F family of proteins from the leucocidin and γ-haemolysin (Prevost *et al.*, 1995) of *S. aureus* but is unrelated to aerolysin, CTX and *C. septicum* α-toxin. Aerolysin and *C. septicum* α-toxin form a unique subclass within this family of toxins since they exhibit significant similarity in their primary structures as well as their mechanisms which is discussed at length later in this chapter.

PURIFICATION OF α-TOXIN

α-Toxin was partially purified by Bernheimer in 1944 (Bernheimer, 1944), at which time he discovered that the lethal and haemolytic activities of *C. septicum* co-purified to the extent possible at that time. This finding was in contrast to earlier reports by Menke (1931) and Robertson (1929) that suggested that the haemolytic and lethal activities were the result of two separate proteins. Nearly 50 years later, Ballard *et al.* (1992) purified α-toxin to homogeneity and confirmed that Bernheimer's original observation was correct, that α-toxin was both haemolytic and lethal. The difficult purification of α-toxin from *C. septicum* described by Ballard *et al.* (1992) is no longer necessary since α-toxin was cloned and expressed in the periplasm of *Escherichia coli* (Ballard *et al.*, 1995). It is expressed with a carboxy-terminal polyhistidine tag, which does not affect its cytolytic function (Sellman *et al.*, 1997). The polyhistidine tag facilitates the affinity purification of the toxin on a metal chelate resin that is charged with Ni^{2+} or Co^{2+}. Although expression of the α-toxin gene in *E. coli* can be achieved with reasonable yields, once its expression is induced the *E. coli* cells quit growing within about an hour, suggesting that the toxin is probably killing the *E. coli*. Purified α-toxin (native or recom-

binant) exhibits a fairly low solubility in its protoxin form if concentrated above 3 mg ml^{-1}. Thus, when storing α-toxin its concentration is best kept at less than 2 mg ml^{-1}.

PRIMARY STRUCTURE OF α-TOXIN

The primary structure of α-toxin, as determined from the cloned gene sequence (Imagawa *et al.*, 1994; Ballard *et al.*, 1995), provided several intriguing insights into the evolution and function of α-toxin. The secreted protoxin exhibits a molecular mass of 46 450 Da and the proteolytically activated toxin has a molecular mass of approximately 41 000 Da. No other sequenced toxin from the clostridia exhibits sequence similarity with α-toxin and so to date α-toxin is unique among the clostridial lethal factors. *Clostridium chauvoei* is a close relative of *C. septicum* and causes a histotoxic disease in cattle, termed blackleg, similar to the non-traumatic gas gangrene caused by *C. septicum* in humans. These two organisms have often been mistakenly identified for one another and it has been determined that 99.3% of their genes are identical (Kuhnert *et al.*, 1996). However, a protein from *C. chauvoei* that was cross-reactive with α-toxin could not be identified using anti-α-toxin antibody (Ballard *et al.*, 1992) and polymerase chain reaction (PCR) analysis using α-toxin-specific primers did not generate a gene product from *C. chauvoei* chromosomal DNA (G. Songer, personal communication). Thus, it is likely that *C. chauvoei* does not produce a protein similar to α-toxin, even though the two organisms are closely related and produce similar disease syndromes.

As mentioned above, the primary structure of α-toxin exhibited sequence similarity with the primary structure of *A. hydrophila* aerolysin (Ballard *et al.*, 1995). The extent of this similarity was a remarkable 72%, with 27% identity. This was the first, and to date, the only example (that we know of) of toxins from a Gram-positive and Gram-negative bacterial species to show this extent of similarity in their primary structures. This observation suggests that these two species probably exchanged genetic material some time in the past. The similarities between these two toxins extend beyond just a simple sequence similarity, as will become evident in the subsequent sections of this chapter. The crystal structure of aerolysin has been solved (Parker *et al.*, 1994) and appears to be divided into one small lobe (domain 1) and one large lobe (domains 2–4). The similarity between α-toxin and aerolysin begins at D56 of α-toxin and D97 of aerolysin. What is evident from the alignment of these two sequences is that α-toxin is homologous with the large lobe of aerolysin and lacks the coding region for the small lobe. The function of the small lobe in aerolysin has not been completely elucidated. However, it appears that it may participate in receptor binding by aerolysin (Rossjohn et al., 1997).

The secreted form of α-toxin contains a single cysteine at position 86 whose function is unknown, but appears to be at a structurally sensitive site that does not tolerate changes (Tweten, unpublished observations). No apparent regions of hydrophobicity exist in the structure, but there are several regions which exhibit amphipathic characteristics and may be candidates for a membrane-spanning amphipathic β-sheet. The only other remarkable regions are the proteolytic activation site and propeptide itself, both of which are discussed in detail below.

α-TOXIN CELLULAR RECEPTORS

The first step in the cytolytic process is receptor binding by the protoxin. It has been shown that the related toxin, aerolysin, binds to glycosylphosphatidylinositol (GPI)-anchored membrane proteins (Cowell et al., 1997; Nelson et al., 1997) and that the GPI anchor itself is part of the binding determinant (Diep et al., 1998). Similarly, α-toxin also appears to bind to GPI-anchored membrane proteins, although in general α-toxin binds a different repertoire of GPI-anchored proteins than does aerolysin (Gordon et al., 1998). The difference in receptor specificity between these two toxins is not surprising, owing to the differences in the disease syndromes caused by each organism. Each toxin has probably evolved a receptor specificity that best fits the pathogenic traits of each bacterial species, since it is likely that each toxin may target different tissues in different hosts.

ACTIVATION OF α-TOXIN

α-Toxin is secreted by *C. septicum* as an inactive protoxin (Ballard et al., 1993). The protoxin can be activated by a variety of proteases *in vitro* which includes trypsin, chymotrypsin, subtilisin and proteinase K, with trypsin providing the best activation (Ballard et al., 1993). Trypsin was found to cleave the protein on the carboxy side of Arg398, thus generating a 367-residue, 41 327-Da activated toxin and a 45-residue, 5140-Da propeptide fragment. The site at which activation takes place is contained within the sequence of P_{388}LPD-KKRRGKR$_{398}$SVD. As is evident from the sequence in this region, it is rich in basic residues, which explains its *in vitro* susceptibility to trypsin and other serine proteases. More importantly, this region contains a furin consensus sequence of RGKR. The cleavage and activation of the protoxin has been determined to be carried out *in vivo* primarily by furin on the surface of eukaryotic cells, but there are clearly other cell surface proteases that recognize the furin consensus sequence which can also activate the protoxin (Gordon et al., 1997). When the furin recognition sequence was changed from RGKR to SGSR, by *in vitro* mutagenesis of the α-toxin gene, the mutated toxin was inactive on furin expressing CHO cells, but could still be activated *in vitro* by trypsin. Therefore, the furin consensus site is the primary, and possibly the only, site at which α-toxin can be activated *in vivo* by furin or other proteases.

THE PROPEPTIDE OF α-TOXIN, AN INTRAMOLECULAR CHAPERONE

Many eukaryotic and prokaryotic proteins are produced in an inactive state and require a post-translational modification for activation. These modifications can range from glycosylation in the eukaryotic endoplasmic reticulum (Strous, 1979), addition of a lipid moiety to the *E. coli* haemolysin (Wagner et al., 1988) to proteolytic cleavage. A number of both eukaryotic and prokaryotic proteins requires proteolytic cleavage because they are produced as pro-proteins (Eder and Fersht, 1995). This means that the proteins contain a pro-region (propeptide) attached to either the amino- and/or carboxy-terminus of the functional portion of the protein. Propeptides usually act in *cis* as intramolecular inhibitors of the protein on which they are present (Eder and Fersht, 1995). However, the affinity of the propeptide for its respective protein is frequently sufficient to act in *trans* as an intermolecular inhibitor, as is the case with propeptides from porcine carboxypeptidase *A*, *Pseudomonas aeruginosa* elastase, cathepsin B and cathepsin L (San Segundo et al., 1982; Fox et al., 1992; Kessler and Safrin, 1994; Carmona et al., 1996). For cathepsin L, Carmona et al. (1996) demonstrated that propeptides can be highly specific inhibitors for the protein with which they are associated.

Hendrick and Hartl (1993) have stated 'Presently, we define a molecular chaperone as a protein that binds to and stabilizes an otherwise unstable conformer...'. They go on to state that a chaperone 'facilitates its [the target protein] correct fate *in vivo*: be it folding, oligomeric assembly, transport to a particular subcellular compartment...'. By definition, the propeptide of α-toxin corresponds to an intramolecular chaperone since it clearly stabilizes an inherently unstable intermediate (activated α-toxin) until its final assembly into a multimeric complex (α-toxin pore) on the cell membrane (Sellman and Tweten, 1997). Sellman and Tweten (1997) found that when the covalent link between the toxin and propeptide was cleaved by protease activation the propeptide remained associated with the toxin. This observation indicated that the propeptide exhibited a reasonable degree of affinity for its site on the toxin, even after its peptide bond with the toxin was cleaved. Two sets of findings by Sellman and Tweten (1997) were critical to the definition of the propeptide as an intramolecular chaperone. First, proteolytic activation of the protoxin while it was in solution resulted in the aggregation of the toxin into non-functional complexes. Even though the remaining soluble activated toxin was fully active the majority of the toxin aggregated into inactive soluble and insoluble complexes. Aggregation of solution activated α-toxin could be minimized by activating very dilute solutions of α-toxin (<100 μg ml^{-1}), which presumably minimized intermolecular interactions that lead to the formation of the inactive complexes. In contrast, when α-toxin was activated after it bound to the membrane (its proper assembly site) it retained 10-fold more activity than when the protoxin was activated in solution. The difference in activity was due to the fact that toxin activated in solution quickly aggregated into an inactive complex in the absence of a membrane, whereas if it was activated after binding to the membrane it followed its proper assembly sequence into an oligomeric, pore-forming complex. Therefore, the propeptide 'stabilizes an otherwise unstable conformer' and 'facilitates its [α-toxin] correct fate *in vivo*' (Hendrick and Hartl, 1993).

The second set of findings that helped to confirm the chaperone role of the α-toxin propeptide was that: (a) the propeptide could not be part of the final complex since it was a potent inhibitor of the toxin when added in *trans*; and (b) even though it was unstable in solution propeptide-free α-toxin was cytolytically active (Sellman and Tweten, 1997). The former observation was consistent with the reports mentioned above that described the inhibitory activity of propeptides of various proteases. When added in molar excess (10–20-fold) over-activated α-toxin the purified propeptide of α-toxin was shown to inhibit the oligomerization step and thus inhibited cytolysis. Therefore, the propeptide inhibited the oligomerization of α-toxin when it was covalently bound, as in the protoxin, or when it was added in molar excess in *trans* to the activated toxin.

When does the propeptide dissociate from the toxin? The current hypothesis is that once the propeptide-toxin peptide bond is cleaved the propeptide is displaced by oligomerization of the toxin (Sellman and Tweten, 1997). One could envisage the process beginning with weak interactions between monomers that proceed to form strong interactions which, in the absence of a covalent link between the propeptide and the toxin, would displace the propeptide. Thus, prior to proteolytic activation only weak intermolecular interactions could occur since the propeptide could not be displaced. An alternative hypothesis is that membrane-bound protoxin might rapidly displace the propeptide once the protoxin is activated by cell-surface proteases such as furin. However, the membrane-bound toxin remained in a monomeric state in the presence of excess propeptide. Therefore, the receptor-bound toxin does not appear to displace the propeptide (Sellman and Tweten, 1997).

FORMATION OF THE PRE-PORE COMPLEX AND ITS MEMBRANE INSERTION

In order for α-toxin and other pore-forming toxins of this family to form channels in a membrane they must insert into the lipid bilayer of the membrane. The insertion event presents a problem since these toxins are predominantly hydrophilic proteins with no obvious transmembrane domains. The question therefore arises as to how these hydrophilic molecules interact with and insert into the hydrophobic core of the lipid bilayer. Recent advances in the study of the *S. aureus* α-haemolysin membrane oligomer crystal structure (Song *et al.*, 1996) coupled with a fluorescence-based analysis of its membrane-penetrating domain (Valeva *et al.*, 1996) confirmed a model originally proposed by Howard *et al.* (1987) for aerolysin. Howard *et al.* proposed that the interaction of these oligomerizing toxins may be similar to the amphipathic β-barrel utilized by porins which forms the transmembrane domain of these proteins in the outer membrane of Gram-negative bacteria (Cowan *et al.*, 1992). In contrast to a porin these toxins do not form a complete β-barrel until the toxin monomers oligomerize on the membrane. In this model each monomer within the oligomeric structure would form a stave in the β-barrel, thereby contributing only a fraction of the overall hydrophobic character to the transmembrane β-barrel. The hydrophobic amino acid side-chains of each β-strand would be located on the outside of the β-barrel and would interact with the lipid bilayer, while the alternating hydrophilic amino acid side chains would line the fluid-filled channel. Thus, the monomers could retain their hydrophilic qualities prior to their interaction with the membrane, but would still be capable of forming a membrane-penetrating complex.

The size of the oligomeric α-toxin complex has not yet been characterized in detail but is likely to be comprised of six or seven monomers based on the migration of the oligomer on denaturing gels (Ballard et al., 1993) and the fact that the related toxin, aerolysin, has been shown to form heptameric complexes on the membrane (Wilmsen et al., 1992). Interestingly, the crystal structure of the oligomeric pore complex of S. aureus α-haemolysin exhibited a heptameric structure (Song et al., 1996), while atomic force microscopy analysis of membrane-bound complexes (Fang et al., 1997; Czajkowsky et al., 1998) indicated that S. aureus α-haemolysin could form a hexamer or a heptamer. These data suggest that perhaps S. aureus α-haemolysin, as well as other toxins in this family, such as the C. septicum α-toxin, can form more than one size of oligomer.

How does this β-barrel form and when does it insert into the membrane? The available data (Van der Goot et al., 1993; Valeva et al., 1996; Fang et al., 1997; Sellman et al., 1997) suggest that oligomerization of the toxin monomers into a pre-pore complex occurs prior to the insertion of the β-barrel into the membrane. Clostridium septicum α-toxin has been shown to form a pre-pore complex in which the toxin monomers oligomerized prior to membrane insertion (Sellman et al., 1997). The present authors were able to take advantage of an observation in the laboratory of the fact that C. septicum α-toxin was haemolytically inactive at 4°C. It was determined that at 4°C membrane binding and oligomerization of α-toxin were unaffected and so a subsequent step in the mechanism must be blocked at low temperature. This step turned out to be insertion of the oligomerized complex (Sellman et al., 1997). It was determined that channels did not form either on erythrocytes or in artificial lipid bilayers in a planar membrane system at 4°C. In addition, low temperature could not close previously opened pores in osmotically stabilized erythrocytes. These studies suggested that the oligomer remained in a pre-insertion state at low temperature and only inserted after the temperature block was removed. The conversion of the pre-pore complex of α-toxin to a fully inserted, pore-forming complex occurred rapidly when the temperature was increased to 25°C, although the precise temperature at which insertion could proceed was not determined. Why low temperature blocked the insertion of the oligomer is currently unknown. It is unlikely that a thermal transition in the lipid bilayer was responsible for blocking insertion, at least in the erythrocytes, since

the presence of cholesterol typically abolishes the lipid thermal phase transitions in membranes (Keough and Davis, 1984).

The physical characteristics of the α-toxin pore are fairly typical for this class of toxins. The pore exhibited a weak anion selectivity and was estimated to be 1–2 nm in diameter (Ballard et al., 1993). Unlike aerolysin, the α-toxin pore was not affected by the addition of Zn^{2+}, which is often found to close the channels of these toxins.

THE ROLE OF α-TOXIN IN PATHOGENESIS

The involvement of α-toxin in C. septicum pathogenesis has not been proven directly by gene knockout experiments as has been done recently for C. perfringens θ-toxin (perfringolysin O) and α-toxin (sphingomyelinase/phospholipase C) (Awad et al., 1995) since genetic transformation has not been achieved in C. septicum. However, some studies and observations suggest that α-toxin is the main lethal factor in C. septicum disease. Ballard et al. (1992) demonstrated that immunization of mice with α-toxin affords partial protection when the mice are challenged with an LD_{100} of viable C. septicum. Also, during the intial purification of α-toxin, the various fractions from the crude culture supernatant from C. septicum BX96 were assayed for both haemolytic activity on human erythrocytes and lethality in mice. α-Toxin was the only haemolytic and/or lethal factor detected during its purification (Ballard et al., 1992), which was consistent with the much earlier report by Bernheimer in which he suggested that the two were the same (Bernheimer, 1944). Mice exhibit shock-like effects within minutes of intravenous administration of pure α-toxin (Ballard et al., 1992). These observations are consistent with the description of the events in human non-traumatic gas gangrene cases (Stevens et al., 1990) in which a severe toxaemia and shock occur within a few hours of diagnosis of the disease. In addition, in survivors of non-traumatic gas gangrene, α-toxin was the immunodominant extracellular antigen (Johnson et al., 1994) and oligomerized α-toxin was detected in the blister fluid from these patients (Tweten, unpublished observations). Thus, it is likely that α-toxin plays a major role in the development of the more serious aspects of non-traumatic gas gangrene, particularly the shock.

One aspect of α-toxin needs to be clarified when addressing its role in pathogenesis, that of its haemolytic characteristic. That it has been described as a haemolytic toxin does not necessarily mean that it causes haemolysis *in vivo*. Cytolytic toxins are often assayed for their cytolytic activity on erythrocytes simply because erythrocytes serve as a convenient substrate for routine analysis of cytolytic activity. In the case of α-toxin the human erythrocyte serves as a convenient method to assay its cytolytic activity; however, it is necessary to add trypsin to the assay mixture to activate the toxin (Sellman *et al.*, 1997). Although human erythrocytes can bind protoxin they do not efficiently activate it, probably because they lack an activating protease. Thus, the specificity of α-toxin for a cell is not only tied to the presence of a receptor, but also requires the presence of an activating protease. Also, cases of *C. septicum* non-traumatic gas gangrene in which intravascular haemolysis was reported as an associated clinical syndrome are infrequent (no reported cases of non-traumatic gas gangrene-associated haemolysis were found in a search of case studies in the National Library of Medicine).

CONCLUDING REMARKS AND FUTURE DIRECTIONS

When one considers the information available for α-toxin *in toto*, a reasonable schematic can be presented which describes the transition of α-toxin from the *C. septicum* cell, where it is secreted, to its ultimate destination as a oligomeric, pore-forming complex on a target membrane (Figure 23.1). It is from the point of secretion by the bacterial cell to the initial interaction with the membrane that the propeptide functions as an intramolecular chaperone to protect the toxin from premature oligomerization and inactivation. Once the toxin encounters a target cell of the host it binds to its receptor(s). In Figure 23.1 the receptor is shown to be associated with the toxin until the oligomer is formed. Whether or not the toxin dissociates from the receptor during some stage in the mechanism is currently unknown for *C. septicum* α-toxin as well as for other oligomerizing, pore-forming toxins. It has been shown that a purified system of lipid and receptor can be reconstituted for aerolysin (Gruber *et al.*, 1994), thus the possibility is open to examine the receptor–toxin interaction for aerolysin, as well as for α-toxin. The next step, the oligomerization of α-toxin into the pre-pore complex, is the penultimate step to insertion of the amphipathic β-barrel. The regions of α-toxin that are involved in the formation of the β-barrel have not yet been defined but there are several candidate regions for amphipathic β-strands in its primary structure.

Why is α-toxin produced as a protoxin? It is clear that propeptide plays an essential role as an intramolecular chaperone that stabilizes the protoxin until it reaches the membrane. This stabilization may have a more practical aspect with regard to *C. septicum* pathogenesis. It is possible that the protoxin structure extends the half-life of the toxin so that it can be disseminated throughout the body from the site of the infection, which is typically localized to muscle tissue. Indirect studies of the action of purified α-toxin *in vivo* suggest that it is responsible for the major clinical syndrome, shock, of non-traumatic gas gangrene. It is not known whether α-toxin also contributes to the massive tissue necrosis which occurs in the muscle during gas gangrene. Since it is a cytolytic toxin it is not unreasonable

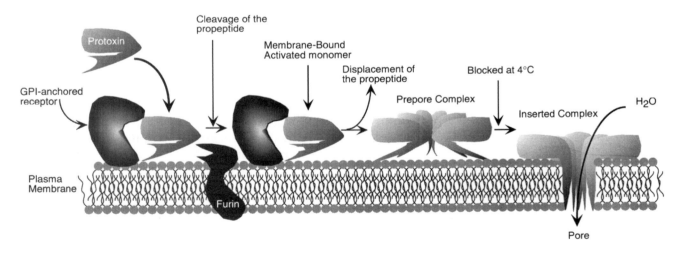

FIGURE 23.1 Proposed steps in the cytolytic mechanism of *C. septicum* α-toxin.

to assume that it does. The full spectrum of its contribution to pathogenesis cannot be assessed until isogenic isolates of *C. septicum* which only differ in the production of α-toxin can be generated and tested in the mouse model (Ballard *et al.*, 1992).

The study of the α-toxin of *C. septicum* has provided several intriguing results which have extended our knowledge about proteolytically activated cytolytic proteins as well as some surprises with respect to the origin of the α-toxin gene. In the future, we expect that the study of α-toxin will help to reveal other secrets of this family of toxins.

REFERENCES

Alpern, R. and Dowell, V. (1969). *Clostridium septicum* infections and malignancy. *JAMA* **209**, 385–8.

Arenas, R., Seaman, D., McLaughlin, C. and Ciardiello, K. (1988). *Clostridium septicum* myonecrosis in association with colonic malignancy. *Connect. Med.* **52**, 709–10.

Awad, M.M., Bryant, A.E., Stevens, D.L. and Rood, J.I. (1995). Virulence studies on chromosomal α-toxin and θ-toxin mutants constructed by allelic exchange provide genetic evidence for the essential role of α-toxin in *Clostridium perfringens*-mediated gas gangrene. *Molec. Microbiol.* **15**, 191–202.

Ballard, J., Bryant, A., Stevens, D. and Tweten, R.K. (1992). Purification and characterization of the lethal toxin (alpha-toxin) of *Clostridium septicum*. *Infect. Immun.* **60**, 784–90.

Ballard, J., Sokolov, Y., Yuan, W.-L., Kagan, B.L. and Tweten, R.K. (1993). Activation and mechanism of *Clostridium septicum* alpha toxin. *Molec. Microbiol.* **10**, 627–34.

Ballard, J., Crabtree, J., Roe, B.A. and Tweten, R.K. (1995). The primary structure of *Clostridium septicum* alpha-toxin exhibits similarity with *Aeromonas hydrophila* aerolysin. *Infect. Immun.* **63**, 340–4.

Bar-Joseph, G., Halberthal, M., Sweed, Y., Bialik, V., Shoshani, O. and Etzioni, A. (1997). *Clostridium septicum* infection in children with cyclic neutropenia. *J. Pediat.* **131**, 317–19.

Bernheimer, A.W. (1944). Parallelism in the lethal and haemolytic activity of the toxin of *Clostridium septicum*. *J. Exp. Med.* **80**, 309–20.

Carmona, E., Dufuor, E., Plouffe, C., Takebe, S., Mason, P., Mort, J.S. and Menard, R. (1996). Potency and selectivity of the cathepsin L propeptide as an inhibitor of cysteine proteases. *Biochemistry* **35**, 8149–57.

Cheng, Y., Huang, C., Leu, H., Chen, J. and Kiu, M. (1997). Central nervous system infection due to *Clostridium septicum*. A case report and review of the literature. *Infection* **25**, 171–4.

Corey, E. (1991). Non-traumatic gas gangrene: case report and review of emergency therapeutics. *J. Emerg. Med.* **9**, 431–6.

Cowan, S.W., Schirmer, T., Rummel, G., Steiert, S., Ghosh, R., Paupit, R.A., Jansonius, J.N. and Rosenbusch, J.P. (1992). Crystal structures explain functional properties of two *Escherichia coli* porins. *Nature* **358**, 727–33.

Cowell, S., Aschauer, W., Gruber, H.J., Nelson, K.L. and Buckley, J.T. (1997). The erythrocyte receptor for the channel-forming toxin aerolysin is a novel glycosylphosphatidylinositol-anchored protein. *Mol. Microbiol.* **25**, 343–50.

Czajkowsky, D.M., Sheng, S. and Shao, Z. (1998). Staphylococcal alpha-haemolysin can form hexamers in phospholipid bilayers. *J. Mol. Biol.* **276**, 325–30.

Diep, D.B., Nelson, K.L., Raja, S.M., Pleshak, E.N. and Buckley, J.T. (1998). Glycosylphosphatidylinositol anchors of membrane glycoprotaeins are binding determinants for the channel-forming toxin aerolysin. *J. Biol. Chem.* **273**, 2355–60.

Eder, J. and Fersht, A.R. (1995). Pro-sequence-assisted protein folding. *Molec. Microbiol.* **16**, 609–14.

Fang, Y., Cheley, S., Bayley, H. and Yang, J. (1997). The heptameric pre-pore of a staphylococcal alpha-haemolysin mutant in lipid bilayers imaged by atomic force microscopy. *Biochemistry* **36**, 9518–22.

Fontes, W., Sousa, M.V., Aragao, J.B. and Morhy, L. (1997). Determination of the amino acid sequence of the plant cytolysin enterolobin. *Arch. Biochem. Biophys.* **347**, 201–7.

Fox, T., de Miguel, E., Mort, J.S. and Storer, A.C. (1992). Potent slow-binding of cathepsin B by its propeptide. *Biochemistry* **31**, 12571–6.

Gordon, V.M., Benz, R., Fujii, K., Leppla, S.H. and Tweten, R.K. (1997). *Clostridium septicum* alpha toxin is proteolytically activated by furin. *Infect. Immun.* **65**, 4130–4.

Gordon, V.M., Nelson, K.L., Buckley, J.T., Stevens, V.L., Tweten, R.K., Elwood, P.C. and Leppla, S.H. (1999). *Clostridium septicum* alpha-toxin uses GPI-anchored proteins receptors. *J. Biol. Chem.* submitted.

Gouaux, E., Hobaugh, M. and Song, L. (1997). Alpha-Haemolysin, gamma-haemolysin, and leukocidin from *Staphylococcus aureus*, distant in sequence but similar in structure. *Protein Sci.* **6**, 2631–5.

Gruber, H.J., Wilmsen, H.U., Cowell, S., Schindler, H. and Buckley, J.T. (1994). Partial purification of the rat erythrocyte receptor for the channel-forming toxin aerolysin and reconstitution into planar lipid bilayers. *Mol. Microbiol.* **14**, 1093–101.

Hendrick, J.P. and Hartl, F.U. (1993). Molecular chaperone functions of heat shock proteins. In: *Annual Review of Biochemistry*, Vol. 62 (eds C.C. Richardson, J.N. Abelson, A. Meistaer and C.T. Walsh), pp. 349–84. Annual Reviews, Palo Alto, CA.

Howard, S.P. and Buckley, J.T. (1985). Activation of the hole-forming toxin aerolysin by extracellular processing. *J. Bacteriol.* **163**, 336–40.

Howard, S.P., Garland, W.J., Green, M.J. and Buckley, J.T. (1987). Nucleotide sequence of the gene for the hole-forming toxin aerolysin of *Aeromonas hydrophila*. *J. Bacteriol.* **169**, 2869–71.

Imagawa, T., Dohi, Y. and Higashi, Y. (1994). Cloning nucleotide sequence and expression of a haemolysin gene of *Clostridium septicum*. *FEMS Microbiol. Lett.* **117**, 287–92.

Jendrzejewski, J., Jones, S., Newcombe, R. and Gilbert, D. (1978). Non-traumatic clostridial myonecrosis. *Am. J. Med.* **65**, 542–5.

Johnson, S., Driks, M.R., Tweten, R.K., Ballard, J., Stevens, D.L., Anderson, D.J. and Janoff, E.N. (1994). Clinical courses of seven survivors of *Clostridium septicum* infection and their immunological responses to α-toxin. *Clin. Infect. Dis.* **19**, 761–4.

Keough, K.M.W. and Davis, P.J. (1984). Thermal analysis of membranes. In: *Membrane Fluidity*, Vol. 12 (eds M. Kates and L.A. Manson), pp. 55–97. Plenum Press, New York.

Kessler, E. and Safrin, M. (1994). The propeptide of *Pseudomonas aeruginosa* elastase acts as an elastase inhibitor. *J. Biol. Chem.* **269**, 22726–31.

Kizer, K. and Ogle, L. (1981). Occult clostridial myonecrosis. *Ann. Emerg. Med.* **10**, 307–11.

Kornbluth, A., Danzig, J. and Bernstein, L. (1989). *Clostridium septicum* infection associated with malignancy. *Medicine* **68**, 30–7.

Kuhnert, P., Capaul, S.E., Nicolet, J. and Frey, J. (1996). Phylogenetic positions of *Clostridium chauvoei* and *Clostridium septicum* based on 16S rRNA gene sequences. *Int. J. Syst. Bact.* **46**, 1174–6.

Menke, W. (1931). Zur Wirkung des Toxin des novyschen Bazillus. *Znt. Bakteriol. Parasitenk. Abt.* **123**, 49.

Nelson, K.L., Raja, S.M. and Buckley, J.T. (1997). The glycosylphosphatidylinositol-anchored surface glycoprotein Thy-1 is a receptor for the channel-forming toxin aerolysin. *J. Biol. Chem.* **272**, 12170–4.

Parker, M.W., Buckley, J.T., Postma, J.P.M., Tucker, A.D., Leonard, K., Pattus, F. and Tsernoglou. D. (1994). Structure of the *Aeromonas* toxin proaerolysin in its water-soluble and membrane-channel states. *Nature* **367**, 292–5.

Pasteur, L. and Joubert, P.A. (1877). Charbon et saepticémie. *Bull. Acad. Med.* **6**, 781.

Prevost, G., Cribier, B., Couppie, P., Petiau, P., Supersac, G., Finck-Barbançon, V., Monteil, H. and Piemont, Y. (1995). Panton–Valentine leucocidin and gamma-haemolysin from *Staphylococcus aureus* ATCC 49775 are encoded by distinct genetic loci and have different biological activities. *Infect. Immun.* **63**, 4121–9.

Robertson, M. (1929). The organisms associated with gas gangrene. Medical Research Council System. *Bacteriology* **3**, 225.

Rossjohn, J., Buckley, J.T., Hazes, B., Murzin, A.G., Read, R.J. and Parker, M.W. (1997). Aerolysin and pertussis toxin share a common receptor-binding domain. *EMBO J.* **16**, 3426–34.

San Segundo, B., Martinez, M.C., Vilanova, M., Cuchillo, C.M. and Aviles, F.X. (1982). The severed activation segment of porcine pancreatic procarboxypeptidase A is a powerful inhibitor of the active enzyme, isolation and characterization of the activation peptide. *Biochim. Biophys. Acta* **707**, 74–80.

Sellman, B.R. and Tweten, R.K. (1997). The propeptide of *Clostridium septicum* alpha toxin functions as an intramolecular chaperone and is a potent inhibitor of alpha toxin-dependent cytolysis. *Molec. Microbiol.* **25**, 429–40.

Sellman, B.R., Kagan, B.L. and Tweten, R.K. (1997). Generation of a membrane-bound, oligomerized pre-pore complex is necessary for pore formation by *Clostridium septicum* alpha toxin. *Molec. Microbiol.* **23**, 551–8.

Smith, L.D. (1975). *The Pathogenic Anaerobic Bacteria*. Charles C. Thomas, Springfield, IL.

Song, L.Z., Hobaugh, M.R., Shustak, C., Cheley, S., Bayley H. and Gouaux, J.E. (1996). Structure of staphylococcal alpha-haemolysin, a heptameric transmembrane pore. *Science* **274**, 1859–66.

Stevens, D.L., Musher, D.M., Watson, D.A., Eddy, H., Hamill, R.J., Gyorkey, F., Rosen, H. and Mader, J. (1990). Spontaneous, non-traumatic gangrene due to *Clostridium septicum*. *Rev. Infect. Dis.* **12**, 286-96.

Strous, G.J. (1979). Initial glycosylation of proteins with acetyl-galactosaminylserine linkages. *Proc. Natl Acad. Sci. USA* **76**, 2694–8.

Valeva, A., Weisser, A., Walker, B., Kehoe, M., Bayley, H., Bhakdi, S. and Palmer, M. (1996). Molecular architecture of a toxin pore: a 15-residue sequence lines the transmembrane channel of staphylococcal alpha-toxin. *EMBO J.* **15**, 1857–64.

Van der Goot, F.G., Pattus, F., Wong, K.R. and Buckley, J.T. (1993). Oligomerization of the channel-forming toxin aerolysin precedes insertion into lipid bilayers. *Biochemistry* **21**, 2636–42.

Wagner, W., Kuhn, M. and Goebel, W. (1988). Active and inactive forms of haemolysin (HlyA) from *Escherichia coli*. *Biol. Chem. Hoppe-Seyler* **369**, 39.

Wilmsen, H.U., Leonard, K.R., Tichelaar, W., Buckley, J.T. and Pattus, F. (1992). The aerolysin membrane channel is formed by heptamerization of the monomer. *EMBO J.* **11**, 2457–63.

Xiong, G., Struckmeier, M. and Lutz, F. (1994). Pore-forming *Pseudomonas aeruginosa* cytotoxin. *Toxicology* **87**, 69–83.

Introduction to the family of the structurally related cholesterol-binding cytolysins ('sulfhydryl-activated' toxins)

Joseph E. Alouf

INTRODUCTION

The so-called sulfhydryl (SH)-activated (also formerly known as oxygen-labile) toxins are a family of 50- to 60-kDa single-chain amphipathic bacterial proteins produced by 23 taxonomically different species of Gram-positive (aerobic or anaerobic, sporulating or non-sporulating) bacteria from the genera *Streptococcus* (six species), *Bacillus* (four species), *Clostridium* (nine species), *Listeria* (three species) and *Arcanobacterium* (Table 24.1).

These toxins, at least those investigated in detail over many years, are known to share the following features: (a) lethal (cardiotoxic) to animals (and probably humans); (b) highly potent lytic agents towards eukaryotic cells including erythrocytes (hence the name 'haemolysins' often formerly used for these toxins); (c) lytic and lethal properties are suppressed by sulfhydryl-group blocking agents and restored by thiols or other reducing agents; (d) these properties are irreversibly abrogated by very low (nanomolar) concentrations of cholesterol and other related 3β-hydroxysterols; (e) membrane cholesterol is thought to be the toxin-binding site at the surface of eukaryotic cells; (f) the toxins are antigenically related (common epitopes) as evidenced by the elicitation in humans or immunized animals of neutralizing and precipitating cross-reacting antibodies; (g) toxin molecules bind as monomers to the membrane surface with subsequent oligomerization into large arc- and ring-shaped structures surrounding large pores generated by this process; (h) the 13 structural genes of the toxins cloned and sequenced to date (Table 24.1) are all chromosomal. The primary structure of the proteins deduced from the nucleotide sequence of the encoding genes shows obvious sequence homology and a common characteristic consensus 11-amino-acid sequence containing a Cys residue near the *C*-terminus of the molecule, critical for biological activity. However, some variations have been observed in the case of the recently discovered toxins intermedilysin and pyolysin from *Streptococcus intermedius* (Nagamune *et al.*, 1996) and *Arcanobacterium* (*Actinomyces*) *pyogenes* (Billington *et al.*, 1997), respectively.

No enzymic activity of this family of toxins has been identified.

Apart from pneumolysin, which is an intracytoplasmic toxin (Johnson, 1977), all of the other toxins are secreted to the extracellular medium during bacterial growth. In contrast, among the species producing this group of toxins, only *L. monocytogenes*, *L. ivanovii* (but not the apathogenic *L. seeligeri*) are intracellular pathogens which grow and release their toxins in the phagocytic cells of the host. Interestingly the 80-kDa cholesterol-inhibitable membrane-damaging metridiolysin produced by the eukaryotic organism *Metridium senile* (sea anemone) shows important similarities with the bacterial toxins reviewed here, particularly in the formation of arc- and ring-shaped structures on lysed cell membranes (Bernheimer *et al.*, 1979).

TABLE 24.1 The family of the cholesterol-binding 'sulfhydryl-activated' cytolytic toxins

Bacterial genus	Species	Toxin name[a]	Gene acronym[a]
Streptococcus	S. pyogenes	Streptolysin O (SLO)	slo
	S. equisimilis[b]	Streptolysin O (SLO)	slo
	S. canis[c]	Streptolysin O (SLO)[d]	slo
	S. pneumoniae	Pneumolysin (PLY)	ply
	S. suis	Suilysin (SLY)	sly
	S. intermedius	Intermedilysin (ILY)	ily
Bacillus	B. cereus	Cereolysin O (CLO)	clo
	B. alvei	Alveolysin (ALV)	alv
	B. thuringiensis	Thuringiolysin O (TLO)	tlo
	B. laterosporus	Laterosporolysin (LSL)	lsl
Clostridium	C. tetani	Tetanolysin (TLY)	tly
	C. botulinum	Botulinolysin (BLY)	bly
	C. perfringens	Perfringolysin O (PFO)	pfo
	C. septicum	Septicolysin O (SPL)	spl
	C. histolyticum	Histolyticolysin O (HLO)	hlo
	C. novyi A (oedematiens)	Novyilysin (NVL)	nvl
	C. chauvoei	Chauveolysin (CVL)	cvl
	C. bifermentans	Bifermentolysin (BFL)	bfl
	C. sordellii	Sordellilysin (SDL)	sdl
Listeria	L. monocytogenes	Listeriolysin O (LLO)	llo
	L. ivanovii	Ivanolysin (ILO)	ilo
	L. seeligeri	Seeligerolysin (LSO)	lso
Arcanobacterium (Actinomyces)	A. pyogenes	Pyolysin (PLO)	plo

[a]The abbreviation of toxin names (in parentheses) and gene acronyms are those reported in the literature or suggested by the author of the chapter.
[b]Group C streptococcus; [c]group G streptococcus; [d]also called canilysin.
Underlined: toxins purified to apparent homogeneity and cloned and sequenced genes.

The family of cholesterol-binding cytolysins (SH-activated toxins) has been described in various reviews (Bernheimer, 1976; Alouf, 1977; Smyth and Duncan, 1978; Bernheimer and Rudy, 1986; Bhakdi and Tranum-Jensen, 1988; Alouf and Geoffroy, 1991; Braun and Focareta, 1991; Morgan et al., 1996). This introductory chapter will briefly summarize the general features of this group of cytolysins and will deal more specifically with the newly discovered members of the family. The recent achievements in the biochemistry, molecular genetics and cell-damaging properties of the toxins are also briefly reviewed. The reader is referred to the following specific chapters in this volume concerning streptolysin O, perfringolysin O, pneumolysin and related toxins from Listeria species (Chapters 25–28).

THE PROBLEMS OF TOXIN NOMENCLATURE

The toxins of the family described here are known in the literature under various denominations: oxygen-labile, (oxygen-sensitive), thiol- or sulfhydryl-activated (or -dependent) toxins, lysins, cytolysins or haemolysins ('haemotoxins' in earlier publications). On the basis of

our present knowledge, most of these denominations are partially inappropriate and somewhat misleading. For example, the term haemolysin has been and is still often used because toxin identification or detection is in most cases based on the lysis of red blood cells. However, the toxins are not only lytic to erythrocytes but lyse or damage many other eukaryotic cells (all cells so far tested). Therefore, as suggested by Bernheimer and Rudy (1986), the term cytolysin, cytolytic toxin or membrane-damaging toxin should be used instead of the restrictive term haemolysin.

Oxygen-labile/oxygen-sensitive toxins

These denominations, which were formerly used, stem from early observations that the haemolytic activities of crude or partially purified preparations of many toxins of the group (pneumolysin, tetanolysin, perfringolysin O, streptolysin O) progressively disappeared on standing in air and less rapidly in the absence of oxygen, but were immediately restored to the original level (or to higher levels) by reducing agents (see Neill, 1926a–c; Neill and Mallory, 1926; Neill and Fleming, 1927; Todd, 1934; Smythe and Harris, 1940; Herbert and Todd, 1941). The loss of the lytic activity was attributed (without direct chemical proof) to the oxidation of toxin molecules by O_2. However, this hypothesis was not supported by the finding that highly purified preparations of streptolysin O (SLO), alveolysin (ALV) and perfringolysin O (PFO) did not lose their haemolytic activities by flushing toxin solutions with air or O_2 for several hours (see Alouf, 1980; Sato, 1986). The loss observed in crude preparations is probably due to a kinetically slow O_2-favoured blockade of toxin SH-group by undefined molecules from culture media or bacterial metabolism. Therefore, the term oxygen-labile should be abandoned.

Sulfhydryl-activated toxins

This denomination or its equivalents (thiol-activated/thiol-dependent toxins) are based on the observations that various thiols and other reducing agents restore the lytic and lethal activities of the toxins previously suppressed by interaction with sulfhydryl-group blocking reagents (see Alouf and Geoffroy, 1991, for a review and Chapter 25). The presence of a Cys residue deduced from the nucleotide sequences of the structural genes of SLO, PFO, ALV, pneumolysin (PLY), listeriolysin O (LLO) and, recently, ivanolysin (ILO) and pyolysin (PLO) appeared to support the concept of a functionally 'essential' sulfhydryl group postulated over three decades (Alouf, 1980; Walker et al., 1987; Alouf and Geoffroy, 1991). However, this contention was disproved by the finding that the replacement, by site-directed mutagenesis, of the Cys residue in SLO, PLY, LLO and PFO by other amino acids generated haemolytically active recombinant molecules (Pinkney et al., 1989). Moreover, the two recently investigated toxins, pyolysin (Billington et al. 1997) and intermedilysin (Nagamune, personal communication), do not contain the cysteine residue common to the consensus undecapeptide (the so-called cysteine motif) present in the other toxins of the group, despite sharing antigenic relatedness and susceptibility to inhibition of their lytic activity by cholesterol. On these grounds, the widely used denomination 'SH-activated toxins' also appears inappropriate.

Proposal for a new generic denomination

A generic name reflecting a common, specific, unchallengeable intrinsic property of these membrane-damaging toxins should be used. In a previous edition of this book, Alouf and Geoffroy (1991) proposed the name 'cholesterol-binding cytolysins (or cytolytic toxins)', abbreviated CBC. The term 'antigenically related' could be added when necessary for differentiation from other possible cholesterol-inhibitable cytolysins such as

Bulkholderia pseudomallei haemolysin (Ashdown and Koehler, 1990). The specific names of each member of the toxin family (Table 24.1) coined by or according to Bernheimer (1976) should be maintained, including the appended letter O (reflecting the supposed O_2-lability) due to its wide usage and for differentiation from other homonymic toxins.

CHARACTERIZATION AND PURIFICATION

Among the toxins listed in Table 24.1, only 11 (those underlined) have been purified to apparent homogeneity to allow a full description of their properties. The other toxins of the group have been largely neglected since their discovery (see Oakley et al., 1947; Guillaumie, 1950; Moussa, 1958; Bernheimer and Grushoff, 1967a; Rutter and Collee, 1969; Hatheway, 1990). Since they are still poorly characterized, these toxins will not be considered here. The purification procedures, characterization, physicochemical, molecular and biological properties of streptolysin O, pneumolysin, perfringolysin O and the listeriolysins are described in separate chapters. Here the main features of other toxins of the group are reviewed.

Cereolysin O (CLO)

The haemolytic activity of Bacillus cereus cultures was known by 1930 and further work pointed to the production of more than one lytic protein. Presently, three different extracellular cytolysins are known: cereolysin first characterized named by Bernheimer and Grushoff (1967b) as an SLO-like cholesterol-inhibitable toxin; cereolysin AB, a bifactorial enzymic component; and another bicomponent haemolysin designated BL. The last two lytic proteins (see Chapter 22) are unrelated to cereolysin and to the other toxins reviewed here. On the basis of the nomenclature proposed by Bernheimer (1976) the author proposes the term cereolysin O (CLO) for its differentiation from the other cereolysins. CLO purification to homogeneity was reported by Cowell et al. (1976). A 'secondary' culture supernatant resulting from a bacterial suspension obtained in a preliminary culture was precipitated by ammonium sulfate and the solubilized material obtained was submitted to isoelectric focusing and to gel filtration. The purified CLO (specific activity ca. $4-5 \times 10^5$ HU mg^{-1} of protein) was characterized as a single-chain polypeptide with a pI of 6.5–6.6 in the 'reduced' form and a relative molecular mass of ca. 55 000 as found by sodium dodecyl sulfate–polyacrylamide gel electrophoresis (SDS–PAGE) and gel filtration. Amino acid analysis showed that the toxin contained 518 residues. The calculated molecular weight (M_r) was 55 636. The *clo* gene was cloned and expressed in *Escherichia coli* and *Bacillus subtilis* (Kreft *et al.*, 1983). A partial nucleotide sequence of *clo* and the deduced amino acid sequence of the toxin has been reported (Yutsudo, 1994). The sequence comprised the undecapeptide containing the Cys residue common to almost all CBCs (Figure 24.1). Toxin interaction with cholesterol and membranes has been widely investigated by Bernheimer and his co-workers (Shany *et al.*, 1974; Cowell and Bernheimer, 1978; Cowell *et al.*, 1978).

Alveolysin (ALV)

This name has been attributed by Bernheimer (1976) to the 'SH-activated' toxin produced by *Bacillus alvei* as first reported by Bernheimer and Grushoff (1967a). No further characterization was done until the systematic study of this toxin by Alouf and his co-workers. *Bacillus alvei* is a soil organism which is often found in honey-bee larvae suffering from the so-called European foul brood. It is not considered as pathogenic in vertebrates. However, rare cases of human opportunistic infection have been reported to date (Reboli *et al.*, 1989; Coudron *et al.*, 1991). Alveolysin was purified to homogeneity by Alouf *et al.* (1977) by ammonium sulfate precipitation, gel filtration, isoelectric focusing and ion-exchange chromatography. Toxin production was enhanced by *ca.* 10-fold (2×10^4 HU ml^{-1}) in *B. alvei* shaken cultures in the presence of activated charcoal and purified by a method based on thiol-disulfide exchange chromatography (Geoffroy and Alouf, 1983). The purified toxin (specific activity 10^6 HU mg^{-1} of protein) had an apparent relative molecular mass of 63 000 and a p*I* of 5.1. The sequence of the *N*-terminal domain (Edman degradation) determined by Alouf *et al.* (1986) was in agreement (except for residue 24) with that of the primary structure of the toxin deduced later from the nucleotide sequence of the gene (Geoffroy and Alouf, 1988; Geoffroy *et al.*, 1990). The mature form com-

prised 469 amino acid residues (51 766 Da) and contained a unique Cys residue in the consensus undecapeptide. The membrane-damaging activities of alveolysin were investigated on cultured human diploid lung fibroblasts (Thelestam *et al.*, 1981). Toxin interaction with thiol group reagents, particularly tosyl lysine chloromethyl ketone (TLCK) investigated with iodine-labelled toxin (Geoffroy and Alouf, 1982) prevented binding to erythrocytes and thereby haemolysis. Alveolysin was also used to permeabilize Raji cells and purified human peripheral blood lymphocytes. This procedure allowed the investigation of the phosphoinositide pathway in these cells (Berthou *et al.*, 1992).

Thuringiolysin O (TLO)

This toxin produced by *Bacillus thuringiensis* was first characterized as an SLO-like toxin by Pendleton *et al.* (1973), who obtained a partially purified preparation by precipitation of culture supernatants by ammonium sulfate followed by isoelectric focusing. Thuringiolysin O focused into two distinct peaks of haemolytic activity at p*I* 6.0 and p*I* 6.5, respectively. The estimated relative molecular mass was 47 000. This toxin has been purified to homogeneity by Rakotobé and Alouf (1984) by ultrafiltration of the culture filtrates of strain H1-30 serotype 1 followed by column chromatography on thiopropyl-Sepharose 6B and ion-exchange chromatography on DEAE-cellulose. The specific activity of the purified toxin was 7×10^5 HU mg^{-1} of protein. A single band was observed by SDS–PAGE corresponding to a relative molecular mass of 58 000 (Alouf *et al.*, 1986). The toxin shared all the properties of the other 'classical' antigenically related cholesterol-binding cytolysins.

Tetanolysin (TLY)

This toxin was first recognized in 1898 in the culture filtrates of *Clostridium tetani* by Ehrlich (1898) (see also

						Consensus 'Cys motif' undecapeptide				
PLY	423	V	K	I	R	E C T G L A W E W W R	T	V	Y	E
SLO	525	I	M	A	R	E C T G L A W E W W R	K	V	I	D
PFO	452	I	K	A	R	E C T G L A W E W W R	D	V	I	S
ALV	456	I	F	A	R	E C T G L A W E W W R	T	V	V	D
LLO	479	V	Y	A	K	E C T G L A W E W W R	T	V	I	D
ILO	479	I	H	A	K	E C T G L A W E W W R	T	V	V	D
LSO	479	I	Y	A	R	E C T G **F** A W E W W R	T	V	I	D
PLO	487	V	E	A	G	E **A** T G L A W **D P** W W -	T	V	I	N
CLO	*	I	V	A	R	E C T G L A W E W W R	T	I	I	K
ILY	*	?	?	?	?	G **A** T G L A W E P W R	?	?	?	?

FIGURE 24.1 Alignment of the amino acid sequences of pneumolysin, streptolysin O, perfringolysin O, alveolysin, listeriolysin O, ivanolysin O, seeligerolysin O, pyolysin O, cereolysin O and intermedilysin within and around the consensus 'Cys motif' undecapeptide.

Neill, 1926c) who clearly differentiated it from the 'true killing toxin' of *C. tetani* (tetanus neurotoxin) as established later in more detail by Fleming (1927) and Hardegree (1965). Little information is available on tetanolysin production by *C. tetani* strains. Lucain and Piffaretti (1977) reported a comparative study of the tetanolysins produced by nine different Tulloch serotypes of *C. tetani* as regards relative molecular mass, p*I* and electrophoretic mobilities. All tetanolysin preparations were similar in these respects. The relative molecular mass values, measured by gel filtration, were 47 000 ± 3000 and the p*I* values, determined by isoelectric focusing in sucrose gradient, were 6.54 for a major form and 6.15 and 5.84 for minor forms, respectively. Tetanolysin purification from culture filtrates of Massachusetts C2 strain and Harvard A-47 strain was reported by Alving *et al.* (1979) and Mitsui *et al.* (1980), respectively. The procedure used by the former group involved ammonium sulphate precipitation, gel filtration and hydroxylapatite chromatography. This procedure was modified by Rottem *et al.* (1990), who passed on a Mono-Q column the material eluted from the Sephadex G-100 column. In both procedures, the specific activity of the purified toxin was 10^6 HU mg^{-1}. The relative molecular mass was 45 000. Two forms of p*I* 6.1 and 6.4 were found (Blumenthal and Habig, 1984). Tetanolysin purification by Mitsui *et al.* (1980) yielded a preparation (specific activity 5×10^5 HU mg^{-1} of protein) consiting of a mixture of four haemolytic entities (48–53 kDa). Whether this heterogeneity corresponds to proteolytic nicking is not known. A high relative molecular mass tetanolysin (≥100 kDa) was later characterized by Mitsui *et al.* (1982). This form eluted from the void volume of Sepharose 6B and was shown by electron microscopy as a particulate material which was thought to be the 50-kDa tetanolysin complexed to cytoplasmic membrane fragments. Cell damage and sterol-binding by tetanolysin are discussed in another section.

Botulinolysin (BLY)

This cytolysin produced by *Clostridium botulinum* constitutes with eight other clostridial membrane-damaging toxins, the largest subgroup of the family of the cholesterol-binding cytolysins (Table 24.1). A comparative study by Haque *et al.* (1992) of toxin production in culture supernatants of six, three and five strains from *C. botulinum* type C, D and E, respectively, showed important variation in toxin production (140–3400 haemolytic units ml^{-1}). The high toxin-producing strain C-340 was used by these investigators for toxin purification to electrophoretic homogeneity. Culture supernatants were fractionated by ammonium sulfate

precipitation, ion-exchange and gel-filtration procedures and SP Toyopearl 650 M ion-exchange column chromatography. The M_r of purified BLY was *ca.* 58 kDa with a p*I* of 8.4. The 50% lethal dose in mice was 310 ng ml^{-1} and the 50% cytotoxic dose for Vero cells 120 ng ml^{-1}.

Surprisingly, BLY was reported to contain four cysteine residues by amino acid analysis, in contrast to the single Cys residues found in SLO, PLY, ALV, PFO, LLO and LSO molecules (two residues occur in ivanolysin). BLY lysed the erythrocytes of many animal species (higher lytic potency for rabbit, human and guinea pig cells).

Toxin damage on rabbit erythrocytes was investiged by Sekiya *et al.* (1998) by electron microscopy. Numerous ring and semicircular (arc) structures were observed, as previously found by the authors for SLO (Sekiya *et al.*, 1993, 1996). These structures reflect toxin oligomerization on the erythrocyte surface. BLY rings (47 nm) appeared larger than those formed by SLO (36 nm) and were supposed to be composed of more than 50 BLY molecules. BLY bound to membranes at 0°C but subsequent treatment with glutaraldehyde prevented ring formation. Zn^{2+} ions inhibited ring formation but not binding of BLY to membranes.

The pathophysiological properties of BLY investigated by Sugimoto *et al.* (1995, 1997) showed an increase in the contractility of rat aortic ring due to inhibition of endothelium-dependent relaxation of blood vessels. The cardiotoxicity elicited by the toxin, similar to that shared by tetanolysin (Hardegree *et al.*, 1971), SLO (see Chapter 25) and other CBCs, probably reflects toxin effects on blood vessels.

Pyolysin (PLO)

This newly discovered cholesterol-binding cytolysin (Ding and Lämmler, 1996; Funk *et al.*, 1996; Billington *et al.*, 1997) is produced by *Arcanobacterium* (*Actinomyces*) *pyogenes*, a commensal bacterium in animals which may become pathogenic in cattle, sheep, swine and other economically important animals. The toxin is lytic for the erythrocytes of various animal species, dermonecrotic and lethal for laboratory animals, and cytotoxic to animal polymorphonuclear leucocytes and kidney cells. As shown by Billington *et al.* (1997), who investigated the molecular genetics and biochemical aspects of PLO, this toxin belongs without doubt to the CBC family in spite of certain unusual molecular features.

The *plo* gene was cloned and sequence analysis revealed an open reading frame of 1605 bp encoding a 57.9-kDa protein comprising 534 amino acids with a consensus signal peptidase cleavage site between residues 27 and 28 leading to a predicted 55.1-kDa mature form of 507 amino acid residues with a pI of 9.3. This form, as discussed in another section, had only 31.2–40.6% identity (48.0–55.8% similarity) with the primary structure of the seven other CBCs so far recognized. Moreover, in contrast to the latter, the invariant undecapeptide consensus sequence did not contain a Cys residue in position 2; this amino acid was replaced by Ala (Figure 24.1). This finding is consistent with the insensitivity of PLO to reducing agents. Mutagenesis of the Ala residue to Cys did not confer thiol activation of PLO, suggesting a conformational difference in the consensus sequence which could be (at least partly) due to the substitution of the Glu residue in position 7 by Asp and the insertion of a supernumerary Pro residue between Asp and Trp residues.

Intermedilysin (ILY)

Streptococcus intermedius (one of the *S. milleri* group of streptococci) is a constituent of normal flora of the human oral cavity. This micro-organism, which may also cause purulent infections of internal organs, produces a 54-kDa cholesterol-binding protein designated intermedilysin (Nagamune *et al.*, 1996). This newly discovered toxin was purified by gel filtration and hydrophobic chromatography from culture medium of the strain UNS 46 isolated from a human liver abscess.

The sequencing of five internal peptide fragments of ILY showed 42–71% homology with pneumolysin. A further comparison of a 150- amino acid residue sequence of ILY showed obvious homology with PLY, SLO, PFO, LLO, ILO and LSO (Nagamune *et al.*, 1997), including the undecapeptide consensus region (Figure 24.1) indicating that ILY belongs to the CBC family. This affiliation is also supported by the inhibition of the haemolytic activity by cholesterol. However, the inhibitory concentration of cholesterol was two orders of magnitude higher than that required for SLO. ILY was highly lytic to human erythrocytes, 100-fold less effective on monkey erythrocytes and had no lytic effect on the erythrocytes of many animal species tested. Moreover, the lytic effect of ILY was not neutralized by anti-SLO antibodies. Similar to PLO, ILY does not contain a Cys residue in the consensus undecapeptide and consequently its lytic activity is not affected by thiol-blocking agents.

Suilysin (SLY)

This toxin was isolated, purified and identified from the culture fluids of *Streptococcus suis* type 2 (Jacobs *et al.*, 1994) which is a major pathogen in pigs. Occasionnally, this micro-organism is associated with disease (meningitis, septicaemia) in humans (Feder *et al.*, 1994). Toxin purification by Superose-12 column chromatography and ammonium sulfate precipitation yielded a protein of 54 kDa exhibiting a specific activity of 0.7×10^6 haemolytic units per mg, comparable to the values found for SLO, PLY or LLO. The lytic activity was tested on erythrocytes from a variety of animal species. All were equally susceptible to lysis. Toxin activity was inhibited by cholesterol. It was blocked by oxidation or by treatment with thiol-blocking agents and restored after addition of reducing agents, suggesting that SLY belongs to the CBC family. Moreover, the sequence of the *N*-terminal 16 amino acids of SLY showed many similarities to parts of the *N*-terminal amino acid sequences of PFO, SLO, LLO, ALV and PLY. A study of the kinetics of SLY production and the effects of various reagents on the lytic activity of crude preparations of this cytolysin has been reported by Feder *et al.* (1994).

Purification of other toxins

Bifermentolysin, sordellilysin and oedematolysin have been purified from culture supernatants by single-step immunoaffinity chromatography through columns of anti-pneumolysin monoclonal antibodies coupled to 2C5 or 3H10 Sepharose (Sato, 1986).

STEROL-BINDING PROPERTIES

The most characteristic biochemical property of the CBC toxins is the specific and irreversible inhibition of their lytic, lethal and cardiotoxic properties by very low (nanomolar) amounts of cholesterol and certain structurally related sterols (reviewed by Bernheimer, 1976; Alouf, 1977, 1980; Cowell and Bernheimer, 1978; Smyth and Duncan, 1978). The inactivation by cholesterol of the haemolytic activity of tetanolysin was shown as early as 1902 and that of pneumolysin and SLO in 1914 and 1939, respectively (see Bernheimer, 1976). The inactivation of the other toxins by cholesterol and other sterols was subsequently demonstrated (see the above-mentioned references and Geoffroy and Alouf, 1983; Sato, 1986; Geoffroy *et al.*, 1987; Ohno-Iwashita *et al.*, 1988; Vazquez-Boland *et al.*, 1989a).

Structural requirements

The structural and stereospecific features necessary for inhibition were investigated in detail on quantitative bases in the case of PLY, SLO and PFO (see Watson and

Kerr, 1974; Prigent and Alouf, 1976; and the references mentioned above). All inhibitory sterols possessed (a) an OH group in β-configuration on carbon-3 of ring A of the cyclopentanoperhydrophenanthrene ring of cholesterol and kindred sterols; (b) a lateral aliphatic side-chain of suitable size (isooctyl chain in the case of cholesterol) attached to carbon-17 on the D-ring; (c) the presence of a methyl group at C-20; (d) an intact B-ring. Neither the saturation state of the B-ring, the positions of the double bonds, nor the stereochemical relationships of rings A and B were critical factors. The presence of an α-OH group on C-3 (epicholesterol), esterification of the β-OH group or its substitution with a keto group or with 3β-SH group (thiocholesterol) rendered the sterol inactive (Alouf and Geoffroy, 1979) against the lytic effect, presumably because such substitutions or orientations preclude correct presentation of the reactive 3β-OH group to the toxins. The usual sterols fulfilling the appropriate criteria for the inhibition of toxin activity are: cholesterol, 7-dehydrocholesterol, dihydrocholesterol, stigmasterol, ergosterol, lathosterol and β-sitosterol, respectively.

Characteristics of toxin–sterol interaction

Early experiments in the 1970s in the liquid phase suggested the formation of complexes between toxin and sterol molecules (Cowell and Bernheimer, 1978; Smyth and Duncan, 1978). The first direct demonstration of the occurrence of these complexes and thereafter their separation were provided by Alouf and his co-workers. The complexes were visualized by allowing SLO or alveolysin to diffuse from wells in sterol-containing agar gels (Prigent and Alouf, 1976; Alouf and Geoffroy, 1979; Geoffroy and Alouf, 1983). The complexes formed appeared as white opaque halos (around toxin wells) constituted by insoluble (hydrophobic) precipitates of irreversibly bound toxin–sterol material stainable by protein dyes. Only inhibitory sterols showed such a pattern.

The separation of toxin–sterol complexes formed in liquid phase was reported by Johnson et al. (1980), who mixed pure pneumolysin, SLO or alveolysin preparations with [^{3}H]cholesterol dispersions at concentrations in which this sterol was present in a micellar form. Advantage was taken of the fact that when dispersions of cholesterol in phosphate buffer are added to Sephadex on Sephacryl columns, the free cholesterol molecules stick to the gel and only toxin–sterol complexes elute with buffer. The separated complexes were haemolytically inactive and of high molecular weight. The amounts of bound sterol increased linearly with toxin concentration. The reaction was rapid and temperature independent, similar to the binding step

during toxin interaction with target cells. The specificity of cholesterol binding was assessed by adding unlabelled inhibitory or non-inhibitory sterols to toxin before adding [^{3}H]cholesterol. Epicholesterol caused only a small decrease in binding, whereas 7-dehydrocholesterol inhibited radiolabelled binding to an extent equal to that observed with unlabelled cholesterol. Haemolytically inactive SLO obtained by treatment with oxidized dithiothreitol or by reaction with parahydroxymercuribenzoate showed no decrease in cholesterol-binding activity, whereas the ability of the toxins to bind to erythrocytes was modified by such treatment. This result is in agreement with the further finding that the SH group of cysteine is not involved in toxin interaction with its binding 'receptor' (cholesterol). The blockade of the SH group probably produced a steric hindrance to toxin fixation to the cholesterol embedded in the erythrocyte membrane; such a hindrance does not take place when the toxin interacts with free cholesterol. The isolation of toxin–sterol complexes in liquid phase was also reported by Geoffroy and Alouf (1983), who separated and identified [^{3}H]cholesterol–alveolysin complexes by sucrose gradient ultracentrifugation. The complexes formed were relatively heterogeneous in size. Heat-denatured toxin did not bind cholesterol, indicating that the native structure is essential for binding. Ivanolysin O–cholesterol complexes were also separated by centrifugation at 25 000 g by Vazquez-Boland et al. (1989a, b).

Stoichiometry

The limited solubility of cholesterol and other sterols in aqueous phase, the non-uniform nature of their dispersions in water and their tendency to stick to solid surfaces, among other reasons, has made it difficult to obtain information on the stoichiometry and thermodynamics of toxin–sterol interaction. Cholesterol has a maximum solubility in aqueous solutions of 1.8 μg ml^{-1} (4.7 μM) and undergoes a reversible self-association at the critical micellar concentration (CMC) of approximately 27–44 nM at 25°C (see Geoffroy and Alouf, 1983, for references). Such cholesterol aggregates are heterogeneous in size and rod-shaped, each containing ca. 260–360 cholesterol molecules (Smyth and Duncan, 1978). Ideally, the study of toxin–cholesterol interactions should be performed at concentrations below the CMC to ensure proper stoichiometry with cholesterol molecules dispersed as monomers rather than aggregates. Aggregate status appears to have been the case in the studies on SLO (Prigent and Alouf, 1976; Badin and Denne, 1978), CLO (Cowell and Bernheimer, 1978) and PFO (Hase et al., 1976) as cholesterol concentrations found to inhibit 1 HU ranged from 12 to 25 nmol, corresponding to a molar cholesterol–toxin ratio of *ca.* 500–1000. The data of the literature compiled by Smyth and Duncan (1978) showed that the number of cholesterol molecules required to neutralize a single molecule of toxin ranged from 170 to 1×10^6. By using a dilution technique involving sterol solutions in absolute ethanol, Geoffroy and Alouf (1983) avoided dealing with micellar solutions and were able to determine a linear inhibition 'titration' curve of alveolysin by cholesterol. The stoichiometry was practically equimolar (*ca.* 1.6 molecule of cholesterol neutralized by 1 molecule of alveolysin).

Nature of the binding forces

The physical forces involved in sterol–toxin interaction are poorly understood. The formation of a covalent linkage between the SH group and the 3β-OH of cholesterol is unlikely. London–van der Waals forces and/or hydrophobic interaction probably occur (but not exclusively?) in the interaction. The requirement of both the A-ring and isooctyl chain of the sterol molecules for inhibition of toxin activity suggest that appropriate aromatic and/or side-chains of amino acid residues of cholesterol-binding domain of the toxins interact with these two separate regions of the cholesterol molecule to create the complex. This hypothesis may be validated if it becomes possible to determine the three-dimensional structure of co-crystallized toxin–cholesterol complexes. Only the crystal structure of PFO has been established so far (Rossjohn *et al.*, 1997).

CYTOLYTIC AND MEMBRANE-DAMAGING EFFECTS

The most striking biological property of the toxins reviewed here is their potent lytic activity towards all mammalian and other eukaryotic cells so far tested. Lysis results from the disorganization or disruption of the cytoplasmic membrane of target cells. The intracellular organelles are also disrupted. The membrane-damaging effects are also reflected by alterations in the permeability and integrity of artificial model membranes.

Cell lysis markers

The lytic process triggered by the toxins reviewed here and other cytolysins is classically investigated and monitored by the measurement of the released intracellular components (from cell cytoplasm or from the disrupted organelles) to the incubation medium by the toxin-damaged cells (Alouf, 1977; Smyth and Duncan,

1978). The release of radiolabelled, coloured or fluorescent markers entrapped in resealed erythrocyte membrane vesicles (Buckingham and Duncan, 1983) or in liposomes (Duncan, 1984; Menestrina *et al.*, 1990) has been used in studies of toxin-induced damage of these vesicles.

The spectrophotometric determination of haemoglobin (Hb) is the most usual method employed in the monitoring of erythrocyte lysis (haemolysis) and its kinetics. The release of K^+ or ^{86}Rb from toxin-treated erythrocytes was also used as a marker of cell lysis by tetanolysin (Blumenthal and Habig, 1984) or by SLO (Smyth and Duncan, 1978). A highly sensitive approach is the measurement by a bioluminescence method of the release of adenosine triphosphate (ATP) from erythrocytes and other cells, as devised by Fehrenbach *et al.* (1980) and also used for the study of the kinetics of erythrocyte lysis by SLO (Fehrenbach *et al.*, 1982; Niedermeyer, 1985). A comparative study of Hb and ATP release showed that the latter was a much more sensitive indicator of SLO-induced lysis (10^{-11}–10^{-12} M ATP).

Platelet lysis by SLO and alveolysin was followed by the assay of released serotonin, lactate dehydrogenase and other enzymes (Launay and Alouf, 1979; Launay *et al.*, 1984). Membrane damage of human lung fibroblasts by SLO, PFO and ALV was assessed by the leakage of three different-sized cytoplasmic markers (Thelestam and Möllby, 1980; Thelestam *et al.*, 1981). The cells were first preloaded with [^{14}C]amino isobutyric acid (M_r 103) or treated with radiolabelled uridine to obtain either a labelled nucleotide ($M_r < 1000$) or ribonucleic acid (RNA) ($M_r < 200\,000$) before toxin challenge. The time course of release of these markers was monitored to study the lytic process and estimate the size of the functional 'pores' (whatever their physical reality) elicited in the damaged membrane (Thelestam and Möllby, 1980). In this respect, references to 'pores', 'channels' or 'lesions' are meant to indicate functional entities with the potential to allow passage of molecules up to a certain size (Buckingham and Duncan, 1983).

Preloaded [^3H]choline or [^{35}S]methionine Lettre cells (murine tumour cells) were used as precursors for labelled intracellular markers for the study of the mechanism of lysis of these cells by PFO and other toxins (Menestrina *et al.*, 1990).

Membrane cholesterol as the toxin-binding site and target

Several lines of experimental evidence suggest that cholesterol molecules embedded in the lipid bilayer of the cytoplasmic membrane of eukaryotic cells constitute the binding site of the toxins.

1. The toxins are inactivated by cholesterol and can no longer bind target cells (Alouf, 1980; Ohno-Iwashita *et al.*, 1986).
2. The toxins have no lytic effects on prokaryotic cells (bacterial protoplasts and spheroplasts) which lack cholesterol in their cytoplasmic membranes. However, toxin binds to parasitic mycoplasma cells that contain cholesterol and are damaged by the toxins but not saprophytic mycoplasma cells in which carotenol replaces cholesterol in the membrane (Rottem *et al.*, 1976, 1990; Smyth and Duncan, 1978). The incubation of these cells with a cholesterol–Tween mixture re-established their ability to bind CLO. Furthermore, the treatment with cholesterol oxidase of human erythrocyte membranes and *A. laidlawii* cells containing cholesterol abolished their ability to bind CLO and to inhibit haemolysis (Cowell and Bernheimer, 1978).
3. The lytic activity of the toxins is abrogated by incubation with erythrocyte membranes; in membrane extracts, only those lipid fractions that contain cholesterol inactivate the toxins (Hase *et al.*, 1976; Smyth and Duncan, 1978). Similarly, *Acholeplasma laidlawii* cells grown in the presence of cholesterol inhibited the haemolytic activity of CLO but not *A. laidlawii* cells grown in the absence of cholesterol.
4. The pretreatment of erythrocyte membranes with agents that bind cholesterol, such as polyene antibiotics and alfalfa saponins, prevented the inhibition of the haemolytic activity of CLO and SLO (Shany *et al.*, 1974). In addition, the treatment of nucleated mammalian cells (L-cells and HeLa cells) with inhibitors of cholesterol synthesis, such as 20α-hydroxysterol or 25-hydroxysterol, significantly reduced SLO binding and abolished cell susceptibility to the lytic effect of the toxin; the incubation of refractory cells with serum or cholesterol restored their sensitivity to SLO (Duncan and Buckingham, 1980).
5. The partial extraction of cholesterol (*ca.* 30%) from human erythrocytes decreased perfringolysin O binding and cell susceptibility to lysis (Mitsui *et al.*, 1982), whereas experimentally cholesterol-enriched erythrocytes exhibited increased sensitivity to lysis by SLO (Linder and Bernheimer, 1984).
6. Only cholesterol-containing phospholipid (but not pure phospholipid) vesicles or films are bound and disrupted by the toxins (Alouf *et al.*, 1984).

MOLECULAR GENETICS OF CHOLESTEROL-BINDING TOXINS

As mentioned above, the nucleotide sequences of the genes encoding SLO, PLY, PFO, ALV, LLO, ILO, LSO and PLO (and thereby the amino acid sequences of these toxins) have been established (Table 24.2). Moreover, partial sequences of the genes encoding CLO and ILY are available (Yutsudo *et al.*, 1994; Nagamune, personal communication). The relative molecular mass of the mature forms of the toxins established so far ranges from 60 to 51 kDa. The longest amino acid (AA) sequence is that of streptolysin O (high M_r form, see Chapter 25) followed by that of pyolysin, seeligerolysin, ivanolysin, listeriolysin O, pneumolysin, perfringolysin O and alveolysin, respectively.

TABLE 24.2 Characteristics of toxin molecules deduced from nucleotide sequences of encoding genes

Toxins	M_r of mature protein	aa[a]	Cys position[b]	References
Streptolysin O[e]	60 151	33 + 538	497	Kehoe *et al.* (1984, 1987), Okumura *et al.* (1994)
Pneumolysin	52 800[c]	0 + 471	428	Walker *et al.* (1987)
Cereolysin[d]	–	–	–	Yutsudo *et al.* (1994)
Perfringolysin O	52 469	27 + 472	429	Tweten (1988)
Alveolysin	51 766	32 + 469	429	Geoffroy *et al.* (1990)
Listeriolysin O	55 842	25 + 504	459	Mengaud *et al.* (1987, 1988), Domann and Chakraborty (1989)
Ivanolysin	55 966	23 + 505	461	Haas *et al.* (1992)
Seeligerolysin	56 371	25 + 505	459	Haas *et al.* (1992)
Pyolysin	57 900	27 + 507	–	Billington *et al.* (1997)

[a]Number of amino acid residues of signal peptide + that of secreted (mature) form (except for pneumolysin which is intracellular).
[b]Position of the unique cysteine residue of the polypeptide chain starting from the *N*-terminal residue of the mature form. Pyolysin does not contain any Cys residue in the conserved consensus undecapeptide.
[c]Low M_r active form of SLO resulting from proteolytic cleavage of the mature toxin form subsequent to secretion.
[d]Only a partial nucleotide sequence of cereolysin is available. The gene was cloned by Kreft *et al.* (1983).
[e]SLO from group A streptococci (Kehoe *et al.*, 1987) and from groups C (*Streptococcus equisimilis*) and G streptococci (*Streptococcus canis*) (Okumura *et al.*, 1994).

Sequence homology

Comparison of the sequences of the toxins revealed homologous stretches along the whole molecules (see Chapter 27). However, homology was clearly stronger in the C-terminal part of the molecules. The amino acid sequence identity ranged from 40% to 65% (Morgan *et al.*, 1996), except for pyolysin (30–41%), which is the most divergent member of the family (Billington *et al.*, 1997). The primary structure of SLO showed a higher level of homology (*ca.* 65%) with PFO than does PLY with respect to SLO (42% only).

The largest continuous region of identity among the toxins is a conserved 11-amino acid (undecapeptide) sequence near the C-terminus: Glu-Cys-Thr-Gly-Leu-Ala-Trp-Glu-Trp-Trp-Arg   (E-C-T-G-L-A-W-E-W-W-R) identically shared by SLO, PLY, PFO, ALV, LLO, ILO and CLO (Figure 24.1). Interestingly, the cysteine residue which lies in this consensus ('signature') sequence is unique in all of these toxins, with the exception of ILO which contains a second cysteine residue outside the undecapeptide and closer to the C-terminus. The presence of a unique cysteine residue in most of the toxins of the family precluded the possibility of intramolecular disulfide bridge formation as the basis for reversible oxidation and reduction of these molecules, as postulated since the 1940s (see Alouf and Geoffroy, 1991).

The corresponding undecapeptides in LSO, PLO and ILY show some differences. LSO contains a single Cys residue in position two but the residue in the fifth position is phenylalanine instead of leucine. The Cys residue is not present in either PLO or ILY and is replaced by an Ala residue. Moreover, in the latter, the glutamic acid residue is replaced by glycine and tryptophan by proline. In PLO, the Glu residue in position eight is replaced by aspartic acid and a supernumerary Pro residue is inserted.

A phylogenetic tree of the CBCs so far sequenced is shown in Figure 24.2.

Role of the common cysteine-containing region

Genetic replacement (by site-directed mutagenesis) of the Cys residue in the consensus undecapeptide by different amino acids was investigated in SLO, PLY and LLO (Pinkney *et al.*, 1989; Saunders *et al.*, 1989; Boulnois *et al.*, 1990; Michel *et al.*, 1990; Hill *et al.*, 1994). These studies showed that this residue was not essential for toxin function, as discussed in the other specific chapters on streptolysin O, pneumolysin, perfringolysin and listeriolysin. Similarly, other residues in the undecapeptide have been mutagenized, particularly the Trp residues, and in these cases, some of the three Trp occurring in this region appeared critical (Michel *et al.*, 1990; Sekino-Suzuki *et al.*, 1996).

Trp has the largest aromatic side-chain of the amino acids and therefore may confer a local hydrophobic character to the undecapeptide, the structure of which is typically amphiphilic. It is very likely, as shown below, that at least one Trp residue is essential for toxin insertion into the cholesterol/phospholipid bilayer (through hydrophobic interactions). This insertion is probably the mechanism by which the cell membrane is disrupted by the toxins.

The individual substitution by site-directed mutagenesis of Trp^{433}, Trp^{435} and Trp^{436} residues of pneumolysin by phenylalanine and the Trp^{466} and Trp^{467} residues of listeriolysin O by alanine was reported by Boulnois *et al.* (1990) and Michel *et al.* (1990), respectively. Compared with the haemolytic activity of the recombinant wild-type PLY, that of PLY $Trp^{433} \rightarrow$ Phe, $Trp^{435} \rightarrow$ Phe and $Trp^{436} \rightarrow$ Phe derivatives was 0.6%, 13% and 100%, indicating that Trp^{433} was critical for activity, whereas Trp^{436} was apparently not. Interestingly, the cholesterol-binding capacity of the $Trp^{433} \rightarrow$ Phe remained identical to that of the wild-type protein. The haemolytic activity of the $Trp^{466} \rightarrow$ Ala and $Trp^{467} \rightarrow$ Ala derivatives of listeriolysin O was severely reduced by 95% and 99%, respectively. However, these derivatives retained their capacity to bind to cell membranes and to combine with cholesterol similar to the wild-type toxins. The differences in activity between PLY and LLO mutated at the same Trp positions may be due to steric factors since alanine is a much smaller molecule and less hydrophobic than the bulky Phe or Trp molecules,

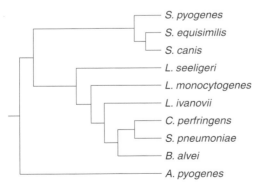

S. pyogenes
S. equisimilis
S. canis
L. seeligeri
L. monocytogenes
L. ivanovii
C. perfringens
S. pneumoniae
B. alvei
A. pyogenes

FIGURE 24.2 Phylogenetic analysis of protein sequences showing the relationship between the cholesterol-binding cytolysins from *A. pyogenes*, *B. alvei*, *C. perfringens*, *Listeria ivanovii* and *Listeria seeligeri*, *L. monocytogenes*, *Streptococcus canis*, and *Streptococcus equisimilis*, *S. pneumoniae* and *S. pyogenes* compiled using the protein parsimony program in the PHYLIP package. (Reproduced from Billington *et al.* and by permission of the authors and of the *Journal of Bacteriology*, American Society for Microbiology, Washington, DC.)

with their phenyl or indole rings. This may confer to the undecapeptide a different conformation and higher hydrophobicity than that resulting from alanine substitution.

Another modification of the residues of the common peptide was the substitution of Glu[434] by asparagine in pneumolysin. The activity of the resulting derivative was 25% that of the unsubstituted recombinant toxin.

On the basis of the data hitherto available it appears that the cytolytic activity depends rather on an appropriate overall structure of the undecapeptide. In this respect, single and double amino acid substitutions were created in PFO to investigate the role of individual tryptophan residues in lytic activity (Sekino-Suzuki, 1996). Wild-type and mutant toxins were overproduced in *E. coli* and purified. The relative lytic activities of four mutants, each with a Trp to Phe substitution outside the common Cys-containing region, were more than 60% that of wild-type PFO. In contrast, mutant toxins with Phe replacements within the Cys-containing region (at Trp[436], Trp[438] or Trp[439]) showed significantly reduced haemolytic and erythrocyte membrane-binding activities. The largest reduction in binding affinity, more than 100-fold, was observed for Trp[438] mutant toxins. However, the mutants retained binding specificity for cholesterol and the ability to form arc-shaped and ring-shaped structures on membranes. These findings indicate that the low haemolytic activities of these mutant toxins can be ascribed, at least in part, to reduced binding activities. With respect to protease susceptibility and far-ultraviolet circular-dichroism spectra, only the W436→F mutant toxin showed any considerable difference from wild-type toxin in secondary or higher-order structures, indicating that Trp[436] is essential for maintenance of toxin structure.

This view is in agreement with the two-site hypothesis of Alouf and Raynaud, who showed by biochemical and immunochemical approaches that two antigenically and topologically distinct sites called f (for fixation) and l (for lytic) are involved in the SLO-induced cytolysis (Alouf, 1977, 1980). The occurrence of these two predicted domains was confirmed immunologically by Watson and Kerr (1985) and is presently supported by the genetic and biochemical approaches mentioned above.

ACKNOWLEDGEMENTS

The author would like to thank colleagues who have made results of their studies available to him. The invaluable secretarial assistance of Mrs Patricia Paul in the preparation of this manuscript is greatly appreciated.

REFERENCES

Alouf, J.E. (1977). Cell membranes and cytolytic bacterial toxins. In: *Specificity and Action of Animal, Bacterial and Plant Toxins* (ed. P. Cuatrecasas), pp. 211–70. Chapman & Hall, London.

Alouf, J.E. (1980). Streptococcal toxins (streptolysin O, streptolysin S, erythrogenic toxin). *Pharmacol. Ther.* **11**, 661–717.

Alouf, J.E. and Geoffroy, C. (1979). Comparative effects of cholesterol and thiocholesterol on streptolysin O. *FEMS Microbiol. Lett.* **6**, 413–16.

Alouf, J.E. and Geoffroy, C. (1991). The family of the antigenically-related cholesterol-binding (sulphydryl-activated) cytolytic toxins. In: *Sourcebook of Bacterial Toxins* (eds J.E. Alouf and J.H. Freer), pp. 147–86. Academic Press, London.

Alouf, J.E., Kiredjian, M. and Geoffroy, C. (1977). Purification de l'hémolysine thiol-dépendante extracellulaire de *Bacillus alvei*. *Biochimie* **59**, 329–36.

Alouf, J.E., Geoffroy, C., Pattus, F. and Verger, R. (1984). Surface properties of bacterial sulphydryl-activated cytolytic toxins. Interaction with monomolecular films of phosphatidylcholine and various sterols. *Eur. J. Biochem.* **141**, 205–10.

Alouf, J.E., Geoffroy, C., Gilles, A.-M. and Falmagne, P. (1986). Structural relatedness between five sulphydryl activated toxins: streptolysin O, perfringolysin O, alveolysin, pneumolysin and thuringiolysin. *Ztbl. Bakt. Microbiol. Hygiene I Abteilung*, Suppl. 15, 49–50.

Alving, C.R., Habig, W.H., Urban, K.A. and Hardegree, M.C. (1979). Cholesterol-dependent tetanolysin damage to liposomes. *Biochim. Biophys. Acta* **551**, 224–8.

Ashdown, L.R. and Koehler, J.M. (1990). Production of haemolysin and other extracellular enzymes by clinical isolates of *Pseudomonas pseudomallei*. *J. Clin. Microbiol.* **28**, 2331–4.

Badin, J. and Denne, M.-A. (1978). A cholesterol fraction for streptolysin O binding on cell membranes and lipoproteins – I. Optimal conditions for the determination. *Cell. Molec. Biol.* **22**, 133–43.

Bernheimer, A.W. (1976). Sulfhydryl activated toxins. In: *Mechanisms in Bacterial Toxinology* (ed. A.W. Bernheimer), pp. 85–97. John Wiley and Sons, New York.

Bernheimer, A.W. and Grushoff, P. (1967a). Extracellular hemolysins of aerobic sporogenic bacilli. *J. Bacteriol.* **93**, 1541–3.

Bernheimer, A.W. and Grushoff, P. (1967b). Cereolysin: production, purification and partial characterization. *J. Gen. Microbiol.* **46**, 143–50.

Bernheimer, A.W. and Rudy, B. (1986). Interaction between membranes and cytolytic peptides. *Biochim. Biophys. Acta* **864**, 123–41.

Bernheimer, A.W., Avigad, L.S. and Kim, K.-S. (1979). Comparison of metridiolysin from the sea anemone with thiol-activated cytolysins from bacteria. *Toxicon* **17**, 69–75.

Berthou, L., Corvaia, N., Geoffroy, C., Mutel, V., Launay, J.-M. and Alouf, J.E. (1992). The phosphoinositide pathway of lymphoid cells: labelling after permeabilization by alveolysin, a bacterial sulfhydryl-activated cytolysin. *Eur. J. Cell. Biol.* **58**, 377–82.

Bhakdi, S. and Tranum-Jensen, J. (1988). Damage to cell membranes by pore-forming bacterial cytolysins. *Progr. Allergy* **40**, 1–43.

Billington, S.J., Jost, B.H., Cuevas, W.A., Bright, K.R. and Songer, J.G. (1997). The *Arcanobacterium* (*Actinomyces*) *pyogenes* hemolysin, pyolysin, is a novel member of the thiol-activated cytolysin family. *J. Bacteriol.* **179**, 6100–6.

Blumenthal, R. and Habig, W.H. (1984). Mechanism of tetanolysin-induced membrane damage: studies with black lipid membranes. *J. Bacteriol.* **157**, 321–3.

Boulnois, G.J., Mitchell, T.J., Saunders, F.K., Mendez, F.J. and Andrew, P.W. (1990). Structure and function of pneumolysin, the thiol-activated toxin of *Streptococcus pneumoniae*. In: *Bacterial Protein Toxins* (eds R. Rappuoli, J.E. Alouf, P. Falmagne *et al.*), *Zbl. Bakt.* Suppl. 19, pp. 43-51. Gustav Fischer, Stuttgart.

Braun, V. and Focareta, T. (1991). Pore-forming bacterial protein hemolysins (cytolysins). *Crit. Rev. Microbiol.* **18**, 115–58.

Buckingham, L. and Duncan, J.L. (1983). Approximate dimensions of membrane lesions produced by streptolysin S and streptolysin O. *Biochim. Biophys. Acta* **729**, 115–22.

Coudron, P.E., Payne, J.M. and Markowitz, S.M. (1991). Pneumonia and empyema infection associated with a *Bacillus* species that resembles *B. alvei. J. Clin. Microbiol.* **29**, 1777–9.

Cowell, J.L. and Bernheimer, A.W. (1978). Role of cholesterol in the action of cereolysin membranes. *Arch. Biochem. Biophys.* **190**, 306–10.

Cowell, J.L., Grushoff-Kosyk, P. and Bernheimer, A.W. (1976). Purification of cereolysin and the electrophoretic separation of the active (reduced) and inactive (oxidized) forms of the purified toxin. *Infect. Immun.* **14**, 144–54.

Cowell, J.L., Kim, K. and Bernheimer, A.W. (1978). Alteration by cereolysin of the structure of cholesterol-containing membranes. *Biochim. Biophys. Acta* **507**, 230–41.

Ding, H. and Lämmler, C. (1996). Purification and further characterization of a hemolysin of *Actinomyces pyogenes. J. Vet. Med. Ser.* **43**, 179–88.

Domann, E. and Chakraborty, T. (1989). Nucleotide sequence of the listeriolysin gene from a *Listeria monocytogenes* serotype1/2a strain. *Nucleic Acids Res.* **17**, 6406.

Duncan, J.L. (1984). Liposomes as membrane models in studies of bacterial toxins. *J. Toxicol. Toxin. Rev.* **3**, 1–51.

Duncan, J.L. and Buckingham, L. (1980). Resistance to streptolysin O in mammalian cells treated with oxygenated derivatives of cholesterol: cholesterol content of resistant cells and recovery of streptolysin O sensitivity. *Biochim. Biophys. Acta* **603**, 278–87.

Ehrlich, P. (1898).Diskusston Wahrend der Geselleschoft der Charite-Ärzte. *Berliner Klinische Wochenschr.* (Berlin) **35**, 273.

Feder, I., Chengappa, M.M., Fenwick, B., Rider, M. and Staats, J. (1994). Partial characterization of *Streptococcus suis* type 2 hemolysin. *J. Clin. Microbiol.* **32**, 1256–60.

Fehrenbach, F.J., Huser, H. and Jaschinski, C. (1980). Measurement of bacterial cytolysins with a highly sensitive kinetic method. *FEMS Microbiol. Lett.* **7**, 285–6.

Fehrenbach, F.J., Schmidt, C.M. and Huser, H. (1982). Early and long events in streptolysin O-inducing hemolysis. *Toxicon* **20**, 233–8.

Fleming, W.L. (1927). Studies on the oxidation and reduction of immunological substances the differentiation of tetanolysin and tetanospasmin. *J. Exp. Med.* **46**, 279–90.

Funk, P.G., Staats, J.J., Howe, M., Nagaraja, T.G. and Chengappa, M.M. (1996). Identification and partial characterization of an *Actinomyces pyogenes* hemolysin. *Vet. Microbiol.* **50**, 129–42.

Geoffroy, C. and Alouf, J.E. (1982). Interaction of alveolysin, a sulfhydryl-activated bacterial cytolytic toxin, with thiol group reagents and cholesterol. *Toxicon* **20**, 239–41.

Geoffroy, C. and Alouf, J.E. (1983). Selective purification by thiol-disulfide interchange chromatography of alveolysin, a sulfhydryl-activated toxin of *B. alvei. J. Biol. Chem.* **258**, 9968–72.

Geoffroy, C. and Alouf, J.E. (1988). Molecular cloning and characterization of *Bacillus alvei* thiol-dependent cytolytic toxin expressed in *E. coli. J. Gen. Microbiol.* **134**, 1961–70.

Geoffroy, C., Gaillard, J.-L., Alouf, J.E. and Berche, P. (1987). Purification, characterization and toxicity of the sulfhydryl-activated hemolysin listeriolysin O from *Listeria monocytogenes. Infect. Immun.* **55**, 1641–6.

Geoffroy, C., Mengaud, J. Alouf, J.E. and Cossart, P. (1990). Alveolysin, the thiol-activated toxin of *Bacillus cereus*, is homologous to listeriolysin O, perfringolysin O, pneumolysin and streptolysin O and contains a single cysteine. *J. Bacteriol.* **172**, 7301–5.

Guillaumie, M. (1950). Hémolysines bactériennes et anti-hémolysines. *Ann. Inst. Pasteur* **79**, 661–71.

Haas, A., Dumbsky, M. and Kreft, J. (1992). Listeriolysin genes: complete nucleotide sequence of *ilo* ivanolysin from *Listeria ivanovii* and of listeriolysin from *Listeria seeligeri. Biochim. Biophys. Acta* **1130**, 81–4.

Hase, J., Mitsui, K. and Shonaka, E. (1976). *Clostridium perfringens* exotoxins. IV. Inhibition of θ-toxin induced haemolysis by steroids and related compounds. *Jpn. J. Exp. Med.* **46**, 45–50.

Haque, A., Sugimoto, N., Horiguchi, Y., Okabe, T., Miyata, T., Iwanaga, S. and Matsuda, M. (1992). Production, purification and characterization of botunolysin, a thiol-activated hemolysin of *Clostridium botulinum. Infect. Immun.* **60**, 71–8.

Hardegree, M.C., Palmer, A.E. and Duffin, N. (1971). Tetanolysin: *in vivo* effects in animals. *J. Infect. Dis.* **123**, 51–60.

Hardegree, M.E. (1965). Separation of neurotoxin and hemolysin of *Clostridium tetani. Proc. Soc. Exp. Biol. Med.* **119**, 405–8.

Hatheway, C.L. (1990). Toxigenic clostridia. *Clin. Microbiol. Rev.* **3**, 66–98.

Herbert, D. and Todd, E.W. (1941). Purification and properties of a haemolysin produced by group A hemolytic streptococci (streptolysin O). *Biochem. J.* **35**, 1124–39.

Hill, J., Andrew, P.W. and Mitchell, T.J. (1994). Amino acids in pneumolysin important for haemolytic activity identified by random mutagenesis. *Infect. Immun.* **62**, 757–8.

Jacobs, A.A.C., Loeffen, P.L.W., Van Den Berg, A.J.G. and Storm, P.K. (1994). Identification, purification and characterization of a thiol-activated hemolysin (suilysin) of *Streptococcus suis. Infect. Immun.* **62**, 1742–8.

Johnson, M.K. (1977). Cellular location of pneumolysin. *FEMS Microbiol. Lett.* **2**, 243–52.

Johnson, M.K., Geoffroy, C. and Alouf, J.E. (1980). Binding of cholesterol by sulfhydryl-activated cytolysins. *Infect. Immun.* **27**, 97–101.

Kehoe, M.A. and Timmis, K.N. (1984). Cloning and expression in *Escherichia coli* of the streptolysin O determinant from *Streptococcus pyogenes*: characterization of the cloned streptolysin O determinant and demonstration of the absence of substantial homology with determinants of other thiol-activated toxins. *Infect. Immun.* **43**, 804–10.

Kehoe, M.A., Miller, L., Walker, J.A. and Boulnois, G.J. (1987). Nucleotide sequence of the streptolysin O (SLO) gene: structural homologies of the between SLO and other membrane-damaging, thiol-activated toxins. *Infect. Immun.* **55**, 3228–32.

Kreft, J., Berger, H., Hartlein, M., Muller, B., Weidinger, G. and Goebel, W. (1983). Cloning and expression in *Escherichia coli* and *Bacillus subtilis* of the hemolysin (cereolysin) determinant from *Bacillus cereus. J. Bacteriol.* **155**, 681–9.

Launay, J.M. and Alouf, J.E. (1979). Biochemical and ultrastructural study of the disruption of blood platelets by streptolysin O. *Biochim. Biophys. Acta* **556**, 278–91.

Launay, J.M., Geoffroy, C., Costa, J.L. and Alouf, J.E. (1984). Purified SH-activated toxins (streptolysin O, alveolysin): new tools for determination of platelet enzyme activities. *Thromb. Res.* **33**, 189–96.

Linder, R. and Bernheimer, A.W. (1984). Action of bacterial cytotoxins on normal mammalian cells and cells with altered membrane lipid composition. *Toxicon* **22**, 641–51.

Lucain, C. and Piffaretti, J.-C. (1977). Characterization of the haemolysins of different serotypes of *Clostridium tetani. FEMS Microbiol. Lett.* **1**, 231–4.

Menestrina, G., Bashford, C.L. and Pasternak, C.A. (1990). Pore-forming toxins: experiments with *S. aureus*, α-toxin, *C. perfringens*, θ-toxin and *E. coli* hemolysin in lipid bilayers, liposomes and intact cells. *Toxicon* **28**, 477–91.

Mengaud, J., Chenevert, J., Geoffroy, C., Gaillard, J.-L. and Cossart, P. (1987). Identification of the structural gene encoding the SH-activated hemolysin of *Listeria monocytogenes*: listeriolysin O is homologous to streptolysin O and pneumolysin. *Infect. Immun.* **55**, 3225–7.

Mengaud, J., Vicente, M.F., Chenevert, J., Moniz-Periera, J., Geoffroy, C., Gicquel-Sanzey, B., Baquero, F., Pérez-Diaz, J.C. and Cossart, P. (1988). Expression in *Escherichia coli* and sequence analysis of the listeriolysin O determinant of *Listeria monocytogenes*. *Infect. Immun.* **56**, 766–72.

Michel, E., Reich, K.A., Favier, R., Berche, P. and Cossart, P. (1990). Attenuated mutants of the intracellular bacterium *Listeria monocytogenes* obtained by a single amino acid substitution in listeriolysin O. *Molec. Microbiol.* **4**, 2167–78.

Mitsui, N., Mitsui, K. and Hase, J. (1980). Purification and some properties of tetanolysin. *Microbiol. Immunol.* **24**, 575–84.

Mitsui, K., Saeki, Y. and Hase, J. (1982). Effects of cholesterol evulsion on susceptibility of perfringolysin O of human erythrocytes. *Biochim. Biophys. Acta* **686**, 307–13.

Morgan, P.J., Andrew, P.W. and Mitchell, T.J. (1996). Thiol-activated cytolysins. *Rev. Med. Microbiol.* **7**, 221–9.

Moussa, R.S. (1958). Complexity of toxins from *Clostridium septicum* and *Clostridium chauvoei*. *J. Bacteriol.* **76**, 538–45.

Nagamune, H., Ohnishi, C., Katsuura, A., Fushitani, K., Whiley, R.A., Tsuji, A. and Matsuda, Y. (1996). Intermedilysin, a novel cytotoxin specific for human cells, secreted by *Streptococcus intermedius* UNS46 isolated from a human liver abscess. *Infect. Immun.* **64**, 3093–100.

Nagamune, H., Ohnishi, C., Katsuura, A., Taoka, Y., Fushitani, K., Whiley, R. A., Yamashita, K., Tsuji, A., Matsuda, Y., Maeda, T., Korai, H. and Kitamura, S. (1997). Intermedilysin: a cytolytic toxin specific for human cells of a *Streptococcus intermedius* isolated from human liver abscess. In: *Streptococci and the Host* (eds T. Horaud *et al.*), pp. 773–5. Plenum Press, New York.

Neill, J.M. (1926a). Studies on the oxidation and reduction of immunological substances. I. *Pneumococcus* hemotoxin. *J. Exp. Med.* **44**, 199–213.

Neill, J.M. (1926b). Studies on the oxidation and reduction of immunological substances. II. The hemotoxin of *B. welchii*. *J. Exp. Med.* **44**, 215–26.

Neill, J.M. (1926c). Studies on the oxidation and reduction of immunological substances. III. Tetanolysin. *J. Exp. Med.* **44**, 227–40.

Neill, J.M. and Fleming, W.L. (1927). Studies on the oxidation of immunological substances. VI. The 'reactivation' of the bacteriolytic activity of oxidized *pneumococcus* extracts. *J. Exp. Med.* **46**, 263–77.

Neill, J.M. and Mallory, J.B. (1926). Studies on the oxidation and reduction of immunological substances. IV. Streptolysin. *J. Exp. Med.* **44**, 241–60.

Niedermeyer, W. (1985). Interaction of streptolysin O with membranes: kinetic and morphological studies on erythrocytes membranes. *Toxicon* **23**, 425–35.

Oakley, C.L., Warrack, G.H. and Clarke, P.H. (1947). The toxins of *Clostridium oedematiens* (*Cl. novyi*). *J. Gen. Microbiol.* **1**, 91–107.

Ohno-Iwashita, Y., Iwamoto, M., Mitsui, K., Kawasaki, H. and Ando, S. (1986). Cold-labile haemolysin produced by limited proteolysis of θ-toxin from *Clostridium perfringens*. *Biochemistry* **25**, 6048–53.

Ohno-Iwashita, Y., Iwamoto, M., Mitsui, K., Ando, S. and Nagai, K. (1988). Protease-nicked θ-toxin of *Clostridium perfringens* a new membrane probe with no catalytic effects reveals two classes of cholesterol as toxin-binding sites on sheep erythrocytes. *Eur. J. Biochem.* **176**, 95–101.

Okumura, K., Hara, A., Tanaka, T., Nishiguchi, I., Minamide, W., Igarashi, H. and Yutsudo, T. (1994). Cloning and sequencing the streptolysin O genes of group C and group G streptococci. *DNA Seq.* **4**, 325–8.

Pendleton, I.R., Bernheimer, A.W. and Grushoff, P. (1973). Purification and characterization of hemolysin from *Bacillus thuringiensis*. *J. Inverteb. Pathol.* **21**, 131–5.

Pinkney, M., Beachey, E. and Kehoe, M. (1989). The thiol-activated toxin streptolysin O does not require a thiol group for cytolytic activity. *Infect. Immun.* **57**, 2553–8.

Prigent, D. and Alouf, J.E. (1976). Interaction of streptolysin O with sterols. *Biochim. Biophys. Acta* **443**, 288–300.

Rakotobé, F. and Alouf, J.E. (1984). Purification and properties of thuringiolysin a sulfhydryl-activated cytolytic toxin of *Bacillus thuringiensis*. In: *Bacterial Protein Toxins* (eds J.E. Alouf *et al.*), p. 265. Academic Press, London.

Reboli, A.C., Bryan, S.C. and Farrar, W.E. (1989). Bacteremia and hip prosthesis caused by *Bacillus alvei*. *J. Clin. Microbiol.* **27**, 1395–6.

Rossjohn, J., Feil, S.C., McKinstrey, W.J., Tweten, R.K. and Parker, M.W. (1997). Structure of a cholesterol-binding, thiol-activated cytolysin and a model of its membrane form. *Cell* **89**, 685–92.

Rottem, S., Hardegree, M.C., Grabowski, M.R., Fornwald, R. and Barile, M.F. (1976). Interaction between tetanolysin and mycoplasma cell membrane. *Biochim. Biophys. Acta* **455**, 879–88.

Rottem, S., Groover, K., Habig, W.H., Barile, M.F. and Hardegree, M.C. (1990). Transmembrane diffusion channels in *Mycoplasma gallisepticum* induced by tetanolysin. *Infect. Immun.* **58**, 598–602.

Rutter, J.M. and Collee, J.F. (1969). Studies on the soluble antigens of *Clostridium oedematiens* (*Cl. novyi*). *J. Med. Microbiol.* **2**, 395–421.

Sato, H. (1986). Monoclonal antibodies against *Clostridium perfringens* θ-toxin (perfringolysin O). In: *Monoclonal Antibodies Against Bacteria*, Vol. III (eds A.J.L. Macario and E. Conway de Macario), pp. 203–28. Academic Press, New York.

Saunders, F.K., Mitchell, T.J., Walker, J.A., Andrew, P.W. and Boulnois, G.J. (1989). Pneumolysin, the thiol-activated toxin of *Streptococcus pneumoniae* does not require a thiol group for *in vivo* activity. *Infect. Immun.* **57**, 2547–52.

Sekino-Suzuki, N., Nakamura, M., Mitsui, K. and Ohno-Iwashita, Y. (1996). Contribution of individual tryptophan residues to the structure and activity of θ-toxin (perfringolysin O), a cholesterol-binding cytolysin. *Eur. J. Biochem.* **241**, 941–7.

Sekiya, K., Satoh, R., Danbara, H. and Futaesaku, Y. (1993). A ring-shaped structure with a crown formed by streptolysin O on the erthrocyte membrane. *J. Bacteriol.* **175**, 5953–61.

Sekiya, K., Satoh, R., Danbara, H. and Futaesaku, Y. (1996). Electron microscopic evaluation of a two-step theory of pore formation by streptolysin O. *J. Bacteriol.* **178**, 6998–7002.

Sekiya, K., Danbara, H., Futaesaku, Y., Haque, A., Sugimoto, N. and Matsuda, M. (1998). Formation of ring-shaped structures on erythrocyte membranes after treatment with botulinolysin a thiol-activated haemolysin from *Clostridium botulinum*. *Infect. Immun.* **66**, 2987–90.

Shany, S., Bernheimer, A.W., Grushoff, P.S. and Kim, K.-S. (1974). Evidence for membrane cholesterol as the common binding site for cereolysin, streptolysin O and saponin. *Cell. Mol. Biochem.* **3**, 179–86.

Smyth, C.J. and Duncan, J.L. (1978). Thiol-activated (oxygen-labile) cytolysins. In: *Bacterial Toxins and Cell Membranes* (eds J. Jeljaszewicz and T. Wadström), pp. 129–83. Academic Press, London.

Smythe, C.V. and Harris, T.N. (1940). Some properties of a haemolysin produced by group A hemolytic streptococci. *J. Immunol.* **38**, 283–300.

Sugimoto, N., Haque, A., Horiguchi, Y. and Matsuda, M. (1995). Coronary vasoconstriction is the most probable cause of death of rats intoxicated with botulinolysin, a hemolysin produced by *Clostridium perfringens*. *Toxicon* **33**, 1215–30.

Sugimoto, N., Haque, A., Horiguchi, Y. and Matsuda, M. (1997). Botulinolysin, a thiol-activated hemolysin produced by *Clostridium botulinum*, inhibits endothelium-dependent relaxation of rat aortic ring. *Toxicon* **35**, 1011–23.

Thelestam, M. and Möllby, R. (1980). Interaction of streptolysin O from *Streptococcus pyogenes* and theta toxin from *Clostridium perfringens* with human fibroblasts. *Infect. Immun.* **29**, 863–77.

Thelestam, M., Alouf, J.E., Geoffroy, C. and Möllby, R. (1981). Membrane-damaging action of alveolysin from *Bacillus alvei*. *Infect. Immun.* **32**, 1187–92.

Todd, E.W. (1934). A comparative serological study of streptolysins derived from human and from animal infections, with notes on pneumococcal haemolysin, tetanolysin and *staphylococcus* toxin. *J. Pathol. Bacteriol.* **39**, 299–321.

Tweten, R.K. (1988). Cloning and expression in *Escherichia coli* of the perfringolysin O (theta-toxin) gene from *Clostridium perfringens* and characterization of the gene product. *Infect. Immun.* **56**, 3228–34.

Vazquez-Boland, J.A., Dominguez, L., Rodriguez-Ferri, E.F. and Suarez, G. (1989a). Purification and characterization of two *Listeria ivanovii* cytolysins, a sphingomyelinase C and a thiol-activated toxin (ivanolysin O). *Infect. Immun.* **57**, 3928–35.

Vazquez-Boland, J.A., Dominguez, L., Rodriguez-Ferri, E.F., Fernandez-Garayzabal, J.F. and Suarez, G. (1989b). Preliminary evidence that different domains are involved in cytolytic activity and receptor (cholesterol) binding in listeriolysin O, the *Listeria monocytogenes* thiol-activated toxin. *FEMS Microbiol. Lett.* **65**, 95–100.

Walker, J.A., Allen, R.L., Falmagne, P., Johnson, M.K. and Boulnois, G. (1987). Molecular cloning, characterization and complete nucleotide sequence of the gene for pneumolysin, the sulfhydryl-activated toxin of *Streptococcus pneumoniae*. *Infect. Immun.* **55**, 1184–9.

Watson, K.C. and Kerr, E.J.C. (1974). Sterol requirements for inhibition of streptolysin O activity. *Biochem. J.* **140**, 95–8.

Watson, K.C. and Kerr, E.J.C. (1985). Specificity of antibodies for T sites and F sites of streptolysin O. *Med. Microbiol.* **19**, 1–7.

Yutsudo, T. (1994). Amino acid sequence of cereolysin (derived from nucleotide sequence PID: g600252). Genbank Accession D21270.

25

Streptolysin O

Joseph E. Alouf and Michael W. Palmer

INTRODUCTION

Among the family of the antigenically related cholesterol-binding cytolysins (the classical 'sulfhydryl-activated' toxins) characterized so far (see Chapter 24, 'The family of the structurally related cholesterol-binding cytolysins'), streptolysin O (SLO) has been the most intensively investigated membrane-damaging toxin over the past 50 years. In this respect, this potent haemolytic and, more generally, cytolytic protein, produced by groups A, C and G streptococci, is considered to be the prototype of the 'SH-activated' toxins (Weller *et al.*, 1996).

Besides the general reviews devoted to this group of toxins (Bernheimer, 1976; Alouf, 1977; Smyth and Duncan, 1978; Alouf and Geoffroy, 1991; Morgan *et al.*, 1996), several specific reviews on SLO have been published over the past 25 years (Halbert, 1970; Bernheimer, 1972, 1976; Jeljaszewicz *et al.*, 1978; Alouf, 1980; Wannamaker, 1983; Wannamaker and Schlievert, 1988). This chapter will briefly summarize the main well-established properties and features of SLO and will deal more specifically with the recent achievements concerning the genetics, biochemistry, mechanisms of action and pathophysiological properties of the toxin. The reader is referred to the preceding chapter for the many common properties shared by SLO and homologous toxins.

HISTORICAL BACKGROUND

The potent lytic activity towards red blood cells in the culture filtrates of certain streptococci (Lancefield group A streptococci, as established later) discovered by Marmorek (1902) and confirmed by many reports in the early decades of this century (see Neill and Mallory, 1926; Köhler, 1963) led to the pioneering investigations by Todd (1932, 1938) and Weld (1934), who attributed the lytic effect to two different 'haemolysins'. These lytic agents are the so-called 'oxygen-labile' streptolysin O (SLO, to indicate that this factor becomes progressively haemolytically inactive upon exposure to air) and the 'oxygen-stable' streptolysin S (SLS), which was found to be stable in air and the production of which was inducible by serum.

Prior to the differentiation of the two streptolysins, Neill and Mallory (1926) observed the 'spontaneous deterioration' of the lytic activity of culture filtrates exposed to air and reported the restoration of haemolytic potency of the filtrates (very often to higher titres than the initial ones) by treatment with a reducing agent (hydrosulfite). This effect was confirmed and more widely investigated by Herbert and Todd (1941), who demonstrated that thiols were the more potent agents to 'activate' oxidized crude or partially purified SLO preparations. Since then, the effects of various organic and inorganic reducing agents, as well as the kinetics of reactivation, have been extensively investigated (Rahman *et al.*, 1969; Bernheimer, 1976; Smyth and Duncan, 1978; Alouf, 1980).

Historically, this oxidation–reduction process led to the concept of 'sulfhydryl or thiol-activated toxins', requiring the addition of reducing agents for the manifestation of the lytic potency of SLO and other cytolysins released by certain Gram-positive bacteria. As described in the preceding chapter, the concept of

457

activation by thiols remains important but not crucial since the discovery in the 1990s that the unique cysteine residue of the toxins could be genetically replaced by other amino acid residues without significant loss of their lytic properties (see SLO genetics section).

Another, absolutely essential, feature of SLO and related toxins is the inhibition of their haemolytic, cytolytic and lethal effects by minute amounts of cholesterol (Hewitt and Todd, 1939) and other sterols with a 3β-hydroxy group on carbon-3 of the A ring of the cyclopentanoperhydrophenanthrene nucleus, and a lateral hydrophobic side-chain at carbon-17, as first shown by Howard et al. (1953). Toxin interaction with sterols and its relevance to the mode of action of SLO at the cellular level are discussed in the section on Molecular mechanism of oligomerization and pore-formation by streptolysin O.

BACTERIOLOGICAL ASPECTS OF STREPTOLYSIN O

Production by group A streptococci (*Streptococcus pyogenes*)

SLO is released in culture media during the exponential and stationary growth phases (Halbert, 1970; Dassy and Alouf, 1973) by virtually all strains investigated in earlier works (Herbert and Todd, 1941; Bernheimer, 1954; Halbert, 1970) or in more recent systematic evaluations on 143, 42 and 212 isolates, respectively (Tiesler and Trinks, 1979; Suzuki et al., 1988; Müller-Alouf et al., 1997). Toxin production is unrelated to the M and T serotypes of strains and of clinical status of the individuals from whom they were collected. SLO appears as a constitutive (permanent) exoprotein released by all strains of the *S. pyogenes* species (Pinkney et al., 1995). This contention is supported by the finding of Tyler et al. (1992) that 100% of 152 strains collected in Canada over five decades (1940–91) possessed the *slo* gene detected by polymerase chain reaction. On these grounds, it is possible that the rare strains, which phenotypically failed to produce detectable haemolytic activity (Bernheimer, 1954; Tiesler and Trinks, 1979), may have released very small amounts of SLO below the detection threshold. Indeed, individual strains found by the above-mentioned authors vary quantitatively in their ability to produce SLO within a range of 1 to more than 1000 haemolytic units (HU) per millilitre, as also shown in other studies (Halbert, 1970; Alouf, 1980; Wannamaker and Schlievert, 1988; Müller-Alouf et al., 1997).

SLO is also produced, but in moderate amounts, in appropriate chemically defined (synthetic) media (Bernheimer, 1954; Dassy and Alouf, 1973). However, better yields are obtained in culture media containing peptone and yeast extract dialysates (see Köhler, 1963; Alouf, 1980; Bhakdi et al., 1984; Alouf and Geoffroy, 1988). Complex media such as Todd–Hewitt broth should not be used for purification purposes because of the presence of high molecular weight peptides. Glucose is necessary (0.5–1%) for optimal toxin production; however, as lactic acid is produced during growth, the pH should be always maintained between 6.8 and 7.4 to avoid toxin destruction by acidity and (or) by proteolysis by the streptococcal proteinase that is released into the medium when the pH drops below 6.7. The best yields (500–1000 HU ml^{-1}) in optimal culture conditions have been obtained with the following strains: C203 S type 3 (Halbert, 1970; Smyth and Fehrenbach, 1974; Gerlach et al., 1993), Richards type 3 (Duncan and Schlegel, 1975; Bhakdi et al., 1984; Pinkney et al., 1995) and Kalback's S 84 type 3 (Alouf and Raynaud, 1973). SLO may be produced by bacterial cultivation in fermenters under pH control (see Halbert, 1970; Dassy and Alouf, 1973; Alouf, 1980).

The variability of SLO production by individual strains suggests the involvement of regulation mechanisms of toxin synthesis and excretion at both genetic and environmental levels, which still await investigation. It has been shown, for example, that at concentrations of antibiotics too low to affect bacterial growth, the production of SLO and erythrogenic A toxin remained at significant levels, and similarly with subinhibitory and even inhibitory concentrations of antibiotics which impair cell-wall synthesis (e.g. amoxicillin, vancomycin). In contrast, toxin production was abrogated in the presence of antibiotics that affect protein synthesis (erythromycin, gentamicin), used at concentrations subinhibitory for bacterial growth (Müller-Alouf et al., 1996).

Release of streptolysin O *in vivo*

Toxin production *in vivo* following streptococcal infections in humans, as reflected by the occurrence of anti-SLO (and other anti-exoprotein) antibodies in blood plasma, is widely documented (Köhler, 1963; Halbert, 1970; Watson and Kerr, 1985). The determination of these antibodies is of great clinical importance for diagnostic purposes (Ayoub and Harden, 1992). However, it is likely that SLO produced *in vivo* rapidly combines with antibody as it is formed, establishing an equilibrium of constant formation and slow dissociation of antigen–antibody complexes. This may establish a slow-release reservoir of active SLO during chronic infections, with progressive accumulation to toxic levels in susceptible tissues such as the heart (Alouf, 1980). Direct experimental evidence of toxin production *in vivo* was reported by Duncan (1983), who implanted in mice peritoneal cavities micropore filter chambers, in which various group A streptococcal strains were grown. Both toxin production and antibody response in the mice were detected.

Production by group C and G streptococci

A number of strains from group G streptococci and from human (but not animal) group C streptococci produces SLO in contrast to other streptococcal groups and species, as first reported by Todd (1939) and, later, other investigators (Halbert, 1970; Tiesler and Trinks, 1982; Suzuki *et al.*, 1988). The purification procedure for SLO from group C streptococci (Gerlach *et al.*, 1993) and the cloning and sequencing of SLO genes from groups A, C and G (Okumura *et al.*, 1994) showed that the SLO molecules from the three groups are practically immunochemically and structurally identical. These findings confirm the identity of SLO from group A and C streptococci deduced in the 1960s on the basis of electrophoretic and immunochemical criteria (Halbert, 1970).

Purification of streptolysin O

The purification of SLO to molecular homogeneity from culture fluids is difficult, owing to the multiplicity of the extracellular proteins (more than 25) released along with SLO (Alouf, 1980). In this respect, only partially purified preparations were obtained (see Halbert, 1970; Alouf, 1980; Bhakdi *et al.*, 1984; Gerlach *et al.*, 1993) before the introduction of novel techniques of protein separation at the turn of the 1980s. Since then, several procedures for SLO purification have been reported. Immunoaffinity chromatography (Linder, 1979), taking advantage of the antigenic relationship between SLO and the other related toxins, involved precipitation of culture supernatant by 80% ammonium sulfate, AH-Sepharose chromatography followed by immunochromatography on anti-tetanolysin coupled to CNBr-activated Sepharose 4B and acid elution. The purified SLO material was 80% homogeneous. Sodium dodecyl sulfate–polyacrylamide gel electrophoresis (SDS–PAGE) showed a major band corresponding to a relative molecular mass of 56 000 and two minor bands. According to the author, better homogeneity could be obtained by using additional steps.

Bhakdi *et al.* (1984) described a procedure involving ammonium sulphate and polyethylene glycol precipitation, DEAE-ion-exchange chromatography, prepar-

ative isoelectric focusing and chromatography on Sephacryl S-300. Two forms of the toxin possessing similar haemolytic activity were isolated: a native form of 69 kDa with a p*I* at pH 6.0–6.4 and a proteolysed form of 57 kDa (p*I* 7.0–7.5). The two forms appeared homogeneous by SDS–PAGE, and their specific activity was 8×10^5 HU mg^{-1} protein according to the definition proposed by Alouf (1980).

A procedure based on the selective separation of SH-proteins by thiol-disulfide covalent chromatography on thiopropyl-Sepharose (Alouf, 1980; Alouf and Geoffroy, 1988) also allowed for the purification of SLO to apparent electrophoretic homogeneity (specific activity 5×10^5 HU mg^{-1} protein). The purification process involved concentration of the culture fluid by ultrafiltration, batchwise absorption of the retentate on calcium phosphate gel, elution with sodium phosphate, thiol-Sepharose 6B column chromatography, gel filtration on Sephacryl S-200, preparative isoelectric focusing (if necessary) and gel filtration on Biogel P100. The apparent size of SLO determined by SDS–PAGE ranged from 55 to 60 kDa, depending on the batches processed. The determination by Edman degradation (Alouf *et al.*, 1986) of the sequence of the first amino acids of the purified proteins (NH$_2$-Ser-Asp-Glu-Asp) was in agreement with that deduced by Kehoe *et al.* (1987) from the nucleotide sequence of the *slo* gene. The purification of alveolysin, perfringolysin O, thuringiolysin O, pneumolysin, listeriolysin O and ivanolysin also involved disulfide chromatography on thiopropyl-Sepharose 6B (see preceding chapter). An original protocol of the purification of SLO from the group C *Streptococcus equisimilis* H 46 A strain was reported by Gerlach *et al.* (1993). The process involved culture supernatant absorption on a silica gel column, washing and elution with triethanolamine/HCl of various molarities, followed by ammonium sulfate precipitation, chromatography on hydroxylapatite, gel filtration on Silica 300 polyol and preparative isoelectric focusing. Similar to SLO from *S. pyogenes*, two slightly larger molecular forms of SLO (75 and 68 kDa) purified from *S. equisimilis* were identified.

GENE CLONING AND SEQUENCING OF STREPTOLYSIN O

The *slo* gene encoding SLO from *Streptococcus pyogenes* was cloned and expressed in *Escherichia coli* by Kehoe and Timmis (1984). Two forms of the SLO gene product (61 and 68 kDa) were detected in *E. coli* minicells. Hybridization experiments performed at a stringency of *ca.* 80% showed no homology between the cloned *slo* DNA sequences and the DNA isolated from bacteria expressing the homologous toxins, cereolysin, pneumolysin, listeriolysin O, perfringolysin O, histolyticolysin O and oedematolysin O (novilysin).

The nucleotide sequence of the *slo* gene and the deduced amino acid sequence of the toxin were reported by Kehoe *et al.* (1987). The primary *slo* gene product (Figure 25.1) consists of 571 amino acid residues (M_r = 63 645). The 33 N-terminal residues possess the features of a signal peptide, suggesting that the toxin is secreted from the streptococcal cell as a polypeptide of 538 amino acids (M_r 60 151). However, in streptococcal liquid cultures, the native polypeptide appeared to undergo a proteolytic cleavage subsequent to secretion, removing a *ca.* 7000-Da segment leading to a predominant low relative molecular mass form of

```
                          10         20         30         40         50
Group A SLO    1 MKDMSNKKTF KKYSRVAGLL TAALIIGNLV TANAESNKQN TASTETTTTN
group C SLO    1 MKDMSNKKIF KKYSRVAGLL TAALIVGNLV TANADSNKQN TANTETTTTN
Group G SLO    1 MKDMSNKKIF KKYSRVAGLL TAALIVGNLV TANADSNKQN TANTETTTTN
                                              *          *          *

                          60         70         80         90        100
Group A SLO   51 EQPKPESSEL TTEKAGQKTD DMLNSNDMIK LAPKEMPLES AEKEEKKSED
Group C SLO   51 EQPKPESSEL TTEKAGQKMD DMLNSNDMIK LAPKEMPLES AEKEEKKSED
Group G SLO   51 EQPKPESSEL TTEKAGQKMD DMLNSNDMIK LAPKEMPLES AEKEEKKSED
                                    *

                         110        120        130        140        150
Group A SLO  101 KKKSEEDHTE EINDKIYSLN YNELEVLAKN GETIENFVPK EGVKKADKFI
Group C SLO  101 NKKSEEDHTE EINDKIYSLN YNELEVLAKN GETIENFVPK EGVKKADKFI
Group G SLO  101 NKKSEEDHTE EINDKIYSLN YNELEVLAKN GETIENFVPK EGVKKADKFI
                  *

                         160        170        180        190        200
Group A SLO  151 VIERKKKNIN TTPVDISIID SVTDRTYPAA LQLANKGFTE NKPDAVVTKR
Group C SLO  151 VIERKKKNIN TTPVDISIID SVTDRTYPAA LQLANKGFTE NKPDAVVTKR
Group G SLO  151 VIERKKKNIN TTPVDISIID SVTDRTYPAA LQLANKGFTE NKPDAVVTKR

                         210        220        230        240        250
Group A SLO  201 NPQKIHIDLP GMGDKATVEV NDPTYANVST AIDNLVNQWH DNYSGGNTLP
Group C SLO  201 NPQKIHIDLP GMGDKATVEV NDPTYANVST AIDNLVNQWH DNYSGGNTLP
Group G SLO  201 NPQKIHIDLP GMGDKATVEV NDPTYANVST AIDNLVNQWH DNYSGGNTLP

                         260        270        280        290        300
Group A SLO  251 ARTQYTESMV YSKSQIEAAL NVNSKILDGT LGIDFKSISK GEKKVMIAAY
Group C SLO  251 ARTQYTESMV YSKSQIEAAL NVNSKILDGT LGIDFKSISK GEKKVMIAAY
Group G SLO  251 ARTQYTESMV YSKSQIEAAL NVNSKILDGT LGIDFKSISK GEKKVMIAAY

                         310        320        330        340        350
Group A SLO  301 KQIFYTVSAN LPNNPADVFD KSVTFKELQR KGVSNEAPPL FVSNVAYGRT
Group C SLO  301 KQIFYTVSAN LPNNPADVFD KSVTFKELQR KGVSNEAPPL FVSNVAYGRT
Group G SLO  301 KQIFYTVSAN LPNNPADVFD KSVTFKELQA KGVSNEAPPL FVSNVAYGRT
                                              *         *

                         360        370        380        390        400
Group A SLO  351 VFVKLETSSK SNDVEAAFSA ALKGTDVKTN GKYSDILENS SFTAVVLGGD
Group C SLO  351 VFVKLETSSK SNDVEAAFSA ALKGTDVKTN GKYSDILENS SFTAVVLGGD
Group G SLO  351 VFVKLETSSK SNDVEAAFSA ALKGTDVKTN GKYSDILENS SFTAVVLGAD
                                                                  *

                         410        420        430        440        450
Group A SLO  401 AAEHNKVVTK DFDVIRNVIK DNATFSRKNP AYPISYTSVF LKNNKIAGVN
Group C SLO  401 AAEHNKVVTK DFDVIRNVIK DNATFSRKNP AYPISYTSVF LKNNKIAGVN
Group G SLO  401 AAEHNKVVTK DFDVIRNVIK ANATFSRKNP AYPISYTSVF LKNNKIAGVN
                                         *

                         460        470        480        490        500
Group A SLO  451 NRTEYVETTS TEYTSGKINL SHQGAYVAQY EILWDEINYD DKGKEVITKR
Group C SLO  451 NRSEYVETTS TEYTSGKINL SHQGAYVAQY EILWDEINYD DKGKEVITKR
Group G SLO  451 NRSEYVETTS TEYTSGKINL SHQGAYVAQY EILWDEINYD DKGKEVITKR
                   *

                         510        520        530        540        550
Group A SLO  501 RWDNNWYSKT SPFSTVIPLG ANSRNIRIMA RECTGLAWEW WRKVIDERDV
Group C SLO  501 RWDNNWYSKT SPFSTVIPLG ANSRNIRIMA RECTGLAWEW WRKVIDERDV
Group G SLO  501 RWDNNWYSKT SPFSTVIPLG ANSRNIRIMA RECTGLAWEW WRKVIDERDV

                         560        570
Group A SLO  551 KLSKEINVNI SGSTLSPYGS ITYK...... .......... ..........
Group C SLO  551 KLSKEINVNI SGSTLSPYGS ITYK...... .......... ..........
Group G SLO  551 KLSKEINVNI SGSTLSPYGS ITYK...... .......... ..........
```

FIGURE 25.1 Alignment of the deduced amino acid sequences of SLO from group A, C and G streptococci (*S. pyogenes*, *S. equisimilis* and *S. canis*), deduced from the nucleotide sequence of their respective cloned genes. According to the sequence published by Kehoe *et al.* (1987) the N-terminus (Met) of the 33-amino acid signal peptide is in position 4 in this figure and that of the mature form (Asn) in position 37. The star marks at the bottom of the columns indicate the position at which different amino acid residues are observed. (Reproduced from Okumura *et al.*, 1994, and by permission of the *Journal of Sequencing and Mapping*, Harwood Academic Publishers.)

about 53 000, in agreement with the M_r of 55 000 and 56 000 reported by Alouf and Raynaud (1973) and Linder (1979), respectively. These data deduced from gene sequencing support the biochemical identification of both high and low molecular mass forms of *S. pyogenes* SLO by Bhakdi *et al.* (1984), as also found for the SLO purified from *S. equisimilis* (Gerlach *et al.*, 1993) and designated as nSLO (n for native) and fSLO (f for fragment).

The first four *N*-terminal residues found by Alouf *et al.* (1986) on a purified preparation corresponded to the residues 101 and 104 of the predicted SLO sequence. As a Lys residue occurs at position 100, Kehoe *et al.* (1987) suggested that the low M_r form (471 residues) was generated by the proteolytic removal of the 67 amino acids between the end of the predicted signal sequence and the Ser residue at position 101. Interestingly, the predicted sequence of the native haemolytically active pneumolysin (minus the *N*-terminal methionine) is 470 residues in length (Walker *et al.*, 1987). The *ca.* 70 kDa peptide cleaved from the native high M_r SLO is unique for this toxin and not found in the other related toxins of the group. According to Braun and Focareta (1991), the cleaved sequence is interesting from an evolutionary point of view, since it seems to have been added after the divergence from a common ancestor of the SH-activated cytolysins had occurred. Also, the signal sequences of these toxins have little in common. They either were added later or evolved as the mature proteins from a common ancestor; but, without a functional selection pressure, they diverged considerably. There is very little sequence constraint on signal peptides with the exception that they must have a positive charge which is followed by a sequence of uncharged residues close to the *N*-terminal end.

Suvorov *et al.* (1992) analysed by polymerase chain reaction the *slo* gene from 44 different *S. pyogenes* M serotypes in order to detect possible nucleotide sequence heterogeneity of the gene. Minor variations (6%) in sequence were observed in a few serotypes, suggesting a high conservation in *slo* genes among the different M serotypes.

The *slo* gene of *S. pyogenes* has been expressed in *Bacillus subtilis* (Yamada, 1995). A hybrid gene, consisting of the promoter and signal sequence fused to the region encoding the mature sequence of the *slo* gene, was constructed for the secretion of SLO from this micro-organism. To increase secretion of the toxin into the culture supernatant, several SLO expression vectors containing various combinations of promoters and pre-pro sequences were constructed. *Bacillus subtilis*, harbouring pPA consisting of the P-43 promoter and the coding sequence of the pre-region of the alka-

line protease gene that was fused to the pro-mature region of the *slo* gene, secreted SLO into the medium. The haemolytic activity was about 40-fold higher in the genetically engineered *B. subtilis* strain than in *S. pyogenes*. Recombinant SLO (rSLO) reacted with anti-SLO antibodies.

The *slo* gene as well as the genes of perfringolysin O (*pfo*) and listeriolysin O (*llo*) have been expressed in *B. subtilis* by Portnoy *et al.* (1992). These authors investigated the capacity of the recombinant strains to grow within a murine macrophage cell line and their ability, through the release of the relevant toxins, to mediate the lysis of the host vacuole membrane, as established for *Listeria monocytogenes* (Gaillard *et al.*, 1987). The recombinant strains expressing LLO and PFO promoted intracellular growth, whereas the strain expressing SLO did not, indicating that other conditions were necessary for growth and lysis (for discussion, see Portnoy *et al.*, 1992; Sheehan *et al.*, 1994; Goebel and Kreft, 1997).

In contrast to the SLO purified from *S. pyogenes*, the *N*-terminus of the low M_r form of SLO purified from *S. equisimilis* corresponded to residues Ala79 (Pinkney *et al.* 1995) or Leu78 (Dobereiner *et al.* 1996) rather than residue 101 of the deduced *S. pyogenes* SLO sequence. It was suggested that this reflects a species difference in proteolytic cleavage of secreted SLO or a discrepancy in published amino acid sequencing data. To clarify this issue, Pinkney *et al.* (1995) investigated the *N*-terminal sequences of both high and low M_r forms of SLO. Either form was purified from its native host cell (*S. pyogenes*, strain Richards) and also recombinantly expressed in *E. coli* LE 392. The high M_r forms were identical, and removal of a 31-residue signal peptide (instead of the 33-peptide reported by Kehoe *et al.*) occurred during nSLO and rSLO translocation across the cytoplasmic membrane of *S. pyogenes* and *E. coli*, respectively. The *N*-terminal residue of the mature toxins was Glu32 (instead of Asn34 as found by Kehoe *et al.*). The *N*-termini of the low M_r nSLO and rSLO were identical, and corresponded to Leu78. Interestingly, SLO cleavage between Lys77 and Leu78 corresponded to a highly sensitive cleavage site for the streptococcal cysteine proteinase (erythrogenic/pyrogenic exotoxin B), which is produced in parallel with SLO by *S. pyogenes*, suggesting that the low form of nSLO could be generated by this proteinase. However, the similarity in size of the low M_r rSLO was coincidental. Indeed, the separated low form was inactive owing to cleavage in the *C*-terminal part of the molecule by *E. coli* periplasmic proteases (Pinkney *et al.*, 1995; Dobereiner *et al.*, 1996). In contrast, the active low M_r rSLO co-purified with the high M_r form.

Interestingly, by fusion of the SLO gene to the *E. coli* maltose-binding protein (MBP) gene, SLO could be expressed to high levels and in a soluble form in the cytoplasm of *E. coli* (Dobereiner *et al.*, 1996; Weller *et al.*, 1996). The MBP–SLO fusion protein was haemolytic and cytotoxic to human fibroblasts and keratinocytes, and displayed equal activity to nSLO or rSLO (32–571). This fusion–protein complex was used as a tool to investigate structure–activity relationships.

Gene cloning from group C and G streptococci

The cloning and nucleotide sequences of the *slo* gene in various strains of group C (*S. equisimilis*) and group G (*S. canis*) streptococci were determined by Okumura *et al.* (1994). As shown in Figure 25.1, the nucleotide sequences of the SLO from group C and group G streptococci were both highly homologous to, but not completely identical to, that of *S. pyogenes*, indicating that the *slo* gene is highly conserved among the three streptococcal groups. Figure 25.2 shows a phylogenic tree constructed by the authors. The *S. pyogenes slo* gene is presumed to have diverged from a common ancestral gene of *S. equisimilis* and *S. canis*. Thereafter, the genes of these species diverged.

Relatedness of *slo* gene to the genes of homologous cholesterol-binding toxins

Besides the *slo* genes of group A, C and G streptococci, the complete nucleotide sequences of the genes and derived amino acid sequences of the following homologous toxins are known: pneumolysin (PLY), perfringolysin O (PFO), alveolysin (ALV), listeriolysin (LLO), ivanolysin (ILO), seeligerolysin (LSO) and pyolysin (PLO) from *Arcanobacterium pyogenes* (Kehoe *et al.*, 1987; Tweten, 1988; Geoffroy *et al.*, 1990; see Haas *et al.*, 1992, for the three *Listeria* toxins; Billington *et al.*, 1997). Moreover, partial gene sequences encoding

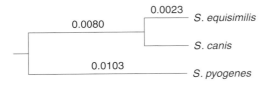

FIGURE 25.2 Phylogenic tree of the SLO gene of *S. pyogenes* (group A), *S. equisimilis* (group C) and *S. canis* (group G). The numbers indicate the genetic distances. (Reproduced from Okumura *et al.*, 1994, and by permission of the *Journal of Sequencing and Mapping*, Harwood Academic Publishers.)

cereolysin (CLO, Yutsudo *et al.*, 1994, database accession D21270 NID g418066) and intermedilysin (ILY, Nagamune, personal communication and in Japanese, *Seikagaku*, 1997, **69**, 343–8) are currently available. Apart from PLY, which is intracytoplasmic, all of these toxins possess conventional *N*-terminal leader sequences.

The longest sequence is that of SLO (538 amino acids for the mature secreted high M_r form) followed by pyolysin (507 aa), ivanolysin and seeligerolysin (505 aa), listeriolysin O (504 aa), perfringolysin O (472 aa), pneumolysin (471 aa) and alveolysin (469 aa). The comparison of the complete sequences of the toxins revealed homologous stretches along the whole molecules (see preceding chapter and Chapter 27 on perfringolysin O). However, homology was clearly stronger in the *C*-terminal part of the molecule.

Amino acid sequence identity ranged from 40 to 65% (Morgan *et al.*, 1996) except for pyolysin (30–41%), which appears to be the most divergent member of the family (Billington *et al.*, 1997). Surprisingly, the primary structure of SLO shows a higher level of homology (*ca.* 65%) with *Clostridium perfringens* PFO than does *Streptococcus pneumoniae* PLY with respect to SLO (42% only).

The largest continuous region of identity among the toxins is a conserved 11-amino acid (undecapeptide) sequence near the *C*-terminus:

Glu – Cys – Thr – Gly – Leu – Ala – Trp – Glu – Trp – Trp – Arg
E – C – T – G – L – A – W – E – W – W – R

shared by SLO, PLY, PFO, ALV, LLO, ILO and CLO. Interestingly, the cysteine residue that lies in this consensus sequence is unique (single) in all of these toxins (position 497 in the secreted high M_r form of SLO and position 530 in the translated product with signal peptide), with the exception of ivanolysin (ILO), which contains a second cysteine residue outside the undecapeptide and closer to the *C*-terminus. The presence of a single cysteine residue in most of the toxins of the family precluded the possibility of intramolecular disulfide bridge formation as the basis for reversible oxidation and reduction of these molecules, as suggested earlier (see Harris, 1940; Herbert and Todd, 1941; Cowell *et al.*, 1976; Smyth and Duncan, 1978; Alouf *et al.*, 1984). No data are available on whether or not such a bridge occurs in ivanolysin.

The corresponding undecapeptides in LSO, PLO and ILY show some differences. LSO contains a single Cys residue in position 2, but the residue in the fifth position is phenylalanine instead of leucine. The Cys residue is not present in either PLO or ILY, but is replaced by an alanine residue. Moreover, in the latter, the glutamic acid residue is replaced by glycine.

Structure–activity relationship

A truncate of SLO with five residues (Ser-Ile-Thr-Tyr-Lys) deleted from the *C*-terminus of the molecule reduced the haemolytic activity of the wild-type form by 99.5% (Kehoe *et al.*, 1988). A similar finding was reported by Owen *et al.* (1994) after deletion of the six *C*-terminal residues of pneumolysin (Asp-Lys-Val-Glu-Asn-Asp), who showed the loss of both haemolytic and binding activity.

Dobereiner *et al.* (1996) showed that the deletion of even the single *C*-terminal Lys571 residue of SLO was sufficient clearly to reduce cell-binding capacity. Deletion of the two *C*-terminal Tyr-Lys residues resulted in *ca.* 50% reduction in binding, suggesting that the extreme *C*-terminal end of SLO makes a significant contribution to the cell-binding domain of the toxin. This finding is supported by other experiments by these authors, who generated a great number of site-directed mutants in the *C*-terminal end of SLO. The mutations greatly reduced (<10% of wild-type) activity. For example, mutations changing SLO-Ile568 to SLO-Asn568 or upstream SLO- Gly166 to SLO-Val166 resulted in significant reduction in SLO activity. In contrast, the first 68 or even 103 amino acids (one-eighth or one-fifth of the total size of the toxin molecule) appear not to be required for pore formation/toxin oligomerization and, thereby, cell lysis. It is noteworthy that a large *N*-terminal part of SLO and related toxins is not or is poorly conserved.

As regards toxin binding, Hugo *et al.* (1986) generated a panel of monoclonal antibodies, some of which were able to neutralize the haemolytic activity of SLO. These antibodies inhibited toxin oligomerization on the target membrane but failed to block the binding of SLO to the membrane.

Role of the cysteine-containing region

The 11-amino acid sequence common to SLO and almost all related toxins has the potential to form an amphiphilic α-helix, making it a candidate for a membrane-insertion domain. Moreover, this undecapeptide and possibly surrounding amino acids may at least partly be involved in toxin binding to target cells, since the blockade of the cysteine SH-group by bulky reagents such as 5,5′-dithiobis(2-nitrobenzoic acid) reduced membrane-bound PFO to 1/50th that of native toxin (Iwamoto *et al.*, 1990).

To explore the putative functional role of the Cys residue (and of other residues), genetic replacement of the Cys residue by alanine in SLO was achieved through site-directed mutagenesis. The modified toxin showed haemolytic and cell-binding activities equiva-

lent to those in controls. Cys substitution by a Ser residue reduced by *ca.* 25% the lytic and binding capacity of the toxin (Pinkney *et al.*, 1989; Saunders *et al.*, 1989). Clearly, these results indicate that the Cys residue is not critical for toxin activity. The reagents that chemically modify free SH groups caused a marked inhibition of the cytolytic activity of wild-type SLO but, as expected, did not change that of the Cys-free mutants. Similar results were observed upon cysteine (or other undecapeptide residues) replacement in other cholesterol-binding toxins, as fully discussed in the preceding chapter.

TOXIN INTERACTION WITH STEROLS AND OTHER LIPIDS

As mentioned in the 'Historical background' section, SLO is inactivated by nanomolecular concentrations of cholesterol and other related sterols, fulfilling strict structural features (Watson and Kerr, 1974; Prigent and Alouf, 1976; Alouf and Geoffroy, 1979). This property, which is shared by the other toxins of the family, is discussed in detail in the preceding chapter. SLO was shown by Alouf and co-workers to generate micellar hydrophobic complexes with sterols (see above-mentioned references and Johnson *et al.*, 1980; Geoffroy and Alouf, 1983), leading, with other lines of experimental evidence, to the establishment of the widely accepted concept that membrane cholesterol, which occurs in all cytoplasmic membranes of eukaryotic cells, is the binding site of SLO and related toxins (Shany *et al.*, 1974; Duncan and Buckingham, 1980). The binding of SLO by cholesterol–phospholipid liposomes was investigated by Duncan and Schlegel (1975) and by Rosenqvist *et al.* (1980). The latter observed clear damage to the vesicles. SLO and other related toxins were also shown by Alouf *et al.* (1984) to penetrate and disrupt monomolecular phospholipid–cholesterol films. No penetration was observed with pure phospholipid films.

DAMAGING EFFECTS *IN VITRO* ON EUKARYOTIC CELLS AND TISSUES

All mammalian and other eukaryotic cells tested are damaged by SLO. In most cases, cell death is accompanied by cell lysis and the release of various intracytoplasmic components (e.g. haemoglobin in RBCs), indicating major damage to the cytoplasmic membrane of target cells. As shown in another section, membrane cholesterol is the binding site of the toxin. This property explains the insensitivity to SLO of prokaryotic cells (bacterial protoplasts and spheroplasts and saprophytic mycoplasma), which, unlike eukaryotic cells, lack cholesterol in their cell membranes (see Bernheimer, 1976). Primitive cells such as free-living protozoa and *Arbaccia* eggs are unaffected (Bernheimer, 1954).

Toxin-induced damage and lysis of erythrocytes

The lytic process triggered by toxin interaction with erythrocytes has been intensively investigated since the pioneering studies of Neill and Malory (1926) and Herbert and Todd (1941), and thereafter by many investigators (see Alouf and Raynaud, 1968; Bernheimer, 1976; Smyth and Duncan, 1978; Alouf, 1980; Niedermeyer, 1985; Alouf and Geoffroy, 1991).

The choice of erythrocytes (generally sheep, rabbit, human) for the study of SLO-elicited lysis is due to the convenient supply, manipulation and standardization of these cells. In addition, red blood cells are very convenient systems for quantitative studies, particularly for the investigation of the parameters of toxin binding, cell lysis and toxin-induced kinetics (monitored by the measurement of the haemoglobin released). Another advantage over other cells is our extensive knowledge of erythrocyte membrane composition and structure.

Cytolytic effects on other cells and tissues

Besides erythrocytes, SLO damages and lyses a wide variety of cells, among them human and rabbit leukocytes and macrophages, amnion cells, Ehrlich ascites tumour cells, human spermatozoa, platelets from various animal species, as well as a variety of cells cultured *in vitro*, such as rabbit kidney cells, HeLa and KB cells and human fibroblasts (see Halbert, 1970; Ginsburg, 1972; Bernheimer, 1976; Smyth and Duncan, 1978; Launay and Alouf, 1979; Alouf, 1980; Thelestam and Möllby, 1980; Launay *et al.*, 1984).

Optical and electron microscopy of SLO-challenged cells shows striking damage, as evidenced by the almost instantaneous swelling of the cell, with numerous spherical cytoplasmic blebs, vacuolization, leakage of intracellular substances, disruption of cytoplasmic organelles (mitochondria, lysosomes) and other bodies, and impairment of the transport of various metabolites. These changes support other indications that SLO acts primarily through physical disruption of cytoplasmic membrane and membranes of surrounding intracellular organelles (Buckingham and Duncan, 1983; Launay *et al.*, 1984).

MOLECULAR MECHANISM OF OLIGOMERIZATION AND PORE FORMATION BY STREPTOLYSIN O

The first example of a pore-forming protein was the terminal complement complex, which inserts into the target lipid bilayers as an oligomeric cylinder surrounding a discrete aqueous channel. Subsequent to this fundamental discovery, the same mechanism of cell damage has been attributed to a large number of proteins elaborated by both microbes and higher organisms, including the cholesterol-binding (SH-activated) toxins.

This wide distribution of pore-forming proteins might suggest that the task of making a hole in the host cell membrane is a very simple one. There are, however, several problems that a pore-forming molecule must deal with to fulfil its task efficiently.

Usually, the cytolytic activity is associated with the toxin monomer as opposed to the oligomer; isolated oligomers no longer attack membranes. This means that premature oligomerization in the absence of a target membrane has to be avoided.

Another problem is peculiar to SLO and related cholesterol-binding cytolysins. The membrane lesions created by SLO have been examined in numerous excellent electron microscopic studies, as first reported in the pioneering work of Dourmashkin and Rosse (1966) and by many other investigators since then (Duncan and Schlegel, 1975; Launay and Alouf, 1979; Bhakdi *et al.*, 1985; Sekiya *et al.*, 1993, 1996). Most authors agree that the individual oligomer consists of some 50 subunits and surrounds a pore about 30 nm in diameter. This extraordinarily large pore size requires that, following binding of the toxin monomers to the membrane, oligomerization occurs in a controlled manner. If too many oligomers were initiated simultaneously, these would run short of available monomers before reaching maturity. It is thus necessary that completion of a growing oligomer is kinetically favoured over initiation of a new one.

Finally, in the formation of a transmembrane pore, lipid molecules must be separated from each other. In the case of the huge SLO pore, this will require a considerable amount of energy, which must be raised by the process of protein oligomerization. Pore formation must, therefore, be intimately linked to oligomerization. The complete oligomeric pore can hardly be envisaged to arise in an instantaneous fashion by simultaneous assembly of all of its subunits; rather, monomers will successively be added to an incomplete, growing oligomer. This consideration suggests that pore formation will likewise occur in a successive fashion rather than in a single, sudden event. Consequently, immature SLO oligomers must exist, which should be associated with some sort of incomplete membrane lesion.

The present purpose is to provide a comprehensible idea of the molecular mechanism of SLO oligomerization and pore formation. Beyond the above functional requirements, any hypothesis on the molecular mechanism of SLO will have to accommodate the existing knowledge of the structure of the protein molecule. Owing to the high degree of homology between the thiol-activated toxins, the conformation of the SLO monomer can be modelled according to that of PFO, which has been elucidated by crystallography (Rossjohn et al., 1997). The molecule is divided into four domains (Figure 25.3). Three domains are arranged in a row, giving the molecule an elongated shape, which has also been inferred previously from hydrodynamic data (Morgan et al., 1993) for the homologous pneumolysin. Domain 3 is covalently connected to N-terminal domain 1 and packed laterally against domain 2. All available data indicate that membrane interaction of the monomer is mediated by domain 4. In contrast, oligomerization involves several sites scattered throughout the sequence, which can now be assigned to domains 1 and 3, respectively.

Owing to their large and somewhat non-homogeneous size, the oligomers of the cholesterol-binding toxins have resisted crystallographic analysis, and their structure is therefore unknown and subject to speculation and some debate. Along with the X-ray analysis of the perfringolysin monomer, a hypothetical structure of the membrane-associated oligomer has been proposed (Rossjohn et al., 1997). In that model, it is assumed that only minor changes in conformation are imposed on the monomeric molecule when it becomes part of the pore. The tryptophan-rich region around the conserved cysteine residue within domain 4 is assumed to adapt conformationally to cholesterol, and domain 3 is envisaged to move across the 'hinge' by which it is connected to domain 1. Consequently, in both the oligomer and the monomer, domain 4 is the only part of the toxin molecule in immediate contact with the lipid bilayer.

A different conclusion has been arrived at in a spectrofluorometric study on the structure of the SLO oligomer (Palmer et al., 1996). In those experiments, single cysteine residues were introduced into a cysteine-less active mutant of SLO (C530A) by site-directed mutagenesis to provide unique sites of attachment for the fluorescent reporter molecule acrylodan (6-acryloyl-2-dimethyl-amino-naphthalene). The fluorescence emission of thiol-derivatized acrylodan is strongly dependent on solvent polarity; characteristic values for the emission maximum are 540 nm with water and 435 nm with dioxane (Prendergast et al., 1983). Consequently, if a water-exposed mutant cysteine penetrates the hydrophobic membrane core during oligomerization, this will result in a distinct blue-shift in the fluorescence spectrum of the acrylodan attached. This approach was previously applied to the identification of the transmembrane part of the pore of staphylococcal alpha-toxin, where it has been fully validated by subsequent crystallographic analysis of the toxin oligomer. With several mutant cysteines within domain 3 of SLO (e.g. residue S286C), the acrylodan emission spectra of the membrane-associated oligomers closely resembled those of lipid-embedded residues within the α-toxin pore. It has thus been suggested that, during oligomerization, domain 3 participates in membrane penetration, which implies a more thorough conformational reorganization of the toxin molecule than that suggested by the above perfringolysin model.

We shall now consider a model for events during SLO oligomerization and pore formation, in the light of structural knowledge of the molecule (see Figure 25.4).

Initially, the toxin monomer binds reversibly to the membrane. Binding imposes a conformational change to the SLO molecule, which involves not only domain 4 but also domains 1 and 3, which participate in the

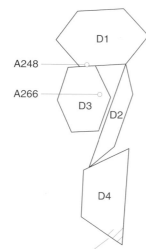

FIGURE 25.3 Domain structure of monomeric SLO (based on the crystal structure of the homologous molecule perfringolysin O, Rossjohn et al., 1997). The molecule has an elongated shape and consists of four domains (D1–D4). The tryptophan-rich motif within domain 4 also contains the single residue of the wild-type molecule and is involved in primary binding to membrane cholesterol. A change in conformation during membrane binding of the SLO monomer has been detected at residues A248 and A266, the approximate locations of which are also indicated.

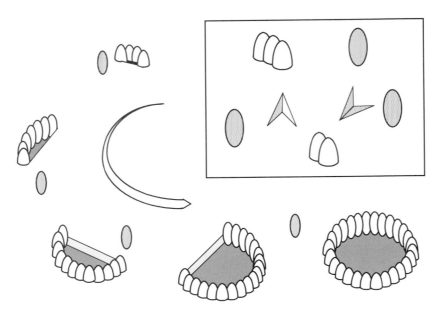

FIGURE 25.4 Hypothetical molecular mechanism of SLO oligomerization and pore formation. Inset: the initial stages of oligomerization. In the initiation step, two reversibly membrane-bound monomers (stippled ovals) join into a dimer. Concomitantly, the two monomers again change conformation and insert into the membrane. Subsequently, the dimer is elongated by association of another monomer. Here, only the additional monomer is required to change conformation and to penetrate the lipid bilayer. The elongation step is therefore kinetically favoured over initiation. The elongation reaction is serially repeated upon the trimer, which results in a successively growing arc-shaped oligomerization intermediate and, ultimately, in the circular pore; the complete oligomer consists of about 50 subunits. The membrane lesion arises at an early stage of the oligomer: its concave surface, owing to its hydrophilic character, repels the adjacent lipid molecules, which retract to form a straight line across the two ends of the arc. The lesion then grows in size concomitantly with the oligomer up to its final size.

subsequent steps of toxin action. The conformational transition thereby acts as a trigger of oligomerization, so that the latter is confined to the membrane-bound state of the SLO monomer. The membrane thus acts as an allosteric effector in eliciting its own permeabilization. The oligomerization of bound SLO is governed by a two-step kinetic mechanism, which resembles the polymerization of actin. The first (nucleation) step involves two membrane-bound SLO monomers, which combine into a small oligomer. In the second step, this small oligomer is elongated by sequential addition of further monomers. The elongation reaction is considerably faster than nucleation, which thus limits the overall rate of oligomerization. These kinetic characteristics result in quick completion of an oligomer once it has been initiated.

The intermediates of oligomer growth form arc-shaped structures in the membrane plane. At a very early stage, the arc-shaped oligomer inserts into the lipid bilayer. Similar to the inner circumference of a fully assembled pore, the concave surface of an arc-shaped oligomer must be ascribed a hydrophilic character. It will therefore deprive the adjacent lipid molecules of hydrophobic adhesion; these, in turn, will become subject to surface tension and hence arrange into a straight

line spanning the two ends of the arc. In conclusion, a membrane-inserted arc will evoke a crescent-shaped membrane defect lined on one side by the protein oligomer and on the other by a free edge of the lipid membrane. Initially, the arc pore will be small, but it will grow with each successive addition of a further monomer to one of its free ends. Concomitantly, the free membrane edge will be expanded against the resistance of its surface tension. However, it will contract again once the arc has grown to more than half a ring, and it may finally guide the ends of the oligomer to come close and merge, so that the ring-shaped SLO pore is completed.

Experimental approaches

The proposed molecular mechanism of SLO pore formation has a series of implications amenable to experimental testing. These include: (a) the reversibility of monomer binding; (b) the change in conformation that accompanies monomer binding to provide the allosteric switch of oligomerization; (c) the two-step kinetic mechanism of oligomerization; (d) the detection of arc-shaped oligomerization intermediates; (e) the association with these arcs of pores reduced in size.

The reversibility of monomer binding was first demonstrated by the following simple experiment: erythrocytes were exposed to SLO at low temperature in order to allow for monomer binding (which readily occurs under these conditions) but to prevent lysis (which is inhibited at low temperature), as shown in earlier works on SLO–erythrocyte interaction (Herbert and Todd, 1941; Alouf and Raynaud, 1968; Oberley and Duncan, 1971; Fehrenbach et al., 1982). The toxin-treated cells were then washed to remove unbound SLO and mixed with an excess of untreated cells. After an incubation for some time, the temperature was raised in order to induce cell lysis. By photometric quantitation of the haemoglobin released, it was found that the cells lysed outnumbered those initially treated with the toxin, indicating that monomeric SLO had migrated between cells before lysis was initiated and, therefore, had been reversibly associated with the membrane prior to oligomerization (Kanbayashi et al., 1972).

The reversal of monomer binding to erythrocyte membranes has also been observed in a more direct way by use of a radio-iodinated SLO preparation (Palmer et al., 1995). In that experiment, erythrocytes were first incubated with the radiolabelled toxin, which was employed at a concentration sufficiently low to prevent oligomerization at all. The cells were washed by centrifugation to remove unbound SLO and then suspended in a solution of the thiol-specific reagent, dithio-bis-nitrobenzoic acid (DTNB; Ellmans reagent). From the mixture, samples were drawn after various time intervals and centrifuged to quantitate the proportion of cell-bound radioactivity in the pellet. Within a few minutes, the labelled SLO accumulated in the supernatant, whereas, in a control sample without DTNB, the label remained associated with the cells. This indicated that SLO molecules, upon spontaneous dissociation from the membrane, were trapped in solution by reaction with DTNB prior to re-binding to the cells. The rate of SLO dissociation did not depend on the concentration of DTNB, and it stopped immediately when DTNB was removed from the sample. These findings ruled out the possibility that DTNB had acted directly upon cell-bound SLO, and they reinforce the notion that the essential sulfhydryl group of SLO is directly involved in membrane binding of the monomer.

The second prediction of the oligomerization model is that a conformational difference exists between the water-soluble and the membrane-associated state of the SLO monomer. This issue was experimentally addressed by spectrofluorometric characterization of acrylodan-labelled single cysteine replacement mutants of SLO, according to the method detailed above for the screening of membrane-inserted residues. The rationale of these experiments was that, apart from insertion of a given labelled residue into the lipid bilayer, a change in conformation may also induce a local change in environmental polarity.

A total of 23 single cysteine replacement mutants of SLO has been examined by acrylodan fluorescence assay (Palmer et al., 1996, 1998a). Among these, four mutant cysteine residues (S218C, A248C, A266C, T277C) were detected that exhibited a distinct emission shift associated with binding of the monomer to erythrocyte membranes. Strikingly, with two of these mutants (S218C and A266C), the emission maximum of the membrane-bound monomer was red-shifted with respect to the water-soluble monomer. This indicated an increased ambient polarity upon binding, so that the labelled residue had definitely not penetrated the lipid membrane. The fluorescence spectra of mutants A248C and T277C were blue-shifted upon binding, but the respective emission maxima remained well above those typical of membrane-inserted amino acid residues. It is thus probable that with all four mutants the spectral shifts were due to a local change in conformation. The residues affected are all located within domains 1 or 3 and, therefore, are remote from the membrane binding site of the monomer. Covalent modification of one of these residues (A248C) as well as an adjacent one (T250C) interferes with oligomerization (see below). Moreover, three of the four residues are subject to further changes in environment during oligomerization, as are two residues in the vicinity of T277C (S274C and S286C, respectively). In conclusion, membrane binding conformationally affects a distant part of the molecule that is intimately involved in the subsequent steps of pore formation, thereby qualifying as a prime candidate for the allosteric switch in question.

Kinetics of toxin oligomerization

The third element of the proposed model of SLO pore formation is concerned with the kinetic mechanism of oligomer assembly. This subject has been quantitatively analysed by use of a radio-iodinated SLO tracer preparation (Palmer et al., 1995). The toxin was first bound to erythrocytes on ice; following removal of excess unbound SLO, oligomerization was suddenly started by rapid dilution of the sample with warm buffer. After various time intervals, samples were drawn from the reaction mixture and immediately solubilized by addition of the detergent sodium deoxycholate.

This step served both to terminate oligomerization and to facilitate subsequent separation of monomers from oligomers by sucrose density centrifugation

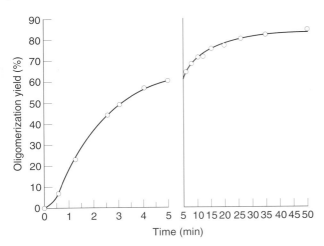

FIGURE 25.5 Kinetics of SLO oligomerization. Rabbit erythrocytes were incubated with ^{125}I-labelled wild type SLO at 0°C. After centrifugation to remove unbound label, oligomerization was started by resuspending the cells in warm buffer (34°C), and the reaction was allowed to proceed at the same temperature. Aliquots were drawn from the reaction at various time intervals, and the amount of label incorporated into oligomers was quantitated. The curve represents the best numerical fit of the initiation–elongation kinetics model (see text) to the experimental values, whereby the two panels depict the early and late phases of the reaction at different time scales. (Reproduced with permission from Palmer *et al.*, 1995.)

(Bhakdi *et al.*, 1985). The fraction of SLO incorporated into oligomers was plotted and analysed as a function of time. An example of such a kinetic experiment is illustrated in Figure 25.5. The time course of oligomerization is evidently sigmoidal in its early stage, which indicates that the reaction is kinetically influenced by at least two successive steps. A quantitative kinetics model was developed, which assumed oligomerization to occur through two separate successive reactions. The initial nucleation event would produce a dimer from two membrane-bound monomers. In the second reaction, the dimer would subsequently be extended by addition of a further monomer to yield a trimer, and this elongation event would be iterated on the trimer and so on up to the final size of the oligomer. By fitting the collective experimental data to this kinetics model, it was deduced that elongation, although it has to be repeated some 50 times on each growing oligomer, is much faster than nucleation. The latter thus has a rate-limiting role in the whole oligomerization reaction. This kinetic pattern ensures that an oligomer, once initiated, will be rapidly completed and not run out of available monomers.

Ring- and arc-shaped structures on target cells

Apart from typical circular membrane lesions, arc-shaped oligomers have been repeatedly observed by electron microscopy of erythrocyte membranes treated with wild-type SLO, as mentioned above. These arcs have been proposed to create functional lesions in the membrane (Bhakdi *et al.*, 1985). While an arc pore would be lined on one side by the protein oligomer, it would be faced by a free edge of the lipid membrane on the other side. However, owing to the concomitant presence of circular lesions in excess, the arc pore could not be substantiated by functional evidence, so that this concept was not widely accepted.

The arc pore is an essential element in the model of SLO pore formation presented here, since it represents the nascent state of the membrane lesion. An important piece of evidence in support of this hypothesis has been obtained recently by the use of a functionally variant derivative of SLO (Palmer *et al.*, 1998b). This protein was generated by thiol-specific modification of the single cysteine mutant, T250C, with the charged reagent *N*-(iodoacetaminoethyl)-1-naphthylamine-5-sulfonic acid (IAEDANS). The modified protein (T250C-AEDANS) was capable of binding to membranes but then failed to initiate oligomerization. However, when admixed with functionally intact SLO, this mutant could be incorporated into hybrid oligomers. Characterization by density gradient centrifugation indicated that T250C-AEDANS caused the hybrid oligomers to be reduced in size in a dose-dependent manner. The small hybrid oligomers appeared as typical arcs in the electron microscope, and again the free membrane edges were clearly visible. Importantly, with a sufficient excess of T250C-AEDANS over the active SLO species, arcs could be observed in the virtual absence of rings, so that their ability to permeabilize the target membrane could be functionally tested. To this end, rabbit erythrocytes were incubated with various mixtures of active SLO and T250C-AEDANS. Colloid-osmotic lysis occurred readily, indicating that the arcs indeed effected permeabilization. When the same experiment was repeated in the presence of dextran at an osmotically balanced concentration, the cells were partially protected from lysis. The largest dextran species that could be employed in these experiments was dextran 60 (M_r 60 000). The molecular diameter of dextran 60 has been estimated as 10 nm (Scherrer and Gerhardt, 1971), and it should therefore readily permeate the complete ring-shaped pore. In keeping with this expectation, dextran 60 exhibited hardly any protection of red cells treated with wild-type SLO. In contrast, when T250C-AEDANS was employed at a threefold excess over active SLO, dextran 60 virtually abolished haemolysis. With this toxin mixture, no complete inhibition but a marked delay of cell lysis was observed even with dextran 4 (molecular diameter: 3 nm). Intermediate protective effects were observed with toxin mixtures containing lower amounts of T250C-AEDANS. Evidently, the functional diameter of the arc pores is reduced in relation to the size of the arc, which conforms to the prediction of the pore-formation model presented above.

Entirely consistent results were obtained with experiments that relied on the release of marker molecules of different size from resealed erythrocyte ghosts. Rhodamine-labelled dextrans were included inside the ghosts, along with the small marker calcein. With mixtures of active SLO and T250C-AEDANS, a differential release of the two markers was observed, whereby the degree of discrimination again depended on both the relative molecular mass of the dextran and the toxin ratio. An important additional finding was that, with T250C-AEDANS in large excess, pores could be detected that effected discrimination between calcein ($M_r=622$) and a small dextran species ($M_r=4400$). These very small SLO pores should correspond to very small oligomers. Pore formation therefore probably commences at a very early stage of oligomerization. In conclusion, incomplete SLO oligomers are endowed with size-selective permeability characteristics that are in full agreement with the concept of the progressively growing arc pore.

The behaviour of the T250C-AEDANS derivative of SLO is strikingly similar to that of actin molecules following ADP-ribosylation (due to the clostridial C3 enzyme). Both molecules are incapable of initiating oligomerization, but they may join growing oligomers of the respective functional species. With SLO, this confirms the existence of distinct nucleation and elongation steps in the oligomerization reaction. Incorporation of both T250C-AEDANS and the ADP-ribosylated actin into the homologous oligomers causes premature termination of the latter. This mode of interfering with polymerization has been aptly termed 'capping' in the case of actin. Circumstantial evidence indicates that the capping effect of T250C-AEDANS is incomplete, so that addition of further functional subunits subsequent to a T250C-AEDANS monomer is delayed rather than abrogated. Nevertheless, the data clearly indicate that the region around residue T250 has a crucial role in oligomerization. As already stated, chemical derivatization of the mutant cysteine A248C yields very much the same findings as noted here for T250C-AEDANS. The corresponding molecular region is thus both conformationally affected by binding and functionally involved in oligomerization, which ties in with the idea that binding acts as a trigger of oligomer and pore formation.

Although available experimental evidence is in good agreement with the presented model of pore-formation, significant areas remain unclear. One of these unsettled issues is the critical size that the oligomer must attain in order to insert into the membrane and thereby initiate pore formation. The observation of very small arc pores (as evident from selective permeability for molecules of 600 as opposed to 4000 Da) suggests that membrane insertion occurs very early during oligomerization, possibly even at the dimer stage. However, this question awaits more definite experimental clarification. Similarly, the hypothesis that the target membrane acts as an allosteric effector that induces oligomerization of the SLO molecule requires additional corroboration. For example, it should be possible to obtain point mutations that abrogate the conformational transition of domains 1 and 3 along with the ability of the molecule to oligomerize.

As the major challenge, the structure of the oligomer remains to be elucidated. Most of the available information on sites of contact between its subunits and of interaction with the target lipid bilayer is either preliminary or presumptive, and this subject deserves both more exacting and more thorough analysis. A detailed comparison of the oligomer structure to that of the monomer would certainly yield new insights into the molecular mechanism of SLO. The comparative analysis of monomeric and oligomeric molecular structures presently represents a major goal in the whole field of pore-forming toxins, but so far has met with limited success. The only high-resolution oligomer structure available is that of staphylococcal alpha-haemolysin (Song et al., 1996). The alpha-haemolysin oligomer is much smaller and uniformly consists of seven subunits. With this toxin, the mechanism of pore formation deviates substantially from that of SLO. The oligomerization of alpha-haemolysin does not accompany but rather precedes membrane permeabilization; only when all subunits have assembled into the heptamer do they cooperatively insert into the membrane to form the pore (Walker et al., 1992; Valeva et al., 1997). Accordingly, no pores associated with incomplete oligomers have been detected. This may serve to illustrate that there is variation in the field of pore-forming toxins and, therefore, certainly room for discovery.

PATHOPHYSIOLOGICAL EFFECTS OF STREPTOLYSIN O

Since the early observations on the lethal, cardiotoxic tissue-damaging cytotoxic and cytolytic properties of SLO, a large volume of literature had accumulated by the end of the 1970s (Bernheimer, 1954, 1974; Halbert, 1970; Ginsburg, 1972; Alouf, 1980). We will briefly summarize the earlier works in this section and will focus on the recent achievements at the turn of the 1990s.

Lethal toxicity

In earlier reports, intravenous injection of partially purified SLO preparations (see Bernheimer, 1954) or highly purified SLO (Halbert, 1970; Alouf and Raynaud, 1973) was rapidly lethal for mice, rats, guinea pigs, rabbits and cats. These animals differed in their susceptibility to the toxin. The average LD_{50} doses by the i.v. route were 0.2 µg, 6 µg and 2 µg in the mouse, guinea pig and rabbit, respectively. The minimal lethal dose for 9-day embryonated eggs was 20 ng (Alouf and Raynaud, 1973). The titrations of lethal activity tend to show sharp endpoints. Death occurred in mice or rabbits within seconds or minutes, depending on doses, and was preceded by convulsions, nasal frothing, pulmonary oedema and haemorrhage. Intravascular haemolysis occurred, but to a degree insufficient to account for death. With several multiples of a lethal dose, death was almost instantaneous. At dose levels that kill only a portion of the animals, death generally occurred within minutes or survival was indefinite (Bernheimer, 1954; Halbert, 1970).

Cardiotoxicity

The almost instantaneous lethal effects of SLO on laboratory animals (rabbits, mice, rats) can be largely attributed to the direct cardiotoxicity of the toxin, as shown *in vivo* and *in vitro* on isolated heart and derived tissues (see Bernheimer, 1954, 1974; Halbert *et al.*, 1961; Coraboeuf and Goullet, 1963; Halpern and Rahman, 1968; Halbert, 1970; Gupta and Gupta, 1979). Electrocardiogram tracings in rabbits and mice given lethal doses of SLO revealed profound disorganization of heart function, characterized by conduction defects and ventricular disturbances. The possible mechanism(s) of the cardiotoxicity of SLO are still not entirely clear. On the basis of the *in vivo* observations on heart functions, the extremely rapid cardiac alterations elicited by SLO might result from the release of vasoactive substances rather than from the direct action of the protein itself, as substantiated by the observation that smaller lethal or sublethal doses immediately produced a pronounced but transient sinus bradycardia followed by temporary or permanent recovery.

Tissue damage and inflammatory effects

Several lines of evidence suggest that SLO may be involved alone or in combination with other streptococcal exotoxins or enzymes in various types of tissue damage that could be relevant to streptococcal diseases. On the basis of autopsy data from cases of streptococcal toxic shock syndrome (STSS), showing the accumulation of polymorphonuclear leucocytes (PMNL) within lung and soft tissue microvasculature, SLO and the erythrogenic (pyrogenic) exotoxin A (SPE A) were investigated as to their potential capacity to stimulate *in vitro* PMNL-dependent adherence on gelatin matrices (Bryant *et al.*, 1992). SPE A modestly enhanced adherence, in contrast to the strong adherence elicited by SLO and mediated by CD11/CD18 adherence glycoprotein.

Intratracheal administration of SLO caused lung injury in rats, as measured by pulmonary vascular leak (Shanley *et al.*, 1996). This injury was synergistically augmented by the streptococcal cysteine proteinase (erythrogenic exotoxin B) paralleled by neutrophil accumulation and production of the pro-inflammatory cytokines interleukin-1β (IL-1β) and tumour necrosis factor (TNF-α). Interestingly, SLO was previously shown to elicit *in vitro* high amounts of these cytokines by human monocytes (Hackett and Stevens, 1992). This effect was potentiated by SPE A, which is considered as a major effector in STSS pathogenesis (see Chapter 32 on '*S. pyogenes* erythrogenic toxins'). Recently, SLO, SPE A, cysteine proteinase and viable or killed group A streptococci, including SLO-deficient strains, were tested *in vitro* on human mononuclear cells for TNF-α production in response to these stimulants (Stevens and Bryant, 1997).

SLO and other related toxins elicited important modulation of lipid effectors of the inflammatory cascade in immune system cells (Bremm *et al.*, 1985, 1987; Köller *et al.*, 1993). An increase in the release of leukotrienes (LT), particularly the chemotactic factor LTB4, the spasmogenic factors, LTC4, LTD4 and LTE4, as well as leukotriene-metabolizing enzymes was observed. This is probably related to the inflammatory responses in chronic and acute streptococcal and clostridial diseases. The inflammatory potential of SLO was shown in the context of SLO-induced complement activation and attack on autologous cell membranes (Bhakdi and Tranum-Jensen, 1985).

SLO-deficient mutants have been constructed by Ruiz *et al.* (1998) by insertional inactivation of the *slo* gene in various *S. pyogenes* strains. Inability to produce SLO caused a considerable reduction in the expression of various pro-inflammatory cytokines and prostaglandin E_2 by human keratinocytes infected with the strains lacking the gene. The failure of the mutants to induce these pharmacological responses was attributed to the absence of SLO-induced permeabilization of target cell membrane. Isogenic insertion inactivation of *slo* gene of an M1 serotype *S. pyogenes* strain by Schmidt *et al.* (1998) resulted in a reduced virulence of the strain in the chicken embryo model, suggesting that SLO contributes to the pathogenicity of group A streptococci.

APPLICATIONS OF STREPTOLYSIN O IN CELL BIOLOGY

The extraordinary size of the SLO pore readily allows for passage of proteins and other macromolecules into and out of the target cell cytosol. SLO has therefore become popular for the permeabilization of cells in cell biological experiments. Although this procedure will ultimately result in cell death, several elements of intracellular metabolism and trafficking remain amenable to experimental study in the permeabilized cells if these are immersed in a 'cytosolic' milieu. A thorough coverage of the many different applications is beyond the scope of this chapter, so just a few recent and instructive examples will be mentioned here. In a study on the effect of actin disassembly upon exocytosis, various actin-binding proteins were introduced into SLO-permeabilized pancreatic acinar cells (Muallem et al., 1995). The dependence of insulin secretion on SNAP-25 (a protein involved in neuronal transmitter release) was probed by selective proteolytic cleavage with botulinum toxin; the latter would not have entered unpermeabilized pancreatic B-cells. Cytosolic components required in peroxisomal protein import were examined with permeabilized cells using model proteins consisting of albumin or luciferase fused to the peroxisomal signal sequence (Wendland and Subramani, 1993). In several elegant studies on polarized protein sorting in epithelial cells, these were grown as monolayers on a filter support dividing a two-chamber system, so that the apical and the basolateral membranes could be accessed separately. Following SLO permeabilization of the cells, tagged proteins were introduced that selectively interfered with either apical or basolateral protein transport. Components of the respective sorting machinery were captured by interaction with the tags (Pimplikar et al., 1994; Ikonen et al., 1995; Yoshimori et al., 1996). Finally, SLO has been applied to the study of intracellular parasites. The parasitic cells are commonly contained in a parasitophorous vacuole. With toxoplasmas (Beckers et al., 1994) and malarial plasmodia (Ansorge et al., 1996, 1997), the vacuolar membrane becomes exposed upon destruction of the cytoplasmic membrane with SLO but, owing to its low content of cholesterol, resists attack by the toxin, so that parasite metabolism may continue and become amenable to analysis in the absence of interfering host cell activities. Selective permeabilization of the host cell therefore holds great promise for future research on these increasingly important subjects.

Apart from its permeabilizing activity, there is a second potentially useful feature of SLO, which so far, however, has met with limited interest by cell biologists. It is quite well established that cholesterol is the ligand of the thiol-activated toxins in the cell membrane, and also that there is a direct (though not linear) relation between the membrane cholesterol content and the number of toxin binding sites (Ohno Iwashita et al., 1992; Iwamoto et al., 1987). SLO derivatives (labelled fluorescently or otherwise) should thus be suitable tools for visualizing the distribution and movement of cholesterol within biological membranes. Among the various mutant derivatives available, those devoid of cytolytic activity should be particularly useful, because they may be combined with non-destructive imaging methods, such as confocal laser scanning microscopy. Serious concerns have been raised about the use of filipin for labelling of cellular cholesterol (Behnke et al., 1984). It thus appears worthwhile, or even overdue, to explore SLO as an alternative cholesterol marker.

CONCLUDING REMARKS AND TRENDS FOR FUTURE RESEARCH

Our knowledge on SLO and other related toxins has progressed considerably over the past 10 years. The greatest achievements in the research area of SLO concerned the molecular genetics and the mechanisms of toxin-induced membrane damage. Techniques of gene cloning, sequencing and site-directed mutagenesis allowed a detailed understanding of the structure of this molecule and led to important conclusions about structure–activity relationships. Among the most important findings is the demonstration that the unique cysteine residue is not critical per se through its SH group for activity, as thought for over 60 years. Another important achievement is the considerable advance in our knowledge of the mechanism of pore formation through toxin oligomerization on target cells.

Remarkable developments were also realized in the domain of the pathophysiological properties of SLO, particularly the inflammatory aspects in animal models or in immune system cells in vitro. The progress of our knowledge in the field of basic and clinical cytokine research was a major step in our understanding of the possible involvement of SLO in streptococcal pathogenesis.

Research on SLO allowed wide applications of this toxin in cell biology for the permeabilization of eukaryotic cell membranes and as a potent probe in the study of their structural and biochemical architecture. Future research should involve the study of genetic regulation of toxin expression, which is still largely unknown. Toxin crystallization for the determination of its three-dimensional structure, as already realized for perfringolysin O, will progress the study of structure–activity relationships, particularly in the case of co-crystallization of SLO with cholesterol.

The precise evaluation of the role of SLO in disease, particularly as a virulence factor (similarly to listeriolysin O) in the pathogenesis of acute and chronic infections by *S. pyogenes* and their non-suppurative sequelae (rheumatic fever, post-streptococcal glomerulonephritis and Sydenham's chorea) still awaits investigation.

ACKNOWLEDGEMENTS

The authors thank all colleagues who have made the results of their studies available to them. The invaluable secretarial assistance of Mrs Patricia Paul in the preparation of this manuscript is greatly appreciated.

REFERENCES

Alouf, J.E. (1977). Cell membranes and cytolytic bacterial toxins. In: *Specificity and Action of Animal, Bacterial and Plant Toxins* (ed. P. Cuatrecasas), pp. 211–70. Chapman & Hall, London.

Alouf, J.E. (1980). Streptococcal toxins (streptolysin O, streptolysin S, erythrogenic toxin). *Pharmacol. Ther.* **11**, 661–717.

Alouf, J.E. and Geoffroy, C. (1979). Comparative effects of cholesterol and thiocholesterol on streptolysin O. *FEMS Microbiol. Lett.* **6**, 413–16.

Alouf, J.E. and Geoffroy, C. (1988). Production, purification and assay of streptolysin O. In: *Microbial Toxins; Tools in Enzymology. Methods in Enzymology*, Vol. 165 (ed. S. Harshman), pp. 52–9. Academic Press, San Diego.

Alouf, J.E. and Geoffroy, C. (1991). The family of the antigenically-related cholesterol-binding (sulphydryl-activated) cytolytic toxins. In: *Sourcebook of Bacterial Toxins* (eds J.E. Alouf and J.H. Freer), pp. 147–86. Academic Press, London.

Alouf, J.E. and Raynaud, M. (1968). Some aspects of the mechanisms of lysis of rabbit erythrocytes by streptolysin O. In: *Current Research on Group A Streptococcus* (ed. R. Caravano), pp. 192–206. Excerpta Medica Foundation, Amsterdam.

Alouf, J.E. and Raynaud, M. (1973). Purification and some properties of streptolysin O. *Biochimie* **55**, 1187–93.

Alouf, J.E., Geoffroy, C., Pattus, F. and Verger, R. (1984). Surface properties of bacterial sulfhydryl-activated cytolytic toxins. Interaction with monomolecular films of phosphatidylcholine and various sterols. *Eur. J. Biochem.* **141**, 205–10.

Alouf, J.E., Geoffroy, C., Gilles, A.M. and Falmagne, P. (1986). Structural relatedness between five bacterial sulphydryl activated toxins; streptolysin O, perfringolysin O, alveolysin, pneumolysin and thuringiolysin. In: *Bacterial Protein Toxins* (eds P. Falmagne, J.E. Alouf, F.J. Fehrenbach, J. Jeljaszewicz and M. Thekstam), pp. 49–50. Gustav Fischer, Stuttgart.

Ansorge, I., Benting, J., Bhakdi, S. and Lingelbach, K. (1996). Protein sorting in *Plasmodium falciparum*-infected red blood cells permeabilized with the pore-forming protein streptolysin O. *Biochem. J.* **315**, 307–14.

Ansorge, I., Paprotka, K., Bhakdi, S. and Lingelbach, K. (1997). Permeabilization of the erythrocyte membrane with streptolysin O allows access to the vacuolar membrane of *Plasmodium falciparum* and a molecular analysis of membrane topology. *Mol. Biochem. Parasitol.* **84**, 259–61.

Ayoub, E.M. and Harden, E. (1992). Immune responses to streptococcal antigens; diagnostic methods. In: *Manual of Clinical Laboratory Immunology* (eds N.R. Rose, E. Conway de Macario, J.L. Fahey, H. Friedman and G.M. Fenn), pp. 427–34. American Society for Microbiology, Washington, DC.

Beckers, C.J., Dubremetz, J.F., Mercereau Puijalon, O. and Joiner, K.A. (1994). The *Toxoplasma gondii* rhoptry protein ROP 2 is inserted into the parasitophorous vacuole membrane, surrounding the intracellular parasite, and is exposed to the host cell cytoplasm. *J. Cell Biol.* **127**, 947–61.

Behnke, O., Tranum Jensen, J. and van Deurs, B. (1984). Filipin as a cholesterol probe. II. Filipin-cholesterol interaction in red blood cell membranes. *Eur. J. Cell Biol.* **35**, 200–15.

Bernheimer, A.W. (1954). Streptolysins and their inhibitors. In: *Streptococcal Infections* (ed. M. McCarty), pp. 19–38. Columbia University Press, New York.

Bernheimer, A.W. (1972). Haemolysins of streptococci; characterization and effects on biological membranes. In: *Streptococci and Streptococcal Diseases* (eds L.W. Wannamaker and J.M. Matsen), pp. 19–31. Academic Press, New York.

Bernheimer, A.W. (1976). Sulfhydryl activated toxins. In: *Mechanisms in Bacterial Toxinology* (ed. A.W. Bernheimer), pp. 85–97. John Wiley, New York.

Bhakdi, S. and Tranum-Jensen, J. (1985). Complement activation and attack on autologous cell membranes induced by streptolysin O. *Infect. Immun.* **48**, 713–19.

Bhakdi, S., Roth, M., Sziegoleit, A. and Tranum-Jensen, J. (1984). Isolation and identification of two hemolytic forms of streptolysin O. *Infect. Immun.* **46**, 394–400.

Bhakdi, S., Tranum-Jensen, J. and Sziegoleit, A. (1985). Mechanism of membrane damage by streptolysin O. *Infect. Immun.* **47**, 52–60.

Billington, S.J., Jost, B.H., Cuevas, W.A., Bright, K.R. and Songer, J.G. (1997). The *Arcanobacterium (Actinomyces) pyogenes* hemolysin, pyolysin, is a novel member of the thiol-activated cytolysin family. *J. Bacteriol.* **179**, 6100–6.

Braun, V. and Focareta, T. (1991). Pore-forming bacterial protein haemolysins (cytolysins). *Crit. Rev. Microbiol.* **18**, 115–58.

Bremm, K.D., König, W., Pfeiffer, P., Rauschen, I., Theobald, K., Thelestam, M. and Alouf, J.E. (1985). Effect of thiol-activated toxins (streptolysin O, aveolysin and theta toxin) on the generation of leukotrienes and leukotriene-inducing and -metabolizing enzymes from human polymorphonuclear granulocytes. *Infect. Immun.* **50**, 844–51.

Bremm, K.D., König, W., Thelestam, M. and Alouf, J.E. (1987). Modulation of granulocyte functions by bacterial exotoxins and endotoxins. *Immunology* **62**, 363–71.

Bryant, A.E., Kehoe, M.A. and Stevens, D.L. (1992). Streptococcal pyrogenic exotoxin A and streptolysin O enhances polymorphonuclear leukocyte binding to gelatin matrixes. *J. Infect. Dis.* **166**, 165–9.

Buckingham, L. and Duncan, J.L. (1983). Approximate dimensions of membrane lesions produced by streptolysin S and streptolysin O. *Biochim. Biophys. Acta* **729**, 115–22.

Coraboeuf, E. and Goullet, E. (1963). Quelques aspects de l'action de la streptolysine O sur le coeur isolé du rat. *J. Physiol. Paris* **55**, 232–3.

Cowell, J.L., Grushoff-Kosyk, P. and Bernheimer, A.W. (1976). Purification of cereolysin and the electrophoretic separation of the active (reduced) and inactive (oxidized) forms of the purified toxin. *Infect. Immun.* **14**, 144–54.

Dassy, B. and Alouf, J.E. (1983). Growth of *Streptococcus pyogenes* and streptolysin O production in complex and synthetic media. *J. Gen. Microbiol.* **129**, 643–51.

Dobereiner, A.H., Agrawal, P., Pinkney, M., Glanville, M., Palmer, M., Weller, U., Messner, N., Kapur, V., Musser, J., Bhakdi, S. and Kehoe, M.A. (1996). Structure–function studies on streptolysin O (SLO). In: *Bacterial Protein Toxins* (eds P. Frandsen, J.E. Alouf, P. Falmagne, F.J. Fehrenboch, J.H. Freer, C. Montecucco, S. Olsnes, R. Rappuoli and T. Wadström), pp. 103–4. *Zentralbl. Bakt. Suppl.* 28. Gustav Fischer, Jena.

Dourmashkin, R.R. and Rosse, W.F. (1966). Morphology changes in the membranes of red blood cells undergoing hemolysis. *Am. J. Med.* **41**, 699–710.

Duncan, J.L. (1983). Streptococcal growth and toxin production *in vivo*. *Infect. Immun.* **40**, 501–5.

Duncan, J.L. and Buckingham, L. (1980). Resistance to streptolysin O in mammalian cells treated with oxygenated derivatives of cholesterol; cholesterol content of resistant cells and recovery of streptolysin O sensitivity. *Biochim. Biophys. Acta* **603**, 278–87.

Duncan, J.L. and Schlegel, R. (1975). Effect of streptolysin O on erythrogenic membranes, liposomes and lipid dispersions. *J. Cell. Biol.* **67**, 160–73.

Fehrenbach, F.-J., Schmidt, C.-M. and Huser, H. (1982). Early and long events in streptolysin O-inducing hemolysis. *Toxicon* **20**, 233–8.

Gaillard, J.-L., Berche, P., Mounier, J., Richard, S. and Sansonetti, P. (1987). *In vitro* model of penetration and intracellular growth of *Listeria monocytogenes* in the human enterocyte-like cell line Caco-2. *Infect. Immun.* **55**, 2822–9.

Gerlach, D., Kohler, W., Gunther, E. and Mann, K. (1993). Purification and characterization of streptolysin O secreted by *Streptococcus equisimilis* (group C). *Infect. Immun.* **61**, 2727–31.

Ginsburg, I. (1972) Mechanisms of cell and tissue injury induced by group A streptococci; relation to post-streptococcal sequelae. *J. Infect. Dis.* **126**, 294–340.

Goebel, W. and Kreft, J. (1997). Cytolysins and the intracellular life of bacteria. *Trends Microbiol.* **5**, 86–8.

Gupta, S.O. and Gupta, R.K. (1979). Disruption of electrocardiographic activity by streptolysin O in rats. *Toxicon* **17**, 167–9.

Haas, A., Dumbsky, M. and Kreft, J. (1992). Listeriolysin genes; complete nucleotide sequence of *ilo* ivanolysin from *Listeria ivanovii* and of listeriolysin from *Listeria seeligeri*. *Biochim. Biophys. Acta* **1130**, 81–4.

Halbert, S.P. (1970). Streptolysin O. In: *Microbial Toxins*, Vol. III (eds T.C. Montie, S. Kadis and S.J. Ajl), pp. 69–98. Academic Press, New York.

Halbert, S.P., Bircher, R. and Dahle, E. (1961). The analysis of streptococcal infections. V. Cardiotoxicity of streptolysin O for rabbits *in vivo*. *J. Exp. Med.* **113**, 759–84.

Halpern, B.N. and Rahman, S. (1968). Studies on the cardiotoxicity of streptolysin O. *Br. J. Pharmacol.* **32**, 441–52.

Herbert, D. and Todd, E.W. (1941). Purification and properties of a haemolysin produced by group A haemolytic streptococci (streptolysin O). *Biochem. J.* **35**, 1124–39.

Hewitt, L.F. and Todd, E.W. (1939). The effect of cholesterol and of sera contaminated with bacteria on the haemolysins produced by haemolytic streptococci. *J. Pathol. Bacteriol.* **49**, 45–51.

Howard, J.G., Wallace, K.R. and Wright, G.P. (1953). The inhibitory effects of cholesterol and related sterols on haemolysis by streptolysin O. *Br. J. Exp. Pathol.* **34**, 174–80.

Hugo, F., Reichwein, J., Arvand, J., Kramer, S. and Bhakdi, S. (1986). Use of a monoclonal antibody to determine the mode of transmembrane pore-formation by streptolysin O. *Infect. Immun.* **54**, 641–5.

Ikonen, E., Tagaya, M., Ullrich, O., Montecucco, C. and Simons, K. (1995). Different requirements for NSF, SNAP and Rab proteins in apical and basolateral transport in MDCK cells. *Cell* **81**, 571–80.

Iwamoto, M., Ohno Iwashita, Y. and Ando, S. (1987). Role of the essential thiol group in the thiol-activated cytolysin from *Clostridium perfringens*. *Eur. J. Biochem.* **167**, 425–30.

Iwamoto, M., Ohno Iwashita, Y. and Ando, S. (1990). Effect of isolated C-terminal fragment of θ-toxin (perfringolysin O) on toxin assembly and membrane lysis. *Eur. J. Biochem.* **194**, 25–31.

Jeljaszewicz, J., Szmigielski, S. and Hryniewicz, W. (1978). Biological effects of staphylococcal and streptococcal toxins. In: *Bacterial Toxins and Cell Membranes* (eds J. Jeljaszewicz and T. Wadström), pp. 185–227. Academic Press, London.

Johnson, M.K., Geoffroy, C. and Alouf, J.E. (1980). Binding of cholesterol by sulfhydryl-activated cytolysins. *Infect. Immun.* **27**, 97–101.

Kanbayashi, Y., Hotta, M. and Koyama, J. (1972). Kinetic study of streptolysin O. *J. Biochem. (Tokyo)* **71**, 227–37.

Kehoe, M.A. and Timmis, K.N. (1984). Cloning and expression in *Escherichia coli* of the streptolysin O determinant from *Streptococcus pyogenes*; characterization of the cloned streptolysin O determinant and demonstration of the absence of substantial homology with determinants of other thiol-activated toxins. *Infect. Immun.* **43**, 804–10.

Kehoe, M.A., Miller, L., Walker, J.A. and Boulnois, G.J. (1987). Nucleotide sequence of the streptolysin O (SLO) gene; structural homologies between SLO and other membrane-damaging, thiol-activated toxins. *Infect. Immun.* **55**, 3228–32.

Kehoe, K., Walker, J.A., Boulnois, G.J., Shields, J. and Miller, L. (1988). Genetics and structure of thiol-activated toxins. In: *Bacterial Protein Toxins* (eds F.J. Fehrenbach, J.E. Alouf, P. Falmagne, W. Goebel, J. Jaljaszewicz, D. Jürgens and R. Rappuoli), pp. 201–5. *Zbl. Bakt. Suppl.* 17. Gustav Fischer, Stuttgart.

Köhler, W. (1963). *Die Serologie des Rheumatismus und der Streptokokkeninfektionen*, pp. 13–37. J.H. Barth, Leipzig.

Köller, M., Hensler, T. König, B., Prevost, G., Alouf, J. and König, W. (1993). Induction of heat-shock proteins by bacterial toxins, lipid mediators and cytokines in human leukocytes. *Zentralbl. Bakt.* **278**, 365–76.

Launay, J.M. and Alouf, J.E. (1979). Biochemical and ultrastructural study of the disruption of blood platelets by streptolysin O. *Biochim. Biophys. Acta* **556**, 278–91.

Launay, J.M., Geoffroy, C., Costa, J.L. and Alouf, J.E. (1984). Purified SH-activated toxins (streptolysin O, alveolysin); new tools for determination of platelet enzyme activities. *Thromb. Res.* **33**, 189–96.

Linder, R. (1979). Heterologous immunoaffinity chromatography in the purification of streptolysin O. *FEMS Microbiol. Lett.* **5**, 339–42.

Marmorek, A. (1902). La toxine streptococcique. *Ann. Inst. Pasteur* **16**, 169–77.

Morgan, P.J., Varley, P.G., Rowe, A.J., Andrew, P.W. and Mitchell, T.J. (1993). Characterization of the solution properties and conformation of pneumolysin, the membrane-damaging toxin of *Streptococcus pneumoniae*. *Biochem. J.* **296**, 671–4.

Morgan, P.J., Andrew, P.W. and Mitchell, T.J. (1996). Thiol-activated cytolysins. *Rev. Med. Microbiol.* **7**, 221–9.

Muallem, S., Kwiatkowska, K., Xu, X. and Yin, H.L. (1995). Actin filament disassembly is a sufficient final trigger for exocytosis in nonexcitable cells. *J. Cell Biol.* **128**, 589–98.

Müller-Alouf, H., Geoffroy, C. and Alouf, J.E. (1996). Comparative study of the inhibitory effects of amoxicillin and macrolides on the production of erythrogenic (pyrogenic) exotoxin A and streptolysin O by *Streptococcus pyogenes*. In: *Bacterial Protein Toxins* (eds P. Frandsen *et al.*), pp. 230–31. *Zentralbl. Bakt. Suppl.* 28. Gustav Fischer, Jena.

Müller-Alouf, H., Geoffroy, C., Geslin, P., Bouvet, A., Felten, A., Günther, E., Ozegowski, J.-H. and Alouf, J.E. (1997). Streptococcal pyrogenic, exotoxin A, streptolysin O, exoenzymes, serotype and biotype profiles of *Streptococcus pyogenes* isolates from patients with toxic shock syndrome and other severe infections. *Zentralbl. Bakt.* **286**, 421–33.

Neill, J.M. and Mallory, T.B. (1926). Studies on the oxidation and reduction of immunological substances. IV. Streptolysin. *J. Exp. Med.* **44**, 241–60.

Niedermeyer, W. (1985). Interaction of streptolysin O with membranes; kinetic and morphological studies on erythrocytes membranes. *Toxicon* **23**, 425–35.

Oberley, T.D. and Duncan, J.L. (1971). Characteristics of streptolysin O action. *Infect. Immun.* **4**, 683–7.

Ohno Iwashita, Y., Iwamoto, M., Ando, S. and Iwashita, S. (1992). Effect of lipidic factors on membrane cholesterol topology-mode of binding of theta-toxin to cholesterol in liposomes. *Biochim. Biophys. Acta* **1109**, 81–90.

Okumura, K., Hara, A., Tanaka, T., Nishiguchi, I., Minamide, W., Igarashi, H. and Yutsudo, T. (1994). Cloning and sequencing the streptolysin O genes of group C and group G streptococci. *DNA Seq.* **4**, 325–8.

Owen, R.H.G., Boulnois, G.J., Andrew, P.W. and Mitchell, T.J. (1994). A role in cell-binding for the C-terminus of pneumolysin, thiol-activated toxin of *Streptococcus pneumoniae*. *FEMS Microbiol. Lett.* **121**, 217–22.

Palmer, M., Valeva, A., Kehoe, M. and Bhakdi, S. (1995). Kinetics of streptolysin O self-assembly. *Eur. J. Biochem.* **231**, 388–95.

Palmer, M., Saweljew, P., Vulicevic, I., Valeva, A., Kehoe, M. and Bhakdi, S. (1996). Membrane-penetrating domain of streptolysin O identified by cysteine scanning mutagenesis. *J. Biol. Chem.* **271**, 26664–7.

Palmer, M., Vulicevic, I., Saweljew, P., Valeva, A., Kehoe, M. and Bhakdi, S. (1998a). Streptolysin O; a proposed model of allosteric interaction between a pore-forming protein and its target lipid bilayer. *Biochemistry* **37**, 2378–83.

Palmer, M., Harris, R., Freytag, C., Kehoe, M., Tranum-Jensen, J. and Bhakdi, S. (1998b). Assembly mechanism of the oligomeric streptolysin O pore; the early membrane lesion is lined by a free edge of the lipid membrane and is gradually extended during oligomerization. *EMBO J.* **17**, 1598–605.

Pimplikar, S.W., Ikonen, E. and Simons, K. (1994). Basolateral protein transport in streptolysin O-permeabilized MDCK cells. *J. Cell Biol.* **125**, 1025–35.

Pinkney, M., Beachey, E. and Kehoe, M. (1989). The thiol-activated toxin streptolysin O does not require a thiol group for cytolytic activity. *Infect. Immun.* **57**, 2553–8.

Pinkney, M., Kapur, V., Smith, J., Weller, U., Palmer, M., Glanville, M., Messner, M., Musser, J.M., Bhakdi, S. and Kehoe, M.A. (1995). Different forms of streptolysin O produced by *Streptococcus pyogenes* and by *Escherichia coli* expressing recombinant toxin: cleavage by streptococcal cysteine protease. *Infect. Immun.* **63**, 2776–9.

Portnoy, D.A., Tweten, R.K., Kehoe, M. and Bielecki, J. (1992). Capacity of listeriolysin O, streptolysin O and perfringolysin O to mediate growth of *Bacillus subtilis* with mammalian cells. *Infect. Immun.* **60**, 2710–17.

Prendergast, F.G., Meyer, M., Carlson, C.L., Iida, S. and Potter, J.D. (1983). Synthesis, spectral properties and use of 6-Acryloyl-2-dimethylaminonaphthalene (Acrylodan). *J. Biol. Chem.* **258**, 7541–4.

Prigent, D. and Alouf, J.E. (1976). Interaction of streptolysin O with sterols. *Biochim. Biophys. Acta* **443**, 288–300.

Rahman, S., Rebeyrotte, P., Halpern, B. and Beslman, D. (1969). Kinetics of streptolysin O reactivation by different thiols. *Biochim. Biophys. Acta* **443**, 288–300.

Rosenqvist, E., Michaelsen, T.E. and Vistnes, A.I. (1980). Effect of streptolysin O and digitonin on egg lecithin/cholesterol vesicles. *Biochim. Biophys. Acta* **600**, 91–102.

Rossjohn, J., Feil, S.C., McKinstrey, W.J., Tweten, R.K. and Parker, M.W. (1997). Structure of a cholesterol-binding, thiol-activated cytolysin and a model of its membrane form. *Cell* **89**, 685–92.

Ruiz, N., Wang, B., Pentland, A. and Caparon, M. (1998). Streptolysin O and adherence synergistically modulate proinflammatory responses of keratinocytes to group A streptococci. *Mol. Microbiol.* **27**, 337–46.

Saunders, F. K., Mitchell, T.J., Walker, J.A., Andrew, P.W. and Boulnois, G.J. (1989). Pneumolysin, the thiol-activated toxin of *Streptococcus pneumoniae*, does not require a thiol group for *in vitro* activity. *Infect. Immun.* **57**, 2547–52.

Scherrer, R. and Gerhardt, P. (1971). Molecular sieving by the *Bacillus megaterium* cell wall and protoplast. *J. Bacteriol.* **107**, 718–35.

Schmidt, K.-H., Podbielski, A. and Gerlach, D. (1998). Isogenic insertion inactivation of the streptolysin O gene (slo) in *Streptococcus pyogenes* M type 1 strain 38541. In: *Bacterial Protein Toxins* (eds J. Hacker, J.E. Alouf, B.C. Brand, P. Falmagne, F.J. Fehrenbach, J.H. Freer, W. Goebel, R. Gross, J. Helcemann, C. Locht, C. Montecucco, S. Olsnes, R. Rappuoli, J. Reidl and T. Wadström), pp. 293–4. *Zentralbl. Bakt. Suppl.* 30. Gustav Fischer, Jena.

Sekiya, K., Satoh, R., Danbara, H. and Futaesaku, Y. (1993). A ring-shaped structure with a crown formed by streptolysin O on the erthrocyte membrane. *J. Bacteriol.* **175**, 5953–61.

Sekiya, K., Satoh, R., Danbara, H. and Futaesaku, Y. (1996). Electron microscopic evaluation of a two-step theory of pore formation by streptolysin O. *J. Bacteriol.* **178**, 6998–7002.

Shanley, T.P., Schrier, D., Kapur, V., Kehoe, M., Musser, M. and Ward, P.A. (1996). Streptococcal cysteine protease augments lung injury induced by products of group A streptococci. *Infect. Immun.* **64**, 870–7.

Shany, S., Bernheimer, A.W., Grushoff, P.S. and Kim, K.-S. (1974). Evidence for membrane cholesterol as the common binding site for cereolysin, streptolysin O and saponin. *Cell. Molec. Biochem.* **3**, 179–86.

Sheehan, B., Kocks, C., Dramsi, S., Gouin, E., Klarsfeld, A.D. and Mengand, J. (1994). Molecular and genetic determinants of the *Listeria monocytogenes* infections process. In: *Current Topics in Microbiology and Immunology*, Vol. 192 (ed. J. Dangi), pp. 187–216. Springer, Berlin.

Smyth, C.J. and Duncan, J.L. (1978). Thiol-activated (oxygen-labile) cytolysins. In: *Bacterial Toxins and Cell Membranes* (eds J. Jeljaszewicz and T. Wadström), pp. 129–83. Academic Press, London.

Smyth, C.J. and Fehrenbach, F.J. (1974). Isoelectric analysis of haemolysins and enzymes from streptococci of groups A, C and G. *Acta Pathol. Microbiol. Scand.* **B82**, 860–74.

Song, L., Hobaugh, M.R., Shustak, C., Cheley, S., Bayley, H. and Gouaux, J.E. (1996). Structure of staphylococcal alpha-hemolysin, a heptameric transmembrane pore. *Science* **274**, 1859–66.

Stevens, D.L. and Bryant, A.E. (1997). Streptolysin O modulates cytokine synthesis in human peripheral blood mononuclear cells. *Adv. Exp. Med. Biol.* **418**, 925–7.

Suvorov, A., Tang, Y., Yu, C.E. and Ferretti, J.J. (1992). PER analysis of streptolysin O (slo) gene of *Streptococcus pyogenes*. In: *New Perpsectives on Streptococci and Streptococcal Infections* (ed. G. Orefici), pp. 348–9. Gustav Fischer, Stuttgart.

Suzuki, J., Kobayashi, S., Kagaya, K. and Fukazawa, Y. (1988). Heterogeneity of hemolytic efficiency and isoelectric point of streptolysin O. *Infect. Immun.* **56**, 2474–8.

Thelestam, M. and Möllby, R. (1980). Interaction of streptolysin O from *Streptococcus pyogenes* and theta-toxin from *Clostridium perfringens* with human fibroblasts. *Infect. Immun.* **29**, 863–77.

Tiesler, E. and Trinks, U. (1979). Die Streptolysin O - Bildung beta-hämolysierender Streptokokken der Gruppe A [The production of streptolysin O by beta-hemolytic streptococci of group A]. *Bakt. Hyg. I Abt. Orig. A* **245**, 17–24.

Tiesler, E. and Trinks, C. (1982). Das Vorkommen extrazellulärer Stoffwechselprodukte bei Streptokokken der gruppen C und G. [Release of extracellular products by groups C and G streptococci]. *Zentralbl. Bakt. Hyg. I. Abt. Orig. A* **253**, 81–7.

Todd, E.W. (1932). Antigenic streptococcal hemolysin. *J. Exp. Med.* **55**, 267–80.

Todd, E.W. (1938). The differentiation of two distinct serological varieties of streptolysin, streptolysin O and streptolysin S. *J. Pathol. Bacteriol.* **47**, 423–45.

Todd, E.W. (1939). The streptolysins of haemolytic streptococci of various groups and types. *J. Hygiene* **39**, 1–11.

Tweten, R.K. (1988). Nucleotide sequence of the gene for perfringolysin O (theta toxin) from *Clostridium perfringens*; significant homology with the genes for streptolysin O and pneumolysin. *Infect. Immun.* **56**, 3235–43.

Tyler, S.D., Johnson, W.M., Huang, J.C., Ashton, F.E., Wang, G., Low, D.E. and Rozee, K.R. (1992). Streptococcal erythrogenic toxin genes; detection by polymerase chain reaction and association with diseases in strains isolated in Canada from 1940 to 1991. *J. Clin. Microbiol.* **30**, 3127–31.

Valeva, A., Palmer, M. and Bhakdi, S. (1997). Staphylococcal α-toxin: formation of the heptameric pore is partially cooperative and proceeds through multiple intermediate stages. *Biochemistry* **36**, 13298–304.

Walev, I., Palmer, M., Valeva, A., Weller, U. and Bhakdi, S. (1995). Binding, oligomerization and pore formation by streptolysin O in erythrocytes and fibroblast membranes; detection of nonlytic polymers. *Infect. Immun.* **63**, 1188–94.

Walker, J.A., Allen, R.L., Falmagne, P., Johnson, M.K. and Boulnois, G.J. (1987). Molecular cloning, characterization and complete nucleotide sequence of the gene for pneumolysin, the sulphydryl-activated toxin of *Streptococcus pneumoniae*. *Infect. Immun.* **55**, 1184–9.

Walker, B., Krishnasastry, M., Zorn, L. and Bayley, H. (1992). Assembly of the oligomeric membrane pore formed by staphylococcal alpha-hemolysin examined by truncation mutagenesis. *J. Biol. Chem.* **267**, 21782–6.

Wannamaker, L.W. (1983). Streptococcal toxins. *Rev. Infect. Dis.* **5** (Suppl. 4), S723–32.

Wannamaker, L.W. and Schlievert, P.M. (1988). Exotoxins of group A streptococci. In: *Bacterial Toxins, Handbook of Natural Toxins* (eds M.C. Hardegree and A.T. Tu), pp. 267–95. Marcel Dekker, New York.

Watson, K.C. and Kerr, E.J.C. (1974). Sterol requirements for inhibition of streptolysin O activity. *Biochem. J.* **140**, 95–8.

Watson, K.C. and Kerr, E.J.C. (1985). Specificity of antibodies for T sites and F sites of streptolysin O. *Med. Microbiol.* **19**, 1–7.

Weld, J.T. (1934). The toxic properties of serum extracts of hemolytic streptococci. *J. Exp. Med.* **59**, 83–95.

Weller, U., Müller, L., Messner, M., Palmer, M., Valeva, A., Tranum-Jensen, J., Agrawal, P., Biermann, C., Döbereiner, A., Kehoe, M. and Bhakdi, S. (1996). Expression of active streptolysin O in *Escherichia coli* as a maltose-binding-protein-streptolysin O fusion protein. The N-terminal to amino acids are not required for hemolytic activity. *Eur. J. Biochem.* **236**, 34–9.

Wendland, M. and Subramani, S. (1993). Cytosol-dependent peroxisomal protein import in a permeabilized cell system. *J. Cell Biol.* **120**, 675–85.

Yamada, S. (1995). Expression of streptolysin O gene in *Bacillus subtilis*. *Biosci. Biotechnol. Biochem.* **59**, 363–6.

Yoshimori, T., Keller, P., Roth, M.G. and Simons, K. (1996). Different biosynthetic transport routes to the plasma membrane in BHK and CHO cells. *J. Cell. Biol.* **133**, 247–56.

26

Pneumolysin: structure, function and role in disease

Timothy J. Mitchell

INTRODUCTION

Pneumolysin is a haemolytic protein toxin produced by *Streptococcus pneumoniae* (the pneumococcus). The pneumococcus causes several important diseases of humans, including pneumonia, bacteraemia, meningitis and otitis media. The organism produces several potential virulence factors, including a polysaccharide capsule, a range of enzymes (hyaluronidase, neuraminidases and superoxide dismutase), surface proteins (PspA and PsaA) and at least two haemolysins (Paton *et al.*, 1993a; Canvin *et al.*, 1997). All of these factors have been investigated with regard to their role in the pathogenesis of pneumococcal infection, and some can be considered as possible antigens for use in protective vaccination. This chapter will concentrate on pneumolysin and will consider its mode of action, its role in the disease process and its possible role in pneumococcal vaccines.

Pneumococci were first reported to produce a haemolysin in 1905 (Libman, 1905). Since this report, numerous studies have been carried out using crude preparations of the toxin (Cole, 1914; Neill, 1926; Neill, 1927; Cohen *et al.*, 1940; Halbert *et al.*, 1946), which demonstrated that the protein was toxic, susceptible to oxidation (a process which could be reversed by treatment with thiol-reducing agents), antigenic and irreversibly inactivated by cholesterol. The sensitivity of the toxin to oxidation and re-activation with thiol-reducing agents led to the toxin's being named 'a thiol-activated toxin', a name that may not be appropriate in the light of more recent findings discussed below. The toxin belongs to a large family of proteins known as the thiol-activated toxins (Smyth and Duncan, 1978), which are produced by four genera of Gram-positive bacteria. Recent advances in molecular biology and protein purification have now allowed large amounts of highly purified pneumolysin to be produced and evaluated in a range of biological systems both *in vitro* and *in vivo*.

STRUCTURE AND FUNCTION STUDIES OF PNEUMOLYSIN

Mechanism of lytic action – the dogma

Pneumolysin is a lytic toxin, able to lyse all cells that have cholesterol in their membranes. The mechanisms of action of the members of this group of toxins are believed to be similar and have been reviewed (Smyth and Duncan, 1978; Morgan *et al.*, 1996).

Two of the major steps in the lytic process are membrane binding and oligomerization of the toxin to form pores. These two steps show differences in their temperature dependency. The binding of the toxin to cells is temperature independent, while oligomerization only occurs at higher temperatures. The lytic activity of the toxin can be inhibited by preincubation with cholesterol. This has led to the suggestion that membrane cholesterol is the receptor for this toxin. Once bound to the cell membrane, monomeric toxin inserts into the membrane and undergoes oligomerization, to give high molecular mass pore-like structures composed of up to 50 monomeric units. These pores are believed to mediate the lysis of the cells by osmotic mechanisms. Inhibition of the lytic process by cholesterol is assumed to represent occupancy of the cholesterol binding site by free cholesterol, preventing binding of the toxin to target cell membranes. However, studies with the related protein listeriolysin O suggest that inhibition of toxin action by cholesterol may be due not to a blockage of binding, but to interference with the oligomerization process (Jacobs et al., 1998). The exact mechanism of cell lysis by pneumolysin (and other members of the thiol-activated toxin family) still remains to be defined. The latest theories for the way in which this process works will be discussed below.

Analysis and modification of the primary amino acid sequence of pneumolysin

Pneumolysin consists of a single 53-kDa polypeptide chain and is produced by virtually all clinical isolates of the pneumococcus (Paton et al., 1983; Kanclerski and Mollby, 1987). Pneumolysin is the only member of this family of proteins that is not secreted from the cell but remains within the bacterial cytoplasm (Johnson, 1977).

The genes encoding pneumolysin from serotype-1 and serotype-2 pneumococci have been cloned, sequenced and expressed in *Escherichia coli* (Paton et al., 1986; Walker et al., 1987; Mitchell et al., 1989). Genes for pneumolysin have also been sequenced from several other serotypes, including from the genome sequence of a type-4 pneumococcus. Comparison of the derived amino acid sequences of these proteins shows a very high level of conservation, with only a single amino acid change in the type-1 or type-4 sequence when compared with the type-2 sequence. (At the time of writing the available genome sequence of the type-4 pneumococcus is not complete, and this change could represent a sequencing error.)

Alignment of the amino acid sequence of pneumolysin with the other members of the family shows that there is a high degree of similarity between the proteins (Table 26.1). Analysis of the primary amino acid sequence of pneumolysin shows no major areas of hydrophobicity which could be involved in membrane insertion. The hydrophobic nature of pneumolysin (Johnson et al., 1982) must therefore reflect the generation of hydrophobic areas during the folding of the primary sequence (see below on the folded form of pneumolysin). In agreement with the cytoplasmic location of the toxin, the predicted amino acid sequence does not contain a signal sequence for secretion.

As pneumolysin is thiol activated, it was thought for many years that the activation process involved the reduction of an intramolecular disulfide bond (Smyth and Duncan, 1978). However, the primary amino acid sequence shows that pneumolysin contains only a single cysteine residue. This residue lies at the C-terminal end of the molecule (amino acid position 428) in a region of 11 amino acids that is conserved amongst many of the family of toxins (Morgan et al., 1996). This conserved sequence (ECTGLAWEWWR) plays an important role in the activity of the toxin, as judged by studies using site-directed mutagenesis. Interestingly, the cysteine residue is not essential for activity and can be replaced with alanine with no effect on the lytic activity of the toxin *in vitro* (Saunders et al., 1989). However, the nature of the residue at this position is

TABLE 26.1 Sequence pair distances of the thiol-activated cytolysins calculated using the Clustal method from DNAstar analysis package (DNAstar Inc., Madison, USA)

Percentage similarity										
	1	2	3	4	5	6	7	8		
1		39.9	36.5	42.3	42.5	43.7	42.3	40.8	1	Pneumolysin
2	58.5		65.4	39.4	38.3	66.8	38.3	58.2	2	Alveolysin
3	60.2	32.2		37.4	36.6	62.1	37.4	55.6	3	Cereolysin
4	56.9	56.3	59.6		79.0	41.0	74.8	38.3	4	Ivanolysin
5	56.7	58.4	60.9	20.3		39.2	81.1	37.4	5	Listeriolysin
6	55.4	31.2	35.4	56.7	58.9		40.0	60.6	6	Perfringolysin
7	56.9	58.5	60.0	24.4	18.3	58.0		38.9	7	Seeligolysin
8	58.0	39.9	43.4	59.4	61.6	38.7	59.0		8	Streptolysin
	1	2	3	4	5	6	7	8		
Percentage divergence										

important, as substitution with glycine or serine results in a reduction in the activity of the toxin. Substitutions in other residues within the motif also affect the activity of the toxin, with changes in the tryptophan at position 433 having a dramatic effect (Table 26.2).

Attempts to define the nature of the defect in the mutants with reduced activity have been unsuccessful and have only determined that the effect is not due to a gross defect in the ability of the toxins to bind to cells or oligomerize (Saunders *et al.*, 1989). It seems that the cysteine motif may play an important part in mediating a conformational change in the toxin when it interacts with membranes and allows the insertion of the toxin monomer into the lipid bilayer (see below).

Naturally occurring mutations in the amino acid sequence of pneumolysin can also be informative. Pneumolysin derived from some serotype-8 pneumococci contains several alterations in amino acid sequence compared with the protein from other serotypes (Lock *et al.*, 1996). The type-8 protein has substitutions at positions 172, 224 and 265, and has a two amino acid deletion (residues 270–271). Interestingly, these changes make the protein run with an apparent M_r on sodium dodecyl sulfate–polyacrylamide gel electrophoresis (SDS–PAGE) greater than that from other more typical strains of the pneumococcus (Lock *et al.*, 1996). Construction of chimeric molecules between the type-8 and the type-2 proteins demonstrated that it is the substitution at position 172 (threonine to isoleucine) that is responsible for the decreased activity of this version of the protein (Lock *et al.*, 1996).

Histidine residues are also important in the lytic mechanism of pneumolysin. Modification of histidine residues within pneumolysin by treatment with diethyl pyrocarbonate abolishes the lytic activity. Studies using site-directed mutagenesis have shown that histidine-367 is crucial in the activity of the toxin. Substitution of this residue with an arginine abolishes the ability of the toxin to form pores but does not affect its ability to bind to cells (Mitchell *et al.*, 1994).

The mechanism by which pneumolysin binds to cells is still unclear. The evidence that cholesterol is the receptor for the toxin is based on studies of the inhibition of toxin activity by prior incubation of the toxin with cholesterol in solution. The finding that the related toxin listeriolysin O is inactivated by cholesterol but is still able to bind to cells suggests that binding of these toxins to cells may involve a receptor other than cholesterol (Jacobs *et al.*, 1998). Interaction with cholesterol may play a role in the assembly of the functional oligomer. The region of the pneumolysin molecule involved in cell binding has been inferred from studies using truncated versions of the protein (Owen *et al.*, 1994). Removal of the five C-terminal amino acids of the toxin was sufficient to prevent cell lysis by the protein. A substitution of the proline residue at position 462 with serine causes a 90% reduction in the ability of the toxin to bind to red blood cells.

Studies with monoclonal antibodies have also been used to investigate the functional regions of pneumolysin (de los Toyos *et al.*, 1996). A panel of monoclonal antibodies was raised against pneumolysin, all of which recognized linear epitopes of the protein. Three of these antibodies (termed PLY-4, PLY-5 and PLY-7) were capable of inhibiting the haemolytic activity of the toxin. One of the neutralizing antibodies (PLY-4) did not prevent binding of the toxin to cells,

TABLE 26.2 Amino acid substitutions in the cysteine-containing region of pneumolysin. The effect of substitutions on the ability of the toxin to lyse red cells, to induce choline leakage from Lettre cells and to activate the complement pathway is shown. The sensitivity of toxin-induced leakage to divalent cations is also given

Variant	Haemolytic activity (% of wild-type)	Leakage from Lettre cells (% of wild-type)	Concentration (M) of cation causing 50% inhibition of leakage from Lettre cells			Complement activation (% of wild-type)
			Zn^{2+}	Ca^{2+}	Mg^{2+}	
Wild-type	100	100	3×10^{-5}	10^{-3}	$>10^{-2}$	100
Cys428 → Gly	3	6	6×10^{-5}	$>10^{-2}$	$>10^{-2}$	100
Cys428 → Ser	25	62	3×10^{-5}	10^{-3}	$>10^{-2}$	100
Cys428 → Ala	100	25	10^{-4}	$>10^{-2}$	$>10^{-2}$	100
Trp433 → Phe	1	3	6×10^{-4}	$>10^{-2}$	$>10^{-2}$	nd
Glu434 → Gln	20	12	8×10^{-5}	2×10^{-3}	$>10^{-2}$	100
Glu434 → Asp	50	6	10^{-4}	10^{-3}	$>10^{-2}$	100
Trp435 → Phe	20	5	6×10^{-5}	$>10^{-2}$	$>10^{-2}$	nd
Trp436 → Phe	50	10	6×10^{-5}	$>10^{-2}$	$>10^{-2}$	nd

Data from Korchev *et al.* (1998).
nd: not determined.

but did prevent the subsequent oligomerization of the toxin to form pores (as determined by electron microscopy), while the other two (PLY-5 and PLY-7) prevented cell binding by the toxin. Proteolytic treatment of pneumolysin with proteinase K generates two fragments of 37 kDa and 15 kDa (Morgan *et al.*, 1997). Purification and characterization of these fragments showed that the 15-kDa fragment represented a *N*-terminal portion of the protein (K1), while the 37-kDa fragment contained the remaining *C*-terminal part of the protein. The site of cleavage as determined by amino acid sequencing was between amino acids 142 and 143 (Morgan *et al.*, 1997). Isolated K2 was still able to bind to liposomes and to form oligomer-like structures.

Antibodies PLY-5 and PLY-7 both recognized the K2 portion of the protein, whereas PLY-4 recognized neither fragment. PLY-4 therefore recognizes at or near to the proteolytic cleavage site and prevents oligomerization. Further analysis of the binding site for PLY-5 and PLY-7 using *C*-terminal truncated versions of the pneumolysin molecule showed that PLY-5 recognized the extreme *C*-terminus of the molecule (amino acids 464–470), while PLY-7 recognized around residue 419. As both antibodies block cell binding, this would suggest two possibilities: (a) there are two binding sites; (b) the linear epitopes recognized by the antibodies are involved in a single binding site in the folded protein. The findings with monoclonal antibodies support the findings, using truncated pneumolysin, that the extreme *C*-terminus of the molecule is involved in cell binding.

The mechanism of pore formation by pneumolysin is also still unclear. Site-directed mutations can be made in the conserved cysteine-containing region that have a dramatic effect on the lytic activity of the toxin, but no effect on the ability of the toxin to generate oligomers as observed by electron microscopy. In attempts to define the differences in the nature of the pore formed by wild-type pneumolysin and a lytic deficient mutant in which the tryptophan at position 433 has been replaced with phenylalanine (F433), the effects of the proteins on cells and on planar lipid bilayers have been studied (Korchev *et al.*, 1998).

The mutation at position 433 has no effect on the ability of the toxin to bind to cells or to form oligomeric structures, as observed by electron microscopy. The mutation does affect the ability of the toxin to induce leakage of markers from Lettre cells and to induce conductance channels in planar lipid bilayers. Pneumolysin-induced leakage from Lettre cells is sensitive to inhibition by bivalent cations. The sensitivity to inhibition by bivalent cations was much reduced in the mutant form of the toxin (Table 26.2). When inserted into planar lipid bilayers, pneumolysin induces a range of conductance channels exhibiting small (less than 30 pS), medium (30 pS–1 nS) and large (greater than 1 nS) conductance steps. Small and medium channels are preferentially closed by bivalent cations. Wild-type toxin forms mainly small channels, whereas the mutant formed mostly large channels, which are insensitive to closure by bivalent cations. Osmotic protection studies show that cells treated with wild-type toxin can be protected from lysis by polysaccharides with a size of more than 15 kDa, whereas cells treated with the mutant form of the toxin were not protected by polysaccharides with relative molecular mass over 40 000. This also suggests that the functional pores are larger in the membranes treated with the F433 mutant. The cysteine motif therefore probably plays a role in the functionality of the pores generated by pneumolysin. Since no differences were observed in the structure of the pores, as observed by electron microscopy, the structural appearance of the pores does not reflect the functional state of the channels.

Purified pneumolysin is also able to activate the classical complement pathway (Paton *et al.*, 1984), and this effect has been linked to the ability of the toxin to bind to the Fc portion of human immunoglobulin (Mitchell *et al.*, 1991). Analysis of the amino acid sequence of pneumolysin showed it did not contain any homology to known IgG binding proteins, but did show some limited homology to the human acute phase protein, C-reactive protein (CRP) (Mitchell *et al.*, 1991). CRP is also able to activate the classical complement pathway, although this effect is mediated by a direct binding of complement component C1q (Volanakis and Kaplan, 1974). Mutagenesis of the homologous region in pneumolysin revealed that residues 384 and 385 are involved in antibody binding and complement activation by the toxin (Mitchell *et al.*, 1991).

Structural models of the pneumolysin molecule

Initial studies of the structure of pneumolysin relied on relatively low-resolution techniques in attempts to model the structure of the monomer and predict how these might fit together to form the oligomer. Electron-microscope images of metal shadowed pneumolysin monomers show it to be an asymmetric molecule composed of four domains (Morgan *et al.*, 1994). The molecules appear to be 5 nm in width and to have a total length of approximately 13 nm. Each of the four domains appears to be about 3 nm. Three of the domains are linearly associated, with the fourth observed at a range of angles, indicating a degree of flexibility. The ring-shaped oligomer has an external diameter of 30–45 nm and a ring width of 6.5 nm. These dimensions are very similar to those reported for other members of this toxin family. Metal shadowing of isolated oligomers enabled the height of the oligomer to be estimated at 9.3 nm. Hydrodynamic bead modelling of the monomer allowed a model of the oligomer to be proposed. The oligomer was proposed to consist of about 30 subunits, arranged so that three of the domains lie adjacent to each other to form a cylindrical column and the fourth flexible domain forms a flange on the outside of the cylinder (Morgan *et al.*, 1994).

In order to understand fully the mechanism of pore formation by pneumolysin, it would be useful to have a three-dimensional structure for the protein monomer. Attempts to crystallize pneumolysin have so far been unsuccessful. A homology model of

pneumolysin has been constructed, based on the crystal structure of another pore-forming protein, aerolysin (Sowdhamini *et al.*, 1997). The sequence identity between aerolysin and pneumolysin is only 16%. Modelling was undertaken on the basis that the two proteins seem to be composed of four domains and have some biological properties in common. Also, the homology score for proteins of known structure was highest for aerolysin. However, the 2.7-Å crystal structure of perfringolysin has recently been solved by Rossjohn *et al.* (1997). As the pneumolysin and perfringolysin sequences share 48% sequence identity and 60% sequence similarity extending over the full length of the molecule, it has been possible to model a structure for the pneumolysin monomer by homology with the perfringolysin structure (Rossjohn *et al.*, 1998) (Figure 26.1). It is now apparent that, although perfringolysin and aerolysin are similarly shaped molecules, possessing the same number of domains, there is no structural homology between the two toxins. On the basis of the homology model with perfringolysin, the pneumolysin molecule is long and rod-shaped, with overall dimensions of 11 nm × 5 nm × 3 nm. These values are in good agreement with those derived for pneumolysin from metal shadowed electron microscopy (as discussed above). Again, in agreement with the electron-microscope data, the molecule consists of four domains. The overall structural composition is 41% sheet and 21% helix. The conserved cysteine motif forms a loop at the bottom of domain 4.

Structural aspects of pore formation

Availability of the three-dimensional structure of perfringolysin and the model for pneumolysin has allowed a mechanism of pore formation to be proposed for these toxins (Rossjohn *et al.*, 1997, 1998). There are no long runs of hydrophobic amino acids within the primary sequence of pneumolysin and no large hydrophobic patches on the surface of the molecule. There are no helices long enough to span a cell membrane, and there does not appear to be a large change in the structure of the protein upon membrane insertion, as judged by circular dichroism (Rossjohn *et al.*, 1998). Thus, it appears to be impossible for the pneumolysin molecule to insert into the membrane.

A mechanism can be proposed based on that suggested for perfringolysin O (Rossjohn *et al.*, 1997). Individual domains of the toxin enter the lipid bilayer where membrane cholesterol is proposed to interact with the protein near the cysteine motif. The monomers then aggregate to form the oligomer,

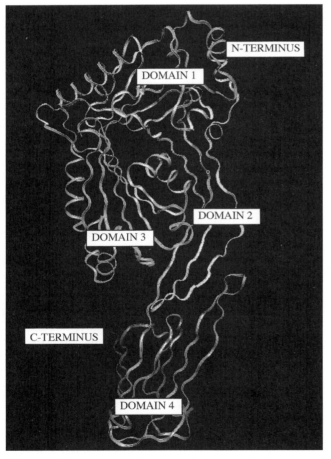

FIGURE 26.1 A model of the pneumolysin molecule showing the four-domain structure of the protein. This model was generated from the crystal co-ordinates of perfringolysin (Rossjohn *et al.*, 1997, 1998). It is believed that the link between domain 3 and the rest of the molecule is flexible, such that domain 3 can move away from domain 2 on membrane insertion.

causing the cysteine region to be displaced, resulting in the formation of a so-called 'hydrophobic dagger', which penetrates into the lipid bilayer. This model predicts that oligomerization occurs prior to membrane insertion and suggests that there must be a significant bending of the molecule at the interface between domains 1 and 3. The oligomer is held by strong interactions between domain 1 outside the membrane, and only domain 4 enters the membrane. Domain 4 is suggested to undergo a reorganization to allow it to bind to cholesterol to protect the hydrophilic face of domain 4 within the bilayer and to generate a continuous intermolecular β-sheet around the pore structure by one face of domain 4.

While this model does not explain all of the observations with regard to pneumolysin, it does serve as a useful tool for designing further studies. The proposed model of the oligomer does not agree with dimensions obtained by electron microscopy. The oligomers have a height of approximately 10 nm as measured by electron microscopy, while the model suggests that this dimension is closer to 14 nm. Studies with the related toxin streptolysin O (SLO) suggest that membrane insertion is a relatively early event in oligomer assembly (Palmer *et al.*, 1995). Also, membrane leakage can be induced by small oligomers of pneumolysin (Korchev *et al.*, 1992) . The mechanism based on the crystal structure suggests that extensive oligomerization occurs before insertion into the membrane. Fluorescence emission studies with SLO (Palmer *et al.*, 1996) also suggest that at least part of domain 3 enters the lipid bilayer, which again does not agree with the proposed model. This model therefore serves as a useful tool, but needs to be refined to take account of some of the anomalies described.

Interpretation of mutagenesis data

Lytic activity

The homology model of pneumolysin can be used to explain some of the biological data generated from biochemical and mutagenesis studies of the protein. The single cysteine residue of the cysteine motif is present in domain 4, sandwiched between Val377 and Trp435, and forms a hydrogen bond with Gln374. Substitution of the cysteine to alanine has no effect on the activity of the toxin, as this is a conservative mutation. However, the model explains why substitution with other residues has a dramatic effect on activity. Replacing the cysteine with a glycine would destabilize the conserved loop, while a serine residue would form a stronger hydrogen bond to Gln374 and reduce the mobility of the loop, and thus affect the formation of the hydrophobic dagger described above.

The monoclonal antibody data discussed above suggest that residues around position 142 are essential for pore formation (de los Toyos *et al.*, 1996). This supports the idea that domain 1 is involved in the oligomerization process. Analysis of site-directed mutagenesis data suggests a vital role for the domain 2/domain 3 interface. Mutations in this region (Hill *et al.*, 1994; Lock *et al.*, 1996) cause a large reduction in haemolytic activity.

The histidine residue at position 367 has been substituted with an arginine residue, to construct a version of the protein that does not form oligomeric structures (Mitchell *et al.*, 1994). Analysis of the structural model suggests that His367 is located within the core of domain 4 with its aromatic side-chain forming interactions with Tyr375 and Tyr371. His367 is hydrogen bonded to the carbonyl moieties of residues 371 and 368. It is predicted that mutation of His367 would disrupt the hydrophobic core of domain 4. The difficulty in purifying this mutant and its tendency to aggregate support the idea that this mutation causes a large disruption in the structure of domain 4 rather than a specific effect on an interaction required for oligomerization.

Complement-activating activity

Pneumolysin activates the classical complement pathway in the absence of specific antitoxin antibodies (Paton *et al.*, 1984; Mitchell *et al.*, 1991). This activation in the fluid phase could divert complement away from intact bacteria and consume complement components. Activation of the pathway would lead to inflammation. Pneumolysin can bind to the Fc portion of antibodies, and it has been proposed that this aggregation of antibody is responsible for the activation of complement (Mitchell *et al.*, 1991). Pneumolysin also has some limited sequence homology to CRP, a human acute phase protein, which can also activate the classical complement pathway in the absence of antibody by binding to the C1q component of the complement pathway (Volanakis and Kaplan, 1974). Mutations of amino acid residues within the region of pneumolysin homologous to CRP (residues 384 and 385) cause a decrease in the ability of the toxin to bind antibody and activate the complement pathway (Mitchell *et al.*, 1991). Also, a monoclonal antibody has been produced that recognizes both pneumolysin and CRP (Mendez *et al.*, 1992).

These data suggest there may be some structural similarity between pneumolysin and CRP. Analysis of the model shows this is apparently not the case. The structural basis for the common reaction of pneumolysin and CRP with a monoclonal antibody is unclear. There is, however, a structural similarity between domain 4 of pneumolysin and the Fc portion of antibody. As complement activation via the classical pathway is dependent on C1q binding to multimers of Fc, it is possible to suggest several mechanisms by which activation could occur.

Aggregates of pneumolysin could act to bind C1q in a similar manner to native Fc. Alternatively, the toxin may form mixed aggregates with immunoglobulins and activate complement. Certainly, the ability of the toxin to activate complement seems to be linked to its ability to interact with antibody (Mitchell *et al.*, 1991). However, if pneumolysin itself has structural similarities to immunoglobulin, it is possible that the original experiments measuring antibody binding are flawed. The two mutations that reduce complement activation by the toxin lie on the extreme end of domain 4, a region that would be accessible in the membrane-bound form of the toxin. The molecular mechanism of complement activation by pneumolysin requires further study.

BIOLOGICAL EFFECTS OF PNEUMOLYSIN

The biological effects of pneumolysin are summarized in Table 26.3 (Mitchell and Andrew, 1997). Pneumolysin has been studied in a range of systems in attempts to define those properties that are important in the pathogenesis of infection. These systems include the study of the interaction of the purified protein with isolated cells, with organ cultures and with whole animals. The use of molecular genetic approaches has also allowed the role of the toxin to be addressed within the context of the whole organism via the construction of isogenic mutants that either do not express the protein or express a version that has an activity altered by modification of the toxin gene. These approaches have allowed a detailed model to be built up of the role played by pneumolysin in a range of experimental infections.

Effect of pneumolysin on isolated cells

The major effect of relatively high concentrations of pneumolysin on cells is cell lysis. The mechanisms of pore formation in red blood cells and planar lipid bilayers have been discussed above. Pneumolysin is also able to damage a range of other eukaryotic cell types, including alveolar epithelial cells (Rubins *et al.*, 1993), pulmonary arterial epithelial cells (Rubins *et al.*, 1992) and bronchial epithelial cells (Steinfort *et al.*, 1989). These cell types are involved in the lung–capillary interface, and damage to these cell types *in vivo* may contribute to the

TABLE 26.3 A summary of the biological properties of purified pneumolysin

Activity	Reference
Lysis of red blood cells	Mitchell *et al.* (1989)
Inhibition of cilial beat of respiratory mucosa	Feldman *et al.* (1990)
Increased alveolar permeability in isolated rat lung	Rubins *et al.* (1993)
Toxic to bovine artery endothelial cells	Rubins *et al.* (1992)
Toxic to pulmonary alveolar epithelial cells	Rubins *et al.* (1993)
Inhibition of mitogen-induced proliferation and antibody production by human lymphocytes	Ferrante *et al.* (1984)
Inhibition of PMNL respiratory burst, random migration and chemotaxis	Paton and Ferrante (1983)
Activation of classical complement pathway by binding to the Fc portion of antibody	Mitchell *et al.* (1991), Paton *et al.* (1984)
Stimulation of TNF-α and IL-1β production from human monocytes	Houldsworth *et al.* (1994)
Activation of phospholipase A$_2$	Rubins *et al.* (1994)
Induction of CSF leucocytosis and increased TNF in rabbit brain	Friedland *et al.* (1995)
Electrophysiological and histological damage when perfused into the cochlea of guinea pigs	Comis *et al.* (1993)

histopathology of early pneumococcal pneumonia, including alveolar flooding and haemorrhage.

As well as its lytic activity, the toxin has a range of other activities that may play a role in the virulence of the pneumococcus. Very low concentrations of the toxin interfere with the function of cells of the immune system and inhibit processes such as respiratory burst of phagocytic cells (Paton and Ferrante, 1983) and antibody production by lymphocytes (Ferrante *et al.*, 1984). The toxin can also stimulate other activities, such as cytokine production by monocytes (Houldsworth *et al.*, 1994) and phospholipase A activity in endothelial cells (Rubins *et al.*, 1994). The effects of the toxin on phagocytic cells are mediated by extremely low concentrations of the toxin. Femtomolar concentrations of pneumolysin can both inhibit the respiratory burst of human polymorphonuclear leucocytes and promote inflammatory cytokine production by human monocytes.

Pneumolysin is more efficacious than bacterial lipopolysaccharide in promoting inflammatory cytokine production (Figure 26.2). The effect of the toxin on inhibition of respiratory burst of macrophages appears to be linked to the lytic activity of the toxin (Saunders *et al.*, 1989). Mutations of the single cysteine residue that interfere with the lytic activity of the toxin also reduce the potency of the toxin to inhibit respiratory burst (Table 26.4). However, the reduction in respiratory burst appears not to be due to membrane damage owing to the low amounts of toxin used and the continued viability of the cells, as judged by trypan blue dye exclusion.

As well as promoting inflammation by stimulating the release of cytokines and activation of the complement pathway, pneumolysin can activate phospholipase A in pulmonary endothelial cells (Rubins *et al.*, 1994). Activation of phospholipase activity is linked to the ability of the toxin to form transmembrane pores. The activated phospholipase has a broad substrate specificity for cellular membrane phospholipid. Activation of phospholipase during an infection could contribute directly to lung damage by the release of free fatty acids and lysophosphatides. Release of arachidonic acid by the activated phospholipase could promote chemotaxis and respiratory burst of neutrophils (Badwey *et al.*, 1984; Curnutte *et al.*, 1984). Arachidanic acid could be metabolized through the eicosanoid pathway, leading to the production of leukotrienes and platelet activating precursor (Holtzman, 1991). Products of the eicosanoid pathway are major chemotaxins for neutrophils (Cabellos *et al.*, 1992). Recruitment of large numbers of neutrophils is a feature of pneumococcal disease, and release of toxic components from these cells may play a role in injury to pulmonary tissue. Thus, activation of phospholipase A could provide a link between the lytic activity of the toxin and its abil-

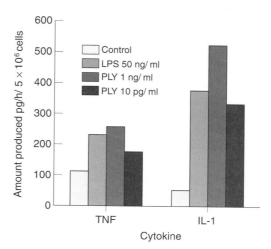

FIGURE 26.2 Stimulation of cytokine production from human monocytes by pneumolysin. 10^6 cells were exposed to the toxin and the amount of cytokine released was measured by ELISA. Pneumolysin in femtomolar concentrations is more active than lipopolysaccharide at promoting cytokine release (Houldsworth *et al.*, 1994).

ity to induce inflammation in models of pneumonitis (Feldman *et al.*, 1991).

Interaction of pneumolysin with animal tissue

Purified pneumolysin has been evaluated for its effects on a range of tissues. When introduced into a ligated lobe of rat lung, the toxin induces histological features that are very similar to those observed in pneumonia (Feldman *et al.*, 1991). When altered forms of the toxin are used, with decreased lytic or complement-activating activity, the histological changes observed are much less than those induced by the wild-type toxin. This is an *in vivo* indication that the ability of

TABLE 26.4 Effect of amino acid substitutions at cysteine 428 on interaction with human polymorphonuclear leucocytes. Inhibition of respiratory burst (induced by treatment with phorbol ester) occurs at a concentration of toxin 1000-fold lower than those required to affect viability

Variant	Toxin to induce 50% decrease in viability of 10^6 human PMNL (pg)	Toxin to induce 50% inhibition of respiratory burst of 10^6 human PMNL (pg)
Wild-type	1800	3.2
Cys428 → Gly	3200	>5.0
Cys428 → Gly	2700	4.1
Cys428 → Gly	1800	3.2

Data taken from Saunders *et al.* (1989).

pneumolysin to lyse cells and to activate the complement pathway contributes to the pathogenesis of the inflammatory disease observed in the lung.

The effects of pneumolysin on human respiratory mucosa have been investigated in an organ culture model. The toxin causes a slowing of the cilial beat of human nasal epithelium and, at higher concentrations, disrupts the integrity of the tissue (Feldman et al., 1990). This effect may be important in compromising the physical barrier of the mucociliary escalator during the development of an infection. Pneumolysin increases the alveolar permeability of perfused rat lungs and is toxic to type-II rat alveolar epithelial cells (Rubins et al., 1993). The alveolar capillary barrier is essential in alveolar water and solute transport and may also provide a physical barrier to infection. Perturbation of this barrier by the toxin may play an important role in the development of pneumococcal pneumonia.

As well as causing pneumonia, the pneumococcus can also cause meningitis. A common complication of pneumococcal meningitis is hearing loss (Fortnum, 1992). The role of pneumolysin in hearing loss has been investigated by direct perfusion of the toxin through the guinea pig cochlea. When the toxin is perfused through the scala tympani, widespread histological and electrophysiological damage was induced (Comis et al., 1993). Electrophysiological changes include reduced amplitude of both the compound action potential and the cochlear microphonic potential. Histological damage observed by scanning electron microscopy includes damage to both inner and outer hair cells and supporting cells. Inner hair cells and outer hair cells of row three were most susceptible to the toxin. Damage to hair cells includes disruption and splaying of stereocilia, loss of stereocilia and complete dissolution of hair bundles. The effects seen were dependent on the toxin concentration used and, at lower concentrations, the effects were reversible.

The toxic effects of pneumolysin on the guinea pig cochlea may be mediated, at least in part, by nitric oxide (Amee et al., 1995). Pretreatment of the cochlea with N^G-methyl-L-arginine, a known inhibitor of nitric oxide synthesis, blocked the effect of the toxin. Pretreatment of the cochlea with MK-801, an N-methyl-D-aspartate (NMDA) receptor antagonist, also confers protection against pneumolysin. This latter finding is consistent with the idea that excessive stimulation of NMDA receptors within the cochlea is responsible for the production of nitric oxide. The role of pneumolysin in hearing loss has also been confirmed in an infection model of meningitis, and this will be discussed below. Pneumolysin may also perturb the membrane of the round window during pneumococcal otitis media and allow access of the toxin from the middle ear to the cochlea (Engel et al., 1995).

Pneumolysin has dramatic effects on the ciliated ependymal cells from the brain of rats. Brain cilia were more sensitive to the toxin than respiratory cilia and as little as 100 ng ml^{-1} of toxin caused cilial stasis within 15 min (Mohammed et al., unpublished data). This concentration of toxin had no effect on respiratory cilia (Feldman et al., 1990). Ciliated cells line the ventricular surface of the brain and cerebral aqueducts and form a barrier between the cerebrospinal fluid (CSF), which is infected during meningitis, and the brain tissue (Alfzelius, 1979). These cilia may protect the neuronal tissue from damage during infection by allowing continual movement of CSF and preventing migration of bacteria during meningitis. Perturbation of brain cilia by pneumolysin could therefore compromise this defence mechanism.

The role of pneumolysin in ocular infections has also been investigated. When the purified toxin is instilled into the eye of a rabbit, an inflammatory response is generated (Johnson and Allen, 1975). The toxin induces similar pathology to that caused by natural infections with S. pneumoniae. If rabbits were made neutropenic prior to challenge with the toxin, the pathology induced was reduced, indicating that leucocytes may be a source of tissue-damaging enzymes such as collagenase (Harrison et al., 1993). The role of pneumolysin in ocular infections has been confirmed in a rat model of endopthalmitis (Ng et al., 1997). In this study, purified pneumolysin was injected intravitreally and induced many of the clinical and histopathological features of pneumococcal endopthalmitis. The toxin therefore plays an important role in the inflammation and tissue damage that occurs in pneumococcal endopthalmitis. The role of pneumolysin in ocular infection has also been investigated, using infection models with genetically modified pneumococci. These studies will be considered below.

Purified pneumolysin affects a variety of cells and tissues. Major effects include compromise of host defence mechanisms and production of a potent inflammatory response. However, extrapolation from studies with a purified virulence factor to the role played by this virulence factor in the pathogenesis of an infection is not always valid. The generation of defined isogenic mutants of the pneumococcus that either do not produce the toxin or synthesize versions of the protein with altered biological activity allows the role played by the toxin in the pathogenesis of disease to be investigated in animal models of the diseases caused by S. pneumoniae.

THE ROLE OF PNEUMOLYSIN IN PATHOGENESIS

Pneumolysin-negative mutant

Although the existence of a toxin produced by the pneumococcus has been known about for more than 90 years, it was not until the studies in James Paton's laboratory in the 1980s and early 1990s that the toxin was shown to play an important role in the virulence of *S. pneumoniae*. The Paton group showed that immunization with a partially inactivated form of pneumolysin could partially protect mice from challenge with virulent pneumococci (Paton *et al.*, 1983). These studies formed the basis of a continuing evaluation of pneumolysin toxoids as vaccine candidates (see below).

The cloning and sequencing of the gene for pneumolysin (Paton *et al.*, 1986; Walker *et al.*, 1987) allowed the construction of isogenic pneumolysin-negative mutants of serotype-2 and serotype-3 pneumococci (Berry *et al.*, 1989, 1992). In the case of serotype-2 pneumococci, the virulence of the organism was reduced by 100-fold when given intranasally to mice (Berry *et al.*, 1989). When injected intravenously into mice, the pneumolysin-negative mutant survived less well (Berry *et al.*, 1989). The serotype-2 pneumolysin-negative mutant was designated PLN-A, and this mutant has been key in studies of the role of pneumolysin in the pathogenesis of disease in laboratories around the world. PLN-A has been used to define the role played by pneumolysin in a mouse model of bronchopneumonia (Canvin *et al.*, 1995). PLN-A induced much less inflammation when instilled into the mouse lung. This was an intriguing finding, as it had long been stated that the inflammation induced by pneumococci was a result of the release of inflammatory cell-wall components (Tuomanen *et al.*, 1987). Studies with PLN-A suggest that most of the inflammation was induced by the toxin. The toxin-negative mutant also showed a reduced ability to replicate in the lung. Invasion from the lung into the bloodstream was also delayed in the infection with PLN-A.

PLN-A has also been compared with wild-type pneumococci in a bacteraemia model (Benton *et al.*, 1995). These workers showed that pneumolysin protected the pneumococcus from infection-induced host resistance. The wild-type parent organism exhibited exponential growth after intravenous injection until numbers reached as high as 10^{10} colony-forming units (CFU) per ml of blood when the animals died. When PLN-A was used, the initial growth rate was the same as wild-type until numbers of approximately 10^7 CFU ml^{-1} were reached, when the increase in CFU

per ml ceased and the numbers of organisms in the blood remained constant for several days. If PLN-A was co-injected with wild-type organisms, it demonstrated wild-type growth characteristics. This observation suggests that pneumolysin exerts its effect at a distance. In mice infected with PLN-A, there was evidence of an inflammatory response, including a rise in plasma levels of interleukin-6 (IL-6), a cessation of the net growth of PLN-A and control of the net growth of wild-type pneumococci if given as a subsequent challenge. These data suggest that pneumolysin allows the pneumococcus to cause acute sepsis rather than a chronic bacteraemia. This effect seems to be limited to early in the infection and, once chronic bacteraemia is established, pneumolysin can no longer act as a virulence factor. Mice actively infected with wild-type pneumococcus produced more IL-6 per CFU in plasma than mice infected with comparable levels of PLN-A, suggesting that the increased resistance in PLN-A-infected mice is not mediated by IL-6 (Benton *et al.*, 1995). Gamma interferon was only detected in mice infected with wild organisms that were near death from sepsis, suggesting that this cytokine does not mediate the resistance. As pneumolysin mediates the production of tumour necrosis factor alpha (TNF-α) (Houldsworth *et al.*, 1994), a study was undertaken to investigate the role of pro-inflammatory cytokines in the resistance to pneumococcal infection (Benton *et al.*, 1998). These studies showed that the host resistance developed early during infection with PLN-A dependent on TNF-α but independent of IL-1β or IL-6. The ability to cause acute sepsis in this model is therefore related to the ability to produce pneumolysin. An inability to produce pneumolysin is associated with chronic bacteraemia and development of resistance to wild-type bacteria. This resistance is mediated by the production of TNF-α. The importance of the ability of pneumolysin to stimulate TNF-α production is unclear in this model.

When evaluating the role played by pneumolysin in infection, it is clear that the role played by the toxin differs according to the type of infection model used. In the pneumonia model described above, in which infection is established via the intranasal route of inoculation, the infection is typical of a bronchopneumonia. It is also possible to use a model of lobar pneumonia, in which the infection is established by intratracheal inoculation of mice (Rubins *et al.*, 1995). When evaluated in this lobar pneumonia model, PLN-A was 10 times less virulent than its wild-type parent. This model has been used to evaluate the possible role of pneumolysin in pulmonary infection. Infection of animals with wild-type pneumococci caused an increase in leakage of serum albumin into the air spaces, indicating that the permeability of the alveolar capillary barrier had been compromised. Pneumolysin-negative pneumococci showed a decreased ability to grow in the air spaces of the lung and had much less effect on the permeability of the air/blood interface. Once in the lung tissue, the mutant organisms showed a decreased ability to grow within the lung tissue and to cause bacteraemia. The virulence properties of PLN-A could be reverted to wild-type by the co-administration of a bolus of pneumolysin.

A mechanism by which pneumolysin might affect the permeability of the alveolar capillary junction has been suggested in studies using human respiratory mucosa grown in organ culture (Rayner et al., 1995). The interaction of wild-type and PLN-A pneumococci with human respiratory mucosa in an organ culture with an air interface were studied for up to 48 h. Beating of the cilia on the epithelium and adherence to and invasion of the epithelium by bacteria were monitored by scanning electron microscopy. Both wild-type and PLN-A caused a decrease in cilial beat frequency, although the onset of the inhibition was delayed in the case of PLN-A. This suggests that pneumolysin is involved in causing cilial slowing, but is not the only factor involved. Histological damage was induced by both wild-type and PLN-A, but again the effect was delayed and less severe when the organisms did not produce pneumolysin. Only the wild-type organism caused separation of tight junctions between epithelial cells and at 48-h post-infection, wild-type pneumococci were adherent to the separated edges of otherwise healthy unciliated cells. This separation of tight junctions and adherence to exposed areas may account for the changes in permeability and tissue invasion in the lobar pneumonia model described above.

The interaction of a toxin with the host during pathogenesis of a bacterial infection is also influenced by the genetic status of that host. The interaction of pneumolysin with mice genetically deficient in complement has been investigated (Rubins et al., 1995). Lobar pneumonia in complement-deficient mice showed several different features to those seen in complement-sufficient animals. Lack of a complete complement system was associated with increased numbers of bacteria in the lungs and an earlier and greater level of bacteraemia. The clearance of PLN-A and especially wild-type pneumococci from lungs was reduced in complement-deficient animals. The total bacterial load increased in the complement-deficient lungs of PLN-A-infected mice, whereas complement-sufficient mice were able to clear this organism. The effect of the removal of pneumolysin expression can therefore be partially reversed by the use of complement-deficient mice, illustrating

that the interaction of pneumolysin with the complement system is important in these model systems.

The role of pneumolysin in ocular infections with the pneumococcus has been investigated. When a deletion mutant lacking pneumolysin was constructed and used in a rabbit model of ocular infection, it showed greatly reduced virulence (Johnson et al., 1990). A non-haemolytic strain of the pneumococcus produced by chemical mutagenesis (probably involving the generation of a point mutation in the pneumolysin gene) showed the same virulence characteristics as the parent strain. This suggested that an activity of pneumolysin other than its lytic function was important in the role of the toxin during pathogenesis of ocular infection. This was confirmed in studies using mutants of the toxin devoid of different activities (see below).

From the above, it is clear that pneumolysin makes a contribution to the virulence of the pneumococcus in models of pulmonary, systemic and ocular infections. Its role in some other diseases is less clear. Use of toxin-positive and toxin-negative isogenic pneumococci in the chinchilla otitis media model showed that the amount of inflammation induced was similar (Sato et al., 1996).

It should be pointed out that these studies were done with a serotype-3 organism, whereas the studies described above with PLN-A are serotype-2. It may be that the contribution of pneumolysin to virulence differs between serotypes. In a study of the role of pneumolysin in pathogenesis of pneumococcal meningitis, it was found that although direct intracisternal injection of pneumolysin into rabbits caused a rapid inflammatory response, there was no evidence of a contribution of the toxin to the inflammation caused by the whole organism, as determined by infection with wild-type and PLN-A pneumococci (Friedland et al., 1995). Therefore, although pneumolysin can stimulate the inflammatory cascade in the central nervous system, it is not necessary for the pathogenesis of meningeal inflammation. It has also been shown that the toxin plays no role in post-antibiotic enhancement of meningeal inflammation (Friedland et al., 1995).

These findings with regard to inflammation were largely confirmed in the study of Winter et al. (1997), who used a guinea pig model of pneumococcal meningitis. Use of wild-type and PLN-A pneumococci in this model showed there was no difference in the amount of inflammation induced. There was, however, less protein influx into the CSF of animals infected with PLN-A. This may be related to the decreased effect that PLN-A is known to have on other cell barriers (discussed above for the alveolar/capillary barrier). Although no difference in inflammation was detected, PLN-A induced substantially less ultrastructural damage to the cochlea of infected animals, and there was less associated hearing loss (Figure 26.3). When meningitis was induced with PLN-A, there was almost no ultrastructural cochlear damage and significantly less hearing loss at all frequencies tested. The ultrastructural damage observed following infection with wild-type pneumococci was very similar to that induced by microperfusion of the cochlea with purified pneumolysin (Comis *et al.*, 1993). The ultrastructural changes induced by cochlear perfusion with pneumolysin are those that were absent after meningitis due to PLN-A.

Thus, in this experimental model, pneumolysin appears not to contribute to inflammation but the majority of cochlear damage and hearing loss is due to pneumolysin. This also challenges the dogma that damage to the cochlea is due to bystander damage as a result of the inflammatory response.

The above describes some extensive studies on the role of pneumolysin, using simple isogenic knockout mutants. With the availability of molecular techniques, it has also proved possible to evaluate the contribution of individual activities or amino acids of the toxin to the virulence of the pneumococcus. These studies are described below.

Administration of altered versions of purified pneumolysin

As considered above, pneumolysin is a multifunctional protein and these various activities can be allocated to various parts of the protein. This process has been facilitated by the generation of a homology model of the three-dimensional structure of the toxin (Rossjohn *et al.*, 1998). Structure–function data for the protein can be used to define the roles played by the various activities of the toxin *in vivo*. Most studies have concerned the relative contribution of complement activating activity and lytic activity to the pathogenesis of infection.

The first evidence that both complement activation and lytic activity contributed separately to the activity of the toxin *in vivo* was provided by Feldman *et al.* (1991), who showed by direct instillation of purified pneumolysin into ligated lobes of the rat lung that the molecule induced an inflammatory response similar to that seen in pneumonia. Use of mutant forms of pneumolysin that lacked either lytic or complement-activating activity demonstrated that both of these activities contributed to the inflammatory process.

Studies by Rubins *et al.* (1995), using the rat lobar pneumonia model, showed that full virulence could be restored to a pneumolysin-negative mutant of the pneumococcus (PLN-A) by co-administration of a bolus of the wild-type toxin. Co-instillation of mutant forms of the toxin showed that the lytic activity of the toxin was important for allowing growth of bacteria in the lung and invasion into the lung tissue during the early times post-infection (up to 12 h). Complement activation by the toxin only appeared to be important at later times.

Isogenic mutants expressing altered versions of pneumolysin

In order to understand how the various activities of pneumolysin are involved in the virulence of the whole organism, a series of isogenic mutants of a type-2 pneumococcus was constructed, expressing versions of the toxin that carry various amino acid substitutions, which affect the activity of the toxin (Berry *et al.*, 1995). This panel of mutants has been used in several models of pneumococcal infection. The contribution of the various activities of the toxin to virulence varies according to the model used. A summary of the mutation used in these studies is given in Table 26.5 (the strain designations are taken from Alexander *et al.*, 1998).

Assay of isogenic strains in vitro

Strains of pneumococci-producing pneumolysin that differed in their ability to activate the complement pathway or cause cell lysis have been used in assays of complement-mediated killing by human phagocytes to show that complement activation by the toxin is involved in modulating this process (Rubins *et al.*, 1996). Interaction with the complement system may well be important, as it is involved in opsonophagocytosis and clearance of pneumococci from the lung (Winkelstein, 1981). Although the pneumococcal cell wall can activate the alternative complement pathway, the pneumococcal polysaccharide capsule is believed to prevent the access of phagocytes to these cell-wall bound opsonins (Winkelstein, 1981). It has been proposed that deposition of opsonins on the surface of the capsule is essential for effective opsonophagocytosis (Brown *et al.*, 1983). The experiments with isogenic pneumococci suggest that pneumolysin may be able to protect pneumococci from the classical complement pathway by consuming the components of the classical complement factors in solution. Evasion of the complement system in this way would allow increased bacterial survival, especially when the concentrations of complement factors are already low, such as in the lung.

Assay of isogenic strains in vivo: systemic infection

The virulence of the strains described in Table 26.5 was compared in a model of intraperitoneal inoculation into

(a)

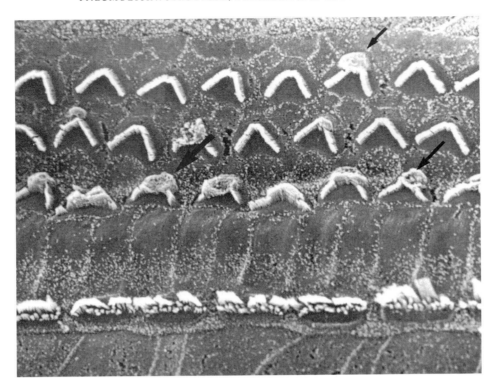

(b)

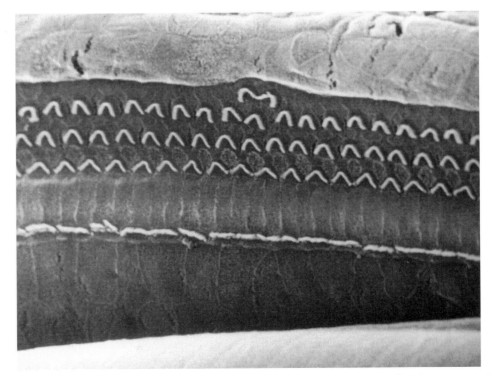

FIGURE 26.3 Comparison of the effect of wild-type and pneumolysin-negative pneumococci on the ultrastructure of the organ of Corti from the cochleas of guinea pigs with pneumococcal meningitis. (a) Organ of Corti from animal infected with wild-type pneumococci: ballooning and disruption of the apical surface of outer hair cells is apparent (indicated by arrows). Original magnification ×1500. (b) Organ of Corti from animal infected with pneumolysin-negative pneumococci (PLN-A): organ of Corti looks normal. Original magnification ×500.

TABLE 26.5 Isogenic strains of *Streptococcus pneumoniae* used in the study of the role of the different activities of pneumolysin in the pathogenesis of infection

Strain designation	Mutation in pneumolysin gene	Haemolytic activity (% of wild-type)	Complement-activating activity (% of wild-type)
H+/C+	Parent	100	100
H1–/C+	Trp433 → Phe	0.1	100
H2–/C+	His367 → Arg	0.02	100
H3–/C+	Cys428 → Gly + Trp433 → Phe	0.001	100
H+/C–	Asp385 → Asn	100	0
H3-/C–	Cys428 → Gly + Trp433 → Phe + Asp385 → Asn	0.001	0

Stain designations are taken from Alexander *et al.* (1998).

mice (Berry *et al.*, 1995). There was no significant difference in either median survival time or overall survival rate between mice challenged with wild-type or those with a deficiency in complement-activation (strain H+/C–). Strains with reduced haemolytic activity (H2–/C+ and H3–/C+) that reduce the activity of the toxin produced to 0.02 and 0.0001%, respectively, had greater median survival times and overall survival rate than mice challenged with wild-type organisms. In the intraperitoneal model, the contribution of pneumolysin to virulence is largely related to its cytotoxic properties rather than its ability to activate complement. It is interesting to note that very small amounts of haemolytic activity may be required for virulence as strain H1–/C+, in which the lytic activity of the toxin is only 0.1% of wild-type, had intermediate virulence.

The same isogenic strains have also been used in an intravenous challenge model (Benton *et al.*, 1997). In this study, strains with decreased complement activation (H+/C–) or decreased lytic activity (H1–/C+, H3–/C+) had virulence characteristics similar to wild-type organisms. Only when mutations in the two activities were combined (H3–/C–) was there any effect on virulence. Moreover, the effect of these combined point mutations on virulence was small compared with that seen in the insertion duplication mutant PLN-A (pneumolysin-negative mutant). The PLN-A mutation increased survival time by 6 days, whereas the H3–/C– mutation increased survival by only 0.5 days. This suggests that pneumolysin has another activity other than lytic or complement activity that can contribute to virulence in this model.

Pneumonia models

In bronchopneumonia following intranasal inoculation of pneumococci, both lytic activity and complement activity are important, as reductions in these activities in the isogenic mutants reduced virulence (Alexander *et al.*, 1998). However, it was the ability to activate complement that most affected the behaviour of pneumo-

cocci in the lungs and associated bacteraemia in the first 24 h following infection.

These findings are similar to those using a lobar pneumonia model (Rubins *et al.*, 1996), where it was confirmed using strains H3–/C+, H+/C– and H3–/C– that absence of either lytic activity or complement-activating activity from pneumolysin rendered mutants less virulent. In contrast to the bronchopneumonia model, these two mutations were not additive in their effect. The lack of additive effect in this system was proposed to show that the lytic activity of the toxin is involved in several steps during pathogenesis, whereas complement activity has a more limited role in reducing bacterial clearance in the lung. The lytic activity of the toxin correlated with acute lung injury and bacterial growth for up to 6 h after inoculation. The complement-activating activity of the toxin correlated with bacterial growth and bacteraemia at 24 h post-inoculation. Complement activation by pneumolysin in lobar pneumonia plays a singular role in pulmonary infection by promoting survival of pneumococci in the lung tissues and facilitating invasion in the blood (Rubins *et al.*, 1996).

Summary of the role of pneumolysin

A summary of the role played by pneumolysin in the different infections is given in Figure 26.4. It is clear that pneumolysin has different roles in different forms of pneumococcal disease. Thus, the toxin has no role in inflammation in meningitis (Friedland *et al.*, 1995; Winter *et al.*, 1997), but has a distinct role in meningitis-associated deafness (Winter *et al.*, 1997) and bacteraemia (Benton *et al.*, 1995) as well as pneumonia. It also appears that, where pneumolysin has a role, the contributions of the haemolytic and complement-activating abilities differ. In bronchopneumonia, the haemolytic activity and complement activation by pneumolysin each contribute to the virulence of pneumococci and these activities have a distinct role in the disease (Alexander *et al.*, 1998). This is also true in lobar pneumonia (Rubins *et al.*, 1996), but with no additive effect on virulence when both activities were decreased, while in bronchopneumonia the survival time increased if both activities were reduced. After intraperitoneal infection, it was only changes that affected the lytic activity of the toxin that changed virulence (Berry *et al.*, 1995), and the amount of haemolytic activity required for virulence was low. In the lung, haemolytic and complement-activating activities have distinct effects on the behaviour of pneumococci, and the time at which each activity contributes is different. In bronchopneumonia, complement activation is important throughout the first 24 h post-infection. Reduction in haemolytic activity had an effect only after 6 h. In contrast, in lobar pneumonia haemolytic activity was important in the first 6 h, with complement activation only playing a role afterwards. It is probably significant that the inoculum entering the lungs is much smaller after intranasal inoculation (to establish bronchopneumonia) and the complement-activating ability of the toxin may be important in protecting this relatively small inoculum. When a large inoculum is used to establish lobar pneumonia, it is the ability to damage the alveolar capillary barrier to release nutrients that becomes important.

During bronchopneumonia, the level of bacteraemia is closely related to the level of growth of bacteria in the lung. Mutations affecting growth in the lung (removal of complement-activating ability) therefore caused a corresponding decrease in the amount of bacteraemia. Reduction of haemolytic activity had little effect on growth in the lung and, therefore, a small effect on numbers of pneumococci in the blood. Studies using intravenous instillation of pneumococci into the blood suggest that all mutants should grow at the same rate once in the blood (Benton *et al.*, 1995, 1997; Berry *et al.*, 1995). It therefore appears that continual seeding of bacteria from the lungs contributes to the level of bacteria in the blood rather than being due to a growth of an inoculum in the blood. An alternative explanation is that the wild-type and H–/C+ strain may be affected by passage through the lungs, i.e. entry to the blood via the lungs is different from entry to the blood via a syringe. Wild-type organisms injected into the blood reach levels of up to 10^{10} CFU ml^{-1} (Benton *et al.*, 1995), whereas those entering the blood from the lung reach maximal levels of 10^8 (Alexander *et al.*, 1998). This difference may be host mediated. Entry via the lungs may induce a host response that controls subsequent bacteraemia, whereas direct injection bypasses the induction of this response and allows greater growth of pneumococci. Pneumolysin may contribute to the down-regulation of this response when the organisms gain access to the lungs.

Pneumococci carrying a three-point mutation in the pneumolysin gene were not as avirulent as those mak-

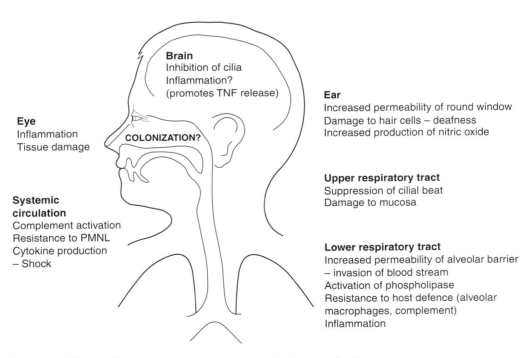

FIGURE 26.4 Summary of the possible roles that pneumolysin may play during infection.

ing no toxin. This indicates that some further unidentified activity of pneumolysin plays a role in infection (Benton et al., 1997; Alexander et al., 1998). This could be related to the ability of strain H3–/C– to bind to the Fc portion of antibody (Mitchell et al., 1991) or could be related to the anticellular activity of sublytic concentrations of the toxin. It will be important to determine the nature of the 'unidentified' activity, if pneumolysin genetic toxoids are to be considered as vaccine candidates.

USE OF PNEUMOLYSIN AS A VACCINE

The currently used pneumococcal vaccine for use in humans is based on a mixture of 23 capsular polysaccharides selected from the 90 known types. This vaccine has two major shortcomings. The first is that it is serotype specific, and the second is that protection against the types included may be very poor in high-risk groups who have a poorer antibody response to polysaccharide vaccines than healthy adults (Broome, 1981; Forrester et al., 1987). It has been proposed that a cross-serotype protective protein could overcome these problems. Protective pneumococcal antigens would also be attractive candidates for use as carriers for polysaccharides in a conjugate vaccine. Conjugation of polysaccharides to proteins increases their immunogenicity and also converts them to antigens capable of generating immunological memory (Snippe et al., 1983). Studies with pneumococcal polysaccharides conjugated to a range of proteins have demonstrated enhanced immunogenicity of the polysaccharide (Andrew et al., 1994). Early trials of conjugate vaccines look promising but, as these vaccines have limited serotype specificity, care needs to be taken, since large-scale vaccination might over time lead to a shift in the capsular types towards those that are not included in the vaccine. Such a shift may be promoted by the frequent horizontal exchange of capsule genes (Barnes et al., 1995; Hermans et al., 1997). Also, the new generation of conjugate vaccines may be too expensive to be used in all parts of the world. A protein-based vaccine would therefore have attractions.

Pneumolysin is a good candidate as a vaccine antigen, as it is produced by virtually all clinical isolates of the pneumococcus (Kanclerski and Mollby, 1987) and its primary structure varies little (Mitchell et al., 1990). Evidence for the production of pneumolysin during invasive pneumococcal disease in humans is based on serological analysis. Bacteraemic pneumococcal pneumonia is associated with a rise in anti-pneumolysin titres in 50–85% of adults, especially in those with more severe pneumonia (Kalin et al., 1987; Kanclerski et al., 1988; Jalonen et al., 1989). Although native pneumolysin is too toxic to be proposed for inclusion in a vaccine, genetically altered pneumolysin with reduced toxicity could be used (Boulnois, 1992). Structure–function data and information on pathogenesis can be used for the rational design of candidate vaccine molecules.

Immunization of mice with such a genetically engineered toxoid confers protection against at least nine serotypes of the pneumococcus (Alexander et al., 1994). In this study the amount of protection conferred varied with the serotype of pneumococcus used and the route of challenge. No protection was afforded against the serotype-3 strain GBO5 when given intranasally. The absence of protection is not related solely to serotype, as some protection was afforded against another serotype-3 strain. The lack of protection against some pneumococcal isolates conferred by the pneumolysin toxoid suggests that additional pneumococcal immunogens may be required in conjunction with the pneumolysin toxoid. Pneumolysin could be used as a component of a multivalent protein vaccine, as an addition to a conventional conjugate vaccine or as a carrier for pneumococcal polysaccharide. A conjugate of pneumolysin toxoid and 19F polysaccharide has been produced and generates a protective immune response in infant mice (Lee et al., 1994). Further protection may be achieved by alternative delivery routes for the pneumolysin toxoid.

The experiments described above have considered only systemic immunity raised to the pneumolysin toxoid. Since colonization of the nasopharyngeal mucosa by pneumococci is a prerequisite for the development of pneumococcal disease, it may be that mucosal immunity will give better protection from disease. Polysaccharides induce little secretory IgA at the mucosal surface and do not affect the carriage rate of pneumococci (Douglas et al., 1986). It is possible that conjugation of the polysaccharide may prove more effective at reducing carriage. In the case of Haemophilus influenzae, the use of a conjugate vaccine eliminated carriage of the organism in children (Takala et al., 1991). It may also be possible to generate a secretory IgA response by mucosal immunization and thus influence colonization by virulent pneumococci. One mechanism for mucosal immunization is the use of live delivery vehicles. Pneumolysin has been expressed in such a delivery vehicle (Paton et al., 1993b). An attenuated Salmonella strain expressing pneumolysin toxoid was able to elicit both IgG and IgA antibodies to the toxoid when given by the oral route (Paton et al., 1993b). New developments in DNA vaccine technology may also provide alternative routes and methods of vaccination with pneumolysin.

CONCLUSION

Many years' work on pneumolysin mean that much is now known about its activity. Detailed structure–function analysis has revealed where some of the biological properties of the toxin lie within the molecule. A good homology model of the toxin now exists, although we still await a determination of the three-dimensional crystal structure of this toxin. The mechanism of pore formation by pneumolysin is partially understood but there are still many aspects to this intriguing process that remain to be clarified.

Availability of structure–function data and molecular biology technology in combination with traditional methods of studying pathogenesis has allowed a detailed analysis of the role of this toxin and its various activities in the pathogenesis of pneumococcal infection to be dissected. The role of the toxin is complex and varies according to which disease is examined. These studies highlight the difficulties in defining the role played by bacterial toxins in the disease process. Studies *in vitro* can only go some way to understanding the role of a toxin. Ultimate understanding can only be achieved by the study of *in vivo* systems.

ACKNOWLEDGEMENTS

I would like to acknowledge the many people who have been involved in our studies on pneumolysin. Much of their data is included in this review. In particular, I would like to thank Peter Andrew, Graham Boulnois, Jeff Rubins, Rob Wison and James Paton for many years of fruitful collaboration and discussion. I thank Mike Parker for providing the crystal co-ordinates of perfringolysin and the model of pneumolysin and Nithin Rai for producing Figure 26.1. I thank Andy Winter for Figure 26.3. Work in the author's laboratory has been supported by the Medical Research Council, the Royal Society and the Wellcome Trust, to whom I am grateful.

REFERENCES

Alexander, J., Lock, R.A., Peeters, C.C.A.M., Poolman, J.T., Andrew, P.W., Mitchell, T.J., Hansman, D. and Paton, J.C. (1994). Immunization of mice with pneumolysin toxoid confers a significant degree of protection against at least nine serotypes of *Streptococcus pneumoniae*. *Infect. Immun.* **62**, 5683–8.

Alexander, J.E., Berry, A.M., Paton, J.C., Rubins, J.B., Andrew, P.W. and Mitchell, T.J. (1998). The course of pneumococcal pneumonia is altered by amino acid changes affecting the activity of pneumolysin. *Microb. Pathog.* **24**, 167–74.

Alfzelius, B.A. (1979). The immotile-cilia syndrome and other ciliary diseases. *Int. Rev. Exp. Pathol.* **19**, 1–43.

Amee, F.Z., Comis, S.D. and Osbourne, M.P. (1995). N^G-Methyl-L-arginine protects the guinea pig cochlea from the cytotoxic effects of pneumolysin. *Acta Otolaryngol.* **115**, 386–91.

Andrew, P.W., Boulnois, G.J., Mitchell, T.J., Lee, C.-J., Paton, J.C., Poolman, J.T. and Peteers, C.C.A.M. (1994). Pneumococcal vaccines. *Zentralbl. Bakteriol.* **S24**, 453–66.

Badwey, J.A., Curnutte, J.T., Robinson, J.M., Berde, C.B., Karnovsky, M.J. and Karnovsky, M.L. (1984). Effects of free fatty acids on release of superoxide and on change of shape by human neutrophils. Reversibility by albumin. *J. Biol. Chem.* **259**, 7870–7.

Barnes, D.M., Whittier, S., Gilligan, P.H., Soares, S., Tomasz, A. and Henderson, F.W. (1995). Transmission of multidrug-resistant serotype 23F *Streptococcus pneumoniae* in group day care: evidence suggesting capsular transformation of the resistant strain *in vivo*. *J. Infect. Dis.* **171**, 890–6.

Benton, K.A., Everson, M.P. and Briles, D.E. (1995). A pneumolysin-negative mutant of *Streptococcus pneumoniae* causes chronic bacteremia rather than acute sepsis in mice. *Infect. Immun.* **63**, 448–55.

Benton, K.A., Paton, J.C. and Briles, D.E. (1997). The hemolytic and complement-activating properties of pneumolysin do not contribute individually to virulence in a pneumococcal bacteramia model. *Microb. Pathog.* **23**, 201–9.

Benton, K.A., VanCott, J.L. and Briles, D.E. (1998). Role of tumor necrosis factor alpha in the host response of mice to bacteremia caused by pneumolysin-deficient *Streptococcus pneumoniae*. *Infect. Immun.* **66**, 839–42.

Berry, A.M., Yother, J., Briles, D.E., Hansman, D. and Paton, J.C. (1989). Reduced virulence of a defined pneumolysin-negative mutant of *Streptococcus pneumoniae*. *Infect. Immun.* **57**, 2037–42.

Berry, A.M., Paton, J.C. and Hansman, D. (1992). Effect of insertional inactivation of the genes encoding pneumolysin and autolysin on the virulence of *Streptococcus pneumoniae* type 3. *Microb. Pathog.* **12**, 87–93.

Berry, A.M., Alexander, J.E., Mitchell, T.J., Andrew, P.W., Hansman, D. and Paton, J.C. (1995). Effect of defined point mutations in the pneumolysin gene on the virulence of *Streptococcus pneumoniae*. *Infect. Immun.* **63**, 1969–74.

Boulnois, G.J. (1992). Pneumococcal proteins and the pathogenesis of disease caused by *Streptococcus pneumoniae*. *J. Gen. Microbiol.* **138**, 249–59.

Broome, C.V. (1981). Efficacy of pneumococcal polysaccharide vaccines. *Rev. Infect. Dis.* **3** (Suppl.), S82–8.

Brown, E.J., Joiner, K.A., Cole, R.M. and Berger, M. (1983). Localisation of complement component C3 on *Streptococcus pneumoniae*: anti-capsular antibody causes complement deposition on the pneumococcal capsule. *Infect. Immun.* **39**, 403–9.

Cabellos, C., MacIntyre, D.E., Forrest, M., Burroughs, M., Prasad, S. and Tuomanen, E. (1992). Differing roles for platelet-activating factor during inflammation of the lung and sub-arachnoid space: the special case of *Streptococcus pneumoniae*. *J. Clin. Invest.* **90**, 612–18.

Canvin, J.R., Marvin, A.P., Sivakumaran, M., Paton, J.C., Boulnois, G.J., Andrew, P.W. and Mitchell, T.J. (1995). The role of pneumolysin and autolysin in the pathology of pneumonia and septicemia in mice infected with a type 2 pneumococcus. *J. Infect. Dis* **172**, 119–23.

Canvin, J.R., Paton, J.C., Boulnois, G.J., Andrew, P.W. and Mitchell, T.J. (1997). *Streptococcus pneumoniae* produces a second haemolysin that is distinct from pneumolysin. *Microb. Pathog.* **22**, 129–32.

Cohen, B., Perkins, M.E. and Putterman, S. (1940). The reaction between hemolysin and cholesterol. *J. Bacteriol.* **39**, 59–60.

Cole, R. (1914). Pneumococcus hemotoxin. *J. Exp. Med.* **20**, 346–62.

Comis, S. D., Osborne, M.P., Stephen, J., Tarlow, M.J., Hayward, T.L., Mitchell, T.J., Andrew, P.W. and Boulnois, G.J. (1993). Cytotoxic effects on hair cells of guinea pig cochlea produced by pneumolysin, the thiol activated toxin of *Streptococcus pneumoniae*. *Acta Otolaryngol.* **113**, 152–9.

Curnutte, J.T., Badwey, J.M., Robinson, J.M., Karnovsky, M.J. and Karnovsky, M.L. (1984). Studies on the mechanism of superoxide release from human neutrophils stimulated with arachidonate. *J. Biol. Chem.* **259**, 11851–7.

de los Toyos, J.R., Mendez, F.J., Aparicio, J.F., Vazquez, F., del Mar Garcia Suarez, M., Fleites, A., Hardisson, C., Morgan, P.J., Andrew, P.J. and Mitchell, T.J. (1996). Functional analysis of pneumolysin by use of monoclonal antibodies. *Infect. Immun.* **64**, 480–4.

Douglas, R.M., Hansman, D., Miles, H. and Paton, J.C. (1986). Pneumococcal carriage and type specific antibody. Failure of a 14-valent vaccine to reduce carriage in healthy children. *Am. J. Dis. Child.* **140**, 1183–5.

Engel, F., Blatz, R., Kellner, J., Palmer, M., Weller, U. and Bhakdi, S. (1995). Breakdown of the round window permeability barrier evoked by streptolysin O: possible etiologic role in the development of sensorineural hearing loss in acute otitis media. *Infect. Immun.* **63**, 1305–10.

Feldman, C., Mitchell, T.J., Andrew, P.W., Boulnois, G.J., Read, R.C., Todd, H.C., Cole, P.J. and Wilson, R. (1990). The effect of *Streptococcus pneumoniae* pneumolysin on human respiratory epithelium *in vitro*. *Microb. Pathog.* **9**, 275–84.

Feldman, C., Munro, N.C., Jeffrey, D.K., Mitchell, T.J., Andrew, P.W., Boulnois, G.J., Gueirreiro, D., Rohde, J.A.L., Todd, H.C., Cole, P.J. and Wilson, R. (1991). Pneumolysin induces the salient features of pneumococcal infection in the rat lung *in vivo*. *Am. J. Respir. Cell. Mol. Biol.* **5**, 416–23.

Ferrante, A., RowanKelly, B. and Paton, J.C. (1984). Inhibition of *in vitro* human lymphocyte response by the pneumococcal toxin pneumolysin. *Infect. Immun.* **46**, 585–9.

Forrester, H.L., Jahigen, D.W. and LaForce, F.M. (1987). Inefficacy of pneumococcal vaccine in high-risk population. *Am. J. Med.* **83**, 425–30.

Fortnum, H.M. (1992). Hearing impairment after bacterial meningitis: a review. *Arch. Dis. Child.* **67**, 1128–33.

Friedland, I.R., Paris, M.M., Hickey, S., Shelton, S., Olsen, K., Paton, J.C. and McCracken, G.H. (1995). The limited role of pneumolysin in the pathogenesis of pneumococcal meningitis. *J. Infect. Dis.* **172**, 805–9.

Halbert, S.P., Cohen, B. and Perkins, M.E. (1946). Toxic and immunological properties of pneumococcal hemolysin. *Bull. Johns Hopkins Hosp.* **78**, 340–59.

Harrison, J.C., Karcioglu, Z.A. and Johnson, M.K. (1993). Response of leukopenic rabbits to pneumococcal toxin. *Curr. Eye Res.* **2**, 705–10.

Hermans, P.W., Sluijter, M., Dejsirilert, S., Lemmens, N., Elzennaar, K., Veen, A.v., Gossens, W.H. and Groot, R.d. (1997). Molecular epidemiology of drug-resistant pneumococci: toward an international approach. *Microb. Drug Resist.* **3**, 243–51.

Hill, J., Andrew, P.W. and Mitchell, T.J. (1994). Amino acids in pneumolysin important for hemolytic activity identified by random mutagenesis. *Infect. Immun.* **62**, 757–8.

Holtzman, M.J. (1991). Arachidonic acid metabolism. Implications of biological chemistry for lung function and disease. *Am. Rev. Respir. Dis.* **143**, 188–203.

Houldsworth, S., Andrew, P.W. and Mitchell, T.J. (1994). Pneumolysin stimulates production of tumor necrosis factor alpha and interleukin-1beta by human mononuclear phagocytes. *Infect. Immun.* **62**, 1501–3.

Jacobs, T., Darji, A., Frahm, N., Rohde, M., Wehland, J., Chakraborty, T. and Weiss, S. (1998). Listeriolysin O: cholesterol inhibits cytolysis but not binding to cellular membranes. *Mol. Microbiol.* **28**, 1081–9.

Jalonen, E., Paton, J.C., Koskela, M., Kerttula, Y. and Leinonen, M. (1989). Measurement of antibody resonses to pneumolysin – a promising diagnostic method for the presumptive aetiological diagnosis of pneumococcal pneumonia. *J. Infect.* **19**, 127–34.

Johnson, M.K. (1977). Cellular location of pneumolysin. *FEMS Microbiol. Lett.* **2**, 243–5.

Johnson, M.K. and Allen, J.H. (1975). The role of cytolysin in pneumococcal ocular infection. *Am. J. Opthalmol.* **80**, 518–20.

Johnson, M.K., Knight, R.J. and Drew, G.K. (1982). The hydrophobic nature of thiol-activated cytolysins. *Biochem. J.* **207**, 557–60.

Johnson, M.K., Hobden, J.A., Hagenah, M., OCallaghan, R.J., Hill, J.M. and Chen, S. (1990). The role of pneumolysin in ocular infections with *Streptococcus pneumoniae*. *Curr. Eye Res.* **9**, 1107–14.

Kalin, M., Kanclerski, K., Granstrom, M. and Mollby, R. (1987). Diagnosis of pneumococcal pneumonia by enzyme-linked immunosorbent assay of antibodies to pneumococcal hemolysin (pneumolysin). *J. Clin. Microbiol.* **25**, 226–9.

Kanclerski, K. and Mollby, R. (1987). Production and purification of *Streptococcus pneumoniae* hemolysin (pneumolysin). *J. Clin. Microbiol.* **25**, 222–5.

Kanclerski, K., Blomquist, S., Granstrom, M. and Mollby, R. (1988). Serum antibodies to pneumolysin in patients with pneumonia. *J. Clin. Microbiol.* **26**, 96–100.

Korchev, Y.E., Bashford, C.L. and Pasternak, C.A. (1992). Differential sensitivity of pneumolysin-iduced channels to gating by divalent cations. *J. Membr. Biol.* **127**, 195–3.

Korchev, Y.E., Bashford, C.L., Pederzolli, C., Pasternack, C.A., Morgan, P.J., Andrew, P.W. and Mitchell, T.J. (1998). A conserved tryptophan in pneumolysin is a determinant of the characteristics of channels formed by pneumolysin in cells and planar lipid bilayers. *Biochem. J.* **329**, 571–7.

Lee, C.J., Lock, R.A., Andrew, P.W., Mitchell, T.J., Hansman, D. and Paton, J.C. (1994). Protection of infant mice from challenge with *Streptococcus pneumoniae* type 19F by immunization with a type 19F polysaccharide–pneumolysoid conjugate. *Vaccine*, **12**, 875–8.

Libman, E. (1905). A pneumococcus producing a peculiar form of hemolysis. *Proc. NY Pathol. Soc.* **5**, 168.

Lock, R.A., Zhang, Q.Y., Berry, A.M. and Paton, J.C. (1996). Sequence variation in the *Streptococcus pneumoniae* pneumolysin gene affecting haemolytic activity and electrophoretic mobility of the toxin. *Microb. Pathog.* **21**, 71–83.

Mendez, F.J., de los Toyos, J.R., Fleites, A., Aparico, J.F., Vazquez, F., Hardisson, C., Mitchell, T.J., Andrew, P.W. and Boulnois, G.J. (1992). Anti-pneumolysin antibodies as diagnostical and therapeutical tools. In: *Profiles on Biotechnology* (eds T.G. Villa and J. Abaide), pp. 107–14. Universidade de Santiago, Santiago.

Mitchell, T.J. and Andrew, P.W. (1997). Biological properties of pneumolysin. *Microb. Drug Resist.* **3**, 19–26.

Mitchell, T.J., Walker, J.A., Saunders, F.K., Andrew, P.A. and Boulnois, G.J. (1989). Expression of the pneumolysin gene in *Escherichia coli*: rapid purification and biological properties. *Biochim. Biophys. Acta* **1007**, 67–72.

Mitchell, T.J., Paton, J.C., Andrew, P.W. and Boulnois, G.J. (1990). Comparison of pneumolysin genes and proteins from *Streptococcus pneumoniae* types 1 and 2. *Nucleic Acids Res.* **18**, 4010.

Mitchell, T.J., Andrew, P.W., Saunders, F.K., Smith, A.N. and Boulnois, G.J. (1991). Complement activation and antibody binding by pneumolysin via a region of the toxin homologous to a human acute-phase protein. *Mol. Microbiol.* **5**, 1883–8.

Mitchell, T.J., Hill, J. and Andrew, P.W. (1994). The role of histidine residues in the cytolytic action of pneumolysin. In: *Bacterial Protein Toxins* (eds J. Freer, R. Aitken, J.E. Alouf, G.J. Boulnois, P.G. Falmagne, F. Fehrenbach, C. Montecucco, Y. Piemont, R. Rappuoli, T. Wadstrom and B. Witholt), pp. 335–6. Gustav Fisher, Stuttgart.

Morgan, P.J., Hyman, S.C., Byron, O., Andrew, P.W., Mitchell, T.J. and Rowe, A.J. (1994). Modeling the bacterial protein toxin, pneumolysin, in its monomeric and oligomeric form. *J. Biol. Chem.* **269**, 25315–20.

Morgan, P.J., Andrew, P.W. and Mitchell, T.J. (1996). Thiol-activated cytolysins. *Rev. Med. Micro.* **7**, 221–9.

Morgan, P.J., Harrison, G., Freestone, P.P.E., Crane, D., Rowe, A.J., Mitchell, T.J., Andrew, P.W. and Gilbert, R.J.C. (1997). Structural and functional characterisation of two proteolytic fragments of the bacterial protein toxin, pneumolysin. *FEBS Lett.* **412**, 563–7.

Neill, J.M. (1926). Studies on the oxidation and reduction of immunological substances. I. Pneumococcus hemotoxin. *J. Exp. Med.* **44**, 199–213.

Neill, J.M. (1927). Studies on the oxidation and reduction of immunological substances. V. Production of anti-hemotoxin by immunization with oxidized pneumococcus hemotoxin. *J. Exp. Med.* **45**, 105–13.

Ng, E.W.M., Samily, N., Rubins, J.B., Cousins, F.V., Ruoff, K.L., Baker, A.S. and DAmico, D.J. (1997). Implication of pneumolysin as a virulence factor in *Streptococcus pneumoniae* endopthalmitis. *Retina* **17**, 521–9.

Owen, R.H.G.O., Boulnois, G.J., Andrew, P.W. and Mitchell, T.J. (1994). A role in cell-binding for the C-terminus of pneumolysin, the thiol-activated toxin of *Streptococcus pneumoniae. FEMS Lett.* **121**, 217–22.

Palmer, M., Valeva, A., Kehoe, M. and Bhakdi, S. (1995). Kinetics of streptolysin O self-assembly. *Eur. J. Biochem.* **231**, 388–95.

Palmer, M., Saweljew, P., Vulicevic, I., Valeva, A., Kehoe, M. and Bhakdi, S. (1996). Membrane-penetrating domain of streptolysin O identified by scanning mutagenesis. *J. Biol. Chem.* **271**, 26662–7.

Paton, J.C. and Ferrante, A. (1983). Inhibition of human polymorphonuclear leukocyte respiratory burst, bactericidal activity, and migration by pneumolysin. *Infect. Immun.* **41**, 1212–16.

Paton, J.C., Lock, R.A. and Hansman, D.J. (1983). Effect of immunization with pneumolysin on survival time of mice challenged with *Streptococcus pneumoniae. Infect. Immun.* **40**, 548–52.

Paton, J.C., Rowan-Kelly, B. and Ferrante, A. (1984). Activation of human complement by the pneumococcal toxin pneumolysin. *Infect. Immun.* **43**, 1085–7.

Paton, J.C., Berry, A.M., Lock, R.A., Hansman, D. and Manning, P.A. (1986). Cloning and expression in *Escherichia coli* of the *Streptococcus pneumoniae* gene encoding pneumolysin. *Infect. Immun.* **54**, 50–5.

Paton, J.C., Andrew, P.W., Boulnois, G.J. and Mitchell, T.J. (1993a). Molecular analysis of the pathogenicity of *Streptococcus pneumoniae*: the role of pneumococcal proteins. *Annu. Rev. Microbiol.* **47**, 89–115.

Paton, J.C., Morona, J.K., Harrer, S., Hansman, D. and Morona, R. (1993b). Immunization of mice with *Salmonella typhimurium* C5 aroA expressing a genetically toxoided derivative of the pneumococcal toxin pneumolysin. *Microb. Pathog.* **14**, 95–102.

Rayner, C., Jackson, A.D., Rutman, A., Dewar, A., Mitchell, T.J., Andrew, P.W., Cole, P.J. and Wilson, R. (1995). Interaction of pneumolysin-sufficient and -deficient isogenic variants of *Streptococcus pneumoniae* with human respiratory mucosa. *Infect. Immun.* **63**, 442–7.

Rossjohn, J., Feil, S.C., McKinstry, W.J., Tweten, R.K. and Parker, M.W. (1997). Structure of a cholesterol-binding, thiol-activated cytolysin and a model of its membrane form. *Cell* **89**, 685–92.

Rossjohn, J., Gilbert, R.J.C., Crane, D., Morgan, P.J., Mitchell, T.J., Rowe, A.J., Andrew, P.W., Paton, J.C., Tweten, R.K. and Parker, M.W. (1998). The molecular mechanism of pneumolysin, a virulence factor from *Streptococcus pneumoniae. J. Mol. Biol.* **248**, 449–61.

Rubins, J.B., Duane, P.G., Charboneau, D. and Janoff, E.N. (1992). Toxicity of pneumolysin to pulmonary endothelial cells *in vitro. Infect. Immun.* **60**, 1740–6.

Rubins, J.B., Duane, P.G., Clawson, D., Charboneau, D., Young, J. and Niewoehner, D.E. (1993). Toxicity of pneumolysin to pulmonary alveolar epithelial cells. *Infect. Immun.* **61**, 1352–8.

Rubins, J.B., Mitchell, T.J., Andrew, P.W. and Niewoehner, D.E. (1994). Pneumolysin activates phospholipase A in pulmonary artery endothelial cells. *Infect. Immun.* **62**, 3829–36.

Rubins, J.B., Charboneau, D., Paton, J.C., Mitchell, T.J., Andrew, P.W. and Janoff, E.N. (1995). Dual function of pneumolysin in the early pathogenesis of murine pneumococcal pneumonia. *J. Clin. Invest.* **95, 62**, 142–50.

Rubins, J.B., Charboneau, D., Fasching, C., Berry, A.M., Paton, J.C., Alexander, J.E., Andrew, P.W., Mitchell, T.J. and Janoff, E.N. (1996). Distinct role for pneumolysin's cytotoxic and complement activities in the pathogenesis of pneumococcal pneumonia. *Am. J. Respir. Crit. Care Med.* **153**, 1339–46.

Sato, K., Quartey, M.K., Liebeler, C.L., Le, C.T. and Giebink, G.S. (1996). Roles of autolysin and pneumolysin in middle ear inflammation caused by a type 3 *Streptococcus pneumoniae* strain in the chinchilla otitis media model. *Infect. Immun.* **64**, 1140–5.

Saunders, F.K., Mitchell, T.J., Walker, J.A., Andrew, P.W. and Boulnois, G.J. (1989). Pneumolysin, the thiol-activated toxin of *Streptococcus pneumoniae*, does not require a thiol group for *in vitro* activity. *Infect. Immun.* **57**, 2547–52.

Smyth, C.J. and Duncan, J.L. (1978). Thiol-activated (oxygen-labile) cytolysins. In: *Bacterial Toxins and Cell Membranes* (eds J. Jeljaszewicz and T. Wadstrom), pp. 129–83. Academic Press, London.

Snippe, H., Houte, A.J. v., Dam, J.E.G. v., Reuver, M.J.d., Jansze, M. and Williers, J.M.N. (1983). Immunogenic properties in mice of hexasaccharide from the capsular polysaccharide of *Streptococcus pneumoniae I type 3. Infect. Immun.* **40**, 856–61.

Sowdhamini, R., Mitchell, T.J., Andrew, P.W. and Morgan, P.J. (1997). Structural and functional analogy between pneumolysin and proaerolysin. *Protein Eng.* **10**, 207–15.

Steinfort, C., Wilson, R., Mitchell, T., Feldman, C., Rutman, A., Todd, H., Sykes, D., Walker, J., Saunders, K., Andrew, P.W., Boulnois, G.J. and Cole, P.J. (1989). Effect of *Streptococcus pneumoniae* on human respiratory epithelium *in vitro. Infect. Immun.* **57**, 2006–13.

Takala, A.K., Eskola, J., Leinonen, M., Kayhty, H., Nissinen, A., Pekkanen, E. and Makela, P.H. (1991). Reduction of oropharyngeal carriage of haemophilus-influenzae type-b (hib) in children immunized with an hib conjugate vaccine. *J. Infect. Dis.* **164**, 982–6.

Tuomanen, E., Rich, R. and Zak, O. (1987). Induction of pulmonary inflammation by components of the pneumococcal cell surface. *Am. Rev. Respir. Dis.* **135**, 868–74.

Volanakis, J.E. and Kaplan, M.H. (1974). Interaction of C-reactive protein complexes with the complement system. II. Consumption of guinea pig complement by CRP complexes: requirement for human C1q. *J. Immunol.* **113**, 9–17.

Walker, J.A., Allen, R.L., Falmagne, P., Johnson, M.K. and Boulnois, G.J. (1987). Molecular cloning, charcaterization, and complete nucleotide sequence of the gene for pneumolysin, the sulfhydryl-activated toxin of *Streptococcus pneumoniae*. *Infect. Immun.* **55**, 1184–9.

Winkelstein, J.A. (1981). Complement and the host's defence against the pneumococcus. *CRC Crit. Rev. Microbiol.* **11**, 187–208.

Winter, A.J., Comis, S.D., Osborne, M.P., Tarlow, M.J., Stephen, J., Andrew, P.W., Hill, J. and Mitchell, T.J. (1997). A role for pneumolysin but not neuraminidase in the hearing loss and cochlear damage induced by experimental pneumococcal meningitis in guinea pigs. *Infect. Immun.* **65**, 4411–18.

27

Perfringolysin O

Jamie Rossjohn, Rodney K. Tweten, Julian I. Rood and Michael W. Parker

INTRODUCTION

Clostridium perfringens is responsible for a variety of human and animal diseases including gas gangrene, food poisoning, necrotic enteritis and enterotoxemia (for a review, see Rood *et al.*, 1997). Identification of the major virulence factors of the bacterium has been complicated by the fact that more than 16 different toxic factors have been identified (see Rood and Cole, 1991, for a review). One of these is perfringolysin O (PFO or *θ*-toxin). PFO is a member of a large family of Gram-positive bacterial protein toxins with more than 20 members so far identified. They share many properties in common, which is not surprising considering they share greater than 40% amino acid sequence identity. A defining feature of the family is the requirement of cholesterol in the membrane of the target cell for toxicity to occur. The toxic effect of these proteins is thought to be based on the formation of large oligomeric arcs and rings of up to 15 nm diameter on the target cell membrane. These structures lead to permeabilization of the membrane but it is not clear whether this occurs through local disruption of the bilayer or via the assembly of protein pores. Either way, the structures cause membrane damage leading to rapid cell death through the uncontrolled efflux of essential nutrients into the surrounding milieu.

The family of toxins to which PFO belongs has been the subject of a number of reviews (for example, Bernheimer, 1976; Smyth and Duncan, 1978; Alouf and

Geoffrey, 1984; Bernheimer and Rudy, 1986; Bhakdi and Tranum-Jensen, 1988; Alouf and Geoffrey, 1991; Braun and Focareta, 1991; Tweten, 1995). Although PFO itself has not been the sole subject of a review, a notable number of recent advances suggests that such a review is now warranted (Table 27.1). Chief amongst these advances is the recent determination of the three-dimensional atomic structure of PFO resolved by X-ray crystallography (Rossjohn *et al.*, 1997). This advance has allowed the plethora of biochemical and biophysical data that exists for the toxin to be explained at the molecular level.

Nomenclature

PFO was originally referred to as *θ*-toxin but a more descriptive name was later adopted whereby the toxin was named after the organism from which it derived and the appended letter 'O' reflected the observed oxygen lability found in impure preparations of the toxin (Bernheimer, 1976). This property also led to the naming of the toxin family as the 'oxygen-labile' haemolysins (van Heyningen, 1950). However, the oxygen lability was latter found to be an artefact of the early impure toxin preparations. The family was subsequently renamed the 'thiol-activated' cytolysins, reflecting their ability to be activated by thiol-reducing agents. However, mutagenesis studies have since indicated that the term 'thiol-activated' is misleading (see below). Alouf and Geoffroy (1991) have proposed the

TABLE 27.1 Key events in the study of perfringolysin O

Year	Event	Reference
1923	Activity first described	Wuth (1923)
1947	Pores shown to allow passage of large macromolecules	Bernheimer (1947)
1970s	First purifications reported	Mitsui *et al.* (1973), Nord *et al.* (1974), Smyth (1975), Möllby *et al.* (1976)
1975	Visualization of arcs and rings in membranes by EM	Smyth *et al.* (1975)
1980s–90s	Series of proteolysis and other biochemical studies	See text
1988	Cloned and expressed in *E. coli*	Tweten (1988a, b)
1990s	Discovery of regulatory genes	Shimizu *et al.* (1991), Lyristis *et al.* (1994), Shimizu *et al.* (1994), Ba-Thein *et al.* (1996)
1993	Detailed EM studies of oligomer	Olofsson *et al.* (1993)
1995	Virulence studies on *pfoA* mutants	Awad *et al.* (1995)
1997	Crystal structure published	Rossjohn *et al.* (1997)

alternative name: cholesterol-binding cytolysins (CBCs), reflecting another common property of each toxin. The authors have chosen to use this terminology in this review.

CHARACTERIZATION AND PURIFICATION

PFO was first described by Wuth (1923) and purification procedures for the toxin subsequently were reported in the early 1970s (Mitsui *et al.*, 1973; Nord *et al.*, 1974; Smyth, 1975; Möllby *et al.*, 1976). These purifications were performed on culture fluids of *C. perfringens* type A strains and involved standard chromatographic procedures. With the advent of recombinant PFO expression in *E. coli*, milligram quantities could be purified by gel filtration and chromatography (Tweten, 1988a, b).

BIOCHEMICAL AND MUTAGENESIS STUDIES

Early studies showed that PFO could be inactivated by cholesterol solutions. The toxin appeared to be highly specific for the sterol since the presence of the 3β-hydroxyl group was found critical for the inhibitory activity. This and subsequent studies also indicated the importance of a long alkyl tail on the C-17 position of the sterol ring (Alouf *et al.*, 1984). Efforts to establish the stoichiometry of interaction between cholesterol and toxin have been fraught with difficulties because of the limited solubility of cholesterol in water and its tendency to form micelles at nanomolar concentrations. The most reliable estimate is between one and two mol-ecules of cholesterol per molecule of CBC (Geoffroy and Alouf, 1983).

Early chemical modification studies showed the CBCs could be inactivated by thiol-reactive agents (reviewed in Alouf and Geoffroy, 1991). For example, upon treatment with 5,5′-dithiobis(2-nitrobenzoic acid), the binding affinity of PFO for membranes was reduced 100-fold (Iwamoto *et al.*, 1987). Sequencing of CBCs indicated the site of inactivation was a cysteine residue (cysteine 459 in PFO) located within an almost strictly conserved undecapeptide sequence near the C-terminus (Figure 27.1). However, subsequent site-directed mutagenesis studies of various CBCs demonstrated that this residue was not essential for toxin function (Pinkney *et al.*, 1989; Saunders *et al.*, 1989; Boulnois *et al.*, 1990; Michel *et al.*, 1990; Hill *et al.*, 1994). These paradoxical results were explained by assuming that the membrane-binding site was close to, but did not include, the 'essential' cysteine. This hypothesis explained the observation that the effect on binding activity was proportional to the size of the thiol reagent used to inactivate PFO (Iwamoto *et al.*, 1990). The bulkier thiol reagents would cause greater steric disturbance to the nearby membrane-binding site.

As well as the 'essential' cysteine, other residues in the almost strictly conserved undecapeptide sequence (Figure 27.1) have been the targets for mutagenesis studies (Sekino-Suzuki *et al.*, 1996). Mutations of either Trp[464], Trp[466] or Trp[467] to phenylalanine in PFO lead to greatly reduced membrane binding and haemolytic activities although all mutants retained specificity for cholesterol and still were capable of oligomerization. The most sensitive site for mutation was at Trp[466] with a 100-fold drop in binding affinity. Only mutation of Trp[464] seemed to cause structural changes, as judged by increased protease susceptibility and changes in the

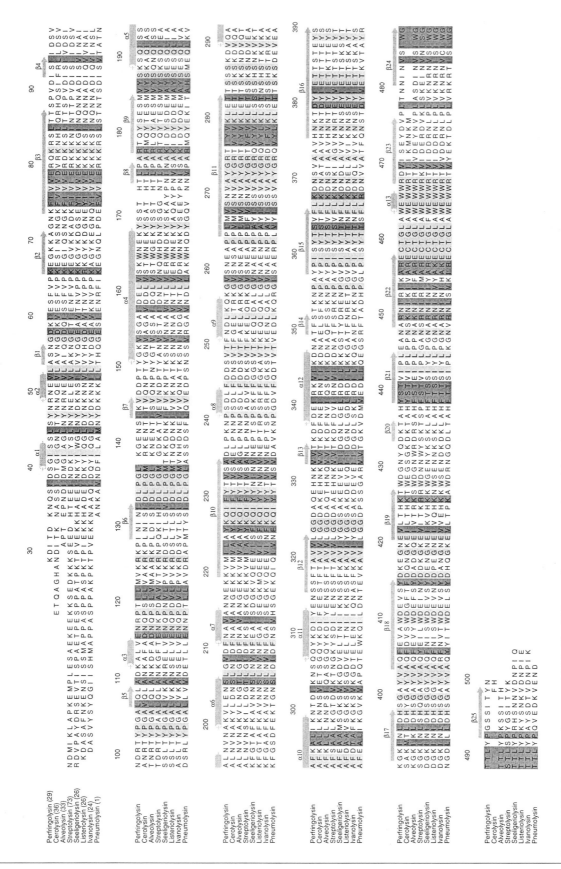

FIGURE 27.1 Sequence alignment of CBCs. The PFO secondary structure, derived from the crystal structure, is indicated above the alignment. Helices are represented by cylinders and β-strands by arrows. Residue numbering is for PFO and the number of the first residue shown in the alignment is indicated between the parentheses. Light shade indicates identical residues and darker shade indicates conservatively substituted residues. This figure was produced using the program ALSCRIPT (Barton, 1993).

far-ultraviolet circular dichroism spectra. A conserved threonine residue at position 460 was also a target for mutation. Mutation to valine caused a large drop in haemolytic activity and membrane binding.

Early observations of CBC activity suggested that they were capable of oligomerizing to form very large pores in membranes. For example, PFO was shown to produce pores large enough to allow the release of large macromolecules such as haemoglobin (Bernheimer, 1947). The first kinetic studies indicated that pore formation was a multi-hit process and the process of pore formation could be divided into two steps. In the first step, PFO binds to membranes in a process that is independent of pH, ionic strength and temperature. In a second temperature-dependent step, cell lysis occurs (Hase et al., 1975). These studies also appeared to rule out the involvement of other proteins in the insertion step since erythrocytes treated with proteases were still susceptible to PFO. The lytic activity was also found to be pH dependent with the optimal pH range between 5.8 and 6.5 (Alouf et al., 1984; Menestrina et al., 1990). Kinetic studies, based on fluorescence energy transfer measurements, indicated a distinct lag phase between binding and haemolysis which was interpreted to reflect the time required for membrane-bound monomer to oligomer conversion (Harris et al., 1991). These studies also suggested that the oligomer grew in a linear fashion. Although most studies of PFO activity have been performed on erythrocytes, other cells such as fibroblasts (Thelestam and Mölby, 1980), hepatocytes (Zs.-Nagy et al., 1988), Hela cells (Duncan and Buckingham, 1980) and myocardial cells (Fisher et al., 1981) are also susceptible to CBCs.

A Japanese group led by Ohno-Iwashita have performed a number of elegant studies on the behaviour of proteolytic fragments of PFO (Ohno-Iwashita et al., 1986, 1988, 1990, 1991, 1992; Iwamoto et al., 1990; Nakamura et al., 1995). Limited proteolysis of PFO with subtilisin produced a nicked toxin composed of two fragments: a 15-kDa N-terminal fragment and a 39-kDa C-terminal fragment (Ohno-Iwashita et al., 1986). The nicked toxin was found to be haemolytically inactive but still capable of binding to cholesterol below 20°C. Experiments with the nicked toxin also revealed the presence of both high-affinity (dissociation constant of 2 nM) and low-affinity (dissociation constant of 0.3 μM) binding sites for cholesterol (Ohno-Iwashita et al., 1988). The Japanese group also isolated a C-terminal fragment of 25 kDa produced by limited proteolysis with trypsin (Iwamoto et al., 1990). This fragment bound to erythrocytes with the same affinity and specificity as intact toxin, indicating that the cholesterol-binding sites were located in the C-terminal half of the molecule. Although the fragment could bind to membranes, it could not oligomerize or cause haemolysis, suggesting a direct connection between the two events. Further studies showed that the fragment could inhibit wild-type toxin activity by preventing polymerization. However, the fragment itself does possess a site of aggregation (Iwamoto et al., 1990; Tweten et al., 1991). Spectroscopic analysis of a nicked toxin, which cannot oligomerize, indicated that large changes in the environment of tryptophan residues occurred upon membrane binding, although the β-sheet-rich secondary structure did not change. Quenching of the intrinsic tryptophan fluorescence by brominated phospholipids suggested that the nicked toxin at least partially inserted into membranes (Nakamura et al., 1995).

MOLECULAR GENETICS

The PFO structural gene, pfoA, is located on the C. perfringens chromosome in a hypervariable 250-kb region that is known to contain several other extracellular toxin genes, including the plc gene, which encodes the α-toxin or phospholipase C, and the colA gene, which encodes the κ-toxin or collagenase (Katayama et al., 1995). The pfoA and colA genes are located within a 10-kb region, but the intervening DNA contains housekeeping genes that are not involved in virulence or toxin production (Ohtani et al., 1997).

Immediately upstream of the pfoA gene there is a putative regulatory gene, pfoR, which encodes a 343 amino acid protein that has a possible helix–turn–helix domain (Shimizu et al., 1991). Studies carried out exclusively in E. coli suggest that the PfoR protein is a positive activator of pfoA expression. In these studies, deletion of the pfoR gene, or an internal pfoR fragment, from a recombinant plasmid that contained the pfoRpfoA gene region led to a 20-fold reduction in PFO production (Shimizu et al., 1991). Confirmation of the role of PfoR in regulating PFO production will be dependent upon the isolation and analysis of pfoR mutants from C. perfringens.

Other studies have clearly shown that in C. perfringens, PFO production is regulated at the transcriptional level by a two-component signal transduction system (Lyristis et al., 1994; Shimizu et al., 1994; Ba-Thein et al., 1996). The virRS operon encodes the sensor histidine kinase, VirS, and its cognate response regulator, VirR (Lyristis et al., 1994). Mutations in either virR or virS lead to complete loss of pfoA expression and PFO production, and reduced production of α-toxin, κ-toxin, sialidase and protease. The precise physiological mechanisms which regulate this system have not been elucidated (Rood and Lyristis, 1995). However, it appears that an environmental or growth phase signal, which may involve the production of the uncharacterized dialysable compound previously identified as substance A (Imagawa and Higashi, 1992), leads to the autophosphorylation of the membrane-associated VirS protein. Activated VirS then acts as a phosphodonor and leads to the phosphorylation and activation of VirR. Recent studies suggest that VirR directly activates the transcription of the *pfoA* gene by binding to a site located just upstream of the *pfoA* promoter (J.K. Cheung and J.I. Rood, unpublished results). However, the mechanism by which VirR partially activates the expression of the extracellular toxin genes, such as the *plc* and *colA* genes, is not known.

GENE CLONING AND SEQUENCING

The gene encoding PFO has been expressed in *E. coli* (Tweten, 1988a) and its nucleotide sequence determined (Tweten, 1988b). The gene product was found to pos-

sess a 27 amino acid residue N-terminal signal sequence with characteristics typically found in other bacterial signal sequences. The mature toxin consists of 500 amino acid residues with a molecular weight close to 53 kDa. Extensive amino acid sequence similarity of between 40% and 83% was observed with other CBCs (Figure 27.1). Of particular interest was the strict conservation of a undecapeptide stretch near the C-terminus of each toxin, which included the 'essential' cysteine, Cys[459]. Analysis of the gene product indicated no significant stretches of hydrophobic amino acid residues to explain how the toxin might interact with membranes.

THREE-DIMENSIONAL STRUCTURE

The three-dimensional structure of PFO has recently been determined by X-ray crystallography to a resolution of 2.7 Å (Rossjohn *et al.*, 1997). Its structure superficially resembles the structures of a growing list of toxins called the hydrophilic channel-forming proteins (HCFs, Parker *et al.*, 1996). This diverse range of toxins includes aerolysin (Parker *et al.*, 1994), α-haemolysin from *Staphylococcus aureus* (Song *et al.*, 1996) and the protective antigen of anthrax toxin (Petosa *et al.*, 1997). They share the property of being elongated molecules rich in β-sheet structure. In the case of PFO, the molecule is 11.5 nm long and contains 40% β-sheet (Figure 27.2). One of the 25 β-strands spans two-thirds of the long axis of the molecule and measures about 7 nm. Despite the superficial resemblance, the detailed structure of each HCF is markedly different.

PFO consists of four discontinuous domains: domain 1 (residues 37–53, 90–178, 229–274, 350–373), domain 2 (residues 54–89, 374–390), domain 3 (residues 179–228, 275–349) and domain 4 (residues 391–500). Searches for similar three-dimensional folds to PFO in the Brookhaven Protein Database yielded only hits to domain 4. These matches reflected the fact that the β-sandwich fold of domain 4 is a common fold found in many proteins. There was one hit with a HCF: to the C-terminal domain of aerolysin which also adopts a β-sandwich fold. However, no functional significance could be attached to this match.

The arrangement of domains within the PFO molecule is worthy of a few comments (Figure 27.2). First, there is a highly pronounced curvature of the five-stranded β-sheet that connects domain 1 to domain 3. In contrast, there is a single connection made up of a glycine linker between domains 2 and 4. Second, the domains pack poorly against one another and interdomain interactions are predominantly polar in

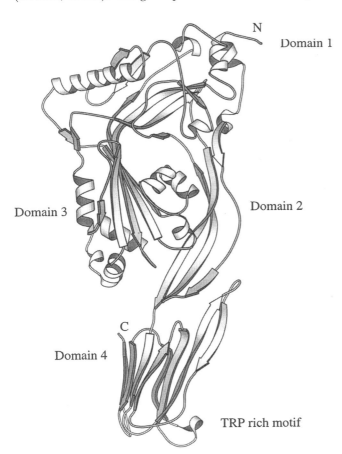

FIGURE 27.2 Structure of the PFO molecule. The structure is shown in ribbon representation. This figure was generated using MOLSCRIPT (Kraulis, 1991).

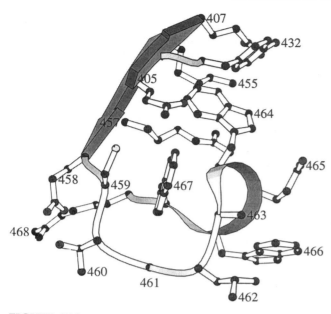

FIGURE 27.3 View of the conserved Trp-rich motif region. The 'essential' cysteine is located at residue 459. This figure was generated using MOLSCRIPT (Kraulis, 1991).

nature. These observations are suggestive that the domains may be free to rotate with respect to one another in solution.

CHOLESTEROL-BINDING SITES

Cholesterol, the PFO receptor, is thought to act by concentrating toxin molecules in cholesterol-rich areas of the target cell membrane. This, in turn, would promote oligomerization (Alouf and Geoffroy, 1991; Tweten, 1995). Numerous chemical modification and mutagenesis studies have indicated that the cholesterol-binding site is close to the Trp-rich motif and, in particular, the conserved cysteine residue, Cys^{459} (see above). Cys^{459} is sandwiched between a β-sheet and Trp^{467} near the bottom of domain 4 (Figure 27.3). The tryptophan residue forms part of the Trp-rich motif which adopts an elongated loop conformation that folds back on to the sheet (Figures 27.2 and 27.3). At the very tip of the loop, Trp^{464} points directly into the sheet. This tryptophan has an unusual microenvironment in that it is surrounded completely by long surface side-chains which have the effect of completely burying Trp^{464} in a hydrophobic environment. It was suggested that the same hydrophobic pocket could be the cholesterol-binding site (Rossjohn *et al.*, 1997). Using computer graphics a convincing fit for the cholesterol molecule was found in this site where the sterol rings pack against the aliphatic portions of the long surface side-chains and the 3β-hydroxyl forms hydrogen bonds

with Glu^{407} and Arg^{457}. Of course, in order for cholesterol to bind, the elongated loop must be displaced. Since the loop is predominantly hydrophobic, it is conceivable that cholesterol binding could initiate membrane insertion via the loop and the aliphatic tail of the cholesterol molecule. This hypothesis is supported by the experiments that demonstrate cholesterol binding induces partial membrane insertion, triggers conformational changes leading to changes in the environment of some tryptophan residues without change in secondary structure (Nakamura *et al.*, 1995) and explains why the maintenance of the hydrophobicity of the loop is critical for lysis (Pinkney *et al.*, 1989; Saunders *et al.*, 1989; Boulnois *et al.*, 1990; Michel *et al.*, 1990; Hill *et al.*, 1994). The hypothesis also provides a molecular explanation for the large number of apparently contradictory experiments that demonstrate that the Trp-rich motif is close to, but not directly involved in cholesterol binding. At this stage the possibility that the hypothesized site is incorrect or that other cholesterol-binding sites are present cannot be excluded.

MOLECULAR BASIS OF THIOL ACTIVATION

Cys^{459} is the site for the observed oxidative inactivation of the toxin. As already discussed, this 'essential' cysteine is sandwiched between one of the β-sheets of domain 4 and the Trp-rich loop that folds back on to the sheet (Figures 27.2 and 27.3). The tightly packed microenvironment around Cys^{459} explains why modification of the residue with a bulky thiol-blocking reagent interferes with cholesterol and membrane binding. Since the Trp-rich loop forms one wall of the pocket in which Cys^{459} is located, chemical modification would adversely impact on the conformation of the loop and on conformational changes that occur to the loop on binding to cholesterol and subsequent membrane interaction.

MECHANISM OF MEMBRANE INSERTION

Based on analogy with other well-studied pore-forming toxins, there must be some hydrophobic region on the toxin that interacts with biological membranes. One candidate region is the Trp-rich loop, which has been proposed to spring out on cholesterol binding to form a hydrophobic dagger, which could interact with membranes. However, this loop is too short to span fully a membrane and there is some evidence that CBCs completely insert into membranes (Alouf and

Geoffroy, 1991; Tweten, 1995). Besides the Trp-rich loop, there are no hydrophobic patches on the surface of the PFO molecule or in the domain interfaces that are large enough to span a membrane. Furthermore, there are no helices long enough to span a membrane. Two different insertion mechanisms are considered, both of which assume the PFO molecule does completely span the bilayer (Figure 27.4, in colour plate section).

The first mechanism is based on the assumption that domain 4 appears to be the membrane-insertion domain, based on overwhelming biochemical and mutagenesis data together with the hypothesis that it houses the cholesterol-binding site (Rossjohn et al., 1997). The sheet in domain 4 is 29 Å long and thus sufficiently long to span the bilayer core. A striking and unusual feature of this domain is the high proportion of hydrophobic residues in the loops at its tip; the high hydrophobicity of the Trp-rich loop has already been mentioned but it is also true for the other loops. This distribution is consistent with the hypothesis that the tip of domain 4 leads the insertion into the bilayer. There are only two charged groups in these loops: Glu[465] in the Trp-rich loop and Asp[434] in one of the other loops. There is a low pH environment on the surface of biological membranes due to a high concentration of protons caused by surface potential. Furthermore, PFO requires a low pH optimum for toxic activity. Partitioning into the bilayer could occur since the low pH environment would neutralize the carboxylates. Such a mechanism has been previously proposed for the membrane penetration of diphtheria toxin (Choe et al., 1992). There is also a striking paucity of charged residues on the β-sheet face of domain 4 that is opposite to the sheet that packs against the Trp-rich loop (Figure 27.2). In fact, there are only two charged residues. The first, Lys[393], is positioned close to the interface between domains 2 and 4. If domain 4 spanned the bilayer, this residue would interact with the polar headgroups. The second residue, Asp[397], is involved in multiple hydrogen bonding interactions with a number of polar residues including Ser[485], Ser[440] and His[438]. Unlike the loops at the tip, this face is not hydrophobic but polar. It has been suggested that domain 4 could only incorporate itself fully into membranes if this face could interact directly with the 3β-hydroxy moieties of cholesterol molecules in the membrane, shielding the face from direct interaction with the hydrophobic bilayer core (Rossjohn et al., 1997). The face of the other β-sheet in domain 4 would form part of the water channel of a pore.

An alternative mechanism of membrane insertion comes from the observation that domain 3 possesses three helices located at the domain 3–domain 2 interface, which resembles a series of wound-up springs.

Analysis of the sequence in this region shows an alternating pattern of hydrophobic and hydrophilic residues. Such a pattern has previously been observed in loop regions of S. aureus α-haemolysin (Song et al., 1996) and in the protective antigen of anthrax toxin (Petosa et al., 1997). In these toxins, the loop regions are thought to flip out on membrane insertion to form extended transmembrane β-sheets with neighbouring molecules. Recent mutagenesis studies support the hypothesis that the wound-up helices of PFO do form an extended β-hairpin on membrane insertion and the resultant hydrophobic face of the newly formed β-sheet interacts with the membrane (Shepard et al., 1998). The experimental data suggesting that the Trp-rich loop inserts into membranes appear overwhelming (see above). In a new model of membrane insertion the authors hypothesized that both the Trp-rich loop and the wound-up helices of domain 3 insert into the membrane (Figure 27.4). However, distinct from our previously published model, domain 4 is thought to sit on the membrane surface rather than inserting and spanning the bilayer. The predominantly hydrophobic loops at the bottom of domain 4 would be nestled into the bilayer, whereas the unusual, uncharged β-sheet face of this domain would be interacting with the polar headgroups of the bilayer surface. It may be envisaged that domain 4 makes the initial interaction with membrane by binding to its receptor, cholesterol, since pores will not form in the absence of the receptor. The partial insertion of domain 4 would bring domain 3 close to the membrane surface and perhaps force it to rotate away from the major lobe of the molecule. The helical springs of domain 3 would then be free to unwind to form the β-hairpin that punches into the membrane core.

ELECTRON MICROSCOPY STUDIES

Early electron microscopic (EM) investigations of PFO interactions with membranes indicated the presence of arc- and ring-shaped features of varying sizes (Smyth et al., 1975; Mitsui et al., 1979a, b). Rings appeared only at high toxin concentrations. On average, the outer diameter of the rings was approximately 35 nm with a width of about 6 nm. These early studies also suggested that the presence of membrane-bound cholesterol was not always essential for the oligomers to form. The size of the 'pores' appears to be directly related to the concentration of toxin and the amount of cholesterol in the membranes (Ohno-Iwashita et al., 1992; R.K. Tweten, unpublished results). PFO 'pores' have been characterized in cholesterol-containing planar lipid

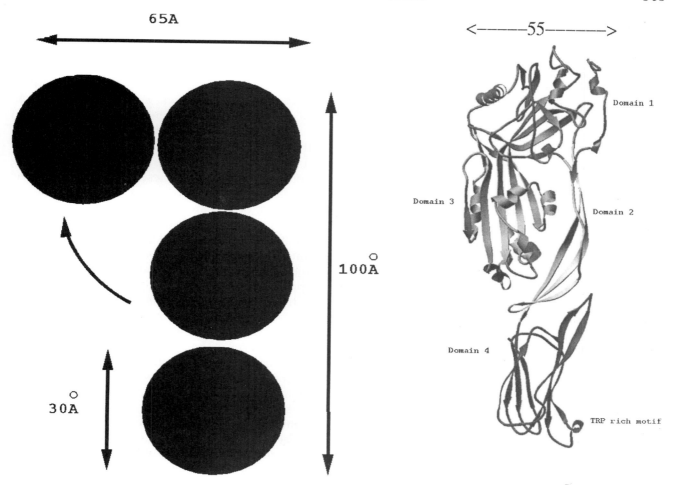

FIGURE 27.5 Fitting the crystal structure to the electron microscopy data. On the left is a sphere model derived from published electron microscopy data and on the right side is a ribbon picture of the crystal structure. The crystal structure can be fitted into the electron microscopy data by appropriate rotation of domain 3. This figure was generated using the program MidasPlus (Ferrin *et al.*, 1988).

bilayers and shown to be very heterogeneous in their conductance (Menestrina *et al.*, 1990). Olofsson and co-workers (1993) performed the most recent published electron microscopy study of PFO. This study allowed a more detailed view of the PFO oligomer. The mono-meric repeating unit was observed with a periodicity of 2.4 nm, suggesting that the rings contained up to 50 monomers and each monomer appeared to span the width of the ring.

The PFO results are in good agreement with those obtained from electron microscopy studies of other CBC oligomers (Duncan and Schlegel, 1975; Cowell *et al.*, 1978; Rottem *et al.*, 1982; Niedermeyer *et al.*, 1985; Bhakdi and Tranum-Jensen, 1988; Sekiya *et al.*, 1993; Morgan *et al.*, 1994; Morgan *et al.*, 1995). A consensus picture of the oligomer has evolved from these studies (Figures 27.5 and 27.6, see colour plate section). When viewed down the channel axis, the outer diameter of the oligomer is between 30 and 4.5 nm depending on the number of monomers contributing to the ring

(between 40 and 50 monomers based on rotational averaging of the images) and the ring width is 6.5 nm. The ring consists of two domains: a compact inner domain forming close contacts with the equivalent domain in the next monomer and an elongated outer domain extending radially outwards from the ring with much less contact with neighbouring mono-mers. The inner domain has a periodic repeat of 2.4 nm around the ring. When viewed side-on, the oligomer is mushroom-shaped with a height of between 9 and 10 nm. The bottom half of the mushroom stalk spans the membrane bilayer (Sekiya *et al.*, 1993).

MODELLING OF THE OLIGOMER

Two very different models of the oligomer have been built using the crystal structure of the monomer and the consensus structure of the CBC oligomer obtained by electron microscopy. The modelling was guided by the considerable amount of biochemical and mutagenesis data that have been published on CBCs. First, there are overwhelming experimental data showing that domain 4, and in particular the Trp-rich motif, interacts with the membrane-bound cholesterol (see above). These data indicate that the domain-4 end of the oligomer must interact with the membrane. Second, antibody data are consistent with regions of domain 1 being involved in oligomerization contacts (Darji *et al.*, 1996; De Los Toyos *et al.*, 1996). This observation was explored further by examining whether this domain could be docked on to itself in convincing fashion using a protein-docking program (Rossjohn *et al.*, 1997). The resulting solution was very convincing. The concave surface of one domain 1 fitted neatly on to the convex surface of the other domain 1. The buried surface area between the two domains was approximately 1500 Å^2, which is within the range normally observed in protein–protein complexes. A detailed analysis of the domain interface indicated many complimentary interactions. These included 26 hydrogen-bonding interactions and six salt bridges (Arg^{49} to Asp^{111}, Lys^{112} to Glu^{51}, Lys^{125} to Asp^{243}, Asp^{148} to Lys^{239}, Asp^{159} to Lys^{139}, Lys^{257} to Asp^{39}). There was only one hydrophobic patch of residues on each interacting surface and these patches were found to interact with each other (Met^{123} and Ile^{121} of one monomer and Leu^{237}, Val^{365} and Val^{372} of the other monomer). Finally, biophysical studies have indicated that one of the domains is flexibly linked to the rest of the toxin molecule (Morgan *et al.*, 1994). Two candidates for the flexibly linked domain were obvious from inspection of the crystal structure. Domain 4 has only a single linker between it and the rest of the molecule. Although domain 3 has many links with domain 1, it was noted that the β-sheet connecting the two domains was markedly bent and that the interface between domains 3 and 2 was poorly packed and very polar. These observations suggested that domain 3 might also be capable of flexing away from the rest of the molecule. The presence of possible hinge points at conserved glycine residues in the domain boundary (positions 274, 324 and 325) supported this hypothesis.

In the first model, the L-shaped crystal structure was fitted into the L-shaped repeating unit of the oligomer with domains 1, 2 and 4 fitting into the cylindrical stem and domain 3 placed in the 'flange' of the mushroom-shaped oligomer (Figures 27.5 and 27.6a, colour plate section; Rossjohn *et al.*, 1997). Domain 4 was positioned in the membrane-spanning base of the oligomer, in agreement with the biochemical data, with one face of the domain forming a continual β-sheet with the same domain in the other monomers. The only required alteration to the crystal structure was a rotation of domain 3 by 35° from domain 2 in order that the width of the ring agreed with the electron microscopy data (Figure 27.5). The rotation had the effect of straightening out the very pronounced bent sheet that extended from domains 1 to 3 and was essential to avoid clashes of domain 3 with the membrane upon insertion of the oligomer. This model was in excellent agreement with the electron microscopy data of CBC oligomers and provided a convincing framework for explaining the plethora of biochemical results at a molecular level (Rossjohn *et al.*, 1997).

However, doubts arose with this model when site-directed mutagenesis results suggested that regions in domain 3 might be involved in membrane interaction (Palmer *et al.*, 1996; Shepard *et al.*, 1998). This has led to the proposal that the wound-up helices in domain 3 spring out to form a transmembrane β-hairpin which would generate an extended β-sheet with other monomers upon oligomerization (see above). In the case of ring formation, the extended β-sheet would circularize to form a β-barrel. In an alternative oligomer model, domain 3 and the extended β-hairpin form the stalk of the mushroomed-shaped oligomer and domain 4 would form the flange (Figure 27.6b, in colour plate section). The domain-1 positioning at the top of the 'mushroom head' is the same in both models.

Whether the oligomer forms on the surface of the membrane or only when inserted remains controversial. In support of the first proposal is the observation that CBC oligomers can form in solution in the absence of cholesterol or detergent (Smyth *et al.*, 1975; Cowell *et al.*, 1978; Mitsui *et al.*, 1979a, b; Niedermeyer, 1985). However, there is a conceptual problem with this surface model in that upon insertion and pore formation, the oligomer would have to displace a large amount of lipid from the membrane. Furthermore, there is no evidence to indicate that oligomers which form in solution are active. Indeed, there is evidence for the existence of non-lytic forms of CBC oligomers (Walev *et al.*, 1995). Bhakdi and co-workers (Palmer *et al.*, 1998) have recently demonstrated that a poorly oligomerizing mutant of SLO forms small arcs in membranes which are nevertheless capable of allowing the passage of small molecules. This result would be consistent with the gradual growth of oligomers in the membrane and suggests that pores can form which consist of part protein and part lipid.

RELEVANCE TO OTHER CBCS

The pronounced serological cross-reactivity of CBCs (Smyth and Duncan, 1978) and closely related primary structures indicates that they also share very similar three-dimensional structures. We have modelled other members on the family based on the PFO crystal structure (Rossjohn and Parker, unpublished results). As expected, the core regions are strongly conserved, whereas the *N*-terminal and some loop regions are the least conserved. In contrast, the Trp-rich loop and other loops at the tip of domain 4 appear well conserved and retain their predominant hydrophobicity. Other regions that have been identified as functionally important, such as the wound-up helices in domain 3 and the domain-1 oligomeric interfaces, are not strictly conserved but are characterized by either conservative substitutions or compensating mutations. These common features are in keeping with the models of membrane insertion described above. An intriguing feature of the modelling is the observation that the electrostatic surface of each toxin can vary markedly, particularly at the top of domain 1. For example, whereas this region is neutral in PFO, it is electronegative in pneumolysin and electropositive in listeriolysin. The differences may reflect the requirement of each toxin to remain soluble in different compartments of either the host or target cell. For example, pneumolysin is the only CBC that is not secreted but remains in the cytosol of the *S. pneumoniae* cell. At one stage in its toxic life, listeriolysin finds itself engulfed in the low pH environment of a phagosomal compartment. A possible reason for the location of these highly charged surfaces on the top of domain 1 can be understood by reference to the oligomer models (Figure 27.6, in colour plate section). Most of the other surfaces of the toxin molecule are either involved in membrane interaction or oligomeric contacts, which would prohibit the development of such toxin-dependent surfaces.

RELEVANCE TO OTHER PORE-FORMING PROTEINS

A number of models has been put forward in recent years to explain how some proteins, which are perfectly water soluble, are capable of inserting into biological membranes. All of these models are based on the premise that insertion and pore formation require the exposure of hydrophobic protein patches which could interact with the hydrophobic core of the bilayer. Perhaps the simplest example is the case where a long stretch of hydrophobic amino acid sequence is buried in the water-soluble form of the protein but becomes exposed on membrane insertion. This mechanism was first put forward for the pore-forming colicins (Parker *et al.*, 1989) and has subsequently been proposed to

operate in a number of pore-forming proteins including diphtheria toxin (Choe *et al.*, 1992), insecticidal δ-endotoxin (Li *et al.*, 1991), apolipoprotein (Breiter *et al.*, 1991), bacteriophage Pf1 (Nambudripad *et al.*, 1991), apocytochrome c (De Jongh *et al.*, 1994), the cell-death inhibitor Bcl-x_L (Muchmore *et al.*, 1996) and uteroglobin (De La Cruz and Lee, 1996). An alternative model of membrane insertion can be envisaged in which the hydrophobic surface is not formed by a linear stretch of sequence but instead is formed by a three-dimensional array of hydrophobic side-chains. Such a surface could be buried in the water-soluble form of the protein owing to the presence of a masking propeptide, or by virtue of its location in a domain or oligomeric interface. A typical example of this model is aerolysin, which exists as a water-soluble dimer, but upon removal of a propeptide converts via monomers into a heptameric membrane pore (Parker *et al.*, 1994). A very different model has been proposed for the annexins in which the protein sits on the surface of the membrane and induces lysis by electroporation owing to an electrostatic field generated by the molecule (Berendes *et al.*, 1993).

The model we have put forward to explain PFO membrane insertion is reminiscent of a model suggested for *S. aureus* α-haemolysin (Song *et al.*, 1996). The structure of α-haemolysin was determined from crystals of the heptameric, membrane-bound form of the toxin. The membrane-spanning stalk of the toxin was found to consist of a long β-hairpin which assembled together with its monomeric neighbours to form a fully circularized β-barrel. The hairpin formed favourable interactions with the bilayer due to an alternating sequence of hydrophobic and hydrophilic residues along its length. The hydrophobic residues projected into the bilayer core and the hydrophilic residues projected into the aqueous pore. It was thought that the long hairpin would not be stable by itself in the water-soluble monomeric form of the toxin. Instead, it was suggested that the loop would form a highly mobile, irregular structure in an aqueous environment. This proposal was supported by the crystal structure determinations of anthrax protective antigen in both its monomeric and pre-pore, oligomeric states (Petosa *et al.*, 1997). These structures showed that a similar long loop of alternating hydrophobicity adopted a very mobile, irregular structure in the water-soluble form of the toxin and suggested that it sprang out into the bilayer on membrane insertion. We have proposed a similar large-scale movement of such a loop in PFO membrane insertion. A significant difference between PFO and the other two toxins is that the membrane-inserting loop in PFO adopts a large amount of secondary structure in the water-soluble form of the toxin. The transition of the membrane-inserting region of PFO from a tightly packed helical structure in the monomeric form to an extended β-sheet in the oligomer is the first description of this type of structural transistion for a membrane-spanning domain of a pore-forming toxin. The authors expect to see other examples of this membrane-insertion mechanism in future years.

ROLE IN PATHOGENESIS OF GAS GANGRENE

Both genetic and immunological studies have shown that α-toxin is the major toxin involved in the pathogenesis of *C. perfringens*-mediated gas gangrene or clostridial myonecrosis. Mice can be protected from the disease by immunization with a non-functional but immunogenic α-toxin domain (Williamson and Titball, 1993) and do not show normal signs of disease when infected with a mutant strain that has an insertionally activated *plc* gene (Awad *et al.*, 1995).

The role of PFO in the disease process is less clearly defined. One of the key features of gas gangrene lesions is the absence of acute inflammatory cells and it has been suggested that both α-toxin and PFO are involved in inhibiting the migration of such cells to the site of infection (Bryant and Stevens, 1997). Although PFO is less toxic than α-toxin, it is still lethal when injected intravenously into mice (Stevens *et al.*, 1988), is leucocytoxic, and decreases polymorphonuclear leucocyte (PMNL) mobility (Stevens *et al.*, 1987). Mice passively immunized with monoclonal antibodies specific for PFO exhibit a delayed onset of *C. perfringens* infection (Bryant *et al.*, 1993; Stevens *et al.*, 1997). In addition, purified PFO is able to decrease significantly PMNL influx into the lesion, causing a concomitant increase in PMNL accumulation in the adjacent blood vessels. Further studies have shown that in endothelial cells α-toxin upregulates expression of the adherence molecules endothelial cell leucocyte adhesion molecule-1 (ELAM-1) and intercellular adhesion molecule-1 (ICAM-1) and the neutrophil chemoattractant-activator IL-8 (Bryant and Stevens, 1996). By contrast, PFO only modulates ICAM-1 expression. These results suggest that α-toxin may be the major toxin involved in increasing the ability of endothelial cells lining the blood vessels to bind inflammatory cells and therefore to impair the delivery of phagocytes to the site of infection. However, it also appears that PFO plays a role in the process, perhaps in synergy with α-toxin.

PFO mutants have been constructed by homologous recombination and have been virulence tested in mice (Awad *et al.*, 1995; Stevens *et al.*, 1997). Mice infected with *pfoA* mutants still exhibit the clinical signs of clostridial myonecrosis and comparable muscle destruction to that observed in mice infected with the isogenic wild-type strain (Awad *et al.*, 1995). By contrast, the *plc* mutants are avirulent. These studies showed that α-toxin, not PFO, is the major toxin involved in the necrotic disease process. Additional studies have confirmed that α-toxin is also the major toxin involved in preventing the influx of PMNL into the myonecrotic lesion (Stevens *et al.*, 1997). However, it appears that PFO does play a role in *C. perfringens*-mediated gas gangrene since a somewhat increased PMNL influx is observed in mice infected with a *pfoA* mutant. This increase is not observed when the mutation is complemented with a recombinant plasmid carrying a wild-type *pfoA* gene (Stevens *et al.*, 1997). In conclusion, the precise role of PFO in the disease remains to be determined but it appears to involve at least some anti-inflammatory effects.

OTHER TOXIC EFFECTS IN VIVO

The molecular basis for PFO toxicity remains uncertain. Some workers suggest that the membrane is sufficiently disrupted by the binding and removal of membrane-bound cholesterol (for example, Smyth and Duncan, 1978) whilst others have argued that it is the eventual formation of large pores in the membranes which leads to cellular damage (for example, Bhakdi and Tranum-Jensen, 1988). The possibility that toxic infection arises from sub-lytic doses of PFO cannot be ruled out. Sublethal doses of PFO exert an inhibitory effect on neutrophil migration (Wilkinson, 1975), decrease leukotriene production and cause morphological changes to polymorphonuclear leucocytes (Bremm *et al.*, 1985; Stevens *et al.*, 1987). Two CBCs, pneumolysin and streptolysin O, have been shown to activate the classical pathway of complement in the absence or presence of toxin-directed antibodies (Paton *et al.*, 1984; Bhakdi and Tranum-Jensen, 1985). The molecular basis of action is thought to involve binding of the CBC to the Fc portion of immunoglobulin G (Mitchell *et al.*, 1991). The similarity of the fold of domain 4 of PFO (and domain 4 of the homology models of pneumolysin and streptolysin O) to Fc and our proposal that PFO forms oligomers via domain 4 interactions suggests a model for complement activation. In this model domain 4 of the CBC would interact with Fc to form an aggregated array resulting in activation of complement (Rossjohn *et al.*, 1998). Although toxin-generated complement activation has not been reported for PFO, the high sequence similarities in domain 4 between the CBCs suggest that PFO will behave in a similar fashion.

CONCLUDING REMARKS AND FUTURE DIRECTIONS

A number of major advances in the study of PFO has been made in recent years. First, the crystal structure determination has led to a number of plausible models for how the toxin interacts with membranes. These models are clearly testable and should lead to a detailed insight into the processes of membrane insertion, oligomerization and membrane lysis. Because of the quite high sequence similarity between members of the CBC family, it is expected that the PFO crystal structure will also stimulate a wide range of experiments directed towards other members of the family. In the longer term, the crystal structure might also facilitate the design of novel therapeutic agents, by the well-established method of structure-based drug design, for the treatment of a variety of infectious diseases caused by bacteria that produce CBCs. Second, there has been some progress towards understanding the role of PFO in disease, particularly its possible involvement in anti-inflammatory responses. However, further studies are required to pinpoint the precise role of PFO, not only in gas gangrene, but also in other diseases caused by *C. perfringens*.

ACKNOWLEDGEMENTS

We would like to thank Susanne Feil, William McKinstry and Galina Polekhina for their various important contributions towards our structural studies of CBCs. Jamie Rossjohn is an Australian Research Council Postdoctoral Fellow. This work was supported in part by a Provost's Award from the University of Oklahoma Health Sciences Center and a grant from the National Institutes of Health to Rodney K. Tweten. Work in Rood's laboratory was supported by grants from the Australian National Health and Medical Research Council. Michael W. Parker is an Australian Research Council Senior Research Fellow and acknowledges support from the Australian National Health and Medical Research Council.

REFERENCES

Alouf, J.E. and Geoffroy, C. (1984). Structure activity relationships in the sulfhydryl-activated toxins. In: *Bacterial Protein Toxins* (eds J. Alouf, F.J. Fehrenbach, J.H. Freer and J. Jeljaszewicz), pp. 165–71. Academic Press, London.

Alouf, J.E. and Geoffroy, C. (1991). The family of the antigenically-related cholesterol-binding ('sulphydryl-activated') cytolytic toxins. In: *Sourcebook of Bacterial Toxins* (eds J.E. Alouf and J.H. Freer), pp. 147–86. Academic Press, London.

Alouf, J.E., Geoffroy, C., Pattus, F. and Verger, R. (1984). Surface properties of bacterial sulfhydryl-activated cytolytic toxins. Interaction with monomolecular films of phosphotidylcholine and various sterols. *Eur. J. Biochem.* **141**, 205–10.

Awad, M.M., Bryant, A.E., Stevens, D.L. and Rood, J.I. (1995). Virulence studies on chromosomal α-toxin and θ-toxin mutants constructed by allelic exchange provide genetic evidence for the essential role of α-toxin in *Clostridium perfringens*-mediated gas gangrene. *Molec. Microbiol.* **15**, 191–202.

Barton, G.J. (1993). ALSCRIPT – a tool to format multiple sequence alignments. *Protein Engng* **6**, 37–40.

Ba-Thein, W., Lyristis, M., Ohtani, K., Nisbet, I.T., Hayashi, H., Rood, J.I. and Shimizu, T. (1996). The *virR/virS* locus regulates the transcription of genes encoding extracellular toxin production in *Clostridium perfringens*. *J. Bacteriol.* **178**, 2514–20.

Berendes, R., Voges, D., Demange, P., Huber, R. and Burger, A. (1993). Structure–function analysis of the ion channel selectivity filter in human annexin V. *Science* **262**, 427–30.

Bernheimer, A.W. (1947). Comparative kinetics of haemolysis induced by bacterial and other haemolysins. *J. Gen. Physiol.* **30**, 337–53.

Bernheimer, A.W. (1976). Sulfhydryl activated toxins. In: *Mechanisms in Bacterial Toxinology* (ed. A.W. Bernheimer), pp. 85–97. John Wiley, New York.

Bernheimer, A.W. and Rudy, B. (1986). Interactions between membranes and cytolytic peptides. *Biochim. Biophys. Acta* **864**, 123–41.

Bhakdi, S. and Tranum-Jensen, J. (1985). Complement activation and attack on autologous cell membranes induced by streptolysin O. *Infect. Immun.* **48**, 713–19.

Bhakdi, S. and Tranum-Jensen, J. (1988). Damage to cell membranes by pore-forming bacterial cytolysins. *Prog. Allergy* **40**, 1–43.

Boulnois, G.J., Mitchell, T.J., Saunders, F.K., Mendez, F.J. and Andrew, P.W. (1990). Structure and function of pneumolysin, the thiol-activated toxin from *Streptococcus pneumoniae*. In: *Bacterial Protein Toxins* (eds R. Rappuoli, J.E. Alouf, F. Falmagne, J. Fehrenbach, J. Freer, R. Gross, J. Jeljaszewicz, C. Montecucco, T. Tomasi, T. Wadström and B. Witholt), pp. 43–51. Gustav Fischer, Stuttgart.

Braun, V. and Focareta, T. (1991). Pore-forming bacterial protein haemolysins (cytolysins). *Crit. Rev. Microbiol.* **18**, 115–58.

Breiter, D.R., Kanost, M.R., Benning, M.M., Wesenberg, G., Law, J.H., Wells, M.A., Rayment, I. and Holden, H.M. (1991). Molecular structure of an apolipoprotein determined at 2.5 Å resolution. *Biochemistry* **30**, 603–8.

Bremm, K.D., König, W., Pfeiffer, P., Rauschen, I., Theobold, K., Thelestam, M. and Alouf, J.E. (1985). Effect of thiol-activated toxins (streptolysin O, aveolysin, and theta toxin) on the generation of leukotrienes and leukotriene-inducing and -metabolizing enzymes from human polymorphonuclear granulocytes. *Infect. Immun.* **50**, 844–51.

Bryant, A.E. and Stevens, D.L. (1996). Phospholipase C and perfringolysin O from *Clostridium perfringens* upregulate endothelial cell–leukocyte adherence molecule 1 and intercellular leukocyte adherence molecule 1 expression and induce interleukin-8 synthesis in cultured human umbilical vein endothelial cells. *Infect. Immun.* **64**, 358–62.

Bryant, A.E. and Stevens, D.L. (1997). The pathogenesis of gas gangrene. In: *The Clostridia: Molecular Biology and Pathogenesis* (eds J.I. Rood, B.A. McClane, J.G. Songer and R.W. Titball), pp. 185–96. Academic Press, London.

Bryant, A.E., Bergstrom, R., Zimmermann, G.A., Salyer, J.L., Hill, H.R., Tweten, R.K., Sato, H. and Stevens, D.L. (1993). *Clostridium perfringens* invasiveness is enhanced by the effects of theta toxin upon PMNL structure and function: the roles of leukocytotoxicity and expression of CD11/CD18 adherence glycoprotein. *FEMS Immunol. Med. Microbiol.* **7**, 321–36.

Choe, S., Bennett, M.J., Fujii, G., Curmi, P.M.G., Kantardjieff, K.A., Collier, R.J. and Eisenberg, D. (1992). The crystal structure of diphtheria toxin. *Nature* **357**, 216–22.

Cowell, J.L., Kim, K–S. and Bernheimer, A.W. (1978). Alteration by cereolysin of the structure of cholesterol-containing membranes. *Biochim. Biophys. Acta* **507**, 230–41.

Darji, A., Niebuhr, K., Hense, M., Wehland, J., Chakraborty, T. and Weiss, S. (1996). Neutralizing monoclonal antibodies against listeriolysin: mapping of epitopes involved in pore formation. *Infect. Immun.* **64**, 2356–8.

De Jongh, H.H., Brasseur, R. and Killian, J.A. (1994). Orientation of the α-helices of apocytochrome c and derived fragments at membrane interfaces, as studied by circular dichroism. *Biochemistry* **33**, 114529–35.

De La Cruz, X. and Lee, B. (1996). The structural homology between uteroglobin and the pore-forming domain of colicin A suggests a possible mechanism of action of uteroglobin. *Protein Sci.* **5**, 857–61.

De Los Toyos, J.R., Méndez, F.J., Aparicio, J.F., Vázquez, F., Del Mar García Suárez, M., Fleites, A., Hardisson, C., Morgan, P.J., Andrew, P.W. and Mitchell, T.J. (1996). Functional analysis of pneumolysin by use of monoclonal antibodies. *Infect. Immun.* **64**, 480–4.

Duncan, J.L. and Buckingham, L. (1980). Resistance to streptolysin O in mammalian cells treated with oxygenated derivatives of cholesterol: cholesterol content of resistant cells and recovery of streptolysin O sensitivity. *Biochim. Biophys. Acta* **603**, 278–87.

Duncan, J.L. and Schlegel, R. (1975). Effect of streptolysin O on erythrocyte membranes, liposomes, and lipid dispersions. A protein–cholesterol interaction. *J. Cell Biol.* **67**, 160–73.

Ferrin, T.E., Huang, C.C., Jarvis, L.E. and Langridge, R. (1988). The MIDAS display system. *J. Mol. Graphics* **6**, 12–27.

Fisher, M.H., Kaplan, E.L. and Wannamaker, L.M. (1981). Cholesterol inhibition of streptolysin O for myocardial cells in tissue culture. *Proc. Soc. Exp. Biol. Med.* **163**, 233–7.

Geoffroy, C. and Alouf, J.E. (1983). Selective purification by thiol-disulfide interchange chromatography of alveolysin, a sulfhydryl-activated toxin of *B. alvei. J. Biol. Chem.* **258**, 9886–972.

Harris, R.W., Sims, P.J. and Tweten, R.K. (1991). Kinetic aspects of the aggregation of *Clostridium perfringens* θ-toxin on erythrocyte membranes. *J. Biol. Chem.* **266**, 6936–41.

Hase, J., Mitsui, K. and Shonaka, E. (1975). *Clostridium perfringens* exotoxins. III. Binding of θ-toxin to erythrocyte membrane. *Jpn. J. Exp. Med.* **45**, 433–8.

Hill, J., Andrew, P.W. and Mitchell, T.J. (1994). Amino acids in pneumolysin important for haemolytic activity identified by random mutagenesis. *Infect. Immun.* **62**, 757–8.

Imagawa, T. and Higashi, Y. (1992). An activity which restores theta-toxin activity in some theta toxin-deficient mutants of *Clostridium perfringens. Microbiol. Immunol.* **36**, 523–7.

Iwamoto, M., Ohno-Iwashita, Y. and Ando, S. (1987). Role of the essential thiol group in the thiol-activated cytolysin from *Clostridium perfringens. Eur. J. Biochem.* **167**, 425–30.

Iwamoto, M., Ohno-Iwashita, Y. and Ando, S. (1990). Effect of isolated C-terminal fragment of θ-toxin (perfringolysin O) on toxin assembly and membrane lysis. *Eur. J. Biochem.* **194**, 25–31.

Katayama, S.I., Dupuy, B., Garnier, T. and Cole, S.T. (1995). Rapid expansion of the physical and genetic map of the chromosome of *Clostridium perfringens. J. Bacteriol.* **177**, 5680–5.

Kraulis, P. (1991). MOLSCRIPT: a program to produce both detailed and schematic plots of proteins. *J. Appl. Crystallogr.* **24**, 946–50.

Li, J., Carroll, J. and Ellar, D.J. (1991). Crystal structure of insecticidal δ-endotoxin from *Bacillus thuringiensis* at 2.5 Å resolution. *Nature* **353**, 815–21.

Lyristis, M., Bryant, A.E., Sloan, J., Awad, M.M., Nisbet, I.T., Stevens, D.L. and Rood, J.I. (1994). Identification and molecular analysis of a locus that regulates extracellular toxin production in *Clostridium perfringens. Molec. Microbiol.* **12**, 761–77.

Menestrina, G., Bashford, C.L. and Pasternak, C.A. (1990). Pore-forming toxins: experiments with *S. aureus* α-toxin, *C. perfringens* θ-toxin and *E. coli* haemolysin in lipid bilayers, liposomes and intact cells. *Toxicon* **28**, 477–91.

Michel, E., Reich, K.A., Favier, R., Berche, P. and Cossart, P. (1990). Attentuated mutants of the intracellular bacterium *Listeria monocytogenes* obtained by single amino acid substitutions in listeriolysin O. *Molec. Microbiol.* **4**, 2167–78.

Mitchell, T.J., Andrew, P.W., Saunders, F.K., Smith, A.N. and Boulnois, G.J. (1991). Complement activation and antibody binding by pneumolysin via a region of the toxin homologous to a human acute-phase protein. *Molec. Microbiol.* **5**, 1883–8.

Mitsui, K., Mitsui, N. and Hase, J. (1973). *Clostridium perfringens* exotoxins. II. Purification and some properties of θ-toxin. *Jpn. J. Exp. Med.* **43**, 377–91.

Mitsui, K., Sekiya, T., Nozawa, Y. and Hase, J. (1979a). Alteration of human erythrocyte plasma membranes by perfringolysin O as revealed by freeze-fracture electron microscopy. *Biochim. Biophys. Acta* **554**, 68–75.

Mitsui, K., Sekiya, T., Okamura, S., Nozawa, Y. and Hase, J. (1979b). Ring formation of perfringolysin O as revealed by negative stain electron microscopy. *Biochim. Biophys. Acta* **558**, 307–13.

Möllby, R., Holme, T., Nord, C.-E., Smyth, C.J. and Wadström, T (1976). Production of phospholipase C (alpha-toxin), haemolysins and lethal toxins by *Clostridium perfringens* types A to D. *J. Gen. Microbiol.* **96**, 137–44.

Morgan, P.J., Hyman, S.C., Byron, O., Andrew, P.W., Mitchell, T.J. and Rowe, A.J. (1994). Modeling the bacterial protein toxin, pneumolysin, in its monomeric and oligomeric form. *J. Biol. Chem.* **269**, 25315–20.

Morgan, P.J., Hyman, S.C., Rowe, A.J., Mitchell, T.J., Andrew, P.W. and Saibil, H.R. (1995). Subunit organization and symmetry of poreforming, oligomeric pneumolysin. *FEBS Lett.* **371**, 77–80.

Muchmore, S.W., Sattler, M., Liang, H., Meadows, R.P., Harlan, J.E., Yoon, H.S., Nettesheim, D., Chang, B.S., Thompson, C.B., Wong, S.-L., Ng, S.-C. and Fesik, S.W. (1996). X-ray and NMR-structure of human Bcl-x$_L$, an inhibitor of programmed cell death. *Nature* **381**, 335–41.

Nakamura, M., Sekino, N., Iwamoto, M. and Ohno-Iwashita, Y. (1995). Interaction of θ-toxin (perfringolysin O), a cholesterol-binding cytolysin, with liposomal membranes: change in the aromatic side chains upon binding and insertion. *Biochemistry* **34**, 6513–20.

Namudripad, R., Stark, W., Opella, S.J. and Makowski, L. (1991). Membrane-mediated assembly of filamentous bacteriophage Pf1 coat protein. *Science* **252**, 1305–8.

Niedermeyer, W. (1985). Interaction of streptolysin-O with biomembranes: kinetic and morphological studies on erythrocyte membranes. *Toxicon* **23**, 425–39.

Nord, C.-E., Möllby, R., Smyth, C.J. and Wadstrom, T. (1974). Formation of phospholipase C and theta-haemolysin in prereduced media in batch and continuous culture of *Clostridium perfringens* type A. *J. Gen. Microbiol.* **84**, 117–27.

Ohno-Iwashita, Y., Iwamoto, M., Mitsui, K., Kawasaki, H. and Ando, S. (1986). Cold-labile haemolysin produced by limited proteolysis of theta-toxin from *Clostridium perfringens. Biochemistry* **25**, 6048–53.

Ohno-Iwashita, Y., Iwamoto, M., Mitsui, K., Ando, S. and Nagai, K. (1988). Protease-nicked θ-toxin of *Clostridium perfringens*, a new membrane probe with no catalytic effect reveals two classes of cholesterol as toxin-binding sites on sheep erythrocytes. *Eur. J. Biochem.* **176**, 95–101.

Ohno-Iwashita, Y., Iwamoto, M., Ando, S., Mitsui, K. and Iwashita, S. (1990). A modified θ-toxin produced by limited proteolysis and methylation: a probe for the functional study of membrane cholesterol. *Biochim. Biophys. Acta* **1023**, 441–8.

Ohno-Iwashita, Y., Iwamoto, M., Mitsui, K., Ando, S. and Iwashita, S. (1991). A cytolysin, theta-toxin, preferentially binds to membrane cholesterol surrounded by phospholipids with 18-carbon hydrocarbon chains in cholesterol-rich regions. *J. Biochem. (Tokyo)* **110**, 369–75.

Ohno-Iwashita, Y., Iwamoto, M., Ando, S. and Iwashita, S. (1992). Effect of lipidic factors on membrane cholesterol topology – mode of binding of θ-toxin to cholesterol in liposomes. *Biochim. Biophys. Acta* **1109**, 81–90.

Ohtani, K., Bando, M., Swe, T., Banu, S., Oe, M., Hayashi, H. & Shimizu, T. (1997). Collagenase gene (*colA*) is located in the 3′-flanking region of the perfringolysin O (*pfoA*) locus in *Clostridium perfringens. FEMS Microbiol. Lett.* **146**, 155–9.

Olofsson, A., Hebert, H. and Thelestam, M. (1993). The projection structure of perfringolysin O. *FEBS Lett.* **319**, 125–7.

Palmer, M., Saweljew, P., Vulicevic, I., Valeva, A., Kehoe, M. and Bhakdi, S. (1996). Membrane-inserting domain of streptolysin O identified by cysteine scanning mutagenesis. *J. Biol. Chem.* **271**, 26664–7.

Palmer, M., Harris, R., Freyfag, C., Kehoe, M., Tranum-Jensen, J. and Bhakdi, S. (1998). Assembly mechanism of the oligomeric streptolysin O pore: the early membrane lesion is lined by a free edge of the lipid membrane and is extended gradually during oligomerization. *EMBO J.* **17**, 1598–605.

Parker, M.W., Pattus, F., Tucker, A.D. and Tsernoglou, D. (1989). Structure of the membrane pore-forming fragment of colicin A. *Nature* **337**, 93–6.

Parker, M.W., Buckley, J.T., Postma, J.P.M., Tucker, A.D., Leonard, K., Pattus, F. and Tsernoglou, D. (1994). Structure of the *Aeromonas* toxin proaerolysin in its water-soluble and membrane-channel states. *Nature* **367**, 292–5.

Parker, M.W., van der Goot, F.G. and Buckley, J.T. (1996). Aerolysin – the ins and outs of a prototypical channel-forming toxin. *Molec. Microbiol.* **19**, 205–12.

Paton, J.C., Rowan-Kelly, B. and Ferrante, A. (1984). Activation of human complement by the pneumococcal toxin pneumolysin. *Infect. Immun.* **43**, 1085–7.

Petosa, C., Collier, R.J., Klimpel, K.R., Leppla, S.H. and Liddington, R.C. (1997). Crystal structure of the anthrax toxin protective antigen. *Nature* **385**, 833–8.

Pinkney, M., Beachey, E. and Kehoe, M. (1989). The thiol-activated toxin streptolysin O does not require a thiol group for cytolytic activity. *Infect. Immun.* **57**, 2553–8.

Rood, J.I. and Cole, S.T. (1991). Molecular genetics and pathogenesis of *Clostridium perfringens. Microbiol. Rev.* **55**, 621–48.

Rood, J.I. and Lyristis, M. (1995). Regulation of extracellular toxin production in *Clostridium perfringens. Trends Microbiol.* **3**, 192–6.

Rood, J.I. McClane, B.A., Songer, J.G. and Titball, R.W. (1997). *The Clostridia. Molecular Biology and Pathogenesis of the Clostridia*, Academic Press, London.

Rossjohn, J., Feil, S.C., McKinstry, W.J., Tweten, R.K. and Parker, M.W. (1997). Structure of a cholesterol-binding, thiol-activated cytolysin and a model of its membrane form. *Cell* **89**, 685–92.

Rossjohn, J., Gilbert, R.J.C., Crane, D., Morgan, P.J., Mitchell, T.J., Rowe, A.J., Andrew, P.W., Paton, J.C., Tweten, R.K. and Parker, M.W. (1998). The molecular mechanism of pneumolysin, a virulence factor from *Streptococcus pneumoniae. J. Mol. Biol.* **284**, 449–61.

Rottem, S., Cole, R.M., Habig, W.H., Barile, M.F. and Hardegree, M.C. (1982). Structural characteristics of tetanolysin and its binding to lipid vesicles. *J. Bacteriol.* **152**, 888–92.

Saunders, F.K., Mitchell, T.J., Walker, J.A., Andrew, P.W. and Boulnois, G.J. (1989). Pneumolysin, the thiol-activated toxin of *Streptococcus pneumoniae*, does not require a thiol group for *in vitro* activity. *Infect. Immun.* **57**, 2547–52.

Sekino-Suzuki, N., Nakamura, M., Mitsui, K. and Ohno-Iwashita, Y. (1996). Contribution of individual tryptophan residues to the structure and activity of θ-toxin (perfringolysin O), a cholesterol-binding cytolysin. *Eur. J. Biochem.* **241**, 941–7.

Sekiya, K., Satoh, R., Danbara, H. and Futaesaku, Y. (1993). A ring-shaped structure with a crown formed by streptolysin O on the erthrocyte membrane. *J. Bacteriol.* **175**, 5953–61.

Shepard, L.A., Heuck, A.P., Hamman, B.D., Rossjohn, J., Parker, M.W., Ryan, K.R., Johnson, A.E. and Tweten, R.K. (1998). Identification of a membrane-spanning domain of the thiol-activated pore-forming toxin *Clostridium perfringens* perfringolysin O: an α-helical to β-sheet transition identified by fluorescence spectroscopy. *Biochemistry* **37**, 14563–74.

Shimizu, T., Okabe, A., Minami, J. and Hayashi, H. (1991). An upstream regulatory sequence stimulates expression of the perfringolysin O gene of *Clostridium perfringens. Infect. Immun.* **59**, 137–42.

Shimizu, T., Ba-Thein, W., Tamaki, M. and Hayashi, H. (1994). The *virR* gene, a member of a class of two-component response regulators, regulates the production of perfringolysin O, collagenase, and haemagglutinin in *Clostridium perfringens. J. Bacteriol.* **176**, 1616–23.

Smyth, C.J. (1975). The identification and purification of multiple forms of θ-haemolysin (θ-toxin) of *Clostridium perfringens* type A. *J. Gen. Microbiol.* **87**, 219–38.

Smyth, C.J. and Duncan, J.L. (1978). Thiol-activated (oxygen-labile) cytolysins. In: *Bacterial Toxins and Cell Membranes* (eds J. Jeljaszewicz and T. Wadstrom), pp. 129–83. Academic Press, London.

Smyth, C.J., Freer, J.H. and Arbuthnot, J.P. (1975). Interaction of *Clostridium perfringens* theta-haemolysin, a contaminant of commercial phospholipase C, with erythrocyte ghost membranes and lipid dispersions. A morphological study. *Biochim. Biophys. Acta* **382**, 479–93.

Song, L., Hobaugh, M.R., Shustak, C., Cheley, S., Bayley, H. and Gouaux, J.E. (1996). Structure of Staphylococcal α-haemolysin, a heptameric transmembrane pore. *Science* **274**, 1859–66.

Stevens. D.L., Mitten, J. and Henry, C. (1987). Effects of α- and θ-toxins from *Clostridium perfringens* on human polymorphonuclear leukocytes. *J. Infect. Dis.* **156**, 324–33.

Stevens, D.L., Troyer, B.E., Merrick, D.T., Mitten, J.E. and Olson, R.D. (1988). Lethal effects and cardiovascular effects of purified α- and θ-toxins from *Clostridium perfringens. J. Infect. Dis.* **157**, 272–9.

Stevens, D.L., Tweten, R.K., Awad, M.M., Rood, J.I. and Bryant, A.E. (1997). Clostridial gas gangrene: evidence that α- and θ-toxins differentially modulate the immune response and induce acute tissue necrosis. *J. Infect. Dis.* **176**, 189–95.

Thelestam, M. and Möllby, R. (1980). Interaction of streptolysin O from *Streptococcus pyogenes* and theta-toxin from *Clostridium perfringens* with human fibroblasts. *Infect. Immun.* **29**, 863–72.

Tweten, R.K. (1988a). Cloning and expression in *Escherichia coli* of the perfringolysin O (theta toxin) gene from *Clostridium perfringens* and characterization of the gene product. *Infect. Immun.* **56**, 468–76.

Tweten, R.K. (1988b). Nucleotide sequence of the gene for per-fringolysin O (theta-toxin) from *Clostridium perfringens*: significant homology with the genes for streptolysin O and pneumolysin. *Infect. Immun.* **56**, 3235–40.

Tweten, R.K. (1995). Pore-forming toxins in Gram-positive bacteria. In: *Virulence Mechanisms of Bacterial Pathogens* (eds J.A. Roth, C.A. Bolin, K.A. Brogden, C. Minion and M.J. Wannemuehler), pp. 207–29. American Society for Microbiology, Washington, DC.

Tweten, R.K., Harris, R.W. and Sims, P.J. (1991). Isolation of a tryptic fragment from *Clostridium perfringens* theta-toxin that contains sites for membrane binding and self-aggregation. *J. Biol. Chem.* **266**, 12449–54.

Van Heyningen, W.E. (1950). *Bacterial Toxins.* C.C. Thomas, Springfield, IL.

Walev, I., Palmer, M., Valeva, A., Weller, U. and Bhakdi, S. (1995). Binding, oligomerization, and pore formation by streptolysin O

in erythrocytes and fibroblast membranes: detection of nonlytic polymers. *Infect. Immun.* **63**, 1188–94.

Wilkinson, P.C. (1975). Inhibition of leukocyte locomotion and chemotaxis by lipid-specific bacterial toxins. *Nature* **255**, 485–7.

Williamson, E.D. and Titball, R.W. (1993). A genetically engineered vaccine against the alpha-toxin of *Clostridium perfringens* protects against experimental gas gangrene. *Vaccine* **11**, 1253–8.

Wuth, O. (1923). Serologische und biochemische Studien über das Hämolysin des Fränkelschen Gasbrandbazillus. *Biochem. Zeit.* **142**, 19–28.

Zs.-Nagy, I., Ohno-Iwashita, Y., Ohta, M., Zs.-Nagy, V., Kitani, K., Ando, S. and Imahori, K. (1988). Effect of perfringolysin O on the lateral diffusion constant of membrane proteins of hepatocytes as revealed by fluorescence recovery after photobleaching. *Biochim. Biophys. Acta* **939**, 551–60.

28

Listeriolysin, the thiol-activated haemolysin of *Listeria monocytogenes*

Thomas Jacobs, Ayub Darji, Siegfried Weiss and Trinad Chakraborty

INTRODUCTION

Listeriolysin O (LLO) is the most important virulence factor of the Gram-positive, facultative, intracellular pathogen *Listeria monocytogenes* (Portnoy *et al.*, 1988; Cossart *et al.*, 1989). The bacterium is a ubiquitously occurring soil micro-organism which can cause bacterial meningitis in infants, the immunocompromised and elderly people world-wide. Natural infection occurs by uptake of contaminated food via the gastrointestinal tract, where *L. monocytogenes* enters intestinal epithelial cells, traverses the mucosa and disseminates via lymphatics and blood to deeper tissues (Racz *et al.*, 1973; Schlech *et al.*, 1983; Farber and Peterkin, 1991; Dalton *et al.*, 1997). The ability of *Listeria* to invade host parenchymal cells, such as fibroblasts and hepatocytes, and the relative ease of reproducing this aspect of infection under laboratory conditions have sparked great interest in the intracellular lifestyle of this bacterium (Gaillard *et al.*, 1987; Cossart and Mengaud, 1989; Portnoy *et al.*, 1992b; Cossart *et al.*, 1996; Chakraborty and Wehland, 1997; Finlay and Cossart, 1997). This has led to the identification of several virulence factors that are essential for the various steps of the infection cycle. Following adhesion and internalization, the bacterium lyses the phagosomal membrane to gain access to the cytosol of the cell where it can replicate (Portnoy *et al.*, 1988; Sun *et al.*, 1990). By recruiting elements of the host-cell cytoskeleton, *Listeria* can move within the infected cell and spread intercellularly (Tilney and Portnoy, 1989; Smith and Portnoy, 1997).

This strategy accelerates propagation of infection and effectively circumvents the effect of the humoral immune response of the host. As will be shown in this chapter, listeriolysin is a key molecule for the intracellular lifestyle of this bacterium with many faceted effects on the host cell.

GENETICS AND VIRULENCE

Listeriolysin is a member of a closely related family of toxins, the so-called thiol-activated toxins, produced by various Gram-positive pathogenic bacteria. These toxins have 40–70% sequence homology to one another and share three hallmark traits: loss of activity following oxidation or modification of a thiol moiety, requirement of cholesterol-containing membranes for cytolytic activity, and oligomerization of toxin monomers to generate pores in the membrane. Most of these toxins are actively secreted by the producing bacteria and have been shown to be cytolytic for most types of erythrocytes and nucleated cells (Tweten, 1995).

LLO was first identified as a factor essential for growth of pathogenic *Listeria* in the livers and spleens of infected mice, where it represents one of the major antigens against which protective immunity is generated (Bouwer *et al.*, 1997). Also, in animal and human infections, the humoral response to listeriolysin is a highly specific diagnostic indicator of prior infection (Berche *et al.*, 1990; L'Hopital *et al.*, 1993; Grenningloh *et al.*, 1997). LLO comprises 529 amino acids and harbours the prototype hallmark sequence motif ECT-GLAWEWWR undecapeptide present in all of the thiol-activated toxins. Its overall sequence is also highly homologous to the haemolysins of *Listeria ivanovii* and *Listeria seeligerii*, ivanolysin and seeligeriolysin, respectively (see Figure 28.1; Haas *et al.*, 1992).

The synthesis of listeriolysin is determined by the *hly* gene which is embedded in a cluster of virulence genes uniquely present on the chromosome of pathogenic *Listeria* spp. (Leimeister-Wächter and Chakraborty, 1989; Mengaud *et al.*, 1989). Expression is regulated by the global virulence transcriptional activator protein PrfA, which binds to the palindromic sequence TTAACAttTGTTAA overlapping the −35 promotor binding sequence of RNA polymerase (Portnoy *et al.*, 1992a). A number of external stimuli including growth temperature and sugar availability has been shown to regulate the production of the toxin. LLO is secreted into culture supernatants as a protein of 58 kDa (Goeffroy *et al.*, 1987; Leimeister-Wächter and Chakraborty, 1989). It is assumed that this transport is a Sec-dependent process initiated by its signal peptide comprising 24 amino acids.

Studies using the mouse infection model in conjunction with tissue culture cells have revealed that this toxin is necessary for the bacterium to escape from the phagolysosome of most host cells, thus enabling them to leave the aggressive mileu of these vesicles and to grow in the cytosol of the infected cell (Portnoy *et al.*, 1988). Epitope-tagged recombinant listeriolysin was generated to follow its expression and study its localization following bacterial infection of the macrophage-like cell line J774. In Figure 28.2 (see colour plate section), a confocal image of infected J774 cells is shown depicting strong expression of listeriolysin within vacuoles containing entrapped bacteria. However, following release from the vacuole, listeriolysin expression is detected as clouds around bacteria or in distinct organelle structures within the host-cell cytoplasm.

The importance of LLO is emphasized by the fact that mutant strains lacking LLO are cleared rapidly from the host without even inducing a specific immune response (Berche *et al.*, 1988). However, introduction of LLO into other bacteria such as *Bacillus subtilis* and *Salmonella typhimurium* (Bielecki *et al.*, 1990; Portnoy *et al.*, 1992a) confers upon these bacteria the ability to escape from the phagolysosomal vacuole into the host cytosol. Interestingly, the toxicity of LLO appears well balanced with regard to damaging the host cell. Cells infected with LLO producing *L. monocytogenes* are usually able to survive for a considerable length of time, while cells infected with recombinant *L. monocytogenes* that expressed the haemolytically more active thiol-activated toxin, perfringolysin (PFO)

instead of LLO, were rapidly and severely damaged. Selection of attenuated strains expressing mutant PFO allowed these bacteria to escape from the vacuole without killing the host cells (Jones *et al.*, 1996). However, such *L. monocytogenes* strains were avirulent in mice, which indicates that PFO cannot fully compensate the lack of LLO. Thus, listeriolysin possesses other essential functions for the survival of the pathogen in the host other than just allowing the bacterium to enter the cytosol of invaded cells.

MECHANISM OF CYTOXICITY

Membrane binding

One common feature of this family of toxins is that cholesterol functions as a receptor in the membrane of the host cell. Direct binding to cholesterol has been demonstrated for many of these haemolysins (Johnson *et al.*, 1980). Depletion of cholesterol from target cells resulted in an almost complete absence of cytolysis and binding of LLO to such cells was considerably reduced (Jacobs *et al.*, manuscript in preparation). Molecular modelling using the published structure of PFO showed that cholesterol would fit into a pocket formed by the region around the conserved undecapeptide (Rossjohn *et al.*, 1997). Consistently, lytic activity of LLO and other members of this family of haemolysins is inhibited by the addition of low amounts of cholesterol (Duncan and Schlegel, 1975; Prigent and Alouf, 1976). Thus, inhibition of cytolysis could be assumed to be due to occupancy of the cholesterol-binding site by free cholesterol. Therefore, pretreatment of LLO with cholesterol should result in a loss of binding of the toxin to membranes.

However, using flow cytometry and immunoblot techniques it was demonstrated that binding of LLO to membranes was not impaired by the pretreatment of LLO with cholesterol, indicating that an interference with the oligomerization step may account for the cholesterol inactivation that is observed (Jacobs *et al.*, 1998). To investigate this hypothesis further, polymerization of LLO on erythrocyte ghosts was visualized directly using electron microscopy. Samples treated with LLO show the characteristic ring- and arc-shaped structures as for other toxins of this family (Bhakdi *et al.*, 1985; Jacobs *et al.*, 1998). These structures were not visible in samples treated with cholesterol-inactivated LLO.

These findings have been corroborated by studying the influence of cholesterol on LLO binding to artificial membranes. LLO binds to liposomes consisting of phosphatidylcholine and cholesterine, whereas binding to pure phosphatidylcholine liposomes is strongly

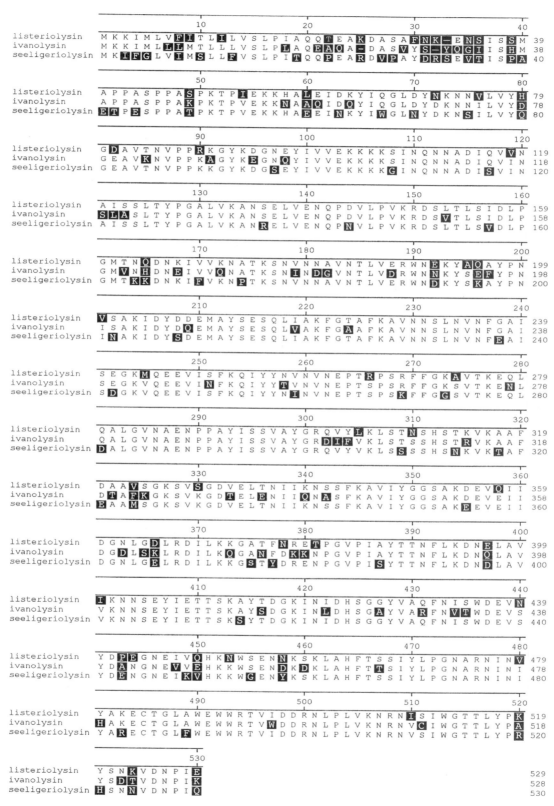

FIGURE 28.1 Sequence alignments of the thiol-activated toxins listeriolysin, ivanolysin and seeligeriolysin of *L. monocytogenes*, *L. ivanovii* and *L. seeligeri* produced using the Clustal V algorithm. Differences in the primary sequences are boxed.

reduced. Pretreatment of LLO with cholesterol enables the toxin to bind to cholesterol-free membranes. To interpret these findings, it was postulated that the interaction of cholesterol with LLO in solution induces a conformational change in the protein which allows the toxin to bind to a target membrane. This is consistent with the structural modelling by Rossjohn *et al.*, who postulate a conformational change in PFO after binding of cholesterol that would render the molecule more hydrophobic and membrane-seeking (Rossjohn *et al.*, 1997).

The inhibition of LLO by cholesterol could thus be explained by assuming an induction of a change to the more hydrophobic conformation that would induce membrane binding but would no longer allow the membrane intercalation. Data consistent with this interpretation could be obtained using cytofluorometry and polyclonal antibodies specific for LLO. A stronger signal was observed with cholesterol-inactivated LLO than with listeriolysin alone. This indicates that functionally active LLO inserts deeper into the membrane than cholesterol-inactivated LLO. The unaltered binding of inactivated LLO to cellular membranes also explains why non-lytic LLO–cholesterol complexes could trigger cytokine expression (Nishibori *et al.*, 1996), release of interleukin-1 (IL-1) (Yoshikawa *et al.*, 1993) and stimulation of the phosphatidylinositol cascade (Sibelius *et al.*, 1996) in target cells.

Experimental evidence that the undecapeptide is involved in cholesterol binding has been obtained by mutagenesis studies. Exchanging the tryptophan residues of this peptide in PFO with phenylalanine revealed reduced binding to cholesterol (Sekino-Suzuki *et al.*, 1996). Similar data were obtained with LLO in which tryptophan 491 or 492 was replaced by alanine (Jacobs *et al.*, unpublished). In addition, cytofluorometry revealed decreased binding of variant LLO molecules to nucleated host cells (Darji *et al.*, unpublished). Thus, the gradual decrease in lytic activity of the respective LLO variant could be correlated with a sequential loss in binding and with the reduced induction of apoptosis in murine dendritic cells by variant LLO (Guzmann *et al.*, 1996).

However, this is in clear contrast to the results obtained using membranes of erythrocytes. Similar binding capacity was obtained for all LLO mutants including a variant in which the cysteine 484 had been exchanged to a serine residue (Michel *et al.*, 1990). Similarly, induction of mucin secretion by a human colon epithelial cell line was achieved with wild-type LLO as well as mutants of tryptophan 491 and 492 (Coconnier *et al.*, 1998). It seems therefore that the process of membrane binding by LLO is more than mere binding of LLO to cholesterol followed by a sub-

sequent conformational change. Clearly, post-binding effects, including the generation of signal transduction events, contribute towards the overall toxicity mediated by this molecule.

Oligomerization

Most of the work regarding the structure–function relationship of pore-forming toxins has been done with streptolysin O (SLO) (Bhakdi *et al.*, 1985; Sekiya *et al.*, 1993; Palmer *et al.*, 1995, 1996, 1998a, b) and perfringolysin O (PFO) (Mitsui *et al.*, 1979; Morgan *et al.*, 1994; Nakamura *et al.*, 1995; Rossjohn *et al.*, 1997) and these data are included here to present a coherent picture on the mechanism of pore formation and structure–function relationship of listeriolysin. It must be pointed out here that apart from the unique properties described previously, LLO is expected to exhibit special additional features since it is the only one of these toxins which is most active at acidic pH (Geoffroy *et al.*, 1987).

As indicated above, listeriolysin belongs to the family of pore-forming, thiol-activated haemolysins which include SLO, PFO and pneumolysin (Tweten, 1995). These toxins share common features. They are activated by thiols and are able to form pores of 20–30 nm in the membrane of the host cell. These have experimentally been visualized using high concentrations of SLO, PFO and LLO. However, it is questionable as to whether it is necessary and/or possible to form fully grown pores under physiological conditions. For SLO it has recently been found that even dimers of the growing pore already exhibit lytic effects (Palmer *et al.*, 1998a).

Recently, the three-dimensional structure of PFO was established (Rossjohn *et al.*, 1997). Molecular simulation using these data allowed the prediction of a structure for LLO. According to this model LLO consists of four domains exhibiting an elongated wedge-like shape that is rich in β-pleated sheets. Domains 1–3 are thought to be involved in polymerization while domain 4 seems to harbour the membrane-binding portion of the toxin. The hydrophobic undecapeptide located close to its C-terminal end and conserved in all members of the pore-forming toxin family is exposed at the very end of domain 4. This model is consistent with data obtained using neutralizing monoclonal antibodies with known epitope specificities. Neutralizing antibodies that bound to epitopes within domain 1 did not inhibit the binding of LLO to cellular membranes but interfered with the oligomerization of toxin monomers (Darji *et al.*, 1996). In contrast, a neutralizing monoclonal antibody against pneumolysin that cross-reacted with LLO and recognized part of the conserved undecapeptide in domain 4 inhibited the binding of both toxins to cellular membranes (de los Toyos *et al.*, 1996; de los Toyos, personal communication).

Based on these data and assumptions a picture of pore formation starts to emerge. The hydrophobic undecapeptide that is present in all members of this toxin family is exposed at the very end of domain 4 and could act as a hydrophobic dagger mediating the initial contact to hydrophobic membranes. This step is reversible and independent of the temperature, while the subsequent steps such as polymerization and lysis of sheep red blood cells (SRBC) are temperature dependent, i.e. they are blocked at 4°C (Palmer *et al.*, 1995; Sekiya *et al.*, 1996). After binding, the toxin could interact with cholesterol which is buried in the membrane. This contact would initiate refolding of the toxin thus exposing the tryptophan residues of the conserved undecapeptide, and rendering the structure more hydrophobic (Nakamura *et al.*, 1995; Jacobs *et al.*, 1998; Palmer *et al.*, 1998b). The molecules would now be entrapped in the membrane and could serve as a starting point for pore formation by acquiring either more monomeric toxin from solution or other membrane-bound monomeric toxin molecules by lateral diffusion. Since membrane dynamics are more restricted at low temperatures, the kinetic energy required for polymerization would be blocked in the frozen membrane. In support of this idea, a conformational change after treatment with cholesterol has been demonstrated for PFO using cluster of differentiation (CD) spectroscopy in the near and far ultraviolet (UV) (Nakamura *et al.*, 1995). For SLO a co-operative effect of bound toxin was also demonstrated (Weller *et al.*, 1996; Palmer *et al.*, 1998b).

The undecapeptide contains the unique cysteine residue of LLO at position 484. Surprisingly, the exchange of this amino acid by alanine does not interfere with the haemolytic activity and the exchange with the more bulky amino acid serine reduces the activity only to 20%. In SLO, changing the cysteine to alanine renders the toxin insensitive to oxidation (Pinkney *et al.*, 1989). Thus, the cysteine is most probably responsible for the sensitivity of these toxins to oxidation. Despite this, it is not essential for haemolytic activity. The selective pressure to retain the cysteine at an identical position in all of these toxins is not yet known.

Differential sensitivity of cell lines against the toxicity of pore-forming, cholesterol-binding haemolysins has been observed (Walev *et al.*, 1995; Greningloh, personal communication). In particular, Epstein–Barr virus-transformed human B-cell lines and primary human dendritic cells were extremely resistant against LLO (Greningloh, personal communication). The underlying mechanism is unclear since cholesterol as the receptor should be found in the membrane of every mammalian cell in roughly similar quantities. However, a number of overlapping mechanisms could contribute to this phenotype. The efficiency of the membrane-repair mechanism might be different in the various cells. Also, formation of non-lytic polymers as described for SLO (Walev *et al.*, 1995) or shedding of toxin-containing vesicles as described for α-toxin and SLO could be invoked as an explanation for the differential resistance (Walev *et al.*, 1994). In the case of dendritic cells the membrane fluidity and dynamics might also have an impact on toxin activity. Differences in the amounts and types of proteases, in particular membrane-associated proteases, could also contribute to this phenomenon. The authors observed degradation of LLO that was tagged with an epitope of haemagglutinin (HA) following incubation with nucleated cells. Since the HA-tag that was used for detection was fused to the *N*-terminus it was estimated that a 5-kDa fragment was removed from the *C*-terminus. This would remove the undecapeptide that was shown to be essential for haemolytic activity. Similarly, deletion of a *C*-terminal fragment was shown to inactivate pneumolysin (Boulnois *et al.*, 1990; Owen *et al.*, 1994).

LLO might exhibit a shorter half-life in comparison to other toxins of this family (Demuth *et al.*, 1994; Goebel and Kreft, 1997). Although there are no direct comparisons of half-lives of such toxins it could be shown that LLO is degraded rapidly in the cytosol of the host cell by proteasomes since epitopes of LLO were presented by major histocompatibility complex (MHC) class-I molecules shortly after infection (Villanueva *et al.*, 1995). Appropriate regulation of LLO expression seems to be an additional prerequisite for host cell survival since *L. monocytogenes* that hyperexpress LLO lyse host cells even when infected at a very low multiplicity of infection (Nicole Frahm, personal communication). These results indicate that LLO is more than just another member of cholesterol-binding toxins with an acidic pH optimum and maybe a short half-life. Rather, it represents a toxin that is well balanced to allow sufficient time for survival of the pathogen within the invaded cell to facilitate further spreading of the bacteria in the infected host.

CELLULAR REACTIONS

Host-cell signalling

Pathogenic bacteria frequently subvert host-cell functions such as signal transduction pathways, cytoskeleton rearrangement and vacuolar trafficking (Finlay and

Cossart, 1997; Finlay and Falkow, 1997). In recent years, data have been accumulating to suggest that at very low, sublytic concentrations, listeriolysin is an important factor in inducing host-cell signal molecules both prior to and following entry of the bacterium. A rapid, but transient, activation of the kinase cascade of the Raf-MEK (mitogen-activated protein kinase kinase)–MAPK (mitogen-activated protein kinase) signal transduction pathway has been documented in a number of cell types including epithelial cell lines and macrophages following incubation of cells with LLO-producing bacteria (Weiglein *et al.*, 1997; Tang *et al.*, 1998). These data are consistent with the observation that NIH3T3 cells transformed with a eukaryotic expression plasmid-producing listeriolysin grow more rapidly and induce focus formation (Demuth *et al.*, 1994).

LLO is also a potent trigger of host signalling molecules such as inositol triphosphate (IP3) and diacylglycerol (DAG). In addition, two lipid inflammatory mediators, platelet activating factor (PAF) and prostglandin I_2 (PGI$_2$), that have diverse biological activities such as inflammatory and chemotatic effects (PAF) and increased vascular permeability and neutrophil chemotaxis (PGI$_2$) are also increased in human umbilical vein endothelial cells following listeriolysin addition (Sibelius *et al.*, 1996). In keeping with these findings, it has recently been demonstrated that the expression of the endothelial adhesion molecule, P-selectin, is also directly induced by LLO (Krull *et al.*, 1997). Furthermore, release of IL-1, IL-6 and tumour necrosis factor (TNF-α) is observed when murine macrophages are exposed to bacteria-producing listeriolysin (Kretschmar *et al.*, 1993). All of these responses can be reproduced using purified toxin. LLO has also been identified as a molecule critical for the induction of apoptosis in mouse dendritic cells (Guzmann *et al.*, 1996).

Many of these responses are a result of transition of the host-cell stress-inducible transcription factor, NF-κB, from the cytoplasm to the nucleus, where it regulates the activity of different genes involved in the immune response (Whitley *et al.*, 1994). NF-κB is a member of the Rel family of transcriptional activator proteins (Israel, 1995). It is sequestered in the cytoplasm as a p50-p65 heterodimer that is associated with two major inhibitory proteins, I-κBα and I-κBβ (Gilmore and Morin, 1993). Stimulation of cells leads to phosphorylation, polyubiquitination and degradation of the inhibitory proteins resulting in NF-κB activation (Chen, *et al.*, 1995). *Listeria monocytogenes* wild-type related NF-κB translocation into the nucleus of endothelial cells was demonstrated recently by immunofluorescent microscopy (Drevets, 1997). In P388D1 macrophages, listeriolysin

and the gene products of the lecithinase operon are required for long-lasting activation of NF-κB and are concomitant with the complete proteolytic degradation of I-κBβ (Hauf *et al.*, 1997).

Vacuolar trafficking

Compared to most other members of the family of thiol-activated haemolysins, LLO allows escape of the bacterium to the cytosol of the host cell but does not harm the host cell under these circumstances (Goebel and Kreft, 1997). This could be due to the fact that its activity is preferentially exhibited intracellularly in vesicles. One potential reason for this selectivity was assumed to be due to the pH dependence of LLO. In an elegant study Beauregard and colleagues have shown that lysis of macrophage phagosomes is pH dependent (Beauregard *et al.*, 1997). By using a membrane-impermeant fluorophore they measured the pH of the phagosome and its integrity. *Listeria monocytogenes*-containing vacuoles began to acidify soon after uptake of the bacteria. Subsequently, the pH increased again, accompanied by a loss of the included dye. This loss of dye could be drastically reduced by using lysosomotropic agents such as ammonium chloride. From these data the authors draw the following conclusion: *L. monocytogenes*-containing phagosomes begin to acidify soon after uptake. The lowered pH facilitates the activity of the accumulating LLO. This event destabilizes the phagosomal membrane leading to pH equilibration across the vacuolar membrane, followed by release of the dye and deactivation of LLO. Although this scenario seems to be conclusive it is not compelling that the pH dependence is the only reason for the preferential action of LLO on the phagosomal membrane. Degradation and repair of functional pores by the host cells could be other reasons for the balanced toxicity of LLO. Fast recovery from the activity of LLO of some vesicles that are involved in antigen presentation has been demonstrated (see below).

Immunological properties of listeriolysin O

LLO is essential for the induction of protective immunity against *L. monocytogenes*. Strains that lack LLO and that are unable of escape from the phagolysosome of macrophages are cleared quickly from the host by the innate immune system (Berche *et al.*, 1987a, b). The time of infection might be too short to attract specific immunocytes to sites of invasion. Efficacious induction of protective immunity is probably the result of the various properties described for this toxin above. Thus, the host-cell signalling that is induced by LLO in macrophages and other host cells might be required for an enhanced immune response by increasing local inflammation. The ability of LLO to induce pore formation at higher toxin concentrations is paramount for the introduction of the invading bacterium and other protein molecules into the cytosol of the host cell. Antigens from this cellular compartment are usually presented by MHC class-I molecules and induce CD8$^+$ cytotoxic T-cells. Such T-cells are known to provide protection against *L. monocytogenes*.

LLO represents a protective antigen of *L. monocytogenes*. CD4$^+$ and CD8$^+$ T-cells are induced against the toxin when mice are infected with sublethal doses of *L. monocytogenes* (Kaufmann, 1993; Mielke *et al.*, 1993; Pamer, 1997). Propagation *in vitro* allowed the assignment of epitopes that are recognized by such T-cells. Interestingly, in mice of the H2^d haplotype the CD8$^+$ T-cell response is exclusively directed against a single epitope LLO 91–99 that is presented by the K^d (Hess *et al.*, 1996; Pamer, 1997). Adoptively transferred CD8$^+$ T-cell clones against LLO 91–99 have been shown to provide protection against lethal challenges by *L. monocytogenes* (Pamer, 1997). Similarly protective responses were induced with recombinant *Salmonella typhimurium* that either secreted LLO (Hess *et al.*, 1996) or were used as a shuttle to induce a genetic immunization against LLO (Darji *et al.*, 1997a). In the former study CD8$^+$ as well as CD4$^+$ T-cells were shown to provide protection.

The existence of a single epitope for CD8$^+$ T-cells in H2^d mice allows the generation of variant LLO that no longer stimulates CD8$^+$ T-cells but allows the bacteria to escape from the vacuole. With such variant strains it could be shown that mice were protected without a CD8$^+$ T-cell response against LLO, thus demonstrating that the pore-forming property is the essential feature of LLO for induction of protective immunity (Bouwer *et al.*, 1996).

LLO also serves as the major protein antigen for the humoral response in humans and animals that were infected with *L. monocytogenes*. Antibodies have been shown to play no role in protection in experimental listeriosis (Kaufmann, 1993). Nevertheless, screening of sera of healthy donors that had a high chance of infection by *L. monocytogenes* owing to their profession or listeriosis patients revealed LLO as one of the two major antigens against which antibodies were directed (Berche *et al.*, 1990; Grenningloh *et al.*, 1998). It is likely that the immunomodulatory and stimulatory properties of LLO induce the production of antibodies. Whether local antibodies play a role in natural infection through the intestinal barrier remains to be determined.

When purified LLO was injected into mice in Freund's adjuvant, CD4$^+$ as well as CD8$^+$ T-cells were induced. In addition, LLO-specific CD8$^+$ T-cells could be maintained in culture with soluble active LLO (Darji *et al.*, 1995b, 1997a). CD8$^+$ T-cells usually recognize epitopes derived from intracellular sources presented by MHC class-I molecules. Thus, it was concluded that the pore-forming activity of LLO led to introduction of soluble LLO into the cytosol of antigen-presenting cells (APC), a conclusion confirmed by using heat-inactivated LLO or LLO that had been inactivated by cholesterol (Darji *et al.*, 1995b). The pore-forming activity could also be used to introduce passenger proteins into the cytosol of APC *in vitro* as well as *in vivo*. The ease with which these passenger proteins can be introduced into the MHC class-I processing pathway emphasizes the suitability of this system in screening for potential vaccine candidates amongst complex antigen mixtures for intracellular pathogens. The presentation of passenger proteins by MHC class-I molecules was found to be dependent on TAP (Darji *et al.*, 1997b), demonstrating that the conventional class-I pathway is involved.

Several years ago *L. monocytogenes* was shown to interfere with the presentation of MHC class-II-dependent antigens (Cluff and Ziegler, 1987). *In vitro*, stimulation of CD4$^+$ T-cells was inhibited when APC were infected with *L. monocytogenes* and *in vivo* antibody responses were seriously hampered during the early phase of infection (Hage-Chahine *et al.*, 1992). By using mutant strains that lacked LLO or partially purified LLO it could be shown that the inhibitory effect was caused by listeriolysin (Cluff *et al.*, 1990). Recently, using LLO that was purified from the supernatants of hyperexpressing recombinant *L. inoccua* (Darji *et al.*, 1995a), it was discovered (Darji *et al.*, 1997c) that the inactivation was due to a T-cell receptor antagonism. T-cell receptor antagonism was originally described for synthetic peptides in which single amino acids had been replaced. Such class II/peptide complexes induce only a partial signal through the T-cell receptor and are inhibitory even when agonistic peptides are added at the same time to the cultures (Lancaster and Allen, 1996). Similar effects have been observed on T-cells by LLO-treated APC. Adding agonistic synthetic peptides to cultures in which LLO-treated APC present antigen does not result in T-cell stimulation (Darji *et al.*, 1997c), indicating that only a partial signal is induced via the T-cell receptor in such cells. However, such cells can be rescued by the addition of ionomycin but not by PMA or IL-2, suggesting that the LLO-induced class II/peptide complexes could still activate the ras-dependent signalling pathway but not the pathway that includes the phosphoinositol cascade resulting in calcium release from intracellular stores. Further experiments demonstrated that LLO-treated APC proliferation and IL-2 secretion of T-cells are inhibited while IFN-γ and IL-3 secretion is not affected (Darji, unpublished data). This could explain why the antibody response but not granuloma formation is inhibited during the early phase of an infection with *L. monocytogenes*. Taken together, these data show that the inhibiton of MHC class-II restricted CD4 T-cells is due to a partial antagonism similar to the antagonism induced by altered peptide ligands. Additional experiments showed that this inhibition of T-cells is only transient. Within 3–5 days T-cells or T-cell hybridomas recover and become fully active again.

APPLICATIONS OF LISTERIOLYSIN O

Numerous studies are available in which pore-forming proteins are used to introduce soluble factors into cells. Cholesterol-binding toxins are widely used since the diameter of the pore is estimated to be around 30 nm and appears to be large enough to introduce complex biological molecules such as proteins DNA into the cytosol of cells (Ahnert-Hilger *et al.*, 1989; Barry *et al.*, 1993; Bhakdi *et al.*, 1993; Liu *et al.*, 1993; Gottschalk *et al.*, 1995). To date, there is no estimation of the exclusion limit of the LLO pore. Antigenic proteins such as hen-egg lysozyme, ovalbumin and complement C5 are presented via MHC class-I. However, the amount of material that is taken up is difficult to estimate since the cytotoxic assay used is extremely sensitive and even a single MHC class-I/peptide complex presented by a cell might be sufficient to trigger a killing event. Attempts to introduce eukaryotic expression plasmids encoding β-galactosidase into cells via LLO have failed so far (Jacobs and Darji, unpublished). However, phosphorothioate antisense oligonucleotides which resulted in the downregulation of targeted cell-surface markers (Jacobs, unpublished) could be achieved using LLO. Thus, there may be limitations in the introduction of molecules into target cells via LLO with regard to their size or physicochemical properties such as charge.

As indicated above, mixtures of LLO and passenger molecules usually induces CD8+ T-cell-dependent responses against the passenger protein as well as against LLO. However, the fact that only a simple immune dominant epitope is recognized by CD8+ T-cells of the H2d haplotype (BALB/c) could be used to generate LLO mutants that would no longer stimulate LLO-specific CD8+ T-cells in such mice (Bouwer *et al.*, 1997; Bruder *et al.*, 1998). The variant LLO was successfully used to induce and to characterize a CD8+ cytotoxic T-cell line against the listerial virulence factor ActA (Darji *et al.*, 1998). The mutation of the MHC class-I epitope of LLO avoids the concomitant induction of LLO-specific CD8+ T-cells while the inhibitory potency of LLO for CD4+ T-cell response is retained. Thus, epitopes generated under such circumstances will undergo the same selection procedures as naturally occurring epitopes, for example during viral infections, and may be used in cancer therapy (Paterson and Ikonomidis, 1996).

The introduction of passenger proteins could probably not be explained by simple perforation of cellular membranes thereby allowing antigenic material access to the cytosol. Endocytosis of LLO and antigen probably takes place, and subsequent acidification of endosomes would result in activation of LLO and in the perforation of their membranes, which would in turn generate a route into the cytosol (Beauregard *et al.*, 1997). This pathway could be the reason for the efficacy of LLO in generating cytotoxic T-cells. Such an approach was used by Lee *et al.*, who used acid-labile liposomes containing antigen and LLO for antigen delivery. These liposomes were taken up by phagocytic cells and disintegrated upon acidification of the endosomes. The liberated LLO was thus able to deliver the antigen from the vesicles into the cytosol (Lee *et al.*, 1996).

SUMMARY AND PERSPECTIVES

Listeriolysin, the pore-forming haemolysin of *L. monocytogenes* is a remarkable molecule with diverse effects on the physiology and metabolism of many different cell types. Despite a considerable increase in the understanding of the biological properties displayed by this molecule, many aspects of structure and function remain unaddressed and have been derived from studies done on comparative toxins, in particular perfringolysin. Thus, the mechanism of secretion of toxin by the bacteria has not been studied and its transition from a water-soluble protein to a membrane-bound complex remains a mystery. Evidence presented in this chapter suggests that although cholesterol is important for binding of the toxin to the membrane many of the biological effects are probably a result of interaction of the toxin with other molecules. Where LLO interacts with the host-cell signalling pathway is not known. In this respect, nothing is known about the localization of the toxin following its interaction with the eukaryotic membrane nor of its location following escape of the bacterium into the host cytosol. Clearly, one general consequence of its interaction with cholesterol is to increase the membrane affinity of this protein but this could also be important in the regulation of protein–protein interactions. At sublytic concentrations,

LLO acts as a pseudo-chemokine, with diverse biological effects on the affected cell. High LLO concentrations produce cell lysis by the induction of oligomeric pores in target cell membranes.

REFERENCES

Ahnert-Hilger, G., Mach, W., Fohr, K.J. and Gratzl, M. (1989). Poration by alpha-toxin and streptolysin O: an approach to analyse intracellular processes. *Methods Cell. Biol.* **31**, 63–90.

Barry, E.L., Gesek, F.A. and Friedman, P.A. (1993). Introduction of antisense oligonucleotides into cells by permeabilization with streptolysin O. *BioTechniques* **15**, 1018–20.

Beauregard, K.E., Lee, K.D., Collier, R.J. and Swanson J.A. (1997). pH-dependent perforation of macrophage phagosomes by listeriolysin O from *Listeria monocytogenes*. *J. Exp. Med.* **186**, 1159–63.

Berche, P., Gaillard, J.L., Geoffroy, C. and Alouf, J.E. (1987). T-cell recognition of listeriolysin O is induced during infection with *Listeria monocytogenes*. *J. Immunol.* **139**, 3813–21.

Berche, P., Gaillard, J.L., Geffroy, C. and Alouf, J.E. (1987a). T-cell recognition of listeriolysin O is induced during infection with *Listeria monocytogenes*. *J. Immunol.* **139**, 3813–21.

Berche, P., Gaillard, J.L. and Sansonetti, P.J. (1987b). Intracellular growth of *L. monocytogenes* as a prerequisite for *in vivo* induction of T-cell-mediated immunity. *J. Immunol.* **138**, 2266–71.

Berche, P., Reich, K.A., Bonnichon, M., Beretti, J.-L., Geoffroy, C., Raveneau, J., Cossart, P., Gaillard, J.-L., Geslin, P., Kreis, H. and Veron, M. (1990). Detection of anti-listeriolysin O for serodignosis of human listeriosis. *Lancet* **335**, 624–27.

Bhakdi, S., Tranum-Jensen, J. and Sziegoleit, A. (1985). Mechanism of membrane damage by streptolysin O. *Infect. Immun.* **47**, 52–60.

Bhakdi, S., Weller, U., Walev, I., Martin, E., Jonas, D. and Palmer, M. (1993). A guide to the use of pore-forming toxins for controlled permeabilization of cell membranes. *Med. Microbiol. Immunol.* **182**, 167–75.

Bielecki, J., Youngman, P., Connelly, P. and Portnoy, D.A. (1990). *Bacillus subtilis* expressing a haemolysin gene from *Listeria monocytogenes* can grow in mammalian cells. *Nature* **435**, 175–6.

Boulnois, G.J., Mitchell, T.J., Saunders, F.K., Mendez, F.J. and Andrew, P.W. (1990). In: *Structure and Function of Pneumolysin, the Thiol-activated Toxin of* Streptococcus pneumoniae (eds B. Withold, *et al.*), pp. 43–51. Gustav Fischer, Stuttgart.

Bouwer, H.G.A., Moors, M. and Hinrichs, D.J. (1996). Elimination of the listeriolysin O-directed immune response by conservative alteration of the immunodominant listeriolysin O amino acid 91 to 99 epitope. *Infect. Immun.* **64**, 3728–35.

Bouwer, H.G.A., Barry, R.A. and Hinrichs, D.J. (1997). Acquired immunity to an intracellular pathogen: immunologic recognition of *L. monocytogenes*-infected cells. *Immunol. Rev.* **158**, 137–46.

Bruder, D., Darji, A., Gakamsky, D.M., Chakraborty, T., Pecht, I., Wehland, J. and Weiss, S. (1998). Efficient induction of cytotoxic CD8[+] T-cells against exogenous proteins: establishment and characterization of a T-cell line specific for the membrane protein ActA of *Listeria monocytogenes Eur. J. Immunol.* **28**, 2630–9.

Chakraborty, T. and Wehland, J. (1997). The host cell infected with *Listeria monocytogenes*. In: *Host Response to Intracellular Pathogens* (ed. S.H.E. Kaufmann), pp. 271–90. R.G. Landes, Austin, TX.

Chen, Z., Hagler, J., Palombella, V.J., Melandri, F., Scherer, D., Ballard, D., Maniatis, T., Cluff, C.W. and Ziegler, H.K. (1995). Signal-induced site-specific phosphorylation targets I kappa B alpha to the ubiquitin-proteasome pathway. *Genes Dev.* **9**, 1586–9.

Cluff, C.W. and Ziegler, H.K. (1987). Inhibition of macrophage-mediated antigen presentation by haemolysin-producing *Listeria monocytogenes*. *J. Immunol.* **139**, 3808–12.

Cluff, C.W., Garcia M. and Ziegler, H.K. (1990). Intracellular haemolysin-producing *Listeria monocytogenes* strains inhibit macrophage-mediated antigen processing. *Infect. Immun.* **58**, 3601–12.

Coconnier, M.H., Dlissi, E., Robard, M., Laboisse, C.L., Gaillard, J.L. and Servin, A.L. (1998). *Listeria monocytogenes* stimulates mucus exocytosis in cultured human polarized mucosecreting intestinal cells through action of listeriolysin O. *Infect. Immun.* **66**, 3673–81.

Cossart, P. and Mengaud, J. (1989). *Listeria monocytogenes*: a model system for the molecular study of intracellular parasitism. *Mol. Biol. Med.* **6**, 463–74.

Cossart, P., Vicente, M.F., Mengaud, J., Bawuero, F., Perez, D.J.C. and Berche, P. (1989). Listeriolysin is essential for virulence of *Listeria monocytogenes*, direct evidence obtained by gene complementation. *Infect. Immun.* **57**, 3629–36.

Cossart, P., Boquet, S., Normark, S. and Rappuoli, R. (1996). Cellular microbiology emerging. *Science* **271**, 315–16.

Dalton, C.B., Austin, C.C., Sobel, J., Hayes, P.S., Bibb, W.F., Graves, L.M., Swaminathan, B., Proctor, M.E. and Griffin, P.M. (1997). An outbreak of gastroenteritis and fever due to *Listeria monocytogenes* in milk. *N. Eng. J. Med.* **336**, 100–5.

Darji, A., Chakraborty, T., Niebuhr, K., Tsonis, N., Wehland, J. and Weiss, S. (1995a). Hyperexpression of listeriolysin in the non-pathogenic species *Listeria innocua* and high yield purification. *J. Biotechnol.* **43**, 205–12.

Darji, A., Chakraborty, T., Wehland, J. and Weiss, S. (1995b). Listeriolysin generates a route for the presentation of exogenous antigens by major histocompatibility complex class I. *Eur. J. Immunol.* **25**, 2967–71.

Darji, A., Niebuhr, K., Hense, M.T., Wehland, J., Chakraborty, T. and Weiss, S. (1996). Neutralizing monoclonal antibodies against listeriolysin: mapping of epitopes involved in pore formation. *Infect. Immun.* **64**, 2356–8.

Darji, A., Guzmann, C.A., Gerstel, B., Wachholz, P., Timmis, K.N., Wehland, J., Chakraborty, T. and Weiss, S. (1997a). Oral somatic transgene vaccination using attenuated *S. typhimurium*. *Cell* **91**, 765–75.

Darji, A., Chakraborty, T., Wehland, J. and Weiss, S. (1997b). TAP-dependent major histocompatibility complex class I presentation of soluble proteins using listeriolysin. *Eur. J. Immunol.* **27**, 1353–9.

Darji, A., Stockinger, J., Wehland, T., Chakraborty, T. and Weiss, S. (1997c). Antigen-specific T-cell receptor antagonism by antigen-presenting cells treated with the haemolysin of *Listeria monocytogenes*: a novel type of immune escape. *Eur. J. Immunol.* **27**, 1696–703.

Darji, A., Bruder, D., zur Lage, S., Gerstel, B., Chakraborty, T., Wehland, J. and Weiss, S. (1998). The role of the bacterial membrane protein ActA in immunity and protection against *Listeria monocytogenes*. *J. Immunol.* **161**, 2414–20.

Demuth, A., Chakraborty, T., Krohne, G. and Goebel, W. (1994). Mammalian cells transfected with the listeriolysin gene exhibit enhanced proliferation and focus formation. *Infect. Immun.* **62**, 5102–11.

Drevets, D.A. (1997). *Listeria monocytogenes* infection of cultured endothelial cells stimulates neutrophil adhesion and adhesion molecule expression. *J. Immunol.* **158**, 5305–13.

Drevets, D.A. (1998). *Listeria monocytogenes* virulence factors that stimulate endothelial cells. *Infect. Immun.* **66**, 232–8.

Duncan, J.L. and Schlegel, R. (1975). Effect of streptolysin O on erythrocyte membranes, liposomes, and lipid dispersions. A protein–cholesterol interaction. *J. Cell Biol.* **67**, 160–73.

Finlay, B.B. and Cossart, P. (1997). Exploitation of mammalian host-cell functions by bacterial pathogens. *Science* **276**, 718–25.

Finlay, B.B. and Falkow, S. (1997). Common themes in microbial pathogenicity revisited. *Microbiol. Mol. Biol. Rev.* **61**, 136–9.

Farber, J. and Peterkin, P.I. (1991). *Listeria monocytogenes*, a food-borne pathogen *Microbiol. Rev.* **55**, 476–511.

Gaillard, J.L., Berche, P., Mounier, J., Richard, S. and Sansonetti, P.J. (1987). *In vitro* model of penetration and intracellluar growth of *Listeria monocytogenes* in the human enterocyte-like cell line caco-2. *Infect. Immun.* **55**, 2822–9.

Geoffroy, C., Gaillard, J.-L., Alouf, J.E. and Berche, P. (1987). Purification, characterization, and toxicity of the sulfhydryl-activated haemolysin listeriolysin O from *Listeria monozytogenes*. *Infect. Immun.* **55**, 1641–6.

Gilmore, T.D. and Morin, P.J. (1993). The I kappa B proteins: members of a multifunctional family. *Trends Genet.* **9**, 427–33.

Goebel, W. and Kreft, J. (1997). Cytolysins and the intracellular life of bacteria. *Trends Microbiol.* **5**, 86–8.

Gottschalk, S., Tweten, R.K., Smith L.C. and Woo, S.L.C. (1995). Efficient gene delivery and expression in mammalian cells using DNA coupled with perfringolysin O. *Gene Therapy* **2**, 498–503.

Grenningloh, R., Darji, A., Wehland, J., Chakraborty, T. and Weiss, S. (1997). Listeriolysin and IrpA are major protein targets of the human humoral response against *Listeria monocytogenes*. *Infect. Immun.* **65**, 3976–80.

Guzmann, C.A., Domann, E., Rohde, M., Bruder, D., Darji, A., Weiss, S., Wehland, J., Chakraborty, T. and Timmis, K.N. (1996). Apoptosis of mouse dendritic cells is triggered by listeriolysin, the major virulence determinant of *Listeria monocytogenes*. *Mol. Microbiol.* **20**, 119–26.

Haas, A., Dumsky, M. and Kreft, J. (1992). Listeriolysin genes: complete sequence of *ilo* from *listeria seeligeri*. *Biochim. Biophys. Acta* **1130**, 81–4.

Hage-Chahine, C.M., Del Giudice, G., Lambert, P.H. and Pechere, J.C. (1992). Haemolysin-producing *Listeria monocytogenes* affects the immune response to T-cell-dependent and T-cell-independent antigens. *Infect. Immun.* **60**, 1415–21.

Hauf, N., Goebel, W., Fiedler, F., Sokolovic, Z. and Kuhn, M. (1997). *Listeria monocytogenes* infection of P388D1 macrophages results in a biphasic NF-kappaB (RelA/p50) activation induced by lipoteichoic acid and bacterial phospholipases and mediated by IkappaBalpha and IkappaBbeta degradation. *Proc. Natl Acad. Sci. USA* **94**, 9394–9.

Hess, J., Gentschev, I., Miko, D., Wetzel, M., Ladel, C. and Goebel, W. (1996). Superior efficacy of secreted over somatic antigen display in *Salmonella* vaccine induced protection against listeriosis. *Proc. Natl Acad. Sci. USA* **93**, 1458–63.

Israel, A. (1995). A role for phosphorylation and degradation in the control of NF-kappa B activity. *Trends Genet.* **11**, 203–5.

Jacobs, T., Darji, A., Frahm, N., Rohde, M., Wehland, J., Chakraborty, T. and Weiss, S. (1998). Listeriolysin O: cholesterol inhibits cytolysis but not binding to cellular membranes. *Mol. Microbiol.* **28**, 1081–9.

Johnson, M.K., Geoffroy, C. and Alouf, J.E. (1980). Binding of cholesterol by sulfhydryl-activated cytolysins. *Infect. Immun.* **27**, 97–101.

Jones, S., Preiter, K. and Portnoy, D.A. (1996). Conversion of an extracellular cytolysin into a phagosome-specific lysin which supports the growth of an intracellular pathogen. *Mol. Microbiol.* **21**, 1219–25.

Kaufmann, S.H.E. (1993). Immunity to intracellular bacteria. *Annu. Rev. Immunol.* **11**, 129–63.

Kretschmar, M., Nichterlein, T., Chakraborty, T., Aufenanger, J. and Hof, H. (1993). Evidence that listeriolysin is a major inducer of early IL-6 production in *Listeria monocytogenes* infected mice. *Med. Microbiol. Lett.* **2**, 95–101.

Krull, M., Nost, R., Hippenstiel, S., Domann, E., Chakraborty, T. and Suttorp, N. (1997). *Listeria monocytogenes* potently induces up-regulation of endothelial adhesion molecules and neutrophil adhesion to cultured human endothelial cells. *J. Immunol.* **159**, 1970–6.

Lancaster, J.S. and Allen, P.M. (1996). Altered peptide ligand-induced partial T-cell activation: molecular mechanisms and role in T-cell biology. *Annu. Rev. Immunol.* **14**, 1–27.

Lee, K.D., Oh, Y.K., Portnoy, D.A. and Swanson, J.A. (1996). Delivery of macromolecules into cytosol using liposomes containing haemolysin from *Listeria monocytogenes*. *J. Biol. Chem.* **271**, 7249–52.

Leimeister-Wachter, M. and Chakraborty, T. (1989). Detection of listeriolysin, the thiol-dependent haemolysin in *Listeria monocytogenes*, *Listeria ivanovii*, and *Listeria seeligeri*. *Infect. Immun.* **57**, 2350–7.

L'Hopital, S., Marly, J., Pardon, P. and Berche, P. (1993). Kinetics of antibody production against listeriolysin O in sheep with listeriosis. *J. Clin. Microbiol.* **31**, 1537–40.

Liu, S.H., Chu, J.C.J. and Ng, S.Y. (1993). Streptolysin O-permeabilized cell system for studying *trans*-acting activities of exogenous nuclear proteins. *Nucleic Acid Res.* **21**, 4005–10.

Mengaud, J., Vicente, M.F. and Cossart, P. (1989). Transcriptional mapping and nucleotide sequence of the *Listeria monocytogenes* hlyA region reveal structural features that may be involved in regulation. *Infect. Immun.* **57**, 3695–701.

Michel, E., Reich, K.A., Favier, R., Berche, P. and Cossart, P. (1990). Attenuated mutants of the intracellular bacterium *Listeria monocytogenes* by single amino acid substitutions in listeriolysin O. *Mol. Microbiol.* **4**, 3609–19.

Mielke, M.E.A., Ehlers, S. and Hahn, H. (1993). The role of cytokines in experimental listeriosis. *Immunobiology* **189**, 285–315.

Mitsui, K., Sekiya, T., Okamura, Y., Nozawa, Y. and Hase, J. (1979). Ring formation of perfringolysin O as revealed by negative stain electron microscopy. *Biochim. Biophys. Acta* **558**, 307–13.

Morgan, P.J., Hyman, S.C., Byron, O., Andrew, P.W., Mitchell, T. J. and Rowe, A.J. (1994). Modelling the bacterial protein toxin, pneumolysin, in its monomeric and oligomeric form. *J. Biol. Chem.* **269**, 25315–20.

Nakamura, M., Sekino, N., Iwamoto, M. and Ohno Iwashita, Y. (1995). Interaction of theta-toxin (perfringolysin O), a cholesterol-binding cytolysin, with liposomal membranes: change in the aromatic side-chains upon binding and insertion. *Biochemistry* **34**, 6513–20.

Nishibori, T., Xiong, H., Kawamura, I., Arakawa, M. and Mitsuyama, M. (1996). Induction of cytokine gene expression by listeriolysin O and roles of macrophages and NK cells. *Infect. Immun.* **64**, 3188–95.

Owen, R.G.H., Boulnois, G.J., Andrew, P.W. and Mitchell, T.J. (1994). A role in cell-binding for the *C*-terminus of pneumolysin, the thiol-activated toxin of *Streptococcus pneumoniae*. *FEMS Microbiol. Lett.* **121**, 217–22.

Palmer, M., Valeva, A., Kehoe, M. and Bhakdi, S. (1995). Kinetics of streptolysin O self-assembly. *Eur. J. Biochem.* **231**, 388–95.

Palmer, M., Saweljew, P., Vulicevic, I., Valeva, A., Kehoe, M. and Bhakdi, S. (1996). Membrane-penetrating domain of streptolysin O identified by cysteine mutagenesis. *J. Biol. Chem.* **37**, 2378–83.

Palmer, M., Harris, R., Freytag, C., Kehoe, M., Tranum-Jensen, J. and Bhakdi, S. (1998a). Assembly mechanism of the oligomeric streptolysin O pore: the early membrane lesion is lined by a free edge of the lipid membrane and is extended gradually during oligomerization. *EMBO J.* **6**, 1598–605.

Palmer, M., Vulicevic, I., Saweljew, P., Valeva, P., Kehoe, M. and Bhakdi, S. (1998b) Streptolysin O: a proposed model of allosteric interaction between a pore-forming protein and its target lipid bilayer. *Biochemistry* **37**, 2378–83.

Pamer, E.G. (1997). The host cell infected with *L. monocytogenes*. In: *Host Response to Intracellular Pathogens* (ed. S.H.E. Kaufmann), pp. 131–8. R.G. Landes, Austin, TX.

Paterson, Y. and Ikonomidis, G. (1996). Recombinant *Listeria monocytogenes* cancer vaccines. *Curr. Opin. Immunol.* **8**, 664–9.

Pinkney, M., Beachey, E. and Kehoe, M. (1989). The thiol-activted toxin streptolysin O does not require a thiol group for cytolytic activity. *Infect. Immun.* **57**, 2553–8.

Portnoy, D.A., Jacks, P.S. and Hinrichs, D.J. (1988). Role of haemolysin for the intracellular growth of *Listeria monocytogenes*. *J. Exp. Med.* **167**, 1459–71.

Portnoy, D.A., Tweten, R.K., Kehoe, M. and Bielecki, J. (1992a). Capacity of listeriolysin O, streptolysin O, and perfringolysin O to mediate growth of *Bacillus subtilis* within mammalin cells. *Infect. Immun.* **60**, 2710–17.

Portnoy, D.A., Chakraborty, T., Goebel, W. and Cossart, P. (1992b) Molecular determinants of *Listeria monocytogenes* pathogenesis. *Infect. Immun.* **60**, 1263–7.

Prigent, D. and Alouf, J.E. (1976). Interaction of streptolysin O with sterols. *Biochim. Biophys. Acta* **443**, 288–300.

Racz, P., Tenner, K. and Mero, E. (1973). Experimental *Listeria enteritis*. I. An electron microscopic study of the epithelial phase in experimental *Listeria* infection. *Lab. Invest.* **26**, 694–700.

Rossjohn, J., Feil, S.C., McKinstry, W.J., Tweten, R.K. and Parker, M.W. (1997). Structure of a cholesterol-binding, thiol-activated cytolysin and a model of its membrane form. *Cell* **89**, 685–92.

Schlech, W.F., Lavigne, P.M., Bortolussi, R.A., Allen, A.C., Haldane, E.V., Wort, A.J., Hightower, A.W., Johnson, S.E., King, S.H., Nicholls, E.S. and Broome, C.V. (1983). Epidemic listeriosis – evidence for transmission by food. *N. Engl. J. Med.* **308**, 203–6.

Sekino-Suzuki, N., Nakamura, M., Mitsui, K.I. and Ohno Iwashita, Y. (1996). Contribution of individual tryptophane residues to the structure and activity of theta-toxin (perfringolysin O), a cholesterol-binding cytolysin. *Eur. J. Biochem.* **241**, 941–7.

Sekiya, K., Satoh, R., Danbara, H. and Futaesaku, Y. (1993). A ring-shaped structure with a crown formed by streptolysin O on the erythrocyte membrane. *J. Bacteriol.* **175**, 5953–61.

Sekiya, K., Danbara, H., Yase, K. and Futaesaku, Y. (1996). Electron microscopic evaluation of a two-step theory of pore formation by streptolysin O. *J. Bacteriol.* **178**, 6998–7002.

Sibelius, U., Rose, F., Chakraborty, T., Darji, A., Wehland, J., Weiss, S., Seeger, W. and Grimminger, F. (1996). Listeriolysin is a potent inducer of the phosphatidylinositol response and lipid mediator generation in human endothelial cells. *Infect. Immun.* **64**, 674–6.

Smith, G.A. and Portnoy, D.A. (1997). How the *Listeria monocytogenes* ActA protein converts actin polymerization into a motile force. *Trends Microbiol.* **5**, 272–6.

Sun, A.N., Camilli, A. and Portnoy, D.A. (1990). Isolation of *Listeria monocytogenes* small plaque mutants defective for intracelluar growth and cell-to-cell spread. *Infect. Immun.* **58**, 3770–8.

Tang, P., Sutherland, C.L., Gold, M.R. and Finlay, B.B. (1998). *Listeria monocytogenes* invasion of epithelial cells requires the MEK-1/ERK-2 mitogen-activated protein kinase pathway. *Infect. Immun.* **66**, 1106–12.

Tilney, L.G. and Portnoy, D.A. (1989). Actin filaments and the growth, movement and spread of the intracellular bacterial parasite, *Listeria monocytogenes*. *J. Cell Biol.* **109**, 1597–608.

de los Toyos, J.R., Méndez, F.J., Aparicio, J.F., Vázquez, F., del Mar García Suárez, M., Fleites, A., Hardisson, C., Morgan, P.J., Andrew, P.W. and Mitchell, T.J. (1996). Functional analysis of pneumolysin by use of monoclonal antibodies. *Infect. Immun.* **64**, 480–4.

Tweten, R.K. (1995). In: *Pore-forming Toxins of Gram-positive Bacteria*, Vol. 2 (ed. J. Roth), pp. 208–14. American Society for Microbiology, Washington, DC.

Villanueva, M.S., Sijts, A.J.A.M. and Pamer, E.G. (1995). Listeriolysin is processed efficiently into an MHC Class I-associated epitope in *Listeria monocytogenes*-infected cells. *J. Immunol.* **155**, 5227–33.

Walev, I., Palmer, M., Martin, E., Jonas, D., Weller, U., Höhn-Bentz, H., Husmann, M. and Bhakdi, S. (1994). Recovery of human fibroblasts from attack by the pore-forming alpha-toxin of *Staphylococcus aureus*. *Microb. Pathog.* **17**, 187–201.

Walev, I., Palmer, M., Valeva, A., Weller, U. and Bhakdi, S. (1995). Binding, oligomerization, and pore formation by streptolysin O in erythrocytes and fibroblasts membranes: detection of nonlytic polymers. *Infect. Immun.* **63**, 1188–94.

Weiglein, I., Goebel, W., Troppmair, J., Rapp, U.R., Demuth, A. and Kuhn, M. (1997). *Listeria monocytogenes* infection of HeLa cells results in listeriolysin O-mediated transient activation of the Raf-MEK-MAP kinase pathway. *FEMS Microbiol. Lett.* **148**, 189–95.

Weller, U., Müller, L., Messner, M., Palmer, M., Valeva, A., Tranum-Jensen, J., Agrawal, P., Biermann, C., Döbereiner, A., Kehoe, M.A. and Bhakdi, S. (1996). Expression of active streptolysin O in *Escherichia coli* as a maltose-binding-protein streptolysin O fusion protein. *Eur. J. Biochem.* **236**, 34–9.

Whitley, M.Z., Thanos, D., Read, M.A., Maniatis, T. and Collins, T. (1994). A striking similarity in the organization of the E-selectin and beta interferon gene promoters. *Mol. Cell Biol.* **14**, 6464–75.

Yoshikawa, H., Kawamura, I., Fujita, M., Tsukada, H., Arakawa, M. and Mitsuyama, M. (1993). Membrane damage and interleukin-1 production in murine macrophages exposed to listeriolysin O. *Infect. Immun.* **61**, 1334–9.

SECTION III

OTHER TOXINS OF CLINICAL, PHARMACOLOGICAL, IMMUNOLOGICAL AND THERAPEUTIC INTEREST

29

Enterotoxigenic *Escherichia coli* heat-stable toxins

J. Daniel Dubreuil

INTRODUCTION

Although most *Escherichia coli* strains live as symbiotic organisms in the intestine of animals including humans, certain pathogenic strains belonging to distinct groups are now defined on the basis of specific virulence factors that they produce. Pathogenic strains belonging to the different groups cause diarrhoea via diverse mechanisms. Enterotoxigenic *Escherichia coli* (ETEC) represents one of these groups. ETEC strains are an important cause of traveller's diarrhoea and diarrhoeal illnesses in children in developing countries. In addition to causing human diseases, specific ETEC strains are also responsible for severe diarrhoea in domesticated and in some wild animals (Caprioli *et al.*, 1991; Sears and Kaper, 1996).

ETEC pathogenesis comprises two steps. Firstly, the micro-organism has to colonize the small intestine by means of specific adherence factors. Fimbrial or non-fimbrial adhesins that permit attachment to enterocytes of specific animal hosts and hence contribute to intestinal colonization are produced by ETEC. These adhesins enable ETEC to attach to receptors on the intestinal epithelium of susceptible hosts, to overcome physiological defences such as peristalsis. Once established, ETEC strains produce one or more enterotoxins. Following interaction with specific receptors on the surface of the enterocytes, the enterotoxins will cause secretory diarrhoea. Adherence of ETEC to intestinal epithelial cells is crucial for establishment of the

pathogen and also for efficient delivery of toxins to their targets. Minimal histopathological lesions are observed in the intestinal mucosa of animals infected by ETEC. In fact, microscopic observations reveal a layer of adherent bacteria covering uniformly the brush border epithelium of the small intestine without visible alteration of the tissue.

ETEC cause diarrhoea by production of two distinct types of enterotoxin, heat-labile (LT) toxin with two subtypes (LTI and LTII) and a family of heat-stable (ST) enterotoxins. LT is a high-molecular-mass toxin (85 kDa) functionally and structurally related to *Vibrio cholerae* enterotoxin (Sprangler, 1992). STs are low-molecular-mass toxins that share the property of retaining toxic activity after incubation at 100°C for 30 min, whereas the activity of LT is abrogated under the same conditions. Two types of ST, STa and STb (also known as STI and STII), based on solubility in methanol and activity in infant mice, were described. STa is soluble in methanol and active in infant mice, whereas STb is insoluble in methanol and not active in infant mice. STa has further been subdivided into STaH (or STIb) and STaP (or STIa), so named as they were isolated originally from a human or a porcine ETEC strain, respectively. These toxins are slightly different with respect to their sequence and number of amino acids. STb has no homology with STa enterotoxin at either the gene or the protein level. This chapter will be devoted to enterotoxigenic *E. coli* heat-stable toxins as knowledge now stands.

STa TOXIN

STa polypeptide

STa represents a family of toxins of approximately 2000 Da composed of a single peptide chain (Dreyfus *et al.*, 1983; Thompson and Gianella, 1985). Toxins produced by human and porcine strains differ slightly. The STaH primary structure comprises 19 amino acid residues compared with 18 for STaP. The amino acid sequences of the two toxin subtypes are not identical (Figure 29.1). To date, STaP polypeptide has been observed in isolates from animal species including pigs, calves, lambs, chickens and horses and also from humans. In contrast, STaH is produced solely by human isolates. These toxins share a highly conserved sequence of 15 amino-acid residues which resides in the carboxy-terminus. This sequence represents a common antigenic determinant. The amino-terminal amino acids to the first cysteine can be cleaved from the molecule and the remaining peptide is fully active (Staples *et al.*, 1980).

Both toxins are synthesized as larger precursors (pre-pro-STa) that are subsequently cleaved to the active mature toxin. Six cysteine residues involved in disulfide bond formation are present at the same position in both STaH and STaP (Figure 29.2). The tertiary structure formed by the disulfide bonds is critical and required for full biological activity (Okamoto *et al.*, 1987). Chemically synthesized STa behaves in a similar manner to ETEC-produced STa (Takeda *et al.*, 1991; Gyles, 1994).

Native STa toxins are poorly immunogenic. However, they can be conjugated to carrier proteins in order to produce polyclonal antisera and monoclonal antibodies (Brandwein *et al.*, 1985; Aitken and Hirst, 1993). The toxicity of STaH and STaP can be neutralized by homologous but also by heterologous antisera (Takeda *et al.*, 1983).

	Number of amino acids	NH₂-terminus	COOH-terminus	References
***Escherichia coli* STaH**	19	N - S - S - N - Y - **C** - **C** - E - L - **C** - **C** - N - P - A - C - T - G - **C** - Y		Aimoto *et al.* (1982)
***Escherichia coli* STaP**	18	N - T - F - Y - **C** - **C** - E - L - **C** - **C** - N - P - A - C - A - G - **C** - Y		Takao *et al.* (1983)
Citrobacter freundii	18	N - T - F - Y - **C** - **C** - E - L - **C** - **C** - N - P - A - C - A - G - **C** - Y		Guarino *et al.* (1989)
***Yersinia enterolitica* STa**	30	...S S D Y D - **C** - **C** - D - Y - **C** - **C** - N - P - A - C - A - G - **C**		Takao *et al.* (1985)
***Vibrio cholerae* non-01**	17	I - D - **C** - **C** - E - I - **C** - **C** - N - P - A - C - F - G - **C** - L - N		Yoshimura *et al.* (1986)
***Vibrio cholerae* non-01** Hataka strain	18	L - I - D - **C** - **C** - E - I - **C** - **C** - N - P - A - C - F - G - **C** - L - N		Arita *et al.* (1991ᵃ)
Vibrio mimicus	17	I - D - **C** - **C** - E - I - **C** - **C** - N - P - A - C - F - G - **C** - L - N		Arita *et al.* (1991ᵇ)
***Escherichia coli* EAST-1**	38	...A - S - S - Y - A - S - **C** - I - W - **C** - T - - - T - A - C - A - S - **C** - H - G...		Savarino *et al.* (1993)
Conus geographus α-Conotoxin Gᵢ	13	E - **C** - **C** - N - P - A - C - G - R - H - Y - S - **C**		Gray *et al.* (1981)
Guanylin (Human)	15	P - G - T - **C** - E - I - **C** - A - Y - A - A - C - T - G - **C**		Greenberg *et al.* (1997)
(Rat)	15	P - N - T - **C** - E - I - **C** - A - Y - A - A - C - T - G - **C**		"
Uroguanylin (Human)	15	N - D - D - **C** - E - L - **C** - V - N - V - A - C - T - G - **C** - L		"
(Rat)	15	T - D - E - **C** - E - L - **C** - I - N - V - A - C - T - G - **C**		"

FIGURE 29.1 Primary sequences of STaH and STaP compared to other heat-stable enterotoxins of bacterial origin. The sequence of *Conus geographus*, α-conotoxin Gᵢ, is represented as it demonstrates some homology to STa. Guanylin and uroguanylin sequences of human and rat origin are also aligned for comparison. The number of amino acids in the mature molecules is indicated. In bold are represented the conserved sequence found in most heat-stable toxins. In the box is the consensus N-P-A-C sequence conserved at least partly between all the molecules appearing in the figure.

(a)

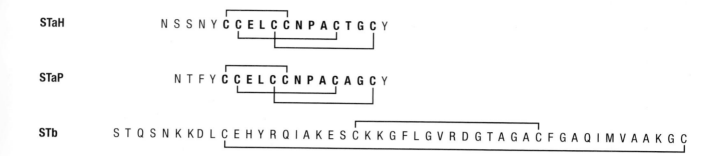

(b)

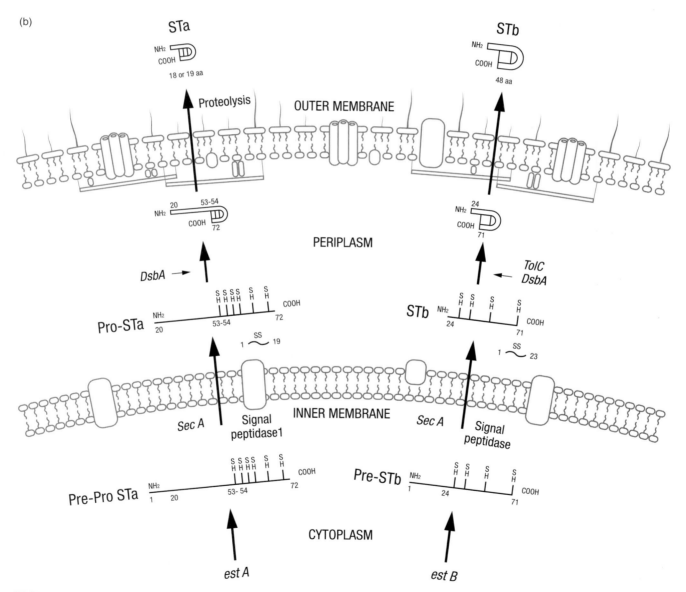

FIGURE 29.2 (a) Primary sequences of STaH, STaP and STb showing the disulfide bond structure. In bold are the 13 amino acid residues constituting the toxic domain of STaH and STaP. (b) Disulfide bond formation and secretory pathways as determined for STa and STb toxins.

Biochemical characteristics

STa enterotoxin is a heat-stable molecule and its small size is responsible for this biochemical characteristic. It is an acidic peptide with a pI of 3.98 (Dreyfus *et al.*, 1983). The molecule is soluble in water and organic solvents including methanol and resists several proteases such as pronase, trypsin and chymotrypsin. It is resistant to acidic but not to basic pH. The toxin is completely inactivated by reducing and oxidizing agents which disrupt disulfide bonds (Dreyfus *et al.*, 1983).

estA gene

The genes encoding STa are found on plasmids of varying molecular sizes (Harnett and Gyles, 1985). For animal ETEC isolates, it is common to find the gene coding for STa, colonization factors, drug resistance and production of colicin on the same plasmid. For human isolates, the same plasmid carries *estA* and an antibiotic resistance gene (Gyles *et al.*, 1977; Echeverria *et al.*, 1978). Genetic studies demonstrated the existence of two types of *estA*. The gene was first cloned from a bovine isolate (STaP) and found to be part of a transposon (Tn *1681*) which is flanked by inverted repeats of IS*1* (So and McCarthy, 1980). STaP genes from ETEC isolated from other animal species (including humans) are also part of the same transposon (Sekizaki *et al.*, 1985). Finally, genes encoding STaH and STaP may be carried by a single human ETEC strain. Synthesis of STa by *E. coli* is subject to catabolite repression and optimal yields of toxin are obtained in glucose-free media (Alderete and Robertson, 1977; Stieglitz *et al.*, 1988). A study by Sommerfelt *et al.* (1989) revealed that loss of STa production is the result of deletions of DNA fragments harbouring the toxin gene rather than loss of plasmids.

Secretion of STa and formation of disulfide bonds

STa is produced as a 72 amino acid precursor molecule referred to as pre-pro-STa (Stieglitz *et al.*, 1988; Okamoto and Takahara, 1990). The 72 amino acid polypeptide consists of a 19 amino acid signal peptide (pre-STa), a 35 amino acid pro sequence and then an 18 or 19 amino acid mature STa (Figure 29.2b). Whereas the pro-STa is translocated across the cytoplasmic membrane and seems to require *secA*-dependent transport, the signal sequence is cleaved by signal peptidase 1 (Rasheed *et al.*, 1990). Yamanaka and Okamoto (1996) substituted the charged amino acids at positions 29 to 31 of the pro-region of STa. Hydrophobic or basic

residues significantly reduced STa translocation across the inner membrane, indicating that a net negative charge near amino acid 30 is important for efficient translocation. The pro-region was shown to guide STa into the periplasmic space (Yamanaka *et al.*, 1993). However, this region does not seem to be involved in transport of the peptide extracellularly (Okamoto and Takahara, 1990).

In the periplasm, the three intramolecular disulfide bonds, important for toxicity, are formed by a protein (DsbA) prior to secretion (Yang *et al.*, 1992; Sanchez *et al.*, 1993). In the pro-region the cysteine-39 residue is important for the correct recognition by DsbA and proteases during maturation of STa (Yamanaka *et al.*, 1994). A negative charge at position 7 in mature STaP (Glu) was shown to be required for toxin interaction with DsbA in the periplasm and formation of the intramolecular disulfide bonds *in vivo* (Yamanaka *et al.*, 1998).

A second proteolysis event occurs extracellularly owing to an undefined protease to produce biologically active 18 and 19 amino acids STa (Rasheed *et al.*, 1990). STa toxin is secreted from the cell as it is synthesized and is not cell associated. Mature STa diffuses across the outer membrane (Yang *et al.*, 1992; Yamanaka *et al.*, 1994). Yamanaka *et al.* (1997) demonstrated, by mutating every cysteine residue of STa, that the formation of the three intramolecular disulfide bonds is not absolutely necessary for the mature toxin to pass through the outer membrane.

The three intramolecular disulfide bonds in STaH link cysteines 6 and 11, 7 and 15, and 10 and 18; in STaP disulfide bonds link cysteines 5 and 10, 6 and 14 and 9 and 17 (Figure 29.2a). The use of STa analogues demonstrated that the second disulfide bond was essential for toxicity whereas analogues lacking the first or the third disulfide bond showed only a reduced toxicity (Hidaka *et al.*, 1991). In conclusion, the disulfide bonds between cysteines 7 and 15 of STaH or 6 and 14 of STaP are essential for biological activity.

Structure of STa and identification of the toxic domain

The three-dimensional structure of STa was determined using nuclear magnetic resonance (NMR) spectroscopy and X-ray crystallography. The tertiary structure consists of a folded peptide backbone assembled as a right-handed spiral from the first cysteine at the NH$_2$-terminus to the last cysteine residue at the COOH-terminus (Ozaki *et al.*, 1991). Three β-turns are located along this spiral and are stabilized by the three intramolecular disulfide bonds (Gariépy *et al.*, 1986, 1987). The disulfide bonds are responsible for the spatial structure of STa and this 3D structure is required for toxicity. The STa molecule appears to be hydrophobic, but there are three hydrophilic areas (i.e. Cys^6-Leu^8, Cys^{10}-Asp^{11} and Pro^{12}-Cys^{17}) in STaP (Ozaki et al., 1991).

Overall, a 13 amino acid sequence from the amino-terminal cysteine to the carboxyl-terminal cysteine is essential for toxic activity of both STaH and STaP (Yoshimura et al., 1985; Waldman and O'Hanley, 1989; Carpick and Gariépy, 1991). Thus, this segment constitutes the toxic domain of STa (Figure 29.2a). For this region, a striking identity is shared with other heat-stable enterotoxins secreted by other enteric pathogens such as *Citrobacter freundii, Yersinia enterocolitica, Vibrio cholerae* non-01 strains *and Vibrio mimicus* (Figure 29.1). The second β-turn at residues 11 to 14 (STaP) is proposed to be the most important for toxicity of STa as this region of the molecule interacts directly with its receptor (Ozaki et al., 1991; Sato et al., 1994). More specifically, alanine at position 13 in STaP and at position 14 in STaH is essential for toxicity. The methyl side-chain of this alanine is directed outward of the molecule and could interact directly with a hydrophobic area located on the surface of the receptor.

The four amino acids (N-P-A-C) are conserved in STaP and STaH and other heat-stable enterotoxins produced by other bacterial species. This sequence is also conserved in the marine snail *Conus geographus* α-conotoxin G_I. In guanylin and uroguanylin and *E. coli* EAST-1 toxin, an alanine followed by a cysteine residue is partially conserved. It is interesting to note that in guanylin, the endogenous ligand for STa receptor, these amino acids interact with the receptor (Greenberg et al., 1997).

Receptor identity

Since the mid-1980s, numerous studies have demonstrated that STa can bind non-covalently, with high or low affinity, to several proteins in the plasma membrane of susceptible eukaryotic cells (Dreyfus and Robertson, 1984; Waldman et al., 1986).

STa stimulates particulate guanylate cyclase in intestinal epithelial cells resulting in elevation of cyclic GMP (cGMP) (Field et al., 1978; Hughes et al., 1978). The same studies also demonstrated that 8-bromo-cGMP mimics the effect of STa in animal systems. Thus, the activity of STa is mediated by the intracellular second messenger cGMP and research has focused on evaluation of the binding of STa to cellular proteins and on demonstration of the possible guanylate cyclase activity of the molecule.

The STa receptor has been the subject of intensive studies. Several solubilization and cross-linking experiments using ^{125}I-STa and intestinal brush border membranes, aimed at identifying the STa receptor were

carried out. For example, de Sauvage et al. (1992) have identified proteins of 49, 56, 68, 81, 133 and 153 kDa, whereas Hirayama et al. (1992) identified a 200-kDa protein cross-linked to radioiodinated STa. Purification of the receptor by ligand-affinity chromatography identified a protein subunit of 74 kDa to which ^{125}I-STa was binding (Hughes et al., 1992). The molecules identified in these studies were glycoproteins but none showed guanylate cyclase activity as expected. The heterogeneity of the proposed intestinal receptors for STa is exemplified by recent additional studies (Thompson and Gianella, 1990; Ivens et al., 1990; Hughes et al., 1991; Cohen et al., 1993; Hakki et al., 1993).

Conclusive identification of STa receptor was achieved when the STa receptor was cloned from cDNA libraries of rat (Schulz et al., 1990), pig (Wada et al., 1994) and human intestine (de Sauvage et al., 1991). New members of the guanylyl cyclase family were looked for in the small intestinal mucosa using degenerated oligonucleotide primers based on conserved sequences in the catalytic domains of membrane and soluble guanylyl cyclases. When the cDNA was transfected into naive COS cells, specific STa binding activity and guanylate cyclase activity were expressed (Schulz et al., 1990). The deduced amino acid sequence and functional expression in mammalian cells indicated that the STa receptor is guanylyl cyclase C (GC-C) belonging to the atrial natriuretic peptide receptor family that includes GC-A and GC-B (Chang et al., 1989; Hirayama et al., 1992; Vaandrager et al., 1993). For these receptors, the binding site and the catalytic activity of guanylate cyclase reside on the same protein.

The GC-C unglycosylated protein has a M_r of 120 000 and is found as a 140–160-kDa glycoprotein after N-linked glycosylation (Mann et al., 1993). At the NH_2-terminus, it consists of an extracellular receptor domain, a transmembrane domain, and a cytoplasmic domain including a kinase homology domain and a guanylyl cyclase catalytic domain at the COOH-terminus. GC-C null mice are refractory to the secretory action of STa, proving that the GC-C receptor is necessary for the diarrhoeal response induced by this toxin (Mann et al., 1997).

Thus, STa activates particulate guanylate cyclase. Particulate guanylate cyclases are brush border membrane glycoproteins. In the human intestine, approximately 75–80% of the total guanylate cyclase activity is particulate, while the rest is soluble (Garbers, 1991; Schulz et al., 1991).

GC-C also contains a carboxy-terminal tail of about 60 amino acids. The function of this segment is uncertain but it possibly links the molecule to the cytoskeleton. The endogenous agonist for GC-C was later found to be a 15 amino acid hormone called guanylin

(Greenberg *et al.*, 1997). This hormone appears to play a role in the regulation of fluid and electrolyte absorption in the gut. Guanylin is 50% homologous to STa and it contains four cysteine residues that are involved in disulfide bond formation and are essential for biological activity. This hormone is less potent than STa in activating GC-C and in stimulating chloride (Cl⁻) secretion (Currie *et al.*, 1992; Wiegarrd *et al.*, 1992; Carpick and Gariépy, 1993; Forte *et al.*, 1993; Kuhn *et al.*, 1994; Greenberg *et al.*, 1997). It thus appears that STa opportunistically utilizes GC-C to alter the basal gut homeostasis.

A form of guanylin circulates in the blood of patients with chronic kidney failure (Kuhn *et al.*, 1993). There also exists in animal and human urine a guanylin-like molecule called uroguanylin (Hamra *et al.*, 1993; Kita *et al.*, 1994). It is clear from their sequences that guanylin and uroguanylin are the products of separate genes. It thus appears that guanylin but also uroguanylin are endogenous ligands for GC-C and could be modulators of Cl⁻ secretion not only in the intestine but also in kidneys and other organs (Giannella, 1995; Greenberg *et al.*, 1997). The receptor for these molecules is located on enterocytes, colonocytes and cells of various extraintestinal tissues.

In vitro studies involving cell lines and *in vivo* analysis of the receptor distribution

Intestinal cell lines were used to study the *in vitro* binding of STa and its biological action. For example, both human cancer cell lines Caco-2 and T84 possess the specific receptor for STa (Guarino *et al.*, 1987a; Cohen *et al.*, 1993). Binding of STa to these cell lines is coupled to cGMP production as was previously observed in the rat. When T84 polarized cells are mounted in a Ussing chamber they can be induced to secrete Cl⁻ by addition of STa (Huott *et al.*, 1988). Other intestinal cell lines such as IEC-6 (rat cell line) can also bind STa; however, the receptor is not coupled to guanylate cyclase (Thompson and Giannella, 1990). Nevertheless, the relevance for humans of the data first obtained with rats was confirmed by studying the normal human intestinal mucosa (Cohen *et al.*, 1986; 1988).

These studies revealed that STa receptors are found throughout the human small intestine and colon with a decreasing number along the longitudinal axis of the gut (Krause *et al.*, 1994). Specific binding of radioiodinated STa was noted in both crypts and villi of the small intestine and in crypts and on the surface epithelium of the colon (Field *et al.*, 1989a, b). Binding was also maximal in the villus preparations and decreased along the villus-to-crypt axis (Cohen *et al.*, 1992).

A relationship was found between the number of STa receptors and age. The number was highest at birth, decreased in the first 1–3 weeks of life and was stable thereafter (Cohen *et al.*, 1988). The large number of receptors observed in infants could account for the severity of diarrhoea due to STa in young children (Guarino *et al.*, 1987b). A similar age-related difference in the number of STa receptors had already been observed in rats (Cohen *et al.*, 1986).

Recently, a study by Swenson *et al.* (1996) indicated that GC-C was present in mouse intestinal crypts and villi of adult small intestine and surface epithelium of the colon. The presence of GC-C in mouse intestinal crypts supports the putative role of GC-C in fluid secretion and electrolyte homeostasis and resembles the pattern seen with human tissues.

Alternative STa receptor

A number of smaller STa-binding proteins previously identified cross-reacts with anti-GC-C antibodies, indicating that these polypeptides could be the result of GC-C extracellular binding domain proteolysis (Cohen *et al.*, 1993; Vaandrager *et al.*, 1993; Scheving and Chong, 1997). This observation can most probably account for numerous proteins with varying M_r previously described as STa receptor. In fact, similar lower molecular mass STa-binding proteins were also observed in cells transfected to express GC-C, supporting the occurrence of processing and/or proteolysis of a single STa binding molecule (de Sauvage *et al.*, 1992; Hirayama *et al.*, 1992; Vaandrager *et al.*, 1993).

Now, at least one receptor for STa has been clearly identified as being GC-C. Nevertheless, the existence of other receptors for STa was suggested by different studies. Non-GC-C-linked STa receptors may also exist and these could be linked to more than a single signal transduction system, as will be discussed later.

Mechanism of action

STa is a highly potent toxin with a rapid action (attaining maximal levels of cGMP within 5 min) but of short duration. For example, 6 ng of STa results in a positive fluid response in mouse intestine compared with 200 ng of STb or CT in the same model (Hitotsubashi *et al.*, 1992a). STa (and guanylin) binds to particulate guanylate cyclase in the jejunum and ileum brush border leading to elevation of cGMP levels. GC-C is found primarily, but not exclusively, in intestinal cells lining the small intestine (Krause *et al.*, 1994).

STa enterotoxin utilizes GC-C to alter ion transport in the gut. Following binding of STa to its receptor a cascade of events ensues (Figure 29.3). The end result of STa action is the inhibition of Na⁺-coupled-Cl⁻ absorption in villus tips and stimulation of electrogenic Cl⁻ secretion in crypt cells, resulting in a net fluid secretion in the lumen of the intestine (Field *et al.*, 1978; Dreyfus *et al.*, 1984; Huott *et al.*, 1988; Forte *et al.*, 1992).

Inhibition of NaCl absorption could be attributed to binding of STa to enterocytes villi responsible for NaCl absorption (Cohen *et al.*, 1992; Almenoff *et al.*, 1993). Treatment of Caco-2 cells with STa prevented uptake of taurine which is coupled to Na⁺ absorption. This mechanism could also contribute to decreasing the absorption of sodium (Brandsch *et al.*, 1995).

GC-C is composed of a receptor, a transmembrane domain, a kinase-like domain and a catalytic domain. In the basal state, GC-C is a homotrimer. After binding of a STa molecule, a bond between two of the three subunits is stabilized, as activation of GC-C may require the interaction of STa with two cyclase domains. Thus, internal dimerization within the homotrimeric GC-C appears to be a key step in activation of guanylate cyclase by STa (Vaandrager *et al.*, 1994). Rudner *et al.* (1995) recently proposed that the binding of STa to GC-C removes the inhibitory effects of the kinase-homology domain and promotes an interaction bet-

ween the GC catalytic domains. Adenine nucleotides were shown to regulate the binding of STa to its receptor. They actually decrease binding but stimulate guanylate cyclase activity (Gazzano *et al.*, 1991; Katwa *et al.*, 1992). Activation of GC-C results in an increased level of cGMP leading to a net fluid secretion involving Cl⁻.

An interaction between protein kinase C and the STa receptor was suggested by Weikel *et al.* (1990), who showed that treatment with phorbol ester, an active analogue of PKC, doubled STa-stimulated cGMP production in cultured intestinal cells. This increase was also observed by Crane *et al.* (1990) in STa-stimulated T84 cells. Crane *et al.* (1992) later demonstrated that treatment of T84 cells with purified PKC similarly increased STa-stimulated guanylyl cyclase activity. This increased response was blocked by PKC inhibitors and by ATP-S, a biologically non-hydrolysable ATP analogue, indicating that PKC regulated the STa receptor through a phosphorylation step. They also presented evidence that PKC could phosphorylate and activate the STa receptor both *in vivo* and *in vitro*. In addition, using synthetic peptides, they have demonstrated that Ser¹⁰²⁹ of the STa receptor is most probably a phosphorylation site for PKC. A study by Wada *et al.* (1996)

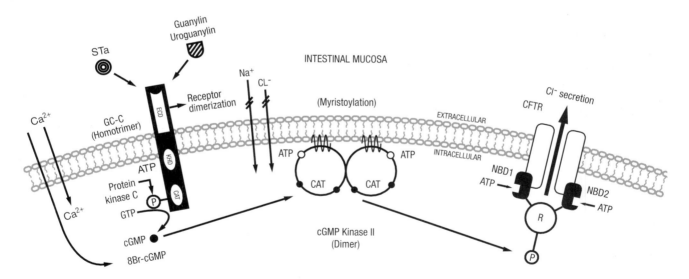

FIGURE 29.3 Molecular model for the induction of a secretory response by STa toxin. STa and guanylin (uroguanylin) interact with guanylate cyclase C. GC-C is a homotrimer composed of a receptor domain (ECD), a transmembrane domain, a kinase-homology domain (KHD) and a catalytic domain (CAT). Interaction of STa with the receptor domain results in dimerization of two of the three CAT domains. Following STa-receptor interaction, Ca²⁺ influx occurs and this event activates protein kinase C which phosphorylates CAT. Conversion of GTP to cGMP takes place. cGMP or 8-Br-cGMP but not 8-Br-cAMP can then activate a cGMP-dependent protein kinase II (cGMP kinase II) which is a dimer molecule linked to the membrane via myristoyl molecules. The catalytic (CAT) domain of cGMP kinase II phosphorylates the R (Regulatory) domain of CFTR (cystic fibrosis transmembrane conductance regulator), a Cl⁻ conductory pore. The R domain interacts with NDB1 and NDB2 (nucleotide binding domains). Then, Cl⁻ secretion is achieved via the CFTR channel. The Na⁺–Cl⁻-coupled influx is blocked. The cGMP-signalling enzyme (GC-C), the signal transducer (cGMP kinase II) and the effector (CFTR) are co-localized on the membrane of intestinal epithelial cells.

supported this hypothesis. It can thus be concluded that PKC activity in target cells modulates the cellular sensitivity to *E. coli* STa enterotoxin.

In vivo, the increase in cGMP results in a cGMP-dependent protein kinase activation of a linear chloride channel known as CFTR (cystic fibrosis transmembrane conductance regulator). CFTR is a Cl⁻ channel encoded by the cystic fibrosis gene. The role of CFTR in secretion was confirmed by the observation of a lack of potential difference response but not of a cGMP response to STa in the colon of cystic fibrosis (CF) patients (Goldstein *et al.*, 1994). In addition, CFTR-knockout mice were insensitive to STa (Pfeifer *et al.*, 1996).

The CFTR protein contains two transmembrane domains forming the Cl⁻ conducting pore, two nucleotide-binding domains, and a regulatory domain where multiple phosphorylation sites for cAK (Gadsby and Nairm, 1994) and cGK (French *et al.*, 1995), which are cAMP- and cGMP-dependent kinases, respectively, are present.

In vivo, following binding of STa to its receptor, only cyclic GMP-dependent kinase type II (cGKII) can effectively phosphorylate the R domain and gate the CFTR Cl⁻ channel (French *et al.*, 1995; Vaandrager *et al.*, 1997a). cGKII is a membrane-associated cGMP-dependent kinase whereas cGKI is a soluble kinase. cGMP cross-activation of cAK does not occur in native intestinal epithelium but is observed *in vitro* in cGK-deficient T84 intestinal tumour cell lines (Forte *et al.*, 1992; Lin *et al.*, 1992; Tien *et al.*, 1994).

Recent studies strongly support a central role of cGKII and not cGKI or cAK as the signal transducer in STa- and cGMP-induced Cl⁻ secretion (Vaandrager *et al.*, 1997b). In fact, synthesis of endogenous cGKII in crypt cells correlates well with the ability of STa and 8-Br-cGMP (and not 8-Br-cAMP) to stimulate Cl⁻ secretion (Markert *et al.*, 1995). STa could not stimulate Cl⁻ secretion in the colon, where cGKII is not found. Homozygous cGKII knockout mice, and rat intestine pretreated with cGKII inhibitors have lost the potential for intestinal Cl⁻ secretion induced by STa and 8-Br-cGMP. A native membrane environment appears to be crucial for cGKII-selective phosphorylation of CFTR. The amino-terminal myristoyl groups present in cGKII dimer could be responsible for bringing cGKII and CFTR in close apposition so that the dimer can effectively phosphorylate and activate the channel, ultimately resulting in Cl⁻ secretion (Vaandrager *et al.*, 1996, 1997b).

The cGMP signalling enzyme (GC-C), the signal transducer (cGMP-dependent protein kinase II) and the effector (CFTR) are co-localized in the apical membrane of intestinal cells (Markert *et al.*, 1995).

STa action via alternative pathways

Non-GC-C STa receptor, if present, may be linked to alternative signalling pathways. Although the CFTR Cl⁻ channel accounts for the main secretory pathway stimulated by STa, there is still controversy as to whether cGMP is the sole signal transducer. In fact, as different pharmaceutical agents could alter the secretory response of STa, alternative pathways responsible for fluid secretion have been proposed (Table 29.1). Overall, to date, important major contradictory results have been reported from different research teams, adding to the perplexity of the situation. For example, as STa-secretory response was completely abolished by 5-hydroxytryptamine (5-HT or serotonin) receptor antagonists, this secretagogue has been proposed as an important mediator of secretion (Beubler *et al.*, 1992). Interestingly, in these experiments, the cGMP response was not influenced. Thus, 5-HT could possibly mediate STa secretion by affecting prostaglandin synthesis (5HT-2 receptors) and/or neurons (5HT-3 receptors). In support of this mechanism of action, it was shown that the cyclooxygenase inhibitor indomethacin caused an inhibition of STa-mediated fluid secretion in suckling mice (Madsen and Knoop, 1978). However, this inhibition was not observed by Kuhn *et al.* (1994) in the human intestine. Thus, the potential role of 5-HT and prostaglandin or leukotrienes remains controversial.

TABLE 29.1 STa proposed alternative mechanisms of action

	References
5-Hydroxytryptamine (serotonin)	Beubler *et al.* (1992)
	Madsen and Knoop (1978)
Prostaglandins/leukotrienes	Greenberg *et al.* (1982)
	Hayden *et al.* (1996)
	Thomas and Knoop (1982)
	Greenberg *et al.* (1982)
Calcium/PKC	Thompson and Knoop (1982)
	Banik and Ganguly (1988, 1989)
	Chaudhuri *et al.* (1993)
	Chaudhuri and Ganguli (1995)
	Guanguly and Talukder (1985)
	Knoop *et al.* (1991)
	Chaudhuri *et al.* (1993)
Neural activity	Eklund *et al.* (1985)
	Mathias *et al.* (1982)
	Rolfe and Levin (1994)
	Giannella *et al.* (1983)
	Roussel *et al.* (1992)
F-actin/C fibre	Matthews *et al.* (1993)
	Nzegwu and Levin (1996)
	Rolfe and Levin (1994)

In fact, studies involving prostaglandin synthesis inhibitors and isolated rat intestinal epithelial cells suggested that STa secretion could result from phosphatidylinositol and diacylglycerol release coupled with elevation of intracellular calcium levels and activation of protein kinase C (Greenberg *et al.*, 1982; Thompson and Knoop, 1982; Ganguly and Talukder, 1985; Banik and Ganguly, 1988, 1989; Knoop *et al.*, 1991; Chaudhuri *et al.*, 1993; Chaudhuri and Ganguly, 1995).

In contrast, the results of the study by Dreyfus *et al.* (1984) refuted the evidence for a role of prostaglandins and leukotrienes, calcium and/or calmodulin in STa secretory action. Supporting this study it was shown that STa increased cGMP levels in T84 cells resulting in Cl$^-$ secretion, although no elevation of calcium levels and no hydrolysis of phospholipids was noted (Huott *et al.*, 1988; Vajanaphanich *et al.*, 1993).

STa was shown to cause changes in the myoelectric activity of the small intestine, indicating that the enteric nervous system could play a role in STa secretion. Following the use of neuronal inhibitors, a reduction in STa secretory response was observed *in vivo* and *in vitro* (Mathias *et al.*, 1982; Eklund *et al.*, 1985; Rolfe and Levin, 1994). The observed changes in myoelectric activity could result in the loss of normal peristalsis (Gianella *et al.*, 1983; Roussel *et al.*, 1992).

Finally, the secretory response of T84 cells to STa was reported to involve microfilament (F-actin) rearrangement at the basal pole of these polarized cells (Matthews *et al.*, 1993). More recently, Rolfe and Levin (1994) showed that STa activated electrogenic Cl$^-$ secretion through a myenteric secretory reflex with afferent C fibre component. A study by Nzegwu and Levin (1996) corroborated this observation.

For now, no definitive consensus can be reached with respect to the alternative pathways that STa could use to stimulate fluid secretion.

STb TOXIN

STb polypeptide

In contrast to STa, no variant of the toxin STb has been found. To date, the enterotoxins isolated from different animal species show the same nucleotide and amino acid sequences. Also, only one antigenic type exists. STb toxin has be observed in *E. coli* strains isolated principally from pigs but also from cattle (including water buffaloes), humans and chickens (Dubreuil, 1997).

The mature STb polypeptide comprises 48 amino acids containing four cysteine residues involved in disulfide bridge formation. The enterotoxin has an M_r

of 5200. As already stated, STb bears no homology to STa enterotoxin (Dreyfus *et al.*, 1992). The STb polypeptide is synthesized as a 71 amino acid precursor comprising a 23 amino acid signal sequence (Lee *et al.*, 1983; Picken *et al.*, 1983). Sukumar *et al.* (1995) indicated that the peptide spanning from Cys^{33} to Cys^{71} has full biological activity and that the first seven amino acids at the NH$_2$-terminus of the mature toxin are not involved in either the structure or toxicity of STb.

STb is a poorly immunogenic molecule but antibodies can be produced against the mature toxin provided numerous booster injections are given. However, a better serological response can be obtained following immunization by either fusion proteins or proteins chemically coupled to STb (Urban *et al.*, 1990, 1991; Dubreuil *et al.*, 1996). In the latter case, neutralizing antibodies can be obtained but this occurs rarely when the native toxin is used for immunization. The anti-STb antibodies can neutralize STb toxicity but are unable to neutralize STa or CT toxins (Hitotsubashi *et al.*, 1992a).

Biochemical characteristics

For STb, the determined isolelectric point of 9.6 corresponds to a highly basic protein (Handl *et al.*, 1993). STb is insoluble in methanol and the toxin loses biological activity following β-mercaptoethanol or trypsin treatment (Burgess *et al.*, 1978; Dubreuil *et al.*, 1991; Fujii *et al.*, 1991). However, it resists acid (pH 2), alkaline (pH 12) and 8 M urea treatments (Dubreuil *et al.*, 1991). STb is very susceptible to protease degradation, particularly by trypsin-like enzymes (Whipp, 1987). Out of 48 residues, the mature toxin contains one tyrosine, two phenylalanines and no tryptophan; this results in a low absorbance at 280 nm (Handl *et al.*, 1993).

estB gene

The *estB* gene that encodes STb is found on heterogeneous plasmids that may also code for other properties including other enterotoxins (i.e. LT, STa), colonization factors, drug resistance, colicin production and transfer functions (Harnett and Gyles, 1985; Echeverria *et al.*, 1985). The *estB* gene is part of a transposon of approximately 9 kb designated Tn*4521* (Lee *et al.*, 1985; Hu *et al.*, 1987; Hu and Lee, 1988). This transposon is flanked by defective IS2 elements but is nevertheless functional as the STb gene can transpose from one plasmid to another (Lee *et al.*, 1985; Hu and Lee, 1988). The structural gene for STb from different clinical isolates appears to be uniform in size but the flanking sequences are heterogeneous, suggesting that *estB* could be found on different transposons (Lee *et al.*, 1985). Thus it seems that transposition of *estB* is a mechanism by which this virulence factor is disseminated among ETEC. Spandau and Lee (1987) reported that the promoter for *estB* expression was weak. The promoter did not conform to the observed consensus sequence, as one important base in the Pribnow box (−10 region), the final invariant T, is replaced by a G. However, the −35 region is highly homologous to the −35 consensus sequence. Thus, the STb promoter is capable of binding RNA polymerase, but seems to be a poor transcription initiator, hence very little STb is produced. Lawrence *et al.* (1990) indicated that cloning the *estB* gene into a high-expression vector downstream to the strong bacteriophage lambda p_L promoter increased by 10–20-fold the mRNA produced, but the amount of STb enterotoxin was not increased.

STb synthesis by wild-type *E. coli* strains varies with the composition of the culture medium used (Busque *et al.*, 1995). A repressive effect of glucose on STb production and a reversal of this effect upon addition of cAMP were noted. Catabolite repression of STb was confirmed using mutant strains for adenylate cyclase and the catabolite activator protein. A DNA homology search revealed a sequence with 72% identity with the cAMP receptor protein (CRP)-binding site located 26 bp upstream of the −35 region of the transcriptional start site. The *estB* gene appears to be quite stable as laboratory strains can be studied for many years (>10 years) without loss of the genetic trait (Dubreuil, unpublished data).

Secretion of STb and formation of disulfide bonds

STb intramolecular disulfide bonds must be correctly formed in order to produce an active molecule. The pathway by which these bonds are formed has been elucidated. STb polypeptide is synthesized as a 71 amino acid precursor (Lee *et al.*, 1983; Picken *et al.*, 1983). The NH$_2$-terminus of pre-STb (residues 1–23) has characteristics of a signal sequence that is cleaved by a signal peptidase during export to the periplasm (Kupersztoch *et al.*, 1990). Thus, an 8.1-kDa precursor (pre-STb) is converted to a transiently cell-associated 5.2-kDa form consisting of 48 amino acids. Conversion of pre-STb to cellular STb depends on the *secA* gene product. For STb, like STa, translocation of the precursor to the periplasm requires energy (Kupersztoch *et al.*, 1990). These data indicate that export of STb relies on the general export pathway of *E. coli*. After STb is detected as a cell-associated molecule, an indistinguishable extracellular form becomes apparent, indicating that no proteolytic processing occurs during mobilization of STb from the periplasm to the culture supernatant. Conversion of cellular STb to extracellular STb does not depend on membrane potential or oxidative phosphorylation.

Foreman *et al.* (1995) obtained secretion-deficient mutants using a synthetic transposon. In *dsbA* and *tolC* defective mutants, STb was absent from the culture supernatant, indicating that these genes were required for secretion of STb. Dreyfus *et al.* (1992) explored the role of the four cysteine residues in STb secretion. Cysteines were separately substituted with serine. The resulting peptides were exported to and degraded in the periplasm, suggesting that formation of disulfide bonds protected STb from protease activity. Therefore, as observed by Foreman *et al.* (1995), a *dsbA* mutant which formed disulfide bonds at a slower rate yielded a STb-negative phenotype since the reduced form of the toxin was degraded. Similarly, Okamoto *et al.* (1995) transformed a *dsbA* mutant with a plasmid harbouring *estB*. STb was not detected either in the cells or in the culture supernatant. STb production was restored by introducing the wild-type *dsbA* gene into the mutant strain, thus confirming that DsbA is involved in forming the disulfide bonds in STb and that its absence results in degradation during the secretory process. Using oligonucleotide-directed site-specific mutagenesis on the four cysteine residues, it was established by Arriaga *et al.* (1995) that the two intramolecular disulfide bonds must be formed for the efficient secretion of STb. Elimination of either one of the bonds renders the toxin susceptible to periplasmic proteolysis. Circular dichroism studies have also indicated that the integrity of the disulfide bonds is crucial for the structure and function of the toxin as reduced STb adopted a random coiled conformation (Sukumar *et al.*, 1995). Mature STb is not associated with the cellular fraction but found preferentially in the culture supernatant (Kupersztoch *et al.*, 1990).

Structure of STb and identification of the toxic domain

Sukumar *et al.* (1995) determined the solution structure of STb by two- and three-dimensional NMR methods. The NMR-derived structure showed that STb is helical between residues 33 and 45 and residues 61 and 67. The helical structure in the region 33–45 is amphipathic, exposing several polar residues to the solvent. The loop region between residues 44 and 59 contains a cluster of hydrophobic residues. Circular dichroism studies confirmed that the integrity of the disulfide bonds is crucial for the structure and function of the toxin. STb is a highly organized molecule with 73 ± 2% α-helix, 4 ± 2% β-structure and 22% remainder.

As the isoelectric point of 9.6 indicates, the side-chains of some of the basic amino acid residues project outside the molecule. Some studies have been conducted to determine the role of selected amino acids, and in particular of the basic residues, on STb toxicity. It is peculiar that, among the 48 amino acids composing mature STb, nine are basic (one histidine, two arginines and six lysines).

A loop defined by the disulfide bond between Cys^{44} and Cys^{59} and containing 14 amino acids (Figure 29.2a), including four glycine residues and four of the nine charged residues, suggested the presence of an extended coil region based on secondary structure predictions (Dreyfus et al., 1992). The location of a Arg^{52} and Asp^{53} charged pair inside this loop is such that it is highly exposed in a hydrophilic environment. The authors speculated that these two amino acid residues would probably be involved in receptor recognition. Using site-directed mutagenesis, amino-acid substitutions were performed on these two residues. When Arg^{52} was changed to serine to eliminate the charge, a significant reduction in specific activity of the mutant molecule was noted, whereas a smaller reduction in toxicity was associated with the substitution of Asp^{53}. No alteration in the stability of the mutated molecules was noted in the intestinal loop model. The mutated toxins (serine replacing Arg^{52} or Asp^{53}) did not interfere with the toxic activity of native STb, suggesting that they were not competing for the putative STb receptor. This result suggested that these amino acids might be responsible for receptor recognition and binding. Fujii et al. (1994) similarly investigated the role of basic amino-acid residues on STb enterotoxicity. Studies involving chemically modified STb indicated that lysine residues play an important role in STb toxicity and that the contribution of other basic residues to toxicity is relatively low. These results were confirmed using oligonucleotide-directed mutagenesis. In fact, when lysine residues at positions 41, 45, 46 and 69 were replaced by neutral amino acids, the toxicity of the molecule was reduced. Mutations of Lys^{45} and Lys^{46} resulted in a substantial alteration of the biological activity.

Receptor identity

The presence and nature of the STb receptor on different tissues of various animals have been investigated but, to date, only the results of a few studies have been reported. For example, a study by Weikel et al. (1986) examining the response of human adult ileal mucosa to STb in an Ussing chamber, in contrast to the piglet jejunum which responded electrogenically to STb, indicated that human tissue showed no response. These authors inferred that the adult human ileum could lack the receptor for STb. Dreyfus

et al. (1993) suggested, based on an in vitro study, that a STb receptor was present on cell types of both intestinal and non-intestinal origin, as Madin-Darby canine kidney, $HT29/C_1$ human intestinal epithelial cells and primary rat pituitary cells all responded to STb. They showed that human intestinal epithelial cells possessed a receptor for STb, as a dose-dependent calcium increase was observed due to the action of STb. Hitotsubashi et al. (1994) identified a 25-kDa protein from the membranes of mouse intestine to which STb bound. The specificity of the interaction was corroborated by competition experiments between radiolabelled (^{125}I) and unlabelled toxin. Preliminary characterization of the molecule suggested that it was not a glycoprotein. The study also indicated tissue specificity, as the STb-binding protein was not found in certain mouse tissues such as the liver, lung, spleen and kidney. More recently, Chao and Dreyfus (1997) studied the interaction of STb with cultured human intestinal cell lines. ^{125}I-STb bound specifically to T84 and HT-29 cells with low affinity ($\leqslant 10^5 M^{-1}$) to a high number of binding sites ($> 10^6$ per cell), reaching equilibrium within 5–10 min at either 4, 22 or 37°C. Approximately half of the bound toxin molecules were stably associated with the plasma membrane and/or internalized into the cytoplasm, as an acidic saline treatment could not remove them. Binding and subsequent internalization were not affected by treatment of cells with trypsin, endoglycosidase F/peptide N-glycosidase F, Vibrio cholerae neuraminidase, tunicamycin or sodium chlorate, indicating that protein, glycoprotein or sulfated proteoglycans were not involved in the process. It was also shown that STb internalization was independent of temperature, cytoskeleton rearrangement, energy or hypertonic conditions. Overall, their results indicated that STb was probably binding to membrane lipids of the plasma membrane. Nevertheless, STb appeared to interact selectively with plasma membranes since it bound with lower affinity to CHO cells and fibroblasts than to intestinal epithelial cells. From their study, Chao and Dreyfus (1997) inferred that STb becomes stably associated with the lipid bilayer and possibly penetrates the bilayer or disrupts it sufficiently to activate directly the protein G_{i3} that was proposed to be implicated in the mode of action of STb. This hypothesis is based on the fact that the amphipathic NH_2-terminal helix of STb possesses a strong membrane association potential (Segrest et al., 1990; Sukumar et al., 1995). In addition, the study of Chao and Dreyfus (1997) shed new light on the 25-kDa protein previously identified by Hitotsubashi et al. (1994) as the STb receptor. Based on their experimental data, the 25-kDa protein represented not a surface-specific receptor for the enterotoxin but rather a cytoplasmic protein or a protein associated with the inner leaflet of the plasma membrane to which STb binds.

Finally, Rousset *et al.* (1998) developed a semi-quantitative binding assay based on indirect fluorescence microscopy and using biotinylated biologically active STb. The binding characteristics and the chemical nature of the molecule responsible for STb binding to the pig jejunum, and the natural host tissue, were determined. The cellular complexity of sections of pig intestinal tissue, although presenting some experimental limitations, was taken into consideration as it could lead to significantly different results from those of *in vitro* cultured intestinal epithelial cells, especially if they originate from other animal species. Using frozen sections of pig jejunum, it was demonstrated that biotinylated STb attached to microvilli. As in the study of Chao and Dreyfus (1997), binding was rapid, reaching saturation in 10 min, and the process was temperature independent. However, binding was pH dependent, being optimum at a pH of 5.8, which is comparable with the situation found *in vivo* in the small intestine. All pig tissues tested (including duodenum, ileum, caecum, colon, liver, lung spleen and kidney) showed binding towards STb, suggesting that a molecule common to these tissues could be acting as an STb receptor.

The chemical nature of the molecule involved in STb binding was determined using chemical and enzymic treatments of the jejunal sections. Together, the data suggested that the molecule is composed of a ceramide moiety comprising a terminal neuraminic acid and/or alpha-linked terminal glucose residue(s). Thus, a glycosphingolipid present in high number in the plasma membrane could specifically be recognized by STb. A high number of receptors was inferred from the fact that a 100-molar fold excess of STb could not effectively compete for binding with biotinylated STb. Chao and Dreyfus (1997) had also indicated that competition was limited, reaching approximately 50% in the presence of either 5, 100 or 1000-fold excess of STb. These two studies suggested that a molecule of lipidic nature present in high numbers in plasma membrane could represent the STb receptor.

Pursuing the investigation (Rousset and Dubreuil, unpublished data), the present author evaluated the binding of STb to commercially available glycosphingolipids using a microplate binding assay. STb binding varied greatly depending on the molecule tested and, unexpectedly, sulfatide ($SO_4$3-galactosyl-ceramide), a molecule that does not contain neuraminic acid or alpha-linked glucose residues, was the molecule to which STb bound with greatest affinity. GM_3 also bound STb at approximately 55% of the level observed for sulfatide. Total lipid extraction of pig jejunum and thin-layer chromatographic analysis revealed that sulfatide is present in this tissue. The band to which STb bound was revealed with an anti-sulfatide monoclonal antibody recognizing the $SO_4$3-galactose epitope and laminin, which specifically binds sulfated glycolipids. It thus appears that STb could bind to the pig jejunum through interaction with sulfatide and/or a neuraminic acid containing molecules such as GM_3.

Mechanism of action

STb is a rapidly acting toxin but of moderate activity. In mouse intestinal loops, purified toxin elicited a response in 30 min and fluid accumulation reached a maximum after about 3 h (Hitotsubashi *et al.*, 1992b). Kennedy *et al.* (1984) first reported, using crude culture filtrates of STb-positive ETEC strains, that STb induced fluid secretion after 3–6 h and that the toxin did not disrupt intestinal histology. STb stimulated a cyclic nucleotide-independent secretion. Thus, STb appeared to be a cytotonic toxin with properties and a mechanism of action different from that of STa.

Weikel and Guerrant (1985) showed that, relative to controls, significant accumulation of Na^+ and Cl^- occurred intraluminally *in vivo*. Measurements of the electrolyte content of ligated intestinal segments *in vivo* further suggested that STb stimulated bicarbonate (HCO_3^-) secretion. This HCO_3^- secretion was also observed by other researchers (Argenzio *et al.*, 1984; Weikel *et al.*, 1986).

Using light microscopy, Hitotsubashi *et al.* (1992a) observed that exposure of mouse jejunum to purified STb for 3 h caused a dilation of capillaries of the submucosa and a decrease in the thickness of the lamina propria. No cellular damage or inflammation was observed. Thus, it appears that damage to the epithelium does not occur with pure toxin but only when STb-containing supernatants are used (Rose *et al.*, 1987; Whipp *et al.*, 1987).

Using the mouse intestinal loop assay model and purified toxin, Hitotsubashi *et al.* (1992a) confirmed that STb did not alter cGMP or cAMP levels in intestinal mucosal cells, thus indicating that the mechanism of action of STb in inducing fluid secretion differs from that of STa and of the cholera toxin (CT) (Figure 29.4). The level of prostaglandin E_2 (PGE_2) in the intestinal intraluminal fluid increased as a result of STb action and prostaglandin synthesis inhibitors significantly reduced the response to STb. This report was the first to involve PGE_2 in the mechanism of action of STb. More recently, Fujii *et al.* (1995) confirmed that the quantity of PGE_2 produced by intestinal cells was directly related to the dose of STb administered to the mouse. In addition, the quantity of PGE_2 correlated

INTESTINAL MUCOSA

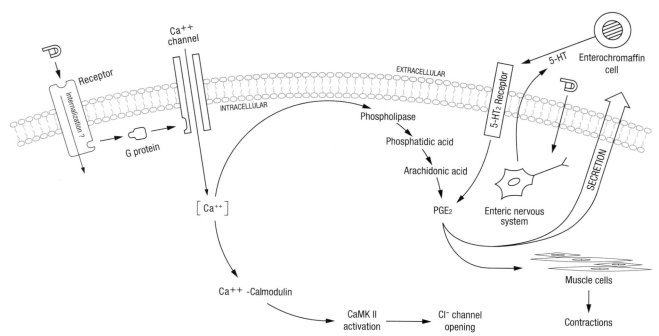

FIGURE 29.4 Model of the molecular mechanism by which STb induces secretion. STb toxin interacts with a receptor on the brush border membrane and transduces a signal to a G-protein (G_{i3}). This G-protein opens a Ca^{2+}-gated channel and the intracellular Ca^{2+} concentration increases. This activates phospholipases which act on membrane phospholipids leading to phosphatidic acid and prostaglandin E_2 (PGE_2) production. PGE_2 can either act on muscle cells to affect muscle contractions or directly result in secretion. Some studies also show that STb can probably, after internalization, directly interact with the enteric nervous system and induce 5-HT (serotonin) production. Specialized cells (enterochromaffin cells) also produce 5-HT in response to STb and activate PGE_2 synthesis via phosphatidyl inositol. Alternatively, the intracellular Ca^{2+} influx could activate CaMKII, through the calmodulin pathway, resulting in opening of a Cl^- channel leading to fluid secretion. Secretion of Cl^- and/or bicarbonate result from STb action.

with the volume of fluid released into the intestinal lumen. Levels of arachidonic acid and phosphatidic acid were also elevated by STb, indicating that arachidonic acid metabolism is stimulated by STb, possibly through phospholipase A_2 activity. Phosphatidic acid is also produced, from membrane phospholipids, through phospholipase C activity.

Independently, two groups have reported that both PGE_2 and 5-HT, the latter being regarded as another secretagogue, were released into the intestinal fluid (Harville and Dreyfus, 1995; Peterson and Whipp, 1995). Peterson and Whipp (1995) compared the secretory effects of CT, STa and STb using the pig intestinal loop model and measured the effects of these toxins on the synthesis of cAMP, cGMP and PGE_2, as well as the release of 5-HT from intestinal enterochromaffin cells. A combination of maximal doses of STa and STb yielded additive effects on fluid accumulation, suggesting different mechanisms of action. A similar additive effect on fluid accumulation and luminal release of 5-HT was noted with a combination of CT and STb. A cAMP and cGMP response to STb could not be demonstrated in either mucosal tissue or luminal fluid. Treatment of rats

with ketanserin, a 5-HT receptor antagonist, reduced the intestinal secretion induced by STb (Harville and Dreyfus, 1995). None the less, the mode of action of STb may be somewhat similar to that of CT since the latter toxin stimulates the release of both PGE_2 and 5-HT, suggesting a potential effect on the enteric nervous system. Interestingly, a study by Eklund et al. (1985) using an E. coli strain producing both STa and STb showed that, at least in rats and cats, these heat-stable enterotoxins evoke secretion, in part, via activation of the enteric nervous system, as drugs influencing nervous activity significantly diminished the secretory response. Furthermore, a study using isolated mouse ileum indicated that STb could also act directly on the muscle cells of the ileal serosa, increasing the spontaneous motility of the intestine and resulting in contractions (Hitotsubashi et al., 1992b). These contractions were not induced when the toxin was applied to the mucosa. The spontaneous motility was not inhibited by atropine, indicating that it was not the result of excitation of cholinergic nerves. In addition, papaverine, which causes relaxation of smooth muscle, had an inhibitory effect on STb, implying that STb acts directly on muscle cells.

In vitro experiments suggested that STb functions by opening a G-protein-linked (G_{i3}, a pertussis toxin-sensitive G-protein), receptor-operated calcium channel in the plasma membrane (Dreyfus *et al.*, 1993). A dose-dependent increase in intracellular Ca^{2+}, linked to extracellular Ca^{2+}, was observed. This process could be blocked by agents that impair GTP-binding regulatory function. Elevated intracellular Ca^{2+} activates phospholipases and the release of arachidonic acid from membrane lipids. It thus appears that the initial action of STb in the induction of diarrhoea is the uptake of Ca^{2+} into cells. Subsequently, synthesis of PGE_2 and other secretagogues is stimulated, leading to diarrhoea.

A recent study by Fujii *et al.* (1997) suggested the involvement of a Ca^{2+}-calmodulin-dependent protein kinase II (CaMKII) in the intestinal secretory action of STb. An increase in CaMKII activity in mouse intestinal cells was noted after addition of STb in intestinal loops. A calmodulin antagonist and an inhibitor of CaMKII both reduced the observed activity of STb. Worrell and Frizzell (1991) have previously shown that Cl^- secretion by Ca^{2+} is mediated by CaMKII in cultured colonic cell lines (T84). Thus, it was proposed that after an influx of Ca^{2+} in the cells and activation of CaMKII, through the calmodulin pathway, there would be stimulation and opening of a Cl^- channel followed by subsequent fluid secretion.

At this time, which Cl^- channel is stimulated by CaMKII remains unresolved. A Ca^{2+} pathway in which a receptor-generated increase in intracellular Ca^{2+} activates multifunctional CaMKII leading to phosphorylation and opening of a Ca^{2+}-dependent Cl^- channel could be implicated (Wagner *et al.*, 1991, 1992; Worrell and Frizzell, 1991). These Cl^- channels are entirely independent of CFTR involved in the STa mechanism of action (Wagner *et al.*, 1991). Interestingly, Seidler *et al.* (1997) reported that CFTR is also involved in electrogenic HCO_3^- secretion in the small intestine of mice. Three intracellular signal transduction pathways, including cAMP, cGMP and Ca^{2+}, were shown to stimulate HCO_3^- secretion. Since HCO_3^- secretion was reported as a result of STb action, it is tempting to relate the observed Ca^{2+} influx as the first step in STb action, leading to HCO_3^- secretion via CFTR.

CONCLUDING REMARKS

The heat-stable enterotoxins STa and STb produced by ETEC induce intestinal fluid accumulation, result-ing in diarrhoea in humans and other animals. Almost 20 years of research have been devoted to the study of STa, establishing it as a prototype for the later discovered heat-stable toxins. These studies have provided a precise knowledge of the structure, toxic domain, and more recently the identification of the STa receptor. This information culminated rapidly in a general understanding of the secretory pathway exploited by STa. Following the identification of CG-C as the STa receptor, cGKII as the signal transducer and CFTR as the secretory effector, drugs that interfere either with the processes leading to secretion or directly with the Cl^- channel could be used to abolish or limit STa action. Future studies on STa should now be directed at determining whether the alternative secretory pathways are truly exploited and, if so, to what extent compared with the CFTR Cl^- channel.

In comparison, research on STb toxin has been neglected and thus many questions still remain concerning the toxin itself and its interaction with cellular targets. In fact, for the STb toxin, crystal structure determination will soon be required as the receptor identity could be confirmed in the near future. These data should provide us with delineation of the binding and toxic domains. As knowledge on the mechanism of action of STb accumulates, it may become evident that the apparently different toxins, STa and STb, possibly share similarities in their secretory pathway. Manifestly, some of the secretagogues identified for both toxins are identical. As CFTR was recently shown also to be responsible for HCO_3^- secretion, it would be interesting to determine whether STb would be active in CF^- mice. Future studies should provide us with exciting revelations concerning the yet unsolved questions.

ACKNOWLEDGEMENTS

The author is supported by grants from the Natural Sciences and Engineering Research Council of Canada (OGPO139070) and 'Fonds pour la formation de chercheurs et l'Aide à la recherche' (93-ER-0214). The author acknowledges S. Radacovici, L.-A. Lortie, J. Harel, M. Bossé, C.E. Handl, A. Letellier, P. Busque, É. Rousset, N. Pressé, H.-E. Beausoleil, V. Labrie and M. Côté who worked on STb under his supervision; P. Sasseville from DITER (University of Montréal) for the graphic work; E. Rousset, J. Harel and and J.M. Fairbrother who provided critical readings of the manuscript; J. Brais for typing the manuscript; and, finally, Pascal-Olivier, my son, for teaching me the important things in life.

REFERENCES

Aimoto, S., Takao, T., Shimonishi, Y., Hara, S., Takeda, T., Takeda, Y. and Miwatani, T. (1982). Amino-acid sequence of heat-stable enterotoxin produced by human enterotoxigenic *Escherichia coli*. *Eur. J. Biochem.* **129**, 257–63.

Aitken, R. and Hirst, T.R. (1993). Recombinant enterotoxins as vaccines against *Escherichia coli*-mediated diarrhoea. *Vaccine,* **11**, 227–33.

Alderete, J.F. and Robertson, D.C. (1977). Repression of heat-stable enterotoxin synthesis in enterotoxigenic *Escherichia coli*. *Infect. Immun.* **17**, 629–33.

Almenoff, J.S., Williams, S.I., Scheving, L.A., Judd, A.K. and Schoolnik, G.K. (1993). Ligand-based histochemical localization and capture of cells expressing heat-stable receptors. *Mol. Microbiol.* **8**, 865–73.

Argenzio, R.A., Liacos, J., Berscheneider, H.M., Whipp, S.C. and Robertson, D.C. (1984). Effect of heat-stable enterotoxin of *Escherichia coli* and theophylline on ion transport in porcine small intestine. *Can. J. Comp. Med.* **48**, 14–22.

Arita, M., Honda, T., Miwatani, T., Ohmori, K., Takao, T. and Shimonishi, Y. (1991a). Purification and characterization of a new heat-stable enterotoxin produced by *Vibrio cholerae* non-O1 serogroup Hakata. *Infect. Immun.* **59**, 2186–8.

Arita, M., Honda, T., Miwatani, T., Takeda, T., Takao, T. and Shimonishi, Y. (1991b). Purification and characterization of a heat-stable enterotoxin of *Vibrio mimicus*. *FEMS Microbiol. Lett.* **79**, 105–10.

Arriaga, Y.L., Harville, B.A. and Dreyfus, L.A. (1995). Contribution of individual disulfide bonds to biological action of *Escherichia coli* heat-stable enterotoxin B. *Infect. Immun.* **63**, 4715–20.

Banik, N. and Ganguly, U. (1988). Stimulation of phosphoinositides breakdown by the heat-stable *E. coli* enterotoxin in rat intestinal epithelial cells. *FEBS Lett.* **236**, 489–92.

Banik, N.D. and Ganguly, U. (1989). Diacylglycerol breakdown in plasma membrane of rat intestinal epithelial cells. *FEBS Lett.* **250**, 201–4.

Beubler, E., Badhri, P. and Schirgi-Degen, A. (1992). 5-HT receptor antagonists and heat-stable *Escherichia coli* enterotoxin-induced effects in the rat. *Eur. J. Pharmacol.* **219**, 445–50.

Brandsch, M., Ramamorthy, S., Marczin, N., Catravas, J.D., Leibach, J.W., Ganapathy, V. and Leibach, F.H. (1995). Regulation of taurine transport by *Escherichia coli* heat-stable enterotoxin and guanylin in human intestinal cell lines. *J. Clin. Invest.* **96**, 361–9.

Brandwein, H., Deutsch, A., Thompson, M. and Giannella, R. (1985). Production of neutralizing monoclonal antibodies to *Escherichia coli* heat-stable enterotoxin. *Infect. Immun.* **47**, 242–6.

Burgess, M.N., Bywater, R.J., Cowley, C.M., Mullan, N.A. and Newsome, P.M. (1978). Biological evaluation of a methanol-soluble, heat-stable *Escherichia coli* enterotoxin in infant mice, pigs, rabbits, and calves. *Infect. Immun.* **21**, 526–31.

Busque, P., Letellier, A., Harel, J. and Dubreuil, J.D. (1995). Production of *Escherichia coli* STb enterotoxin is subject to catabolite repression. *Microbiology* **141**, 1621–7.

Caprioli, A., Donelli, G., Falbo, V., Passi, C., Pagano, A. and Mantovani, A. (1991). Antimicrobial resistance and production of toxins in *Escherichia coli* strains from wild ruminants and the alpine marmot. *J. Wildlife Dis.* **27**, 324–7.

Carpick, B.W. and Gariépy, J. (1991). Structural characterization of functionally important regions of the *Escherichia coli* heat-stable enterotoxin STIb. *Biochemistry* **30**, 4803–9.

Carpick, B.W. and Gariépy, J. (1993). The *Escherichia coli* heat-stable enterotoxin is a long-lived superagonist of guanylin. *Infect. Immun.* **61**, 4710–15.

Chang, M.-S., Lowe, D.G., Lewis, M., Hellmiss, R., Chen, E. and Goeddel, D.V. (1989). Differential activation by atrial and brain natriuretic peptides of two different receptor guanylate cyclases. *Nature* **341**, 68–72.

Chao, K.L. and Dreyfus, L.A. (1997). Interaction of *Escherichia coli* heat-stable enterotoxin B with cultured human intestinal epithelial cells. *Infect. Immun.* **65**, 3209–17.

Chaudhuri, A.G. and Ganguly, U. (1995). Evidence for stimulation of the inositol triphosphate-Ca^{2+} signalling system in rat enterocytes by heat-stable enterotoxin of *Escherichia coli*. *Biochim. Biophys. Acta* **1267**, 131–3.

Chaudhuri, A.G., Sen, P.C. and Ganguly, U. (1993). Evidence for protein kinase C stimulation in rat enterocytes pretreated with heat-stable enterotoxin of *Escherichia coli*. *FEMS Microbiol. Lett.* **110**, 185–90.

Cohen, M.B., Moyer, M.S., Luttrell, M. and Giannella, R.A. (1986). The immature rat small intestine exhibits an increased sensitivity and response to *Escherichia coli* heat-stable enterotoxin. *Pediatr. Res.* **20**, 555–60.

Cohen, M.B., Guarino, A., Shukla, R. and Giannella, R.A. (1988). Age-related differences in receptors for *Escherichia coli* heat-stable enterotoxin in the small and large intestine of children. *Gastroenterology* **94**, 367–73.

Cohen, M.B., Mann, E.A., Lau, C., Henning, S.J. and Giannella, R.A. (1992). A gradient in expression of the *Escherichia coli* heat-stable enterotoxin receptor exists along the villus-to-crypt axis of rat small intestine. *Biochem. Biophys. Res. Commun.* **186**, 483–90.

Cohen, M.B., Jensen, N.J., Hawkins, J.A., Mann, E.A., Thompson, M.R., Lentze, M.J. and Giannella, R.A. (1993). Receptors for *Escherichia coli* heat-stable enterotoxin in human intestine and in a human intestinal cell line (Caco-2). *J. Cell Physiol.* **156**, 138–44.

Crane, J.K., Burrell, L.L., Weikel, C.S. and Guerrant, R.L. (1990). Carbachol mimics phorbol esters in its ability to enhance cyclic GMP production by STa, the heat-stable toxin of *Escherichia coli*. *FEBS Lett.* **274**, 199–202.

Crane, J.K., Wehner, M.S., Bolen, E.J., Sando, J.J., Linden, J., Guerrant, R.L. and Sears, C.L. (1992). Regulation of intestinal guanylate cyclase by the heat-stable enterotoxin of *Escherichia coli* (STa) and protein kinase C. *Infect. Immun.* **60**, 5004–12.

Currie, M.G., Fok, K.F., Kato, J., Moore, R.J., Hamra, F.K., Duffin, K.L. and Smith, C.E. (1992). Guanylin: an endogenous activator of intestinal guanylate cyclase. *Proc. Natl Acad. Sci. USA* **89**, 947–51.

de Sauvage, F.J., Camerato, T.R. and Goeddel, D.V. (1991). Primary structure and functional expression of the human receptor for *Escherichia coli* heat-stable enterotoxin. *J. Biol. Chem.* **266**, 17912–18.

de Sauvage, F.J., Horuk, R., Bennett, G., Quan, C., Burnier, J.P. and Goeddel, D.V. (1992). Characterization of the recombinant human receptor for *Escherichia coli* heat-stable enterotoxin. *J. Biol. Chem.* **267**, 6479–82.

Dreyfus, L.A. and Robertson, D.C. (1984). Solubilization and partial characterization of the intestinal receptor for *Escherichia coli* heat-stable enterotoxin. *Infect. Immun.* **46**, 537–43.

Dreyfus, L.,A., Frantz, J.C. and Robertson, D.C. (1983). Chemical properties of heat-stable enterotoxins produced by enterotoxigenic *Escherichia coli* of different host origins. *Infect. Immun.* **42**, 539–48.

Dreyfus, L.A., Jaso-Friedman, L. and Robertson, D.C. (1984). Characterization of the mechanism of action of *Escherichia coli* heat-stable enterotoxin. *Infect. Immun.* **44**, 493–501.

Dreyfus, L.A., Urban, R.G., Whipp, S.C., Slaughter, C., Tachias, K., and Kupersztoch, Y.M. (1992). Purification of the ST_B enterotoxin of *Escherichia coli* and the role of selected amino-acids on its secretion, stability and toxicity. *Mol. Microbiol.* **6**, 2397–406.

Dreyfus, L.A., Harville, B., Howard, D.E., Shaban, R., Beatty, D.M. and Morris, S.J. (1993). Calcium influx mediated by the *Escherichia coli* heat-stable enterotoxin B (ST$_B$). *Proc. Natl Acad. Sci. USA* **90**, 3202–6.

Dubreuil, J.D. (1997). *Escherichia coli* STb enterotoxin (a review). *Microbiology* **143**, 1783–95.

Dubreuil, J.D., Fairbrother, J.M., Lallier, R. and Larivière, S. (1991). Production and purification of heat-stable enterotoxin b from a porcine *Escherichia coli* strain. *Infect. Immun.* **59**, 198–203.

Dubreuil, J.D., Letellier, A. and Harel, J. (1996). A recombinant *Escherichia coli* heat-stable enterotoxin b (STb) fusion protein eliciting neutralizing antibodies. *FEMS Immunol. Med. Microbiol.* **13**, 317–23.

Echeverria, P., Verhaert, L. and Uylangco, V. (1978). Antimicrobial resistance and enterotoxin production among isolates of *Escherichia coli* in the Far East. *Lancet* **ii**, 589–92.

Echeverria, P., Seriwatana, J., Taylor, D.N., Tirapat, C. and Rowe, B. (1985). *Escherichia coli* contains plasmids coding for heat-stable b, other enterotoxins, and antibiotic resistance. *Infect. Immun.* **48**, 843–6.

Eklund, S., Jodal, M. and Lundgren, O. (1985). The enteric nervous system participates in the secretory response to the heat stable enterotoxins of *Escherichia coli* in rats and cats. *Neuroscience* **14**, 673–81.

Field, M., Graf, L.H., Laird, W.J. and Smith, P.L. (1978). Heat-stable enterotoxin of *Escherichia coli*: *in vitro* effects on guanylate cyclase activity, cyclic GMP concentration, and ion transport in small intestine. *Proc. Natl Acad. Sci. USA* **75**, 2800–4.

Field, M., Rao, M.C. and Chang, E.B. (1989a). Intestinal electrolyte transport and diarrheal disease. I. *N. Engl. J. Med.* **321**, 800–6.

Field, M., Rao, M.C. and Chang, E.B. (1989b). Intestinal electrolyte transport and diarrheal disease. II. *N. Engl. J. Med.* **321**, 879–83.

Foreman, D.T., Martinez, Y., Coombs, G., Torres, A. and Kupersztoch, Y.M. (1995). TolC and DsbA are needed for the secretion of ST$_B$, a heat-stable enterotoxin of *Escherichia coli*. *Mol. Microbiol.* **18**, 237–45.

Forte, L.R., Thorne, P.K., Eber, S.L., Krause, W.J., Freeman, R.H., Francis, S.H. and Corbin, J.D. (1992). Stimulation of intestinal Cl⁻ transport by heat-stable enterotoxin: activation of cAMP-dependent protein kinase by cGMP. *Am. J. Physiol.* **263**, C607–15.

Forte, L.R., Eber, S.L., Turner, J.T., Freeman, R.H., Fok, K.F. and Currie, M.G. (1993). Guanylin stimulation of Cl⁻ secretion in human intestinal T$_{84}$ cells via cyclic guanosine monophosphate. *J. Clin. Invest.* **91**, 2423–8.

French, P.J., Bijman, J., Edixhoven, M., Vaandrager, A.B., Scholte, B.J., Lohmann, S.M., Nairm, A.C. and de Jonge, H.R. (1995). Isotype-specific activation of cystic fibrosis transmembrane conductance regulator-chloride channels by cGMP-dependent protein kinase II. *J. Biol. Chem.* **270**, 26626–31.

Fujii, Y., Hayashi, M., Hitotsubashi, S., Fuke, Y., Yamanaka, H. and Okamoto, K. (1991). Purification and characterization of *Escherichia coli* heat-stable enterotoxin II. *J. Bacteriol.* **173**, 5516–22.

Fujii, Y., Okamuro, Y., Hitotsubashi, S., Saito, A., Akashi, N. and Okamoto, K. (1994). Effects of alterations of basic amino acid residues of *Escherichia coli* heat-stable enterotoxin II on enterotoxicity. *Infect. Immun.* **62**, 2295–301.

Fujii, Y., Kondo, Y. and Okamoto, K. (1995). Involvement of prostaglandin E2 synthesis in the intestinal secretory action of *Escherichia coli* heat-stable enterotoxin II. *FEMS Microbiol. Lett.* **130**, 259–66.

Fujii, Y., Nomura, T., Yamanaka, H. and Okamoto, K. (1997). Involvement of Ca^{2+}-calmodulin-dependent protein kinase II in the intestinal secretory action of *Escherichia coli* heat-stable enterotoxin II. *Microbiol. Immunol.* **41**, 633–6.

Gadsby, D.C. and Nairm, A.C. (1994). Regulation of CFTR channel gating. *Trends Biochem. Sci.* **19**, 513–18.

Ganguly, U. and Talukder, S. (1985). Effect of *Escherichia coli* heat-stable enterotoxin (ST) on calcium uptake by rat intestinal brush border membrane vesicles (BBMV). *FEMS Microbiol. Lett.* **26**, 255–7.

Garbers, D. (1991). The guanylyl cyclase-receptor family. *Can. J. Physiol. Pharmacol.* **69**, 1618–21.

Gariépy, J., Lane, A., Frayman, F., Wilbur, D., Robien, W., Schoolnik, G.K. and Jardetzki, O. (1986). Structure of the toxic domain of *Escherichia coli* heat-stable enterotoxin STI. *Biochemistry* **25**, 7854–66.

Gariépy, J., Judd, A.K. and Schoolnik, G.K. (1987). Importance of disulfide bridges in the structure and activity of *Escherichia coli* enterotoxin ST1b. *Proc. Natl Acad. Sci. USA* **84**, 8907–11.

Gazzano, H., Wu, H.I. and Waldman, S.A. (1991). Activation of particulate guanylate cyclase by *Escherichia coli* heat-stable enterotoxin is regulated by adenine nucleotides. *Infect. Immun.* **59**, 1552–7.

Giannella, R.A. (1995). *Escherichia coli* heat-stable enterotoxins, guanylins, and their receptors: what are they and what do they do? *J. Lab. Clin. Med.* **125**, 173–81.

Giannella, R.A., Luttrell, M. and Thompson, M. (1983). Binding of *Escherichia coli* heat-stable enterotoxin to receptors on rat intestinal cells. *Am. J. Physiol.* **245**, G492–8.

Goldstein, J.L., Sahi, J., Bhuva, M., Layden, T.J. and Rao, M.C. (1994). *Escherichia coli* heat-stable enterotoxin-mediated colonic Cl⁻ secretion is absent in cystic fibrosis. *Gastroenterology* **107**, 950–6.

Gray, W.R., Luque, A., Olivera, B.M., Barrett, J. and Cruz, L.J. (1981). Peptide toxins from *Conus geographus* venom. *J. Biol. Chem.* **256**, 4734–40.

Greenberg, R.N., Guerrant, R.L., Chang, B., Robertson, D.C. and Murad, F. (1982). Inhibition of *Escherichia coli* heat-stable enterotoxin effects on intestinal guanylate cyclase and fluid secretion by quinacrine. *Biochem. Pharmacol.* **31**, 2005–9.

Greenberg, R.N., Hill, M., Crytzer, J., Krause, W.J., Eber, S.L., Hamra, F.K. and Forte, L.R. (1997). Comparison of effects of uroguanylin, guanylin and *Escherichia coli* heat-stable enterotoxin STa in mouse intestine and kidney: evidence that uroguanylin is an intestinal natriuretic hormone. *J. Invest. Med.* **45**, 276–83.

Guarino, A., Cohen, M., Thompson, M., Charmsathaphorn, K. and Giannella, R. (1987a). T84 cell receptor binding and guanylate cyclase activation by *Escherichia coli* heat-stable toxin. *Am. J. Physiol.* **253**, G775–80.

Guarino, A., Cohen, M.B. and Giannella, R.A. (1987b). Small and large intestinal guanylate cyclase activity in children: effect of age and stimulation by *Escherichia coli* heat-stable enterotoxin. *Pediatr. Res.* **20**, 551–5.

Guarino, A., Giannella, R. and Thompson, M.R. (1989). *Citrobacter freundii* produces an 18-amino-acid heat-stable enterotoxin identical to the 18-amino-acid *Escherichia coli* heat-stable enterotoxin (STI$_a$). *Infect. Immun.* **57**, 649–52.

Gyles, C.L. (1994). *Escherichia coli* enterotoxins. In: Escherichia coli *in Domestic Animals and Human* (ed. C.L. Gyles), pp. 337–64. CAB International, Wallingford, UK.

Gyles, C.L., Palchaudhuri, S. and Maas, W. (1977). Naturally occuring plasmid carrying genes for enterotoxin production and drug resistance. *Science* **198**, 198–9.

Hakki, S., Robertson, D.C. and Waldman, S.A. (1993). A 56 kDa binding protein for *Escherichia coli* heat-stable enterotoxin isolated from the cytoskeleton of rat intestinal membranes does not possess guanylate cyclase activity. *Biochim. Biophys. Acta* **1152**, 1–8.

Hamra, F.K. Forte, L.R. Eber, S.L., Pidhorodeckyj, N.V., Krause, W.J. Freeman, R.H., Chin, D.T., Tompkins, J.A., Fok, K.F., Smith, C.E., Duffin, K.L., Siegel, N.R. and Currie, M.G. (1993). Uroguanylin:

structure and activity of a second endogenous peptide that stimulates intestinal guanylate cyclase. *Proc. Natl Acad. Sci. USA* **90**, 10464–8.

Handl, C.E., Harel, J., Flock, J.-I. and Dubreuil, J.D. (1993). High yield of active STb enterotoxin from a fusion protein (MBP-STb) expressed in *Escherichia coli*. *Protein Expr. Purif.* **4**, 275–81.

Harnett, N.M. and Gyles, C.L. (1985). Linkage of genes for heat-stable enterotoxin, drug resistance, K99 antigen, and colicin in bovine and porcine strains of enterotoxigenic *Escherichia coli. Am. J. Vet. Res.* **46**, 428–33.

Harville, B.A. and Dreyfus, L.A. (1995). Involvement of 5-hydroxytryptamine and prostaglandin E$_2$ in intestinal secretory action of *Escherichia coli* heat-stable enterotoxin B. *Infect. Immun.* **63**, 745–50.

Hayden, U.L., Richard, M.A., Greenberg, N., Hannah, M.D. and Carey, V. (1996). Role of prostaglandins and enteric nerves in *Escherichia coli* heat-stable enterotoxin (STa)-induced intestinal secretion in pigs. *Am. J. Vet. Res.* **57**, 211–15.

Hidaka, Y., Ohmori, K., Wada, A., Ozaki, H., Ito, H., Hirayama, T., Takeda, Y. and Shimonishi, Y. (1991). Synthesis and biological properties of carba-analogs of heat-stable enterotoxin (ST) produced by enterotoxigenic *Escherichia coli. Biochem. Biophys. Res. Commun.* **176**, 958–65.

Hirayama, T., Wada, A., Iwata, N., Takasaki, S., Shimonishi, Y. and Takeda, Y. (1992). Glycoprotein receptors for a heat-stable enterotoxin (STh) produced by enterotoxigenic *Escherichia coli. Infect. Immun.* **60**, 4213–20.

Hitotsubashi, S., Fujii, Y., Yamanaka, H. and Okamoto, K. (1992a). Some properties of purified *Escherichia coli* heat-stable enterotoxin II. *Infect. Immun.* **60**, 4468–74.

Hitotsubashi, S., Akagi, M., Saitou, A., Yamanaka, H., Fujii, Y., and Okamoto, I. (1992b). Action of *Escherichia coli* heat-stable enterotoxin II on isolated section of mouse ileum. *FEMS Microbiol Lett.* **90**, 249–52.

Hitotsubashi, S., Fujii, Y. and Okamoto, K. (1994). Binding protein for *Escherichia coli* heat-stable enterotoxin II in mouse intestinal membrane. *FEMS Microbiol. Lett.* **122**, 297–302.

Hu, S.T. and Lee, C.H. (1988). Characterization of the transposon carrying the STII gene of enterotoxigenic *Escherichia coli. Mol. Gen. Genet.* **214**, 490–5.

Hu, S.T., Yang, M.K., Spandau, D.F., and Lee, C.H. (1987). Characterization of the terminal sequences flanking the transposon that carries the *Escherichia coli* enterotoxin STII gene. *Gene* **55**, 157–67.

Hughes, J.M., Murad, F., Chang, B. and Guerrant, R.L. (1978). Role of cyclic GMP in the action of heat-stable enterotoxin of *Escherichia coli. Nature* **271**, 755–6.

Hughes, M., Crane, M., Hakki, S., O'Hanley, P. and Waldman, S.A. (1991). Identification and characterization of a new family of high-affinity receptors for *Escherichia coli* heat-stable enterotoxin in rat intestinal membranes. *Biochemistry* **30**, 10738–45.

Hughes, M., Crane, M.R., Thomas, B.R., Robertson, D., Gazzano, D., O'Hanley, P. and Waldman, S.A. (1992). Affinity purification of functional receptors for *Escherichia coli* heat-stable enterotoxin from rat intestine. *Biochemistry* **31**, 12–26.

Huott, P.A., Liu, W., McRoberts, J.A., Giannella, R.A. and Dharmsathaphorn, K. (1988). Mechanism of action of *Escherichia coli* heat-stable enterotoxin in a human colonic cell line. *J. Clin. Invest.* **82**, 514–23.

Ivens, K., Gazzano, H., O'Hanley, P. and Waldman, S.A. (1990). Heterogeneity of intestinal receptors for *Escherichia coli* heat-stable enterotoxin. *Infect. Immun.* **58**, 1817–20.

Katwa, L.C., Parker, C.D., DyBing, J.K. and White, A.A. (1992). Nucleotide regulation of heat-stable enterotoxin receptor binding and of guanylate cyclase activation. *Biochem. J.* **283**, 727–35.

Kennedy, D.J., Greenberg, R.N., Dunn, J.A., Abernathy, R., Ryerse, J.S. and Guerrant, R.L. (1984). Effects of *Escherichia coli* heat-stable enterotoxin STb on intestines of mice, rats, rabbits, and piglets. *Infect. Immun.* **46**, 639–43.

Kita, T., Smith, C.E., Fok, K.F., Duffin, K.L., Moore, W.M., Karabotsos, P.J., Kachur, J.F., Hamra, F.K., Pidhorodeckyj, N.V., Forte, L.R. and Currie, M.G. (1994). Characterization of human uroguaylin: a member of the guanylin peptide family. *Am. J. Physiol.* **266**, F342–8.

Knoop, F.C., Owens, M., Marcus, J.N. and Murphy, B. (1991). Elevation of calcium in rat enterocytes by *Escherichia coli* heat-stable (STa) enterotoxin. *Curr. Microbiol.* **23**, 291–6.

Krause, W.J., Cullingford, G.L., Freeman, R.H., Eber, S.L., Richardson, K.C., Fok, K.F., Currie, M.G. and Forte, L.R. (1994). Distribution of heat-stable enterotoxin/guanylin receptors in the intestinal tract of man and other mammals. *J. Anat.* **184**, 407–17.

Kuhn, M., Raida, M., Adermann, K., Schulz-Knappe, P., Gerzer, R., Helm, J.-M. and Forssmann, W.G. (1993). The circulating bioactive form of human guanylin is a high molecular weight peptide (10.3 kDa). *FEBS Lett.* **18**, 205–9.

Kuhn, M., Adermann, K., Jähne, J., Forssmann, W.G. and Rechkemmer, G. (1994). Segmental differences in the effects of guanylin and *Escherichia coli* heat-stable enterotoxin on Cl$^-$ secretion in human gut. *J. Physiol. (London)* **479**, 433–40.

Kupersztoch, Y.M., Tachias, K., Moomaw, C.R., Dreyfus, L.A., Urban, R., Slaughter, C. and Whipp, S.C. (1990). Secretion of a methanol-insoluble heat-stable enterotoxin (ST$_B$): energy- and *sec*-A dependent conversion of pre-ST$_B$ to an intermediate indistinguishable from the extracellular toxin. *J. Bacteriol.* **172**, 2427–32.

Lawrence, R.M., Huang, P.-T., Glick, J., Oppenheim, J.D. and Maas, W.K. (1990). Expression of the cloned gene for enterotoxin STb of *Escherichia coli. Infect. Immun.* **58**, 970–7.

Lee, C.H., Moseley, S.L., Moon, H.W., Whipp, S.C., Gyles, C.L. and So, M. (1983). Characterization of the gene encoding heat-stable toxin II and preliminary molecular epidemiological studies of enterotoxigenic *Escherichia coli* heat-stable toxin II producers. *Infect. Immun.* **42**, 264–8.

Lee, C.H., Hu, S.T., Swiatek, P.J., Moseley, S.L., Allen, S.D. and So, M. (1985). Isolation of a novel transposon which carries the *Escherichia coli* enterotoxin STII gene. *J. Bacteriol.* **162**, 615–20.

Lin, M., Nairm, A.C. and Guggino, S.E. (1992). cGMP-dependent protein kinase regulation of a chloride channel in T84 cells. *Am. J. Physiol.* **262**, C1304–12.

Madsen, G.L. and Knoop, F.C. (1978). Inhibition of the secretory activity of *Escherichia coli* heat-stable enterotoxin by indomethacin. *Infect. Immun.* **22**, 143–7.

Mann, E.A., Cohen, M.B. and Giannella, R.A. (1993). Comparison of receptors for *Escherichia coli* heat-stable enterotoxin: novel receptor present in IEC-6 cells. *Am. J. Physiol.* **264**, G172–8.

Mann, E.A., Jump, M.L., Wu, J., Yee, E. and Giannella, R.A. (1997). Mice lacking the guanylyl cyclase C receptor are resistant to STa-induced intestinal secretion. *Biochem. Biophys. Res. Commun.* **239**, 463–6.

Markert, T., Vaandrager, A.B., Gambaryan, S., Pöhler, D., Häusler, C., Walter, U., deJonge, H.R., Jarchau, T. and Lohmann, S.M. (1995). Endogenous expression of type II cGMP-dependent protein kinase mRNA and protein in rat intestine. *J. Clin. Invest.* **96**, 822–30.

Mathias, J.R., Nogueira, J., Martin, J.L., Carlson, G.M. and Giannella, R.A. (1982). *Escherichia coli* heat-stable toxin: its effect on motility of the small intestine. *Am. J. Physiol.* **242**, G360–3.

Matthews, J.B., Awtrey, C.S., Thompson, R., Hung, T., Tally, K.J. and Madara, J.L. (1993). Na$^+$-K$^+$-2Cl$^-$ cotransport and Cl$^-$ secretion evoked by heat-stable enterotoxin is microfilament dependent in T84 cells. *Am. J. Physiol.* **265**, G370–8.

Nzegwu, H.C. and Levin, R.J. (1996). Luminal capsaicin inhibits fluid secretion induced by enterotoxigenic *E. coli* STa, but not by carbachol, *in vivo* in rat small and large intestine. *Exp. Med.* **81**, 313–15.

Okamoto, K. and Takahara, M. (1990). Synthesis of *Escherichia coli* heat-stable enterotoxin STp as a pre-pro form and role of the pro sequence in secretion. *J. Bacteriol.* **172**, 5260–5.

Okamoto, K., Okamoto, Yukitake, J., Kawamoto, Y. and Muyama, A. (1987). Substitutions of cysteine residues of *Escherichia coli* heat-stable enterotoxin by oligonucleotide-directed mutagenesis. *Infect. Immun.* **55**, 2121–5.

Okamoto, K., Baba, T., Yamanaka, H., Akashi, N. and Fujii, Y. (1995). Disulfide bond formation and secretion of *Escherichia coli* heat-stable enterotoxin II. *J. Bacteriol.* **177**, 4579–86.

Ozaki, H., Sato, T., Kubota, H., Hata, Y., Katsube, Y. and Shimonishi, Y. (1991). Molecular structure of the toxic domain of heat-stable enterotoxin produced by a pathogenic strain of *Escherichia coli*. *J. Biol. Chem.* **266**, 5934–41.

Peterson, J.W. and Whipp, S.C. (1995). Comparison of the mechanisms of action of cholera toxin and the heat-stable enterotoxins of *Escherichia coli*. *Infect. Immun.* **63**, 1452–61.

Pfeifer, A., Aszódi, A., Seidler, U., Ruth, P., Hofmann, F. and Fässler, R. (1996). Intestinal secretory defects and dwarfism in mice lacking cGMP-dependent protein kinase II. *Science* **274**, 2082–6.

Picken, R.N., Mazaitis, A.J., Maas, W.K., Rey, M. and Heyneker, H. (1983). Nucleotide sequence of the gene for heat-stable enterotoxin II of *Escherichia coli*. *Infect. Immun.* **42**, 269–75.

Rasheed, J.K., Guzmán-Verduzco, L.-M. and Kupersztoch, Y.M. (1990). Two precursors of the heat-stable enterotoxin of *Escherichia coli*: evidence of extracellular processing. *Mol. Microbiol.* **4**, 265–73.

Rolfe, V. and Levin, R.J. (1994). Enterotoxin *Escherichia coli* STa activates a nitric oxide-dependent myenteric plexus secretory reflex in the rat ileum. *J. Physiol.* **475**, 531–7.

Rose, R., Whipp, S.C. and Moon, H.W. (1987). Effects of *Escherichia coli* heat-stable enterotoxin b on small intestinal villi in pigs, rabbits, and lambs. *Vet. Pathol.* **24**, 71–9.

Roussel, A.J., Woode, G.N., Waldron, R.C., Sriranganathan, N. and Jones, M.K. (1992). Myoelectric activity of the small intestine in enterotoxin-induced diarrhea of calves. *Am. J. Vet. Res.* **53**, 1145–8.

Rousset, E., Harel, J. and Dubreuil, J.D. (1998). Binding characteristics of *Escherichia coli* enterotoxin b (STb) to the pig jejunum and partial characterization of the molecule involved. *Microb. Pathogen.* **24**, 277–88.

Rudner, X.L., Mandal, K.K., deSauvage, F.J., Kindman, L.A. and Almenoff, J.S. (1995). Regulation of cell signaling by the cytoplasmic domains of the heat-stable enterotoxin receptor: identification of auto-inhibitory and activating motifs. *Proc. Natl Acad. Sci. USA* **92**, 5169–73.

Sanchez, J., Solorzano, R.M. and Holmgren, J. (1993). Extracellular secretion of STa heat-stable enterotoxin by *Escherichia coli* after fusion to a heterologous leader peptide. *FEBS Lett.* **330**, 265–9.

Sato, T., Ozaki, H., Hata, Y., Kitagawa, Y., Katsube, Y. and Shimonishi, Y. (1994). Structural characteristics for biological activity of heat-stable enterotoxin produced by enterotoxigenic *Escherichia coli*: X-ray crystallography of weakly toxic and nontoxic analogues. *Biochemistry* **33**, 8641–50.

Savarino, S.J., Fasano, A., Watson, J., Martin, B.M., Levine, M.M., Guandalini, S. and Guerry, P. (1993). Enteroaggregative *Escherichia coli* heat-stable enterotoxin 1 represents another family of *E. coli* heat-stable toxin. *Proc. Natl Acad. Sci. USA* **90**, 3093–7.

Scheving L.A. and Chong, K.M. (1997). Differential processing of guanylyl cyclase C along villus-crypt axis of rat small intestine. *Am. J. Physiol.* **272**, C1995–2004.

Schulz, S., Green, C.K., Yuen, P.S.T. and Garbers, D.L. (1990). Guanylyl cyclase is a heat-stable enterotoxin receptor. *Cell* **63**, 941–8.

Schulz, S., Yuen, P. and Garbers, D. (1991). The expanding family of guanylyl cyclases. *Trends Pharmacol. Sci.* **12**, 116–20.

Sears, C.L. and Kaper, J.B. (1996). Enteric bacterial toxins: mechanisms of action and linkage to intestinal secretion. *Microbiol. Rev.* **60**, 167–215.

Segrest, J.P., DeLoof, H., Dohlman, J.G., Brouillette, C.G. and Ananthaaramaiah, G.M. (1990). Amphiphatic helix motif: classes and properties. *Proteins* **8**, 103–17.

Seidler, U., Blumenstein, I., Kretz, A., Viellard-Baron, D., Rossmann, H., Colledge, W.H., Evans, M., Ratcliff, R. and Gregor, M. (1997). A functional CFTR protein is required for mouse intestinal cAMP-, cGMP- and Ca(2+)-dependent HCO_3^- secretion. *J. Physiol.* **505**, 411–23.

Sekizaki, T., Akashi, H. and Terakado, N. (1985). Nucleotide sequences of the genes for *Escherichia coli* heat-stable enterotoxin I of bovine, avian, and porcine origins. *Am. J. Vet. Res.*, **46**, 909–12.

So, M. and McCarthy, B.J. (1980). Nucleotide sequence of the bacterial transposon Tn*1681* encoding a heat-stable (ST) toxin and its identification in enterotoxigenic *Escherichia coli* stains. *Proc. Natl Acad. Sci. USA* **77**, 4011–15.

Sommerfelt, H., Haukanes, B.-I., Kalland, K.H., Svennerholm, A.-M., Sanchéz, J. and Bjorvatn, B. (1989). Mechanism of spontaneous loss of heat-stable toxin (STa) production in enterotoxigenic *Escherichia coli*. *APMIS* **97**, 436–40.

Spandau, D.F. and Lee, C.-H. (1987). Determination of the promoter strength of the gene encoding *Escherichia coli* heat-stable enterotoxin II. *J. Bacteriol.* **169**, 1740–4.

Sprangler, B.D. (1992). Structure and function of cholera toxin and the related *Escherichia coli* heat-labile enterotoxin. *Microbiol. Rev.* **56**, 622–47.

Staples, S.J., Asher, S.E. and Giannella, R.A. (1980). Purification and characterization of heat-stable enterotoxin produced by a strain of *E. coli* pathogenic for man. *J. Biol. Chem.* **255**, 4716–21.

Stieglitz, H., Cervantes, L., Robledo, R., Covarrubias, L., Bolivar, F. and Kupersztoch, Y.M. (1988). Cloning, sequencing and expression in Ficoll generated minicells of an *Escherichia coli* heat-stable enterotoxin gene. *Plasmid* **20**, 42–53.

Sukumar, M., Rizo, J., Wall, M., Dreyfus, L.A., Kupersztoch, Y.M. and Gierasch, L.M. (1995). The structure of *Escherichia coli* heat-stable enterotoxin b by nuclear magnetic resonance and circular dichroism. *Protein Sci.* **4**, 1718–29.

Swenson, E.S., Mann, E.A., Jump, M.L., Witte, D.P. and Giannella, R.A. (1996). The guanylin/STa receptor is expressed in crypts and apical epithelium throughout the mouse intestine. *Biochem. Biophys. Res. Commun.* **225**, 10009–14.

Takao, T., Hitouji, T., Aimoto, S., Shimonishi, Y., Hara, S., Takeda, T., Takeda, Y. and Miwatani, T. (1983). Amino-acid sequence of a heat-stable enterotoxin isolated from enterotoxigenic *Escherichia coli* strain 18D. *FEBS Lett.* **152**, 1–5.

Takao, T., Tominaga, N., Yoshimura, S., Shimonishi, Y., Hara, S., Inoue, T. and Miyama, A. (1985). Isolation, primary structure and synthesis of heat-stable enterotoxin produced by *Yersinia enterolitica*. *Eur. J. Biochem.* **152**, 199–206.

Takeda, T., Takeda, Y., Aimoto, S., Takao, T., Ikemura, H., Shimonishi, Y. and Miwatani, T. (1983). Neutralization of activity of two different heat-stable enterotoxins (ST$_h$ and ST$_p$) of enterotoxigenic *Escherichia coli* by homologous and heterologous antisera. *FEMS Microbiol. Lett.* **20**, 357–9.

Takeda, Y., Yamasaki, S., Hirayama, T. and Shimonishi, Y. (1991). Heat-stable enterotoxins produced by enteric bacteria. In: *Molecular Pathogenesis of Gastrointestinal Infections* (eds T. Wädstrom, P.H. Mäkela, A.-M. Svennerholm and H. Wolf-Watz), pp. 125–8, FEMS Symposium No. 58. Plenum Press, New York.

Thomas, D.D. and Knoop, F.C. (1982). The effect of calcium and prostaglandin inhibitors on the intestinal fluid response to heat-stable enterotoxin of *Escherichia coli*. *J. Infect. Dis.* **145**, 141–7.

Thompson, M.R. and Giannella, R.A. (1985). Revised amino acid sequence for a heat-stable enterotoxin produced by an *Escherichia coli* strain (18D) that is pathogenic for humans. *Infect. Immun.* **47**, 834–6.

Thompson, M.R. and Giannella, R.A. (1990). Different crosslinking agents identify distinctly different putative *Escherichia coli* heat-stable enterotoxin rat intestinal cell receptor proteins. *J. Receptor Res.* **10**, 97–117.

Tien, X.-Y., Brasitus, T.A., Kaetzel, M.A., Dedman, J.R. and Nelson, D.J. (1994). Activation of the cystic fibrosis transmembrane conductance regulator by cGMP in the human colonic cancer cell line CaCo-2. *J. Biol. Chem.* **269**, 51–4.

Urban, R.G, Dreyfus, L.A. and Whipp, S.C. (1990). Construction of a bifunctional *Escherichia coli* heat-stable enterotoxin (STb)-alkaline phosphatase fusion protein. *Infect. Immun.* **58**, 3645–52.

Urban, R.G., Pipper, E.M. and Dreyfus, L.A. (1991). Monoclonal antibodies specific for the *Escherichia coli* heat-stable enterotoxin STb. *J. Clin. Microbiol.* **29**, 1963–8.

Vaandrager, A.B., van der Wiel, E. and de Jonge, H.R. (1993). Heat-stable enterotoxin activation of immunopurified guanylyl cyclase C. *J. Biol. Chem.* **268**, 19598–603.

Vaandrager, A.B., van der Wiel, E., Hom, M.L., Luthjens, L.H. and de Jonge, H.R. (1994). Heat-stable enterotoxin receptor/guanylyl cyclase C is an oligomer consisting of functionally distinct subunits, which are non-covalently linked in the intestine. *J. Biol. Chem.* **269**, 16409–15.

Vaandrager, A.B., Ehlert, E.M.E., Jarchau, J., Lohmann, S.M. and de Jonge, H.R. (1996). N-terminal myristoylation is required for membrane localization of cGMP-dependent protein kinase type II. *J. Biol. Chem.* **271**, 7025–9.

Vaandrager, A.B., Bot, A.G.M. and de Jonge, H.R. (1997a). Guanosine 3′, 5′-cyclic monophosphate-dependent protein kinase II mediates heat-stable enterotoxin-provoked chloride secretion in rat intestine. *Gastroenterology* **112**, 437–43.

Vaandrager, A.B., Edixhoven, M., Bot, A.G.M., Kroos, M.A., Jarchau, T., Lohmann, S., Genieser, H.-G. and de Jonge, H.R. (1997b). Endogenous type II cGMP-dependent protein kinase exists as a dimer in membranes and can be functionally distinguished from the type I isoforms. *J. Biol. Chem.* **272**, 11816–23.

Vajanaphanich, M., Kachintorn, U., Barrett, K.E., Cohn, J.A., Dharmsathaphorn, K. and Traynor-Kaplan, A. (1993). Phosphatidic acid modulates Cl secretion in T84 cells: varying effects depending on mode of stimulation. *Am. J. Physiol.* **264**, C1210–18.

Wada, A., Hirayama, T., Kitao, S., Fujisawa, J.-I., Hidaka, Y. and Shimonishi, Y. (1994). Pig intestinal membrane-bound receptor (Guanylyl cyclase) for heat-stable enterotoxin: cDNA cloning, functional expression, and characterization. *Microbiol. Immunol.* **38**, 535–41.

Wada, A., Hasegawa, M., Matsumoto, K., Niidone, T., Kawano, Y., Hidaka, Y., Padilla, P.I., Kurazono, H., Shimonishi, Y. and Hirayama, T. (1996). The significance of Ser[1029] of the heat-stable enterotoxin receptor (STaR): relation of STa-mediated guanylyl cyclase activation and signaling by phorbol myristate acetate. *FEBS Lett.* **384**, 75–7.

Wagner, J.A., Cozens, A.L., Schulman, H., Gruenert, D.C., Stryer, L. and Gardner, P. (1991). Activation of chloride channels in normal and cystic fibrosis airway epithelial cells multifunctional calcium/calmodulin-dependent protein kinase. *Nature* **349**, 793–6.

Wagner, J.A., McDonald, T.V., Nghiem, P.T., Lowe, A.W., Shulman, H., Gruenert, D.C., Stryer, L. and Gardner, P. (1992). Antisense oligonucleotides to the cystic fibrosis transmembrane conductance regulator inhibit cAMP-activated but not calcium-activated chloride currents. *Proc. Natl Acad. Sci. USA* **89**, 6785–9.

Waldman, S.A. and O'Hanley, P. (1989). Influence of a glycine or proline substitution on the functional properties of a 14-amino-acid analogue of *Escherichia coli* heat-stable enterotoxin. *Infect. Immun.* **57**, 2420–4.

Waldman, S.A., Kuno, T., Kamisaki, Y., Chang, L.Y. Gariépy, J., O'Hanley, P., Schoolnik, G. and Murad, F. (1986). Intestinal receptor for heat-stable enterotoxin of *Escherichia coli* is tightly coupled to a novel form of particulate guanylate cyclase. *Infect. Immun.* **51**, 320–6.

Weikel, C.S. and Guerrant, R.L. (1985). STb enterotoxin of *Escherichia coli*: cyclic nucleotide-independent secretion. In: *Microbial Toxins and Diarrhoeal Disease* (Ciba Foundation Symposium 112), pp. 94–115. Pitman, London.

Weikel, C.S., Nellans, H.N. and Guerrant, R.L. (1986). *In vivo* and *in vitro* effects of a novel enterotoxin, STb, produced by *Escherichia coli*. *J. Infect. Dis.* **153**, 893–901.

Weikel, C.S., Spann, C.L., Chambers, C.P., Crane, J.K., Linden, J. and Hewlett, E.L. (1990). Phorbol esters enhance the cyclic GMP response of T84 cells to the heat-stable enterotoxin of *Escherichia coli* (STa). *Infect. Immun.* **58**, 1402–7.

Whipp, S.C. (1987). Protease degradation of *Escherichia coli* heat-stable, mouse-negative, pig-positive enterotoxin. *Infect. Immun.* **55**, 2057–60.

Whipp, S.C., Kokue, E., Morgan, R.W., Rose, R. and Moon, H.W. (1987). Functional significance of histologic alterations induced by *Escherichia coli* pig-specific, mouse-negative, heat-stable enterotoxin (STb). *Vet. Res. Commun.* **11**, 41–55.

Wiegarrd, R.C., Kato, J., Huang, M.D., Fok, K.F., Kachur, J.-F. and Currie, M.G. (1992). Human guanylin: cDNA isolation, structure, and activity. *FEBS Lett.* **311**, 150–4.

Worrell, R.T. and Frizzell, R.A. (1991). CaMKII mediates stimulation of chloride conductance by calcium in T84 cells. *Am. J. Physiol.* **260**, C877–82.

Yamanaka, H. and Okamoto, K. (1996). Amino acid residues in the pro region of *Escherichia coli* heat-stable enterotoxin I that affect efficiency of translocation across the inner membrane. *Infect. Immun.* **64**, 2700–8.

Yamanaka, H., Fuke, Y., Hitotsubashi, S., Fujii, Y. and Okamoto, K. (1993). Functional properties of pro region of *Escherichia coli* heat-stable enterotoxin. *Microbiol. Immunol.* **37**, 195–205.

Yamanaka, H., Kameyama, M., Baba, T., Fujii, Y. and Okamoto, K. (1994). Maturation pathway of *Escherichia coli* heat-stable enterotoxin I: Requirement of *DsbA* for disulfide bond formation. *J. Bacteriol.* **176**, 2906–13.

Yamanaka, H., Nomura, T., Fujii, Y. and Okamoto, K. (1997). Extracellular secretion of *Escherichia coli* heat-stable enterotoxin I across the outer membrane. *J. Bacteriol.* **179**, 3383–90.

Yamanaka, H., Nomura, T. and Okamoto, K. (1998). Involvement of glutamic acid residue at position 7 in the formation of the intramolecular disulfide bond of *Escherichia coli* heat-stable enterotoxin Ip *in vivo*. *Microb. Pathogen.* **24**, 145–54.

Yang, Y., Gao, Z., Guzmán-Verduzco, L.-M., Tachias, K. and Kupersztoch, Y.M. (1992). Secretion of the ST$_{A3}$ heat-stable enterotoxin of *Escherichia coli*: extracellular delivery of pro-ST$_A$ is accomplished by either pro or ST$_A$. *Mol. Microbiol.* **6**, 3521–9.

Yoshimura, S., Ikemura, H., Watanabe, H., Aimoto, S., Shimonishi, Y., Hara, S., Takeda, T., Miwatani, T. and Takeda, Y. (1985). Essential structure for full enterotoxigenic activity of heat-stable enterotoxin produced by enterotoxigenic *Escherichia coli*. *FEBS Lett.* **181**, 138–41.

Yoshimura, S., Takao, T., Shimonishi, Y., Hara, S., Arita, M., Takeda, T., Imaishi, H., Honda, T. and Miwatani, T. (1986). A heat-stable enterotoxin of *Vibrio cholerae* non-01: chemical synthesis, and biological and physiochemical properties. *Biopolymers* **25**, S69–83.

Heat-stable enterotoxins of *Vibrio* and *Yersinia* species

Tae Takeda, Ken-ichi Yoshino, Thandavarayan Ramamurthy and G. Balakrish Nair

INTRODUCTION

Diarrhoeal diseases caused by bacterial pathogens are raising concerns in industrialized countries and remain an important cause of morbidity and mortality in developing countries. During the past decade important advances in the epidemiology, transmission and pathogenesis of bacterial-mediated diarrhoeas have been made. Improved microbiological and molecular methods have resulted in more frequent detection of pathogens in association with diarrhoea and in detection of novel virulence factors resulting in better understanding of pathogenesis of these diseases.

The mechanism of pathogenesis of enteric bacteria is a labyrinthine event which requires co-ordinated expression of several virulence factors to colonize the intestine, invade intestinal epithelial cells and/or produce one or more toxins. Production of toxins is one of the mechanisms by which enteric pathogens precipitate diarrhoea. The heat-stable enterotoxin (ST) produced by enteric bacteria is a small peptide toxin (molecular weight of approximately 2000) (Dreyfus *et al.*, 1983; Thompson and Giannella, 1985), which belongs to a family of cysteine-rich peptides, that is capable of inducing a secretory response by a mechanism still not completely understood.

This chapter reviews our current understanding of the STs produced by diarrhoeagenic *Vibrio* and *Yersinia* species. Emphasis is placed on the genetic and chemical characteristics of ST produced by species belonging to these two genera and the development of diagnostic techniques for detection of the toxin.

CLASSIFICATION OF ST PRODUCED BY ENTERIC BACTERIA

ST was first described in enterotoxigenic *Escherichia coli* (ETEC). Based on structure and function, the *E. coli* ST is divided into two types, namely STa or STI and STb or STII (Burgess *et al.*, 1978; Weikel and Guerrant, 1985). STI is a methanol-soluble suckling mouse-active protease-resistant extracellular peptide toxin, while STII is methanol insoluble, protease sensitive and active in weaned pigs and was earlier thought to be inactive in the suckling mouse assay. A recent study using a trypsin inhibitor to block protease activity showed an intestinal response to STII in mice, rats, rabbits and calves (Whipp, 1990). Another ST produced by enteroaggregative *E. coli* (EAggEC) is genetically and immunologically different from *E. coli* STI or STII, as shown by STh-ELISA, suckling mouse assay and STh DNA probe hybridization tests (Savarino *et al.*, 1991).

Apart from strains of ETEC or EAggEC which produce ST-like toxins, other enteric bacteria are capable of producing ST (Table 30.1). These include strains of

TABLE 30.1 Amino acid sequence of STa toxins

Abbreviation toxin	Length of mature toxin	Amino acid sequence	Minimum effective dose (pmol)
EC-STh	19	Asn – Ser – Ser – Asn – Tyr – Cys – Cys – Glu – Leu – Cys – Cys – Asn – Pro – Ala – Cys – Thr – Gly – Cys – Tyr	0.4
EC-STp	18	Asn – Thr – Phe – Tyr – Cys – Cys – Glu – Leu – Cys – Cys – Asn – Pro – Ala – Cys – Ala – Gly – Cys – Tyr	0.5
C-ST	18	Asn – Thr – Phe – Tyr – Cys – Cys – Glu – Leu – Cys – Cys – Asn – Pro – Ala – Cys – Ala – Gly – Cys – Tyr	0.5
NAG-ST	17	Ile – Asp – Cys – Cys – Glu – Ile – Cys – Cys – Asn – Pro – Ala – Cys – Phe – Gly – Cys – Leu – Asn	2.8
H-ST	18	Leu – Ile – Asp – Cys – Cys – Glu – Ile – Cys – Cys – Asn – Pro – Ala – Cys – Phe – Gly – Cys – Leu – Asn	2.8
M-ST	17	Ile – Asp – Cys – Cys – Glu – Ile – Cys – Cys – Asn – Pro – Ala – Cys – Phe – Gly – Cys – Leu – Asn	2.8
O1-ST	28	Asn – Leu – Ile – Asp – Cys – Cys – Glu – Ile – Cys – Cys – Asn – Pro – Ala – Cys – Phe – Gly – Cys – Leu – Asn	2.8
Y-STa	30	Ser – Ser – Asp – Trp – Asp – Cys – Cys – Asp – Val – Cys – Cys – Asn – Pro – Ala – Cys – Ala – Gly – Cys	7.8
Y-STb	30	Glu – Glu – Asp – Asp – Trp – Cys – Cys – Glu – Val – Cys – Cys – Asn – Pro – Ala – Cys – Ala – Gly – Cys	0.4
Y-STc	53	Asn – Asp – Trp – Cys – Cys – Glu – Leu – Cys – Cys – Asn – Pro – Ala – Cys – Phe – Gly – Cys	0.1

Residue numbering (1, 5, 10, 15, 20, 25, 30, 40, 45, 50) is indicated beneath the sequences.

Vibrio cholerae non-O1 non-O139 which produce NAG-ST (Takao *et al.*, 1985a; Arita *et al.*, 1986), Hakata strains of *V. cholerae* non-O1 which produce H-ST (Arita *et al.*, 1991a), *V. cholerae* O1 which produce O1-ST (Takeda *et al.*, 1991; Mallard and Desmarchelier, 1995) and *V. mimicus* which produce M-ST (Arita *et al.*, 1991b; Yuan *et al.*, 1994; Ramamurthy *et al.*, 1994). Among the other species, Y-ST and its subtypes are produced by *Yersinia enterocolitica* (Delor *et al.*, 1990; Huang *et al.*, 1997; Ramamurthy *et al.*, 1997), Yk-ST from *Y. kristensenii* (Delor *et al.*, 1990; Ibrahim *et al.*, 1992), C-ST from *Citrobacter freundii* (Guarino *et al.*, 1987, 1989a) and ST-like toxin from *Klebsiella pneumoniae* (Guarino *et al.*, 1989b). ST capable of inducing fluid accumulation in rabbit ligated loops without altering intestinal histology has been reported in *Plesiomonas shigelloides* but DNA probes specific for STh and STp show no homologous sequences (Matthews *et al.*, 1988).

DISEASES CAUSED BY VIBRIOS AND YERSINAE

Vibrio cholerae is the causative agent of the disease cholera. However, only two serotypes, namely O1 and O139, are associated with cholera. The other serotypes of *V. cholerae*, collectively known as non-O1 non-O139, are now recognized as the causative agents of gastroenteritis and sometimes extraintestinal infections (Aldova *et al.*, 1968; El-Shawi and Thewaini, 1969; Shehabi *et al.*, 1980). At times, the clinical features of the disease caused by non-O1 non-O139 serotypes are indistinguishable from cholera (Morris *et al.*, 1990). During cholera epidemics, more than 10% of patients with cholera-like diarrhoea are infected with non-O1 non-O139 vibrios. Virulence factors of non-O1 vibrios such as cholera toxin-like enterotoxin (Levine *et al.*, 1988), El Tor haemolysin (Honda and Finkelstein, 1979; Brown and Manning, 1985; Ichinose *et al.*, 1987), Kanagawa haemolysin (Honda *et al.*, 1986; Nishibuchi *et al.*, 1992), Shiga-like toxin (O'Brien *et al.*, 1984) and heat-stable enterotoxin known as NAG-ST (Hoge *et al.*, 1990; Morris *et al.*, 1990; Ogawa *et al.*, 1990) have been proposed to explain the clinical manifestation of *V. cholerae* non-O1 gastroenteritis. Some strains of *V. cholerae* O1, Hakata strains of *V. cholerae* non-O1 and *V. mimicus* also produce ST and have been implicated in diarrhoeal disease (Takeda *et al.*, 1991; Arita *et al.*, 1991a, b; Ramamurthy *et al.*, 1994).

The genus *Yersinia* defines a diverse range of micro-organisms including the three major pathogenic species *Y. pestis*, *Y. pseudotuberculosis* and *Y. enterocolitica*. Since it was first described in 1939, *Y enterocolitica*

has become well established as an enteric pathogen and is capable of causing a variety of clinical disorders in humans (Bottone, 1981, 1983; Cornelis *et al.*, 1987). The severity of the disease caused by *Y. enterocolitica* varies considerably and includes diarrhoea, mesenteric lymphadenitis and ileitis resembling appendicitis (Cover and Aber, 1989). The organism has been isolated most frequently in temperate areas of the world, with a majority of cases reported from northern Europe and North America (Cover and Aber, 1989). While diarrhoea is the most commonly recognized clinical manifestation, *Y. enterocolitica* can also cause invasive diseases (mesenteric adenitis, septicaemia) and has been associated with a variety of autoimmune manifestations, including arthritis and erythema nodosum (Cover and Aber, 1989).

Yersinia enterocolitica produces a heat-stable toxin, known as Y-ST (Robins-Browne *et al.*, 1979; Okamoto *et al.*, 1981), and the biological expression of this toxin can be detected by the suckling mice assay (SMA) (Pai and Mors, 1978). Some strains of *Y. intermedia* (Kwaga *et al.*, 1992; Ramamurthy *et al.*, 1997) and *Y. kristensenii* (Ibrahim *et al.*, 1992) harbour the *yst* gene responsible for ST production. To date, the *yst* gene has not been found among strains classified as *Y. pestis* and *Y. pseudotuberculosis*. However, the fundamental question about the clinical significance of Y-ST produced by *Y. enterocolitica* has been raised by several workers (Boyce *et al.*, 1979; Amirmozafari and Robertson, 1993), mainly because Y-ST is produced *in vitro* only at temperatures below 30°C. Several studies were unable to show a clear correlation between production of Y-ST and the occurrence of diarrhoeal illness (Noble *et al.*, 1987; Bissett *et al.*, 1990; Morris *et al.*, 1991). However, recent examination of *yst* deletion mutants in a rabbit model indicated that the *yst* gene is expressed *in vivo* and is associated with virulence (Delor and Cornelis, 1992). In addition, *yst* transcription can be induced at 37°C by increasing the osmolarity and pH of the culture medium to values normally present in the ileum lumen (Mikulskis *et al.*, 1994).

NUCLEOTIDE AND AMINO ACID SEQUENCE ANALYSIS

Significant divergence has been observed among the genes encoding NAG-ST and *E. coli* STh and STp, although the three nucleotide sequences share a similar sequence at the C-terminal region (Ogawa *et al.*, 1990). Despite the nucleotide sequence divergence, the predicted amino acid sequence shows considerable structural similarity, that is, 39 or 36 of the 78 amino acid residues of NAG-ST are identical to those of *E. coli* STh or STp, respectively (Ogawa *et al.*, 1990). Cluster identity was found only at the *C*-terminus, but throughout the sequence small identities were found, suggesting that the precursors may have structural similarity. The hydropathy plot analysis for each precursor revealed similar profiles between the plots, suggesting that NAG-ST and *E. coli* STs have a common evolutionary origin (Ogawa *et al.*, 1990).

The nucleotide sequence of *sto* (gene encoding O1 ST) is very similar to that of *stn* (gene encoding non-O1 ST and NAG-ST, respectively). Three transitions and three transversions between *sto* and *stn* in their open reading frames (ORFs) give rise to four amino acid changes without affecting the mature ST amino acid sequences (Ogawa and Takeda, 1993). Both genes are flanked by 123-bp direct repeats (DRs) which have at least 93% homology to one another, which also included some inverted repeats (IR) (Ogawa and Takeda, 1993). These DRs appear to be important for the evolutionary process by mediating chromosomal rearrangements such as transpositions, duplications and deletions. It has been reported that the *E. coli* STp is encoded within a transposon (Tn 1681) flanked by IRs of IS1 (So *et al.*, 1979) but in the case of the *sto* and *stn*, Ogawa and Takeda (1993) have reported that there is no typical transposon or insertion sequence in the flanking region. The IR at the junction of the DR and *sto* or *stn* segment might have functional significance and might explain why the ST genes had been encoded between the DRs. The haemagglutinin gene *hag* of *V. cholerae* O1 is also flanked by 123-bp DRs which show 95% homology to DRs flanking the ST genes, indicating that the DR is a repetitive extragenic sequence interspersed in the chromosome of *V. cholerae*. However, not all strains of vibrios have DR sequences, indicating that the DR sequence is a transposable element or is genetically unstable in the chromosome (Ogawa and Takeda, 1993).

Comparison of the sequences of *ystA* and the *estA1* (encoding STh) showed that there is a sequence divergence of 58% and only 23 amino acids are conserved (Delor *et al.*, 1990). A high degree of homology (80%) between *estA1* and *ystA* occurs at the 3′ end of the gene, encoding 11 identical amino acids out of 13 (85%). The length of the active extracellular polypetide differs between Y-STa and STh, as described by Delor *et al.* (1990). The similarity within the *yst* gene family as well as their deduced amino acid sequences are shown in Table 30.2. Among this group of genes, the highest degree of homology exists between *ystB* and *ykst* genes with 76.9% nucleotide similarity, while both *ystA* and *ystC* share 73.5% nucleotide similarity with *ystB* (Ramamurthy *et al.*, 1997). Comparison of the deduced

TABLE 30.2 Nucleotide and amino acid sequence comparison between structural genes and precursor proteins among *Yersinia*-ST toxins

Toxin gene	Percentage homology of		
	ystA	*ystC*	*ykst*
ystB	73.5	73.5	76.9
ystA		73.1	80.6
ystC			70.3

Precursor protein	Percentage homology of		
	Y-STa	Y-STc	Yk-ST
Y-STb	57	60	68
Y-STa		57	76
Y-STc			61

amino acid sequence of precursor protein of Y-STb showed more homology with Yk-ST (68%) compared with Y-STa (57%) and Y-STc sequences (Ramamurthy *et al.*, 1997).

TRANSCRIPTION OF Y-ST

The transcription of Y-ST is a complex process depending on culture medium, temperature and growth phase (Boyce *et al.*, 1979; Delor *et al.*, 1990; Amirmozafari and Robertson, 1993). In commonly used culture media, Y-ST can be detected only at temperatures below 30°C (Amirmozafari and Robertson, 1993), which argued against the role of Y-ST in prolonged diarrhoea at body temperature. Recently, it was shown that the Y-ST gene expression could indeed be induced at 37°C if the osmolarity and pH of the culture medium are close to those of the intestinal environment (Fordtran, 1973; Mikulskis *et al.*, 1994). These findings reconciled the observation regarding Y-ST expression in a host environment and in bacterial cultures, thus reinforcing the role of Y-ST in diarrhoea. Transcription analysis of a *yst*-lacZ operon fusion in a *ymoA* mutant suggested that the negative regulator YmoA participated in *yst* silencing, temperature repression and growth-phase regulation of *yst* (Mikulskis *et al.*, 1994). It has been demonstrated that Yrp (*Yersinia* regulator for pleiotrophic) protein is essential for *yst* expression, and spontaneous mutations in the *yrp* locus might cause a Y-ST silent phenotype (Nakao *et al.*, 1995).

During the growth phase, the *rpoS* gene (Lange and Hengge-Aronis, 1991; McCann *et al.*, 1991), coding for an alternative sigma subunit of RNA polymerase (Tanaka *et al.*, 1993) controls the expression of stationary-phase genes in enteric bacteria, especially the virulence gene *spv* in *Salmonella typhimurium* (Kowarz *et al.*, 1994) and acid resistance in *Shigella flexneri* (Small *et al.*, 1994). An intact *rpoS* gene is necessary for full expression of *yst* in the stationary phase. In *Y. enterocolitica*, Iriatre *et al.* (1995) have characterized the homology of *rpoS* to determine its role in the expression *of yst*. In *rpoS* mutant studies, they showed that RpoS is required for *yst* expression. This suggests that, as for many *E. coli rpoS*-regulated genes, additional control mechanisms that also respond to growth phase are involved in the regulation of *yst*. Interestingly, *yst* expression is modulated by increasing osmolarity, and growth phase regulation of *yst* is also affected by the absence of the histone-like protein, YmoA (Mikulskis *et al.*, 1994). The exact role of *rpoS* in *yst* regulation appears, therefore, to be quite complex.

TOXICITY AND TOXIC DOMAINS

STs produced by various enteropathogenic bacteria share a highly conserved sequence which consists of the following 13 amino acid residues with six cysteine residues linked intramolecularly by three disulfide bonds as follows: Cys-Cys-[Glu/Asp]-[Leu/Ile/Val]-Cys-Cys-Asn-Pro-Ala-Cys-[Ala/Thr/Phe]-Gly-Cys (Shimonishi *et al.*, 1987; Hidaka *et al.*, 1988; Nair and Takeda, 1998). This tridecapeptide constitutes the minimal structure essential for toxicity and has been designated as the 'toxic domain' or the 'core sequence' (Yoshimura *et al.*, 1985). Four amino acids (residues 11–14), Asn-Pro-Ala-Cys, are conserved in all ST family members, as shown in Table 30.1.

Shimonishi *et al.* (1987) have documented that STh molecules in which Glu and Leu residues take the positions of acidic and aliphatic amino acid residues between two Cys-Cys sequences in the toxic domain, respectively, appear to possess high toxicity and also a Trp residue at a position just preceding the toxic domain functions to enhance the toxicity, as in the case of Y-STb (Yoshino *et al.*, 1994). The biological potency of STs depends to a great extent on correct formation of the three disulfide bridges. Peptides with only one disulfide bond show no biological activity. These disulfide bonds are necessary for toxicity as well as for stabilization of the spatial structure of STs required for biological activity, as shown in *E. coli* STh (Yamasaki *et al.*, 1988).

In O1-ST produced by *V. cholerae* O1, the toxin molecule with shorter N-terminal sequences showed more potent toxicity and the minimum effective dose (MED) of the largest one with 28 amino acid residues was 10 times that of the smallest one with 17 amino acid

residues (Yoshino *et al.*, 1993). The purified Y-STb showed an MED of 0.35 pmol in the SMA, indicating that the Y-STb is 20-fold more potent than Y-STa [MED 7.8 pmol (Okamoto *et al.*, 1982)] and is as potent as *E. coli* STh [MED 0.39 pmol (Ikemura *et al.*, 1984)]. It appears that one amino acid replacement of Asp[20] by Glu enhanced toxic activity of Y-STb as produced by 'chimera-peptide' analysis (Yoshino *et al.*, 1994). The higher toxicity of Y-STb compared with Y-STa is due to substitution of Trp for Asp[17] in addition to that of Glu for Asp[20] (Yoshino *et al.*, 1994). Purified Y-STc showed an MED of 0.6 ng (0.1 pmol) in the SMA. This indicates that Y-STc is fourfold more potent than Y-STb (Yoshino *et al.*, 1995) and *E. coli* STh, which was previously thought to be the most toxic among the ST family. STs produced by members of the genus *Vibrio* (O1-ST, NAG-ST, H-ST and M-ST) are less potent (MED 2.8 to 28 pmol) than *E. coli* STh and Y-STb (Takao *et al.*, 1985b; Arita *et al.*, 1991a, b; Yoshino *et al.*, 1993).

Recent studies have shown that Y-STc is the most toxic in the ST family despite having the largest sequence. ST molecules in which Glu and Leu residues take the positions of acidic and aliphatic amino acid residues between two Cys-Cys sequences in the toxic domain, respectively, are known to possess high toxicity (Shimonishi *et al.*, 1988). It has been reported that a Trp residue at a position just preceding the toxic domain functions to enhance the toxicity, as shown in Y-ST (Yoshino *et al.*, 1994). The highest enterotoxicity of Y-STc has been ascribed to a synergism of both of the above properties which are present in Y-STc.

In Y-STc, the toxic domain is also conserved in its C-terminal portion. Furthermore, an amino acid residue at the position just preceding the toxic domain plays an important role in the enterotoxicity but also confers the protease-resistant property to the toxic domain and a few amino acid residues nearby (Ikemura *et al.*, 1984). In other words, the C-terminal rear portion is susceptible to various proteases in the host's intestine.

While the pro-region (see below) plays an important role in the intramolecular disulfide linkage formation and for secretion of the toxin, it does not contribute towards the toxicity of the molecule. Therefore, the boundary between the pro-region and the mature toxin, and the number of amino acid residues at the N-terminus of the toxic domain do not appear to have significance for the biological activity of ST since most of the N-terminal portion of the pro-toxin is degraded, keeping the toxic domain and a few amino acid residues nearby. The mature toxin is produced after processing by various extracellular proteases. Therefore, the cleavage site between the pro-region and mature toxin region of ST depends on both the amino acid sequence of the pro-toxin and the protease system of bacteria responsible for maturation of the toxin.

MATURATION PROCESS AND SECRETION

It has been demonstrated that the ST precursor proteins are comprised of three regions, namely the pre-region, the pro-region and the mature toxin region (Okamoto and Takahara, 1990; Rasheed *et al.*, 1990). The pre-region comprises a conventional leader peptide of the *N*-terminal 18 or 19 amino acid residues. The pre-region is cleaved by a signal peptidase. This cleavage allows the resulting polypeptide (pro-form of ST) to translocate through the inner membrane to the periplasm. The ST pro-form undergoes a cleavage that results in the mature toxin, but where the second proteolysis takes place has remained controversial. Therefore, the pro-region is defined as a sequence between the signal peptide and the mature toxin. The role of the pro-sequence in ST maturation has been examined by several investigators (Okamoto and Takahara, 1990; Rasheed *et al.*, 1990; Yang *et al.*, 1992; Yamanaka *et al.*, 1993, 1994).

Like other STs biosynthesized in the cells as pre-protoxin, *V. cholerae* O1-ST typically consists of three regions (Ogawa and Takeda, 1993). The potential signal sequence, consisting of 18 amino acid residues, is similar to that of other ST-producing bacteria (von Heijne, 1984, 1985), and is cleaved upon export to the periplasm (Sjostrom *et al.*, 1987). By comparing four synthetic peptides, Yoshino *et al.* (1993) documented that the amino acid sequence of Cys-Cys-Glu-Ile-Cys-Cys-Asn-Pro-Ala-Cys-Phe-Gly-Cys in O1-ST is essential for expression of the biological activity and their toxicity becomes more potent with the loss of the *N*-terminal amino acids.

The nucleotide sequence of the *sto* gene is highly homologous to that typical of the *stn* gene (Ogawa *et al.*, 1990; Takeda *et al.*, 1991). However, only the toxin product of NAG-ST could be recovered from the culture supernatant of *V. cholerae* non-O1 (Takao *et al.*, 1985a). The difference in gene products between *V. cholerae* O1 and non-O1 may be ascribed to the specificity of proteases secreted into the periplasmic space and extracellular media by these bacteria (Dohi *et al.*, 1993). There is evidence that abnormally short peptides are easily degraded in *E. coli* (Bukhari and Zipser, 1973; Langley *et al.*, 1975). Similar degradation might occur in *V. cholerae* and the precursor form might prevent the folding of ST, as the leader sequences do (Pork *et al.*, 1988; Weiss *et al.*, 1989).

Y-STa is a 71-amino acid preprotoxin that is processed similarly to *E. coli* STh, in two proteolytic steps to a biologically active 30 amino acid protein (Delor *et al.*, 1990). The Y-STa is cleaved by signal peptidase I to a 53 amino acid peptide (Delor *et al.*, 1990). This peptide is translocated to the periplasm, where three intermolecular disulfide bonds crucial to toxin activity are formed by a protein termed DsbA prior to secretion by the bacteria (Yamanaka *et al.*, 1994).

To elucidate the effect of 13 amino acid substitutions in Y-STb on enhancing the toxicity, several short analogues of Y-STb were compared in the SMA (Yoshino *et al.*, 1994). The enhanced enterotoxicity could be ascribed to the substitution of Trp for Asp[11] in addition to that of Glu for Asp[20]. The amino acid residues adjacent to the *N*-terminus of the Y-STb toxic domain play an important role in the biological activity of ST, as shown by Yoshino *et al.* (1994). In Y-STc, it was shown that the long polypeptide consisting of 53 amino acid residues, including the putative pro-sequence, is secreted as the mature toxin with a highly potent enterotoxicity. This implies that, in *Y. enterocolitica*, the proteolytic processing of a putative pro-toxin appearing in the periplasm is not mandatory for the extracellular secretion of the Y-ST through the other outer membrane and this supports the model proposed by Cruzman-Verduzco and Kupersztoch (Rasheed *et al.*, 1990; Yang *et al.*, 1992). Since Y-STc includes the pro-region-like sequence in its mature toxin region (Yoshino *et al.*, 1995; Huang *et al.*, 1997), the amino acid sequence of the Y-STc precursor protein was useful in understanding the mechanism of maturation and secretion of toxins of the ST family.

The *N*-terminal sequence (1–19) of Y-STa precursor protein has 79% homology with the pre-region (1–19) of Y-STa precursor protein. However, the mature toxin region of Y-STc was encoded following the pre-region, indicating that the precursor protein of Y-STc does not possess a pro-region like other ST precursors (Yoshino *et al.*, 1995; Huang *et al.*, 1997). This was shown by alignment of the *N*-terminal 22 amino acid residues of the native Y-STc and the pro-toxin of Y-STa with 50% similarity (11 out of 22 residues) suggesting that the *N*-terminal 22 amino acid residues of Y-STc correspond to the pro-region of Y-STa. Thus, in the case of Y-STc, a long polypeptide corresponding to progenitor toxin of other STs was secreted as the mature toxin. As further proof, Huang *et al.* (1997) showed that the recombinant Y-ST secreted by an *E. coli* strain was shorter and composed of the *C*-terminal 24 amino acid residues of the native Y-STc. Thus, the length of polypeptide chain of Y-STc from the wild-type *Y. enterocolitica* and an *E. coli* host (JM-109) were different. Likewise, Dohi *et al.* (1993) reported that the NAG-ST gene product in *E. coli* JM-

109 was released to the extracellular compartment as the *C*-terminal 23 or 26 amino acid residues of NAG-ST precursor protein, while native NAG-ST produced by *V. cholerae* non-O1) was 17 amino acid residues long (Takao *et al.*, 1985a). Thus, the peptide chain length of the final products was different, despite being products of the same gene. This indicates that the protease system of the recombinant *E. coli* strain is different from that of the wild-type *V. cholerae* and *Y. enterocolitica* strains. The existence of ST and its subtypes among *V. cholerae* and *Y. enterocolitica* indicates that multiple enzymes take part in the maturation of the toxins. On the basis of these facts, it was proposed that the distinction between the pro-region and the mature ST is not as hard and fast as was previously thought.

RECEPTOR BINDING

In *E. coli* STh, amino acid residues 5–17 or 6–18 confer full binding and enterotoxic activities (Waldman and O'Hanley, 1989). This region shares a striking identity with STs secreted by other enteric bacteria including *Y. enterocolitica* and *V. cholerae* (Takao *et al.*, 1985a, b; Yoshimura *et al.*, 1985). The STs act by binding to an intestinal epithelial receptor protein in the brush border membrane. ST/guanylin receptors in the intestinal tract of humans are found throughout the small intestine and colon, with decreasing receptors along the longitudinal axis of the gut (Krause *et al.*, 1994). For *E. coli* STh, guanylate cyclase type C (GC-C), located in the apical membrane of the intestinal epithelial cells, was identified as the receptor (Schulz *et al.*, 1990; Thompson and Giannella, 1990; Cohen *et al.*, 1993). GC-C is a 120-kDa unglycosylated protein and a 140–160-kDa protein after *N*-linked glycosylation and belongs to a family of receptor cyclases that include the atrial natriuretic peptide receptors, GC-A and GC-B (Chang *et al.*, 1989; Hirayama *et al.*, 1992; Vaandrager *et al.*, 1993). STh of *E. coli* is a super agonist of guanylin GC-C and contains four cysteine residues to activate GC-C and stimulate chloride secretion (Carpick and Gariepy, 1993; Forte *et al.*, 1993; Kuhn *et al.*, 1994). However, it is unclear whether or not the other bacterial STs that elevate intracellular cyclic GMP levels activate GC-C similar to *E. coli* STh (Sears and Kaper, 1996). The active site form of Y-ST stimulates intestinal particulate GC-C (Takao *et al.*, 1985b) and the purified Y-ST increases cGMP levels in mouse intestinal and cultured cells (Inoue *et al.*, 1983). Activation of GC-C results in increased levels of intracellular cGMP that stimulate chloride secretion and/or inhibit absorption, resulting in net intestinal fluid secretion (Field *et al.*, 1978; Hughes *et al.*, 1978; Rao *et al.*, 1980; Guandalini *et al.*, 1982; Huott *et al.*, 1988). Thus, the mechanism by which Y-ST stimulates secretion is thought to be similar or identical to that used by *E. coli* STh.

IMMUNOLOGICAL INTER-RELATIONSHIPS

There appear to be differences in the immunological properties of NAG-ST and various subtypes of Y-ST, as clearly demonstrated by their pattern of neutralization with monoclonal antibodies (MAbs) specific for *E. coli* STh (Takeda *et al.*, 1993) and NAG-ST (Takeda *et al.*, 1990). The biological activity of Y-STa could be completely neutralized by NAG-ST MAb-2F, weakly neutralized by *E. coli* STh MAb SH-1 and not neutralized by any other MAbs (Takeda *et al.*, 1990). In contrast, Y-STb could be neutralized by both 2F and SH-1 while Y-STc could be neutralized by the *E. coli* STh MAbs 53–4 and SH-1 but not neutralized by 2F. Therefore, Y-STb and Y-STc appear to be more immunologically related to *E. coli* STh than the archetype Y-STa.

EPIDEMIOLOGY OF ST-MEDIATED DIARRHOEA

Even though members of the ST family of toxins are detected in different bacterial groups, the links between the presence of ST gene, ST production and the global problem of diarrhoeal disease are not clearly understood.

Among vibrios, *V. mimicus* isolates from clinical and environmental sources are the main reservoir of the ST gene (*stn*) (Ramamurthy *et al.*, 1994; Yuan *et al.*, 1994). The importance of NAG-ST was strengthened by a volunteer study using a CT-negative *V. cholerae* non-O1 non-O139 strain producing NAG-ST (Morris *et al.*, 1990). The *stn* gene was detected in clinical and environmental strains from different geographical locations such as Thailand (Bagchi *et al.*, 1993) and Calcutta, India (Pal *et al.*, 1992). The distribution of the NAG-ST gene (*stn*) among *V. cholerae* non-O1 non-O139 was more predominant in Cuba (18.2%) (Guglielmetti *et al.*, 1994) than in Thailand (6.8%) (Hoge *et al.*, 1990) and Calcutta (2.3%) (Pal *et al.*, 1992). However, in the USA and Mexico, *V. cholerae* non-O1 isolates harbouring *stn* were not detected (Hoge *et al.*, 1990). Takeda *et al.* (1991) described a *V. cholerae* O1 strain, which produced ST toxin (O1-ST) in addition to cholera toxin. The *sto* gene showed high homology with the *stn* gene of *V. cholerae* non-O1 (Takeda *et al.*, 1991). It appears that the distribution of *V. cholerae* O1 strains harbouring *sto* is restricted to environmental isolates of the serotype

Inaba from Australia (Mallard and Desmarchelier, 1995) with the same frequency as the *stn* gene found among *V. cholerae* non-O1 non-O139 strains. Interestingly, as found in the archetype *V. cholerae* O1 strain GP-156 (Takeda *et al.*, 1991), in all 13 Australian strains *sto* was located on the chromosome and found in CT-producing *V. cholerae* (Mallard and Desmarchelier, 1995).

In the genus *Yersinia*, *Y. enterocolitica*, *Y. kristensenii* and *Y. intermedia* are the three species that produce ST (Delor *et al.*, 1990; Ibrahim *et al.*, 1992; Kwaga *et al.*, 1992; Ramamurthy *et al.*, 1997). Production of Y-ST by *Y. enterocolitica* (Delor *et al.*, 1990) is postulated to be important in the pathogenesis of the watery diarrhoea syndrome. Three distinct molecular subtypes of *yst* have been documented, namely *ystA* (Delor *et al.*, 1990), *ystB* (Ramamurthy *et al.*, 1997) and *ystC* (Huang *et al.*, 1997).

Yersinia enterocolitica comprises a group of organisms with 57 serotypes and six biotypes (Wauters *et al.*, 1987, 1991). Based on epidemiological studies, *Y. enterocolitica* was categorized into pathogenic and non-pathogenic serotypes (Cornelis *et al.*, 1987). The serotypes commonly associated with human yersiniosis are limited to O:1,3; O:3; O:9; O:5,27 (European strains) and O:4,32; O:8; O:13a; O:13b; O:18; O:20; O:21 (American strains), which are widely referred to as pathogenic strains (Kay *et al.*, 1983; Cornelis *et al.*, 1987). The pathogenic potential of biotype 1A was reported to be low (Pai and Mors, 1978; Pai *et al.*, 1978; de Boer, 1995) compared with biotypes 2–4 (Wauters *et al.*, 1991), which are generally common among the pathogenic serotypes. Based on DNA–DNA hybridization results, Delor *et al.* (1990) reported that *ystA* is present in all of the tested pathogenic strains of *Y. enterocolitica* but not harboured in the non-pathogenic counterparts. However, several studies in the past have shown that there is strain-to-strain variation in the expression of virulence properties (Bottone, 1981; Kay *et al.*, 1983; Cornelis *et al.*, 1987) and some strains lack the established virulence traits (Noble *et al.*, 1987; Bissett *et al.*, 1990) but are significantly associated with sporadic

cases of diarrhoea (Morris *et al.*, 1991). In addition, several recent epidemiological investigations have raised doubts about the link between traditional criteria for pathogenicity and diarrhoeal disease (Noble *et al.*, 1987; Bissett *et al.*, 1990; Morris *et al.*, 1991; Robins-Browne *et al.*, 1993). Recently, Ramamurthy *et al.* (1997) have shown that the *ystB* gene is harboured in a significant proportion of *Y. enterocolitica* strains belonging to the so-called non-pathogenic bio-serotypes. Even though in *Y. enterocolitica* non-pathogenic serotypes, biotype 1A-associated diarrhoea is uncommon, the presence of the novel subtype gene *ystB* as a virulence trait among these strains might be responsible for causing sporadic cases of diarrhoea (Ramamurthy *et al.*, 1997). Prevalence of this *ystB* subtype is recorded in 18 countries, predominantly from clinical sources in Europe and Chile. Generally, *ystC* gene is not frequent (1% among *Y. enterocolitica*) and is usually observed in both pathogenic and non-pathogenic serotypes (Ramamurthy *et al.*, 1997).

DETECTION AND DIAGNOSIS OF ST PRODUCING VIBRIOS AND YERSINIA ENTEROCOLITICA

A competitive enzyme-linked immunosorbent assay (ELISA) has been developed for screening Y-ST and NAG-ST-producing strains of *Y. enterocolitica* and *V. cholerae* non-O1 and non-O139, respectively, using high-affinity MAb (Nair *et al.*, 1993). In this study there was good concordance between the Y-ST ELISA and the SMA but the evaluation of NAG-ST ELISA and SMA showed discordant results. The inherent low production of NAG-ST by the wild strains of *V. cholerae* non-O1 (Arita *et al.*, 1986) and the production of haemolysin appear to be the two main reasons for the discordant results. However, the NAG-ST ELISA can be used to authenticate NAG-ST-producing strains of *V. cholerae*. Apart from the traditional SMA, ST producing vibrios and sub-

TABLE 30.3 DNA probes, nucleotide sequences of PCR primers and oligonucleotide probes for the detection of different genes encoding ST of vibrios and Yersiniae

Gene	DNA probe/PCR primer pair	Size	Reference
stn/sto	pAO111	0.27 kb	Ogawa *et al.* (1990)
ystA	pID	1.8 kb	Delor *et al.* (1990)
ystA	5'-AAAGATAGTTTTTCTTGT-3' 5'-GCAGCCAGCACACGCGGG-3'	208 bp	Ibrahim *et al.* (1992)
ystB	5'-AAAGCGTGCGATACTCAGAC-3' 5'-CAACATACCTCACAACACCA-3'	68 bp	Ramamurthy *et al.* (1997)
stn	5'-CCTATTCATTGCATTAATG-3' 5'-CCAAAGCAAGCTGGATTGC-3'	215 bp	Ogawa *et al.* (1990)

types of *Y. enterocolitica* could be identified employing several DNA probes, oligionucleotide probes and by PCR amplification. Nucleotide sequences of PCR oligonucleotide primers, poly- and oligonucleotide probes used for identification ST producing vibrios and *Y. enterocolitica* are shown in Table 30.3.

REFERENCES

Aldova, E., Lazonikova, K., Stepankova, E. and Lietova, J. (1968). Isolation of nonagglutinable vibrios from an enteritis outbreak in Czechoslovakia. *J. Infect. Dis.* **118**, 25–31.

Amirmozafari, N. and Robertson, D.C. (1993). Nutritional requirements for synthesis of heat-stable enterotoxin by *Yersinia enterocolitica*. *Appl. Environ. Microbiol.* **59**, 3314–20.

Arita, M., Takeda, T., Honda, T. and Miwatani, T. (1986). Purification and characterization of *Vibrio cholerae* non O1 heat-stable enterotoxin. *Infect. Immun.* **52**, 45–9.

Arita, M., Honda, T., Miwatani, T., Ohmori, K., Takao, T. and Shimonishi, Y. (1991a). Purification and characterization of a new heat-stable enterotoxin produced by *Vibrio cholerae* non O1 serogroup Hakata. *Infect. Immun.* **59**, 2186–8.

Arita, M., Honda, T., Miwatani, T., Takeda, T., Takao, T. and Shimonishi, Y. (1991b). Purification and characterization of a heat-stable enterotoxin of *Vibrio mimicus*. *FEMS Microbiol. Lett.* **79**, 105–10.

Bagchi, K., Echeverria, P., Arthur, J.D., Sethabutr, O., Serichantalergs, O. and Hoge, C.W. (1993). Epidemic diarrhoea caused by *Vibrio cholera* non O1 that produced heat-stable toxin among Khmers in a camp in Thailand. *J. Clin. Microbiol.* **31**, 1315–17.

Bissett, M.L., Powers, C., Abbott, S.L. and Janda, J.M. (1990). Epidemiological investigations of *Yersinia enterocolitica* and related species: sources, frequency, and serogroup distribution. *J. Clin. Microbiol.* **28**, 910–12.

Bottone, E.J. (ed.) (1981). Yersinia enterocolitica. CRC Press, Boca Raton, FL.

Bottone, E.J. (1983). Current trends of *Yersinia enterocolitica* isolates in the New York City area. *J. Clin. Microbiol.* **17**, 63–7.

Boyce, J.M., Evans, D.J., Jr, Evans, D.G. and DuPont, H.L. (1979). Production of heat-stable, methanol-soluble enterotoxin by *Yersinia enterocolitica*. *Infect. Immun.* **25**, 532–7.

Brown, M.H. and Manning P.A. (1985). Haemolysin genes of *Vibrio cholerae*: presence of homologous DNA in non-haemolytic O1 and haemolytic non O1 strains. *FEMS Microbiol. Lett.* **30**, 197–201.

Bukhari, A.I. and Zipser, D. (1973). Mutants of *Escherichia coli* with a defect in the degradation of nonsense fragments. *Nature New Biol.* **243**, 238–41.

Burgess, M.N., Bywater, R.J., Corley, C.M. Mullan, M.N. and Newsome, P.M. (1978). Biological evaluation of a methanol soluble, heat-stable *Escherichia coli* enterotoxin in infant mice, pigs, rabbits, and calves. *Infect. Immun.* **21**, 526–31.

Carpick, B.W. and Gariepy, J. (1993). The *Escherichia coli* heat-stable enterotoxin is a long-lived superagonist of guanylin. *Infect. Immun.* **61**, 4710–15.

Chang, M., Lowe, D.G., Lewis, M., Hellamiss, R., Chen, E. and Goeddel, D.V. (1989). Differential activation by atrial and brain natriuretic peptides of two different receptor guanylate cyclases. *Nature (London)* **341**, 68–72.

Cohen, M.B., Jensen, N.J., Hawkins, J.A., Mann, E.A., Thompson, M.R., Lentze, M.J. and Giannella, R.A. (1993). Receptors for *Escherichia coli* heat-stable enterotoxin in human intestine and in a human intestinal cell line (Caco-2). *J. Cell. Physiol.* **156**, 138–44.

Cornelis, G., Laroche, Y., Ballingad, G., Sory, M-P. and Wauters, G. (1987). *Yersinia enterocolitica*, a primary model for bacterial invasiveness. *Rev. Infect. Dis.* **9**, 64–87.

Cover, T.L. and Aber, R.C. (1989). *Yersinia enterocolitica. N. Engl. J. Med.* **321**, 16–24.

de Boer, E. (1995). Isolation of *Yersinia enterocolitica* from foods. *Contrib. Microbiol. Immunol.* **13**, 71–3.

Delor, I. and Cornelis, G.R. (1992). Role of *Yersinia enterocolitica* Yst toxin in experimental infection of young rabbits. *Infect. Immun.* **60**, 4269–77.

Delor, I., Kaeckenbeeck, A., Wauters, G. and Cornelis, G.R. (1990). Nucleotide sequence of *yst*, the *Yesinia enterocolitica* gene encoding the heat-stable enterotoxin, and prevalence of the gene among pathogenic and non-pathogenic *Yersina*. *Infec. Immun.* **60**, 4269–77.

Dohi, S., Kasuga, H., Nakao, H., Ogawa, A., Nair, G.B. and Takeda, T. (1993). Heterogeneity in the molecular species of heat-stable enterotoxin of *Vibrio cholerae* non O1 expressed by *Escherichia coli* carrying the cloned toxin gene. *FEMS Microbiol. Lett.* **106**, 223–8.

Dreyfus, L.A., Frantz, J.C. and Robertson, D.C. (1983). Chemical properties of heat-stable enterotoxins produced by enterotoxigenic *Escherichia coli* of different host origins. *Infect. Immun.* **42**, 539–48.

El-Shawi, N. and Thewaini, A.J. (1969). Non-agglutinable vibrios isolated in the 1966 epidemic of cholera in Iraq. *Bull. WHO* **40**, 163–6.

Field, M., Graf, L.H., Laird, W.J. and Smith, P.L. (1978). Heat-stable enterotoxin of *Escherichia coli*: in vivo effects on guanylate cyclase activity, cyclic GMP concentration, and ion transport in small intestine. *Proc. Natl Acad. Sci. USA* **75**, 2800–4.

Fordtran, J.S. (1973). Diarrhoea. In: *Gastrointestinal Disease* (eds M.H. Sleisenger and J.E. Fordtran), pp. 291–301. W.B. Saunders, Philadelphia, PA.

Forte, L.R., Eber, S.L., Turner, J.T., Freeman, R.H., Fok, K.F. and Currie, M.G. (1993). Guanylin stimulation of Cl⁻ secretion in human intestinal T84 cells via cyclic guanosine monophosphate. *J. Clin. Invest.* **91**, 2423–8.

Guandalini, S., Rao, M.C., Smith, P.L. and Field, M. (1982). cGMP modulation of ileal ion transport: in vitro effects of *Escherichia coli* heat-stable enterotoxin. *Am. J. Physiol.* **243**, G36–41.

Guarino, A., Capano G., Malamisura, B., Alessio, M., Guandalini, S. and Rubino, A. (1987). Production of *Escherichia coli* STa-like heat-stable enterotoxin by *Citrobacter freundii* isolated from humans. *J. Clin. Microbiol.* **25**, 110–14.

Guarino, A., Giannella, R. and Thompson, M.R. (1989a). *Citrobacter freundii* an 18-amino-acid heat-stable enterotoxin identical to the 18-amino-acid *Escherichia coli* heat-stable enterotoxin (STa). *Infect. Immun.* **57**, 649–52.

Guarino, A., Guandalini, S., Akssio, M., Gentile, F., Tarallo, L., Capano, G., Migliavacca, M. and Rubsino, A. (1989b). Characteristics and mechanism of action of a heat-stable enterotoxin produced by *Klebsiella pneumoniae* from infants with secretory diarrhoea. *Pediatr. Res.* **25**, 514–18.

Guglielmetti, P., Bravo, L., Zanchi, A., Monte, R., Lombardi, G. and Rossolini, G.M. (1994). Detection of the *Vibrio cholerae* heat-stable enterotoxin gene by polymerase chain reaction. *Mol. Cell. Probes* **8**, 39–44.

Hidaka, Y., Kubota, H., Yoshimura, S., Ito, H., Takeda, Y. and Shimonishi, Y. (1988). Disulphide linkages in a heat-stable enterotoxin (STp) produced by a porcine strain of enterotoxigenic *Escherichia coli*. *Bull. Chem. Soc. Jpn* **61**, 1265–71.

Hirayama, T., Wada, A., Iwata, N., Takasaki, S., Shimonsishi, Y. and Takeda, Y. (1992). Glycoprotein receptors for a heat-stable enterotoxin (STh) produced by enterotoxigenic *Escherichia coli*. *Infect. Immun.* **60**, 4213–20.

Hoge, C.W., Sethabutr, O., Bodhidatta, L., Echeverria, P., Robertson, D.C. and Morris, J.G., Jr (1990). Use of a synthetic oligonucleotide probe to detect strains of non-serovar O1 *Vibrio cholerae* carrying the gene for heat-stable enterotoxin (NAG-ST). *J. Clin. Microbiol.* **28**, 1473–6.

Honda, T. and Finkelstein, R.A. (1979). Purification and characterization of hemolysin produced by *Vibrio cholerae* biotype El Tor: another toxic substance produced by cholera vibrios. *Infect. Immun.* **26**, 1020–7.

Honda, T., Nishibuchi, M., Miwatani, T. and Kaper, J.B. (1986). Demonstration of a plasmid-borne gene encoding a thermonstable direct haemolysin in *Vibrio cholerae* non O1 strains. *Appl. Environ. Microbiol.* **52**, 1218–20.

Huang, X., Yoshino, K., Nakao, H. and Takeda, T. (1997). Nucleotide sequence of a gene encoding the novel *Yersinia enterocolitica* heat-stable enterotoxin that includes a pro-region-like sequence in its mature toxin molecule. *Microb. Pathog.* **22**, 89–97.

Hughes, J.M., Murad, F., Chang, B. and Guerrant, R.L. (1978). Role of cyclic GMP in the action of heat-stable enterotoxin of *Escherichia coli*. *Nature* (*London*) **271**, 755–6.

Huott, P.A., Liu, W., McRoberts, J.A., Giannella, R.A. and Dharmsathaphorn, K. (1988). Mechanism of action of *Escherichia coli* heat-stable enterotoxin in a human colonic cell line. *J. Clin. Invest.* **82**, 514–23.

Ibrahim, A., Liesack, W. and Stackebrandt, E. (1992) Differentiation between pathogenic and non-pathogenic *Yersinia enterocolitica* strains by colony hybridization with a PCR-mediated digoxigenin-dUTP-labelled probe. *Mol. Cell. Probes.* **6**, 163–71.

Ichinose, Y., Yamamoto, K., Nakasone, N., Tanabe, M.J., Takeda, T., Miwatani, T. and Iwanaga, M. (1987). Enterotoxicity of El Tor-like haemolysin of non O1 *Vibrio cholerae*. *Infect. Immun.* **55**, 1090–3.

Ikemura, H., Yoshimura, S., Aimoto, S., Shimonishi, Y., Hara, S., Takeda, T., Takeda, Y. and Miwatani, T. (1984). Synthesis of heat-stable enterotoxin (STh) produced by a human strain SK-1 of enterotoxigenic *Escherichia coli*. *Bull. Chem. Soc. Jpn* **57**, 2543–9.

Inoue, R., Okamoto, K., Moriyama, T., Takahashi, T., Shimuzu, K. and Miyama, A. (1983). Effect of *Yersinia enterocolitica* ST on cyclic guanosine 3′, 5′-monophosphate levels in mouse intestines and cultured cells. *Microbiol. Immunol.* **27**, 159–66.

Iriarte, M., Stainier, I. and Cornelis, G.R. (1995). The *rpoS* gene from *Yersinia enterocolitica* and its influence on expression of virulence factors. *Infect. Immun.* **63**, 1840–7.

Kay, B.A., Wachsmuth, K., Gemski, P., Feeley J.C., Quan, T.J. and Brenner, D.J. (1983). Virulence and phenotypic characterization of *Yersinia enterocolitica* isolated from humans in the United States. *J. Clin. Microbiol.* **17**, 128–38.

Kowarz, L., Coynault, C., Robbe-Saule, V. and Norel, F. (1994). The *Salmonella typhimurium katF (rpoS)* gene: cloning nucleotide sequence, and regulation of *spvR* and *spvABCD* virulence plasmid genes. *J. Bacteriol.* **176**, 6852–60.

Krause, W.J., Cullingford, G.L., Freeman, R.H., Eber, S.L., Richardson, K.C., Fok, K.F., Currie, M.G. and Forte, L.R. (1994). Distribution of heat-stable enterotoxin/guanylin receptors in the intestinal tract of man and other mammals. *J. Anat.* **184**, 407–17.

Kuhn, M., Adermann, K., Jahne, J., Forssmann, W.G. and Rechkemmer, G. (1994). Segmental differences in the effects of guanylin and *Escherichia coli* heat-stable enterotoxin on Cl⁻ secretion in human gut. *J. Physiol.* (*London*) **479**, 433–40.

Kwaga, J., Iversen, J.O. and Misra. V. (1992). Detection of pathogenic *Yersinia enterocolitica* by polymerase chain reaction and digoxigenin-labelled polynucleotide probes. *J. Clin. Microbiol.* **30**, 2668–73.

Lange, R. and Hengge-Aronsis, R. (1991). Growth phase-regulated expression of *bolA* and morphology of stationary-phase *Escherichia coli* cells are controlled by the novel sigma factor. *J. Bacteriol.* **173**, 4474–81.

Langley, K.E., Fowler, A.V. and Zabin, I. (1975). Amino acid sequence of β-galactosidase. *J. Biol. Chem.* **250**, 2587–92.

Levine, M.M., Kaper, J.B., Herrington, D., Losonsky, G., Morris, J.G., Clements, M.L., Black, R.E., Tall, B. and Hall, R. (1988). Volunteer studies of deletion mutants of *Vibrio cholerae* O1 prepared by recombinant techniques. *Infec. Immun.* **56**, 161–7.

Mallard, K.E. and Desmarchelier, P.M. (1995). Detection of heat-stable enterotoxin genes among Australian *Vibrio cholerae* O1 strains. *FEMS Microbiol. Lett.* **127**, 111–15.

Matthews, B.G., Douglas, H. and Guiney, D.G. (1988). Production of a heat-stable enterotoxin by *Plesiomonas shigelloides*. *Microb. Pathog.* **5**, 207–13.

McCann, M.P., Kidwell, J.P. and Matin, A. (1991). The putative s factor KatF has a central role in development of starvation-mediated general resistance in *Escherichia coli*. *J. Bacteriol.* **173**, 4188–94.

Mikulskis, A.V., Delor, I., Thi, V.H. and Cornelis, G.R. (1994). Regulation of the *Yersinia enterocolitica* enterotoxin *yst* gene. Influence of growth phase, temperature, osmolarity, pH and bacterial host factors. *Mol. Microbiol.* **14**, 905–15.

Morris, J.G., Jr, Takeda, T., Tall, B.D., Losonsky, G.A., Bhattacharya, S.K., Forrest, B.D., Kay, B.A. and Nishibuchi, M. (1990). Experimental non-O group 1 *Vibrio cholerae* gastroenteritis in humans. *J. Clin. Invest.* **85**, 697–705.

Morris, J.G. Jr, Prado, V., Ferreccio, C., Robins-Brown, R.M., Bordun, A.-M., Cayazzo, M., Kay, B.A. and Levine, M.M. (1991). *Yersinia enterocolitica* isolated from two cohorts of young children in Santiago, Chile: incidence of and lack of correlation between illness and proposed virulence factors. *J. Clin. Microbiol.* **29**, 2784–8.

Nair, G.B. and Takeda, Y. (1998). The heat-stable enterotoxins. *Microb. Pathog.* **24**, 123–31.

Nair, G.B., Bhattacharya, S.K. and Takeda, T. (1993). Identification of heat-stable enterotoxin-producing strains of *Yersinia enterocolitica* and *Vibrio cholerae* non O1 by a monoclonal antibody-based enzyme-linked immunosorbent assay. *Microbiol. Immunol.* **37**, 181–6.

Nakao, H., Watanabe, H., Nakayama, K. and Takeda, T. (1995). *yst* gene expression in *Yersinia enterocolitica* is positively regulated by a chromosomal region that is highly homologous to *Escherichia coli* host factor 1 gene (*hfq*). *Mol. Microbiol.* **18**, 859–65.

Nishibuchi, M., Fasano, A., Russell, R.G. and Kaper, J.B. (1992). Enterotoxigenicity of *Vibrio parahaemolyticus* with and without genes encoding thermostable direct haemolysin. *Infect. Immun.* **60**, 3539–45.

Noble, M.A., Barteluk, R.L., Freeman, H.J., Subrarnaniam, R. and Hudson, J.B. (1987). Clinical significance of virulence-related assay of *Yersinia* species. *J. Clin. Microbiol.* **25**, 802–7.

O'Brien, A.D., Chen, M.E., Holmes, R.K., Kaper, J.B. and Levine, M. (1984). Environmental and human isolates of *Vibrio cholerae* and *Vibrio parahaemolyticus* produce *Shigella dysenteriae* 1 (Shiga)-like cytotoxin. *Lancet* **i**, 77–8.

Ogawa, A. and Takeda, T. (1993). The gene encoding the heat-stable enterotoxin of *Vibrio cholerae* is flanked by 123-base pair direct repeats. *Microbiol. Immunol.* **37**, 607–16.

Ogawa, A., Kato, J.-I., Watanabe, H., Nair, G.B. and Takeda, T. (1990). Cloning and nucleotide sequence of a heat-stable enterotoxin gene from *Vibrio cholerae* non O1 isolated from a patient with traveller's diarrhoea. *Infect. Immun.* **58**, 3325–9.

Okamoto, K. and Takahara, M. (1990). Synthesis of *Escherichia coli* heat-stable enterotoxin STp as pre-pro form and role of the pro sequence in secretion. *J. Bacteriol.* **172**, 5260–5.

Okamoto, K., Inove, T., Ichikawa, H., Kawamoto, Y. and Miyama, A. (1981). Partial purification and characterization of heat-stable enterotoxin produced by *Yersinia enterocolitica. Infect. Immun.* **31**, 554–9.

Okamoto, K., Inoue, T., Shimizu, K., Hara, S. and Miyama, A. (1982). Further purification and characterization of heat-stable enterotoxin produced by *Yersinia enterocolitica. Infect. Immun.* **35**, 958–64.

Pai, C.H. and Mors, V. (1978). Production of enterotoxin by *Yersinia enterocolitica* gastroenteritis. *Infect. Immun.* **9**, 908–11.

Pai, C.H., Mors, V. and Toma, S. (1978). Prevalence of enterotoxigenicity in human and nonhuman isolates of *Yersinia enterocolitica. Infect. Immun.* **22**, 334–8.

Pal, A., Ramamurthy, T., Bhadra, R., Takeda, T., Shimada, T., Takeda, Y., Nair, G.B., Pal, S.C. and Chakrabarti, S. (1992). Reassessment of the prevalence of heat-stable enterotoxin (NAG-ST) among environmental *Vibrio cholerae* non O1 strains isolated from Calcutta, India by using a NAG-ST probe. *Appl. Environ. Microbiol.* **58**, 2485–9.

Pork, S., Lie, G., Topping, T.B., Cover, W.H. and Randall, L.L. (1988). Modulation of folding pathways of exported proteins by the leader sequence. *Science* **239**, 1033–5.

Ramamurthy, T., Albert, M.J., Huq, A., Colwell, R.R., Takeda, Y., Takeda, T., Shimada, T., Mandal, B.K. and Nair, G.B. (1994). *Vibrio mimicus* with multiple toxin types isolated from human and environmental sources. *J. Med. Microbiol.* **40**, 194–6.

Ramamurthy, T., Yoshino, K., Huang, X., Nair, G.B., Carniel, F., Maruyama, T., Fukushima, H. and Takeda, T. (1997). The novel heat-stable enterotoxin subtype gene *(ystB)* of *Yersinia enterocolitica*: nucleotide sequence and distribution of the *yst* genes. *Microb. Pathog.* **23**, 189–200.

Rao, M.C., Guandalini, S., Smith, P.L. and Field, M. (1980). Mode of action of heat-stable *Escherichia coli* enterotoxin: tissue and subcellular specificities and role of cyclic GMP. *Biochim. Biophys. Acta* **632**, 35–46.

Rasheed, J.K., Guzuman-Verduzco, L-M. and Kupersztoch, Y.M. (1990). Two precursors of the heat-stable enterotoxin of *Escherichia coli*: evidence of extracellular processing. *Mol. Microbiol.* **4**, 265–73.

Robins-Browne, R.M., Jansen van Vuuren, C.J., Still, C.S., Miliotis, M.D. and Koornhof, H. (1979). The pathogenesis of *Yersinia enterocolitica* gastroenteritis. *Contrib. Microbiol. Immunol.* **5**, 324–8.

Robins-Browne, R.M., Takeda, T., Fasano, A., Bordun, A.M., Dohi, S., Kasuga, H., Fong, G., Prado, V., Guerrant, R.L. and Morris, J.G., Jr. (1993). Assessment of enterotoxin production of *Yersinia enterocolitica* and identification of a novel heat-stable enterotoxin produced by a noninvasive *Y. enterocolitica* strain isolated from clinical material. *Infect. Immun.* **61**, 764–7.

Savarino, S.J., Fasano, A., Robertson, D.C. and Levine, M.M. (1991). Enteroaggregative *Escherichia coli* elaborate a heat-stable enterotoxin demonstrable in an *in vitro* rabbit intestinal model. *J. Clin. Invest.* **87**, 1450–5.

Schulz, S., Green, C.K., Yuen, P.S.T. and Garbers, D.L. (1990). Guanylyl cyclase is a heat-stable enterotoxin receptor. *Cell.* **63**, 941–8.

Sears, C.L. and Kaper, J.B. (1996). Enteric bacterial toxins: Mechanisms of action and linkage to intestinal secretion. *Microbiol. Rev.* **60**, 167–215.

Shehabi, A.A., Rajab, A.B.A. and Shaker, A.A. (1980). Observation on the emergence of non-cholera vibrios during an outbreak of cholera. *Jordan Med. J.* **14**, 123–7.

Shimonishi, Y., Hidaka, Y., Koizumi, M., Hane, M., Aimoto, T., Takeda, T., Miwatani, T. and Takeda, Y. (1987). Mode of disulphide bond formation of a heat-stable enterotoxin (STh) produced by a human strain of enterotoxigenic *Escherichia coli. FEBS Lett.* **215**, 165–70.

Shimonishi, Y., Aimoto, S., Yoshimura, S., Hidaka, Y., Torishima, H., Takeda, T., Miwatani, T. and Takeda, Y. (1988). In: *Advances in Research on Cholera and Related Diarrhoeas* (eds S. Kuwahara and N.F. Pierce), Vol. 5, pp. 241–52. KTK Scientific, Tokyo.

Sjostrom, M., Wold, S., Wieslander, A. and Rilfors, L. (1987). Signal peptide amino acid sequences in *Escherichia coli* contain information related to final protein localization. A multivariate data analysis. *EMBO J.* **6**, 823–31.

Small, P., Blankenhorn, D., Welty, D., Zinser, E. and Slonczewski. (1994). Acid and base resistance in *Escherichia coli* and *Shigella flexneri*: role of *rpoS* and growth pH. *J. Bacteriol.* **176**, 1729–37.

So, M., Heffron, F. and McCarthy, B.J. (1979). The *Escherichia coli* gene encoding heat stable toxin is a bacterial transposon flanked by inverted repeats of ISI. *Nature.* **277**, 453–6.

Takao, T., Shimonishi, Y., Kobayashi, M., Nishimura, O., Arita, M., Takeda, T., Honda, T. and Miwatani, T. (1985a). Amino acid sequence of heat-stable enterotoxin produced by *Vibrio cholerae* non O1. *FEBS Lett.* **193**, 250–4.

Takao, T., Tominaga, N., Yoshimura, S., Shimonishi, Y., Hara, S., Inoue, T. and Miyama, A. (1985b). Isolation, primary structure and synthesis of heat-stable enterotoxin produced by *Yersinia enterocolitica. Eur. J. Biochem.* **152**, 199–206.

Takeda, T., Nair, G.B., Suzuki, K. and Shimonishi, Y. (1990). Production of a monoclonal antibody to *Vibrio cholerae* non O1 heat-stable entertoxin (ST) which is cross-reactive with *Yersinia enterocolitica* ST. *Infect. Immun.* **58**, 2755–9.

Takeda, T., Peina, Y., Ogawa, A., Dohi, S., Abe, H., Nair, G.B. and Pal, S.C. (1991). Detection of heat-stable enterotoxin in a cholera toxin gene-positive strain of *Vibrio cholerae* 01. *FEMS Microbiol. Lett.* **80**, 23–8.

Takeda, T., Nair, G.B., Suzuki, K., Huang, X., Yokoo, Y., De Mol, P., Hemelhof, W., Butzler, J.-P., Takeda, Y. and Shimonishi, Y. (1993). Epitope mapping and characterization of antigenic determinants of heat-stable enterotoxin (STh) of enterotoxigenic *Escherichia coli* by using monoclonal antibodies. *Infect. Immun.* **61**, 289–94.

Tanaka, K., Takayanagi, Y., Fujita, N., Ishihama, A. and Takahashi, H. (1993). Heterogeneity of the principal σ factor in *Escherichia coli*: the *rpoS* gene product σ^{38} is a second principal σ factor of RNA polymerase in stationary-phase *Escherichia coli. Proc. Natl Acad. Sci. USA* **90**, 3511–15.

Thompson, M.R. and Giannella, R.A. (1985). Revised amino acid sequence for a heat-stable enterotoxin produced by an *Escherichia coli* strain (18D) that is pathogenic for humans. *Infect. Immun.* **47**, 834–6.

Thompson, M.R. and Giannella, R.A. (1990). Different cross linking agents identify distinctly different putative *Escherichia coli* heat-stable enterotoxin rat intestinal cell receptor proteins. *J. Receptor Res.* **10**, 97–117.

Vaandrager, A.B., Schulz, S., de Jonge, H.R. and Garbers, D.L. (1993). Guanylyl cyclase C is an *N*-linked glycoprotein receptor that accounts for multiple heat-stable enterotoxin-binding proteins in the intestine. *J. Biol. Chem.* **268**, 2174–9.

von Heijne, G. (1984). How signal sequences maintain cleavage specificity. *J. Mol. Biol.* **173**, 243–51.

von Heijne, G. (1985). Signal sequence: the limits of variation. *J. Mol. Biol.* **184**, 99–105.

Waldaman, S.A. and O'Hanley, P. (1989). Influence of a glycine or proline substitution on the functional properties of a 14-amino-acid analogue of *Escherichia coli* heat-stable enterotoxin. *Infect. Immu.* **57**, 2420-4.

Wauters, G., Kandolo, K. and Janssens, M. (1987). Revised biogrouping scheme of *Yersinia enterocolitica. Contrib. Microbiol. Immunol.* **9**, 14–21.

Wauters, G., Aleksic, S., Charlier, J. and Schulze, G. (1991). Somatic and flagellar antigens of *Yersinia enterocolitica* and related species. *Contrib. Microbiol. Immunol.* **12**, 239–43.

Weikel, C.S. and Guerrant, R.L. (1985). STb enterotoxin of *Escherichia coli:* cyclic nucleotide independent secretion. *CIBA Foundation Symp.* **112**, 94–114.

Weiss, J.B., MacGregor, C.H., Collier, D.N., Fikes, J.D., Ray, P.H. and Bassford, P.J., Jr (1989). Factors influencing the *in vitro* translocation on the *Escherichia coli* maltose-binding protein. *J. Biol. Chem.* **264**, 3021–7.

Whipp, S.C. (1990). Assay for enterotoxigenic *Escherichia coli* heat-stable enterotoxin b in rats and mice. *Infect. Immun.* **58**, 930–4.

Yamanaka, H., Fuke, Y., Hitotsubashi, S., Fuji, Y. and Okamoto, K. (1993). Functional properties of pro region of *Escherichia coli* heat-stable enterotoxin. *Microbiol. Immunol.* **37**, 195–205.

Yamanaka, H., Kameyama, M., Baba, T., Fuji, Y. and Okamoto, T. (1994). Maturation pathway of *Escherichia coli* heat-stable enterotoxin I: requirement of DsbA for disulphide bond formation. *J. Bacteriol.* **176**, 2906–13.

Yamasaki, S., Hidaka, Y., Ito, H., Takeda, Y. and Shimonishi, Y. (1988). Structure requirements for the spatial structure and toxicity of heat-stable enterotoxin (STh) of enterotoxigenic *Escherichia coli*. *Bull. Chem. Soc. Jpn* **61**, 1701–6.

Yang, Y., Gao, Z., Guzuman-Verduzco, L.-M., Tachias, K. and Kupersztoch, Y.M. (1992). Secretion of the STa3 heat-stable enterotoxin of *Escherichia coli:* extracellular delivery of Pro-STa is accomplished by either Pro or STa. *Mol. Microbiol.* **6**, 3521–9.

Yoshimura, S., Ikemura, H., Watanabe, H., Aimoto, S., Shiamonishi, Y., Hara, S., Takeda, T., Miwatani, T. and Takeda, Y. (1985). Essential structure for full entrotoxigenic activity of heat-stable enterotoxin produced by enterotoxigenic *Escherichia coli*. *FEBS Lett.* **181**, 138–42.

Yoshino, K., Miyachi, M., Takao, T., Bag, P.K., Huang, X., Nair, G.B., Takeda, T. and Shimonishi, Y. (1993). Purification and sequence determination of heat-stable enterotoxin elaborated by a cholera toxin-producing strain of *Vibrio cholerae* O1. *FEBS Lett.* **326**, 83–6.

Yoshino, K., Huang, X., Miyachi, M., Hong, Y.-M., Tako, T., Nakao, H., Takeda, T. and Shimonishi, Y. (1994). Amino acid sequence of a novel heat-stable enterotoxin produced by a *yst* gene-negative strain of *Yersinia enterocolitica*. *Lett. Peptide Sci.* **1**, 95–105.

Yoshino K., Takao, T., Huang, X., Murata, H., Nakao, H., Takeda, T. and Shimonishi, Y. (1995). Characterization of a highly toxic, large molecular size heat-stable enterotoxin produced by a clinical isolate of *Yersinia enterocolitica*. *FEBS Lett.* **362**, 319–22.

Yuan, P., Ogawa, A., Ramamurthy, T., Nair, G.B., Shimada, T., Shinoda, S. and Takeda, T. (1994). *Vibrio mimicus* are the reservoirs of the heat-stable enterotoxin gene (nag-st) among species of the genus *Vibrio*. *World J. Micriobiol. Bitechnol.* **10**, 59–63.

31

Bacteroides fragilis toxins

Cynthia L. Sears

INTRODUCTION

Myers *et al.* (1984) first described strains of *Bacteroides fragilis* associated with diarrhoeal illnesses in young lambs (termed enterotoxigenic *B. fragilis* or ETBF). Their subsequent studies revealed that culture supernatants of ETBF strains stimulated intestinal secretion in lamb ligated intestinal loops and that the biologically active factor was a heat-labile protein toxin of approximately 20 kDa (Myers *et al.*, 1984, 1985).

Over the previous 15 years, *B. fragilis* had been identified as the leading pathogenic anaerobe in systemic infections (Polk and Kasper, 1977) and experimental data indicated that the extraintestinal virulence of *B. fragilis* was primarily due to the capsule of the organisms (Onderdonk *et al.*, 1977). However, no toxin or virulence protein of *B. fragilis* had been identified. Thus, the landmark experiments of Myers *et al.* (1984, 1985) both expanded knowledge on the pathogenic potential of *B. fragilis* and identified the first *B. fragilis* toxin (BFT).

GENETIC STRUCTURE AND CLASSIFICATION OF THE *BACTEROIDES FRAGILIS* TOXINS

Putative *B. fragilis* toxins fall into two categories which are described in detail below. The first category (class-

I toxins) contains the most established toxins and is the best studied. This category includes the BFT first described by Myers *et al.* (1984). The second category (class-II toxin), at present, contains only a single putative toxin termed metalloprotease II (MPII).

Class I toxins

Two prototype intestinal ETBF strains have been studied in the greatest depth, 86-5443-2-2 (piglet isolate) and VPI13784 (lamb isolate). The BFTs secreted by these strains are termed 86-BFT and VPI-BFT (Wu *et al.*, 1997). More recently, there has been increasing recognition that extraintestinal *B. fragilis* strains are also toxigenic, but little information is presently available on the BFT(s) secreted by these strains (Pantosti *et al.*, 1994; Kato *et al.*, 1995, 1996, 1998; Mundy and Sears, 1996; Chung *et al.*, 1998). The highest percentages of toxigenic *B. fragilis* from extraintestinal sites have been reported from Asia, particularly Korea (Chung *et al.*, 1998) and Japan (Kato *et al.*, 1995, 1996).

To date, the *B. fragilis* toxin (*bft*) gene has been cloned and sequenced from three ETBF strains, 86-5443-2-2 (86-*bft*), VPI13874 (VPI-*bft* or *bftP*) and Korea 419 (Korea-*bft*) (Moncrief *et al.*, 1995; Franco *et al.*, 1997a; Kling *et al.*, 1997; Chung *et al.*, 1998). Using a collection of 45 intestinal or extraintestinal ETBF strains isolated in the USA, the 86-*bft* and VPI-*bft* alleles were identified in 42 and 58% of strains, respectively (Franco *et al.*, 1997a). In contrast, none of these strains is positive for the Korea-*bft* allele (Franco and Sears, unpublished data), which has only been identified in strains from Korea to date (Chung *et al.*, 1998). Although each *bft* allele sequenced so far is distinct, all *bft* genes identified are chromosomal, have a G + C content of 39% and are predicted to encode a 397-residue holotoxin with a calculated molecular weight of approximately 44.5 kDa. These data, combined with identification of the *N*-terminal amino acid sequence of purified BFT, indicate that ETBF strains probably synthesize BFT as a pre-proprotein toxin which is then processed by ETBF prior to secretion of the biologically active toxin (Figure 31.1) (Van Tassell *et al.*, 1992; Franco *et al.*, 1997a; Kling *et al.*, 1997).

The initial 18 amino acids of the predicted BFT comprise a signal peptide with a pattern typical of lipoprotein signal peptides, although there is no experimental evidence indicating that BFT is a lipoprotein. The 'pro' region is predicted to be 211 residues in length. By analogy with data analysing the function of the proprotein region of the *Pseudomonas aeruginosa* elastase (McIver *et al.* 1995), the BFT proprotein region is hypothesized to be instrumental in proper protein folding and secretion of biologically active BFT (Franco *et al.*, 1997a). The high degree of homology of the pre- and proprotein domains of 86-BFT and VPI-BFT (see below and Figure 31.2) is consistent with the interpretation of conserved functions for these protein domains. Cleavage at the Arg-Ala site at the carboxy-terminus of the proprotein domain releases the mature toxin (Moncrief *et al.*, 1995; Franco *et al.*, 1997a).

The predicted mature toxin domain of each *bft* allele contains the HEXXH motif, suggesting that the encoded proteins belong to the family of zinc-dependent metalloprotease toxins which include, for example, tetanus toxin and botulinum toxin (Häse and Finkelstein, 1993). The presence of a predicted Met-turn region and threonine residue immediately following the zinc-binding motif suggests that BFT is most close-ly related to the family of metalloproteases termed metzincins, subfamily matrixins (matrix metalloproteases) (Moncrief *et al.*, 1995). Consistent with these observations, BFT has been shown to be a protease *in vitro* (Moncrief *et al.*, 1995) and *in vivo* (see below and Wu *et al.*, 1998a) and to contain 1 g-atom of Zn^{2+} per toxin molecule (Moncrief *et al.*, 1995). In addition, the biological activity of BFT is reduced by approximately 90% by zinc chelation of the purified protein (Moncrief *et al.*, 1995; Obiso *et al.*, 1995; Koshy *et al.* 1996). *In vitro* substrates of BFT include G (monomeric)-actin, gelatin, azocoll, tropomyosin, collagen IV, human complement C3 and fibrinogen, although no biological significance of these substrates has been identified (Moncrief *et al.*, 1995; Obiso *et al.*, 1997a; Saidi *et al.*, 1997).

BFT also undergoes autodigestion but it is not known whether this is the key mechanism responsible for release of mature BFT (Moncrief *et al.*, 1995). The carboxy-terminus of the 86-*bft* (but not the VPI-*bft*) also encodes an amphipathic domain, suggesting that the 86-BFT protein may oligomerize and insert into cell membranes (Franco *et al.*, 1997a). However, there is no direct evidence for this potential mechanism of action as yet.

Although only preliminary information is presently available, the Korea-*bft* allele, cloned from a systemic isolate of *B. fragilis*, is most closely related to the 86-*bft* allele but lacks the amphipathic domain (Chung *et al.*, 1998). Kato *et al.* (1998) have also recently reported the identification of a unique *bft* allele from systemic isolates of *B. fragilis* in Japan. The relationship of this gene to the Korea-*bft* is not yet known.

A direct comparison of the predicted amino acid sequences of the 86-*bft* and the VPI-*bft* reveals that the predicted proteins are 95% similar and 92% identical (Figure 31.2) (Franco *et al.*, 1997a). However, the degree of identity varies with the predicted protein domains of the toxins. Namely, comparison of the pre-proprotein domains indicate that these regions are highly conserved with only two amino acid changes. In contrast, 29 amino acid changes are predicted in the mature toxin protein domain.

Consistent with the protein diversity predicted by the gene sequence, the 86-BFT and VPI-BFT purify with distinct biochemical profiles and migrate differently when analysed by sodium dodecyl sulfate–polyacrylamide gel electrophoresis (SDS–PAGE), confirming that the protein products of these *bft* alleles are unique proteins (Table 31.1) (Wu *et al.*, 1998a, b). The VPI-BFT has been shown to be trypsin and chymotrypsin resistant (the 86-BFT has not been specifically studied). Both the 86- and VPI-BFTs are stable over a wide pH range (i.e. pHs 5–10) (Van Tassell *et al.*, 1992; Saidi *et al.*, 1997).

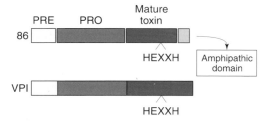

FIGURE 31.1 Schematic of the protein structures of BFT-1 (VPI-BFT) and BFT-2 (86-BFT). The amino acid sequences of these proteins predict that they are pre-proprotein toxins. Each protein contains a HEXXH motif, suggesting that they are zinc-dependent metalloprotease toxins. Zinc-dependent proteolytic activity has been confirmed for BFT-1 and BFT-2. BFT-2 is also predicted to have an amphipathic carboxy-terminus.

Signal Peptide Domain

```
MKNVKLLLMLGTAALLAA CSNEADSLTTSIDAPVTASIDLQSVSYTDLATQLNDVSDFGK   VPI
::::::::::::::::::: ::::::::::::.::::::::::::::::::::::::::::::
MKNVKLLLMLGTAALLAA CSNEADSLTTSIDTPVTASIDLQSVSYTDLATQLNDVSDFGK   086
::::.:  :......:::...  .  .  . .     ..:.:..: ::: :. ..:.  .
MKNIKYYLFFAATALFTA CADDLLHVEETASPQLEHVLNLRSMDYEDLAGVLSKISNTEH   MPII
```

```
MIILKDNGFNRQVHVSMDKRTKIQLDNENVRLFNGRDKDSTSFILGDEFAVLRFYRNGES
::::::::::::::::::::::::::::::::::::::::::::::::::::::::::::::
MIILKDNGFNRQVHVSMDKRTKIQLDNENVRLFNGRDKDSTSFILGDEFAVLRFYRNGES
 :.:....    .  ..   . ..:. .:. : ::.:.:::: . ::. .:..:. :.:.
TIMLQEGSELWTTSIKAIHGVEIEESNRPVYLFEGQDKDSINAILSQSYATIRLQRGGDL
```

```
ISYIAYKEAQMMNEIAEFYAAPFKKTRAINEKEAFECIYDSRTRSAGKDIVSVKINIDKA
:::::::::::::::::::::::::::::::::::::::::::::::::.:::::::::::
ISYIAYKEAQMMNEIAEFYAAPFKKTRAINEKEAFECIYDSRTRSAGKDLVSVKINIDKA
:.::.::. .: : :::..:   .  .  .  :       ::::..... :  :: ::
IDYIVYKDKERMAEIANYYQNHYLSASSDTSDKIVVCNTGEDTRSGNSDIKNIRVDITKA
```

Mature Toxin Domain

```
---KKILNLPECDYINDYIKTPQVPHGITESQTR VPSEPKTVYVICLRENGSTIYPNEV
   :::::::::::::::::::::::::::::::: :::::::::::::.::.:::::::
---KKILNLPECDYINDYIKTPQVPHGITESQTR VPSEPKTVYVICLRESGSTVYPNEV
   .  .:: :: .. .:.  ..:   :.  .:: :  . : . ..:.  ...
IGNNPFKGLPIKDYPTEKLSTID-KNSILSLSSRA--TYPATLEFMLIKEKDGGSLEHDI
```

```
SAQMQDAANSVYAVHGLKRYVNFHFVL--YTTEYSCPSGDAKE----GLEGFTASLKSNP
:::::::::::::::::::::.::.:::  :::::::::::.: :    ::.:::::::.::
SAQMQDAANSVYAVHGLKRFVNLHFVL--YTTEYSCPSGNADE----GLDGFTASLKANP
..:.:  ...:       :: ...  :.   :: :  .:.:.   ...:.  :..
TSQIQAVTTS------LKFLIDSGFITVKYTIKDSSHKGGASDYEVSALESFQNYLRSWD
```

```
                                                           *
KAEGYDDQIYFLIRWGTWDNK-ILGM-SWFNSYNVNTA-SDFEASGMSTTQLMYPGVMAH
::::::::::::::::::::.  :::.  ::..::::::  :::.:::::::::::::::::
KAEGYDDQIYFLIRWGTWDNN-ILGI-SWLDSYNVNTA-SDFKASGMSTTQLMYPGVMAH
 ...:  : . :.:.:  ::::.   .:   :  . ..:.   ..:......:::.   .: ..::
EVKGQDKKPYILLRD GTWDSGKT FGYASGIGVIHLNNPRGNFEVAAISTTSSSHPYTLAH
         ATP/GTP Binding Site Motif A
```

```
*  *  *  *      *
ELGHILGAEHTDNSKDLMYATFTGYLS--HLSEKNMDIIAKNLGWEAADGD
:::::::::.:.:. :::::. .::::   :::::.:: ::::::::: ::::
ELGHILGARHADDPKDLMYSKYTGYLF--HLS EENMYRIAKNLGWEIADGD   Amphipathic Domain
:.::.::.:. :::::. :.  .   ::: .:. .:  .::
EIGHLLGAEHVDNEQDLMYTWYSPQVTPNHLSADNWVRMLECIQ-----K
```

FIGURE 31.2 Comparison of the amino acid sequences of BFT-1, BFT-2 and MPII. BFT-1 (VPI-BFT) and BFT-2 (86-BFT) are highly related proteins with overall 95% similarity and 92% identity in amino acid sequence. Note, however, that only two amino acid changes between the two proteins are present in the protoxin domain whereas 29 amino acid differences are clustered in the mature toxin domain. In contrast, MPII is only approximately 56% similar and 28% identical to BFT-2. MPII (sequence from strain 86-5443-2-2, GenBank Accessions No. AFO56297) contains a lipoprotein signal peptide similar to BFT-1 and BFT-2 but contains a unique ATP/GTP binding motif, suggesting that it may bind a nucleotide. No biological activity has yet been identified for MPII. *Residues forming the Zn^{2+}-binding signature motif.

TABLE 31.1 Comparison of BFT-1 (VPI-BFT) and BFT-2 (86-BFT)

	BFT-I or VPI-BFT (ETBF strains VPI 13784)	BFT-2 or 86-BFT (ETBF strain 86-5443-2-2)
Protein characterization		
MW (SDS–PAGE)	22 kDa	20 kDa
NaCl elution, mono Q column	0.22 ± 0.005 M	0.18 ± 0.001 M[a]
End-point titre, HT29/C1 cells	0.475 ± 0.14 ng ml^{-1}	0.125 ± 0.02 ng ml^{-1}[b]
Heat-labile protein	Yes	Yes
Pathophysiological outcomes[c]		
↑ HT29/C1 cell volume	Yes	Yes
↑ protein synthesis	Yes	Yes
↓ monolayer resistance	Yes	Yes
Actin rearrangement	Yes	Yes
Cleaves E-cadherin	Yes	Yes
Cell effects reversible	Yes	Yes
Cytotoxic	No	No
Enters cells	No	No
Toxin gene characterization		
Predicted proteolytic domain	Present	Present
Predicted amphipathic domain	Absent	Present
Human ETBF epidemiology ($n = 139$)	48.6% strains positive	51.4% strains positive
Strains simultaneously positive with specific *bft-1* and *bft-2* oligonucleotide probes	None	None
B. fragilis Pai[d]	Present	Present

[a]Student's *t*-test, $P < 0.001$.
[b]Student's *t*-test, $P \leq 0.05$.
[c]Results are primarily from studies of BFT action on human intestinal cell lines. However, direct comparisons of BFT-1 and BFT-2 potency in most assays are not yet available.
[d]Pai: Pathogenicity island.

Class-II toxin

Recent data indicate that the chromosome of ETBF strains possesses a 6-kb region of DNA found in ETBF strains but not in non-toxigenic *B. fragilis* strains (Franco *et al.*, 1997b, 1998; Moncrief *et al.*, 1998a). This DNA region has a guanosine plus cytosine content of 35%, which differs from that of the native *B. fragilis* chromosome (*ca.* 42%) (Smith *et al.*, 1998). Because of the strict association of this 6-kb region with ETBF strains and the G + C content which differs from the native *B. fragilis* chromosome, this DNA region has been proposed to be a <u>B</u>. *fragilis* <u>p</u>athogenicity <u>i</u>sland or <u>i</u>slet (BfPai). Sequence analysis of this 6-kb region revealed an open reading frame with homology to other zinc-dependent metalloprotease genes. This gene (termed *mpII* for <u>m</u>etallo<u>p</u>rotease <u>II</u>) (Moncrief *et al.*, 1998a) is predicted to encode a 44.4-kDa protein (termed MPII) which is only *ca.* 56% similar and 28% identical to the other BFTs described to date (Figure 31.2). Similarly to 86-BFT and VPI-BFT, the initial 18 amino acids comprise a domain typical of a lipoprotein signal peptide and preliminary data suggest that MPII is also processed to an approximately 20-kDa protein (A.A. Franco and C.L. Sears, unpublished data). In addition, MPII is predicted to contain an ATP/GTP-binding

site motif A (P-loop) (Franco *et al.*, 1998). This motif may indicate that MPII binds a nucleotide (Walker *et al.*, 1982; Saraste *et al.*, 1990). However, no biological activity has yet been described for this protein, nor has it been purified (Franco *et al.*, 1997b, 1998; Moncrief *et al.*, 1998b).

Given the rapid evolution of our knowledge of the established and putative toxins of *B. fragilis*, the terminology for the toxin genes and proteins is in flux. It has been proposed that the biochemically discrete toxins secreted by ETBF strains VPI13784 and 86-5443-2-2 be termed BFT-1 and BFT-2 and their corresponding genes, *bft-1* and *bft-2*, respectively, consistent with the order in which the proteins were purified and genetic sequences identified (Van Tassell *et al.*, 1992; Franco *et al.*, 1997a; Kling *et al.*, 1997; Wu *et al.*, 1997). Fragilysin has been proposed as an alternative name for BFT (Obiso *et al.*, 1997b). Classification of the Korea-*bft* and Japan-*bft* awaits further genetic, biochemical and biological data.

Biological and physiological activities of *Bacteroides fragilis* toxins

Only the activities of BFT-1 (VPI-BFT) and BFT-2 (86-BFT) have been examined in any detail in animal models, cultured cells and/or monolayers of polarized epithelial cells. Direct 'head-to-head' comparisons of these proteins in each experimental condition are not available. However, to date, all identified biological and physiological activities for BFT-1 and BFT-2 have been very similar, if not identical (Table 31.1). One potential difference between BFT-1 and BFT-2 is that BFT-2 exhibits modest, but consistently greater biological activity than BFT-1 when tested on the cloned human colonic epithelial cell line, HT29/C1. However, given the parallel biological activities of these two proteins (Table 31.1 and see below), the observed difference in specific activity of these two proteins (S. Wu and C.L. Sears, data not shown) may not be pathogenetically important. Thus, in this section, the biological and physiological activities of these proteins will be described without distinguishing between the two isotypes of BFT.

The activity of purified BFT has been demonstrated in ligated intestinal segments or loops of rats, rabbits and lambs (Obiso *et al.*, 1995). These experiments revealed that: (a) histological changes precede the detection of intestinal secretion after treatment with BFT; and (b) BFT stimulates dose-dependent secretion in both ileal and colonic loops in all species examined, although there were species-specific differences in BFT potency in ileal vs colonic loops (Obiso *et al.*, 1995). Consistent with these results, previous data in which ETBF organisms were fed to intact animals have suggested that the primary sites of pathology in ETBF intestinal disease were the distal ileum and colon (Collins *et al.*, 1989; Duimstra *et al.*, 1991). Fluid secreted into the intestinal lumen after treatment with BFT revealed accumulations of sodium and chloride as well as albumen and protein (Obiso *et al.*, 1995). At a higher dose of BFT, mildly haemorrhagic fluid and patchy mucosal wall haemorrhage were observed.

Histological examination of ileal and colonic tissue treated with BFT in all species revealed villus blunting, crypt elongation and neutrophil infiltration consistent with earlier reports of the histopathology of ETBF disease or intestinal tissue exposed to BFT-containing sterile culture filtrates (reviewed in Sears *et al.*, 1995). Extensive detachment and rounding of surface epithelial cells resulting in necrosis of villus tips were observed. The ileal and colonic tissues of rats were the most severely damaged by BFT. The secretory and histological effects of purified BFT were partially inhibited by zinc chelation of the toxin (Obiso *et al.*, 1995). Given that BFT is not a lethal toxin when epithelial cells are treated *in vitro* (see below and Donelli *et al.*, 1996; Koshy *et al.*, 1996; Wells *et al.*, 1996; Chambers *et al.*, 1997; Obiso *et al.*, 1997a), the mechanism(s) leading to tissue damage and haemorrhage are unknown.

Early experiments attempting to define the biological activity of BFT on tissue culture cells failed to identify any activity when crude BFT was tested on Chinese hamster ovary (CHO), Y-1 adrenal and Vero cells (R.B. Sack, personal communication; Sears *et al.*, 1995). However, in 1992, BFT was first observed to have biological activity on continuous human colonic carcinoma cell lines, HT29/C1, Caco-2 and T84 (Weikel *et al.*, 1992). Subconfluent uncloned HT29 or cloned HT29/C1 cell lines have been studied in the greatest detail and, in particular, cloned HT29/C1 cells have been shown to be exquisitely sensitive to BFT with a half-maximal effective concentration of approximately 12.5 pM and as little as 0.5 pM altering cell morphology (Saidi and Sears, 1996). Using the HT29/C1 cell line, the key observation is that BFT (BFT-1, BFT-2 and BFT-Korea) causes rounding of the cells and disruption of cell-to-cell contacts leading to dispersion of the cells from their normally tight cluster morphology (Figure 31.3a) (Weikel *et al.*, 1992; Mundy and Sears, 1996; S. Wu, G.-T. Chung and C.L. Sears, unpublished observations). These changes in the light microscopic appearance of the cells are accompanied by time- and concentration-dependent changes in the F (filamentous)-actin structure of the cells, although the total F-actin content of the cells is unaltered (Donelli *et al.*, 1996; Koshy *et al.*, 1996; Saidi and Sears, 1996). Decreased stress fibres, cell surface blebs and floccular F-actin staining with peripheral F-actin condensation are observed (Donelli *et al.*, 1996; Koshy *et al.*, 1996).

In studies of subconfluent HT29/C1 cells, BFT is a non-cytolethal toxin which alters the morphology of HT29/C1 cells in a time-, temperature- and concentration-dependent manner (Saidi and Sears, 1996; Chambers *et al.*, 1997; Wu *et al.*, 1998). BFT is inactive at 4°C, partially active at 20°C and fully active at 37°C (Moncrief *et al.*, 1995; Saidi and Sears, 1996). At 37°C, the onset of action of BFT is rapidly irreversible. After treatment of HT29/C1 cells with 0.25 nM BFT for only 7 min or 2.5 nM for 2 min (followed by washing to remove the toxin), cells treated with BFT are destined to change shape. Namely, HT29/C1 cells will begin to develop altered morphology as early as 9 min after BFT treatment, with 100% of subconfluent HT29/C1 cells exhibiting rounding by 30–60 min (Saidi and Sears, 1996; Wu *et al.*, 1998). Changes in cell morphology become even more dramatic over the next 3–6 h (Weikel *et al.*, 1992; Saidi and Sears, 1996).

Despite the dramatic changes in morphology, HT29/C1 cells regain a normal appearance by light microscopy by 2–3 days after toxin treatment, indicating that the biological activity of BFT on HT29/C1 cells is reversible (Weikel *et al.*, 1992; Saidi and Sears, 1996; Wu *et al.*, 1998). Consistent with the non-lethal and non-

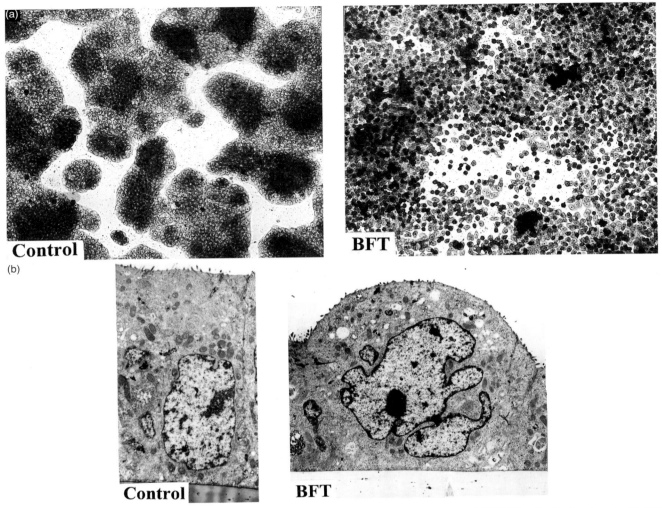

FIGURE 31.3 Effect of BFT on intestinal epithelial cells *in vitro*. (a) Light microscopic appearance of HT29/C1 cells after treatment with BFT (5 nM, 3 h). BFT-treated subconfluent HT29/C1 cells are rounded with loss of the tight cluster morphology when compared with untreated control cells (original magnification ×100). (b) Transmission electron micrograph of a BFT-treated (5 nM, 2 h) polarized T84 cell. The BFT-treated cell reveals a dramatic increase in cell size and development of a domed apical membrane with loss of intact microvilli (original magnification ×3000). (Reproduced with permission from Chambers *et al.*, 1997.)

cytotoxic activity of biologically active concentrations of BFT, protein synthesis is not diminished in BFT-treated cells and, in fact, is stimulated beginning approximately 5 h after BFT treatment (Koshy *et al.*, 1996; Chambers *et al.*, 1997); LDH release is not observed (Koshy *et al.*, 1996; Chambers *et al.*, 1997); BFT-treated cells exclude trypan blue and propidium iodide (Wells *et al.*, 1996; Obiso *et al.*, 1997b); DNA synthesis continues normally (Donelli *et al.*, 1996) and no ^{51}Cr release from BFT-treated cells is observed (Obiso *et al.*, 1997b). Inhibitors of microtubules, and endosomal or Golgi trafficking do not alter BFT's activity (Donelli *et al.*, 1996; Obiso *et al.*, 1997b; Saidi *et al.*, 1997).

When polarized monolayers of epithelial cells (T84, MDCK, HT29, HT29/C1, Caco-2) are treated with BFT *in vitro*, the most striking observations are that: (a) BFT

decreases the resistance of epithelial monolayers in a dose- and time-dependent manner; and (b) the activity of the toxin is polar; namely, BFT decreases resistance more rapidly and at lower toxin concentrations when placed on the basolateral membranes of the cell monolayers than when placed on the apical membranes of the monolayers (Sears *et al.*, 1995; Wells *et al.*, 1996; Chambers *et al.*, 1997; Obiso *et al.*, 1997b; C.L. Sears, unpublished data). Apical treatment of T84, MDCK or HT29 monolayers with BFT (5 nM) results in an approximate 50% decrease in monolayer resistance after 3–6 h (Chambers *et al.*, 1997; Obiso *et al.*, 1997b). In contrast, treatment of the basolateral membranes of T84 monolayers with BFT decreases monolayer resistance by 90% within 2 h (Chambers *et al.*, 1997). In addition, basolateral BFT rapidly increases the short circuit current (Isc; indicative of chloride secretion) in a concentration-dependent manner, a finding which is not observed with apical BFT (Chambers *et al.*, 1997). The increase in Isc stimulated by BFT is self-limited, returning to baseline as monolayer resistance decreases further.

Consistent with these physiological results, apical BFT causes focal changes in the morphology of T84 monolayers, whereas basolateral BFT alters the morphology of all cells in the monolayer with development of a striking domed apical membrane on the cells (Chambers *et al.*, 1997). Transmission and/or scanning electron microscopy reveals that cell swelling occurs with a loss of the microvilli of the apical membrane resulting in the domed membrane structure (Figure 31.3b). Consistent with these observations, BFT stimulates a protracted increase in HT29/C1 cell volume, although the mechanism of this effect is unknown (Koshy *et al.*, 1996). At higher power by transmission electron microscopy, the zonula occludens (tight junction) and zonula adherens, electron-dense structures which determine the 'barrier function' of epithelial monolayers, are observed to develop concentration-dependent changes. Initially, these structures become more dense and compact with subsequent complete loss of these structures between some cells. Sites where the zonula occludens and zonula adherens are noted to dissolve are typically associated with a complete loss of the apical membrane microvillus structure. These morphological changes explain the measured decrease in monolayer resistance ('barrier function'). Interestingly, the development of gaping junctions between HT29 cells treated with BFT has been shown to permit increased association and invasion of the basolateral membranes with pathogenic enteric bacteria (Wells *et al.*, 1996). Despite the striking changes observed in the zonula occludens and zonula adherens, desmosomes remain intact after BFT treatment (Sears *et al.*, 1995; Chambers *et al.*, 1997).

When F-actin is examined by confocal microscopy in T84 monolayers treated with basolateral BFT, the actin of the tight junctional ring and microvilli is nearly absent (Chambers *et al.*, 1997). However, increased F-actin is demonstrated at the basal pole of the cells, consistent with the observations in HT29/C1 cells that total F-actin is unchanged in BFT-treated cells (Saidi *et al.*, 1997). These results suggest that treatment of epithelial cells with BFT stimulates dissociation and reassociation of F-actin, albeit in an entirely new cellular location. Interestingly, recent data examining the action of BFT on human colon *in vitro* confirm the polar activity of BFT on barrier function (tissue resistance) and the altered cellular F-actin architecture stimulated by BFT (Riegler *et al.*, 1998).

MOLECULAR MECHANISM OF ACTION OF *BACTEROIDES FRAGILIS* TOXIN

Because BFT does not appear to enter cells (Donelli *et al.*, 1996; Saidi and Sears, 1996; Obiso *et al.*, 1997b) and its biological activity on intestinal epithelial cells is strikingly polar (Current and Haynes, 1984; Chambers *et al.*, 1997; Obiso *et al.*, 1997b), Wu *et al.* (1998) hypothesized that the cellular substrate for BFT was a cell surface protein present on the basolateral membrane of intestinal epithelial cells. Experiments to address this hypothesis have revealed that BFT does not cleave the zonula occludens protein, occludin, or the basal membrane protein, β_1-integrin, although both of these proteins have extracellular domains presumably accessible to BFT's proteolytic activity. Similarly, consistent with its proposed site of action at the cell surface, no quantitative changes in the intracellular proteins, β-catenin, α-catenin, zonula occludens protein-1 (ZO-1) or actin, are identified by Western blot analysis during the first hour of treatment of intestinal epithelial cells with BFT (Saidi *et al.*, 1997; Wu *et al.*, 1998). However, although these proteins are not BFT substrates, analysis of the cellular distribution of ZO-1 or occluden, for example, by confocal microscopy reveals redistribution of these proteins in the cells by 1 h after treatment with BFT (Obiso *et al.*, 1997b; Wu *et al.*, 1998).

In contrast, the zonula adherens protein, E-cadherin, is rapidly cleaved by BFT, as demonstrated by Western blot and confocal immunofluorescence analysis (Wu *et al.*, 1998). This activity of BFT has been examined in greatest detail using HT29/C1 cells, although E-cadherin is also cleaved on polarized MDCK cells. This proteolytic activity of BFT is time and concentration dependent, with onset of E-cadherin cleavage detectable by 1 min after treatment of intestinal epithelial cells (HT29/C1) with BFT (5 nM) and complete loss of intact cellular E-cadherin by 60–120 min after treatment with BFT. Cleavage of E-cadherin by BFT yields two proteolytic fragments (33 and 28 kDa) detectable by 1 and 10 min after BFT treatment, respectively, with an antibody to the intracellular domain of E-cadherin. Note that detection of the 28-kDa fragment of E-cadherin correlates with the onset of visible morphological changes in HT29/C1 cells. The size of the proteolytic degradation products of E-cadherin combined with data indicating that BFT does not enter cells (Donelli *et al.*, 1996; Saidi and Sears, 1996; Obiso *et al.*, 1997b) suggest that BFT cleaves E-cadherin in its extracellular domain near the cellular plasma membrane. However, the exact site of cleavage has not yet been identified. This likely cell-surface proteolytic activity of BFT has been shown to be independent of cellular ATP stores, consistent with the hypothesis that BFT acts as a specific cell-surface protease toxin. Consistent with these data indicating that BFT cleaves E-cadherin, Wells *et al.* (1996) previously reported that treatment of HT29 cells with BFT resulted in decreased internalization of *Listeria monocytogenes* (but not other pathogenic enteric bacteria). E-cadherin is one ligand utilized by *L. monocytogenes* to enter epithelial cells (Mengaud *et al.*, 1996).

Importantly, after BFT treatment of intestinal epithelial cells for at least 60 min, further degradation of the 33 and 28-kDa E-cadherin fragments occurs. In contrast to the initial ATP-independent proteolytic activity of BFT, this subsequent proteolytic activity requires cellular ATP, suggesting that the complete cleavage of E-cadherin is a two-step event (Figure 31.4a). It is proposed that BFT first directly cleaves the extracellular domain of E-cadherin in an ATP-independent manner and that this cleavage of E-cadherin then triggers the cell to destroy the intracellular domain of this protein, presumably through the action of ATP-dependent cellular proteases. Note that the cellular morphological changes stimulated by BFT require cellular ATP indicating, not unexpectedly, that the BFT-stimulated changes in cell shape and physiology are energy-dependent events. Cellular recovery after BFT treatment correlates with resynthesis of E-cadherin. However, transcriptional upregulation of E-cadherin synthesis has not been demonstrated (Wu *et al.*, 1998).

Although several approaches were attempted to demonstrate cleavage of E-cadherin *in vitro* by BFT, none has been successful, suggesting that the native conformation of E-cadherin on epithelial cells may be critical to observe E-cadherin cleavage by BFT.

Figure 31.4(b) shows a proposed model of the pathogenesis of BFT-mediated intestinal secretion. Although this model builds on the data on the biochemical, biological and physiological sequelae of treatment of intestinal epithelial cells with BFT discussed in this chapter, many of the proposed steps require additional investigation for validation. Data confirming that BFT is a key virulence factor of ETBF strains (for example, creation of an isogenic mutant) in intestinal models of disease are still lacking.

SUMMARY

In recent years, the spectrum of established and putative toxins of *B. fragilis* has evolved rapidly. Much remains to be learned about the biological and physiological activities, the molecular mechanism(s) of action and the role of these proteins in animal and disease human. None the less, these early data on the cellular activities of BFT-1 and BFT-2 suggest that, at a minimum, these proteins will serve as investigative tools to dissect further how the actin cytoskeleton regulates intestinal epithelial cell structure and function.

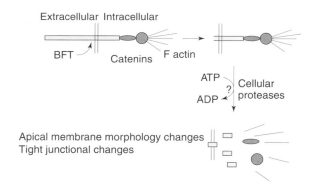
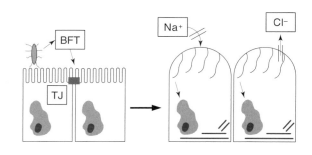

Outcome: Non-inflammatory diarrhoea

FIGURE 31.4 Proposed models of BFT activity. (a) Initial steps in the mechanism of action of BFT. BFT cleaves the extracellular domain of the zonula adherens protein, E-cadherin, in an ATP-independent manner. Modification of the extracellular domain of E-cadherin triggers the ATP-dependent proteolysis of the intracellular domain of E-cadherin, probably by cellular proteases. Loss of intact E-cadherin is predicted to disrupt its linkages with the catenin proteins and secondarily F-actin. These events may lead to the characteristic disruption of the apical cytoskeleton of polarized epithelial cells (Chambers *et al.*, 1997, and Figure 31.1b). (Reproduced with permission from Wu *et al.*, 1998a.) (b) Pathogenesis of BFT-mediated intestinal secretion. ETBF organisms attach to the apical membrane of intestinal epithelial cells and secrete BFT. BFT may diffuse through the zonula occludens to reach its target protein E-cadherin. Cleavage of E-cadherin may precipitate focal morphological changes and diminished barrier function. These events may facilitate the delivery of BFT to the basolateral membranes of intestinal epithelial cells where further cleavage of E-cadherin may augment the apical morphological changes initially stimulated by BFT. The dramatic changes in the apical morphology of BFT-treated intestinal epithelial cells are hypothesized to alter the function of one or more ion transporters, resulting in net intestinal secretion.

The cleavage of E-cadherin by BFT-1 and BFT-2 represents the first time that E-cadherin has been described as a bacterial toxin substrate. In addition, BFT is the first cytoskeletal-altering toxin predicted to act solely at the cell surface and without covalent enzymic modification of an intracellular substrate. The concept that certain bacterial proteases are, in fact, exquisitely specific toxins yielding unique cellular responses requires further investigation. Tetanus toxin and botulinum toxin are examples of well-defined substrate-specific toxins which act intracellularly (Tonello *et al.*, 1996; Williamson *et al.*, 1996), whereas BFT may be the first recognized substrate-specific cell-surface protease toxin. The biological importance of remodelling of cell-surface proteins by extracellular proteases has recently been highlighted as a mechanism yielding specific cellular responses and triggering specific cellular signal transduction pathways (Werb, 1997; Werb and Yan, 1998). Future investigations will seek to determine whether cellular signal transduction pathways are altered by BFT and to identify the mechanism(s) resulting in the dynamic rearrangement of F-actin stimulated by BFT.

REFERENCES

Chambers, F.G., Koshy, S.S., Saidi, R.F., Clark, D.P., Moore, R.D. and Sears, C.L. (1997). *Bacteroides fragilis* toxin exhibits polar activity on monolayers of human intestinal epithelial cells (T84 Cells) *in vitro*. *Infect Immun.* **65**, 3561–70.

Chung, G.-T., Franco, A.A., Oh, H.-B., Wu, S. and Sears, C.L. (1998). Identification of a third metalloprotease toxin gene in systemic isolates of *Bacteroides fragilis*. *98th General Meeting, American Society of Microbiology*, Atlanta, GA, B-272 (Abstract).

Collins, J.H., Bergeland, M.E., Myers, L.L. and Shoop, D.S. (1989). Exfoliating colitis associated with enterotoxigenic *Bacteroides fragilis* in a piglet. *J. Vet. Diagn. Invest.* **1**, 349–51.

Donelli, G., Fabbri, A. and Fiorentini, C. (1996). *Bacteroides fragilis* enterotoxin induces cytoskeletal changes and surface blebbing in HT-29 cells. *Infect. Immun.* **64**, 113–19.

Duimstra, J.R., Myers, L.L., Collins, J.E., Benfield, D.A., Shoop, D.S. and Bradbury, W.C. (1991). Enterovirulence of enterotoxigenic *Bacteroides fragilis* in gnotobiotic pigs. *Vet. Pathol.* **28**, 514–18.

Franco, A.A., Mundy, L.M., Trucksis, M., Wu, S., Kaper, J.B. and Sears, C.L. (1997a). Cloning and characterization of the *Bacteroides fragilis* metalloprotease toxin gene. *Infect Immun.* **65**, 1007–13.

Franco, A.A., Kaper, J.B., Wu, S. and Sears, C.L. (1997b). Enterotoxigenic *Bacteroides fragilis* strains contain a potential pathogenicity island. *97th General Meeting, American Society of Microbiology*, Miami Beach, FL, B-426 (Abstract).

Franco, A.A., Wu, S., Cheng, R., Chung, G.-T. and Sears, C.L. (1998). Molecular evolution of the pathogenicity island of enterotoxigenic *Bacteroides fragilis* strains. Submitted.

Häse, C.C. and Finkelstein, R.A. (1993). Bacterial extracellular zinc-containing metalloproteases. *Microbiol. Rev.* **57**, 823–37.

Kato, N. (1998). Prevalence of enterotoxigenic *Bacteroides fragilis* in children, adults and elderly people in Japan. *2nd World Congress on Anaerobic Bacteria and Infections*, Nice, France (Abstract).

Kato, N., Karuniawati, A., Jotwani, R., Kato, H., Watanabe, K. and Ueno, K. (1995). Isolation of enterotoxigenic *Bacteroides fragilis* from extraintestinal sites by cell culture assay. *Clin. Infect. Dis.* **20**, (Suppl. 2), S141.

Kato, N., Kato, H., Watanabe, K. and Ueno, K. (1996). Association of enterotoxigenic *Bacteroides fragilis* with bacteremia. *Clin. Infect. Dis.* **23**, S83–6.

Kling, J.J., Wright, R.L., Moncrief, J.S. and Wilkens, T.D. (1997). Cloning and characterization of the gene for the metalloprotease enterotoxin of *Bacteroides fragilis*. *FEMS Microbiol. Lett.* **146**, 279–84.

Koshy, S.S., Montrose, M.H. and Sears, C.L. (1996). Human intestinal epithelial cells swell and demonstrate actin rearrangement in response to the metalloprotease toxin of *Bacteroides fragilis*. *Infect. Immun.* **64**, 5022–8.

McIver, K.S., Kessler, E., Olson, J.C. and Ohman, D.E. (1995). The elastase propeptide functions as an intramolecular chaperone required for elastase activity and secretion in *Pseudomonas aeruginosa*. *Molec. Microbiol.* **18**, 877–89.

Mengaud, J., Ohayon, H., Gounon, P., Mege, R. and Cossart, P. (1996). E-cadherin is the receptor for internalin, a surface protein required for entry of *L. monocytogenes* into epithelial cells. *Cell* **84**, 923–32.

Moncrief, J.S., Obiso, R., Barroso, L.A., Kling, J.J., Wright, R.L., Van Tassell, R.L., Lyerly, D.M. and Wilkins, T.D. (1995). The enterotoxin of *Bacteroides fragilis* is a metalloprotease. *Infect. Immun.* **63**, 175–81.

Moncrief, J.S., Duncan, A.J., Wright, R.L., Barroso, L.A. and Wilkins, T.D. (1998a). Molecular characterization of the fragilysin pathogenicity islet of enterotoxigenic *Bacteroides fragilis*. *Infect. Immun.* **66**, 1735–9.

Moncrief, J.S., Barroso, L.A. and Wilkins, T.D. (1998b). Purification and characterization of recombinant metalloprotease II from enterotoxigenic *Bacteroides fragilis*. *98th General Meeting, American Society of Microbiology*, Atlanta, GA, B-274 (Abstract).

Mundy, L.M. and Sears, C.L. (1996). Detection of toxin production by *Bacteroides fragilis*: assay development and screening of extraintestinal clinical isolates. *Clin. Infect. Dis.* **23**, 269–76.

Myers, L.L., Firehammer, B.D., Shoop, D.S. and Border, M.M. (1984). *Bacteroides fragilis*: a possible cause of acute diarrheal disease in newborn lambs. *Infect. Immun.* **44**, 241–4.

Myers, L.L., Shoop, D.S., Firehammer, B.D. and Border, M.M. (1985). Association of enterotoxigenic *Bacteroides fragilis* with diarrheal disease in calves. *J. Infect. Dis.* **152**, 1344–7.

Obiso, R.J., Jr, Lyerly, D.M., Van Tassell, R.L. and Wilkins, T.D. (1995). Proteolytic activity of the *Bacteroides fragilis* enterotoxin causes fluid secretion and intestinal damage *in vivo*. *Infect. Immun.* **63**, 3820–6.

Obiso, R.J., Jr, Bevan, D. and Wilkins, T.D. (1997a). Molecular modeling and analysis of fragilysin, the *Bacteroides fragilis* toxin. *Clin. Infect. Dis.* **25**, S153–5.

Obiso, R.J., Jr, Azghani, A.O. and Wilkins, T.D. (1997b). The *Bacteroides fragilis* toxin fragilysin disrupts the paracellular barrier of epithelial cells. *Infect. Immun.* **65**, 1431–9.

Onderdonk, A., Kasper, D.L., Cisneros, R.L. and Bartlett, J.G. (1977). The capsular polysaccharide of *Bacteroides fragilis* as a virulence factor: comparison of the pathogenic potential of encapsulated and unencapsulated strains. *J. Infect. Dis.* **136**, 82–9.

Pantosti, A., Cerquetti, M., Colangeli, R. and D'Ambrosio, F. (1994). Detection of intestinal and extra-intestinal strains of enterotoxigenic *Bacteroides fragilis* by the HT-29 cytotoxicity assay. *J. Med. Microbiol.* **41**, 191–6.

Polk, B.F. and Kasper, D.L. (1977). *Bacteroides fragilis* subspecies in clinical isolates. *Ann. Intern. Med.* **86**, 569–71.

Riegler, M., Lotz, M., Sears, C.L., Pothoulakis, C., Castagliuolo, I., Wang, C.C., Sedivy, R., Sogukoglu, T., Cosentini, E., Bischof, G., Feil, W., Teleky, B., Hamilton, G., LaMont, J.T. and Wenzl, E. (1998). *Bacteroides fragilis* toxin-2 damages human colonic mucosa *in vitro*. *Gut* (in press).

Saidi, R.F. and Sears, C.L. (1996). *Bacteroides fragilis* toxin rapidly intoxicates human intestinal epithelial cells (HT29/C$_1$) *in vitro*. *Infect. Immun.* **64**, 5029–34.

Saidi, R.F., Jaeger, K., Montrose, M.H., Wu, S. and Sears, C.L. (1997). *Bacteroides fragilis* toxin alters the actin cytoskeleton of HT29/C1 cells *in vivo* qualitatively but not quantitatively. *Cell Motil. Cytoskeleton* **37**, 159–65.

Saraste, M., Sibbald, P.R. and Wittinghofer, A. (1990). The P-loop – a common motif in ATP- and GTP-binding proteins. *Trends Biochem. Sci.* **15**, 430–4.

Sears, C.L., Myers, L.L., Lazenby, A. and Van Tassell, R.L. (1995). Enterotoxigenic *Bacteroides fragilis*. *Clin. Infect. Dis.* **20** (Suppl. 2), S142–8.

Smith, C.J., Tribble, G.D. and Bayley, D.P. (1998). Genetic elements of *Bacteroides* species: a moving story. *Plasmid* **40**, 12–29.

Tonello, F., Montecucco, C., Schiavo, G. and Rossetto, O. (1996). Tetanus and botulism neurotoxins: a novel group of zinc-endopeptidases. *Adv. Exp. Med. Biol.* **389**, 251–60.

Van Tassell, R.L., Lyerly, D.M. and Wilkins, T.D. (1992). Purification and characterization of an enterotoxin from *Bacteroides fragilis*. *Infect. Immun.* **60**, 1343–50.

Walker, J.E., Saraste, M., Runswick, M.J. and Gay, N.J. (1982). Distantly related sequences in the alpha- and beta-subunits of ATP synthase, myosin, kinases and other ATP-requiring enzymes and a common nucleotide binding fold. *EMBO J.* **1**, 945–51.

Weikel, C.S., Grieco, F.D., Reuben, J., Myers, L.L. and Sack, R.B. (1992). Human colonic epithelial cells, HT29/C$_1$, treated with crude *Bacteroides fragilis* enterotoxin dramatically alter their morphology. *Infect. Immun.* **60**, 321–7.

Wells, C.L., Van De Westerlo, E.M.A., Jechorek, R.P., Feltis, B.A., Wilkins, T.D. and Erlandsen, S.L. (1996). *Bacteroides fragilis* enterotoxin modulates epithelial permeability and bacterial internalization by HT-29 enterocytes. *Gastroenterology* **110**, 1429–37.

Werb, Z. (1997). ECM and cell surface proteolysis: regulating cellular ecology. *Cell* **91**, 439–42.

Werb, Z. and Yan, Y (1998). A cellular striptease act. *Science* **282**, 1279–80.

Williamson, L.C., Halpern, J.L., Montecucco, C., Brown, J.E. and Neale, E.A. (1996). Clostridial neurotoxins and substrate proteolysis in intact neurons. *J. Biol. Chem.* **271**, 7694–9.

Wu, S., Dreyfus, L.A., Tzianabos, A.O., Franco, A.A., Hayashi, C. and Sears, C.L. (1998a). Diversity of the metalloprotease toxin produced by enterotoxigenic *Bacteroides fragilis*. *98th General Meeting, American Society for Microbiology*, Atlanta, GA, B-270 (Abstract).

Wu, S., Lim, K.-C., Huang, J., Saidi, R.F. and Sears, C.L. (1998b). *Bacteroides fragilis* enterotoxin cleaves the zonula adherens protein, E-cadherin. *Proc. Natl Acad. Sci. USA* **95**, 14979–84.

32

Superantigenic *Streptococcus pyogenes* erythrogenic/pyrogenic exotoxins

Joseph E. Alouf, Heide Müller-Alouf and Werner Köhler

INTRODUCTION

Streptococcus pyogenes (group A *Streptococcus* in the Lancefield classification) is a pathogen strictly associated with the human species. This Gram-positive micro-organism is responsible for a wide spectrum of diseases ranging from simple skin infections and uncomplicated tonsillopharyngitis to more severe diseases such as scarlet fever, acute rheumatic fever, glomerulonephritis (Hallas, 1985; Quinn, 1989; Knöll *et al.*, 1991; Katz and Morens, 1992), invasive deep tissue infections (cellulitis, myositis, necrotizing fasciitis, and so on), bacteraemia (Stevens, 1992; Bronze and Dale, 1996; Geslin *et al.*, 1996) and the new clinical entity called streptococcal toxic shock syndrome (STSS).

STSS is a life-threatening disease (30–50% mortality) recognized in 1983 (Willoughby and Greenberg, 1983) which has been observed world-wide since 1987 (Cone *et al.*, 1987; Stevens *et al.*, 1989; Köhler, 1990; Stevens, 1992, 1995; Hoge *et al.*, 1993; Reichardt *et al.*, 1992; Bronze and Dale, 1996; Müller-Alouf *et al.*, 1997a). STSS is characterized by hypotension, shock, fever, multiorgan failure, skin desquamation and other biological perturbations (Working Group on Severe Streptococcal Infections, 1993). It corresponds very likely (Köhler *et al.*, 1987; Stevens, 1992; Holm *et al.*, 1992) to the cases of scarlatina maligna ('toxic scarlatina' or 'scarlatina fulminans') with high fatality rates described in older literature (Wretlind, 1874; Osler, 1895; Holt, 1898) during outbreaks of scarlet fever.

STREPTOCOCCUS PYOGENES EXTRACELLULAR PROTEINS

Several lines of clinical, epidemiological and experimental evidence suggest that the pathogenesis of many of these diseases is (at least partially) mediated by certain toxins and enzymes produced by group A streptococci. These bacteria release more than 25 extracellular proteins in culture media and in the infected host, as reflected by the elicitation of specific antibodies (see Alouf, 1980). The secreted proteins comprise the membrane-damaging toxins, streptolysin O (see Chapter 25) and streptolysin S, various hydrolytic enzymes, the chromosomally encoded cysteine proteinase previously referred to as erythrogenic (pyrogenic) exotoxin B, and the classical bacteriophage-encoded erythrogenic toxins A and C, also designated by certain authors as pyrogenic exotoxins A and C (SPE A, SPE C). Toxins A and C belong to the family of bacterial superantigens, described at the turn of the 1990s (Fleischer, 1989; Imanishi *et al.*, 1990; Abe *et al.*, 1991; Fleischer *et al.*, 1991; Leonard *et al.*, 1991). Furthermore, various newly discovered extracellular superantigenic mitogens have been identified (at least eight to date) in the supernatants of clinical *S. pyogenes* isolates from various geographical areas (Table 32.1).

Many aspects of the pathophysiological features and clinical effects of streptococcal (and also staphylococcal) superantigenic toxins result (see Chapter 36) from the overproduction and release of various cytokines and other inflammatory factors.

TABLE 32.1 The family of extracellular superantigens (SAgs) produced by group A streptococci (*Streptococcus pyogenes*)

	Year of discovery
Erythrogenic/pyrogenic exotoxins	
Toxin A (SPE A)	Dick and Dick (1924a, b)
Toxin B (?)[a] (SPE B)	Hooker and Follensy (1934)
Toxin C (SPE C)	Watson (1960)
Novel mitogenic proteins	
30-kDa (p*I* 8.5) mitogen	Cavaillon *et al.* (1979)
Low mol. wt proteins	Gerlach *et al.* (1986/1992)
'Mitogenic factor' (MF/SPE F)	Yutsudo *et al.* (1992, 1994), Norrby-Teglund *et al.* (1994a, b), Toyosaki *et al.* (1996), Iwasaki *et al.* (1997), Norgren *et al.* (1998)
Streptococcal superantigen (SSA)	Mollick *et al.* (1992, 1993), Reda *et al.* (1994), Stevens *et al.* (1996a, b)
Low mol. wt SAg (LMWS)	Geoffroy *et al.* (1994)
'New' mitogens AX and BX	Gerlach *et al.* (1995)
S. pyogenes mitogen (SPM)	Nemoto *et al.* (1996)
Helper T-cell mitogen (SPM-2)	Rikiishi *et al.* (1997)
Exotoxin Z (SMEZ)	Kamezawa *et al.* (1997)

[a]Immunological (Gerlach *et al.*, 1983) and genetic studies (Hauser and Schlievert, 1990; Chaussee *et al.*, 1993; Musser, 1997) established that SPE B is identical to or a variant of the well-known streptococcal cysteine proteinase (SCP) also referred to as interleukin-1β convertase (Musser, 1997). The contention that SPE B/SPC is a superantigen (Leonard *et al.*, 1991; Abe *et al.*, 1991; Imanishi *et al.*, 1992; Norgren and Eriksson, 1997) is disputable according to Braun *et al.* (1993) and Gerlach *et al.* (1994), who attribute the T-cell proliferation activity to contaminants of SPE B preparations. The superantigenic character of SPE B is also ruled out by Rink *et al.* (1996) on the basis of comparative *in vitro* study of cytokine release by human PBMC stimulated with SPE C and SPE B. Recombinant SPE B was also found to be devoid of mitogenic activity (Toyosaki *et al.*, 1996).

What is a superantigen?

The term *superantigen* (SAg) was coined by White *et al.* (1989) to designate particular properties of certain viral and bacterial proteins exhibiting highly potent mitogenic properties toward mammalian (including human) T-lymphocytes. Unlike conventional antigens, SAgs bind to major histocompatibility (MHC) class-II molecules outside the antigen-binding grove and are presented as unprocessed proteins to certain T-lymphocytes expressing specific T-cell (TcR) receptors motifs located on the variable domain of the β-chain (Vβ) of the TcR. As a consequence, SAgs stimulate at nano- to picogram concentrations up to 20% of T-cells, while only one in 10^5–10^6 T-cells (0.001–0.0001%) is activated upon MHC-restricted conventional antigenic peptide presentation to TcR during the immune response process.

Historical background, discovery and classification of erythrogenic/pyrogenic toxins

Classically, three antigenically distinct proteins, historically designated erythrogenic toxins A, B and C, were identified in 1924, 1934 and 1960, respectively (see below). Toxin A was produced by *S. pyogenes* strains of serotypes M28, 12, 17 and 10 (NY5 strain). Toxin B was produced by the NY5 strain and a strain of serotype M19. Toxin type C was produced by a strain of serotype 18.

Erythrogenic/pyrogenic exotoxin A

The pioneering work of George and Gladys Dick (1924b), who provided the proof that streptococci (later identified as Lancefield's group A streptococci) are the causative agents of scarlet fever, constitutes an important hallmark in the microbiology of infectious diseases. The Dicks were also the discoverers of the so-called 'scarlet fever toxin', further designated 'erythrogenic toxin' (later shown as exotoxin A) in broth culture filtrates by haemolytic streptococci of scarlatinal origin (Dick and Dick, 1924a; Dick and Boor, 1935). Toxin detection by these investigators was based on the intradermal injection of culture filtrates (0.1 ml of a 1/1000 dilution) into certain healthy persons. Erythematous and oedematous skin reactions developed in the susceptible persons by 24 h after injection and it was concluded that the resulting reaction was a local 'model' for the skin rash of scarlet fever. This so-called positive 'Dick reaction' was held to be the manifestation of the primary toxicity of erythrogenic toxin. The injection of larger doses induced a generalized reaction with fever, nausea, vomiting and a transient scarlatiniform rash (so-called miniature scarlet fever). Antitoxin neutralized the skin reactions. Persons with sufficient circulating antitoxin did not develop a positive skin response (negative Dick reaction) and they were therefore not considered susceptible to scarlet fever. Antitoxin injected into the rash may neutralize the toxin within the skin and cause a local blanching of the rash (Schultz-Charlton test). An analogy was drawn with the Schick test for susceptibility to diphtheria toxin and the steady rise in the percentage of negatives during childhood was taken as evidence of the acquisition of active antitoxic immunity, either by passing through an attack of scarlet fever or by having an infection that is not apparent.

However, since 1927 many authors have suggested that the scarlatinal rash and the positive Dick skin test were not the manifestation of a direct 'primary' toxic effect but the result of a delayed-type hypersensitivity reaction (see Watson, 1960; Kim and Watson, 1970;

Schlievert *et al.*, 1979a; Schlievert, 1981; Alouf, 1980; Wannamaker and Schlievert, 1988). This contention was based on the observation that scarlet fever does not occur in infants and becomes more prevalent only after repeated streptococcal infections. Conversely, positive Dick tests were found infrequently in individuals (e.g. 1280 men of military age) from geographical areas with a low incidence of streptococcal infections (Rantz *et al.*, 1946).

By the 1970s, the common consensus was that the skin rash results from a combination of direct primary (intrinsic) toxicity and a hypersensitivity reaction not only to erythrogenic toxin A, but also to other extracellular streptococcal products (Kim and Watson, 1970; Hallas, 1985). In this respect, both skin activity and the immune status of the host should be considered in the interpretation of the dermal reaction to the toxin.

Skin reactivity in laboratory animals is variable; young rabbits and guinea pigs are usually not susceptible but become reactive after the injection of a small dose of toxin (Kim and Watson, 1970; Schlievert *et al.*, 1979a). However, Yamaoka *et al.* (1998) reported that intradermal (i.d.) injection of toxin C induced erythema in unsensitized rabbits with as little as 5 ng of toxin. Increased vascular permeability, erythema and leucocyte migration were elicited by erythrogenic toxin A (Kamezawa *et al.*, 1990).

Erythrogenic/pyrogenic exotoxin B

This toxin was identified by Hooker and Follensby (1934) as a protein immunologically distinct from toxin A on the basis of a flocculation reaction with anti-toxin A immune sera. Toxin B was biochemically characterized and purified by Stock and Lynn (1961). In 1983, Gerlach and co-workers demonstrated that SPE B was immunologically identical to the streptococcal cysteine proteinase (SCP), confirmed by the cloning of the *spe* B gene (Hauser and Schlievert, 1990) and its inactivation (Chaussee *et al.*, 1993). The functional and biochemical identity of SPE B and SCP is currently documented (Ohara-Nemoto *et al.*, 1994; Musser, 1997). However, the superantigen nature of SPE B/SCP is questioned by some authors (see Table 32.1 and the reviews by Musser, 1997, and Norgren and Eriksson, 1997). SPE B will not be discussed further in this chapter. Valuable overviews are provided in the above-mentioned references.

Erythrogenic/pyrogenic exotoxin C

This new serotype of erythrogenic/pyrogenic exotoxins (toxin C) was isolated by Watson (1960) from the culture filtrate of a strain (serotype M18) associated with scarlet fever. The toxin was immunologically distinct

TABLE 32.2 *Streptococcus pyogenes* reference strains used for the production and purification of erythrogenic (pyrogenic) toxins A and C

Strains	Reference
Toxin A	
NY5	Gerlach *et al.* (1980, 1981), Houston and Ferretti (1981), Bohach *et al.* (1988), Kamezawa and Nakahara (1989)
594	Nauciel *et al.* (1969), Hallas (1985)
T25₃	Lee and Schlievert (1989)
C203 S	Gray (1979), Schlievert and Gray (1989)
S 84	Cavaillon *et al.* (1979), Geoffroy and Alouf (1988)
ROS	Klein and Collins (1996)
Toxin C	
NY5	Ozegowski *et al.* (1984), Bohach *et al.* (1988)
T 18	Ozegowski *et al.* (1984), Bohach *et al.* (1988), Lee and Schlievert (1989)
AT 13	Ozegowski *et al.* (1984)
86–104	Lee and Schlievert (1989)

from toxins A and B. It induced fever (pyrogenic effect) by i.v. injection in the rabbit (Kim and Watson, 1970) and was purified by Schlievert *et al.* (1977) and thereafter by Ozegowski *et al.* (1984) and other investigators (Table 32.2).

TOXIN NOMENCLATURE

At the beginning of the 1970s the three classical toxins A, B and C were shown to exhibit a multiplicity of biological activities, particularly pyrogenicity and enhancement of animal susceptibility to lethal shock by Gram-negative bacterial endotoxins (Kim and Watson, 1970). The contention that the erythematous Dick reaction was not a primary effect of the toxins but rather the manifestation of an acquired hypersensitivity to streptococcal antigens led Kim and Watson to designate the toxins as 'streptococcal pyrogenic' exotoxins A, B, C (SPE A/B/C), since the toxin-induced fever was thought to be an intrinsic (primary) effect of these proteins. Although this terminology was subsequently frequently used in the literature, the 'historical' Dick's denomination 'erythrogenic toxins' still remains widely used to date (see Köhler, 1990; Yu and Ferretti, 1991; Knöll *et al.*, 1991; Abe *et al.*, 1991; Bahr *et al.*, 1991; Holm *et al.*, 1992; Tyler *et al.*, 1992; Reichardt *et al.*, 1992; Michie *et al.*, 1994; Yamamoto and Ferretti, 1995; Arvand *et al.*, 1996; Chaussee *et al.*, 1996; Müller-Alouf *et al.*, 1996, 1997a; McShan and Ferretti, 1997; Nakano *et al.*, 1997; Cleary *et al.*, 1998).

Pyrogenicity is but one among many 'primary' effects of the toxins (Table 32.4) and since most of these effects are a consequence of the superantigenic nature of these molecules, which is beyond doubt the basic characteristic of the SPEs, a consensual acceptable denomination could be streptococcal superantigenic toxins A, C (SSTA/C). This nomenclature may also be appropriately extended to the novel streptococcal superantigens (Table 32.1).

PRODUCTION OF EXOTOXINS A AND C

These toxins as well as SCP (toxin B) are produced exclusively by group A streptococci, separately, simultaneously or in various combinations (Schlievert *et al.*, 1979b; Hallas, 1985; Knöll *et al.*, 1991; Musser *et al.*, 1991; Reichardt *et al.*, 1992) depending on culture media and strains (whether associated with scarlet fever, STSS or not). However, it should be emphasized, that in contrast to *spe A* and *spe C* genes, *spe B* gene is constitutive and present in virtually all *S. pyogenes* isolates and phenotypically expressed by most strains (Yu and Ferretti, 1991; Tyler *et al.*, 1992; Black *et al.*, 1993; Talkington *et al.*, 1993; Stanley *et al.*, 1996; Hsueh *et al.*, 1997). This gene is chromosomal, in contrast to the genes encoding toxins A and C, which are located on bacteriophage (as described in the section on molecular biology in this chapter).

The reference strains commonly used for toxin production and purification are listed in Table 32.2. Eleven strains from various M and T serotypes investigated by Gerlach *et al.* (1981) produced toxin A in quantities ranging from 16 to 0.03 mg l^{-1}. The amounts of toxin C produced by strains T18, NY5 and AT13 were 1.2, 0.9 and 1 mg l^{-1}, respectively. The mean amount of toxin A produced by 52 strains isolated in 1987/1988 in Germany from children suffering from scarlet fever was approximately 1 µg l^{-1} (Knöll *et al.*, 1991) whereas the average titre was 68 and 8 µg l^{-1} from the strains isolated in the same country during scarlet fever epidemics in 1972/1973 and 1982/1983 (Köhler *et al.*, 1987). In contrast, much higher amounts were produced by 26 strains isolated from patients with STSS in 1987/1988 (Lee and Schlievert, 1989). The average concentrations of toxins A and C released per litre of culture were 3.2 and 0.6 mg, respectively. Interestingly, a strain isolated from a patient with a pharyngitis-associated STSS was reported to produce 67.4 mg of toxin A per litre [as measured by enzyme-linked immunosorbent assay (ELISA)], while no detectable amounts of toxin C and streptococcal proteinase were found (Arvand *et al.*, 1996).

However, many strains isolated from patients with STSS in France, Sweden and Chile produced very low or no detectable amounts of both toxins (Köhler, 1990; Reichardt *et al.*, 1992). In a recent study on culture supernatants from 49 isolates carrying the *spe A* gene,

the concentrations of toxin A varied significantly among the strains (Chaussee *et al.*, 1996). Most strains produced 2–20 ng ml^{-1}, others 28–90 ng ml^{-1}. The common laboratory strain NY-5 produced more than 650 ng/ml.

The reasons for the variable toxin production are unclear. They are probably due to regulatory rather than gene dosage effects (Yu and Ferretti, 1989). The cyclic trends in the qualitative and quantitative pattern of the erythrogenic toxins since the 1930s and their relation to fluctuations in the severity of the diseases and mortality rates have been discussed by Hallas (1985), Köhler *et al.* (1987), Lee and Schlievert (1989) and Yu and Ferretti (1989).

The *in vivo* production of toxins A, B and C in rabbits experimentally infected in a tissue-cage model with various *S. pyogenes* strains of different M and T serotypes was reported by Knöll *et al.* (1982). The toxins released were detected by immunoprecipitation and by their mitogenic properties on human blood lymphocytes during the first week of animal infection. The average amounts of toxins released by the various strains ranged from 5 to 10 µg ml^{-1} of cage fluid. This procedure allowed the detection of toxin A production *in vivo* by strains that did not show toxin release *in vitro* (Knöll *et al.*, 1985), suggesting that biological factors in the host may promote toxin production. Furthermore, the rabbits responded to the infection by eliciting serologically specific antibody response to the type of toxin produced. Using the cage model, Knöll *et al.* (1988) showed that immunization of rabbits with erythrogenic toxin A (ETA) toxoid evoked remarkable protection against death from infection by ETA-producing strains introduced in the cage.

PURIFICATION AND ASSAY OF TOXINS A AND C

Various protocols for the purification of these toxins have been published and the reader is referred to the publications of the authors listed in Table 32.2 and to the review of Wannamaker and Schlievert (1988).

Valuable immunological and biological assays of the toxins have been reported. A highly sensitive method of detection and quantification of toxin A by an ELISA was devised by Houston and Ferretti (1981) and Köhler *et al.* (1987). Microgram amounts of toxin are detectable. A recently described sandwich ELISA can detect as little as 2 ng ml^{-1} of toxin A in culture supernatants (Chaussee *et al.*, 1996).

Two tests based on the erythematous skin reaction and mitogenic activity are used for the bioassay of the toxins (see Geoffroy and Alouf, 1988).

The skin test is performed in rabbits or guinea pigs (see Schlievert *et al.*, 1979a, and Houston and Ferretti, 1981). Toxin samples (100 µl) are injected intradermally. Positive reaction is demonstrated by an erythematous response. The skin is observed at 24 and 48 h after injection. One skin test dose (STD) is that amount of toxin which elicits an erythema of 10 mm in diameter. It corresponds approximately to 100 ng of toxin in rabbit and guinea pig systems according to Hrìbalová *et al.* (1980) and to 1 ng in rabbit according to Kamezawa and Nakahara (1989). This great discrepancy is not explained but may be related to both the complex nature of the reaction and the host, since the skin reactivity appears to reflect Arthus and delayed hypersensitivity reactions in addition to the toxic effects of the toxin (Schlievert *et al.*, 1979a). The specificity of this reactivity is assessed by its neutralization by preincubation of the toxins with corresponding specific antibodies. A skin test in rabbits based on blueing reaction has been described (Kamezawa *et al.*, 1990). This test appears to offer advantages over the classical erythematous skin reaction.

The lymphocyte blast transformation test performed on human or rabbit lymphocytes separated from peripheral blood is based on the mitogenic effect of the toxins. The results are expressed as the stimulation index, which is the ratio of the counts per minute (cpm) of the toxin-stimulated culture pulsed with [^{3}H]thymidine to the cpm of the non-stimulated control cultures (Cavaillon *et al.*, 1979; Knöll *et al.*, 1983). Inhibition of the lymphocyte mitogenicity of toxin A has been used for the assay of antitoxin antibodies in human sera (Abe *et al.*, 1990; Bahr *et al.*, 1991). The production of monoclonal antibodies against toxins A and C has been reported (Wollweber *et al.*, 1994).

THE MOLECULAR BIOLOGY OF ERYTHROGENIC/PYROGENIC EXOTOXINS A AND C

Exotoxin A

As early as 1927, a filterable agent from strains of *S. pyogenes* causing scarlet fever was reported by Frobisher and Brown (1927) to induce erythrogenic toxin production by non-scarlatinal strains. Bingel (1949) confirmed this observation and Zabriskie (1964) showed the conversion of the SPE A non-lysogenic strain T25$_3$ to a SPE A$^+$ strain by a bacteriophage (φT12) from the *S. pyogenes* strain T12g1. Zabriskie used the rabbit skin test to demonstrate the presence of SPE A in the culture filtrates of the converted strain and its absence in the original strain. Subsequently, several investigators confirmed that SPE A production by various strains was under bacteriophage control (Johnson *et al.*, 1980; McKane and Ferretti, 1981; Nida and Ferretti, 1982; Johnson and Schlievert, 1983).

The *spe* A gene encoding SPE A was cloned from bacteriophage T12 and expressed in *E. coli* (Johnson and Schlievert, 1984). The gene consists of 756 base pairs (bps) encoding 252 amino acids, the first 30 of which (signal peptide) are removed during the secretion of the toxin (Table 32.3). Two putative promoter regions have been located, as well as the ribosomal binding site.

The recombinant product was biologically and immunologically related to SPE A, as demonstrated by Weeks and Ferretti (1984), who also determined the nucleotide sequence of the *spe* A gene (Weeks and Ferretti, 1986). This sequence was also established by Johnson *et al.* (1986a).

TABLE 32.3 Molecular characteristics and Vβ (TCR) specificities of *S. pyogenes* erythrogenic/pyrogenic exotoxins A and C and other extracellular superantigens

Toxins/mitogens	aa	Molecular weight (Da)	p*I*	Human T-cell receptor Vβ motif(s) specificity	References
Erythrogenic/pyrogenic exotoxin A (SPE A)	222	25 787	5.2	2, 12, 14, 15	Weeks and Ferretti (1986), Norgren and Eriksson (1997)
Erythrogenic/pyrogenic exotoxin C (SPE C)	208	24 354	6.8	1, 2, 5.1, 10	Goshorn *et al.* (1988), Norgren and Eriksson (1997)
Streptococcal superantigen (SSA)	234	26 892	ND	1, 3, 15	Mollick *et al.* (1993), Reda *et al.* (1994)
Mitogen F (SPE F)/DNase	228	25 363	7.6–8.6	2, 4, 8, 15, 19	Iwasaki *et al.* (1993), Norrby-Teglund *et al.* (1994b)
Low mol. wt SAg (LMWZ)	ND	*ca.* 10 000	7.2	4, 7, 8	Gerlach *et al.* (1992)
Low mol. wt SAg (LMP)	ND	12 400	4.3–4.7	ND	Geoffroy *et al.* (1994)
Mitogen AX, BX	ND	27 000	8.3–7.3	ND	Gerlach *et al.* (1995)
SPM	ND	28 000	9.2	13	Nemoto *et al.* (1996)
SPM-2	ND	29 000	ND	4,7, 8	Rikiishi *et al.* (1997)
Exotoxin MZ (SMEZ)	234	25 524	5.3	2, 8	Kamezawa *et al.* (1997)

The *spe A* gene is adjacent to the chromosomal insertion site of the bacteriophage (Johnson *et al.*, 1986b). Recently, McShan *et al.* (1997) discovered that the *spe A* gene carried by phage T12 integrates into *S. pyogenes* chromosome at gene *att B* for serine tRNA. This phage is a member of a group of temperate bacteriophages which do not all carry the *spe A* gene (e.g. φ 436) but use the same serine tRNA gene attachment site as φ 12 (McShan and Ferretti, 1997). Interestingly, other toxinogenic strains harbour a *spe A* gene on a phage (φ 49) bearing a larger genome than T 12 (40 kb versus 36 kb) and having a physical map distinct from that of T 12. The *spe A*-positive φ 49 did not integrate into the *att B* serine tRNA gene into the bacterial chromosome, indicating that this gene does not invariably correlate with the carriage of *spe A* (McShan and Ferretti, 1997). It was suggested that recombination between phage genomes has been an important means of generating diversity and disseminating virulence-associated genes such as *spe A*.

Toxin A regulation

The regulation of the expression of the *spe A* gene is poorly defined (Chaussee *et al.*, 1996). According to Kim and Schlievert (1997), *spe A* expression varies in a clonal manner generally with respect to the M serotype of the strains. M3 strains appear to be higher producers of toxin A, in contrast to the recently isolated M1 type strain. This variability appears to be proportional to the amount of *spe A* messenger RNA made and may be due to an element external to the chromosomal insertion site of the bacteriophage. A *cis*-acting element near this site may be controlling SPE A production, as known for several *S. pyogenes* virulence factors expressed at the bacterial surface. Indeed, many of these factors including *emm* genes are controlled by a global regulatory genomic system known as multigene regulator of group A *Streptococcus* (*mga*) regulon (see Kihilberg *et*

al., 1995). Interestingly, when *spe A* is cloned in *S. aureus*, the gene becomes partially under the control of the global regulatory genomic system *agr* (Kim and Schlievert, 1997). Recently, Cleary *et al.* (1998) showed that an insertion mutation in *mga*, the multigene activator, downregulated epithelial cell invasion by *S. pyogenes* but did not significantly affect the expression of the *spe A* gene.

The effect of temperature and other environmental factors (pH, osmolarity, various atmospheric concentrations of CO_2 and O_2) on toxin A production in relation to bacterial growth in culture medium was investigated by Xu and Collins (1996). Toxin production was not affected by environmental factors except for temperature. A fourfold greater level of toxin was found at 37°C compared with the cells grown at 26°C. Toxin expression was temperature regulated at the level of transcription (sixfold increase in specific RNA at 37°C). This increase in *spe* A expression was not seen with *spe* B and *emm* 3 encoding M-serotype expression. This finding suggests an increase in toxin A production in severe *S. pyogenes* infections in soft tissue and the bloodstream as opposed to *S. pyogenes* that have not breached the epithelial layer and are living on the surface of the skin.

Expression of spe A *in other bacteria*

The *spe A* gene has been successfully expressed in other bacteria. The gene was cloned on a high copy-number plasmid in *Bacillus subtilis*. This micro-organism produced 32-fold more toxin than the native streptococcal strain. However, difficulties were encountered for the solubilization of the recombinant protein (Kreiswirth *et al.*, 1987). The *spe A* gene was also cloned and expressed in *Streptococcus sanguis* (Gerlach *et al.*, 1987), which does not naturally produce the toxin. The recombinant toxin produced was one-quarter of the amount produced by *S. pyogenes*. It was processed identically in

TABLE 32.4 Major biological and pathophysiological activities of *S. pyogenes* superantigenic toxins[a]

- Superantigenicity: polyclonal proliferation of T-lymphocytes following binding to Vβ motifs of T-cell receptor and to non-polymorphic domains of MHC class-II molecules on the surface of antigen-presenting cells (APC).
- Induction and release of APC and Th1/Th2-derived, pro- and anti-inflammatory, immunoregulatory and haematopoetic cytokines – induction of certain cytokine receptors. Production of inducible nitric oxide synthase.
- Elicitation under certain conditions of T-cell anergy or apoptosis.
- Mimicry of cognate T/B-cell interaction, proliferation/differentiation of resting B-cells into immunoglobulin-secreting cells or suppression of Ig classes.
- Pyrogenicity in animals (direct action on hypothalamus, IL-1 and TNF-α release).
- Blockade of reticuloendothelial system (impairment of clearance functions).
- Enhancement of host susceptibility (experimental animals) to lethal shock by endotoxins.
- Lethality, shock in experimental animals, direct capillary leak, skin rash.
- Involvement in the pathogenesis of severe diseases including toxic shock syndrome.

[a]These activities have been established for toxin A and for most of them for toxin C. They are shared by staphylococcal superantigenic toxins (see Chapter 33 of this book). Some of these activities have also been evaluated for the other streptococcal superantigens listed in Table 32.1.

both *S. sanguis* and *S. pyogenes* and showed total antigenic identity with native toxin A. This toxin has also been expressed in *S. aureus* and the recombinant protein conserved full biological activity (Roggiani *et al.*, 1997). However, Yamamoto and Ferretti (1995) reported overexpression of toxin A in *E. coli*. A two-step high-performance liquid chromatographic (HPLC) purification of the protein accumulated in the periplasm and extracted with 0.5 M sucrose resulted in a highly purified, biologically active recombinant toxin. Recombinant SPE A as well as SPE C and the streptococcal superantigens SSA and MF (Table 32.1) were also expressed in *E. coli* by Morita *et al.* (1997). The recombinant superantigens were purified from bacterial lysates. They also proved biologically active, as confirmed by their mitogenic activity on rabbit peripheral blood lymphocytes.

Mutagenesis of the spe A gene

Single- and double-site mutations of toxin A were generated by Roggiani *et al.* (1997) at residues K16, N20, C87, C90, C98, K157, S195, N20/C98 and N20/K157. The SPE A mutants were analysed *in vivo* for their lethal activity toward American Dutch belted rabbits and *in vitro* for their superantigenic mitogenicity. This study indicated that the ability of toxin A to induce lethality and endotoxin enhancement does not require superantigenicity. Conversely, superantigenicity does not necessarily lead to lethality. Thus, these properties and their relative contributions to the onset of hypotension and shock may be separable. It was suggested that certain mutant toxins may be suitable for use as vaccine toxoids.

Sriskandan *et al.* (1998) replaced the *spe A* gene in an M1T1 scarlet fever *S. pyogenes* isolate (strain H305) with a centrally disrupted *spe A* construct though a double recombination event (strain H326). They also created a number of *spe A*-negative mutants through insertion duplication mutagenesis. The wild-type strain H305 supernatant contained toxin A whereas the mutant strain H326 did not. *Spe A*-negative mutants strains and *spe A*-positive strain H305 were then compared in a mouse model of streptococcal thigh muscle infection, which progresses to bacteraemia and multi-organ failure. Preliminary experiments showed no difference in mortality between the two groups of animals, suggesting that *spe A* is not a major determinant of lethality in murine streptococcal fasciitis. This finding appears to be supported by the report of Hsueh *et al.* (1997), who found that invasive group A streptococcal disease in Taiwan was not associated with the presence of *spe* s genes. Similarly, very low or no detectable production of toxins A and C *in vitro* was reported for clinical strains isolated from patients with STSS (Köhler *et al.*, 1990; Reichardt *et al.*, 1992).

Allelic forms of spe A

Four naturally occurring *spe A* alleles have been demonstrated on 20 *S. pyogenes* strains recovered from patients with severe invasive diseases from the USA and various European countries (Nelson *et al.*, 1991). Three of these, *spe A*1, *spe A*2 and *spe A*3, encode toxins differing by a single amino acid. In SPE A2, a serine at position 110 replaces the glycine residue in SPE A1. In SPE A3, an isoleucine is found at position 106 in place of valine in SPE A1. The toxin encoded by *spe A*4 was 9% divergent from the other three with 26 amino acid changes. Almost all changes occurred in regions that are not highly conserved when toxin A is aligned with toxin C and *S. aureus* enterotoxins (SEA to SEE). The *spe A* gene cloned for the first time (Johnson and Schlievert, 1984) expressed the A1 allele.

Musser *et al.* (1995) analysed by multilocus enzyme electrophoresis and restriction fragment profiling by pulse-field gel electrophoresis 125 *S. pyogenes* strains expressing M1 protein serotype collected from 13 countries on five continents, recovered from individuals with a variety of severe and mild streptococcal infections. The isolates were studied for various alleles and its *spe A* and for allelic polymorphism of *emm* 1, *spe B*, *ska* (streptokinase) and *sep* (C5a peptidase) genes. The *spe A* gene was found in 94 of the 126 isolates (75%). Only *spe A*1 and *spe A*2 genes were detected among the strains.

Interestingly, *spe A*1 allele was found in many distinct electrophoretic types and its presence in diverse phylogenetic lineages means that the gene has been distributed horizontally among divergent strains. Moreover, this wider distribution of *spe A*1 allele suggests (Kehoe *et al.*, 1996) that it may be evolutionary older than *spe A*2 and *spe A*3, each of which was restricted to individual clones. However, the absence of synonymous (silent) nucleotide differences between *spe A*1, *spe A*2 and *spe A*3 is unusual and raises the possibility that the allelic variation observed may not be selectively neutral.

Strains expressing *spe A*2 and *spe A*3 have caused the majority of STSS episodes in the past 12 years, suggesting, according to Kline and Collins (1996), that the gene products *spe A*2 and *spe A*3 may be more bioactive toxic forms of SPE A. These authors analysed the products of alleles 1, 2 and 3 for mitogenic stimulation of human peripheral blood mononuclear cells (PBMC) and affinity for class-II major histocompatibility complex (MHC) molecules DQ. It was shown that the product of allele 2 (*spe A*2) had slightly higher affinity for class-II MHC molecule compared with *spe A*1, but not significantly greater mitogenic activity. *Spe A*3, however, was significantly increased in mitogenic activity and affinity for class-II MHC compared with *spe A*1, providing evidence that the toxin encoded by some of the highly virulent *S. pyogenes* STSS-associated isolates is a more active form of *spe A* since STSS is thought to result, at least partly, from the overproduction of cytokines released by the stimulated PBMC.

Exotoxin C

The *spe C* gene encoding this toxin is carried similarly to the *spe A* gene by a bacteriophage isolated by Goshorn and Schlievert (1989) from *S. pyogenes* strain CS 112. The gene lies near the attachment site for the phage insertion into the bacterial chromosome. The cloning and the determination of the nucleotide sequence of *spe C* were reported by Goshorn *et al.* (1988) and Goshorn and Schlievert (1988). The gene encodes 235 amino acids, the first 27 of which are removed as the signal peptide (Table 32.3). *Spe A* and *spe C* genes may occur simultaneously in certain strains (Reichardt *et al.*, 1992).

Similarly to toxin A, toxin C shows variability in the amounts released from strain to strain, generally dependent on M serotype (Kim and Schlievert, 1997). Evidence for two alleles *spe C1* and *spe C2* has been reported by Kapur *et al.* (1992); *spe C2* differed from *spe C1* by nucleotide changes at position 438 and 452, both of which are synonymous (silent) A-to-G transitions in the third position of codons for lysine. Two new alleles designated *spe C3* and *spe C4* were found by nested polymerase chain reaction (PCR) in clinical Swedish isolates by Norrby-Teglund *et al.* (1994c).

Expression of spe C in other bacteria

The mature form of toxin C possessing Asp28 to Cys235 was expressed in *E. coli* as a C-terminal fusion protein of glutathione-*S*-transferase (GST). Recombinant toxin C was separated from the GST moiety and HPLC-purified to homogeneity (Ohara-Nemoto and Kaneko, 1996).

Recombinant SPE C (and also *S. aureus*, SEA, SEB and TSST-1) was also expressed from the GEX vector in *E. coli* as glutathione-*S*-transferase fusion protein and purified from bacterial lysates by glutathione chromatography (Li *et al.*, 1997).

Mutagenesis of spe C gene

Three single-residue mutants have been raised by Yamaoka *et al.* (1998) and used as described in a further section to analyse the superantigenicity of this toxin.

Frequency of spe A *and* spe C *genes among* Streptococcus pyogenes *clinical isolates*

The detection of these genes for epidemiological and pathophysiological purposes has been determined by many investigators since 1989 in North America,

Europe, Asia and Australia. The first report concerned 448 strains from different countries (Yu and Ferretti, 1989). Among 300 isolates from individuals with a variety of streptococcal diseases except for scarlet fever, 15% harboured the *spe A* gene, whereas this gene was found in 45% of 145 scarlet fever strains, suggesting a more likely association of toxin A with this disease. The gene occurred in certain serotypes (M1, M3, M49, T1, T2, T3) but not in others (M4, M12, M22, T2/22, T14). It was suggested that this situation reflects the host range of encoding bacteriophages rather than clinical significance. As concerns *spe C*, Yu and Ferretti (1991) reported a frequency of 50% of this gene among 512 isolates from 13 countries, whatever their clinical origin (scarlet fever, other diseases).

The gene was found more frequently associated with M2, M4 and M6 serotype strains and less frequently with M1, M3 and M49 strains.

The detection of *spe A* and *spe C* on a series of 34, 108, 62, 117 and 62 strains from patients with STSS and other severe diseases in the USA was undertaken by Hauser and Schlievert (1991), Musser *et al.* (1991), Talkington *et al.* (1993), Black *et al.* (1993) and Chaussee *et al.* (1993), respectively.

The following frequency of *spe A* was found: 85%, 51%, 56% and 44%. That of *spe C* ranged from 21% to 27%. In the isolates investigated by Talkington and co-workers, 5% carried both *spe A* and *spe C* and 21% neither gene. The study by Reichardt *et al.* (1992) of 53 strains from patients with STSS isolated in Europe and Chile (1988–1991) showed that *spe A* and *spe C* were carried by 64% and 28% of the strains. Three contained both *spe A* and *spe C* (5–7%) and 13 (24.5%) neither gene.

Tyler *et al.* (1992) examined 152 clinical strains isolated in Canada between 1940 and 1991. The overall frequency of *spe A* was lower than that reported in the preceding studies, namely 33.6% for this gene and 28.9% for *spe C*. However, the isolates associated with scarlet fever, pharyngitis and severe invasive infections showed statistically significant differences in the presence of *spe A*, with scarlet fever strains having the highest association (81.3%), severe infections the next highest association (42.9%) and pharyngitis the lowest association (18.4%). In contrast, significant differences were observed in *spe C* frequencies in isolates associated with the three disease categories. In a series of 13 strains isolated in Scotland from patients with severe streptococcal infections (Upton *et al.*, 1996), eight tested positive for *spe A* and *ssa*, the gene encoding a novel superantigen (Table 32.1).

A genotypic study of 56 invasive and non-invasive strains isolated from three epidemiologically distinct outbreaks in the UK was reported by Stanley *et al.* (1996). The *spe A* and *spe C* genes were carried by 34%

and 68% of the strains, respectively. Neither *spe A* nor *spe C* was found in the isolates from localized sepsis.

A high frequency of *spe A* (80%) was found in the isolates from patients with STSS in Australia (Carapetis *et al.*, 1995). Recently, Hsueh *et al.* (1998) examined 44 invasive and 28 non-invasive strains from patients in Taiwan. The frequency of *spe A* was 39% for the whole invasive strains and surprisingly 36% for the non-invasive strains. Furthermore, among the invasive isolates, the STSS-mortality strains showed a very low frequency of *spe A* (13%). Interestingly, *spe C* and *spe F* (MF mitogen gene) were carried by 90% and 64% of both invasive and non-invasive strains. This led Hsueh *et al.* (1998) to the contention that the invasive strains isolated in Taiwan are not associated with *spe A*, *spe C* and *spe F* genes. In contrast, serotype M1, which was prevalent among these isolates and the high protease activity appeared significantly associated with STSS.

The data from the various studies reported in this section reveal great variations in the occurrence of *spe A* and *spe C* among clinical isolates as well as the absence of these genes in many strains isolated from severe infections, leading to the contention that other superantigens and non-superantigenic toxins may influence the clinical outcome of infection (Reichardt *et al.*, 1992; Stevens, 1995; Musser *et al.*, 1995; Bronze and Dale, 1996; Chaussee *et al.*, 1996; Watanabe-Ohnishi, 1996). However, the same studies showed the predominant association of certain M serotypes, particularly M1 and M3 is (but not always) associated with toxin A and C genes in the streptococcal strains isolated from patients with STSS and other severe infections.

MOLECULAR STRUCTURE OF TOXINS A AND C

The mature forms of these toxins comprise 222 and 208 amino acid residues, resulting in relative molecular masses of 25 787 (Weeks and Ferretti, 1986) and 24 354 (Goshorn *et al.*, 1988), respectively (Table 32.3). All allelic forms of toxin A comprise three cysteine residues at positions 87, 90 and 98 in the mature form. It has been postulated, before the establishment of the primary structure, that the toxin may have a disulfide bond (Nauciel *et al.*, 1969) but to the authors' knowledge no direct evidence of this bridge has been demonstrated. However, according to Roggiani *et al.* (1997), such a disulfide bridge may connect two of the Cys residues, Cys87 and Cys98 but not Cys90. The cysteine loop but not the Cys residues themselves may contribute to biological activity. Interestingly, the two Cys residues of the staphylococcal superantigenic enterotoxins are also linked by a disulfide bridge (see Alouf *et al.*, 1991; Munson *et al.*, 1998, and Chapter 33 of this book).

The analysis of toxin A3 isolated from a strain harbouring the *spe A3* allele and recombinant toxin A1 by sodium dodecyl sulfate–polyacrylamide gel electrophoresis (SDS–PAGE) under non-reducing and reducing conditions showed a shift in electrophoretic mobility, suggesting that the protein contains a disulfide bond (Kline and Collins, 1996).

When the cysteine residue at position 90 was changed to serine, the protein displayed the same migration shift as observed with *r*Spe A1, indicating that Cys90 was not involved in disulfide bond formation. In contrast, when Cys98 was changed to serine, the shift was no longer observed. Therefore, removal of this cysteine residue prevented the formation of the disulfide bond. Moreover, substituting a serine residue for Cys87 caused decreased stability of the toxin. SDS–PAGE analysis revealed several degradation products. One possible explanation for this instability is that the remaining cysteine residues at positions 90 and 98 formed an aberrant disulfide bond that changed the tertiary structure of the toxin and this conformational change resulted in a protein that was more susceptible to proteolysis. This modified protein had little or no mitogenic activity. The mutation of Cys90 (to Ser90) which was not involved in the disulfide bond caused an approximate 50% decrease in toxin mitogenicity, indicating that the disulfide bond and surrounding residues play a critical role in the superantigenicity of toxin A.

The exotoxin C contains a unique cysteine residue at position 27, whereas the 26 892-kDa streptococcal superantigen (SSA) contains five cysteine residues. A disulfide bond may be formed by residues Cys93 and C108 (Reda *et al.*, 1994).

MOLECULAR AND IMMUNOLOGICAL RELATIONSHIP BETWEEN STREPTOCOCCAL AND STAPHYLOCOCCAL SUPERANTIGENS

Early immunological comparisons of these toxins showed important cross-reactivity between streptococcal toxin A and staphylococcal enterotoxin B and C1 (SEB, SEC 1), as shown by Ouchterlony double diffusion, Western immunoblot and immunodot analysis (Hynes *et al.*, 1987).

Anti-toxin A immune serum reacted more strongly with SEB than with SEC 1. Similar findings with polyclonal and monoclonal antibodies also showed cross-reactivity by ELISA (Bohach *et al.*, 1988).

The homology has also been confirmed directly by comparison of the amino acid sequences of all superantigens. The alignment of these sequences showed that toxin A is most closely related to SEB, the SECs and SEG (44–42% of sequence identity) and much less related to the other enterotoxins (29–20% of sequence identity) (Munson *et al.*, 1998) and to streptococcal toxin C (20%), which share little similarity to any other streptococcal and staphylococcal superantigens (Roussel *et al.*, 1997; Kim and Schlievert, 1997). According to Kim and Schlievert it is possible that toxin A arose as a result of gene transfer from *S. aureus* to *S. pyogenes*.

Crystal structure of exotoxin C

The three-dimensional (3D) structure of this toxin at 2.4 Å has been solved recently (Roussel *et al.*, 1997). That of toxin A and the other streptococcal superantigens is still believed to be unknown.

The toxin exhibits the following features: it occurs in a dimeric form in solution at pH > 6.0 and binds zinc, a property also shared by the staphylococcal enterotoxins A, D and E (SEA, SED, SEE), which allows efficient interaction with the MHC class-II molecule.

The crystals of toxin C contained one monomer in the asymmetric unit with an unusually high solvent (75%) content. The monomer forms a flattened ellipsoid of approximately $50 \times 45 \times 30$ Å. Like the other bacterial SAgs, the toxin molecule is folded into two closely packed domains. Domain 1 comprises most of the *N*-terminal half of the molecule residues (20–95) and domain 2 the *C*-terminal half residues (96–208), with the *N*-terminal residues 5–18 forming a helix that packs against domain 2. Domain 1 is folded as a β-barrel made up of two antiparallel sheets that curve around to enclose the hydrophobic core. Domain 2 has a 'β-grasp' fold in which a long amphipathic α-helix (α4) is packed against a mixed β-sheet of five strands (β6, β7, β9, β10, β12).

When aligned according to the 3D structures, toxin C shows only a low level of sequence identity with SEA (23%), SEB (21%) and TSST-1 (18%). In the *N*-terminal half, toxin C most closely resembles TSST-1 in overall structure, although it has the least sequence identity with these toxins. Like the latter, toxin C lacks the disulfide loop present in toxin A and staphylococcal enterotoxins. In the *C*-terminal domain, toxin C closely resembles SEA. This may be linked to the fact that both molecules share a common Zn-dependent MHC-binding site (called Zn-A) on this domain. Indeed, toxin C binds zinc with an affinity similar to SEE and SEA. Sequence comparisons also identified His201 and Asp203 as corresponding to the zinc ligands His225 and Asp227 in SEA. Another Zn-binding site (on His35 and

Glu54) of lower affinity (Zn-B) was also identified and found located within the dimer interface, although the zinc atom was not essential for dimerization.

Very recently the crystal structure of toxin A (Spe A1) at 2.6Å resolution has been determined (Papageorgiou *et al.*, 1999).

BIOLOGICAL AND PATHO-PHYSIOLOGICAL PROPERTIES OF EXOTOXINS A AND C

These toxins exhibit a remarkable spectrum of biological and pharmacological activities (Table 32.4) which are shared for most of them by the superantigenic staphylococcal enterotoxins and toxic shock toxin-1 (TSST-1) and, to the authors' knowledge, by the recently discovered streptococcal superantigens (see Wannamaker and Schlievert, 1988; Alouf *et al.*, 1991; Kim and Schlievert, 1997; Norgren and Eriksson, 1997; and Chapter 33 of this book). Except for the superantigenicity of toxins A and C and the other related effectors described at the beginning of the nineteenth century, the other 'classical' properties were defined in the pioneering investigations of Watson (1960), Schuh (1965), Hanna and Watson (1965, 1968), Hrìbalová and Pospíšil (1973), Seravalli and Taranta (1974). Over the following 25 years, a great number of investigations was undertaken for a better understanding of these properties. Currently, these properties clearly appear to derive (for most of them at least) from the superantigenicity of these streptococcal and staphylococcal proteins. The major biological activities of toxins A and C are summarized below.

Pyrogenicity

Fever induction in experimental animals is one of the most important properties reported first by Watson (1960) and later by many other authors (Schuh, 1965; Schuh and Hrìbalová, 1966; Schuh *et al.*, 1970; Kim and Watson, 1970). The i.v. injection of the toxins in rabbits is characterized in most reports by a latency of 30–60 min and elevated temperature up to 4–5 h (Kim and Watson, 1970; Hrìbalová *et al.*, 1980; Murai *et al.*, 1987). The pyrogenic effect was directly proportional to the logarithm of toxin dose. A biphasic response was reported by Schuh *et al.* (1970) with culture filtrates from various toxin-producing strains. The minimal pyrogenic dose of toxin A or C determined either 3 or 4 h after i.v. injection ranged from 0.7 to 0.1 μg kg^{-1} except for the toxin used by Murai *et al.* (1987), who found a value of 1 μg kg^{-1}. A study of toxin C pyrogenicity in mice was investigated by Schlievert and Watson (1978), who also demonstrated the high diffusibility of this toxin since similar response peaking at 4 h post-injection was observed whatever the injection route (i.v., i.m., i.d. or s.c.). These investigators provided evidence that toxin C was capable of crossing the blood–brain barrier to produce fever by direct stimulation of the hypothalmic fever response control centre rather than indirectly through endogenous pyrogen (Schuh and Hrìbalová, 1966). Cerebrospinal fluid from normal rabbits, obtained 3 h after i.v. administration of toxin C induced fever in normal rabbits. The mechanism of fever elicitation by the toxins is very likely to be attributable to the release of interleukin-1 and tumour necrosis factor-α from macrophages, whether peripheral or within the central nervous system and through direct effects on the hypothalamus (Fast et al., 1989; Kim and Schlievert, 1997).

Reticuloendothelial system blockade

Toxins A and C depressed significantly the clearance function of the reticuloendothelial system (RES) in rabbits and mice, as measured by the reduction in the rate of clearance of colloidal carbon (Hanna and Watson, 1965; Hrìbalová, 1979) or ^{51}Cr-labelled sheep red blood cells (Cunningham and Watson, 1978a) from blood after i.v. injection of the toxins. In addition, toxin treatment failed to clear endotoxin from the circulation in parallel with the inhibition of RNA synthesis in Kupffer cells (see Wannamaker and Schlievert, 1988).

Enhancement of host susceptibility to lethal shock by endotoxins

The susceptibility of rabbits and other animals to the lethal shock induced by Gram-negative bacterial endotoxins [lipopolysaccharides (LPS)] was considerably enhanced by i.v. injection of toxins A and C (0.1 to 10 µg), as first reported by Watson (1960) and thereafter by Kim and Watson (1970) and Hrìbalová (1979). This synergistic effect is certainly a key property which may be relevant to the pathogenesis of severe streptococcal infections, particularly STSS (see Kim and Schlievert, 1997).

Toxin A greatly enhanced the susceptibility to endotoxin shock in American Dutch belted (Kim and Watson, 1970) rabbits, monkeys and mice by as much as 100 000-fold. Myocardial and liver damage were observed in surviving animals. Deaths were recorded over 48 h. A similar effect was observed with toxin C by Schlievert and Watson (1978), who showed that the ability to produce fever and to enhance lethal endotoxin shock were separate functions of the toxin. The mechanism of potentiation effect was investigated by Murai et al. (1987), who observed severe pathophysiological changes in Japanese white rabbits. These changes included transient hyperglycaemia followed by profound hypoglycaemia, elevation of the blood lipoperoxide level and an acute increase in plasma β-glucuronidase activity, suggesting a general potentiation of physiological failures. Murai et al. (1990) showed that the macrophages from toxin-treated rabbits exhibited hyper-reactivity to endotoxin, as assessed by their increased consumption of glucose. This enhancing effect was also reflected by potentiation of the febrile response to endotoxin in rabbits pretreated with toxin A, suggesting an increased production of endogenous pyrogens. It is likely that other cytokines are massively released, which may contribute to the lethal enhancing effect.

The mechanism underlying the enhancement phenomenon is still not clear. The reduced clearance function of the liver due to the RES blockade suggests that the endotoxin may enter the circulation, bind LPS-receptor(s) and trigger the release of tumour necrosis factor-α (TNF-α) from macrophages, as well as capillary leak. However, toxin A binds to endotoxin through the KDO moiety in the core region of the LPS. This complex has the ability to bind to lymphocyte (but not by interaction with the TcR) and causes cell death (Kim and Schlievert, 1997).

Alteration and modulation of humoral and cell-mediated immune functions

Toxins A and C (and the staphylococcal superantigens) depress the immune response (see Wannamaker and Schlievert, 1988; Alouf et al., 1991). In a series of investigations (Hanna and Watson, 1968; Cunningham and Watson, 1978b; Cavaillon et al., 1983), the toxins were shown to suppress the in vitro IgM response against sheep erythrocytes in rabbits and mice in a direct Jerne-plaque forming cells (PFC) assay. The immunosuppression was followed by a late delayed PFC burst of IgG synthesis at 10–12 days. This immuno-enhancement of antibody response to the erythrocytes (called deregulation by Hanna et al., 1980) was a T-cell-dependent property of the toxins (Cunningham and Watson, 1978b). The in vivo suppression of antibody response to sheep erythrocytes by the toxins was established by Hanna and Watson (1968) in the rabbit and by Cunningham and Watson (1978a) in the mouse. It is very likely that the activation of CD4$^+$ T-cells by the toxins (see below) elicits the release of interferon-γ, which results in antibody suppression.

Superantigenicity and mediator release by toxin-stimulated immune system cells

The mitogenic (lymphocyte-transforming) activity of toxins A and C, that of other streptococcal mitogens and of staphylococcal enterotoxins and TSST-1 on the T-lymphocytes of humans and other animal species is the most important property of this group of proteins and currently the main focus of research in the field of the immunocytotropic bacterial toxins. Before the demonstration of the superantigenic feature of these pathogenic streptococcal and staphylococcal products, these effectors were originally viewed as non-specific mitogens similar to lectins.

The potent mitogenicity of the streptococcal toxins A and C was first reported by Kim and Watson (1972), Nauciel (1973), Hřibalová and Pospíšil (1973) and Abe and Alouf (1976), as measured by [³H]thymidine incorporation in target cells. Kim and Watson (1972) found that 50% transformation of human peripheral blood lymphocytes required only 0.00005 μg ml^{-1} of streptococcal culture filtrate. Cavaillon et al. (1979) isolated and purified two distinct, non-specific, extracellular mitogenic proteins from culture fluid of a *S. pyogenes*. The mitogen with a pI of 4.8 showed complete immunological cross-reaction with toxin A. The other, of pI 8.5, was immunologically different. Both were highly potent on human and rabbit peripheral blood lymphocytes as well as on rabbit thymocytes and splenocytes.

These initial findings were supported and expanded by other investigators (Barsumian et al., 1978; Petermann et al., 1978; Gray, 1979; Abe et al., 1980; Hřibalová et al., 1980; Rasmussen and Wuepper, 1981; Regelman et al., 1982; Knöll et al., 1983). The mitogenic effect was also found *in vivo* in the rabbit subcutaneous tissue cage model of Knöll et al. (1981). Only T-lymphocytes were transformed by the toxins A and C. Most data indicated that the toxins are very potent mitogens since as low as 10^{-4} μg ml^{-1} (~3 × 10^{-12} M) of toxin and sometimes less was sufficient to elicit human lymphocyte proliferation. Mouse lymphocytes were much less sensitive. It was demonstrated (Cavaillon et al., 1983; Regelmann et al., 1982; Alouf et al., 1986) that adherent cells (even irradiated) were required for T-cell mitogenesis, in contrast to other mitogens (concanavalin, phytohaemaglutinin). This finding was supported by the further establishment of the superantigenic nature of the toxins which indeed require presentation by antigen-presenting cells (APCs).

In vitro infection of human CD4$^+$ cells with human immunodeficiency virus-1 (HIV-1) in the presence of toxin A as mitogen led to a six- to 10-fold higher yield of virus compared with similar lymphocyte stimulation with haemagglutinin (Alouf et al., 1986). This finding may be of clinical significance for subjects contaminated by HIV-1 who become infected with mitogen-producing streptococci. However, the mechanisms by which these toxins (and the other bacterial toxins and mitogens sharing the same polyclonal lymphocyte-

transforming activity) stimulate immune systems cells remained poorly understood until their characterization as 'superantigens'.

Superantigenic behaviour of the toxins

At the beginning of the 1990s toxins A and C were characterized as members of the family of the bacterial superantigens (see Introduction) active on T-lymphocytes bearing appropriate Vβ motifs (Table 32.3) at concentration of 1–5 ng (on human T-cells) and even at lower concentration for toxin C (Tomai et al., 1992; Li et al., 1997). Both wild-type and recombinant toxins were highly active. As low as 1 ng of recombinant toxin A elicited significant stimulation of PBMCs (Yamamoto and Ferretti, 1995). However, the most potent of the recombinant toxins was toxin C, which gave a half-maximal response of 5 fg ml^{-1} and was still active at below 0.1 fg ml^{-1} (Li et al., 1997). In contrast, toxin C lacked proliferative activity for mouse T-cells but bound with high affinity to both human HLA-DR and murine I-E molecules (but not to murine I-A molecules) in a zinc-dependent fashion. Competition binding studies with other recombinant toxins revealed that toxin C lacked the generic low-affinity MHC class-II alpha-chain binding site common to all other bacterial superantigens. Non-denaturing SDS electrophoresis and size-exclusion chromatography revealed that both wild-type and recombinant toxin C exist in a stable dimer state at neutral or alkaline pH. These data support the recent crystal structure of SPE-C as discussed in another section and reveal yet another mechanism by which bacterial superantigens ligate and cross-link MHC class II (Li et al., 1997).

Three single-residue mutants (Y151, A16E, Y171) of toxin C in which Tyr15, Ala16 and Tyr17 were replaced by Ile, Glu and Ile, respectively, have been raised by Yamaoka et al. (1998). These single mutations significantly reduce mitogenic activity (1000-fold less for Y151 mutant) toward Vβ2 human T-lymphocytes *in vitro*. However, these mutants retained their ability to bind to MHC class-II antigen, indicating that the mutated residues were critically important for toxin C interaction with the TcR beta-chain.

Interestingly, the erythema induced by intradermal injection of wild-type toxin in unsensitized rabbits was abrogated by the pretreatment of these animals by cyclosporine. Moreover, the Y151 mutant injected into the rabbits showed a 1000-fold decrease in erythema. These results suggest that this skin reaction is attributable to T-cell stimulatory activity of this superantigenic toxin.

Over the 1990–98 period a great number of investigations was conducted on toxins A, C and the interaction of other superantigenic mitogens with APCs, T-lymphocytes and the formation of the trimolecular complex: MHC class-II molecules/the superantigen/the T-cell receptor (TcR).

The studies concerned the *in vitro* stimulation and expansion of both human and murine CD4 and rabbit CD8 T-cells, the T-lymphocytes bearing the γ/δ TcR, the characterization of the Vβ repertoire recognized by the streptococcal SAgs (Table 32.3), the determination of the regions of the toxins involved in the formation of the ternary complexes and the subsequent release of cytokines and other biological effectors.

Toxin–immune cell interactions

Both human CD4 and CD8 T-lymphocytes bearing the appropriate Vβ motifs were stimulated by toxin A and C (Abe *et al.*, 1991; Imanishi *et al.*, 1992; Tomai *et al.*, 1992; Gougeon *et al.*, 1993; Braun *et al.*, 1993; Dadaglio *et al.*, 1994; Uchiyama *et al.*, 1994). According to Abe *et al.*, the CD4 T-cell subset was proportionately more expanded than the CD8 T-cell subset. In contrast, the latter was not expanded (Alouf *et al.*, 1986). Vβ specific anergy affecting both CD4 and CD8 T-cells of HIV patients was observed (Gougeon *et al.*, 1993; Dadaglio *et al.*, 1994). Imanishi *et al.* (1992) demonstrated that toxin A binds to HLA-DQ molecules with a greater affinity than to either HLA-DR or HLA-DP. Toxin A interaction with MHC class-II molecules was also investigated by Hartwig *et al.* (1994).

Toxin A-induced proliferation of murine T-cells bearing specific Vβ motifs was reported by Leonard *et al.* (1991) and Imanishi *et al.* (1990).

A comparative study of the stimulation of human PBMC and rabbit and murine splenocytes with toxin A and various mutants of this toxin was reported by Roggiani *et al.* (1997). These authors examined in detail the lymphocytic proliferative and lethal activities of the mutants compared with the wild-type toxin, as mentioned in the section on molecular biology in this chapter. Kline and Collins (1996) generated 20 mutants of toxin A1 encoded by the *Spe* A1 allele. The mutants were analysed for mitogenic stimulation of human PBMC and affinity for MHC class-II DQ molecules. Residues necessary for each of these functions were identified.

In a further study, Kline and Collins (1997) investigated the recombinant toxins encoded by *Spe* A alleles 1, 2, 3 and 19 toxins resulting from distinct point mutations. The analysis indicated that the residues of toxin A needed for a productive TcR interaction differ for each Vβ-chain examined. An amino acid substitution at only one site significantly affected the toxin's ability to stimulate Vβ 2.1-expressing T-cells, three individual amino acid substitutions resulted in significant loss of ability to stimulate Vβ 12.2-expressing T-cells and substitution at

13 individual sites significantly affected the ability to stimulate Vβ 14.1-expressing T-cells. To elucidate the regions of the Vβ-chains that interacted with toxin A, synthetic peptides representative of the human Vβ 12.2 complementary-determining regions (CDRs) 1, 2 and 4 were used to block the toxin A-mediated proliferation of human PBMCs. The CDR1, CDR2 and CDR4 peptides were each able to block proliferation with the activity of CDR1 > CDR2 > CDR4. Combinations of CDR1 peptide with CDR2 or CDR4 peptides allosterically enhanced the ability of each to block proliferation, suggesting that toxin A has distinct binding sites for the CDR loops.

A detailed study of toxin C interaction with APCs and T-cells was also reported by several investigators who all stressed the highly potent mitogenic activity of toxin C compared to toxin A towards human T-lymphocytes bearing the TcR Vβ2-chain. As low as 1 pg ml^{-1} of toxin C elicited lymphocyte transformation and specifically increased the amounts of Vβ2 mRNA transcripts (Ohara-Nemoto and Kaneko, 1996).

Stimulation of T-lymphocytes bearing the γ/δ T-cell receptor

The supernatant fluid from a culture of an *S. pyogenes* strain isolated from an adult patient with pharyngitis-associated STSS was shown by Arvand *et al.* (1996) to expand markedly the $\gamma\delta$ T-cell fraction of PBMC of two healthy donors bearing the Vδ1 cell subset.

Stimulation of other cells

Besides interaction with T-lymphocytes and monocytes/macrophages, toxin A interacts with other cells. The interaction with human neutrophils elicited the induction of heat shock proteins (Köller *et al.*, 1993), the modulation of the release of the inflammatory mediator leukotriene B4 (LTB4) and the expression of the receptors for this mediator, for formyl-methionyl-leucyl-phenylalanine (fMLP) and G-protein activity of these cells (Hensler *et al.*, 1993).

Using colloidal gold-labelled toxin A, Wagner *et al.* (1993) showed by transmission electron microscopy the binding of the toxin to certain human skin cells, such as Langherans cells and dermal fibroblasts. Since *S. pyogenes* is an important skin pathogen, the demonstration of toxin binding to target skin cells appears clinically important.

Toxin A and staphylococcal superantigens were also shown to stimulate the human colonic epithelial cell line HT-29 to stimulate T-cell proliferation. Cells of this line behave as APC to these superantigens. The response was a HLA-DR- and ICAM-1-dependent, but β7-2- and LFA-3-independent process and thereby different from professional APC monocytes (Liu *et al.*,

1997). Buslau *et al.* (1993) investigated toxin A and SEB presentation by human epidermal cells and the induction of autologous T-cells.

Inhibition of toxin-induced T-cell proliferation by immune sera

The *in vitro* lymphoproliferative responses to toxin A and to a streptococcal mitogen produced by the same strain purified by Cavaillon *et al.* (1979) were investigated by Bahr *et al.* (1992) on the PBMC from children with acute rheumatic fever (ARF) and patients with chronic rheumatic heart disease (CRHD). Antibody levels to the streptococcal products were also analysed in the sera of those with ARF or CRHD as well as in the sera of children with streptococcal pharyngitis or post-streptococcal glomerulonephritis. Depressed lymphoproliferative responses during the active stage of rheumatic fever were observed. The depressed responses were neither induced by mitogen-specific suppressor cells nor related to a dose–response phenomenon. However, antibody levels to the extracellular mitogens were significantly elevated in the sera of children with ARF compared with the levels in the rest of the groups.

Non-neutralizing IgG antibodies to toxin A were recovered by Nakano *et al.* (1997) in the sera of patients with *S. pyogenes* infection or Kawasaki disease. These antibodies are postulated to play an important role in the protection against toxin A by binding to non-mitogenic epitopes on the toxin molecule, enabling the host to handle the molecule not as a SAg but as a conventional peptide antigen.

Expression of cytokines and other mediators by immune cells stimulated by streptococcal exotoxins A and C and other mitogens

Owing to the superantigenic character of these bacterial effectors, the activation of an unusually high pro-portion of T-lymphocytes and APCS *in vivo* and *in vitro* upon binding the SAg molecules triggers an initial production of a variety of cytokines and other mediators by the stimulated cells. This process leads to the elicitation (through a complex of upregulated and down-regulated immunological network, activation signals and the co-operation of adhesion molecules on target cells) of a cascade of events including further release of a wide array of cytokines and other pharmacologically active products.

Interferon-γ (IFN-γ) was the first cytokine reported to be elicited by human PBMC stimulated with toxin A (Cavaillon *et al.*, 1982; Tonew *et al.*, 1982). Since then, the production of toxin-induced cytokines has been widely investigated through three different experimental approaches.

Identification and assay of the cytokines released *in vitro* by PBMC or other cells stimulated with the toxins

This approach was undertaken by a great number of authors. As summarized in Table 32.5, the challenge of target cells by toxins A, C and other streptococcal super-antigens elicited the release of substantial amounts of cytokines, among them the mainly monocyte-derived cytokines IL-1α and β, IL-6, IL-8 and TNF-α, as well as IL-2, TNF-β and interferon-γ produced by type 1 helper T-lymphocytes (Th1). Besides these pro-inflammatory cytokines, the release of the anti-inflammatory cytokines IL-4 and IL-10 which derive from type 2 helper T-cells (Th2) and the monocyte-derived soluble IL-1 receptor antagonist (IL-1ra) were released in response to toxins A and C. IL-12, IL-13 and the haematopoietic cytokines IL-3, IL-5 and granulocyte macrophage–colony-stimulating factor (GM-CSF) were also elicited *in vitro* by human PBMC stimulated with toxin A. IL-5 and IL-13 are typical Th2-derived cytokines. IL-3 is produced by both Th1 and Th2 cells, whereas a variety of immune

TABLE 32.5 Cytokine release by human peripheral blood mononuclear cells stimulated *in vitro* by the superantigenic streptococcal erythrogenic/pyrogenic toxins A, C

Authors	Toxin	Cytokines
Cavaillon *et al.* (1982)	A	IFN
Tonew *et al.* (1982)	A	IFN
Fast *et al.* (1989)	A	TNF
Hackett and Stevens (1992, 1993)	A	IL-1β, TNF-α, TNF-β
Müller-Alouf *et al.* (1994, 1996)	A, C	IL-1α, IL-1β, IL-6, IL-8, TNF-α, TNF-β, IFN-γ, IL-2, IL-3, IL-10, IL-1ra
König *et al.* (1994)	A	IL-8
Rink *et al.* (1996)		IL-1α, IL-1β, IL-2, IL-4, IL-6, IFN-γ, TNF-α
Sriskandan *et al.* (1996b)		IFN-γ, IL-12, TNF-β
Müller-Alouf *et al.* (1997b)	A	IL-3, IL-4, IL-5, GM-CSF, IL-12p40, IL-12p70, IL-13
Müller-Alouf *et al.* (1997c)	A, C	Above-mentioned cytokines and IL-1α, IL-1β, IL-2, IL-6, IL-8, TNF-α, TNF-β, sIL-1ra, MIP-1, RANTES, TGF-β1

and non-immune cells produces GM-CSF. This cytokine as well as IL-3 and IL-5 activate early haematopoietic stem cells. IL-12, mainly produced by monocytes and macrophages, plays an important role in immune regulation and primes human T-cells for high production of both IFN-γ and IL-10.

König et al. (1994) demonstrated the transcription of IL-8 (expression of mRNA) and its release by PBMC and lymphocyte-monocyte-basophil cell population stimulated with toxin A. IL-8 synthesis was not influenced by the protein tyrosine kinase inhibitors lavandustin A and tyrphostin 25. In contrast, these inhibitors abrogated IL-8 induction by the membrane-damaging toxins staphylococcal leucocidins and Bacillus alvei cytolysin, indicating that various pathways may lead to IL-8 induction.

Interestingly, heat-killed streptococcal cells proved highly potent in vitro inducers of a wide array of inflammatory and other cytokines by PBMC, suggesting that the bacteria may jointly contribute with streptococcal exotoxins to cytokine-mediated development of severe streptococcal diseases (Müller-Alouf et al., 1994, 1996, 1997b, c).

Besides the release of cytokines, toxin A and TSST-1 elicited the release of nitric oxide by human PBMC (Sriskandan et al., 1996b). Cytokine release was also reported for murine splenocytes challenged with toxin A (Müller-Alouf et al., 1992). Toxins A and C induced nitric oxide synthase in murine macrophages (Christ et al., 1997).

Cytokine production at the single cell level

This elegant procedure was described by Andersson et al. (1992), who detected the intracellular production of various cytokines by PBMC challenged with toxin A and other superantigens by indirect immunofluorescence staining with cytokine specific fluorescent-tagged monoclonal antibodies. A similar analysis of superantigen-induced Th1- and Th2-derived cytokines was reported by Norrby-Teglund et al. (1994a, 1997).

The in vitro stimulation of PBMC by toxins A and C and staphylococcal SEB and TSST-1 was shown by Leung et al. (1995b) specifically to induce IL-12 production and a significant increase in the cutaneous-associated antigen (CLA) of T-cell blasts by immunofluorescence staining. CLA induction was blocked by anti-IL-12 antibodies, suggesting that the expansion of skin-homing CLA$^+$ T-cells takes place in an IL-12-dependent manner and thus may contribute to the development of skin rashes in SAg-mediated diseases.

Cytokine mRNA expression in toxin-stimulated cells

This procedure was also used to investigate cytokine elicitation in PBMC stimulated with toxin A or C by determination of cytokine mRNA by reverse transcription of RNA followed by PCR (König et al., 1994; Rink et al., 1996; Ohara-Nemoto and Kaneko, 1996; Müller-Alouf et al., 1997b).

SUPERANTIGENS AND THE PATHOGENESIS OF TOXIC SHOCK SYNDROME AND OTHER SEVERE DISEASES

Several lines of experimental, epidemiological and clinical observations lead to the concept that the superantigenic toxins A and C and most probably the other streptococcal SAgs play a pivotal role in the pathogenesis of STSS and other severe diseases (see Introduction).

First of all, historical strains of S. pyogenes from patients with severe scarlet fever make toxin A (Schlievert et al., 1996). This observation is widely supported by the many epidemiological studies on toxin A-expressing strains, as described in another section of this chapter.

Toxin C was also found epidemiologically associated with development of severe invasive streptococcal diseases including STSS (Leggiadro et al., 1993). Interestingly, S. pyogenes strains associated with guttate psoriasis consistently produce the toxin either alone or together with toxin A and cysteine proteinase (SPE B) (Mollick et al., 1992).

Experiments with isogenic S. pyogenes strains harbouring or lacking the spe A gene administered in subcutaneous implanted Wiffle balls in rabbits showed that the toxin A-producing strain caused STSS, whereas the toxin-negative strain did not (Schlievert et al., 1996). Purified toxin A administered to rabbits in subcutaneous mini-osmotic pumps also provoked STSS (Lee and Schlievert, 1989). A murine model of toxin-induced fasciitis and multi-organ failure was developed by Sriskandan et al. (1996a). In addition, two researchers injected themselves with toxin A and induced clinical STSS (see Kim and Schlievert, 1997).

The massive overproduction of cytokines by immune systems cells is pathogenic in infected patients and generates large regulatory disturbances and thereby severe imbalance of the cytokine network of the host (Kotb, 1997). This imbalance is thought to contribute to the disease and systematic lethality, as experimentally supported in the animal models studies mentioned above and in a primate (baboon) model of *S. pyogenes* bacteraemia that mimics human STSS developed by Stevens *et al.* (1996). These authors investigated the role of TNF-α and the dynamics of cardiovascular and laboratory abnormalities. Profound hypotension, leucopenia, metabolic acidosis, renal impairment, thrombocytopenia and disseminated coagulopathy similar to those observed in patients with STSS (see Working Group on Severe Streptococcal Infections, 1993) developed within 3 h after i.v. injection of a strain isolated from a patient with STSS. The strain produced toxin A and the superantigens SSA and MF. Serum TNF-α peaked at 3 h and returned to baseline by 10 h. Mortality was 100% and anti-TNF-α monoclonal antibody markedly improved arterial pressure and survival, suggesting that TNF-α plays an important role in the induction of shock, organ failure and subsequent death.

The excessive endogenous release of TNF-α and -β, IL-1 and IFN-γ is considered to be the primary cause of capillary leak hypotension and shock, the most severe manifestations of STSS (Hackett and Stevens, 1992, 1993; Uchiyama *et al.*, 1994; Schafer and Sheil, 1995; Roggiani *et al.*, 1997; Kotb, 1997). Christ *et al.* (1997) suggested that nitric oxide release also plays an important role in STSS pathogenesis. These contentions are strongly supported by clinical observations showing high levels of cytokines in the plasma and/or cerebrospinal fluid in patients with severe streptococcal infections, particularly STSS (Hackett and Stevens, 1992; Nadal *et al.*, 1993; Norrby-Teglund *et al.*, 1995; Cavaillon *et al.*, 1997). Interestingly, streptococcal SAg-induced T-lymphocyte proliferation and cytokine production was inhibited by plasma from patients with severe invasive *S. pyogenes* infections treated with human polyspecific IgG (Norrby-Teglund *et al.*, 1996a). These IgG contained neutralizing antibodies against streptococcal SAgs (Norrby-Teglund *et al.*, 1996b).

Direct evidence for the role of streptococcal SAgs in disease is also derived from studies that examined the TcR Vβ repertoire of patients during illness (Kotb, 1997). Specific changes in the peripheric repertoire are reflected by either Vβ expansion or Vβ depletion. As reported by Michie *et al.* (1994), lymphocytes collected from two patients with STSS during the acute phase of their illness showed a marked decrease in CD 4+T (helper) and CD 45 RA+ (naibe T-lymphocytes) expressing the Vβ2 chain and an increase in those expressing CD 8+ and CD 45 RO markers. Similarly, a pattern of depletion of Vβ1, Vβ5.1 and Vβ12 was observed in patients with severe streptococcal infections (Watanabe-Ohnishi *et al.*, 1995).

Streptococcal and/or staphylococcal SAgs are thought to be involved in the pathogenesis of Kawasaki disease (KD), an acute multisystem vasculitis of unknown aetiology characterized by marked activation of T-lymphocytes and monocytes and elevated levels of IL-1, IL-6, IFN-γ and TNFα in the plasma of patients in the acute stage of the disease (Abe *et al.*, 1990; Leung *et al.*, 1995a; Nakano *et al.*, 1997; Rowley, 1998). Toxin A-induced angiocutaneous lymph node lesions and lymphocytic arteritis in rabbits similar to those of KD (Abe *et al.*, 1995, 1998) appear to support the role of toxin A in KD. However, according to Morita *et al.* (1997), this hypothesis appears unlikely on the basis of the low frequency of detection of anti-SAg antibodies in patients' sera. A valuable survey of KD has been reported by Rowley (1998).

Finally, several lines of clinical evidence support the involvement of streptococcal superantigens in acute guttate psoriasis (Leung *et al.*, 1995c; Baker *et al.*, 1997).

ACKNOWLEDGEMENTS

The authors thank Dr J.M. Cavaillon (Institut Pasteur, Paris) and Dr D. Gerlach (University of Jena), who have made results of their studies available to them, and the Fondation de la Recherche Médicale (Paris) for their grant (1995) to H. Müller-Alouf. The invaluable secretarial assistance of Mrs Patricia Paul in the preparation of this manuscript is greatly appreciated.

REFERENCES

Abe, J., Forrester, J., Nakahara, T. and Lafferty, J.A. (1991). Selective stimulation of human T cells with streptococcal erythrogenic toxins A and B. *J. Immunol.* **146**, 3747–50.

Abe, Y. and Alouf, J.E. (1976). Partial purification of exocellular lymphocyte mitogen from *Streptococcus pyogenes*. *Jpn. J. Exp. Med.* **46**, 363–9.

Abe, Y., Alouf, J.E., Kuirhara, T. and Kawashima, H. (1980). Species-dependent response to streptococcal lymphocyte mitogens in rabbits, guinea pig and mice. *Infect. Immun.* **29**, 814–18.

Abe, Y., Nakano, S., Nakahara, T., Kamezawa, Y., Kato, I., Ushijima, H., Yoshino, K., Ito, S., Noma, S., Okitsu, S. and Tajima, M. (1990). Detection of serum antibody by the anti mitogen assay against streptococcal erythrogenic toxins. Age distribution in children and the relation to Kawasaki disease. *Pediatr. Res.* **27**, 11–16.

Abe, Y., Nakano, S., Sagishima, M. and Aita, K. (1995). Erythrogenic-toxin induced angiocutaneous lymph node lesions as an experimental model. In: *Kawasaki Disease* (ed. H. Kato), pp. 163–9. Elsevier Science, Amsterdam.

Abe, Y., Nakano, S., Aita, K. and Sagishima, M. (1998). Streptococcal and staphylococcal superantigen induced lymphocytic arteritis in a local type experimental model: comparison with acute vasculitis in the Arthus reaction. *J. Lab. Clin. Med.* **131**, 93–102.

Alouf, J.E. (1980). Streptococcal toxins (streptolysin O, streptolysin S, erythrogenic toxin). *Pharmacol. Ther.* **11**, 661–718.

Alouf, J.E., Geoffroy, C., Klatzmann, D., Gluckman, J.C., Gruest, J. and Montagnier, L. (1986). High production of the AIDS virus (LAV) by human T lymphocytes stimulated by streptococcal mitogenic toxins. *J. Clin. Microbiol.* **24**, 639–41.

Alouf, J.E., Knöll, H. and Köhler, W. (1991). The family of mitogenic, shock-inducing and superantigenic toxins from staphylococci and streptococci. In: *Sourcebook of Bacterial Protein Toxins* (eds J.E. Alouf and J.H. Freer), pp. 367–414. Academic Press, London.

Andersson, J., Nagy, S., Björk, L., Abrams, J., Holm, S. and Andersson, U. (1992). Bacterial-toxin induced cytokine production studied at the single-cell level. *Immunol. Rev.* **127**, 69–96.

Arvand, M., Schneider, T., Jahn, H.-U. and Hahn, H. (1996). Streptococcal toxic shock syndrome associated with marked γδ T cell expansion: case report. *Clin. Infect. Dis.* **22**, 362–5.

Bahr, G.M., Yousof, A.H., Behbehani, K., Majeed, H.A., Sakkalok, S., Sonan, K., Jarrad, I., Geoffroy, C. and Alouf, J.E. (1991). Antibody levels and *in vitro* lymphoproliferative responses to *Streptococcus pyogenes* erythrogenic toxin A and mitogen of patients with rheumatic fever. *J. Clin. Microbiol.* **29**, 1789–94.

Baker, B.S., Garioch, J.J., Hardman, C., Powles, A. and Fry, L. (1997). Induction of cutaneous lymphocyte-associated antigen expression by group A streptococcal antigens in psoriasis. *Arch. Dermatol. Res.* **289**, 671–6.

Barsumian, E.L., Schlievert, P.M. and Watson, D.W. (1978). Non specific and specific immunological mitogenicity by group A streptococcal pyrogenic exotoxins. *Infect. Immun.* **22**, 681–8.

Bingel, K.F. (1949). Neue Untersuchungen zur Scharlachätiologie. *Deutsch. Med. Wochensch.* **127**, 703–6.

Black, C.M., Talkington, D.F., Messmer, T.O., Facklam, R.R., Hornes, E. and Olsvik, Ø. (1993). Detection of streptococcal pyrogenic exotoxins genes by a nested polymerase chain reaction. *Molec. Cell. Probes* **7**, 255–9.

Bohach, G.A., Hovde, C.J., Handley, J.P. and Schlievert, P.M. (1988). Cross-neutralization of staphylococcal and streptococcal pyrogenic toxins by monoclonal and polyclonal antibodies. *Infect. Immun.* **56**, 400–4.

Braun, M.A., Gerlach, D., Hartwig, U.F., Ozegowski, J.-H, Romagne, F., Carrel, S., Köhler, W. and Fleischer, B. (1993). Stimulation of human T cells by streptococcal 'superantigens' erythrogenic toxins (scarlet fever toxins). *J. Immunol.* **150**, 2457–66.

Bronze, M.S. and Dale, J.D. (1996). The reemergence of serious group A streptococcal infections and acute rheumatic fever. *Am. J. Med. Sci.* **311**, 41–54.

Buslau, M., Kappus, R., Gerlach, D., Köhler, W., Diel, S. and Holzmann, H. (1993). Streptococcal and staphylococcal superantigens (ETA, SEB): presentation by human epidermal cells and induction of autologous T cell *in vitro*. *Acta. Derm. Venereol.* **73**, 94–6.

Carapetis, J., Robins-Brown, R., Martin, D., Shelby-James, T. and Hogg, G. (1995). Increasing seventy of invasive group A streptococcal disease in Australia: clinical and molecular epidemiological features and identification of a virulent M-non typeable clone. *Clin. Infect. Dis.* **21**, 1220–7.

Cavaillon, J.-M., Geoffroy, C. and Alouf, J.E. (1979). Purification of two extracellular streptococcal mitogens and their effect on human, rabbit and mouse lymphocytes. *J. Clin. Lab. Immunol.* **2**, 155–63.

Cavaillon, J.-M., Rivière, Y., Svab, J., Montagnier, L. and Alouf, J.E. (1982). Induction of interferon by *Streptococcus pyogenes* extracellular products. *Immunol. Lett.* **5**, 323–6.

Cavaillon, J.-M., Leclerc, C. and Alouf, J.E. (1983). Polyclonal antibody-forming cell activation and immunomodulation of the *in vitro* immune response induced by streptococcal extracellular products. *Cell. Immunol.* **76**, 200–6.

Cavaillon, J.-M., Müller-Alouf, H. and Alouf, J.E. (1997). Cytokines in streptococcal infections. An opening lecture. In: *Streptococci and the Host* (eds T. Horaud *et al.*), pp. 869–79. Plenum Press, New York.

Chaussee, M.S., Gerlach, D., Yu, C.-E. and Ferretti, J.J. (1993). Inactivation of the streptococcal erythrogenic toxin B gene (*spe B*) in *Streptococcus pyogenes*. *Infect. Immun.* **61**, 3719–23.

Chaussee, M.S., Liu, J., Stevens, D.L. and Ferretti, J.J. (1996). Genetic and phenotypic diversity among isolates of *Streptococcus pyogenes* from invasive infections. *J. Infect. Dis.* **173**, 901–8.

Christ, E.A., Meals, E. and English, B.K. (1997). Streptococcal pyrogenic exotoxins A (Spe A) and C (Spe C) stimulate the production of inducible nitric oxide synthase (iNOS) protein in RAW 264.7 macrophages. *Shock* **8**, 450–3.

Cleary, P.P., McLandsborough, L., Ikeda, L., Cue, D., Krawczak, J. and Lam, H. (1998). High-frequency intracellular infection and erythrogenic toxin A expression undergo phase variation in M1 group A streptococci. *Mol. Microbiol.* **28**, 157–67.

Cone, L.A., Voodard, D.R., Schlievert, P.M. and Tomory, G.S. (1987). Clinical and bacteriologic observations of a toxic shock-like syndrome due to *Streptococcus pyogenes*. *N. Engl. J. Med.* **317**, 146–9.

Cunningham, C.M. and Watson, D.W. (1978a). Alteration of clearance function by group A streptococcal pyrogenic exotoxin and its relation to suppression of the antibody response. *Infect. Immun.* **19**, 51–7.

Cunningham, C.M. and Watson, D.W. (1978b). Suppression of antibody response by group A streptococcal pyrogenic exotoxin and characterization of the cells involved. *Infect. Immun.* **19**, 470–6.

Dadaglio, G., Garcia, S., Montagnier, L. and Gougeon, M.L. (1994). Selective anergy of Vβ 8⁺ T cells in human immuno-deficiency virus-infected individuals. *J. Exp. Med.* **179**, 413–24.

Dick, G.F. and Boor, A.K. (1935). Scarlet fever toxin I. A method of purification and concentration. *J. Infect. Dis.* **57**, 164–73.

Dick, G.F. and Dick, G.H. (1924a). A skin test for susceptibility to scarlet fever. *JAMA* **82**, 265–6.

Dick, G.F. and Dick, G.H. (1924b). The etiology of scarlet fever. *JAMA* **82**, 301–2.

Fast, D.J., Schlievert, P.M. and Nelson, R.D. (1989). Toxic shock syndrome-associated staphylococcal and streptococcal pyrogenic toxins are potent inducers of tumor necrosis factor production. *Infect. Immun.* **57**, 291–4.

Fleischer, B. (1989). Bacterial toxins as probes for the T-cell antigen receptor. *Immunol. Today* **10**, 262–4.

Fleischer, B., Gerardy-Schahu, R., Carrel, S., Gerlach, D. and Köhler, W. (1991). An evolutionary conserved mechanism of T cell activation by microbial toxins. *J. Immunol.* **146**, 11–17.

Frobisher, M., Jr and Brown, J.H. (1927). Transmissible toxigenicity of streptococci. *Bull. Johns Hopkins Hosp.* **41**, 167–73.

Geoffroy, C. and Alouf, J.E. (1988). Production, purification and assay of streptococcal erythrogenic toxin. In: *Methods in Enzymology*, Vol. 165, *Microbial Toxins: Tools in Enzymology* (ed. S. Harshman), pp. 64–7. Academic Press, San Diego, CA.

Geoffroy, C., Müller-Alouf, H., Champagne, E., Cavaillon, J.-M. and Alouf, J.E. (1994). Identification of a new extracellular mitogen from group A streptococci. *Zbl. Bakt. Suppl.* **24**, 90–1.

Gerlach, D., Knöll, H. and Köhler, W. (1980). Purification and characterization of erythrogenic toxins. 1. investigation of erythrogenic toxin A produced by *Streptococcus pyogenes* strain NY-5. *Ztbl. Bakteriol. Hyg. Abt. I orig. A* **247**, 177–91.

Gerlach, D., Knöll, H., Köhler, W. and Ozegowski, J.H. (1981). Isolation and characterization of erythrogenic toxins of *Streptococcus pyogenes*. 3. Comparative studies of type A erythrogenic toxins (in German). *Zbl. Bakt. Hyg. I. Abt. Orig. A* **250**, 277–86.

Gerlach, D., Knöll, H., Köhler, W., Ozegowski, J.H. and Hřibalová, V. (1983). Isolation and characterization of erythrogenic toxins. V. Identity of erythrogenic toxin type B and streptococcal proteinase precursor. *Zbl. Bakt. Hyg. I. Abt. Orig. A* **255**, 221–33.

Gerlach, D., Ozegowski, J.H., Knöll, H. and Köhler, W. (1986). Isolation and characterization of erythrogenic toxins.VIII Purification of a biological active low molecular weight 10 000 (LMP-10 K) protein from filtrates of *Streptococcus pyogenes*, strain NY-5. Relations to erythrogenic toxin type A. *Zbl. Bakt. Hyg. A* **261**, 75–84.

Gerlach, D., Köhler, W., Knöll, H., Moravek, L., Weeks, C.R. and Ferretti, J.J. (1987). Purification and characterization of *Streptococcus pyogenes* erythrogenic toxin type A produced by a cloned gene in *Streptococcus sanguis. Zbl. Bakt. Hyg.* **266**, 347–58.

Gerlach, D., Alouf, H., Moravek, L., Pavlick, M. and Köhler, W. (1992). The characterization of two new low molecular weight proteins (LMPs) from *Streptococcus pyogenes. Zbl. Bakt.* **277**, 1–9.

Gerlach, D., Reichardt, W., Fleischer, B. and Schmidt, K.-H. (1994). Separation of mitogenic and pyrogenic activities from so-called erythrogenic toxin type B (streptococcal proteinase). *Zbl. Bakt.* **280**, 507–14.

Gerlach, D., Günther, E., Köhler, W., Vettermann, S., Fleischer, B. and Schmidt, K.-H. (1995). Isolation and characterization of a mitogen characteristic for group A streptococci (*Streptococcus pyogenes*). *Zbl. Bakt.* **282**, 67–82.

Geslin, P., Kriz-Kuzemenska, P., Frémaux, A., Havlickova, H., Dublanchet, A. and Bouvet, A. (1996). Septicaemia caused by phenotypic variants of group A streptococci. *Res. Microbiol.* **147**, 273–7.

Goshorn, S.C. and Schlievert, P.M. (1988). Nucleotide sequence of streptococcal pyrogenic exotoxin type C. *Infect. Immun.* **56**, 2518–20.

Goshorn, S.C., Bohach, G.A. and Schlievert, P.M. (1988). Cloning and characterization of the gene, speC, for pyrogenic exotoxin type C from *Streptococcus pyogenes. Molec. Gen. Genet.* **212**, 66–70.

Goshorn, S.C. and Schlievert, P.M. (1989). Bacteriophage association of streptococcal pyrogenic exotoxin type C. *J. Bacteriol.* **171**, 3068–73.

Gougeon, M.L., Dadaglio, G., Garcia, S. and Müller-Alouf, H. (1993). Is a dominant superantigen involved in AIDS pathogenicity? *Lancet* **142**, 50–1.

Gray, E.D. (1979). Purification and properties of an extracellular blastogen produced by group A streptococci. *J. Exp. Med.* **149**, 1438–49.

Hackett, S.P. and Stevens, D.L. (1992). Streptococcal toxic shock syndrome synthesis of tumor necrosis factor and IL-1 by monocytes stimulated with pyrogenic exotoxin A and streptolysin O. *J. Infect. Dis.* **165**, 879–85.

Hackett, S.P. and Stevens, D.L. (1993). Superantigens associated with staphylococcal and streptococcal toxic shock syndrome are potent inducers of tumor necrosis factor β synthesis. *J. Infect. Dis.* **168**, 232–5.

Hallas, G. (1985). The production of pyrogenic exotoxins by group A streptococci. *J. Hyg. Camb.* **95**, 47–57.

Hanna, E.E. and Watson, D.W. (1965). Host–parasite relationships among group A streptococci. III. Depression of reticuloendothelial function by streptococcal pyrogenic exotoxins. *J. Bacteriol.* **89**, 154–8.

Hanna, E.E. and Watson, D.W. (1968). Host–parasite relationships among group A streptococci. IV. Suppression of antibody response by streptococcal pyrogenic exotoxins. *J. Bacteriol.* **95**, 14–21.

Hanna, E.E., Hale, M.L. and Misfeldt, M.L. (1980). Deregulation of mouse antibody forming cells by streptococcal pyrogenic exotoxin (SPE). III. Modification of T-cell-dependent plaque-forming cell responses of mouse immunocytes is a common property of highly purified and crude preparation of SPE. *Cell Immunol.* **56**, 247–57.

Hartwig, U.F., Gerlach, D. and Fleischer, B. (1994). Major histocompatibility complex class II binding site for streptococcal pyrogenic (erythrogenic) toxin A. *Med. Microbiol. Immunol.* **183**, 257–64.

Hauser, A.R. and Schlievert, P.M. (1990). Nucleotide sequence of the streptococcal pyrogenic exotoxin type B gene and relationship between the toxin and the streptococcal proteinase precursor. *J. Bacteriol.* **172**, 4536–42.

Hauser, A.R., Stevens, D.L., Kaplan, E.L. and Schlievert, P.M. (1991). Molecular analysis of pyrogenic exotoxins from *Streptococcus pyogenes* isolates associated with toxic shock-like syndrome. *J. Clin. Microbiol.* **29**, 1562–7.

Hensler, T., Köller, M., Geoffroy, C., Alouf, J.E. and König, W. (1993). *Staphylococcus aureus* toxic shock syndrome toxin 1 and *Streptococcus pyogenes* erythrogenic toxin A modulate inflammatory mediator release from human neutrophils. *Infect. Immun.* **61**, 1055–61.

Hoge, C.W., Schwarz, B., Talkington, D.F., Breiman, R.F., MacNeill, E.M. and Englender, J.J. (1993). The changing epidemiology of invasive group A streptococcal infections and the emergence of streptococcal toxic shock-like syndrome. A retrospective population-based study. *JAMA.* **269**, 384–9.

Holm, S.E., Norrby, A., Bergholm, A.-M. and Norgren, M. (1992). Aspects of pathogenesis in serious group A streptococcal infections in Sweden 1988–1989. *J. Infect. Dis.* **166**, 31–7.

Holt, L.E. (1898). *The Diseases of Infancy and Childhood*, pp. 888–910. Appleton, New York.

Hooker, S.B. and Follensby, E.M. (1934). Studies on scarlet fever. II. Different toxins produced by hemolytic streptococci of scarlatinal origin. *J. Immunol.* **27**, 177–93.

Houston, C.W. and Ferretti, J.J. (1981). Enzymelinked immunosorbent assay for detection of type A streptococcal exotoxin. Kinetics and regulation during growth of *Streptococcus pyogenes. Infect. Immun.* **33**, 862–9.

Hřibalová, V. (1979). Effect of scarlet fever toxin on the phagocytic activity of the reticuloendothelial system. *Folia Microbiol.* **24**, 415–27.

Hřibalová, V. and Pospíšil, M. (1973). Lymphocyte-stimulating activity of scarlet fever toxin. *Experientia* **29**, 704–5.

Hřibalová, V., Knöll, H., Gerlach, D. and Köhler, W. (1980). Purification and characterization of erythrogenic toxins. II. Communications: *in vivo* biological activities of erythrogenic toxin produced by *Streptococcus pyogenes* strain NY-5. *Zbl. Bakt. Hyg.* **248**, 314–22.

Hsueh, P.-R., Wu, J.-J., Tsai, P.-J., Liu, J.-W., Chuang, Y.-C. and Luh, K.-T. (1998). Invasive group A streptococcal disease in Taiwan is not associated with the presence of streptococcal pyrogenic exotoxin genes. *Clin. Infect. Dis.* **26**, 584–9.

Hynes, W.L., Weeks, C.R., Iandolo, J.J. and Ferretti, J.J. (1987). Immunologic cross-reactivity of type A streptococcal exotoxin (erythrogenic toxin) and staphylococcal enterotoxins B and C1. *Infect. Immun.* **55**, 837–8.

Imanishi, K., Igarashi, H. and Uchiyama, T. (1990). Activation of murine T cells by streptococcal pyrogenic exotoxin type A. Requirement for MHC class II molecules on accessory cells and identification of Vβ elements in T cell receptor of toxin-reactive T cells. *J. Immunol.* **145**, 3170–6.

Imanishi, K., Igarashi, H. and Uchiyama, T. (1992). Relative abilities of distinct isotypes of human major histocompatibility complex class II molecules to bind streptococcal pyrogenic exotoxins A and B. *Infect. Immun.* **60**, 5025–9.

Iwasaki, M., Igarashi, H., Hinuma, Y. and Yutsudo, T. (1993). Cloning, characterization and overexpression of a *Streptococcus pyogenes* gene encoding a new type of mitogenic factor. *FEBS Lett.* **331**, 187–92.

Iwasaki, M., Igarashi, H. and Yutsudo, T. (1997). Mitogenic factor secreted by *Streptococcus pyogenes* is a heat-stable nuclease requiring his[122] for activity. *Microbiology* **143**, 2449–55.

Johnson, L.P. and Schlievert, P.M. (1983). A physical map of the group A streptococcal pyrogenic exotoxin bacteriophage T12gl genome. *Mol. Gen. Genet.* **189**, 251–5.

Johnson, L.P. and Schlievert, P.M. (1984). Group A streptococcal phage T12 carries the structural gene for pyrogenic exotoxin type A. *Mol. Gen. Genet.* **194**, 52–6.

Johnson, L.P., Schlievert, P.M. and Watson, D.W. (1980). Transfer of group A streptococcal pyrogenic exotoxin production to non-toxigenic strains by lysogenic conversion. *Infect. Immun.* **28**, 254–7.

Johnson, L.P., L'Italien, J.J. and Schlievert, P.M. (1986a). Streptococcal pyrogenic exotoxin type A (scarlet fever toxin) is related to *Staphylococcus aureus* enterotoxin B. *Mol. Gen. Genet.* **203**, 354–6.

Johnson, L.P., Tonnai, M.A. and Schlievert, P.M. (1986b). Bacteriophage involvement in group A streptococcal pyrogenic exotoxin A production. *J. Bacteriol.* **166**, 623–7.

Kamezawa, Y. and Nakahara, T. (1989). Purification and characterization of streptococcal erythrogenic toxin type A produced by *Streptococcus pyogenes* strain NY-5 cultured in the synthetic medium NCTC-135. *Microbiol. Immunol.* **33**, 183–94.

Kamezawa, Y., Nakahara, T., Abe, Y. and Kato, I. (1990). Increased vascular permeability, erythema and leukocyte emigration induced in rabbit skin by streptococcal erythrogenic toxin type A. *FEMS Microbiol. Lett.* **68**, 159–62.

Kamezawa, Y., Nakahara, T., Nakano, S., Abe, Y., Nozaki-Renard, J. and Isono, T. (1997). Streptococcal mitogenic exotoxin Z, a novel acidic superantigenic toxin produced by a T1 strain of *Streptococcus pyogenes*. *Infect. Immun.* **65**, 3828–33.

Kapur, V., Nelson, K., Schlievert, P.M., Selander, R.K. and Musser, J.M. (1992). Molecular population genetic evidence of horizontal spread of two alleles of the pyrogenic exotoxin C gene (*speC*) among pathogenic clones of *Streptococcus pyogenes*. *Infect. Immun.* **60**, 3513–17.

Katz, A.R. and Morens, D.M. (1992). Severe streptococcal infections in historical perspective. *Clin. Infect. Dis.* **14**, 298–307.

Kehoe, M.A., Kapur, V., Whatmore, A.M. and Musser, J.J. (1996). Horizontal gene transfer among group A streptococci: implications for pathogenesis and epidemiology. *Trends Microbiol.* **4**, 436–43.

Kihilberg, B.-M., Looney, J., Caparon, M., Olsén, A. and Björk, L. (1995). Biological properties of *Streptococcus pyogenes* mutant generated by Tn 916 insertion in *mga*. *Microb. Pathog.* **19**, 299–315.

Kim, Y.B. and Watson, D.W. (1970). A purified group A streptococcal pyrogenic exotoxin. Physiochemical and biological properties including the enhancement of susceptibility to endotoxin lethal shock. *J. Exp. Med.* **131**, 611–28.

Kim, Y.B. and Watson, D.W. (1972). Streptococcal exotoxins: biological and pathological properties. In: *Streptococci and Streptococcal Diseases* (eds L.W. Wannamaker and J.H. Matsen), pp. 33–50. Academic Press, New York.

Kim, M.M. and Schlievert, P.M. (1997). Molecular genetics, structure and immunobiology of streptococcal pyrogenic exotoxins A and C. In: *Superantigens. Molecular Biology, Immunobiology and Relevance to Human Diseases* (eds D.Y.M. Leung, B. Huber and P.M. Schlievert), pp. 257–79. Marcel Dekker, New York.

Kline, J.B. and Collins, C.M. (1996). Analysis of the superantigenic activity of mutant and allelic forms of streptococcal pyrogenic exotoxin A. *Infect. Immun.* **64**, 861–9.

Kline, J.B. and Collins, C.M. (1997). Analysis of the interaction between the bacterial superantigen streptococcal pyrogenic exotoxin A (*SpeA*) and the human T-cell receptor. *Mol. Microbiol.* **24**, 191–202.

Knöll, H., Hrìbalová, V., Gerlach, D. and Köhler, W. (1981). Purification and characterization of erythrogenic toxins. IV. Commun mitogenic activity of erythrogenic toxin produced by *Streptococcus pyogenes* strain NY-5. *Zbl. Bakt. Hyg.* **A251**, 15–26.

Knöll, H., Holm, J.E., Gerlach, D. and Köhler, W. (1982). Tissue cages for study of experimental streptococcal infection in rabbits. I. Production of erythrogenic toxins *in vivo*. *Immunobiology* **162**, 128–40.

Knöll, H., Gerlach, D., Ozegowski, J.-H., Hrìbalová, V. and Köhler, W. (1983). Mitogenic activity of isoelectrically focused erythrogenic toxin preparations and culture supernatants of group A streptococci. *Zbl. Bakt. Hyg. A.* **256**, 49–60.

Knöll, H., Holm, S.E., Gerlach, D., Kühnemund, O. and Köhler, W. (1985). Tissue cages for study of experimental streptococcal infection of rabbits. II. Humoral and cell mediated immune response to erythrogenic toxins. *Immunobiol.* **169**, 116–27.

Knöll, H., Holm, S.E., Gerlach, D., Ozegowski, J.H. and Köhler, W. (1988). Tissue cages for study of experimental streptococcal infection in rabbits. III. Influence of immunization with erythrogenic toxin A (ET A) and its toxoid on subsequent infection with an ET A producing strain. *Zbl. Bakt. Hyg. A* **269**, 366–76.

Knöll, H., Srámek, J., Vrbová, K., Gerlach, D., Reichardt, W. and Köhler, W. (1991). Scarlet fever and types of erythrogenic toxins produced by the infecting streptococcal strains. *Zbl. Bakt.* **276**, 94–106.

Köhler, W. (1990). Streptococcal toxic syndrome. *Zbl. Bakt.* **272**, 257–64.

Köhler, W., Gerlach, D. and Knöll, H. (1987). Streptococcal outbreaks and erythrogenic toxin type A. *Zbl. Bakt. Hyg. A.* **266**, 104–15.

Köller, M., Hensler, T., König, B., Prévost, G., Alouf, J.E. and König, W. (1993). Induction of heat-shock proteins by bacterial toxins, lipid mediators and cytokines in human leukocytes. *Zbl. Bakt.* **278**, 365–76.

König, B., Köller, M., Prévost, G., Piémont, Y., Geoffroy, C., Alouf, J.E., Schreiner, A. and König, W. (1994). Activation of human effector cells by different bacterial toxins (leukocidin, alveolysin, erythrogenic toxin A): generation of interleukin-8. *Infect. Immun.* **62**, 4831–7.

Kotb, M. (1997). Superantigens in human diseases. *Clin. Microbiol. Newletter* **19**, 145–52.

Kreiswirth, B.N., Handley, J.P., Schlievert, P.M. and Novick, R.P. (1987). Cloning and expression of streptococcal pyrogenic exotoxin A and staphylococcal toxic-shock syndrome toxin-1 in *Bacillus subtilis*. *Molec. Gen. Genet.* **208**, 84–7.

Lee, P.K. and Schlievert, P.M. (1989). Quantification and toxicity of group A streptococcal pyrogenic exotoxins in an animal model of toxic shock syndrome-illness. *J. Clin. Microbiol.* **27**, 1890–2.

Leggiadro, R.J., Bugnitz, M.C., Peck, B.A., Luedtke, G.S., Kim, M.H., Kaplan, E.L. and Schlievert, P.M. (1993). Group A streptococcal bacteremia in a mid-south children's hospital. *South Med. J.* **86**, 615–18.

Leonard, B.A., Lee, P.K., Jenkins, M.K. and Schlievert, M. (1991). Cell and receptor requirements for streptococcal pyrogenic exotoxin T-cell mitogenicity. *Infect. Immun.* **59**, 1210–14.

Leung, D.Y.M., Meissner, H.C., Fulton, D.R., Quimby, F. and Schlievert, P.M. (1995a). Superantigens in Kawasaki syndrome. *Clin. Immunol. Immunopathol.* **77**, 119–26.

Leung, D.Y.M., Gately, M., Trumble, A., Ferguson-Darnell, B., Schlievert, P.M. and Picker, L.J. (1995b). Bacterial superantigens induce T-cell expression of the skin-selective homing receptor, the cutaneous lymphocyte-associated antigen via stimulation of interleukin-12 production. *J. Exp. Med.* **181**, 747–53.

Leung, D.Y.M., Travers, J.B., Giorno, R., Norris, D.A., Skinner, R., Aelion, J., Xaremi, L.V., Kim, M.H., Trumble, A.E., Kotb, M. and Schlievert, P.M. (1995c). Evidence for a streptococcal superantigen-driven process in acute guttate psoriasis. *J. Clin. Investig.* **96**, 2106–12.

Li, P.L., Tiedemann, R.E., Moffat, S.L. and Fraser, J.D. (1997). The superantigen streptococcal pyrogenic exotoxin C (SpeC) exhibits a novel mode of action. *J. Exp. Med.* **186**, 375–83.

Liu, Z.X., Sugawara, S., Hiwatashi, N., Noguchi, M., Rikiishi, H., Kumagai, K. and Toyota, T. (1997). Accessory cell function of a human colonic epithelial cell line HT-29 for bacterial superantigens. *Clin. Exp. Immunol.* **108**, 384–91.

McKane, L. and Ferretti, J.J. (1981). Phage–host interactions and the production of type A streptococcal exotoxin in group A streptococci. *Infect. Immun.* **34**, 915–19.

McShan, W.M. and Ferretti, J.J. (1997). Genetic diversity in temperate bacteriophages of *Streptococcus pyogenes*: identification of a second attachment site for phages carrying the erythrogenic toxin gene. *J. Bacteriol.* **179**, 6509–11.

McShan, W.M., Tang, Y.-P. and Ferretti, J.J. (1997). Bacteriophage T12 of *Streptococcus pyogenes* integrates into the gene for a serine tRNA. *Mol. Microbiol.* **23**, 719–28.

Michie, C., Scott, A., Cheesborough, J., Beverley, P. and Pasvol, G. (1994). Streptococcal toxic shock-like syndrome: evidence of superantigen activity and its effects on T lymphocyte subsets *in vivo*. *Clin. Exp. Immunol.* **98**, 140–4.

Mollick, J.A., Miller, G.G., Musser, J.M., Cook, R.G. and Rich, R.R. (1992). Isolation and characterization of a novel streptococcal superantigen. *Trans. Assoc. Am. Phys.* **60**, 110–22.

Mollick, J.A., Miller, G.G., Musser, J.M., Cook, R.G., Grossman, D. and Rich, R.R. (1993). A novel superantigen isolated from pathogenic strains of *Streptococcus pyogenes* with aminoterminal homology to staphylococcal enterotoxins B and C. *J. Clin. Invest.* **92**, 710–19.

Morita, A., Imada, Y., Igarashi, H. and Yutsudo, T. (1997). Serologic evidence that streptococcal superantigens are not involved in the pathogenesis of Kawasaki disease. *Microbiol. Immunol.* **41**, 895–900.

Müller-Alouf, H., Alouf, J.E., Gerlach, D., Fitting, C. and Cavaillon, J.-M. (1992). Cytokine production by murine cells activated by erythrogenic toxin type A superantigen of *Streptococcus pyogenes*. *Immunobiology* **186**, 435–48.

Müller-Alouf, H., Alouf, J.E., Gerlach, D., Ozegowski, J.-H., Fitting, C. and Cavaillon, J.-M. (1994). Comparative study of cytokine release by human peripheral blood mononuclear cells stimulated with *Streptococcus pyogenes* superantigenic erythrogenic toxins, heat-killed streptococci and lipopolysaccharide. *Infect. Immun.* **62**, 4915–21.

Müller-Alouf, H., Alouf, J.E., Gerlach, D., Ozegowski, J.-H., Fitting, C. and Cavaillon, J.-M. (1996). Human pro- and anti-inflammatory cytokine pattern induced by *Streptococcus pyogenes* erythrogenic (pyrogenic) exotoxins A and C superantigens. *Infect. Immun.* **64**, 1450–3.

Müller-Alouf, H., Geoffroy, C., Geslin, P., Bouvet, A., Felten, A., Günther, E., Ozegowski, J.-H. and Alouf, J.E. (1997a). Streptococcal pyrogenic exotoxin A, streptolysin O, exoenzymes, serotype and biotype profiles of *Streptococcus pyogenes* isolates from patients with toxic shock syndrome and other severe infections. *Zbl. Bakt.* **286**, 421–33.

Müller-Alouf, H., Gerlach, D., Desreumaux, P., Leportier, C., Alouf, J.E. and Capron, M. (1997b). Streptococcal pyrogenic exotoxin A (SPE A) superantigen induced production of hematopoietic cytokines, IL-12 and IL-13 by human peripheral blood mononuclear cells. *Microb. Pathogen.* **23**, 265–72.

Müller-Alouf, H., Capron, M., Alouf, J.E., Geoffroy, C., Gerlach, D., Ozegowski, J.-H., Fitting, C. and Cavaillon, J.-M. (1997c). Cytokine profile of human peripheral blood mononucleated cells stimulated with a novel streptococcal superantigen, SPE A, SPE C and streptococcal cells. In: *Streptococci and the Host* (eds T. Horaud *et al*.), pp. 922–31. Plenum Press, New York.

Munson, S.H., Tremaine, M.T., Betley, M.T. and Welch, R.A. (1998). Identification and characterization of staphylococcal enterotoxin types G and I from *Staphylococcus aureus*. *Infect. Immun.* **66**, 3337–48.

Murai, T., Ogawa, Y., Kawasaki, H. and Kanoh, S. (1987). Physiology of the potentiation of lethal endotoxin shock by streptococcal pyrogenic exotoxin in rabbits. *Infect. Immun.* **55**, 2456–60.

Murai, T., Ogawa, Y. and Kawasaki, H. (1990). Macrophage hyperreactivity to endotoxin induced by streptococcal pyrogenic exotoxin in rabbits. *FEMS Microbiol. Lett.* **68**, 61–4.

Musser, J.M. (1997). Streptococcal superantigen, mitogenic factor and pyrogenic exotoxin B expressed by *Streptococcus pyogenes*. *Preparat. Biochem. Biotechnol.* **27**, 143–72.

Musser, J.M., Hauser, A.R., Kim, M.H., Schlievert, P.M., Nelson, K. and Selander, R.K. (1991). *Streptococcus pyogenes* causing toxic shock-like syndrome and other invasive disease: clonal diversity and pyrogenic exotoxin expression. *Proc. Natl Acad. Sci. USA* **88**, 2668–72.

Musser, J.M., Kapur, V., Szeto, J., Pan, X., Swanson, D.S. and Martin, D.R. (1995). Genetic diversity and relationships among *Streptococcus pyogenes* strains expressing serotype M1 protein: recent intercontinental spread of a subclone causing episodes of invasive disease. *Infect. Immun.* **63**, 994–1003.

Nadal, D., Lauener, R.P., Braegger, C.P., Kaufhold, A., Simma, B., Lütticken, R. and Seger, A. (1993). T cell activation and cytokine release in streptococcal toxic shock-like syndrome. *J. Pediatr.* **122**, 727–9.

Nakano, S., Abe, Y., Kamezawa, Y. and Nakahara, T. (1997). Nonneutralizing antibody to erythrogenic toxin A alpha (NY5 ETA) in human sera. *Adv. Exp. Med. Biol.* **418**, 923–4.

Nauciel, C. (1973). Mitogenic activity of purified streptococcal erythrogenic toxin on lymphocytes. *Ann. Immunol. (Inst. Pasteur)* **114**, 796–811.

Nauciel, C., Blass, J., Maugalo, R. and Raynaud, M. (1969). Evidence for two molecular forms of streptococcal erythrogenic toxin. *Eur. J. Biochem.* **11**, 160–4.

Nelson, K., Schlievert, P.M., Selander, R.K. and Musser, J.M. (1991). Characterization and clonal distribution of four alleles of the *speA* gene encoding pyrogenic exotoxin A (scarlet fever toxin) in *Streptococcus pyogenes*. *J. Exp. Med.* **174**, 1271–4.

Nemoto, E., Rikiishi, H., Sugawara, S., Okamoto, S., Tamura, K., Maruyama, Y. and Kumagai, K. (1996). Isolation of a new superantigen with potent mitogenic activity to murine T cells from *Streptococcus pyogenes*. *FEMS Immunol. Med. Microbiol.* **15**, 81–91.

Nida, S.K. and Ferretti, J.J. (1982). Phage influence in the synthesis of extracellular toxins in group A streptococci. *Infect. Immun.* **36**, 746–50.

Norgren, M. and Eriksson, A. (1997). Streptococcal superantigens and their role in the pathogenesis of severe infections. *J. Toxicol. – Toxin Rev.* **16**, 1–32.

Norgren, M.S., Holm, S.E. and Eriksson, A. (1998). Identification of superantigen and DNase related epitopes of streptococcal pyrogenic exotoxin F (*Spe* F). *Zbl. Bakt. Suppl.* **29**, 210–11.

Norrby-Teglund, A., Norgren, M., Holm, S.E., Andersson, U. and Andersson, J. (1994a). Similar cytokine induction profiles of a novel streptococcal exotoxin, MF, and pyrogenic exotoxins A and B. *Infect. Immun.* **62**, 3731–8.

Norrby-Teglund, A., Newton, D., Kotb, M., Holm, S.E. and Norgren, M. (1994b). Superantigenic properties of the group A streptococcal exotoxin SPE F (MF). *Infect. Immun.* **62**, 5227–33.

Norrby-Teglund, A., Holm, S.E. and Norgren, M. (1994c). Detection and nucleotide sequence analysis of the *speC* gene in swedish clinical group A streptococcal isolates. *J. Clin. Microb.* **32**, 705–9.

Norrby-Teglund, A., Pauksens, K., Norgren, M. and Holm, S.E. (1995). Correlation between serum TNF alpha and IL6 levels and severity of group A streptococcal infections. *Scand. J. Infect. Dis.* **27**, 125–30.

Norrby-Teglund, A., Kaul, R., Low, D.E., McGeer, A., Newton, D.W., Anderson, J., Anderson, U. and Kotb, M. (1996a). Plasma from patients with severe invasive group A streptococcal infections treated with normal polyspecific IgG inhibits streptococcal superantigen-induced T cell proliferation and cytokine production. *J. Immunol.* **156**, 3057–64.

Norrby-Teglund, A., Kaul, R., Low, D.E., McGeer, A., Anderson, J., Anderson, U. and Kotb, M. (1996b). Evidence for the presence of streptococcal superantigen neutralizing antibodies in normal polyspecific immunoglobulin G. *Infect. Immun.* **64**, 5395–8.

Norrby-Teglund, A., Lustig. R. and Kotb, M. (1997). Differential induction of Th1 versus Th2 cytokines by group A streptococcal toxic shock syndrome isolates. *Infect. Immun.* **65**, 5209–15.

Ohara-Nemoto, Y. and Kaneko, M. (1996). Expression of T-cell receptor Vβ2 and type 1 helper T-cell – related cytokine mRNA in streptococcal pyrogenic exotoxin C activated human peripheral blood mononuclear cells. *Can. J. Microbiol.* **42**, 1104–11.

Ohara-Nemoto, Y., Sasaki, M., Kaneko, M., Nemoto, T. and Ota, M. (1994). Cysteine protease activity of streptococcal pyrogenic exotoxin B. *Can. J. Microbiol.* **46**, 930–6.

Osler, W. (1895). *The Principles and Practice of Medicine*, 2nd edn, pp. 71–80. Appleton, New York.

Ozegowski, J.H., Knöll, H., Gerlach, D. and Köhler, W. (1984). Isolierung und Characterizierung von Erythrogenen Toxinen. VII. Bestimmung des von *Streptococcus Pyogenes* gebildeten Erythrogenen Toxins typ C. *Zbl. Bakt. Hyg. A* **257**, 38–50.

Papageorgiou, A.C., Willins, C.M., Gutman, D.M., Kline, J.B., O'Brien, S.M., Trauten, H.S. and Achanya, K.R. (1999). Structural basis for the recognition of superantigen streptococcal pyrogenic exotoxin A (Spe A1) by MHC class II molecules and T-cell receptors. *EMBO J.* **18**, 9–21.

Petermann, F., Knöll, H. and Köhler, W. (1978). Untersuchungen zur Mitogenität erythrogener Toxin. I. Mitteilung: Typenspezifische Hemmung der mitogenen Aktivät erythrogener Toxin durch antitoxische Seren von Kaninchen. *Zbl. Bakt. Hyg. I. Abt. Orig. A* **240**, 366–79.

Quinn, R.W. (1989). Comprehensive review of morbidity and mortality trends for rheumatic fever, streptococcal disease and scarlet fever: the decline of rheumatic fever. *Rev. Infect. Dis.* **11**, 928–53.

Rantz, L.A., Paul, M.D.J., Boisvert, P.J. and Spink, W.W. (1946). The Dick test in military personnel. With special reference to the pathogenesis of the skin reaction. *N. Engl. J. Med.* **235**, 39–43.

Rasmussen, E.O. and Wuepper, K.D. (1981). Purification and characterization of streptococcal proliferation factor. *J. Invest. Dermatol.* **77**, 246–9.

Reda, K.B., Kapur, V., Mollick, J.A., Lamphear, J.G., Musser, J.M. and Rich, R.R. (1994). Molecular characterization and phylogenetic distribution of the streptococcal superantigen gene (*ssa*) from *Streptococcus pyogenes*. *Infect. Immun.* **62**, 1867–74.

Regelmann, W.E., Gray, E.D. and Wannamaker, L.W. (1982). Characterization of the human cellular immune response to purified group A streptococcal blastogen A. *J. Immunol.* **128**, 1631–6.

Reichardt, W., Müller-Alouf, H., Alouf, J.E. and Köhler, W. (1992). Erythrogenic toxins A, B and C: occurrence of the genes and exotoxin formation from clinical *Streptococcus pyogenes* strains associated with streptococcal toxic shock like syndrome. *FEMS Microbiol. Lett.* **100**, 313–22.

Rikiishi, H., Okamoto, S., Sugawara, S., Tamura, K., Liu, Z.X. and Kumagai, K. (1997). Superantigenicity of helper T-cell mitogen (SPM-2) isolated from culture supernatants of *Streptococcus pyogenes*. *Immunology* **91**, 406–13.

Rink, L., Luhm, J., Koestler, M. and Kirchner, H. (1996). Induction of a cytokine network by superantigens with parallel TH1 and TH2 stimulation. *J. Interferon Cytokine Res.* **16**, 41–7.

Roggiani, M., Stoehr, J.A., Leonard, B.A.B. and Schlievert, P.M. (1997). Analysis of toxicity of streptococcal pyrogenic exotoxin A mutants. *Infect. Immun.* **65**, 2868–75.

Roussel, A., Anderson, B.F., Baker, H.M., Fraser, J.D. and Baker, E.N. (1997). Crystal structure of the streptococcal superantigen SPE-C: dimerization and zinc binding suggest a novel mode of interaction with MHC class II molecules. *Nature Struct. Biol.* **4**, 635–43.

Rowley, A.H. (1998). Controversies in Kawasaki syndrome. *Adv. Pediatr. Infect. Dis.* **13**, 127–41.

Schafer, R. and Sheil, J.M. (1995). Superantigens and their role in infectious disease. *Adv. Pediatr. Infect. Dis.* **10**, 369–90.

Schlievert, P.M. (1981). Scarlet fever: role of pyrogenic exotoxins. *Infect. Immun.* **31**, 732–6.

Schlievert, P.M and Watson, D.W. (1978). Group A streptococcal pyrogenic exotoxin: pyrogenicity, alteration of blood–brain barrier and separation of sites for pyrogenicity and enhancement of lethal endotoxic shock. *Infect. Immun.* **21**, 753–63.

Schlievert, P.M. and Gray, E.D. (1989). Group A streptococcal pyrogenic exotoxin (scarlet fever toxin) type A and blastogen A are the same protein. *Infect. Immun.* **57**, 1865–7.

Schlievert, P.M., Bettin, K.M. and Watson, D.W. (1977). Purification and characterisation of group A streptococcal pyrogenic exotoxin type C. *Infect. Immun.* **16**, 673–9.

Schlievert, P.M., Bettin, K.M. and Watson, D.W. (1979a). Reinterpretation of the Dick test: role of group A streptococcal pyrogenic exotoxin. *Infect. Immun.* **26**, 467–72.

Schlievert, P.M., Bettin, K.M. and Watson, D.W. (1979b). Production of pyrogenic exotoxin by groups of streptococci, association with group A. *J. Infect. Dis.* **140**, 676–81.

Schlievert, P.M., Assimacopoulos, A.P. and Cleary, P.P. (1996). Severe group A streptococcal disease: clinical description and mechanisms of pathogenesis. *J. Lab. Clin. Med.* **127**, 13–22.

Schuh, V. (1965). The pyrogenic effect of scarlet fever toxin. I. Neutralization with antitoxin; the nature of tolerance. *Folia Microbiol.* **10**, 156–62.

Schuh, V. and Hřibalová, V. (1966). The pyrogenic effect of scarlet fever toxin. II. Leukocytic pyrogen formation induced by scarlet fever toxin or *Salmonella paratyphi B* endotoxin. *Folia Microbiol.* **11**, 112–22.

Schuh, V., Hřibalová, V. and Atkins, E. (1970). The pyrogenic effect of scarlet fever toxin. IV. Pyrogenicity of strain C 203 U filtrate: comparison with some basic characteristics of the known types of scarlet fever toxin. *Yale J. Biol. Med.* **43**, 31–42.

Seravalli, E. and Taranta, A. (1974). Lymphocyte transformation and macrophage migration inhibition by electrofocused and gel-filtered fractions of group A streptococcal filtrate. *Cell. Immunol.* **14**, 366–75.

Sriskandan, S., Moyes, D., Buttery, L. *et al.* (1996a). Streptococcal pyrogenic exotoxin A release, distribution and role in a murine model of fasciitis and multiorgan failure due to *Streptococcus pyogenes*. *J. Infect. Dis.* **173**, 119–25.

Sriskandan, S., Evans, T.J. and Cohen, J. (1996b). Bacterial superantigen-induced human lymphocyte responses are nitric oxide dependent and mediated by IL-12 and IFN-γ. *J. Immunol.* **156**, 2430–5.

Sriskandan, S., Unnkrishnan, M. and Cohen, J. (1998). Invasive infection due to *Streptococcus pyogenes* with a disruption of the gene encoding streptococcal pyrogenic exotoxin A. *Abstract B 67, 38th Interscience Conference on Antimicrobial Agents and Chemotherapy*, San Diego, CA.

Stanley, J., Desai, M., Xerry, J., Tanna, A., Efstratiou, A. and George, R. (1996). High-resolution genotyping elucidates the epidemiology of group A *Streptococcus* outbreaks. *J. Infect. Dis.* **174**, 505–6.

Stevens, D.L. (1992). Invasive group A *streptococcus* infections. *Clin. Infect. Dis.* **14**, 2–13.

Stevens, D.L. (1995). Streptococcal toxic shock syndrome: spectrum of disease, pathogenesis and new concepts in treatment. *Emerg. Infect. Dis.* **1**, 69–78.

Stevens, D.L., Tanner, M.H., Winship, J., Swarts, R., Ries, K.M., Schlievert, P.M. and Kaplan, E. (1989). Severe group A streptococcal infections associated with toxic shock-like syndrome and scarlet fever toxin A. *N. Engl. J. Med.* **321**, 1–7.

Stevens, D.L., Bryant, A.E., Hackett, S.P., Chang, A., Peer, G., Kosanke, S., Emerson, T. and Hinshaw, L. (1996). Group A streptococcal bacteremia: the role of tumor necrosis factor in shock and organ failure. *J. Infect. Dis.* **173**, 619–26.

Stevens, K.R., Van, M., Lamphear, J.G., Orkiszewski, R.S., Ballard, K.D., Cook, R.G. and Rich, R.R. (1996a). Species-dependent post-translational modification and position 2 allelism: effects on streptococcal superantigen SSA structure and V beta specificity. *J. Immunol.* **157**, 2479–87.

Stevens, K.R., Van, M., Lamphear, J.G. and Rich, R.R. (1996b). Altered orientation of streptococcal superantigen (SSA) on HLA-DR1 allows unconventional regions to contribute to SSA Vβ specificity. *J. Immunol.* **157**, 4970–8.

Stock, A.H. and Lynn, R.J. (1961). Preparation and properties of partially purified erythrogenic toxin B of group A streptococci. *J. Immunol.* **86**, 561–6.

Talkington, D.F., Schwartz, B., Black, C.M., Todd, J.K., Elliott, J., Breiman, R.F. and Facklam, R.R. (1993). Association of phenotypic and genotypic characteristics of invasive *Streptococcus pyogenes* isolated with clinical components of streptococcal toxic shock syndrome. *Infect. Immun.* **61**, 3369–74.

Tomai, M.A., Schlievert, P.M. and Kotb, M. (1992). Distinct T-cell receptor Vβ gene usage by human T lymphocytes stimulated with the streptococcal pyrogenic exotoxins and pep M5 protein. *Infect. Immun.* **60**, 701–5.

Tonew, E., Gerlach, D., Tonew, M. and Köhler, W. (1982). Induktion von Immuninterferon durch Erythrogene Toxine A und B des *Streptococcus pyogenes*. *Zbl. Bakt. Hyg. Orig.* **A 252**, 463–71.

Toyosaki, T., Yoshioka, T., Tsuruta, Y., Yutsudo, T., Iwasaki, M. and Suzuki, R. (1996). Definition of the mitogenic factor (MF) as a novel streptococcal superantigen that is different from pyrogenic exotoxins A, B and C. *Eur. J. Immunol.* **26**, 2693–701.

Tyler, S.D., Johnson, W.M., Huang, J.C., Ashton, F.E., Wang, G., Low, D.E. and Rozee, K.R. (1992). Streptococcal erythrogenic toxin genes: detection by polymerase chain reaction and association with disease in strains isolated in Canada from 1940 to 1991. *J. Clin. Microb.* **30**, 3127–31.

Uchiyama, T., Yan, X.-J., Imanishi, K. and Yagi, Y. (1994). Bacterial superantigens: mechanism of T cell activation by the superantigens and their role in the pathogenesis of infections diseases. *Microbiol. Immunol.* **38**, 245–56.

Upton, M., Carter, P.E., Orange, G. and Pennington, T.H. (1996). Genetic heterogeneity of M type 3 group A streptococci causing severe infections in Tayside, Scotland. *J. Clin. Microbiol.* **34**, 196–8.

Wagner, B., Buslau, M., Gerlach, D., Ramirez-Bosca, A., Stracke, R. and Wagner, M. (1993). Binding of *Streptococcus pyogenes* erythrogenic toxin A to CD1 a-positive and CD1 a-negative human epidermal cells and to dermal fibroblasts. *Eur. J. Dermatol.* **3**, 704–8.

Wannamaker, L.W. and Schlievert, P.M. (1988). Exotoxins of group A streptococci. In: *Bacterial Toxins* (eds M.C. Hardegree and A.T. Tu), Vol. 4, pp. 267–95. Marcel Dekker, New York.

Watanabe-Ohnishi, R., Low, D.E., McGeer, A., Stevens, D.L., Schlievert, P.M., Newton, D., Schwartz, B., Kreiswirth, B., Ontario Streptococcal study project and Kotb, M. (1995). Selective depletion of Vβ-bearing T cells in patients with severe invasive group A streptococcal infections and streptococcal toxic shock syndrome. *J. Infect. Dis.* **171**, 74–84.

Watson, D.W. (1960). Host parasite factors in group A streptococcal infections: pyrogenic and other effects on immunologic distinct exotoxins related to scarlet fever toxins. *J. Exp. Med.* **111**, 255–83.

Weeks, C. R. and Ferretti, J.J. (1984). The gene for type A streptococcal exotoxin (erythrogenic toxin) is located in bacteriophage T12. *Infect. Immun.* **46**, 531–6.

Weeks, C.R. and Ferretti, J.J. (1986). Nucleotide sequence of the type A streptococcal exotoxin (erythrogenic toxin) gene from *Streptococcus pyogenes* bacteriophage T12. *Infect. Immun.* **52**, 144–50.

White, J., Herman, A., Pullen, A.M., Kubo, R., Kappler, J.W. and Marrack, P. (1989). The Vβ specific superantigen staphylococcal enterotoxin toxin B stimulation of mature T cells and clonal deletion in neonatal mice. *Cell* **56**, 27–35.

Willoughby, R. and Greenberg, R.N. (1983). Toxic shock syndrome and streptococcal pyrogenic exotoxins. *Ann. Intern. Med.* **98**, 559.

Wollweber, L., Fritzke, H., Ozegowski, J.-H., Gerlach, D. and Köhler, W. (1994). Production and partial characterization of monoclonal antibodies against erythrogenic toxins type A and C from *Streptococcus pyogenes*. *Hybridoma* **13**, 403–8.

Working Group on Severe Streptococcal Infections (1993). Defining the group A streptococcal toxic shock syndrome. *JAMA* **269**, 390–1.

Wretlind, E.W. (1874). *Scharlakansfebern, dess orsker, kännetecken och behandling*. Carlssons Förlag, Stockholm.

Xu, S. and Collins, C.M. (1996). Temperature regulation of the streptococcal pyrogenic exotoxin A-encoding gene (*spe A*). *Infect. Immun.* **64**, 5399–402.

Yamaoka, J., Nakamura, E., Takeda, Y., Imamura, S. and Minato, N. (1998). Mutational analysis of superantigen activity responsible for the induction of skin erythema by streptococcal pyrogenic exotoxin C. *Infect. Immun.* **66**, 5020–6.

Yamamoto, M. and Ferretti, J.J. (1995). High level expression of *Streptococcus pyogenes* erythrogenic toxin A (*SPE A*) in *Escherichia coli* and its rapid purification by HPLC. *FEMS Microb.* **132**, 209–13.

Yu, C.-E. and Ferretti, J.J. (1989). Molecular epidemiologic analysis of the type A streptococcal exotoxin (erythrogenic toxin) gene (*spe A*) in clinical *Streptococcus pyogenes* strains. *Infect. Immun.* **57**, 3715–19.

Yu, C.-E. and Ferretti, J.J. (1991). Frequency of the erythrogenic toxin B and C genes (*spe B* and *spe C*) among clinical isolates of group A streptococci. *Infect. Immun.* **59**, 211–15.

Yutsudo, T., Murai, H., Gonzalez, J., Takao, T., Shimonishi, Y., Takeda, Y., Igarashi, H. and Hinuma, Y. (1992). A new type of mitogenic factor produced by *Streptococcus pyogenes*. *FEBS Lett.* **308**, 30–4.

Yutsudo, T., Okumura, K., Iwasaki, M., Hara, A., Kamitani, S., Minamide, W., Igarashi, H. and Hinuma, Y. (1994). The gene encoding a new mitogenic factor in a *Streptococcus pyogenes* strain is distributed only in group A streptococci. *Infect. Immun.* **62**, 4000–4.

Zabriskie, J.B. (1964). The role of temperate bacteriophage in the production of erythrogenic toxin by group A streptococci. *J. Exp. Med.* **119**, 761–79.

33

Properties of *Staphylococcus aureus* enterotoxins and toxic shock syndrome toxin-1

Steven R. Monday and Gregory A. Bohach

INTRODUCTION, HISTORICAL ASPECTS AND NOMENCLATURE

Staphylococcus aureus strains may express a plethora of exoproteins facilitating their ability to colonize and persist within their broad range of hosts. These proteins, often enzymic or cytotoxic in nature, include several membrane-active haemolysins, nucleases, proteases, leucocidins, lipases, collagenases, cell-surface proteins and various superantigens (SAgs), including members of the pyrogenic toxin (PT) family. Collectively, these exoproteins serve the bacteria by promoting adherence or providing nutrients for the microbes. In addition, the PT SAgs directly, or indirectly, avert or alter the function of various cells of the host immune system and, in combination, these processes may result in disease.

The PT family consists of moderately sized proteins, ranging from approximately 22 000 to 30 000 Da. These toxins are expressed by two genera of Gram-positive cocci, *Staphylococcus* sp. or *Streptococcus* sp., and are grouped together because of their shared biological and physicochemical properties. This family of proteins, which initially included the staphylococcal enterotoxins (SEs) and toxic shock syndrome toxin-1 (TSST-1) of *S. aureus* and the streptococcal pyrogenic exotoxins (SPEs) of *S. pyogenes*, continues to grow as more proteins with the same or similar properties are isolated and characterized. Recently, the PT family has been

expanded to include several new SE variants (Munson *et al.*, 1998), as well as toxins expressed by groups A, B, C, F and G streptococci (Bohach *et al.*, 1990; Iwasaki *et al.*, 1993; Mollick *et al.*, 1993; Schlievert *et al.*, 1993; Reda *et al.*, 1994). We suspect that newly described PTs will continue to be discovered as further investigation of these pathogens is conducted.

All PTs have the shared ability to interact with several host cellular targets to induce a variety of biological consequences which may ultimately lead to disease. Collectively, these activities can result in the life-threatening staphylococcal or streptococcal toxic shock syndromes (TSS). In addition to these shared biological activities, the SEs have the unique ability to induce symptoms in the gastrointestinal tract following oral ingestion, a property that distinguishes them from the other PTs. Consequently, in addition to inducing TSS, the SEs are the causative agent of staphylococcal food poisoning (SFP), a very common form of food-associated gastroenteritis world-wide.

Discovery and classification of the SEs

Although not the first food-borne toxins recognized, the SEs, which induce their effects in the gastrointestinal tract, were the first true enterotoxins described. These toxins have long been recognized as key mediators of the pathogenesis of SFP. This common illness is usually a self-limiting disease resulting from the ingestion of contaminated food containing preformed toxins. Barber (1914) reported that consuming milk contaminated with staphylococci can result in symptoms known to be typical of SFP. He provided the first strong evidence that a soluble toxin was responsible for SFP symptoms (Holmberg and Blake, 1984; Bergdoll, 1989). Dack *et al.* (1930) extended Barber's work by demonstrating that voluntary ingestion of culture supernatants prepared from staphylococcal agents isolated from other sources also generated SFP-like symptoms in the volunteers.

Conclusive identification of the SEs required several decades of investigation. The main reason for the difficulties encountered centred around interpretation of immunological methods used. Antisera directed against an SE isolated from one strain often failed to react with SEs produced by other strains. It was eventually determined that *S. aureus* expresses several antigenically different SEs with similar biological activities. The classification scheme used currently is still based largely on antigenic differences (Table 33.1). Presently, there are eight (A, B, C, D, E, G, H and I) known major SE types. Each toxin type is designated according to standard nomenclature in which the SE is assigned a letter in the order of its discovery. The amino acid sequence of each of the SE types has been determined by either protein or nucleic acid sequencing and the degree of immunological relatedness roughly parallels the molecular relatedness at the sequence level. Except for SEC, which has at least seven distinct molecular and antigenic variants, sequence heterogeneity has not been reported for the other SEs.

Association of SEs and TSST-1 with TSS

Compared to SFP, TSS is more serious and life-threatening and results from colonization or infection of susceptible hosts by staphylococci- or streptococci-expressing PTs *in vivo*. A description of TSS was first formally published in 1978 (Todd and Fishaut, 1978), although a comparable illness, staphylococcal scarlet fever, had been reported as early as the 1920s (Stevens, 1927; Aranow and Wood, 1942; Dunnett and Schallibaum, 1960). In 1981, an exotoxin identified by Schlievert *et al.* (1981) as staphylococcal pyrogenic exotoxin C and Bergdoll *et al.* (1981) as staphylococcal enterotoxin F (SEF) was proposed to be the key staphylococcal virulence factor responsible for TSS. This toxin induced most TSS symptoms in animals. However, as it did not induce emesis in primates following ingestion, a key feature of SEs, it was subsequently renamed toxic shock syndrome toxin-1 (TSST-1). To minimize potential confusion in future nomenclature, the SEF designation has been retired (Betley *et al.*, 1990). Only two molecular variants of TSST, TSST-1 and TSST$_{ovine}$, expressed by isolates of human and ovine staphylococci, respectively, display a minor degree of sequence heterogeneity. However, unlike the SEs, only a single immunological serotype has been described thus far (Bohach *et al.*, 1997).

While TSST-1 is responsible for nearly all menstrual TSS cases, only approximately 60% of non-menstrual staphylococcal TSS cases are caused by TSST-1-expressing isolates of *S. aureus* (Deresiewicz, 1997). The remaining cases are attributed to SE production, primarily SEB and less often to other SEs, especially SEC

TABLE 33.1 Biophysical characteristics and reported Vβ specificities of the staphylococcal enterotoxins (SEs) and toxic shock syndrome toxin-1 (TSST-1)

Toxin	Molecular mass (Da)	p*I*	Human Vβ specificity	Reference
SEA	27 100	6.8–7.3	1.1, 5.3, 6.3, 6.4, 6.9, 7.3, 7.4, 9.1, 18	Betley and Mekalanos (1988), Svensson *et al.* (1997)
SEB	28 336	8.6	3, 12, 13.2, 14, 15, 17, 20	Jones and Khan (1986), Deringer *et al.* (1996)
SEC1	27 496	8.6	3, 12, 13.2, 14, 15, 17, 20	Bohach and Schlievert (1987), Deringer *et al.* (1996)
SEC2	27 531	7.0	3[a], 12, 13.1, 13.2, 14, 15, 17, 20	Bohach and Schlievert (1989), Deringer *et al.* (1996)
SEC3$_{FRI909}$	27 588	8.0	3[a], 12, 13.1, 13.2, 14, 15, 17, 20	Marr *et al.* (1993), Deringer *et al.* (1996)
SEC3$_{FRI913}$	27 563	8.1	3[a], 12, 13.1, 13.2, 14, 15, 17, 20	Hovde *et al.* (1990), Deringer *et al.* (1996)
SEC$_{bovine}$	27 618	7.6	3[a], 12, 13.2, 14, 15, 17, 20	Deringer *et al.* (1997)
SEC$_{ovine}$	27 517	7.6	3[a], 12, 13.2, 14, 15, 17, 20	Deringer *et al.* (1997)
SEC$_{canine}$	27 600	7.0	3[a], 12, 13.2, 14, 15, 17, 20[a]	Edwards *et al.* (1997)
SED	26 360	7.4	5, 12	Bayles and Iandolo (1989), Svensson *et al.* (1997)
SEE	26 425	8.5	5.1, 6.3, 6.4, 6.9, 8.1, 18	Couch *et al.* (1988), Svensson *et al.* (1997)
SEG	27 042	ND[b]	ND[b]	Munson *et al.* (1998)
SEH	25 145	5.7	ND[b]	Ren *et al.* (1994)
SEI	24 928	ND[b]	ND[b]	Munson *et al.* (1998)
TSST-1	22 000	7.2	2	Blomster-Hautamaa *et al.* (1986), Svensson *et al.* (1997)
TSST$_{ovine}$	22 000	8.5	ND[b]	Lee *et al.* (1992)

[a]Markedly reduced compared to SEC1 and SEB. [b]Not determined.

and SEA (Garbe *et al.*, 1985; Crass and Bergdoll, 1986; Whiting *et al.*, 1989; Schlievert *et al.*, 1990). More recently, a nearly identical disease associated with toxigenic group A streptococci has been reported (Stevens *et al.*, 1989). The fact that the PTs produced by both organisms cause TSS further highlights the shared biological properties of this family of toxins.

BIOLOGICAL ACTIVITIES OF THE SE AND TSST-1

Classically, a rather long list of biological properties has been demonstrable with these toxins, either experimentally or in patients with toxigenic illnesses (Table 33.2). For example, fever, hypotension, changes in lymphocyte number and activation status, and transient immunosuppression are common features of TSS. Recently, with the elucidation of the SAg concept, it became clear that many of the shared properties in this list could be attributed, in part, to a single underlying mechanism. As SAgs, the SEs and TSST-1 induce the aberrant release of cytokines and other mediators that ultimately induce the symptoms in the host. However, other studies have shown, in addition, that super-antigenicity cannot explain all of the toxic effects observed. Finally, some PTs have unique activities that are not properties of all toxins in the family. When this is the case, as with the emetic ability of the SEs, not all of the toxins may be expected to be associated with a particular illness, such as SFP.

Stimulation of T-cells and antigen-presenting cells as superantigens

The SEs and TSST-1 are regarded as prototype bacterial SAgs (Jablonski and Bohach, 1997). Although their ability to cause T-cell proliferation and activation of antigen-presenting cells (APCs) has been long recognized, the mechanism by which they stimulate cells of the immune system remained elusive until recently. Marrack and Kapler's group (White *et al.*, 1989; Marrack and Kappler, 1990) described the concept of 'superantigens' to explain the unique molecular and cellular mechanisms by which the PTs induce T-cell proliferation. In work performed initially with SEB, SAgs were characterized as molecules with certain unique properties including:

(a) polyclonal Vβ-specific proliferation of certain T-cell subpopulations

(b) a requirement for the presence of APCs to induce maximum T-cell stimulation

(c) deletion of the stimulated Vβ-expressing T-cell subpopulations.

Subsequent investigations elucidated the mechanisms responsible for these observations. Consistent with a T-cell receptor (TCR) and major histocompatability complex (MHC) class-II-dependent stimulation, maximum T-cell activation requires the formation of a trimolecular complex consisting of MHC class II, SAg and the TCR (Figure 33.1). The molecular interactions of TSST-1, and some SEs, with MHC class II have been well characterized following the demonstration by several groups that

TABLE 33.2 Biological properties of various pyrogenic toxin (PTs)

Biological activity	Type of PT	Confirmed or proposed mechanism(s)
Pyrogenicity	All	Direct action on hypothalamus
Induction of endogenous pyrogens		
Enhancement of:		
endotoxic shock	All	RES[a] impairment
		PT–endotoxin complex formation
		Cytotoxicity
cardiotoxicity	Streptococcal pyrogenic exotoxins (SPEs)	Unknown
Immune cell stimulation	All	Superantigenicity
Cytokine release	All	Superantigenicity
Immunosuppression	All	Superantigenicity
Lethality/shock	All	Superantigenicity
Cytotoxicity		
Direct capillary leak	All	Superantigenicity
Cytotoxicity		
Emesis (upon ingestion)	Staphylococcal enterotoxins (SEs)	Vagus nerve stimulation
		Mast cell stimulation
		Neuropeptides
		Leucocyte inflammatory mediators
Hind-limb paralysis	Group B SPE	Unknown

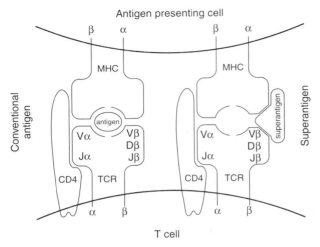

FIGURE 33.1 Schematic depiction comparing the manners in which conventional antigens and SAgs engage the MHC class-II molecule and the TCR. Following endocytosis and proteolytic processing by the APC, conventional antigen peptides are presented in the binding cleft of MHC class II for recognition by multiple elements in the TCR (left). Unlike conventional antigens, TSST-1, the SEs and most other SAg require no processing by the APC, but associate directly with MHC class II and the TCR outside the typical binding cleft (right). This Vβ-dependent association with the TCR and specific domains of MHC class II gives the toxins a wedge-like nature which could potentially displace the conventional antigenic peptide from within the binding cleft.

SEs bind specifically to MHC class-II molecules (Fraser, 1989; Rich *et al.*, 1989). Interestingly, although all currently known PT SAgs are 'presented' to the T-cell via MHC class II, the molecular interactions between the MHC molecule and various PTs vary greatly from one toxin to another (discussed below). One feature that is consistent among the toxins covered in this chapter and most SAgs is that they are not endocytosed or proteolysed by APCs (Jablonski and Bohach, 1997). Unlike conventional antigens, PT SAgs bind directly to invariant regions of the MHC class-II receptor on the cell surface of the APC. Although they recognize multiple MHC class-II alleles, some allelic preference is seen among the various toxins which also could have widely different affinities for a particular MHC class-II molecule (discussed below). Although other factors are involved, the affinities of these toxins for the MHC molecule correlate with their ability to stimulate T-cell proliferation (Chintagumpala *et al.*, 1991).

TCR binding requires only residues within the variable region of the β-chain (Vβ-), independent of the antigen specificity of that receptor. As the amino acids recognized varies from one toxin to another, a unique repertoire of T-cell subpopulations can be activated by each toxin (Table 33.1). Because the binding of SAgs to the TCR requires only residues in the Vβ element, the

specificity is only moderate, resulting in a polyclonal activation of T-cells. Unlike conventional antigen-mediated immune responses, in which approximately one in 10^4–10^6 T-cells is activated, SAgs may stimulate and activate up to 10–50% of the host T-cells (Choi *et al.*, 1990).

Pyrogenicity and lethal shock: role of cytokines and other inflammatory mediators

By acting as SAgs, the SEs and TSST-1 directly activate an unusually high percentage of T-cells and APCs, resulting in an initial release of lymphokines and monokines. This first burst of cytokines causes a cascade of events leading to further release of cytokines and other inflammatory mediators. Induction of numerous cytokines has been noted following SAg stimulation *in vitro* and *in vivo*. Those receiving most attention include the monokines tumour necrosis factor-α (TNF-α) and interleukin-1 (IL-1), in addition to numerous lymphokines, especially TNF-α, TNF-β, interferon-γ (IFN-γ) and IL-2 (Ikejima *et al.*, 1984; Parsonnet *et al.*, 1985; Uchiyama *et al.*, 1986; Jupin *et al.*, 1988; Parsonnet and Gillis, 1988; Fast *et al.*, 1989; Micusan *et al.*, 1989; Hackett and Stevens, 1993). The presence of these cytokines, primarily TNF, induces the expression of the pro-inflammatory mediators IL-6, and IL-8, as well as nitric oxide, platelet-activating factor, leukotrienes, thromboxane A_2 and prostaglandins (Bone, 1991; Boyle *et al.*, 1994; Florquin *et al.*, 1994; Hasko *et al.*, 1998). TNF also enhances leucocyte and endothelial cell adhesion and activation of the complement and coagulation cascades. TNF may also function as an endogenous pyrogen, acting in conjunction with IL-1 directly on the hypothalamus to produce fever (Bone, 1991). Finally, TNF, which is directly toxic to endothelial cells, may act in conjunction with other pro-inflammatory mediators to produce the extensive endothelial damage, capillary leakage and decrease in peripheral vascular resistance, culminating in hypotension and shock (Chesney, 1997).

Several lines of evidence suggest a crucial role for T-cell-derived cytokines in systemic lethality. Miethke *et al.* (1992) proposed that the T-cell-dependent TNF-α and TNF-β, although possibly working in combination with numerous other agents, are the prime effectors in staphylococcal TSS. Passive immunization with an anti-TNF monoclonal antibody resulted in effective resistance to SEB-induced T-cell-mediated shock. Moreover, as TNF is the key mediator in lethal shock and there are distinctive differences in the kinetics of lymphokine-derived TNF induction by PTs and monokine TNF induction by endotoxin, it is likely that differences in TSS and Gram-negative septic shock are due to the differences in the characteristics of these cytokines. TNF-β shares many biological properties with TNF-α (Paul and Ruddle, 1988). As such, this cytokine shares the ability to induce many of the same biological consequences associated with TNF-α in Gram-negative septic shock.

Immunosuppression

Despite the fact that TSST-1 and SEs are potent T-cell stimulants, numerous examples of their immunosuppressive features have been shown. One property which may be demonstrated *in vitro* and *in vivo* is a non-specific suppression of immunoglobulin synthesis. The fact that recurrent TSS is more common than most other exotoxin-associated systemic diseases has been attributed to failure of many TSS patients to generate neutralizing antitoxin antibodies. In addition, systemic exposure to TSST-1 or SEs, either experimentally or naturally through disease, results in initial elevation of certain Vβ-expressing T-cells. Eventually, however, the levels and function of these same subpopulations decline (White *et al.*, 1989; Mahlknecht *et al.*, 1996).

The mechanisms by which these effects are mediated by SAgs have been extensively investigated. The factors involved are complex, and the outcome of SAg engagement with the TCR depends not only upon the initial interaction but also on mediators elucidated by the activated cell populations. Depending on the types and pathways of signal transduction, positive or negative, at least three outcomes (activation, apoptosis or anergy) may result. Apoptosis is one method by which a large population of T-cells, activated by an SAg, can be down-regulated or deleted. Apoptosis is initiated by a cascade of events that can be mediated by Fas (CD95), a receptor present on numerous cell types, and Fas ligand (FasL), which is predominantly expressed on T cells (Nagata, 1997; Lincz, 1998). Apoptosis is characterized by the breakdown of individual cells to form discrete, membrane-bound, bundles of varying size. These bundles fail to elicit an immune response, they dissipate through intercellular spaces and are phagocytosed without causing inflammation in the surrounding tissues (Lincz, 1998).

In the presence of APCs the SEs and TSST-1 activate naive, mature CD4$^+$ and CD8$^+$ T-cells in a typical SAg fashion, triggering tremendous proliferation of cells and cytokine expression. Although Fas and FasL expression is increased, the end result is not solely apoptosis, but can be proliferation depending upon the balance between positive and negative signals present (Boshell *et al.*, 1996). CD8$^+$ cells suppress CD4$^+$ cell responses by inducing apoptosis of CD4$^+$ cells through ligation with Fas. This suppression may be due in part to

the increase in FasL expression on activated CD8$^+$, but not on CD4$^+$, target cells; it is more likely to be related to other factors such as the Th profiles of the cells, a time-dependent variable in SAg-stimulated cultures (Ferens *et al.*, 1998). Although CD4$^+$ cells expressing Th2 cytokines are more resistant to CD8$^+$-mediated apoptosis (Noble *et al.*, 1998), most primary SAg-stimulated CD4$^+$ cells express Th1 cytokines, most notably IL-2, and are consequently susceptible to apoptosis. This also supports the findings that IL-2, a Th1 cytokine, mediates apoptosis (Miethke *et al.*, 1996) while IL-10, a Th2 cytokine, protects cells against apoptosis (Hasko *et al.*, 1998). Activation by SAg and IL-2 is a prerequisite for apoptosis. IL-2 increases the transcription and expression of FasL, while it suppresses transcription of FLIP, an inhibitor of apoptosis (Refaeli *et al.*, 1998). CD8$^+$ cells are more Fas resistant, and Fas may actually enhance CD8$^+$-cell proliferation during SAg activation. It is possible that most CD8$^+$-cell deletion occurs through an alternative apoptosis pathway triggered by TNF (Noble *et al.*, 1998).

CD8$^+$ and CD4$^+$ cells are more resistant to apoptosis upon secondary activation with the same SAg. This is due in part to the deletion of Fas-sensitive cells during the primary activation, and also to the preferential expression of Th2-type cytokines upon repeated SAg stimulation (Florquin and Aaldering, 1997). The preferential apoptosis of Th1-cells, mainly CD4$^+$ cells, upon initial SAg engagement alters the Th profile of the lymphocytes and promotes the non-responsive or non-proliferative state seen during subsequent exposure to the same SAg. SEB-reactive T-cells do not proliferate or secrete IL-2 upon a second exposure to SEB (Florquin and Aaldering, 1997); this phenomenon is defined as anergy. In effect, the cells are not actually in a state of non-responsiveness, they are simply exhibiting a different response (McLeod *et al.*, 1998). Also, they are capable of mounting a proliferative reaction to other less related SAgs such as SEA (Florquin and Aaldering, 1997). Secondary SAg responder T-cells lose their IL-2 receptors. As IL-2 is known to maintain the proliferation of cells which express IL-2 receptors, the loss of the IL-2 receptor, in essence IL-2, mediates anergy.

High doses of SAg also promote CD4$^+$-cell-mediated B-cell apoptosis and the down-regulation of activated Ig-secreting B-cells. Cell-to-cell contact is required, and Fas is most the likely mediator (Stohl *et al.*, 1998). As with most biological systems, there are many conflicting views related to SAg-induced immunosuppression. However, it seems likely that SAg-induced immunosuppression provides a selective advantage to the organism by averting the host immune system. *Staphylococcus aureus* is an organism with a broad host range. The heterogeneity of molecular forms of SEs, and to a lesser extent TSST, may represent adaptability of the organism in its attempt to modify its SAgs to interact more effectively with the immune system of its numerous potential hosts. There are many cellular and mediator effects induced by these toxins which could facilitate staphylococcal colonization and perhaps infection. For example, the mixed cytokine profile generated may induce an initial effect on the cellular composition with the eventual expression of Th2 cytokines. Interestingly, Th2 cytokines are generally less effective at clearing intracellular infections. While *S. aureus* is classically considered to be an extracellular pathogen, it is now clear that this organism may occupy an intracellular niche that may be important in the pathogenesis (Bayles *et al.*, 1998).

Miscellaneous shared biological activities

PT SAgs also produce other biological activities, distinct from their action as SAgs, which could contribute significantly to the pathogenesis of disease. For example, it is possible to construct TSST-1 mutants which retain the ability to function as SAgs, but are not lethal in experimental models (Murray *et al.*, 1994). The SEs and TSST-1 may enhance host susceptibility to lethal endotoxin shock by a factor of greater than 100 000-fold (Schlievert, 1982; Bohach *et al.*, 1990). This effect may be the consequence of the interference with liver function, possibly through inhibition of hepatocyte RNA synthesis by the toxins (Schlievert *et al.*, 1980). The inability of these cells effectively to clear endotoxin presumably results in excess levels of endotoxin within the circulation further enhancing TNF-α production, subsequently causing capillary leakage and hypotension. TSST-1 also interacts with endothelial cells (Kushnaryov *et al.*, 1989; Lee *et al.*, 1991) and thereby may also directly cause additional capillary leakage contributing to hypotension. The SEs and TSST-1 may also induce fever by acting directly on the hypothalamus (Schlievert and Watson, 1978).

Enterotoxic activity of the SEs

Ingestion of SEs induces emesis and diarrhoea in humans and non-human primates (Bergdoll, 1988), and is associated with SFP. TSST-1 and other PT SAgs do not have this ability (Bohach *et al.*, 1998). A dose of 20–35 µg of purified SE is sufficient to produce vomiting in human volunteers (Merson, 1973), although the amount required if ingested with food is somewhat lower. Upon ingestion, these inherently stable proteins produce symptoms by indirectly stimulating the medullary emetic reflex centre. Earlier work suggested a stimulation of neural receptors and impulse

transmission along the vagus nerve, although the exact nature of the cellular and molecular events involved remains to be elucidated (Bayliss, 1940; Clark, 1962; Sugiyama and Hayama, 1965; Elwell *et al.*, 1975).

There is mounting evidence that inflammatory mediators (Scheuber *et al.*, 1987a; Boyle *et al.*, 1994) contribute significantly to SE activity in the gastrointestinal tract following ingestion. Jett *et al.* (1994) demonstrated elevated levels of the arachidonic acid metabolites, prostaglandin E$_2$, leukotriene B$_4$ and 5-hydroxyeicosatetraenoic acid following oral ingestion of SEB. Scheuber *et al.* (1987b) linked cysteinyl leukotriene production, namely leukotriene E4, to the gastrointestinal symptoms in SFP and suggested that mast cells play a significant role in SFP pathogenesis. Studies by other investigators have provided additional evidence that mast cells may be involved in the production of SFP-associated symptoms (Reck *et al.*, 1988; Komisar *et al.*, 1992). Alber *et al.* (1989) were not able to induce directly the release of inflammatory mediators with SEB, but implicated the neuropeptide substance P in mast-cell release of compounds. Based on these combined studies, it has been proposed that activation of the mast cells through a non-immunological pathway, and involvement of the vagus nervous system, may mediate SFP pathogenesis. The involvement of specific cells in the emetic response, and identification of the putative receptor, remain to be determined conclusively.

CLINICAL ASPECTS

SEs and TSST-1, having confirmed roles in two human diseases and being suspected in several others, can cause a diverse set of clinical symptoms. Classically, the SEs had been implicated in SFP, a food-borne illness acquired by ingestion of preformed toxin in contaminated food. In the 1980s their role in non-menstrual TSS became evident. TSST-1 is not involved in food-borne illnesses, but was the initial toxin implicated in TSS. These toxins may also be expressed by veterinary isolates, especially those from bovine mastitis and canine pyoderma (Marr *et al.*, 1993; Edwards *et al.*, 1997). The role of these toxins in symptoms of animal diseases is unclear, although cases of TSS-like illness in domesticated animals are documented.

Staphylococcal food poisoning

World-wide, SFP is the leading cause of food-borne illness caused by microbial intoxication and also constitutes a major source of all food-borne illnesses. Although the incidence varies considerably according to geographical location, approximately 14–40% of all microbial food-borne illnesses have been attributed to SFP (Bergdoll, 1979; Holmberg and Blake, 1984). Because SFP is usually self-limiting, it is likely to be largely under-reported. In addition, reporting is highly skewed toward well-publicized outbreaks, whereas most cases occurring in the home are not reported. The highest incidence is typically in the late summer when temperatures are warm, and a second peak in winter is associated with leftover holiday food. Both are attributed to improper storage of food which allows bacterial growth and production of SE.

Typical cases of SFP present as self-limiting gastrointestinal illnesses with emesis following a short incubation period (mean of 4.4 h). Although vomiting is the hallmark symptom, its occurrence may be variable. The frequency of the most common symptoms are as follows: vomiting (82%), nausea (74%), diarrhoea (usually watery) (68%) and abdominal pain (64%) (Holmberg and Blake, 1984). The duration of symptoms is approximately 1 day (mean of 26.3 h), although ranges of 1–88 h have been reported. The fatality rate is low for the general public (0.03%), but may reach 4.4% for certain populations such as children and the elderly (Holmberg and Blake, 1984). Approximately 10% of patients seek medical attention. Treatment is usually minimal, although fluids should be administered when diarrhoea and vomiting are severe. The most important factors contributing to the likelihood of developing symptoms and their severity include susceptibility of the individual to the toxin, the total amount of food ingested and the overall health of the affected person. The toxin type also contributes to the

likelihood of disease. Although SFP outbreaks attributed to ingestion of SEA are much more common, individuals exposed to SEB exhibit more severe symptoms (Holmberg and Blake, 1984; Jablonski and Bohach, 1997).

Toxic shock syndrome

Patients lacking neutralizing antibody are susceptible to TSS when infected or colonized by *S. aureus* expressing TSST-1, SEs or both. TSS is an acute illness which can affect several organ systems. Although the symptoms observed can vary, several criteria established in 1981 (Reingold *et al.*, 1982) are presently required to define firmly a case of TSS (Table 33.3). Nearly all patients present with hypotension, fever, rash and desquamation during convalescence. Involvement of at least three additional organ systems produces several of the variable symptoms which may include mucous membrane hyperaemia and vomiting/diarrhoea, or neurological affects manifesting as confusion or combativeness. Patients may also have electrolyte imbalances, especially hypocalcaemia and hypophosphataemia. As TSS mimics other illnesses including scalded skin syndrome and Kawasaki disease, careful differential diagnosis is needed. Some have recommended revision of the current case definition to include patients with less severe disease resulting from early treatment (Parsonnet, 1998). In such cases, demonstration of toxigenic and/or serological conversion to the putative toxin might be used to confirm the aetiology.

Staphylococcal TSS may manifest in either of two major forms, menstrual or non-menstrual. Menstrual

TABLE 33.3 Toxic shock syndrome (TSS) – clinical case definition[a]

1. Fever with temperature ≥38.9°C
2. Diffuse macular erythroderma rash
3. Desquamation 1–2 weeks after onset of illness, particularly on palms, soles, fingers and toes
4. Hypotension with systolic blood pressure ≤90 mmHg for adults; < 5th percentile for children < 16 years of age; orthostatic drop in diastolic pressure ≥15 mmHg from lying to sitting; orthostatic syncope, or orthostatic dizziness
5. Involvement of three or more of the following organ systems:
   (a) gastrointestinal: (vomiting or diarrhoea usually at onset)
   (b) muscular: severe myalgia or elevated creatine phosphokinase (CPK) levels
   (c) mucous membrane hyperaemia: usually vaginal, oropharyngeal or conjunctival
   (d) renal: elevated blood urea nitrogen (BUN) or creatinine; or at least five white blood cells per high-power field in the absence of a urinary tract infection
   (e) hepatic: elevated total bilirubin, serum glutamic oxaloacetic transaminase (SGOT) or serum glutamic pyruvic transaminase (SGPT)
   (f) haematological: platelets ≤ 100 000 mm^{-3}
   (g) central nervous system: disorientation or alterations in consciousness without focal neurological signs when fever and hypotension are absent

Negative results on the following tests, if obtained:
   (a) Blood, throat or cerebrospinal fluid cultures (blood culture may be positive for *S. aureus*)
   (b) Serological tests for Rocky Mountain spotted fever, leptospirosis or measles

[a]TSS is confirmed if all six of the clinical findings described above occur, including desquamation, unless the patient dies before desquamation. Cases with five of the six clinical findings are defined as probable TSS.

TSS generally has an onset within 2 days of menstruation in women whose vaginal/cervical mucosa is colonized by TSST-1-producing *S. aureus*. SEs may be co-expressed by these strains but it is unlikely that they contribute to menstrual cases. First, strains lacking the ability to express TSST-1 are rarely, if ever, solely isolated from genital sites from women with menstrual TSS. Also, unlike TSST-1, the SEs are inefficient in inducing systemic symptoms if administered intravaginally, presumably owing to their inability to cross mucosal barriers (Bohach *et al.*, 1998). In addition to vaginal colonization by TSST-1-expressing *S. aureus*, tampon use is the other risk factor in menstrual TSS. A direct correlation between tampon absorbency and risk of developing TSS has been clearly established.

The work by Schlievert and Blomster (1983) suggested that introduction of oxygen into the vagina upon insertion of the tampon allows staphylococci to express increased amounts of TSST-1, as the toxin is not expressed in significant levels under anaerobic conditions of the vagina. Many environmental conditions, including divalent cations, also affect toxin production and could be affected by tampon use (Bohach *et al.*, 1990). For example, some brands of tampon bind significant levels of Mg, a divalent cation that reduces levels of toxin in the environment. This results not from a direct effect on toxin expression but from altering the growth of *S. aureus* (Bohach *et al.*, 1990).

Non-menstrual TSS may result from *S. aureus* infection or colonization elsewhere in the body. Examples of some types of non-menstrual TSS have included post-surgical, post-partum and influenza-associated tracheal colonization by *S. aureus* (Deresiewicz, 1997). The case definitions for menstrual and non-menstrual TSS are the same. However, in regard to toxins involved, the aetiology of the two forms can be different as either SEs or TSST-1 may mediate the non-menstrual form. In most non-menstrual cases the toxins are expressed at a body site which would not require them to cross vaginal or other mucosa, a property that SEs apparently lack.

Although data have been somewhat limited, one large-scale active surveillance project estimated the average incidence of TSS to be 0.53 per 100 000, although the distribution of TSS cases seems to be changing (Gaventa *et al.*, 1998). Coinciding with the removal of high-absorbency tampons from the market, the majority of staphylococcal TSS cases reported at present are non-menstrual in nature (Schlievert, 1998). Patients with TSS should receive supportive therapy for life-threatening symptoms, chemistry imbalances and intravascular volume reduction caused by capillary leakage. Also critical is identification of and, if possible, sterilization of the site of infection. Aggressive antimicrobial therapy for

10–14 days is recommended. Intravenous penicillinase-resistant β-lactam antibiotics are generally used, although there is growing evidence that concurrent administration of protein synthesis inhibitors should also be employed to block toxin synthesis by the organism. Several studies have demonstrated the efficacy of administering antitoxin in the form of intravenous immunoglobulin. Because a significant percentage of TSS patients do not develop protective immunity to TSST-1, and elimination of *S. aureus* is difficult or temporary, recurrent TSS is possible. Likewise, AIDS patients are predisposed to a unique form of TSS, termed recalcitrant erythaematous desquamating disorder, which has a prolonged course and high mortality.

Role of TSST-1 and SEs in other clinical situations

In addition to recurrences of TSS, patients may experience neurological or allergic sequelae. There is growing evidence that the SAgs also have key roles in several other human illnesses. Regarding a potential role for TSST-1 and SEs, some rather convincing evidence has been provided for atopic dermatitis (AD) and Kawasaki syndrome (KS). The majority of AD patients harbour staphylococci-expressing SEs or TSST-1 and often produce antitoxin IgE, capable of activating basophils and mast cells.

KS is another illness linked to staphylococcal, and some streptococcal, SAgs. KS is a febrile vascular illness usually seen in young children. While the vasculitis in KS patients, and especially coronary artery aneurysms, are not part of the required criteria for TSS, both illnesses have considerable clinical similarities. Although the significance is somewhat controversial at the present time, isolation of toxigenic organisms from patients with KS and selective Vβ skewing of lymphocyte subpopulations in KS patients further suggest that PT SAgs may be involved (Leung and Schlievert, 1997). Several additional human diseases affecting the skin, and some cases of sudden infant death syndrome, have been linked to TSST-1 and the SEs. At present, these associations are quite controversial and it remains to be determined whether a conclusive link or mechanism of pathogenesis will be established.

THE MOLECULAR BIOLOGY OF TSST-1 AND THE SE

Cloning and sequence analysis

The structural genes for TSST-1 and at least 15 SEs have been cloned and sequenced. At the nucleotide level, the

SE structural genes share between 27.5 and > 98% homology (Marr *et al.*, 1993; Munson *et al.*, 1998), and the degree of relatedness closely parallels the relatedness at the protein level (Figure 33.2). The gene-encoding TSST-1, *tst*, shares little, if any, sequence identity with the SE genes and cannot be confidently aligned with any member of the PT family (Van den Bussche *et al.*, 1993). The transcripts of most SEs are monocistronic and slightly larger than the open reading frames (ORFs) encoding these toxins. Each transcript contains a single promoter and accessory transcriptional elements (Iandolo, 1989; Betley *et al.*, 1992). Both *sea* and *seb* promoters have been characterized (Mahmood and Khan, 1990; Borst and Betley, 1994). Interestingly, Northern blot analysis revealed that the *seg* mRNA is 6.7 kb in size, suggesting that this gene is unique as it is contained within a polycistronic transcript (Munson *et al.*, 1998). Also, DNA flanking the *seg* ORF contains additional SE-like elements, suggesting that other uncharacterized SEs may be located within the large transcript.

There is substantial evidence that the SE and other PT genes originated from a common ancestral gene. Several SE genes are located on mobile genetic elements that could disseminate the genes within or between species. Betley and Mekalanos (1985) demonstrated that *sea* is associated with the bacteriophage PS42-D. The toxin gene is located on the phage genome near the phage–bacteria DNA junction. In addition, *see* is

associated with bacteriophage DNA (Couch *et al.*, 1988). An extensive analysis of the DNA isolated from Seb+ strains has demonstrated that *seb* is associated with a discrete genetic determinant at least 26.8 kb in size. This element is believed to have originated from either bacteriophage or plasmid DNA (Johns and Khan, 1988). Consistent with a possible phage origin is the finding that the *seb* genetic determinant is often located at the same chromosomal locus of Seb+ strains, suggesting a mode of integration similar to that of the bacteriophages pS42-D and L54a, which also typically integrate into the chromosome at the same locus (Betley and Mekalanos, 1985; Lee and Iandolo, 1986). Additionally, *seb* is located near the junction of the genetic unit, much like that of other toxin genes carried on bacteriophage genomes (Laird and Groman, 1976; Weeks and Ferretti, 1984; Betley and Mekalanos, 1985; Johnson *et al.*, 1986).

Altboum *et al.* (1985) reported that a 56.2-kb penicillin-resistant plasmid pZA10, isolated from one clinical strain of *S. aureus* could transform several non-toxigenic *S. aureus* strains to either an Seb+, Sec1+ or Seb+/Sec1+ phenotype, providing evidence that both of these elements may be at least transiently harboured on plasmids. Interestingly, the plasmid frequently integrated, re-excised and rearranged activities that could lead to permanent integration of these SE genes into the chromosome. Another toxin gene, *sed*, is carried on a 27.6-kb penicillinase-producing plasmid in all Sed+ strains evaluated (Bayles and Iandolo, 1989).

The evolutionary relatedness between TSST-1 and the SEs is uncertain. However, like most of the SE genes, *tst* is a chromosomally located gene and is part of a larger mobile genetic unit that was originally suspected to be a site-specific transposon (Kreiswirth *et al.*, 1989). Lindsay *et al.* (1998) recently characterized the 15.2-kb genetic element harbouring *tst* and showed that it possesses many of characteristics used to describe pathogenicity islands in other organisms. Specifically, this element, designated SaPI, contains at least one known virulence gene (*tst* and possibly two others), is dispensable and is flanked by directly repeating sequences 17 nucleotides in length. Unlike typical pathogenicity islands, SaPI is efficiently mobilized by a generalized transducing phage and readily transferred to *recA*– recipients.

The SE molecular diversity is proposed to be the result of adaptation, by staphylococci, allowing the most efficient interactions with immune cell receptors in its broad range of potential hosts. Several mechanisms have generated SE diversity, including recombination. Although SEC1 shares only 69% sequence homology with SEB, its *N*-terminal residues are more homologous with SEB than with either SEC2 or SEC3.

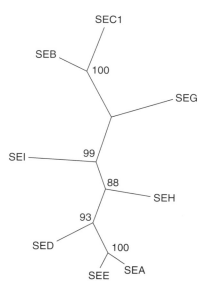

FIGURE 33.2 Phylogenetic analysis of SE mature proteins using parsimony. Branch lengths are proportional to distance. Numbers on some branches represent percentage support of branching based on bootstrap (100 replications) analysis. Bootstrap percentages of less than 50% are not shown (Swofford, 1993).

This suggests that the gene encoding SEC1 originated through recombination between the genes for SEB and SEC2 (or SEC3) (Couch and Betley, 1989). Diversity has also been generated by point mutation. For example, the host-specific SECs, SEC$_{bovine}$ and SEC$_{ovine}$ share 236 of 239 residues. The three divergent residues are located at positions 94, 127 and 165, and determine the distinct biological properties distinguishing these two toxins (Deringer *et al.*, 1997).

Genetic regulation of TSST-1 and SEs

Except for SEG, SEH and SEI, expression of the other SEs and TSST-1 has been thoroughly investigated. With the exception of SEA (Tremaine *et al.*, 1993), regulation of these toxins is influenced by at least two global regulators, *agr* and *sar*. Consistent with the observation that SEA expression is not influenced by either of these *trans*-acting regulators, while the other toxins are, is the finding that no portion of *sea* DNA upstream of the canonical transcriptional promoter site is required for efficient transcription (Borst and Betley, 1994). In contrast, an analysis of the *seb* gene has determined that an upstream portion of DNA, located between −59 and −93 above the transcriptional start site, is required for efficient expression of the toxin (Mahmood and Khan, 1990). Presumably, each of the other toxins regulated by these *trans*-acting factors contains similar regions essential for expression.

The best-characterized regulon is the pleiotropic staphylococcal accessory gene regulator (*agr*). Generally speaking, the *Agr* locus regulates secreted post-exponential phase-produced exotoxins in a positive fashion, while simultaneously negatively regulating many cell-surface-associated virulence factors (Bohach *et al.*, 1997). The degree to which *agr* regulates its potential targets varies and a classification system reflecting the influence of this global regulon has been established. Class-I proteins, which include TSST-1, are tightly regulated by *agr* and are expressed in high quantities in *agr*$^+$ stains and only minimally in *agr*$^−$ strains. Class-II proteins, which are not as tightly influenced by *agr*, are produced in high quantities in *agr*$^+$ strains and moderately in *agr*$^−$ strains. SEB, SEC and SED are regulated by *agr* as class-II proteins. Class-III proteins, which include protein A, coagulase and fibronectin binding proteins, are negatively regulated by *agr* and thus are made in higher amounts in *agr*$^−$ stains than in *agr*$^+$ strains.

The *agr* locus consists of five genes contained in two divergent operons (Figure 33.3). One operon contains *agrA*, *agrB*, *agrC* and *agrD*. The second operon contains only the open reading frame for the staphylococcal haemolysin (*hld*), as well as a large amount of untran-

scribed sequence (Peng *et al.*, 1988). Transcription of the *agr* locus is initiated from three different promoters; P1, P2 and P3. P1 is weakly constitutive and expresses AgrA. The P2 transcript, designated RNAII, spans the entire operon containing the *agr* genes, and is transcribed during the late exponential and early stationary phase. This transcript is responsible for the expression of AgrA, AgrB, AgrC and AgrD (Kornblum *et al.*, 1990). AgrA and AgrC share many properties with members of bacterial two-component signal transduction systems and, in *S. aureus*, are necessary for efficient transcription from the P2 and P3 promoter. Following translation, AgrC becomes localized to the cell membrane and is the environmental sensory component which ultimately serves as the phosphate donor for the regulatory component of the system, AgrA (Lina *et al.*, 1998). AgrA, which lacks DNA binding features of other response regulators known to activate transcription, has not been shown to bind directly to the *agr* promoters but possibly interacts with proteins (i.e. SarA) to activate transcription of RNAII and RNAIII (Morfeldt *et al.*, 1996).

Ji *et al.* (1995) reported that transcriptional activation through the AgrA/AgrC signal transduction system is

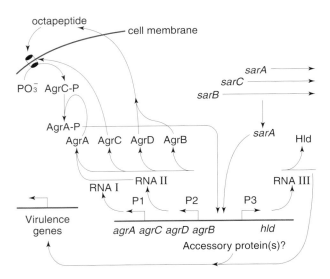

FIGURE 33.3 Schematic representation of Agr- and Sar-mediated regulation of staphylococcal virulence genes. Transcription of the *agr* operon from the P2 promoter produces the gene products AgrA, AgrB, AgrC and AgrD. AgrA, the transducing component of a typical bacterial two-component signal transduction pathway becomes phosphorylated by AgrC (sensor component) and a portion of AgrD (AgrC activation-inducing octapeptide) is required for efficient transcription of RNAIII. Currently, the exact mechanism by which AgrA mediates its influence in uncertain. In addition to AgrA, SarA binds DNA between the P2 and P3 promoters, and is essential for optimal transcription of RNAIII. RNAIII, possibly complexed with additional accessory proteins, is believed to be the effector that regulates, in a negative or positive fashion, various staphylococcal virulence genes.

modulated in a density-dependent fashion by an octapeptide that is synthesized from the two remaining P2 operon genes, *agrB* and *agrD*. These investigators (Ji *et al.*, 1997) subsequently determined that this octapeptide is derived from sequences within the *agrD* gene. Currently, there is emerging evidence that processing and transport of the octapeptide is mediated by AgrB. This system not only serves as a positive regulator of *agr*, but may be utilized as a bacterial interference mechanism by negatively affecting the *agr* homologues of other staphylococci.

Also important, transcription from the *agr* P3 promoter occurs during the late exponential–early stationary phase to generate the RNAIII transcript. This 514 nucleotide transcript, which includes much untranscribed sequence, encodes the 26-residue staphylococcal haemolysin. Given that the Agr phenotype can be restored with a plasmid-transcribing RNAIII from an inducible promoter, this transcript acts as the diffusible effector of the staphylococcal virulence factors. While it is not entirely clear how RNAIII-mediated regulation is achieved, one proposal is that formation of a RNA–protein complex is required. Morfeldt *et al.* (1995) determined that RNAIII binds the 5′ untranslated region of the *S. aureus* α-toxin mRNA transcript, suggesting that RNAIII acts not only at the level of transcription, as originally believed, but at the level of translation as well.

The regulation of toxin expression by RNAIII is directly influenced by the another global regulon. The pleiotropic staphylococcal accessory regulator (*sar*) positively regulates, again at the transcriptional level, expression of certain virulence factors through a mechanism which has only recently been elucidated (Cheung *et al.*, 1994; Cheung *et al.*, 1997). The *sar* locus consists of three overlapping ORFs, *sarA*, *sarB* and *sarC*, each transcribed from a distinct promoter, but sharing a common 3′ end. Studies using isogenic *sar* mutants demonstrated that expression of RNAIII and exoproteins is largely influenced by the presence of SarA (Cheung and Projan, 1994). However, these studies also revealed that complementation of the *sar* mutant with SarA could only partially restore the parental levels of RNAIII transcription. Full restoration to wild-type levels required complementation using a DNA fragment encompassing not only the *sarA* gene, but also elements within the larger *sarB* transcriptional unit. Two ORFs that could potentially encode short peptides, upstream of *sarA*, seem to be involved in SarA-mediated transcriptional activation.

Cheung *et al.* (1997) recently demonstrated by gel-shift analysis that the SarA protein regulates RNAIII (and RNAII) transcription by directly interacting with the *agr* locus. More recently, Chien and Cheung (1998) showed that SarA binds a 29-bp sequence located in the intergenic region located between the P2 and P3 promoter of *agr* in a dose-dependent fashion. Except for the most *N*-terminal residues (residues 1–15), full-level transcription of RNAII and RNAIII requires an intact SarA protein.

The effect of various environmental conditions on expression of TSST-1 and SEs, and their influence on global regulation, have received a significant amount of attention. TSST-1 is most efficiently expressed when the organism is grown aerobically at 37°C and pH 7–8 (Schlievert and Blomster, 1983). The effects are not attributed solely to changes in growth rate, suggesting that these environmental conditions influence *tst* gene expression. Regassa and Betley (1992) demonstrated that SEC expression is down-regulated in bacteria grown in an alkaline pH. Given that these bacteria are found to have reduced RNAIII levels, the reduction in toxin expression was probably due to an effect of pH at the level of *agr*. Regassa *et al.* (1992) also showed that glucose down-regulates SEC expression independent of *agr*. Interestingly, this work revealed that the effect of glucose on toxin expression is further enhanced in conditions that negatively influence *agr* expression (i.e. alkaline or unmaintained pH). Finally, osmotic concentrations of the culture media also affect toxin expression. Growth of *S. aureus* in conditions of high osmolarity (1.2 M NaCl) results in 8–16-fold less toxin than in control cultures. The effect of NaCl appears to be independent of *agr* as an identical response was observed in an *agr⁻* strain (Regassa and Betley, 1993).

SE AND TSST-1 STRUCTURAL BIOLOGY

General characteristics

As a group, the SEs and TSST-1 are heat stable and resistant to inactivation by proteases, although the extent of their stability varies, depending on the toxin. Despite their physical and chemical relatedness, the structures of some members of this group of toxins have diverged to a degree which manifests in unique biological properties. The most notable example is the well-known and unique ability of the SEs to cause emesis, a property that seems to depend on an area of their molecular structure near their cysteine loops. As another example, heterogeneity exists even among the SEs themselves. Some, but not all, of the SEs require Zn^{2+} for maximal induction of SAg activity. Although many of the toxins bind Zn^{2+}, the mechanisms by which this cation is bound and the physiological consequences vary (Bohach, 1997; Svensson *et al.*, 1997).

In part, similarities among the structural features of the SEs roughly parallel the extent of their sequence homology. Because of this, it has become customary to discuss the structural biology of these toxins as groups of proteins rather than considering them individually. TSST-1 is generally considered unique among the SEs because it lacks sequence relatedness with the enterotoxins and has no association with SFP. One group of SEs is comprised of SEB and all subtypes and molecular variants of SEC. Each SEC molecular variant shares greater than 90% amino acid identity with other SECs and at least 65% amino acid identity with SEB. The second group of related SEs contains the prototypic SEA, as well as SEE and SED which share 82% and 52% sequence identity with SEA, respectively. It is possible that SEH, which shares approximately 30% identity with SEA (Betley *et al.*, 1992), will be placed in this group when its structure is elucidated. With the possible exception of SEH, which remains to be thoroughly characterized, each of the other proteins in this group has a demonstrated need for Zn^{2+} (discussed below) to bind efficiently MHC class II (Fraser *et al.*, 1992; Abrahmsen *et al.*, 1995). Other than sequence comparisons, little structure–function information is available for two recently reported SEs (G and I). At the amino acid level, SEG is equally similar (39%) to both SEB and SEC, and shares at least one antigenic epitope with SEC. SEI, a more divergent member of the SE family, appears to be most closely related to SEA, SED and SEE (26–28%) (Munson *et al.*, 1998), but may be sufficiently different to warrant its separation into a separate group.

Many strategies have been employed to date in efforts to study the physicochemical properties of TSST-1 and the SEs. A complete survey of these is beyond the scope of this chapter, but an overview of the earlier work in this area may be obtained in several previously pub-lished reviews (Bohach *et al.*, 1990, 1997). In this chapter, work performed during the past decade to define structural features involved in SAg and enterotoxic biological activities will be highlighted. To date, crystal structures of TSST-1, SEA, SEB, SEC2, SEC3 and SED have been solved (Swaminathan *et al.*, 1992; Prasad *et al.*, 1993; Hoffmann *et al.*, 1994; Schad *et al.*, 1995; Sundstrom *et al.*, 1996). The first crystal structure reported for a SAg was that of SEB, published by Swaminathan *et al.* in 1992. Comparison of the SEB crystal structure with a structure of SEC3, subsequently reported in 1994 (Hoffmann *et al.*, 1994), and several other SEs thereafter, demonstrated that the overall conformations of these toxins are very similar. Interestingly, several structural features seem to be conserved among all SEs. SE molecules may be described as tightly compacted ellipsoidal proteins. All known SE structures are folded into two unequal sized domains comprised of mixed α–β structures. Even TSST-1, which shares little or no sequence homology with the SEs, has similar structural features, consistent with its shared set of biological properties.

SEB and SEC crystal structures

The smaller domain, domain 1, contains (in SEC3) residues 35–120, but not the *N*-terminal residues (Figure 33.4). The β-strands of domain 1 and the single α-helix which caps one end of the domain form a characteristic 'Greek key β-barrel' structural conformation. This domain resembles the oligonucleotide–oligosaccharide binding (OB) motif, reported by Murzin (1993). Although several other bacterial exotoxins possess this same fold, which is responsible for their binding to oligosaccharide receptors, there is presently no evidence for an analogous binding function of this domain in the SEs. Evidence suggests that the SEs do not interact with glycosylated moieties of their two known receptors nec-

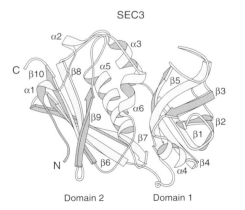
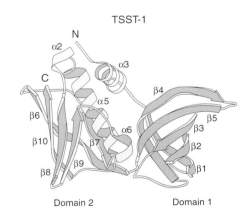

FIGURE 33.4 A comparison of the three-dimensional crystal structures of TSST-1 and SEC3. The outline is orientated to demonstrate the major structural similarities and differences. Labelling of key structural elements in TSST-1 is based on analogous structures in SEC3.

essary for SAg function, MHC class II or the TCR. Domain 1 also contains a disulfide loop, formed by covalent linkage of cysteine residue 110 in the β5-strand with cysteine residue 93 located within the loop itself. This loop, which is found in all SEs, lies at the top of domain 1. The SE loops are very flexible and difficult to model with confidence. The degree of flexibility and resistance to proteolysis appear to correlate directly with the length of the loop. SEB and SEC, having loops of 19 and 16 residues, respectively, are more susceptible to proteolysis than other SEs with shorter loops.

The larger domain 2 is comprised of the N-terminus (residues 1–33) and C-terminal residues 123–239 of SEC3. Residues 123–239 form five β-strands, which fold into a β-grasp motif, and the α-helices, which overlay the β-strands. The N-terminal 20 residues of SEC3 form a loosely attached structure which extends over the top of this domain. Residues 21–26, immediately downstream of the first 20 N-terminal amino acids, form an important α-helix (α3 in SEC3) of domain 2. This helix contains the residues that define the SEC subtype-specific antigenic epitopes (Turner et al., 1992). Moreover, this helix, and the analogous α2-helix in SEB, delineate part of the interface between the two domains of these toxins, and form a shallow cavity located at the top of the SE molecule important for TCR interactions (see below). The remainder of the interface between the two domains is defined by a large groove present in both SEB and SEC. This groove, formed by the α4 and α5 helices of SEB and SEC3, respectively, runs the length of the molecule and contains most of the highly conserved residues of the PT family (Hoffmann et al., 1994).

SEC3 binds a Zn^{2+} atom on the back side of the molecule at the base of the α5 groove (Bohach et al., 1995). Residues D83, H118 and H122 of SEC3 are involved in the co-ordination of Zn^{2+} by this protein. Several features of this SEC3 Zn^{2+} binding environment closely resemble those of thermolysin, the prototype of one class of zinc metalloenzymes. SEC2 also contains a virtually identical Zn^{2+} binding site and, based on sequence data, this may be a feature shared by all SEC variants. Currently, there is no evidence suggesting that SEB, which lacks the SEC-type Zn^{2+} binding motif, has any bound Zn^{2+}.

Crystal structures of SEA and SEE

The SEA and SED crystal structures (Schad et al., 1995; Sundstrom et al., 1996) and a molecular model of SEE, constructed using shared-sequence homology with SEA (82% identity), indicated that these SEs have molecular topologies similar to those of SEB and SEC. All three toxins have a smaller domain 1 comprised of a β-barrel motif capped at one end by an α-helix, which may be described as an OB-like motif described above. Each

toxin also has a characteristic disulfide loop in the same region of domain 1 as seen above for SEC3. Domain 2 of these toxins is also highly structurally homologous to its SEC3 counterpart. In SEA, this domain contains the typical β-grasp motif which consists of a β-sheet, formed by β6–β12-strands, and the α4-helix. It is within this domain that the highest degree of dissimilarity between SEA and SEE exists (Svensson et al., 1997).

It is now clear that the SEs possess at least two mechanisms of Zn^{2+} co-ordination. SEA, SED and SEE require Zn^{2+} to interact efficiently with the MHC class-II molecule (Fraser et al., 1992; Sundstrom et al., 1996). A high-affinity binding site in SEA, in contrast to the SEC3 Zn^{2+} binding site, is located on the external portion of the β-grasp motif of domain 2, very near the N-terminus. Mutagenesis and crystallographic analysis of SEA and SED mapped this site to residues H187, H225 and D227 of SEA (D182, H220 and D222 of SED). SED has a second low-affinity binding site located in a position similar to that of SEC2 and SEC3. This site results in interdomain binding of Zn^{2+} by His residues 109 and 113 of one SED molecule. Two additional co-ordination sites are provided by His8 and Glu12 of a second SED protein. The role of this second site remains unclear; however, it has been postulated to stabilize a particular structural configuration locally (Sundstrom et al., 1996).

TSST-1 crystal structure

The crystal structure of TSST-1 reported initially by Prasad et al. (1993) revealed that this toxin also consists of two separate tightly associated domains. Domain 1, formed by residues 1–17 and 90–194, contains a long central α5-helix (residues 125–140) surrounded by a β-sheet consisting of five β-strands. In addition, the short N-terminal α3-helix is located at the top of this domain. The second domain, containing residues 18–89, forms a five β-stranded claw motif. The α5-helix of domain 1 lies at the base of two grooves separating the two domains. These grooves, referred to as the front and back-side grooves, are further developed by the small N-terminal α-helix and loop structures present on opposite sides of the α5-helix. Despite the lack of sequence homology with the SEs, the TSST-1 crystal structure shares remarkable similarities with the SEs. However, TSST-1 lacks several features shared by the SEs. The most obvious of these include:

(a) an elongated N-terminus which folds back over the top of domain 2
(b) an α-helix in the loop at the base of domain 1 β-sheet
(c) a second α-helix positioned in the groove between both domains
(d) the SE-associated disulfide loop.

Interaction of the SEs and TSST-1 with MHC class II

MHC class-II binding is a property of all SAgs and, in general, HLA-DR1 is most efficient in promoting the activity of SAgs (Svensson *et al.*, 1997). Despite this, the SEs and TSST-1 vary largely in regard to binding to MHC class II. Each toxin binds to MHC class-II receptors with its own unique affinity and also recognizes a particular set of MHC alleles (Herman *et al.*, 1990). The affinities of SEB, SEC1 and TSST-1 for HLA-DR1 are approximately three–nine-fold weaker than that of SEA, but significantly higher than those of either SED or SEE (Chintagumpala *et al.*, 1991; Mollick *et al.*, 1991). Moreover, individual SEs may bind to different regions of the MHC class-II molecule, and two individual SEs may bind efficiently to entirely different MHC chains. For instance, SEA binds the receptor at both the α- and β-chains, while SEB has only been shown conclusively to interact efficiently with the α-chain (Jardetzky *et al.*, 1994; Abrahmsen *et al.*, 1995; Hudson *et al.*, 1995). These differences result from divergent regions within the toxin molecules and also depend on the presence or absence of Zn^{2+} bound in a manner that can promote interaction with other toxin or MHC molecules.

An SEB–HLA-DR1 crystal structure reported by Jardetzky *et al.* (1994) demonstrated that SEB interacts with the residues of the HLA-DR1 α1-chain. The association occurs at a concave surface on the receptor which lies adjacent to, but outside, the peptide binding groove. This interaction requires 19 residues of SEB (residues 43, 44, 45, 46, 47, 65, 67, 69, 71, 78, 89, 92, 94, 95, 96, 115, 211, 212 and 215) in four main stretches of sequence on the outside and top of domain 1 within and adjacent to the cysteine loop. Binding to the receptor in this manner orientates domain 2 away from the HLA-DR1 α-chain. Twenty-one residues of the HLA-DR1 α1-chain are involved in binding to SEB. Unlike TSST-1, SEB does not extend over the top of the receptor α1-chain to interact with the bound peptide. Several lines of evidence support the existence of at least one alternative MHC class-II binding site for SEB. Mutagenesis and peptide studies implicated several *N*-terminal residues of the toxin, and other amino acids on the outside surface of domain 2, in affecting binding to MHC class II (Kappler *et al.*, 1992; Komisar *et al.*, 1994; Soos and Johnson, 1994).

Kim *et al.* (1994) solved the crystal structure of HLA-DR1 complexed with TSST-1. With a few significant exceptions, TSST-1 binding to this receptor was very similar to that of SEB. Residues within three regions of domain 1 bind primarily to the α-chain of HLA-DR1 and secondarily to the β-chain. Region 1 centres around L30 of TSST-1, which hydrogen bonds with two loops of the HLA-DR1 α-chain, and TSST-1 residues D27, K58, S53 and P50. Region II involves four β-strands (β2, β3, β5 and β4) of TSST-1 which associate with the external face of the HLA-DR1 α-domain. Region III is involved in two interactions. The first of these is an association of T75 and S76 in the loop region between β4 and β5 of TSST-1, with the *C*-terminus of the HLA-DR1-bound peptide. Binding to the peptide is a significant difference distinguishing TSST-1 and SEB. The other interaction of this region of the toxin involves an association with TSST-1 Q73 with the β-chain of HLA-DR1.

Although crystallographic data for an SEA:MHC class-II complex have not been reported, MHC class-II binding by this toxin may be more complex than that observed for either SEB or TSST-1. Competition studies demonstrate that SEA competes with SEB and TSST-1 for binding to HLA-DR1 (Thibodeau *et al.*, 1994). In contrast, excess SEB does not inhibit MHC class-II binding by SEA (Fraser, 1989; Chintagumpala *et al.*, 1991; Purdie *et al.*, 1991), suggesting that SEA has not only an MHC class-II binding domain, which overlaps that of SEB/TSST-1, but an additional binding site unique to SEA. Evidence in support of this prediction was provided by Hudson *et al.* (1995), who proposed the existence of two non-overlapping MHC binding sites in SEA. The first high-affinity site involves the tetrahedral co-ordination of Zn^{2+} via SEA residues H187, H225, D227 and H81 of the MHC class-II β-chain. The second low-affinity site, located on the opposite side of SEA, is a Zn^{2+}-independent site analogous to that of SEB which is proposed to bind to the MHC class-II α-chain. Key SEA residues in the low-affinity site were F47 and L48 (analogous to SEB F44 and L45) (Abrahmsen *et al.*, 1995).

The likely existence of two MHC binding sites on SEA suggests several possible mechanisms for induction of SEA-mediated biological activities. SEA binding to MHC class-II molecules was originally noted to be a co-operative process in which the binding of SEA to the β-chain of the HLA-DR1 molecule assists the binding of a second toxin molecule to the α-chain of the receptor to form a DR1–SEA$_2$ trimer (Hudson *et al.*, 1995; Tiedemann *et al.*, 1995). Subsequent studies with toxins altered in either HLA-DR1 α- or β-chain binding sites (Tiedemann and Fraser, 1996) demonstrated that SEA can also function bivalently and cross-link two HLA-DR1 molecules (forming a DR1$_\alpha$–SEA–DR1$_\beta$ trimer) through both the α- and β-chains. Formation of this trimer was essential for T-cell activation and in inducing expression of proinflammatory cytokines and cell adhesion by monocytes and B-cells, respectively (Trede *et al.*, 1991; Mehindate *et al.*, 1995; Tiedemann *et al.*, 1996).

SED, SEE and SEH possess high-affinity Zn^{2+} binding sites (Fraser *et al.*, 1992; Ren *et al.*, 1994) and could potentially interact with the MHC class-II molecules in a fashion similar to that of SEA. SED and SEE have both been shown to contain two MHC class-II binding sites and to form the MHC class-II α–SE–MHC Class II β trimers (Al-Daccak *et al.*, 1998). Although SED may actively participate in the formation of these trimers, this toxin may also form Zn^{2+}-dependent homodimers utilizing the high-affinity Zn^{2+} binding site (Thibodeau *et al.*, 1997). The crystal structure of the SED homodimer revealed that the molecules pack against one another in such a fashion that forms two symmetrical Zn^{2+} co-ordination sites involving the high-affinity Zn^{2+} binding residues (D182, H220 and D222) of one molecule and H218 of the other molecule (Sundstrom *et al.*, 1996). Moreover, this SED homodimer could potentially interact with MHC class-II molecules to produce three different types of tetrameric complexes; an α-chain–dimer–α-chain complex, a β-chain–dimer–β-chain complex or an α-chain–dimer–β-chain complex, each having the capacity to induce cytokine expression in a monocytic cell line. Interestingly, while all of the possible SED complexes result in monocyte-derived cytokine expression, only complexes simultaneously involving the binding of two MHC class-II molecules through the α-chain of one and the β-chain of the other have the capacity to activate T-cells bearing the human Vβ3 element (Al-Daccak *et al.*, 1998).

The binding of the SECs to MHC class II is comparatively weak, and much less is known about the interaction of these toxins with the receptor. As the primary sequences and crystal structures of the SECs are similar to those of SEB, both groups of toxins probably bind in a manner similar to that elucidated for the SEB–HLA-DR1 complex. Consistent with this notion is the fact that the residues involved in binding of SEB to MHC are largely conserved with the SECs (Jardetzky *et al.*, 1994). The proposal (Hoffmann *et al.*, 1994) that synthetic oligopeptides derived from residues located within the SEC1 α5 groove inhibit T-cell proliferation by binding to MHC class II implied that SEC has a second binding site unique to itself, and not shared with SEB. Although SECs bind Zn^{2+} trivalently at the base of the α5 groove, leaving an empty potential liganding site, site-directed mutagenesis showed that Zn^{2+} binding is not required for induction of T-cell proliferation (Stauffacher *et al.*, 1998). Therefore, if Zn^{2+} mediates SEC–MHC class-II complex formation, it probably serves as an alternative mechanism to that provided by a site analogous to that of SEB.

Interactions of SEs and TSST-1 with the TCR

The SEs and TSST-1 each interact with a defined TCR repertoire (Table 33.1), determined in part by sequences of the Vβ region of the receptor. The TCR repertoire recognized also depends on residues in the TCR binding regions of each toxin. Although large differences are noted among individual toxin–Vβ pairs, K_ds reported for various toxins binding to TCRs with compatible Vβs range from 0.82×10^{-6} to 4.5×10^{-6} M, and thus are similar to, but generally somewhat weaker than, affinities for MHC class II (Seth *et al.*, 1994; Malchiodi *et al.*, 1995; Leder *et al.*, 1998). In general, highly homologous toxins are more likely to interact with related Vβs. For example, SEB and SEC variants consistently stimulate T-cells bearing human Vβs 12, 13.2, 14, 15 and 17, which are highly related Vβs falling into subgroup 4 as classified by Chothia *et al.* (1988). In contrast, SEs in the SEA–SEE group predominantly stimulate T-cells bearing Vβs in the distantly related subgroup 1; Vβ2 stimulated by TSST-1 is likewise uniquely in subgroup 3. Even highly related toxins may have slight differences in Vβ preferences. For example, SEC1 strongly induces proliferation of T-cells bearing Vβ3 but not Vβ13.1, while SEC2 and SEC3 have the opposite pattern in regard to these two Vβs.

Through mutagenesis experiments, Deringer *et al.* (1996) showed that the differences in Vβ specificity between SEC1 and SEC2 (or SEC3) were determined by residue 26 located at the base of a shallow cavity at the top of the SEC molecule. Residue 26 is located at the end of the α3-helix in SEC3 (Figure 33.4). Work by other investigators implicated residues on analogous structures in SEA, SEB, SEE and TSST-1 in TCR binding. Also, residues further away in the primary sequence, but in close proximity to this α-helix in the three-dimensional structure, were shown to be involved in interactions with the TCR (Harris *et al.*, 1993; Hudson *et al.*, 1993; Hayball *et al.*, 1994). Together, these studies defined the general boundaries of the shallow cavity responsible for TCR binding by all known PT SAgs. The crystal structure of SEC3 bound to the extracellular portion of the β-chain of a murine TCR (14.3d) extended the above results (Fields *et al.*, 1996). The SEC3:14.3.d structure revealed that binding of SEC3 to the TCR was dependent on contacts formed through amino acid residues in the complementarity determining region 2 (CDR2) and, to a lesser extent, CDR1 and the hypervariable region 4 (HV4) of the Vβ domain. In the crystal structure, these loops interact with residues in the small interdomain cavity at the top of the toxin molecule described above. Specifically, N60, Y90, V91, G102, K103, V104 and G106 of the small SEC3 domain and G19, T20, N23, Y26, F176 and Q210 of the large domain associate with N28 and N30 of CDR1, Y50, G51, A52, G53, S54 and T55 of CDR2, and P70 and Q72 of the HV4. Leder *et al.* (1998) determined the relative contribution of each toxin residue to the SEC3–Vβ complex. Based on the properties of site-specific toxin mutants, N23, Y90 and Q210 were the most important toxin residues. In addition, significant contributions were noted for the residues T20, Y26, N60, V91 and F126. With the exception of F126, all of the most energetically important residues (N23, Y26, V91 Y90 and Q210) are clustered at the centre of the α3 cavity, a finding that confirmed the importance of this region in TCR binding. Probably because of the intrinsic flexibility of this region, mutation of the contact sites located within the disulfide loop (G102, K103, G106) had little effect on binding.

Properties of the MHC–toxin–TCR trimolecular complex

Superimposition of crystal structures from (a) a Vβ–SEC3 complex; (b) the SEB–HLA-DR1 complex; and (c) a TCR Vα homodimer by Fields *et al.* (1996) led to a model representing the toxin-induced trimolecular complex at the interface of the T-cell and APC. The model provides some interesting insights into the nature of this complex. First, it demonstrates that these toxins could bridge both the TCR and the MHC class-II molecule outside the peptide binding groove, acting as a 'wedge' between the TCR β-chain and the MHC class-II α-chain. Formation of the complex in this manner could dislodge the peptide binding cleft away from the TCR, circumventing any association of this region with the TCR. Furthermore, consistent with the notion that SAg interaction with the TCR occurs in a Vβ-dependent manner, this model conclusively demonstrates that the Vα-chain of the TCR does not directly engage the toxin, but instead associates with the β1-chain of MHC class II. Finally, this model showed that the Vβ region of the TCR does not interact with the MHC class-II α-chain.

Leder *et al.* (1998) also demonstrated that the ability of various SEs and SE mutants to induce T-cell proliferation is directly related to the efficiency with which they bind the TCR Vβ domain. As a similar effect on T-cell proliferation is determined by affinities for MHC class II (see above), and because toxin stimulation of T-cells requires simultaneous interactions with both the TCR and MHC molecules, it is likely that the overall affinity of the entire complex determines the effectiveness of the stimulation. Consistent with this proposal, Leder *et al.* (1998) showed that low-affinity binding of a particular toxin to the TCR can be compensated for by stronger binding to MHC class II, and vice versa. For example, SEB, which binds to HLA-DR1 with a three-fold higher affinity than SEC3, induces proliferation of T-cells

bearing murine Vβ8.2 more effectively than SEC3 despite the fact that SEB has a 35-fold lower affinity for this Vβ.

Structural aspects involved in emesis

Induction of emesis following oral ingestion by primates is a key feature distinguishing the SEs from TSST-1 and other members of the PT family. Several lines of evidence, derived from mutational analysis or chemical modification of various SEs, strongly suggest that no direct correlation between emetic activity and function of the SEs as SAgs exists and that these functions are separable using appropriate techniques. For instance, substitution of a key residue (N25) in the TCR binding region of SEA with alanine dramatically reduced TCR binding and subsequent T-cell proliferation despite having no demonstrable effect on induction of the emetic response (Harris and Betley, 1995). Similar conclusions have been made by Alber *et al.* (1990) and Hovde *et al.* (1994) through related work with SEB and SEC, respectively.

Because all SEs possess an intramolecular disulfide bond and cysteine loop, this feature has been suspected to be involved in their emetic response. However, elucidation of the SE sequences and structural information revealed that the cysteine loop itself is highly flexible and not well conserved in regard to content or number of amino acid residues. In addition, another related streptococcal PTs, SPEA, is believed to form a similar cysteine loop, but is non-emetic (Bohach *et al.*, 1998). Therefore, it is unlikely that the cysteines themselves or the cysteine loop residues play a key direct role in induction of emesis. Mutagenesis studies demonstrated that when the cysteines of SEC1 (positions 93 and 110) are replaced with alanine, the recombinant mutant toxin lacks the ability to induce an emetic response (Hovde *et al.*, 1994). In contrast, substitution of the cysteine residues with serine generated an emetic mutant. Given that serine is unable to form a disulfide bond, this study demonstrated that the bond itself was not required, but that the loop region is indirectly involved in the ability of SEs to induce emesis. Because serine substitution apparently stabilized the conformational integrity of the toxin, presumably through the formation of hydrogen bonds, the contribution of the disulfide bond to the induction of emesis could be to maintain a correct spatial orientation of other critical residues (Hovde *et al.*, 1994; Bohach *et al.*, 1995). The results obtained with SEC, plus the properties of SEA mutants near the cysteine loop (Harris and Betley, 1995), are suggestive that a highly conserved stretch of residues in the SE small domain (i.e. the

SEC3 β5-strand) could determine two overlapping but separable biological activities, T-cell stimulation and emesis.

Other potentially important molecular regions

Despite the fact that T-cells and T-cell cytokines clearly play a central role in the pathogenesis of TSS, a complete link between superantigenicity and lethality induced by TSST-1 and SEs is not conclusively established. Mutagenesis experiments, mostly using TSST-1 and the non-lethal TSST$_{ovine}$, have shown that it is possible at least partially to separate T-cell stimulatory activity from lethal effects *in vivo*. By constructing hybrid TSST toxins, in which residues of TSST-1 and TSST$_{ovine}$ were switched, Murray *et al.* (1994, 1996) showed that alteration of two residues (132 and 136) on the TSST central α-helix (see above and Figure 33.4) generated mutants with opposite effects, suggesting that these two activities could be separable. For example, substitution of Q136 of TSST-1 with alanine generated a mutant that, even in very high amounts, was non-lethal in a rabbit TSS model. The Q136A mutant was only moderately reduced in T-cell proliferative ability. Conversely, substitution of K132 in TSST$_{ovine}$ with its analogous residue (Glu) in TSST-1 produced a lethal hybrid mutant which had only minimal ability to induce T-cell proliferation. The cellular target and corresponding receptors involved in the proposed lethal activity of TSST-1 and other PTs have not been defined and their nature remains somewhat controversial. One proposal is that lethality is dependent upon cytotoxicity for certain cells, possibly in the kidneys, liver or vascular endothelium. This remains to be determined.

ACKNOWLEDGEMENTS

This work was supported by US PHS grant AI28401 and the United Dairymen of Idaho. The authors wish to thank Claudia Deobald for assistance in preparation of the manuscript. The assistance of Phil Berger, Cathy Earhart, Matt Marshall, Douglas Ohlendorf and Gregory Vath in artwork and preparation of figures is greatly appreciated.

REFERENCES

Abrahmsen, L., Dohlsten, M., Segren, S., Bjork, P., Jonsson, E. and Kalland, T. (1995). Characterization of two distinct MHC class II binding sites in the superantigen staphylococcal enterotoxin A. *EMBO J.* **14**, 2978-86.

Al-Daccak, R., Mehindate, K., Damdoumi, F., Etongue-Mayer, P., Nilsson, H., Antonsson, P., Sundstrom, M., Dohlsten, M., Sekaly, R.P. and Mourad, W. (1998). Staphylococcal enterotoxin D is a promiscuous superantigen offering multiple modes of interactions with the MHC class II receptors. *J. Immunol.* **160**, 225–32.

Alber, G., Scheuber, P.H., Reck, B., Sailer-Kramer, B., Hartmann, A. and Hammer, D.K. (1989). Role of substance P in immediate-type skin reactions induced by staphylococcal enterotoxin B in unsensitized monkeys. *J. Allergy Clin. Immunol.* **84**, 880–5.

Alber, G., Hammer, D.K. and Fleischer, B. (1990). Relationship between enterotoxic- and T lymphocyte-stimulating activity of staphylococcal enterotoxin B. *J. Immunol.* **144**, 4501–6.

Altboum, Z., Hertman, I. and Sarid, S. (1985). Penicillinase plasmid-linked genetic determinants for enterotoxins B and C1 production in *Staphylococcus aureus. Infect. Immun.* **47**, 514–21.

Aranow, H., Jr and Wood, W.B. (1942). Staphylococcal infection simulating scarlet fever. *JAMA* **119**, 1495.

Barber, M.A. (1914). Milk poisoning due to a type of *Straphylococcus albus* occurring in the udder of a healthy cow. *Philipp. J. Sci.* **B9**, 515.

Bayles, K.W. and Iandolo, J.J. (1989). Genetic and molecular analyses of the gene encoding staphylococcal enterotoxin D. *J. Bacteriol.* **171**, 4799–806.

Bayles, K.W., Wesson, C.A., Liou, L.E., Fox, L.K., Bohach, G.A. and Trumble, W.R. (1998). Intracellular *Staphylococcus aureus* escapes the endosome and induces apoptosis in epithelial cells. *Infect. Immun.* **66**, 336–42.

Bayliss, M. (1940). Studies on the mechanism of vomiting produced by staphylococcal enterotoxin. *J. Exp. Med.* **72**, 669–84.

Bergdoll, M.S. (1979). Staphylococcal intoxications. In: *Food-borne Infections and Intoxications* (eds H. Riemann and F.L. Bryan), pp. 443–94. Academic Press, New York.

Bergdoll, M.S. (1988). Monkey feeding test for staphylococcal enterotoxin. *Methods Enzymol.* **165**, 324–33.

Bergdoll, M.S. (1989). *Staphylococcus aureus*. In: *Foodborne Bacterial Pathogens* (ed. M.P. Doyle), pp. 463–523. Marcel Dekker, New York.

Bergdoll, M.S., Crass, B.A., Reiser, R.F., Robbins, R.N. and Davis, J.P. (1981). A new staphylococcal enterotoxin, enterotoxin F, associated with toxic-shock-syndrome *Staphylococcus aureus* isolates. *Lancet* **i**, 1017–21.

Betley, M.J. and Mekalanos, J.J. (1985). Staphylococcal enterotoxin A is encoded by phage. *Science* **229**, 185–7.

Betley, M.J. and Mekalanos, J.J. (1988). Nucleotide sequence of the type A staphylococcal enterotoxin gene. *J. Bacteriol.* **170**, 34–41.

Betley, M.J., Schlievert, P.M., Bergdoll, M.S., Bohach, G.A., Iandolo, J.J., Khan, S.A., Pattee, P.A. and Reiser, R.R. (1990). Staphylococcal gene nomenclature. *ASM News* **56**, 182.

Betley, M.J., Borst, D.W. and Regassa, L.B. (1992). Staphylococcal enterotoxins, toxic shock syndrome toxin and streptococcal pyrogenic exotoxins: a comparative study of their molecular biology. *Chem. Immunol.* **55**, 1–35.

Blomster-Hautamaa, D.A., Kreiswirth, B.N., Kornblum, J.S., Novick, R.P. and Schlievert, P.M. (1986). The nucleotide and partial amino acid sequence of toxic shock syndrome toxin-1. *J. Biol. Chem.* **261**, 15783–6.

Bohach, G.A. (1997). Staphylococcal enterotoxins B and C. Structural requirements for superantigenic and entertoxigenic activities. *Prep. Biochem. Biotechnol.* **27**, 79–110.

Bohach, G.A. and Schlievert, P.M. (1987). Nucleotide sequence of the staphylococcal enterotoxin C1 gene and relatedness to other pyrogenic toxins. *Mol. Gen. Genet.* **209**, 15–20.

Bohach, G.A. and Schlievert, P.M. (1989). Conservation of the biologically active portions of staphylococcal enterotoxins C1 and C2. *Infect. Immun.* **57**, 2249–52.

Bohach, G.A., Dinges, M.M., Mitchell, D.T., Ohlendorf, D.H. and Schlievert, P. (1997). Exotoxins. In: *The Staphylococci in Human Disease* (eds K.B. Crossley and G.A. Archer), pp. 83–111. Churchill Livingstone, New York.

Bohach, G.A., Fast, D.J., Nelson, R.D. and Schlievert, P.M. (1990). Staphylococcal and streptococcal pyrogenic toxins involved in toxic shock syndrome and related illnesses. *Crit. Rev. Microbiol.* **17**, 251–72.

Bohach, G.A., Jablonski, L.M., Deobald, C.F., Chi, Y.I. and Stauffacher, C.V. (1995). Functional domains of staphylococcal enterotoxins. In: *Molecular Approaches to Food Safety: Issues Involving Toxic Microorganisms* (eds M. Ecklund, J.L. Richard and K. Mise), pp. 339–56. Alaken, Fort Collins, CO.

Bohach, G., Jablonski, L., Roggiani, M., Sadler, I., Schlievert, P., Mitchell, D. and Ohlendorf, D. (1998). Biological activity of pyrogenic toxins delivered at the mucosal surface. In: *European Conference on Toxic Shock Syndrome*. (eds F. Arbuthnott and B. Furman), pp. 170–2. Royal Society of Medicine, London.

Bone, R.C. (1991). The pathogenesis of sepsis. *Ann. Intern. Med.* **115**, 457–69.

Borst, D.W. and Betley, M.J. (1994). Promoter analysis of the staphylococcal enterotoxin A gene. *J. Biol. Chem.* **269**, 1883–8.

Boshell, M., McLeod, J., Walker, L., Hall, N., Patel, Y. and Sansom, D. (1996). Effects of antigen presentation on superantigen-induced apoptosis mediated by Fas/Fas ligand interactions in human T cells. *Immunology* **87**, 586–92.

Boyle, T., Lancaster, V., Hunt, R., Gemski, P. and Jett, M. (1994). Method for simultaneous isolation and quantitation of platelet activating factor and multiple arachidonate metabolites from small samples: analysis of effects of *Staphylococcus aureus* enterotoxin B in mice. *Anal. Biochem.* **216**, 373–82.

Chesney, J.P. (1997). Toxic shock syndrome. In: *The Staphylococci in Human Disease* (eds K.B. Crossley and G.A. Archer), pp. 509–25. Churchill Livingstone, New York.

Cheung, A.L. and Projan, S.J. (1994). Cloning and sequencing of *sarA* of *Staphylococcus aureus*, a gene required for the expression of *agr*. *J. Bacteriol.* **176**, 4168–72.

Cheung, A.L., Eberhardt, K.J., Chung, E., Yeaman, M.R., Sullam, P.M., Ramos, M. and Bayer, A.S. (1994). Diminished virulence of a *sar*/*agr*-mutant of *Staphylococcus aureus* in the rabbit model of endocarditis. *J. Clin. Invest.* **94**, 1815–22.

Cheung, A.L., Bayer, M.G. and Heinrichs, J.H. (1997). *sar* genetic determinants necessary for transcription of RNAII and RNAIII in the *agr* locus of *Staphylococcus aureus*. *J. Bacteriol.* **179**, 3963–71.

Chien, Y. and Cheung, A.L. (1998). Molecular interactions between two global regulators, *sar* and *agr*, in *Staphylococcus aureus*. *J. Biol. Chem.* **273**, 2645–52.

Chintagumpala, M.M., Mollick, J.A. and Rich, R.R. (1991). Staphylococcal toxins bind to different sites on HLA-DR. *J. Immunol.* **147**, 3876–81.

Choi, Y.W., Herman, A., DiGiusto, D., Wade, T., Marrack, P. and Kappler, J. (1990). Residues of the variable region of the T-cell-receptor beta-chain that interact with *S. aureus* toxin superantigens. *Nature* **346**, 471–3.

Chothia, C., Boswell, D.R. and Lesk, A.M. (1988). The outline structure of the T-cell alpha beta receptor. *EMBO J.* **7**, 3745–55.

Clark, W.G. (1962). Emetic effect of purified staphylococcal enterotoxin in cats. *Proc. Soc. Exp. Biol. Med.* **111**, 205–7.

Couch, J.L. and Betley, M.J. (1989). Nucleotide sequence of the type C3 staphylococcal enterotoxin gene suggests that intergenic recombination causes antigenic variation. *J. Bacteriol.* **171**, 4507–10.

Couch, J.L., Soltis, M.T. and Betley, M.J. (1988). Cloning and nucleotide sequence of the type E staphylococcal enterotoxin gene. *J. Bacteriol.* **170**, 2954–60.

Crass, B.A. and Bergdoll, M.S. (1986). Involvement of staphylococcal enterotoxins in nonmenstrual toxic shock syndrome. *J. Clin. Microbiol.* **23**, 1138–9.

Dack, G.M., Cary, W.E., Woolper, O. and Wiggers, H. (1930). An outbreak of food poisoning proved to be due to a yellow hemolytic *Staphylococcus*. *J. Prev. Med.* **4**, 167–75.

Deresiewicz, R.L. (1997). Staphylococcal toxic shock syndrome. In: *Superantigens: Molecular Biology, Immunology, and Relevance to Human Disease* (eds D.Y. Leung, B.T. Huber and P. Schlievert.), pp. 435–79. Marcel Dekker, New York.

Deringer, J.R., Ely, R.J., Stauffacher, C.V. and Bohach, G.A. (1996). Subtype-specific interactions of type C staphylococcal enterotoxins with the T-cell receptor. *Mol. Microbiol.* **22**, 523–34.

Deringer, J.R., Ely, R.J., Monday, S.R., Stauffacher, C.V. and Bohach, G.A. (1997). Vbeta-dependent stimulation of bovine and human T cells by host-specific staphylococcal enterotoxins. *Infect. Immun.* **65**, 4048–54.

Dunnett, W.N. and Schallibaum, E.M. (1960). Scarlet fever-like illness due to staphylococcal infection. *Lancet* **ii**, 1229.

Edwards, V.M., Deringer, J.R., Callantine, S.D., Deobald, C.F., Berger, P.H., Kapur, V., Stauffacher, C.V. and Bohach, G.A. (1997). Characterization of the canine type C enterotoxin produced by *Staphylococcus intermedius* pyoderma isolates. *Infect. Immun.* **65**, 2346–52.

Elwell, M.R., Liu, C.T., Spertzel, R.O. and Beisel, W.R. (1975). Mechanisms of oral staphylococcal enterotoxin B-induced emesis in the monkey (38553). *Proc. Soc. Exp. Biol. Med.* **148**, 424–7.

Fast, D.J., Schlievert, P.M. and Nelson, R.D. (1989). Toxic shock syndrome-associated staphylococcal and streptococcal pyrogenic toxins are potent inducers of tumor necrosis factor production. *Infect. Immun.* **57**, 291–4.

Ferens, W.A., Davis, W.C., Hamilton, M.J., Park, Y.H., Deobald, C.F., Fox, L. and Bohach, G. (1998). Activation of bovine lymphocyte subpopulations by staphylococcal enterotoxin C. *Infect. Immun.* **66**, 573–80.

Fields, B.A., Malchiodi, E.L., Li, H., Ysern, X., Stauffacher, C.V., Schlievert, P.M., Karjalainen, K. and Mariuzza, R.A. (1996). Crystal structure of a T-cell receptor beta-chain complexed with a superantigen. *Nature* **384**, 188–92.

Florquin, S. and Aaldering, L. (1997). Superantigens: a tool to gain new insight into cellular immunity. *Res. Immunol.* **148**, 373–86.

Florquin, S., Amraoui, Z., Dubois, C., Decuyper, J. and Goldman, M. (1994). The protective role of endogenously synthesized nitric oxide in staphylococcal enterotoxin B-induced shock in mice. *J. Exp. Med.* **180**, 1153–8.

Fraser, J.D. (1989). High-affinity binding of staphylococcal enterotoxins A and B to HLA-DR. *Nature* **339**, 221–3.

Fraser, J.D., Urban, R.G., Strominger, J.L. and Robinson, H. (1992). Zinc regulates the function of two superantigens. *Proc. Natl Acad. Sci. USA* **89**, 5507–11.

Garbe, P.L., Arko, R.J., Reingold, A.L., Graves, L.M., Hayes, P.S., Hightower, A.W., Chandler, F.W. and Broome, C.V. (1985). *Staphylococcus aureus* isolates from patients with nonmenstrual toxic shock syndrome. Evidence for additional toxins. *JAMA* **253**, 2538–42.

Gaventa, S., Reingold, A.L., Hightower, A.W., Broome, C.V., Schwartz, B., Hoppe, C., Harwell, J., Lefkowitz, L.K., Makintubee, S. and Cundiff, D.R. (1989). Active surveillance for toxic shock syndrome in the United States, 1986. *Rev. Infect. Dis.* **11**, Suppl. 1, S28–34.

Hackett, S.P. and Stevens, D.L. (1993). Superantigens associated with staphylococcal and streptococcal toxic shock syndrome are potent inducers of tumor necrosis factor-beta synthesis. *J. Infect. Dis.* **168**, 232–5.

Harris, T.O. and Betley, M.J. (1995). Biological activities of staphylococcal enterotoxin type A mutants with *N*-terminal substitutions. *Infect. Immun.* **63**, 2133–40.

Harris, T.O., Grossman, D., Kappler, J.W., Marrack, P., Rich, R.R. and Betley, M.J. (1993). Lack of complete correlation between emetic and T-cell-stimulatory activities of staphylococcal enterotoxins. *Infect. Immun.* **61**, 3175–83.

Hasko, G., Virag, L., Egnaczyk, G., Salzman, A.L. and Szabo, C. (1998). The crucial role of IL-10 in the suppression of the immunological response in mice exposed to staphylococcal enterotoxin B. *Eur. J. Immunol.* **28**, 1417–25.

Hayball, J.D., Robinson, J.H., O'Hehir, R.E., Verhoef, A., Lamb, J.R. and Lake, R.A. (1994). Identification of two binding sites in staphylococcal enterotoxin B that confer specificity for TCR V beta gene products. *Int. Immunol.* **6**, 199–211.

Herman, A., Croteau, G., Sekaly, R.P., Kappler, J. and Marrack, P. (1990). HLA-DR alleles differ in their ability to present staphylococcal enterotoxins to T cells. *J. Exp. Med.* **172**, 709–17.

Hoffmann, M.L., Jablonski, L.M., Crum, K.K., Hackett, S.P., Chi, Y.I., Stauffacher, C.V., Stevens, D.L. and Bohach, G.A. (1994). Predictions of T-cell receptor- and major histocompatibility complex- binding sites on staphylococcal enterotoxin C1. *Infect. Immun.* **62**, 3396–407.

Holmberg, S.D. and Blake, P.A. (1984). Staphylococcal food poisoning in the United States. New facts and old misconceptions. *JAMA* **251**, 487–9.

Hovde, C.J., Hackett, S.P. and Bohach, G.A. (1990). Nucleotide sequence of the staphylococcal enterotoxin C3 gene: sequence comparison of all three type C staphylococcal enterotoxins. *Mol. Gen. Genet.* **220**, 329–33.

Hovde, C.J., Marr, J.C., Hoffmann, M.L., Hackett, S.P., Chi, Y.I., Crum, K.K., Stevens, D.L., Stauffacher, C.V. and Bohach, G.A. (1994). Investigation of the role of the disulphide bond in the activity and structure of staphylococcal enterotoxin C1. *Mol. Microbiol.* **13**, 897–909.

Hudson, K.R., Robinson, H. and Fraser, J.D. (1993). Two adjacent residues in staphylococcal enterotoxins A and E determine T cell receptor V beta specificity. *J. Exp. Med.* **177**, 175–84.

Hudson, K.R., Tiedemann, R.E., Urban, R.G., Lowe, S.C., Strominger, J.L. and Fraser, J.D. (1995). Staphylococcal enterotoxin A has two cooperative binding sites on major histocompatibility complex class II. *J. Exp. Med.* **182**, 711–20.

Iandolo, J.J. (1989). Genetic analysis of extracellular toxins of *Staphylococcus aureus*. *Annu. Rev. Microbiol.* **43**, 375–402.

Ikejima, T., Dinarello, C.A., Gill, D.M. and Wolff, S.M. (1984). Induction of human interleukin-1 by a product of *Staphylococcus aureus* associated with toxic shock syndrome. *J. Clin. Invest.* **73**, 1312–20.

Iwasaki, M., Igarashi, H., Hinuma, Y. and Yutsudo, T. (1993). Cloning, characterization and overexpression of a *Streptococcus pyogenes* gene encoding a new type of mitogenic factor. *FEBS Lett.* **331**, 187–92.

Jablonski, L.M. and Bohach, G.A. (1997). *Staphylococcus aureus*. In: *Food Microbiology: Fundamentals and Frontiers* (eds M.P. Doyle, L.P. Beuchat and T.J. Montville), pp. 353–75. ASM Press, Washington, DC.

Jardetzky, T.S., Brown, J.H., Gorga, J.C., Stern, L.J., Urban, R.G., Chi, Y.I., Stauffacher, C., Strominger, J.L. and Wiley, D.C. (1994). Three-dimensional structure of a human class II histocompatibility molecule complexed with superantigen. *Nature* **368**, 711–18.

Jett, M., Neill, R., Welch, C., Boyle, T., Bernton, E., Hoover, D., Lowell, G., Hunt, R.E., Chatterjee, S. and Gemski, P. (1994). Identification of staphylococcal enterotoxin B sequences important for induction of lymphocyte proliferation by using synthetic peptide fragments of the toxin. *Infect. Immun.* **62**, 3408–15.

Ji, G., Beavis, R.C. and Novick, R.P. (1995). Cell density control of staphylococcal virulence mediated by an octapeptide pheromone. *Proc. Natl Acad. Sci. USA* **92**, 12055–9.

Ji, G., Beavis, R. and Novick, R.P. (1997). Bacterial interference caused by autoinducing peptide variants. *Science* **276**, 2027–30.

Johns, M.B.J. and Khan, S.A. (1988). Staphylococcal enterotoxin B gene is associated with a discrete genetic element. *J. Bacteriol.* **170**, 4033–9.

Johnson, L.P., Tomai, M.A. and Schlievert, P.M. (1986). Bacteriophage involvement in group A streptococcal pyrogenic exotoxin A production. *J. Bacteriol.* **166**, 623–7.

Jones, C.L. and Khan, S.A. (1986). Nucleotide sequence of the enterotoxin B gene from *Staphylococcus aureus*. *J. Bacteriol.* **166**, 29–33.

Jupin, C., Anderson, S., Damais, C., Alouf, J.E. and Parant, M. (1988). Toxic shock syndrome toxin 1 as an inducer of human tumor necrosis factors and gamma interferon. *J. Exp. Med.* **167**, 752–61.

Kappler, J.W., Herman, A., Clements, J. and Marrack, P. (1992). Mutations defining functional regions of the superantigen staphylococcal enterotoxin B. *J. Exp. Med.* **175**, 387–96.

Kim, J., Urban, R.G., Strominger, J.L. and Wiley, D.C. (1994). Toxic shock syndrome toxin-1 complexed with a class II major histocompatibility molecule HLA-DR1. *Science* **266**, 1870–4.

Komisar, J., Rivera, J., Vega, A. and Tseng, J. (1992). Effects of staphylococcal enterotoxin B on rodent mast cells. *Infect. Immun.* **60**, 2969–75. Published erratum appears in *Infect. Immun.* **60**, 4976.

Komisar, J.L., Small-Harris, S. and Tseng, J. (1994). Localization of binding sites of staphylococcal enterotoxin B (SEB), a superantigen, for HLA-DR by inhibition with synthetic peptides of SEB. *Infect. Immun.* **62**, 4775–80.

Kornblum, J., Kreiswirth, B.N., Projan, S.J., Ross, H. and Novick, R.P. (1990). Agr: a polycistronic locus regulating exoprotein synthesis in *Staphylococcus aureus*. In: *Molecular Biology of the Staphylococci* (ed. R.P. Novick), pp. 373–402. VCH, New York.

Kreiswirth, B.N., Projan, S.J., Schlievert, P.M. and Novick, R.P. (1989). Toxic shock syndrome toxin 1 is encoded by a variable genetic element. *Rev. Infect. Dis.* **11**, Suppl. 1, S83–8.

Kushnaryov, V.M., MacDonald, H.S., Reiser, R.F. and Bergdoll, M.S. (1989). Reaction of toxic shock syndrome toxin 1 with endothelium of human umbilical cord vein. *Rev. Infect. Dis.* **11**, Suppl. 1, S282–7.

Laird, W. and Groman, N. (1976). Orientation of the *tox* gene in the prophage of corynebacteriophage beta. *J. Virol.* **19**, 228–31.

Leder, L., Llera, A., Lavoie, P.M., Lebedeva, M.I., Li, H., Sekaly, R.P., Bohach, G.A., Gahr, P.J., Schlievert, P.M., Karjalainen, K. and Mariuzza, R.A. (1998). A mutational analysis of the binding of staphylococcal enterotoxins B and C3 to the T cell receptor beta chain and major histocompatibility complex class II. *J. Exp. Med.* **187**, 823–33.

Lee, C.Y. and Iandolo, J.J. (1986). Integration of staphylococcal phage L54a occurs by site-specific recombination: structural analysis of the attachment sites. *Proc. Natl Acad. Sci. USA* **83**, 5474–8.

Lee, P.K., Vercellotti, G.M., Deringer, J.R. and Schlievert, P.M. (1991). Effects of staphylococcal toxic shock syndrome toxin 1 on aortic endothelial cells. *J. Infect. Dis.* **164**, 711–19.

Lee, P.K., Kreiswirth, B.N., Deringer, J.R., Projan, S.J., Eisner, W., Smith, B.L., Carlson, E., Novick, R.P. and Schlievert, P.M. (1992). Nucleotide sequences and biologic properties of toxic shock syndrome toxin 1 from ovine- and bovine-associated *Staphylococcus aureus*. *J. Infect. Dis.* **165**, 1056–63.

Leung, D.Y. and Schlievert, P.M. (1997). Superantigens in human disease. In: *Superantigens: Molecular Biology, Immunology and Relevance to Human Disease* (eds D.Y. Leung, B.T. Huber and P.M. Schlievert), pp. 581–601. Marcel Dekker, New York.

Lina, G., Jarraud, S., Ji, G., Greenland, T., Pedraza, A., Etienne, J., Novick, R.P. and Vandenesch, F. (1998). Transmembrane topology and histidine protein kinase activity of AgrC, the *agr* signal receptor in *Staphylococcus aureus*. *Mol. Microbiol.* **28**, 655–62.

Lincz, L.F. (1998). Deciphering the apoptotic pathway: all roads lead to death. *Immunol. Cell Biol.* **76**, 1–19.

Lindsay, J.A., Ruzin, A., Ross, H.F., Kurepina, N. and Novick, R.P. (1998). The gene for toxic shock is carried by a family of mobile pathogenicity islands in *Staphylococcus aureus*. *Mol. Microbiol.* **29**, 527–43.

Mahlknecht, U., Herter, M., Hoffmann, M.K., Niethammer, D. and Dannecker, G.E. (1996). The toxic shock syndrome toxin-1 induces anergy in human T cells *in vivo*. *Hum. Immunol.* **45**, 42–5.

Mahmood, R. and Khan, S.A. (1990). Role of upstream sequences in the expression of the staphylococcal enterotoxin B gene. *J. Biol. Chem.* **265**, 4652–6.

Malchiodi, E.L., Eisenstein, E., Fields, B.A., Ohlendorf, D.H., Schlievert, P.M., Karjalainen, K. and Mariuzza, R.A. (1995). Superantigen binding to a T cell receptor beta chain of known three-dimensional structure. *J. Exp. Med.* **182**, 1833–45.

Marr, J.C., Lyon, J.D., Roberson, J.R., Lupher, M., Davis, W.C. and Bohach, G.A. (1993). Characterization of novel type C staphylococcal enterotoxins: biological and evolutionary implications. *Infect. Immun.* **61**, 4254–62.

Marrack, P. and Kappler, J. (1990). The staphylococcal enterotoxins and their relatives. *Science* **248**, 1066.

McLeod, J.D., Walker, L.S., Patel, Y.I., Boulougouris, G. and Sansom, D.M. (1998). Activation of human T cells with superantigen (staphylococcal enterotoxin B) and CD28 confers resistance to apoptosis via CD95. *J. Immunol.* **160**, 2072–9.

Mehindate, K., Thibodeau, J., Dohlsten, M., Kalland, T., Sekaly, R.P. and Mourad, W. (1995). Cross-linking of major histocompatibility complex class II molecules by staphylococcal enterotoxin A superantigen is a requirement for inflammatory cytokine gene expression. *J. Exp. Med.* **182**, 1573–7.

Merson, M.H. (1973). The epidemiology of staphylococci foodborne disease. In: *Proceedings Staphylococci in Foods (Conference)*, pp. 20–37. Pennsylvania State University Press, University Park.

Micusan, V.V., Desrosiers, M., Gosselin, J., Mercier, G., Oth, D., Bhatti, A.R., Heremans, H. and Billiau, A. (1989). Stimulation of T cells and induction of interferon by toxic shock syndrome toxin 1. *Rev. Infect. Dis.* **11**, Suppl. 1, S305–12.

Miethke, T., Wahl, C., Heeg, K., Echtenacher, B., Krammer, P.H. and Wagner, H. (1992). T cell-mediated lethal shock triggered in mice by the superantigen staphylococcal enterotoxin B: critical role of tumor necrosis factor. *J. Exp. Med.* **175**, 91–8.

Miethke, T., Vabulas, R., Bittlingmaier, R., Heeg, K. and Wagner, H. (1996). Mechanisms of peripheral T cell deletion: anergized T cells are Fas resistant but undergo proliferation-associated apoptosis. *Eur. J. Immunol.* **26**, 1459–67.

Mollick, J.A., Chintagumpala, M., Cook, R.G. and Rich, R.R. (1991). Staphylococcal exotoxin activation of T cells. Role of exotoxin-MHC class II binding affinity and class II isotype. *J. Immunol.* **146**, 463–8.

Mollick, J.A., Miller, G.G., Musser, J.M., Cook, R.G., Grossman, D. and Rich, R.R. (1993). A novel superantigen isolated from pathogenic strains of *Streptococcus pyogenes* with amino terminal homology to staphylococcal enterotoxins B and C. *J. Clin. Invest.* **92**, 710–19.

Morfeldt, E., Taylor, D., von Gabain, A. and Arvidson, S. (1995). Activation of alpha-toxin translation in *Staphylococcus aureus* by the trans-encoded antisense RNA, RNAIII. *EMBO J.* **14**, 4569–77.

Morfeldt, E., Tegmark, K. and Arvidson, S. (1996). Transcriptional control of the *agr*-dependent virulence gene regulator, RNAIII, in *Staphylococcus aureus*. *Mol. Microbiol.* **21**, 1227–37.

Munson, S.H., Tremaine, M.T., Betley, M.J. and Welch, R.A. (1998). Identification and characterization of staphylococcal enterotoxin types G and I from *Staphylococcus aureus*. *Infect. Immun.* **66**, 3337–48.

Murray, D.L., Prasad, G.S., Earhart, C.A., Leonard, B.A., Kreiswirth, B.N., Novick, R.P., Ohlendorf, D.H. and Schlievert, P.M. (1994). Immunobiologic and biochemical properties of mutants of toxic shock syndrome toxin-1. *J. Immunol.* **152**, 87–95.

Murray, D.L., Earhart, C.A., Mitchell, D.T., Ohlendorf, D.H., Novick, R.P. and Schlievert, P.M. (1996). Localization of biologically important regions on toxic shock syndrome toxin 1. *Infect. Immun.* **64**, 371–4.

Murzin, A.G. (1993). OB(oligonucleotide/oligosaccharide binding)-fold: common structural and functional solution for non-homologous sequences. *EMBO J.* **12**, 861–7.

Nagata, S. (1997). Apoptosis by death factor. *Cell* **88**, 355–65.

Noble, A., Pestano, G.A. and Cantor, H. (1998). Suppression of immune responses by CD8 cells. I. Superantigen-activated CD8 cells induce unidirectional Fas-mediated apoptosis of antigen-activated CD4 cells. *J. Immunol.* **160**, 559–65.

Parsonnet, J. (1998). Case definition of staphylococcal TSS: a proposed revision incorporating laboratory findings. In: *European Conference on Toxic Shock Syndrome* (eds F. Arbuthnott and B. Furman), p. 15. Royal Society of Medicine, London.

Parsonnet, J. and Gillis, Z.A. (1988). Production of tumor necrosis factor by human monocytes in response to toxic-shock-syndrome toxin-1. *J. Infect. Dis.* **158**, 1026–33.

Parsonnet, J., Hickman, R.K., Eardley, D.D. and Pier, G.B. (1985). Induction of human interleukin-1 by toxic-shock-syndrome toxin-1. *J. Infect. Dis.* **151**, 514–22.

Paul, N.L. and Ruddle, N.H. (1988). Lymphotoxin. *Annu. Rev. Immunol.* **6**, 407–38.

Peng, H.L., Novick, R.P., Kreiswirth, B., Kornblum, J. and Schlievert, P. (1988). Cloning, characterization, and sequencing of an accessory gene regulator (*agr*) in *Staphylococcus aureus*. *J. Bacteriol.* **170**, 4365–72.

Prasad, G.S., Earhart, C.A., Murray, D.L., Novick, R.P., Schlievert, P.M. and Ohlendorf, D.H. (1993). Structure of toxic shock syndrome toxin 1. *Biochemistry* **32**, 13761–6.

Purdie, K., Hudson, K.R. and Fraser, J.D. (1991). Bacterial superantigens. In: *Antigen Processing and Presentation* (ed. J. MacCluskey), pp. 193–211. CRC Press, Boca Raton, FL.

Reck, B., Scheuber, P.H., Londong, W., Sailer-Kramer, B., Bartsch, K. and Hammer, D.K. (1988). Protection against the staphylococcal enterotoxin-induced intestinal disorder in the monkey by anti-idiotypic antibodies. *Proc. Natl Acad. Sci. USA* **85**, 3170–4.

Reda, K.B., Kapur, V., Mollick, J.A., Lamphear, J.G., Musser, J.M. and Rich, R.R. (1994). Molecular characterization and phylogenetic distribution of the streptococcal superantigen gene (*ssa*) from *Streptococcus pyogenes*. *Infect. Immun.* **62**, 1867–74.

Refaeli, Y., Van Parijs, L., London, C.A., Tschopp, J. and Abbas, A.K. (1998). Biochemical mechanisms of IL-2-regulated Fas-mediated T cell apoptosis. *Immunity* **8**, 615–23.

Regassa, L.B. and Betley, M.J. (1992). Alkaline pH decreases expression of the accessory gene regulator (*agr*) in *Staphylococcus aureus*. *J. Bacteriol.* **174**, 5095–100.

Regassa, L.B. and Betley, M.J. (1993). High sodium chloride concentrations inhibit staphylococcal enterotoxin C gene (*sec*) expression at the level of *sec* mRNA. *Infect. Immun.* **61**, 1581–5.

Regassa, L.B., Novick, R.P. and Betley, M.J. (1992). Glucose and nonmaintained pH decrease expression of the accessory gene regulator (*agr*) in *Staphylococcus aureus*. *Infect. Immun.* **60**, 3381–8.

Reingold, A.L., Hargrett, N.T., Shands, K.N., Dan, B.B., Schmid, G.P., Strickland, B.Y. and Broome, C.V. (1982). Toxic shock syndrome surveillance in the United States, 1980 to 1981. *Ann. Intern. Med.* **96**, 875–80.

Ren, K., Bannan, J.D., Pancholi, V., Cheung, A.L., Robbins, J.C., Fischetti, V.A. and Zabriskie, J.B. (1994). Characterization and biological properties of a new staphylococcal exotoxin. *J. Exp. Med.* **180**, 1675–83.

Rich, R.R., Mollick, J.A. and Cook, R.G. (1989). Superantigens: interaction of staphylococcal enterotoxins with MHC class II molecules. *Trans. Am. Clin. Climatol. Assoc.* **101**, 195–204.

Schad, E.M., Zaitseva, I., Zaitsev, V.N., Dohlsten, M., Kalland, T., Schlievert, P.M., Ohlendorf, D.H. and Svensson, L.A. (1995). Crystal structure of the superantigen staphylococcal enterotoxin type A. *EMBO J.* **14**, 3292–301.

Scheuber, P.H., Denzlinger, C., Wilker, D., Beck, G., Keppler, D. and Hammer, D.K. (1987a). Cysteinyl leukotrienes as mediators of staphylococcal enterotoxin B in the monkey. *Eur. J. Clin. Invest.* **17**, 455–9.

Scheuber, P.H., Denzlinger, C., Wilker, D., Beck, G., Keppler, D. and Hammer, D.K. (1987b). Staphylococcal enterotoxin B as a non-immunological mast cell stimulus in primates: the role of endogenous cysteinyl leukotrienes. *Int. Arch. Allergy Appl. Immunol.* **82**, 289–91.

Schlievert, P. (1998). Incidence studies of toxic shock syndrome. In: *European Conference on Toxic Shock Syndrome* (eds F. Arbuthnott and B. Furman), pp. 34–6. Royal Society of Medicine, London.

Schlievert, P.M. (1982). Enhancement of host susceptibility to lethal endotoxin shock by staphylococcal pyrogenic exotoxin type C. *Infect. Immun.* **36**, 123-8.

Schlievert, P.M. and Blomster, D.A. (1983). Production of staphylococcal pyrogenic exotoxin type C: influence of physical and chemical factors. *J. Infect. Dis.* **147**, 236–42.

Schlievert, P.M. and Watson, D.W. (1978). Group A streptococcal pyrogenic exotoxin: pyrogenicity, alteration of blood-brain barrier, and separation of sites for pyrogenicity and enhancement of lethal endotoxin shock. *Infect. Immun.* **21**, 753–63.

Schlievert, P.M., Bettin, K.M. and Watson, D.W. (1980). Inhibition of ribonucleic acid synthesis by group A streptococcal pyrogenic exotoxin. *Infect. Immun.* **27**, 542–8.

Schlievert, P.M., Shands, K.N., Dan, B.B., Schmid, G.P. and Nishimura, R.D. (1981). Identification and characterization of an exotoxin from *Staphylococcus aureus* associated with toxic-shock syndrome. *J. Infect. Dis.* **143**, 509–16.

Schlievert, P.M., Bohach, G.A., Hovde, C.J., Kreiswirth, B.N. and Novick, R.P. (1990). Molecular studies of toxic shock syndrome-associated staphylococcal and streptococcal toxins. In: *Molecular Biology of the Staphylococci* (ed. R.P. Novick), pp. 311–25. VCH, New York.

Schlievert, P.M., Gocke, J.E. and Deringer, J.R. (1993). Group B streptococcal toxic shock-like syndrome: report of a case and purification of an associated pyrogenic toxin. *Clin. Infect. Dis.* **17**, 26–31.

Seth, A., Stern, L.J., Ottenhoff, T.H., Engel, I., Owen, M.J., Lamb, J.R., Klausner, R.D. and Wiley, D.C. (1994). Binary and ternary complexes between T-cell receptor, class II MHC and superantigen *in vitro*. *Nature* **369**, 324–7.

Soos, J.M. and Johnson, H.M. (1994). Multiple binding sites on the superantigen, staphylococcal enterotoxin B, imparts versatility in binding to MHC class II molecules. *Biochem. Biophys. Res. Commun.* **201**, 596–602.

Stauffacher, C., Chi, Y., Deobald, C. and Bohach, G. (1998). The structure and function of relationships revealed by mutagenesis in MHCII binding face of staphylococcal enterotoxin C3. In: *European Conference on Toxic Shock Syndrome* (eds F. Arbuthnott and B. Furman), p. 92. Royal Society of Medicine, London.

Stevens, D.L., Tanner, M.H., Winship, J., Swarts, R., Ries, K.M., Schlievert, P.M. and Kaplan, E. (1989). Severe group A streptococcal infections associated with a toxic shock- like syndrome and scarlet fever toxin A. *N. Engl. J. Med.* **321**, 1–7.

Stevens, F.A. (1927). The occurrence of *Staphylococcus aureus* infection with a scarlatiniform rash. *JAMA* **88**, 1958.

Stohl, W., Elliott, J.E., Lynch, D.H. and Kiener, P.A. (1998). CD95 (Fas)-based, superantigen-dependent, CD4$^+$ T cell-mediated down-regulation of human *in vitro* immunoglobulin responses. *J. Immunol.* **160**, 5231–8.

Sugiyama, H.T. and Hayama, T. (1965). Abdominal viscera as site of emetic action for staphylococcal enterotoxin in the monkey. *J. Infect. Dis.* **115**, 330–6.

Sundstrom, M., Abrahmsen, L., Antonsson, P., Mehindate, K., Mourad, W. and Dohlsten, M. (1996). The crystal structure of staphylococcal enterotoxin type D reveals Zn^{2+}-mediated homodimerization. *EMBO J.* **15**, 6832–40.

Svensson, L.A., Schad, E.M., Sundstrom, M., Antonsson, P., Kalland, T. and Dohlsten, M. (1997). Staphylococcal enterotoxins A, D, and E. Structure and function, including mechanism of T-cell superantigenicity. *Prep. Biochem. Biotechnol.* **27**, 111–41.

Swaminathan, S., Furey, W., Pletcher, J. and Sax, M. (1992). Crystal structure of staphylococcal enterotoxin B, a superantigen. *Nature* **359**, 801–6.

Swofford, D.L. (1993). *PAUP: Phylogenetic Analysis Using Parsimony*, Version 3.1. Computer program distributed by the Illinois Natural History Survey, Champaign, IL.

Thibodeau, J., Cloutier, I., Lavoie, P.M., Labrecque, N., Mourad, W., Jardetzky, T. and Sekaly, R.P. (1994). Subsets of HLA-DR1 molecules defined by SEB and TSST-1 binding. *Science* **266**, 1874–8.

Thibodeau, J., Dohlsten, M., Cloutier, I., Lavoie, P.M., Bjork, P., Michel, F., Leveille, C., Mourad, W., Kalland, T. and Sekaly, R.P. (1997). Molecular characterization and role in T cell activation of staphylococcal enterotoxin A binding to the HLA-DR alpha-chain. *J. Immunol.* **158**, 3698–704.

Tiedemann, R.E. and Fraser, J.D. (1996). Cross-linking of MHC class II molecules by staphylococcal enterotoxin A is essential for antigen-presenting cell and T cell activation. *J. Immunol.* **157**, 3958–66.

Tiedemann, R.E., Urban, R.J., Strominger, J.L. and Fraser, J.D. (1995). Isolation of HLA-DR1 (staphylococcal enterotoxin A)2 trimers in solution. *Proc. Natl Acad. Sci. USA* **92**, 12156–9.

Todd, J. and Fishaut, M. (1978). Toxic-shock syndrome associated with phage-group-I Staphylococci. *Lancet* **ii**, 1116–18.

Trede, N.S., Geha, R.S. and Chatila, T. (1991). Transcriptional activation of IL-1 beta and tumor necrosis factor-alpha genes by MHC class II ligands. *J. Immunol.* **146**, 2310–15.

Tremaine, M.T., Brockman, D.K. and Betley, M.J. (1993). Staphylococcal enterotoxin A gene (sea) expression is not affected by the accessory gene regulator (agr). *Infect. Immun.* **61**, 356–9.

Turner, T.N., Smith, C.L. and Bohach, G.A. (1992). Residues 20, 22, and 26 determine the subtype specificities of staphylococcal enterotoxins C1 and C2. *Infect. Immun.* **60**, 694–7.

Uchiyama, T., Kamagata, Y., Wakai, M., Yoshioka, M., Fujikawa, H. and Igarashi, H. (1986). Study of the biological activities of toxic shock syndrome toxin-1. I. Proliferative response and interleukin 2 production by T cells stimulated with the toxin. *Microbiol. Immunol.* **30**, 469–83.

Van den Bussche, R.A., Lyon, J.D. and Bohach, G.A. (1993). Molecular evolution of the staphylococcal and streptococcal pyrogenic toxin gene family. *Mol. Phylogenet. Evol.* **2**, 281–92.

Weeks, C.R. and Ferretti, J.J. (1984). The gene for type A streptococcal exotoxin (erythrogenic toxin) is located in bacteriophage T12. *Infect. Immun.* **46**, 531–6.

White, J., Herman, A., Pullen, A.M., Kubo, R., Kappler, J.W. and Marrack, P. (1989). The V beta-specific superantigen staphylococcal enterotoxin B: stimulation of mature T cells and clonal deletion in neonatal mice. *Cell* **56**, 27–35.

Whiting, J.L., Rosten, P.M. and Chow, A.W. (1989). Determination by western blot (immunoblot) of seroconversions to toxic shock syndrome (TSS) toxin 1 and enterotoxin A, B, or C during infection with TSS- and non-TSS-associated *Staphylococcus aureus*. *Infect. Immun.* **57**, 231–4.

34

Yersinia pseudotuberculosis superantigenic toxins

Christophe Carnoy and Michel Simonet

INTRODUCTION

Superantigens are toxins produced by various pathogenic bacteria and viruses, and elicit powerful immune responses by an unconventional mechanism. Superantigens are presented to T-cells by direct binding to major histocompatibility (MHC) molecules present on antigen-presenting cell (APC) surfaces, without MHC restriction, and are specifically recognized by the variable region Vβ of T-cell receptors (TCR) (Marrack and Kappler, 1990). As a consequence of this interaction, T-cells and APCs release large amounts of inflammatory cytokines which can cause shock and tissue damage (Kotb, 1995).

Gram-positive *Staphylococcus aureus* and *Streptococcus pyogenes* are considered major sources of bacterial superantigens (see Chapters 24 and 32 of this book) but recently a mitogen with superantigenic features has been characterized in the Gram-negative bacillus *Yersinia pseudotuberculosis*.

YERSINIA PSEUDOTUBERCULOSIS, A PATHOGEN RESPONSIBLE FOR A BROAD SPECTRUM OF CLINICAL MANIFESTATIONS IN HUMANS

Malassez and Vignal (1883) first described *Y. pseudotuberculosis* as a causative agent of tuberculosis-like (pseudotuberculosis) lesions in guinea pigs. Seventy years later, Knapp (1958) isolated the bacterium from enlarged ileocaecal lymph nodes of two German children who had had surgery for suspected appendicitis. There have been many subsequent reports of human infections with *Y. pseudotuberculosis* from Europe, North America, Japan and eastern Russia. Except for a few outbreaks which occurred in Finland, eastern Russia and Japan (Somov and Martinevsky, 1973; Inoue *et al.*, 1984, 1988; Tertti *et al.*, 1984, 1989; Nakano *et al.*, 1989; Pebody, 1997), most cases are sporadic. In Eurasia and North America, *Y. pseudotuberculosis* is enzootic in various mammal and bird species which may harbour this micro-organism as healthy carriers (Fukushima and Gomyoda, 1991).

Human infection is most probably established by ingestion of food or water contaminated with the excreta of infected animals. It is noteworthy that asymptomatic infection is not uncommon in humans. In normal hosts, infection is self-limited and causes relatively mild illness. The most frequent clinical symptoms are fever and abdominal pain rather than diarrhoea and vomiting usually associated with gastrointestinal infection. Abdominal pain in the right fossa mimics acute appendicitis (pseudo-appendicular syndrome) leading to surgery but per-operatively the appendix appears normal, although considerable mesenteric adenitis and/or acute terminal ileitis are observed. Several cases of Crohn's ileitis following *Y. pseudotuberculosis* infection have also been reported in the literature, but whether this bacterium triggers this inflammatory bowel disease is questionable (Delchier *et al.*, 1983; Blaser *et al.*, 1984; Treacher and Jewell, 1985). In debilitated patients, especially those with cirrhosis due to alcoholism or haemochromatosis as well as other liver diseases, micro-organisms have the propensity to spread from the digestive tract to deep organs via blood, and septicaemia is frequently fatal (Butler, 1983).

A typical feature of *Y. pseudotuberculosis* infection is the common occurrence of post-infection complications such as reactive arthritis and erythema nodosum (Butler, 1983; Tertti *et al.*, 1984, 1989). Aseptic reactive arthritis consists of mononuclear cell infiltration of synovial fluid and membrane. Most patients (75–80%) have the MHC class-I gene HLA-B27. The delay between onset of digestive symptoms and onset of arthritis ranges from days to months but is typically less than 3 weeks. An average of three joints is commonly affected, mainly knees, ankles and metatarsophalangeal joints. Urethritis and conjunctivitis (Reiter's syndrome) may be associated. Commonly, arthritis lasts for 6 months and the clinical course of illness varies between an acute self-limiting form in most cases and intermittent relapsing or chronic arthritis in a limited number of cases (Simonet, 1999). Erythema nodosum, which is sometimes associated with reactive arthritis, is characterized by typical aseptic and inflammatory nodules on the anterior aspect of legs of patients. In contrast with reactive arthritis, erythema nodosum is not linked to the presence of the HLA-B27 antigen in patients.

In Japan, where pseudotuberculosis is often epidemic, infection in children frequently associates fever, conjunctival injection, a biphasic erythematous rash, lip changes (fissure, redness or crust), induration of hands or feet, erythema of palms or soles, desquamation of peripheral extremities, strawberry tongue, inflamed oral mucosa, enlarged lymph nodes, erythema nodosum and arthritis (Sato *et al.*, 1983). These clinical manifestations, rarely reported in European countries, resemble Kawasaki syndrome and Far East scarlet fever disease reported in eastern Russia during the 1960s (Somov and Martinevsky, 1973). It has been suggested that *Y. pseudotuberculosis* might be a possible but non-exclusive aetiological agent of these illnesses.

Finally, approximately 10% of the infected patients develop renal complications. The most frequent one (80% of cases) is acute renal failure probably due to tubulointerstitial nephritis, but acute nephritic syndrome, IgA nephropathy and haemolytic uraemic syndrome have also been described (Takeda *et al.*, 1991).

YERSINIA PSEUDOTUBERCULOSIS PRODUCES VARIOUS VIRULENCE FACTORS

Once ingested, *Y. pseudotuberculosis* reaches the small intestine, binds to intestinal mucus and subsequently crosses the gut mucosa. As for other entero-invasive pathogens, intestinal translocation occurs primarily through M cells, a cell type found in the epithelial lining of the Peyer's patches which are part of the gut-associated lymphoid tissue (Fujimura *et al.*, 1992; Marra and Isberg, 1997). Electron microscopic studies of intestine sections of experimentally infected rodents showed that *Y. pseudotuberculosis* binds preferentially to the apical surface of M cells and not to enterocytes (Marra and Isberg, 1997). *Yersinia* infection causes a recruitment of phagocytes in Peyer's patches and subsequently the formation of abscesses and ultimately lymphoid follicle destruction. Then, micro-organisms migrate via lymphatic vessels to the mesenteric lymph nodes where they proliferate causing adenitis. In lymphoid tissues, *Y. pseudotuberculosis* organisms are predominantly, if not exclusively, extracellular (Simonet *et al.*, 1990).

Genes encoding for bacterial factors involved in *Y. pseudotuberculosis* pathogenesis are located on the chromosome and on a 70-kb plasmid, called pYV and present in all virulent strains (Brubaker, 1991). *Yersinia pseudotuberculosis* makes two products that are important for colonization and entry into the intestinal mucosa: invasin and YadA. Invasin is a surface-exposed protein encoded by a chromosomal gene (*inv*), that allows both attachment to and internalization into cultured epithelial cells (Isberg *et al.*, 1987). *In vitro*, invasin attaches to $\beta 1$ integrins which are expressed in the intestinal epithelia layer only by M cells (Isberg and Leong, 1990; Clark *et al.*, 1998). YadA, an outer membrane fibrillar structure encoded by *yadA* present on pYV, links to mucus as well as to collagens, fibronectin and laminin and, like invasin, binds to $\beta 1$ integrins (Emödy *et al.*, 1989; Paerregaard *et al.*, 1991; Schulze-Koops *et al.*, 1992; Tertti *et al.*, 1992; Bliska *et al.*, 1993). Moreover, YadA also mediates *in vitro* uptake of *Y. pseudotuberculosis* by cultured epithelial cells but to a lesser extent than invasin (Bliska *et al.*, 1993). *In vivo*, it has been clearly established by testing an *inv* or *yadA* mutant in a murine model of infection that, unlike YadA, invasin is required for efficient bacterial translocation from the intestinal epithelium surface to the submucosa (Simonet and Falkow, 1992; Marra and Isberg, 1997). By contrast, the fimbrial adhesin called pH6 antigen, encoded by the chromosomal *psaA* gene, contributes to *Y. pseudotuberculosis* adhesion to cultured epithelial cells but is not required for the bacterial translocation (Yang *et al.*, 1996; Marra and Isberg, 1997).

In addition to YadA, plasmid pYV governs the massive release of a dozen proteins called *Yersinia* outer proteins (Yops). The export pathway of these exotoxins involves 22 proteins [Yop secretion proteins (Ysc)] which constitutes a type-III or contact-dependent secretion system (Cornelis and Wolf-Watz, 1997). YopH, YopO, YopM, YopE and, probably, YopJ are injected directly inside the cytosol of macrophages by translocation of these molecules through the eukaryotic cell plasma membrane. This translocation requires an invasin and YadA-dependent contact between the bacterium and the host cell, and two translocators, YopD and YopB, the latter forming a pore in the eukaryotic cell plasma membrane (Cornelis and Wolf-Watz, 1997). Some of the Yops are essential effectors in bacterial virulence and allow *Y. pseudotuberculosis* to escape the infected host's defence mechanisms, contributing to extracellular location of the pathogen:

(a) YopH is a protein tyrosine phosphatase acting on tyrosine-phosphorylated proteins of macrophages contributing to the inhibition of bacterial uptake and oxidative burst in these cells (Bliska *et al.*, 1991; Bliska and Black, 1995)

(b) YopO (previously YpkA) is a serine/threonine kinase which probably interferes with some signal transduction pathway of the eukaryotic cell (Galyov *et al.*, 1993)

(c) YopE is a cytotoxin that causes disruption of actin microfilaments of the cytoskeleton, possibly by modification of small G-proteins (Rosqvist *et al.*, 1990)

(d) YopM binds human α-thrombin and competes with human platelets for thrombin, and potentially interferes with platelet-mediated events of the inflammatory response (Leung *et al.*, 1990)

(e) YopJ induces macrophage apoptosis by interfering with cell death inhibiting signals, and blocks macrophage TNF-α production (Monack *et al.*, 1997; Palmer *et al.*, 1998; Schesser *et al.*, 1998).

YERSINIA PSEUDOTUBERCULOSIS PRODUCES SUPERANTIGENS

The typical post-infection immunopathological complications which may occur after infection with *Y. pseudotuberculosis* (reactive arthritis, erythema nodosum, Kawasaki syndrome, etc.) suggested the involvement of a superantigen-like molecule. In 1993, the production of a mitogenic activity by *Y. pseudotuberculosis* was simultaneously described from *Y. pseudotuberculosis* strains involved in a mass outbreak in Japan (Uchiyama *et al.*, 1993) and from a strain isolated from a patient manifesting Kawasaki-like symptoms (Abe *et al.*, 1993; Yoshino *et al.*, 1994). The substance, purified from a lysate of *Y. pseudotuberculosis* and exhibiting a mitogenic activity on human peripheral blood mononuclear cells (PBMC), was first designated YPM for *Y. pseudotuberculosis*-derived mitogen (Miyoshi-Akiyama *et al.*, 1993), and later YPMa after the discovery of a variant (Ramamurthy *et al.*, 1997).

The characterization of the biological activity of YPMa revealed that it induces interleukin-2 (IL-2) production from PBMC but not from T-cell-depleted PBMC, and expands T-lymphocytes bearing Vβ3, Vβ9, Vβ13.1 and Vβ13.2 (Abe *et al.*, 1993; Uchiyama *et al.*, 1993). YPMa has the ability to stimulate both CD4$^+$ and CD8$^+$ T-cells with a predominant proliferation of CD4$^+$ lymphocytes (Ito *et al.*, 1995). Furthermore, YPMa also requires the presence of MHC class-II molecules to induce T-cell proliferation. This is demonstrated by the observation that YPMa stimulates T-cells in the presence of fibroblasts transfected with HLA-DP, HLA-DQ or HLA-DR class-II molecules, although the degree of expansion was higher in presence of HLA-DR (DR4 subtype). Lymphocyte stimulation by YPMa can be inhibited by monoclonal antibodies directed to HLA-DR, and paraformaldehyde fixation of fibroblasts transfected with HLA-DR does not alter the stimulation properties of YPMa, suggesting that it binds directly to HLA class-II molecules without being processed (Abe *et al.*, 1993; Uchiyama *et al.*, 1993). Together, these data demonstrate that the mitogen YPMa produced by *Y. pseudotuberculosis* displays all the criteria of a superantigenic toxin as proposed by Marrack and Kappler (1990).

YPMa activity is not limited to human cells, as YPMa is also able to induce proliferation of murine T-cells in an MHC class-II-dependent manner and to stimulate IL-2 production of murine C57BL/6 splenic T-cells bearing the Vβ7, Vβ8 and, possibly, Vβ5 valuable regions (Miyoshi-Akiyama *et al.*, 1997).

YPMa, which is released in the supernatant of a culture of *Y. pseudotuberculosis*, was first described as a 21-kDa protein (Uchiyama *et al.*, 1993), but in fact has a relative molecular mass of 14 500, as demonstrated by electrospray ionization mass spectrometry (Yoshino *et al.*, 1994; Abe and Takeda, 1997). YPMa is a small superantigen when compared with the known bacterial superantigens which range from 20 to 30 kDa (Kotb, 1995). Cloning and sequencing of the YPMa gene revealed a 456-base pair (bp) open reading frame (ORF) designated *ypmA* (Figure 34.1) and encoding a 151 amino acid precursor protein which is processed into

FIGURE 34.1 Complete nucleotide sequences of *ypmA*, *ypmB* and *ypmC*. The −10 and −35 regions (solid lines above the nucleotide sequence) correspond to the promoter of *ypmA* according to Ito *et al.* (1995), whereas −10′ and −35′ positions (underlined nucleotides) correspond to the promoter of *ypmA* described by Miyoshi-Akiyama *et al.* (1995). The Shine–Dalgarno sequence (SD) and the stop codon are underlined and indicated by bold characters. Dashes (−) indicate identical nucleotides compared to *ypmA*, whereas dots (·) correspond to deletions of nucleotides.

```
YPMa   M K N K L L S L L L T F T L F S G V A L A T D Y D N T L N S I P S L R I P N I A T     40
YPMb   - - K - F - - - - - L - F - - - L - - - A - - - - - - - - - - - - - - - - - - - E -
YPMc   - - - - - - - - - - - - - - - - - - - - - - - - - - - - - - - - - - - - - - - - - -

YPMa   Y T G T I Q G K G E V C I I G N K E G K T R G G E L Y A V L H S T N V N A D M T     80
YPMb   - - - - - - - - - - - - R - - - - - S - - - - - - - - - R - - - A - - - - -
YPMc   - - - - - - - - - - - - - - - - - - - - - - - - - - - Y - - - - - - - -

YPMa   L I L L R N V G G N G W G E I K R N D I D K P L K Y E D Y Y T S G . L S W I W K     119
YPMb   - - - - C S I R . D - - K - V - - S - - - R - - R - - - - - - P - A - - - - E
YPMc   - - - - - - - - - - - - - - - - - - - - - - - - - - - - - - - - - - - - - - -

YPMa   I K N N S S E T S N Y S L D A T V H D D K E D S D V L T K C P V     151
YPMb   - - - - - - - A - D - - - S - - - - - - - - - - - - - - - M - - -
YPMc   - - - - - - - - - - - - - - - - - - - - - - - - - - - - - - - -
```

FIGURE 34.2 Deduced amino acid sequences of YPMa, YPMb and YPMc. Amino acids are represented by a single-letter code. The signal sequence is underlined. Identical amino acids compared to YPMa are indicated with dashes (–) and dots (·) correspond to deletions.

a mature form of 131 amino acids after proteolytic cleavage of a 20 amino acid hydrophobic signal sequence (Figure 34.2) (Ito *et al.*, 1995; Miyoshi-Akiyama *et al.*, 1995). The presence of a signal peptide suggests that YPMa is secreted via the general secretory pathway found in most Gram-negative bacteria. A p*I* value for YPMa was calculated from the deduced amino acid sequence as 4.95 (Ito *et al.*, 1995). A significant proliferative response of PBMC by purified or recombinant YPMa was detectable at concentrations as low as 1 pg ml^{-1}, a value comparable to the known bacterial superantigenic toxins (Ito *et al.*, 1995).

Even if the translation start of the *ypmA* gene has been precisely located, the promoter region is still uncharacterized. Indeed, there is a discrepancy between authors on the location of the −10 and −35 promoter regions (Figure 34.1) (Ito *et al.*, 1995; Miyoshi-Akiyama *et al.*, 1995). A primer extension experiment will be necessary to position clearly the promoter of *ypmA*. A mRNA transcript around 500 bases was detected after Northern blot with a *ypmA*-specific probe, demonstrating the monocistronic organization of *ypmA* and eliminating possible organization as an operon (personal communication).

YERSINIA PSEUDOTUBERCULOSIS EXPRESSES SEVERAL VARIANTS OF YPMa

Among strains of *Y. pseudotuberculosis* tested for the presence of *ypmA*, Ramamurthy *et al.* (1997) found that 20% of their isolates expressing a mitogenic activity failed to hybridize with a specific probe for *ypmA*. This new mitogen produced by *Y. pseudotuberculosis* was characterized and designated YPMb. The correspond-ing gene, *ypmB*, which displays 88.9% homology with *ypmA*, encodes a 150 amino acid protein with 83% homology with YPMa. YPMb, which is one amino acid shorter than YPMa, has the lowest amino acid homology with YPMa in the central region between amino acid 65 and 86 of the mature proteins (Figure 34.2). YPMa has two cysteine residues (position 32 and 129 of mature protein), whereas YPMb has an additional residue at position 66. All of these differences do not alter the TCR recognition because YPMb, like YPMa, stimulates T-cells bearing Vβ3, Vβ9, Vβ13.1 and Vβ13.2, suggesting that Vβ recognition could be attributed to common amino acids between YPMa and YPMb (Takeda *et al.*, 1996). Analysis of nucleotide sequences upstream and downstream of *ypmA* and *ypmB* ORFs revealed a much lower homology for the downstream region than for the upstream region (Figure 34.1). This may indicate that *ypmA* and *ypmB* have a distinct genetic environment. Genes flanking *ypmA* or *ypmB* are so far uncharacterized.

In a study performed in their laboratory, the authors screened for the presence of the *ypmA* gene among 118 strains obtained from various strain collections. Thirty strains (25%) were positive by colony hybridization under mild stringency conditions with a probe specific to *ypmA*. To evaluate the polymorphism of *ypm* genes, the authors amplified by polymerase chain reaction (PCR) with oligonucleotides sup1 and sup2 (Figure 34.1) and sequenced a 418-bp internal region of the superantigen encoding genes. Out of 29 sequenced genes, 11 were identical to *ypmA*, four were identical to *ypmB* and 14 had one nucleotide difference with *ypmA* which turned the histidine residue at position 51 into a tyrosine residue. This last gene was designated *ypmC* (Figure 34.1). The determination of its Vβ specificity is in progress in the laboratory.

ARE OTHER *YERSINIA* SPECIES SUPERANTIGEN-PRODUCING MICRO-ORGANISMS?

In addition to *Y. pseudotuberculosis*, two other *Yersinia* species cause disease in humans: *Y. enterocolitica*, which causes diarrhoea and abdominal pain, and *Y. pestis*, the bacterial agent responsible for the plague. *Yersinia enterocolitica* was in fact the first *Yersinia* described as a mitogen-producing bacterium. Early in 1992, Stuart and Woodward (1992) found a superantigenic substance from lysate of *Y. enterocolitica* (serotype O:8) able to stimulate BALB/c murine T-cells bearing Vβ3, Vβ6 and Vβ11, and possibly Vβ7 and Vβ9. This protease-sensitive substance, which can be membrane associated, is also active on human T-cells as it is able to stimulate human lymphocytes bearing Vβ3, Vβ12, Vβ14 and Vβ17 in an MHC class-II-dependent manner (Stuart *et al.*, 1995). To date, the mitogen produced by *Y. enterocolitica* has been partially purified, but a DNA sequence is not available yet to ascertain whether there are any homologies between the mitogen produced by *Y. enterocolitica* and the superantigens of *Y. pseudotuberculosis*. Nevertheless, several arguments suggest that the two superantigens are distinct. First, the Vβ specificity found for the mitogenic activity produced by *Y. enterocolitica* is different from the T-cell specificities of YPMa. Secondly, Yoshino *et al.* (1995) did not detect the presence of the *ypmA* gene among 225 strains of *Y. enterocolitica* tested. Finally, the present authors tested the strain of *Y. enterocolitica* described as mitogen-producing and did not detect any *ypmA*-specific signal after low-stringency hybridization on genomic DNA (personal communication).

Although DNA–DNA hybridization indicated 90% homology between the genome of *Y. pseudotuberculosis* and *Y. pestis*, the question of production of a mitogenic activity by *Y. pestis* remains unanswered. Miyoshi-Akiyama *et al.* (1997) could not detect the presence of YPMa in several strains of *Y. pestis*, but a more complete study is needed before concluding that a superantigen is absent in *Y. pestis*. However, there are no arguments from the pathophysiology of *Y. pestis* to suggest the presence of a superantigen (Perry and Fetherston, 1997).

YERSINIA PSEUDOTUBERCULOSIS SUPERANTIGENS SHOW NO SIGNIFICANT HOMOLOGY WITH GRAM-POSITIVE SUPERANTIGENS

Before the discovery of YPMs in *Y. pseudotuberculosis*, it was thought that bacterial superantigens were exclusively expressed by Gram-positive bacteria. Except for *Yersinia*, Gram-negative bacteria producing mitogen activity are rather uncommon. Gram-negative *Pseudomonas aeruginosa* produces exotoxin A, which displays the same biological mechanism as diphtheria toxin and was described as a mitogen inducing the proliferation of Vβ8 murine T-cells (Legaard *et al.*, 1991, 1992). Although exotoxin A was referred to as a superantigen, it does not fulfil the criteria for a superantigenic toxin as proposed by Marrack and Kappler (1990) as exotoxin A requires processing by APC proteases prior to the expression of mitogen activity. Exoenzyme S, a virulence factor of *P. aeruginosa* with an ADP-ribosylating activity, also has mitogenic activity for human T-cells but the absence of Vβ-dependent T-cell activation clearly demonstrates that exoenzyme S is a mito-

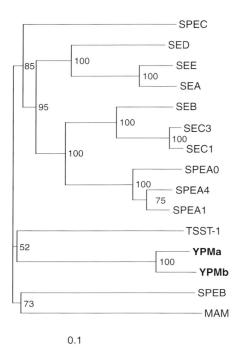

0.1
Fixed mutation per position

FIGURE 34.3 Phylogenetic analysis of amino acid sequences of various bacterial superantigenic toxins. Sequences were aligned according to the neighbour-joining method (Saitou and Nei, 1987). The numbers adjacent to nodes represent the proportion of 1000 bootstrap trees. Evolutionary distance is read only on the horizontal component of the tree. Amino acid sequences were obtained from the following accession numbers: MAM, U33151 (Cole *et al,.* 1996); SEA, M18970 (Betley and Mekalanos, 1988); SEB, M11118 (Ranelli *et al.*, 1985); SEC1, X05815 (Bohach and Schlievert, 1987); SEC3, X51661 (Hovde *et al.*, 1990); SED, M28521 (Bayles and Iandolo, 1989); SEE, M21319 (Couch *et al.*, 1988); SPEA0, X03929 (Johnson *et al.*, 1986); SPEA1 and SPEA4, X61560 and X61573, respectively (Nelson *et al.*, 1991); SPEB, L26150 (Kapur *et al.*, 1993); SPEC, M33514 (Goshorn and Schlievert 1988); TSST-1, J02615 (Blomster-Hautamaa *et al.*, 1986); YPMa, D38523 (Ito *et al.*, 1995) and D38638 (Miyoshi-Akiyama *et al.*, 1995); YPMb, D88144 (Ramamurthy *et al.*, 1997).

gen rather than a superantigen (*Bruno* et al., 1998). To date, it seems that *Y. pseudotuberculosis* is the only Gram-negative bacterium producing a superantigen *sensu stricto*.

Although YPMs share some of the TCR Vβ specificity with other staphylococcal superantigens such as enterotoxins SEB and SEC, there is no significant homology at the amino acid level between YPMa or YPMb and the superantigens described so far, especially SEB and SEC (*Ranelli* et al., 1985; Bohach and Schlievert, 1987; Ito *et al.*, 1995; Miyoshi-Akiyama *et al.*, 1995). A phylogenetic analysis based on protein sequence identity of various superantigens clearly demonstrates that YPMs diverge from all Gram-positive superantigens and from the superantigen MAM (*Mycoplasma arthritidis* mitogen) produced by *Mycoplasma arthritidis*, a pathogen involved in arthritis in mice (*Cole et al.*, 1996) (Figure 34.3). Although staphylococcal toxic shock syndrome toxin-1 (TSST-1) and YPMs are located on the same branch of the phylogenetic tree, they only share 23% identity at the amino acid level, confirming that YPMs belong to a new type of bacterial superantigen (Figure 34.3).

SUPERANTIGENS ARE NOT PRODUCED BY ALL STRAINS OF *YERSINIA PSEUDOTUBERCULOSIS*

Strains of *Y. pseudotuberculosis* containing YPM encoding genes are not human restricted as many strains are isolated from various animal and environmental sources (Yoshino *et al.*, 1995). In our study, two-thirds of the *ypm*-containing strains were recovered from non-human sources and were found in a wide range of hosts (rodents, livestock, etc.) (Figure 34.4). It is unknown whether these strains are pathogens in their hosts.

Although the *ypmA* gene is often associated with strains from the Far East, it has also been found in a few European strains. Yoshino *et al.* (1995) detected *ypmA* in all strains from eastern Russia, in 95% of the clinical isolates from Japan but in only 17% of European clinical isolates. Furthermore, clinical manifestations due to *Y. pseudotuberculosis* infections are frequently more severe in the Far East than in Europe. It seems that there is a correlation between these clinical manifestations

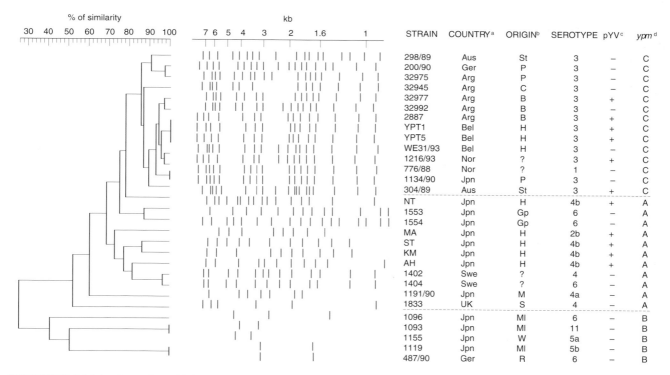

FIGURE 34.4 Molecular typing with insertion sequence IS*1541* of superantigen expressing *Y. pseudotuberculosis*. Genomic DNA was digested with *Hinc*II, and DNA fragments were separated by electrophoresis and transferred on to a Nylon membrane. DNA was hybridized with a digoxigenin-labelled IS*1541*-specific probe. The dendrogram, based on Dice similarity index, was established using GelCompar® software. (a) Arg, Argentina; Aus, Australia; Bel, Belgium; Ger, Germany; Jpn, Japan; Nor, Norway; Swe, Sweden; UK, United Kingdom. (b) B, Bovine; C, calf; Gp, guinea pig; H, human; M, mouse; MI, mole; P, pig; R, rabbit; S, sheep; St, stag; W, water. (c) The presence of the virulence plasmid pYV was PCR-detected with *yopH*-specific primers. (d) A, *ypmA*; B, *ypmB*; C, *ypmC*.

(Far East scarlet fever, Izumi fever, Kawasaki syndrome) and the presence of superantigen genes, but so far there is no direct evidence of the involvement of a superantigen in these pathologies.

A relationship between the O-serotype and the presence of a superantigen has also been suggested to explain the prevalence of superantigen-producing strains in the Far East. Sixteen serotypes have been described for *Y. pseudotuberculosis* but the geographical distribution of these serotypes is rather heterogeneous. Human isolates from Europe and North America belong mostly to serotype 1, whereas those from the Far East belong to serotypes 4 and 5 (Tsubokura *et al.*, 1989). Yoshino *et al.* (1995), who tested the presence of *ypmA* among 204 strains, found that strains from the Far East containing *ypmA* belonged to 10 different serotypes, whereas serotype 3 was predominant among strains producing YPMs isolated from Europe. From these data it appears that there is not a strict association between the O-serotype of *Y. pseudotuberculosis* and the presence of superantigen encoding genes.

An epidemiological study of superantigen-producing *Y. pseudotuberculosis* collected from different Japanese and European laboratories was performed by molecular typing with insertion sequence IS*1541*, a molecular method used to discriminate pathogenic *Y. pseudotuberculosis* strains (Odaert *et al.*, 1996). This study, presented in Figure 34.4, revealed that superantigen-producing strains could be separated into three clusters of strains. The first group, which seems anterior to the other two, contained *ypmB*-expressing strains all originating from Japan except for one from Germany (strain 487/90). A second cluster of superantigen-expressing *Y. pseudotuberculosis* contained the *ypmA* gene, whereas the third group contained *ypmC*. Interestingly, all strains but one from this last group belonged to serotype 3. This epidemiological study seems to suggest a Far Eastern origin for the superantigen expressing strains of *Y. pseudotuberculosis*. Furthermore, *ypmB* appears as the ancestral gene when compared to *ypmA* and *ypmC*. This study, which tested 30 superantigen-expressing *Y. pseudotuberculosis*, deserves extension to a more complete epidemiological work involving all superantigenic strains.

ARE YPM GENES LOCATED ON A MOBILE GENETIC ELEMENT?

At the present time little information is available on the genetic location of *ypm* genes. It is now established that *ypmA* is not located on the virulence plasmid pYV because it is present in strains which do not harbour the virulence plasmid or which are cured from pYV

(Carnoy *et al.*, 1998). As they are not present on pYV, *ypm* genes are likely to be located on the chromosome, although no experimental evidence is available to confirm this hypothesis (Miyoshi-Akiyama *et al.*, 1995). Nevertheless, some arguments suggest that *ypm* genes might be located on a mobile genetic element. First, *ypm* genes are not present in all *Y. pseudotuberculosis* strains, raising the question of the transmission of the superantigen genes among the *Y. pseudotuberculosis* population. Furthermore, analysis of the nucleotide sequence of *ypm* genes revealed a guanine and cytosine (G + C) content of 35%, whereas the genome of *Y. pseudotuberculosis* has a higher G + C content (47%). This would suggest that *Y. pseudotuberculosis* obtained *ypm* genes from low G + C percentage micro-organisms. Interestingly, *S. aureus*, *S. pyogenes* and *M. arthritidis*, which are superantigen-producing bacteria, display a low G + C percentage. This is an attractive hypothesis to explain the origin of *ypm* genes but so far there is no experimental evidence demonstrating that non-*Yersinia* micro-organisms are the source of *ypm* genes.

YPMS, A NOVEL CLASS OF SUPERANTIGEN?

The molecular mechanism of interaction of YPMa with its ligands, TCR and MHC class-II molecules, is still unclear. The absence of homology at the amino acid level between YPMa and other bacterial superantigens strongly suggests a unique molecular structure of YPMa. To date, two preliminary studies have raised the question of the structure–function of YPMs. Yoshino *et al.* (1996) tested the inhibitory effect of synthetic peptides covering YPMa on YPMa-induced proliferation of PBMC and found that the synthetic peptide corresponding to the first 23 amino acids of the mature protein was the best inhibitor. However, this peptide inhibited only 50% of the proliferation induced by YPMa, indicating that the biological activity was not strictly located at the *N*-terminal region and that other structural domains might be involved. Furthermore, site-directed mutagenesis on cysteine residues at positions 32 and 129 of mature YPMa did not affect the biological activity, questioning the importance of the disulfide bond in the TCR recognition (Ito *et al.*, 1998). A complete mutational analysis is now needed to map the *N*-terminal residues directly involved in the biological activity.

Some information on the structure–function of the superantigens of *Y. pseudotuberculosis* may also be collected from a comparison of the amino acid sequences of the different variants of YPMs (Figure 34.2). Superantigens YPMa and YPMb share 83% homology and the same Vβ specificity (Takeda *et al.*, 1996; Ramamurthy *et al.*, 1997). From that comparison it may be deduced that the 28 amino acids which differ in YPMa and YPMb, and especially the most heterogeneous region between amino acid residues 65 and 86 of the mature protein, do not affect the TCR recognition.

Although sharing few homologies at the amino acid level, TSST-1 and enterotoxin SEB display analogous tridimensional structural features since they are both organized in β-strands within two distinct domains (Ulrich *et al.*, 1995). Therefore, the unique amino acid sequence of YPMs does not predict a tridimensional structure distinct from other superantigens. Characterization of the tridimensional structure of YPMs should be determined in order to resolve the question of whether YPMs have unique interactions with their ligands.

YERSINIA PSEUDOTUBERCULOSIS SUPERANTIGEN AND DISEASES

Superantigens have been extensively characterized at the molecular level but their relationship to specific diseases is still uncertain, especially for *Y. pseudotuberculosis* superantigens (Michie and Cohen, 1998). Since their characterization in 1995, very little experimental data have been available on the role of YPMs in diseases. A recent study showed that sera from 61% of acutely infected patients with *Y. pseudotuberculosis* had elevated anti-YPM IgG levels (with a higher titre in patients with systemic complications) demonstrating the expression of YPMa *in vivo* (Abe *et al.*, 1997). Furthermore, Vβ3-bearing T-cells were increased in the acute-phase patients compared with control subjects, whereas the amount of T-cells bearing Vβ9, Vβ13.1 and Vβ13.2 was unchanged. It is unclear why the level of Vβ9, Vβ13.1 and Vβ13.2 T-cells is not increased as these cells are stimulated *in vitro* by YPMs. Abe *et al.* (1997) suggested that expansion of given Vβ T cells occurred in lymph nodes after entry of *Y. pseudotuberculosis* through M cells and that only the major subset (Vβ3) stimulated can be detected in PBMC. This was partially demonstrated by examination in the Vβs expressed by mesenteric lymph node T-cells from a patient infected with *Y. pseudotuberculosis* and undergoing surgery after severe abdominal pain. In the mesenteric lymph node of this *Y. pseudotuberculosis*-infected patient an increase of Vβ3- and Vβ13.2-bearing T cells was observed. These experimental data strengthen the hypothesis of Abe *et al.* (1997), but the study should be extended to include a greater number of cases. However, this approach, which requires surgery, is difficult to reproduce.

Yersinia pseudotuberculosis and Kawasaki syndrome

The Kawasaki syndrome, which was first described in the 1960s, is an acute multisystem vasculitis complicated by the development of coronary abnormalities and primarily affecting children (Kawasaki, 1967; Kawasaki *et al.*, 1974). The aetiology of the illness is still unknown but it has now been firmly established from clinical and epidemiological features that an infectious agent is responsible for the syndrome (Rowley, 1998). The most appealing hypothesis is the involvement of a superantigen-producing micro-organism but there is still controversy about the agent involved. A report showed an increase in Vβ2 and Vβ8.1 T-cells in PBMC of patients with acute Kawasaki syndrome, which suggested the involvement of a superantigen like the staphylococcal TSST-1 (Abe *et al.*, 1992), but this study was not confirmed in other cases of Kawasaki syndrome (Terai *et al.*, 1995). *Yersinia pseudotuberculosis* was also cited as a possible aetiological agent for Kawasaki syndrome as this bacterium was isolated from patients with clinical manifestations resembling Kawasaki disease (Sato *et al.*, 1983; Baba *et al.*, 1998). YPMa was actually purified from a strain of *Y. pseudotuberculosis* isolated from a patient suffering of this illness (Yoshino *et al.*, 1994). However, an increase or a depletion of the T-cells bearing the Vβ, which are recognized by YPMs (Vβ3, Vβ9, Vβ13.1 and Vβ13.2), as well as the presence of anti-YPM antibodies, has never been described in patients suffering from Kawasaki syndrome.

YPMa induces lethal shock in mice

The effect of YPMa *in vivo* was tested in a murine experimental model. Miyoshi-Akiyama *et al.* (1997) found that 50 µg of YPMa induced lethal shock in D-galactosamine presensitized C57BL/6 mice and that the induction of the shock could be blocked by anti-CD4, anti-Vβ7 or anti-Vβ8 monoclonal antibodies, but not by anti-CD8, anti-Vβ9 or anti-Vβ11 antibodies, confirming the Vβ specificity found *in vitro* (Vβ7 and Vβ8) and that the targets of YPMa are the CD4$^+$ T-cells. Monoclonal antibodies to tumour necrosis factor-α (TNF-α) and to interferon-γ (IFN-γ) also blocked the YPMa-induced shock. This indicates that these two lymphokines are involved in the shock. Although a toxic shock is usually not observed in patients infected with *Y. pseudotuberculosis*, these experimental data demonstrate the toxicity of YPMa *in vivo* and the potential virulence properties of the superantigenic toxin of *Y. pseudotuberculosis*.

Several studies demonstrated that *Y. pseudotuberculosis* YopJ, a protein translocated inside the macrophage, suppressed the production of TNF-α by macrophages by inhibiting NF-κB activation (Palmer *et al.*, 1998; Schesser *et al.*, 1998). There is apparently an opposite effect between, on the one hand, the induction by YPMs of TNF-α production in mice and, on the other hand, inhibition of TNF-α production by YopJ. It would be interesting to investigate whether *in vivo* YopJ reduces the potential mitogenic activity of YPMa. It now appears that the role of the superantigen of *Y. pseudotuberculosis* should be included in the context of the other virulence factors produced by *Y. pseudotuberculosis*.

CONCLUSION

A feature of *Y. pseudotuberculosis* infection is the frequent immunopathological post-infection complications which can appear in some patients, but the reasons for such complications are still obscure. The recent discovery of the production of a superantigenic activity by *Y. pseudotuberculosis* might bring new insight to these post-infection complications. However, although YPMa is highly mitogenic for human and mouse T-cells, induces production of anti-YPM antibodies in patients and can be toxic to mice, a role in *Y. pseudotuberculosis* pathologies, especially reactive arthritis and Kawasaki syndrome, remains to be demonstrated. The construction of an isogenic mutant deficient in superantigen and the measurement of its virulence *in vivo* in various experimental models might shed some light on the role of YPMs in *Y. pseudotuberculosis* infections.

How YPMs interact with their ligands, TCR and MHC class-II molecules, is still unknown but the question deserves serious consideration. Indeed, the low relative molecular mass of these proteins and the absence of homologies with other superantigens makes them potential novel superantigens. Extensive structure–function analysis is now required to understand the molecular mechanism of interaction of these superantigens and to confirm their unique structure.

Furthermore, it will be important to evaluate the relationship of these superantigens with other virulence factors of *Y. pseudotuberculosis* involved in its pathophysiology, in order to have a global view of the pathogenesis of *Y. pseudotuberculosis*.

ACKNOWLEDGEMENTS

This work was supported by the Conseil Régional du Nord-Pas de Calais, and Christophe Carnoy was supported by a grant from the Délégation à la Recherche du Centre Hospitalier et Universitaire de Lille.

REFERENCES

Abe, J. and Takeda, T. (1997). Characterization of a superantigen produced by *Yersinia pseudotuberculosis*. *Prep. Biochem. Biotechnol.* **27**, 173–208.

Abe, J., Kotzin, B.L., Jujo, K., Melish, M.E., Glode, M.P., Kohsaka, T. and Leung D.Y.M. (1992). Selective expansion of T cells expressing T-cell receptor variable regions Vβ2 and Vβ8 in Kawasaki disease. *Proc. Natl Acad. Sci. USA* **89**, 4066–70.

Abe, J., Takeda, T., Watanabe, Y., Nakao, H., Kobayashi, N., Leung, D.Y.M. and Kohsaka, T. (1993). Evidence for superantigen production by *Yersinia pseudotuberculosis*. *J. Immunol.* **151**, 4183–8.

Abe, J., Onimaru, M., Matsumoto, S., Noma, S., Baba, K., Ito, Y., Kohsaka, T. and Takeda, T. (1997). Clinical role for a superantigen in *Yersinia pseudotuberculosis* infection. *J. Clin. Invest.* **99**, 1823–30.

Baba, K., Konishi, N., Oonishi, H., Maruko, T., Waki, K., Abe, J. and Tanaka, M. (1998). Cases of Kawasaki disease with coronary arterial lesions documented with *Yersinia pseudotuberculosis* infection. In: *7th International Congress on* Yersinia, Nijmegen, The Netherlands, 14–16 June, Abstract O-35. Netherlands Tijdschrift voor Medische Microbiologie, The Netherlands.

Bayles, K.W. and Iandolo, J.J. (1989). Genetic and molecular analyses of the gene encoding staphylococcal enterotoxin D. *J. Bacteriol.* **171**, 4799–806.

Betley, M.J. and Mekalanos, J.J. (1988). Nucleotide sequence of the type A staphylococcal enterotoxin gene. *J. Bacteriol.* **170**, 34–41.

Blaser, M.J., Miller, R.A., Lacher, J. and Singleton, J.W. (1984). Patients with active Crohn's disease have elevated serum antibodies to antigens of seven enteric bacterial pathogens. *Gastroenterology* **87**, 888–94.

Bliska, J.B. and Black, D.S. (1995). Inhibition of the Fc receptor-mediated oxidative burst in macrophages by the *Yersinia pseudotuberculosis* tyrosine phosphatase. *Infect. Immun.* **63**, 681–5.

Bliska, J.B., Guan, K., Dixon, J.E. and Falkow, S. (1991). Tyrosine phosphate hydrolysis of host proteins by an essential *Yersinia* virulence determinant. *Proc. Natl Acad. Sci. USA* **88**, 1187–91.

Bliska, J.B., Copass, M.C. and Falkow, S. (1993). The *Yersinia pseudotuberculosis* adhesin YadA mediates intimate bacterial attachment to and entry into HEp-2 cells. *Infect. Immun.* **61**, 3914–21.

Blomster-Hautamaa, D.A., Kreiswirth, B.N., Kornblum, J.S., Novick, R.P. and Schlievert, P.M. (1986). The nucleotide and partial amino acid sequence of toxic shock syndrome toxin-1. *J. Biol. Chem.* **261**, 15783–6.

Bohach, G.A. and Schlievert, P.M. (1987). Nucleotide sequence of the staphylococcal enterotoxin C1 gene and relatedness to other pyrogenic toxins. *Mol. Gen. Genet.* **209**, 15–20.

Brubaker, R.R. (1991). Factors promoting acute and chronic disease caused by yersiniae. *Clin. Microbiol. Rev.* **4**, 309–24.

Bruno, T.F., Buser, D.E., Syme, R.M., Woods, D.E. and Mody, C.H. (1998). *Pseudomonas aeruginosa* exoenzyme S is a mitogen but not a superantigen for human T lymphocytes. *Infect. Immun.*, **66**, 3072–9.

Butler, T. (1983). *Plague and Other* Yersinia *Infection*. Plenum Press, New York.

Carnoy, C., Müller-Alouf, H., Haentjens, S. and Simonet, M. (1998). Polymorphism of *ypm*, *Yersinia pseudotuberculosis* superantigen encoding gene. In: *Bacterial Protein Toxins*. Zentralbl. Bakteriol. Suppl. 29, 397–8.

Clark, M.A., Hirst, B.H. and Jepson, M.A. (1998). M-cell surface β1 integrin expression and invasin-mediated targeting of *Yersinia pseudotuberculosis* to mouse Peyers patch M cells. *Infect. Immun.* **66**, 1237–43.

Cole, B.C., Knudtson, K.L., Oliphant, A., Sawitzke, A.D., Pole, A., Manohar, M., Benson, L.S., Ahmed, E. and Atkin, C.L. (1996). The sequence of the *Mycoplasma arthritidis* superantigen, MAM: identification of functional domains and comparison with microbial superantigens and plant lectin mitogens. *J. Exp. Med.* **183**, 1105–10.

Cornelis, G.R. and Wolf-Watz, H. (1997). The *Yersinia* Yop virulon: a bacterial system for subverting eukaryotic cells. *Mol. Microbiol.* **23**, 861–7.

Couch, J.L., Soltis, M.T. and Betley, M.J. (1988). Cloning and nucleotide sequence of the type E staphylococcal enterotoxin gene. *J. Bacteriol.* **170**, 2954–60.

Delchier, J.C., Constantini, D. and Soule, J.C. (1983). Présence d'agglutinines anti-*Yersinia pseudotuberculosis* lors d'une poussée de maladie de Crohn iléale. A propos de trois cas. *Gastroenterol. Clin. Biol.* **7**, 580–4.

Emödy, L., Heesemann, J., Wolf-Watz, H., Skurnik, M., Kapperud, G., O'Toole, P. and Wadström, T. (1989). Binding to collagen by *Yersinia enterocolitica* and *Yersinia pseudotuberculosis*: evidence for *yopA*-mediated and chromosomally encoded mechanisms. *J. Bacteriol.* **171**, 6674–9.

Fujimura, Y., Kihara, T. and Mine, H. (1992). Membranous cells as a portal of *Yersinia pseudotuberculosis* entry into rabbit ileum. *J. Clin. Electron. Microsc.* **25**, 35–45.

Fukushima, H. and Gomyoda, M. (1991). Intestinal carriage of *Yersinia pseudotuberculosis* by wild birds and mammals in Japan. *Appl. Environ. Microbiol.* **57**, 1152–5.

Galyov, E.E., Håkansson, S., Forsberg, Å. and Wolf-Watz, H. (1993). A secreted protein kinase of *Yersinia pseudotuberculosis* is an indispensable virulence determinant. *Nature* **361**, 730–2.

Goshorn, S.C. and Schlievert, P.M. (1988). Nucleotide sequence of streptococcal pyrogenic exotoxin type C. *Infect. Immun.* **56**, 2518–20.

Hovde, C.J., Hackett, S.P. and Bohach, G.A. (1990). Nucleotide sequence of the staphylococcal enterotoxin C3 gene: sequence comparison of all three type C staphylococcal enterotoxins. *Mol. Gen. Genet.* **220**, 329–33.

Inoue, M., Nakashima, H., Ueba, O., Ishida, T., Date, H., Kobashi, S., Takagi, K., Nishu, T. and Tsubokura, M. (1984). Community outbreak of *Yersinia pseudotuberculosis*. *Microbiol. Immunol.* **28**, 883–91.

Inoue, M., Nakashima, H., Ishida, T. and Tsubokura, M. (1988). Three outbreaks of *Yersinia pseudotuberculosis* infection. *Zentralbl. Bakteriol. Hyg.* **186**, 504–11.

Isberg, R.R. and Leong, J.M. (1990). Multiple β1 chain integrins are receptors for invasin, a protein that promotes bacterial penetration into mammalian cells. *Cell* **60**, 861–71.

Isberg, R.R., Voorhis, D.L. and Falkow, S. (1987). Identification of invasin: a protein that allows enteric bacteria to penetrate cultured mammalian cells. *Cell* **50**, 769–78.

Ito, Y., Abe, J., Yoshino, K.-I., Takeda, T. and Kohsaka T. (1995). Sequence analysis of the gene for a novel superantigen produced by *Yersinia pseudotuberculosis* and expression of the recombinant protein. *J. Immunol.* **154**, 5896–906.

Ito, Y., Seprényi, G., Abe, J. and Kohsaka T. (1998). Structure and function analysis of a unique superantigen YPM using a gene mutagenesis technique. In: *7th International Congress on* Yersinia, Nijmegen, The Netherlands, 14–16 June, Abstract P-38. Nederlands Tijdschrift voor Medische Microbiologie, The Netherlands.

Johnson, L.P., L'Italien, J.J. and Schlievert, P.M. (1986). Streptococcal pyrogenic exotoxin type A (scarlet fever toxin) is related to *Staphylococcus aureus* enterotoxin B. *Mol. Gen. Genet.* **203**, 354–6.

Kapur V., Topouzis, S., Majesky, M.W., Li, L.-L., Hamrick, M.R., Hamill, R.J., Patti, J.M. and Musser, J.M. (1993). A conserved *Streptococcus pyogenes* extracellular cysteine protease cleaves human fibronectin and degrades vitronectin. *Microb. Pathogen.* **15**, 327–46.

Kawasaki, T. (1967). Acute febrile mucocutaneous syndrome with lymphoid involvement with specific desquamation of the fingers and toes in children. *Jpn. J. Allergy* **16**, 178–222. (In Japanese.)

Kawasaki, T., Kosaki, F., Okawa, S., Shigematsu, I. and Yanagawa, H. (1974). A new infantile acute febrile mucocutaneous lymph node syndrome (MLNS) prevailing in Japan. *Pediatrics* **54**, 271–6.

Knapp, W. (1958). Mesenteric adenitis due to *Pasteurella pseudotuberculosis* in young people. *N. Engl. J. Med.* **259**, 776–8.

Kotb, M. (1995). Bacterial pyrogenic exotoxins as superantigens. *Clin. Microbiol. Rev.* **8**, 411–26.

Legaard, P.K., LeGrand, R.D. and Misfeldt, M.L. (1991). The superantigen *Pseudomonas* exotoxin A requires additional functions from accessory cells for T lymphocyte proliferation. *Cell Immunol.* **135**, 372–82.

Legaard, P.K., LeGrand, R.D. and Misfeldt, M.L. (1992). Lymphoproliferative activity of *Pseudomonas* exotoxin A is dependent on intracellular processing and is associated with the carboxyl-terminal portion. *Infect. Immun.* **60**, 1273–8.

Leung, K.Y., Reisner, B.S. and Straley, S. (1990). YopM inhibits platelet aggregation and is necessary for virulence of *Yersinia pestis* in mice. *Infect. Immun.* **58**, 3262–71.

Malassez, L. and Vignal, W. (1883). Tuberculose zooléique (forme ou espèce du tuberculose sans bacilles). *Arch. Physiol. No.*

Marra, A. and Isberg, R.R. (1997). Invasin-dependent and invasin-independent pathways for translocation of *Yersinia pseudotuberculosis* across the Peyer's patch intestinal epithelium. *Infect. Immun.* **65**, 3412–21.

Marrack, P. and Kappler, J. (1990). The staphylococcal enterotoxins and their relatives. *Science* **248**, 705–11.

Michie, C.A. and Cohen, J. (1998). The clinical significance of T-cell superantigens. *Trends Microbiol.* **6**, 61–5.

Miyoshi-Akiyama, T., Imanishi, K. and Uchiyama T. (1993). Purification and partial characterization of a product from *Yersinia pseudotuberculosis* with the ability to activate human T cells. *Infect. Immun.* **61**, 3922–7.

Miyoshi-Akiyama, T., Abe, A., Kato, H., Kawahara, K., Narimatsu, H. and Uchiyama, T. (1995). DNA sequencing of the gene encoding a bacterial superantigen, *Yersinia pseudotuberculosis*-derived mitogen (YPM), and characterization of the gene product, cloned YPM. *J. Immunol.* **154**, 5228–34.

Miyoshi-Akiyama, T., Fujimaki, W., Yan, X.J., Yagi, J., Imanishi, K., Kato, H., Tomonari, K. and Uchiyama, T. (1997). Identification of murine T cells reactive with the bacterial superantigen *Yersinia pseudotuberculosis*-derived mitogen (YPM) and factors involved in YPM-induced toxicity in mice. *Microbiol. Immunol.* **41**, 345–52.

Monack, D.M., Mecsas, J., Ghori, N. and Falkow, S. (1997). *Yersinia* signals macrophages to undergo apoptosis and YopJ is necessary for this cell death. *Proc. Natl Acad. Sci. USA* **94**, 10385–90.

Nakano, T., Kawaguchi, H., Nakao, K., Maruyama, T., Kamiya, H and Sakurai, M. (1989). Two outbreaks of *Yersinia pseudotuberculosis* 5a infection in Japan. *Scand. J. Infect. Dis.* **21**, 175–9.

Nelson, K., Schlievert, P.M., Selander, R.K. and Musser, J.M. (1991). Characterization and clonal distribution of four alleles of the *speA* gene encoding pyrogenic exotoxin A (scarlet fever toxin) in *Streptococcus pyogenes*. *J. Exp. Med.* **174**, 1271–4.

Odaert, M., Berche, P. and Simonet, M. (1996). Molecular typing of *Yersinia pseudotuberculosis* by using an IS*200*-like element. *J. Clin. Microbiol.* **34**, 2231–5.

Paerregaard, A., Espersen, F. and Skurnik, M. (1991). Adhesion of *Yersinia* to rabbit intestinal constituents: role of outer membrane protein YadA and modulation by intestinal mucous. *Contrib. Microbiol. Immunol.* **12**, 171–5.

Palmer, L.E., Hobbie, S., Galan, J.E. and Bliska, J.B. (1998). YopJ of *Yersinia pseudotuberculosis* is required for the inhibition of macrophage TNF-α production and downregulation of the MAP kinases p38 and JNK. *Mol. Microbiol.* **27**, 953–65.

Pebody, R. (1997). Outbreak of *Yersinia pseudotuberculosis* in central Finland. *Eurosurveillance Weekly* **1**, 970918. (http://www.eurosurv.org)

Perry, R.D. and Fetherston, J.D. (1997). *Yersinia pestis* – etiologic agent of plague. *Clin. Microbiol. Rev.* **10**, 35–66.

Ramamurthy, T., Yoshino, K.-I., Abe, J., Ikeda, N. and Takeda, T. (1997). Purification, characterization and cloning of a novel variant of the superantigen *Yersinia pseudotuberculosis*-derived mitogen. *FEBS Lett.* **413**, 174–6.

Ranelli, D.M., Jones, C.L., Johns, M.B., Mussey, G.J. and Khan, S.A. (1985). Molecular cloning of staphylococcal enterotoxin B gene in *Escherichia coli* and *Staphylococcus aureus*. *Proc. Natl Acad. Sci. USA* **82**, 5850–4.

Rosqvist, R., Forsberg, Å., Rimpiläinen, M., Bergman, T. and Wolf-Watz, H.H. (1990). The cytotoxic protein YopE of *Yersinia* obstructs the primary host defence. *Mol. Microbiol.* **4**, 657–67.

Rowley, A.H. (1998). Controversies in Kawasaki syndrome. *Adv. Pediatr. Infect. Dis.* **13**, 127–41.

Saitou, N. and Nei, M. (1987). The neighbor-joining method: a new method for reconstructing phylogenetic trees. *Mol. Biol. Evol.* **4**, 406–25.

Sato, K., Ouchi, K. and Taki, M. (1983). *Yersinia pseudotuberculosis* infection in children, resembling Izumi fever and Kawasaki syndrome. *Pediatr. Infect. Dis.* **2**, 123–6.

Schesser, K., Spiik, A.-K., Dukuzumuremyi, J.-M., Neurath, M.F., Pettersson, S. and Wolf-Watz, H. (1998). The *yopJ* locus is required for *Yersinia*-mediated inhibition of NF-*k*B activation and cytokine expression: YopJ contains a eukaryotic SH2-like domain that is essential for its repressive activity. *Mol. Microbiol.* **28**, 1067–79.

Schulze-Koops, H., Burkhardt, H., Heesemann, J., Von der Mark, K. and Emmrich, F. (1992). Plasmid-encoded outer membrane protein YadA mediates specific binding of enteropathogenic yersiniae to various types of collagen. *Infect. Immun.* **60**, 2153–9.

Simonet, M. (1999). Enterobacteria in reactive arthritis: *Yersinia*, *Shigella* and *Salmonella*. *Rev. Rhum.* (*Engl. Ed.*) **66**, 145–95.

Simonet, M. and Falkow, S. (1992). Invasin production in *Yersinia pseudotuberculosis*. *Infect. Immun.* **60**, 4414–17.

Simonet, M., Richard, S. and Berche, P. (1990). Electron microscopic evidence for *in vivo* extracellular localization of *Yersinia pseudotuberculosis* harboring the pYV plasmid. *Infect. Immun.* **58**, 841–5.

Somov, G.P. and Martinevsky, I.L. (1973). New facts about pseudotuberculosis in the USSR. *Contr. Microbiol. Immunol.* **2**, 214–16.

Stuart, P.M. and Woodward, J.G. (1992). *Yersinia enterocolitica* produces superantigenic activity. *J. Immunol.* **148**, 225–33.

Stuart, P.M., Munn, R.K., DeMoll, E. and Woodward, J.G. (1995). Characterization of human T-cell responses to *Yersinia enterocolitica* superantigen. *Hum. Immunol.* **43**, 269–75.

Takeda, N., Usami, I., Fujita A., Baba K. and Tanaka M. (1991). Renal complications of *Yersinia pseudotuberculosis* infection in children. *Contrib. Microbiol. Immunol.* **12**, 301–6.

Takeda, T., Ramamurthy, T., Yoshino, K.-I., Ikeda, N. and Kato, Y. (1996) Purification and characterization of a novel variant of the superantigen *Yersinia pseudotuberculosis*-derived mitogen. *Jpn. J. Med. Sci. Biol.* **49**, 252.

Terai, M., Miwa, K., Williams, T., Kabat, W., Fukuyama, M., Okajima, Y., Igarashi, H. and Shulman, S.T. (1995). The absence of evidence of staphylococcal toxin involvement in the pathogenesis of Kawasaki disease. *J. Infect. Dis.* **172**, 558–61.

Tertti, R., Granfors, K., Lehtonen, O.-P., Mertsola, J., Mäkelä, A.-L., Välimäki, I., Hänninen, P. and Toivanen, A. (1984). An outbreak of *Yersinia pseudotuberculosis* infection. *J. Infect. Dis.* **149**, 245–50.

Tertti, R., Skurnik, M., Vartio, T. and Kuusela, P. (1992). Adhesion protein YadA of *Yersinia* species mediates binding of bacteria to fibronectin. *Infect. Immun.* **60**, 3021–4.

Tertti, R., Vuento, R., Mikkola, P., Granfors, K., Mäkelä, A.-L. and Toivanen A. (1989). Clinical manifestations of *Yersinia pseudotuberculosis* infection in children. *Eur. J. Clin. Microbiol. Infect. Dis.* **8**, 587–91.

Treacher, D.F. and Jewell, D.P. (1985). *Yersinia* colitis associated with Crohn's disease. *Postgrad. Med. J.* **61**, 173–4.

Tsubokura, M., Otsuki, K., Sato, K., Tanaka, M., Hongo, T., Fukushima, H., Maruyama, T. and Inoue, M. (1989). Special features of distribution of *Yersinia pseudotuberculosis* in Japan. *J. Clin. Microbiol.* **27**, 790–1.

Uchiyama, T., Miyoshi-Akiyama, T., Kato, H., Fujimaki, W., Imanishi, K. and Yan, X.-J. (1993). Superantigenic properties of a novel mitogenic substance produced by *Yersinia pseudotuberculosis* isolated from patients manifesting acute and systemic symptoms. *J. Immunol.* **151**, 4407–13.

Ulrich, R.G., Bavari, S. and Olson, M.A. (1995). Bacterial superantigens in human disease: structure, function and diversity. *Trends Microbiol.* **3**, 463–8.

Yang, Y., Merriam, J.J., Mueller, J.P. and Isberg, R.R. (1996). The *psa* locus is responsible for thermoinducible binding of *Yersinia pseudotuberculosis* to cultured cells. *Infect. Immun.* **64**, 2483–9.

Yoshino, K.-I., Abe, J., Murata, H., Takao, T., Kohsaka, T., Shimonishi, Y. and Takeda, T. (1994). Purification and characterization of a novel superantigen produced by a clinical isolate of *Yersinia pseudotuberculosis*. *FEBS Lett.* **356**, 141–4.

Yoshino, K.-I., Ramamurthy, T., Nair, B.G., Fukushima, H., Ohtomo, Y., Takeda, N., Kaneko, S. and Takeda, T. (1995). Geographical heterogeneity between Far East and Europe in prevalence of *ypm* gene encoding the novel superantigen among *Yersinia pseudotuberculosis* strains. *J. Clin. Microbiol.* **33**, 3356–8.

Yoshino, K.-I., Takao, T., Ishibashi, M., Samejima, Y., Shimonishi, Y. and Takeda, T. (1996). Identification of the functional region on the superantigen *Yersinia pseudotuberculosis*-derived mitogen responsible for induction of lymphocyte proliferation by using synthetic peptides. *FEBS Lett.* **390**, 196–8.

35

The pathogenesis of shock and tissue injury in clostridial gas gangrene

Dennis L. Stevens and Amy E. Bryant

INTRODUCTION

The genus *Clostridium* encompasses over 60 species of Gram-positive anaerobic spore-forming rods that cause a variety of infections in humans and animals by virtue of a myriad of proteinaceous exotoxins. *Clostridium tetani* and *C. botulinum* manifest specific clinical disease by elaborating single, but highly potent, toxins. Although botulism is usually the result of ingestion of preformed toxin, tetanus requires the bacteria to proliferate at the site of penetrating injury. Frequently, signs of infection are not apparent even with lethal exotoxinaemia. In contrast, other strains of clostridia, such as *C. perfringens* and *C. septicum*, cause aggressive necrotizing infections of the soft tissues, attributable, in part, to the elaboration of bacterial proteases, phospholipases and cytotoxins.

Histotoxic clostridial organisms can be isolated from soft tissues in three distinct settings (Smith, 1975a). First is simple wound contamination. Infection *per se* does not develop because there is insufficient devitalized tissue. This is a very common occurrence; MacLennan (1962) reported that 30–80% of open traumatic wounds are contaminated with clostridial species. The second type, anaerobic cellulitis, occurs when there is modest devitalized tissue in a wound, sufficient for growth of *C. perfringens* or other strains. Although gas is produced locally and may extend along fascial planes, invasion of healthy tissue and bacteraemia do not occur. Appropriate medical and surgical management including prompt removal of the devitalized tissue is all that is necessary and mortality is generally nil (Smith, 1975a).

The third type, clostridial gas gangrene or myonecrosis, occurs in four different settings (Table 35.1), and is characterized by the rapid and progressive invasion and destruction of healthy living muscle (MacLennan, 1962). The first and most common form is traumatic gas gangrene which develops after deep, penetrating injury that compromises the blood supply (e.g. knife or gunshot wound, crush injury or automobile accident) creating an anaerobic environment ideal for clostridial proliferation. This type of trauma accounts for about 70% of cases of gas gangrene and *C. perfringens* is found in about 80% of such infections (Smith, 1975a). The remaining cases are caused by *C. septicum, C. novyi, C. histolyticum, C. bifermentans, C. tertium* and *C. fallax*. Other conditions associated with traumatic gas gangrene include bowel and biliary tract surgery, intramuscular injection of epinephrine, criminal abortion, retained placenta, prolonged rupture of the membranes, intrauterine fetal demise, missed abortion in postpartum patients and, most recently, intradermal injection ('skin popping') of black tar heroin. Second, spontaneous or non-traumatic gas gangrene is most commonly caused by the more aerotolerant *C. septicum*. As described later, most of these cases occur in patients with gastrointestinal portals of entry such as adenocarcinoma but with no antecedent trauma. *Clostridium tertium* has also been associated

623

TABLE 35.1 Clinical settings of clostridial myonecrosis

- Traumatic gas gangrene
  Caused by *C. perfringens*, *C. septicum*, *C. histolyticum* and *C. novyii*
  Trauma is usually crush injury or associated with compromised blood supply
- Spontaneous gas gangrene
  More commonly due to *C. septicum*
  Often associated with metastatic seeding from bowel portal
  Predisposing factors are intra-abdominal tumour, acute leukaemia, neutropenia, cancer chemotherapy or radiation therapy
- Recurrent gas gangrene
  More than one episode of gas gangrene
- Gastrointestinal necrosis
  Due to ingestion of meats contaminated with *C. perfringens* type C
  Symptoms range from mild abdominal pain to life-threatening rupture of the bowel
  Associated with β-toxin production by the organism
  Also associated with high levels of trypsin inhibitors in the human host

with spontaneous myonecrosis; however, it more commonly causes bacteraemia in compromised hosts who have received long courses of antibiotics. Third, recurrent gas gangrene caused by *C. perfringens* has been described in individuals with non-penetrating injuries at sites of previous gas gangrene where spores of *C. perfringens*, which may have remained quiescent in tissue for periods of 10–20 years, germinate when minor trauma provides conditions suitable for growth (Stevens *et al.*, 1988a). Fourth, necrotizing infections of the gastrointestinal tract largely due to *C. perfringens* type C have been described in New Guinea and in prisoners of war in Europe following World War II. The epidemiology and clinical course of this disease will be described later in this chapter.

Lastly, *C. sordellii* may be associated with deep soft tissue infections, most commonly in postpartum women. Unlike the previous condition where gas and necrosis of tissue occurs, *C. sordellii* infection is characterized by oedema, shock, haemoconcentration, absence of fever and a leukaemoid reaction.

This chapter discusses four types of necrotizing clostridial infections: spontaneous, non-traumatic gas gangrene due to *C. septicum*; necrotizing enteritis; *C. sordellii* infections and traumatic gas gangrene.

SPONTANEOUS, NON-TRAUMATIC GAS GANGRENE DUE TO *CLOSTRIDIUM SEPTICUM*

Clinical manifestations

The onset of disease is abrupt, often with excruciating pain, although the patient may sense only heaviness or numbness (MacLennan, 1962; Alpern and Dowell, 1969; Smith, 1975a; Stevens *et al.*, 1990; Johnson *et al.*, 1994).

The first symptom may be confusion or malaise. Extremely rapid progression of gangrene follows with demonstrable gas in the tissues (Figure 35.1). Later, swelling increases and bullae appear filled with clear, cloudy, haemorrhagic or purplish fluid (Figure 35.2). The skin around such bullae also has a purple hue, perhaps reflecting vascular compromise resulting from bacterial toxins diffusing into surrounding tissues (Stevens *et al.*, 1990). The advancing margin of tissue necrosis is well demarcated (Figure 35.2). Histopathology of muscle and connective tissues includes cell lysis and gas formation; inflammatory cells are remarkably absent (Stevens *et al.*, 1990).

Predisposing factors include colonic carcinoma, diverticulitis, gastrointestinal surgery, leukaemia, lymphoproliferative disorders, cancer chemotherapy, radiation therapy and, more recently, AIDS (Alpern and

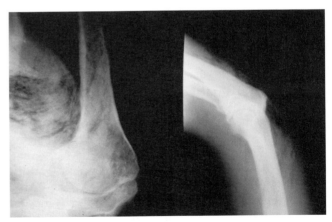

FIGURE 35.1 Spontaneous gas gangrene. Radiograph of the upper thorax and shoulder of a patient who developed spontaneous gas gangrene of the hand, which spread rapidly up the arm and on to the thorax. *Clostridium septicum* was grown from blood and necrotic tissue.

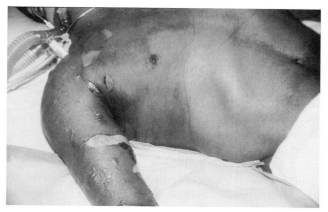

FIGURE 35.2 The line of demarcation in the patient shown in Figure 35.1. The patient underwent extensive surgical debridement, including amputation of the arm, and received hyperbaric oxygen therapy and antibiotics. Two months after recovery, carcinoma of the colon was diagnosed.

Dowell, 1969; Stevens *et al.*, 1990; Johnson *et al.*, 1994). Cyclic or other neutropenia is also associated with spontaneous gas gangrene due to *C. septicum*, and in such cases necrotizing enterocolitis, caecitis or distal ileitis is commonly found. These gastrointestinal pathologies permit bacterial access to the bloodstream; consequently, the aerotolerant *C. septicum* can proliferate in normal tissues (Smith, 1975a). Patients surviving bacteraemia or spontaneous gangrene due to *C. septicum* should have appropriate diagnostic studies to rule out gastrointestinal pathology.

Diagnosis

Unlike traumatic gas gangrene, a transient bacteraemia precedes cutaneous manifestations by several hours. In the absence of the usual cutaneous manifestations of gas gangrene, other causes of fever and extremity pain such as deep vein thrombophlebitis (DVT) are frequently entertained. Diagnostic evaluations for DVT tend to prolong establishment of the appropriate diagnosis, delay treatment and, as a consequence, increase mortality.

Pathogenesis

Clostridium septicum produces four toxins (Table 35.2): alpha (α)-toxin (lethal, haemolytic, necrotizing activity), beta (β)-toxin (DNase), gamma (γ)-toxin (hyaluronidase) and delta (Δ)-toxin (septicolysin, an oxygen-labile haemolysin), as well as a protease and a neuraminidase (Smith, 1975a). The *C. septicum* α-toxin does not possess phospholipase activity and is thus distinct from the α-toxin of *C. perfringens*. Active immunization against α-toxin significantly protects against challenge with viable *C. septicum* (Ballard *et al.*, 1992). α-Toxin, a pore-forming toxin, requires proteolytic cleavage for activation and probably contributes to *C. septicum* pathogenesis by lytic destruction of host cells (see Chapter 23 of this book).

TABLE 35.2 Major virulence factors of clostridial species causing necrotizing infections in humans

Organism	Clinical infection	Major virulence factor(s)	Mechanism of action
C. perfringens type A	Traumatic gas gangrene	α-Toxin	Phospholipase C
		θ-Toxin	Thiol-activated cytolysin
C. perfringens type C	Enteritis necroticans	β-Toxin	Cytolytic for intestinal microvilli
C. septicum	Spontaneous and traumatic gas gangrene	α-Toxin	Cytotoxic, lethal, haemolytic, antigenically related to α-toxin of *C. histolyticum*
		Δ-Toxin	Thiol-activated cytolysin
C. histolyticum	Traumatic gas gangrene	α-Toxin	Cytotoxic, lethal, haemolytic, antigenically related to α-toxin of *C. septicum*
		β-Toxin	Collagenase
		γ-Toxin	Thiol-activated protease
		Δ-Toxin	Elastase
		ε-Toxin	Thiol-activated cytolysin
C. novyii	Traumatic gas gangrene	α-Toxin	Dermonecrotic, causes gelatinous oedema
		γ-Toxin	Phospholipase C
		Δ-Toxin	Thiol-activated cytolysin

Treatment, prevention and prognosis

No human trials have compared the efficacy of antibiotics with and without hyperbaric oxygen treatment (HBO) for treating clinical cases of spontaneous gas gangrene. *In vitro* data indicate that *C. septicum* is uniformly susceptible to penicillin, tetracycline, erythromycin, clindamycin, chloramphenicol and metronidazole. The aerotolerance of *C. septicum* may reduce the efficacy of HBO therapy (Hill and Osterhout, 1972a).

The mortality of spontaneous clinical gangrene ranges from 67% to 100%, with the majority of deaths occurring within 24 h of onset. Risk factors include underlying malignancy and compromised immune status.

NECROTIZING ENTERITIS

Clinical manifestations

Neutropenic enterocolitis is a fulminant form of necrotizing enteritis that occurs in neutropenic patients (Farnell, 1987). The degree of neutropenia is often profound, and may be related to cyclic neutropenia, leukaemia, aplastic anaemia or chemotherapy (Farnell, 1987; Bartlett, 1990; Gorbach, 1992). Symptoms include abdominal pain, chills and malaise. Copious watery diarrhoea, abdominal distention and pain localizing to the right lower quadrant develop, followed rapidly by signs of toxicity such as tachycardia, fever and delirium.

Diagnosis

Radiographic examinations may reveal thickening of the wall of the colon or caecum and, in advanced cases, gas in the wall of the colon. Anecdotal reports suggest that computerized tomographic (CT) scanning may be a superior means of diagnosing this condition. Rupture of the bowel with peritonitis and bacteraemia results in death in 100% of cases.

Post-mortem examinations reveal that among children dying of leukaemia, localized infection of the ileocaecal region (typhilitis) is extremely common and may have contributed to death in nearly 40% of patients (Farnell, 1987). Most of these children have profound neutropenia. *Clostridium septicum* is the most common organism isolated from the blood of such patients, and Gram stain and immunofluorescence studies demonstrate that these bacteria invade the bowel wall in most cases.

Pathogenesis

Other forms of necrotizing enteritis have occurred endemically in New Guinea (pigbel; Lawrence and Walker, 1976), in epidemic proportions in Germany following World War II (Darmbrand; reviewed in Guerrant, 1990), and sporadically in Africa, south-east Asia and the USA (Bartlett, 1990; Gorbach, 1992). All cases are associated with the ingestion of meats contaminated with *C. perfringens* type C. Clinical courses vary from abdominal pain, fever and diarrhoea, which resolves spontaneously, to bloody diarrhoea, ruptured bowel and death. β-Toxin from *C. perfringens* type C has been implicated as causing these infections. β-Toxin paralyses the intestinal villi, and causes friability and necrosis of the bowel wall. Predisposing factors include malnutrition, specifically in those with diets low in protein and rich in trypsin inhibitors such as sweet potato or soy bean (Lawrence and Walker, 1976; Guerrant, 1990). In addition, *Ascaris lumbricoides* is commonly found in such patients and it, too, secretes a trypsin inhibitor. These protease inhibitors protect β-toxin from intraluminal proteolysis.

Treatment, prognosis and prevention

Aggressive supportive measures, surgical intervention and appropriate antibiotics (see the subsection on 'Spontaneous, non-traumatic gas gangrene due to *C. septicum*') have reduced the mortality to 25% (Farnell, 1987). Medical management should include aggressive fluid and electrolyte replacement, bowel decompression and antibiotic treatment with penicillin or chloramphenicol. Surgical resection of necrotic bowel is necessary in 50% of patients, and mortality rates as high as 40% have been described. If peritonitis develops, broader antibiotic coverage may be necessary. Immunization of children in New Guinea with a β-toxoid vaccine has dramatically reduced the incidence of this disease (Lawrence *et al.*, 1979).

CLOSTRIDIUM SORDELLII INFECTIONS

Patients with *C. sordellii* infection present with unique clinical features including oedema, absence of fever, leukaemoid reaction, haemoconcentration and, later, shock and multiorgan failure (Bartlett, 1990). Often, *C. sordellii* infections develop after childbirth or after gynaecological procedures (McGregor *et al.*, 1989), and most represent endometrial infection. Rarely, other cases have occurred at sites of minor trauma such as lacerations of the soft tissues of an extremity. Unlike *C. per-*

fringens and *C. septicum* infections, pain may not be a prominent feature. The absence of fever and paucity of signs and symptoms of local infection make early diagnosis difficult (Bartlett, 1990). The mechanisms of diffuse capillary leak, massive oedema and haemoconcentration are not well established, but clearly are related to elaboration of a potent toxin. Haematocrits of 75–80 have been described and leucocytosis of 50 000–100 000 cells mm^{-3} with a left shift is common (McGregor *et al.*, 1989; Stevens, 1995).

TRAUMATIC GAS GANGRENE

Clinical manifestations

The first symptom is usually the sudden onset of severe pain at the site of surgery or trauma (MacLennan, 1962; Weinstein and Barza, 1972). The mean incubation period is less than 24 h, but ranges from 6–8 h to several days, probably depending on the degree of soil contamination or bowel spillage and the extent of vascular compromise. The skin may initially appear pale, but quickly changes to bronze then purplish red, and becomes tense and exquisitely tender. Bullae develop; they may be clear, red, blue or purple. Gas present in tissue may be obvious by physical examination, soft tissue radiographs, CT scan or magnetic resonance imaging (MRI). Interestingly, none of these radiographic procedures has been more specific or more sensitive than the physical finding of crepitus in the soft tissue (Gozal *et al.*, 1986). However, radiographic procedures are particularly helpful to demonstrate gas in deeper tissue such as the uterus. Signs of systemic toxicity develop rapidly, including tachycardia, low-grade fever and diaphoresis, followed by shock and multiorgan failure. Shock was present in 50% of patients at the time they presented to the hospital (Hart *et al.*, 1983). Of those who developed shock at some point during their time in hospital, 40% died compared with 20% mortality in the group as a whole (Hart *et al.*, 1983). Bacteraemia occurs in 15% of patients and may be associated with brisk haemolysis. One patient has been described with a decrease in haematocrit from 37% to 0% over a 24-h period (Terebelo *et al.*, 1982). Subsequently, despite transfusion with 10 units of packed red blood cells over a 4-h period, the haematocrit never exceeded 7.2% (Terebelo *et al.*, 1982). Based on studies with recombinant α- and theta (θ)-toxins, it is quite clear that both toxins contribute to this marked intravascular haemolysis. Not all cases of *C. perfringens* bacteraemia have been associated with gas gangrene; thus in this setting, the presence of *C. perfringens* in the

blood may be merely a transient phenomenon (Gorbach and Thadepalli, 1975). In contrast, in a recent study 90% of *C. perfringens* isolates and 100% of *C. septicum* blood isolates were associated with clinically significant infection (Brook, 1989). Additional complications of clostridial myonecrosis include jaundice, renal failure, hypotension and liver necrosis. Renal failure is largely due to haemoglobinuria and myoglobinuria, but is complicated by acute tubular necrosis following hypotension. Renal tubular cells are probably directly affected by toxins, but this has not been proven.

Diagnosis

Increasing pain at the site of prior injury or surgery, together with signs of systemic toxicity and gas in the tissue support the diagnosis. Definitive diagnosis rests on demonstrating large, Gram-variable rods at the injury site. Note that although clostridia stain Gram-positive when obtained from bacteriological media, when visualized from infected tissues, they appear both Gram-positive and Gram-negative (Figure 35.3). In fresh material *C. perfringens* may appear to be encapsulated (Butler, 1943), although this was not corroborated in gas gangrene associated with war-time trauma (Keppie and Robertson, 1944). Surgical exploration is essential and demonstrates muscle that does not bleed or contract when stimulated. Grossly, muscle tissue is oedematous and may have a reddish-blue to black coloration. Usually, necrotizing fasciitis and cutaneous necrosis are also present. Microscopic evaluation of biopsy material invariably demonstrates organisms among degenerating muscle bundles and, characteristically, an absence of acute inflammatory cells (MacLennan, 1962; Stevens, 1995) (Figure 35.3).

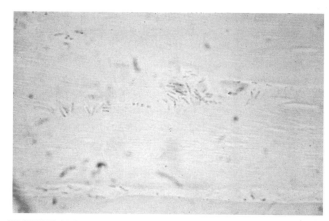

FIGURE 35.3 Tissue Gram stain showing slender rods with subterminal spores. Note that *in vivo C. perfringens* are Gram-variable and may be confused with Gram-negative rods.

Histopathology

Descriptions of gangrenous infections with gas in the tissues date back to the Middle Ages. World War I surgeons stationed at the allied casualty clearing stations recognized that wounds containing severed main blood vessels were associated with massive or 'group' gangrene, meaning that bacterial proliferation and muscle destruction easily and rapidly ensued throughout all muscle groups that had been cut off from the blood supply (McNee and Dunn, 1917). However, it was the careful observations of McNee and Dunn that provided an early insight regarding the mechanisms whereby gas gangrene advanced into healthy, viable tissue. Impressed by the natural course of this infection, they wrote: 'The rapidity of spread of gas gangrene into living muscle, once the disease has begun, is so remarkable as to demand some definite explanation' (McNee and Dunn, 1917). Their 1917 report in the *British Medical Journal* attempted to provide that explanation based on the histopathology of serial sections of single muscle bundles from freshly amputated limbs of soldiers with gas gangrene (McNee and Dunn, 1917). At the advancing edge of the infection they noted that few organisms were present, that the fibres appeared ischaemic and that a toxic fluid, formed in the gangrenous tissue behind, spread between the fibres, breaking down the tissue and providing an optimal environment for the organisms to proliferate. They also noted that in these areas 'leucocytes are generally conspicuous by their absence' from involved tissues but were seen in the interfascial planes. These gross and microscopic observations have stood the test of time (Stevens, 1995).

Subsequent experimental histopathology studies demonstrated that the tissue pathologies characteristic of clostridial gas gangrene could be reproduced by injection of either a crude clostridial toxin preparation (Robb-Smith, 1945) or recombinant clostridial α- (Bryant *et al.*, 1995) or θ-toxin (Bryant *et al.*, 1993) into healthy muscle (Figure 35.4).

Thus, the histopathology of gas gangrene is unique, and distinctly different from infections caused by bacteria such as *Staphylococcus aureus*, *Haemophilus influenzae* or *Streptococcus pneumoniae*. In these latter cases, a luxuriant pyogenic response occurs at the site of infection. With clostridial myonecrosis, leucocytes, when present, are localized between fascial planes (McNee and Dunn, 1917) and are often amassed within small vessels near the demarcation between healthy and necrotic tissues (Bryant *et al.*, 1993). Leucocytes in these areas exhibit altered morphology and karyolysis (Robb-Smith, 1945), suggesting that they are directly affected *in vivo* by the presence of clostridial exotoxins.

FIGURE 35.4 Histopathology of mouse thigh muscle 8 h after injection of recombinant α-toxin. Note extensive muscle destruction, accumulation of leucocytes within the adjacent vessel and an absence of leucocytes in the tissue.

Pathogenesis

The organism and its toxins

Clostridium perfringens is a Gram-positive, spore-forming, non-motile, rod-shaped organism commonly found in soil and in the intestines of humans and other animal species. Although classified as an anaerobe, *C. perfringens* is somewhat aerotolerant. Under optimal conditions, the generation time for *C. perfringens* can be as little as 8–10 min and growth is accompanied by abundant gas production (Smith, 1975b). The species produces 12 extracellular toxins, four of which are considered major lethal toxins: α-toxin, β-toxin, epsilon (ε)-toxin and iota (ι)-toxin. Based on the presence of these major toxins, the species has been divided into five distinct types, A–E (Smith, 1975b).

Of these subgroups, *C. perfringens* type A causes the majority of human infections and, out of the lethal toxins listed above, this organism produces only the α-toxin, a phospholipase C (Table 35.2). This exotoxin has the distinction of being the first bacterial toxin to which an enzymic activity (i.e. lecithinase) was ascribed. α-Toxin is haemolytic, destroys platelets and leucocytes, and increases capillary permeability, effects which are probably related to its ability to cleave sphingomyelin and the phosphoglycerides of choline, ethanolamine and serine present in eukaryotic cell membranes (Ispolatovskaya, 1972). α-Toxin requires divalent cations such as calcium and magnesium for optimal activity (Ispolatovskaya, 1972). Zinc enhances α-toxin production in culture (Murata *et al.*, 1969) and is essential for its activity *in vivo*. Histidine residues have been shown to be essential for the binding of zinc ions (Titball and Rubidge, 1990). Basak *et al.* have crystallized α-toxin and have provided preliminary X-ray diffraction analysis of the protein (Basak *et al.*, 1994). Titball *et al.* have determined that the protein is composed of two domains; the *N*-terminal domain possesses the phospholipase C activity and the *C*-terminal domain confers the cytolytic properties (see Chapter 15 in this book).

α-Toxin's role as the major lethal factor in *C. perfringens* infections is supported by numerous studies utilizing a variety of approaches. Both active and passive immunization of animals against α-toxin (Pober *et al.*, 1986; Williamson and Titball, 1993) or its *C*-terminal fragment (Williamson and Titball, 1993) are protective in experimental wild-type infections. Similarly, experimental infections established with genetic mutants of *C. perfringens* lacking α-toxin (Awad *et al.*, 1995) or with strains which produce less α-toxin (Ninomiya *et al.*, 1994) were markedly less fulminant and mortality was significantly reduced.

The importance of θ-toxin in the pathogenesis of gas gangrene has been largely controversial despite the early knowledge of its haemolytic nature (Bernheimer, 1947), and it serological and antigenic relationships to the cholesterol-binding, thiol-activated cytolysins (TAC) from *Streptococcus pyogenes*, *Streptococcus pneumoniae* and *Listeria monocytogenes* (Todd, 1941; Bernheimer, 1976; Cowell and Bernheimer, 1977; Geoffroy and Alouf, 1984). Recently, the amino acid and nucleotide sequences of pneumolysin, streptolysin O and θ-toxin have been determined (Paton *et al.*, 1986; Kehoe *et al.*, 1987; Tweten, 1988a, b). Impressive homology exists among the amino acid sequences of these TAC toxins, particularly in the region containing the cysteine residue near the amino-terminus, where a highly conserved segment of 12 amino acids is identical for all three toxins. Some of these TAC toxins, including θ-toxin, have been shown to facilitate the growth of TAC-toxin-producing organisms within mammalian phagocytic cells (Portnoy *et al.*, 1992). Further, experimental animal studies have demonstrated protective efficacy of several antibody preparations against TAC toxins (Paton *et al.*, 1983; Bailey *et al.*, 1987; Bryant *et al.*, 1993). These studies support a principal role for thiol-activated cytolysins in the pathogenesis of their respective diseases.

Other exotoxins produced by *C. perfringens* type A include collagenase and hyaluronidase, and two possible leucocidins, the nu antigen and the 'non-alpha–delta–theta' haemolysin (Smith, 1975b). Investigations into the roles of collagenase and hyaluronidase in the pathogenesis of gas gangrene are only now being undertaken. Similarly, little is known regarding the roles of the nu antigen or the 'non-alpha–delta–theta' haemolysin, as the former is without lethal or necrotizing activity and the latter is rare in *C. perfringens* type A (Smith, 1975b).

Mechanisms of shock

Haemodynamic collapse is a common occurrence in patients with gas gangrene caused by *C. perfringens*. Two exotoxins, α-toxin (phospholipase C) and θ-toxin, which have been extensively characterized *in vitro*, contribute to the dramatic clinical course of patients with this infection (Smith, 1979).

Despite its frequency, the course of shock in gas gangrene has not been extensively studied in humans, but studies in experimental animals provide some important clues regarding the mechanisms of septic shock associated with *C. perfringens* infection. For example, a prompt reduction in cardiac index (CI) occurred in rabbits receiving either α-toxin or a crude toxin preparation (Asmuth *et al.*, 1995). Although many physiological mechanisms could contribute to such a reduction, a direct reduction in myocardial contractility (dF/dt) was subsequently demonstrated in isolated atrial strips bathed with r-α toxin (Asmuth *et al.*, 1995) (Table 35.3). As reflected by the increased mortality in the rabbits receiving r-α toxin and crude toxin, a greater reduction in CI was also measured in these groups compared with those receiving r-θ toxin or normal saline (Asmuth *et al.*, 1995) (Figure 35.5a). Similarly, a marked decline in mean arterial pressure (MAP) was observed in r-α toxin- and crude-toxin-treated rabbits, although these effects were delayed until the later stages of the experiment (Figure 35.5b) (Asmuth *et al.*, 1995). Thus, rabbits receiving α-toxin-containing toxin preparations maintained MAP in the face of a falling CI by an as yet uncharacterized compensatory mechanism for a brief period before hypotension ultimately occurred. The physiological mechanisms which maintained MAP did not include significant changes in heart rate or central venous pressure until the terminal stages of the experiments, if at all (Asmuth *et al.*, 1995).

In contrast, r-θ toxin-treated rabbits demonstrated changes most characteristic of 'warm shock', with profound falls in peripheral vascular resistance after approximately 1 h of toxin infusion (Stevens *et al.*, 1988b; Asmuth *et al.*, 1995). Interestingly, rabbits

TABLE 35.3 The effects of extracellular toxins of *Clostridium perfringens* on myocardial contractility

Toxin preparation	dF/dt at concentration		
	1	2	3
Crude exotoxin	136 ± 10	87 ± 5	36 ± 5
Purified α-toxin	141 ± 18	85 ± 24	22 ± 5
Purified θ-toxin	137 ± 14		157 ± 26
Purified θ-toxin plus L-cysteine	95 ± 7	85 ± 6	80 ± 6

Adapted from Stevens *et al.* (1988b).

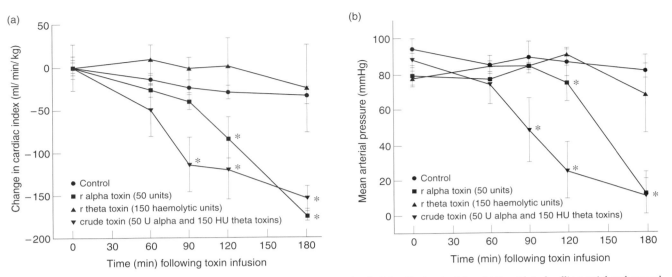

FIGURE 35.5 Effects of clostridial toxins on haemodynamic parameters. Awake New Zealand white rabbits with indwelling atrial and ascending aortic catheters were monitored for 1 h prior to obtaining baseline haemodynamic measurements. Animals were then infused either with sterile normal saline (control), or with crude or recombinant toxin preparations. Following toxin infusion, cardiac index (a) and mean arterial pressure (b) were measured every 15 min over a 3-h period. *$P < 0.05$ compared to control.

receiving crude toxin (containing both α- and θ-toxin activity) demonstrated peripheral vascular resistance (PVR) pressures intermediary to the other toxin groups. This latter observation provides insights into possible antagonistic interactions between r-θ and r-α toxin toxins in terms of PVR. Support for such antagonism is derived from experiments performed by Sakurai *et al.* (1985, 1991), who demonstrated that when sublethal doses of α-toxin purified from crude toxin were administered to rats, increases in blood pressure were observed over the course of the experiment (Sakurai *et al.*, 1985). However, in an *ex vivo* model of isolated rat skeletal muscle, purified α-toxin inhibited electrically stimulated muscle contraction after prolonged incubation in a dose-dependent manner (Sakurai *et al.*, 1991). In aggregate, these experiments suggest that α-toxin initially augments cardiac function, counteracting the vasodilatory effect of r-θ toxin. Later, decreased cardiac output and death occur.

Haemolysis and/or hypoxaemia have been described clinically, yet in experimental studies these factors alone were not responsible for mortality in either the r-α toxin- or crude-toxin-treated rabbits because rabbits that survived r-θ toxin treatment had similar degrees of haemolysis and hypoxaemia to those that did not survive.

These data suggest that the cardiovascular dysfunction of r-α toxin may be mediated by direct myocardial toxicity. This mechanism, by which r-α toxin contributes to haemodynamic changes, is supported by *in vitro* data (Regal and Shigeman, 1980) demonstrating inhibition of

the inotropic cardiac response in isolated embryonic chick heart preparations through a calcium-dependent enzyme. In addition, α-toxin may contribute indirectly to shock by stimulating production of endogenous mediators such as tumour necrosis factor (TNF) (Stevens and Bryant, 1997) and platelet-activating factor (Bunting *et al.*, 1997).

θ-Toxin probably contributes to septic shock through indirect routes including augmented release of TNF, interleukin-1 (IL-1) and IL-6 (Hackett and Stevens, 1992; Houldsworth *et al.*, 1994), platelet activating factor (PAF) and prostaglandin I_2 (PGI_2) (Whatley *et al.*, 1989). Perhaps θ-toxin-induced synthesis of nitric oxide by host cells such as macrophages or endothelial cells could also play a role in early hypotension. In addition, α- and θ-toxins may act synergistically in inducing hypotension, hypoxia and reduced cardiac output (Figure 35.5). This latter point should be considered when interpreting results from experiments that utilize isogenic mutants or single toxins.

Thus, shock associated with gas gangrene may be attributable, in part, to direct and indirect effects of toxins. α-Toxin directly suppresses myocardial contractility (Stevens *et al.*, 1988b), thereby contributing to profound hypotension via a sudden reduction in cardiac output (Asmuth *et al.*, 1995). θ-Toxin reduces systemic vascular resistance and markedly increases cardiac output (Stevens *et al.*, 1988b; Asmuth *et al.*, 1995). This afterload reduction occurs undoubtedly through induction of endogenous mediators that cause relaxation of blood vessel wall tension (Whatley *et al.*, 1989). Reduced vascular tone develops rapidly and, in order to maintain adequate tissue perfusion, a compensatory host response is required either to increase cardiac output or rapidly expand the intravascular blood volume. In contrast, patients with Gram-negative sepsis compensate for hypotension by markedly increasing cardiac output; however, this adaptive mechanism may not be possible in *C. perfringens*-induced shock owing to direct suppression of myocardial contractility by α-toxin (Stevens *et al.*, 1988b). The roles of other endogenous mediators such as IL-1 and IL-6, as well as the potent endogenous vasodilator, bradykinin, have not been elucidated.

The pathogenesis of tissue necrosis

The initiating trauma introduces organisms (either vegetative forms or spores) into the deep tissues, and produces an anaerobic niche with a sufficiently low redox potential and acid pH for optimal clostridial growth (MacLennan, 1962; Smith, 1975a). The rapid progression of infection and tissue necrosis is related to the absence of an acute tissue inflammatory response (Bryant *et al.*, 1993, 1995), to tissue perfusion deficits resulting from toxin-mediated vascular dysfunction and injury (Bryant *et al.*, 1998; Nagata *et al.*, 1998), and to the elaboration of potent cytotoxins and proteases.

Mechanisms of toxin-induced suppression of the acute inflammatory response

Several plausible mechanisms exist to explain the lack of a tissue inflammatory response in clostridial gas gangrene. First, an absence of bacterial- or host-derived chemoattractants could account for the paucity of leucocytes in the tissues; however, studies have shown that both an extracellular component of bacterial culture and serum incubated with killed bacilli were potent chemoattractants (Bryant *et al.*, 1993). Second, both α- and θ-toxins are cytolytic for leucocytes in high concentrations (Stevens *et al.*, 1987a, 1989), and destruction of any infiltrating phagocytes at the site of active bacterial proliferation and toxin elaboration probably contributes to the marked reduction of inflammatory cells in these areas. However, were this the only mechanism responsible for the absence of a tissue inflammatory response, one would expect to observe abundant phagocytes in tissues approaching the nidus of infection but abruptly halted at the point in the diffusion gradient where toxin concentrations reached cytolytic proportions.

Furthermore, leucocyte extravasation from vessels distal to the focus of infection would proceed unimpaired. Thus, cytotoxicity *per se* does not fully account for what is classically observed in both human and experimental cases of gas gangrene – the paucity of inflammatory cells in the tissues and the marked leucostasis in adjacent vasculature.

A third possible mechanism involves the effects of sublytic amounts of α- and θ-toxins on the function and interaction of leucocytes and endothelial cells. It was hypothesized that toxin-induced dysregulation of the normal, physiological mechanisms of leucocyte accumulation, adherence and extravasation, which orchestrate the pyogenic responses with other infections, could, in part, explain the leucostasis and anti-inflammatory response characteristic of clostridial gas gangrene. Further, these dysregulated events could lead to local and regional ischaemia, thereby extending the region for optimal clostridial proliferation (see the following sections).

Dose-dependent effects of clostridial toxins on adherence molecule expression and chemokine production by endothelial cells

Successful transmigration of leucocytes through the vessel and to the site of infection is the culmination of a complex cascade of both leucocyte- and endothelial cell (EC)-dependent adherence and activational events

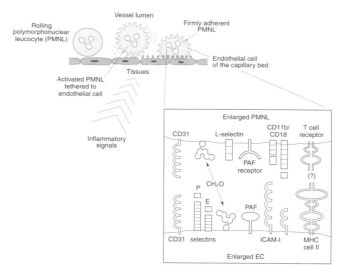

FIGURE 35.6 Schematic of proinflammmatory leucocyte/endothelial cell receptor–ligand interactions. Endothelial cells, activated by host- or bacterial-derived inflammatory signals, tether leucocytes from the circulating pool. This loose, selectin-dependent, adherence activates the leucocyte causing functional upregulation of CD11b–CD18 and tight adhesion to its ligand, ICAM-1, that is constitutively expressed on the endothelium. Local production of inflammatory mediators (cytokines and lipid autocoids) augments the inflammatory response by stimulating EC to increase expression of ICAM-1 and E-selectin. Activated leucocytes move to endothelial cell junctions and migrate between the cells and out into the tissue.

(reviewed in Bevilacqua, 1993) (Figure 35.6). Initially, the circulating, unactivated leucocyte is tethered to the activated EC and rolls along the vessel's luminal surface, processes that are mediated by selectins. Tethering results in juxtacrine activation of the leucocyte by PAF, functional upregulation of leucocyte CDllb–CD18 (MAC-1; CR3), and firm adhesion to intercellular adhesion molecule-1 (ICAM-1; CD54) constitutively expressed on the EC (Patel *et al.*, 1993). Local production of cytokines (e.g. TNF; IL-1) augments the inflammatory response by stimulating EC to produce the neutrophil chemoattractant/activator, IL-8, to increase ICAM-1 expression, and to express transiently endothelial leucocyte adhesion molecule-1 (ELAM-1; E-selectin; CD62E). Strongly adherent, activated leucocytes move to EC junctions and emigrate between endothelial cells, aided by platelet–endothelial cell adhesion molecule-1 (PECAM-1; CD31).

Work carried out in the authors' laboratory has shown that α-toxin strongly induces the expression of E-selectin and ICAM-1 on cultured human umbilical vein endothelial cells (Bryant and Stevens, 1996). The magnitude and duration of these responses were similar to those reported in other studies using either TNF or IL-1 (Pober *et al.*, 1986), or lipopolysaccharide (LPS) from Gram-negative organisms (Bevilacqua *et al.*, 1987). In addition, α-toxin stimulated production of endothelial cell-derived IL-8 (Bryant and Stevens, 1996). As with E-selectin expression, the dynamics of PLC-induced IL-8 synthesis was comparable to that induced by TNF or LPS (Whatley *et al.*, 1988). The local production of IL-8 in gas gangrene could amplify the recruitment of leucocytes and prime them for enhanced respiratory burst activity. Alternatively, Smith *et al.* have shown that neutrophils exposed to 100 mM IL-8 for 30 min lost the ability to transmigrate through an endothelial cell monolayer in response to an IL-8, but not an *N*-formyl-methionyl-leucyl-phenylalanine (fMLP), chemoattractant gradient, a process termed 'homologous desensitization' (Smith *et al.*, 1993). Exposure of neutrophils to higher IL-8 concentrations (>1000 nM) resulted in heterologous desensitization (Smith *et al.*, 1993). Similarly, Kitayama *et al.* showed that stimulation of neutrophils with fMLP prevented neutrophil transendothelial migration in response to fMLP or C5a, and that this cross-desensitization correlated with inhibition of neutrophil F-actin polymerization in response to chemotactic factor stimulation (Kitayama *et al.*, 1997).

Dose-dependent effects of clostridial toxins on neutrophil function

In contrast to α-toxin, θ-toxin caused a modest, but significant, increase in ICAM-1, had no effect on

E-selectin expression and did not induce detectable IL-8 synthesis (Bryant and Stevens, 1996). However, intramuscular injection of θ-toxin in mice produces marked vascular leucostasis adjacent to the site of toxin injection (Bryant *et al.*, 1993), suggesting that θ-toxin may impair the inflammatory response by primarily affecting neutrophil, rather than endothelial cell, function. For instance, it was previously shown that sublytic concentrations of θ-toxin dose-dependently stimulated random migration of neutrophils but decreased directed migration towards fMLP or a complement-derived chemoattractant (Stevens *et al.*, 1987a) and prevented F-actin polymerization by leucocytes in response to chemotactic factor stimulation (Bryant *et al.*, 1993) Whether homologous and/or heterologous desensitization of neutrophils contributes to the lack of phagocyte emigration into tissues infected with *C. perfringens* remains to be determined.

Proposed model for the role of α- and θ-toxins in the progression of tissue destruction and shock

Toxin-induced hyperadhesion of leucocytes (see the previous section) with enhanced respiratory burst activity, due to toxins directly (Stevens *et al.*, 1987a; Bryant *et al.*, 1993), to toxin-induced IL-8 (Bryant and Stevens, 1996) or PAF (Whatley *et al.*, 1989) synthesis by host cells, and toxin-induced chemotaxis deficits (Stevens *et al.*, 1987a; Bryant *et al.*, 1993), could result in neutrophil-mediated vascular injury. Direct toxin-induced cytopathic effects on EC may also contribute to vascular abnormalities associated with gas gangrene. Over prolonged incubation periods α-toxin, at sublytic concentrations, causes EC to undergo profound shape changes (Bryant and Stevens, 1996) similar to those described following prolonged TNF or interferon-γ (IFN-γ) exposure (Stolpen *et al.*, 1986). *In vivo*, conversion of EC to this fibroblastoid morphology could contribute to the localized vascular leakage and massive swelling observed clinically with this infection. Similarly, the direct cytotoxicity of θ-toxin could disrupt endothelial integrity and contribute to progressive oedema both locally and systemically.

Thus, via the mechanisms outlined above, both α- and θ-toxins may cause local, regional and systemic vascular dysfunction. For instance, local absorption of exotoxins within the capillary beds could affect the physiological function of the endothelium lining the post-capillary venules, resulting in impairment of phagocyte delivery at the site of infection. Toxin-induced endothelial dysfunction and microvascular injury could also cause loss of albumin, electrolytes and water into the interstitial space, resulting in marked localized oedema. These events, combined with leucostasis within the venule, would increase venous pressures and favour further loss of fluid and protein in the distal capillary bed. Ultimately, a reduced arteriolar flow would impair oxygen delivery, thereby attenuating phagocyte oxidative killing and facilitating anaerobic glycolysis of muscle tissue. The resultant drop in tissue pH, together with reduced oxygen tension, might further decrease the redox potential of viable tissues to a point suitable for growth of this anaerobic bacillus (Smith, 1975b). As infection progresses and additional toxin is absorbed, larger venous channels would become affected, causing regional vascular compromise, increased compartment pressures and rapid anoxic necrosis of large muscle groups. When toxins reach the arterial circulation, systemic shock and multiorgan failure rapidly ensue, and death is common.

Treatment

Aggressive debridement of devitalized tissue, as well as rapid repair of the compromised vascular supply and prophylactic antibiotics, greatly reduce the frequency of gas gangrene in contaminated deep wounds (Bartlett, 1990; Gorbach, 1992). Intramuscular epinephrine, prolonged application of tourniquets and surgical closure of traumatic wounds should be avoided. Patients presenting with gas gangrene of an extremity have a better prognosis than those with truncal or intra-abdominal gas gangrene, largely because it is difficult adequately to debride such lesions (Hart *et al.*, 1983; Bartlett, 1990; Gorbach, 1992). In addition, patients with associated bacteraemia and intravascular haemolysis have the greatest likelihood of progressing to shock and death. Patients who are in shock at the time that diagnosis is made have the highest mortality (Hart *et al.*, 1983).

Penicillin, clindamycin, tetracycline, chloramphenicol, metronidazole and a number of cephalosporins have excellent *in vitro* activity against *C. perfringens* and other clostridia. No controlled clinical trials have been conducted to compare the efficacy of these agents in humans. Based strictly on *in vitro* susceptibility data, most textbooks state that penicillin is the drug of choice (Bartlett, 1990; Gorbach, 1992). However, experimental studies in mice suggest that clindamycin has the greatest efficacy, and penicillin the least (Stevens *et al.*, 1987b, c). Other agents with greater efficacy than penicillin included erythromycin, rifampin, tetracycline, chloramphenicol and metronidazole (Stevens *et al.*, 1987b, c). Slightly greater survival was observed in animals receiving both clindamycin and penicillin; in contrast, antagonism was observed with penicillin plus metronidazole (Stevens *et al.*, 1987c). Because some strains (2–5%) are resistant to clindamycin, a combination of penicillin and clindamycin is warranted. Based

on his experimental studies and his vast clinical experience with gas gangrene, the late Dr William Altemeier recommended tetracycline and penicillin (Altemeier and Fullen, 1971). Thus, given an absence of efficacy data from a clinical trial in humans, the best treatment would appear to be clindamycin or tetracycline combined with penicillin.

The failure of penicillin in experimental clostridial myonecrosis may be related to continued toxin production by filamentous forms of the organism induced

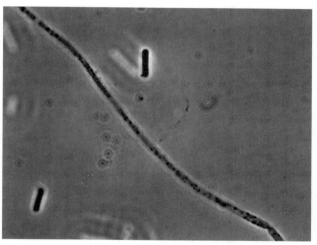

FIGURE 35.7 Penicillin-induced filament formation in *C. perfringens*. Fluid cultures of log phase *C. perfringens* (ATCC 13124) treated with penicillin at concentrations below its minimal inhibitory concentration demonstrated filamentous bacterial forms in approximately one-third of the bacteria. Some bacteria, like the one pictured here, had lengths 20–30 times untreated bacteria. No morphological abnormalities were observed in bacteria treated with other antibiotics.

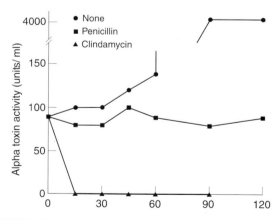

FIGURE 35.8 Dynamics of α-toxin suppression by antibiotics. Antibiotics were added to 10^7 log-phase *C. perfringens* (ATCC 13124) per ml to provide a concentration of drug that was 10-fold greater than the minimal inhibitory concentration. α-Toxin activity was measured in duplicate by radial diffusion in wells in blood agar plates. A zone diameter of 5 mm represents 100 units ml^{-1}.

by this cell wall active agent (Figure 35.7) (Stevens *et al.*, 1987d). In contrast, the efficacy of clindamycin and tetracycline may be related to their ability rapidly to inhibit toxin synthesis (Stevens *et al.*, 1987d) (Figure 35.8).

The use of hyperbaric oxygen (HBO) is controversial, although some non-randomized studies have reported excellent results with HBO therapy when combined with antibiotics and surgical debridement (Heimbach *et al.*, 1977; Hart *et al.*, 1983; Bakker, 1988). Some experimental studies demonstrate that HBO alone was an effective treatment if the inoculum was small and treatment was begun immediately (Hill and Osterhout, 1972b). In contrast, other studies have demonstrated that HBO was only of slight benefit when combined with penicillin (Stevens *et al.*, 1993). In these studies, however, survival rates were greater with clindamycin alone than with either HBO alone, penicillin alone or HBO plus penicillin together (Stevens *et al.*, 1993). The benefit of HBO, at least theoretically, is to inhibit bacterial growth (Hill and Osterhout, 1972a), preserve marginally perfused tissue and inhibit toxin production (van Unnik, 1965). Interestingly, Altemeier did not use HBO and was able to realize a mortality rate of less than 15% using surgical debridement and antibiotics (tetracycline plus penicillin) alone (Altemeier and Fullen, 1971).

Therapeutic strategies directed against toxin expression *in vivo*, such as neutralization with specific antitoxin antibody or inhibition of toxin synthesis, may be valuable adjuncts to traditional antimicrobial regimens. Currently, antitoxin is no longer available. Future strategies may target endogenous proadhesive molecules such that toxin-induced vascular leucostasis and resultant tissue injury are attenuated.

REFERENCES

Alpern, R.J. and Dowell, V.R., Jr (1969). *Clostridium septicum* infections and malignancy. *JAMA* **209**, 385–8.

Altemeier, W.A. and Fullen, W.D. (1971). Prevention and treatment of gas gangrene. *JAMA* **217**, 806–13.

Asmuth, D.A., Olson, R.D., Hackett, S.P., Bryant, A.E., Tweten, R.K., Tso, J.Y., Zollman, T. and Stevens, D.L. (1995). Effects of *Clostridium perfringens* recombinant and crude phospholipase C and theta toxins on rabbit hemodynamic parameters. *J. Infect. Dis.* **172**, 1317–23.

Awad, M.M., Bryant, A.E., Stevens, D.L. and Rood, J.I. (1995). Virulence studies on chromosomal α-toxin and θ-toxin mutants constructed by allelic exchange provide genetic evidence for the essential role of α-toxin in *Clostridium perfringens*-mediated gas gangrene. *Mol. Microbiol.* **15**, 191–202.

Bailey, J.T., Halsey, N.A. and Johnson, M.K. (1987). Prevention of *S. pneumoniae* bacteremia and mortality in rats by immunization with pneumolysin toxoid. In: *Interscience Conference on Antimicrobial Agents and Chemotherapy*, Abstract 895. American Society for Microbiology, Washington, DC.

Bakker, D.J. (1988). Clostridial myonecrosis. In: *Problem Wounds: The Role of Oxygen* (eds J.C. Davis and T.K. Hunt), pp. 153–72. Elsevier, New York.

Ballard, J., Bryant, A., Stevens, D. and Tweten, R.K. (1992). Purification and characterization of the lethal toxin (alpha-toxin) of *Clostridium septicum*. *Infect. Immun.* **60**, 784–90.

Bartlett, J.G. (1990). Gas gangrene (other clostridium-associated diseases). In: *Principles and Practice of Infectious Diseases* (eds G.L. Mandell, R.G. Douglas and J.E. Bennett), pp. 1851–60. Churchill Livingstone, New York.

Basak, A.K., Stuart, D.I., Nikura, T., Bishop, D.H., Kelly, D.C., Feam, A. and Titball, R.W. (1994). Purification, crystallization and preliminary X-ray diffraction studies of alpha-toxin of *Clostridium perfringens*. *J. Mol. Biol.* **244**, 648–50.

Bernheimer, A.W. (1947). Comparative kinetics of hemolysis induced by bacterial and other hemolysins. *J. Gen. Physiol.* **30**, 337–53.

Bernheimer, A.W. (1976). Sulfhydryl activated toxins. In: *Mechanisms in Bacterial Toxinology* (ed. A.W. Bernheimer), pp. 85–97. John Wiley, New York.

Bevilacqua, M.P. (1993). Endothelial-leukocyte adhesion molecules. *Annu. Rev. Immunol.* **11**, 767–804.

Bevilacqua, M.P., Pober, J.S., Mendrick, D.L., Cotran, R.S. and Gimbrone, M.A. (1987). Identification of an inducible endothelial-leukocyte adhesion molecule. *Proc. Natl Acad. Sci. USA* **84**, 9238–42.

Brook, I. (1989). Anaerobic bacterial bacteremia: 12-year experience in two military hospitals. *J. Infect. Dis.* **160**, 1071–5.

Bryant, A.E. and Stevens, D.L. (1996). Phospholipase C and perfringolysin O from *Clostridium perfringens* upregulate endothelial cell–leukocyte adherence molecule 1 and intercellular leukocyte adherence molecule 1 expression and induce interleukin-8 synthesis in cultured human umbilical vein endothelial cells. *Infect. Immun.* **64**, 358–62.

Bryant, A.E., Bergstrom, R., Zimmerman, G.A., Salyer, J.L., Hill, H.R., Tweten, R.K., Sato, H. and Stevens, D.L. (1993). *Clostridium perfringens* invasiveness is enhanced by effects of theta toxin upon PMNL structure and function: the roles of leukocytotoxicity and expression of CD11/CD18 adherence glycoprotein. *FEMS Immunol. Med. Microbiol.* **7**, 321–36.

Bryant, A.E., Awad, M.M., Lyristis, M., Rood, J.I. and Stevens, D.L. (1995). Perfringolysin O and phospholipase C from *Clostridium perfringens* impair host inflammatory cell defense mechanisms. In: *First International Conference on Molecular Biology and Pathogenesis of Clostridia*, Rio Rico, Arizona, Abstract B6, p. 21. American Society for Microbiology, Washington, DC.

Bryant, A.E., Chen, R., Nagata, Y., Bunting, M., Prescott, S., Zimmerman, G. and Stevens, D.L. (1998). Microvascular dysfunction and leukocyte adhesion are induced by *Clostridium perfringens* exotoxins *in vivo*. In: *First ASM Conference on Microbes, Haemostasis, and Vascular Biology*, Abstract 09.

Bunting, M., Lorant, D.E., Bryant, A.E., Zimmerman, G.A., McIntyre, T.M., Stevens, D.L. and Prescott, S.M. (1997). Alpha toxin from *Clostridium perfringens* induces proinflammatory changes in endothelial cells. *J. Clin. Invest.* **100**, 565–74.

Butler, H.M. (1943). Pathogenicity of washed *Cl. welchii* and mode of development of *Cl. welchii* infections in man. *Med. J. Aust.* **2**, 224–6.

Cowell, J.L. and Bernheimer, A.W. (1977). Antigenic relationships among thiol-activated cytolysins. *Infect. Immun.* **16**, 397–9.

Farnell, M.B. (1987). Neutropenic enterocolitis: a surgical disease? *Infect. Surg.* **6**, 120–32.

Geoffroy, C. and Alouf, J.E. (1984). Antigenic relationship between sulfhydryl-activated toxins. In: *Bacterial Protein Toxins* (ed. J.E. Alouf), pp. 241–3. Academic Press, London.

Gorbach, S.L. (1992). *Clostridium perfringens* and other clostridia. In: *Infectious Diseases* (eds S.L. Gorbach, J.G. Bartlett and N.R. Blacklow), pp. 1587–96. W.B. Saunders, Philadelphia, PA.

Gorbach, S.L. and Thadepalli, H. (1975). Isolation of *Clostridium* in human infections: evaluation of 114 cases. *J. Infect. Dis.* **131**, S81–5.

Gozal, D., Ziser, A., Shupak, A., Ariel, A. and Melamed, Y. (1986). Necrotizing fasciitis. *Arch. Surg.* **121**, 233–5.

Guerrant, R.L. (1990). Inflammatory enteritides. In: *Principles and Practice of Infectious Diseases*, Vol. 3 (eds G.L. Mandell, R.G. Douglas and J.E. Bennett), pp. 870–80. Churchill Livingstone, New York.

Hackett, S.P. and Stevens, D.L. (1992). Streptococcal toxic shock syndrome: synthesis of tumor necrosis factor and interleukin-1 by monocytes stimulated with pyrogenic exotoxin A and streptolysin O. *J. Infect. Dis.* **165**, 879–85.

Hart, G.B., Lamb, R.C. and Strauss, M.B. (1983). Gas gangrene: I. A collective review. *J. Trauma* **23**, 991–1000.

Heimbach, R.D., Boerema, I., Brummelkamp, W.H. and Wolfe, W.G. (1977). Current therapy of gas gangrene. In: *Hyperbaric Oxygen Therapy* (eds J.C. Davis and T.K. Hunt), pp. 153–76. Undersea Medical Society, Bethesda, MD.

Hill, G.B. and Osterhout, S. (1972a). Experimental effects of hyperbaric oxygen on selected clostridial species: I. *In vitro* studies. *J. Infect. Dis.* **125**, 17–25.

Hill, G.B. and Osterhout, S. (1972b). Experimental effects of hyperbaric oxygen on selected clostridial species: II. *In vivo* studies on mice. *J. Infect. Dis.* **125**, 26–35.

Houldsworth, S., Andrew, P.W. and Mitchell, T.J. (1994). Pneumolysin stimulates production of tumor necrosis factor alpha and interleukin-1 beta by human mononuclear phagocytes. *Infect. Immun.* **62**, 1501–3.

Ispolatovskaya, M.V. (1972). Type A *Clostridium perfringens* toxin. In: *Microbial Toxins* (eds S. Kadis, T.C. Montie and S.J. Ajl), pp. 109–58. Academic Press, New York.

Johnson, S., Driks, M.R., Tweten, R.K., Ballard, J., Stevens, D.L., Anderson, D.J. and Janoff, E.N. (1994). Clinical courses of seven survivors of *Clostridium septicum* infection and their immunologic responses to α toxin. *Clin. Infect. Dis.* **19**, 761–4.

Kehoe, M.A., Miller, L., Walker, J.A. and Boulnois, G.J. (1987). Nucleotide sequence of the streptolysin O (SLO) gene: structural homologies between SLO and other membrane-damaging, thiol-activated toxins. *Infect. Immun.* **55**, 3228–32.

Keppie, J. and Robertson, M. (1944). The *in vitro* toxigenicity and other characters of strains of *Cl. welchii* type A from various sources. *J. Pathol. Bacteriol.* **56**, 123–6.

Kitayama, J., Carr, M.W., Roth, S.J., Buccola, J. and Springer, T.A. (1997). Contrasting responses to multiple chemotactic stimuli in transendothelial migration. Heterologous desensitization in neutrophils and augmentation of migration in eosinophils. *J. Immunol.* **158**, 2340–9.

Lawrence, G. and Walker, P.D. (1976). Pathogenesis of enteritis necroticans in Papua New Guinea. *Lancet* **i**, 125–8.

Lawrence, G., Shann, F., Freestone, D.S. and Walker, P.D. (1979). Prevention of necrotizing enteritis in Papua New Guinea by active immunization. *Lancet* **i**, 227–30.

MacLennan, J.D. (1962). The histotoxic clostridial infections of man. *Bacteriol. Rev.* **26**, 177–76.

McGregor, J.A., Soper, D.E., Lovell, G. and Todd, J.K. (1989). Maternal deaths associated with *Clostridium sordellii* infection. *Am. J. Obstet. Gynecol.* **161**, 987–95.

McNee, J.W. and Dunn, J.S. (1917). The method of spread of gas gangrene into living muscle. *BMJ.* **i**, 727–9.

Murata, R., Soda, S., Yamamoto, A., Sato, H. and Ito, A. (1969). The effect of zinc on the production of various toxins of *Clostridium perfringens*. *Jpn. J. Med. Sci. Biol.* **22**, 133–48.

Nagata, Y., Stevens, D.L., Bryant, A.E., Guth, P.H., Finegold, S. and Chen, R.Y.Z. (1998). Microcirculatory derangements induced by clostridial exotoxins in rat skeletal muscle. In: *Program and Abstracts of the American Federation for Clinical Research*, Abstract. American Federation for Clinical Research, Thorofare, NJ.

Ninomiya, M., Matsushita, O., Minami, J., Sakamoto, H., Nakano, M. and Okabe, A. (1994). Role of alpha-toxin in *Clostridium perfringens* infection determined by using recombinants of *C. perfringens* and *Bacillus subtilis*. *Infect. Immun.* **62**, 5032–9.

Patel, K.D., Lorant, E., Jones, D.A., Prescott, M., McIntyre, T.M. and Zimmerman, G.A. (1993). Juxtacrine interactions of endothelial cells with leukocytes: tethering and signaling molecules. *Behring Inst. Mitt.* **92**, 144–64.

Paton, J.C., Lock, R.A. and Hansman, D.J. (1983). Effect of immunization with pneumolysin on survival time of mice challenged with *Streptococcus pneumoniae*. *Infect. Immun.* **40**, 548–52.

Paton, J.C., Merry, A.M., Lock, R.A., Hansman, D. and Manning, P.A. (1986). Cloning and expression in *Escherichia coli* of the *Streptococcus pneumoniae* gene encoding pneumolysin. *Infect. Immun.* **54**, 50–5.

Pober, J.S., Gimbrone, M.A., Jr, Lapierre, L.A., Mendrick, D.L., Fiers, W., Rothlein, R. and Springer, T.A. (1986). Overlapping patterns of activation of human endothelial cells by interleukin 1, tumor necrosis factor, and immune interferon. *J. Immunol.* **137**, 1893–6.

Portnoy, D.A., Tweten, R.K., Kehoe, M. and Bielecki, J. (1992). Capacity of lysteriolysin O, streptolysin O, and perfringolysin O to mediate growth of *Bacillus subtilis* within mammalian cells. *Infect. Immun.* **60**, 2710–17.

Regal, J.F. and Shigeman, F.E. (1980). The effect of phospholipase C on the responsiveness of cardiac receptors. I. Inhibition of the adrenergic inotropic response. *J. Pharmacol. Exp. Ther.* **214**, 282–90.

Robb-Smith, A.H.T. (1945). Tissues changes induced by *C. welchii* type a filtrates. *Lancet* **ii**, 362–8.

Sakurai, J., Oshita, Y. and Fujii, Y. (1985). Effect of *Clostridium perfringens* α-toxin on the cardiovascular system of rats. *Toxicon* **23**, 905–12.

Sakurai, J., Tsuchiya, Y., Ochi, S. and Fujii, Y. (1991). Effect of *Clostridium perfringens* α-toxin on contraction of isolated guinea-pig diaphragm. *Microbiol. Immunol.* **35**, 481–6.

Smith, L.D.S. (1975a). Clostridial wound infections. In: *The Pathogenic Anaerobic Bacteria* (ed. L.D.S. Smith), pp. 321–4. Charles C. Thomas, Springfield, IL.

Smith, L.D.S. (1975b). *Clostridium perfringens*. In: *The Pathogenic Anaerobic Bacteria* (ed. L.D.S. Smith), pp. 115–76. Charles C. Thomas, Springfield, IL.

Smith, L.D.S. (1979). Virulence factors of *Clostridium perfringens*. *Rev. Infect. Dis.* **1**, 254–60.

Smith, W.B., Gamble, J.R., Clark-Lewis, I. and Vadas, M.A. (1993). Chemotactic desensitization of neutrophils demonstrates interleukin-8 (IL-8)-dependent and IL-8-independent mechanisms of transmigration through cytokine-activated endothelium. *Immunology* **78**, 491–7.

Stevens, D.L. (1995). Clostridial infections. In: *Atlas of Infectious Diseases* (eds D.L. Stevens and G.L. Mandell), pp. 13.1–9. Churchill Livingstone, Philadelphia, PA.

Stevens, D.L. and Bryant, A.E. (1997). Pathogenesis of *Clostridium perfringens* infection: mechanisms and mediators of shock. *Clin. Infect. Dis.* **25**, S160–4.

Stevens, D.L., Mitten, J. and Henry, C. (1987a). Effects of alpha and theta toxins from *Clostridium perfringens* on human polymorphonuclear leukocytes. *J. Infect. Dis.* **156**, 324–33.

Stevens, D.L., Maier, K.A., Laine, B.M. and Mitten, J.E. (1987b). Comparison of clindamycin, rifampin, tetracycline, metronidazole, and penicillin for efficacy in prevention of experimental gas gangrene due to *Clostridium perfringens*. *J. Infect. Dis.* **155**, 220–8.

Stevens, D.L., Laine, B.M. and Mitten, J.E. (1987c). Comparison of single and combination antimicrobial agents for prevention of experimental gas gangrene caused by *Clostridium perfringens*. *Antimicrob. Agents Chemother.* **31**, 312–16.

Stevens, D.L., Maier, K.A. and Mitten, J.E. (1987d). Effect of antibiotics on toxin production and viability of *Clostridium perfringens*. *Antimicrob. Agents Chemother.* **31**, 213–18.

Stevens, D.L., Laposky, L.L., Montgomery, P. and Harris, I. (1988a). Recurrent gas gangrene at a site of remote injury: localization due to circulating antitoxin. *West. J. Med.* **148**, 204–5.

Stevens, D.L., Troyer, B.E., Merrick, D.T., Mitten, J.E. and Olson, R.D. (1988b). Lethal effects and cardiovascular effects of purified alpha- and theta-toxins from *Clostridium perfringens*. *J. Infect. Dis.* **157**, 272–9.

Stevens, D.L., Gibbons, A.E. and Bergstrom, R.A. (1989). Ultrastructural changes in human granulocytes induced by purified exotoxins from *Clostridium perfringens*. In: *Program and Abstracts of the American Society for Microbiology*, Abstract J17. American Society for Microbiology, Washington, DC.

Stevens, D.L., Musher, D.M., Watson, D.A., Eddy, H., Hamill, R.J., Gyorkey, F., Rosen, H. and Mader, J. (1990). Spontaneous, nontraumatic gangrene due to *Clostridium septicum*. *Rev. Infect. Dis.* **12**, 286–96.

Stevens, D.L., Bryant, A.E., Adams, K. and Mader, J.T. (1993). Evaluation of hyperbaric oxygen therapy for treatment of experimental *Clostridium perfringens* infection. *Clin. Infect. Dis.* **17**, 231–7.

Stolpen, A.H., Guinan, E.C., Fiers, W. and Pober, J.S. (1986). Recombinant tumor necrosis factor and immune interferon act singly and in combination to reorganize human vascular endothelial cell monolayers. *Am. J. Pathol.* **123**, 16–24.

Terebelo, H.R., McCue, R.L. and Lenneville, M.S. (1982). Implication of plasma free hemoglobin in massive clostridial hemolysis. *JAMA* **248**, 2028–9.

Titball, R.W. and Rubidge, T. (1990). The role of histidine residues in the alpha toxin of *Clostridium perfringens*. *FEMS Microbiol. Lett.* **56**, 261–5.

Todd, E.W. (1941). The oxygen-labile haemolysin for θ toxin of *Clostridium welchii*. *Br. J. Exp. Pathol.* **22**, 172–8.

Tweten, R.K. (1988a). Cloning and expression in *Escherichia coli* of the perfringolysin O (theta-toxin) gene from *Clostridium perfringens* and characterization of the gene product. *Infect. Immun.* **56**, 3228–34.

Tweten, R.K. (1988b). Nucleotide sequence of the gene for perfringolysin O (theta-toxin) from *Clostridium perfringens*: significant homology with the genes for streptolysin O and pneumolysin. *Infect. Immun.* **56**, 3235–40.

van Unnik, A.J.M. (1965). Inhibition of toxin production in *Clostridium perfringens in vitro* by hyperbaric oxygen. *Antonie von Leeuwenhoek* **31**, 181–6.

Weinstein, L. and Barza, M. (1972). Gas gangrene. *N. Engl. J. Med.* **289**, 1129–31.

Whatley, R.E., Zimmerman, G.A., McIntyre, T.M. and Prescott, S.M. (1988). Endothelium from diverse vascular sources synthesize platelet activating factor. *Arteriosclerosis* **8**, 321–31.

Whatley, R.E., Zimmerman, G.A., Stevens, D.L., Parker, C.J., McIntyre, T.M. and Prescott, S.M. (1989). The regulation of platelet activating factor production in endothelial cells – the role of calcium and protein kinase C. *J. Biol. Chem.* **264**, 6325–33.

Williamson, E.D. and Titball, R.W. (1993). A genetically engineered vaccine against alpha-toxin of *Clostridium perfringens* protects against experimental gas gangrene. *Vaccine* **11**, 1253–8.

36

Toxin-induced modulation of inflammatory processes

Brigitte König, Andreas Drynda, Andreas Ambrosch and Wolfgang König

INTRODUCTION

Microbial toxins and pathogenicity factors – role in immunity and inflammation

Molecular biological tools, as well as the advancement in the understanding of microbial pathogenicity factors and immune effector functions, have broadened our scope for detailed studies on microbial cell interaction (König, W. *et al.* 1991a, b; König, B. *et al.*, 1994c, 1996b, c; Kotwal, 1997; Rappuoli and Montecucco, 1997).

The interaction of bacterial protein toxins, and more generally, pathogenicity factors with immune effector cells leads to a cascade of inflammatory host reactions with the generation of pro- and anti-inflammatory molecules. These mediators are either preformed (e.g. proteases, enzymes, histamine) within cells or newly generated upon activation of inflammatory effector cells (Marrack and Kappler, 1990; Gröner *et al.*, 1992; König, W. *et al.*, 1992a; Arnold et al., 1993; Arnold and König, 1996; Lucey *et al.*, 1996; Mantovani *et al.*, 1997; Trinchieri, 1998).

To the latter belong arachidonic acid metabolites, for example the prostanoids and among them the leukotrienes with chemotactic (LTB$_4$) as well as spasmogenic properties (LTC$_4$, LTD$_4$, LTE$_4$) (König, W. *et al.*, 1990a). Basal activities of cytokines, for example interleukins (IL-) 4 and 8, may be preformed in cells, but overall cytokines are synthesized during cell activation. They may attract inflammatory cells by binding to spe-cific receptors, induce a cellular network of communi-cation by the upregulation or solubilization of adhesion molecules and may determine T-, as well as B-cell activ-ities including isotype regulation (Baggiolini and Clark-Lewis, 1992; Del Prete *et al.*, 1993; Höpken *et al.*, 1996; Clerici *et al.*, 1997).

Immunological reactions seem to be regulated by T-helper (Th-1 or Th-2) cells expressing and releasing a different set of cytokines on activation: Th-1 cells with the release of IL-2, interferon-γ (IFN-γ) and Th2-cells with the release of IL-4 and IL-5; the latter cytokines are involved in the induction of allergic and inflam-matory reactions. Th-1/Th-2-derived cytokines favour T-cell-mediated immunity and defence against intracellular micro-organisms or antibody-mediated immunity against extracellular pathogens (De Waal Malefyt *et al.*, 1991; O'Garra and Murphy, 1994; Lamont and Adorini, 1996; Dai *et al.*, 1997; Estaquier and Ameisen, 1997; Jeannin *et al.*, 1997).

In discussing human diseases, the nomenclature of type-1 (Th-1-like) and type-2 (Th-2-like) cytokines is used, which includes all cell types producing these cytokines rather than only CD4$^+$ T-cells. Type-1 cytokines include IFN-γ, IL-12 and tumour necrosis fac-tor-beta (TNF-β), while type-2 cytokines include IL-4, IL-5, IL-6, IL-10 and IL-13. In general, type-1 cytokines favour the development of a strong cellular immune response, whereas type-2 cytokines facilitate a humoral immune response. Some of these type-1 and type-2 cytokines are cross-regulatory. For example, IFN-γ and

IL-12 decrease the levels of type-2 cytokines, whereas IL-4 and IL-10 decrease the levels of type-1 cytokines (Punnonen et al., 1991; de Vries and Yssel, 1996; Romani et al., 1997; Trinchieri, 1998).

IL-12, a heterodimeric cytokine comprising p35 and p40 chains, originally described as a factor that promotes both natural killer (NK)-cell and cytotoxic T-lymphocyte (CTL) activity, plays a major role in microbial immunity. Furthermore, this cytokine has been found to induce IFN-γ secretion, promote growth of activated T and NK cells, modulate IgE synthesis and induce commitment from the T-helper$_O$ (Th-O) to the Th-1 phenotype. More recently, cells other than CD4$^+$ T-cells, including CD8$^+$ T-cells, monocytes, NK cells, B-cells, eosinophils, mast cells, basophils and other cells, have been shown to be capable of producing Th-1 and Th-2 cytokines.

The activity of Th-1 and Th-2-cells is also regulated by cytokines which modulate during infection the rate of apoptosis. In sepsis, an imbalance of Th-1/Th-2 is observed and this may be due to a Th-2 preponderence to elevation of IgE and eosinophilia. Apoptosis may thus represent a subtle and finely tuned system to initiate and prolong, as well as determine, immunological and inflammatory reactions.

T-cell–B-cell interaction is facilitated by adhesion molecules, for example CD40–CD40-ligand (CD40L) interactions (Armitage et al., 1993; Bonnefoy et al., 1996). In addition, CD23, the low-affinity receptor for IgE, exerts the role of an adhesion as well as an antigen-presenting molecule (König, W. et al., 1991b). Its expression and solubilization are regulated by IL-4 and IFN-γ. Surprisingly, during activation of inflammatory cells (e.g. mast cells) cytokines which are involved in IgE-synthesis can also be released from B-cells. In addition, the CD40L-induced activation of B-cells can be carried out by mast cells in the absence of T-cells.

Thus, microbial pathogenicity factors are able to induce multiple steps of cellular activation leading to Th-1- or Th-2-regulated immune effector functions. Direct inflammatory mediator release by these factors may thus bypass T-cell recognition and activation, and directly influence B-cell activation. This may lead to immunosuppression or even hypersensitivity reactions with the generation of IgE (König, W. et al., 1994a; Jabara and Geha, 1996; Galli, 1997). The events described above are mediated via a network of defined cytokines and cell–cell interactions, as well as signal transduction cascades (Hensler et al., 1991; Arnold et al., 1995b; Huang et al., 1996; Kagnoff and Eckmam, 1997).

It has become evident in the past years that not a single factor but a complex mixture of pathogenicity factors including bacterial toxins may be released and therefore determine the immunopathological reactions.

Furthermore, micro-organisms may secrete pathogenicity factors in relation to the cellular environment or to the stage of the disease (Köller et al., 1993, 1996, 1997; Köller and König, 1995; Keel et al., 1997). In order to clarify these processes the following are described in this chapter, as representative examples.

1. *Staphylococcus aureus* leucotoxins which are bicomponent complexes consisting of S and F proteins expressed by different genetic loci (Hensler et al., 1994a; Prévost et al., 1994; see also Chapter 21 in this book).
2. *Staphylococcus aureus* native and mutant toxic shock syndrome toxin-1 (TSST-1) and enterotoxins which belong to the family of superantigens (Hensler et al., 1993; Drynda et al., 1995; Schlievert, 1997).
3. *Helicobacter pylori* expressing vacuolating toxin (VacA$^+$) or lacking the toxin (VacA$^-$).
4. *Pseudomonas aeruginosa* haemolytic and non-haemolytic phospholipase C and lipase.

Pseudomonas aeruginosa is involved, for example, in burns, sepsis and cystic fibrosis, and releases multiple toxins and pathogenicity factors with different actions on target cells (Bergmann et al., 1989; Friedl et al., 1992).

The predominance of one or the other pathogenicity factor may determine the pathophysiological outcome (König, W. et al., 1990b; Schlüter and König, 1990; Köller and König, 1995). The in vitro analysis of pathogenicity factors with regard to the release of inflammatory mediators has created a powerful tool with which to understand the cell biological reaction, as well as the regulation of various cytokines and cell–cell interaction. In this regard, many toxins are primarily powerful inducers of Th-1-cytokines (IL-2, IFN-γ, IL-12). They may activate and recruit a defined set of inflammatory mediators from defined target cells or may deactivate cells by interacting with various elements of the cellular signal transduction cascade (Brom and König, 1992a, b; Brom, C. et al., 1992; Brom, J. et al., 1993; Hensler et al., 1994b).

In the course of activation, the resulting deactivation for a subsequent response may occur, which leads to an unresponsiveness of the cell for a secondary stimulus. Thus, the outcome of a microbial cell interaction is the result of the nature of the pathogenicity factors, as well as the responsiveness and the given environment of target cells. The latter may by greatly influenced by the released inflammatory mediators including cytokines.

Clearly, our understanding as to the biological role of various toxins has been aided by the availability of mutant strains and mutant toxins (Ostroff et al., 1989; Bonventre et al., 1995). They represent powerful tools to analyse the toxin action as a single molecule or in a combined interaction, or may represent novel therapeutic strategies to combat toxin-induced deleterious effects.

In this chapter it is demonstrated that microbial toxins and their mutants may up- or downregulate immune effector functions leading to inflammation, allergy and immunosuppression with tissue destruction (Köller *et al.*, 1993).

BICOMPONENT TOXINS OF *STAPHYLOCOCCUS AUREUS*

Panton–Valentine leucocidin

Among the bicomponent toxins of *S. aureus* the Panton–Valentine leucocidin (Luk-PV) and γ-haemolysin consist of type S and F proteins. The secretion of the Panton–Valentine leucocidin (Luk-PV) but not of another leucocidin (Luk-R) from S. *aureus* strains is correlated with severe pyodermic infections (dermonecrosis). The effects of both Luk-PV and Luk-R (0–5000 ng) on inflammatory mediator release from human leucocytes were investigated (Hensler *et al.*, 1994a, b; König, B. *et al.*, 1994a, 1995a).

Luk-PV, but not Luk-R, induced a pronounced release of the vasodilator histamine from human basophilic granulocytes (up to 55%) and of enzymes (β-glucuronidase, up to 45%; lysozyme, up to 35%), chemotactic components leukotriene B_4 (42 ng/10^7 cells) and IL-8 (up to 33 ng/10^7 cells), and oxygen metabolites from human neutrophilic granulocytes. The results indicate that granulocytes play a central role in dermonecrosis; these *in vitro* data account for the histological picture of Luk-PV infections, characterized by local vasodilation, infiltration of granulocytes and a central necrotic area.

A potent chemotactic factor for neutrophils is the 5-lipoxygenase product, leukotriene B_4 (LTB_4), in addition to various cytokines (e.g. IL-8) and related peptides. The effect of leucocidin from *S. aureus* V8 strains (Luk-PV) on the generation of LTB_4 and its metabolites from human polymorphonuclear neutrophils (PMNs) was analysed. In this regard, the S and F components of leucocidin acted synergistically.

The calcium ionophore A23187 induced LTB_4 generation, and the metabolism of exogenously added LTB_4 into biologically less active omega-oxidized compounds was significantly decreased after leucocidin exposure. Activation of immune effector cells by leucocidin is accompanied by apoptosis of the cells. Priming of PMNs with granulocyte-macrophage colony-stimulating factor (GM-CSF), or G-CSF prior to leucocidin exposure, substantially increased toxin- and calcium ionophore A23187-induced LTB_4 formation.

The inhibitory effects of leucocidin on mediator release were accompanied by membrane damage and DNA fragmentation, which were both restored after pretreatment with GM-CSF, suggesting that the presence of co-stimulatory priming factors such as GM-CSF or G-CSF in the microenvironment of an inflammatory focus determines the pathophysiological effects induced by the leucocidin, and thus may prolong a chronic inflammatory signal.

For the induction of inflammatory mediators, GTP-binding proteins (G-proteins) are involved in the Luk-PV-activated signal transduction of PMNs. ADP-ribosylation of heterotrimeric G-proteins by cholera and pertussis toxins decreased the Luk-PV-induced LTB_4 generation. In contrast, ADP-ribosylation of the low-molecular weight G-proteins rho and rac by *Clostridium botulinum* exoenzyme C3 increased the Luk-PV-induced LTB_4 synthesis. The subsequent stimulation of Luk-PV-treated PMNs by either calcium ionophore A23187, sodium fluoride or formylmethionyl-leucyl-phenylalanine (fMLP) was significantly inhibited (Hensler *et al.*, 1994a, b). This decrease was paralleled by a loss of G-protein functions, including GTPase activity and GTP-binding capacity. An increase in G-protein functions was obtained with small amounts in Luk-PV.

In addition to the modulated G-protein functions, ADP-ribosylation of 24-, 40- and 45-kDa proteins by Luk-PV was detected. As shown in control experiments, the ADP-ribosylated 24-kDa proteins were not substrates for this exoenzyme C3. Introduction of ras p21 into digitonin-permeabilized PMNs was without effect on subsequent Luk-PV stimulation. In addition, the translocation of ras p21, ras GAP and 5-lipoxygenase into the membrane of Luk-PV-treated PMNs, as well as the expression of chemotactic membrane receptors for LTB_4 and fMLP, was significantly diminished.

Panton–Valentine leucocidin and γ-haemolysin

Clinical isolates harbour not only the two genes coding for Luk-PVL (S-protein: LukS-PVL; F-protein: LukF-PVL) but also the three genes encoding γ-haemolysin (S protein, HlgA, HlgB; F protein, HlgC) (Cooney *et al.*, 1993; Kamio *et al.*, 1993; Supersac *et al.*, 1993; Prévost *et al.*, 1994, 1995; Gouaux *et al.*, 1997).

Class S components were regarded as binding specifically membranes of target cells prior to the secondary interaction of class F components. Class S and class F components are interchangeable and yield toxins with genuine biological activity. Clinical isolates were obtained that harbour and express not only the two genes encoding for Luk-PVL, but also the three genes encoding the γ-haemolysin (Tomita and Kamio, 1997).

Two parameters of inflammation are presented by *in vitro* experiments: the release of histamine and of IL-8, respectively. Histamine released by mast cell and basophils has been implicated as a potentially important mediator of inflammatory diseases. In addition to its vasodilatatory action, histamine can affect the release of soluble mediators from other cell types. The various toxin combinations clearly differed with regard to histamine release. The potency was: LukS-PVL/LukF-PVL = LukS-PVL/HlgB > HlgC/LukF-PVL = HlgC/HlgB > HlgA/LukF-PVL = HlgA/HlgB (König, B. *et al.*, 1997a).

Effect of toxins on inflammatory effector cells

Our current view as to the role of inflammatory cells has been significantly changed by the recent increase in knowledge of cytokines, chemokines and chemokine receptors (Baggiolini and Dahinden, 1994).

Bacteria (e.g. *S. aureus*), their exoproducts, as well as viruses (e.g. RSV), have been implicated in the induction of allergic diseases (Arnold *et al.*, 1994, 1995a). Obviously, the mechanisms involved may be very different. During allergic disease, leucocytes infiltrate the affected tissues and release their mediators and cytokines and, thereby, the local inflammatory process is induced and maintained. The induction of cytokines (e.g. GM-CSF, IL-3, IL-5) from inflammatory cells by bacterial exoproducts (e.g. from *S. aureus*) will prime, for example, basophils, in a sense, that a secondary stimulus will lead to an upregulated and exaggerated inflammatory response. Furthermore, the inflammatory environment also influences cell adhesion events which induce binding of basophils via integrins with the subsequent release of preformed and newly generated mediators.

This reaction is not exclusive for the basophils and mast cells; in addition, eosinophils, neutrophils, as well as endothelial and epithelial cells, can be triggered by the various toxins in a very defined way (Hensler *et al.*, 1994b; Knol *et al.*, 1996; Harris *et al.*, 1997; Hachicha *et al.*, 1998).

In the course of cellular activation arachidonic acid and defined metabolites are released which interact and stabilize cytokine synthesis (e.g. IL-8). While each individual cell releases a specific pattern of leukotrienes (neutrophil = LTB_4, mast cells – LTC4/PGD2, eosinophil – LTC4, platelets – 12-hydroxyeicosatetraenoic acid) there is an additional potent amplification loop. In this regard an inflammatory stimulus may induce LTB_4 and the epoxide LTA4 – this metabolite is transformed by platelets into LTC4, by red blood cells to LTB_4 and by endothelial cells to LTC4. Clearly, this amplification loop

also modulates a cytokine and adhesion molecule expression (e.g. ICAM-1, VLA-4), as well as the potency of inflammatory mediator release and chemotactic receptor expression. Free arachidonic acid also promotes the apoptosis of cells.

Cytokines such as GM-CSF and IL-3 suppress apoptosis, enhance cell survival and, in case of inflammatory mediator release, they are supportive of continuous mediator release (Hilger *et al.*, 1991, 1992; Brom and König, 1992a, b; Brom, C. *et al.*, 1992; Hensler *et al.*, 1993, 1994a; Brom, J. *et al.*, 1995).

These observations suggest that toxins and related molecules in an individual and cell-specific manner may contribute to a continuous activation, as well as specific regulation, within the immune network. The resulting cascade of events may be specific for the individual toxin in a primary way and then proceed via common pathophysiological and cell biological pathways.

Monocyte endothelial–epithelial cell interaction

Cytokines released from monocytes by bacterial toxins may be crucial for the immunological and inflammatory response of endothelial and epithelial cells (Mantovani *et al.*, 1997; Rasmussen *et al.*, 1997). The effects of different *Helicobacter pylori* strains on the immune response of THP1 monocytes were investigated. *Helicobacter* strains were obtained from gastric specimens and were subsequently characterized by their potency of producing vacuolating cytotoxin A (VacA) and their expression of specific mRNA encoding *vacA* and *cagA* gene. Mono-chamber incubation of THP1 monocytes (10^6 cells ml^{-1}) with toxin-positive *Helicobacter* strains (10^7 bacteria ml^{-1}) induced a nearly 10-fold higher increase in TNF-α secretion in comparison to toxin-negative strains. Similar results were

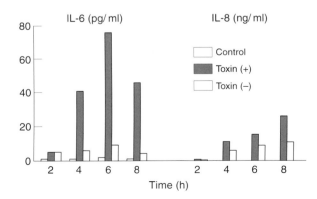

FIGURE 36.1 Cytokine (IL-6, IL-8) release from monocytes induced by toxin-positive/toxin-negative *Helicobacter pylori* strains.

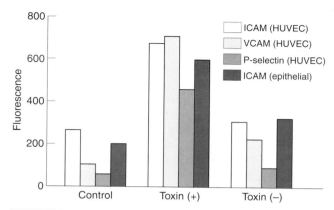

FIGURE 36.2 Expression of adhesion molecules on endothelial cells after activation of monocytes on interaction with toxin-positive vs toxin-negative *Helicobacter pylori* strains.

obtained with regard to IL-6 and IL-8. Double-chamber incubation with *Helicobacter* induced TNF-α secretion from monocytes only with toxin-positive strains, therefore indicating an exclusive effect of the vacuolating toxin (Figure 36.1).

As cytokines play a critical role in the pathogenesis of mucosal and vascular inflammation, dysregulation of monocyte cytokine production induced by bacteria might contribute to mucosa epithelial and vascular endothelial activation and dysfunction (Ye *et al.*, 1997). In a double-chamber incubation model, THP1 monocytes – co-stimulated with toxin-positive *Helicobacter* strains – enhanced ICAM-1 and induced VCAM and P-selectin expression on human endothelial cell, and also enhanced ICAM-1 expression on gastric mucosa epithelial cells. These results indicate that epithelial and endothelial cells may be not only the primary target but also the secondary target of infection (Figure 36.2).

IL-8 generation – a prototype of chemokine

IL-8, a prototype of C–X–C chemokines with a potent neutrophil chemotactic activity, exhibits multiple functions against non-leucocytic as well as leucocytic cells, thereby contributing to the establishment of inflammatory reactions. Increased IL-8 levels have been observed in microbial infection, for example in sputa of cystic fibrosis patients (Baggiolini and Clark-Lewis, 1992; Baggiolini and Dahinden, 1994; Schwartz *et al.*, 1997).

Accumulating evidence indicates that not only bacteria but also various types of viruses and viral products can induce IL-8 production (Arnold *et al.*, 1994, 1995a). Infection and microbial activation of inflammatory cells may also increase gene expression of a specific receptor for IL-8, the C–X–C R1. IL-8, in addition to its chemotactic and granulocytic activating and secretory properties, has been implicated in reducing the potency of IFN-α *in vivo*, thus aggravating viral infections. The upregulation and expression of chemokine receptors may thus facilitate viral entry, as is observed in HIV. Thus, bacterial exoproducts and toxins (e.g. leucocidins) may also contribute via IL-8 to viral infection and entry (König, W. *et al.*, 1992b; König, B. *et al.*, 1994a).

The inductive capacity of various micro-organisms for IL-8 is different. Recently, it was shown that *Salmonella typhimurium*, *P. aeruginosa* and *S. aureus* strongly induced both IL-8 and the macrophage inflammatory protein (MIP-1α), whereas *S. pneumoniae*, *S. epidermidis* and the opportunistic yeast *Candida albicans* were less potent. *Saccharomyces cerevisiae* induced IL-8 secretion but failed to stimulate MIP-1α. Depending on the presence of the proinflammatory cytokines (e.g. TNF-α), either both cytokines or only IL-8 was released. Thus, various micro-organisms by virtue of distinct pathogenicity factors are able selectively to recruit inflammatory cells by the production of various chemokines and depending on the inflammatory environment. Chemokines bind and signal through seven G-protein-coupled transmembrane receptors (Brom, J. and König, W., 1992a, b; Brom, C. *et al.*, 1992).

IL-8 generation by *Staphylococcus aureus* leucocidins

IL-8 represents one important chemotactic factor for PMNs. The effects of *S. aureus* leucocidins on the release of this cytokine from human LMB were investigated. It is evident that LukS-PVL–LukF-PVL and LukS-PVL–HlgB, already at low concentrations, were the most potent mediators of IL-8 release, again followed by the toxins HlgC–LukF-PVL and HlgC–HlgB.

The toxins containing HlgA as S components were significantly less active. The results indicate that the individual subunits (S, F) differ in their activities. The following biological activities were obtained for the S components: LukS-PVL > HlgC > HlgA. The F components LukF-PVL and HLgB had similar activities (König, B. *et al.*, 1997a).

The potency of the toxin complex to induce inflammatory mediator release is predominantly determined by the S component. This component is responsible for the membrane interaction and binding. By analogy with α-haemolysin, γ-haemolysin and leucocidin may also form oligomeric, transmembrane channels in which an antiparallel β-sheet constitutes the primary membrane-embedded domain.

From several cell models it is concluded that the leucotoxins bind to a receptor linked to a divalent cation-selective channel or to the channel itself which is activated. Then, the leucotoxins open a second pathway by insertion into the membrane and subsequent formation of specific pores allowing an influx of ethidium bromide (Staali *et al.*, 1998). Thus, these toxins exert broad cellular activation with a diverse array of inflammatory mediators, and thereby modulate signal transduction of the activated cells. These complex events may lead to an overexaggerated response or finally to a paralysed immune system which is observed in shock and sepsis (Köller and König, 1995).

To identify the pivotal region responsible for the haemolytic function of LukS towards rabbit erythrocytes, a series of chimeric genes (lukS-PV–lukS) and mutant genes of lukS-PV was created and expressed in *Escherichia coli*. The results indicate that a two-residue segment (D12I13) of lukS is the minimum region essential for the haemolytic function of LukS towards rabbit erythrocytes (Nariya *et al.*, 1997). Such an approach may be helpful to develop novel strategies to treat staphylococcal infections which are due to bicomponent toxin generation.

SUPERANTIGENS AS IMMUNOLOGICAL EFFECTORS AND MODULATORS

T-lymphocytes recognize a wide variety of antigens through highly diverse cell-surface glycoproteins known as T-cell receptors (TCRs). These disulfide-linked heterodimers are comprised of α and β (or γ and δ) chains that have variable (V) and constant (C) regions homologous to those of immunoglobulins. However, unlike antibodies, which recognize the antigen alone, $\alpha\beta$TCRs recognize antigen only in the form of peptides bound to major histocompatibility complex (MHC) molecules. In addition, TCRs interact with a class of viral or bacterial proteins known as superantigens (SAgs) which stimulate T-cells bearing particular Vβ elements, leading to the massive release of T-cell-derived lymphokines such as IL-2 and tumour necrosis factor, usually followed by the eventual disappearance or inactivation of responding T-cells (Abrahmsén, 1995; Schlievert, 1997).

The structurally and immunologically best-characterized group of SAgs is *S. aureus* enterotoxins, which cause both toxic shock syndrome and food poisoning. Some *S. aureus* isolates also produce toxin shock syndrome toxin-1 (TSST-1), which has been implicated in the majority of cases of menstrual toxic shock. Other microbial proteins with superantigenic properties include the *Staphylococcal* exfoliative toxins, the *Streptococcal* pyrogenic exotoxins (Müller-Alouf *et al.*, 1996, 1997), *Mycoplasma arthritidis* mitogen (MAM) and *Yersinia pseudotuberculosis*-derived mitogen (YPM). In addition, mouse mammary tumour viruses (MMTVs) encode endogenous SAgs that enable these retroviruses to exploit the host immune system for their transmission.

It has also been proposed that SAgs derived from bacteria, mycoplasma or viruses may initiate autoimmune disease by activating T-cells specific for self-antigens. In addition to the classical 'T-cell superantigens' B-cell superantigens have recently been described (Silverman, 1997).

Toxic shock syndrome toxin (TSST-1) and mutant proteins in immunological and allergic reactions

Staphylococcus aureus TSST-1 plays a crucial role in TSS and, possibly, in Kawasaki syndrome and atopic dermatitis. In this regard, it seems that the capacity of TSST-1 to induce massive proliferation of T-cells and concomitant release of proinflammatory cytokines, particularly TNF-α, is an essential precursor of lethal shock.

TSST-1 is a 22-kDa single-chain polypeptide including 194 amino acid residues. It belongs to the group of superantigens, bacterial exotoxins, that strongly and specifically stimulate CD4$^+$ and CD8$^+$ T-lymphocytes by cross-linking the TCR with MHC class-II molecules on accessory or target cells. The three-dimensional structure of TSST-1 has recently been determined (Acharya *et al.*, 1994; Papageorgiou *et al.*, 1996). Structural domains responsible for specific biological activities have been studied. In this regard, it was shown that modification of one or two of the nine tyrosine residues leads to an 85% decrease in the mitogenic activity in a murine model.

The C-terminal region of TSST-1 (residues 88–194) implicated in mitogenic activity may be class-II binding sites, but are considered more likely to be TCR binding sites. Mutations at residues 132, 135 and 140 result in partial or total loss of mitogenic activity for murine lymphocytes. Amino acid exchange at position 135 abolished the lethality of TSST-1, as measured in a rabbit infection model of TSS (Bonventre *et al.*, 1995; Drynda *et al.*, 1995).

It has been shown that activation of peripheral blood mononuclear cells (PBMC) by TSST-1-results in the release of several cytokines. Among them, TNF-α is prominent in sepsis, IL-1α exhibits pyrogenic activity and IL-6 is a potent inducer of acute phase responses.

Recently, Bonventre *et al.* (1995) studied the effect of recombinant TSST-1 (p17), mutant toxin Y115A (tyrosine residue modified to alanine) and toxin H135A (histidine residue modified to alanine) in a murine system for mitogenic activity. The mutant H135A was nearly devoid of mitogenic activity for murine splenocytes, whereas Y115A showed a reduction in mitogenic activity of approximately 50% compared with p17. These findings were supported and extended in a human *in vitro* system. We investigated the role of the C-terminal structural unit of TSST-1 with regard to proliferation, cytokine release (TNF-α, IL-6 and IL-8), mRNA expression for IL-6, IL-8, IL-10, TNF-α and CD40L, synthesis of immunoglobulin E (IgE), IgA, IgG and IgM, CD23 expression, and soluble CD23 (sCD23) release from human PBMC. For this purpose, the recombinant wild-type TSST-1 (p17), and mutant toxins Y115A and H135A were analysed. The H135A toxin was inactive in a thymidine proliferation assay, whereas mutant Y115A was less active than p17 at low concentrations (0.5–0.05 ng ml^{-1}) (Figure 36.3).

B-cells can be stimulated to proliferate in response to cell contact-dependent signals provided by activated but not resting T-cells. In this regard, the CD40–CD40L system plays an important role (Armitage *et al.*, 1993; Bonnefoy *et al.*, 1996).

In the human system, antibodies specific for the surface antigen CD40, as well as the soluble CD40L, may substitute for the B- and T-cell contact. Our data show that mRNA for CD40L was strongly induced following activation with TSST-1 (p17) and to a lesser degree with the mutant toxin Y115A. In contrast, the H135A mutant toxin did not elicit mRNA expression for CD40L compared with the unstimulated control. Thus, the capacity of the various toxins to induce proliferation paralleled the capacity to induce CD40L mRNA expression. In addition to its role in proliferation, the CD40–CD40L system, in combination with IL-4 or IL-13, is involved in IgE synthesis and CD23 expression as well (Fischer and König, 1990; Bonnefoy *et al.*, 1996; de Vries and Yssel, 1996; Fasler *et al.*, 1998).

IFN-γ and IL-12 are known to counteract IL-4 effects. Indeed, it was shown that the staphylococcal superantigens (e.g. enterotoxins A and B, as well as TSST-1) induce significant amounts of IFN-γ and IL-12 (mRNA expression and protein secretion). IL-12 is a potent immunoregulatory cytokine. In several experimental models of bacterial, parasitic, viral and fungal infection, endogenous IL-12 is required for early control of infection and for generation, and perhaps maintenance, of acquired protective immunity, directed by T-helper type-1 (Th-1) cells and mediated by phagocytes. Treatment of animals with IL-12, either alone or as a vaccine adjuvant, has been shown to prevent disease by many of the same infectious agents, by stimulating innate resistance or promoting specific reactivity. Continued investigation into the possible application of IL-12 therapy to human infections is warranted by the role of the cytokine in inflammation, immuno-

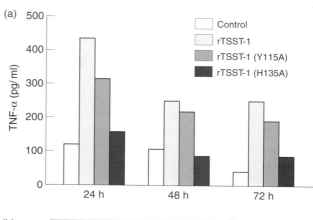

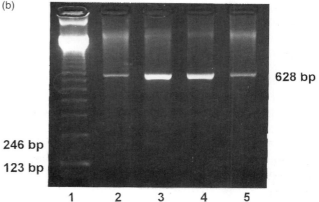

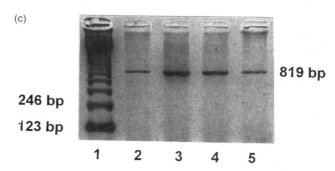

FIGURE 36.3 (a) TNF-α secretion elicited by recombinant TSST-1 and mutants. Human LMBs (1×10^6 ml^{-1}) were incubated with the toxins rTSST-1, Y115A and H135A (10 ng ml^{-1}) for 24, 48 and 72 h. The cell supernatant was assessed for TNF-α release (mean values of four independent experiments). (b) mRNA analysis for IL-6. Human LMBs (1×10^6 ml^{-1}) were activated with rTSST-1, Y115A and H135A (10 ng ml^{-1}) for 24 h. Total RNA of 10^6 cells was reverse transcribed and 10 μl was used for amplication (1, marker; 2, cells in medium; 3, rTSST-1 activation; 4, Y115A activation; 5, H135A activation) of cells with either 4 (Y115A) or 5 (H135A). (c) mRNA analysis for CD40 ligand. Human LMBs (1×10^6 ml^{-1}) were activated with rTSST-1, Y115A and H135A (10 ng ml^{-1}) for 24 h. Total RNA of 10^6 cells was reverse transcribed and 10 μl was used for amplication (1, marker; 2, cells in medium; activation of cells with: 3, rTSST-1; 4, Y115A; 5, H135A).

pathology and autoimmunity (Vercelli and Geha, 1993; Takatsu, 1997; Yoshimoto *et al.*, 1997; Trinchieri, 1998).

IL-10 possesses a wide range of activities on a number of cell types, and as a potent mediator of host response is released during injury and infection. It exerts significant anti-inflammatory activities as a cytokine synthesis inhibitor and is able to promote allergic responses with the activation of mast cells and eosinophils, as well as promoting Th-2-type cytokine release. A predominance of Th-2-type cytokines is found in allergy and sepsis, as well as in HIV-infected patients (Yamaguchi *et al.*, 1997; Ying *et al.*, 1997).

With regard to monocytes/macrophages, reported *in vitro* effects of IL-10 include the ability to downregulate class-II MHC antigens or monocytes, to suppress TNF-α and additional proinflammatory cytokine (IL-1, IL-6, IL-8) release. Moreover, IL-10 prevents IL-4-induced IgE synthesis by inhibiting accessory cell functions of monocytes (De Waal Malefyt *et al.*, 1991; König, W. *et al.*, 1991b, 1994a, b; Del Prete *et al.*, 1993).

Our data clearly show that toxin p17 and, to a lesser degree, toxin Y115, but not toxin H135A, induced mRNA expression for IL-10. Thus, it may be concluded that Ig synthesis, CD23 expression and sCD23 release are influenced by CD40L and T-helper cell (Th-1/Th-2) subset-specific cytokines.

The monokine TNF-α is a prominent factor in the onset of TSS and septic shock. In this regard, the injection of purified TNF-α either alone or with IL-1 into rabbits reproduced many of the features of TSS. Like Parsonnet and Gillis, we observed secretion and mRNA expression for TNF-α and IL-6 in human PBMC challenged with TSST-1.

Mutation of amino acid 115 diminished, and mutation of amino acid 135 abolished, the capacity of TSST-1 to induce TNF-α and IL-6 expression (mRNA and protein). TNF-α, IL-6 and IL-10 are produced by lymphocytes and monocytes as well. As a PBMC suspension (a mixture of T- and B-lymphocytes and monocytes) was used in our experiments, we cannot distinguish the source of expressed mRNA. However, the induction of TNF-α and IL-1β by purified staphylococcal TSST-1 requires the presence of both monocytes and T-lymphocytes.

Recombinant TSST-1 (p17) and mutant toxins Y115A and H135A bind equally well to MHC class-II-expressing antigen-presenting cells (murine B-cells and human lymphomas). The binding to T-cells can be inferred only from the functional assays (e.g. T-cell proliferation). As residues 132–140 are located within the second α-helix and are exposed and therefore solvent accessible, Acharya *et al.* (1994) speculated that these residues interact with the TCR. Because toxin H135A is the only mutant which demonstrates total absence of T-cell activation potential, histidine 135 would appear to be crucial for effective interaction between TSST-1 and the TCR (Drynda *et al.*, 1995). The structural change in the mutant, however, is minor, as H135A and p17 are antigenically similar and H135A elicits anti-TSST-1 neutralizing antibodies.

Superantigens in allergic and inflammatory responses

IL-4 and the recently described cytokine IL-13 have been described as switch factors for IgE synthesis. In addition to elevation of IgE levels, IL-4 has several other effects on PBMC. In this regard. CD23 expression is observed after stimulation with IL-4. CD23 is the low-affinity receptor for IgE (Fc$_\varepsilon$RII), represents a marker for activated B-cells and is an important differentiation molecule (König, B. *et al.*, 1995c, d).

CD23 is a glycosylated type-II membrane protein with a molecular weight of 45 kDa. It consists of a short intracellular cytoplasmic domain with 23 amino acids, a transmembrane domain and an extracellular domain of 277 amino acids. CD23 belongs to the lectin receptor familiy. It is expressed on B-cells, T-cells, monocytes, eosinophils, platelets and epidermal Langerhans cells. In humans two nearly identical receptors are present (e.g. CD23A and CD23B) which differ in seven amino acids of the intracellular amino-terminal end. CD23A is a stage-specific B-cell antigen marker, while CD23B is regulated by IL-4 and its role in allergic diseases and parasite host defence has been elucidated in the past. Recently, it has been shown that the staphylococcal enterotoxins A and B affect the regulation of IgE synthesis and CD23 expression on human PBMC depending on the presence of Th-1 and Th-2 cells. Activation of Th-1-cells leads to suppression of IgE synthesis and CD23 expression, unlike the activation of Th-2 cells (König, W. *et al.*, 1991b; Neuber *et al.*, 1991a; Neuber and König, 1992; Jabara and Geha, 1996).

As it had been reported that *S. aureus* toxins are involved in allergic and inflammatory disease processes (Neuber *et al.*, 1991b, 1992; Gelfand *et al.*, 1995), we analysed the influence of the recombinant wild-type p17, as well as recombinant mutant toxins Y115A and H135A on human PBMC with regard to proliferation, Ig synthesis, CD23 expression, sCD23 release, and cytokine expression and release.

Although toxin p17 induced IL-4 mRNA expression (data not shown), toxin p17 downregulated IL-4 as well as IL-4–aCD40-induced Ig synthesis and IL-4-induced CD23 expression and sCD23 release; mutant toxin Y115A was less active. Mutant toxin H135A did not modulate Ig synthesis, CD23 expression or sCD23 release. Thus, we may conclude that additional factors released from toxin (p17, Y115A or H135A)-treated PBMC influence Ig synthesis, CD23 expression and sCD23 release.

In the presence of TSST-1, Y115A suppressed IgE synthesis, unlike H135A. The effects on CD23 expression were less pronounced. A 50% reduction was observed with TSST-1, less with Y115A and no effect was obtained with H135A. Thus, our data suggest that the exchange of histidine by alanine at position 135 led to a loss in biological activity of the toxin. Therefore, histidine obviously plays a prominent role in the induction of immunological reactions by TSST-1 (Figure 36.4).

IgE regulation by toxins

Patients with atopic dermatitis are frequently colonized with *S. aureus* (Neuber *et al.*, 1991a, b; 1992; Neuber and König, 1992; König, W. *et al.*, 1994b; Hofer *et al.*, 1995;

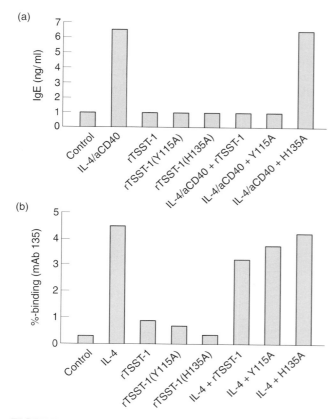

FIGURE 36.4 (a) IgE synthesis. Human LMBs (1×10^6 ml^{-1}) were incubated with the toxins rTSST-1, Y115A and H135A (10 ng ml^{-1}) in the absence and presence of rhIL-4 (1 nm)/aCD40 (100 ng ml^{-1}) for 10 days. IgE content was assessed in the cell culture supernatants by RIA (mean value of four independent experiments). (b) CD23 expression. Human LMBs (1×10^6 ml^{-1}) were activated with rTSST-1, Y115A and H135A (10 ng ml^{-1}) in the absence and presence of rhIL-4 (1 nm). The cell cultures were assessed after 5 days. Percentage binding (%-binding) of mAb 135 is indicated (mean value of four independent experiments).

Saloga *et al.*, 1996; Campbell and Kemp, 1997; Klubal *et al.*, 1997; MacGlashan *et al.*, 1998).

Superantigens could induce and promote allergic inflammatory reactions in the following way: the process is initiated following interaction of a native Ag with an Ig receptor on the surface of a B-cell, internalization and processing of the Ag–Ig complex, and presentation of an Ag fragment via a specific MHC class-II molecule on the cell surface (de Bentzmann *et al.*, 1996; Linehan *et al.*, 1998; Leung, 1998; Maggi, 1998).

The class-II–Ag complex is presented to a specific TCR on a Th-2 (or Th-O) cell, and signals transduced through the TCR–CD3 complex lead to CD40L expression on the surface of the T-cell. CD40L combines with CD40 present on the interacting B-cell, and transduced signals from the T-cell–B-cell interaction result in transcription and translation of the IL-4 gene. IL-4, released from the T-cell, interacts with IL-4R on Ag-specific B-cells and induces class-switching to IgE (producing a sterile ε transcript). Expression of the mature IgE transcript requires a 'second signal' derived from the CD40L–CD40 interaction. IgE can also be produced following non-cognate recognition via basophil–B-cell interaction (Park *et al.*, 1997; Punnonen *et al.*, 1997; Emson *et al.*, 1998; Hasbold *et al.*, 1998; Jeannin *et al.*, 1998).

This process is thought to be initiated following basophil activation, for example by Ag cross-linking of specific IgE molecules bound to Fc$_\varepsilon$RI and by bacterial peptides interacting with fMLP peptide receptors on the basophil. In addition to Fc$_\varepsilon$RI, IgE binds via Fc$_\varepsilon$RII (CD23) and IgE binding structures to inflammatory effector cells, thus leading to acute and chronic inflammatory diseases.

The presence of IgE enhances Fc$_\varepsilon$RI expression (Dahinden *et al.*, 1997). IgE receptors (Fc$_\varepsilon$RI and Fc$_\varepsilon$RII) are also present on Langerhans cells in the skin and are linked to the pathophysiology of atopic dermatitis (Kraft *et al.*, 1998; Rajakulasingam *et al.*, 1998). Isotype switching to IgE requires two signals. The first signal is provided by the cytokines IL-4 or IL-13, and the second signal is delivered by the interaction between the B-cell antigen CD40 and its ligand (CD40L) which is expressed on activated T-cells. In addition, co-stimulatory signals, such as CD23–CD21 interaction, contribute further, ensuring a selective control over this production. Recently, CD28, expressed on T-cells, has been reported to be involved in this process. The CD28 ligands, CD80 (B7-1) and CD86 (B7-2), are expressed on human tonsillar B-cells, and their expression is upregulated by IL-4, IL-13 and/or an anti-CD40 monoclonal antibody (mAb) (Armitage *et al.*, 1993; Bonnefoy *et al.*, 1996; Kimata *et al.*, 1996a, b; Barnes and Marsh, 1998; Fasler *et al.*, 1998).

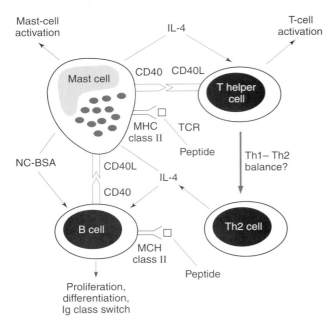

FIGURE 36.5 T-cell/B-cell regulation by mast cells. CD40–CD40L interaction, mast-cell-derived B-cell stimulatory activity (MC-BSA).

Engagement of peptide–MHC class-II complexes with the TCR leads to mast-cell activation and the induction of cytokine genes. The production of IL-4 by mast cells results in differentiation of T-cells into the Th-2 phenotype. Mast cells also produce a soluble factor (MC-BSA) which induces B-cell proliferation and differentiation. In humans, mast cells have been shown to express CD40L and can physically interact with B-cells. This interaction, together with production of IL-4, induces B-cells to differentiate into IgE-producing plasma cells. The presence of MHC class-I and class-II molecules on mast cells allows binding to, and clonal expansion of, CD8$^+$ or CD4$^+$ T-cells, respectively (Figure 36.5).

T-cell expansion is regulated by the finely tuned expression of B7 (CD80 and CD86) on mast cells (Wu *et al.*, 1997). MHC class-II-dependent Ag presentation in mast cells is strictly limited to exogenous Ags and SAgs, as endogenous self-peptides and SAgs cannot be presented (Mécheri and David, 1997).

Mast-cell mediators

Mast cells can be activated for mediator release (histamine, PG, LT and cytokines) by bacterial components, parasites and complement components through conserved receptors (Frandji *et al.*, 1996; Kimata *et al.*, 1996a, b; Galli, 1997; Harris *et al.*, 1997; He and Walls, 1997; Levy *et al.*, 1997).

Among the diverse mediators, tryptase, a protease unique to the mast-cell secretory granule, is released in

substantial quantities into the respiratory tract of patients with inflammatory disease of airways. Tryptase stimulated a catalytic site-dependent release of IL-8 from epithelial cells and this was associated with upregulation of ICAM-1 expression and in the recruitment of granulocytes following mast-cell activation. Eosinophils, through the release of IL-1β and TNF-α, might participate in the amplification of the inflammatory reaction by activating the vascular endothelium.

Staphylococcal enterotoxin B and mutants – effects of cytokines and IgE regulation

Mutants were obtained by site-directed mutagenesis. The SEB–T-cell receptor mutant (rSEB–TCR-mut) shows at positions 23 and 60 an exchange of asparagine, while the SEB–MHCII mutant shows an exchange at position 45 of lysine and at position 89 of tyrosine. The SEA T-cell receptor binding mutant (rSEA–TCR-mut = Y64A) shows at position 64 an exchange of tyrosine against alanine.

Human LMBs were stimulated with the various toxins for 24, 48 and 72 h, and the supernatant of the stimulated cells was assessed for TNF-α, IFN-γ, IL-6, IL-10, IL-12 and IL-13 release. For the various cytokines,

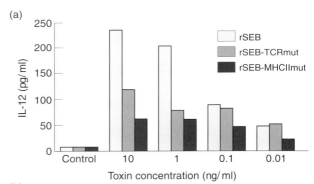
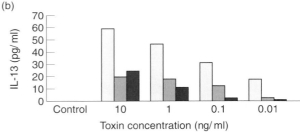

FIGURE 36.6 (a) Release of IL-12. Human LMBs (1×10^6 ml^{-1}) were incubated with the toxins rSEB, rSEB–TCR-mut and rSEB–MHCll-mut (a), (b) (0.01–10 ng ml^{-1}). The cell culture supernatants were assessed after 48 h for IL-12 release (mean value of four independent experiments). (b) Release of IL-13. Human LMBs (1×10^6 ml^{-1}) were incubated with the toxins rSEB, rSEB–TCR-mut and rSEB–MHCII-mut (0.01–10 ng ml^{-1}). The cell culture supernatants were assessed after 48 h for IL-13 release (mean value of four independent experiments).

differences in the optimum of induction were observed. However, the rSEB proved to be more active than the rSEB–TCR mutant. The rSEB–MHCII mutant proved to be less active (Figure 36.6). Thus, these results clearly indicate that mutated toxin molecules are less biologically active. The TCR mutant showed a 50% remainder activity compared with the rSEB.

Synthetic peptides from staphylococcal enterotoxin B

Synthetic peptides were obtained from SEB, and comprised the regions 21–31 (peptide I), 93–107 (peptide II) and 202–212 (peptide III). With regard to proliferation, peptide III showed 50% remainder activity compared with the native SEB; no activity was obtained for peptides I and II. In the aCD40–IL-4-induced IgE synthesis model, the native SEB reduced IgE synthesis by 80% while peptide III at similar concentrations showed a reduction of 50%, unlike peptide I and peptide II which were ineffective.

Superantigens and mutant proteins in therapy

The studies on superantigens have facilitated our knowledge on inflammatory, as well as anergic, disease processes. Structural considerations as well as mutated proteins are excellent approaches to study T-cell receptor and MHC class-II binding prerequisites for their biological activities (Salgame *et al.*, 1991; Swaminathan *et al.*, 1992; Acharya *et al.*, 1994; Abrahmsén, 1995; Wen *et al.*, 1996; Papageorgiou and Acharya, 1997; Woody *et al.*, 1997). Clearly, superantigens can bridge lymphocyte–lymphocyte, as well as non-lymphocyte–lymphocyte, interactions.

The various inflammatory mediators derived from, for example, mast cells, eosinophils during superantigen or bacterial exotoxin activation may then skew the immune system depending on the inflammatory environment (Neuber *et al.*, 1992; Lamkhioued *et al.*, 1996; Nakajima *et al.*, 1996; Patella *et al.*, 1996). While superantigens predominantly induce a Th-1 response and IL-12 induction, the presence of IL-4 and IL-5 may lead to a predominance of Th-2 cells which are then further amplified by superantigens as Th-2 cytokine-producing cells. This happens in allergy but also in shock and immunosuppression.

Allergic diseases proceed via two phases:

(a) a Th-2 cytokine response is responsible for allergic inflammation

(b) the chronicity of disease is governed by Th-1 cytokine responses.

Thus, SAgs contribute to the regulation of normal immunity as well as the progression of acute to chronic diseases, as observed in allergy and atopic dermatitis: Th-2 (acute inflammation) → Th-1 (chronic disease).

The concomitant release of exotoxins and superantigens, for example from *S. aureus*, may amplify this response as leucocidins directly induce the release of various cytokines (stored or newly generated) from inflammatory cells. The availability of mutant toxins as therapeutic strategy to counteract inflammatory and cell biological reactivity has to await further studies (Swaminathan *et al.*, 1992; Prasad *et al.*, 1993; Bonventre *et al.*, 1995; Drynda *et al.*, 1995; Köller and König, 1995; Papageorgiou *et al.*, 1996; Papageorgiou and Acharya, 1997; Silverman, 1997; Woody *et al.*, 1997; Li *et al.*, 1998).

Although, conversely, one could argue that superantigens may also present therapeutic bullets to boost or redirect an immune response prior to anergy and immunosuppression.

PSEUDOMONAS AERUGINOSA AND CYSTIC FIBROSIS

The role of microbial pathogenicity factors for disease

Pseudomonas aeruginosa is an opportunistic pathogen which causes either localized infections, such as pneumonia in patients with cystic fibrosis (CF), or generalized septicaemia in immunocompromised hosts suffering from severe burns, cancer or recurring immunosuppressive therapy (Bonfield *et al.*, 1995). Basically, the occurrence of such diverse clinical diseases may derive from either an altered host defence in these patients and/or a different virulence of taxonomically identical *P. aeruginosa* strains (König, W. *et al.*, 1991a, 1992b; Henderson *et al.*, 1996).

Mucosal surfaces are highly adapted to mediate interactions between the host and pathogenic microorganisms. Multiple innate and immunologically based mechanisms exist to prevent inadvertently inspired bacteria from reaching the epithelial surface. A massive accumulation of neutrophils mainly due to enhanced IL-8 levels is believed to contribute to the deleterious effects of *P. aeruginosa* lung infection in, for example, cystic fibrosis (König, B. *et al.*, 1995b).

Pseudomonas aeruginosa induces CD11/CD18-dependent emigration during acute pneumonia and recurrent pneumonia at previously uninflamed sites. However, adhesion pathways are altered in regions of chronic inflammation, and a greater proportion of neutrophil emigration occurs through CD11/CD18-independent pathways (Saiman and Prince, 1993; Moser *et al.*, 1997; Pier *et al.*, 1997; Saiman and Price, 1997; Yu *et al.*, 1998).

Pseudomonas aeruginosa produces multiple virulence factors that seem to contribute to its pathogenic properties. Exotoxin A modulates lymphocyte functions. The heat-labile phospholipase C (PLC) and the heat-stable (glycolipid) haemolysin have been shown to induce inflammatory mediator release from human granulocytes as well as mast cells. Alginate, the mucoid exopolysaccharide of *P. aeruginosa*, which is mainly present in isolates from patients with cystic fibrosis interferes with granulocyte functions, for example phagocytosis. Phospholipase C, an exoenzyme of *P. aeruginosa*, has been identified as a critical component in the pathogenesis of infection (Scharfman *et al.*, 1996). In addition, intratracheal instillation of highly purified CF sputum DNA caused acute inflammation similar to that induced by bacterial DNA. These findings suggest that bacterial DNA, and unmethylated CpG motifs in particular, may play an important pathogenic role in inflammatory lung disease (Schwartz *et al.*, 1997).

Bronchial secretions of cystic fibrosis patients, collected on the first day of admission for antibiotic treatment, showed a high chemotactic index. Fractionation by gel filtration of bronchial secretions resulted in three chemotactic fractions. The first factor corresponded to IL-8, and the second activated neutrophils via the fMLP receptor. The third factor, which was of lower molecular weight, did not activate fFMLP or leukotriene B_4 receptors, and its nature is still under investigation.

Human whole-blood cultures were incubated with several concentrations of purified *P. aeruginosa* products, including porins, exomucopolysaccharide, lipopolysaccharide and toxin A. All of the *P. aeruginosa* components, except for toxin A, were able to stimulate the release of TNF-α and IL-6. In addition, we showed that mucoid, unlike non-mucoid strains, led to a defined modulation of the neutrophil responses, i.e. decrease in chemiluminescence, a reduced leukotriene B_4 formation and differences in signal transduction activation. Mucoid bacteria induced a two-fold enhanced GTPase activity but activated the protein kinase C (PKC) to a lesser degree than non-mucoid *P. aeruginosa* bacteria. The interaction of pathogenicity factors from *P. aeruginosa* with platelets revealed that the purified glycolipid (heat-stable haemolysin) induced the generation of significant amounts of 12-hydroxyeicosatetraenoic acid (12-HETE) and serotonin from platelets (König, B. and König, 1993a; König, B. *et al.*, 1994b). These mediators are potent chemotactic factors, cause neutrophil degranulation, initiate mucus secretion and mediate leucocyte diapedesis through the vascular endothelium and inflammatory mediator release.

Stimulation of platelets with glycolipid was accompanied by significant changes in the signal transduction cascade, for example calcium influx translocation of PKC and increased GTPase activity).

These components may be responsible for the chronically overactive inflammatory response associated with persistent lung infection in cystic fibrosis patients.

Haemolytic and non-haemolytic phospholipase C

Pseudomonas aeruginosa produces at least two types of phospholipase C: haemolytic (PLC-H) and non-haemolytic (PLC-N). PLC-H (78 kDa) haemolyses human and sheep erythrocytes, and degrades not only phosphatidylcholine but also sphingomyelin, which are key components of eukaryotic cell membranes. PLC-N (78 kDa) is non-haemolytic and does not degrade sphingomyelin.

Human neutrophilic granulocytes (PMNL) as well as human peripheral blood mononuclear cells (PBMC) were treated with culture supernatants from clinical *P. aeruginosa* strains isolated at different disease stages of CF, with culture supernatants from *P. aeruginosa* PAO1 and mutants, expressing only haemolytic, non-

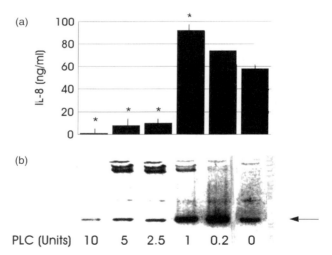

FIGURE 36.7 Effect of purified haemolytic phospholipase C (PLC) of *P. aeruginosa* on IL-8 release (ng/ml^{-1}) from human monocytes. Human monocytes (4×10^5/ml^{-1}) were incubated with purified PLC of *P. aeruginosa* at the indicated concentration (units) for 24 h. Values represent mean values ± SD of three independent experiments: (a) cytokine release and (b) cytokine expression. *Significant difference compared with unstimulated human monocytes (OU PLC) ($P < 0.05$).

haemolytic or not PLC, as well as with purified haemolytic *P. aeruginosa* PLC. Experiments were performed to evaluate the potential contribution of *P. aeruginosa* PLC to neutrophil accumulation during infection. Therefore, PLC-H and PLC-N were compared with regard to IL-8 generation from human monocytes and stimulatory effects of PLC-H on IL-8 release (Ostroff *et al.*, 1989, 1990; König, B. *et al.*, 1997b) (Figure 36.7).

The effects on IL-8 release induced by PLC-H were observed at concentrations that induce negligible contents of leukotriene B$_4$ (LTB$_4$). Conversely, concentrations that induce large amounts of LTB$_4$ showed small amounts of IL-8 release.

These data were supported by the use of mutant *P. aeruginosa* strains capable of producing only one kind of PLC or none at all. *Pseudomonas aeruginosa* PAOI mutant strains with a deletion of haemolytic PLC exhibited a marked reduction in virulence. PLC-N failed to modulate IL-8 release. In addition to IL-8, the haemolytic *P. aeruginosa* PLC has potent effects on immunomodulatory cytokines (e.g. IL-10). *Pseudomonas aeruginosa* culture supernatants expressing haemolytic PLC, as well as purified haemolytic PLC, induced IL-10 release from human PBMC in a dose-dependent manner.

Role of lipase and phospholipase C – co-operative properties of pathogenicity factors

We have previously shown that *P. aeruginosa* lipase and PLC, two extracellular lipolytic enzymes, interact with each other during 12-hydroxyeicosatetraenoic acid (HETE) generation from human platelets. In this regard, the addition of purified *P. aeruginosa* lipase to PLC-containing crude *P. aeruginosa* culture supernatants enhanced the generation of the chemotactically active 12-HETE from human platelets. Therefore, we analysed the interaction of purified *P. aeruginosa* lipase and purified haemolytic *P. aeruginosa* PLC with regard to inflammatory mediator release from human platelets, neutrophilic and basophilic granulocytes, and monocytes (Figure 36.8).

Purified *P. aeruginosa* PLC, but not purified lipase, induced in a dose-dependent manner 12-HETE generation from human platelets, leukotriene B$_4$ (LTB$_4$), oxygen metabolites and enzyme release from human neutrophils, histamine release from basophils and diminished IL-8 release from human monocytes. The addition of purified lipase enhanced PLC-induced 12-HETE and LTB$_4$ generation, did not influence enzyme, histamine or IL-8 release, but diminished the PLC-induced chemiluminescence response.

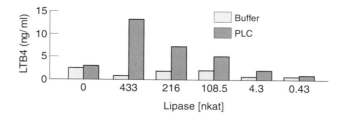

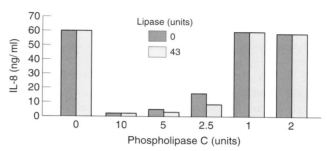

FIGURE 36.8 Co-operative effects of *P. aeruginosa* lipase and phospholipase C (PLC) on inflammatory mediator release. PMNL (1 × 10^7/500 µl) were left unstimulated with different concentrations of lipase and PLC. Release of LTB$_4$ and IL-8 was studied.

Similar results were obtained when the haemolytic PLC from *Clostridium perfringens* was used instead of *P. aeruginosa* PLC. For further comparison, we used the well-known calcium ionophore A23187 and phorbol-12-myristate-13-acetate (PMA) as stimuli. Lipase enhanced calcium ionophore-induced LTB$_4$ generation and β-glucuronidase release, but reduced calcium ionophore-induced and PMA-induced chemiluminescence. In parallel, we analysed the role of lipase in a crude *P. aeruginosa* culture supernatant containing PLC and lipase (Figure 36.8). Our results suggest that the simultaneous secretion of lipase and PLC by *P. aeruginosa* residing in an infected host may result in severe pathological effects which cannot be explained by the sole action of the individual virulence factor on inflammatory effector cells (Jaeger *et al.*, 1992; König, B. *et al.*, 1996a).

Clinical *Pseudomonas aeruginosa* isolates from cystic fibrosis – role of chronic colonization in inflammation

In the CF lung colonized with *P. aeruginosa* a complex mixture of inflammatory mediators is observed (Kammouri *et al.*, 1997a, b). Up to now the chronological order of inflammatory mediator generation in response to *P. aeruginosa* pathogenicity factors has not been clear (Li *et al.*, 1997). Two clones of *P. aeruginosa* and another species co-existed in four samples. Genomically homogeneous populations of *P. aeruginosa* are characteristic of chronically colonized lungs in most cases of cystic fibrosis (Schmidt *et al.*, 1996).

The distribution of bacterial populations in the airways of 13 patients with cystic fibrosis, who were colonized for 6–23 years with *P. aeruginosa*, was investigated by genotyping of bacterial chromosomes directly isolated from 21 sputa. *Pseudomonas aeruginosa* isolates of six CF patients in the initial early, chronic medium and late chronic disease phase were studied for the induction of inflammatory mediator release from human effector cells. *Pseudomonas aeruginosa* resulting from the onset of colonization and infection showed highest expression for haemolysins, PLC and lipase. *Pseudomonas aeruginosa* from this stage of disease induced significant amounts of LTB$_4$, lysosomal enzyme release and histamine release (König, B. *et al.*, 1994a, b, 1996a). *Pseudomonas aeruginosa* of late-stage disease was the most pronounced inducer of IL-8 but no longer of leukotriene B$_4$ and enzyme release (Figure 36.9). Our data thus indicate that the strategy of *P. aeruginosa* to change the expression pattern for pathogenicity factors is coupled to a change in the pattern of inflammatory mediators generated from human inflammatory effector cells. A precise knowledge of bacteria–cell interaction may contribute to the development of new strategies against chronic *P. aeruginosa* lung infection.

CONCLUSION

Bacteria have been able to develop a variety of sophisticated strategies of survival and of modification of host physiology in order to promote their own multiplication and spread. Micro-organisms interact with the immune system in multiple ways. In an interaction between a micro-organism and its host the defence of the host does not go unchallenged. This aggressive mechanism of interaction with the components of the host immune system allows the microbe not only to block the normal function of immune components on the surface of immune cells from functions, but also to obliterate a vital immune function, i.e. cellular as well as humoral immunity. Toxins are believed to increase the chance of survival and/or proliferation and/or spreading of a particular organism.

It is clear from our and other authors' results that the various microbial pathogenicity factors described here induce inflammatory mediator release to a different degree (König and König, 1993a, b). Obviously, not a single cell but the cellular interaction may decide the pathophysiological outcome. Clearly, the *in vitro* experiments permit a judgement to be made on the efficacy of the various pathogenicity factors in immunoregulation and inflammation. They also allow development of novel therapeutic strategies which after *in vitro* studies can be probed under *in vivo* conditions.

The inflammatory scenario is now envisaged as a result of a multitude of mediators, among which cytokines and low-molecular-weight mediators act together. Cytokines (IL-3, GM-CSF, IL-8) may prime the target cells or contribute to a differential release of low-molecular-weight mediators (Horie *et al.*, 1996).

In this regard, microbial exotoxins and superantigens as released effector molecules or expressed with

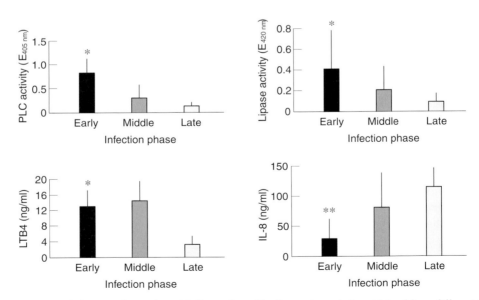

FIGURE 36.9 Cystic fibrosis and chronic colonization with *P. aeruginosa*. The *P. aeruginosa* strains obtained from different time intervals (early, middle, late in years) were characterized with regard to the phospholipase C and lipase activities. These strains were incubated with human peripheral cells, and their inductive capacity to generate leukotriene B$_4$ and IL-8 was analysed from human cells.

additional microbial components (e.g. adhesins or haemagglutinins) are powerful inducers and modulators of the inflammatory response (König *et al.*, 1991). They are powerful tools to activate as well as counteract host defence mechanisms. While in the past major emphasis was directed towards their cytotoxic role, it is increasingly evident that toxins at non-cytolytic concentrations modulate cellular functions.

Mediators serve as local hormones of the microenvironment. Toxins also modulate cellular functions by interacting with the signal-transduction cascade (e.g. protein kinase C, phosphoinositol turnover, G-proteins). Thus, toxins are powerful probes in cell biology and membrane biochemistry and also interfere with cell-mediated immune processes, such as antigen presentation and cytokine formation.

In a selective way microbial toxins and exoproducts may prime inflammatory effector cells and lymphocytes to a varying degree, and also initiate selective inflammatory mediator formation. An exotoxin which induces leukotriene formation and histamine release does not necessarily initiate cytokine production. Among the various cytokines the inductive capacity of microbial products differs to a great extent, as shown for the staphylococcal leucocidin, the superantigens and *Pseudomonas* ssp. exoproducts. Differences in cellular reactivity are also observed when different cells are studied. Thus, to a variable extent, inflammatory cells combine, or selective cell populations are primed, stimulated or deactivated.

Crucial, in this regard, is the microbial- or cytokine-induced upregulation of adhesion molecules on endothelial cells, the products of which initiate immunity, angiogenesis and inflammation.

The directional response of inflammatory cells by microbial products is facilitated by the generation and release of various chemokines, and the expression of receptors and adhesion molecules on the target cells. These chemokines prime inflammatory cells for a subsequent response. A downregulation of chemotactic factor responses and generation is obtained by the concomitant release of enzymes, for example elastase which cleaves IL-8. Conversely, the upregulation of chemotaxin receptors may additionally facilitate viral entry, and thus perpetuate and amplify infection.

Of utmost interest is the notion that microbial factors may skew the Th-O, Th-1 and Th-2 responses by inducing Th-1 or Th-2 cytokines from inflammatory effector cells. Cytokines produced by toxins from various cells also regulate the apoptosis of effector cells, thus contributing to acute and chronic inflammatory reactions or anergy. In this regard mast cells and eosinophils and non-T-cells can be regarded as IL-4-producing cells which on activation by, for example, superantigens or

microbial enzymes may facilitate allergic and inflammatory reactions as well as immunosuppression. The primary micro-organisms involved in the disease process may change their pathogenicity profile, as shown for cystic fibrosis. The *P. aeruginosa* strains revealed, over time, differences as to the pathogenicity factor expression which may serve as an additional explanation for the disease outcome.

In cystic fibrosis, a complex inflammatory reaction of acute and chronic inflammatory mediator generation concomitantly occurs and is decisive for inflammation, immunosuppression and tissue distruction.

Of major interest is the question of to what extent microbial pathogenicity factors may serve as therapeutic bullets to initiate an immune response in life-threatening immune paralysis and anergy or dampen an overexaggerated inflammatory reaction leading to acute and chronic diseases. Clearly, major progress requires precise knowledge of the major pathogenicity factors and their mode of action on immune effector cells, as well as the effect on cell biological signal transduction elements. In this regard, microbial factors may also deviate an immune response by interaction with cellular signalling and transduction system in a sense that a secondary cellular activation is impaired or abolished. Evidently, our knowledge of microbial toxins and exoproducts relies on work that is being carried out in multidisciplinary scientific fields, and further progress demands more interdisciplinary research and activities.

REFERENCES

Abrahmsén, L. (1995). Superantigen engineering. *Curr. Opin. Struct. Biol.* **5**, 464–70.

Acharya, K.R., Passalaqua, E.F., Jones, E.Y., Harlos, K., Stuart, D.I., Brehm, R.E. and Tranter, H.S. (1994). Structural basis of superantigen action inferred from crystal structure of toxic-shock syndrome toxin-1. *Nature* **367**, 94–7.

Armitage, R.J., Maliszewski, C.R., Alderson, M.R., Grabstein, K.H., Springs, M.K. and Fanslow, W. (1993). CD40 L: a multi-functional ligand. *Semin. Immunol.* **5**, 401–12.

Arnold, R. and König, W. (1996). ICAM-1 expression and low-molecular-weight G-protein activation of human bronchial epithelial cells (A549) infected with RSV. *J. Leukoc. Biol.* **60**, 766–71.

Arnold, R., Scheffer, J., König, B. and König, W. (1993). Effect of *Listeria monocytogenes* and *Yersinia enterocolitica* on cytokine gene expression and release from polymorphonuclear granulocytes and human epithelial. *Infect. Immun.* **60**, 2545–52.

Arnold, R., Humbert, B., Werchau, H., Gallati, H. and König, W. (1994). Interleukin-8, IL-6 and sTNF-RI release from a human pulmonary epithelial cell line (A549) exposed to respiratory syncytial virus (RSV). *Immunology* **82**, 126–33.

Arnold, R., König, B., Gallati, H., Werchau, H. and König, W. (1995a). Cytokine (IL-8, IL-6, TNF-α) and sTNFR-I release from human peripheral blood mononuclear cells after respiratory syncytial virus (RSV) infection. *Immunology* **85**, 324–72.

Arnold, R., Werchau, H. and König, W. (1995b). Increase of intercellular adhesion molecule-1 on human pulmonnary epithelial cells (A549) after respiratory syncytial virus (RSV) infection. In: *Effectors and Pathogenesis of Allergic Diseases. Int. Arch. Allergy Appl. Immunol.* **107**, 392–3.

Baggiolini, M. and Clark-Lewis, I. (1992). Interleukin-8, a chemotactic and inflammatory cytokine. *FEBS Lett.* **307**, 97–101.

Baggiolini, M. and Dahinden, C.A. (1994). CC chemokines in allergic inflammation. *Immunol. Today* **15**, 127–33.

Barnes, K.C. and Marsh, D.G. (1998). The genetics and complexity of allergy and asthma. *Immunol. Today* **19**, 325–32.

Bergmann, U. Scheffer, J., Köller, M., Schönfeld, W., Erbs, G., Müller, F.E. and König, W. (1989). Induction of inflammatory mediators (histamine and leukotrienes) from rat peritoneal mast cells and human granulocytes by *Pseudomonas aeruginosa* strains from burn patients. *Infect. Immun.* **57**, 2187–95.

Bonfield, T.L., Panuska, J.R., Konstan, M.W., Killiard, K.A., Hilliard, J.B., Ghnaim, H. and Berger, M. (1995). Inflammatory cytokines in cystic fibrosis lungs. *Am. J. Respir. Crit. Care Med.* **152**, 2111–18.

Bonnefoy, J.Y., Gauchat, J.F., Life, P., Graber, P., Mazzei, G. and Aubry, J.P. (1996). Pairs of surface molecules involved in human IgE regulation: CD23–CD21 and CD40–CD40L. *Eur. Respir. J. Suppl.* **22**, 63–6s.

Bonventre, P.F., Heeg, H., Edwards, C.K. and Cullen, C. (1995). A mutation at histidine 135 of toxic shock syndrome toxin 1 yields an immunogenic protein with minimal toxicity. *Infect. Immun.* **63**, 509–15.

Brom, C., Brom, J. and König, W. (1992). GTPases and low molecular weight G-proteins during cell–cell-interaction between neutrophils and platelets. *Int. Arch. Allergy Appl. Immunol.* **99**, 397–9.

Brom, J. and König, W. (1992a). Signal transduction and priming of human neutrophils. *Int. Arch. Allergy Appl. Immunol.* **99**, 387–9.

Brom, J. and König, W. (1992b). Cytokine-induced (IL-3, IL-6, IL-8 and TNF-β) activation and deactivation of human neutrophils. *Immunology* **75**, 281–5.

Brom, J., Köller, M., Müller-Lange, P.M., Steinau, H.U. and König, W. (1993). GTP-binding proteins in polymorphonuclear granulocytes of severely burned patients. *J. Leukoc. Biol.* **53**, 268–72.

Brom, J., Köller, M., Schlüter, B., Steinau, H.U. and König, W. (1995). Expression of the adhesion molecule CD11b and polymerization of action by polymorphonuclear granulocytes of burned and septic patients. *Burns* **21**, 427–31.

Campbell, D.E. and Kemp, A.S. (1997). Proliferation and production of interferon-gamm (IFN-gamma) and IL-4 in response to *Staphylococcus aureus* and staphylococcal superantigen in childhood atopic dermatitis. *Clin. Exp. Immunol.* **107**, 392–7.

Clerici, M., Fusi, M.L., Ruzzante, S., Piconi, S., Biasin, M., Arienti, D., Trabattoni, D. and Villa, M.L. (1997). Type 1 and type 2 cytokines in HIV infection – a possible role in apoptosis and disease progression. (Editorial.) *Ann. Med.* **29**, 185–8.

Cooney, J., Kienle, Z.l., Foster, T.J. and O`Toole, P.W. (1993). The gamma-hemoysin locus of *Staphylococcus aureus* comprises three linked genes, two of which are identical to the genes for the F and S components of leukocidin. *Infect. Immun.* **61**, 768–71.

Dahinden, C.A., Rhihs, S. and Ochsenberger, B. (1997). Regulation of cytokine expression by human blood basophils. *Int. Arch. Allergy Immunol.* **113**, 134–7.

Dai, W.J., Köhler, G. and Brombacher, F. (1997). Both innate and acquired immunity to *Listeria monocytogenes* infection are increased in IL-10-deficient mice. *J. Immunol.* **158**, 2259–67.

de Bentzmann, S., Plotkowski, C. and Puchelle, E. (1996). Receptors in the *Pseudomonas aeruginosa* adherence to injured and repairing airway epithelium. *Am. J. Respir. Crit. Care Med.* **154**, 155–62.

Del Prete, G., DeCarli, M., Almerigogna, F., Giudizi, M.G., Biagiotti, R. and Romagnani, S. (1993). Human IL-10 is produced by both type 1 helper (Th1) and type 2 helper (Th2) T cell clones and inhibits their antigen-specific proliferation and cytokine production. *J. Immunol.* **150**, 353–60.

de Vries, J.E. and Yssel, H. (1996). Modulation of the human IgE response. *Eur. Respir. J. Suppl.* **22**, 582–602.

De Waal Malefyt, R. and de Vries, J.E. (1993). Interleukin 13 induces interleukin 4 independent IgG4 and IgE synthesis and CD23 expression by human B cells. *Immunology* **90**, 3730–4.

Drynda, A., König, B., Bonventre, P.F. and König, W. (1995). Role of a carboxy-terminal site of toxic shock syndrome toxin 1 in eliciting immune responses of human peripheral blood mononuclear cells. *Infect. Immun.* **63**, 1095–101.

Emson, C.L., Bell, S.E., Jones, A., Wisden, W. and McKenzie, A.N.J. (1998). Interleukin (IL)-4-independent induction of immunoglobulin (Ig)E, and perturbation of T cell development in transgenic mice expressing IL-13. *Exp. Med.* **188**, 399–404.

Estaquier, J. and Ameisen, J.C. (1997). A role for T-helper type-1 and type-2 cytokines in the regulation of human monocyte. *Blood* **90**, 1618–25.

Fasler, S., Aversa, G., de Vries, J.E. and Yssel, H. (1998). Antagonistic peptides specifically inhibit proliferation, cytokine production, CD40L expression, and help for IgE synthesis by Der p 1-specific human T-cell clones. *J. Allergy Clin. Immunol.* **102**, 521–30.

Fischer, A. and König, W. (1990). Regulation of CD23 expression, soluble CD23 release and immunoglobulin synthesis of peripheral blood lymphocytes by glucocorticoids. *Immunology* **71**, 473–9.

Frandji, P., Tkaczyk, C., Oskeritzian, C., David, B., Desaymard, C. and Mécheri, S. (1996). Exogenous and endogenous antigens are differentially presented by mast cells to CD4[+] T lymphocytes. *Eur. J. Immunol.* **26**, 2517–28.

Friedl, P., König, B. and König, W. (1992). Effects of mucoid and nonmucoid *Pseudomonas aeruginosa* isolates from cystic fibrosis patients on inflammatory mediator release from human polymorphonuclear granulocytes and rat mast cells. *Immunology* **76**, 86–94.

Galli, S.J. (1997). The Paul Kallos Memorial Lecture. The mast cell: a versatile effector cell for a challenging world. *Int. Arch. Allergy Immunol.* **113**, 14–22.

Gelfand, E.W., Saloga, J. and Lack, G. (1995). Modification of immediate hypersensitivity responses by staphylococcal enterotoxin B. *J. Clin. Immunol.* **15**, Suppl., 37S–41S.

Gouaux, E., Hobaugh, M. and Song. L. (1997). Alpha-hemolysin, gamma-hemolysin, and leukocidin from *Staphylococcus aureus*: distant in sequence but similar in structure. *Protein Sci.* **6**, 2631–5.

Gröner, M., Scheffer, J. and König, W. (1992). Modulation of leukotriene generation by invasive bacteria. *Immunology* **77**, 400–7.

Hachicha, M., Rathanaswami, P., Naccache, P.H. and McColl, R.S., (1998). Regulation of chemokine gene expression in human peripheral blood neutrophils phagocytosing microbial pathogens. *J. Immunol.* **160**, 449–54.

Harris, R.R., Komater, V.A., Marett, R.A., Wilcox, D.M. and Bell, R.L. (1997). Effect of mast cell deficiency and leukotriene inhibition on the influx of eosinophils induced by eotaxin. *J. Leukoc. Biol.* **62**, 688–91.

Hasbold, J., Lyons, A.B., Kehry, M.R. and Hodgkin, P.D. (1998). Cell division number regulates IgG1 and IgE switching of B cells following stimulation by CD40 ligand and IL-4. *Eur. J. Immunol.* **28**, 1040–51.

He, S. and Walls, A.F. (1997). Human mast cell tryptase: a stimulus of microvascular leakage and mast cell activation. *Eur. J. Pharmacol.* **328**, 89–97.

Henderson, B., Poole, S. and Wilson, M. (1996). Bacterial modulins: a novel class of virulence factors which cause host tissue pathology by inducing cytokine synthesis. *Microbiol. Rev.* **60**, 316–41.

Hensler, T., Köller, M., Alouf, J.E. and König, W. (1993). Toxic shock syndrome toxin 1 and the erythrogenic toxin A modulate the chemotactic response of human neutrophils. *Infect. Immun.* **58**, 1055–61.

Hensler, T., König, B., Prévost, G., Piémont, Y., Köller, M. and König, W. (1994a). Leukotriene B4-generation and DNA fragmentation induced by leukocidin from *Staphylococcus aureus*: protective role of granulocyte-macrophage colony-stimulating factor (GMCSF) and G-CSF on human neutrophils. *Infect. Immun.* **62**, 2529–35.

Hensler, T., Köller, M., Prévost, G., Piémont, Y. and König, W. (1994b). GTP-binding proteins are involved in the modulated activity of human neutrophils treated with the Panton–Valentine leukocidin from *Staphylococcus aureus*. *Infect. Immun.* **62**, 5281–9.

Hensler, Th., Köller, M., Alouf, J.E. and König, W. (1991). Bacterial toxins induce heat shock proteins in human neutrophils. *Biochem. Biophys. Res. Commun.* **179**, 872–9.

Hilger, R., Knöller, J. and König, W. (1991). Modulation of leukotriene formation by cellular composition and exogenous leukotriene A4. *Int. Arch. Allergy Appl. Immunol.* **94**, 254–6.

Hilger, R.A., Knöller, J. and König, W. (1992). Leukotriene A4 modulates generation of leukotriene B4 and sulphidopeptide leukotrienes by human neutrophils. *Immunology* **77**, 408–15.

Hofer, M.F., Lester, M.R., Schlievert, P.M. and Leung, D.Y. (1995). Upregulation of IgE synthesis by staphylococcal toxic shock syndrome toxin-1 in peripheral blood mononuclear cells from patients with atopic dermatitis. *Clin. Exp. Allergy* **25**, 1218–27.

Höpken, U.E., Lu. B., Gerard, N.P. and Gerard, C. (1996). The C5a chemoattractant receptor mediates mucosal defence to infection. *Nature* **383**, 86–9.

Horie, S., Gleich, G.J. and Kita, H. (1996). Cytokines directly induce degranulation and superoxide production from human eosinophils. *J. Allergy Clin. Immunol.* **98**, 371–81.

Huang, G.T., Eckmann, L., Savidge, T.C. and Kagnoff, M.F. (1996). Infection of human intestinal epithelial cells with invasive bacteria upregulates apical intercellular adhesion molecule-1 (ICAM)-1) expression and neutrophil adhesion. *J. Clin. Invest.* **98**, 572–83.

Jabara, H.H. and Geha, R.S. (1996). The superantigen toxic shock syndrome toxin-1 induces CD40 ligand expression and modulates IgE isotype switching. *Int. Immunol.* **8**, 1503–10.

Jaeger, K.E., Kinscher, D.A., König, B. and König, W. (1992). Extracellular lipase of *Pseudomonas aeruginosa*: biochemistry and potential role as a virulence factor. In: *European Congress of Cystic Fibrosis* (eds S.S. Pedersen and N. Hoiby), pp. 113–19.

Jeannin, P., Delneste, Y., Lecoanet-Henchoz, S., Gauchat, J.F., Ellis, J. and Bonnefoy, J.Y. (1997). CD86 (B7-2) on human B cells. A functional role in proliferation and selective differentiation into IgE- and IgG4-producing cells. *J. Biol. Chem.* **272**, 15613–19.

Jeannin, P., Lecoanet-Henchoz, S., Delneste, Y., Gauchat, J.F. and Bonnefoy, J.Y. (1998). Alpha-1 antitrypsin up-regulates human B cell differentiation selectively into IgE- and IgG4-secreting cells. *Eur. J. Immunol.* **28**, 1815–22.

Kagnoff, M.F. and Eckmann, L. (1997). Epithelial cells as sensors for microbial infection. *J. Clin. Invest.* **100**, 6–10.

Kamio, Y., Rahmann, A., Nariya, H., Ozawa, T. and Izaki, K. (1993). The two staphylococcal bi-component toxins, leukocidin and gamma-hemolysin, share one component in common. *FEBS Lett.* **321**, 15–18.

Kammouni, W., Figarella, C., Baeza, N., Marchland, S. and Merten, M.D. (1997a). *Pseudomonas aeruginosa* lipopolysaccharide induces CF-like alteration of protein secretion by human tracheal gland cells. *Biochem. Biophys. Res. Commun.* **241**, 305–11.

Kammouni, W., Figarella, C., Marchand, S. and Merten, M. (1997b). Altered cytokine production by cystic fibrosis tracheal gland serous cells. *Infect. Immun.* **65**, 5176–83.

Keel, M., Ungethüm, U., Steckholzer, U., Niederer, E., Hartung, T., Trentz, O. and Ertel, W. (1997). Interleukin-10 counterregulates proinflammatory cytokine-induced inhibition of neutrophil apoptosis during severe sepsis. *Blood* **90**, 3356–63.

Kimata, H., Fujimoto, M., Ishioka, C. and Yoshida, A. (1996a). Histamine selectively enhances human immunoglobulin E (IgE) and IgG4 production induced by anti-CD58 monoclonal antibody. *J. Exp. Med.* **184**, 357–64.

Kimata, H. Yoshida, A., Ishioka, C., Fujimoto, M. Lindley, I. and Furusho, K. (1996b). RANTES and macrophage inflammatory protein 1 alpha selectively enhance immunoglobulin (IgE) and IgG4 production by human B cells. *J. Exp. Med.* **183**, 2397–402.

Klubal, R., Osterhoff, B., Wang, B., Kinet, J.P., Maurer, D. and Stingl, G. (1997). The high-affinity receptor for IgE is the predominant IgE-binding structure in lesional skin of atopic dermatitis patients. *J. Invest. Dermatol.* **108**, 336–42.

Knol, E.F., Mul, F.P., Lie, W.J., Verhoeven, A.J. and Ross, D. (1996). The role of basophils in allergic disease. *Eur. Respir. J. Suppl.* **22**, 126–31s.

Köller, M. and König, W. (1995). Interaction of bacteria with host defense cells: mechanisms of burn wound sepsis. In: *Wound Healing and Skin Physiology* (eds P. Altmeyer *et al.*), pp. 413–21. Springer, Berlin.

Köller, M., Brom, J. and König, W. (1993). Analysis of 5-lipoxygenase and 5-lipoxygenase activating protein in neutrophil granulocytes of severely burned patients. In: *Eicosanoids and Other Bioactive Lipids in Cancer, Inflammation and Radiation Injury* (eds S. Nigam, C.W. Honn and T. Walden, Jr), pp. 759–61. Academic Press, Boston, MA.

Köller, M., Hilger, R.A. and König, W. (1996). Effect of the PAF-receptor antagonist SM-12502 on human pletelets. *Inflammation* **20**, 71–85.

Köller, M., Wachtler, P., Dávid, A., Muhr, G. and König, W. (1997). Arachidonic acid induces DNA-fragmentation in human polymorphonuclear neutrophil granulocytes. *Inflammation* **21**, 463–74.

König, B. and König, W. (1993a). Induction and suppression of cytokine release (tumour necrosis factor-alpha; interleukin-6; interleukin 1β) by *Escherichia coli* pathogenicity factors (adhesins, alpha-haemoylsin). *Immunology* **78**, 526–33.

König, B. and König, W. (1993b). The role of the phosphatidylinositolturnover in 12-hydroxyeicosatetraenoic acid (12-HETE) generation from human platelets by the *E.coli*, α-hemolysin, thrombin and fluoride. *Immunology* **80**, 633–9.

König, B., Jaeger, K.E., Sage, A.E., Vasil, M.L. and König, W. (1996a). Role of *Pseudomonas aeruginosa* lipase in inflammatory mediator release from human inflammatory effector cells (platelets, granulocytes, and monocytes). *Infect. Immun.* **64**, 3252–8.

König, B., Streckert, H.-J., Krusat, T. and König, W. (1996b). Respiratory Syncytial-Virus G-protein modulates cytokine release from human peripheral blood mononuclear cells. *J. Leukoc. Biol.* **59**, 403–8.

König, B., Krusat, T., Streckert, H.J. and König, W. (1996c). Interleukin-8 release from human neutrophils by the Respiratory Syncytial Virus (RSV) is independent from viral replication. *J. Leukoc. Biol.* **60**, 253–60.

König, B., Köller, M., Prévost, G., Piémont, Y., Geoffrey, C., Alouf, J.E. and König, W. (1994a). Activation of effector cells by different bacterial toxins (leukocidin, alveolysin, erythrogenic toxin A), generation of interleukin-8. *Infect. Immun.* **62**, 4831–7.

König, B., Jaeger, K.E. and König, W. (1994b). Induction of inflammatory mediator release (12-hydroxyeicosatetraenoic acid) from human platelets by *Pseudomonas aeruginosa. Int. Arch. Allergy Appl. Immunol.* **104**, 33–41.

König, B., Ludwig, A., Goebel, W. and König, W. (1994c). Pore formation by the *Escherichia coli* α-hemolysin: role for mediator release from human inflammatory cells. *Infect. Immun.* **63**, 4612–17.

König, B., Prévost, G., Piémont, Y. and König, W. (1995a). Effects of *Staphylococcus aureus* leukocidins on inflammatory mediator release from human granulocytes. *J.Infect. Dis.* **171**, 607–13.

König, B., Ceska, M. and König, W. (1995b). Effect of *Pseudomonas aeruginosa* on interleukin-8 release from human phagocytes. *Int. Arch. Allergy Immunol.* **106**, 357–65.

König, B., Neuber, K. and König, W. (1995c). Responsiveness of peripheral blood mononuclear cells from normal and atopic donors to microbial superantigens. *Int. Arch. Allergy Immunol.* **106**, 124–33.

König, B., Fischer, A. and König, W. (1995d). Modulation of cell bound and soluble (s)CD23, spontaneous and ongoing IgE synthesis of human peripheral blood mononuclear cells by soluble IL-4 receptors (sIL-4R) and the partial antagonistic IL-4 mutant protein IL-4(Y124D). *Immunology* **85**, 604–10.

König, B., Prévost, G. and König, W. (1997a). Composition of staphylococcal bi-component toxins determines pathophysiological reactions. *J. Med. Microbiol.* **46**, 479–85.

König, B., Vasil, M. L. and König, W. (1997b). Role of haemolytic and non-haemolytic phospholipase C from *Pseudomonas aeruginosa* in interleukin-8 release from human monocytes. *J. Med. Microbiol.* **46**, 471–8.

König, W., Schönfeld, W., Raulf, M., Köller, M., Knöller, J., Scheffer, J. and Brom, J. (1990a). The neutrophil and leukotrienes – role in health and disease. *Eicosanoids* **3**, 1–22.

König, W., Köller, M., Brom, J., Knöller, J., Schönfeld, W. (1990b). Parameters of host resistance in acute and chronic infections. In: *Natural Resistance to Infection*, Vol. 6, *Local Immunity* (ed. C. Sorg), pp. 11–28. Gustav Fischer, Stuttgart.

König, W., Scheffer, J., Knöller, J., Schönfeld, W., Brom, J. and Köller, M. (1991a). *Effects of Bacterial Toxins of Activity and Release of Immunomediators* (eds J.E. Alouf and J.H. Freer), pp. 461–90. Academic Press, London.

König, W., Fischer, U., Stephan, Th. and Bujanowski-Weber, J. (1991b). Regulation of CD23 in allergic disease. In: *Monographs in Allergy – CD23, a novel multifunctional regulator of the immune system*, Vol. 29 (ed. J. Gordon), pp. 94–123. Karger, Basel.

König, W., Köller, M. and Brom, J. (1992a). Priming mechanisms and induction of heat shock proteins in human polymorphonuclear granulocytes induced by eicosanoids and cytokines. *Eisosanoids* **5**, 539–41.

König, W., Kasimir, S., Hensler, Th., Scheffer, J., König, B., Hilger, R. and Brom, J. (1992b). Release of inflammatory mediators by toxin stimulated immune system cells and platelets. In: *Bacterial Protein Toxins (Fifth European Workshop on Bacterial Protein Toxins, 1991)* (eds E. Witholt *et al.*), pp. 385–94. Gustav Fischer, Jena.

König, W., Bujanowski-Weber, J., Stephan, U. and Fischer, A. (1994a). Molekulare und zellbiologische Mechanismen der IgE-Regulation. (Molecular and cell biological mechanisms of IgE regulation.) In: *Pädiatrische Immunologie* (eds U. Wahn, R. Seger and V. Wahn), pp. 65–81. Gustav Fischer, Jena.

König, W., Fischer, A., Hilger, R., König, B. and Köller, M. (1994b). *Grundlagen und Mechanismen der allergischen Reaktion* (*Basis and Mechanisms of Allergic Reaction*), pp. 1–62. Thieme, Stuttgart.

Kotwal, G.J. (1997). Microorganisms and their interaction with the immune system. *J. Leukoc. Biol.* **62**, 415–29.

Kraft, S., Wessensdorf, J.H., Hanau, D. and Bieber, T. (1998). Regulation of the high affinity receptor for IgE on human epidermal Langerhans cells. *J. Immunol.* **161**, 1000–6.

Lamkhioued, B., Gounni, A.S., Aldebert, D., Delaporte, E., Prin, L., Capron, A. and Capron, M. (1996). Synthesis of type 1 (IFN gamma) and type 2 (IL-4, IL-5, and IL-19) cytokines by human eosinophils. *Ann. NY Acad. Sci.* **796**, 203–8.

Lamont, A.G. and Adorini, L. (1996). IL-12: a key cytokine in immune regulation. *Immunol. Today* **17**, 214–17.

Leung, D.Y. (1998). Molecular basis of allergic diseases. *Mol. Genet. Metab.* **63**, 157–67.

Levy, F., Kristofic, C., Heusser, C. and Brinkmann, V. (1997). Role of IL-13 in CD4 T cell-dependent IgE production in atopy, *Int. Arch. Allergy Immunol.* **112**, 49–58.

Li, H., Llera, A. and Mariuzza, R.A. (1998). Structure–function studies of T-cell receptor–superantigen interactions. *Immunol. Rev.* **163**, 177–86.

Li, J.-D., Dohrman, A.F., Gallup, M., Miyata, S., Gum, J.R., Kim, Y.S., Nadel, J.A., Prince, A. and Basbaum, C.B. (1997). Transcriptional activation of mucin in *Pseudomonas aeruginosa* lipopolysaccharide in the pathogenesis of cystic fibrosis lung disease. *Proc. Natl Acad. Sci. USA* **94**, 967–72.

Linehan, L.A., Warren, W.D., Thompson, P.A., Grusby, M.J. and Berton, M.T. (1998). STAT6 is required for IL-4-induced germline Ig gene transcription and switch recombination. *J. Immunol.* **161**, 302–10.

Lucey, D.R., Clerici, M., and Shearer, G.M. (1996). Type 1 and type 2 cytokine dysregulation in human infectious, neoplastic, and inflammatory diseases. *Clin. Microbiol. Rev.* **9**, 532–62.

MacGlashan, D.J., Lavens-Phillips, S. and Katsushi, M. (1998). IgE-mediated desensitization in human basophils and mast cells. *Front. Biosci.* **3**, D746–56.

Maggi, E. (1998). The Th1/Th2 paradigm in allergy. *Immunotechnology* **3**, 233–44.

Mantovani, A., Bussolino, F. and Introna, M. (1997). Cytokine regulation of endothelial cell function: from molecular level to the bedside. *Immunol. Today* **18**, 231–9.

Marrack, P. and Kappler, J. (1990). The staphylococcal enterotoxins and their relatives. *Science* **248**, 705–11.

Mécheri, S. and David, B. (1997). Unravelling the mast cell dilemma: culprit or victim of its generosity? *Immunol. Today* **18**, 212–15.

Moser, C., Johansen, H.K., Song, Z., Hougen, H.P., Rygaard, J. and Hoiby, N. (1997). Chronic *Pseudomonas aeruginosa* lung infection is more severe in Th2 responding BALB/c mice compared to Th1 responding C3H/HeN mice. *APMIS* **105**, 838–42.

Müller-Alouf, H., Alouf, J.E., Gerlach, D., Ozegowski, J.H., Fitting, C. and Cavaillon, J.M. (1996). Human pro- and anti-inflammatory cytokine patterns induced by *Streptococcus pyogenes* erythrogenic (pyrogenic) exotoxin A and C superantigens. *Infect. Immun.* **64**, 1450–3.

Müller-Alouf, H., Gerlach, D., Desreumaux, P., Leportier, C., Alouf, J.E. and Capron, M. (1997). Streptococcal pyrogenic exotoxin A (SPE A) superantigen induced production of hematopoietic cytokines, IL-12 and IL-13 by human peripheral blood mononuclear cells. *Microb. Pathogen.* **23**, 265–72.

Nakajima, H., Gleich, G.J. and Kita, H. (1996). Constitutive production of IL-4 and IL-10 and stimulated production of IL-8 by normal peripheral blood eosinophils. *J. Immunol.* **156**, 4859–66.

Nariya, H., Shimatani, A., Tomita, T. and Kamio, Y. (1997). Identification of the essential amino acid residues in lukS for the hemolytic activity of staphylococcal leukocidin towards rabbit erythrocytes. *Biosci. Biotechnol. Biochem.* **61**, 2095–9.

Neuber, K. and König, W. (1992). Effects of *Staphylococcus aureus* cell wall products (teichoic acid, peptidoglycan) and enterotoxin B on immunoglobulin (IgE, IgA, IgE) synthesis and CD23 expression in patients with atopic dermatitis. *Immunology* **75**, 23–8.

Neuber, K., Stephan, U., Fränken, J. and König, W. (1991a). *Staphylococcus aureus* modifies the cytokine-induced immunoglobulin synthesis and CD23 expression in patients with atopic dermatitis. *Immunology* **73**, 197–204.

Neuber, K., Hilger, R.A. and König, W. (1991b). Interleukin-3, Interleukin-8, FMLP and C5a enhance the release of leukotrienes from neutrophils of patients with atopic dermatitis. *Immunology* **73**, 83–7.

Neuber, K., Hilger, R.A. and König, W. (1992). Differential increase in 12-HETE release and CD29/CD49f expression of platelets from normal donors and from patients with atopic dermatitis by *Staphylococcus aureus*. *Int. Arch. Allergy Appl. Immunol.* **98**, 339–42.

O'Garra, A. and Murphy, K. (1994). Role of cytokines in determining T-lymphocyte function. *Curr. Opin. Immunol.* **6**, 458–66.

Ostroff, R.M., Wretlind, B. and Vasil, M.L. (1989). Mutations in the hemolytic-phospholipase C operon result in decreased virulende of *Pseudomonas aeruginosa* PAO1 grown under phosphate-limiting conditions. *Infect. Immun.* **57**, 1369–73.

Ostroff, R.M., Vasil, A.I. and Vasil, M.L. (1990). Molecular comparison of a nonhemolytic and a hemolytic phospholipase C from *Pseudomonas aeruginosa*. *J. Bacteriol.* **172**, 5915–23.

Papageorgiou, A.C. and Acharya, K.R. (1997). Superantigens as immunomodulators: recent structural insights. *Structure* **5**, 991–6.

Papageorgiou, A.C., Quinn, C.P., Beer, D., Brehm, R.D., Tranter, H.S. and Bonventre, P.F. (1996). Crystal structure of a biologically inactive mutant of toxic shock syndrome toxin-1 at 2.5 Å resolution. *Protein Sci.* **5**, 1737–41.

Park, H.J., Choi, Y.S. and Lee, C.E. (1997). Indentification and activation mechanism of the interleukin-4-induced nuclear factor binding to the CD23(b) promoter in human B lymphocytes. *Mol. Cells* **7**, 755–61.

Parsonnet, J. and Gillis, Z.A. (1988). Production of tumor necrosis factor by human monocytes in response to toxic shock syndrome toxin-1. *J. Infect. Dis.* **158**, 1026–33.

Patella, V., de Crescenzo, G., Marino, I., Genovese, A., Adt, M., Gleich, G.J. and Marone, G. (1996). Eosinophil granule proteins activate human heart mast cells. *J. Immunol.* **157**, 1219–25.

Pier, G.B., Grout, M. and Zaidi, T.S. (1997). Cystic fibrosis transmembrane conductance regulator is an epithelial cell receptor for clearance of *Pseudomonas aeruginosa* from the lung. *Proc. Natl Acad. Sci. USA* **94**, 12088–93.

Prasad, G.S., Earhart, C.A., Murray, D.L., Novick, R.P., Schlievert, P.M. and Ohlendorf, D.H. (1993). Structure of toxic shock syndrome toxin 1. *Biochemistry* **32**, 13761–6.

Prévost, G., Supersac, G., Colin, D.A., Couppié, P., Sire, S., Hensler, T., Petiau, P., Meunier, O., Cribiert, B., König, W. and Piemont, Y. (1994). In: *Bacterial Protein Toxins* (eds J. Freer *et al.*). *Zentralbl. Bakteriol.* **24**, 284–93.

Prévost, G., Cribier, B., Couppié, P, Petiau, P., Supersac, G., Finck-Barbarcon, V., Monteil, H. and Viemont, Y. (1995). Panton–Valentine leucocidin and gamma-hemolysin from *Staphylococcus aureus* ATCC 49775 are encoded by distinct genetic loci and have different biological activities. *Infect. Immun.* **63**, 4121–9.

Punnonen, J., Aversa, G., Cocks, B.G., McKenzie, A.N.J., Menon, S., Zurawski, G., De Waal Malefyt, R., Abrams, J., Bennett, B., Figdor, C.G. and Vries, J.E. (1991). Interleukin 10 (IL-10) inhibits cytokine synthesis by human monocytes: an autoregulatory role of IL-10 produced by monocytes. *J. Exp. Med.* **174**, 1209–20.

Punnonen, J., Yssel, H. and de Vries, J.E. (1997). The relative contribution of IL-4 and IL-13 to human IgE synthesis induced by activated CD4[+] or CD8[+] T cells. *J. Allergy Clin. Immunol.* **100**, 792–801.

Rajakulasingam, K., Till, S., Ying, S., Humbert, M., Barkans, J., Sullivan, M., Meng, Q., Corrigan, C.J., Bungre, J., Grant, J.A., Kay, A.B. and Durham, S.R. (1998). Increased expression of high affinity IgE (FcepsilonRI) receptor-alpha chain mRNA and protein-bearing eosinophils in human allergen-induced atopic asthma. *Am. J. Respir. Crit. Care Med.* **158**, 233–40.

Rappuoli, R. and Montecucco, C. (1997). *Guidebook to Protein Toxins and Their Use in Cell Biology*. Sambrook & Tooze Publication at Oxford University Press, Oxford.

Rasmussen, S.J., Eckmann, L., Quayle, A.J., Shen, L., Zhang, Y.X., Anderson, D.J., Fierer, J., Stephens, R.S. and Kagnoff, M.F. (1997). Secretion of proinflammatory cytokines by epithelial cells in response to *Chlamydia* infection suggests a central role for epithelial cells in chlamydial pathogenesis. *J. Clin. Invest.* **99**, 77–87.

Romani, L., Puccetti, P. and Bistoni, F. (1997). Interleukin-12 in infectious diseases. *Clin. Microbiol. Rev.* **10**, 611–36.

Saiman, L. and Prince, A. (1993). *Pseudomonas aeruginosa* pili bind to asialoGMI which is increased on the surface of cystic fibrosis epithelial cell. *J. Clin. Invest.* **92**, 1875–80.

Salgame, P., Abrams, J.S., Clayberger, C., Goldstein, H., Convit, J., Modlin, R.J. and Bloom, B.R. (1991). Differing lymphokine profiles of functional subsets of human CD4 and CD8 T cell clones. *Science* **254**, 279–82.

Saloga, J., Leung, D.Y., Reardon, C., Giorno, R.C., Born, W. and Gelfand, E.W. (1996). Cutaneous exposure to the superantigen staphylococcal enterotoxin B elicits a T-cell-dependent inflammatory response. *J. Invest. Dermatol.* **106**, 982–8.

Scharfmann, A., Van Brussel, E., Houdret, N., Lamblin, G. and Roussel, P. (1996). Interactions between glycoconjugates from human respiratory airways and *Pseudomonas aeruginosa*. *Am. J. Respir. Crit. Care Med.*, **154**, 163–9.

Schlievert, P.M. (1997). Searching for superantigens. *Immunol. Invest.* **26**, 283–90.

Schlüter, B. and König, W. (1990). Microbial pathogenicity and host defense mechanisms – crucial parameters of posttraumatic infections. In: *Deutsche Gesellschaft für Throax-, Herz- und Gefäßchirurgie (The Thoraxic and Cardiovascular Surgeon)*, Vol. 38, pp. 339–47. Georg Thieme, Stuttgart.

Schmidt, K.D., Tümmler, B. and Römling, U. (1996). Comparative genome mapping of *Pseudomonas aeruginosa* PAO with *P. aeruginosa* C, which belongs to a major clone in cystic fibrosis patients and aquatic habitats. *J. Bacteriol.* **178**, 85–93.

Schwartz, D.A., Quinn, T.J., Thorne, P.S., Sayeed, S., Yi, A.K. and Krieg, A.M. (1997). CpG motifs in bacterial DNA cause inflammation in the lower respiratory tract. *J. Clin. Invest.* **100**, 68–73.

Silverman, G.J. (1997). B-cell superantigens. *Immunol. Today* **18**, 379–86.

Staali, L., Monteil, H. and Colin, D.A. (1998). The staphylococcal pore-forming leukotoxins open Ca^{2+} channels in the membrane of human polymorphonuclear neutrophils. *J. Membr. Biol.* **162**, 209–16.

Supersac, G., Prévost, G. and Piemont, Y. (1993). Sequencing of leucocidin from *Staphylococcus aureus* O83 suggests that staphylococcal leucocidins and gamma-hemolysin are members of a single, two-component family of toxins. *Infect. Immun.* **61**, 580–7.

Swaminathan, S., Furey, W., Pletcher, J. and Sax, M. (1992). Crystal structure of staphylococcal enterotoxin B, a superantigen. *Nature* **359**, 801–6.

Takatsu, K. (1997). Cytokines involved in B-cell differentiation and their sites of action. *Proc. Soc. Exp. Biol. Med.* **215**, 121–33.

Tomita, T. and Kamio, Y. (1997). Molecular biology of the pore-forming cytolysins from *Staphylococcus aureus*, alpha- and gamma-hemolysins and leukocidin. *Biosci. Biotechnol. Biochem.* **61**, 565–72.

Trinchieri, G. (1998). Immunobiology of interleukin-12. *Immunol. Res.* **17**, 269–78.

Vercelli, D. and Geha, R.S. (1993). Regulation of IgE synthesis: from the membrane to the genes. *Springer Semin. Immunopathol.* **15**, 5–16.

Wen, R., Cole, G.A., Surman, S. Blackman, M.A. and Woodland, D.L. (1996). Major histocompatibility complex class II-associated peptides control the presentation of bacterial superantigens to T cells. *J. Exp. Med.* **183**, 1083–92.

Woody, M.A., Krakauer, T. and Stiles, B.G. (1997). Staphylococcal enterotoxin B mutants (N23K and F44S): biological effects and vaccine potential in a mouse model. *Vaccine* **15**, 133–9.

Wu, Y., Cuo, Y., Huang, A., Zheng, P. and Liu, Y. (1997). CTLA-4-B7 interaction is sufficient to costimulate T cell clonal expansion. *J. Exp. Med.* **185**, 1327–35.

Yamaguchi, M., Lantz, C.S., Oettgen, H.C., Katona, I.M., Fleming, T., Miyajima, I., Kinet, J.P. and Galli, S.J. (1997). IgE enhances mouse mast cell Fc(epsilon)RI expression *in vitro* and *in vivo*: evidence for a novel amplification mechanism in IgE-dependent reactions. *J. Exp. Med.* **185**, 663–72.

Ye, G., Barrera, C., Fan, X., Gourley, W.K., Crowe, S.E., Ernst, P.B. and Reyes, V.E. (1997). Expression of B7-1 and B7-2 costimulatory molecules by human gastric epithelial cells: potential role in CD4$^+$ T cell activation during *Helicobacter pylori* infection. *J. Clin. Invest.* **99**, 1628–36.

Ying, S., Meng, Q., Barata, L.T., Robinson, D.S., Durham, S.R. and Kay, A.B. (1997). Associations between IL-13 and IL-4 (mRNA and protein), vascular cell adhesion molecule-1 expression, and the infiltration of eosinophils, macrophages, and T cells in allergen-induced late-phase cutaneous reactions in atopic subjects. *J. Immunol.* **158**, 5050–7.

Yoshimoto, T., Okamura, H., Tagawa, Y.I., Iwakura, Y. and Nakanishi, K. (1997). Interleukin 18 together with interleukin 12 inhibits IgE production by induction of interferon-gamma production from activated B cells. *Proc. Natl Acad. Sci. USA* **94**, 3948–53.

Yu, H., Hanes, M., Chrisp, C.E., Boucher, J.C. and Deretic, V. (1998). Microbial pathogenesis in cystic fibrosis: pulmonary clearance of mucoid *Pseudomonas aeruginosa* and inflammation in a mouse model of repeated respiratory challenge. *Infect. Immun.* **66**, 280–8.

Staphylococcal epidermolytic toxins: structure, biological and pathophysiological properties

Yves Piémont

HISTORY

In 1878, Gottfried Ritter von Rittershain reported 297 cases of 'dermatitis exfoliativa neonatorum' observed for over 10 years in the Royal and Imperial Maternity Hospital of Prague (Ritter von Rittershain, 1880). After purulent conjunctivitis or infection of the upper respiratory tract, the disease began by a perioral scarlet-like erythema, followed by cutaneous blisters and generalized exfoliation in human newborns aged 2–5 weeks. Half of the patients recovered spontaneously without scars 7–10 days after the beginning of the erythema and the other half died. At that time, no cause of the disease was evident. A mouth infection was supposed to be the origin of the disease, but skin lesions did not seem to be infectious. In this maternity hospital, mothers and nurses were never sick, so the disease was considered to be non-infectious.

In 1898, in the same maternity hospital, Rudolf Winternitz (Winternitz, 1898) cultivated yellow-pigmented staphylococci in a patient suffering from the 'dermatitis exfoliativa neonatorum'. Gradually, the disease was associated to the presence of *Staphylococcus aureus* (Koblenzer, 1967), although culture of the clear liquid obtained from bullae was usually negative. By its clinical presentation (bullae, large skin exfoliation) this syndrome closely resembles a scald caused by boiling water (Figure 37.1 in the colour plate section).

The possible confusion between burns and this syndrome could have led to awful consequences as exem-

plified by the article in *The Daily Mail* of 29 December 1904 (Jackson, 1974):

> A Lewisham child which was supposed to have been scalded to death was yesterday declared by Dr Toogood, of the Lewisham Infirmary to have fallen a victim to a mysterious and fatal disease unnamed in the medical textbooks. [...]. The child, aged two weeks, died from what a local doctor believed to be scald and he refused to give a certificate, with the result that an inquest was held. The doctor said he thought death was due to scalding, and his opinion was supported by two other medical men.

In the vicinity of scald-like lesions, the skin could appear intact, but when it was tangentially rubbed, it remained wrinkled because the upper part of the epidermis was no longer adherent to the underlying part of the dermis containing elastic fibres. So, the upper part of the epidermis, devoid of any elastic fibre, remained wrinkled. This feature is known as the sign of Nikolsky, who described it in 1898 for another histopathologically related disease called foliaceous pemphigus (Nikolsky, 1898).

A differential diagnosis of the Ritter von Rittershain syndrome is the so-called Lyell syndrome or toxic epidermal necrolysis (TEN) described by Lyell (1956) and also by Lang and Walker (1956) which also includes a scalded appearance of the skin, and which can be caused by drugs such as pyrazolone, antibiotics, barbiturates, sulfamides or other factors which are more or less well known. Both clinical entities had often been confused for several years because of their terminal

---

aspect mimicking scald. In fact, both are very distinct syndromes by their causes, the age of the patients, the duration of the disease, the prognosis and, particularly, the histopathological features. The Ritter von Rittershain syndrome is caused by *S. aureus*, it occurs mostly in newborn or young infants and its evolution is linked to the development or the control of the infection; there is no cytotoxic lesion in the staphylococcal syndrome: the only lesions are constituted by an intraepidermal splitting between the stratum granulosum and the stratum spinosum (Wuepper *et al.*, 1975) which are two of the cellular layers that constitute the human epidermis. Few acanthosic cells can be present within the cleavage space. By contrast, in the Lyell syndrome, the splitting occurs not within the epidermis, but at the junction between the dermis and the epidermis. All of these features allow a clear distinction between the staphylococcal syndrome and the toxic epidermal necrolysis. Therefore, the Ritter von Rittershain syndrome is now clearly defined and also generally known as 'staphylococcal scalded skin syndrome' (SSSS) (Lowney *et al.*, 1967).

In 1955, it appeared that most staphylococcal isolates obtained from SSSS belonged to the phage group 2 (Parker *et al.*, 1955). This result was confirmed by several other authors (Howells and Jones, 1961; Benson *et al.*, 1962; Lowney *et al.*, 1967; Melish and Glasgow, 1970; Kapral and Miller, 1971; Anthony *et al.*, 1972; Dajani, 1972). The production of epidermolytic toxins (ETs) is not under the control of bacteriophage(s) (Rogolsky *et al.*, 1974). However, *S. aureus* strains belonging to other phage groups (such as the phage groups 1 and 3) are also occasionally responsible for SSSS (McClosky, 1973; Kondo *et al.*, 1975; Rasmussen, 1975; Arbuthnott and Billcliffe, 1976; Sarai *et al.*, 1977). Conversely, only 31–40% of *S. aureus* isolates that belong to phage group 2 produce epidermolytic toxin type A (ETA) or ETB (Kapral, 1975; Fleurette and Ritter, 1980; Willard *et al.* 1984). A milestone for the study of the pathophysiology of SSSS was the discovery by Melish and Glasgow in 1970 that subcutaneous or intraperitoneal injection of mice with *S. aureus* strains obtained from patients with SSSS, was responsible for the Nikolsky sign (Melish and Glasgow, 1970, 1971). In this animal model, splitting occurred, as in humans, between the stratum spinosum and the stratum granulosum.

From this major breakthrough, the staphylococcal products responsible for SSSS could be purified and characterized. They were called exfoliative toxins or epidermolysins or exfoliatins.

Apart from *S. aureus*, another staphylococcal species, *S. hyicus*, a common bacterium of the porcine flora, is able to promote in piglets cutaneous lesions known as exudative epidermitis of piglets. Exudative epidermi-

tis is a generalized infection of the skin, characterized by greasy exudation, vesicle formation, exfoliation and formation of a crust (Sompolinski, 1950). These lesions can be accompanied by ulcerative glossitis and stomatitis (Amtsberg *et al.*, 1973; Devriese, 1977; Phillips *et al.*, 1980). Not all strains of *S. hyicus* are responsible for the piglet disease. *S. hyicus* is merely a skin pathogen, as it can also be isolated from arthritic lesions in pigs. Skin lesions caused by *S. hyicus* ssp. hyicus may also be found in cattle (Devriese and Derycke, 1979) and in horses (Devriese *et al.*, 1983). In cattle, the bacterium usually complicates mange lesions but it is probably also the primary cause of uncomplicated superficial epidermolysis and alopecia in cattle in winter quarters (Devriese and Nuytten, cited by Devriese, 1986).

ANIMAL MODELS

Melish and Glasgow (1970, 1971) reported in 1970 that injection of *S. aureus* strains originating from SSSS to newborn mice resulted in cutaneous lesions mimicking, clinically and histopathologically, the human Ritter von Rittershain disease. A Nikolsky sign similar to the one observed in humans was clearly seen in the injected animals by gently rubbing the skin (Figures 37.2 and 37.3); the epidermis irreversibly wrinkled because after the action of the toxin the upper part of the epidermis was separated from the underlying part of the skin (particularly the dermis) containing elastic fibres. So, the upper part of the epidermis, devoid of any elastic fibre, remains wrinkled because of the loss of adherence between the elements of the skin. There is minimal inflammation in the underlying epidermis despite the

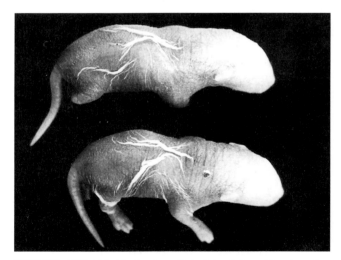

FIGURE 37.2 Nikolsky's sign elicited in 1-day-old mice after intraperitoneal injection of ETA.

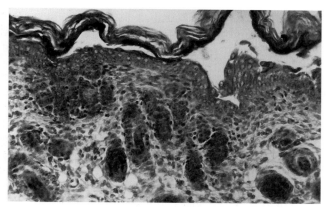

FIGURE 37.3 Histological section of newborn mouse skin after subcutaneous injection of ETA. The cleavage plane is between the stratum granulosum and the stratum apinosum.

presence of bacteria seen in the animal model (Rogolsky, 1979). This discovery was the starting point of renewed studies on this disease because previously there had been no means of exploring the question of why some staphylococcal isolates were able to cause this syndrome. So, Melish and Glasgow were able to fulfil Koch's postulates for the role of *S. aureus* in the disease (Magner, 1992). The experimental model worked only if the number of bacteria injected intraperitoneally or subcutaneously did not exceed one-tenth of the lethal dose, because the staphylococcal α-toxin (mostly responsible for the death of the animals) acts more rapidly than the exfoliative toxin and because the experimental SSSS requires live animals in order to occur. However, the intra-epidermal blisters obtained after intraperitoneal or subcutaneous injection of *S. aureus* strains remained sterile. Similarly, the blisters observed in the generalized SSSS were also usually sterile. Therefore, Jefferson (1967), Lowney *et al.* (1967) and Melish and Glasgow (1970, 1971) hypothesized that the occurrence of SSSS was due to one or several soluble staphylococcal substances.

Concerning *S. hyicus*, experimental infections have shown that subcutaneous injection of sterile concentrated culture supernatant from certain strains of *S. hyicus* (referred to as virulent) might induce exudative epidermitis in pigs (Amtsberg *et al.*, 1973; Wegener *et al.*, 1993). About 80% of *S. hyicus* strains secrete that virulence factor which is currently not well characterized. The microscopic alterations produced after injection of piglets with sterile concentrated supernatant were acanthosis, epidermal exocytosis and crust formation (Andresen *et al.*, 1993), comparable to lesions seen in natural and experimental infections. As with exfoliative toxin from *Staphylococcus aureus*, a cleavage plane was obtained between the stratum corneum and stratum

granulosum and at the stratum granulosum (Sato *et al.*, 1991a). Because its action on the skin is similar to that of the exfoliative toxin of *S. aureus*, this supernatant factor is also called exfoliative toxin from *S. hyicus* or shET, although there is no serological cross-reactivity between the toxins produced by *S. aureus* and those produced by *S. hyicus*. Sato *et al.* (1991b) discovered that not only piglets but also young chickens were sensitive to the action of shETs. Moreover, in contrast to purified ETs of *S. aureus*, partially purified shETs caused a rounding effect without cell death in cultured cells (NCTC 2544, HeLa S/3, HmLu-1, CHO and BHK-21) (Sato *et al.*, 1991b).

PURIFICATION OF TOXINS

Since 1971, Arbuthnott *et al.* (1971, 1974a, b), Kapral and Miller (1971), Melish *et al.* (1972), Kondo *et al.* (1973, 1974), Johnson *et al.* (1975), Dimond and Wuepper (1976), Wiley *et al.* (1976a), Piémont and Monteil (1983), Piémont *et al.* (1984), Cavarelli *et al.* (1997) and Vath *et al.* (1997) have published numerous techniques to purify to homogeneity the factors responsible for the Nikolsky sign in the newborn mouse. Useful culture media include beef heart broth without the addition of carbon dioxide in the atmosphere of the culture (Wiley *et al.*, 1976a), TY medium (Kapral and Miller, 1971; Kondo *et al.*, 1973; Johnson *et al.*, 1975) consisting of a saline-buffered solution of yeast extract and biotrypcase which should be used with an atmosphere of 10% carbon dioxide.

The toxins can also be produced *in vivo* (Melish *et al.*, 1972; Dimond and Wuepper, 1976) by inserting in the peritoneal cavity of rats or rabbits a dialysis bag containing cell-culture medium previously inoculated with ET-producing staphylococci; 5 days after insertion into the peritoneum, the dialysis bag is removed and the bacterial-culture supernatant appears to contain exfoliative toxin. The purification techniques generally involve a preliminary purification step (ammonium sulfate precipitation or ethanol precipitation or gel filtration or ultrafiltration) and a further step (mostly hydroxyapatite chromatography, ion-exchange chromatography, chromatofocusing, thin-layer isoelectric focusing, hydrophobic chromatography). These protein factors were called epidermolytic toxin (Arbuthnott *et al.*, 1971), epidermolysin (Dimond and Wuepper, 1976), exfoliatin (Kapral and Miller, 1971) and exfoliative toxins or ET (Melish *et al.*, 1972).

Three serological types of ET produced by *S. aureus* are currently known: type A (ETA), type B (ETB), the latter described for the first time by Kondo *et al.* (1974), and a type C (ETC) recently reported from infection of a horse with phlegmon by Sato *et al.* (1994). Only the sequences of ETA (O'Toole and Foster, 1987) and ETB (Lee *et al.*, 1987) are known as their respective genes have been cloned and sequenced. By contrast, ETC is not well characterized and its sequence remains unknown. Antibodies directed to the exfoliatins are protective, which confirms the role of these toxins in pathogenesis.

ETA and ETB are synthesized as precursor proteins with a signal sequence which is proteolytically cleaved during the secretion process. In both ETA and ETB, the cleavage occurs after an alanine residue. The ETA precursor and the mature protein contain 280 and 242 amino acid residues, respectively (Lee *et al.*, 1987; O'Toole and Foster, 1987). The ETB precursor and the mature protein contain 277 and 246 amino acid residues, respectively (Lee *et al.*, 1987). The calculated relative molecular mass of ETA and ETB is 26 951 and 27 318, respectively. Secreted ETA and ETB share 55% sequence identity. They are chromosome- and plasmid-encoded, respectively (Lee *et al.*, 1987). The gene coding for ETB is located on a 37-kb plasmid, the restriction map of which has been published by Warren (1981). This plasmid also possesses genes coding for cadmium resistance, for a bacteriocin called staphylococcin and for the ability to resist the action of staphylococcin (Jackson and Iandolo, 1986; Lee *et al.*, 1987).

Thermal denaturation of ETA occurs at 57–59°C as observed by measurement of intrinsic fluorescence techniques; that of ETB occurs at 52.5–54°C (Piémont *et al.*, 1986). These data correlate with the biological activity of the heat-treated toxins.

ETA and ETB are antigenic molecules. In rabbits, precipiting ETA-antibodies are obtained easily after five weekly intradermal injections of ETA (1 mg ml^{-1}) plus Freund's complete adjuvant in multiple sites. For ETB, precipiting antibodies are obtained in the same way, but 16–20 weekly injections are necessary (Rogolsky, 1979); however, neutralizing antibodies are obtained earlier, after 2 months.

The data concerning exfoliative toxins from *S. hyicus* are less detailed. Sato *et al.* (1996) described two shET: one is chromosomally encoded and the second is controlled by a 42-kb plasmid. Both toxins have a relative molecular mass of about 27 000–30 000, and there is no serological cross-reaction between these two toxins. A DNA probe synthesized on the basis of the conservative nucleotide sequences of the ETA and ETB genes hybridized with DNA of the plasmid- and chromosome-encoded shETs, but not with the DNA of strains not producing shETs. One serotype of exfoliative toxin produced by *S. hyicus* has been purified by Tanabe *et al.* (1993) from strain P-1. It has a relative molecular mass of 27 000 and does react with either ETA or ETB pro-

duced by *S. aureus*. Andresen *et al.* also purified shETs by ammonium sulfate precipitation, hydrophobic interaction chromatography and anion-exchange chromatography (Andresen *et al.*, 1993, 1997; Andresen, 1998). These proteins of approximately 30 kDa have antigenic diversity, depending on the shET-producing strain. The antigenic variants produced by strains NCTC 10350, 1289D-88 and 842A-88 are called ExhA, ExhB and ExhC, respectively (Andresen, 1998). These three serotype-producing strains apparently account for only 71% of the epidemics of exudative epidermitis in piglets. The genes coding for the epidermolytic toxins of *S. hyicus* have not been cloned.

STRUCTURE OF EXFOLIATIVE TOXINS

Sequence analysis of ETA and ETB revealed about 40% identity with each other and no apparent sequence homology to other bacterial toxins. The ETs proved to have some homology to serine proteases (Dancer *et al.*, 1990; Bailey and Redpath, 1992) such as the staphylococcal V8 protease and the Glu–SGP protease from *Streptomyces griseus*, both of which are specific for a glutamate residue. The ETA sequence includes the conserved catalytic triad Ser195, His72 and Asp120, and the ETB sequence the conserved catalytic triad Ser186, His65 and Asp114, which is common to serine proteases, esterases and lipases. The functional role of this triad was confirmed in ETA, as site-directed mutagenesis of these amino acid residues led to biologically inactive toxin in the newborn mouse model (Prévost *et al.*, 1991, 1992; Redpath *et al.*, 1991). The physiological target of ETs remains unknown and no enzymic activity has ever been demonstrated. Only an esterolytic activity with quite poor kinetic constants was found for both ETA and ETB with the Boc-L-Glu *o*-phenyl synthetic substrate (Bailey and Redpath, 1992).

Both ETA and ETB have been crystallized. ETA crystals belong to space group P2$_1$, with unit cell dimensions $a = 49.09$ Å, $b = 66.40$ Å, $c = 81.77$ Å, $\beta = 93.92°$ (Yoo *et al.*, 1978; Cavarelli *et al.*, 1997; Vath *et al.*, 1997); ETB crystals also belong to space group P2$_1$, with unit cell dimensions $a = 55.9$ Å, $b = 107.9$ Å, $c = 42.8$ Å, $\beta = 90.9°$ (Moras *et al.*, 1984). The crystal structure of ETA was determined by multiple isomorphous replacement to 1.7 Å resolution with a crystallographic *R* factor of 0.184 by Cavarelli *et al.* (1997), and to 2.1 and 2.3 Å resolution with *R* factors of 0.17 and 0.19, respectively, by Vath *et al.* (1997). The ETA protein contains two domains of similar structure, which are built around a six-stranded antiparallel β-sheet folded into a β-barrel. The catalytic site lies at the interface between the two domains. There is no metal ion binding site in the ETA structure. ETA is a new and unique member of the trypsin- or chymotrypsin-like serine protease family and cleaves substrate(s) after acidic residues. In contrast to other serine protease folds, ETA can be characterized by ETA-specific surface loops, a lack of cysteine bridges, an oxyanion hole which is not preformed, an S1-specific pocket designed for a negatively charged amino acid and an ETA-specific *N*-terminal helix which is crucial for substrate hydrolysis. The conformation of a loop adjacent to the catalytic site is thought to be crucial in regulating the proteolytic activity of ETA through controlling whether the main chain carbonyl group of Pro192 occupies the oxyanion hole. A unique amino-terminal domain containing a 15-residue amphipathic α-helix may also be involved in protease activation through binding a specific receptor. Despite very low sequence homology between ETA and other trypsin-like serine proteases, the ETA crystal structure, together with biochemical data and site-directed mutagenesis studies, strongly confirms the classification of ETA in the Glu-endopeptidase family (see Figure 37.4 in the colour plate section). Direct links can be made between the protease architecture of ETA and its biological activity.

MODE OF ACTION OF EXFOLIATIVE TOXINS

After subcutaneous injection of purified ETA or ETB to 1-day-old mice, the time required for a Nikolsky sign to appear was dose-dependent. *In vivo*, the lowest biologically active dose in this model was about 0.2 μg injected subcutaneously; it resulted in a Nikolsky sign 180 min after injection. By contrast, injection of 200 μg led to a Nikolsky sign less than 10 min after injection (unpublished results). *In vitro*, skin fragments can be maintained alive in cell-culture medium; a Nikolsky sign can be elicited if the medium contains large amounts (5 mg ml^{-1}) of exfoliative toxin (Elias *et al.*, 1976; Nishioka *et al.*, 1981). Non-differentiated keratinocytes in cell culture are not susceptible to the action of the epidermolysin; when *in vitro*-differentiated keratinocytes obtained by Gentilhomme *et al.* (1990) in cell culture were exposed to ETA, they underwent intra-epithelial splitting.

The mechanism of action of exfoliative toxins is still unknown. The hypothetical mode of action is based on the histological modifications observed by electron microscopy on newborn mouse skin and on the recently determined ETA structure. ETs have a very narrow species specificity as they are active only on the epidermis of newborn mice, adult hamsters (slow and

localized exfoliation), newborn hamsters (generalized exfoliation), adult *Macaca rhesus* monkeys, and newborn and adult humans (Elias and Levy, 1976). Most studies have been performed on the newborn mouse model. The toxin is also tissue specific, as it is active only on keratinized epithelia: the skin is particularly susceptible and other epithelia can display a desquamation increase of the horny cells, without intra-epithelial splitting. Although the cause of the epidermal separation has remained elusive, investigators have explored at least four mechanisms: protease activity; modification of cell metabolism; ability to bind to cell or tissue structures; and involvement of the immune system.

Protease activity

Electron microscopy was used by Lillibridge *et al.* (1972), Elias *et al.* (1975), McLay *et al.* (1975), Wuepper *et al.* (1975) and Dimond *et al.* (1977) to study the mode of action of the toxin. Lillibridge *et al.* (1972) hypothesized, according to their observations, that the target of ETs consisted of desmosomes. The cells of the stratum spinosum are normally linked together by desmosomes. In the absence of treatment by ET, there are small vesicles located between the cells of the upper stratum granulosum and the cells of the lower stratum corneum. These vesicles would be identified, according to Dimond *et al.* (1977), as extracellular keratinosomes. Lillibridge *et al.* (1972) noted that 20 min after subcutaneous injection of ET, these vesicles widened, lost their oval appearance and disappeared, and the intercellular spaces enlarged. When the Nikolsky's sign appeared, at the 25th minute, the desmosomes were split into two hemidesmosomes. Later, at the 150th minute, a cleavage plane was apparent between the cells within the stratum granulosum, and the cells conserved their full integrity. All desmosomes located along the cleavage plane were split into two hemidesmosomes. Lillibridge *et al.* (1972) hypothesized that the intercellular vesicles may contain proteolytic enzyme or proenzyme that would be activated by ET. This activated enzyme would separate the nearest desmosomes and, as a result, the enzyme would diffuse to other desmosomes. So, the cells would be split and the cleavage plane would be achieved. The hypothesis of a proteolytic action of ETs is strengthened by the data deduced from the tridimensional structure of ETA (Cavarelli *et al.*, 1997; Vath *et al.*, 1997) and by site-directed mutagenesis (Prévost *et al.*, 1991, 1992).

According to other authors, the target of the toxin would be the intercellular cement rather than the desmosomes. After application of toxin on to the skin of newborn mice, Elias *et al.* (1975) saw neither early modification of the intercellular vesicles (referred to as extracellular keratinosomes) nor early desmosome splitting. The first modifications observed consisted of an enlargement of the inter-desmosomal extracellular regions. In other locations, a separation occurred simultaneously along desmosomal and inter-desmosomal surfaces. These authors believed that extracellular components located along the inter-desmosomal regions were initially involved in the action of exfoliative toxin. These elements could include structures containing specific proteins, because of the structure-deduced hypothesis of proteolytic activity of ETs. McLay *et al.* (1975) observed the same phenemenon and hypothesized that desmosomal splitting was probably a secondary event following the enlargement of the intercellular spaces.

The efforts undertaken to unravel the precise mode of action of the ET have hitherto been unsuccessful. No proteolytic activity has been detected either directly on several natural polypeptide substrates or indirectly by using protease inhibitors (Lillibridge *et al.*, 1972; McLay *et al.*, 1975; Wuepper *et al.*, 1975; Elias *et al.*, 1977; Nishioka *et al.*, 1981). Only Dancer *et al.* (1990) reported that the serine protease inhibitors, DFP and PMSF, increased the time needed for ETA to show skin separation in newborn mice *in vivo*.

Another aspect which remains to be explored more extensively was identified by Takiuchi *et al.* (1987). These authors described a proteolytic activity arising only when an epidermal fraction was mixed *in vitro* with ETA. These data indicate that an epidermal factor may be activated by ET or that ET may be activated by an epidermal factor, so determining a protease-induced intra-epidermal splitting. These results remain to be confirmed.

Ability to modify the cell metabolism

Wuepper *et al.* (1975) and Dimond *et al.* (1977) observed by electron microscopy an intracellular lightening along the cell membranes limiting the cleavage plane. This observation was interpreted either as a localized disappearance of cytoplasmic organelles or as a localized cell oedema.

Dimond *et al.* (1977) observed no detectable histochemical modification for the following enzymes after action of ET on the skin: glucose-phosphate dehydrogenase, glyceraldehyde-3 phosphate dehydrogenase, lactate dehydrogenase, non-specific esterase, acid and alkaline phosphatases, leucine aminopeptidase, α-naphthylesterase, naphthol-AS-D-acetate esterase and naphthol-AS-D-chloroacetate esterase. Elias *et al.* (1976) showed that the resistance of rat to exfoliative toxin was not linked to plasma or tissue factors. In another paper, Elias *et al.* (1977) demonstrated that the

exfoliation process occurred at a maximum speed at 37°C and that it stopped at 4°C. This process was not inhibited either by protein synthesis inhibitors, such as cycloheximide or puromycine, or by 2-deoxyglucose, or in an atmosphere of 100% nitrogen. There was no interference with the adenylate cyclase activity or with a system of translation *in vitro* (personal unpublished results).

Ability to bind to cell or tissue structures

The exfoliative toxin did not modify the pemphigus antigen and did not interfere with the binding of pemphigus antibodies originating from patients suffering of pemphigus vulgaris (nor with foliaceous pemphigus which better mimics the histopathological lesions of SSSS). So, the target of ET should be different to the pemphigus vulgaris antigen (Elias *et al.*, 1975). The exfoliative toxin does not interfere either with HLA antigens of the keratinocyte surface or with concanavalin A-reactive sugars. The exfoliative toxin did not bind to cells or other tissue structures (Baker *et al.*, 1978a). Smith and Bailey (1986) described the binding of ETB to intracellular profilaggrins. The relationship between the keratohyalin granules, the intracellular filaments (keratins) and the desmosomes could explain the cell acanthosis often observed in the ET-induced cleavage plane. However, this binding could not easily explain the occurrence of the intra-epidermal splitting. Only Tanabe *et al.* (1995) showed that shET bound to the GM4-like glycolipid extracted from the skin of 1-day-old chickens, but did not bind to glycolipid from adult chickens or suckling mice. ET produced by *S. aureus* bound the GM4-like glycolipid extracted from the skin of suckling mice, but not to glycolipid from 1-day-old or adult chickens. Preincubation of shET and *S. aureus* ETs with GM4-like glycolipid from newborn and suckling mice, respectively, resulted in a loss of their toxic activity. So, skin-extracted GM4-like lipid is a candidate for the cutaneous receptor of ETs.

Involvement of the immune system

The cutaneous lesions associated with the exfoliative toxin did not seem to be mediated by the immune system. Morlock *et al.* (1980) demonstrated that ETA had mitogenic properties on murine T-lymphocytes, but this action was not necessary to trigger skin lesions as the latter were still observed in 'nude' athymic newborn mice. Complement is not required either, because mice genetically deficient in C5, or those in which the complement system has been lysed by cobra venom, remain susceptible to the cutaneous action of exfoliative toxins (Wuepper *et al.*, 1975).

Other studies of Marrack and Kapler (1990) have shown that ETs belongs to the family of proteins referred to as 'superantigens'. ETA induces Vβ2$^+$ T-lymphocyte proliferation if bound to the MHC2 receptor of Langerhans cells or keratinocytes. The role of ETB as a superantigen has been less convincingly established. The superantigenic activity of ETA is not responsible for the action of the toxin on the skin, as substitution of the active site serine residue with cysteine abolished the ability of ETA to produce the characteristic separation of epidermal layers, but not its ability to induce T-cell proliferation (Vath et al., 1997). ETs also stimulate keratinocytes to produce cytokines that activate T-cells (Tokura et al., 1992, 1994); they trigger murine macrophages to secrete TNF-α and IL-6 (Fleming et al., 1991) and also stimulate murine splenic B cells.

Concerning the mode of action of S. hyicus exfoliative toxin, few data are available. Only Andresen et al. (1997) noticed that the activity of the toxin purified from strain 1289D-88 was dependent on Cu^{2+} ions.

To summarize, the exfoliative toxins determine splitting between the stratum spinosum and the stratum granulosum of the epidermis. In most reports, the first morphological change observed is an enlargement of the intercellular spaces in the area where the splitting will occur. The separation of the desmosomes would occur later. No cell lesion was induced by the toxin. The target of the toxin remains hitherto unknown and no enzymic activity on a natural substrate has been described. No interference by the toxin with enzymic or metabolic activities has been reproducibly demonstrated. The immune system does not seem to be involved in the occurrence of the skin lesions, except, perhaps, the skin flush occurring in some patients before generalized exfoliation.

EPIDEMIOLOGY AND CLINICAL DATA

Exfoliative toxins are produced by about 5.5% of S. aureus isolates obtained in a clinical laboratory attached to a general hospital (Piémont et al., 1984). Eighty-eight per cent of the ET-producing isolates produce only serotype A, 4% only serotype B and 8% both serotypes (Piémont et al., 1984); in that study, there was no screening for serotype C. There is currently no description of coagulase-negative staphylococci naturally producing exfoliative toxin.

Before the work of Lina et al. (1997) it was thought that ET-producing strains were responsible for three clinical presentations: bullous impetigo, generalized scalded skin syndrome and staphylococcal scarlet fever. Lina et al. (1997), as well as Brook and Bannister (1991), demonstrated that staphylococcal scarlet fever was not related to ET-producing strains; this syndrome should be classed rather as an abortive form of toxic shock syndrome due to staphylococcal superantigens (TSST-1 or enterotoxins) or as a separate syndrome.

Bullous impetigo is the most common manifestation of SSSS (Figure 37.5). It most frequently affects newborn and children, but can be found in patients of any age (Todd, 1985). It represents about 10% of all impetigo cases. The disease is first seen as small vesicles that become flaccid bullae containing S. aureus in a clear fluid. Staphylococci may be found only in the skin lesions and not in other parts of the body. The bullae then rupture and heal quickly, forming brown crusts (Rogolsky, 1979). Lesions are not tender and constitutional symptoms are usually lacking. During the course of the disease, ET-antibodies can be detected (Gemmell, 1995), and this may prevent such patients from developing generalized scalded skin syndrome. These antibodies are protective, and neutralize ETA and ETB in humans and in mice (Miller and Kapral, 1972; Elias et al., 1974a; McLay et al., 1975; Wuepper et al., 1975; Wiley et al., 1976b). The Nikolsky sign cannot be elicited in this mild manifestation of SSSS.

Generalized scalded skin syndrome is also called Ritter's disease or Ritter von Rittershain's disease. It usually occurs in children, particularly in nurseries as epidemics. Cases complicating varicella have been reported by several authors (Melish, 1973; Wald et al., 1973; Brook and Bannister, 1991; Lina et al., 1997). At the beginning, children often have a trivial cutaneous infection due to an ET-producing strain. Strains from patients with generalized exfoliative syndrome are isolated not only from sites of focal S. aureus infection but

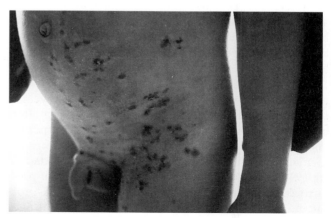

FIGURE 37.5 Bullous impetigo due to an ETA-producing strain of S. aureus.

also from colonized nasopharyngeal and ocular sites (Lina *et al.*, 1997). This disease develops in non-immune patients in whom the toxin should be absorbed and transported systemically to subepidermal layers of the skin, resulting in pathological effects (Melish and Glasgow, 1971).

Although systemic diffusion of the toxin has never been shown in humans, it was demonstrated in newborn and adult mice by Fritsch *et al.* (1976) with radio-labelled toxin, which is excreted by the kidneys. This illness may also develop in adults who are generally immunocompromised (Khuong *et al.*, 1993; Cribier *et al.*, 1994). This syndrome is the more severe of the two SSSS presentations. Cutaneous involvement is widespread and may begin abruptly by an erythroderma which is accentuated in perioral and flexural areas, accompanied by skin tenderness and fever. This flushing of the skin spreads over the entire body within a few days and a Nikolsky's sign can be elicited. Erythroderma may be due to the superantigenic activity of ETA (and maybe ETB) (Lina *et al.*, 1997). Within a short time, large, flaccid bullae appear, exudative lesions with a yellow crusty eschar are observed. The disseminated bullous lesions are sterile, in contrast to what is usually observed in bullous impetigo. When desquamation occurs, it is markedly painful. The exfoliated areas dry out and develop seborrhoea-like flakes (Rogolsky, 1979; Barg and Harris, 1997). The entire sequence from disease onset to recovery is 7–10 days. There is no permanent scarring on areas of the skin. Although the causative toxins are not themselves lethal to the host, the disease may be fatal in 1–10% of cases. Antibodies do not appear until convalescence (Gemmell, 1995), which indicates that the generalized form of SSSS is probably due to a lack of ET-specific neutralizing maternal antibodies.

The occurrence of SSSS is not reported frequently in a normal population, although one isolate among 20 is likely to produce the toxin. Host factors play an important role in determining skin injuries. Three host factors, which are often related, have been identified: patient age, immune status and renal maturity level.

In humans, SSSS often occurs in newborns and infants; the disease is rare after the age of 5 years. Infants are usually colonized several hours or days after birth by ET-producing strains from the environment and particularly from healthy nasal carriers (family, medical staff). There are rare observations (Haddad *et al.*, 1991) of newborns infected at birth because of pre-birth perineal colonization of the mother by ET-producing *S. aureus*. The rarity of the disease after the age of 5 years has been related to the presence of ET-antibodies. ET-antibodies injected into mice (Miller and

Kapral, 1972; McLay *et al.*, 1975) or humans (Elias *et al.*, 1974a; Wiley *et al.*, 1976b) before challenge with ET suppress any cutaneous activity of the toxin. It is probable that neutralizing antibodies towards ETA or ETB prevent the occurrence of SSSS in adults. Small doses of purified ET (0.3–6.0 µg), to which most human adults are insensitive, can induce the formation of bullae in adults without detectable ET-antibodies (Kapral, 1974). Among 64 people randomly selected with different ages, 47 had detectable ET-antibodies (Wiley *et al.*, 1976b). Other studies from Baker and Wuepper (1977) and Todd (1985) show that ET-antibodies are rarely present in infants with SSSS and that ET-antibodies appear in 70% of these infants during convalescence. So, Baker *et al.* (1978b) hypothesize that SSSS occurs only in immunocompromised subjects unable to respond to the bacterial production of toxin by neutralizing antibodies.

In some instances, immunocompetent adults can suffer from ET-induced skin lesions (Opal *et al.*, 1988; Cribier *et al.*, 1994), as can HIV-1 seropositive patients (Farell *et al.*, 1996). Moreover, human adults receiving cytostatic or immunosuppressive treatments are more prone to SSSS (Peyramond *et al.*, 1980). So, the immune system, by means of antibodies, plays a major role in the protection against the action of the epidermolysin. The therapeutic use of pooled immunoglobulins obtained from collected human sera is not useful in the prevention or cure of SSSS as Kapral (1979) demonstrated that from 16 commercial batches, only four contained ET-neutralizing antibodies at low concentration.

However, the ability to create skin lesions in mice aged more than 5 days, even with high doses (Elias *et al.*, 1974b), and the rarity of the lesions in adults led Elias *et al.* (1977) to suggest that other factors may be involved. For example, biochemical and biological modifications occurring during the maturation of the epidermis could be responsible for protection against the action of ET. Elias *et al.* (1977) suggested and Fritsch *et al.* (1976) demonstrated that adult animals excrete the toxin in urine better than young ones. This observation is supported by rare clinical observations of SSSS in patients with renal function impairment (Elias *et al.*, 1976b).

In hospital patients (infants and adults) and in nurses, nasal carriage of ET-producing strains is not unusual. So, SSSS often appears as small outbreaks (Anthony *et al.*, 1972), particularly in infant care units. Sporadic cases are also often observed (Willard *et al.*, 1984). Nasal carriage constitutes an important reservoir for future epidemics. Kaplan *et al.* (1986) showed that ETA-producing strains contaminate the environment of infant care units and that these strains are identical to those responsible for outbreaks in the same care unit.

These data demonstrate, once more, the importance of stringent hygiene regulations, particularly in hospital care units with newborn and young infants. Outbreaks can be controlled by the use of both strict handwashing and masks. Another question is to decide whether the whole medical staff should be regularly tested for the presence of ET-producing *S. aureus* in anterior nares.

The action of ET is superimposed on to that of numerous other virulence factors of *S. aureus* (adhesion factors, toxins and enzymes). The diagnosis of SSSS is confirmed by the recovery of the organism and by the identification of the toxin (immunological methods or biological method with the newborn mouse test) or the toxin gene (Johnson *et al.*, 1991). Therefore, ET-induced lesions should be treated by adapted systemic antibiotherapy in order to limit the development of the staphylococcal infection and, hence, the production of epidermolysin. A supportive treatment, including fluid replacement in severe cases and wound care, is not sufficient but often remains necessary. After treatment, the patients usually recover without scars as the ET-induced lesions never involve the dermis.

REFERENCES

Amtsberg, G., Bollwahn, W., Hazem, S., Jordan, B. and Schmidt, U. (1973). Bacteriologische, serologische und tierexperimentelle Untersuchungen zur ätiologischen Bedeutung von *Staphylococcus hyicus* beim Nässenden Ekzem des Schweines. *Dtsch. Tierarztl. Wochenschr.* **80**, 493–516.

Andresen L.O. (1998). Differentiation and distribution of three types of exfoliative toxin produced by *Staphylococcus hyicus* from pigs with exudative epidermitis. *FEMS Immunol. Med. Microbiol.* **20**, 301–10.

Andresen, L.O., Wegener, H.C. and Bille-Hansen, V. (1993). *Staphylococcus hyicus*-skin reactions in piglets caused by crude extracellular products and by partially purified exfoliative toxin. *Microb. Pathogen.* **15**, 217–25.

Andresen, L.O., Bille-Hansen, V. and Wegener, H.C. (1997). *Staphylococcus hyicus* exfoliative toxin: purification and demonstration of antigenic diversity among toxins from virulent strains. *Microb. Pathogen.* **22**, 113–22.

Anthony, B.F., Giuliano, D.M. and Oh, W. (1972). Nursery outbreak of staphylococcal scalded skin syndrome. *Am. J. Dis. Child.* **124**, 41–4.

Arbuthnott, J.P. and Billcliffe, B. (1976). Qualitative and quantitative methods for detecting staphylococcal epidermolytic toxin. *J. Med. Microbiol.* **9**, 191–201.

Arbuthnott, J.P., Kent, J., Lyell, A. and Gemmell, C.G. (1971). Toxic epidermal necrolysis produced by an extracellular product of *Staphylococcus aureus*. *Br. J. Dermatol.* **85**, 145–9.

Arbuthnott, J.P., Billcliffe, B. and Thompson, W. D. (1974a). Isoelectric focusing studies of staphylococcal epidermolytic toxin. *FEBS Lett.* **46**, 92–5.

Arbuthnott, J.P., Kent, J., Lyell, A. and Gemmell, C.G. (1974b). Studies of staphylococcal toxins in relation to toxic epidermal necrolysis (the scalded skin syndrome). *Br. J. Dermatol.* **86**, 35–59.

Bailey, C.J. and Redpath, M.B. (1992). The esterolytic activity of epidermolytic toxins. *Biochem. J.* **284**, 177–80.

Baker, D.H. and Wuepper, K.D. (1977). Measurement of the epidermolytic toxin of *Staphylococcus aureus* in blister fluid from patients with bullous impetigo. *Clin. Res.* **25**, 197.

Baker, D.H., Dimond, R.L. and Wuepper, K.D. (1978a). The epidermolytic toxin of *Staphylococcus aureus*. Its failure to bind to cells and its detection in blister fluids of patients with bullous impetigo. *J. Invest. Dermatol.* **71**, 274–5.

Baker, D.H., Wuepper, K.D. and Rasmussen, J.E. (1978b). Staphylococcal skin syndrome: detection of antibody to epidermolytic toxin by a primary binding assay. *Clin. Exp. Dermatol.* **3**, 17–23.

Barg, N.L. and Harris, T. (1997). Toxin-mediated syndromes. In: *The Staphylococci in Human Disease* (eds K.B. Crossley and G.L. Archer), pp. 527–43. Churchill Livingstone, New York.

Benson, P.F., Rankin, G.L.S. and Ripper, J.J. (1962). An outbreak of exfoliative dermatitis of the newborn (Ritter's disease) due to *Staphylococcus aureus* phage type 55/71. *Lancet* **i**, 999–1002.

Brook, N.G. and Bannister, B.A. (1991). Staphylococcal enterotoxins in scarlet fever complicating chickenpox. *Postgrad. Med. J.* **67**, 1013–14.

Cavarelli, J., Prévost, G., Bourguet, W., Moulinier, L., Chevrier, B., Delagoutte, B., Bilwes, A., Mourey, L., Rifai, S., Piémont, Y. and Moras, D. (1997). The structure of *Staphylococcus aureus* epidermolytic toxin A, an atypic serine protease, at 1.7 Å resolution. *Structure* **5**, 813–24.

Cribier, B., Piémont, Y. and Grosshans, E. (1994). Staphylococcal scalded skin syndrome in adults. A clinical review illustrated with a new case. *J. Am. Acad. Dermatol.* **30**, 319–24.

Dajani, A. (1972). The scalded skin syndrome: relationship of phage group 2 staphylococci. *J. Infect. Dis.* **125**, 548–51.

Dancer, C.J., Garatt, R., Saldnha, J., Jhotti, H. and Evans, R. (1990). The epidermolytic toxins are proteases. *FEBS Lett.* **268**, 129–32.

Devriese, L.A. (1977). Isolation and identification of *Staphylococcus hyicus*. *Am. J. Vet. Res.* **38**, 787–92.

Devriese, L.A. (1986). Coagulase-negative staphylococci in animals. In: *Coagulase-negative Staphylococci* (eds P.A. Mårdh and K.H. Schleifer), pp. 51–7. Almqvist & Wiksell International, Stockholm.

Devriese, L.A. and Derycke, J. (1979). *Staphylococcus hyicus* in cattle. *Res. Vet. Sci.* **26**, 356–8.

Devriese, L.A., Vlaminck, K., Nuytten, J. and De Keersmaecker, Ph. (1983). *Staphylococcus hyicus* in skin lesions of horses. *Equine Vet. J.* **15**, 263–5.

Dimond, R.L. and Wuepper, K.D. (1976). Purification and characterization of a staphylococcal epidermolytic toxin. *Infect. Immun.* **13**, 627–33.

Dimond, R.L., Wolff, H.H. and Braun-Falco, O. (1977). The staphylococcal scalded skin syndrome. An experimental histochemical and electron microscopic study. *Br. J. Dermatol.* **96**, 483–92.

Elias, P.M. and Levy, S.W. (1976). Bullous impetigo: occurrence of localized scalded skin syndrome in an adult. *Arch. Dermatol.* **112**, 856.

Elias, P.M., Fritsch, P., Tuppeiner, G., Mittermayer, H. and Wolff, K. (1974a). Experimental staphylococcal toxic epidermal necrolysis (TEN) in adult humans and mice. *J. Lab. Clin. Med.* **84**, 414–24.

Elias, P.M., Mittermayer, H., Tuppeiner, G., Fritsch, P. and Wolff, K. (1974b). Staphylococcal toxic epidermal necrolysis (TEN): the expanded mouse model. *J. Invest. Dermatol.* **63**, 467–75.

Elias, P.M., Fritsch, P.F., Dahl, M.V. and Wolff, K. (1975). Staphylococcal toxic epidermal necrolysis: pathogenesis and studies on the subcellular site of action of exfoliatin. *J. Invest. Dermatol.* **65**, 501–12.

Elias, P.M., Fritsch, P.F. and Mittermayer, H. (1976). Staphylococcal toxic epidermal necrolysis: species and tissue susceptibility and resistance. *J. Invest. Dermatol.* **66**, 80–9.

Elias, P.M., Fritsch, P. and Epstein, E.H. (1977). Staphylococcal scalded skin syndrome. *Arch. Dermatol.* **113**, 207–19.

Farell, A.M., Ross, J.S., Unasankar, S. and Bunker, C.B. (1996). Staphylococcal scalded skin syndrome in an HIV-1 seropositive man. *Br. J. Dermatol.* **134**, 962–5.

Fleming, S.D., Iandolo, J.J. and Chapes, S.K. (1991). Murine macrophage activation by staphylococcal exotoxins. *Infect. Immun.* **59**, 4049–55.

Fleurette, J. and Ritter, J. (1980). Prévalence des souches de *Staphylococcus aureus* productrices d'exfoliatine dans le groupe bactériophagique II. *Ann. Microbiol. (Inst. Pasteur)* **131B**, 175–83.

Fritsch, P., Elias, P.M. and Varga, J. (1976). The fate of staphylococcal exfoliatin in newborn and adult mice. *Br. J. Dermatol.* **95**, 275–84.

Gemmell, C.G. (1995). Staphylococcal scalded skin syndrome. *J. Med. Microbiol.* **43**, 318–27.

Gentilhomme, E., Faure, M., Piémont, Y., Binder, P. and Thivolet, J. (1990). Action of staphylococcal exfoliative toxins on epidermal cell cultures and organotropic skin. *J. Dermatol.* **17**, 526–32.

Haddad, J., Piémont, Y., Barrinuevo, G., Baldauf, J.J., Monteil, H., Ritter, J. and Willard, D. (1991). Syndrome de Lyell et infection staphylococcique materno-foetale. *Arch. Fr. Pediatr.* **48**, 227–8.

Howells, C.H.L. and Jones, E.H. (1961). Two outbreaks of neonatal skin sepsis caused by *Staphylococcus aureus* phage type 71. *Arch. Dis. Child.* **36**, 214–16.

Jackson, R. (1974). Sir Jonathan Hutchinson on toxic epidermal necrolysis – an historical note. *Br. J. Dermatol.* **91**, 469–72.

Jackson, M.P. and Iandolo, J.J. (1986). Cloning and expression of the exfoliative toxin B gene from *Staphylococcus aureus*. *J. Bacteriol.* **166**, 574–80.

Jefferson, J. (1967). Lyell's toxic epidermal necrolysis: a staphylococcal etiology. *BMJ.* **ii**, 802–4.

Johnson, A.D., Metzger, J.F. and Spero, L. (1975). Production, purification and chemical characterization of *Staphylococcus aureus* exfoliative toxin. *Infect. Immun.* **12**, 1206–10.

Johnson, W.M., Tyler, S.D., Ewan, E.P., Ashton, F.E., Pollard, D.R. and Rozee, K.R. (1991). Detection of genes for enterotoxins, exfoliative toxins, and toxic shock syndrome toxin-1 in *Staphylococcus aureus* by the polymerase chain reaction. *J. Clin. Microbiol.* **29**, 426–30.

Kaplan, M.H., Chmel, H., Hsieh, H.C., Stephens, A. and Brinsko, V. (1986). Importance of exfoliatin toxin A production by *Staphylococcus aureus* strains isolated from clustered epidemics of neonatal pustulosis. *J. Clin. Microbiol.* **23**, 83–91.

Kapral, F.A. (1974). *Staphylococcus aureus*: some host–parasite interactions. *Ann. NY Acad. Sci.* **236**, 267–76.

Kapral, F.A. (1975). *Staphylococcus aureus* exfoliatin. In: *Microbiology* (ed. D. Schlessinger), pp. 263–66. American Society for Microbiology, Washington, DC.

Kapral, F.A. (1979). Levels of exfoliatin antitoxin in pooled human serum globulin. *J. Infect. Dis.* **139**, 209–10.

Kapral, F.A. and Miller, M.M. (1971). Product of *Staphylococcus aureus* responsible for the scalded skin syndrome. *Infect. Immun.* **4**, 541–5.

Khuong, M.A., Chosidow, O., El Sohl, N., Wechsler J., Piémont, Y., Roujeau J.C. and Revuz, J. (1993). Staphylococcal scalded skin syndrome in an adult: possible influence of non-steroidal anti-inflammatory drugs. *Dermatology* **186**, 153–4.

Koblenzer, P.J. (1967). Acute epidermal necrolysis (Ritter von Rittershain-Lyell). A clinicopathologic study. *Arch. Dermatol.* **95**, 608–17.

Kondo, I., Sakurai, S. and Sarai, Y. (1973). Purification of exfoliatin produced by *Staphylococcus aureus* of bacteriophage group 2 and its physicochemical properties. *Infect. Immun.* **8**, 156–64.

Kondo, I., Sakurai, S. and, Sarai, Y. (1974). New type of exfoliatin obtained from staphylococcal strains, belonging to phage groups other than group II, isolated from patients with impetigo and Ritter's disease. *Infect. Immun.* **10**, 851–61.

Kondo, I., Sakurai, S., Sari, Y. and Futaki, S. (1975). Two serotypes of exfoliatin and their distribution in staphylococcal strains isolated from patients with scalded skin syndrome. *J. Clin. Microbiol.* **1**, 397–400.

Lang, R. and Walker, J. (1956). An unusual bullous eruption. *S. Afr. Med. J.* **30**, 97–8.

Lee, C.Y., Schmidt, J.J., Johnson-Winegar, A.D., Spero, L. and Iandolo, J.J. (1987). Sequence determination and comparison of exfoliative toxin A and toxin B from *Staphylococcus aureus*. *J. Bacteriol.* **169**, 3904–9.

Lillibridge, C.B., Melish, M.E. and Glasgow, L.A. (1972). Site of action of exfoliative toxin in the staphylococcal scalded skin syndrome. *Pediatrics* **50**, 728–38.

Lina, G., Gillet, Y., Vandenesch, F., Jones, M.E., Floret, D. and Etienne, J. (1997). Toxin involvement in staphylococcal scalded skin syndrome. *Clin. Infect. Dis.* **25**, 1369–73.

Lowney, E.D., Baublis, J.V., Kreyer, G.M., Harrell, E.R. and McKenzie, A.R. (1967). The scalded skin syndrome in small children. *Arch. Dermatol.* **95**, 359–69.

Lyell, A. (1956). Toxic epidermal necrolysis: an eruption resembling scalding of the skin. *Br. J. Dermatol.* **68**, 355–61.

Magner, L.N. (1992). The germ theory of disease: medical microbiology. In: *A History of Medicine*, pp. 305–33. Marcel Dekker, New York.

Marrack, P. and Kappler, J. (1990). The staphylococcal enterotoxins and their relatives. *Science* **248**, 705–11.

McClosky, R.V. (1973). Scarlet fever and necrotizing vasculitis caused by coagulase-positive hemolytic *Staphylococcus aureus*, phage type 85. *Ann. Intern. Med.* **75**, 85–7.

McLay, A.L.C., Arbuthnott, J.P. and Lyell, A. (1975). Action of staphylococcal epidermolytic toxin on mouse skin: an electron microscopic study. *J. Invest. Dermatol.* **65**, 423–8.

Melish, M.E. (1973). Bullous varicella: its association with staphylococcal scalded skin syndrome. *J. Pediatr.* **83**, 1019–21.

Melish, M.E. and Glasgow, L.A. (1970). The staphylococcal scalded skin syndrome: development of an experimental model. *N. Engl. J. Med.* **282**, 1114–19.

Melish, M.E. and Glasgow, L.A. (1971). Staphylococcal scalded skin syndrome: the expanded clinical syndrome. *J. Pediatr.* **78**, 958–67.

Melish, M.E., Glasgow, L.A. and Turner, M.D. (1972). The staphylococcal scalded skin syndrome: isolation and partial characterization of the exfoliative toxin. *J. Infect. Dis.* **125**, 129–40.

Miller, M.M. and Kapral, F.A. (1972). Neutralization of *Staphylococcus aureus* exfoliation by antibody. *Infect. Immun.* **6**, 561–3.

Moras, D., Thierry, J.C., Cavarelli, J. and Piémont, Y. (1984). Preliminary crystallographic data for exfoliative toxin B from *Staphylococcus aureus*. *J. Mol. Biol.* **175**, 89–91.

Morlock, B.A., Spero, L. and Johnson, A.D. (1980). Mitogenic activity of staphylococcal exfoliative toxin. *Infect. Immun.* **30**, 381–4.

Nikolski (sic), P.W. (1898). Trois nouveaux cas de pemphigus foliacé étudiés au point de vue de la symptomatologie. *Ann. Dermatol. Syphiligr.* **9**, 1026–31. (Article erroneously signed by Lindstroem. Erratum published in *Ann. Dermatol. Syphiligr.* **9**, 1184.)

Nishioka, K., Katayama, I. and, Sano, S. (1981). Possible binding of epidermolytic toxin to a subcellular fraction of the epidermis. *J. Dermatol.* **8**, 7–12.

Opal, S.M., Johnson-Winegar, A.D. and Gross, A.S. (1988). Staphylococcal scalded skin syndrome in two immunocompetent adults caused by exfoliatin B-producing *Staphylococcus aureus*. *J. Clin. Microbiol.* **26**, 1283–6.

O'Toole, P.W. and Foster, T.J. (1987). Nucleotide sequence of the epidermolytic toxin A gene of *Staphylococcus aureus*. *J. Bacteriol.* **169**, 3910–15.

Parker, M.T., Tomlinson, A.J.H. and Williams, R.E.O. (1955). Impetigo contagiosa. The association of certain types of *Staphylococcus aureus* and *Streptococcus pyogenes* in superficial skin infections. *J. Hyg. Cambr.* **53**, 458–73.

Peyramond, D., Bensahkoun, D. and Bertrand, J.L. (1980). Syndromes cutanés aigus induits par la toxine exfoliante staphylococcique chez l'enfant. *Arch. Fr. Pediat.* **37**, 219–25.

Phillips, W.E., King, R.E. and Kloos, W.E. (1980). Isolation of *Staphylococcus hyicus* subsp. *hyicus* from a pig with septic polyarthritis. *Am. J. Vet. Res.* **41**, 274–6.

Piémont, Y. and Monteil, H. (1983). New approach in the separation of two exfoliative toxins from *Staphylococcus aureus*. *FEMS Microbiol. Lett.* **17**, 191–5.

Piémont, Y., Rasoamananjara, D., Fouace, J.M. and Bruce, T. (1984). Epidemiological investigation of exfoliative toxin-producing *Staphylococcus aureus* strains in hospitalized patients. *J. Clin. Microbiol.* **19**, 417–20.

Piémont, Y., Piémont, E. and Gérard, D. (1986). Fluorescence studies of thermal stability of staphylococcal exfoliative toxins A and B. *FEMS Microbiol. Lett.* **36**, 245–9.

Prévost, G., Rifai, S., Chaix, M.L. and Piémont, Y. (1991). Functional evidence that the Ser-195 residue of staphylococcal exfoliative toxin A is essential for biological activity. *Infect. Immun.* **59**, 3337–9.

Prévost, G., Rifai, S., Chaix, M.L., Meyer, S. and Piémont, Y. (1992). Is the His72, Asp120, Ser195 triad constitutive of the catalytic site of staphylococcal exfoliative toxin A? In: *Bacterial Protein Toxins, Fifth European Workshop on Bacterial Protein Toxins, Veldhoven, The Netherlands* (eds B. Witholt, J.E. Alouf, G.J. Boulnois, P. Cossart, B.W. Dijkstra, P. Falmagne, F.J. Fehrenbach, J. Freer, H. Niemann, R. Rappuoli and T. Wadström), pp. 488–9. Gustav Fischer, Stuttgart.

Rasmussen, J.E. (1975). Toxic epidermal necrolysis: a review of 75 cases in children. *Arch. Dermatol.* **111**, 1135–9.

Redpath, M.B., Foster, T.J. and Bailey, C.J. (1991). The role of the serine protease active site in the mode of action of epidermolytic toxin of *Staphylococcus aureus*. *FEMS Microbiol. Lett.* **81**, 151–5.

Ritter von Rittershain, G. (1880). Die exfoliative Dermatitis jüngerer Säuglinge und Cazenave's pemphigus foliaceus. *Arch. Kinderheilk.* **1**, 53–62.

Rogolsky, M. (1979). Nonenteric toxins of *Staphylococcus aureus*. *Microbiol. Rev.* **43**, 320–60.

Rogolsky, M., Warren, R., Wiley, B.B., Nakamura, H.T. and Glasgow, L.A. (1974). Nature of the genetic determinant controlling exfoliative toxin production in *Staphylococcus aureus*. *J. Bacteriol.* **117**, 157–65.

Sarai, Y., Nakahara, H., Ishikawa, T., Kondo, I., Futaki, S. and Hirayama, K. (1977). A bacteriological study on children with staphylococcal toxic epidermal necrolysis in Japan. *Dermatology* **154**, 161–7.

Sato, H., Tanabe, T., Kuramoto, M., Tanaka, K., Hashimoto, T. and Saito, H. (1991a). Isolation of exfoliative toxin from *Staphylococcus hyicus* subsp. *hyicus* and its exfoliative activity in the piglet. *Vet. Microbiol.* **27**, 263–75.

Sato, H., Kuramoto, M., Tanabe, T. and Saito, H. (1991b). Susceptibility of various animals and cultured cells to exfoliative toxin produced by *Staphylococcus hyicus* subsp. *hyicus*. *Vet. Microbiol.* **28**, 157–69.

Sato, H., Matsumori, Y., Tanabe, T., Saito, H., Shimizu, A. and Kawano, J. (1994). A new type of staphylococcal exfoliative toxin from a *Staphylococcus aureus* strain isolated from a horse with phegmon. *Infect. Immun.* **62**, 3780–5.

Sato, H., Tanabe, T., Watanabe, T., Teruya, K., Ohtaka, A., Saito, H. and Maehara, N. (1996). Chromosomal and extrachromosomal synthesis of exfoliative toxin from *Staphylococcus hyicus*. In: *Abstracts of the 8th International Symposium on Staphylococci and Staphylococcal Infections, Aix-les-Bains, France*, Abstract O-33, p. 83. Societé Française de Microbiologie, Paris.

Smith, T.P. and Bailey, C.J. (1986). Epidermolytic toxin from *Staphylococcus aureus* binds to filaggrins. *FEBS Lett.* **194**, 309–13.

Sompolinski, D. (1950). Impetigo contagiosa suis. *Maanedsskr. Dyrlæg.* **61**, 401–53.

Takiuchi, I., Kawamura, M., Teramoto, T. and Higuchi, D. (1987). Staphylococcal exfoliative toxin induces caseinolytic activity. *J. Infect. Dis.* **156**, 508–9.

Tanabe, T., Sato, H., Kuramoto, M. and Saito, H. (1993). Purification of exfoliative toxin produced by *Staphylococcus hyicus* and its antigenicity. *Infect. Immun.* **61**, 2973–7.

Tanabe, T., Sato, H., Ueda, K., Chihara, H., Watanabe, T., Nakano, K., Saito, H. and Maehara, N. (1995). Possible receptor for exfoliative toxins produced by *Staphylococcus hyicus* and *Staphylococcus aureus*. *Infect. Immun.* **63**, 1591–4.

Todd, J.K. (1985). Staphylococcal toxin syndromes. *Annu. Rev. Med.* **36**, 337–47.

Tokura, Y., Heald, P.W., Yan, S.L. and Edelson, R.L. (1992). Stimulation of cutaneous T-cell lymphoma cells with superantigenic staphylococcal toxins. *J. Invest. Dermatol.* **98**, 33–7.

Tokura, Y., Yagi, J., O'Malley, M., Lewis, J.M., Takigawa, M., Edelson, R.L., Tigelaar, R.E. (1994). Superantigenic staphylococcal exotoxins induce T-cell proliferation in the presence of Langerhans cells or class II-bearing keratinocytes and stimulate keratinocytes to produce T-cell-activating cytokines. *J. Invest. Dermatol.* **102**, 31–8.

Vath, G.M., Earhart, C.A., Rago, J.V., Kim, M.H., Bohach, G.A., Schlievert, P.M. and Ohlendorf, D.H. (1997). The structure of the superantigen exfoliative toxin A suggests a novel regulation as a serine protease. *Biochemistry* **36**, 1559–66.

Wald, E.R., Levine, M.M. and Togo, Y. (1973). Concomitant varicella and staphylococcal scalded skin syndrome. *J. Pediatr.* **83**, 1017–19.

Warren, R.L. (1981). Restriction endonuclease map of phage group 2 *Staphylococcus aureus* exfoliative toxin plasmid. *Infect. Immun.* **33**, 7–10.

Wegener, H.C., Andresen, L.O. and Bille-Hansen, V. (1993). *Staphylococcus hyicus* virulence in relation to exudative epidermitis in pigs. *Can. J. Vet. Res.* **57**, 119–25.

Wiley, B.B., Glasgow, L.A. and Rogolsky, M. (1976a). Studies on staphylococcal scalded skin syndrome (SSSS): isolation and purification of toxin and development of a radioimmuno-binding assay for antibodies to exfoliative toxin (ET). In: *Staphylococci and Staphylococcal Diseases* (ed J. Jeljaszewicz), pp. 499–516. Gustav Fischer, Stuttgart.

Wiley, B.B., Glasgow, L.A. and Rogolsky, M. (1976b). Staphylococcal scalded skin syndrome: development of a primary binding assay for human antibody to the exfoliative toxin. *Infect. Immun.* **13**, 513–20.

Willard, D., Monteil, H., Piémont, Y., Assi, R., Messer, I., Lavillaureix, J., Minck, R. and Gandar, R. (1984). L'exfoliatine dans les staphylococcies néonatales. *Nouv. Presse Med. (Paris)* **11**, 3769–71.

Winternitz, R. (1898). Ein Beitrag zur Kenntnis der Dermatitis exfoliativa neonatorum (Ritter). *Arch. Dermatol. Syph.* **44**, 397–416.

Wuepper, K.D., Dimond, R.L. and Knutson, D. (1975). Studies of the mechanism of epidermal injury by a staphylococcal epidermolytic toxin. *J. Invest. Dermatol.* **65**, 191–200.

Yoo, C.S., Wang, B.C., Sax, M. and Johnson, A.D. (1978). Preliminary crystallographic data for *Staphylococcus aureus* exfoliative toxin. *J. Mol. Biol.* **124**, 421–3.

38

Bacterial toxins as food poisons

Per Einar Granum and Sigrid Brynestad

INTRODUCTION

Food- and waterborne illnesses are common, distressing and sometimes life-threatening problems for people world-wide. People in developing countries suffer the most from these diseases. A significant proportion of cases of the estimated 1.5 billion episodes per year of diarrhoea in children under 5 years, including 3 million deaths, are of food- and waterborne origin. In many industrialized countries, despite all precautions, the incidence of foodborne infection has increased in recent years. Surveys indicate that 5–10% of the population are affected annually, and in addition to the human suffering, food poisoning causes substantial economic loss. In the USA costs are estimated to be US$6.5–34.9 billion/year, and an outbreak of cholera in Peru cost US$700 million in lost fish exports and another US$70 million in decreased tourism and local food sales (Todd, 1989; WHO, 1997).

The bacteria responsible for causing food poisoning have managed to adapt to the new niches presented to them with modern food processing. Much of the adaptation is through assimilation of new genes, including toxin and invasion-related genes. The toxin that is responsible for the direct action on host tissues, and thereby development of disease, is dependent on the physical and genetic background of both the bacteria and the host. The toxins are, for some of the food-poisoning diseases, the only cause of the symptoms but may also only increase the severity of the disease. Many of the toxins described here are on mobile elements (plasmids, phages and transposons), and can conceivably be transferred to bacteria that have not yet been identified as potential food pathogens. The presence of the same toxin gene in different species of bacteria has caused confusion in terminology of the proteins and of the bacteria.

THE FOOD-POISONING BACTERIA

There is a limited number of bacteria that are able to cause food poisoning, and they act through intoxications or infections. For some of these species the toxin(s) may be the only cause of the symptoms, while for others the toxin(s) plays a moderate to minor role in pathogenicity. The role of the toxins along with the mechanisms of pathogenicity make it possible to divide the food-poisoning bacteria into five different groups according to the criteria given in Table 38.1. Group 1 contains the bacteria causing foodborne intoxications (toxins preformed in foods), while the remaining four groups comprise the species that have the ability to give us foodborne infections. For foodborne intoxications the bacteria may or may not be present in the food at the time of consumption, and the bacteria can be killed by heat while the toxin survives. Taking a closer look at the species belonging to groups 2 and 3 (Table 38.2), we see that although they are by

TABLE 38.1 The different groups of bacteria causing food poisoning divided according to pathogenicity mechanisms

Group[a]	Mechanisms of pathogenicity
1	Enterotoxin or neurotoxins produced in foods (preformed toxin)
2	Enterotoxin produced in the intestine without bacterial adherence to epithelial cells
3	Enterotoxin produced after bacterial adherence to epithelial cells, no bacterial invasion
4	Bacterial invasion localized to the epithelium and the intestinal immune system, with or without toxin production
5	Systemic infection

[a]Bacterial intoxications: group 1, bacterial infections: group 2–5.

ria are able to invade the hosts cells, and the bacteria in group 5 cause systemic infections. From our present knowledge it appears that protein toxins do not play a direct role in virulence/pathogenesis for the bacteria in group 5, and these bacteria will not be discussed further in this chapter.

Most of the foodborne diseases have been well characterized, and it is possible to summarize the knowledge on infective dose, incubation time, symptoms and duration as seen in Table 38.2. For the intoxications described in group 1, the toxins are always preformed in the food. For the next group the toxins (enterotoxins) are produced in the host, but without any interaction with the host cells. Even for group 3 the bacteria have little effect on the host, although colonization of the epithelial cells is usually necessary. For these first three groups, although the diseases might be severe (i.e. *C. botulinum*, EHEC and *V. cholera*), fever is not a common symptom and is never high. In contrast, fever is usually a major symptom for groups 4 and 5 because of the bacteria's direct interaction and invasion of the host.

definition infections, the symptoms are almost entirely caused by toxin production in the host. In contrast, for the members of groups 4 and 5, which usually cause the most severe infections, the toxins play a less important role in pathogenicity, mainly because these bacte-

TABLE 38.2 The most important food-poisoning organisms (Granum, 1996)

Species	Infective dose	Incubation time	Symptoms[a] in the order they usually appear	Duration
Intoxications				
Group 1				
Staphylococcus aureus	Toxin	1–6 h	NAV (DF)	8–24 h
Bacillus cereus (emetic)	Toxin	1–6 h	NV	6–24 h
Clostridium botulinum	Toxin	12–72 h	Neurological	days–months
Infections				
Group 2				
Bacillus cereus (diarrhoeal type)	10^5–10^7	6–12 h	AD	12–24 h
Clostridium perfringens	10^7–10^8	8–16 h	ADN(F)	16–24 h
Group 3				
Aeromonas spp.	10^6–10^8	6–48 h	DA (F)	24–28 h
Escherichia coli				
ETEC (ST)	10^5–10^8	16–48 h	D (AVF)	1–2 days
ETEC (LT)	10^5–10^7	16–48 h	D (AVF)	1–3 days
EHEC (O157:H7)	10	1–7 days	DAB (H)	days–weeks
Vibrio cholerae	10^8	2–5 days	DA (V)	4–6 days
Vibrio parahaemolyticus	10^5–10^7	3–76 h	DA (N V F)	3–7 days
Group 4				
Campylobacter jejuni/coli	$\geqslant 10^3$	3–8 days	FADB	weeks
Salmonella spp (non-typhoid)	10^3–10^6	6–72 h	DAF (V H)	2–7 days
Shigella spp.	10^2–10^5	1–7 days	AFDB (H N V)	days–weeks
Yersinia enterocolitica	10^6–10^7	3–5 days	FDA (V H)	weeks
Group 5				
Listeria monocytogenes	10^7–10^8	days	Systemic	weeks
Salmonella typhi	1–10^2	10–21 days	Systemic	weeks
Salmonella paratyphi	1–10^2	10–21 days	Systemic	weeks

[a]A, abdominal pain; H, Headache; B, bloody diarrhoea; N, nausea; D, diarrhoea; V, vomiting; F, fever.

THE TOXINS

The toxins involved in bacterial food poisoning mainly belong to three different groups of toxins: emetic toxins, neurotoxins and enterotoxins. Emetic toxins cause vomiting by binding specific receptors in the duodenum. Neurotoxins are proteins or small chemical substances (not relevant in this context) which act on the nervous system. Enterotoxins are defined as protein toxins that are active in the intestine, and cause diarrhoea (fluid accumulation in loop tests). Enterotoxins can be divided into two groups: cytotoxic enterotoxins, which disrupt the cell membrane or other vital functions in the cell, thus causing cell death, and cytotonic enterotoxins, which enter the epithelial cell and cause diarrhoea without direct membrane disruption or cell death.

Table 38.3 shows some characteristics of the toxins of the different food-poisoning bacteria. Several of these toxins are described in detail in other chapters in this book, and will only be treated briefly here. The information in this chapter concentrates on bacterial toxins involved in food poisoning that are not treated elsewhere in this book.

Staphylococcus aureus enterotoxin (see Chapters 21, 33 and 37)

Staphylococcus aureus enterotoxins (SE) are the cause of one of the most common types of food poisoning in the developed world. The enterotoxin is preformed in foods stored above 15–18°C, and it usually takes about 3 h to produce enough enterotoxin to cause food poisoning. As little as 100–200 ng may cause emesis in humans (Jablonski and Bohach, 1997). The enterotoxins are specific to primates and are also potent mitogenic superantigens (Marrack and Kappler, 1990; and Chapter 33). They consist of a family of seven different but related toxins (Alouf et al., 1991). Similar toxins are not found in other bacterial species. These toxins cause vomiting, yet they are called enterotoxins because of their ability to produce fluid accumulation in ileal loop tests. These toxins may also cause diarrhoea, but not a severe type and probably only at high concentrations. Although these toxins are proteins (26–28 kDa) the active part of the molecule is not influenced by heat treatment, so even autoclavation is not sufficient to reduce activity of preformed SE. Thus

TABLE 38.3 Characteristics of toxins involved in food poisoning produced by different bacteria

Species	Number of toxins	Toxin type[a]	Heat labile(L)/ or stable (S)	Receptor	Mode of action	Enzyme activity
Staphylococcus aureus	Seven (type A–E) closely related	Emetic and CnEnt, protein	S	TCRVβ T-cells	Emesis: via nervus vagus and sympaticus	?
Bacillus cereus (emetic)	One	Emetic cyclic peptide	S	5-HT$_3$	Emesis: via nervus vagus	No
Clostridium botulinum	Seven (type A–G) closely related	Neurotoxin, protein	L	80–116-kDa glycoprotein	Split presynaptic membrane proteins	Zn^{2+}-dependent endopeptidase
Bacillus cereus (diarrhoeal type)	Two three-component	CtEnt protein	L	Unknown	Membrane disruption	See text
Clostridium perfringens type A	One	CtEnt protein	L	22-kDa protein	Membrane disruption	No
Aeromonas spp.	Several types	CtEnt and CnEnt	L and S			?
Escherichia coli						
ETEC (ST)	Several	CnEnt, peptide	S		cGMP accum.	?
ETEC (LT)	One	CnEnt, protein	L	GM1	cAMP accum.	ADP-ribosylase
EHEC (O157:H7)	Two	CtEnt, protein	L	Gb3	Inhib. of protein synthesis	N-glycosidase
Vibrio cholerae	One	CnEnt, protein	L	GM1	cAMP accum.	ADP-ribosylase
Vibrio parahaemolyticus	One (additional types)	CnEnt, protein	S	GM2/GM1	Ca^{2+} signal transduction	?
Campylobacter jejuni/coli	One (additional types)	CnEnt, protein		GM1	cAMP accum.	ADP-ribosylase
Salmonella spp	One	CnEnt, protein	L	GM1	cAMP accum.	ADP-ribosylase
Shigella spp.	One	CtEnt, protein	L	Gb3	Inhib. of protein synthesis	N-glycosidase
Yersinia enterocolitica	One (additional types)	CnEnt, peptide	S	M-cells (uncertain)	cGMP accum.	?

[a]Ent, enterotoxin; Cn, cytotonic; Ct, cytotoxic.

SE may be found in sterile heat-treated foods where it has been produced prior to heat treatment, i.e. canned foods.

Clostridium botulinum neurotoxin (see Chapters 9 and 10)

Botulism is a rare but often fatal disease. The different types of *C. botulinum* vary in potency and geographical locations. In Argentina, Brazil, China and western parts of the USA, where the highly potent type A is the most frequently isolated type, fatality rates are higher than in countries where other types are prevalent (Dodds and Austin, 1997). *Clostridium botulinum* produces seven different neurotoxins (Hauschild, 1989). *Clostridium tetani* neurotoxin is related, but not involved in food poisoning. Also, other *Clostridium* species may produce neurotoxins of the type found in *C. botulinum*. These toxins are now known to be Zn^{2+}-dependent endopeptidases, that split three different neuronal proteins at specific target sites (Montecucco and Schiavo, 1994). In contrast to *S. aureus* enterotoxins the botulinum toxins are heat labile and are rapidly reduced in activity at 60°C (Hauschild, 1989).

Bacillus cereus toxins

Bacillus cereus is a Gram-positive, spore-forming, motile, facultative anaerobic rod. It causes two different types of food poisoning: the diarrhoeal type and the emetic type. The diarrhoeal type of food poisoning is caused by a complex of enterotoxins (Beecher and Wong, 1997; Lund and Granum, 1997) produced during vegetative growth of *B. cereus* in the small intestine (Granum, 1994), while the emetic toxin is produced by growing cells in foods (Kramer and Gilbert, 1989). The closely related *B. thuringiensis* is reported to produce enterotoxins (Ray, 1991; Jackson *et al.*, 1995), and this could potentially cause serious problems, as spraying of this organism to protect crops against insect attacks has become common in several countries. There has been a confirmed *B. thuringiensis* outbreak of food poisoning (Jackson *et al.*, 1995). However, since the procedures normally used for confirmation of *B. cereus* would not differentiate between the two organisms, undetected outbreaks may have occurred. To assure safe spraying with *B. thuringiensis*, the organism in use should not carry enterotoxin genes and be unable to produce food-poisoning toxins.

Foodborne outbreaks

The dominating type of disease caused by *B. cereus* differs from country to country. In Japan the emetic type is reported about 10 times more frequently than the diarrhoeal type (Kramer and Gilbert, 1989), while in Europe and North America the diarrhoeal type is the most frequently reported (Kramer and Gilbert, 1989). Since *B. cereus* food poisoning is not a reportable disease in any country, there are very few figures given of the total number of these kinds of food poisoning. Even if the two syndromes were reportable one would expect dramatic under-reportation, since few seek medical help during the active phase of the disease, and the patients recover quickly thereafter.

Characteristics of disease

The emetic toxin results in vomiting, while the diarrhoeal type, caused by enterotoxins, gives diarrhoea (Kramer and Gilbert, 1989). In a small number of cases both types of symptom are recorded (Kramer and Gilbert, 1989), probably due to production of both types of toxin. There has been some debate about whether or not the enterotoxin(s) can be preformed in foods, and cause an intoxication. By reviewing the literature it is obvious that the incubation time is a little too long for an intoxication (>6 h; average 12 h) (Kramer and Gilbert, 1989), and in model experiments it has been shown that the enterotoxin(s) is degraded on its way to the ileum (Granum, 1994). Although the enterotoxin(s) can be preformed, the number of *B. cereus* cells in the food would be at least two orders of magnitude higher than that necessary for causing food poisoning (Granum, 1997), and such products would no longer be acceptable to the consumer.

Counts ranging from 10^3 to 10^9 g^{-1} (or ml) *B. cereus* (Granum, 1997) have been reported in the incriminated foods after food poisoning, giving total infective doses ranging from about 10^5 to 10^{11}. Partly owing to the large differences in the amount of enterotoxin produced by different strains (Granum, 1997), the total infective dose seems to vary between about 10^5 and 10^8 viable cells or spores. Thus, any food containing more than 10^3 *B. cereus* g^{-1} cannot be considered completely safe for consumption.

The emetic toxin

The emetic toxin causes emesis (vomiting) only and its structure remained a mystery as long as the only detection system involved living primates (Kramer and Gilbert, 1989). The recent discovery that the toxin could be detected (vacuolation activity) by the use of HEp-2 cells (Hughes *et al.*, 1988) has led to its isolation and determination of its structure (Agata *et al.*, 1994). Although there has been some doubt as to whether the emetic toxin and the vacuolating factor are the same component (Kramer and Gilbert, 1989) there is no doubt that it is the same toxin (Agata *et al.*, 1995a; Shinagawa *et al.*, 1995). The emetic toxin, cereulide, consists of a ring structure of three repeats of four amino-

and/or oxy-acids: [D-*O*-Leu-D-Ala-L-*O*-Val-L-Val]$_3$. This ring structure (dodecadepsipeptide) has a molecular mass of 1.2 kDa, and is chemically closely related to the potassium ionophore valinomycin (Agata *et al.*, 1994). The emetic toxin is resistant to heat, pH and proteolysis but is not antigenic (Kramer and Gilbert, 1989) (Table 38.1). The biosynthetic pathway and mechanism of action of the emetic toxin still have to be elucidated, although it has just been shown that it stimulates the vagus afferent through binding to the 5-HT$_3$ receptor (Agata *et al.*, 1995a). It is not clear whether the toxin is a modified gene product or whether it is enzymically produced. However, with such an unusual structure it is most likely that cereulide is an enzymically synthesized peptide and not a genetic product.

Enterotoxins

The number of enterotoxins and their properties have also been the subject of debate for some time (Kramer and Gilbert, 1989; Granum, 1997), and at least three different enterotoxins have been characterized (Beecher and Wong, 1994; Agata *et al.*, 1995b; Beecher *et al.*, 1995; Lund and Granum, 1996). The early studies on the enterotoxin (Kramer and Gilbert, 1989; Granum, 1997) suggested a single or a multi-component enterotoxin. Further work has now shown that *B. cereus* produces two different three-component enterotoxins (Beecher and Wong, 1994, 1997; Beecher *et al.*, 1995; Lund and Granum, 1996, 1997;) which are considered to be the main virulence factors of the diarrhoeal type *B. cereus* food poisoning.

A three-component haemolysin (HBL; consisting of three proteins: B, L$_1$ and L$_2$: see also Chapter 22) with enterotoxin activity has been purified and characterized (Beecher and Wong, 1994, 1997; Beecher *et al.*, 1995). This toxin also has dermonecrotic and vascular permeability activities, and causes fluid accumulation in ligated rabbit ileal loops. HBL has therefore been suggested to be a primary virulence factor in *B. cereus* diarrhoea (Beecher *et al.*, 1995). Convincing evidence has shown that all three components are necessary for maximal enterotoxin activity (Beecher *et al.*, 1995). A non-haemolytic three-component enterotoxin (NHE) was recently characterized (Lund and Granum, 1996). The three components of this toxin were different from the components of HBL. The characteristics of the two three-component enterotoxins are given in Table 38.4.

The three components of NHE enterotoxin were first purified from a *B. cereus* strain isolated after a large food-poisoning outbreak in Norway in 1995. This strain was originally chosen mainly because it did not produce the L$_2$-component (Granum, 1994; Granum *et al.*, 1996; Lund and Granum, 1996). The authors expected to show that the L$_2$-component was unnecessary for biological activity of the HBL enterotoxin, but instead

TABLE 38.4 Characteristics of the three enterotoxins from *B. cereus*

	Enterotoxin HBL	Enterotoxin NHE
Number of components	Three	Three
Size of active component(s)	L$_2$: 46 kDa	45 kDa
	L$_1$: 38 kDa	105 kDa
	B: 37 kDa	39 kDa
Haemolytic	Yes	No
Toxicity in cell tests	80 ng	70 ng
Shown to be involved in food poisoning	Yes	Yes
Cloned and sequenced	Yes	Yes[a]

[a]Granum *et al.* (manuscript in preparation).

came across another three-component enterotoxin. Binary combination of the components of this enterotoxin possesses some biological activity, but not nearly at as high a level as when all of the components are present (Lund and Granum, 1997).

Some strains produce both of the three-component enterotoxins, while other stains only contain genes for one of them (Granum *et al.*, 1996; Lund and Granum, 1996, 1997). Currently, we do not know the distribution of the two enterotoxin complexes among strains, or how important each of them is in relation to food poisoning. However, it seems as if they are both involved in food poisoning.

It has been suggested, from studies of interactions with erythrocytes, that the B-protein is the component that binds HBL to the target cells, and that L$_1$ and L$_2$ have lytic functions (Beecher and Macmillan, 1991). Recently, another model for the action of HBL has been proposed, suggesting that the components of HBL bind to target cells independently and then constitute a membrane-attacking complex resulting in a colloid osmotic lysis mechanism (Beecher and Wong, 1997). Studies of interactions between NHE and Vero cells have shown that the 105-kDa protein might be the binding component of that complex (Lund and Granum, 1997). It has just been shown that the 105-kDa protein has gelatinolytic and collagenolytic activities (Lund and Granum, manuscript in preparation). The two other components are probably not able to bind to Vero or epithelial cells alone. It is still not clear whether the two different toxin complexes only damage the plasma membranes of the targets cells or whether some of the components are translocated to the cytosol and have lytic activity inside the cell.

There is a high degree of identity between the *N*-terminal part of L$_1$ and the corresponding part of the 39-kDa protein, and high identity also exists between parts of L$_2$ and the 45-kDa protein. Together, these observations indicate some similarities between HBL and NHE.

Gene organization of the enterotoxins

All three proteins of the HBL are transcribed from one operon (*hbl*) (Ryan *et al.*, 1997) and Northern blot analysis has shown an RNA transcript of 5.5 kb (Figure 38.1). *hblC* (transcribing L₂) and *hblD* (transcribing L₁) are only separated by 37 bp and encode proteins of 447 amino acids (aa) and 384 aa. L₂ has a signal peptide of 32 aa and L₁ a signal peptide of 30 aa. The B-protein, transcribed from *hblA*, consists of 375 aa, with a signal peptide of 31 aa (Heinrichs *et al.*, 1993) The exact spacing between *hblD* and *hblA* is at least 100 bp (overlapping sequence not published), but is claimed to be approximately 115 bp (Ryan *et al.*, 1997). The spacing between *hblA* and *hblB* is 381 bp, and the length of *hblB* is not known (Heinrichs *et al.*, 1993). However, based on the length of the Northern blot it is tempting to suggest a similar size to *hblA*. The B and the putative B'-protein are very similar in the first 158 aa (the known sequence of B'-protein based on the DNA sequence). The function of this putative protein is not yet known, but it is possible that it may substitute for the B-protein. It was, however, shown by polymerase chain reaction (PCR) analysis that not all strains containing the *hbl* operon have *hblB* present with the known sequence. The *hbl* operon is mapped to the unstable part of the *B. cereus* chromosome (Carlson *et al.*, 1996).

The two smallest proteins of the NHE complex are similar to the L₁ and L₂-proteins of HBL. We have just sequenced the operon containing the 45-kDa protein and the 39-kDa protein gene. This operon also carries a second 39-kDa protein gene with a stop codon two-thirds into the gene (Figure 38.2). The gene of the collagenase (105-kDa protein) is not directly included in the operon, based on PCR analysis using primers constructed from the *N*-terminal sequences of the proteins of NHE. The authors have now sequenced more than half of the collagenase gene (Granum *et al.*, manuscript in preparation).

Regulation of enterotoxin production

Very little is yet known about the regulation of enterotoxin transcription. It seems as if maximal enterotoxin activity is found during late exponential or early stationary phase, and indeed this has been shown for the 45-kDa protein of the NHE that it is under regulation of *plcR* (Agaisse *et al.*, 1997), a gene first described in connection with the regulation of *plcA* (phospholipase C) (Lereculus *et al.*, 1996). The binding sequence for this regulatory protein (TATG8NCATG) is not found upstream of the *hbl* operon.

Clostridium perfringens enterotoxin

Clostridium perfringens are ubiquitous spore-forming, anaerobic, Gram-positive, non-motile rods which grow well between 20 and 50°C (Hatheway, 1990). They are commonly found as a part of the intestinal flora of mammals. *Clostridium perfringens* can produce at least 13 different toxins, although each individual isolate only produces a subset of these. The production of the four major lethal toxins is used to type isolates (A–E) (Table 38.5). The enterotoxin gene (*cpe*) has been found in all types. There are two major types of food poisoning, a diarrhoeal type caused by the *C. perfringens* enterotoxin (CPE) and necrotic enteritis mainly caused by the β-toxin in type C strains.

Clostridium perfringens *type C food poisoning*

Clostridium perfringens type C can be involved in necrotic enteritis known in Germany as 'Darmbrand' and in New Guinea as 'pig-bel'. The symptoms are

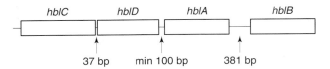

FIGURE 38.1 The map of the *hbl* operon (Heinrichs *et al.*, 1993; Ryan *et al.*, 1997). The operon codes for the three known proteins of the HBL: L2-protein (*hblC*), L1-protein (*hblD*), and B-protein b (*hblA*) and the B-protein (*hylB*), with 73% identity to the B-protein in the first 158 amino acids [5].

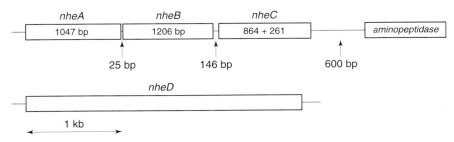

FIGURE 38.2 The map of the *nhe* operon and the 105-kDa protein (collagenase) (Granum *et al.*, 1998). The *nhe* operon codes for: 45-kDa protein (*nheA*), 39-kDa protein (*nheB*) and a 39-kDa-like protein (*nheC*) with a stop codon after 864 bp. The last protein in NHE complex is a 105-kDa collagenase (*nheD*).

TABLE 38.5 The distribution of the major lethal toxins used for typing of *Clostridium perfringens*

| Type | Major toxins | | | |
	Alpha (α)	Beta (β)	Epsilon (ϵ)	Iota (ι)
A	+	–	–	–
B	+	+	+	–
C	+	+	–	–
D	+	–	+	–
E	+	–	–	+

mainly caused by the β-toxin (phospholipase), although δ-toxin (haemolysin) and θ-toxin (perfringolysin O; see Chapter 27) contribute to the disease, which can be quite serious and has a mortality rate of 15–25% even with treatment. The toxin is trypsin sensitive and is normally inactivated by stomach enzymes. The disease is associated with individuals who have a low level of proteolytic enzymes in their intestinal tracts most often caused by low intake of protein. In New Guinea pig-bel outbreaks coincided with traditional festivals where large amounts of spit-grilled pork were consumed, while their staple diet, sweet potatoes, contains a trypsin inhibitor (Granum, 1990). The risk of *C. perfringens* type C food poisoning is minimal for healthy individuals with normal levels of proteolytic enzymes. Although the disease is not common, the danger of this type of food poisoning is real in immunocompromised individuals on special (vegetarian) diets.

Food poisoning caused by Clostridium perfringens type A enterotoxin

A relatively mild type of diarrhoea is the most common form of food poisoning caused by *C. perfringens* and is one of the most common forms of human gastrointestinal illness in industrialized nations (Granum, 1990). Although the enterotoxin gene (*cpe*) has been found in all types of *C. perfringens*, only type A has been associated with CPE-caused food poisoning. After ingestion of food contaminated with large numbers (approximately 10^8) of cells, CPE is produced by the bacterium during sporulation in the intestine and released upon lysis of the mother cell. CPE then binds to a protein receptor, after which a pore is formed, resulting in altered membrane permeability, which causes diarrhoea.

Characteristics of the type A disease

The symptoms are diarrhoea and abdominal pain; nausea and fever are less common with vomiting being rare. The symptoms usually appear 6–16 h after ingestion of the contaminated food, with a duration that is usually less than 24 h (Granum, 1990). The disease is usually associated with protein-rich foods, and in 75% of the cases meat and meat products are accountable for the outbreaks (Johnson and Gerding, 1997). The ability to sporulate, the optimum growth temperature of 45°C and a generation time of down to 8 min are important traits that allow the bacterium to survive and multiply in food (McClane, 1997). Most cases are identified from institutions where hygiene routines have been broken, generally too slow cooling and/or inadequate reheating of food. 'Kitchen strains' where repeated warming and sporulation have occurred could select for strains which make heat-resistant spores and have acquired the enterotoxin gene.

The Clostridium perfringens enterotoxin

CPE is a single, 319 amino acid polypeptide with a molecular weight of 35 317 Da and an isoelectric point of 4.3, and is heat and pH labile (Kokai-Kun *et al.*, 1997a). There is no significant homology to any known proteins. A two-domain structure has been proposed (Granum and Stewart, 1993) where the *N*-terminal is necessary for insertion/cytotoxicity and the *C*-terminal end (aa 290–319) contains the binding region, with specific binding necessary for toxicity (Kokai-Kun and McClane, 1997b). The binding also causes a change in secondary structure from mainly β-sheet to mainly α-helix (Granum and Harbitz, 1985). The removal of the first 25 to 34 aa increases activity (Granum *et al.*, 1981; Granum and Richardson, 1991) and the first 44 aa can be removed without loss of activity (Kokai-Kun and McClane, 1997b).

Two types of protein receptor for CPE have been cloned, CPE-R and RVP1, both *ca.* 22 030 Da with over 90% similarity. The CPE-R protein receptor binds CPE with high affinity while RVP1 has a lower affinity, and seems to be expressed at a lower level in cells. The receptors have been found to be present in many cell types, including lungs, heart, skeletal muscle, liver, kidneys and the intestine (Katahira *et al.*, 1997a, b). The exact mechanism of action is still under investigation, and although there are discrepancies in the reports in the sizes of complexes, some steps in formation of an active complex are clear (Kokai-Kun and McClane, 1997a; Katahira *et al.*, 1997a, b). When CPE binds to the specific receptors a complex is formed. After a physical change, this complex interacts with other proteins and forms a larger (160–200 kDa) very hydrophobic complex, which causes small molecule membrane permeabilities to develop. Evidence points towards the large complex serving as a pore, but permeability could conceivably be caused by some other mechanism (Kokai-Kun and McClane, 1997a). CPE seems to differ substantially from other known pore-forming toxins.

Gene organization of the enterotoxin

The observation that strains could lose the ability to produce enterotoxin, and the fact that only 5–10% of naturally isolated strains carried *cpe* enterotoxin suggested that *cpe* was carried on some sort of mobile element (Granum and Stewart, 1993). This possibility was confirmed when it was found that *cpe* is on a transposon, Tn*5565*, integrated between two housekeeping genes in human food-poisoning strains (Brynestad *et al.*, 1997). In animal isolates, the gene is carried on large plasmids with some strain to strain variation in the genetic structure (Cornillot *et al.*, 1995). The transposon contains two different IS elements, flanking copies of IS*1470*, which is a member of the IS*30* family and an IS*200*-like element, IS*1469* (see Figure 38.3). IS*1470* is found in multiple copies in many of the *cpe*-positive strains, a type C *cpe*-negative strain was positive for IS*1470*, while many of the strains where *cpe* is plasmid borne do not carry IS*1470*, but another IS element, IS*1151*. All type A *cpe*-positive *C. perfringens* tested have IS*1469* in the same position upstream regardless of the genetic placement of *cpe*. The actual movement of *cpe* has not yet been shown, so which of the IS elements are essential in the transfer of *cpe* is yet to be established.

Regulation of enterotoxin production

Enterotoxin production is associated with sporulation, and while little to no CPE is produced during vegetative growth, up to 15% and possibly up to 30% of the total protein produced under sporulation can be CPE (McClane, 1997). The genetic background of *C. perfringens* is essential for production of large amounts of CPE (Melville *et al.*, 1994). The *cpe* mRNA seems to be very stable with a half-life of up to 58 min, which could help to explain the large amounts of CPE produced (Labbé and Duncan, 1977). The mapped promoters have sequence homology to the promoters regulated by SigE and SigK in *B. subtilis*, although the *C. perfringens* promoter is not recognized by the *B. subtilis* sigma factors. Isolation of the clostridial sigma factors should help to elucidate the sporulation specificity of expression of CPE (Melville and Zhao, 1998). Binding sites for the transition state regulator Hpr were found up- and downstream of *cpe* (Brynestad *et al.*, 1994), and a Hpr-like protein could provide regulation that agrees with the observed regulation pattern.

Aeromonas spp. enterotoxin

Although *Aeromonas* spp. have been suspected to be the cause of water- and foodborne (mainly seafood) gastroenteritis (Kirov, 1997), no definite proof for their involvement has been supplied so far. However, outbreaks have recently been reported by several authors (for a review see Kirov, 1997). Three species, *A. hydrophila*, *A. caviae* and *A. veronii* biovar sobria, have been suggested as a cause of human gastroenteritis (Deohar *et al.*, 1991; Kirov, 1997). *Aeromonas* spp. represent a potentially serious food problem, as many of them can grow at refrigeration temperatures, at pH values from 4 to 10, under most atmospheres used to prevent bacterial growth in foods and in the presence of high salt concentration (Kirov, 1993, 1997). A variety of virulence factors and potential enterotoxins has been characterized (Kirov, 1997), and some strains have been shown to invade epithelial cells (Lawson *et al.*, 1985; Watson *et al.*, 1985).

Aerolysin (see Chapter 18) and other haemolysins with closely related sequences seem to be the best candidates for the major enterotoxin (Kirov, 1997; Granum *et al.*, 1998). These haemolysins are able to disrupt the membranes in epithelial cells and thereby cause diarrhoea (Granum *et al.*, 1998). It is not unlikely that more than one enterotoxin may be involved in food poisoning of humans, but addressing this question requires much more research, preferably on

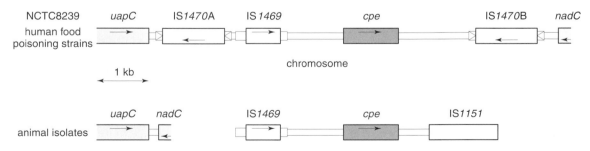

FIGURE 38.3 The map of the transposon Tn5565, containing *cpe* in type A human food-poisoning strains and the configuration in animal isolates. The genetic location is specified. The direction of transcription is indicated by arrows, inverted repeats with boxed arrows, and direct repeats with solid boxes. Specifics of each gene: IS1470: 1210-bp IS element with a 346 aa potential transposase with 30% homology to IS30. IS1469: 718-bp IS element with a 151 aa putative transposase, 60% identity to IS200. *uapC*: putative purine permease, *nadC*: putative quinolate phosphoribosyl transferase. IS1151: 1696 bp with a 474 aa putative transposase.

strains that have been involved in food poisoning. Although involvement of *Aeromonas* spp. in food poisoning is still controversial, an increasing number of papers points in the direction of direct involvement. The number of potential enterotoxins that has been suggested during the past few years (Kirov, 1997) may show the diversity of the different *Aeromonas* spp. strains. They are pathogenic to a variety of different animal species living under very different conditions (from fish to humans).

Escherichia coli enterotoxins (see Chapters 6, 29 and 40)

There are at least five different types of *E. coli* that can cause food poisoning: enterotoxic (ETEC; two toxin types, ST and LT), enterohaemorrhagic (EHEC), enteroinvasive (EIEC), enteropathogenic (EPEC) and enteroaggregative (EAaggEC) (Nataro and Kaper, 1998). The two first types produce enterotoxins (Table 38.3). Enterotoxic *E. coli* are probably the most common cause of travellers' diarrhoea, and are an important cause of food poisoning throughout the world. The main *E. coli* concern for the food industry is EHEC, because of the low infective dose (Table 38.2) and the severity of the disease, especially for the infirm and children, who are susceptible to the development of kidney failure (haemolytic–uraemic syndrome: HUS).

Both types of ETEC are relatively mild diseases and the symptoms are caused only by the enterotoxins (see Chapters 6 and 29). The Shiga-like toxins produced by EHEC (O157:H7, amongst other serotypes) which are also found in other related bacteria are closely related to the Shigella toxin, and with identical mechanism of action (Table 38.3). There are two types of *E. coli* Shigella-like toxins (STI and STII) and strains can produce both or only one of them. The genes are phage mediated and seem to be transferred from one bacterium to another. It is almost impossible for the food industry to control the problem of the presence of these types of *E. coli,* as all meat is a potential source of infection, and thus a specific problem in ground-meat products.

Vibrio spp. enterotoxins (see Chapters 6 and 19)

Vibrio spp. are motile, faculatively anaerobic, Gram-negative rods, where eight of the species are known food-associated pathogens. *Vibrio cholera* and *V. para-haemolyticus* are the most common and have the best characterized virulence mechanisms, and are the only two that are included here. *Vibrio* spp. are found in estuarine waters throughout the world. They can be found in essentially all seafood, and raw, undercooked or recontaminated seafood are the major sources of food poisoning. *Vibrio* has been shown to survive for extended periods at refrigeration temperatures and can grow rapidly at room temperature (non-refrigeration) (Oliver and Kaper, 1997).

Vibrio cholera (see Chapter 6) can cause an explosive, potentially fatal dehydrating diarrhoea caused by ingestion of serotypes (01/0139) which carry the cholera enterotoxin (CT). The majority of infections involving *V. cholera* 01 are mild and volunteer studies showed large differences in the susceptibility between individuals. Strains without CT can, in some cases, still cause diarrhoea, but without the typical 'rice water' stools. Additional toxins such as Zot, Ace, haemolysin/cytolysin are suspected to be involved in the pathogenesis of the disease. Genetic factors other than CT would seem to be important, as other bacteria, such as *E. coli*, carry a gene coding for a toxin comparable to CT, but these bacteria have not been shown to give cholera-like symptoms (Oliver and Kaper, 1997). The main source of infection is contaminated water, but seafood has also been involved.

Vibrio parahaemolyticus can cause a disease, mainly involving diarrhoea, which is linked to the ability to produce Kanagawa haemolysin, TDH (KP$^+$ strains). This protein is only partially inactivated at 100°C for 30 min at pH 6.0. TDH produces oedema, erythema and induration in skin, and lyses erythrocytes from many animals, except horses. More relevant for food poisoning is the ability to alter transport in the intestinal tract and thereby induce diarrhoea. Studies using Ussing chambers demonstrated that TDH induces intestinal chloride ion secretion with the trisialoganglioside GT$_{1b}$ as the cellular receptor. TDH uses Ca^{2+} as an intercellular second messenger, instead of cAMP or cGMP, and is the first bacterial enterotoxin where changes in intercellular calcium and secretory activity have been shown to be directly linked (Oliver and Kaper, 1997).

KP$^+$ isolates usually contain two non-identical copies of genes coding for TDH, *tdh*1 and *tdh*2. Both gene products contribute to the KP$^+$ phenotype, but 90% of the TDH produced comes from the high-level expression of *tdh*2. A homologue of ToxRS, which regulates toxin production in *V. cholera*, is found in *V. parahaemolyticus* (ToxR). ToxR promoters regulate the expression of *tdh*2 but not *tdh*1. Another similar haemolysin, TRH, is found in KP$^-$ strains, and could explain cases where only KP$^-$ strains were isolated from stools of sick individuals. In general, individual strains produce only one of the haemolysins, TDH or TRH (Nishibuchi and Kaper, 1995). The main source of food poisoning is seafood from warm seawater.

Campylobacter jejuni/coli enterotoxins

Campylobacter jejuni and *C. coli* are Gram-negative microaerophilic S-shaped motile spiral rods which only grow between 30 and 47°C. They can survive at 4°C for extended periods, and have a viable but not culturable coccoid form. Reservoirs include water, domestic animals (notably poultry) and pets, and wild birds. Most outbreaks occur in the summer months. Campylobacteriosis has been the most frequent cause of foodborne outbreaks in some European countries in the 1990s (Wassenaar, 1997). Suspected food sources include raw milk, poultry, eggs, beef and water. Even though *Campylobacter* spp. are susceptible to low pH and heat and do not grow at low temperatures, the low dosage needed to cause sickness makes it difficult to avoid in food, and it will probably continue to be one of the most common food-poisoning agents in industrialized nations. Patients can be asymptomatic to severely ill. Symptoms include fever, cramping and diarrhoea, with or without leucocytes, that last for several days to over a week. Infections are usually self-limiting, but extraintestinal infections and sequelae do occur, including bacteraemia, reactive arthritis and Guillain-Burré syndrome (GBS) (Nachamkin, 1997).

Campylobacter jejuni has been reported to produce a cholera-like enterotoxin which elevated cAMP levels in rabbit ileal loop tests, but genetic probing using CT probes proved negative. Shiga toxins, cytolethal distending toxin and hepatoxins have also been described. Genetic studies have been difficult to perform in *Campylobacter*, and the lack of suitable animal models has hindered the study of the contribution of the different toxins in pathogenicity (Wassenaar, 1997). The wide range of symptoms seen during *Camplyobacter* spp. infections suggests that different strains of *Campylobacter* spp. carry one or a number of different virulence genes, and the severity of the sickness is a result of the expression of the combination of genes found in the infecting bacteria.

Salmonella spp. enterotoxin

Salmonella is one of the most common food-poisoning organisms in industrialized nations. They are resilient organisms that adapt to extreme environmental conditions, and there are strains that can grow at 54°C, and others that can grow at 2–4°C and tolerate pH values from 6.0 to 9.0. These bacteria are a major problem in poultry and egg products and other meat products. *Salmonella* infections can cause sickness ranging from uncomplicated diarrhoea to serious systemic infections. The bacterium has a number of genes which facilitate the attachment to and invasion of host cells, and attach-

ment is necessary for pathogenesis. Although many of the details of invasion of the host cells are known, the contribution of toxins produced by the bacteria in the total pathogenicity picture has not been completely elucidated (D'Aoust, 1997).

Although the enterotoxin is not necessary for development of salmonellosis it contributes to the disease. The enterotoxin is a thermolabile protein encoded by a 6.3-kb operon (*stx*) which encodes three genes. The toxin increases the level of cAMP in cells, which leads to fluid exsorption and to diarrhoea. Although genetic studies show that STX and cholera toxin are not closely related, functional studies show that the toxins share a similar mode of action with A and B components, and probably use the same GM_1 ganglioside receptor. *Salmonella* spp. also produce a thermolabile cytotoxic protein localized in the bacterial outer membrane. This cytotoxin inhibits protein synthesis, apparently using the same mechanisms as Shiga toxins, although genetic studies show that the toxins are not identical (D'Aoust, 1997).

Shigella spp. enterotoxin (see Chapter 11)

Shigella are Gram-negative non-motile rods which are the cause of bacillary dysentery. The clinical picture ranges from watery diarrhoea to severe dysentery (bloody diarrhoea), and complications such as reactive arthritis and kidney failure (HUS) are seen. The infective dose is very low, and that facilitates the infectious spread of the bacteria through food and water. Most of the foodborne incidents are caused by infected food handlers, but the largest outbreaks are after natural or artificial disasters (war) where sanitation and water supply lines have broken down. Attachment and invasion genes are the primary virulence factors, while the production of Shiga toxins seems to play a role in the progression of the disease, as dysentery is not seen in strains without the toxin (Maurelli and Lampel, 1997).

Yersinia enterocolitica enterotoxin (see Chapter 34)

Yersinia is a Gram-negative, facultative anaerobe rod that can grow between 0 and 44°C. Fever and watery to muccoid diarrhoea are the most common symptoms, while bacteremia and reactive arthritis are the most common complications. It is not a common form of food poisoning, and is most often seen in Northern climates. Pigs are often carriers of *Yersinia* but raw milk is the most commonly implicated food source. As with *Salmonella* and *Shigella*, invasiveness is the major virulence factor, and these genes are homologous in all three bacteria. Most strains of *Yersinia* secrete a heat-stable enterotoxin, Yst, which is homologous to the heat-stable toxins from enterotoxigenic *E. coli* and *V. cholera* non-01. The mechanism of action is elevation of intercellular cGMP levels, which affects the fluid transport pathways and results in diarrhoea. The role of the toxin, here as in other invasive bacteria, is uncertain, but it appears that toxins can either worsen the clinical symptoms or assist in the attachment to the host cell (Robins-Browne, 1997).

CONCLUDING REMARKS

Only a few species of all known bacteria are involved in foodborne illness. It is not sufficient for bacteria to possess genes that can encode for toxins or other virulence factors, the bacteria must also be able to survive one or a number of detrimental and changing environmental factors, including dehydration, heat, cold, low pH values (stomach), competing flora and other intestinal defence mechanisms. There are many bacteria which carry the same or equivalent virulence genes as bacteria known to cause foodborne illness which have not yet been confirmed in human illness, e.g. *Bacteroides fragilis* (see Chapter 31) and *Citrobacter* spp.

The modes of action for different groups of toxins are becoming clear, but the full importance of these toxins in the clinical picture is still somewhat unclear in all but the first two groups of toxins and some members of group 3 (Tables 38.1–38.3). Even in these groups genetic characterization of the different toxins and their prevalence among the strains of the species has just started. Equivalent toxins in different, and even the same, bacterial species can give very different symptoms, apparently due to the genetic background in which the toxins are found. It would be beneficial to the field if the nomenclature of the protein toxins (Granum *et al.*, 1995) and the tests used to determine their action, preferably using available cell lines such as Vero, Caco and HeLa, were standardized. Research on the role of toxins, their movement between bacteria, and regulation of expression is needed in order effectively to monitor and prevent unwanted bacteria in our food, and possibly prevent the rise of new pathogens.

REFERENCES

Agaisse, H., Salamitou, S., Gominet, M. and Lereclus, D. (1997). PlcR, the transcriptional activator of phospholipase expression in *Bt*, is a pleiotropic regulator. *First International Workshop on the Molecular Biology of B. cereus, B anthracis and B. thuringiensis*, Oslo, 23–25 May 1997, Abstract 8.

Agata, N., Mori, M., Ohta, M., Suwan, S., Ohtani, I. and Isobe, M. (1994). A novel dodecadepsipeptide, cereulide, isolated from *Bacillus cereus* causes vacuole formation in HEp-2 cells. *FEMS Microbiol. Lett.* **121**, 31–4.

Agata, N., Ohta, M., Mori, M., and Isobe, M. (1995a). A novel dodecadepsipeptide, cereulide, is an emetic toxin of *Bacillus cereus*. *FEMS Microbiol. Lett.* **129**, 17–20.

Agata, N., Ohta, M., Arakawa, Y. and Mori, M. (1995b). The *bceT* gene of *Bacillus cereus* encodes an enterotoxic protein. *Microbiology* **141**, 983–8.

Alouf, J.E., Knöll, H. and Köhler, W. (1991). The family of mitogenic, shock-inducing and superantigenic toxins from staphlococci and streptococci. In: *Sourcebook of Bacterial Protein Toxins* (eds J.E. Alouf and J.H. Freer), pp. 367–414. Academic Press, London.

Beecher, D.J. and Macmillan, J.D. (1991). Characterization of the components of hemolysin BL from *Bacillus cereus*. *Infect. Immun.* **59**, 1778–84.

Beecher, D.J. and Wong, A.C.L. (1994). Improved purification and characterization of hemolysin BL, a hemolytic dermonecrotic vascular permeability factor from *Bacillus cereus*. *Infect. Immun.* **62**, 980–6.

Beecher, D.J. and Wong, A.C.L. (1997). Tripartite hemolysin BL from *Bacillus cereus*. Hemolytic analysis of component interaction and model for its characteristic paradoxical zone phenomenon. *J. Biol. Chem.* **272**, 233–9.

Beecher, D.J., Schoeni, J.L. and Wong, A.C.L. (1995). Enterotoxin activity of hemolysin BL from *Bacillus cereus*. *Infect. Immun.* **63**, 4423–8.

Brynestad, S., Iwanejko, L.A., Stewart, G.S.A.B and Granum, P.E. (1994). A complex array of Hpr consensus DNA recognition sequences proximal to the enterotoxin gene in *Clostridium perfringens* Type A. *Microbiology* **140**, 97–104.

Brynestad, S., Synstad, B. and Granum, P.E. (1997). The *Clostridium perfringens* enterotoxin gene is on a transposable element in type A human food poisoning strains. *Microbiology* **143**, 2109–15.

Carlson, C.R., Johansen, T. and Kolstø, A.-B. (1996). The chromosome map of *Bacillus thuringiensis* subsp. *canadensis* HD224 is highly similar to that of the *Bacillus cereus* type strain ATCC 14579. *FEMS Microbiol. Lett.* **141**, 163–7.

Cornillot, E., Saintjoanis, B., Daube, G., Katayama, S., Granum, P.E., Canard, B. and Cole, S.T. (1995). The enterotoxin gene (*cpe*) of *Clostridium perfringens* can be chromosomal or plasmid-borne. *Mol. Microbiol.* **15**, 639–47.

D'Aoust, J.-V. (1997). *Salmonella* In: *Food Microbiology: Fundamentals and Frontiers* (eds M. Doyle, L. Beuchat and T. Montville), pp. 129–58. ASM Press, Washington, DC.

Deodhar L.P., Saraswathi, K. and Varudkar, A. (1991). *Aeromonas* spp. and their association with human diarrheal disease. *J. Clin. Microbiol.* **29**, 853–6.

Dodds, K.L. and Austin, J.W. (1997). *Clostridium botulinum*. In: *Food Microbiology: Fundamentals and Frontiers* (eds M. Doyle, L. Beuchat and T. Montville), pp. 288–304. ASM Press, Washington, DC.

Granum, P.E. (1990). *Clostridium perfringens* toxins involved in food poisoning. *Int. J. Food Microbiol.* **10**, 101–12.

Granum, P.E. (1994). *Bacillus cereus* and its toxins. *J. Appl. Bacteriol. Symp. Suppl.* **76**, 61–6S.

Granum, P.E. (1996). Smittsomme sykdommer fra mat: *Næringsmiddelbårne infeksjoner og intoksikasjoner*. Kristiansand, Norwegian Academic Press.

Granum, P.E. (1997). *Bacillus cereus* In: *Food Microbiology: Fundamentals and Frontiers* (eds M. Doyle, L. Beuchat and T. Montville), pp. 327–36. ASM Press, Washington, DC.

Granum, P.E. and Harbitz, O. (1985). A circular dichroism study of the enterotoxin from *Clostridium perfringens* type A. *J. Food Biochem.* **9**, 137–46.

Granum, P.E. and Richardson, M. (1991). Chymotrypsin treatment increases the activity of *Clostridium perfringens* enterotoxin. *Toxicon* **29**, 898–900.

Granum, P.E. and Stewart, G.S.A.B. (1993). Molecular biology of *Clostridium perfringens* enterotoxin. In: *Genetics and Molecular Biology of Anaerobic Bacteria* (ed. M. Sebald), pp. 235–47. Springer, New York.

Granum, P.E., Whitaker, J.R. and Skjelkvåle, R. (1981). Trypsin activation of enterotoxin from *Clostridium perfringens* type A. *Biochim. Biophys. Acta* **688**, 325–32.

Granum, P.E., Tomas, J.M. and Alouf, J.E. (1995). A survey of bacterial toxins involved in food poisoning: a suggestion for bacterial food poisoning toxin nomenclature. *Int. J. Food Microbiol.* **28**, 129–44.

Granum, P.E., Andersson, A., Gayther, C., te Giffel, M.C., Larsen, H., Lund, T. and O´Sullivan, K. (1996). Evidence for a further enterotoxin complex produced by *Bacillus cereus. FEMS Microb. Lett.* **141**, 145–9.

Granum, P.E., O´Sullivan, K., Tomas, J.M. and Ørmen, Ø. (1998). Possible virulence factors of *Aeromonas* spp isolated from food and water. *FEMS Immun. Med. Microbiol.* **21**, 131–7.

Hatheway, C.L. (1990). Toxigenic Clostridia. *Clin. Microbiol. Rev.* **3**, 68–98.

Hauschild, A.H.W. (1989). *Clostridium botulinum* In: *Foodborne Bacterial Pathogens* (ed. M.P. Doyle), pp. 111–89. Marcel Dekker, New York.

Heinrichs, J.H., Beecher, D.J., MacMillan, J.M. and Zilinskas, B.A. (1993). Molecular cloning and characterization of the *hblA* gene encoding the B component of hemolysin BL from *Bacillus cereus. J. Bacteriol.* **175**, 6760–6.

Hughes, S., Bartholomew, B., Hardy, J.C. and Kramer, J.M. (1988). Potential application of a HEp-2 cell assay in the investigation of *Bacillus cereus* emetic-syndrome food poisoning. *FEMS Microbiol. Lett.* **52**, 7–12.

Jablonski, K.M. and Bohach, G.A. (1997). *Staphylococcus aureus* In: *Food Microbiology: Fundamentals and Frontiers* (eds M. Doyle, L. Beuchat and T. Montville), pp. 327–36. ASM Press, Washington, DC.

Jackson, S.G., Goodbrand, R.B., Ahmed, R. and Kasatiya, S. (1995). *Bacillus cereus* and *Bacillus thuringiensis* isolated in a gastroenteritis outbreak investigation. *Lett. Appl. Microbiol.* **21**, 103–5.

Johnson, S. and Gerding, D. (1997). Enterotoxemic infections. In: *The Clostridia: Molecular Biology and Pathogenesis* (eds J. Rood, B.A. McClane, J.G. Songer and R.W. Titball), pp. 117–40. Academic Press, London.

Katahira, J., Inoue, N., Horigushi, Y., Matsuda, M. and Sugimoto, N. (1997a). Molecular cloning and functional characterization of the receptor for *Clostridium perfringens* enterotoxin. *J. Cell Biol.* **136**, 1239–47.

Katahira, J., Sugiyama, H., Inoue, N., Horigushi, Y., Matsuda, M. and Sugimoto, N. (1997b). *Clostridium perfringens* utilizes two structurally related membrane proteins as functional receptors *in vivo. J. Biol. Chem.* **272**, 26652–8.

Kirov, S.M. (1993). The public health significance of *Aeromonas* ssp in foods. *Int. J. Food Microbiol.* **20**, 179–98.

Kirov, S.M. (1997). *Aeromonas* and *Plesiomonas* species In: *Food Microbiology: Fundamentals and Frontiers* (eds M. Doyle, L. Beuchat and Montville, T.), pp. 265–87. ASM Press, Washington, DC.

Kokai-Kun, J.F. and McClane, B.A. (1997a). The *Clostridium perfringens* enterotoxin. In: *The Clostridia: Molecular Biology and Pathogenesis* (eds J. Rood, B.A. McClane, J.G. Songer and R.W. Titball), pp. 325–57. Academic Press, London.

Kokai-Kun, J.F. and McClane, B.A. (1997b). Deletion analysis of the *Clostridium perfringens* enterotoxin. *Infect. Immun.* **65**, 1014–22.

Kramer, J.M. and Gilbert, R.J. (1989). *Bacillus cereus* and other *Bacillus* species. In: *Foodborne Bacterial Pathogens* (ed. M.P. Doyle), pp. 21–70. Marcel Dekker, New York.

Labbé, R.G. and Duncan, C.L. (1977). Evidence for stable messenger ribonucleic acid during sporulation and enterotoxin synthesis by *Clostridium perfringens* type A. *J. Bacteriol.* **129**, 843–9.

Lawson, M.A., Burke, V. and Chang, B.J. (1985). Invasion of HEp-2 cells by fecal isolates of *Aeromonas hydrophila. Infect. Immun.* **47**, 1531–7.

Lereclus, D., Agaisse, H., Gominet, M., Salamitou, S. and Sanchis, V. (1996). Identification of a *Bacillus thuringiensis* gene that positively regulates transcription of the phosphatidylinositol-specific phospholipase C gene at the onset of the stationary phase. *J. Bacteriol.* **178**, 2749–56.

Lund, T. and Granum, P.E. (1996). Characterisation of a non-haemolytic enterotoxin complex from *Bacillus cereus* isolated after a foodborne outbreak. *FEMS Microbiol. Lett.* **141**, 151–6.

Lund, T. and Granum, P.E. (1997). Comparison of biological effect of the two different enterotoxin complexes isolated from three different strains of *Bacillus cereus. Microbiology* **143**, 3329–36.

Marrack, P. and Kappler, J. (1990). The staphylococcal enterotoxins and their relatives. *Science* **248**, 705–11.

Maurelli, A.T. and Lampel, K.A. (1997). *Shigella* species. In: *Food Microbiology: Fundamentals and Frontiers* (eds M. Doyle, L. Beuchat and T. Montville), pp. 216–27. ASM Press, Washington, DC.

McClane, B.A. (1997). *Clostridium perfringens*, In: *Food Microbiology: Fundamentals and Frontiers* (eds M. Doyle, L. Beuchat and T. Montville), pp. 305–26. ASM Press, Washington, DC.

Melville, S.B. and Zhao, Y. (1998). Identification and characterization of sporulation-dependent promoters upstream of the enterotoxin gene (*cpe*) of *Clostridium perfringens. J. Bacteriol.* **189**, 136–42.

Melville, S.B., Labbé, R. and Sonenshein, A.L. (1994). Expression from the *Clostridium perfringens cpe* promoter in *Clostridium perfringens* and *Bacillus subtilis. Infect. Immun.* **62**, 5550–8.

Montecucco, C. and Schiavo, G. (1994). Microreview: Mechanism of action of tetanus and botulinum neurotoxins. *Mol. Microbiol.* **13**, 1–8.

Nachamkin, I. (1997). *Campylobacter jejuni.* In: *Food Microbiology: Fundamentals and Frontiers* (eds M. Doyle, L. Beuchat and T. Montville), pp. 159–70. ASM Press, Washington, DC.

Nataro, J.P. and Kaper, J.B. (1998). Diarrheagenic *Escherichia coli. Clin. Microbiol. Rev.* **11**, 142–201.

Nishibuchi, M. and Kaper, J.B. (1995). Thermostable direct hemolysin gene of *Vibrio parahaemolyticus*: a virulence gene acquired by a marine bacterium. *Infect. Immun.* **63**, 2093–9.

Oliver, J.D. and Kaper, J.B. (1997). *Vibrio* species. In: *Food Microbiology: Fundamentals and Frontiers* (eds M. Doyle, L. Beuchat and T. Montville), pp. 228–64. ASM Press, Washington, DC.

Ray, D.E. (1991). Pesticides derived from plants and other organisms. In: *Handbook of Pesticide Toxicology* (eds W.J. Hayes and E.R. Laws, Jr), pp. 585–636. Academic Press, New York.

Robins-Browne, R.M. (1997). *Yersinia enterocolitica.* In: *Food Microbiology: Fundamentals and Frontiers* (eds M. Doyle, L. Beuchat and T. Montville), pp. 192–215. ASM Press, Washington, DC.

Ryan, P.A., Macmillan, J.M. and Zilinskas, B.A. (1997). Molecular cloning and characterization of the genes encoding the L_1 and L_2 components of hemolysin BL from *Bacillus cereus. J. Bacteriol.* **179**, 2551–6.

Shinagawa, K., Konuma, H., Sekita, H. and Sugii, S. (1995). Emesis of rhesus monkeys induced by intragastric administration with the HEp-2 vacuolation factor (cereulide) produced by *Bacillus cereus. FEMS Microbiol. Lett.* **130**, 87–90.

Todd, E.C.D. (1989). Costs of acute bacterial foodborne disease in Canada and the United States. *Int. J. Food Microbiol.* **9**, 313–26.

Wassenaar, T.M. (1997). Toxin production by *Campylobacter* spp. *Clin. Microbiol. Rev.* **10**, 466–76.

Watson, I.M., Robinson, J.O., Burke, V. and Gracey, M. (1985). Invasiveness of *Aeromonas* spp. in relation to biotype, virulence factors and clinical features. *J. Clin. Microbiol.* **22**, 48–51.

WHO Press Release (1997). htttp//www.who.ch/programmes/fsf

CHAPTER
39

Diphtheria toxin-based interleukin-2 fusion proteins

Johanna C. vanderSpek and John R. Murphy

INTRODUCTION

Diphtheria toxin (DT) is one of the most toxic substances known and will inhibit protein synthesis in sensitive cells at concentrations as low as 10^{-13} M. The catalytic (C) domain of DT catalyses the NAD^+-dependent ADP-ribosylation of a post-translationally modified His residue (diphthamide) present in elongation factor 2 (EF2) (Gill and Pappenheimer, 1971; van Ness *et al.*, 1980). Once ADP-ribosylated, EF2 can no longer function and cellular protein synthesis is irreversibly inhibited, leading to the death of the cell (Yamaizumi *et al.*, 1978). The transmembrane (T) domain of DT is required for cellular processing and facilitating delivery of the C-domain to the target cell cytosol. Finally, the receptor-binding (R) domain of DT targets DT to sensitive, receptor-expressing cells (Choe *et al.*, 1992; Murphy and vanderSpek, 1995). The cell surface receptor for native DT is a heparin-binding epidermal growth factor-like precursor, and in sensitive species the receptor is widely distributed throughout all organ systems (Pappenheimer, 1977; Naglich *et al.*, 1992).

In order to target differentially the ADP-ribosyltransferase activity of the C-domain towards specific receptors on the surface of cancer cells, in the 1980s we envisioned the construction of fusion protein toxins in which the native R-domain of diphtheria toxin would be replaced with either a polypeptide hormone or growth factor (Murphy *et al.*, 1986). With careful selection, we anticipated the development of unique

reagents that might be useful as experimental therapeutics and might serve as useful probes in the study of toxin–target cell interactions. The selection of a given surrogate R-domain required three criteria: limited distribution, an internalization pathway that included acidified early endosomes, and enrichment on the surface of target malignant cells. The first of the fusion protein toxins has recently received Food and Drug Administration approval for the treatment of refractory cutaneous T-cell lymphoma (CTCL) and is being clinically used. In addition, this fusion protein toxin and related mutants have proven to be extremely useful probes in furthering the understanding of structure–function relationships within the fusion toxin and mechanism by which the C-domain enters the cytosol of interleukin (IL)-2 receptor-positive eukaryotic cells.

INTERLEUKIN-2 RECEPTOR TARGETED FUSION PROTEIN TOXINS AS REAGENTS FOR MOLECULAR CELL BIOLOGY

The first DT-based, genetically constructed, fusion protein toxin to be reported was $DAB_{486}\alpha MSH$ (Murphy *et al.*, 1986). In this instance oligonucleotides encoding the 13 amino acid α-melanocyte stimulating hormone were cloned into a unique *Sph*I restriction endonuclease site at the 3'-end of the diphtheria toxin structural gene replacing the C-terminal 50 amino acids of the native toxin. While subject to marked proteolytic degradation, this fusion protein was found to be specifically toxic for cells that expressed α-MSH receptors. In order to overcome difficulties associated with degradation in recombinant *Escherichia coli*, the second surrogate R-domain that was employed was that of human IL-2 (Williams *et al.*, 1987). This fusion protein toxin, DAB$_{486}$IL-2, proved to be resistant to proteolytic degradation and was found to be remarkably potent towards only those cells which displayed the high-affinity form of the IL-2 receptor (Bacha *et al.*, 1988; Waters *et al.*, 1990). Subsequent in-frame internal deletion analysis genetically defined the optimal fusion junction between the diphtheria toxin-related sequences and IL-2 sequences to be Thr387 (Williams *et al.*, 1990a); DAB$_{389}$IL-2 was generated as the second-generation form of the IL-2 receptor targeted fusion toxin. As shown in Table 39.1, Thr387 is a universal fusion junction in the construction of a number of fusion protein toxins directed towards a variety of cell surface receptors. In each instance, the substitution of the native diphtheria toxin R-domain with a polypeptide hormone or growth factor resulted in the formation of a fusion protein toxin that was highly selective and potent against cells which expressed the targeted cell surface receptor. The subsequent elucidation of the X-ray crystal structure of DT by Choe *et al.* (1992) clearly defined the C, T and R domains of DT and confirmed that the boundary between the T and R domains was composed of a random coil that included Thr387.

Largely because of its ease of purification and remarkable receptor binding affinity and specificity, the fusion protein toxin DAB$_{389}$IL-2 has been used extensively to study the structure–function relationships within the C-, T- and IL-2R binding domains. As is well known, the mechanism of DT intoxication involves several steps subsequent to binding its receptor. These steps include: (a) proteolytic 'nicking' of the toxin in a sensitive 14 amino acid residue loop subtended by a disulfide bond between Cys186 and Cys201; (b) internalization of the toxin–receptor complex by receptor-mediated endocytosis and, upon acidification of the early endosome; (c) the insertion of the T-domain in the early endosomal membrane which forms a channel and facilitates delivery of the C-domain to the cell cytosol. Except for their target cell specificity, the diphtheria toxin-based fusion protein toxins appear to follow an identical route of entry into the cell (Murphy *et al.*, 1995).

Site-directed mutagenesis has been used to study extensively the structure of DAB$_{486}$IL-2 and DAB$_{389}$IL-2 necessary for the delivery of its C-domain into the cytosol of target cells. Importantly, these studies have provided additional insight into the precise molecular mechanism of target cell intoxication. Williams *et al.* (1990b) used DAB$_{486}$IL-2 to determine the structural requirements for proteolytic processing between the C and T-domains. There are three arginine residues in the protease sensitive loop (Cys-Ala-Gly-Asn-Arg-Val-Arg-Arg-Ser-Val-Gly-Ser-Ser-Leu-Ser-Cys) and site-directed mutagenesis was employed to mutate each independently. The results of this analysis conclusively demonstrated that Arg194 in the fusion toxin (i.e. Arg193 in native DT) was required for processing. Mutation of either Arg191 or Arg193 to Gly resulted in the formation of mutants with a 50-fold lower cytotoxic potency, but the mutation of Arg193 to Lys maintained full cytotoxic potency. It was later demonstrated that the endoprotease furin was directly involved in the proteolytic processing of both native diphtheria toxin

TABLE 39.1 DT-Related fusion protein toxins

Fusion protein	Targeting ligand	IC$_{50}$ (M)[a]	Reference
DAB$_{486}$ α-MSH	α-Melanocyte stimulating hormone	Not done	Murphy *et al.* (1986)
DAB$_{486}$ IL-2	Interleukin-2	10^{-11}	Williams *et al.* (1987)
DAB$_{389}$ IL-2	Interleukin-2	10^{-12}	Williams *et al.* (1990a, b)
DAB$_{389}$ α-MSH	α-Melanocyte stimulating hormone	10^{-9}	Wen *et al.* (1991)
DAB$_{389}$ EGF	Epidermal growth factor	10^{-13}	Shaw *et al.* (1991)
DAB$_{389}$ mIL-4	Murine interleukin-4	10^{-9}	Lakkis *et al.* (1991)
DAB$_{389}$ IL-6	Interleukin-6	10^{-11}	Jean and Murphy (1992)
DAB$_{389}$ CD4	Cell determinant 4	10^{-9}	Aullo *et al.* (1992)
DAB$_{389}$ IL-7	Interleukin-7	10^{-10}	Sweeney *et al.* (1998)
DAB$_{389}$ sIL-15	Simian interleukin-15	10^{-8}	vanderSpek *et al.* (1995)
DAB$_{389}$ SP	Substance P	10^{-11}	Fisher *et al.* (1996)
DAB$_{389}$ GMCSF	Granular monocyte stimulating factor	10^{-10}	Bendel *et al.* (1997)
DAB$_{389}$ GRP	Gastrin releasing peptide	10^{-11}	vanderSpek *et al.* (1997)
DAB$_{389}$ mIL-3	Murine interleukin-3	10^{-10}	Liger *et al.* (1997)
DAB$_{389}$ NT4	Neurotrophin 4	10^{-10}	Negro and Skaper (1997)

[a]Approximate IC$_{50}$ as determined on target eukaryotic cells *in vitro*. Assay **protocols varied.**

and the diphtheria toxin-based fusion protein toxins (Tsuneoka *et al.*, 1993; Murphy, unpublished observations).

While internal in-frame deletion studies demonstrated that the Thr389 was the optimal junction for the construction of fusion toxins, additional analysis by both insertion and deletion mutagenesis demonstrated that this region of the fusion protein toxins was critical both in determining the relative receptor binding affinity, and in providing structure required for the efficient delivery of the catalytic domain across the endosomal membrane and into the cytosol of target cells. Kiyokawa *et al.* (1991) demonstrated that the introduction of a flexible 'spacer' between the diphtheria toxin transmembrane domain and IL-2 sequences increased both the relative cytotoxic potency and receptor binding affinity of the IL-2R targeted fusion proteins. The first 10 amino acid residues of IL-2 were chosen for insertion based on the FLEXPRO analysis and its predicted flexibility. The insertion mutant variants proved to be two- to fivefold more potent, and their increased potency was directly related to their more avid binding to the high-affinity form of the IL-2 receptor.

The precise role played by the T-domain in the efficient delivery of the C-domain to the cytosol of target cells is only partially defined. The T-domain is composed of nine α-helices and their connecting loops. Helices 1–3 are amphipathic in nature, containing a hydrophobic and a hydrophilic face, whereas helices 5–9 are hydrophobic in nature. The T-domain of native DT has been shown to form channels of characteristic conductance in artificial bilayer membranes (Boquet *et al.*, 1976; Donovan *et al.*, 1981; Kagan *et al.*, 1981). In a series of studies, the minimal helix requirement for channel formation in planar lipid bilayers indicated that only helices 8 and 9 were required (Silverman *et al.*, 1994a, b; Mindell *et al.*, 1994a, b). While these studies clearly demonstrate that only helices 8 and 9 are necessary to form ion conductive channels, it was not clear whether these channels could support the productive delivery of the C-domain to the cytosol and result in the intoxication of target cells. Accordingly, a series of mutational studies was performed with DAB$_{389}$IL-2 to determine whether channel formation alone was sufficient for effective C-domain delivery and to determine the functions of the helices in the T-domain. A mutant was created in which the first three amphipathic helices were deleted from the T-domain of DAB$_{389}$IL-2. The resulting mutant was not cytotoxic. However, this mutant still formed channels in planar lipid bilayers with a characteristic conductance, and retained the wild-type IL-2 receptor-binding affinity (vanderSpek *et al.*, 1993). The first α-helix of the DAB$_{389}$IL-2 T-domain was subsequently replaced such that the overall

charge distribution and hydrophobicity were maintained. Interestingly, the mutant fusion toxin was cytotoxic with an IC$_{50}$ of 1×10^{-10} M; however, the kinetics of intoxication appeared to have been affected. Furthermore, the introduction of a charged residue (Glu) in the hydrophobic face of this resulted in the complete loss of cytotoxic potency (IC$_{50}$ > 10^{-7} M) (vanderSpek *et al.*, 1994a). These results suggest that the first three helices of the T-domain may be required for 'non-specific' membrane surface interactions that result in the stabilization of the fusion protein on the cell, thereby anchoring the T-domain in an orientation for subsequent insertion of the channel-forming helices. Internal in-frame deletions of the carboxy-terminal end of the T-domain of DAB$_{389}$IL-2 demonstrated that an intact helix 9 was required for both membrane insertion and formation of channels of nominal conductance and efficient delivery of the C-domain to the cytosol (vanderSpek *et al.*, 1994b). This result indicated that helix 9 was involved with channel formation.

X-ray crystallographic analysis of native DT showed that the nine α-helices of the T-domain are arranged in three helical layers. Recent studies by Hu *et al.* (1998) have shown that the introduction of helix-breaking proline substitution mutations into each α-helix layer results in a complete loss of cytotoxicity. Helix-breaking mutations were independently introduced into the middle of helices 1 and 3 of the first helix layer. Both mutants still formed channels in planar lipid bilayers and possessed binding affinities similar to wild-type DAB$_{389}$ IL-2; however, these mutants were no longer cytotoxic. These results are in agreement with those of the previous studies involving the amphipathic region, indicating that although the region is not within the channel, it is still required for effective C-domain delivery. Mutations in the second helical layer, within helices 5, 6 and 7 of the T-domain, affected conformation of the fusion protein, as shown by fluorescence studies. The proteins were not cytotoxic and channels formed possessed abnormal conductances. These results were of interest as previous studies had indicated that only helices 8 and 9, in the third helix layer, were necessary for channel formation (Silverman *et al.*, 1994a, b). The results from the introduction of helix-breaking mutations into layer 2 indicate that this helix layer is also required for effective C-domain delivery. The first helix layer probably functions as a membrane surface anchor and is involved in the kinetics of channel formation, but does not affect the channel conductance once it has formed, whereas helix layer 2 appears to be more intimately involved with channel formation. As expected, insertion of a helix-breaking mutation into helix 8 of helix layer 3 resulted in a fusion protein that was not cytotoxic and did not form channels. In contrast, the introduction of a helix-breaking mutation into helix 9 resulted in a mutant which retained the ability to form characteristic channels; however, this mutant was also devoid of cytotoxic activity. From this and previous studies it is clear that helices 8 and 9 are indeed required for channel formation. Despite a large number of studies, the precise architecture of the T-domain channel within the membrane of the early endosome is still poorly understood. Moreover, the precise mechanism by which the C-domain is delivered across the membrane and into the cytosol remains largely unknown.

DAB$_{389}$IL-2 has also been used as a cytotoxic probe to study the interactions between its IL-2 receptor binding domain and the high-affinity form of the IL-2 receptor (vanderSpek *et al.*, 1996). Various point mutations were generated in the IL-2 receptor targeting domain and the effects on cytotoxicity, the kinetics of cytotoxicity, DNA stimulation and binding were examined. A number of interesting mutants in this region of the fusion protein toxin was isolated and characterized. For example, the Thr439 to Pro mutation in the IL-2 portion of the fusion toxin (T51P in IL-2) resulted in a mutant that was 300-fold less cytotoxic than wild-type DAB$_{389}$IL-2, had a decreased binding affinity, but stimulated DNA synthesis to a greater extent than expected. The intoxication kinetics indicated that these results were due to an increased contact time between the mutant and the IL-2 receptor, perhaps as a result of a decreased internalization rate. The Gln514 to Asp mutation (Gln126 in IL-2) resulted in a mutant that was 2000-fold less cytotoxic, had a decreased binding affinity and did not stimulate DNA synthesis. Taken together, these results suggest that the Thr439 was involved with binding and signalling internalization of the ligand–receptor complex, whereas Gln514 is involved with receptor binding and activation of DNA synthesis. These results confirm and extend the observations of Walz *et al.* (1989) and further suggest that activation of DNA synthesis results in an increased rate of cytotoxicity (i.e. activated cells are more readily intoxicated). If this is indeed the case, this property of the fusion protein toxins is a useful characteristic as cancer cells often proliferate more rapidly than their normal counterparts.

Little is known of the role the C-domain plays in its own delivery to the cytosol. Falnes *et al.* (1994) prepared a series of double cysteine mutants in the C-domain which they expected to form disulfide bonds. Four of their mutants formed disulfide bonds and in all instances, the translocation process was inhibited. These results clearly suggest that the C-domain must be completely unfolded in order to be delivered through the channel formed by the T-domain. Similar conclu-

sions had previously been put forth by Wiedlocha *et al.* (1992). These investigators modified the C-domain by addition of acidic fibroblast growth factor (a-FGF) to the N-terminus and determined that the C-domain was still delivered to the cytosol. In the presence of heparin, which induces tight folding of a-FGF, the fusion protein toxin was no longer cytotoxic. These results suggest that under these conditions the a-FGF component of the fusion protein could no longer be denatured and passage of the modified C-domain through the endosomal membrane was blocked.

In order to study further C-domain delivery into the eukaryotic cell cytosol, Lemichez *et al.* (1997) have described an important *in vitro* protocol in which Vero cell endosomes were preloaded with DT in the presence of bafilomycin A1. The early endosomal compartment was then isolated by sucrose gradient centrifugation. Following removal of bafilomycin A1, the early endosomes were allowed to acidify and ADP-ribosyltransferase activity could then be detected in the external medium. The present authors have recently modified this protocol in order to examine the transmembrane delivery of the C-domain of DAB$_{389}$IL-2 from purified early endosomes isolated from HUT102/6TG cells (Zeng, vanderSpek and Murphy, unpublished). Preliminary observations strongly suggest that upon binding to the IL-2 receptor the ADP-ribosyltransferase activity of fusion protein toxin becomes denatured rapidly. Following internalization and purification of early endosomes in the presence of bafilomycin A1, C-domain catalytic activity was measured in endosomal lysates. ADP-ribosyltransferase activity could only be detected in these lysates following the addition of cytosolic factors and ATP. These experiments further suggest that the fusion toxin C-domain is denatured in the lumen of the early endosome and requires re-folding into an active conformation. Initial experiments designed to study the translocation of the C-domain to the external medium also suggest that the presence of cytosolic factors is required. Time-course experiments demonstrate that the export of ADP-ribosyltransferase activity is linear with time, and up to 80% of the total activity may be exported within 45 min at 37°C. Since fusion protein toxin mutants which contain internal disulfide bonds within the C-domain analogous to those described by Falnes *et al.* (1994) remain within the purified early endosomes, it can be concluded that export of the C-domain is specific.

The authors are currently exploring the hypothesis that the export of ADP-ribosyltransferase activity from an early endosomal compartment is analogous to that of the import of proteins into the mitochondria. In this system, it is clear that chaparonins facilitate import, perhaps by the sequential binding and refolding of the denatured protein as it emerges from the membrane surface. In this way, it remains possible that the entry of the C-domain into the cytosol of target cells employs cellular factors to facilitate the entry process, and that refolding into an active conformation is a concomitant process.

The working model of the intoxication of high-affinity IL-2 receptor bearing cells by $DAB_{389}IL$-2 parallels that of native diphtheria toxin: $DAB_{389}IL$-2 binds to the high-affinity form of the IL-2 receptor, signalling internalization of the receptor–ligand complex and stimulation of DNA synthesis. Soon after binding, the C-domain becomes denatured and loses its ADP-ribosyltransferase activity, and the cellular protease furin 'nicks' the α-carbon backbone of the fusion toxin at Arg194. As internalization proceeds, the T-domain of the fusion toxin is stabilized on the endosomal membrane by the amphipathic, first helical layer (helices 1–3). As the early endosomal compartment is acidified the decrease in pH causes spontaneous insertion of the T-domain and channel formation through the endosome membrane. The insertion of the third helix layer, channel-forming helices 8 and 9, is stabilized conformationally by the insertion of second helix layer, composed of helices 5, 6 and 7. As the channel is forming, the denatured C-domain is translocated through the channel by a mechanism that might involve positioning of the C-domain by T-domain helix layer 1. The disulfide bond between the C- and T-domains is reduced and the carboxy-terminal end of the C-domain is inserted into the channel. Once the first few hydrophobic residues of the C-domain reach the cytosol, it is possible that cellular chaparonins facilitate the entry process by sequentially binding and refolding the denatured polypeptide into an active conformation as it emerges from the lumen of the early endosome.

It is important to note, however, that there is much to learn about the precise molecular events that take place during the entry process. None the less, studies of $DAB_{389}IL$-2 continue to provide a unique perspective on this fundamental biological process.

INTERLEUKIN-2 RECEPTOR TARGETED FUSION PROTEIN TOXINS AS CLINICAL REAGENTS FOR REFRACTORY LYMPHOMAS

As initially envisioned (Murphy *et al.*, 1986), the diphtheria toxin-based fusion proteins had the potential to form a new class of biological therapeutic agent.

While conceptually analogous to the immunotoxins (i.e. monoclonal antibody/toxin chimeras assembled through chemical cross-linking), the fusion protein toxins differ significantly in that they are assembled at the level of the gene. Since the fusion protein toxins are expressed in recombinant *E. coli* as single-chain proteins, their large-scale production is relatively simple. Moreover, as discrete chemical entities the fusion protein toxins are readily purified and characterized to Food and Drug Administration (FDA) standards.

Based on the potential of $DAB_{486}IL$-2 to intoxicate selectively cells that displayed the high-affinity form of the IL-2 receptor, a series of phase-I/II clinical trials was conducted. The first of these studies was conducted with $DAB_{486}IL$-2 and established the 'proof-of-principle' for biological efficacy of IL-2 receptor targeted cytotoxic therapy in lymphoma (LeMaistre *et al.*, 1992, 1993; Schwartz *et al.*, 1992; Hesketh *et al.*, 1993; Foss *et al.*, 1995, 1998). All patients who were enrolled in these studies presented with refractory disease and had failed at least two prior chemotherapy regimens. Patients were initially treated to determine the safety, tolerability and pharmacokinetics of DAB_{486} IL-2. Detection of IL-2 receptors on tumour tissues was not a prerequisite for enrolment in these early studies as fresh tissue samples were not available from all patients. Single and multiple doses of the fusion protein toxin were administered by intravenous injection as a bolus, or by 90-min infusions. Dose escalations were performed on groups of three patients with a starting dose of 700 $\mu g\ kg^{-1}$ per day, which was gradually increased to 400 $\mu g\ kg^{-1}$ per day.

Adverse effects associated with the intravenous administration of $DAB_{486}IL$-2 were transient and included fever, malaise, hypersensitivity, nausea/vomiting and increased levels of serum hepatic transaminases. The maximum tolerated dose was determined to be 400 $\mu g\ kg^{-1}$ per day; above this level, renal insufficiency occurred. Time-course studies indicated that this fusion protein toxin cleared from the serum with a $t_{1/2}$ of 11 min in dose ranges of 200–400 $\mu g\ kg^{-1}$ (LeMaistre *et al.*, 1993). Increased levels of soluble IL-2 receptor were detected in the serum of many patients, but had no effect on either clearance rates or potential efficacy. Furthermore, antibodies to IL-2 developed in 50% of the patients during the course of the study, but seemed to have no effect on anti-tumour response. Clinical responses occurred in 8% (4/51) of the patients with low and intermediate grade non-Hodgkin's lymphoma, 7% (1/14) of the patients with Hodgkin's disease and 17% (6/36) of patients with CTCL. One patient with tumour stage CTCL had a complete remission and has remained disease-free for over 5 years (Hesketh *et al.*, 1993) (at the time of writing this patient has continued to remain free of disease). As anticipated from *in vitro* studies, it was notable that all patients who responded to $DAB_{486}IL$-2 therapy had demonstrable IL-2 receptor expression as shown by immunoreactivity with anti-Tac (p55) antibody.

Because of the increase in both ease of purification and biological activity, all additional human clinical trials were conducted with $DAB_{389}IL$-2. Because of the developmental need to focus on a single disease entity for potential regulatory approval, clinical trials with $DAB_{389}IL$-2, which began in 1992, were limited to patients with CTCL and non-Hodgkin's lymphoma and Hodgkin's disease. Moreover, only those patients who presented with tumours that expressed IL-2 receptor as assessed by immunohistochemical staining using anti-p55 and anti-p75 antibodies were enrolled (LeMaistre *et al.*, 1998). The patient population had a mean of five previous therapies and included 25 patients who had received bone marrow transplants. A cohort dose-escalation trial was once again employed and patients received doses ranging from 3 to 31 µg kg^{-1} per day for 5 days. Treatment cycles were repeated every 3 weeks. Side-effects once again included fever/chills, nausea/vomiting, malaise and reversible elevation of serum transaminases. Eight of the patients with CTCL experienced hypoalbuminaemia, hypotension and oedema and these were thought to be symptoms of mild vascular leak syndrome. Dose-limiting toxicity was determined to be 31 µg kg^{-1} per day based on malaise.

Importantly, all observed toxicities were reversible and were not cumulative. Thirty-nine (53%) patients discontinued the study because of disease progression and 12 (16%) stopped treatment because of toxicity. Of the 73 patients who began the treatment 52 completed two courses, 39 completed three courses of treatment and 18 completed all six cycles. As in the earlier study, pre-existing antibodies and antibody developing after treatment with $DAB_{389}IL$-2, did not appear to interfere with the clinical response or contribute to adverse reactions. Clinical responses were observed in 16 of the 73 patients (22%) with effects occurring in 13 out of 35 patients (37%) with CTCL and three of the 17 (18%) non-Hodgkin's lymphoma patients. $DAB_{389}IL$-2 did not elicit response in Hodgkin's disease patients. The median response was found to occur after two courses of treatment and the median duration of response was 10 months. Since four out of five complete responses were found in those patients who presented with CTCL, phase-III studies of $DAB_{389}IL$-2 were initiated in this patient population.

Cutaneous T-cell lymphoma

As a clinical entity, CTCL describes a series of low-grade non-Hodgkin's lymphomas in which malignant T-cells invade the skin (Broder and Bunn, 1980). This malignancy is also known as either mycosis fungoides, or in its erythrodermic leukaemic variant, Sezary syndrome. While in its early stages mycosis fungoides is a disease of the skin; in later stages lymph nodes, spleen, liver and other organs may also be involved. Skin lesions associated with this disease usually progress through patch, plaque and tumour phases. Patch lesions are flat, scaly, erythrematous macules which may itch, whereas plaques are generally raised, red to purple in colour, and may be thick and scaly and usually intensely itchy. Reddened nodular tumours may arise from plaque lesions and tend to predominate on the face and intertriginous areas of the body. With time, tumorous lesions become ulcerated and secondarily infected, and may spread to regional lymph nodes and visceral organs. In the case of Sezary syndrome, atypical lymphocytes are found in peripheral blood, and in severe cases the skin appears to be scalded.

In general, patients who present with CTCL become symptomatic even in early stages of their disease. Both the breakdown of the normal skin barrier and a depression in cell-mediated immunity compromise this patient population and predispose them towards infection (Axelrod *et al.*, 1992). While the 8–10-year overall survival rate for these patients is similar to that for other non-Hodgkin's lymphomas, once disease has progressed to lymph node or organ system involvement the median survival is less than 3 years. While there are many therapeutic regimens that have shown efficacy in early stage disease, patients with advanced disease are refractory and complete clinical responses are rare. Once the disease progresses beyond more than 10% of the total body surface area involvement, spontaneous remission from disease does not occur and the disease is usually fatal.

$DAB_{389}IL$-2: phase-III clinical evaluation

Given the promising results from the open-label phase-II clinical trials with $DAB_{389}IL$-2, a phase-III clinical trial was designed to test rigorously the potential efficacy of treating CTCL patients with this fusion protein toxin. This phase-III trial was conducted as two randomized, double-blind studies in mutually exclusive patient populations. The first arm of the trial evaluated the intravenous administration of $DAB_{389}IL$-2 for up to eight courses of therapy at either 9 or 18 µg kg^{-1} per day in patients with advanced refractory disease, whereas the second arm of the trial evaluated $DAB_{389}IL-2$ in patients with less advanced disease at 9 and 18 $\mu g\,kg^{-1}$ per day and included a placebo control group. In the latter study, patients whose disease progressed were unblinded and if they had received placebo were allowed to enrol in an open-label study in which they received $DAB_{389}IL-2$ at 18 $\mu g\,kg^{-1}$ per day.

Upon completion of the first arm of the phase-III trial, the overall response rate in patients that met the inclusion criteria demonstrated that 30% of the patients had a 50% or greater reduction in their tumour burden for at least 6 weeks following treatment with $DAB_{389}IL-2$. Ten per cent of the total patients that could be evaluated had either a complete or complete clinical response (i.e. complete responders were histologically free of disease). While there was a trend towards a higher response rate in patients treated at the 18 $\mu g\,kg^{-1}$ per day level (36%) compared with those treated at 9 $\mu g\,kg^{-1}$ per day (23%), the total number of patients in each group is sufficiently small to allow statistical separation of the two groups.

In this study the most common adverse events experienced by this patient population was chills/fever, malaise and nausea/vomiting. Less frequent adverse events were hypotension, oedema, rash and a capillary leak syndrome. It is clear that CTCL is a devastating malignancy and, unlike many other non-Hodgkin's lymphomas, CTCL patients suffer a substantial and often disfiguring disability during the course of their illness. The rigorous analysis of the phase-III clinical trial that has been completed has demonstrated that $DAB_{389}IL-2$ therapy has the ability to offer substantial reduction in tumour burden and relief from constitutional symptoms in a large percentage of patients who are otherwise refractory. As a result of these clinical findings, $DAB_{389}IL-2$ (ONTAK®) has recently received approval from the Food and Drug Administration to be used for the treatment of patients presenting with refractory CTCL. In 1986, Murphy *et al.* 'envisioned that these chimeric molecules might serve as ... targeted toxins for the treatment of human malignancies'. $DAB_{389}IL-2$ is the first recombinant receptor-targeted biologic to achieve this status.

ACKNOWLEDGEMENTS

J.vdS. and J.R.M. are supported by Public Health Service Grants CA48626 and CA60934 from the National Cancer Institute, and AI21628 from the National Institute of Allergy and Infectious Diseases. We wish to thank John Love, David Perlman and Jay Sutherland for their critical reading of the manuscript and many helpful suggestions.

REFERENCES

Aullo, P., Alcani, J., Popoff, M.R., Klatzmann, D.R., Murphy, J.R. and Boquet, P. (1992). *In vitro* effects of a recombinant diphtheria-human CD4 fusion toxin on acute and chronically HIV-1 infected cells. *EMBO J.* **11**, 575–83.

Axelrod, P.I., Lorber, B. and Vonderheid, E.C. (1992). Infections complicating mycosis fungoides and Sezary syndrome. *JAMA* **267**, 1354–8.

Bacha, P., Waters, C., Williams, J., Murphy, J.R. and Strom, T.B. (1988). Interleukin-2 targeted cytotoxicity: selective action of a diphtheria toxin-related interleukin-2 fusion protein. *J. Exp. Med.* **167**, 612–22.

Bendel, A.E., Shao, Y., Davies, S.M., Warman, B., Yang, C.H., Waddick, K.G., Uckun, F.M. and Perentesis, J.P. (1997). A recombinant fusion toxin targeted to the granulocyte-macrophage colony-stimulating factor receptor. *Leukaemia Lymphoma* **25**, 257–70.

Boquet, P., Silverman, M.S., Pappenheimer, A.M., Jr and Vernon, W.B. (1976). Binding of Triton X-100 to diphtheria toxin, crossreacting material 45, and their fragments. *Proc. Natl Acad. Sci. USA* **73**, 4449–53.

Broder, S. and Bunn, P.A., Jr (1980). Cutaneous T cell lymphomas. *Semin. Oncol.* **7**, 310–31.

Choe, S., Bennett, M.J., Fujii, G., Curmi, P.M.G., Kantardjieff, K.A., Collier, R.J. and Eisenberg, D. (1992). The crystal structure of diphtheria toxin. *Nature* **357**, 216–22.

Donovan, J.J., Simon, M.I., Draper, R.K. and Montal, M. (1981). Diphtheria toxin forms transmembrane channels in planar lipid bilayers. *Proc. Natl Acad. Sci. USA* **78**, 172–6.

Falnes, P.O., Choe, S., Madshus, I.H., Wilson, B.A. and Olsnes, S. (1994). Inhibition of membrane translocation of diphtheria toxin A-fragment by internal disulfide bridges. *J. Biol. Chem.* **269**, 8402–7.

Fisher, C.E., Sutherland, J.A., Krause, J.E., Murphy, J.R., Leeman, S. and vanderSpek, J.C. (1996). Genetic construction and properties of a diphtheria toxin-related substance P fusion protein: *in vitro* elimination of substance P receptor bearing cells. *Proc. Natl Acad. Sci. USA* **93**, 7341–5.

Foss, F.M. and Kuzel, T.M. (1995). Experimental therapies in the treatment of cutaneous T-cell lymphoma. *Haematol.–Oncol. Clin. N. Am.* **9**, 1127–37.

Foss, F.M., Saleh, M.N., Krueger, J.G., Nichols, J.C. and Murphy, J.R. (1998). Diphtheria toxin fusion proteins. In: *Current Topics in Microbiology and Immunology, Clinical Applications of Immunotoxins*, Vol. 234 (ed. A.E. Frankel). Springer, Berlin.

Gill, D.M. and Pappenheimer, A.M., Jr (1971). Structure–activity relationships in diphtheria toxin. *J. Biol. Chem.* **246**, 1492–5.

Hesketh, P., Caguioa, P., Koh, H., Dewey, H., Facada, A., McCaffrey, R., Parker, K., Nylen, P. and Woodworth, T. (1993). Clinical activity of a cytotoxic fusion protein in the treatment of cutaneous T cell lymphoma. *J. Clin. Oncol.* **11**, 1682–90.

Hu, H-Y., Hunth, P.D., Murphy, J.R. and vanderSpek, J.C. (1998). The effects of helix breaking mutations in the diphtheria toxin transmembrane domain helix layers of the fusion toxin $DAB_{389}IL-2$. *Protein Engng* **11**, 101–7.

Jean, L.-F. and Murphy, J.R. (1992). Diphtheria toxin receptor binding domain substitution with interleukin-6: genetic construction and interleukin-6 receptor specific action of a diphtheria toxin-related interleukin-6 fusion protein. *Protein Engng* **4**, 989–94.

Kagan, B.L., Finkelstein, A. and Colombini, M. (1981). Diphtheria toxin fragment forms large pores in phospholipid bilayer membranes. *Proc. Natl Acad. Sci. USA* **78**, 4950–4.

Kiyokawa, T., Williams, D.P., Snider, C.E., Strom, T.B. and Murphy, J.R. (1991). Protein engineering of diphtheria toxin-related interleukin-2 fusion toxins to increase biologic potency for high affinity interleukin-2 receptor bearing target cells. *Protein Engng* **4**, 463–8.

Lakkis, F., Steele, A., Pacheco-Silva, A., Kelley, V.E., Strom, T.B. and Murphy, J.R. (1991). Interleukin-4 receptor targeted cytotoxicity: genetic construction and properties of diphtheria toxin-related interleukin-4 fusion toxins. *Eur. J. Immunol.* **21**, 2253–8.

LeMaistre, C.F., Meneghetti, C., Rosenblum, M., Reuben, J., Parker, K., Shaw, J., Woodworth, T. and Parkinson, D. (1992). Phase I trial of an interleukin-2 receptor (IL-2R) fusion toxin (DAB$_{486}$IL-2) in haematologic malignancies expressing the IL-2 receptor. *Blood* **79**, 2547–54.

LeMaistre, C.F., Craig, F.E., Meneghetti, C., McMullin, B., Parker, K., Reuben, J., Boldt, D.H., Rosenblum, M. and Woodworth, T. (1993). Phase I trial of a 90-minute infusion of the fusion toxin DAB$_{486}$IL-2 in haematological cancers. *Cancer Res.* **53**, 3930–4.

LeMaistre, C.F., Saleh, M.N., Kuzel, T.M., Foss, F., Platanias, L.C., Schwartz, G., Ratain, M., Rook, A., Freytes, C.O., Craig, F., Reuben, J. and Nichols, J.C. (1998). Phase I trial of a ligand fusion-protein (DAB$_{389}$IL-2) in lymphomas expressing the receptor for interleukin-2. *Blood* **91**, 399–405.

Lemichez, E., Bomsel, M., vanderSpek, J., Lukianov, E.V., Murphy, J.R., Olsnes, S. and Boquet, P. (1997). Membrane translocation of diphtheria toxin fragment A exploits early to late endosome trafficking machinery. *Mol. Microbiol.* **23**, 445–57.

Liger, D., vanderSpek, J., Gaillard, C., Cansier, C., Murphy, J.R., Leboulch, P. and Gillet, D. (1997). Characterization and receptor specific toxicity of two diphtheria toxin-related interleukin 3 fusion proteins DAB$_{389}$IL-3 and DAB$_{389}$(Gly$_4$Ser)$_2$-mIL-3. *FEBS Lett.* **406**, 157–61.

Mindell, J.A., Silverman, J.A., Collier, R.J. and Finkelstein, A. (1994a). Structure–function relationships in diphtheria toxin channels: III. Residues which affect the cis pH dependence of channel conductance. *J Membrane Biol.* **137**, 45–57.

Mindell, J.A., Silverman, J.A., Collier, R.J. and Finkelstein, A. (1994b). Structure–function relationships in diphtheria toxin channels: II. A residue responsible for the channel's dependence on trans pH. *J Membrane Biol.* **137**, 29–44.

Murphy, J.R. and vanderSpek, J.C. (1995). Targeting diphtheria toxin to growth factor receptors. *Semin. Cancer Biol.* **6**, 259–67.

Murphy, J.R., Bishai, W., Borowski, M., Miyanohara, A., Boyd, J. and Nagle, S. (1986). Genetic construction, expression, and melanoma selective cytotoxicity of a diphtheria toxin α-melanocyte stimulating hormone fusion protein. *Proc. Natl Acad. Sci. USA* **83**, 8258–62.

Murphy, J.R., vanderSpek, J.C., Lemichez, E. and Boquet, P. (1995). In: *Bacterial Toxins and Virulence Factors in Disease, Handbook of Natural Toxins*, Vol. 8 (eds J. Moss, B. Iglewski, M. Vaughan and A.T. Tu), pp. 23–45. Marcel Dekker, New York.

Naglich, J.G., Rolf, J.M. and Eidels, L. (1992). Expression cloning of a diphtheria toxin receptor: identity with a heparin-binding EGF-like growth factor precursor. *Cell* **69**, 1051–61.

Negro, A. and Skaper, S.D. (1997). Synthesis and cytotoxic profile of a diphtheria toxin-neurotrophin-4 chimera. *J. Neurochem.* **68**, 554–63.

Pappenheimer, A.M., Jr (1977). Diphtheria A. toxin. *Annu. Rev. Biochem.* **46**, 69–94.

Schwartz, G., Tepler, I., Charette, J., Kadin, L., Parker, K., Woodworth, T. and Schnipper, L. (1992). Complete response of a Hodgkin's lymphoma in a phase I trial of DAB$_{486}$IL-2. *Blood* **79**, 175a.

Shaw, J.P., Akiyoshi, D.E., Arrigo, D.A., Rhoad, A.E., Sullivan, B., Thomas, J., Genbauffe, F.S., Bacha, P. and Nichols, J.C. (1991). Cytotoxic properties of DAB486EGF and DAB389EGF, epidermal growth factor (EGF) receptor-targeted fusion toxins. *J Biol. Chem.* **266**, 21118–24.

Silverman, J.A., Mindell, J.A., Finkelstein, A., Shen, W.H. and Collier, R.J. (1994a). Mutational analysis of the helical hairpin region of diphtheria toxin transmembrane domain. *J. Biol. Chem.* **269**, 22524–32.

Silverman, J.A., Mindell, J.A., Zhan, H., Finkelstein, A. and Collier, R.J. (1994b). Structure–function relationships in diphtheria toxin channels: I. Determining a minimal channel-forming domain. *J. Membrane Biol.* **137**, 17–28.

Sweeney, E.B., vanderSpek, J.C., Foss, F. and Murphy, J.R. (1998). Genetic construction and characterization of DAB$_{389}$IL-7: a novel agent for the elimination of IL-7 receptor positive cells. *Bioconjugate Chem.* **9**, 201–7.

Tsuneoka, M., Nakagama, K., Hatsuzawa, K., Komada, M., Kitamura, N. and Mekada, E. (1993). Evidence for involvement of furin in cleavage and activation of diphtheria toxin. *J. Biol. Chem.* **268**, 26461–5.

vanderSpek, J.C., Mindel, J., Finkelstein, A. and Murphy, J.R. (1993). Structure function analysis of the transmembrane domain of the interleukin-2 receptor target fusion toxin DAB$_{389}$-IL-2: the amphipathic helical region of the transmembrane domain is essential for the efficient delivery of the catalytic domain to the cytosol of target cells. *J. Biol. Chem.* **263**, 12077–82.

vanderSpek, J., Howland, K., Friedman, T. and Murphy, J.R. (1994a). Maintenance of the hydrophobic face of the diphtheria toxin amphipathic transmembrane helix 1 is essential for the efficient delivery of the catalytic domain to the cytosol of target cells. *Protein Engng* **7**, 985–9.

vanderSpek, J., Cassidy, D., Genbauffe, F., Huynh, P. and Murphy, J.R. (1994b). An intact transmembrane helix 9 is essential for the efficient delivery of the diphtheria toxin catalytic domain to the cytosol of target cells. *J. Biol. Chem.* **269**, 21455–9.

vanderSpek, J.C., Sutherland, J., Sampson, E. and Murphy, J.R. (1995). Genetic construction and characterization of the diphtheria toxin-related interleukin 15 fusion protein DAB$_{389}$IL-15. *Protein Engng* **8**, 1317–21.

vanderSpek, J.C., Sutherland, J., Ratnarathorn, M., Howland, K., Ciardelli, T.L. and Murphy, J.R. (1996). DAB$_{389}$IL-2 receptor binding domain mutations: cytotoxic probes for studies of ligand/receptor interactions. *J Biol. Chem.* **271**, 12145–9.

vanderSpek, J.C., Sutherland, J.A., Zeng, H., Battey, J.F., Jensen, R.T. and Murphy, J.R. (1997). Inhibition of protein synthesis in small cell lung cancer cells induced by the diphtheria toxin-related fusion proteins DAB$_{389}$GRP and DAB$_{389}$SP. *Cancer Res.* **57**, 290–4.

van Ness, B.G., Howard, J.B. and Bodley, J.W. (1980). ADP-ribosylation of elongation factor 2 by diphtheria toxin. NMR spectra and proposed structure of ribosyl diphtamide and its hydrolysis products. *J. Biol. Chem.* **255**, 10710–16.

Walz, G., Zanker, B., Brand, K., Swanlund, D., Genbauffe, F., Zeldis, J.C., Murphy, J.R. and Strom, T.B. (1989). Sequential effects of interleukin-2/diphtheria toxin fusion protein on T-cell activation. *Proc. Natl Acad. Sci. USA* **86**, 9485–8.

Waters, C.A., Schimke, P., Snider, C.E., Itoh, K., Smith, K.A., Nichols, J.C., Strom, T.B. and Murphy, J.R. (1990). Interleukin-2 receptor targeted cytotoxicity: receptor binding requirements for entry of IL-2 toxin into cells. *Eur. J. Immunol.* **20**, 785–91.

Wen, Z., Tao, X., Lakkis, F., Kiyokawa, T. and Murphy, J.R. (1991). Expression, purification, and α-melanocyte stimulating hormone receptor specific toxicity of DAB–MSH fusion toxins. *J. Biol. Chem.* **266**, 12289–93.

Wiedlocha, A., Madhus, I.H., Mach, H., Middaugh, C.R. and Olsnes, S. (1992). Tight folding of acidic fibroblast growth factor prevents its translation to the cytosol with diphtheria toxin as vector. *EMBO J.* **11**, 4835–42.

Williams, D., Parker, K., Bishai, W., Borowski, M., Genbauffe, F., Strom, T.B. and Murphy, J.R. (1987). Diphtheria toxin receptor binding domain substitution with interleukin-2: genetic construction and properties of a diphtheria toxin-related interleukin-2 fusion protein. *Protein Engng* **1**, 493–8.

Williams, D.P., Snider, C.E., Strom, T.B. and Murphy, J.R. (1990a). Structure function analysis of IL-2-toxin (DAB$_{486}$-IL-2): fragment B sequences required for the delivery of fragment A to the cytosol of target cells. *J. Biol. Chem.* **65**, 11885–9.

Williams, D.P., Wen, Z., Watson, R.S., Boyd, J., Strom, T.B. and Murphy, J.R. (1990b). Cellular processing of the fusion toxin DAB$_{486}$-IL-2 and efficient delivery of diphtheria toxin fragment A to the cytosol of target cells requires Arg194. *J. Biol. Chem.* **265**, 20673–7.

Yamaizumi, M., Mekada, E., Uchida, T. and Okada, Y. (1978). One molecule of diphtheria toxin fragment A introduced into a cell can kill the cell. *Cell* **15**, 245–50.

Native and genetically engineered toxin-based vaccines for cholera and *Escherichia coli* diarrhoeas

Ann-Mari Svennerholm and Jan Holmgren

INTRODUCTION

New techniques and novel concepts have given great hopes for future vaccine development, and this is especially true for toxin-based vaccines, for which the genetic and molecular information basis is often more advanced than for many other types of vaccine. The new techniques give prospects for, for example:

(a) improved vaccine delivery and presentation systems, by, for example, biodegradable microspheres
(b) development of useful adjuvants for human use
(c) recombinant vaccine production in plants
(d) use of 'naked DNA' vaccines.

Conceptually, the prospects for the development of immunotherapeutic vaccines against important autoimmune, allergic and other chronic immunopathological disorders have also dramatically increased based on recent new insights into the regulation of immunity including immunopathological observations. Interestingly, in these latter applications, which are outside the established area of infectious disease, bacterial toxins and toxin derivatives have also been found to have a special role, not as specific antigens but instead as promising immunomodulating agents.

Mucosal immunization, especially by the oral route, has attracted special interest recently in vaccinology, both for its capacity to elicit protective immunity against infectious diseases and as a promising approach to be used for immunological treatment of various diseases caused by an aberrant immune response associated with tissue-damaging inflammation. Again, important roles have been identified for bacterial toxins and toxin-derived subunits both as specific immunogens and as carriers and/or immunomodulating agents for various antigens.

The majority of infections occur at or begin at a mucosal surface, and topical application of vaccine is usually needed or is beneficial for inducing a protective immune response. As discussed in the main part of this chapter, the recent detailed definition of the pathogenesis and the critical virulence proteins of important causative agents of enteric infections, together with an improved understanding of protective mucosal immune mechanisms and how to stimulate protective mucosal immunity by topical vaccine application, have now resulted in the successful development of several important new oral vaccines against cholera, enterotoxigenic *Escherichia coli* (ETEC) and rotavirus diarrhoea, with additional vaccines against enteric infections being developed (World Health Organization, 1998). In both cholera and ETEC vaccines, the stimulation of local antitoxic immunity by vaccination with a toxin-derived B subunit protein is important and, as will also be discussed in this chapter, a similar situation may hold true for vaccine development against the severe disease caused by Verotoxin-producing *E. coli* (VTEC) (the latter organism

691

also being known as enterohaemorrhagic *E. coli*, EHEC). Likewise, there are good prospects for developing novel mucosal vaccines against infections in the respiratory and genital tracts using toxin B subunits or mutant toxins as carrier proteins and/or immunostimulants (Holmgren *et al.*, 1996).

Mucosal immunization may suppress systemic immune responses in a specific manner, including inflammatory hypersensitivity (DTH) reactions caused by T-cells, often evident in chronic infections associated with immunopathology and in various autoimmune and allergic diseases (Weiner *et al.*, 1994; Czerkinsky and Holmgren, 1995). Particularly promising results, with regard to both induction of mucosal immune responses and suppression of systemic inflammation, have been obtained using conjugates between specific antigens and the B subunit of cholera toxin (CTB) as mucosal immunogens. Thus, oral administration of conjugates of various model antigens and CTB has suppressed antigen-specific DTH reactions by 90–100% (Sun *et al.*, 1994). Different CTB–antigen conjugates, when given before or after immune sensitization, have effectively suppressed various DTH-associated autoimmune and inflammatory disorders in animal models:

(a) myelin basic protein (MBP)–CTB suppressed experimental allergic encephalomyelitis (EAE) in Lewis rats (Sun *et al.*, 1996)
(b) insulin–CTB suppressed diabetes in adult non-obese diabetic (NOD) mice (Bergerot *et al.*, 1997)
(c) collagen II–CTB suppressed arthritis in DBA mice (Tarkowski *et al.*, 1999)
(d) *Schistosoma mansoni* egg antigen–CTB reduced liver granuloma formation and immunopathology-associated mortality in experimental schistosomiasis in mice (Sun *et al.*, 1999).

On this basis it may be possible to develop CTB-based mucosal anti-infectious (preventive), as well as anti-inflammatory (immunotherapeutic), 'vaccines' for use in selected infections, autoimmune and allergic disorders.

The chapter describes the development of toxin-based vaccines for use in humans against diarrhoea caused by enterotoxin-producing bacteria based on current knowledge of mechanisms of disease and immunity in these infections. Vaccines that are already available or under development, are presented in Table 40.1. A brief summary will be given of the current intense efforts to use native and especially genetically engineered mutant enterotoxins or their receptor-binding B subunits as carriers and/or immunomodulators, either for inducing mucosal anti-infectious immunity or for giving rise to an anti-inflammatory immunotherapeutic response by inducing a state of 'oral tolerance' at the systemic level.

TABLE 40.1 Toxin-based vaccines for cholera and *E. coli* diarrhoeas – available and under development

Pathogen	Type of vaccine	Status
Vibrio cholerae O1	Oral inactivated CTB-WC (Dukoral®)	Licensed
	Oral live CVD103HgR (Orochol®)	Licensed
Vibrio cholerae O139	Oral inactivated CTB-O1 + O139 WC	Phase II
	Oral live O139	Phase III
Non-cholera vibrios	CTB	Available in, for example, Dukoral®
Enterotoxigenic *E. coli*	Oral inactivated rCTB CF ETEC	Phase III
Verotoxin-producing *E. coli*	VT-O157-LPS conjugates	Not available experimental
	Parenteral VT toxoid	Not available experimental

ENTEROTOXIN-INDUCED DIARRHOEAL DISEASES

Diarrhoeal disease remains one of the leading global health problems. It has been estimated that between three billion and five billion episodes of diarrhoea, resulting in approximately three million deaths, occur annually in developing countries, with the highest incidence and severity in children below the age of 5 years (Black, 1986; World Health Organization, 1996). The majority of these diseases are caused by enteric infections; the aetiological agents include a large number of bacteria, some viral agents and parasites. About half of all diarrhoeas are due to bacteria that cause watery stools by producing one or more enterotoxins. Cholera, which results from infection with *Vibrio cholerae*, is the most severe form of these 'enterotoxic enteropathies' (Craig, 1980), whereas disease caused by ETEC is the most common one.

Diarrhoea is also the leading cause of illness among international travellers visiting countries with a high incidence of enteric infections. It has been estimated that approximately one-third of travellers to developing countries experience diarrhoeal disease, which accounts for at least 10 million cases per year (Black, 1990; Levine and Svennerholm, 1997).

Based on the great health impact of infections with enterotoxin-producing bacteria, both in children in developing countries and in travellers to these areas, there has been great interest in developing effective vaccines, particularly against cholera and ETEC disease. During recent years, effective cholera vaccines have been developed and studies to evaluate ETEC vaccines for protective efficacy in humans have been initiated.

Cholera

Cholera may be caused by *V. cholerae* of two different O groups, i.e. O1 and O139. *Vibrio cholerae* of serogroup O1 is the prototype for the enterotoxin-producing bacteria. Until the beginning of the twentieth century all *V. cholerae* O1 isolates examined were of the classical, non-haemolytic biotype. At the beginning of the twentieth century, however, vibrios of a haemolytic biotype, El Tor, were identified and for many years vibrios of either the classical or El Tor biotype were isolated from cholera cases (Blake, 1994). Both classical and El Tor *V. cholerae* O1 can be subdivided into two main serotypes, Inaba and Ogawa.

At the beginning of the nineteenth century, cholera started to spread from Bengal and since then seven cholera pandemics have been described. The last pandemic spread to a large number of countries in Asia and Africa, and since 1991 it has also become endemic in South and Central America (Blake, 1994; Tauxe *et al.*, 1994). In addition, *V. cholerae* of a previously unknown serogroup, i.e. O139 Bengal, was first identified as an additional, new cause of cholera in India and Bangladesh in 1992 (Morris, 1994). Cholera caused by *V. cholerae* O139 Bengal has subsequently been identified in a number of neighbouring countries (e.g. Thailand, Nepal, Burma and Indonesia), but so far has been restricted to south-east Asia. Although there has been a decline in the incidence during recent years, *V. cholerae* O139 still accounts for approximately 10% of all cholera in, for example, Bangladesh. The overall incidence of cholera is uncertain owing to under-reporting of the disease from many countries. Therefore, recent figures from the World Health Organization (1996) of *ca.* 5.5 million cases and 120 000 deaths from cholera annually may be substantial underestimates of the present situation.

The disease caused by *V. cholerae* is characterized by watery stools without blood and mucus; the clinical spectrum is broad, ranging from severe dehydrating cholera to inapparent infection. In the most severe cases patients may purge as much as 15–25 litres of water and electrolytes per day and the mortality rate in such untreated cases may be as high as 30–50%. In addition to the often large volumes of rice water-like stools, severely affected patients may manifest different signs of dehydration, including vomiting, poor skin turgor, sunken eyes, intense peripheral vasoconstriction, and eventually shock and death.

Enterotoxigenic *Escherichia coli* diarrhoea

ETEC is the most common cause of diarrhoea in developing countries, accounting for *ca.* 400 million cases a year (World Health Organization, 1996). Although the disease is generally mild, ETEC infections are responsible for 300 000–700 000 deaths annually in children below 5 years of age in developing countries, which corresponds to almost 10–20% of the global total deaths from diarrhoeal disease in this age group. ETEC is also the most common cause of traveller's diarrhoea, causing one-third to one-half of all diarrhoeal episodes in travellers to endemic areas in Africa, Asia and Latin America (Black, 1990).

ETEC diarrhoea may be caused by a wide variety of *E. coli* of different serotypes. The bacteria, which are non-invasive, cause a secretory type of diarrhoea by producing a heat-labile enterotoxin (LT) or a heat-stable enterotoxin (ST) or both toxins. These toxins cause fluid and electrolyte secretion in the bowel, primarily from the intestine crypt cells (Guerrant, 1985). In order to cause disease, the bacteria must colonize the small intestine. To do so they have to overcome a number of host defences, including peristalsis, mucus production and epithelial shedding.

While both geographical and age-related variations in ETEC infections may occur, it has previously been estimated that approximately one-third of all clinical ETEC isolates produce LT alone, one-third ST alone and one-third LT in combination with ST (Svennerholm and Holmgren, 1995). However, a higher proportion of ETEC strains producing ST alone was recently reported (Wolf, 1997).

Although most episodes of ETEC disease are relatively mild, ETEC infection may result in moderate to severe dehydration that is sometimes fatal. Indeed, ETEC disease may vary from mild diarrhoea to a severe cholera-like disease, and is often accompanied by nausea, vomiting, abdominal cramps and anorexia, often with significant fever (Black, 1986).

Disease caused by Verotoxin-producing *Escherichia coli*

Verotoxin (VT)-producing *E. coli* (VTEC), often termed enterohaemorrhagic *E. coli* (EHEC), causes diarrhoeal disease in both humans and other animals. The clinical importance of VTEC was first recognized in 1982 when *E. coli* strains of a certain serotype, O157:H7, that produced VT were isolated from stools of patients during outbreaks of haemorrhagic colitis associated with the consumption of beef (Riley *et al.*, 1983). Since then O157 VTEC have been identified in many outbreaks and in sporadic cases of bloody diarrhoea in several countries, and VTEC has emerged as the leading cause of diarrhoea in North America (Noël and Boedeker, 1997). Also, a close association has been established between VTEC and haemolytic uraemic syndrome (HUS), which is characterized by kidney failure and microangiopathic anaemia, as initially described by Karmali (1989). Indeed, VTEC-induced diarrhoeas have the potential for progression to HUS in 5–10% of patients (Noël and Boedeker, 1997).

Although VTEC of many different serotypes have been isolated from human infections, VTEC strains of serotypes other than O157:H7 have not been identified as a major cause of diarrhoeal illness. The second most commonly reported serotype is O26:H11, which has also been associated with haemorrahgic colitis; other relatively common serotypes are O103:H2 and O113:H21.

VTEC may produce either or both of two major classes of VT, identified by toxin neutralization and DNA hybridization tests (Scotland *et al.*, 1988). The first, VT1, is neutralized by antibodies to Shiga toxin whereas the other, VT2, is not. The VT1 class is very homogeneous, whereas there are several variants of VT2 (e.g. VT2c and VT2e). Examination of VTEC for the type of toxin they produce has revealed geographical differences. Thus, whereas strains producing both VT1 and VT2 predominated in Canada, strains producing only VT2 have predominated in studies, for example, in the UK. Among non-O157 VTEC, the VT-types produced are serotype related (Willshaw *et al.*, 1997).

MECHANISM OF DISEASE AND IMMUNITY IN ENTEROTOXIC BACTERIAL INFECTIONS

Enterotoxins and antitoxic immunity

The major pathogenic mechanisms of enterotoxigenic bacteria include initial bacterial colonization of the small intestine, followed by production of one or more enterotoxins that, through various mechanisms, can induce electrolyte and water secretion resulting in diarrhoea (Guerrant, 1985). These enterotoxins stimulate secretion, probably mainly by inducing increased formation of cyclic AMP and/or cyclic GMP in the epithelial cells. However, there is increasing evidence that in addition to the fluid secretion mediated via these pathways, there may be a local neurogenic component in the diarrhoeal secretory process, triggered primarily via the action of enterotoxins on intestinal enterochromaffin cells (Jodal and Lundgren, 1995).

The prototype enterotoxin is cholera toxin (CT), which is produced by most *V. cholerae* O1 bacteria of either the classical or El Tor biotype (Holmgren, 1981). CT with identical structure to O1 El Tor CT is also produced by *V. cholerae* O139 bacteria (Waldor and Mekalanos, 1994). CT consists of five identical binding (B) subunits associated in a ring into which a single toxin-active (A) subunit is non-covalently inserted. (For a more detailed review see Chapter 6 in this book.)

ETEC bacteria may produce LT alone, ST alone or a combination of both toxins. LT is structurally, functionally and immunologically closely related – although not identical – to CT. Thus, LT also consists of five B subunits and one A subunit, and both of these proteins cross-react immunologically with the corresponding CT subunits, even though there are also specific epitopes on the A, as well as the B, subunits of both toxins (Guerrant, 1985). Both anti-CT and anti-LT immune responses are mainly directed against the B subunit portions of the molecules and prevent the toxins from binding to the intestinal cell receptors (Holmgren and Svennerholm, 1992).

Studies in experimental animals have shown a direct correlation between protection against CT-induced fluid secretion and intestinal synthesis of secretory IgA (SIgA) antibodies, and also between protection and the number of IgA antitoxin-producing cells in the intestine (Holmgren and Svennerholm, 1992). A protective role of SIgA antitoxin has also been indicated by the observation in breast-fed children in Bangladesh of a reduced risk for cholera when having elevated levels of antitoxin antibodies in the ingested milk (Glass *et al.*, 1983).

The identification of the subunit structure of CT and LT, and the roles of the different subunits in pathogenesis and immunity, have indicated that the purified CT or LT B subunits (CTB or LTB) are suitable toxoid candidates for inclusion in a vaccine against enterotoxic enteropathies (Holmgren *et al.*, 1977; Svennerholm *et al.*, 1989). Thus, antitoxic immunity in both cholera and *E. coli* LT diarrhoea is mainly, if not exclusively, mediated by locally produced antibodies directed against the B subunit portion of the toxoid molecules. Furthermore, the B subunit pentamers are particularly well suited as oral immunogens, because they are stable in the small intestine and are capable of binding to intestinal epithelium, including M-cells of Peyer's patches; these properties are important for stimulating mucosal immunity, including local immunological memory (Neutra and Kraehenbühl, 1992). However, the pentameric forms of CTB or LTB are sensitive to the acid pH in the stomach.

Antibodies against CTB may cross-protect against *E. coli* LT diarrhoea and, vice versa, anti-LT immunity may be effective against experimental cholera, although the protection observed against the homologous toxin may be somewhat stronger (Svennerholm and Holmgren, 1995). As a consequence of this substantial degree of immunological cross-reactivity between CT and LT, studies both in endemic areas and in travellers have shown that peroral vaccination with CTB-containing cholera vaccine gives rise to significant protection also against diarrhoea caused by LT-producing *E. coli* (Clemens *et al.*, 1988; Peltola *et al.*, 1991).

Heat-stable enterotoxins (ST) are a family of closely related peptides that induce diarrhoea in both humans and other animals. *Escherichia coli* STs are classified into two structurally, functionally and immunologically unrelated types, named STa and STb (Burgess *et al.*, 1978; Guerrant, 1985). STa includes methanol-soluble, infant mouse-active peptide toxins, while STb is methanol-insoluble and active in weaned pigs. The only type of ST that is important for diarrhoea in humans is STa, which includes STh produced by *Escherichia coli* strains of human origin and STp by *E. coli* strains of porcine origin. Most human strains produce STh, but a few clinical isolates have been shown to produce STp.

STa is a small molecule consisting of 18 (STh) or 19 (STp) amino acids. The molecules have a common, highly conserved region with 10 amino acids, including six cysteine residues, that are linked intramolecularly by three disulfide bonds (for a review see Chapter 30 of this book, and Nair and Takeda, 1997). *Escherichia coli* STa is poorly immunogenic, but high titre polyclonal anti-ST sera, as well as specific monoclonal antibodies, can be obtained by immunizing with conjugates consisting of ST coupled to appropriate protein carriers (e.g. bovine serum albumin or CTB). Antisera raised against such ST–carrier protein conjugates have had strong ST-neutralizing capacity. However, these conjugates have retained toxic activity, and numerous attempts to immunize with non-toxic ST–carrier protein conjugates, either chemically modified or recombinantly produced, have resulted in antibody preparations with low to high anti-ST antibody activity but with no neutralizing capacity (Sanchez *et al.*, 1988; Svennerholm *et al.*, 1988; Aitken and Hirst, 1993). Thus, it is still unclear whether sufficiently effective anti-ST immunity can be induced by vaccination with non-toxic STa–carrier protein conjugates to provide practically significant protection against disease caused by ST-producing *E. coli* in humans (Svennerholm and Holmgren, 1995).

Antibacterial immunity in cholera and ETEC infections

Colonization of the small intestine is a prerequisite for enterotoxin-producing bacteria to cause diarrhoea. The colonization is dependent on receptor–ligand interactions between the bacteria and the host cells. Colonization is mediated by so-called adhesins or colonization factors (CFs), that may be fimbrial or fibrillar in nature (Evans and Evans, 1989; Gaastra and Svennerholm, 1996).

Vibrio cholerae bacteria may express a number of fimbrial structures, for example toxin-coregulated pilus (TCP) and mannose-sensitive haemagglutinin (MSHA), which mediate bacterial attachment to the small intestine or epithelial cells (Jonsson *et al.*, 1991; Voss *et al.*, 1996). Antibodies against both these fimbriae protect against infection and disease with *V. cholerae* expressing the corresponding structures in experimental animals (Osek *et al.*, 1992). It remains to be shown, however, whether immunity against either or both of these fimbriae can add significantly to the antibacterial protection afforded by immunity against *V. cholerae* lipopolysaccharide (LPS). Thus, antibacterial immunity against experimental O1 classical or El Tor cholera, as well as against *V. cholerae* O139, is to a large extent directed against *V. cholerae* O1 and O139 LPS, respectively (Svennerholm, 1980; Jonsson *et al.*, 1996). Thus, immunity against cholera seems to be predominantly directed against LPS and CTB, respectively (Holmgren *et al.*, 1977).

These types of antibodies may confer strong protection against disease by inhibiting bacterial colonization and toxin-binding and, when present together, to co-operate synergistically in providing cholera immunity (Svennerholm and Holmgren, 1976; Jonsson *et al.*, 1996).

At variance with the situation in cholera, protective antibacterial immunity against experimental ETEC infection seems to be directed against different colonization factors rather than the LPS O-antigens (Svennerholm *et al.*, 1989). A problem for vaccine developmental work is, however, that more than 20 different colonization factors (CSs) have been identified to date (Gaastra and Svennerholm, 1996), although epidemiological studies have suggested that 50–80% of the clinical ETEC isolates express one of three different types of colonization factor antigens (CFAs), i.e. CFA/I, CFA/II or CFA/IV (Evans and Evans, 1989). Whereas CFA/I is a homogeneous protein, CFA/II consists of three different subcomponents, i.e. CS3 alone or in combination with CS1 or CS2, and CFA/IV of CS6 alone or together with CS4 or CS5. All of these different CFs have been shown to promote colonization of ETEC in the gut and to induce protective immune responses following infection (Gaastra and Svennerholm, 1996).

The various CFs are associated with a limited number of serogroups, and associations between the toxin type and expression of certain CFs or a specific serotype also exist (Gaastra and Svennerholm, 1996). Thus, most LT-only strains belong to serogroups that occur infrequently and some CFs occur exclusively on strains producing ST, whereas others are present exclusively on strains producing LT. Furthermore, 90% of the strains that produce both toxins, 60% of the strains producing ST alone and 10% of the strains producing LT only express one or more of the CFs identified to date.

It has been suggested that the most important antigens to be included in a future vaccine against ETEC diarrhoea are a suitable toxoid in combination with the CFs that occur with the highest frequency in the target population.

Pathogenic and immune mechanisms in VTEC disease

The association of VTEC with the clinical symptoms observed in haemorrhagic colitis and HUS suggests that VTs are major virulence factors in these diseases. Thus, VT1 is capable of causing fluid accumulation in rabbit ileal loops, and subsequently death when injected parentally into mice; monoclonal antibodies against the B subunits of the toxin may prevent these biological effects (Strockbine *et al.*, 1986). Injection of VT2 into the peritoneal cavity of rabbits results in diarrhoea and neurological symptoms. However, studies in animals have shown that VT-negative variants of O157 strains may still cause diarrhoea.

The virulence of VTEC is determined by several factors. They include the Verotoxins VT1, VT2 and VT2 variants (also named Shiga-like toxins SLT-I, SLT-II and SLT-II variants), but also fimbriae or other adhesins as well as the gene products of the chromasomal locus for enterocyte effacement, that produce the attaching/effacing (A/E) lesions (Noël and Boedeker, 1997). Both the enteroadhesive and toxigenic properties are required for full virulence of VTEC.

Both VT1 and VT2 are composed of one toxin-active A subunit and five B subunits which form a pentameric ring that mediates receptor-binding of the holotoxin to the epithelial cells in analogy with CT and LT (Willshaw *et al.*, 1997). The genetic information for the Verotoxins is encoded on two distinct bacteriophages (Strockbine *et al.*, 1986). The two types of VT are similar in function but differ immunologically. VT1 is clearly related to Shiga toxin produced, for example, by *Shigella dysenteriae*; the two toxins cross-react immunologically and they only differ by one amino acid in the A subunit. The A subunits of VT1 and VT2, however, are immunologically distinct and have only 57% homology (as reviewed by Noël and Boedeker, 1997). The B subunits of VT1 and VT2 are also only about 60% homologous at the amino acid level, whereas the different variants of VT2 are 95% homologous. Antibodies to the B subunit of VT1 and VT2 do not cross-neutralize each other, whereas polyclonal antisera against VT2 can neutralize VT2 and all variants of VT2 (Noël and Boedeker, 1997).

VTs are among the most potent toxins that cause disease in humans. However, the toxins are poor immunogens, at least during infection, probably because the amount of toxin coming into the circulation during infection is too small to be immunogenic. However, it is likely that small amounts of toxin gain access to the circulation and cause the extraintestinal renal and CNS vascular lesions seen in HUS (Noël and Boedeker, 1997).

DEVELOPMENT OF VACCINES AGAINST CHOLERA

Until the mid-1980s the only cholera vaccines available were injectable killed whole-cell preparations that contained vibrios of the two serotypes Inaba and Ogawa, but no toxin antigen. Such vaccines only provided limited protection of short duration, at most up to 50% protection for 3–6 months (Holmgren and Svennerholm, 1992). However, following the identification of CT as a key virulence factor in cholera and the insight that protective immunity is afforded by mucosal rather than systemic immunity (Holmgren and Svennerholm, 1992), work was initiated in many laboratories to develop new cholera vaccines that induced antitoxic in addition to antibacterial immunity locally in the gut. This work has resulted in the development of both oral inactivated as well as live genetically engineered toxin-based vaccines against cholera.

Oral inactivated cholera vaccines

An oral cholera vaccine consisting of the highly immunogenic CTB protein in combination with heat and formalin-killed *V. cholerae* O1 vibrios of the classical as well as El Tor biotypes was developed in Sweden in the early 1980s (Holmgren and Svennerholm, 1992). This CTB-whole cell (CTB-WC) vaccine, which is produced by SBL Vaccin, Stockholm, Sweden, was licensed in 1991, and in a second-generation form based on the use of recombinantly produced CTB (rCTB; Sanchez and Holmgren, 1989) licensed in 1993 (Table 40.2). The vaccine, which is given together with a bicarbonate buffer to preserve the pentameric structure of CTB, necessary for protective oral mucosal immunogenicity, is taken as a drink. In numerous extensive clinical trials in different countries, including

TABLE 40.2 Oral B subunit (CTB) – whole cell (WC) cholera and ETEC vaccines

Composition per dose	Clinical evaluation
Cholera	
(A) CTB-O1 WC (Bangladesh field trial formulation)	
1 mg CTB (purified from CT) + 1×10^{11} killed bacteria	Safe, immunogenic and protective both in volunteers and in large
$\quad 2.5 \times 10^{10}$ heat-killed Inaba vibrios (strain Cairo 48)	field trials in Bangladesh: 85% efficacy in the first 6 months; *ca.* 60%
$\quad 2.5 \times 10^{10}$ heat-killed Ogawa vibrios (strain Cairo 50)	in the first 3 years. Cross-protection against ETEC: *ca.* 70%
$\quad 2.5 \times 10^{10}$ formalin-killed classical vibrios (strain Cairo 50)	short-term efficacy. Licensed in 1991. Replaced by (B) in 1993
$\quad 2.5 \times 10^{10}$ formalin-killed El Tor vibrios (strain Phil 6973)	
(B) rCTB-O1 WC (currently licensed formulation)	
1 mg recombinant CTB + same WC composition as in A	Same safety, immunogenicity and protective efficacy as for (A). Field trial in Peru showed 86% protection against O1 El Tor cholera. Licensed in 1993
ETEC	
(C) rCTB-CF ETEC vaccine	
1 mg rCTB + 1×10^{11} formalin-killed ETEC	Safe and immunogenic in phase-I and phase-II clinical trials. Studies
(five different *E. coli* strains expressing CFA/I and CS1-CS6)	for efficacy initiated in travellers and in children in an endemic country

a large field trial in Bangladesh, the CTB-WC vaccine was shown to be safe and to provide strong protection against cholera and also partial protection against *E. coli* LT diarrhoea.

The oral CTB-WC vaccine has been designed to evoke antitoxic as well as antibacterial immune responses locally in the gut, as both of these types of immunity are protective and cooperate synergistically, when present together, in providing protection against experimental cholera (Svennerholm and Holmgren, 1976). Phase-I and phase-II clinical studies, in non-endemic and endemic areas, have established that the vaccine is safe and does not cause any significant side-effects (Holmgren *et al.*, 1997). Furthermore, oral administration of two or three doses of the vaccine results in gut mucosal IgA antitoxic and antibacterial immune responses comparable to those induced by cholera disease (Svennerholm *et al.*, 1984; Quiding *et al.*, 1991). Also, the vaccine provided highly significant protection in American volunteers given a challenge with cholera vibrios that caused disease in 100% of concurrently tested unvaccinated controls (Black *et al.*, 1987).

Based on the promising results obtained in the various phase-I and phase-II trials in Sweden, the USA and Bangladesh, a large double-blind, placebo-controlled field trial was initiated in Bangladesh. In this trial groups of volunteers (2–65 years of age), with approximately 30 000 subjects in each, were given either the CTB-WC, the WC-component alone or an *E. coli* K12 placebo. When comparing the incidence of cholera in the three different study groups it was found that the combined CTB-WC vaccine had a very high, 85%, protective efficacy, that was the same in all age groups studied during the initial 4–6 months (Clemens *et al.*, 1986). Indeed, the protection induced by the CTB-WC

vaccine was 64% higher ($P < 0.05$) than that provided by the WC-component alone, which gave a protective effect of 58% during the same study period. Direct comparison of the two vaccine preparations revealed that the CTB-WC vaccine continued to be significantly more protective than the WC-component alone during the first 8 months after vaccination. Thereafter, the efficacy was similar, i.e. approximately 60%, for both vaccines for a 3-year follow-up period (Clemens *et al.*, 1990). These results strongly support an important contribution of antitoxic immunity for protection against cholera, particularly during the first 6–8 months after vaccination.

The CTB-WC vaccine tested in the large field trial in Bangladesh contained CTB that was prepared by chemical isolation, using gel filtration under acidic conditions, from CT produced by a high-level expression wild-type strain (Tayot *et al.*, 1981), which made the preparation of CTB relatively laborious and expensive. Subsequently, however, an efficient recombinant overexpression system was developed that allows large-scale production of CTB at low cost. Thus, Sanchez and Holmgren (1989) constructed a recombinant *V. cholerae* O1 strain in which the gene for the toxic A subunit was deleted and a plasmid containing the gene for CTB coupled to the *tac*-promoter introduced, which allows overexpression of CTB in the absence of the toxic A subunit. Indeed, fermentor culturing of this strain, followed by a simple purification step of the CTB subunit, results in production in the order of 1000 g of rCTB per 1000-litre culture, which corresponds to approximately one million doses of the oral cholera vaccine (Holmgren and SBL Vaccin, unpublished data). As extensive clinical testing has shown that oral cholera vaccine containing rCTB is equally as safe and immunogenic as the old formulation used in the field trial in Bangladesh (Jertborn *et al.*, 1992; Sanchez *et al.*, 1993; Begue *et al.*, 1995), rCTB-WC has become the currently produced vaccine formulation. This vaccine has been licensed and is marketed under the trade name of Dukoral® by SBL Vaccin AB, Stockholm, Sweden. In a recent field trial in Peruvian militaries, the rCTB-WC vaccine was as effective as the Bangladeshi formulation in providing short-term protection against cholera. Thus, the rCTB-WC conferred 86% protection against cholera in Peru, in spite of the fact that it was only given in two doses (Sanchez *et al.*, 1994).

Together with SBL Vaccin, we have also developed an oral bivalent rCTB-O1/O139 WC cholera vaccine by adding formalin-killed O139 vibrios to the oral rCTB-O1 WC vaccine. When tested in Swedish volunteers, this rCTB-O1/O139 WC vaccine was safe and immunogenic (Jertborn *et al.*, 1996). Two vaccine doses given 2 weeks apart induced strong intestinal-mucosal IgA antibody responses to CT (100%), as well as to O1 vibrios (78%) and O139 vibrios (78%) as tested in pre- and post-vaccination intestinal lavage or faecal extract specimens.

Oral live vaccines

Recombinant DNA techniques have also been applied to construct various attenuated *V. cholerae* O1 strains for possible use as live oral vaccines. In the early 1980s several vaccine candidates were constructed from wild-type *V. cholerae* O1 strains by introducing deletions in the chomosomal genes encoding both the A and B subunits of CT (Kaper *et al.*, 1984, 1997). This first generation of recombinant vaccine strains, for example JBK70 (derived from the El Tor strain N16961) and CVD101 (derived from the classical Ogawa strain 395), were markedly attenuated compared with the wild-type parents. However, they still caused unacceptable adverse reactions in the form of mild to moderate diarrhoea, which was often associated with additional reactions, such as malaise, abdominal cramps and vomiting in up to 50% of the immunized volunteers. These findings suggested that the removal of even the whole CT gene from pathogenic *V. cholerae* O1 strains was not sufficient to provide safe, non-reactogenic vaccine strains. These findings could subsequently partly be explained by the identification of accessory toxins produced by *V. cholerae* [e.g. Zona occludens toxin (Zot) and accessory CT (Ace)], which may have accounted for the pathogenicity of, for example, JBK70 and CVD101 (Tacket *et al.*, 1993). Thus, Zot is a toxin that may be produced in the supernates of *V. cholerae* cultures and that affects intracellular tight junctions of intestinal epithelial cells that render them permeable to solute (Fasano *et al.*,

1991). Ace is a distinct toxin that causes an increase in potential difference in Ussing chambers and secretion in rabbit ligated ileal loops (Trucksis *et al.*, 1993). Indeed, CT, Zot and Ace comprise a genetic virulence cassette that may contribute to the full pathogenic potential of *V. cholerae*. Against this background a recombinant strain, CVD110 (derived from El Tor Ogawa E7946), that lacks CT, Zot and Ace, was developed and tested in clinical trials. However, this strain, when given in doses of 10^8 colony-forming units (cfu) to adult healthy American volunteers, also induced diarrhoea in a majority of the vaccinees, suggesting that yet another unrecognized toxin may be responsible for the diarrhoea induced after ingestion of this strain (Tacket *et al.*, 1993). Although all of these recombinant cholera vaccines were too reactogenic to be used as oral vaccines, they all induced signficant protection against challenge with fully virulent cholera in human volunteers (Kaper *et al.*, 1997).

The first attenuated *V. cholerae* O1 vaccine strain to be well tolerated and yet highly immunogenic and protective in human volunteers was CVD103 (Levine *et al.*, 1988). This strain was derived from the wild-type *V. cholerae* O1 classical Inaba strain, 569B, which is pathogenic in volunteers, by deletion of the gene for the CT A subunit, *ctxA*. The parent strain was reported to lack Shiga-like toxin, thought to be involved in vaccine reactogenicity of previous recombinant *V. cholerae* vaccine strains (Kaper *et al.*, 1997). In addition, this parent strain also colonized the intestine at lower levels than other toxigenic *V. cholerae* strains. Immunization with CVD103 elicited seroconversion in a majority of the vaccinees, both against the bacteria and against CT, after a single oral dose of the vaccine. However, before CVD103 was tested in outpatient studies, genes encoding resistance to mercury were added to the strain to provide a marker that allows differentiation of the vaccine strain from wild-type *V. cholerae*. The resulting strain, CVD103 HgR, was considered to exhibit many of the characteristics of an ideal cholera vaccine and has subsequently been extensively studied in numerous phase-I, phase-II and, recently, phase-III clinical trials.

At the beginning of the 1990s several thousand subjects (aged 7 months–65 years) participated in placebo-controlled clinical trials of CVD103 HgR both in industrialized areas and in developing countries with epidemic or endemic cholera (Simanjuntak *et al.*, 1993; Kaper *et al.*, 1997). All of these studies confirmed that the vaccine was safe and caused no adverse reactions other than those observed in concurrently tested placebo recipients. In the different studies in industrialized countries a single dose of 5×10^8 cfu caused antibacterial antibody responses in approximately

90% of vaccinees. However, when this dose was used in endemic countries, very low seroconversion rates were observed. By increasing the vaccine dose 10-fold a marked increase in response rate was seen, with a single dose eliciting seroconversions in 75–85% of the vaccinees.

CVD103 HgR was also extensively tested in adult American volunteers. These studies showed very good protection against cholera induced by classical *V. cholerae*, i.e. 100% protection against severe disease and 76% against any diarrhoea. Highly significant, although somewhat lower protection, was observed after challenge with *V. cholerae* El Tor strains (49–67% irrespective of serotype). Based on these results, CVD103 HgR, which is manufactured by the Swiss Serum and Vaccine Institut, Berne, Switzerland, has been licensed for use in travellers in some European and Latin American countries under the trade name Orochol® (Levine and Svennerholm, 1997).

In 1993, a large, randomized, double-blind, placebo-controlled field trial was initiated in more than 67 000 volunteers (aged 2–42 years) in Jakarta, Indonesia to assess the efficacy of a single oral dose of CVD103 HgR. However, the results from this study unfortunately did not support the promising results from challenge studies in American volunteers during the four-year follow-up. Thus, the vaccine did not confer significant protection in the endemic population (Levine and Svennerholm, to be published).

The same genetic manipulations that were used to construct CVD110 were applied to another toxigenic El Tor strain N16117, as this parent strain was less virulent than other wild-type El Tor challenge strains (Kaper *et al.*, 1997). The resulting vaccine candidate, CVD111, which lacked the *ctxA*, *ace* and *zot*-genes, expressed the CTB subunit and mercury resistence. When tested in human volunteers this strain caused mild diarrhoea in only a few (12%) cases, and at the same time induced strong antibacterial and antitoxin responses in more than 90% of the immunized volunteers. Also, the vaccine strain afforded highly significant protection against cholera challenge (81% protective efficacy).

Mekalanos and co-workers have prepared a series of interesting live vaccine candidates generated from wild-type El Tor strains, which have been tested for safety and immunogenicity in volunteers (Taylor *et al.*, 1994; Coster *et al.*, 1995). These various vaccine constructs have in common the deletion of the whole virulence cassette containing the genes encoding CT, zot, ace as well as the RS1-element and attRS1; the latter sequences flank one core element, and are used in site-specific and homologous recombination. Similar to previous experience with attenuation of classical strains and the first-generation El Tor vaccine candidate strains, they gave rise to unacceptable adverse reactions with diarrhoea and usually additional gastrointestinal symptoms. More recently, however, a motility-deficient mutant strain which colonized less well, designated Peru15, was constructed and this has given much more promising results in clinical studies, with little reactogenicity and yet good protective immunogenicity (Taylor *et al.*, 1994; Coster *et al.*, 1995).

DEVELOPMENT OF VACCINES AGAINST ETEC DIARRHOEA

The findings of drastically decreased rates of ETEC diarrhoea in children in developing countries with increasing age (Black, 1986) and the decreased disease-to-infection rates observed in highly endemic areas (Cravioto *et al.*, 1988; López-Vidal *et al.*, 1990) have suggested that protective immunity against ETEC may develop naturally. Furthermore, travellers from industrialized countries who remain in less-developed countries for at least 1 year, and who are therefore likely to be repeatedly exposed to ETEC infections, exhibit significally lower incidence rates of ETEC disesase than newly arrived travellers, supporting the notion of protection by acquired immunity (Levine and Svennerholm, 1997).

A strong indication for the potential of inducing effective antitoxic immunity against ETEC disease in humans by vaccination is the finding that oral CTB-WC cholera vaccine through its CTB component, which cross-reacts immunologically with *E. coli* LTB, afforded significant protection against diarrhoea caused by LT-producing ETEC in the cholera vaccine trial in Bangladesh (Clemens *et al.*, 1988). The protection observed was about 67% (protective efficacy) for 3 months and was similar against ETEC-producing LT and against ETEC-producing LT in combination with ST. The oral CTB-WC cholera vaccine affords significant protection (50–60%) against LT-producing *E. coli* (LT or LT + ST strains) in placebo-controlled trials in Finnish travellers going to Morocco (Peltola *et al.*, 1991) and in American college students studying in Mexico for limited periods (Scerpella *et al.*, 1995).

Thus, there is strong support for the potential of developing effective ETEC vaccines for use in humans. A broad and strong protective efficacy is to be expected if the mucosal immunity against LT can be combined with antibacterial immunity directed against the predominant CFs on human ETEC strains. Different ETEC vaccine candidates have recently been considered based on these premises (e.g. inactivated or live vaccines that may provide both anti-colonization and antitoxic immunities).

Oral inactivated vaccines

Enterotoxoids

As purified CTB and LTB have both been shown to be strongly immunogenic and lack toxicity, it has been suggested that they may be suitable candidate antigens for providing anti-LT immunity. Furthermore, both types of B subunit are particularly well suited as oral immunogens, because they are stable in the gastrointestinal milieu and bind efficiently to the intestinal epithelium (Holmgren and Svennerholm, 1992). Immunization with purified LT or LTB protected rats and rabbits against challenge in ligated loops (Klipstein *et al.*, 1981; Svennerholm *et al.*, 1989; Tacket and Levine, 1997). Even though CTB provided significant protection against *E. coli* LT disease in animals as well as in humans (Clemens *et al.*, 1988; Peltola *et al.*, 1991; Holmgren and Svennerholm, 1992), it cannot be excluded that an LT toxoid may be slightly more effective than CTB in inducing protective anti-LT immunity. However, suitable methods for large-scale production of LTB are not yet available. Therefore, rCTB is currently used for immunoprophylaxis against ETEC, but may later be replaced by a more LT-like B subunit. For example, Lebens *et al.* (1996) have constructed hybrid proteins between the B subunits of LT and CT by substituting CTB amino acids with those at corresponding positions in LTB in order to obtain B subunits that display both cross-reactive and toxin-specific epitopes. Sera raised against such CTB/LTB hybrid proteins neutralize the toxic effects of both LT and CT better than sera raised against either CTB or LTB. Such hybrid B subunits are promising candidates for inclusion in an ETEC vaccine or a combined cholera and ETEC vaccine.

The significance of anti-ST immunity for protection against ST-producing *E. coli* has not been well documented. By coupling *E. coli* STa, either chemically or recombinantly, to different carrier proteins (e.g. CTB or bovine serum albumin), chimeric proteins that are capable of inducing ST-neutralizing antibodies have been achieved (Franz and Robertson, 1981; Svennerholm *et al.*, 1986; Sanchez *et al.*, 1988; Aitken and Hirst, 1993), although all of these ST conjugates have retained toxic activity. The possibility of preparing a non-toxic ST-peptide is supported by the finding that monoclonal antibodies that recognize the amino-terminal regions of STa, that are not essential for toxic activity, have a potent protective capacity (Takeda *et al.*, 1993). Although it is possible to prepare non-toxic ST-peptides, either by protein synthesis or by recombinant methods (Sanchez *et al.*, 1988; Svennerholm *et al.*, 1988; Aitken and Hirst 1993), immunization with such peptides coupled to different carrier proteins has usually failed to induce ST-neutralizing antibodies. However, Clements (1990) has prepared a genetically engineered LTB-ST fusion peptide that he reported was non-toxic and capable of inducing specific antibodies that recognize and neutralize native toxins. Even though it may be possible to induce high levels of ST-neutralizing antibodies, it is uncertain whether or not such a toxoid might play a significant role in an ETEC vaccine, as comparatively large amounts of anti-ST antibodies would be required to provide neutralization of the small ST-molecule.

CTB-CF ETEC vaccines

Based on such considerations, we have concluded that a practical way to construct an inactivated ETEC vaccine is to prepare killed ETEC bacteria that express the most important CFs on their surface and combine these organisms with an appropriate B subunit component, i.e. rCTB. In collaboration with SBL Vaccin, we have developed a CTB-containing CF whole-cell ETEC (CTB-CF ETEC) vaccine with the potential for providing broad protective coverage against ETEC diseases in different countries.

In initial studies a prototype vaccine consisting of a mixture of CTB and killed *E. coli* expressing CFA/I and the different CS components (CS1–CS3) of CFA/II was evaluated. The B subunit component was provided as conventionally purified CTB in the oral CTB-WC cholera vaccine. Strains that express the different CFs in high concentrations had been selected for preparation of the ETEC whole-cell component. The bacteria were inactivated by mild formalin treatment, which resulted in complete killing without significant loss of antigenicity of the different CFs (Svennerholm *et al.*, 1989).

The safety and immunogenicity of the prototype CTB-CF ETEC vaccine were demonstrated in Swedish volunteers given two or three oral doses of the vaccine at 2-week intervals. Surveillance for side-effects revealed that the vaccine did not give rise to any significant side-effects as tested in more than 100 adult Swedish volunteers (Wennerås *et al.*, 1992; Åhrén *et al.*, 1993). Most importantly, the vaccine induced significant mucosal IgA antibody responses locally in the small intestine against CFs and CTB in a majority of vaccinees (Åhrén *et al.*, 1993). The prototype ETEC vaccine also gave rise to significant increases in peripheral blood antibody-secreting cells (ASCs), with specificities for CFA/I, CFA/II and CTB in 85–100% of the volunteers (Wennerås *et al.*, 1992). Responses were predominantly in IgA-producing cells, but high frequencies of IgG ASCs against CTB and of IgM ASC responses against CFs were also seen. Two oral immunizations seemed to be optimal in inducing specicific mucosal immune responses (Wennerås et al., 1992; Åhrén et al., 1993).

Based on results from studies of the prototype ETEC vaccine in Sweden, a modified, more definitive, formulation of the ETEC vaccine was developed (Table 40.2). This vaccine contains recombinantly produced CTB (the same as in the rCTB-WC cholera vaccine) in combination with five different *E. coli* strains expressing CFA/I and the different subcomponents of CFA/II and CFA/IV, i.e. CS1–CS6 (Svennerholm et al., 1997). Based on a large number of epidemiological studies of CF and toxin profiles of ETEC in different geographical areas, this modified rCTB-CF ETEC vaccine has a potential protective coverage of at least 70–80%.

The rCTB-CF ETEC vaccine has been evaluated for safety and immunogenicity in several phase-I and phase-II trials in different countries. Peroral immunization with one, or in most instances two, doses of the vaccine 2 weeks apart of more than 1000 Swedish, Bangladeshi, Egyptian, Israeli, Swiss, Austrian or American adult volunteers, and recently of more than 150 Egyptian children in the age group 6 months–10 years, has shown that the vaccine is safe. The capacity of the rCTB-CF vaccine to induce a mucosal immune response has been assessed by determining ASC responses in peripheral blood against CTB, as well as the different vaccine CFs, as our recent studies have suggested that peripheral blood ASC responses may be good proxy measures of intestinal immune responses (Åhrén et al., 1998). The rCTB-CF ETEC vaccine induced high frequencies of ASC responses against CTB as well as the different CFs of the vaccine in a majority of the immunized Swedish, Egyptian and Bangladeshi adult volunteers (Jertborn et al., 1998; Savarino et al., 1998; Qadri et al., 1999). Interestingly, frequencies of ASC responses in peripheral blood to CTB and the different CFs after one or two vaccine doses were very similar, with 70–100% against the different antigens in the different groups of adult volunteers.

The promising results obtained from the studies of the rCTB-CF ETEC vaccine have encouraged the initiation of different phase-III trials of the vaccine in travellers to ETEC-endemic areas. Thus, a study has been undertaken in European travellers to different countries in Asia, Africa and Latin America with very promising preliminary results (Wiedermann et al., 1999) and another one has been initiated in Swiss and German travellers going to Kenya (R. Steffen et al., Zürich, Switzerland). Yet another study was recently initiated in American students going to Guatemala for summer courses (D. Sack et al., Baltimore, USA). In these different trials adult volunteers are given two doses of the vaccine or an *E. coli* K12 placebo in a double-blind fashion, and the incidence of ETEC disease in the two study groups is evaluated during follow-up. A study

was also initiated in Egypt in early 1999 (S. Savarino et al., Cairo, Egypt) to test the vaccine for protective efficacy against ETEC diarrhoea in children aged 6–18 months. This phase-III trial has been preceded by extensive phase-II trials in Egypt, which have confirmed that the vaccine is safe and immunogenic in children aged 2–10 years (Savarino et al., 1999), and recently also in infants aged 6–18 months (Savarino et al., to be published).

The results from these different phase-III trials may support the use of an inactivated toxoid containing ETEC vaccine for immunoprophylaxis against traveller's diarrhoea caused by ETEC, as well as for use as a public health tool to control the most prevalent form of diarrhoea in children in developing countries.

Oral live vaccines

Live bacteria producing LTB and expressing the major CFs may also be considered as ETEC vaccine candidates. If such strains could effectively colonize and multiply in the gut and produce substantial levels of LTB while multiplying in the small intestine, they might provide a sustained antigen stimulation for the local intestinal immune system (Levine, 1990). Such strains could only provide antibacterial protection against ST-only producing strains, as recombinant strains producing non-toxic immunogenic ST conjugates are not available. Thus, any live vaccine against *E. coli* ST disease should ideally provide immunity against the most prevalent CFs. However, as the different CFs are normally not expressed in the same strains, and it has not yet been possible to clone successfully the genes for more than two or three different CFs in the same host organisms, such vaccines must, at least for the time being, be based on a cocktail of several different strains. With any mixed vaccine, however, there is a risk of overgrowth of one of the included vaccine strains with suppression of the others. Other potential drawbacks of live ETEC vaccines may be the risk of reversion to toxicity by uptake of toxin-encoding plasmids/genes, or that they would produce insufficient LTB during growth *in vivo* to induce significant antitoxic immunity.

Attenuated live ETEC

The potential of live ETEC vaccines was demonstrated when a prototype live vaccine strain expressing CS1 and CS3 fimbriae, but lacking genes that encode LT and ST, was given to human volunteers (Levine, 1990). By providing a single dose of this attenuated strain, SIgA CF antibodies were induced locally in the intestine and 75% protection against experimental challenge with wild-type ETEC-producing CS1 and CS3 fimbriae, as well as LT and ST, was induced. Based on these findings, one strategy may be to administer a collection of attenuated *E. coli* strains expressing the major fimbrial CFs and an LT antigen such as LTB or mutant LT (Levine and Svennerholm, 1997).

Recombinant strains

Another approach for the preparation of a live ETEC vaccine may be to introduce LTB as well as CF-encoding plasmids into heterologous bacteria, such as attenuated Salmonellae which, because of their invasive properties, can reside in the bowel for long periods (Tacket and Levine, 1997). Recently developed attenuated *S. typhi* strains, which are immunogenic when administered as a single dose, may be used. Thus, cloned genes for the expression of CFA/I and CS3 have been introduced on stable plasmids into such an *S. typhi* strain (CVD 908), and a high level of co-expression of these fimbriae has been demonstrated (Girón *et al.*, 1995). Attempts to clone LTB into such strains have not been reported. Other attenuated bacterial strains, such as Shigellae or *V. cholerae*, have also been considered as live vectors for the expression of ETEC CFs and LTB or CTB. Thus, expression of CFA/I and CS3 fimbriae in an attenuated *S. flexneri* 2a live vaccine candidate has been reported (Noriega *et al.*, 1996). However, a problem with these various approaches appears to be the exposure of several CFs together with LTB in the same host organism.

PROSPECTS FOR VACCINES AGAINST VTEC DISEASE

The possibility for immunoprophylaxis against VTEC is suggested by experimental studies showing that parenteral immunization of rabbits with the B subunits of VT1 protected the animals from intravenous challenge with VT1 toxin (Boyd *et al.*, 1991). Furthermore, parenteral immunization of rabbits with VT1 B subunits protected them against haemorrhagic colitis induced by a VT1-producing *E. coli* strain (Noël *et al.*, 1994).

Based on these results it may be suggested that parenteral immunization with a VT toxoid may be useful for preventing EHEC disease in high-risk groups. Another possibility may be an oral vaccine containing a mixture of VTEC bacterial antigens (e.g. fimbriae), O157 LPS and attaching/effacing factors in combination with a VT toxoid, consisting of both VT1 and VT2 antigens. Such a vaccine could be given either orally or parenterally. Another approach may be to prepare a conjugate between 0157 LPS and a VT toxoid for parenteral use (SZU, NIH, Bethesda, personal communication).

ENTEROTOXINS AND B SUBUNITS AS IMMUNOMODULATING AGENTS: PROSPECTS FOR DEVELOPMENT OF NOVEL MUCOSAL ANTI-INFECTIOUS AND ANTI-INFLAMMATORY (TOLEROGENIC) VACCINES

As outlined in the introduction to this chapter, and partly as a result of the promising development of oral vaccines against cholera and related enteric toxigenic diseases described above, mucosal immunization (especially by the oral but also to a lesser extent by the intranasal routes) has recently attracted much interest as a means of eliciting protective immunity against other mucosal infections, including gastrointestinal, respiratory and urogenital infections. Furthermore, a kind of mucosal immunization, now with the intention to induce tolerance rather than an immune response, has also emerged as a possible approach for immunological treatment of various diseases caused by an aberrant immune response associated with tissue-damaging inflammation. Therefore, mucosal vaccination and mucosal-tolerance induction represent interesting approaches to protect an individual against, respectively, mucosal infectious agents and inflammatory 'immunopathologies' as seen, for example, in selected chronic infections, autoimmune disorders and allergies, respectively (Czerkinsky and Holmgren, 1995). Interestingly enough, CT and LT, and their B subunits CTB and LTB, have emerged as key immunomodulating molecules in relation to these objectives and serve as both models and promising vehicles for achieving the goals of either stimulating a strong mucosal immune response to an infectious agent or inducing an anti-inflammatory 'oral tolerance' immunotherapeutic response (Holmgren *et al.*, 1996).

IMMUNOMODULATION BY CHOLERA TOXIN AND CTB

In contrast to most other soluble proteins, both cholera toxin (CT) and its non-toxic binding subunit CTB have proved to be excellent mucosal immunogens, as well as efficient mucosal carrier–delivery systems for linked foreign antigens. The corresponding *E. coli* enterotoxin proteins, LT and LTB, have analogous immunomodulating properties. CT and LT, in contrast to CTB and LTB, can markedly increase the mucosal immunogenicity of admixed rather than linked antigens. However, whilst both CT/LT and CTB/LTB work to increase immune responses at

TABLE 40.3 Examples of antigens for which conjugation to CTB has markedly increased their mucosal immunogenicity and led to enhanced local IgA antibody responses (Holmgren *et al.*, 1996)

- Model proteins, e.g. horseradish peroxidase (HRP); human gamma globulin (HGG); ovalbumin (OVA)
- Model polysaccharides, e.g. dextrans
- Candidate vaccine antigens:
  Bacterial, e.g. *Streptococcus mutans* antigen I/II; *Chlamydia trachomatis* major outer membrane protein antigen A8-VDIV peptide; *Haemophilus influenzae* type B capsular polysaccharide
  Viral, e.g. Sendai virus, hepatitis B virus peptide; human immunodeficiency virus-1 gp120 peptides; rotavirus peptides
  Parasitic, e.g. *Schistosoma mansoni* GST antigen; *Entamoeba histolytica*

mucosal surfaces, they have opposite effects with regard to the induction of systemic oral tolerance: the holotoxins inhibit the development of oral tolerance, whereas the B subunits induce oral tolerance for linked antigens (Holmgren *et al.*, 1996).

The strong immunomodulating properties of these various proteins can be attributed to the fact that both CT/LT and CTB/LTB are resistant to degradation by mucosal enzymes, can bind with high affinity to mucosal epithelial and lymphoid cells including M-cells, antigen presenting cells (APCs) and B- and T-cells, and have inherent immunomodulating activity on APCs, B- and T-cells (Holmgren *et al.*, 1993; Lycke, 1997). The partly differential immunomodulating properties of CT/LT vs CTB/LTB in their turn can most probably be explained by the inherent biological activity of the A subunit of the holotoxins (CTA and LTA). However, it remains unclear whether it is only the ADP-ribosylating activity or also other features of CTA/LTA that mediate the additional and/or opposite immuno-modulating effects of the toxins compared with the B subunits.

Indeed, a very active area of research recently has been to prepare mutant holotoxins of LT or CT with defined modifications in the A subunits with the objective of eliminating toxic activity without removing the adjuvant activity. A number of LT mutants retains at least partial adjuvant activity for admixed foreign antigens when given intranasally or orally to mice (De Magistris *et al.*, 1998). It has not yet been possible to retain good adjuvant activity of such mutated toxins when given by the oral route: there seems to exist a more stringent dependence on the ADP-ribosylating/cyclic AMP-inducing activity of the toxins when they are given orally compared with nasally. To overcome this problem, an alternative approach has been to link the ADP-ribosylating intact CTA subunit to a different cell-binding molecule to avoid the enterotoxicity while binding to and presenting the A subunit to appropriate antigen-presenting and antigen-responding cells in the mucosa (Ågren *et al.*, 1998). To date, however, it seems that these molecules are active

mainly, or exclusively, when given nasally rather than orally, most probably due to the more leaky nasal mucosa which allows easier transmucosal penetration of these molecules.

CTB AS A CARRIER FOR FOREIGN VACCINE ANTIGENS

CTB is an important protective antigen component of recently developed oral vaccines against cholera and enterotoxigenic *E. coli* diarrhoea. The coupling of immunogens to CTB can dramatically enhance their ability to induce mucosal immune responses at mucosal surfaces. McKenzie and Halsey (1984) first reported that covalent coupling of horseradish perox-idase to CTB markedly enhanced gut immune responses to the peroxidase antigen after oral admin-istration. This observation has since been confirmed and extended to a large number of antigens, as exemplified in Table 40.3.

By recombinant DNA techniques, efficient systems are now in place for the construction and large-scale production of hybrid mucosal vaccine candidate pro-teins based on CTB to which various foreign antigens are fused in different locations, either directly to CTB or indirectly via a modified CTA subunit (Lebens and Holmgren, 1994; Bäckström *et al.*, 1995).

CTB AND ORAL TOLERANCE

Mucosally induced immunological tolerance has been proposed as a strategy to prevent or to reduce the inten-sity of allergic reactions or autoimmune diseases, and some promising clinical results have been reported (Weiner *et al.*, 1994). However, its therapeutic potential has remained limited, mainly due to the fact that cur-rent protocols of mucosally induced tolerance have had limited success in suppressing the expression of an already established state of systemic immunological sensitization.

Oral administration of microgram amounts of different prototype antigens conjugated to CTB, the non-toxic receptor-binding moiety of cholera toxin, can readily induce tolerance in primarily the peripheral T-cell compartment and is effective in both naive and systemically sensitized animals (Sun *et al.*, 1994). Furthermore, oral administration of minute amounts of an autoantigen, myelin basic protein (MBP), coupled to CTB can prevent experimental allergic encephalomyelitis (EAE) in Lewis rats and, most importantly, the oral CTB-MBP was effective when given after the induction of EAE (Sun *et al.*, 1996). In another auto-immune disease model, spontaneously developing diabetes in adult non-obese diabetic (NOD) mice, it was found that feeding a small amount of insulin conjugated to CTB (insulin–CTB) significantly delayed the onset of clinical disease. The protective effect could be adoptively transferred with T-cells from animals fed insulin–CTB in syngenic recipients co-injected with diabetogenic T-cells, and correlated with reduced lesions of insulitis (Bergerot *et al.*, 1997). Promising results against type-II collagen-induced arthritis in DBA mice have also been obtained by intranasal administration of low doses of type-II collagen–CTB conjugate. Both the incidence and severity of disease development were much reduced, and even when treatment was delayed until disease manifestations were evident, a significant effect was achieved (Tarkowski *et al.*, 1998). These results indicate that mucosal administration of autoantigens conjugated to CTB may provide a future treatment approach for several important autoimmune diseases such as multiple sclerosis, type-I diabetes and rheumatoid arthritis. Furthermore, in other systems we have also been able partially to suppress liver granuloma formation and mortality in experimental murine schistosomiasis (Sun *et al.*, 1999). Based on this, the development of therapeutic anti-inflammatory mucosal vaccines is a feasible adjunct to the more conventional concept of largely preventive anti-infectious mucosal vaccines.

ACKNOWLEDGEMENTS

Financial support for these studies was provided by the Swedish Medical Research Council, SIDA-SAREC (Sweden) and the World Health Organization.

REFERENCES

Ågren, L., Löwenadler, B. and Lycke, N. (1998). A novel concept in mucosal adjuvanticity: the CTA1-DD adjuvant is a B cell-targeted fusion protein that incorporates the enzymatically active cholera toxin A1 subunit. *Immunol. Cell. Biol.* **76**, 280–7.

Åhrén, C., Wennerås, C., Holmgren, J. and Svennerholm, A.-M. (1993). Intestinal antibody response after oral immunization with a prototype enterotoxigenic *Escherichia coli* vaccine. *Vaccine* **11**, 929–34.

Åhrén, C., Jertborn, M., Holmgren, J. and Svennerholm, A.-M. (1998). Intestinal immune responses and associated immunological A responses to an oral inactivated enterotoxigenic *Escherichia coli* vaccine. *Infect. Immun.* **66**, 3311–16.

Aitken, R. and Hirst, T.R. (1993). Recombinant enterotoxin as vaccines against *Escherichia coli*-modified diarrhoea. *Vaccine* **11**, 227–33.

Bäckström M., Holmgren, J., Schödel, F. and Lebens, M. (1995). Characterization of an internal permissive site in the cholera toxin B-subunit and insertion of epitopes from human immunodeficiency virus-1, hepatitis B virus and enterotoxigenic *Escherichia coli*. *Gene* **165**, 163–71.

Begue, R.E., Castellares, G., Ruiz, R., Hayashi, K.E., Sanchez, J.L., Gotuzzo, E., Bourgeois, A.L., Oberst, R.B., Taylor, D.N. and Svennerholm, A.-M. (1995). Community-based assessment of safety and immunogenicity of the whole cell plus recombinant B subunit (WC/rBS) oral cholera vaccine in Peru. *Vaccine* **23**, 25–33.

Bergerot, I., Ploix, C., Petersen, J., Moulin, V., Rask, C., Fabien, N., Lindblad, M., Mayer, A., Czerkinsky, C., Holmgren, J. and Thivolet, C. (1997). A cholera toxoid-insulin conjugate as an oral vaccine against spontaneous autoimmune diabetes. *Proc. Natl Acad. Sci. USA* **94**, 4610–14.

Black, R.E. (1986). The epidemiology of cholera and enterotoxigenic *E. coli* diarrheal disease. In: *Development of Vaccines and Drugs Against Diarrhea* (eds J. Holmgren, A. Lindberg and R. Möllby), 11th Nobel Conference, pp. 23–32. Studentlitteratur, Stockholm.

Black, R.E. (1990). Epidemiology of traveller's diarrhea and relative importance of various pathogens. *Rev. Infect. Dis.* **12**, S73–9.

Black, R.E., Levine, M.M., Clemens, M.L., Young, C.R., Svennerholm, A.-M. and Holmgren, J. (1987). Protective efficacy in man of killed whole vibrio oral cholera vaccine with and without the B subunit of cholera toxin. *Infect. Immun.* **77**, 1116–29.

Blake, P.A. (1994). Historical perspectives on pandemic cholera. In: *Vibrio cholerae and Cholera: Molecular to Global Perspectives* (eds I.K.Wachsmuth, P.A. Blake and Ø. Olsvik), pp. 293–5. American Society for Microbiology, Washington, DC.

Boyd, B., Richardson, S. and Gariépy, J. (1991). Serologic response to the B-subunit of Shiga-like toxin I and its peptide fragments indicate that the B-subunit is a vaccine candidate to counter the action of toxin. *Infect. Immun.* **59**, 750–7.

Burgess, M.N., Bywater, R.J., Corley, C.M., Mullan, M.N. and Newsome, P.M. (1978). Biological evaluation of a methanol soluble, heat-stable *Escherichia coli* enterotoxin in infant mice, pigs, rabbits, and calves. *Infect. Immun.* **21**, 526–31.

Clemens, J., Sack, D.A., Harris, J.R., Chakraborty, J., Khan, M.R., Stanton, B.F., Kay, B.A., Khan, M.U., Yunus, M.D., Atkinson, W., Svennerholm, A.-M. and Holmgren, J. (1986). Field trial of oral cholera vaccines in Bangladesh. *Lancet* **i**, 124–7.

Clemens, J., Sack, D.A., Harris, J.R., Chakraborty, J., Neogy, P.K., Stanton, F., Huda, N., Khan, M.U., Kay, B.A., Khan, M.R., Ansurazzaman, M., Yunus, M., Rao, M.R., Svennerholm, A.-M. and Holmgren, J. (1988). Cross-protection by B subunit-whole cell cholera vaccine against diarrhea associated with heat-labile toxin-producing enterotoxigenic *Escherichia coli*: results of a large-scale field trial. *J. Infect. Dis.* **158**, 372–7.

Clemens, J.D., Sack, D.A., Harris, J.R., van Loon, F., Chakraborty, J., Ahmed, F., Rao, M.R., Khan, M.R., Yunus, M.D., Huda, N., Stanton, B.F., Kay, B.A., Walter, S., Eeckels, R., Svennerholm, A.-M. and Holmgren, J. (1990). Field trial of oral cholera vaccines in Bangladesh: results from three-year follow-up. *Lancet* **355**, 270–3.

Clements, J.D. (1990). Construction of a nontoxic fusion peptide for immunization against *Escherichia coli* strains that produce heat-labile and heat-stable enterotoxins. *Infect. Immun.* **58**, 1159–66.

Coster, T.S., Killeen, K.P., Waldor, M.K., Beattie, D.T., Spriggs, D.R., Kenner, J.R., Trofa, A., Sadoff, J.C., Mekalanos, J.J. and Taylor, D.N. (1995). Safety, immunogenicity, and efficacy of live attenuated *Vibrio cholerae* O139 vaccine prototype. *Lancet* **345**, 949–52.

Craig, J.P. (1980). A survey of the enterotoxic enteropathies. In: *Cholera and Related Diarrheas, Molecular Aspects of a Global Health Problem* (eds Ö. Ouchterlony and J. Holmgren), 43rd Nobel Symposium, pp. 15–25. Karger, Basel.

Cravioto, A., Reyes, R.E., Ortega, R., Fernandéz, G., Hernandez R. and López, D. (1988). Prospective study of diarrhoeal diseases in a cohort of rural Mexican children: incidence and isolated pathogens during the first two years of life. *Epidemiol. Infect.* **101**, 123–34.

Czerkinsky, C. and Holmgren, J. (1995). The mucosal immune system and prospects for anti-infectious and anti-inflammatory vaccines. *Immunologist* **3**, 97–103.

De Magistris, M.T., Pizza, M., Douce, G., Ghiara, P., Dougan, G. and Pappuoli, R. (1998). Adjuvant effect of non-toxic mutants of *E. coli* heat-labile enterotoxin following intranasal, oral and intravaginal immunization. *Dev. Biol. Stand.* **92**, 123–6.

Evans, D.J. and Evans, D.G. (1989). Determinants of microbial attachment and their genetic control. In: *Enteric Infection. Mechanisms, Manifestations and Management* (eds M.J.G. Farthing and G.T. Keusch), pp. 31–40. Chapman & Hall, London.

Fasano, A., Baudry, B., Pumplin, D.W., Wasserman, S.S., Tall, B.D., Ketley, J.M. and Kaper, J.B. (1991). *Vibrio cholerae* produces a second enterotoxin, which affects intestinal tight junctions. *Proc. Natl Acad. Sci. USA* **88**, 5242–6.

Frantz, J.C. and Robertson, D.C. (1981). Immunological properties of *Escherichia coli* heat-stable enterotoxins: development of a radioimmunoassay specific for heat-stable enterotoxins with suckling mouse activity. *Infect. Immun.* **33**, 193–8.

Gaastra, W. and Svennerholm, A.-M. (1996). Colonization factors of human enterotoxigenic *Escherichia coli* (ETEC). *Trends Microbiol.* **4**, 444–52.

Girón, J.A., Xu, J.-G., González, C.R., Hone, D.M., Kaper, J.B. and Levine, M.M. (1995). Simultaneous expression of CFA/I and CS3 colonization factor antigens of enterotoxigenic *Escherichia coli* by ?aroC, ?aroD *Salmonella typhi* vaccine strain CVD 908. *Vaccine* **13**, 939–46.

Glass, R., Svennerholm, A.-M., Stoll, B.J., Khan, M.R., Hossain, K.M.B., Huq, M.I. and Holmgren, J. (1983). Protection against cholera in breast-fed children by antibodies in breast milk. *N. Engl. J. Med.* **308**, 1389–92.

Guerrant, R.L. (1985). Microbial toxins and diarrhoeal disease: introduction and overview. In: *Microbial Toxins and Diarrhoeal Disease* (eds D. Evered and J. Whelan), Ciba Foundation Symposium 112, pp. 1–13. Pitman, London.

Holmgren, J. (1981). Actions of cholera toxin and the prevention and treatment of cholera. *Nature* **292**, 413–17.

Holmgren, J. and Svennerholm, A.-M. (1992). Bacterial enteric infections and vaccine development. In: *Mucosal Immunology, Gastroenterology Clinics of North America* (eds R.P. McDermott and C.O. Elson), pp. 283–302. W.B. Saunders, Philadelphia, PA.

Holmgren, J., Svennerholm, A.-M., Lönnroth, I., Fall-Persson, M., Markman, B. and Lundbäck, H. (1977). Development of improved cholera vaccine based on subunit toxoid. *Nature* **269**, 602–4.

Holmgren, J., Lycke, N. and Czerkinsky, C. (1993). Cholera toxin and cholera B subunits as oral-mucosal adjuvant and antigen vector systems. *Vaccine* **11**, 1179–84.

Holmgren, J., Czerkinsky, C., Sun, J.-B. and Svennerholm, A.-M. (1996). Oral vaccination, mucosal immunity and oral tolerance with special reference to cholera toxin. In: *Concepts in Vaccine Design* (ed. S. Kaufmann), pp. 437–58. Berlin.

Holmgren, J., Jertborn, M. and Svennerholm, A.-M. (1997). New and improved vaccines against cholera: oral B subunit killed whole-cell cholera vaccine. In: *New Generation Vaccines*, 2nd edn (eds M.M. Levine, G.C. Woodrow, J.B. Kaper and G.S. Cobon), pp. 459–68. Marcel Dekker, New York.

Jertborn, M., Svennerholm, A.-M. and Holmgren, J. (1992). Safety and immunogenicity of an oral recombinant cholera B subunit-whole cell vaccine in Swedish volunteers. *Vaccine* **10**, 130–2.

Jertborn, M., Svennerholm, A.-M. and Holmgren, J. (1996). Intestinal and systemic immune responses in humans after oral immunization with a bivalent B subunit-O1/O139 whole cell cholera vaccine. *Vaccine* **14**, 1459–65.

Jertborn, M., Åhrén, C., Holmgren J. and Svennerholm, A.-M. (1998). Safety and immunogenicity of an oral inactivated enterotoxigenic *Escherichia coli* vaccine. *Vaccine* **16**, 255–60.

Jodal, M. and Lundgren, O. (1995). Neural reflex modulation of intestinal epithelial transport. In: *Regulatory Mechanisms in Gastrointestinal Function* (ed. T.S. Gaginella), pp. 99–144. CRC Press, Boca Raton, FL.

Jonsson, G., Holmgren, J. and Svennerholm, A.M. (1991). Identification of a mannose-binding pilus on *Vibrio cholerae* El Tor. *Microb. Pathogen.* **11**, 433–41.

Jonsson, G., Osek, J., Svennerholm, A.-M. and Holmgren, J. (1996). Immune mechanisms and protective antigens of *Vibrio cholerae* serogroup O139 as a basis for vaccine development. *Infect. Immun.* **64**, 3778–85.

Kaper, J.B., Lockman, H., Baldini, M.M. and Levine, M.M. (1984). Recombinant nontoxigenic *Vibrio cholerae* strains as attenuated cholera vaccine candidates. *Nature* **308**, 655–8.

Kaper, J.B., Tacket, C.O. and Levine, M.M. (1997). New and improved vaccines against cholera: attenuated *Vibrio cholerae* O1 and O139 strains as live oral cholera vaccines. In: *New Generation Vaccines*, 2nd edn (eds M.M. Levine, G.C. Woodrow, J.B. Kaper and G.S. Cobon), pp. 447–58. Marcel Dekker, New York.

Karmali, M.A. (1989). Infection by Verocytotoxin-producing *Escherichia coli*. *Clin. Microbiol. Rev.* **2**, 15–38.

Klipstein, F.A., Engert, R.E. and Clements, J.D. (1981). Immunization of rats with heat-labile enterotoxin provides uniform protection against heterologous serotypes of enterotoxigenic *Escherichia coli*. *Infect. Immun.* **32**, 1100–4.

Lebens, M. and Holmgren, J. (1994). Mucosal vaccines based on the use of cholera toxin B subunit as immunogen and antigen carrier. Symposium on Recombinant vectors in vaccine development, Albany, NY. *Dev. Biol. Stand.* **82**, 215–27.

Lebens, M., Shahabi, V., Bäckström, M., Houze, T., Lindblad, M. and Holmgren, J. (1996). Synthesis of hybrid molecules between heat-labile entertoxin and cholera toxin B subunits: potential for use in broad spectrum vaccine. *Infect. Immun.* **64**, 2144–50.

Levine, M.M. (1990). Vaccines against enterotoxigenic *Escherichia coli* infections. In: *New Generation Vaccines* (eds G.C. Woodrow and M.M. Levine), pp. 649–60. Marcel Dekker, New York.

Levine, M.M. and Svennerholm, A.-M. (1997). Future enteric vaccines. In: *Travel Medicine* (eds H. DuPont and R. Steffen), pp. 169–77. Marcel Dekker, New York.

Levine, M.M., Kaper, J.B., Herrington, D., Losonsky, G., Morris, J.G., Clements, M., Black, R.E., Tall, B. and Hall, R. (1988). Safety, immunogenicity and efficacy of recombinant live oral cholera vaccine CVD103 and CVD103-HgR. *Lancet* **ii**, 467–70.

López-Vidal, Y., Calva, J.J., Trujillo, A., de León, A.P., Ramos, A., Svennerholm, A.-M. and Ruiz-Palacios, S. (1990). Enterotoxins and adhesins of enterotoxigenic *Escherichia coli*: are they risk factors for acute diarrhea in the community? *J. Infect. Dis.* **162**, 442–4.

Lycke, N. (1997). The mechanism of cholera toxin adjuvanticity. *Res. Immunol.* **148**, 504–20.

McKenzie, S.J. and Halsey, J.F. (1984). Cholera toxin B subunit as a carrier protein to stimulate a mucosal immune response. *J. Immunol.* **133**, 1818–25.

Morris, J.G. (1994). *Vibrio cholerae* O139 Bengal. In: Vibrio cholerae *and* Cholera: Molecular to Global Perspectives (eds I.K. Wachsmuth, P.A. Blake and Ø. Olsvik), pp. 95–102. American Society for Microbiology, Washington, DC.

Nair, G.B. and Takeda, Y. (1997). The heat-labile and heat-stable enterotoxins of *Escherichia coli*. In: Escherichia coli: *Mechanisms of Virulence* (ed. M. Sussman), pp. 237–56. Cambridge University Press, Cambridge.

Neutra, M.R. and Kraehenbühl, J.-P. (1992). Transepithelial transport and mucosal defence. The role of M cells. *Trends Cell Biol.* **2**, 134–8.

Noël, J.M. and Boedeker, E.C. (1997). Enterohemorrhagic *Escherichia coli*: a family of emerging pathogens. *Dig. Dis.* **15**, 67–91.

Noël, J.M., Mullett, C., Acheson, D., Lowell, G., Tseng, L., Fleming, E., Boedeker, E. and Latimer, J. (1994). Parenteral immunization with Shiga-like toxin B subunit in proteosomes protects in a model of enterohemorrhagic *E. coli* colitis. *J. Pediatr. Gastroenterol. Nutr.* **19**, A89.

Noriega, F.R., Losonsky, G., Wang, J.Y., Formal, S.B. and Levine, H.M. (1996). Further characterization of ΔaroA ΔvirG *Shigella flexneri* 2a strain CVD 1203 as a mucosal *Shigella* vaccine and as a live-vector vaccine for delivering antigens of enterotoxigenic *Escherichia coli*. *Infect. Immun.* **64**, 23–7.

Osek, J., Svennerholm, A.-M. and Holmgren, J. (1992). Protection against *Vibrio cholerae* El Tor infection by specific antibodies against mannose-binding hemagglutinin pili. *Infect. Immun.* **60**, 4961–4.

Peltola, H., Siitonen, A., Kyrönseppä, H., Simula, I., Mattila, L., Oksanen, P., Kataja, M.J. and Cadoz, M. (1991). Prevention of travellers' diarrhoea by oral B-subunit/whole cell cholera vaccine. *Lancet* **338**, 1285–9.

Qadri, F., Wennerås, C., Bardhan, P.K., Hossain, J., Sack, R.B. and Svennerholm, A.-M. (1999). B cell responses to enterotoxigenic *Escherichia coli* (ETEC) induced by oral vaccination and natural disease. Submitted for publication.

Quiding, M., Nordström, I., Kilander, A., Andersson, G., Hanson, L.-Å., Holmgren, J. and Czerkinsky, C. (1991). Intestinal immune responses in humans. Oral cholera vaccination induces strong intestinal antibody responses, gamma-interferon production, and evokes local immunological memory. *J. Clin. Invest.* **88**, 143–8.

Riley, L.W., Remis, R.S., Helgerson, S.D., McGee, H.B., Wells, J.G., Davis, B.R., Hebert, R.J., Olcott, E.S., Johnson, L.M., Hargrett, N.T., Blake, P.A. and Cohen, M.L. (1983). Hemorrhagic colitis associated with a rare *Escherichia coli* serotype. *N. Engl. J. Med.* **308**, 681–5.

Sanchez, J. and Holmgren, J. (1989). Recombinant system for overexpression of cholera toxin B subunit in *Vibrio cholerae* as a basis for vaccine development. *Proc. Natl Acad. Sci. USA* **86**, 481–5.

Sanchez, J., Svennerholm, A.-M. and Holmgren, J. (1988). Genetic fusion of a non-toxic heat-stable enterotoxin-related decapeptide antigen to cholera toxin B-subunit. *FEBS Lett.* **241**, 110–14.

Sanchez, J.L., Trofa, A.F., Taylor, D.N., Kuschner, R.A., DeFraites, R.F., Craig, S.C., Rao, M.R., Clemens, J.D., Svennerholm, A.-M., Sadoff, J.C. and Holmgren, J. (1993). Safety and immunogenicity of the oral, whole cell/recombinant B subunit cholera vaccine in North American volunteers. *J. Infect. Dis.* **167**, 1446–9.

Sanchez, J.L., Vasques, B., Begue, R.E., Meza, R., Castellares, G., Cabezas, C., Watts, D.M., Svennerholm, A.-M., Sadoff, J.C. and Taylor, D.N. (1994). Protective efficacy of the oral, whole cell/recombinant B subunit cholera vaccine in Peruvian military recruits. *Lancet* **344**, 1273–6.

Savarino, S.J., Brown, F.M., Hall, E., Bassily, S., Youssef, F., Wierzba, T., Peruskil, El-Masry, N.A., Safwat, M., Rao, M., Bourgeois, A.L., Jertborn, M., Svennerholm, A.-M., Lee, Y. and Clemens, J.D. (1998). Safety and immunogenicity of an oral, killed enterotoxigenic *Escherichia coli*-cholera toxin B subunit vaccine in Egyptian adults. *J. Infect. Dis.* **177**, 796–9.

Savarino, S.J., Hall, E.R., Bassily, S., Brown, M., Youssef, F., Wierzba, T.F., Peruskil, El-Masry, N.A., Safwat, M., Rao, M., Eng, M., Svennerholm, A.-M., Lee, Y.J. and Clemens, J.D. (1999). Oral, whole cell enterotoxigenic *Escherichia coli* plus cholera toxin B subunit vaccine: results of the initial evaluation in children. *J. Infect. Dis.* **179**, 107–14.

Scerpella, E.G., Sanchez, J.L., Mathweson, J.J., Torres-Cordero, J.V., Sadoff, J.C., Svennerholm, A.-M., DuPont, H.L., Taylor, D.N. and Ericsson, C.D. (1995). Safety, immunogenicity and protective efficacy of the whole-cell/recombinant B subunit (WC/rBS) oral cholera vaccine against travelers' diarrhea. *J. Travel Med.* **2**, 22–7.

Scotland, S.M., Rowe, B., Smith, H.R., Willshaw, G.A. and Gross, R.J. (1988). Vero cytotoxin-producing strains of *Escherichia coli* from children with haemolytic uramic syndrome and their detection by specific DNA probes. *J. Med. Microbiol.* **25**, 237–43.

Simanjuntak, C.H., O'Hanley, P., Punjabi, N.H., Noriega, F., Pazzaglia, G., Dykstra, P., Kay, B., Suharyono, Budiarso, A., Rifai, A.R., Wasserman, S.S., Losonsky, G., Kaper, J., Cryz, S. and Levine, M.M. (1993). Safety, immunogenicity, and transmissibility of single-dose live oral cholera vaccine strain CVD 103-HgR in 24- to 59-month-old Indonesian children. *J. Infect. Dis.* **168**, 1169–76.

Strockbine, N.A., Marques, L.R.M., Holmes, R.K. and O'Brien, A.D. (1986). Two toxin-converting phages from *Escherichia coli* O157:H7 strain 933 encode antigenically distinct toxins with similar biological activities. *Infect. Immun.* **53**, 135–40.

Sun, J.-B., Holmgren, J. and Czerkinsky, C. (1994). Cholera toxin B subunit. An efficient transmucosal carrier delivery system for induction of peripheral immunological tolerance. *Proc. Natl Acad. Sci. USA* **91**, 10795–9.

Sun, J.-B., Rask, C., Olsson, T., Holmgren, J. and Czerkinsky, C. (1996). Treatment of experimental autoimmune encephalomyelitis by feeding myelin basic protein conjugated to cholera toxin B subunit. *Proc. Natl Acad. Sci. USA* **93**, 7196–201.

Sun, J.-B., Mielcarek, N., Lakew, M., Grzych, J.-M., Capron, A., Holmgren, J. and Czerkinsky, C (1999). Intranasal administration of a *Schistosoma mansoni* glutathione S transferase-cholera B subunit conjugate vaccine evokes anti-parasitic pathology immunity in mice. Submitted for publication.

Svennerholm, A.-M. (1980). The nature of protective immunity in cholera. In: *Cholera and Related Diarrheal Disease* (eds Ö. Ouchterlony and J. Holmgren), 43rd Nobel Symposium, pp. 171–84. Karger, Basel.

Svennerholm, A.-M. and Holmgren, J. (1976). Synergistic protective effect in rabbits of immunization with *Vibrio cholerae* lipopolysaccharide and toxin/toxoid. *Infect. Immun.* **13**, 735–40.

Svennerholm, A.-M., Jertborn, M., Gothefors, L., Karim, M., Sack, D.A. and Holmgren, J. (1984). Mucosal antitoxic and antibacterial immunity after cholera disease and after immunization with a combined B subunit-whole cell vaccine. *J. Infect. Dis.* **149**, 884–93.

Svennerholm, A.-M., Wikström, M., Lindblad, M. and Holmgren, J. (1986). Monoclonal antibodies against *E. coli* heat-stable toxin (STa) and their use in diagnostic ST ganglioside GM1-enzyme-linked immunosorbent assay. *J. Clin. Microbiol.* **24**, 585–90.

Svennerholm, A.-M., Lindblad, M., Svennerholm, B. and Holmgren, J. (1988). Synthesis of nontoxic, antibody-binding *Escherichia coli* heat-stable enterotoxin (STa) peptides. *FEMS Microbiol. Lett.* **55**, 23–8.

Svennerholm, A.-M., Holmgren, J. and Sack, D.A. (1989). Development of oral vaccines against enterotoxigenic *Escherichia coli* diarrhoea. *Vaccine* **7**, 196–8.

Svennerholm, A.-M. and Holmgren, J. (1995). Oral B-subunit whole-cell vaccines against cholera and enterotoxigenic *Escherichia coli* diarrhoea. In: *Molecular and Clinical Aspects of Bacterial Vaccine Development* (eds D.A.A. Ala'Aldeen and C.E. Hormaeche), pp. 205–32. John Wiley, Chichester.

Svennerholm, A.-M., Åhrén, C. and Jertborn, M. (1997). Oral inactivated vaccines against enterotoxigenic *Escherichia coli*. In: *New Generation Vaccines*, 2nd edn (eds M.M. Levine, G.C. Woodrow, J.B. Kaper and G.S. Gabon), pp. 865–74. Marcel Dekker,. New York.

Tacket, C. and Levine, M.M. (1997). Vaccines against enterotoxigenic *Escherichia coli* infections. Part ii: Live oral vaccines and subunit (purified fimbriae and toxin subunit vaccines. In: *New Generation Vaccines*, 2nd edn (eds M.M. Levine, G.C. Woodrow, J.B. Kaper and G.S. Cobon), pp. 875–83. Marcel Dekker, New York.

Tacket, C.O., Losonsky, G., Nataro, J.P., Cryz, S.J., Edelman, R., Fasano, A., Michalski, J., Kaper, J.B. and Levine, M.M. (1993). Safety and immunogenicity of live oral cholera vaccine candidate CVD110, a ΔctxA Δzot Δace derivative of El Tor Ogawa *Vibrio cholerae. J. Infect. Dis.* **168**, 1536–40.

Takeda, T., Nair, G.B., Suzuki, K., Zhe, H.X., Yokoo, Y., De Mol, P., Hemelhof, W., Butzler, J.P., Takeda,Y. and Shimonishi, Y. (1993). Epitope mapping and characterization of antigenic determinants of heat-stable enterotoxin (STh) of enterotoxigenic *Escherichia coli* by using monoclonal antibodies. *Infect. Immun.* **61**, 289–94.

Tarkowski, A., Sun, J.-B., Holmdahl, R., Holmgren J. and Czerkinsky, C. (1999). Successful prophylaxis and treatment of experimental arthritis by intranasal administration of collagen II conjugated to cholera toxin B subunit. *Eur. J. Immunol.*, in press.

Tauxe, R., Seminario, L., Tapia, R. and Libel, M. (1994). The Latin American epidemic. In: Vibrio cholerae *and Cholera: Molecular to Global Perspectives* (eds I.K.Wachsmuth, P.A. Blake and Ö. Olsvik), pp. 321–44. American Society for Microbiology, Washington, DC.

Taylor, D.N., Killeen, K.P., Hack, D.C., Kenner, J.K., Coster, T.S., Beattie, D.T., Ezzell, J., Hyman, T., Trofa, A., Sjogren, M.H., Friedlander, A., Mekalanos, J.J. and Sadoff, J.C. (1994). Development of a live, oral and attenuated vaccine against El Tor cholera. *J. Infect. Dis.* **170**, 1518–23.

Tayot, J.-L., Holmgren, J., Svennerholm, L., Lindblad, M. and Tardy, M. (1981). Receptor-specific large scale purification of cholera toxin on silica beads derivatized with lyso-GM1 ganglioside. *Eur. J. Biochem.* **113**, 249–58.

Trucksis, M., Galen, J.E., Michalski, J., Fasano, A. and Kaper, J.B. (1993). Accessory cholera enterotoxin (Ace), the third member of a *Vibrio cholerae* virulence cassette. *Proc. Natl Acad. Sci. USA* **90**, 5267–71.

Voss, E., Manning, P.A. and Attridge, S.R. (1996). The toxin-coregulated pilus is a colonization factor and protective antigen of *Vibrio cholerae* El Tor. *Microb. Pathogen.* **20**, 141–53.

Waldor, M.K. and Mekalanos, J.J. (1994). Emergence of a new cholera pandemic: molecular analysis of virulence determinants in *Vibrio cholerae* O139 and development of a live vaccine prototype. *J. Infect. Dis.* **170**, 278–83.

Weiner, H.L., Friedman, A., Miller, A., Khoury, S.J., Al-Sabbagh, A., Santos, L., Sayegh, M., Nussenblatt, R.B., Trentham, D.E. and Hafler, D.A. (1994). Oral tolerance: immunologic mechanisms and treatment of animal and human organ-specific autoimmune diseases by oral administration of autoantigens. *Annu. Rev. Immunol.* **12**, 809–37.

Wennerås, C., Svennerholm, A.-M., Åhrén, C. and Czerkinsky, C. (1992). Antibody-secreting cells in human peripheral blood after oral immunization with an inactivated enterotoxigenic *Escherichia coli* vaccine. *Infect. Immun.* **60**, 2605–11.

Wiedermann, G., Kollaritsch, H., Kundi, M., Svennerholm, A.-M. and Bjare, U. (1999). Double blind, randomized, placebo controlled pilot study evaluating efficacy and reactogenicity of an oral ETEC B-subunit-inactivated whole cell vaccine against travelers diarrhea (Preliminary report). *J. Travel Medicine* (in press).

Willshaw, G.A., Scotland, S.M. and Rowe, B. (1997). Vero-cytotoxin-producing *Escherichia coli*. In: Escherichia coli: *Mechanisms of Virulence* (ed. M. Sussman), pp. 421–48. Cambridge University Press, Cambridge.

Wolf, M.K. (1999). Occurence, distribution and association of O and H serogroups, colonization factors, antigens, and toxins of enterotoxigenic *Escherichia coli. Clin. Microbiol. Rev.* **10**, 569–84.

World Health Organization (1996). *State of the World's Vaccines and Immunization* (ed. S. Davey). World Health Organization, Geneva.

World Health Organization (1998). 1998 Report on diarrheal disease vaccines. GPV/VRD/98.06. World Health Organization, Geneva.

Index